INSTRUCTOR'S SOLUTIONS MANUAL

Mark A. McKibben
Goucher College

to accompany

Precalculus & Precalculus with Limits

Cynthia Y. Young
University of Central Florida

JOHN WILEY & SONS, INC.

Cover Photo Credit ©John Kelly/Getty Images, Inc.

Copyright © 2010 John Wiley & Sons, Inc. All rights reserved.

No part of this publication may be reproduced, stored in a retrieval system or transmitted in any form or by any means, electronic, mechanical, photocopying, recording, scanning or otherwise, except as permitted under Section 107 or 108 of the 1976 United States Copyright Act, without either the prior written permission of the Publisher or authorization through payment of the appropriate per-copy fee to the Copyright Clearance Center, Inc., 222 Rosewood Drive, Danvers, MA 01923, website www.copyright.com. Requests to the Publisher for permission should be addressed to the Permissions Department, John Wiley & Sons, Inc., 111 River Street, Hoboken, NJ 07030-5774, (201) 748-6011, fax (201) 748-6008, website www.wiley.com/go/permissions.

Evaluation copies are provided to qualified academics and professionals for review purposes only, for use in their courses during the next academic year. These copies are licensed and may not be sold or transferred to a third party. Upon completion of the review period, please return the evaluation copy to Wiley. Return instructions and a free of charge return shipping label are available at www.wiley.com/go/returnlabel. Outside of the United States, please contact your local representative.

ISBN 978-0-470-53206-5

Printed in the United States of America

10 9 8 7 6 5 4 3 2 1

Printed and bound by Hamilton Printing

TABLE OF CONTENTS

CHAPTER 0

Section 0.1 Solutions ---

1.

$$9m-7=11$$
$$9m=18$$
$$\boxed{m=2}$$

2.

$$2x+4=5$$
$$2x=1$$
$$\boxed{x=1/2}$$

3.

$$5t+11=18$$
$$5t=7$$
$$\boxed{t=7/5}$$

4.

$$7x+4=21+24x$$
$$7x=17+24x$$
$$-17x=17$$
$$\boxed{x=-1}$$

5.

$$3x-5=25+6x$$
$$3x=30+6x$$
$$-3x=30$$
$$\boxed{x=-10}$$

6.

$$5x+10=25+2x$$
$$5x=15+2x$$
$$3x=15$$
$$\boxed{x=5}$$

7.

$$20n-30=20-5n$$
$$20n=50-5n$$
$$25n=50$$
$$\boxed{n=2}$$

8.

$$14c+15=43+7c$$
$$14c=28+7c$$
$$7c=28$$
$$\boxed{c=4}$$

9.

$$4(x-3)=2(x+6)$$
$$4x-12=2x+12$$
$$2x=24$$
$$\boxed{x=12}$$

10.

$$5(2y-1)=2(4y-3)$$
$$10y-5=8y-6$$
$$2y=-1$$
$$\boxed{y=-\tfrac{1}{2}}$$

11.

$$-3(4t-5)=5(6-2t)$$
$$-12t+15=30-10t$$
$$-15=2t$$
$$\boxed{-\tfrac{15}{2}=t}$$

12. $2(3n+4)=-(n+2)$ $6n+8=-n-2$ $7n=-10$ $\boxed{n=-10/7}$	**13.** $2(x-1)+3=x-3(x+1)$ $2x-2+3=x-3x-3$ $2x+1=-2x-3$ $4x=-4$ $\boxed{x=-1}$
14. $4(y+6)-8=2y-4(y+2)$ $4y+24-8=2y-4y-8$ $4y+16=-2y-8$ $6y=-24$ $\boxed{y=-4}$	**15.** $5p+6(p+7)=3(p+2)$ $5p+6p+42=3p+6$ $11p+42=3p+6$ $8p=-36$ $\boxed{p=-\frac{36}{8}=-\frac{9}{2}}$
16. $3(z+5)-5=4z+7(z-2)$ $3z+15-5=4z+7z-14$ $3z+10=11z-14$ $-8z=-24$ $\boxed{z=3}$	**17.** $7x-(2x+3)=x-2$ $7x-2x-3=x-2$ $5x-3=x-2$ $4x=1$ $\boxed{x=1/4}$
18. $3x-(4x+2)=x-5$ $3x-4x-2=x-5$ $-x-2=x-5$ $3=2x$ $\boxed{3/2=x}$	**19.** $2-(4x+1)=3-(2x-1)$ $2-4x-1=3-2x+1$ $1-4x=4-2x$ $-3=2x$ $\boxed{-3/2=x}$
20. $5-(2x-3)=7-(3x+5)$ $5-2x+3=7-3x-5$ $8-2x=2-3x$ $\boxed{x=-6}$	**21.** $2a-9(a+6)=6(a+3)-4a$ $-7a-54=6a+18-4a$ $-7a-54=2a+18$ $-9a=72$ $\boxed{a=-8}$

22.	$25-\left[2+5y-3\left(y+2\right)\right]=-3\left(2y-5\right)-\left[5\left(y-1\right)-3y+3\right]$ $25-[2+5y-3y-6]=-6y+15-[5y-5-3y+3]$ $25-2-5y+3y+6=-6y+15-5y+5+3y-3$ $29-2y=-8y+17$ $6y=-12$ $\boxed{y=-2}$
23.	$32-\left[4+6x-5\left(x+4\right)\right]=4\left(3x+4\right)-\left[6\left(3x-4\right)+7-4x\right]$ $32-[4+6x-5x-20]=12x+16-[18x-24+7-4x]$ $32-4-6x+5x+20=12x+16-18x+24-7+4x$ $48-x=-2x+33$ $\boxed{x=-15}$
24.	$12-\left[3+4m-6\left(3m-2\right)\right]=-7\left(2m-8\right)-3\left[\left(m-2\right)+3m-5\right]$ $12-[3+4m-18m+12]=-14m+56-3[m-2+3m-5]$ $12-3-4m+18m-12=-14m+56-3m+6-9m+15$ $-3+14m=-26m+77$ $40m=80$ $\boxed{m=2}$
25.	$20-4\left[c-3-6\left(2c+3\right)\right]=5\left(3c-2\right)-\left[2\left(7c-8\right)-4c+7\right]$ $20-4[c-3-12c-18]=15c-10-[14c-16-4c+7]$ $20-4c+12+48c+72=15c-10-14c+16+4c-7$ $44c+104=5c-1$ $39c=-105$ $\boxed{c=\frac{-105}{39}=\frac{-35}{13}}$

26.

$$46-\left[7-8y+9(6y-2)\right]=-7(4y-7)-2\left[6(2y-3)-4+6y\right]$$
$$46-[7-8y+54y-18]=-28y+49-2[12y-18-4+6y]$$
$$46-7+8y-54y+18=-28y+49-24y+36+8-12y$$
$$-46y+57=-64y+93$$
$$18y=36$$
$$\boxed{y=2}$$

27.

$$60\left(\frac{1}{5}m\right)=60\left(\frac{1}{60}m+1\right)$$
$$12m=m+60$$
$$11m=60$$
$$\boxed{m=\frac{60}{11}}$$

28.

$$24\left(\frac{1}{12}z\right)=24\left(\frac{1}{24}z+3\right)$$
$$2z=z+72$$
$$\boxed{z=72}$$

29.

$$63\left(\frac{x}{7}\right)=63\left(\frac{2x}{63}+4\right)$$
$$9x=2x+252$$
$$7x=252$$
$$\boxed{x=36}$$

30.

$$22\left(\frac{a}{11}\right)=22\left(\frac{a}{22}+9\right)$$
$$2a=a+198$$
$$\boxed{a=198}$$

31.

$$24\left(\frac{1}{3}p\right)=24\left(3-\frac{1}{24}p\right)$$
$$8p=72-p$$
$$9p=72$$
$$\boxed{p=8}$$

32.

$$10\left(\frac{3x}{5}-x\right)=10\left(\frac{x}{10}-\frac{5}{2}\right)$$
$$6x-10x=x-25$$
$$-5x=-25$$
$$\boxed{x=5}$$

33.

$$84\left(\frac{5y}{3}-2y\right)=84\left(\frac{2y}{84}+\frac{5}{7}\right)$$
$$140y-168y=2y+60$$
$$-30y=60$$
$$\boxed{y=\frac{60}{-30}=-2}$$

34.

$$72\left(2m-\frac{5m}{8}\right)=72\left(\frac{3m}{72}+\frac{4}{3}\right)$$
$$144m-45m=3m+96$$
$$96m=96$$
$$\boxed{m=1}$$

35. $8\left(p+\dfrac{p}{4}\right)=8\left(\dfrac{5}{2}\right)$ $8p+2p=20$ $10p=20$ $\boxed{p=2}$	**36.** $4\left(\dfrac{c}{4}-2c\right)=4\left(\dfrac{5}{4}-\dfrac{c}{2}\right)$ $c-8c=5-2c$ $-5c=5$ $\boxed{c=-1}$
37. $\dfrac{x-3}{3}-\dfrac{x-4}{2}=1-\dfrac{x-6}{6}$ $6\cdot\left[\dfrac{x-3}{3}-\dfrac{x-4}{2}\right]=6\cdot\left[1-\dfrac{x-6}{6}\right]$ $2(x-3)-3(x-4)=6-(x-6)$ $2x-6-3x+12=6-x+6$ $-x+6=-x+12$ $6=12$, which is false. Hence, $\boxed{\text{no solution}}$.	**38.** $1-\dfrac{x-5}{3}=\dfrac{x+2}{5}-\dfrac{6x-1}{15}$ $15\cdot\left[1-\dfrac{x-5}{3}\right]=15\cdot\left[\dfrac{x+2}{5}-\dfrac{6x-1}{15}\right]$ $15-5(x-5)=3(x+2)-(6x-1)$ $15-5x+25=3x+6-6x+1$ $40-5x=-3x+7$ $33=2x$ $\boxed{33/2=x}$
39 Let $x=$ distance from Angela's home to the restaurant. Home $\rightarrow$ Train station $=1$ mile On train $\rightarrow\frac{3}{4}x$ In taxi $\rightarrow\frac{1}{6}x$ $1+\frac{3}{4}x+\frac{1}{6}x=x$ LCD $=12$ $12+9x+2x=12x$ $12+11x=12x$ $x=12$ Angela travels $\boxed{\text{12 miles}}$ to the restaurant.	**40.** Let $x=$ distance from her house to VAB House $\rightarrow$ Park & Ride $=7$ miles Park & Ride $\rightarrow$ H.Q. $=\frac{5}{6}x$ H.Q. $\rightarrow$ VAB $=\frac{1}{20}x$ $7+\frac{5}{6}x+\frac{1}{20}x=x$ LCD $=60$ $420+50x+3x=60x$ $420=7x$ $x=60$ She travels $\boxed{\text{60 miles}}$ to the VAB.

41. Fixed costs = 15,000 Variable costs = 18.50x Total costs = 20,000 $18.50x + 15,000 = 20,000$ $18.50x = 5000$ $x = 270.27$ Approximately $\boxed{\text{270 units}}$ can be produced.	**42.** Let x = number of sets of napkins Fixed monthly costs = 1329.50 Variable costs = 3.70x Total budget = 1870 $1329.50 + 3.70x = 1870$ $3.70x = 540.5$ $x = 146.08$ She can afford approximately $\boxed{146}$ sets of napkins.
43. r_1 = radius of smaller circle r_2 = radius of larger circle $r_2 = r_1 + 3$ Circumference of smaller circle $= 2\pi r_1$ Circumference of larger circle $= 2\pi r_2$ Ratio of circumferences $= \dfrac{2\pi r_2}{2\pi r_1} = \dfrac{r_2}{r_1} = \dfrac{2}{1}$ $r_2 = 2r_1$ $2r_1 = r_1 + 3$ $r_1 = 3$ $\boxed{r_1 = 3 \text{ feet} \;\; r_2 = 6 \text{ feet}}$	**44.** w = width l = length $= 3w + 2$ $p = 2l + 2w$ $28 = 2(3w + 2) + 2w$ $28 = 6w + 4 + 2w$ $24 = 8w$ $w = 3$ $\boxed{\begin{array}{l}\text{width} = 3 \text{ inches} \\ \text{length} = 11 \text{ inches}\end{array}}$
45. Let x = length of alligator in feet. Solve: $\dfrac{3.5}{0.5} = \dfrac{x}{0.75}$ $0.5x = 2.625$ $x = 5.25$ The alligator is about $\boxed{\text{5.25 feet}}$.	**46.** Let x = length of snake in inches. $\dfrac{\text{Fang}}{\text{Body}} = \dfrac{2}{36} = \dfrac{2.6}{x}$ Solve: $2x = 93.6$ $x = 46.8$ The snake is about 3.9 feet $= \boxed{\text{46.8inches}}$.

47. Let $x =$ amount invested at 4%.

$120{,}000 - x =$ amount invested at 7%

Solve:

$$0.04x + 0.07(120{,}000 - x) = 7{,}800$$
$$0.04x + 8400 - 0.07x = 7{,}800$$
$$-0.03x = -600$$
$$x = 20{,}000$$

$\boxed{\$20{,}000 \text{ at } 4\% \text{ and } \$100{,}000 \text{ at } 7\%}$

48. Let $x =$ amount invested at 10%.

$13{,}000 - x =$ amount invested at 14%

Solve:

$$0.10x + 0.14(13{,}000 - x) = 1580$$
$$0.10x + 1820 - 0.14x = 1580$$
$$-0.04x = -240$$
$$x = 6000$$

$\boxed{\$6{,}000 \text{ at } 10\% \text{ and } \$7{,}000 \text{ at } 14\%}$

49. Let $x =$ amount invested at 10%.

$\frac{14{,}000 - x}{2} =$ amount invested at 2%

$\frac{14{,}000 - x}{2} =$ amount invested at 40%

Interest earned $= 16{,}610 - 14{,}000 = 2{,}610$

Solve:

$$0.1x + 0.02\left(\frac{14{,}000 - x}{2}\right) + 0.4\left(\frac{14{,}000 - x}{2}\right) = 2610$$
$$0.1x + 140 - 0.01x + 2800 - 0.2x = 2610$$
$$-0.11x = -330$$
$$x = 3{,}000$$

$\boxed{\begin{array}{l} \$3{,}000 \text{ at } 10\% \\ \$5{,}500 \text{ at } 2\% \\ \$5{,}500 \text{ at } 40\% \end{array}}$

50. \$2500 in money market $(\text{rate} = x)$

\$2500 in stock market $(\text{rate} = 3x)$

Interest earned = \$150

Solve:

$$2500x + 2500(3x) = 150$$
$$2500x + 7500x = 150$$
$$10{,}000x = 150$$
$$x = 0.015$$

$\boxed{\begin{array}{l} \text{Money market: } 1.5\% \\ \text{Stock market: } 4.5\% \end{array}}$

51. Let $x =$ ml of 5% HCl

Solve:

$100 - x =$ ml of 15% HCl

$$0.05x + 0.15(100 - x) = 0.08(100)$$
$$0.05x + 15 - 0.15x = 8$$
$$-0.1x = -7$$
$$x = 70$$

$\boxed{\begin{array}{l} \text{70ml of 5\% HCl} \\ \text{30ml of 15\% HCl} \end{array}}$

52. Let $x =$ gallons of 100% alcohol

Solve:

$$0.20(5) + 1.00x = 0.50(5 + x)$$
$$1 + x = 2.5 + 0.50x$$
$$0.50x = 1.5$$
$$x = \frac{1.5}{0.5} = 3$$

$\boxed{\text{3 gallons}}$.

53. distance = rate · time

distance = 100,000,000 miles

rate = 670,616,629 mph

$$\text{time} = \frac{\text{distance}}{\text{rate}}$$

= 0.15 hours ≅ **9 minutes**

54. distance = rate · time

distance = 0.5 miles

rate = 760 mph

$$\text{time} = \frac{\text{distance}}{\text{rate}}$$

= 0.0006579 hours ≅ **2.4 seconds**

55. rate (r) = boat speed (s) ± current speed (c)

boat speed: s = 16 mph

upstream: $r = s - c$, $t = 1/3$ hours

downstream: $r = s + c$, $t = 1/4$ hours

Distance is the same both ways (rate · time)

Solve:

$$(16-c)\left(\frac{1}{3}\right)=(16+c)\left(\frac{1}{4}\right)$$

$$4(16-c)=3(16+c)$$

$$64-4c=48+3c$$

$$7c=16$$

$$\boxed{c=\frac{16}{7}\cong 2.3 \text{ mph}}$$

56. rate (r) = plane speed (s) ± wind speed (w)

plane speed: s = 130 mph

upwind: $r = s - w$, $t = 2$ hours

downwind: $r = s + w$, $t = 1.25$ hours

Distance is the same both ways (rate · time)

Solve:

$$2(130-w)=1.25(130+w)$$

$$260-2w=162.5+1.25w$$

$$97.5=3.25w$$

$$\boxed{w=30 \text{ mph}}$$

57. rate of walker = r_w

rate of jogger = $r_w + 2$

time of walker = 1 hour

time of jogger = $\frac{2}{3}$ hour

$$r_w(1)=(r_w+2)(2/3)$$

$$r_w=\tfrac{2}{3}r_w+4/3$$

$$\tfrac{1}{3}r_w=\tfrac{4}{3}$$

$$r_w=4$$

walker: 4 mph
jogger: 6 mph

58. dist. of Southern route = d

dist. of Northern route = $d + 300$

time of S route = 45 hours

time of N route = 50 hours

$$\frac{d}{45}=\frac{d+300}{50}$$

$$50d=45(d+300)$$

$$50d=45d+13,500$$

$$5d=13,500$$

$$d=2700$$

S route = 2,700 miles
N route = 3,000 miles

59. Let x = number of minutes it takes a rider to get to class
Using Distance = Rate × Time, and the fact that since they use the same path, their distances are the same, we must solve the equation:

$$2(12+x)=6(x)$$
$$24+2x=6x$$
$$24=4x$$
$$x=6$$

So, it takes the bicyclist 6 minutes to get to class, and the walker 18 minutes.

60. Let x = number of minutes a car travels before catching the truck
When it catches the truck, the car and truck will have traveled the same distance. So, we must solve:

$$70x=50(x+30)$$
$$70x=50x+1500$$
$$x=75$$

It takes the car 75 minutes to catch the truck.

61. Let x = hours it takes Cynthia to paint house alone. Christopher can paint 1/15 house per hour. Cynthia can paint $1/x$ house per hour. Together they paint $\left(\frac{1}{15}+\frac{1}{x}\right)$ house per hour.

$$\frac{1}{15}+\frac{1}{x}=\frac{1}{9} \Rightarrow 3x+45=5x \Rightarrow 2x=45 \Rightarrow x=22.5$$

Cynthia can paint the house alone in 22.5 hours.

62. Let x = number of hours it takes Morgan to complete the yard alone.
Jay: 1/3 of the yard per hour
Morgan: $1/x$ of the yard per hour
Together: 1 hour for entire yard
Solve:

$$\frac{1}{3}+\frac{1}{x}=1$$
$$3x\cdot\left(\frac{1}{3}+\frac{1}{x}\right)=3x$$
$$x+3=3x$$
$$x=1.5$$

It takes 1.5 hours for Morgan to do the yard herself.

63. Tracey can do 1/4 of a delivery per hour, and Robin can do 1/6 of a delivery per hour. Together, they complete 1/4 + 1/6 = 1/(12/5) of the delivery in an hour. So, together, they complete the job in $\boxed{2.4 \text{ hours}}$.

64. Joshua can do the job at a rate of 1/30 job per minute. Amber can do it in 1/20 job per minute. Together, they work at $(1/30 + 1/20) = 1/12$ job per minute. They will finish in $\boxed{12 \text{ minutes}}$

65. Let x = number of field goals. Then $8 - x$ = number of touchdowns. So,

$$3x + 7(8 - x) = 48$$
$$3x + 56 - 7x = 48$$
$$-4x = -8$$
$$x = 2$$

$\boxed{\text{2 field goals, 6 touchdowns}}$

66. TE: $\frac{100\text{yds}}{12\text{secs}}$ DB: $\frac{100\text{yds}}{10\text{secs}}$

Let d be the distance the TE runs.

Then $d + 5$ is the distance the DB runs.

Time spent running is the same for both.

dist=rate · time, time=dist/rate

$$\text{TE}: t = \frac{d}{100/12}$$
$$\text{DB}: t = \frac{d+5}{100/10}$$

$$\frac{d}{100/12} = \frac{d+5}{100/10}$$
$$\frac{12}{100}d = \frac{10}{100}(d+5)$$
$$12d = 10d + 50$$
$$2d = 50$$
$$d = 25$$

$\boxed{\text{TE catches ball at 20 yard line and is tackled 25 yards later at the 45.}}$

67.

$$(42)(5) = (60)(x)$$
$$210 = 60x$$
$$x = 3.5$$

$\boxed{\text{Maria should sit 3.5 feet from the center.}}$

68.

$$(33 + 42)(4) = 60x$$
$$(75)(4) = 60x$$
$$x = 5$$

$\boxed{\text{Maria should sit 5 feet from the center.}}$

69. Let the board be 1 unit long.

Let x = distance from Maria to fulcrum.

$1-x$ = distance from Max to fulcrum.

$$60x = 42(1-x)$$
$$60x = 42-42x$$
$$102x = 42$$
$$x \cong 0.4$$

Fulcrum is 0.4 units from Maria and 0.6 units from Max.

70. Let the board be 1 unit long.

Let x = distance from Maria to fulcrum.

$1-x$ = distance from Max/Martin to fulcrum.

$$60x = (42+33)(1-x)$$
$$60x = 75-75x$$
$$135x = 75$$
$$x = 0.5\overline{5}$$

Fulcrum is 0.56 units from Maria and 0.44 units from Max/Martin.

71. Should have subtracted $4x$ and added 7 to both sides. The correct answer is $x = 5$.

72. Forgot to distribute the negative sign through the parentheses.

73.

$$ax + b = c \quad \boxed{a \neq 0}$$
$$ax = c - b$$
$$\boxed{x = \frac{c-b}{a}}$$

74.

$$2 - a = 2 + 5 - 3a(2)$$
$$2 - a = 7 - 6a$$
$$-5 = -5a$$
$$\boxed{a = 1}$$

75.

$$P = 2l + 2w$$
$$P - 2l = 2w$$
$$\frac{P-2l}{2} = w$$

76.

$$P = 2l + 2w$$
$$P - 2w = 2l$$
$$\frac{P-2w}{2} = l$$

77.

$$A = \tfrac{1}{2}bh$$
$$2A = bh$$
$$\frac{2A}{b} = h$$

78.

$$C = 2\pi r$$
$$\frac{C}{2\pi} = r$$

79.

$$A = lw$$
$$\frac{A}{l} = w$$

80.

$$d = rt$$
$$\frac{d}{r} = t$$

81.

$$V = lwh$$
$$\frac{V}{lw} = h$$

82.

$$V = \pi r^2 h$$
$$\frac{V}{\pi r^2} = h$$

83. Let x = Janine's average speed (in mph).
Then, Tricia's speed = $(12 + x)$ mph. We must solve the equation:

$$2.5(12+x) + 2.5x = 320$$
$$30 + 2.5x + 2.5x = 320$$
$$5x = 290$$
$$x = 58$$

So, Janine's average speed is 58 mph and Tricia's average speed is 70 mph.

84. Let x = Rick's average speed (in mph).
Then, Mike's speed = $(8 + x)$ mph. We must solve the equation:

$$1.5(8+x)+1.5x=210$$
$$12+1.5x+1.5x=210$$
$$3x=198$$
$$x=66$$

So, Rick's average speed is 66 mph and Mike's average speed is 74 mph.

85. $y1=3(x+2)-5x$

$y2=3x-4$

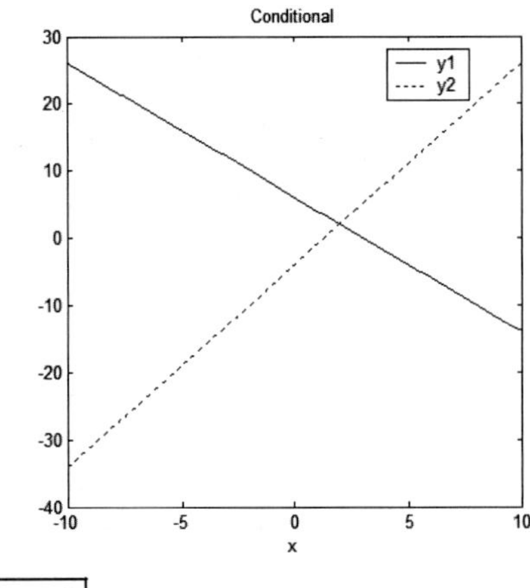

$x=2$

86. $y1=-5(x-1)-7$

$y2=10-9x$

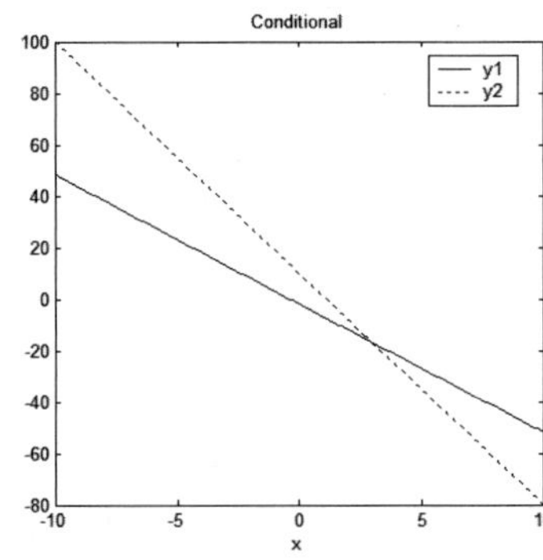

$x=3$

87. $y1=2x+6$

$y2=4x-2x+8-2$

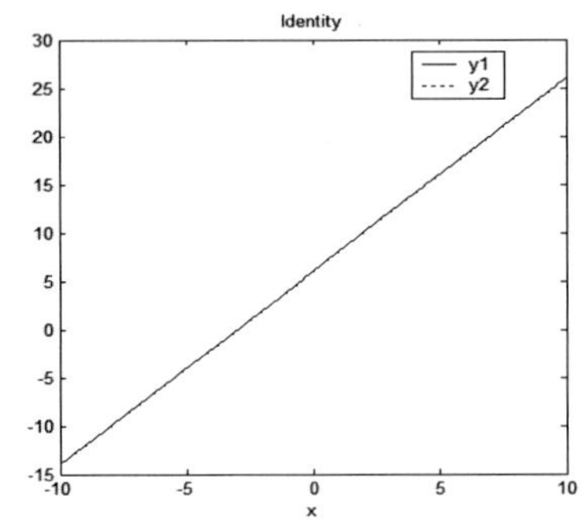

All real numbers

88. $y1=10-20x$

$y2=10x-30x+20-10$

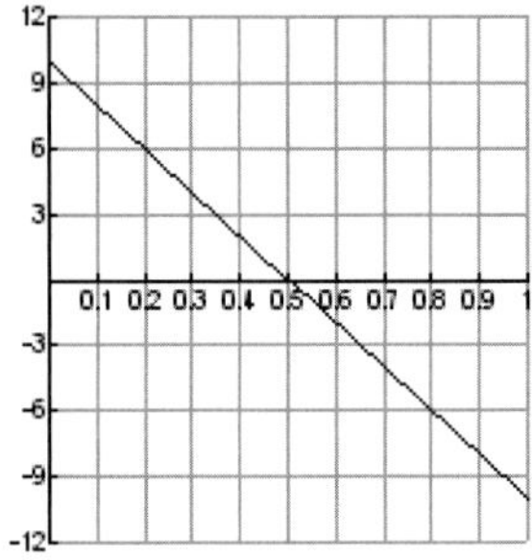

All real numbers

89. $y = 11896.67x + 132500$

$\boxed{\$191,983.35}$

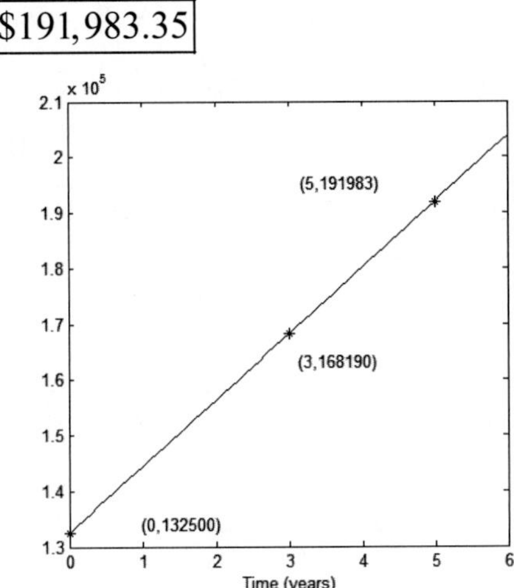

90. $y = 11896.67x + 132500$

$\boxed{\$144,397}$

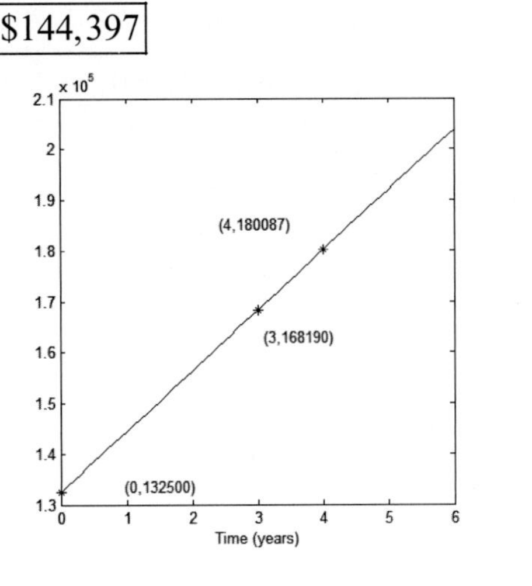

91. Let $x =$ number of times you play.

Option A: $y_1 = 300 + 15x$

Option B: $y_2 = 150 + 42x$

Option B is better if you play about 5 times or less per month. Option A is better if you play 6 times or more per month.

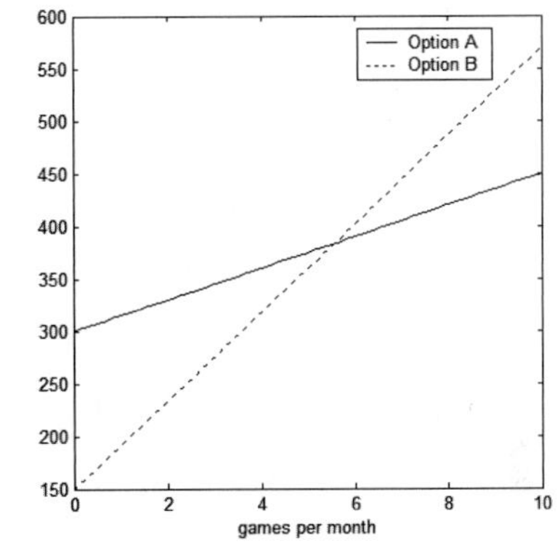

92. Let $x =$ number of minutes used.

Plan A: $y_1 = 30 + 0.1x$

Plan B: $y_2 = 50 + 0.03x$

Plan A is better if you use about 285 minutes or less per month. Plan B is better if you use more than 285 minutes per month.

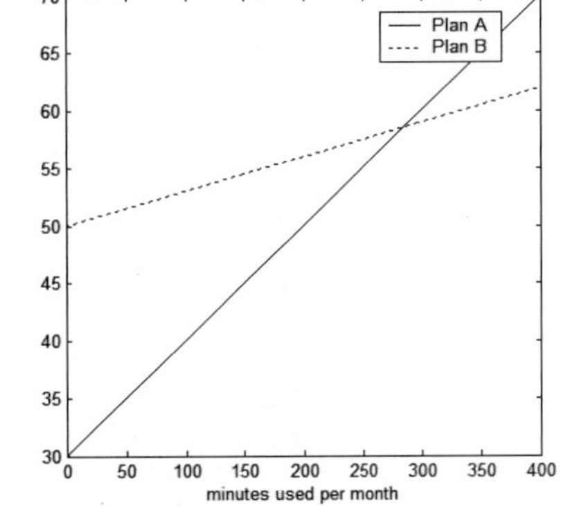

Section 0.2 Solutions

1.	$x^2-5x+6=0$ $(x-3)(x-2)=0$ $x-3=0$ or $x-2=0$ $\boxed{x=3 \text{ or } x=2}$	**2.**	$v^2+7v+6=0$ $(v+6)(v+1)=0$ $v+6=0$ or $v+1=0$ $\boxed{v=-6 \text{ or } v=-1}$
3.	$p^2-8p+15=0$ $(p-5)(p-3)=0$ $\boxed{p=5 \text{ or } p=3}$	**4.**	$u^2-2u-24=0$ $(u-6)(u+4)=0$ $\boxed{u=6 \text{ or } u=-4}$
5.	$x^2=12-x$ $x^2+x-12=0$ $(x+4)(x-3)=0$ $x+4=0$ or $x-3=0$ $\boxed{x=-4 \text{ or } x=3}$	**6.**	$11x=2x^2+12$ $2x^2-11x+12=0$ $(2x-3)(x-4)=0$ $2x-3=0$ or $x-4=0$ $\boxed{x=\frac{3}{2} \text{ or } x=4}$
7.	$16x^2+8x=-1$ $16x^2+8x+1=0$ $(4x+1)(4x+1)=0$ $4x+1=0$ $\boxed{x=-1/4}$	**8.**	$3x^2+10x-8=0$ $(3x-2)(x+4)=0$ $3x-2=0$ or $x+4=0$ $\boxed{x=\frac{2}{3} \text{ or } x=-4}$
9.	$9y^2+1=6y$ $9y^2-6y+1=0$ $(3y-1)(3y-1)=0$ $\boxed{y=\frac{1}{3}}$	**10.**	$4x=4x^2+1$ $4x^2-4x+1=0$ $(2x-1)(2x-1)=0$ $\boxed{x=\frac{1}{2}}$
11.	$8y^2-16y=0$ $8y(y-2)=0$ $8y=0$ or $y-2=0$ $\boxed{y=0 \text{ or } y=2}$	**12.**	$3A^2=-12A$ $3A^2+12A=0$ $3A(A+4)=0$ $3A=0$ or $A+4=0$ $\boxed{A=0 \text{ or } A=-4}$

13.	$9p^2 = 12p - 4$ $9p^2 - 12p + 4 = 0$ $(3p-2)(3p-2) = 0$ $3p - 2 = 0$ $\boxed{p = \frac{2}{3}}$	**14.**	$4u^2 = 20u - 25$ $4u^2 - 20u + 25 = 0$ $(2u-5)(2u-5) = 0$ $\boxed{u = 5/2}$
15.	$x^2 - 9 = 0$ $(x+3)(x-3) = 0$ $x + 3 = 0$ or $x - 3 = 0$ $\boxed{x = -3 \text{ or } x = 3}$	**16.**	$16v^2 - 25 = 0$ $(4v-5)(4v+5) = 0$ $\boxed{v = 5/4 \text{ or } v = -5/4}$
17.	$x(x+4) = 12$ $x^2 + 4x = 12$ $x^2 + 4x - 12 = 0$ $(x+6)(x-2) = 0$ $x + 6 = 0$ or $x - 2 = 0$ $\boxed{x = -6 \text{ or } x = 2}$	**18.**	$3t^2 - 48 = 0$ $3(t^2 - 16) = 0$ $3(t+4)(t-4) = 0$ $t + 4 = 0$ or $t - 4 = 0$ $\boxed{t = -4 \text{ or } t = 4}$
19.	$2p^2 - 50 = 0$ $2(p^2 - 25) = 0$ $2(p-5)(p+5) = 0$ $\boxed{p = -5 \text{ or } p = 5}$	**20.**	$5y^2 - 45 = 0$ $5(y^2 - 9) = 0$ $5(y-3)(y+3) = 0$ $\boxed{y = -3 \text{ or } y = 3}$
21.	$3x^2 = 12$ $3x^2 - 12 = 0$ $3(x^2 - 4) = 0$ $3(x-2)(x+2) = 0$ $\boxed{x = -2 \text{ or } x = 2}$	**22.**	$7v^2 = 28$ $7v^2 - 28 = 0$ $7(v^2 - 4) = 0$ $7(v-2)(v+2) = 0$ $\boxed{v = -2 \text{ or } v = 2}$

23.	$p^2 - 8 = 0$ $p^2 = 8$ $p = \pm\sqrt{8}$ $\boxed{p = \pm 2\sqrt{2}}$	**24.**	$y^2 - 72 = 0$ $y^2 = 72$ $y = \pm\sqrt{72}$ $\left(72 = 2^3 \cdot 3^2\right)$ $\boxed{y = \pm 6\sqrt{2}}$
25.	$x^2 + 9 = 0$ $x^2 = -9$ $\boxed{x = \pm 3i}$	**26.**	$v^2 + 16 = 0$ $v^2 = -16$ $\boxed{v = \pm 4i}$
27.	$(x-3)^2 = 36$ $x - 3 = \pm 6$ $x = 3 \pm 6$ $\boxed{x = -3, 9}$	**28.**	$(x-1)^2 = 25$ $x - 1 = \pm 5$ $x = \pm 5 + 1$ $\boxed{x = -4, 6}$
29.	$(2x+3)^2 = -4$ $2x + 3 = \pm 2i$ $2x = -3 \pm 2i$ $\boxed{x = \frac{-3 \pm 2i}{2}}$	**30.**	$(4x-1)^2 = -16$ $4x - 1 = \pm 4i$ $4x = 1 \pm 4i$ $\boxed{x = \frac{1}{4} \pm i}$
31.	$(5x-2)^2 = 27$ $5x - 2 = \pm\sqrt{27}$ $5x = 2 \pm 3\sqrt{3}$ $\boxed{x = \frac{2 \pm 3\sqrt{3}}{5}}$	**32.**	$(3x+8)^2 = 12$ $3x + 8 = \pm\sqrt{12}$ $3x = -8 \pm 2\sqrt{3}$ $\boxed{x = \frac{-8 \pm 2\sqrt{3}}{3}}$
33.	$(1-x)^2 = 9$ $1 - x = \pm 3$ $-x = -1 \pm 3$ $\boxed{x = 1 \pm 3 = -2, 4}$	**34.**	$(1-x)^2 = -9$ $1 - x = \pm 3i$ $-x = -1 \pm 3i$ $\boxed{x = 1 \pm 3i}$

35. $x^2+2x=3$ $x^2+2x+1=3+1$ $(x+1)^2=4$ $x+1=\pm 2$ $x=-1\pm 2$ $\boxed{x=-3,1}$	**36.** $y^2+8y-2=0$ $y^2+8y+16=2+16$ $(y+4)^2=18$ $y+4=\pm\sqrt{18}$ $\boxed{y=-4\pm 3\sqrt{2}}$
37. $t^2-6t=-5$ $t^2-6t+9=-5+9$ $(t-3)^2=4$ $t-3=\pm 2$ $\boxed{t=3\pm 2=1,5}$	**38.** $x^2+10x=-21$ $x^2+10x+25=-21+25$ $(x+5)^2=4$ $x+5=\pm 2$ $\boxed{x=-5\pm 2=-7,-3}$
39. $y^2-4y=-3$ $y^2-4y+4=-3+4$ $(y-2)^2=1$ $y-2=\pm 1$ $\boxed{y=\pm 1+2=1,3}$	**40.** $x^2-7x=-12$ $x^2-7x+\left(\frac{7}{2}\right)^2=-12+\left(\frac{7}{2}\right)^2$ $\left(x-\frac{7}{2}\right)^2=-12+\frac{49}{4}=\frac{1}{4}$ $x-\frac{7}{2}=\pm\frac{1}{2}$ $\boxed{x=\frac{7}{2}\pm\frac{1}{2}=3,4}$
41. $2p^2+8p=-3$ $2(p^2+4p)=-3$ $2(p^2+4p+4)=-3+8$ $2(p+2)^2=5$ $(p+2)^2=\frac{5}{2}$ $p+2=\pm\sqrt{\frac{5}{2}}$ $\boxed{p=-2\pm\sqrt{\frac{5}{2}}=\frac{-4\pm\sqrt{10}}{2}}$	**42.** $2x^2-4x=-3$ $2(x^2-2x)=-3$ $2(x^2-2x+1)=-3+2$ $2(x-1)^2=-1$ $(x-1)^2=\frac{-1}{2}$ $x-1=\pm i\frac{1}{\sqrt{2}}$ $\boxed{x=1\pm i\frac{\sqrt{2}}{2}}$

43.

$$2x^2 - 7x = -3$$

$$2\left(x^2 - \frac{7}{2}x\right) = -3$$

$$2\left(x^2 - \frac{7}{2}x + \left(\frac{7}{4}\right)^2\right) = -3 + 2\left(\frac{7}{4}\right)^2$$

$$2\left(x - \frac{7}{4}\right)^2 = -3 + 2\left(\frac{49}{16}\right)$$

$$\left(x - \frac{7}{4}\right)^2 = \frac{-3}{2} + \frac{49}{16} = \frac{25}{16}$$

$$x - \frac{7}{4} = \pm\frac{5}{4}$$

$$\boxed{x = \frac{7}{4} \pm \frac{5}{4} = \frac{1}{2}, 3}$$

44.

$$3x^2 - 5x = 10$$

$$3\left(x^2 - \frac{5}{3}x\right) = 10$$

$$3\left(x^2 - \frac{5}{3}x + \left(\frac{5}{6}\right)^2\right) = 10 + 3\left(\frac{5}{6}\right)^2$$

$$3\left(x - \frac{5}{6}\right)^2 = 10 + 3\left(\frac{25}{36}\right)$$

$$\left(x - \frac{5}{6}\right)^2 = \frac{10}{3} + \frac{25}{36} = \frac{145}{36}$$

$$x - \frac{5}{6} = \pm\sqrt{\frac{145}{36}} = \pm\frac{\sqrt{145}}{6}$$

$(145 = 5 \cdot 29)$ so the radical can't be reduced.

$$\boxed{x = \frac{5}{6} \pm \frac{\sqrt{145}}{6} = \frac{5 \pm \sqrt{145}}{6}}$$

45.

$$\frac{x^2}{2} - 2x = \frac{1}{4}$$

$$x^2 - 4x = \frac{1}{2}$$

$$x^2 - 4x + 4 = \frac{1}{2} + 4$$

$$(x-2)^2 = \frac{9}{2}$$

$$x - 2 = \pm\frac{3}{\sqrt{2}}$$

$$\boxed{x = 2 \pm \frac{3}{\sqrt{2}} = \frac{4 \pm 3\sqrt{2}}{2}}$$

46.

$$\frac{t^2}{3} + \frac{2t}{3} + \frac{5}{6} = 0$$

$$t^2 + 2t = -\frac{5}{2}$$

$$t^2 + 2t + 1 = -\frac{5}{2} + 1$$

$$(t+1)^2 = \frac{-3}{2}$$

$$t + 1 = \pm i\sqrt{\frac{3}{2}}$$

$$\boxed{t = -1 \pm i\sqrt{\frac{3}{2}} = -1 \pm i\frac{\sqrt{6}}{2}}$$

47.

$$t^2 + 3t - 1 = 0$$

$$t = \frac{-3 \pm \sqrt{9+4}}{2}$$

$$\boxed{t = \frac{-3 \pm \sqrt{13}}{2}}$$

48.

$$t^2 + 2t - 1 = 0$$

$$t = \frac{-2 \pm \sqrt{4+4}}{2} = -1 \pm \frac{1}{2}\sqrt{8}$$

$$\boxed{t = -1 \pm \sqrt{2}}$$

49. $s^2+s+1=0$ $s=\frac{-1\pm\sqrt{1-4}}{2}=\frac{-1\pm\sqrt{-3}}{2}$ $\boxed{s=\frac{-1\pm i\sqrt{3}}{2}}$	**50.** $2s^2+5s+2=0$ $s=\frac{-5\pm\sqrt{25-16}}{4}=\frac{-5\pm\sqrt{9}}{4}$ $\boxed{s=\frac{-5\pm 3}{4}=-2,\ \frac{-1}{2}}$
51. $3x^2-3x-4=0$ $x=\frac{3\pm\sqrt{9+48}}{6}=\frac{1}{2}\pm\frac{\sqrt{57}}{6}$ $\boxed{x=\frac{3\pm\sqrt{57}}{6}}$	**52.** $4x^2-2x-7=0$ $x=\frac{2\pm\sqrt{4+4\cdot 28}}{8}=\frac{2\pm\sqrt{116}}{8}$ $(116=2^2\cdot 29)$ $\boxed{x=\frac{2\pm 2\sqrt{29}}{8}=\frac{1\pm\sqrt{29}}{4}}$
53. $x^2-2x+17=0$ $x=\frac{2\pm\sqrt{4-4\cdot 17}}{2}=\frac{2\pm\sqrt{-64}}{2}$ $\boxed{x=\frac{2\pm 8i}{2}=1\pm 4i}$	**54.** $4m^2+7m+8=0$ $m=\frac{-7\pm\sqrt{49-4\cdot 32}}{8}=\frac{-7\pm\sqrt{-79}}{8}$ $\boxed{m=-\frac{7}{8}\pm i\frac{\sqrt{79}}{8}}$
55. $5x^2+7x-3=0$ $x=\frac{-7\pm\sqrt{49+60}}{10}$ $\boxed{x=\frac{-7\pm\sqrt{109}}{10}}$	**56.** $3x^2+5x+11=0$ $x=\frac{-5\pm\sqrt{25-132}}{6}=\frac{-5\pm\sqrt{-107}}{6}$ $\boxed{x=-\frac{5}{6}\pm i\frac{\sqrt{107}}{6}}$
57. $\frac{1}{4}x^2+\frac{2}{3}x-\frac{1}{2}=0$ $3x^2+8x-6=0$ $x=\frac{-8\pm\sqrt{64-4(3)(-6)}}{2(3)}=\frac{-8\pm 2\sqrt{34}}{2(3)}$ $\boxed{x=\frac{-4\pm\sqrt{34}}{3}}$	**58.** $\frac{1}{4}x^2-\frac{2}{3}x-\frac{1}{3}=0$ $3x^2-8x-4=0$ $x=\frac{8\pm\sqrt{64-4(3)(-4)}}{2(3)}=\frac{8\pm 4\sqrt{7}}{2(3)}$ $\boxed{x=\frac{4\pm 2\sqrt{7}}{3}}$

59. $v^2-8v-20=0$ $(v-10)(v+2)=0$ $\boxed{v=-2,10}$	**60.** $v^2-8v+20=0$ $v=\frac{8\pm\sqrt{64-80}}{2}=\frac{8\pm\sqrt{-16}}{2}$ $\boxed{v=\frac{8\pm 4i}{2}=4\pm 2i}$
61. $t^2+5t-6=0$ $(t+6)(t-1)=0$ $\boxed{t=-6,1}$	**62.** $t^2+5t+6=0$ $(t+2)(t+3)=0$ $\boxed{t=-2,-3}$
63. $(x+3)^2=16$ $x+3=\pm 4$ $\boxed{x=-3\pm 4=-7,1}$	**64.** $(x+3)^2=-16$ $x+3=\pm 4i$ $\boxed{x=-3\pm 4i}$
65. $(p-2)^2=4p$ $p^2-4p+4=4p$ $p^2-8p+4=0$ $p=\frac{8\pm\sqrt{64-4(1)(4)}}{2(1)}=\frac{8\pm 4\sqrt{3}}{2}$ $\boxed{p=4\pm 2\sqrt{3}}$	**66.** $(u+5)^2=16u$ $u^2+10u+25=16u$ $u^2-6u+25=0$ $u=\frac{6\pm\sqrt{36-4(25)}}{2}=\frac{6\pm 8i}{2}$ $\boxed{u=3\pm 4i}$
67. $8w^2+2w+21=0$ $w=\frac{-2\pm\sqrt{4-4\cdot 8\cdot 21}}{16}$ $w=\frac{-2\pm\sqrt{-668}}{16}=\frac{-2\pm 2i\sqrt{167}}{16}$ $\boxed{w=\frac{-1\pm i\sqrt{167}}{8}}$	**68.** $8w^2+2w-21=0$ $(4w+7)(2w-3)=0$ $4w+7=0$ or $2w-3=0$ $\boxed{w=\frac{-7}{4},\frac{3}{2}}$
69. $3p^2-9p+1=0$ $p=\frac{9\pm\sqrt{81-12}}{6}$ $\boxed{p=\frac{9\pm\sqrt{69}}{6}}$	**70.** $3p^2-9p-1=0$ $p=\frac{9\pm\sqrt{81+12}}{6}$ $\boxed{p=\frac{9\pm\sqrt{93}}{6}}$

71.
$$\frac{2}{3}t^2 - \frac{4}{3}t - \frac{1}{5} = 0$$
$$\text{LCD} = 15$$
$$10t^2 - 20t - 3 = 0$$
$$t = \frac{20 \pm \sqrt{400+120}}{20}$$
$$t = \frac{20 \pm \sqrt{520}}{20} = \frac{20 \pm 2\sqrt{130}}{20}$$
$$\boxed{t = \frac{10 \pm \sqrt{130}}{10}}$$

72.
$$\frac{1}{2}x^2 + \frac{2}{3}x - \frac{2}{5} = 0$$
$$\text{LCD} = 30$$
$$15x^2 + 20x - 12 = 0$$
$$x = \frac{-20 \pm \sqrt{400 + 4 \cdot 15 \cdot 12}}{30}$$
$$x = \frac{-20 \pm \sqrt{1120}}{30}$$
$$x = \frac{-20 \pm 4\sqrt{70}}{30}$$
$$\boxed{x = \frac{-10 \pm 2\sqrt{70}}{15}}$$

73.
$$x^2 - 0.1x - 0.12 = 0$$
$$(x - 0.4)(x + 0.3) = 0$$
$$\boxed{x = -0.3,\ 0.4}$$

74.
$$y^2 - 0.5y + 0.06 = 0$$
$$(y - 0.2)(y - 0.3) = 0$$
$$\boxed{y = 0.2,\ 0.3}$$

75.
$$-4t^2 + 80t - 360 = 24$$
$$t^2 - 20t + 90 = -6$$
$$t^2 - 20t + 96 = 0$$
$$(t-8)(t-12) = 0$$
$$\boxed{t = 8 \text{ (August 2003) and } 12 \text{ (Dec. 2003)}}$$

76.
$$2t^2 - 12t + 70 = 60$$
$$t^2 - 6t + 35 = 30$$
$$t^2 - 6t + 5 = 0$$
$$(t-1)(t-5) = 0$$
$$\boxed{t = 1 \text{ (Nov. 2003) and } 5 \text{ (March 2004)}}$$

77. Form a right triangle with legs of length x and 25in. and hypotenuse of length 32in. Then, by the Pythagorean Theorem, we solve:
$$x^2 + 25^2 = 32^2$$
$$x^2 = 399$$
$$x = \pm\sqrt{399} \approx \pm 20$$
So, the TV is approximately $\boxed{\text{20 inches}}$ high.

78. Form a right triangle with legs of length x and 20in. and hypotenuse of length 42in. Then, by the Pythagorean Theorem, we solve:
$$x^2 + 20^2 = 42^2$$
$$x^2 = 1364$$
$$x = \pm\sqrt{1364} \approx \pm 37$$
So, the TV is approximately $\boxed{\text{37 inches}}$ wide.

79. Let the numbers be x, $x+1$. $x+(x+1)=35$ $2x=34 \Rightarrow x=17$ $x(x+1)=306$ $x^2+x=306$ $x^2+x-306=0$ $(x+18)(x-17)=0$ $x=\cancel{-18},17$ So, $\boxed{\text{the numbers are 17 and 18}}$.	**80.** Let the numbers be x, $x+2$. $x+(x+2)=24$ $2x=22 \Rightarrow x=11$ $x(x+2)=143$ $x^2+2x=143$ $x^2+2x-143=0$ $(x+13)(x-11)=0$ $x=\cancel{-13},11$ So, $\boxed{\text{the numbers are 11 and 13}}$.
81. Let l = length of the rectangle (in ft.) Then, the width $w=l-6$ (in ft.) We must solve: $135=lw$ $135=l(l-6)$ $l^2-6l-135=0$ $(l-15)(l+9)=0$ $l=15, \cancel{-9}$ So, the rectangle has: $\boxed{\text{length 15ft. and width 9ft.}}$	**82.** $\text{Area} = \text{length}\cdot\text{width} = (2w+2)(w)=31.5$ $2w^2+2w-31.5=0$ $w=\dfrac{-2\pm\sqrt{4+4\cdot2\cdot31.5}}{4}$ $w=\dfrac{-2\pm\sqrt{256}}{4}=\dfrac{-2\pm16}{4}$ Widths are lengths which are always positive, so $\boxed{w=\dfrac{14}{4}=\dfrac{7}{2}}$
83. $\text{Area} = \dfrac{1}{2}b\cdot h=60$ $h=3b+2$ $\dfrac{1}{2}b(3b+2)=60$ $\dfrac{3}{2}b^2+b=60$ $3b^2+2b-120=0$ $(3b+20)(b-6)=0$ $\boxed{b=\cancel{\dfrac{-20}{3}},6;\ h=20}$	**84.** $s^2=A$ $(s+3)^2=A+69$ $(s+3)^2=s^2+69$ $s^2+6s+9=s^2+69$ $6s-60=0$ $\boxed{s=10}$

85. $h = -16t^2 + 100$ Ground $\rightarrow h = 0$ $-16t^2 + 100 = 0$ $t^2 = \frac{100}{16}$ $t = \pm\frac{10}{4}$ (Time must be ≥ 0) Impact with ground in 2.5 sec	**86.** $h = -16t^2 - 5t + 100$ Ground $\rightarrow h = 0$ $16t^2 + 5t - 100 = 0$ $t = \frac{-5 \pm \sqrt{25 + 4 \cdot 16 \cdot 100}}{32}$ $t = \frac{-5 \pm \sqrt{6425}}{32} = \frac{-5 \pm 5\sqrt{257}}{32}$ $t \cong 2.3$ sec
87. $15^2 + 15^2 = r^2$ $r^2 = 450$ $r = \pm\sqrt{450} = \pm 15\sqrt{2}$ $r \approx 21.2$ feet	**88.** $d^2 = 90^2 + 90^2$ $d^2 = 16200$ $d = \pm\sqrt{16200} = \pm 90\sqrt{2}$ $d \cong 127$ feet
89. volume $= l \cdot w \cdot h$ $v = (x-2)(x-2)(1)$ $9 = (x-2)^2$ $x - 2 = \pm 3$ $x = 2 \pm 3 = -1, 5$ $x = 5$ Original square was 5ft $\times$ 5ft	**90.** volume $= l \cdot w \cdot h = 12$ $l = 2w,\ h = 1$ $v = (2w-2)(w-2)(1) = 12$ $(2w-2)(w-2) = 12$ $2w^2 - 6w + 4 = 12$ $2w^2 - 6w - 8 = 0$ $w^2 - 3w - 4 = 0$ $(w+1)(w-4) = 0$ Original rectangle was 4ft $\times$ 8ft

91.	Let $w =$ width of border Total area of garden + border $= (8+2w)(5+2w) = 4w^2 + 26w + 40$ Area of garden $= 8 \cdot 5 = 40$ Area of border $= \underbrace{(4w^2 + 26w + 40)}_{total} - \underbrace{40}_{garden} = 4w^2 + 26w$ Volume of border = Area $\cdot$ depth (depth = 4 in. = 1/3 ft) $= (4w^2 + 26w)(1/3)$ Volume = 27 ft^3 $\frac{1}{3}(4w^2 + 26w) = 27$ $4w^2 + 26w = 81$ $4w^2 + 26w - 81 = 0$ $w = \frac{-26 \pm \sqrt{26^2 + 4 \cdot 4 \cdot 81}}{2 \cdot 4} = \frac{-26 \pm \sqrt{1972}}{8}$ $w \cong \cancel{-8.8}, 2.3$ Width of border is 2.3 feet.
92.	Area of Rose Garden $= \frac{\pi r^2}{2} = \frac{\pi 6^2}{2} = 18\pi$ Volume of mulch = Area $\cdot$ depth $= 18\pi \cdot d$ Volume of mulch = 54 ft^3 $18\pi d = 54$ $d = \frac{54}{18\pi} \cong 0.95$ Mulch will be about 1 foot deep

93.

Let x = days for Kimmie to complete job herself.

$x-5$ = days for Lindsey to complete job herself.

$\frac{1}{x}$ = % of job Kimmie can do per day.

$\frac{1}{x-5}$ = % of job Lindsey can do per day.

$\frac{1}{\text{x}}+\frac{1}{x-5}=\frac{1}{6}$ (Together they can do it in 6 days.)

$\text{LCD}=x(x-5)6$ $\boxed{x\neq 0,5}$

$6(x-5)+6x=x(x-5)$

$6x-30+6x=x^2-5x$

$x^2-17x+30=0$

$(x-15)(x-2)=0$

$x=\not{2},15$

Kimmie alone: 15 days

$\boxed{\text{Lindsey alone: 10 days}}$

94.

Jack can clean $\frac{1}{4}$ house per hour.

Ryan can clean $\frac{1}{6}$ house per hour.

Together they can clean $\frac{1}{4}+\frac{1}{6}=\frac{10}{24}=\frac{5}{12}$ house per hour.

$\boxed{\text{They can clean the house in } \frac{12}{5}=2.4 \text{ hours.}}$

95.

Factored incorrectly

$\text{t}^2-5t-6=0$

$(t+1)(t-6)=0$

$\boxed{t=-1,\,6}$

96.

Forgot ±

$(2y-3)^2=25$

$2y-3=\pm 5$

$2y=3\pm 5$

$y=\frac{3\pm 5}{2}=-1,4$

97. $\sqrt{-a}$ is imaginary for positive a $a^2 = -\frac{9}{16}$, so $\boxed{a = \pm\sqrt{\frac{9}{16}} = \pm\frac{3}{4}i}$	**98.** In completing the square we should add 2 (not 1) to the right side. So, $2\left(x^2 - 2x + 1\right) = 3 + 2$. The solution is $1 \pm \sqrt{\frac{5}{2}}$
99. False $x = -5/3$ satisfies 1st equation but not 2nd	**100.** True
101. True	**102.** True
103. If $x = a$ is a repeated root for a quadratic equation, then $(x-a)^2 = 0$. Simplifying yields: $\boxed{x^2 - 2ax + a^2 = 0}$	**104.** If $x = bi$ is a root for a quadratic equation, then so is $x = -bi$. $(x+bi)(x-bi) = 0$ Then, $x^2 - b^2i^2 = 0$ $\boxed{x^2 + b^2 = 0}$
105. $(x-2)(x-5) = 0$ $\boxed{x^2 - 7x + 10 = 0}$	**106.** $x(x+3) = 0$ $\boxed{x^2 + 3x = 0}$
107. $s = \frac{1}{2}gt^2$ $t^2 = \frac{2s}{g}$ $\boxed{t = \pm\sqrt{\frac{2s}{g}}}$	**108.** $A = P(1+r)^2$ $\frac{A}{P} = (1+r)^2$ $1 + r = \pm\sqrt{\frac{A}{P}}$ $\boxed{r = -1 \pm \sqrt{\frac{A}{P}}}$
109. $a^2 + b^2 = c^2$ $\boxed{c = \pm\sqrt{a^2 + b^2}}$	**110.** $P = EI - RI^2$ $RI^2 - EI + P = 0$ This equation is quadratic in I. $\boxed{I = \frac{E \pm \sqrt{E^2 - 4RP}}{2R}}$
111. $x^4 - 4x^2 = 0$ $x^2\left(x^2 - 4\right) = 0$ $x^2(x-2)(x+2) = 0$ $\boxed{x = 0, \pm 2}$	**112.** $3x - 6x^2 = 0$ $3x(1-2x) = 0$ $\boxed{x = 0, \tfrac{1}{2}}$

113.

$$x^3 + x^2 - 4x - 4 = 0$$
$$\left(x^3 + x^2\right) - 4\left(x+1\right) = 0$$
$$x^2\left(x+1\right) - 4\left(x+1\right) = 0$$
$$\left(x^2 - 4\right)\left(x+1\right) = 0$$
$$(x-2)(x+2)(x+1) = 0$$
$$\boxed{x = -1,\ \pm 2}$$

114.

$$x^3 + 2x^2 - x - 2 = 0$$
$$\left(x^3 + 2x^2\right) - \left(x+2\right) = 0$$
$$x^2\left(x+2\right) - \left(x+2\right) = 0$$
$$\left(x^2 - 1\right)\left(x+2\right) = 0$$
$$(x-1)(x+1)(x+2) = 0$$
$$\boxed{x = \pm 1,\ -2}$$

115.

$$x_1 = \frac{-b + \sqrt{b^2 - 4ac}}{2a} \qquad x_2 = \frac{-b - \sqrt{b^2 - 4ac}}{2a}$$

$$x_1 + x_2 = \frac{-b}{2a} + \frac{\sqrt{b^2 - 4ac}}{2a} - \frac{b}{2a} - \frac{\sqrt{b^2 - 4ac}}{2a}$$
$$= \frac{-2b}{2a} = \boxed{\frac{-b}{a}}$$

116.

$$x_1 = \frac{-b + \sqrt{b^2 - 4ac}}{2a} \qquad x_2 = \frac{-b - \sqrt{b^2 - 4ac}}{2a}$$

$$x_1 \cdot x_2 = \frac{\left(-b + \sqrt{b^2 - 4ac}\right)\left(-b - \sqrt{b^2 - 4ac}\right)}{2a \cdot 2a}$$
$$= \frac{b^2 - \left(b^2 - 4ac\right)}{4a^2} = \frac{4ac}{4a^2} = \boxed{\frac{c}{a}}$$

117.

$$\left[x - \left(3 + \sqrt{5}\right)\right]\left[x - \left(3 - \sqrt{5}\right)\right] = 0$$
$$\left[\left(x-3\right) - \sqrt{5}\right]\left[\left(x-3\right) + \sqrt{5}\right] = 0$$
$$\left(x-3\right)^2 - 5 = 0$$
$$x^2 - 6x + 9 - 5 = 0$$
$$\boxed{x^2 - 6x + 4 = 0}$$

118.

$$\left[x - \left(2 - i\right)\right]\left[x - \left(2 + i\right)\right] = 0$$
$$\left[\left(x-2\right) + i\right]\left[\left(x-2\right) - i\right] = 0$$
$$\left(x-2\right)^2 - i^2 = 0$$
$$x^2 - 4x + 4 + 1 = 0$$
$$\boxed{x^2 - 4x + 5 = 0}$$

119.
Let x = speed in still air and y = time to make the trip with a tail wind.
Using Distance = Rate × Time, we obtain the following two equations:
With tail wind: $(x+50)y = 600$ **(1)**
Against head wind:
$(x-50)(y+1) = 600$ **(2)**

Solve **(1)** for y: $y = \dfrac{600}{x+50}$

Substitute this into **(2)** and solve for x:

$$(x-50)\left(\frac{600}{x+50}+1\right) = 600$$
$$(x-50)\left(\frac{600+x+50}{x+50}\right) = 600$$
$$(x-50)(650+x) = 600(x+50)$$
$$650x - 32{,}500 - 50x + x^2 = 600x + 30{,}000$$
$$x^2 - 62{,}500 = 0$$
$$(x-250)(x+250) = 0$$
$$x = 250, \cancel{-250}$$

So, the plane in still air travels at $\boxed{250\text{mph}}$.

120.
Let c = Current rate.
Down river: rate $= 10 + c$ in t hours
Up river: rate $= 10 - c$ in $t+1$ hours.
Using Distance = Rate × Time, we obtain:
Distance down river: $d = (10+c)(t) = 24$ **(1)**
Distance up river: $d = (10-c)(t+1) = 24$ **(2)**
Equating **(1)** and **(2)** yields:

$$(10+c)(t) = (10-c)(t+1)$$
$$10t + ct = 10t + 10 - ct - c$$
$$2ct = 10 - c$$
$$t = \frac{10-c}{2c} \quad \textbf{(3)}$$

Substitute **(3)** into **(1)** to solve for c:

$$(10+c)\left(\frac{10-c}{2c}\right) = 24$$
$$(10+c)(10-c) = 2c(24)$$
$$100 - c^2 = 48c$$
$$-c^2 - 48c + 100 = 0$$
$$c^2 + 48c - 100 = 0$$
$$c = \frac{-48 \pm \sqrt{48^2 + 400}}{2}$$
$$= \frac{-48 \pm \sqrt{2704}}{2} = \frac{-48 \pm 52}{2}$$

The rate must be postive, so $\boxed{c = 2 \text{ mph}}$.

121.
2 distinct real roots of $ax^2 + bx + c = 0$ are: $x_1 = \dfrac{-b+\sqrt{b^2-4ac}}{2a}$ $\quad x_2 = \dfrac{-b-\sqrt{b^2-4ac}}{2a}$

If real roots are negatives of x_1, x_2, then $x_1^* = \dfrac{b-\sqrt{b^2-4ac}}{2a}$ $\quad x_2^* = \dfrac{b+\sqrt{b^2-4ac}}{2a}$

Replace b with $-b$. So, $\boxed{ax^2 - bx + c = 0}$.

122.
$$ax^2+bx+c=0 \;\Rightarrow\; x=\frac{-b\pm\sqrt{b^2-4ac}}{2a}$$

Label the roots as $x_1=\dfrac{-b+\sqrt{b^2-4ac}}{2a}$, $x_2=\dfrac{-b-\sqrt{b^2-4ac}}{2a}$

$$\boxed{x_1^*=\frac{1}{x_1}=\frac{2a}{-b+\sqrt{b^2-4ac}} \qquad x_2^*=\frac{1}{x_2}=\frac{-2a}{b+\sqrt{b^2-4ac}}}$$

Using these roots in a new quadratic equation,

$$\left[x-\frac{2a}{-b+\sqrt{b^2-4ac}}\right]\left[x-\frac{-2a}{b+\sqrt{b^2-4ac}}\right]=0$$

$$\left[x+\frac{2a}{b-\sqrt{b^2-4ac}}\right]\left[x+\frac{2a}{b+\sqrt{b^2-4ac}}\right]=0$$

$$x^2+x\left(\frac{2a}{b+\sqrt{b^2-4ac}}\right)+x\left(\frac{2a}{b-\sqrt{b^2-4ac}}\right)+\left(\frac{2a}{b-\sqrt{b^2-4ac}}\right)\left(\frac{2a}{b+\sqrt{b^2-4ac}}\right)=0$$

$$x^2+\frac{2ax\left(b+\sqrt{b^2-4ac}\right)+2ax\left(b-\sqrt{b^2-4ac}\right)}{\left(b-\sqrt{b^2-4ac}\right)\left(b+\sqrt{b^2-4ac}\right)}+\frac{4a^2}{4ac}=x^2+\frac{4abx+4a^2}{4ac}=0$$

$$x^2+\frac{b}{c}x+\frac{a}{c}=0$$

$$\boxed{cx^2+bx+a=0}$$

123. Let x = speed of small jet (in mph). Then, the speed of the 757-jet = x+100 (mph) Form a right triangle depicting the relative position of the jets after two hours of flight. Using Distance = Rate × time, this triangle will have legs of length $2x$ and $2(x+100)$, and hypotenuse of length 1000 miles. Using the Pythagorean Theorem then yields

$$(2x)^2+(2(x+100))^2=1000^2$$
$$4x^2+4x^2+800x+40{,}000=1{,}000{,}000$$
$$x^2+100x-120{,}000=0$$

$$x=\frac{-100\pm\sqrt{100^2+4(120{,}000)}}{2}=\frac{-100\pm700}{2}=\cancel{-400},\ 300$$

So, the speed of the small jet is 300mph and the speed of the 757-jet is 400mph.

124. Let x = speed of small boat (in mph).
Then, the speed of the large boat = x+10 (mph).
Form a right triangle depicting the relative position of the jets after three hours. Using Distance = Rate × time, this triangle will have legs of length $3x$ and $3(x+10)$, and hypotenuse of length 150 miles. Using the Pythagorean Theorem then yields

$$(3x)^2 + (3(x+10))^2 = 150^2$$
$$9x^2 + 9x^2 + 180x + 900 = 22,500$$
$$x^2 + 10x - 1200 = 0$$
$$(x+40)(x-30) = 0$$
$$x = \cancel{-40}, 30$$

So, $\boxed{\text{the speed of the small boat is 30mph and the speed of the large boat is 40mph}}$.

125. $x^2 - x - 2 = 0$

$(x-2)(x+1) = 0$

$\boxed{x = -1, 2}$

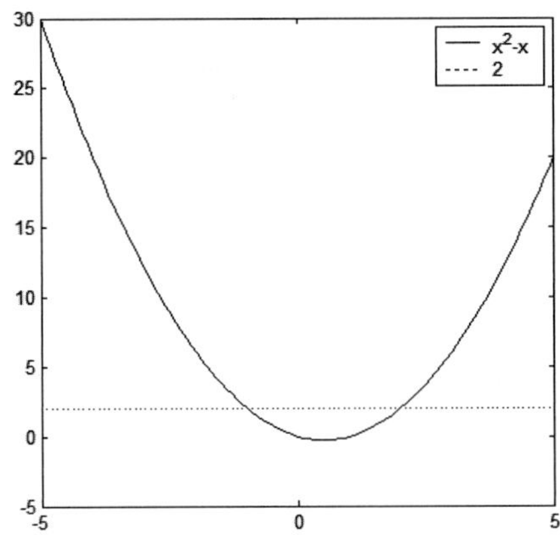

126. $x^2 - 2x + 2 = 0$

$$x = \frac{2 \pm \sqrt{4 - 4 \cdot 2}}{2} = \frac{2 \pm \sqrt{-4}}{2}$$

$\boxed{x = 1 \pm i}$

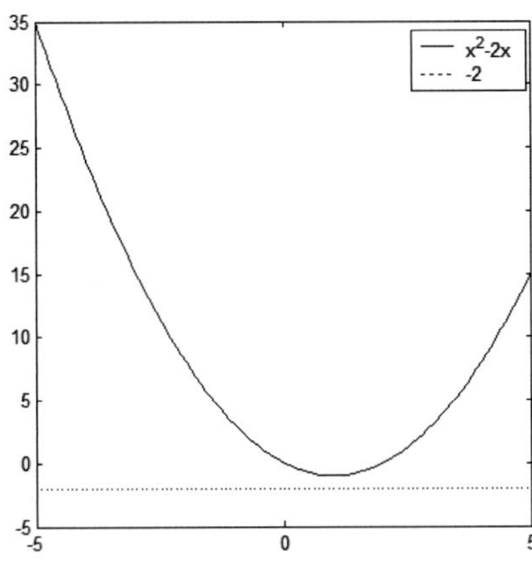

127. (a) Consider $x^2 - 2x - b = 0$. **(1)**
For $b = 8$, **(1)** factors as $(x-4)(x+2) = 0$, so that $x = -2, 4$.
Graphically, we let $y1 = x^2 - 2x,\ y2 = 8$ and look for the intersection points of the graphs:

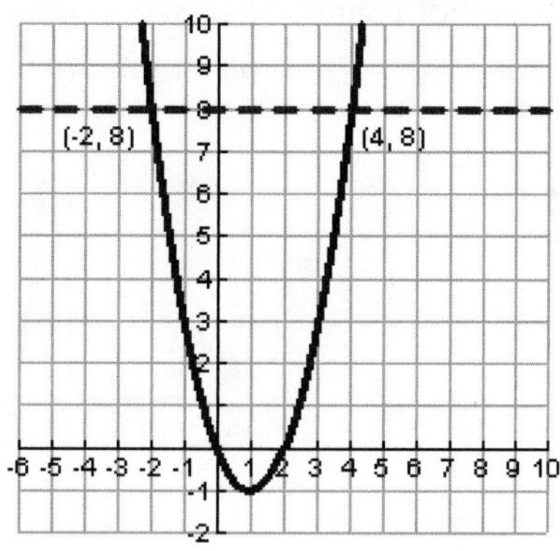

Note that they intersect at precisely the x-values obtained algebraically. So, yes, these values agree with the points of intersections.

(b) We do the same thing now for different values of b.

$\underline{b = -3}$:

$$x^2 - 2x + 3 = 0$$

$$x = \frac{2 \pm \sqrt{4 - 4(3)}}{2} = 1 \pm i\sqrt{2}$$

So, we don't expect the graphs to intersect. Indeed, we have:

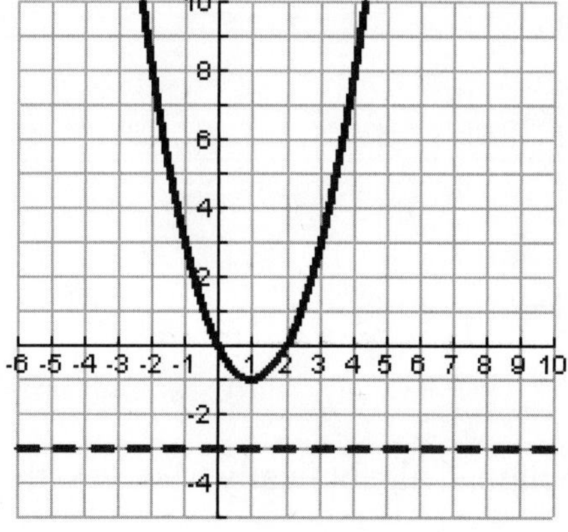

$\underline{b = -1}$:

$$x^2 - 2x + 1 = 0$$

$$(x-1)^2 = 0$$

$$x = 1$$

So, we expect the graphs to intersect once. Indeed, we have:

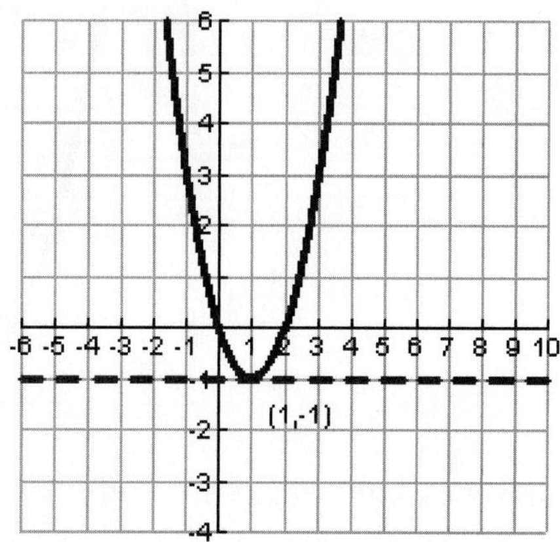

$\underline{b=0}$:

$$x^2-2x=0$$
$$x(x-2)=0$$
$$x=0,2$$

So, we expect the graphs to intersect twice as in part **(a)**. Indeed, we have:

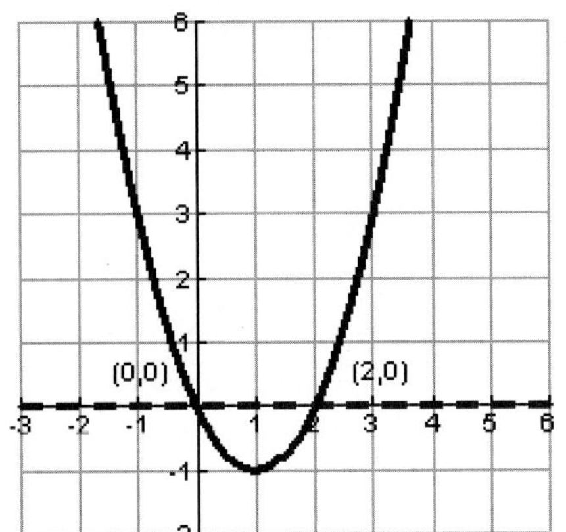

$\underline{b=5}$:

$$x^2-2x-5=0$$
$$x=\frac{2\pm\sqrt{4+4(5)}}{2}=1\pm\sqrt{6}$$

So, we expect the graphs to intersect twice as in part **(a)**. Indeed, we have:

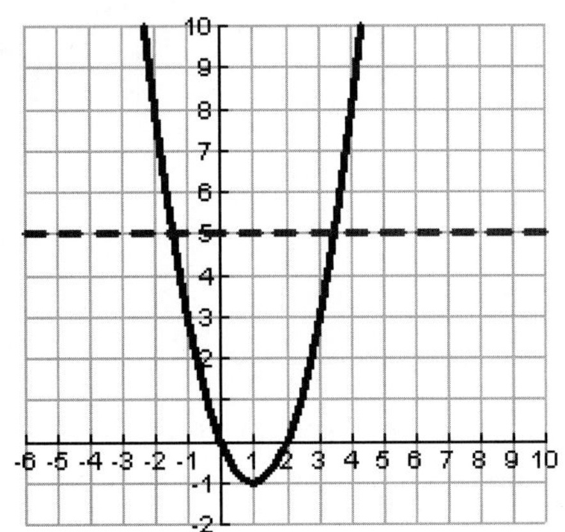

128. **(a)** Consider $x^2+2x-b=0$. **(1)**
For $b=8$, **(1)** factors as $(x+4)(x-2)=0$, so that $x=2,-4$.
Graphically, we let $y1=x^2+2x,\ y2=8$ and look for the intersection points of the graphs:

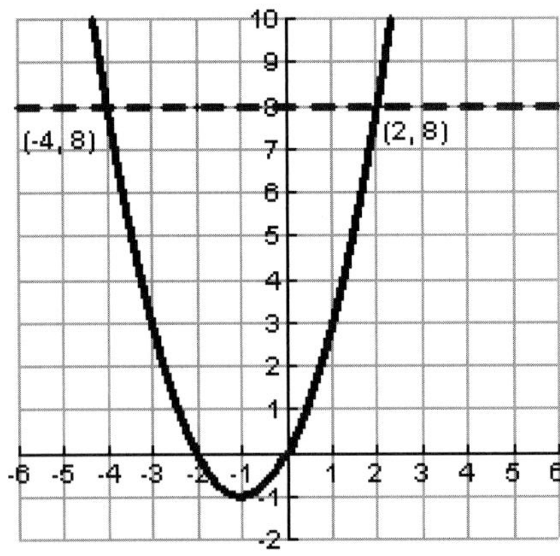

Note that they intersect at precisely the x-values obtained algebraically. So, yes, these values agree with the points of intersections.

(b) We do the same thing now for different values of b.

$\underline{b=-3}$:

$x^2+2x+3=0$

$$x=\frac{-2\pm\sqrt{4-4(3)}}{2}=-1\pm i\sqrt{2}$$

So, we don't expect the graphs to intersect. Indeed, we have:

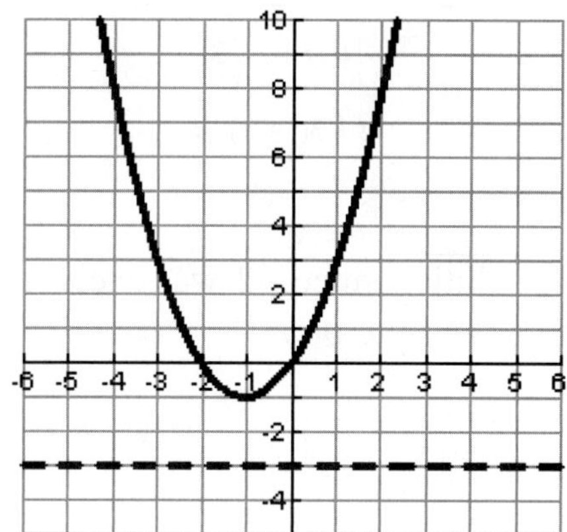

$\underline{b=-1}$:

$$x^2+2x+1=0$$

$$(x+1)^2=0$$

$$x=-1$$

So, we expect the graphs to intersect once. Indeed, we have:

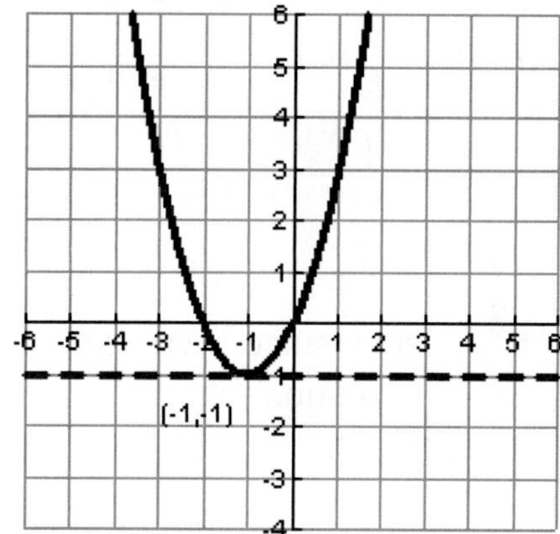

$\underline{b=0}$:

$$x^2+2x=0$$

$$x(x+2)=0$$

$$x=0,-2$$

So, we expect the graphs to intersect twice as in part **(a)**. Indeed, we have:

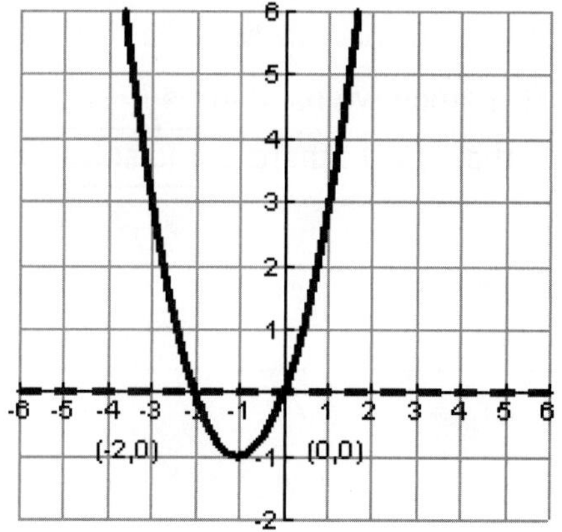

$\underline{b=5}$:

$x^2+2x-5=0$

$$x=\frac{-2\pm\sqrt{4+4(5)}}{2}=-1\pm\sqrt{6}$$

So, we expect the graphs to intersect twice as in part **(a)**. Indeed, we have:

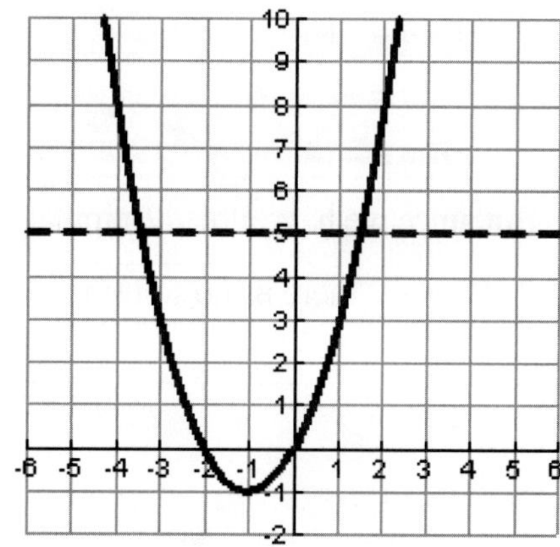

Section 0.3 Solutions

1. $(x-2)\left(\frac{x}{x-2}+5\right)=(x-2)\left(\frac{2}{x-2}\right)$ $\boxed{x\neq 2}$ $x+5(x-2)=2$ $x+5x-10=2$ $6x=12$ $x=2$ $\boxed{\text{No solution}}$ since 2 was excluded from the solution set.	**2.** $(n-5)\left(\frac{n}{n-5}+2\right)=(n-5)\left(\frac{n}{n-5}\right)$ $\boxed{n\neq 5}$ $n+2(n-5)=n$ $3n-10=n$ $2n=10$ $n=5$ $\boxed{\text{No}}$ $\boxed{\text{solution}}$ since 5 was excluded from the solution set.
3. $(p-1)\left(\frac{2p}{p-1}\right)=(p-1)\left(3+\frac{2}{p-1}\right)$ $\boxed{p\neq 1}$ $2p=3(p-1)+2$ $2p=3p-3+2$ $2p=3p-1$ $p=1$ $\boxed{\text{No solution}}$ since 1 was excluded from the solution set.	**4.** $(t+2)\left(\frac{4t}{t+2}\right)=(t+2)\left(3-\frac{8}{t+2}\right)$ $\boxed{t\neq -2}$ $4t=3(t+2)-8$ $4t=3t+6-8$ $t=-2$ $\boxed{\text{No solution}}$ since -2 was excluded from the solution set.
5. $(x+2)\left(\frac{3x}{x+2}-4\right)=(x+2)\left(\frac{2}{x+2}\right)$ $\boxed{x\neq -2}$ $3x-4(x+2)=2$ $-x-8=2$ $\boxed{x=-10}$	**6.** $(2y-1)\left(\frac{5y}{2y-1}-3\right)=(2y-1)\left(\frac{12}{2y-1}\right)$ $\boxed{y\neq \frac{1}{2}}$ $5y-3(2y-1)=12$ $5y-6y+3=12$ $-y+3=12$ $\boxed{y=-9}$
7. $\frac{1}{n}+\frac{1}{n+1}=\frac{-1}{n(n+1)}$ $\boxed{n\neq -1,0}$ LCD is $n(n+1)$. So, $(n+1)+n=-1$ $n+1+n=-1$ $2n=-2$ $n=-1$ But since we have already stipulated that $n\neq -1$, there is $\boxed{\text{no solution.}}$	**8.** $x(x-1)\left(\frac{1}{x}+\frac{1}{x-1}\right)=x(x-1)\left(\frac{1}{x(x-1)}\right)$ First notice that $\boxed{x\neq 0,1}$ $(x-1)+x=1$ $2x=2$ $x=1$ But since we have already stipulated that $x\neq 1$, there is $\boxed{\text{no solution.}}$

9. $\frac{3}{a}-\frac{2}{a+3}=\frac{9}{a(a+3)}$ $\boxed{a\neq 0,-3}$

LCD is $a(a+3)$. So,

$$3(a+3)-2a=9$$
$$3a+9-2a=9$$
$$a=0$$

But since we have already stipulated that $a\neq 0$, there is $\boxed{\text{no solution.}}$

10. $\frac{1}{c-2}+\frac{1}{c}=\frac{2}{c(c-2)}$ $\boxed{c\neq 0,2}$

LCD is $c(c-2)$. So,

$$c+(c-2)=2$$
$$2c-2=2$$
$$2c=4$$
$$c=2$$

But since we have already stipulated that $c\neq 2$, there is $\boxed{\text{no solution.}}$

11.

$$\frac{n-5}{6(n-1)}=\frac{1}{9}-\frac{n-3}{4(n-1)} \quad \boxed{n\neq 1}$$

LCD is $36(n-1)$. So,

$$\frac{(n-5)(36)(n-1)}{6(n-1)}=\frac{36(n-1)}{9}-\frac{(n-3)(36)(n-1)}{4(n-1)}$$
$$6(n-5)=4(n-1)-9(n-3)$$
$$6n-30=4n-4-9n+27$$
$$6n-30=-5n+23$$
$$11n=53$$

So, the final solution is: $\boxed{n=\frac{53}{11}}$

12.

$$\frac{5}{m}+\frac{3}{m-2}=\frac{6}{m(m-2)} \quad \boxed{m\neq 0,2}$$

LCD is $m(m-2)$. So,

$$5(m-2)+3(m)=6$$
$$5m-10+3m=6$$
$$8m-10=6$$
$$8m=16$$
$$m=2$$

Hence, $\boxed{\text{no solution}}$ since we have already stipulated that $m\neq 2$.

13. $\dfrac{2}{5x+1}=\dfrac{1}{2x-1}$ $\boxed{x\neq -\frac{1}{5},\frac{1}{2}}$ $2(2x-1)=1(5x+1)$ $4x-2=5x+1$ $\boxed{x=-3}$	**14.** $\dfrac{3}{4n-1}=\dfrac{2}{2n-5}$ $\boxed{n\neq \frac{1}{4},\frac{5}{2}}$ $3(2n-5)=2(4n-1)$ $6n-15=8n-2$ $-13=2n$ $\boxed{n=-\frac{13}{2}}$
15. $\dfrac{t-1}{1-t}=\dfrac{3}{2}$ $\boxed{t\neq 1}$ $3(1-t)=2(t-1)$ $3-3t=2t-2$ $-5t=-5$ $t=1$ No solution (boxed) since 1 was excluded from the solution set.	**16.** $\dfrac{2-x}{x-2}=\dfrac{3}{4}$ $\boxed{x\neq 2}$ $4(2-x)=3(x-2)$ $8-4x=3x-6$ $14=7x$ $x=2$ No solution (boxed) since 2 is excluded from the solution set.
17. $x+\dfrac{12}{x}=7$ $\boxed{x\neq 0}$ $x^2+12=7x$ $x^2-7x+12=0$ $(x-3)(x-4)=0$ $x-3=0$ or $x-4=0$ $\boxed{x=3 \text{ or } x=4}$	**18.** $x-\dfrac{10}{x}=-3$ $\boxed{x\neq 0}$ $x^2-10=-3x$ $x^2+3x-10=0$ $(x+5)(x-2)=0$ $x+5=0$ or $x-2=0$ $\boxed{x=-5 \text{ or } x=2}$

19. $\dfrac{4(x-2)}{x-3}+\dfrac{3}{x}=\dfrac{-3}{x(x-3)}$ $\boxed{x \neq 0,3}$

$\text{LCD}=x(x-3)$

$4x(x-2)+3(x-3)=-3$

$4x^2-8x+3x-9=-3$

$4x^2-5x-6=0$

$(4x+3)(x-2)=0$

$4x+3=0$ or $x-2=0$

$\boxed{x=-3/4 \text{ or } x=2}$

20. $\dfrac{5}{y+4}=4+\dfrac{3}{y-2}$ $\boxed{y \neq -4,2}$

$\text{LCD}=(y+4)(y-2)$

$5(y-2)=4(y+4)(y-2)+3(y+4)$

$5y-10=4(y^2+2y-8)+3y+12$

$5y-10=4y^2+8y-32+3y+12$

$-4y^2-6y+10=0$

$2y^2+3y-5=0$

$(2y+5)(y-1)=0$

$2y+5=0$ or $y-1=0$

$\boxed{y=-\tfrac{5}{2} \text{ or } y=1}$

21. $\sqrt{u+1}=-4$

$\boxed{\text{no solution}}$

$u+1=16$

$u=15$

Check: $\sqrt{15+1}$

$=\sqrt{16}=4$

22. $-\sqrt{3-2u}=9$

$\sqrt{3-2u}=-9$

$\boxed{\text{no solution}}$

$3-2u=81$

$2u=-78$

$u=-39$

$-\sqrt{3+2\cdot 39}=-9$

23. $\sqrt[3]{5x+2}=3$

$5x+2=3^3=27$

$5x=25$

$\boxed{x=5}$

24. $\sqrt[3]{1-x}=-2$

$1-x=-8$

$\boxed{x=9}$

25. $(4y+1)^{\frac{1}{3}}=-1$

$4y+1=-1$

$4y=-2$

$\boxed{y=-\dfrac{1}{2}}$

26. $(5x-1)^{\frac{1}{3}}=4$

$5x-1=64$

$5x=65$

$\boxed{x=13}$

27. $\sqrt{12+x}=x$

$12+x=x^2$

$x^2-x-12=0$

$(x+3)(x-4)=0$

$x=-3, \boxed{4}$

Check -3:

$\sqrt{12-3}=\sqrt{9}\neq -3$

Check 4:

$\sqrt{12+4}=\sqrt{16}=4$

28. $x=\sqrt{56-x}$

$x^2=56-x$

$x^2+x-56=0$

$(x+8)(x-7)=0$

$x=-8, \boxed{7}$

Check -8:

$\sqrt{56+8}=\sqrt{64}\neq -8$

Check 7:

$\sqrt{56-7}=\sqrt{49}=7$

29. $y=5\sqrt{y}$ $y^2=25y$ $y^2-25y=0$ $y(y-25)=0$ $\boxed{y=0,25}$ Check 0: $0=5\sqrt{0}$ Check 25: $25=5\sqrt{25}$	**30.** $\sqrt{y}=\frac{y}{4}$ $y=\frac{y^2}{16}$ $y^2-16y=0$ $y(y-16)=0$ $\boxed{y=0,16}$ Check 0: $\sqrt{0}=0/4$ Check 16: $\sqrt{16}=16/4$	**31.** $s=3\sqrt{s-2}$ $s^2=9(s-2)$ $s^2=9s-18$ $s^2-9s+18=0$ $(s-3)(s-6)=0$ $\boxed{s=3,6}$ Check 3: $3=3\sqrt{3-2}=3\sqrt{1}$ Check 6: $6=3\sqrt{6-2}=3\sqrt{4}$	**32.** $-2s=\sqrt{3-s}$ $4s^2=3-s$ $4s^2+s-3=0$ $(s+1)(4s-3)=0$ $s=\boxed{-1},\frac{3}{4}$ Check -1: $(-2)(-1)=\sqrt{3+1}$ $2=2$ Check 3/4: $-2\left(\frac{3}{4}\right)=\sqrt{3-\frac{3}{4}}$ $-\frac{3}{2}=\sqrt{\frac{9}{4}}\neq\frac{3}{2}$

33. $\sqrt{2x+6}=x+3$ $2x+6=(x+3)^2$ $x^2+4x+3=0$ $(x+3)(x+1)=0$ $x=\boxed{-3,-1}$ Check -3: $\sqrt{2(-3)+6}=-3+3$ $\sqrt{0}=0$ Check -1: $\sqrt{2(-1)+6}=-1+3$ $\sqrt{4}=2$	**34.** $\sqrt{8-2x}=2x-2$ $8-2x=4x^2-8x+4$ $4x^2-6x-4=0$ $2x^2-3x-2=0$ $(2x+1)(x-2)=0$ $x=\frac{-1}{2},\boxed{2}$ Check $\frac{-1}{2}$: $\sqrt{8-2\left(\frac{-1}{2}\right)}=2\left(\frac{-1}{2}\right)-2$ $\sqrt{9}\neq-3$ Check 2: $\sqrt{8-4}=2(2)-2$ $\sqrt{4}=2$	**35.** $\sqrt{1-3x}=x+1$ $1-3x=x^2+2x+1$ $x^2+5x=0$ $x(x+5)=0$ $x=-5,\boxed{0}$ Check -5: $\sqrt{1+15}\neq-4$ Check 0: $\sqrt{1}=1$
36. $\sqrt{2-x}=x-2$ $2-x=(x-2)^2$ $2-x=x^2-4x+4$ $x^2-3x+2=0$ $(x-2)(x-1)=0$ $x=\not{1},\boxed{2}$	**37.** $\sqrt{x^2-4}=x-1$ $x^2-4=(x-1)^2$ $x^2-4=x^2-2x+1$ $2x=5$ $\boxed{x=\frac{5}{2}}$	**38.** $\sqrt{25-x^2}=x+1$ $25-x^2=(x+1)^2$ $25-x^2=x^2+2x+1$ $2x^2+2x-24=0$ $x^2+x-12=0$ $(x+4)(x-3)=0$ $x=\not{-4},\boxed{3}$

39.
$$\sqrt{x^2-2x-5}=x+1$$
$$x^2-2x-5=(x+1)^2$$
$$x^2-2x-5=x^2+2x+1$$
$$-6=4x$$
$$\cancel{-\tfrac{3}{2}}=x$$
$\boxed{\text{No solution}}$.

40.
$$\sqrt{2x^2-8x+1}=x-3$$
$$2x^2-8x+1=(x-3)^2$$
$$2x^2-8x+1=x^2-6x+9$$
$$x^2-2x-8=0$$
$$(x-4)(x+2)=0$$
$$\boxed{x=\cancel{-2},4}$$

41.
$$\sqrt{2x-1}=1+\sqrt{x-1}$$
$$2x-1=1+2\sqrt{x-1}+x-1$$
$$x-1=2\sqrt{x-1}$$
$$x^2-2x+1=4(x-1)$$
$$x^2-2x+1=4x-4$$
$$x^2-6x+5=0$$
$$(x-5)(x-1)=0$$
$$\boxed{x=1,5}$$

42.
$$\sqrt{8-x}=2+\sqrt{2x+3}$$
$$8-x=4+4\sqrt{2x+3}+2x+3$$
$$-3x+1=4\sqrt{2x+3}$$
$$9x^2-6x+1=16(2x+3)$$
$$9x^2-6x+1=32x+48$$
$$9x^2-38x-47=0$$
$$(9x-47)(x+1)=0$$
$$x=\tfrac{47}{9},\boxed{-1}$$

43.
$$\sqrt{3x-5}=7-\sqrt{x+2}$$
$$3x-5=49-14\sqrt{x+2}+x+2$$
$$2x-56=-14\sqrt{x+2}$$
$$x-28=-7\sqrt{x+2}$$
$$x^2-56x+784=49(x+2)$$
$$x^2-56x+784=49x+98$$
$$x^2-105x+686=0$$
$$(x-98)(x-7)=0$$
$$x=\boxed{7},\cancel{98}$$

44.
$$\sqrt{x+5}=1+\sqrt{x-2}$$
$$x+5=1+2\sqrt{x-2}+x-2$$
$$6=2\sqrt{x-2}$$
$$9=x-2$$
$$\boxed{x=11}$$

45.
$$\sqrt{2+\sqrt{x}}=\sqrt{x}$$
$$2+\sqrt{x}=x$$
$$\sqrt{x}=x-2$$
$$x=x^2-4x+4$$
$$x^2-5x+4=0$$
$$(x-4)(x-1)=0$$
$$x=\cancel{1},\boxed{4}$$

46.
$$\sqrt{2-\sqrt{x}}=\sqrt{x}$$
$$2-\sqrt{x}=x$$
$$\sqrt{x}=2-x$$
$$x=4-4x+x^2$$
$$x^2-5x+4=0$$
$$(x-1)(x-4)=0$$
$$x=\boxed{1},\cancel{4}$$

47. Let $u=x^{1/3}$
$$u^2+2u=0$$
$$u(u+2)=0$$
$$u=-2,0$$
$$x^{1/3}=0\rightarrow\boxed{x=0}$$
$$x^{1/3}=-2\rightarrow\boxed{x=-8}$$

48. Let $u=x^{1/4}$
$$u^2-2u=0$$
$$u(u-2)=0$$
$$u=0,2$$
$$x^{1/4}=0\rightarrow\boxed{x=0}$$
$$x^{1/4}=2\rightarrow\boxed{x=16}$$

49. Let $u=x^2$
$$u^2-3u+2=0$$
$$(u-1)(u-2)=0$$
$$u=1,2$$
$$x^2=1\rightarrow\boxed{x=\pm1}$$
$$x^2=2\rightarrow\boxed{x=\pm\sqrt{2}}$$

50. Let $u=x^2$
$$u^2-8u+16=0$$
$$(u-4)^2=0$$
$$u=4$$
$$x^2=4$$
$$\boxed{x=\pm2}$$

51. Let $u = x^2$

$$2u^2 + 7u + 6 = 0$$
$$(2u+3)(u+2) = 0$$

$u = -3/2$, $x^2 = -3/2$, $x = \pm i\sqrt{3/2}$, $\boxed{x = \dfrac{\pm i\sqrt{6}}{2}}$

$u = -2$, $x^2 = -2$, $x = \pm i\sqrt{2}$, $\boxed{x = \pm i\sqrt{2}}$

52. Let $u = x^4$

$$u^2 - 17u + 16 = 0$$
$$(u-16)(u-1) = 0$$

$u = 1$, $x^4 = 1$, $x^2 = \pm 1$; if $x^2 = 1$, $\boxed{x = \pm 1}$; if $x^2 = -1$, $\boxed{x = \pm i}$

$u = 16$, $x^4 = 16$, $x^2 = \pm 4$; if $x^2 = 4$, $\boxed{x = \pm 2}$; if $x^2 = -4$, $\boxed{x = \pm 2i}$

53. Let $u = t - 1$

$$4u^2 - 9u + 2 = 0$$
$$(4u-1)(u-2) = 0$$

$u = 1/4$, $t - 1 = 1/4$, $\boxed{t = 5/4}$

$u = 2$, $t - 1 = 2$, $\boxed{t = 3}$

54. Let $u = 1 - y$

$$2u^2 + 5u - 12 = 0$$
$$(2u-3)(u+4) = 0$$

$u = 3/2$, $1 - y = 3/2$, $\boxed{y = -1/2}$

$u = -4$, $1 - y = -4$, $\boxed{y = 5}$

55. Let $u = x^{-4}$

$$u^2 - 17u + 16 = 0$$
$$(u-16)(u-1) = 0$$

$u = 1$, $x^{-4} = 1$, $x^2 = \pm 1$, $\boxed{x = \pm 1, \pm i}$

$u = 16$, $x^{-4} = 16$, $x^2 = \pm 1/4$, $\boxed{x = \pm\frac{1}{2}, \pm\frac{1}{2}i}$

56. Let $x = u^{-1}$

$$2x^2 + 5x - 12 = 0$$
$$(2x-3)(x+4) = 0$$

$x = 3/2$, $u^{-1} = 3/2$, $\boxed{u = 2/3}$

$x = -4$, $u^{-1} = -4$, $\boxed{u = -1/4}$

57. Let $u = y^{-1}$

$$3u^2 + u - 4 = 0$$
$$(3u+4)(u-1) = 0$$

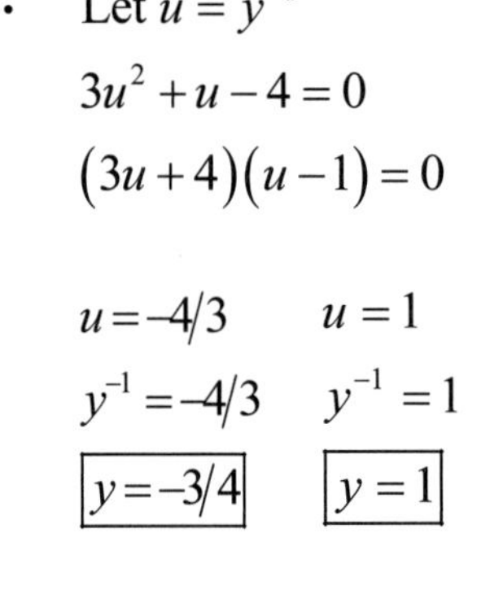

$u = -4/3$, $y^{-1} = -4/3$, $\boxed{y = -3/4}$

$u = 1$, $y^{-1} = 1$, $\boxed{y = 1}$

58. Let $u = a^{-1}$

$$5u^2 + 11u + 2 = 0$$
$$(5u+1)(u+2) = 0$$

$u = -1/5 \Rightarrow a^{-1} = -1/5 \Rightarrow \boxed{a = -5}$

$u = -2 \Rightarrow a^{-1} = -2 \Rightarrow \boxed{a = -1/2}$

59. Let $u = z^{1/5}$

$$u^2 - 2u + 1 = 0$$
$$(u-1)^2 = 0$$

$u = 1 \Rightarrow z^{1/5} = 1 \Rightarrow \boxed{z = 1}$

60.

Let $u = x^{1/4}$

$2u^2 + u - 1 = 0 \Rightarrow (2u-1)(u+1) = 0 \Rightarrow u = -1, \frac{1}{2}$

$x^{1/4} = \frac{1}{2}$ or ~~$x^{1/4} = -1$~~ $\Rightarrow \boxed{x = \frac{1}{16}}$

61. Let $u = t^{-1/3}$

$6u^2 - u - 1 = 0$

$(3u+1)(2u-1) = 0$

$u = -1/3 \quad u = 1/2$

$t^{-1/3} = -1/3 \quad t^{-1/3} = 1/2$

$t = (-1/3)^{-3} \quad t = (1/2)^{-3}$

$\boxed{t = -27} \quad \boxed{t = 8}$

62. $u = t^{-1/3}$

$u^2 - u - 6 = 0$

$(u-3)(u+2) = 0$

$u = 3 \quad u = -2$

$t^{-1/3} = 3 \quad t^{-1/3} = -2$

$t = 3^{-3} \quad t = (-2)^{-3}$

$\boxed{t = 1/27} \quad \boxed{t = -1/8}$

63. $3 = \frac{1}{(x+1)^2} + \frac{2}{(x+1)}$ $\boxed{x \neq -1}$

$3(x+1)^2 = 1 + 2(x+1)$

$3(x+1)^2 - 2(x+1) - 1 = 0$

Let $u = x+1$

$3u^2 - 2u - 1 = 0$

$(3u+1)(u-1) = 0$

$u = -1/3 \quad u = 1$

$x + 1 = -1/3 \quad x + 1 = 1$

$\boxed{x = -4/3} \quad \boxed{x = 0}$

64. $\frac{1}{(x+1)^2} + \frac{4}{x+1} + 4 = 0$

$\boxed{x \neq -1}$

$\text{LCD} = (x+1)^2$

$1 + 4(x+1) + 4(x+1)^2 = 0$

Let $u = x+1$

$4u^2 + 4u + 1 = 0$

$(2u+1)^2 = 0$

$u = -1/2$

$x + 1 = -1/2$

$\boxed{u = -3/2}$

65. Let $x = u^{2/3}$

$x^2 - 5x + 4 = 0$

$(x-4)(x-1) = 0$

$x = 4 \quad x = 1$

$u^{2/3} = 4 \quad u^{2/3} = 1$

$u = \pm 4^{3/2} \quad u = \pm 1^{3/2}$

$\boxed{u = \pm 8} \quad \boxed{u = \pm 1}$

66. Let $x = u^{2/3}$

$x^2 + 5x + 4 = 0$

$(x+4)(x+1) = 0$

$x = -4 \quad x = -1$

$u^{2/3} = -4 \quad u^{2/3} = -1$

$u = (-4)^{3/2} \quad u = (-1)^{3/2}$

$u = \left[(-4)^{1/2}\right]^3 \quad u = \left[(-1)^{1/2}\right]^3$

$u = [2i]^3 \quad u = (\pm i)^3$

$\boxed{u = \pm 8i} \quad \boxed{u = \pm i}$

67.

$$x^3 - x^2 - 12x = 0$$
$$x\left(x^2 - x - 12\right) = 0$$
$$x(x-4)(x+3) = 0$$
$$\boxed{x = 0, -3, 4}$$

68.

$$2y^3 - 11y^2 = 12y = 0$$
$$y\left(2y^2 - 11y + 12\right) = 0$$
$$y(2y-3)(y-4) = 0$$
$$\boxed{y = 0, 4, 3/2}$$

69.

$$4p^3 - 9p = 0$$
$$p\left(4p^2 - 9\right) = 0$$
$$p(2p-3)(2p+3) = 0$$
$$\boxed{p = 0, \pm 3/2}$$

70.

$$25x^3 = 4x$$
$$25x^3 - 4x = 0$$
$$x\left(25x^2 - 4\right) = 0$$
$$x(5x-2)(5x+2) = 0$$
$$\boxed{x = 0, \pm 2/5}$$

71.

$$u^5 - 16u = 0$$
$$u\left(u^4 - 16\right) = 0$$
$$u\left(u^2 - 4\right)\left(u^2 + 4\right) = 0$$
$$u(u-2)(u+2)(u-2i)(u+2i) = 0$$
$$\boxed{u = 0, \pm 2, \pm 2i}$$

72.

$$t^5 - 81t = 0$$
$$t\left(t^4 - 81\right) = 0$$
$$t\left(t^2 - 9\right)\left(t^2 + 9\right) = 0$$
$$t(t-3)(t+3)(t-3i)(t+3i) = 0$$
$$\boxed{t = 0, \pm 3, \pm 3i}$$

73.

$$x^3 - 5x^2 - 9x + 45 = 0$$
$$\left(x^3 - 5x^2\right) - \left(9x - 45\right) = 0$$
$$x^2(x-5) - 9(x-5) = 0$$
$$\left(x^2 - 9\right)(x-5) = 0$$
$$(x-3)(x+3)(x-5) = 0$$
$$\boxed{x = \pm 3, 5}$$

74.

$$2p^3 - 3p^2 - 8p + 12 = 0$$
$$\left(2p^3 - 3p^2\right) - \left(8p - 12\right) = 0$$
$$p^2(2p-3) - 4(2p-3) = 0$$
$$\left(p^2 - 4\right)(2p-3) = 0$$
$$(p-2)(p+2)(2p-3) = 0$$
$$\boxed{p = \pm 2, 3/2}$$

<table>
<tr>
<td>75
$y(y-5)^3-14(y-5)^2=0$
$(y-5)^2\left[y(y-5)-14\right]=0$
$(y-5)^2\left(y^2-5y-14\right)=0$
$(y-5)^2(y-7)(y+2)=0$
$\boxed{y=-2,5,7}$</td>
<td>76.
$v(v+3)^3-40(v+3)^2=0$
$(v+3)^2\left[v(v+3)-40\right]=0$
$(v+3)^2\left(v^2+3v-40\right)=0$
$(v+3)^2(v-5)(v+8)=0$
$\boxed{v=-8,-3,5}$</td>
</tr>
<tr>
<td>77.
$x^{9/4}-2x^{5/4}-3x^{1/4}=0$
$x^{1/4}\left[x^2-2x-3\right]=0$
$x^{1/4}(x-3)(x+1)=0$
$\boxed{x=0,3,\cancel{-1}}$</td>
<td>78.
$u^{7/3}+u^{4/3}-20u^{1/3}=0$
$u^{1/3}\left[u^2+u-20\right]=0$
$u^{1/3}(u+5)(u-4)=0$
$\boxed{u=-5,0,4}$</td>
</tr>
<tr>
<td>79.
$t^{5/3}-25t^{-1/3}=0$
$t^{-1/3}\left[t^2-25\right]=0$
$t^{-1/3}(t-5)(t+5)=0$
$\boxed{t=\pm 5}$
(Note: $t^{-1/3}=0$ has no solution.)</td>
<td>80.
$4x^{9/5}-9x^{-1/5}=0$
$x^{-1/5}\left[4x^2-9\right]=0$
$x^{-1/5}(2x-3)(2x+3)=0$
$\boxed{x=\pm 3/2}$
(Note: $x^{-1/5}=0$ has no solution.)</td>
</tr>
<tr>
<td>81.
$y^{3/2}-5y^{1/2}+6y^{-1/2}=0$
$y^{-1/2}\left[y^2-5y+6\right]=0$
$y^{-1/2}(y-3)(y-2)=0$
$\boxed{y=2,3}$
(Note: $y^{-1/2}=0$ has no solution.)</td>
<td>82.
$4p^{5/3}-5p^{2/3}-6p^{-1/3}=0$
$p^{-1/3}\left(4p^2-5p-6\right)=0$
$p^{-1/3}(4p+3)(p-2)=0$
$\boxed{p=-3/4,2}$
(Note: $p^{-1/3}=0$ has no solution.)</td>
</tr>
<tr>
<td>83. $p-7=3$ $p-7=-3$
$\boxed{p=10}$ $\boxed{p=4}$</td>
<td>84. $p+7=3$ $p+7=-3$
$\boxed{p=-4}$ $\boxed{p=-10}$</td>
</tr>
<tr>
<td>85. $4-y=-1$ $4-y=1$
$\boxed{y=5}$ $\boxed{y=3}$</td>
<td>86. $2-y=-11$ $2-y=11$
$\boxed{y=13}$ $\boxed{y=-9}$</td>
</tr>
</table>

87.	$3t-9=3$ $3t=12$ $\boxed{t=4}$	$3t-9=-3$ $3t=6$ $\boxed{t=2}$
88.	$4t+2=2$ $4t=0$ $\boxed{t=0}$	$4t+2=-2$ $4t=-4$ $\boxed{t=-1}$
89.	$7-2x=-9$ $2x=16$ $\boxed{x=8}$	$7-2x=9$ $2x=-2$ $\boxed{x=-1}$
90.	$6-3y=12$ $-3y=6$ $\boxed{y=-2}$	$6-3y=-12$ $-3y=-18$ $\boxed{y=6}$
91.	$1-3y=1$ $-3y=0$ $\boxed{y=0}$	$1-3y=-1$ $-3y=-2$ $\boxed{y=2/3}$
92.	$5-x=2$ $-x=-3$ $\boxed{x=3}$	$5-x=-2$ $-x=-7$ $\boxed{x=7}$
93.	$\frac{2}{3}x-\frac{4}{7}=-\frac{5}{3}$ LCD = 21 $14x-12=-35$ $14x=-23$ $\boxed{x=-23/14}$	$\frac{2}{3}x-\frac{4}{7}=\frac{5}{3}$ LCD = 21 $14x-12=35$ $14x=47$ $\boxed{x=47/14}$
94.	$\frac{1}{2}x+\frac{3}{4}=-\frac{1}{16}$ $8x+12=-1$ $8x=-13$ $\boxed{x=-13/8}$	$\frac{1}{2}x+\frac{3}{4}=\frac{1}{16}$ $8x+12=1$ $8x=-11$ $\boxed{x=-11/8}$
95.	$\|x-5\|=8$ $x-5=8$ $\boxed{x=13}$	$x-5=-8$ $\boxed{x=-3}$
96.	$\|x+3\|=11$ $x+3=11$ $\boxed{x=8}$	$x+3=-11$ $\boxed{x=-14}$
97.	$2\|p+3\|=20$ $\|p+3\|=10$ $p+3=10$ $\boxed{p=7}$	$p+3=-10$ $\boxed{p=-13}$
98.	$-3\|p-4\|=-6$ $\|p-4\|=2$ $p-4=2$ $\boxed{p=6}$	$p-4=-2$ $\boxed{p=2}$
99.	$5\|y-2\|-10=4\|y-2\|-3$ $\|y-2\|=7$ $y-2=7$ $\boxed{y=9}$	$y-2=-7$ $\boxed{y=-5}$
100.	$3-\|y+9\|=11-3\|y+9\|$ $2\|y+9\|=8$ $\|y+9\|=4$ $y+9=4$ $\boxed{y=-5}$	$y+9=-4$ $\boxed{y=-13}$

101. $4 - x^2 = -1$ $\quad$ $4 - x^2 = 1$ $x^2 = 5$ $\quad$ $x^2 = 3$ $\boxed{x = \pm\sqrt{5}}$ $\quad$ $\boxed{x = \pm\sqrt{3}}$	**102.** $7 - x^2 = -3$ $\quad$ $7 - x^2 = 3$ $x^2 = 10$ $\quad$ $x^2 = 4$ $\boxed{x = \pm\sqrt{10}}$ $\quad$ $\boxed{x = \pm 2}$
103. $x^2 + 1 = -5$ $\quad$ $x^2 + 1 = 5$ $x^2 = -6$ $\quad$ $x^2 = 4$ $x = \pm i\sqrt{6}$ $\quad$ $\boxed{x = \pm 2}$	**104.** $x^2 - 1 = -5$ $\quad$ $x^2 - 1 = 5$ $x^2 = -4$ $\quad$ $x^2 = 6$ $x = \pm 2i$ $\quad$ $\boxed{x = \pm\sqrt{6}}$
105. $\frac{1}{f} = \frac{1}{d_0} + \frac{1}{d_i}$ $f = 3,\ d_i = 5$ $\frac{1}{3} = \frac{1}{d_0} + \frac{1}{5}$ $\text{LCD} = 15d_0$ $5d_0 = 15 + 3d_0$ $2d_0 = 15$ Object is $\boxed{d_0 = 7.5}$ cm from lens.	**106.** $\frac{1}{f} = \frac{1}{s_0} + \frac{1}{s_i}$ $f = 8,\ s_i = 2$ $\frac{1}{8} = \frac{1}{s_0} + \frac{1}{2}$ $\text{LCD} = 8s_0$ $s_0 = 8 + 4s_0$ $-3s_0 = 8$ $s_0 = \frac{-8}{3} \approx -2.67$ Object is $\boxed{\text{2.67 cm behind lens}}$.
107. $\frac{1}{f} = \frac{1}{s_0} + \frac{1}{s_i}$ $f = 2,\ s_i = \frac{1}{2}s_0$ $\frac{1}{2} = \frac{1}{s_0} + \frac{1}{\frac{1}{2}s_0}$ Since $\frac{1}{\frac{1}{2}s_0} = \frac{2}{s_0}$, $\frac{1}{2} = \frac{1}{s_0} + \frac{2}{s_0} = \frac{3}{s_0} \Rightarrow s_0 = 6$ $\boxed{\begin{array}{l}\text{Object distance} = 6\,\text{cm}\\ \text{Image distance} = 3\text{cm}\end{array}}$	**108.** $\frac{1}{f} = \frac{1}{s_0} + \frac{1}{s_i}$ $f = 8,\ s_i = \frac{1}{2}s_0$ $\frac{1}{8} = \frac{1}{s_0} + \frac{1}{\left(\frac{1}{2}s_0\right)}$ $\frac{1}{8} = \frac{1}{s_0} + \frac{2}{s_0}$ $\frac{1}{8} = \frac{3}{s_0}$ $\boxed{s_0 = 24\text{ cm}}$

109.

$$T=\frac{\sqrt{d}}{4}+\frac{d}{1100},\ T=3$$

$$3=\frac{\sqrt{d}}{4}+\frac{d}{1100}$$

$$\text{LCD}=1100$$

$$3300=275\sqrt{d}+d$$

$$d+275\sqrt{d}-3300=0$$

$$\text{Let } u=\sqrt{d}$$

$$u^2+275u-3300=0$$

$$u=\frac{-275\pm\sqrt{275^2+4\cdot1\cdot3300}}{2(1)}$$

$$u=-286.5, 11.5$$

$$\sqrt{d}=11.5$$

$$\boxed{d=132\text{ ft}}$$

110.

$$\frac{\sqrt{d}}{4}=3\ \Rightarrow\ \sqrt{d}=12\ \Rightarrow\ \boxed{d=144\text{ feet}}$$

111.

$$1=2\pi\sqrt{\frac{L}{9.8}}$$

$$\left(\frac{1}{2\pi}\right)^2=\frac{L}{9.8}$$

$$0.24824\text{ m}\approx\frac{9.8}{4\pi^2}=L$$

Convert to centimeters:

$$\frac{0.24824\ \cancel{\text{m}}\ \big|\ 100\text{ cm}}{\big|\ 1\ \cancel{\text{m}}}\approx\boxed{25\text{ cm}}$$

112.

$$1=2\pi\sqrt{\frac{L}{32}}$$

$$\left(\frac{1}{2\pi}\right)^2=\frac{L}{32}$$

$$0.81057\text{ ft}\approx\frac{32}{4\pi^2}=L$$

Convert to inches:

$$\frac{0.81057\ \cancel{\text{ft}}\ \big|\ 12\text{ in}}{\big|\ 1\ \cancel{\text{ft}}}\approx\boxed{10\text{ in}}$$

113.

$$18 = 30\sqrt{1-\frac{v^2}{c^2}}$$

$$\frac{3}{5} = \frac{18}{30} = \sqrt{1-\frac{v^2}{c^2}}$$

$$\left(\frac{3}{5}\right)^2 = 1-\frac{v^2}{c^2}$$

$$\frac{16}{25} = \frac{v^2}{c^2}$$

$$v^2 = \frac{16}{25}c^2$$

$$v = \frac{4}{5}c$$

So, $\boxed{\text{80\% of the speed of light}}$.

114.

$$5 = 30\sqrt{1-\frac{v^2}{c^2}}$$

$$\frac{1}{6} = \frac{5}{30} = \sqrt{1-\frac{v^2}{c^2}}$$

$$\left(\frac{1}{6}\right)^2 = 1-\frac{v^2}{c^2}$$

$$\frac{35}{36} = \frac{v^2}{c^2}$$

$$v^2 = \frac{35}{36}c^2$$

$$v = \frac{\sqrt{35}}{6}c$$

So, $\boxed{\text{about 98.6\% of the speed of light}}$.

115. $t = 5$ is extraneous; there is no solution.

116. $x = -1$ is extraneous. $\boxed{x = 2}$

117. Cannot cross multiply- must multiply by LCD first. The correct answer is $p = 6/5$.

118. Should have eliminated $x = 0,\ x = 1$ from the domain first.

119. False $\boxed{x \neq 0}$

120. False $\boxed{x \neq 1, -2}$

121.

$$x \cdot \left(\frac{a}{x} - \frac{b}{x}\right) = x \cdot c \quad \boxed{x \neq 0,\ c \neq 0} \quad \Rightarrow \quad a - b = cx \quad \Rightarrow \quad \boxed{x = \frac{a-b}{c}}$$

122.

$$\frac{1}{y-a} + \frac{1}{y+a} = \frac{2}{y-1} \quad \boxed{y \neq -a, a, 1}$$

LCD is $(y-a)(y+a)(y-1)$. So,

$$(y+a)(y-1) + (y-a)(y-1) = 2(y-a)(y+a)$$

$$y^2 - y + ay - a + y^2 - y - ay + a = 2\left(y^2 + ay - ay - a^2\right)$$

$$y^2 - y + \cancel{ay} \cancel{-a} + y^2 - y \cancel{-ay} \cancel{+a} = 2y^2 \cancel{+2ay} \cancel{-2ay} - 2a^2$$

$$-2y = -2a^2$$

$$-y = -a^2$$

$$y = a^2 \quad \boxed{y \geq 0}$$

123.

$$\sqrt{x+6}+\sqrt{11+x}=5\sqrt{3+x}$$
$$(x+6)+2\sqrt{x+6}\sqrt{11+x}+(11+x)=25(3+x)$$
$$2x+17+2\sqrt{x+6}\sqrt{11+x}=75+25x$$
$$2\sqrt{x+6}\sqrt{11+x}=58+23x$$
$$4(x+6)(11+x)=529x^2+2668x+3364$$
$$4\left(x^2+17x+66\right)=529x^2+2668x+3364$$
$$4x^2+68x+264=529x^2+2668x+3364$$
$$525x^2+2600x+3100=0$$
$$21x^2+104x+124=0$$
$$(21x+62)(x+2)=0$$
$$\cancel{x=\frac{-62}{21}},\ \boxed{x=-2}$$

124.

Factor out $x^{1/3}$

$$x^{1/3}\left(3x^{1/4}-x^{1/2}-2\right)=0$$

Let $u=x^{1/4}$.

$$3u-u^2-2=0$$
$$u^2-3u+2=0$$
$$(u-2)(u-1)=0$$

$$u=2 \qquad u=1 \qquad x^{1/3}=0$$
$$x^{1/4}=2 \qquad x^{1/4}=1 \qquad x=0$$
$$\boxed{x=16} \qquad \boxed{x=1}$$

125.

$$y=\frac{a}{1+\frac{b}{x}+c} \qquad \boxed{x\neq 0,\ -\frac{b}{c+1}}$$
$$y=\frac{a}{\frac{x+b+cx}{x}}$$
$$y=\frac{ax}{b+x(c+1)}$$
$$y\left(b+x(c+1)\right)=ax$$
$$yb+xy(c+1)-ax=0$$
$$x\left[y(c+1)-a\right]=-yb$$
$$\boxed{x=\frac{by}{a-y-cy}}$$

126.

$$\frac{t+\frac{1}{t}}{\frac{1}{t}-1}=1 \qquad \boxed{t\neq 0,1}$$
$$\cancel{\frac{1}{t}}-1=t+\cancel{\frac{1}{t}} \quad \Rightarrow \quad \boxed{t=-1}$$

127.

$$\sqrt{x-3} = 4 - \sqrt{x+2}$$

$$x - 3 = 16 - 8\sqrt{x+2} + x + 2$$

$$-21 = -8\sqrt{x+2}$$

$$441 = 64(x+2) = 64x + 128$$

$$313 = 64x$$

$$\boxed{x = \frac{313}{64} \cong 4.891}$$

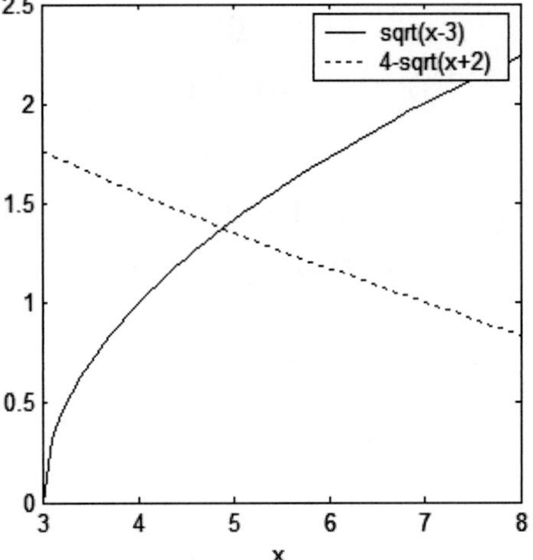

128.

$$2\sqrt{x+1} = 1 + \sqrt{3-x}$$

$$4(x+1) = 1 + 2\sqrt{3-x} + 3 - x$$

$$4x + 4 = 4 - x + 2\sqrt{3-x}$$

$$5x = 2\sqrt{3-x}$$

$$25x^2 = 4(3-x) = 12 - 4x$$

$$25x^2 + 4x - 12 = 0$$

$$x = \frac{-4 \pm \sqrt{4^2 - 4(25)(-12)}}{2(25)}$$

$$x \cong \frac{-4 \pm 34.9}{50} \cong \cancel{-0.778}, \boxed{0.62}$$

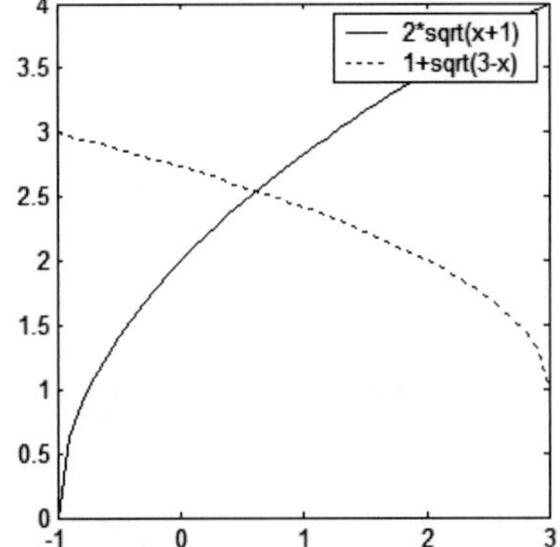

129.

$$x^{1/2} = -4x^{1/4} + 21$$

$$x^{1/2} + 4x^{1/4} - 21 = 0$$

Let $u = x^{1/4}$ to obtain

$$u^2 + 4u - 21 = 0$$

$$(u+7)(u-3) = 0$$

$$u = -7, 3$$

$x^{1/4} = -7$ $\qquad$ $x^{1/4} = 3$

$\boxed{\text{no solution}}$ $\qquad$ $\boxed{x = 81}$

Graphically, let:

$y1 = x^{1/2}, \ y2 = -4x^{1/4} + 21.$

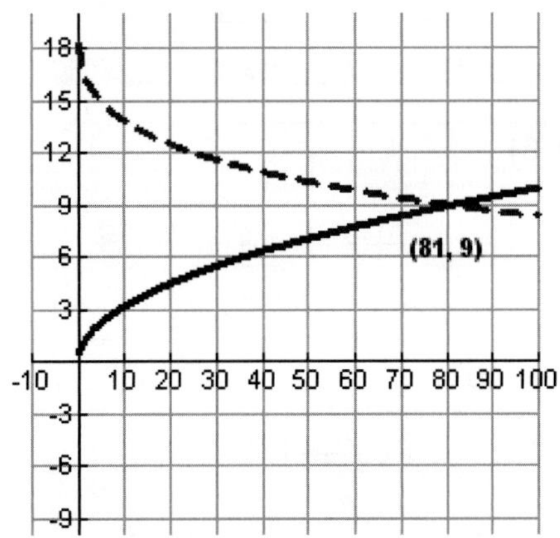

Yes, the two solutions agree.

130.

$$x^{-1} = 3x^{-2} - 10$$
$$3x^{-2} - x^{-1} - 10 = 0$$
Let $u = x^{-1}$ to obtain
$$3u^2 - u - 10 = 0$$
$$(3u+5)(u-2) = 0$$
$$u = -\tfrac{5}{3},\ 2$$
$$x^{-1} = -\tfrac{5}{3} \qquad x^{-1} = 2$$
$$\boxed{x = -\tfrac{3}{5}} \qquad \boxed{x = \tfrac{1}{2}}$$

Yes, the two solutions agree.

Graphically, let:
$y1 = x^{-1},\ y2 = 3x^{-2} - 10$.

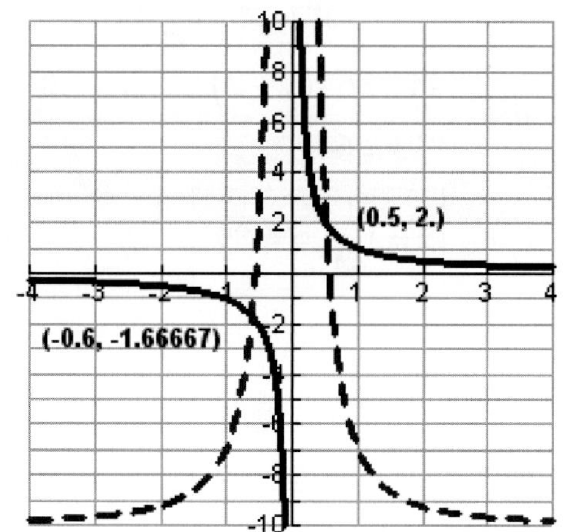

Section 0.4 Solutions --

1. $[-2,3)$ 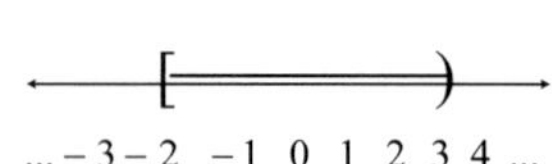	**2.** $[-4,-1]$ 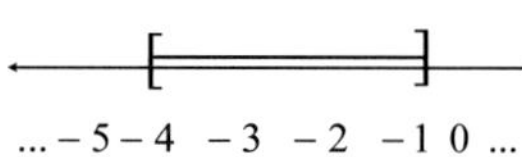
3. $(-3,5]$ 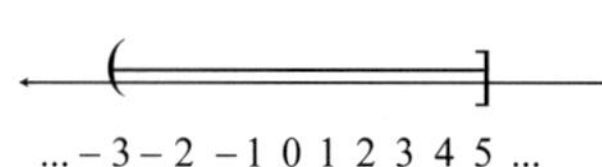	**4.** $(0,6)$ 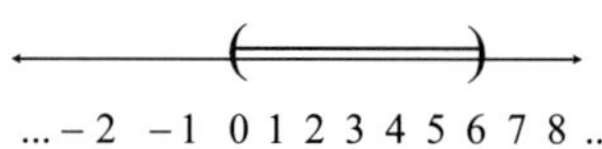
5. $[4,6]$ ...3 4 5 6 7 ...	**6.** $(-3,2]$ 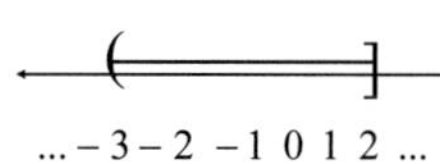
7. $[-8,-6]$ 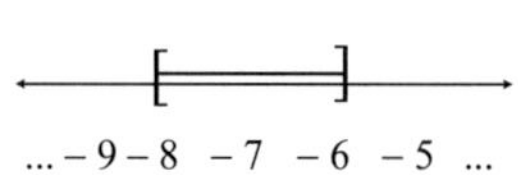	**8.** $(-\infty,2)$ 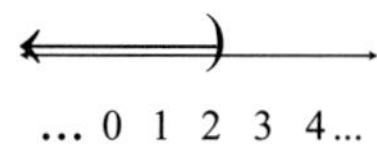

9. $\varnothing$ ... −3 −2 −1 0 1 2 3 ...	**10.** $\varnothing$ ... −3 −2 −1 0 1 2 3 ...
11. $[1,4)$ ... 0 1 2 3 4 ...	**12.** $[-3,\infty)$ ... −3 −2 −1 0 1 2 3 4 ...
13. $[-1,2)$ ... −2 −1 0 1 2 3...	**14.** $[-2,5)$ ... −3 −2 −1 0 1 2 3 4 5...
15. $(-\infty,4)\cup(4,\infty)$ ...2 3 4 5 6 7 ...	**16.** $(-\infty,\infty)$ ... −3 −2 −1 0 1 2 3 ...
17. $(-\infty,-3]\cup[3,\infty)$ ... −3 −2 −1 0 1 2 3 ...	**18.** $(-2,1]$ ... −3 −2 −1 0 1 2 3 4 ...
19. $(-3,2]$ ... −4 −3 −2 −1 0 1 2 3 4 ...	**20.** $(-\infty,\infty)$ ... −3 −2 −1 0 1 2 3 ...
21. $3(t+1)>2t$ $3t+3>2t$ $t+3>0$ $t>-3$ $\boxed{(-3,\infty)}$	**22.** $2(y+5)\le 3(y-4)$ $2y+10\le 3y-12$ $10\le y-12$ $22\le y$ $\boxed{[22,\infty)}$

23. $7-2(1-x)>5+3(x-2)$ $7-2+2x>5+3x-6$ $5+2x>3x-1$ $5>x-1$ $x<6$ $\boxed{(-\infty,6)}$	**24.** $4-3(2+x)<5$ $4-6-3x<5$ $-2-3x<5$ $-3x<7$ $x>-7/3$ $\boxed{(-7/3,\infty)}$
25. Multiply by LCD = 6 $4y-3(5-y)<10y-6(2+y)$ $4y-15+3y<10y-12-6y$ $7y-15<4y-12$ $3y-15<-12$ $3y<3$ $y<1$ $\boxed{(-\infty,1)}$	**26.** $\frac{s}{2}-\frac{s-3}{3}>\frac{s}{4}-\frac{1}{12}$ LCD = 12 $6s-4(s-3)>3s-1$ $6s-4s+12>3s-1$ $2s+12>3s-1$ $s<13$ $\boxed{(-\infty,13)}$
27. $-3<1-x\le 9$ $-4<-x\le 8$ Divide by -1 Flip the signs $-8\le x<4$ $\boxed{[-8,4)}$	**28.** $3\le -2-5x\le 13$ $5\le -5x\le 15$ Divide by -5 Flip the signs $-1\ge x\ge -3$ $\boxed{[-3,-1]}$
29. $0<2-\frac{1}{3}y<4$ $-2<-\frac{1}{3}y<2$ Multiply by -3 Flip the signs $-6<y<6$ $\boxed{(-6,6)}$	**30.** $3<\frac{1}{2}A-3<7$ $6<\frac{1}{2}A<10$ Multiply by 2 $12<A<20$ $\boxed{(12,20)}$

31.

$$\frac{1}{2} \le \frac{1+y}{3} \le \frac{3}{4}$$

Multiply by 3

$$\frac{3}{2} \le 1+y \le \frac{9}{4}$$

$$\frac{1}{2} \le y \le \frac{5}{4}$$

$$\boxed{\left[\frac{1}{2}, \frac{5}{4}\right]}$$

32.

$-1 < \frac{2-z}{4} \le \frac{1}{5}$ (Multiply by 4)

$-4 < 2-z \le \frac{4}{5}$

$-6 < -z \le -\frac{6}{5}$ (Multiply by -1 and Flip the signs)

$6 > z \ge \frac{6}{5}$

$$\boxed{\left[\frac{6}{5}, 6\right)}$$

33. $2t^2 - t - 3 \le 0$

$(2t-3)(t+1) \le 0$

CP's: $t = -1, 3/2$

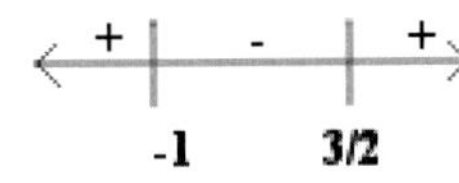

$\boxed{[-1, 3/2]}$

34. $3t^2 + 5t - 2 \ge 0$

$(3t-1)(t+2) \ge 0$

CP's: $t = -2, 1/3$

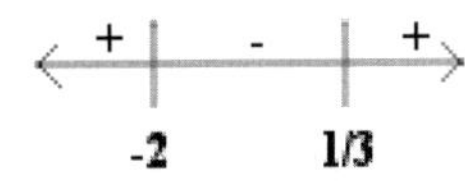

$\boxed{(-\infty, -2] \cup [1/3, \infty)}$

35. $6v^2 - 5v + 1 < 0$

$(3v-1)(2v-1) < 0$

CP's: $v = 1/3, 1/2$

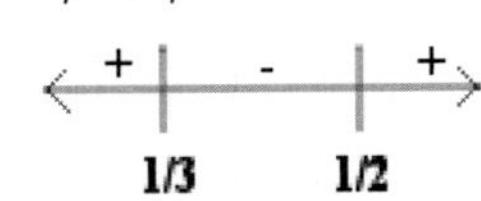

$\boxed{(1/3, 1/2)}$

36. $12t^2 - 37t - 10 < 0$

$(3t-10)(4t+1) < 0$

CP's: $t = -1/4, 10/3$

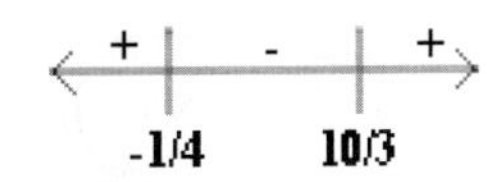

$\boxed{(-1/4, 10/3)}$

37. $2s^2 - 5s - 3 \ge 0$

$(2s+1)(s-3) \ge 0$

CP's: $s = -1/2, 3$

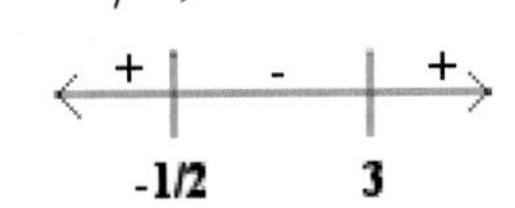

$\boxed{(-\infty, -1/2] \cup [3, \infty)}$

38. $s^2 + 8s + 12 \le 0$

$(s+2)(s+6) \le 0$

CP's: $s = -6, -2$

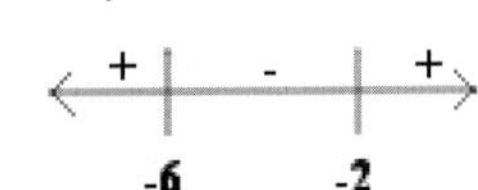

$\boxed{[-6, -2]}$

39. $y^2+2y-4\geq 0$ Note: Can't factor

To find CP's solve $y^2+2y-4=0$

$$y=\frac{-2\pm\sqrt{2^2-4(1)(-4)}}{2(1)}$$

$$y=\frac{-2\pm\sqrt{20}}{2}$$

$$y=\frac{-2\pm2\sqrt{5}}{2}=-1\pm\sqrt{5}$$

+ | - | +

-1-√5   -1+√5

$$\boxed{\left(-\infty,-1-\sqrt{5}\right]\cup\left[-1+\sqrt{5},\infty\right)}$$

40. $y^2+3y-1\leq 0$ Note: can't factor

To find CP's solve $y^2+3y-1=0$

$$y=\frac{-3\pm\sqrt{3^2-4(1)(-1)}}{2(1)}$$

$$y=\frac{-3\pm\sqrt{13}}{2}$$

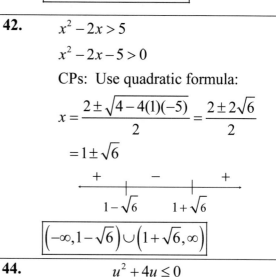

+ | - | +

(-3-√13)/2   (-3+√13)/2

$$\boxed{\left[\frac{-3-\sqrt{13}}{2},\frac{-3+\sqrt{13}}{2}\right]}$$

41. $x^2-4x<6$

$x^2-4x-6<0$

CPs: Use quadratic formula:

$$x=\frac{4\pm\sqrt{16-4(1)(-6)}}{2}=\frac{4\pm2\sqrt{10}}{2}$$

$$=2\pm\sqrt{10}$$

+ | − | +

$2-\sqrt{10}$ $2+\sqrt{10}$

$$\boxed{\left(2-\sqrt{10},2+\sqrt{10}\right)}$$

42. $x^2-2x>5$

$x^2-2x-5>0$

CPs: Use quadratic formula:

$$x=\frac{2\pm\sqrt{4-4(1)(-5)}}{2}=\frac{2\pm2\sqrt{6}}{2}$$

$$=1\pm\sqrt{6}$$

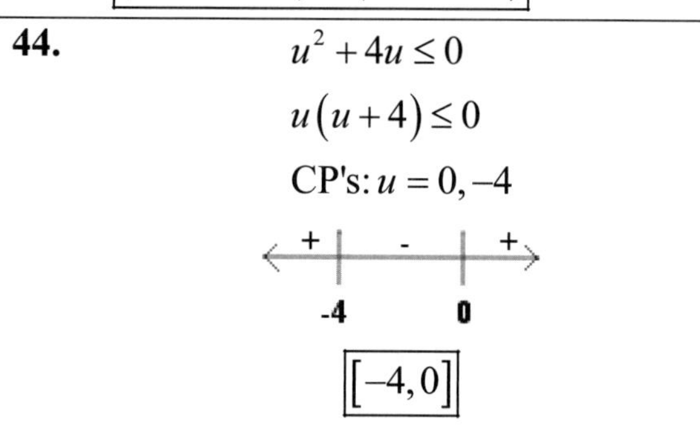

+ | − | +

$1-\sqrt{6}$ $1+\sqrt{6}$

$$\boxed{\left(-\infty,1-\sqrt{6}\right)\cup\left(1+\sqrt{6},\infty\right)}$$

43. $u^2-3u\geq 0$

$u(u-3)\geq 0$

CP's: $u=0, 3$

+ | - | +

0   3

$$\boxed{(-\infty,0]\cup[3,\infty)}$$

44. $u^2+4u\leq 0$

$u(u+4)\leq 0$

CP's: $u=0,-4$

+ | - | +

-4   0

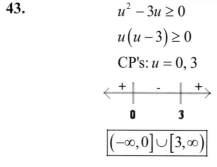

$$\boxed{[-4,0]}$$

45. $x^2-9>0$ $(x-3)(x+3)>0$ CP's: $x=-3,3$ + - + -3 3 $\boxed{(-\infty,-3)\cup(3,\infty)}$	**46.** $t^2-49\le 0$ $(t-7)(t+7)\le 0$ CP's: $t=-7,7$ + − + −7 7 $\boxed{[-7,7]}$
47. $x^3+x^2-2x\le 0$ $x(x^2+x-2)\le 0$ $x(x+2)(x-1)\le 0$ − + − + −2 0 1 $\boxed{(-\infty,-2]\cup[0,1]}$	**48.** $x^3+2x^2-3x>0$ $x(x^2+2x-3)>0$ $x(x+3)(x-1)>0$ − + − + −3 0 1 $\boxed{(-3,0)\cup(1,\infty)}$
49. $x^3+x>2x^2$ $x^3-2x^2+x>0$ $x(x^2-2x+1)>0$ $x(x-1)^2>0$ − + + 0 1 $\boxed{(0,1)\cup(1,\infty)}$	**50.** $x^3+4x\le 4x^2$ $x^3-4x^2+4x\le 0$ $x(x^2-4x+4)\le 0$ $x(x-2)^2\le 0$ − + + 0 2 $\boxed{(-\infty,0]}$
51. $\dfrac{s+1}{(2-s)(2+s)}\ge 0$ CP's: $s=-2,-1,2$ + - + - -2 -1 2 $\boxed{(-\infty,-2)\cup[-1,2)}$	**52.** $\dfrac{s+5}{(2-s)(2+s)}\le 0$ CP's: $s=-5,-2,2$ + - + - -5 -2 2 $\boxed{[-5,-2)\cup(2,\infty)}$

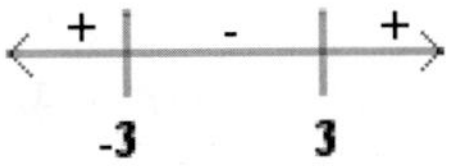

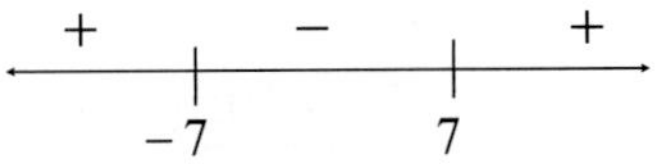

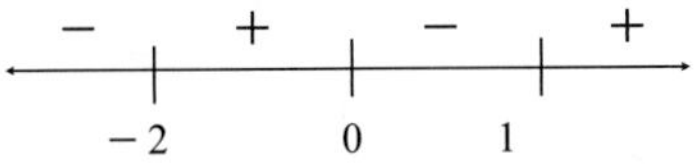

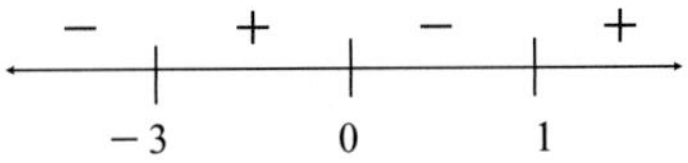

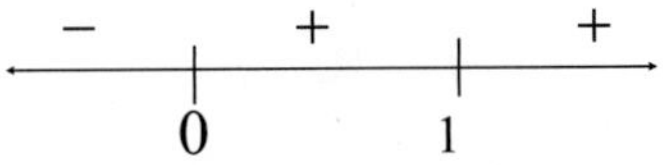

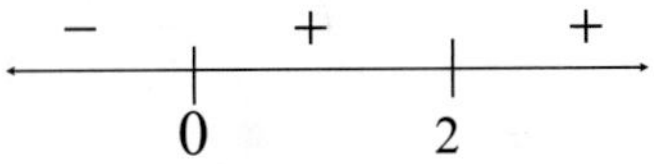

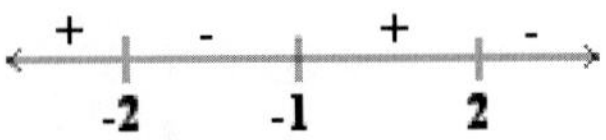

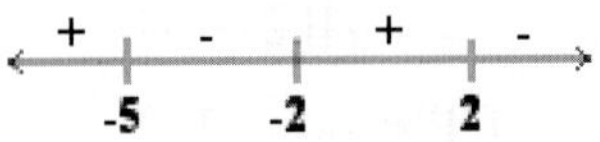

53.

$$\frac{3t^2}{t+2}-5t\ge 0$$

$$\frac{3t^2-5t(t+2)}{t+2}\ge 0$$

$$\frac{3t^2-5t^2-10t}{t+2}\ge 0$$

$$\frac{-2t^2-10t}{t+2}\ge 0$$

$$\frac{-2t(t+5)}{t+2}\ge 0 \quad \text{CP's: } t=-5,-2,0$$

\+   -   +   -

-5   -2   0

$$\boxed{(-\infty,-5]\cup(-2,0]}$$

54.

$$\frac{-2t-t^2}{4-t}-t\ge 0$$

$$\frac{-2t-t^2-t(4-t)}{4-t}\ge 0$$

$$\frac{-2t-t^2-4t+t^2}{4-t}\ge 0$$

$$\frac{-6t}{4-t}\ge 0$$

CP's: $t=0, 4$

\+   -   +

0   4

$$\boxed{(-\infty,0]\cup(4,\infty)}$$

55.

$$\frac{3p-2p^2}{4-p^2}-\frac{(3+p)}{(2-p)}<0$$

$$\frac{p(3-2p)}{(2-p)(2+p)}-\frac{(3+p)}{(2-p)}<0$$

$$\frac{p(3-2p)-(3+p)(2+p)}{(2-p)(2+p)}<0$$

$$\frac{3p-2p^2-6-5p-p^2}{(2-p)(2+p)}>0$$

$$\frac{-3p^2-2p-6}{(2-p)(2+p)}<0$$

$$\frac{3p^2+2p+6}{(2-p)(2+p)}>0$$

CP's: $p=-2, 2$

\-   +   -

-2   2

$$\boxed{(-2,2)}$$

56.

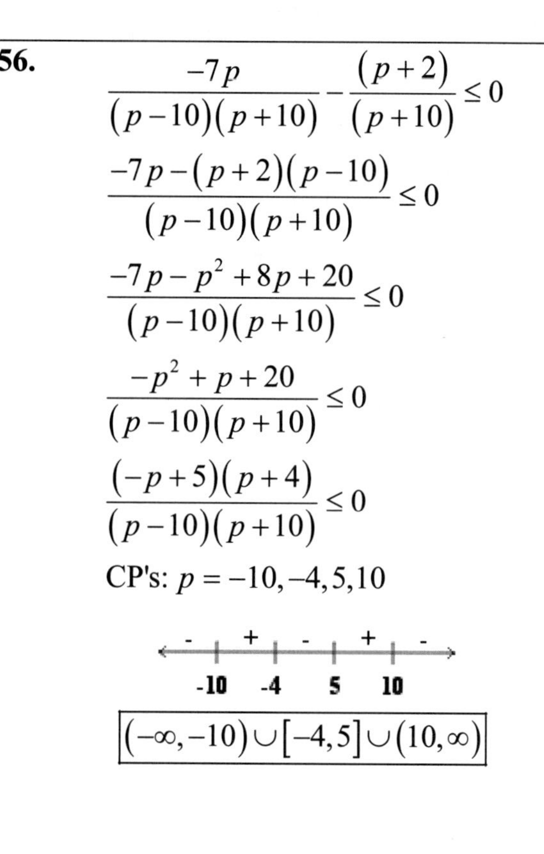

$$\frac{-7p}{(p-10)(p+10)}-\frac{(p+2)}{(p+10)}\le 0$$

$$\frac{-7p-(p+2)(p-10)}{(p-10)(p+10)}\le 0$$

$$\frac{-7p-p^2+8p+20}{(p-10)(p+10)}\le 0$$

$$\frac{-p^2+p+20}{(p-10)(p+10)}\le 0$$

$$\frac{(-p+5)(p+4)}{(p-10)(p+10)}\le 0$$

CP's: $p=-10,-4,5,10$

\-   +   -   +   -

-10   -4   5   10

$$\boxed{(-\infty,-10)\cup[-4,5]\cup(10,\infty)}$$

57.

$$\frac{x^2+10}{x^2+16}>0$$

$\boxed{\mathbb{R}}$ (consistent)

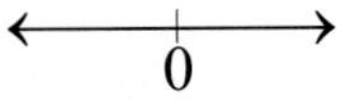

58.

$$-\left(\frac{x^2+2}{x^2+4}\right)<0$$

$\boxed{\mathbb{R}}$ (consistent)

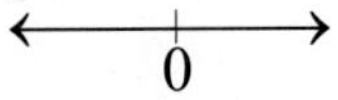

59.

$$\frac{(v-3)(v+3)}{(v-3)}\geq 0 \quad \boxed{v\neq 3}$$

$$v+3\geq 0$$

$$v\geq -3$$

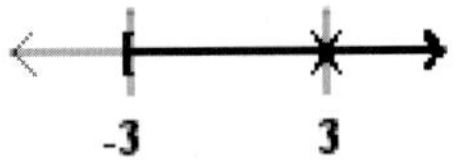

$$\boxed{[-3,3)\cup(3,\infty)}$$

60.

$$(v-1)\frac{(v+1)}{(v+1)}\leq 0 \quad \boxed{v\neq -1}$$

$$(v-1)\leq 0$$

$$v\leq 1$$

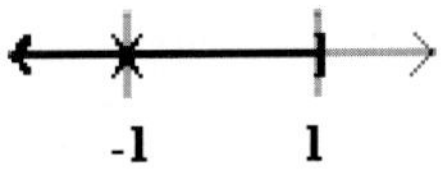

$$\boxed{(-\infty,-1)\cup(-1,-1]}$$

61.

$$\frac{2}{t-3}+\frac{1}{t+3}\geq 0$$

$$\frac{2(t+3)+(t-3)}{(t-3)(t+3)}\geq 0$$

$$\frac{3t+3}{(t-3)(t+3)}\geq 0$$

$$\frac{3(t+1)}{(t-3)(t+3)}\geq 0$$

CPs: $-1, \pm 3$

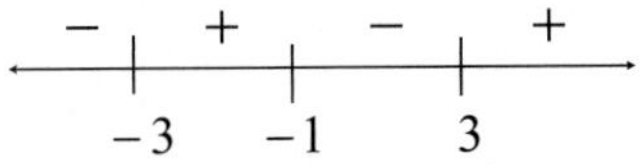

$$\boxed{(-3,-1]\cup(3,\infty)}$$

62.

$$\frac{1}{t-2}+\frac{1}{t+2}\leq 0$$

$$\frac{(t+2)+(t-2)}{(t-2)(t+2)}\leq 0$$

$$\frac{2t}{(t-2)(t+2)}\leq 0$$

CPs: $0, \pm 2$

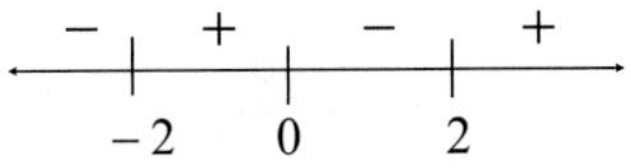

$$\boxed{(-\infty,-2)\cup[0,2)}$$

63.

$$\frac{3}{x+4}-\frac{1}{x-2}\le 0$$
$$\frac{3(x-2)-(x+4)}{(x+4)(x-2)}\le 0$$
$$\frac{2x-10}{(x+4)(x-2)}\le 0$$
$$\frac{2(x-5)}{(x+4)(x-2)}\le 0$$

CPs: $-4,2,5$

Sign chart: $-$ on $(-\infty,-4)$, $+$ on $(-4,2)$, $-$ on $(2,5)$, $+$ on $(5,\infty)$; tick marks at -4, 2, 5

$\boxed{(-\infty,-4)\cup(2,5]}$

64.

$$\frac{2}{x-5}-\frac{1}{x-1}\ge 0$$
$$\frac{2(x-1)-(x-5)}{(x-5)(x-1)}\ge 0$$
$$\frac{x+3}{(x-5)(x-1)}\ge 0$$

CPs: $-3,1,5$

Sign chart: $-$ on $(-\infty,-3)$, $+$ on $(-3,1)$, $-$ on $(1,5)$, $+$ on $(5,\infty)$; tick marks at -3, 1, 5

$\boxed{[-3,1)\cup(5,\infty)}$

65.

$$\frac{1}{p-2}-\frac{1}{p+2}-\frac{3}{p^2-4}\ge 0$$
$$\frac{(p+2)-(p-2)-3}{(p+2)(p-2)}\ge 0$$
$$\frac{1}{(p+2)(p-2)}\ge 0$$

CPs: ± 2

Sign chart: $+$ on $(-\infty,-2)$, $-$ on $(-2,2)$, $+$ on $(2,\infty)$; tick marks at -2, 2

$\boxed{(-\infty,-2)\cup(2,\infty)}$

66.

$$\frac{2}{2p-3}-\frac{1}{p+1}-\frac{1}{2p^2-p-3}\le 0$$
$$\frac{2}{2p-3}-\frac{1}{p+1}-\frac{1}{(2p-3)(p+1)}\le 0$$
$$\frac{2(p+1)-(2p-3)-1}{(2p-3)(p+1)}\le 0$$
$$\frac{4}{(2p-3)(p+1)}\le 0$$

CPs: $-1, 3/2$

Sign chart: $+$ on $(-\infty,-1)$, $-$ on $(-1,3/2)$, $+$ on $(3/2,\infty)$; tick marks at -1, $3/2$

$\boxed{(-1, 3/2)}$

67.

$$x-4<-2 \quad \text{or} \quad x-4>2$$
$$x<2 \qquad\qquad x>6$$
$$(-\infty,2)\cup(6,\infty)$$

68.

$$-3<x-1<3$$
$$-2<x<4$$
$$(-2,4)$$

69. $-1\le 4-x\le 1$ $-5\le -x\le -3$ $3\le x\le 5$ $[3,5]$	**70.** $-3<1-y<3$ $-4<-y<2$ $4>y>-2$ $(-2,4)$
71. $\mathbb{R}$	**72.** No solution
73. $\lvert 7-2y\rvert\ge 3$ $7-2y\ge 3$ or $7-2y\le -3$ $-2y\ge -4$ or $-2y\le -10$ $y\le 2$ or $y\ge 5$ $\boxed{(-\infty,2]\cup[5,\infty)}$	**74.** $\lvert 6-5y\rvert\le 1$ $-1\le 6-5y\le 1$ $-7\le -5y\le -5$ $7/5\ge y\ge 1$ $\boxed{[1,7/5]}$
75. $\mathbb{R}$	
76. $4-3x\le -1$ or $4-3x\ge 1$ $-3x\le -5$ or $-3x\ge -3$ $x\ge 5/3$ $\quad x\le 1$ $\boxed{(-\infty,1]\cup[5/3,\infty)}$	**77.** $2\lvert 4x\rvert-9\ge 3$ $2\lvert 4x\rvert\ge 12$ $\lvert 4x\rvert\ge 6$ $4x\ge 6$ or $4x\le -6$ $x\ge 3/2$ or $x\le -3/2$ $\boxed{(-\infty,-3/2]\cup[3/2,\infty)}$
78. $5\lvert x-1\rvert+2\le 7$ $5\lvert x-1\rvert\le 5$ $\lvert x-1\rvert\le 1$ $-1\le x-1\le 1$ $0\le x\le 2$ $\boxed{[0,2]}$	**79.** $9-\lvert 2x\rvert<3$ $-\lvert 2x\rvert<-6$ $\lvert 2x\rvert>6$ $2x>6$ or $2x<-6$ $x>3$ or $x<-3$ $\boxed{(-\infty,-3)\cup(3,\infty)}$

80.	**81.**
$4-\lvert x+1\rvert>1$ $-\lvert x+1\rvert>-3$ $\lvert x+1\rvert<3$ $-3<x+1<3$ $-4<x<2$ $\boxed{(-4,2)}$	$x^2-1\le 8$ $x^2-9\le 0$ $(x-3)(x+3)\le 0$ CP's: $x=-3,3$ + -3 - 3 + $-3\le x\le 3$ $\boxed{[-3,3]}$

82.

$$x^2+4\ge 29$$
$$x^2-25\ge 0$$
$$(x-5)(x+5)\ge 0$$

CP's: – 5, 5

+ -5 - 5 +

$$\boxed{(-\infty,-5]\cup[5,\infty)}$$

83.	**84.**
$0.9\,r_T\le r_R\le 1.1\,r_T$	$\frac{S}{N}\ge 2$ if N fluctuates by 10% $\frac{S}{1.1\,N}>2$ or $\boxed{S>2.2\,N}$

85. Let T = amount of tax paid. Least amount of tax = \$4,386.25
Greatest amount of tax = \$15,698.75
So, the range of taxes is: $\boxed{4{,}386.25\le T\le 15{,}698.75}$

86. Let T = amount of tax paid. Least amount of tax = \$15,698.75
Greatest amount of tax = \$39,148.75
So, the range of taxes is: $\boxed{15{,}698.75\le T\le 39{,}148.75}$

87. $-x^2+130x-3000>0$

$x^2-130x+3000<0$

$(x-30)(x-100)<0$

CP's: $x=30,100$

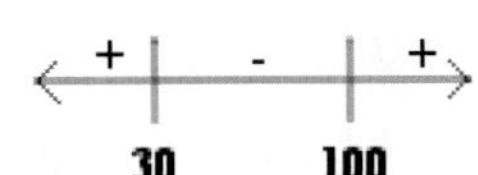

Between 30 and 100 orders

88. $x^2-130x+3600>0$

$(x-40)(x-90)>0$

CP's: $x=40,90$

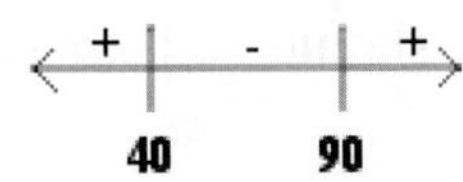

Less than 40 or more than 90 orders

89. Car is worth more than you owe:

$\frac{t}{t-3}>0$ CP's: $t=0,3$

+ - +

0 3

$(3,\infty)$ Greater than 3 years

You owe more than it's worth:

$\frac{t}{t-3}<0$ CP's: $t=0,3$

+ - +

0 3

$(0,3)$ First 3 years

90. Car is worth more than you owe:

$-\left(\frac{2-t}{4-t}\right)>0$

- + -

2 4

$[2,4)$ Between 2 and 4 years

You owe more than it's worth:

$-\left(\frac{2-t}{4-t}\right)<0$

$(0,2)\cup(4,\infty)$

91. $h=-16t^2+1200t$

bullet is in the air if $h>0$

$-16t^2+1200t>0$

$-16t(t-75)>0$

CP's: $t=0,75$

$(0,75)$

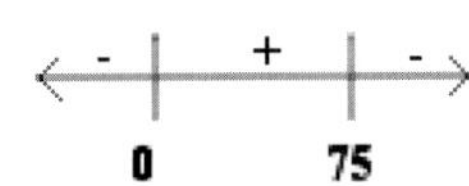

Bullet is in the air for 75 sec

92. $h=-16t^2+600t$

bullet is in the air if $h>0$

$-16t^2+600t>0$

$-8t(2t-75)>0$

CP's: $t=0,75/2$

$(0,75/2)$

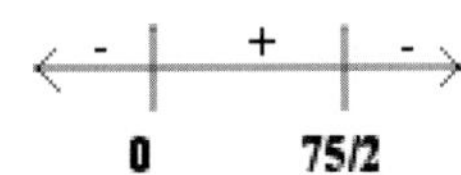

Bullet is in the air for 37.5 sec

<table>
<tr><td>93. In order to win the hole, $d < 4$.
In order to have a tie, $d = 4$.</td><td>94. $|f - f_C| \le 15$</td></tr>
<tr><td>95. Forgot to flip the sign when dividing by -3. Answer should be $[2, \infty)$.</td><td>96.
Cannot take square root.
$u^2 - 25 < 0$
$(u-5)(u+5) < 0$
$(-5, 5)$</td></tr>
<tr><td>97.
Cannot divide by x.
$x^2 - 3x > 0$
$x(x-3) > 0$
$(-\infty, 0) \cup (3, \infty)$</td><td>98.
Can't cross-multiply
$\frac{x+4}{x} + \frac{1}{3} < 0$
$\frac{3(x+4)+x}{3x} < 0$
$\frac{4x+12}{3x} < 0$
$\frac{4(x+3)}{3x} < 0$
$x = 0, -3$ are CP's
$(-3, 0)$</td></tr>
<tr><td>99. True. In fact, the two inequalities are equivalent.</td><td>100. False. Need to switch the sign.</td></tr>
<tr><td>101. False $(-a, a)$</td><td>102. False $(-\infty, -a] \cup [a, \infty)$</td></tr>
<tr><td>103.
$\frac{x^2 + a^2}{x^2 + b^2} \ge 0$
$\boxed{\mathbb{R}}$</td><td>104.
$\frac{x^2 - b^2}{x + b} < 0$ $\boxed{x \ne -b}$
$\frac{(x-b)(x+b)}{x+b} < 0$
$x - b < 0$
$x < b$ $\boxed{(-\infty, -b) \cup (-b, b)}$</td></tr>
<tr><td colspan="2">105. No solution</td></tr>
</table>

106.

$$3x^2 - 7x + 2 < -8$$
$$3x^2 - 7x + 10 < 0$$
no solution

$$3x^2 - 7x + 2 > 8$$
$$3x^2 - 7x - 6 > 0$$
$$(3x+2)(x-3) > 0$$
CP's: $x = -2/3, 3$

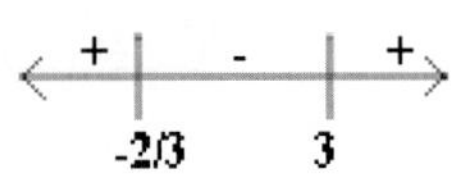

$$\boxed{(-\infty, -2/3) \cup (3, \infty)}$$

107.

a)

$$x - 3 < 2x - 1 < x + 4$$
$$-3 < x - 1 < 4$$
$$-2 < x < 5$$
$$(-2, 5)$$

b)

c) Agree

108. **a)**

$$x - 2 < 3x + 4 \le 2x + 6$$
$$x - 2 < 3x + 4 \quad \text{and} \quad 3x + 4 \le 2x + 6$$
$$-6 < 2x \quad \text{and} \quad x \le 2$$
$$-3 < x$$
$$(-3, 2]$$

c) Agree

b) Graphically, let
$y1 = x - 2,\ y2 = 3x + 4,\ y3 = 2x + 6$

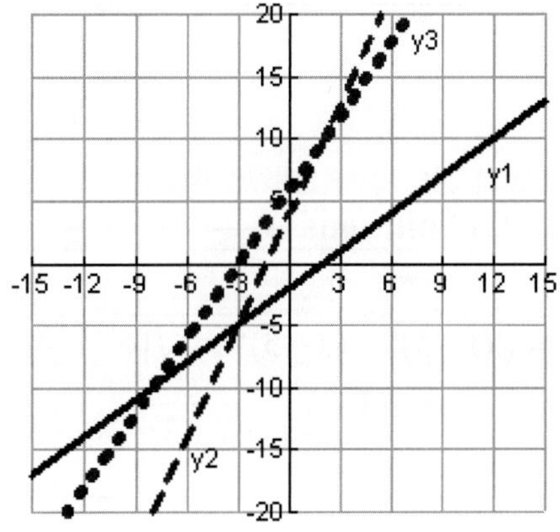

109.

$$y_1 = \left| \frac{x}{x+1} \right|$$

$$y_2 = 1$$

Find when $y_1 < y_2$.

$$\boxed{\left(-\tfrac{1}{2}, \infty\right)}$$

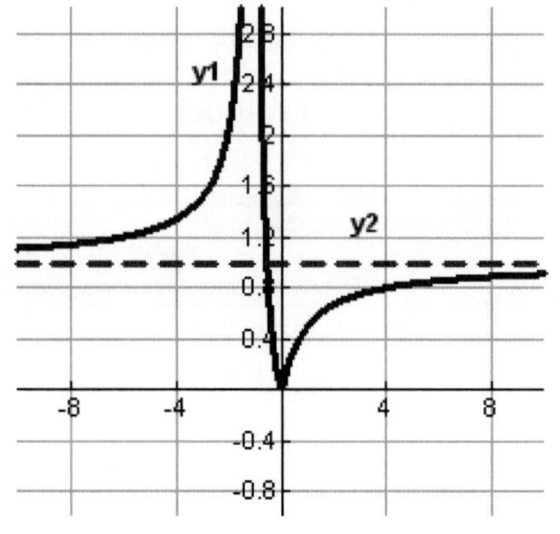

110.

$$y_1 = \left| \frac{x}{x+1} \right|$$

$$y_2 = 2$$

Find when $y_1 < y_2$.

$$\boxed{(-\infty, -2) \cup \left(-\tfrac{2}{3}, \infty\right)}$$

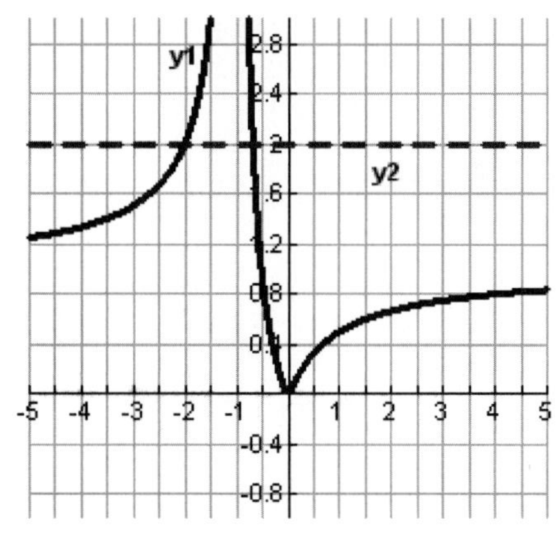

Section 0.5 Solutions --

1.

$$d = \sqrt{(1-5)^2 + (3-3)^2} = \sqrt{16} = \boxed{4}$$

$$M = \left(\frac{1+5}{2}, \frac{3+3}{2} \right) = \boxed{(3,3)}$$

2.

$$d = \sqrt{(-2-(-2))^2 + (4-(-4))^2} = \sqrt{64} = \boxed{8}$$

$$M = \left(\frac{-2+(-2)}{2}, \frac{4+(-4)}{2} \right) = \boxed{(-2,0)}$$

3.

$$d = \sqrt{(-1-3)^2 + (4-0)^2} = \sqrt{32} = \boxed{4\sqrt{2}}$$

$$M = \left(\frac{-1+3}{2}, \frac{4+0}{2} \right) = \boxed{(1,2)}$$

4.

$$d = \sqrt{(-3-1)^2 + (-1-3)^2} = \sqrt{32} = \boxed{4\sqrt{2}}$$

$$M = \left(\frac{-3+1}{2}, \frac{-1+3}{2} \right) = \boxed{(-1,1)}$$

5.

$$d=\sqrt{(-10-(-7))^2+(8-(-1))^2}$$
$$=\sqrt{90}=\boxed{3\sqrt{10}}$$
$$M=\left(\frac{-10+(-7)}{2},\ \frac{8+(-1)}{2}\right)=\boxed{\left(\frac{-17}{2},\ \frac{7}{2}\right)}$$

6.

$$d=\sqrt{(-2-7)^2+(12-15)^2}$$
$$=\sqrt{90}=\boxed{3\sqrt{10}}$$
$$M=\left(\frac{-2+7}{2},\ \frac{12+15}{2}\right)=\boxed{\left(\frac{5}{2},\ \frac{27}{2}\right)}$$

7.

$$d=\sqrt{(-3-(-7))^2+(-1-2)^2}=\sqrt{25}=\boxed{5}$$
$$M=\left(\frac{-3+(-7)}{2},\ \frac{-1+2}{2}\right)=\boxed{\left(-5,\ \frac{1}{2}\right)}$$

8.

$$d=\sqrt{(-4-(-9))^2+(5-(-7))^2}=\sqrt{169}=\boxed{13}$$
$$M=\left(\frac{-4+(-9)}{2},\ \frac{5+(-7)}{2}\right)=\boxed{\left(-\frac{13}{2},\ -1\right)}$$

9.

$$d=\sqrt{(-6-(-2))^2+(-4-(-8))^2}=\sqrt{32}=\boxed{4\sqrt{2}}$$
$$M=\left(\frac{-6+(-2)}{2},\ \frac{-4+(-8)}{2}\right)=\boxed{(-4,-6)}$$

10.

$$d=\sqrt{(0-(-4))^2+(-7-(-5))^2}=\sqrt{20}=\boxed{2\sqrt{5}}$$
$$M=\left(\frac{0+(-4)}{2},\ \frac{-7+(-5)}{2}\right)=\boxed{(-2,-6)}$$

11.

$$d=\sqrt{\left(-\frac{1}{2}-\frac{7}{2}\right)^2+\left(\frac{1}{3}-\frac{10}{3}\right)^2}=\sqrt{25}=\boxed{5}$$
$$M=\left(\frac{-\frac{1}{2}+\frac{7}{2}}{2},\ \frac{\frac{1}{3}+\frac{10}{3}}{2}\right)=\boxed{\left(\frac{3}{2},\ \frac{11}{6}\right)}$$

12.

$$d=\sqrt{\left(\frac{1}{5}-\frac{9}{5}\right)^2+\left(\frac{7}{3}-\left(-\frac{2}{3}\right)\right)^2}$$
$$=\sqrt{\frac{289}{25}}=\boxed{\frac{17}{5}}$$
$$M=\left(\frac{\frac{1}{5}+\frac{9}{5}}{2},\ \frac{\frac{7}{3}-\frac{2}{3}}{2}\right)=\boxed{\left(1,\ \frac{5}{6}\right)}$$

13.

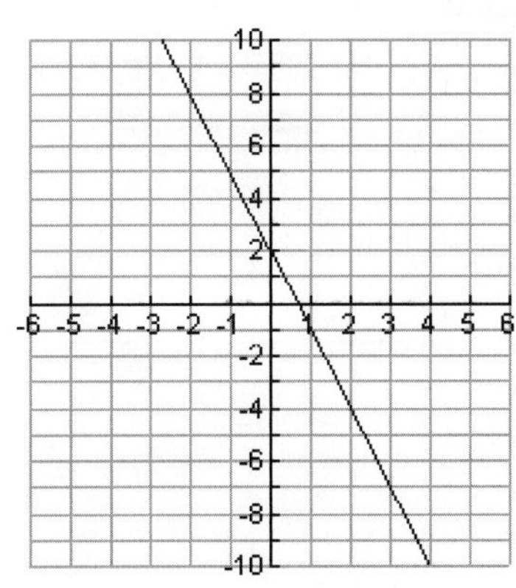

14.

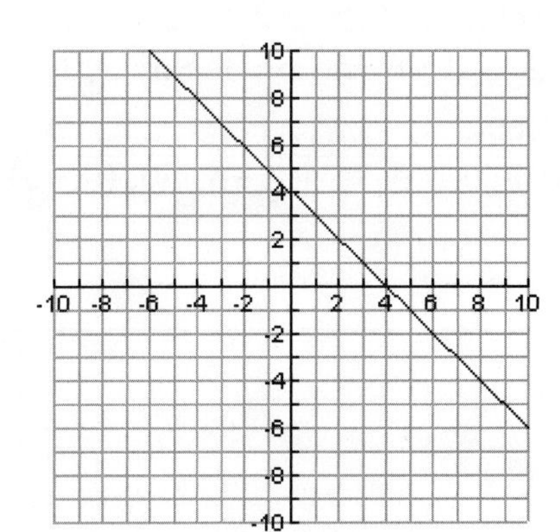

15.

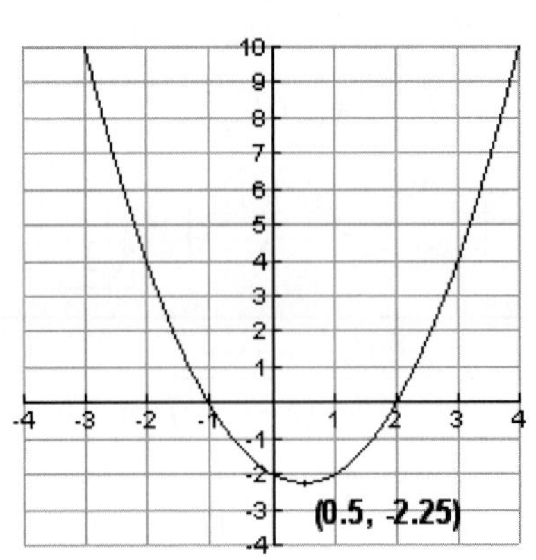

16.

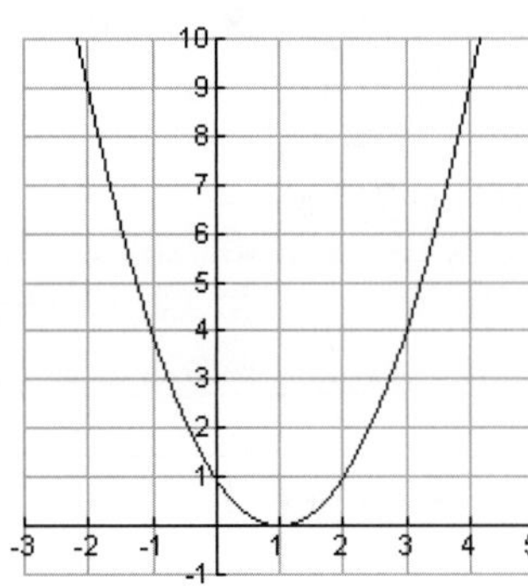

17.

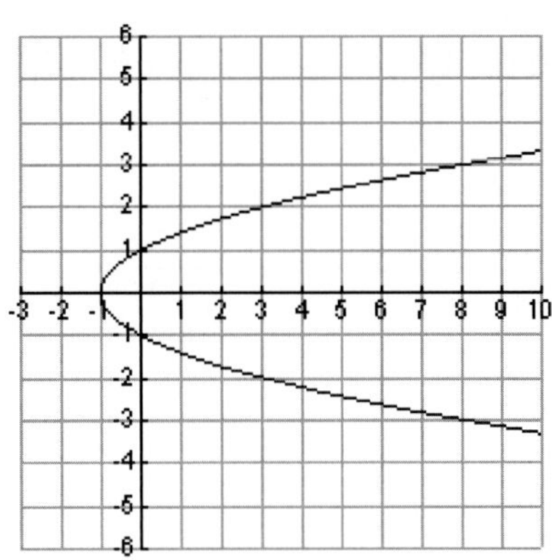

18.

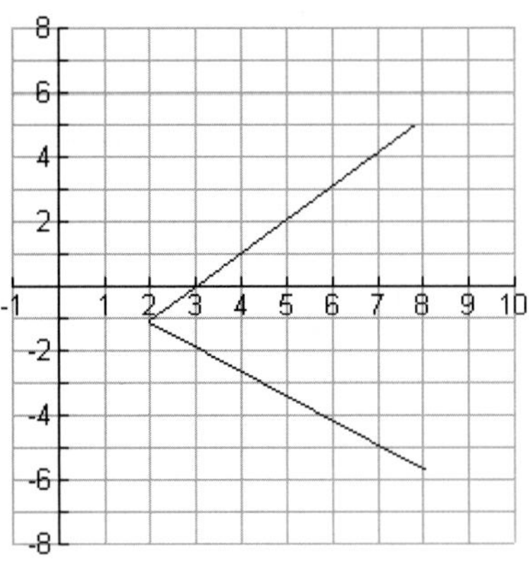

19.

x-intercept:

$$2x - 0 = 6 \Rightarrow x = 3 \quad \text{So, } (3,0).$$

y-intercept:

$$2(0) - y = 6 \Rightarrow y = -6 \quad \text{So, } (0,-6).$$

20.

x-intercepts:

$$4x^2 - 1 = 0 \Rightarrow (2x-1)(2x+1) = 0 \Rightarrow x = \pm\tfrac{1}{2}.$$

So, $(\pm\frac{1}{2}, 0)$.

y-intercept:

$$4(0)^2 - 1 = y \Rightarrow y = -1 \quad \text{So, } (0,-1).$$

21.

x-intercept:

$$\sqrt{x-4}=0 \Rightarrow x=4 \quad \text{So, } (4,0).$$

y-intercept:

$$\underbrace{\sqrt{0-4}}_{\text{undefined}}=y. \quad \text{So, no } y-\text{intercept.}$$

22.

x-intercepts:

$$\frac{x^2-x-12}{x}=0 \Rightarrow x^2-x-12=(x-4)(x+3)=0$$
$$\Rightarrow x=4,-3 \quad \text{So, } (4,0),\ (-3,0).$$

y-intercept:

$$\frac{0^2-0-12}{0}=y$$ is undefined. So, no y-int.

23.

x-intercepts:

$$4x^2+0=16 \Rightarrow x^2-4=0 \Rightarrow (x-2)(x+2)=0$$
$$\Rightarrow x=\pm 2. \quad \text{So, } (\pm 2,0).$$

y-intercept:

$$4(0)^2+y^2=16 \Rightarrow y^2-16=0 \Rightarrow (y-4)(y+4)=0$$
$$\Rightarrow y=\pm 4. \quad \text{So, } (0,\pm 4).$$

24.

x-intercepts:

$$x^2-0=9 \Rightarrow x^2-9=0 \Rightarrow (x-3)(x+3)=0$$
$$\Rightarrow x=\pm 3. \quad \text{So, } (\pm 3,0).$$

y-intercept:

$0-y^2=9$ which has no solution. So, no $y-$intercept.

25. x-axis symmetry (Replace y by $-y$):

$$x=(-y)^2+4$$
$$x=y^2+4$$

Yes, since equivalent to the original.

y-axis symmetry (Replace x by $-x$):

$$-x=y^2+4$$
$$x=-\left(y^2+4\right)$$

No, since not equivalent to the original.

symmetry about origin (Replace y by $-y$ and x by $-x$):

$$-x=(-y)^2+4=y^2+4$$
$$x=-\left(y^2+4\right)$$

No, since not equivalent to the original.

26. x-axis symmetry (Replace y by $-y$):

$$-y=x^5+1$$
$$y=-\left(x^5+1\right)$$

No, since not equivalent to the original.

y-axis symmetry (Replace x by $-x$):

$$y=(-x)^5+1$$
$$y=-x^5+1$$

No, since not equivalent to the original.

symmetry about origin (Replace y by $-y$ and x by $-x$):

$$-y=(-x)^5+1=-x^5+1$$
$$y=x^5-1$$

No, since not equivalent to the original.

27. <u>x-axis symmetry</u> (Replace y by $-y$):

$$x = |-y| = |-1||y|$$
$$x = |y|$$

Yes, since equivalent to the original.

<u>y-axis symmetry</u> (Replace x by $-x$):

$$-x = |y|$$
$$x = -|y|$$

No, since not equivalent to the original.

<u>symmetry about origin</u> (Replace y by $-y$ and x by $-x$):

$$-x = |-y| = |-1||y| = |y|$$
$$x = -|y|$$

No, since not equivalent to the original.

28. <u>x-axis symmetry</u> (Replace y by $-y$):

$$x^2 + 2(-y)^2 = 30$$
$$x^2 + 2y^2 = 30$$

Yes, since equivalent to the original.

<u>y-axis symmetry</u> (Replace x by $-x$):

$$(-x)^2 + 2y^2 = 30$$
$$x^2 + 2y^2 = 30$$

Yes, since equivalent to the original.

<u>symmetry about origin</u> (Replace y by $-y$ and x by $-x$):

$$(-x)^2 + 2(-y)^2 = 30$$
$$x^2 + 2y^2 = 30$$

Yes, since equivalent to the original.

29. <u>x-axis symmetry</u> (Replace y by $-y$):

$$-y = x^{2/3}$$
$$y = -x^{2/3}$$

No, since not equivalent to the original.

<u>y-axis symmetry</u> (Replace x by $-x$):

$$y = (-x)^{2/3}$$
$$y = x^{2/3}$$

Yes, since equivalent to the original.

<u>symmetry about origin</u> (Replace y by $-y$ and x by $-x$):

$$-y = (-x)^{2/3} = x^{2/3}$$
$$y = -x^{2/3}$$

No, since not equivalent to the original.

30. <u>x-axis symmetry</u> (Replace y by $-y$):

$$x(-y) = 1$$
$$-xy = 1$$

No, since not equivalent to the original.

<u>y-axis symmetry</u> (Replace x by $-x$):

$$(-x)y = 1$$
$$-xy = 1$$

No, since not equivalent to the original.

<u>symmetry about origin</u> (Replace y by $-y$ and x by $-x$):

$$(-x)(-y) = 1$$
$$xy = 1$$

Yes, since equivalent to the original.

31.

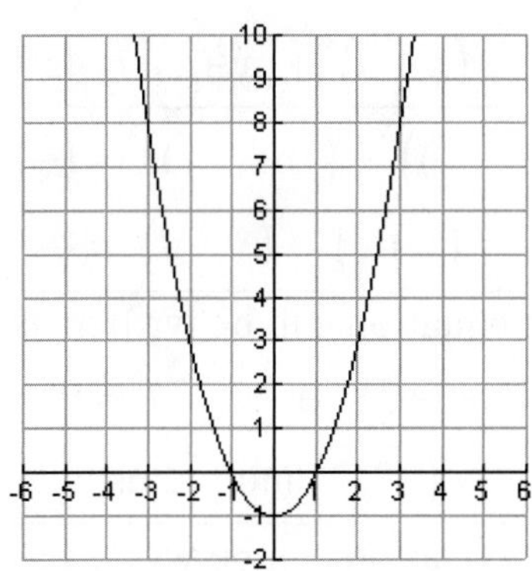

32.

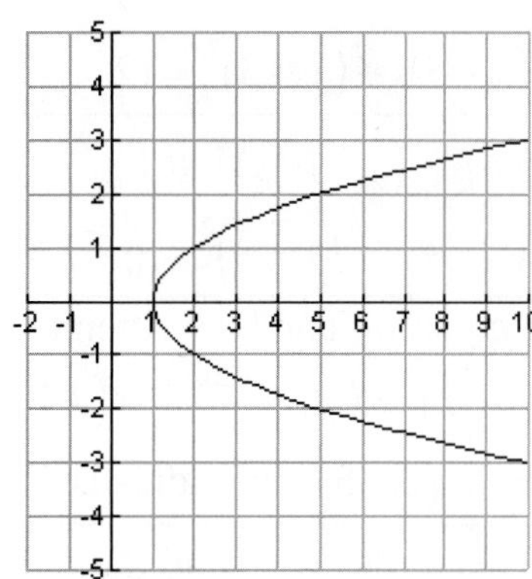

33.

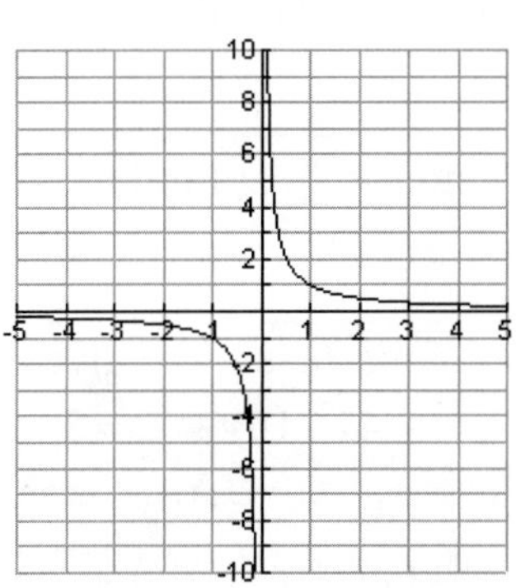

34.

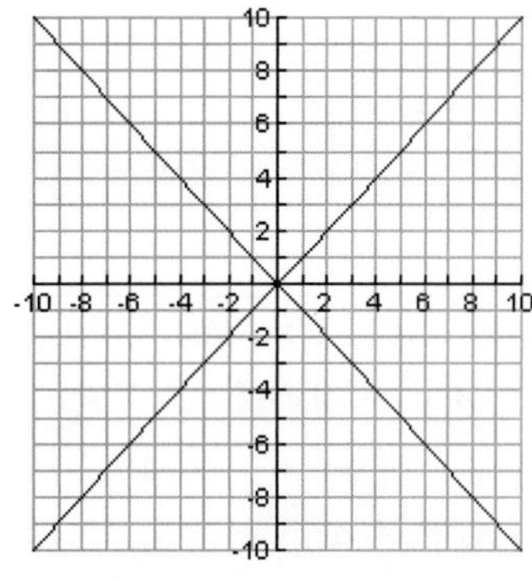

35.

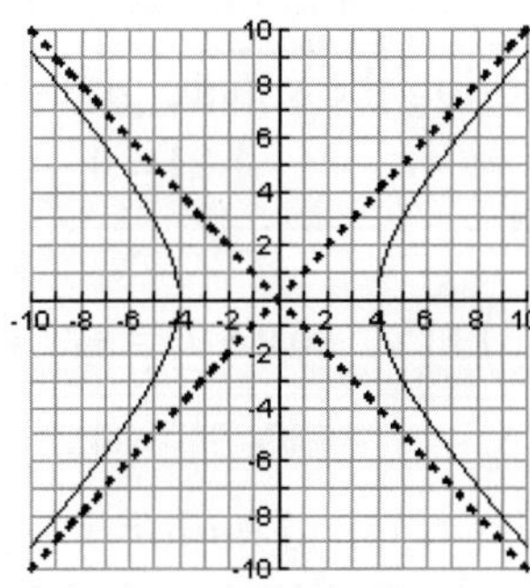

36.

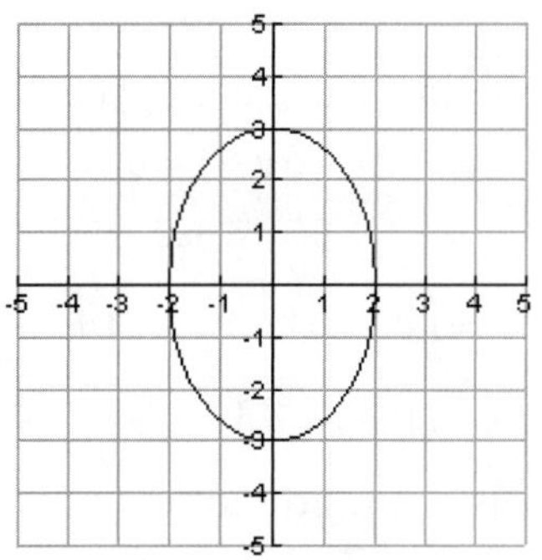

37. $(x-5)^2+(y-7)^2=81$

38. $(x-2)^2+(y-8)^2=36$

39. $(x-(-11))^2+(y-12)^2=13^2$

$(x+11)^2+(y-12)^2=169$

40. $(x-6)^2+(y-(-7))^2=64$

$(x-6)^2+(y+7)^2=64$

41. $(x-5)^2+(y-(-3))^2=(2\sqrt{3})^2$ $(x-5)^2+(y+3)^2=12$	**42.** $(x-(-4))^2+(y-(-1))^2=(3\sqrt{5})^2$ $(x+4)^2+(y+1)^2=45$
43. $(x-\frac{2}{3})^2+(y-(-\frac{3}{5}))^2=(\frac{1}{4})^2$ $(x-\frac{2}{3})^2+(y+\frac{3}{5})^2=\frac{1}{16}$	**44.** $(x-(-\frac{1}{3}))^2+(y-(-\frac{2}{7}))^2=(\frac{2}{5})^2$ $(x+\frac{1}{3})^2+(y+\frac{2}{7})^2=\frac{4}{25}$
45. The equation can be written as $(x-2)^2+(y-(-5))^2=(7)^2$. So, center is $(2,-5)$ and radius is 7.	**46.** The equation can be written as $(x-(-3))^2+(y-7)^2=9^2$. So, center is $(-3,7)$ and radius is 9.
47. The center is $(4,9)$ and the radius is $\sqrt{20}=2\sqrt{5}$.	**48.** The equation can be written as $(x-(-1))^2+(y-(-2))^2=(\sqrt{8})^2$. So, center is $(-1,-2)$ and radius is $\sqrt{8}=2\sqrt{2}$.
49. The center is $(\frac{2}{5},\frac{1}{7})$ and the radius is $\sqrt{\frac{4}{9}}=\frac{2}{3}$.	**50.** The center is $(\frac{1}{2},\frac{1}{3})$ and the radius is $\sqrt{\frac{9}{25}}=\frac{3}{5}$.
51. Completing the square gives us: $x^2+y^2-10x-14y-7=0$ $x^2-10x \quad +y^2-14y=7$ $(x^2-10x+25)+(y^2-14y+49)=7+25+49$ $(x-5)^2+(y-7)^2=81$ So, center is $(5,7)$ and radius is 9.	**52.** Completing the square gives us: $x^2+y^2-4x-16y+32=0$ $x^2-4x \quad +y^2-16y=-32$ $(x^2-4x+4)+(y^2-16y+64)=-32+4+64$ $(x-2)^2+(y-8)^2=36$ So, center is $(2,8)$ and radius is 6.
53. Completing the square gives us: $x^2+y^2-2x-6y+1=0$ $x^2-2x \quad +y^2-6y=-1$ $(x^2-2x+1)+(y^2-6y+9)=-1+1+9$ $(x-1)^2+(y-3)^2=9$ So, center is $(1,3)$ and radius is 3.	**54.** Completing the square gives us: $x^2+y^2-8x-6y+21=0$ $x^2-8x \quad +y^2-6y=-21$ $(x^2-8x+16)+(y^2-6y+9)=-21+16+9$ $(x-4)^2+(y-3)^2=4$ So, center is $(4,3)$ and radius is 2.

55. Completing the square gives us:

$$x^2+y^2-10x+6y+22=0$$
$$x^2-10x \quad +y^2+6y=-22$$
$$\left(x^2-10x+25\right)+\left(y^2+6y+9\right)=-22+25+9$$
$$(x-5)^2+(y+3)^2=12$$
$$(x-5)^2+(y-(-3))^2=12$$

So, center is $(5,-3)$ and radius is $\sqrt{12}=2\sqrt{3}$.

56. Completing the square gives us:

$$x^2+y^2+8x+2y-28=0$$
$$x^2+8x \quad +y^2+2y=28$$
$$\left(x^2+8x+16\right)+\left(y^2+2y+1\right)=28+16+1$$
$$(x+4)^2+(y+1)^2=45$$
$$(x-(-4))^2+(y-(-1))^2=45$$

So, center is $(-4,-1)$ and radius is $\sqrt{45}=3\sqrt{5}$.

57. Completing the square gives us:

$$x^2+y^2-6x-4y+1=0$$
$$x^2-6x \quad +y^2-4y=-1$$
$$\left(x^2-6x+9\right)+\left(y^2-4y+4\right)=-1+9+4$$
$$(x-3)^2+(y-2)^2=12$$

So, center is $(3,2)$ and radius is $\sqrt{12}=2\sqrt{3}$.

58. Completing the square gives us:

$$x^2+y^2-2x-10y+2=0$$
$$x^2-2x \quad +y^2-10y=-2$$
$$\left(x^2-2x+1\right)+\left(y^2-10y+25\right)=-2+1+25$$
$$(x-1)^2+(y-5)^2=24$$

So, center is $(1,5)$ and radius is $\sqrt{24}=2\sqrt{6}$.

59. Completing the square gives us:

$$x^2+y^2-x+y+\tfrac{1}{4}=0$$
$$x^2-x \quad +y^2+y=-\tfrac{1}{4}$$
$$\left(x^2-x+\tfrac{1}{4}\right)+\left(y^2+y+\tfrac{1}{4}\right)=-\tfrac{1}{4}+\tfrac{1}{4}+\tfrac{1}{4}$$
$$\left(x-\tfrac{1}{2}\right)^2+\left(y+\tfrac{1}{2}\right)^2=\tfrac{1}{4}$$
$$\left(x-\tfrac{1}{2}\right)^2+\left(y-(-\tfrac{1}{2})\right)^2=\tfrac{1}{4}$$

So, center is $\left(\tfrac{1}{2},-\tfrac{1}{2}\right)$ and radius is $\sqrt{\tfrac{1}{4}}=\tfrac{1}{2}$.

60. Completing the square gives us:

$$x^2+y^2-\tfrac{1}{2}x-\tfrac{3}{2}y+\tfrac{3}{8}=0$$
$$x^2-\tfrac{1}{2}x \quad +y^2-\tfrac{3}{2}y=-\tfrac{3}{8}$$
$$\left(x^2-\tfrac{1}{2}x+\tfrac{1}{16}\right)+\left(y^2-\tfrac{3}{2}y+\tfrac{9}{16}\right)=-\tfrac{3}{8}+\tfrac{1}{16}+\tfrac{9}{16}$$
$$\left(x-\tfrac{1}{4}\right)^2+\left(y-\tfrac{3}{4}\right)^2=\tfrac{1}{4}$$

So, center is $\left(\tfrac{1}{4},\tfrac{3}{4}\right)$ and radius is $\sqrt{\tfrac{1}{4}}=\tfrac{1}{2}$.

61. Assume that Columbia is located at (0,0). Then, Atlanta is located at (–215,0) and Savannah is located at $(0,-160)$. So, d(Atlanta, Savannah) is:

$$\sqrt{(-215-0)^2+(0-(-160))^2}\cong \boxed{268 \text{ miles}}$$

62. $d=\sqrt{(5-(-10))^2+(50-2)^2}\cong 50.29$, so the distance is approximately $\boxed{50.29 \text{ yards}}$.

63. $M=\left(\dfrac{2002+2004}{2},\dfrac{400+260}{2}\right)$

$$=(2003,330)$$

So, the estimated revenue in 2003 is $\boxed{\$330 \text{ million}}$.

64. $M = \left(\dfrac{1993+2001}{2}, \dfrac{56+28}{2} \right)$

$= (1997, 42)$

So, the average price of a ticket in 1997 was $\boxed{\$42}$.

65. The center of the inner circle is $(0,0)$. Since the diameter is 3000 ft., the radius is 1500 ft. So, the equation of the inner circle is $(x-0)^2 + (y-0)^2 = 1500^2$, that is $x^2 + y^2 = 2,250,000$.

66. The center of the outer circle is $(0,0)$. Since the diameter is 6000 ft., the radius is 3000 ft. So, the equation of the outer circle is $(x-0)^2 + (y-0)^2 = 3000^2$, that is $x^2 + y^2 = 9 \times 10^6$.

67. Location of tower = center of reception area = (0,0). We also know the radius is 200 miles. Hence, the equation of the circle is $\boxed{x^2 + y^2 = 40,000}$.

68. Assuming the fire station is located at the origin, we know that since the radius is 2 miles, the equation of the circle where fire has spread is $\boxed{x^2 + y^2 = 4}$

69. The equation is not linear – you need more than two points to plot the graph.

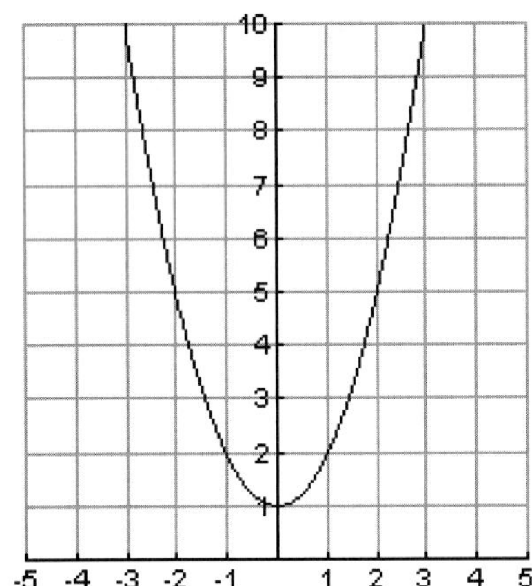

70.

To test for symmetry about the *x*-axis, replace *y* by $-y$, NOT *x* by $-x$. The conclusion should be that the graph is symmetric about the *y*-axis. The correct graph is given to the right:

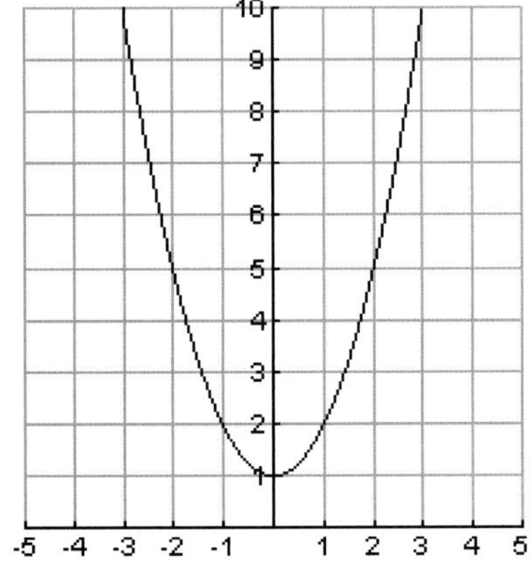

71. The center should be $(4, -3)$.

72. The radius should be $\sqrt{2}$.

73. False The correct conclusion would be that the point $(a,-b)$ must also be on the graph. For instance, $y^2 = x$ is symmetric about the x-axis and $(1,1)$ is on the graph, but $(-1,1)$ is not.

74. True
By definition of symmetry about the y-axis.

75. True

76. False. This is true only for linear equations.

77. Completing the square gives us:

$$x^2+y^2+10x-6y+34=0$$
$$x^2+10x \quad +y^2-6y=-34$$
$$\left(x^2+10x+25\right)+\left(y^2-6y+9\right)=-34+25+9$$
$$\left(x+5\right)^2+\left(y-3\right)^2=0$$

Thus, the graph consists of the single point $(-5,3)$.

78. Completing the square gives us:

$$x^2+y^2-4x+6y+49=0$$
$$x^2-4x \quad +y^2+6y=-49$$
$$\left(x^2-4x+4\right)+\left(y^2+6y+9\right)=-49+4+9$$
$$\left(x-2\right)^2+\left(y+3\right)^2=-36$$

Since no ordered pair of real numbers (x, y) can satisfy this equation, there is no graph.

79. <u>x-axis symmetry</u> (Replace y by $-y$):

$$-y=\frac{ax^2+b}{cx^3}$$
$$y=-\left(\frac{ax^2+b}{cx^3}\right)$$

No, since not equivalent to the original.

<u>y-axis symmetry</u> (Replace x by $-x$):

$$y=\frac{a(-x)^2+b}{c(-x)^3}=\frac{ax^2+b}{-cx^3}$$
$$y=-\left(\frac{ax^2+b}{cx^3}\right)$$

No, since not equivalent to the original.

<u>symmetry about origin</u> (Replace y by $-y$ and x by $-x$):

$$-y=\frac{a(-x)^2+b}{c(-x)^3}=\frac{ax^2+b}{-cx^3}=-\left(\frac{ax^2+b}{cx^3}\right) \text{ so that } y=\frac{ax^2+b}{cx^3}$$

Yes, since equivalent to the original.

80. Consider $y=(x-a)^2-b^2$.

<u>x-intercepts</u>: $0=(x-a)^2-b^2 \Rightarrow 0=\left(x-a-b\right)\left(x-a+b\right) \Rightarrow x=a+b,\, a-b$. So, $(a+b,0)$, $(a-b,0)$.

<u>y-intercept</u>: $y=(0-a)^2-b^2=a^2-b^2$. So, $\left(0,a^2-b^2\right)$.

81. The diameter is equal to the distance between $(5,2)$ and $(1,-6)$, which is given by $\sqrt{(5-1)^2+(2-(-6))^2}=\sqrt{80}=4\sqrt{5}$. So, the radius is $\frac{1}{2}(4\sqrt{5})=2\sqrt{5}$.
The center is the midpoint of the segment joining the points $(5,2)$ and $(1,-6)$, which is given by $\left(\frac{5+1}{2},\frac{2+(-6)}{2}\right)=(3,-2)$. Therefore, the equation is then given by:

$$(x-3)^2+(y-(-2))^2=(2\sqrt{5})^2$$
$$(x-3)^2+(y+2)^2=20$$

82. The diameter is equal to the distance between $(3,0)$ and $(-1,-4)$, which is given by $\sqrt{(3-(-1))^2+(0-(-4))^2}=\sqrt{32}=4\sqrt{2}$. So, the radius is $\frac{1}{2}(4\sqrt{2})=2\sqrt{2}$.
The center is the midpoint of the segment joining the points $(3,0)$ and $(-1,-4)$, which is given by $\left(\frac{3+(-1)}{2},\frac{0+(-4)}{2}\right)=(1,-2)$. Therefore, the equation is then given by:

$$(x-1)^2+(y+2)^2=\underbrace{(2\sqrt{2})^2}_{8}$$

83. Completing the square gives us:

$$x^2+y^2+ax+by+c=0$$
$$x^2+ax \quad +y^2+by=-c$$
$$\left(x^2+ax+\tfrac{a^2}{4}\right)+\left(y^2+by+\tfrac{b^2}{4}\right)=-c+\tfrac{a^2}{4}+\tfrac{b^2}{4}$$
$$\left(x+\tfrac{a}{2}\right)^2+\left(y+\tfrac{b}{2}\right)^2=-c+\tfrac{a^2}{4}+\tfrac{b^2}{4}$$

If $-c+\frac{a^2}{4}+\frac{b^2}{4}=0$, which is equivalent to $4c=a^2+b^2$, then the graph will consist of the single point $\left(-\frac{a}{2},-\frac{b}{2}\right)$.

84. Completing the square gives us:

$$x^2+y^2+ax+by+c=0$$
$$x^2+ax \quad +y^2+by=-c$$
$$\left(x^2+ax+\tfrac{a^2}{4}\right)+\left(y^2+by+\tfrac{b^2}{4}\right)=-c+\tfrac{a^2}{4}+\tfrac{b^2}{4}$$
$$\left(x+\tfrac{a}{2}\right)^2+\left(y+\tfrac{b}{2}\right)^2=-c+\tfrac{a^2}{4}+\tfrac{b^2}{4}$$

If $-c+\frac{a^2}{4}+\frac{b^2}{4}<0$, which is equivalent to $4c>a^2+b^2$, then there is no graph.

85.

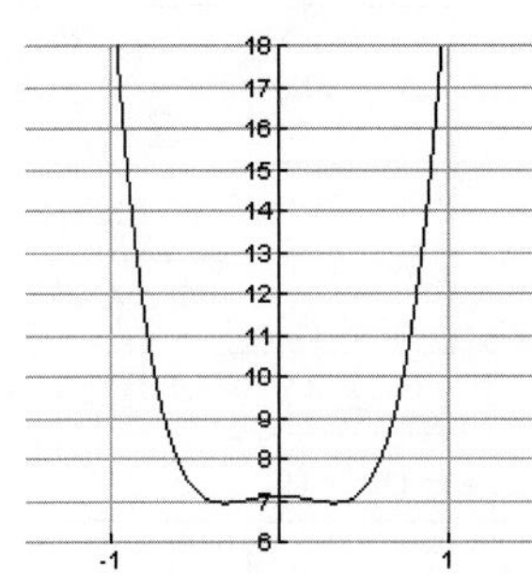

Symmetric with respect to the y-axis.

86.

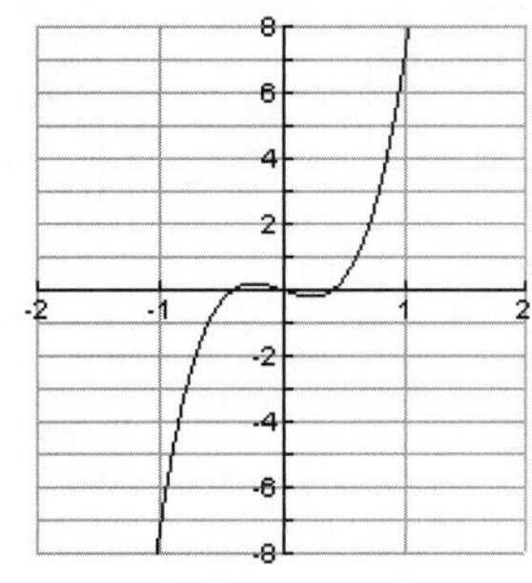

Symmetric with respect to the origin.

87. a. Completing the square yields

$$\left(x^2-11x\quad\right)+\left(y^2+3y\quad\right)=7.19$$

$$\left(x^2-11x+30.25\right)+\left(y^2+3y+2.25\right)=7.19+30.25+2.25=39.69$$

$$(x-5.5)^2+(y+1.5)^2=39.69 \quad \textbf{(1)}$$

b. Solving **(1)** for y yields:

$$(y+1.5)^2=39.69-(x-5.5)^2$$

$$y+1.5=\pm\sqrt{39.69-(x-5.5)^2}$$

$$y=-1.5\pm\sqrt{39.69-(x-5.5)^2}$$

c. and d. The graph in both parts **a** and **b** are the same, and are as follows:

88. a. Completing the square yields

$$\left(x^2+1.2x\quad\right)+\left(y^2-3.2y\quad\right)=-2.11$$

$$\left(x^2+1.2x+0.36\right)+\left(y^2-3.2y+2.56\right)=-2.11+0.36+2.56=0.81$$

$$(x+0.6)^2+(y-1.6)^2=0.81 \quad \textbf{(1)}$$

b. Solving **(1)** for y yields:

$$(y-1.6)^2=0.81-(x+0.6)^2$$

$$y-1.6=\pm\sqrt{0.81-(x+0.6)^2}$$

$$y=1.6\pm\sqrt{0.81-(x+0.6)^2}$$

c. and d. The graph in both parts **a** and **b** are the same, and are as follows:

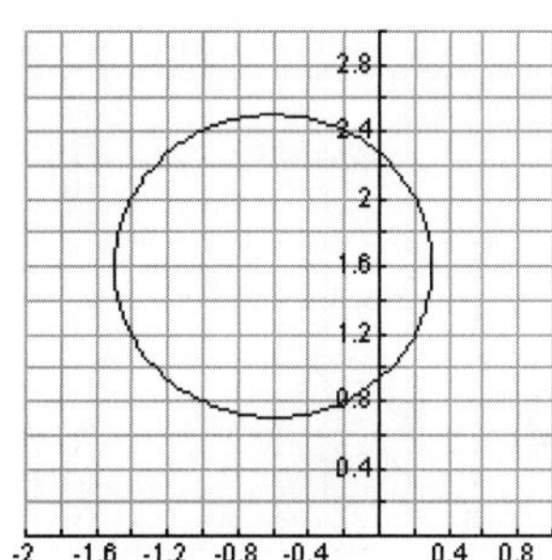

Section 0.6 Solutions

1. $m=\frac{3-6}{1-2}=\boxed{3}$	**2.** $m=\frac{1-9}{2-4}=\boxed{4}$	**3.** $m=\frac{5-(-3)}{-2-2}=\boxed{-2}$
4. $m=\frac{-4-6}{-1-4}=\boxed{2}$	**5.** $m=\frac{9-(-10)}{-7-3}=\boxed{-\frac{19}{10}}$	**6.** $m=\frac{-3-6}{11-(-2)}=\boxed{-\frac{9}{13}}$

7. $m=\frac{-1.7-5.2}{0.2-3.1}=\frac{6.9}{2.9}\cong\boxed{2.379}$

8. $m=\frac{1.7-(-2.3)}{-2.4-(-5.6)}=\frac{4.0}{3.2}=\boxed{1.25}$

9. $m=\frac{-\frac{1}{4}-\left(-\frac{3}{4}\right)}{\frac{2}{3}-\frac{5}{6}}=-\frac{\frac{1}{2}}{\frac{1}{6}}=\boxed{-3}$

10. $m=\frac{\frac{3}{5}-\frac{7}{5}}{\frac{1}{2}-\left(-\frac{3}{4}\right)}=-\frac{\frac{4}{5}}{\frac{5}{4}}=\boxed{-\frac{16}{25}}$

11. x-intercept: $(0.5,0)$
y-intercept: $(0,-1)$
slope: $m=\frac{-3-3}{-1-2}=2$
rising

12. x-intercept: $(-2,0)$
y-intercept: $(0,3)$
slope: $m=\frac{-3-3}{-4-0}=\frac{3}{2}$
rising

13. x-intercept: $(1,0)$
y-intercept: $(0,1)$
slope:
$m=\frac{3-(-1)}{-2-2}=-1$
falling

14. x-intercept: $(0,0)$
y-intercept: $(0,0)$
slope: $m=\frac{3-(-2)}{-3-2}=-1$
falling

15. x-intercept: None
y-intercept: $(0,1)$
slope: $m=0$
horizontal

16. x-intercept: $(-4,0)$
y-intercept: None
slope: Undefined
vertical

17.
x-intercept:
$0=2x-3$
$\frac{3}{2}=x$ So, $\left(\frac{3}{2},0\right)$.
y-intercept: $y=2(0)-3=-3$. So, $(0,-3)$.

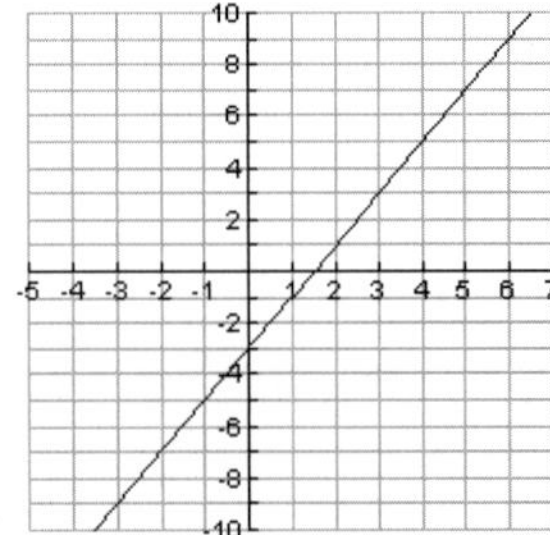

18.
x-intercept:
$0=-3x+2$
$\frac{2}{3}=x$ So, $\left(\frac{2}{3},0\right)$.
y-intercept: $y=-3(0)+2=2$. So, $(0,2)$.

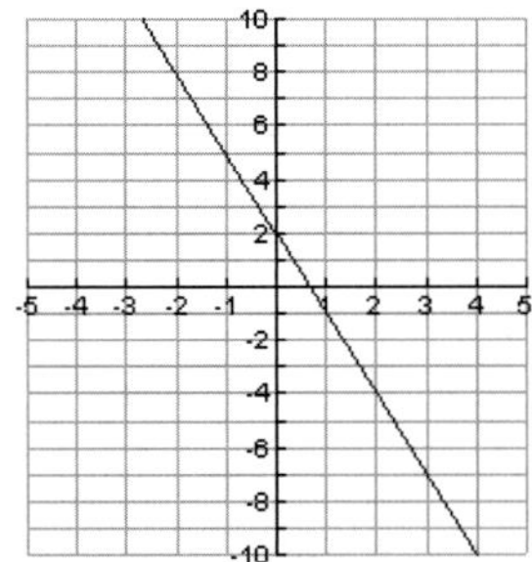

19.

x-intercept:

$$0 = -\tfrac{1}{2}x + 2 \quad \text{So, } (4,0).$$
$$4 = x$$

y-intercept: $y = -\tfrac{1}{2}(0) + 2 = 2$. So, $(0,2)$.

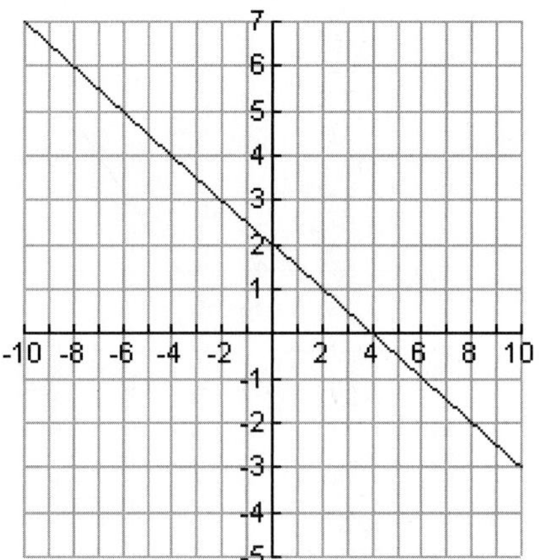

20.

x-intercept:

$$0 = \tfrac{1}{3}x - 1 \quad \text{So, } (3,0).$$
$$3 = x$$

y-intercept: $y = \tfrac{1}{3}(0) - 1 = -1$. So, $(0,-1)$.

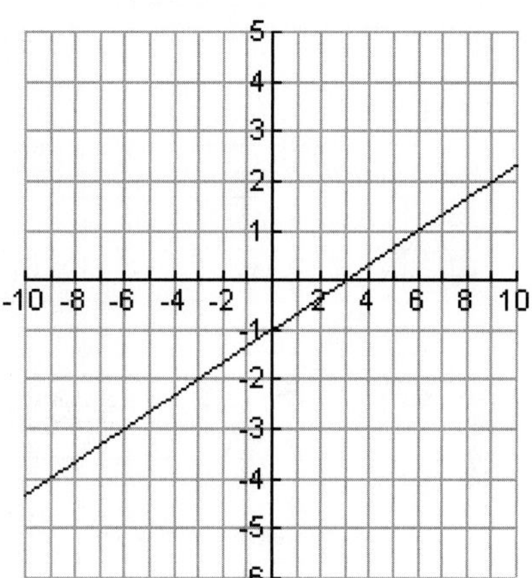

21.

x-intercept:

$$4 = 2x - 3(0) \quad \text{So, } (2,0).$$
$$2 = x$$

y-intercept:

$$4 = 2(0) - 3y$$
$$-\tfrac{4}{3} = y \quad \text{. So, } \left(0,-\tfrac{4}{3}\right).$$

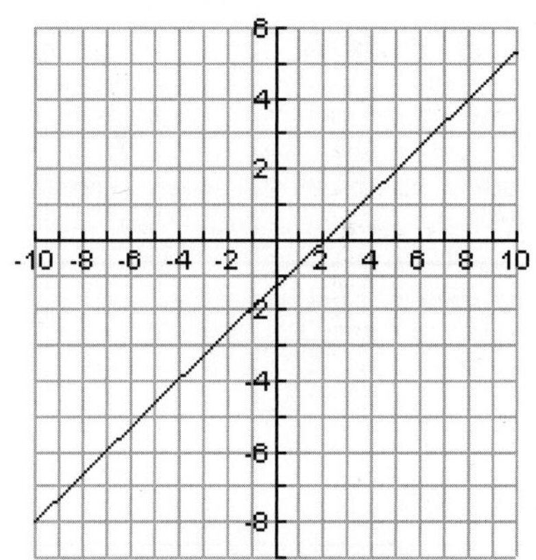

22.

x-intercept:

$$-1 = -x + 0 \quad \text{So, } (1,0).$$
$$1 = x$$

y-intercept:

$$-1 = 0 + y$$
$$-1 = y \quad \text{. So, } (0,-1).$$

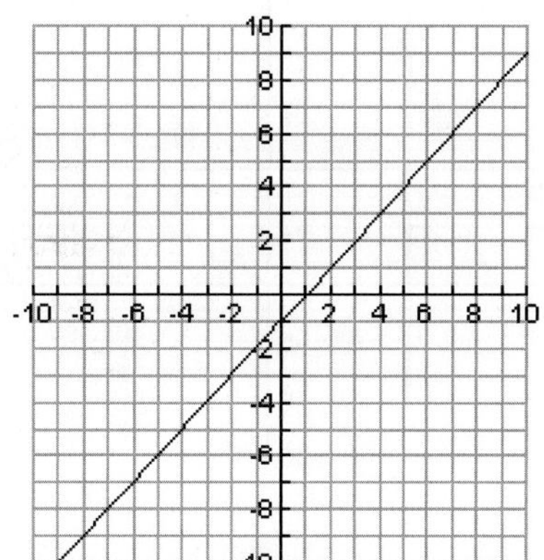

23.

<u>*x*-intercept</u>:

$$-1 = \tfrac{1}{2}x + \tfrac{1}{2}(0) \qquad \text{So, } (-2,0).$$
$$-2 = x$$

<u>*y*-intercept</u>:

$$-1 = \tfrac{1}{2}(0) + \tfrac{1}{2}y \quad \text{. So, } (0,-2).$$
$$-2 = y$$

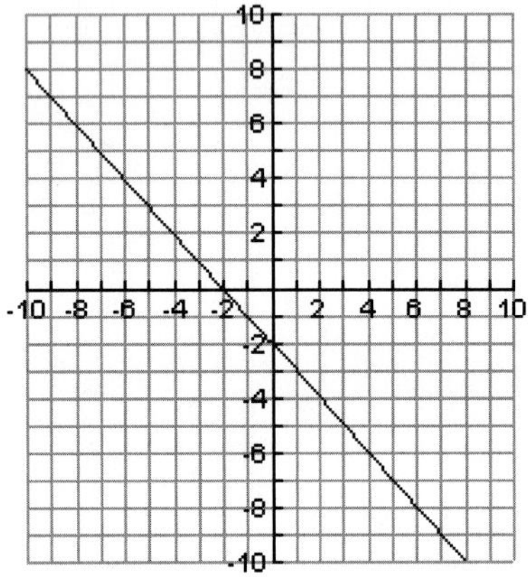

24.

<u>*x*-intercept</u>:

$$\tfrac{1}{12} = \tfrac{1}{3}x - \tfrac{1}{4}(0) \qquad \text{So, } \left(\tfrac{1}{4},0\right).$$
$$\tfrac{1}{4} = x$$

<u>*y*-intercept</u>:

$$\tfrac{1}{12} = \tfrac{1}{3}(0) - \tfrac{1}{4}y \quad \text{. So, } \left(0,-\tfrac{1}{3}\right).$$
$$-\tfrac{1}{3} = y$$

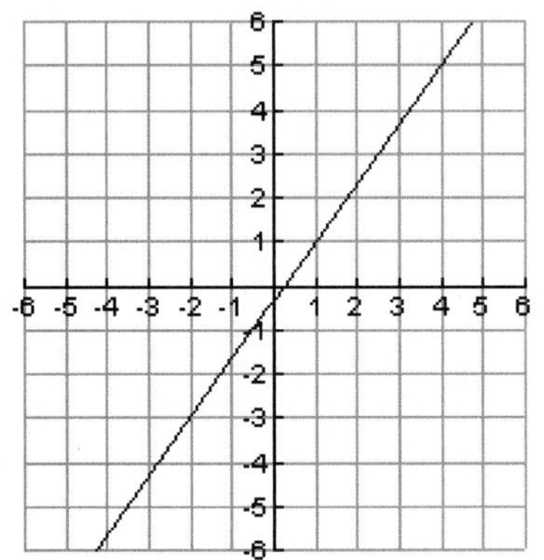

25.

x-intercept: $(-1,0)$

y-intercept: None

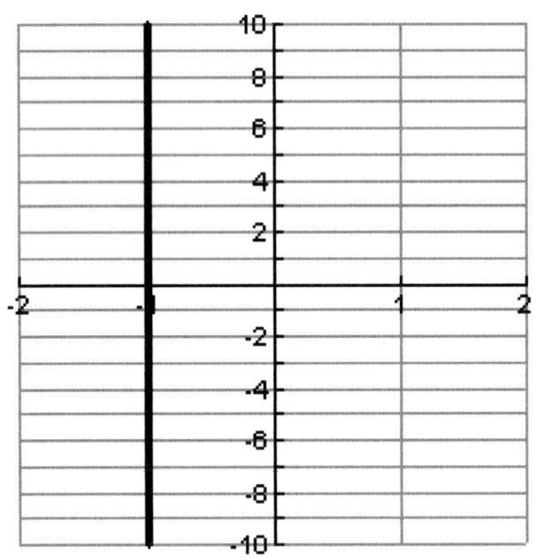

26.

x-intercept: None

y-intercept: $(0,-3)$

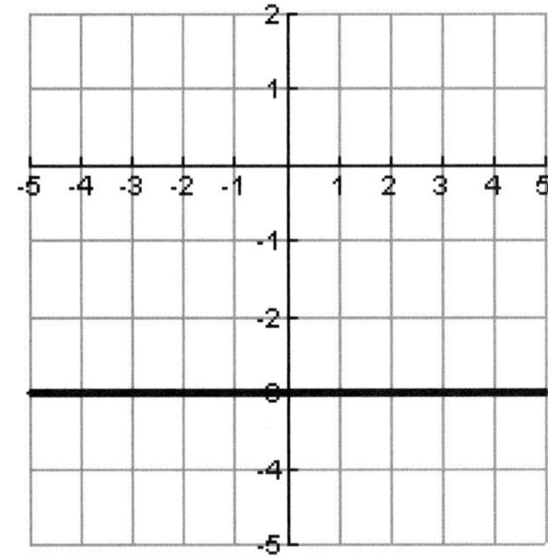

27. x-intercept: None y-intercept: $(0,1.5)$ 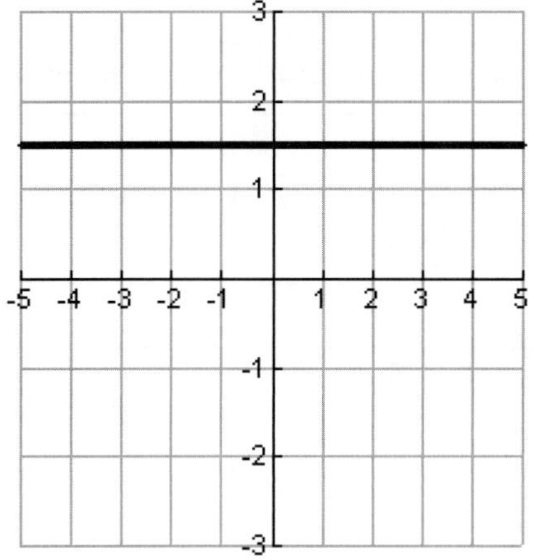	**28.** x-intercept: $(-7.5,0)$ y-intercept: None 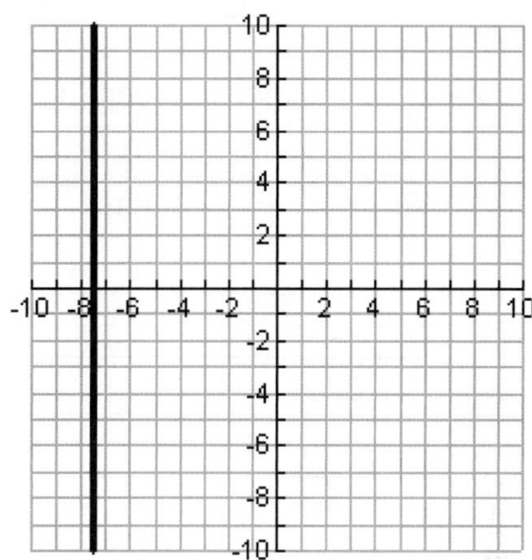
29. x-intercept: $\left(-\frac{7}{2},0\right)$ y-intercept: None 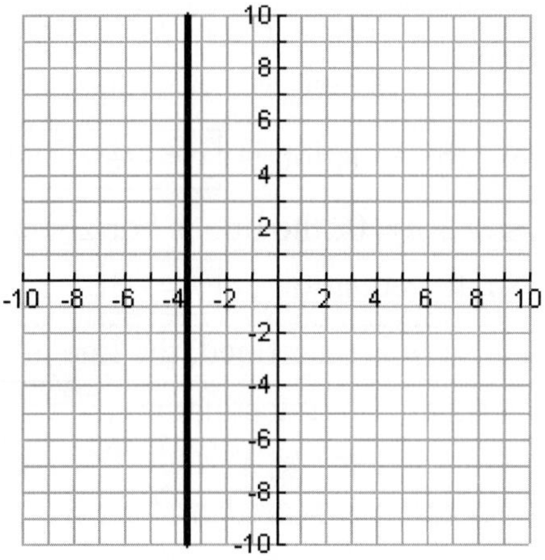	**30.** x-intercept: None y-intercept: $\left(0,\frac{5}{3}\right)$ 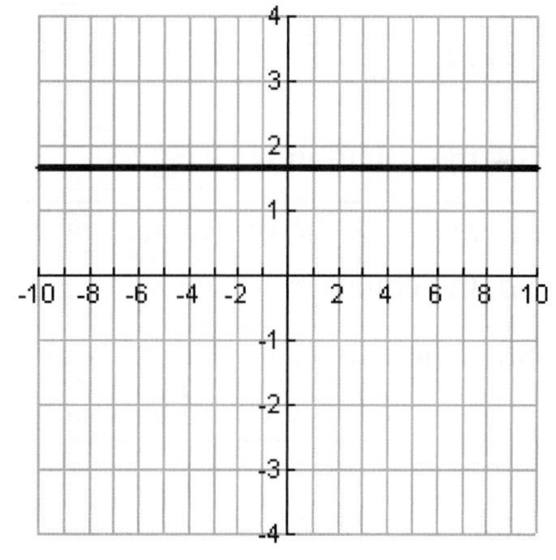
31. $y=\frac{2}{5}x-2 \quad m=\frac{2}{5} \quad y-$intercept: $(0,-2)$	**32.** $y=\frac{3}{4}x-3 \quad m=\frac{3}{4} \quad y-$intercept: $(0,-3)$
33. $y=-\frac{1}{3}x+2 \quad m=-\frac{1}{3} \quad y-$intercept: $(0,2)$	**34.** $y=-\frac{1}{2}x+4 \quad m=-\frac{1}{2} \quad y-$intercept: $(0,4)$

35. $y = 4x - 3 \quad m = 4 \quad y-$intercept: $(0, -3)$	**36.** $y = x - 5 \quad m = 1 \quad y-$intercept: $(0, -5)$
37. $y = -2x + 4 \quad m = -2 \quad y-$intercept: $(0, 4)$	**38.** $y = \frac{1}{4}x - \frac{1}{2} \quad m = \frac{1}{4} \quad y-$intercept: $\left(0, -\frac{1}{2}\right)$
39. $y = \frac{2}{3}x - 2 \quad m = \frac{2}{3} \quad y-$intercept: $(0, -2)$	**40.** $y = -4x + 3 \quad m = -4 \quad y-$intercept: $(0, 3)$
41. $y = -\frac{3}{4}x + 6 \quad m = -\frac{3}{4} \quad y-$intercept: $(0, 6)$	**42.** $y = -\frac{5}{8}x + 5 \quad m = -\frac{5}{8} \quad y-$intercept: $(0, 5)$
43. $y = 2x + 3$	**44.** $y = -2x + 1$
45. $y = -\frac{1}{3}x + 0 = -\frac{1}{3}x$	**46.** $y = \frac{1}{2}x - 3$
47. $y = 2$	**48.** $y = -1.5$
49. $x = \frac{3}{2}$	**50.** $x = -3.5$
51. Find b: $-3 = 5(-1) + b$, $2 = b$. So, the equation is $\boxed{y = 5x + 2}$.	**52.** Find b: $-1 = 2(1) + b$, $-3 = b$. So, the equation is $\boxed{y = 2x - 3}$.
53. Find b: $2 = -3(-2) + b$, $-4 = b$. So, the equation is $\boxed{y = -3x - 4}$.	**54.** Find b: $-4 = -1(3) + b$, $-1 = b$. So, the equation is $\boxed{y = -x - 1}$.
55. Find b: $-1 = \frac{3}{4}(1) + b$, $-\frac{7}{4} = b$. So, the equation is $\boxed{y = \frac{3}{4}x - \frac{7}{4}}$.	**56.** Find b: $3 = -\frac{1}{7}(-5) + b$, $\frac{16}{7} = b$. So, the equation is $\boxed{y = -\frac{1}{7}x + \frac{16}{7}}$.
57. Since $m = 0$, the line is horizontal. So, the equation is $\boxed{y = 4}$.	**58.** Since $m = 0$, the line is horizontal. So, the equation is $\boxed{y = -3}$.
59. Since m is undefined, the line is vertical. So, the equation is $\boxed{x = -1}$.	**60.** Since m is undefined, the line is vertical. So, the equation is $\boxed{x = 4}$.

61.
slope:
$$m = \frac{-1-2}{-2-3} = \frac{3}{5}$$
y-intercept:
Use the point $(-2,-1)$ to find b:
$$-1 = \tfrac{3}{5}(-2)+b$$
$$\tfrac{1}{5} = b$$
So, the equation is $\boxed{y = \tfrac{3}{5}x + \tfrac{1}{5}}$.

62.
slope:
$$m = \frac{-3-1}{-4-5} = \frac{4}{9}$$
y-intercept:
Use the point $(-4,-3)$ to find b:
$$-3 = \tfrac{4}{9}(-4)+b$$
$$-\tfrac{11}{9} = b$$
So, the equation is $\boxed{y = \tfrac{4}{9}x - \tfrac{11}{9}}$.

63.
slope:
$$m = \frac{-1-(-6)}{-3-(-2)} = -5$$
y-intercept:
Use the point $(-3,-1)$ to find b:
$$-1 = -5(-3)+b$$
$$-16 = b$$
So, the equation is $\boxed{y = -5x - 16}$.

64.
slope:
$$m = \frac{-8-(-2)}{-5-7} = \frac{1}{2}$$
y-intercept:
Use the point $(-5,-8)$ to find b:
$$-8 = \tfrac{1}{2}(-5)+b$$
$$-\tfrac{11}{2} = b$$
So, the equation is $\boxed{y = \tfrac{1}{2}x - \tfrac{11}{2}}$.

65.
slope:
$$m = \frac{-37-(-42)}{20-(-10)} = \frac{1}{6}$$
y-intercept:
Use the point $(20,-37)$ to find b:
$$-37 = \tfrac{1}{6}(20)+b$$
$$-\tfrac{121}{3} = b$$
So, the equation is $\boxed{y = \tfrac{1}{6}x - \tfrac{121}{3}}$.

66.
slope:
$$m = \frac{12-(-12)}{-8-(-20)} = 2$$
y-intercept:
Use the point $(-8,12)$ to find b:
$$12 = 2(-8)+b$$
$$28 = b$$
So, the equation is $\boxed{y = 2x + 28}$.

67. slope: $m=\frac{4-(-5)}{-1-2}=-3$ y-intercept: Use the point $(-1,4)$ to find b: $4=-3(-1)+b$ $1=b$ So, the equation is $\boxed{y=-3x+1}$.	**68.** slope: $m=\frac{3-(-3)}{-2-2}=-\frac{3}{2}$ y-intercept: Use the point $(-2,3)$ to find b: $3=-\frac{3}{2}(-2)+b$ $0=b$ So, the equation is $\boxed{y=-\frac{3}{2}x}$.
69. slope: $m=\frac{\frac{3}{4}-\frac{9}{4}}{\frac{1}{2}-\frac{3}{2}}=\frac{3}{2}$ y-intercept: Use the point $\left(\frac{1}{2},\frac{3}{4}\right)$ to find b: $\frac{3}{4}=\frac{3}{2}\left(\frac{1}{2}\right)+b$ $0=b$ So, the equation is $\boxed{y=\frac{3}{2}x}$.	**70.** slope: $m=\frac{-\frac{1}{2}-\frac{1}{2}}{-\frac{2}{3}-\frac{7}{3}}=\frac{1}{3}$ y-intercept: Use the point $\left(-\frac{2}{3},-\frac{1}{2}\right)$ to find b: $-\frac{1}{2}=\frac{1}{3}\left(-\frac{2}{3}\right)+b$ $-\frac{5}{18}=b$ So, the equation is $\boxed{y=\frac{1}{3}x-\frac{5}{18}}$.
71. Since m is undefined, the line is vertical. So, the equation is $\boxed{x=3}$.	**72.** Since m is undefined, the line is vertical. So, the equation is $\boxed{x=-5}$.
73. Since $m=\frac{7-7}{3-9}=0$, the line is horizontal. So, the equation is $\boxed{y=7}$.	**74.** Since $m=\frac{-1-(-1)}{-2-3}=0$, the line is horizontal. So, the equation is $\boxed{y=-1}$.
75. The slope is $m=\frac{6-0}{0-(-5)}=\frac{6}{5}$. Since $(0,6)$ is the y-intercept, $b=6$. So, the equation is $\boxed{y=\frac{6}{5}x+6}$.	**76.** Since m is undefined, the line is vertical. So, the equation is $\boxed{x=0}$.
77. Since m is undefined, the line is vertical. So, the equation is $\boxed{x=-6}$.	**78.** Since m is undefined, the line is vertical. So, the equation is $\boxed{x=-9}$.
79. The slope is undefined. So, the equation of the line passing through the points $\left(\frac{2}{5},-\frac{3}{4}\right)$ and $\left(\frac{2}{5},\frac{1}{2}\right)$ is $\boxed{x=\frac{2}{5}}$.	**80.** Since m is undefined, the line is vertical. So, the equation is $\boxed{x=\frac{1}{3}}$.

81. We identify the slope and y-intercept, and then express the equation in slope-intercept form as $\boxed{y = x - 1}$.	**82.** We identify the slope and y-intercept, and then express the equation in slope-intercept form as $\boxed{y = x + 1}$.
83. We identify the slope and y-intercept, and then express the equation in slope-intercept form as $\boxed{y = -2x + 3}$.	**84.** We identify the slope and y-intercept, and then express the equation in slope-intercept form as $\boxed{y = -4x + 2}$.
85. We identify the slope and y-intercept, and then express the equation in slope-intercept form as $\boxed{y = -\frac{1}{2}x + 1}$.	**86.** We identify the slope and y-intercept, and then express the equation in slope-intercept form as $\boxed{y = -\frac{1}{3}x + 2}$.
87. slope: Since parallel to $y = 2x - 1$, $m = 2$. y-intercept: Use the point $(-3,1)$ to find b: $1 = 2(-3) + b$ $7 = b$ So, the equation is $\boxed{y = 2x + 7}$.	**88.** slope: Since parallel to $y = -x + 2$, $m = -1$. y-intercept: Use the point $(1,3)$ to find b: $3 = -1(1) + b$ $4 = b$ So, the equation is $\boxed{y = -x + 4}$.
89. slope: Since perpendicular to $2x + 3y = 12$ (whose slope is $-\frac{2}{3}$), $m = \frac{3}{2}$. y-intercept: Since $(0,0)$ is assumed to be on the line and is the y-intercept, $b = 0$. So, the equation is $\boxed{y = \frac{3}{2}x}$.	**90.** slope: Since perpendicular to $x - y = 7$ (whose slope is 1), $m = -1$. y-intercept: Since $(0,6)$ is assumed to be on the line and is the y-intercept, $b = 6$. So, the equation is $\boxed{y = -x + 6}$.
91. Since parallel to x-axis, the line is horizontal. So, its equation is of the form $y = b$. Since $(3,5)$ is assumed to be on the line, the equation must be $\boxed{y = 5}$.	**92.** Since parallel to y-axis, the line is vertical. So, its equation is of the form $x = a$. Since $(3,5)$ is assumed to be on the line, the equation must be $\boxed{x = 3}$.
93. Since perpendicular to y-axis, the line is horizontal. So, its equation is of the form $y = b$. Since $(-1,2)$ is assumed to be on the line, the equation must be $\boxed{y = 2}$.	**94.** Since perpendicular to x-axis, the line is vertical. So, its equation is of the form $x = a$. Since $(-1,2)$ is assumed to be on the line, the equation must be $\boxed{x = -1}$.

95.
slope:
Since parallel to $\frac{1}{2}x - \frac{1}{3}y = 5$ (whose slope is $\frac{3}{2}$), $m = \frac{3}{2}$.
y-intercept:
Use the point $(-2,-7)$ to find b:

$$-7 = \tfrac{3}{2}(-2) + b$$
$$-4 = b$$

So, the equation is $\boxed{y = \frac{3}{2}x - 4}$.

96.
slope:
Since perpendicular to $-\frac{2}{3}x + \frac{3}{2}y = -2$ (whose slope is $\frac{4}{9}$), $m = -\frac{9}{4}$.
y-intercept:
Use the point $(1,4)$ to find b:

$$4 = -\tfrac{9}{4}(1) + b$$
$$\tfrac{25}{4} = b$$

So, the equation is $\boxed{y = -\frac{9}{4}x + \frac{25}{4}}$.

97. Let h = number of hours a job lasts. The cost (in dollars) is given by $C(h) = 1200 + 25h$. So, for $h = 32$, $C = 2000$. This means that a $\boxed{\text{32-hour job will cost \$2,000}}$.

98. Let x = number of miles. The cost (in dollars) is given by $\boxed{C = 50 + 0.39x}$.

99. Let L = monthly loan payment and x = # times filled up gas tank.
Then, $500 = L + 25x$. So, if $x = 5$, then $500 = L + 125$, so that $L = 375$.
The monthly payment is $\boxed{\$375}$.

100. Let B = monthly loan payment
M = cost of filling gas tank once
x = # times filled up gas tank

Then, the corresponding total monthly cost is given by $C = B + Mx$.
We are given that $(3,520)$ and $(5,600)$ satisfy the equation.
We compute M and B as follows:

slope: $M = \dfrac{520 - 600}{3 - 5} = 40$

y-intercept: Use $(3,520)$ to find B:

$$520 = B + 40(3)$$
$$400 = B$$

Thus, the monthly loan payment is \$400 and it costs \$40 every time you fill up with gasoline.

101. Assume that F and C are related by $F = mC + b$.
We are given that $(25,77)$ and $(20,68)$ (where the points are listed in the form (C,F)) satisfy the equation. We compute m and b as follows:

slope: $m = \dfrac{77-68}{25-20} = \dfrac{9}{5}$ y-intercept: Use $(25,77)$ to find b:

$$77 = \tfrac{9}{5}(25) + b$$
$$32 = B$$

So, the equation relating F and C is given by $\boxed{F = \tfrac{9}{5}C + 32}$.

The temperature for which $F = C$ can be found by substituting $F = C$ into the equation:

$$C = \tfrac{9}{5}C + 32$$
$$-\tfrac{4}{5}C = 32$$
$$-40 = C$$

Thus, $\boxed{-40 \text{ degrees Celsius} = -40 \text{ degrees Fahrenheit}}$.

102. Assume $C = mx + b$, where C is temperature (in degrees Celsius) and x is elevation (in feet). We are given that $(0,15)$ and $(500,14)$ satisfy the equation. So, We compute m and $b$as follows:

slope: $m = \dfrac{15-14}{0-500} = -\dfrac{1}{500}$

y-intercept: Since $(0,15)$ is assumed to be on the line and is the y-intercept, $b = 15$.

So, the equation is $\boxed{C = -\dfrac{1}{500}x + 15}$.

The expected temperature at 2500 ft. is $\boxed{10 \text{ degrees Celsius}}$.

103. Since $(1900,67)$ and $(2000,69)$ are assumed to be on the line, the slope of the line is $m = \dfrac{69-67}{2000-1900} = \dfrac{2}{100} = \dfrac{1}{50}$. So, the rate of change (in inches per year) is $\boxed{\dfrac{1}{50}}$.

104. Since $(1900,64)$ and $(2000,67)$ are assumed to be on the line, the slope of the line is $m = \dfrac{67-64}{2000-1900} = 0.03$. So, the rate of change (in inches per year) is $\boxed{0.03}$.

105. Note: 6 pounds 4 ounces = 100 ounces 6 pounds 10 ounces = 106 ounces

Since $(1900,100)$ and $(2000,106)$ are assumed to be on the line, the slope of the line is $m=\frac{106-100}{2000-1900}=0.06$. So, the rate of change (in ounces per year) is $\boxed{0.06}$.

To determine the expected weight in the year 2040, we need the equation of the line. So far, we know the form of the equation is $y=0.06x+b$. Use $(1900,100)$ to find b:

$$100=0.06(1900)+b$$
$$-14=b$$

So, the equation is $y=0.06x-14$. Therefore, in 2040, $y=0.06(2040)-14=108.4$ oz. Since 108.4 oz. = 6.775 pounds, and 0.775 pounds = 12.4 oz., we expect a baby to weigh $\boxed{\text{6 pounds 12.4 oz}}$ in 2040.

106. Since $(1906,4.5)$ and $(1957,4)$ are assumed to be on the line, the slope of the line is $m=\frac{4.5-4}{1906-1957}\cong-0.01$. So, the rate of change is $\boxed{-0.01}$.

107. The y-intercept represents the flat monthly fee of $\boxed{\$35}$ since $x=0$ implies that no long distance minutes were used.

108. x-intercept: Represents the age at which the car is worth \$0. It is as follows:

$$0=11,100-1850x \Rightarrow x=6.$$

So, the x-intercept is (6,0).

y-intercept: Represents the initial value of the car when it is brand new. It is (0, 11,100).

109. The rate of change is

$$\frac{3.8-5.2}{4-0}=-\frac{1.4}{4}=-0.35 \text{ in./yr.}$$

Using the point (0, 5.2), and assuming that $x=0$ corresponds to 2003, the equation that governs rainfall per year is given by

$$y-5.2=-0.35(x-0) \Rightarrow y=-0.35x+5.2.$$

Since 2010 corresponds to $x=7$, we see that $y=-0.35(7)+5.2=2.75$. So, the expected average rainfall for July in 2010 is $\boxed{\text{2.75 inches}}$.

110. The rate of change is

$$\frac{44.5-43}{2-0}=\frac{1.5}{2}=0.75\ \circ\text{F/yr.}$$

Using the point (0, 43), and assuming that $x = 0$ corresponds to Jan. 2005, the equation that governs this trend is given by

$$y-43=0.75(x-0)\ \Rightarrow\ y=0.75x+43.$$

Hence, in Jan. 2010, you can expect the average temperature to be $0.75(5)+43=$ $\boxed{46.75^{\circ}\text{F}}$.

111. The rate of change is

$$\frac{392-380}{5-0}=2.4\ \text{plastic bags per year (in billions).}$$

Using the point (0, 380), and assuming that $x = 0$ corresponds to 2000, the equation that governs this trend is given by

$$y-380=2.4(x-0)\ \Rightarrow\ y=2.4x+380.$$

Hence, in 2010, you can expect the average number of plastic bags used to be $380+2.4(10)=\boxed{404\text{ billion}}$.

112. The rate of change is

$$\frac{788-744}{2-0}=22\ \text{dollars per year}$$

Using the point (0, 744), and assuming that $x = 0$ corresponds to 2004, the equation that governs this trend is given by

$$y-744=22(x-0)\ \Rightarrow\ y=22x+744.$$

Hence, in 2008, you can expect the average number of dollars owed to be $744+22(4)=\boxed{832\text{ dollars}}$.

113. The computations used to calculate the x- and y-intercepts should be reversed. So, the x-intercept is $(3,0)$ and the y-intercept is $(0,-2)$.

114. The denominator of the slope should be $4-(-2)$, resulting in $m=-\frac{1}{3}$.

115. The denominator and numerator in the slope computation should be switched, resulting in the slope being undefined.

116. These two are listed incorrectly: **a.** horizontal **b.** vertical

117. True. Since the equation $mx+b=0$ has at most one solution.

118. False. Any vertical line $x=a\ (a\neq 0)$ does not have a y-intercept.

119. False. The lines are perpendicular.

120. True.

121. Any vertical line is perpendicular to a line with slope 0.

122. Any vertical line is parallel to a line with no slope.

123. Since $B \neq 0$, $Ax + By = C$ can be written as $y = -\frac{A}{B}x + \frac{C}{B}$. Since the desired line is parallel to this line, its slope $m = -\frac{A}{B}$. Use the point $(-B, A+1)$ on the line to find the y-intercept:

$$\begin{aligned} A+1 &= -\tfrac{A}{B}(-B) + b \\ A+1 &= A+b \\ b &= 1 \end{aligned}$$

So, the equation of the line in this case is $y = -\frac{A}{B}x + 1$.

124. We know that the point $(B, A-1)$ is on the line. Since it is parallel to $Ax + By = C$, and the slope of this line is $-\frac{A}{B}$ (assuming $B \neq 0$), the equation of the desired line is

$$\begin{aligned} y-(A-1) &= -\tfrac{A}{B}(x-B) \\ y &= (A-1) - \tfrac{A}{B}x + A \\ &\boxed{y = -\tfrac{A}{B}x + (2A-1)} \end{aligned}$$

125. We know that the point $(-A, B-1)$ is on the line. Since it is perpendicular to $Ax + By = C$, and the slope of this line is $-\frac{A}{B}$ (assuming $B \neq 0$), the slope of the desired line is $\frac{B}{A}$ (assuming $A \neq 0$). So, the equation of the desired line is

$$\begin{aligned} y-(B-1) &= \tfrac{B}{A}(x+A) \\ y &= (B-1) + \tfrac{B}{A}x + B \\ &\boxed{y = \tfrac{B}{A}x + (2B-1)} \end{aligned}$$

126. Case 1: $A \neq 0$ and $B \neq 0$

In such case, $Ax + By = C$ can be written as $y = -\frac{A}{B}x + \frac{C}{B}$. Since the desired line is perpendicular to this line, its slope $m = \frac{B}{A}$. Use the point $(A, B+1)$ on the line to find the y-intercept:

$$\begin{aligned} B+1 &= \tfrac{B}{A}(A) + b \\ b &= 1 \end{aligned}$$

So, the equation of the line in this case is $y = \frac{B}{A}x + 1$.

Case 2: $A = 0$ and $B \neq 0$

In such case, the line $Ax + By = C$ can be written as $y = \frac{C}{B}$, which is horizontal. Since the desired line is perpendicular to this one, it must be vertical. So, since $(A, B+1)$ is on the line, its equation is $x = A$, which is $x = 0$.

Case 3: $A \neq 0$ and $B = 0$

In such case, the line $Ax + By = C$ can be written as $x = \frac{C}{A}$, which is vertical. Since the desired line is perpendicular to this one, it must be horizontal. So, since $(A, B+1)$ is on the line, its equation is $y = B+1$, which is $y = 1$.

127. Let $y_1 = mx + b_1$ and $y_2 = mx + b_2$, assuming that $b_1 \neq b_2$. At a point of intersection of these two lines, $y_1 = y_2$. This is equivalent to $mx + b_1 = mx + b_2$, which implies $b_1 = b_2$, which contradicts our assumption. Hence, there are no points of intersection.

128. Solving $m_1 x + b_1 = m_2 x + b_2$ yields

$$(m_1 - m_2)x = b_2 - b_1 \Rightarrow \boxed{x = \frac{b_2 - b_1}{m_1 - m_2}}.$$

This is the x-coordinate of the intersection point.

129. Slope of $y_1 = 17x + 22$ is 17.

Slope of $y_2 = -\frac{1}{17}x - 13$ is $-\frac{1}{17}$.

Since $17\left(-\frac{1}{17}\right) = -1$, $\boxed{\text{perpendicular}}$.

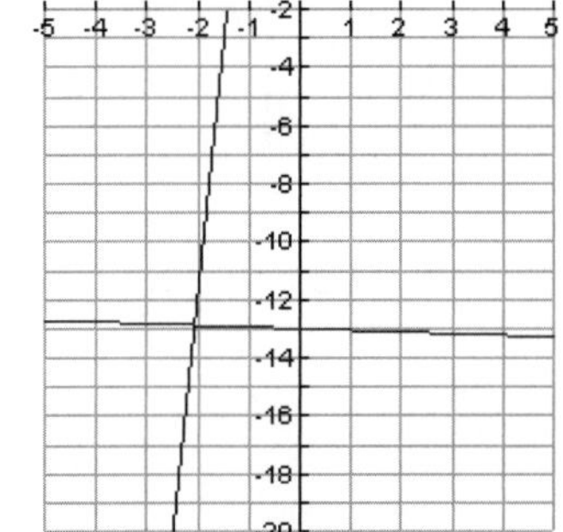

130. Slope of $y_1 = 0.35x + 2.7$ is 0.35.

Slope of $y_2 = 0.35x - 1.2$ is 0.35.

Since they have the same slope, $\boxed{\text{parallel}}$.

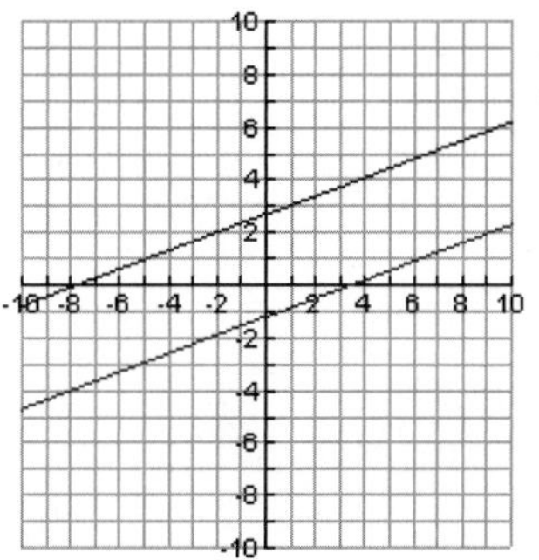

131. Slope of $y_1 = 0.25x + 3.3$ is 0.25.

Slope of $y_2 = -4x + 2$ is -4.

Since $0.25(-4) = -1$, they are $\boxed{\text{perpendicular}}$.

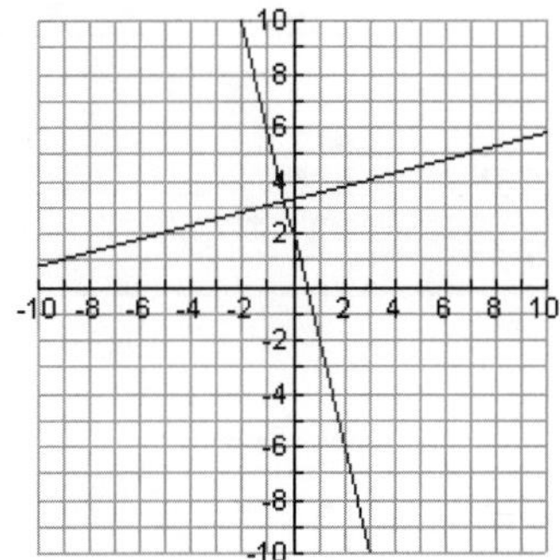

132. Slope of $y_1 = \frac{1}{2}x + 5$ is $\frac{1}{2}$.

Slope of $y_2 = 2x - 3$ is 2.

Since $\frac{1}{2}(2) \neq -1$ and $\frac{1}{2} \neq 2$, they are $\boxed{\text{neither}}$ parallel nor perpendicular.

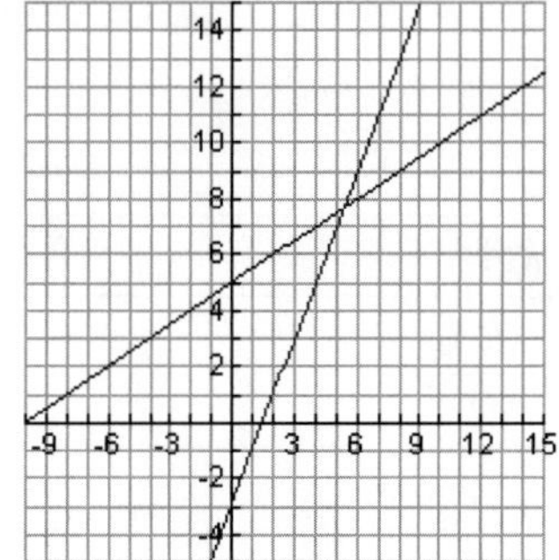

133. Slope of $y_1 = 0.16x + 2.7$ is 0.16. Slope of $y_2 = 6.25x - 1.4$ is 6.25. Since $(0.16)(6.25) \neq -1$ and $0.16 \neq 6.25$, they are neither parallel nor perpendicular. 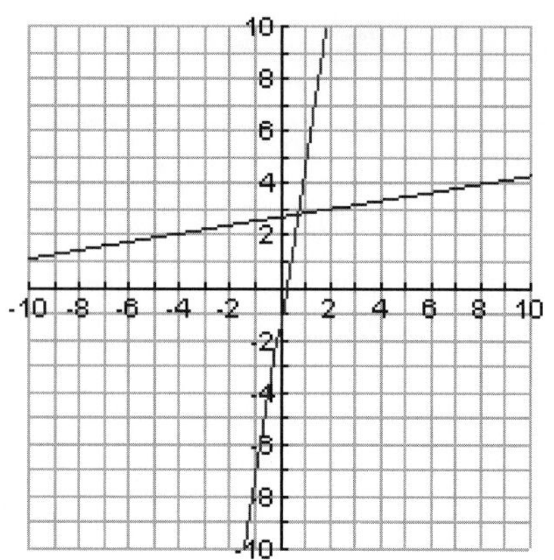	**134.** Slope of $y_1 = -3.75x + 8.2$ is -3.75. Slope of $y_2 = \frac{4}{15}x + \frac{5}{6}$ is $\frac{4}{15}$. Since $(-3.75)(\frac{4}{15}) = -1$ they are perpendicular. 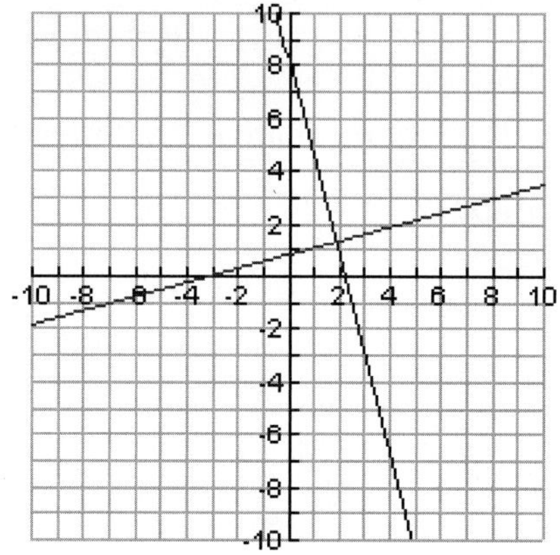

Section 0.7 Solutions --

1. $y = kx$	**2.** $s = kt$
3. $V = kx^3$	**4.** $A = kx^2$
5. $z = km$	**6.** $h = k\sqrt{t}$
7. $f = \dfrac{k}{\lambda}$	**8.** $P = \dfrac{k}{r^2}$
9. $F = \dfrac{kw}{L}$	**10.** $V = \dfrac{kT}{P}$
11. $v = kgt$	**12.** $S = ktd$
13. $R = \dfrac{k}{PT}$	**14.** $y = \dfrac{k}{xz}$
15. $y = k\sqrt{x}$	**16.** $y = \dfrac{k}{t^3}$

17. The general equation is $d = kt$. Using the fact that $d = k(1) = r$, we see that $k = r$. So, $\boxed{d = rt}$.	**18.** The general equation is $F = km$. Using the fact that $F = k(1) = a$, we see that $k = a$. So, $\boxed{F = ma}$.
19. The general equation is $V = klw$. Using the fact that $V = k(2)(3) = 6h$ we see that $k = h$. So, $\boxed{V = lwh}$.	**20.** The general equation is $A = kbh$. Using the fact that $A = k(5)(4) = 10$ we see that $k = \frac{1}{2}$. So, $\boxed{A = \frac{1}{2}bh}$.
21. The general equation is $d = kr^2$. Using the fact that $A = k(3)^2 = 9\pi$, we see that $k = \pi$. So, $\boxed{A = \pi r^2}$.	**22.** The general equation is $V = kr^3$. Using the fact that $V = k(3)^3 = 36\pi$, we see that $k = \frac{4}{3}\pi$. So, $\boxed{V = \frac{4}{3}\pi r^3}$.

23. The general equation is $V = khr^2$. Using the fact that $V = k\left(\frac{4}{\pi}\right)(2)^2 = 1$, we see that $k = \frac{\pi}{16}$. So, $\boxed{V = \frac{\pi}{16}hr^2}$.

24. The general equation is $W = kRI^2$. Using the fact that $W = k(100)(0.25)^2 = 4$, we see that $\frac{25k}{4} = 4$, so that $k = \frac{16}{25}$. Hence, $\boxed{W = \frac{16}{25}RI^2}$.

25. The general equation is $V = \frac{k}{P}$. Using the fact that $V = \frac{k}{400} = 1000$, we see that $k = 400,000$. Hence, $\boxed{V = \frac{400,000}{P}}$.

26. The general equation is $I = \frac{k}{d^2}$. Using the fact that $I = \frac{k}{16^2} = 42$, we see that $k = 10,752$. Hence, $\boxed{I = \frac{10,752}{d^2}}$.

27. The general equation is $F = \frac{k}{\lambda L}$. Using the fact that $F = \frac{k}{(10^{-6}m)(10^5 m)} = 20\pi$, we see that $k = 2\pi$. So, $\boxed{F = \frac{2\pi}{\lambda L}}$.

28. The general equation is $y = \frac{k}{xz}$. Using the fact that $y = \frac{k}{4(0.05)} = 32$, we see that $k = 4(0.05)(32) = 6.4$. Hence, $\boxed{y = \frac{6.4}{xz}}$.
29. The general equation is $t = \frac{k}{s}$. Using the fact that $t = \frac{k}{8} = 2.4$, we see that $k = 2.4(8) = 19.2$. Hence, $\boxed{t = \frac{19.2}{s}}$.
30. The general equation is $W = \frac{k}{d^2}$. Using the fact that $W = \frac{k}{(0.2)^2} = 180$, we see that $k = 180(0.2)^2 = 7.2$. Hence, $\boxed{W = \frac{7.2}{d^2}}$.
31. The general equation is $R = \frac{k}{I^2}$. Using the fact that $R = \frac{k}{(3.5)^2} = 0.4$, we see that $k = (3.5)^2(0.4) = 4.9$. Hence, $\boxed{R = \frac{4.9}{I^2}}$.
32. The general equation is $y = \frac{k}{x\sqrt{z}}$. Using the fact that $y = \frac{k}{(0.2)\sqrt{4}} = 12$, we see that $k = 12(0.2)(2) = 4.8$. Hence, $\boxed{y = \frac{4.8}{x\sqrt{z}}}$.
33. The general equation is $R = \frac{kL}{A}$. Using the fact that $R = \frac{k(20)}{(0.4)} = 0.5$, we see that $50k = 0.5$, so that $k = 0.01$. Hence, $\boxed{R = \frac{0.01L}{A}}$.
34. The general equation is $F = \frac{km}{d}$. Using the fact that $F = \frac{k(20)}{8} = 32$, we see that $k = 32\left(\frac{8}{20}\right) = 12.8$. Hence, $\boxed{F = \frac{12.8m}{d}}$.

35. The general equation is $F = \frac{km_1m_2}{d^2}$. Using the fact that $F = \frac{k(8)(16)}{(0.4)^2} = 20$, we see that $k = \frac{20(0.4)^2}{(8)(16)} = 0.025$. Hence, $\boxed{F = \frac{0.025m_1m_2}{d^2}}$.

36. The general equation is $w = \frac{k\sqrt{g}}{t^2}$. Using the fact that $w = \frac{k\sqrt{16}}{(0.5)^2} = 20$, we see that $k = \frac{20(0.5)^2}{4} = 1.25$. Hence, $\boxed{w = \frac{1.25\sqrt{g}}{t^2}}$.

37. Assume that $W = kH$. We need to determine *k*. Using Jason's data, we see that $172.50 = 23k$, so that $k = 7.5$. So, $\boxed{W = 7.5H}$. (Note that Valerie's data also satisfies this equation.)

38.

Orange County: Assume that $T = kP$. We need to determine *k*. Using the data, we see that $2.60 = 40k$, so that $k = 0.065$. So, $\boxed{T = 0.065P}$.

Seminole County: Assume that $T = kP$. We need to determine *k*. Using the data, we see that $0.84 = 12k$, so that $k = 0.07$. So, $\boxed{T = 0.07P}$.

39. Let S = speed of the object, and M = Mach number. We are given that $S = kM$. We also know that when $S = 760$ mph (at sea level), $M = 1$. As such, $k = 760$. Hence, for U.S. Navy Blue Angels, $S = 760(1.7) = \boxed{1{,}292 \text{ mph}}$.

40. Using the same model as in Exercise 39, we see that for the F-22A Raptor, $S = 760(1.5) = \boxed{1{,}140 \text{ mph}}$.

41. We are given that $F = kH$. Using the fact that $F = 11$ when $H = 6.8$, we see that $11 = 6.8k$, so that $k = 1.618$. Hence, $\boxed{F = 1.618H}$.

42. Let S_1 and S_2 denote two abutting sections of a finger. It is known that $S_1 = kS_2$. Using the fact that $S_2 = 5$ when $S_1 = 8$, we see that $k = 1.6$. So, $\boxed{S_1 = 1.6S_2}$.

43. Assume Hooke's law holds: $F = kx$. Using the fact that $F = 30$N when $x = 10$ cm, we see that $k = 3 \text{ N/cm}$. So, $F = 3x$. As such, $72\text{N} = \left(3 \text{ N/cm}\right)x$, so that $\boxed{x = 24 \text{ cm}}$.

44. Using the same information as in Exercise 43, we see that the given information yields

$$F = \left(3 \text{ N/cm}\right)(18 \text{ cm}) = \boxed{54\text{N}}.$$

45. Let D = demand for Levi's jeans
P = price for Levi's jeans.

We are told that $D = \frac{k}{P}$. Using the given information for Flare 519 jeans (namely that $P = 20$ when $D = 300{,}000$) yields $300{,}000 = \frac{k}{20}$ so that $k = 6{,}000{,}000$.

Thus, for Vintage Flare jeans, the demand D is given by: $D = \frac{6{,}000{,}000}{300} = \boxed{20{,}000}$.

46. Let D = demand for Levi's jeans
P = price for Levi's jeans.

We are told that $D = \frac{k}{P}$. Using the given information for Silver Tab baggy jeans (namely that $P = 30$ when $D = 400{,}000$) yields $400{,}000 = \frac{k}{30}$ so that $k = 12{,}000{,}000$. Thus, for Offender jeans, the demand D is given by:

$D = \frac{12{,}000{,}000}{160} = \boxed{75{,}000}$.

47. Use the formula $I = \frac{k}{D^2}$. Using the data for Earth, we obtain:

$$1400 \ {}^{w}\!/_{m^2} = \frac{k}{(150{,}000 \text{ km})^2} \text{ so that}$$

$$k = \left(1400 \ {}^{w}\!/_{m^2}\right)(150{,}000 \text{ km})^2 = \left(1400 \ {}^{w}\!/_{m^2}\right)(150{,}000{,}000 \ m)^2 = 3.15 \times 10^{19} \, w$$

Hence, the intensity for Mars is given by:

$$I = \frac{3.15 \times 10^{19} \, w}{(228{,}000 \ km)^2} = \frac{3.15 \times 10^{19} \, w}{(228{,}000{,}000 \, m)^2} \approx \boxed{600 \ {}^{w}\!/_{m^2}}$$

48. Use the formula $I = \frac{k}{D^2}$. Using the data for Earth, we obtain:

$$1400 \ {}^{w}\!/_{m^2} = \frac{k}{(150{,}000 \text{ km})^2} \text{ so that}$$

$$k = \left(1400 \ {}^{w}\!/_{m^2}\right)(150{,}000 \text{ km})^2 = \left(1400 \ {}^{w}\!/_{m^2}\right)(150{,}000{,}000 \ m)^2 = 3.15 \times 10^{19} \, w$$

Hence, the intensity for Mercury is given by:

$$I = \frac{3.15 \times 10^{19} \, w}{(58{,}000 \ km)^2} = \frac{3.15 \times 10^{19} \, w}{(58{,}000{,}000 \, m)^2} \approx \boxed{9400 \ {}^{w}\!/_{m^2}}$$

49. Use the formula $I = kPt$.

Bank of America:
$750 = k(25{,}000)(2)$, so that $k = 0.015$, which corresponds to 1.5%.

Navy Federal Credit Union:
$1500 = k(25{,}000)(2)$, so that $k = 0.03$, which corresponds to 3%.

50. Use the formula $I = kPt$. Observe that $3250 = k(130{,}000)\left(\frac{1}{2}\right)$, so that $k = 0.05$. So, $\boxed{5\% \text{ interest rate}}$ is needed to make \$3250 in interest in 6 months.

51. Use the formula $P = \frac{kT}{V}$ with $T = 300\text{K}$, $P = 1$ atm., and $V = 4$ ml to obtain

$k = \frac{PV}{T} = \frac{(1\text{ atm})(4)}{300} = \frac{4}{300}$. Thus, $P = \frac{4}{300}\left(\frac{275}{4}\right) = \boxed{\frac{11}{12} \text{ or } 0.92 \text{ atm}}$.

52. Use the formula $P = \frac{kT}{V}$ with $T = 300\text{K}$, $P = 1$ atm., and $V = 4$ ml to obtain

$k = \frac{PV}{T} = \frac{(1\text{ atm})(4)}{300} = \frac{4}{300}$. Thus, $P = \frac{4}{300}\left(\frac{300}{3}\right) = \boxed{\frac{4}{3} \text{ or } 1.33 \text{ atm}}$.

53. Should be y is <u>inversely</u> proportional to x.

54. y varies directly with the <u>square</u> of x (x^2), NOT the square root ($\sqrt{x}$).

55. True. $A = \frac{1}{2}bh$, so area is directly proportional to both base and height.

56. False. Since $d = rt$, it follows that $r = \frac{d}{t}$. So, r is directly proportional to d, but inversely proportional to t.

57. b

58. a

59. Use the equation $\sigma_{p_1}^2 = \alpha C_n^2 k^{7/6} L^{11/6}$ with the following information (all converted to meters):

$C_n^2 = 1.0\times10^{-13}$, $L = 2000\text{m}$, $\lambda = 1.55\times10^{-6}\text{m}$ (so that $k = \frac{2\pi}{1.55\times10^{-6}\text{m}}$), and $\sigma_{p_1}^2 = 7.1$.

Substituting this information into the equation yields α:

$$7.1 = \alpha\left(1.0\times10^{-13}\right)\left(\frac{2\pi}{1.55\times10^{-6}}\right)^{7/6} 2000^{11/6}$$

so that

$$\alpha = \frac{7.1}{\left(1.0\times10^{-13}\right)\left(\frac{2\pi}{1.55\times10^{-6}}\right)^{7/6} 2000^{11/6}} \approx 1.23.$$

Thus, the equation is given by $\boxed{\sigma_{p_1}^2 = 1.23 C_n^2 k^{7/6} L^{11/6}}$.

60. Use the equation $\sigma_{s_p}^2 = \alpha C_n^2 k^{7/6} L^{11/6}$ with the following information (all converted to meters):

$C_n^2 = 1.0\times10^{-13}$, $L = 2000\text{m}$, $\lambda = 1.55\times10^{-6}\text{m}$ (so that $k = \dfrac{2\pi}{1.55\times10^{-6}\text{m}}$), and $\sigma_{s_p}^2 = 2.3$.

Substituting this information into the equation yields α:

$$2.3 = \alpha\left(1.0\times10^{-13}\right)\left(\frac{2\pi}{1.55\times10^{-6}}\right)^{7/6} 2000^{11/6}$$

so that

$$\alpha = \frac{2.3}{\left(1.0\times10^{-13}\right)\left(\dfrac{2\pi}{1.55\times10^{-6}}\right)^{7/6} 2000^{11/6}} \approx 0.399.$$

Thus, the equation is given by $\boxed{\sigma_{s_p}^2 = 0.399 C_n^2 k^{7/6} L^{11/6}}$.

61. (a) The least squares regression line is $y = 2.93x + 201.72$, and is plotted as seen below:

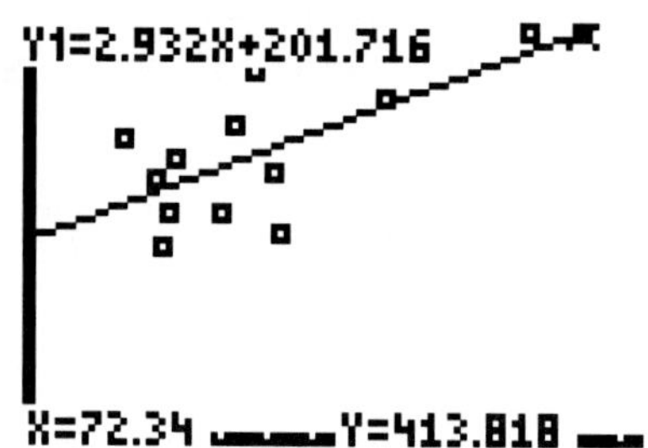

(b) The variation constant is 120.07 and the equation of the direct variation is $y = 120.074x^{0.259}$.

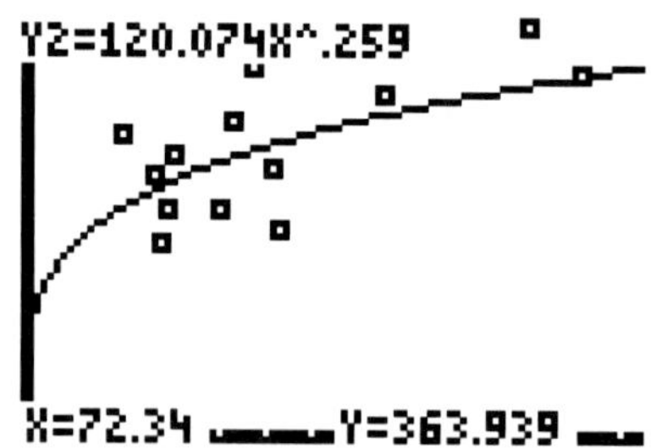

(c) When the oil price is \$72.70 per barrel in September 2006, the predicted stock index obtained from the least squares regression line is 415, and the value from the equation of direct variation is 364. In this case, the least squares regression line provides a closer approximation to the actual value, 417.

62. (a) The least squares regression line is $y = -0.04x + 6.08$.

The following is the sequence of commands, and screen captures, to use on the TI-8*.

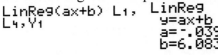

```
LinReg(ax+b) L1,
L4,Y1
```

```
LinReg
 y=ax+b
 a=-.039
 b=6.083
```

```
WINDOW
 Xmin=0
 Xmax=80
 Xscl=1
 Ymin=0
 Ymax=10
 Yscl=1
 Xres=1
```

Then, the graph of the least squares regression line, with the scatterplot, is given by:

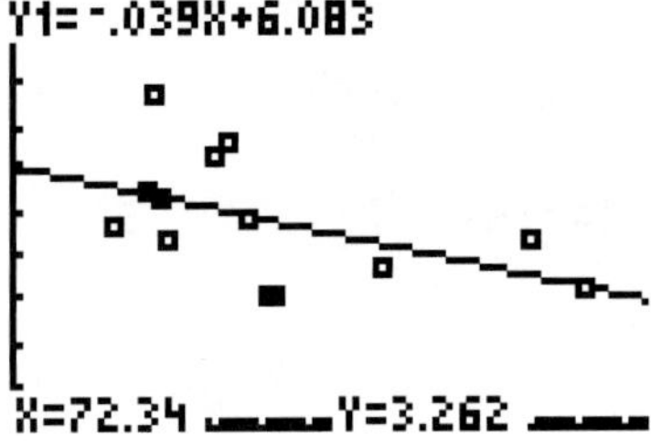

(b) The variation constant is 11.53 and the equation of the inverse variation is $y = \dfrac{11.53}{x^{0.27}}$. The following is the sequence of commands, and screen captures, to use on the TI-8*.

```
PwrReg L1,L4,Y2
```

```
PwrReg
 y=a*x^b
 a=11.528
 b=-.271
```

(c) When oil price is \$72.70 per barrel in September 2006, the predicted 5-year maturity rate obtained from the least squares regression line is 3.25, and the equation of inverse variation is 3.61. The equation of the inverse variation provides a closer approximation to the actual value, 5.02. Here are the corresponding screen captures from the TI-8*.

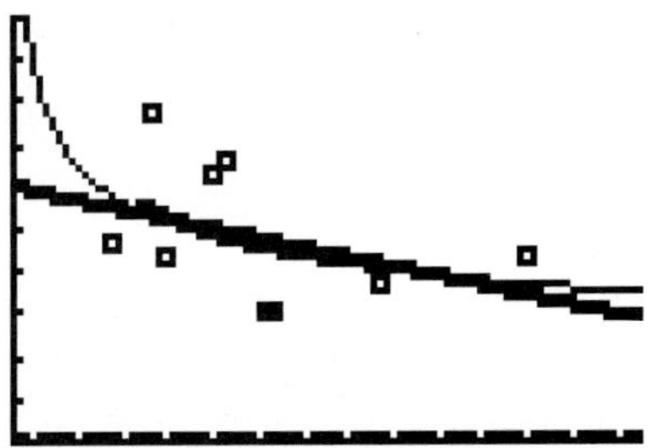

```
72.7→X
                72.700
Y1
                 3.248
Y2
                 3.608
```

63. (a) The least squares regression line is $y = -141.73x + 2,419.35$. The following is the sequence of commands, and screen captures, to use on the TI-8*.

```
LinReg(ax+b) L4, L3,Y1

LinReg
 y=ax+b
 a=-141.732
 b=2419.351

Plot1 Plot2 Plot3
On Off
Type:
Xlist:L4
Ylist:L3
Mark: ▪ + ·

WINDOW
 Xmin=0
 Xmax=10
 Xscl=1
 Ymin=0
 Ymax=2500
 Yscl=1
 Xres=1
```

Then, the graph of the least squares regression line, with the scatterplot, is given by:

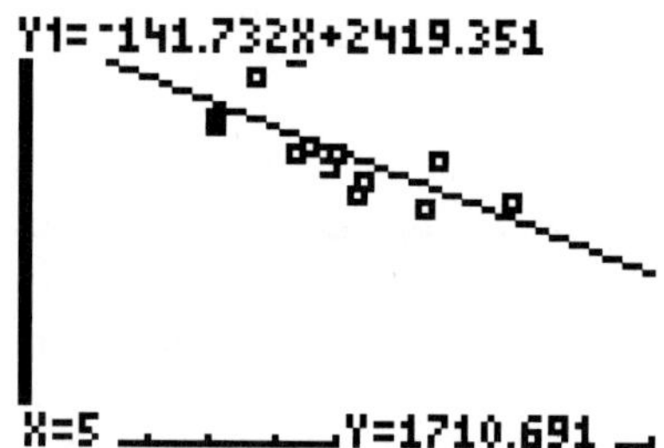

(b) The variation constant is 3,217.69 and the equation of the inverse variation is $y = \dfrac{3217.69}{x^{0.41}}$. The following is the sequence of commands, and screen captures, to use on the TI-8*.

```
PwrReg L4,L3,Y2

PwrReg
 y=a*x^b
 a=3217.691
 b=-.410
```

The graph of the curve of inverse variation, with the scatter plot, is given by:

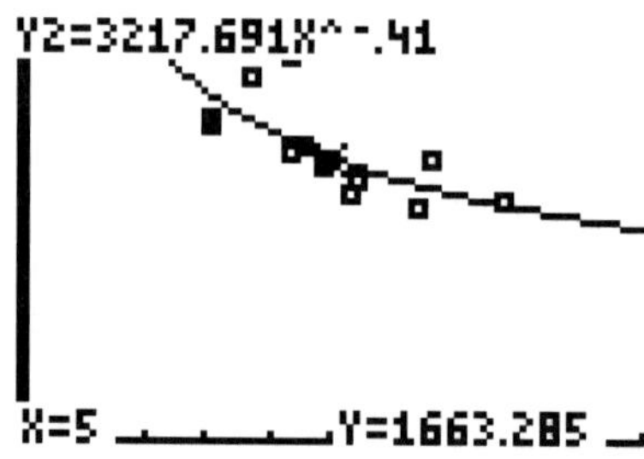

(c) When the 5-year maturity rate is 5.02% in September 2006, the predicted number of housing units obtained from the least squares regression line is 1708, and the equation of inverse variation is 1661. The equation of the least squares regression line provides a closer approximation to the actual value, 1861. The picture of the least squares line, with the scatter plot, as well as the computations using the TI-8* are as follows:

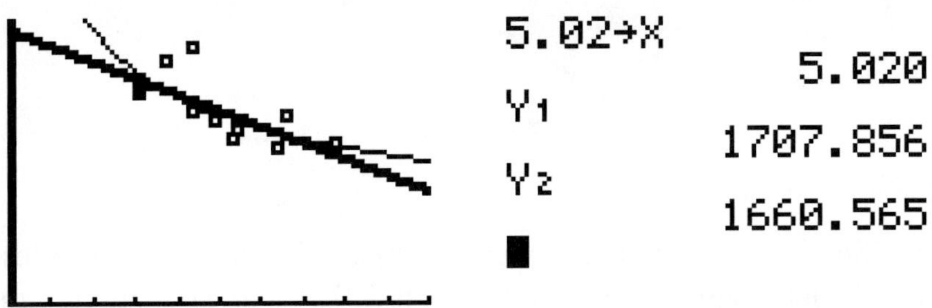

64. (a) The least squares regression line is $y = 0.15x + 22.60$. The following is the sequence of commands, and screen captures, to use on the TI-8*.

```
LinReg(ax+b) L3,
L2,Y1
```

```
LinReg
 y=ax+b
 a=.153
 b=22.601
```

```
Plot1 Plot2 Plot3
On Off
Type:
Xlist:L3
Ylist:L2
Mark: ▫ + ·
```

```
WINDOW
 Xmin=0
 Xmax=2500
 Xscl=1
 Ymin=0
 Ymax=420
 Yscl=1
 Xres=1
```

Then, the graph of the least squares regression line, with the scatterplot, is given by:

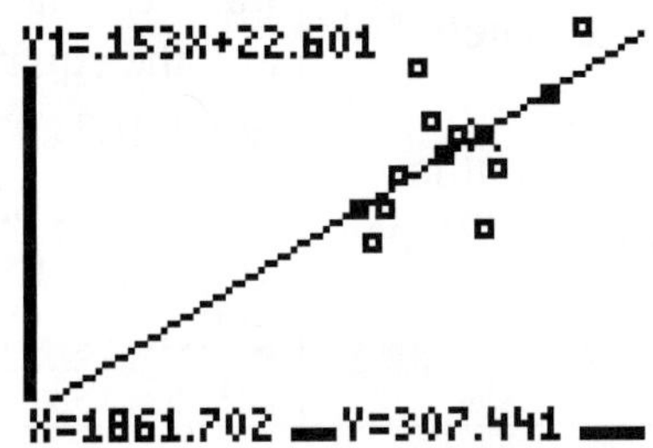

(b) The variation constant is 0.32 and the equation of the inverse variation is $y = 0.32x^{0.91}$.

```
PwrReg L3,L2,Y2
```

```
PwrReg
 y=a*x^b
 a=.317
 b=.912
```

Then, the graph of the curve of inverse variation, with the scatterplot, is given by:

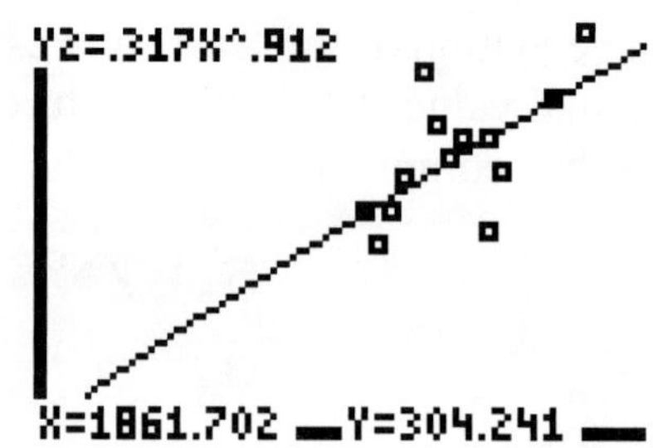

(c) There are 1861 housing units in September 2006. The predicted utilities stock index obtained from the least squares regression line is 307, and the equation of direct variation is 304. The equation of the least squares regression line provides a closer approximation to the actual value, 417. The screen captures from the TI-8* are as follows:

$1861 \rightarrow X$

1861.000

$Y1$

308.186

$Y2$

303.583

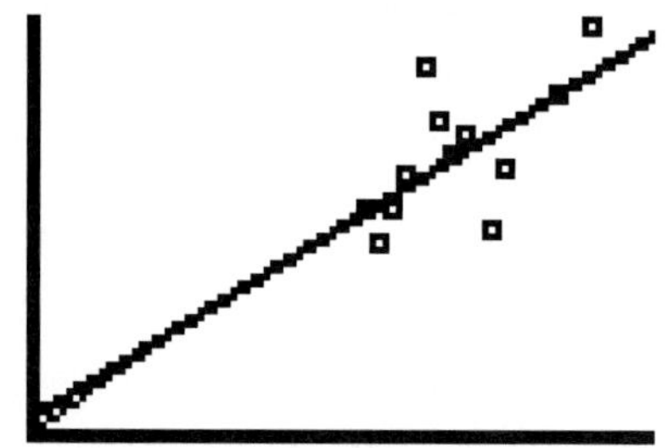

65. (a) The equation is approximately $y = 0.218x + 0.898$

(b) About \$2.427 per gallon. Yes, it is very close to the actual price at \$2.425 per gallon.

(c) \$3.083

66. (a) The variation constant is 0.346, and the equation is approximately $y = 1.163x^{0.346}$.

(b) About \$2.283 per gallon. No, it is very close to the actual price at \$2.425 per gallon.

(c) \$2.583

Chapter 0 Review Solutions --

1. $7x-4=12$ $7x=16$ $\boxed{x=16/7}$	**2.** $13d+12=7d+6$ $6d=-6$ $\boxed{d=-1}$
3. $20p+14=6-5p$ $25p=-8$ $\boxed{p=-8/25}$	**4.** $4x-28-4=4$ $4x=36$ $\boxed{x=9}$
5. $3x+21-2=4x-8$ $\boxed{x=27}$	**6.** $7c+3c-15=2c+6-14$ $10c-15=2c-8$ $8c=7$ $\boxed{c=7/8}$
7. $14-[-3y+12+9]=8y+12-6+4$ $14+3y-21=8y+10$ $-17=5y$ $\boxed{y=-17/5}$	**8.** $6-4x+2x-14-52=6x-12+6[6x-9+6]$ $-2x-60=6x-12+36x-18$ $-2x-60=42x-30$ $-30=44x$ $\boxed{x=\frac{-30}{44}=\frac{-15}{22}}$
9. $b\neq 0$ $12-3b=6+4b$ $6=7b$ $\boxed{b=6/7}$	**10.** $3g+9g=7$ $12g=7$ $\boxed{g=7/12}$
11. $\text{LCD}=28$ $4(13x)-28x=7x-2(3)$ $52x-28x=7x-6$ $17x=-6$ $\boxed{x=-6/17}$	**12.** $\text{LCD}=6$ $30b+b=2b-29$ $29b=-29$ $\boxed{b=-1}$

13. $x = $ amount invested @ 20% $25000 - x = $ amount invested @ 8% Earned interest $= 27600 - 25000 = 2600$ $0.2x + 0.08(25000 - x) = 2600$ $0.2x + 2000 - 0.08x = 2600$ $0.12x = 600$ $x = 5000$ $\boxed{\begin{array}{l}\$5,000 @ 20\% \\ \$20,000 @ 8\%\end{array}}$	**14.** \$2500 in mutual funds \$2500 in stock $x = $ rate of mutual fund $4x = $ rate of stock $x(2500) + 4x(2500) = 250$ $2500x + 10000x = 250$ $12500x = 250$ $x = 0.02$ $\boxed{\begin{array}{l}\text{Mutual Fund: } 2\% \\ \text{Stock: } 8\%\end{array}}$
15. $x = $ ml of 5% $150 - x = $ ml of 10% $0.05x + 0.10(150 - x) = 0.08(150)$ $0.05x + 15 - 0.10x = 12$ $-0.05x = -3$ $x = 60$ $\boxed{\begin{array}{l}\text{60 ml of 5\%} \\ \text{90 ml of 10\%}\end{array}}$	**16.** $x = $ ounces of 8% 4 ounces of 20% Desired: 12% $0.08x + 0.20(4) = 0.12(x + 4)$ Multiply by 100 $8x + 80 = 12x + 48$ $32 = 4x$ $\boxed{x = 8}$
17. $b^2 - 4b - 21 = 0$ $(b - 7)(b + 3) = 0$ $\boxed{b = -3, 7}$	**18.** $x^2 - 3x - 54 = 0$ $(x - 9)(x + 6) = 0$ $\boxed{x = -6, 9}$
19. $x^2 - 8x = 0$ $x(x - 8) = 0$ $\boxed{x = 0, 8}$	**20.** $(3y - 5)(2y + 1) = 0$ $3y - 5 = 0 \quad 2y + 1 = 0$ $\boxed{y = 5/3 \text{ or } y = -1/2}$
21. $q^2 = 169$ $q = \pm\sqrt{169}$ $\boxed{q = \pm 13}$	**22.** $c^2 = -36$ $c = \pm\sqrt{-36}$ $\boxed{c = \pm 6i}$

23. $2x-4=\pm\sqrt{-64}$ $2x-4=\pm 8i$ $2x=4\pm 8i$ $\boxed{x=2\pm 4i}$	**24.** $d+7=\pm\sqrt{4}$ $d=-7\pm 2$ $\boxed{d=-9,-5}$
25. $x^2-4x=12$ $x^2-4x+4=12+4$ $(x-2)^2=16$ $x-2=\pm 4$ $x=2\pm 4$ $\boxed{x=-2,6}$	**26.** $2x^2-5x=7$ $2\left(x^2-\frac{5}{2}x\right)=7$ $2\left(x^2-\frac{5}{2}x+\frac{25}{16}\right)=7+\frac{25}{8}$ $2\left(x-\frac{5}{4}\right)^2=\frac{81}{8}$ $\left(x-\frac{5}{4}\right)^2=\frac{81}{16}$ $x-\frac{5}{4}=\pm\frac{9}{4}$ $x=\frac{5}{4}\pm\frac{9}{4}$ $\boxed{x=-1,\frac{7}{2}}$
27. $x^2-x=8$ $x^2-x+\frac{1}{4}=8+\frac{1}{4}$ $\left(x-\frac{1}{2}\right)^2=\frac{33}{4}$ $x-\frac{1}{2}=\pm\sqrt{\frac{33}{4}}$ $\boxed{x=\frac{1\pm\sqrt{33}}{2}}$	**28.** $m^2-8m=-15$ $m^2-8m+16=-15+16$ $(m-4)^2=1$ $m-4=\pm 1$ $m=4\pm 1$ $\boxed{m=3,5}$

29. $3t^2-4t-7=0$ $a=3, b=-4, c=-7$ $t=\frac{-(-4)\pm\sqrt{(-4)^2-4(3)(-7)}}{2(3)}$ $t=\frac{4\pm\sqrt{100}}{6}=\frac{4\pm10}{6}$ $\boxed{t=-1, \frac{7}{3}}$	**30.** $4x^2+5x+7=0$ $a=4, b=5, c=7$ $x=\frac{-5\pm\sqrt{5^2-4(4)(7)}}{2(4)}$ $x=\frac{-5\pm\sqrt{-87}}{8}$ $\boxed{x=\frac{-5\pm i\sqrt{87}}{8}}$
31. $8f^2-\frac{1}{3}f-\frac{7}{6}=0$ LCD = 6 $48f^2-2f-7=0$ $a=48, b=-2, c=-7$ $f=\frac{-(-2)\pm\sqrt{(-2)^2-4(48)(-7)}}{2(48)}$ $f=\frac{2\pm\sqrt{1348}}{96}$ $f=\frac{2\pm2\sqrt{337}}{96}$ $\boxed{f=\frac{1\pm\sqrt{337}}{48}}$	**32.** $x^2+6x-6=0$ $a=1, b=6, c=-6$ $x=\frac{-6\pm\sqrt{6^2-4(1)(-6)}}{2(1)}$ $x=\frac{-6\pm\sqrt{60}}{2}$ $x=\frac{-6\pm2\sqrt{15}}{2}$ $\boxed{x=-3\pm\sqrt{15}}$
33. $a=5, b=-3, c=-3$ $q=\frac{-(-3)\pm\sqrt{(-3)^2-4(5)(-3)}}{2(5)}$ $\boxed{q=\frac{3\pm\sqrt{69}}{10}}$	**34.** $x-7=\pm\sqrt{-12}$ $\boxed{x=7\pm2i\sqrt{3}}$

35. $(2x-5)(x+1)=0$ $\boxed{x=-1,\frac{5}{2}}$	**36.** $g^2+3g-3=0$ $a=1, b=3, c=-3$ $g=\frac{-3\pm\sqrt{3^2-4(1)(-3)}}{2(1)}$ $\boxed{g=\frac{-3\pm\sqrt{21}}{2}}$
37. $7x^2+19x-6=0$ $(7x-2)(x+3)=0$ $\boxed{x=-3,2/7}$	**38.** $2b^2+2=7$ $2b^2=5$ $b^2=\frac{5}{2}$ $b=\pm\sqrt{\frac{5}{2}}$ $\boxed{b=\pm\frac{\sqrt{10}}{2}}$
39. $A=\frac{1}{2}bh$ $b=h+3 \quad A=2$ $2=\frac{1}{2}(h+3)h$ $4=h^2+3h$ $h^2+3h-4=0$ $(h+4)(h-1)=0$ $h=-4,1$ (height must be positive) $\boxed{h=1\text{ ft}, b=4\text{ ft}}$	**40.** $-16t^2+500=0$ $16t^2=500$ $t^2=\frac{500}{16}$ $t=\pm\sqrt{\frac{500}{16}}=\pm\frac{10\sqrt{5}}{4}=\pm\frac{5\sqrt{5}}{2}$ Approximately 5.6 seconds

41. $x \neq 0$ $\text{LCD} = x$ $1-4x = 3x(x-7)+5x$ $1-4x = 3x^2-21x+5x$ $3x^2-12x-1=0$ $x = \frac{-(-12)\pm\sqrt{(-12)^2-4(3)(-1)}}{2(3)}$ $x = \frac{12\pm\sqrt{156}}{6} = \frac{12\pm 2\sqrt{39}}{6}$ $\boxed{x = \frac{6\pm\sqrt{39}}{3}}$	**42.** $x \neq -1, 1$ $\text{LCD} = (x+1)(x-1)$ $4(x-1)-8(x+1) = 3(x+1)(x-1)$ $4x-4-8x-8 = 3(x^2-1)$ $-4x-12 = 3x^2-3$ $3x^2+4x+9=0$ $x = \frac{-4\pm\sqrt{4^2-4(3)(9)}}{2(3)}$ $x = \frac{-4\pm\sqrt{-92}}{6} = \frac{-4\pm 2i\sqrt{23}}{6}$ $\boxed{x = -\frac{2}{3}\pm i\frac{\sqrt{23}}{3}}$
43. $t \neq -4, 0$ $\text{LCD} = t(t+4)$ $2t-7(t+4)=6$ $2t-7t-28=6$ $-5t=34$ $\boxed{t=-34/5}$	**44.** $x \neq -\frac{1}{3}, \frac{7}{2}$ $-2(2x-7) = 3(3x+1)$ $-4x+14 = 9x+3$ $13x = 11$ $\boxed{x = 11/13}$
45. $x \neq 0$ $\text{LCD} = 2x$ $3-12 = 18x$ $-9 = 18x$ $\boxed{x = -\frac{1}{2}}$	**46.** $m \neq -\frac{5}{2}, 0$ $3-\frac{5}{m} = 2+\frac{5}{m}$ $3m-5 = 2m+5$ $\boxed{m=10}$
47. $2x-4 = 2^3 = 8$ $2x = 12$ $\boxed{x=6}$	**48.** no solution (boxed) because square roots cannot be negative.

<table>
<tr><td>

49.

$$2x-7=3^5$$
$$2x=7+243=250$$
$$\boxed{x=125}$$

</td><td>

50.

$$x^2=7x-10$$
$$x^2-7x+10=0$$
$$(x-2)(x-5)=0$$
$$\boxed{x=2,5}$$

</td></tr>
<tr><td>

51.

$$(x-4)^2=x^2+5x+6$$
$$x^2-8x+16=x^2+5x+6$$
$$13x=10$$
$$\cancel{x=\frac{10}{13}} \quad \left(\begin{array}{l}\text{This answer would make}\\ \text{the first } \sqrt{\ } \text{ equal to a}\\ \text{negative number}\end{array}\right)$$
$$\boxed{\text{no solution}}$$

</td><td>

52.

$$2x-7=x+3$$
$$\boxed{x=10}$$

</td></tr>
<tr><td>

53.

$$x+3=4-4\sqrt{3x+2}+3x+2$$
$$-2x-3=-4\sqrt{3x+2}$$
$$2x+3=4\sqrt{3x+2}$$
$$(2x+3)^2=16(3x+2)$$
$$4x^2+12x+9=48x+32$$
$$4x^2-36x-23=0$$
$$x=\frac{36\pm\sqrt{36^2-4(4)(-23)}}{2(4)}$$
$$x=\frac{36\pm\sqrt{1664}}{8}\cong-0.6, 9.6$$
$$\boxed{x\cong-0.6} \text{ (9.6 doesn't check)}$$

</td><td>

54.

$$16+8\sqrt{x-3}+x-3=x-5$$
$$8\sqrt{x-3}=-18$$
$$\boxed{\text{no solution}}$$

</td></tr>
<tr><td>

55.

$$y^{-2}-5y^{-1}+4=0$$
Let $u=y^{-1}$
$$u^2-5u+4=0$$
$$(u-4)(u-1)=0$$
$$u=4,1$$
So, we have:
$$y^{-1}=4 \Rightarrow \boxed{y=\tfrac{1}{4}}$$
$$y^{-1}=1 \Rightarrow \boxed{y=1}$$

</td><td>

56.

$$p^{-2}+4p^{-1}-12=0$$
Let $u=p^{-1}$
$$u^2+4u-12=0$$
$$(u+6)(u-2)=0$$
$$u=-6,2$$
So, we have:
$$p^{-1}=-6 \Rightarrow \boxed{p=-\tfrac{1}{6}}$$
$$p^{-1}=2 \Rightarrow \boxed{p=\tfrac{1}{2}}$$

</td></tr>
</table>

57. $2x^{2/3}+3x^{1/3}-5=0$ Let $u=x^{1/3}$ $2u^2+3u-5=0$ $(2u+5)(u-1)=0$ $u=-\frac{5}{2},1$ $x^{1/3}=-\frac{5}{2}$ $x^{1/3}=1$ $x=\left(-\frac{5}{2}\right)^3$ $\boxed{x=1}$ $\boxed{x=-\frac{125}{8}}$	**58.** Let $u=x^{1/3}$ $2u^2-3u-5=0$ $(2u-5)(u+1)=0$ $u=-1,\frac{5}{2}$ $x^{1/3}=\frac{5}{2}$ $x^{1/3}=-1$ $\boxed{x=\frac{125}{8}}$ $\boxed{x=-1}$
59. $x^{-2/3}+3x^{-1/3}+2=0$ Let $u=x^{-1/3}$. $u^2+3u+2=0$ $(u+2)(u+1)=0$ $u=-2,-1$ So, we have: $x^{-1/3}=-2 \Rightarrow x=(-2)^{-3}=\boxed{-1/8}$ $x^{-1/3}=-1 \Rightarrow x=(-1)^{-3}=\boxed{-1}$	**60.** $y^{-1/2}-2y^{-1/4}+1=0$ Let $u=y^{-1/4}$. $u^2-2u+1=0$ $(u-1)^2=0$ $u=1$ So, we have: $y^{-1/4}=1 \Rightarrow \boxed{y=1}$
61. Let $u=x^2$ $u^2+5u-36=0$ $(u+9)(u-4)=0$ $u=-9,4$ $-9=x^2$ $4=x^2$ $\boxed{x=\pm 3i}$ $\boxed{x=\pm 2}$	**62.** Let $u=x^{-1/2}$ $u^2-4u+3=0$ $(u-1)(u-3)=0$ $u=1,3$ $1=x^{-1/2}$ $3=x^{-1/2}$ $\boxed{x=1^{-2}=1}$ $\boxed{x=3^{-2}=\frac{1}{9}}$

63.	64.
$x^3+4x^2-32x=0$ $x(x^2+4x-32)=0$ $x(x+8)(x-4)=0$ $\boxed{x=0,-8,4}$	$9t^3-25t=0$ $t(9t^2-25)=0$ $t(3t-5)(3t+5)=0$ $\boxed{t=0,\pm\frac{5}{3}}$
65.	**66.**
$p^3-3p^2-4p+12=0$ $(p^3-3p^2)-4(p-3)=0$ $p^2(p-3)-4(p-3)=0$ $(p^2-4)(p-3)=0$ $(p-2)(p+2)(p-3)=0$ $\boxed{p=\pm2,3}$	$4x^3-9x^2+4x-9=0$ $(4x^3-9x^2)+(4x-9)=0$ $x^2(4x-9)+(4x-9)=0$ $(x^2+1)(4x-9)=0$ $\boxed{x=\pm i,\ \frac{9}{4}}$
67.	**68.**
$p(2p-5)^2-3(2p-5)=0$ $(2p-5)[p(2p-5)-3]=0$ $(2p-5)(2p^2-5p-3)=0$ $(2p-5)(2p+1)(p-3)=0$ $\boxed{p=-\frac{1}{2},\ \frac{5}{2},\ 3}$	$2(t^2-9)^3-20(t^2-9)^2=0$ $2(t^2-9)^2\underbrace{[t^2-9-10]}_{t^2-19}=0$ $2(t-3)(t+3)(t-\sqrt{19})(t+\sqrt{19})=0$ $\boxed{t=\pm3,\ \pm\sqrt{19}}$
69.	**70.**
$y-81y^{-1}=0$ $y-\frac{81}{y}=0$ $\frac{y^2-81}{y}=0$ $\frac{(y-9)(y+9)}{y}=0$ $\boxed{y=\pm9}$	$9x^{3/2}-37x^{1/2}+4x^{-1/2}=0$ $x^{-1/2}(9x^2-37x+4)=0$ $x^{-1/2}(9x-1)(x-4)=0$ $\boxed{x=\frac{1}{9},\ 4}$
71. $\lvert x-3\rvert=-4$ no solution (boxed)	**72.** $2+x=-5$ or $2+x=5$ $\boxed{x=-7 \text{ or } x=3}$

73.
$3x - 4 = -1.1$
$3x = 2.9$
$\boxed{x = 0.9667}$

$3x - 4 = 1.1$
$3x = 5.1$
$\boxed{x = 1.7}$

74.
$x^2 - 6 = -3$
$x^2 = 3$
$\boxed{x = \pm\sqrt{3}}$

$x^2 - 6 = 3$
$x^2 = 9$
$\boxed{x = \pm 3}$

75. $(4, \infty)$

... 0 1 2 3 4 5 6...

76. $(-\infty, 2]$

... 0 1 2 3 4 ...

77. $[8, 12]$

...7 8 9 10 11 12 ...

78. $\varnothing$

... −3 −2 −1 0 1 2 3 ...

79.
$3x < 5$
$x < 5/3$
$\boxed{(-\infty, 5/3)}$

... 0 $\frac{1}{3}$ $\frac{2}{3}$ 1 $\frac{4}{3}$ $\frac{5}{3}$ ...

80.
$6x \le -2$
$x \le -1/3$
$\boxed{(-\infty, -1/3]}$

... −1 $-\frac{2}{3}$ $-\frac{1}{3}$ 0 $\frac{1}{3}$...

81.
$4x - 4 > 2x - 7$
$2x > -3$
$x > -3/2$
$\boxed{(-3/2, \infty)}$

... −2 $-\frac{3}{2}$ −1 $-\frac{1}{2}$ 0...

82.
$x + 3 \ge 18$
$x \ge 15$
$\boxed{[15, \infty)}$

... 13 14 15 16 ...

83.
$6 < 2 + x \le 11$
$4 < x \le 9$
$\boxed{(4, 9]}$

...3 4 5 6 7 8 9 10...

84.
$-6 \le -4x - 7 \le 16$
$1 \le -4x \le 23$
$-\dfrac{1}{4} \ge x \ge -\dfrac{23}{4}$
$\boxed{\left[-\dfrac{23}{4}, -\dfrac{1}{4}\right]}$

... −6 $-\frac{23}{4}$ $-\frac{22}{4}$ ... $-\frac{1}{4}$ 0...

85.
$\text{LCD} = 12$
$8 \le 2(1 + x) \le 9$
$8 \le 2 + 2x \le 9$
$6 \le 2x \le 7$
$3 \le x \le 7/2$
$\boxed{[3, 7/2]}$

... $\frac{5}{2}$ 3 $\frac{7}{2}$ 4 ...

86. LCD = 18 $6x+2(x+4)>3x-6$ $6x+2x+8>3x-6$ $5x>-14$ $x>-14/5$ $\boxed{(-14/5,\infty)}$ ... -3 $-\frac{14}{5}$ $-\frac{13}{5}$ $-\frac{12}{5}$ ... 0...	**87.** $x^2-36\le 0$ $(x-6)(x+6)\le 0$ CP's: $x=-6,6$ + - + -6 6 $\boxed{[-6,6]}$
88. $6x^2-7x-20<0$ $(3x+4)(2x-5)<0$ CP's: $x=-\frac{4}{3},\frac{5}{2}$ + - + -4/3 5/2 $\boxed{\left(-\frac{4}{3},\frac{5}{2}\right)}$	**89.** $x^2-4x\ge 0$ $x(x-4)\ge 0$ CP's: $x=0, x=4$ + - + 0 4 $\boxed{(-\infty,0]\cup[4,\infty)}$
90. $x^2+9x+14\le 0$ $(x+7)(x+2)\le 0$ CP's: $-7,-2$ + - + -7 -2 $\boxed{[-7,-2]}$	**91.** $4x^2-12>13x$ $4x^2-13x-12>0$ $(4x+3)(x-4)>0$ CPs: $x=-3/4, 4$ + − + $-3/4$ 4 $\boxed{(-\infty,-3/4)\cup(4,\infty)}$
92. $3x\le x^2+2$ $x^2-3x+2\ge 0$ $(x-2)(x-1)\ge 0$ CPs: $x=1,2$ + − + 1 2 $(-\infty,1]\cup[2,\infty)$	**93.** $\frac{x}{x-3}<0$ $\boxed{x\ne 3}$ CP's: $x=0,3$ + - + 0 3 $\boxed{(0,3)}$

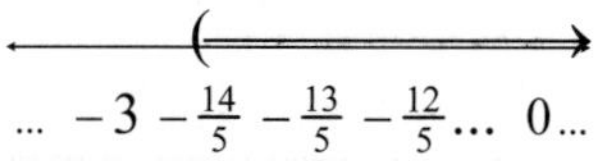

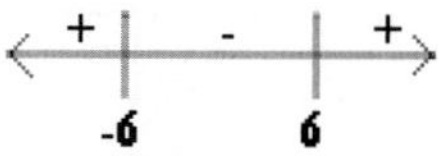

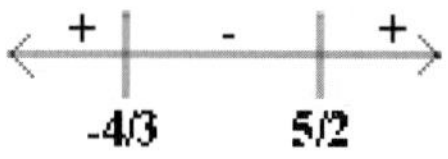

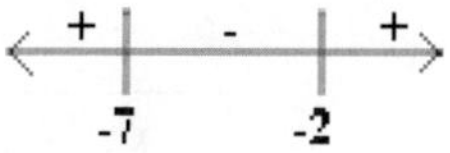

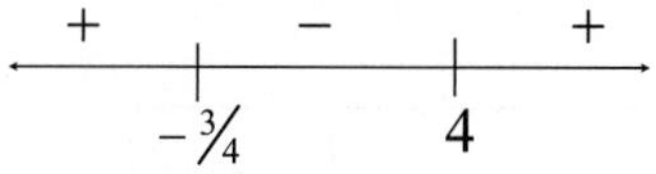

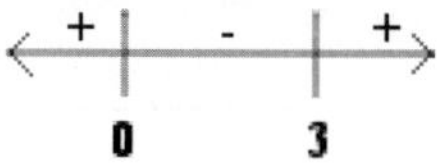

94.

$$\frac{x-1}{x-4} > 0 \quad \boxed{x \neq 4}$$

CP's: 1,4

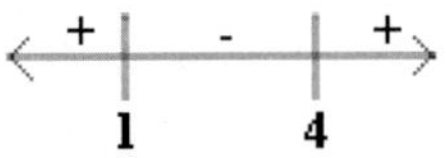

$+$ | 1 | $-$ | 4 | $+$

$$\boxed{(-\infty,1)\cup(4,\infty)}$$

95.

$$\frac{x^2-3x}{3} - \frac{18(3)}{3} \geq 0$$

$$\frac{x^2-3x-54}{3} \geq 0$$

$$\frac{(x-9)(x+6)}{3} \geq 0$$

CP's: $x = -6, 9$

$+$ | -6 | $-$ | 9 | $+$

$$\boxed{(-\infty,-6]\cup[9,\infty)}$$

96.

$$\frac{(x-7)(x+7)}{x-7} \geq 0 \quad \boxed{x \neq 7}$$

$$x+7 \geq 0$$

$$x \geq -7$$

$$\boxed{[-7,7)\cup(7,\infty)}$$

97.

$$\frac{3}{x-2} - \frac{1}{x-4} \leq 0$$

$$\frac{3(x-4)-(x-2)}{(x-2)(x-4)} \leq 0$$

$$\frac{2x-10}{(x-2)(x-4)} \leq 0$$

$$\frac{2(x-5)}{(x-2)(x-4)} \leq 0$$

CPs: $x = 2, 4, 5$

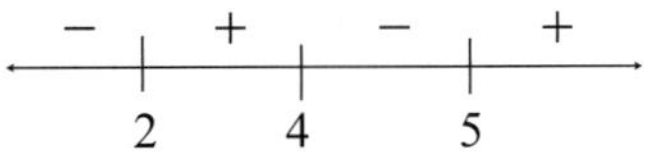

$-$ | 2 | $+$ | 4 | $-$ | 5 | $+$

$$\boxed{(-\infty,2)\cup(4,5]}$$

98.

$$\frac{4}{x-1} \leq \frac{2}{x+3}$$

$$\frac{4}{x-1} - \frac{2}{x+3} \leq 0$$

$$\frac{4(x+3)-2(x-1)}{(x-1)(x+3)} \leq 0$$

$$\frac{2(x+7)}{(x-1)(x+3)} \leq 0$$

CPs: $x = -7, -3, 1$

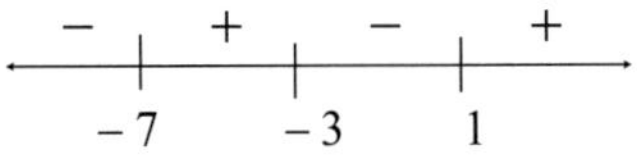

$-$ | −7 | $+$ | −3 | $-$ | 1 | $+$

$$\boxed{(-\infty,-7]\cup(-3,1)}$$

99.

$$x+4 < -7 \qquad x+4 > 7$$

$$x < -11 \qquad x > 3$$

$$\boxed{(-\infty,-11)\cup(3,\infty)}$$

100. $-4 \le -7 + y \le 4$ $3 \le y \le 11$ $\boxed{[3,11]}$	**101.** $\lvert 2x \rvert > 6$ $2x < -6 \qquad 2x > 6$ $x < -3 \qquad x > 3$ $\boxed{(-\infty,-3)\cup(3,\infty)}$
102. $\dfrac{4+2x}{3} \le -\dfrac{1}{7} \qquad \dfrac{4+2x}{3} \ge \dfrac{1}{7}$ $\text{LCD} = 21 \qquad 7(4+2x) \ge 3$ $7(4+2x) \le -3 \qquad 28+14x \ge 3$ $28+14x \le -3 \qquad 14x \ge -25$ $14x \le -31 \qquad x \ge -25/14$ $x \le -31/14$ $\boxed{(-\infty,-31/14]\cup[-25/14,\infty)}$	**103.** $\mathbb{R}$ **104.** $-4 \le 1-2x \le 4$ $-5 \le -2x \le 3$ $\dfrac{5}{2} \ge x \ge -\dfrac{3}{2}$ $\boxed{\left[-\dfrac{3}{2},\dfrac{5}{2}\right]}$
105. $\sqrt{(-2-4)^2+(0-3)^2} = \sqrt{45} = \boxed{3\sqrt{5}}$	**106.** $\sqrt{(1-4)^2+(4-4)^2} = \boxed{3}$
107. $\sqrt{(-4-2)^2+(-6-7)^2} = \boxed{\sqrt{205}}$	**108.** $\sqrt{\left(\frac{1}{4}-\frac{1}{3}\right)^2+\left(\frac{1}{12}-\left(-\frac{7}{3}\right)\right)^2} =$ $\sqrt{\left(-\frac{1}{12}\right)^2+\left(\frac{29}{12}\right)^2} = \dfrac{\sqrt{842}}{12} \cong \boxed{2.418}$
109. $\left(\dfrac{2+3}{2},\dfrac{4+8}{2}\right) = \boxed{\left(\frac{5}{2},6\right)}$	**110.** $\left(\dfrac{-2+5}{2},\dfrac{6+7}{2}\right) = \boxed{\left(\frac{3}{2},\frac{13}{2}\right)}$
111. $\left(\dfrac{2.3+5.4}{2},\dfrac{3.4+7.2}{2}\right) = \boxed{(3.85,5.3)}$	**112.** $\left(\dfrac{-a+a}{2},\dfrac{2+4}{2}\right) = \boxed{(0,3)}$
113. <u>x-intercepts</u>: $x^2+4(0)^2=4 \Rightarrow x=\pm 2$ So, $(\pm 2,0)$. <u>y-intercepts</u>: $0^2+4y^2=4 \Rightarrow y=\pm 1$ So, $(0,\pm 1)$.	**114.** <u>x-intercepts</u>: $x^2-x+2=0 \Rightarrow x=\dfrac{1\pm\sqrt{1-4(2)}}{2}=\dfrac{1\pm i\sqrt{7}}{2}$ Since these solutions are not real, there are no x-intercepts. <u>y-intercept</u>: $y=0^2-0+2 \Rightarrow y=2$ So, $(0,2)$.

115.	**116.**
<u>x-intercepts</u>: $\sqrt{x^2-9}=0 \Rightarrow x=\pm 3$ So, $(\pm 3,0)$. <u>y-intercepts</u>: $3i=\sqrt{0^2-9}=y$. Since this is not real, there is no *y*-intercept.	<u>x-intercepts</u>: $\frac{x^2-x-12}{x-12}=\frac{(x-4)(x+3)}{x-12}=0 \Rightarrow x=-3,4$ So, $(-3,0)$, $(4,0)$. <u>y-intercepts</u>: $y=\frac{0^2-0-12}{0-12}=1$. So, $(0,1)$.

117. <u>x-axis symmetry</u> (Replace y by $-y$):

$$x^2+(-y)^3=4$$
$$x^2-y^3=4$$

No, since not equivalent to the original.

<u>y-axis symmetry</u> (Replace x by $-x$):

$$(-x)^2+y^3=4$$
$$x^2+y^3=4$$

Yes, since equivalent to the original.

<u>symmetry about origin</u> (Replace y by $-y$ and x by $-x$):

$$(-x)^2+(-y)^3=4$$
$$x^2-y^3=4$$

No, since not equivalent to the original.

118. <u>x-axis symmetry</u> (Replace y by $-y$):

$$-y=x^2-2$$
$$y=-x^2+2$$

No, since not equivalent to the original.

<u>y-axis symmetry</u> (Replace x by $-x$):

$$y=(-x)^2-2$$
$$y=x^2+2$$

Yes, since equivalent to the original.

<u>symmetry about origin</u> (Replace y by $-y$ and x by $-x$):

$$-y=(-x)^2-2=x^2-2$$
$$y=-x^2+2$$

No, since not equivalent to the original.

119. <u>x-axis symmetry</u> (Replace y by $-y$):

$$x(-y)=4$$
$$-xy=4$$

No, since not equivalent to the original.

<u>y-axis symmetry</u> (Replace x by $-x$):

$$(-x)y=4$$
$$-xy=4$$

No, since not equivalent to the original.

<u>symmetry about origin</u> (Replace y by $-y$ and x by $-x$):

$$(-x)(-y)=4$$
$$xy=4$$

Yes, since equivalent to the original.

120. <u>x-axis symmetry</u> (Replace y by $-y$):

$$(-y)^2 = 5 + x$$
$$y^2 = 5 + x$$

Yes, since equivalent to the original.

<u>y-axis symmetry</u> (Replace x by $-x$):

$$y^2 = 5 + (-x)$$
$$y^2 = 5 - x$$

No, since not equivalent to the original.

<u>symmetry about origin</u> (Replace y by $-y$ and x by $-x$):

$$(-y)^2 = 5 + (-x)$$
$$y^2 = 5 - x$$

No, since not equivalent to the original.

121.

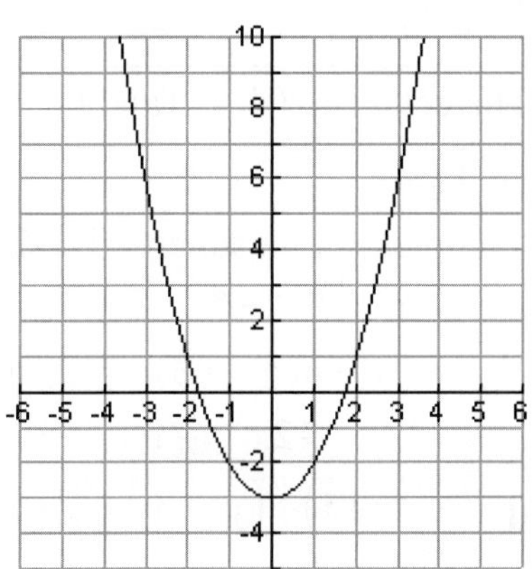

122.

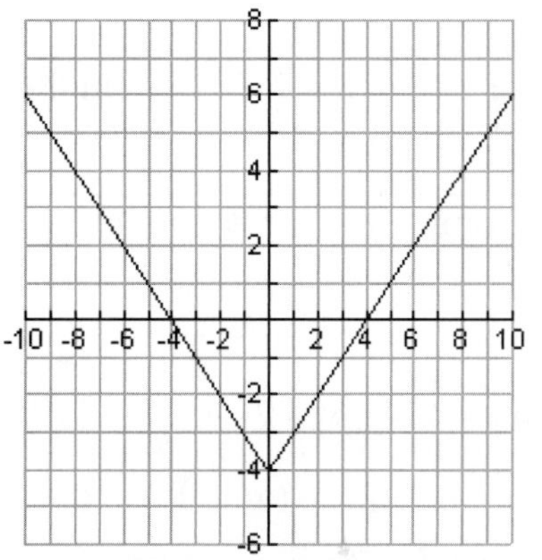

123.

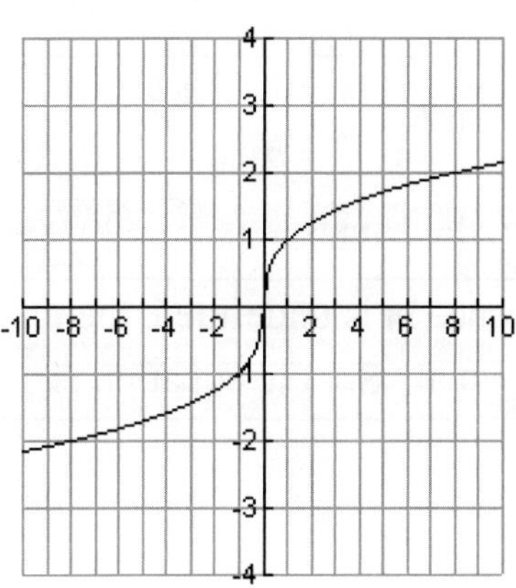

124.

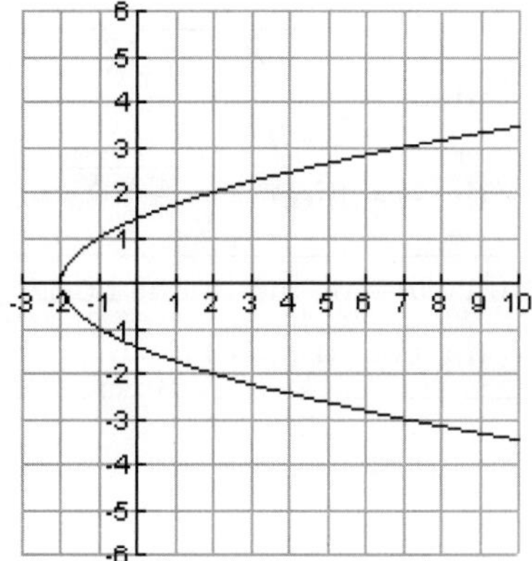

125. 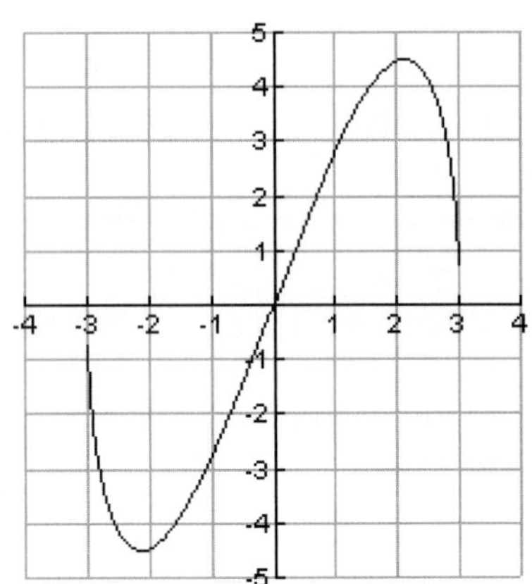	**126.** 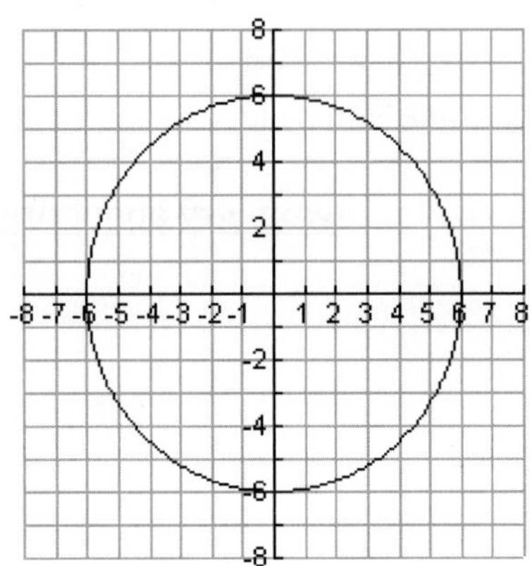
127. The center is $(-2,-3)$ and the radius is $\sqrt{81}=9$.	**128.** The center is $(4,-2)$ and the radius is $\sqrt{32}=4\sqrt{2}$.
129. Completing the square gives us: $x^2+y^2-4x+2y+11=0$ $x^2-4x \quad +y^2+2y=-11$ $(x^2-4x+4)+(y^2+2y+1)=-11+4+1$ $(x-2)^2+(y+1)^2=-6$ Since the left-side is always non-negative, there is no graph. Hence, this is not a circle and so there is no center or radius in this case.	**130.** Completing the square gives us: $3x^2+3y^2-6x-7=0$ $3x^2-6x \quad +3y^2=7$ $x^2-2x \quad +y^2=\frac{7}{3}$ $(x^2-2x+1)+(y^2+0)=\frac{7}{3}+1$ $(x-1)^2+(y-0)^2=\frac{10}{3}$ So, center is $(1,0)$ and radius is $\sqrt{\frac{10}{3}}$.
131. Find b: $4=-2(-3)+b$ $-2=b$ So, the equation is $\boxed{y=-2x-2}$.	**132.** Find b: $16=\frac{3}{4}(2)+b$ $\frac{29}{2}=b$ So, the equation is $\boxed{y=\frac{3}{4}x+\frac{29}{2}}$.
133. Since $m=0$, the line is horizontal. So, the equation is $\boxed{y=6}$.	**134.** Since m is undefined, the line is vertical. So, the equation is $\boxed{x=2}$.

135. slope: $m = \frac{-2-3}{-4-2} = \frac{5}{6}$ y-intercept: Use the point $(-4,-2)$ to find b: $-2 = \frac{5}{6}(-4) + b$ $\frac{4}{3} = b$ So, the equation is $\boxed{y = \frac{5}{6}x + \frac{4}{3}}$.	**136.** slope: $m = \frac{4-5}{-1-(-2)} = -1$ y-intercept: Use the point $(-1,4)$ to find b: $4 = (-1)(-1) + b$ $3 = b$ So, the equation is $\boxed{y = -x + 3}$.
137. slope: $m = \frac{\frac{1}{2}-\frac{5}{2}}{-\frac{3}{4}-\left(-\frac{7}{4}\right)} = -2$ y-intercept: Use the point $\left(-\frac{3}{4},\frac{1}{2}\right)$ to find b: $\frac{1}{2} = -2\left(-\frac{3}{4}\right) + b$ $-1 = b$ So, the equation is $\boxed{y = -2x - 1}$.	**138.** slope: $m = \frac{-2-2}{3-(-9)} = -\frac{1}{3}$ y-intercept: Use the point $(3,-2)$ to find b: $-2 = \left(-\frac{1}{3}\right)(3) + b$ $-1 = b$ So, the equation is $\boxed{y = -\frac{1}{3}x - 1}$.
139. slope: Since parallel to $2x - 3y = 6$ (whose slope is $\frac{2}{3}$), $m = \frac{2}{3}$. y-intercept: Use the point $(-2,-1)$ to find b: $-1 = \frac{2}{3}(-2) + b$ $\frac{1}{3} = b$ So, the equation is $\boxed{y = \frac{2}{3}x + \frac{1}{3}}$.	**140.** slope: Since perpendicular to $5x - 3y = 0$ (whose slope is $\frac{5}{3}$), $m = -\frac{3}{5}$. y-intercept: Use the point $(5,6)$ to find b: $6 = -\frac{3}{5}(5) + b$ $9 = b$ So, the equation is $\boxed{y = -\frac{3}{5}x + 9}$.
141. The general equation is $C = kr$. Using the fact that $C = k(1) = 2\pi$, we see that $k = 2\pi$. So, $\boxed{C = 2\pi r}$.	**142.** The general equation is $V = klw$. Using the fact that $V = k(2)(6) = 12h$, we see that $k = h$. So, $\boxed{V = lwh}$.

143. The general equation is $A = kr^2$. Using the fact that $A = k5^2 = 25\pi$, we see that $k = \pi$. So, $\boxed{A = \pi r^2}$.	**144.** The general equation is $F = \dfrac{k}{\lambda L}$. Using the fact that $F = \dfrac{k}{(10^{-6}m)(10^3 m)} = 20\pi$ we see that $k = \pi/50$. So, $\boxed{F = \dfrac{\pi}{50\lambda L}}$.

Chapter 0 Practice Test Solutions ---

1. $4p - 7 = 6p - 1$ $-6 = 2p$ $\boxed{-3 = p}$	**2.** $-2z + 2 + 3 = -3z + 3z - 3$ $-2z = -8$ $\boxed{z = 4}$
3. $3t = t^2 - 28$ $t^2 - 3t - 28 = 0$ $(t-7)(t+4) = 0$ $\boxed{t = -4, 7}$	**4.** $8x^2 - 13x = 6$ $8x^2 - 13x - 6 = 0$ $(8x+3)(x-2) = 0$ $\boxed{x = -3/8, 2}$
5. $6x^2 - 13x - 8 = 0$ $(3x-8)(2x+1) = 0$ $\boxed{x = -\frac{1}{2}, \frac{8}{3}}$	**6.** $\frac{3}{x-1} - \frac{5}{x+2} = 0$ $\frac{3(x+2) - 5(x-1)}{(x-1)(x+2)} = 0$ $\frac{-2x+11}{(x-1)(x+2)} = 0$ $\boxed{x = 11/2}$

7. $\frac{5}{y-3}+1-\frac{30}{y^2-9}=0$ $\frac{5(y+3)+(y^2-9)-30}{(y-3)(y+3)}=0$ $\frac{y^2+5y-24}{(y-3)(y+3)}=0$ $\frac{(y+8)\cancel{(y-3)}}{\cancel{(y-3)}(y+3)}=0$ $\boxed{y=-8}$	**8.** $x^4-5x^2-36=0$ $(x^2+4)(x^2-9)=0$ $(x-2i)(x+2i)(x-3)(x+3)=0$ $\boxed{x=\pm 3,\ \pm 2i}$
9. $\sqrt{2x+1}+x=7$ $\sqrt{2x+1}=7-x$ $2x+1=(7-x)^2$ $2x+1=49-14x+x^2$ $x^2-16x+48=0$ $(x-12)(x-4)=0$ $\boxed{x=4},\ \cancel{12}$	**10.** Let $u=x^{1/3}$ $2u^2+3u-2=0$ $(2u-1)(u+2)=0$ $u=-2,1/2$ So, $\boxed{x=-8,\ \frac{1}{8}}$
11. $3y-2=9-6\sqrt{3y+1}+3y+1$ $-12=-6\sqrt{3y+1}$ $\sqrt{3y+1}=2$ $3y+1=4$ $3y=3$ $\boxed{y=1}$	**12.** $x(3x-5)^3-2(3x-5)^2=0$ $(3x-5)^2[x(3x-5)-2]=0$ $(3x-5)^2(3x^2-5x-2)=0$ $(3x-5)^2(3x+1)(x-2)=0$ $\boxed{x=5/3,\ -1/3,\ 2}$
13. $x^{7/3}-8x^{4/3}+12x^{1/3}=0$ $x^{1/3}(x^2-8x+12)=0$ $x^{1/3}(x-6)(x-2)=0$ $\boxed{x=0,2,6}$	**14.** $\left\lvert\frac{1}{5}x+\frac{2}{3}\right\rvert=\frac{7}{15}$ $\frac{\lvert 3x+10\rvert}{15}=\frac{7}{15}$ $\lvert 3x+10\rvert=7$ $3x+10=7$ or $3x+10=-7$ $3x=-3$ $3x=-17$ $\boxed{x=-1}$ $\boxed{x=-\frac{17}{3}}$

15.
$$3x+19\ge 5x-15$$
$$34\ge 2x$$
$$17\ge x$$
$$\boxed{(-\infty,17]}$$

16.
$$-1\le 3x+5<26$$
$$-6\le 3x<21$$
$$-2\le x<7$$
$$\boxed{[-2,7)}$$

17.
$$\frac{2}{5}<\frac{x+8}{4}\le\frac{1}{2}$$
$$8<5(x+8)\le 10$$
$$-32<5x\le -30$$
$$-\frac{32}{5}<x\le -6$$
$$\boxed{\left(-\frac{32}{5},-6\right]}$$

18.
$$3x\ge 2x^2$$
$$0\ge 2x^2-3x$$
$$0\ge x(2x-3)$$
CPs: $x=0,\frac{3}{2}$

$+$ | 0 | $-$ | $\frac{3}{2}$ | $+$

$$\boxed{[0,\tfrac{3}{2}]}$$

19.
$$3p^2-p-4\ge 0$$
$$(3p-4)(p+1)\ge 0$$
CP's: $p=\frac{4}{3},-1$
$$\boxed{(-\infty,-1]\cup\left[\tfrac{4}{3},\infty\right)}$$

20.
$$|5-2x|>1$$
$5-2x>1$ or $5-2x<-1$

$-2x>-4$ or $-2x<-6$

$x<2$ or $x>3$
$$\boxed{(-\infty,2)\cup(3,\infty)}$$

21.
$$\frac{x-3}{2x+1}\le 0$$
CPs: $x=-\frac{1}{2},3$

$+$ | $-\frac{1}{2}$ | $-$ | 3 | $+$

$$\boxed{(-\tfrac{1}{2},3]}$$

22.
$$\frac{x+4}{x^2-9}\ge 0$$
$$\frac{x+4}{(x-3)(x+3)}\ge 0$$
CPs: $x=-4,\ \pm 3$

$-$ | -4 | $+$ | -3 | $-$ | 3 | $+$

$$\boxed{[-4,-3)\cup(3,\infty)}$$

23. $d=\sqrt{(-7-2)^2+(-3-(-2))^2}=\boxed{\sqrt{82}}$

24. $M=\left(\frac{-3+5}{2},\ \frac{5+(-1)}{2}\right)=\boxed{(1,2)}$

25.

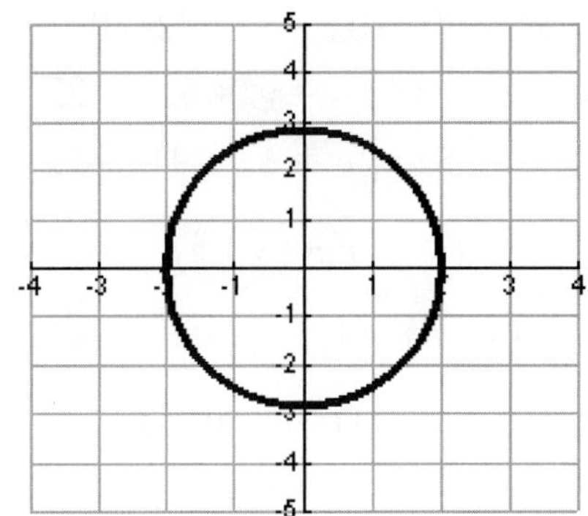

26.

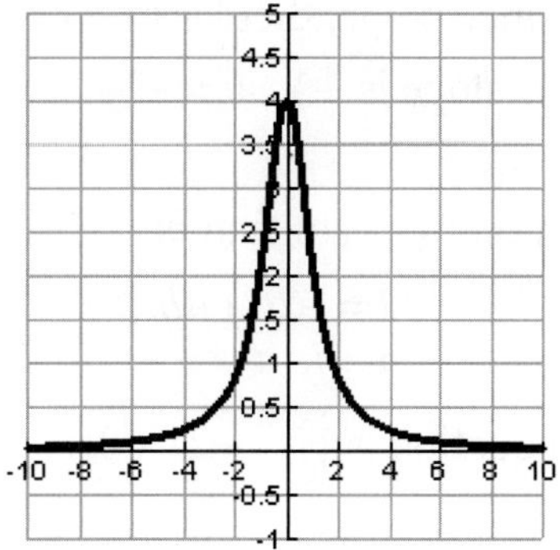

27.

<u>x-intercept</u>: $x-3(0)=6 \Rightarrow x=6$. So, (6,0).

<u>y-intercept</u>: $0-3y=6 \Rightarrow y=-2$.

So, $(0,-2)$.

28. <u>x-intercepts</u>:

$4x^2-9(0)^2=36 \Rightarrow x^2=9 \Rightarrow x=\pm 3$

So, $(\pm 3, 0)$.

<u>y-intercepts</u>:

$4(0)^2-9y^2=36$, which has no real solution.

So, no y-intercepts.

29.

$\frac{2}{3}x-\frac{1}{4}y=2 \Rightarrow -\frac{1}{4}y=2-\frac{2}{3}x \Rightarrow \boxed{y=\frac{8}{3}x-8}$

30.

$4x-6y=12 \Rightarrow -6y=-4x+12 \Rightarrow \boxed{y=\frac{2}{3}x-2}$

31.

<u>slope</u>: $m=\dfrac{9-2}{4-(-3)}=1$

<u>y-intercept</u>: Use the point $(-3,2)$ to find b:

$$2=1(-3)+b$$
$$5=b$$

So, the equation is $\boxed{y=x+5}$.

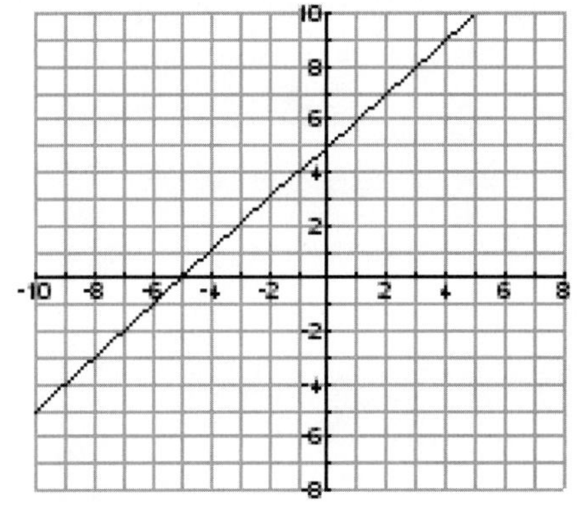

32.

<u>slope</u>: Since parallel to $y=4x+3$ (whose slope is 4), $m=4$.

<u>y-intercept</u>: Use the point $(1,7)$ to find b:

$$7=4(1)+b$$
$$3=b$$

So, the equation is $\boxed{y=4x+3}$.

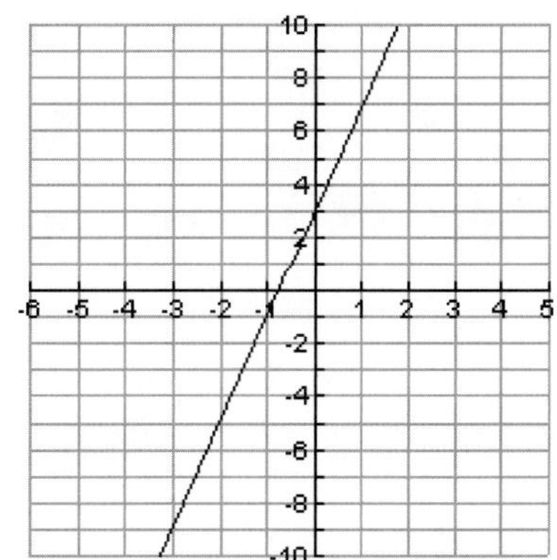

33.
slope: Since perpendicular to $2x-4y=5$ (whose slope is $\frac{1}{2}$), $m=-2$.
y-intercept:
Use the point $(1,1)$ to find b:

$$1=-2(1)+b$$
$$3=b$$

So, the equation is $\boxed{y=-2x+3}$.

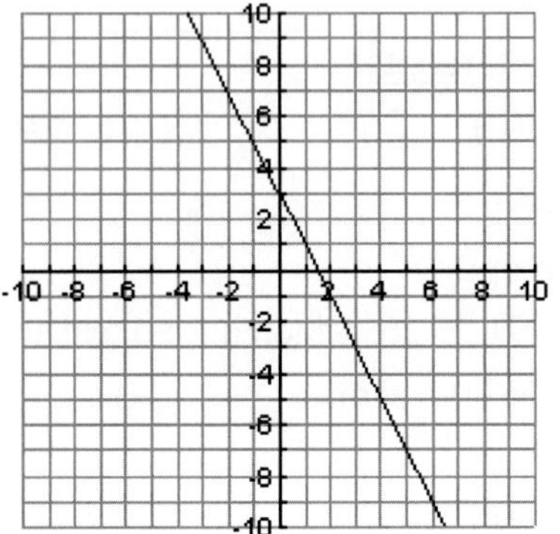

34. Completing the square gives us:

$$x^2+y^2-10x+6y+22=0$$
$$x^2-10x \quad +y^2+6y=-22$$
$$\left(x^2-10x+25\right)+\left(y^2+6y+9\right)=-22+25+9$$
$$\left(x-5\right)^2+\left(y+3\right)^2=12$$
$$\left(x-5\right)^2+\left(y-(-3)\right)^2=\left(\sqrt{12}\right)^2$$

So, center is $(5,-3)$ and radius is $\sqrt{12}=2\sqrt{3}$.

35. The general equation is $F=\dfrac{km}{P}$.

Using the fact that $F=\dfrac{k(2)}{3}=20$, we see that $k=20\left(\frac{3}{2}\right)=30$. So, $F=\dfrac{30m}{P}$.

36. The general equation is $y=kx^2$.
Using the fact that $y=k(5)^2=8$, we see that $\boxed{k=\frac{8}{25}}$. So, $y=\frac{8}{25}x^2$

CHAPTER 1

Section 1.1 Solutions ---

1. Not a function – 0 maps to both -3 and 3.	**2.** Not a function – 2 maps to both -2 and 2, and 5 maps to both -5 and 5.
3. Not a function – 4 maps to both -2 and 2, and 9 maps to both -3 and 3.	**4.** Function
5. Function	**6.** Function
7. Not a function – Since $\left(1,-2\sqrt{2}\right)$ and $\left(1,2\sqrt{2}\right)$ are both on the graph, it does not pass vertical line test.	**8.** Not a function – Since $(1,-1)$ and $(1,1)$ are both on the graph, it does not pass the vertical line test.
9. Not a function – Since $(1,-1)$ and $(1,1)$ are both on the graph, it does not pass the vertical line test.	**10.** Function
11. Function	**12.** Function
13. Not a function – Since $(0,5)$ and $(0,-5)$ are both on the graph, it does not pass the vertical line test.	**14.** Not a function – Since $(0,4)$ and $(0,-4)$ are both on the graph, it does not pass the vertical line test.
15. Function	**16.** Function
17. Not a function – Since $(0,-1)$ and $(0,-3)$ are both on the graph, it does not pass the vertical line test.	**18.** Function
19. **a)** 5 **b)** 1 **c)** -3	**20.** **a)** 1 **b)** -5 **c)** 0
21. **a)** 3 **b)** 2 **c)** 5	**22.** **a)** 0 **b)** 4 **c)** -5
23. **a)** -5 **b)** -5 **c)** -5	**24.** **a)** -2 **b)** -6 **c)** -4
25. **a)** 2 **b)** -8 **c)** -5	**26.** **a)** DNE **b)** 0 **c)** 3
27. 1	**28.** -1.5 and 3
29. 1 and -3	**30.** -7
31. For all x in the interval $[-4,4]$	**32.** For all x in the set $[-4,0)\cup[4]$

33. 6	**34.** −3
35. $f(-2)=2(-2)-3=\boxed{-7}$	**36.** $G(-3)=(-3)^2+2(-3)-7=\boxed{-4}$
37. $g(1)=5+1=\boxed{6}$	**38.** $F(-1)=4-(-1)^2=\boxed{3}$
39. Using #35 and #37, we see that $f(-2)+g(1)=-7+6=\boxed{-1}$.	**40.** Using #36 and #38, we see that $G(-3)-F(-1)=-4-3=\boxed{-7}$.
41. Using #35 and #37, we see that $3f(-2)-2g(1)=3(-7)-2(6)=\boxed{-33}$.	**42.** Using #36 and #38, we see that $2F(-1)-2G(-3)=2(3)-2(-4)=\boxed{14}$.
43. Using #35 and #37, we see that $\frac{f(-2)}{g(1)}=\boxed{-\frac{7}{6}}$.	**44.** Using #36 and #38, we see that $\frac{G(-3)}{F(-1)}=\boxed{-\frac{4}{3}}$.
45. $\frac{f(0)-f(-2)}{g(1)}=\frac{(2(0)-3)-(-7)}{6}=\frac{-3+7}{6}=\boxed{\frac{2}{3}}$	**46.** $\frac{G(0)-G(-3)}{F(-1)}=\frac{(0^2+2(0)-7)-(-4)}{3}=\frac{-7+4}{3}=\boxed{-1}$

47.

$$\begin{aligned} f(x+1)-f(x-1) &= [2(x+1)-3]-[2(x-1)-3] \\ &= [2x+2-3]-[2x-2-3] \\ &= [2x-1]-[2x-5] \\ &= 2x-1-2x+5 \\ &= \boxed{4} \end{aligned}$$

48.

$$\begin{aligned} F(t+1)-F(t-1) &= \left[4-(t+1)^2\right]-\left[4-(t-1)^2\right] \\ &= \left[4-\left(t^2+2t+1\right)\right]-\left[4-\left(t^2-2t+1\right)\right] \\ &= \left[4-t^2-2t-1\right]-\left[4-t^2+2t-1\right] \\ &= 4-t^2-2t-1-4+t^2-2t+1 \\ &= \boxed{-4t} \end{aligned}$$

49.

$$\begin{aligned} g(x+a)-f(x+a) &= [5+(x+a)]-[2(x+a)-3] \\ &= [5+x+a]-[2x+2a-3] \\ &= 5+x+a-2x-2a+3 \\ &= \boxed{8-x-a} \end{aligned}$$

50. $$G(x+b)+F(b)=\left[(x+b)^2+2(x+b)-7\right]+\left[4-b^2\right]$$ $$=x^2+2bx+b^2+2x+2b-7+4-b^2$$ $$=\boxed{x^2+2bx+2x+2b-3}$$
51. $$\frac{f(x+h)-f(x)}{h}=\frac{[2(x+h)-3]-[2x-3]}{h}$$ $$=\frac{2x+2h-3-2x+3}{h}$$ $$=\frac{2h}{h}=\boxed{2}$$
52. $$\frac{F(t+h)-F(t)}{h}=\frac{\left[4-(t+h)^2\right]-\left[4-t^2\right]}{h}$$ $$=\frac{\left[4-\left(t^2+2ht+h^2\right)\right]-\left[4-t^2\right]}{h}$$ $$=\frac{\left[4-t^2-2ht-h^2\right]-\left[4-t^2\right]}{h}$$ $$=\frac{4-t^2-2ht-h^2-4+t^2}{h}$$ $$=\frac{-2ht-h^2}{h}=\frac{-h(2t+h)}{h}=\boxed{-(2t+h)}$$
53. $$\frac{g(t+h)-g(t)}{h}=\frac{[5+(t+h)]-[5+t]}{h}$$ $$=\frac{5+t+h-5-t}{h}=\frac{h}{h}=\boxed{1}$$
54. $$\frac{G(x+h)-G(x)}{h}=\frac{\left[(x+h)^2+2(x+h)-7\right]-\left[x^2+2x-7\right]}{h}$$ $$=\frac{x^2+2hx+h^2+2x+2h-7-x^2-2x+7}{h}$$ $$=\frac{2hx+h^2+2h}{h}=\frac{h(2x+h+2)}{h}=\boxed{2x+h+2}$$

55. It follows directly from the computation in #51 with $x = -2$ that this equals $\boxed{2}$.	**56.** It follows directly from the computation in #52 with $t = -1$ that this equals $\boxed{2-h}$.
57. It follows directly from the computation in #53 with $t = 1$ that this equals $\boxed{1}$.	**58.** It follows directly from the computation in #54 with $x = -3$ that this equals $\boxed{h-4}$.
59. The domain is $\mathbb{R}$. This is written using interval notation as $\boxed{(-\infty,\infty)}$.	**60.** The domain is $\mathbb{R}$. This is written using interval notation as $\boxed{(-\infty,\infty)}$.
61. The domain is $\mathbb{R}$. This is written using interval notation as $\boxed{(-\infty,\infty)}$.	**62.** The domain is $\mathbb{R}$. This is written using interval notation as $\boxed{(-\infty,\infty)}$.
63. The domain is the set of all real numbers x such that $x-5 \neq 0$, that is $x \neq 5$. This is written using interval notation as $\boxed{(-\infty,5)\cup(5,\infty)}$.	**64.** The domain is the set of all real numbers t such that $t+3 \neq 0$, that is $t \neq -3$. This is written using interval notation as $\boxed{(-\infty,-3)\cup(-3,\infty)}$.
65. The domain is the set of all real numbers x such that $x^2-4=(x-2)(x+2)\neq 0$, that is $x \neq -2,2$. This is written using interval notation as $\boxed{(-\infty,-2)\cup(-2,2)\cup(2,\infty)}$.	**66.** The domain is the set of all real numbers x such that $x^2-1=(x-1)(x+1)\neq 0$, that is $x \neq -1,1$. This is written using interval notation as $\boxed{(-\infty,-1)\cup(-1,1)\cup(1,\infty)}$.
67. Since $x^2+1\neq 0$, for every real number x, the domain is $\mathbb{R}$. This is written using interval notation as $\boxed{(-\infty,\infty)}$.	**68.** Since $x^2+4\neq 0$, for every real number x, the domain is $\mathbb{R}$. This is written using interval notation as $\boxed{(-\infty,\infty)}$.
69. The domain is the set of all real numbers x such that $7-x\geq 0$, that is $7 \geq x$. This is written using interval notation as $\boxed{(-\infty,7]}$.	**70.** The domain is the set of all real numbers t such that $t-7\geq 0$, that is $t \geq 7$. This is written using interval notation as $\boxed{[7,\infty)}$.

71. The domain is the set of all real numbers x such that

$$2x+5\geq 0,$$

that is $x\geq -\frac{5}{2}$.

This is written using interval notation as $\boxed{\left[-\frac{5}{2},\infty\right)}$.

72. The domain is the set of all real numbers x such that

$$5-2x\geq 0,$$

that is $\frac{5}{2}\geq x$.

This is written using interval notation as $\boxed{\left(-\infty,\frac{5}{2}\right]}$.

73. The domain is the set of all real numbers t such that $t^2-4\geq 0$, which is equivalent to $(t-2)(t+2)\geq 0$.

CPs are -2, 2

+ − +

-2 2

This is written using interval notation as $\boxed{(-\infty,-2]\cup[2,\infty)}$.

74. The domain is the set of all real numbers x such that $x^2-25\geq 0$, which is equivalent to $(x-5)(x+5)\geq 0$.

CPs are -5, 5

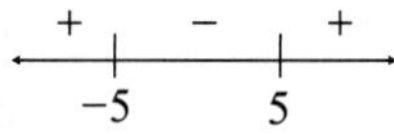

This is written using interval notation as $\boxed{(-\infty,-5]\cup[5,\infty)}$.

75. The domain is the set of all real numbers x such that

$$x-3>0,$$

that is $x>3$.

This is written using interval notation as $\boxed{(3,\infty)}$.

76. The domain is the set of all real numbers x such that

$$5-x>0,$$

that is $5>x$.

This is written using interval notation as $\boxed{(-\infty,5)}$.

77. Since $1-2x$ can be any real number, there is no restriction on x, so that the domain is $\boxed{(-\infty,\infty)}$.

78. Since $7-5x$ can be any real number, there is no restriction on x, so that the domain is $\boxed{(-\infty,\infty)}$.

79. The only restriction is that $x+4\neq 0$, so that $x\neq -4$. So, the domain is $\boxed{(-\infty,-4)\cup(-4,\infty)}$.

80. The only restriction is that $x^2-9=(x-3)(x+3)\neq 0$, so that $x\neq \pm 3$. So, the domain is $\boxed{(-\infty,-3)\cup(-3,3)\cup(3,\infty)}$.

81. The domain is the set of all real numbers x such that

$$3-2x>0,$$

that is $\frac{3}{2}>x$.

This is written using interval notation as $\boxed{\left(-\infty,\frac{3}{2}\right)}$.

82. The domain is the set of all real numbers t such that $25-x^2>0$, which is equivalent to $(5-x)(5+x)>0$.

CPs are -5, 5

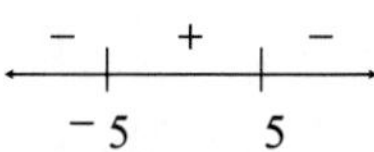

This is written using interval notation as $\boxed{(-5,5)}$.

83. The domain is the set of all real numbers t such that $t^2 - t - 6 > 0$, which is equivalent to $(t-3)(t+2) > 0$. CPs are –2, 3 + – + on the number line with marks at −2 and 3 This is written using interval notation as $\boxed{(-\infty,-2)\cup(3,\infty)}$.	**84.** Since $t^2 + 9 > 0$, for all real numbers t, there is no restriction. So, the domain is $\boxed{(-\infty,\infty)}$.
85. The domain is the set of all real numbers t such that $x^2 - 16 \geq 0$, which is equivalent to $(x-4)(x+4) \geq 0$. CPs are –4, 4 + – + on the number line with marks at −4 and 4 This is written using interval notation as $\boxed{(-\infty,-4]\cup[4,\infty)}$.	**86.** There is no restriction on x. So, the domain is $\boxed{(-\infty,\infty)}$.
87. The function can be written as $r(x) = \dfrac{x^2}{\sqrt{3-2x}}$. So, the domain is the set of real numbers x such that $3 - 2x > 0$, that is $\frac{3}{2} > x$. This is written using interval notation as $\boxed{(-\infty,\frac{3}{2})}$.	**88.** The function can be written as $p(x) = \dfrac{(x-1)^2}{(x^2-9)^{3/5}}$. So, the domain is the set of real numbers x such that $x^2 - 9 = (x-3)(x+3) \neq 0$, so that $x \neq \pm 3$. So, the domain is $\boxed{(-\infty,-3)\cup(-3,3)\cup(3,\infty)}$.
89. The domain of any linear function is $\boxed{(-\infty,\infty)}$.	**90.** The domain of any quadratic function is $\boxed{(-\infty,\infty)}$.
91. Solve $x^2 - 2x - 5 = 3$. $x^2 - 2x - 8 = 0$ $(x-4)(x+2) = 0$ $\boxed{x = -2, 4}$	**92.** Solve $\frac{5}{6}x - \frac{3}{4} = \frac{2}{3}$. $10x - 9 = 8$ $10x = 17$ $\boxed{x = \frac{17}{10}}$

93.

$$2x(x-5)^3 - 12(x-5)^2 = 0$$
$$2(x-5)^2\left[x(x-5)-6\right] = 0$$
$$2(x-5)^2(x^2-5x-6) = 0$$
$$2(x-5)^2(x-6)(x+1) = 0$$
$$\boxed{x = -1, 5, 6}$$

94.

$$3x(x+3)^2 - 6(x+3)^3 = 0$$
$$3(x+3)^2\left[x-2(x+3)\right] = 0$$
$$3(x+3)^2(-x-6) = 0$$
$$\boxed{x = -3, -6}$$

95. Assume: 6am corresponds to $x = 6$
noon corresponds to $x = 12$

Then, the temperature at 6am is:

$$T(6) = -0.7(6)^2 + 16.8(6) - 10.8 = 64.8^\circ F$$

The temperature at noon is:

$$T(12) = -0.7(12)^2 + 16.8(12) - 10.8$$
$$= 90^\circ F$$

96. 9am corresponds to $x = 9$ and 3pm corresponds to $x = 15$. So,

$$T(9) = -0.5(9)^2 + 14.2(9) - 2.8 = 84.5^\circ F$$
$$T(15) = -0.5(15)^2 + 14.2(15) - 2.8 = 97.7^\circ F$$

97. $h(2) = -16(2)^2 + 45(2) + 1 = 27$

Since height must be nonnegative, only those values of t for which $h(t) \geq 0$ should be included in the domain. As such, we must solve $-16t^2 + 45t + 1 \geq 0$. Graphically, we see that

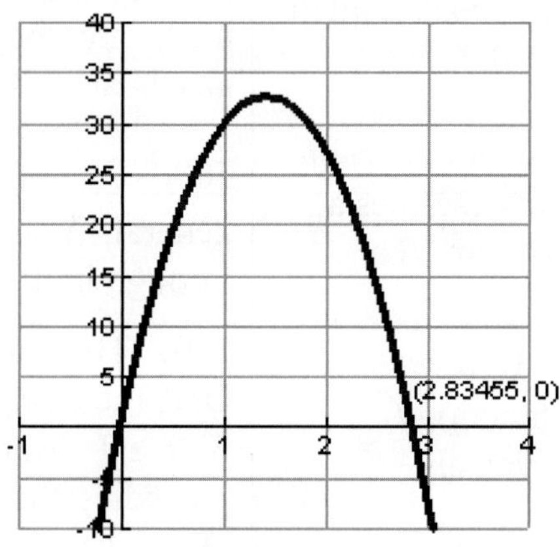

Hence, the domain of h is approximately [0, 2.8].

98. $h(14) = -16(4)^2 + 128(4) = \boxed{256\text{ ft}}$. The domain is $[0,\infty)$ since we are starting at time $t = 0$ sec.

99. Start with a square piece of cardboard with dimensions 10 in. $\times$ 10 in.. Then, cut out 4 square corners with dimensions x in. $\times$ x in., as shown in the diagram:

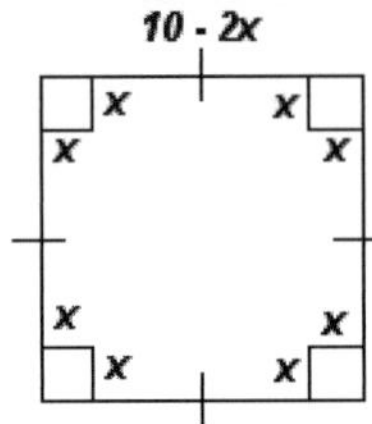

Upon bending all four corners up, a box of height x is formed. Notice that all four sides of the base of the resulting box have length $10-2x$. The volume of the box, $V(x)$, is given by:

$$\begin{aligned} V(x) &= (\text{Length})\cdot(\text{Width})\cdot(\text{Height}) \\ &= (10-2x)(10-2x)(x) \\ &= x(10-2x)^2 \end{aligned}$$

The domain is $(0,5)$. (For any other values of x, one cannot form a box.)

100. The volume of a right circular cylindrical tank whose base radius is 10 ft and whose height is h is given by $V(h) = \pi(10)^2 h = 100\pi h$. If the height is increased by 2 ft, the corresponding volume would be:

$$V(h+2) = \pi(10)^2(h+2) = 100\pi h + 200\pi$$

So, the volume increased by 200π cubic ft, which corresponds to $200\pi \cdot 7.48\,\text{gal} \cong \boxed{4700\,\text{gal}}$.

101. Yes. For because every input (year), there corresponds to exactly one output (federal funds rate).

102. (2000, 5.45), (2002, 1.73), (2004, 1.00), (2006, 4.50), (2008, 3.50)

103. (1989, 4000), (1993, 6000), (1997, 6000), (2001, 8000), (2005,11000),

104. Yes, for every input there corresponds a unique output.

105. **a)** $F(50)$ = number of tons of carbon emitted by natural gas in 1950 = 0
b) $g(50)$ = number of tons of coal emitted by natural gas in 1950 = 1000
c) $h(5) = 2000$

106. $F(100) + g(100) + G(100)$ represents the total amount (in millions of metric tons) of carbon emitted in 2000 by natural gas, coal, and petroleum.

107. Should apply the vertical line test to determine if the relationship describes a function. The given relationship IS a function in this case.

108. $H(3) - H(-1) \neq H(3) + H(1)$, in general. You cannot distribute -1 through in this manner.

109. $f(x+1) \neq f(x) + f(1)$, in general. You cannot distribute the function f through the input at which you are evaluating it.

110. There are two mistakes. One, the computation $3-t>0$ should be $3-t \geq 0$. And two, the statement directly preceding the computation should be, "What can $3-t$ be?"
The domain should be $(-\infty,3]$.

111. $G(-1+h) \neq G(-1)+G(h)$, in general.

112. $f(1)=-1$ means the point $(1,-1)$ must satisfy the function. So,

$$-1=|1-A|-1$$
$$0=|1-A|$$
$$\boxed{1=A}$$

113. False. Consider the function $f(x)=\sqrt{9-x^2}$ on its domain $[-3,3]$. The vertical line test $x=4$ doesn't intersect the graph, but it still defines a function.

114. False. Consider the function $f(x)=x^2$ on its domain $\mathbb{R}$.

115. True

116. True

117.

$$f(1)=A(1)^2-3(1)=-1$$
$$A-3=-1$$
$$A=2$$

118. $g(3)=\dfrac{1}{b-3}$ is undefined only if $b=3$.

119. $F(-2)=\dfrac{C-(-2)}{D-(-2)}=\dfrac{C+2}{D+2}$ is undefined only if $\boxed{D=-2}$. So,

$F(-1)=\dfrac{C-(-1)}{D-(-1)}=\dfrac{C+1}{D+1}=\dfrac{C+1}{-2+1}=-(C+1)=4$ implies that $\boxed{C=-5}$.

120. Many functions will work here. The easiest ones to construct are of the form $g(x)=\dfrac{b}{x-5}$. For such a function, certainly $g(5)$ is undefined. In order for $(1,-1)$ to be on the graph, it must be the case that $-1=\dfrac{b}{1-5}=\dfrac{b}{-4}$, so that $b=4$. So, one function that works is $g(x)=\dfrac{4}{x-5}$.

121. The domain is the set of all real numbers x such that $x^2-a^2=(x-a)(x+a)\neq 0$, which is equivalent to $x \neq \pm a$. So, the domain is $\boxed{(-\infty,-a)\cup(-a,a)\cup(a,\infty)}$.

122. The domain is the set of all real numbers x such that $x^2-a^2=(x-a)(x+a)\geq 0$.

CPs: $x=\pm a$

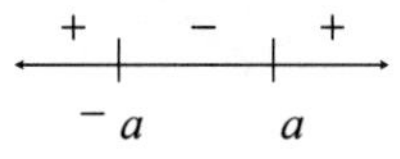

So, the domain is $\boxed{(-\infty,-a]\cup[a,\infty)}$.

123.

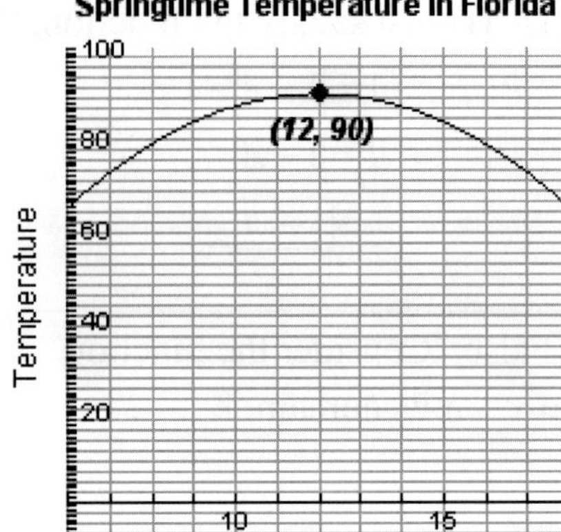

The time of day when it is warmest is Noon ($x = 12$) and the temperature is approximately 90 degrees.

This model is only valid on the interval $[6,18]$ since the values of T outside the interval $[6,18]$ are too small to be considered temperatures in Florida.

124.

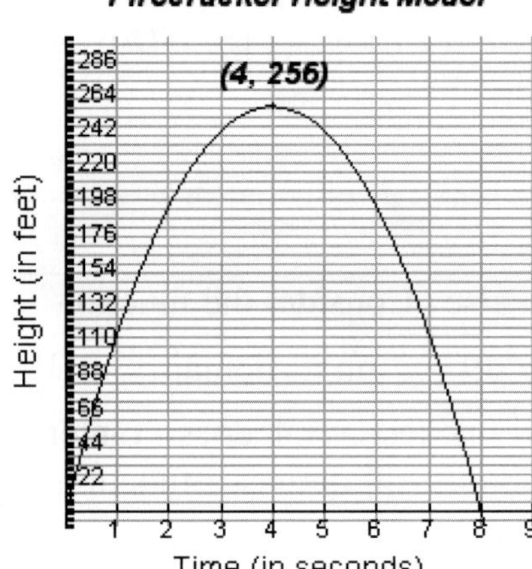

The firecracker is airborne for 8 seconds after liftoff – this corresponds to the x-intercept.

The firecracker maintains a maximum height of 256 ft within 4 seconds of liftoff.

The model only applies for 8 seconds since the firecracker will land after 8 seconds, and presumably will not travel through the ground afterwards.

125.

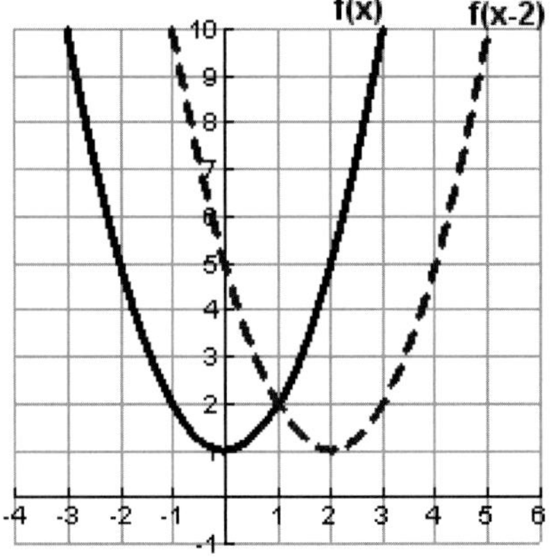

The graph of y2 can be obtained by shifting the graph of y1 two units to the right.

126.

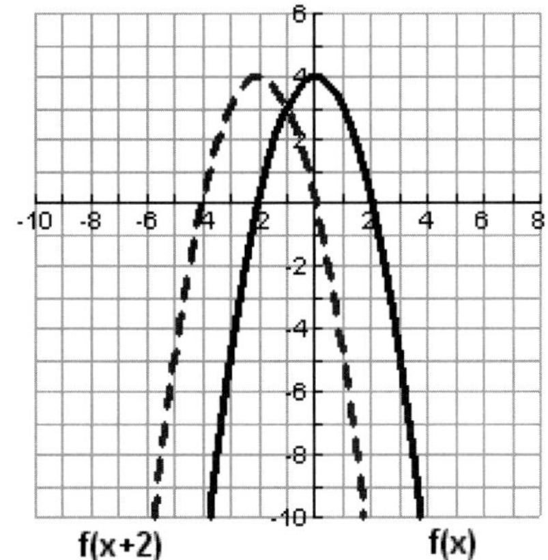

The graph of y2 can be obtained by shifting the graph of y1 two units to the left.

127.

$$\frac{f(x+h)-f(x)}{h}=\frac{\left[(x+h)^3+(x+h)\right]-\left[x^3+x\right]}{h}$$

$$=\frac{x^3+3x^2h+3xh^2+h^3+x+h-x^3-x}{h}$$

$$=\frac{h\left(3x^2+3xh+h^2+1\right)}{h}=\boxed{3x^2+3xh+h^2+1}$$

So, at $h=0$ we get $\boxed{f'(x)=3x^2+1}$.

128.

$$\frac{f(x+h)-f(x)}{h}=\frac{\left[6(x+h)+\sqrt{x+h}\right]-\left[6x+\sqrt{x}\right]}{h}$$

$$=\frac{6h}{h}+\frac{\sqrt{x+h}-\sqrt{x}}{h}$$

$$=6+\frac{\left(\sqrt{x+h}-\sqrt{x}\right)\left(\sqrt{x+h}+\sqrt{x}\right)}{h\left(\sqrt{x+h}+\sqrt{x}\right)}$$

$$=6+\frac{x+h-x}{h\left(\sqrt{x+h}+\sqrt{x}\right)}=6+\frac{1}{\sqrt{x+h}+\sqrt{x}}$$

So, at $h=0$ we get $\boxed{f'(x)=6+\frac{1}{2\sqrt{x}}}$.

129.

$$\frac{f(x+h)-f(x)}{h}=\frac{\frac{x+h-5}{x+h+3}-\frac{x-5}{x+3}}{h}$$

$$=\frac{(x+h-5)(x+3)-(x-5)(x+h+3)}{h(x+h+3)(x+3)}$$

$$=\frac{\left(x^2+xh-5x+3x+3h-15\right)-\left(x^2+hx+3x-5x-5h-15\right)}{h(x+h+3)(x+3)}$$

$$=\frac{8h}{h(x+h+3)(x+3)}=\frac{8}{(x+h+3)(x+3)}$$

So, at $h=0$ we get $\boxed{f'(x)=\frac{8}{(x+3)^2}}$

130.

$$\frac{f(x+h)-f(x)}{h}=\frac{\sqrt{\frac{x+h+7}{5-(x+h)}}-\sqrt{\frac{x+7}{5-x}}}{h}$$

$$=\frac{\sqrt{x+h+7}\sqrt{5-x}-\sqrt{x+7}\sqrt{5-x-h}}{h\sqrt{5-x-h}\sqrt{5-x}}$$

$$=\frac{\sqrt{(x+h+7)(5-x)}-\sqrt{(x+7)(5-x-h)}}{h\sqrt{5-x-h}\sqrt{5-x}}$$

$$=\frac{\sqrt{(x+h+7)(5-x)}-\sqrt{(x+7)(5-x-h)}}{h\sqrt{5-x-h}\sqrt{5-x}}\cdot\frac{\sqrt{(x+h+7)(5-x)}+\sqrt{(x+7)(5-x-h)}}{\sqrt{(x+h+7)(5-x)}+\sqrt{(x+7)(5-x-h)}}$$

$$=\frac{\left(5h+35-x^2-hx-2x\right)-\left(-2x-x^2-hx+35-7h\right)}{h\sqrt{5-x-h}\sqrt{5-x}\left(\sqrt{(x+h+7)(5-x)}+\sqrt{(x+7)(5-x-h)}\right)}$$

$$=\frac{12}{\sqrt{5-x-h}\sqrt{5-x}\left(\sqrt{(x+h+7)(5-x)}+\sqrt{(x+7)(5-x-h)}\right)}$$

So, at $h=0$ we get

$$f'(x)=\frac{12}{\sqrt{5-x}\sqrt{5-x}\left(\sqrt{(x+7)(5-x)}+\sqrt{(x+7)(5-x)}\right)}$$

$$=\frac{6}{(5-x)\sqrt{(x+7)(5-x)}}=\boxed{\frac{6}{(x+7)^{1/2}(5-x)^{3/2}}}$$

Section 1.2 Solutions ---

1. $h(-x)=(-x)^2+2(-x)=x^2-2x\neq h(x)$

So, not even.

$-h(-x)=-\left(x^2-2x\right)=-x^2+2x\neq h(x)$

So, not odd. Thus, $\boxed{\text{neither}}$.

2. $G(-x)=2(-x)^4+3(-x)^3$

$=2x^4-3x^3\neq G(x)$

So, not even.

$-G(-x)=-\left(2x^4-3x^3\right)\neq G(x)$

So, not odd. Thus, $\boxed{\text{neither}}$.

3. $h(-x)=(-x)^{1/3}-(-x)$ $=-\left(x^{1/3}-x\right)\neq h(x)$ So, not even. $-h(-x)=-\left(-\left(x^{1/3}-x\right)\right)=x^{1/3}-x=h(x)$ So, odd.	**4.** $g(-x)=(-x)^{-1}+(-x)$ $=-\left(x^{-1}+x\right)\neq g(x)$ So, not even. $-g(-x)=-\left(-\left(x^{-1}+x\right)\right)$ $=x^{-1}+x=g(x)$ So, odd. (<u>Note</u>: $(-x)^{-1}=\frac{1}{-x}=-\frac{1}{x}=-(x)^{-1}$)
5. $f(-x)=\lvert -x\rvert+5=\lvert -1\rvert\lvert x\rvert+5$ $=\lvert x\rvert+5=f(x)$ So, even. Thus, f cannot be odd.	**6.** $f(-x)=\lvert -x\rvert+(-x)^2$ $=\lvert -1\rvert\lvert x\rvert+x^2=f(x)$ So, even. Thus, f cannot be odd.
7. $f(-x)=\lvert -x\rvert=\lvert -1\rvert\lvert x\rvert=f(x)$ So, even. Thus, f cannot be odd.	**8.** $f(-x)=\left\lvert (-x)^3\right\rvert=\left\lvert -x^3\right\rvert=\lvert -1\rvert\left\lvert x^3\right\rvert=f(x)$ So, even. Thus, f cannot be odd.
9. $G(-t)=\lvert(-t)-3\rvert=\lvert-(t+3)\rvert$ $=\lvert t+3\rvert\neq G(t)$ So, not even. $-G(-t)=-\lvert t+3\rvert\neq G(t)$ So, not odd. Thus, neither.	**10.** $G(-t)=\lvert(-t)+2\rvert\neq G(t)$ So, not even. $-G(-t)=-\lvert(-t)+2\rvert\neq G(t)$ So, not odd. Thus, neither.
11. $G(-t)=\sqrt{-t-3}=\sqrt{-(t+3)}\neq G(t)$ So, not even. $-G(-t)=\underbrace{-\sqrt{-(t+3)}}_{\text{Note: Cannot distribute } -1 \text{ here}}\neq G(t)$ So, not odd. Thus, neither.	**12.** $f(-x)=\sqrt{2-(-x)}=\sqrt{2+x}\neq f(x)$ So, not even. $-f(-x)=-\sqrt{2+x}\neq f(x)$ So, not odd. Thus, neither.
13. $g(-x)=\sqrt{(-x)^2+(-x)}$ $=\sqrt{x^2-x}\neq g(x)$ So, not even. $-g(-x)=-\sqrt{x^2-x}\neq g(x)$ So, not odd. Thus, neither.	**14.** $f(-x)=\sqrt{(-x)^2+2}=\sqrt{x^2+2}=f(x)$ So, even. Thus, f cannot be odd.

15. $h(-x) = \frac{1}{-x} + 3 \neq h(x)$

So, not even.

$-h(-x) = -\left(\frac{1}{-x} + 3\right) = \frac{1}{x} - 3 \neq h(x)$

So, not odd. Thus, neither.

16. $h(-x) = \frac{1}{-x} - 2(-x) = -\left(\frac{1}{x} - 2x\right) \neq h(x)$

So, not even.

$-h(-x) = -\left(-\left(\frac{1}{x} - 2x\right)\right) = \frac{1}{x} - 2x = h(x)$

So, odd.

17.

Domain	$(-\infty, \infty)$
Range	$[-1, \infty)$
Increasing	$(-1, \infty)$
Decreasing	$(-3, -2)$
Constant	$(-\infty, -3) \cup (-2, -1)$

d) 0
e) -1
f) 2

18.

Domain	$[-4, \infty)$
Range	$(-\infty, 3]$
Increasing	$(1, 2)$
Decreasing	$(-3, 0) \cup (2, \infty)$
Constant	$[-4, -3) \cup (0, 1)$

d) -1
e) approximately 1.8
f) 1

19.

Domain	$[-7, 2]$
Range	$[-5, 4]$
Increasing	$(-4, 0)$
Decreasing	$(-7, -4) \cup (0, 2)$
Constant	nowhere

d) 4
e) 1
f) -5

20.

Domain	$(-\infty,\infty)$
Range	$(-\infty,\infty)$
Increasing	$(-\infty,-3)\cup(3,\infty)$
Decreasing	$(-3,3)$
Constant	nowhere

d) 0
e) 3.5
f) approximately -3.3

21.

Domain	$(-\infty,\infty)$
Range	$(-\infty,\infty)$
Increasing	$(-\infty,-3)\cup(4,\infty)$
Decreasing	nowhere
Constant	$(-3,4)$

d) 2
e) 2
f) 2

22.

Domain	$(-\infty,\infty)$
Range	$(-\infty,\infty)$
Increasing	nowhere
Decreasing	$(-\infty,\infty)$
Constant	nowhere

d) 2
e) 1
f) -1

23.

Domain	$(-\infty,\infty)$
Range	$[-4,\infty)$
Increasing	$(0,\infty)$
Decreasing	$(-\infty,0)$
Constant	nowhere

d) -4
e) 0
f) 0

24.

Domain	$(-\infty, \infty)$
Range	$[0, \infty)$
Increasing	$(3, \infty)$
Decreasing	$(-\infty, -3)$
Constant	$(-3, 3)$

d) 0
e) 0
f) 0

25.

Domain	$(-\infty, 0) \cup (0, \infty)$
Range	$(-\infty, 0) \cup (0, \infty)$
Increasing	$(-\infty, 0) \cup (0, \infty)$
Decreasing	nowhere
Constant	nowhere

d) undefined
e) 3
f) –3

26.

Domain	$(-\infty, 4) \cup (4, \infty)$
Range	$(-\infty, \infty)$
Increasing	$(-\infty, 0) \cup (4, \infty)$
Decreasing	$(0, 4)$
Constant	nowhere

d) 4
e) approximately 3.5
f) approximately 2.5

27.

Domain	$(-\infty, 0) \cup (0, \infty)$
Range	$(-\infty, 5) \cup [7]$
Increasing	$(-\infty, 0)$
Decreasing	$(5, \infty)$
Constant	$(0, 5)$

d) undefined
e) 3
f) 7

28.

Domain	$(-8,0)\cup(0,4]$
Range	$(-4,3]$
Increasing	$(-8,-5)\cup(0,4)$
Decreasing	$(-5,0)$
Constant	nowhere

d) undefined
e) approximately –0.8
f) 0

29.

$$\frac{\left[(x+h)^2-(x+h)\right]-\left[x^2-x\right]}{h}=$$

$$\frac{x^2+2hx+h^2-x-h-x^2+x}{h}=$$

$$\frac{\cancel{h}(2x+h-1)}{\cancel{h}}=\boxed{2x+h-1}$$

30.

$$\frac{\left[(x+h)^2+2(x+h)\right]-\left[x^2+2x\right]}{h}=$$

$$\frac{x^2+2hx+h^2+2x+2h-x^2-2x}{h}=$$

$$\frac{\cancel{h}(2x+h+2)}{\cancel{h}}=\boxed{2x+h+2}$$

31.

$$\frac{\left[(x+h)^2+3(x+h)\right]-\left[x^2+3x\right]}{h}=$$

$$\frac{x^2+2hx+h^2+3x+3h-x^2-3x}{h}=$$

$$\frac{\cancel{h}(2x+h+3)}{\cancel{h}}=\boxed{2x+h+3}$$

32.

$$\frac{\left[-(x+h)^2+5(x+h)\right]-\left[-x^2+5x\right]}{h}=$$

$$\frac{-x^2-2hx-h^2+5x+5h+x^2-5x}{h}=$$

$$\frac{\cancel{h}(-2x-h+5)}{\cancel{h}}=\boxed{-2x-h+5}$$

33.

$$\frac{\left[(x+h)^2-3(x+h)+2\right]-\left[x^2-3x+2\right]}{h}=$$

$$\frac{x^2+2hx+h^2-3x-3h+2-x^2+3x-2}{h}=$$

$$\frac{\cancel{h}(2x+h-3)}{\cancel{h}}=\boxed{2x+h-3}$$

34.

$$\frac{\left[(x+h)^2-2(x+h)+5\right]-\left[x^2-2x+5\right]}{h}=$$

$$\frac{x^2+2hx+h^2-2x-2h+5-x^2+2x-5}{h}=$$

$$\frac{\cancel{h}(2x+h-2)}{\cancel{h}}=\boxed{2x+h-2}$$

35.

$$\frac{\left[-3(x+h)^2+5(x+h)-4\right]-\left[-3x^2+5x-4\right]}{h}=$$

$$\frac{-3x^2-6hx-3h^2+5x+5h-4+3x^2-5x+4}{h}=$$

$$\frac{\not{h}(-6x-3h+5)}{\not{h}}=\boxed{-6x-3h+5}$$

36.

$$\frac{\left[-4(x+h)^2+2(x+h)-3\right]-\left[-4x^2+2x-3\right]}{h}=$$

$$\frac{-4x^2-8hx-4h^2+2x+2h-3+4x^2-2x+3}{h}=$$

$$\frac{\not{h}(-8x-4h+2)}{\not{h}}=\boxed{-8x-4h+2}$$

37. Note that $(a+b)^3=a^3+3a^2b+3ab^2+b^3$. As such, we have

$$\frac{f(x+h)-f(x)}{h}=\frac{\left[(x+h)^3+(x+h)^2\right]-\left[x^3+x^2\right]}{h}$$

$$=\frac{x^3+3x^2h+3xh^2+h^3+x^2+2xh+h^2-x^3-x^2}{h}$$

$$=\frac{h\left(3x^2+3xh+h^2+2x+h\right)}{h}=\boxed{3x^2+3xh+h^2+2x+h}$$

38. Note that $(a-b)^4=a^4-4ab^3+6a^2b^2-4a^3b+b^4$. As such, we have

$$\frac{f(x+h)-f(x)}{h}=\frac{\left[(x+h)-1\right]^4-\left[x-1\right]^4}{h}$$

$$=\frac{\left[(x+h)^4-4(x+h)^3+6(x+h)^2-4(x+h)+1\right]-\left[x^4-4x^3+6x^2-4x+1\right]}{h}$$

$$=\frac{\left[x^4-4x^3+6x^2-4x+1+h\left(4x^3+6x^2h+4xh^2+h^3-12x^2-12xh-4h^2+12x+6h-4\right)\right]-x^4+4x^3-6x^2+4x-1}{h}$$

$$=\frac{h\left(4x^3+6x^2h+4xh^2+h^3-12x^2-12xh-4h^2+12x+6h-4\right)}{h}$$

$$=\boxed{4x^3+6x^2h+4xh^2+h^3-12x^2-12xh-4h^2+12x+6h-4}$$

39.

$$\frac{f(x+h)-f(x)}{h}=\frac{\frac{2}{x+h-2}-\frac{2}{x-2}}{h}=\frac{2(x-2)-2(x+h-2)}{h(x+h-2)(x-2)}=\frac{2x-4-2x-2h+4}{h(x+h-2)(x-2)}$$

$$=\frac{-2h}{h(x+h-2)(x-2)}=\boxed{\frac{-2}{(x+h-2)(x-2)}}$$

40.

$$\frac{f(x+h)-f(x)}{h}=\frac{\frac{x+h+5}{x+h-7}-\frac{x+5}{x-7}}{h}=\frac{(x+h+5)(x-7)-(x+5)(x+h-7)}{h(x+h-7)(x-7)}$$

$$=\frac{\left(x^2+xh+5x-7x-7h-35\right)-\left(x^2+hx-7x+5x+5h-35\right)}{h(x+h-7)(x-7)}$$

$$=\frac{-12h}{h(x+h-7)(x-7)}=\boxed{\frac{-12}{h(x+h-7)(x-7)}}$$

41.

$$\frac{f(x+h)-f(x)}{h}=\frac{\sqrt{1-2(x+h)}-\sqrt{1-2x}}{h}$$

$$=\frac{\sqrt{1-2(x+h)}-\sqrt{1-2x}}{h}\cdot\frac{\sqrt{1-2(x+h)}+\sqrt{1-2x}}{\sqrt{1-2(x+h)}+\sqrt{1-2x}}$$

$$=\frac{(1-2x-2h)-(1-2x)}{h\left(\sqrt{1-2(x+h)}+\sqrt{1-2x}\right)}=\frac{-2h}{h\left(\sqrt{1-2(x+h)}+\sqrt{1-2x}\right)}$$

$$=\boxed{\frac{-2}{\sqrt{1-2(x+h)}+\sqrt{1-2x}}}$$

42.

$$\frac{f(x+h)-f(x)}{h}=\frac{\sqrt{(x+h)^2+(x+h)+1}-\sqrt{x^2+x+1}}{h}$$

$$=\frac{\sqrt{(x+h)^2+(x+h)+1}-\sqrt{x^2+x+1}}{h}\cdot\frac{\sqrt{(x+h)^2+(x+h)+1}+\sqrt{x^2+x+1}}{\sqrt{(x+h)^2+(x+h)+1}+\sqrt{x^2+x+1}}$$

$$=\frac{\left[(x+h)^2+(x+h)+1\right]-\left[x^2+x+1\right]}{h\left(\sqrt{(x+h)^2+(x+h)+1}+\sqrt{x^2+x+1}\right)}=\frac{\left[x^2+2hx+h^2+x+h+1\right]-\left[x^2+x+1\right]}{h\left(\sqrt{(x+h)^2+(x+h)+1}+\sqrt{x^2+x+1}\right)}$$

$$=\frac{h(2x+h+1)}{h\left(\sqrt{(x+h)^2+(x+h)+1}+\sqrt{x^2+x+1}\right)}=\boxed{\frac{2x+h+1}{\sqrt{(x+h)^2+(x+h)+1}+\sqrt{x^2+x+1}}}$$

43.

$$\frac{f(x+h)-f(x)}{h}=\frac{\frac{4}{\sqrt{x+h}}-\frac{4}{\sqrt{x}}}{h}=\frac{4\sqrt{x}-4\sqrt{x+h}}{h\sqrt{x}\sqrt{x+h}}=\frac{4\left(\sqrt{x}-\sqrt{x+h}\right)}{h\sqrt{x}\sqrt{x+h}}\cdot\frac{\sqrt{x}+\sqrt{x+h}}{\sqrt{x}+\sqrt{x+h}}$$

$$=\frac{4\left(x-(x+h)\right)}{h\sqrt{x}\sqrt{x+h}\left(\sqrt{x}+\sqrt{x+h}\right)}=\boxed{\frac{-4}{\sqrt{x(x+h)}\left(\sqrt{x}+\sqrt{x+h}\right)}}$$

44.

$$\frac{f(x+h)-f(x)}{h}=\frac{\sqrt{\frac{x+h}{x+h+1}}-\sqrt{\frac{x}{x+1}}}{h}=\frac{\sqrt{x+h}\sqrt{x+1}-\sqrt{x}\sqrt{x+h+1}}{h\sqrt{x+1}\sqrt{x+h+1}}$$

$$=\frac{\sqrt{(x+h)(x+1)}-\sqrt{x(x+h+1)}}{h\sqrt{x+1}\sqrt{x+h+1}}$$

$$=\frac{\sqrt{(x+h)(x+1)}-\sqrt{x(x+h+1)}}{h\sqrt{x+1}\sqrt{x+h+1}}\cdot\frac{\sqrt{(x+h)(x+1)}+\sqrt{x(x+h+1)}}{\sqrt{(x+h)(x+1)}+\sqrt{x(x+h+1)}}$$

$$=\frac{\left(x^2+hx+x+h\right)-\left(x^2+xh+x\right)}{h\sqrt{x+1}\sqrt{x+h+1}\left(\sqrt{(x+h)(x+1)}+\sqrt{x(x+h+1)}\right)}$$

$$=\frac{h}{h\sqrt{x+1}\sqrt{x+h+1}\left(\sqrt{(x+h)(x+1)}+\sqrt{x(x+h+1)}\right)}$$

$$=\boxed{\frac{1}{\sqrt{x+1}\sqrt{x+h+1}\left(\sqrt{(x+h)(x+1)}+\sqrt{x(x+h+1)}\right)}}$$

45. $\frac{3^3-1^3}{3-1}=\frac{27-1}{2}=\boxed{13}$	**46.** $\frac{\frac{1}{3}-\frac{1}{1}}{3-1}=\frac{-\frac{2}{3}}{2}=\boxed{-\frac{1}{3}}$
47. $\frac{\lvert 3\rvert-\lvert 1\rvert}{3-1}=\boxed{1}$	**48.** $\frac{2(3)-2(1)}{3-1}=\frac{4}{2}=\boxed{2}$
49. $\frac{(1-2(3))-(1-2(1))}{3-1}=\frac{-5-(-1)}{2}=\boxed{-2}$	**50.** $\frac{\left(9-3^2\right)-\left(9-1^2\right)}{3-1}=\frac{0-8}{2}=\boxed{-4}$
51. $\frac{\lvert 5-2(3)\rvert-\lvert 5-2(1)\rvert}{3-1}=\frac{\lvert -1\rvert-3}{2}=\boxed{-1}$	**52.** $\frac{\sqrt[3]{3^2-1}-\sqrt[3]{1^2-1}}{3-1}=\frac{\sqrt[3]{8}}{2}=\boxed{1}$

53.

Domain	$(-\infty,\infty)$
Range	$(-\infty,2]$
Increasing	$(-\infty,2)$
Decreasing	nowhere
Constant	$(2,\infty)$

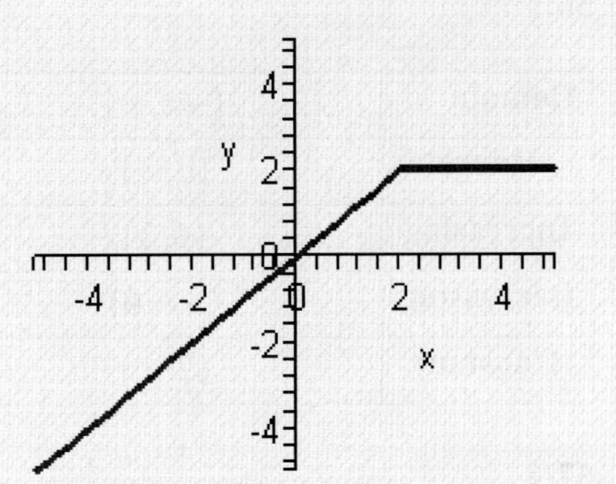

54.

Domain	$(-\infty,\infty)$
Range	$\{-1\}\cup(1,\infty)$
Increasing	nowhere
Decreasing	$(-\infty,-1)$
Constant	$(-1,\infty)$

Notes on Graph: There should be an open hole at $(-1,1)$, and a closed hole at $(-1,-1)$.

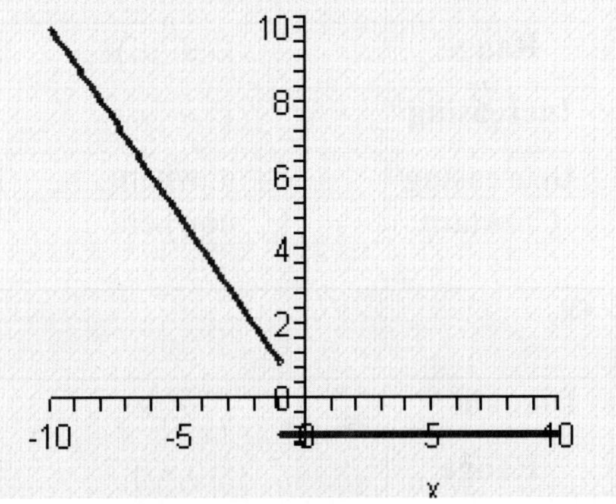

55.

Domain	$(-\infty,\infty)$
Range	$[0,\infty)$
Increasing	$(0,\infty)$
Decreasing	$(-1,0)$
Constant	$(-\infty,-1)$

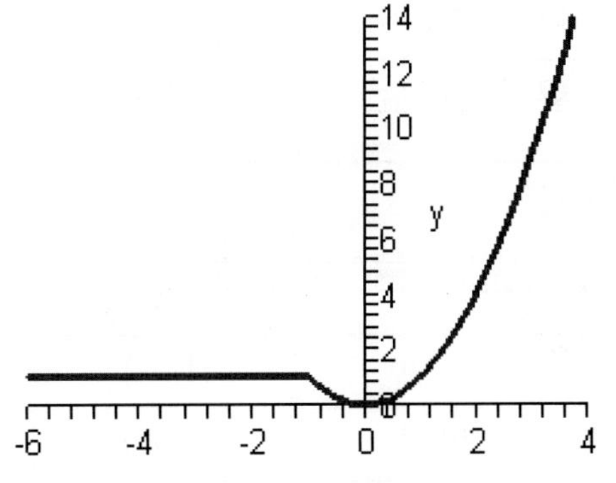

56.

Domain	$(-\infty,\infty)$
Range	$[0,\infty)$
Increasing	$(0,2)$
Decreasing	$(-\infty,0)$
Constant	$(2,\infty)$

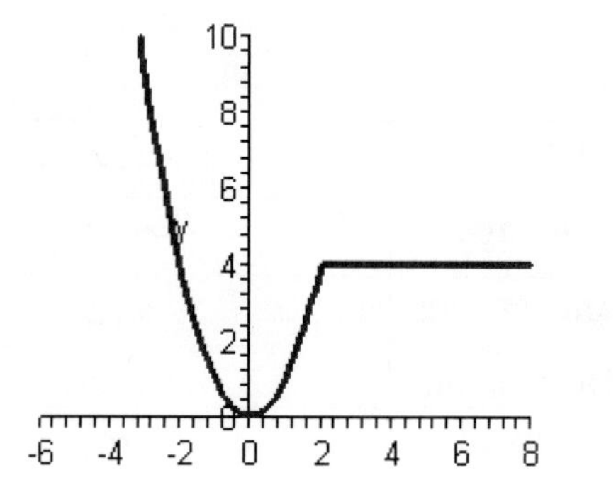

57.

Domain	$(-\infty,\infty)$
Range	$(-\infty,\infty)$
Increasing	$(-\infty,\infty)$
Decreasing	nowhere
Constant	nowhere

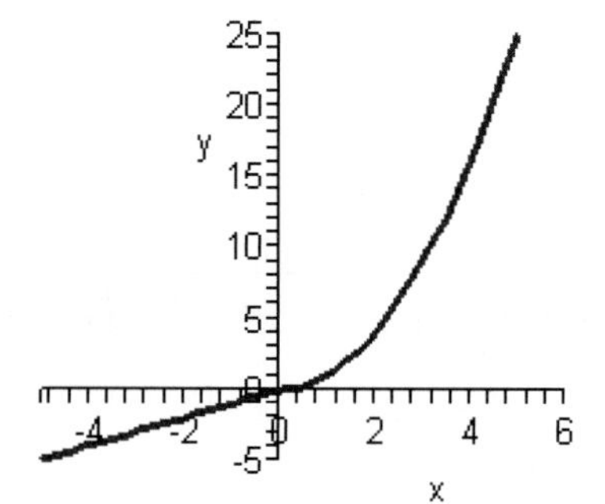

58.

Domain	$(-\infty,\infty)$
Range	$[0,\infty)$
Increasing	$(0,\infty)$
Decreasing	$(-\infty,0)$
Constant	nowhere

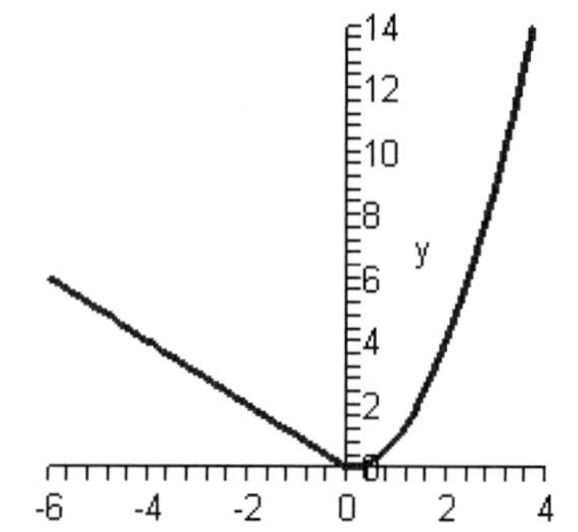

59.

Domain	$(-\infty,\infty)$
Range	$[1,\infty)$
Increasing	$(1,\infty)$
Decreasing	$(-\infty,1)$
Constant	nowhere

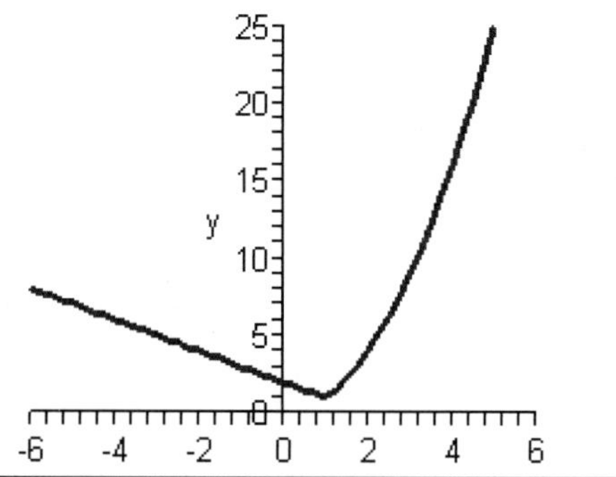

60.

Domain	$(-\infty,\infty)$
Range	$(-\infty,\infty)$
Increasing	$(-\infty,-1)\cup(0,\infty)$
Decreasing	$(-1,0)$
Constant	nowhere

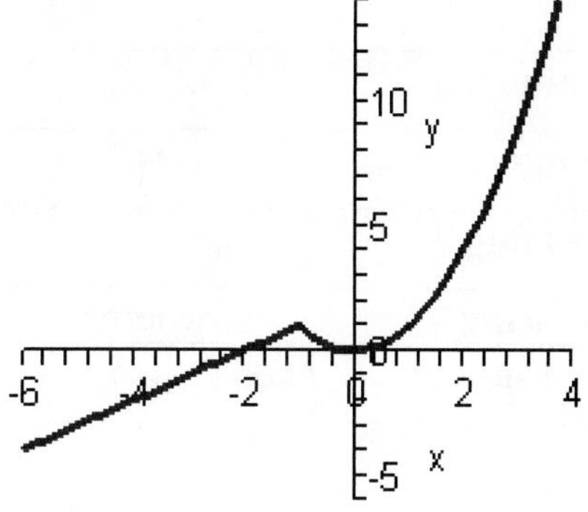

61.

Domain	$(-\infty,\infty)$
Range	$[-1,3]$
Increasing	$(-1,3)$
Decreasing	nowhere
Constant	$(-\infty,-1)\cup(3,\infty)$

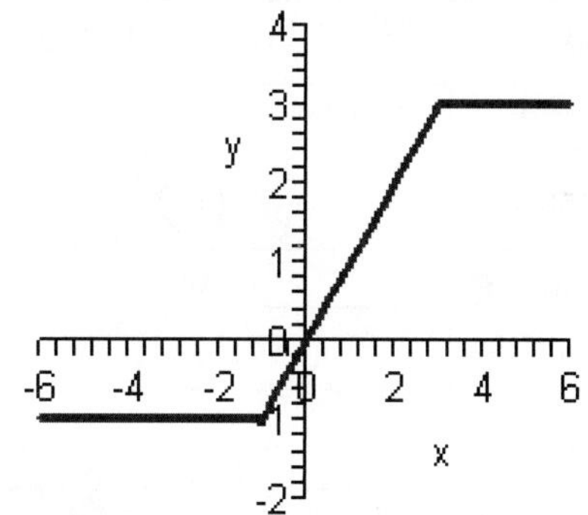

62.

Domain	$(-\infty,-1)\cup(-1,3)\cup(3,\infty)$
Range	$[-1,3]$
Increasing	$(-1,3)$
Decreasing	nowhere
Constant	$(-\infty,-1)\cup(3,\infty)$

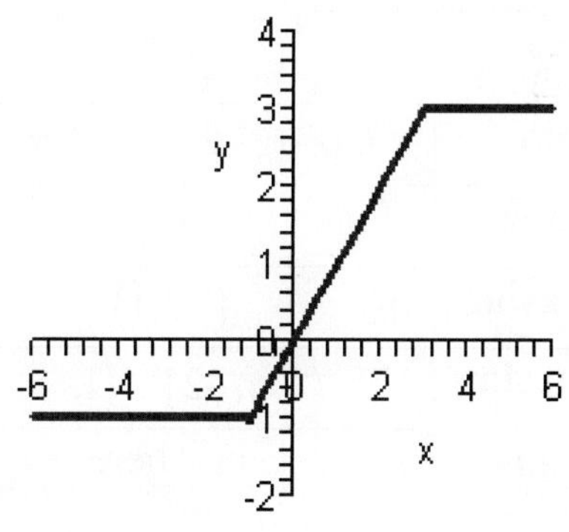

Notes on Graph: There should be open holes at $(-1,-1)$ and $(3,3)$.

63.

Domain	$(-\infty,\infty)$
Range	$[1,4]$
Increasing	$(1,2)$
Decreasing	nowhere
Constant	$(-\infty,1)\cup(2,\infty)$

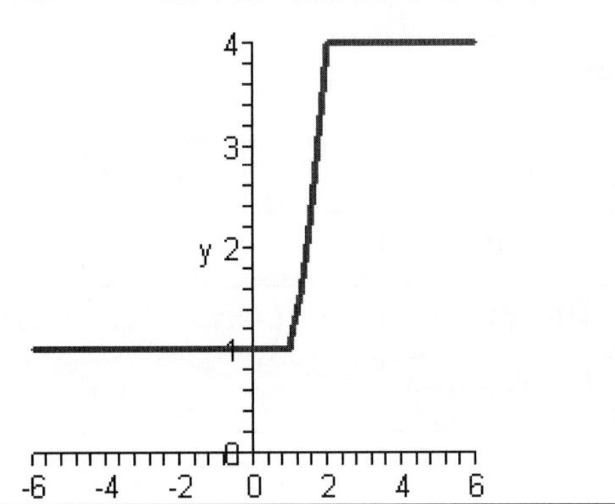

64.

Domain	$(-\infty,1)\cup(1,2)\cup(2,\infty)$
Range	$[1,4]$
Increasing	$(1,2)$
Decreasing	nowhere
Constant	$(-\infty,1)\cup(2,\infty)$

Notes on Graph: There should be open holes at $(1,1)$ and $(2,4)$.

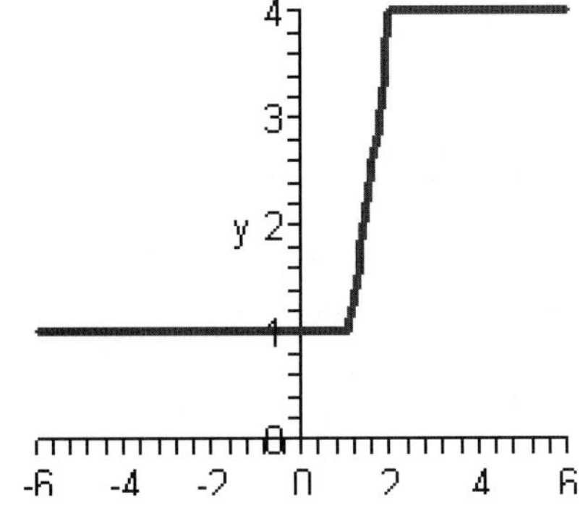

65.

Domain	$(-\infty,-2)\cup(-2,\infty)$
Range	$(-\infty,\infty)$
Increasing	$(-2,1)$
Decreasing	$(-\infty,-2)\cup(1,\infty)$
Constant	nowhere

Notes on Graph: There should be open holes at $(-2,1)$, $(-2,-1)$, and $(1,2)$, and a closed hole at $(1,0)$.

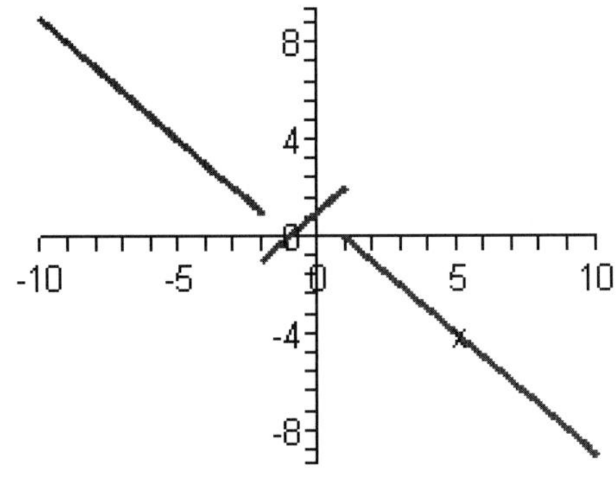

66.

Domain	$(-\infty,1)\cup(1,\infty)$
Range	$(-\infty,\infty)$
Increasing	$(-2,1)$
Decreasing	$(-\infty,-2)\cup(1,\infty)$
Constant	nowhere

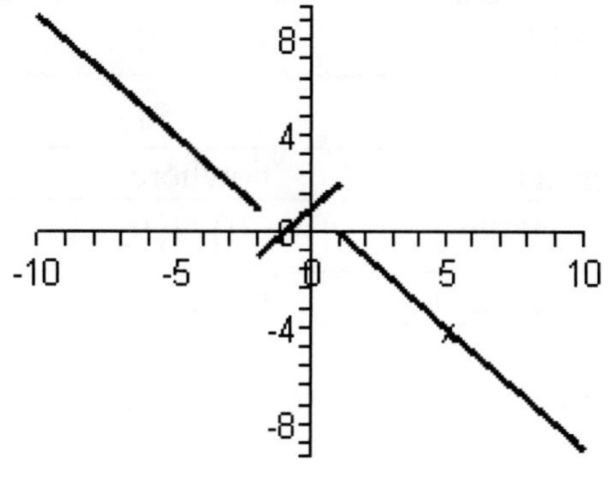

Notes on Graph: There should be open holes at $(-2,-1)$, $(1,2)$, and $(1,0)$, and a closed hole at $(-2,1)$.

67.

Domain	$(-\infty,\infty)$
Range	$[0,\infty)$
Increasing	$(0,\infty)$
Decreasing	nowhere
Constant	$(-\infty,0)$

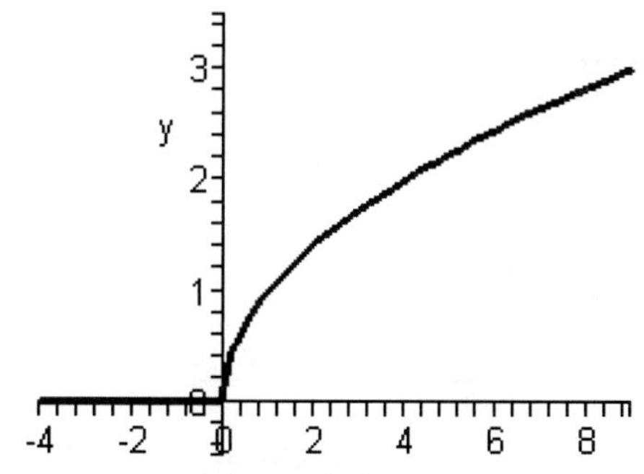

68.

Domain	$(-\infty,1)\cup(1,\infty)$
Range	$[1,\infty)$
Increasing	$(1,\infty)$
Decreasing	nowhere
Constant	$(-\infty,1)$

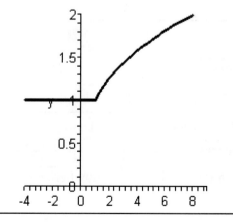

Notes on Graph: There should be an open hole at $(1,1)$.

69.

Domain	$(-\infty,\infty)$
Range	$(-\infty,\infty)$
Increasing	nowhere
Decreasing	$(-\infty,0)\cup(0,\infty)$
Constant	nowhere

Notes on Graph: There should be a closed hole at (0,0).

70.

Domain	$(-\infty,\infty)$
Range	$(-\infty,\infty)$
Increasing	$(-\infty,0)\cup(0,\infty)$
Decreasing	nowhere
Constant	nowhere

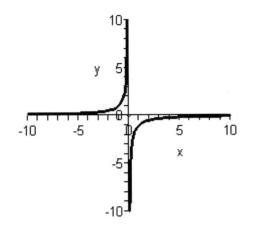

Notes on Graph: There should be a closed hole at (0,0).

71.

Domain	$(-\infty,1)\cup(1,\infty)$
Range	$(-\infty,-1)\cup(-1,\infty)$
Increasing	$(-1,1)$
Decreasing	$(-\infty,-1)\cup(1,\infty)$
Constant	nowhere

Notes on Graph: There should be open holes at $(-1,-1)$, $(1,1)$ and $(1,-1)$. Also, the graph of $-\sqrt[3]{x}$ should appear on the interval $(-\infty,-1)$ with a closed hole at (-1,1).

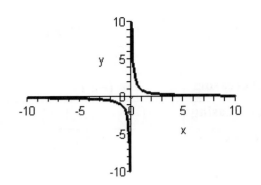

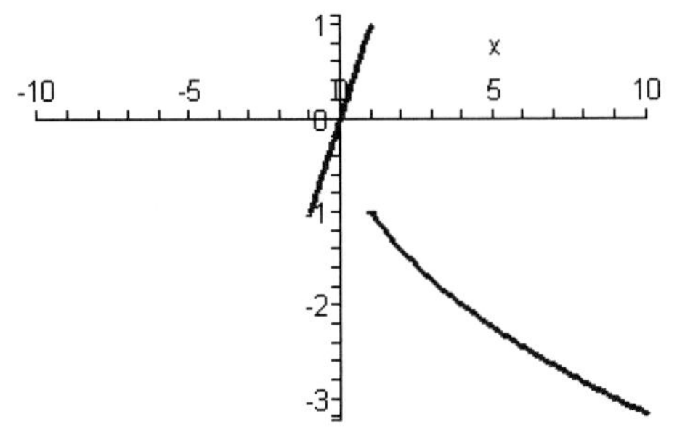

72.

Domain	$(-\infty,1)\cup(1,\infty)$
Range	$(-\infty,1)\cup(1,\infty)$
Increasing	$(1,\infty)$
Decreasing	$(-\infty,-1)$
Constant	nowhere

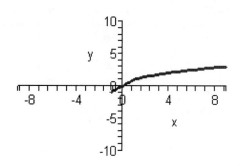

Notes on Graph: The graph of $-\sqrt[3]{x}$ should appear on the interval $(-\infty,-1)$.

73.

Domain	$(-\infty,\infty)$
Range	$(-\infty,2)\cup[4,\infty)$
Increasing	$(-\infty,-2)\cup(0,2)\cup(2,\infty)$
Decreasing	$(-2,0)$
Constant	nowhere

Notes on Graph: There should be open holes at $(-2,2)$, $(2,2)$ and closed holes at $(-2,1)$, $(2,4)$.

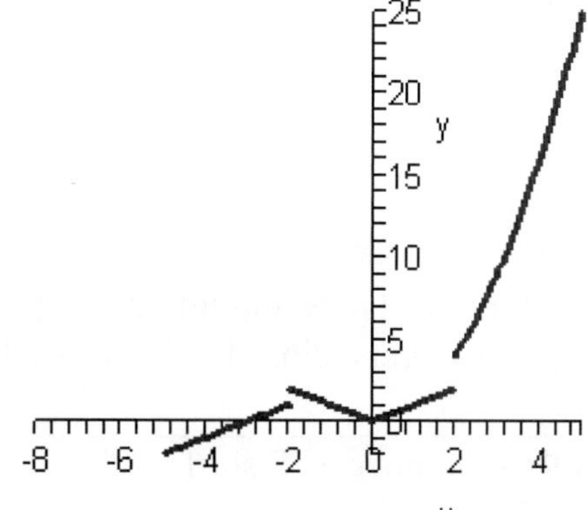

74.

Domain	$(-\infty,-1)\cup(-1,1)\cup(1,\infty)$
Range	$[1,\infty)$
Increasing	$(1,\infty)$
Decreasing	$(-\infty,-1)$
Constant	$(-1,1)$

Notes on Graph: There should be open holes at $(-1,1)$, $(1,1)$.

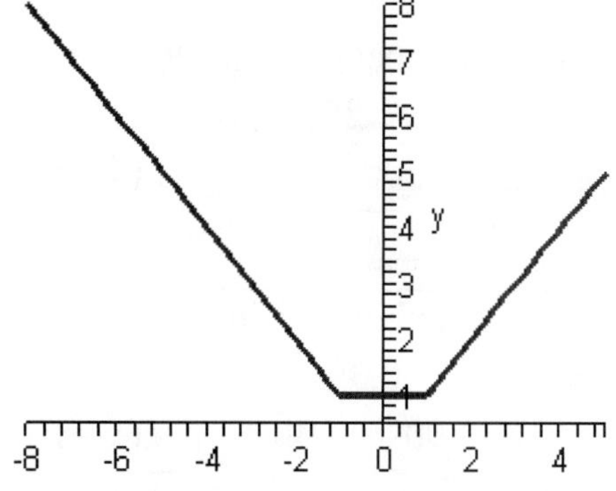

75.

Domain	$(-\infty,1)\cup(1,\infty)$
Range	$(-\infty,1)\cup(1,\infty)$
Increasing	$(-\infty,1)\cup(1,\infty)$
Decreasing	nowhere
Constant	nowhere

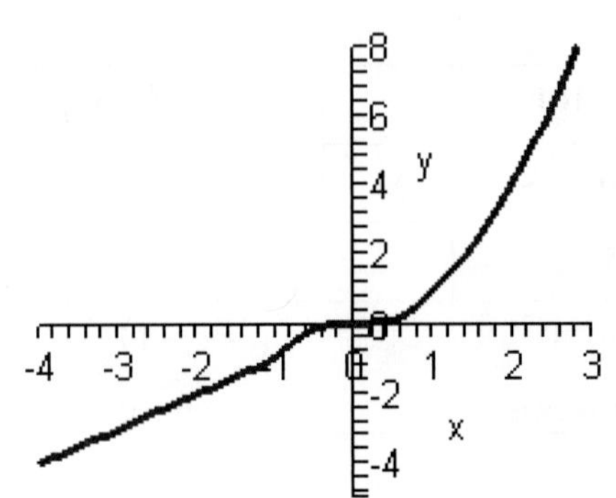

Notes on Graph: There should be an open hole at $(1,1)$.

76.

Domain	$(-\infty,\infty)$
Range	$(-1,\infty)$
Increasing	$(-1,\infty)$
Decreasing	$(-\infty,-1)$
Constant	nowhere

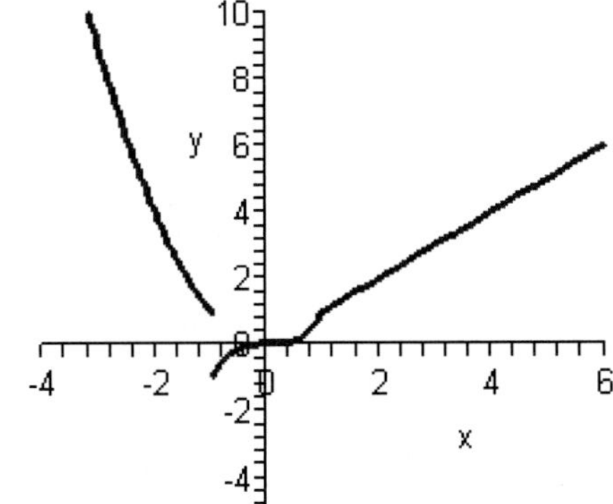

Notes on Graph: There should be an open hole at $(-1,-1)$ and a closed hole at $(-1,1)$.

77. Let x = number of T-shirts ordered. The cost function is given by

$$C(x)=\begin{cases}10x, & 0\le x\le 50\\ 9x, & 50<x<100\\ 8x, & x\ge 100\end{cases}$$

78. Let x = number of new uniforms ordered. The cost function is given by

$$C(x)=\begin{cases}176.12x, & 0\le x\le 50\\ 159.73x, & 50<x\le 100\end{cases}$$

79. Let x = number of boats entered. The cost function is given by

$$C(x)=\begin{cases}250x, & 0\le x\le 10\\ \underbrace{2500}_{\text{Cost for first 10 boats}}+175\cdot\underbrace{(x-10)}_{\text{\# of boats beyond first 10}}, & x>10\end{cases}=\begin{cases}250x, & 0\le x\le 10\\ 175x+750, & x>10\end{cases}$$

80. Let x = number of minutes. The cost function is given by

$$C(x)=\begin{cases}0.39x, & 0\le x\le 10\\ \underbrace{3.90}_{\text{Cost for first 10 minutes}}+0.12\cdot\underbrace{(x-10)}_{\text{\# of minutes beyond first 10}}, & x>10\end{cases}=\begin{cases}0.39x, & 0\le x\le 10\\ 0.12x+2.7, & x>10\end{cases}$$

81. Let x = number of people attending the reception. The cost function is given by

$$C(x)=\begin{cases}\underbrace{1000}_{\text{Fee for reserving dining room}} + 35x, & 0\le x\le 100\\ 1000+\underbrace{3500}_{\text{Cost for first 100 guests}} + 25\cdot\underbrace{(x-100)}_{\text{\# of guests beyond first 100}}, & x>100\end{cases}$$

Simplifying the terms above yields

$$C(x)=\begin{cases}1000+35x, & 0\le x\le 100\\ 2000+25x, & x>100\end{cases}$$

82. Let x = number of hours of labor. The cost function is

$$C(x)=\underbrace{1400}_{\text{Cost for parts}}+\underbrace{25x}_{\text{Cost for labor}}.$$

83. Let x = number of books sold.
Since a single book sells for \$20, the amount of money earned for x books is $20x$.
Then, the amount of royalties due to the author (as a function of x) is given by:

$$R(x)=\begin{cases}\underbrace{50,000}_{\text{Amount upfront}}+\underbrace{0.15(20x)}_{\text{Amount from 15\% royalties}}, & 0\le x\le 100,000\\ 50,000+\underbrace{0.15(2,000,000)}_{\text{Royalties from first 100,000 books}}+\underbrace{0.20(20)(x-100,000)}_{\text{20\% royalties on books beyond initial 100,000}}, & 100,000<x\end{cases}$$

Simplifying the terms above yields

$$R(x)=\begin{cases}50,000+3x, & 0\le x\le 100,000\\ -50,000+4x, & x>100,000\end{cases}$$

84. Let x = number of books sold.
Since a single book sells for \$20, the amount of money earned for x books is $20x$.
Then, the amount of royalties due to the author (as a function of x) is given by:

$$R(x)=\begin{cases}\underbrace{35,000}_{\text{Amount upfront}}+\underbrace{0.15(20x)}_{\text{Amount from 15\% royalties}}, & 0\le x\le 100,000\\ 35,000+\underbrace{0.15(2,000,000)}_{\text{Royalties from first 100,000 books}}+\underbrace{0.25(20)(x-100,000)}_{\text{25\% royalties on books beyond initial 100,000}}, & x>100,000\end{cases}$$

Simplifying the terms above yields

$$R(x)=\begin{cases}35,000+3x, & 0\le x\le 100,000\\ -165,000+5x, & x>100,000\end{cases}$$

85. Let x = number of stained glass units sold.
Total monthly cost is given by: $C(x) = \underbrace{100}_{\substack{\text{Business}\\\text{Costs}}} + \underbrace{700}_{\substack{\text{Studio}\\\text{Rent}}} + \underbrace{35x}_{\substack{\text{Cost of materials}\\\text{for } x \text{ units}}} = 800 + 35x$.

Revenue for x units sold is given by: $R(x) = 100x$

So, the total profit is given by: $P(x) = R(x) - C(x) = 100x - (800 + 35x) = 65x - 800$

86. Let x = number of people who attend.
Since it is assumed that each person eats 1 lb. of shrimp, it will cost $5x$ dollars for x people for the shrimp.
So, the total cost is given by: $C(x) = 30 + 5x$
The total revenue is given by: $R(x) = 10x$
So, the total profit is given by: $P(x) = R(x) - C(x) = 10x - (30 + 5x) = 5x - 30$

87. Observe that

$$f(x) = \begin{cases} 0.88, & 0 \le x < 1 \\ 0.88 + 0.17, & 1 \le x < 2 \\ 0.88 + 0.17(2), & 2 \le x < 3 \\ \vdots & \\ 0.88 + 0.17(n), & n \le x < n+1 \end{cases}$$

Using the greatest integer function, we have $\boxed{f(x) = 0.88 + 0.17[\![x]\!],\ x \ge 0}$.

88. Observe that

$$f(x) = \begin{cases} 1.22, & 0 \le x < 1 \\ 1.22 + 0.17, & 1 \le x < 2 \\ 1.22 + 0.17(2), & 2 \le x < 3 \\ \vdots & \\ 1.22 + 0.17(n), & n \le x < n+1 \end{cases}$$

Using the greatest integer function, we have $\boxed{f(x) = 1.22 + 0.17[\![x]\!],\ x \ge 0}$.

89. $f(t) = 3(-1)^{[\![t]\!]},\ t \ge 0$

90. $f(x) = (-1)^{\left(1+\left[\!\left[\frac{x}{100}\right]\!\right]\right)},\ x \ge 0$

91. a) $\dfrac{1500-500}{1950-1900} = \boxed{20 \text{ per year}}$

b) $\dfrac{7000-1500}{2000-1950} = \boxed{110 \text{ per year}}$

92. a) $\dfrac{5000-1500}{1975-1950} = \boxed{140 \text{ per year}}$

b) $\dfrac{7000-5000}{2000-1975} = \boxed{80 \text{ per year}}$

93.

$$\frac{h(2)-h(1)}{2-1} = \frac{\left(-16(2)^2 + 48(2)\right) - \left(-16(1)^2 + 48(1)\right)}{2-1}$$

$\boxed{= 0 \text{ ft/sec}}$

94.

$$\frac{h(3)-h(1)}{3-1} = \frac{\left(-16(3)^2 + 48(3)\right) - \left(-16(1)^2 + 48(1)\right)}{2-1}$$

$\boxed{= -16 \text{ ft/sec}}$

95. Should exclude the origin since $x = 0$ is not in the domain. The range should be $(0, \infty)$.

96. The open and closed holes at $x = 1$ should be switched, and then the range should be changed to $[-1, \infty)$.

97. The portion of $C(x)$ for $x > 30$ should be: $15 + \underbrace{x - 30}_{\substack{\text{Number miles}\\\text{beyond first 30}}}$

98. The portion of $C(x)$ for $x > 10{,}000$ should be:
$0.02(10{,}000) + 0.04(x - 10{,}000)$

99. False. For instance, $f(x)=x^3$ is always increasing.	**100.** True.

101. The individual pieces used to form *f*, namely ax, bx^2, are continuous on $\mathbb{R}$. So, the only *x*-value with which we need to be concerned regarding the continuity of *f* is $x=2$. For *f* to be continuous at 2, we need $a(2)=b(2)^2$, which is the same as $\boxed{a=2b}$.

102. Both $\frac{1}{x}$ and $-\frac{1}{x}$ are undefined at $x=0$. So, for every value of *a*, either $a>0$ or $a\le 0$. Hence, we would need to evaluate either $\frac{1}{x}$ or $-\frac{1}{x}$ at 0, which is not possible. So, this function cannot be continuous, for any value of *a*.

103. Since *f* is already continuous on $(-\infty,-2]\cup[1,\infty)$ (being defined in terms of continuous functions), we need only to focus our attention on the interval $[-2,1]$. In order for *f* to be continuous at $x=-2$, we need $f(-2)=a(-2)+b$. This is equivalent to

$$\underbrace{-(-2)^2-10(-2)-13}_{=3}=-2a+b$$

In order for *f* to be continuous at $x = 1$, we need $f(1)=a(1)+b$. This is equivalent to

$$\underbrace{\sqrt{1-1}-9}_{=-9}=a+b.$$

As such, we must solve the system

$$\begin{cases} -2a+b=3 & \textbf{(1)} \\ a+b=-9 & \textbf{(2)} \end{cases}$$

Subtract **(1)** – **(2)** to eliminate *b*: $-3a=12$, so that $\boxed{a=-4}$.

Substitute this into **(2)** to find *b*: $\boxed{b=-5}$.

104. The first two expressions in the definition of *f* must agree at $x=-2$, and the last two expressions must agree at $x=2$. This yields the system:

$$\begin{cases} -2(-2)-a+2b=\sqrt{-2+a} \\ \sqrt{2+a}=2^2-4(2)+a+4 \end{cases} \text{ which is equivalent to } \begin{cases} 4-a+2b=\sqrt{a-2} & \textbf{(1)} \\ \sqrt{a+2}=a & \textbf{(2)} \end{cases}$$

Solve **(2)** for *a*:

$$a+2=a^2$$

$$a^2-a-2=0$$

$$(a-2)(a+1)=0$$

$$\boxed{a=2},\ \cancel{-1}$$

Substitute this into **(1)** to find *b*: $4-2+2b=0$, so that $\boxed{b=-1}$.

105. The graph is odd.

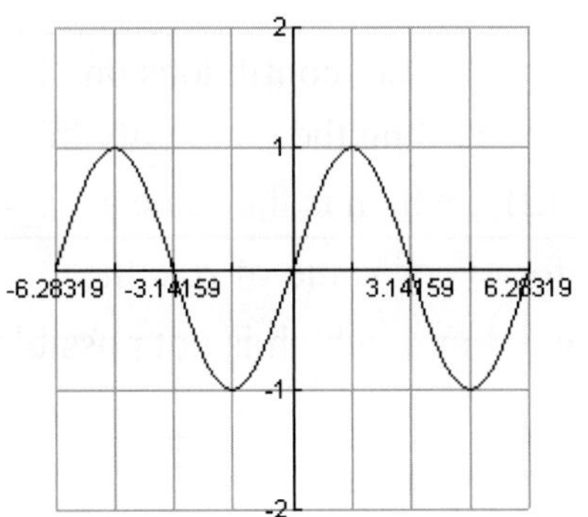

106. The graph is even.

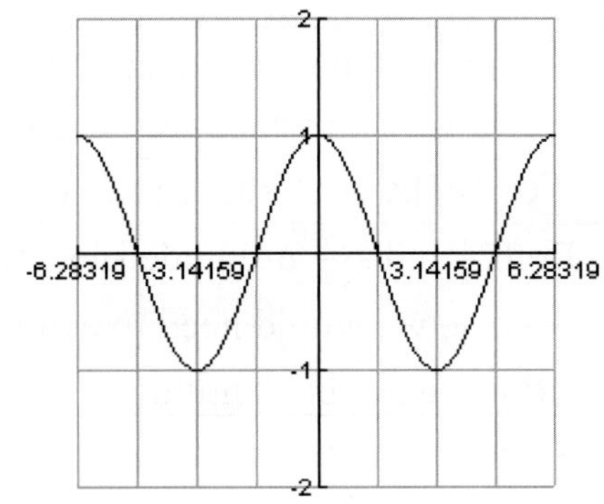

107. The graph is odd.

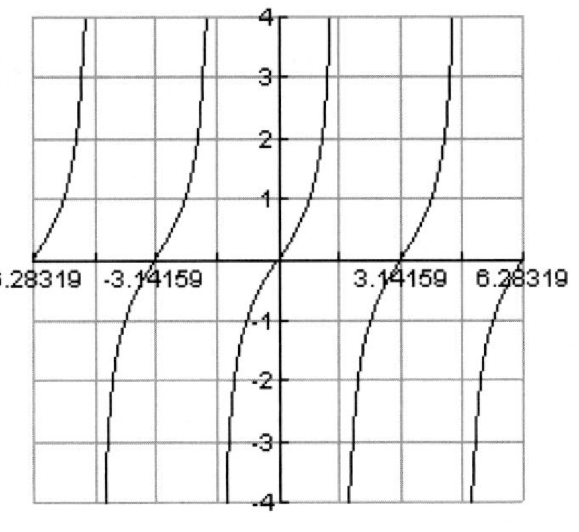

108. This is the graph of $\tan x$ (in #107).

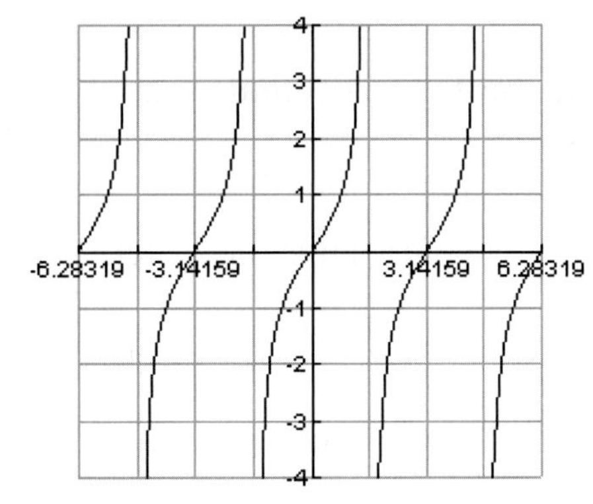

109.

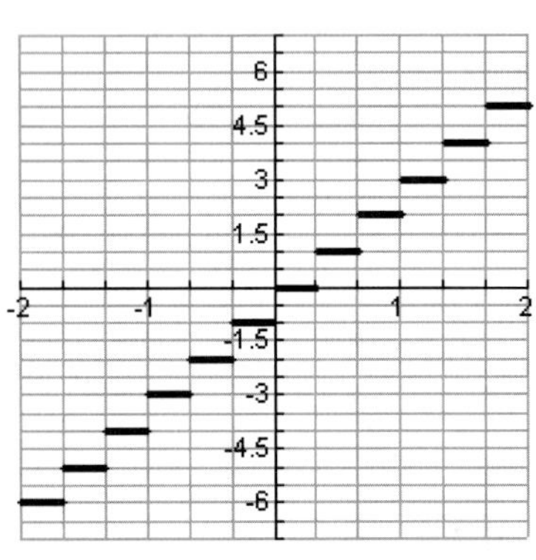

Domain: $\mathbb{R}$ Range: The set of integers

110.

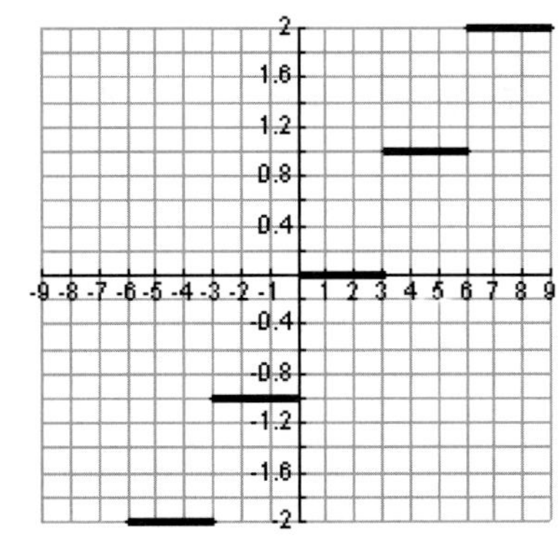

Domain: $\mathbb{R}$ Range: The set of integers

111. $\frac{f(x+h)-f(x)}{h}=\frac{k-k}{h}=0$. So, at $h=0$ we get $\boxed{f'(x)=0}$.
112. $\frac{f(x+h)-f(x)}{h}=\frac{[m(x+h)+b]-[mx+b]}{h}=\frac{mx+mh+b-mx-b}{h}=\frac{mh}{h}=m$ So, at $h=0$ we get $\boxed{f'(x)=m}$.
113. $$\frac{f(x+h)-f(x)}{h}=\frac{\left[a(x+h)^2+b(x+h)+c\right]-\left[ax^2+bx+c\right]}{h}$$ $$=\frac{ax^2+2axh+ah^2+bx+bh+c-ax^2-bx-c}{h}$$ $$=\frac{h(2ax+ah+b)}{h}=2ax+ah+b$$ So, at $h=0$ we get $\boxed{f'(x)=2ax+b}$.
114. Apply the results from problems #111 – 113 on each individual expression (on the OPEN intervals) to obtain $$f'(x)=\begin{cases}0, & x<0,\\ -3, & 0<x<4.\\ 2x+4, & x>4\end{cases}$$

Section 1.3 Solutions --

1. $y=\|x\|+3$	**2.** $y=\|x+4\|$
3. $y=\|-x\|=\|x\|$ (since $\|-x\|=\|-1\|\|x\|=\|x\|$)	**4.** $y=-\|x\|$
5. $y=3\|x\|$	**6.** $y=\frac{1}{3}\|x\|$
7. $y=x^3-4$	**8.** $y=(x-3)^3$
9. $y=(x+1)^3+3$	**10.** $y=-x^3$
11. $y=-x^3$	**12.** $y=x^3$

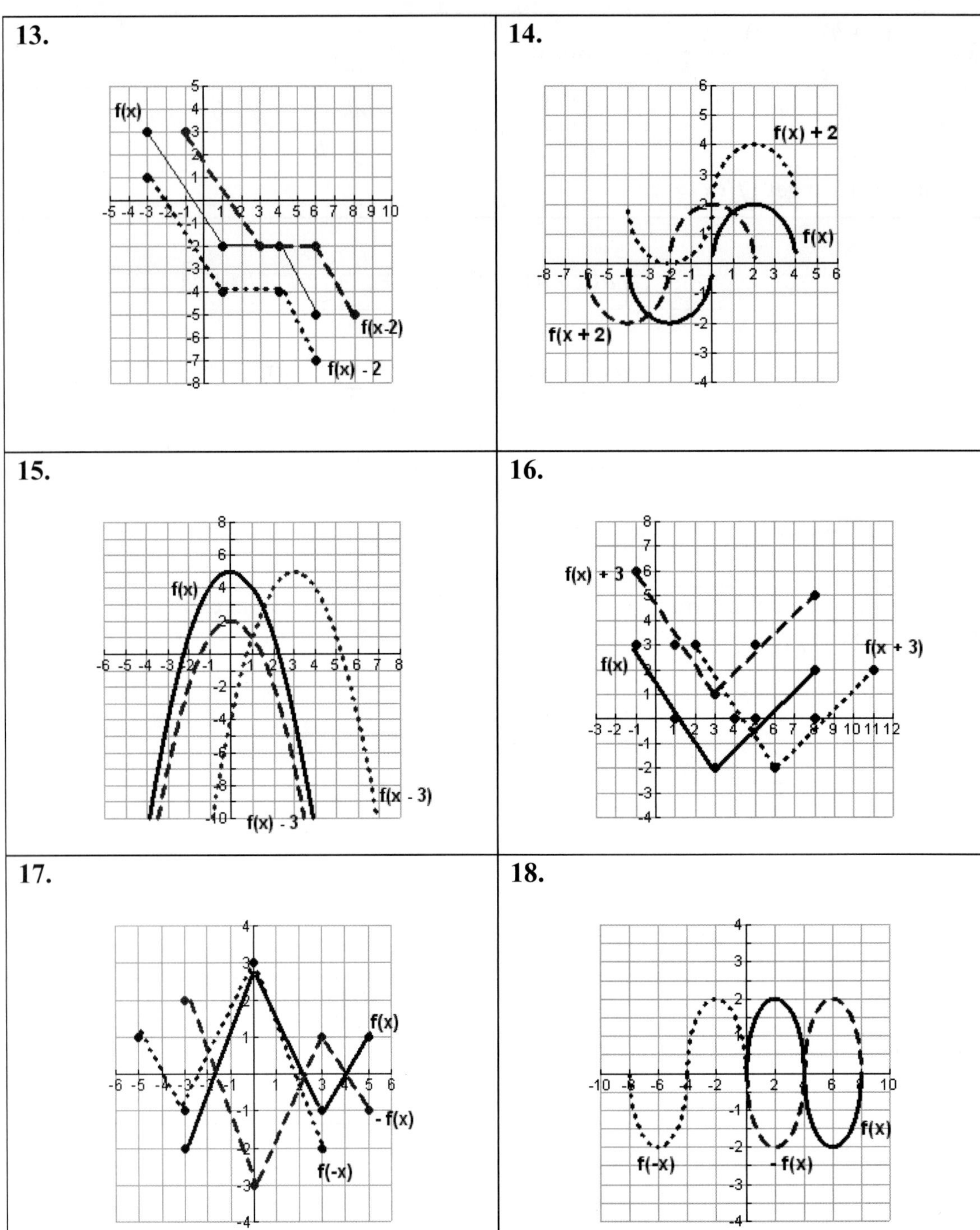
13.
f(x)
f(x-2)
f(x) - 2
14.
f(x) + 2
f(x)
f(x + 2)
15.
f(x)
f(x - 3)
f(x) - 3
16.
f(x) + 3
f(x)
f(x + 3)
17.
f(x)
- f(x)
f(-x)
18.
f(-x)
- f(x)
f(x)

19.

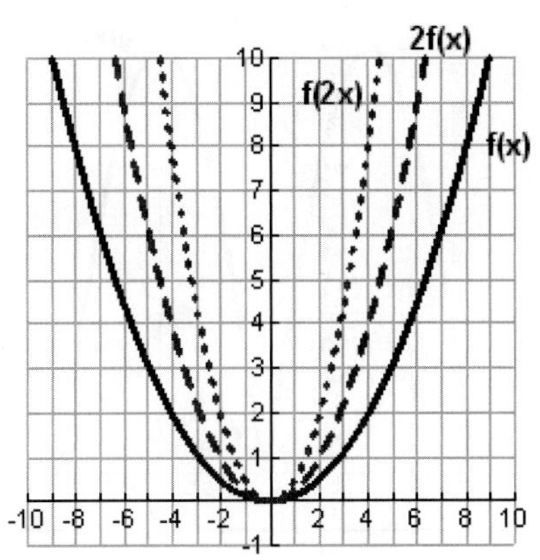

20.

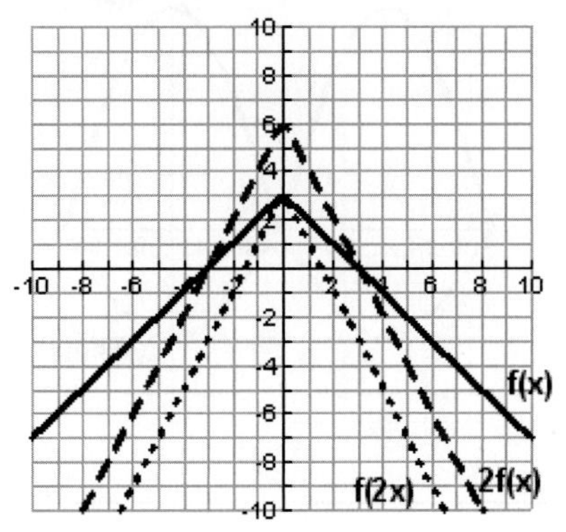

21.

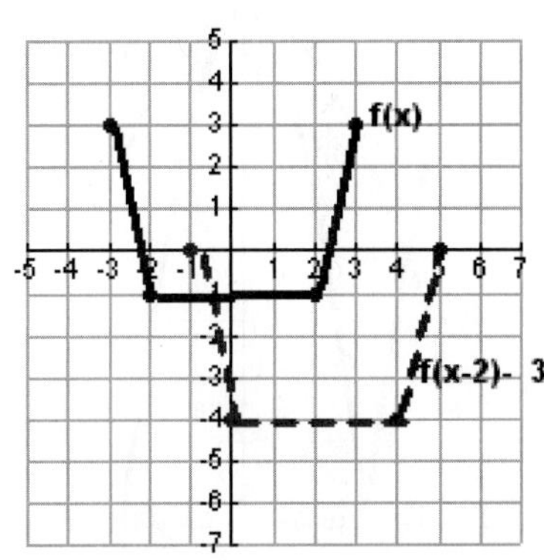

22.

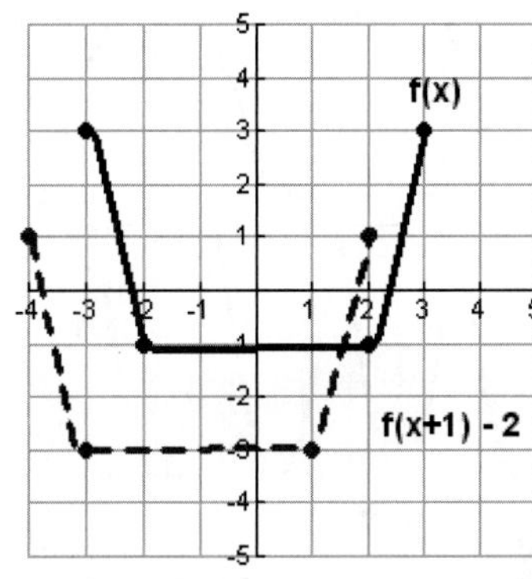

23.

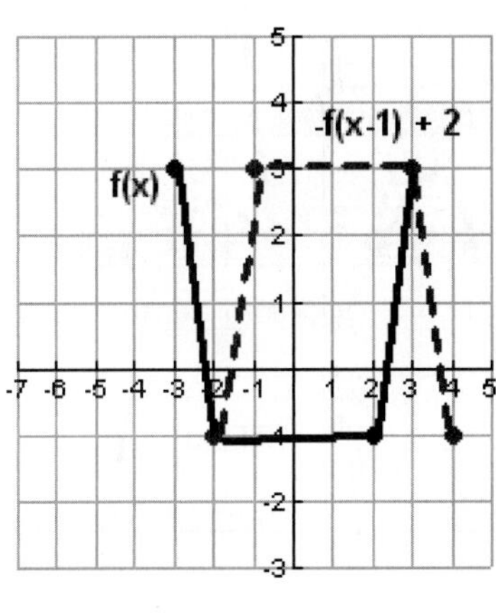

24.

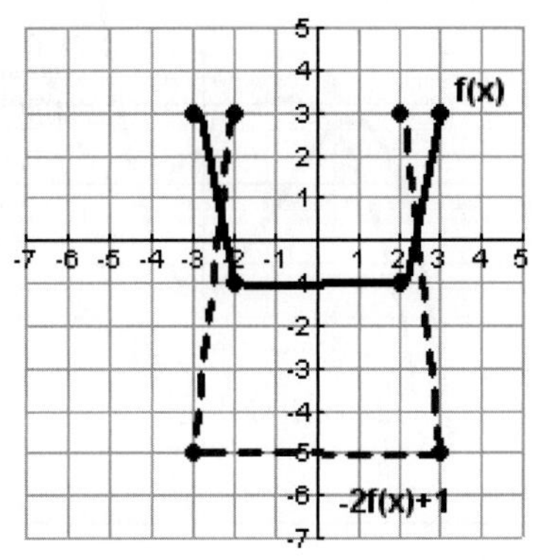

25.

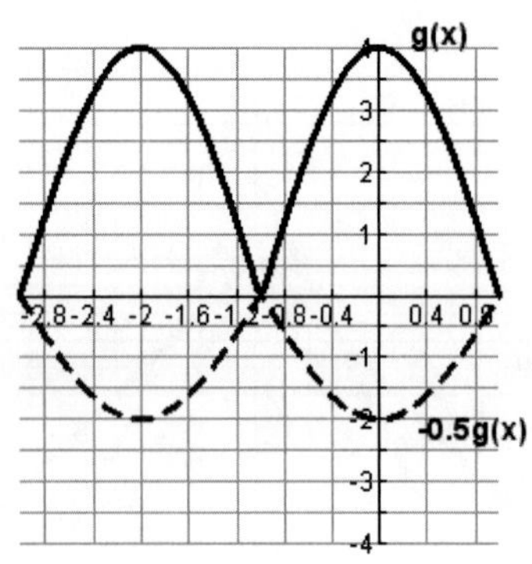

26.

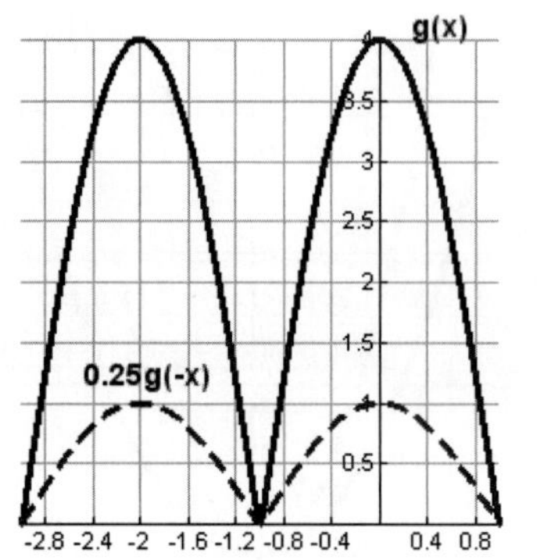

27.

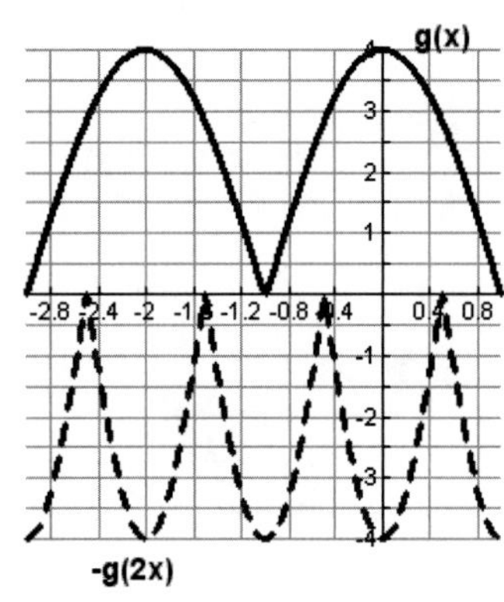

28.

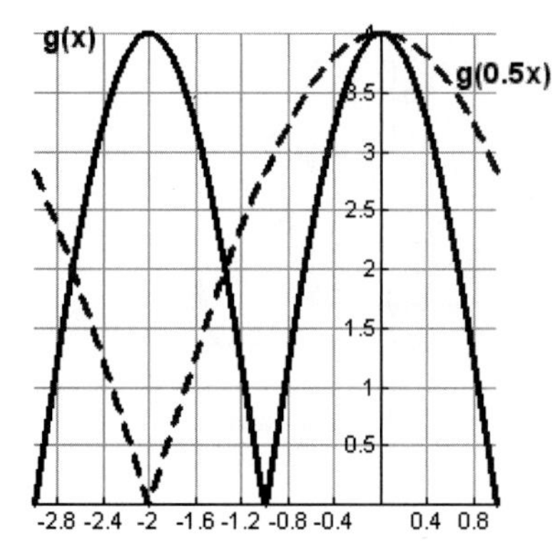

29.

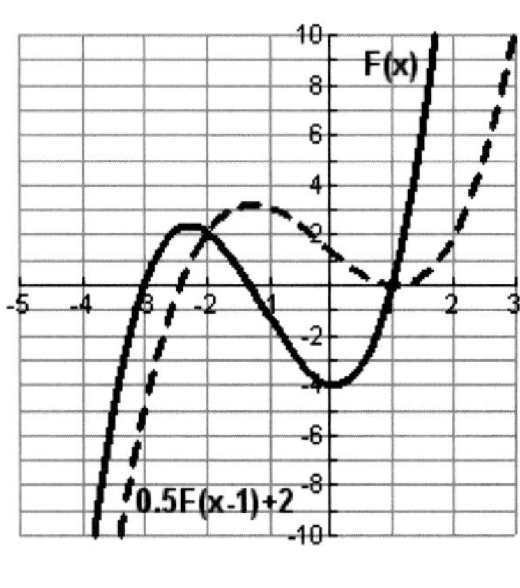

30.

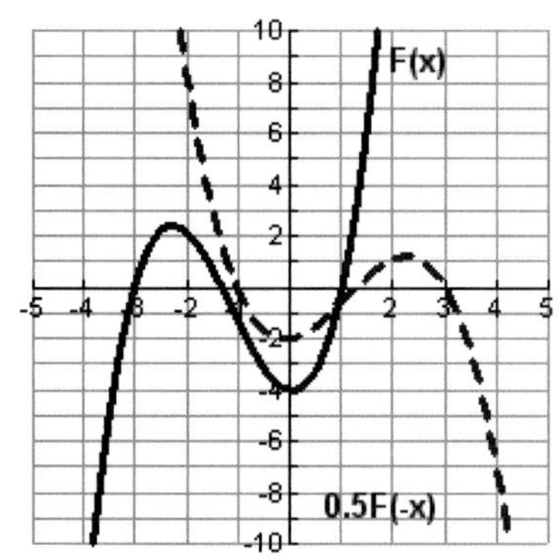

31.

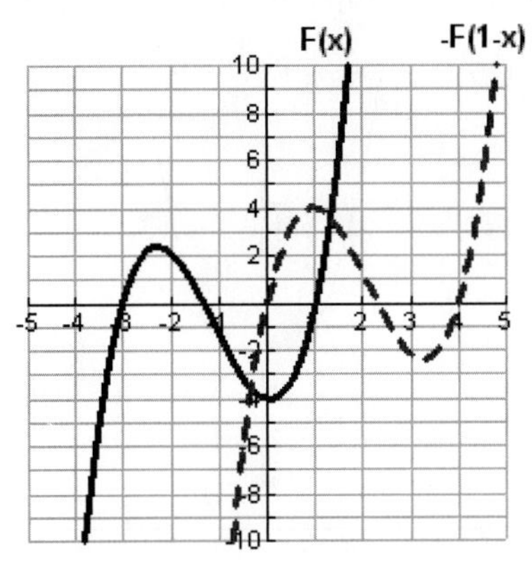

32.

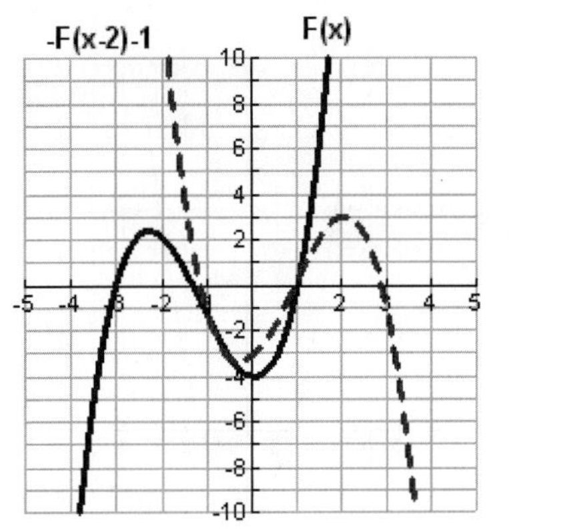

33.

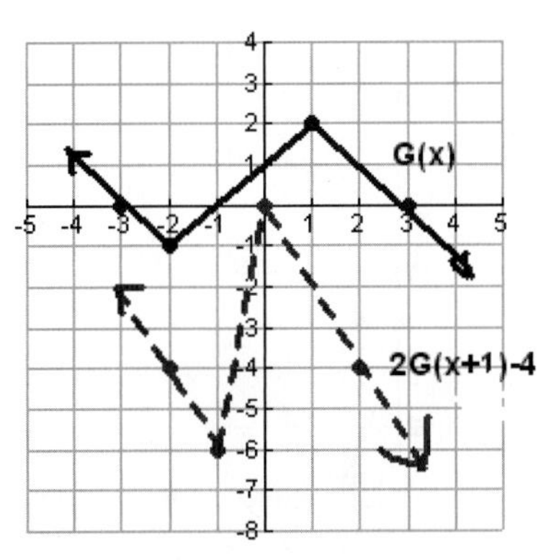

34.

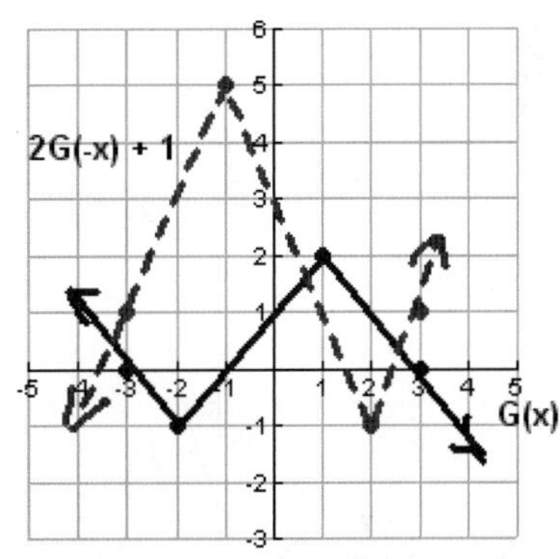

35.

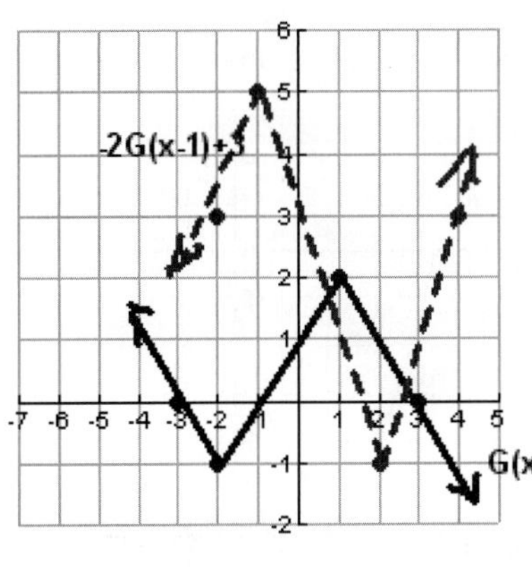

36.

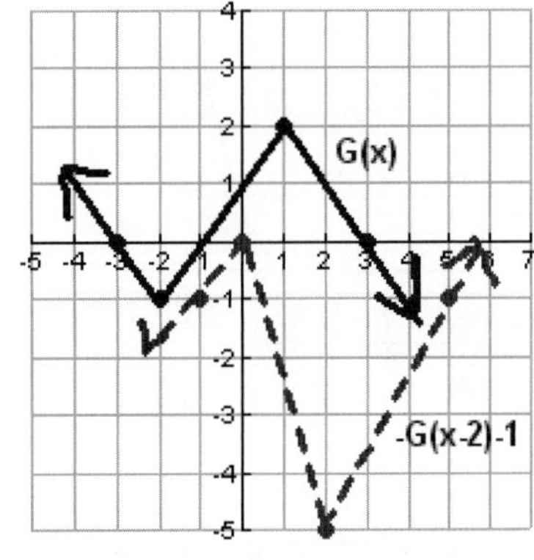

37. Shift the graph of x^2 down 2 units.

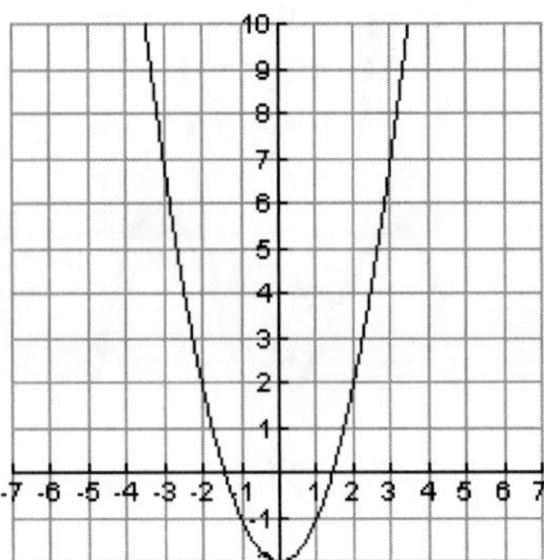

38. Shift the graph of x^2 up 3 units.

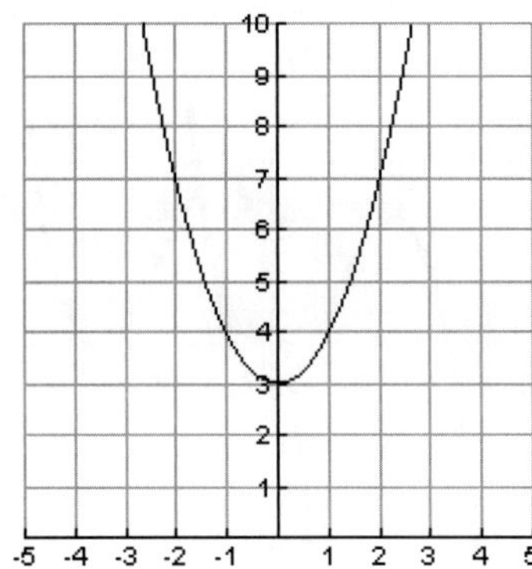

39. Shift the graph of x^2 left 1 unit.

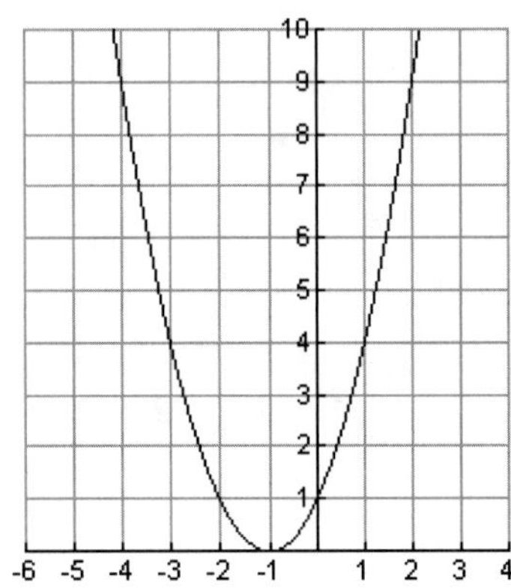

40. Shift the graph of x^2 right 2 units.

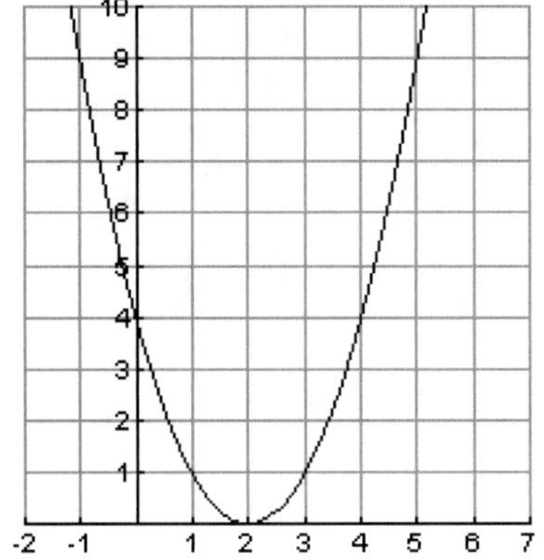

41. Shift the graph of x^2 right 3 units, and up 2 units.

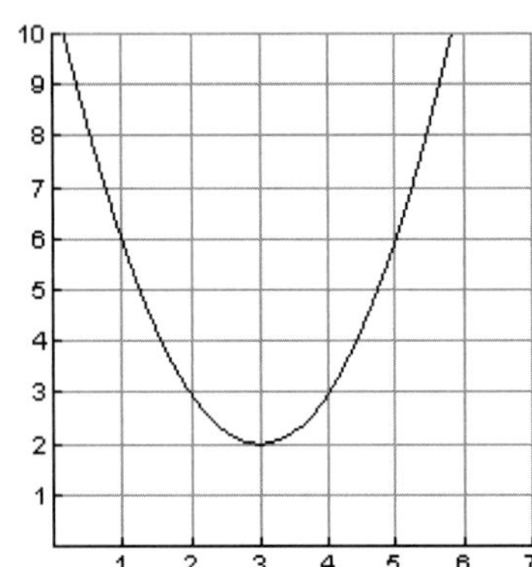

42. Shift the graph of x^2 left 2 units, and up 1 unit.

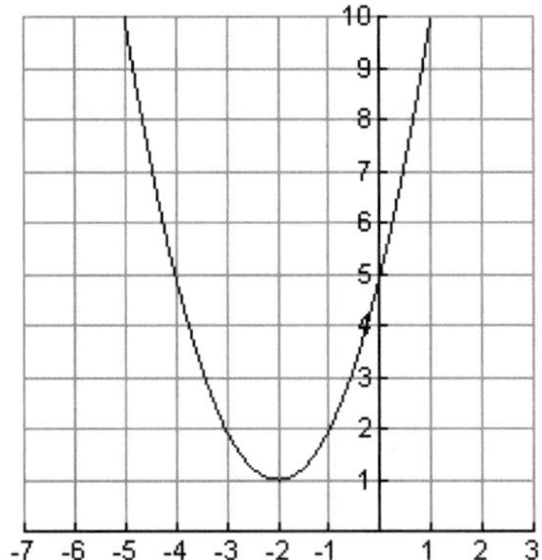

43. Shift the graph of x^2 right 1 unit, and then reflect over x-axis.

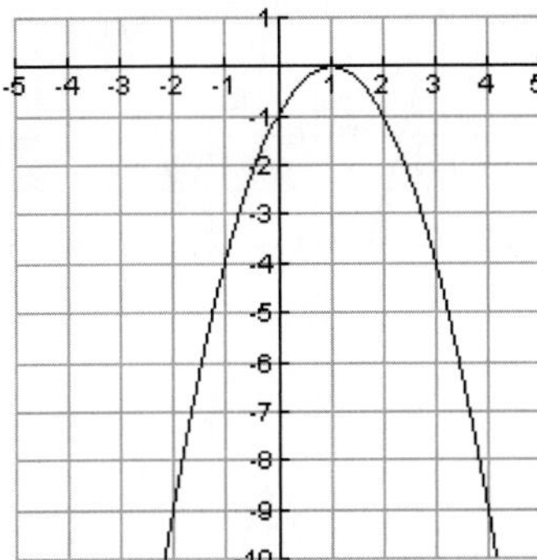

44. Shift the graph of x^2 left 2 units, and then reflect over x-axis.

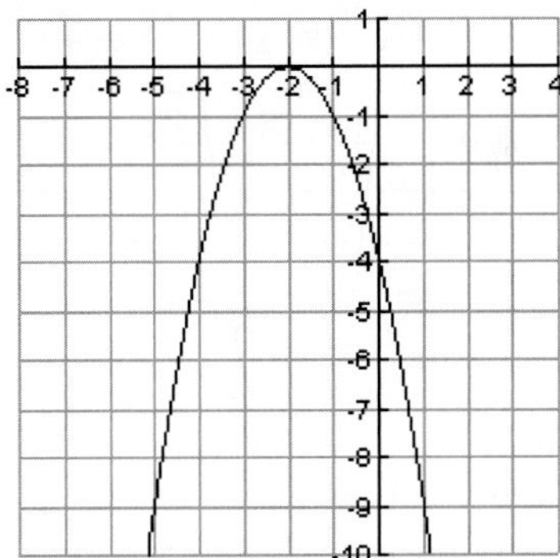

45. Reflect the graph of $|x|$ over y-axis. (This yields the same graph as $|x|$ since $|-x| = |-1||x| = |x|$.)

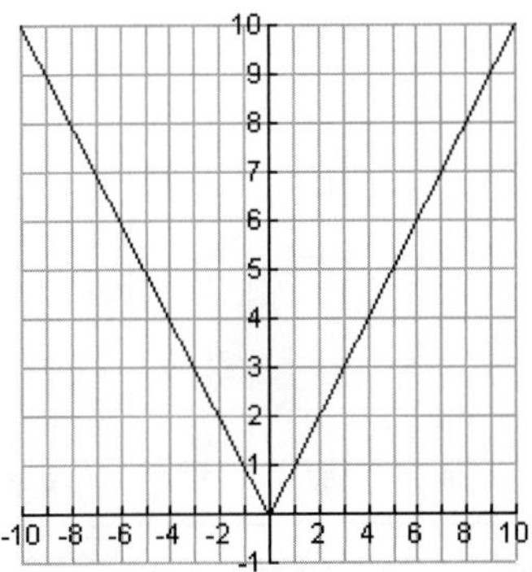

46. Reflect the graph of $|x|$ over x-axis.

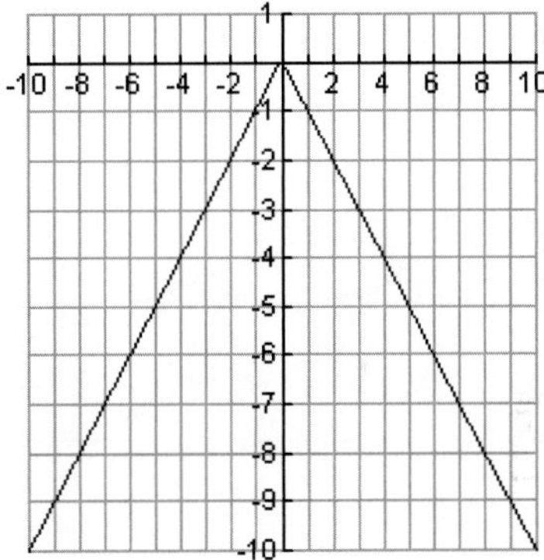

47. Reflect the graph of $|x|$ over x-axis, then shift left 2 units and down 1 unit.

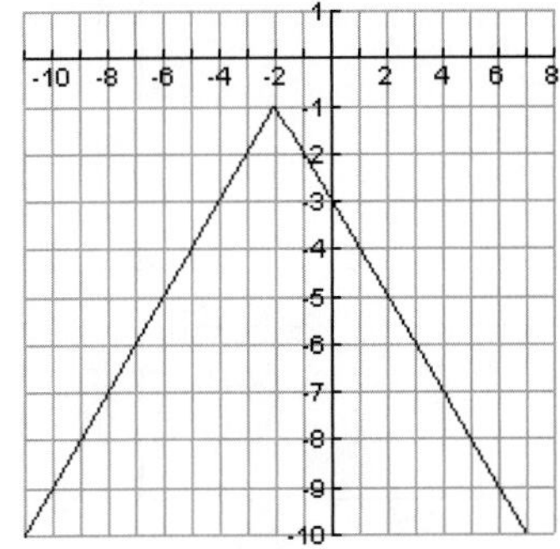

48. Since $|1-x|+2 = |x-1|+2$, shift the graph of $|x|$ right 1 unit, and up 2 units.

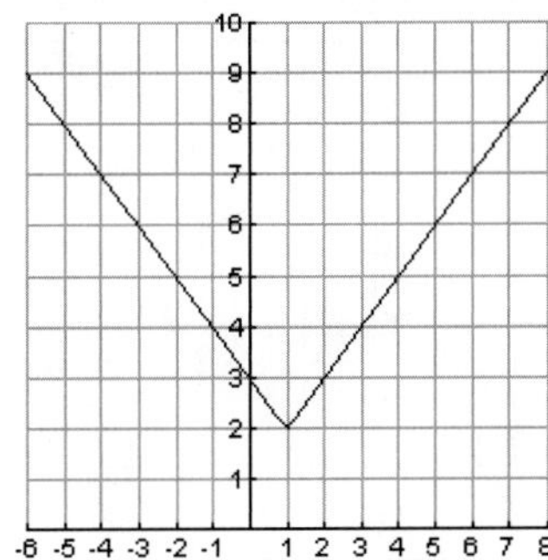

49. Vertically stretch the graph of x^2 by a factor of 2, then shift up 1unit.

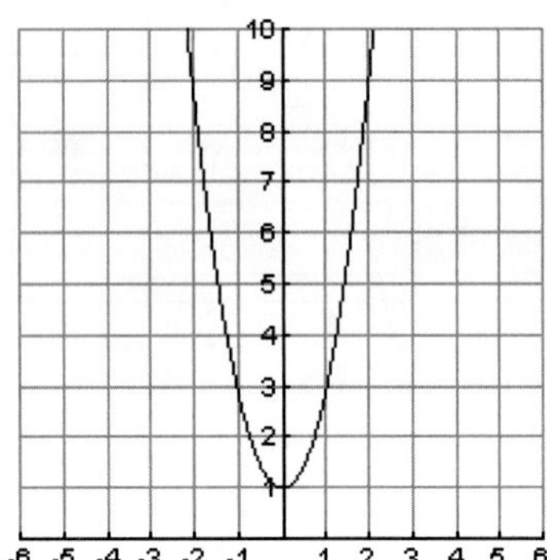

50. Vertically stretch the graph of $|x|$ by a factor of 2, then shift up 1unit.

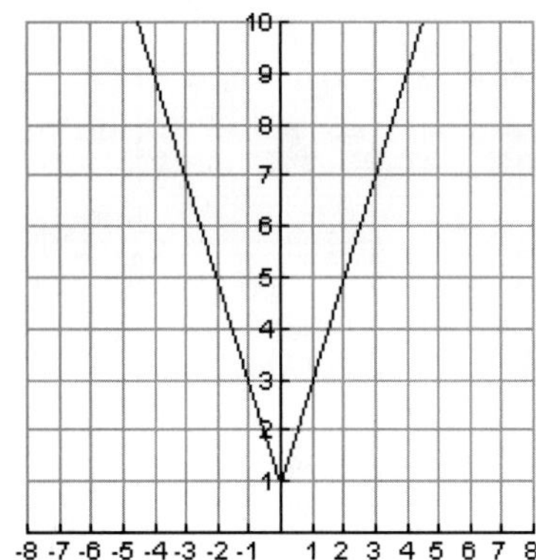

51. Shift the graph of $\sqrt{x}$ right 2 units, then reflect over x-axis.

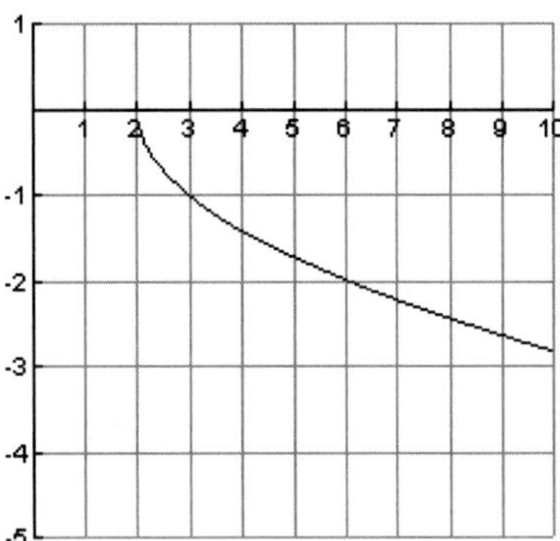

52. Since $\sqrt{2-x} = \sqrt{-(x-2)}$, reflect the graph of $\sqrt{x}$ over y-axis, then shift right 2 units.

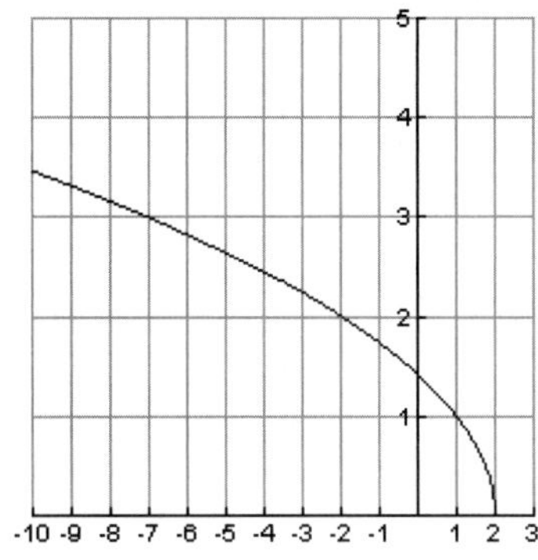

53. Reflect the graph of $\sqrt{x}$ over x-axis, then shift left 2 units and down 1 unit.

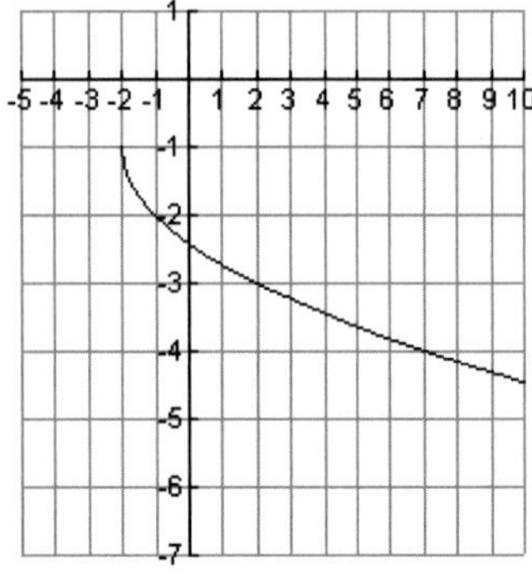

54. Since $\sqrt{2-x}+3 = \sqrt{-(x-2)}+3$, reflect the graph of $\sqrt{x}$ over y-axis, then right 2 units and up 3 units.

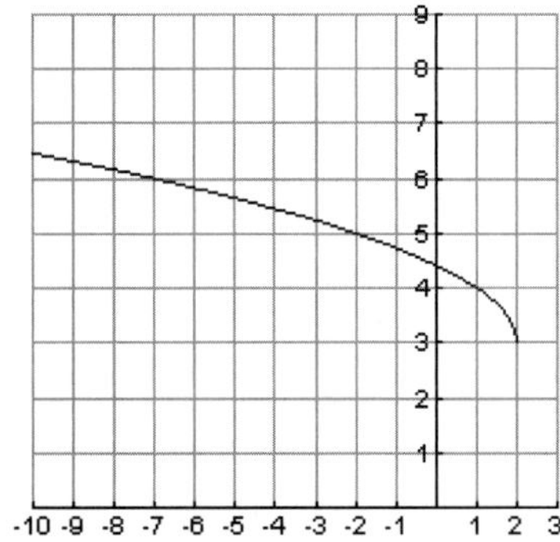

55. Shift the graph of $\sqrt[3]{x}$ right 1 unit, then up 2 units.

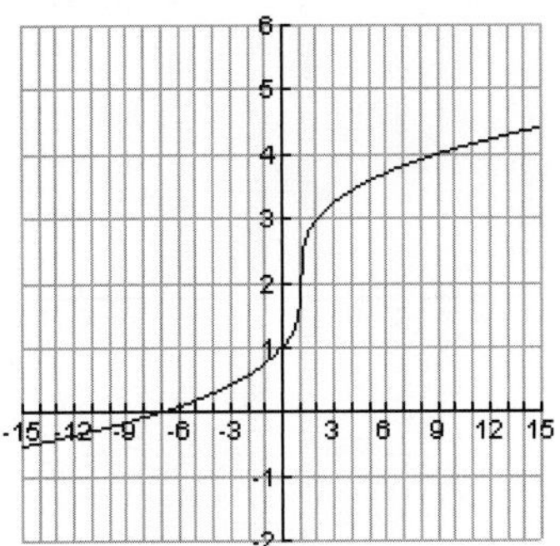

56. Shift the graph of $\sqrt[3]{x}$ left 2 units, then down 1 unit.

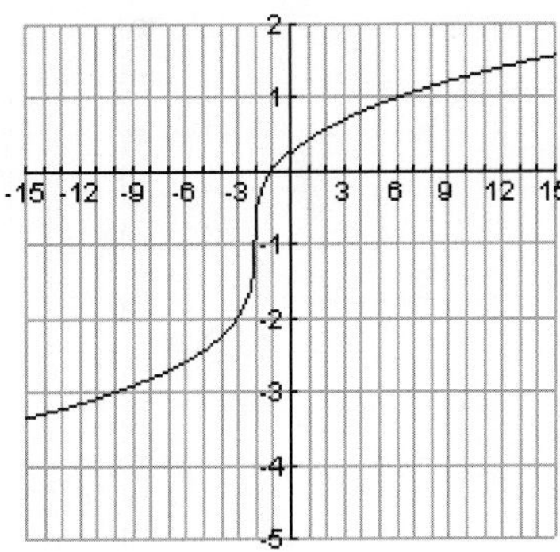

57. Shift the graph of $\frac{1}{x}$ left 3 units, then up 2 units.

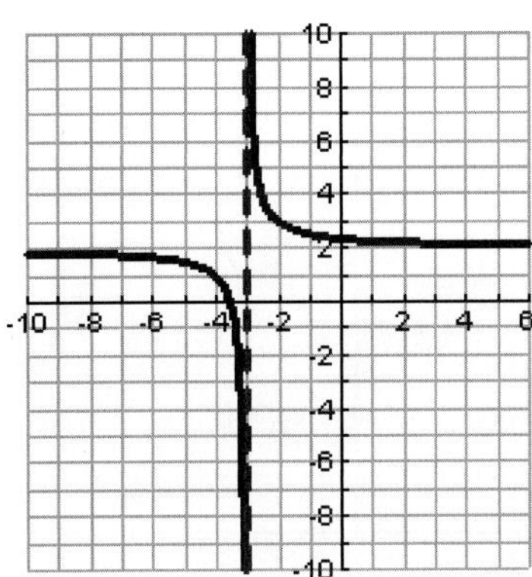

58. Since $\frac{1}{3-x} = -\frac{1}{x-3}$, shift the graph of $\frac{1}{x}$ right 3 units, and then reflect over x-axis.

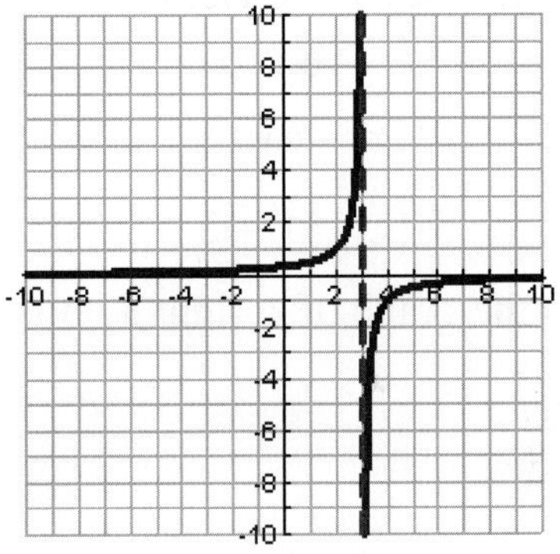

59. Shift the graph of $\frac{1}{x}$ left 2 units, then reflect over x-axis, and then shift up 2 units.

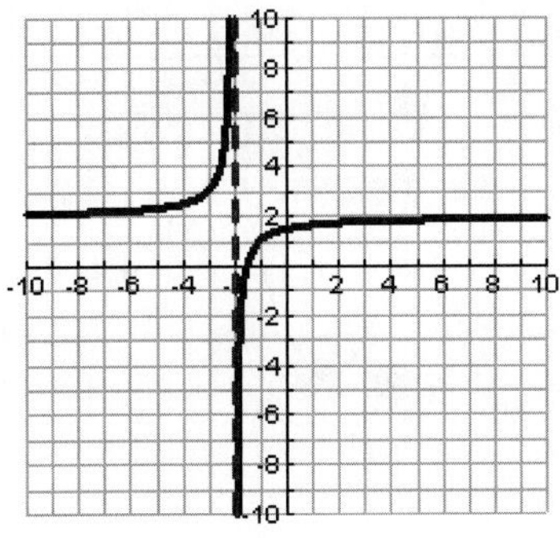

60. Since $2 - \frac{1}{-(x-1)} = 2 + \frac{1}{x-1}$, shift the graph of $\frac{1}{x}$ right 1 unit, then up 2 units.

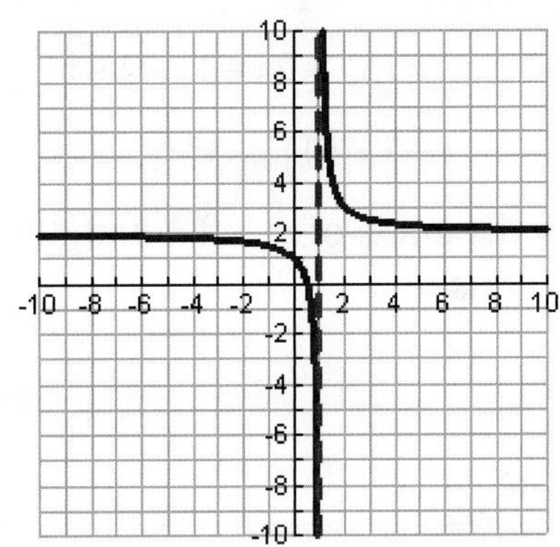

61. Reflect the graph of $\sqrt{x}$ over y-axis, then expand vertically by a factor of 5.

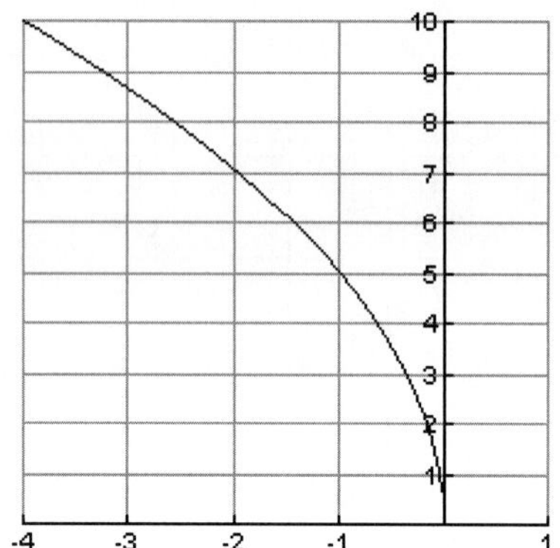

62. Reflect the graph of $\sqrt{x}$ over x-axis, then contract vertically by a factor of 5.

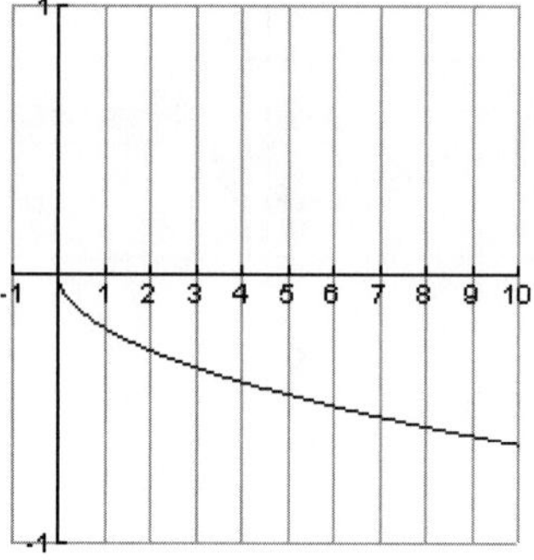

63.

Completing the square yields

$$\begin{aligned} f(x) &= x^2 - 6x + 11 \\ &= \left(x^2 - 6x + 9\right) + 11 - 9 \\ &= (x-3)^2 + 2 \end{aligned}$$

So, shift the graph of x^2 right 3 units, then up 2 units.

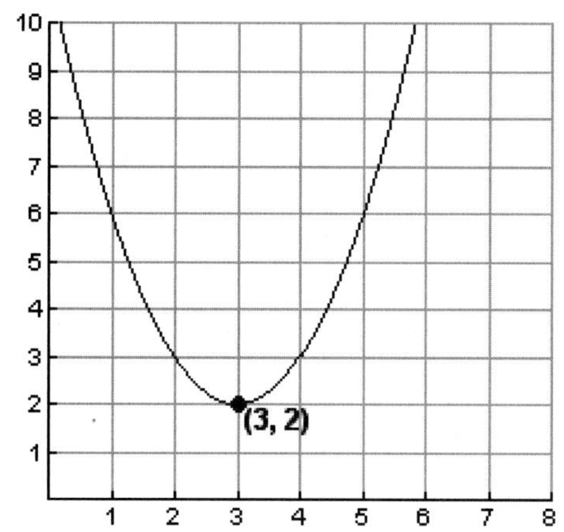

64.

Completing the square yields

$$\begin{aligned} f(x) &= x^2 + 2x - 2 \\ &= \left(x^2 + 2x + 1\right) - 2 - 1 \\ &= (x+1)^2 - 3 \end{aligned}$$

So, shift the graph of x^2 left 1 unit, then down 3 units.

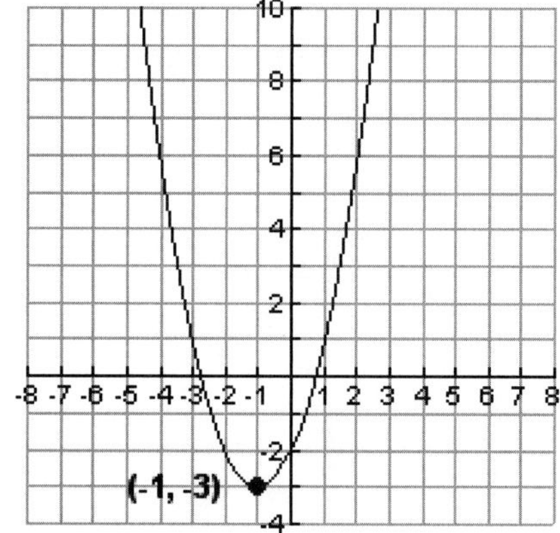

65.

Completing the square yields

$$f(x) = -\left(x^2 + 2x\right)$$
$$= -\left(x^2 + 2x + 1\right) + 1$$
$$= -\left(x + 1\right)^2 + 1$$

So, reflect the graph of x^2 over x-axis, then shift left 1 unit, then up 1 unit.

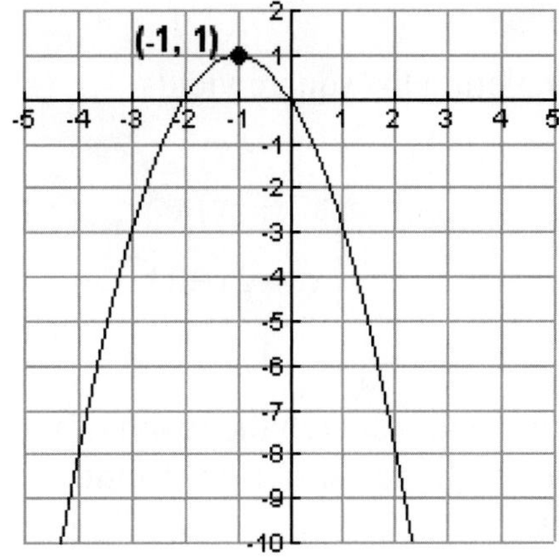

66.

Completing the square yields

$$f(x) = -x^2 + 6x - 7$$
$$= -\left(x^2 - 6x\right) - 7$$
$$= -\left(x^2 - 6x + 9\right) - 7 + 9$$
$$= -\left(x - 3\right)^2 + 2$$

So, reflect the graph of x^2 over x-axis, then shift right 3 units, then up 2 units.

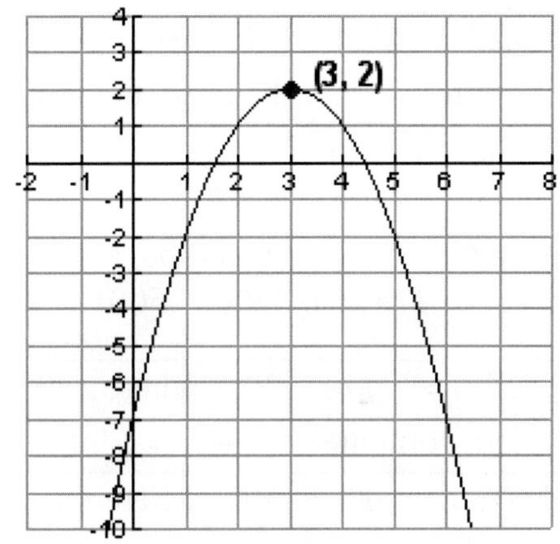

67.

Completing the square yields

$$f(x) = 2x^2 - 8x + 3$$
$$= 2\left(x^2 - 4x\right) + 3$$
$$= 2\left(x^2 - 4x + 4\right) + 3 - 8$$
$$= 2\left(x - 2\right)^2 - 5$$

So, vertically stretch the graph of x^2 by a factor of 2, then shift right 2 units, then down 5 units.

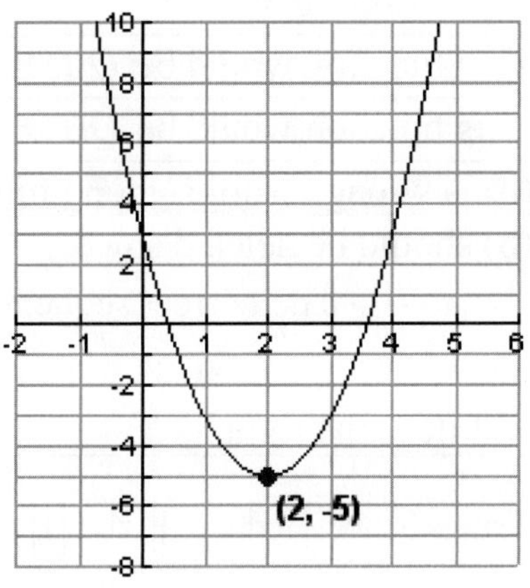

68.

Completing the square yields

$$\begin{aligned} f(x) &= 3x^2 - 6x + 5 \\ &= 3\left(x^2 - 2x\right) + 5 \\ &= 3\left(x^2 - 2x + 1\right) + 5 - 3 \\ &= 3\left(x - 1\right)^2 + 2 \end{aligned}$$

So, vertically stretch the graph of x^2 by a factor of 3, then shift right 1 unit, then up 2 units.

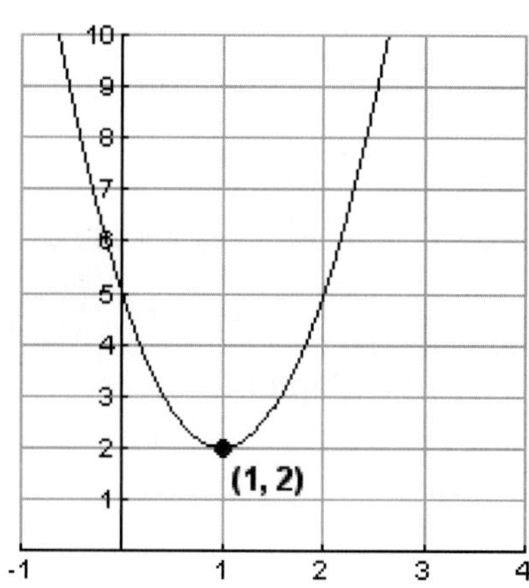

69. Let x = number of hours worked per week. Then, the salary is given by $\boxed{S(x) = 10x}$ (in dollars). After 1 year, taking into account the raise, the new salary is $\boxed{\overline{S}(x) = 10x + 50}$.

70. The profit in a rainy year is given by $P(x) - 10(\text{Cost of } 1)$, where x is the number of pallets sold.
Since they are giving away 10 pallets in a rainy year, they don't make a profit on the first 10. So, the profit would be $P(x-10)$.

71. The 2006 taxes would be:

$$T(x) = 0.33(x - 6500)$$

72. The actual amount administered if the weight is overestimated by 3 ounces is $A(x+3) = \sqrt{x+3} + 2$.

73. $f(x) = 7.00 + 0.30x$, where x is in miles and $f(x)$ is in dollars. If 2 more miles are traveled, compute $f(x+2) = 7.00 + 0.30(x+2) = 7.60 + 0.30x$

74. $f(x) = 7.00 + 0.30x$, where x is in miles and $f(x)$ is in dollars. Since the restaurant is between the bank and home (which are 1 miles apart), and the restaurant is x miles south of the bank, from home her taxi bill would be
$f(1-x) = 7.00 + 0.30(1-x) = 7.30 - 0.30x$

75. This function would be $Q(t) = P(t + 50)$.

76. This function would be $Q(t) = P(t + 60)$.

77. (b) is wrong – shift right 3 units.

78. (c) is wrong – reflect over x-axis.

79. (b) should be deleted since $|3 - x| = |x - 3|$. The correct sequence of steps would be: $(a) \to (c)^* \to (d)$, where (c)* : Shift to the right 3

80. (b) is wrong and (d) is misplaced. The correct sequence of steps would be:

$$(a) \to (d) \to (*) \to (c),$$

where $(*)$ = reflect over x-axis.

81. True. Since $|-x| = |-1||x| = |x|$.

82. False. $y = \sqrt{-x}$ is the reflection of $y = \sqrt{x}$ over the y-axis.

83. True.

84. True.

85. True

86. False. The shift is a units to the left.

87. The graph of $y = f(x-3)+2$ is the graph of $y = f(x)$ shifted right 3 units, then up 2 units. So, if the point (a,b) is on the graph of $y = f(x)$, then the point $(a+3,b+2)$ is on the graph of the translation $y = f(x-3)+2$.

88. The graph of $y = -f(-x)+1$ is the graph of $y = f(x)$ reflected over y-axis, then over x-axis, and then shifted up 1 unit. So, if the point (a,b) is on the graph of $y = f(x)$, then the point $(-a,-b+1)$ is on the graph of the translation $y = -f(-x)+1$.

89. We do this in three steps:

$f(x)$		(a,b)
$f(x+1)$	Shift left 1 unit	$(a-1,b)$
$2f(x+1)$	Multiply all outputs by 2	$(a-1,2b)$
$2f(x+1)-1$	Shift vertically down 1 unit	$(a-1,2b-1)$

So, the point $(a-1,2b-1)$ is guaranteed to lie on the graph.

90. We do this in three steps:

$f(x)$		(a,b)
$f(x-3)$	Shift right 3 units	$(a+3,b)$
$-2f(x-3)$	Multiply all outputs by -2	$(a+3,-2b)$
$-2f(x-3)+4$	Shift vertically up 4 units	$(a+3,-2b+4)$

So, the point $(a+3,-2b+4)$ is guaranteed to lie on the graph.

91. a.

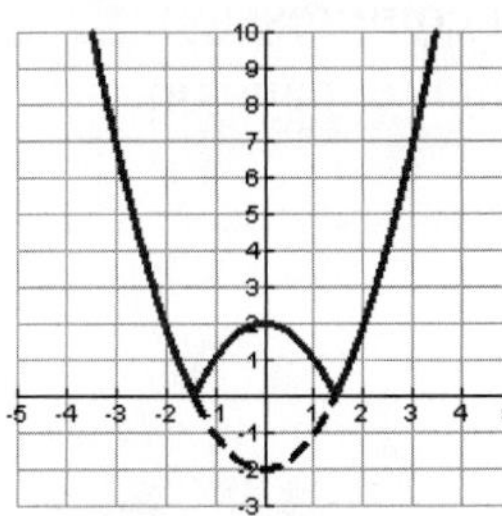

b.

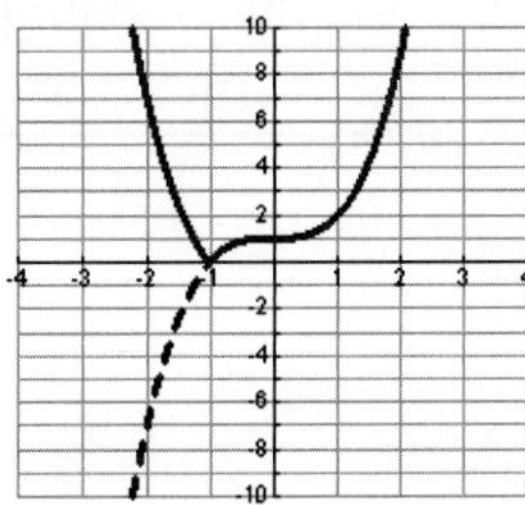

Any part of the graph of $y = f(x)$ that is below the x-axis is reflected above it for the graph of $y = |f(x)|$.

92. a.

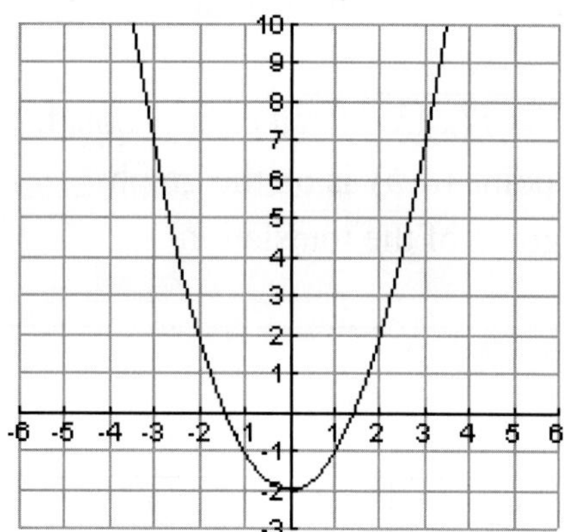

b.

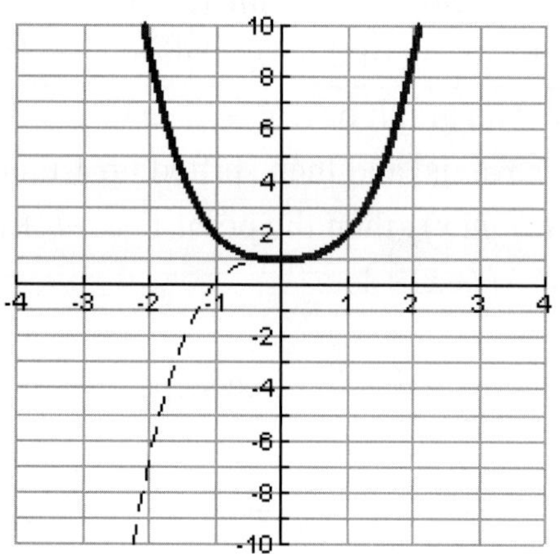

The relationship is described by: $f\left(|x|\right)=\begin{cases} f(x), & x\geq 0 \\ f(-x), & x<0 \end{cases}$

93. a.

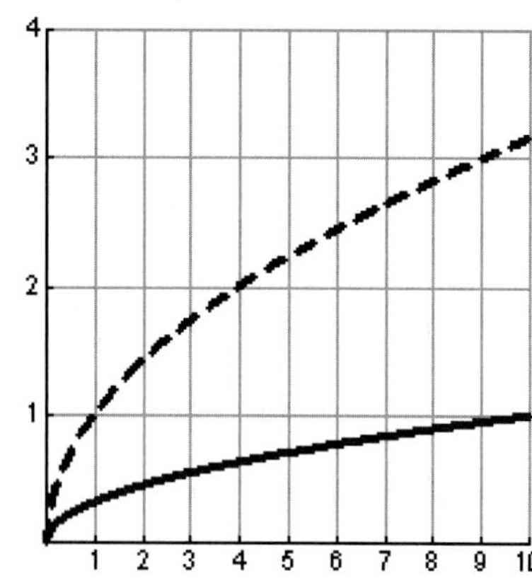

b.

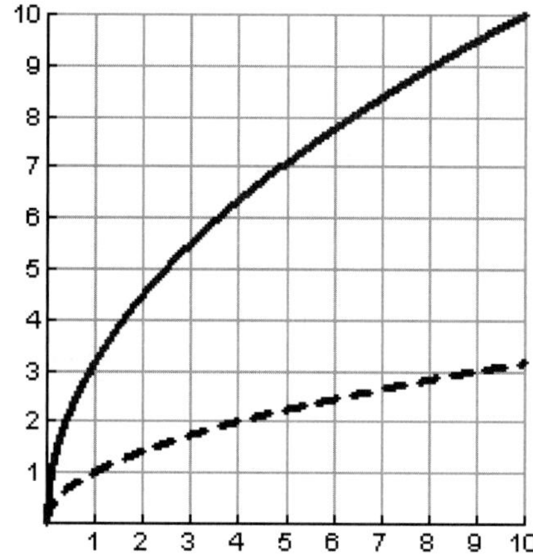

If $a>1$, then the graph is a horizontal compression.
If $0<a<1$, then the graph is a horizontal expansion.

94. a.

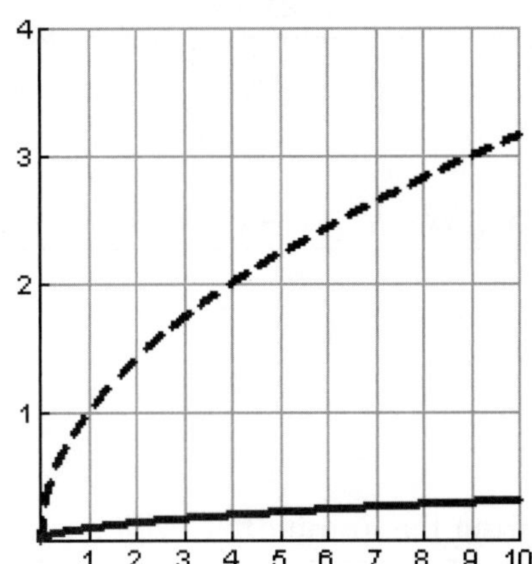

b.

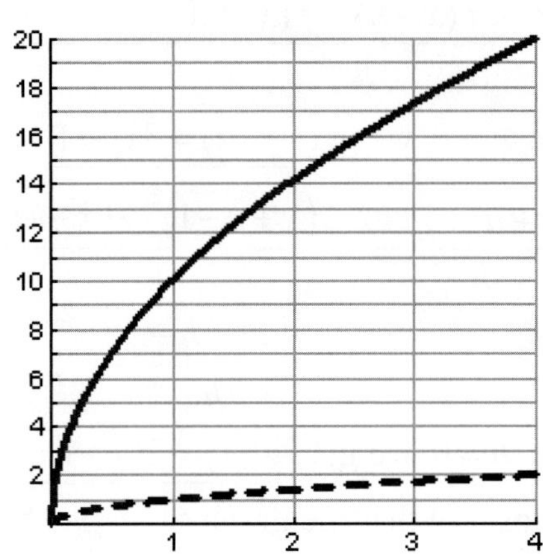

If $0 < a < 1$, then the graph is a vertical compression.
If $a > 1$, then the graph is a vertical expansion.

95.
The graph of *f* is as follows:

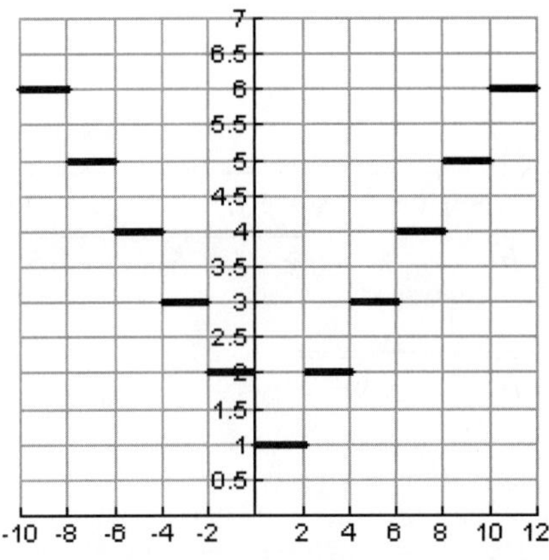

Each horizontal line in the graph of $y = [\![x]\!]$ is stretched by a factor of 2. Any portion of the graph that is below the *x*-axis is reflected above it. Also, there is a vertical shift up of one unit.

96.
The graph of *g* is as follows:

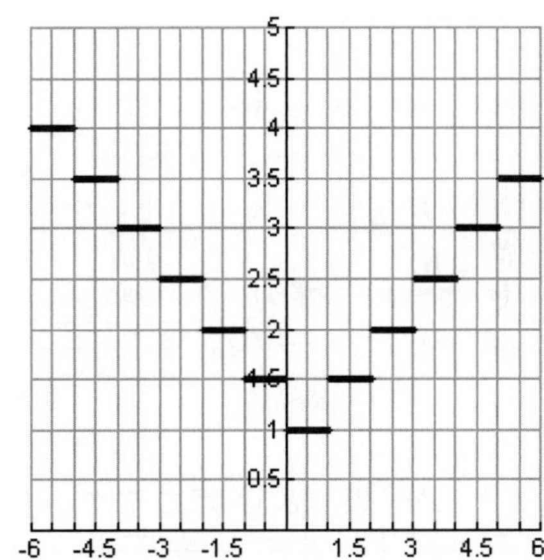

Any portion of the graph of $y = [\![x]\!]$ that is below the *x*-axis is reflected above it. Also, there is a vertical shift up of one unit and a vertical compression by a factor of $\frac{1}{2}$.

97.

$$\frac{f(x+h)-f(x)}{h}=\frac{(x+h)^2-x^2}{h}=\frac{x^2+2hx+h^2-x^2}{h}=\frac{h(2x+h)}{h}=2x+h$$

So, at $h=0$ we get $\boxed{f'(x)=3x^2+1}$.

$$\frac{g(x+h)-g(x)}{h}=\frac{(x+h-1)^2-(x-1)^2}{h}=\frac{x^2+hx-x+hx+h^2-h-x-h+1-x^2+2x-1}{h}$$

$$=\frac{h(2x+h-2)}{h}=2x+h-2=2(x-1)+h$$

So, at $h=0$ we get $\boxed{g'(x)=2(x-1)}$.

We observe that the graph of g' is obtained by shifting the graph of f' right 1 unit.

98.

$$\frac{f(x+h)-f(x)}{h}=\frac{\sqrt{x+h}-\sqrt{x}}{h}=\frac{\sqrt{x+h}-\sqrt{x}}{h}\cdot\frac{\sqrt{x+h}+\sqrt{x}}{\sqrt{x+h}+\sqrt{x}}$$

$$=\frac{x+h-x}{h\left(\sqrt{x+h}+\sqrt{x}\right)}=\frac{1}{\sqrt{x+h}+\sqrt{x}}$$

So, at $h=0$ we get $\boxed{f'(x)=\frac{1}{2\sqrt{x}}}$.

$$\frac{g(x+h)-g(x)}{h}=\frac{\sqrt{x+h+5}-\sqrt{x+5}}{h}=\frac{\sqrt{x+h+5}-\sqrt{x+5}}{h}\cdot\frac{\sqrt{x+h+5}+\sqrt{x+5}}{\sqrt{x+h+5}+\sqrt{x+5}}$$

$$=\frac{x+h+5-x-5}{h\left(\sqrt{x+h+5}+\sqrt{x+5}\right)}=\frac{1}{\sqrt{x+h+5}+\sqrt{x+5}}$$

So, at $h=0$ we get $\boxed{g'(x)=\frac{1}{2\sqrt{x+5}}}$.

We observe that the graph of g' is obtained by shifting the graph of f' left 5 units.

99. $\frac{f(x+h)-f(x)}{h}=\frac{2(x+h)-2x}{h}=\frac{2h}{h}=2$. So, at $h=0$ we get $\boxed{f'(x)=2}$.

$\frac{g(x+h)-g(x)}{h}=\frac{[2(x+h)+7]-[2x+7]}{h}=\frac{2h}{h}=2$. So, at $h=0$ we get $\boxed{g'(x)=2}$.

We observe that the graphs of f' and g' are the same.

100.

$$\frac{f(x+h)-f(x)}{h}=\frac{\left[(x+h)^3\right]-\left[x^3\right]}{h}=\frac{x^3+3x^2h+3xh^2+h^3-x^3}{h}$$

$$=\frac{h\left(3x^2+3xh+h^2\right)}{h}=\boxed{3x^2+3xh+h^2}$$

So, at $h=0$ we get $\boxed{f'(x)=3x^2}$.

$$\frac{g(x+h)-g(x)}{h}=\frac{\left[(x+h)^3-4\right]-\left[x^3-4\right]}{h}=\frac{x^3+3x^2h+3xh^2+h^3-4-x^3}{h}+4$$

$$=\frac{h\left(3x^2+3xh+h^2\right)}{h}=\boxed{3x^2+3xh+h^2}$$

So, at $h=0$ we get $\boxed{g'(x)=3x^2}$.

We observe that the graphs of f' and g' are the same.

Section 1.4 Solutions ---

1.

$$f(x)+g(x)=(2x+1)+(1-x)$$
$$=x+2$$
$$f(x)-g(x)=(2x+1)-(1-x)$$
$$=2x+1-1+x$$
$$=3x$$
$$f(x)\cdot g(x)=(2x+1)(1-x)$$
$$=2x+1-2x^2-x$$
$$=-2x^2+x+1$$
$$\frac{f(x)}{g(x)}=\frac{2x+1}{1-x}$$

Domains:

$$\left.\begin{array}{r}dom(f+g)\\ dom(f-g)\\ dom(fg)\end{array}\right\}=(-\infty,\infty)$$

$$dom\left(\frac{f}{g}\right)=(-\infty,1)\cup(1,\infty)$$

2.

$$f(x)+g(x)=(3x+2)+(2x-4)$$
$$=5x-2$$
$$f(x)-g(x)=(3x+2)-(2x-4)$$
$$=3x+2-2x+4$$
$$=x+6$$
$$f(x)\cdot g(x)=(3x+2)\cdot(2x-4)$$
$$=6x^2-12x+4x-8$$
$$=6x^2-8x-8$$
$$\frac{f(x)}{g(x)}=\frac{3x+2}{2x-4}$$

Domains:

$$\left.\begin{array}{r}dom(f+g)\\ dom(f-g)\\ dom(fg)\end{array}\right\}=(-\infty,\infty)$$

$$dom\left(\frac{f}{g}\right)=(-\infty,2)\cup(2,\infty)$$

3.

$$f(x)+g(x)=\left(2x^2-x\right)+\left(x^2-4\right)$$
$$=3x^2-x-4$$
$$f(x)-g(x)=\left(2x^2-x\right)-\left(x^2-4\right)$$
$$=2x^2-x-x^2+4$$
$$=x^2-x+4$$
$$f(x)\cdot g(x)=\left(2x^2-x\right)\cdot\left(x^2-4\right)$$
$$=2x^4-x^3-8x^2+4x$$
$$\frac{f(x)}{g(x)}=\frac{2x^2-x}{x^2-4}$$

Domains:

$dom(f+g),\ dom(f-g),\ dom(fg)\}=(-\infty,\infty)$

$dom\left(\frac{f}{g}\right)=(-\infty,-2)\cup(-2,2)\cup(2,\infty)$

4.

$$f(x)+g(x)=\left(3x+2\right)+\left(x^2-25\right)$$
$$=x^2+3x-23$$
$$f(x)-g(x)=\left(3x+2\right)-\left(x^2-25\right)$$
$$=3x+2-x^2+25$$
$$=-x^2+3x+27$$
$$f(x)\cdot g(x)=\left(3x+2\right)\cdot\left(x^2-25\right)$$
$$=3x^3+2x^2-75x-50$$
$$\frac{f(x)}{g(x)}=\frac{3x+2}{x^2-25}$$

Domains:

$dom(f+g),\ dom(f-g),\ dom(fg)\}=(-\infty,\infty)$

$dom\left(\frac{f}{g}\right)=(-\infty,-5)\cup(-5,5)\cup(5,\infty)$

5.

$$f(x)+g(x)=\tfrac{1}{x}+x=\frac{1+x^2}{x}$$
$$f(x)-g(x)=\tfrac{1}{x}-x=\frac{1-x^2}{x}$$
$$f(x)\cdot g(x)=\tfrac{1}{x}\cdot x=1$$
$$\frac{f(x)}{g(x)}=\frac{\frac{1}{x}}{x}=\frac{1}{x^2}$$

Domains:

$$\left.\begin{array}{c} dom(f+g)\\ dom(f-g)\\ dom(fg)\\ dom\left(\frac{f}{g}\right)\end{array}\right\}=(-\infty,0)\cup(0,\infty)$$

6.

$$f(x)+g(x)=\frac{2x+3}{x-4}+\frac{x-4}{3x+2}$$
$$=\frac{(2x+3)(3x+2)+(x-4)^2}{(x-4)(3x+2)}$$
$$=\frac{6x^2+9x+4x+6+x^2-8x+16}{(x-4)(3x+2)}$$
$$=\frac{7x^2+5x+22}{(x-4)(3x+2)}$$

$$f(x)-g(x)=\frac{2x+3}{x-4}-\frac{x-4}{3x+2}$$
$$=\frac{(2x+3)(3x+2)-(x-4)^2}{(x-4)(3x+2)}$$
$$=\frac{6x^2+9x+4x+6-x^2+8x-16}{(x-4)(3x+2)}$$
$$=\frac{5x^2+21x-10}{(x-4)(3x+2)}$$

$$f(x)\cdot g(x)=\frac{2x+3}{\cancel{x-4}}\cdot\frac{\cancel{x-4}}{3x+2}=\frac{2x+3}{3x+2}$$

$$\frac{f(x)}{g(x)}=\frac{\frac{2x+3}{x-4}}{\frac{x-4}{3x+2}}=\frac{2x+3}{x-4}\cdot\frac{3x+2}{x-4}$$
$$=\frac{(2x+3)(3x+2)}{(x-4)^2}$$

<u>Domains</u>:

$$\left.\begin{array}{c} dom(f+g) \\ dom(f-g) \\ dom(fg) \\ dom\left(\frac{f}{g}\right) \end{array}\right\}=\left(-\infty,-\tfrac{2}{3}\right)\cup\left(-\tfrac{2}{3},4\right)\cup(4,\infty)$$

7.

$$f(x)+g(x)=\sqrt{x}+2\sqrt{x}=3\sqrt{x}$$
$$f(x)-g(x)=\sqrt{x}-2\sqrt{x}=-\sqrt{x}$$
$$f(x)\cdot g(x)=\sqrt{x}\cdot 2\sqrt{x}=2x$$
$$\frac{f(x)}{g(x)}=\frac{\sqrt{x}}{2\sqrt{x}}=\frac{1}{2}$$

<u>Domains</u>:

$$\left.\begin{array}{c} dom(f+g) \\ dom(f-g) \\ dom(fg) \end{array}\right\}=[0,\infty)$$

$$dom\left(\frac{f}{g}\right)=(0,\infty)$$

8.

$$f(x)+g(x)=\sqrt{x-1}+2x^2$$
$$f(x)-g(x)=\sqrt{x-1}-2x^2$$
$$f(x)\cdot g(x)=2x^2\sqrt{x-1}$$
$$\frac{f(x)}{g(x)}=\frac{\sqrt{x-1}}{2x^2}$$

<u>Domains</u>:

Must have both $x-1\geq 0$ and $2x^2\neq 0$. So,

$$\left.\begin{array}{c} dom(f+g) \\ dom(f-g) \\ dom(fg) \\ dom\left(\frac{f}{g}\right) \end{array}\right\}=[1,\infty)$$

9.

$$f(x)+g(x)=\sqrt{4-x}+\sqrt{x+3}$$
$$f(x)-g(x)=\sqrt{4-x}-\sqrt{x+3}$$
$$f(x)\cdot g(x)=\sqrt{4-x}\cdot\sqrt{x+3}$$
$$\frac{f(x)}{g(x)}=\frac{\sqrt{4-x}}{\sqrt{x+3}}=\frac{\sqrt{4-x}\sqrt{x+3}}{x+3}$$

Domains:
Must have both $4-x\geq 0$ and $x+3\geq 0$. So,

$$\left.\begin{array}{r} dom(f+g) \\ dom(f-g) \\ dom(fg) \end{array}\right\}=[-3,4].$$

For the quotient, must have both $4-x\geq 0$ and $x+3>0$. So, $dom\left(\frac{f}{g}\right)=(-3,4]$.

10.

$$f(x)+g(x)=\sqrt{1-2x}+\tfrac{1}{x}$$
$$f(x)-g(x)=\sqrt{1-2x}-\tfrac{1}{x}$$
$$f(x)\cdot g(x)=\sqrt{1-2x}\cdot\tfrac{1}{x}$$
$$\frac{f(x)}{g(x)}=\frac{\sqrt{1-2x}}{\frac{1}{x}}=x\sqrt{1-2x}$$

Domains:
Must have both $1-2x\geq 0$ and $x\neq 0$. So,

$$\left.\begin{array}{r} dom(f+g) \\ dom(f-g) \\ dom(fg) \\ dom\left(\frac{f}{g}\right) \end{array}\right\}=(-\infty,0)\cup\left(0,\tfrac{1}{2}\right]$$

11.

$$(f\circ g)(x)=2\left(x^2-3\right)+1=2x^2-6+1=2x^2-5$$
$$(g\circ f)(x)=(2x+1)^2-3=4x^2+4x+1-3=4x^2+4x-2$$

Domains:
$dom(f\circ g)=(-\infty,\infty)=dom(g\circ f)$

12.

$$(f\circ g)(x)=(2-x)^2-1=4-4x+x^2-1=x^2-4x+3$$
$$(g\circ f)(x)=2-\left(x^2-1\right)=2-x^2+1=-x^2+3$$

Domains:
$dom(f\circ g)=(-\infty,\infty)=dom(g\circ f)$

13.

$$(f\circ g)(x)=\frac{1}{(x+2)-1}=\frac{1}{x+1}$$
$$(g\circ f)(x)=\frac{1}{x-1}+2=\frac{1+2(x-1)}{x-1}=\frac{1+2x-2}{x-1}=\frac{2x-1}{x-1}$$

Domains:
$dom(f\circ g)=(-\infty,-1)\cup(-1,\infty),\quad dom(g\circ f)=(-\infty,1)\cup(1,\infty)$

14.

$$(f \circ g)(x) = \frac{2}{(2+x)-3} = \frac{2}{x-1}$$

$$(g \circ f)(x) = 2 + \frac{2}{x-3} = \frac{2(x-3)+2}{x-3} = \frac{2x-6+2}{x-3} = \frac{2x-4}{x-3}$$

Domains:

$dom(f \circ g) = (-\infty,1) \cup (1,3) \cup (3,\infty)$, $dom(g \circ f) = (-\infty,3) \cup (3,\infty)$

15.

$$(f \circ g)(x) = \left|\frac{1}{x-1}\right| = \frac{1}{|x-1|}$$

$$(g \circ f)(x) = \frac{1}{|x|-1}$$

Domains:

$dom(f \circ g) = (-\infty,1) \cup (1,\infty)$

$dom(g \circ f) = (-\infty,-1) \cup (-1,1) \cup (1,\infty)$

16.

$$(f \circ g)(x) = \left|\frac{1}{x} - 1\right| = \left|\frac{1-x}{x}\right|$$

$$(g \circ f)(x) = \frac{1}{|x-1|}$$

Domains:

$dom(f \circ g) = (-\infty,0) \cup (0,\infty)$

$dom(g \circ f) = (-\infty,0) \cup (0,1) \cup (1,\infty)$

17.

$$(f \circ g)(x) = \sqrt{(x+5)-1} = \sqrt{x+4}$$

$$(g \circ f)(x) = \sqrt{x-1} + 5$$

Domains:

$dom(f \circ g)$: Must have $x+4 \geq 0$. So, $dom(f \circ g) = [-4,\infty)$.

$dom(g \circ f)$: Must have $x-1 \geq 0$. So, $dom(g \circ f) = [1,\infty)$.

18.

$$(f \circ g)(x) = \sqrt{2-\left(x^2+2\right)} = \sqrt{2-x^2-2} = \sqrt{-x^2}$$

$$(g \circ f)(x) = \left(\sqrt{2-x}\right)^2 + 2 = 2-x+2 = 4-x$$

Domains:

$dom(f \circ g) = [0]$ since $-x^2 \geq 0$ only when $x = 0$. $dom(g \circ f) = (-\infty,2]$

19.

$$(f \circ g)(x) = \left[(x-4)^{1/3}\right]^3 + 4 = x-4+4 = x$$

$$(g \circ f)(x) = \left[\left(x^3+4\right)-4\right]^{1/3} = \left[x^3\right]^{1/3} = x$$

Domains:

$dom(f \circ g) = (-\infty,\infty) = dom(g \circ f)$

20. $(f\circ g)(x)=\sqrt[3]{\left(x^{2/3}+1\right)^2-1}=\sqrt[3]{x^{4/3}+2x^{2/3}+1-1}=\sqrt[3]{x^{4/3}+2x^{2/3}}=\sqrt[3]{x^{2/3}\left(x^{2/3}+2\right)}$ $(g\circ f)(x)=\left(\sqrt[3]{x^2-1}\right)^{2/3}+1=\left(x^2-1\right)^2+1=x^4-2x^2+1+1=x^4-2x^2+2$ Domains: $dom(f\circ g)=(-\infty,\infty)=dom(g\circ f)$	
21. $(f+g)(2)=f(2)+g(2)$ $=\left[2^2+10\right]+\sqrt{2-1}$ $=14+1=\boxed{15}$	**22.** $(f+g)(10)=f(10)+g(10)$ $=\left[10^2+10\right]+\sqrt{10-1}$ $=110+3=\boxed{113}$
23. $(f-g)(2)=f(2)-g(2)$ $=\left[2^2+10\right]-\sqrt{2-1}$ $=14-1=\boxed{13}$	**24.** $(f-g)(5)=f(5)-g(5)$ $=\left[5^2+10\right]-\sqrt{5-1}$ $=35-2=\boxed{33}$
25. $(f\cdot g)(4)=f(4)\cdot g(4)$ $=\left[4^2+10\right]\cdot\sqrt{4-1}$ $=\boxed{26\sqrt{3}}$	**26.** $(f\cdot g)(5)=f(5)\cdot g(5)$ $=\left[5^2+10\right]\cdot\sqrt{5-1}$ $=35(2)=\boxed{70}$
27. $\left(\frac{f}{g}\right)(10)=\frac{f(10)}{g(10)}=\frac{10^2+10}{\sqrt{10-1}}=\boxed{\frac{110}{3}}$	**28.** $\left(\frac{f}{g}\right)(2)=\frac{f(2)}{g(2)}=\frac{2^2+10}{\sqrt{2-1}}=\boxed{14}$
29. $f(g(2))=f\left(\underbrace{\sqrt{2-1}}_{=1}\right)=1^2+10=\boxed{11}$	**30.** $f(g(1))=f\left(\underbrace{\sqrt{1-1}}_{=0}\right)=0^2+10=\boxed{10}$
31. $g(f(-3))=g\left(\underbrace{(-3)^2+10}_{=19}\right)=\sqrt{19-1}=\boxed{3\sqrt{2}}$	**32.** $g(f(4))=g\left(\underbrace{4^2+10}_{=26}\right)=\sqrt{26-1}=\boxed{5}$
33. 0 is not in the domain of g, so that $g(0)$ is not defined. Hence, $f(g(0))$ is $\boxed{\text{undefined}}$.	**34.** $g(f(0))=g\left(\underbrace{0^2+10}_{=10}\right)=\sqrt{10-1}=\boxed{3}$

35. $f(g(-3))$ is not defined since $g(-3)$ in not defined.	**36.** $g\left(f\left(\sqrt{7}\right)\right)=g\left(\left(\sqrt{7}\right)^2+10\right)$ $=g(17)=\sqrt{17-1}=\boxed{4}$
37. $(f\circ g)(4)=f(g(4))=f\left(\sqrt{4-1}\right)$ $=f\left(\sqrt{3}\right)=\left(\sqrt{3}\right)^2+10=\boxed{13}$	**38.** $(g\circ f)(-3)=g(f(-3))=g\left((-3)^2+10\right)$ $=g(19)=\sqrt{19-1}=\boxed{3\sqrt{2}}$
39. $f(g(1))=f\left(\underbrace{2(1)+1}_{=3}\right)=\boxed{\frac{1}{3}}$ $g(f(2))=g\left(\frac{1}{2}\right)=2\left(\frac{1}{2}\right)+1=\boxed{2}$	**40.** $f(g(1))=f\left(\frac{1}{\underbrace{2-1}_{=1}}\right)=1^2+1=\boxed{2}$ $g(f(2))=g\left(\underbrace{2^2+1}_{=5}\right)=\frac{1}{2-5}=\boxed{-\frac{1}{3}}$
41. $f(g(1))=f\left(\underbrace{1^2+2}_{=3}\right)$ Since 3 is not in the domain of *f*, this is $\boxed{\text{undefined}}$. Likewise, $g(f(2))$ is $\boxed{\text{undefined}}$ since 2 is not in the domain of *f*.	**42.** $f(g(1))=f\left(\underbrace{1^2+1}_{=2}\right)=\sqrt{3-2}=\boxed{1}$ $g(f(2))=g\left(\underbrace{\sqrt{3-2}}_{=1}\right)=1^2+1=\boxed{2}$
43. $f(g(1))=f\left(\underbrace{1+3}_{=4}\right)=\frac{1}{\lvert 4-1\rvert}=\boxed{\frac{1}{3}}$ $g(f(2))=g\left(\frac{1}{\underbrace{\lvert 2-1\rvert}_{=1}}\right)=1+3=\boxed{4}$	**44.** $f(g(1))=f\left(\underbrace{\lvert 2(1)-3\rvert}_{=1}\right)=\frac{1}{1}=\boxed{1}$ $g(f(2))=g\left(\frac{1}{2}\right)=\left\lvert 2\left(\frac{1}{2}\right)-3\right\rvert=\boxed{2}$
45. $f(g(1))=f\left(\underbrace{1^2+5}_{=6}\right)=\sqrt{6-1}=\boxed{\sqrt{5}}$ $g(f(2))=g\left(\underbrace{\sqrt{2-1}}_{=1}\right)=1^2+5=\boxed{6}$	**46.** $f(g(1))=f\left(\frac{1}{\underbrace{1-3}_{=-2}}\right)=\sqrt[3]{-\frac{1}{2}-3}=\boxed{\sqrt[3]{-\frac{7}{2}}}$ $g(f(2))=g\left(\underbrace{\sqrt[3]{2-3}}_{=-1}\right)=\frac{1}{-1-3}=\boxed{-\frac{1}{4}}$
47. $f(g(1))$ is $\boxed{\text{undefined}}$ since $g(1)$ is not defined. $g(f(2))=g\left(\frac{1}{2^2-3}\right)=g(1)$, which is not defined. So, this is also $\boxed{\text{undefined}}$.	**48.** $f(g(1))=f\left(4-1^2\right)=f(3)=\frac{3}{2-3}=\boxed{-3}$ $g(f(2))$ is $\boxed{\text{undefined}}$ since $f(2)$ is not defined.

49. $f(g(1)) = f\left(1^2+2(1)+1\right) = f(4)$ $= (4-1)^{1/3} = \boxed{\sqrt[3]{3}}$ $g(f(2)) = g\left((2-1)^{1/3}\right) = g(1)$ $= 1^2+2(1)+1 = \boxed{4}$	**50.** $f(g(1)) = f\left((1-3)^{1/3}\right) = f\left((-2)^{1/3}\right)$ $= \left(1-\left((-2)^{1/3}\right)^2\right)^{1/2} = \left(\underbrace{1-2^{2/3}}_{<0}\right)^{1/2}$, which is $\boxed{\text{undefined}}$ $g(f(2))$ is $\boxed{\text{undefined}}$ since $f(2)$ is not defined.
51. $f(g(x)) = \not{2}\left(\dfrac{x-1}{\not{2}}\right)+1 = x-1+1 = x$ $g(f(x)) = \dfrac{(2x+1)-1}{2} = \dfrac{2x}{2} = x$	**52.** $f(g(x)) = \dfrac{(3x+2)-2}{3} = \dfrac{3x}{3} = x$ $g(f(x)) = \not{3}\left(\dfrac{x-2}{\not{3}}\right)+2 = x-2+2 = x$
53. $f(g(x)) = \sqrt{\left(x^2+1\right)-1} = \sqrt{x^2} = \underbrace{\lvert x\rvert = x}_{\text{Since } x\ge 1}$ $g(f(x)) = \left(\sqrt{x-1}\right)^2+1 = (x-1)+1 = x$	**54.** $f(g(x)) = 2-\left(\sqrt{2-x}\right)^2 = 2-(2-x)$ $= 2-2+x = x$ $g(f(x)) = \sqrt{2-\left(2-x^2\right)} = \sqrt{2-2+x^2} = \sqrt{x^2} = x$
55. $f(g(x)) = \dfrac{1}{\frac{1}{x}} = x \quad g(f(x)) = \dfrac{1}{\frac{1}{x}} = x$	**56.** $f(g(x)) = \left[5-\left(5-x^3\right)\right]^{1/3} = \left[5-5+x^3\right]^{1/3}$ $= \left[x^3\right]^{1/3} = x$ $g(f(x)) = 5-\left[(5-x)^{1/3}\right]^3 = 5-(5-x)$ $= 5-5+x = x$

57.

$$f(g(x)) = 4\left(\frac{\sqrt{x+9}}{2}\right)^2 - 9 = 4\left(\frac{x+9}{4}\right) - 9 = x$$

$$g(f(x)) = \frac{\sqrt{\left(4x^2-9\right)+9}}{2} = \frac{\sqrt{4x^2}}{2} = \frac{2x}{2} = x$$

58. $$f(g(x)) = \sqrt[3]{8\left(\frac{x^3+1}{8}\right)-1} = \sqrt[3]{x^3} = x$$ $$g(f(x)) = \frac{\left(\sqrt[3]{8x-1}\right)^3+1}{8} = \frac{8x-1+1}{8} = x$$	
59. $f(g(x)) = \frac{1}{\frac{x+1}{x}-1} = \frac{1}{\frac{x+1-x}{x}} = \frac{1}{\frac{1}{x}} = x \quad g(f(x)) = \frac{\frac{1}{x-1}+1}{\frac{1}{x-1}} = \frac{\frac{1+x-1}{x-1}}{\frac{1}{x-1}} = \frac{\frac{x}{\cancel{x-1}}}{\frac{1}{\cancel{x-1}}} = x$	
60. $f(g(x)) = g(f(x)) = \sqrt{25-\left(\sqrt{25-x^2}\right)^2} = \sqrt{25-\left(25-x^2\right)} = \sqrt{x^2} = x$ since $x \geq 0$.	
61. $f(x) = 2x^2+5x \quad g(x) = 3x-1$	**62.** The most natural pairs are: $f(x) = \frac{1}{x} \quad g(x) = x^2+1$; $f(x) = \frac{1}{x+1} \quad g(x) = x^2$
63. $f(x) = \frac{2}{\lvert x\rvert} \quad g(x) = x-3$	**64.** $f(x) = \sqrt{x} \quad g(x) = 1-x^2$
65. $f(x) = \frac{3}{\sqrt{x}-2} \quad g(x) = x+1$	**66.** $f(x) = \frac{x}{3x+2} \quad g(x) = \sqrt{x}$
67. $F(C(K)) = \frac{9}{5}\left(K-273.15\right)+32$	

68. We need to calculate the composition function $(K \circ C)(F)$.

Solve $F = \frac{9}{5}C+32$ for C: $C = \frac{5}{9}(F-32)$

Solve $C = K-273.15$ for K: $K = C+273.15$

So, $(K \circ C)(F) = K(C(F)) = K\left(\frac{5}{9}(F-32)\right) = \frac{5}{9}(F-32)+273.15 = \boxed{\frac{5}{9}F+255.37}$.

Thus, $32^\circ F$ corresponds to $\frac{5}{9}(32)+255.37 = 273.15K$, and

$212^\circ F$ corresponds to $\frac{5}{9}(212)+255.37 = 373.15K$.

69. Let x = number of linear feet of fence purchased.

a. Let l = length of each side of the square pen $\left(=\frac{x}{4}\right)$

A = area of the square pen $= l^2$.

So, $(A \circ l)(x) = A\left(\frac{x}{4}\right) = \left(\frac{x}{4}\right)^2$.

b. $A(100) = \left(\frac{100}{4}\right)^2 = 625\text{ ft}^2$

c. $A(200) = \left(\frac{200}{4}\right)^2 = 2500\text{ ft}^2$

70. Let x = number of linear feet of fence purchased. **a.** Let l = length of a radius of the circular pen $\underbrace{2\pi l}_{\text{Circumference}} = x$ so that $l = \frac{x}{2\pi}$ A = area of the square pen $= \pi l^2$. So, $(A \circ l)(x) = A\left(\frac{x}{2\pi}\right) = \pi\left(\frac{x}{2\pi}\right)^2 = \frac{x^2}{4\pi}$. **b.** $A(100) = \frac{100^2}{4\pi} = \frac{2500}{\pi}$ **c.** $A(200) = \frac{200^2}{4\pi} = \frac{10,000}{\pi}$
71. First, solve $p = 3000 - \frac{1}{2}x$ for x: $x = 2(3000 - p) = 6000 - 2p$ **a.** $C(x(p)) = C(6000 - 2p) = 2000 + 10(6000 - 2p) = 62,000 - 20p$ **b.** $R(x(p)) = 100(6000 - 2p) = 600,000 - 200p$ **c.** Profit $P = R - C$. So, $P(x(p)) = R(x(p)) - C(x(p))$ $= (600,000 - 200p) - (62,000 - 20p)$ $= 538,000 - 180p$
72. First, solve $p = 10,000 - \frac{1}{4}x$ for x: $x = 4(10,000 - p) = 40,000 - 4p$ **a.** $C(x(p)) = C(40,000 - 4p) = 30,000 + 5(40,000 - 4p) = 230,000 - 20p$ **b.** $R(x(p)) = 1000(40,000 - 4p) = 40,000,000 - 4000p$ **c.** Profit $P = R - C$. So, $P(x(p)) = R(x(p)) - C(x(p))$ $= (40,000,000 - 4000p) - (230,000 - 20p)$ $= 39,770,000 - 3,980p$
73. $A(t) = \pi\left[150\sqrt{t}\right]^2 = 22,500\pi t$ ft^2
74. Volume of rectangular pool = length × width × height = $20 \times 10 \times h = 200h$. Let t = number of hours water has been pumped into the pool. Then, from the information we are given, volume = $50t$. Thus, solving $50t = 200h$ for h yields: $\underbrace{h = \frac{50t}{200} = \frac{t}{4}}_{\text{Height of water as a function of time } t}$

75. Let h = height of the fireworks above ground.

Then, the distance between the family and the fireworks is given by

$$d(h) = \sqrt{(h-0)^2 + (0-2)^2} = \sqrt{h^2 + 4}.$$

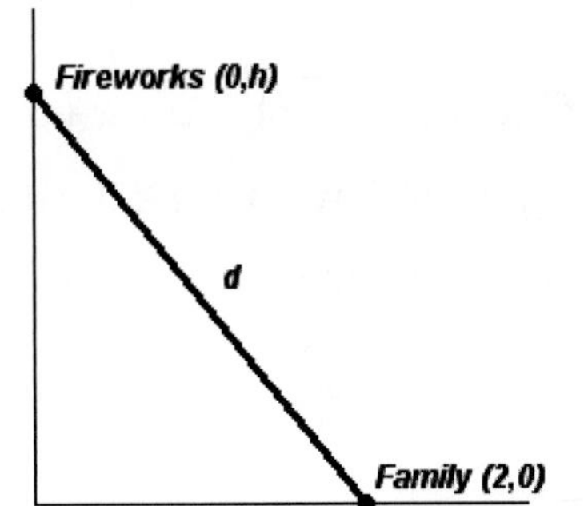

76. Let p = asking price for a home. Then, the profit is given by

$$R(p) = p - \Bigg(\underbrace{172{,}000}_{\text{Amount initially paid for home}} + \overbrace{0.6p}^{\text{Realtor Commission}} \Bigg)$$

77. Must exclude -2 from the domain.

78. Must also exclude -2 from the domain.

79. $(f \circ g)(x) = f(g(x))$, not $f(x) \cdot g(x)$

80. Domain is $[3, \infty)$

81. The mistake made was that $(f+g)$ was multiplied by 2 when it ought to have been evaluated at 2.

82. Didn't distribute " – " to all parts of $g(x)$. Should have been:

$$\begin{aligned} f(x) - g(x) &= (x+2) - \left(x^2 - 4\right) \\ &= x + 2 - x^2 + 4 \\ &= -x^2 + x + 6 \end{aligned}$$

83. False.
The domain of the sum, difference, or product of two functions is the <u>intersection</u> of their domains; the domain of the quotient is the set obtained by intersecting the two domains and then excluding all values where the denominator equals 0.

84. False.
For example, consider the functions

$$f(x) = x+1, \ g(x) = 3.$$

Observe that

$$\begin{aligned} f(g(4)) &= f(3) = 4 \\ g(f(4)) &= g(5) = 3 \end{aligned}$$

85. True

86. False

87.

$(g \circ f)(x) = \dfrac{1}{(x+a)-a} = \dfrac{1}{x}$ <u>Domain</u>: $x \neq 0, a$

88.

$$\begin{aligned} (g \circ f)(x) &= \frac{1}{\left(ax^2 + bx + c\right) - c} = \frac{1}{ax^2 + bx} \\ &= \frac{1}{x(ax+b)} \end{aligned}$$

<u>Domain</u>: $x \neq 0, -\frac{b}{a}, c$

89.

$(g \circ f)(x) = \left(\sqrt{x+a}\right)^2 - a = x + a - a = x$

Domain: Must have $x + a \geq 0$, so that $x \geq -a$. So, domain is $[-a, \infty)$.

90. $(g \circ f)(x) = \dfrac{1}{\left(\dfrac{1}{x^a}\right)^b} = \dfrac{1}{\dfrac{1}{x^{ab}}} = x^{ab}$

Domain: The domain depends on the values of a and b. For instance, if $a = 1$ and $b = 3$, then domain is $(-\infty, 0) \cup (0, \infty)$. If $a = \frac{1}{2}$ and $b = 1$, then domain is $(0, \infty)$.

91.

Notes on the graph:
The dotted curve is the graph of y_1, while the thick, solid curve is the graph of $y_1 + y_2$.

Domain of y_2 is [-7, 9].

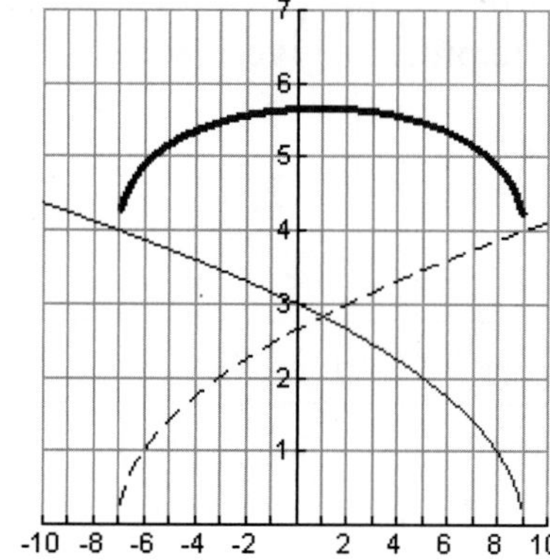

92.

$y1 = \sqrt[3]{x+5}$

$y2 = \dfrac{1}{\sqrt{3-x}}$

$y3 = \dfrac{y1}{y2}$

Domain of $y3 = \boxed{(-\infty, 3)}$

The graph of $y3$ is as follows:

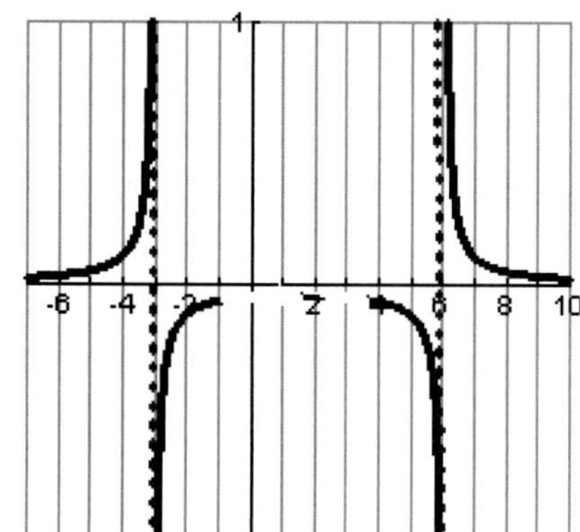

93.

$y1 = \sqrt{x^2 - 3x - 4}$

$y2 = \dfrac{1}{x^2 - 14}$

$y3 = \dfrac{1}{(y1)^2 - 14}$

Domain of $y3 = \boxed{(-\infty, 3) \cup (-3, -1] \cup [4, 6) \cup (6, \infty)}$

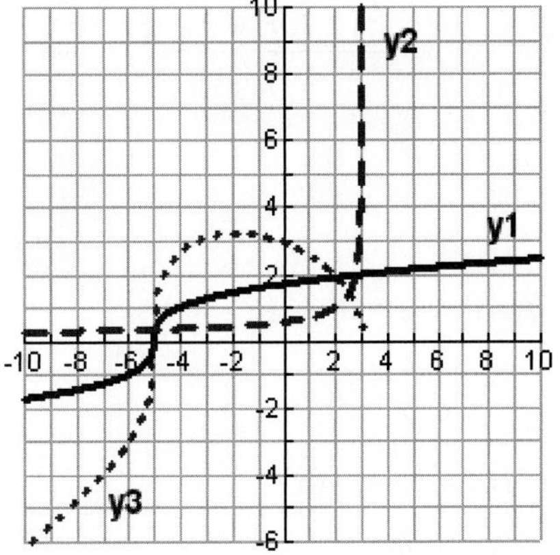

94.

<u>Notes on the graph</u>:

The dotted curve is the graph of y_1, while the thick, solid curve is the graph of $y_3 = y_1^2 + 2$.

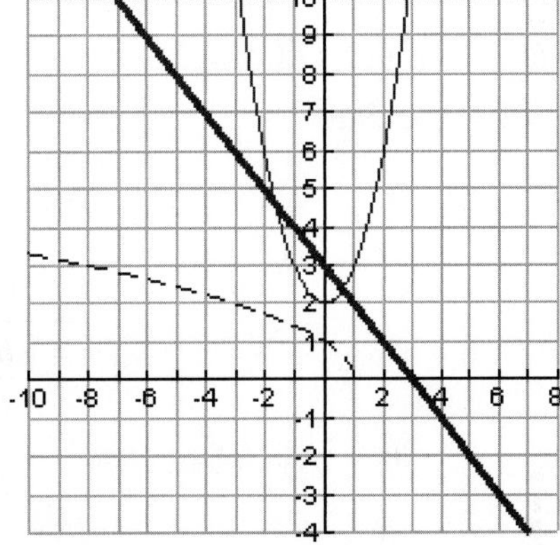

95. Observe that $\dfrac{F(x+h)-F(x)}{h} = \dfrac{(x+h)-x}{h} = \dfrac{h}{h} = 1$. So, at $h = 0$ we get $F'(x) = 1$.

Also, $\dfrac{G(x+h)-G(x)}{h} = \dfrac{(x+h)^2 - x^2}{h} = \dfrac{x^2 + 2hx + h^2 - x^2}{h} = \dfrac{h(2x+h)}{h} = 2x + h$

So, at $h = 0$ we get $G'(x) = 2x$.

Finally, observe that

$$\frac{H(x+h)-H(x)}{h} = \frac{\left[(x+h)+(x+h)^2\right]-\left[x+x^2\right]}{h}$$

$$= \frac{(x+h)-x}{h} + \frac{x^2+2hx+h^2-x^2}{h} = \frac{h}{h} + \frac{h(2x+h)}{h} = 1 + 2x + h$$

So, at $h = 0$ we get $H'(x) = 1 + 2x = F'(x) + G'(x)$. As such, we conclude that it appears as though $H'(x) = F'(x) + G'(x)$.

96. Observe that

$$\frac{F(x+h)-F(x)}{h}=\frac{\sqrt{x+h}-\sqrt{x}}{h}=\frac{\sqrt{x+h}-\sqrt{x}}{h}\cdot\frac{\sqrt{x+h}+\sqrt{x}}{\sqrt{x+h}+\sqrt{x}}$$
$$=\frac{x+h-x}{h\left(\sqrt{x+h}+\sqrt{x}\right)}=\frac{1}{\sqrt{x+h}+\sqrt{x}}$$

So, at $h=0$ we get $F'(x)=\frac{1}{2\sqrt{x}}$.

Also,

$$\frac{G(x+h)-G(x)}{h}=\frac{\left[(x+h)^3+1\right]-\left[x^3+1\right]}{h}=\frac{x^3+3x^2h+3xh^2+1+h^3-x^3-1}{h}$$
$$=\frac{h\left(3x^2+3xh+h^2\right)}{h}=\boxed{3x^2+3xh+h^2}$$

So, at $h=0$ we get $G'(x)=3x^2$.

Finally, observe that

$$\frac{H(x+h)-H(x)}{h}=\frac{\left(\sqrt{x+h}-\left[(x+h)^3+1\right]\right)-\left(\sqrt{x}-\left[x^3+1\right]\right)}{h}$$
$$=\frac{\sqrt{x+h}-\sqrt{x}}{h}-\frac{\left[(x+h)^3+1\right]-\left[x^3+1\right]}{h}$$
$$=\frac{1}{\sqrt{x+h}+\sqrt{x}}-\left(3x^2+3xh+h^2\right)$$

So, at $h=0$, we get $H'(x)=\frac{1}{2\sqrt{x}}-3x^2=F'(x)-G'(x)$. As such, we conclude that it appears as though $H'(x)=F'(x)-G'(x)$.

97. Observe that $\frac{F(x+h)-F(x)}{h}=\frac{5-5}{h}=0$. So, at $h=0$ we get $F'(x)=0$. Also,

$$\frac{G(x+h)-G(x)}{h}=\frac{\sqrt{x+h-1}-\sqrt{x-1}}{h}=\frac{\sqrt{x+h-1}-\sqrt{x-1}}{h}\cdot\frac{\sqrt{x+h-1}+\sqrt{x-1}}{\sqrt{x+h-1}+\sqrt{x-1}}$$
$$=\frac{x+h-1-x+1}{h\left(\sqrt{x+h-1}+\sqrt{x-1}\right)}=\frac{1}{\sqrt{x+h-1}+\sqrt{x-1}}$$

So, at $h=0$ we get $G'(x)=\dfrac{1}{2\sqrt{x-1}}$. Finally, observe that

$$\frac{H(x+h)-H(x)}{h}=\frac{5\sqrt{x+h-1}-5\sqrt{x-1}}{h}=5\left[\frac{\sqrt{x+h-1}-\sqrt{x-1}}{h}\right]=\frac{5}{\sqrt{x+h-1}+\sqrt{x-1}}.$$

Hence, $H'(x)=\dfrac{5}{2\sqrt{x-1}}\neq F'(x)G'(x)$. So, we conclude that it appears as though $H'(x)\neq F'(x)G'(x)$.

98. Observe that $\dfrac{F(x+h)-F(x)}{h}=\dfrac{(x+h)-x}{h}=\dfrac{h}{h}=1$. So, at $h=0$ we get $F'(x)=1$.

Also,

$$\frac{G(x+h)-G(x)}{h}=\frac{\sqrt{x+h+1}-\sqrt{x+1}}{h}=\frac{\sqrt{x+h+1}-\sqrt{x+1}}{h}\cdot\frac{\sqrt{x+h+1}+\sqrt{x+1}}{\sqrt{x+h+1}+\sqrt{x+1}}$$

$$=\frac{x+h-1-x+1}{h\left(\sqrt{x+h+1}+\sqrt{x+1}\right)}=\frac{1}{\sqrt{x+h+1}+\sqrt{x+1}}$$

So, at $h=0$ we get $G'(x)=\dfrac{1}{2\sqrt{x+1}}$. Finally, observe that

$$\frac{H(x+h)-H(x)}{h}=\frac{\dfrac{x+h}{\sqrt{x+h+1}}-\dfrac{x}{\sqrt{x+1}}}{h}=\frac{(x+h)\sqrt{x+1}-x\sqrt{x+h+1}}{h\sqrt{x+1}\sqrt{x+h+1}}$$

$$=\frac{(x+h)\sqrt{x+1}-x\sqrt{x+h+1}}{h\sqrt{x+1}\sqrt{x+h+1}}\cdot\frac{(x+h)\sqrt{x+1}+x\sqrt{x+h+1}}{(x+h)\sqrt{x+1}+x\sqrt{x+h+1}}$$

$$=\frac{(x+h)^2(x+1)-x^2(x+h+1)}{h\sqrt{x+1}\sqrt{x+h+1}\left[(x+h)\sqrt{x+1}+x\sqrt{x+h+1}\right]}$$

$$=\frac{x^3+2hx^2+h^2x+x^2+2hx+h^2-x^3-x^2h-x^2}{h\sqrt{x+1}\sqrt{x+h+1}\left[(x+h)\sqrt{x+1}+x\sqrt{x+h+1}\right]}$$

$$=\frac{\cancel{h}\left(x^2+hx+2x+h\right)}{\cancel{h}\sqrt{x+1}\sqrt{x+h+1}\left[(x+h)\sqrt{x+1}+x\sqrt{x+h+1}\right]}$$

So, $H'(x)=\dfrac{x^2+2x}{(x+1)\left[2x\sqrt{x+1}\right]}=\dfrac{x(x+2)}{2x(x+1)\sqrt{x+1}}=\dfrac{x+2}{2(x+1)\sqrt{x+1}}\neq\dfrac{F'(x)}{G'(x)}$.

Section 1.5 Solutions ---

1. Not a function since 4 maps to both 2 and -2.	**2.** Is a function. Not one-to-one 0,1,2,3 all map to 1 in the range.	**3.** Is a function. Not one-to-one $0, 2, -2$ all map to 1 in the range, for instance.
4. Is a function. One-to-one	**5.** Is a function. Not one-to-one Doesn't pass the horizontal line test. Both $(-1,1)$, $(0,1)$ are on the graph.	**6.** Is a function. Not one-to-one Doesn't pass the horizontal line test.
7. Is a function. One-to-one	**8.** Is a function. One-to-one	**9.** Is a function. Not one-to-one Doesn't pass the horizontal line test.
10. Is a function. One-to-one		

11. Not one-to-one. Both $(0,3)$, $(6,3)$ lie on the graph.

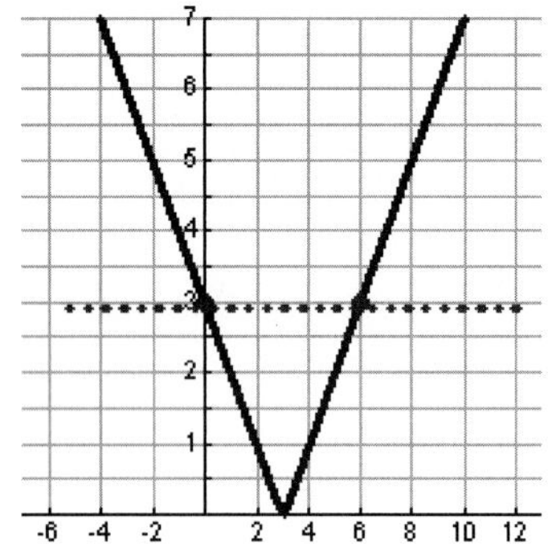

12. Not one-to-one. Both $(0,5)$, $(4,5)$ lie on the graph.

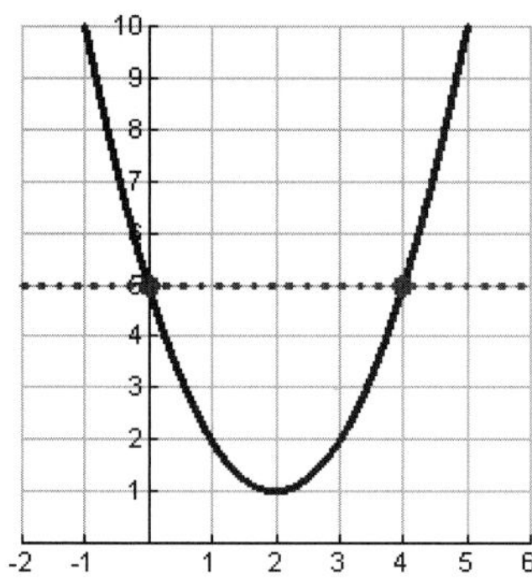

13.

$$f(x_1) = f(x_2) \Rightarrow \frac{1}{x_1 - 1} = \frac{1}{x_2 - 1}$$
$$\Rightarrow x_2 - 1 = x_1 - 1$$
$$\Rightarrow x_2 = x_1$$

One-to-one

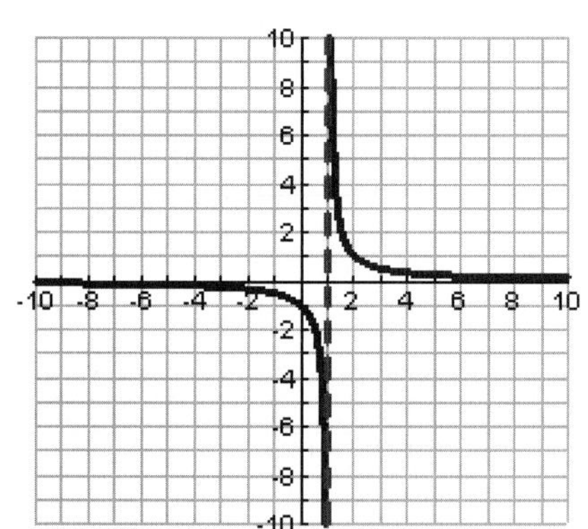

14.

$$f(x_1) = f(x_2) \Rightarrow \sqrt[3]{x_1} = \sqrt[3]{x_2}$$

$$\Rightarrow \left(\sqrt[3]{x_1}\right)^3 = \left(\sqrt[3]{x_2}\right)^3$$

$$\Rightarrow x_1 = x_2$$

One-to-one

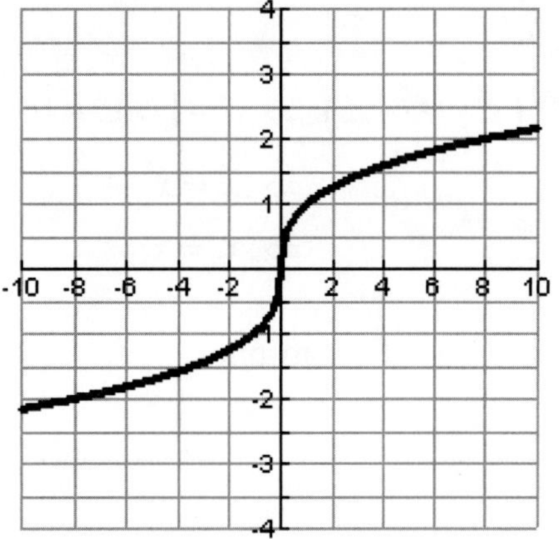

15. f is not one-to-one since, for example, $f(-1) = f(1) = -3$.

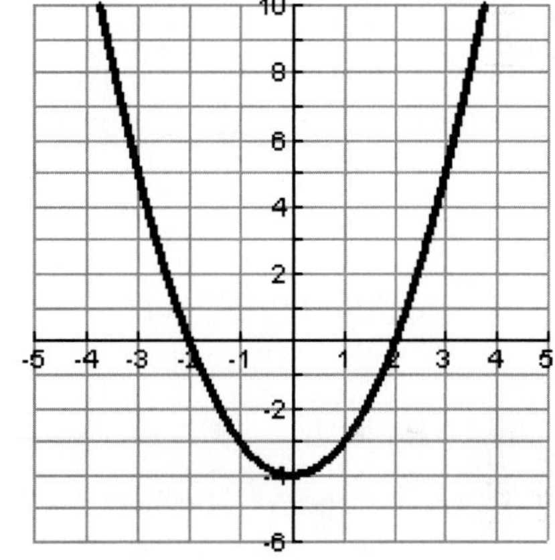

16.

$$f(x_1) = f(x_2) \Rightarrow \sqrt{x_1 + 1} = \sqrt{x_2 + 1}$$

$$\Rightarrow x_1 + 1 = x_2 + 1$$

$$\Rightarrow x_1 = x_2$$

One-to-one

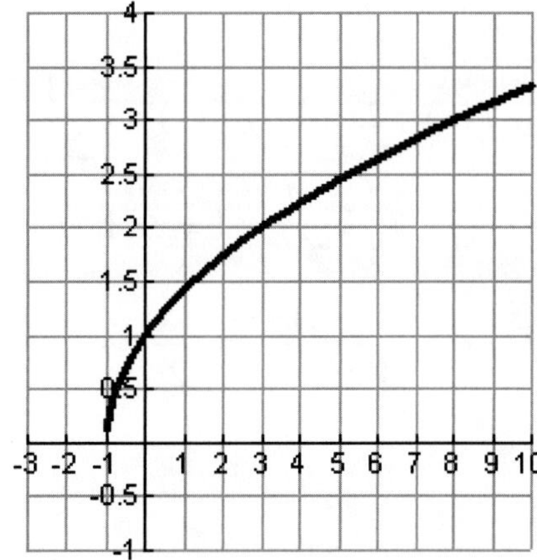

17.

$$f(x_1) = f(x_2) \Rightarrow x_1^3 - 1 = x_2^3 - 1$$
$$\Rightarrow x_1^3 = x_2^3$$
$$\Rightarrow x_1 = x_2$$

One-to-one

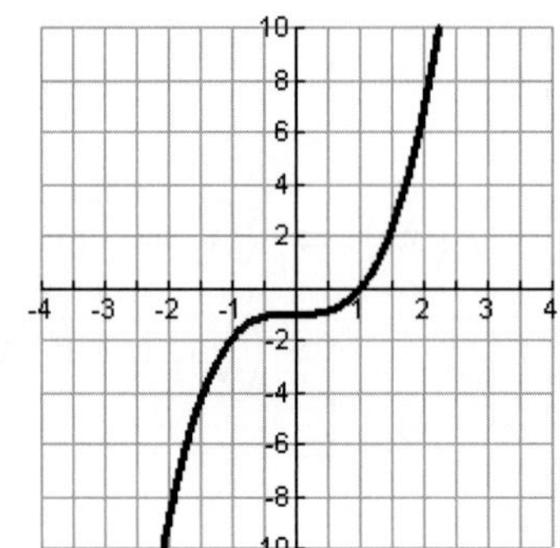

18.

$$f(x_1) = f(x_2) \Rightarrow \frac{1}{x_1+2} = \frac{1}{x_2+2}$$
$$\Rightarrow x_2 + 2 = x_1 + 2$$
$$\Rightarrow x_2 = x_1$$

One-to-one

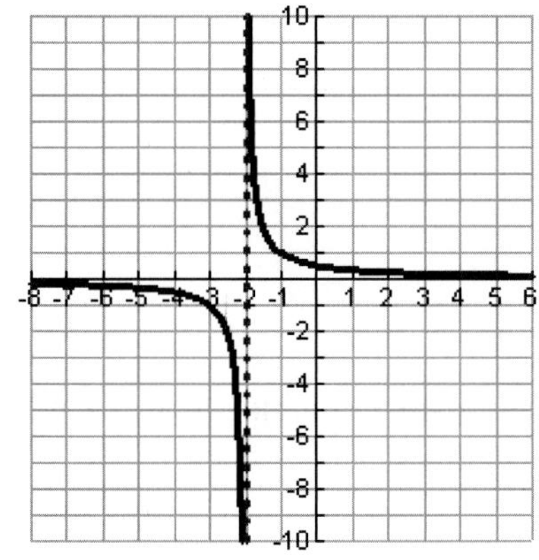

19.

Given: $f(x) = 2x+1,\ f^{-1}(x) = \dfrac{x-1}{2}$

$$f\left(f^{-1}(x)\right) = \not{2}\left(\frac{x-1}{\not{2}}\right) + 1 = x - 1 + 1 = x$$

$$f^{-1}\left(f(x)\right) = \frac{(2x+1)-1}{2} = \frac{2x}{2} = x$$

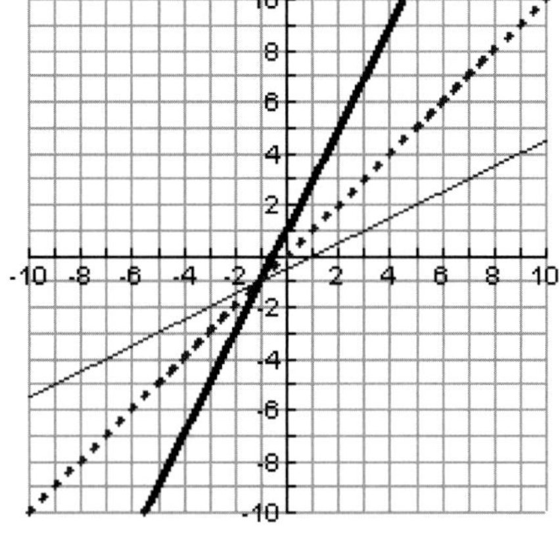

Notes on the Graphs:
Thick, solid curve is the graph of *f*.
Thin, solid curve is the graph of f^{-1}.
Thick, dotted curve is the graph of $y = x$.

20.

<u>Given</u>: $f(x)=\dfrac{x-2}{3},\ f^{-1}(x)=3x+2$

$$f\left(f^{-1}(x)\right)=\frac{(3x+2)-2}{3}=\frac{3x}{3}=x$$

$$f^{-1}\left(f(x)\right)=\cancel{3}\left(\frac{x-2}{\cancel{3}}\right)+2=x-2+2=x$$

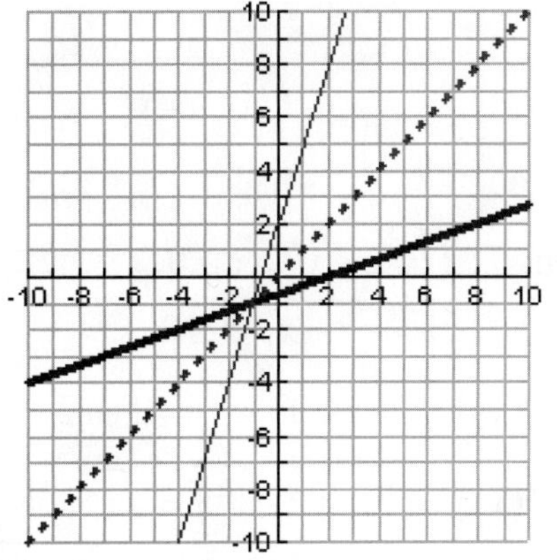

<u>Notes on the Graphs</u>:
Thick, solid curve is the graph of *f*.
Thin, solid curve is the graph of f^{-1}.
Thick, dotted curve is the graph of $y=x$.

21.

<u>Given</u>: $f(x)=\sqrt{x-1},\ x\geq 1$

$f^{-1}(x)=x^2+1,\ x\geq 0$

$$f\left(f^{-1}(x)\right)=\sqrt{\left(x^2+1\right)-1}=\sqrt{x^2}=\underbrace{|x|}_{\text{Since } x\geq 0}=x$$

$$f^{-1}\left(f(x)\right)=\left(\sqrt{x-1}\right)^2+1=x-1+1=x$$

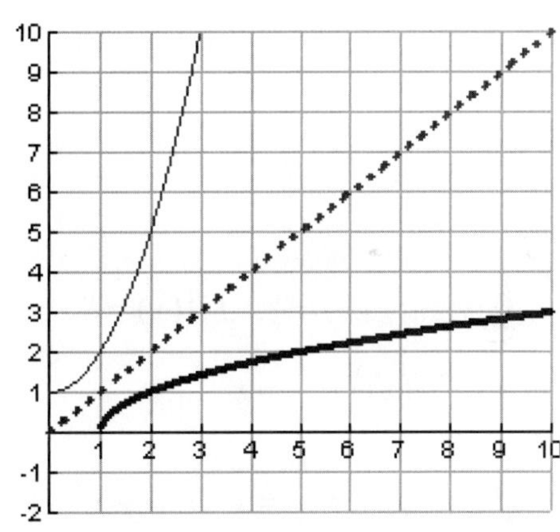

<u>Notes on the Graphs</u>:
Thick, solid curve is the graph of *f*.
Thin, solid curve is the graph of f^{-1}.
Thick, dotted curve is the graph of $y=x$.

22.

<u>Given</u>: $f(x)=2-x^2,\ x\ge 0$

$f^{-1}(x)=\sqrt{2-x},\ x\le 2$

$$f\left(f^{-1}(x)\right)=2-\left(\sqrt{2-x}\right)^2=2-(2-x)=x$$

$$f^{-1}\left(f(x)\right)=\sqrt{2-\left(2-x^2\right)}=\sqrt{x^2}=\underbrace{|x|=x}_{\text{Since } x\ge 0}$$

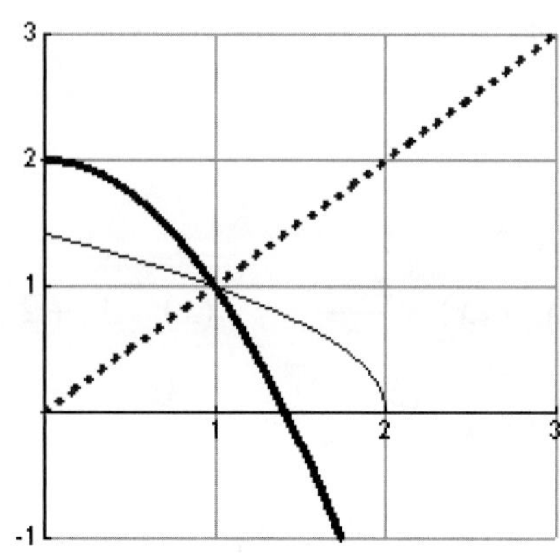

<u>Notes on the Graphs</u>:
Thick, solid curve is the graph of *f*.
Thin, solid curve is the graph of f^{-1}.
Thick, dotted curve is the graph of $y=x$.

23.

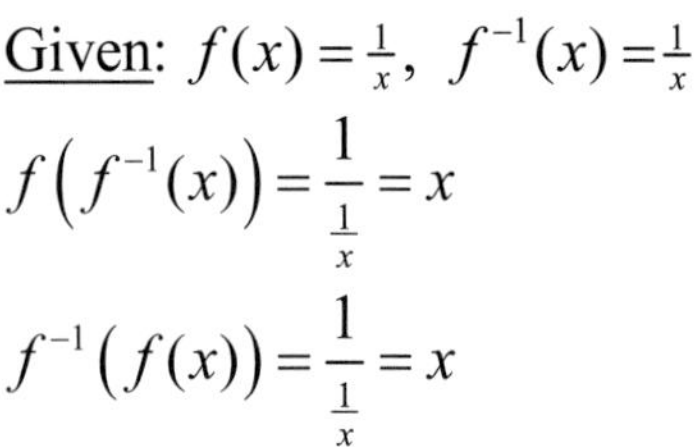

<u>Given</u>: $f(x)=\frac{1}{x},\ f^{-1}(x)=\frac{1}{x}$

$$f\left(f^{-1}(x)\right)=\frac{1}{\frac{1}{x}}=x$$

$$f^{-1}\left(f(x)\right)=\frac{1}{\frac{1}{x}}=x$$

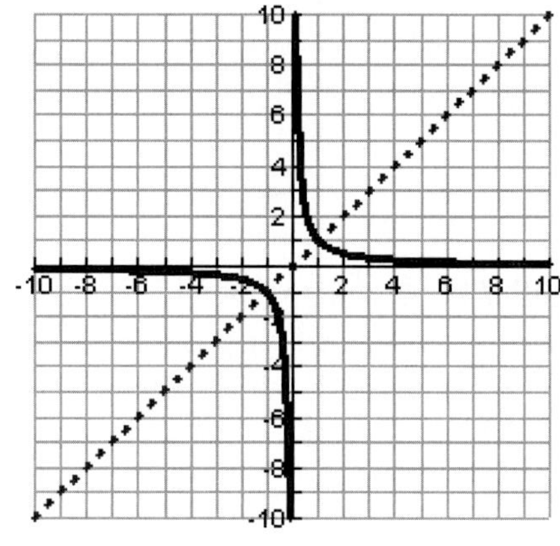

<u>Notes on the Graphs</u>:
Thick, solid curve is the graph of *f*.
Thin, solid curve is the graph of f^{-1}.
(Note: These curves overlap since the functions are the same.)
Thick, dotted curve is the graph of $y=x$.

24.

<u>Given</u>: $f(x)=(5-x)^{1/3}$, $f^{-1}(x)=5-x^3$

$$f\left(f^{-1}(x)\right)=(5-(5-x^3))^{1/3}=(5-5+x^3)^{1/3}$$
$$=(x^3)^{1/3}=x$$
$$f^{-1}\left(f(x)\right)=5-\left[(5-x)^{1/3}\right]^3=5-(5-x)=x$$

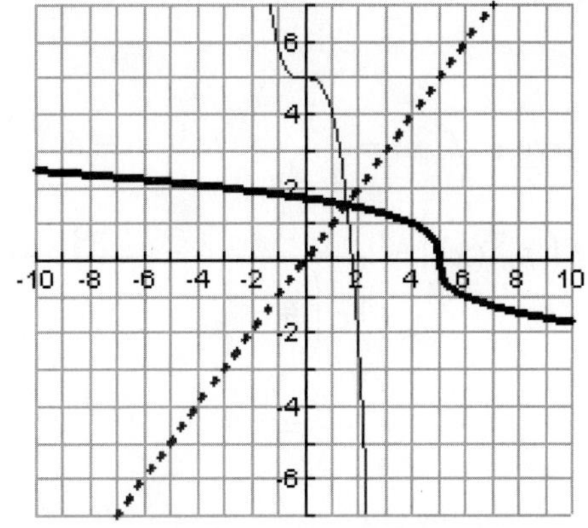

<u>Notes on the Graphs</u>:
Thick, solid curve is the graph of *f*.
Thin, solid curve is the graph of f^{-1}.
Thick, dotted curve is the graph of $y=x$.

25.

<u>Given</u>: $f(x)=\dfrac{1}{2x+6}$, $f^{-1}(x)=\dfrac{1}{2x}-3$

$$f\left(f^{-1}(x)\right)=\frac{1}{2\left(\frac{1}{2x}-3\right)+6}=\frac{1}{\frac{1}{x}-6+6}$$
$$=\frac{1}{\frac{1}{x}}=x$$
$$f^{-1}\left(f(x)\right)=\frac{1}{2\left[\frac{1}{2x+6}\right]}-3=\frac{1}{\frac{1}{x+3}}-3$$
$$=x+3-3=x$$

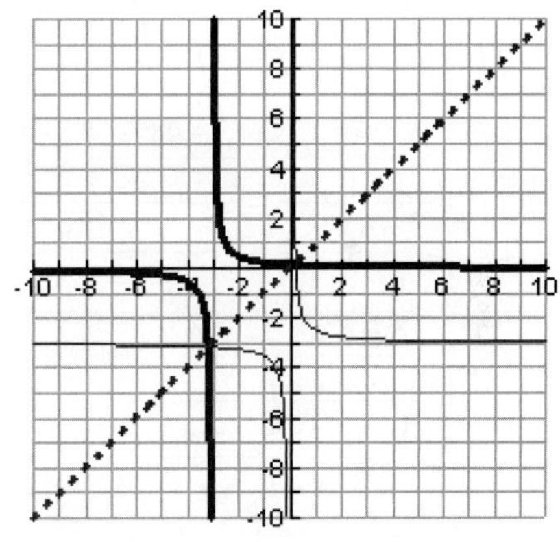

<u>Notes on the Graphs</u>:
Thick, solid curve is the graph of *f*.
Thin, solid curve is the graph of f^{-1}.
Thick, dotted curve is the graph of $y=x$.

26.

<u>Given</u>: $f(x)=\dfrac{3}{4-x}$, $f^{-1}(x)=4-\dfrac{3}{x}$

$$f\left(f^{-1}(x)\right)=\frac{3}{4-\left(4-\frac{3}{x}\right)}=\frac{3}{4-4+\frac{3}{x}}$$

$$=3\cdot\frac{x}{3}=x$$

$$f^{-1}\left(f(x)\right)=4-\frac{3}{\frac{3}{4-x}}=4-\cancel{3}\cdot\frac{4-x}{\cancel{3}}$$

$$=4-4+x=x$$

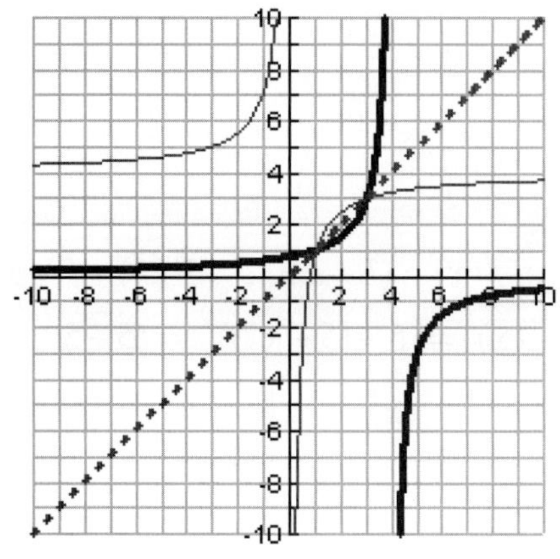

<u>Notes on the Graphs</u>:
Thick, solid curve is the graph of *f*.
Thin, solid curve is the graph of f^{-1}.
Thick, dotted curve is the graph of $y=x$.

27.

<u>Given</u>: $f(x)=\dfrac{x+3}{x+4}$, $f^{-1}(x)=\dfrac{3-4x}{x-1}$

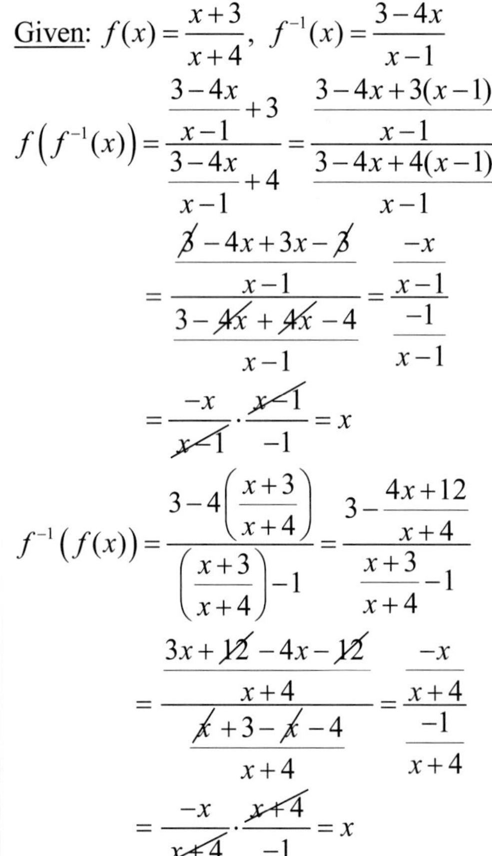

$$f\left(f^{-1}(x)\right)=\frac{\frac{3-4x}{x-1}+3}{\frac{3-4x}{x-1}+4}=\frac{\frac{3-4x+3(x-1)}{x-1}}{\frac{3-4x+4(x-1)}{x-1}}$$

$$=\frac{\frac{\cancel{3}-4x+3x-\cancel{3}}{x-1}}{\frac{3-\cancel{4x}+\cancel{4x}-4}{x-1}}=\frac{\frac{-x}{x-1}}{\frac{-1}{x-1}}$$

$$=\frac{-x}{\cancel{x-1}}\cdot\frac{\cancel{x-1}}{-1}=x$$

$$f^{-1}\left(f(x)\right)=\frac{3-4\left(\frac{x+3}{x+4}\right)}{\left(\frac{x+3}{x+4}\right)-1}=\frac{3-\frac{4x+12}{x+4}}{\frac{x+3}{x+4}-1}$$

$$=\frac{\frac{3x+\cancel{12}-4x-\cancel{12}}{x+4}}{\frac{\cancel{x}+3-\cancel{x}-4}{x+4}}=\frac{\frac{-x}{x+4}}{\frac{-1}{x+4}}$$

$$=\frac{-x}{\cancel{x+4}}\cdot\frac{\cancel{x+4}}{-1}=x$$

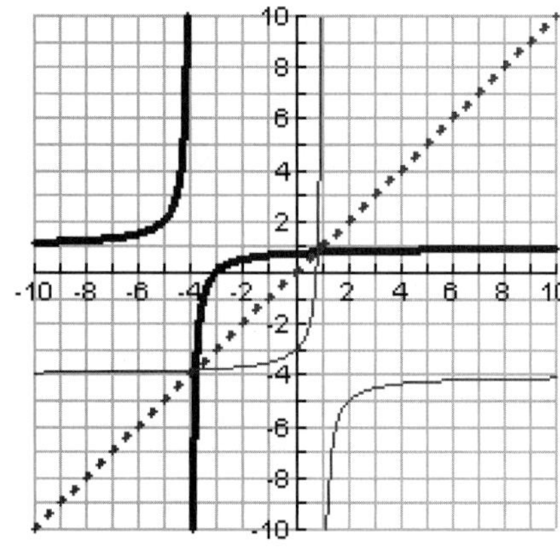

<u>Notes on the Graphs</u>:
Thick, solid curve is the graph of *f*.
Thin, solid curve is the graph of f^{-1}.
Thick, dotted curve is the graph of $y=x$.

28.

Given: $f(x)=\dfrac{x-5}{3-x},\ f^{-1}(x)=\dfrac{3x+5}{x+1}$

$$f\left(f^{-1}(x)\right)=\frac{\frac{3x+5}{x+1}-5}{3-\frac{3x+5}{x+1}}=\frac{\frac{3x+5-5(x+1)}{x+1}}{\frac{3x+3-(3x+5)}{x+1}}$$

$$=\frac{\frac{3x+\cancel{5}-5x-\cancel{5}}{x+1}}{\frac{\cancel{3x}+3-\cancel{3x}-5}{x+1}}=\frac{\frac{-2x}{x+1}}{\frac{-2}{x+1}}$$

$$=\frac{-2x}{\cancel{x+1}}\cdot\frac{\cancel{x+1}}{-2}=x$$

$$f^{-1}\left(f(x)\right)=\frac{3\left(\frac{x-5}{3-x}\right)+5}{\left(\frac{x-5}{3-x}\right)+1}=\frac{\frac{3x-15}{3-x}+5}{\frac{x-5}{3-x}+1}$$

$$=\frac{\frac{3x-15+5(3-x)}{3-x}}{\frac{x-5+3-x}{3-x}}=\frac{\frac{3x-\cancel{15}+\cancel{15}-5x}{3-x}}{\frac{\cancel{x}-5+3-\cancel{x}}{3-x}}$$

$$=\frac{-2x}{\cancel{3-x}}\cdot\frac{\cancel{3-x}}{-2}=x$$

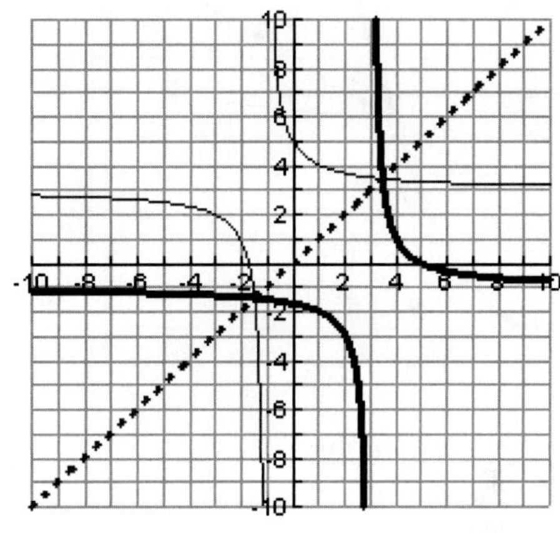

Notes on the Graphs:
Thick, solid curve is the graph of *f*.
Thin, solid curve is the graph of f^{-1}.
Thick, dotted curve is the graph of $y=x$.

29.

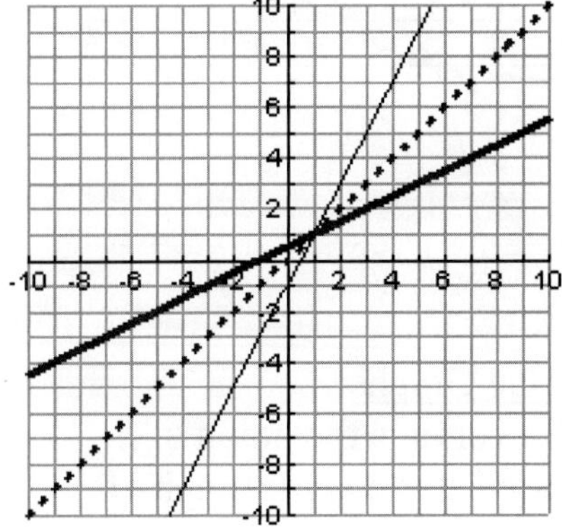

Notes on the Graphs:
Thin, solid curve is the graph of *f*.
Thick, dotted curve is the graph of $y=x$.
Thick, solid curve is the graph of f^{-1}.

30.

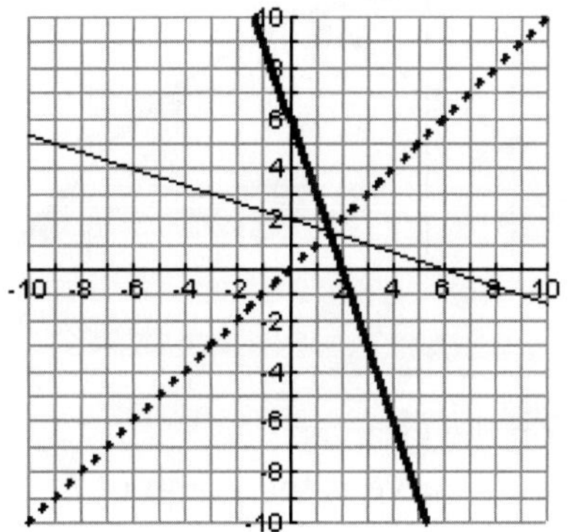

Notes on the Graphs:
Thin, solid curve is the graph of *f*.
Thick, dotted curve is the graph of $y=x$.
Thick, solid curve is the graph of f^{-1}.

31.

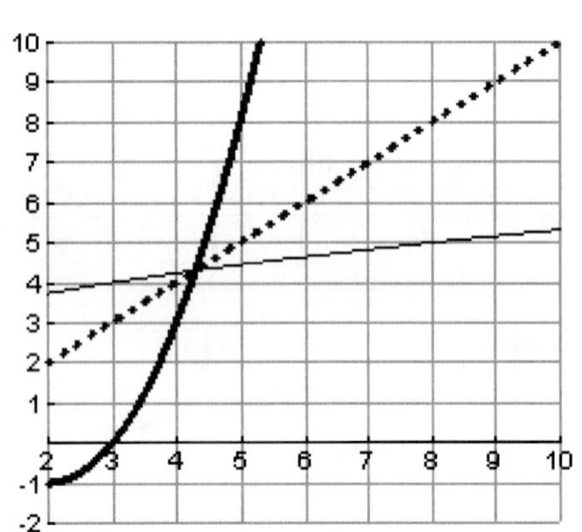

Notes on the Graphs:
Thin, solid curve is the graph of *f*.
Thick, dotted curve is the graph of $y = x$.
Thick, solid curve is the graph of f^{-1}.

32.

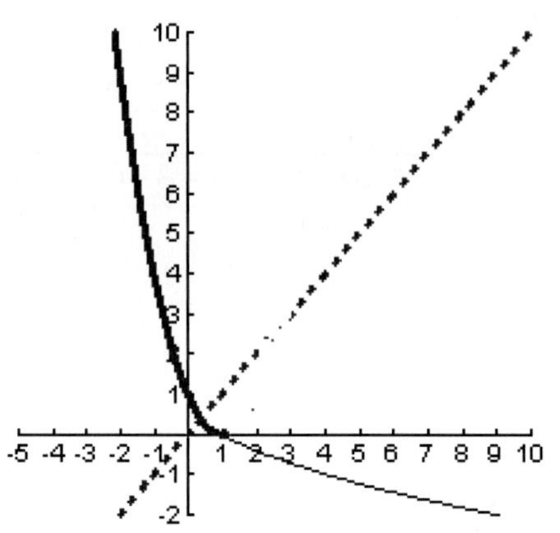

Notes on the Graphs:
Thin, solid curve is the graph of *f*.
Thick, dotted curve is the graph of $y = x$.
Thick, solid curve is the graph of f^{-1}.

33.

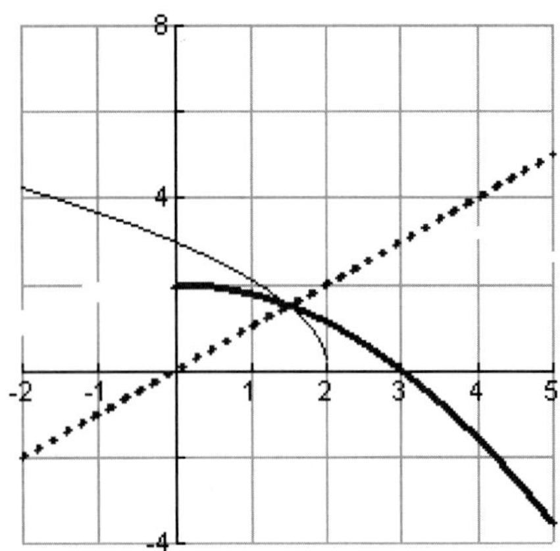

Notes on the Graphs:
Thin, solid curve is the graph of *f*.
Thick, dotted curve is the graph of $y = x$.
Thick, solid curve is the graph of f^{-1}.

34.

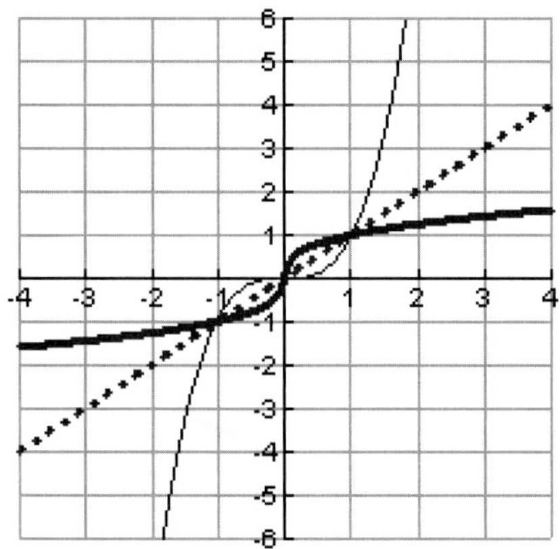

Notes on the Graphs:
Thin, solid curve is the graph of *f*.
Thick, dotted curve is the graph of $y = x$.
Thick, solid curve is the graph of f^{-1}.

35.	**36.**
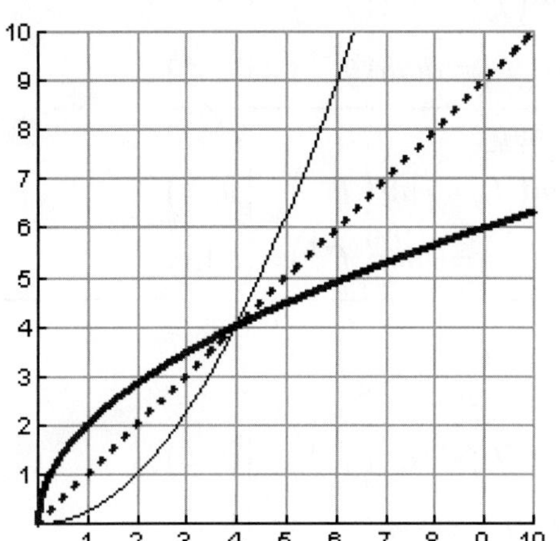	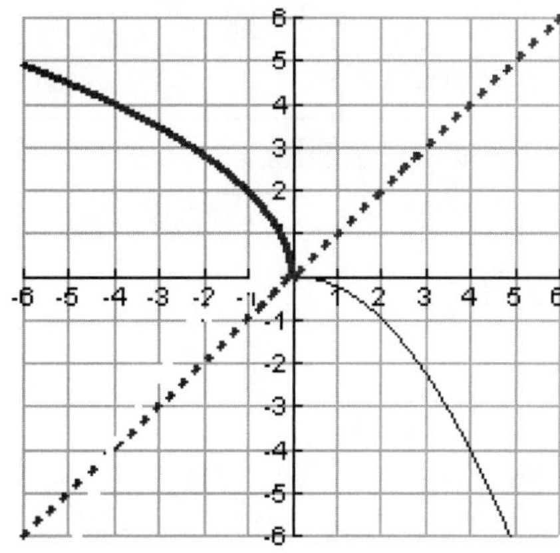
Notes on the Graphs: Thin, solid curve is the graph of *f*. Thick, dotted curve is the graph of $y = x$. Thick, solid curve is the graph of f^{-1}.	Notes on the Graphs: Thin, solid curve is the graph of *f*. Thick, dotted curve is the graph of $y = x$. Thick, solid curve is the graph of f^{-1}.
37. Solve $y = -3x + 2$ for *x*: $x = -\frac{1}{3}(y-2)$ Thus, $f^{-1}(x) = -\frac{1}{3}(x-2) = -\frac{1}{3}x + \frac{2}{3}$.	Domains: $dom(f) = rng(f^{-1}) = (-\infty, \infty)$ $rng(f) = dom(f^{-1}) = (-\infty, \infty)$
38. Solve $y = 2x + 3$ for *x*: $x = \frac{1}{2}(y-3)$ Thus, $f^{-1}(x) = \frac{1}{2}(x-3)$.	Domains: $dom(f) = rng(f^{-1}) = (-\infty, \infty)$ $rng(f) = dom(f^{-1}) = (-\infty, \infty)$
39. Solve $y = x^3 + 1$ for *x*: $x = \sqrt[3]{y-1}$ Thus, $f^{-1}(x) = \sqrt[3]{x-1}$.	Domains: $dom(f) = rng(f^{-1}) = (-\infty, \infty)$ $rng(f) = dom(f^{-1}) = (-\infty, \infty)$
40. Solve $y = x^3 - 1$ for *x*: $x = \sqrt[3]{y+1}$ Thus, $f^{-1}(x) = \sqrt[3]{x+1}$.	Domains: $dom(f) = rng(f^{-1}) = (-\infty, \infty)$ $rng(f) = dom(f^{-1}) = (-\infty, \infty)$
41. Solve $y = \sqrt{x-3}$ for *x*: $x = y^2 + 3$ Thus, $f^{-1}(x) = x^2 + 3$.	Domains: $dom(f) = rng(f^{-1}) = [3, \infty)$ $rng(f) = dom(f^{-1}) = [0, \infty)$

42. Solve $y=\sqrt{3-x}$ for x: $$x=3-y^2$$ Thus, $f^{-1}(x)=3-x^2$.	<u>Domains</u>: $dom(f)=rng(f^{-1})=(-\infty,3]$ $rng(f)=dom(f^{-1})=[0,\infty)$
43. Solve $y=x^2-1$ for x: $$x=\sqrt{y+1}$$ Thus, $f^{-1}(x)=\sqrt{x+1}$.	<u>Domains</u>: $dom(f)=rng(f^{-1})=[0,\infty)$ $rng(f)=dom(f^{-1})=[-1,\infty)$
44. Solve $y=2x^2+1$ for x: $$2x^2=y-1$$ $$x=+\sqrt{\frac{y-1}{2}}\text{ (since } x\ge 0)$$ Thus, $f^{-1}(x)=\sqrt{\frac{x-1}{2}}$.	<u>Domains</u>: $dom(f)=rng(f^{-1})=[0,\infty)$ $rng(f)=dom(f^{-1})=[1,\infty)$
45. Solve $y=(x+2)^2-3$ for x: $$y+3=(x+2)^2$$ $$\sqrt{y+3}=x+2\text{ (since } x\ge -2)$$ $$-2+\sqrt{y+3}=x$$ Thus, $f^{-1}(x)=-2+\sqrt{x+3}$.	<u>Domains</u>: $dom(f)=rng(f^{-1})=[-2,\infty)$ $rng(f)=dom(f^{-1})=[-3,\infty)$
46. Solve $y=(x-3)^2-2$ for x: $$y+2=(x-3)^2$$ $$\sqrt{y+2}=x-3\text{ (since } x\ge 3)$$ $$3+\sqrt{y+2}=x$$ Thus, $f^{-1}(x)=3+\sqrt{x+2}$.	<u>Domains</u>: $dom(f)=rng(f^{-1})=[3,\infty)$ $rng(f)=dom(f^{-1})=[-2,\infty)$
47. Solve $y=\frac{2}{x}$ for x: $$xy=2$$ $$x=\frac{2}{y}$$ Thus, $f^{-1}(x)=\frac{2}{x}$.	<u>Domains</u>: $dom(f)=rng(f^{-1})=(-\infty,0)\cup(0,\infty)$ $rng(f)=dom(f^{-1})=(-\infty,0)\cup(0,\infty)$
48. Solve $y=-\frac{3}{x}$ for x: $$yx=-3$$ $$x=-\frac{3}{y}$$ Thus, $f^{-1}(x)=-\frac{3}{x}$.	<u>Domains</u>: $dom(f)=rng(f^{-1})=(-\infty,0)\cup(0,\infty)$ $rng(f)=dom(f^{-1})=(-\infty,0)\cup(0,\infty)$

49. Solve $y = \frac{2}{3-x}$ for x: $$(3-x)y = 2$$ $$3y - xy = 2$$ $$xy = 3y - 2$$ $$x = \frac{3y-2}{y}$$ Thus, $f^{-1}(x) = \frac{3x-2}{x} = 3 - \frac{2}{x}$.	Domains: $dom(f) = rng(f^{-1}) = (-\infty,3) \cup (3,\infty)$ $rng(f) = dom(f^{-1}) = (-\infty,0) \cup (0,\infty)$
50. Solve $y = \frac{7}{x+2}$ for x: $$(x+2)y = 7$$ $$2y + xy = 7$$ $$xy = 7 - 2y$$ $$x = \frac{7-2y}{y}$$ Thus, $f^{-1}(x) = \frac{7-2x}{x}$.	Domains: $dom(f) = rng(f^{-1}) = (-\infty,-2) \cup (-2,\infty)$ $rng(f) = dom(f^{-1}) = (-\infty,0) \cup (0,\infty)$
51. Solve $y = \frac{7x+1}{5-x}$ for x: $$y(5-x) = 7x+1$$ $$5y - xy = 7x+1$$ $$-7x - xy = 1 - 5y$$ $$-x(7+y) = 1 - 5y$$ $$x = \frac{5y-1}{7+y}$$ Thus, $f^{-1}(x) = \frac{5x-1}{x+7}$.	Domains: $dom(f) = rng(f^{-1}) = (-\infty,5) \cup (5,\infty)$ $rng(f) = dom(f^{-1}) = (-\infty,-7) \cup (-7,\infty)$
52. Solve $y = \frac{2x+5}{7+x}$ for x: $$y(7+x) = 2x+5$$ $$xy + 7y = 2x+5$$ $$-2x + xy = 5 - 7y$$ $$x(y-2) = 5 - 7y$$ $$x = \frac{5-7y}{y-2}$$ Thus, $f^{-1}(x) = \frac{5-7x}{x-2}$.	Domains: $dom(f) = rng(f^{-1}) = (-\infty,-7) \cup (-7,\infty)$ $rng(f) = dom(f^{-1}) = (-\infty,2) \cup (2,\infty)$

53. Solve $y = \frac{1}{\sqrt{x}}$ for x:

$$y = \frac{1}{\sqrt{x}}$$
$$y\sqrt{x} = 1$$
$$\sqrt{x} = \frac{1}{y},\ y > 0$$
$$x = \frac{1}{y^2},\ y > 0$$

Thus, $f^{-1}(x) = \frac{1}{x^2}$

Domains:
$dom(f) = rng(f^{-1}) = (0, \infty)$
$rng(f) = dom(f^{-1}) = (0, \infty)$

54. Solve $y = \frac{x}{\sqrt{x+1}}$, $x > -1$, for x:

$$y = \frac{x}{\sqrt{x+1}}$$
$$y\sqrt{x+1} = x$$
$$y^2(x+1) = x^2$$
$$y^2x + y^2 = x^2$$
$$x^2 - xy^2 - y^2 = 0$$
$$x = \frac{y^2 \pm \sqrt{y^4 + 4y^2}}{2}$$
$$x = \frac{y^2 \pm y\sqrt{y^2 + 4}}{2}$$

Since $y\sqrt{y^2+4} > y^2$, we know that eventually $y^2 - y\sqrt{y^2+4} \le -1$, which cannot occur because of the initial restriction on x. Hence,

$$f^{-1}(x) = \frac{x^2 \pm x\sqrt{x^2+4}}{2}$$

Domains:
$dom(f) = rng(f^{-1}) = (-1, \infty)$
$rng(f) = dom(f^{-1}) = (-\infty, \infty)$

55. Solve $y=\sqrt{\frac{x+1}{x-2}}$ for *x*: $$y=\sqrt{\frac{x+1}{x-2}}$$ $$y^2=\frac{x+1}{x-2}$$ $$y^2(x-2)=x+1$$ $$x(y^2-1)=1+2y^2$$ $$x=\frac{2y^2+1}{y^2-1}$$ So, $f^{-1}(x)=\frac{2x^2+1}{x^2-1}$.	Note: We must have $\frac{x+1}{x-2}\geq 0$ for domain of *f*. Domains: $dom(f)=rng(f^{-1})=(-\infty,-1]\cup(2,\infty)$ $rng(f)=dom(f^{-1})=(-\infty,-1)\cup(-1,1)\cup(1,\infty)$
56. Solve $y=\sqrt{x^2-1},\ x\geq 1$ for *x*: $$y=\sqrt{x^2-1},\ x\geq 1$$ $$y^2=x^2-1$$ $$y^2+1=x^2$$ $$x=\pm\sqrt{y^2+1}$$ Since $y\geq 0$, we conclude that $f^{-1}(x)=\sqrt{x^2+1}$.	Domains: $dom(f)=rng(f^{-1})=[1,\infty)$ $rng(f)=dom(f^{-1})=[0,\infty)$

57. Not one-to-one

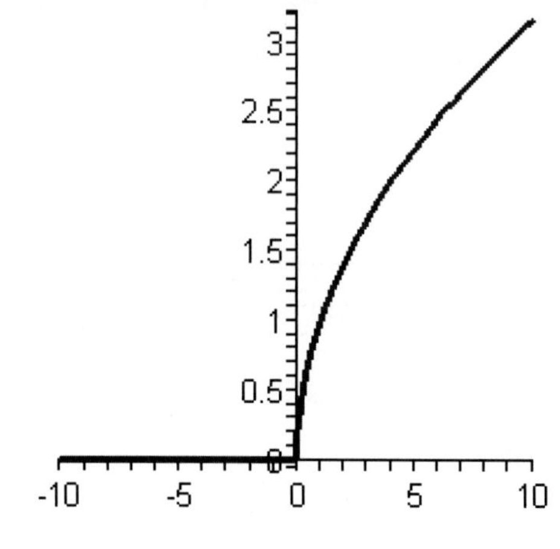

58. One-to-one

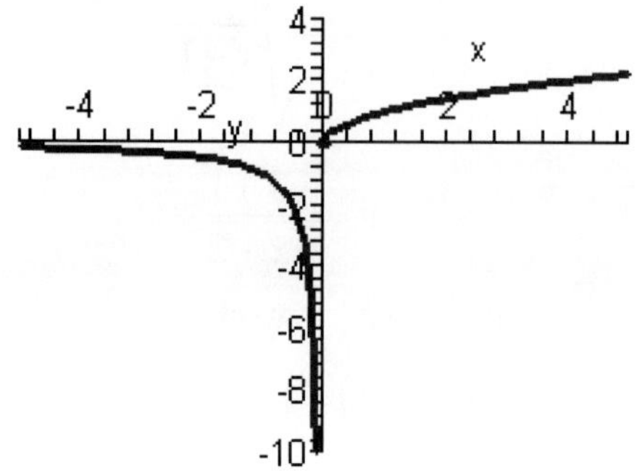

Calculate the inverse function piecewise:

For $x<0$: Solve $y=\frac{1}{x}$ for x:

$$y=\tfrac{1}{x}$$
$$xy=1$$
$$x=\tfrac{1}{y}$$

So, $G^{-1}(x)=\frac{1}{x}$ on $(-\infty,0)$.

For $x\geq 0$: Solve $y=\sqrt{x}$ for x:

$$y=\sqrt{x}$$
$$x=y^2$$

So, $G^{-1}(x)=x^2$ on $(0,\infty)$.

Thus, the inverse function is given by:

$$G^{-1}(x)=\begin{cases}\frac{1}{x}, & x<0\\ x^2, & x\geq 0\end{cases}$$

59. One-to-one

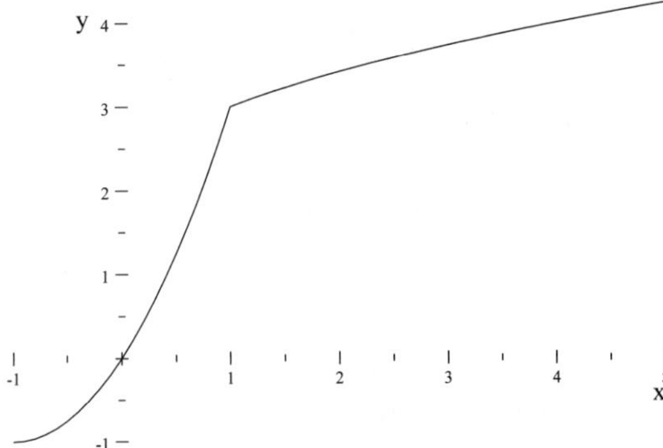

Calculate the inverse piecewise:

For $x \le -1$: Solve $y = \sqrt[3]{x}$ for x.

$$y = \sqrt[3]{x}$$
$$x = y^3$$

So, $f^{-1}(x) = x^3$ on $(-\infty, -1)$.

For $-1 < x \le 1$: Solve $y = x^2 + 2x$ for x.

$$y = x^2 + 2x$$
$$y = x^2 + 2x + 1 - 1$$
$$y = (x+1)^2 - 1$$
$$y + 1 = (x+1)^2$$
$$\pm\sqrt{y+1} = x + 1$$
$$x = -1 \pm \sqrt{y+1}$$

Since *rng*(*f*) = dom(f^{-1}) = (-1, 3) on this portion, we see that $f^{-1}(x) = -1 + \sqrt{x+1}$ on (-1, 1).

For $x > 1$: Solve $y = \sqrt{x} + 2$ for x.

$$y = \sqrt{x} + 2$$
$$y - 2 = \sqrt{x}$$
$$x = (y-2)^2$$

So, $f^{-1}(x) = (x-2)^2$ on $(1, \infty)$.

Thus,

$$f^{-1}(x) = \begin{cases} x^3, & x \le -1, \\ -1 + \sqrt{x+1}, & -1 < x \le 1, \\ (x-2)^2, & x > 1 \end{cases}$$

60.

One-to-one

Calculate the inverse piecewise:

For $x < -2$: Solve $y = -x$ for x.

$$y = -x$$
$$x = -y$$

So, $f^{-1}(x) = -x$ on $(2, \infty)$.

For $-2 < x \leq 0$: Solve $y = \sqrt{4 - x^2}$ for x.

$$y = \sqrt{4 - x^2}$$
$$y^2 = 4 - x^2$$
$$x^2 = 4 - y^2$$
$$x = \pm\sqrt{4 - y^2}$$

Since *rng*(*f*) = dom(f^{-1}) = [-2, 0] on this portion, we see that $f^{-1}(x) = -\sqrt{4 - x^2}$ on [0,2].

For $x > 0$: Solve $y = -\frac{1}{x}$ for x.

$$y = -\tfrac{1}{x}$$
$$x = -\tfrac{1}{y}.$$

So, $f^{-1}(x) = -\frac{1}{x}$ on $(-\infty, 0)$.

Thus,

$$f^{-1}(x) = \begin{cases} -x^3, \ x > 2, \\ -\sqrt{4 - x^2}, \ 0 \leq x \leq 2, \\ -\frac{1}{x}, \ x < 0 \end{cases}$$

61. One-to-one

The portion of the graph for non-negative *x* values is as follows. The graph for negative *x*-values is merely a reflection of this graph over the origin.

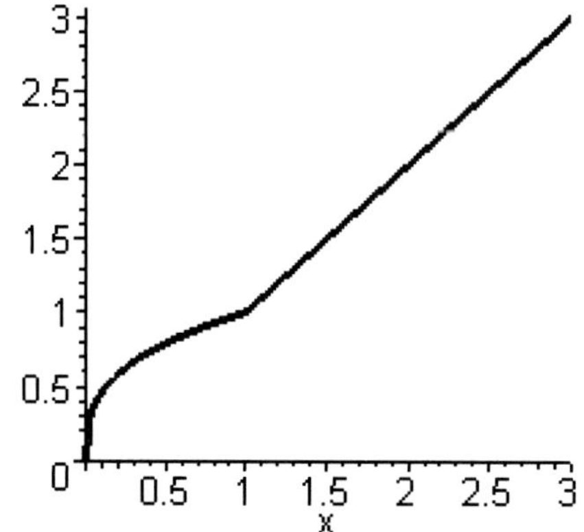

62. Not one-to-one

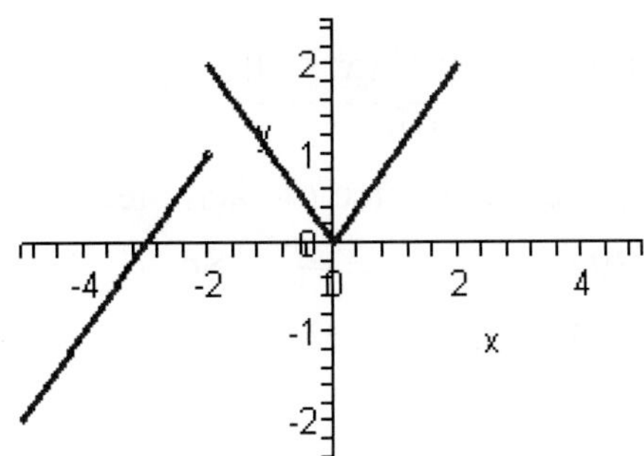

(Note: Should also have the graph of $y = x^2, x \geq 2$, above. The present curve would have an open hole at (2,2), and this newly-added piece would have a closed hole at (2,4) and extend upward to the right quadratically.)

63. Solve $y = \frac{9}{5}x + 32$ for x:

$$y - 32 = \frac{9}{5}x$$
$$\frac{5}{9}(y - 32) = x$$

So, $f^{-1}(x) = \frac{5}{9}(x - 32)$.
The inverse function represents the conversion from degrees Fahrenheit to degrees Celsius.

64. Solve $y = \frac{5}{9}(x - 32)$ for x:

$$\frac{9}{5}y = x - 32$$
$$\frac{9}{5}y + 32 = x$$

So, $C^{-1}(x) = \frac{9}{5}x + 32$. The inverse function represents the conversion from degrees Celsius to degrees Fahrenheit.

65. Let x = number of boats entered. The cost function is

$$C(x) = \begin{cases} 250x, & 0 \leq x \leq 10 \\ \underbrace{2500 + 175(x - 10)}_{= 175x + 750}, & x > 10 \end{cases}$$

To calculate $C^{-1}(x)$, we calculate the inverse of each piece separately:
For $0 \leq x \leq 10$: Solve $y = 250x$ for x: $x = \frac{y}{250}$. So, $C^{-1}(x) = \frac{x}{250}$, for $0 \leq x \leq 2500$.
For $x > 10$: Solve $y = 175x + 750$ for x: $x = \frac{y-750}{175}$. So, $C^{-1}(x) = \frac{x-750}{175}$, for $x > 2500$.
Thus, the inverse function is given by:

$$C^{-1}(x) = \begin{cases} \frac{x}{250}, & 0 \leq x \leq 2500 \\ \frac{x-750}{175}, & x > 2500 \end{cases}$$

66. Let x = number of long-distance minutes. The cost function is

$$C(x)=\begin{cases} 0.39x, & 0\le x\le 10 \\ \underbrace{3.9+0.12(x-10)}_{=0.12x+2.7}, & x>10 \end{cases}$$

To calculate $C^{-1}(x)$, we calculate the inverse of each piece separately:

For $0\le x\le 10$: Solve $y=0.39x$ for x: $x=\frac{y}{0.39}$. So, $C^{-1}(x)=\frac{x}{0.39}$, for $0\le x\le 3.9$.

For $x>10$: Solve $y=0.12x+2.7$ for x: $x=\frac{y-2.7}{0.12}$. So, $C^{-1}(x)=\frac{x-2.7}{0.12}$, for $x>3.9$.

Thus, the inverse function is given by:

$$C^{-1}(x)=\begin{cases} \frac{x}{0.39}, & 0\le x\le 3.9 \\ \frac{x-2.7}{0.12}, & x>3.9 \end{cases}$$

67. Let x = number of hours worked. Then, the take home pay is given by

$$E(x)=\underbrace{7x}_{\substack{\$7\text{ per hour,}\\ \text{for } x \text{ hours}}}-\underbrace{0.25(7x)}_{\substack{\text{Amount withheld}\\ \text{for taxes}}}=5.25x.$$

To calculate E^{-1}, solve $y=5.25x$ for x: $x=\frac{y}{5.25}$. So, $E^{-1}(x)=\frac{x}{5.25}$, $x\ge 0$.

The inverse function tells you how many hours you need to work to attain a certain take home pay.

68. Let x = number of hours worked.

Since the hourly rate for overtime pay is 1.5(8) = 12 dollars per hour, we see that the weekly earnings are described by the following function:

$$E(x)=\begin{cases} 8x, & 0\le x\le 40 \\ \underbrace{320}_{\substack{\text{Pay for first}\\ \text{40 hours}}}+\underbrace{12(x-40)}_{\substack{\text{Amount of overtime}\\ \text{pay}}}, & x>40 \end{cases} = \begin{cases} 8x, & 0\le x\le 40 \\ 12x-160, & x>40 \end{cases}$$

To calculate $E^{-1}(x)$, we calculate the inverse of each piece separately:

For $0\le x\le 40$: Solve $y=8x$ for x: $x=\frac{y}{8}$. So, $E^{-1}(x)=\frac{x}{8}$, for $0\le x\le 320$.

For $x>40$: Solve $y=12x-160$ for x: $x=\frac{y+160}{12}$. So, $E^{-1}(x)=\frac{x+160}{12}$, for $x>320$.

Thus, the inverse function is given by:

$$E^{-1}(x)=\begin{cases} \frac{x}{8}, & 0\le x\le 320 \\ \frac{x+160}{12}, & x>320 \end{cases}$$

The inverse function tells you how many hours you need to work to attain a certain take home pay.

69.

$$M(x)=\begin{cases} 0.60x, & 0\le x\le 15 \\ \underbrace{0.60(15)+0.90(x-15)}_{=0.90x-4.5}, & x>15 \end{cases}$$

To calculate $M^{-1}(x)$, we calculate the inverse of each piece separately:

For $0\le x\le 15$: Solve $y=0.60x$ for x: $x=\frac{5y}{3}$. So, $M^{-1}(x)=\frac{5x}{3}$, for $0\le x\le 9$.

For $x>15$: Solve $y=0.90x-4.5$ for x: $x=\frac{10y+45}{9}$. So, $M^{-1}(x)=\frac{10x+45}{9}$, for $x>9$.

Thus, the inverse function is given by:

$$M^{-1}(x)=\begin{cases} \frac{5x}{3}, & 0\le x\le 9 \\ \frac{10x}{9}+5, & x>9 \end{cases}$$

70. Let x = number of miles.

$$C(x)=\underbrace{19.95(2)}_{\text{Cost per day}}+\underbrace{0.80x}_{\text{Cost per mile}}+\underbrace{0.10[19.95(2)+0.80x]}_{\text{Taxes}}=43.89+0.88x,\ x\ge 0.$$

To compute $C^{-1}(x)$, solve $y=43.89+0.88x$ for x: $x=\dfrac{y-43.89}{0.88}$

Thus, $C^{-1}(x)=\dfrac{x-43.89}{0.88}$.

71.

$$V(x)=\begin{cases} 20{,}000-600x, \ 0\le x\le 5, \\ 20{,}000-600(5)-900(x-5), \ x>5 \end{cases} = \begin{cases} 20{,}000-600x, \ 0\le x\le 5, \\ 21{,}500-900x, \ x>5 \end{cases}$$

To calculate $V^{-1}(x)$, we calculate the inverse of each piece separately:

For $0\le x\le 5$: Solve $y=20{,}000-600x$ for x: $x=\frac{20{,}000-y}{600}$. So, $V^{-1}(x)=\frac{20{,}000-x}{600}$, for $17{,}000\le x\le 20{,}000$.

For $x>5$: Solve $y=21{,}500-900x$ for x: $x=\frac{21{,}500-y}{900}$. So, $V^{-1}(x)=\frac{21{,}500-x}{900}$, for $0\le x\le 17{,}000$.

Thus, the inverse function is given by:

$$V^{-1}(x)=\begin{cases} \frac{21{,}500-x}{900}, & 0\le x\le 17{,}000 \\ \frac{20{,}000-x}{600}, & 17{,}000<x\le 20{,}000 \end{cases}$$

72. $P(x) = x\left(\frac{50-x}{5}\right) = \frac{50x-x^2}{5}$, $x < 25$.

To calculate $P^{-1}(x)$, solve $y = \frac{50x-x^2}{5}$ for x:

$$y = \frac{50x-x^2}{5} \Rightarrow 5y = 50x - x^2 \Rightarrow x^2 - 50x + 5y = 0$$

$$\Rightarrow x = \frac{50 \pm \sqrt{2500-20y}}{2} \Rightarrow x = \frac{50 \pm \sqrt{4(625-5y)}}{2} = 25 \pm \sqrt{625-5y}$$

Since dom(P) = rng(P^{-1}) = [0, 25], we use $P^{-1}(x) = 25 - \sqrt{625-5x}$. This tells you the number of plants required to produce x pounds of strawberries per square yard.

73. Not a function since the graph does not pass the vertical line test.

74. To determine the points on the graph of the inverse of f, switch the order of the x and y in the ordered pairs rather than multiplying by -1. So, in this case, since $(3,3)$, $(0,-4)$ are on the graph of f, the points $(3,3)$, $(-4,0)$ are on the graph of f^{-1}.

75. Must restrict the domain to a portion on which f is one-to-one, say $x \geq 0$. Then, the calculation will be valid.

76. $dom\left(f^{-1}\right) = rng(f) = [0,\infty)$, not $[2,\infty)$.

77. False. In fact, no even function can be one-to-one since the condition $f(x) = f(-x)$ implies that the horizontal line test is violated.

78. False. The function $f(x) = 0$ is odd, but not one-to-one.

79. False. Consider $f(x) = x$. Then, $f^{-1}(x) = x$ also.

80. True. $dom(f)$ is inside $(-\infty,0)$ and $rng(f)$ is inside $(0,\infty)$. Since they are switched for f^{-1}, $dom(f^{-1})$ is inside $(0,\infty)$ and $rng(f^{-1})$ is inside $(-\infty,0)$. Thus, the graph of f^{-1} is in Quadrant IV.

81. $(b,0)$ since the x and y coordinates of all points on the graph of f are switched to get the corresponding points on the graph of f^{-1}.

82. If $(a,0)$ is on the graph of f, then $(0,a)$ is on the graph of f^{-1}. This is its y-intercept.

83. The equation of the unit circle is $x^2 + y^2 = 1$. The portion in Quadrant I is given by

$$y = \sqrt{1-x^2},\ 0 \leq x \leq 1,\ 0 \leq y \leq 1.$$

To calculate the inverse of this function, solve for x:

$$y^2 = 1 - x^2, \text{ which gives us } x = \sqrt{1-y^2}$$

So, $f^{-1}(x) = \sqrt{1-x^2}$, $0 \leq x \leq 1$. The domain and range of both are $[0,1]$.

84. Let $f(x)=\frac{c}{x}$, $c \neq 0$. To calculate the inverse of this function, solve for x:

$$y=\tfrac{c}{x} \Rightarrow x=\tfrac{c}{y} \Rightarrow yx=c \Rightarrow y=\tfrac{c}{x}.$$

Thus, $f(x)=f^{-1}(x),\ x\neq 0$.

85. As long as $m \neq 0$ (that is, while the graph of f is not a horizontal line), it is one-to-one.

86. Assume $m \neq 0$. Then, solving $y=mx+b$ for x yields: $x=\frac{y-b}{m}$. So, the inverse of $f(x)=mx+b$ is $f^{-1}(x)=\frac{x-b}{m}$.

87. We know from earlier problems that rational functions of the form $g(x)=\dfrac{1}{x+b}$ are one-to-one. Further, if a rational function possesses two vertical asymptotes, it is impossible for it to be one-to-one. As such, we choose a such that the numerator cancels with one of the factors in the denominator. The choice of a that does this is $a=4$. Then, $f(x)=\dfrac{x-2}{x^2-4}=\dfrac{x-2}{(x-2)(x+2)}=\dfrac{1}{x+2}$. Observe also that

$$dom(f)=rng(f^{-1})=(-\infty,-2)\cup(-2,\infty)$$
$$dom(f^{-1})=rng(f)=(-\infty,0)\cup(0,\infty)$$

To calculate $f^{-1}(x)$, solve $y=\dfrac{1}{x+2}$ for x:

$$y=\frac{1}{x+2} \Rightarrow y(x+2)=1 \Rightarrow yx+2y=1 \Rightarrow x=\frac{1-2y}{y}$$

So, we conclude that $f^{-1}(x)=\dfrac{1-2x}{x}$.

88. The only point guaranteed to be on the graph of $f^{-1}(x)$ is (b, a).

89. Not one-to-one

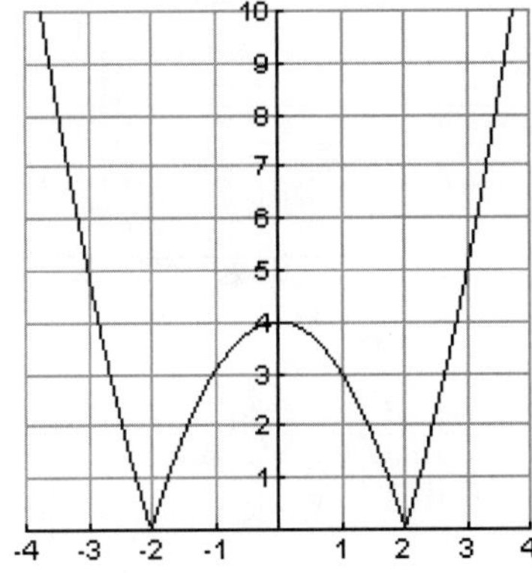

90. One-to-one

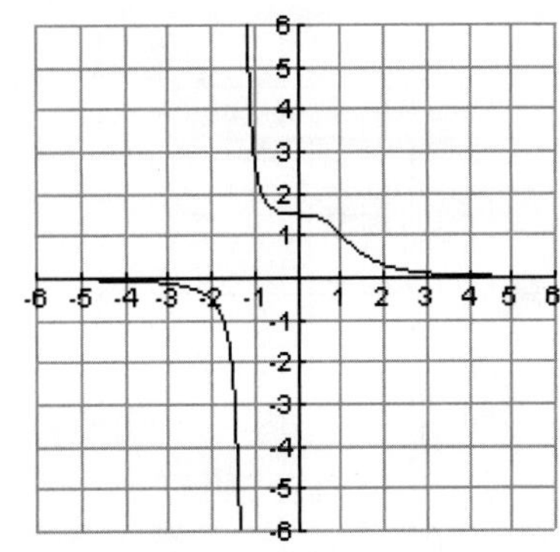

91. Not one-to-one

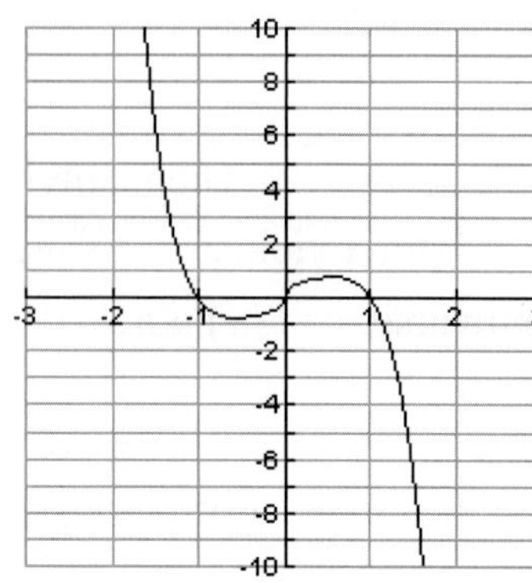

92. One-to-one

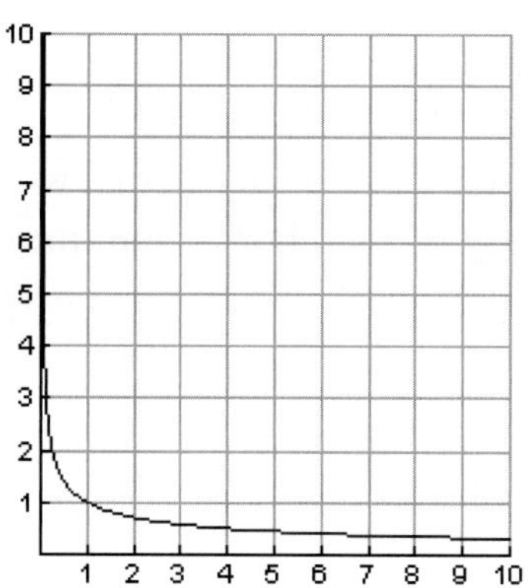

93. Not inverses. Had we restricted the domain of the parabola to $[0, \infty)$, then they would have been.

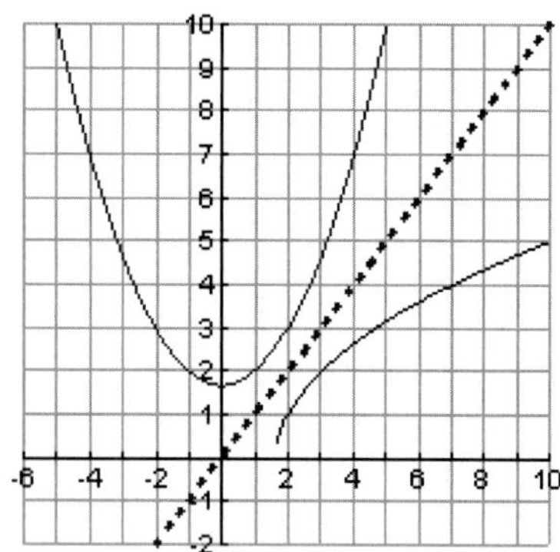

94. Yes, it appears as though the given functions are inverses of each other.

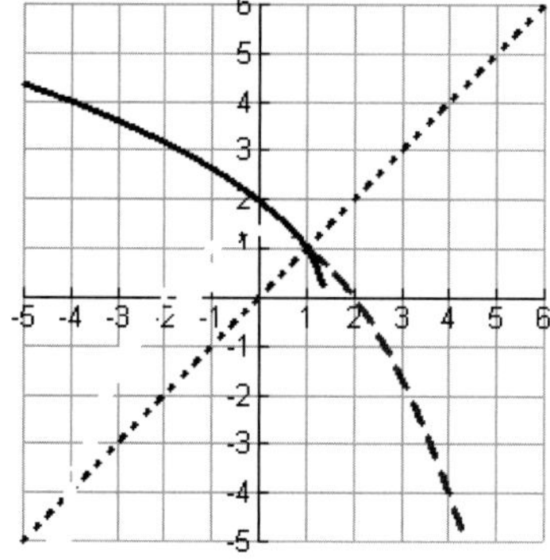

95. Yes, it appears as though the given functions are inverses of each other.

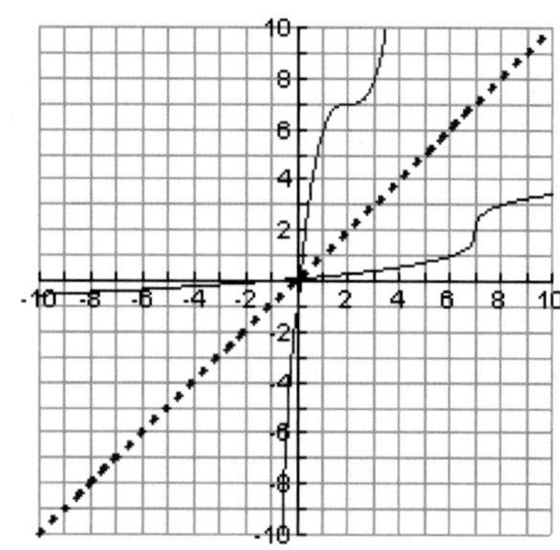

96. No, they are not inverses.

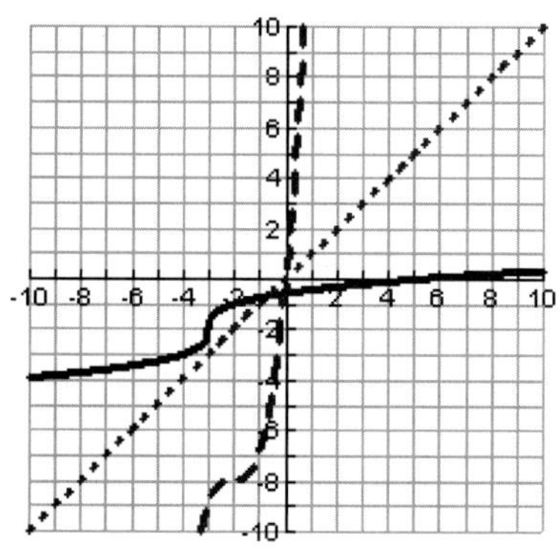

97. **a.** Solve $y = 2x+1$ for x: $y = 2x+1 \Rightarrow x = \frac{y-1}{2}$. So, $f^{-1}(x) = \frac{x-1}{2}$.

b. $\frac{f(x+h)-f(x)}{h} = \frac{(2(x+h)+1)-(2x+1)}{h} = \frac{2x+2h+1-2x-1}{h} = 2$.

So, evaluating at $h=0$ yields $f'(x) = 2$.

c. $\frac{f^{-1}(x+h)-f^{-1}(x)}{h} = \frac{\frac{1}{2}(x+h-1)-\frac{1}{2}(x-1)}{h} = \frac{\frac{1}{2}x+\frac{1}{2}h-\frac{1}{2}-\frac{1}{2}x+\frac{1}{2}}{h} = \frac{1}{2}$.

So, evaluating at $h=0$ yields $\left(f^{-1}\right)'(x) = \frac{1}{2}$.

d. $\frac{1}{f'\left(f^{-1}(x)\right)} = \frac{1}{f'(\frac{1}{2})} = \frac{1}{2} = \left(f^{-1}\right)'(x)$

98. **a.** Solve $y = x^2,\ x > 0$ for x: $x = y^2 \Rightarrow x = \pm\sqrt{y}$. So, $f^{-1}(x) = \sqrt{x}$.

b. $\frac{f(x+h)-f(x)}{h} = \frac{(x+h)^2 - x^2}{h} = \frac{x^2+2hx+h^2-x^2}{h} = \frac{h(2x+h)}{h} = 2x+h$.

So, evaluating at $h=0$ yields $f'(x) = 2x$.

c. $\frac{f^{-1}(x+h)-f^{-1}(x)}{h} = \frac{\sqrt{x+h}-\sqrt{x}}{h} = \frac{\sqrt{x+h}-\sqrt{x}}{h}\cdot\frac{\sqrt{x+h}+\sqrt{x}}{\sqrt{x+h}+\sqrt{x}} = \frac{x+h-x}{h\left(\sqrt{x+h}+\sqrt{x}\right)} = \frac{1}{\sqrt{x+h}+\sqrt{x}}$.

So, evaluating at $h=0$ yields $\left(f^{-1}\right)'(x) = \frac{1}{2\sqrt{x}}$.

d. $\frac{1}{f'\left(f^{-1}(x)\right)} = \frac{1}{f'(\sqrt{x})} = \frac{1}{2\sqrt{x}} = \left(f^{-1}\right)'(x)$

99. **a.** Solve $y = \sqrt{x+2},\ x > -2$ for x: $y = \sqrt{x+2} \Rightarrow y^2 = x+2 \Rightarrow x = y^2-2$. So, $f^{-1}(x) = x^2-2,\ x \geq 0$.

b. $\frac{f(x+h)-f(x)}{h} = \frac{\sqrt{x+h+2}-\sqrt{x+2}}{h} = \frac{\sqrt{x+h+2}-\sqrt{x+2}}{h}\cdot\frac{\sqrt{x+h+2}+\sqrt{x+2}}{\sqrt{x+h+2}+\sqrt{x+2}}$

$= \frac{x+h+2-x-2}{h\left(\sqrt{x+h+2}+\sqrt{x+2}\right)} = \frac{1}{\sqrt{x+h+2}+\sqrt{x+2}}$

So, evaluating at $h=0$ yields $f'(x) = \frac{1}{2\sqrt{x+2}}$.

c. $$\frac{(f^{-1})(x+h)-(f^{-1})(x)}{h}=\frac{\left[(x+h)^2-2\right]-(x^2-2)}{h}=\frac{x^2+2hx+h^2-2-x^2+2}{h}$$
$$=\frac{h(2x+h)}{h}=2x+h$$

So, evaluating at $h=0$ yields $(f^{-1})'(x)=2x$.

d. $$\frac{1}{f'(f^{-1}(x))}=\frac{1}{f'(x^2-2)}=\frac{1}{2\sqrt{(x^2-2)+2}}=\frac{1}{\frac{1}{2x}}=2x=(f^{-1})'(x)$$

100. a. Solve $y=\frac{1}{x+1}$, $x>-1$ for x:

$$y=\frac{1}{x+1}\Rightarrow y(x+1)=1\Rightarrow yx+y=1\Rightarrow x=\frac{1-y}{y},\ y>0.\text{ So, } f^{-1}(x)=\frac{1-x}{x},\ x>0.$$

b.

$$\frac{f(x+h)-f(x)}{h}=\frac{\frac{1}{x+h+1}-\frac{1}{x+1}}{h}=\frac{(x+1)-(x+h+1)}{h(x+h+1)(x+1)}=\frac{-h}{h(x+h+1)(x+1)}=\frac{-1}{(x+h+1)(x+1)}$$

So, evaluating at $h=0$ yields $f'(x)=-\frac{1}{(x+1)^2}$.

c. $$\frac{(f^{-1})(x+h)-(f^{-1})(x)}{h}=\frac{\frac{1-(x+h)}{x+h}-\frac{1-x}{x}}{h}=\frac{[1-(x+h)]x-(1-x)(x+h)}{hx(x+h)}$$
$$=\frac{x-x^2-hx-x-h+x^2+xh}{hx(x+h)}=-\frac{1}{x(x+h)}$$

So, evaluating at $h=0$ yields $(f^{-1})'(x)=\frac{-1}{x^2}$.

d. $$\frac{1}{f'(f^{-1}(x))}=\frac{1}{f'\left(\frac{1-x}{x}\right)}=\frac{1}{\frac{-1}{\left(\frac{1-x}{x}+1\right)^2}}=-\left(\frac{1-x}{x}+1\right)^2=-\left(\frac{1}{x}\right)^2=(f^{-1})'(x)$$

Chapter 1 Review Solutions ---

1. Yes	**2.** Yes
3. No, since both $(0,6)$ and $(0,-6)$ satisfy the equation, so that the graph fails the vertical line test.	**4.** No, since the graph fails the vertical line test.
5. Yes **6.** Yes	**7.** No, since the graph fails the vertical line test. **8.** Yes
9. **(a)** 2 **(b)** 4 **(c)** when $x=-3,\ 4$	**10.** **(a)** 0 **(b)** -4 **(c)** when $x\approx -2, 3.2$
11. **(a)** 0 **(b)** -2 **(c)** when $x\approx -5, 2$	**12.** **(a)** 7 **(b)** -1.5 **(c)** never
13. $f(3)=4(3)-7=\boxed{5}$	**14.** $F(4)=4^2+4(4)-3=\boxed{29}$
15. $f(-7)\cdot g(3)=\left(4(-7)-7\right)\cdot\left\|3^2+2(3)+4\right\| = -35\left\|19\right\| = \boxed{-665}$	**16.** $\dfrac{F(0)}{g(0)}=\boxed{-\dfrac{3}{4}}$
17. $\dfrac{f(2)-F(2)}{g(0)}=\dfrac{\left(4(2)-7\right)-\left(2^2+4(2)-3\right)}{4}$ $\dfrac{1-9}{4}=\boxed{-2}$	**18.** $f(3+h)=4(3+h)-7=\boxed{5+4h}$

19.

$$\frac{f(3+h)-f(3)}{h}=\frac{\left(4(3+h)-7\right)-\left(4(3)-7\right)}{h} = \frac{5+4h-5}{h}=\boxed{4}$$

20.

$$\begin{aligned}\frac{F(t+h)-F(t)}{h}&=\frac{\left((t+h)^2+4(t+h)-3\right)-\left(t^2+4t-3\right)}{h}\\&=\frac{t^2+2ht+h^2+4t+4h-3-t^2-4t+3}{h}\\&=\frac{2ht+h^2+4h}{h}=\frac{h\left(2t+h+4\right)}{h}=\boxed{2t+h+4}\end{aligned}$$

21. $(-\infty,\infty)$	**22.** $(-\infty,\infty)$	**23.** $(-\infty,-4)\cup(-4,\infty)$

24. $(-\infty,\infty)$	**25.** We need $x-4\geq 0$, so the domain is $[4,\infty)$.	**26.** We need $2x-6>0$, so the domain is $(3,\infty)$.

27. Solve $2=f(5)=\dfrac{D}{5^2-16}$ for D: $2=\frac{D}{9}$, so that $\boxed{D=18}$.

28. There are many such functions. The most natural one to construct has the form $f(x)=\dfrac{D}{(x+3)(x-2)}$. Since $(0,-4)$ is to lie on the graph of *f*, we substitute this point into the equation for the function to find the corresponding value of *D* that will ensure this: $-4=\dfrac{D}{(0+3)(0-2)}=\dfrac{D}{-6}$, so that $D=24$. Hence, one such function is given by:

$$f(x)=\frac{24}{(x+3)(x-2)}.$$

29.

$h(-x)=(-x)^3-7(-x)=-\left(x^3-7x\right)\neq h(x)$

So, not even.

$-h(-x)=-\left(-\left(x^3-7x\right)\right)=x^3-7x=h(x)$

So, $\boxed{\text{odd}}$.

30.

$f(-x)=(-x)^4+3(-x)^2=x^4+3x^2=f(x)$

So, $\boxed{\text{even}}$.

Hence, cannot be odd.

31.

$f(-x)=\dfrac{1}{(-x)^3}+3(-x)=-\left(\dfrac{1}{x^3}+3x\right)\neq f(x)$

So, not even.

$-f(-x)=-\left(-\left(\dfrac{1}{x^3}+3x\right)\right)=\dfrac{1}{x^3}+3x=f(x)$

So, $\boxed{\text{odd}}$.

32.

$f(-x)=\dfrac{1}{(-x)^2}+3(-x)^4+|-x|=f(x)$

So, $\boxed{\text{even}}$.

Hence, *f* cannot be odd.

33.

Domain	$[-5,\infty)$
Range	$[-3,\infty)$
Increasing	$(-5,-3)\cup(3,\infty)$
Decreasing	$(-1,1)$
Constant	$(-3,1)\cup(1,3)$

d) 2
e) 3
f) 1

34.

Domain	$(-\infty,\infty)$
Range	$[-4,\infty)$
Increasing	$(-2,\infty)$
Decreasing	$(-\infty,-2)$
Constant	none

d) 0
e) -3
f) 3

35.

Domain	$[-6,6]$
Range	$[0,3]\cup\{-3,-2,-1\}$
Increasing	$(0,3)$
Decreasing	$(3,6)$
Constant	$(-6,-4)\cup(-4,-2)\cup(-2,0)$

d) -1
e) -2
f) 3

36.

Domain	$[-6,6]$
Range	$[-3,1]$
Increasing	$(-3,1)\cup(3,5)$
Decreasing	$(-6,-3)\cup(1,3)\cup(5,6)$
Constant	none

d) 0
e) -3
f) 1

37.

$$\frac{f(x+h)-f(x)}{h}=\frac{\left[(x+h)^3-1\right]-\left[x^3-1\right]}{h}=\frac{x^3+3x^2h+3xh^2+h^3-1-x^3+1}{h}$$

$$=\frac{h\left(3x^2+3xh+h^2\right)}{h}=\boxed{3x^2+3xh+h^2}$$

38.

$$\frac{f(x+h)-f(x)}{h}=\frac{\frac{x+h-1}{x+h+2}-\frac{x-1}{x+2}}{h}=\frac{(x+h-1)(x+2)-(x-1)(x+h+2)}{h(x+h+2)(x+2)}$$

$$=\frac{(x^2+xh-x+2x+2h-2)-(x^2+hx+2x-x-h-2)}{h(x+h+2)(x+2)}$$

$$=\frac{3h}{h(x+h+2)(x+2)}=\boxed{\frac{3}{(x+h+2)(x+2)}}$$

39.

$$\frac{f(x+h)-f(x)}{h}=\frac{\left(x+h+\frac{1}{x+h}\right)-\left(x+\frac{1}{x}\right)}{h}=\frac{h+\frac{1}{x+h}-\frac{1}{x}}{h}$$

$$=1+\frac{x-(x+h)}{hx(x+h)}=\boxed{1-\frac{1}{x(x+h)}}$$

40.

$$\frac{f(x+h)-f(x)}{h}=\frac{\sqrt{\frac{x+h}{x+h+1}}-\sqrt{\frac{x}{x+1}}}{h}$$

$$=\frac{\sqrt{x+h}\sqrt{x+1}-\sqrt{x}\sqrt{x+h+1}}{h\sqrt{x+h+1}\sqrt{x+1}}$$

$$=\frac{\sqrt{(x+h)(x+1)}-\sqrt{x(x+h+1)}}{h\sqrt{x+h+1}\sqrt{x+1}}$$

$$=\frac{\sqrt{(x+h)(x+1)}-\sqrt{x(x+h+1)}}{h\sqrt{x+h+1}\sqrt{x+1}}\cdot\frac{\sqrt{(x+h)(x+1)}+\sqrt{x(x+h+1)}}{\sqrt{(x+h)(x+1)}+\sqrt{x(x+h+1)}}$$

$$=\frac{x^2+x+h+xh-x^2-xh-x}{h\sqrt{x+h+1}\sqrt{x+1}\left(\sqrt{(x+h)(x+1)}+\sqrt{x(x+h+1)}\right)}$$

$$=\boxed{\frac{1}{\sqrt{x+h+1}\sqrt{x+1}\left(\sqrt{(x+h)(x+1)}+\sqrt{x(x+h+1)}\right)}}$$

41. $\frac{(4-2^2)-(4-0^2)}{2}=\boxed{-2}$

42. $\frac{|2(5)-1|-|2(1)-1|}{5-1}=\frac{9-1}{4}=\boxed{2}$

43. Domain: $(-\infty, \infty)$ Range: $(0, \infty)$

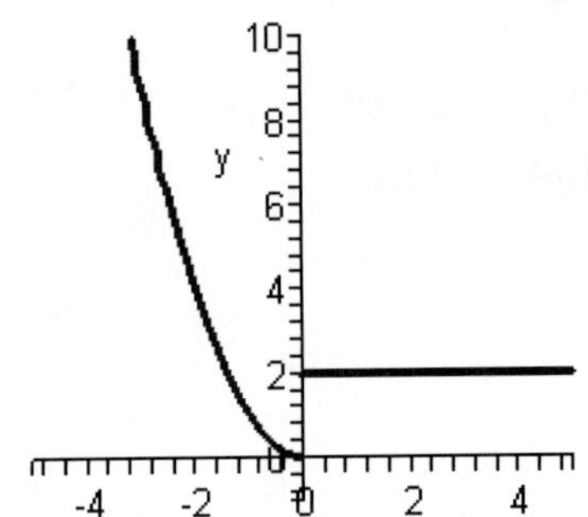

Notes on the graph: There is an open hole at (0,0), and a closed hole at (0,2).

44. Domain: $(-\infty, \infty)$ Range: $[-3, \infty)$

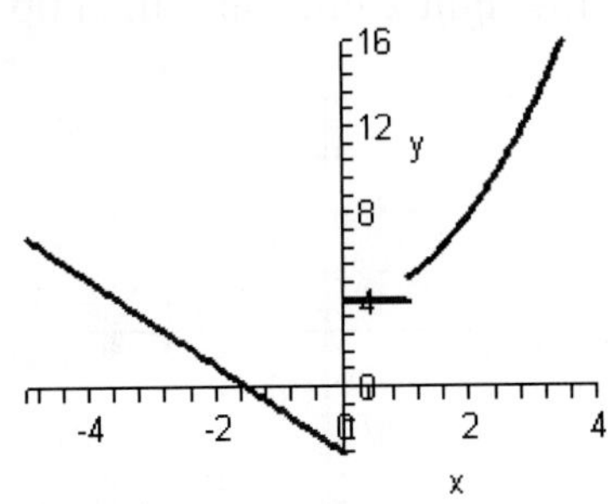

Notes on the graph: There are open holes at (0,4) and (1,5), and closed holes at (1,4) and $(0,-3)$.

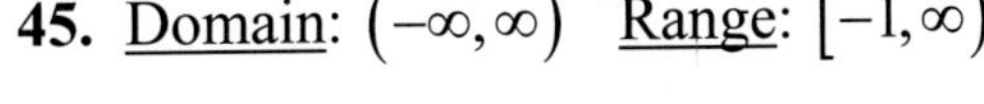

45. Domain: $(-\infty, \infty)$ Range: $[-1, \infty)$

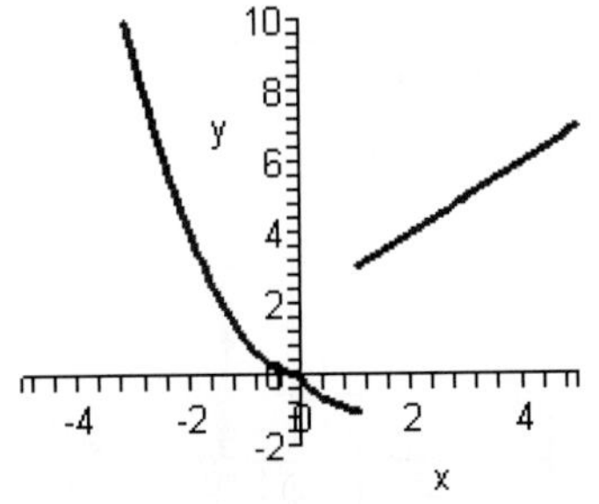

Notes on the graph: There is an open hole at (1,3), and a closed hole at $(1,-1)$.

46. Domain: $(-\infty, 0) \cup (0, \infty)$

Range: $(-\infty, -2] \cup (0, \infty)$

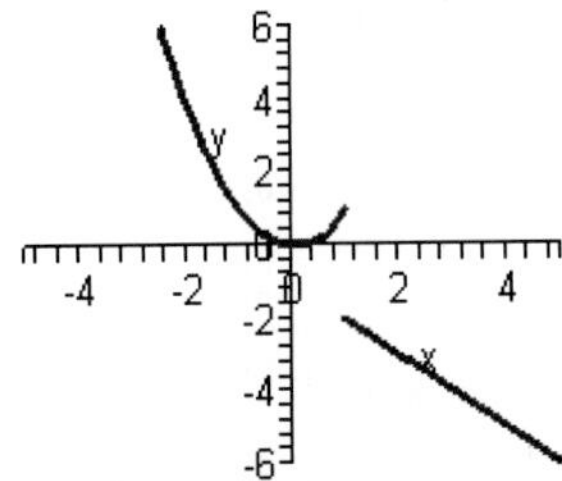

Notes on the graph: There are open holes at (0,0), (1,1), and a closed hole at $(1,-2)$.

47. $\dfrac{280{,}000 - 135{,}000}{5 - 0} = \dfrac{145{,}000}{5} = \boxed{\$29{,}000 \text{ per year}}$

48. $\dfrac{64 - 38}{10 - 0} = \dfrac{26}{10} = \boxed{2.6\%}$

49. Reflect the graph of x^2 over x-axis, then shift right 2 units and then up 4 units.

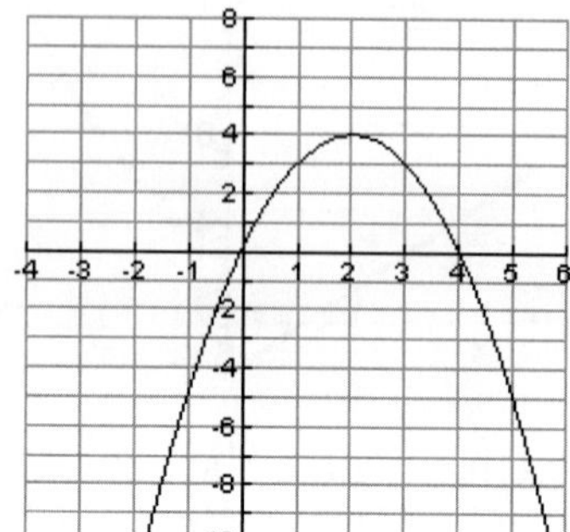

50. First, note that

$$|-x+5|-7=|-(x-5)|-7=|x-5|-7$$

So, shift the graph of $|x|$ right 5 units and then down 7 units.

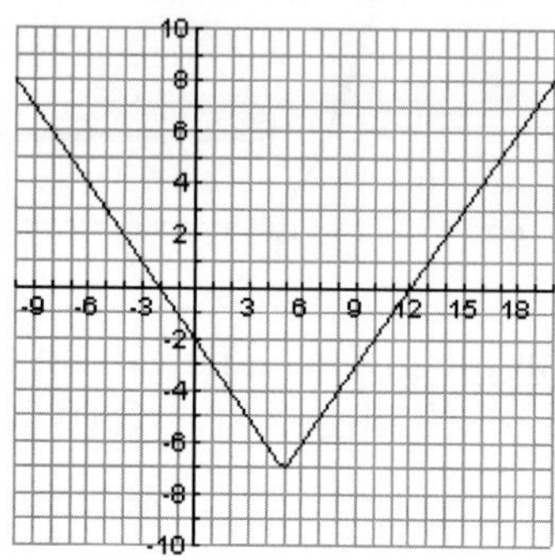

51. Shift the graph of $\sqrt[3]{x}$ right 3 units, and then up 2 units.

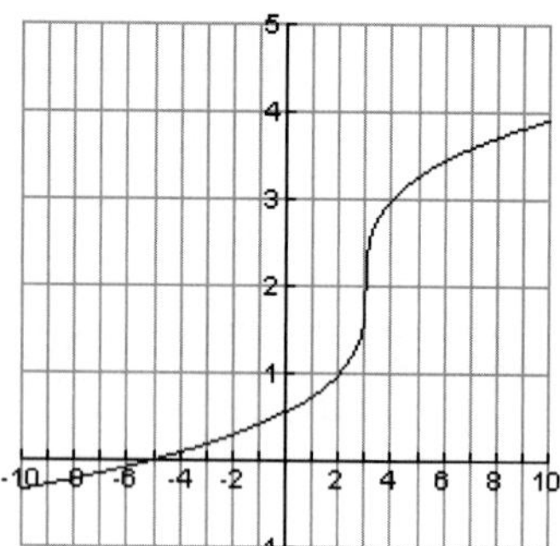

52. Shift the graph of $\frac{1}{x}$ right 2 units, and then down 4 units.

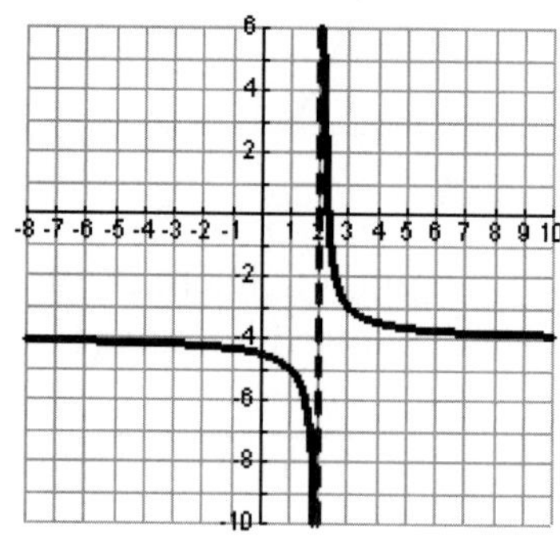

53. Reflect the graph of x^3 over x-axis, and then contract vertically by a factor of 2.

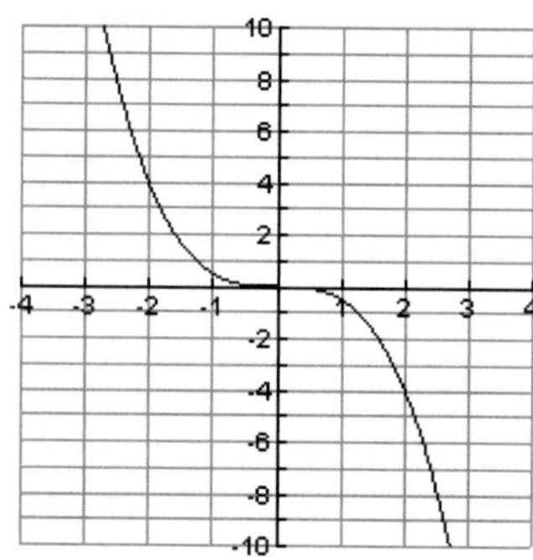

54. Expand the graph of x^2 vertically by a factor of 2, and then shift up 3 units.

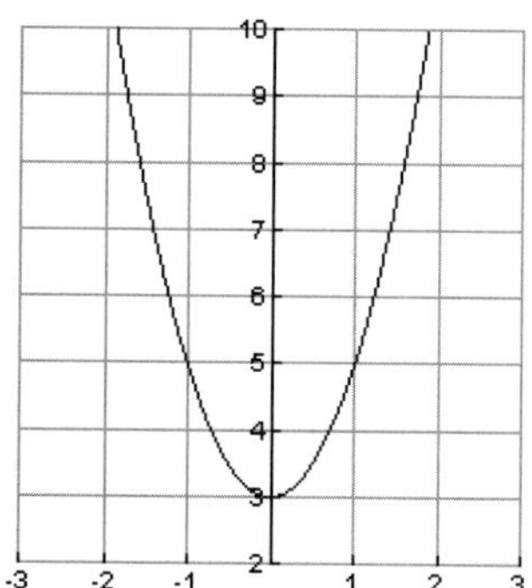

55.

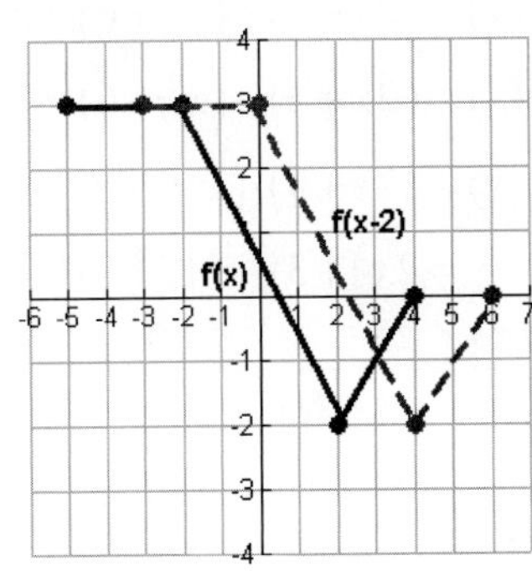

56.

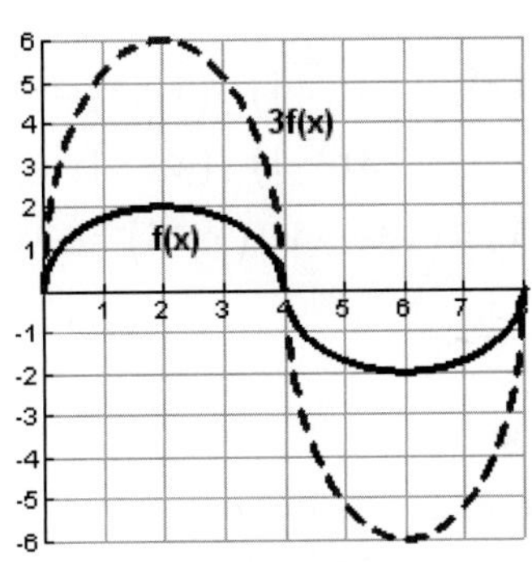

57.

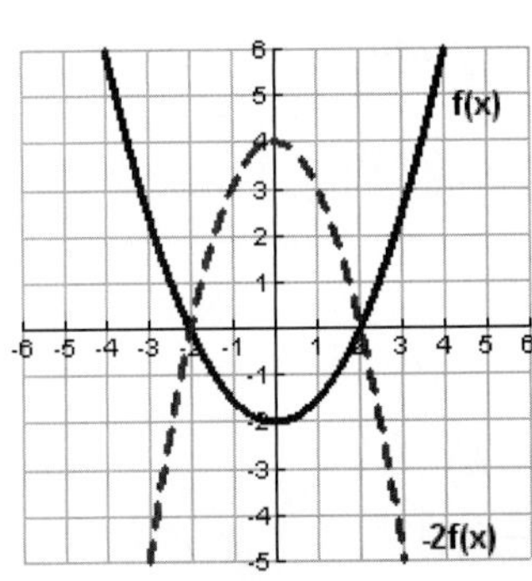

58.

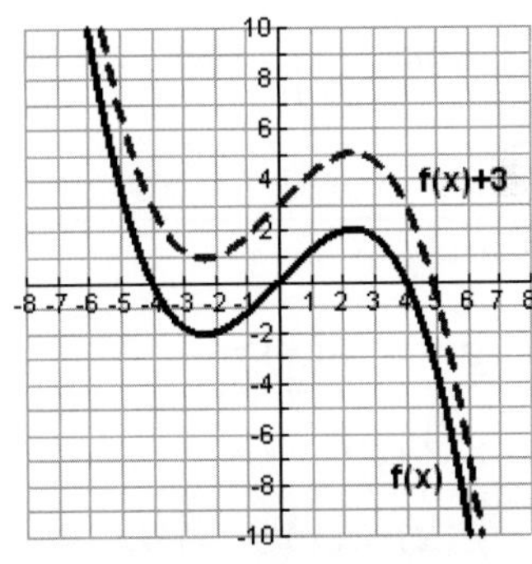

59. $y=\sqrt{x+3}$ Domain: $[-3,\infty)$

60. $y=\sqrt{x}-4$ Domain: $[0,\infty)$

61. $y=\sqrt{x-2}+3$ Domain: $[2,\infty)$

62. $y=\sqrt{-x}$ Domain: $(-\infty,0]$

63. $y=5\sqrt{x}-6$ Domain: $[0,\infty)$

64. $y=\frac{1}{2}\sqrt{x}+3$ Domain: $[0,\infty)$

65. $y=\left(x^2+4x+4\right)-8-4=\left(x+2\right)^2-12$

Domain: $\mathbb{R}$ or $(-\infty,\infty)$

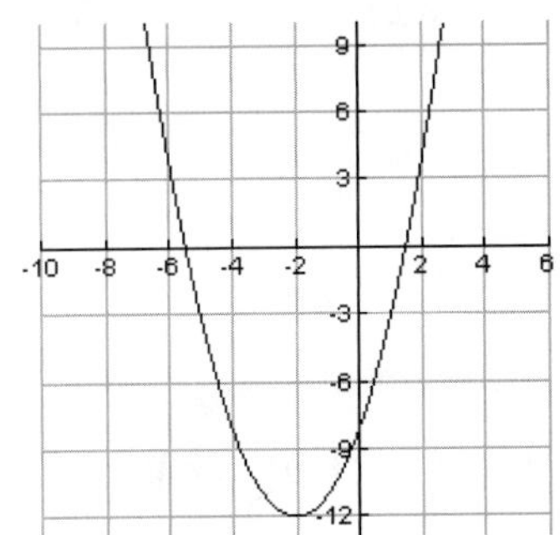

66.

$y=2\left(x^2+3x\right)-5=2\left(x^2+3x+\frac{9}{4}\right)-5-\frac{9}{2}$

$=2\left(x+\frac{3}{2}\right)^2-\frac{19}{2}$

Domain: $\mathbb{R}$ or $(-\infty,\infty)$

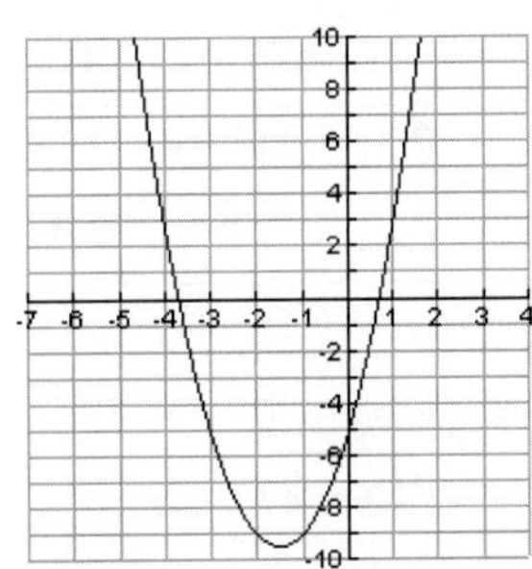

67.

$$g(x)+h(x)=(-3x-4)+(x-3)=-2x-7$$
$$g(x)-h(x)=(-3x-4)-(x-3)=-4x-1$$
$$g(x)\cdot h(x)=(-3x-4)\cdot(x-3)=-3x^2+5x+12$$
$$\frac{g(x)}{h(x)}=\frac{-3x-4}{x-3}$$

Domains:

$$\left.\begin{array}{c} dom(g+h) \\ dom(g-h) \\ dom(gh) \end{array}\right\}=(-\infty,\infty)$$
$$dom\left(\frac{g}{h}\right)=(-\infty,3)\cup(3,\infty)$$

68.

$$g(x)+h(x)=(2x+3)+(x^2+6)=x^2+2x+9$$
$$g(x)-h(x)=(2x+3)-(x^2+6)=-x^2+2x-3$$
$$g(x)\cdot h(x)=(2x+3)\cdot(x^2+6)$$
$$=2x^3+3x^2+12x+18$$
$$\frac{g(x)}{h(x)}=\frac{2x+3}{x^2+6}$$

Domains:

$$\left.\begin{array}{c} dom(g+h) \\ dom(g-h) \\ dom(gh) \\ dom\left(\frac{g}{h}\right) \end{array}\right\}=(-\infty,\infty)$$

69.

$$g(x)+h(x)=\tfrac{1}{x^2}+\sqrt{x}$$
$$g(x)-h(x)=\tfrac{1}{x^2}-\sqrt{x}$$
$$g(x)\cdot h(x)=\tfrac{1}{x^2}\cdot\sqrt{x}=\tfrac{1}{x^{3/2}}$$
$$\frac{g(x)}{h(x)}=\frac{\frac{1}{x^2}}{\sqrt{x}}=\frac{1}{x^{5/2}}$$

Domains:

$$\left.\begin{array}{c} dom(g+h) \\ dom(g-h) \\ dom(gh) \\ dom\left(\frac{g}{h}\right) \end{array}\right\}=(0,\infty)$$

70.

$$g(x)+h(x)=\frac{x+3}{2(x-2)}+\frac{3x-1}{x-2}$$
$$=\frac{(x+3)+2(3x-1)}{(2x-4)}$$
$$=\frac{7x+1}{2(x-2)}$$

$$g(x)-h(x)=\frac{x+3}{2(x-2)}-\frac{3x-1}{x-2}$$
$$=\frac{(x+3)-2(3x-1)}{(2x-4)}$$
$$=\frac{-5x+5}{2(x-2)}$$

$$g(x)\cdot h(x)=\frac{x+3}{2(x-2)}\cdot\frac{3x-1}{x-2}=\frac{(x+3)\cdot(3x-1)}{2(x-2)^2}$$

$$\frac{g(x)}{h(x)}=\frac{\frac{x+3}{2(x-2)}}{\frac{3x-1}{x-2}}=\frac{x+3}{2\cancel{(x-2)}}\cdot\frac{\cancel{x-2}}{3x-1}=\frac{x+3}{2(3x-1)}$$

Domains:

$$\left.\begin{array}{c} dom(g+h) \\ dom(g-h) \\ dom(gh) \end{array}\right\}=(-\infty,2)\cup(2,\infty)$$
$$dom\left(\frac{f}{g}\right)=(-\infty,\tfrac{1}{3})\cup(\tfrac{1}{3},2)\cup(2,\infty)$$

71.

$$g(x)+h(x)=\sqrt{x-4}+\sqrt{2x+1}$$
$$g(x)-h(x)=\sqrt{x-4}-\sqrt{2x+1}$$
$$g(x)\cdot h(x)=\sqrt{x-4}\cdot\sqrt{2x+1}$$
$$\frac{g(x)}{h(x)}=\frac{\sqrt{x-4}}{\sqrt{2x+1}}$$

<u>Domains</u>:
Must have both $x-4\geq 0$ and $2x+1\geq 0$. So,

$$\left.\begin{array}{r} dom(f+g) \\ dom(f-g) \\ dom(fg) \end{array}\right\}=[4,\infty).$$

For the quotient, must have both $x-4\geq 0$ and $2x+1>0$. So,

$$dom\left(\frac{f}{g}\right)=[4,\infty).$$

72.

$$g(x)+h(x)=\left(x^2-4\right)+\left(x+2\right)=x^2+x-2$$
$$g(x)-h(x)=\left(x^2-4\right)-\left(x+2\right)=x^2-x-6$$
$$g(x)\cdot h(x)=\left(x^2-4\right)\cdot\left(x+2\right)$$
$$=x^3+2x^2-4x-8$$
$$\frac{g(x)}{h(x)}=\frac{x^2-4}{x+2}=x-2,\quad x\neq -2$$

<u>Domains</u>:

$$\left.\begin{array}{r} dom(g+h) \\ dom(g-h) \\ dom(gh) \end{array}\right\}=(-\infty,\infty)$$

$$dom\left(\frac{g}{h}\right)=(-\infty,-2)\cup(-2,\infty)$$

73.

$$(f\circ g)(x)=3\left(2x+1\right)-4=6x-1$$
$$(g\circ f)(x)=2\left(3x-4\right)+1=6x-7$$

<u>Domains</u>:
$dom(f\circ g)=(-\infty,\infty)=dom(g\circ f)$

74.

$$(f\circ g)(x)=\left(x+3\right)^3+2\left(x+3\right)-1$$
$$=x^3+9x^2+29x+3$$
$$(g\circ f)(x)=\left(x^3+2x-1\right)+3=x^3+2x+2$$

<u>Domains</u>:
$dom(f\circ g)=(-\infty,\infty)=dom(g\circ f)$

75.

$$(f\circ g)(x)=\frac{2}{\frac{1}{4-x}+3}=\frac{2}{\frac{1+3(4-x)}{4-x}}=\frac{2(4-x)}{13-3x}=\frac{8-2x}{13-3x}$$

$$(g\circ f)(x)=\frac{1}{4-\frac{2}{x+3}}=\frac{1}{\frac{4(x+3)-2}{x+3}}=\frac{x+3}{4x+10}$$

Domains:

$dom(f\circ g)=(-\infty,4)\cup\left(4,\tfrac{13}{3}\right)\cup\left(\tfrac{13}{3},\infty\right)$

$dom(g\circ f)=(-\infty,-3)\cup\left(-3,-\tfrac{5}{2}\right)\cup\left(-\tfrac{5}{2},\infty\right)$

76.

$$(f\circ g)(x)=\sqrt{2\left(\sqrt{x+6}\right)^2-5}=\sqrt{2(x+6)-5}=\sqrt{2x+7}$$

$$(g\circ f)(x)=\sqrt{\sqrt{2x^2-5}+6}$$

Domains:

$dom(f\circ g)$: Need both $x+6\geq 0$ and $2x+7\geq 0$. Thus, $dom(f\circ g)=\left[-\tfrac{7}{2},\infty\right)$.

$dom(g\circ f)$: Note $\sqrt{2x^2-5}+6\geq 0$, for all values of x for which $\sqrt{2x^2-5}$ is defined. This is true when $2x^2-5\geq 0$. So, solving this inequality yields:

$$2x^2-5\geq 0$$
$$x^2-\tfrac{5}{2}\geq 0$$
$$\left(x-\sqrt{\tfrac{5}{2}}\right)\left(x+\sqrt{\tfrac{5}{2}}\right)\geq 0$$

CPs: $\pm\sqrt{\frac{5}{2}}$ 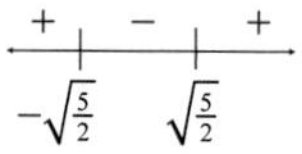

So, $dom(g\circ f)=\left(-\infty,-\sqrt{\tfrac{5}{2}}\right)\cup\left(\sqrt{\tfrac{5}{2}},\infty\right)$.

77.

$$(f\circ g)(x)=\sqrt{x^2-4-5}=\sqrt{(x-3)(x+3)}$$

$$(g\circ f)(x)=\left(\sqrt{x-5}\right)^2-4=x-9$$

Domains:

$dom(f\circ g)$: Need $(x-3)(x+3)\geq 0$.

CPs: ± 3 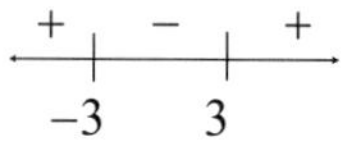

So, $dom(g\circ f)=(-\infty,-3]\cup[3,\infty)$.

$dom(g\circ f)$: Need $x-5\geq 0$. Thus, $dom(g\circ f)=[5,\infty)$.

78.

$$(f\circ g)(x)=\frac{1}{\sqrt{\frac{1}{x^2-4}}}=\sqrt{x^2-4}$$

$$(g\circ f)(x)=\frac{1}{\left(\frac{1}{\sqrt{x}}\right)^2-4}=\frac{1}{\frac{1}{x}-4}=\frac{1}{\frac{1-4x}{x}}=\frac{x}{1-4x}$$

Domains:

$dom(f\circ g)$: Need $(x-2)(x+2)>0$.

CPs: ± 2 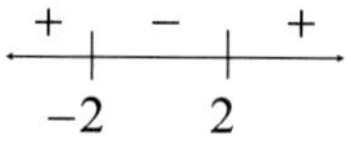

So, $dom(g\circ f)=(-\infty,-2)\cup(2,\infty)$.

$dom(g\circ f)$: Need $1-4x\neq 0$, so that $x\neq\frac{1}{4}$. So,

$dom(g\circ f)=(-\infty,0)\cup\left(0,\tfrac{1}{4}\right)\cup\left(\tfrac{1}{4},\infty\right)$.

79. $g(3)=6(3)-3=15$ $f(g(3))=f(15)=4(15)^2-3(15)+2=857$ $f(-1)=4(-1)^2-3(-1)+2=9$ $g(f(-1))=g(9)=6(9)-3=51$	**80.** $g(3)=3^2+5=14$, but $f(g(3))=f(14)$ is not defined. $f(-1)=\sqrt{4-(-1)}=\sqrt{5}$ $g(f(-1))=g\left(\sqrt{5}\right)=\left(\sqrt{5}\right)^2+5=10$
81. $g(3)=\lvert 5(3)+2\rvert=17$ $f(g(3))=f(17)=\frac{17}{\lvert 2(17)-3\rvert}=\frac{17}{31}$ $f(-1)=\frac{-1}{\lvert 2(-1)-3\rvert}=-\frac{1}{5}$ $g(f(-1))=g\left(-\frac{1}{5}\right)=\left\lvert 5\left(-\frac{1}{5}\right)+2\right\rvert=1$	**82.** $g(3)=3^2-1=8$ $f(g(3))=f(8)=\frac{1}{8-1}=\frac{1}{7}$ $f(-1)=\frac{1}{-1-1}=-\frac{1}{2}$ $g(f(-1))=g\left(-\frac{1}{2}\right)=\left(-\frac{1}{2}\right)^2-1=-\frac{3}{4}$
83. $f(g(3))=\left(\sqrt[3]{3-4}\right)^2-\left(\sqrt[3]{3-4}\right)+10$ $=(-1)^2-(-1)+10=\boxed{12}$ $g(f(-1))=g\left((-1)^2+1+10\right)$ $=\sqrt[3]{12-4}=\boxed{2}$	**84.** $f(g(3))$ is $\boxed{\text{undefined}}$ since $g(3)$ is not defined. $g(f(-1))=g\left(\frac{4}{(-1)^2-2}\right)=g(-4)=\boxed{\frac{1}{7}}$
85. Let $f(x)=3x^2+4x+7,\ \ g(x)=x-2$. Then, $h(x)=f(g(x))$.	**86.** Let $f(x)=\frac{x}{1-x},\ \ g(x)=\sqrt[3]{x}$. Then, $h(x)=f(g(x))$.
87. Let $f(x)=\frac{1}{\sqrt{x}},\ \ g(x)=x^2+7$. Then, $h(x)=f(g(x))$.	**88.** Let $f(x)=\sqrt{x},\ \ g(x)=\lvert 3x+4\rvert$. Then, $h(x)=f(g(x))$.
89. The area of a circle with radius $r(t)$ is given by: $A(t)=\pi\left(r(t)\right)^2=\pi\left(25\sqrt{t+2}\right)^2$ $=625\pi(t+2)\text{ in}^2$	**90.** Since $42=lw,\ \ l=\frac{42}{w}$. So, the perimeter formula becomes: $36=2l+2w=2\left(\frac{42}{w}\right)+2w=\frac{84+2w^2}{w}$ so that $2w^2-36w+84=0$ $w^2-18w+42=0$
91. Yes	**92.** Yes
93. Yes	**94.** No, since both (1,1) and (−1,1) satisfy the equation.

95. One-to-one

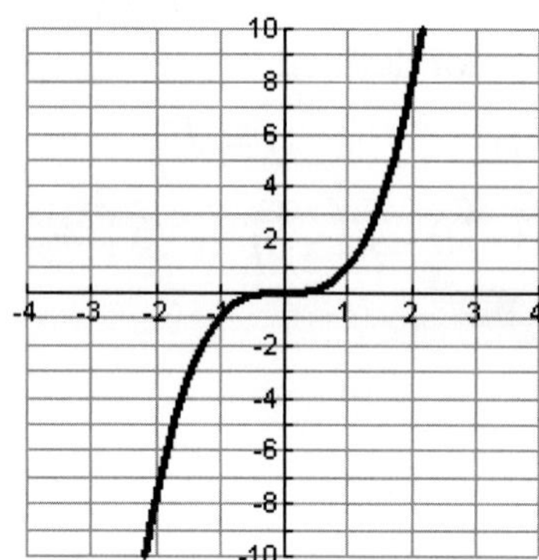

96. Not one-to-one, since $f(-1) = f(1) = 1$, for instance.

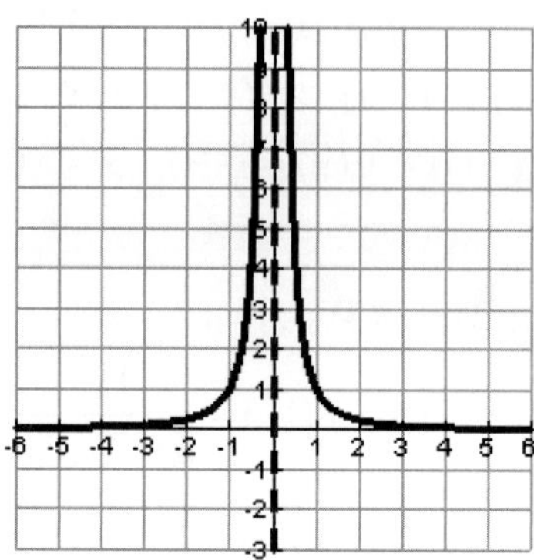

97. Not one-to-one. No function that is a constant value on an entire value can be one-to-one.

98. Not one-to-one. For instance, $f(4) = f(-1) = 0$.

99. One-to-one

100. One-to-one

101.

$$f\left(f^{-1}(x)\right) = \not{3}\left(\frac{x-4}{\not{3}}\right) + 4 = x - 4 + 4 = x$$

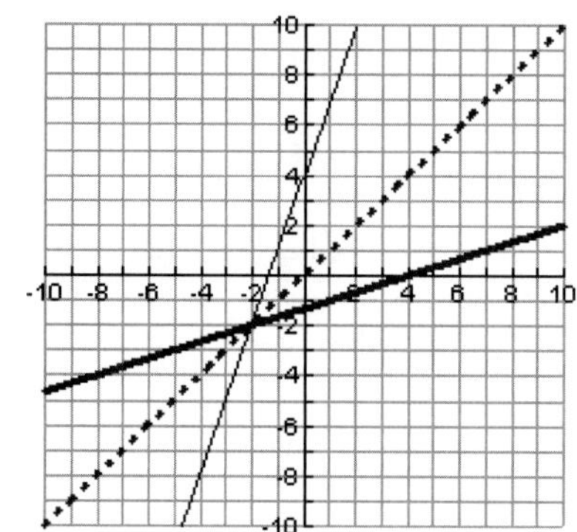

102.

$$f\left(f^{-1}(x)\right) = \frac{1}{4\left(\frac{1+7x}{4x}\right) - 7} = \frac{1}{\frac{1+7x}{x} - 7}$$

$$= \frac{1}{\frac{1}{x} + 7 - 7} = \frac{1}{\frac{1}{x}} = x$$

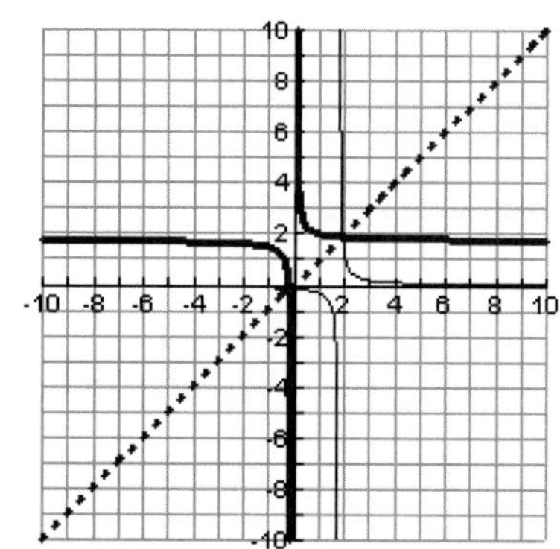

103.

$f\left(f^{-1}(x)\right)=\sqrt{\left(x^2-4\right)+4}=\sqrt{x^2}=x$,
since $x \geq 0$.

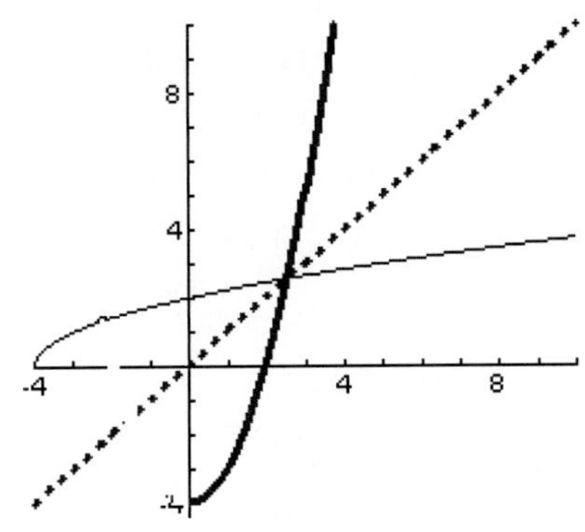

104.

$$f\left(f^{-1}(x)\right)=\frac{\frac{7x+2}{x-1}+2}{\frac{7x+2}{x-1}-7}=\frac{\frac{7x+2+2(x-1)}{x-1}}{\frac{7x+2-7(x-1)}{x-1}}$$

$$=\frac{7x+2+2x-2}{7x+2-7x+7}=\frac{9x}{9}=x$$

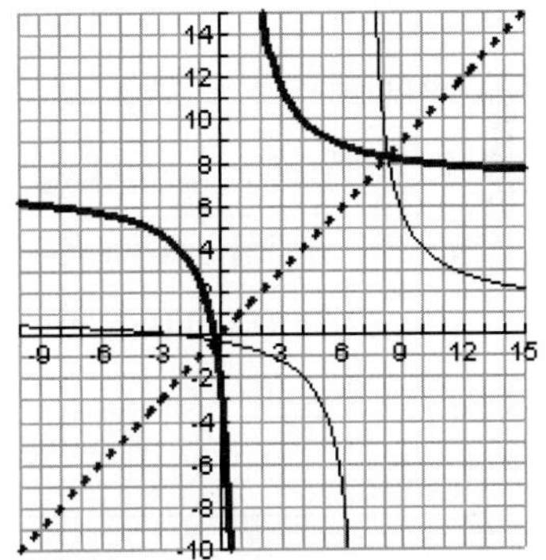

105. Solve $y=2x+1$ for x:

$$x=\tfrac{1}{2}(y-1)$$

Thus, $f^{-1}(x)=\tfrac{1}{2}(x-1)=\dfrac{x-1}{2}$.

Domains:

$dom(f)=rng\left(f^{-1}\right)=(-\infty,\infty)$

$rng(f)=dom\left(f^{-1}\right)=(-\infty,\infty)$

106. Solve $y=x^5+2$ for x:

$$x=\sqrt[5]{y-2}$$

Thus, $f^{-1}(x)=\sqrt[5]{x-2}$.

Domains:

$dom(f)=rng\left(f^{-1}\right)=(-\infty,\infty)$

$rng(f)=dom\left(f^{-1}\right)=(-\infty,\infty)$

107. Solve $y=\sqrt{x+4}$ for x:

$$x=y^2-4$$

Thus, $f^{-1}(x)=x^2-4$.

Domains:

$dom(f)=rng\left(f^{-1}\right)=[-4,\infty)$

$rng(f)=dom\left(f^{-1}\right)=[0,\infty)$

108. Solve $y=(x+4)^2+3$ for x:

$$\sqrt{y-3}=x+4$$

$$-4+\sqrt{y-3}=x$$

Thus, $f^{-1}(x)=-4+\sqrt{x-3}$.

Domains:

$dom(f)=rng\left(f^{-1}\right)=[-4,\infty)$

$rng(f)=dom\left(f^{-1}\right)=[3,\infty)$

109. Solve $y = \frac{x+6}{x+3}$ for x:

$$(x+3)y = x+6$$
$$xy + 3y = x+6$$
$$xy - x = 6 - 3y$$
$$x(y-1) = 6 - 3y$$
$$x = \frac{6-3y}{y-1}$$

Thus, $f^{-1}(x) = \frac{6-3x}{x-1}$.

Domains:
$dom(f) = rng(f^{-1}) = (-\infty,-3)\cup(-3,\infty)$
$rng(f) = dom(f^{-1}) = (-\infty,1)\cup(1,\infty)$

110. Solve $y = 2\sqrt[3]{x-5} - 8$ for x:

$$y + 8 = 2\sqrt[3]{x-5}$$
$$\left(\tfrac{1}{2}(y+8)\right)^3 = x - 5$$
$$5 + \left(\tfrac{1}{2}(y+8)\right)^3 = x$$

Thus, $f^{-1}(x) = 5 + \left(\frac{1}{2}(x+8)\right)^3$.

Domains:
$dom(f) = rng(f^{-1}) = (-\infty,\infty)$
$rng(f) = dom(f^{-1}) = (-\infty,\infty)$

111. Let x = total dollars worth of products sold. Then, $S(x) = 22{,}000 + 0.08x$. Solving $y = 22{,}000 + 0.08x$ for x yields: $x = \frac{1}{0.08}(y - 22{,}000)$

Thus, $S^{-1}(x) = \frac{x - 22{,}000}{0.08}$. This inverse function tells you the sales required to earn a desired income.

112. $V(s) = 3s^2$, $s \ge 0$. Solving $y = 3s^2$ for s yields: $s = \sqrt{\frac{1}{3}y}$.

So, $V^{-1}(s) = \sqrt{\frac{1}{3}s}$. This inverse function tells you the length s of a side of a base required to get a desired volume.

113. The graph of f is as follows:

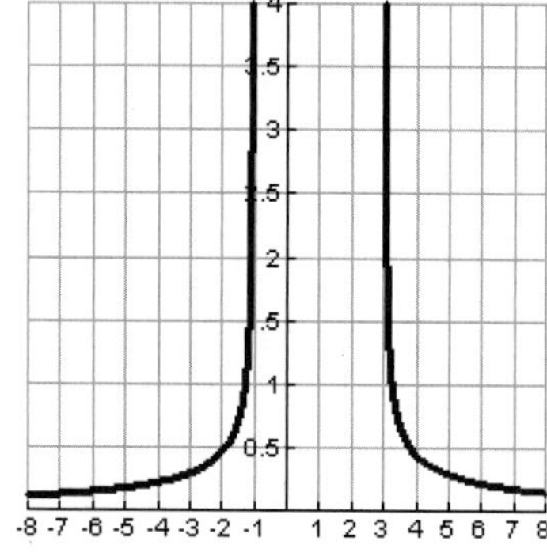

Domain of f: $\boxed{(-\infty,-1)\cup(3,\infty)}$

114. The graph of f is as follows:

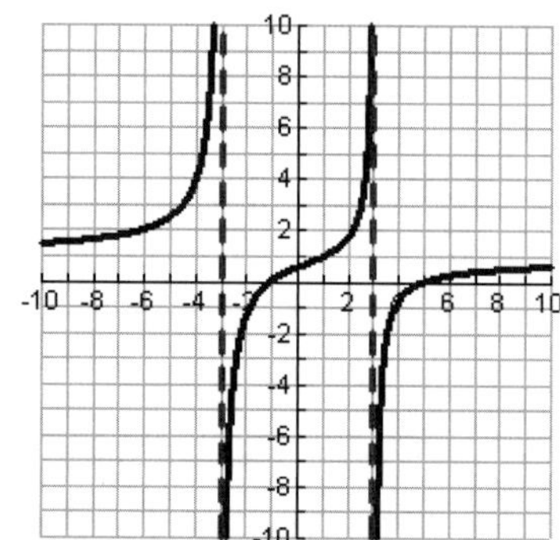

Domain of f: $\boxed{(-\infty,-3)\cup(-3,3)\cup(3,\infty)}$

115.

Domain	$(-\infty,2)\cup(2,\infty)$
Range	$\{-1,0,1\}\cup(2,\infty)$
Increasing	$(2,\infty)$
Decreasing	$(-\infty,-1)$
Constant	$(-1,0)\cup(0,1)\cup(1,2)$

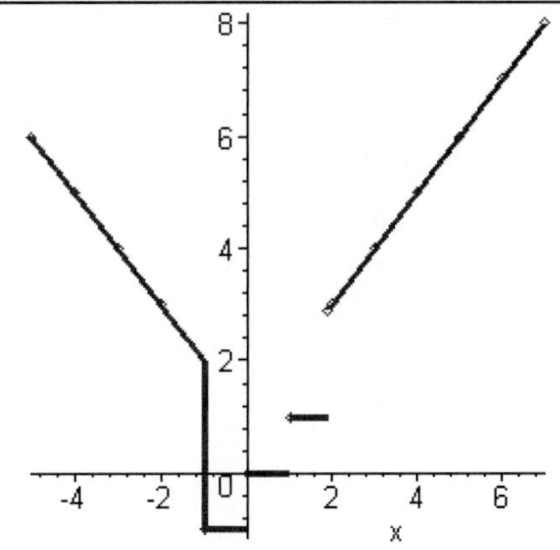

Note on the graph: The vertical line at $x=-1$ should NOT be a part of the graph.

116.

Domain	$(-\infty,2)\cup(2,\infty)$
Range	$[0,3)\cup(4,\infty)$
Increasing	$(-1,0)\cup(1,2)\cup(2,\infty)$
Decreasing	$(-2,-1)\cup(0,1)$
Constant	nowhere

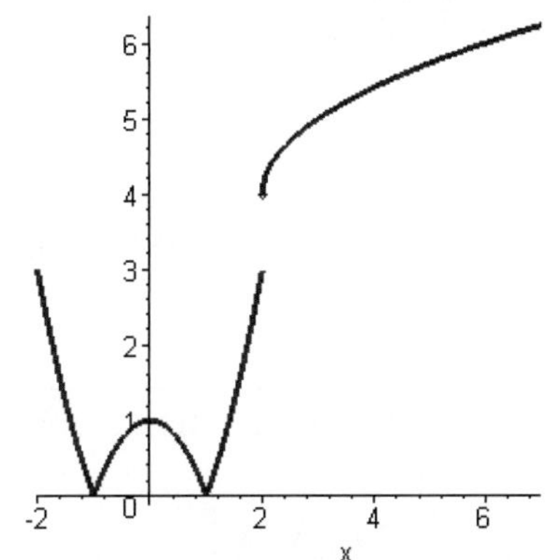

117. The graphs of f and g are as follows:

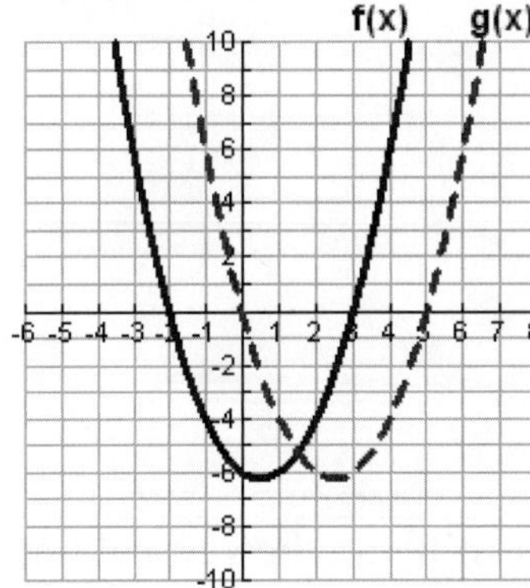

The graph of f can be obtained by shifting the graph of g two units to the left. That is, $f(x)=g(x+2)$.

118. The graphs of f and g are as follows:

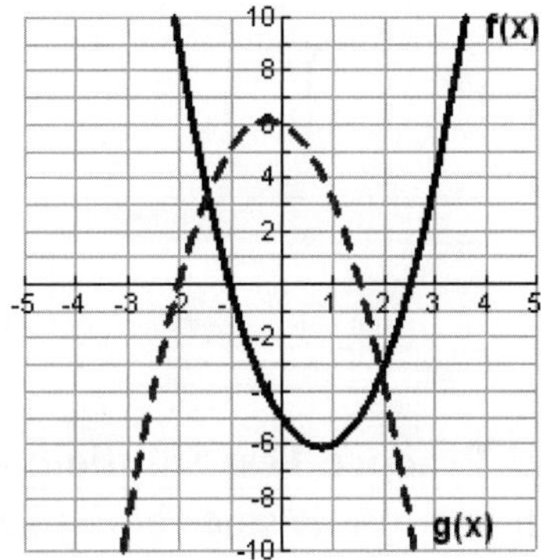

The graph of f can be obtained by shifting the graph of g one unit to the right and then reflecting it over the x-axis. That is, $f(x)=-g(x-1)$.

119.

$y1 = \sqrt{2x+3}$

$y2 = \sqrt{4-x}$

$y3 = \dfrac{y1}{y2}$

Domain of $y3 = \boxed{[-1.5, 4)}$

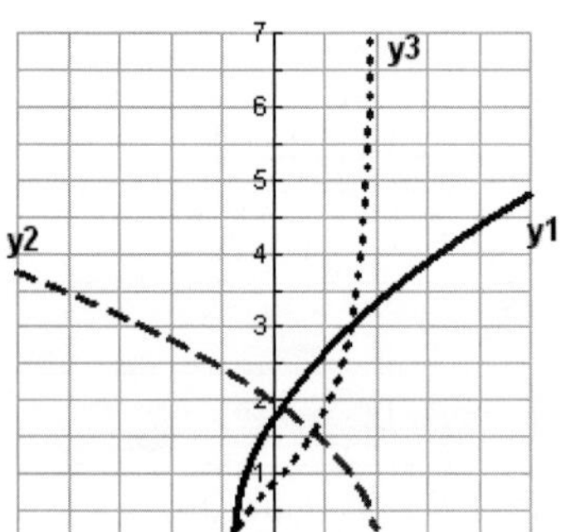

120.

$y1 = \sqrt{x^2 - 4}$

$y2 = x^2 - 5$

$y3 = (y1)^2 - 5$

Domain of $y3 = \boxed{(-\infty, -2] \cup [2, \infty)}$

121. Yes, the function is one-to-one.

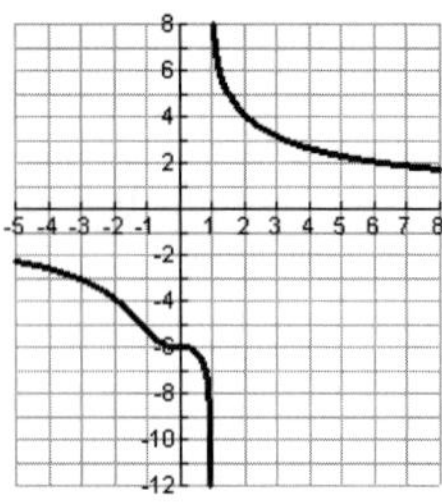

122. Yes, the functions are inverses.

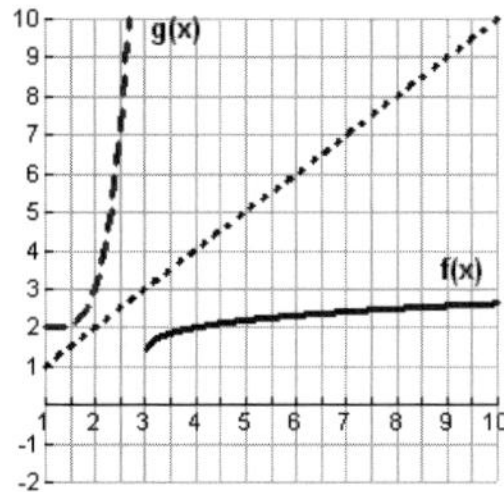

Chapter 1 Practice Test Solutions

1. b (Not one-to-one since both (0,3) and $(-3,3)$ lie on the graph.)	**2. a** (Doesn't pass the vertical line test.)
3. c	**4.** Observe that $f(11) = \sqrt{11-2} = \sqrt{9} = 3$ $g(-1) = (-1)^2 + 11 = 12$ So, $f(11) - 2g(-1) = 3 - 2(12) = -21$.

5. $\left(\frac{f}{g}\right)(x)=\frac{\sqrt{x-2}}{x^2+11}$ <u>Domain</u>: $[2,\infty)$	**6.** $\left(\frac{g}{f}\right)(x)=\frac{x^2+11}{\sqrt{x-2}}$ <u>Domain</u>: $(2,\infty)$
7. $g(f(x))=\left(\sqrt{x-2}\right)^2+11=x-2+11=x+9$ Domain: $(2,\infty)$	**8.** $(f+g)(6)=f(6)+g(6)$ $=\sqrt{6-2}+\left(6^2+11\right)=2+47=\boxed{49}$
9. $f\left(g\left(\sqrt{7}\right)\right)=f\left(\left(\sqrt{7}\right)^2+11\right)=f(18)$ $=\sqrt{18-2}=\boxed{4}$	**10.** $f(-x)=\lvert -x\rvert-(-x)^2=\lvert x\rvert-x^2=f(x)$ So, $\boxed{\text{even}}$. Therefore, f cannot be odd.
11. $f(-x)=9(-x)^3+5(-x)-3$ $=-\left[9x^3+5x+3\right]\neq f(x)$ So, not even. $-f(-x)=-\left(-\left[9x^3+5x+3\right]\right)$ $=9x^3+5x+3\neq f(x)$ So, not odd. Thus, $\boxed{\text{neither}}$.	**12.** $f(-x)=\frac{2}{-x}=-\frac{2}{x}=-f(x)$ So, $\boxed{\text{odd}}$. Therefore, f cannot be even.
13. $f(x)=-\sqrt{x-3}+2$ Reflect the graph of $\sqrt{x}$ over the x-axis, then shift right 3 units, and then up 2 units. <u>Domain</u>: $[3,\infty)$ <u>Range</u>: $(-\infty,2]$ 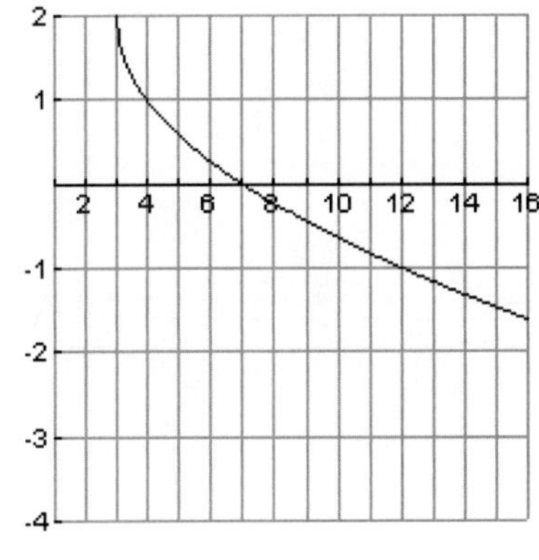	**14.** $f(x)=-2(x-1)^2$ Reflect the graph of x^2 over the x-axis, then expand vertically by a factor of 2, and then shift right 1 unit. <u>Domain</u>: $(-\infty,\infty)$ <u>Range</u>: $(-\infty,0]$ 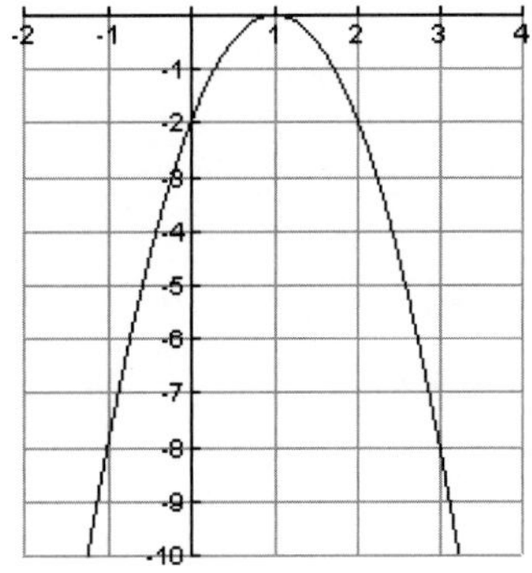

15. $f(x)=\begin{cases} -x, & x<-1 \\ 1, & -1<x<2 \\ x^2, & x\geq 2 \end{cases}$

Domain: $(-\infty,-1)\cup(-1,\infty)$

Range: $[1,\infty)$

Open holes at (-1,1) and (2,1); closed hole at (2,4)

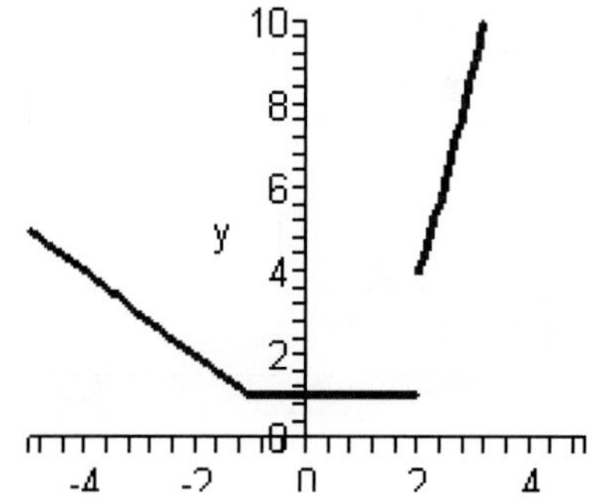

16. **(a)** 5 **(b)** 2 **(c)** 7 **(d)** when $x=-1,1,5.5$
(e) when $x=7$

17.
(a) -2
(b) 4
(c) -3
(d) when $x=-3,2$

18.
(a) -3
(b) never
(c) -1
(d) 2

19.

$$\frac{\left(3(x+h)^2-4(x+h)+1\right)-\left(3x^2-4x+1\right)}{h}=\frac{3x^2+6xh+3h^2-4x-4h+1-3x^2+4x-1}{h}$$

$$=\frac{h\left(6x+3h-4\right)}{h}=\boxed{6x+3h-4}$$

20.

$$\frac{f(x+h)-f(x)}{h}=\frac{\left[(x+h)^3-\frac{1}{\sqrt{x+h}}\right]-\left[x^3-\frac{1}{\sqrt{x}}\right]}{h}$$

$$=\frac{x^3+3x^2h+3xh^2+h^3-x^3}{h}-\frac{\frac{1}{\sqrt{x+h}}-\frac{1}{\sqrt{x}}}{h}$$

$$=\frac{h\left(3x^2+3xh+h^2\right)}{h}-\frac{\sqrt{x}-\sqrt{x+h}}{h\sqrt{x}\sqrt{x+h}}$$

$$=\left(3x^2+3xh+h^2\right)-\frac{\sqrt{x}-\sqrt{x+h}}{h\sqrt{x}\sqrt{x+h}}\cdot\frac{\sqrt{x}+\sqrt{x+h}}{\sqrt{x}+\sqrt{x+h}}$$

$$=\left(3x^2+3xh+h^2\right)-\frac{x-(x+h)}{h\sqrt{x}\sqrt{x+h}\left(\sqrt{x}+\sqrt{x+h}\right)}$$

$$=\boxed{\left(3x^2+3xh+h^2\right)+\frac{1}{\sqrt{x}\sqrt{x+h}\left(\sqrt{x}+\sqrt{x+h}\right)}}$$

21.

$$\frac{\left(64-16(2)^2\right)-\left(64-16(0)^2\right)}{2}=\frac{0-64}{2}=\boxed{-32}$$

22.

$$\frac{\sqrt{10-1}-\sqrt{2-1}}{10-2}=\frac{3-1}{8}=\boxed{\frac{1}{4}}$$

23. Solve $y=\sqrt{x-5}$ for x:

$$y^2=x-5$$

$$y^2+5=x$$

Thus, $f^{-1}(x)=x^2+5$.

<u>Domains</u>:

$dom(f)=rng\left(f^{-1}\right)=[5,\infty)$

$rng(f)=dom\left(f^{-1}\right)=[0,\infty)$

24. Solve $y=x^2+5$ for x:

$$y=x^2+5$$

$$\sqrt{y-5}=x,\ \text{ since } x\geq 0.$$

Thus, $f^{-1}(x)=\sqrt{x-5}$.

<u>Domains</u>:

$dom(f)=rng\left(f^{-1}\right)=[0,\infty)$

$rng(f)=dom\left(f^{-1}\right)=[5,\infty)$

25. Solve $y = \dfrac{2x+1}{5-x}$ for x:

$$\begin{aligned}(5-x)y &= 2x+1\\ 5y - xy &= 2x+1\\ 5y-1 &= x(y+2)\\ x &= \frac{5y-1}{y+2}\end{aligned}$$

Thus, $f^{-1}(x) = \dfrac{5x-1}{x+2}$.

Domains:

$dom(f) = rng(f^{-1}) = (-\infty,5)\cup(5,\infty)$

$rng(f) = dom(f^{-1}) = (-\infty,-2)\cup(-2,\infty)$

26. We compute the inverse of f piecewise:

For $x \le 0$: Solve $y = -x$ for x: $x = -y$. So, $f^{-1}(x) = -x$ on $(-\infty, 0]$.

For $x > 0$: Solve $y = -x^2\ (\le 0)$ for x: $x = -\sqrt{-y}$. So, $f^{-1}(x) = -\sqrt{-x}$ on $(0,\infty)$.

Thus, the inverse function is given by

$$f^{-1}(x) = \begin{cases} -x, & x \ge 0 \\ -\sqrt{-x}, & x < 0 \end{cases}$$

27. Can restrict to $[0,\infty)$ so that f will have an inverse. Also, one could restrict to any interval of the form $[a,\infty)$ or $(-\infty,-a]$, where a is a positive real number, to ensure f is one-to-one.

28. The point $(5,-2)$ (switch x and y coordinates to get a point on the inverse.)

29. $\dfrac{\Delta P}{\Delta d} = \dfrac{28-10}{100-0} = 0.18\ \text{psi/ft.}$ $\quad \dfrac{\Delta d}{\Delta t} = \dfrac{5}{1} = 5\ \text{ft/sec.}$

So, $\dfrac{\Delta P}{\Delta t} = \dfrac{\Delta P}{\Delta d}\cdot\dfrac{\Delta d}{\Delta t} = (0.18)(5)\ \text{psi/sec.} = 0.9\ \text{psi/sec.}$ is the slope of the line.

Using (0, 10) as a point on the line, we see that the equation is $\boxed{P(t) = \dfrac{9}{10}t + 10}$.

30. Recall that $V = \dfrac{4}{3}\pi R^3$ **(1)** and $S = 4\pi R^2$ **(2)**

Solve **(2)** for R: $R = \sqrt{\dfrac{S}{4\pi}}$ Then, substitute this into **(1)**:

$$V = \frac{4}{3}\pi\left(\sqrt{\frac{S}{4\pi}}\right)^3 = \frac{S}{3}\sqrt{\frac{S}{4\pi}} = \frac{S}{6}\sqrt{\frac{S}{\pi}}$$

31. Consider $f(x) = -\sqrt{1-x^2}$, $-1 \le x \le 0$. (The graph of *f* is the quarter unit circle in the third quadrant.) To find its inverse, solve $y = -\sqrt{1-x^2}$ for *x*:

$$y = -\sqrt{1-x^2} \;\Rightarrow\; (-y)^2 = 1-x^2 \;\Rightarrow\; x^2 = 1-y^2 \;\Rightarrow\; x = -\sqrt{1-y^2} \text{ since } -1 \le x \le 0$$

So, $f^{-1}(x) = -\sqrt{1-x^2}$ (The graph looks identical to that of *f*.)

32. Solve $r(t) = 15$ (At this point, the puddle just touches the sidelines.)

$$10\sqrt{t} = 15$$
$$\sqrt{t} = 1.5$$
$$t = (1.5)^2 = 2.25$$

So, after 2.25 hours, the puddle will reach the sidelines.

33. Let *x* = number of minutes. Then,

$$C(x) = \begin{cases} 15, & 0 \le x \le 30 \\ 15 + \underbrace{1(x-30)}_{\text{Amount for minutes beyond the initial 30.}}, & x > 30 \end{cases}$$

$$= \begin{cases} 15, & 0 \le x \le 30 \\ x-15, & x > 30 \end{cases}$$

34. slope $= \dfrac{\Delta T}{\Delta CO_2} = \dfrac{46.23-48.86}{379.7-369.4} = \dfrac{0.37}{10.3} \approx 0.036$

Using the point (369.4, 45.86), we find that the equation of the line is

$$T(x) = 0.036(x-369.4) + 45.86.$$

As such, $T(375) = 0.036(375-369.4) + 45.86 = 46.1^\circ F$.

CHAPTER 2

Section 2.1 Solutions ---

1. b Vertex $(-2,-5)$ and opens up	**2. d** Vertex (1,3) and opens up
3. a Vertex $(-3,2)$ and opens down	**4. c** Vertex (2,3) and opens down

5. Since the coefficient of x^2 is positive, the parabola opens up, so it must be *b* or *d*. Since the coefficient of *x* is positive, the *x*-coordinate of the vertex is negative. So, the graph is **b.**

6. As in #5, the parabola opens up, so it must be *b* or *d*. Since the coefficient of *x* is negative, the *x*-coordinate of the vertex is positive. So, the graph is **d**.

7. Since the coefficient of x^2 is negative, the parabola opens down, so it must be *a* or *c*. In comparison to #8, this one will grow more slowly in the negative *y*-direction. So, the graph is **c**.

8. Since the coefficient of x^2 is -2, the parabola opens down and twice as quickly as the graph of $y=-x^2$. So, the graph is **a**.

9.

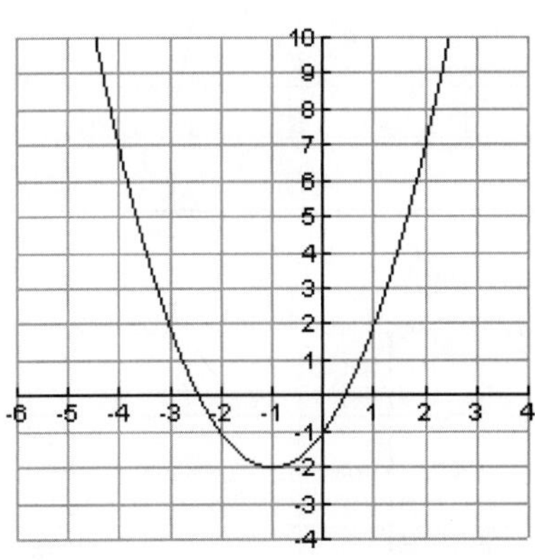

10.

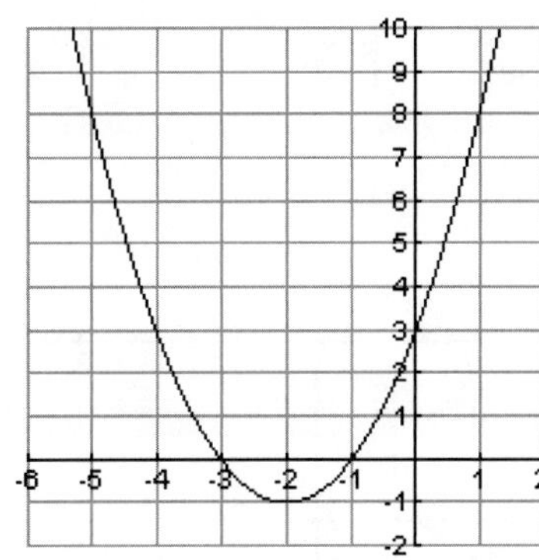

11.

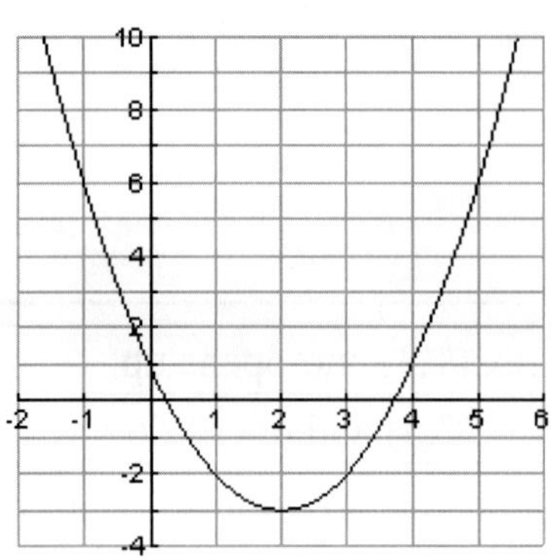

12.

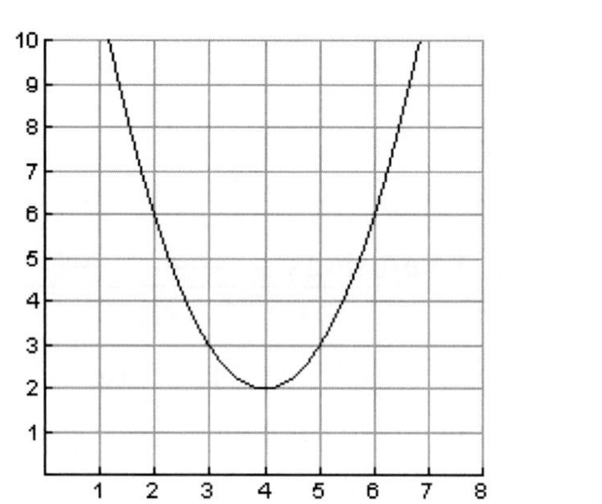

13.

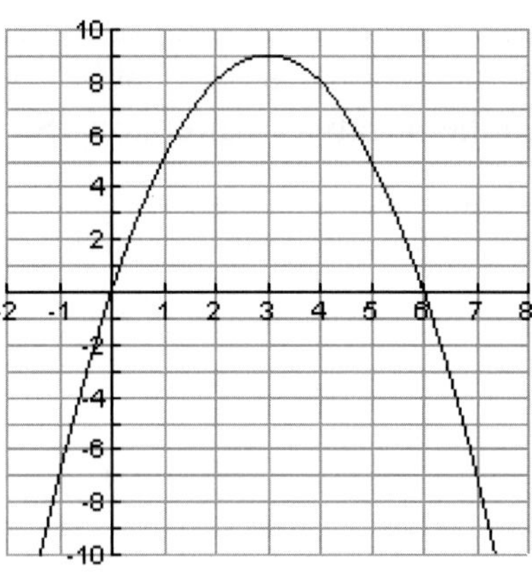

14.

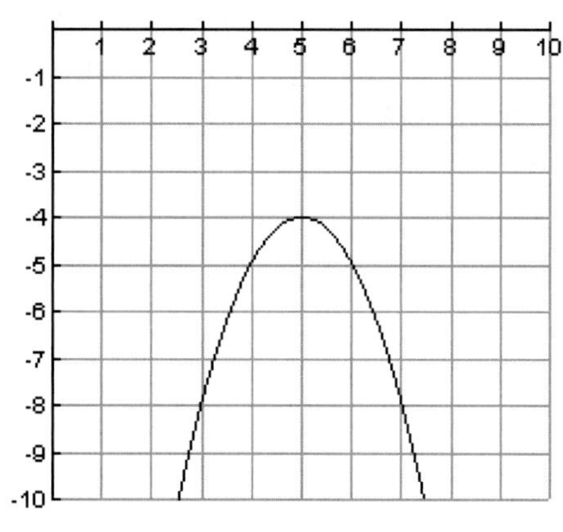

15.

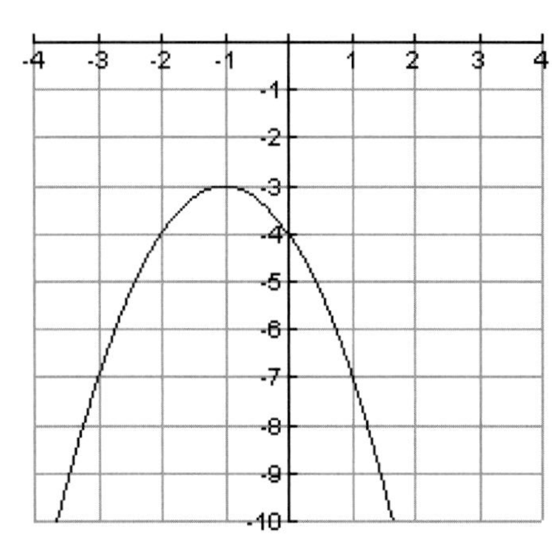

16.

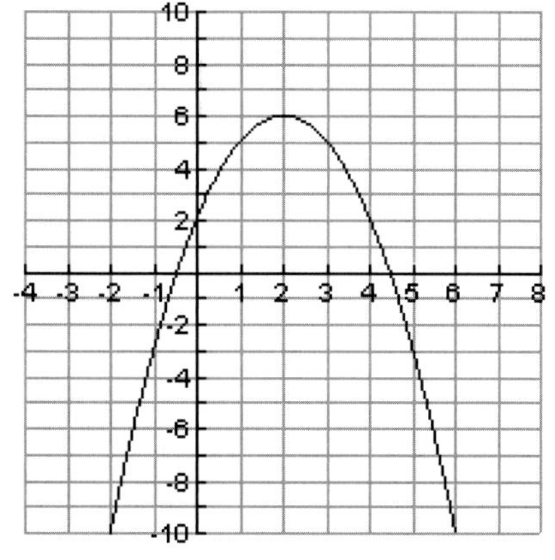

17.

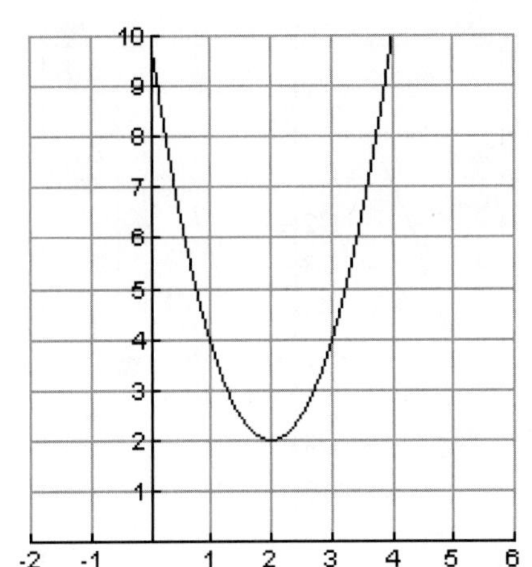
10
9
8
7
6
5
4
3
2
1
-2
-1
1
2
3
4
5
6

18.

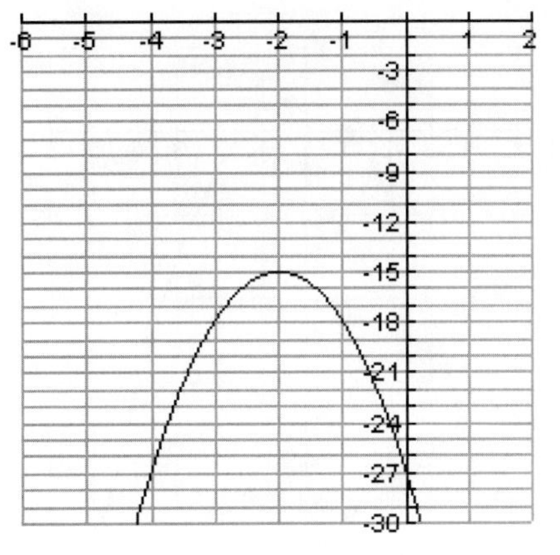
-6
-5
-4
-3
-2
-1
1
2
-3
-6
-9
-12
-15
-18
-21
-24
-27
-30

19.

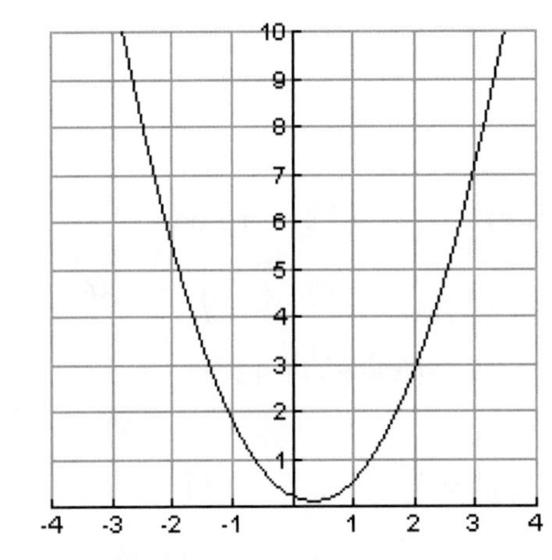
10
9
8
7
6
5
4
3
2
1
-4
-3
-2
-1
1
2
3
4

20.

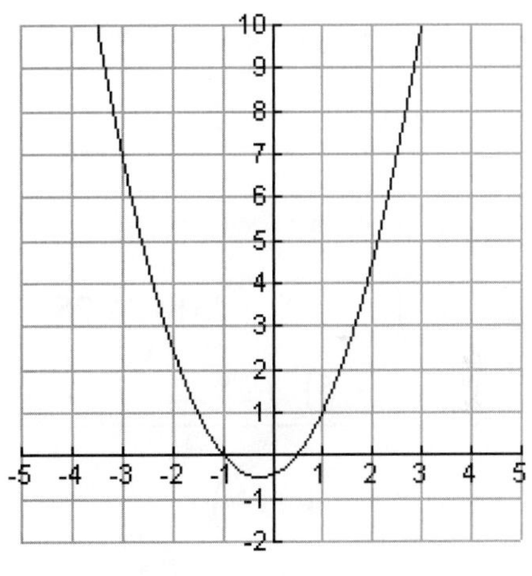
10
9
8
7
6
5
4
3
2
1
-1
-2
-5
-4
-3
-2
-1
1
2
3
4
5

21. 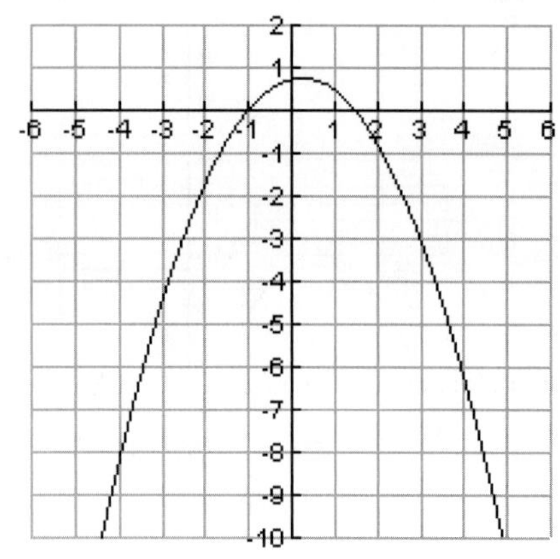	**22.** 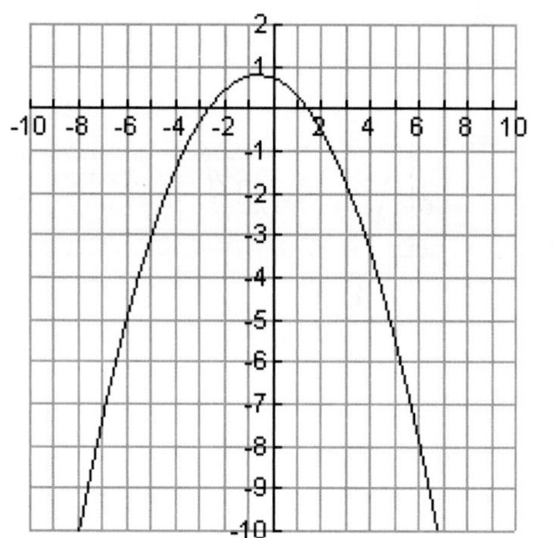
23. $f(x)=(x^2+6x+9)-3-9$ $=(x+3)^2-12$	**24.** $f(x)=(x^2+8x+16)+2-16$ $=(x+4)^2-14$
25. $f(x)=-(x^2+10x\quad)+3$ $=-(x^2+10x+25)+3+25$ $=-(x+5)^2+28$	**26.** $f(x)=-(x^2+12x\quad)+6$ $=-(x^2+12x+36)+6+36$ $=-(x+6)^2+42$
27. $f(x)=2(x^2+4x\quad)-2$ $=2(x^2+4x+4)-2-8$ $=2(x+2)^2-10$	**28.** $f(x)=3(x^2-3x\quad)+11$ $=3(x^2-3x+\frac{9}{4})+11-\frac{27}{4}$ $=3(x-\frac{3}{2})^2+\frac{17}{4}$
29. $f(x)=-4(x^2-4x\quad)-7$ $=-4(x^2-4x+4)-7+16$ $=-4(x-2)^2+9$	**30.** $f(x)=-5(x^2-20x\quad)-36$ $=-5(x^2-20x+100)-36+500$ $=-5(x-10)^2+464$

31.

$$f(x) = \left(x^2 + 10x + 25\right) - 25$$
$$= (x+5)^2 - 25$$

32.

$$f(x) = -4\left(x^2 - 3x \quad\right) - 2$$
$$= -4\left(x^2 - 3x + \tfrac{9}{4}\right) - 2 + 9$$
$$= -4\left(x - \tfrac{3}{2}\right)^2 + 7$$

33.

$$f(x) = \tfrac{1}{2}\left(x^2 - 8x \quad\right) + 3$$
$$= \tfrac{1}{2}\left(x^2 - 8x + 16\right) + 3 - 8$$
$$= \tfrac{1}{2}(x-4)^2 - 5$$

34.

$$f(x) = -\tfrac{1}{3}\left(x^2 - 18x \quad\right) + 4$$
$$= -\tfrac{1}{3}\left(x^2 - 18x + 81\right) + 4 + 27$$
$$= -\tfrac{1}{3}(x-9)^2 + 31$$

35.

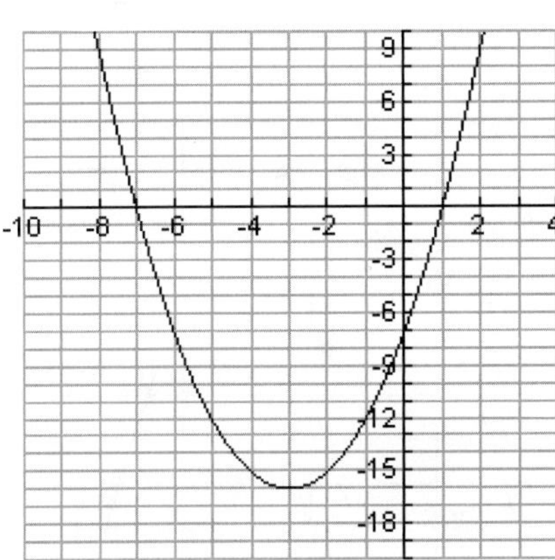

36.

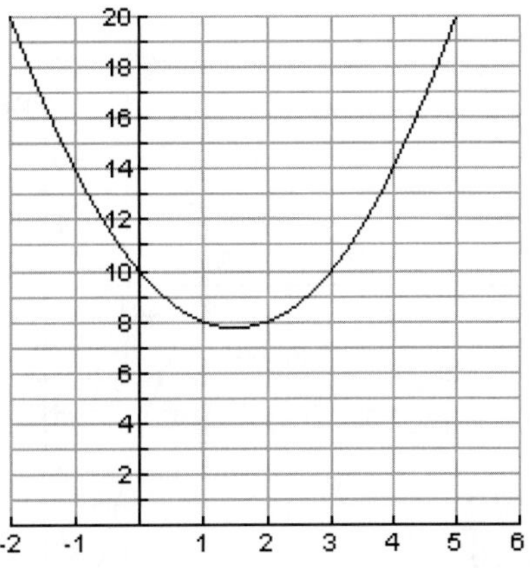

37.

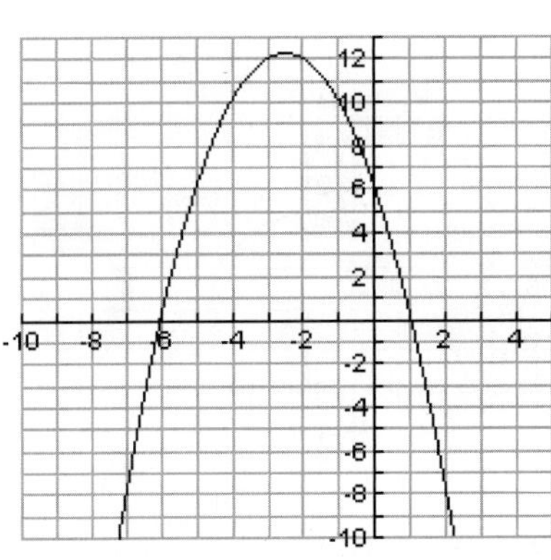

38.

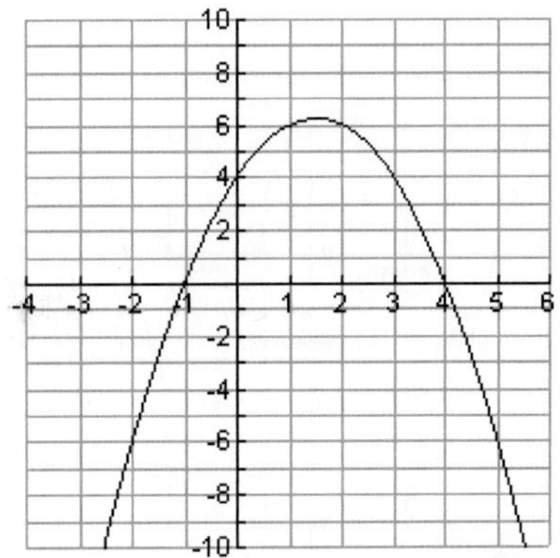

39.

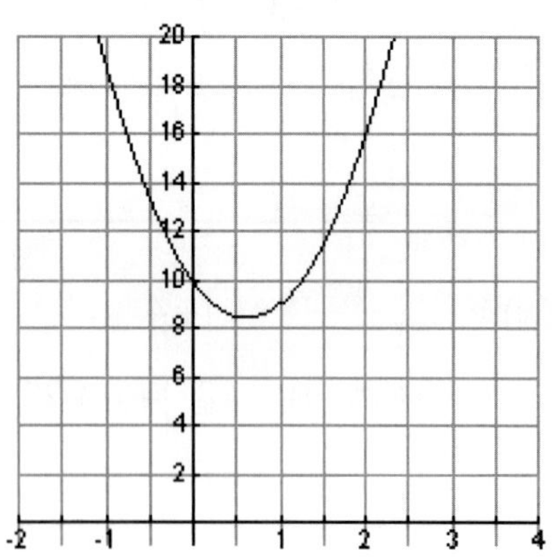

40.

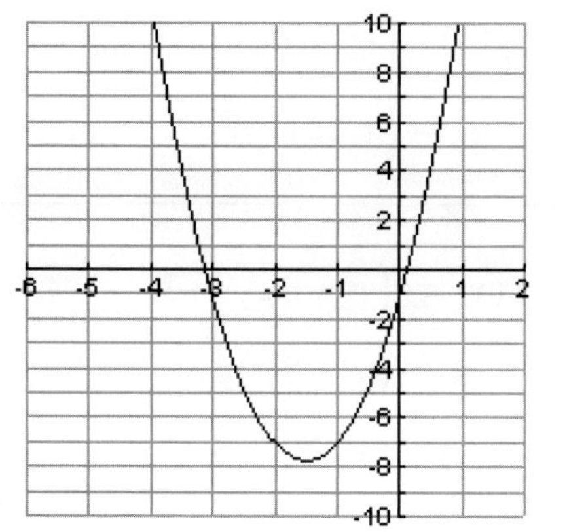

41.

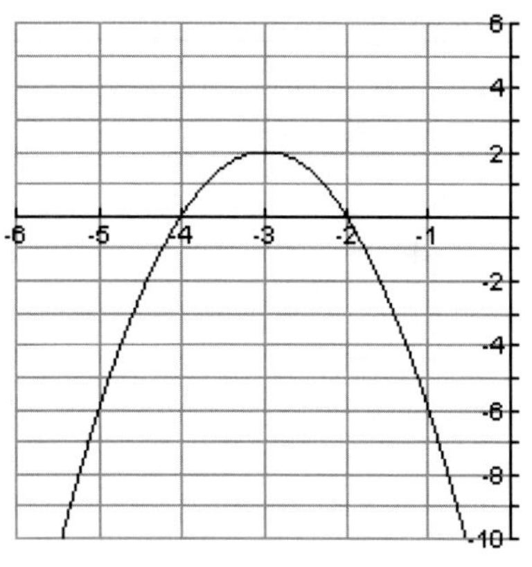

42.

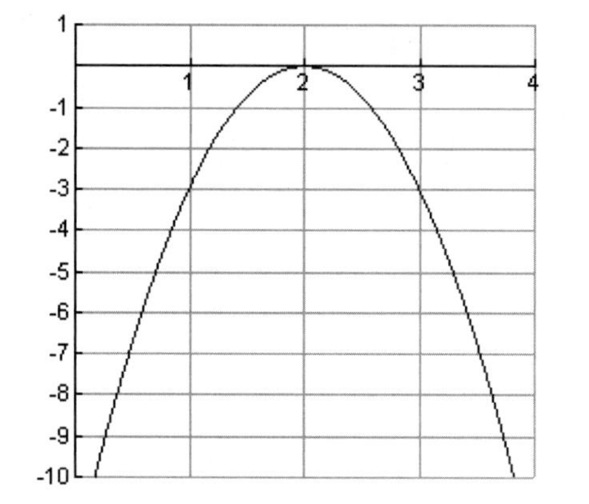

43.

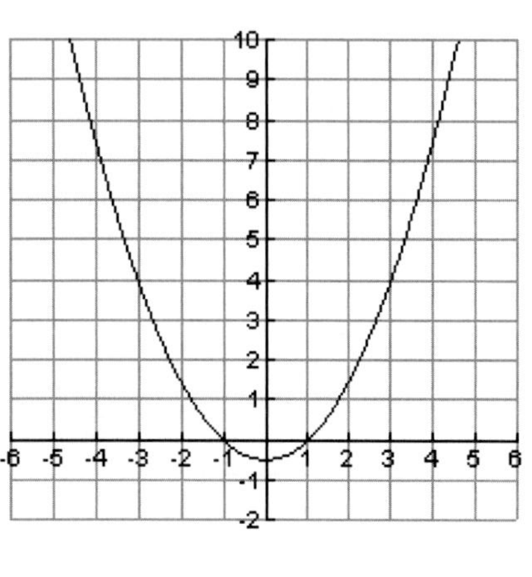

44.

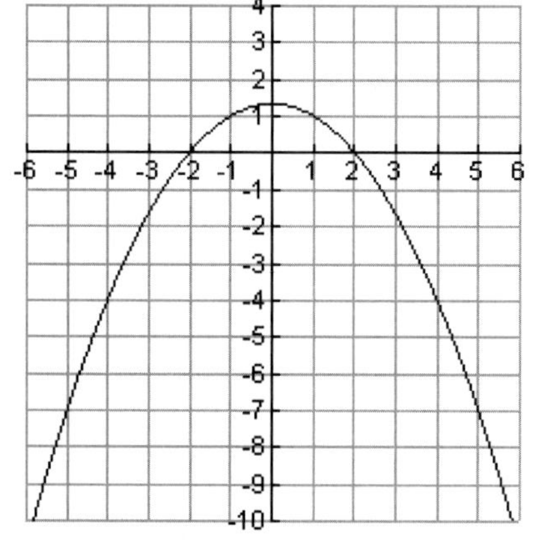

45.

$$\left(-\frac{b}{2a}, c-\frac{b^2}{4a}\right)=\left(-\frac{-2}{2(33)}, 15-\frac{(-2)^2}{4(33)}\right)$$
$$=\left(\frac{1}{33}, \frac{494}{33}\right)$$

46.

$$\left(-\frac{b}{2a}, c-\frac{b^2}{4a}\right)=\left(-\frac{4}{2(17)}, -3-\frac{(4)^2}{4(17)}\right)$$
$$=\left(-\frac{2}{17}, -\frac{55}{17}\right)$$

47.

$$\left(-\frac{b}{2a}, c-\frac{b^2}{4a}\right)=\left(-\frac{-7}{2(\frac{1}{2})}, 5-\frac{(-7)^2}{4(\frac{1}{2})}\right)$$
$$=\left(7, -\frac{39}{2}\right)$$

48.

$$\left(-\frac{b}{2a}, c-\frac{b^2}{4a}\right)=\left(-\frac{\frac{2}{5}}{2(-\frac{1}{3})}, 4-\frac{(\frac{2}{5})^2}{4(-\frac{1}{3})}\right)$$
$$=\left(\frac{3}{5}, \frac{103}{25}\right)$$

49.

$$\left(-\frac{b}{2a}, c-\frac{b^2}{4a}\right)=\left(-\frac{\frac{3}{7}}{2(-\frac{2}{5})}, 2-\frac{(\frac{3}{7})^2}{4(-\frac{2}{5})}\right)$$
$$=\left(\frac{15}{28}, \frac{829}{392}\right)$$

50.

$$\left(-\frac{b}{2a}, c-\frac{b^2}{4a}\right)=\left(-\frac{(-\frac{2}{3})}{2(-\frac{1}{7})}, \frac{1}{9}-\frac{(-\frac{2}{3})^2}{4(-\frac{1}{7})}\right)$$
$$=\left(-\frac{7}{3}, \frac{8}{9}\right)$$

51.

$$\left(-\frac{b}{2a}, c-\frac{b^2}{4a}\right)=\left(-\frac{-0.3}{2(-0.002)}, 1.7-\frac{(-0.3)^2}{4(-0.002)}\right)$$
$$=(-75, 12.95)$$

52.

$$\left(-\frac{b}{2a}, c-\frac{b^2}{4a}\right)=\left(-\frac{2.5}{2(0.05)}, -1.5-\frac{(2.5)^2}{4(0.05)}\right)$$
$$=(-25, -32.75)$$

53.

$$\left(-\frac{b}{2a}, c-\frac{b^2}{4a}\right)=\left(-\frac{-2.6}{2(0.06)}, 3.52-\frac{(-2.6)^2}{4(0.06)}\right)$$
$$=(21.67, -24.65)$$

54.

$$\left(-\frac{b}{2a}, c-\frac{b^2}{4a}\right)=\left(-\frac{0.8}{2(-3.2)}, -0.14-\frac{(0.8)^2}{4(-3.2)}\right)$$
$$=(0.125, -0.09)$$

55. Since the vertex is $(-1,4)$, the function has the form $y=a(x+1)^2+4$. To find *a*, use the fact that the point (0,2) is on the graph:

$$2=a(0+1)^2+4$$
$$-2=a$$

So, the function is $y=-2(x+1)^2+4$.

56. Since the vertex is $(2,-3)$, the function has the form $y=a(x-2)^2-3$. To find *a*, use the fact that the point (0,1) is on the graph:

$$1=a(0-2)^2-3$$
$$1=4a-3$$
$$1=a$$

So, the function is $y=(x-2)^2-3$.

57. Since the vertex is $(2,5)$, the function has the form $y=a(x-2)^2+5$. To find *a*, use the fact that the point (3,0) is on the graph: $$0=a(3-2)^2+5$$ $$-5=a$$ So, the function is $y=-5(x-2)^2+5$.	**58.** Since the vertex is $(1,3)$, the function has the form $y=a(x-1)^2+3$. To find *a*, use the fact that the point $(-2,0)$ is on the graph: $$0=a(-2-1)^2+3$$ $$0=9a+3$$ $$-\tfrac{1}{3}=a$$ So, the function is $y=-\tfrac{1}{3}(x-1)^2+3$.
59. Since the vertex is $(-1,-3)$, the function has the form $y=a(x+1)^2-3$. To find *a*, use the fact that the point $(-4,2)$ is on the graph: $$2=a(-4+1)^2-3$$ $$2=9a-3$$ $$\tfrac{5}{9}=a$$ So, the function is $y=\tfrac{5}{9}(x+1)^2-3$.	**60.** Since the vertex is $(0,-2)$, the function has the form $y=a(x-0)^2-2$. To find *a*, use the fact that the point $(3,10)$ is on the graph: $$10=a(3-0)^2-2$$ $$10=9a-2$$ $$\tfrac{12}{9}=\tfrac{4}{3}=a$$ So, the function is $y=\tfrac{4}{3}(x-0)^2-2$.
61. Since the vertex is $(-2,-4)$, the function has the form $y=a(x+2)^2-4$. To find *a*, use the fact that the point $(-1,6)$ is on the graph: $$6=a(-1+2)^2-4$$ $$6=a-4$$ $$10=a$$ So, the function is $y=10(x+2)^2-4$.	**62.** Since the vertex is $(5,4)$, the function has the form $y=a(x-5)^2+4$. To find *a*, use the fact that the point $(2,-5)$ is on the graph: $$-5=a(2-5)^2+4$$ $$-5=9a+4$$ $$-1=a$$ So, the function is $y=-(x-5)^2+4$.
63. Since the vertex is $(\tfrac{1}{2},\tfrac{-3}{4})$, the function has the form $y=a(x-\tfrac{1}{2})^2-\tfrac{3}{4}$. To find *a*, use the fact that the point $(\tfrac{3}{4},0)$ is on the graph: $$0=a(\tfrac{3}{4}-\tfrac{1}{2})^2-\tfrac{3}{4}$$ $$0=\tfrac{1}{16}a-\tfrac{3}{4}$$ $$12=a$$ So, the function is $y=12(x-\tfrac{1}{2})^2-\tfrac{3}{4}$.	**64.** Since the vertex is $(-\tfrac{5}{6},\tfrac{2}{3})$, the function has the form $y=a(x+\tfrac{5}{6})^2+\tfrac{2}{3}$. To find *a*, use the fact that the point $(0,0)$ is on the graph: $$0=a(0+\tfrac{5}{6})^2+\tfrac{2}{3}$$ $$0=\tfrac{25}{36}a+\tfrac{2}{3}$$ $$-\tfrac{24}{25}=a$$ So, the function is $y=-\tfrac{24}{25}(x+\tfrac{5}{6})^2+\tfrac{2}{3}$.

65. Since the vertex is $(2.5,-3.5)$, the function has the form $y=a(x-2.5)^2-3.5$. To find a, use the fact that the point $(4.5,1.5)$ is on the graph:

$$1.5=a(4.5-2.5)^2-3.5$$
$$1.5=4a-3.5$$
$$\tfrac{5}{4}=a$$

So, the function is $y=\tfrac{5}{4}(x-2.5)^2-3.5$.

66. Since the vertex is $(1.8,2.7)$, the function has the form $y=a(x-1.8)^2+2.7$. To find a, use the fact that the point $(-2.2,-2.1)$ is on the graph:

$$-2.1=a(-2.2-1.8)^2+2.7$$
$$-2.1=16a+2.7$$
$$-0.3=a$$

So, the function is $y=-0.3(x-1.8)^2+2.7$.

67. a. The maximum occurs at the vertex, which is $(-5,40)$. So, the maximum height is $\boxed{\text{120 feet}}$.

b. If the height of the ball is assumed to be zero when the ball is kicked, and is zero when it lands, then we need to simply compute the x-intercepts of h and determine the distance between them. To this end, solve

$$0=-\tfrac{8}{125}(x+5)^2+40 \Rightarrow \underbrace{40\left(\tfrac{125}{8}\right)}_{=625}=(x+5)^2 \Rightarrow \pm 25=x+5 \Rightarrow x=-30,20$$

So, the distance the ball covers is $\boxed{\text{50 yards}}$.

68. a. The maximum occurs at the vertex, which is $(30,50)$. So, the maximum height is $\boxed{\text{150 feet}}$.

b. If the height of the ball is assumed to be zero when the ball is kicked, and is zero when it lands, then we need to simply compute the x-intercepts of h and determine the distance between them. To this end, solve

$$0=-\tfrac{5}{40}(x-30)^2+50 \Rightarrow \underbrace{50\left(\tfrac{40}{5}\right)}_{=400}=(x-30)^2 \Rightarrow \pm 20=x-30 \Rightarrow x=10,50$$

So, the distance the ball covers is $\boxed{\text{40 yards}}$.

69. Let x = length and y = width.

The total amount of fence is given by: $4x+3y=10,000$ so that $y=\dfrac{10,000-4x}{3}$ **(1)**.

The combined area of the two identical pens is $2xy$. Substituting **(1)** in for y, we see that the area is described by the function:

$$A(x)=2x\left(\frac{10,000-4x}{3}\right)=-\tfrac{8}{3}x^2+\tfrac{20,000}{3}x$$

Since this parabola opens downward (since the coefficient of x^2 is negative), the maximum occurs at the x-coordinate of the vertex, namely

$$x=-\tfrac{b}{2a}=\frac{-\frac{20,000}{3}}{2\left(-\frac{8}{3}\right)}=1250.$$

The corresponding width of the pen (from **(1)**) is $y=\dfrac{10,000-4(1250)}{3}\approx 1666.67$

So, each of the two pens would have area $\boxed{\cong 2,083,333 \text{ sq. ft.}}$

70. Let x = length of one of the four pastures, y = width of one of the four pastures.

The total amount of fence is given by: $8x+5y=30,000$ so that $y=\dfrac{30,000-8x}{5}$ **(1)**.

The combined area of the four identical pastures is $4xy$. Substituting **(1)** in for y, we see that the area is described by the function:

$$A(x)=4x\left(\frac{30,000-8x}{5}\right)=-\tfrac{32}{5}x^2+24,000x$$

Since this parabola opens downward (since the coefficient of x^2 is negative), the maximum occurs at the x-coordinate of the vertex, namely

$$x=-\tfrac{b}{2a}=\frac{-24,000}{2\left(-\frac{32}{5}\right)}=1875$$

The corresponding width of the pen (from **(1)**) is $y=\dfrac{30,000-8(1875)}{5}=3000$

So, each of the four pastures would have area $\boxed{5,625,000 \text{ sq. ft.}}$

71. a. Completing the square on h yields

$$h(t) = -16\left(t^2 - 2t\right) + 100$$
$$= -16\left(t^2 - 2t + 1\right) + 100 + 16$$
$$= -16(t-1)^2 + 116$$

So, it takes 1 second to reach maximum height of 116 ft.

b. Solve $h(t) = 0$.

$$-16(t-1)^2 + 116 = 0$$
$$(t-1)^2 = \tfrac{116}{16}$$
$$t - 1 = \pm\sqrt{\tfrac{116}{16}}$$
$$t = 1 \pm \sqrt{\tfrac{116}{16}}$$

Since time must be positive, we conclude that the rock hits the water after about $t = 1 + \sqrt{\tfrac{116}{16}} \cong 3.69$ seconds (assuming the time started at $t = 0$).

c. The rock is above the cliff between 0 and 2 seconds (0<t<2).

72. Solve $-16t^2 + 1200t = 0$:

$$-16t^2 + 1200t = 0$$
$$t(-16t + 1200) = 0$$
$$t = 0, 75$$

So, the person has 75 seconds to get out of the way of the bullet.

73.

$$A(x) = -0.0003\left(x^2 - 31{,}000x\right) - 46{,}075$$
$$= -0.0003\left(x - 15{,}500\right)^2 - 46{,}075 + 72{,}075$$
$$= -0.0003(x - 15{,}500)^2 + 26{,}000$$

Also, we need the x-intercepts to determine the horizontal distance. Observe

$$-0.0003(x - 15{,}500)^2 + 26{,}000 = 0$$
$$(x - 15{,}500)^2 = \frac{26{,}000}{0.0003} \approx 86{,}666{,}667$$
$$x - 15{,}500 \approx \pm 9309.49$$
$$x \approx 15{,}500 \pm 9309.49$$
$$= 6{,}309.49 \text{ and } 24{,}809.49$$

So, the maximum altitude is 26,000 ft. over a horizontal distance of 8,944 ft.

74. a. Maximum height is the y-coordinate of the vertex of $H(x)$. Completing the square yields

$$H(x) = -0.0128\left(x^2 - 78.125x\right) = -0.0128\left(x^2 - 78.125x + 1525.87891\right) + 19.53125$$
$$= -0.0128(x - 39.06250)^2 + 19.53125$$

So, the maximum height is $\boxed{19.53125 \text{ ft}}$.

b. Need the x-intercepts. To find them, solve

$$-0.0128x^2 + x = 0 \;\Rightarrow\; x(-0.0128x + 1) = 0 \;\Rightarrow\; x = 0, 78.125$$

So, the ball travels $\boxed{78.125 \text{ ft.}}$

75. a. Maximum profit occurs at the x-coordinate of the vertex of $P(x)$. Completing the square yields

$$P(x) = -0.4\left(x^2 + 200x \quad\right) + 20{,}000$$
$$= -0.4\left(x^2 + 200x + 10{,}000\right) + 20{,}000 + 4{,}000 = -0.4(x-100)^2 + 24{,}000$$

So, the profit is maximized when 100 boards are sold.

b. The maximum profit is the y-coordinate of the vertex, namely $\boxed{\$24{,}000}$.

76. a. Since the vertex is $(50,30)$, the function has the form $y = a(x-50)^2 + 30$. To find a, use the fact that the point $(70,25)$ is on the graph:

$$25 = a(75-50)^2 + 30$$
$$-5 = 400a \text{ so that } -0.0125 = a$$

So, the function is $y(x) = -0.0125(x-50)^2 + 30$.

b. Since $y(90) = -0.0125(90-50)^2 + 30 = 10$, you would expect 10 mpg.

77. First, completing the square yields

$$P(x) = (100-x)x - 1000 - 20x$$
$$= -x^2 + 80x - 1000 = -\left(x^2 - 80x\right) - 1000 = -\left(x^2 - 80x + 1600\right) - 1000 + 1600$$
$$= -(x-40)^2 + 600$$

Now, solve $-(x-40)^2 + 600 = 0$:

$$-(x-40)^2 + 600 = 0$$
$$(x-40)^2 = 600$$
$$x - 40 = \pm\sqrt{600} \text{ so that } x = 40 \pm \sqrt{600} \approx 15.5,\ 64.5$$

So, $\boxed{\text{15 to 16 units to break even, or 64 or 65 units to break even.}}$

78. Using #69, we see that the maximum profit occurs at the y-coordinate of the vertex, namely \$600.

79. We are given that the vertex is $(h,k) = (0, 16)$ and that (9, 100) is on the graph.

a. We need to find a such that the equation governing the situation is

$$y = a(x-0)^2 + 16. \quad \textbf{(1)}$$

To do this, we use the fact that (9, 100) satisfies **(1)**: $100 = a(9-0)^2 + 16 \Rightarrow a = \frac{28}{27}$

Thus, the equation is $\boxed{y = \frac{28}{27}x^2 + 16}$, where y is measured in millions.

b. Note that the year 2010 corresponds to $t = 14$. Observe that

$$y(14) = \tfrac{28}{27}(14)^2 + 16 \approx \boxed{\text{219 million}}.$$

80. a. We are given that the vertex is $(h,k)=(0, 49)$, that (8, 36) is on the graph (since 2006 corresponds to $t = 8$), and that the graph opens down ($a < 0$) since the peak occurs at the vertex. We need to find a such that the equation governing the situation is

$$y = a(x-0)^2 + 49. \quad \textbf{(1)}$$

To do this, we use the fact that (8, 36) satisfies **(1)**: $36 = a(8-0)^2 + 49 \Rightarrow a = -\frac{13}{64}$

Thus, the equation is $\boxed{y = -\frac{13}{64}x^2 + 49}$.

b. Note that the year 2010 corresponds to $x = 12$, so that $y(12) = -\frac{13}{64}(12)^2 + 49 \approx 19.75$. So, approximately $\boxed{19.75\%.}$

81. a. We are given that the vertex is $(h,k)=(225, 400)$, that (50, 93.75) is on the graph, and that the graph opens down ($a < 0$) since the peak occurs at the vertex. We need to find a such that the equation governing the situation is

$$y = a(t-225)^2 + 400. \quad \textbf{(1)}$$

To do this, we use the fact that (50, 93.75) satisfies **(1)**:

$$93.75 = a(50-225)^2 + 400 \Rightarrow a = -0.01$$

Thus, the equation is $\boxed{y = -0.01(t-225)^2 + 400}$.

b. We must find the value(s) of t for which $0 = -0.01(t-225)^2 + 400$. To this end,

$$0 = -0.01(t-225)^2 + 400 \Rightarrow (t-225)^2 = 40{,}000$$
$$\Rightarrow t - 225 = \pm\sqrt{40{,}000} = \pm 200$$
$$\Rightarrow t = 225 \pm 200 = 25, 425$$

So, it takes $\boxed{\text{425 minutes}}$ for the drug to be eliminated from the bloodstream.

82. a. We know that the points (70, 20) and (50, 25) lie on this line. Hence, the slope is $m = \frac{25-20}{50-70} = -\frac{1}{4}$. Using point-slope form, the price function is:

$$p - 20 = -\tfrac{1}{4}(x-70) \Rightarrow \boxed{p(x) = -\tfrac{1}{4}x + \tfrac{75}{2}}$$

b. $R(x) = xp(x) = -\frac{1}{4}x^2 + \frac{75}{2}x$

c. The maximum revenue occurs at the vertex of $R(x)$. Completing the square yields

$$-\tfrac{1}{4}x^2 + \tfrac{75}{2}x = -\tfrac{1}{4}\left(x^2 - 150x\right)$$
$$= -\tfrac{1}{4}\left(x^2 - 150x + 5{,}625\right) + \tfrac{5{,}625}{4}$$
$$= -\tfrac{1}{4}\left(x - 75\right)^2 + \tfrac{5{,}625}{4}$$

The vertex is $\left(75, \frac{5{,}625}{4}\right)$. So, he needs to wash $\boxed{\text{75 cars}}$ in order to maximize revenue.

d. To maximize revenue, he should charge $p(x) = -\frac{1}{4}(75) + \frac{75}{2} = \frac{75}{4} = \boxed{\$18.75}$.

83. Step 2 is wrong: Vertex is $(-3,-1)$ Step 4 is wrong: The x-intercepts are $(-2,0)$, $(-4,0)$. So, should graph $y=(x+3)^2-1$.	
84. $b=-6$, not 6	**85.** Step 2 is wrong: $\left(-x^2+2x\right)=-\left(x^2-2x\right)$
86. Step 3 is wrong: $f(9)=a(9-2)^2-3=0$ $49a-3=0$ $a=\frac{3}{49}$ So, $f(x)=\frac{3}{49}(x-2)^2-3$.	**87.** True. $f(x)=a(x-h)^2+k$, so that $f(0)=ah^2+k$. **88.** False. Consider $f(x)=-1-x^2$.
89. False. The graph would not pass the vertical line test in such case, and hence wouldn't define a function.	**90.** True. Consider $f(x)=x^2-1$.

91. Completing the square yields

$$\begin{aligned} f(x) &= ax^2+bx+c \\ &= a\left(x^2+\tfrac{b}{a}x\right)+c \\ &= a\left(x^2+\tfrac{b}{a}x+\left(\tfrac{b}{2a}\right)^2\right)+c-a\left(\tfrac{b}{2a}\right)^2 \\ &= a\left(x+\tfrac{b}{2a}\right)^2+c-\tfrac{b^2}{4a}=a\left(x+\tfrac{b}{2a}\right)^2+\tfrac{4ac-b^2}{4a} \end{aligned}$$

So, the vertex is $\left(-\frac{b}{2a},c-\frac{b^2}{4a}\right)$. Observe that $f\left(-\frac{b}{2a}\right)=a\underbrace{\left(-\frac{b}{2a}+\frac{b}{2a}\right)^2}_{=0}+c-\frac{b^2}{4a}=c-\frac{b^2}{4a}$.

92. Given that $f(x)=a(x-h)^2+k$, we have:

y-intercept: $a(0-h)^2+k=ah^2+k$, so that the y-intercept is $\left(0,ah^2+k\right)$.

x-intercepts: Solve $a(x-h)^2+k=0$.

$$\begin{aligned} a(x-h)^2+k &= 0 \\ (x-h)^2 &= -\tfrac{k}{a} \\ x-h &= \pm\sqrt{-\tfrac{k}{a}} \\ x &= h\pm\sqrt{-\tfrac{k}{a}} \end{aligned}$$

So, the x-intercepts are $\left(h+\sqrt{-\frac{k}{a}},0\right)$, $\left(h-\sqrt{-\frac{k}{a}},0\right)$.

93. a. Let x = width of rectangular pasture, y = length of rectangular pasture.
Then, the total amount of fence is described by $2x+2y=1000$ (so that $y=500-x$ **(1)**).

The area of the pasture is xy. Substituting in **(1)** yields $x(500-x)=-x^2+500x$.

The maximum area occurs at the y-coordinate of the vertex (since the coefficient of x^2 is negative); in this case this value is $c-\frac{b^2}{4a}=0-\frac{500^2}{4(-1)}=62,500$.

So the maximum area is 62,500 sq. ft.

b. Let x = radius of circular pasture.
Then, the total amount of fence is described by $2\pi x=1000$ (so that $x=\frac{1000}{2\pi}=\frac{500}{\pi}$ **(2)**).

The area of the pasture is πx^2, which in this case must be (by **(2)**) $\pi\left(\frac{500}{\pi}\right)^2=\frac{500^2}{\pi}$.

So, the area of the pasture must be approximately 79,577 sq. ft.

94. Let x = number of increases in room rate. Then, the monthly income for the hotel is

$$I(x)=(90+5x)(600-10x)=-50x^2+2100x+54,000$$

Completing the square then yields

$$\begin{aligned}-50x^2+2100x+54,000&=-50\left(x^2-42x\right)+54,000\\&=-50\left(x^2-42x+441\right)+54,000+22,050\\&=-50(x-21)^2+76,050\end{aligned}$$

Since the coefficient of x^2 is negative, the maximum income must be \$76,050, which occurs when there are 21 increases in room rate (i.e., at the vertex). Hence, the room rate that yields the maximum profit is $90+21(5)=\$195$.

95. Observe that

$$\begin{aligned}&\frac{1}{x+11}+\frac{1}{x+4}=\frac{25}{144}\\&\frac{x+4+x+11}{(x+11)(x+4)}=\frac{25}{144}\\&144(2x+15)=25(x+11)(x+4)\\&288x+2160=25x^2+375x+1100\\&25x^2+87x-1060=0\\&x=\frac{-87\pm\sqrt{87^2+4925)(1060)}}{2(25)}=5,\ \cancel{-8.48}\end{aligned}$$

Since speed cannot be negative, we conclude that $x=5$.

96. Let x = length and y = width of the original rectangle.
The 25% reduction results in a rectangle whose length is 0.25x, which equals y by assumption, and whose width is 0.25y. As such, using the information about the area yields

$$y(0.25y)=36 \Rightarrow y^2=144 \Rightarrow y=12.$$

So, the original rectangle has dimensions 12 ft by 12/0.25 = 44 ft.

97. a. Vertex:

$$\left(-\tfrac{5.7}{2(-0.002)}, -23-\tfrac{(5.7)^2}{4(-0.002)}\right)=(1425,4038.25)$$

b. y-intercept: $f(0)=-23$, so $(0,-23)$

c. x-intercepts: Solve

$$-0.002x^2+5.7x-23=0.$$

Using part **a.**, the standard form is $-0.002(x-1425)^2+4038.25=0$. So, solving this equation yields:

$$(x-1425)^2=2{,}019{,}125$$

$$x=1425\pm\sqrt{2{,}019{,}125}$$

$$\cong 4.04,\ 2845.96$$

So, the x-intercepts are

$$(4.04,0),\ (2845.96,0).$$

d. Axis of symmetry: $x=1425$

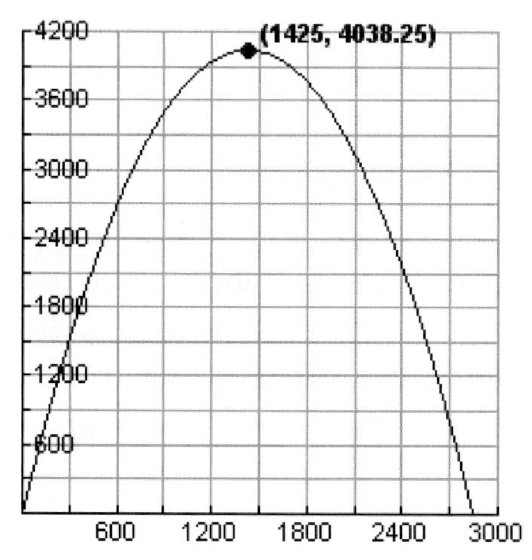

b.

98. a. Since the vertex is $(-0.5,1.7)$, the function has the form $y=a(x+0.5)^2+1.7$. To find a, use the fact that the point $(0,4)$ is on the graph:

$$4=a(0+0.5)^2+1.7$$

$$4=0.25a+1.7$$

$$9.2=a$$

So, the function is $y=9.2(x+0.5)^2+1.7$.

c. Yes, they agree.

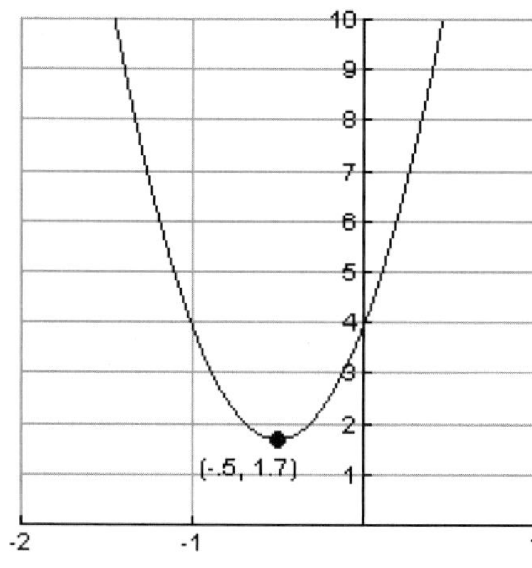

99. a. The equation obtained using this particular calculator feature is

$$f(x) = -2x^2 + 12.8x + 4.32.$$

b. Completing the square yields

$$\begin{aligned} f(x) &= -2x^2 + 12.8x + 4.32 \\ &= -2\left(x^2 - 6.4x\right) + 4.32 \\ &= -2\left(x^2 - 6.4x + 10.24\right) + 4.32 + 20.48 \\ &= -2(x - 3.2)^2 + 24.8 \end{aligned}$$

So, the vertex is $(3.2, 24.8)$.

c. The graph is to the right. Yes, they agree with the given values.

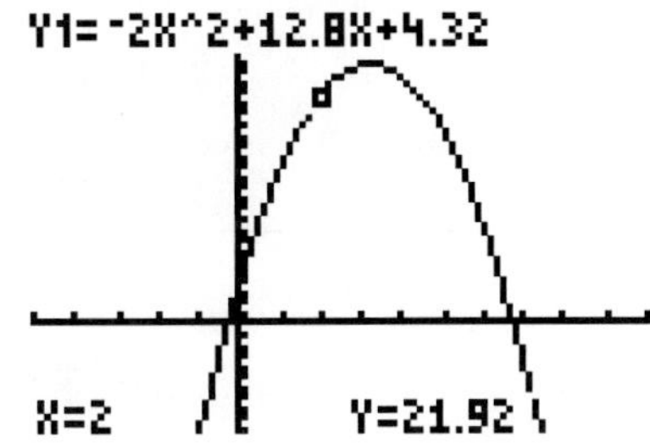

100. a. The equation obtained using this particular calculator feature is

$$f(x) = 0.44x^2 + 2.92x - 12.10.$$

b. Completing the square yields

$$\begin{aligned} f(x) &= 0.44x^2 + 2.92x - 12.10 \\ &= 0.44\left(x^2 + 6.64x\right) - 12.10 \\ &= 0.44\left(x^2 + 6.64x + 11.0224\right) - 12.10 - 4.85 \\ &= 0.44(x + 3.32)^2 - 16.95 \end{aligned}$$

So, the vertex is $(-3.32, -16.95)$.

c. The graph is to the right. Yes, they agree with the given values.

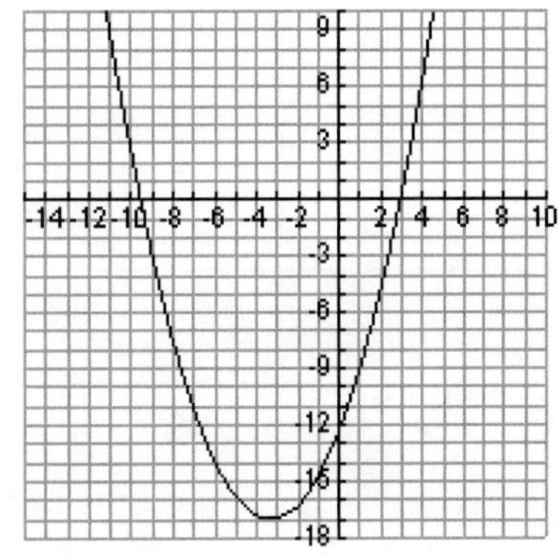

101.

$$\begin{aligned} 4(x-0)^2 + 9\left(y^2 - 4y\right) &= 0 \\ 4(x-0)^2 + 9\left(y^2 - 4y + 4\right) &= 36 \\ 4(x-0)^2 + 9(y-2)^2 &= 36 \\ \frac{(x-0)^2}{9} + \frac{(y-2)^2}{4} &= 1 \quad \text{ellipse} \end{aligned}$$

102.	$$x^2-2x-4y^2+16y=19$$ $$\left(x^2-2x+1\right)-4\left(y^2-4y+4\right)=19+1-16$$ $$(x-1)^2-4(y-2)^2=4$$ $$\frac{(x-1)^2}{4}-(y-2)^2=1 \quad \text{hyperbola}$$
103.	$$x^2+6x=20y-5$$ $$x^2+6x+9=20y-5+9$$ $$(x+3)^2=20y+4$$ $$(x+3)^2=20(y+\tfrac{1}{5}) \quad \text{parabola}$$
104.	$$x^2-6x+4y^2+40y=-105$$ $$\left(x^2-6x+9\right)+4\left(y^2+10y+25\right)=-105+9+100$$ $$(x-3)^2+4(y+5)^2=4$$ $$\frac{(x-3)^2}{4}+(y+5)^2=1 \quad \text{ellipse}$$

Section 2.2 Solutions ---

1. Polynomial with degree 5	**2.** Polynomial with degree 6
3. Polynomial with degree 7	**4.** Polynomial with degree 9
5. Not a polynomial (due to the term $x^{1/2}$)	**6.** Not a polynomial (due to the term $x^{1/2}$)
7. Not a polynomial (due to the term $x^{1/3}$)	**8.** Not a polynomial (due to the term $\frac{2}{3x}$)
9. Not a polynomial (due to the terms $\frac{1}{x}, \frac{1}{x^2}$)	**10.** Polynomial with degree 2
11. h linear function	**12. g** Parabola that opens down
13. b Parabola that opens up	**14. f** Note that $-2x^3+4x^2-6x=-2x\left(x^2-2x+3\right)$. Since x^2-2x+3 is a parabola opening up with vertex (1,2), it has no real roots. So, this polynomial has only 1 *x*-intercept at which it crosses.

15. e $x^3 - x^2 = x^2(x-1)$ So, there are two x-intercepts: the graph is tangent at 0 and crosses at 1.

16. d Note that

$$2x^4 - 18x^2 = 2x^2\left(x^2 - 9\right) = 2x^2(x-3)(x+3)$$

There are three x-intercepts (0, 3, -3) and it crosses at each of them.

17. c $-x^4 + 5x^3 = -x^3(x-5)$ So, there are two x-intercepts (0, 5) and the graph crosses at each of them.

18. a

$$\begin{aligned} x^5 - 5x^3 + 4x &= x\left(x^4 - 5x^2 + 4\right) \\ &= x\left(x^2 - 4\right)\left(x^2 - 1\right) \\ &= x(x-1)(x+1)(x-2)(x+2) \end{aligned}$$

So, there are five x-intercepts (0, 1, 2, -1, -2) and the graph crosses at each of them.

19.

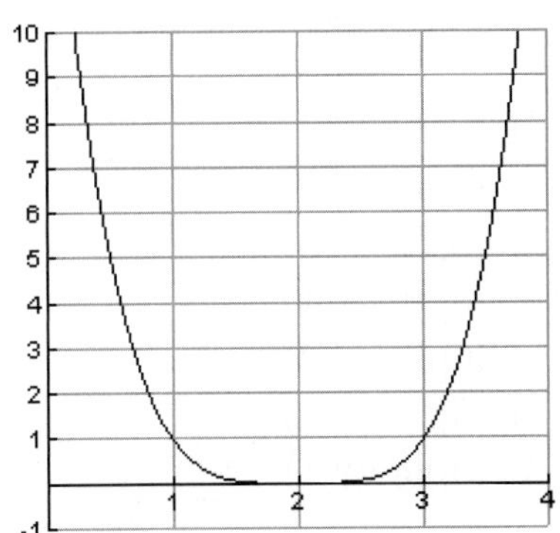

20.

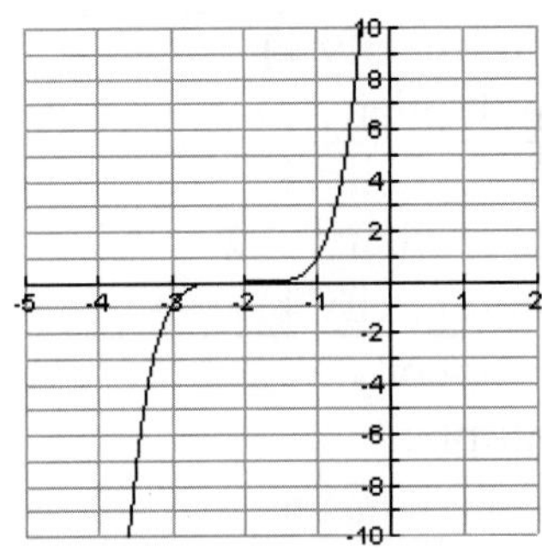

21.

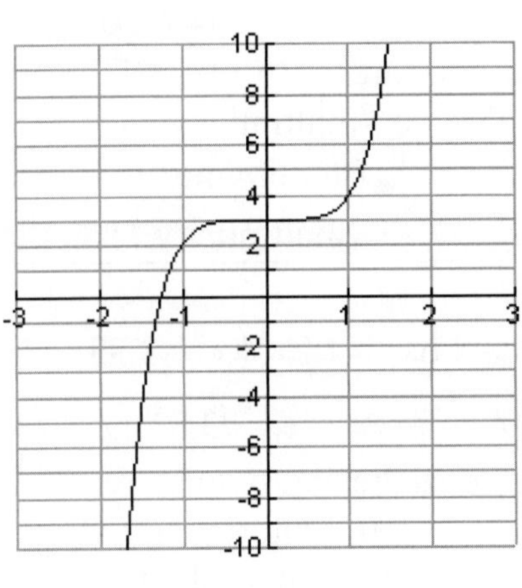

22.

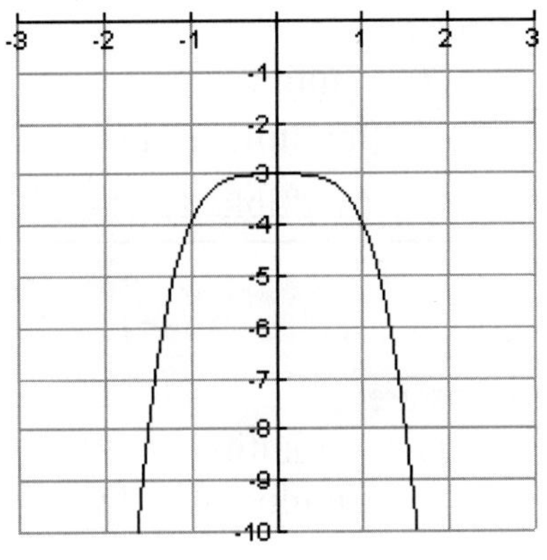

23. 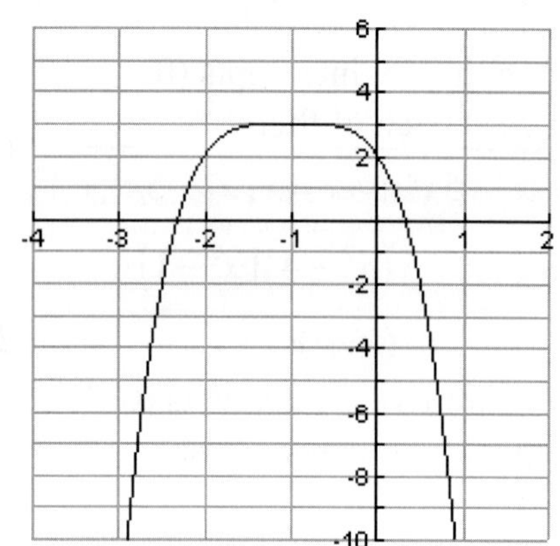	**24.** 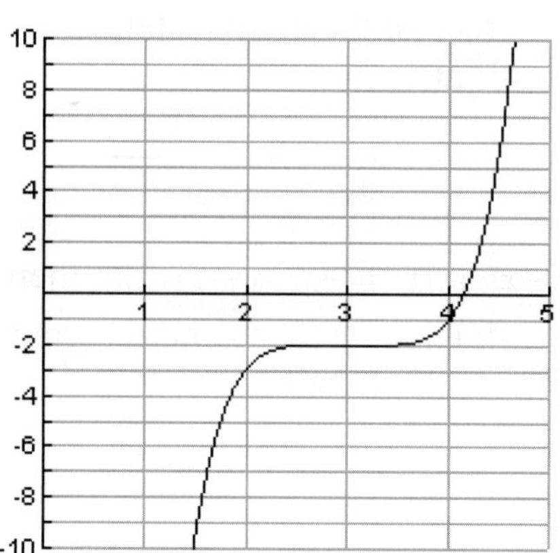
25. 3 (multiplicity 1) –4 (multiplicity 3)	**26.** –2 (multiplicity 3) 1 (multiplicity 2)
27. 0 (multiplicity 2) 7 (multiplicity 2) –4 (multiplicity 1)	**28.** 0 (multiplicity 3) –1 (multiplicity 4) 6 (multiplicity 1)
29. 0 (multiplicity 2) 1 (multiplicity 2) Note: $x^2+4=0$ has no real solutions.	**30.** 0 (multiplicity 2) –1 (multiplicity 1) 1 (multiplicity 1)
31. $8x^3+6x^2-27x=x\left(8x^2+6x-27\right)$ $=x(2x-3)(4x+9)$ So, the zeros are: 0 (multiplicity 1) $\frac{3}{2}$ (multiplicity 1) $-\frac{9}{4}$ (multiplicity 1)	**32.** $2x^4+5x^3-3x^2=x^2\left(2x^2+5x-3\right)$ $=x^2(2x-1)(x+3)$ So, the zeros are: 0 (multiplicity 2) $\frac{1}{2}$ (multiplicity 1) –3 (multiplicity 1)
33. $-2.7x^3-8.1x^2=-2.7x^2\left(x+3\right)$ So, the zeros are: 0 (multiplicity 2) –3 (multiplicity 1) **35.** $\frac{1}{3}x^6+\frac{2}{5}x^4=\frac{1}{3}x^4\underbrace{\left(x^2+\frac{6}{5}\right)}_{\text{Always positive}}$ So, the only zero is: 0 (multiplicity 4)	**34.** $1.2x^6-4.6x^4\cong1.2x^4\left(x^2-3.83\right)$ $=x(x-\sqrt{3.83})(x+\sqrt{3.83})$ So, the zeros are approximately: 0 (multiplicity 1) 1.957 (multiplicity 1) –1.957 (multiplicity 1)

36. Note that $\frac{2}{7}x^5 - \frac{3}{4}x^4 + \frac{1}{2}x^3 = \frac{2}{7}x^3\left(x^2 - \frac{21}{8}x + \frac{7}{4}\right)$. To determine if the second term is factorable, we complete the square:

$$x^2 - \frac{21}{8}x + \frac{7}{4} = \left(x^2 - \frac{21}{8}x + \left(\frac{21}{16}\right)^2\right) + \frac{7}{4} - \left(\frac{21}{16}\right)^2$$
$$= \underbrace{\left(x - \frac{21}{16}\right)^2 + \frac{7}{256}}_{\text{Always positive}}$$

Since $\left(x - \frac{21}{16}\right)^2 + \frac{7}{256} = 0$ has no real solutions, we conclude that the only real zero is:

0 (multiplicity 3)

37. $P(x) = x(x+3)(x-1)(x-2)$	**38.** $P(x) = x(x+2)(x-2)$
39. $P(x) = x(x+5)(x+3)(x-2)(x-6)$	**40.** $P(x) = x(x-1)(x-3)(x-5)(x-10)$
41. $P(x) = (2x+1)(3x-2)(4x-3)$	**42.** $P(x) = x(4x+3)(3x+1)(2x-1)$
43. $P(x) = \left(x-(1-\sqrt{2})\right)\left(x-(1+\sqrt{2})\right) = x^2 - 2x - 1$	**44.** $P(x) = \left(x-(1-\sqrt{3})\right)\left(x-(1+\sqrt{3})\right) = x^2 - 2x - 2$
45. $P(x) = x^2(x+2)^3$	**46.** $P(x) = (x+4)^2(x-5)^3$
47. $P(x) = (x+3)^2(x-7)^5$	**48.** $P(x) = x(x-10)^3$
49. $P(x) = x^2(x+1)(x+\sqrt{3})^2(x-\sqrt{3})^2$	**50.** $P(x) = x(x-1)^2(x+\sqrt{5})^2(x-\sqrt{5})^2$

51. $f(x) = (x-2)^3$

a. Zeros: 2 (multiplicity 3)

b. Crosses at 2

c. y-intercept: $f(0) = -8$, so $(0, -8)$

d. End behavior: Behaves like $y = x^3$.

Odd degree and leading coefficient positive, so graph falls without bound to the left and rises to the right.

e.

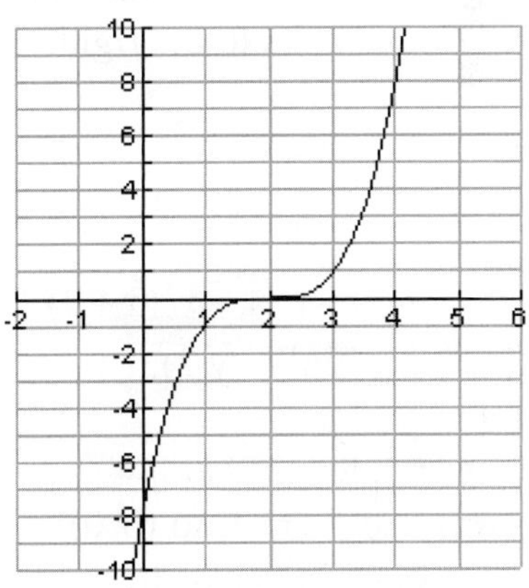

52. $f(x) = -(x+3)^3$

a. Zeros: -3 (multiplicity 3)

b. Crosses at -3

c. y-intercept: $f(0) = -27$, so $(0,-27)$

d. End behavior: Behaves like $y = -x^3$.

Odd degree and leading coefficient negative, so graph falls without bound to the right and rises to the left.

e.

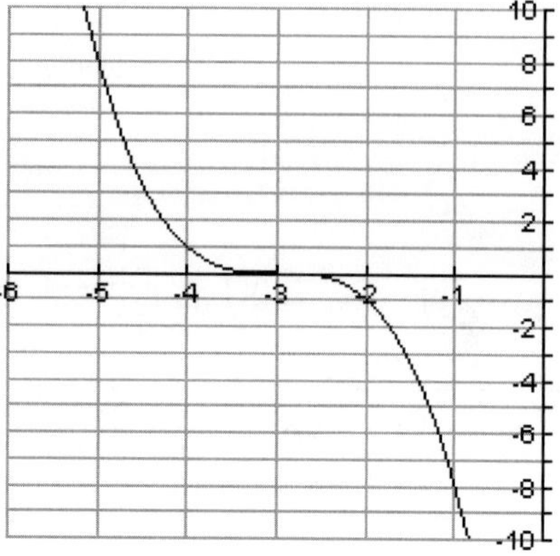

53. $f(x) = x^3 - 9x = x(x-3)(x+3)$

a. Zeros: 0, 3, -3 (multiplicity 1)

b. Crosses at each zero

c. y-intercept: $f(0) = 0$, so $(0,0)$

d. End behavior: Behaves like $y = x^3$.

Odd degree and leading coefficient positive, so graph falls without bound to the left and rises to the right.

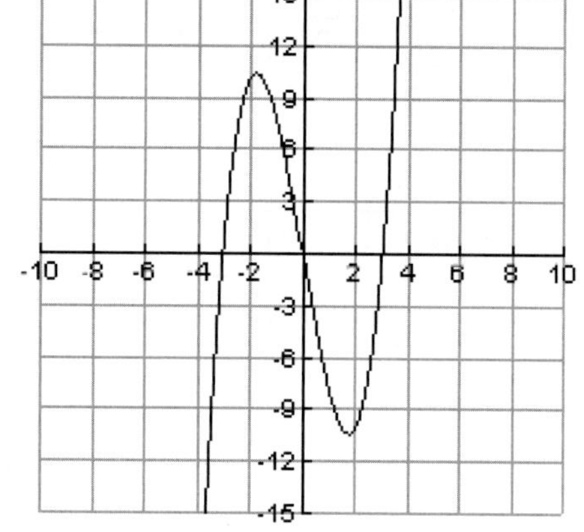

54. $f(x) = -x^3 + 4x^2 = -x^2(x-4)$

a. Zeros: 0 (multiplicity 2), 4 (multiplicity 1)

b. Crosses at 4, touches at 0

c. y-intercept: $f(0) = 0$, so $(0,0)$

d. End behavior: Behaves like $y = -x^3$.

Odd degree and leading coefficient negative, so graph falls without bound to the right and rises to the left.

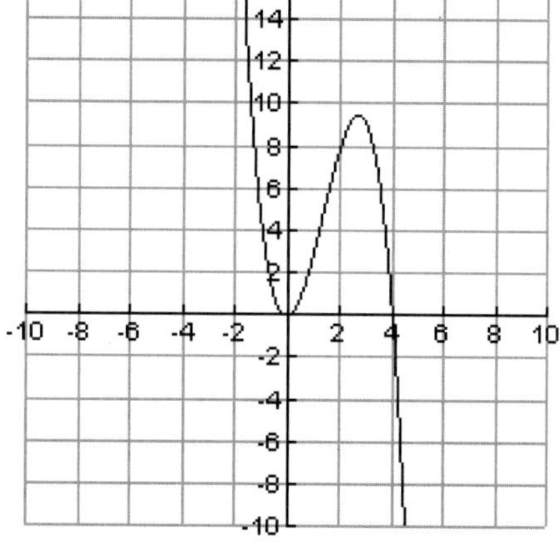

55.
$f(x) = -x^3 + x^2 + 2x = -x(x-2)(x+1)$

a. Zeros: 0, 2, -1 (multiplicity 1)
b. Crosses at each zero
c. y-intercept: $f(0) = 0$, so $(0,0)$
d. End behavior: Behaves like
$y = -x^3$.
Odd degree and leading coefficient negative, so graph falls without bound to the right and rises to the left.

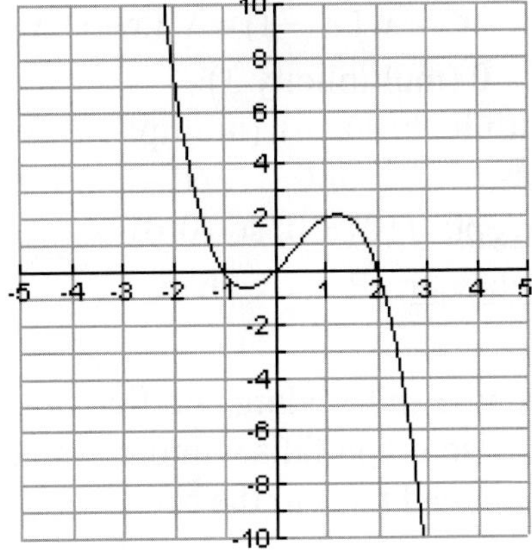

56. $f(x) = x^3 - 6x^2 + 9x = x(x-3)^2$

a. Zeros: 0 (multiplicity 1)
3 (multiplicity 2)
b. Crosses at 0, touches at 3
c. y-intercept: $f(0) = 0$, so $(0,0)$
d. End behavior: Behaves like
$y = x^3$.
Odd degree and leading coefficient positive, so graph falls without bound to the left and rises to the right.

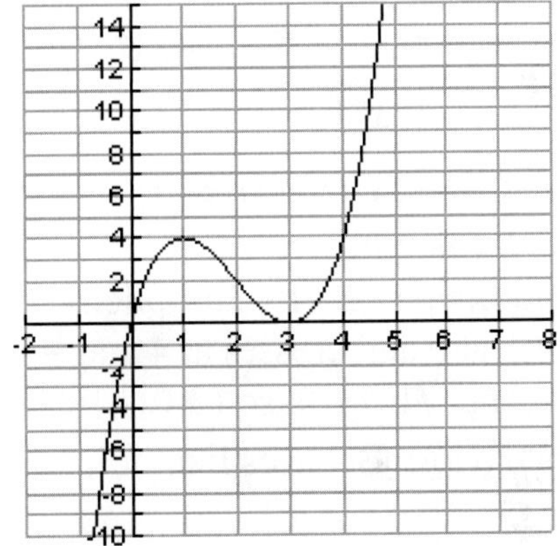

e.

57. $f(x) = -x^4 - 3x^3 = -x^3(x+3)$

a. Zeros: 0 (multiplicity 3)
-3 (multiplicity 1)
b. Crosses at both 0 and -3
c. y-intercept: $f(0) = 0$, so $(0,0)$
d. End behavior: Behaves like
$y = -x^4$.
Even degree and leading coefficient negative, so graph falls without bound to left and right.

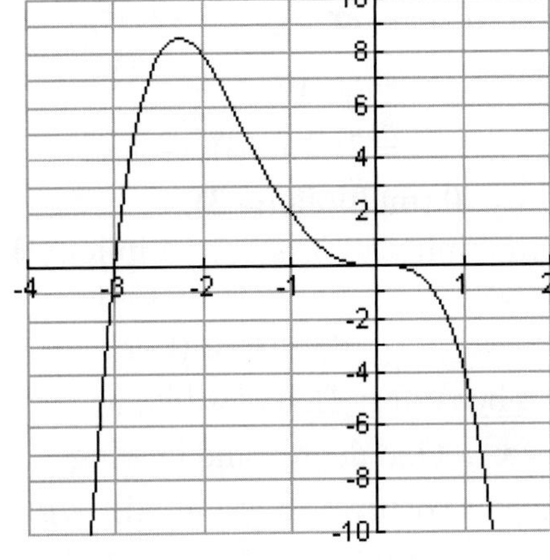

58.

e.

$f(x) = x^5 - x^3 = x^3(x^2 - 1) = x^3(x-1)(x+1)$

a. <u>Zeros</u>: 0 (multiplicity 3)
1 (multiplicity 1), -1 (multiplicity 1)

b. Crosses at each of 0, 1, and -1

c. <u>y-intercept</u>: $f(0) = 0$, so $(0,0)$

d. <u>End behavior</u>: Behaves like $y = x^5$.
Odd degree and leading coefficient positive, so graph falls without bound to the left and rises to the right.

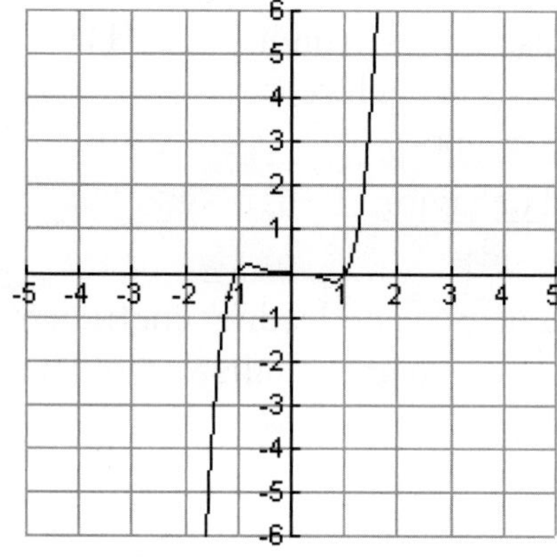

59.

e.

$$\begin{aligned} f(x) &= 12x^6 - 36x^5 - 48x^4 \\ &= 12x^4\left(x^2 - 3x - 4\right) \\ &= 12x^4(x-4)(x+1) \end{aligned}$$

a. <u>Zeros</u>: 0 (multiplicity 4),
4 (multiplicity 1), -1 (multiplicity 1)

b. Touches at 0 and crosses at 4 and -1.

c. <u>y-intercept</u>: $f(0) = 0$, so $(0,0)$

d. <u>End behavior</u>: Behaves like $y = x^6$.
Even degree and leading coefficient positive, so graph rises without bound to the left and right.

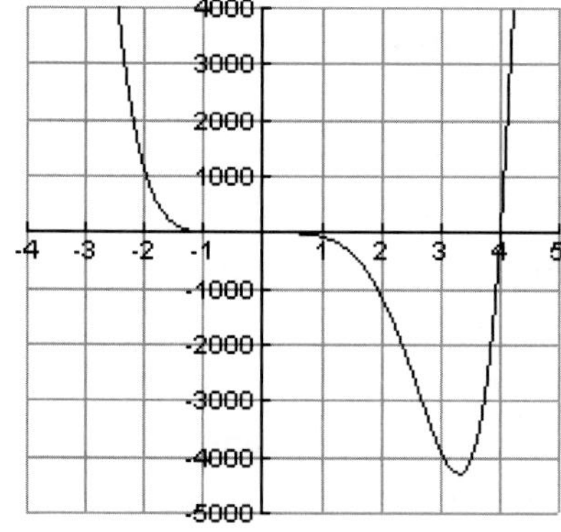

60.

e.

$$\begin{aligned} f(x) &= 7x^5 - 14x^4 - 21x^3 \\ &= 7x^3\left(x^2 - 2x - 3\right) \\ &= 7x^3(x-3)(x+1) \end{aligned}$$

a. <u>Zeros</u>: 0 (multiplicity 3),
3 (multiplicity 1), -1 (multiplicity 1)

b. Crosses at each of 0, 3, and -1

c. <u>y-intercept</u>: $f(0) = 0$, so $(0,0)$

d. <u>End behavior</u>: Behaves like $y = x^5$. Odd degree and leading coefficient positive, so graph falls without bound to the left and rises to the right.

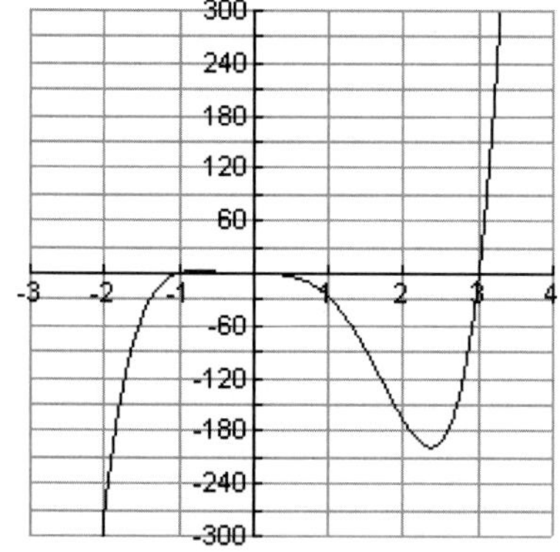

61.

$f(x) = 2x^5 - 6x^4 - 8x^3$
$= 2x^3(x-4)(x+1)$

a. Zeros: 0 (multiplicity 3), 4(multiplicity 1), −1 (multiplicity 1)

b. Crosses at each zero

c. y-intercept: $f(0) = 0$, so $(0,0)$

d. End behavior: Behaves like $y = x^5$. Odd degree and leading coefficient positive, so graph falls without bound to the left and rises to the right.

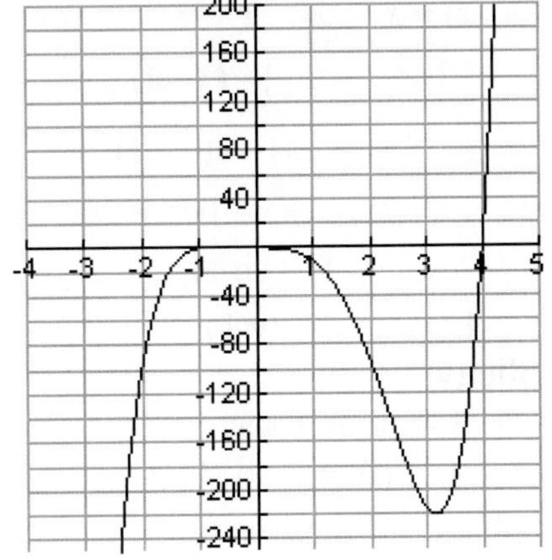

62.

$f(x) = -5x^4 + 10x^3 - 5x^2 = -5x^2(x-1)^2$

a. Zeros: 0 (multiplicity 2)
1 (multiplicity 2)

b. Touches at each zero

c. y-intercept: $f(0) = 0$, so $(0,0)$

d. End behavior: Behaves like $y = -x^4$.
Even degree and leading coefficient negative, so graph falls without bound to left and right.

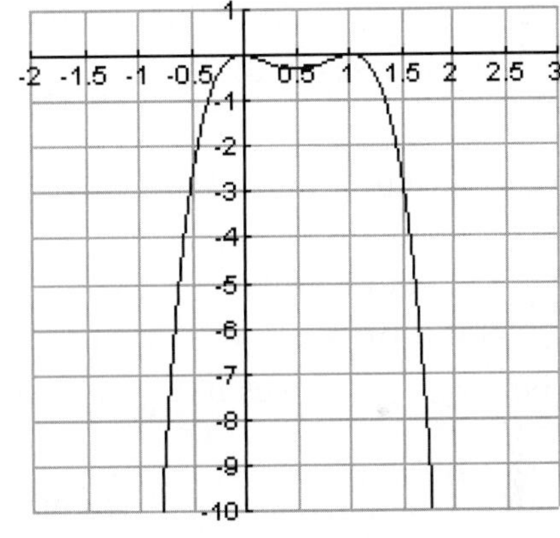

63.

$$f(x) = x^3 - x^2 - 4x + 4$$
$$= (x^3 - x^2) - 4(x-1)$$
$$= x^2(x-1) - 4(x-1)$$
$$= (x-2)(x+2)(x-1)$$

a. Zeros: 1, 2, -2 (multiplicity 1)

b. Crosses at each zero

c. y-intercept: $f(0) = 4$, so $(0,4)$

d. End behavior: Behaves like $y = x^3$.
Odd degree and leading coefficient positive, so graph falls without bound to the left and rises to the right.

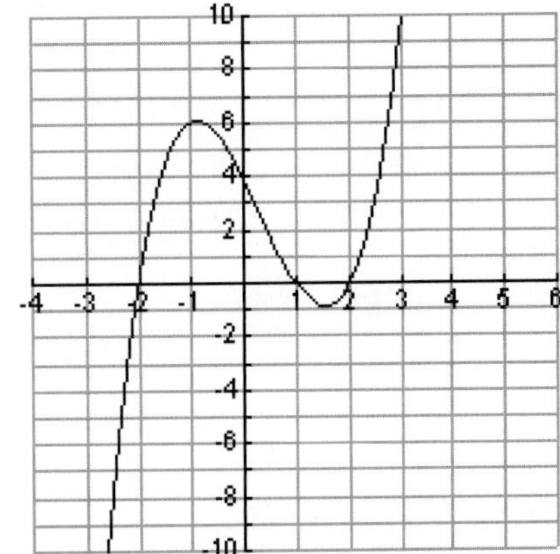

64.

$$f(x) = x^3 - x^2 - x + 1$$
$$= \left(x^3 - x^2\right) - (x - 1)$$
$$= x^2(x-1) - 4(x-1)$$
$$= (x^2 - 1)(x - 1)$$
$$= (x-1)^2(x+1)$$

a. Zeros: -1 (multiplicity 1), 1 (multiplicity 2)

b. Crosses at -1, touches at 1

c. y-intercept: $f(0) = 1$, so $(0,1)$

d. End behavior: Behaves like $y = x^3$.

Odd degree and leading coefficient positive, so graph falls without bound to the left and rises to the right.

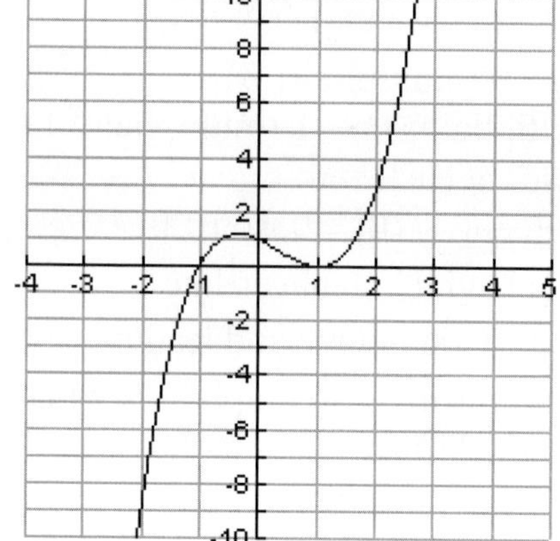

65. $f(x) = -(x+2)^2(x-1)^2$

a. Zeros: -2 (multiplicity 2)
1 (multiplicity 2)

b. Touches at both -2 and 1

c. y-intercept: $f(0) = -4$, so $(0,-4)$

d. End behavior: Behaves like $y = -x^4$.

Even degree and leading coefficient negative, so graph falls without bound to left and right.

e.

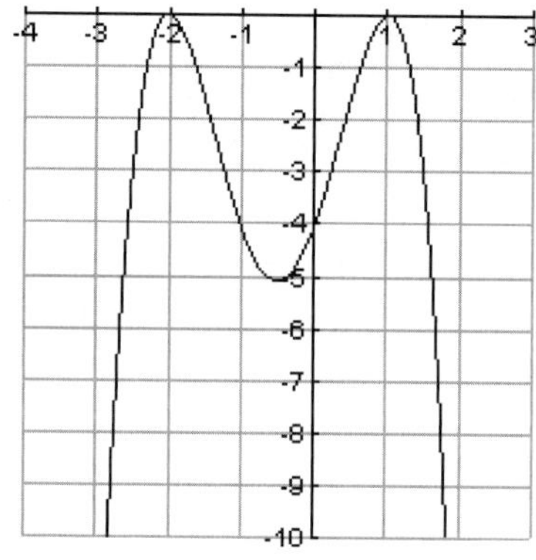

66. $f(x) = (x-2)^3(x+1)^3$

a. Zeros: -1 (multiplicity 3)
2 (multiplicity 3)

b. Crosses at both -1 and 2

c. y-intercept: $f(0) = -8$, so $(0,-8)$

d. End behavior: Behaves like $y = x^6$.
Even degree and leading coefficient positive, so graph rises without bound to the left and right.

e.

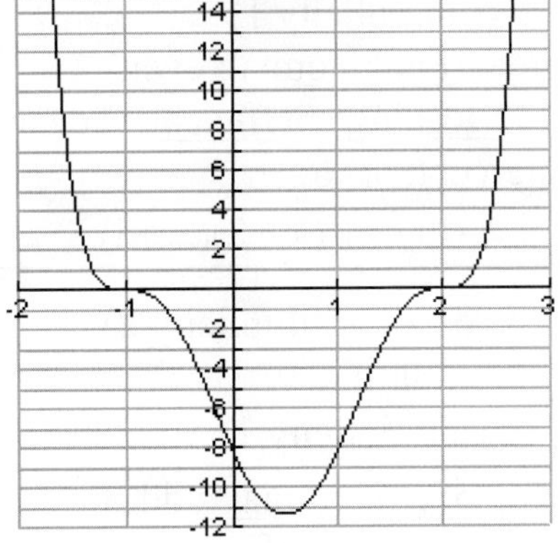

67. $f(x) = x^2(x-2)^3(x+3)^2$

a. Zeros: 0 (multiplicity 2)
2 (multiplicity 3)
-3 (multiplicity 2)

b. Touches at both 0 and -3, and crosses at 2.

c. y-intercept: $f(0) = 0$, so $(0,0)$

d. End behavior: Behaves like $y = x^7$.
Odd degree and leading coefficient positive, so graph falls without bound to the left and rises to the right.

e.

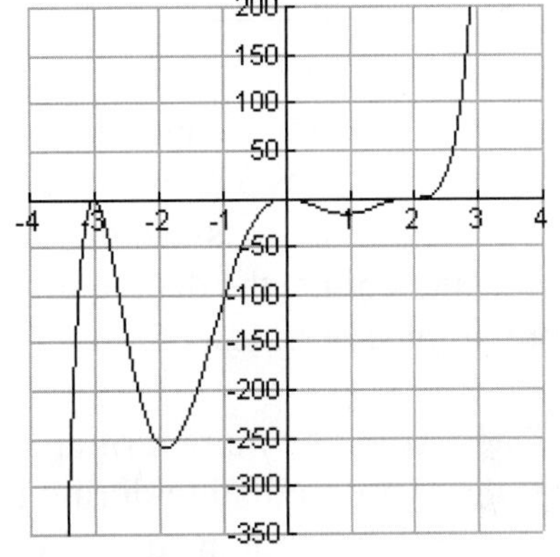

68. $f(x) = -x^3(x-4)^2(x+2)^2$

a. Zeros: 0 (multiplicity 3)
4 (multiplicity 2)
-2 (multiplicity 2)

b. Touches at 4 and -2, and crosses at 0.

c. y-intercept: $f(0) = 0$, so $(0,0)$

d. Long-term behavior: Behaves like $y = -x^7$.
Odd degree and leading coefficient negative, so graph falls without bound to the right and rises to the left.

e.

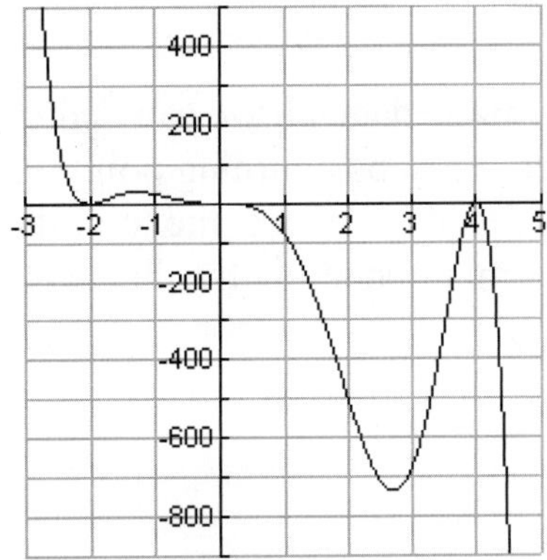

69. a. <u>zeros</u>: -3 (multiplicity 1) -1 (multiplicity 2) 2 (multiplicity 1) **b.** <u>degree of polynomial</u>: even **c.** <u>sign of leading coefficient</u>: negative **d.** <u>y-intercept</u>: (0,6) **e.** $f(x)=-(x+1)^2(x-2)(x+3)$.	**70. a.** <u>zeros</u>: -2 (multiplicity 1) 2 (multiplicity 2) 0 (multiplicity 1) **b.** <u>degree of polynomial</u>: even **c.** <u>sign of leading coefficient</u>: positive **d.** <u>y-intercept</u>: (0,0) **e.** $f(x)=x(x+2)(x-2)^2$.
71. a. <u>zeros</u>: 0 (multiplicity 2) -2 (multiplicity 2) $\frac{3}{2}$ (multiplicity 1) **b.** <u>degree of polynomial</u>: odd **c.** <u>sign of leading coefficient</u>: positive **d.** <u>y-intercept</u>: (0,0) **e.** $f(x)=x^2(2x-3)(x+2)^2$.	**72. a.** <u>zeros</u>: -3 (multiplicity 1) 0 (multiplicity 1) $-\frac{3}{2}$ (multiplicity 1) 1 (multiplicity 2) **b.** <u>degree of polynomial</u>: odd **c.** <u>sign of leading coefficient</u>: negative **d.** <u>y-intercept</u>: (0,0) **e.** $f(x)=-x(2x+3)(x+2)(x-1)^2$.

73. From the data, the turning points occur between months 3 & 4, 4 & 5, 5 & 6, 6 & 7, and 7 & 8. So, since there are at least 5 turning points, the degree of the polynomial must be at least 6.

<u>Note</u>: There could be more turning points if there was more oscillation during these month-long periods, but the data doesn't reveal such behavior.

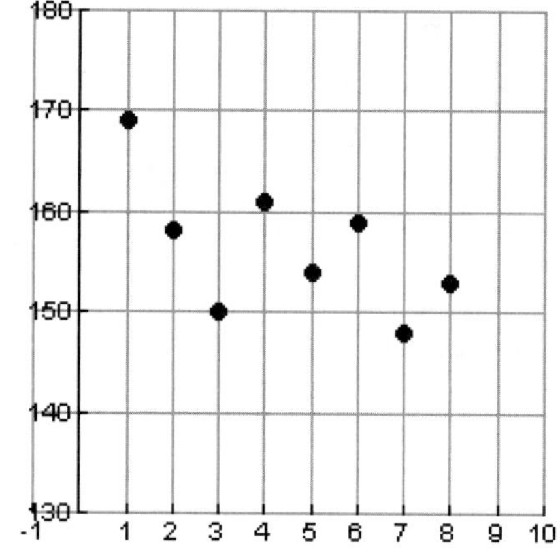

74. From the data, all we can deduce is that there must be a turning point between periods 2 and 3. Hence, the lowest degree a polynomial modeling this behavior could be is 2.

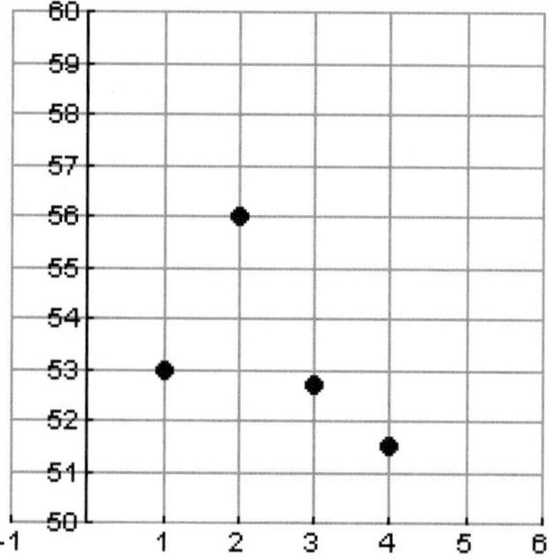

75. From the data, there is a turning point at 2. Since it is a third degree polynomial, one would expect the price to go down.

76. From the data, we know that there is a turning point at 3. Since a third degree polynomial is used to model the stock, we expect there to be a turning point at 4, so that the stock should go up in the fifth period.

77. The given graph has three turning points, and is assumed to be defined only on the interval [2000,2008]. Since it seems to touch the axis at 2004, the lowest degree polynomial that would work is 4.

78. Judging from the fact that both ends point upward at the endpoints, the end behavior is best described by $y = x^{2n}$, for some positive integer n. So, the coefficient of the leading term must be positive.

79. Let Monday correspond to $x = 1$, Tuesday $x = 2$, etc… We have the following data:

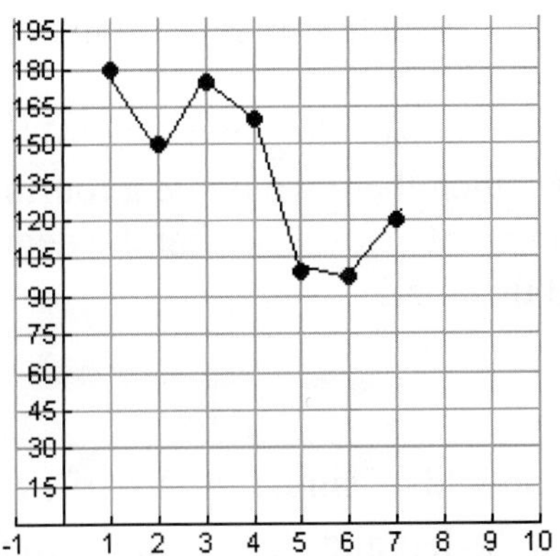

Note that a continuous curve passing through these points would have 3 turning points. Hence, its minimum degree is 4.

80. Let Monday correspond to $x = 1$, Tuesday $x = 2$, etc… We have the following data:

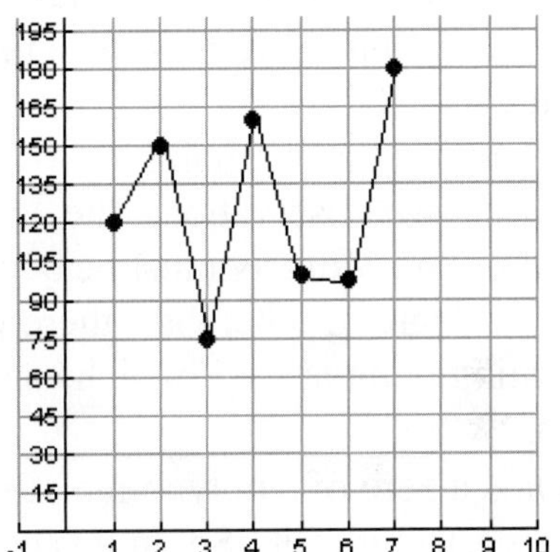

Note that a continuous curve passing through these points would have 2 turning points. Hence, its minimum degree is 3.

81. Let Monday correspond to $x = 1$, Tuesday $x = 2$, etc... We have the following data: 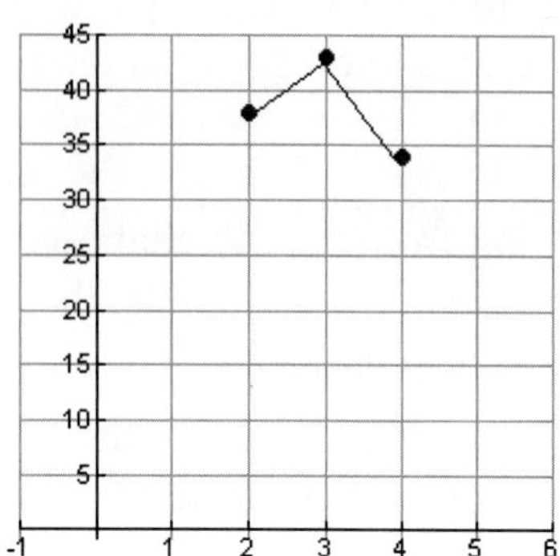 Observe: 1.) Since Thursday has the lowest temperature, the y-values at $x = 5, 6, 7$ are all ≥ 35. Since a cubic polynomial can be nowhere constant, there must be a turning point at $x = 4$. 2.) There is a turning point at $x = 3$. From these two observations, the temperature on Monday must be between 35 and 39 degrees; otherwise, there would be a turning point at $x = 2$, which cannot happen.	**82.** Let x = game #. (a) We have the following data: 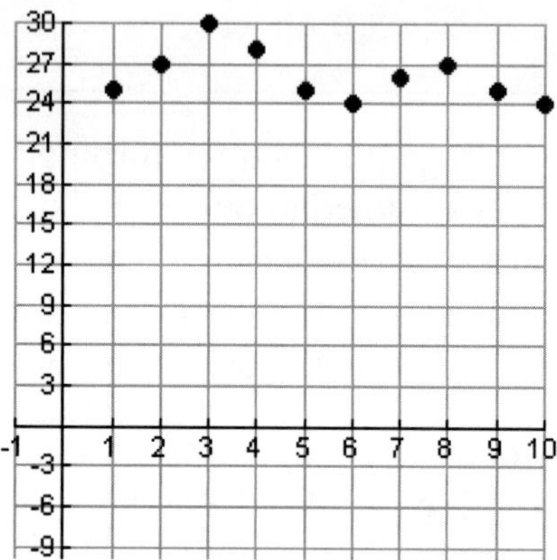 Since the graph has three turning points, its degree is 4. (b) If we replace the point (10,24) by (10,26), then there would be a fourth turning point at $x = 9$, so that the degree would then be 5.
83. If h is a zero of a polynomial, then $(x-h)$ is a factor of it. So, in this case the function would be: $f(x) = (x+2)(x+1)(x-3)(x-4)$	**84.** It should be similar to $y = x^4$, which rises to the left and right without bound.

85. The zeros are correct. But, it is a fifth degree polynomial. The graph should touch at 1 (since even multiplicity) and cross at -2. The graph should look like:	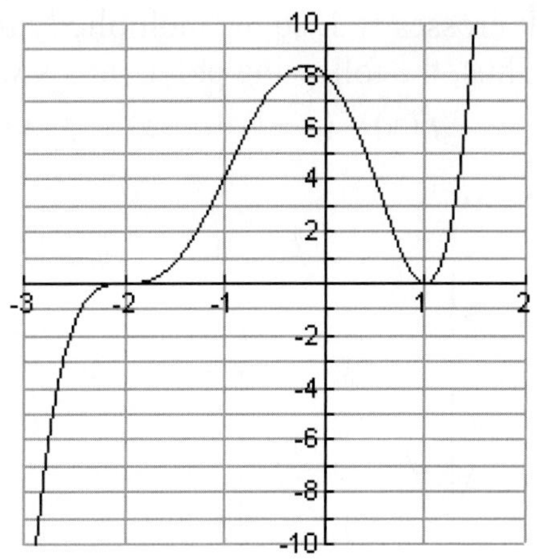
86. Should <u>touch</u>, not cross, at 1 and -1 since both have even multiplicity. The graph should look like:	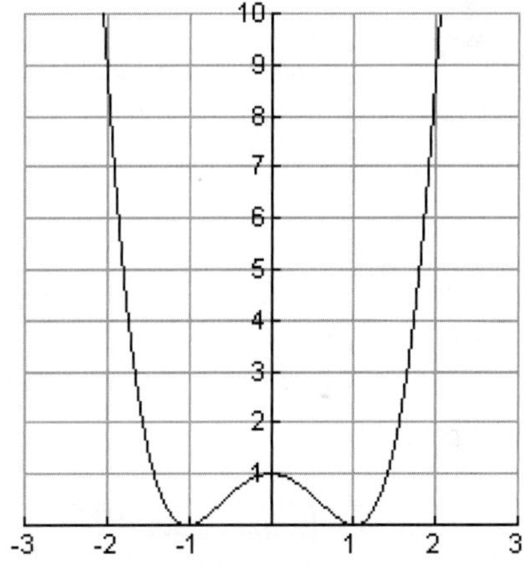
87. False. A polynomial has general form $f(x)=a_n x^n + a_{n-1}x^{n-1}+...+a_1 x + a_0$. So, $f(0)=a_0$.	**88.** True. Consider $f(x)=x^2+1$ **89.** True. **90.** False. Consider $f(x)=x^2+1$. Its range is $[1,\infty)$.

91. A polynomial of degree n can have at most n zeros.

92. An nth degree polynomial can have at most $(n-1)$ turning points, and this would occur if it had n distinct real zeros, each with multiplicity 1.

93. It touches at -1, so the multiplicity of this zero must be 2, 4, or 6.
It crosses at 3, so the multiplicity of this zero must be 1, 3, or 5.
Thus, the following polynomials would work:

$$f(x)=(x+1)^2(x-3)^5,\quad g(x)=(x+1)^4(x-3)^3,\quad h(x)=(x+1)^6(x-3)$$

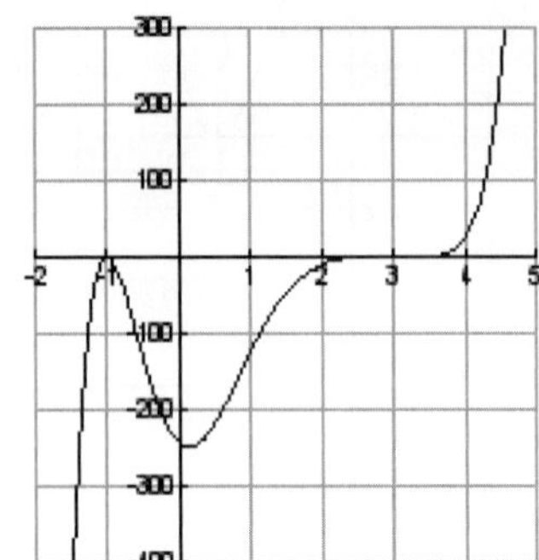

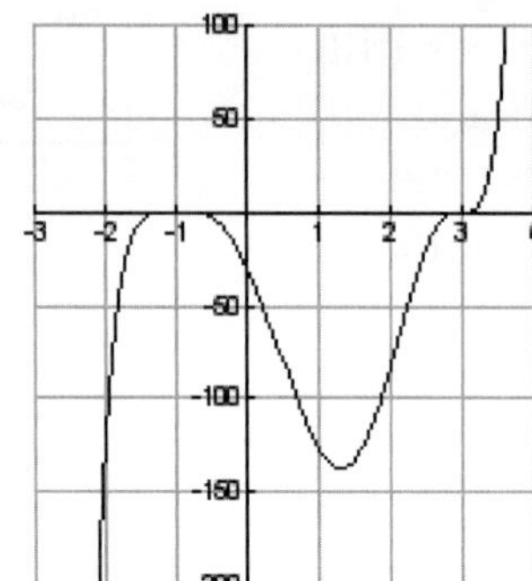

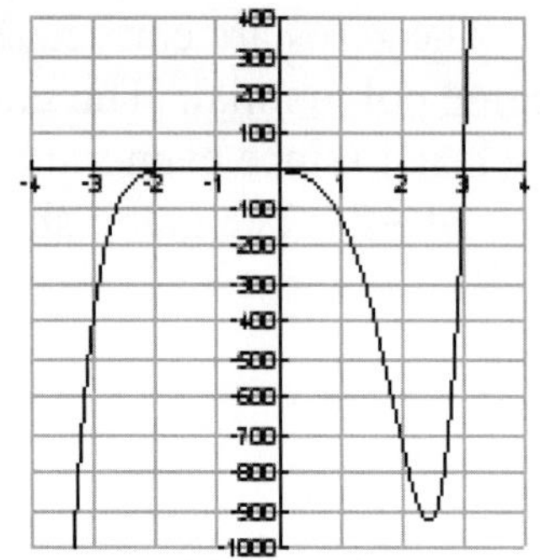

94. A 5th degree polynomial satisfying these properties is of one of the following two forms. You can multiply each by a nonzero real constant to obtain others.

$P_1(x)=(x-0)^4(x-4)$

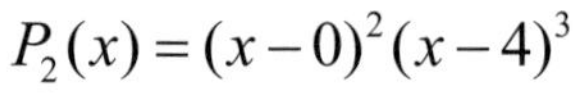

$P_2(x)=(x-0)^2(x-4)^3$

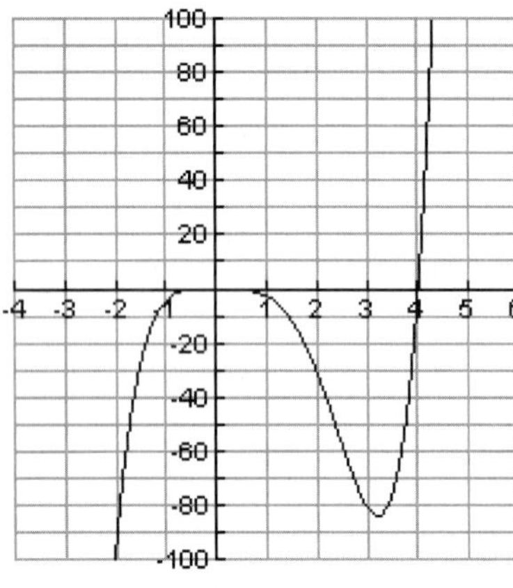

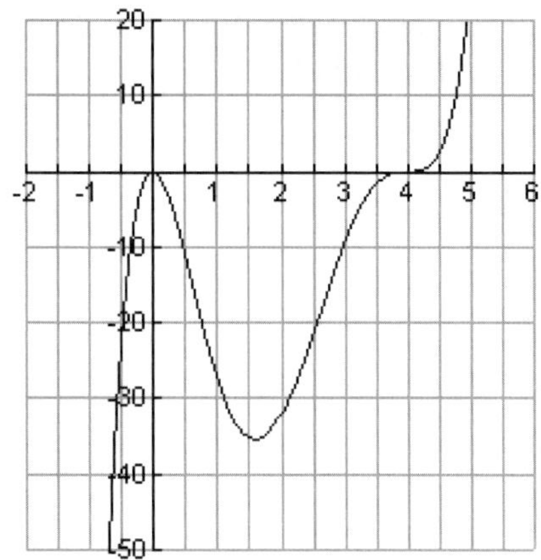

95. Observe that $x^3+(b-a)x^2-abx=x\left(x^2+(b-a)x-ab\right)=x(x-a)(x+b)$. So, the zeros are 0, a, and $-b$.

96. The function $f(x) = x^2(x-a)^2(x-b)^2$ has zeros of 0, a, and b, all of which touch the x-axis. The graph is as follows:

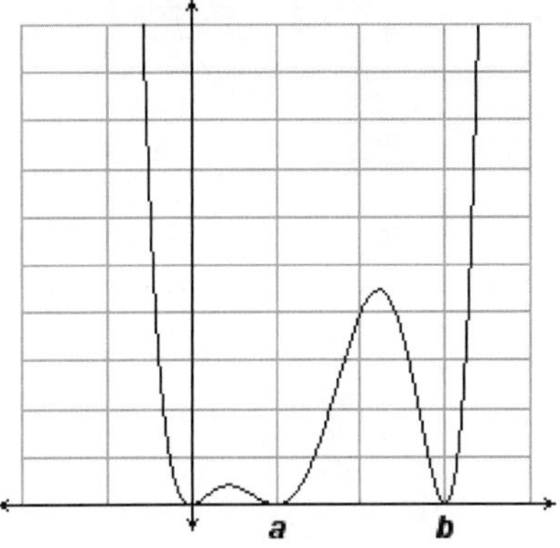

97. Note that $x^4 + 2x^2 + 1 = \underbrace{(x^2+1)^2}_{\text{Always positive}}$. So, there are no real zeros, and hence no x-intercepts. The graph is as follows:

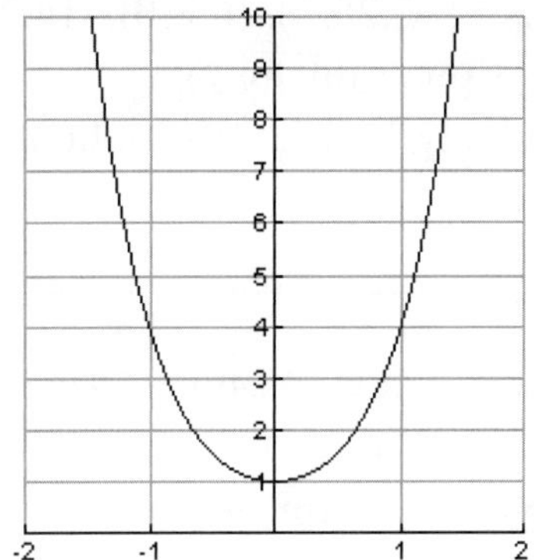

98. First, note that $1.1x^3 - 2.4x^2 + 5.2x = x\left(1.1x^2 - 2.4x + 5.2\right)$.

Further, using the quadratic formula, the solutions of $1.1x^2 - 2.4x + 5.2 = 0$ are:

$$\frac{2.4 \pm \sqrt{-17.12}}{2.2}$$

Hence, the only real zero (and hence, x-intercept) of the above function is 0. The graph is as follows:

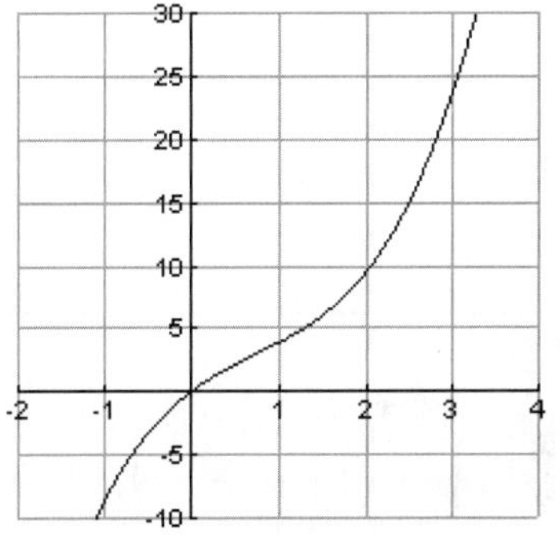

99. From the following graph of both $f(x) = -2x^5 - 5x^4 - 3x^3$ (solid curve) and $y = -2x^5$ (dotted curve), we do conclude that they have the same end behavior.

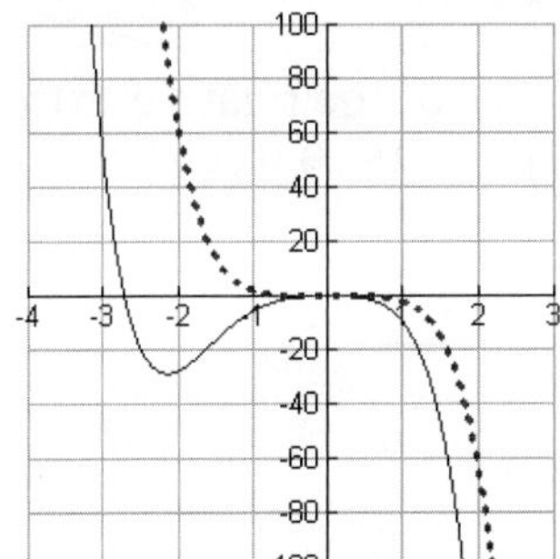

100. From the following graph of both $f(x) = x^4 - 6x^2 + 9$ and $y = x^4$, we do conclude that they have the same end behavior.

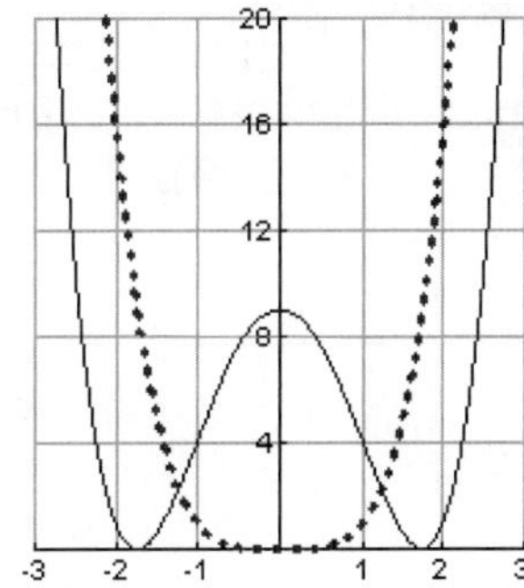

101.
$f(x)=x^4-15.9x^3+1.31x^2+292.905x+445.7025$

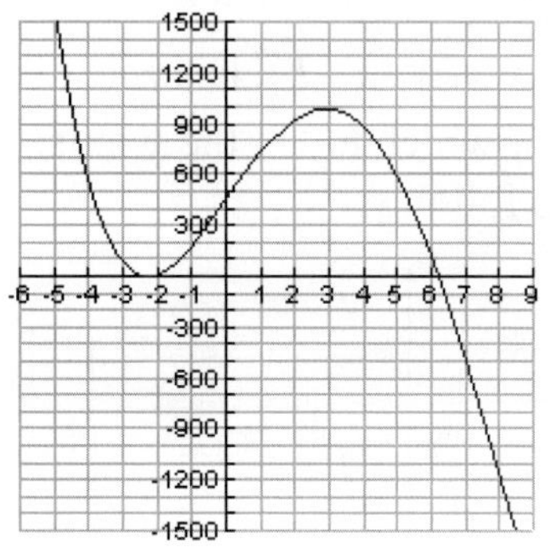

<u>The following are estimates:</u>
x-intercepts: $(-2.25,0)$, $(6.2,0)$, $(14.2,0)$
zeros: -2.25 (multiplicity 2),
6.2 (multiplicity 1), 14.2 (multiplicity 1)

102.
$$f(x)=-x^5+2.2x^4+18.49x^3-29.878x^2 \\ -76.5x+100.8$$

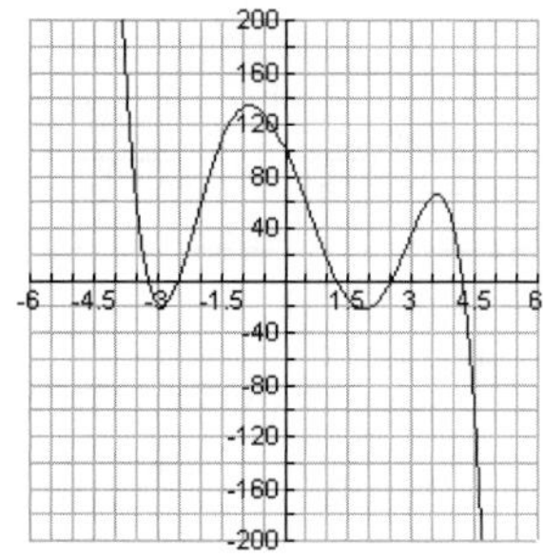

<u>The following are estimates:</u>
x-intercepts:
$(-3.2,0)$, $(-2.5,0)$, $(1.2,0)$, $(2.5,0)$, $(4.2,0)$

zeros: -3.2, -2.5, 1.2, 2.5, 4.2 (all have multiplicity 1)

103. The graph is:

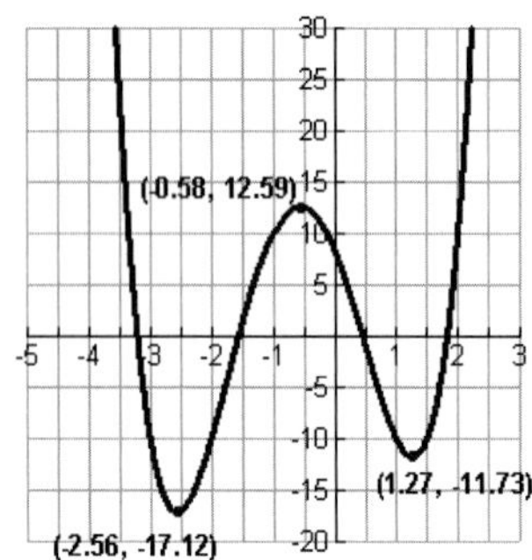

The coordinates of the relative extrema are (-2.56,-17.12), (-0.58,12.59), (1.27,-11.73).

104. The graph is:

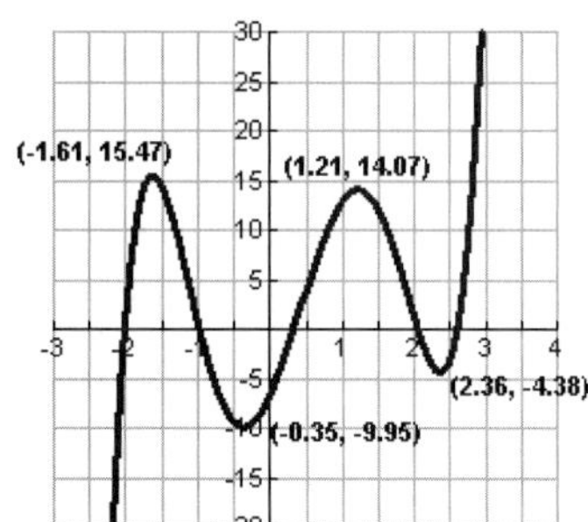

The coordinates of the relative extrema are (-1.61,15.47), (-0.35,-9.95), (1.21,14.07), (2.36,-4.38).

105. The graph is:

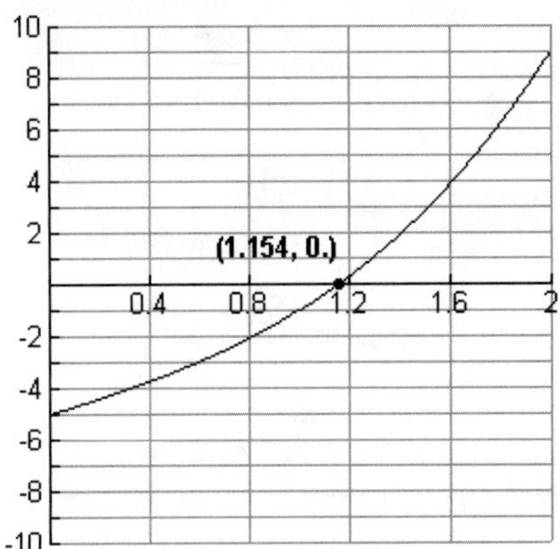

There is one root of f in this interval, namely approximately $x = 1.154$.

106. The graph is:

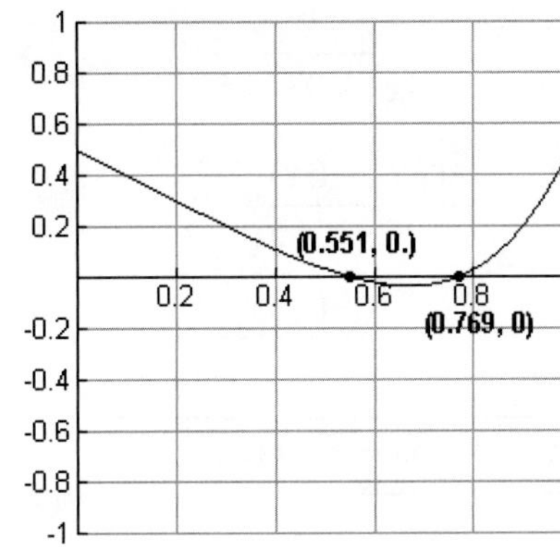

There are two rootsof f in this interval, namely approximately $x = 0.551$ and $x=$ 0.769.

107. The graph is :

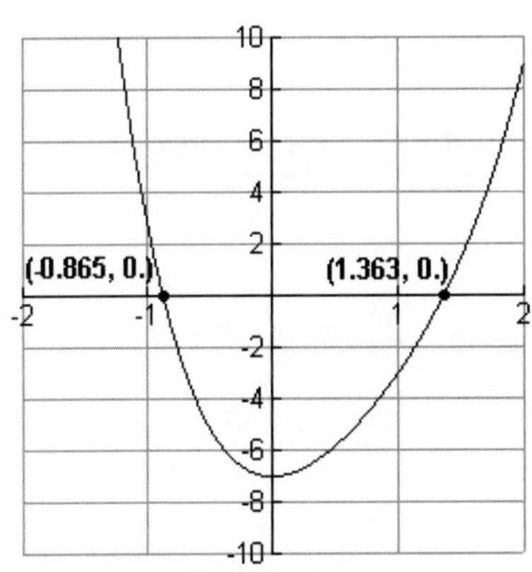

There are two rootsof f in this interval, namely approximately $x = -0.865$ and $x=$ 1.363.

108. The graph is:

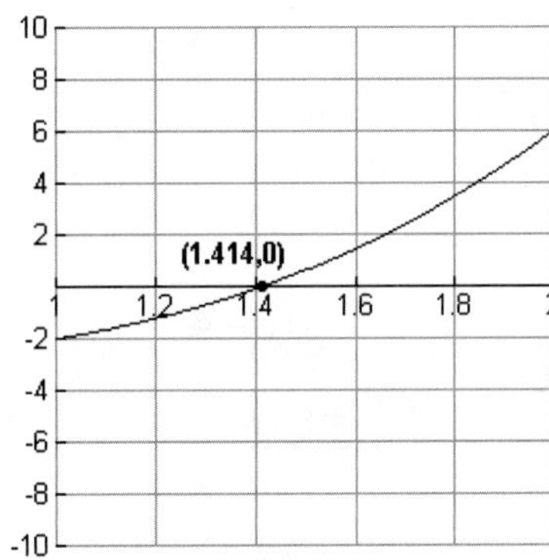

There is one root of f in this interval, namely approximately $x = 1.414$.

Section 2.3 Solutions ---

1.

$$\begin{array}{r|l} & 3x-3 \\ \hline x-2 & 3x^2-9x-5 \\ & -(3x^2-6x) \\ \hline & -3x-5 \\ & -(-3x+6) \\ \hline & -11 \end{array}$$

So, $\boxed{Q(x)=3x-3,\ \ r(x)=-11}$.

2.

$$\begin{array}{r|l} & x+5 \\ \hline x-1 & x^2+4x-3 \\ & -(x^2-x) \\ \hline & 5x-3 \\ & -(5x-5) \\ \hline & 2 \end{array}$$

So, $\boxed{Q(x)=x+5,\ \ r(x)=2}$.

3.

$$\begin{array}{r|l} & 3x-28 \\ \hline x+5 & 3x^2-13x-10 \\ & -(3x^2+15x) \\ \hline & -28x-10 \\ & -(-28x-140) \\ \hline & 130 \end{array}$$

So, $\boxed{Q(x)=3x-28,\ \ r(x)=130}$.

4.

$$\begin{array}{r|l} & 3x+2 \\ \hline x-5 & 3x^2-13x-10 \\ & -(3x^2-15x) \\ \hline & 2x-10 \\ & -(2x-10) \\ \hline & 0 \end{array}$$

So, $\boxed{Q(x)=3x+2,\ \ r(x)=0}$.

5.

$$\begin{array}{r|l} & x-4 \\ \hline x+4 & x^2+0x-4 \\ & -(x^2+4x) \\ \hline & -4x-4 \\ & -(-4x-16) \\ \hline & 12 \end{array}$$

So, $\boxed{Q(x)=x-4,\ \ r(x)=12}$.

6.

$$\begin{array}{r|l} & x+2 \\ \hline x-2 & x^2+0x-9 \\ & -(x^2-2x) \\ \hline & 2x-9 \\ & -(2x-4) \\ \hline & -5 \end{array}$$

So, $\boxed{Q(x)=x+2,\ \ r(x)=-5}$.

7.

$$\begin{array}{r|l} & 3x+5 \\ \hline 3x-5 & 9x^2+0x-25 \\ & -(9x^2-15x) \\ \hline & 15x-25 \\ & -(15x-25) \\ \hline & 0 \end{array}$$

So, $\boxed{Q(x)=3x+5,\ \ r(x)=0}$.

8.

$$\begin{array}{r|l} & 5x-5 \\ \hline x+1 & 5x^2+0x-3 \\ & -(5x^2+5x) \\ \hline & -5x-3 \\ & -(-5x-5) \\ \hline & 2 \end{array}$$

So, $\boxed{Q(x)=5x-5,\ \ r(x)=2}$.

9.

$$\begin{array}{r} 2x-3 \\ 2x+3 \overline{)\, 4x^2+0x-9} \\ \underline{-(4x^2+6x)} \\ -6x-9 \\ \underline{-(-6x-9)} \\ 0 \end{array}$$

So, $\boxed{Q(x)=2x-3, \quad r(x)=0}$.

10.

$$\begin{array}{r} 4x^2-6x+9 \\ 2x+3 \overline{)\, 8x^3+0x^2+0x+27} \\ \underline{-(8x^3+12x^2)} \\ -12x^2+0x \\ \underline{-(-12x^2-18x)} \\ 18x+27 \\ \underline{-(18x+27)} \\ 0 \end{array}$$

So, $\boxed{Q(x)=4x^2-6x+9, \quad r(x)=0}$.

11.

$$\begin{array}{r} 4x^2+4x+1 \\ 3x+2 \overline{)\, 12x^3+20x^2+11x+2} \\ \underline{-(12x^3+8x^2)} \\ 12x^2+11x \\ \underline{-(12x^2+8x)} \\ 3x+2 \\ \underline{-(3x+2)} \\ 0 \end{array}$$

So, $\boxed{Q(x)=4x^2+4x+1, \quad r(x)=0}$.

12.

$$\begin{array}{r} 6x^2+7x+2 \\ 2x+1 \overline{)\, 12x^3+20x^2+11x+2} \\ \underline{-(12x^3+6x^2)} \\ 14x^2+11x \\ \underline{-(14x^2+7x)} \\ 4x+2 \\ \underline{-(4x+2)} \\ 0 \end{array}$$

So, $\boxed{Q(x)=6x^2+7x+2, \quad r(x)=0}$.

13.

$$\begin{array}{r} 2x^2-x-\frac{1}{2} \\ 2x+1 \overline{)\, 4x^3+0x^2-2x+7} \\ \underline{-(4x^3+2x^2)} \\ -2x^2-2x \\ \underline{-(-2x^2-x)} \\ -x+7 \\ \underline{-(-x-\frac{1}{2})} \\ \frac{15}{2} \end{array}$$

So, $\boxed{Q(x)=2x^2-x-\frac{1}{2}, \quad r(x)=\frac{15}{2}}$.

14.

$$\begin{array}{r} -2x^3-\frac{4}{3}x^2-\frac{2}{9}x-\frac{4}{27} \\ -3x+2 \overline{)\, 6x^4+0x^3-2x^2+0x+5} \\ \underline{-(6x^4-4x^3)} \\ 4x^3-2x^2 \\ \underline{-(4x^3-\frac{8}{3}x^2)} \\ \frac{2}{3}x^2+0x \\ \underline{-(\frac{2}{3}x^2-\frac{4}{9}x)} \\ \frac{4}{9}x+5 \\ \underline{-(\frac{4}{9}x-\frac{8}{27})} \\ \frac{143}{27} \end{array}$$

So,

$\boxed{Q(x)=-2x^3-\frac{4}{3}x^2-\frac{2}{9}x-\frac{4}{27}, \quad r(x)=\frac{143}{27}}$.

15.

$$\begin{array}{r} 4x^2-10x-6 \\ x-\tfrac{1}{2}\overline{\big)\ 4x^3-12x^2-x+3} \\ \underline{-(4x^3-2x^2)} \\ -10x^2-x \\ \underline{-(-10x^2+5x)} \\ -6x+3 \\ \underline{-(-6x+3)} \\ 0 \end{array}$$

So, $\boxed{Q(x)=4x^2-10x-6,\quad r(x)=0}$.

16.

$$\begin{array}{r} 12x^2+12x+3 \\ x+\tfrac{1}{3}\overline{\big)\ 12x^3+16x^2+7x+1} \\ \underline{-(12x^3+4x^2)} \\ 12x^2+7x \\ \underline{-(12x^2+4x)} \\ 3x+1 \\ \underline{-(3x+1)} \\ 0 \end{array}$$

So, $\boxed{Q(x)=12x^2+12x+3,\quad r(x)=0}$.

17.

$$\begin{array}{r} -2x^2-3x-9 \\ x^3-3x^2+0x+1\overline{\big)-2x^5+3x^4+0x^3-2x^2+0x+0} \\ \underline{-(-2x^5+6x^4+0x^3-2x^2)} \\ -3x^4+0x^3+0x^2+0x \\ \underline{-(-3x^4+9x^3+0x^2-3x)} \\ -9x^3+0x^2+3x+0 \\ \underline{-(-9x^3+27x^2+0x-9)} \\ -27x^2+3x+9 \end{array}$$

So, $\boxed{Q(x)=-2x^2-3x-9,\quad r(x)=-27x^2+3x+9}$.

18.

$$\begin{array}{r} -3x^2+0x+\tfrac{7}{3} \\ 3x^4+0x^3+0x^2-2x+1\overline{\big)-9x^6+0x^5+7x^4-2x^3+0x^2+0x+5} \\ \underline{-(-9x^6+0x^5+0x^4+6x^3-3x^2)} \\ 7x^4-8x^3+3x^2+0x+5 \\ \underline{-(7x^4+0x^3+0x^2-\tfrac{14}{3}x+\tfrac{7}{3})} \\ -8x^3+3x^2+\tfrac{14}{3}x+\tfrac{8}{3} \end{array}$$

So, $\boxed{Q(x)=-3x^2+\tfrac{7}{3},\quad r(x)=-8x^3+3x^2+\tfrac{14}{3}x+\tfrac{8}{3}}$.

19.

$$\begin{array}{r} x^2+0x+1 \\ x^2+0x-1\overline{\big)\, x^4+0x^3+0x^2+0x-1} \\ \underline{-(x^4+0x^3-x^2)} \\ x^2+0x-1 \\ \underline{-(x^2+0x-1)} \\ 0 \end{array}$$

So, $\boxed{Q(x)=x^2+1,\quad r(x)=0}$.

20.

$$\begin{array}{r} x^2+0x-3 \\ x^2+0x+3\overline{\big)\, x^4+0x^3+0x^2+0x-9} \\ \underline{-(x^4+0x^3+3x^2)} \\ -3x^2+0x-9 \\ \underline{-(-3x^2+0x-9)} \\ 0 \end{array}$$

So, $\boxed{Q(x)=x^2-3,\quad r(x)=0}$.

21.

$$\begin{array}{r} x^2+x+\frac{1}{6} \\ 6x^2+x-2\overline{\big)\, 6x^4+7x^3+0x^2-22x+40} \\ \underline{-(6x^4+x^3-2x^2)} \\ 6x^3+2x^2-22x \\ \underline{-(6x^3+x^2-2x)} \\ x^2-20x+40 \\ \underline{-(x^2+\frac{1}{6}x-\frac{1}{3})} \\ -\frac{121}{6}x+\frac{121}{3} \end{array}$$

So, $\boxed{Q(x)=x^2+x+\frac{1}{6},\quad r(x)=-\frac{121}{6}x+\frac{121}{3}}$.

22.

$$\begin{array}{r} x^2+0x-1 \\ 4x^2+0x-9\overline{\big)\, 4x^4+0x^3-13x^2+0x+9} \\ \underline{-(4x^4+0x^3-9x^2)} \\ -4x^2+0x+9 \\ \underline{-(-4x^2+0x+9)} \\ 0 \end{array}$$

So, $\boxed{Q(x)=x^2-1,\quad r(x)=0}$.

23.

$$\begin{array}{r} -3x^3+5.2x^2+3.12x-0.128 \\ x-0.6\overline{\big)\, -3x^4+7x^3+0x^2-2x+1} \\ \underline{-(-3x^4+1.8x^3)} \\ 5.2x^3+0x^2 \\ \underline{-(5.2x^3-3.12x^2)} \\ 3.12x^2-2x \\ \underline{-(3.12x^2-1.872x)} \\ -0.128x+1 \\ \underline{-(-0.128x+0.0768)} \\ 0.9232 \end{array}$$

So, $\boxed{Q(x)=-3x^3+5.2x^2+3.12x-0.128,\quad r(x)=0.9232}$.

24.

$$
\begin{array}{r|l}
 & 2x^4 + 1.8x^3 - 2.38x^2 + 0.858x + 0.7722 \\
\hline
x - 0.9 & 2x^5 + 0x^4 - 4x^3 + 3x^2 + 0x + 5 \\
 & \underline{-(2x^5 - 1.8x^4)} \\
 & 1.8x^4 - 4x^3 \\
 & \underline{-(1.8x^4 - 1.62x^3)} \\
 & -2.38x^3 + 3x^2 \\
 & \underline{-(-2.38x^3 + 2.142x^2)} \\
 & 0.858x^2 + 0x \\
 & \underline{-(0.858x^2 - 0.7722x)} \\
 & 0.7722x + 5 \\
 & \underline{-(0.7722x - 0.69498)} \\
 & 5.69498
\end{array}
$$

So, $\boxed{Q(x) = 2x^4 + 1.8x^3 - 2.38x^2 + 0.858x + 0.7722, \quad r(x) = \frac{5.69498}{x-0.9}}$.

25.

$$
\begin{array}{r|l}
 & x^2 - 0.6x + 0.09 \\
\hline
x^2 + 1.4x + 0.49 & x^4 + 0.8x^3 - 0.26x^2 - 0.168x + 0.0441 \\
 & \underline{-(x^4 + 1.4x^3 + 0.49x^2)} \\
 & -0.6x^3 - 0.75x^2 - 0.168x \\
 & \underline{-(-0.6x^3 - 0.84x^2 - 0.294x)} \\
 & 0.09x^2 + 0.126x + 0.0441 \\
 & \underline{-(0.09x^2 + 0.126x + 0.0441)} \\
 & 0
\end{array}
$$

So, $\boxed{Q(x) = x^2 - 0.6x + 0.09, \quad r(x) = 0}$.

26.

$$\begin{array}{r} x^3 + 3.4x^2 + 3.29x + 0.98 \\ x^2-0.6x+0.09 \overline{\big)\, x^5+2.8x^4+1.34x^3-0.688x^2-0.2919x+0.0882} \\ \underline{-(x^5-0.6x^4+0.09x^3)} \\ 3.4x^4+1.25x^3-0.688x^2 \\ \underline{-(3.4x^4-2.04x^3+0.306x^2)} \\ 3.29x^3-0.994x^2-0.2919x \\ \underline{-(3.29x^3-1.974x^2+0.2961x)} \\ 0.98x^2-0.588x+0.0882 \\ \underline{-(0.98x^2-0.588x+0.0882)} \\ 0 \end{array}$$

So, $\boxed{Q(x)=x^3+3.4x^2+3.29x+0.98, \quad r(x)=0}$.

27.

$$\begin{array}{r|rrr} -2 & 3 & 7 & 2 \\ & & -6 & -2 \\ \hline & 3 & 1 & 0 \end{array}$$

So, $\boxed{Q(x)=3x+1, \quad r(x)=0}$.

28.

$$\begin{array}{r|rrr} -5 & 2 & 7 & -15 \\ & & -10 & 15 \\ \hline & 2 & -3 & 0 \end{array}$$

So, $\boxed{Q(x)=2x-3, \quad r(x)=0}$.

29.

$$\begin{array}{r|rrr} -1 & 7 & -3 & 5 \\ & & -7 & 10 \\ \hline & 7 & -10 & 15 \end{array}$$

So, $\boxed{Q(x)=7x-10, \quad r(x)=15}$.

30.

$$\begin{array}{r|rrr} 2 & 4 & 1 & 1 \\ & & 8 & 18 \\ \hline & 4 & 9 & 19 \end{array}$$

So, $\boxed{Q(x)=4x+9, \quad r(x)=19}$.

31.

$$\begin{array}{r|rrrrr} -2 & -1 & -2 & 3 & 4 & -4 \\ & & 2 & 0 & -6 & 4 \\ \hline & -1 & 0 & 3 & -2 & 0 \end{array}$$

So, $\boxed{Q(x)=-x^3+3x-2, \quad r(x)=0}$.

32.

$$\begin{array}{r|rrrr} 1 & 1 & 3 & 0 & -4 \\ & & 1 & 4 & 4 \\ \hline & 1 & 4 & 4 & 0 \end{array}$$

So, $\boxed{Q(x)=x^2+4x+4, \quad r(x)=0}$.

33.

$$\begin{array}{r|rrrrr} -1 & 1 & 0 & 0 & 0 & 1 \\ & & -1 & 1 & -1 & 1 \\ \hline & 1 & -1 & 1 & -1 & 2 \end{array}$$

So, $\boxed{Q(x)=x^3-x^2+x-1, \quad r(x)=2}$.

34.

$$\begin{array}{r|rrrrr} -3 & 1 & 0 & 0 & 0 & 9 \\ & & -3 & 9 & -27 & 81 \\ \hline & 1 & -3 & 9 & -27 & 90 \end{array}$$

So, $\boxed{Q(x)=x^3-3x^2+9x-27, \quad r(x)=90}$.

35.

$$\begin{array}{r|rrrrr} -2 & 1 & 0 & 0 & 0 & -16 \\ & & -2 & 4 & -8 & 16 \\ \hline & 1 & -2 & 4 & -8 & 0 \end{array}$$

So, $\boxed{Q(x) = x^3 - 2x^2 + 4x - 8,\quad r(x) = 0}$.

36.

$$\begin{array}{r|rrrrr} 3 & 1 & 0 & 0 & 0 & -81 \\ & & 3 & 9 & 27 & 81 \\ \hline & 1 & 3 & 9 & 27 & 0 \end{array}$$

So, $\boxed{Q(x) = x^3 + 3x^2 + 9x + 27,\quad r(x) = 0}$.

37.

$$\begin{array}{r|rrrr} -\frac{1}{2} & 2 & -5 & -1 & 1 \\ & & -1 & 3 & -1 \\ \hline & 2 & -6 & 2 & 0 \end{array}$$

So, $\boxed{Q(x) = 2x^2 - 6x + 2,\quad r(x) = 0}$.

38.

$$\begin{array}{r|rrrr} -\frac{1}{3} & 3 & -8 & 0 & 1 \\ & & -1 & 3 & -1 \\ \hline & 3 & -9 & 3 & 0 \end{array}$$

So, $\boxed{Q(x) = 3x^2 - 9x + 3,\quad r(x) = 0}$.

39.

$$\begin{array}{r|rrrrr} \frac{2}{3} & 2 & -3 & 7 & 0 & -4 \\ & & \frac{4}{3} & -\frac{10}{9} & \frac{106}{27} & \frac{212}{81} \\ \hline & 2 & -\frac{5}{3} & \frac{53}{9} & \frac{106}{27} & -\frac{112}{81} \end{array}$$

So,

$\boxed{Q(x) = 2x^3 - \frac{5}{3}x^2 + \frac{53}{9}x + \frac{106}{27},\quad r(x) = -\frac{112}{81}}$.

40.

$$\begin{array}{r|rrrrr} \frac{3}{4} & 3 & 1 & 0 & 2 & -3 \\ & & \frac{9}{4} & \frac{39}{16} & \frac{117}{64} & \frac{735}{256} \\ \hline & 3 & \frac{13}{4} & \frac{39}{16} & \frac{245}{64} & -\frac{33}{256} \end{array}$$

So,

$\boxed{Q(x) = 3x^3 + \frac{13}{4}x^2 + \frac{39}{16}x + \frac{245}{64},\quad r(x) = -\frac{33}{256}}$

41.

$$\begin{array}{r|rrrrr} -1.5 & 2 & 9 & -9 & -81 & -81 \\ & & -3 & -9 & 27 & 81 \\ \hline & 2 & 6 & -18 & -54 & 0 \end{array}$$

So, $\boxed{Q(x) = 2x^3 + 6x^2 - 18x - 54,\quad r(x) = 0}$.

42.

$$\begin{array}{r|rrrr} -0.8 & 5 & -1 & 6 & 8 \\ & & -4 & 4 & -8 \\ \hline & 5 & -5 & 10 & 0 \end{array}$$

So, $\boxed{Q(x) = 5x^2 - 5x + 10,\quad r(x) = 0}$.

43.

$$\begin{array}{r|rrrrrrrr} 1 & 1 & 0 & 0 & -8 & 0 & 3 & 0 & 1 \\ & & 1 & 1 & 1 & -7 & -7 & -4 & -4 \\ \hline & 1 & 1 & 1 & -7 & -7 & -4 & -4 & -3 \end{array}$$

So,

$\boxed{\begin{array}{l} Q(x) = x^6 + x^5 + x^4 - 7x^3 - 7x^2 - 4x - 4, \\ r(x) = -3 \end{array}}$.

44.

$$\begin{array}{r|rrrrrrr} -1 & 1 & 4 & 0 & -2 & 0 & 0 & 7 \\ & & -1 & -3 & 3 & -1 & 1 & -1 \\ \hline & 1 & 3 & -3 & 1 & -1 & 1 & 6 \end{array}$$

So,

$\boxed{\begin{array}{l} Q(x) = x^5 + 3x^4 - 3x^3 + x^2 - x + 1, \\ r(x) = 6 \end{array}}$.

45.

$$\begin{array}{r|rrrrrrr} \sqrt{5} & 1 & 0 & -49 & 0 & -25 & 0 & 1225 \\ & & \sqrt{5} & 5 & -44\sqrt{5} & -220 & -245\sqrt{5} & -1225 \\ \hline & 1 & \sqrt{5} & -44 & -44\sqrt{5} & -245 & -245\sqrt{5} & 0 \end{array}$$

So, $\boxed{Q(x) = x^5 + \sqrt{5}x^4 - 44x^3 - 44\sqrt{5}x^2 - 245x - 245\sqrt{5}, \quad r(x) = 0}$.

46.

$$\begin{array}{r|rrrrrrr} \sqrt{3} & 1 & 0 & -4 & 0 & -9 & 0 & 36 \\ & & \sqrt{3} & 3 & -\sqrt{3} & -3 & -12\sqrt{3} & -36 \\ \hline & 1 & \sqrt{3} & -1 & -\sqrt{3} & -12 & -12\sqrt{3} & 0 \end{array}$$

So, $\boxed{Q(x) = x^5 + \sqrt{3}x^4 - x^3 - \sqrt{3}x^2 - 12x - 12\sqrt{3}, \quad r(x) = 0}$.

47.

$$\begin{array}{r l} & 2x-7 \\ 3x-1 \overline{)} & 6x^2 - 23x + 7 \\ & \underline{-(6x^2 - 2x)} \\ & \quad -21x + 7 \\ & \quad \underline{-(-21x + 7)} \\ & \qquad 0 \end{array}$$

So, $\boxed{Q(x) = 2x - 7, \quad r(x) = 0}$.

48.

$$\begin{array}{r l} & 3x+2 \\ 2x-1 \overline{)} & 6x^2 + x - 2 \\ & \underline{-(6x^2 - 3x)} \\ & \quad 4x - 2 \\ & \quad \underline{-(4x - 2)} \\ & \qquad 0 \end{array}$$

So, $\boxed{Q(x) = 3x + 2 \quad r(x) = 0}$.

49.

$$\begin{array}{r|rrrr} 1 & 1 & -1 & -9 & 9 \\ & & 1 & 0 & -9 \\ \hline & 1 & 0 & -9 & 0 \end{array}$$

So, $\boxed{Q(x) = x^2 - 9, \quad r(x) = 0}$.

50.

$$\begin{array}{r|rrrr} -2 & 1 & 2 & -6 & -12 \\ & & -2 & 0 & 12 \\ \hline & 1 & 0 & -6 & 0 \end{array}$$

So, $\boxed{Q(x) = x^2 - 6, \quad r(x) = 0}$.

51.

$$\begin{array}{r l} & x+6 \\ x^2 + 0x - 1 \overline{)} & x^3 + 6x^2 - 2x - 5 \\ & \underline{-(x^3 + 0x - x)} \\ & \quad 6x^2 - x - 5 \\ & \quad \underline{-(6x^2 + 0x - 6)} \\ & \qquad -x + 1 \end{array}$$

So, $\boxed{Q(x) = x + 6, \quad r(x) = -x + 1}$.

52.

$$\begin{array}{r l} & 3x^2 - 3x + 5 \\ x^3 + x^2 - x + 1 \overline{)} & 3x^5 + 0x^4 - x^3 + 2x^2 + 0x - 1 \\ & \underline{-(3x^5 + 3x^4 - 3x^3 + 3x^2)} \\ & \quad -3x^4 + 2x^3 - x^2 + 0x \\ & \quad \underline{-(-3x^4 - 3x^3 + 3x^2 - 3x)} \\ & \qquad 5x^3 - 4x^2 + 3x - 1 \\ & \qquad \underline{-(5x^3 + 5x^2 - 5x + 5)} \\ & \qquad\quad -9x^2 + 8x - 6 \end{array}$$

So,

$\boxed{Q(x) = 3x^2 - 3x + 5, \quad r(x) = -9x^2 + 8x - 6}$

53.

$$\begin{array}{r} x^4-2x^3-4x+7 \\ x^2+0x+1\overline{)x^6-2x^5+x^4-6x^3+7x^2-4x+7} \\ -(x^6+0x^5+x^4) \\ \hline -2x^5+0x^4-6x^3 \\ -(-2x^5+0x^4-2x^3) \\ \hline -4x^3+7x^2-4x \\ -(-4x^3+0x^2-4x) \\ \hline 7x^2+0x+7 \\ -(7x^2+0x+7) \\ \hline 0 \end{array}$$

So, $\boxed{Q(x)=x^4-2x^3-4x+7,\quad r(x)=0}$

54.

$$\begin{array}{r} x^4-x^3+x+7 \\ x^2+x+1\overline{)x^6+0x^5+0x^4+0x^3+0x^2+0x-1} \\ -(x^6+x^5+x^4) \\ \hline -x^5-x^4+0x^3 \\ -(-x^5-x^4-x^3) \\ \hline x^3+0x^2+0x \\ -(x^3+x^2+x) \\ \hline -x^2-x-1 \\ -(-x^2-x-1) \\ \hline 0 \end{array}$$

So, $\boxed{Q(x)=x^4-x^3+x+7,\quad r(x)=0}$

55.

$$\begin{array}{r|rrrrrr} 2 & 1 & 0 & 4 & 2 & 0 & -1 \\ & & 2 & 4 & 16 & 36 & 72 \\ \hline & 1 & 2 & 8 & 18 & 36 & 71 \end{array}$$

So,

$\boxed{Q(x)=x^4+2x^3+8x^2+18x+36,\; r(x)=71}$.

56.

$$\begin{array}{r|rrrrr} -5 & 1 & 0 & -1 & 3 & -10 \\ & & -5 & 25 & -120 & 585 \\ \hline & 1 & -5 & 24 & -117 & 575 \end{array}$$

So,

$\boxed{Q(x)=x^3-5x^2+24x-117,\; r(x)=575}$.

57.

$$\begin{array}{r} x^2+0x+1 \\ x^2+0x-1\overline{)x^4+0x^3+0x^2+0x-25} \\ -(x^4+0x^3-x^2) \\ \hline x^2+0x-25 \\ -(x^2+0x-1) \\ \hline -24 \end{array}$$

So, $\boxed{Q(x)=x^2+1,\quad r(x)=-24}$.

58.

$$\begin{array}{r} x \\ x^2+0x-2\overline{)x^3+0x^2+0x-8} \\ -(x^3+0x^2-2x) \\ \hline 2x-8 \end{array}$$

So, $\boxed{Q(x)=x,\quad r(x)=2x-8}$.

59.

$$\begin{array}{r|rrrrrrrr} 1 & 1 & 0 & 0 & 0 & 0 & 0 & 0 & -1 \\ & & 1 & 1 & 1 & 1 & 1 & 1 & 1 \\ \hline & 1 & 1 & 1 & 1 & 1 & 1 & 1 & 0 \end{array}$$

So,

$\boxed{Q(x)=x^6+x^5+x^4+x^3+x^2+x+1,\; r(x)=0}$.

60.

$$\begin{array}{r|rrrrrrr} 3 & 1 & 0 & 0 & 0 & 0 & 0 & -27 \\ & & 3 & 9 & 27 & 81 & 243 & 729 \\ \hline & 1 & 3 & 9 & 27 & 81 & 243 & 702 \end{array}$$

So,

$\boxed{Q(x)=x^5+3x^4+9x^3+27x^2+81x+243,\; r(x)=702}$.

61. Area = length × width. So, solving for width, we see that width = Area ÷ length. So, we have:

$$\begin{array}{r} 3x^2+2x+1 \\ 2x^2+0x-1 \overline{\big) 6x^4+4x^3-x^2-2x-1} \\ \underline{-(6x^4+0x^3-3x^2)} \\ 4x^3+2x^2-2x \\ \underline{-(4x^3+0x^2-2x)} \\ 2x^2+0x-1 \\ \underline{-(2x^2+0x-1)} \\ 0 \end{array}$$

Thus, the width (in terms of x) is $\boxed{3x^2+2x+1 \text{ feet}}$.

62. Volume = (Area of base) × height. So, solving for height, we see that height = Volume ÷ (Area of base). So, we have:

$$\begin{array}{r} 3x+1 \\ 6x^4+4x^3-x^2-2x-1 \overline{\big) 18x^5+18x^4+x^3-7x^2-5x-1} \\ \underline{-(18x^5+12x^4-3x^3-6x^2-3x)} \\ 6x^4+4x^3-x^2-2x-1 \\ \underline{-(6x^4+4x^3-x^2-2x-1)} \\ 0 \end{array}$$

So, the height (in terms of x) is $\boxed{3x+1 \text{ feet}}$.

63. Distance = Rate × Time. So, solving for Time, we have: Time = Distance ÷ Rate. So, we calculate $\left(x^3+60x^2+x+60\right) \div (x+60)$ using synthetic division:

$$\begin{array}{r|rrrr} -60 & 1 & 60 & 1 & 60 \\ & & -60 & 0 & -60 \\ \hline & 1 & 0 & 1 & 0 \end{array}$$

So, the time is $\boxed{x^2+1 \text{ hours}}$.

64. Distance = Rate × Time. So, solving for Rate, we have: Rate = Distance ÷ Time. So, we calculate $\left(-x^2-5x+50\right)\div(5-x)$ using long division:

$$\begin{array}{r|l} & \quad x+10 \\ \hline -x+5 & -x^2-5x+50 \\ & \underline{-(-x^2+5x)} \\ & \qquad -10x+50 \\ & \qquad \underline{-(-10x+50)} \\ & \qquad\qquad 0 \end{array}$$

So, the rate is $\boxed{x+10 \text{ yards per second}}$.

65. Should have subtracted each term in the long division rather than adding them.

66. The zero of the divisor is used in synthetic division. So, 2 should replace -2 as the divisor.

67. Forgot the "0" placeholder.

68. Cannot use synthetic division with a quadratic divisor. Use long division instead.

69. True.

70. False. For instance,

$$\left(x^3-x^2+x-1\right)\div\left(x-1\right)=x^2+1.$$

71. False. Only use when the divisor has degree 1.

72. True.

73. False. For example, $\dfrac{x+2}{x+1}\neq 1$.

74. True.

75.

$$\begin{array}{r|rrrr} -b & 1 & 2b-a & b^2-2ab & -ab^2 \\ & & -b & -b^2+ab & ab^2 \\ \hline & 1 & b-a & -ab & 0 \end{array}$$

Since the remainder is 0 upon using synthetic division, YES, $(x+b)$ is a factor of $x^3+(2b-a)x^2+\left(b^2-2ab\right)x-ab^2$.

76.

$$\begin{array}{r|rrrrr} -b & 1 & 0 & b^2-a^2 & 0 & -a^2b^2 \\ & & -b & b^2 & -2b^3+a^2b & 2b^4-a^2b^2 \\ \hline & 1 & -b & 2b^2-a^2 & -2b^3+a^2b & 2b^4-2a^2b^2 \end{array}$$

Since the remainder is not 0 upon using synthetic division, NO, $(x+b)$ is not a factor of the given polynomial, <u>unless</u> $b=0$, in which case the above simplifies to saying $x-0$ is a factor of $x^4-a^2x^2$.

77.

$$\begin{array}{r} x^{2n}+2x^n+1 \\ x^n-1\overline{)x^{3n}+x^{2n}-x^n-1} \\ \underline{-(x^{3n}-x^{2n})} \\ 2x^{2n}-x^n \\ \underline{-(2x^{2n}-2x^n)} \\ x^n-1 \\ \underline{-(x^n-1)} \\ 0 \end{array}$$

So, $\boxed{Q(x)=x^{2n}+2x^n+1,\quad r(x)=0}$

78. First, we rewrite the polynomial in a more familiar form using the substitution $y=x^n$. Doing so yields

$x^{3n}+5x^{2n}+8x^n+4=y^3+5y^2+8y+4$.

Now, apply synthetic division:

$$\begin{array}{r|rrrr} -1 & 1 & 5 & 8 & 4 \\ & & -1 & -4 & -4 \\ \hline & 1 & 4 & 4 & 0 \end{array}$$

Thus,

$$\begin{aligned} y^3+5y^2+8y+4 &= (y+1)(y^2+4y+4) \\ &= (y+1)(y+2)^2. \end{aligned}$$

Going back to the original polynomial, this says:

$x^{3n}+5x^{2n}+8x^n+4=(x^n+1)(x^n+2)^2$.

79. Long division gives us

$$\begin{array}{r} 2x-1 \\ x^2+0x+5\overline{)\,2x^3-x^2+10x-5} \\ \underline{-(2x^3+0x^2+10x)} \\ -x^2+0x-5 \\ \underline{-(-x^2+0x-5)} \\ 0 \end{array}$$

So, the graph is a line, as shown:

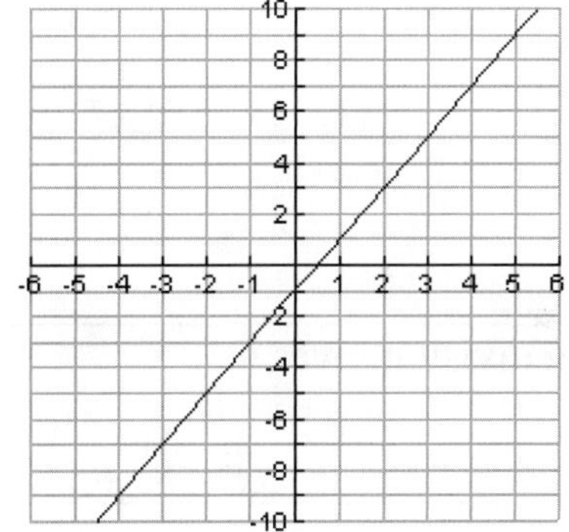

80. Synthetic division gives us

$$\begin{array}{r|rrrr} 3 & 1 & -3 & 4 & -12 \\ & & 3 & 0 & 12 \\ \hline & 1 & 0 & 4 & 0 \end{array}$$

So, the quotient is the quadratic function $y=x^2+4$ with a hole at 3, shown to the right:

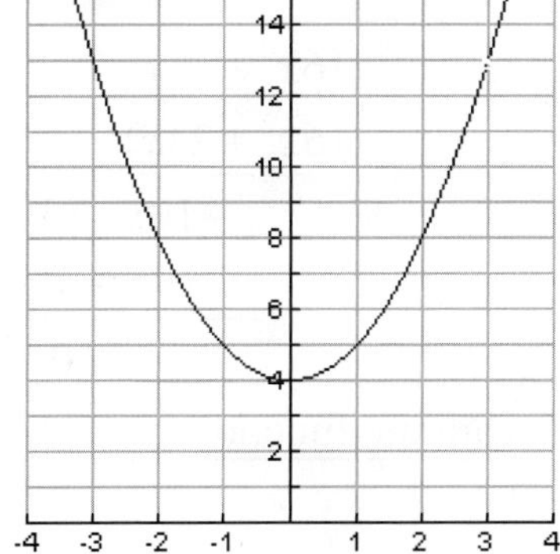

81. Using synthetic division gives us:

$$\begin{array}{r|rrrrr} -2 & 1 & 2 & 0 & -1 & -2 \\ & & -2 & 0 & 0 & 2 \\ \hline & 1 & 0 & 0 & -1 & 0 \end{array}$$

So, the quotient is the cubic $y = x^3 - 1$, as shown to the right:

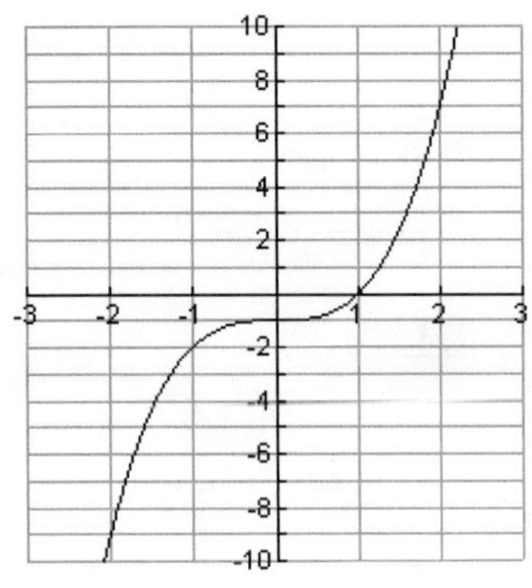

82. Long division yields:

$$\begin{array}{r} x-3 \\ x^4-6x^3+0x^2+2x+1 \overline{\smash{\big)}\, x^5-9x^4+18x^3+2x^2-5x-3} \\ \underline{-(x^5-6x^4+0x^3+2x^2+x)} \\ -3x^4+18x^3+0x^2-6x-3 \\ \underline{-(-3x^4+18x^3+0x^2-6x-3)} \\ 0 \end{array}$$

So, the graph is the line $y = x - 3$, as shown below:

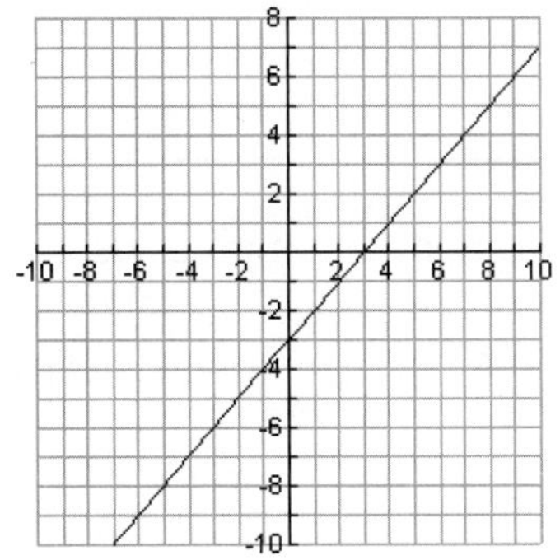

83. Long division gives us

$$\begin{array}{r} -3x^2+8x-5 \\ 2x+3 \overline{\smash{\big)}\, -6x^3+7x^2+14x-15} \\ \underline{-(-6x^3-9x^2)} \\ 16x^2+14x \\ \underline{-(16x^2+24x)} \\ -10x-15 \\ \underline{-(-10x-15)} \\ 0 \end{array}$$

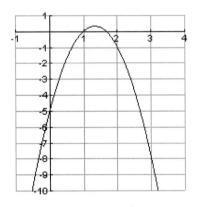

So, it is a quadratic function.

84. Long division gives us

$$\begin{array}{r} -x^3+9x \\ 3x^2+4x-2\overline{\big)-3x^5-4x^4+29x^3+36x^2-18x} \\ \underline{-(-3x^5-4x^4+2x^3)} \\ 27x^3+36x^2-18x \\ \underline{-(27x^3+36x^2-18x)} \\ 0 \end{array}$$

So, it is a third-degree polynomial.

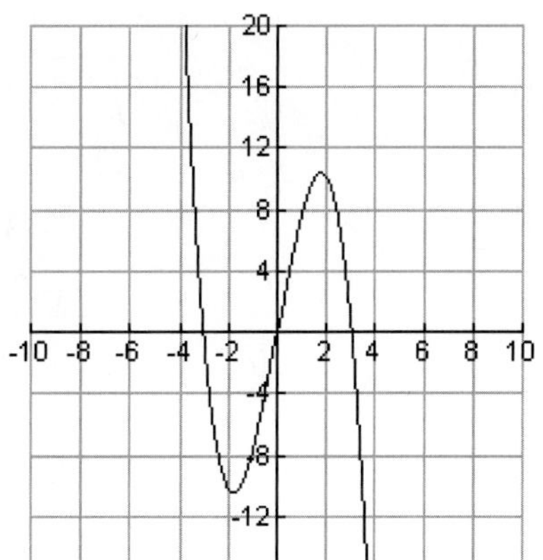

85. Using synthetic division gives us:

$$\begin{array}{r|rrr} -2 & 2 & -1 & 0 \\ & & -4 & 10 \\ \hline & 2 & -5 & 10 \end{array}$$

So, $\dfrac{2x^2-x}{x+2}=(2x-5)+\dfrac{10}{x+2}$.

86. Using synthetic division gives us:

$$\begin{array}{r|rrrr} 3 & 5 & 2 & -3 & 0 \\ & & 15 & 51 & 144 \\ \hline & 5 & 17 & 48 & 144 \end{array}$$

So,

$$\frac{5x^3+2x^2-3x}{x-3}=\left(5x^2+17x+48\right)+\frac{144}{x-3}$$

87.

$$\begin{array}{r} 2x^2-2x+3 \\ x^2+x+1\overline{\big)\,2x^4+0x^3+3x^2+0x+6} \\ \underline{-(2x^4+2x^3+2x^2)} \\ -2x^3+x^2+0x \\ \underline{-\left(-2x^3-2x^2-2x\right)} \\ 3x^2+2x+6 \\ \underline{-\left(3x^2+3x+3\right)} \\ -x+3 \end{array}$$

So, $\dfrac{2x^4+3x^2+6}{x^2+x+1}=\left(2x^2-2x+3\right)-\dfrac{x-3}{x^2+x+1}$.

88.

$$\begin{array}{r} 3x^3 - 3x^2 - 14x + 30 \\ x^2 + x + 5 \overline{) 3x^5 + 0x^4 - 2x^3 + x^2 + x - 6} \\ \underline{-(3x^5 + 3x^4 + 15x^3)} \\ -3x^4 - 17x^3 + x^2 \\ \underline{-(-3x^4 - 17x^3 + x^2)} \\ -14x^3 + 16x^2 + x \\ \underline{-(-14x^3 - 14x^2 - 70x)} \\ 30x^2 + 71x - 6 \\ \underline{-(30x^2 + 30x + 150)} \\ 41x - 156 \end{array}$$

So, $\dfrac{3x^5 - 2x^3 + x^2 + x - 6}{x^2 + x + 5} = \left(3x^3 - 3x^2 - 14x + 30\right) + \dfrac{41x - 156}{x^2 + x + 5}$

Section 2.4 Solutions ---

1.

$$\begin{array}{r|rrrr} 1 & 1 & 0 & -13 & 12 \\ & & 1 & 1 & -12 \\ \hline & 1 & 1 & -12 & 0 \end{array}$$

So,

$$\begin{aligned} P(x) &= (x-1)(x^2 + x - 12) \\ &= (x-1)(x+4)(x-3) \end{aligned}.$$

The zeros are 1, 3 and -4.

2.

$$\begin{array}{r|rrrr} 3 & 1 & 3 & -10 & -24 \\ & & 3 & 18 & 24 \\ \hline & 1 & 6 & 8 & 0 \end{array}$$

So,

$$\begin{aligned} P(x) &= (x-3)(x^2 + 6x + 8) \\ &= (x-3)(x+4)(x+2) \end{aligned}.$$

The zeros are 3, -2 and -4.

3.

$$\begin{array}{r|rrrr} \frac{1}{2} & 2 & 1 & -13 & 6 \\ & & 1 & 1 & -6 \\ \hline & 2 & 2 & -12 & 0 \end{array}$$

So,

$$\begin{aligned} P(x) &= (x - \tfrac{1}{2})(2x^2 + 2x - 12) \\ &= 2(x - \tfrac{1}{2})(x^2 + x - 6) \\ &= 2(x - \tfrac{1}{2})(x+3)(x-2) \\ &= (2x-1)(x+3)(x-2) \end{aligned}.$$

The zeros are $\frac{1}{2}, -3$ and 2.

4.

$$\begin{array}{r|rrrr} -\frac{1}{3} & 3 & -14 & 7 & 4 \\ & & -1 & 5 & -4 \\ \hline & 3 & -15 & 12 & 0 \end{array}$$

So,

$$\begin{aligned} P(x) &= (x + \tfrac{1}{3})(3x^2 - 15x + 12) \\ &= 3(x + \tfrac{1}{3})(x^2 - 5x + 4) \\ &= 3(x + \tfrac{1}{3})(x-4)(x-1) \\ &= (3x+1)(x-4)(x-1) \end{aligned}.$$

The zeros are 1, 4 and $-\frac{1}{3}$.

5. Since -3 and 5 are both zeros, we know that $(x+3)$ and $(x-5)$ are factors of $P(x)$ and hence, must divide $P(x)$ evenly. (Note: $(x+3)(x-5)=x^2-2x-15$.)

$$\begin{array}{r} x^2+4 \\ x^2-2x-15\,\big)\,\overline{x^4-2x^3-11x^2-8x-60} \\ \underline{-(x^4-2x^3-15x^2)} \\ 4x^2-8x-60 \\ \underline{-(4x^2-8x-60)} \\ 0 \end{array}$$

So, $P(x)=(x^2+4)(x^2-2x-15)=\boxed{(x^2+4)(x-5)(x+3)}$. The real zeros are 5 and -3.

6. Since -1 and 2 are both zeros, we know that $(x+1)$ and $(x-2)$ are factors of $P(x)$ and hence, must divide $P(x)$ evenly. (Note: $(x+1)(x-2)=x^2-x-2$.)

$$\begin{array}{r} x^2+9 \\ x^2-x-2\,\big)\,\overline{x^4-x^3+7x^2-9x-18} \\ \underline{-(x^4-x^3-2x^2)} \\ 9x^2-9x-18 \\ \underline{-(9x^2-9x-18)} \\ 0 \end{array}$$

So, $P(x)=(x^2+9)(x^2-x-2)=(x^2+9)(x-2)(x+1)$. The real zeros are 2 and -1.

7. Since –3 and 1 are both zeros, we know that $(x+3)$ and $(x-1)$ are factors of $P(x)$ and hence, must divide $P(x)$ evenly. (Note: $(x+3)(x-1)=x^2+2x-3$.)

$$\begin{array}{r} x^2-2x+2 \\ x^2+2x-3\,\big)\,\overline{x^4+0x^3-5x^2+10x-6} \\ \underline{-(x^4+2x^3-3x^2)} \\ -2x^3-2x^2+10x \\ \underline{-(-2x^3-4x^2+6x)} \\ 2x^2+4x-6 \\ \underline{-(2x^2+4x-6)} \\ 0 \end{array}$$

So, $P(x)=(x^2+2x-3)(x^2-2x+2)=(x-1)(x+3)(x^2-2x+2)$. The real zeros are 1 and -3.

8. Since –2 and 4 are both zeros, we know that $(x+2)$ and $(x-4)$ are factors of $P(x)$ and hence, must divide $P(x)$ evenly. (Note: $(x+2)(x-4)=x^2-2x-8$.)

$$
\begin{array}{r}
x^2-2x+5 \\
x^2-2x-8\overline{\big)\, x^4-4x^3+x^2+6x-40} \\
\underline{-(x^4-2x^3-8x^2)} \\
-2x^3+9x^2+6x \\
\underline{-(-2x^3+4x^2+16x)} \\
5x^2-10x-40 \\
\underline{-(5x^2-10x-40)} \\
0
\end{array}
$$

So, $P(x)=(x^2-2x+5)(x^2-2x-8)=(x^2-2x+5)(x-4)(x+2)$. The real zeros are 4 and -2.

9.

$$
\begin{array}{r|rrrrr}
-2 & 1 & 6 & 13 & 12 & 4 \\
 & & -2 & -8 & -10 & -4 \\ \hline
-2 & 1 & 4 & 5 & 2 & 0 \\
 & & -2 & -4 & -2 & \\ \hline
 & 1 & 2 & 1 & 0 &
\end{array}
$$

So,
$P(x)=(x+2)^2(x^2+2x+1)=(x+2)^2(x+1)^2$

The zeros are -2 and -1, both with multiplicity 2.

10.

$$
\begin{array}{r|rrrrr}
1 & 1 & 4 & -2 & -12 & 9 \\
 & & 1 & 5 & 3 & -9 \\ \hline
1 & 1 & 5 & 3 & -9 & 0 \\
 & & 1 & 6 & 9 & \\ \hline
 & 1 & 6 & 9 & 0 &
\end{array}
$$

So,
$P(x)=(x-1)^2(x^2+6x+9)=(x-1)^2(x+3)^2$
The zeros are -3 and 1, both with multiplicity 2.

11.
Factors of 4: $\pm1, \pm2, \pm4$
Factors of 1: ±1
Possible rational zeros: $\pm1, \pm2, \pm4$

12.
Factors of 4: $\pm1, \pm2, \pm4$
Factors of -1: ±1
Possible rational zeros: $\pm1, \pm2, \pm4$

13.
Factors of 12: $\pm1, \pm2, \pm3, \pm4, \pm6, \pm12$
Factors of 1: ±1
Possible rational zeros:
$\pm1, \pm2, \pm3, \pm4, \pm6, \pm12$

14.
Factors of 9: $\pm1, \pm3, \pm9$
Factors of 1: ±1
Possible rational zeros: $\pm1, \pm3, \pm9$

15. Factors of 8: $\pm 1, \pm 2, \pm 4, \pm 8$ Factors of 2: $\pm 1, \pm 2$ Possible rational zeros: $\pm \frac{1}{2}, \pm 1, \pm 2, \pm 4, \pm 8$	**16.** Factors of -10: $\pm 1, \pm 2, \pm 5, \pm 10$ Factors of 3: $\pm 1, \pm 3$ Possible rational zeros: $\pm 1, \pm 2, \pm 5, \pm 10, \pm \frac{1}{3}, \pm \frac{2}{3}, \pm \frac{5}{3}, \pm \frac{10}{3}$
17. Factors of -20: $\pm 1, \pm 2, \pm 4, \pm 5, \pm 10, \pm 20$ Factors of 5: $\pm 1, \pm 5$ Possible rational zeros: $\pm 1, \pm 2, \pm 4, \pm 5, \pm 10, \pm 20, \pm \frac{1}{5}, \pm \frac{2}{5}, \pm \frac{4}{5}$	**18.** Factors of -21: $\pm 1, \pm 3, \pm 7, \pm 21$ Factors of 4: $\pm 1, \pm 2, \pm 4$ Possible rational zeros: $\pm 1, \pm 3, \pm 7, \pm 21, \pm \frac{1}{2}, \pm \frac{1}{4}, \pm \frac{3}{2},$ $\pm \frac{7}{2}, \pm \frac{21}{2}, \pm \frac{3}{4}, \pm \frac{7}{4}, \pm \frac{21}{4}$
19. Factors of 8: $\pm 1, \pm 2, \pm 4, \pm 8$ Factors of 1: ± 1 Possible rational zeros: $\pm 1, \pm 2, \pm 4, \pm 8$ Testing: $P(1) = P(-1) = P(2) = P(-4) = 0$	**20.** Factors of 3: $\pm 1, \pm 3$ Factors of 1: ± 1 Possible rational zeros: $\pm 1, \pm 3$ Testing: $P(1) = P(-1) = P(-3) = 0$

21.
Factors of -3: $\pm 1, \pm 3$
Factors of 2: $\pm 1, \pm 2$
Possible rational zeros: $\pm 1, \pm 3, \pm \frac{1}{2}, \pm \frac{3}{2}$
Testing: $P(1) = P(3) = P\left(\frac{1}{2}\right) = 0$

22.
Factors of -8: $\pm 1, \pm 2, \pm 4, \pm 8$
Factors of 3: $\pm 1, \pm 3$
Possible rational zeros: $\pm 1, \pm 2, \pm 4, \pm 8, \pm \frac{1}{3}, \pm \frac{2}{3}, \pm \frac{4}{3}, \pm \frac{8}{3}$
Testing: $P(-2) = P(4) = P\left(-\frac{1}{3}\right) = 0$

23.
Number of sign variations for $P(x)$: 1
$P(-x) = P(x)$, so
Number of sign variations for $P(-x)$: 1
Since $P(x)$ is degree 4, there are 4 zeros, the real ones of which are classified to the right:

Positive Real Zeros	**Negative Real Zeros**
1	1

24.
Number of sign variations for $P(x)$: 0
$P(-x)=P(x)$, so
Number of sign variations for $P(-x)$: 0
Since $P(x)$ is degree 4, there are 4 zeros, the real ones of which are classified to the right:

Positive Real Zeros	Negative Real Zeros
0	0

25.
Number of sign variations for $P(x)$: 1
$P(-x)=(-x)^5-1=-x^5-1$, so
Number of sign variations for $P(-x)$: 0
Since $P(x)$ is degree 5, there are 5 zeros, the real ones of which are classified to the right:

Positive Real Zeros	Negative Real Zeros
1	0

26.
Number of sign variations for $P(x)$: 0
$P(-x)=-x^5+1$, so
Number of sign variations for $P(-x)$: 1
Since $P(x)$ is degree 5, there are 5 zeros, the real ones of which are classified to the right:

Positive Real Zeros	Negative Real Zeros
0	1

27.
Number of sign variations for $P(x)$: 2
$P(-x)=-x^5+3x^3+x+2$, so
Number of sign variations for $P(-x)$: 1
Since $P(x)$ is degree 5, there are 5 zeros, the real ones of which are classified to the right:

Positive Real Zeros	Negative Real Zeros
2	1
0	1

28.
Number of sign variations for $P(x)$: 1
$P(-x)=P(x)$, so
Number of sign variations for $P(-x)$: 1
Since $P(x)$ is degree 4, there are 4 zeros, the real ones of which are classified to the right:

Positive Real Zeros	Negative Real Zeros
1	1

29.

Number of sign variations for $P(x)$: 1

$P(-x) = -9x^7 - 2x^5 + x^3 + x$, so

Number of sign variations for $P(-x)$: 1

Since $P(x)$ is degree 7, there are 7 zeros. But, 0 is also a zero. So, we classify the remaining real zeros to the right:

Positive Real Zeros	Negative Real Zeros
1	1

30.

Number of sign variations for $P(x)$: 3

$P(-x) = -16x^7 - 3x^4 - 2x - 1$, so

Number of sign variations for $P(-x)$: 0

Since $P(x)$ is degree 7, there are 7 zeros, the real ones of which are classified to the right:

Positive Real Zeros	Negative Real Zeros
3	0
1	0

31.

Number of sign variations for $P(x)$: 2

$P(-x) = P(x)$, so

Number of sign variations for $P(-x)$: 2

Since $P(x)$ is degree 6, there are 6 zeros, the real ones of which are classified to the right:

Positive Real Zeros	Negative Real Zeros
2	2
0	2
2	0
0	0

32.

Number of sign variations for $P(x)$: 1

$P(-x) = -7x^6 - 5x^4 - x^2 - 2x + 1$, so

Number of sign variations for $P(-x)$: 1

Since $P(x)$ is degree 6, there are 6 zeros, the real ones of whicha are classified to the right:

Positive Real Zeros	Negative Real Zeros
1	1

33.

Number of sign variations for $P(x)$: 4

$P(-x) = -3x^4 - 2x^3 - 4x^2 - x - 11$, so

Number of sign variations for $P(-x)$: 0

Since $P(x)$ is degree 4, there are 4 zeros, the real ones of which are classified to the right:

Positive Real Zeros	Negative Real Zeros
4	0
2	0
0	0

34.

Number of sign variations for $P(x)$: 2

$P(-x) = 2x^4 + 3x^3 + 7x^2 - 3x + 2$, so

Number of sign variations for $P(-x)$: 2

Since $P(x)$ is degree 4, there are 4 zeros, the real ones of which classified to the right:

Positive Real Zeros	Negative Real Zeros
2	2
2	0
0	2
0	0

35.

a. Number of sign variations for $P(x)$: 0

$P(-x) = -x^3 + 6x^2 - 11x + 6$, so

Number of sign variations for $P(-x)$: 3

Since $P(x)$ is degree 3, there are zeros, the real ones of which are classified as:

Positive Real Zeros	Negative Real Zeros
0	3
0	1

b. Factors of 6: $\pm1, \pm2, \pm3, \pm6$

Factors of 1: ±1

Possible rational zeros: $\pm1, \pm2, \pm3, \pm6$

c. Note that $P(-1) = P(-2) = P(-3) = 0$. So, the rational zeros are $-1, -2, -3$. These are the only zeros since P has degree 3.

d. $P(x) = (x+1)(x+2)(x+3)$

36.

a. Number of sign variations for $P(x)$: 3

$P(-x) = -x^3 - 6x^2 - 11x - 6$, so

Number of sign variations for $P(-x)$: 0

Since $P(x)$ is degree 3, there are 3 zeros, the real ones of which are classified as:

Positive Real Zeros	Negative Real Zeros
3	0
1	0

b. Factors of -6: $\pm1, \pm2, \pm3, \pm6$

Factors of 1: ±1

Possible rational zeros: $\pm1, \pm2, \pm3, \pm6$

c. Note that $P(1) = P(2) = P(3) = 0$. So, the rational zeros are $1, 2, 3$. These are the only zeros since P has degree 3.

d. $P(x) = (x-1)(x-2)(x-3)$

37.

a. Number of sign variations for $P(x)$: 2

$P(-x) = -x^3 - 7x^2 + x + 7$, so

Number of sign variations for $P(-x)$: 1

Since $P(x)$ is degree 3, there are 3 zeros, the real ones of which are classified as:

Positive Real Zeros	Negative Real Zeros
2	1
0	1

b. Factors of 7: $\pm 1, \pm 7$

Factors of 1: ± 1

Possible rational zeros: $\pm 1, \pm 7$

c. Note that $P(-1) = P(1) = P(7) = 0$. So, the rational zeros are $-1, 1, 7$. These are the only zeros since P has degree 3.

d. $P(x) = (x+1)(x-1)(x-7)$

38.

a. Number of sign variations for $P(x)$: 2

$P(-x) = -x^3 - 5x^2 + 4x + 20$, so

Number of sign variations for $P(-x)$: 1

Since $P(x)$ is degree 3, there are 3 zeros, the real ones of which are classified as:

Positive Real Zeros	Negative Real Zeros
2	1
0	1

b. Factors of 20:

$\pm 1, \pm 2, \pm 4, \pm 5, \pm 10, \pm 20$

Factors of 1: ± 1

Possible rational zeros:

$\pm 1, \pm 2, \pm 4, \pm 5, \pm 10, \pm 20$

c. Note that $P(-2) = P(2) = P(5) = 0$. So, the rational zeros are $-2, 2, 5$. These are the only zeros since P has degree 3.

d. $P(x) = (x+2)(x-2)(x-5)$

39.

a. Number of sign variations for $P(x)$: 1

$P(-x) = x^4 - 6x^3 + 3x^2 + 10x$, so

Number of sign variations for $P(-x)$: 2

Since $P(x)$ is degree 4, there are 4 zeros, one of which is 0. We classify the remaining real zeros below:

Positive Real Zeros	Negative Real Zeros
1	2
1	0

b. $P(x) = x\left(x^3 + 6x^2 + 3x - 10\right)$ We list the possible nonzero rational zeros below:

Factors of -10: $\pm 1, \pm 2, \pm 5, \pm 10$

Factors of 1: ± 1

Possible rational zeros: $\pm 1, \pm 2, \pm 5, \pm 10$

c. Note that

$$P(0) = P(1) = P(-2) = P(-5) = 0.$$

So, the rational zeros are $0, 1, -2, -5$. These are the only zeros since P has degree 4.

d. $P(x) = x(x-1)(x+2)(x+5)$

40.

a. Number of sign variations for $P(x)$: 2

$P(-x) = x^4 + x^3 - 14x^2 - 24x$, so

Number of sign variations for $P(-x)$: 1

Since $P(x)$ is degree 4, there are 4 zeros, one of which is 0. We classify the remaining real zeros below:

Positive Real Zeros	**Negative Real Zeros**
2	1
0	1

b. $P(x) = x\left(x^3 - x^2 - 14x + 24\right)$ We list the possible nonzero rational zeros below:

Factors of 24:

$\pm 1, \pm 2, \pm 3, \pm 4, \pm 6, \pm 8, \pm 12, \pm 24$

Factors of 1: ± 1

Possible rational zeros:

$\pm 1, \pm 2, \pm 3, \pm 4, \pm 6, \pm 8, \pm 12, \pm 24$

c. Note that

$$P(0) = P(-4) = P(2) = P(3) = 0.$$

So, the rational zeros are $0, -4, 2, 3$. These are the only zeros since P has degree 4.

d. $P(x) = x(x+4)(x-2)(x-3)$

41.

a. Number of sign variations for $P(x)$: 4

$P(-x) = x^4 + 7x^3 + 27x^2 + 47x + 26$, so

Number of sign variations for $P(-x)$: 0

Since $P(x)$ is degree 4, there are 4 zeros, the real ones of which are classified as:

Positive Real Zeros	**Negative Real Zeros**
4	0
2	0
0	0

b. Factors of 26: $\pm 1, \pm 2, \pm 13, \pm 26$

Factors of 1: ± 1

Possible rational zeros:

$\pm 1, \pm 2, \pm 13, \pm 26$

c. Note that $P(1) = P(2) = 0$. After testing the others, it is found that the only rational zeros are 1, 2. So, we at least know that $(x-1)(x-2) = x^2 - 3x + 2$ divides $P(x)$ evenly. To find the remaining zeros, we long divide:

$$\begin{array}{r}
x^2 - 4x + 13 \\
x^2 - 3x + 2 \,\big)\, \overline{x^4 - 7x^3 + 27x^2 - 47x + 26} \\
\underline{-(x^4 - 3x^3 + 2x^2)} \qquad\qquad \\
-4x^3 + 25x^2 - 47x \qquad \\
\underline{-(-4x^3 + 12x^2 - 8x)} \qquad \\
13x^2 - 39x + 26 \\
\underline{-(13x^2 - 39x + 26)} \\
0
\end{array}$$

Since $x^2 - 4x + 13$ is irreducible, the real zeros are 1 and 2.

d. $P(x) = (x-1)(x-2)\left(x^2 - 4x + 13\right)$

42.

a. Number of sign variations for $P(x)$: 3

$P(-x) = x^4 + 5x^3 + 5x^2 - 25x - 26$, so

Number of sign variations for $P(-x)$: 1

Since $P(x)$ is degree 4, there are 4 zeros, the real ones of which are classified as:

Positive Real Zeros	Negative Real Zeros
3	1
1	1

b. Factors of -26: $\pm1, \pm2, \pm13, \pm26$

Factors of 1: ±1

Possible rational zeros:

$\pm1, \pm2, \pm13, \pm26$

c. Note that $P(1) = P(-2) = 0$. After testing the others, it is found that the only rational zeros are 1, -2. So, we at least know that $(x-1)(x+2) = x^2 + x - 2$ divides $P(x)$ evenly. To find the remaining zeros, we long divide:

$$\begin{array}{r} x^2 - 6x + 13 \\ x^2 + x - 2 \overline{\big) x^4 - 5x^3 + 5x^2 + 25x - 26} \\ \underline{-(x^4 + x^3 - 2x^2)} \\ -6x^3 + 7x^2 + 25x \\ \underline{-(-6x^3 - 6x^2 + 12x)} \\ 13x^2 + 13x - 26 \\ \underline{-(13x^2 + 13x - 26)} \\ 0 \end{array}$$

Since $x^2 - 6x + 13$ is irreducible, the real zeros are 1 and -2.

d. $P(x) = (x-1)(x+2)\left(x^2 - 6x + 13\right)$

43.

a. Number of sign variations for $P(x)$: 2

$P(-x) = -10x^3 - 7x^2 + 4x + 1$, so

Number of sign variations for $P(-x)$: 1

Since $P(x)$ is degree 3, there are 3 zeros, the real ones of which are classified as:

Positive Real Zeros	Negative Real Zeros
2	1
0	1

b. Factors of 1: ±1

Factors of 10: $\pm1, \pm2, \pm5, \pm10$

Possible rational zeros: $\pm1, \pm\frac{1}{2}, \pm\frac{1}{5}, \pm\frac{1}{10}$

c. Note that $P(1) = P(-\frac{1}{2}) = P(\frac{1}{5}) = 0$. So, the rational zeros are $-1, -\frac{1}{2}, \frac{1}{5}$. These are the only zeros since P has degree 3.

d. $P(x) = (x-1)(2x+1)(5x-1)$

44.

a. Number of sign variations for $P(x)$: 3

$P(-x) = -12x^3 - 13x^2 - 2x - 1$, so

Number of sign variations for $P(-x)$: 0

Since $P(x)$ is degree 3, there are 3 zeros, the real ones of which are classified as:

Positive Real Zeros	Negative Real Zeros
3	0
1	0

b. Factors of 1: ± 1

Factors of 12:

$\pm 1, \pm 2, \pm 3, \pm 4, \pm 6, \pm 12$

Possible rational zeros:

$\pm 1, \pm \frac{1}{2}, \pm \frac{1}{3}, \pm \frac{1}{4}, \pm \frac{1}{6}, \pm \frac{1}{12}$

c. After testing, we conclude that the only rational zero is 1. To determine the other zeros, we use synthetic division:

$$\begin{array}{r|rrrr} 1 & 12 & -13 & 2 & -1 \\ & & 12 & -1 & 1 \\ \hline & 12 & -1 & 1 & 0 \end{array}$$

Since $12x^2 - x + 1$ is irreducible, there are no other rational zeros.

d. $P(x) = (x-1)\left(12x^2 - x + 1\right)$

45.

a. Number of sign variations for $P(x)$: 1

$P(-x) = -6x^3 + 17x^2 - x - 10$, so

Number of sign variations for $P(-x)$: 2

Since $P(x)$ is degree 3, there are 3 zeros, the real ones of which are classified as:

Positive Real Zeros	Negative Real Zeros
1	2
1	0

b. Factors of -10: $\pm 1, \pm 2, \pm 5, \pm 10$

Factors of 6: $\pm 1, \pm 2, \pm 3, \pm 6$

Possible rational zeros:

$\pm 1, \pm 2, \pm 5, \pm 10, \pm \frac{1}{2}, \pm \frac{1}{3}, \pm \frac{1}{6}, \pm \frac{2}{3}, \pm \frac{5}{2}, \pm \frac{5}{3}, \pm \frac{5}{6}, \pm \frac{10}{3}$

c. Note that $P(-1) = P(-\frac{5}{2}) = P(\frac{2}{3}) = 0$.

So, the rational zeros are $-1, -\frac{5}{2}, \frac{2}{3}$. These are the only zeros since P has degree 3.

d.
$$\begin{aligned} P(x) &= (x+1)(2x+5)(3x-2) \\ &= 6(x+1)\left(x+\tfrac{5}{2}\right)\left(x-\tfrac{2}{3}\right) \end{aligned}$$

46.

a. Number of sign variations for $P(x)$: 1

$P(-x) = -6x^3 + x^2 + 5x - 2$, so

Number of sign variations for $P(-x)$: 2

Since $P(x)$ is degree 3, there are 3 zeros, the real ones of which are classified as:

Positive Real Zeros	Negative Real Zeros
1	2
1	0

b. Factors of -2: $\pm 1, \pm 2$

Factors of 6: $\pm 1, \pm 2, \pm 3, \pm 6$

Possible rational zeros:

$\pm 1, \pm 2, \pm \frac{1}{2}, \pm \frac{1}{3}, \pm \frac{1}{6}, \pm \frac{2}{3}$

c. Note that $P(1) = P(-\frac{1}{2}) = P(-\frac{2}{3}) = 0$. So, the rational zeros are $1, -\frac{1}{2}, -\frac{2}{3}$. These are the only zeros since P has degree 3.

d.
$$\begin{aligned} P(x) &= (x-1)(3x+2)(2x+1) \\ &= 6(x-1)\left(x+\tfrac{2}{3}\right)\left(x+\tfrac{1}{2}\right) \end{aligned}$$

47.

a. Number of sign variations for $P(x)$: 4

$P(-x) = x^4 + 2x^3 + 5x^2 + 8x + 4$, so

Number of sign variations for $P(-x)$: 0

Since $P(x)$ is degree 4, there are 4 zeros, the real ones of which are classified as:

Positive Real Zeros	Negative Real Zeros
4	0
2	0
0	0

b. Factors of 4: $\pm 1, \pm 2, \pm 4$

Factors of 1: ± 1

Possible rational zeros: $\pm 1, \pm 2, \pm 4$

c. Note that $P(1) = 0$. After testing the others, it is found that the only rational zeros is 1. Hence, by **a**, 1 must have multiplicity 2 or 4. So, we know that at least $(x-1)^2 = x^2 - 2x + 1$ divides $P(x)$ evenly. To find the remaining zeros, we long divide:

$$\begin{array}{r} x^2 + 4 \\ x^2 - 2x + 1 \overline{\smash{)}\, x^4 - 2x^3 + 5x^2 - 8x + 4} \\ \underline{-(x^4 - 2x^3 + x^2)} \\ 4x^2 - 8x + 4 \\ \underline{-(4x^2 - 8x + 4)} \\ 0 \end{array}$$

Since $x^2 + 4$ is irreducible, the only real zero is 1 (multiplicity 2).

d. $P(x) = (x-1)^2(x^2+4)$

48.

a. Number of sign variations for $P(x)$: 0

$P(-x) = x^4 - 2x^3 + 10x^2 - 18x + 9$, so

Number of sign variations for $P(-x)$: 4

Since $P(x)$ is degree 4, there are 4 zeros, the real ones of which are classified as:

Positive Real Zeros	Negative Real Zeros
0	4
0	2
0	0

b. Factors of 9: $\pm 1, \pm 3, \pm 9$

Factors of 1: ± 1

Possible rational zeros: $\pm 1, \pm 3, \pm 9$

c. Note that $P(-1) = 0$. After testing the others, it is found that the only rational zero is -1. Hence, by **a**, -1 must have at least multiplicity 2. So, we know that $(x+1)^2 = x^2 + 2x + 1$ divides $P(x)$ evenly. To find the remaining zeros, we long divide:

$$\begin{array}{r} x^2 + 9 \\ x^2 + 2x + 1 \overline{\smash{)}\, x^4 + 2x^3 + 10x^2 + 18x + 9} \\ \underline{-(x^4 + 2x^3 + x^2)} \\ 9x^2 + 18x + 9 \\ \underline{-(9x^2 + 18x + 9)} \\ 0 \end{array}$$

Since $x^2 + 9$ is irreducible, the only real zero is -1 (multiplicity 2).

d. $P(x) = (x+1)^2(x^2+9)$

49.

a. Number of sign variations for $P(x)$: 1

$P(-x) = P(x)$, so

Number of sign variations for $P(-x)$: 1

Since $P(x)$ is degree 6, there are 6 zeros, the real ones of which are classified as:

Positive Real Zeros	Negative Real Zeros
1	1

b. Factors of -36:

$\pm 1, \pm 2, \pm 3, \pm 4, \pm 6, \pm 9, \pm 12, \pm 18, \pm 36$

Factors of 1: ± 1

Possible rational zeros:

$\pm 1, \pm 2, \pm 3, \pm 4, \pm 6, \pm 9, \pm 12, \pm 18, \pm 36$

c. Note that $P(-1) = P(1) = 0$. From **a**, there can be no other rational zeros for P.

We know that $(x+1)(x-1) = x^2 - 1$ divides $P(x)$ evenly. To find the remaining zeros, we long divide:

$$\begin{array}{r} x^4 + 13x^2 + 36 \\ x^2 - 1 \overline{\big) x^6 + 0x^5 + 12x^4 + 0x^3 + 23x^2 + 0x - 36} \\ \underline{-(x^6 + 0x^5 - x^4)} \\ 13x^4 + 0x^3 + 23x^2 \\ \underline{-(13x^4 + 0x^3 - 13x^2)} \\ 36x^2 + 0x - 36 \\ \underline{-(36x^2 + 0x - 36)} \\ 0 \end{array}$$

Observe that

$$x^4 + 13x^2 + 36 = (x^2 + 9)(x^2 + 4),$$

both of which are irreducible. So, the real zeros are: -1 and 1

d. $P(x) = (x+1)(x-1)(x^2+9)(x^2+4)$

50.

a. Number of sign variations for $P(x)$: 2

$P(-x) = x^4 + x^3 - 16x^2 + 16$, so

Number of sign variations for $P(-x)$: 2

Since $P(x)$ is degree 4, there are 4 zeros, the real ones of which are classified as:

Positive Real Zeros	Negative Real Zeros
2	2
2	0
0	2
0	0

b. Factors of 16: $\pm 1, \pm 2, \pm 4, \pm 8, \pm 16$

Factors of 1: ± 1

Possible rational zeros:

$\pm 1, \pm 2, \pm 4, \pm 8, \pm 16$

c. Note that $P(1) = 0$. After testing, we conclude that the only rational zero is 1. The best we can do is estimate the remaining zeros graphically:

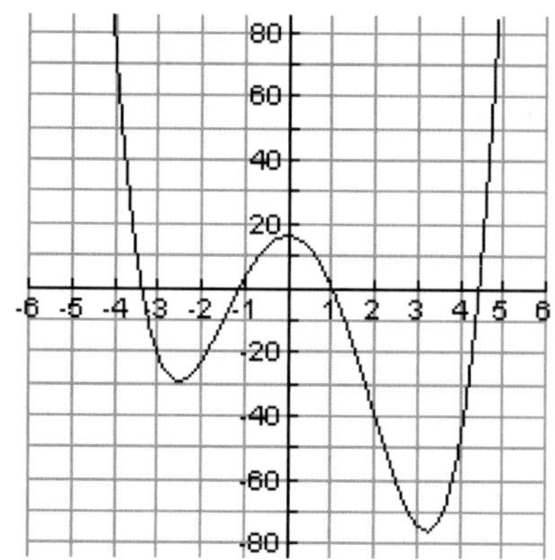

From the graph, we see that the other real zeros are approximately -3.35026, -1.07838, and 4.42864.

d. An approximate factorization of $P(x)$ is:

$P(x) = (x-1)(x+3.35026)(x+1.07838)(x-4.2864)$

51.

a. Number of sign variations for $P(x)$: 4

$$P(-x) = 4x^4 + 20x^3 + 37x^2 + 24x + 5,$$

so

Number of sign variations for $P(-x)$: 0

Since $P(x)$ is degree 4, there are 4 zeros, the real ones of which are classified as:

Positive Real Zeros	Negative Real Zeros
4	0
2	0
0	0

b. Factors of 5: $\pm 1, \pm 5$

Factors of 4: $\pm 1, \pm 2, \pm 4$

Possible rational zeros:

$$\pm 1, \pm 5, \pm \tfrac{1}{2}, \pm \tfrac{1}{4}, \pm \tfrac{5}{2}, \pm \tfrac{5}{4}$$

c. Note that $P(\frac{1}{2}) = 0$. After testing, we conclude that there the only rational zero is $\frac{1}{2}$, which has multiplicity 2 or 4. So, we know that at least $(x - \frac{1}{2})^2$ divides $P(x)$ evenly:

$$\begin{array}{r|rrrrr} \frac{1}{2} & 4 & -20 & 37 & -24 & 5 \\ & & 2 & -9 & 14 & -5 \\ \hline \frac{1}{2} & 4 & -18 & 28 & -10 & 0 \\ & & 2 & -8 & 10 & \\ \hline & 4 & -16 & 20 & 0 & \end{array}$$

So,

$$\begin{aligned} P(x) &= (x - \tfrac{1}{2})^2 (4x^2 - 16x + 20) \\ &= 4(x - \tfrac{1}{2})^2 (x^2 - 4x + 5) \\ &= (2x - 1)^2 (x^2 - 4x + 5) \end{aligned}$$

52.

a. Number of sign variations for $P(x)$: 2

$$P(-x) = 4x^4 + 8x^3 + 7x^2 - 30x + 50,$$ so

Number of sign variations for $P(-x)$: 2

Since $P(x)$ is degree 4, there are 4 zeros, the real ones of which are classified as:

Positive Real Zeros	Negative Real Zeros
2	2
0	2
0	0
2	0

b. Factors of 50:

$$\pm 1, \pm 2, \pm 5, \pm 10, \pm 25, \pm 50$$

Factors of 4: $\pm 1, \pm 2, \pm 4$

Possible rational zeros:

$$\pm 1, \pm 2, \pm 5, \pm 10, \pm 25, \pm 50, \pm \tfrac{1}{2}, \pm \tfrac{5}{2}, \pm \tfrac{25}{2}, \pm \tfrac{1}{4}, \pm \tfrac{5}{4}, \pm \tfrac{25}{4}$$

c. After testing, we conclude that there are no rational zeros. The best we can do is graph the polynomial to locate any real zeros:

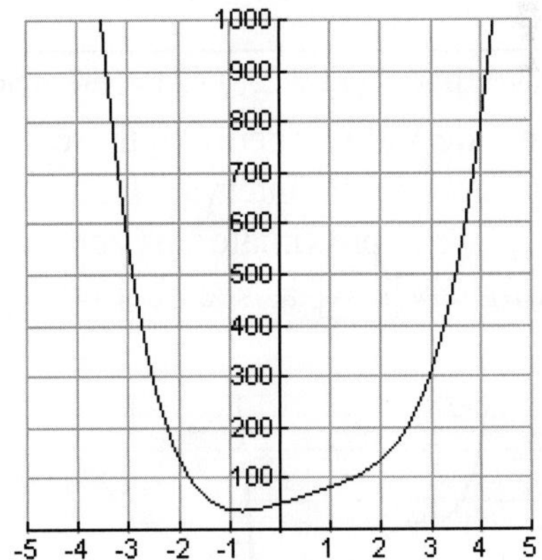

As seen from the graph, there are no real zeros.

d. An accurate factorization of $P(x)$ is not possible since we don't have values for the zeros.

53.

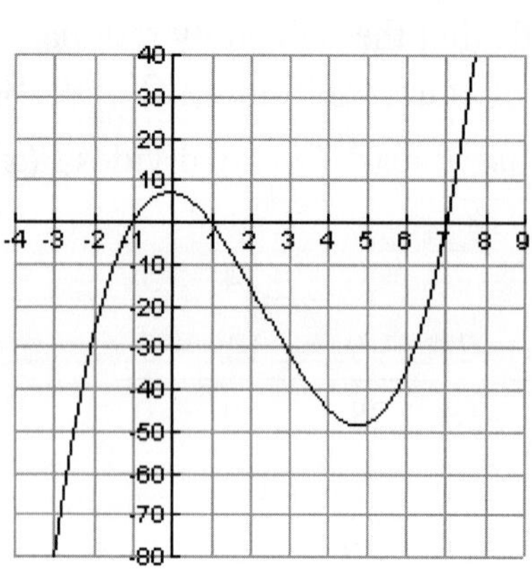

54.

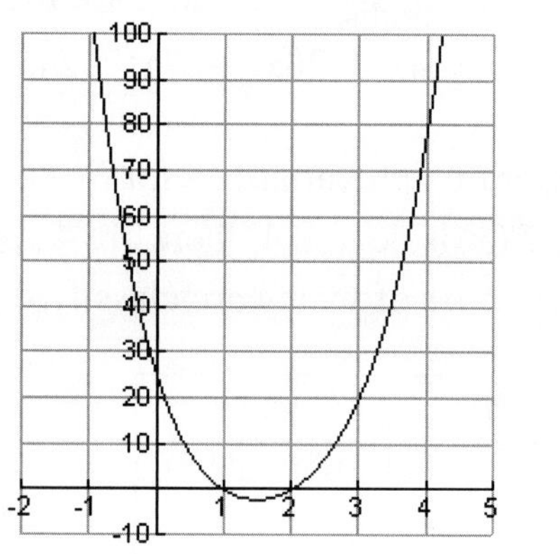

55.

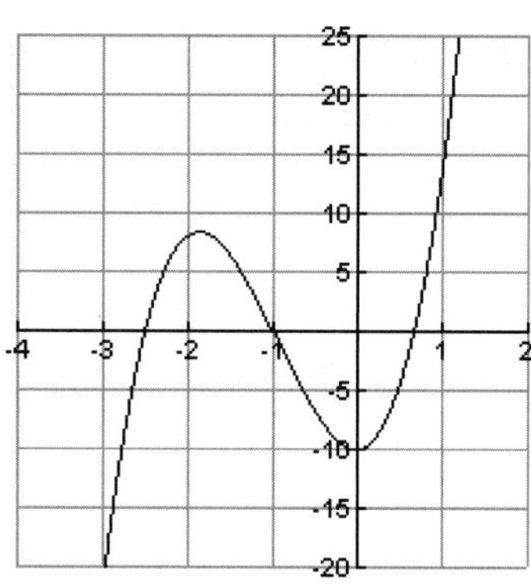

56.

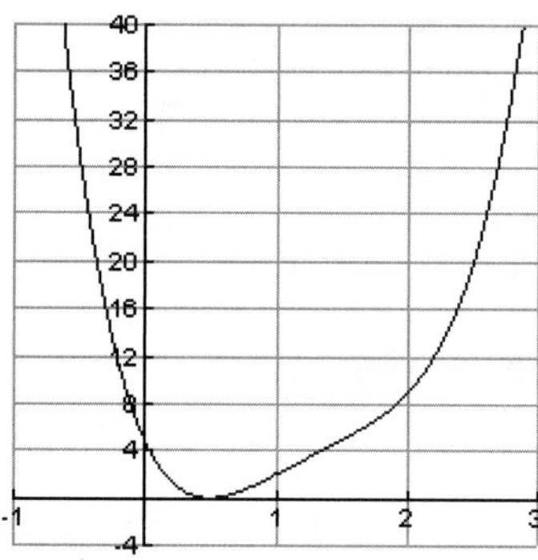

57. Observe that $f(1)=2,\ f(2)=-4$. So, by Intermediate Value Theorem, there must exist a zero in the interval (1,2). Graphically, we approximate this zero to be approximately $\boxed{1.34}$, as seen below:

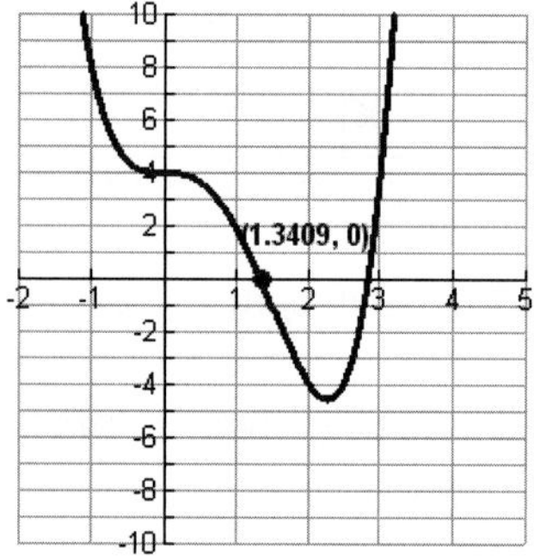

58. Observe that $f(0)=1,\ f(1)=-1$. So, by Intermediate Value Theorem, there must exist a zero in the interval (0,1). Graphically, we approximate this zero to be approximately $\boxed{0.74}$, as seen below:

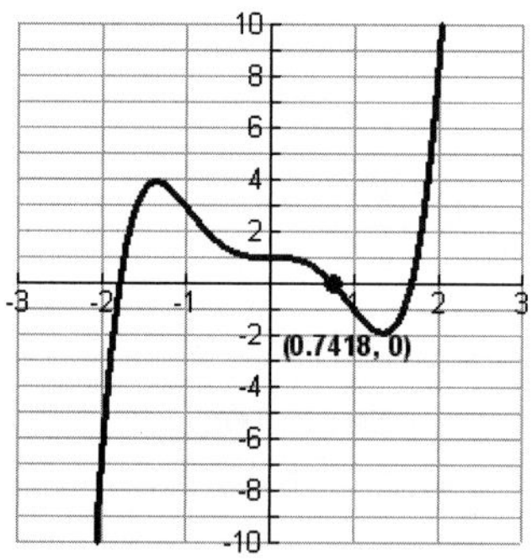

59. Observe that $f(0)=-1,\ f(1)=9$. So, by Intermediate Value Theorem, there must exist a zero in the interval (0,1). Graphically, we approximate this zero to be approximately $\boxed{0.22}$, as seen below:

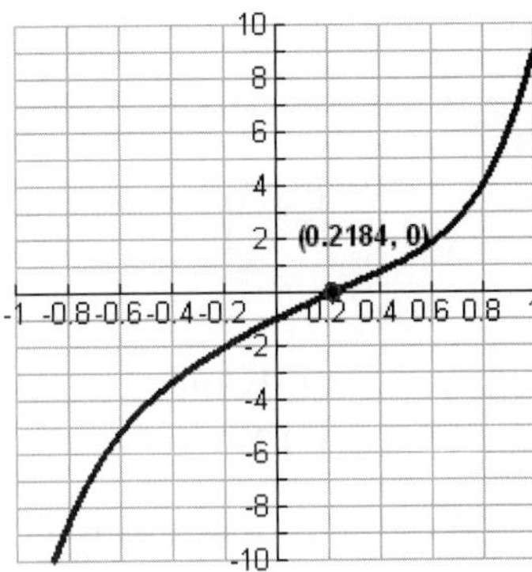

60. Observe that $f(-2)=9,\ f(-1)=-8$. So, by Intermediate Value Theorem, there must exist a zero in the interval $(-2,-1)$. Graphically, we approximate this zero to be approximately $\boxed{-1.64}$, as seen below:

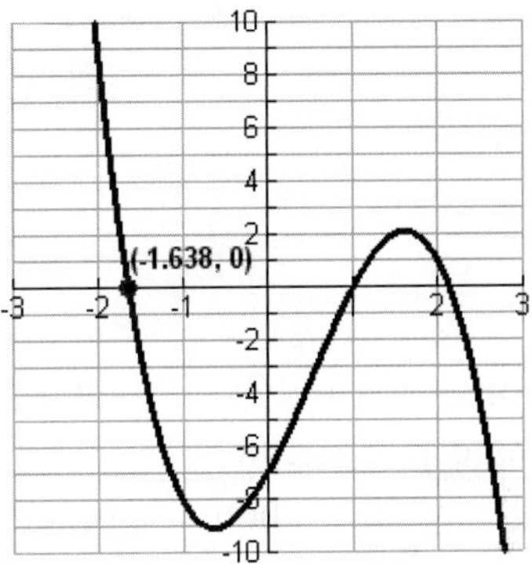

61. Observe that $f(-1)=2,\ f(0)=-3$. So, by Intermediate Value Theorem, there must exist a zero in the interval (-1, 0). Graphically, we approximate this zero to be approximately $\boxed{-0.43}$, as seen below:

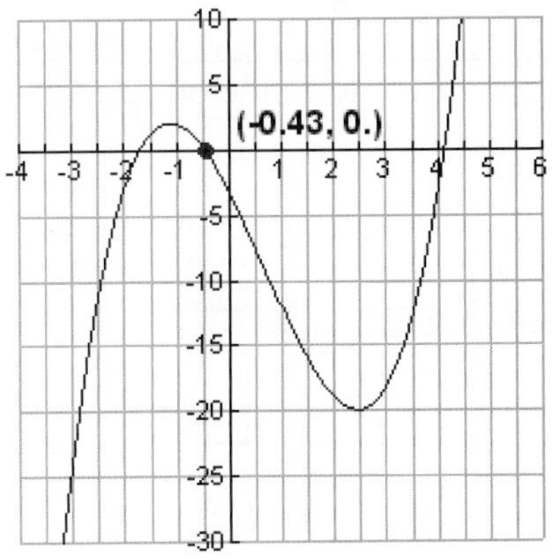

62. Observe that $f(-2)=33,\ f(-1)=-1$. So, by Intermediate Value Theorem, there must exist a zero in the interval (-2,-1). Graphically, we approximate this zero to be approximately -1.05, as seen below:

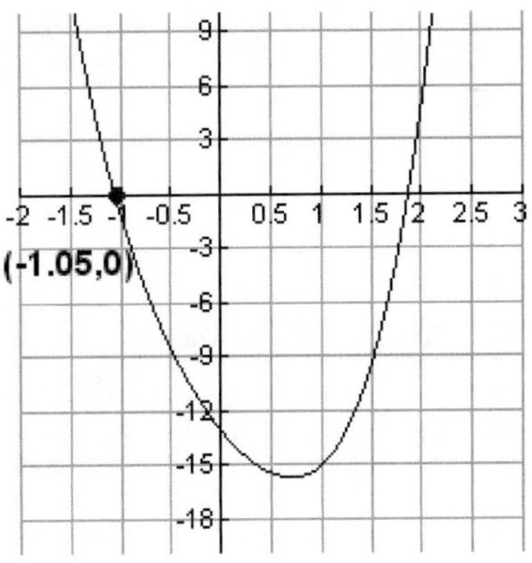

63.

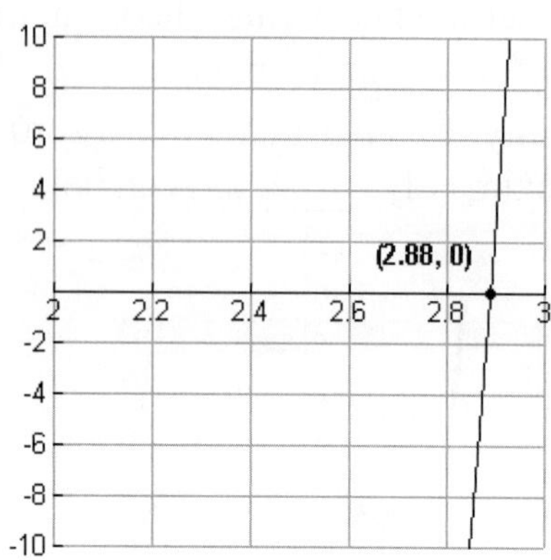

There is one root in this interval, namely approximately $x = 2.88$.

64.

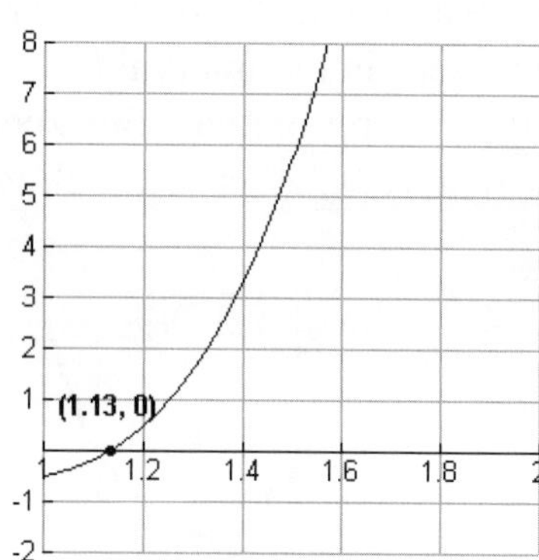

There is one root in this interval, namely approximately $x = 1.13$.

65. Let x = width. Then, the other leg has length $x+2$ and the diagonal has length $x+4$ (since it must be the longest side). By the Pythagorean theorem,

$$x^2 + (x+2)^2 = (x+4)^2$$

$$x^2 + x^2 + 4x + 4 = x^2 + 8x + 16$$

$$x^2 - 4x - 12 = 0$$

$$(x-6)(x+2) = 0$$

$$x = 6, \cancel{-2}$$

So, the width is 6 in. and length is 8 in.

66. Let x = width of the base.
Then, the length is $x + 3.5$ and the height is $x + 4.0$. The volume of the box is then given by

$$x(x+3.5)(x+4) = 97.5$$

$$x^3 + 7.5x^2 + 14x - 97.5 = 0$$

The graph of the left-side on [2,3] is given by

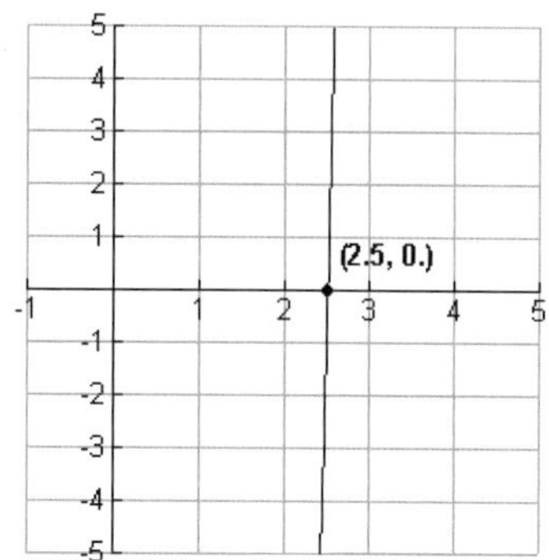

So, $x = 2.5$. Thus, the dimensions of the box are 2.5 in x 6 in. x 6.5 in.

67. We must solve $x^3+21x^2-1480x-1500=0$. We attempt to factor the left-side using synthetic division:

$$\begin{array}{r|rrrr} 30 & 1 & 21 & -1480 & -1500 \\ & & 30 & 1530 & 1500 \\ \hline & 1 & 51 & 50 & 0 \end{array}$$

Hence,

$$x^3+21x^2-1480x-1500=(x-30)\left(x^2+51x+50\right)=(x-30)(x+50)(x+1)=0.$$

As such, $x=30, \cancel{-50}, \cancel{-1}$, so that there are 30 cows.

68. We must solve $2x^4-7x^3+3x^2+8x-4=0$. We attempt to factor the left-side using synthetic division:

$$\begin{array}{r|rrrrr} 2 & 2 & -7 & 3 & 8 & -4 \\ & & 4 & -6 & -6 & 4 \\ \hline 2 & 2 & -3 & -3 & 2 & 0 \\ & & 4 & 2 & -2 & \\ \hline & 2 & 1 & -1 & 0 & \end{array}$$

Hence,

$$2x^4-7x^3+3x^2+8x-4=(x-2)^2\left(2x^2+x-1\right)=(x-2)^2(2x-1)(x+1)=0.$$

As such, $x=2, \cancel{\tfrac{1}{2}}, \cancel{-1}$. So, there are 2 loaves.

69. It is true that one can get 5 negative zeros here, but there may be just 1 or 3.

Positive Real Zeros	Negative Real Zeros
0	5
0	3
	1

70. Use 2, not -2.

71. True

72. False. Consider $f(x)=x^2+1$. There are no real zeros.

73. False. For instance, $f(x)=\left(x^2+1\right)\left(x^2+2\right)$ cannot be factored over the reals.

74. False. This is one possibility, but if there are 2 or more such sign changes, then there could be fewer.

75. False. For instance, $f(x)=(x-1)^2$ has only one x-intercept.

76. False. For instance, $f(x)=(x-1)^2(x-3)^2(x-4)^2$ has three x-intercepts, but degree 6.

77.

$$\begin{array}{r|rrrr} a & 1 & -(a+b+c) & (ab+ac+bc) & -abc \\ & & a & -ab-ac & abc \\ \hline & 1 & -b-c & bc & 0 \end{array}$$

So, $P(x)=(x-a)(x^2-(b+c)x+bc)=(x-a)(x-b)(x-c)$. So, the other zeros are b, c.

78.

$$\begin{array}{r|rrrr} a & 1 & -a+b-c & -(ab+bc-ac) & abc \\ & & a & ab-ac & -abc \\ \hline & 1 & b-c & -bc & 0 \end{array}$$

So, $P(x)=(x-a)(x^2+(b-c)x-bc)=(x-a)(x+b)(x-c)$.

79. First, note that

$$\begin{array}{r|rrrrr} b & 1 & -(a+b) & (ab-c^2) & (a+b)c^2 & -abc^2 \\ & & b & -ab & -bc^2 & abc^2 \\ \hline & 1 & -a & -c^2 & ac^2 & 0 \end{array}$$

At this point, the possible rational zeros are a, c, and $-c$. Continuing the synthetic division yields

$$\begin{array}{r|rrrr} a & 1 & -a & -c^2 & ac^2 \\ & & a & 0 & -ac^2 \\ \hline & 1 & 0 & -c^2 & \end{array}$$

Thus, we see that

$$P(x)=(x-b)(x-a)\left(x^2-c^2\right)=(x-b)(x-a)(x-c)(x+c).$$

So, the other three zeros are a, c, and $-c$.

80. The possible rational zeros are $\pm a,\pm b$. Using synthetic division yields

$$\begin{array}{r|rrrrr} a & 1 & 2(b-a) & a^2-4ab+b^2 & 2ab(a-b) & a^2b^2 \\ & & a & 2ab-a^2 & ab^2-2a^2b & -a^2b^2 \\ \hline a & 1 & 2b-a & b^2-2ab & -ab^2 & 0 \\ & & a & 2ab & ab^2 & \\ \hline & 1 & 2b & b^2 & 0 & \end{array}$$

Thus, we see that

$$P(x)=(x-a)^2\left(x^2+2bx+b^2\right)=(x-a)^2(x+b)^2.$$

So, the other zeros are a, $-b$ (multiplicity 2).

81.

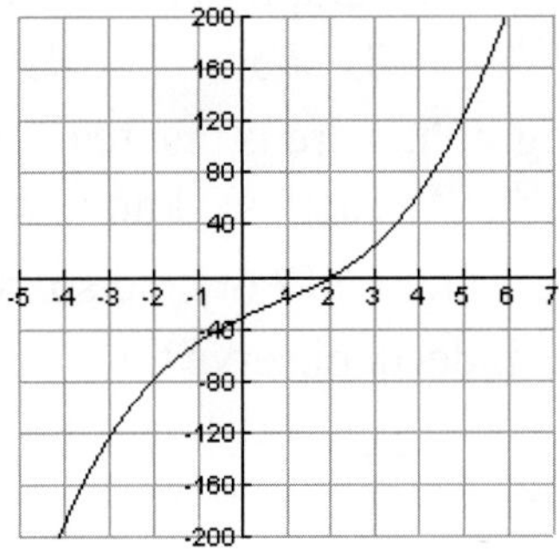

Factors of 32:
$\pm 1, \pm 2, \pm 4, \pm 8, \pm 16, \pm 32$

Factors of 1: ± 1

Possible rational zeros:

$$\pm 1, \pm 2, \pm 4, \pm 8, \pm 16, \pm 32$$

From the graph, we conclude that the only rational zero is 2.

82.

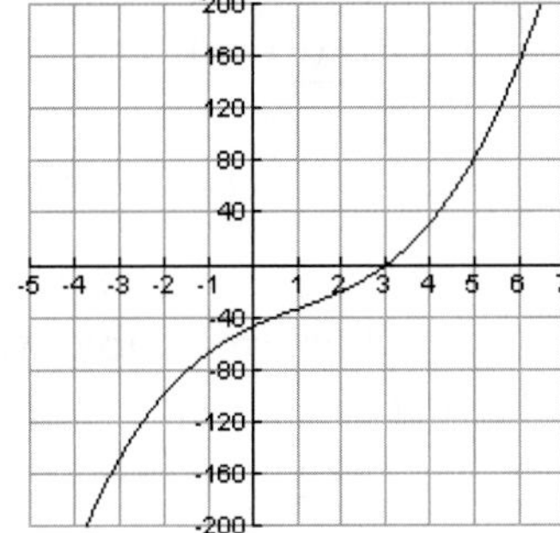

Factors of 48:

$$\pm 1, \pm 2, \pm 3, \pm 4, \pm 6, \pm 8, \pm 12, \pm 16, \pm 24, \pm 48$$

Factors of 1: ± 1

Possible rational zeros:

$$\pm 1, \pm 2, \pm 3, \pm 4, \pm 6, \pm 8, \pm 12, \pm 16, \pm 24, \pm 48$$

From the graph, we conclude that 3 is the only real zero.

83. Consider the graph to the right of

$$P(x) = 12x^4 + 25x^3 + 56x^2 - 7x - 30.$$

From the graph, there are two real zeros, namely $-\frac{3}{4}$ and $\frac{2}{3}$. As such, we know that both $(3x-2)$ and $(4x+3)$ are factors of $P(x)$ and so, $(3x-2)(4x+3) = 12x^2 + x - 6$ divides it evenly. Indeed, observe that

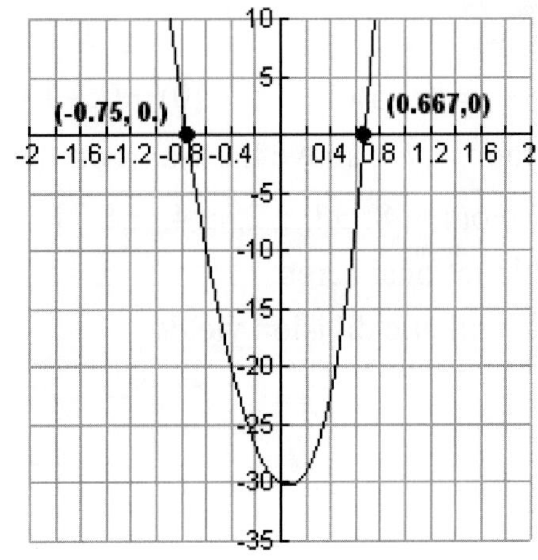

$$\begin{array}{r} x^2 + 2x + 5 \\ 12x^2 + x - 6 \overline{)12x^4 + 25x^3 + 56x^2 - 7x - 30} \\ \underline{-(12x^4 + x^3 - 6x^2)} \\ 24x^3 + 62x^2 - 7x \\ \underline{-(24x^3 + 2x^2 - 12x)} \\ 60x^2 + 5x - 30 \\ \underline{-(60x^2 + 5x - 30)} \\ 0 \end{array}$$

Since $x^2 + 2x + 5$ is irreducible, $P(x)$ factors as $(3x-2)(4x+3)\left(x^2 + 2x + 5\right)$.

84. Consider the graph to the right of

$$P(x) = -3x^3 - x^2 - 7x - 49.$$

From the graph, there is one real zero, namely $-\frac{7}{3}$. As such, we know that $(3x+7)$ is a factor of $P(x)$ and so, divides it evenly. Indeed, observe that

$$\begin{array}{r} -x^2 + 2x - 7 \\ 3x+7\overline{)-3x^3 - x^2 - 7x - 49} \\ \underline{-(-3x^3 - 7x^2)} \\ 6x^2 - 7x \\ \underline{-(6x^2 + 14x)} \\ -21x - 49 \\ \underline{-(-21x - 49)} \\ 0 \end{array}$$

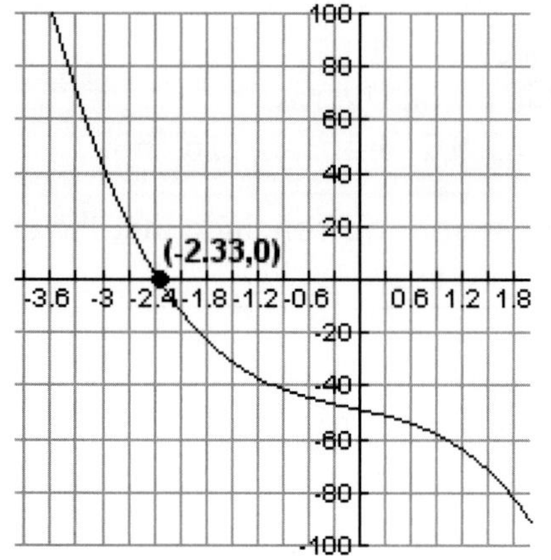

Since $-x^2 + 2x - 7$ is irreducible, we see that $P(x)$ factors as $(3x+7)\left(-x^2 + 2x - 7\right) = -(3x+7)\left(x^2 - 2x + 7\right)$.

85. The possible rational zeros of $f(x) = x^3 - 4x^2 - 7x + 10$ are $\pm 1, \pm 2, \pm 5, \pm 10$. Using synthetic division yields:

$$\begin{array}{r|rrrr} 1 & 1 & -4 & -7 & 10 \\ & & 1 & -3 & -10 \\ \hline & 1 & -3 & -10 & 0 \end{array}$$

So, $f(x) = (x-1)(x^2 - 3x - 10) = (x-1)(x-5)(x+2)$ and the zeros are 1, 5, -2.
The graph of f is above the x-axis on the set: $(-2,-1) \cup (5,\infty)$.

86. Factors of 8: $\pm 1, \pm 2, \pm 4, \pm 8$ Factors of 6: $\pm 1, \pm 2, \pm 3, \pm 6$
Possible rational zeros: $\pm 1, \pm 2, \pm 4, \pm 8, \pm\frac{1}{2}, \pm\frac{1}{3}, \pm\frac{1}{6}, \pm\frac{2}{3}, \pm\frac{4}{3}, \pm\frac{8}{3}$
Using synthetic division yields:

$$\begin{array}{r|rrrr} -1 & 6 & -13 & -11 & 8 \\ & & -6 & 19 & -8 \\ \hline & 6 & -19 & 8 & 0 \end{array}$$

So,

$$f(x) = (x+1)\left(6x^2 - 19x + 8\right) = (x+1)(3x-8)(2x-1),$$

and the zeros are $-1, \frac{8}{3}, \frac{1}{2}$.
The graph of f is above the x-axis on the set: $\left(-1, \frac{1}{2}\right) \cup \left(\frac{8}{3}, \infty\right)$.

87. Factors of 6: $\pm1, \pm2, \pm3, \pm6$ Factors of 2: $\pm1, \pm2$

Possible rational zeros: $\pm1, \pm2, \pm3, \pm6, \pm\frac{1}{2}, \pm\frac{3}{2}$

Using synthetic division yields:

$$\begin{array}{r|rrrrr} 3 & -2 & 5 & 7 & -10 & -6 \\ & & -6 & -3 & 12 & 6 \\ \hline -\frac{1}{2} & -2 & -1 & 4 & 2 & 0 \\ & & 1 & 0 & -2 & \\ \hline & -2 & 0 & 4 & & \end{array}$$

So,

$f(x)=(x-3)(x+\frac{1}{2})(-2x^2+4)=-2(x-3)(x+\frac{1}{2})(x^2-2)=-2(x-3)(x+\frac{1}{2})(x-\sqrt{2})(x+\sqrt{2})$,

and the zeros are $-\frac{1}{2},3,\pm\sqrt{2}$.

The graph of f is above the x-axis on the set: $\left(-\sqrt{2},-\frac{1}{2}\right)\cup\left(\frac{1}{2},3\right)$.

88. Factors of 8: $\pm1, \pm2, \pm4, \pm8$ Factors of 3: $\pm1, \pm3$

Possible rational zeros: $\pm1, \pm2, \pm4, \pm8, \pm\frac{1}{3}, \pm\frac{2}{3}, \pm\frac{4}{3}, \pm\frac{8}{3}$

Using synthetic division yields:

$$\begin{array}{r|rrrrr} 4 & -3 & 14 & -11 & 14 & -8 \\ & & -12 & 8 & -12 & 8 \\ \hline \frac{2}{3} & -3 & 2 & -3 & 2 & 0 \\ & & -2 & 0 & -2 & \\ \hline & -3 & 0 & -3 & & \end{array}$$

So, $f(x)=(x-4)(x-\frac{2}{3})(-3x^2-3)=-3(x-4)(x-\frac{2}{3})(x^2+1)$ and the real zeros are 4, $\frac{2}{3}$.

The graph of f is above the x-axis on the set: $\left(\frac{2}{3},4\right)$.

Section 2.5 Solutions --

1. $P(x)=(x+2i)(x-2i)$. Zeros are $\pm2i$.

2. $P(x)=(x+3i)(x-3i)$. Zeros are $\pm3i$

3. Note that the zeros are

$$x^2-2x+2=0 \Rightarrow$$

$$x=\frac{2\pm\sqrt{4-4(2)}}{2}=\frac{2\pm2i}{2}=1\pm i$$

So, $P(x)=(x-(1-i))(x-(1+i))$.

4. Note that the zeros are

$$x^2-4x+5=0 \Rightarrow$$

$$x=\frac{4\pm\sqrt{16-4(5)}}{2}=\frac{4\pm2i}{2}=2\pm i$$

So, $P(x)=(x-(2-i))(x-(2+i))$.

5. Observe that $P(x)=\left(x^2-4\right)\left(x^2+4\right)$ $=(x-2)(x+2)(x-2i)(x+2i)$ So, the zeros are $\pm 2,\ \pm 2i$.	**6.** Observe that $P(x)=\left(x^2-9\right)\left(x^2+9\right)$ $=(x-3)(x+3)(x-3i)(x+3i)$ So, the zeros are $\pm 3,\ \pm 3i$.
7. Observe that $P(x)=\left(x^2-5\right)\left(x^2+5\right)$ $=\left(x-\sqrt{5}\right)\left(x+\sqrt{5}\right)\left(x-i\sqrt{5}\right)\left(x+i\sqrt{5}\right)$ So, the zeros are $\pm\sqrt{5},\ \pm i\sqrt{5}$.	**8.** Observe that $P(x)=\left(x^2-3\right)\left(x^2+3\right)$ $=\left(x-\sqrt{3}\right)\left(x+\sqrt{3}\right)\left(x-i\sqrt{3}\right)\left(x+i\sqrt{3}\right)$ So, the zeros are $\pm\sqrt{3},\ \pm i\sqrt{3}$.
9. If i is a zero, then so is its conjugate $-i$. Since $P(x)$ has degree 3, this is the only missing zero.	**10.** If $-i$ is a zero, then so is its conjugate i. Since $P(x)$ has degree 3, this is the only missing zero.
11. Since $2i$ and $3-i$ are zeros, so are their conjugates $-2i$ and $3+i$, respectively. Since $P(x)$ has degree 4, these are the only missing zeros.	**12.** Since $3i$ and $2+i$ are zeros, so are their conjugates $-3i$ and $2-i$, respectively. Since $P(x)$ has degree 4, these are the only missing zeros.
13. Since $1-3i$ and $2+5i$ are zeros, so are their conjugates $1+3i$ and $2-5i$, respectively. Since $P(x)$ has degree 6 and 0 is a zero with multiplicity 2, these are the only missing zeros.	**14.** Since $1-5i$ and $2+3i$ are zeros, so are their conjugates $1+5i$ and $2-3i$, respectively. Since $P(x)$ has degree 6 and -2 is a zero with multiplicity 2, these are the only missing zeros.
15. Since $-i$ and $1-i$ are zeros, so are their conjugates i and $1+i$, respectively. Since $1-i$ has multiplicity 2, so does its conjugate. Since $P(x)$ has degree 6, these are the only missing zeros.	**16.** Since $2i$ and $1+i$ are zeros, so are their conjugates $-2i$ and $1-i$, respectively. Since $1+i$ has multiplicity 2, so does its conjugate. Since $P(x)$ has degree 6, these are the only missing zeros.

17. Since -2i is a zero of $P(x)$, so is its conjugate 2i. As such, $(x-2i)(x+2i)=x^2+4$ divides $P(x)$ evenly. Indeed, observe that

$$\begin{array}{r} x^2-2x-15 \\ x^2+0x+4\overline{)x^4-2x^3-11x^2-8x-60} \\ \underline{-(x^4+0x^3+4x^2)} \\ -2x^3-15x^2-8x \\ \underline{-(-2x^3+0x^2-8x)} \\ -15x^2-60 \\ \underline{-(-15x^2-60)} \\ 0 \end{array}$$

So,

$$\begin{aligned} P(x) &= (x-2i)(x+2i)\left(x^2-2x-15\right) \\ &= (x-2i)(x+2i)(x-5)(x+3) \end{aligned}$$

So, the zeros are $\pm 2i,\ -3,\ 5$.

18. Since 3i is a zero of $P(x)$, so is its conjugate -3i. As such, $(x-3i)(x+3i)=x^2+9$ divides $P(x)$ evenly. Indeed, observe that

$$\begin{array}{r} x^2-x-2 \\ x^2+0x+9\overline{)x^4-x^3+7x^2-9x-18} \\ \underline{-(x^4+0x^3+9x^2)} \\ -x^3-2x^2-9x \\ \underline{-(-x^3+0x^2-9x)} \\ -2x^2-18 \\ \underline{-(-2x^2-18)} \\ 0 \end{array}$$

So,

$$\begin{aligned} P(x) &= (x-3i)(x+3i)\left(x^2-x-2\right) \\ &= (x-3i)(x+3i)(x-2)(x+1) \end{aligned}$$

So, the zeros are $\pm 3i,\ -1,\ 2$.

19. Since i is a zero of $P(x)$, so is its conjugate -i. As such, $(x-i)(x+i)=x^2+1$ divides $P(x)$ evenly. Indeed, observe that

$$\begin{array}{r} x^2-4x+3 \\ x^2+0x+1\overline{)x^4-4x^3+4x^2-4x+3} \\ \underline{-(x^4+0x^3+x^2)} \\ -4x^3+3x^2-4x \\ \underline{-(-4x^3+0x^2-4x)} \\ 3x^2+3 \\ \underline{-(3x^2+3)} \\ 0 \end{array}$$

So,

$$\begin{aligned} P(x) &= (x-i)(x+i)\left(x^2-4x+3\right) \\ &= (x-i)(x+i)(x-3)(x-1) \end{aligned}$$

So, the zeros are $\pm i,\ 1,\ 3$.

20. Since -2i is a zero of $P(x)$, so is its conjugate 2i. As such, $(x-2i)(x+2i)=x^2+4$ divides $P(x)$ evenly. Indeed, observe that

$$\begin{array}{r} x^2-x-2 \\ x^2+0x+4\overline{)x^4-x^3+2x^2-4x-8} \\ \underline{-(x^4+0x^3+4x^2)} \\ -x^3-2x^2-4x \\ \underline{-(-x^3+0x^2-4x)} \\ -2x^2-8 \\ \underline{-(-2x^2-8)} \\ 0 \end{array}$$

So,

$$\begin{aligned} P(x) &= (x-2i)(x+2i)\left(x^2-x-2\right) \\ &= (x-2i)(x+2i)(x-2)(x+1) \end{aligned}$$

So, the zeros are $\pm 2i,\ -1,\ 2$.

21. Since -$3i$ is a zero of $P(x)$, so is its conjugate $3i$. As such, $(x-3i)(x+3i) = x^2+9$ divides $P(x)$ evenly. Indeed, observe that

$$\begin{array}{r} x^2-2x+1 \\ x^2+0x+9\overline{\big)x^4-2x^3+10x^2-18x+9} \\ \underline{-(x^4+0x^3+9x^2)} \\ -2x^3+x^2-18x \\ \underline{-(-2x^3+0x^2-18x)} \\ x^2+9 \\ \underline{-(x^2+9)} \\ 0 \end{array}$$

So,

$$P(x) = (x-3i)(x+3i)\left(x^2-2x+1\right)$$
$$= (x-3i)(x+3i)(x-1)^2$$

So, the zeros are $\pm 3i$ and 1 (multiplicity 2).

22. Since $5i$ is a zero of $P(x)$, so is its conjugate -$5i$. As such, $(x-5i)(x+5i) = x^2+25$ divides $P(x)$ evenly. Indeed, observe that

$$\begin{array}{r} x^2-3x-4 \\ x^2+0x+25\overline{\big)x^4-3x^3+21x^2-75x-100} \\ \underline{-(x^4+0x^3+25x^2)} \\ -3x^3-4x^2-75x \\ \underline{-(-3x^3+0x^2-75x)} \\ -4x^2-100 \\ \underline{-(-4x^2-100)} \\ 0 \end{array}$$

So,

$$P(x) = (x-5i)(x+5i)\left(x^2-3x-4\right)$$
$$= (x-5i)(x+5i)(x-4)(x+1)$$

So, the zeros are $\pm 5i$, -1, 4.

23. Since $1+i$ is a zero of $P(x)$, so is its conjugate $1-i$. As such, $(x-(1+i))(x-(1-i)) = x^2-2x+2$ divides $P(x)$ evenly. Indeed, observe that

$$\begin{array}{r} x^2+2x-7 \\ x^2-2x+2\overline{\big)x^4+0x^3-9x^2+18x-14} \\ \underline{-(x^4-2x^3+2x^2)} \\ 2x^3-11x^2+18x \\ \underline{-(2x^3-4x^2+4x)} \\ -7x^2+14x-14 \\ \underline{-(-7x^2+14x-14)} \\ 0 \end{array}$$

Now, we find the roots of x^2+2x-7:

$$x = \frac{-2 \pm \sqrt{4-4(-7)}}{2} = -1 \pm 2\sqrt{2}$$

So,

$$P(x) = (x-(1+i))(x-(1-i))\cdot$$
$$(x-(-1-2\sqrt{2}))(x-(-1+2\sqrt{2}))$$

So, the zeros are $1 \pm i$, $-1 \pm 2\sqrt{2}$.

24. Since $1-2i$ is a zero of $P(x)$, so is its conjugate $1+2i$. As such, $(x-(1+2i))(x-(1-2i)) = x^2-2x+5$ divides $P(x)$ evenly. Indeed, observe that

$$\begin{array}{r} x^2-2x-8 \\ x^2-2x+5\overline{\big)x^4-4x^3+x^2+6x-40} \\ \underline{-(x^4-2x^3+5x^2)} \\ -2x^3-4x^2+6x \\ \underline{-(-2x^3+4x^2-10x)} \\ -8x^2+16x-40 \\ \underline{-(-8x^2+16x-40)} \\ 0 \end{array}$$

So,

$$P(x) = (x-(1+2i))(x-(1-2i))(x-4)(x+2)$$

So, the zeros are $1 \pm 2i$, -2, 4.

25. Since 3 - i is a zero of $P(x)$, so is its conjugate 3 + i. As such, $(x-(3+i))(x-(3-i)) = x^2-6x+10$ divides $P(x)$ evenly. Indeed, observe that

$$\begin{array}{r}
x^2-4 \\
x^2-6x+10\overline{)x^4-6x^3+6x^2+24x-40} \\
\underline{-(x^4-6x^3+10x^2)} \\
-4x^2+24x-40 \\
\underline{-(-4x^2+24x-40)} \\
0
\end{array}$$

So,

$$P(x)=(x-(3+i))(x-(3-i))(x-2)(x+2)$$

So, the zeros are $3\pm i,\ \pm 2$.

26. Since 2+ i is a zero of $P(x)$, so is its conjugate 2- i. As such, $(x-(2+i))(x-(2-i)) = x^2-4x+5$ divides $P(x)$ evenly. Indeed, observe that

$$\begin{array}{r}
x^2-1 \\
x^2-4x+5\overline{)x^4-4x^3+4x^2+4x-5} \\
\underline{-(x^4-4x^3+5x^2)} \\
-x^2+4x-5 \\
\underline{-(-x^2+4x-5)} \\
0
\end{array}$$

So,

$$P(x)=(x-(2+i))(x-(2-i))(x-1)(x+1)$$

So, the zeros are $2\pm i,\ \pm 1$.

27. Since 2-i is a zero of $P(x)$, so is its conjugate 2+i. As such, $(x-(2+i))(x-(2-i)) = x^2-4x+5$ divides $P(x)$ evenly. Indeed, observe that

$$\begin{array}{r}
x^2-5x+4 \\
x^2-4x+5\overline{)x^4-9x^3+29x^2-41x+20} \\
\underline{-(x^4-4x^3+5x^2)} \\
-5x^3+24x^2-41x \\
\underline{-(-5x^3+20x^2-25x)} \\
4x^2-16x+20 \\
\underline{-(4x^2-16x+20)} \\
0
\end{array}$$

So,

$$P(x)=(x-(2+i))(x-(2-i))(x-1)(x-4)$$

So, the zeros are $2\pm i,\ 1, 4$.

28. Since 3+i is a zero of $P(x)$, so is its conjugate 3-i. As such, $(x-(3+i))(x-(3-i)) = x^2-6x+10$ divides $P(x)$ evenly. Indeed, observe that

$$\begin{array}{r}
x^2-x-2 \\
x^2-6x+10\overline{)x^4-7x^3+14x^2+2x-20} \\
\underline{-(x^4-6x^3+10x^2)} \\
-x^3+4x^2+2x \\
\underline{-(-x^3+6x^2-10x)} \\
-2x^2+12x-20 \\
\underline{-(-2x^2+12x-20)} \\
0
\end{array}$$

So,

$$P(x)=(x-(3+i))(x-(3-i))(x-2)(x+1)$$

So, the zeros are $3\pm i,\ -1,\ 2$.

29.

$$\begin{aligned}
x^3-x^2+9x-9 &= \left(x^3-x^2\right)+9\left(x-1\right) \\
&= x^2(x-1)+9\left(x-1\right) \\
&= \left(x^2+9\right)(x-1) \\
&= (x+3i)(x-3i)(x-1)
\end{aligned}$$

30.

$$\begin{aligned}
x^3-2x^2+4x-8 &= \left(x^3-2x^2\right)+4\left(x-2\right) \\
&= x^2(x-2)+4\left(x-2\right) \\
&= \left(x^2+4\right)(x-2) \\
&= (x+2i)(x-2i)(x-2)
\end{aligned}$$

31.

$$\begin{aligned} x^3-5x^2+x-5 &= \left(x^3-5x^2\right)+\left(x-5\right) \\ &= x^2(x-5)+\left(x-5\right) \\ &= \left(x^2+1\right)(x-5) \\ &= (x+i)(x-i)(x-5) \end{aligned}$$

32.

$$\begin{aligned} x^3-7x^2+x-7 &= \left(x^3-7x^2\right)+\left(x-7\right) \\ &= x^2(x-7)+\left(x-7\right) \\ &= \left(x^2+1\right)(x-7) \\ &= (x+i)(x-i)(x-7) \end{aligned}$$

33.

$$\begin{aligned} x^3+x^2+4x+4 &= \left(x^3+x^2\right)+4\left(x+1\right) \\ &= x^2(x+1)+4\left(x+1\right) \\ &= \left(x^2+4\right)(x+1) \\ &= (x+2i)(x-2i)(x+1) \end{aligned}$$

34. Consider $P(x)=x^3+x^2-2$.
By the Rational Zero Theorem, the only possible rational roots are $\pm 1, \pm 2$. Note that

$$\begin{array}{r|rrrr} 1 & 1 & 1 & 0 & -2 \\ & & 1 & 2 & 2 \\ \hline & 1 & 2 & 2 & 0 \end{array}$$

So, $P(x)=(x-1)\left(x^2+2x+2\right)$. Now, we find the roots of x^2+2x+2:

$$x=\frac{-2\pm\sqrt{4-4(2)}}{2}=-1\pm i$$

So,
$P(x)=(x-1)(x-(-1+i))(x-(-1-i))$.

35. Consider $P(x)=x^3-x^2-18$.
By the Rational Zero Theorem, the only possible rational roots are
$\pm 1, \pm 2, \pm 3, \pm 6, \pm 9, \pm 18$. Note that

$$\begin{array}{r|rrrr} 3 & 1 & -1 & 0 & -18 \\ & & 3 & 6 & 18 \\ \hline & 1 & 2 & 6 & 0 \end{array}$$

So, $P(x)=(x-3)\left(x^2+2x+6\right)$. Now, we find the roots of x^2+2x+6:

$$x=\frac{-2\pm\sqrt{4-4(6)}}{2}=-1\pm i\sqrt{5}$$

So,
$P(x)=(x-3)(x-(-1+i\sqrt{5}))(x-(-1-i\sqrt{5}))$.

36. Consider $P(x)=x^4-2x^3-2x^2-2x-3$.
By the Rational Zero Theorem, the only possible rational roots are $\pm 1, \pm 3$. Note that

$$\begin{array}{r|rrrrr} -1 & 1 & -2 & -2 & -2 & -3 \\ & & -1 & 3 & -1 & 3 \\ \hline & 1 & -3 & 1 & -3 & 0 \end{array}$$

So,

$$\begin{aligned} P(x) &= (x+1)\left(x^3-3x^2+x-3\right) \\ &= (x+1)\left[x^2(x-3)+(x-3)\right]. \\ &= (x+1)\left(x^2+1\right)(x-3) \\ &= (x+1)(x-3)(x+i)(x-i) \end{aligned}$$

37. Consider $P(x)=x^4-2x^3-11x^2-8x-60$. By the Rational Zero Theorem, the only possible rational roots are

$$\pm1,\ \pm2,\ \pm3,\ \pm4,\ \pm5,\ \pm6,$$
$$\pm10,\ \pm12,\ \pm15,\ \pm20,\ \pm30,\ \pm60.$$

Note that

$$\begin{array}{r|rrrrr} -3 & 1 & -2 & -11 & -8 & -60 \\ & & -3 & 15 & -12 & 60 \\ \hline & 1 & -5 & 4 & -20 & 0 \end{array}$$

So,

$$\begin{aligned} P(x) &= (x+3)\left(x^3-5x^2+4x-20\right) \\ &= (x+3)\left[x^2(x-5)+4(x-5)\right]. \\ &= (x+3)\left(x^2+4\right)(x-5) \\ &= (x+3)(x-5)(x+2i)(x-2i) \end{aligned}$$

38. Consider $P(x)=x^4-x^3+7x^2-9x-18$. By the Rational Zero Theorem, the only possible rational roots are

$$\pm1,\ \pm2,\ \pm3,\ \pm6,\ \pm9,\ \pm18.$$

Note that

$$\begin{array}{r|rrrrr} -1 & 1 & -1 & 7 & -9 & -18 \\ & & -1 & 2 & -9 & 18 \\ \hline & 1 & -2 & 9 & -18 & 0 \end{array}$$

So,

$$\begin{aligned} P(x) &= (x+1)\left(x^3-2x^2+9x-18\right) \\ &= (x+1)\left[x^2(x-2)+9(x-2)\right]. \\ &= (x+1)\left(x^2+9\right)(x-2) \\ &= (x+1)(x-2)(x+3i)(x-3i) \end{aligned}$$

39. Consider $P(x)=x^4-4x^3-x^2-16x-20$. By the Rational Zero Theorem, the only possible rational roots are

$$\pm1,\ \pm2,\pm4,\ \pm5,\ \pm10,\ \pm20$$

Note that

$$\begin{array}{r|rrrrr} -1 & 1 & -4 & -1 & -16 & -20 \\ & & -1 & 5 & -4 & 20 \\ \hline & 1 & -5 & 4 & -20 & 0 \end{array}$$

So,

$$\begin{aligned} P(x) &= (x+1)\left(x^3-5x^2+4x-20\right) \\ &= (x+1)\left[x^2(x-5)+4(x-5)\right]. \\ &= (x+1)\left(x^2+4\right)(x-5) \\ &= (x+1)(x-5)(x+2i)(x-2i) \end{aligned}$$

40. Consider $P(x)=x^4-3x^3+11x^2-27x+18$. By the Rational Zero Theorem, the only possible rational roots are

$$\pm1,\ \pm2,\pm3,\ \pm6,\ \pm9,\ \pm18$$

Note that

$$\begin{array}{r|rrrrr} 1 & 1 & -3 & 11 & -27 & 18 \\ & & 1 & -2 & 9 & -18 \\ \hline & 1 & -2 & 9 & -18 & 0 \end{array}$$

So,

$$\begin{aligned} P(x) &= (x-1)\left(x^3-2x^2+9x-18\right) \\ &= (x-1)\left[x^2(x-2)+9(x-2)\right]. \\ &= (x-1)\left(x^2+9\right)(x-2) \\ &= (x-1)(x-2)(x+3i)(x-3i) \end{aligned}$$

41. Consider
$P(x)=x^4-7x^3+27x^2-47x+26.$
By the Rational Zero Theorem, the only possible rational roots are

$$\pm 1,\ \pm 2, \pm 13,\ \pm 26$$

Note that

$$\begin{array}{r|rrrrr} 1 & 1 & -7 & 27 & -47 & 26 \\ & & 1 & -6 & 21 & -26 \\ \hline 2 & 1 & -6 & 21 & -26 & 0 \\ & & 2 & -8 & 26 & \\ \hline & 1 & -4 & 13 & 0 & \end{array}$$

So, $P(x)=(x-1)(x-2)\left(x^2-4x+13\right).$

Next, we find the roots of $x^2-4x+13$:

$$x=\frac{4\pm\sqrt{16-4(13)}}{2}=2\pm 3i$$

So,
$P(x)=(x-1)(x-2)(x-(2-3i))(x-(2+3i))$

42. Consider $P(x)=x^4-5x^3+5x^2+25x-26.$
By the Rational Zero Theorem, the only possible rational roots are

$$\pm 1,\ \pm 2, \pm 13,\ \pm 26$$

Note that

$$\begin{array}{r|rrrrr} 1 & 1 & -5 & 5 & 25 & -26 \\ & & 1 & -4 & 1 & 26 \\ \hline -2 & 1 & -4 & 1 & 26 & 0 \\ & & -2 & 12 & -26 & \\ \hline & 1 & -6 & 13 & 0 & \end{array}$$

So, $P(x)=(x-1)(x+2)\left(x^2-6x+13\right).$

Next, we find the roots of $x^2-6x+13$:

$$x=\frac{6\pm\sqrt{36-4(13)}}{2}=3\pm 2i$$

So,
$P(x)=(x-1)(x+2)(x-(3-2i))(x-(3+2i))$

43. Consider $P(x)=-x^4-3x^3+x^2+13x+10.$
By the Rational Zero Theorem, the only possible rational roots are

$$\pm 1,\ \pm 2, \pm 5,\ \pm 10$$

Note that

$$\begin{array}{r|rrrrr} -1 & 1 & 3 & -1 & -13 & -10 \\ & & -1 & -2 & 3 & 10 \\ \hline 2 & 1 & 2 & -3 & -10 & 0 \\ & & 2 & 8 & 10 & \\ \hline & 1 & 4 & 5 & 0 & \end{array}$$

So, $P(x)=-(x+1)(x-2)\left(x^2+4x+5\right).$

Next, we find the roots of x^2+4x+5:

$$x=\frac{-4\pm\sqrt{16-4(5)}}{2}=-2\pm i$$

So,
$P(x)=-(x+1)(x-2)(x-(-2-i))(x-(-2+i))$

44. Consider
$P(x)=-x^4-x^3+12x^2+26x+24.$
By the Rational Zero Theorem, the only possible rational roots are

$$\pm 1,\ \pm 2, \pm 3,\ \pm 4,\ \pm 6,\ \pm 8, \pm 12,\ \pm 24$$

Note that

$$\begin{array}{r|rrrrr} -3 & 1 & 1 & -12 & -26 & -24 \\ & & -3 & 6 & 18 & 24 \\ \hline 4 & 1 & -2 & -6 & -8 & 0 \\ & & 4 & 8 & 8 & \\ \hline & 1 & 2 & 2 & 0 & \end{array}$$

So, $P(x)=-(x+3)(x-4)\left(x^2+2x+2\right).$

Next, we find the roots of x^2+2x+2:

$$x=\frac{-2\pm\sqrt{4-4(2)}}{2}=-1\pm i$$

So,
$P(x)=-(x+3)(x-4)(x-(-1-i))(x-(-1+i))$

45. Consider $P(x)=x^4-2x^3+5x^2-8x+4$.
By the Rational Zero Theorem, the only possible rational roots are $\pm 1, \pm 2, \pm 4$.
Note that

$$\begin{array}{r|rrrrr} 1 & 1 & -2 & 5 & -8 & 4 \\ & & 1 & -1 & 4 & -4 \\ \hline & 1 & -1 & 4 & -4 & 0 \end{array}$$

So,

$$\begin{aligned} P(x) &= (x-1)\left(x^3-x^2+4x-4\right) \\ &= (x-1)\left[x^2(x-1)+4(x-1)\right]. \\ &= (x-1)\left(x^2+4\right)(x-1) \\ &= (x-1)^2(x+2i)(x-2i) \end{aligned}$$

46. Consider $P(x)=x^4+2x^3+10x^2+18x+9$.
By the Rational Zero Theorem, the only possible rational roots are $\pm 1, \pm 3, \pm 9$.
Note that

$$\begin{array}{r|rrrrr} -1 & 1 & 2 & 10 & 18 & 9 \\ & & -1 & -1 & -9 & -9 \\ \hline & 1 & 1 & 9 & 9 & 0 \end{array}$$

So,

$$\begin{aligned} P(x) &= (x+1)\left(x^3+x^2+9x+9\right) \\ &= (x+1)\left[x^2(x+1)+9(x+1)\right]. \\ &= (x+1)\left(x^2+9\right)(x+1) \\ &= (x+1)^2(x+3i)(x-3i) \end{aligned}$$

47. Consider $P(x)=x^6+12x^4+23x^2-36$.
By the Rational Zero Theorem, the only possible rational roots are

$\pm 1, \pm 2, \pm 4, \pm 6, \pm 9, \pm 18, \pm 36$

Note that

$$\begin{array}{r|rrrrrrr} 1 & 1 & 0 & 12 & 0 & 23 & 0 & -36 \\ & & 1 & 1 & 13 & 13 & 36 & 36 \\ \hline -1 & 1 & 1 & 13 & 13 & 36 & 36 & 0 \\ & & -1 & 0 & -13 & 0 & -36 & \\ \hline & 1 & 0 & 13 & 0 & 36 & 0 & \end{array}$$

So,

$$\begin{aligned} P(x) &= (x-1)(x+1)\left(x^4+13x^2+36\right) \\ &= (x-1)(x+1)\left(x^2+4\right)\left(x^2+9\right) \\ &= (x-1)(x+1)(x-2i)(x+2i)(x-3i)(x+3i) \end{aligned}$$

48. Consider

$P(x)=x^6-2x^5+9x^4-16x^3+24x^2-32x+16$.

By the Rational Zero Theorem, the only possible rational roots are

$\pm 1, \pm 2, \pm 4, \pm 8, \pm 16$

Note that

$$\begin{array}{r|rrrrrrr} 1 & 1 & -2 & 9 & -16 & 24 & -32 & 16 \\ & & 1 & -1 & 8 & -8 & 16 & -16 \\ \hline 1 & 1 & -1 & 8 & -8 & 16 & -16 & 0 \\ & & 1 & 0 & 8 & 0 & 16 & \\ \hline & 1 & 0 & 8 & 0 & 16 & 0 & \end{array}$$

So,

$$\begin{aligned} P(x) &= (x-1)^2\left(x^4+8x^2+16\right) \\ &= (x-1)^2\left(x^2+4\right)^2 \\ &= (x-1)^2(x-2i)^2(x+2i)^2 \end{aligned}$$

49. Consider

$P(x) = 4x^4 - 20x^3 + 37x^2 - 24x + 5$.

By the Rational Zero Theorem, the only possible rational roots are

$$\pm 1, \pm 5, \pm \tfrac{1}{2}, \pm \tfrac{5}{2}, \pm \tfrac{1}{4}, \pm \tfrac{5}{4}$$

Note that

$$\begin{array}{r|rrrrr} \tfrac{1}{2} & 4 & -20 & 37 & -24 & 5 \\ & & 2 & -9 & 14 & -5 \\ \hline \tfrac{1}{2} & 4 & -18 & 28 & -10 & 0 \\ & & 2 & -8 & 10 & \\ \hline & 4 & -16 & 20 & 0 & \end{array}$$

So, $P(x) = 4\left(x - \tfrac{1}{2}\right)^2 \left(x^2 - 4x + 5\right)$.

Next, we find the roots of $x^2 - 4x + 5$:

$$x = \frac{4 \pm \sqrt{16 - 4(5)}}{2} = 2 \pm i$$

So, $P(x) = 4\left(x - \tfrac{1}{2}\right)^2 (x - (2 - i))(x - (2 + i))$

$= (2x - 1)^2 (x - (2 - i))(x - (2 + i))$

50. Consider $P(x) = 4x^4 - 44x^3 + 145x^2 - 114x + 26$.

By the Rational Zero Theorem, the only possible rational roots are

$$\pm 1, \pm 2, \pm \tfrac{1}{2}, \pm \tfrac{13}{2}, \pm \tfrac{13}{4}, \pm 13, \pm 26$$

Note that

$$\begin{array}{r|rrrrr} \tfrac{1}{2} & 4 & -44 & 145 & -114 & 26 \\ & & 2 & -21 & 62 & -26 \\ \hline \tfrac{1}{2} & 4 & -42 & 124 & -52 & 0 \\ & & 2 & -20 & 52 & \\ \hline & 4 & -40 & 104 & 0 & \end{array}$$

So, $P(x) = \left(x - \tfrac{1}{2}\right)^2 \left(4x^2 - 40x + 104\right)$

$= 4\left(x - \tfrac{1}{2}\right)^2 \left(x^2 - 10x + 26\right)$.

Next, we find the roots of $x^2 - 10x + 26$:

$$x = \frac{10 \pm \sqrt{100 - 4(26)}}{2} = 5 \pm i$$

So, $P(x) = 4\left(x - \tfrac{1}{2}\right)^2 (x - (5 - i))(x - (5 + i))$

51. Consider

$P(x) = 3x^5 - 2x^4 + 9x^3 - 6x^2 - 12x + 8$.

By the Rational Zero Theorem, the only possible rational roots are

$$\pm 1, \pm 2, \pm 4, \pm 8, \pm \tfrac{1}{3}, \pm \tfrac{2}{3}, \pm \tfrac{4}{3}, \pm \tfrac{8}{3}$$

Note that

$$\begin{array}{r|rrrrrr} -1 & 3 & -2 & 9 & -6 & -12 & 8 \\ & & -3 & 5 & -14 & 20 & -8 \\ \hline 1 & 3 & -5 & 14 & -20 & 8 & 0 \\ & & 3 & -2 & 12 & -8 & \\ \hline \tfrac{2}{3} & 3 & -2 & 12 & -8 & 0 & \\ & & 2 & 0 & 8 & & \\ \hline & 3 & 0 & 12 & 0 & & \end{array}$$

So,

$P(x) = (x - 1)(x + 1)(x - \tfrac{2}{3})(3x^2 + 12)$

$= (x - 1)(x + 1)(3x - 2)(x - 2i)(x + 2i)$

52. Consider

$P(x) = 2x^5 - 5x^4 + 4x^3 - 26x^2 + 50x - 25$.

By the Rational Zero Theorem, the only possible rational roots are

$$\pm 1, \pm 5, \pm 25, \pm \tfrac{1}{2}, \pm \tfrac{5}{2}, \pm \tfrac{25}{2}$$

Note that

$$\begin{array}{r|rrrrrr} 1 & 2 & -5 & 4 & -26 & 50 & -25 \\ & & 2 & -3 & 1 & -25 & 25 \\ \hline 1 & 2 & -3 & 1 & -25 & 25 & 0 \\ & & 2 & -1 & 0 & -25 & \\ \hline \tfrac{5}{2} & 2 & -1 & 0 & -25 & 0 & \\ & & 5 & 10 & 25 & & \\ \hline & 2 & 4 & 10 & 0 & & \end{array}$$

So, $P(x) = (x - 1)^2 (x - \tfrac{5}{2})(2x^2 + 4x + 10)$

$= (x - 1)^2 (2x - 5)(x^2 + 2x + 5)$

Next, we find the roots of $x^2 + 2x + 5$:

$$x = \frac{-2 \pm \sqrt{4 - 4(5)}}{2} = -1 \pm 2i$$

So, $P(x) = (x - 1)^2 (2x - 5)(x + 1 - 2i)(x + 1 + 2i)$.

53. Yes. In such case, $P(x)$ never touches the x-axis (since crossing it would require $P(x)$ to have a real root) and is always above it since the leading coefficient is positive, indicating that the end behavior should resemble that of $y=x^{2n}$, for some positive integer n. So, profit is always positive and increasing.	**54.** No. In such case, $P(x)$ never touches the x-axis (since crossing it would require $P(x)$ to have a real root) and is always below it since the leading coefficient is negative. So, never have a positive profit.
55. No. In such case, it crosses the x-axis and looks like $y=-x^3$. So, profit is decreasing.	**56.** Yes. In such case, it crosses the x-axis and looks like $y=x^3$. So, profit is increasing.
57. Step 2 is an error. In general, the additive inverse of a real root need not be a root. This is being confused with the fact that complex roots occur in conjugate pairs.	**58.** Possible rational roots include $\pm\frac{1}{2}$ from the Rational Zero theorem.
59. False. For example, consider $P(x)=(x-1)(x+3)$. Note that 1 is a zero of P, but -1 is not.	**60.** False. Complex zeros do not correspond to x-intercepts.
61. True. It has n complex zeros.	**62.** True.
63. No. Complex zeros occur in conjugate pairs. So, the collection of complex solutions contribute an even number of zeros, thereby requiring there to be at least one real zero.	**64.** Yes. For example, $P(x)=x^2+4$ has zeros $\pm 2i$, both of which are imaginary.
65. Since bi is a zero of multiplicity 3, its conjugate $-bi$ is also a zero of multiplicity 3. Hence, $P(x)=(x-bi)^3(x+bi)^3$ $=\left[(x-bi)(x+bi)\right]^3$ $=\left(x^2+b^2\right)^3$ $=x^6+3b^2x^4+3b^4x^2+b^6$	**66.** Since $a+bi$ is a zero of multiplicity 2, its conjugate $a-bi$ is also a zero of multiplicity 2. Hence, $P(x)=\left(x-(a+bi)\right)^2\left(x-(a-bi)\right)^2$ $=\left[\left(x-(a+bi)\right)\left(x-(a-bi)\right)\right]^2$ $=x^2-2ax+(a^2+b^2)$

67. Since ai is a zero with multiplicity 2 and bi is a zero, it follows that their conjugates $-ai$ and $-bi$ satisfy the same conditions, so that $\left(x^2+a^2\right)^2$ and $\left(x^2+b^2\right)$ are factors. These must divide $P(x)$ evenly. Hence, a 6th degree polynomial satisfying these conditions is:

$$P(x)=\left(x^2+a^2\right)^2\left(x^2+b^2\right)=\left(x^4+2a^2x^2+a^4\right)\left(x^2+b^2\right)=x^6+\left(2a^2+b^2\right)x^4+\underbrace{\left(a^4+2a^2b^2\right)}_{=a^2\left(a^2+2b^2\right)}x^2+b^2a^4$$

68. Since ai and bi are both zeros, it follows that their conjugates $-ai$ and $-bi$ are also. So, (x^2+a^2) and (x^2+b^2) are factors. These must divide $P(x)$ evenly. Hence, a polynomial with minimal degree satisfying these conditions is

$$P(x)=(x^2+a^2)(x^2+b^2)=x^4+(a^2+b^2)x^2+a^2b^2.$$

69. The possible combinations of zeros of $P(x)$ are:

Real Zeros	Complex Zeros
0	4
2	2
4	0

From the graph of P to the right, we note that all roots are complex.

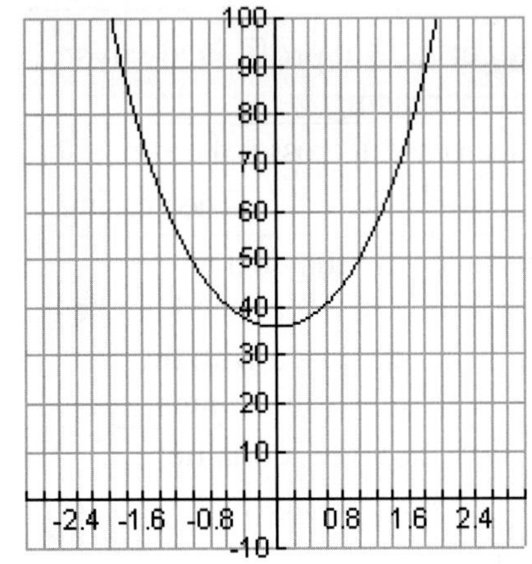

70. The possible combinations of zeros of $P(x)$ are:

Real Zeros	Complex Zeros
0	6
2	4
4	2
6	0

From the graph of P to the right, we note that there are two real roots.

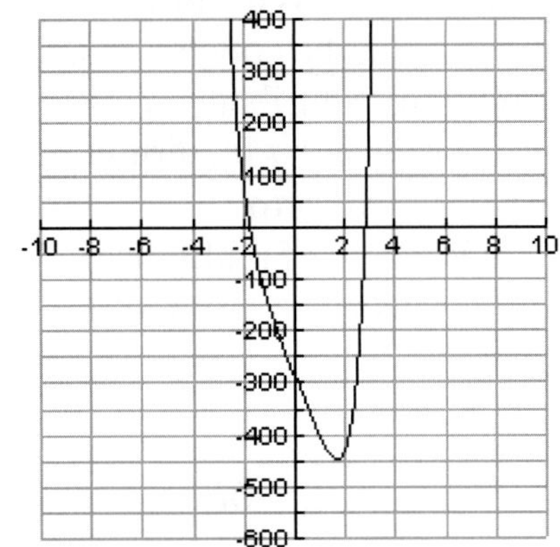

71. Consider the graph of $P(x)$ to the right. Note that the only real zero is $\frac{3}{5}$, so that there must be four complex zeros. Observe that

$$\begin{array}{r|rrrrrr} 0.6 & -5 & 3 & -25 & 15 & -20 & 12 \\ & & -3 & 0 & -15 & 0 & -12 \\ \hline & -5 & 0 & -25 & 0 & -20 & 0 \end{array}$$

So,

$$\begin{aligned} P(x) &= (x-0.6)\left(-5x^4-25x^2-20\right) \\ &= -5(x-0.6)\left(x^4+5x^2+4\right) \\ &= -5(x-0.6)\left(x^2+4\right)\left(x^2+1\right) \\ &= -5(x-0.6)(x-2i)(x+2i)(x-i)(x+i) \end{aligned}$$

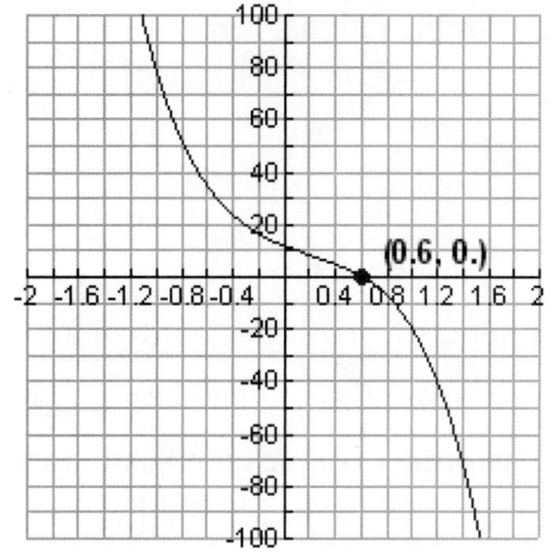

72. Consider the graph of $P(x)$ to the right. Note that both 1.5 and -2.4 are real zeros, and -2.4 has multiplicity of 2 or 4 (since the graph touches the x-axis at this root). As such, we know that at least $(x-1.5)(x+2.4)^2$ divides $P(x)$ evenly. Indeed, dividing $P(x)$ by this yields the quotient $x^2-1.2x+0.40$, the roots of which are $0.6 \pm 0.2i$. Thus,

$$\begin{aligned} P(x) = (x-1.5)(x+2.4)^2 \cdot \\ (x-(0.6+0.2i))(x-(0.6-0.2i)) \end{aligned}$$

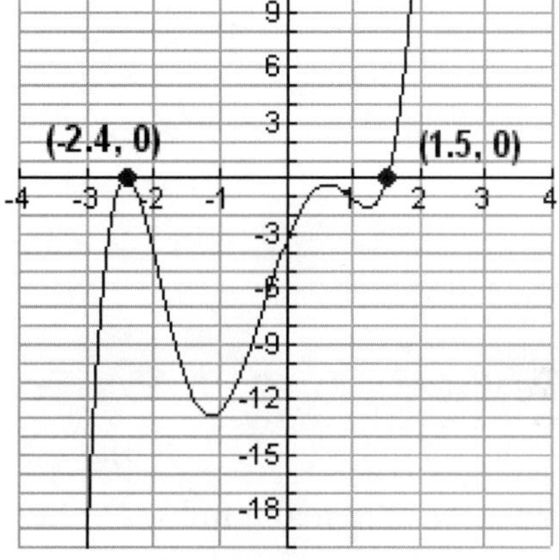

73. The possible rational zeros are ± 1. Using synthetic division yields

$$\begin{array}{r|rrrr} -1 & 1 & 1 & 1 & 1 \\ & & -1 & 0 & -1 \\ \hline & 1 & 0 & 1 & 0 \end{array}$$

a) Factoring over the complex numbers yields $f(x) = (x+1)\left(x^2+1\right) = (x+1)(x-i)(x+i)$.

b) Factoring over the real numbers yields $f(x) = (x+1)\left(x^2+1\right)$.

74. The possible rational zeros are $\pm1, \pm2, \pm13, \pm26$. Using synthetic division yields $$\begin{array}{r\|rrrr} 2 & 1 & -6 & 21 & -26 \\ & & 2 & -8 & 26 \\ \hline & 1 & -4 & 13 & 0 \end{array}$$ **a)** Observe that $x^2-4x+13=0 \Rightarrow x=\frac{4\pm\sqrt{16-4(13)}}{2}=2\pm3i$. So, factoring over the complex numbers yields $f(x)=(x-2)(x-2-3i)(x-2+3i)$. **b)** Factoring over the real numbers yields $f(x)=(x-2)\left(x^2-4x+13\right)$.
75. a) Factoring over the complex numbers yields $$f(x)=\left(x^2+4\right)\left(x^2+1\right)=(x+2i)(x-2i)(x+i)(x-i).$$ **b)** Factoring over the real numbers yields $f(x)=\left(x^2+4\right)\left(x^2+1\right)$.
76. The possible rational zeros are $\pm1, \pm2, \pm3, \pm6, \pm9, \pm18$. Using synthetic division yields $$\begin{array}{r\|rrrrr} 3 & 1 & -2 & -7 & 18 & -18 \\ & & 3 & 3 & -12 & 18 \\ \hline -3 & 1 & 1 & -4 & 6 & \\ & & -3 & 6 & -6 & \\ \hline & 1 & -2 & 2 & & \end{array}$$ **a)** Observe that $x^2-2x+2=0 \Rightarrow x=\frac{2\pm\sqrt{4-4(2)}}{2}=1\pm i$. So, factoring over the complex numbers yields $f(x)=(x-3)(x+3)(x-1-i)(x-1+i)$. **b)** Factoring over the real numbers yields $f(x)=(x-3)(x+3)(x^2-2x+2)$.

Section 2.6 Solutions ---

1. Note that $x^2+x-12=(x+4)(x-3)$. Domain: $(-\infty,-4)\cup(-4,3)\cup(3,\infty)$	**2.** Note that $x^2+2x-3=(x+3)(x-1)$. Domain: $(-\infty,-3)\cup(-3,1)\cup(1,\infty)$
3 Note that $x\neq\pm2$. So, Domain: . $(-\infty,-2)\cup(-2,2)\cup(2,\infty)$	**4.** Note that $x\neq\pm7$. So, Domain: $(-\infty,-7)\cup(-7,7)\cup(7,\infty)$
5. Note that x^2+16 is never zero. Domain: $(-\infty,\infty)$	**6.** Note that x^2+9 is never zero. Domain: $(-\infty,\infty)$

7. Note that $2\left(x^2-x-6\right)=2(x-3)(x+2)$. Domain: $(-\infty,-2)\cup(-2,3)\cup(3,\infty)$	**8.** Note that $x^2-x-6=(x-3)(x+2)$. Domain: $(-\infty,-2)\cup(-2,3)\cup(3,\infty)$
9. Vertical Asymptote: $x+2=0$, so $x=-2$ is the VA. Horizontal Asymptote: Since the degree of the numerator is less than degree of the denominator, $y=0$ is the HA.	**10.** Vertical Asymptote: $5-x=0$, so $x=5$ is the VA. Horizontal Asymptote: Since the degree of the numerator is less than degree of the denominator, $y=0$ is the HA.
11. Vertical Asymptote: $x+5=0$, so $x=-5$ is the VA. Horizontal Asymptote: Since the degree of the numerator is greater than degree of the denominator, there is no HA.	**12.** Vertical Asymptote: $2x-7=0$, so $x=\frac{7}{2}$ is the VA. Horizontal Asymptote: Since the degree of the numerator is greater than degree of the denominator, there is no HA.
13. Vertical Asymptote: $6x^2+5x-4=(2x-1)(3x+4)=0$, so $x=\frac{1}{2}$, $x=-\frac{4}{3}$ are the VAs. Horizontal Asymptote: Since the degree of the numerator is greater than degree of the denominator, there is no HA.	**14.** Vertical Asymptote: $3x^2-5x-2=(3x+1)(x-2)=0$, so $x=2$, $x=-\frac{1}{3}$ are the VAs. Horizontal Asymptote: Since the degree of the numerator equals the degree of the denominator, $y=\frac{6}{3}=2$ is the HA.
15. Vertical Asymptote: $x^2+\frac{1}{9}$ is never 0, so there is no VA. Horizontal Asymptote: Since the degree of the numerator equals the degree of the denominator, $y=\frac{1}{3}$ is the HA.	**16.** Vertical Asymptote: $(2x-1)=0$, so $x=\frac{1}{2}$ is the VA. Horizontal Asymptote: Since the degree of the numerator is greater than degree of the denominator, there is no HA.
17. To find the slant asymptote, we use synthetic division: $\begin{array}{r\|rrr} -4 & 1 & 10 & 25 \\ & & -4 & -24 \\ \hline & 1 & 6 & 1 \end{array}$ So, the slant asymptote is $y=x+6$.	**18.** To find the slant asymptote, we use synthetic division: $\begin{array}{r\|rrr} 3 & 1 & 9 & 20 \\ & & 3 & 36 \\ \hline & 1 & 12 & 56 \end{array}$ So, the slant asymptote is $y=x+12$.

19. To find the slant asymptote, we use synthetic division:

$$\begin{array}{r|rrr} 5 & 2 & 14 & 7 \\ & & 10 & 120 \\ \hline & 2 & 24 & 127 \end{array}$$

So, the slant asymptote is $y = 2x + 24$.

20. To find the slant asymptote, we use long division:

$$\begin{array}{r} 3x+7 \\ x^2 - x - 30 \overline{) \, 3x^3 + 4x^2 - 6x + 1} \\ \underline{-(3x^3 - 3x^2 - 90x)} \\ 7x^2 + 84x + 1 \\ \underline{-(7x^2 - 7x - 210)} \\ 91x + 211 \end{array}$$

So, the slant asymptote is $y = 3x + 7$.

21. To find the slant asymptote, we use long division:

$$\begin{array}{r} 4x + \frac{11}{2} \\ 2x^3 - x^2 + 3x - 1 \overline{) \, 8x^4 + 7x^3 + 0x^2 + 2x - 5} \\ \underline{-(8x^4 - 4x^3 + 12x^2 - 4x)} \\ 11x^3 - 12x^2 + 6x - 5 \\ \underline{-(11x^3 - \frac{11}{2}x^2 + \frac{33}{2}x - \frac{11}{2})} \\ -\frac{13}{2}x^2 - \frac{21}{2}x + \frac{1}{2} \end{array}$$

So, the slant asymptote is $y = 4x + \frac{11}{2}$.

22. To find the slant asymptote, we use long division:

$$\begin{array}{r} 2x \\ x^5 + 0x^4 + 0x^3 + 0x^2 + 0x - 1 \overline{) \, 2x^6 + 0x^5 + 0x^4 + 0x^3 + 0x^2 + 0x + 1} \\ \underline{-(2x^6 + 0x^5 + 0x^4 + 0x^3 + 0x^2 - 2x)} \\ 2x + 1 \end{array}$$

So, the slant asymptote is $y = 2x$.

23. b Vertical Asymptote: $x = 4$
Horizontal Asymptote: $y = 0$

24. d Vertical Asymptote: $x = 4$
Horizontal Asymptote: $y = 3$

25. a Vertical Asymptotes: $x = 2,\ x = -2$
Horizontal Asymptote: $y = 3$

26. f Vertical Asymptote: None
Horizontal Asymptote: $y = -3$
Graph never goes above the x-axis

27. e Vertical Asymptotes: $x = 2,\ x = -2$
Horizontal Asymptote: $y = -3$

28. c Vertical Asymptote: $x = -4$
Horizontal Asymptote: None

29. Vertical Asymptote: $x=-1$ Horizontal Asymptote: $y=0$ 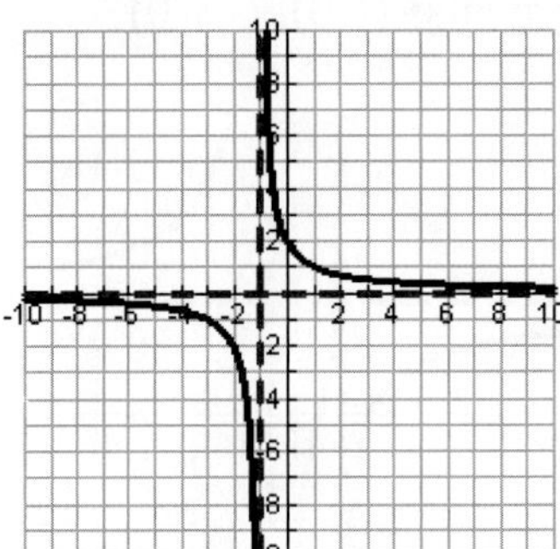	**30.** Vertical Asymptote: $x=2$ Horizontal Asymptote: $y=0$ 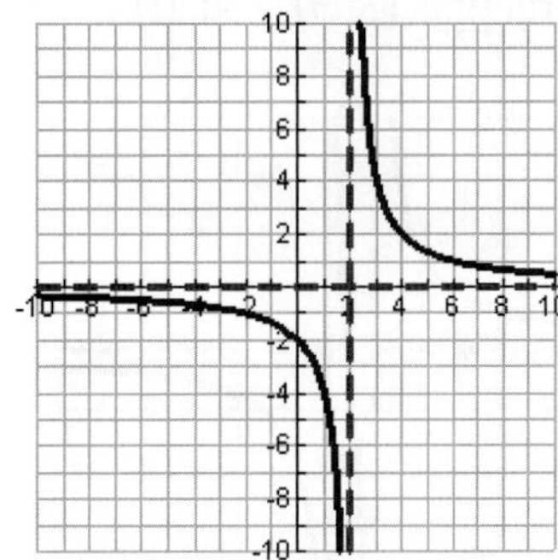
31. Vertical Asymptote: $x=1$ Horizontal Asymptote: $y=2$ Intercept: $(0,0)$ 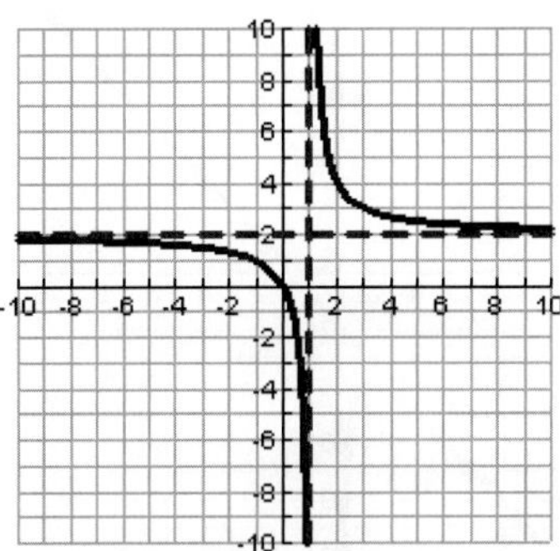	**32.** Vertical Asymptote: $x=-2$ Horizontal Asymptote: $y=4$ Intercept: $(0,0)$ 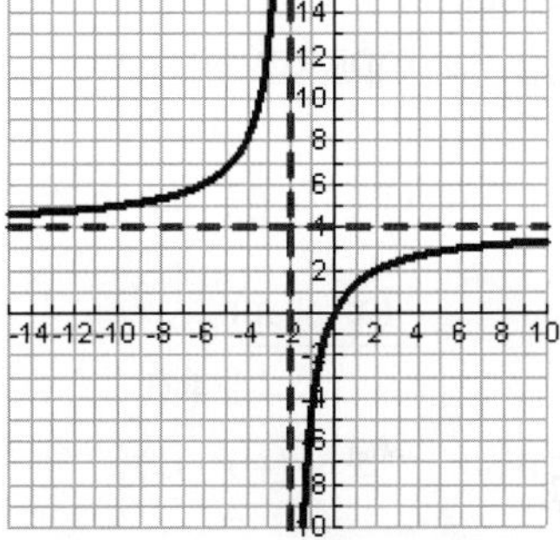
33. Vertical Asymptote: $x=0$ Horizontal Asymptote: $y=1$ Intercept: $(1,0)$ 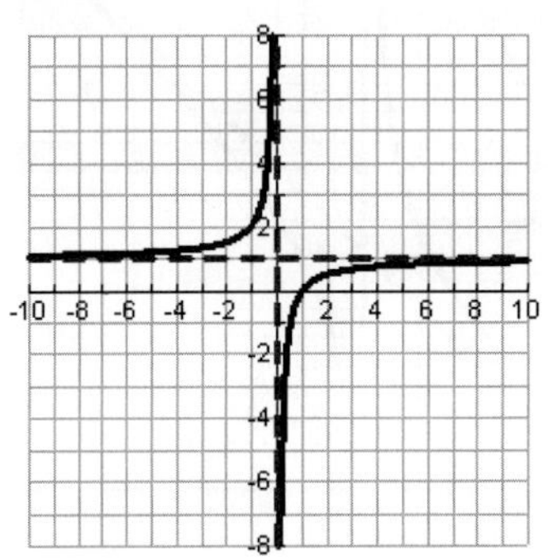	**34.** Vertical Asymptote: $x=1$ Horizontal Asymptote: $y=1$ Intercepts: $(-2,0)$, $(0,-2)$ 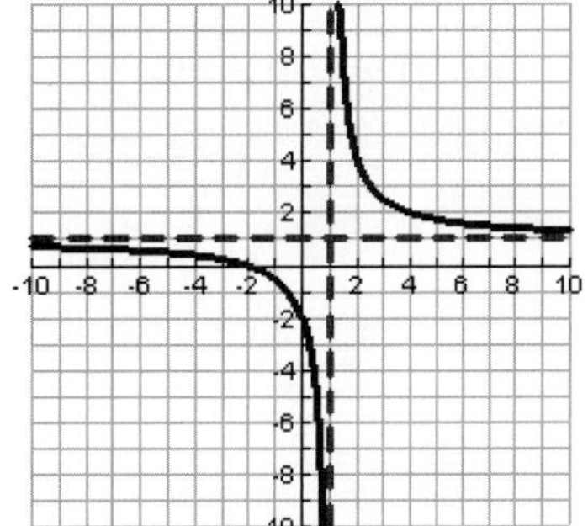

35. Vertical Asymptotes: $x = 0,\ x = -2$
Horizontal Asymptote: $y = 2$
Intercepts: $(3,0),\ (-1,0)$

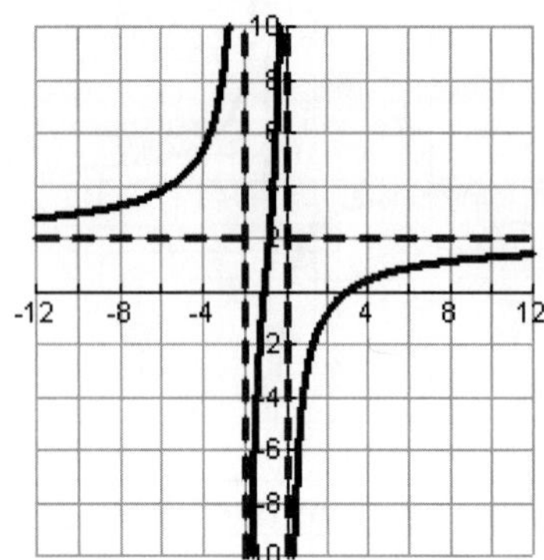

36. Vertical Asymptotes: $x = 0,\ x = 3$
Horizontal Asymptote: $y = 3$
Intercepts: $(1,0),\ (-1,0)$

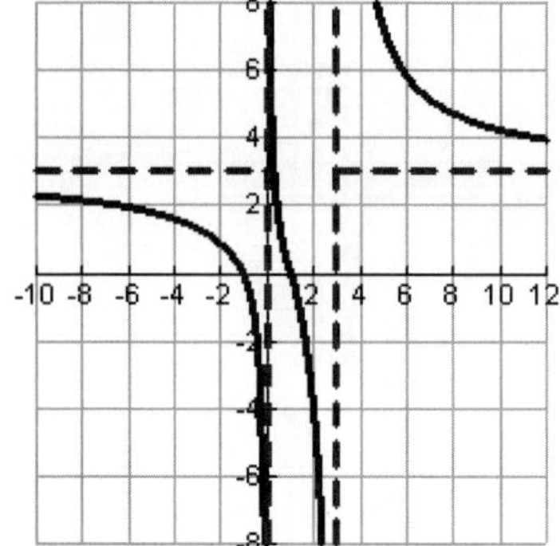

37. Vertical Asymptote: $x = -1$
Intercept: (0,0)
Slant Asymptote: $y = x - 1$

$$\begin{array}{r} x-1 \\ x+1\overline{\smash{)}\,x^2+0x+0} \\ \underline{-(x^2+x)} \\ -x+0 \\ \underline{-(-x-1)} \\ 1 \end{array}$$

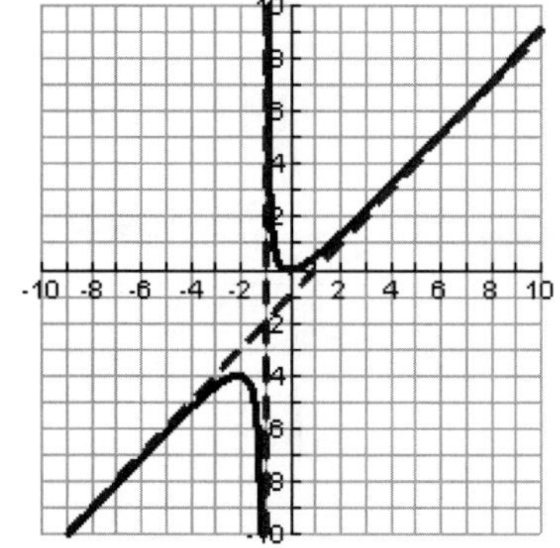

38. Vertical Asymptote: $x = -2$
Intercepts: $(3,0),\ (-3,0)$
Slant Asymptote: $y = x - 2$

$$\begin{array}{r} x-2 \\ x+2\overline{\smash{)}\,x^2+0x-9} \\ \underline{-(x^2+2x)} \\ -2x-9 \\ \underline{-(-2x-4)} \\ -5 \end{array}$$

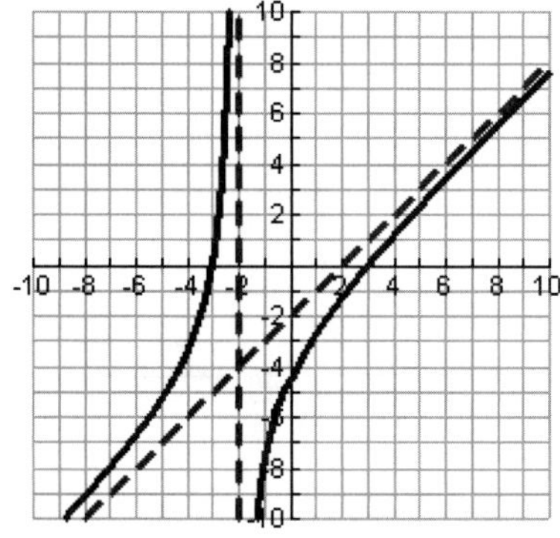

39. <u>Vertical Asymptotes</u>: $x = -2,\ x = 2$
<u>Intercepts</u>: $(0,0),\ (-\frac{1}{2},0),\ (1,0)$
<u>Slant Asymptote</u>: $y = 2x - 1$

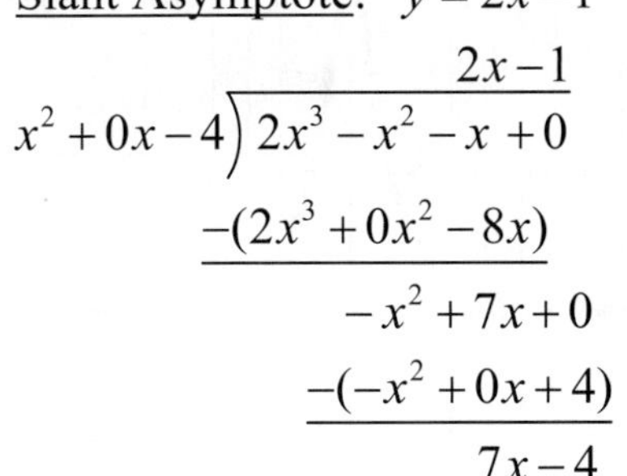

$$\begin{array}{r}
2x-1 \\
x^2+0x-4\overline{\smash{\big)}\,2x^3-x^2-x+0} \\
\underline{-(2x^3+0x^2-8x)} \\
-x^2+7x+0 \\
\underline{-(-x^2+0x+4)} \\
7x-4
\end{array}$$

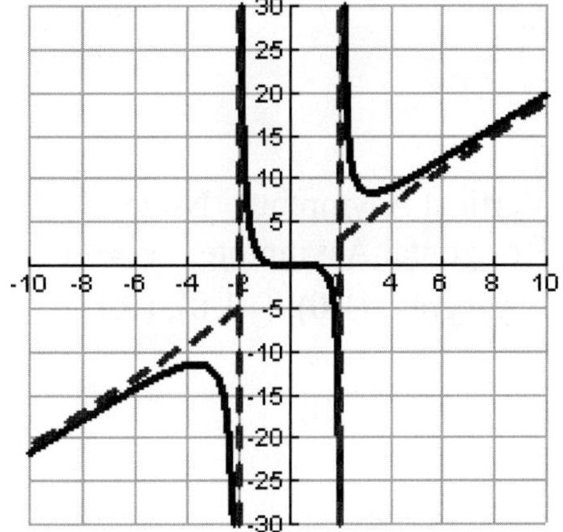

40. <u>Vertical Asymptote</u>: None
<u>Intercepts</u>: $(0,0),\ (\frac{1}{3},0),\ (-2,0)$
<u>Slant Asymptote</u>: $y = 3x + 5$

$$\begin{array}{r}
3x+5 \\
x^2+0x+4\overline{\smash{\big)}\,3x^3+5x^2-2x+0} \\
\underline{-(3x^3+0x^2+12x)} \\
5x^2-14x+0 \\
\underline{-(5x^2+0x+20)} \\
-14x-20
\end{array}$$

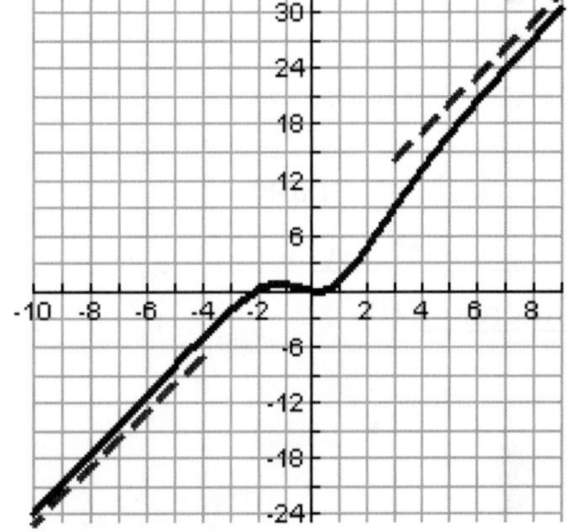

41. <u>Vertical Asymptote</u>: $x = 1,\ x = -1$
<u>Horizontal Asymptote</u>: $y = 1$
<u>Intercept</u>: $(0,-1)$

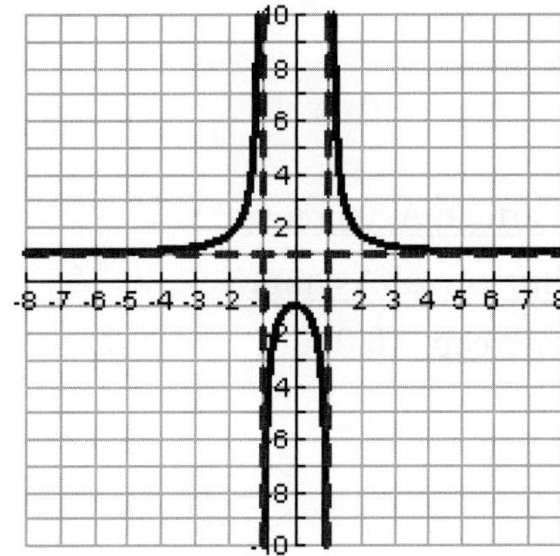

42. Vertical Asymptote: None
Horizontal Asymptote: $y = -1$
Intercepts: $(1,0)$, $(-1,0)$, $(0,1)$

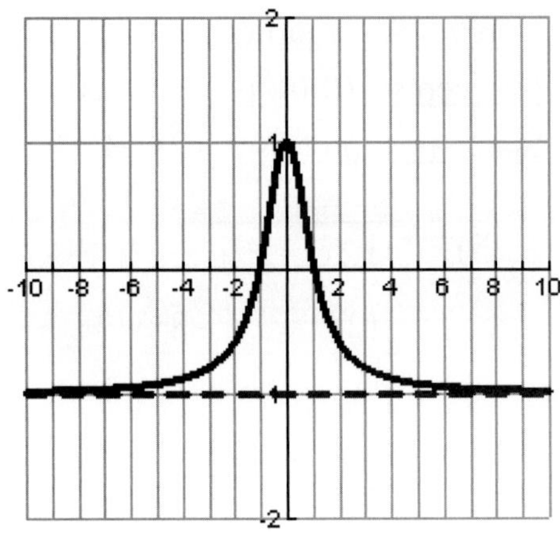

43. Vertical Asymptote: $x = -\frac{1}{2}$
Horizontal Asymptote: $y = \frac{7}{4}$
Intercept: $(0,0)$

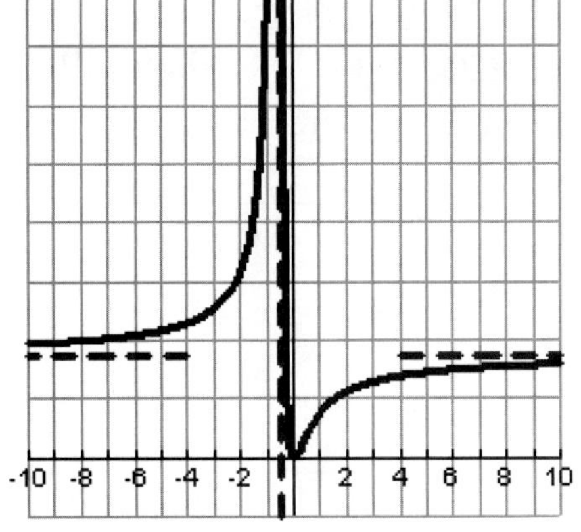

44. Vertical Asymptote: $x = -\frac{1}{3}$
Horizontal Asymptote: $y = \frac{4}{27}$
Intercept: $(0,0)$

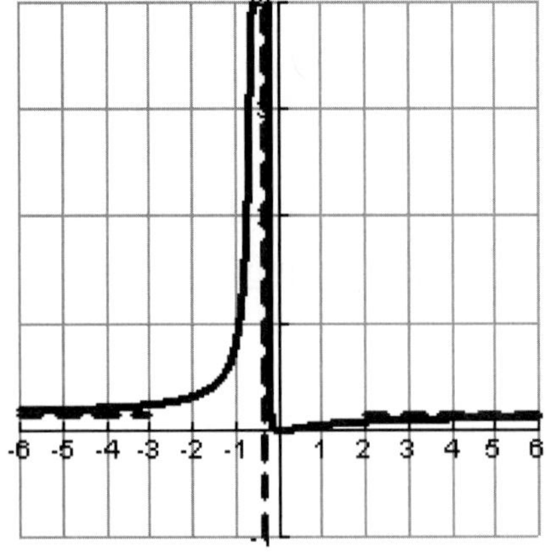

45. Vertical Asymptotes: $x = -\frac{1}{2}$, $x = \frac{1}{2}$
Horizontal Asymptote: $y = 0$
Intercepts: $(0,1)$, $(\frac{1}{3},0)$, $(-\frac{1}{3},0)$

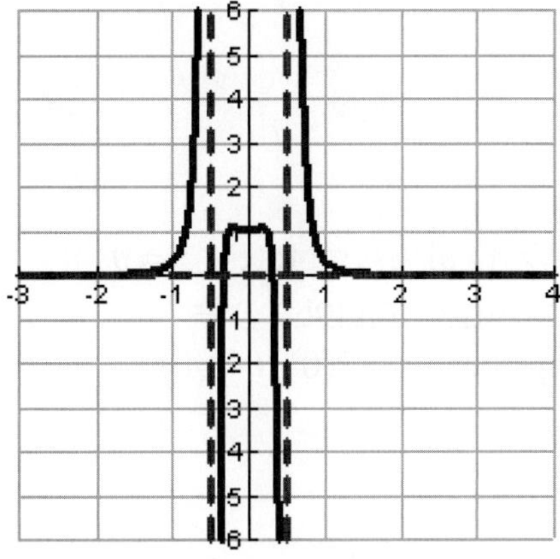

46. Vertical Asymptotes: $x = -\frac{1}{4}$, $x = \frac{1}{4}$
Horizontal Asymptote: $y = 0$
Intercepts: $(0,-1)$, $(\frac{1}{5},0)$, $(-\frac{1}{5},0)$

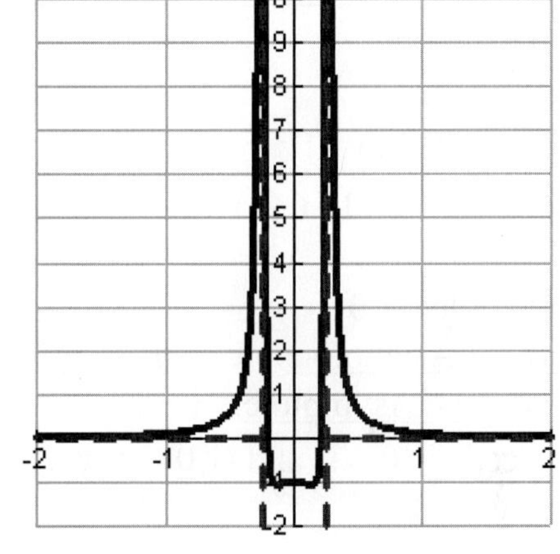

47. Vertical Asymptotes: $x = 0$
Slant Asymptote: $y = 3x$

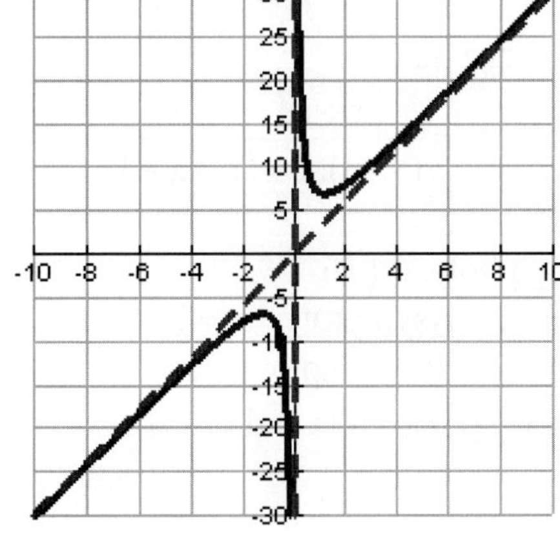

48. Vertical Asymptotes: $x = 0$
Slant Asymptote: $y = x$
Intercepts: $(2,0)$, $(-2,0)$

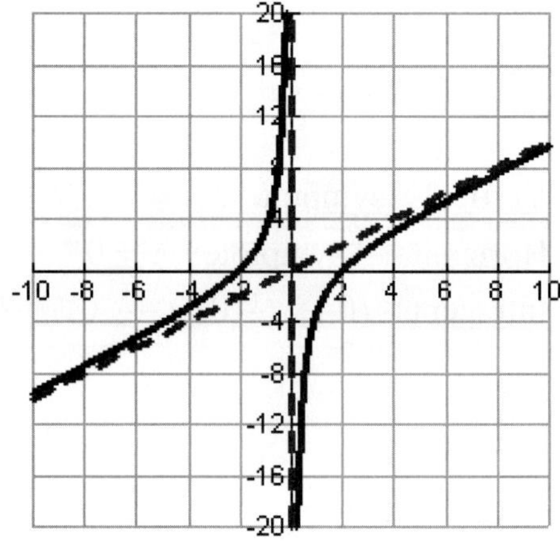

49. Observe that

$$f(x) = \frac{(x-1)^2}{(x-1)(x+1)} = \frac{x-1}{x+1}.$$

Open hole: (1,0)
Vertical Asymptote: $x = -1$
Horizontal Asymptote: $y = 1$
Intercepts: $(0,-1)$ and $(1,0)$

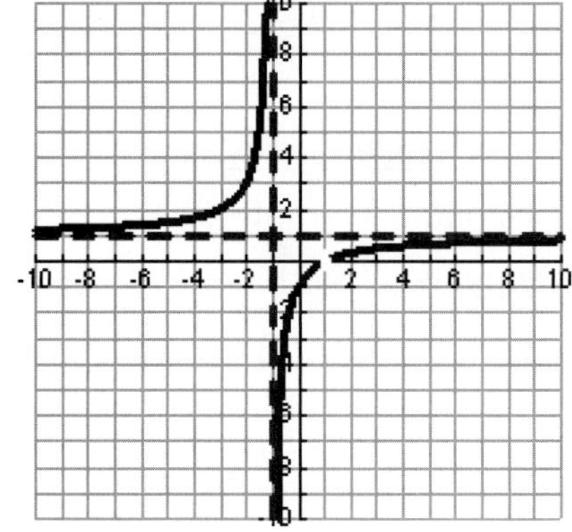

50. Observe that

$$f(x) = \frac{(x+1)^2}{(x-1)(x+1)} = \frac{x+1}{x-1}.$$

Open hole: (-1,0)
Vertical Asymptote: $x = 1$
Horizontal Asymptote: $y = 1$
Intercepts: $(0,-1)$ and $(-1,0)$

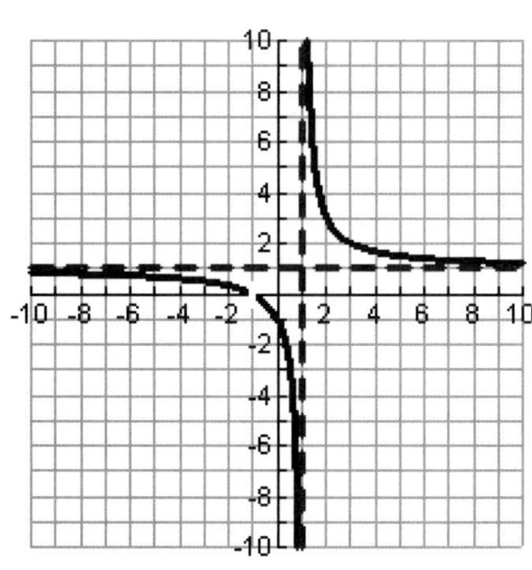

51. Observe that

$$f(x)=\frac{(x-1)(x-2)(x+2)}{(x-2)(x^2+1)}=\frac{(x-1)(x+2)}{x^2+1}$$

.

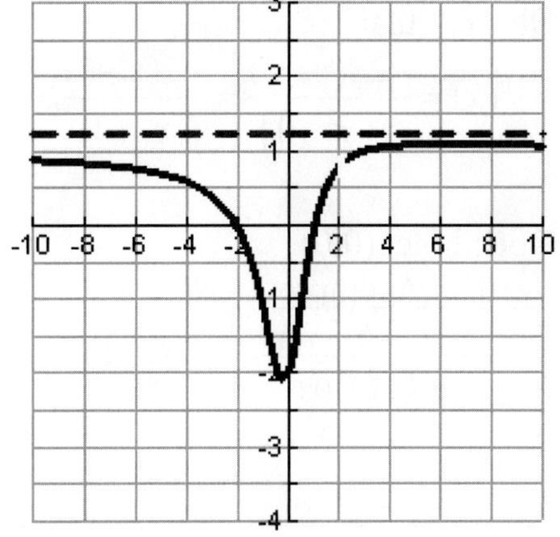

Open hole: (2,0)
Vertical Asymptote: None
Horizontal Asymptote: $y=1$
Intercepts: $(0,-2)$, $(1,0)$, and $(-2,0)$

52. Observe that

$$f(x)=\frac{(x-1)(x-3)(x+3)}{(x-3)(x^2+1)}=\frac{(x-1)(x+3)}{x^2+1}.$$

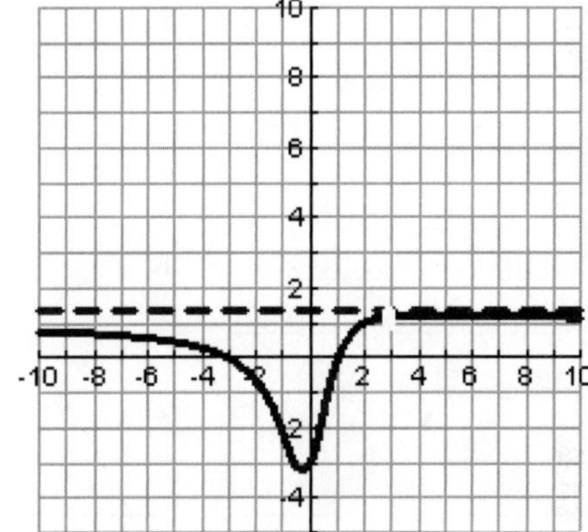

Open hole: (3,0)
Vertical Asymptote: None
Horizontal Asymptote: $y=1$
Intercepts: $(0,-3)$, $(1,0)$, and $(-3,0)$

53. Observe that

$$f(x)=\frac{3x(x-1)}{x(x-2)(x+2)}=\frac{3x-3}{(x-2)(x+2)}.$$

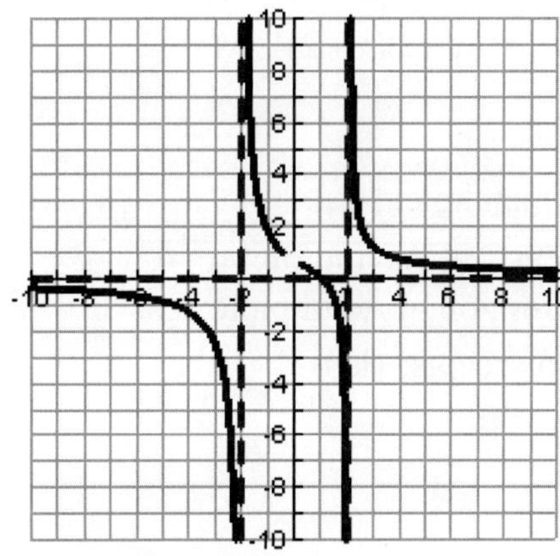

Open hole: (0,0)
Vertical Asymptote: $x=2,\ x=-2$
Horizontal Asymptote: $y=0$
Intercepts: $(1,0)$

54. Observe that

$$f(x)=\frac{-2x(x-3)}{x\left(x^2+1\right)}=\frac{-2(x-3)}{x^2+1}.$$

Open hole: (0,0)
Vertical Asymptote: None
Horizontal Asymptote: $y=0$
Intercepts: $(3,0)$

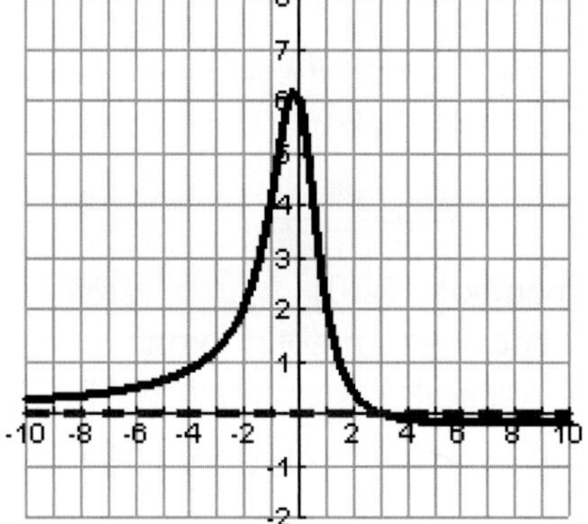

55. Observe that

$$f(x)=\frac{x^2(x+5)}{2x\left(x^2+3\right)}=\frac{x^2+5x}{2\left(x^2+3\right)},\ x\neq 0.$$

Open hole: (0,0)
Vertical Asymptote: None
Horizontal Asymptote: $y=\frac{1}{2}$
Intercepts: $(-5,0)$

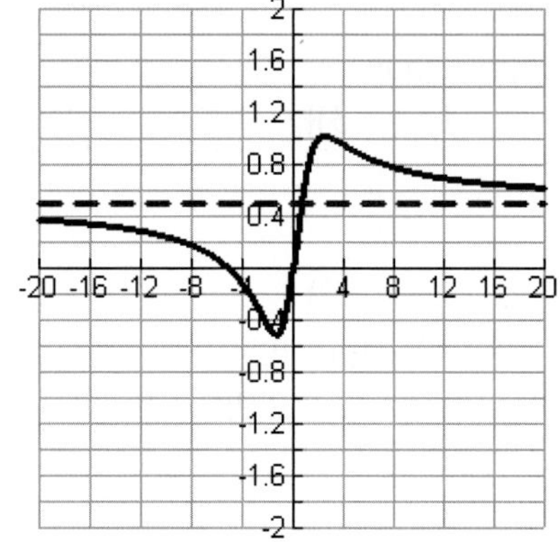

56. Observe that

$$f(x)=\frac{4x(x-1)(x+2)}{x^2(x-2)(x+2)}=\frac{4(x-1)}{x(x-2)},\ x\neq 0,-2$$

Open hole: (-2, $-\frac{3}{2}$)
Vertical Asymptote: x=0, x=2
Horizontal Asymptote: $y=0$
Intercepts: $(1,0)$

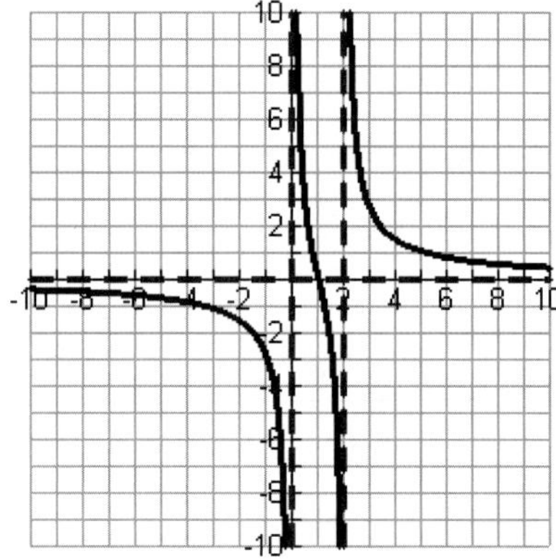

<table>
<tr>
<td>

57. a. <u>x-intercept</u>: (2,0)
<u>y-intercept</u>: (0,0.5)

b. <u>horizontal asymptote</u>: $y = 0$
<u>vertical asymptotes</u>: $x = -1,\ x = 4$

c. $f(x) = \dfrac{x-2}{(x+1)(x-4)}$

</td>
<td>

58. a. <u>x-intercepts</u>: (-0.5,0), (3,0)
<u>y-intercept</u>: (0,0.5)

b. <u>horizontal asymptote</u>: $y = 2$
<u>vertical asymptotes</u>: $x = -4,\ x = 3$

c. $f(x) = \dfrac{(2x+1)(x-3)}{(x+3)(x-2)}$

</td>
</tr>
<tr>
<td>

59. a. <u>x-intercept</u>: (0,0)
<u>y-intercept</u>: (0,0)

b. <u>horizontal asymptote</u>: $y = -3$
<u>vertical asymptotes</u>: $x = -4,\ x = 4$

c. $f(x) = \dfrac{-3x^2}{(x+4)(x-4)}$

</td>
<td>

60. a. <u>x-intercept</u>: (0,0), (2,0)
<u>y-intercept</u>: (0,0)

b. <u>horizontal asymptote</u>: None
<u>vertical asymptote</u>: $x = -1$
<u>slant asymptote</u>: $y = x - 3$

c. $f(x) = \dfrac{x(x-2)}{x+1}$

</td>
</tr>
<tr>
<td>

61. a) $r(4) = 4500$

b) Observe that

$$\frac{9500t-2000}{4+t} = 5500$$
$$9500t - 2000 = 5500(4+t)$$
$$4000t = 24{,}000$$
$$t = 6 \text{ months}$$

c) $n(t)$ gets closer to 9500 as t gets larger. So, the number of infected people stabilizes around 9500.

</td>
<td>

62. a) $r(8) = 3.01$, so about 3%.

b) $r(20) = 3.60$

c) The graph of $r(x) = \dfrac{4x^2}{x^2 + 2x + 5}$ is:

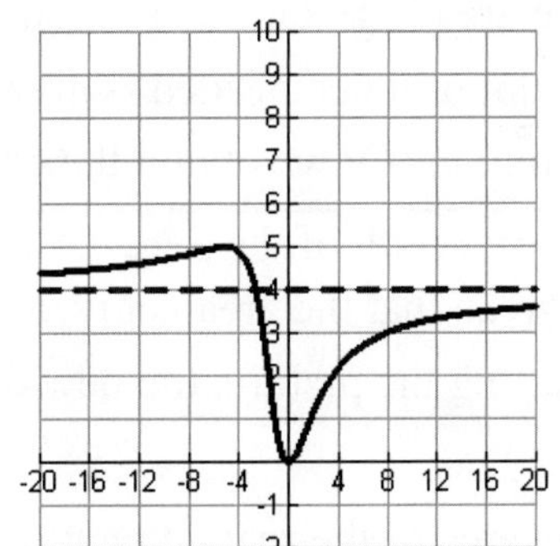

Observe that the graph increases towards 4 as x gets large.

</td>
</tr>
</table>

63. a. $C(1)=\frac{2(1)}{1^2+100}=\frac{2}{101}\cong 0.0198$ **b.** Since 1 hour = 60 minutes, we have $C(60)=\frac{2(60)}{60^2+100}\cong 0.0324$. **c.** Since 5 hours = 300 minutes, we have $C(300)=\frac{2(300)}{300^2+100}\cong 0$. **d.** The horizontal asymptote is $y=0$. So, after several days, C is approximately 0.	**64. a.** $C\left(\frac{1}{2}\right)\cong 0.0124$ **b.** $C(1)\cong 0.0243$ **c.** $C(4)\cong 0.0714$ **d.** Even though the values temporarily get larger, eventually they decrease toward 0. So, the horizontal asymptote is $y=0$. So, after several days, C is approximately 0.
65. a. $N(0)=52$ wpm **b.** $N(12)\cong 107$ wpm **c.** Since 3 years = 36 months, $N(36)\cong 120$wpm **d.** The horizontal asymptote is $y=130$. So, expect to type approximately 130 wpm as time goes on.	**66.** $N(3)\cong 78$, $N(16)\cong 267$ The horizontal asymptote is $y=600$. So, expect to remember at most 600 names.
67. The horizontal asymptote is $y=10$. So most adult cats eventually eat about 10 ounces of food.	**68.** $y(1)\cong 934$. The horizontal asymptote is $y=2800$. So, the maximum number of cards that could be memorized according to this model is 2800.

69. Let w = width of the pen, l = length of the pen.
Then, the area is given by $wl=500$, which is equivalent to $l=\frac{500}{w}$ **(1)**.
The amount of fence needed is given by the perimeter of the pen, namely $2w+2l$.
Substituting in **(1)**, we obtain the equivalent expression $2w+2\left(\frac{500}{w}\right)=\frac{2w^2+1000}{w}$.

70. Let w = width of the pen, l = length of the pen.
We are given that the area is given by $wl=414$, which is equivalent to $w=\frac{414}{l}$ **(2)**.
After matting the picture, the new dimensions become:

$$\text{Width} = 8+w, \quad \text{Length} = 7+l$$

Thus, the area of the matted picture = $(8+w)(7+l)=(8+\frac{414}{l})(7+l)=\dfrac{(8l+414)(7+l)}{l}$
(where we have substituted **(2)** for w).

71. $f(x)=\dfrac{x-1}{x^2-1}=\dfrac{\cancel{x-1}}{(\cancel{x-1})(x+1)}=\dfrac{1}{x+1}$ with a hole at $x=1$. So, $x=1$ is not a vertical asymptote.	**72.** $x^2+1=0$ has no real solution. So, there is no vertical asymptote.

73. In Step 2, the ratio of the leading coefficients should be $\frac{-1}{1}$. So, the horizontal asymptote is $y=-1$.	**74.** "Degree of numerator = Degree of denominator – 1" is not the criterion for the existence of an oblique asymptote. In this case, there is a horizontal asymptote, namely $y=0$, but no oblique asymptote.
75. True. The only way to have a slant asymptote is for the degree of the numerator to be greater than the degree of the denominator (by 1). In such case, there is no horizontal asymptote.	**76.** False. Consider $f(x)=\dfrac{1}{(x-2)(x+2)}$. The vertical asymptotes are $x=-2,\ x=2$.
77. False. This would require the denominator to equal 0, causing the function to be undefined.	**78.** True. Intersections with neither of these types of asymptotes creates a division by 0.
79. <u>Vertical Asymptotes</u>: $x=c,\ x=-d$ <u>Horizontal Asymptote</u>: $y=1$	**80.** There are no vertical asymptotes since $x^2+a^2\neq 0$. The horizontal asymptote is $y=3$. (<u>Note</u>: The actual values of *a* and *b* do not impact this result.)
81. Two such possibilities are: $y=\dfrac{4x^2}{(x+3)(x-1)}$ and $y=\dfrac{4x^5}{(x+3)^3(x-1)^2}$	**82.** $f(x)=\dfrac{x-3}{x^2+1}$ is such a function.

83. There are many different answers. Here is one approach.

Since there is an oblique asymptote, the degree of the numerator equals 1 + degree of the denominator. We are also given that $f(0)=1,\ f(-1)=0$.

The denominator cannot be linear since it <u>would</u> then have a vertical asymptote. So, it must be at least quadratic. Note that $x^2+a^2\neq 0$, for any $a\neq 0$.

Guided by these observations, we assume the general form of *f* is: $f(x)=\frac{x^3+K}{x^2+a^2}$.

(Note that $y=x$ is, in fact, an oblique asymptote for *f*.)
We must find values of *K* and *a* using the two points on the curve – this leads to the following system:

$$\begin{cases} f(0)=\frac{K}{a^2}=1 \Rightarrow K=a^2 \\ f(-1)=\frac{-1+K}{1+a^2}=0 \Rightarrow K=1 \end{cases}$$

From this system, we see that $K=1$ and $a=\pm 1$.
Thus, one function that works is $f(x)=\frac{x^3+1}{x^2+1}$.

84. There are many different answers. Here is one approach.
Since $x=-3, x=1$ are vertical asymptotes, the denominator has factors $(x+3)(x-1)$.
Further, since y=3x is an oblique asymptote for f, we know it has form:

$$f(x)=\frac{3x^3+K}{a(x+3)(x-1)}$$

We must determine values of K and a that satisfy $f(0)=2, f(2)=0$. This leads to the system:

$$\begin{cases} f(0)=\frac{K}{-3a}=2 \Rightarrow K=-6a \\ f(2)=\frac{24+k}{5a}=0 \Rightarrow K=-24 \end{cases}$$

From this system, we see that K = -24 and a = 4.

Thus, one function that works is $f(x)=\dfrac{3x^2-24}{4(x+3)(x-1)}$.

85. $f(x)=\dfrac{x-4}{(x-4)(x+2)}=\dfrac{1}{x+2}$ has a hole at $x=4$ and a vertical asymptote at $x=-2$.

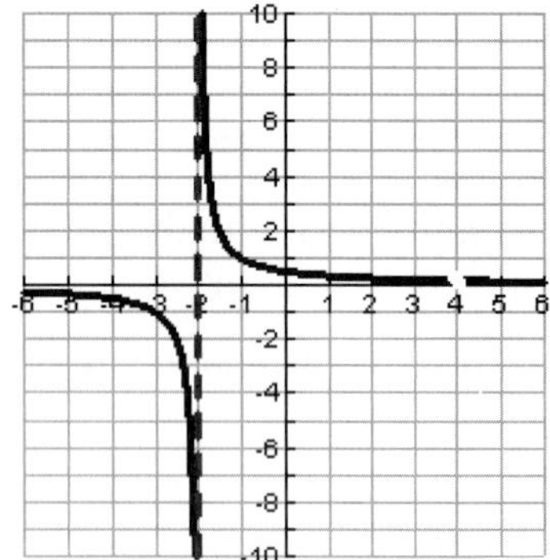

86. $f(x)=\dfrac{2x+1}{(2x+1)(3x-1)}=\dfrac{1}{3x-1}$ has a hole at $x=-\frac{1}{2}$ and a vertical asymptote at $x=\frac{1}{3}$.

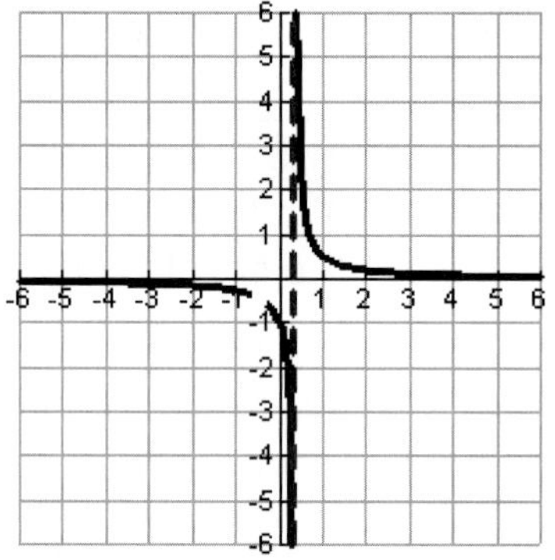

87. $f(x)=\dfrac{x-2(3x+1)}{x(3x+1)}=\dfrac{-5x-2}{x(3x+1)}$

Vertical asymptotes: $x=0,\ x=-\frac{1}{3}$

Horizontal asymptote: $y=0$

Intercepts: $\left(-\frac{2}{5},0\right)$

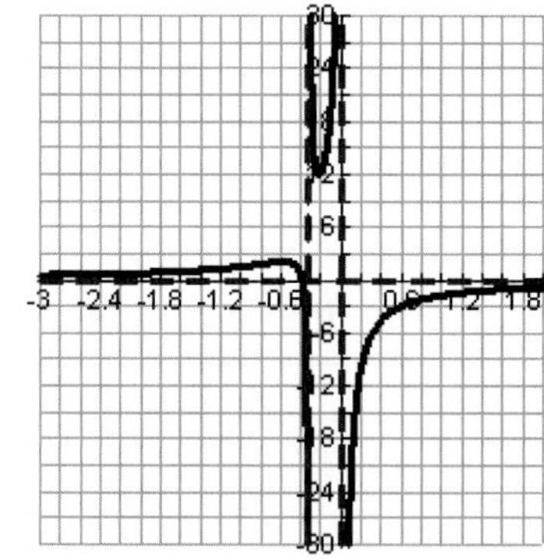

88. $f(x) = \dfrac{-x + x^2 + 1}{x(x^2+1)} = \dfrac{x^2 - x + 1}{x(x^2+1)}$

Vertical asymptotes: $x = 0$
Horizontal asymptote: $y = 0$

Intercepts: None since $x^2 - x + 1 = 0$ has no real solutions.

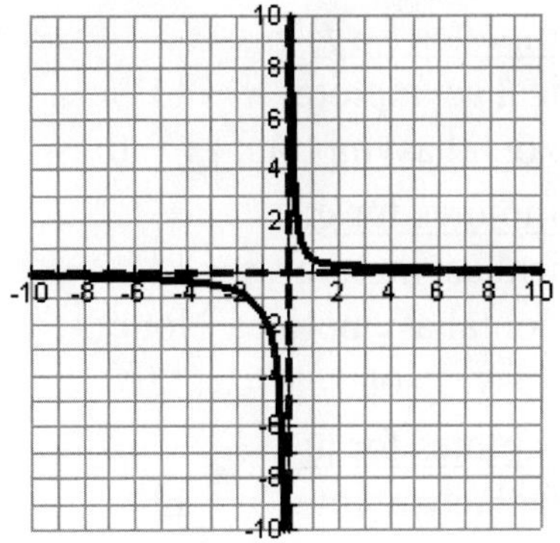

89.

a. Asymptotes for *f*:
vertical asymptote: $x = 3$
horizontal asymptote: $y = 0$
Asymptotes for *g*:
vertical asymptote: $x = 3$
horizontal asymptote: $y = 2$
Asymptotes for h:
vertical asymptote: $x = 3$
horizontal asymptote: $y = -3$

b. The graphs of *f* and *g* are below. As $x \to \pm\infty$, $f(x) \to 0$ and $g(x) \to 2$.

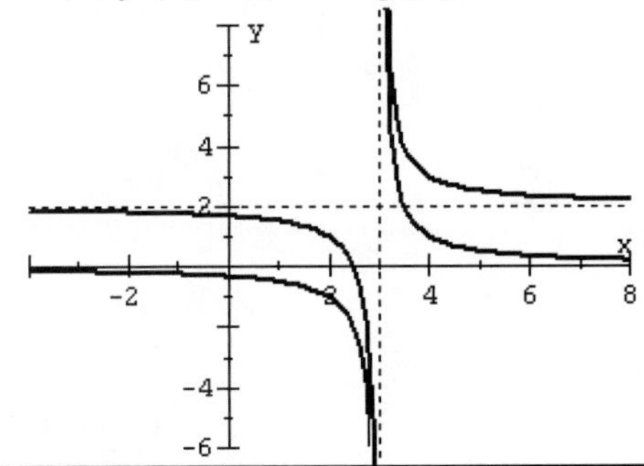

c. The graphs of *g* and *h* are below. As

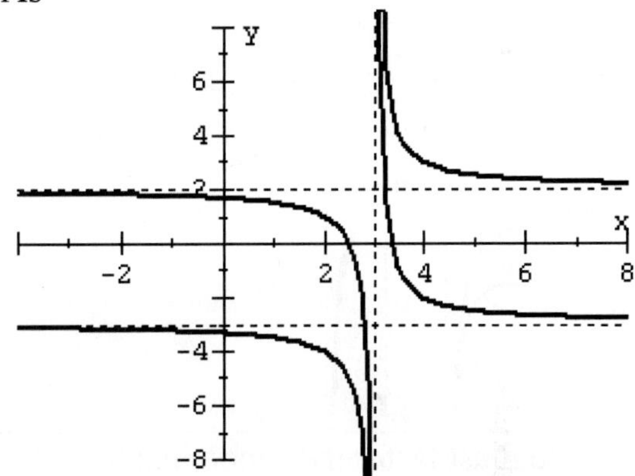

$x \to \pm\infty$, $g(x) \to 2$ and $h(x) \to -3$

d. $g(x) = \dfrac{2x-5}{x-3}$, $h(x) = \dfrac{-3x+10}{x-3}$.

Yes, if the degree of the numerator is the same as the degree of the denominator, then the horizontal asymptote is the ratio of the leading coefficients for both *g* and *h*.

90.

a. Asymptotes for *f*:
vertical asymptote: $x = \pm 1$
horizontal asymptote: $y = 0$

Asymptotes for g:
vertical asymptote: $x = \pm 1$
horizontal asymptote: None
slant asymptote: $y = x$

Asymptotes for h:
vertical asymptote: $x = \pm 1$
horizontal asymptote: None
slant asymptote: $y = x - 3$

b. The graphs of *f* and *g* are below. As $x \to \pm\infty$, $f(x) \to 0$ and $g(x) \to x$.

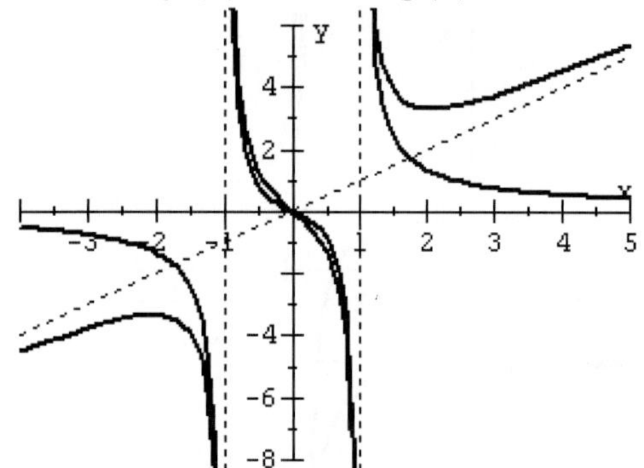

c. The graphs of *g* and *h* are below.
As $x \to \pm\infty$, $g(x) \to x$ and $h(x) \to x - 3$

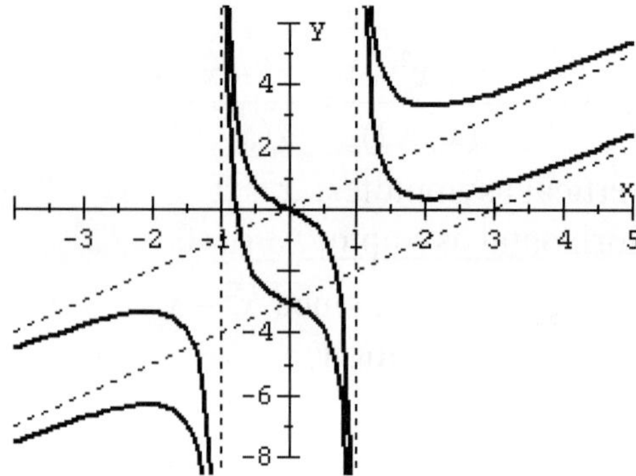

d. $g(x) = \dfrac{x^3 + x}{x^2 - 1}$, $h(x) = \dfrac{x^3 - 3x^2 + x + 3}{x^2 - 1}$.

Yes, if the degree of the numerator is the exactly one more than the degree of the denominator, then the quotient is the slant asymptote.

91. We must factor the denominator. Observe that the possible rational zeros are: $\pm 1, \pm 2, \pm 5, \pm 10$. Using synthetic division yields

$$\begin{array}{r|rrrr} -2 & 1 & -2 & -13 & -10 \\ & & -2 & 8 & 10 \\ \hline -1 & 1 & -4 & -5 & \\ & & -1 & 5 & \\ \hline & 1 & -5 & 0 & \end{array}$$

So, $x^3 - 2x^2 - 13x - 10 = (x+2)(x+1)(x-5)$. None of these factors cancels with one in the numerator. So, the vertical asymptotes are $x = -2, x = -1, x = 5$. So, the integral of *f* exists on [0,3].

92. We must factor the denominator. Observe that the possible rational zeros are: $\pm 1, \pm 2, \pm 5, \pm 10, \pm 25, \pm 50$. Using synthetic division yields $$\begin{array}{r\|rrrr} 5 & 1 & 2 & -25 & -50 \\ & & 5 & 35 & 50 \\ \hline -5 & 1 & 7 & 10 & \\ & & -5 & -10 & \\ \hline & 1 & 2 & 0 & \end{array}$$ So, $x^3+2x^2-25x-50=(x-5)(x+5)(x-2)$. None of these factors cancels with one in the numerator. So, the vertical asymptotes are $x=-5, x=2, x=5$. So, the integral of *f* might not exist on [-3,2].
93. Note that the denominator factors as $6x^2-x-2=(3x-2)(2x+1)$. Neither of these factors cancels with one in the numerator. So, the vertical asymptotes are $x=\frac{2}{3}, x=-\frac{1}{2}$. So, the integral of *f* might not exist on [-2,0].
94. Observe that $f(x)=\dfrac{2x(3-x)}{x\left(x^2+1\right)}=\dfrac{2(3-x)}{x^2+1}$, which has no vertical asymptotes. Hence, the integral of *f* exists on [-1,1].

Chapter 2 Review Solutions ---

1. b Opens down, vertex is $(-6,3)$	**2. c** Opens up, vertex is $(4,2)$
3. a Opens up, vertex is $\left(-\frac{1}{2},-\frac{25}{4}\right)$. $x^2+x-6=(x^2+x+\frac{1}{4})-6-\frac{1}{4}$ $=(x+\frac{1}{2})^2-\frac{25}{4}$	**4. d** Opens down, vertex is $\left(-\frac{5}{3},\frac{49}{3}\right)$ $-3x^2-10x+8=-3(x^2+\frac{10}{3}x\ \ )+8$ $=-3(x^2+\frac{10}{3}x+\frac{25}{9}\ \ )+8+\frac{25}{3}$ $=-3(x+\frac{5}{3})^2+\frac{49}{3}$
5. 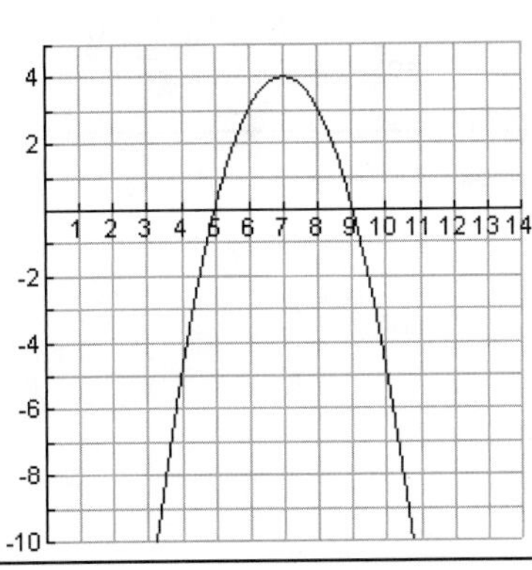	**6.** 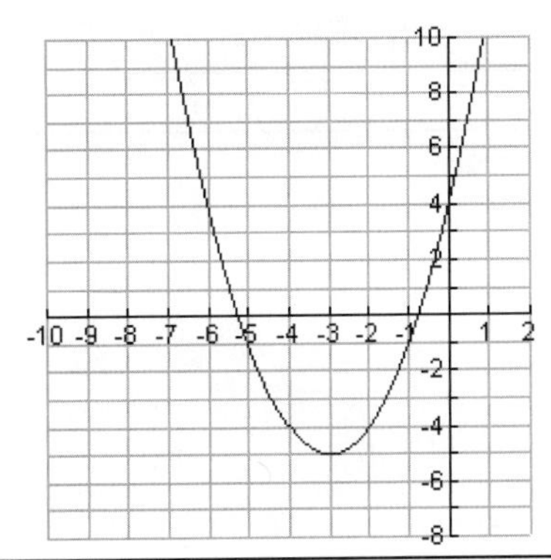

7.

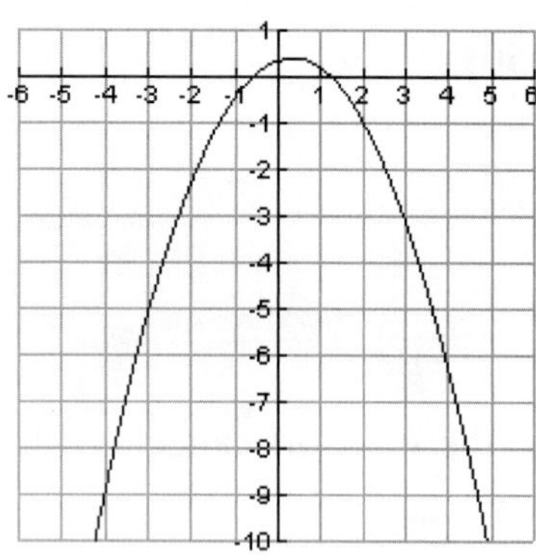

8.

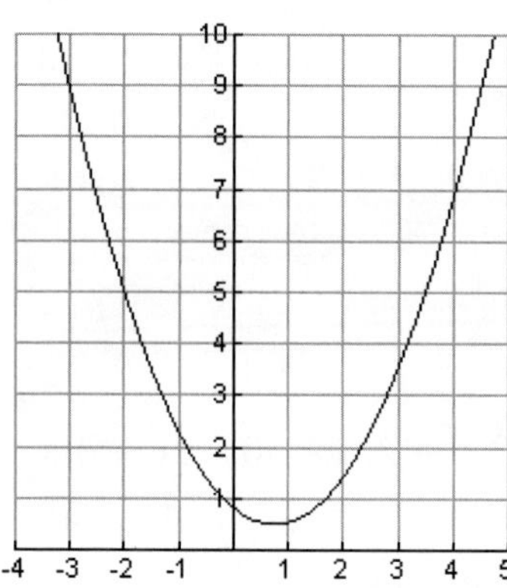

9.

$$x^2 - 3x - 10 = (x^2 - 3x + \tfrac{9}{4}) - 10 - \tfrac{9}{4}$$
$$= (x - \tfrac{3}{2})^2 - \tfrac{49}{4}$$

10.

$$x^2 - 2x - 24 = (x^2 - 2x + 1) - 24 - 1$$
$$= (x-1)^2 - 25$$

11.

$$4x^2 + 8x - 7 = 4(x^2 + 2x \quad) - 7$$
$$= 4(x^2 + 2x + 1) - 7 - 4$$
$$= 4(x+1)^2 - 11$$

12.

$$-\tfrac{1}{4}x^2 + 2x - 4 = -\tfrac{1}{4}(x^2 - 8x \quad) - 4$$
$$= -\tfrac{1}{4}(x^2 - 8x + 16) - 4 + 4$$
$$= -\tfrac{1}{4}(x-4)^2$$

13.

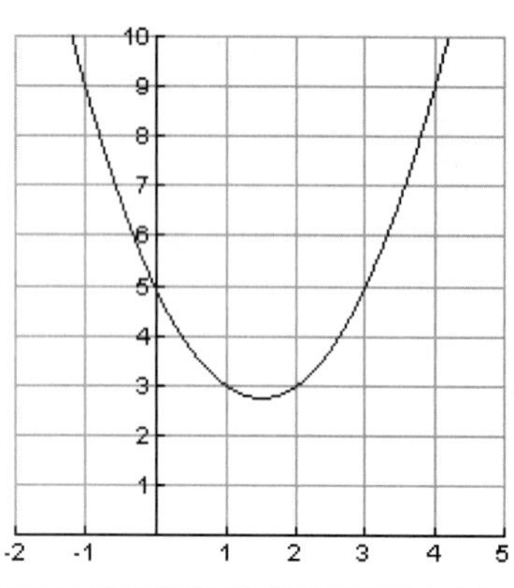

14.

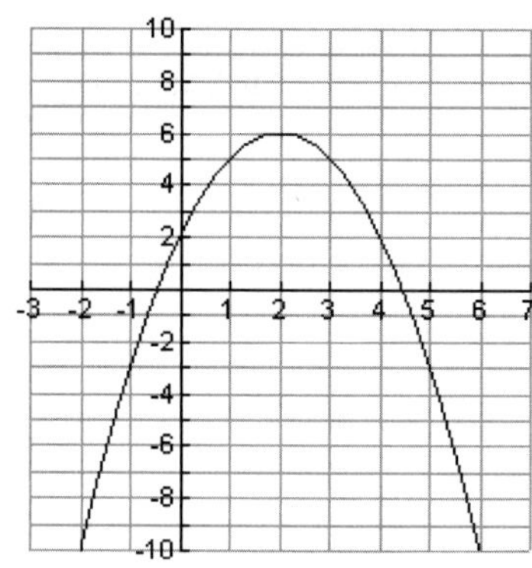

15. 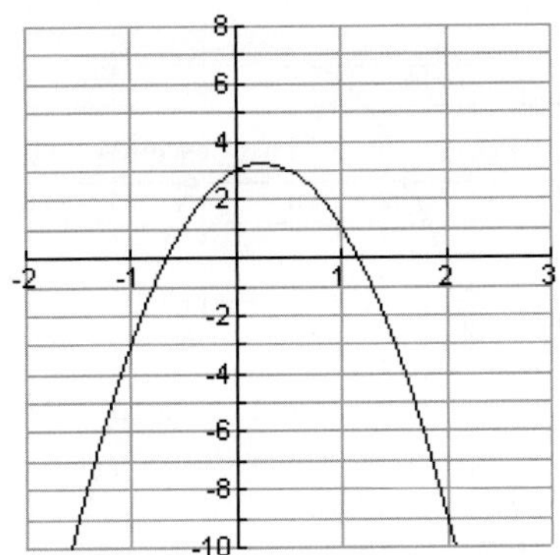	**16.** 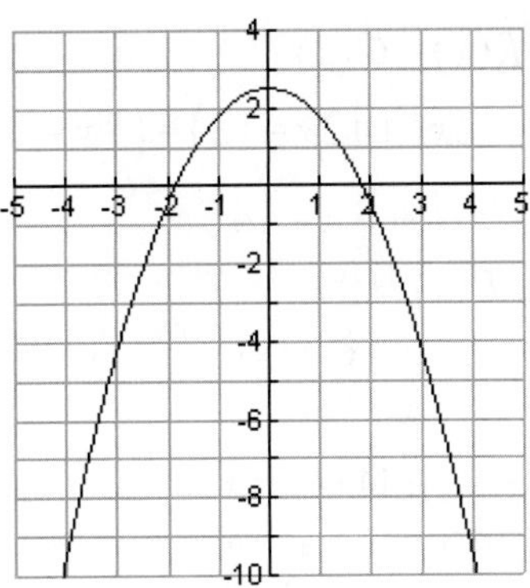
17. Vertex is $\left(-\frac{b}{2a}, c-\frac{b^2}{4a}\right)=\left(-\frac{-5}{2(13)}, 12-\frac{(-5)^2}{4(13)}\right)$ $=\left(\frac{5}{26}, \frac{599}{52}\right)$	**18.** Vertex is $\left(-\frac{b}{2a}, c-\frac{b^2}{4a}\right)=\left(-\frac{-4}{2(\frac{2}{5})}, 3-\frac{(-4)^2}{4(\frac{2}{5})}\right)$ $=(5,-7)$
19. Vertex is $\left(-\frac{b}{2a}, c-\frac{b^2}{4a}\right)=\left(-\frac{-0.12}{2(-0.45)}, 3.6-\frac{(-0.12)^2}{4(-0.45)}\right)$ $=\left(-\frac{2}{15}, \frac{451}{125}\right)$	**20.** Vertex is $\left(-\frac{b}{2a}, c-\frac{b^2}{4a}\right)=\left(-\frac{\frac{2}{5}}{2(-\frac{3}{4})}, 4-\frac{(\frac{2}{5})^2}{4(-\frac{3}{4})}\right)$ $=\left(\frac{4}{15}, \frac{304}{75}\right)$
21. Since the vertex is $(-2,3)$, the function has the form $y=a(x+2)^2+3$. To find *a*, use the fact that the point (1,4) is on the graph: $4=a(1+2)^2+3$ $4=9a+3$ $\frac{1}{9}=a$ So, the function is $y=\frac{1}{9}(x+2)^2+3$.	**22.** Since the vertex is $(4,7)$, the function has the form $y=a(x-4)^2+7$. To find *a*, use the fact that the point $(-3,1)$ is on the graph: $1=a(-3-4)^2+7$ $1=49a+7$ $-\frac{6}{49}=a$ So, the function is $y=-\frac{6}{49}(x-4)^2+7$.
23. Since the vertex is $(2.7,3.4)$, the function has the form $y=a(x-2.7)^2+3.4$. To find *a*, use the fact that the point (3.2, 4.8) is on the graph: $4.8=a(3.2-2.7)^2+3.4$ $4.8=0.25a+3.4$ $5.6=a$ So, the function is $y=5.6(x-2.7)^2+3.4$.	**24.** Since the vertex is $(-\frac{5}{2},\frac{7}{4})$, the function has the form $y=a(x+\frac{5}{2})^2+\frac{7}{4}$. To find *a*, use the fact that the point $(\frac{1}{2},\frac{3}{5})$ is on the graph: $\frac{3}{5}=a(\frac{1}{2}+\frac{5}{2})^2+\frac{7}{4}$ $\frac{3}{5}=9a+\frac{7}{4}$ $-\frac{23}{180}=a$ So, the function is $y=-\frac{23}{180}(x+\frac{5}{2})^2+\frac{7}{4}$.

25.

a.

$$P(x) = R(x) - C(x)$$
$$= \left(-2x^2 + 12x - 12\right) - \left(\tfrac{1}{3}x + 2\right)$$
$$= -2x^2 + \tfrac{35}{3}x - 14$$

b. Solve $P(x) = 0$.

$$-2x^2 + \tfrac{35}{3}x - 14 = 0$$
$$-2\left(x^2 - \tfrac{35}{6}x \quad \right) - 14 = 0$$
$$-2\left(x^2 - \tfrac{35}{6}x + \tfrac{1225}{144}\right) - 14 + \tfrac{1225}{72} = 0$$
$$-2(x - \tfrac{35}{12})^2 + \tfrac{217}{72} = 0$$
$$(x - \tfrac{35}{12})^2 = \tfrac{217}{144}$$
$$x = \tfrac{35}{12} \pm \sqrt{\tfrac{217}{144}}$$
$$= \tfrac{35 \pm \sqrt{217}}{12}$$
$$\cong 4.1442433,\ 1.68909$$

c.

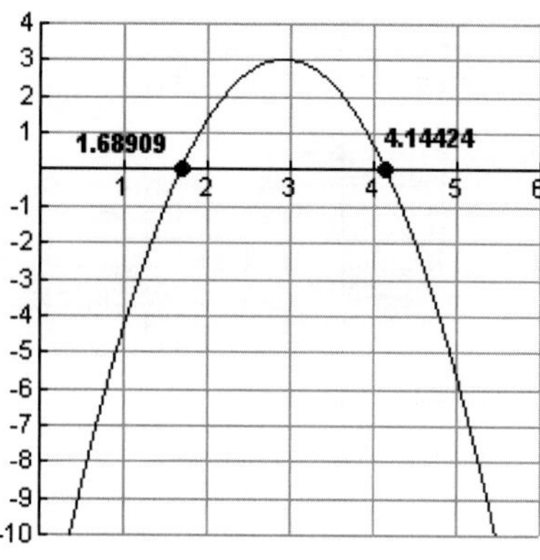

d. The range is approximately $(1.6891, 4.144)$, which corresponds to where the graph is above the x-axis.

26. Area is

$$A(x) = (2x-4)(x+7) = 2x^2 + 10x - 28$$

27. Area is

$$A(x) = \tfrac{1}{2}(x+2)(4-x) = -\tfrac{1}{2}(x^2 - 2x \quad) + 4$$
$$= -\tfrac{1}{2}(x^2 - 2x + 1) + 4 + \tfrac{1}{2}$$
$$= -\tfrac{1}{2}(x-1)^2 + \tfrac{9}{2}$$

Note that $A(x)$ has a maximum at $x = 1$ (since its graph is a parabola that opens down). The corresponding dimensions are both base and height are 3 units.

28. a.

$$h(t) = -12(t^2 - \tfrac{20}{3}t)$$
$$= -12(t^2 - \tfrac{20}{3}t + \tfrac{100}{9}) + 12\left(\tfrac{100}{9}\right)$$
$$= -12(t - \tfrac{10}{3})^2 + \tfrac{400}{3}$$

So, the maximum height is approximately 133.33 units.

b. Solve $h(t) = 0$:

$t(-12t + 80) = 0$

$t = 0,\ \tfrac{80}{12} \cong 6.67$

So, after approximately 6.7 seconds, it will hit the ground.

29. Polynomial with degree 6

30. Polynomial with degree 5

31. Not a polynomial (due to the term $x^{1/4}$)

32. Polynomial with degree 3

33. d linear

34. b Parabola opens down

35. a 4th degree polynomial, looks like $y = x^4$ for x very large.

36. c 7th degree polynomial, looks like $y = x^7$ for x very large.

37. Reflect the graph of x^7 over the x-axis

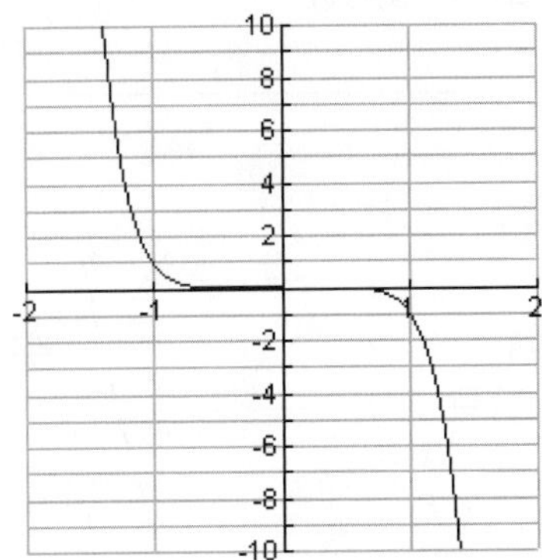

38. Shift the graph of x^3 to the right 3 units.

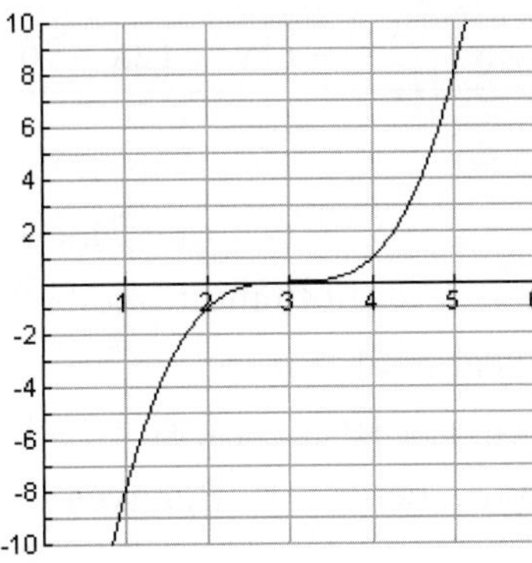

39. Shift the graph of x^4 down 2 units.

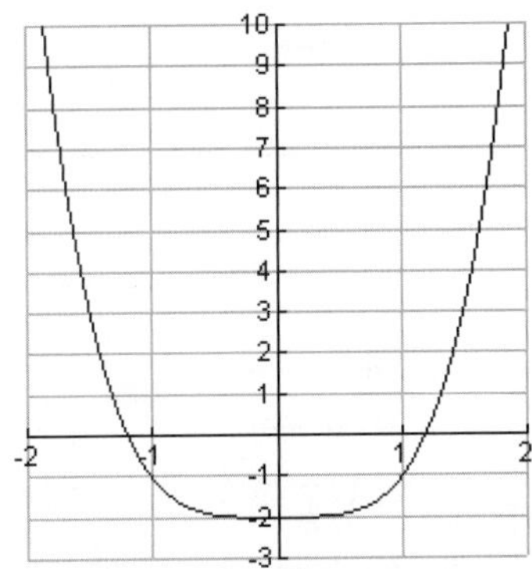

40. Shift the graph of x^5 left 7 units, then reflect over the x-axis, and then move down 6 units and.

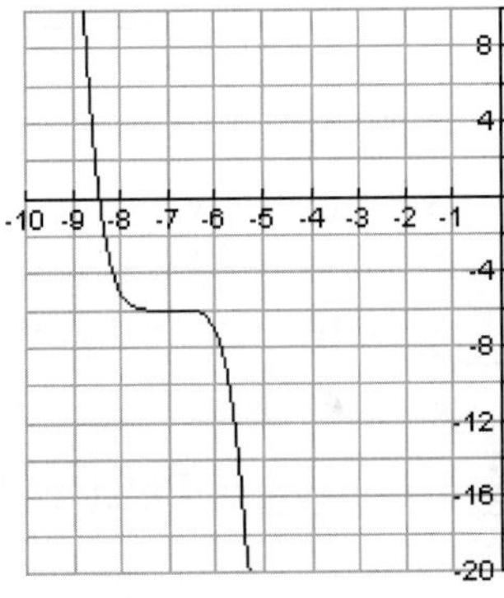

41. 6 (multiplicity 5)
-4 (multiplicity 2)

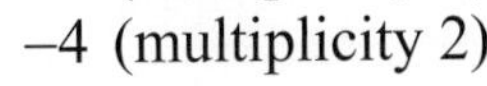

42. 0 (multiplicity 1)
2 (multiplicity 3)
-5 (multiplicity 1)

43.

$$x^5 - 13x^3 + 36x = x(x^2 - 9)(x^2 - 4)$$
$$= x(x+3)(x-3)(x+2)(x-2)$$

So, the zeros are $0, -2, 2, 3, -3$, all with multiplicity 1.

44.

$$4.2x^4 - 2.6x^2 \cong 4.2x^2\left(x^2 - 0.619047\right)$$
$$= 4.2x(x - 0.786795)(x + 0.786795)$$

So, the zeros are approximately:
0 (multiplicity 2)
0.786795, -0.786795 (multiplicity 1)

45. $x(x+3)(x-4)$

46. $(x-2)(x-4)(x-6)(x+8)$

47. $x(x+\frac{2}{5})(x-\frac{3}{4}) = x(5x+2)(4x-3)$

48. $(x-(2-\sqrt{5})(x-(2+\sqrt{5}))$

49. $(x+2)^4(x-3)^2 =$ $x^4-2x^3-11x^2+12x+36$	**50.** $(x-3)^2x^3(x+1)^2$
51. $f(x)=x^2-5x-14=(x-7)(x+2)$ **a.** Zeros: -2, 7 (both multiplicity 1) **b.** Crosses at both -2, 7 **c.** y-intercept: $f(0)=-14$, so $(0,-14)$ **d.** Long-term behavior: Behaves like $y=x^2$. Even degree and leading coefficient positive, so graph rises without bound to the left and right.	**e.** 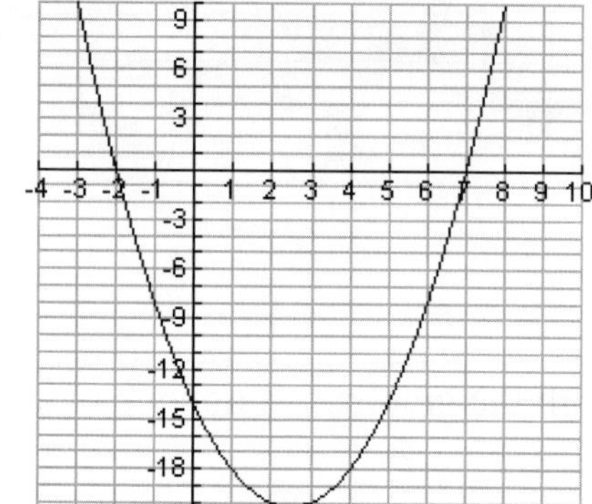
52. $f(x)=-(x-5)^5$ **a.** Zeros: 5 (multiplicity 5) **b.** Crosses at 5 **c.** y-intercept: $f(0)=-(-5)^5=3125$, so $(0,3125)$ **d.** Long-term behavior: Behaves like $y=-x^5$. Odd degree and leading coefficient negative, so graph falls without bound to the right and rises to the left.	**e.** 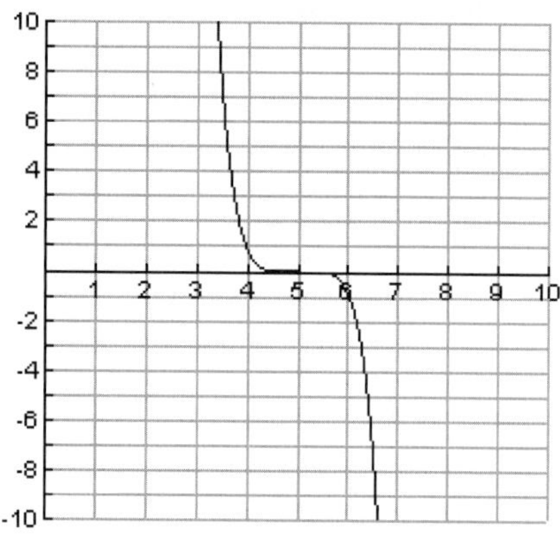

53. $f(x)=6x^7+3x^5-x^2+x-4$

a. Zeros: We first try to apply the Rational Root Test:

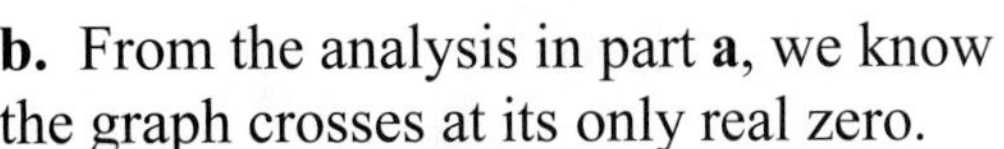

Factors of -4: $\pm1, \pm2, \pm4$

Factors of 6: $\pm1, \pm2, \pm3, \pm6$

Possible rational zeros:

$$\pm1, \pm2, \pm4, \pm\tfrac{1}{2}, \pm\tfrac{1}{3}, \pm\tfrac{2}{3}, \pm\tfrac{4}{3}, \pm\tfrac{1}{6}$$

Unfortunately, it can be shown that none of these are zeros of *f*. To get a feel for the possible number of irrational and complex root, we apply Descartes' Rule of Signs:

Number of sign variations for $f(x)$: 3

$f(-x)=-6x^7-3x^5-x^2-x-4$, so

Number of sign variations for $f(-x)$: 0

Since $f(x)$ is degree 7, there are 7 zeros, classified as:

Positive Real Zeros	Negative Real Zeros	Imaginary Zeros
3	0	4
1	0	6

Need to actually graph the polynomial to determine the approximate zeros. From **e.**, we see there is a zero at approximately (0.8748,0) with multiplicity 1.

b. From the analysis in part **a**, we know the graph crosses at its only real zero.

c. y-intercept: $f(0)=-4$, so $(0,-4)$

d. Long-term behavior: Behaves like $y=x^7$.

Odd degree and leading coefficient positive, so graph falls without bound to the left and rises to the right.

e.

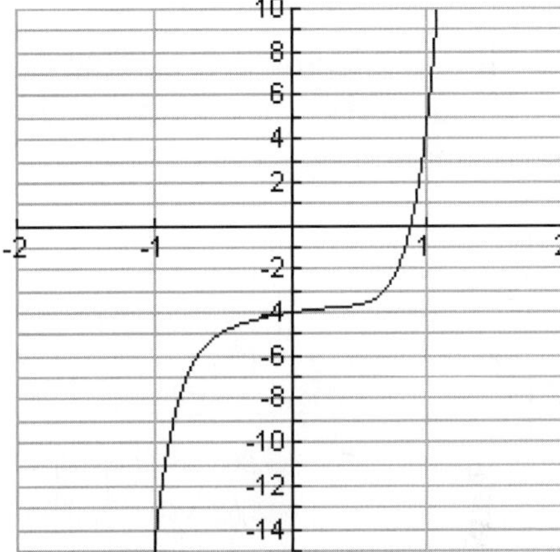

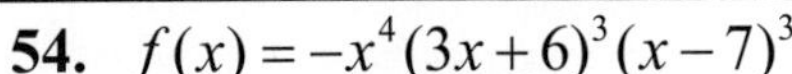

54. $f(x)=-x^4(3x+6)^3(x-7)^3$

a. Zeros: 0 (multiplicity 4)
-2 (multiplicity 3)
7 (multiplicity 3)

b. Touches at 0, and crosses at both -2, 7

c. y-intercept: $f(0)=0$, so $(0,0)$

d. Long-term behavior: Behaves like $y=-x^{10}$.

Even degree and leading coefficient negative, so graph falls without bound to the right and left.

e.

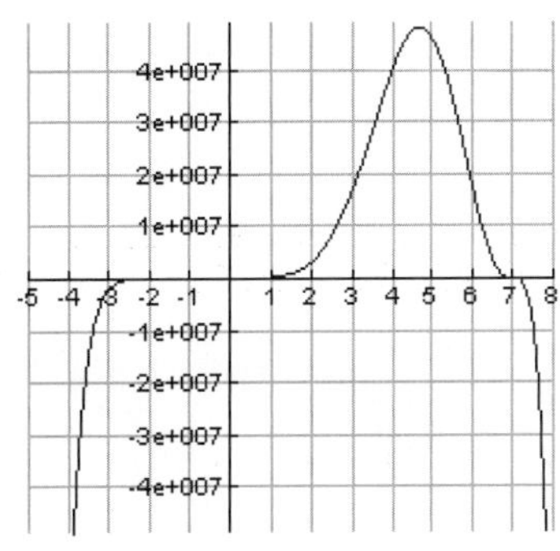

55. $f(x) = (x-1)(x-3)(x-7)$

a.

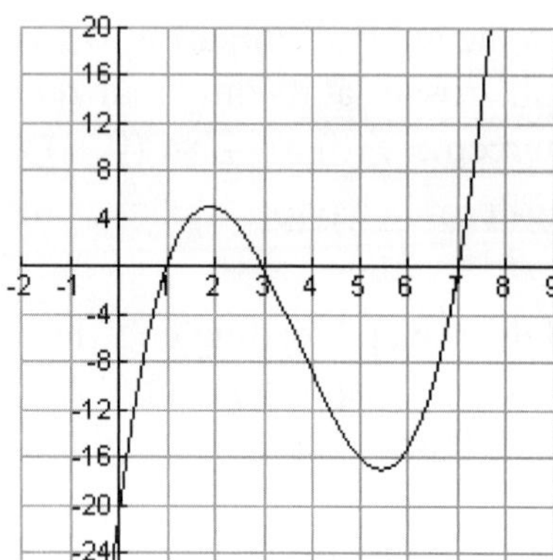

b. Zeros: 1, 3, 7 (all with multiplicity 1)

c. Want intervals where the graph of f is above the x-axis. From **a.**, this occurs on the intervals $(1,3)$ and $(7,\infty)$.

So, between 1 and 3 hours, and more than 7 hours would be financially beneficial.

56. Consider the graph below. The peak seasons occur during the summer and in Nov. and Dec.

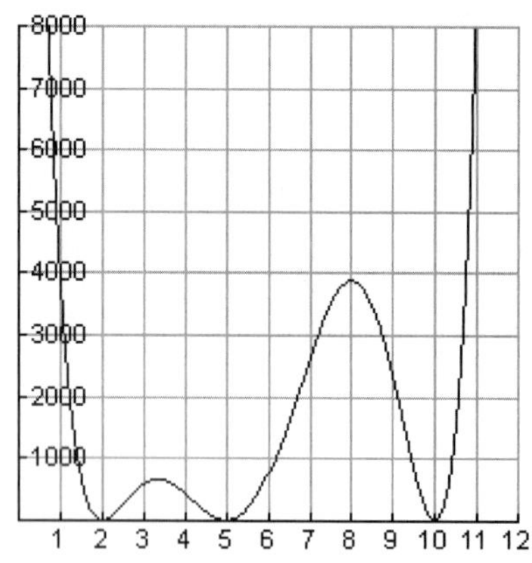

57.

$$\begin{array}{r} x+4 \\ x-2 \overline{\smash{)}\; x^2+2x-6} \\ \underline{-(x^2-2x)} \\ 4x-6 \\ \underline{-(4x-8)} \\ 2 \end{array}$$

So, $\boxed{Q(x) = x+4, \quad r(x) = 2}$.

58.

$$\begin{array}{r} x-1 \\ 2x-3 \overline{\big) \; 2x^2-5x-1} \\ \underline{-(2x^2-3x)} \\ -2x-1 \\ \underline{-(-2x+3)} \\ -4 \end{array}$$

So, $\boxed{Q(x)=x-1, \quad r(x)=-4}$.

59.

$$\begin{array}{r} 2x^3-4x^2-2x-\frac{7}{2} \\ 2x-4 \overline{\big) \; 4x^4-16x^3+12x^2+x-9} \\ \underline{-(4x^4-8x^3)} \\ -8x^3+12x^2 \\ \underline{-(-8x^3+16x^2)} \\ -4x^2+x \\ \underline{-(-4x^2+8x)} \\ -7x-9 \\ \underline{-(-7x+14)} \\ -23 \end{array}$$

So,

$\boxed{Q(x)=2x^3-4x^2-2x-\frac{7}{2}, \quad r(x)=-23}$.

60.

$$\begin{array}{r} -2x^2+2x-2 \\ 2x^2+x-4 \overline{\big) \; -4x^4+2x^3+6x^2-x+2} \\ \underline{-(-4x^4-2x^3+8x^2)} \\ 4x^3-2x^2-x \\ \underline{-(4x^3+2x^2-8x)} \\ -4x^2+7x+2 \\ \underline{-(-4x^2-2x+8)} \\ 9x-6 \end{array}$$

So, $\boxed{Q(x)=-2x^2+2x-2, \quad r(x)=9x-6}$.

61.

$$\begin{array}{r|rrrrr} -2 & 1 & 4 & 5 & -2 & -8 \\ & & -2 & -4 & -2 & 8 \\ \hline & 1 & 2 & 1 & -4 & 0 \end{array}$$

So, $Q(x)=x^3+2x^2+x-4, \quad r(x)=0$.

62.

$$\begin{array}{r|rrrr} -2 & 1 & 0 & -10 & 3 \\ & & -2 & 4 & 12 \\ \hline & 1 & -2 & -6 & 15 \end{array}$$

So, $Q(x)=x^2-2x-6, \quad r(x)=15$.

63.

$$\begin{array}{r|rrrrrrr} -8 & 1 & 0 & 0 & 0 & 0 & 0 & -64 \\ & & -8 & 64 & -512 & 4096 & -32,768 & 262,144 \\ \hline & 1 & -8 & 64 & -512 & 4096 & -32,768 & 262,080 \end{array}$$

So, $Q(x)=x^5-8x^4+64x^3-512x^2+4096x-32,768, \quad r(x)=262,080$.

64.

$$\begin{array}{r|rrrrrr} \frac{3}{4} & 2 & 4 & -2 & 0 & 7 & 5 \\ & & \frac{3}{2} & \frac{33}{8} & \frac{51}{32} & \frac{153}{128} & \frac{3147}{512} \\ \hline & 2 & \frac{11}{2} & \frac{17}{8} & \frac{51}{32} & \frac{1049}{128} & \frac{5707}{512} \end{array}$$

So, $Q(x) = 2x^4 + \frac{11}{2}x^3 + \frac{17}{8}x^2 + \frac{51}{32}x + \frac{1049}{128}$, $r(x) = \frac{5707}{512}$.

65.

$$\begin{array}{r} x+3 \\ 5x^2-7x+3 \overline{\smash{\big)}\ 5x^3+8x^2-22x+1} \\ \underline{-(5x^3-7x^2+3x)} \\ 15x^2-25x+1 \\ \underline{-(15x^2-21x+9)} \\ -4x-8 \end{array}$$

So, $Q(x) = x+3$, $r(x) = -4x-8$.

66.

$$\begin{array}{r|rrrrr} 3 & 1 & 2 & -5 & 4 & 2 \\ & & 3 & 15 & 30 & 102 \\ \hline & 1 & 5 & 10 & 34 & 104 \end{array}$$

So, $Q(x) = x^3 + 5x^2 + 10x + 34$, $r(x) = 104$.

67.

$$\begin{array}{r|rrrr} -1 & 1 & -4 & 2 & -8 \\ & & -1 & 5 & -7 \\ \hline & 1 & -5 & 7 & -15 \end{array}$$

So, $Q(x) = x^2 - 5x + 7$, $r(x) = -15$.

68.

$$\begin{array}{r} x-5 \\ x^2+0x+4 \overline{\smash{\big)}\ x^3-5x^2+4x-20} \\ \underline{-(x^3+0x^2+4x)} \\ -5x^2+0x-20 \\ \underline{-(-5x^2+0x-20)} \\ 0 \end{array}$$

So, $Q(x) = x-5$, $r(x) = 0$.

69. Area = length × width. So, solving for length, we see that length = Area ÷ width. So, in this case,

$$\text{length} = \frac{6x^4 - 8x^3 - 10x^2 + 12x - 16}{2x-4} = \frac{3x^4 - 4x^3 - 5x^2 + 6x - 8}{x-2}.$$

We compute this quotient using synthetic division:

$$\begin{array}{r|rrrrr} 2 & 3 & -4 & -5 & 6 & -8 \\ & & 6 & 4 & -2 & 8 \\ \hline & 3 & 2 & -1 & 4 & 0 \end{array}$$

Thus, the length (in terms of x) is $\boxed{3x^3 + 2x^2 - x + 4 \text{ feet}}$.

70. Let x = width (= length) of corner square.
Then, the dimensions of the box formed are:

$$\text{width} = 10-2x$$
$$\text{length} = 15-2x$$
$$\text{height} = x$$

Thus, the volume of the box is given by

$$V(x) = x(10-2x)(15-2x).$$

71.

$$\begin{array}{r|rrrrrr} -2 & 6 & 1 & 0 & -7 & 1 & -1 \\ & & -12 & 22 & -44 & 102 & -206 \\ \hline & 6 & -11 & 22 & -51 & 103 & -207 \end{array}$$

So, $f(-2) = -207$.

72.

$$\begin{array}{r|rrrrrr} 1 & 6 & 1 & 0 & -7 & 1 & -1 \\ & & 6 & 7 & 7 & 0 & 1 \\ \hline & 6 & 7 & 7 & 0 & 1 & 0 \end{array}$$

So, $f(1) = 0$.

73.

$$\begin{array}{r|rrrr} 1 & 1 & 2 & 0 & -3 \\ & & 1 & 3 & 3 \\ \hline & 1 & 3 & 3 & 0 \end{array}$$

So, $g(1) = 0$.

74.

$$\begin{array}{r|rrrr} -1 & 1 & 2 & 0 & -3 \\ & & -1 & -1 & 1 \\ \hline & 1 & 1 & -1 & -2 \end{array}$$

So, $g(-1) = -2$.

75.
$P(-3) = (-3)^3 - 5(-3)^2 + 4(-3) + 2 = -82$.
So, it is not a zero.

76. $P(-2) = P(2) = 0$
Yes, they are zeros.

77. $P(1) = 2(1)^4 - 2(1) = 0$
Yes, it is a zero.

78. $P(4) = (4)^4 - 2(4)^3 - 8(4) = 96$
No, it is not a zero.

79. $P(x) = x\left(x^3 - 6x^2 + 32\right)$
Observe that since -2 is a zero, synthetic division yields:

$$\begin{array}{r|rrrr} -2 & 1 & -6 & 0 & 32 \\ & & -2 & 16 & -32 \\ \hline & 1 & -8 & 16 & 0 \end{array}$$

So,

$$\begin{aligned} P(x) &= x(x+2)(x^2-8x+16) \\ &= x(x+2)(x-4)^2 \end{aligned}.$$

80. Observe that since 3 is a zero, synthetic division yields:

$$\begin{array}{r|rrrr} 3 & 1 & -7 & 0 & 36 \\ & & 3 & -12 & -36 \\ \hline & 1 & -4 & -12 & 0 \end{array}$$

So,

$$\begin{aligned} P(x) &= (x-3)(x^2-4x-12) \\ &= (x-3)(x-6)(x+2) \end{aligned}.$$

81. $P(x) = x^2\left(x^3 - x^2 - 8x + 12\right)$

We need to factor $x^3 - x^2 - 8x + 12$. To do so, we begin by applying the Rational Root Test:

Factors of 12: $\pm1, \pm2, \pm3, \pm4, \pm6, \pm12$
Factors of 1: ±1
Possible rational zeros:

$$\pm1, \pm2, \pm3, \pm4, \pm6, \pm12$$

Observe that both -3 (multiplicity 1) and 2 (multiplicity 2) are zeros. So,

$$P(x) = x^2(x+3)(x-2)^2.$$

82.

$$x^4 - 32x^2 - 144 = (x^2 - 36)(x^2 + 4)$$
$$= (x-6)(x+6)(x-2i)(x+2i)$$

83.
Number of sign variations for $P(x)$: 1

$$P(-x) = x^4 - 3x^3 - 16\text{, so}$$

Number of sign variations for $P(-x)$: 1

Positive Real Zeros	**Negative Real Zeros**
1	1

84.
Number of sign variations for $P(x)$: 1

$$P(-x) = -x^5 - 6x^3 + 4x - 2\text{, so}$$

Number of sign variations for $P(-x)$: 2

Positive Real Zeros	**Negative Real Zeros**
1	2
1	0

85.
Number of sign variations for $P(x)$: 5

$$P(-x) = -x^9 + 2x^7 + x^4 + 3x^3 - 2x - 1\text{, so}$$

Number of sign variations for $P(-x)$: 2

Positive Real Zeros	**Negative Real Zeros**
5	2
5	0
3	2
3	0
1	2
1	0

86.
Number of sign variations for $P(x)$: 3

$$P(-x) = -2x^5 + 4x^3 + 2x^2 - 7\text{, so}$$

Number of sign variations for $P(-x)$: 2

Positive Real Zeros	**Negative Real Zeros**
3	2
1	2
3	0
1	0

87.
Factors of 6: $\pm1, \pm2, \pm3, \pm6$
Factors of 1: ±1
Possible rational zeros: $\pm1, \pm2, \pm3, \pm6$

88.
Factors of -8: $\pm1, \pm2, \pm4, \pm8$
Factors of 1: ±1
Possible rational zeros: $\pm1, \pm2, \pm4, \pm8$

89.
Factors of 64:
$\pm1, \pm2, \pm4, \pm8, \pm16, \pm32, \pm64$
Factors of 2: $\pm1, \pm2$
Possible rational zeros:
$\pm1, \pm2, \pm4, \pm8, \pm16, \pm32, \pm64, \pm\frac{1}{2}$

90.
Factors of 2: $\pm1, \pm2$
Factors of -4: $\pm1, \pm2, \pm4$
Possible rational zeros: $\pm1, \pm2, \pm\frac{1}{2}, \pm\frac{1}{4}$

91.
Factors of 1: ±1
Factors of 2: $\pm1, \pm2$
Possible rational zeros: $\pm1, \pm\frac{1}{2}$

The only rational zero is $\frac{1}{2}$.

92.
Factors of 3: $\pm1, \pm3$
Factors of 12: $\pm1, \pm2, \pm3, \pm4, \pm6, \pm12$
Possible rational zeros:
$\pm1, \pm\frac{1}{2}, \pm\frac{1}{3}, \pm\frac{1}{4}, \pm\frac{1}{6}, \pm\frac{1}{12}, \pm3, \pm\frac{3}{2}, \pm\frac{3}{4}$

The rational zeros are $-\frac{3}{2}, \frac{1}{3}, \frac{1}{2}$.

93.
Factors of -16: $\pm1, \pm2, \pm4, \pm8, \pm16$
Factors of 1: ±1
Possible rational zeros:
$$\pm1, \pm2, \pm4, \pm8, \pm16$$

The rational zeros are 1, 2, 4, and -2.

94.
Factors of -2: $\pm1, \pm2$
Factors of 24:
$\pm1, \pm2, \pm3, \pm4, \pm6, \pm8, \pm12, \pm24$
Possible rational zeros:
$\pm1, \pm\frac{1}{2}, \pm\frac{1}{3}, \pm\frac{1}{4}, \pm\frac{1}{6}, \pm\frac{1}{8}, \pm\frac{1}{12}, \pm\frac{1}{24}, \pm2, \pm\frac{2}{3}$

There are no rational zeros. Indeed, see the graph to the right:

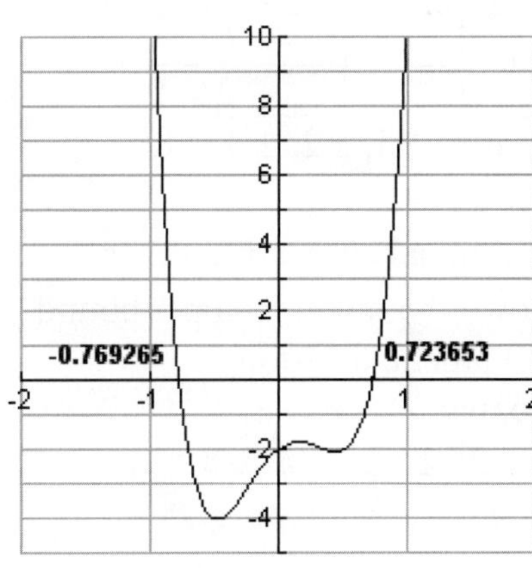

95.

a. Number of sign variations for $P(x)$: 1

$P(-x) = -x^3 - 3x - 5$, so

Number of sign variations for $P(-x)$: 0

Since $P(x)$ is degree 3, there are 3 zeros, the real ones of which are classified as:

Positive Real Zeros	Negative Real Zeros
1	0

b. Factors of -5: $\pm 1, \pm 5$

Factors of 1: ± 1

Possible rational zeros: $\pm 1, \pm 5$

c. -1 is a lower bound, 5 is an upper bound

d. There are no rational zeros.

e. Not possible to accurately factor since we do not have the zeros.

f.

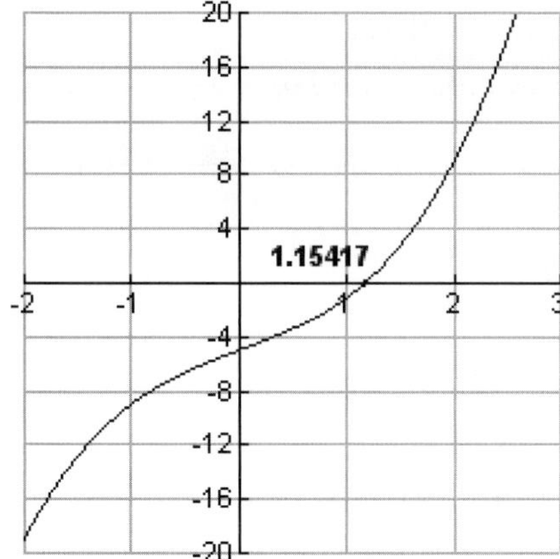

96.

a. Number of sign variations for $P(x)$: 1

$P(-x) = -x^3 + 3x^2 + 6x - 8$, so

Number of sign variations for $P(-x)$: 2

Positive Real Zeros	Negative Real Zeros
1	0
1	2

b. Factors of 8: $\pm 1, \pm 2, \pm 4, \pm 8$

Factors of 1: ± 1

Possible rational zeros: $\pm 1, \pm 2, \pm 4, \pm 8$

c. -8 is a lower bound, 4 is upper bound (check by synthetic division)

d. Observe that

$$\begin{array}{r|rrrr} 2 & 1 & 3 & -6 & -8 \\ & & 2 & 10 & 8 \\ \hline & 1 & 5 & 4 & 0 \end{array}$$

So,

$P(x) = (x-2)(x^2+5x+4) = (x-2)(x+4)(x+1)$

Hence, the zeros are -4, 1, and 2.

e. $P(x) = (x-2)(x+4)(x+1)$

f.

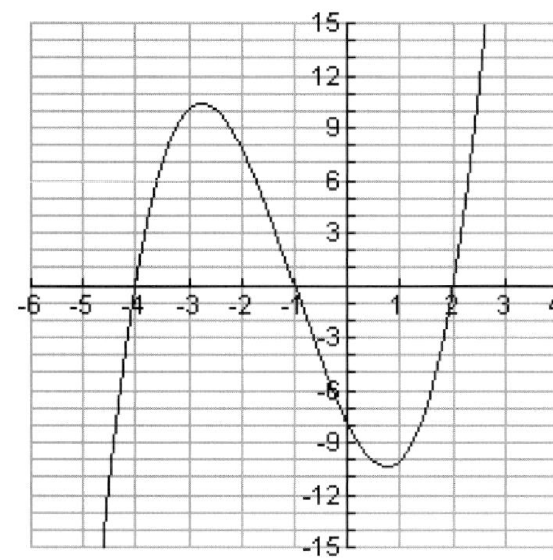

97.

a. Number of sign variations for $P(x)$: 3

$P(-x) = -x^3 - 9x^2 - 20x - 12$, so

Number of sign variations for $P(-x)$: 0

Positive Real Zeros	Negative Real Zeros
3	0
1	0

b. Factors of 12:

$\pm1, \pm2, \pm3, \pm4, \pm6, \pm12$

Factors of 1: ±1

Possible rational zeros:

$\pm1, \pm2, \pm3, \pm4, \pm6, \pm12$

c. -1 is a lower bound, 12 is upper bound (check by synthetic division)

d. Observe that

$$\begin{array}{r|rrrr} 1 & 1 & -9 & 20 & -12 \\ & & 1 & -8 & 12 \\ \hline & 1 & -8 & 12 & 0 \end{array}$$

So,

$P(x) = (x-1)(x^2 - 8x + 12) = (x-1)(x-6)(x-2)$

Hence, the zeros are 1, 2, and 6.

e. $P(x) = (x-1)(x-6)(x-2)$

f.

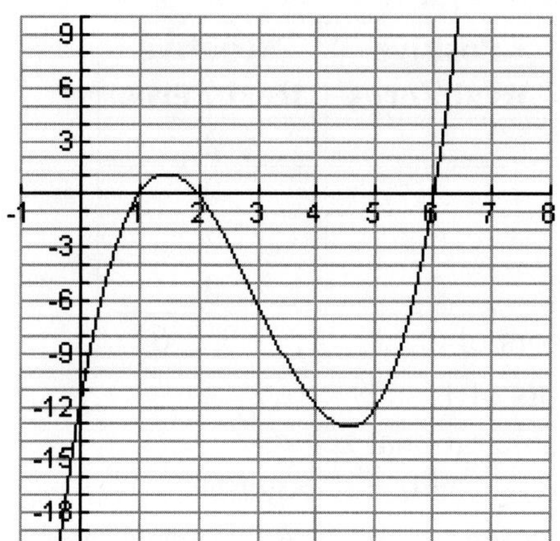

98.

a. Number of sign variations for $P(x)$: 2

$P(-x) = x^4 + x^3 - 7x^2 - x + 6$, so

Number of sign variations for $P(-x)$: 2

Positive Real Zeros	Negative Real Zeros
2	2
2	0
0	2
0	0

b. Factors of 6: $\pm 1, \pm 2, \pm 3, \pm 6$

Factors of 1: ± 1

Possible rational zeros:

$\pm 1, \pm 2, \pm 3, \pm 6$

c. -3 is a lower bound, 6 is upper bound (check by synthetic division)

d. Observe that

$$\begin{array}{r|rrrrr} 1 & 1 & -1 & -7 & 1 & 6 \\ & & 1 & 0 & -7 & -6 \\ \hline -2 & 1 & 0 & -7 & -6 & 0 \\ & & -2 & 4 & 6 & \\ \hline & 1 & -2 & -3 & 0 & \end{array}$$

So,

$$\begin{aligned} P(x) &= (x-1)(x+2)(x^2-2x-3) \\ &= (x-1)(x+2)(x-3)(x+1) \end{aligned}$$

Hence, the zeros are -2, -1, 1, and 3.

e. $P(x) = (x-1)(x+2)(x-3)(x+1)$

f.

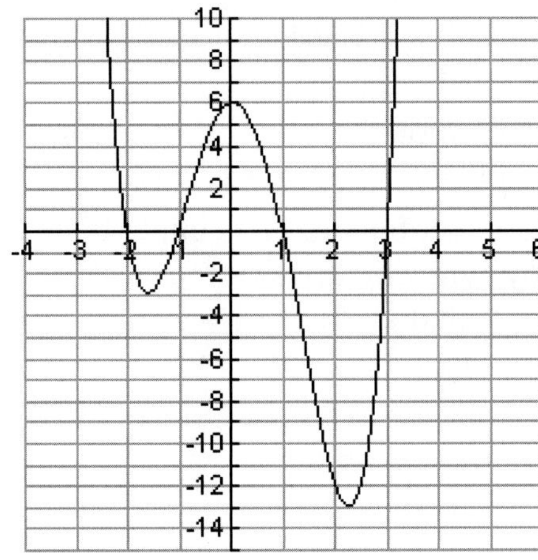

99.

a. Number of sign variations for $P(x)$: 2

$P(-x) = x^4 + 5x^3 - 10x^2 - 20x + 24$, so

Number of sign variations for $P(-x)$: 2

Positive Real Zeros	Negative Real Zeros
0	0
0	2
2	2
2	0

b. Factors of 24:

$\pm 1, \pm 2, \pm 3, \pm 4, \pm 6, \pm 8, \pm 12, \pm 24$

Factors of 1: ± 1

Possible rational zeros:

$\pm 1, \pm 2, \pm 3, \pm 4, \pm 6, \pm 8, \pm 12, \pm 24$

c. -3 is a lower bound, 8 is an upper bound (check by synthetic division)

d. Observe that

$$\begin{array}{r|rrrrr} 2 & 1 & -5 & -10 & 20 & 24 \\ & & 2 & -6 & -32 & -24 \\ \hline -1 & 1 & -3 & -16 & -12 & 0 \\ & & -1 & 4 & 12 & \\ \hline & 1 & -4 & -12 & 0 & \end{array}$$

So,

$$P(x) = (x-2)(x+1)(x^2 - 4x + 12)$$
$$= (x-2)(x+1)(x+2)(x-6)$$

Hence, the zeros are -2, -1, 1, and 6.

e. $P(x) = (x-2)(x+1)(x+2)(x-6)$

f.

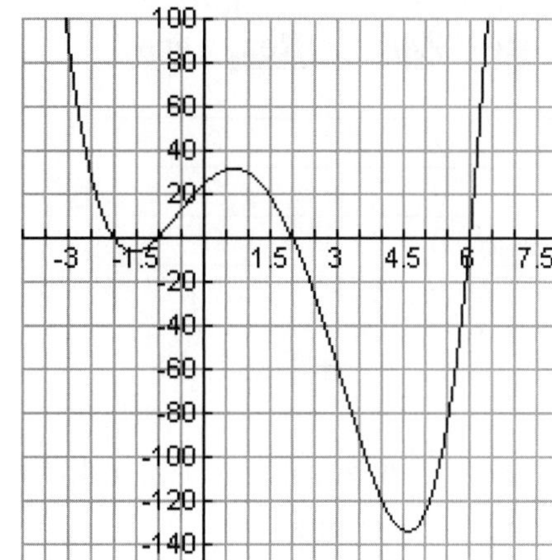

100.

a. Number of sign variations for $P(x)$: 2

$P(-x) = -x^5 + 3x^3 - 6x^2 - 8x$, so

Number of sign variations for $P(-x)$: 2

Since $P(x)$ is degree 5, there are 5 zeros.

Also note that 0 is a zero of P

Positive Real Zeros	Negative Real Zeros
2	2
2	0
0	2
0	0

b. $P(x) = x\left(x^4 - 3x^2 - 6x + 8\right)$

Factors of 8: $\pm 1, \pm 2, \pm 4, \pm 8$

Factors of 1: ± 1

Possible rational zeros: $\pm 1, \pm 2, \pm 4, \pm 8$

c. -2 is a lower bound.

d. The rational zeros are 0, 1, 2.

e. From **d.**, we know that $x(x-1)(x-2)$ divides $P(x)$ evenly. We use synthetic division using 1 and 2 to determine the quotient $P(x) \div x(x-1)(x-2)$:

$$\begin{array}{r|rrrrr} 2 & 1 & 0 & -3 & -6 & 8 \\ & & 2 & 4 & 2 & -8 \\ \hline 1 & 1 & 2 & 1 & -4 & 0 \\ & & 1 & 3 & 4 & \\ \hline & 1 & 3 & 4 & 0 & \end{array}$$

So, $P(x) = x(x-1)(x-2)(x^2 + 3x + 4)$

Now, solve $x^2 + 3x + 4 = 0$ using the quadratic formula: $x = \frac{-3 \pm \sqrt{9-16}}{2} = \frac{-3 \pm i\sqrt{7}}{2}$ So,

$P(x) = x(x-1)(x-2)(x - (\frac{-3+i\sqrt{7}}{2}))(x - (\frac{-3-i\sqrt{7}}{2}))$

f.

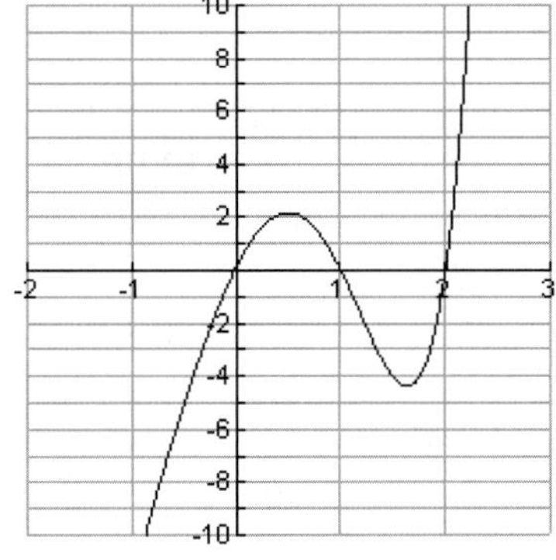

101. $x^2 + 25 = (x - 5i)(x + 5i)$

102. $x^2 + 16 = (x - 4i)(x + 4i)$

103. Note that the zeros are

$x^2 - 2x + 5 = 0 \Rightarrow$

$$x = \frac{2 \pm \sqrt{4 - 4(5)}}{2} = \frac{2 \pm 4i}{2} = 1 \pm 2i$$

So, $P(x) = (x - (1 - 2i))(x - (1 + 2i))$.

104. Note that the zeros are

$x^2 + 4x + 5 = 0 \Rightarrow$

$$x = \frac{-4 \pm \sqrt{16 - 4(5)}}{2} = \frac{-4 \pm 2i}{2} = -2 \pm i$$

So, $P(x) = (x - (-2 - i))(x - (-2 + i))$.

105. Since $-2i$ and $3 + i$ are zeros, so are their conjugates $2i$ and $3 - i$, respectively. Since $P(x)$ has degree 4, these are the only missing zeros.

106. Since $3i$ and $2 - i$ are zeros, so are their conjugates $-3i$ and $2 + i$, respectively. Since $P(x)$ has degree 4, these are the only missing zeros.

107. Since i is a zero, then so is its conjugate $-i$. Also, since $2 - i$ is a zero of multiplicity 2, then its conjugate $2 + i$ is also a zero of multiplicity 2. These are all of the zeros.

108. Since $2i$ is a zero, then so is its conjugate $-2i$. Also, since $1 - i$ is a zero of multiplicity 2, then its conjugate $1 + i$ is also a zero of multiplicity 2. These are all of the zeros.

109. Since i is a zero of $P(x)$, so is its conjugate $-i$. As such, $(x-i)(x+i)=x^2+1$ divides $P(x)$ evenly. Indeed, observe that $$\begin{array}{r} x^2-3x-4 \\ x^2+0x+1\overline{)x^4-3x^3-3x^2-3x-4} \\ \underline{-(x^4+0x^3+x^2)} \\ -3x^3-4x^2-3x \\ \underline{-(-3x^3+0x^2-3x)} \\ -4x^2+0x-4 \\ \underline{-(-4x^2+0x-4)} \\ 0 \end{array}$$ So, $$\begin{aligned} P(x)&=(x-i)(x+i)\left(x^2-3x-4\right) \\ &=(x-i)(x+i)(x-4)(x+1) \end{aligned}$$	**110.** Since $2-i$ is a zero of $P(x)$, so is its conjugate $2+i$. As such, $(x-(2+i))(x-(2-i))=x^2-4x+5$ divides $P(x)$ evenly. Indeed, observe that $$\begin{array}{r} x^2-4 \\ x^2-4x+5\overline{)x^4-4x^3+x^2+16x-20} \\ \underline{-(x^4-4x^3+5x^2)} \\ -4x^2+16x-20 \\ \underline{-(-4x^2+16x-20)} \\ 0 \end{array}$$ So, $P(x)=(x-(2+i))(x-(2-i))(x-2)(x+2)$.
111. Since $-3i$ is a zero of $P(x)$, so is its conjugate $3i$. As such, $(x-3i)(x+3i)=x^2+9$ divides $P(x)$ evenly. Indeed, observe that $$\begin{array}{r} x^2-2x+2 \\ x^2+0x+9\overline{)x^4-2x^3+11x^2-18x+18} \\ \underline{-(x^4+0x^3+9x^2)} \\ -2x^3+2x^2-18x \\ \underline{-(-2x^3+0x^2-18x)} \\ 2x^2+0x+18 \\ \underline{-(2x^2+0x+18)} \\ 0 \end{array}$$ Next, we find the roots of x^2-2x+2: $$x=\frac{2\pm\sqrt{4-4(2)}}{2}=1\pm i$$ So, $P(x)=(x-3i)(x+3i)(x-(1+i))(x-(1-i))$.	**112.** Since $1+i$ is a zero of $P(x)$, so is its conjugate $1-i$. As such, $(x-(1+i))(x-(1-i))=x^2-2x+2$ divides $P(x)$ evenly. Indeed, observe that $$\begin{array}{r} x^2+2x-3 \\ x^2-2x+2\overline{)x^4+0x^3-5x^2+10x-6} \\ \underline{-(x^4-2x^3+2x^2)} \\ 2x^3-7x^2+10x \\ \underline{-(2x^3-4x^2+4x)} \\ -3x^2+6x-6 \\ \underline{-(-3x^2+6x-6)} \\ 0 \end{array}$$ So, $P(x)=(x-(1+i))(x-(1-i))(x-1)(x+3)$.

113.
$$P(x) = x^4 - 81 = (x^2 - 9)(x^2 + 9)$$
$$= (x-3)(x+3)(x-3i)(x+3i)$$

114.
$$P(x) = x^3 - 6x^2 + 12x = x(x^2 - 6x + 12)$$
We need to find the roots of $x^2 - 6x + 12$:
$$x = \frac{6 \pm \sqrt{36 - 4(12)}}{2} = 3 \pm i\sqrt{3}$$
So, $P(x) = x(x - (3 + i\sqrt{3}))(x - (3 - i\sqrt{3}))$.

115. $x^3 - x^2 + 4x - 4 = x^2(x-1) + 4(x-1) = (x^2 + 4)(x-1) = (x-2i)(x+2i)(x-1)$

116. $P(x) = x^4 - 5x^3 + 12x^2 - 2x - 20$

Factors of 20: $\pm 1, \pm 2, \pm 4, \pm 5, \pm 10, \pm 20$

Factors of 1: ± 1

Possible rational zeros:
$$\pm 1, \pm 2, \pm 4, \pm 5, \pm 10, \pm 20$$

$$\begin{array}{r|rrrrr} -1 & 1 & -5 & 12 & -2 & -20 \\ & & -1 & 6 & -18 & 20 \\ \hline 2 & 1 & -6 & 18 & -20 & 0 \\ & & 2 & -8 & 20 & \\ \hline & 1 & -4 & 10 & 0 & \end{array}$$

So, $P(x) = (x+1)(x-2)(x^2 - 4x + 10)$.

Next, we find the roots of $x^2 - 4x + 10$:
$$x = \frac{4 \pm \sqrt{16 - 4(10)}}{2} = 2 \pm i\sqrt{6}$$
So,
$$P(x) = (x+1)(x-2)(x - (2 + i\sqrt{6}))(x - (2 - i\sqrt{6})).$$

117. Vertical Asymptote: $x = -2$

Horizontal Asymptote:

Since the degree of the numerator equals the degree of the denominator,

$y = -1$ is the HA.

118. Vertical Asymptote: $x = 1$

Horizontal Asymptote:

Since the degree of the numerator is less than degree of the denominator,

$y = 0$ is the HA.

119. Vertical Asymptote: $x = -1$

Slant Asymptote: $y = 4x - 4$

To find the slant asymptote, we use long division:

$$\begin{array}{r} 4x - 4 \\ x+1 \overline{\smash{)}\, 4x^2 + 0x + 0} \\ \underline{-(4x^2 + 4x)} \\ -4x + 0 \\ \underline{-(-4x - 4)} \\ 4 \end{array}$$

120. Vertical Asymptote: None since $x^2 + 9 \neq 0$

Horizontal Asymptote:

Since the degree of the numerator equals the degree of the denominator,

$y = 3$ is the HA.

121. No vertical asymptotes,
Horizontal asymptote: $y = 2$

122. vertical asymptotes: $x = -5$
Horizontal asymptote: None
Slant asymptote: $y = -2x + 13$, as seen by the result of this synthetic division:

$$\begin{array}{r|rrr} -5 & -2 & 3 & 5 \\ & & 10 & -65 \\ \hline & -2 & 13 & -60 \end{array}$$

123.

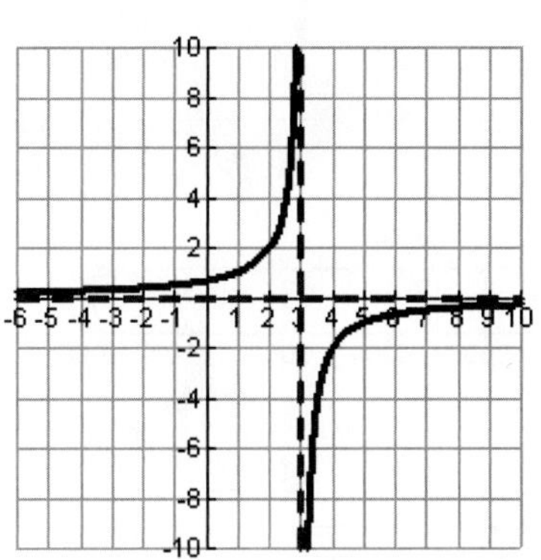

124.

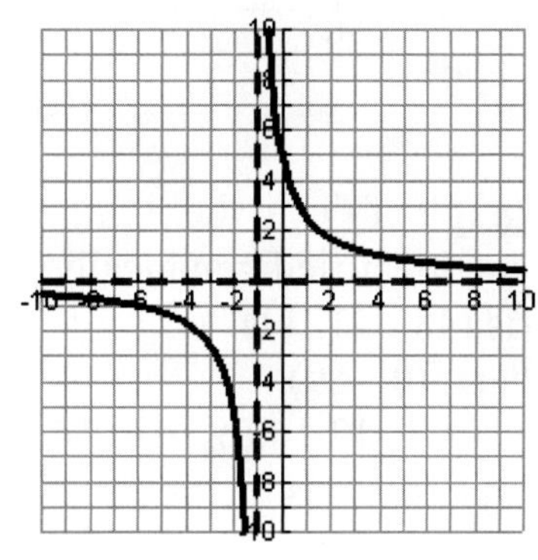

125.

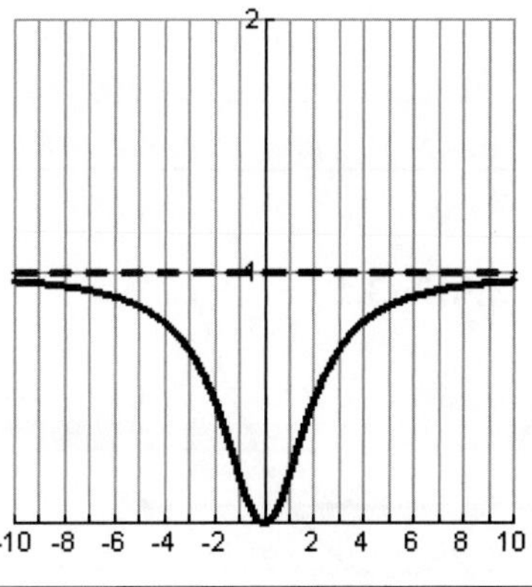

126.

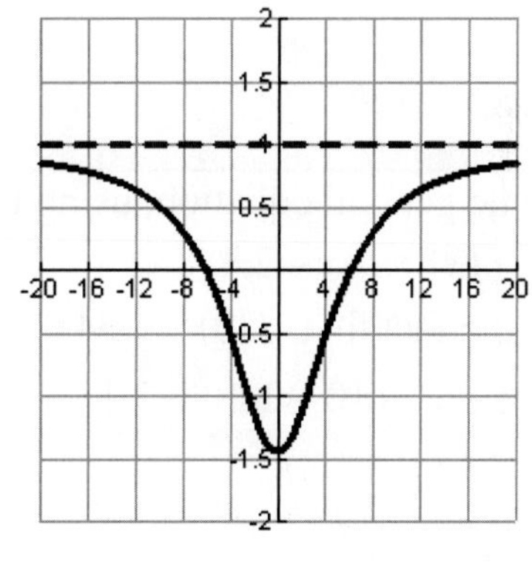

127. Note the hole at $x = -7$.

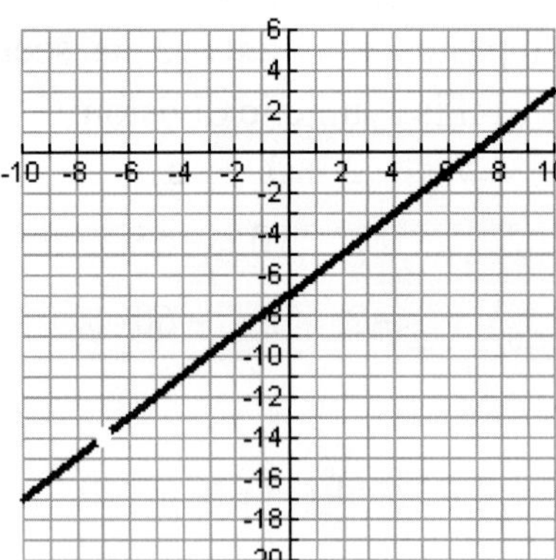

128. Note the hole at $x = -\frac{1}{2}$.

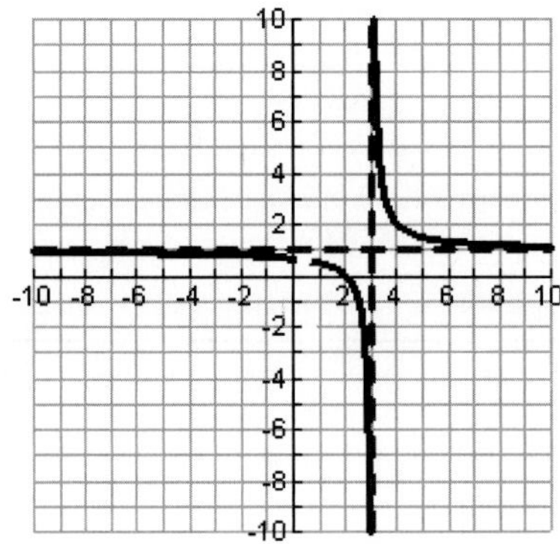

129. The graph of *f* is to the right. The following are identified graphically:
a. Vertex is (480, -1211)
b. *y*-intercept: (0, -59)
c. *x*-intercepts: (-12.14,0), (972.14,0)
d. axis of symmetry is the line $x = 480$.

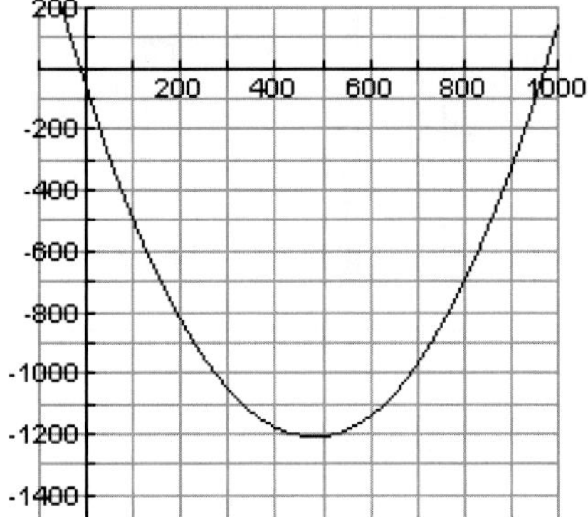

130. a. The general equation, using the vertex, is $f(x) = a(x-2.4)^2 - 3.1$. To find *a*, we use the fact that $f(0) = 5.54$:

$$5.54 = a(0-2.4)^2 - 3.1$$

$$8.64 = 5.76a$$

$$a = 1.5$$

So, $f(x) = 1.5(x-2.4)^2 - 3.1$.

b.

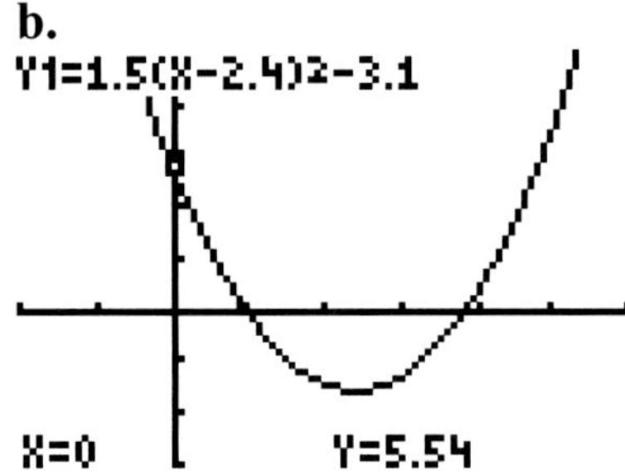

c. Yes, they agree.

131. The graph of *f* is as follows:

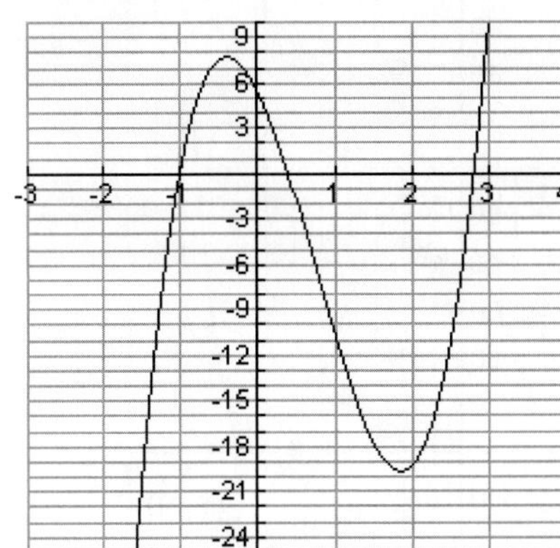

x-intercepts: (-1,0), (0.4, 0), (2.8,0)
zeros: -1, 0.4, 2.8, each with multiplicity 1

132. The graph of *f* is as follows:

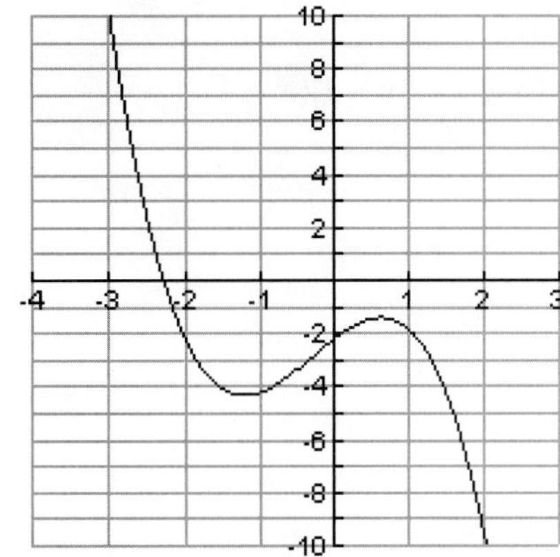

x-intercept: (-2.27,0)
zeros: -2.27 (multiplicity 1)

133. The graph is to the right.
It is a linear function, as the following long division verifies:

$$\begin{array}{r} 5x-4 \\ 3x^2-7x+2\overline{)15x^3-47x^2+38x-8} \\ \underline{-(15x^3-35x^3+10x^2)} \\ -12x^2+28x-8 \\ \underline{-(-12x^2+28x-8)} \\ 0 \end{array}$$

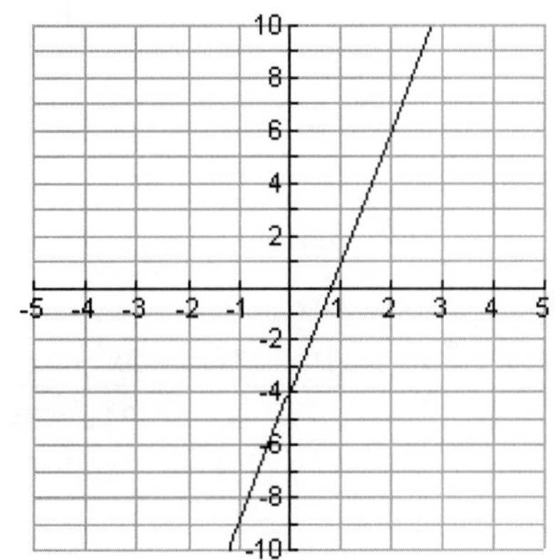

134. The graph is to the right:
It is a quadratic function, as the following long division verifies:

$$\begin{array}{r} -4x^2+2x+5 \\ x-3\overline{)-4x^3+14x^2-x-15} \\ \underline{-(-4x^3+12x^2)} \\ 2x^2-x \\ \underline{-(2x^2-6x)} \\ 5x-15 \\ \underline{-(5x-15)} \\ 0 \end{array}$$

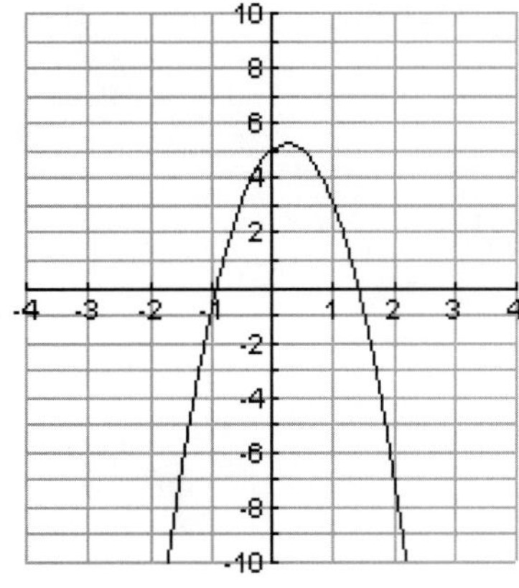

135. The graph is to the right.

a. The zeros are -2 (multiplicity 2), 3, and 4.

b. $P(x) = (x+2)^2(x-3)(x-4)$

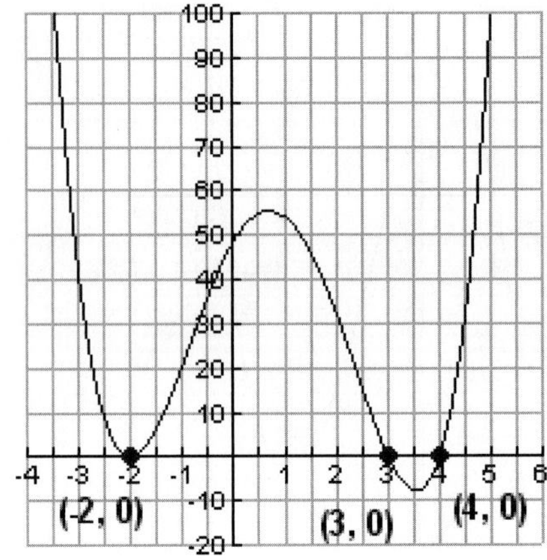

136. The graph is to the right.

a. The zeros are -1 (multiplicity 2) and $\frac{2}{5}$.

As such, we know that

$(x+1)^2(5x-2) = 5x^3 + 8x^2 + x - 2$ divides $P(x)$ evenly. So,

$$\begin{array}{r|l} & -x^2 - 2x - 3 \\ 5x^3 + 8x^2 + x - 2 & -5x^5 - 18x^4 - 32x^3 - 24x^2 + x + 6 \\ & -(-5x^5 - 8x^4 \quad - x^3 \quad + 2x^2) \\ & -10x^4 - 31x^3 - 26x^2 + x \\ & -(-10x^4 - 16x^3 - 2x^2 + 4x) \\ & -15x^3 - 24x^2 - 3x + 6 \\ & -(-15x^3 - 24x^2 - 3x + 6) \\ & 0 \end{array}$$

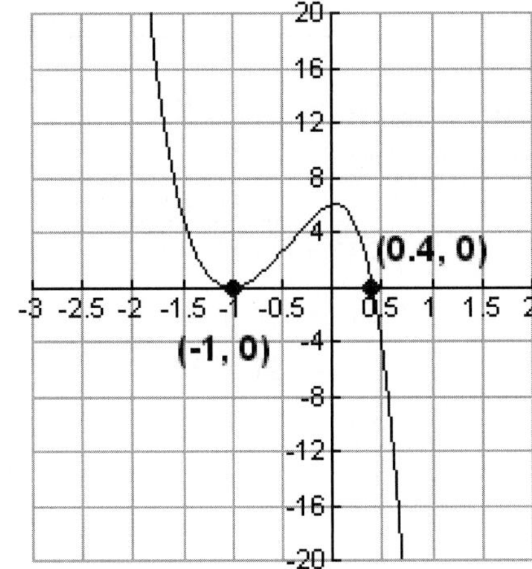

So, $P(x) = -(x+1)^2(5x-2)(x^2+2x+3)$.

137. The graph is to the right.
Note that the only real zero is $\frac{7}{2}$, and so we have:

$$\begin{array}{r|rrrr} \frac{7}{2} & 2 & 1 & -2 & -91 \\ & & 7 & 28 & 91 \\ \hline & 2 & 8 & 26 & 0 \end{array}$$

So,

$$P(x) = \left(x - \tfrac{7}{2}\right)(2x^2 + 8x + 26)$$
$$= (2x - 7)(x^2 + 4x + 13)$$

Need to find the roots of $x^2 + 4x + 13$:

$$x = \frac{-4 \pm \sqrt{16 - 4(13)}}{2} = -2 \pm 3i$$

Thus, $P(x) = (2x - 7)(x + 2 - 3i)(x + 2 + 3i)$.

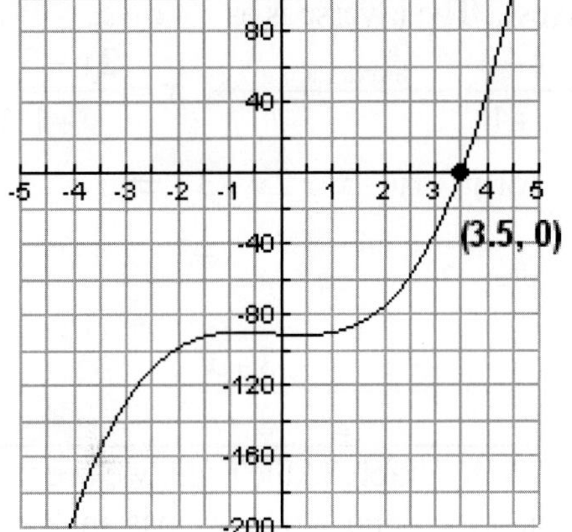

138. The graph is to the right.
Note that the real zeros are -5 and $\frac{3}{2}$, and so $(x+5)(x - \frac{3}{2}) = x^2 + \frac{7}{2}x - \frac{15}{2}$ divides $P(x)$ evenly. So, we have:

$$\begin{array}{r} -2x^2 + 12x - 20 \\ x^2 + \frac{7}{2}x - \frac{15}{2}\overline{\big)-2x^4 + 5x^3 + 37x^2 - 160x + 150} \\ \underline{-(-2x^4 - 7x^3 + 15x^2)} \\ 12x^3 + 22x^2 - 160x \\ \underline{-(12x^3 + 42x^2 - 90x)} \\ -20x^2 - 70x + 150 \\ \underline{-(-20x^2 - 70x + 150)} \\ 0 \end{array}$$

So, $P(x) = -(2x - 3)(x + 5)(x^2 - 6x + 10)$.
Next, we find the roots of $x^2 - 6x + 10$:

$$x = \frac{6 \pm \sqrt{36 - 4(10)}}{2} = 3 \pm i.$$

So,
$P(x) = -(2x - 3)(x + 5)(x - 3 - i)(x - 3 + i)$.

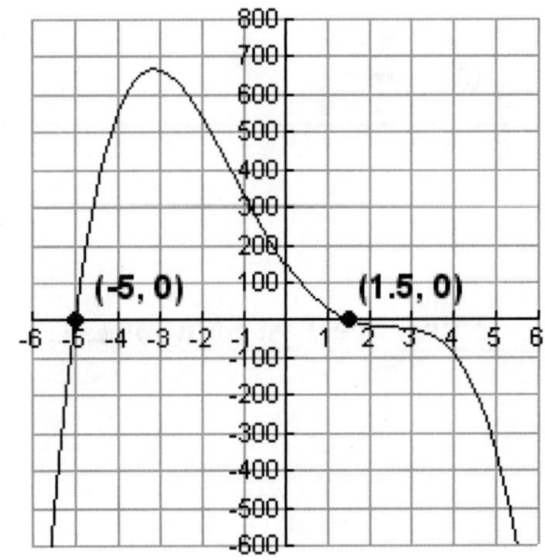

139. From the graph, to the right, we see that indeed f is one-to-one and hence has an inverse (which is also graphed on the same set of axes). The inverse is as follows:

$$y=\frac{2x-3}{x+1}\xrightarrow{\text{Switch } x \text{ and } y} x=\frac{2y-3}{y+1}$$

Solve for y: $xy+x=2y-3 \Rightarrow$

$$xy-2y=-3-x \Rightarrow y=\frac{-3-x}{x-2}$$

Thus, $f^{-1}(x)=\dfrac{-3-x}{x-2}=\dfrac{3+x}{2-x}$.

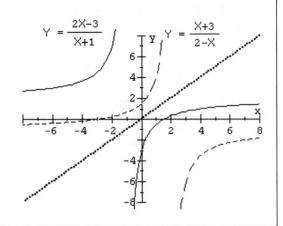

140. From the graph, to the right, we see that indeed f is one-to-one and hence has an inverse (which is also graphed on the same set of axes). The inverse is as follows:

$$y=\frac{4x+7}{x-2}\xrightarrow{\text{Switch } x \text{ and } y} x=\frac{4y+7}{y-2}$$

Solve for y: $xy-2x=4y+7 \Rightarrow$

$$xy-4y=2x+7 \Rightarrow y=\frac{2x+7}{x-4}$$

Thus, $f^{-1}(x)=\dfrac{2x+7}{x-4}$.

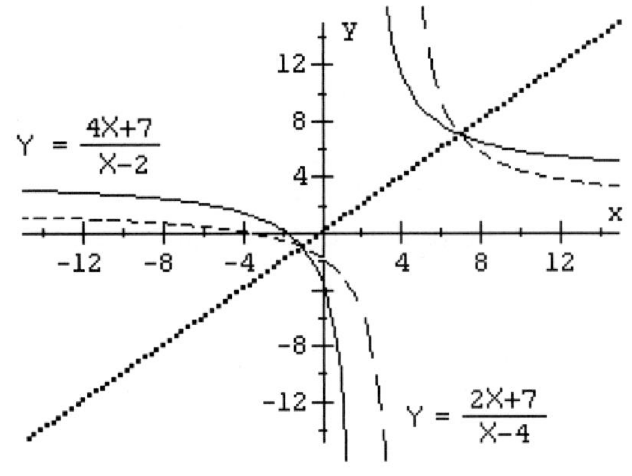

Chapter 2 Practice Test Solutions ---

1.

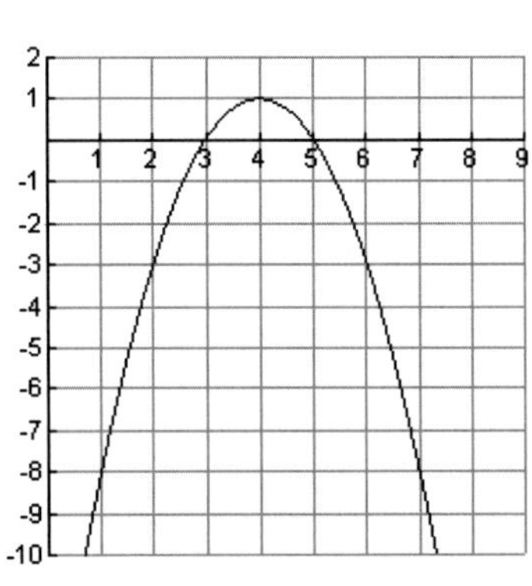

2.

$$\begin{aligned} y &= -\left(x^2-4x\right)-1 \\ &= -(x^2-4x+4)-1+4 \\ &= -(x-2)^2+3 \end{aligned}$$

3.

$$y = -\tfrac{1}{2}\left(x^2 - 6x\right) - 4$$
$$= -\tfrac{1}{2}(x^2 - 6x + 9) - 4 + \tfrac{9}{2}$$
$$= -\tfrac{1}{2}(x-3)^2 + \tfrac{1}{2}$$

So, the vertex is $(3, \tfrac{1}{2})$.

4. Since the vertex is $(-3,-1)$, the function has the form $y = a(x+3)^2 - 1$. To find a, use the fact that the point $(-4,1)$ is on the graph:

$$1 = a(-4+3)^2 - 1$$
$$1 = a - 1$$
$$2 = a$$

So, the function is $y = 2(x+3)^2 - 1$.

5. $f(x) = x(x-2)^3(x-1)^2$

6. $f(x) = x\left(x^3 + 6x^2 - 7\right)$

a. Zeros: Certainly, 0 is a zero. To find the remaining three, we first try to apply the Rational Root Test:

Factors of 7: $\pm 1, \pm 7$

Factors of 1: ± 1

Possible rational zeros: $\pm 1, \pm 7$

Observe that 1 is a zero. So, using synthetic division, we compute the quotient $\left(x^3 + 6x^2 - 7\right) \div (x-1)$:

$$\begin{array}{r|rrrr} 1 & 1 & 6 & 0 & -7 \\ & & 1 & 7 & 7 \\ \hline & 1 & 7 & 7 & 0 \end{array}$$

So, $f(x) = x(x-1)(x^2 + 7x + 7)$. Now, we solve $x^2 + 7x + 7 = 0$ using the quadratic formula: $x = \frac{-7 \pm \sqrt{49 - 4(7)}}{2} = \frac{-7 \pm \sqrt{21}}{2}$

So, there are four distinct x-intercepts.

b. Crosses at all four zeros.

c. y-intercept: $f(0) = 0$, so $(0,0)$

d. Long-term behavior: Behaves like $y = x^4$. Even degree and leading coefficient positive, so graph rises without bound to the left and right.

e.

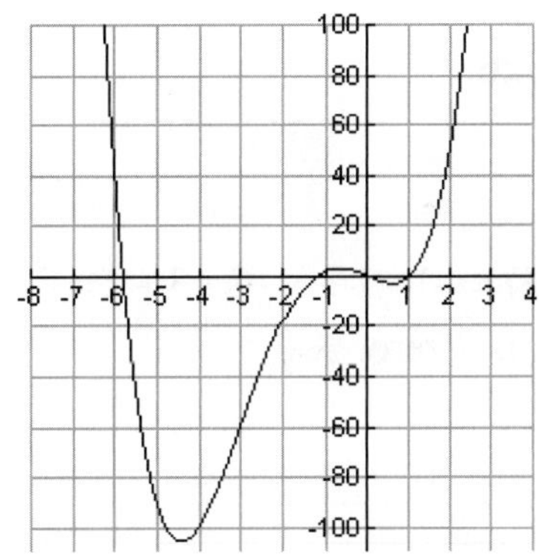

7.

$$
\begin{array}{r}
-2x^2-2x-\frac{11}{2} \\
2x^2-3x+1 \overline{\smash{)}\ -4x^4+2x^3-7x^2+5x-2} \\
\underline{-(-4x^4+6x^3-2x^2)} \\
-4x^3-5x^2+5x \\
\underline{-(-4x^3+6x^2-2x)} \\
-11x^2+7x-2 \\
\underline{-(-11x^2+\frac{33}{2}x-\frac{11}{2})} \\
-\frac{19}{2}x+\frac{7}{2}
\end{array}
$$

So, $\boxed{Q(x)=-2x^2-2x-\frac{11}{2},\ \ r(x)=-\frac{19}{2}x+\frac{7}{2}}$.

8.

$$
\begin{array}{r|rrrrrr}
\underline{-2} & 17 & 0 & -4 & 0 & 2 & -10 \\
 & & -34 & 68 & -128 & 256 & -516 \\
\hline
 & 17 & -34 & 64 & -128 & 258 & -526
\end{array}
$$

So,
$Q(x)=17x^4-34x^3+64x^2-128x+258$,
$r(x)=-526$

9.

$$
\begin{array}{r|rrrrr}
3 & 1 & 1 & -13 & -1 & 12 \\
 & & 3 & 12 & -3 & -12 \\
\hline
 & 1 & 4 & -1 & -4 & 0
\end{array}
$$

Yes, $x-3$ is a factor of $x^4+x^3-13x^2-x+12$.

10. $P(-1)=-1-2+5-7+3+2=0$.
So, yes it is a zero.

11.

$$
\begin{array}{r|rrrr}
7 & 1 & -6 & -9 & 14 \\
 & & 7 & 7 & -14 \\
\hline
 & 1 & 1 & -2 & 0
\end{array}
$$

So,
$P(x)=(x-7)(x^2+x-2)=(x-7)(x+2)(x-1)$

12. Since $3i$ is a zero, its conjugate $-3i$ must also be a zero. So, we know that $(x-3i)(x+3i)=x^2+9$ divides $P(x)$ evenly. This gives us the following, after long division:

$$\begin{array}{r} x^2-3x+10 \\ x^2+9\overline{\big)\,x^4-3x^3+19x^2-27x+90} \\ \underline{-(x^4+0x^3+9x^2)} \\ -3x^3+10x^2-27x \\ \underline{-(-3x^3+0x^2-27x)} \\ 10x^2+0x+90 \\ \underline{-(10x^2+0x+90)} \\ 0 \end{array}$$

So, $P(x)=(x-3i)(x+3i)(x^2-3x+10)=(x-3i)(x+3i)(x-5)(x+2)$

13. Yes, a complex zero cannot be an x-intercept.

14. Number of sign variations for $P(x)$: 4

$P(-x)=-3x^5+2x^4+3x^3+2x^2+x+1$, so

Number of sign variations for $P(-x)$: 1

Since $P(x)$ is degree 5, there are 5 zeros that we classify as:

Positive Real Zeros	Negative Real Zeros	Imaginary Zeros
4	1	0
2	1	2
0	1	4

15.

Factors of 12: $\pm1, \pm2, \pm3, \pm4, \pm6, \pm12$ Factors of 3: $\pm1, \pm3$

Possible rational zeros: $\pm1, \pm2, \pm3, \pm4, \pm6, \pm12, \pm\frac{1}{3}, \pm\frac{2}{3}, \pm\frac{4}{3}$

16. $P(x) = -x^3 + 4x = -x(x-2)(x+2)$

Zeros are $0, \pm 2$.

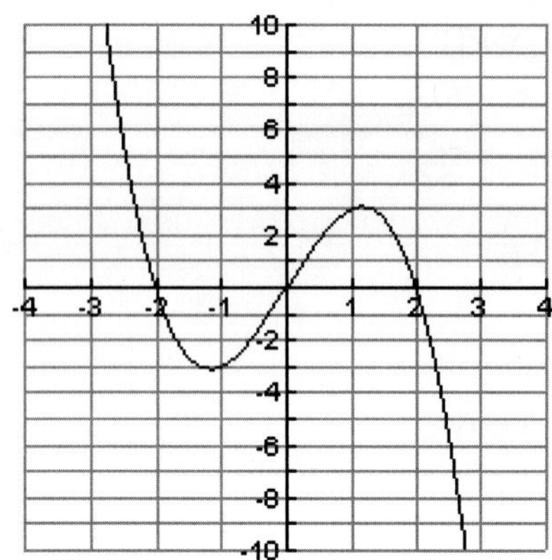

17.

$$\begin{aligned} P(x) &= 2x^3 - 3x^2 + 8x - 12 \\ &= x^2(2x-3) + 4(2x-3) \\ &= (x^2+4)(2x-3) \\ &= (x+2i)(x-2i)(2x-3) \end{aligned}$$

The only real zero is $3/2$, the other two are complex, namely $\pm 2i$.

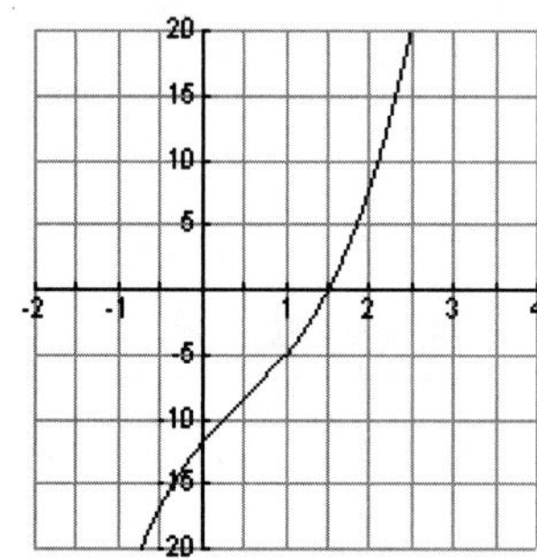

18. Factors of 9: $\pm 1, \pm 3, \pm 9$

Factors of 1: ± 1

Possible rational zeros: $\pm 1, \pm 3, \pm 9$

Observe that

$$\begin{array}{r|rrrrr} 3 & 1 & -6 & 10 & -6 & 9 \\ & & 3 & -9 & 3 & -9 \\ \hline 3 & 1 & -3 & 1 & -3 & 0 \\ & & 3 & 0 & 3 & \\ \hline & 1 & 0 & 1 & 0 & \end{array}$$

So, $P(x) = (x-3)^2(x^2+1)$.

So, the only real zero is 3 (multiplicity 2).

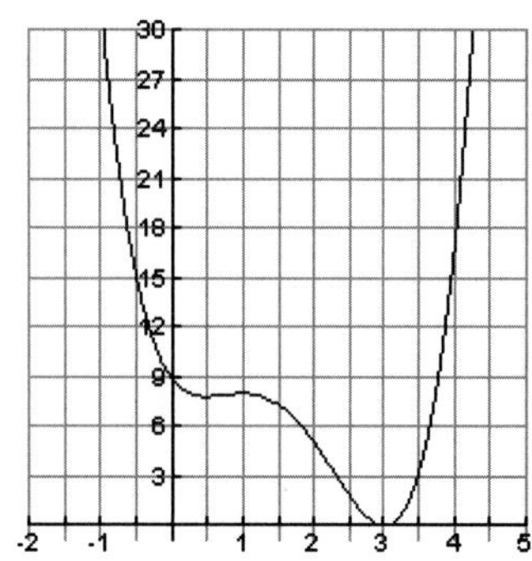

19. The polynomial can have degree 3 since there are 2 turning points. See the graph below:

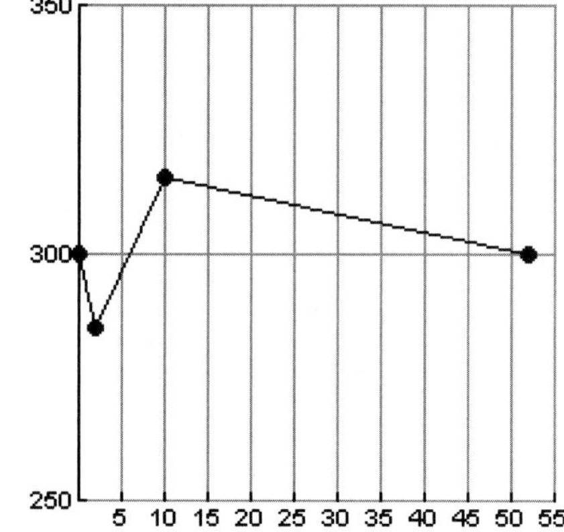

20. We need to solve

$$x^3 - 13x^2 + 47x - 35 = 0.$$

From the graph to the right, we observe that the zeros are 1, 5, and 7. So, you will break even at any of these values.

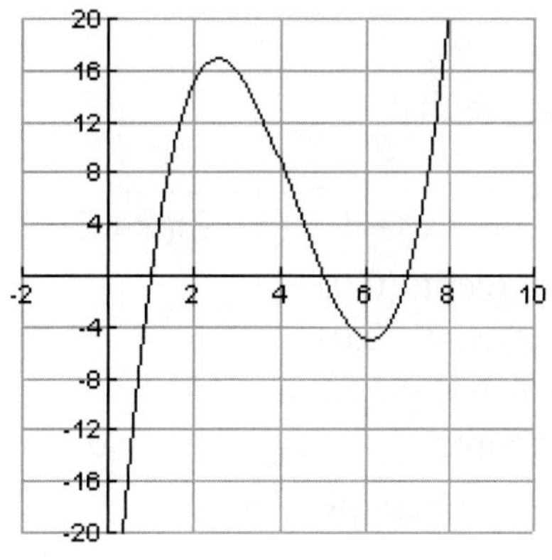

21. Since there are 2 turning points (seen on the graph to the right), the polynomial must be at least degree 3.

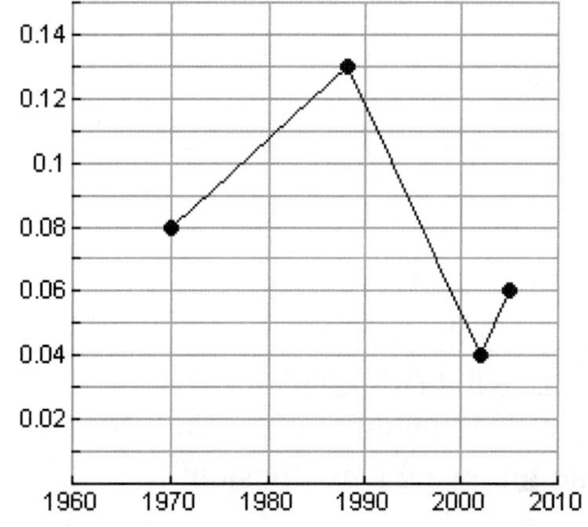

22. x-intercept: $(\frac{9}{2}, 0)$
y-intercept: (0,-3)
Vertical Asymptote: $x = -3$
Horizontal Asymptote: $y = 2$
Slant Asymptote: None

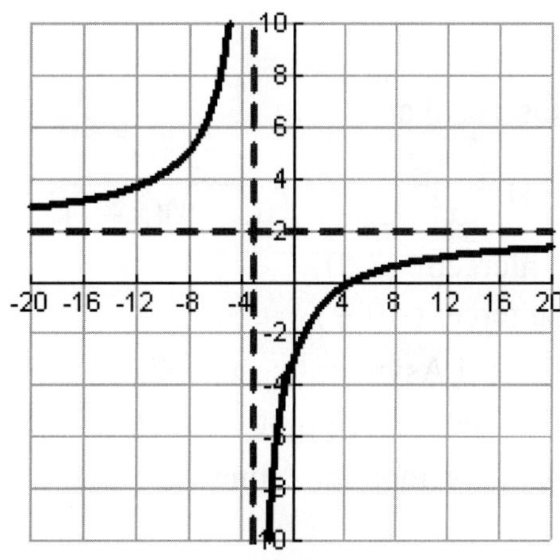

23. Observe that

$$g(x)=\frac{x}{x^2-4}=\frac{x}{(x-2)(x+2)}.$$

<u>x-intercept</u>: (0,0)
<u>y-intercept</u>: (0,0)
<u>Vertical Asymptote</u>: $x=\pm 2$
<u>Horizontal Asymptote</u>: $y=0$
<u>Slant Asymptote</u>: None

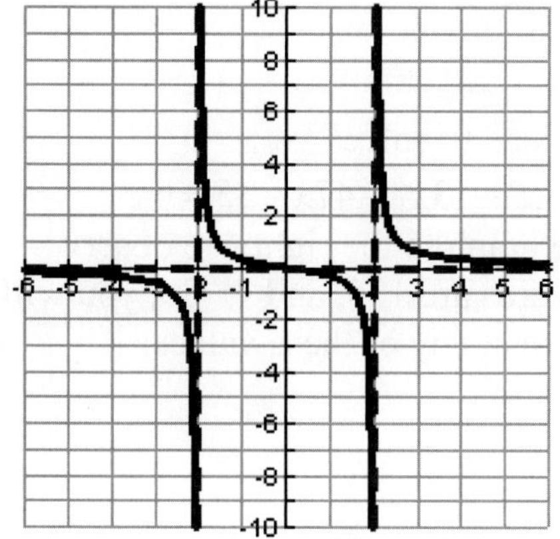

24. Observe that

$$h(x)=\frac{3x^3-3}{x^2-4}=\frac{3(x^3-1)}{(x-2)(x+2)}.$$

<u>x-intercept</u>: (1,0)
<u>y-intercept</u>: $(0,\frac{3}{4})$
<u>Vertical Asymptote</u>: $x=\pm 2$
<u>Horizontal Asymptote</u>: None
<u>Slant Asymptote</u>: $y=3x$, as seen by the following long division:

$$x^2+0x-4\overline{\big)3x^3+0x^2+0x-3}\quad 3x$$

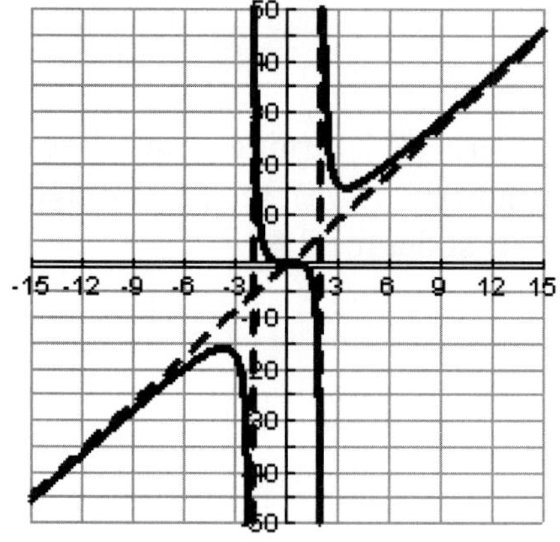

25. Observe that

$$F(x)=\frac{x-3}{x^2-2x-8}=\frac{x-3}{(x-4)(x+2)}.$$

<u>x-intercept</u>: (3,0)
<u>y-intercept</u>: $(0,\frac{3}{8})$
<u>Vertical Asymptote</u>: $x=-2,\ x=4$
<u>Horizontal Asymptote</u>: $y=0$
<u>Slant Asymptote</u>: None

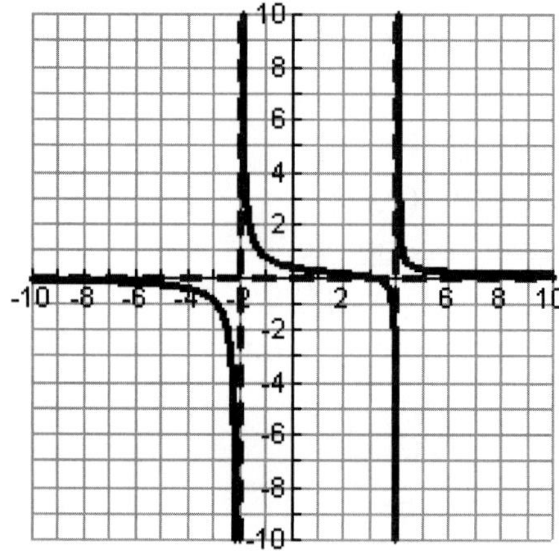

26.

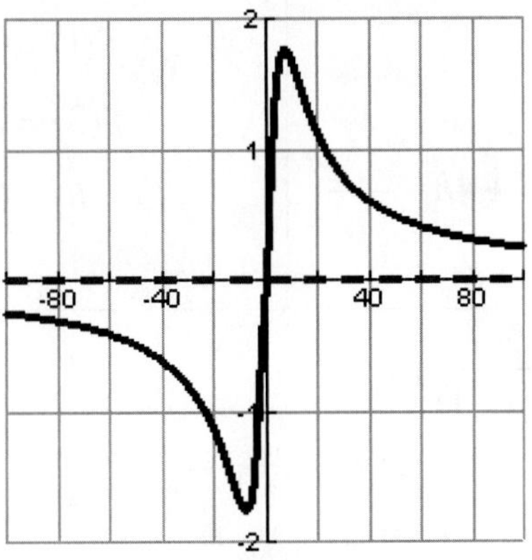

27. a & b. The equation we obtain using the calculator is $y = x^2 - 3x - 7.99$. We complete the square to put this into standard form: $y = (x-1.5)^2 - 10.24$

c. x-intercepts: (-1.7,0) and (4.7,0)

d. The graph is to the right. Yes, they agree.

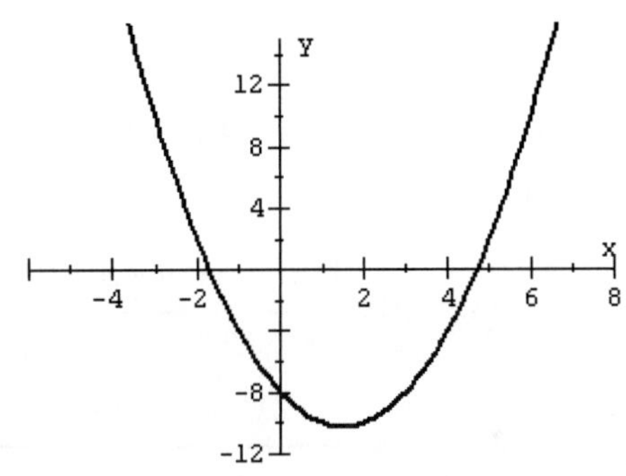

28. Observe that

$$f(x) = \frac{x(2x-3) + x^2 - 3x}{x^2 - 3x} = \frac{3\cancel{x}(x-2)}{\cancel{x}(x-3)}$$

So, there is an open hole at $x = 0$. Also, we have

<u>x-intercept</u>: (2,0)

<u>y-intercept</u>: none

<u>Vertical Asymptote</u>: $x = 3$

<u>Horizontal Asymptote</u>: $y = 3$

The graph of f is to the right. Yes, they agree.

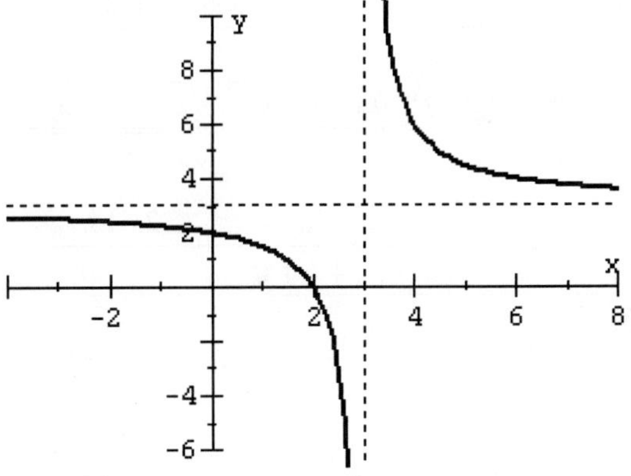

Chapter 2 Cumulative Test --

1.

$$f(2) = 8 - \tfrac{1}{\sqrt{4}} = \tfrac{15}{2}$$
$$f(-1) = -4 - \tfrac{1}{\sqrt{1}} = -5$$
$$f(1+h) = 4(1+h) - \tfrac{1}{\sqrt{1+h+2}} = 4 + 4h - \tfrac{1}{\sqrt{h+3}}$$
$$f(-x) = -4x - \tfrac{1}{\sqrt{-x+2}} = -4x - \tfrac{1}{\sqrt{2-x}}$$

2.

$$f(1) = 0^4 - \sqrt{5} = -\sqrt{5}$$
$$f(3) = 2^4 - \sqrt{9} = 13$$
$$f(x+h) = (x+h-1)^4 - \sqrt{2x+2h+3}$$

3. Note that $f(x) = \dfrac{3x-5}{-(x+2)(x-1)}$.

$$f(-3) = \tfrac{-14}{-(-1)(-4)} = \tfrac{7}{2}$$
$$f(0) = \tfrac{-5}{-(2)(-1)} = -\tfrac{5}{2}$$
$$f(4) = \tfrac{7}{-(6)(3)} = -\tfrac{7}{18}$$

$f(1)$ is undefined

4.

$$\frac{f(x+h)-f(x)}{h} = \frac{\left[4(x+h)^3 - 3(x+h)^2 + 5\right] - \left[4x^3 - 3x^2 + 5\right]}{h}$$
$$= \frac{4\left[x^3 + 3x^2h + 3xh^2 + h^3\right] - 3x^2 - 6xh - 3h^2 + 5 - 4x^3 + 3x^2 + 5}{h}$$
$$= \frac{h\left(12x^2 + 12xh + 4h^2 - 6x - 3h\right)}{h} = \boxed{12x^2 + 12xh + 4h^2 - 6x - 3h}$$

5.

$$\frac{f(x+h)-f(x)}{h} = \frac{\left[\sqrt{x+h} - \frac{1}{(x+h)^2}\right] - \left[\sqrt{x} - \frac{1}{x^2}\right]}{h} = \frac{\sqrt{x+h}-\sqrt{x}}{h} - \frac{\frac{1}{(x+h)^2} - \frac{1}{x^2}}{h}$$
$$= \frac{\sqrt{x+h}-\sqrt{x}}{h} \cdot \frac{\sqrt{x+h}+\sqrt{x}}{\sqrt{x+h}+\sqrt{x}} - \frac{x^2 - (x+h)^2}{hx^2(x+h)^2} = \frac{x+h-x}{h\left(\sqrt{x+h}+\sqrt{x}\right)} - \frac{x^2 - x^2 - 2xh - h^2}{hx^2(x+h)^2}$$
$$= \frac{1}{\sqrt{x+h}+\sqrt{x}} + \frac{2x+h}{x^2(x+h)^2}$$

6.

$$f(-5) = 0, \quad f(0) = 0,$$
$$f(3) = 3(3) + 3^2 = 18$$
$$f(4) = 3(4) + 4^2 = 28$$
$$f(5) = \left|2(5) - 5^3\right| = 115$$

7.

Domain	$(-\infty,10)\cup(10,\infty)$
Range	$[0,\infty)$
Increasing	$(3,8)$
Decreasing	$(-\infty,3)\cup(10,\infty)$
Constant	$(8,10)$

The graph of f is:

8.

Domain	$(-\infty,14]$
Range	$[-6,\infty)$
Increasing	$(-5,10)$
Decreasing	$(-\infty,-5)\cup(10,14)$
Constant	None

The graph of f is:

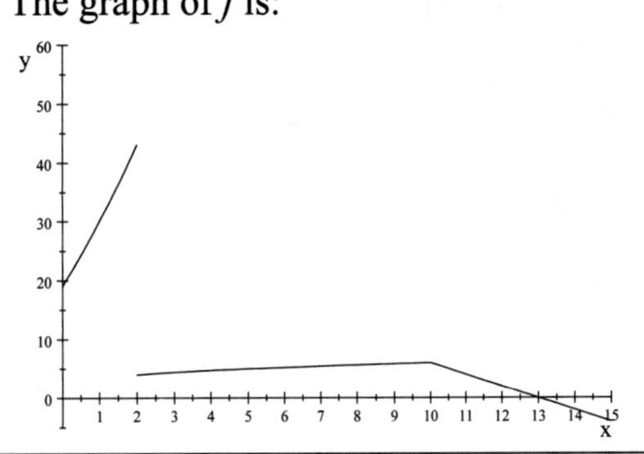

9.

$$\frac{y(9)-y(5)}{9-5}=\frac{\frac{2(9)}{9^2+3}-\frac{2(5)}{5^2+3}}{4}=\frac{\frac{18}{84}-\frac{10}{28}}{4}=\boxed{-\frac{1}{28}}$$

10. Need $6x-7\geq 0$, so that $x\geq\frac{7}{6}$. So, the domain is $\left[\frac{7}{6},\infty\right)$.

11. Observe that

$$g(-x)=\sqrt{-x+10}\neq g(x)$$
$$-g(-x)=-\sqrt{-x+10}\neq g(x)$$

So, neither.

12. Shift the graph of $y=x^2$ left 1 unit, then reflet over the x-axis, and then move up 2 units.

13. Translate the graph of $y=\sqrt{x}$ to the right 1 unit and then up 3 units.

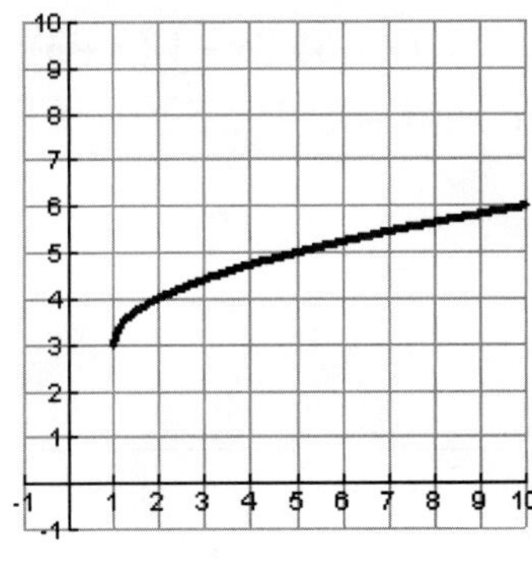

14.

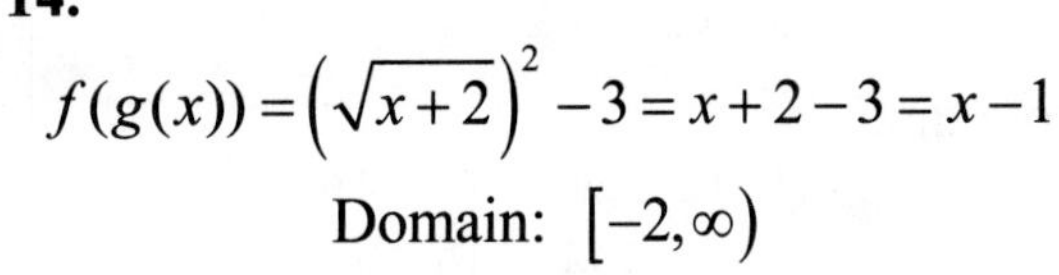

$$f(g(x))=\left(\sqrt{x+2}\right)^2-3=x+2-3=x-1$$

Domain: $[-2,\infty)$

15.

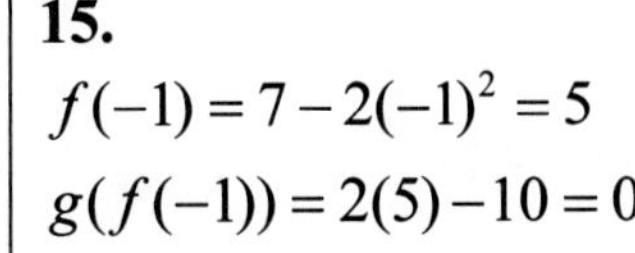

$$f(-1)=7-2(-1)^2=5$$
$$g(f(-1))=2(5)-10=0$$

16. To find the inverse, switch the x and y and solve for y:

$$x=(y-4)^2+2 \quad \Rightarrow \quad x-2=(y-4)^2 \quad \Rightarrow \quad y=4\pm\sqrt{x-2}$$

Since we are assuming $y\geq 4$, use the positive root above. So, $f^{-1}(x)=4+\sqrt{x-2},\ x\geq 2$.

17. Since the vertex is (-2,3), the equation so far is $f(x)=a(x+2)^2+3$. Use (-1,4) to find a:

$$4=a(-1+2)^2+3=a+3 \Rightarrow a=1$$

So, $f(x)=(x+2)^2+3$.

18. $f(x)=-3.7x^3(x+4)$

So, the zeros are 0 (multiplicity 3) and -4 (multiplicity 1)

19. Observe that

$$\begin{array}{r|l} & 4x^2+4x+1 \\ \hline -5x+3 & -20x^3-8x^2+7x-5 \\ & \underline{-(-20x^3+12x^2)} \\ & -20x^2+7x \\ & \underline{-(-20x^2+12x)} \\ & -5x-5 \\ & \underline{-(-5x+3)} \\ & -8 \end{array}$$

So, $Q(x)=4x^2+4x+1,\ r(x)=-8$

20. Observe that

$$\begin{array}{r|rrrr} 3 & 2 & 3 & -11 & 6 \\ & & 6 & 27 & 48 \\ \hline & 2 & 9 & 16 & 54 \end{array}$$

So, $Q(x)=2x^2+9x+16,\ r(x)=54$.

21. $P(x)=12x^3+29x^2+7x-6$

Factors of -6: $\pm1, \pm2, \pm3, \pm6$

Factors of 12: $\pm1, \pm2, \pm3, \pm4, \pm6, \pm12$

Possible rational zeros:

$$\pm1, \pm2, \pm3, \pm6, \pm\tfrac{1}{2}, \pm\tfrac{1}{3}, \pm\tfrac{1}{4}, \pm\tfrac{1}{6}, \pm\tfrac{1}{12}, \pm\tfrac{2}{3}, \pm\tfrac{3}{2}, \pm\tfrac{3}{4}, \pm\tfrac{1}{4}$$

Note that

$$\begin{array}{r|rrrr} -2 & 12 & 29 & 7 & -6 \\ & & -24 & -10 & 6 \\ \hline & 12 & 5 & -3 & 0 \end{array}$$

So, $P(x)=(x+2)(12x^2+5x-3)$
$=(x+2)(4x+3)(3x-1)$

So, the zeros are $-2, -\tfrac{3}{4}, \tfrac{1}{3}$.

22. Since 5 is a zero, we know that $x-5$ divides $P(x)$ evenly. So,

$$\begin{array}{r|rrrr} 5 & 2 & -3 & -32 & -15 \\ & & 10 & 35 & 15 \\ \hline & 2 & 7 & 3 & 0 \end{array}$$

So,

$$P(x)=(x-5)(2x^2+7x+3) = (x-5)(2x+1)(x+3)$$

So, the zeros are $-3, -\tfrac{1}{2}, 5$.

23. Possible rational zeros are:

$\pm 1, \pm 2, \pm 4, \pm 8$

Using synthetic division yields

$$\begin{array}{r|rrrr} -1 & 1 & -5 & 2 & 8 \\ & & -1 & 6 & -8 \\ \hline 2 & 1 & -6 & 8 & \\ & & 2 & -8 & \\ \hline & 1 & -4 & & \end{array}$$

So, $P(x) = (x+1)(x-2)(x-4)$.

24. Possible rational zeros are:

$\pm 1, \pm 2, \pm 3, \pm 4, \pm 6, \pm 12$

Using synthetic division yields

$$\begin{array}{r|rrrrrr} 1 & 1 & 7 & 15 & 5 & -16 & -12 \\ & & 1 & 8 & 23 & 28 & 12 \\ \hline -1 & 1 & 8 & 23 & 28 & 12 & \\ & & -1 & -7 & -16 & -12 & \\ \hline -3 & 1 & 7 & 16 & 12 & & \\ & & -3 & -12 & -12 & & \\ \hline & 1 & 4 & 4 & & & \end{array}$$

So,

$$P(x) = (x-1)(x+1)(x+3)\left(x^2+4x+4\right)$$
$$= (x-1)(x+1)(x+3)(x+2)^2$$

25. Vertical asymptotes: $x = \pm 2$ Horizontal asymptote: $y = 0$

26. Observe that

$$f(x) = \frac{x\left(2x^2-x-1\right)}{(x-1)(x+1)} = \frac{x(2x+1)\cancel{(x-1)}}{\cancel{(x-1)}(x+1)}$$

Open hole: $(1, \frac{3}{2})$

x-intercepts: $(0,0)$, $(-\frac{1}{2},0)$

y-intercept: $(0,0)$

Vertical Asymptote: $x = -1$

Horizontal Asymptote: None

Slant Asymptote: $y = 2x-1$

$$\begin{array}{r|rrr} -1 & 2 & 1 & 0 \\ & & -2 & 1 \\ \hline & 2 & -1 & 1 \end{array}$$

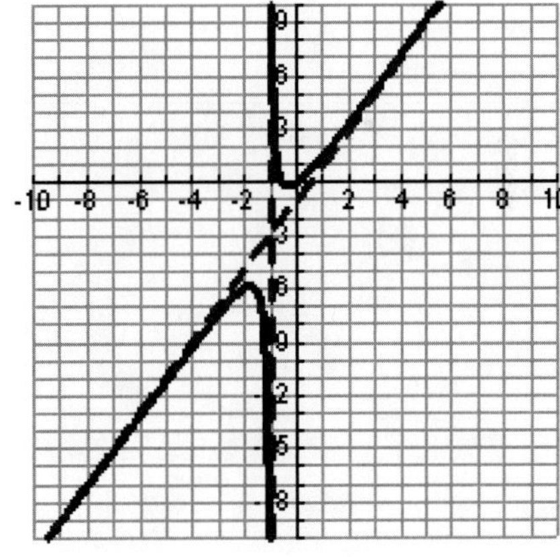

27. Observe that

$$f(x) = \frac{3(x+1)}{x(2x-3)}$$

x-intercepts: (-1,0)

y-intercept: none

Vertical Asymptote: $x = 0$, $x = \frac{3}{2}$

Horizontal Asymptote: $y = 0$ The graph of f is to the right. Yes, they agree.

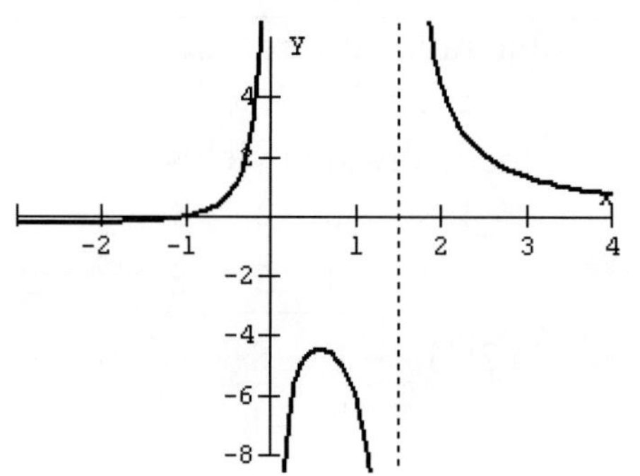

28. Observe that

$$f(x) = \frac{24x^2 - 6x - 18x^2 - 6x}{(3x+1)(4x-1)} = \frac{6x(x-2)}{(3x+1)(4x-1)}$$

x-intercepts: (2,0), (0,0)

y-intercept: (0,0)

Vertical Asymptote: $x = -\frac{1}{3}$, $x = \frac{1}{4}$

Horizontal Asymptote: $y = 0.5$

The graph of f is to the right. Yes, they agree.

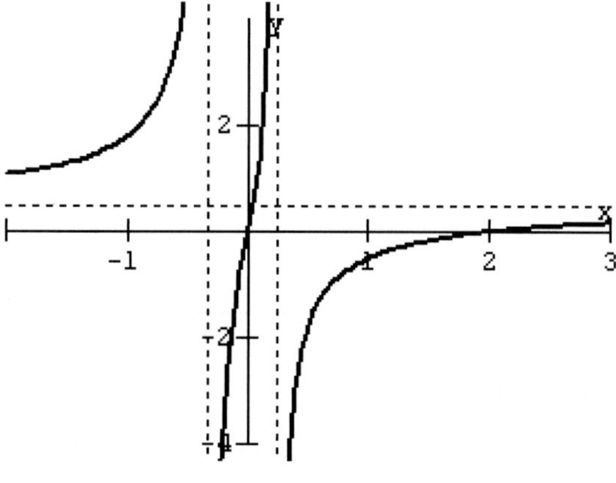

CHAPTER 3

Section 3.1 Solutions ---

1. $\frac{1}{5^2}=\frac{1}{25}$	**2.** $\frac{1}{4^3}=\frac{1}{64}$
3. $8^{2/3}=\left(\sqrt[3]{8}\right)^2=2^2=4$	**4.** $27^{2/3}=\left(\sqrt[3]{27}\right)^2=3^2=9$
5. $\left(\frac{1}{9}\right)^{-3/2}=9^{3/2}=\left(\sqrt{9}\right)^3=3^3=27$	**6.** $\left(\frac{1}{16}\right)^{-3/2}=16^{3/2}=\left(\sqrt{16}\right)^3=4^3=64$
7. 9.7385	**8.** 22.2740
9. 7.3891	**10.** 1.6487
11. 0.0432	**12.** 0.2431
13. $f(3)=3^3=27$	**14.** $h(1)=10^2=100$
15. $g(-1)=\left(\frac{1}{16}\right)^{-1}=16$	**16.** $f(-2)=3^{-2}=\frac{1}{3^2}=\frac{1}{9}$
17. $g\left(-\frac{1}{2}\right)=\left(\frac{1}{16}\right)^{-1/2}=16^{1/2}=4$	**18.** $g\left(-\frac{3}{2}\right)=\left(\frac{1}{16}\right)^{-3/2}=16^{3/2}=\left(\sqrt{16}\right)^3=4^3=64$
19. $f(e)=3^e\approx 19.81$	**20.** $g(\pi)=\left(\frac{1}{16}\right)^{\pi}\approx 0.0002$
21. f $y=\frac{1}{5}\left(5^x\right)$ Rising a factor of 5 slower than 5^x	**22. c** $y=5\left(5^{-x}\right)$ Falling a factor of 5 quicker than 5^{-x}, and y-intercept (0,5)
23. e $y=-\left(5^x\right)$ The reflection of 5^x over the x-axis	**24. d** $y=-\left(5^{-x}\right)$ The reflection of 5^{-x} over the x-axis.
25. b $y=-\left(5^{-x}\right)+1$ The reflection of 5^{-x} over the x-axis and then shifted up 1 unit.	**26. a** $y=5^x-1$ The graph of 5^x shifted down 1 unit.

27. y-intercept: $f(0) = 6^0 = 1$, so (0,1).
HA: $y = 0$
Domain: $(-\infty, \infty)$ Range: $(0, \infty)$
Other points: (-1,1/6), (1,6)

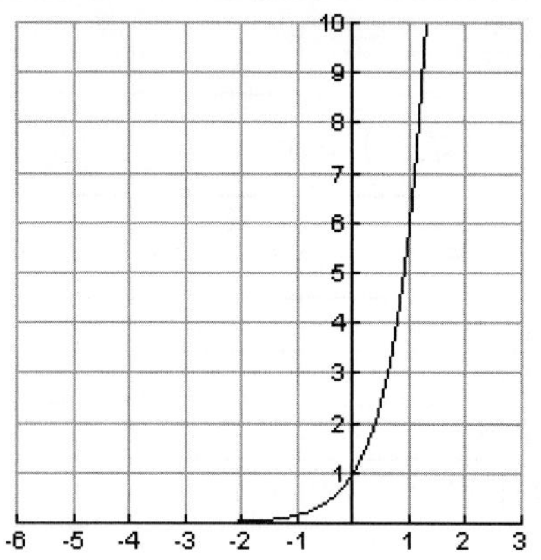

28. y-intercept: $f(0) = 7^0 = 1$, so (0,1).
HA: $y = 0$
Domain: $(-\infty, \infty)$ Range: $(0, \infty)$
Other points: (0,1), (1,7)

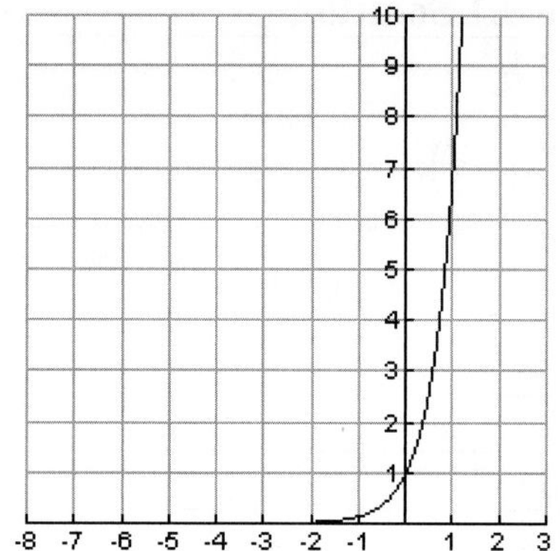

29. y-intercept: $f(0) = 10^0 = 1$, so (0,1).
HA: $y = 0$
Reflect graph of $y = 10^x$ over y-axis.
Domain: $(-\infty, \infty)$ Range: $(0, \infty)$
Other points: (1,0.1), (-1,10)

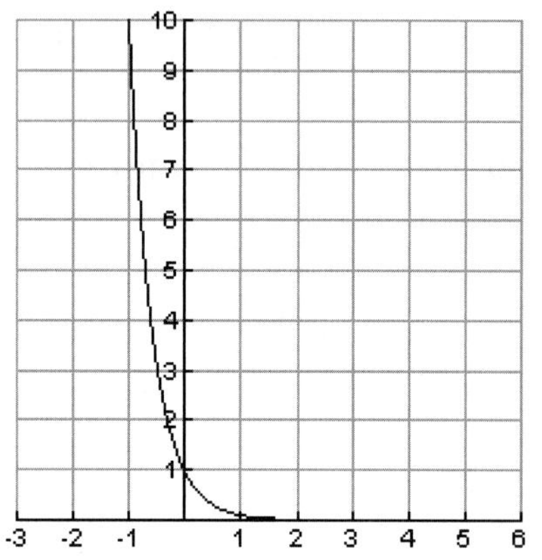

30. y-intercept: $f(0) = 4^0 = 1$, so (0,1).
HA: $y = 0$
Reflect graph of $y = 4^x$ over y-axis.
Domain: $(-\infty, \infty)$ Range: $(0, \infty)$
Other points: (0,1), (-1,4)

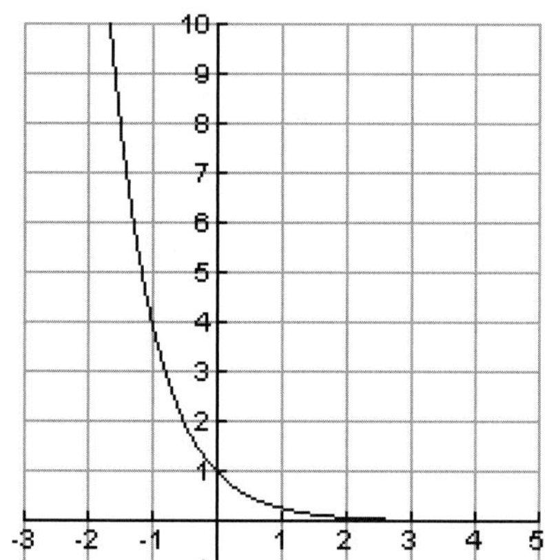

31. <u>y-intercept</u>: $f(0) = e^0 = 1$, so (0,1).

<u>HA</u>: $y = 0$

<u>Domain</u>: $(-\infty, \infty)$ <u>Range</u>: $(0, \infty)$

<u>Other points</u>: (1, e), (2, e^2)

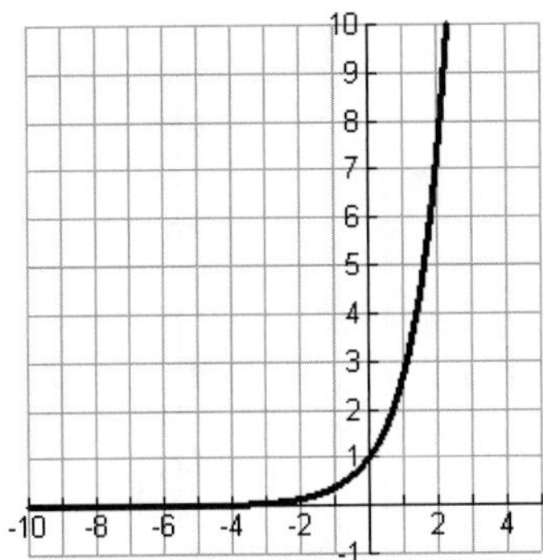

32. <u>y-intercept</u>: $f(0) = -e^0 = -1$, so (0,-1).

<u>HA</u>: $y = 0$

Reflect graph of $y = e^x$ over y-axis, then x-axis.

<u>Domain</u>: $(-\infty, \infty)$ <u>Range</u>: $(-\infty, 0)$

<u>Other points</u>: $(1, -\frac{1}{e})$, (-1, -e)

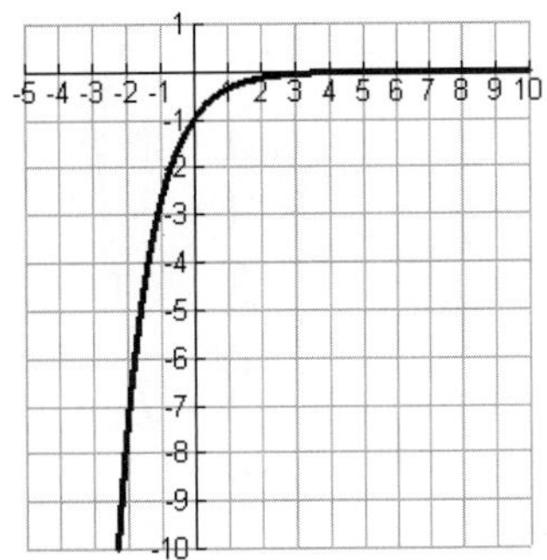

33. <u>y-intercept</u>: $f(0) = e^0 = 1$, so (0,1).

<u>HA</u>: $y = 0$

Reflect graph of $y = e^x$ over y-axis.

<u>Domain</u>: $(-\infty, \infty)$ <u>Range</u>: $(0, \infty)$

<u>Other points</u>: (1,1/e), (-1, e)

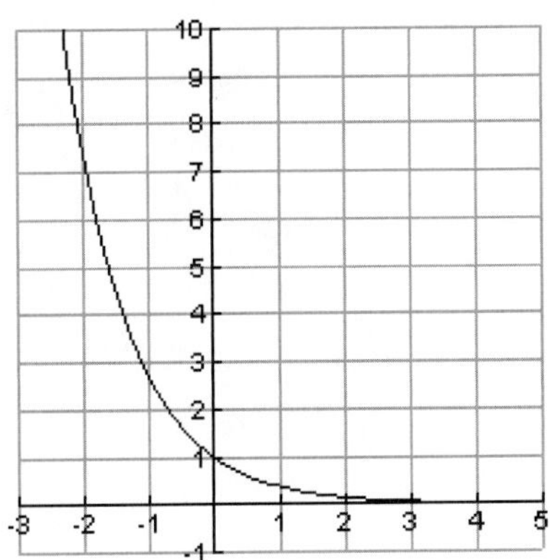

34. <u>y-intercept</u>: $f(0) = -e^0 = -1$, so $(0, -1)$.

<u>HA</u>: $y = 0$

Reflect graph of e^x over the x-axis.

<u>Domain</u>: $(-\infty, \infty)$ <u>Range</u>: $(0, \infty)$

<u>Other points</u>: (0,1), (1, $-e$)

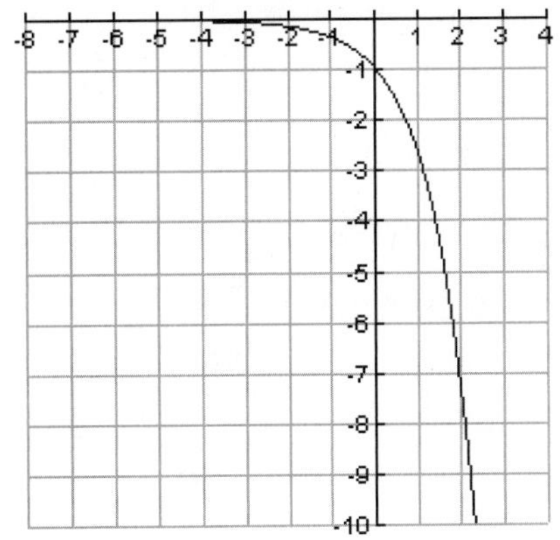

35. <u>y-intercept</u>: $f(0)=2^0-1=0$, so (0,0).

<u>HA</u>: $y=-1$

Shift graph of $y=2^x$ down 1 unit

<u>Domain</u>: $(-\infty,\infty)$ <u>Range</u>: $(-1,\infty)$

<u>Other points</u>: (2,3), (1,1)

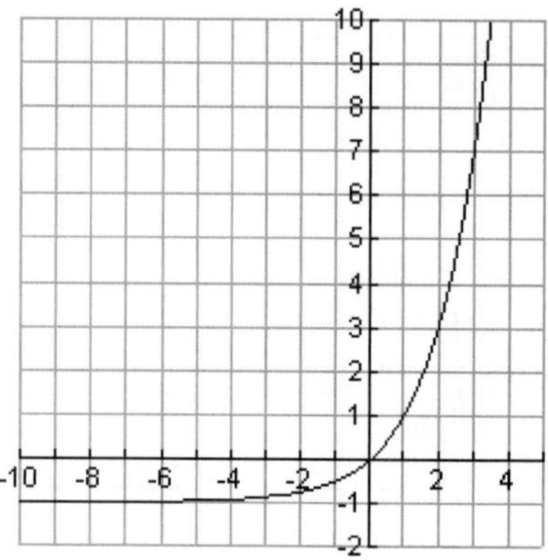

36. <u>y-intercept</u>: $f(0)=3^0-1=0$, so (0,0).

<u>HA</u>: $y=-1$

Shift graph of $y=3^x$ down 1 unit

<u>Domain</u>: $(-\infty,\infty)$ <u>Range</u>: $(-1,\infty)$

<u>Other points</u>: (2,8), (1,2)

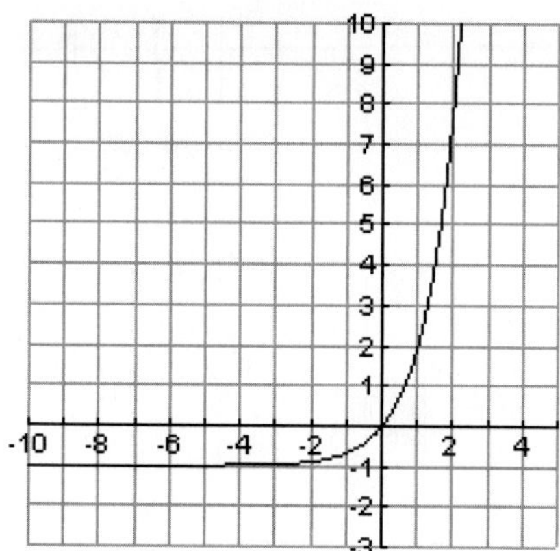

37. <u>y-intercept</u>: $f(0)=2-e^0=1$, so $(0,1)$.

<u>HA</u>: $y=2$

Reflect graph of e^x over the x-axis, then shift up 2 units.

<u>Domain</u>: $(-\infty,\infty)$ <u>Range</u>: $(-\infty,2)$

<u>Other points</u>: (1, 2-e), (-1, 2-1/e)

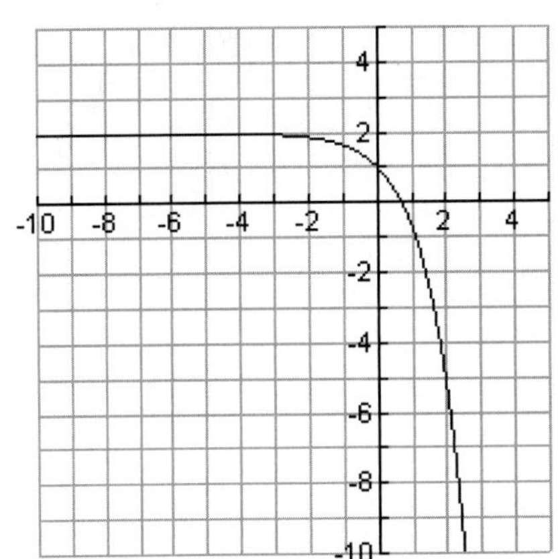

38. <u>y-intercept</u>: $f(0)=1+e^{-0}=2$, so $(0,1)$.

<u>HA</u>: $y=1$

Reflect graph of e^x over the y-axis, then shift up 1 unit.

<u>Domain</u>: $(-\infty,\infty)$ <u>Range</u>: $(1,\infty)$

<u>Other points</u>: (-1, 1+e), $(1,1+\frac{1}{e})$

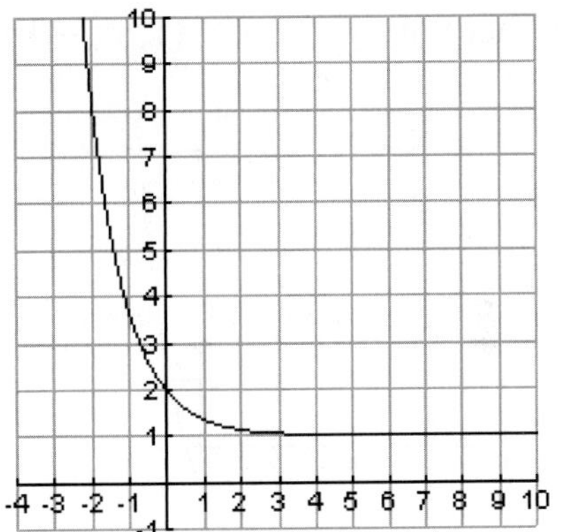

39. <u>y-intercept</u>: $f(0)=5+4^{-0}=6$, so $(0,6)$.

<u>HA</u>: $y=5$

<u>Domain</u>: $(-\infty,\infty)$ <u>Range</u>: $(5,\infty)$

<u>Other points</u>: (1, 5.25), (-1,9)

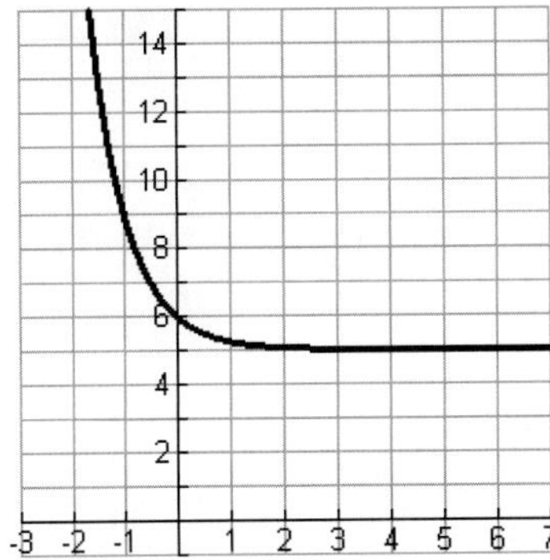

40. <u>y-intercept</u>: $f(0)=5^0-2=-1$, so $(0,-1)$.

<u>HA</u>: $y=-2$

Shift graph of 5^x down 2 units.

<u>Domain</u>: $(-\infty,\infty)$ <u>Range</u>: $(-2,\infty)$

<u>Other points</u>: (1, 3), (2,23)

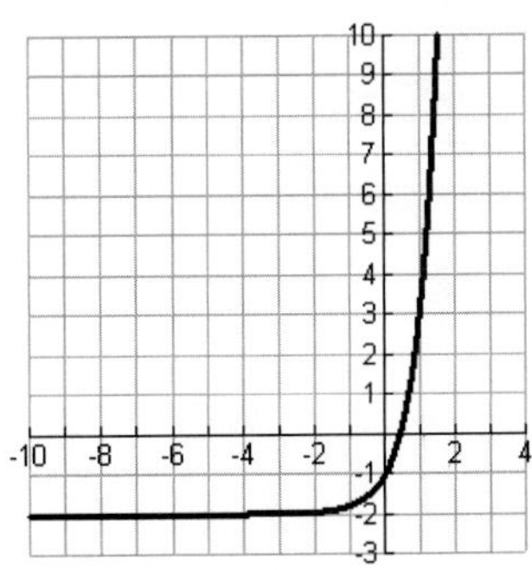

41. <u>y-intercept</u>: $f(0)=e^{0+1}-4=e-4$, so

$(0,e-4)$.

<u>HA</u>: $y=-4$

Shift the graph of e^x left 1 unit, then down 4 units.

<u>Domain</u>: $(-\infty,\infty)$ <u>Range</u>: $(-4,\infty)$

<u>Other points</u>: (-1,-3), (1, e^2-4)

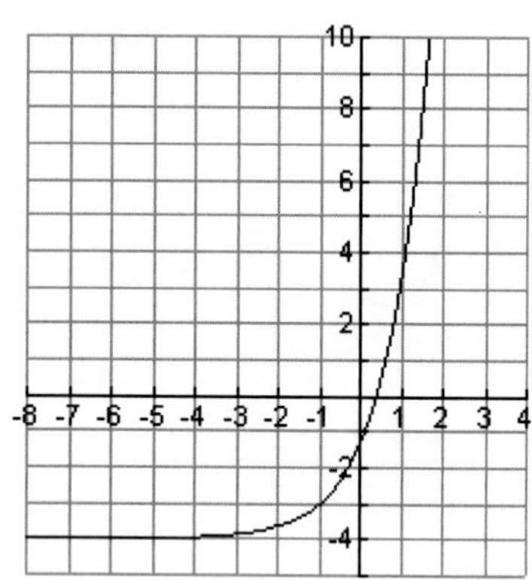

42. <u>y-intercept</u>: $f(0)=2+e^{0-1}=2+\frac{1}{e}$, so

$(0,2+\frac{1}{e})$.

<u>HA</u>: $y=2$

Shift the graph of e^x right 1 unit, then up 2 units.

<u>Domain</u>: $(-\infty,\infty)$ <u>Range</u>: $(2,\infty)$

<u>Other points</u>: (1,3), (2, $e+2$)

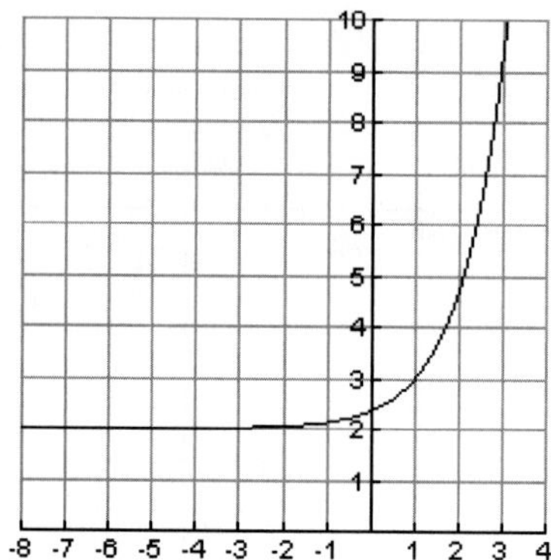

43. y-intercept: $f(0) = 3 \cdot e^0 = 3$, so $(0,3)$.

HA: $y = 0$

Expand the graph of $y = e^x$ horizontally by a factor of 2, then expand vertically by a factor of 3.

Domain: $(-\infty, \infty)$ Range: $(0, \infty)$

Other points: $(2,3e)$, $(1,\ 3\sqrt{e})$

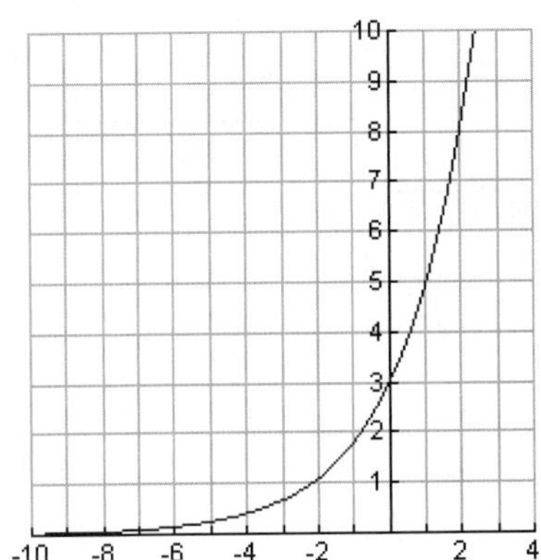

44. y-intercept: $f(0) = 2 \cdot e^{-0} = 2$, so $(0,2)$.

HA: $y = 0$

Reflect the graph of $y = e^x$ about the y-axis, then expand vertically by a factor of 2.

Domain: $(-\infty, \infty)$ Range: $(0, \infty)$

Other points: $(0,2)$, $(-1, 2e)$

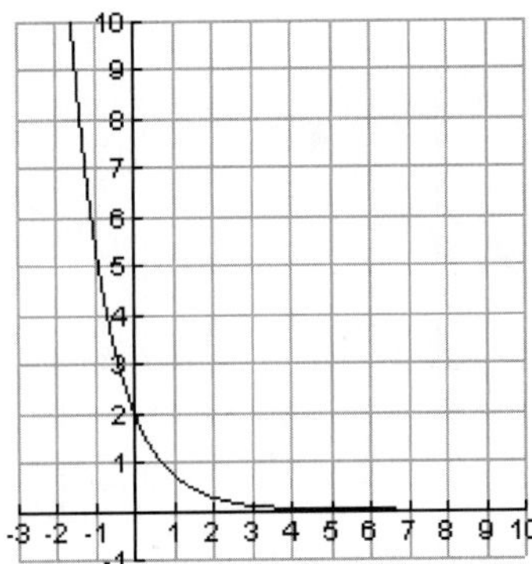

45. <u>y-intercept</u>:
$f(0)=1+\left(\frac{1}{2}\right)^{0-2}=1+2^2=5$,
so $(0,5)$.
<u>HA</u>: $y=1$

Shift the graph of $y=\left(\frac{1}{2}\right)^x$ right 2 units, then up 1 unit.

<u>Domain</u>: $(-\infty,\infty)$ <u>Range</u>: $(1,\infty)$

<u>Other points</u>: (0,5), $(2,2)$

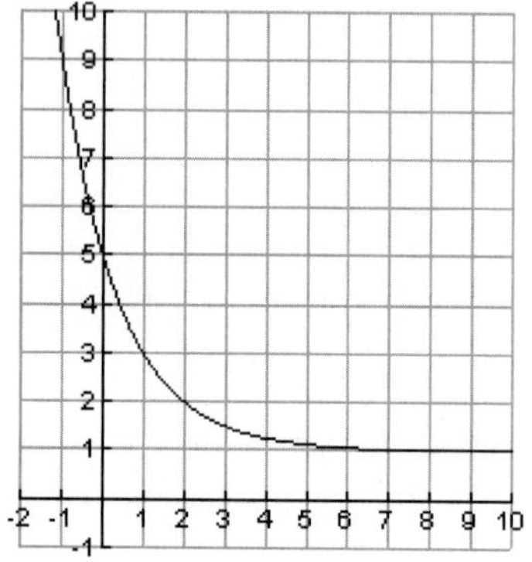

46. <u>y-intercept</u>:
$f(0)=2-\left(\frac{1}{3}\right)^{0+1}=2-\frac{1}{3}=\frac{5}{3}$, so $(0,\frac{5}{3})$.
<u>HA</u>: $y=2$

Shift the graph of $y=\left(\frac{1}{3}\right)^x$ left 1 unit, reflect over x-axis, then shift up 2 units.

<u>Domain</u>: $(-\infty,\infty)$ <u>Range</u>: $(-\infty,2)$

<u>Other points</u>: $(0,\frac{5}{3})$, $(-1,1)$

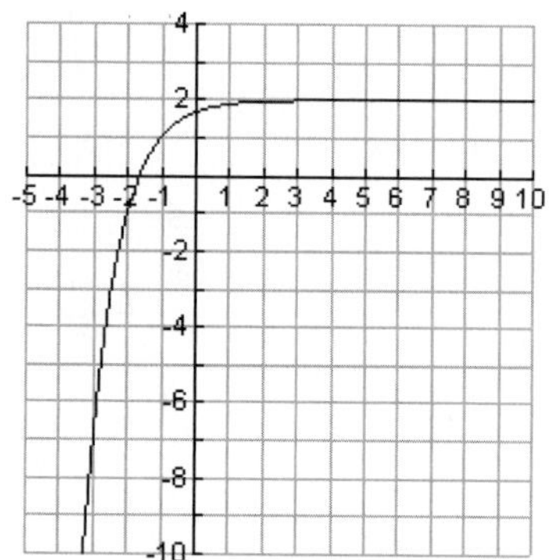

47. Use $P(t)=P_0\left(2^{t/d}\right)$.

Here,
$P_0=7.1,\ d=88,\ t=48\ \ (2090-2002)$.

Thus, $P(48)=7.1\left(2^{48/88}\right)\cong 10.4$.

So, the expected population in 2090 is approximately 10.4 million.

48. Use $P(t)=P_0\left(2^{t/d}\right)$.

Here,
$P_0=43{,}000,\ d=6\ \ (2010-2004)$,
$t=16\ \ (2020-2004)$

Thus, $P(16)=43{,}000\left(2^{16/6}\right)\cong 273{,}033$.

So, the expected population in 2020 is approximately 273,033.

49. Use $P(t)=P_0\left(2^{t/d}\right)$. Here, $P_0=1500,\ d=5$ (doubling time), $t=30$. Thus, $P(30)=1500\left(2^{30/5}\right)\cong 96{,}000$.	**50.** Use $P(t)=P_0\left(2^{t/d}\right)$. Colin: $P_0=55{,}000,\ d=15,\ t=43\ (65-22)$. Thus, $P(43)=55{,}000\left(2^{43/15}\right)\cong 401{,}158$. Cameron: $P_0=35{,}000,\ d=10,\ t=43\ (65-22)$. Thus, $P(43)=35{,}000\left(2^{43/10}\right)\cong 689{,}441$. So, Cameron will have made more money by the time he retires.
51. Use $A(t)=A_0\left(\frac{1}{2}\right)^{t/h}$ Here, $h=119.77$ days, $A_0=200\ mg$, $t=30$ days Thus, $A(30)=200\left(\frac{1}{2}\right)^{30/119.77}\cong 168$. So, 168 mg remain after 30 days.	**52.** Use $A(t)=A_0\left(\frac{1}{2}\right)^{t/h}$ Here, $h=2.807$ days, $A_0=300\ mg$, $t=7$ days T hus, $A(7)=300\left(\frac{1}{2}\right)^{7/2.807}\cong 53$. So, 53 mg remain after 1 week.
53. Use $A(t)=A_0\left(\frac{1}{2}\right)^{t/h}$ Here, $h=13.81\,\text{sec}$, $A_0=800\ mg$, $t=120\,\text{sec}$ Thus, $A(120)=800\left(\frac{1}{2}\right)^{120/13.81}\cong 1.94mg$. So, about 2mg remain after 120 seconds.	**54.** Use $A(t)=A_0\left(\frac{1}{2}\right)^{t/h}$ Here, $h=13.81\,\text{sec}$, $A_0=1{,}000\ mg$, $t=60\,\text{sec}$ Thus, $A(60)=1{,}000\left(\frac{1}{2}\right)^{60/13.81}\cong 49mg$. So, about 49mg remain after 60 seconds.
55. Use $A(t)=A_0\left(\frac{1}{2}\right)^{t/h}$ Here, $h=10$ years, $A_0=8000$, $t=14$ years Thus, $A(14)=8000\left(\frac{1}{2}\right)^{14/10}\cong 3031$. So, the value after 14 years is $3031.	**56.** Use $A(t)=A_0\left(\frac{1}{2}\right)^{t/h}$ Here, $h=1$ year, $A_0=1500$, $t=4$ years Thus, $A(4)=1500\left(\frac{1}{2}\right)^{4/1}\cong 94$. So, the value after 4 years is $94.
57. Use $A(t)=P\left(1+\frac{r}{n}\right)^{nt}$ Here, $P=3200,\ r=0.025,\ n=4,\ t=3$ years. Thus, $A(4)=3200\left(1+\frac{0.025}{4}\right)^{4(3)}\cong 3448.42$. So, the amount in the account after 4 years is $3,448.42.	**58.** Use $A(t)=P\left(1+\frac{r}{n}\right)^{nt}$ Here, $P=10{,}000,\ r=0.035,\ n=1,\ t=5$ years. Thus, $A(5)=10{,}000\left(1+\frac{0.035}{1}\right)^{1(5)}\cong 11{,}876.86$. So, the amount in the account after 5 years is $11,876.86.

59. Use $A(t)=P\left(1+\frac{r}{n}\right)^{nt}$ Here, $A(18)=32{,}000,\ r=0.05,\ n=365,\ t=18$ Thus, solving the above formula for *P*, we have $P=\dfrac{32{,}000}{\left(1+\frac{0.05}{365}\right)^{365(18)}}\cong 13{,}011.03$. So, the initial investment should be \$13,011.03.	**60.** Use $A(t)=P\left(1+\frac{r}{n}\right)^{nt}$ Here, $A(15)=80{,}000,\ r=0.03,\ n=52,\ t=15$. Thus, solving the above formula for *P*, we have $P=\dfrac{80{,}000}{\left(1+\frac{0.03}{52}\right)^{52(15)}}\cong 51{,}016.87$. So, the initial investment should be \$51,016.87.
61. Use $A(t)=Pe^{rt}$. Here, $P=3200,\ r=0.02,\ t=15$. Thus, $A(15)=3200e^{(0.02)(15)}\cong 4319.55$. So, the amount in the account after 15 years is \$4319.55.	**62.** Use $A(t)=Pe^{rt}$. Here, $P=7000,\ r=0.043,\ t=10$. Thus, $A(10)=7000e^{(0.043)(10)}\cong 10{,}760.80$. So, the amount in the account after 10 years is \$10,760.80.
63. Use $A(t)=Pe^{rt}$. Here, $A(20)=38{,}000,\ r=0.05,\ t=20$. Thus, solving the above formula for *P*, we have $P=\dfrac{38{,}000}{e^{(0.05)(20)}}\cong 13{,}979.42$. So, the initial investment should be \$13,979.42.	**64.** Use $A(t)=Pe^{rt}$. Here, $A(18)=80{,}000,\ r=0.06,\ t=18$. Thus, solving the above formula for *P*, we have $P=\dfrac{80{,}000}{e^{(0.06)(18)}}\cong 27{,}167.64$. So, the initial investment should be \$27,167.64.
65. The mistake is that $4^{-1/2}\neq 4^2$. Rather, $4^{-1/2}=\dfrac{1}{4^{1/2}}=\dfrac{1}{2}$.	**66.** The mistake is that $4^{3/2}\neq \dfrac{4^3}{4^2}$. Rather, $4^{3/2}=\left(\sqrt{4}\right)^3=2^3=8$.
67. $r=0.025$ rather than 2.5	**68.** Time is measured in years, so 6 months should be converted to $\frac{1}{2}$. Using $t=\frac{1}{2}$ in the formula yields $A=5075.57$.
69. False. $(0,-1)$ is the *y*-intercept.	**70.** True | **71.** True. $3^{-x}=(3^{-1})^x=\left(\frac{1}{3}\right)^x$
72. False. *e* is irrational, and hence cannot equal a finite decimal.	

73.

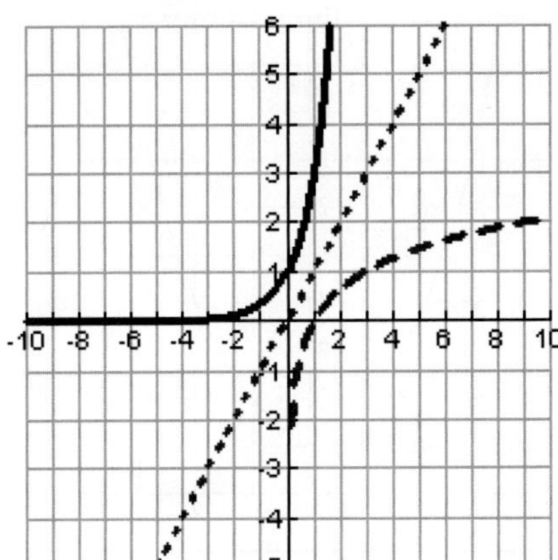

Note on Graphs: Solid curve is $y = 3^x$ and the dashed curve is $y = \log_3 x$.

74.

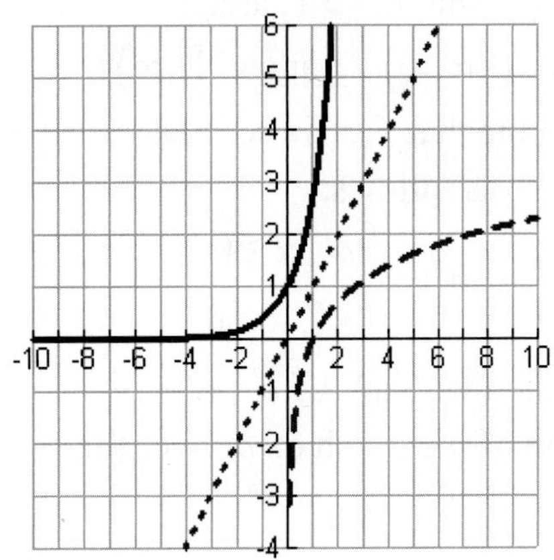

Note on Graphs: Solid curve is $y = e^x$ and the dashed curve is $y = \ln x$.

75.

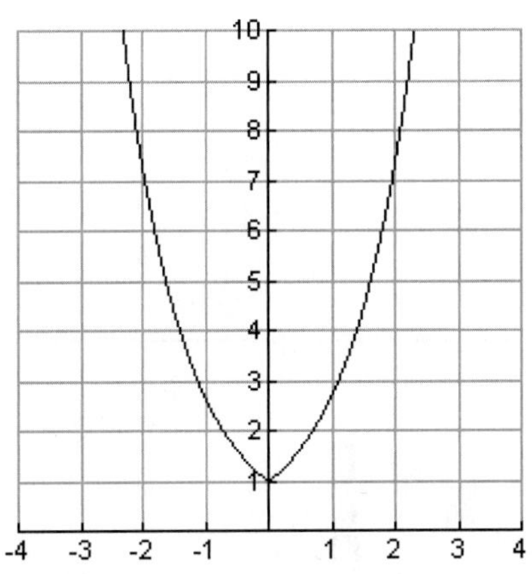

76.

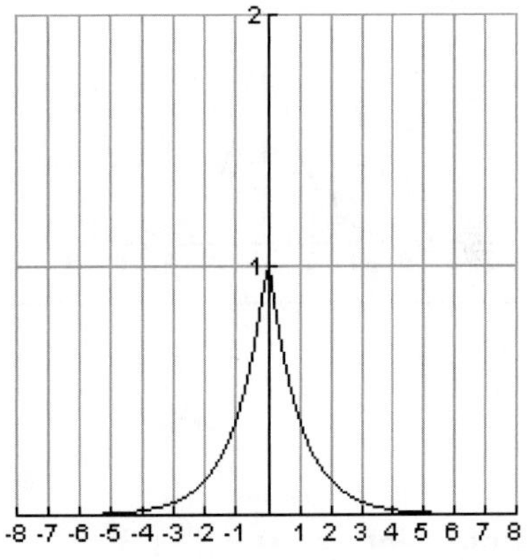

77. y-intercept:
$f(0) = be^{-0+1} - a = be - a$ So, $(0, be - a)$.
Horizontal asymptote: For x very large, $be^{-x+1} \approx 0$, so $y = -a$ is the horizontal asymptote.

78. y-intercept: $f(0) = a + b \cdot e^{0+1} = a + be$, so $(0, a + be)$.
HA: $y = a$

79. The domain and range for $f(x) = b^{|x|}$, where $b > 1$, are:
Domain: $(-\infty, \infty)$ Range: $[1, \infty)$.
Indeed, note that $f(x)$ is defined piecewise, as follows:

$$b^{|x|} = \begin{cases} b^x, & x \geq 0 \\ b^{-x}, & x < 0 \end{cases}$$

Recall that the graph of b^{-x} is the reflection of the graph of b^x over the y-axis.

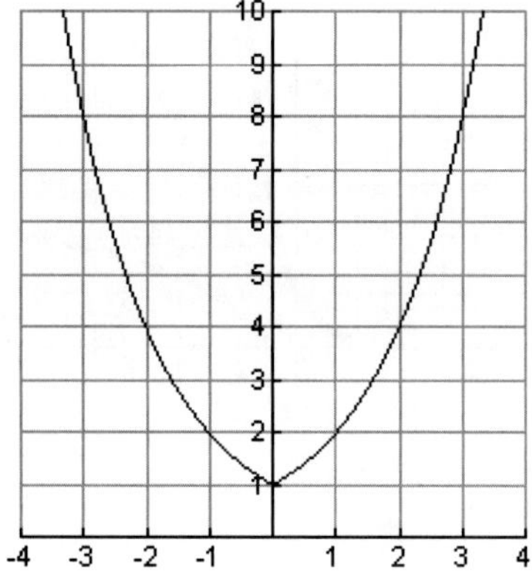

80.

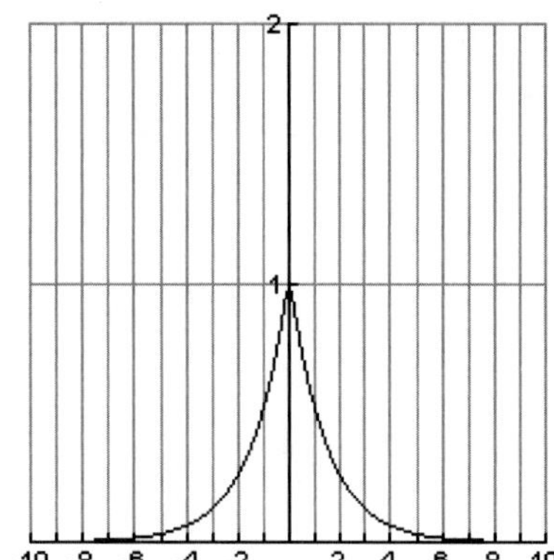

81.

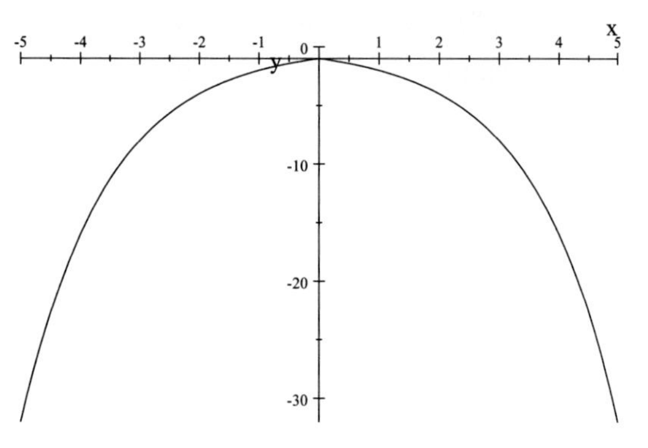

82. y-intercept: $f(0) = 2^0 + 3^0 = 2$.
So, (0,2).

HA: Both 2^x and 3^x approach 0 as x becomes more negative. So, the HA is
$y = 0$.

83.

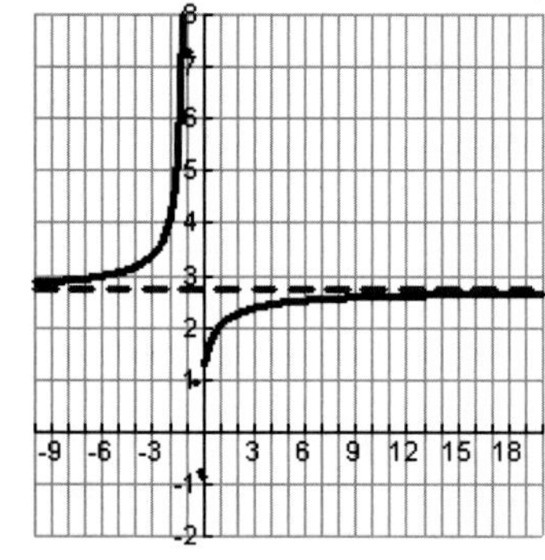

Note on Graphs: Solid curve is $y = (1 + \frac{1}{x})^x$ and dashed curve (the horizontal asymptote) is $y = e$.

84. Since $2<e<3$, we have $2^x<e^x<3^x$, for all *x*, as is seen in the graphs to the right.

Note on Graphs: Solid curve is $y=2^x$, dashed curve is $y=e^x$, and dotted curve is $y=3^x$.

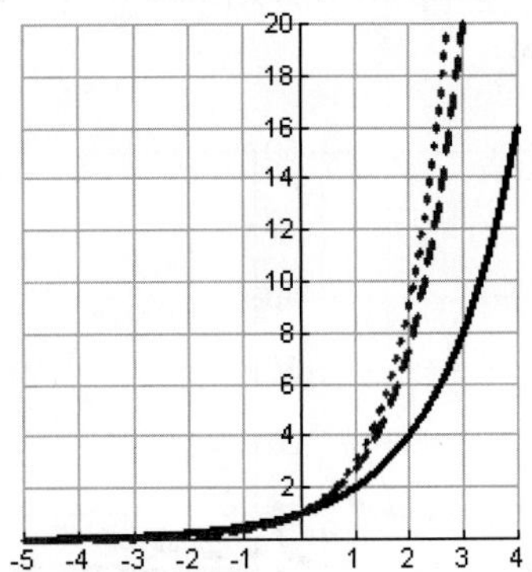

85. The graphs are close on the interval $(-3,3)$.

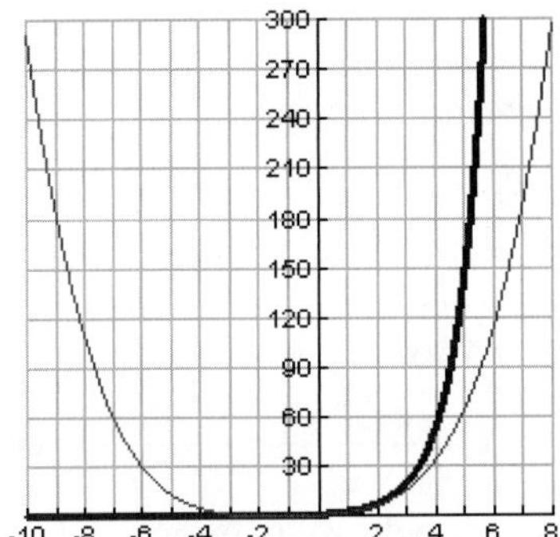

Note on Graphs: Solid curve is $y=e^x$ and thin curve is $y=1+x+\frac{x^2}{2}+\frac{x^3}{6}+\frac{x^4}{24}$.

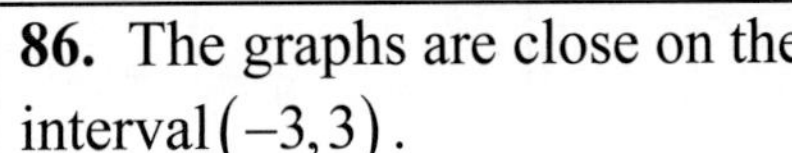

86. The graphs are close on the interval $(-3,3)$.

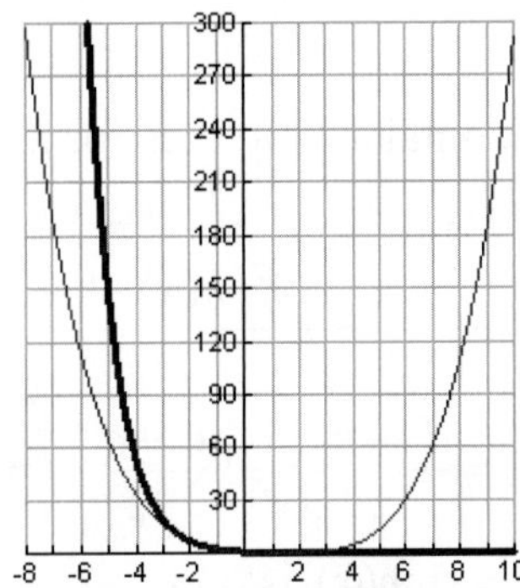

Note on Graphs: Solid curve is $y=e^{-x}$ and thin curve is $y=1-x+\frac{x^2}{2}-\frac{x^3}{6}+\frac{x^4}{24}$.

87. The graphs of f, g, and h are as follows:

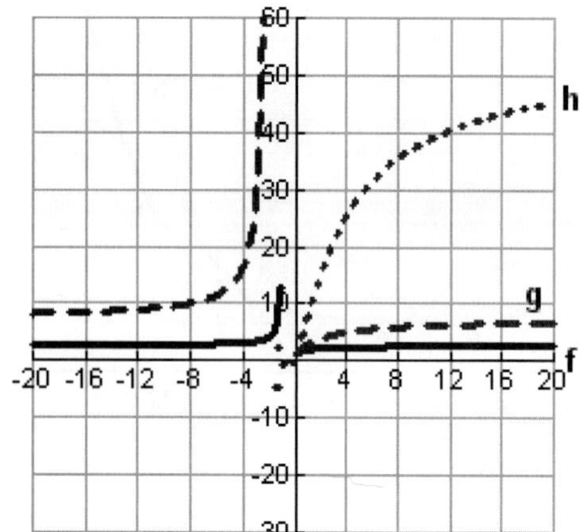

The horizontal asymptotes for f, g, and h are $y = e,\ y = e^2, y = e^4$. As x increases, $f(x) \to e,\ g(x) \to e^2, h(x) \to e^4$.

88. The graphs of f, g, and h are as follows:

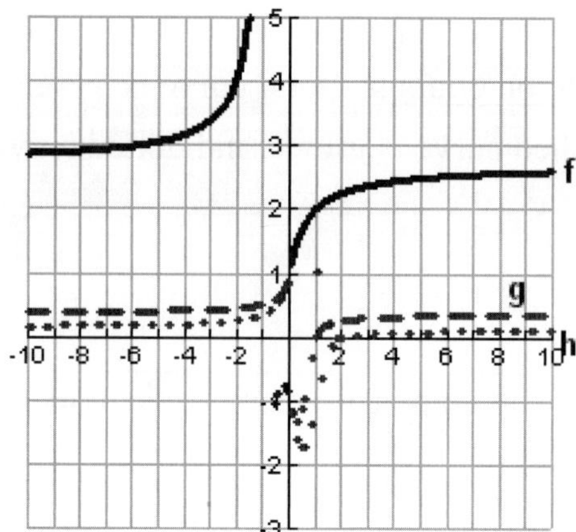

The horizontal asymptotes for f, g, and h are $y = e,\ y = e^{-1}, y = e^{-2}$. As x increases, $f(x) \to e,\ g(x) \to e^{-1}, h(x) \to e^{-2}$.

89. Observe that

$$\sinh(-x) = \frac{e^{-x} - e^{x}}{2} = -\left[\frac{e^{x} - e^{-x}}{2}\right] = -\sinh(x).$$

So, $f(x) = \sinh(x)$ is odd.

90. Observe that

$$\cosh(-x) = \frac{e^{-x} + e^{x}}{2} = \frac{e^{x} + e^{-x}}{2} = \cosh(x).$$

So, $g(x) = \cosh(x)$ is even.

91. Observe that

$$\cosh^2 x - \sinh^2 x = \left[\frac{e^{x} + e^{-x}}{2}\right]^2 - \left[\frac{e^{x} - e^{-x}}{2}\right]^2$$

$$= \frac{e^{2x} + 2e^{x}e^{-x} + e^{-2x}}{4} - \frac{e^{2x} - 2e^{x}e^{-x} + e^{-2x}}{4}$$

$$= \frac{4}{4} = 1$$

92. Observe that

$$\cosh(x) + \sinh(x) = \frac{e^{x} + e^{-x}}{2} + \frac{e^{x} - e^{-x}}{2}$$

$$= \frac{2e^{x}}{2} = e^{x}$$

Section 3.2 Solutions --

1. $81^{\frac{1}{4}} = 3$	**2.** $121^{\frac{1}{2}} = 11$
3. $2^{-5} = \frac{1}{32}$	**4.** $3^{-4} = \frac{1}{81}$
5. $10^{-2} = 0.01$	**6.** $10^{-4} = 0.0001$
7. $10^{4} = 10{,}000$	**8.** $10^{3} = 1{,}000$
9. $\left(\frac{1}{4}\right)^{-3} = 4^{3} = 64$	**10.** $\left(\frac{1}{6}\right)^{-2} = 6^{2} = 36$
11. $e^{-1} = \frac{1}{e}$	**12.** $e^{1} = e$
13. $e^{0} = 1$	**14.** $10^{0} = 1$
15. $e^{x} = 5$	**16.** $e^{y} = 4$
17. $x^{z} = y$	**18.** $x^{y} = z$
19. $x = \log_y(x+y)$ is equivalent to $y^x = x + y$.	**20.** $z = \ln x^y = \log_e x^y \Rightarrow e^z = x^y$
21. $\log(0.00001) = -5$	**22.** $3^6 = 729$ is equivalent to $\log_3 729 = 6$
23. $78{,}125 = 5^7$ is equivalent to $\log_5 78{,}125 = 5^7$.	**24.** $\log(100{,}000) = 5$
25. $\log_{225}(15) = \frac{1}{2}$	**26.** $\log_{343}(7) = \frac{1}{3}$
27. $\log_{2/5}\left(\frac{8}{125}\right) = 3$	**28.** $\log_{2/3}\left(\frac{8}{27}\right) = 3$
29. $\log_{1/27}(3) = -\frac{1}{3}$	**30.** $\log_{1/1024}(4) = -\frac{1}{5}$
31. $\ln 6 = x$	**32.** $\ln 4 = -x$
33. $\log_y x = z$	**34.** $\log_y z = x$
35. $\log_2(1) = 0$	**36.** $\log_5(1) = 0$
37. $\log_5(3125) = x$ $5^x = 3125$ $x = 5$	**38.** $\log_3(729) = x$ $3^x = 729$ $x = 6$
39. $\log_{10}(10^7) = 7$	**40.** $\log_{10}(10^{-2}) = -2$
41. $\log_{1/4}(4096) = x$ $\left(\frac{1}{4}\right)^x = 4096$ $4^{-x} = 4096$ $x = -6$	**42.** $\log_{1/7}(2401) = x$ $\left(\frac{1}{7}\right)^x = 2401$ $7^{-x} = 2401$ $x = -4$
43. undefined	**44.** undefined

45. undefined	**46.** undefined
47. 1.46	**48.** 3.37
49. 5.94	**50.** 2.58
51. undefined	**52.** undefined
53. -8.11	**54.** -3.52
55. Must have $x+5>0$, so that the domain is $(-5,\infty)$.	**56.** Must have $4x-1>0$, so that the domain is $\left(\frac{1}{4},\infty\right)$.
57. Must have $5-2x>0$, so that the domain is $\left(-\infty,\frac{5}{2}\right)$.	**58.** Must have $5-x>0$, so that the domain is $(-\infty,5)$.
59. Must have $7-2x>0$, so that the domain is $\left(-\infty,\frac{7}{2}\right)$.	**60.** Must have $3-x>0$, so that the domain is $(-\infty,3)$.
61. Must have $\lvert x\rvert>0$, so that the domain is $(-\infty,0)\cup(0,\infty)$.	**62.** Must have $\lvert x+1\rvert>0$, so that the domain is $(-\infty,-1)\cup(-1,\infty)$.
63. Must have $x^2+1>0$ (which always occurs), so that the domain is $\mathbb{R}$.	**64.** Must have $1-x^2=(1-x)(1+x)>0$. CPs: -1, 1 − + − (number line: −1, 1) So, the domain is $(-1,1)$.
65. Must have $10+3x-x^2>0$, which is equivalent to $(x-5)(x+2)<0$. CPs: -2, 5 + − + (number line: −2, 5) So, the domain is (-2, 5).	**66.** Must have $x^3-3x^2+3x-1=(x-1)^3>0$. CPs: 1 − + (number line: 1) So, the domain is $(1,\infty)$.
67. b	**68. e** Reflect the graph of $y=\log_5 x$ over the y-axis.
69. c Reflect the graph of $y=\log_5 x$ over the x-axis.	**70. f** Shift the graph of $y=\log_5 x$ left 3 units and down 1 unit.
71. d Since $\log_5(1-x)-2=\log_5(-(x-1))-2$, Reflect the graph of $y=\log_5 x$ over the y-axis, then shift right 1 unit, and then shift down 5 units.	**72. a** Since $-\log_5(3-x)+2=-\log_5(-(x-3))+2$, Reflect the graph of $y=\log_5 x$ over the y-axis, then shift right 3 units, then reflect over the x-axis, an then shift up 2 units.

73. Shift the graph of $y = \log x$ right 1 unit. Domain: $(1, \infty)$ Range: $(-\infty, \infty)$

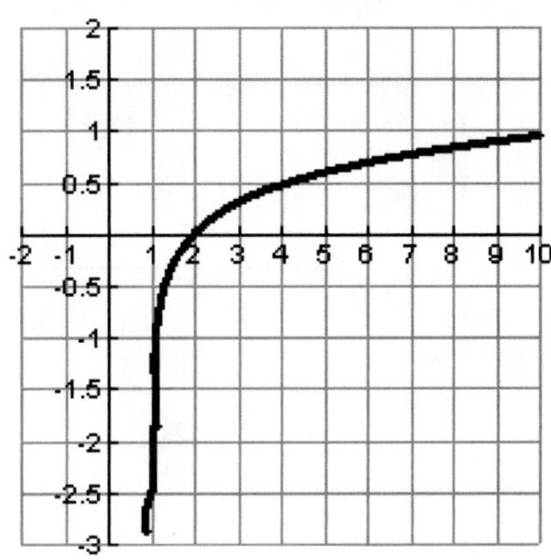

74. Shift the graph of $y = \log x$ left 2 units. Domain: $(-2, \infty)$ Range: $(-\infty, \infty)$

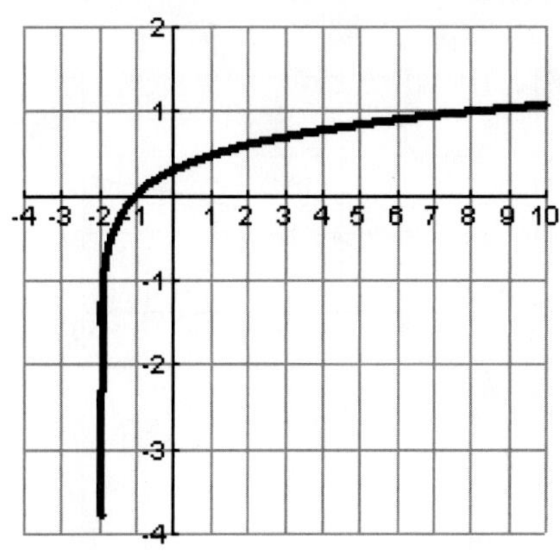

75. Shift the graph of $y = \ln x$ up 2 units. Domain: $(0, \infty)$ Range: $(-\infty, \infty)$

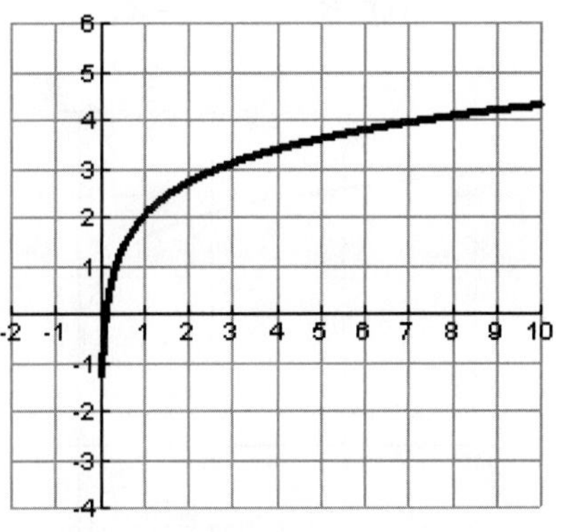

76. Shift the graph of $y = \ln x$ down 1 unit. Domain: $(0, \infty)$ Range: $(-\infty, \infty)$

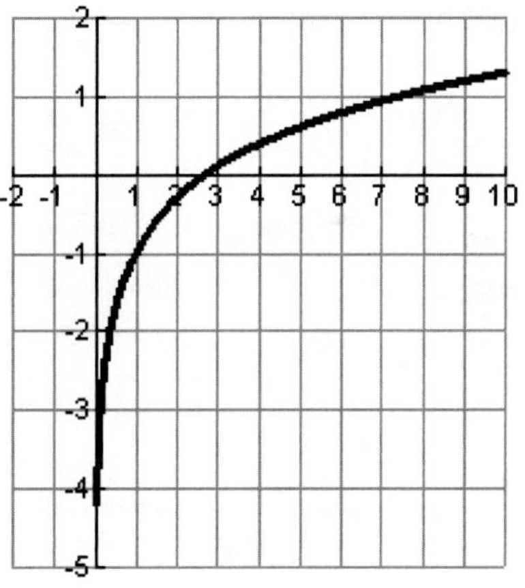

77. Shift the graph of $y = \log_3 x$ left 2 units, then down 1 unit.
Domain: $(-2, \infty)$ Range: $(-\infty, \infty)$

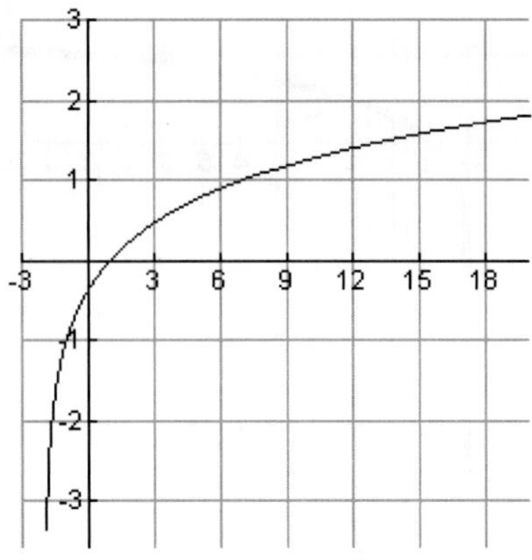

78. Shift the graph of $y = \log_3 x$ left 1 unit, then down 2 units.
Domain: $(-1, \infty)$ Range: $(-\infty, \infty)$

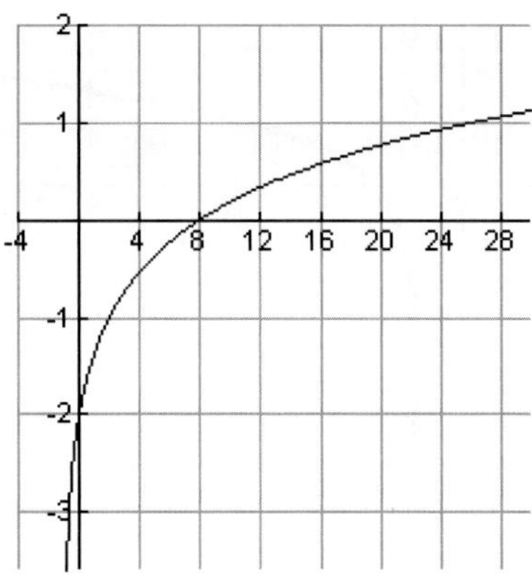

79. Reflect the graph of $y = \log x$ over the x-axis, then shift up 1 unit.
Domain: $(0, \infty)$ Range: $(-\infty, \infty)$

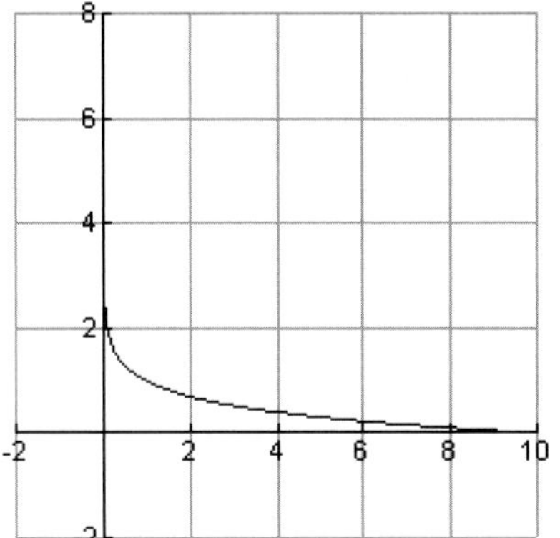

80. Reflect the graph of $y = \log x$ over the y-axis, then shift up 2 units.
Domain: $(-\infty, 0)$ Range: $(-\infty, \infty)$

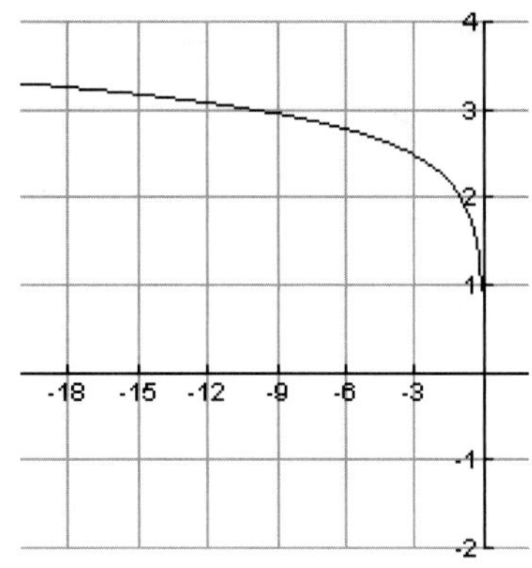

81. Shift the graph of $y = \ln x$ left 4 units.
Domain: $(-4, \infty)$ Range: $(-\infty, \infty)$

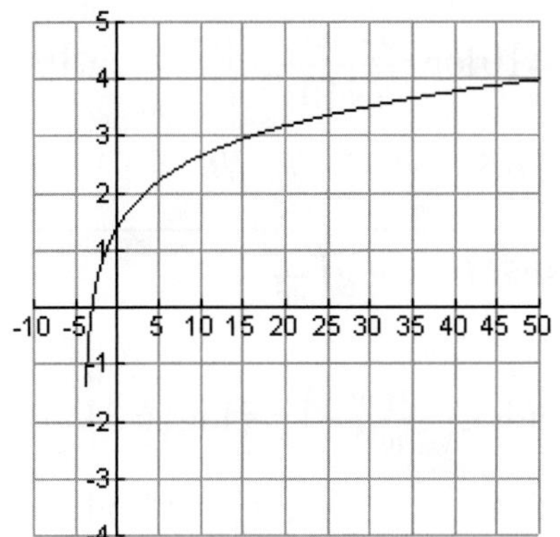

82. Since $\ln(4-x) = \ln(-(x-4))$, then shift the graph of $y = \ln x$ to the right 4 units, then reflect over the *y*-axis.
Domain: $(-\infty, 4)$ Range: $(-\infty, \infty)$

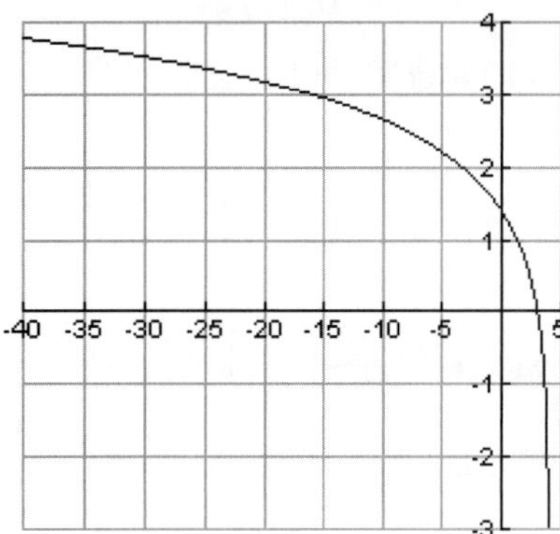

83. Compress the graph of $y = \log x$ horizontally by a factor of 2.
Domain: $(0, \infty)$ Range: $(-\infty, \infty)$

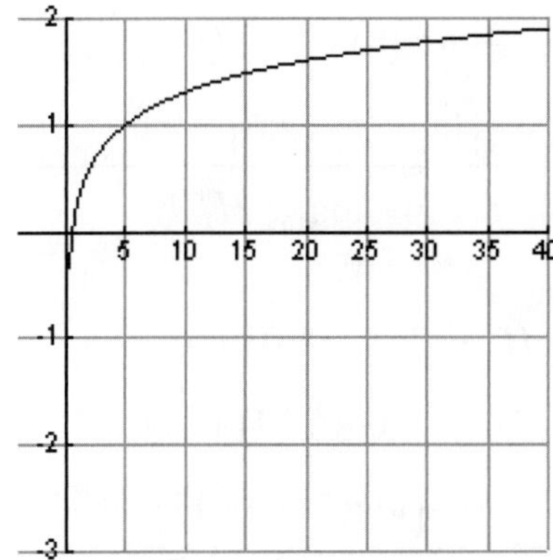

84. Reflect the graph of $y = \ln x$ over the *y*-axis, then expand vertically by a factor of 2. Domain: $(-\infty, 0)$ Range: $(-\infty, \infty)$

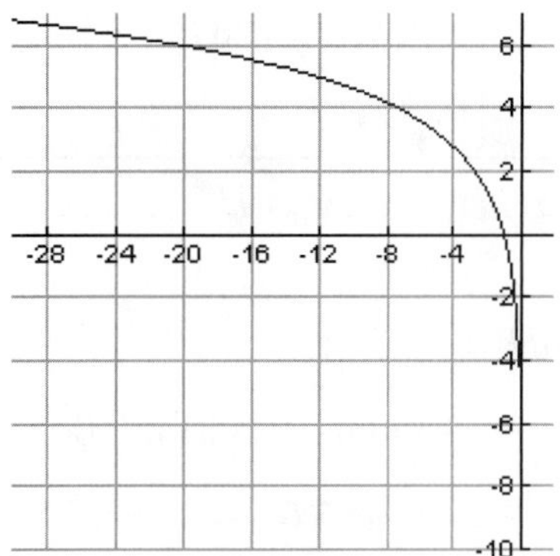

85. Use $D = 10\log\left(\dfrac{I}{1\times 10^{-12}}\right)$.

Here,

$$D = 10\log\left(\frac{1\times 10^{-6}}{1\times 10^{-12}}\right) = 10\log(10^{6})$$
$$= 60\log(10) = 60\,dB$$

86. Use $D = 10\log\left(\dfrac{I}{1\times 10^{-12}}\right)$.

Here,

$$D = 10\log\left(\frac{1\times 10^{1}}{1\times 10^{-12}}\right) = 10\log(10^{13})$$
$$= 130\log(10) = 130\,dB$$

87. Use $D = 10\log\left(\frac{I}{1\times 10^{-12}}\right)$. Here, $D = 10\log\left(\frac{1\times 10^{-0.3}}{1\times 10^{-12}}\right) = 10\log(10^{11.7})$ $= 117\underbrace{\log(10)}_{=1} = 117\,dB$	**88.** Use $D = 10\log\left(\frac{I}{1\times 10^{-12}}\right)$. Here, $D = 10\log\left(\frac{1\times 10^{-4.5}}{1\times 10^{-12}}\right) = 10\log(10^{7.5})$ $= 75\log(10) = 75\,dB$
89. Use $M = \frac{2}{3}\log\left(\frac{E}{10^{4.4}}\right)$. Here, $M = \frac{2}{3}\log\left(\frac{1.41\times 10^{17}}{10^{4.4}}\right) = \frac{2}{3}\log\left(1.41\times 10^{12.6}\right)$ $= \frac{2}{3}\left[\log(1.41) + \log_{10}(10^{12.6})\right]$ $= \frac{2}{3}\left[\log(1.41) + 12.6\right] \cong 8.5$	**90.** Use $M = \frac{2}{3}\log\left(\frac{E}{10^{4.4}}\right)$. Here, $M = \frac{2}{3}\log\left(\frac{6.31\times 10^{15}}{10^{4.4}}\right) = \frac{2}{3}\log\left(6.31\times 10^{10.6}\right)$ $= \frac{2}{3}\left[\log(6.31) + \log_{10}(10^{10.6})\right]$ $= \frac{2}{3}\left[\log(6.31) + 10.6\right] \cong 7.6$
91. Use $M = \frac{2}{3}\log\left(\frac{E}{10^{4.4}}\right)$. Here, $M = \frac{2}{3}\log\left(\frac{2\times 10^{14}}{10^{4.4}}\right) = \frac{2}{3}\log\left(2\times 10^{9.6}\right)$ $= \frac{2}{3}\left[\log(2) + \log_{10}(10^{9.6})\right]$ $= \frac{2}{3}\left[\log(2) + 9.6\right] \cong 6.6$	**92.** Use $M = \frac{2}{3}\log\left(\frac{E}{10^{4.4}}\right)$. Here, $M = \frac{2}{3}\log\left(\frac{8\times 10^{17}}{10^{4.4}}\right) = \frac{2}{3}\log\left(8\times 10^{12.6}\right)$ $= \frac{2}{3}\left[\log(8) + \log_{10}(10^{12.6})\right]$ $= \frac{2}{3}\left[\log(8) + 12.6\right] \cong 9.0$
93. Use $pH = -\log_{10}\left[H^+\right]$. Here, $pH = -\log_{10}(5.01\times 10^{-4})$ $= -\left[\log(5.01) + \log(10^{-4})\right]$ $= -\left[\log(5.01) - 4\right] \cong 3.3$	**94.** Use $pH = -\log_{10}\left[H^+\right]$. Here, $pH = -\log_{10}(5.01\times 10^{-11})$ $= -\left[\log(5.01) + \log(10^{-11})\right]$ $= -\left[\log(5.01) - 11\right] \cong 10.3$
95. Use $pH = -\log_{10}\left[H^+\right]$. Normal Rainwater: $pH = -\log_{10}(10^{-5.6}) = 5.6$ Acid rain/tomato juice: $pH = -\log_{10}(10^{-4}) = 4$	**96.** Use $pH = -\log_{10}\left[H^+\right]$. Here, $pH = -\log_{10}(5.0\times 10^{-13})$ $= -\left[\log(5.0) + \log(10^{-13})\right]$ $= -\left[\log(5.0) - 13\right] \cong 12.3$

97. Use $pH = -\log_{10}\left[H^+\right]$. Here, $pH = -\log_{10}(10^{-3.6})$ $= -(-3.6)\log_{10}10 = 3.6$	**98.** Use $pH = -\log_{10}\left[H^+\right]$. Here, $pH = -\log_{10}(10^{-4.2})$ $= -(-4.2)\log_{10}10 = 4.2$
99. Use $t = -\frac{\ln\left(\frac{C}{500}\right)}{0.0001216}$. Here, $t = -\frac{\ln\left(\frac{100}{500}\right)}{0.0001216} = -\frac{\ln\left(\frac{1}{5}\right)}{0.0001216} \cong 13,236$	**100.** Use $t = -\frac{\ln\left(\frac{C}{500}\right)}{0.0001216}$. Here, $t = -\frac{\ln\left(\frac{40}{500}\right)}{0.0001216} = -\frac{\ln\left(\frac{1}{12.5}\right)}{0.0001216} \cong 20,771$
101. $dB = 10\log\left(\frac{3\times10^{-3}}{1}\right) = 10\left[\log(3) + \underbrace{\log(10^{-3})}_{=-3}\right]$ $= -30 + 10\log(3) \cong -25\,dB$ So, the result is approximately a 25 dB loss.	**102.** $dB = 10\log\left(\frac{2\times10^{-4}}{3}\right) = 10\left[\log\left(\tfrac{2}{3}\right) + \underbrace{\log(10^{-4})}_{=-4}\right]$ $= -40 + 10\log(\tfrac{2}{3}) \cong -42\,dB$
103. $\log_2 4 = x$ is equivalent to $2^x = 4$ (not $x = 2^4$).	**104.** $\log_{100}10 = x$ is equivalent to $100^x = 10$ (not $10^x = 100$).
105. The domain is the set of all real numbers such that $x+5>0$, which is written as $(-5,\infty)$.	**106.** Must exclude $x = 0$ from the domain since $\ln 0$ is not defined.
107. False. The domain is all positive real numbers.	**108.** False. There is no horizontal asymptote for $y = \ln x$. There is, however, a vertical asymptote of $x = 0$ (i.e, the *y*-axis).
109. True.	
110. True.	

111. Consider $f(x) = -\ln(x-a)+b$, where *a*, *b* are real numbers.

Domain: Must have $x-a>0$, so that the domain is (a,∞).

Range: The graph of $f(x)$ is the graph of $\ln x$ shifted *a* units to the right, then reflected over the *x*-axis, and then shifted up *b* units. Through all of this movement, the range of *y*-values remains the same as that of $\ln x$, namely $(-\infty,\infty)$.

x-intercept: Solve $-\ln(x-a)+b=0$:

$$-\ln(x-a)+b=0$$
$$\ln(x-a)=b$$
$$x-a=e^b$$
$$x=a+e^b$$

So, the *x*-intercept is $(a+e^b,0)$.

112. Consider $f(x)=\log(a-x)-b$, where *a*, *b* are real numbers.

Domain: Must have $a-x>0$, so that the domain is $(-\infty,a)$.

Range: Note that $f(x)=\log(-(x-a))-b$. So, the graph of $f(x)$ is the graph of $\log x$ shifted *a* units to the right, then reflected over the *y*-axis, and then shifted down *b* units. Through all of this movement, the range of *y*-values remains the same as that of $\ln x$, namely $(-\infty,\infty)$.

x-intercept: Solve $\log(a-x)-b=0$:

$$\log(a-x)-b=0$$
$$\log(a-x)=b$$
$$a-x=10^b$$
$$x=a-10^b$$

So, the *x*-intercept is $(a-10^b,0)$.

113.

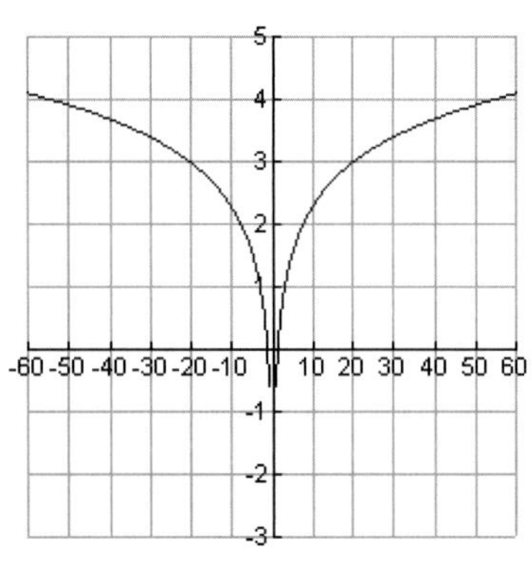

114.

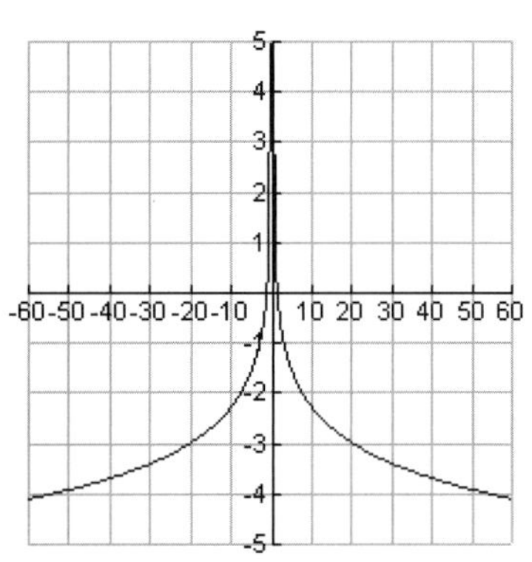

115. The graphs are symmetric about the line $y = x$.

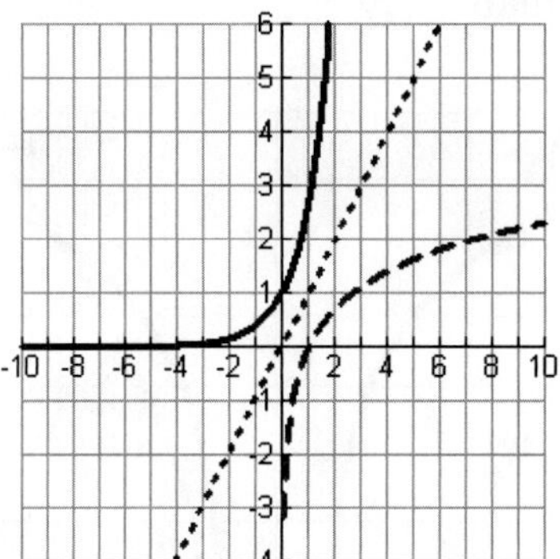

Note on Graphs: Solid curve is $y = e^x$ and dashed curve is $y = \ln x$.

116. The graphs are symmetric about the line $y = x$.

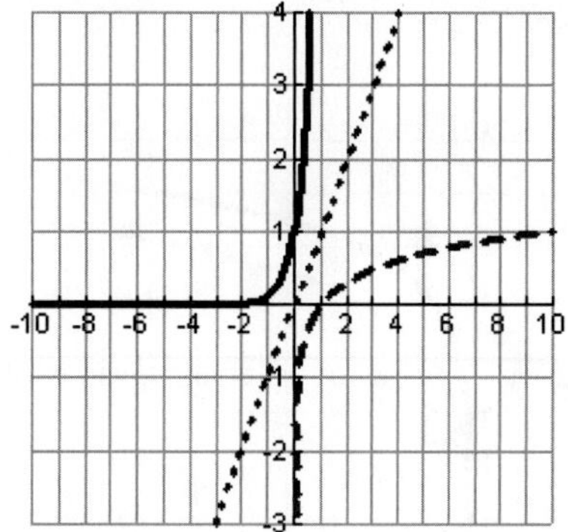

Note on Graphs: Solid curve is $y = 10^x$ and dashed curve is $y = \log x$.

117. The common characteristics are:

- x-intercept for both is $(1,0)$.
- y-axis is the vertical asymptote for both.
- Range is $(-\infty, \infty)$ for both.
- Domain is $(0, \infty)$ for both.

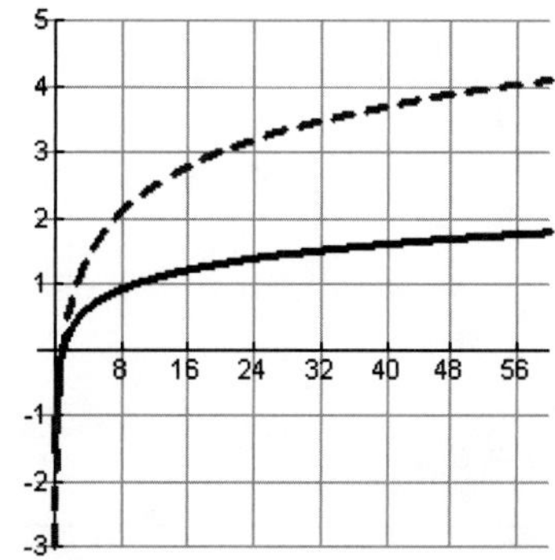

Note on Graphs: Solid curve is $y = \log x$ and dashed curve is $y = \ln x$

118. The function is defined everywhere except at 0.

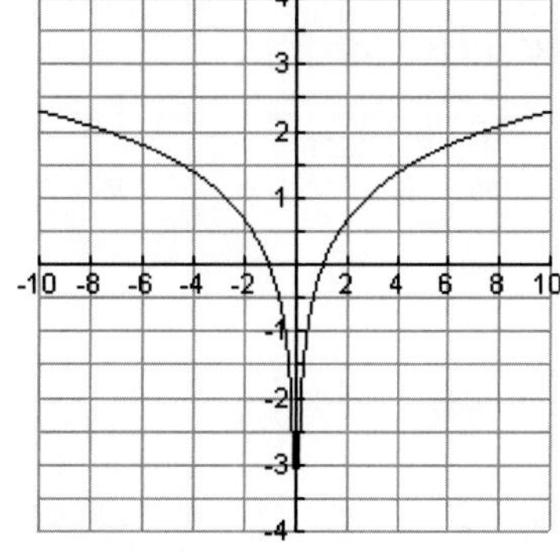

119. The graphs of *f*, *g*, and *h* are below: We note that *f* and *g* have the same graph with domain $(0,\infty)$.

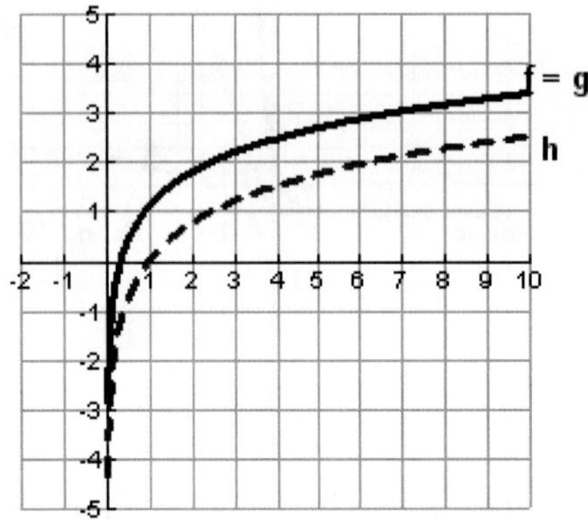

120. The graphs of *f*, *g*, and *h* are below: We note that *f* and *g* have the same graph with domain $(2,\infty)$.

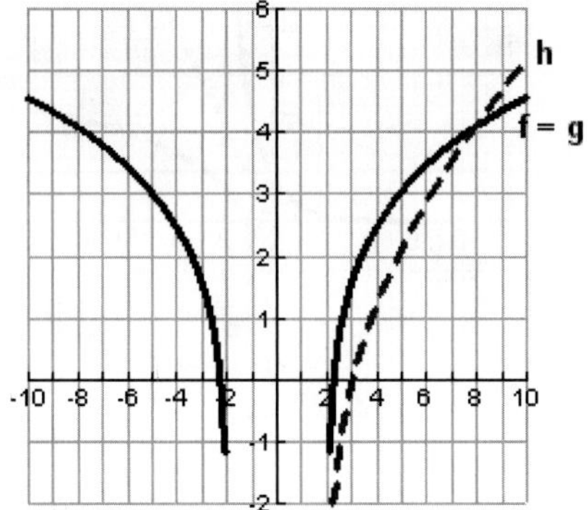

121.

$$\frac{f(x+h)-f(x)}{h}=\frac{e^{(x+h)}-e^{x}}{h}$$

$$=\frac{e^{x}e^{h}-e^{x}}{h}=e^{x}\underbrace{\left[\frac{e^{h}-1}{h}\right]}_{\to 1}$$

$$\to e^{x}$$

So, $f'(x)=e^{x}$.

122.

$$\frac{f(x+h)-f(x)}{h}=\frac{e^{2(x+h)}-e^{2x}}{h}$$

$$=\frac{e^{2x}e^{2h}-e^{2x}}{h}=e^{2x}\left[\frac{e^{2h}-1}{h}\right]$$

$$=e^{2x}\underbrace{\left[e^{h}+1\right]}_{\to 2}\underbrace{\left[\frac{e^{h}-1}{h}\right]}_{\to 1}\to 2e^{2x}$$

So, $f'(x)=2e^{2x}$.

123. a.) Solve $y=e^{x}$ for *x*: $x=\ln y$

Hence, $f^{-1}(x)=\ln x$.

b.) Using Exercise 121, we know that $f'(x)=e^{x}$. Hence,

$$\left(f^{-1}\right)'(x)=\frac{1}{e^{\ln x}}=\frac{1}{x}.$$

124. a.) Solve $y=e^{2x}$ for *x*:

$$2x=\ln y$$

$$x=\tfrac{1}{2}\ln y$$

Hence, $f^{-1}(x)=\tfrac{1}{2}\ln x$.

b.) Using Exercise 122, we know that $f'(x)=2e^{2x}$. Hence,

$$\left(f^{-1}\right)'(x)=\frac{1}{2e^{2\left(\frac{1}{2}\ln x\right)}}=\frac{1}{2x}.$$

Section 3.3 Solutions --

1.

$$\log_9 1 = x$$
$$9^x = 1$$
$$x = 0$$

2.

$$\log_{69} 1 = x$$
$$69^x = 1$$
$$x = 0$$

3.

$$\log_{\frac{1}{2}}\left(\tfrac{1}{2}\right) = x$$
$$\left(\tfrac{1}{2}\right)^x = \tfrac{1}{2}$$
$$x = 1$$

4.

$$\log_{3.3} 3.3 = x$$
$$3.3^x = 3.3$$
$$x = 1$$

5.

$$\log_{10} 10^8 = x$$
$$10^x = 10^8$$
$$x = 8$$

6.

$$\ln e^3 = x$$
$$e^x = e^3$$
$$x = 3$$

7.

$$\log_{10} 0.001 = x$$
$$10^x = 0.001 = 10^{-3}$$
$$x = -3$$

8.

$$\log_3 3^7 = x$$
$$3^x = 3^7$$
$$x = 7$$

9.

$$\log_2 \sqrt{8} = x$$
$$2^x = \sqrt{8} = 2^{\frac{3}{2}}$$
$$x = \tfrac{3}{2}$$

10.

$$\log_5 \sqrt[3]{5} = x$$
$$5^x = \sqrt[3]{5} = 5^{\frac{1}{3}}$$
$$x = \tfrac{1}{3}$$

11. $8^{\log_8 5} = 5$

12. $2^{\log_2 5} = 5$

13. $e^{\ln(x+5)} = x+5$

14.

$$10^{\log\left(3x^2+2x+1\right)} = 3x^2+2x+1$$

15. $5^{3\log_5 2} = 5^{\log_5 2^3} = 2^3 = 8$

16. $7^{2\log_7 5} = 7^{\log_7 5^2} = 5^2 = 25$

17. $7^{-2\log_7 3} = 7^{\log_7 3^{-2}} = 3^{-2} = \frac{1}{9}$

18. $e^{-2\ln 10} = e^{\ln 10^{-2}} = 10^{-2} = \frac{1}{100}$

19. $7e^{-3\ln x} = 7e^{\ln x^{-3}} = 7x^{-3} = \dfrac{7}{x^3}$

20.

$$-19e^{-2\ln x^2} = -19e^{\ln\left(x^2\right)^{-2}} = -19e^{\ln x^{-4}}$$
$$= -19x^{-4} = -\frac{19}{x^4}$$

21.

$$\log_b\left(x^3y^5\right) = \log_b\left(x^3\right) + \log_b\left(y^5\right)$$
$$= 3\log_b(x) + 5\log_b(y)$$

22.

$$\log_b\left(x^{-3}y^{-5}\right) = \log_b\left(x^{-3}\right) + \log_b\left(y^{-5}\right)$$
$$= -3\log_b(x) - 5\log_b(y)$$

23.

$$\log_b\left(x^{\frac{1}{2}}y^{\frac{1}{3}}\right)=\log_b\left(x^{\frac{1}{2}}\right)+\log_b\left(y^{\frac{1}{3}}\right)$$
$$=\tfrac{1}{2}\log_b(x)+\tfrac{1}{3}\log_b(y)$$

24.

$$\log_b\left(\sqrt{r}\sqrt{t}\right)=\log_b\left(r^{\frac{1}{2}}t^{\frac{1}{2}}\right)$$
$$=\log_b\left(r^{\frac{1}{2}}\right)+\log_b\left(t^{\frac{1}{2}}\right)$$
$$=\tfrac{1}{2}\log_b(r)+\tfrac{1}{2}\log_b(t)$$
$$=\tfrac{1}{2}\left(\log_b(r)+\log_b(t)\right)$$

25.

$$\log_b\left(\frac{r^{\frac{1}{3}}}{s^{\frac{1}{2}}}\right)=\log_b\left(r^{\frac{1}{3}}\right)-\log_b\left(s^{\frac{1}{2}}\right)$$
$$=\tfrac{1}{3}\log_b(r)-\tfrac{1}{2}\log_b(s)$$

26.

$$\log_b\left(\frac{r^4}{s^2}\right)=\log_b\left(r^4\right)-\log_b\left(s^2\right)$$
$$=4\log_b(r)-2\log_b(s)$$

27.

$$\log_b\left(\frac{x}{yz}\right)=\log_b(x)-\log_b(yz)$$
$$=\log_b(x)-\left[\log_b(y)+\log_b(z)\right]$$
$$=\log_b(x)-\log_b(y)-\log_b(z)$$

28.

$$\log_b\left(\frac{xy}{z}\right)=\log_b(xy)-\log_b(z)$$
$$=\log_b(x)+\log_b(y)-\log_b(z)$$

29.

$$\log\left(x^2\sqrt{x+5}\right)=\log\left(x^2\right)+\log\left(\sqrt{x+5}\right)$$
$$=\log\left(x^2\right)+\log(x+5)^{\frac{1}{2}}$$
$$=2\log x+\tfrac{1}{2}\log(x+5)$$

30.

$$\log\left((x-3)(x+2)\right)=\log(x-3)+\log(x+2)$$

31.

$$\ln\left(\frac{x^3(x-2)^2}{\sqrt{x^2+5}}\right)=\ln\left(x^3(x-2)^2\right)-\ln\left(\sqrt{x^2+5}\right)$$
$$=\ln\left(x^3\right)+\ln\left((x-2)^2\right)-\ln\left(x^2+5\right)^{\frac{1}{2}}$$
$$=3\ln(x)+2\ln(x-2)-\tfrac{1}{2}\ln\left(x^2+5\right)$$

32.

$$\ln\left(\frac{\sqrt{x+3}\,\sqrt[3]{x-4}}{(x+1)^4}\right)=\ln\left(\sqrt{x+3}\,\sqrt[3]{x-4}\right)-\ln\left((x+1)^4\right)$$
$$=\ln\left(\sqrt{x+3}\right)+\ln\left(\sqrt[3]{x-4}\right)-\ln(x+1)^4$$
$$=\tfrac{1}{2}\ln(x+3)+\tfrac{1}{3}\ln(x-4)-4\ln(x+1)$$

33.

$$\log\left(\frac{x^2-2x+1}{x^2-9}\right)=\log\left(\frac{(x-1)^2}{(x-3)(x+3)}\right)$$
$$=\log\left((x-1)^2\right)-\log\left((x-3)(x+3)\right)$$
$$=2\log(x-1)-\left[\log(x-3)+\log(x+3)\right]$$
$$=2\log(x-1)-\log(x-3)-\log(x+3)$$

34.

$$\log\left(\frac{x^2-x-2}{x^2+3x-4}\right)=\log\left(\frac{(x-2)(x+1)}{(x+4)(x-1)}\right)$$
$$=\log\left((x-2)(x+1)\right)-\log\left((x-1)(x+4)\right)$$
$$=\log(x-2)+\log(x+1)-\left[\log(x-1)+\log(x+4)\right]$$
$$=\log(x-2)+\log(x+1)-\log(x-1)-\log(x+4)$$

35.

$$\ln\sqrt{\frac{x^2+3x-10}{x^2-3x+2}}=\ln\left(\frac{x^2+3x-10}{x^2-3x+2}\right)^{1/2}$$
$$=\tfrac{1}{2}\ln\left(\frac{x^2+3x-10}{x^2-3x+2}\right)$$
$$=\tfrac{1}{2}\ln\left(\frac{(x+5)(x-2)}{(x-2)(x-1)}\right)$$
$$=\tfrac{1}{2}\ln\left(\frac{x+5}{x-1}\right)$$
$$=\tfrac{1}{2}\ln(x+5)-\tfrac{1}{2}\ln(x-1)$$

36.

$$\ln\left[\frac{\sqrt[3]{x-1}(3x-2)^4}{(x+1)\sqrt{x-1}}\right]^2=\ln\left[\frac{(3x-2)^4}{(x+1)(x-1)^{1/6}}\right]^2$$
$$=2\ln\left[\frac{(3x-2)^4}{(x+1)(x-1)^{1/6}}\right]$$
$$=2\ln(3x-2)^4-2\ln\left[(x+1)(x-1)^{1/6}\right]$$
$$=8\ln(3x-2)-2\ln(x+1)-\tfrac{1}{3}\ln(x-1)$$

37.

$$3\log_b x+5\log_b y=\log_b\left(x^3\right)+\log_b y^5$$
$$=\log_b\left(x^3y^5\right)$$

38.

$$2\log_b u+3\log_b v=\log_b u^2+\log_b v^3$$
$$=\log_b\left(u^2v^3\right)$$

39.

$$5\log_b u-2\log_b v=\log_b u^5-\log_b v^2$$
$$=\log_b\left(\frac{u^5}{v^2}\right)$$

40.

$$3\log_b x-\log_b y=\log_b x^3-\log_b y$$
$$=\log_b\left(\frac{x^3}{y}\right)$$

41.	**42.**
$\frac{1}{2}\log_b x + \frac{2}{3}\log_b y = \log_b x^{\frac{1}{2}} + \log_b y^{\frac{2}{3}} = \log_b\left(x^{\frac{1}{2}}y^{\frac{2}{3}}\right)$	$\frac{1}{2}\log_b x - \frac{2}{3}\log_b y = \log_b x^{\frac{1}{2}} - \log_b y^{\frac{2}{3}} = \log_b\left(\frac{x^{\frac{1}{2}}}{y^{\frac{2}{3}}}\right)$

43.

$$\begin{aligned}2\log u - 3\log v - 2\log z &= \log u^2 - \left[\log v^3 + \log z^2\right]\\ &= \log u^2 - \log\left(v^3 z^2\right)\\ &= \log\left(\frac{u^2}{v^3 z^2}\right)\end{aligned}$$

44.

$$\begin{aligned}3\log u - \log 2v - \log z &= \log u^3 - \left[\log 2v + \log z\right]\\ &= \log u^3 - \log(2vz)\\ &= \log\left(\frac{u^3}{2vz}\right)\end{aligned}$$

45.

$$\begin{aligned}\ln(x+1) + \ln(x-1) - 2\ln(x^2+3) &= \ln(x+1) + \ln(x-1) - \ln(x^2+3)^2\\ &= \ln\left[(x+1)(x-1)\right] - \ln(x^2+3)^2\\ &= \ln\left(\frac{x^2-1}{(x^2+3)^2}\right)\end{aligned}$$

46.

$$\begin{aligned}\ln(\sqrt{x-1}) + \ln(\sqrt{x+1}) - 2\ln(x^2-1) &= \ln(x-1)^{\frac{1}{2}} + \ln(x+1)^{\frac{1}{2}} - \ln(x^2-1)^2\\ &= \ln\left[(x-1)^{\frac{1}{2}}(x+1)^{\frac{1}{2}}\right] - \ln(x^2-1)^2\\ &= \ln\left(\frac{(x-1)^{\frac{1}{2}}(x+1)^{\frac{1}{2}}}{(x^2-1)^2}\right)\\ &= \ln\left(\frac{\sqrt{(x-1)(x+1)}}{(x^2-1)^2}\right)\end{aligned}$$

47.

$$\tfrac{1}{2}\ln(x+3)-\tfrac{1}{3}\ln(x+2)-\ln x=\ln(x+3)^{\frac{1}{2}}-\left[\ln(x+2)^{\frac{1}{3}}+\ln x\right]$$
$$=\ln(x+3)^{\frac{1}{2}}-\left[\ln\left(x(x+2)^{\frac{1}{3}}\right)\right]$$
$$=\ln\left(\frac{(x+3)^{\frac{1}{2}}}{x(x+2)^{\frac{1}{3}}}\right)$$

48.

$$\tfrac{1}{3}\ln(x^2+4)-\tfrac{1}{2}\ln(x^2-3)-\ln(x-1)=\ln(x^2+4)^{\frac{1}{3}}-\left[\ln(x^2-3)^{\frac{1}{2}}+\ln(x-1)\right]$$
$$=\ln(x^2+4)^{\frac{1}{3}}-\left[\ln\left((x-1)(x^2-3)^{\frac{1}{2}}\right)\right]$$
$$=\ln\left(\frac{(x^2+4)^{\frac{1}{3}}}{(x-1)(x^2-3)^{\frac{1}{2}}}\right)$$

49. $\log_5 7=\frac{\log 7}{\log 5}\cong 1.2091$	**50.** $\log_4 19=\frac{\log 19}{\log 4}\cong 2.1240$
51. $\log_{\frac{1}{2}} 5=\frac{\log 5}{\log\frac{1}{2}}\cong -2.3219$	**52.** $\log_5\frac{1}{2}=\frac{\log\frac{1}{2}}{\log 5}\cong -0.4307$
53. $\log_{2.7} 5.2=\frac{\log 5.2}{\log 2.7}\cong 1.6599$	**54.** $\log_{7.2} 2.5=\frac{\log 2.5}{\log 7.2}\cong 0.4642$
55. $\log_{\pi} 10=\frac{\log 10}{\log\pi}\cong 2.0115$	**56.** $\log_{\pi} 2.7=\frac{\log 2.7}{\log\pi}\cong 0.8677$
57. $\log_{\sqrt{3}} 8=\frac{\log 8}{\log\sqrt{3}}\cong 3.7856$	**58.** $\log_{\sqrt{2}} 9=\frac{\log 9}{\log\sqrt{2}}\cong 6.3400$

59. Use $D=10\log\left(\frac{I}{1\times10^{-12}}\right)$.

In this case, $I=\underbrace{(1\times10^{-1})\,W/m^2}_{\text{From music}}+\underbrace{(1\times10^{-6})\,W/m^2}_{\text{From conversation}}=\left(1.00001\times10^{-1}\right)W/m^2$.

So, $D=10\log\left(\frac{1.00001\times10^{-1}}{1\times10^{-12}}\right)\cong 110\,dB$ that you are exposed to.

60. Use $D = 10\log\left(\frac{I}{1\times 10^{-12}}\right)$.

In this case, $I = \underbrace{(1\times 10^{-10})\,{}^{W}\!/_{m^2}}_{\text{From whisperer}} + \underbrace{(1\times 10^{-6})\,{}^{W}\!/_{m^2}}_{\text{Normal}} = \left(1.0001\times 10^{-6}\right){}^{W}\!/_{m^2}$.

So, $D = 10\log\left(\frac{1.0001\times 10^{-6}}{1\times 10^{-12}}\right) \cong 60\,dB$ that you are exposed to.

61. Use $M = \frac{2}{3}\log\left(\frac{E}{10^{4.4}}\right)$.

Here, the combined energy is $(4.5\times 10^{12}) + (7.8\times 10^{8})$ joules. The corresponding magnitude on the Richter scale is $M = \frac{2}{3}\log\left(\frac{(4.5\times 10^{12}) + (7.8\times 10^{8})}{10^{4.4}}\right) \cong 5.5$.

62. Use $M = \frac{2}{3}\log\left(\frac{E}{10^{4.4}}\right)$.

Here, the combined energy is $(5.2\times 10^{11}) + (4.1\times 10^{9})$ joules. The corresponding magnitude on the Richter scale is $M = \frac{2}{3}\log\left(\frac{(5.2\times 10^{11}) + (4.1\times 10^{9})}{10^{4.4}}\right) \cong 4.9$.

63. Use $D = \log O = \log\left(\frac{I}{T}\right)$.

If $T = 0.90I$, we see that the density D is $D = \log\left(\frac{I}{0.90I}\right) = \log\left(\frac{10}{9}\right) \approx 0.0458$.

64. Use $pH = -\log\left(\alpha_H\right)$.

Here, $pH = -\log(0.00407) \approx 2.39$.

65. Use $pH = -\log\left(\alpha_H\right)$.

Observe that

$$3.2 = -\log\left(\alpha_H\right) \Rightarrow \alpha_H = 10^{-3.2} \approx 0.000631$$

versus

$$4.4 = -\log\left(\alpha_H\right) \Rightarrow \alpha_H = 104.4 \approx 0.0000398.$$

As such, $\frac{0.000631}{0.0000398} \approx 15.854$, so that about 16 times more acidic.

66. Use $C = W\log_2\left(1+\frac{S}{N}\right)$ and $R = 10\log\left(\frac{S}{N}\right)$.
Here, $W = 3$ and $R = 2$. Determine C.
First, observe that $2 = 10\log\left(\frac{S}{N}\right) \Rightarrow \frac{S}{N} = 10^{1/5}$. Using this in the first equation yields

$$C = W\log_2\left(1+\tfrac{S}{N}\right) = 3\log_2\left(1+10^{-1/5}\right) = 3\left[\frac{\log\left(1+10^{-1/5}\right)}{\log 2}\right] \approx 4.11\,\text{Mbps}.$$

67. Cannot apply the quotient property directly. Observe that

$$3\log 5 - \log 5^2 = 3\log 5 - 2\log 5 = \log 5.$$

68. Step 3 is wrong. Apply the product and quotient properties:

$$\ln 3 + \ln 16 - \ln 8 = \ln(3\cdot 16) - \ln 8 = \ln\left(\frac{48}{8}\right) = \ln 6$$

69. Cannot apply the product and quotient properties to logarithms with different bases. So, you cannot reduce the given expression further without using the change of base formula.

70. Applied the power property incorrectly. Should be $2\log\left(\frac{3}{5}\right) = \log\left(\frac{3}{5}\right)^2$, not $\left(\log\frac{3}{5}\right)^2$.

71. True. $\log e = \dfrac{\ln e}{\ln 10} = \dfrac{1}{\ln 10}$

72. True. $\log 10 = 1$, so $\dfrac{1}{\log 10} = 1$, which does equal $\ln e$.

73. False.
$\ln(xy)^3 = 3\ln(xy) = 3(\ln x + \ln y)$, which does not equal $(\ln x + \ln y)^3$, in general.

74. True. $\dfrac{\log a}{\log b} = \dfrac{\frac{\ln a}{\ln 10}}{\frac{\ln b}{\ln 10}} = \dfrac{\ln a}{\ln b}$

75. False. Note that $36\log x = \log x^{36}$ and $x^{36} \neq 12x^3$ for all x.

76. True. $e^{\ln x^2} = x^2$ because e^x and $\ln x$ are inverses.

77. Claim: $\log_b\left(\frac{M}{N}\right) = \log_b M - \log_b N$

Proof: Let $u = \log_b M$, $v = \log_b N$. Then, $b^u = M$, $b^v = N$.

Observe that $\log_b\left(\frac{M}{N}\right) = \log_b\left(\frac{b^u}{b^v}\right) = \log_b\left(b^{u-v}\right) = u - v = \log_b M - \log_b N$. ▪

78. Claim: $\log_b\left(M^p\right) = p\log_b M$

Proof: Let $u = \log_b M$. Then, $b^u = M$.

Observe that $\log_b\left(M^p\right) = \log_b\left(b^u\right)^p = \log_b\left(b^{u\cdot p}\right) = u\cdot p = \left(\log_b M\right)\cdot p = p\log_b M$. ▪

79.

$$\log_b\left(\sqrt{\frac{x^2}{y^3z^{-5}}}\right)^6 = \log_b\left(\frac{x^2}{y^3z^{-5}}\right)^3 = \log_b\left(\frac{x^6}{y^9z^{-15}}\right)$$

$$= \log_b\left(x^6\right) - \log_b\left(y^9z^{-15}\right)$$

$$= \log_b\left(x^6\right) - \left[\log_b\left(y^9\right) + \log_b\left(z^{-15}\right)\right]$$

$$= 6\log_b x - 9\log_b y + 15\log_b z$$

80. $\log_b\left(\frac{1}{x}\right) = \log_b\left(x^{-1}\right) = -\log_b x$

81.

$$\log_b\left(\frac{a^2}{b^3}\right)^{-3} = -3\log_b\left(\frac{a^2}{b^3}\right) = -3\left[\log_b a^2 - \log_b b^3\right] = -3\left[2\log_b a - 3\log_b b\right]$$

$$= -6\log_b a + 9 = -6\left[\frac{\log a}{\log b}\right] + 9 = 9 - \frac{6}{\frac{\log b}{\log a}} = 9 - \frac{6}{\log_a b}$$

82.

$$\log_b \sqrt{48} = \tfrac{1}{2}\log_b\left(2^4 \cdot 3\right)$$

$$= \tfrac{1}{2}\left[\log_b\left(2^4\right) + \log_b 3\right]$$

$$= \tfrac{1}{2}\left[4\log_b 2 + \log_b 3\right]$$

$$= \tfrac{1}{2}\left[4(0.4307) + 0.6826\right]$$

$$= 1.2027$$

83. Yes, they are the same graph.

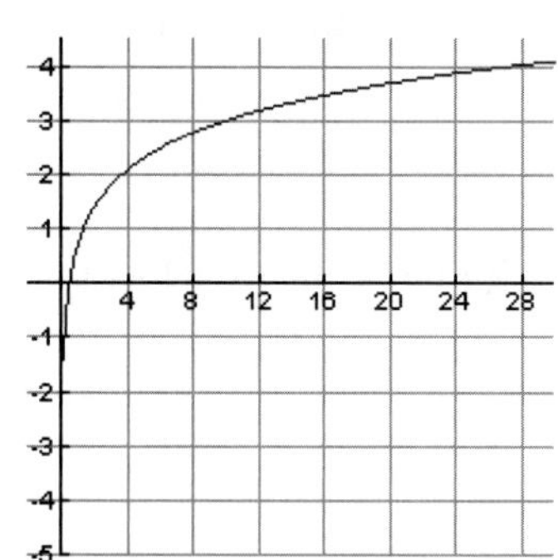

84. No, they are not the same graph.

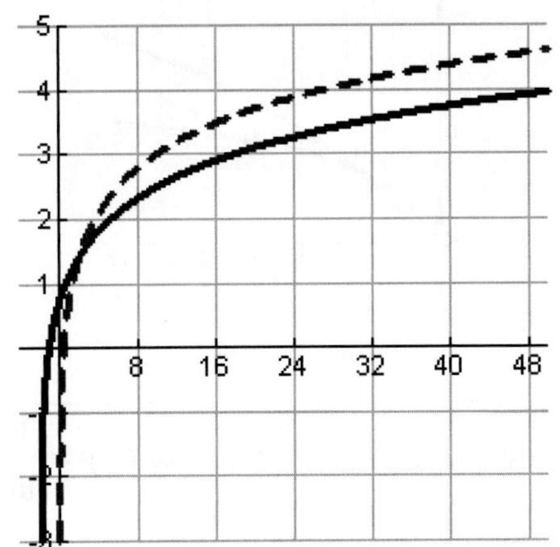

Note on Graphs: Solid curve is $y = \ln(2 + x)$ and dashed curve is $y = \ln 2 + \ln x$

85. No, they are not the same graph.

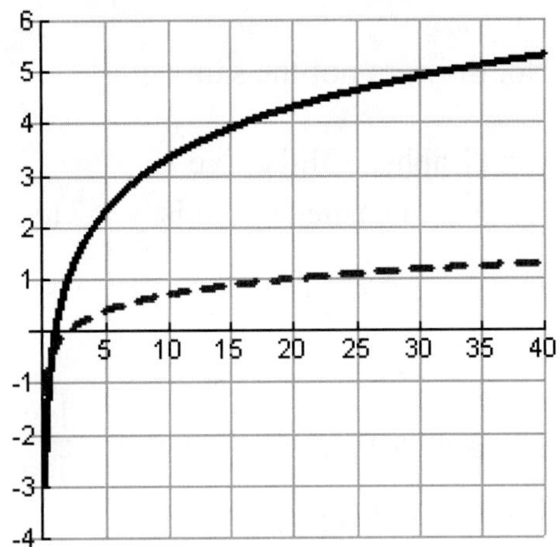

Note on Graphs: Solid curve is $y = \frac{\log x}{\log 2}$ and dashed curve is $y = \log x - \log 2$.

86. Yes, they are the same graph.

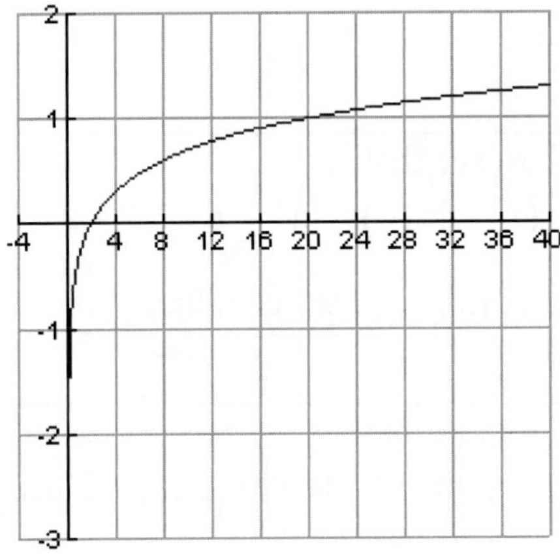

87. No, they are not the same graph, even though the property is true.

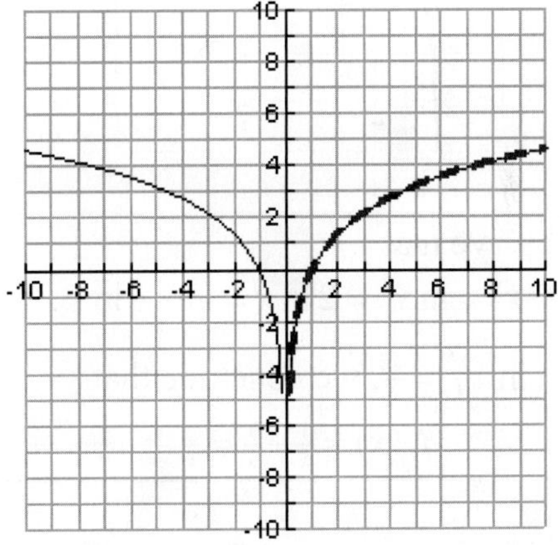

Note on Graphs: The thin curve plus the dashed curve is $y = \ln(x^2)$ (domain all real numbers except 0) and the dashed curve only is $y = 2\ln x$ (domain is $(0, \infty)$).

88. No, they are not the same graph.

Note on Graphs: Solid curve is $y = (\ln x)^2$ and dashed curve is $y = 2\ln x$.

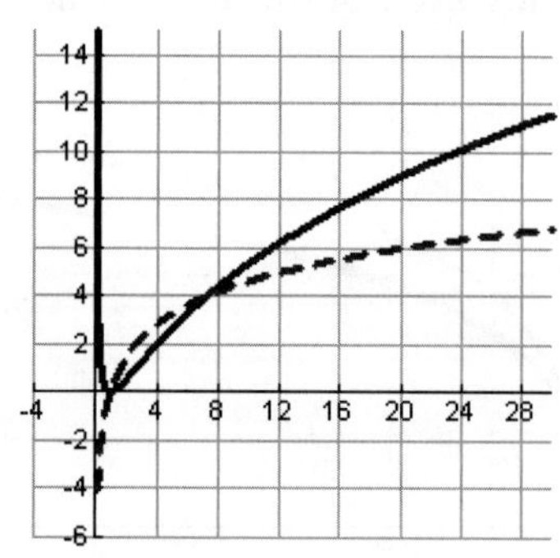

89. The graphs of $y = \ln x$ and $y = \dfrac{\log x}{\log e}$ do coincide, and are as follows:

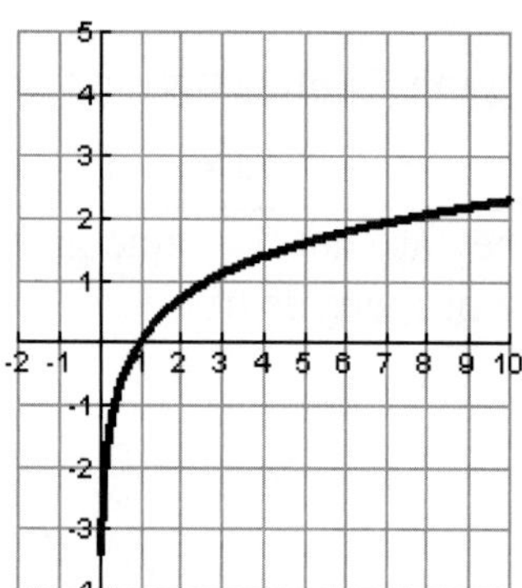

90. The graphs of $y = \log x$ and $y = \dfrac{\ln x}{\ln 10}$ do coincide, and are as follows:

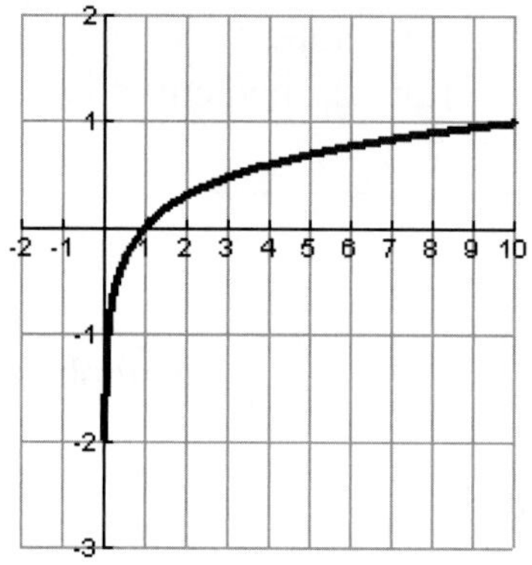

91. Observe that

$$f(x) = \ln x^2 = 2\ln x = \ln x + \ln x.$$

Using $(\ln x)' = \frac{1}{x}$, we conclude that

$$f'(x) = \tfrac{1}{x} + \tfrac{1}{x} = \tfrac{2}{x}$$

92. Observe that

$$f(x) = \ln\left(\tfrac{1}{x}\right) = \ln 1 - \ln x = 0 - \ln x.$$

Using $(\ln x)' = \frac{1}{x}$, we conclude that $f'(x) = -\frac{1}{x}$.

93. Observe that

$$f(x) = \ln\left(\tfrac{1}{x^2}\right) = \ln 1 - \ln x^2$$
$$= 0 - 2\ln x = -\ln x - \ln x$$

Using $(\ln x)' = \frac{1}{x}$, we conclude that

$$f'(x) = -\tfrac{1}{x} - \tfrac{1}{x} = -\tfrac{2}{x}$$

94. Observe that

$$f(x) = \ln x^2 + \ln x^3 = 2\ln x + 3\ln x$$
$$= \underbrace{\ln x + \ldots + \ln x}_{\text{5 times}}$$

Using $(\ln x)' = \frac{1}{x}$, we conclude that

$$f'(x) = \underbrace{\tfrac{1}{x} + \ldots + \tfrac{1}{x}}_{\text{5 times}} = \tfrac{5}{x}.$$

Section 3.4 Solutions --

1.

$$2^{x^2}=16$$
$$2^{x^2}=2^4$$
$$x^2=4$$
$$x=\pm 2$$

2.

$$169^x=13$$
$$\left(13^2\right)^x=13$$
$$13^{2x}=13$$
$$2x=1$$
$$x=\tfrac{1}{2}$$

3.

$$\left(\tfrac{2}{3}\right)^{x+1}=\tfrac{27}{8}$$
$$\left(\tfrac{2}{3}\right)^{x+1}=\left(\tfrac{3}{2}\right)^3=\left(\tfrac{2}{3}\right)^{-3}$$
$$x+1=-3$$
$$x=-4$$

4.

$$\left(\tfrac{3}{5}\right)^{x+1}=\tfrac{25}{9}$$
$$\left(\tfrac{3}{5}\right)^{x+1}=\left(\tfrac{5}{3}\right)^2=\left(\tfrac{3}{5}\right)^{-2}$$
$$x+1=-2$$
$$x=-3$$

5.

$$e^{2x+3}=1=e^0$$
$$2x+3=0$$
$$x=-\tfrac{3}{2}$$

6.

$$10^{x^2-1}=1=10^0$$
$$x^2-1=0$$
$$x=\pm 1$$

7.

$$7^{2x-5}=7^{3x-4}$$
$$2x-5=3x-4$$
$$\boxed{x=-1}$$

8.

$$125^x=5^{2x-3}$$
$$5^{3x}=\left(5^3\right)^x=5^{2x-3}$$
$$3x=2x-3$$
$$\boxed{x=-3}$$

9.

$$2^{x^2+12}=2^{7x}$$
$$x^2+12=7x$$
$$x^2-7x+12=0$$
$$(x-4)(x-3)=0$$
$$\boxed{x=3,4}$$

10.

$$5^{x^2-3}=5^{2x}$$
$$x^2-3=2x$$
$$x^2-2x-3=0$$
$$(x-3)(x+1)=0$$
$$\boxed{x=-1,3}$$

11.

$$9^x=3^{x^2-4x}$$
$$3^{2x}=\left(3^2\right)^x=3^{x^2-4x}$$
$$2x=x^2-4x$$
$$x^2-6x=0$$
$$x(x-6)=0$$
$$\boxed{x=0,6}$$

12. $16^{x-1}=2^{x^2}$ $2^{4x-4}=\left(2^4\right)^{x-1}=2^{x^2}$ $4x-4=x^2$ $x^2-4x+4=0$ $(x-2)^2=0$ $\boxed{x=2}$	**13.** $e^{5x-1}=e^{x^2+3}$ $5x-1=x^2+3$ $x^2-5x+4=0$ $(x-4)(x-1)=0$ $\boxed{x=1,4}$
14. $10^{x^2-8x}=100^x$ $10^{x^2-8x}=\left(10^2\right)^x=10^{2x}$ $x^2-8x=2x$ $x^2-10x=0$ $x(x-10)=0$ $\boxed{x=0,10}$	**15.** $27=2^{3x-1}$ $\log_2(27)=3x-1$ $x=\frac{1}{3}\left[\log_2(27)+1\right]\approx\boxed{1.918}$
16. $15=7^{3-2x}$ $\log_7(15)=3-2x$ $x=-\frac{1}{2}\left[\log_7(15)-3\right]\approx\boxed{0.804}$	**17.** $3e^x-8=7$ $3e^x=15$ $e^x=5$ $x=\ln 5\approx\boxed{1.609}$
18. $5e^x+12=27$ $5e^x=15$ $e^x=3$ $x=\ln 3\approx\boxed{1.100}$	**19** $9-2e^{0.1x}=1$ $2e^{0.1x}=8$ $e^{0.1x}=4$ $0.1x=\ln 4$ $x\approx\boxed{13.863}$

20.

$$\begin{aligned} 21-4e^{0.1x} &= 5 \\ 4e^{0.1x} &= 16 \\ e^{0.1x} &= 4 \\ 0.1x &= \ln 4 \\ x &\approx \boxed{13.863} \end{aligned}$$

21.

$$\begin{aligned} 2\left(3^x\right)-11 &= 9 \\ 2\left(3^x\right) &= 20 \\ 3^x &= 10 \\ x &= \log_3(10) \approx \boxed{2.096} \end{aligned}$$

22.

$$\begin{aligned} 3\left(2^x\right)+8 &= 35 \\ 3\left(2^x\right) &= 27 \\ 2^x &= 9 \\ x &= \log_2(9) \approx \boxed{3.170} \end{aligned}$$

23.

$$\begin{aligned} e^{3x+4} &= 22 \\ \ln\left(e^{3x+4}\right) &= \ln(22) \\ 3x+4 &= \ln(22) \\ 3x &= -4+\ln(22) \\ x &= \tfrac{-4+\ln(22)}{3} \cong -0.303 \end{aligned}$$

24.

$$\begin{aligned} e^{x^2} &= 73 \\ \ln\left(e^{x^2}\right) &= \ln(73) \\ x^2 &= \ln(73) \\ x &= \pm\sqrt{\ln(73)} \approx \pm 2.071 \end{aligned}$$

25.

$$\begin{aligned} 3e^{2x} &= 18 \\ e^{2x} &= 6 \\ \ln\left(e^{2x}\right) &= \ln(6) \\ 2x &= \ln(6) \\ x &= \tfrac{\ln 6}{2} \cong 0.896 \end{aligned}$$

26.

$$\begin{aligned} 4\left(10^{3x}\right) &= 20 \\ 10^{3x} &= 5 \\ \log\left(10^{3x}\right) &= \log 5 \\ 3x &= \log 5 \\ x &\approx 0.233 \end{aligned}$$

27.

$$\begin{aligned} 4e^{2x+1} &= 17 \\ e^{2x+1} &= \tfrac{17}{4} \\ 2x+1 &= \ln\left(\tfrac{17}{4}\right) \\ x &= \frac{\ln\left(\tfrac{17}{4}\right)-1}{2} \approx 0.223 \end{aligned}$$

28.

$$\begin{aligned} 5\left(10^{x^2+2x+1}\right) &= 13 \\ 10^{(x+1)^2} &= \tfrac{13}{5} \\ (x+1)^2 &= \log\left(\tfrac{13}{5}\right) \\ x &= -1 \pm \sqrt{\log\left(\tfrac{13}{5}\right)} \\ &\approx -1.644,\ -0.356 \end{aligned}$$

29.

$$3\left(4^{x^2-4}\right)=16$$
$$4^{x^2-4}=\tfrac{16}{3}$$
$$x^2-4=\log_4\left(\tfrac{16}{3}\right)$$
$$x^2=4+\log_4\left(\tfrac{16}{3}\right)$$
$$x=\pm\sqrt{4+\log_4\left(\tfrac{16}{3}\right)}\approx\pm2.282$$

30.

$$7\left(\tfrac{1}{4}\right)^{6-5x}=3$$
$$\left(\tfrac{1}{4}\right)^{6-5x}=\tfrac{3}{7}$$
$$\log_{\frac{1}{4}}\left(\tfrac{3}{7}\right)=6-5x$$
$$x=-\tfrac{1}{5}\left[-6+\log_{\frac{1}{4}}\left(\tfrac{3}{7}\right)\right]\approx1.078$$

31. Note that $e^{2x}+7e^x-3=0$ is equivalent to $\left(e^x\right)^2+7\left(e^x\right)-3=0$. Let $y=e^x$ and solve $y^2+7y-3=0$ using the quadratic formula:

$$y=\frac{-7\pm\sqrt{7^2-4(1)(-3)}}{2}=\frac{-7\pm\sqrt{61}}{2}$$

So, substituting back in for y, the following two equations must be solved for x:

$$e^x=\frac{-7+\sqrt{61}}{2}\text{ and }e^x=\frac{-7-\sqrt{61}}{2}$$

Since $\frac{-7-\sqrt{61}}{2}<0$, the second equation has no real solution. Solving the first one yields $x=\ln\left(\frac{-7+\sqrt{61}}{2}\right)\cong-0.904$.

32. Note that $e^{2x}-4e^x-5=0$ is equivalent to $\left(e^x\right)^2-4\left(e^x\right)-5=\left(e^x-5\right)\left(e^x+1\right)=0$. So, $e^x-5=0$ or $\underbrace{e^x+1=0}_{\text{No solution}}$. Solving the first equation yields $e^x=5$, so that $x=\ln 5\approx1.609$.

33.

$$\left(3^x-3^{-x}\right)^2=0$$
$$3^x-3^{-x}=0$$
$$3^x=3^{-x}$$
$$x=-x$$
$$x=0$$

34.

$$\left(3^x-3^{-x}\right)\left(3^x+3^{-x}\right)=0$$
$$\left(3^x\right)^2-\left(3^{-x}\right)^2=0$$
$$3^{2x}-3^{-2x}=0$$
$$3^{2x}=3^{-2x}$$
$$2x=-2x$$
$$x=0$$

35.

$$\frac{2}{e^x-5}=1$$
$$2=e^x-5$$
$$e^x=7$$
$$x=\ln(7)\approx1.946$$

36.

$$\begin{aligned} \frac{17}{e^x+4} &= 2 \\ 17 &= 2e^x + 8 \\ 2e^x &= 9 \\ e^x &= \tfrac{9}{2} \\ x &= \ln\left(\tfrac{9}{2}\right) \approx 1.504 \end{aligned}$$

37.

$$\begin{aligned} \frac{20}{6-e^{2x}} &= 4 \\ 20 &= 24 - 4e^{2x} \\ 4e^{2x} &= 4 \\ e^{2x} &= 1 \\ 2x &= 0 \end{aligned}$$

38.

$$\begin{aligned} \frac{4}{3-e^{3x}} &= 8 \\ 4 &= 24 - 8e^{3x} \\ 8e^{3x} &= 20 \\ e^{3x} &= \tfrac{5}{2} \\ 3x &= \ln\left(\tfrac{5}{2}\right) \\ x &\approx 0.305 \end{aligned}$$

39.

$$\begin{aligned} \frac{4}{10^{2x}-7} &= 2 \\ 4 &= 2\left(10^{2x}\right) - 14 \\ 10^{2x} &= 9 \\ 2x &= \log_{10}(9) \\ x &\approx 0.477 \end{aligned}$$

40.

$$\begin{aligned} \frac{28}{10^x+3} &= 4 \\ 28 &= 4\left(10^x\right) + 12 \\ 10^x &= 4 \\ x &= \log_{10}(4) \approx 0.602 \end{aligned}$$

41.

$$\begin{aligned} \log_3(2x+1) &= 4 \\ 2x+1 &= 3^4 \\ 2x &= 80 \\ x &= 40 \end{aligned}$$

42.

$$\begin{aligned} \log_2(3x-1) &= 3 \\ 3x-1 &= 2^3 \\ 3x &= 9 \\ x &= 3 \end{aligned}$$

43.

$$\begin{aligned} \log_2(4x-1) &= -3 \\ 2^{-3} &= 4x-1 \\ 4x &= \tfrac{1}{8} + 1 \\ x &= \tfrac{9}{32} \end{aligned}$$

44.

$$\begin{aligned} \log_4(5-2x) &= -2 \\ 4^{-2} &= 5-2x \\ 2x &= 5 - \tfrac{1}{16} = \tfrac{79}{16} \\ x &= \tfrac{79}{32} \end{aligned}$$

45.

$$\begin{aligned} \ln x^2 - \ln 9 &= 0 \\ \ln x^2 &= \ln 9 \\ x^2 &= 9 \\ x &= \pm 3 \end{aligned}$$

46.

$$\begin{aligned} \log x^2 + \log x &= 3 \\ 2\log x + \log x &= 3 \\ 3\log x &= 3 \\ \log x &= 1 \\ x &= 10 \end{aligned}$$

47.

$$\begin{aligned} \log_5(x-4) + \log_5 x &= 1 \\ \log_5\left(x(x-4)\right) &= 1 \\ x(x-4) &= 5 \\ x^2 - 4x - 5 &= 0 \\ (x-5)(x+1) &= 0 \\ x &= 5, \cancel{-1} \end{aligned}$$

48.

$$\log_2(x-1)+\log_2(x-3)=3$$
$$\log_2(x-1)(x-3)=3$$
$$(x-1)(x-3)=x^2-4x+3=2^3$$
$$x^2-4x-5=0$$
$$(x-5)(x+1)=0$$
$$x=5, \cancel{-1}$$

49.

$$\log(x-3)+\log(x+2)=\log(4x)$$
$$\log(x-3)(x+2)=\log(4x)$$
$$(x-3)(x+2)=4x$$
$$x^2-x-6=4x$$
$$x^2-5x-6=0$$
$$(x-6)(x+1)=0$$
$$x=6, \cancel{-1}$$

50.

$$\log_2(x+1)+\log_2(4-x)=\log_2(6x)$$
$$\log_2\big((x+1)(4-x)\big)=\log_2(6x)$$
$$(x+1)(4-x)=6x$$
$$-x^2+3x+4=6x$$
$$x^2+3x-4=0$$
$$(x+4)(x-1)=0$$
$$x=\cancel{-4}, 1$$

51.

$$\log_4(4x)-\log_4\left(\tfrac{x}{4}\right)=3$$
$$\log_4\left(\frac{4x}{\frac{x}{4}}\right)=3$$
$$\log_4(16)=3$$

Since the last line is a false statement, this equation has no solution.

52.

$$\log_3(7-2x)-\log_3(x+2)=2$$
$$\log_3\left(\frac{7-2x}{x+2}\right)=2$$
$$\frac{7-2x}{x+2}=3^2=9$$
$$9x+18=7-2x$$
$$11x=-11$$
$$x=-1$$

53.

$$\log(2x-5)-\log(x-3)=1$$
$$\log\left(\frac{2x-5}{x-3}\right)=1$$
$$\frac{2x-5}{x-3}=10$$
$$2x-5=10x-30$$
$$8x=25$$
$$x=\tfrac{25}{8}$$

54.

$$\log_3(10-x)-\log_3(x+2)=1$$
$$\log_3\left(\frac{10-x}{x+2}\right)=1$$
$$\frac{10-x}{x+2}=3$$
$$10-x=3(x+2)$$
$$10-x=3x+6$$
$$4=4x$$
$$1=x$$

55.

$$\log_4(x^2+5x+4)-2\log_4(x+1)=2$$
$$\log_4(x+4)(x+1)-\log_4(x+1)^2=2$$
$$\log_4\left[\frac{(x+4)(x+1)}{(x+1)^2}\right]=2$$
$$\log_4\left[\frac{x+4}{x+1}\right]=2$$
$$4^2=\frac{x+4}{x+1}$$
$$16(x+1)=x+4$$
$$15x=-12$$
$$x=-\tfrac{4}{5}$$

56.

$$\log_2(x+1)+\log_2(x+5)-\log_2(2x+5)=2$$
$$\log_2\left[\frac{(x+1)(x+5)}{2x+5}\right]=2$$
$$\frac{(x+1)(x+5)}{2x+5}=4$$
$$8x+20=x^2+6x+5$$
$$x^2-2x-15=0$$
$$(x-5)(x+3)=0$$
$$x=5,\ \cancel{-3}$$

57.

$$\log(2x+5)=2$$
$$2x+5=10^2$$
$$2x=95$$
$$x=47.5$$

58.

$$\ln(4x-7)=3$$
$$4x-7=e^3$$
$$4x=7+e^3$$
$$x=\tfrac{7+e^3}{4}\cong 6.771$$

59.

$$\ln(x^2+1)=4$$
$$x^2+1=e^4$$
$$x^2=e^4-1$$
$$x=\pm\sqrt{e^4-1}\approx\pm 7.321$$

60.

$$\log(x^2+4)=2$$
$$x^2+4=10^2=100$$
$$x^2=96$$
$$x=\pm\sqrt{96}\approx\pm 9.798$$

61.

$$\ln(2x+3) = -2$$
$$2x+3 = e^{-2}$$
$$x = \tfrac{1}{2}\left[-3+e^{-2}\right] \approx -1.432$$

62.

$$\log(3x-5) = -1$$
$$3x-5 = 10^{-1}$$
$$3x = \tfrac{51}{10}$$
$$x = \tfrac{51}{30} = 1.7$$

63.

$$\log(2-3x)+\log(3-2x) = 1.5$$
$$\log\left((2-3x)(3-2x)\right) = 1.5$$
$$(2-3x)(3-2x) = 10^{1.5} \cong 31.622$$
$$6x^2 - 13x + 6 \cong 31.622$$
$$6x^2 - 13x - 25.622 \cong 0$$

Now, use the quadratic formula:

$x = \dfrac{13 \pm \sqrt{783.93}}{12}$, so that $x \cong$ ~~3.42~~, -1.25

64.

$$\log_2(3-x)+\log_2(1-2x) = 5$$
$$\log_2 (3-x)(1-2x) = 5$$
$$(3-x)(1-2x) = 2^5 = 32$$
$$2x^2 - 7x + 3 = 32$$
$$2x^2 - 7x - 29 = 0$$

$x = \dfrac{7 \pm \sqrt{49-4(2)(-29)}}{2(2)} \approx -2.441$, ~~5.941~~

65.

$$\ln x + \ln(x-2) = 4$$
$$\ln x(x-2) = 4$$
$$x(x-2) = e^4$$
$$x^2 - 2x - e^4 = 0$$
$$x = \frac{2 \pm \sqrt{4-4(1)(-e^4)}}{2}$$

$\approx$ ~~-6.456~~, 8.456

66.

$$\ln(4x) + \ln(2+x) = 2$$
$$\ln\left((4x)(2+x)\right) = 2$$
$$4x(2+x) = e^2$$
$$4x^2 + 8x - e^2 = 0$$

Now, use the quadratic formula:

$$x = \frac{-8 \pm \sqrt{64-4(4)(-e^2)}}{2(4)}$$

$\cong 0.6875$, ~~-2.6875~~

67.

$$\log_7(1-x)-\log_7(x+2)=\log_7(x)$$
$$\log_7\left(\frac{1-x}{x+2}\right)=\log_7(x)$$
$$\frac{1-x}{x+2}=x$$
$$1-x=x(x+2)$$
$$x^2+3x-1=0$$

There are no rational solutions since neither 1 nor -1 work. So, graph to find the real roots:

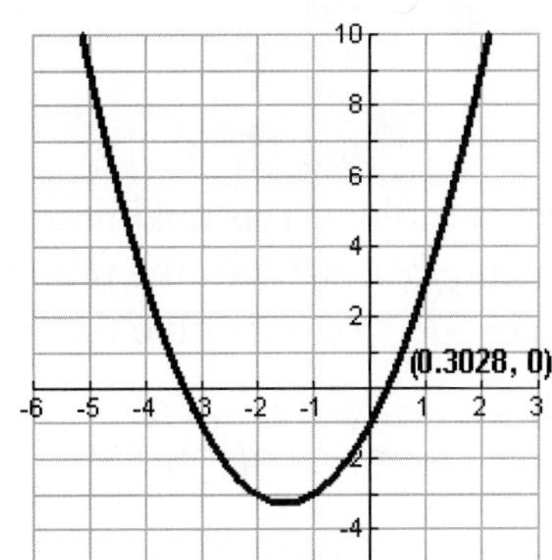

So, the solution is approximately 0.3028.

68.

$$\log_5(x+1)-\log_5(x-1)=\log_5 x$$
$$\log_5\left(\frac{x+1}{x-1}\right)=\log_5 x$$
$$\frac{x+1}{x-1}=x$$
$$x+1=x^2-x$$
$$x^2-2x-1=0$$
$$x=\frac{2\pm\sqrt{4-4(1)(-1)}}{2}=1\pm\sqrt{2}$$
$$\approx \cancel{-0.414},\ 2.414$$

69.

$$\ln\sqrt{x+4}-\ln\sqrt{x-2}=\ln\sqrt{x+1}$$
$$\tfrac{1}{2}\ln(x+4)-\tfrac{1}{2}\ln(x-2)=\tfrac{1}{2}\ln(x+1)$$
$$\ln\left(\frac{x+4}{x-2}\right)=\ln(x+1)$$
$$\frac{x+4}{x-2}=x+1$$
$$x+4=x^2-x-2$$
$$x^2-2x-6=0$$
$$x=\frac{2\pm\sqrt{4-4(1)(-6)}}{2}=1\pm\sqrt{7}$$
$$\approx \cancel{-1.646},\ 3.646$$

70.

$$\log(\sqrt{1-x})-\log(\sqrt{x+2})=\log(x)$$
$$\log\left(\sqrt{\frac{1-x}{x+2}}\right)=\log(x)$$
$$\sqrt{\frac{1-x}{x+2}}=x$$
$$\frac{1-x}{x+2}=x^2$$
$$1-x=x^2(x+2)$$
$$x^3+2x^2+x-1=0$$

Just as in #37, there are no rational solutions since neither 1 nor -1 work. So, graph to find the real roots:

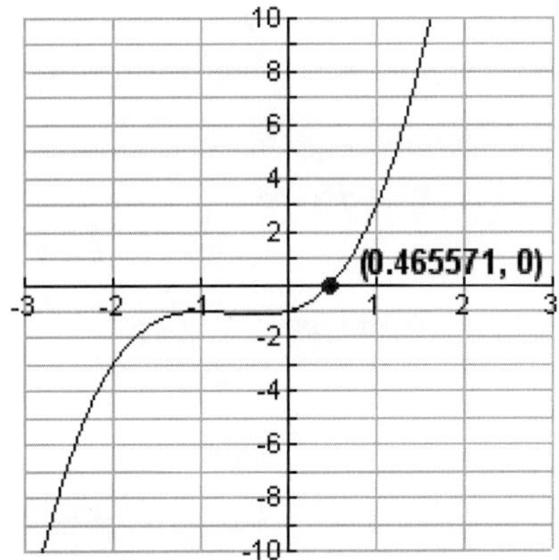

So, the solution is approximately 0.4656.

71. Use $A(t) = P\left(1+\frac{r}{n}\right)^{nt}$.

Here,

$$r = 0.035,\ n = 1.$$

In order to triple, if P is the initial investment, then we seek the time t such that $A(t) = 3P$. So, we solve the following equation:

$$3P = P\left(1+\frac{0.035}{1}\right)^{1(t)}$$

$$3 = (1.035)^t$$

$$\log_{1.035} 3 = t$$

So, it takes approximately 31.9 years to triple.

72. Use $A(t) = P\left(1+\frac{r}{n}\right)^{nt}$.

Here,

$$r = 0.035,\ n = 12.$$

In order to triple, if P is the initial investment, then we seek the time t such that $A(t) = 3P$. So, we solve the following equation:

$$3P = P\left(1+\frac{0.035}{12}\right)^{12(t)}$$

$$3 = (1.0029)^{12t}$$

$$\log_{1.0029} 3 = 12t$$

$$\tfrac{1}{12}\log_{1.0029} 3 = t$$

So, it takes approximately 31.62 years to triple.

73. Use $A(t) = P\left(1+\frac{r}{n}\right)^{nt}$.

Here,

$$A = 20{,}000,\ r = 0.05,\ n = 4,\ P = 7500.$$

So, we solve the following equation:

$$20{,}000 = 7500\left(1+\frac{0.05}{4}\right)^{4(t)}$$

$$2.667 = (1.0125)^{4t}$$

$$\log_{1.0125} 2.667 = 4t$$

$$\tfrac{1}{4}\log_{1.0125} 2.667 = t$$

So, it takes approximately 19.74 years.

74. Use $A(t) = Pe^{rt}$.

Here,

$$A = 15{,}000,\ r = 0.06,\ P = 9000.$$

Substituting into the above equation, we can solve for t:

$$15{,}000 = 9000e^{0.06t}$$

$$1.667 = e^{0.06t}$$

$$\tfrac{1}{0.06}\ln 1.667 = t$$

So, $t \cong 8.514$.

75. Use $M = \frac{2}{3}\log\left(\frac{E}{10^{4.4}}\right)$.

Here, $M = 7.4$. So, substituting in, we can solve for E:

$$7.4 = \tfrac{2}{3}\log\left(\frac{E}{10^{4.4}}\right)$$

$$11.1 = \log\left(\frac{E}{10^{4.4}}\right)$$

$$10^{11.1} = \frac{E}{10^{4.4}}$$

$$\underbrace{10^{11.1}\times 10^{4.4}}_{=10^{15.5}} \approx 3.16\times 10^{15} = E$$

So, it would generate 3.16×10^{15} joules of energy.

76. Use $M = \frac{2}{3}\log\left(\frac{E}{10^{4.4}}\right)$.

Here, $M = 8.3$. So, substituting in, we can solve for E:

$$8.3 = \tfrac{2}{3}\log\left(\frac{E}{10^{4.4}}\right)$$

$$12.45 = \log\left(\frac{E}{10^{4.4}}\right)$$

$$10^{12.45} = \frac{E}{10^{4.4}}$$

$$\underbrace{10^{12.45}\times 10^{4.4}}_{=10^{16.85}} \approx 7.08\times 10^{16} = E$$

So, it would generate $10^{16.85}$ joules of energy.

77. Use $D = 10\log\left(\frac{I}{10^{-12}}\right)$. Here, $D = 120$. So, substituting in, we can solve for *I*: $$120 = 10\log\left(\frac{I}{10^{-12}}\right)$$ $$12 = \log\left(\frac{I}{10^{-12}}\right)$$ $$10^{12} = \frac{I}{10^{-12}}$$ $$1 = 10^{12} \times 10^{-12} = I$$ So, the intensity is $1 \, W/m^2$.	**78.** Use $D = 10\log\left(\frac{I}{10^{-12}}\right)$. Here, $D = 100$. So, substituting in, we can solve for *I*: $$100 = 10\log\left(\frac{I}{10^{-12}}\right)$$ $$10 = \log\left(\frac{I}{10^{-12}}\right)$$ $$10^{10} = \frac{I}{10^{-12}}$$ $$10^{-2} = 10^{10} \times 10^{-12} = I$$ So, the intensity is $10^{-2} \, W/m^2$.
79. Use $A = A_0 e^{-0.5t}$. Here, $A = 0.10A_0$, where A_0 is the initial amount of anesthesia. So, substituting into the above equation, we can solve for *t*: $$0.10A_0 = A_0 e^{-0.5t}$$ $$0.10 = e^{-0.5t}$$ $$\ln(0.10) = -0.5t$$ $$-\tfrac{1}{0.5}\ln(0.10) = t$$ So, it takes about 4.61 hours until 10% of the anesthesia remains in the bloodstream.	**80.** Use $A = A_0 e^{0.03t}$. Here, $A = 2A_0$, where A_0 is the initial investment. So, substituting into the above equation, we can solve for *t*: $$2A_0 = A_0 e^{0.03t}$$ $$2 = e^{0.03t}$$ $$\ln 2 = 0.03t$$ $$\tfrac{1}{0.03}\ln 2 = t$$ So, it takes about 23.105 years until the initial investment doubles.
81. Use $N = \frac{200}{1+24e^{-0.2t}}$. Here, $N = 100$. So, substituting into the above equation, we can solve for *t*: $$100 = \frac{200}{1+24e^{-0.2t}}$$ $$100\left(1+24e^{-0.2t}\right) = 200$$ $$1+24e^{-0.2t} = 2$$ $$24e^{-0.2t} = 1$$ $$e^{-0.2t} = \tfrac{1}{24}$$ $$-0.2t = \ln\left(\tfrac{1}{24}\right)$$ $$t = -\tfrac{1}{0.2}\ln\left(\tfrac{1}{24}\right) \approx 15.89$$ So, it takes about 15.9 years.	**82.** Use $N = \frac{100{,}000}{1+10e^{-2t}}$. Here, $N = 50{,}000$. So, substituting into the above equation, we can solve for *t*: $$50{,}000 = \frac{100{,}000}{1+10e^{-2t}}$$ $$50{,}000\left(1+10e^{-2t}\right) = 100{,}000$$ $$50{,}000 + 500{,}000e^{-2t} = 100{,}000$$ $$e^{-2t} = 0.1$$ $$-2t = \ln(0.1)$$ $$t = -\tfrac{1}{2}\ln(0.1) = \tfrac{1}{2}\ln 10$$ So, it takes $\frac{\ln 10}{2}$ weeks until 50,000 Honda hybrids are on the road.

83. Use $M = \frac{2}{3}\log\left(\frac{E}{10^{4.4}}\right)$.

For P waves: Here, $M = 6.2$. So, substituting in, we can solve for E:

$$6.2 = \tfrac{2}{3}\log\left(\frac{E}{10^{4.4}}\right)$$
$$9.3 = \log\left(\frac{E}{10^{4.4}}\right)$$
$$10^{9.3} = \frac{E}{10^{4.4}}$$
$$\underbrace{10^{9.3} \times 10^{4.4}}_{=10^{13.7}} = E$$

For S waves: Here, $M = 3.3$. So, substituting in, we can solve for E:

$$3.3 = \tfrac{2}{3}\log\left(\frac{E}{10^{4.4}}\right)$$
$$4.95 = \log\left(\frac{E}{10^{4.4}}\right)$$
$$10^{4.95} = \frac{E}{10^{4.4}}$$
$$\underbrace{10^{4.95} \times 10^{4.4}}_{=10^{9.35}} = E$$

So, the combined energy is $\left(10^{13.7} + 10^{9.35}\right)$joules. Hence, the reading on the Richter scale is: $M = \frac{2}{3}\log\left(\frac{10^{9.35} + 10^{13.7}}{10^{4.4}}\right) \approx 6.2$

84. Use $D = 10\log\left(\frac{I}{10^{-12}}\right)$.

Here, $D = \underbrace{100}_{\text{Concert}} + \underbrace{60}_{\text{Converstaion}} = 160$. So, substituting in, we can solve for I:

$$160 = 10\log\left(\frac{I}{10^{-12}}\right)$$
$$16 = \log\left(\frac{I}{10^{-12}}\right)$$
$$10^{16} = \frac{I}{10^{-12}}$$
$$10^{16} \times 10^{-12} = I$$
$$10^{4} = I$$

So, the intensity is 10^4 watts.

85. $\ln(4e^x) \neq 4x$. Should first divide both sides by 4, then take the natural log:

$$4e^x = 9$$
$$e^x = \tfrac{9}{4}$$
$$\ln(e^x) = \ln(\tfrac{9}{4})$$
$$x = \ln(\tfrac{9}{4})$$

86. Step 2 should be:

$$10^{\log(3x)} = 10$$
$$\log(3x) = 1$$
$$3x = 10$$
$$x = \tfrac{10}{3}$$

87. $x = -5$ is not a solution since $\log(-5)$ is not defined.

88. $\log x + \log 2 \neq \log(x+2)$, in general. The computation should be:

$$\log(2x) = \log 5$$
$$2x = 5$$
$$x = \tfrac{5}{2}$$

89. True.

90. False. $\log(x^2) = 2\log(x)$,whereas $(\log x)^2 \neq 2\log x$, in general.

91. False. Since $\log x = \frac{\ln x}{\ln 10}$,

$$e^{\log x} = e^{\frac{\ln x}{\ln 10}} = \left(e^{\ln x}\right)^{\frac{1}{\ln 10}} = x^{\frac{1}{\ln 10}} \neq x.$$

92. True. Since the left side is always positive, and the right side is always negative.

93. False.

$$\log_3\left(x^2+x-6\right)=1 \Rightarrow x^2+x-6=3 \Rightarrow x^2+x-9=0 \Rightarrow x=\frac{-1\pm\sqrt{1+36}}{2}=\frac{-1\pm\sqrt{37}}{2}$$

Checking, we see that $\frac{-1+\sqrt{37}}{2}$ does not satisfy the equation. So, there is only one solution.

94. False. $\frac{\log b}{\log a} \neq \log(b-a)$, in general. For instance, take $a=b=5$ and note that $\frac{\log 5}{\log 5}=1$, but log(0) is undefined.

95.

$$\tfrac{1}{3}\log_b(x^3)+\tfrac{1}{2}\log_b(\underbrace{x^2-2x+1}_{=(x-1)^2})=2$$

$$\log_b(x^3)^{\frac{1}{3}}+\log_b(x-1)^{2\cdot\frac{1}{2}}=2$$

$$\log_b x+\log_b(x-1)=2$$

$$\log_b(x(x-1))=2$$

$$b^2=x(x-1)$$

$$x^2-x-b^2=0$$

Now, use the quadratic formula to find the solutions:

$$x=\frac{1\pm\sqrt{1-4(-b^2)}}{2}$$

$$=\frac{1+\sqrt{1+4b^2}}{2},\quad \underbrace{\frac{1-\sqrt{1+4b^2}}{2}}_{\text{This is negative, for any value of } b\text{. So, it cannot be a solution.}}$$

96.

$$2\log_b(x)+2\log_b(1-x)=4$$

$$\log_b(x)+\log_b(1-x)=2$$

$$\log_b\left(x(1-x)\right)=2$$

$$x(1-x)=b^2$$

$$x^2-x+b^2=0$$

Now, use the quadratic formula to find the solutions:

$$x=\frac{1\pm\sqrt{1-4b^2}}{2}$$

$$=\frac{1+\sqrt{1-4b^2}}{2},\ \frac{1-\sqrt{1-4b^2}}{2}$$

Note that we need to impose a restriction on *b*, namely we can only consider those values of *b* for which $1-4b^2\geq 0$. This occurs for $b\in\left[-\frac{1}{2},\frac{1}{2}\right]$. Now, observe that for all such values of *b*, the integrand is positive and less than 1. Hence, both values of *x* above are positive, and thus are both solutions to the original equation.

97.

$$y = \frac{3000}{1+2e^{-0.2t}}$$

$$y\left(1+2e^{-0.2t}\right) = 3000$$

$$y + 2ye^{-0.2t} = 3000$$

$$2ye^{-0.2t} = 3000 - y$$

$$e^{-0.2t} = \frac{3000-y}{2y}$$

$$-0.2t = \ln\left(\frac{3000-y}{2y}\right)$$

$$t = -5\ln\left(\frac{3000-y}{2y}\right)$$

98. Must have

$x^2 - a^2 = (x-a)(x+a) > 0$.

CPs are $a, -a$.

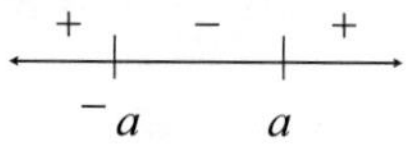

So, the identity is valid for any x in the set $(-\infty, -a) \cup (a, \infty)$.

99. Consider the function $y = \dfrac{e^x + e^{-x}}{2}$, for $x \ge 0,\ y \ge 1$. (Need this restriction in order for the function to be one-to-one, and hence have an inverse.) Solve $y = \dfrac{e^x + e^{-x}}{2}$ for x.

$$y = \frac{e^x + e^{-x}}{2}$$

$$y = \frac{e^x + \frac{1}{e^x}}{2} = \frac{\frac{(e^x)^2+1}{e^x}}{2}$$

$$2y = \frac{(e^x)^2+1}{e^x}$$

$$2ye^x = (e^x)^2 + 1$$

$$(e^x)^2 - 2y\left(e^x\right) + 1 = 0$$

Now, solve using the quadratic formula:

$$e^x = \frac{2y \pm \sqrt{(-2y)^2-4}}{2} = \frac{2y \pm \sqrt{4y^2-4}}{2}$$

$$= \frac{2y \pm 2\sqrt{y^2-1}}{2} = y \pm \sqrt{y^2-1}$$

Since $dom(f) = (0,\infty) = rng(f^{-1})$, we use only $y + \sqrt{y^2-1}$ here. Now to solve for x, take natural log of both sides:

$$e^x = y + \sqrt{y^2-1}$$

$$x = \ln\left(y + \sqrt{y^2-1}\right)$$

Thus, inverse function is given by

$$f^{-1}(x) = \ln\left(x + \sqrt{x^2-1}\right).$$

100. Consider the function $y=\frac{e^x-e^{-x}}{2}$. (There is no need for a restriction on x and y this time since the function is one-to-one on $\mathbb{R}$.) Solve $y=\frac{e^x-e^{-x}}{2}$ for x.

$$y=\frac{e^x-e^{-x}}{2}$$

$$y=\frac{e^x-\frac{1}{e^x}}{2}=\frac{\frac{(e^x)^2-1}{e^x}}{2}$$

$$2y=\frac{(e^x)^2-1}{e^x}$$

$$2ye^x=(e^x)^2-1$$

$$(e^x)^2-2y(e^x)-1=0$$

Now, solve using the quadratic formula:

$$e^x=\frac{2y\pm\sqrt{(-2y)^2+4}}{2}=\frac{2y\pm\sqrt{4y^2+4}}{2}$$

$$=\frac{2y\pm2\sqrt{y^2+1}}{2}=y\pm\sqrt{y^2+1}$$

Since $e^x>0$, we use only $y+\sqrt{y^2+1}$ here. Now to solve for x, take natural log of both sides:

$$e^x=y+\sqrt{y^2+1}$$

$$x=\ln\left(y+\sqrt{y^2+1}\right)$$

Thus, inverse function is given by

$$f^{-1}(x)=\ln\left(x+\sqrt{x^2+1}\right).$$

101. Observe that

$$\ln(3x)=\ln(x^2+1)$$

$$3x=x^2+1$$

$$x^2-3x+1=0$$

$$x=\frac{3\pm\sqrt{9-4}}{2}=\frac{3\pm\sqrt{5}}{2}$$

These solutions agree with the graphical solution seen to the right.

Note on Graphs: Solid curve is $y=\ln(3x)$ and the thin curve is $y=\ln(x^2+1)$.

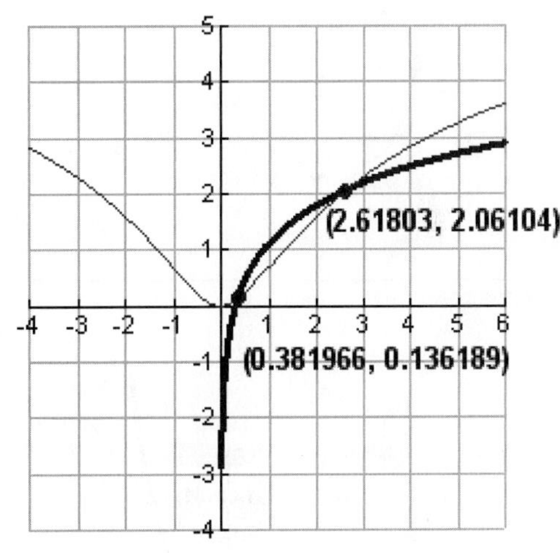

102. Observe that

$$10^{x^2} = 0.001^x$$
$$10^{x^2} = \left(10^{-3}\right)^x$$
$$10^{x^2} = 10^{-3x}$$
$$x^2 = -3x$$
$$x^2 + 3x = x(x+3) = 0, \quad \text{so that } x = 0, -3$$

These solution agrees with the graphical solution seen to the right.

Note on Graphs: Two graphs are provided since the range in values between the two solutions made it difficult to show both on the same graph. In both cases, the solid curve is $y = 10^{x^2}$ and the thin curve is $y = 10^{-3x}$.

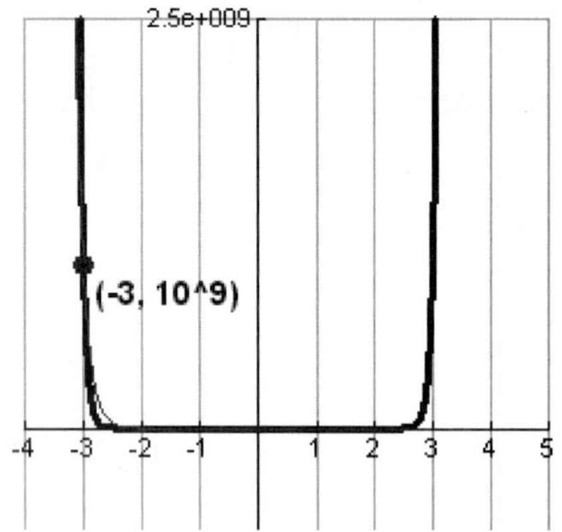

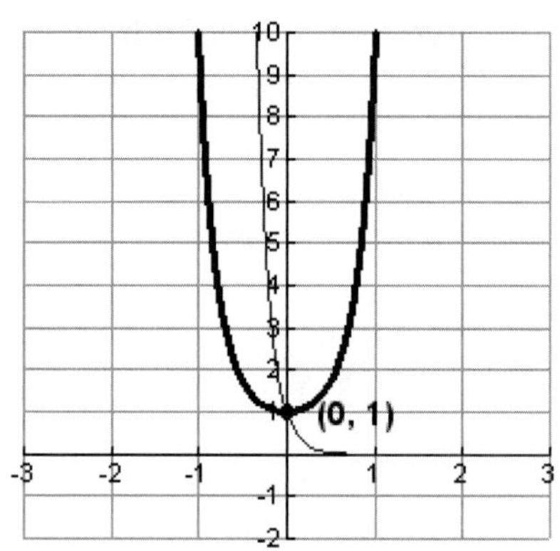

103.

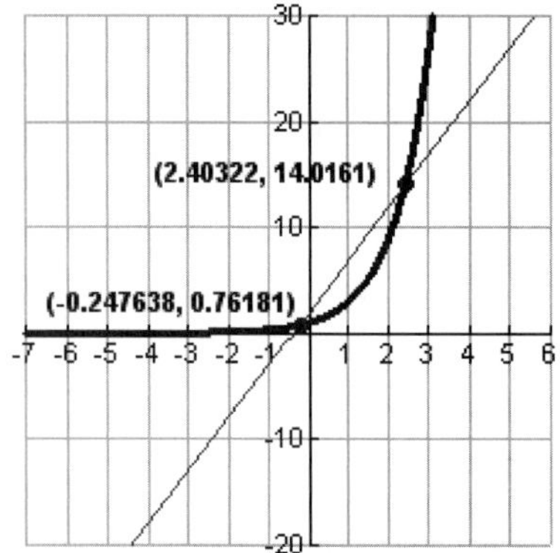

Note on Graphs: Solid curve is $y = 3^x$ and the thin curve is $y = 5x + 2$.

104.

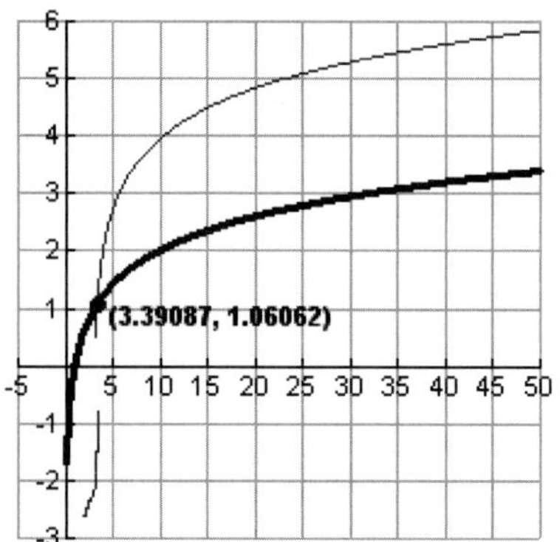

Note on Graphs: Solid curve is $y = 2\log x$ and the thin curve is $y = \ln(x-3) + 2$.

105. The graph of $f(x)=\dfrac{e^x+e^{-x}}{2}$ is given below: 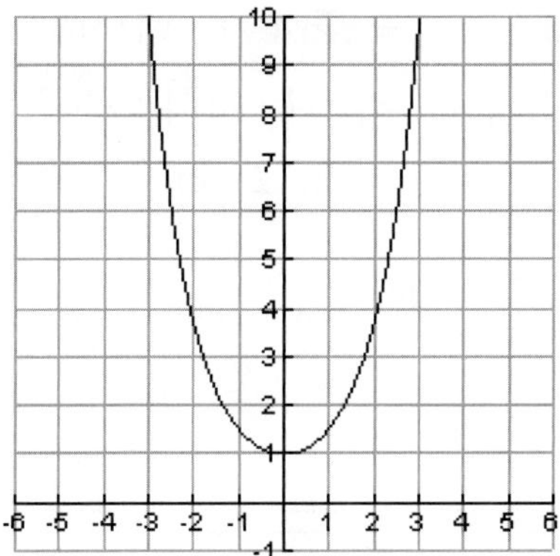 The domain is $(-\infty,\infty)$, and the graph is symmetric about the y-axis. There are no asymptotes.	**106.** The graph of $f(x)=\dfrac{e^x+e^{-x}}{e^x-e^{-x}}$ is: 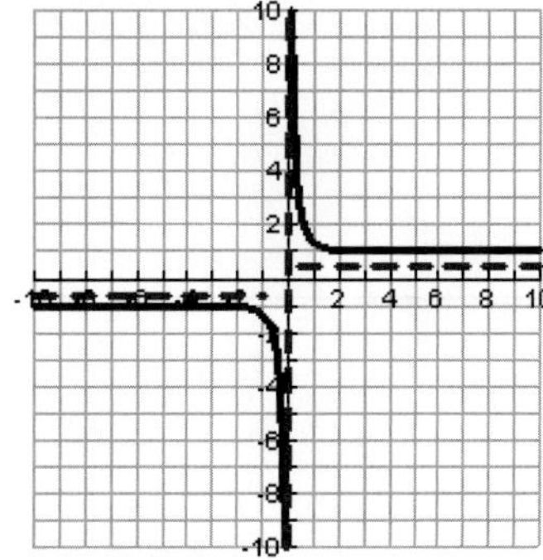 The domain is $(-\infty,0)\cup(0,\infty)$, and the graph is symmetric about the origin. The vertical asymptote is $x=0$, and there are two different horizontal asymptotes (one as $x\to\infty$ and one as $x\to-\infty$). They are $y=1,\ y=-1$ respectively.
107. $y=\frac{e^x-e^{-x}}{2}$ $2y=e^x-e^{-x}$ $2y=\frac{(e^x)^2-1}{e^x}$ $(2y)e^x=(e^x)^2-1$ $(e^x)^2-(2y)e^x-1=0$ $e^x=\dfrac{2y\pm\sqrt{(-2y)^2+4}}{2}$ $e^x=\dfrac{2y\pm2\sqrt{y^2+1}}{2}$ $e^x=y\pm\sqrt{y^2+1}$ $x=\ln\left(y\pm\sqrt{y^2+1}\right)$ Since the input must be positive, we conclude that $f^{-1}(x)=\ln\left(x+\sqrt{x^2+1}\right)$.	**108.** $y=\frac{e^x-e^{-x}}{e^x+e^{-x}}$ $ye^x+ye^{-x}=e^x-e^{-x}$ $ye^x+\frac{y}{e^x}=e^x-\frac{1}{e^x}$ $\frac{y(e^x)^2+y-(e^x)^2+1}{e^x}=0$ $(y-1)(e^x)^2+(y+1)=0$ $(e^x)^2=\frac{y+1}{1-y}$ $e^x=\pm\sqrt{\frac{y+1}{1-y}}$ $x=\ln\left(\pm\sqrt{\frac{y+1}{1-y}}\right)$ Since the input must be positive, we conclude that $f^{-1}(x)=\ln\sqrt{\frac{x+1}{1-x}}$.

109. $\ln y = \ln\left(2^x\right) = x\ln 2$	**110.** $$\begin{aligned} \ln y &= \ln\left(4^x \cdot 3^{x+1}\right) \\ &= \ln\left(4^x\right) + \ln\left(3^{x+1}\right) \\ &= x\ln 4 + (x+1)\ln 3 \\ &= \left[\ln 4 + \ln 3\right]x + \ln 3 \\ &= x\ln 12 + \ln 3 \end{aligned}$$

Section 3.5 Solutions--

1. c (iv)	**2. d (v)**
3. a (iii)	**4. b (ii)**
5. f (i)	**6. e (i)**
7. Use $N = N_0 e^{rt}$. Here, $N_0 = 80$, $r = 0.0236$. Determine N when $t = 7$ (determined by 2010 – 2003): $$N = 80e^{0.0236(7)} \approx 94 \text{ million.}$$	**8.** Use $N = N_0 e^{rt}$. Here, $N_0 = 13.7$, $r = 0.025$. Determine N when $t = 20$ (determined by 2016 – 1996): $$N = 13.7e^{0.025(20)} \approx 23 \text{ million.}$$
9. Use $N = N_0 e^{rt}$. Here, $N_0 = 103{,}800$, $r = 0.12$. Determine t such that $N = 200{,}000$: $$200{,}000 = 103{,}800e^{0.12t}$$ $$\tfrac{200{,}000}{103{,}800} = e^{0.12t}$$ $$t = \tfrac{1}{0.12}\ln\left(\tfrac{200{,}000}{103{,}800}\right) \approx 5.5$$ So, the population would hit 200,000 sometime in the year 2008.	**10.** Use $N = N_0 e^{-rt}$. Here, $N_0 = 776{,}000$, $r = 0.015$. Determine t such that $N = 700{,}000$: $$700{,}000 = 776{,}000e^{-0.015t}$$ $$\tfrac{700{,}000}{776{,}000} = e^{-0.015t}$$ $$t = \tfrac{1}{-0.015}\ln\left(\tfrac{700{,}000}{776{,}000}\right) \approx 6.871$$ So, the population would hit 700,000 sometime in the year 2008.
11. Use $N = N_0 e^{rt}$. Here, $N_0 = 487.4$, $r = 0.165$. Determine N when $t = 3$ (corresponds to the number of cell phone subscribers in 2010): $$N = 487.4e^{0.165(3)} \approx 799.6$$ There are approximately 799.6 subscribers in 2010.	**12.** Use $N = N_0 e^{rt}$ (t is measured in hours). Here, $N_0 = 500$, $r = 0.20$. **a.** $N(12) = 500e^{0.20(12)} \approx 5{,}512$ There are about 5,512 bacteria after 12 hours. **b.** $N(24) = 500e^{0.20(24)} \approx 60{,}756$ There are about 60,756 bacteria after 1 day.

13. Use $N = N_0 e^{rt}$. Here, $N_0 = 185{,}000$, $r = 0.30$. Determine N when $t = 3$: $N = 185{,}000e^{0.30(3)} \approx 455{,}027$ The amount is approximately \$455,000.	**14.** Use $N = N_0 e^{rt}$ **(1)**. Assuming $t = 0$ corresponds to the year 2004, we have $N(0) = N_0 = 230{,}000$ $N(1) = 252{,}000$ **(2)** In order to determine $N(3)$, we need to first determine r. To this end, substitute **(2)** into **(1)** to obtain: $252{,}000 = 230{,}000e^{r} \Rightarrow r = \ln\left(\frac{252{,}000}{230{,}000}\right)$ Thus, $N(3) = 230{,}000e^{3\ln\left(\frac{252{,}000}{230{,}000}\right)} \approx 302{,}500$ The amount is approximately \$302,000.
15. Use $N = N_0 e^{rt}$ (t measured in months). Here, $N_0 = 100$ (million), $r = 0.20$. Determine N when $t = 6$: $N = 100e^{0.20(6)} \approx 332$ million.	**16.** Use $N = N_0 e^{rt}$ (t measured in months). Here, $N_0 = 50$ (million), $r = 0.12$. Determine N when $t = 3$: $N = 50e^{0.12(3)} \approx 72$ million.
17. Use $N = N_0 e^{rt}$ (t measured in months). Here, $N_0 = 1$ (million), $r = 0.025$. Determine N when $t = 7$: $N = 1e^{0.025(7)} \approx 1.19$ million.	**18.** Use $N = N_0 e^{rt}$ (t measured in months). Here, $N_0 = 25$ (million), $r = 0.09$. Determine N when $t = 7$: $N = 25e^{0.09(7)} \approx 46.94$ million.
19. Use $A = 100e^{-0.5t}$. Observe that $A(4) = 100e^{-0.5(4)} \approx 13.53$ ml	**20.** Use $A = 100e^{-0.5t}$. Observe that $A(12) = 100e^{-0.5(12)} \approx 0.25$ ml
21. Use $N = N_0 e^{rt}$ **(1)**. We know that $N(5{,}730) = \frac{1}{2}N_0$ **(2)**. If $N_0 = 5$ (grams), find t such that $N(t) = 2$. To do so, we must first find r. To this end, substitute **(2)** into **(1)** to obtain: $\frac{1}{2} = e^{r(5{,}730)} \Rightarrow \frac{1}{5{,}730}\ln\left(\frac{1}{2}\right) = r$ Now, solve for t: $2 = 5e^{\frac{1}{5{,}730}\ln\left(\frac{1}{2}\right)t}$ $\ln\left(\frac{2}{5}\right) = \frac{1}{5{,}730}\ln\left(\frac{1}{2}\right)t$ $t = \dfrac{\ln\left(\frac{2}{5}\right)}{\frac{1}{5{,}730}\ln\left(\frac{1}{2}\right)} \approx 7575$ years	**22.** Use $N = N_0 e^{rt}$ **(1)**. We know that $N(1{,}600) = \frac{1}{2}N_0$ **(2)**. If $N_0 = 5$ (grams), find t such that $N(t) = 2$. To do so, we must first find r. To this end, substitute **(2)** into **(1)** to obtain: $\frac{1}{2} = e^{r(1{,}600)} \Rightarrow \frac{1}{1{,}600}\ln\left(\frac{1}{2}\right) = r$ Now, solve for t: $2 = 5e^{\frac{1}{1{,}600}\ln\left(\frac{1}{2}\right)t}$ $\ln\left(\frac{2}{5}\right) = \frac{1}{1{,}600}\ln\left(\frac{1}{2}\right)t$ $t = \dfrac{\ln\left(\frac{2}{5}\right)}{\frac{1}{1{,}600}\ln\left(\frac{1}{2}\right)} \approx 2115$ years

23. Use $N = N_0 e^{rt}$ **(1)**. We know that $N(4.5\times10^9) = \frac{1}{2}N_0$ **(2)**. Find *t* such that $N(t) = 0.98N_0$. To do so, we must first find *r*. To this end, substitute **(2)** into **(1)** to obtain:

$$\tfrac{1}{2} = e^{r(4.5\times10^9)} \Rightarrow \tfrac{1}{4.5\times10^9}\ln\left(\tfrac{1}{2}\right) = r$$

Now, solve for *t*:

$$0.98\cancel{N_0} = \cancel{N_0}e^{\frac{1}{4.5\times10^9}\ln\left(\frac{1}{2}\right)t}$$

$$t = \frac{\ln(0.98)}{\frac{1}{4.5\times10^9}\ln\left(\frac{1}{2}\right)} \approx 131{,}158{,}556 \text{ years old}$$

24. Use $N = N_0 e^{rt}$ **(1)**. We know that $N(12) = \frac{1}{2}N_0$ **(2)**. Assuming that $N_0 = 5$, find $N(16)$. To this end, substitute **(2)** into **(1)** to obtain:

$$\tfrac{1}{2} = e^{12r} \Rightarrow \tfrac{1}{12}\ln\left(\tfrac{1}{2}\right) = r$$

Now, compute $N(16)$:

$$N(16) = 5e^{\frac{1}{12}\ln\left(\frac{1}{2}\right)\cdot16} \approx 1.984 \text{ mg}$$

25. Use $T = T_s + \left(T_0 - T_s\right)e^{-kt}$.

Here, $T_0 = 325,\ T_s = 72,$ and $T(10) = 200$. Find $T(30)$. To do so, we must first find *k*. Observe

$$200 = 72 + \left(325 - 72\right)e^{-k(10)}$$

$$128 = 253e^{-10k}$$

$$k = -\tfrac{1}{10}\ln\left(\tfrac{128}{253}\right)$$

Now,

$$T(30) = 72 + 253e^{-\left(-\frac{1}{10}\ln\left(\frac{128}{253}\right)\right)\cdot30} \approx 105^\circ\text{F}.$$

26. Use $T = T_s + \left(T_0 - T_s\right)e^{-kt}$.

Here, $T_0 = 38,\ T_s = 75,$ and $T(5) = 45$. Find $T(20)$. To do so, we must first find *k*. Observe

$$45 = 75 + \left(38 - 75\right)e^{-k(5)}$$

$$-30 = -37e^{-5k}$$

$$k = -\tfrac{1}{5}\ln\left(\tfrac{30}{37}\right)$$

Now,

$$T(20) = 75 - 37e^{-\left(-\frac{1}{5}\ln\left(\frac{30}{37}\right)\right)\cdot20} \approx 59^\circ\text{F}.$$

27. Use $T = T_s + \left(T_0 - T_s\right)e^{-kt}$ **(1)**.

Assume $t = 0$ corresponds to 7am. We know that

$T(0) = 85,\ T(1.5) = 82,\ T_s = 74$. **(2)**

Find *t* such that $T(t) = 98.6$. We first use **(2)** in **(1)** to find *k* and T_0.

$$85 = 74 + \left(T_0 - 74\right)e^0 = T_0 \Rightarrow T_0 = 85$$

$$82 = 74 + \left(85 - 74\right)e^{-1.5k} \Rightarrow 8 = 11e^{-1.5k}$$

$$\Rightarrow k = -\tfrac{1}{1.5}\ln\left(\tfrac{8}{11}\right)$$

28. Use $T = T_s + \left(T_0 - T_s\right)e^{-kt}$ **(1)**.

Assume $t = 0$ corresponds to 4am. We know that

$T(0) = 90,\ T(1) = 86,\ T_s = 60$. **(2)**

Find *t* such that $T(t) = 98.6$. We first use **(2)** in **(1)** to find *k* and T_0.

$$90 = 60 + \left(T_0 - 60\right)e^0 = T_0 \Rightarrow T_0 = 90$$

$$86 = 60 + \left(90 - 60\right)e^{-k} \Rightarrow 26 = 30e^{-k}$$

$$\Rightarrow k = -\ln\left(\tfrac{26}{30}\right)$$

Now, solve for t:

$$98.6 = 74 + (85-74)e^{-\left(-\frac{1}{1.5}\ln\left(\frac{8}{11}\right)\right)t}$$

$$24.6 = 11e^{\frac{1}{1.5}\ln\left(\frac{8}{11}\right)t}$$

$$-3.8 \approx \frac{\ln\left(\frac{24.6}{11}\right)}{\frac{1}{1.5}\ln\left(\frac{8}{11}\right)} = t$$

So, the victim died approximately 3.8 hours before 7am. So, by 8:30am, the victim has been dead for about 5.29 hours.

Now, solve for t:

$$98.6 = 60 + 30e^{-\left(-\ln\left(\frac{26}{30}\right)\right)t}$$

$$38.6 = 30e^{\ln\left(\frac{26}{30}\right)t}$$

$$-1.76 \approx \frac{\ln\left(\frac{38.6}{30}\right)}{\ln\left(\frac{26}{30}\right)} = t$$

So, the victim died approximately 1.76 hours before 4am. So, by 5am, the victim has been dead for about 2.76 hours.

29. Use $N = N_0 e^{-rt}$ **(1)**.

We have

$$\begin{aligned} N(0) &= N_0 = 38{,}000 \\ N(1) &= 32{,}000 \end{aligned} \quad \textbf{(2)}$$

In order to determine $N(4)$, we need to first determine r. To this end, substitute **(2)** into **(1)** to obtain:

$$32{,}000 = 38{,}000e^{-r} \Rightarrow r = -\ln\left(\frac{32{,}000}{38{,}000}\right)$$

Thus, $N(4) = 38{,}000e^{-\left(-\ln\left(\frac{32{,}000}{38{,}000}\right)\right)\cdot 4} \approx 19{,}100$.

The book value after 4 years is approximately $19,100.

30. Use $N = N_0 e^{-rt}$ **(1)**.

We have

$$\begin{aligned} N(0) &= N_0 = 22{,}000 \\ N(2) &= 14{,}000 \end{aligned} \quad \textbf{(2)}$$

In order to determine $N(4)$, we need to first determine r. To this end, substitute **(2)** into **(1)** to obtain:

$$14{,}000 = 22{,}000e^{-r(2)} \Rightarrow r = -\tfrac{1}{2}\ln\left(\frac{14{,}000}{22{,}000}\right)$$

Thus, $N(4) = 22{,}000e^{-\left(-\frac{1}{2}\ln\left(\frac{14{,}000}{22{,}000}\right)\right)\cdot 4} \approx 8{,}900$.

The book value after 4 years is approximately $8,900.

31. Use $N = \dfrac{100{,}000}{1+10e^{-2t}}$.

a. $N(2) = \dfrac{100{,}000}{1+10e^{-2(2)}} \approx 84{,}520$

b. $N(30) = \dfrac{100{,}000}{1+10e^{-2(30)}} \approx 100{,}000$

c. The highest number of new convertibles that will be sold is 100,000 since the smallest that $1+10e^{-2t}$ can be is 1.

32. Use $N = \dfrac{2{,}000{,}000}{1+2e^{-4t}}$.

a. $N(2) = \dfrac{2{,}000{,}000}{1+2e^{-4(2)}} \approx 1{,}998{,}659$

b. $N(4) = \dfrac{2{,}000{,}000}{1+2e^{-4(4)}} \approx 2{,}000{,}000$

33. Use $N = N_0 e^{-rt}$ **(1).**

Assuming that $t = 0$ corresponds to 1997, we have

$$\begin{aligned} N(0) = N_0 = 2{,}422 \\ N(6) = 7{,}684 \end{aligned} \quad \textbf{(2)}$$

In order to determine $N(13)$, we need to first determine r. To this end, substitute **(2)** into **(1)** to obtain:

$$7{,}684 = 2{,}422e^{-r(6)} \quad \Rightarrow \quad r = -\tfrac{1}{6}\ln\left(\tfrac{7{,}684}{2{,}422}\right)$$

Thus,

$$N(13) = 2{,}422e^{-\left(-\frac{1}{6}\ln\left(\frac{7{,}684}{2{,}422}\right)\right)\cdot 13} \approx 29{,}551 \text{ cases.}$$

34. Use $N = N_0 e^{-rt}$ **(1).**

Assuming that $t = 0$ corresponds to 1997, we have

$$\begin{aligned} N(0) = N_0 = 300{,}000 \\ N(3) = 630{,}000 \end{aligned} \quad \textbf{(2)}$$

In order to determine $N(6)$, we need to first determine r. To this end, substitute **(2)** into **(1)** to obtain:

$$630{,}000 = 300{,}000e^{-r(3)} \quad \Rightarrow \quad r = -\tfrac{1}{3}\ln\left(\tfrac{630{,}000}{300{,}000}\right)$$

Thus,

$$\begin{aligned} N(6) &= 300{,}000e^{-\left(-\frac{1}{3}\ln\left(\frac{630{,}000}{300{,}000}\right)\right)\cdot 6} \\ &\approx 1{,}323{,}000 \text{ cases} \end{aligned}$$

35. Find t such that $\dfrac{10{,}000}{1+19e^{-1.56t}} = 5000$.

$$\begin{aligned} 10{,}000 &= 5000\left(1+19e^{-1.56t}\right) \\ 2 &= 1+19e^{-1.56t} \\ 1 &= 19e^{-1.56t} \\ e^{-1.56t} &= \tfrac{1}{19} \\ t &= -\tfrac{1}{1.56}\ln\left(\tfrac{1}{19}\right) \approx 1.89 \text{ years} \end{aligned}$$

36. Find t such that $\dfrac{1600}{1+0.6e^{-0.14t}} = 1200$.

$$\begin{aligned} 1600 &= 1200\left(1+0.6e^{-0.14t}\right) \\ 400 &= 720e^{-0.14t} \\ t &= -\tfrac{1}{0.14}\ln\left(\tfrac{400}{720}\right) \approx 4.20 \text{ years} \end{aligned}$$

37. Consider $I(r) = e^{-r^2}$, whose graph is below. Note that the beam is brightest when $r = 0$.

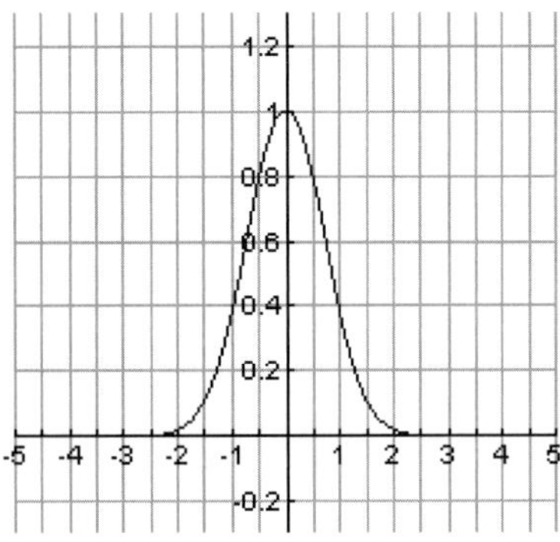

38. Since $I(2) = e^{-2^2} \approx 0.0183$, about 1.83%.

39. Consider $N(x) = 10e^{-\frac{(x-75)^2}{25^2}}$.

a. The graph of N is as follows:

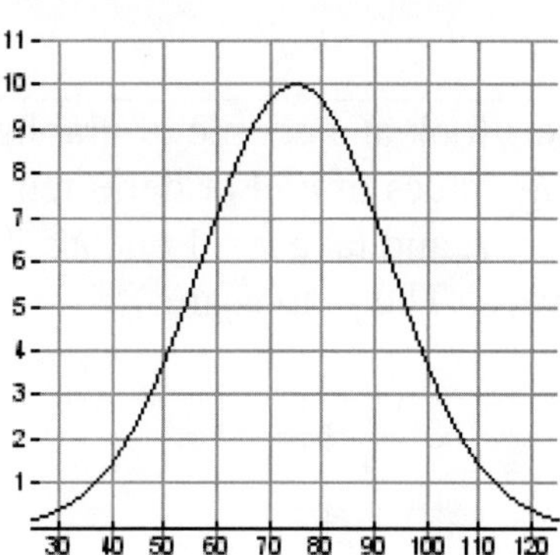

b. Average grade is 75.

c. $N(50) = 10e^{-\frac{(50-75)^2}{25^2}} = 10e^{-1} \approx 4$

d. $N(100) = 10e^{-\frac{(100-75)^2}{25^2}} = 10e^{-1} \approx 4$

40. Consider $N(x) = 10e^{-\frac{(x-80)^2}{16^2}}$.

a. The graph of N is as follows:

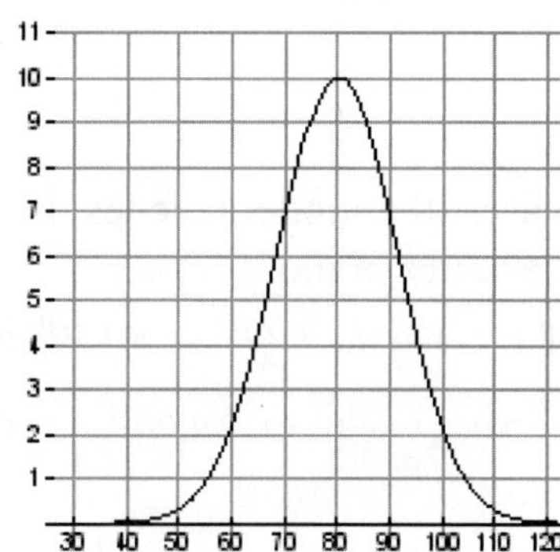

b. Average grade is 80.

c. $N(60) = 10e^{-\frac{(60-80)^2}{16^2}} \approx 2$

d. $N(100) = 10e^{-\frac{(100-80)^2}{16^2}} \approx 2$

41. Use $t = -\dfrac{\ln\left(1 - \frac{Pr}{nR}\right)}{n\ln\left(1 + \frac{r}{n}\right)}$.

a. Here, $P = 80{,}000,\ r = 0.09,\ n = 12, R = 750$.

So, $t = -\dfrac{\ln\left(1 - \frac{80{,}000(0.09)}{12(750)}\right)}{12\ln\left(1 + \frac{0.09}{12}\right)} \approx 18$ years.

b. Here, $P = 80{,}000,\ r = 0.09,\ n = 12, R = 1000$.

So, $t = -\dfrac{\ln\left(1 - \frac{80{,}000(0.09)}{12(1000)}\right)}{12\ln\left(1 + \frac{0.09}{12}\right)} \approx 10$ years.

42. Use $t = -\dfrac{\ln\left(1 - \frac{Pr}{nR}\right)}{n\ln\left(1 + \frac{r}{n}\right)}$.

a. Here, $P = 20{,}000,\ r = 0.17,\ n = 12, R = 300$.

So, $t = -\dfrac{\ln\left(1 - \frac{20{,}000(0.17)}{12(300)}\right)}{12\ln\left(1 + \frac{0.17}{12}\right)} \approx 17.12$ years.

b. Here, $P = 20{,}000,\ r = 0.17,\ n = 12, R = 400$.

So, $t = -\dfrac{\ln\left(1 - \frac{20{,}000(0.17)}{12(400)}\right)}{12\ln\left(1 + \frac{0.17}{12}\right)} \approx 7.30$ years.

43. $r = 0.07$, not 7	**44.** $r = 0.05$, not 5
45. True	**46.** False. Should use logistic curve here.
47. False (Since there is a finite number of students at the school to which the lice can spread.)	**48.** True
49. Take a look at a couple of graphs for increasing values of *c*. For definiteness, let $a = k = 1$, and take $c = 1$ and $c = 5$, respectively. The graphs are: 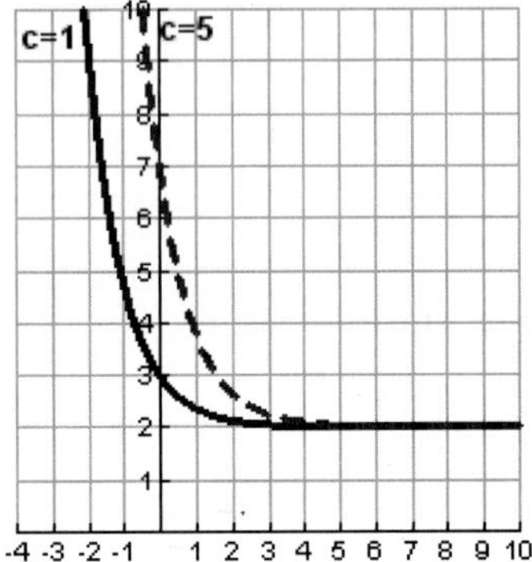 As *c* increases, the model reaches the carrying capacity in more time.	**50.** Take a look at a couple of graphs for increasing values of *k*. For definiteness, let $a = c = 1$, and take $k = 1$ and $k = 5$, respectively. The graphs are: 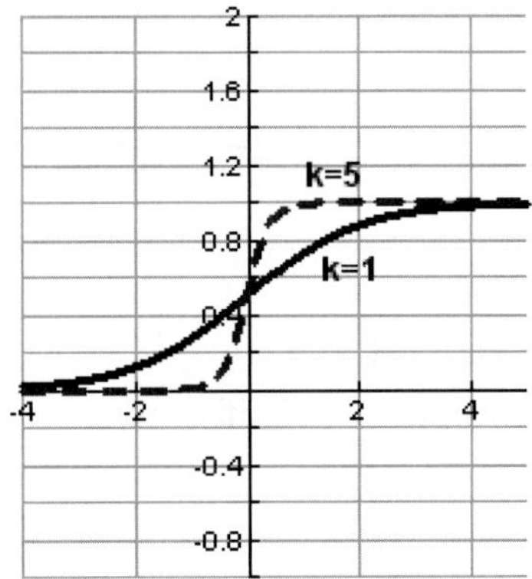 As *k* increases, the model reaches the carrying capacity in more time.
51. Observe that $B(N) = 100(1.20)^N$ and the other bacteria is described by $B^*(N) = 60(1.30)^N$. Solve $100(1.20)^{N+2} = 60(1.30)^N$: $2.4 = \left(\frac{1.30}{1.20}\right)^N \approx 1.0833^N$ $N = \log_{1.0833}(2.4) \approx 10.9$ So, about 10.9 days.	**52. a.)** If $r_1 > r_2$, then $Q = \frac{P_1}{P_2} e^{(r_1 - r_2)t}$ and $r_1 - r_2 > 0$. So, Q is an exponential growth model. **b.)** If $r_1 < r_2$, then $r_1 - r_2 < 0$ and Q is an exponential decay model.
53. Observe that $f(1) = (2+c)e^{-k_1}$ and $g(1) = ce^{-k_2}$. Equating these values yields $(2+c)e^{-k_1} = ce^{-k_2}$ $e^{(k_1 - k_2)} = \frac{2+c}{c}$ $k_1 - k_2 = \ln\left(\frac{2+c}{c}\right)$ $k_1 = k_2 + \ln\left(\frac{2+c}{c}\right)$	**54.** The HA for *f* is $y = a_1$ and the HA for *g* is $y = a_2$. If they coincide, then $a_1 = a_2 = 100$. By definition, these are the carrying capacities.

55. a. The graphs are below:
For the same periodic payment, it will take Wing Shan fewer years to pay off the loan if she can afford to pay biweekly.

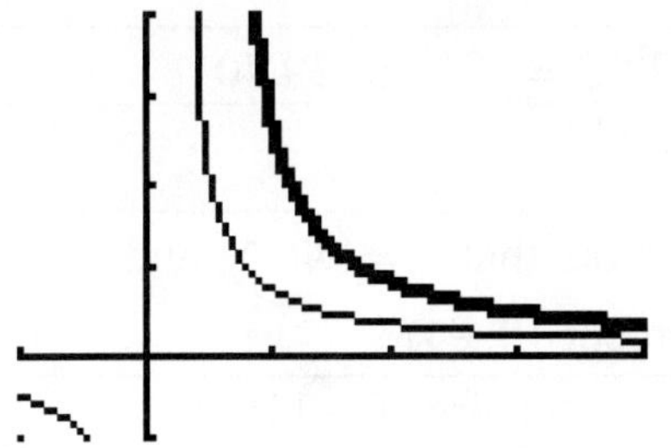

b. 11.58 years
c. 10.33 years
d. 8.54 years, 7.69 years, respectively.

56. The graphs are below:
a. For the same periodic payment, it will take Hong fewer years to pay off the loan if he can afford to pay biweekly.

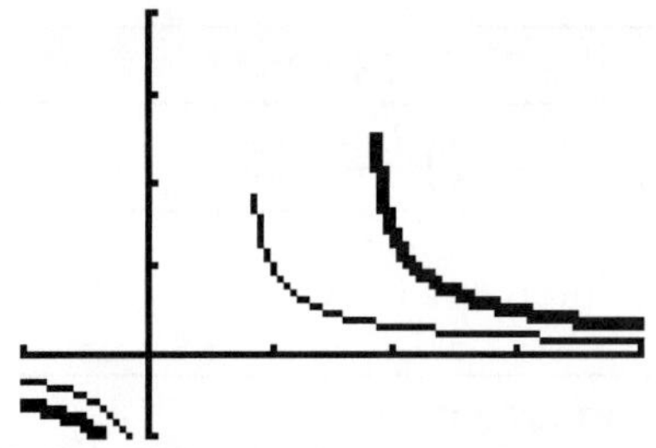

b. 5.12 years
c. 4.50 years
d. 3.35 years, 2.99 years, respectively.

57.

$$\frac{f(x+h)-f(x)}{h}=\frac{Pe^{k(x+h)}-Pe^{kx}}{h}$$
$$=Pe^{kx}\frac{\left(e^{kh}-1\right)}{h}$$

58.

$$\frac{f(x+h)-f(x)}{h}=\frac{Pe^{-k(x+h)}-Pe^{-kx}}{h}$$
$$=Pe^{-kx}\frac{\left(e^{-kh}-1\right)}{h}$$

59.

$$\frac{f(x+h)-f(x)}{h}=\frac{e^{x+h}-e^{x}}{h}+\frac{x+h-x}{h}$$
$$=e^{x}\left(\frac{e^{h}-1}{h}\right)+1$$

Evaluating at h=0 yields $f'(x)=e^{x}+1$.

60. Recall that $f(x)=\cosh(x)=\frac{e^{x}+e^{-x}}{2}$.

So,

$$\frac{f(x+h)-f(x)}{h}=\frac{\frac{e^{x+h}+e^{-(x+h)}}{2}-\frac{e^{x}+e^{-x}}{2}}{h}$$
$$=\frac{e^{x}e^{h}+e^{-x}e^{-h}-e^{x}-e^{-x}}{2h}$$
$$=\frac{e^{x}}{2}\left(\frac{e^{h}-1}{h}\right)+\frac{e^{-x}}{2}\left(\frac{e^{-h}-1}{h}\right)$$

Evaluating at h =0 yields

$$f'(x)=\frac{e^{x}}{2}-\frac{e^{-x}}{2}=\frac{e^{x}-e^{-x}}{2}=\sinh(x).$$

Chapter 3 Review Solutions --

1. 17,559.94	**2.** 1.58
3. 5.52	**4.** 1.24
5. 24.53	**6.** 23.14
7. 5.89	**8.** 0.01
9. $2^{4-(-2.2)} = 2^{6.2} \cong 73.52$	**10.** $-2^{1.3+4} = -2^{5.3} \cong -39.40$
11. $\left(\frac{2}{5}\right)^{1-6(\frac{1}{2})} = \left(\frac{2}{5}\right)^{-2} = 6.25$	**12.** $\left(\frac{4}{7}\right)^{5(\frac{1}{5})+1} = \left(\frac{4}{7}\right)^{2} = \frac{16}{49}$
13. b y-intercept $(0,\frac{1}{4})$	**14. a** Note that $y = -4(\frac{1}{2})^x$ and y-intercept $(0,-4)$
15. c y-intercept $(0,11)$	**16. d** y-intercept $(0,-11)$
17. <u>y-intercept</u>: $(0,-1)$ <u>HA</u>: $y=0$ 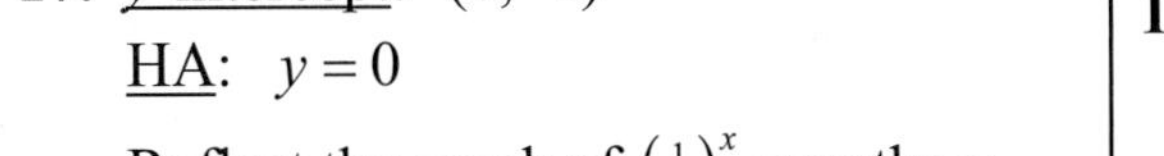 Reflect the graph of $\left(\frac{1}{6}\right)^x$ over the x-axis. 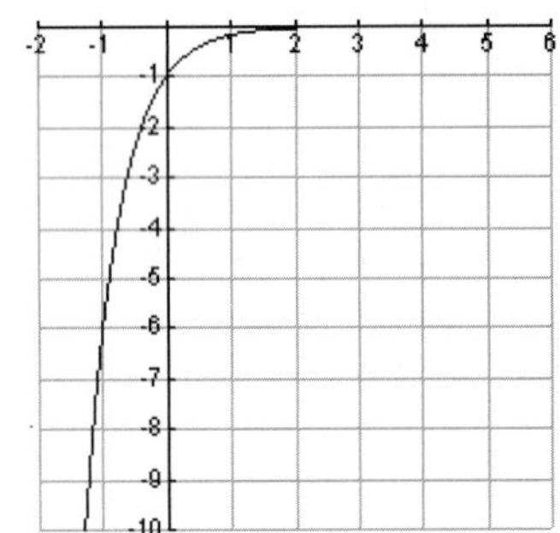	**18.** <u>y-intercept</u>: $(0,3)$. <u>HA</u>: $y=4$ Reflect the graph of 3^x over the x-axis, then shift up 4 units. 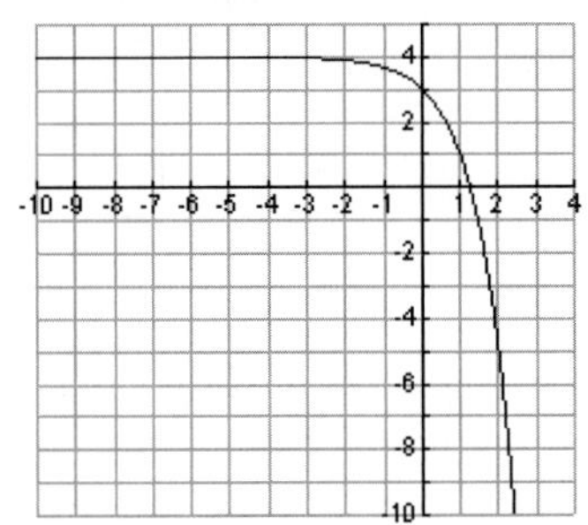
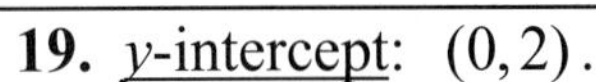 **19.** <u>y-intercept</u>: $(0,2)$. <u>HA</u>: $y=1$ Shift the graph of $\left(\frac{1}{100}\right)^x$ up 1 unit. 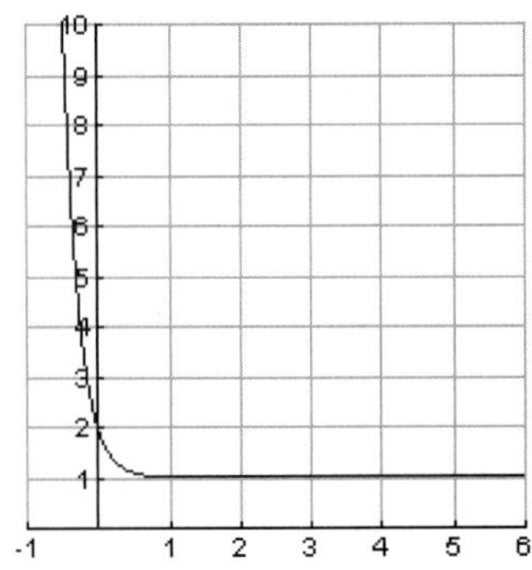	**20.** <u>y-intercept</u>: $(0,-3)$. <u>HA</u>: $y=-4$ Shift the graph of 4^x down 4 units. 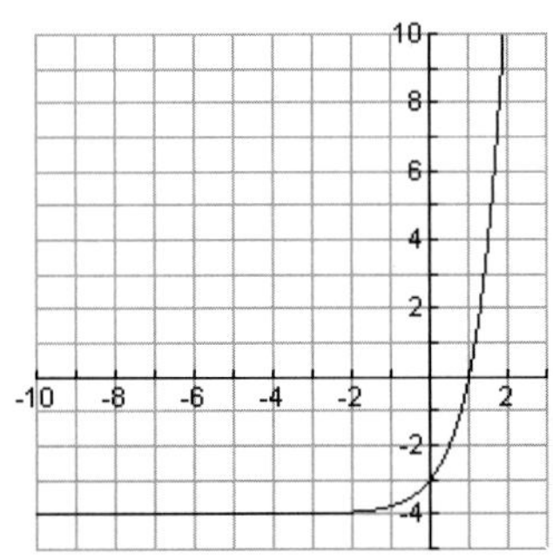

21. y-intercept: $(0,1)$.
HA: $y=0$

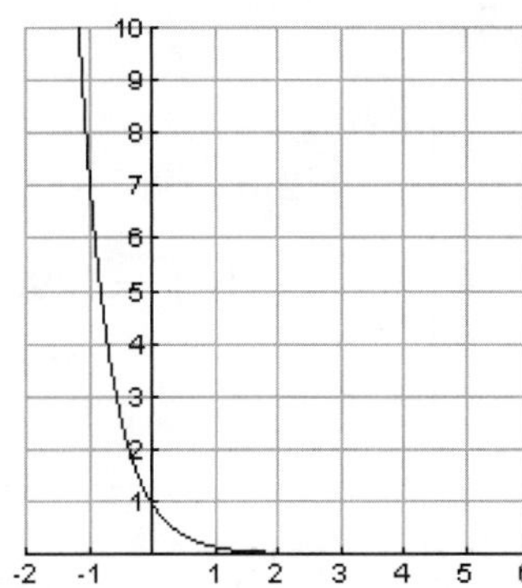

22. y-intercept: $(0,e^{-1})$.
HA: $y=0$

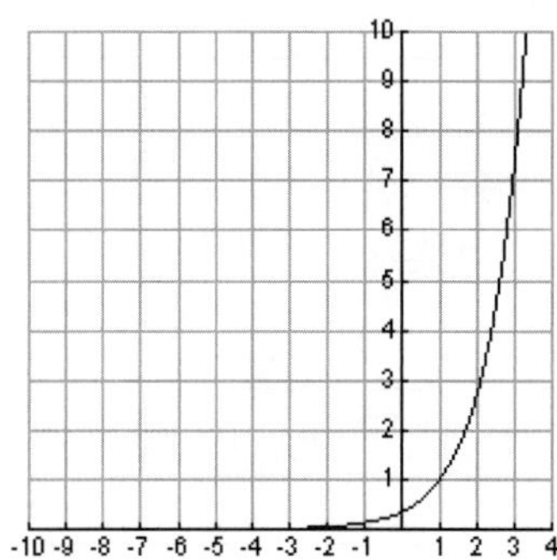

23. y-intercept: $(0,3.2)$.
HA: $y=0$

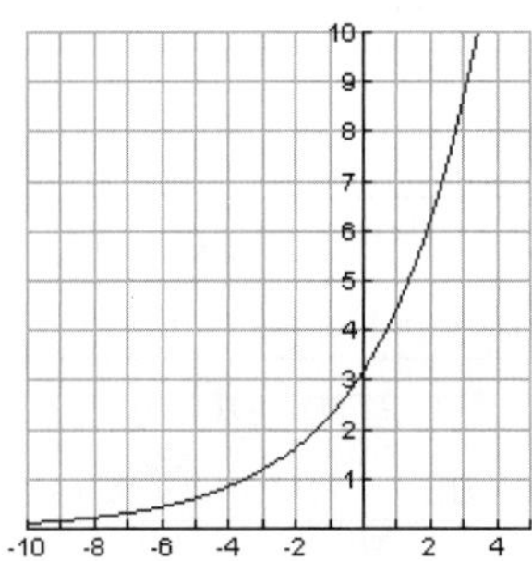

24. y-intercept: $(0,2-e)$.
HA: $y=2$

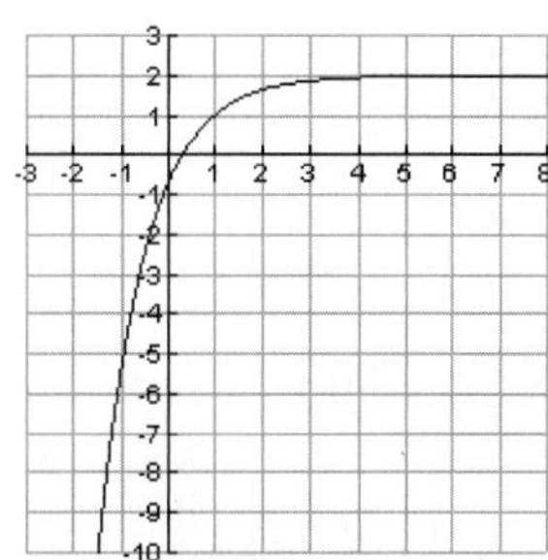

25. Use $A(t)=P\left(1+\frac{r}{n}\right)^{nt}$
Here, $P=4500,\ r=0.045,\ n=2,\ t=7$.
Thus, $A(7)=4500\left(1+\frac{0.045}{2}\right)^{2(7)}\cong 6144.68$.
So, the amount in the account after 7 years is \$6144.68.

26. Use $A(t)=P\left(1+\frac{r}{n}\right)^{nt}$
Here, $A(8)=25{,}000,\ r=0.04,\ n=4,\ t=8$.
Thus, solving the above formula for *P,* we have $P=\dfrac{25{,}000}{\left(1+\frac{0.04}{4}\right)^{4(8)}}\cong 18{,}182.60$.
So, the initial investment should be \$18,182.60.

27. Use $A(t)=Pe^{rt}$.
Here, $P=13{,}450,\ r=0.036,\ t=15$.
Thus,
$A(15)=13{,}450e^{(0.036)(15)}\cong 23{,}080.29$.
So, the amount in the account after 15 years is \$23,080.29.

28. Use $A(t)=Pe^{rt}$.
Here, $A(10)=15{,}000,\ r=0.025,\ t=10$.
Thus, solving the above formula for *P,* we have $P=\dfrac{15{,}000}{e^{(0.025)(10)}}\cong 11{,}682.01$. So, the initial investment should be \$11,682.01.

29. $4^3=64$

30. $4^{\frac{1}{2}}=2$

31. $10^{-2} = \frac{1}{100}$

32. $16^{\frac{1}{2}} = 4$

33. $\log_6 216 = 3$

34. $\log_{10} 0.0001 = -4$

35. $\log_{\frac{2}{13}} \left(\frac{4}{169}\right) = 2$

36. $\log_{512} 8 = \frac{1}{3}$

37.
$$\log_7 1 = x$$
$$7^x = 1$$
$$x = 0$$

38.
$$\log_4 256 = x$$
$$4^x = 256$$
$$x = 4$$

39.
$$\log_{\frac{1}{6}} 1296 = x$$
$$\left(\tfrac{1}{6}\right)^x = 1296 = 6^4$$
$$x = -4$$

40. $\log_{10} 10^{12} = 12\log_{10} 10 = 12$

41. 1.51

42. 3.47

43. –2.08

44. –0.90

45. Must have $x+2>0$, so that the domain is $(-2,\infty)$.

46. Must have $2-x>0$, so that the domain is $(-\infty,2)$.

47. Since $x^2+3>0$, for all values of *x*, the domain is $(-\infty,\infty)$.

48. Must have
$$3-x^2 = (\sqrt{3}-x)(\sqrt{3}+x) > 0.$$
CPs: $-\sqrt{3}$, $\sqrt{3}$

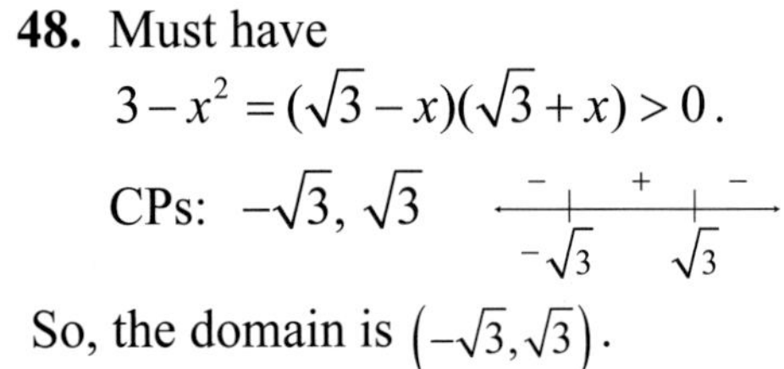

So, the domain is $\left(-\sqrt{3},\sqrt{3}\right)$.

49. b

50. a Reflect the graph of $\log_7 x$ over the *x*-axis, then over the *y*-axis

51. d Shift the graph of $\log_7 x$ left 1 unit, then down 3 units. Also, VA is $x=-1$.

52. c VA is $x=1$

53. Shift the graph of $\log_4 x$ right 4 units, then up 2 units.

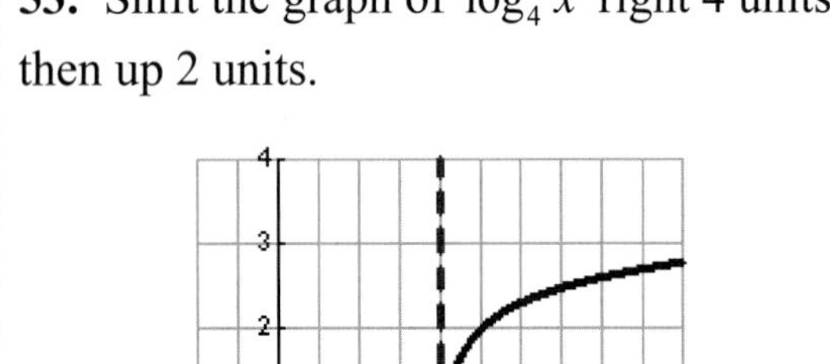

54. Shift the graph of $\log_4 x$ left 4 units, then down 3 units.

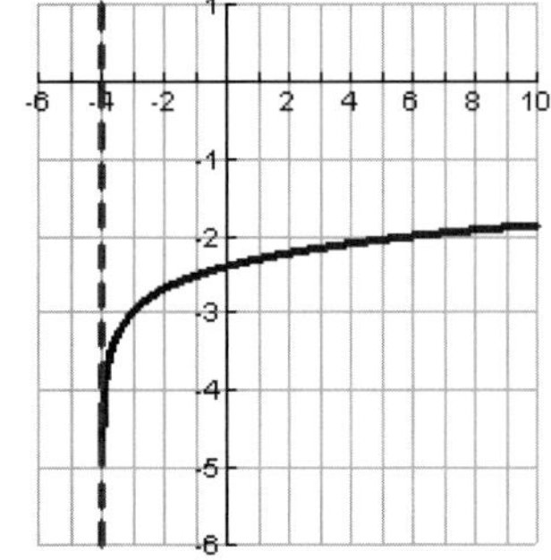

55. Reflect the graph of $\log_4 x$ over the x-axis, then shift down 6 units.

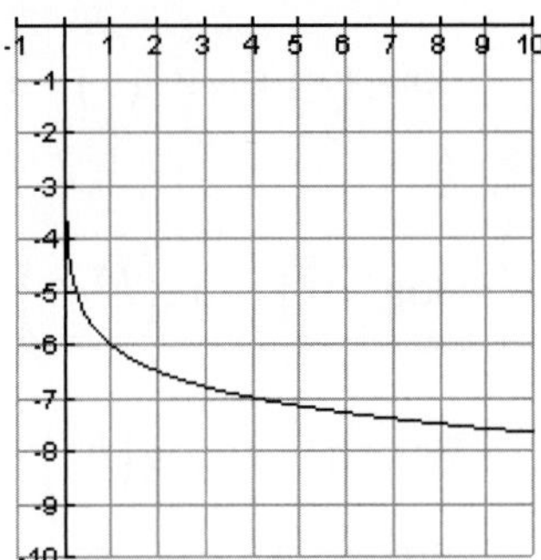

56. Reflect the graph of $\log_4 x$ over the y-axis, then over the x-axis, then expand vertically by a factor of 2, and then shift up 4 units.

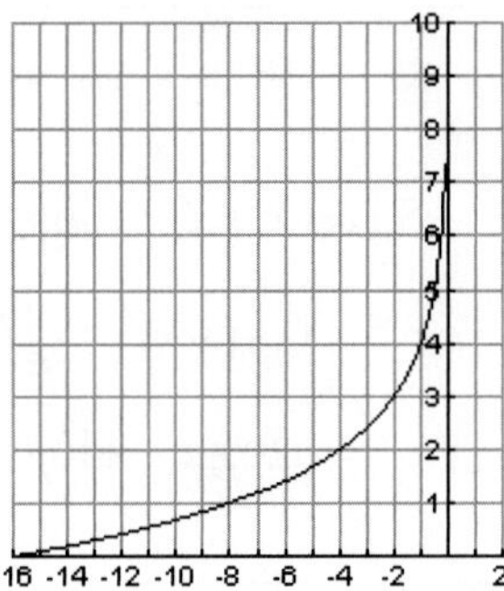

57. Use $pH = -\log_{10}\left[H^+\right]$.

Here,

$$\begin{aligned} pH &= -\log_{10}(3.16\times 10^{-7}) \\ &= -\left[\log(3.16) + \log(10^{-7})\right] \\ &= -\left[\log(3.16) - 7\right] \\ &\cong 6.5 \end{aligned}$$

58. Use $pH = -\log_{10}\left[H^+\right]$.

Here,

$$\begin{aligned} pH &= -\log_{10}(2.0\times 10^{-3}) \\ &= -\left[\log(2.0) + \log(10^{-3})\right] \\ &= -\left[\log(2.0) - 3\right] \\ &\cong 2.7 \end{aligned}$$

59. Use $D = 10\log\left(\dfrac{I}{1\times 10^{-12}}\right)$.

Here,

$$\begin{aligned} D &= 10\log\left(\frac{1\times 10^{-7}}{1\times 10^{-12}}\right) = 10\log(10^5) \\ &= 50\log(10) = 50\,dB \end{aligned}$$

60. Use $D = 10\log\left(\dfrac{I}{1\times 10^{-12}}\right)$.

Here,

$$\begin{aligned} D &= 10\log\left(\frac{1\times 10^{-4}}{1\times 10^{-12}}\right) = 10\log(10^8) \\ &= 80\log(10) = 80\,dB \end{aligned}$$

61. 1

62.

$$\begin{aligned} \log_2 \sqrt{16} &= x \\ 2^x &= \sqrt{16} = 4 \\ x &= 2 \end{aligned}$$

63. 6

64. $e^{-3\ln 6} = e^{\ln 6^{-3}} = 6^{-3} = \frac{1}{216}$

65.

$$\begin{aligned} \log_c\left(x^a y^b\right) &= \log_c\left(x^a\right) + \log_c\left(y^b\right) \\ &= a\log_c(x) + b\log_c(y) \end{aligned}$$

66. $\log_3(x^2y^{-3}) = \log_3(x^2)+\log_3(y^{-3})$ $= 2\log_3(x) - 3\log_3(y)$	**67.** $\log_j\left(\frac{rs}{t^3}\right) = \log_j(rs) - \log_j(t^3)$ $= \log_j(r)+\log_j(s)-3\log_j(t)$
68. $\log\left(x^c\sqrt{x+5}\right) = \log(x^c)+\log\left(\sqrt{x+5}\right)$ $= \log(x^c)+\log(x+5)^{\frac{1}{2}}$ $= c\log x + \frac{1}{2}\log(x+5)$	**69.** $\log\left(\frac{a^{\frac{1}{2}}}{b^{\frac{3}{2}}c^{\frac{2}{5}}}\right) = \log\left(a^{\frac{1}{2}}\right)-\log\left(b^{\frac{3}{2}}c^{\frac{2}{5}}\right)$ $= \log\left(a^{\frac{1}{2}}\right)-\left[\log\left(b^{\frac{3}{2}}\right)+\log\left(c^{\frac{2}{5}}\right)\right]$ $= \frac{1}{2}\log(a)-\frac{3}{2}\log(b)-\frac{2}{5}\log(c)$

70.

$$\log_7\left(\frac{c^3d^{\frac{1}{3}}}{e^6}\right)^{\frac{1}{3}} = \frac{1}{3}\log_7\left(\frac{c^3d^{\frac{1}{3}}}{e^6}\right)$$
$$= \frac{1}{3}\left[\log_7\left(c^3d^{\frac{1}{3}}\right)-\log_7\left(e^6\right)\right]$$
$$= \frac{1}{3}\left[\log_7\left(c^3\right)+\log_7\left(d^{\frac{1}{3}}\right)-\log_7\left(e^6\right)\right]$$
$$= \frac{1}{3}\left[3\log_7(c)+\frac{1}{3}\log_7(d)-6\log_7(e)\right]$$

71. $\log_8 3 = \frac{\log 3}{\log 8}$ $\cong 0.5283$	**72.** $\log_5 \frac{1}{2} = \frac{\log\frac{1}{2}}{\log 5}$ $\cong -0.4307$	**73.** $\log_\pi 1.4 = \frac{\log 1.4}{\log \pi}$ $\cong 0.2939$	**74.** $\log_{\sqrt{3}} 2.5 = \frac{\log 2.5}{\log\sqrt{3}}$ $\cong 1.6681$

75. $4^x = \frac{1}{256}$ $4^x = 4^{-4}$ $x = -4$	**76.** $3^{x^2} = 81 = 3^4$ $x^2 = 4$ $x = \pm 2$
77. $e^{3x-4} = 1 = e^0$ $3x-4=0$ $x = \frac{4}{3}$	**78.** $e^{\sqrt{x}} = e^{4.8}$ $\sqrt{x} = 4.8$ $x = (4.8)^2 = 23.04$

79.

$$\left(\tfrac{1}{3}\right)^{x+2} = 81$$
$$\left(3^{-1}\right)^{x+2} = 3^4$$
$$3^{-x-2} = 3^4$$
$$-x-2 = 4$$
$$x = -6$$

80.

$$100^{x^2-3} = 10$$
$$\left(10^2\right)^{x^2-3} = 10$$
$$10^{2(x^2-3)} = 10$$
$$2(x^2-3) = 1$$
$$2x^2 - 7 = 0$$

So, $x^2 = \frac{7}{2}$, $x = \pm\sqrt{\frac{7}{2}}$

81.

$$e^{2x+3} - 3 = 10$$
$$e^{2x+3} = 13$$
$$2x+3 = \ln 13$$
$$x = \tfrac{-3+\ln 13}{2} \approx -0.218$$

82.

$$2^{2x-1} + 3 = 17$$
$$2^{2x-1} = 14$$
$$\log_2\left(2^{2x-1}\right) = \log_2(14)$$
$$2x-1 = \log_2(14)$$
$$x = \tfrac{1+\log_2(14)}{2} \cong 2.404$$

83. Note that $e^{2x} + 6e^x + 5 = 0$ is equivalent to $\left(e^x\right)^2 + 6\left(e^x\right) + 5 = \left(e^x + 5\right)\left(e^x + 1\right) = 0$. Neither $\underbrace{e^x + 5 = 0}_{\text{No solution}}$ or $\underbrace{e^x + 1 = 0}_{\text{No solution}}$ has a real solution. So, the original equation has no solution.

84.

$$4e^{0.1x} = 64$$
$$e^{0.1x} = 16$$
$$0.1x = \ln(16)$$
$$x \approx 27.726$$

85.

$$\left(2^x - 2^{-x}\right)\left(2^x + 2^{-x}\right) = 0$$
$$\left(2^x\right)^2 - \left(2^{-x}\right)^2 = 0$$
$$2^{2x} - 2^{-2x} = 0$$
$$2^{2x} = 2^{-2x}$$
$$2x = -2x$$
$$x = 0$$

86.

$$5\left(2^x\right) = 25$$
$$2^x = 5$$
$$x = \log_2 5 \cong 2.3219$$

87.

$$\log(3x) = 2$$
$$10^2 = 3x$$
$$100 = 3x$$
$$\tfrac{100}{3} = x$$

88.

$$\log_3(x+2) = 4$$
$$x + 2 = 3^4$$
$$x = 79$$

89.

$$\log_4(x) + \log_4 2x = 8$$
$$\log_4\left(2x^2\right) = 8$$
$$2x^2 = 4^8$$
$$x^2 = \tfrac{1}{2}\left(4^8\right)$$
$$x = \pm\sqrt{\tfrac{1}{2}}\left(4^4\right)$$

$x = 128\sqrt{2}$, ~~$-128\sqrt{2}$~~

90.

$$\log_6(x) + \log_6(2x-1) = \log_6(3)$$
$$\log_6\left(x(2x-1)\right) = \log_6(3)$$
$$x(2x-1) = 3$$
$$2x^2 - x - 3 = 0$$
$$(2x-3)(x+1) = 0$$

$x = \tfrac{3}{2}$, ~~-1~~

91.

$$\ln x^2 = 2.2$$
$$x^2 = e^{2.2}$$
$$x = \pm\sqrt{e^{2.2}} \approx \pm 3.004$$

92.

$$\ln(3x-4) = 7$$
$$3x - 4 = e^7$$
$$3x = 4 + e^7$$
$$x = \tfrac{4+e^7}{3} \cong 366.88$$

93.

$$\log_3(2-x) - \log_3(x+3) = \log_3(x)$$
$$\log_3\left(\tfrac{2-x}{x+3}\right) = \log_3(x)$$
$$\tfrac{2-x}{x+3} = x$$
$$2 - x = x(x+3)$$
$$x^2 + 4x - 2 = 0$$
$$x = \frac{-4 \pm \sqrt{16+8}}{2}$$

$x = -2 + \sqrt{6}$, ~~$-2-\sqrt{6}$~~

≈ 0.449

94.

$$4\log(x+1) - 2\log(x+1) = 1$$
$$\log(x+1) = 0.5$$
$$x + 1 = 10^{0.5}$$
$$x = 10^{0.5} - 1 \approx 2.162$$

95. Use $A(t) = Pe^{rt}$.

Here,

$$A = 30{,}000,\ \ r = 0.05,\ \ t = 1$$

Substituting into the above equation, we can solve for P:

$$30{,}000 = Pe^{0.05(1)}$$
$$28{,}536.88 \cong \tfrac{30{,}000}{e^{0.05}} = P$$

So, the initial investment in 1 year CD should be approximately \$28,536.88.

96. Use $A(t) = Pe^{rt}$.
Here,

$$A = 4000,\ t = 2,\ P = 2800\ (100 \times 28\,\text{shares})$$

Substituting into the above equation, we can solve for r:

$$4000 = 2800e^{r(2)}$$
$$\tfrac{10}{7} = \tfrac{4000}{2800} = e^{2r}$$
$$0.178 \cong \tfrac{1}{2}\ln(\tfrac{10}{7}) = r$$

So, about 17.8%.

97. Use $A(t) = P\left(1+\frac{r}{n}\right)^{nt}$
Here, we know that $r = 0.042,\ n = 4,\ A = 2P$. We can substitute these in to find t.

$$2P = P\left(1+\tfrac{0.042}{4}\right)^{4(t)}$$
$$2 = (1.0105)^{4t}$$
$$\log_{1.0105} 2 = 4t$$
$$16.6 \cong \tfrac{1}{4}\log_{1.0105} 2 = t$$

So, it takes approximately 16.6 years until the initial investment doubles.

98. Use $A(t) = Pe^{rt}$.
Here,

$$A = 22{,}500,\ r = 0.08,\ P = 9000$$

Substituting into the above equation, we can solve for t:

$$22{,}500 = 9000e^{0.08t}$$
$$\tfrac{22{,}500}{9000} = e^{0.08t}$$
$$11.453 \cong \tfrac{1}{0.08}\ln\left(\tfrac{22{,}500}{9000}\right) = t$$

So, it takes approximately 11.453 years.

99. Use $N = N_0 e^{rt}$.
Here, $N_0 = 2.62,\ r = 0.035$. Determine N when $t = 6$:

$$N = 2.62e^{0.035(6)} \approx 3.23 \text{ million}$$

The population in 2010 is about 3.23 million.

100. Use $N = N_0 e^{-rt}$ **(1)**.
We have

$$\begin{aligned} N(0) &= N_0 = 28.3 \\ N(4) &= 32.5 \end{aligned} \quad \textbf{(2)}$$

In order to determine $N(14)$, we need to first determine r. To this end, substitute **(2)** into **(1)** to obtain:

$$32.5 = 28.3e^{r(4)} \ \Rightarrow\ r = \tfrac{1}{4}\ln\left(\tfrac{32.5}{28.3}\right)$$

Thus, $N(14) = 28.3e^{\frac{1}{4}\ln\left(\frac{32.5}{28.3}\right)\cdot 14} \approx 48.93 \text{ million}$.
The population in 2010 is about 48.93 million.

101. Use $N = N_0 e^{rt}$ **(1)**.
We have

$$\left.\begin{aligned} N(0) &= N_0 = 1000 \\ N(3) &= 2500 \end{aligned}\right\} \textbf{(2)}$$

In order to determine $N(6)$, we need to first determine *r*. To this end, substitute **(2)** into **(1)** to obtain:

$$2500 = 1000e^{r(3)} \Rightarrow r = \tfrac{1}{3}\ln\left(\tfrac{2500}{1000}\right)$$

Thus, $N(6) = 1000e^{\frac{1}{3}\ln\left(\frac{2500}{1000}\right)\cdot 6} \approx 6250 \text{ bacteria}$.

102. Use $N = N_0 e^{rt}$ **(1)**.
We have

$$\left.\begin{aligned} N(0) &= N_0 = 1{,}388{,}215 \\ N(1) &= 1{,}418{,}041 \end{aligned}\right\} \textbf{(2)}$$

In order to determine $N(7)$, we need to first determine *r*. To this end, substitute **(2)** into **(1)** to obtain:

$$1{,}418{,}041 = 1{,}388{,}215e^{r(1)} \Rightarrow$$
$$r = \ln\left(\tfrac{1{,}418{,}041}{1{,}388{,}215}\right)$$

Thus,

$$N(7) = 1{,}388{,}215e^{\ln\left(\frac{1{,}418{,}041}{1{,}388{,}215}\right)\cdot 7} \approx 1{,}610{,}947.$$

103. Use $N = N_0 e^{-rt}$.
We know that $N(28) = \frac{1}{2}N_0$. Assuming that $N_0 = 20$, determine *t* such that $N = 5$. To do so, we first find *r*:

$$\tfrac{1}{2}N_0 = N_0 e^{-r(28)} \Rightarrow r = -\tfrac{1}{28}\ln\left(\tfrac{1}{2}\right)$$

Now, solve:

$$5 = 20e^{-\left(-\frac{1}{28}\ln\left(\frac{1}{2}\right)\right)t}$$
$$\tfrac{1}{4} = e^{\frac{1}{28}\ln\left(\frac{1}{2}\right)t}$$
$$t = \frac{\ln\left(\frac{1}{4}\right)}{\frac{1}{28}\ln\left(\frac{1}{2}\right)} \approx 56 \text{ years}$$

104. Use $N = N_0 e^{-rt}$.
We know that $N(25{,}000) = \frac{1}{2}N_0$. Assuming that $N_0 = 100$, determine *t* such that $N = 20$. To do so, we first find *r*:

$$\tfrac{1}{2}N_0 = N_0 e^{-r(25{,}000)} \Rightarrow r = -\tfrac{1}{25{,}000}\ln\left(\tfrac{1}{2}\right)$$

Now, solve:

$$20 = 100e^{-\left(-\frac{1}{25{,}000}\ln\left(\frac{1}{2}\right)\right)t}$$
$$t = \frac{\ln\left(\frac{1}{5}\right)}{\frac{1}{25{,}000}\ln\left(\frac{1}{2}\right)} \approx 58{,}048.2 \text{ years}$$

105. Use $N = N_0 e^{-rt}$.
Assuming that 2003 occurs at $t = 0$, we know that

$$N_0 = 5600, \quad N(1) = 2420.$$

Find $N(7)$. To do so, we first find *r*:

$$2420 = 5600e^{-r(1)} \Rightarrow r = -\ln\left(\tfrac{2420}{5600}\right)$$

Now, solve:

$$N(7) = 5600e^{-\left(-\ln\left(\frac{2420}{5600}\right)\right)(7)} \approx 16 \text{ fish}$$

106. Use $N = N_0 e^{-rt}$.
We know that

$$N_0 = 28{,}200, \quad N(2) = 24{,}500.$$

Find $N(6)$. To do so, we first find *r*:

$$24{,}500 = 28{,}200e^{-r(2)} \Rightarrow r = -\tfrac{1}{2}\ln\left(\tfrac{24{,}500}{28{,}200}\right)$$

Now, solve:

$$N(6) = 28{,}200e^{-\left(-\frac{1}{2}\ln\left(\frac{24{,}500}{28{,}200}\right)\right)(6)}$$
$$\approx \$18{,}492.69$$

107. Use $M = 1000\left(1 - e^{-0.035t}\right)$.
Since $t = 0$ corresponds to 1998, we have
$M(12) = 1000\left(1 - e^{-0.035(12)}\right) \approx 343$ mice.

108. Use $N = N_0 e^{rt}$.
Here, $N_0 = 50{,}000$ $r = 0.023$. Determine N when $t = 60$:

$$N = 50{,}000 e^{0.023(60)} \approx 195{,}745$$

The population in 2030 is about 195,745.

109. The graph is as follows:

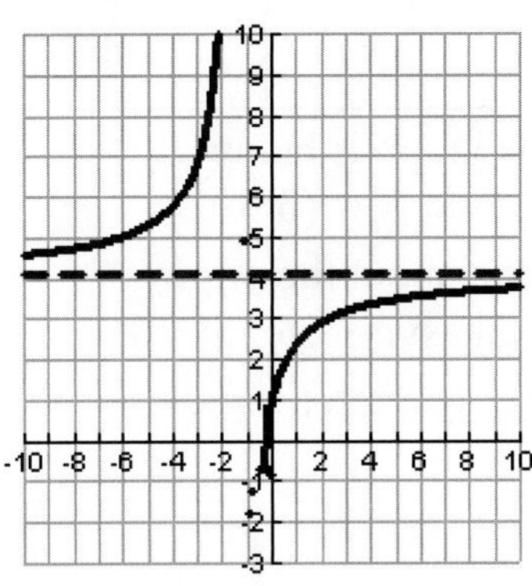

Using the calculator to compute the functional values for large values of x suggests that the HA is about $y = 4.11$. (Note: The exact equation of the HA is $y = e^{\sqrt{2}}$.)

110. Let $y1 = e^{-x+2}$, $y2 = 3^x + 1$. The graphs are as follows:

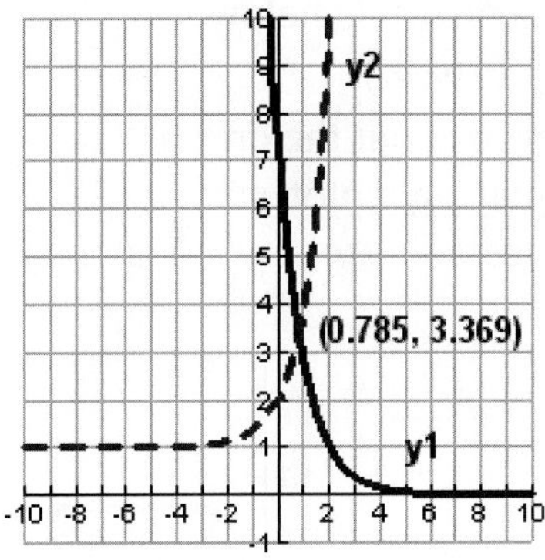

The coordinates of the point of intersection are about (0.785, 3.369).

111. Let
$y1 = \log_{2.4}(3x - 1)$, $y2 = \log_{0.8}(x - 1) + 3.5$.
The graphs are as follows:

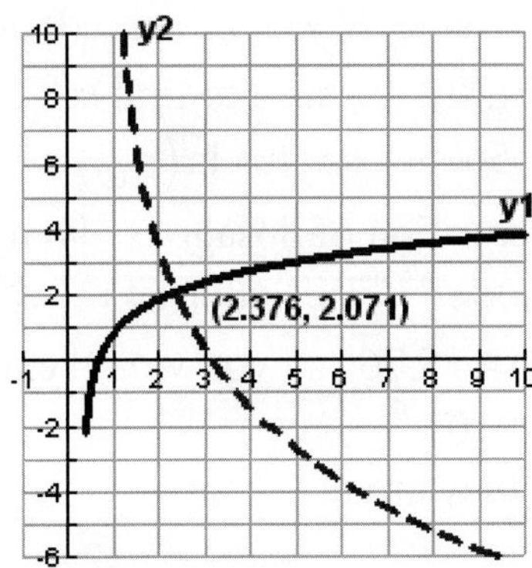

The coordinates of the point of intersection are about (2.376, 2.071).

112. The intersection points occur at (28.09, 2.753) and (1.227, 0.379).

113. The graphs agree on $(0,\infty)$, as seen below.

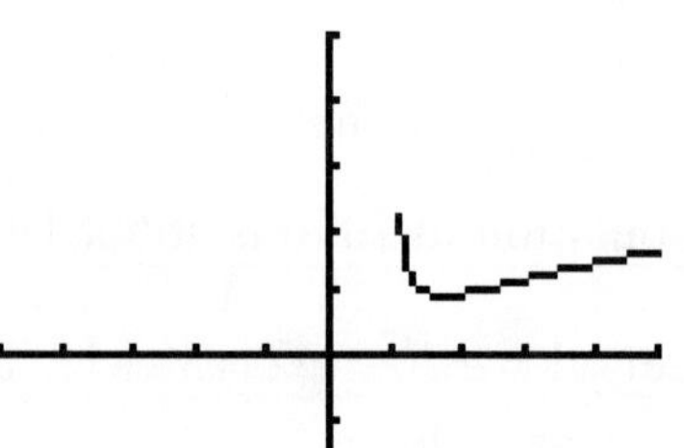

114. The graphs agree on (1, 3), as seen below.

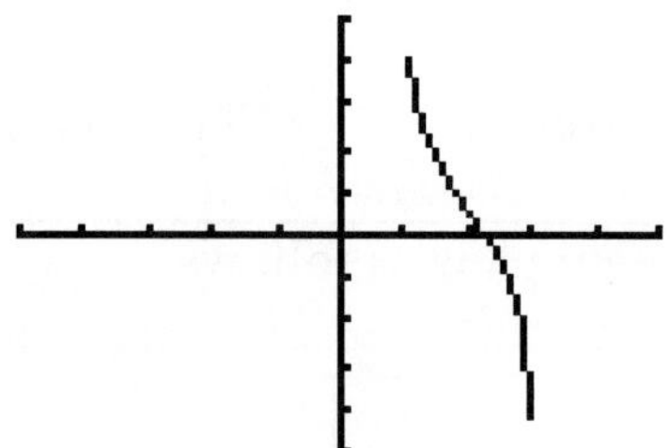

115. The graph is below.
Domain: $(-\infty,\infty)$.
Symmetric about the origin.
Horizontal asymptotes:
$y=-1$ (as $x\to-\infty$), $y=1$ (as $x\to\infty$)

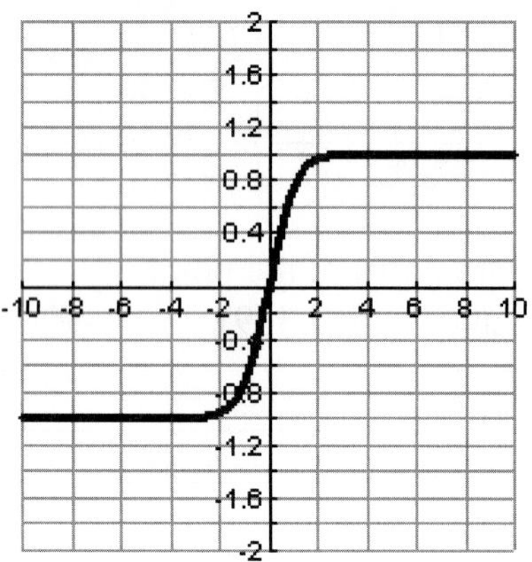

116. The graph is below.
Domain: $(-\infty,0)\cup(0,\infty)$.
Symmetric about the origin.
Horizontal asymptote: $y=0$
Vertical asymptote: $x=0$

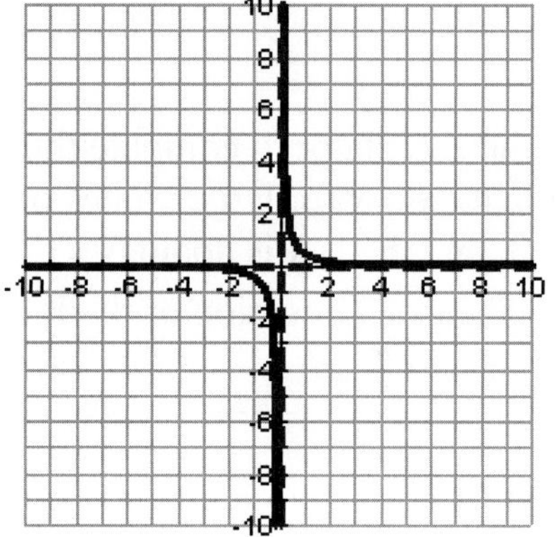

117. a. Using $N=N_0e^{rt}$ with (0,4) and (18,2), we need to find r:

$$2=4e^{18r}\Rightarrow r=\tfrac{1}{18}\ln\left(\tfrac{1}{2}\right)\approx-0.35508$$

So, the equation of dosage is given by

$$N=4e^{-0.038508t}\approx4(0.9622)^t.$$

b. $N=4(0.9622)^t$

c. Yes, they are the same.

118. a. Using $N=N_0e^{rt}$ with (0,5600) and (1,2420), we need to find r:

$$2420=5600e^{r}\Rightarrow r=\ln\left(\tfrac{2420}{5600}\right)\approx-0.8390$$

So, the equation of dosage is given by

$$N=5600e^{-0.8390t}\approx5600(0.4321428571)^t.$$

b. $N(7)\approx15.76$, so you would expect 16 fish in 2010.

c. Yes, they are the same.

Chapter 3 Practice Test Solutions --

1. $\log 10^{x^3} = x^3 \log 10 = x^3$

2. $\log_5 326 = \dfrac{\log 326}{\log 5} \cong 3.60$

3.

$$\log_{1/3} 81 = x$$
$$3^4 = \left(\tfrac{1}{3}\right)^x = 3^{-x}$$
$$-4 = x$$

4.

$$\ln\left(\frac{e^{5x}}{x(x^4+1)}\right) = \ln(e^{5x}) - \ln\left(x(x^4+1)\right)$$
$$= 5x - \left(\ln x + \ln(x^4+1)\right)$$
$$= 5x - \ln x - \ln(x^4+1)$$

5.

$$e^{x^2-1} = 42$$
$$x^2 - 1 = \ln 42$$
$$x^2 = 1 + \ln 42$$
$$x = \pm\sqrt{1+\ln 42} \approx \pm 2.177$$

6. Note that $e^{2x} - 5e^x + 6 = 0$ is equivalent to

$\left(e^x\right)^2 - 5\left(e^x\right) + 6 = \left(e^x - 3\right)\left(e^x - 2\right) = 0$.

Thus,

$$e^x - 3 = 0 \quad \text{or} \quad e^x - 2 = 0$$
$$e^x = 3 \qquad\qquad e^x = 2$$
$$x = \ln 3 \qquad\qquad x = \ln 2$$

7.

$$27e^{0.2x+1} = 300$$
$$e^{0.2x+1} = \tfrac{300}{27}$$
$$0.2x + 1 = \ln\left(\tfrac{300}{27}\right)$$
$$0.2x = -1 + \ln\left(\tfrac{300}{27}\right)$$
$$x = \frac{-1+\ln\left(\frac{300}{27}\right)}{0.2} \cong 7.04$$

8.

$$3^{2x-1} = 15$$
$$2x - 1 = \log_3 15$$
$$x = \tfrac{1}{2}\left(1 + \log_3 15\right) \approx 1.732$$

9.

$$3\ln(x-4) = 6$$
$$\ln(x-4) = 2$$
$$x - 4 = e^2$$
$$x = 4 + e^2 \approx 11.389$$

10.

$$\log(6x+5) - \log(3) = \log(2) - \log(x)$$
$$\log\left(\tfrac{6x+5}{3}\right) = \log\left(\tfrac{2}{x}\right)$$
$$\tfrac{6x+5}{3} = \tfrac{2}{x}$$
$$6x^2 + 5x = 6$$
$$6x^2 + 5x - 6 = 0$$
$$(3x-2)(2x+3) = 0 \;\Rightarrow\; x = \tfrac{2}{3},\ \cancel{-\tfrac{3}{2}}$$

11.

$$\ln(\ln x) = 1$$
$$\ln x = e$$
$$x = e^e \approx 15.154$$

12.

$$\log_2(3x-1)-\log_2(x-1)=\log_2(x+1)$$
$$\log_2\left(\tfrac{3x-1}{x-1}\right)=\log_2(x+1)$$
$$\tfrac{3x-1}{x-1}=x+1$$
$$3x-1=(x+1)(x-1)=x^2-1$$
$$x^2-3x=0 \;\Rightarrow\; x(x-3)=0$$
$$\Rightarrow\; x=\not{0},\,3$$

13.

$$\log_6 x+\log_6(x-5)=2$$
$$\log_6\big(x(x-5)\big)=2$$
$$x^2-5x=36$$
$$x^2-5x-36=0$$
$$(x-9)(x+4)=0$$
$$x=\not{-4},9$$

14.

$$\ln(x+2)-\ln(x-3)=2$$
$$\ln\left(\frac{x+2}{x-3}\right)=2$$
$$\frac{x+2}{x-3}=e^2$$
$$x+2=e^2x-3e^2$$
$$\left(e^2-1\right)x=3e^2+2$$
$$x=\frac{3e^2+2}{e^2-1}\approx 3.783$$

15.

$$\ln x+\ln(x+3)=1$$
$$\ln\big(x(x+3)\big)=1$$
$$x^2+3x=e$$
$$x^2+3x-e=0$$
$$x=\frac{-3\pm\sqrt{9-4(-e)}}{2}$$
$$\approx 0.729,\ \not{-3.729}$$

16.

$$\log_2\left(\frac{2x+3}{x-1}\right)=3$$
$$\frac{2x+3}{x-1}=8$$
$$2x+3=8x-8$$
$$6x=11$$
$$x=\tfrac{11}{6}$$

17.

$$\frac{12}{1+2e^x}=6$$
$$12=6+12e^x$$
$$6=12e^x$$
$$x=\ln\left(\tfrac{1}{2}\right)\approx -0.693$$

18.

$$\ln x+\ln(x-3)=2$$
$$\ln x(x-3)=2$$
$$x^2-3x=e^2$$
$$x^2-3x-e^2=0$$
$$x=\frac{3\pm\sqrt{9-4(1)(-e^2)}}{2}\approx 4.605,\ \not{-1.605}$$

19. Must have $\frac{x}{x^2-1}>0$ and $x^2-1\neq 0$

CPs $-1,0,1$

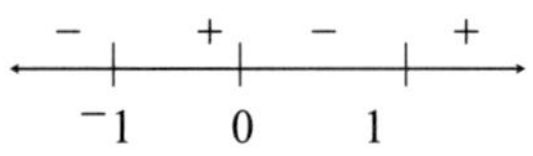

So, the domain is $(-1,0)\cup(1,\infty)$.

20.

In order for the identity $10^{\log(4x-a)}=4x-a$ to hold, we must have $4x-a>0 \;\Rightarrow\; x>\frac{a}{4}$.

21.

<u>y-intercept</u>: $f(0)=3^{-0}+1=2$. So, (0,2).

<u>x-intercept</u>: None

<u>Domain</u>: $(-\infty,\infty)$ <u>Range</u>: $(1,\infty)$

<u>Horizontal Asymptote</u>: $y=1$

The graph is as follows:

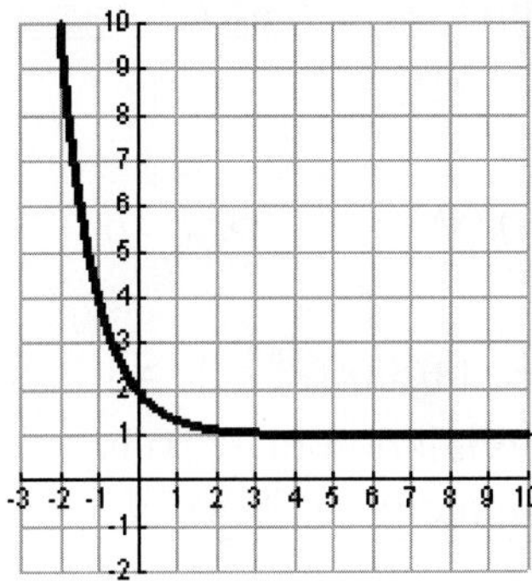

22.

<u>y-intercept</u>: $f(0)=\left(\frac{1}{2}\right)^0-3=-2$. So, (0,-2).

<u>x-intercept</u>: (-1.5850, 0) since

$0=\left(\frac{1}{2}\right)^x-3 \Rightarrow x=\log_{\left(\frac{1}{2}\right)}3\approx-1.5850$

<u>Domain</u>: $(-\infty,\infty)$ <u>Range</u>: $(-3,\infty)$

<u>Horizontal Asymptote</u>: $y=-3$

The graph is as follows:

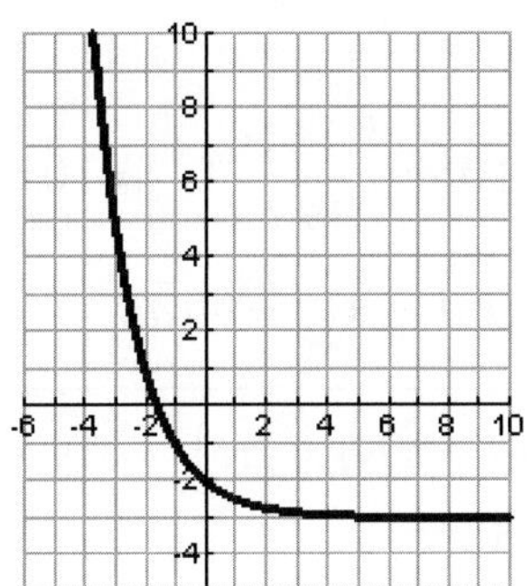

23. <u>y-intercept</u>: None

<u>x-intercept</u>: Must solve $\ln(2x-3)+1=0$.

$$\ln(2x-3)=-1 \Rightarrow 2x-3=\tfrac{1}{e} \Rightarrow x=\frac{3+\frac{1}{e}}{2}$$

So, $\left(\frac{3+\frac{1}{e}}{2},0\right)$.

<u>Domain</u>: $\left(\frac{3}{2},\infty\right)$ <u>Range</u>: $(-\infty,\infty)$

<u>Vertical Asymptote</u>: $x=\frac{3}{2}$

The graph is as follows:

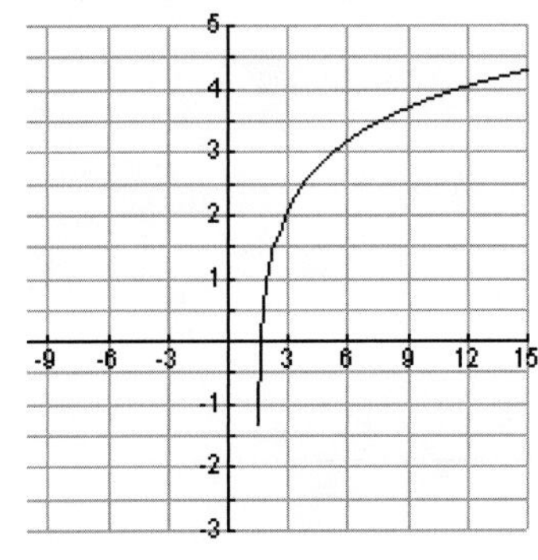

24. <u>y-intercept</u>: $f(0)=\log(1)+2=2$. So, (0,2).

<u>x-intercept</u>: Must solve $\log(1-x)+2=0$.

$$\log(1-x)+2=0 \Rightarrow \log(1-x)=-2$$
$$\Rightarrow 1-x=10^{-2} \Rightarrow x=\tfrac{99}{100}$$

So, $\left(\frac{99}{100},0\right)$.

<u>Domain</u>: $\left(-\infty,\frac{99}{100}\right)$ <u>Range</u>: $(-\infty,\infty)$

<u>Vertical Asymptote</u>: $x=1$

The graph is as follows:

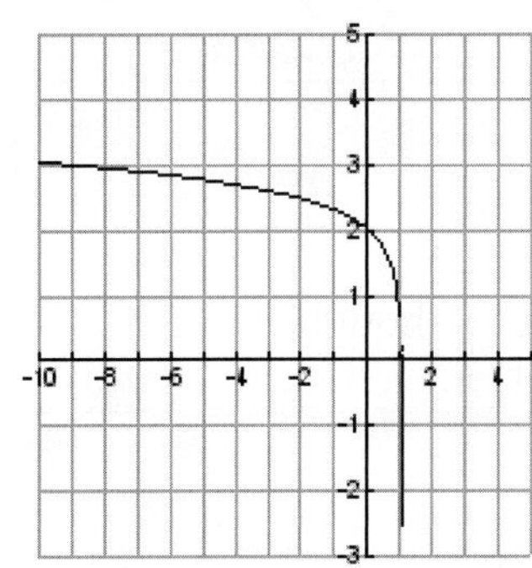

25. Use $A(t) = P\left(1+\frac{r}{n}\right)^{nt}$

Here,

$$P = 5000,\ r = 0.06,\ n = 4,\ t = 8.$$

Thus, $A(8) = 5000\left(1+\frac{0.06}{4}\right)^{4(8)} \cong 8051.62$.

So, the amount in the account after 8 years is \$8051.62.

26. Use $A(t) = Pe^{rt}$.

Here,

$$P = 10,000,\ r = 0.05,\ t = 10$$

Thus,

$$A(10) = 10,000e^{(0.05)(10)} \cong 16,487.21.$$

So, after 10 years, to amount is approximately \$16,487.

27. Use $D = 10\log\left(\dfrac{I}{1\times10^{-12}}\right)$.

Here,

$$D = 10\log\left(\frac{1\times10^{-3}}{1\times10^{-12}}\right) = 10\log(10^9)$$

$$= 90\log(10) = 90\,dB$$

28. Use $N(t) = N_0e^{rt}$.

Here,

$N_0 = 800,000,\ r = 0.05,\ t = 6\ \ (2010-2004)$

Thus,

$$N(6) = 800,000e^{(0.05)(6)} \cong 1,079,887.05.$$

So, to population in 2010 is approximately 1,079,887.05.

29. Use $M = \frac{2}{3}\log\left(\frac{E}{10^{4.4}}\right)$. Here,

For $M = 5$:

$$5 = \tfrac{2}{3}\log\left(\tfrac{E}{10^{4.4}}\right)$$

$$7.5 = \log\left(\tfrac{E}{10^{4.4}}\right)$$

$$10^{7.5} = \tfrac{E}{10^{4.4}}$$

$$\underbrace{10^{7.5}\times10^{4.4}}_{=10^{11.9}} = E$$

So, the energy here is $10^{11.9} \approx 7.9\times10^{11}$ joules.

For $M = 6$:

$$6 = \tfrac{2}{3}\log\left(\tfrac{E}{10^{4.4}}\right)$$

$$9 = \log\left(\tfrac{E}{10^{4.4}}\right)$$

$$10^{9} = \tfrac{E}{10^{4.4}}$$

$$\underbrace{10^{9}\times10^{4.4}}_{=10^{13.4}} = E$$

So, the energy here is $10^{13.4} \approx 2.5\times10^{13}$ joules.

Thus, the range of energy is $7.9\times10^{11} < E < 2.5\times10^{13}$ joules.

30.

$$30 = 50e^{-0.0578t}$$

$$\tfrac{3}{5} = e^{-0.0578t}$$

$$\ln\left(\tfrac{3}{5}\right) = -0.0578t$$

$$8.8 \cong -\tfrac{1}{0.0578}\ln\left(\tfrac{3}{5}\right) = t$$

So, about 8.8 hours.

31. Use $N = N_0 e^{rt}$ **(1)**.
We have

$$N(0) = N_0 = 200 \quad N(2) = 500 \qquad \textbf{(2)}$$

In order to determine $N(8)$, we need to first determine *r*. To this end, substitute **(2)** into **(1)** to obtain:

$$500 = 200e^{r(2)} \Rightarrow r = \tfrac{1}{2}\ln\left(\tfrac{500}{200}\right)$$

Thus, $N(8) = 200e^{\frac{1}{2}\ln\left(\frac{500}{200}\right)\cdot 8} \approx 7800$ bacteria.

32. Use $N = N_0 e^{-rt}$.
We know that $N(5730) = \frac{1}{2}N_0$. Assuming that $N_0 = 100$, determine *t* such that $N = 40$. To do so, we first find *r*:

$$\tfrac{1}{2}N_0 = N_0 e^{-r(5730)} \Rightarrow r = -\tfrac{1}{5730}\ln\left(\tfrac{1}{2}\right)$$

Now, solve:

$$40 = 100e^{-\left(-\frac{1}{5730}\ln\left(\frac{1}{2}\right)\right)t}$$

$$t = \frac{\ln\left(\frac{40}{100}\right)}{\frac{1}{5730}\ln\left(\frac{1}{2}\right)} \approx 7574.65 \text{ years}$$

33. Solve $1000 = \dfrac{2000}{1+3e^{-0.4t}}$.

$$1000(1+3e^{-0.4t}) = 2000$$
$$1+3e^{-0.4t} = 2$$
$$3e^{-0.4t} = 1$$
$$e^{-0.4t} = \tfrac{1}{3}$$
$$-0.4t = \ln\left(\tfrac{1}{3}\right)$$
$$t = -\tfrac{1}{0.4}\ln\left(\tfrac{1}{3}\right) \approx 3 \text{ days}$$

34. Use $N = N_0 e^{rt}$ **(1)**.
We have

$$N(0) = N_0 = 76 \quad N(2) = 83 \qquad \textbf{(2)}$$

In order to determine $N(8)$, we need to first determine *r*. To this end, substitute **(2)** into **(1)** to obtain:

$$83 = 76e^{r(2)} \Rightarrow r = \tfrac{1}{2}\ln\left(\tfrac{83}{76}\right)$$

Thus,
$N(8) = 76e^{\frac{1}{2}\ln\left(\frac{83}{76}\right)\cdot 8} \approx 108.11$ million barrels.

35. The graph is below.
Domain: $(-\infty, \infty)$.
Symmetric about the origin.
No asymptotes

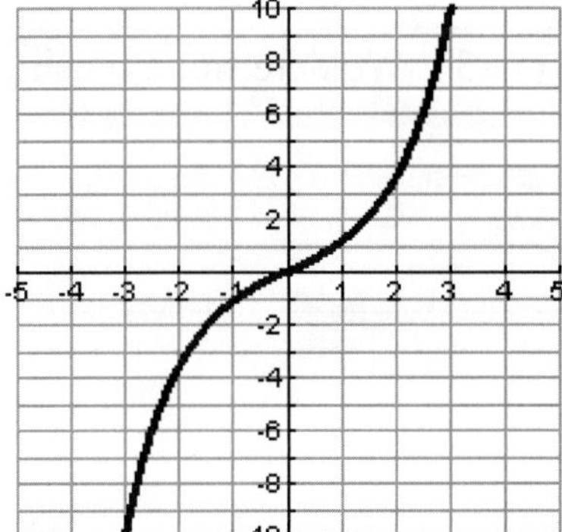

36. Let $y1 = 4^{3-x}$, $y2 = 2x-1$. The graphs are as follows:

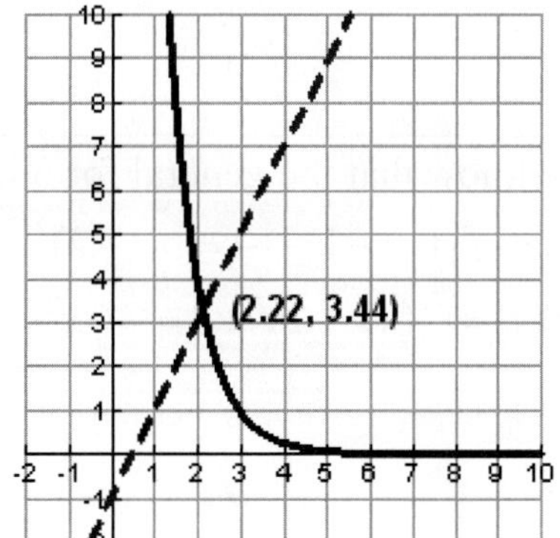

The intersection occurs about when $x = 2.22$.

Chapter 3 Cumulative Test

1. Must have $x^2-9=(x-3)(x+3)>0$. CPs: -3, 3 $+ \quad - \quad +$ on number line with marks at -3, 3 So, the domain is $(-\infty,-3)\cup(3,\infty)$. As for the range, note that $f(x)>0$, for all x and $y=0$ is the HA. Since f has VA at $x=-3$ and $x=3$, we know it attains all values >0. So, the range is $(0,\infty)$.	**2. a)** $f(x)+g(x)=x^2+3x$ Domain: $(-\infty,\infty)$ **b)** $f(x)-g(x)=2+3x-x^2$ Domain: $(-\infty,\infty)$ **c)** $f(x)\cdot g(x)=(1+3x)(x^2-1)$ $=3x^3+x^2-3x-1$ Domain: $(-\infty,\infty)$ **d)** $\dfrac{f(x)}{g(x)}=\dfrac{1+3x}{x^2-1}$ Domain: $(-\infty,-1)\cup(-1,1)\cup(1,\infty)$
3. One possibility is to choose $g(x)=e^{2x}$ and $f(x)=\dfrac{1-x}{1+x}$. Another possibility is $g(x)=e^{x}$ and $f(x)=\dfrac{1-x^2}{1+x^2}$.	**4.** $f(x_1)=f(x_2)\Rightarrow\sqrt[5]{x_1^3+1}=\sqrt[5]{x_2^3+1}$ $\Rightarrow x_1^3+1=x_2^3+1$ $\Rightarrow x_1=x_2$ So, f is one-to-one. Inverse: $y=\sqrt[5]{x^3+1}$ $y^5=x^3+1$ $x=\sqrt[3]{y^5-1}$ Thus, $f^{-1}(x)=\sqrt[3]{x^5-1}$.

5. We know that the general form is $f(x)=a(x+2)^2+3$. We determine a using the fact that $f(1)=-1$: $-1=a(1+2)^2+3\Rightarrow a=-\frac{4}{9}$.
So, $f(x)=-\frac{4}{9}(x+2)^2+3=-\frac{4}{9}x^2-\frac{16}{9}x+\frac{11}{9}$

6. Factors of 18: $\pm 1, \pm 2, \pm 3, \pm 6, \pm 9, \pm 18$ Factors of 3: $\pm 1, \pm 3$

So, the possible rational zeros are: $\pm 1, \pm 2, \pm 3, \pm 6, \pm 9, \pm 18, \pm \frac{1}{3}, \pm \frac{2}{3}$

Using synthetic division yields

$$\begin{array}{r|rrrr} 2 & 3 & 6 & -15 & -18 \\ & & 6 & 24 & 18 \\ \hline -1 & 3 & 12 & 9 & 0 \\ & & -3 & -9 & \\ \hline & 3 & 9 & 0 & \end{array}$$

So, $f(x) = (x-2)(x+1)(3x+9) = 3(x+3)(x+1)(x-2)$.

7.

$$e^x + \sqrt{e^x} - 12 = 0$$
$$e^x - 12 = -\sqrt{e^x}$$
$$\left(e^x - 12\right)^2 = e^x$$
$$\left(e^x\right)^2 - 24\left(e^x\right) + 144 - e^x = 0$$
$$\left(e^x\right)^2 - 25\left(e^x\right) + 144 = 0$$
$$e^x = \frac{25 \pm \sqrt{(-25)^2 - 4(144)}}{2}$$
$$e^x = \frac{25 \pm 7}{2} = 9, 16$$

Thus, $x = \ln 9 \approx 2.197$, ~~$\ln 16$~~

8.

$$\frac{f(x+h) - f(x)}{h} = \frac{4x + 4h - \left(x^2 + 2hx + h^2\right) - 4x + x^2}{h}$$
$$= \frac{h(4 - 2x - h)}{h}$$
$$= 4 - 2x - h$$
$$= \frac{\left(4(x+h) - (x+h)^2\right) - \left(4x - x^2\right)}{h}$$

9. **a.** 1 **b.** 5 **c.** 1 **d.** undefined **e.** Domain: $(-2, \infty)$ Range: $(0, \infty)$

f. Increasing on $(4, \infty)$ Decreasing on (0,4) Constant on (-2,0)

10. Shift the graph of $y = \sqrt{x}$ left 1 unit, and then reflect over y-axis.

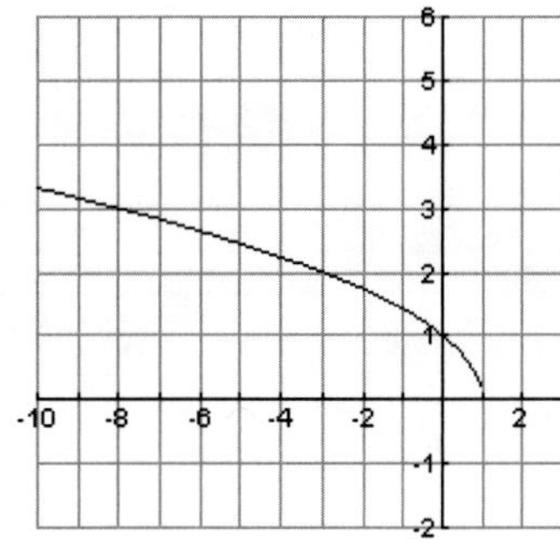

11. Yes, f is one-to-one since

$$\begin{aligned} f(x)=f(y) &\Rightarrow \sqrt{x-4}=\sqrt{y-4} \\ &\Rightarrow x-4=y-4 \\ &\Rightarrow x=y \end{aligned}$$

12. We know that

$$\begin{aligned} 10\pi R^2 &= 400 \\ \pi R^2 &= 40 \\ R^2 &= \tfrac{40}{\pi} \\ R &= \sqrt{\tfrac{40}{\pi}} \approx 3.568 \text{ in.} \end{aligned}$$

13.

$$\begin{aligned} f(x) &= -4x^2+8x-5 \\ &= -4\left(x^2-2x\right)-5 \\ &= -4\left(x^2-2x+1\right)-5+4 \\ &= -4(x-1)^2-1 \end{aligned}$$

So, the vertex is (1,-1).

14. $P(x)=C(x+5)^2(x-9)^4$, where C is any constant.

15. Rewrite as:

$\left(-x^4-4x^3+3x^2+7x-20\right)\div(x-(-4))$

Synthetic division then gives

$$\begin{array}{r|rrrrr} -4 & -1 & -4 & 3 & 7 & -20 \\ & & 4 & 0 & -12 & 20 \\ \hline & -1 & 0 & 3 & -5 & 0 \end{array}$$

So, $Q(x)=-x^3+3x-5,\ r(x)=0$.

16. Since $2+i$ is a zero, so is its conjugate $2-i$. Hence, $(x-(2+i))(x-(2-i))=x^2-4x+5$ divides $P(x)$ evenly. Indeed, note that

$$\begin{array}{r} x^2-3x-4 \\ x^2-4x+5 \overline{\big)\ x^4-7x^3+13x^2+x-20} \\ \underline{-\left(x^4-4x^3+5x^2\right)} \\ -3x^3+8x^2+x \\ \underline{-\left(-3x^3+12x^2-15x\right)} \\ -4x^2+16x-20 \\ \underline{-\left(-4x^2+16x-20\right)} \\ 0 \end{array}$$

So, $P(x)=(x-(2+i))(x-(2-i))(x-4)(x+1)$.

17. Vertical asymptote: $x = 3$
Horizontal asymptote: None
Slant asymptote: $y = x + 3$ since

$$\begin{array}{r} x+3 \\ x-3\overline{\big)\; x^2+0x+7} \\ -\left(x^2+0x-9\right) \\ \hline 16 \end{array}$$

18. The graph is as follows:

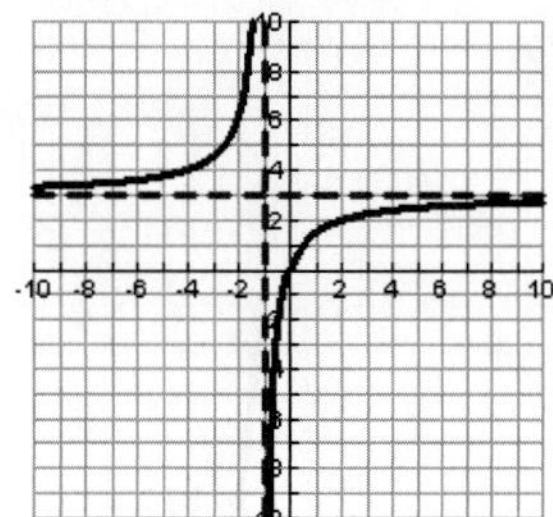

Vertical Asymptote: $x = -1$
Horizontal Asymptote: $y = 3$

19.

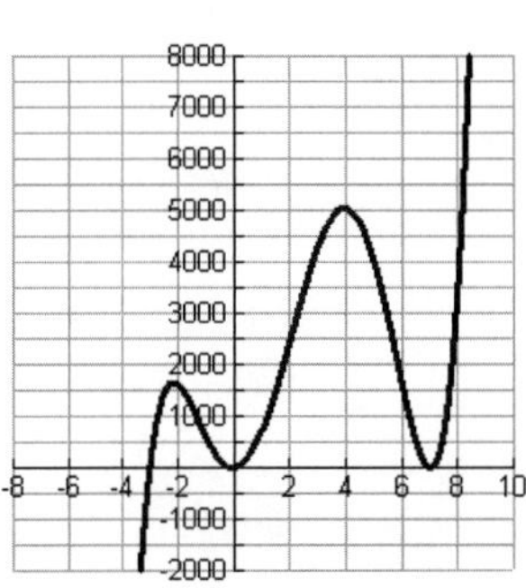

20. Use $A = P\left(1+\frac{r}{n}\right)^{nt}$.

We have

$$P = 5400,\ r = 0.0275,\ n = 12,\ t = 4.$$

Thus, $A \approx 6{,}027.14$.

21. $\log_3 243 = \log_3\left(3^5\right) = 5$

22.

$$\tfrac{1}{2}\ln(x+5) - 2\ln(x+1) - \ln(3x) =$$

$$\ln(x+5)^{1/2} - \left[\ln(x+1)^2 + \ln(3x)\right] =$$

$$\ln(x+5)^{1/2} - \left[\ln\left((x+1)^2(3x)\right)\right] =$$

$$\ln\left(\frac{(x+5)^{1/2}}{(x+1)^2(3x)}\right)$$

23.

$$10^{2\log(4x+9)} = 10^{\log(4x+9)^2} = (4x+9)^2 = 121$$

$$4x+9 = \pm 11$$

$$4x = -9 \pm 11$$

$$x = \tfrac{1}{4}(-9 \pm 11) = \cancel{-5},\ 0.5$$

24.

$$5^{x^2} = 625$$

$$5^{x^2} = 5^4$$

$$x^2 = 4$$

$$x = \pm 2$$

25. Use $A = Pe^{rt}$.
We know that $P = 8500$ and $r = 0.04$.
Determine t such that $A = 12{,}000$.

$$12{,}000 = 8500e^{0.04t}$$

$$t = \tfrac{1}{0.04}\ln\left(\tfrac{12{,}000}{8{,}500}\right) \approx 8.62 \text{ years}$$

26. Let $y1 = e^{3-2x}$, $y2 = 2^{x-1}$. The graphs are below: 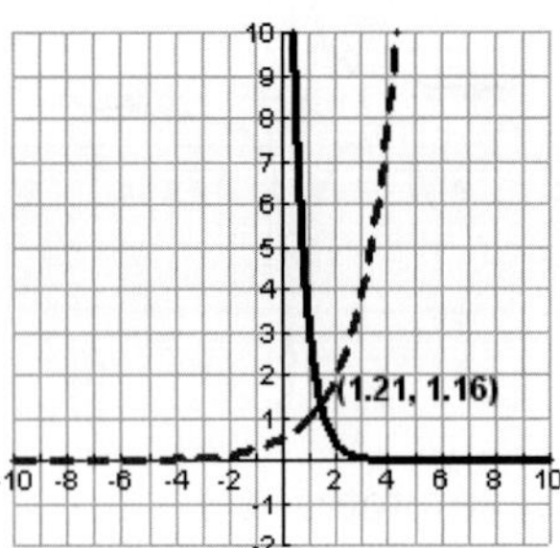 The solution is approximately 1.21.	**27. a.** Using $N = N_0 e^{rt}$ with (0,6) and (28,3), we need to find r: $3 = 6e^{28r} \Rightarrow r = \frac{1}{28}\ln\left(\frac{1}{2}\right) \approx -0.247553$ So, the equation of dosage is given by $N = 6e^{-0.247553t} \approx 6(0.9755486421)^t$. **b.** $N(32) \approx 2.72$ grams

CHAPTER 4

Section 4.1 Solutions ---

1. **a)** <u>complement</u>: $90^\circ - 18^\circ = \boxed{72^\circ}$ **b)** <u>supplement</u>: $180^\circ - 18^\circ = \boxed{162^\circ}$	**2.** **a)** <u>complement</u>: $90^\circ - 39^\circ = \boxed{51^\circ}$ **b)** <u>supplement</u>: $180^\circ - 39^\circ = \boxed{141^\circ}$
3. **a)** <u>complement</u>: $90^\circ - 42^\circ = \boxed{48^\circ}$ **b)** <u>supplement</u>: $180^\circ - 42^\circ = \boxed{138^\circ}$	**4.** **a)** <u>complement</u>: $90^\circ - 57^\circ = \boxed{33^\circ}$ **b)** <u>supplement</u>: $180^\circ - 57^\circ = \boxed{123^\circ}$
5. **a)** <u>complement</u>: $90^\circ - 89^\circ = \boxed{1^\circ}$ **b)** <u>supplement</u>: $180^\circ - 89^\circ = \boxed{91^\circ}$	**6.** **a)** <u>complement</u>: $90^\circ - 75^\circ = \boxed{15^\circ}$ **b)** <u>supplement</u>: $180^\circ - 75^\circ = \boxed{105^\circ}$
7. $\theta = \frac{s}{r} = \frac{4\text{ in.}}{22\text{ in.}} = \frac{2}{11} \approx \boxed{0.18}$	**8.** $\theta = \frac{s}{r} = \frac{1\text{ in.}}{6\text{ in.}} = \frac{1}{6} \approx \boxed{0.17}$
9. $\theta = \frac{s}{r} = \frac{20\text{ mm}}{100\text{ cm}} = \frac{2\text{ cm}}{100\text{ cm}} = \frac{1}{50} = \boxed{0.02}$	**10.** $\theta = \frac{s}{r} = \frac{2\text{ cm}}{1\text{ m}} = \frac{2\text{ cm}}{100\text{ cm}} = \frac{1}{50} = \boxed{0.02}$
11. $\theta = \frac{s}{r} = \frac{\frac{1}{32}\text{ in.}}{\frac{1}{4}\text{ in.}} = \frac{1}{8} = \boxed{0.125}$	**12.** $\theta = \frac{s}{r} = \frac{\frac{3}{14}\text{ cm}}{\frac{3}{4}\text{ cm}} = \frac{4}{14} = \boxed{0.29}$
13. $30^\circ = 30^\circ \cdot \frac{\pi}{180^\circ} = \boxed{\frac{\pi}{6}}$	**14.** $60^\circ = 60^\circ \cdot \frac{\pi}{180^\circ} = \boxed{\frac{\pi}{3}}$
15. $45^\circ = 45^\circ \cdot \frac{\pi}{180^\circ} = \boxed{\frac{\pi}{4}}$	**16.** $90^\circ = 90^\circ \cdot \frac{\pi}{180^\circ} = \boxed{\frac{\pi}{2}}$
17. $315^\circ = 315^\circ \cdot \frac{\pi}{180^\circ} = 7\cancel{(45^\circ)} \cdot \frac{\pi}{4\cancel{(45^\circ)}} = \boxed{\frac{7\pi}{4}}$	**18.** $270^\circ = 270^\circ \cdot \frac{\pi}{180^\circ} = 3\cancel{(90^\circ)} \cdot \frac{\pi}{2\cancel{(90^\circ)}} = \boxed{\frac{3\pi}{2}}$
19. $75^\circ = 75^\circ \cdot \frac{\pi}{180^\circ} = 5\cancel{(15^\circ)} \cdot \frac{\pi}{12\cancel{(15^\circ)}} = \boxed{\frac{5\pi}{12}}$	**20.** $100^\circ = 100^\circ \cdot \frac{\pi}{180^\circ} = 5\cancel{(20^\circ)} \cdot \frac{\pi}{9\cancel{(20^\circ)}} = \boxed{\frac{5\pi}{9}}$

21. $170^\circ = 170^\circ \cdot \frac{\pi}{180^\circ} = 17(10^\circ) \cdot \frac{\pi}{18(10^\circ)} = \boxed{\frac{17\pi}{18}}$	**22.** $340^\circ = 340^\circ \cdot \frac{\pi}{180^\circ} = 17(20^\circ) \cdot \frac{\pi}{9(20^\circ)} = \boxed{\frac{17\pi}{9}}$
23. $780^\circ = 780^\circ \cdot \frac{\pi}{180^\circ} = 13(60^\circ) \cdot \frac{\pi}{3(60^\circ)} = \boxed{\frac{13\pi}{3}}$	**24.** $540^\circ = 540^\circ \cdot \frac{\pi}{180^\circ} = 3(180^\circ) \cdot \frac{\pi}{180^\circ} = \boxed{3\pi}$
25. $-210^\circ = -210^\circ \cdot \frac{\pi}{180^\circ} = -7(30^\circ) \cdot \frac{\pi}{6(30^\circ)} = \boxed{-\frac{7\pi}{6}}$	**26.** $-320^\circ = -320^\circ \cdot \frac{\pi}{180^\circ} = -16(20^\circ) \cdot \frac{\pi}{9(20^\circ)} = \boxed{-\frac{16\pi}{9}}$
27. $-3,600^\circ = -3,600^\circ \cdot \frac{\pi}{180^\circ} = -20(180^\circ) \cdot \frac{\pi}{180^\circ} = \boxed{-20\pi}$	**28.** $1,800^\circ = 1,800^\circ \cdot \frac{\pi}{180^\circ} = 10(180^\circ) \cdot \frac{\pi}{180^\circ} = \boxed{10\pi}$
29. $\frac{\pi}{6} = \frac{\pi}{6} \cdot \frac{180^\circ}{\pi} = \frac{6(30^\circ)}{6} = \boxed{30^\circ}$	**30.** $\frac{\pi}{4} = \frac{\pi}{4} \cdot \frac{180^\circ}{\pi} = \frac{4(45^\circ)}{4} = \boxed{45^\circ}$
31. $\frac{3\pi}{4} = \frac{3\pi}{4} \cdot \frac{180^\circ}{\pi} = \frac{3 \cdot 4(45^\circ)}{4} = \boxed{135^\circ}$	**32.** $\frac{7\pi}{6} = \frac{7\pi}{6} \cdot \frac{180^\circ}{\pi} = \frac{7 \cdot 6(30^\circ)}{6} = \boxed{210^\circ}$
33. $\frac{3\pi}{8} = \frac{3\pi}{8} \cdot \frac{180^\circ}{\pi} = \boxed{67.5^\circ}$	**34.** $\frac{11\pi}{9} = \frac{11\pi}{9} \cdot \frac{180^\circ}{\pi} = \frac{11 \cdot 9(20^\circ)}{9} = \boxed{220^\circ}$
35. $\frac{5\pi}{12} = \frac{5\pi}{12} \cdot \frac{180^\circ}{\pi} = \frac{5 \cdot 12(15^\circ)}{4} = \boxed{75^\circ}$	**36.** $\frac{7\pi}{3} = \frac{7\pi}{3} \cdot \frac{180^\circ}{\pi} = \frac{7 \cdot 3(60^\circ)}{3} = \boxed{420^\circ}$
37. $9\pi = 9\pi \cdot \frac{180^\circ}{\pi} = \boxed{1620^\circ}$	**38.** $-6\pi = -6\pi \cdot \frac{180^\circ}{\pi} = \boxed{-1080^\circ}$
39. $\frac{19\pi}{20} = \frac{19\pi}{20} \cdot \frac{180^\circ}{\pi} = \frac{19 \cdot 20(9^\circ)}{4} = \boxed{171^\circ}$	**40.** $\frac{13\pi}{36} = \frac{13\pi}{36} \cdot \frac{180^\circ}{\pi} = \frac{13 \cdot 36(5^\circ)}{36} = \boxed{65^\circ}$
41. $-\frac{7\pi}{15} = -\frac{7\pi}{15} \cdot \frac{180^\circ}{\pi} = -\frac{7 \cdot 15(12^\circ)}{15} = \boxed{-84^\circ}$	**42.** $-\frac{8\pi}{9} = -\frac{8\pi}{9} \cdot \frac{180^\circ}{\pi} = -\frac{8 \cdot 9(20^\circ)}{9} = \boxed{-160^\circ}$
43. $4 = 4 \cdot \frac{180^\circ}{\pi} = \left(\frac{720}{\pi}\right)^\circ \approx \boxed{229.18^\circ}$	**44.** $3 = 3 \cdot \frac{180^\circ}{\pi} = \left(\frac{540}{\pi}\right)^\circ \approx \boxed{171.89^\circ}$

45. $0.85 = 0.85 \cdot \frac{180^\circ}{\pi} = \left(\frac{153}{\pi}\right)^\circ \approx \boxed{48.70^\circ}$	**46.** $3.27 = 3.27 \cdot \frac{180^\circ}{\pi} = \left(\frac{588.6}{\pi}\right)^\circ \approx \boxed{187.36^\circ}$
47. $-2.7989 = -2.7989 \cdot \frac{180^\circ}{\pi} = \left(\frac{-503.802}{\pi}\right)^\circ$ $\approx \boxed{-160.37^\circ}$	**48.** $-5.9841 = -5.9841 \cdot \frac{180^\circ}{\pi} = \left(\frac{-1,077.138}{\pi}\right)^\circ$ $\approx \boxed{-342.86^\circ}$
49. $2\sqrt{3} = 2\sqrt{3} \cdot \frac{180^\circ}{\pi} = \left(\frac{360\sqrt{3}}{\pi}\right)^\circ \approx \boxed{198.48^\circ}$	**50.** $5\sqrt{7} = 5\sqrt{7} \cdot \frac{180^\circ}{\pi} = \left(\frac{900\sqrt{7}}{\pi}\right)^\circ \approx \boxed{757.95^\circ}$
51. $47^\circ = 47^\circ \cdot \frac{\pi}{180^\circ} = \frac{47\pi}{180} \approx \boxed{0.820}$	**52.** $65^\circ = 65^\circ \cdot \frac{\pi}{180^\circ} = \frac{65\pi}{180} \approx \boxed{1.13}$
53. $112^\circ = 112^\circ \cdot \frac{\pi}{180^\circ} = \frac{112\pi}{180} \approx \boxed{1.95}$	**54.** $172^\circ = 172^\circ \cdot \frac{\pi}{180^\circ} = \frac{172\pi}{180} \approx \boxed{3.00}$
55. $56.5^\circ = 56.5^\circ \cdot \frac{\pi}{180^\circ} = \frac{56.5\pi}{180} \approx \boxed{0.986}$	**56.** $298.7^\circ = 298.7^\circ \cdot \frac{\pi}{180^\circ} = \frac{298.7\pi}{180} \approx \boxed{5.21}$
57. QII	**58.** QII
59. negative y-axis	**60.** negative x-axis
61. Since $-540^\circ = -360^\circ - 180^\circ$, and -360° corresponds to one full revolution clockwise about the circle, the terminal side ends up on the negative x-axis.	**62.** Since $-450^\circ = -360^\circ - 90^\circ$, and -360° corresponds to one full revolution clockwise about the circle, the terminal side ends up on the negative y-axis.
63. Since $0 < \frac{2\pi}{5} < \frac{\pi}{2}$, the terminal side is in Quadrant I.	**64.** Since $\frac{\pi}{2} < \frac{4\pi}{7} < \pi$, the terminal side is in Quadrant II.
65. Since $\frac{3\pi}{2} < \frac{13\pi}{4} < 2\pi$, the terminal side is in Quadrant IV.	**66.** Since $\frac{3\pi}{2} < \frac{18\pi}{11} < 2\pi$, the terminal side is in Quadrant IV.
67. Since $\frac{\pi}{2} < 2.5 < \pi$, the terminal side is in Quadrant II.	**68.** Since $\frac{7\pi}{2} < 11.4 < 4\pi$, the terminal side is in Quadrant IV.

69. Observe that $412^\circ = 360^\circ + 52^\circ$. So, the angle with the smallest positive measure that is coterminal with 412° is $\boxed{52^\circ}$.

70. Observe that $379^\circ = 360^\circ + 19^\circ$. So, the angle with the smallest positive measure that is coterminal with 379° is $\boxed{19^\circ}$.

71. Observe that $268^\circ = 360^\circ - 92^\circ$. So, the angle with the smallest positive measure that is coterminal with -92° is $\boxed{268^\circ}$.	
72. Observe that $173^\circ = 360^\circ - 187^\circ$. So, the angle with the smallest positive measure that is coterminal with -187° is $\boxed{173^\circ}$.	
73. Observe that $-390^\circ = -360^\circ - 30^\circ$. Since 330° and -30° are coterminal, the angle with the smallest positive measure that is coterminal with -390° is $\boxed{330^\circ}$.	
74. Observe that $945^\circ = 2(360^\circ) + 225^\circ$. So, the angle with the smallest positive measure that is coterminal with 945° is $\boxed{225^\circ}$.	
75. Observe that $\frac{29\pi}{3} = 4(2\pi) + \frac{5\pi}{3}$. So, the angle with the smallest positive measure that is coterminal with $\frac{29\pi}{3}$ is $\frac{5\pi}{3}$.	
76. Observe that $\frac{47\pi}{7} = 3(2\pi) + \frac{5\pi}{7}$. So, the angle with the smallest positive measure that is coterminal with $\frac{47\pi}{7}$ is $\frac{5\pi}{7}$.	
77. Observe that $-\frac{313\pi}{9} = -17(2\pi) - \frac{7\pi}{9}$. So, the angle with the smallest positive measure that is coterminal with $-\frac{313\pi}{9}$ is $2\pi - \frac{7\pi}{9} = \frac{11\pi}{9}$.	
78. Observe that $-\frac{217\pi}{4} = -27(2\pi) - \frac{\pi}{4}$. So, the angle with the smallest positive measure that is coterminal with $-\frac{217\pi}{4}$ is $2\pi - \frac{\pi}{4} = \frac{7\pi}{4}$.	
79. Observe that $-\frac{30}{2\pi} = -4.774648 = -4(2\pi) - 0.774648(2\pi)$. So, the angle with the smallest positive measure that is coterminal with -30 is $2\pi - 0.774648(2\pi) \approx 1.42$.	
80. Observe that $\frac{42}{2\pi} = 6.68450761 = 6(2\pi) + 0.68450761(2\pi)$. So, the angle with the smallest positive measure that is coterminal with 42is $0.68450761(2\pi) \approx 4.3$.	
81. Using the formula $s = \theta_r r$, we obtain $s = \frac{\pi}{12}(8 \text{ ft.}) = \boxed{\frac{2\pi}{3}\text{ ft.}}$.	**82.** Using the formula $s = \theta_r r$, we obtain $s = \frac{\pi}{8}(6 \text{ yd.}) = \boxed{\frac{3\pi}{4}\text{ yd.}}$.
83. Using the formula $s = \theta_r r$, we obtain $s = \frac{1}{2}(5 \text{ in.}) = \boxed{\frac{5}{2}\text{ in.}}$.	**84.** Using the formula $s = \theta_r r$, we obtain $s = \frac{3}{4}(20 \text{ m}) = \boxed{15 \text{ m}}$.

85. Using the formula $s=\theta_d\left(\frac{\pi}{180^\circ}\right)r$, we obtain $s=22^\circ\left(\frac{\pi}{180^\circ}\right)(18\ \mu\text{m})=\boxed{\frac{11\pi}{5}\ \mu\text{m}}$.	**86.** Using the formula $s=\theta_d\left(\frac{\pi}{180^\circ}\right)r$, we obtain $s=14^\circ\left(\frac{\pi}{180^\circ}\right)(15\ \mu\text{m})=\boxed{\frac{7\pi}{6}\ \mu\text{m}}$.
87. Using the formula $s=\theta_d\left(\frac{\pi}{180^\circ}\right)r$, we obtain $s=8^\circ\left(\frac{\pi}{180^\circ}\right)(1500\text{ km})=\boxed{\frac{200\pi}{3}\text{ km}}$.	**88.** Using the formula $s=\theta_d\left(\frac{\pi}{180^\circ}\right)r$, we obtain $s=3^\circ\left(\frac{\pi}{180^\circ}\right)(1800\text{ km})=\boxed{30\pi\text{ km}}$.
89. Using the formula $A=\frac{1}{2}r^2\theta_r$, we obtain $A=\frac{1}{2}(2.2\text{ km})^2\left(\frac{3\pi}{8}\right)\approx\boxed{2.85\text{ km}^2}$	**90.** Using the formula $A=\frac{1}{2}r^2\theta_r$, we obtain $A=\frac{1}{2}(13\text{ mi.})^2\left(\frac{5\pi}{6}\right)=\frac{845\pi}{12}\text{ mi.}^2\approx\boxed{221\text{ mi.}^2}$
91. Using the formula $A=\frac{1}{2}r^2\theta_d\left(\frac{\pi}{180^\circ}\right)$, we obtain $A=\frac{1}{2}(4.2\text{ cm})^2(56^\circ)\left(\frac{\pi}{180^\circ}\right)\approx\boxed{8.62\text{ cm}^2}$	**92.** Using the formula $A=\frac{1}{2}r^2\theta_d\left(\frac{\pi}{180^\circ}\right)$, we obtain $A=\frac{1}{2}(2.5\text{ mm})^2(27^\circ)\left(\frac{\pi}{180^\circ}\right)\approx\boxed{1.47\text{ mm}^2}$
93. Using the formula $A=\frac{1}{2}r^2\theta_d\left(\frac{\pi}{180^\circ}\right)$, we obtain $A=\frac{1}{2}(1.5\text{ ft.})^2(1.2^\circ)\left(\frac{\pi}{180^\circ}\right)\approx\boxed{0.0236\text{ ft.}^2}$	**94.** Using the formula $A=\frac{1}{2}r^2\theta_d\left(\frac{\pi}{180^\circ}\right)$, we obtain $A=\frac{1}{2}(3.0\text{ ft.})^2(14^\circ)\left(\frac{\pi}{180^\circ}\right)\approx\boxed{1.10\text{ ft.}^2}$
95. Using the formula $v=\frac{s}{t}$ yields $v=\frac{2\text{ m}}{5\text{ sec.}}=\boxed{\frac{2}{5}\ \text{m}/\text{sec.}}$.	**96.** Using the formula $v=\frac{s}{t}$ yields $v=\frac{12\text{ ft.}}{3\text{ min.}}=\boxed{4\ \text{ft.}/\text{min.}}$.
97. Using the formula $v=\frac{s}{t}$ yields $v=\frac{68{,}000\text{ km}}{250\text{ hr.}}=\boxed{272\ \text{km}/\text{hr.}}$.	**98.** Using the formula $v=\frac{s}{t}$ yields $v=\frac{7{,}524\text{ mi.}}{12\text{ days}}=\boxed{627\ \text{mi.}/\text{day}}$.
99. Using the formula $s=vt$ yields $s=\left(2.8\ \text{m}/\text{sec.}\right)(3.5\text{ sec.})=\boxed{9.8\text{ m}}$.	**100.** Using the formula $s=vt$ yields $s=\left(6.2\ \text{km}/\text{hr.}\right)(4.5\text{ hr.})=\boxed{27.9\text{ km}}$.
101. (Note that 20 min. = $\frac{1}{3}$ hr.) Using the formula $s=vt$ yields $s=\left(4.5\ \text{mi.}/\text{hr.}\right)\left(\frac{1}{3}\text{ hr.}\right)=\boxed{1.5\text{ mi.}}$	**102.** (Note that 2 min. = 120 sec.) Using the formula $s=vt$ yields $s=\left(5.6\ \text{ft.}/\text{sec.}\right)(120\text{ sec.})=\boxed{672\text{ ft.}}$

103. Using the formula $\omega = \frac{\theta}{t}$ (θ measured in radians) yields

$$\omega = \frac{25\pi \text{ rad.}}{10 \text{ sec.}} = \boxed{\frac{5\pi}{2} \text{ rad./sec.}}$$

104. Using the formula $\omega = \frac{\theta}{t}$ (θ measured in radians) yields

$$\omega = \frac{\frac{3\pi}{4} \text{ rad.}}{\frac{1}{6} \text{ sec.}} = \boxed{\frac{9\pi}{2} \text{ rad./sec.}}$$

105. First, note that

$$\theta = 200^\circ\left(\frac{\pi}{180^\circ}\right) = \frac{10\pi}{9}.$$

Using the formula $\omega = \frac{\theta}{t}$ (θ measured in radians) yields

$$\omega = \frac{\frac{10\pi}{9} \text{ rad.}}{5 \text{ sec.}} = \boxed{\frac{2\pi}{9} \text{ rad./sec.}}$$

106. First, note that

$$\theta = 60^\circ\left(\frac{\pi}{180^\circ}\right) = \frac{\pi}{3}.$$

Using the formula $\omega = \frac{\theta}{t}$ (θ measured in radians) yields

$$\omega = \frac{\frac{\pi}{3} \text{ rad.}}{0.2 \text{ sec.}} = \boxed{\frac{5\pi}{3} \text{ rad./sec.}}$$

107. Using the formula $v = r\omega$ yields

$$v = (9 \text{ in.})\left(\frac{2\pi}{3 \text{ sec.}}\right) = \boxed{6\pi \text{ in./sec.}}$$

108. Using the formula $v = r\omega$ yields

$$v = (8 \text{ cm})\left(\frac{3\pi}{4 \text{ sec.}}\right) = \boxed{6\pi \text{ cm/sec.}}$$

109. Using the formula $v = r\omega$ yields

$$v = (5 \text{ mm})\left(\frac{\pi}{20 \text{ sec.}}\right) = \boxed{\frac{\pi}{4} \text{ mm/sec.}}$$

110. Using the formula $v = r\omega$ yields

$$v = (24 \text{ ft.})\left(\frac{5\pi}{16 \text{ sec.}}\right) = \boxed{\frac{15\pi}{2} \text{ ft./sec.}}$$

111. Note that using the two equations $v = \frac{s}{t}$ and $v = r\omega$ together generates the formula: $s = r\omega t$. Using this formula yields

$$s = (5 \text{ cm})\left(\frac{\pi}{6 \text{ sec.}}\right)(10 \text{ sec.})$$

$$= \frac{25\pi}{3} \text{ cm} \approx \boxed{26.2 \text{ cm}}$$

112. Note that using the two equations $v = \frac{s}{t}$ and $v = r\omega$ together generates the formula: $s = r\omega t$. Using this formula yields

$$s = (2 \text{ mm})\left(\frac{6\pi}{1 \text{sec.}}\right)(11 \text{ sec.})$$

$$= 132\pi \text{ mm} \approx \boxed{415 \text{ mm}}$$

113. Note that using the two equations $v = \frac{s}{t}$ and $v = r\omega$ together generates the formula: $s = r\omega t$.

Further, note that

10 min. = 10(60) sec. = 600 sec.

Using this with the above formula yields

$$s = (5.2 \text{ in.})\left(\frac{\pi}{15 \text{ sec.}}\right)(600 \text{ sec.})$$

$$= 208\pi \text{ in.} \approx \boxed{653 \text{ in. or } 54.5 \text{ ft.}}$$

114. Note that using the two equations $v = \frac{s}{t}$ and $v = r\omega$ together generates the formula: $s = r\omega t$.

Further, note that

3 min. = 3(60) sec. = 180 sec.

Using this with the above formula yields

$$s = (3.2 \text{ ft.})\left(\frac{\pi}{4 \text{ sec.}}\right)(180 \text{ sec.})$$

$$= 144\pi \text{ ft.} \approx \boxed{452 \text{ ft.}}$$

115. Note that each multiple of 5 cuts off a sector that corresponds to $1/8$ of the entire circular face. Follow these instructions to obtain the total angle sum that corresponds to the total turning to unlock the lock with combination **35 – 5 – 20 :**

Step 1: 2 full clockwise turns: $2(360°)$

Step 2: Going clockwise, stop at **35**: $\frac{1}{8}(360°)$

Step 3: 1 full counterclockwise turn: $(360°)$

Step 4: From **35**, go to **5** counterclockwise: $\frac{2}{8}(360°)$

Step 5: From **5**, go to **20** clockwise: $\frac{5}{8}(360°)$

Now, the resulting total angle sum is $360°\left[2 + \frac{1}{8} + 1 + \frac{2}{8} + \frac{5}{8}\right] = \boxed{1{,}440°}$.

116. Note that each multiple of 5 cuts off a sector that corresponds to $\frac{1}{8}$ of the entire circular face. Follow these instructions to obtain the total angle sum that corresponds to the total turning to unlock the lock with combination **20 – 15 – 5 :**

Step 1: 2 full clockwise turns: $2\left(360^\circ\right)$

Step 2: Going clockwise, stop at **20**: $\frac{4}{8}\left(360^\circ\right)$

Step 3: 1 full counterclockwise turn: $\left(360^\circ\right)$

Step 4: From **20**, go to **15** counterclockwise: $\frac{7}{8}\left(360^\circ\right)$

Step 5: From **15**, go to **5** clockwise: $\frac{3}{8}\left(360^\circ\right)$

Now, the resulting total angle sum is $360^\circ\left[2+\frac{4}{8}+1+\frac{7}{8}+\frac{3}{8}\right]=\boxed{1,665^\circ}$.

117. First, since 1 mi. = 63,360 in., we see that

$$\omega=\frac{v}{r}=\frac{65\ \text{mi./hr.}}{12.15\text{ in.}}=\frac{65(63,360)\ \text{in./hr.}}{12.15\text{ in.}}\approx 338,962\ \text{rad./hr.}$$

Next, using the formula $v=r\omega$ with $r=13.05$ in. and $\omega\approx 338,962$ rad./hr. yields

$$v=(13.05\text{ in.})\left(338,962\ \text{rad./hr.}\right)=4,423,454\ \text{in./hr.}=\frac{4,423,454\ \text{in./hr.}}{63,360\ \text{in./mi.}}\approx 69.82\ \text{mi./hr.}$$

So, she is actually traveling at a speed of approximately $\boxed{69.82\text{ mph}}$.

118. First, since 1 mi. = 63,360 in., we see that

$$\omega=\frac{v}{r}=\frac{70\ \text{mi./hr.}}{12.4\text{ in.}}=\frac{70(63,360)\ \text{in./hr.}}{12.4\text{ in.}}\approx 357,677\ \text{rad./hr.}$$

Next, using the formula $v=r\omega$ with $r=13.5$ in. and $\omega\approx 357,677$ rad./hr. yields

$$v=(13.5\text{ in.})\left(357,677\ \text{rad./hr.}\right)=4,828,639.5\ \text{in./hr.}=\frac{4,828,639.5\ \text{in./hr.}}{63,360\ \text{in./mi.}}\approx 76.21\ \text{mi./hr.}$$

So, she is actually traveling at a speed of approximately $\boxed{76.21\text{ mph}}$.

119. We are given that $r = 29$ ft. and $v = 200$ mph. We seek ω. Use the formula $\omega = \frac{v}{r}$. Observe that $\omega = \dfrac{200 \text{ mph}}{\left[\frac{29}{5,280}\right] \text{mi.}} \approx 36,413.79 \text{ rad.}/\text{hr.}$ Converting to radians per second, we see that $$\omega \approx \left(\frac{36,413.79 \text{ rad.}}{\text{hr.}}\right)\cdot\left(\frac{1 \text{ hr.}}{3,600 \text{ sec.}}\right) = \boxed{10.11 \text{ rad.}/\text{sec.} = 1.6 \text{ rotations per second}}.$$
120. We are given that $r = 29$ ft. and $\omega = 2\pi \text{ rad.}/\text{sec.}$. We seek v. Use the formula $v = \omega r$. Observe that $$v = \left(2\pi \text{ rad.}/\text{sec.}\right)(29 \text{ ft.}) = 58\pi \text{ ft.} \approx \boxed{182 \text{ ft.}}.$$
121. Angular velocity must be expressed in radians (not degrees) per second. Here, one should use $\pi \text{ rad.}/\text{sec.}$ in place of $180^\circ/\text{sec}$.
122. Angular velocity must be expressed in radians (not degrees) per second. Here, one should use $\pi \text{ rad.}/\text{sec.}$ in place of $180^\circ/\text{sec}$.
123. True. Let θ be a fixed central angle, and let s_r be the arc length corresponding to an arc with central angle θ and radius r. Then, s_{2r} is the arc length corresponding to an arc with central angle θ and radius $2r$. Observe that $$s_r = \theta r \text{ and } s_{2r} = \theta(2r) = 2(\theta r) = 2s_r.$$
124. False. The area, in fact, quadruples (rather than doubles). To see this, let θ be the central angle (in radians), and let A_r be the area of a sector corresponding to an arc with central angle θ and radius r. Then, A_{2r} is the area of a sector corresponding to an arc with central angle θ and radius $2r$. Observe that $$A_{2r} = \frac{1}{2}(2r)^2\,\theta = \frac{1}{2}4r^2\theta = 4\left(\frac{1}{2}r^2\theta\right) = 4A_r.$$
125. True. If the angular speed is $\omega = \frac{\theta}{t}$, then the number of revolutions is $\frac{\omega}{2\pi}$. Hence, if the angular speed is 2ω, then the number of revolutions is $\frac{2\omega}{2\pi} = 2\left(\frac{\omega}{2\pi}\right)$, so that there are twice as many.
126. True. The area of a sector of radius r and central angle θ is $A = \frac{1}{2}r^2\theta$. So, if you double θ, you indeed double the area.

127. First, note that at **12:00** exactly, both the minute and the hour hands are identically on the **12**. Then, for each minute that passes, the minute hand moves $\frac{1}{60}$ the way around the clock face (i.e., 6°). Similarly, for each minute that passes, the hour hand moves $\frac{1}{60}$ the way between the **12** and the **1;** since there are $\frac{1}{12}(360^\circ) = 30^\circ$ between consecutive integers on the clock face, such movement corresponds to $\frac{1}{60}(30^\circ) = 0.5^\circ$.
Now, when the time is **12:20**, we know that the minute hand is on the **4**, but the hour hand has moved $20 \times 0.5^\circ = 10^\circ$ clockwise from the **12** towards the **1**.
The picture is as follows:

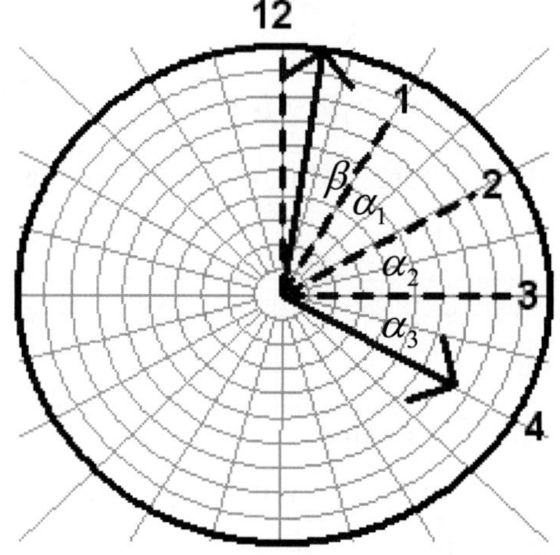

The angle we seek is $\beta + \alpha_1 + \alpha_2 + \alpha_3$. From the above discussion, we know that $\alpha_1 = \alpha_2 = \alpha_3 = 30^\circ$ and $\beta = 20^\circ$. Thus, the angle at time **12:20** is $\boxed{110^\circ}$.

128. First, note that at **9:00** exactly, the minute is identically on the **12** and the hour hand is identically on the **9**. Then, for each minute that passes, the minute hand moves $\frac{1}{60}$ the way around the clock face (i.e., 6°). Similarly, for each minute that passes, the hour hand moves $\frac{1}{60}$ the way between the **9** and the **10;** since there are $\frac{1}{12}(360^\circ) = 30^\circ$ between consecutive integers on the clock face, such movement corresponds to $\frac{1}{60}(30^\circ) = 0.5^\circ$.

Now, when the time is **9:10**, we know that the minute hand is on the **2**, but the hour hand has moved $10 \times 0.5^\circ = 5^\circ$ clockwise from the **9** towards the **10**, thereby leaving an angle of 25° between the hour hand and the **10**. The picture is as follows:

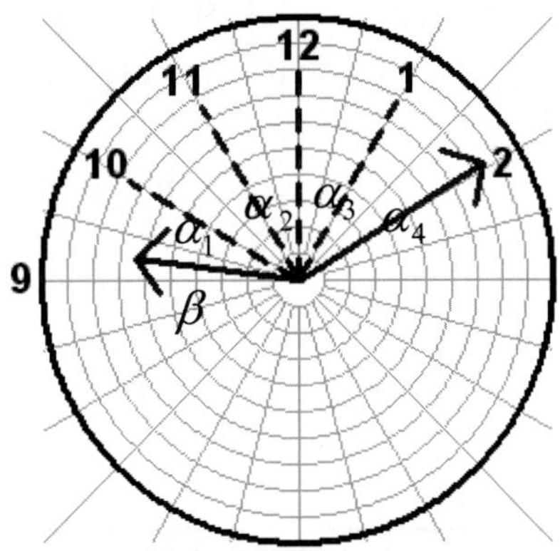

The angle we seek is $\beta + \alpha_1 + \alpha_2 + \alpha_3 + \alpha_4$. From the above discussion, we know that $\alpha_1 = \alpha_2 = \alpha_3 = \alpha_4 = 30^\circ$ and $\beta = 25^\circ$. Thus, the angle at time **9:10** is $\boxed{145^\circ}$.

129. Labeling the sector in QI as #1, the one in QII as #2, and so on, we have:

Area of #1 $= \frac{1}{2}(4)^2\left(\frac{\pi}{2}\right)$ units^2

Area of #2 $= \frac{1}{2}(3)^2\left(\frac{\pi}{2}\right)$ units^2

Area of #3 $= \frac{1}{2}(2)^2\left(\frac{\pi}{2}\right)$ units^2

Area of #4 $= \frac{1}{2}(1)^2\left(\frac{\pi}{2}\right)$ units^2

Hence, the area of the entire shaded region is the sum of the above areas, namely $\frac{15\pi}{2}$ units^2.

130. Labeling the sector in QI as #1, the one in QII as #2, and so on, we have:

Length of #1 $= 4\left(\frac{\pi}{2}\right)$ units

Length of #2 $= 3\left(\frac{\pi}{2}\right)$ units

Length of #3 $= 2\left(\frac{\pi}{2}\right)$ units

Length of #4 $= 1\left(\frac{\pi}{2}\right)$ units

Adding these lengths to the lengths of the linear segments used in the formation of the figure results in a total length of $(6+5\pi)$ units.

131. Since $s = r\theta_r$, we have $94.4 = 78.6\theta_r$, so that $\theta_r \approx 1.201$ rad $\approx 68.812^\circ$. Upon using the command ANGLE/DMS on the TI-calculator, we see that this is equivalent to $68^\circ 48' 44.032''$.	**132.** Since $s = r\theta_r$, we have $23.8 = 14.2\theta_r$, so that $\theta_r \approx 1.676$ rad $\approx 96.031^\circ$. Upon using the command ANGLE/DMS on the TI-calculator, we see that this is equivalent to $96^\circ 1' 51.436''$.
133. Using $s = r\theta_r$, we have $2\pi = 10\theta$, so that $\theta = \frac{\pi}{5}$.	**134.** Observe that $750^\circ = 2\left(360^\circ\right) + 30^\circ$. Since 30° is equivalent to $\frac{\pi}{6}$, we see that the angle with the smallest positive measure that is coterminal with 750° is 30°.
135. Using $S = \frac{1}{2} r^2 \left(\frac{\pi}{180^\circ}\right)\theta_d$, we have $\frac{3\pi}{2} = \frac{1}{2}(3)^2\left(\frac{\pi}{180^\circ}\right)\theta_d \Rightarrow \theta_d = \frac{3\pi(20^\circ)}{\pi} = 60^\circ$ or $\frac{\pi}{3}$	**136.** Using $\omega = \frac{\theta}{t}$, we see that $600^\circ/\text{sec} = \theta_d/3\,\text{sec} \Rightarrow \theta_d = 1800^\circ$ which is equivalent to $1800^\circ\left(\frac{\pi}{180^\circ}\right) = 10\pi$.

Section 4.2 Solutions ---

1. Note that by the Pythagorean Theorem, the hypotenuse has length $\sqrt{5}$. So, $\cos\theta = \frac{1}{\sqrt{5}} = \boxed{\frac{\sqrt{5}}{5}}$.	**2.** Note that by the Pythagorean Theorem, the hypotenuse has length $\sqrt{5}$. So, $\sin\theta = \frac{2}{\sqrt{5}} = \boxed{\frac{2\sqrt{5}}{5}}$.
3. Note that by the Pythagorean Theorem, the hypotenuse has length $\sqrt{5}$. So, $\sec\theta = \frac{1}{\cos\theta} = \frac{1}{\frac{1}{\sqrt{5}}} = \boxed{\sqrt{5}}$.	**4.** Note that by the Pythagorean Theorem, the hypotenuse has length $\sqrt{5}$. So, $\csc\theta = \frac{1}{\sin\theta} = \frac{1}{\frac{2}{\sqrt{5}}} = \boxed{\frac{\sqrt{5}}{2}}$.
5. $\tan\theta = \frac{2}{1} = \boxed{2}$	**6.** $\cot\theta = \frac{1}{\tan\theta} = \boxed{\frac{1}{2}}$
7. $\frac{2\sqrt{10}}{7}$	**8.** $\frac{3}{7}$
9. $\frac{7}{3}$	**10.** $\frac{7\sqrt{10}}{20}$
11. $\frac{3/7}{2\sqrt{10}/7} = \frac{3\sqrt{10}}{20}$	**12.** $\frac{2\sqrt{10}}{3}$

13. a	**14.** b	**15.** b	**16.** a	**17.** c	**18.** c

19. $\tan 30^\circ = \dfrac{\sin 30^\circ}{\cos 30^\circ} = \dfrac{\frac{1}{2}}{\frac{\sqrt{3}}{2}} = \dfrac{1}{\sqrt{3}} = \boxed{\dfrac{\sqrt{3}}{3}}$

20. $\tan\frac{\pi}{4} = \dfrac{\sin\frac{\pi}{4}}{\cos\frac{\pi}{4}} = \dfrac{\frac{\sqrt{2}}{2}}{\frac{\sqrt{2}}{2}} = \boxed{1}$

21. $\tan 60^\circ = \dfrac{\sin 60^\circ}{\cos 60^\circ} = \dfrac{\frac{\sqrt{3}}{2}}{\frac{1}{2}} = \boxed{\sqrt{3}}$

22. $\csc 30^\circ = \dfrac{1}{\sin 30^\circ} = \dfrac{1}{\frac{1}{2}} = \boxed{2}$

23.

$$\sec 30^\circ = \frac{1}{\cos 30^\circ} = \frac{1}{\frac{\sqrt{3}}{2}} = \frac{2}{\sqrt{3}} = \boxed{\frac{2\sqrt{3}}{3}}$$

24. Using Exercise 19, we see that

$$\cot\tfrac{\pi}{6} = \frac{1}{\tan\frac{\pi}{6}} = \frac{1}{\tan 30^\circ} = \frac{1}{\frac{\sqrt{3}}{3}} = \boxed{\sqrt{3}}.$$

25. $\csc\frac{\pi}{3} = \dfrac{1}{\sin\frac{\pi}{3}} = \dfrac{1}{\frac{\sqrt{3}}{2}} = \dfrac{2}{\sqrt{3}} = \boxed{\dfrac{2\sqrt{3}}{3}}$

26. $\sec 60^\circ = \dfrac{1}{\cos 60^\circ} = \dfrac{1}{\frac{1}{2}} = \boxed{2}$

27. Using Exercise 21, we see that

$$\cot 60^\circ = \frac{1}{\tan 60^\circ} = \frac{1}{\sqrt{3}} = \boxed{\frac{\sqrt{3}}{3}}.$$

28. $\csc 45^\circ = \dfrac{1}{\sin 45^\circ} = \dfrac{1}{\frac{\sqrt{2}}{2}} = \boxed{\sqrt{2}}$

29. $\sec\frac{\pi}{4} = \dfrac{1}{\cos\frac{\pi}{4}} = \dfrac{1}{\frac{1}{\sqrt{2}}} = \boxed{\sqrt{2}}$

30. Using Exercise 20, we see that

$$\cot\tfrac{\pi}{4} = \frac{1}{\tan\frac{\pi}{4}} = \frac{1}{1} = \boxed{1}.$$

31. 0.6018	**32.** 0.3057	**33.** 0.1392	**34.** 0.9278
35. 0.2588	**36.** 0.9848	**37.** -0.8090	**38.** 0.9010
39. 1.3764	**40.** 0.9391	**41.** 0.4142	**42.** -0.3249
43. 1.0034	**44.** 5.7588	**45.** 0.7002	**46.** 1.8040

47. Consider this triangle:

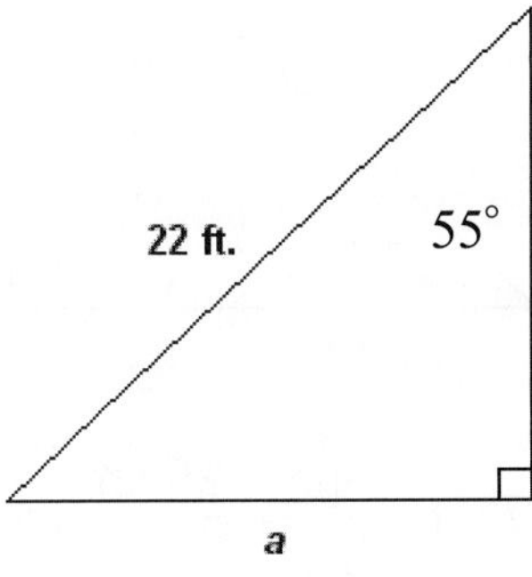

Since $\sin(55^\circ) = \dfrac{\text{opp}}{\text{hyp}} = \dfrac{a}{22 \text{ ft.}}$, we have

$$a = (22 \text{ ft.})\sin(55^\circ) \approx 18.02 \text{ ft.} \approx \boxed{18 \text{ ft.}}$$

48. Consider this triangle:

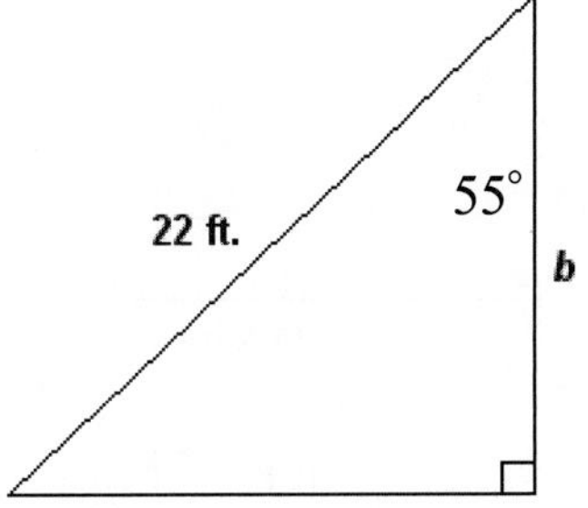

Since $\cos(55^\circ) = \dfrac{\text{adj}}{\text{hyp}} = \dfrac{b}{22 \text{ ft.}}$, we have

$$b = (22 \text{ ft.})\cos(55^\circ) \approx 12.619 \text{ ft.} \approx \boxed{13 \text{ ft.}}$$

49. Consider this triangle: 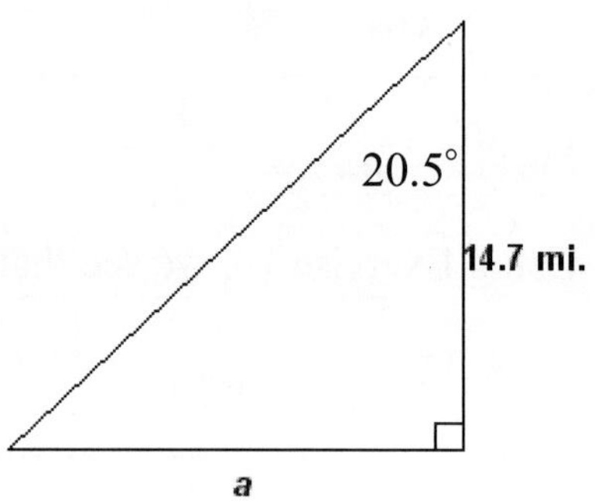 Since $\tan(20.5^\circ) = \frac{\text{opp}}{\text{adj}} = \frac{a}{14.7\text{ mi.}}$, we have $$a = (14.7\text{ mi.})\tan(20.5^\circ) \approx 5.496\text{ mi.} \approx \boxed{5.50\text{ mi.}}$$	**50.** Consider this triangle: 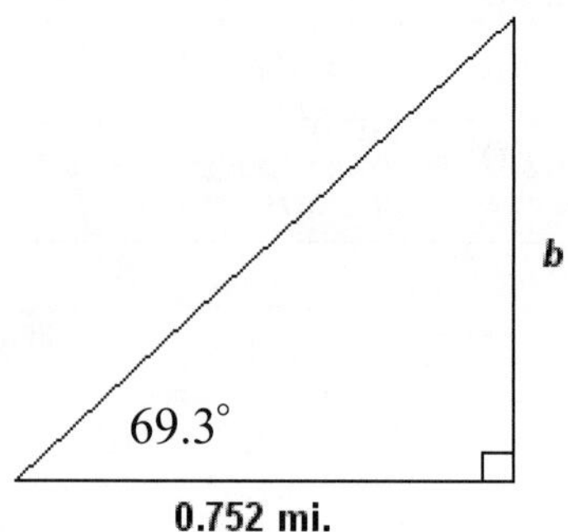 Since $\tan(69.3^\circ) = \frac{\text{opp}}{\text{adj}} = \frac{b}{0.752\text{ mi.}}$, we have $$b = (0.752\text{ mi.})\tan(69.3^\circ) \approx 1.990\text{ mi.} \approx \boxed{1.99\text{ mi.}}$$ (since the number of significant digits is 3).
51. Consider this triangle: 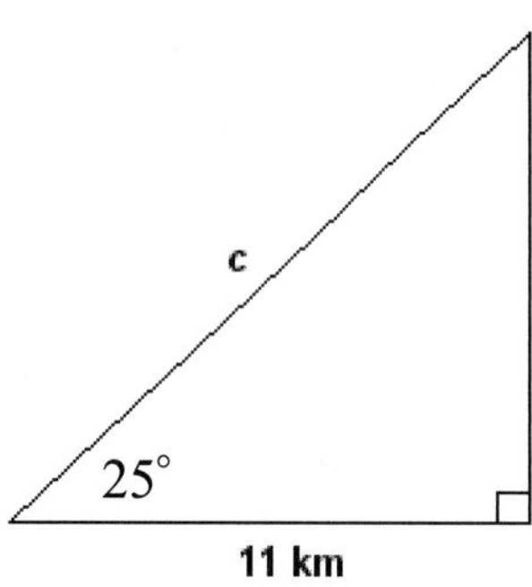 Since $\cos(25^\circ) = \frac{\text{adj}}{\text{hyp}} = \frac{11\text{ km}}{c}$, we have $$c = \frac{11\text{ km}}{\cos(25^\circ)} \approx 12.137\text{ km} \approx \boxed{12\text{ km}}$$ (since the number of significant digits is 2).	**52.** Consider this triangle: 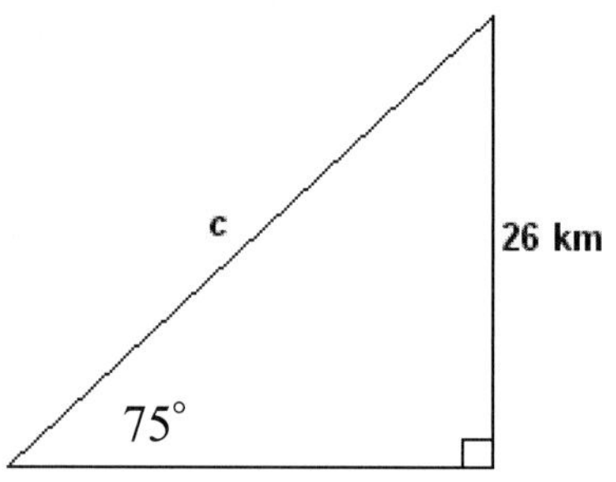 Since $\sin(75^\circ) = \frac{\text{opp}}{\text{hyp}} = \frac{26\text{ km}}{c}$, we have $$c = \frac{26\text{ km}}{\sin(75^\circ)} \approx 26.917\text{ km} \approx \boxed{27\text{ km}}$$ (since the number of significant digits is 2).

53. Consider this triangle:

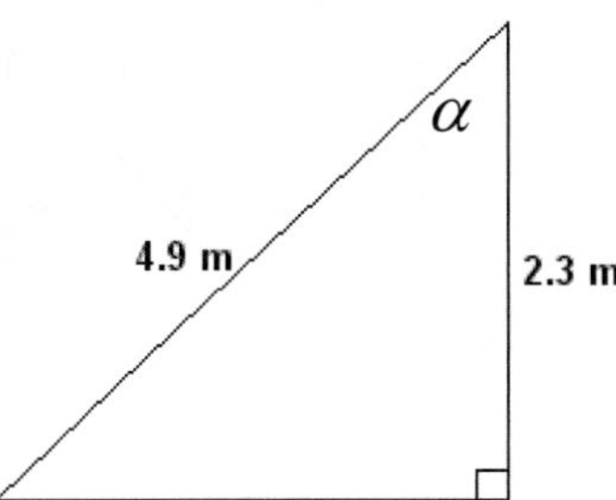

Since $\cos(\alpha) = \frac{\text{adj}}{\text{hyp}} = \frac{2.3 \text{ m}}{4.9 \text{ m}}$, we have

$$\alpha = \cos^{-1}\left(\frac{2.3}{4.9}\right) \approx 62.005 \approx \boxed{62^\circ}$$

(since the number of significant digits is 2).

54. Consider this triangle:

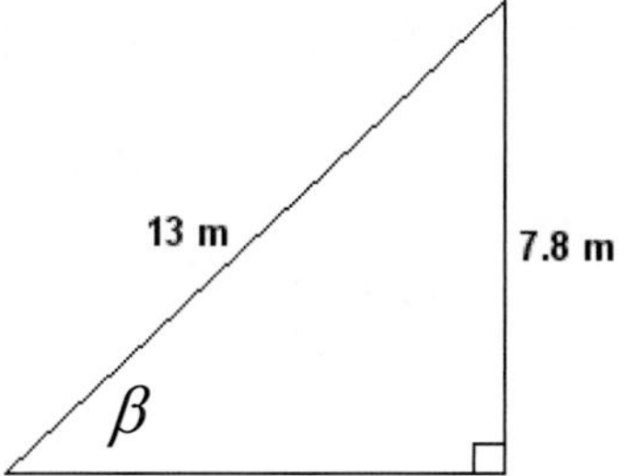

Since $\sin(\beta) = \frac{\text{opp}}{\text{hyp}} = \frac{7.8 \text{ m}}{13 \text{ m}}$, we have

$$\beta = \sin^{-1}\left(\frac{7.8}{13}\right) \approx 36.870 \approx \boxed{37^\circ}$$

(since the number of significant digits is 2).

55. Consider this triangle:

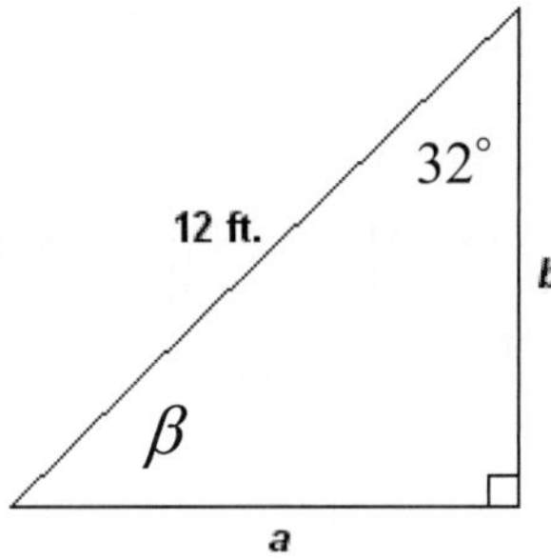

First, observe that

$$\beta = 180^\circ - \left(90^\circ + 32^\circ\right) = 58^\circ.$$

Since $\sin(32^\circ) = \frac{\text{opp}}{\text{hyp}} = \frac{a}{12 \text{ ft.}}$, we have

$$a = (12 \text{ ft.})(\sin 32^\circ) \approx 6.4 \text{ ft.}$$

Similarly, since $\cos(32^\circ) = \frac{\text{adj}}{\text{hyp}} = \frac{b}{12 \text{ ft.}}$, we have

$$b = (12 \text{ ft.})(\cos 32^\circ) \approx 10 \text{ ft.}$$

56. Consider this triangle:

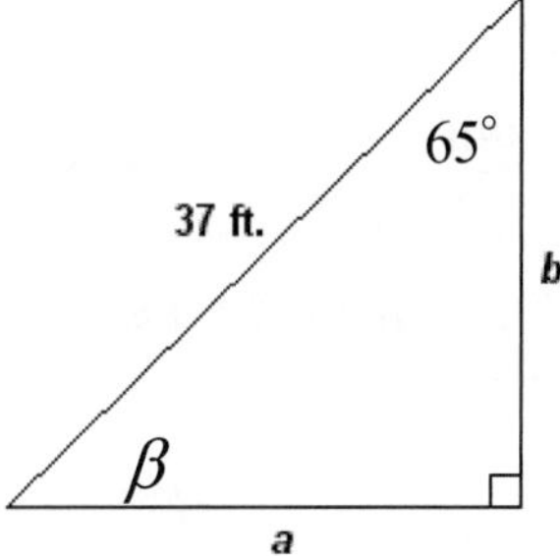

First, observe that

$$\beta = 180^\circ - \left(90^\circ + 65^\circ\right) = 25^\circ.$$

Since $\sin(65^\circ) = \frac{\text{opp}}{\text{hyp}} = \frac{a}{37 \text{ ft.}}$, we have

$$a = (37 \text{ ft.})(\sin 65^\circ) \approx 34 \text{ ft.}$$

Similarly, since $\cos(65^\circ) = \frac{\text{adj}}{\text{hyp}} = \frac{b}{37 \text{ ft.}}$, we have

$$b = (37 \text{ ft.})(\cos 65^\circ) \approx 16 \text{ ft.}$$

57. Consider this triangle:

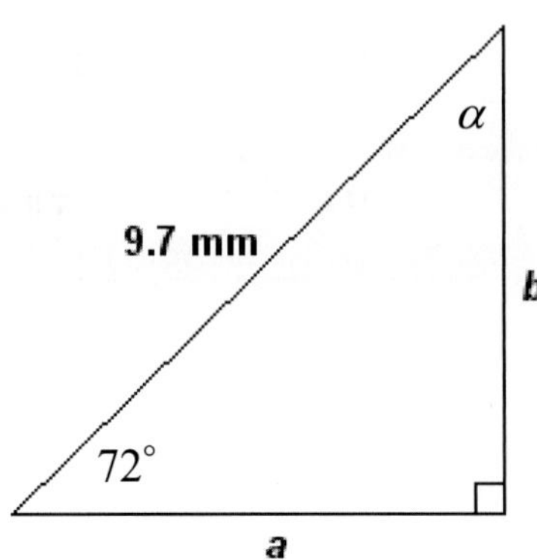

First, observe that

$$\alpha = 180^\circ - \left(90^\circ + 72^\circ\right) = 18^\circ.$$

Since $\sin\left(72^\circ\right) = \frac{\text{opp}}{\text{hyp}} = \frac{b}{9.7 \text{ mm}}$, we have

$$b = \left(9.7 \text{ mm}\right)\left(\sin 72^\circ\right) \approx 9.2 \text{ mm}.$$

Similarly, since $\cos\left(72^\circ\right) = \frac{\text{adj}}{\text{hyp}} = \frac{a}{9.7 \text{ mm}}$, we have

$$a = \left(9.7 \text{ mm}\right)\left(\cos 72^\circ\right) \approx 3.0 \text{ mm}.$$

58. Consider this triangle:

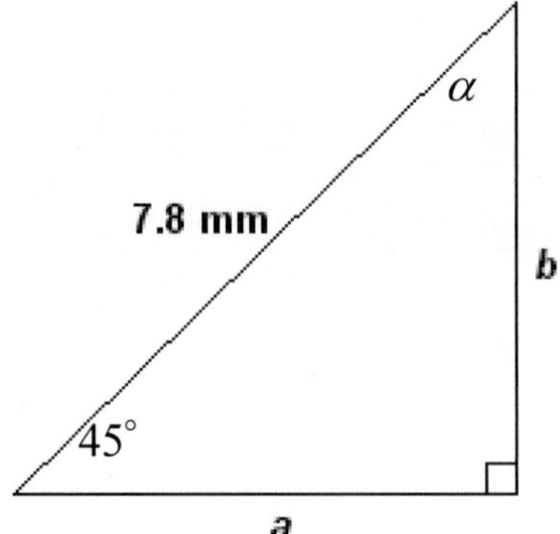

First, observe that

$$\alpha = 180^\circ - \left(90^\circ + 45^\circ\right) = 45^\circ.$$

Since $\sin\left(45^\circ\right) = \frac{\text{opp}}{\text{hyp}} = \frac{b}{7.8 \text{ mm}}$, we have

$$b = \left(7.8 \text{ mm}\right)\left(\sin 45^\circ\right) \approx 5.5 \text{ mm}.$$

Similarly, since $\cos\left(45^\circ\right) = \frac{\text{adj}}{\text{hyp}} = \frac{a}{7.8 \text{ mm}}$, we have

$$a = \left(7.8 \text{ mm}\right)\left(\cos 45^\circ\right) \approx 5.5 \text{ mm}.$$

59. Consider this triangle:

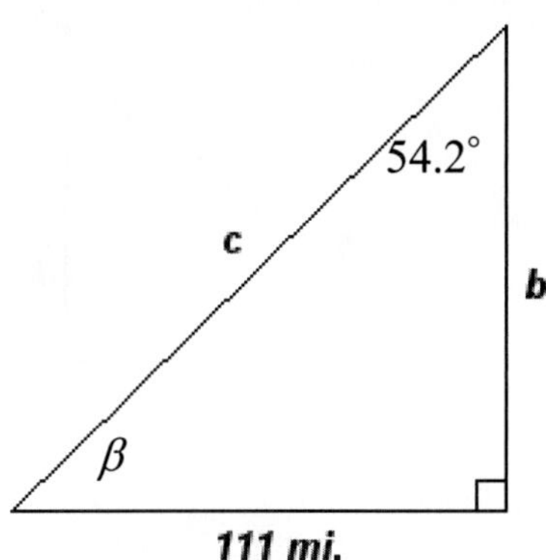

First, observe that

$$\beta = 180^\circ - \left(90^\circ + 54.2^\circ\right) = 35.8^\circ .$$

Since $\sin\left(54.2^\circ\right) = \frac{\text{opp}}{\text{hyp}} = \frac{111 \text{ mi.}}{c}$, we have

$$c = \frac{111 \text{ mi.}}{\sin\left(54.2^\circ\right)} \approx 137 \text{ mi.}$$

Similarly, since

$\tan\left(54.2^\circ\right) = \frac{\text{opp}}{\text{adj}} = \frac{111 \text{ mi.}}{b}$, we have

$$b = \frac{111 \text{ mi.}}{\tan\left(54.2^\circ\right)} \approx 80.1 \text{ mi.}$$

60. Consider this triangle:

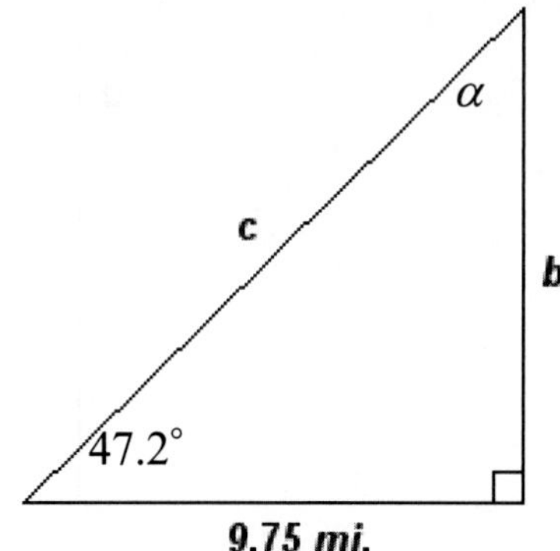

First, observe that

$$\alpha = 180^\circ - \left(90^\circ + 47.2^\circ\right) = 42.8^\circ .$$

Since $\cos\left(47.2^\circ\right) = \frac{\text{adj}}{\text{hyp}} = \frac{9.75 \text{ mi.}}{c}$, we have

$$c = \frac{9.75 \text{ mi.}}{\cos\left(47.2^\circ\right)} \approx 14.4 \text{ mi.}$$

Similarly, since

$\tan\left(47.2^\circ\right) = \frac{\text{opp}}{\text{adj}} = \frac{b}{9.75 \text{ mi.}}$, we have

$$b = \left(9.75 \text{ mi.}\right)\tan\left(47.2^\circ\right) \approx 10.5 \text{ mi.}$$

61. Consider this triangle:

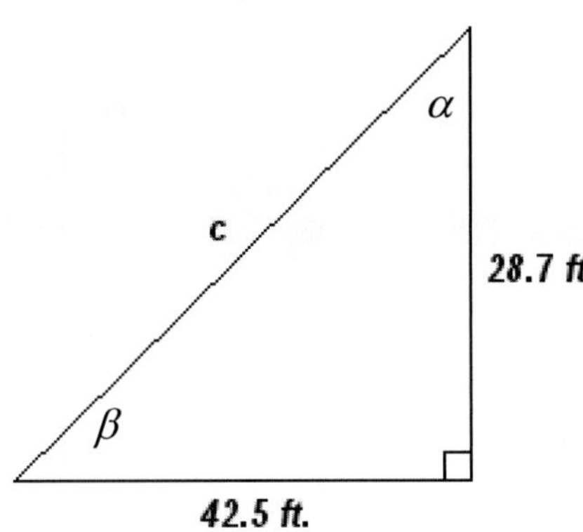

First, since $\tan\alpha = \dfrac{42.5 \text{ ft.}}{28.7 \text{ ft.}}$, we obtain

$$\alpha = \tan^{-1}\left(\frac{42.5}{28.7}\right) \approx 56.0^\circ .$$

So, $\beta \approx 180^\circ - \left(90^\circ + 56.0^\circ\right) \approx 34.0^\circ$.

Next, using the Pythagorean Theorem, we obtain

$$c^2 = (28.7 \text{ ft.})^2 + (42.5 \text{ ft.})^2 = 2{,}629.94 \text{ ft.}^2$$

and so, $c \approx 51.3$ ft.

62. Consider this triangle:

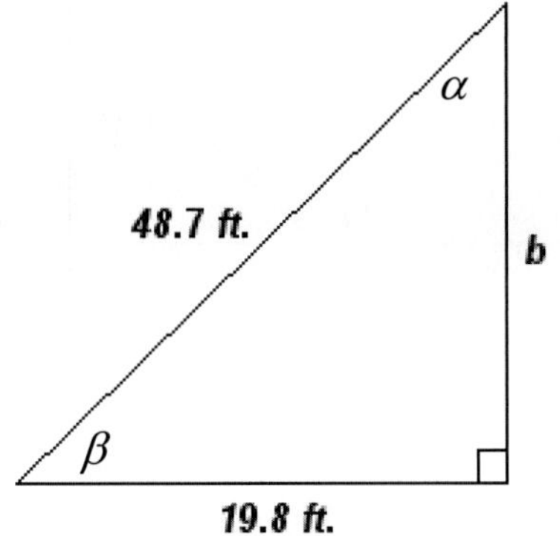

First, since $\sin\alpha = \dfrac{19.8 \text{ ft.}}{48.7 \text{ ft.}}$, we obtain

$$\alpha = \sin^{-1}\left(\frac{19.8}{48.7}\right) \approx 24.0^\circ .$$

Similarly, since $\cos\beta = \dfrac{19.8 \text{ ft.}}{48.7 \text{ ft.}}$, we obtain

$$\beta = \cos^{-1}\left(\frac{19.8}{48.7}\right) \approx 66.0^\circ .$$

Next, using the Pythagorean Theorem, we obtain

$$b^2 + (19.8 \text{ ft.})^2 = (48.7 \text{ ft.})^2$$
$$b^2 = 1{,}979.65 \text{ ft.}^2$$
$$b \approx 44.5 \text{ ft.}$$

63. Consider this triangle:

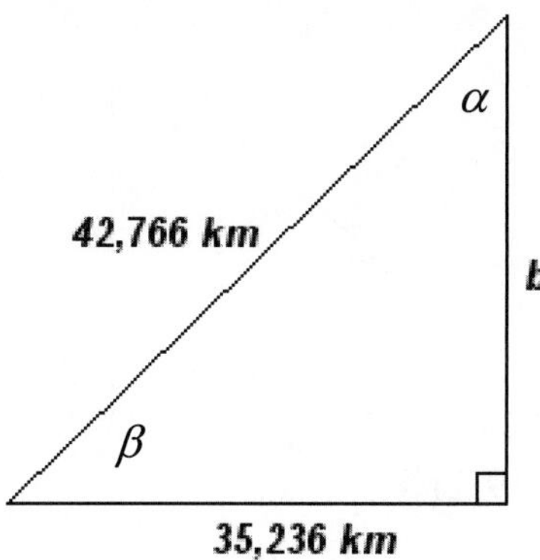

First, since $\sin\alpha = \dfrac{35{,}236\text{ km}}{42{,}766\text{ km}}$, we obtain

$$\alpha = \sin^{-1}\left(\frac{35{,}236}{42{,}766}\right) \approx 55.480^\circ.$$

Similarly, since $\cos\beta = \dfrac{35{,}236\text{ km}}{42{,}766\text{ km}}$, we obtain

$$\beta = \cos^{-1}\left(\frac{35{,}236}{42{,}766}\right) \approx 34.520^\circ.$$

Next, using the Pythagorean Theorem, we obtain

$$b^2 + (35{,}236\text{ km})^2 = (42{,}766\text{ km})^2$$
$$b^2 = 587{,}355{,}060\text{ km}^2$$
$$b \approx 24{,}235\text{ km}$$

64. Consider this triangle:

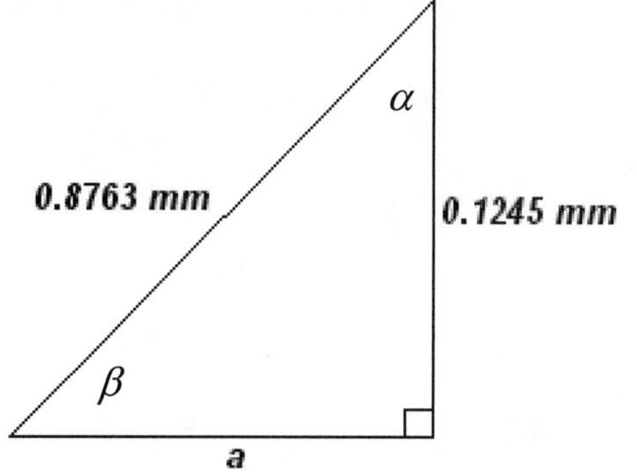

First, since $\cos\alpha = \dfrac{0.1245\text{ mm}}{0.8763\text{ mm}}$, we obtain

$$\alpha = \cos^{-1}\left(\frac{0.1245}{0.8763}\right) \approx 81.83^\circ.$$

Similarly, since $\sin\beta = \dfrac{0.1245\text{ mm}}{0.8763\text{ mm}}$, we obtain

$$\beta = \sin^{-1}\left(\frac{0.1245}{0.8763}\right) \approx 8.17^\circ.$$

Next, using the Pythagorean Theorem, we obtain

$$a^2 + (0.1245\text{ mm})^2 = (0.8763\text{ mm})^2$$
$$a^2 = 0.752401\text{ mm}^2$$
$$a \approx 0.8674\text{ mm}$$

65. Consider this triangle.

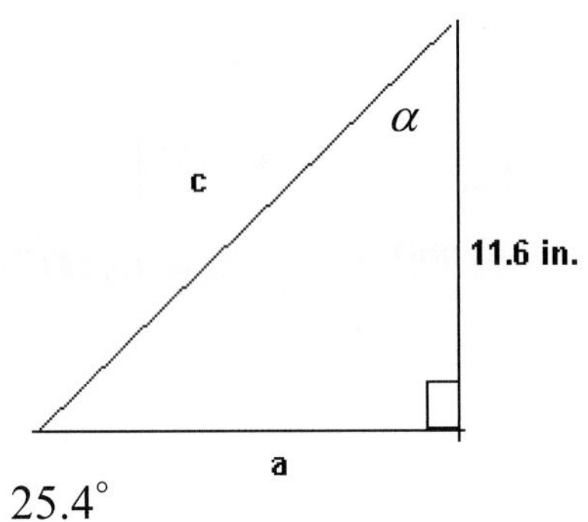

Since $\sin 25.4^\circ = \dfrac{11.6 \text{ in.}}{c}$, we have

$$c = \frac{11.6 \text{ in.}}{\sin 25.4^\circ} \approx 27.0 \text{ in.}$$

Next, since $\tan 25.4^\circ = \dfrac{11.6 \text{ in.}}{a}$, we have

$$a = \frac{11.6 \text{ in.}}{\tan 25.4^\circ} \approx 24.4 \text{ in.}$$

Finally, $\alpha = 180^\circ - (90^\circ + 25.4^\circ) \approx 64.6^\circ$.

66. Consider this triangle.

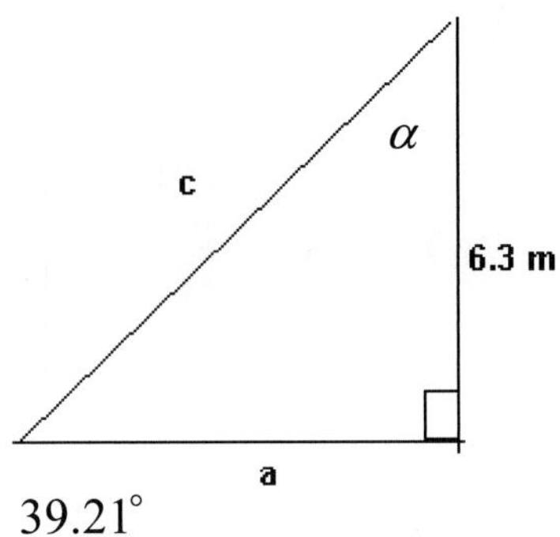

Since $\sin 39.21^\circ = \dfrac{6.3 \text{ m}}{c}$, we have

$$c = \frac{6.3 \text{ m}}{\sin 39.21^\circ} \approx 10.0 \text{ m}$$

Next, since $\tan 39.21^\circ = \dfrac{6.3 \text{ m}}{a}$, we have

$$a = \frac{6.3 \text{ m}}{\tan 39.21^\circ} \approx 7.7 \text{ m}$$

Finally, $\alpha = 180^\circ - (90^\circ + 39.21^\circ) \approx 50.79^\circ$.

67. Consider this triangle:

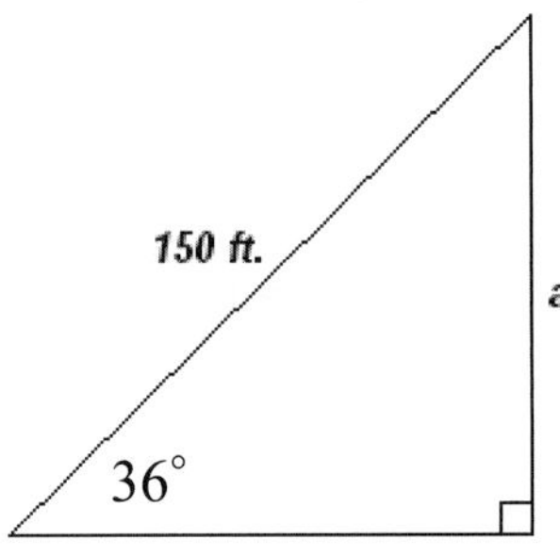

Observe that $\sin 36^\circ = \dfrac{a}{150 \text{ ft.}}$, so that

$$a = (\sin 36^\circ)(150 \text{ ft.}) \approx 88 \text{ ft.}.$$

So, the altitude between the two planes should be approximately $\boxed{88 \text{ ft.}}$

68. Consider this triangle:

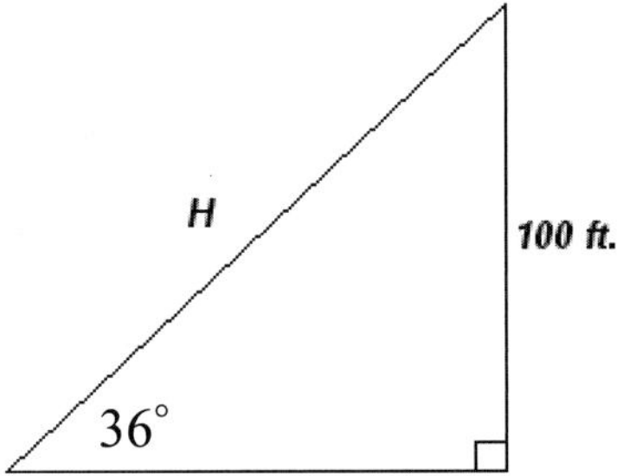

Observe that $\sin 36^\circ = \dfrac{100 \text{ ft.}}{H}$, so that

$$H = \frac{100 \text{ ft.}}{\sin 36^\circ} \approx 170 \text{ ft.}.$$

So, the hose should be approximately $\boxed{170 \text{ ft.}}$ long.

69. Consider this triangle:

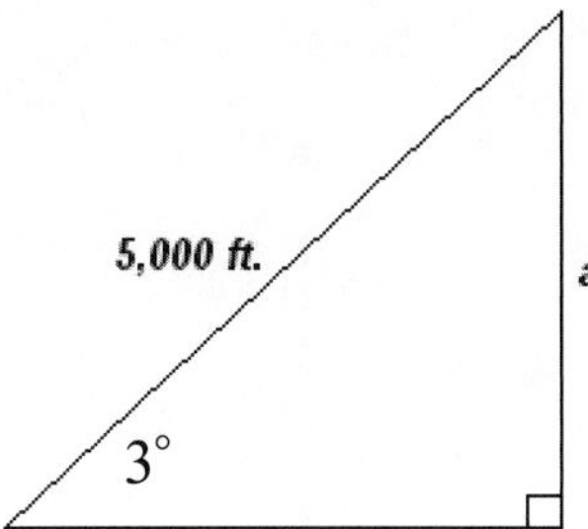

Observe that $\sin 3^\circ = \dfrac{a}{5{,}000 \text{ ft.}}$, so that

$$a = (\sin 3^\circ)(5{,}000 \text{ ft.}) \approx 262 \text{ ft.}$$

So, rounding to two significant digits, we see that the altitude should be approximately $\boxed{260 \text{ ft.}}$

70. Consider this triangle:

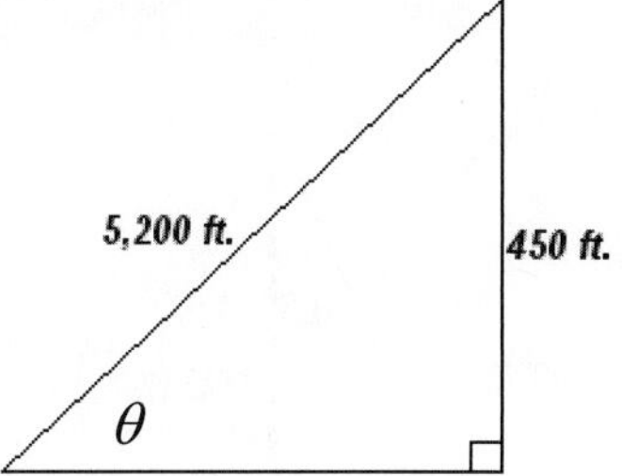

Let θ = glide slope angle. Observe that $\sin\theta = \dfrac{450 \text{ ft.}}{5{,}200 \text{ ft.}}$, so that

$$\theta = \sin^{-1}\left(\frac{450}{5{,}200}\right) \approx \boxed{5^\circ}.$$

Since an angle of 3° will allow the pilot to see both red and white lights while higher angles enable him to see only white lights, we conclude that in this case the pilot will see only white lights.

71. Consider this triangle:

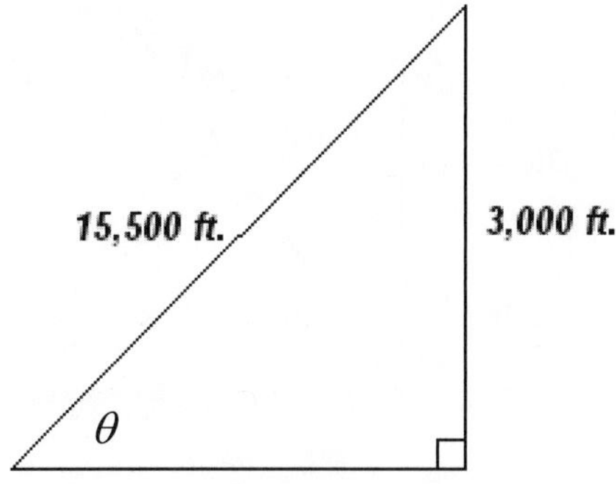

Let θ = glide slope angle. Observe that $\sin\theta = \dfrac{3{,}000 \text{ ft.}}{15{,}500 \text{ ft.}}$, so that

$$\theta = \sin^{-1}\left(\frac{3{,}000}{15{,}500}\right) \approx \boxed{11.0^\circ}.$$

This angle is way too low (since the typical range is $18^\circ - 20^\circ$).

72. Consider this triangle:

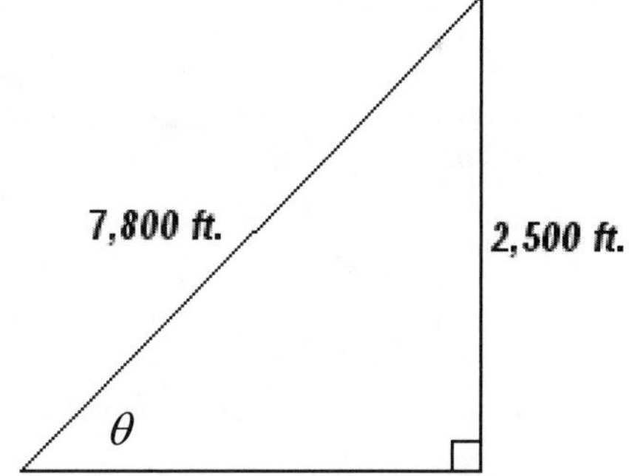

Let θ = glide slope angle. Observe that $\sin\theta = \dfrac{2{,}500 \text{ ft.}}{7{,}800 \text{ ft.}}$, so that

$$\theta = \sin^{-1}\left(\frac{2{,}500}{7{,}800}\right) \approx \boxed{18^\circ}.$$

So, she is within safety specifications to land.

73. Consider the following diagram:

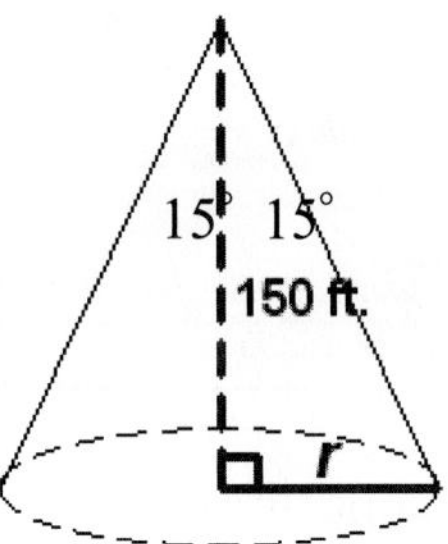

Since $\tan 15^\circ = \dfrac{r}{150 \text{ ft.}}$, we have

$$r = (150 \text{ ft.})(\tan 15^\circ) \approx 40 \text{ ft.}$$

So, the diameter of the circle is $\boxed{80 \text{ ft.}}$

74. Consider the following diagram:

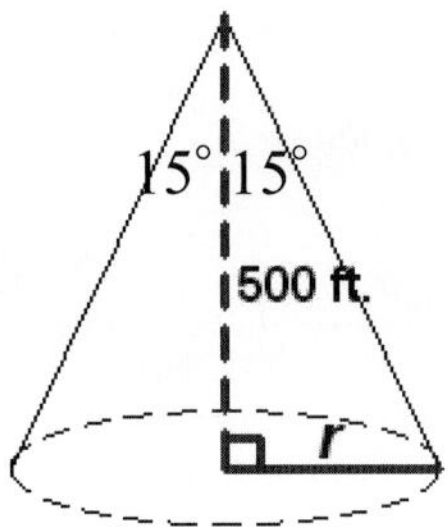

Since $\tan 15^\circ = \dfrac{r}{500 \text{ ft.}}$, we have

$$4 = (500 \text{ ft.})(\tan 15^\circ) \approx 134 \text{ ft.}$$

So, the diameter of the circle is 268 ft., which we round to $\boxed{270 \text{ ft.}}$

75. Consider this triangle:

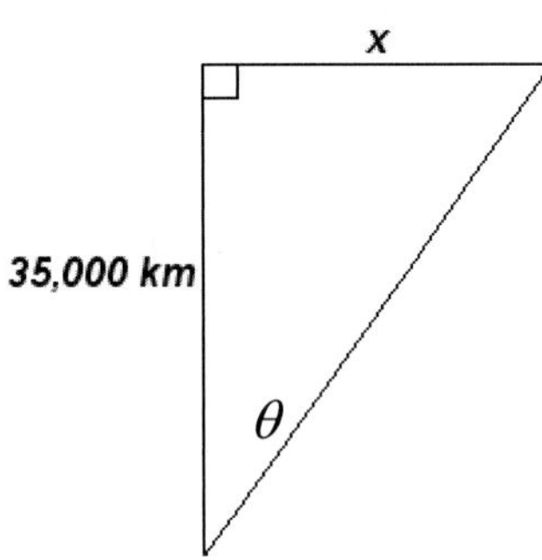

First, note that $\theta = 1'' \approx 0.0002778^\circ$.

Since $\tan\left(0.0002778^\circ\right) = \dfrac{x}{35,000 \text{ km}}$, we have that

$$x = (35,000 \text{ km})\tan\left(0.0002778^\circ\right)$$
$$\approx 0.1697 \text{ km} \approx \boxed{170 \text{ m}}$$

76. Consider this triangle:

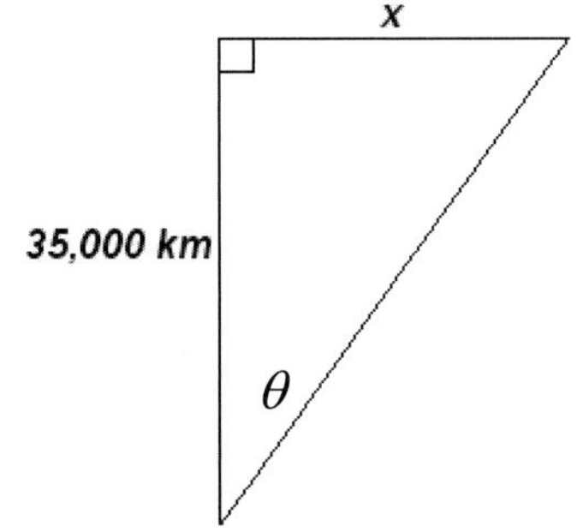

First, note that $\theta = \left(\dfrac{1}{2}\right)'' \approx 0.00013889^\circ$.

Since $\tan\left(0.00013889^\circ\right) = \dfrac{x}{35,000 \text{ km}}$, we have that

$$x = (35,000 \text{ km})\tan\left(0.00013889^\circ\right)$$
$$\approx 0.08484 \text{ km} \approx \boxed{85 \text{ m}}$$

77. Consider this triangle:

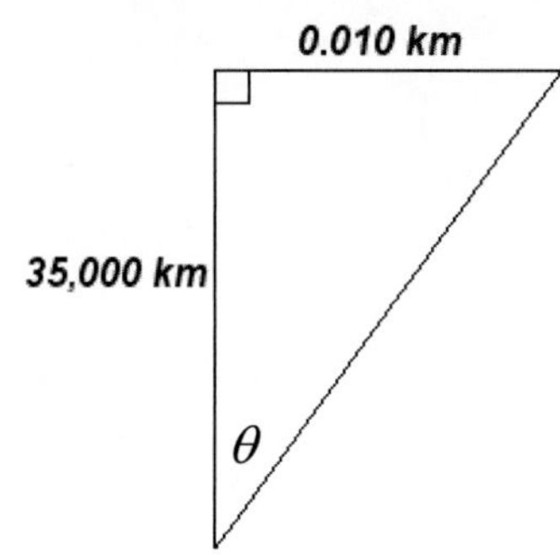

First, note that 0.010 km = 10 m.

Since $\tan\theta = \dfrac{0.010 \text{ km}}{35{,}000 \text{ km}}$, we have that

$$\theta = \tan^{-1}\left(\frac{0.010}{35{,}000}\right) \approx \boxed{0.000016^\circ}.$$

78. Consider this triangle:

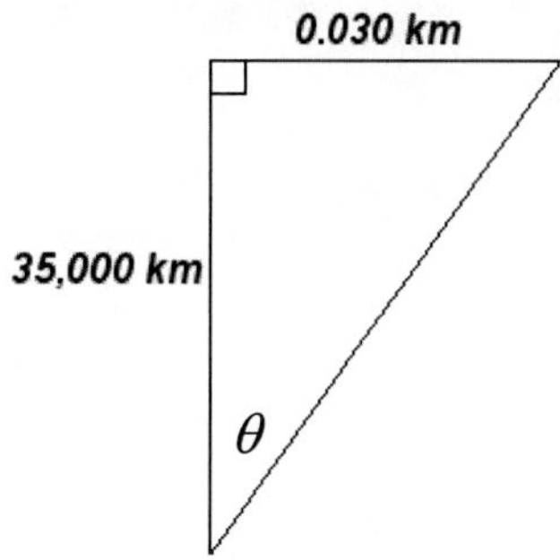

First, note that 0.030 km = 30 m.

Since $\tan\theta = \dfrac{0.030 \text{ km}}{35{,}000 \text{ km}}$, we have that

$$\theta = \tan^{-1}\left(\frac{0.030}{35{,}000}\right) \approx \boxed{0.000049^\circ}.$$

79. Consider this diagram:

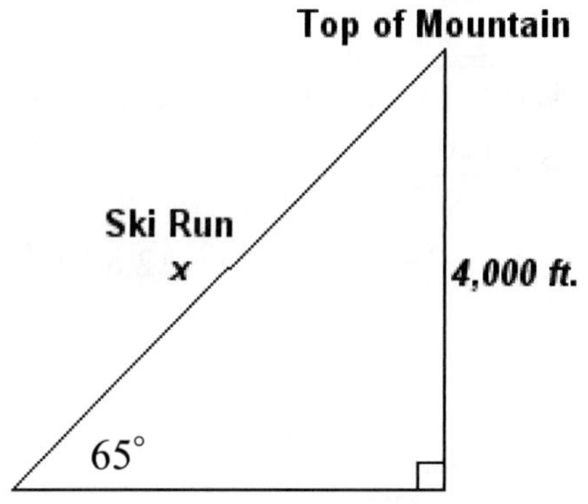

Observe that $\sin 65^\circ = \dfrac{4{,}000 \text{ ft.}}{x}$, so that

$$x = \frac{4{,}000 \text{ ft.}}{\sin 65^\circ} \approx \boxed{4{,}414 \text{ ft.}}.$$

80. Consider this diagram:

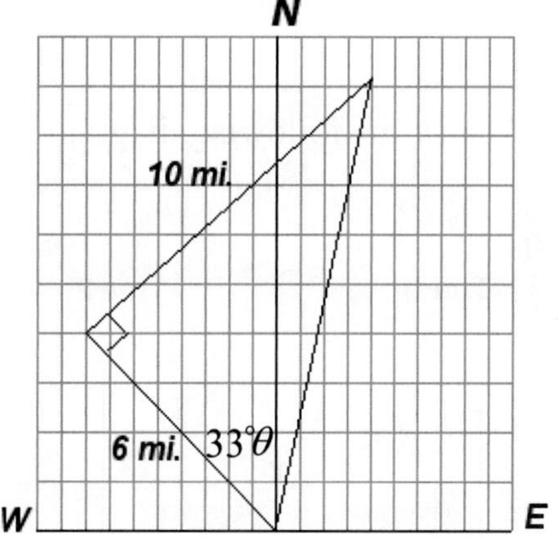

Observe that $\tan\left(33^\circ + \theta\right) = \dfrac{10 \text{ mi.}}{6 \text{ mi.}}$, so that

$$33^\circ + \theta = \tan^{-1}\left(\frac{10}{6}\right)$$

$$\theta = \tan^{-1}\left(\frac{10}{6}\right) - 33^\circ \approx 26^\circ$$

So, the bearing would be $\boxed{N26^\circ E}$.

81. Consider the following diagram:

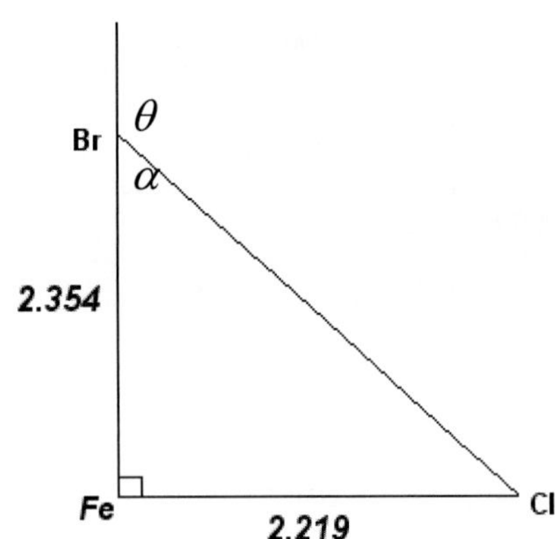

First, observe that $\tan\alpha = \dfrac{2.219}{2.354}$, so that $\alpha = \tan^{-1}\left(\dfrac{2.219}{2.354}\right) \approx 43.309^\circ$. Since $\alpha + \theta = 180^\circ$, we conclude that $\theta \approx \boxed{136.69^\circ}$.

82. Consider the following diagram:

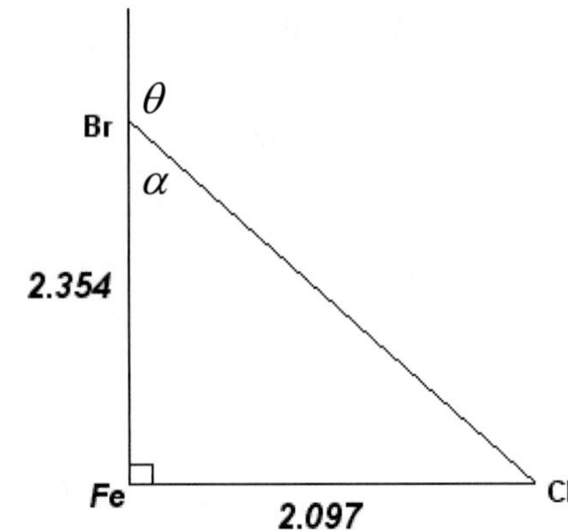

First, observe that $\tan\alpha = \dfrac{2.097}{2.354}$, so that $\alpha = \tan^{-1}\left(\dfrac{2.097}{2.354}\right) \approx 41.695^\circ$. Since $\alpha + \theta = 180^\circ$, we conclude that $\theta \approx \boxed{138.30^\circ}$.

83. Consider the following diagram:

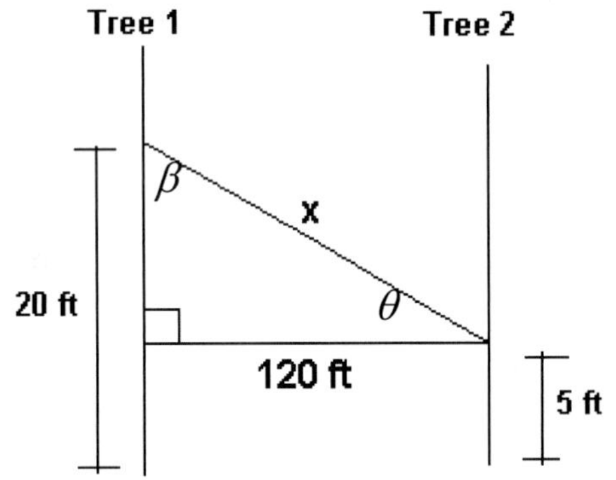

The Pythagorean theorem yields

$$15^2 + 120^2 = x^2$$

$$x \approx 120.93 \text{ ft.}$$

84. Using the diagram in #83, we see that $\tan\beta = \dfrac{120}{15}$, so that

$$\beta = \tan^{-1}\left(\tfrac{120}{15}\right) \approx 82.87^\circ.$$

85. Consider the following diagram:

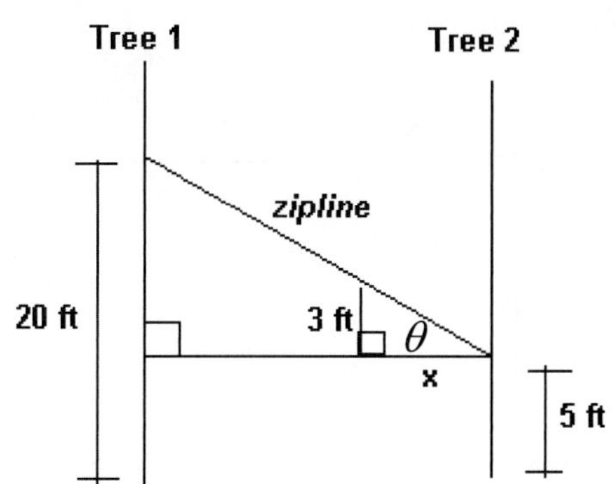

Find x such that $\tan\theta = \frac{3}{x}$. Note from #66 that $\theta = 7.13^{\circ}$. Substituting in yields $x = \frac{3}{\tan 7.13^{\circ}} \approx 24$ ft.

86. We have the following diagram:

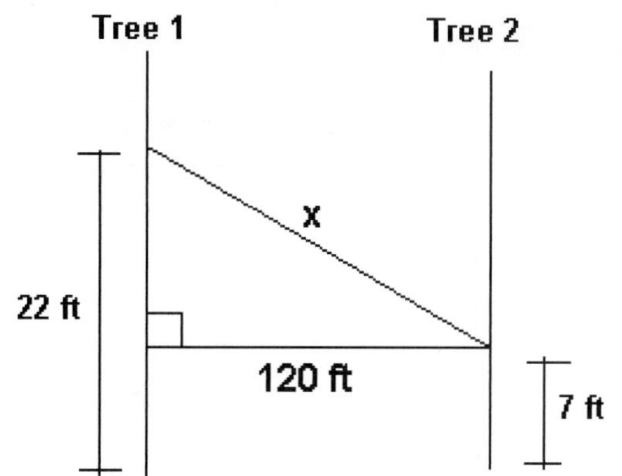

We want the same right triangle. So, the distance from the ground to the point of attachment of the zipline to Tree 2 is 7 ft. So, tack it 4 ft. above the platform.

87. Opposite side has length 3, not 4.

88. $\tan x = \dfrac{\text{opp}}{\text{adj}}$, not $\dfrac{\text{adj}}{\text{opp}}$.

89. $\sec x = \dfrac{1}{\cos x}$, not $\dfrac{1}{\sin x}$.

90. $\csc y = \dfrac{1}{\sin y}$, not $\dfrac{1}{\cos y}$.

91. True

92. True

93. False. Knowing the measures of all angles in a right triangle would not enable you to find any of the side lengths. In fact, you could produce infinitely many similar triangles with these same angles.

94. False. The angle opposite the hypotenuse in a right triangle is 90°. If we only knew this and no other side length, we could not determine the measures of either of the two remaining angles.

95. Consider the following triangle: 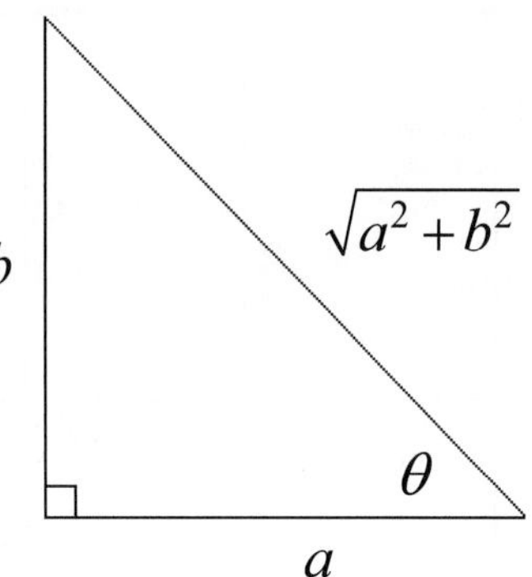 $\sin(0^\circ) \approx \frac{b}{\sqrt{a^2+b^2}} \approx 0$ since $a \gg b$ and so, the denominator is <u>much</u> larger than the numerator.	**96.** Consider the following triangle: 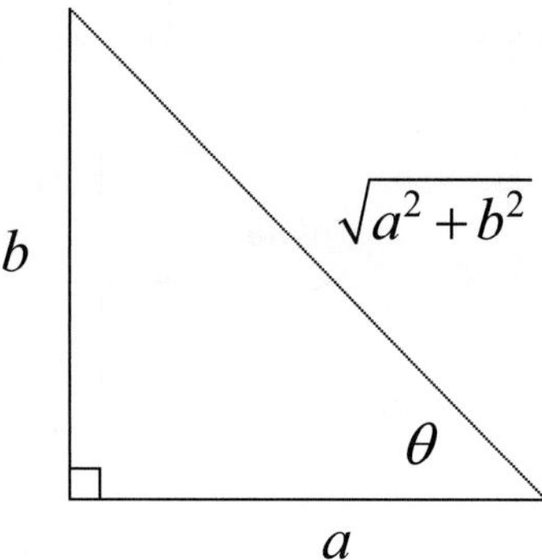 $\cos(0^\circ) \approx \frac{a}{\sqrt{a^2+b^2}} \approx 1$ since $a \gg b$ and so, $a \approx \sqrt{a^2+b^2}$.
97. Using the co-function identity with Exercise 95, we see that $\cos(90^\circ) = \sin(0^\circ) = 0$.	**98.** Using the co-function identity with Exercise 96, we see that $\sin(90^\circ) = \cos(0^\circ) = 1$.
99. Since $3\alpha = 90^\circ$, $\alpha = 30^\circ$. So, $\frac{3x}{4} = \cos 60^\circ = \frac{1}{2} \Rightarrow x = \frac{2}{3}$.	**100.** From the Pythagorean theorem, $y^2 = 1^2 + 4^2 = 17$, so that $y = \sqrt{17}$.
101. a. <u>Step 1</u>: $\cos(70^\circ) \approx 0.342$ <u>Step 2</u>: $\frac{1}{\cos(70^\circ)} \approx 2.92398$ **b.** $\left(\cos(70^\circ)\right)^{-1} \approx 2.92380$ The calculation in part **b** is more accurate since one does not round twice as in part **a**.	**102. a.** <u>Step 1</u>: $\sin(40^\circ) \approx 0.643$ <u>Step 2</u>: $\frac{1}{\sin(40^\circ)} \approx 1.55521$ **b.** $\left(\sin(40^\circ)\right)^{-1} \approx 1.55572$ The calculation in part **b** is more accurate since one does not round twice as in part **a**.
103. a. <u>Step 1</u>: $\tan(54.9^\circ) \approx 1.423$ <u>Step 2</u>: $\frac{1}{\tan(54.9^\circ)} \approx 0.70281$ **b.** 54.9, tan, 1/x: 0.70281	**104. a.** <u>Step 1</u>: $\cos(18.6^\circ) \approx 0.948$ <u>Step 2</u>: $\frac{1}{\cos(18.6^\circ)} \approx 1.05511$ 18.6, cos, 1/x: 1.05511

105.	**106.**
$F\left(\frac{\pi}{3}\right)-F\left(\frac{\pi}{6}\right)=\sec\left(\frac{\pi}{3}\right)-\sec\left(\frac{\pi}{6}\right)$ $=2-\frac{2}{\sqrt{3}}=\frac{6-2\sqrt{3}}{3}$	$F\left(\frac{\pi}{4}\right)-F(0)=\sin^3\left(\frac{\pi}{4}\right)-\sin^3(0)$ $=\left(\frac{\sqrt{2}}{2}\right)^3-0=\frac{\sqrt{2}}{4}$
107.	**108.**
$F\left(\frac{\pi}{3}\right)-F(0)$ $=\left[\tan\left(\frac{\pi}{3}\right)+2\cos\left(\frac{\pi}{3}\right)\right]-\left[\tan(0)+2\cos(0)\right]$ $=\left[\frac{\sqrt{3}/2}{1/2}+2\left(\frac{1}{2}\right)\right]-[0+2]$ $=\sqrt{3}+1-2=\sqrt{3}-1$	$F\left(\frac{\pi}{3}\right)-F\left(\frac{\pi}{4}\right)$ $=\frac{\cot\left(\frac{\pi}{3}\right)-4\sin\left(\frac{\pi}{3}\right)}{\cos\left(\frac{\pi}{3}\right)}-\frac{\cot\left(\frac{\pi}{3}\right)-4\sin\left(\frac{\pi}{3}\right)}{\cos\left(\frac{\pi}{3}\right)}$ $=\left[\frac{\frac{1/2}{\sqrt{3}/2}-4\left(\sqrt{3}/2\right)}{\frac{1}{2}}\right]-\left[\frac{1-4\left(\sqrt{2}/2\right)}{\frac{\sqrt{2}}{2}}\right]$ $=2\left(\frac{1}{\sqrt{3}}-2\sqrt{3}\right)-\frac{\sqrt{2}}{2}\left(1-2\sqrt{2}\right)$ $=\frac{12-3\sqrt{2}-10\sqrt{3}}{3}$ (after some manipulation)

Section 4.3 Solutions ---

1. Consider the following diagram:

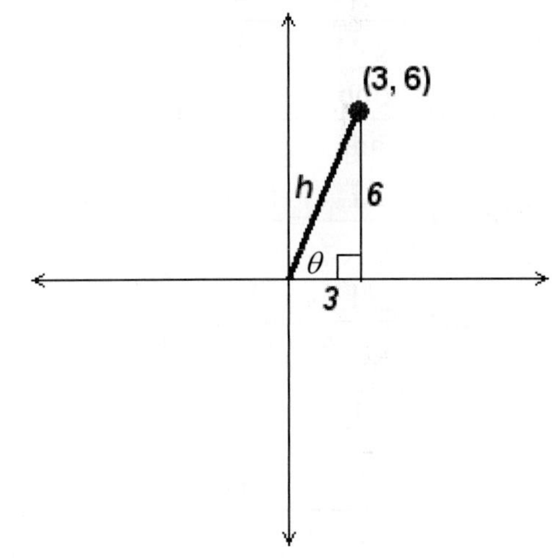

Using the Pythagorean Theorem, we obtain $h=\sqrt{3^2+6^2}=\sqrt{45}=3\sqrt{5}$.

$$\sin\theta=\frac{\text{opp}}{\text{hyp}}=\frac{6}{3\sqrt{5}}=\frac{2}{\sqrt{5}}=\frac{2\sqrt{5}}{5}$$

$$\cos\theta=\frac{\text{adj}}{\text{hyp}}=\frac{3}{3\sqrt{5}}=\frac{1}{\sqrt{5}}=\frac{\sqrt{5}}{5}$$

$$\tan\theta=\frac{\text{opp}}{\text{adj}}=\frac{6}{3}=2$$

$$\cot\theta=\frac{1}{\tan\theta}=\frac{1}{2}$$

$$\sec\theta=\frac{1}{\cos\theta}=\frac{1}{\frac{1}{\sqrt{5}}}=\sqrt{5}$$

$$\csc\theta=\frac{1}{\sin\theta}=\frac{1}{\frac{2}{\sqrt{5}}}=\frac{\sqrt{5}}{2}$$

2. Consider the following diagram:

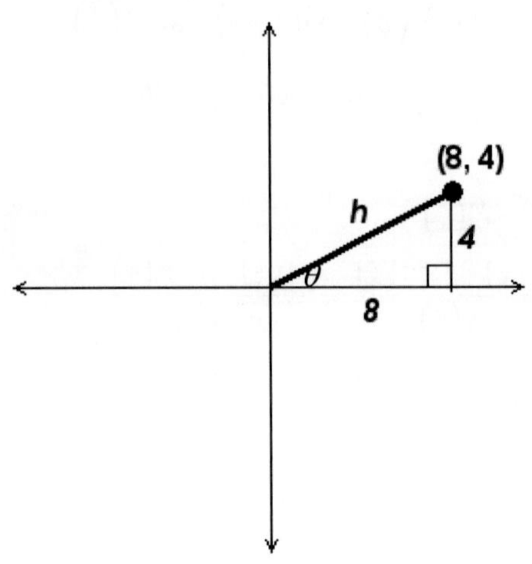

Using the Pythagorean Theorem, we obtain $h=\sqrt{8^2+4^2}=\sqrt{80}=4\sqrt{5}$.

$$\sin\theta=\frac{\text{opp}}{\text{hyp}}=\frac{4}{4\sqrt{5}}=\frac{1}{\sqrt{5}}=\frac{\sqrt{5}}{5}$$

$$\cos\theta=\frac{\text{adj}}{\text{hyp}}=\frac{8}{4\sqrt{5}}=\frac{2}{\sqrt{5}}=\frac{2\sqrt{5}}{5}$$

$$\tan\theta=\frac{\text{opp}}{\text{adj}}=\frac{4}{8}=\frac{1}{2}$$

$$\cot\theta=\frac{1}{\tan\theta}=2$$

$$\sec\theta=\frac{1}{\cos\theta}=\frac{1}{\frac{2}{\sqrt{5}}}=\frac{\sqrt{5}}{2}$$

$$\csc\theta=\frac{1}{\sin\theta}=\frac{1}{\frac{1}{\sqrt{5}}}=\sqrt{5}$$

3. Consider the following diagram:

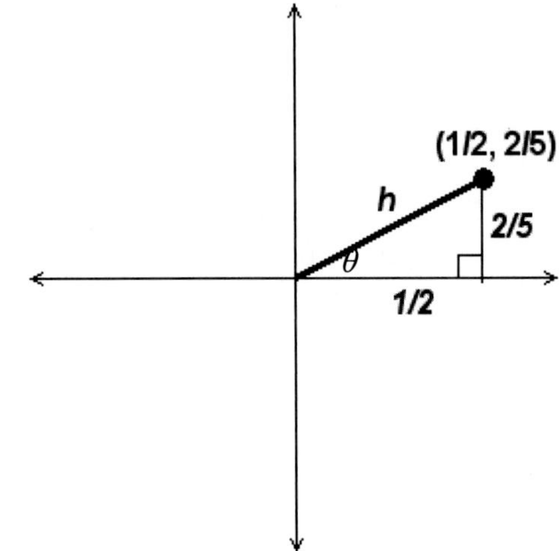

Using the Pythagorean Theorem, we obtain $h=\sqrt{\left(\frac{1}{2}\right)^2+\left(\frac{2}{5}\right)^2}=\sqrt{\frac{1}{4}+\frac{4}{25}}=\frac{\sqrt{41}}{10}$.

$$\sin\theta=\frac{\text{opp}}{\text{hyp}}=\frac{\frac{2}{5}}{\frac{\sqrt{41}}{10}}=\frac{2}{5}\cdot\frac{10}{\sqrt{41}}=\frac{4}{\sqrt{41}}=\frac{4\sqrt{41}}{41}$$

$$\cos\theta=\frac{\text{adj}}{\text{hyp}}=\frac{\frac{1}{2}}{\frac{\sqrt{41}}{10}}=\frac{1}{2}\cdot\frac{10}{\sqrt{41}}=\frac{5}{\sqrt{41}}=\frac{5\sqrt{41}}{41}$$

$$\tan\theta=\frac{\text{opp}}{\text{adj}}=\frac{\frac{2}{5}}{\frac{1}{2}}=\frac{4}{5}$$

$$\cot\theta=\frac{1}{\tan\theta}=\frac{5}{4}$$

$$\sec\theta=\frac{1}{\cos\theta}=\frac{1}{\frac{5}{\sqrt{41}}}=\frac{\sqrt{41}}{5}$$

$$\csc\theta=\frac{1}{\sin\theta}=\frac{1}{\frac{4}{\sqrt{41}}}=\frac{\sqrt{41}}{4}$$

4. Consider the following diagram:

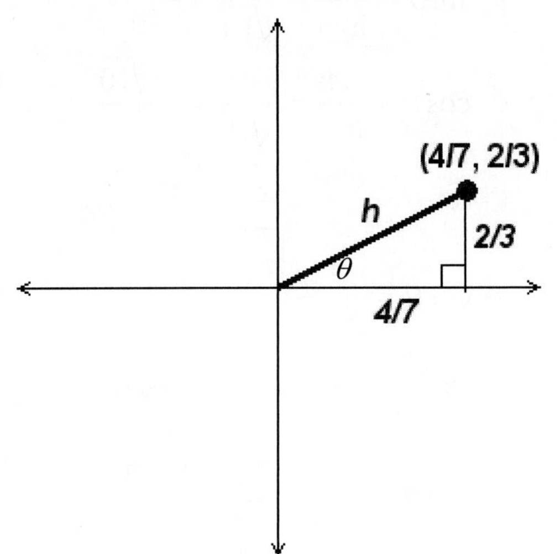

Using the Pythagorean Theorem, we obtain
$h=\sqrt{\left(\frac{4}{7}\right)^2+\left(\frac{2}{3}\right)^2}=\sqrt{\frac{16}{49}+\frac{4}{9}}=\frac{\sqrt{340}}{21}=\frac{2\sqrt{85}}{21}$.

$$\sin\theta=\frac{\text{opp}}{\text{hyp}}=\frac{\frac{2}{3}}{\frac{2\sqrt{85}}{21}}=\frac{2}{3}\cdot\frac{21}{2\sqrt{85}}=\frac{7}{\sqrt{85}}=\frac{7\sqrt{85}}{85}$$

$$\cos\theta=\frac{\text{adj}}{\text{hyp}}=\frac{\frac{4}{7}}{\frac{2\sqrt{85}}{21}}=\frac{4}{7}\cdot\frac{21}{2\sqrt{85}}=\frac{6}{\sqrt{85}}=\frac{6\sqrt{85}}{85}$$

$$\tan\theta=\frac{\text{opp}}{\text{adj}}=\frac{\frac{2}{3}}{\frac{4}{7}}=\frac{7}{6}$$

$$\cot\theta=\frac{1}{\tan\theta}=\frac{6}{7}$$

$$\sec\theta=\frac{1}{\cos\theta}=\frac{1}{\frac{6}{\sqrt{85}}}=\frac{\sqrt{85}}{6}$$

$$\csc\theta=\frac{1}{\sin\theta}=\frac{1}{\frac{7}{\sqrt{85}}}=\frac{\sqrt{85}}{7}$$

5. Consider the following diagram:

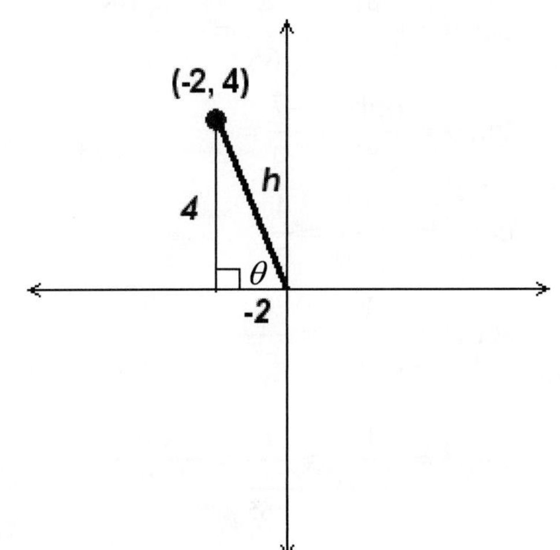

Using the Pythagorean Theorem, we obtain
$h=\sqrt{(-2)^2+4^2}=\sqrt{20}=2\sqrt{5}$.

$$\sin\theta=\frac{\text{opp}}{\text{hyp}}=\frac{4}{2\sqrt{5}}=\frac{2}{\sqrt{5}}=\frac{2\sqrt{5}}{5}$$

$$\cos\theta=\frac{\text{adj}}{\text{hyp}}=\frac{-2}{2\sqrt{5}}=\frac{-1}{\sqrt{5}}=-\frac{\sqrt{5}}{5}$$

$$\tan\theta=\frac{\text{opp}}{\text{adj}}=\frac{4}{-2}=-2$$

$$\cot\theta=\frac{1}{\tan\theta}=-\frac{1}{2}$$

$$\sec\theta=\frac{1}{\cos\theta}=\frac{1}{\frac{-1}{\sqrt{5}}}=-\sqrt{5}$$

$$\csc\theta=\frac{1}{\sin\theta}=\frac{1}{\frac{2}{\sqrt{5}}}=\frac{\sqrt{5}}{2}$$

6. Consider the following diagram:

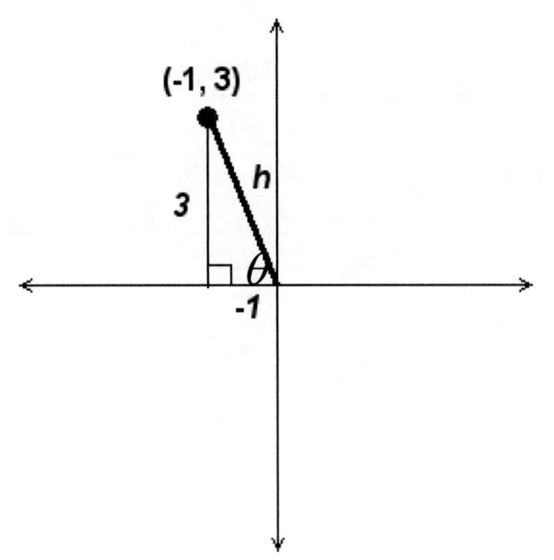

Using the Pythagorean Theorem, we obtain $h = \sqrt{(-1)^2 + 3^2} = \sqrt{10}$.

$$\sin\theta = \frac{\text{opp}}{\text{hyp}} = \frac{3}{\sqrt{10}} = \frac{3\sqrt{10}}{10}$$

$$\cos\theta = \frac{\text{adj}}{\text{hyp}} = \frac{-1}{\sqrt{10}} = -\frac{\sqrt{10}}{10}$$

$$\tan\theta = \frac{\text{opp}}{\text{adj}} = \frac{3}{-1} = -3$$

$$\cot\theta = \frac{1}{\tan\theta} = -\frac{1}{3}$$

$$\sec\theta = \frac{1}{\cos\theta} = \frac{1}{\frac{-1}{\sqrt{10}}} = -\sqrt{10}$$

$$\csc\theta = \frac{1}{\sin\theta} = \frac{1}{\frac{3}{\sqrt{10}}} = \frac{\sqrt{10}}{3}$$

7. Consider the following diagram:

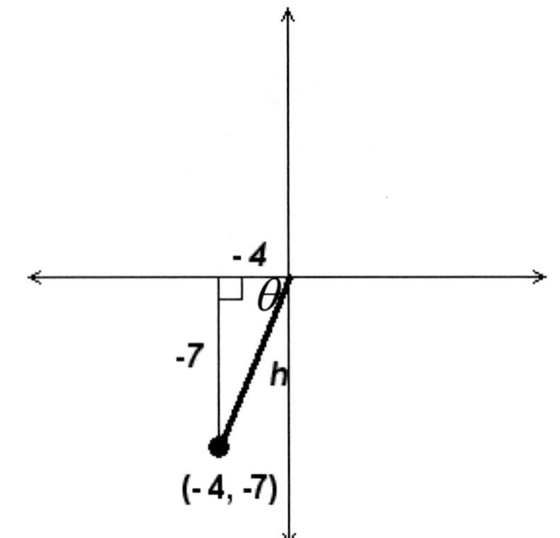

Using the Pythagorean Theorem, we obtain $h = \sqrt{(-4)^2 + (-7)^2} = \sqrt{65}$.

$$\sin\theta = \frac{\text{opp}}{\text{hyp}} = \frac{-7}{\sqrt{65}} = -\frac{7\sqrt{65}}{65}$$

$$\cos\theta = \frac{\text{adj}}{\text{hyp}} = \frac{-4}{\sqrt{65}} = -\frac{4\sqrt{65}}{65}$$

$$\tan\theta = \frac{\text{opp}}{\text{adj}} = \frac{-7}{-4} = \frac{7}{4}$$

$$\cot\theta = \frac{1}{\tan\theta} = \frac{4}{7}$$

$$\sec\theta = \frac{1}{\cos\theta} = \frac{1}{\frac{-4}{\sqrt{65}}} = -\frac{\sqrt{65}}{4}$$

$$\csc\theta = \frac{1}{\sin\theta} = \frac{1}{\frac{-7}{\sqrt{65}}} = -\frac{\sqrt{65}}{7}$$

8. Consider the following diagram:

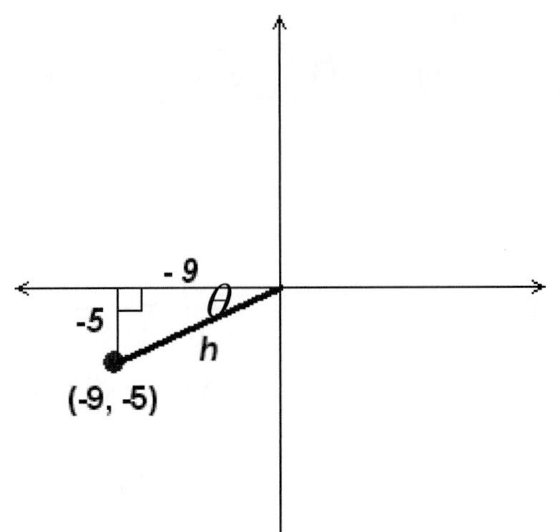

Using the Pythagorean Theorem, we obtain $h = \sqrt{(-5)^2 + (-9)^2} = \sqrt{106}$.

$$\sin\theta = \frac{\text{opp}}{\text{hyp}} = \frac{-5}{\sqrt{106}} = -\frac{5\sqrt{106}}{106}$$

$$\cos\theta = \frac{\text{adj}}{\text{hyp}} = \frac{-9}{\sqrt{106}} = -\frac{9\sqrt{106}}{106}$$

$$\tan\theta = \frac{\text{opp}}{\text{adj}} = \frac{-5}{-9} = \frac{5}{9}$$

$$\cot\theta = \frac{1}{\tan\theta} = \frac{9}{5}$$

$$\sec\theta = \frac{1}{\cos\theta} = \frac{1}{\frac{-9}{\sqrt{106}}} = -\frac{\sqrt{106}}{9}$$

$$\csc\theta = \frac{1}{\sin\theta} = \frac{1}{\frac{-5}{\sqrt{106}}} = -\frac{\sqrt{106}}{5}$$

9. Consider the following diagram:

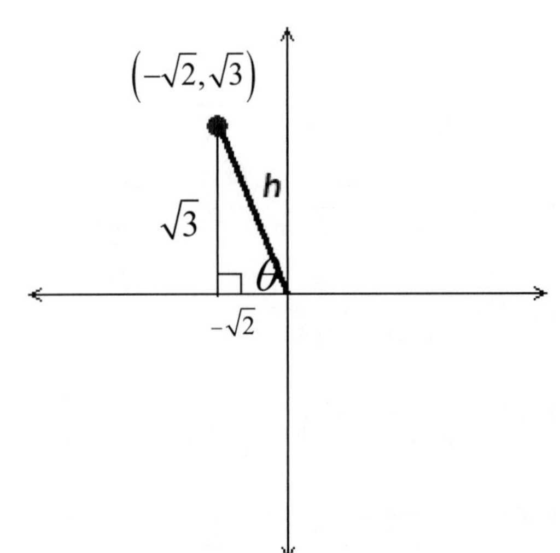

Using the Pythagorean Theorem, we obtain $h = \sqrt{\left(-\sqrt{2}\right)^2 + \left(\sqrt{3}\right)^2} = \sqrt{5}$.

$$\sin\theta = \frac{\text{opp}}{\text{hyp}} = \frac{\sqrt{3}}{\sqrt{5}} = \frac{\sqrt{3}\sqrt{5}}{5} = \frac{\sqrt{15}}{5}$$

$$\cos\theta = \frac{\text{adj}}{\text{hyp}} = \frac{-\sqrt{2}}{\sqrt{5}} = \frac{-\sqrt{2}\sqrt{5}}{5} = -\frac{\sqrt{10}}{5}$$

$$\tan\theta = \frac{\text{opp}}{\text{adj}} = \frac{\sqrt{3}}{-\sqrt{2}} = \frac{-\sqrt{3}\sqrt{2}}{2} = -\frac{\sqrt{6}}{2}$$

$$\cot\theta = \frac{1}{\tan\theta} = -\frac{2}{\sqrt{6}} = -\frac{2\sqrt{6}}{6} = -\frac{\sqrt{6}}{3}$$

$$\sec\theta = \frac{1}{\cos\theta} = \frac{1}{\frac{-\sqrt{10}}{5}} = -\frac{5}{\sqrt{10}} = -\frac{\sqrt{10}}{2}$$

$$\csc\theta = \frac{1}{\sin\theta} = \frac{1}{\frac{\sqrt{15}}{5}} = \frac{5}{\sqrt{15}} = \frac{5\sqrt{15}}{15} = \frac{\sqrt{15}}{3}$$

10. Consider the following diagram:

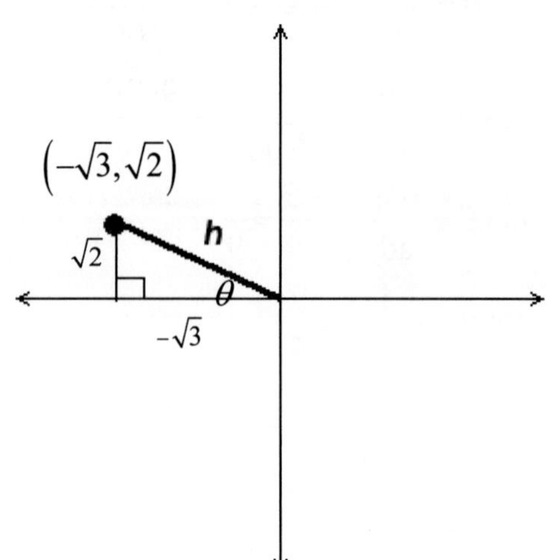

Using the Pythagorean Theorem, we obtain $h=\sqrt{\left(-\sqrt{3}\right)^2+\left(\sqrt{2}\right)^2}=\sqrt{5}$.

$$\sin\theta=\frac{\text{opp}}{\text{hyp}}=\frac{\sqrt{2}}{\sqrt{5}}=\frac{\sqrt{2}\sqrt{5}}{5}=\frac{\sqrt{10}}{5}$$

$$\cos\theta=\frac{\text{adj}}{\text{hyp}}=\frac{-\sqrt{3}}{\sqrt{5}}=\frac{-\sqrt{3}\sqrt{5}}{5}=-\frac{\sqrt{15}}{5}$$

$$\tan\theta=\frac{\text{opp}}{\text{adj}}=\frac{\sqrt{2}}{-\sqrt{3}}=\frac{-\sqrt{2}\sqrt{3}}{3}=-\frac{\sqrt{6}}{3}$$

$$\cot\theta=\frac{1}{\tan\theta}=-\frac{3}{\sqrt{6}}=-\frac{3\sqrt{6}}{6}=-\frac{\sqrt{6}}{2}$$

$$\sec\theta=\frac{1}{\cos\theta}=\frac{1}{\frac{-\sqrt{15}}{5}}=-\frac{5}{\sqrt{15}}=-\frac{\sqrt{15}}{3}$$

$$\csc\theta=\frac{1}{\sin\theta}=\frac{1}{\frac{\sqrt{10}}{5}}=\frac{5}{\sqrt{10}}=\frac{5\sqrt{10}}{10}=\frac{\sqrt{10}}{2}$$

11. Consider the following diagram:

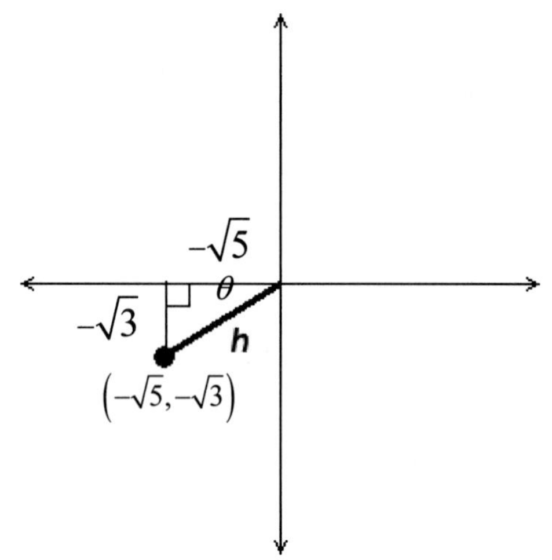

Using the Pythagorean Theorem, we obtain $h=\sqrt{\left(-\sqrt{5}\right)^2+\left(-\sqrt{3}\right)^2}=\sqrt{8}=2\sqrt{2}$.

$$\sin\theta=\frac{\text{opp}}{\text{hyp}}=\frac{-\sqrt{3}}{2\sqrt{2}}=\frac{-\sqrt{3}\sqrt{2}}{4}=-\frac{\sqrt{6}}{4}$$

$$\cos\theta=\frac{\text{adj}}{\text{hyp}}=\frac{-\sqrt{5}}{2\sqrt{2}}=\frac{-\sqrt{5}\sqrt{2}}{4}=-\frac{\sqrt{10}}{4}$$

$$\tan\theta=\frac{\text{opp}}{\text{adj}}=\frac{-\sqrt{3}}{-\sqrt{5}}=\frac{\sqrt{3}\sqrt{5}}{5}=\frac{\sqrt{15}}{5}$$

$$\cot\theta=\frac{1}{\tan\theta}=\frac{5}{\sqrt{15}}=\frac{5\sqrt{15}}{15}=\frac{\sqrt{15}}{3}$$

$$\sec\theta=\frac{1}{\cos\theta}=\frac{1}{\frac{-\sqrt{10}}{4}}=-\frac{4}{\sqrt{10}}=-\frac{2\sqrt{10}}{5}$$

$$\csc\theta=\frac{1}{\sin\theta}=\frac{1}{\frac{-\sqrt{6}}{4}}=-\frac{4}{\sqrt{6}}$$

$$=-\frac{4\sqrt{6}}{6}=-\frac{2\sqrt{6}}{3}$$

12. Consider the following diagram:

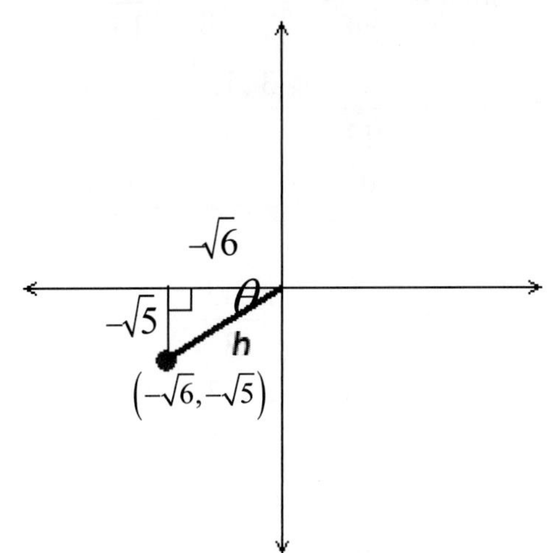

Using the Pythagorean Theorem, we obtain $h=\sqrt{\left(-\sqrt{5}\right)^2+\left(-\sqrt{6}\right)^2}=\sqrt{11}$.

$$\sin\theta=\frac{\text{opp}}{\text{hyp}}=\frac{-\sqrt{5}}{\sqrt{11}}=\frac{-\sqrt{5}\sqrt{11}}{11}=-\frac{\sqrt{55}}{11}$$

$$\cos\theta=\frac{\text{adj}}{\text{hyp}}=\frac{-\sqrt{6}}{\sqrt{11}}=\frac{-\sqrt{6}\sqrt{11}}{11}=-\frac{\sqrt{66}}{11}$$

$$\tan\theta=\frac{\text{opp}}{\text{adj}}=\frac{-\sqrt{5}}{-\sqrt{6}}=\frac{\sqrt{5}\sqrt{6}}{6}=\frac{\sqrt{30}}{6}$$

$$\cot\theta=\frac{1}{\tan\theta}=\frac{6}{\sqrt{30}}=\frac{6\sqrt{30}}{30}=\frac{\sqrt{30}}{5}$$

$$\sec\theta=\frac{1}{\cos\theta}=\frac{1}{\frac{-\sqrt{66}}{11}}=-\frac{11}{\sqrt{66}}=-\frac{\sqrt{66}}{6}$$

$$\csc\theta=\frac{1}{\sin\theta}=\frac{1}{\frac{-\sqrt{55}}{11}}=-\frac{11}{\sqrt{55}}=-\frac{\sqrt{55}}{5}$$

13. Consider the following diagram:

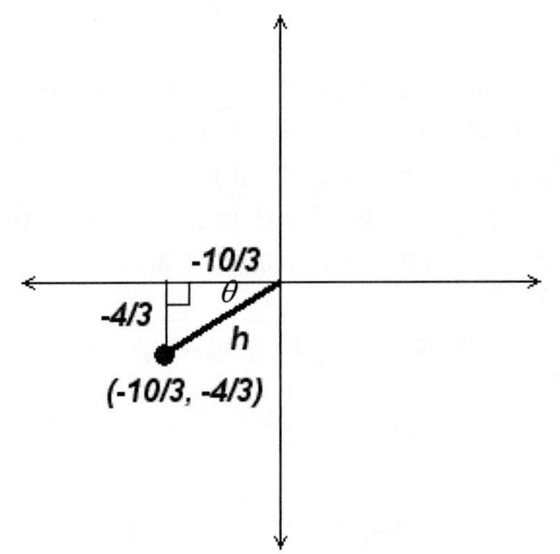

Using the Pythagorean Theorem, we obtain $h=\sqrt{\left(\frac{-10}{3}\right)^2+\left(\frac{-4}{3}\right)^2}=\sqrt{\frac{100}{9}+\frac{16}{9}}=\frac{\sqrt{116}}{3}=\frac{2\sqrt{29}}{3}$.

$$\sin\theta=\frac{\text{opp}}{\text{hyp}}=\frac{-\frac{4}{3}}{\frac{2\sqrt{29}}{3}}=\frac{-4}{3}\cdot\frac{3}{2\sqrt{29}}=\frac{-2}{\sqrt{29}}=-\frac{2\sqrt{29}}{29}$$

$$\cos\theta=\frac{\text{adj}}{\text{hyp}}=\frac{-\frac{10}{3}}{\frac{2\sqrt{29}}{3}}=\frac{-10}{3}\cdot\frac{3}{2\sqrt{29}}=\frac{-5}{\sqrt{29}}=-\frac{5\sqrt{29}}{29}$$

$$\tan\theta=\frac{\text{opp}}{\text{adj}}=\frac{-\frac{4}{3}}{-\frac{10}{3}}=\frac{2}{5}$$

$$\cot\theta=\frac{1}{\tan\theta}=\frac{5}{2}$$

$$\sec\theta=\frac{1}{\cos\theta}=\frac{1}{\frac{-5}{\sqrt{29}}}=-\frac{\sqrt{29}}{5}$$

$$\csc\theta=\frac{1}{\sin\theta}=\frac{1}{\frac{-2}{\sqrt{29}}}=-\frac{\sqrt{29}}{2}$$

14. Consider the following diagram:

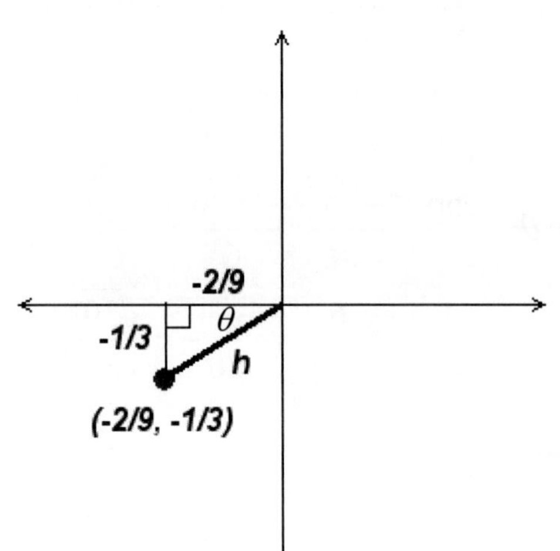

Using the Pythagorean Theorem, we obtain $h=\sqrt{\left(\frac{-2}{9}\right)^2+\left(\frac{-1}{3}\right)^2}=\sqrt{\frac{4}{81}+\frac{1}{9}}=\frac{\sqrt{13}}{9}$.

$$\sin\theta=\frac{\text{opp}}{\text{hyp}}=\frac{-\frac{1}{3}}{\frac{\sqrt{13}}{9}}=\frac{-1}{3}\cdot\frac{9}{\sqrt{13}}=\frac{-3}{\sqrt{13}}=-\frac{3\sqrt{13}}{13}$$

$$\cos\theta=\frac{\text{adj}}{\text{hyp}}=\frac{-\frac{2}{9}}{\frac{\sqrt{13}}{9}}=\frac{-2}{9}\cdot\frac{9}{\sqrt{13}}=\frac{-2}{\sqrt{13}}=-\frac{2\sqrt{13}}{13}$$

$$\tan\theta=\frac{\text{opp}}{\text{adj}}=\frac{-\frac{1}{3}}{-\frac{2}{9}}=\frac{3}{2}$$

$$\cot\theta=\frac{1}{\tan\theta}=\frac{2}{3}$$

$$\sec\theta=\frac{1}{\cos\theta}=\frac{1}{\frac{-2}{\sqrt{13}}}=-\frac{\sqrt{13}}{2}$$

$$\csc\theta=\frac{1}{\sin\theta}=\frac{1}{\frac{-3}{\sqrt{13}}}=-\frac{\sqrt{13}}{3}$$

15. QIV	**16.** QII
17. QII (Need $\cos\theta<0$ and $\sin\theta>0$.)	**18.** QIII (Need $\cos\theta<0$ and $\sin\theta<0$.)
19. QI (Need $\cos\theta>0$ and $\sin\theta>0$.)	**20.** QIII (Need $\cos\theta<0$ and $\sin\theta<0$.)
21. QI (since $\sin\theta>0$)	**22.** QIV (since $\cos\theta>0$)
23. QIII (since $\sec\theta<0$ implies $\cos\theta<0$ and hence, from the given information, $\sin\theta<0$)	**24.** QII (since $\csc\theta>0$ implies $\sin\theta>0$, and hence $\cos\theta<0$ from the given information)
25. Consider the following diagram: 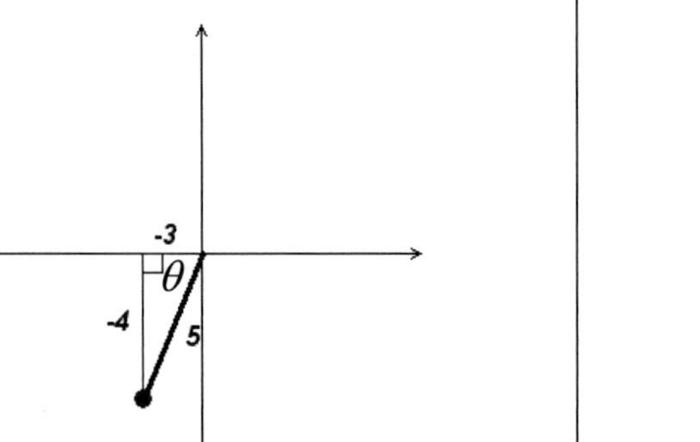 Observe that $\sin\theta=\boxed{-\frac{4}{5}}$.	**26.** Consider the following diagram: 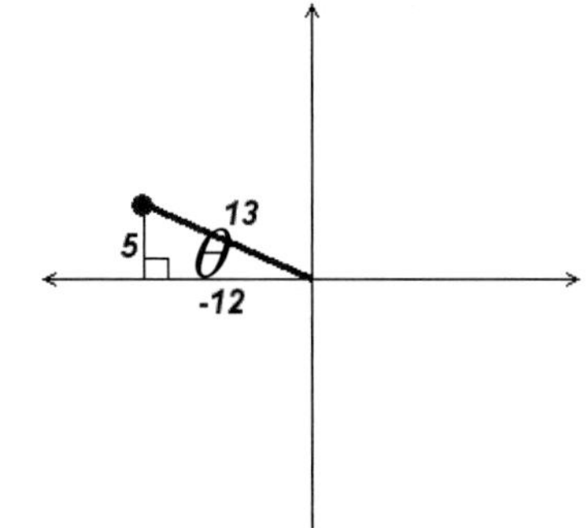 Observe that $\cos\theta=\boxed{-\frac{12}{13}}$.

27. Consider the following diagram:

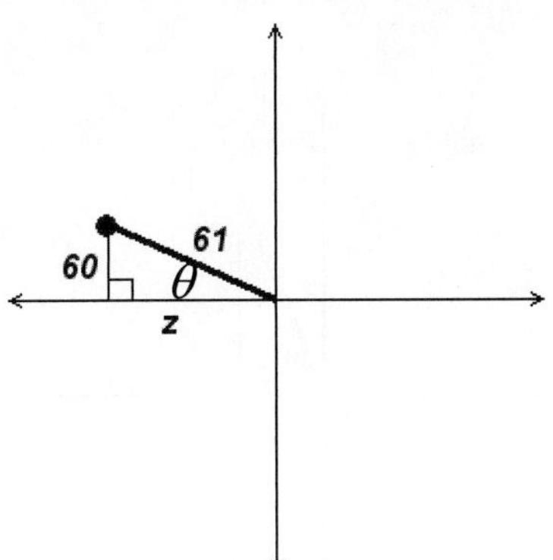

Using the Pythagorean Theorem, we obtain $z = -\sqrt{61^2 - 60^2} = -11$. So, $\tan\theta = \boxed{-\frac{60}{11}}$

28. Consider the following diagram:

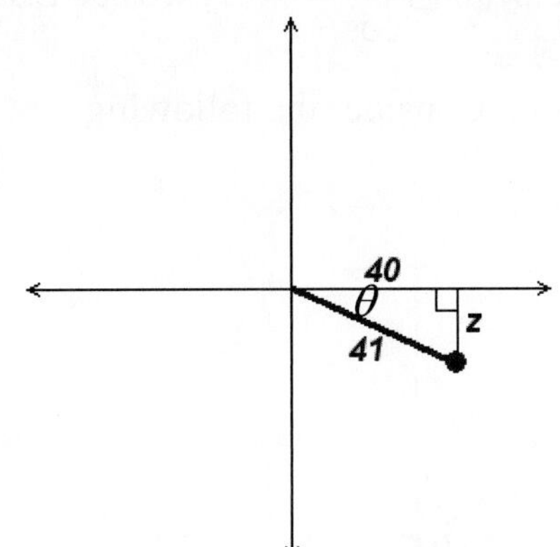

Using the Pythagorean Theorem, we obtain $z = -\sqrt{41^2 - 40^2} = -9$. So, $\tan\theta = \boxed{-\frac{9}{40}}$

29. Consider the following diagram:

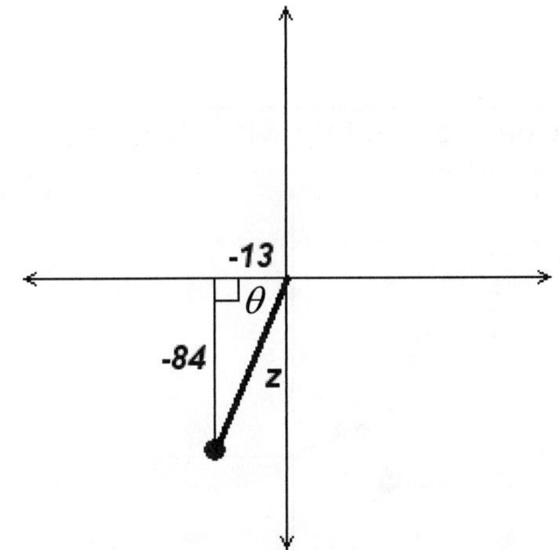

Using the Pythagorean Theorem, we obtain $z = \sqrt{(-13)^2 + (-84)^2} = 85$. So, $\sin\theta = \boxed{-\frac{84}{85}}$.

30. Consider the following diagram:

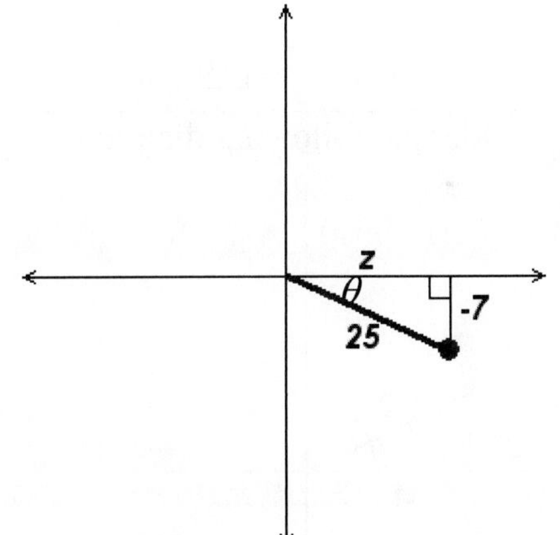

Using the Pythagorean Theorem, we obtain $z = \sqrt{25^2 - (-7)^2} = 24$. So, $\cos\theta = \boxed{\frac{24}{25}}$.

31. Since $\sec\theta = \dfrac{1}{\cos\theta} = -2$, we see that $\cos\theta = -\dfrac{1}{2}$. Consider the following diagram:

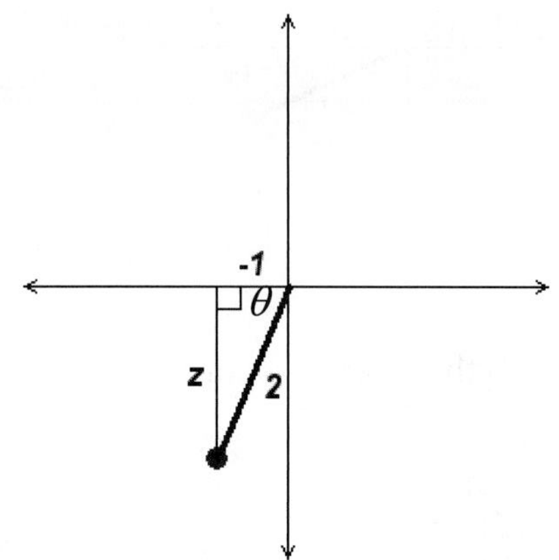

Using the Pythagorean Theorem, we obtain $z = -\sqrt{2^2 - (-1)^2} = -\sqrt{3}$. So,

$$\tan\theta = \frac{-\sqrt{3}}{-1} = \boxed{\sqrt{3}}.$$

32. Consider the following diagram:

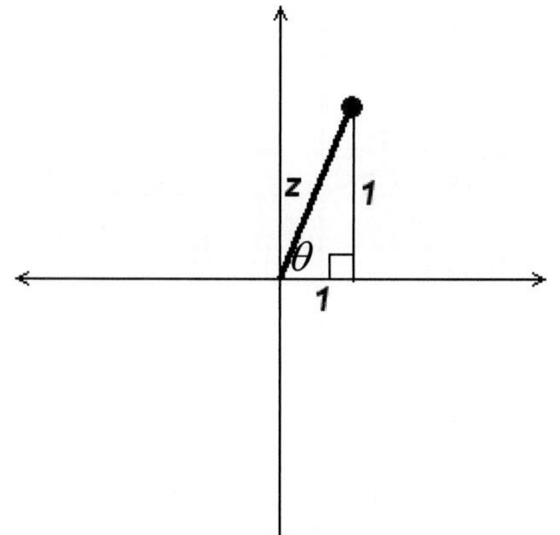

Using the Pythagorean Theorem, we obtain $z = \sqrt{1^2 + 1^2} = \sqrt{2}$. So,

$$\sin\theta = \frac{1}{\sqrt{2}} = \boxed{\frac{\sqrt{2}}{2}}.$$

33. Consider the following diagram:

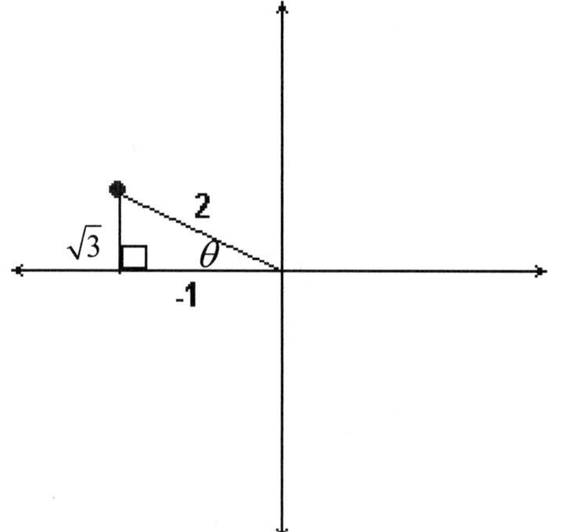

Since $\csc\theta = \dfrac{2}{\sqrt{3}}$, we know that $\sin\theta = \dfrac{\sqrt{3}}{2}$. So, $\cos\theta = -\frac{1}{2}$. As such, $\cot\theta = -\frac{1}{\sqrt{3}} = -\frac{\sqrt{3}}{3}$

34. Consider the following diagram:

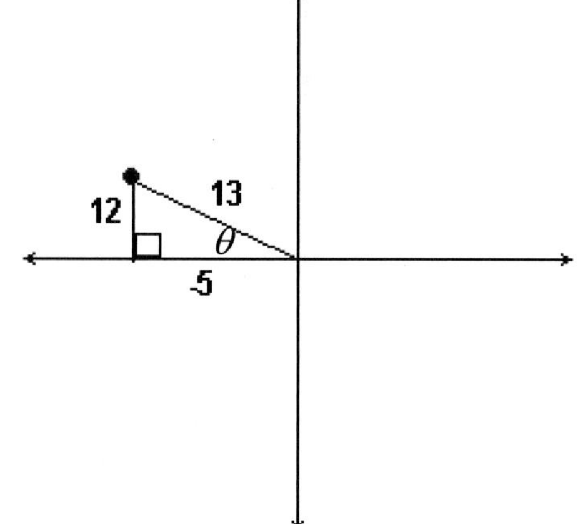

Since $\sec\theta = -\frac{13}{5}$, we know that $\cos\theta = -\frac{5}{13}$. So, $\sin\theta = \frac{12}{13}$. As such, $\csc\theta = \frac{13}{12}$.

35. Consider the following diagram:

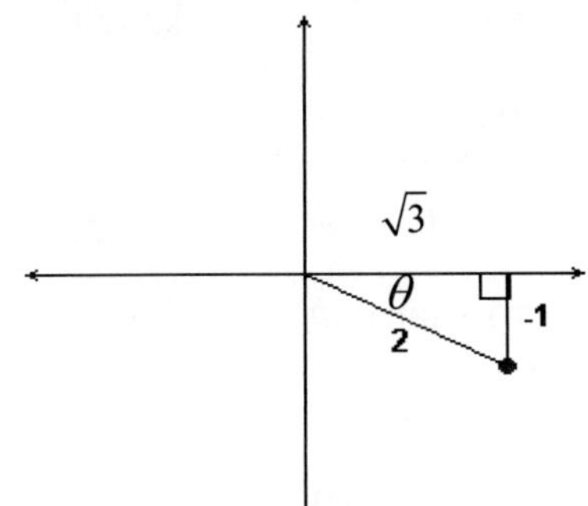

Observe that $\cot\theta = -\sqrt{3} = \frac{\frac{\sqrt{3}}{2}}{-\frac{1}{2}}$. So, $\cos\theta = \frac{\sqrt{3}}{2}$, $\sin\theta = -\frac{1}{2}$. Thus, $\sec\theta = \frac{2\sqrt{3}}{3}$.

36. Consider the following diagram:

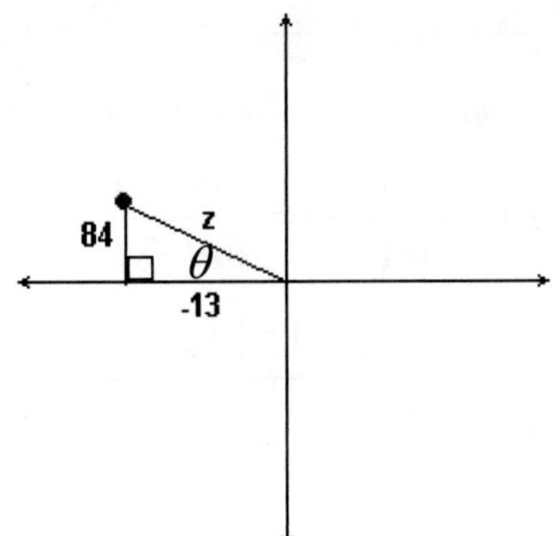

Using the Pythagorean theorem, we obtain $z = \sqrt{84^2 + (-13)^2} = 85$. So,

$$\csc\theta = \frac{1}{\sin\theta} = \frac{1}{\frac{84}{85}} = \frac{85}{84}.$$

37. First, $\cos(-270^\circ) = \cos(90^\circ) = 0$. Also, since $450^\circ = 360^\circ + 90^\circ$, we have $\sin(450^\circ) = \sin(90^\circ) = 1$. Therefore, $\cos(-270^\circ) + \sin(450^\circ) = \boxed{1}$.

38. First, $\sin(-270^\circ) = \sin(90^\circ) = 1$. Also, since $450^\circ = 360^\circ + 90^\circ$, we have $\cos(450^\circ) = \cos(90^\circ) = 0$. Therefore, $\sin(-270^\circ) + \cos(450^\circ) = \boxed{1}$.

39. Since $630^\circ = 360^\circ + 270^\circ$, we have $\sin(630^\circ) = \sin(270^\circ) = -1$. Further, since $-540^\circ = -360^\circ - 180^\circ$, we have

$$\tan(-540^\circ) = \tan(-180^\circ) = \frac{\sin(-180^\circ)}{\cos(-180^\circ)} = 0.$$

So, $\sin(630^\circ) + \tan(-540^\circ) = \boxed{-1}$.

40. Since $-720^\circ = 2(-360^\circ)$, we have $\cos(-720^\circ) = \cos(0^\circ) = 1$. Further, since $720^\circ = 2(360^\circ)$, we have

$$\tan(720^\circ) = \tan(0^\circ) = \frac{\sin(0^\circ)}{\cos(0^\circ)} = 0.$$ So,

$\cos(-720^\circ) + \tan(720^\circ) = \boxed{1}$.

41. Since $3\pi = 2\pi + \pi$, we have $\cos(3\pi) = \cos(\pi) = -1$. Further, since $-3\pi = -2\pi - \pi$, we have $\cos(-3\pi) = \cos(-\pi) = -1$, so that

$$\sec(-3\pi) = \frac{1}{\cos(-3\pi)} = \frac{1}{\cos(-\pi)} = -1.$$ So,

$\cos(3\pi) - \sec(-3\pi) = -1 - (-1) = \boxed{0}$.

42. Since $-\frac{5\pi}{2} = -2\pi - \frac{\pi}{2}$, we have $\sin\left(-\frac{5\pi}{2}\right) = \sin\left(-\frac{\pi}{2}\right) = \sin\left(\frac{3\pi}{2}\right) = -1$.

Further, $\csc\left(\frac{3\pi}{2}\right) = \frac{1}{\sin\left(\frac{3\pi}{2}\right)} = -1$. So,

$\sin\left(-\frac{5\pi}{2}\right) + \csc\left(\frac{3\pi}{2}\right) = -1 + (-1) = \boxed{-2}$.

43. Since $-\frac{7\pi}{2} = -2\pi - \frac{3\pi}{2}$, we have $\csc\left(-\frac{7\pi}{2}\right) = \csc\left(-\frac{3\pi}{2}\right) = \csc\left(\frac{\pi}{2}\right) = 1$. Further, since $\frac{7\pi}{2} = 2\pi + \frac{3\pi}{2}$, we have $\cot\left(\frac{7\pi}{2}\right) = \dfrac{\cos\left(\frac{7\pi}{2}\right)}{\sin\left(\frac{7\pi}{2}\right)} = \dfrac{\cos\left(\frac{3\pi}{2}\right)}{\sin\left(\frac{3\pi}{2}\right)} = 0$. So, $\csc\left(-\frac{7\pi}{2}\right) - \cot\left(\frac{7\pi}{2}\right) = \boxed{1}$.	**44.** Since $-3\pi = -2\pi - \pi$, we have $\cos(-3\pi) = \cos(-\pi) = -1$, so that $\sec(-3\pi) = \dfrac{1}{\cos(-3\pi)} = \dfrac{1}{\cos(-\pi)} = -1$. Further, $\tan(3\pi) = \tan(\pi) = 0$. So, $\sec(-3\pi) + \tan(3\pi) = \boxed{-1}$.
45. Since $\tan(720^\circ) = \tan(0^\circ) = 0$ and $\sec(720^\circ) = \dfrac{1}{\cos(720^\circ)} = \dfrac{1}{\cos(0^\circ)} = 1$, we have $\tan(720^\circ) + \sec(720^\circ) = \boxed{1}$.	**46.** Since $450^\circ = 360^\circ + 90^\circ$, we have $\cos(450^\circ) = \cos(90^\circ) = 0$. Further, since $-450^\circ = -360^\circ - 90^\circ$, we have $\cos(-450^\circ) = \cos(-90^\circ) = 0$. So, $\cos(450^\circ) - \cos(-450^\circ) = \boxed{0}$.
47. $\boxed{\text{Possible}}$ since the range of sine is $[-1,1]$.	**48.** $\boxed{\text{Not possible}}$ since the range of cosine is $[-1,1]$.
49. $\boxed{\text{Not possible}}$ since $\dfrac{2\sqrt{6}}{3} > 1$, and hence out of the range of cosine (which is $[-1,1]$).	**50.** $\boxed{\text{Possible}}$ since $-1 < \dfrac{\sqrt{2}}{10} < 1$.
51. $\boxed{\text{Possible}}$ since the range of tangent is $\mathbb{R}$.	**52.** $\boxed{\text{Possible}}$ since the range of cotangent is $\mathbb{R}$.
53. $\boxed{\text{Possible}}$ since $\frac{-4}{\sqrt{7}} < -1$, and hence in the range of secant (which is $(-\infty,-1] \cup [1,\infty)$).	**54.** $\boxed{\text{Possible}}$ since the range of cosecant is $(-\infty,-1] \cup [1,\infty)$.
55. $\boxed{\text{Possible}}$. $\theta \approx 0.1146^\circ$.	**56.** $\boxed{\text{Not possible}}$, $\sec\theta \in (-\infty,-1] \cup [1,\infty)$.

57.

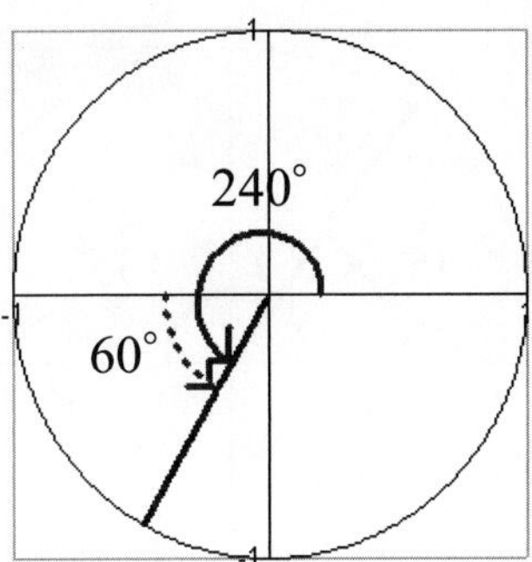

The given angle is shown in **bold**, while the reference angle is *dotted.*

$$\cos\left(240^\circ\right) = -\cos\left(60^\circ\right) = \boxed{-\frac{1}{2}}$$

58.

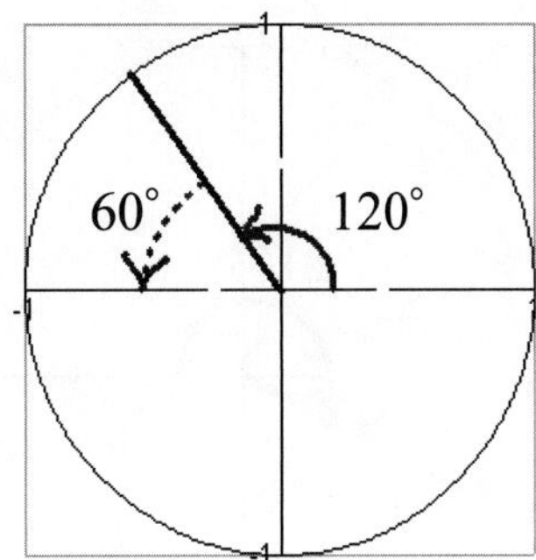

The given angle is shown in **bold**, while the reference angle is *dotted.*

$$\cos\left(120^\circ\right) = -\cos\left(60^\circ\right) = \boxed{-\frac{1}{2}}$$

59.

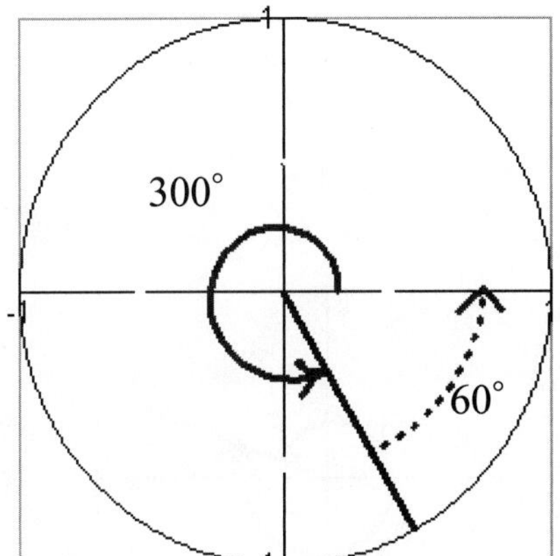

The given angle is shown in **bold**, while the reference angle is *dotted.*

$$\sin\left(300^\circ\right) = -\sin\left(60^\circ\right) = \boxed{-\frac{\sqrt{3}}{2}}$$

60.

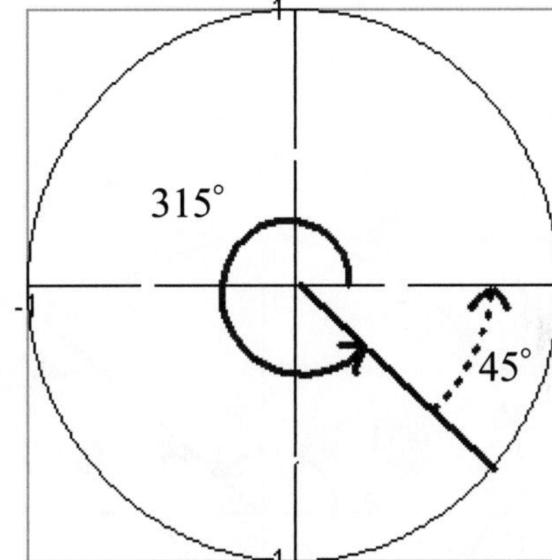

The given angle is shown in **bold**, while the reference angle is *dotted.*

$$\sin\left(315^\circ\right) = -\sin\left(45^\circ\right) = \boxed{-\frac{\sqrt{2}}{2}}$$

61.

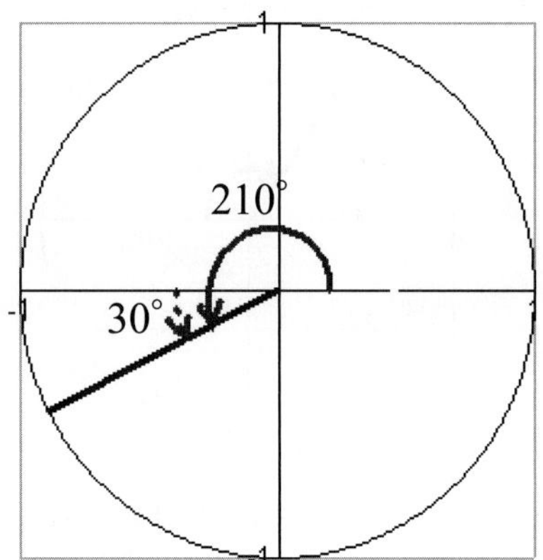

The given angle is shown in **bold**, while the reference angle is *dotted.*

$$\tan(210^\circ) = \tan(30^\circ) = \frac{\sin 30^\circ}{\cos 30^\circ} = \frac{\frac{1}{2}}{\frac{\sqrt{3}}{2}} = \boxed{\frac{\sqrt{3}}{3}}$$

62.

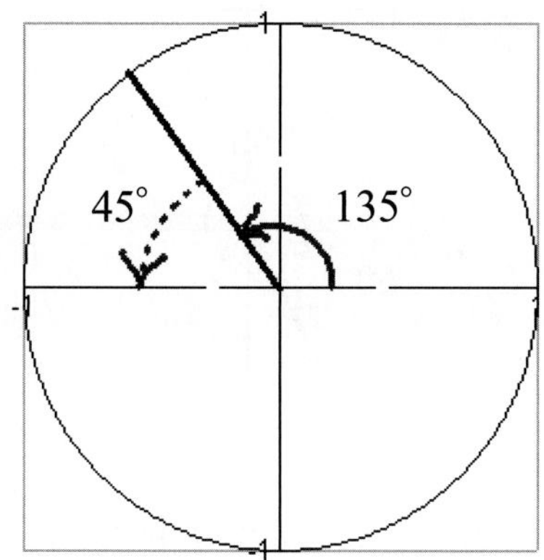

The given angle is shown in **bold**, while the reference angle is *dotted.*

$$\sec(135^\circ) = -\sec(45^\circ) = \frac{-1}{\cos(45^\circ)}$$

$$= -\frac{2}{\sqrt{2}} = \boxed{-\sqrt{2}}$$

63.

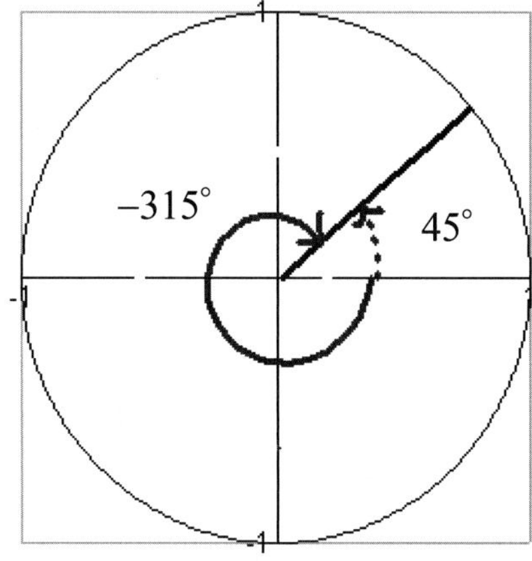

The given angle is shown in **bold**, while the reference angle is *dotted.*

$$\tan(135^\circ) = \tan(45^\circ) = \boxed{1}$$

64.

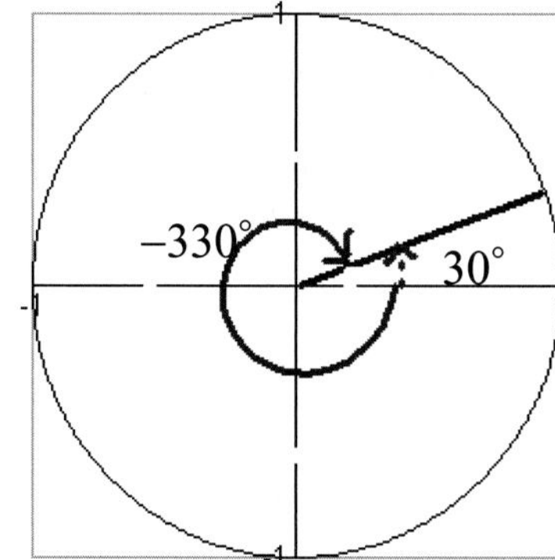

The given angle is shown in **bold**, while the reference angle is *dotted.*

$$\sec(-330^\circ) = \sec(30^\circ) = \frac{1}{\cos(30^\circ)}$$

$$= \frac{2}{\sqrt{3}} = \boxed{\frac{2\sqrt{3}}{3}}$$

65.

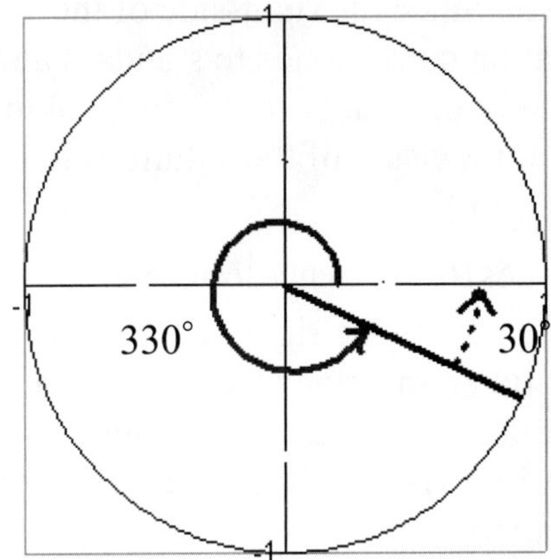

The given angle is shown in **bold**, while the reference angle is *dotted.*

$$\csc(330^\circ) = -\csc(30^\circ) = \frac{-1}{\sin(30^\circ)} = \boxed{-2}$$

66.

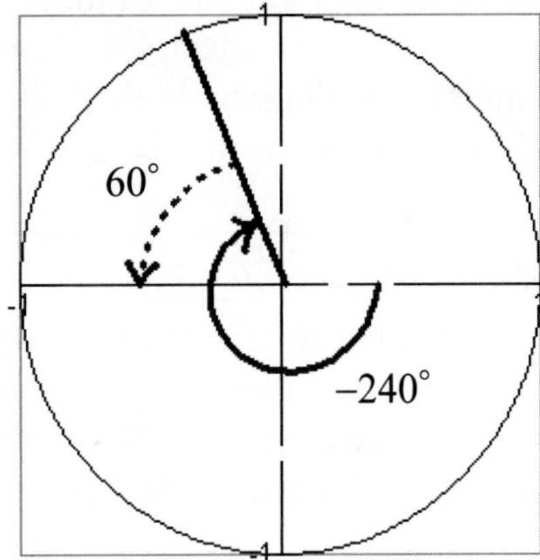

The given angle is shown in **bold**, while the reference angle is *dotted.*

$$\csc(-240^\circ) = \csc(60^\circ) = \frac{1}{\sin(60^\circ)}$$

$$= \frac{2}{\sqrt{3}} = \boxed{\frac{2\sqrt{3}}{3}}$$

67. $\cot(-315^\circ) = \cot(45^\circ) = 1$

68. $\cot(150^\circ) = \dfrac{\cos(150^\circ)}{\sin(150^\circ)} = \dfrac{-\frac{\sqrt{3}}{2}}{\frac{1}{2}} = -\sqrt{3}$

69. $\cos\theta = \dfrac{\sqrt{3}}{2}$ when $\boxed{\theta = 30^\circ \text{ or } 330^\circ}$.

(Note: Since the right-side of the equation corresponds to standard angles, we refer the reader to the diagram in the text for a visual of the solutions to this equation.)

70. $\sin\theta = \dfrac{\sqrt{3}}{2}$ when $\boxed{\theta = 60^\circ \text{ or } 120^\circ}$.

(Note: Since the right-side of the equation corresponds to standard angles, we refer the reader to the diagram in the text for a visual of the solutions to this equation.)

71. $\sin\theta = -\dfrac{1}{2}$ when $\boxed{\theta = 210^\circ \text{ or } 330^\circ}$.

(Note: Since the right-side of the equation corresponds to standard angles, we refer the reader to the diagram in the text for a visual of the solutions to this equation.)

72. $\cos\theta = -\dfrac{1}{2}$ when $\boxed{\theta = 120^\circ \text{ or } 240^\circ}$.

(Note: Since the right-side of the equation corresponds to standard angles, we refer the reader to the diagram in the text for a visual of the solutions to this equation.)

73. $\cos\theta = 0$ when $\boxed{\theta = 90^\circ \text{ or } 270^\circ}$. (Note: Since the right-side of the equation corresponds to standard angles, we refer the reader to the diagram in the text for a visual of the solutions to this equation.)	**74.** $\sin\theta = 0$ when $\boxed{\theta = 0^\circ, 180^\circ, \text{or } 360^\circ}$. (Note: Since the right-side of the equation corresponds to standard angles, we refer the reader to the diagram in the text for a visual of the solutions to this equation.)
75. $\sin\theta = -1$ when $\boxed{\theta = 270^\circ}$. (Note: Since the right-side of the equation corresponds to standard angles, we refer the reader to the diagram in the text for a visual of the solutions to this equation.)	**76.** $\cos\theta = -1$ when $\boxed{\theta = 180^\circ}$. (Note: Since the right-side of the equation corresponds to standard angles, we refer the reader to the diagram in the text for a visual of the solutions to this equation.)
77. Since $\sin\theta = 0.9397$, it follows that $\theta = \sin^{-1}(0.9397) \approx 70^\circ$. This angle has terminal side in QI. Since the desired angle has terminal side in QII, we have the following diagram. **Note:** The desired terminal side is the dashed segment. 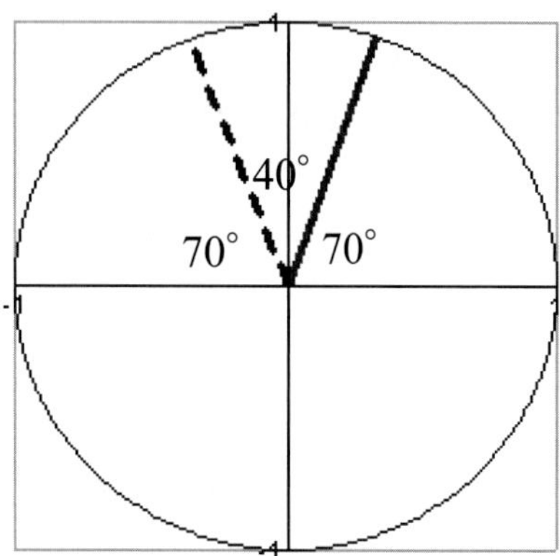 Hence, $\theta \approx 70^\circ + 40^\circ = \boxed{110^\circ}$.	**78.** Since $\cos\theta = 0.7071$, it follows that $\theta = \cos^{-1}(0.7071) \approx 45^\circ$. This angle has terminal side in QI. Since the desired angle has terminal side in QIV, we have the following diagram. **Note:** The desired terminal side is the dashed segment. 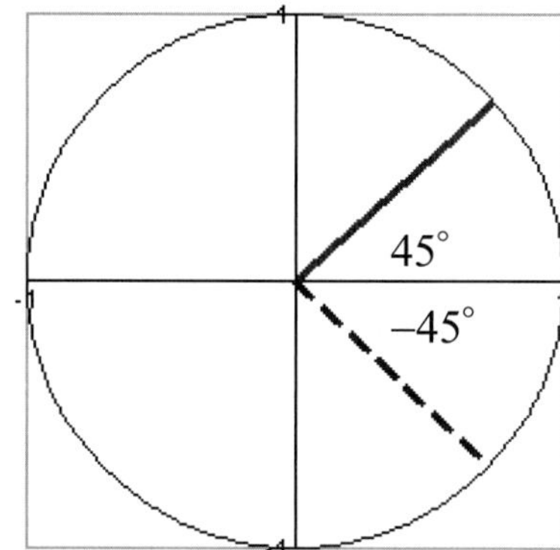 Hence, $\theta \approx 360^\circ - 45^\circ = \boxed{315^\circ}$.

79. Since $\cos\theta = -0.7986$, it follows that $\theta = \cos^{-1}(-0.7986) \approx \boxed{143^\circ}$. This angle has terminal side in QII. Since the desired angle has terminal side also in QII, no adjustment is necessary – this is pictured in the following diagram.
Note: The desired terminal side is the dashed segment.

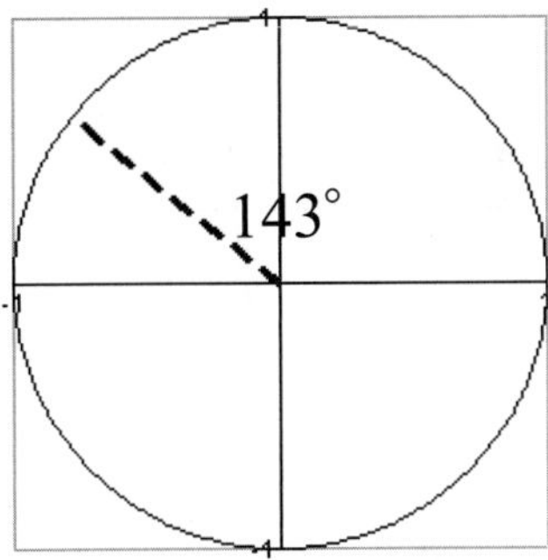

80. Since $\sin\theta = -0.1746$, it follows that $\theta = \sin^{-1}(-0.1746) \approx -10^\circ$. This angle has terminal side in QIV. Since the desired angle has terminal side in QIII, we have the following diagram.
Note: The desired terminal side is the dashed segment.

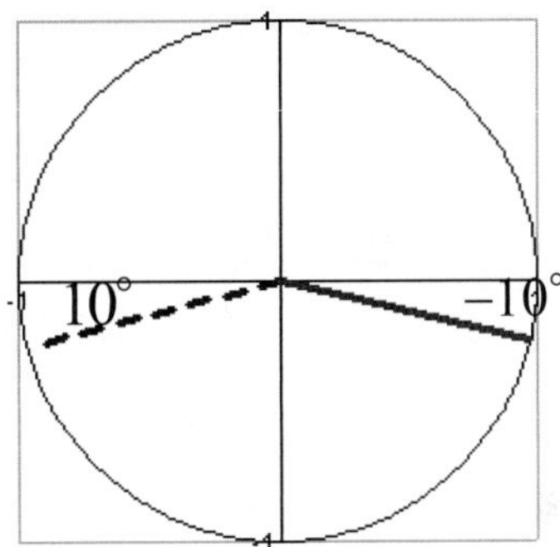

Hence, $\theta \approx 180^\circ + 10^\circ = \boxed{190^\circ}$

81. Since $\tan\theta = -0.7813$, it follows that $\theta = \tan^{-1}(-0.7813) \approx -38^\circ$. This angle has terminal side in QIV. Since the desired angle has terminal side also in QIV, we need only a corresponding angle with positive measure – this is pictured in the following diagram.
Note: The desired terminal side is the dashed segment.

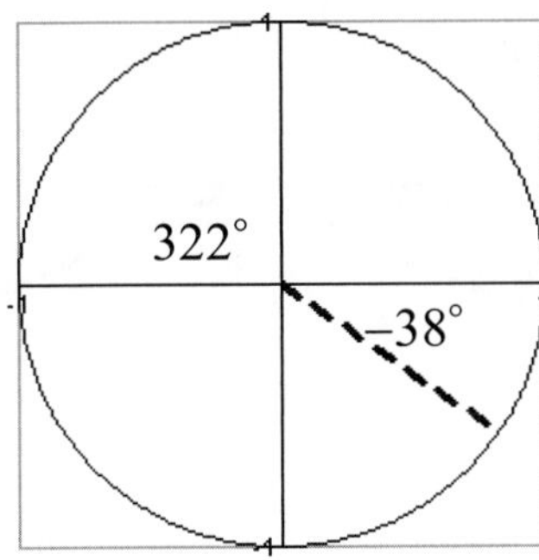

Hence, $\theta \approx 360^\circ - 38^\circ = \boxed{322^\circ}$.

82. Since $\cos\theta = -0.3420$, it follows that $\theta = \cos^{-1}(-0.3420) \approx \boxed{110^\circ}$. This angle has terminal side in QII. Since the desired angle has terminal side in QIII, we have the following diagram.
Note: The desired terminal side is the dashed segment.

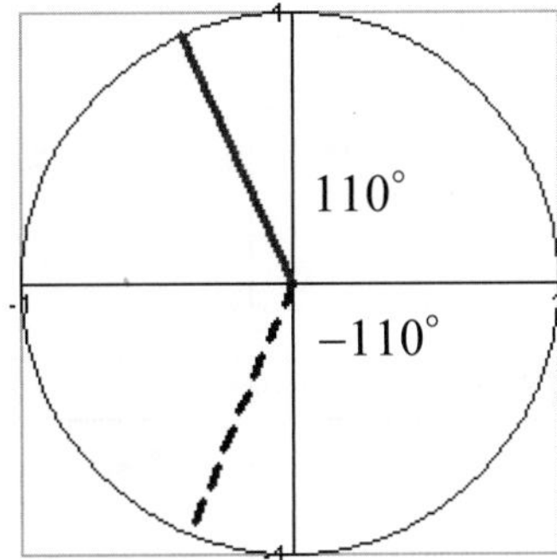

Hence, $\theta \approx 360^\circ - 110^\circ = \boxed{250^\circ}$.

83. Since $\tan\theta = -0.8391$, it follows that $\theta = \tan^{-1}(-0.8391) \approx -40^\circ$. This angle has terminal side in QIV. Since the desired angle has terminal side also in QII, we have the following diagram.
Note: The desired terminal side is the dashed segment.

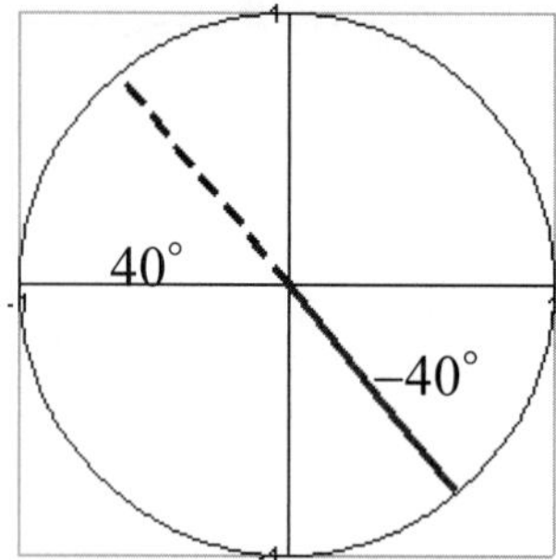

Hence, $\theta \approx 180^\circ - 40^\circ = \boxed{140^\circ}$.

84. Since $\tan\theta = 11.4301$, it follows that $\theta = \tan^{-1}(11.4301) \approx 85^\circ$. This angle has terminal side in QI.. Since the desired angle has terminal side also in QIII, we have the following diagram.
Note: The desired terminal side is the dashed segment.

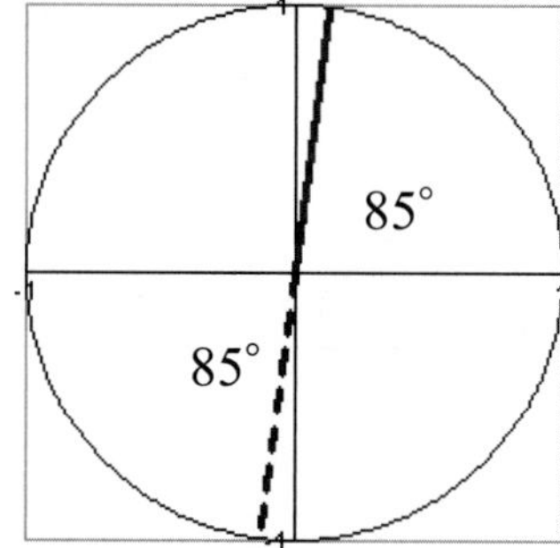

Hence, $\theta \approx 360^\circ - 85^\circ = \boxed{265^\circ}$.

85. Since $\sin\theta = -0.3420$, it follows that $\theta = \sin^{-1}(-0.3420) \approx -20^\circ$. This angle has terminal side in QIV. Since the desired angle has terminal side also in QIV, we need only a corresponding angle with positive measure – this is pictured in the following diagram.
Note: The desired terminal side is the dashed segment.

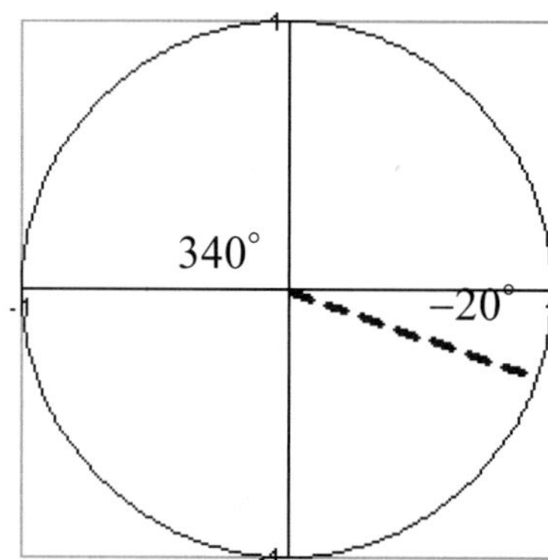

Hence, $\theta \approx 360^\circ - 20^\circ = \boxed{340^\circ}$.

86. Since $\sin\theta = -0.4226$, it follows that $\theta = \sin^{-1}(-0.4226) \approx -25^\circ$. This angle has terminal side in QIV. Since the desired angle has terminal side in QIII, we have the following diagram.
Note: The desired terminal side is the dashed segment.

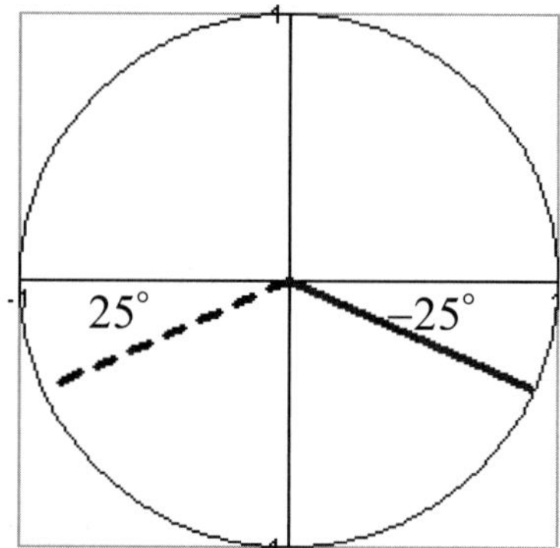

Hence, $\theta \approx 180^\circ + 25^\circ = \boxed{205^\circ}$.

87. $\sec\theta = 1.0001$ implies that $\cos\theta = \frac{1}{1.0001}$. Hence, $\theta = \cos^{-1}\left(\frac{1}{1.0001}\right) \approx 1^\circ$.	**88.** $\sec\theta = -3.1421$ implies that $\cos\theta = -\frac{1}{3.1421}$. Hence, $\theta = \cos^{-1}\left(-\frac{1}{3.1421}\right) \approx 109^\circ$.
89. $\csc\theta = -2.3604$ implies that $\sin\theta = -\frac{1}{2.3604}$. Hence, $\theta = \sin^{-1}\left(-\frac{1}{2.3604}\right) \approx -25^\circ$, which we identify with 335°.	**90.** $\csc\theta = -1.0001$ implies that $\sin\theta = -\frac{1}{1.0001}$. Hence, $\theta = \sin^{-1}\left(-\frac{1}{1.0001}\right) \approx -89^\circ$, which we identify with 271°.
91. Use $n_1\sin(\theta_1) = n_2\sin(\theta_2)$ with $n_1 = 1,\ \theta_1 = 30^\circ,\ \theta_2 = 22^\circ$ to obtain $(1)\sin(30^\circ) = n_2\sin(22^\circ)$ $n_2 = \dfrac{\sin(30^\circ)}{\sin(22^\circ)} \approx \boxed{1.3}$	**92.** Use $n_1\sin(\theta_1) = n_2\sin(\theta_2)$ with $n_1 = 1,\ \theta_1 = 30^\circ,\ \theta_2 = 18^\circ$ to obtain $(1)\sin(30^\circ) = n_2\sin(18^\circ)$ $n_2 = \dfrac{\sin(30^\circ)}{\sin(18^\circ)} \approx \boxed{1.6}$
93. Use $n_1\sin(\theta_1) = n_2\sin(\theta_2)$ with $n_1 = 1,\ n_2 = 2.4,\ \theta_1 = 30$ to obtain $(1)\sin(30^\circ) = (2.4)\sin(\theta_2)$ $\sin(\theta_2) = \dfrac{\sin(30^\circ)}{2.4}$ $\theta_2 = \sin^{-1}\left[\dfrac{\sin(30^\circ)}{2.4}\right] \approx \boxed{12^\circ}$	**94.** Use $n_1\sin(\theta_1) = n_2\sin(\theta_2)$ with $n_1 = 1,\ n_2 = 1.9,\ \theta_1 = 30$ to obtain $(1)\sin(30^\circ) = (1.9)\sin(\theta_2)$ $\sin(\theta_2) = \dfrac{\sin(30^\circ)}{1.9}$ $\theta_2 = \sin^{-1}\left[\dfrac{\sin(30^\circ)}{1.9}\right] \approx \boxed{15^\circ}$

95. The reference angle is measured between the terminal side and the *x*-axis, not the *y*-axis. So, in this case, the reference angle is 60°. Since $\cos 60^\circ = \frac{1}{2}$ and $\cos 120^\circ = -\frac{1}{2}$, we see that $\sec 120^\circ = \dfrac{1}{-\frac{1}{2}} = -2$.

96. Consider the following diagram: 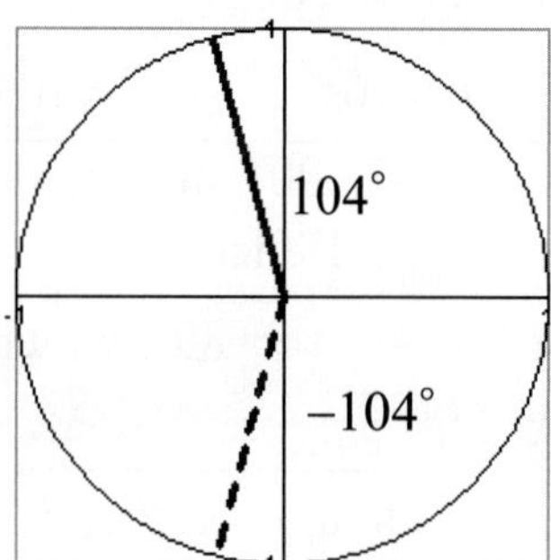	The cosine of angles with terminal sides in QII and QIII is negative. The calculator correctly produces the angle 104°, but since we want the terminal side to lie in QIII, we need to use the correct reference angle. In this case, we angle would be $360^\circ - 104^\circ = 256^\circ$.
97. True. For any $0^\circ < \theta < 90^\circ$, $\sin\theta$ and $\cos\theta$ are both positive. Since all of the trigonometric functions are defined as quotients of sine, cosine, and 1, they are also positive for such values of θ.	
98. False. This requires both $\sin\theta$ and $\cos\theta$ to be negative. But, in such case, $\tan\theta$ and $\cot\theta$ are both positive.	
99. False. For instance, $\cos(-45^\circ) = \frac{\sqrt{2}}{2}$.	**100.** False. For instance, $\cos(120^\circ) = -\frac{1}{2}$.
101. False. Since $1 + \tan^2\theta = \sec^2\theta$, we know that $\tan^2\theta = \sec^2\theta - 1$. Since $\tan^2\theta \geq 0$, it follows that $\sec^2\theta - 1 \geq 0$.	**102.** True. $\sec\theta\csc\theta < 0$ when $\sin\theta$ and $\cos\theta$ have opposite signs. This occurs in QII and QIV.
103. True. The angles θ and $\theta + 360^\circ n$ (where n is an integer) are coterminal, so that $\cos\theta = \cos(\theta + 360^\circ n)$.	**104.** True. This is true since sine is periodic with period 2π (which corresponds to 360°).

105. First, note that it is not possible for $a = 0$ since in such case there would be no triangle. Thus, there are only two cases to consider.

Case 1: $a > 0$

Consider the following diagram:

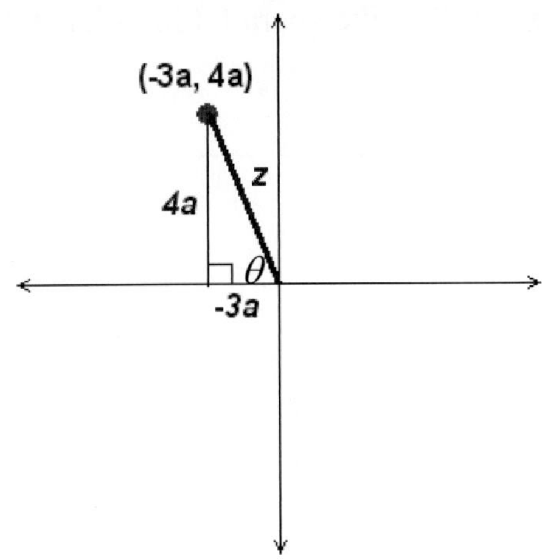

From the Pythagorean Theorem, it follows that $z = \sqrt{(4a)^2 + (-3a)^2} = 5a$.

So, $\cos\theta = \dfrac{-3a}{5a} = -\dfrac{3}{5}$.

Case 2: $a < 0$

Consider the following diagram:

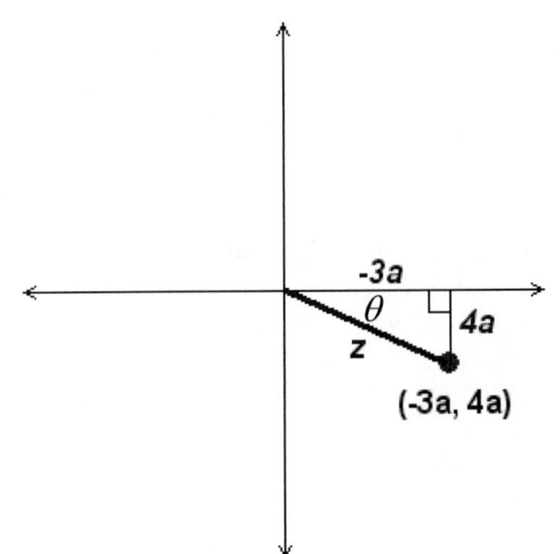

From the Pythagorean Theorem, it follows that $z = \sqrt{(4a)^2 + (-3a)^2} = 5a$.

So, $\cos\theta = \dfrac{-3a}{5a} = -\dfrac{3}{5}$, which is the same conclusion of Case 1.

106. Referring to the diagram in Exercise 105, we observe that in both cases, $\sin\theta = \dfrac{4a}{5a} = \dfrac{4}{5}$.

107. Suppose the form of the equation of the line is given by $y = mx + b$. **First**, substitute the point $(a,0)$ into the equation: $0 = ma + b$ so that $b = -ma$. So, the equation of the line thus far is $y = mx - ma = m(x-a)$.
Next, note that this line is parallel to the corresponding line translate to pass through the origin – the slope of that line is $-\tan\theta$ (by Exercise 104). Since parallel lines have the same slope, we know that $m = -\tan\theta$.
Thus, the equation of the line is $\boxed{y = (-\tan\theta)(x-a)}$.

108. First, note that $\csc\theta = -\frac{\sqrt{29}}{2}$, so that $\sin\theta = -\frac{2}{\sqrt{29}}$.
Consider the following diagram:

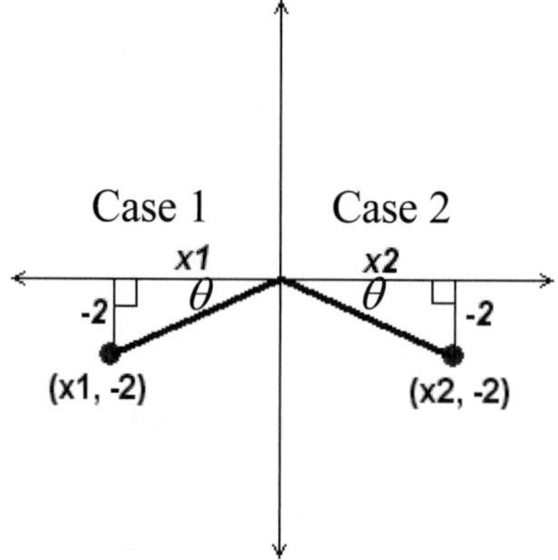

Using the Pythagorean Theorem, we see that

$$x_1^2 + (-2)^2 = \left(\sqrt{29}\right)^2 \text{ and } x_2^2 + (-2)^2 = \left(\sqrt{29}\right)^2.$$

Hence, we have $x_1 = x_2 = \pm 5$. Thus, the two possible x values are $\boxed{\pm 5}$.

109. Assume that $a>0,\ b>0$. Consider the following diagram: 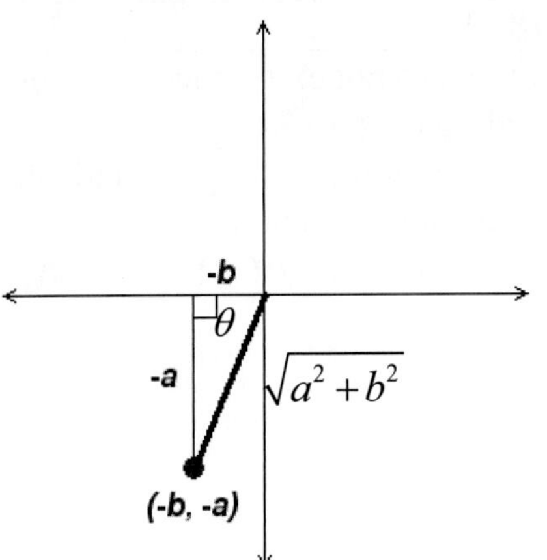 Observe that $\sin\theta = \boxed{-\frac{a}{\sqrt{a^2+b^2}}}$.	**110.** Assume that $a>0,\ b>0$. Consider the following diagram: 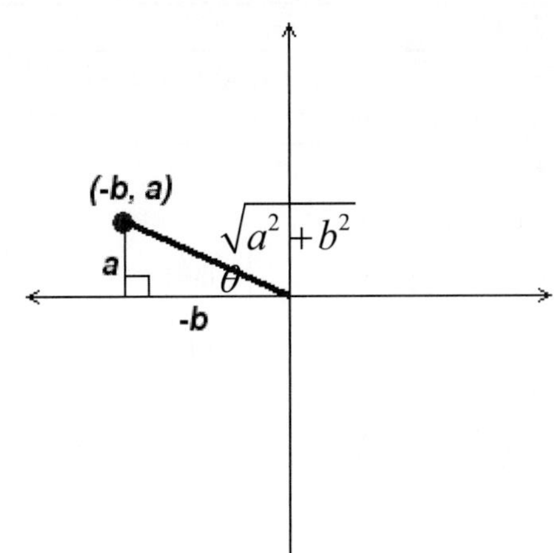 Observe that $\cos\theta = \boxed{-\frac{b}{\sqrt{a^2+b^2}}}$.
111. Consider the following diagram: 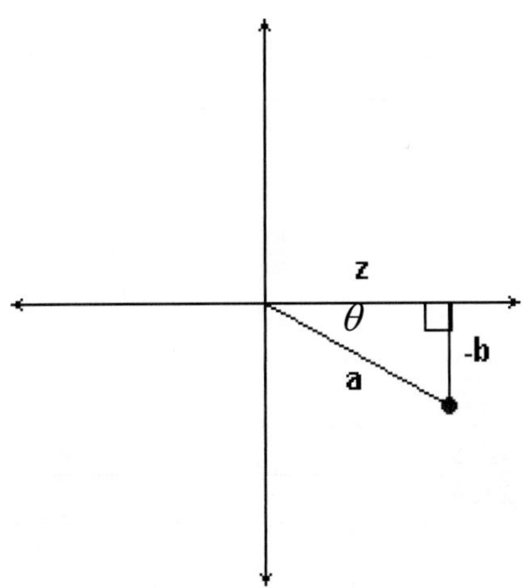 Using the Pythagorean theorem, we obtain $z=\sqrt{a^2-b^2}$. So, $\cot\theta = -\frac{\sqrt{a^2-b^2}}{b}$.	**112.** Consider the following diagram: 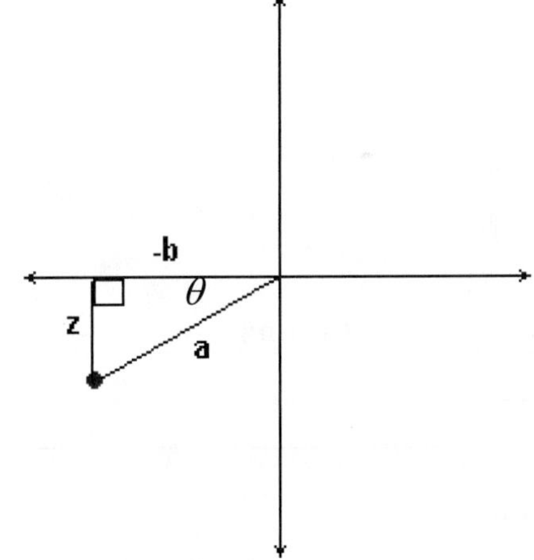 Using the Pythagorean theorem, we obtain $z=-\sqrt{a^2-b^2}$ (since in QIII). So, $\tan\theta = \frac{\sqrt{a^2-b^2}}{b}$.
113. $\cos 270^\circ = 0$	**114.** $\sin 270^\circ = -1$

115. $\cot 270^\circ = \dfrac{\cos 270^\circ}{\sin 270^\circ} = 0$	**116.** $\sin\left(-270^\circ\right) = 1$
117. 0	**118.** 1
119. Does not exist since you would be dividing by zero.	**120.** Does not exist since you would be dividing by zero.
121. $F\left(\frac{\pi}{4}\right) - F\left(-\frac{\pi}{6}\right)$ $= \left[2\tan\left(\frac{\pi}{4}\right) + \cos\left(\frac{\pi}{4}\right)\right] - \left[2\tan\left(-\frac{\pi}{6}\right) + \cos\left(-\frac{\pi}{6}\right)\right]$ $= \left[2(1) + \frac{\sqrt{2}}{2}\right] - \left[2\left(\frac{-\frac{1}{2}}{\frac{\sqrt{3}}{2}}\right) + \frac{\sqrt{3}}{2}\right]$ $= \dfrac{12 + 3\sqrt{2} + \sqrt{3}}{6}$	**122.** Using the Pythagorean identity, we immediately conclude that $F\left(\frac{7\pi}{6}\right) - F\left(\frac{3\pi}{4}\right) = 1 - 1 = 0$.
123. $F\left(\frac{4\pi}{3}\right) - F\left(\frac{5\pi}{6}\right) = \left[\sec^2\left(\frac{4\pi}{3}\right) + 1\right] - \left[\sec^2\left(\frac{5\pi}{6}\right) + 1\right]$ $= \dfrac{1}{\left(-\frac{1}{2}\right)^2} - \dfrac{1}{\left(-\frac{\sqrt{3}}{2}\right)^2}$ $= 4 - \dfrac{4}{3} = \dfrac{8}{3}$	**124.** $F\left(\frac{7\pi}{4}\right) - F\left(\frac{7\pi}{6}\right)$ $= \left[\cot\left(\frac{7\pi}{4}\right) - \csc^2\left(\frac{7\pi}{4}\right)\right] - \left[\cot\left(\frac{7\pi}{6}\right) - \csc^2\left(\frac{7\pi}{6}\right)\right]$ $= \left[-1 - \dfrac{1}{\left(-\frac{\sqrt{2}}{2}\right)^2}\right] - \left[\dfrac{-\frac{\sqrt{3}}{2}}{-\frac{1}{2}} - \dfrac{1}{\left(\frac{1}{2}\right)^2}\right]$ $= 1 - \sqrt{3}$

Section 4.4 Solutions ---

1. SSA	**2.** SSA
3. SSS	**4.** SAS
5. ASA	**6.** ASA

7. Observe that:

$\gamma = 180^\circ - \left(45^\circ + 60^\circ\right) = 75^\circ$

$$\frac{\sin\alpha}{a} = \frac{\sin\beta}{b} \Rightarrow \frac{\sin 45^\circ}{10\,\text{m}} = \frac{\sin 60^\circ}{b} \Rightarrow b = \frac{(10\text{m})(\sin 60^\circ)}{\sin 45^\circ} = \frac{(10\text{m})\left(\frac{\sqrt{3}}{2}\right)}{\frac{\sqrt{2}}{2}} \approx 12.2\,\text{m}$$

$$\frac{\sin\alpha}{a} = \frac{\sin\gamma}{c} \Rightarrow \frac{\sin 45^\circ}{10\,\text{m}} = \frac{\sin 75^\circ}{c} \Rightarrow c = \frac{(10\text{m})(\sin 75^\circ)}{\sin 45^\circ} \approx 13.66\,\text{m}$$

8. Observe that:

$$\alpha = 180^\circ - \left(75^\circ + 60^\circ\right) = 45^\circ$$

$$\frac{\sin\alpha}{a} = \frac{\sin\beta}{b} \Rightarrow \frac{\sin 45^\circ}{a} = \frac{\sin 75^\circ}{25 \text{ in.}} \Rightarrow a = \frac{(25\text{ in.})(\sin 45^\circ)}{\sin 75^\circ} \approx 18\text{ in.}$$

$$\frac{\sin\beta}{b} = \frac{\sin\gamma}{c} \Rightarrow \frac{\sin 75^\circ}{25\text{ in.}} = \frac{\sin 60^\circ}{c} \Rightarrow c = \frac{(25\text{ in.})(\sin 60^\circ)}{\sin 75^\circ} \approx 22\text{ in.}$$

9. Observe that:

$$\beta = 180^\circ - \left(46^\circ + 72^\circ\right) = 62^\circ$$

$$\frac{\sin\alpha}{a} = \frac{\sin\beta}{b} \Rightarrow \frac{\sin 46^\circ}{a} = \frac{\sin 62^\circ}{200 \text{ cm}} \Rightarrow a = \frac{(200\text{ cm})(\sin 46^\circ)}{\sin 62^\circ} \approx 163\text{ cm}$$

$$\frac{\sin\beta}{b} = \frac{\sin\gamma}{c} \Rightarrow \frac{\sin 62^\circ}{200\text{ cm}} = \frac{\sin 72^\circ}{c} \Rightarrow c = \frac{(200\text{ cm})(\sin 72^\circ)}{\sin 62^\circ} \approx 215\text{ cm}$$

10. Observe that:

$$\alpha = 180^\circ - \left(100^\circ + 40^\circ\right) = 40^\circ$$

$$\frac{\sin\alpha}{a} = \frac{\sin\beta}{b} \Rightarrow \frac{\sin 40^\circ}{16 \text{ ft.}} = \frac{\sin 40^\circ}{b} \Rightarrow b = \frac{(16\text{ ft.})(\sin 40^\circ)}{\sin 40^\circ} \approx 16\text{ ft.}$$

$$\frac{\sin\alpha}{a} = \frac{\sin\gamma}{c} \Rightarrow \frac{\sin 40^\circ}{16\text{ ft.}} = \frac{\sin 100^\circ}{c} \Rightarrow c = \frac{(16\text{ ft.})(\sin 100^\circ)}{\sin 40^\circ} \approx 25\text{ ft.}$$

11. Observe that:

$$\beta = 180^\circ - \left(16.3^\circ + 47.6^\circ\right) = 116.1^\circ$$

$$\frac{\sin\alpha}{a} = \frac{\sin\gamma}{c} \Rightarrow \frac{\sin 16.3^\circ}{a} = \frac{\sin 47.6^\circ}{211 \text{ yd.}} \Rightarrow a = \frac{(211\text{ yd.})(\sin 16.3^\circ)}{\sin 47.6^\circ} \approx 80.2\text{ yd.}$$

$$\frac{\sin\beta}{b} = \frac{\sin\gamma}{c} \Rightarrow \frac{\sin 116.1^\circ}{b} = \frac{\sin 47.6^\circ}{211\text{ yd.}} \Rightarrow b = \frac{(211\text{ yd.})(\sin 116.1^\circ)}{\sin 47.6^\circ} \approx 256.6\text{ yd.}$$

12. Observe that:

$$\alpha = 180^\circ - \left(104.2^\circ + 33.6^\circ\right) = 42.2^\circ$$

$$\frac{\sin\alpha}{a} = \frac{\sin\beta}{b} \Rightarrow \frac{\sin 42.2^\circ}{26 \text{ in.}} = \frac{\sin 104.2^\circ}{b} \Rightarrow b = \frac{(26\text{ in.})(\sin 104.2^\circ)}{\sin 42.2^\circ} \approx 37.5\text{ in.}$$

$$\frac{\sin\alpha}{a} = \frac{\sin\gamma}{c} \Rightarrow \frac{\sin 42.2^\circ}{26\text{ in.}} = \frac{\sin 33.6^\circ}{c} \Rightarrow c = \frac{(26\text{ in.})(\sin 33.6^\circ)}{\sin 42.2^\circ} \approx 21.4\text{ in.}$$

13. Observe that:

$$\gamma = 180^\circ - \left(30^\circ + 30^\circ\right) = 120^\circ$$

$$\frac{\sin\alpha}{a} = \frac{\sin\gamma}{c} \Rightarrow \frac{\sin 30^\circ}{a} = \frac{\sin 120^\circ}{12\text{ m}} \Rightarrow a = \frac{(12\text{m})\left(\sin 30^\circ\right)}{\sin 120^\circ} \approx 7\text{ m}$$

$$\frac{\sin\beta}{b} = \frac{\sin\gamma}{c} \Rightarrow \frac{\sin 30^\circ}{b} = \frac{\sin 120^\circ}{12\text{m}} \Rightarrow b = \frac{(12\text{m})\left(\sin 30^\circ\right)}{\sin 120^\circ} \approx 7\text{ m}$$

14. Observe that:

$$\beta = 180^\circ - \left(45^\circ + 75^\circ\right) = 60^\circ$$

$$\frac{\sin\alpha}{a} = \frac{\sin\gamma}{c} \Rightarrow \frac{\sin 45^\circ}{a} = \frac{\sin 75^\circ}{9\text{ in.}} \Rightarrow a = \frac{(9\text{ in.})\left(\sin 45^\circ\right)}{\sin 75^\circ} \approx 7\text{ in.}$$

$$\frac{\sin\beta}{b} = \frac{\sin\gamma}{c} \Rightarrow \frac{\sin 60^\circ}{b} = \frac{\sin 75^\circ}{9\text{ in.}} \Rightarrow b = \frac{(9\text{ in.})\left(\sin 60^\circ\right)}{\sin 75^\circ} \approx 8\text{ in.}$$

15. Observe that:

$$\alpha = 180^\circ - \left(26^\circ + 57^\circ\right) = 97^\circ$$

$$\frac{\sin\alpha}{a} = \frac{\sin\gamma}{c} \Rightarrow \frac{\sin 97^\circ}{a} = \frac{\sin 57^\circ}{100\text{ yd.}} \Rightarrow a = \frac{(100\text{ yd.})\left(\sin 97^\circ\right)}{\sin 57^\circ} \approx 118\text{ yd.}$$

$$\frac{\sin\beta}{b} = \frac{\sin\gamma}{c} \Rightarrow \frac{\sin 26^\circ}{b} = \frac{\sin 57^\circ}{100\text{ yd.}} \Rightarrow b = \frac{(100\text{ yd.})\left(\sin 26^\circ\right)}{\sin 57^\circ} \approx 52\text{ yd.}$$

16. Observe that:

$$\beta = 180^\circ - \left(80^\circ + 30^\circ\right) = 70^\circ$$

$$\frac{\sin\alpha}{a} = \frac{\sin\beta}{b} \Rightarrow \frac{\sin 80^\circ}{a} = \frac{\sin 70^\circ}{3\text{ ft.}} \Rightarrow a = \frac{(3\text{ ft.})\left(\sin 80^\circ\right)}{\sin 70^\circ} \approx 3\text{ ft.}$$

$$\frac{\sin\beta}{b} = \frac{\sin\gamma}{c} \Rightarrow \frac{\sin 70^\circ}{3\text{ft.}} = \frac{\sin 30^\circ}{c} \Rightarrow c = \frac{(3\text{ft.})\left(\sin 30^\circ\right)}{\sin 70^\circ} \approx 2\text{ ft.}$$

17. Step 1: Determine β.

$$\frac{\sin\alpha}{a}=\frac{\sin\beta}{b} \Rightarrow \frac{\sin 16^\circ}{4}=\frac{\sin\beta}{5} \Rightarrow \sin\beta=\frac{5\sin 16^\circ}{4} \Rightarrow \beta=\sin^{-1}\left(\frac{5\sin 16^\circ}{4}\right)\approx 20^\circ$$

This is the solution in QI – label it as β_1. The second solution in QII is given by $\beta_2=180^\circ-\beta_1\approx 160^\circ$. Both are tenable solutions, so we need to solve for two triangles.

Step 2: Solve for both triangles.

QI triangle: $\beta_1\approx 20^\circ$

$$\gamma_1\approx 180^\circ-\left(16^\circ+20^\circ\right)=144^\circ$$

$$\frac{\sin\alpha}{a}=\frac{\sin\gamma_1}{c_1} \Rightarrow \frac{\sin 16^\circ}{4}=\frac{\sin 144^\circ}{c_1} \Rightarrow c_1=\frac{4\sin 144^\circ}{\sin 16^\circ}\approx 9$$

QII triangle: $\beta_2\approx 160^\circ$

$$\gamma_2\approx 180^\circ-\left(16^\circ+160^\circ\right)=4^\circ$$

$$\frac{\sin\alpha}{a}=\frac{\sin\gamma_2}{c_2} \Rightarrow \frac{\sin 16^\circ}{4}=\frac{\sin 4^\circ}{c_2} \Rightarrow c_2=\frac{4\sin 4^\circ}{\sin 16^\circ}\approx 1$$

18. Step 1: Determine γ.

$$\frac{\sin\gamma}{c}=\frac{\sin\beta}{b} \Rightarrow \frac{\sin\gamma}{20}=\frac{\sin 70^\circ}{30} \Rightarrow \sin\gamma=\frac{20\sin 70^\circ}{30} \Rightarrow \gamma=\sin^{-1}\left(\frac{20\sin 70^\circ}{30}\right)\approx 39^\circ$$

Note that there is only one triangle in this case since the angle in QII with the same sine as this value of γ is $180^\circ-39^\circ=141^\circ$. In such case, note that $\beta+\gamma>180^\circ$, therefore preventing the formation of a triangle (since the three interior angles, two of which are β and γ, must sum to 180°).

Step 2: Solve for the triangle.

$$\alpha\approx 180^\circ-\left(70^\circ+39^\circ\right)=71^\circ$$

$$\frac{\sin\alpha}{a}=\frac{\sin\beta}{b} \Rightarrow \frac{\sin 71^\circ}{a}=\frac{\sin 70^\circ}{30} \Rightarrow a=\frac{30\sin 71^\circ}{\sin 70^\circ}\approx 30$$

19. Step 1: Determine α.

$$\frac{\sin\gamma}{c}=\frac{\sin\alpha}{a} \Rightarrow \frac{\sin 40^\circ}{12}=\frac{\sin\alpha}{12} \Rightarrow \sin\alpha=\sin 40^\circ \Rightarrow \alpha=40^\circ$$

Note that there is only one triangle in this case since the angle in QII with the same sine as this value of α is $180^\circ-40^\circ=140^\circ$. In such case, note that $\alpha+\gamma=180^\circ$, therefore preventing the formation of a triangle (since the three interior angles, two of which are α and γ, must sum to 180° - this would only occur if the third angle $\beta=0$, in which case there is no triangle).

Step 2: Solve for the triangle.

$$\beta\approx 180^\circ-\left(40^\circ+40^\circ\right)=100^\circ$$

$$\frac{\sin\gamma}{c}=\frac{\sin\beta}{b} \Rightarrow \frac{\sin 40^\circ}{12}=\frac{\sin 100^\circ}{b} \Rightarrow b=\frac{12\sin 100^\circ}{\sin 40^\circ}\approx 18$$

20. Step 1: Determine β.

$$\frac{\sin\alpha}{a}=\frac{\sin\beta}{b} \Rightarrow \frac{\sin 25^\circ}{80}=\frac{\sin\beta}{111} \Rightarrow \sin\beta=\frac{111\sin 25^\circ}{80} \Rightarrow \beta=\sin^{-1}\left(\frac{111\sin 25^\circ}{80}\right)\approx 36^\circ$$

This is the solution in QI – label it as β_1. The second solution in QII is given by $\beta_2=180^\circ-\beta_1\approx 144^\circ$. Both are tenable solutions, so we need to solve for two triangles.

Step 2: Solve for both triangles.

QI triangle: $\beta_1\approx 36^\circ$

$$\gamma_1\approx 180^\circ-\left(25^\circ+36^\circ\right)=119^\circ$$

$$\frac{\sin\alpha}{a}=\frac{\sin\gamma_1}{c_1} \Rightarrow \frac{\sin 25^\circ}{80}=\frac{\sin 119^\circ}{c_1} \Rightarrow c_1=\frac{80\sin 119^\circ}{\sin 25^\circ}\approx 166$$

QII triangle: $\beta_2\approx 144^\circ$

$$\gamma_2\approx 180^\circ-\left(25^\circ+144^\circ\right)=11^\circ$$

$$\frac{\sin\alpha}{a}=\frac{\sin\gamma_2}{c_2} \Rightarrow \frac{\sin 25^\circ}{80}=\frac{\sin 11^\circ}{c_2} \Rightarrow c_2=\frac{80\sin 11^\circ}{\sin 25^\circ}\approx 36$$

21. Step 1: Determine α.

$$\frac{\sin\alpha}{a}=\frac{\sin\beta}{b} \Rightarrow \frac{\sin\alpha}{21}=\frac{\sin 100^\circ}{14} \Rightarrow \sin\alpha=\frac{21\sin 100^\circ}{14} \Rightarrow \alpha=\sin^{-1}\left(\frac{21\sin 100^\circ}{14}\right),$$

which does not exist. Hence, there is no triangle in this case.

22. Step 1: Determine β.

$$\frac{\sin\alpha}{a}=\frac{\sin\beta}{b} \Rightarrow \frac{\sin 120^\circ}{13}=\frac{\sin\beta}{26} \Rightarrow \sin\beta=\frac{26\sin 120^\circ}{13} \Rightarrow \beta=\sin^{-1}\left(\frac{26\sin 120^\circ}{13}\right),$$

which does not exist. Hence, there is no triangle in this case.

23. Step 1: Determine β.

$$\frac{\sin\alpha}{a}=\frac{\sin\beta}{b} \Rightarrow \frac{\sin 30^\circ}{9}=\frac{\sin\beta}{18} \Rightarrow \sin\beta=\frac{18\left(\frac{1}{2}\right)}{9}=1 \Rightarrow \beta=\sin^{-1}(1)=90^\circ.$$

Step 2: Solve the triangle.

$$\gamma\approx 180^\circ-\left(90^\circ+30^\circ\right)=60^\circ$$

$$\frac{\sin 30^\circ}{9}=\frac{\sin 60^\circ}{c} \Rightarrow c=\frac{9\sin 60^\circ}{\sin 30^\circ}=9\sqrt{3}\approx 16$$

24. Step 1: Determine β.

$$\frac{\sin\alpha}{a}=\frac{\sin\beta}{b} \Rightarrow \frac{\sin 45^\circ}{1}=\frac{\sin\beta}{\sqrt{2}} \Rightarrow \sin\beta=\frac{\sqrt{2}\left(\frac{\sqrt{2}}{2}\right)}{1}=1 \Rightarrow \beta=\sin^{-1}(1)=90^\circ.$$

Hence, there is no triangle in this case since there would be two such angles.

25. First, note that there is only one triangle since $a\geq b$ and $0<\sin\beta<1$.

Step 1: Determine β.

$$\frac{\sin\alpha}{a}=\frac{\sin\beta}{b} \Rightarrow \sin\beta=\frac{7\sin 34^\circ}{10} \Rightarrow \beta=\sin^{-1}\left(\frac{7\sin 34^\circ}{10}\right)\approx 23^\circ$$

Step 2: Solve for the triangle.

$$\gamma\approx 180^\circ-\left(34^\circ+23^\circ\right)=123^\circ$$

$$\frac{\sin\alpha}{a}=\frac{\sin\gamma}{c} \Rightarrow \frac{\sin 34^\circ}{10}=\frac{\sin 123^\circ}{c} \Rightarrow c=\frac{10\sin 123^\circ}{\sin 34^\circ}\approx 15$$

26. First, note that there is only one (acute) triangle since $a\geq b$ and $0<\sin\beta<1$.

Step 1: Determine β.

$$\frac{\sin\alpha}{a}=\frac{\sin\beta}{b} \Rightarrow \sin\beta=\frac{5.2\sin 71^\circ}{5.2} \Rightarrow \beta=\sin^{-1}\left(\frac{5.2\sin 71^\circ}{5.2}\right)\approx 71^\circ$$

Step 2: Solve for the triangle.

$$\gamma\approx 180^\circ-\left(71^\circ+71^\circ\right)=38^\circ$$

$$\frac{\sin\alpha}{a}=\frac{\sin\gamma}{c} \Rightarrow \frac{\sin 71^\circ}{5.2}=\frac{\sin 38^\circ}{c} \Rightarrow c=\frac{5.2\sin 38^\circ}{\sin 71^\circ}\approx 3.4$$

27. First, note that we expect two triangles since α is acute and $a < b$.
Step 1: Determine β.

$$\frac{\sin\alpha}{a}=\frac{\sin\beta}{b} \Rightarrow \frac{\sin 21.3^\circ}{6.03}=\frac{\sin\beta}{6.18} \Rightarrow \sin\beta=\frac{6.18\sin 21.3^\circ}{6.03} \Rightarrow \beta=\sin^{-1}\left(\frac{6.18\sin 21.3^\circ}{6.03}\right)\approx 21.9^\circ$$

This is the solution in QI – label it as β_1. The second solution in QII is given by $\beta_2 = 180^\circ - \beta_1 \approx 158.1^\circ$. Both are tenable solutions, so we need to solve for two triangles.

Step 2: Solve for both triangles.

QI triangle: $\beta_1 \approx 21.9^\circ$

$$\gamma_1 \approx 180^\circ - (21.3^\circ + 21.9^\circ) = 136.8^\circ$$

$$\frac{\sin\alpha}{a}=\frac{\sin\gamma_1}{c_1} \Rightarrow \frac{\sin 21.3^\circ}{6.03}=\frac{\sin 136.8^\circ}{c_1} \Rightarrow c_1=\frac{6.03\sin 136.8}{\sin 21.3^\circ}\approx 11.36$$

QII triangle: $\beta_2 \approx 158.1^\circ$

$$\gamma_2 \approx 180^\circ - (21.3^\circ + 158.1^\circ) = 0.6^\circ$$

$$\frac{\sin\alpha}{a}=\frac{\sin\gamma_2}{c_2} \Rightarrow \frac{\sin 21.3^\circ}{6.03}=\frac{\sin 0.6^\circ}{c_2} \Rightarrow c_2=\frac{6.03\sin 0.6^\circ}{\sin 21.3^\circ}\approx 0.17$$

28. Step 1: Determine β.

$$\frac{\sin\alpha}{a}=\frac{\sin\beta}{b} \Rightarrow \frac{\sin 47.3^\circ}{5.32}=\frac{\sin\beta}{7.3} \Rightarrow \sin\beta=\frac{7.3\sin 47.3^\circ}{5.32}>1.$$

Hence, there is no triangle in this case.

29. First, note that we expect only one triangle since α is obtuse and $a > b$.
Step 1: Determine β.

$$\frac{\sin\alpha}{a}=\frac{\sin\beta}{b} \Rightarrow \sin\beta=\frac{4\sqrt{3}\sin 116^\circ}{5\sqrt{2}}\approx 0.880635 \Rightarrow \beta\approx 62^\circ$$

Step 2: Solve for the triangle.

$$\gamma \approx 180^\circ - (116^\circ + 62^\circ) = 2^\circ$$

$$\frac{\sin\alpha}{a}=\frac{\sin\gamma}{c} \Rightarrow \frac{\sin 116^\circ}{5\sqrt{2}}=\frac{\sin 2^\circ}{c} \Rightarrow c=\frac{5\sqrt{2}\sin 38^\circ}{\sin 116^\circ}\approx 0.275$$

30. First, note that we expect only one triangle since α is acute and $a > b$.

Step 1: Determine β.

$$\frac{\sin\alpha}{a} = \frac{\sin\beta}{b} \Rightarrow \sin\beta = \frac{4\sqrt{3}\sin 51^\circ}{4\sqrt{5}} \approx 0.601975 \Rightarrow \beta \approx 37^\circ$$

Step 2: Solve for the triangle.

$$\gamma \approx 180^\circ - \left(51^\circ + 37^\circ\right) = 92^\circ$$

$$\frac{\sin\alpha}{a} = \frac{\sin\gamma}{c} \Rightarrow \frac{\sin 51^\circ}{4\sqrt{5}} = \frac{\sin 92^\circ}{c} \Rightarrow c = \frac{4\sqrt{5}\sin 92^\circ}{\sin 51^\circ} \approx 11.50$$

31. Step 1: Determine β.

$$\frac{\sin\gamma}{c} = \frac{\sin\beta}{b} \Rightarrow \frac{\sin 40^\circ}{330} = \frac{\sin\beta}{500} \Rightarrow \sin\beta = \frac{500\sin 40^\circ}{330} \Rightarrow \beta = \sin^{-1}\left(\frac{500\sin 40^\circ}{330}\right) \approx 77^\circ$$

This is the solution in QI – label it as β_1. The second solution in QII is given by $\beta_2 = 180^\circ - \beta_1 \approx 103^\circ$. Both are tenable solutions, so we need to solve for two triangles.

Step 2: Solve for both triangles.

QI triangle: $\beta_1 \approx 77^\circ$

$$\alpha_1 \approx 180^\circ - \left(77^\circ + 40^\circ\right) = 63^\circ$$

$$\frac{\sin\alpha_1}{a_1} = \frac{\sin\gamma}{c} \Rightarrow \frac{\sin 63^\circ}{a_1} = \frac{\sin 40^\circ}{330} \Rightarrow a_1 = \frac{330\sin 63^\circ}{\sin 40^\circ} \approx 457$$

QII triangle: $\beta_2 \approx 103^\circ$

$$\alpha_2 \approx 180^\circ - \left(103^\circ + 40^\circ\right) = 37^\circ$$

$$\frac{\sin\alpha_2}{a_2} = \frac{\sin\gamma}{c} \Rightarrow \frac{\sin 37^\circ}{a_2} = \frac{\sin 40^\circ}{330} \Rightarrow a_2 = \frac{330\sin 37^\circ}{\sin 40^\circ} \approx 309$$

32. Step 1: Determine α.

$$\frac{\sin\alpha}{a} = \frac{\sin\beta}{b} \Rightarrow \frac{\sin\alpha}{9} = \frac{\sin 137^\circ}{16} \Rightarrow \sin\alpha = \frac{9\sin 137^\circ}{16} \Rightarrow \alpha = \sin^{-1}\left(\frac{9\sin 137^\circ}{16}\right) \approx 23^\circ$$

Note that there is only one triangle in this case since the angle in QII with the same sine as this value of α is $180^\circ - 23^\circ = 157^\circ$. In such case, note that $\beta + \alpha > 180^\circ$, therefore preventing the formation of a triangle (since the three interior angles, two of which are β and α, must sum to 180°).

Step 2: Solve for the triangle.

$$\gamma \approx 180^\circ - \left(23^\circ + 137^\circ\right) = 20^\circ$$

$$\frac{\sin\gamma}{c} = \frac{\sin\beta}{b} \Rightarrow \frac{\sin 20^\circ}{c} = \frac{\sin 137^\circ}{16} \Rightarrow c = \frac{16\sin 20^\circ}{\sin 137^\circ} \approx 8$$

33. Step 1: Determine α.

$$\frac{\sin\alpha}{a}=\frac{\sin\beta}{b} \Rightarrow \frac{\sin\alpha}{\sqrt{2}}=\frac{\sin 106^\circ}{\sqrt{7}} \Rightarrow \sin\alpha=\frac{\sqrt{2}\sin 106^\circ}{\sqrt{7}} \Rightarrow \alpha=\sin^{-1}\left(\frac{\sqrt{2}\sin 106^\circ}{\sqrt{7}}\right)\approx 31^\circ$$

Note that there is only one triangle in this case since the angle in QII with the same sine as this value of α is $180^\circ-31^\circ=149^\circ$. In such case, note that $\beta+\alpha>180^\circ$, therefore preventing the formation of a triangle (since the three interior angles, two of which are β and α, must sum to 180°).

Step 2: Solve for the triangle.

$$\gamma\approx 180^\circ-\left(31^\circ+106^\circ\right)=43^\circ$$

$$\frac{\sin\gamma}{c}=\frac{\sin\beta}{b} \Rightarrow \frac{\sin 43^\circ}{c}=\frac{\sin 106^\circ}{\sqrt{7}} \Rightarrow c=\frac{\sqrt{7}\sin 43^\circ}{\sin 106^\circ}\approx 2$$

34. Step 1: Determine β.

$$\frac{\sin\gamma}{c}=\frac{\sin\beta}{b} \Rightarrow \frac{\sin 11.6^\circ}{27.2}=\frac{\sin\beta}{15.3} \Rightarrow \sin\beta=\frac{15.3\sin 11.6^\circ}{27.2} \Rightarrow \beta=\sin^{-1}\left(\frac{15.3\sin 11.6^\circ}{27.2}\right)\approx 6.5^\circ$$

Note that there is only one triangle in this case since the angle in QII with the same sine as this value of β is $180^\circ-6.5^\circ=173.5^\circ$. In such case, note that $\beta+\gamma>180^\circ$, therefore preventing the formation of a triangle (since the three interior angles, two of which are β and γ, must sum to 180°).

Step 2: Solve for the triangle.

$$\gamma\approx 180^\circ-\left(11.6^\circ+6.5^\circ\right)=161.9^\circ$$

$$\frac{\sin\gamma}{c}=\frac{\sin\alpha}{a} \Rightarrow \frac{\sin 11.6^\circ}{27.2}=\frac{\sin 161.9^\circ}{a} \Rightarrow a=\frac{27.2\sin 161.9^\circ}{\sin 11.6^\circ}\approx 42.0$$

35. Consider the following diagram:

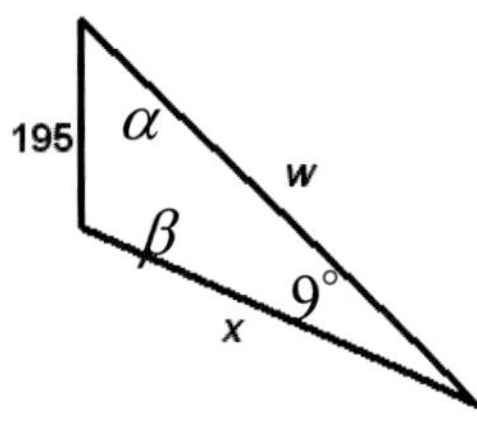

First, using the properties of corresponding and supplementary angles, we see that $\alpha=80^\circ$. As such, the information provided yields an AAS triangle. So, we use Law of Sines to solve the triangle:

$$\beta=180^\circ-\left(9^\circ+80^\circ\right)=91^\circ$$

$$\frac{\sin 9^\circ}{195}=\frac{\sin 91^\circ}{w} \Rightarrow$$

$$w=\frac{195\left(\sin 91^\circ\right)}{\sin 9^\circ}\approx \boxed{1{,}246 \text{ ft.}}$$

36. Using the same triangle and information as in Exercise 27, we compute x as follows:

$$\frac{\sin 80^\circ}{x}=\frac{\sin 9^\circ}{195} \Rightarrow x=\frac{195\left(\sin 80^\circ\right)}{\sin 9^\circ}\approx 1,227.59 \text{ ft.}$$

Now, we seek the length of the base of the smaller dotted right-triangle, the height of which coincides with the height of the launch pad (see diagram in the text) (and the angle opposite of which is 1°). Call the base of this triangle y. Then, from this information, we see that $\cos 1^\circ=\frac{y}{1,227.59}$, so that

$y=1,227.59\left(\cos 1^\circ\right)\approx 1,227.40\approx \boxed{1,227 \text{ ft.}}$

37. Consider the following diagram:

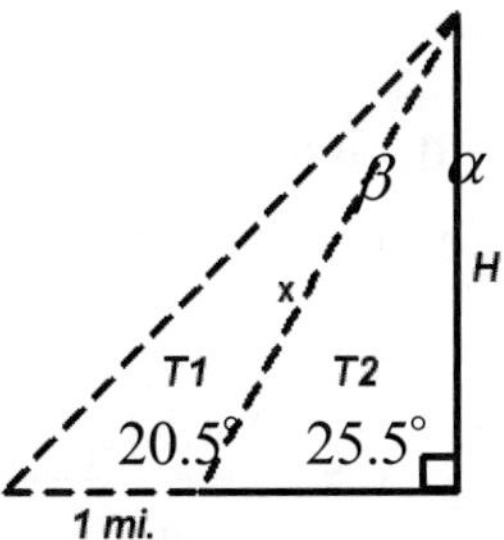

We need to determine H. To this end, first observe that $\alpha=154.5^\circ$, and so $\beta=180^\circ-\left(20.5^\circ+154.5^\circ\right)=5^\circ$. So, the information provided yields an AAS situation for triangle T_1. So, use Law of Sines to find x:

$$\frac{\sin 20.5^\circ}{x}=\frac{\sin 5^\circ}{1 \text{ mi.}} \Rightarrow x=\frac{\sin 20.5^\circ}{\sin 5^\circ}\approx 4.01818 \text{ mi.}$$

Now, to find H, we turn our focus to triangle T_2. Again, from the information we have, the situation is AAS, so we use Law of Sines to find H:

$$\frac{\sin 90^\circ}{x}=\frac{\sin 25.5^\circ}{H} \Rightarrow \frac{1}{4.01818}=\frac{\sin 25.5^\circ}{H}$$

so that $H\approx 4.01818\left(\sin 25.5^\circ\right)\approx \boxed{1.7 \text{ mi.}}$

38. Consider the following diagram:

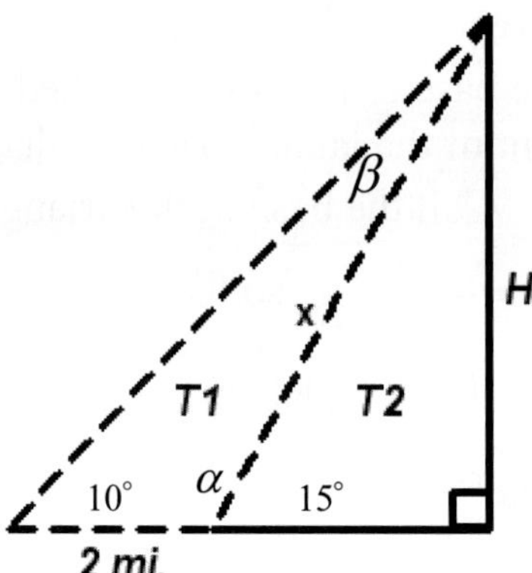

We need to determine H. To this end, first observe that $\alpha = 165^\circ$, and so $\beta = 180^\circ - \left(10^\circ + 165^\circ\right) = 5^\circ$. So, the information provided yields an AAS situation for triangle T_1. So, we use Law of Sines to find x:

$$\frac{\sin 10^\circ}{x} = \frac{\sin 5^\circ}{2 \text{ mi.}} \Rightarrow x = \frac{(2 \text{ mi.})\sin 10^\circ}{\sin 5^\circ} \approx 3.984778 \text{ mi.}$$

Now, to find H, we turn our focus to triangle T_2. Again, from the information we have, the situation is AAS, so we use Law of Sines to find H:

$$\frac{\sin 90^\circ}{x} = \frac{\sin 15^\circ}{H} \Rightarrow \frac{1}{3.984778} = \frac{\sin 15^\circ}{H}$$

so that

$$H \approx 3.984778\left(\sin 15^\circ\right) \approx \boxed{1.0 \text{ mi.}}$$

39. Consider the following diagram:

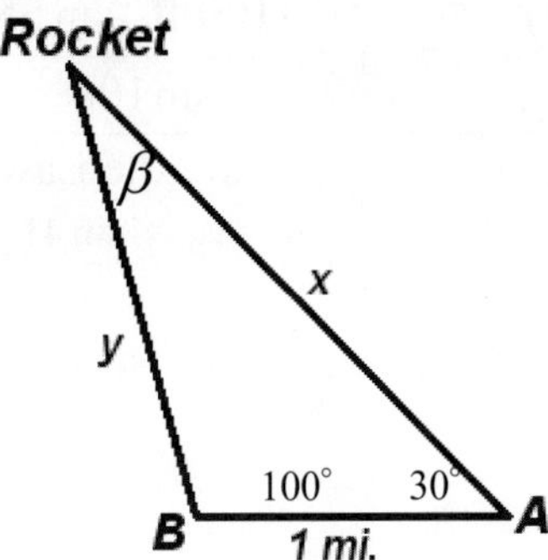

We seek x. To this end, first note that $\beta = 50^\circ$. So, the information provided yields an AAS situation for triangle T_1. So, we use Law of Sines to find x:

$$\frac{\sin 100^\circ}{x} = \frac{\sin 50^\circ}{1 \text{ mi.}} \Rightarrow x = \frac{(1 \text{ mi.})\sin 100^\circ}{\sin 50^\circ} \approx \boxed{1.3 \text{mi.}}$$

40. Refer to the diagram in Exercise 41. We seek y this time. Again, using the Law of Sines yields: $\dfrac{\sin 30^\circ}{y} = \dfrac{\sin 50^\circ}{1 \text{ mi.}} \Rightarrow y = \dfrac{(1 \text{ mi.})\sin 30^\circ}{\sin 50^\circ} \approx \boxed{0.7 \text{mi.}}$

41. Consider the following diagram:

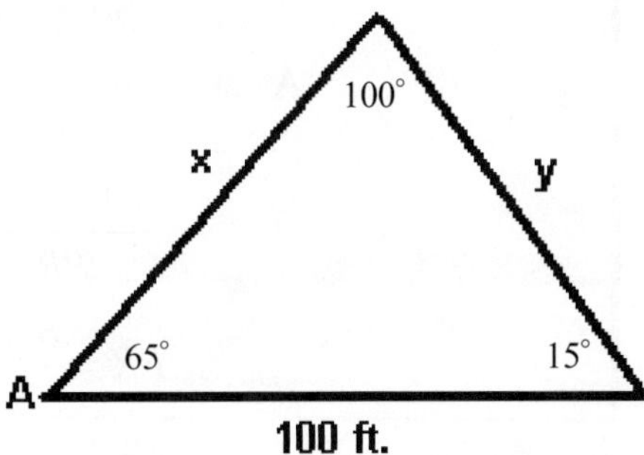

We seek x. The information provided yields an AAS situation for triangle T_1. So, we use Law of Sines to find x:

$$\frac{\sin 15^\circ}{x} = \frac{\sin 100^\circ}{100 \text{ ft.}} \Rightarrow x = \frac{(100 \text{ ft.})\sin 15^\circ}{\sin 100^\circ} \approx \boxed{26 \text{ ft.}}$$

42. Refer to the diagram in Exercise 41. We seek y this time. Again, using the Law of Sines yields:

$$\frac{\sin 65^\circ}{y} = \frac{\sin 100^\circ}{100 \text{ ft.}} \Rightarrow y = \frac{(100 \text{ ft.})\sin 65^\circ}{\sin 100^\circ} \approx \boxed{92 \text{ ft.}}$$

43. Consider the following diagram:

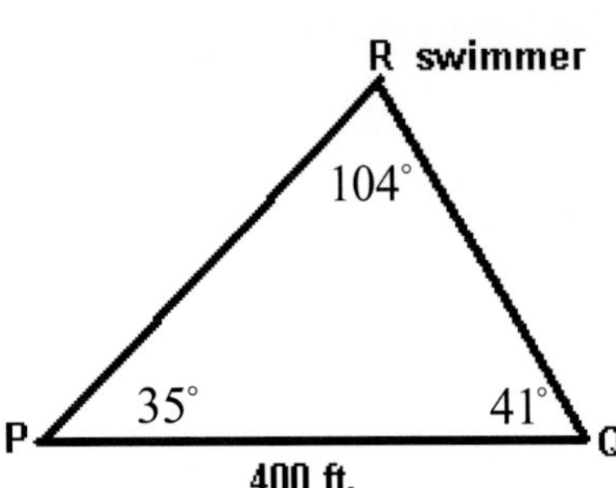

Using the Law of Sines yields

$$\frac{\sin 41^\circ}{PR} = \frac{\sin 104^\circ}{400}$$

$$PR = \frac{400 \sin 41^\circ}{\sin 104^\circ} \approx \boxed{270 \text{ ft.}}$$

44. Consider the following diagram:

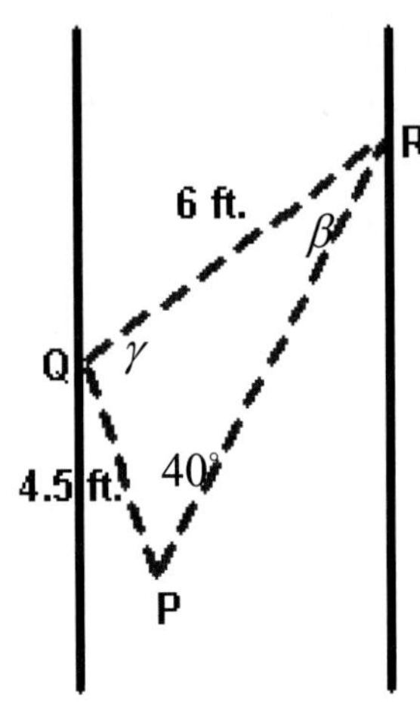

Using the Law of Sines yields

$$\frac{\sin \beta}{4.5} = \frac{\sin 40^\circ}{6} \Rightarrow \sin \beta = \frac{4.5 \sin 40^\circ}{6}$$

$$\Rightarrow \beta = \sin^{-1}\left(\frac{4.5 \sin 40^\circ}{6}\right) \approx 28.822^\circ$$

So, $\gamma = 180^\circ - (40^\circ + 28.822^\circ) = 111.178^\circ$.

As such, using the Law of Sines again yields

$$\frac{\sin 111.178^\circ}{PR} = \frac{\sin 40^\circ}{6} \Rightarrow$$

$$PR = \frac{6 \sin 111.178^\circ}{\sin 40^\circ} \approx \boxed{8.7 \text{ ft.}}$$

45. Consider the following diagram: 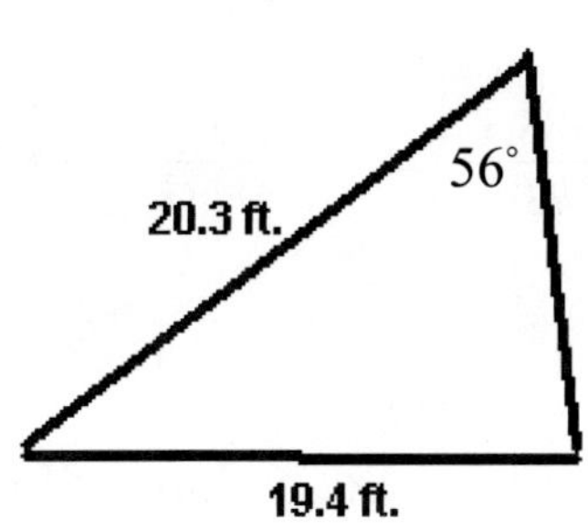	Using the Law of Sines yields $\frac{\sin\beta}{20.3}=\frac{\sin 56^\circ}{19.4} \Rightarrow \sin\beta=\frac{20.3\sin 56^\circ}{19.4}$ $\Rightarrow \beta=\sin^{-1}\left(\frac{20.3\sin 56^\circ}{19.4}\right)\approx 60.169^\circ$ So, $\alpha=180^\circ-\left(56^\circ+60.169^\circ\right)=\boxed{63.83^\circ}$
46. Consider the following diagram: 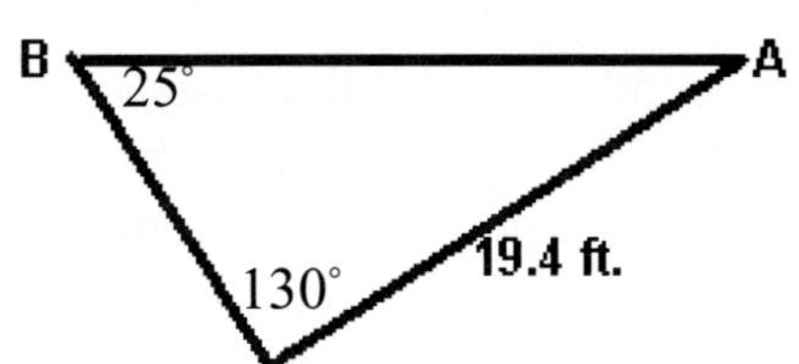	Using the Law of Sines yields $\frac{\sin 130^\circ}{AB}=\frac{\sin 25^\circ}{19.4} \Rightarrow$ $AB=\frac{19.4\sin 130^\circ}{\sin 25^\circ}\approx \boxed{35\text{ ft.}}$
47. Consider the following diagram: 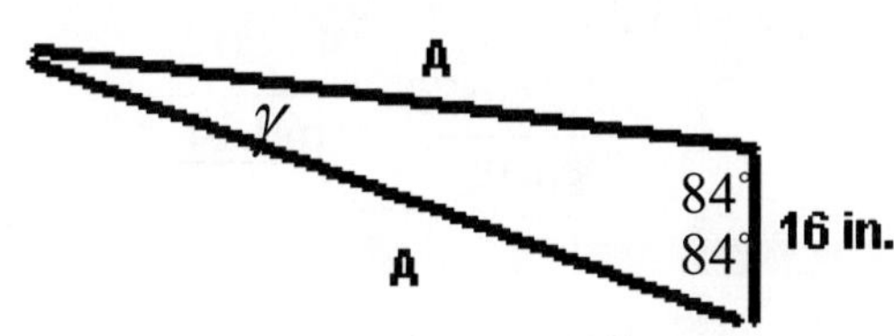	They travel the same distance since the triangle is isosceles. To find this distance, first note that $\gamma=180^\circ-2\left(84^\circ\right)=12^\circ$. So, using the Law of Sines yields $\frac{\sin 12^\circ}{16}=\frac{\sin 84^\circ}{A} \Rightarrow$ $A=\frac{16\sin 84^\circ}{\sin 12^\circ}\approx \boxed{76.5\text{ ft.}}$

48. Consider the following diagram:

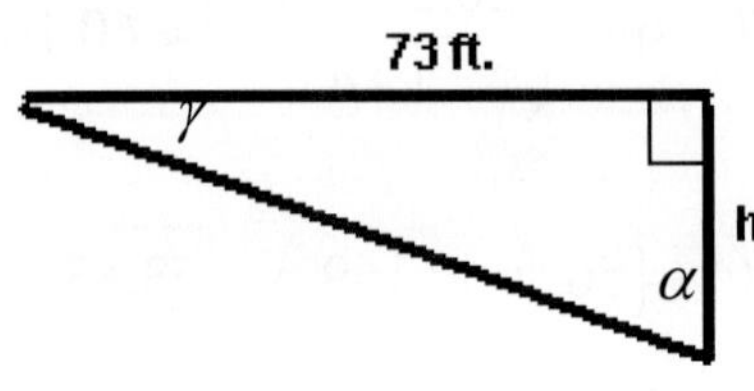

First, note that the triangle is right since the arrow travels horizontally. In order to get the range, we need to determine h in two triangles, namely when $\alpha = 88.3^\circ$ and when $\alpha = 88.7^\circ$. We use Law of Sines in both.

Case I: $\alpha = 88.3^\circ$

Note that $\gamma = 1.7^\circ$. So, by Law of Sines, we have

$$\frac{\sin 1.7^\circ}{h} = \frac{\sin 88.3^\circ}{73} \Rightarrow h = \frac{73 \sin 1.7^\circ}{\sin 88.3^\circ} \approx 2.16 \text{ ft.}$$

Case II: $\alpha = 88.7^\circ$

Note that $\gamma = 1.3^\circ$. So, by Law of Sines, we have

$$\frac{\sin 1.3^\circ}{h} = \frac{\sin 88.7^\circ}{73} \Rightarrow h = \frac{73 \sin 1.3^\circ}{\sin 88.7^\circ} \approx 1.66 \text{ ft.}$$

So, the second arrow hit the target between 19.92 and 25.92 inches below the bullseye, resulting in a score of either 5 points or 10 points. Since the bullseye counts for 50 points, the overall score is either 55 points or 60 points.

49. Consider the following diagram:

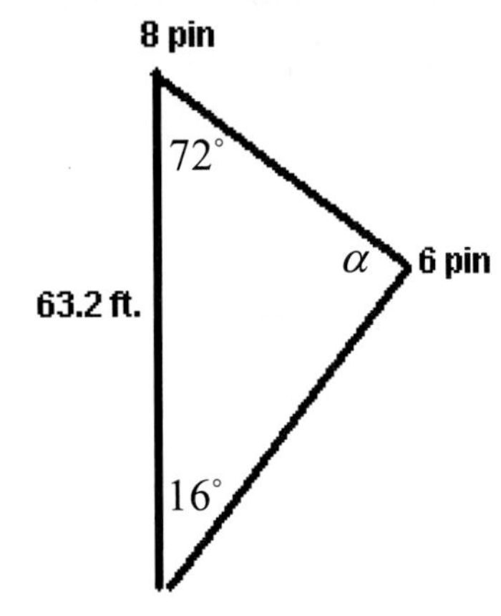

First, note that

$$\alpha = 180^\circ - \left(72^\circ + 16^\circ\right) = 92^\circ.$$

Using the Law of Sines yields

$$\frac{\sin 92^\circ}{63.2} = \frac{\sin 72^\circ}{A} \Rightarrow$$

$$A = \frac{63.2 \sin 72^\circ}{\sin 92^\circ} \approx \boxed{60.14 \text{ ft.}}$$

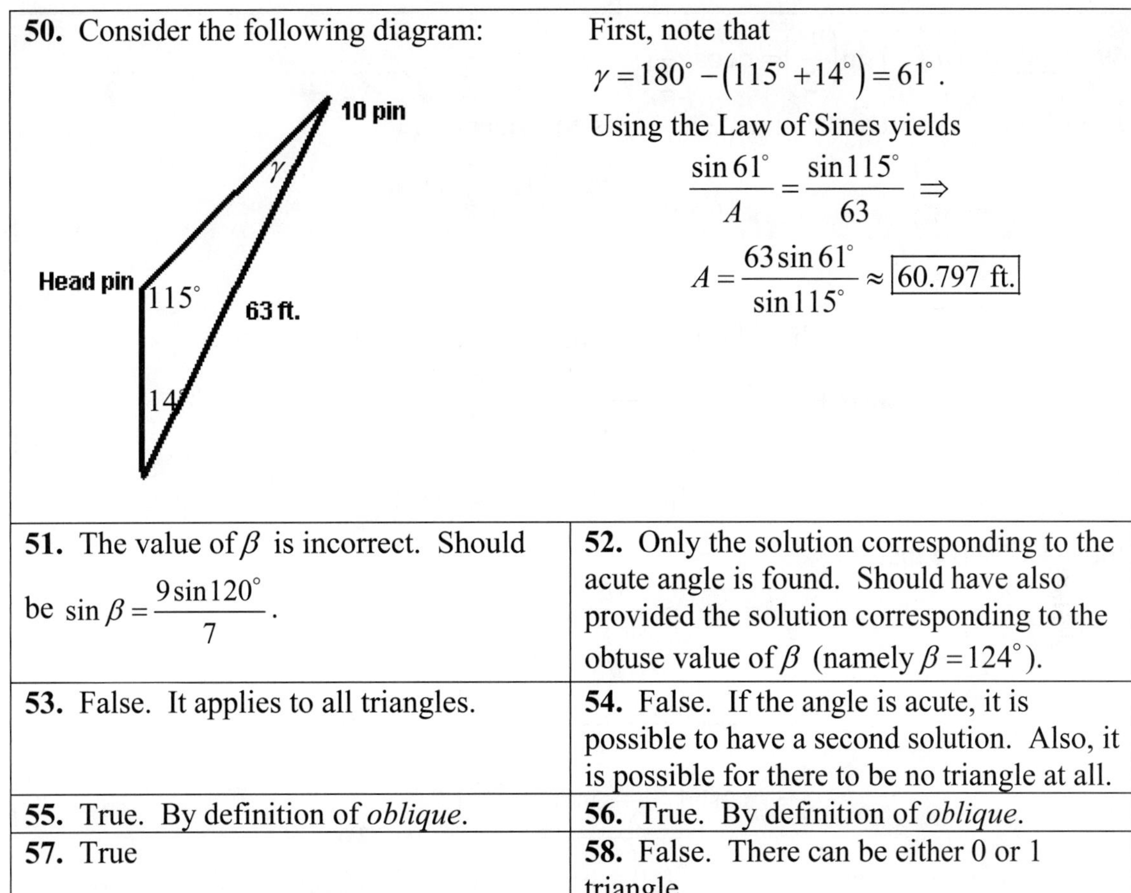

50. Consider the following diagram: Head pin, 10 pin, 115°, 14°, γ, 63 ft.	First, note that $\gamma = 180^\circ - (115^\circ + 14^\circ) = 61^\circ$. Using the Law of Sines yields $\frac{\sin 61^\circ}{A} = \frac{\sin 115^\circ}{63} \Rightarrow$ $A = \frac{63\sin 61^\circ}{\sin 115^\circ} \approx \boxed{60.797 \text{ ft.}}$
51. The value of β is incorrect. Should be $\sin\beta = \frac{9\sin 120^\circ}{7}$.	**52.** Only the solution corresponding to the acute angle is found. Should have also provided the solution corresponding to the obtuse value of β (namely $\beta = 124^\circ$).
53. False. It applies to all triangles.	**54.** False. If the angle is acute, it is possible to have a second solution. Also, it is possible for there to be no triangle at all.
55. True. By definition of *oblique*.	**56.** True. By definition of *oblique*.
57. True	**58.** False. There can be either 0 or 1 triangle.

59. **<u>Claim</u>:** $(a+b)\sin\left(\frac{\gamma}{2}\right)=c\cos\left[\frac{1}{2}(\alpha-\beta)\right]$

***Proof*:** First, note that the given equality is equivalent to:

$$\frac{(a+b)}{c}=\frac{\cos\left[\frac{1}{2}(\alpha-\beta)\right]}{\sin\left(\frac{\gamma}{2}\right)}$$

We shall start with the right-side, and we will need the following identities:

$$\sin\alpha+\sin\beta=2\sin\left(\frac{\alpha+\beta}{2}\right)\cos\left(\frac{\alpha-\beta}{2}\right) \qquad \textbf{(1)}$$

$$\sin\gamma=\sin\left(2\cdot\frac{\gamma}{2}\right)=2\sin\left(\frac{\gamma}{2}\right)\cos\left(\frac{\gamma}{2}\right) \qquad \textbf{(2)}$$

Indeed, observe that

$$\frac{a+b}{c}=\frac{a}{c}+\frac{b}{c}=\frac{\sin\alpha}{\sin\gamma}+\frac{\sin\beta}{\sin\gamma}=\frac{\sin\alpha+\sin\beta}{\sin\gamma}$$

$$=\frac{2\sin\left(\frac{\alpha+\beta}{2}\right)\cos\left(\frac{\alpha-\beta}{2}\right)}{\sin\gamma} \quad \text{(by } \textbf{(1)}\text{)}$$

$$=\frac{2\sin\left(\frac{\alpha+\beta}{2}\right)\cos\left(\frac{\alpha-\beta}{2}\right)}{2\sin\left(\frac{\gamma}{2}\right)\cos\left(\frac{\gamma}{2}\right)} \quad \text{(by } \textbf{(2)}\text{)}$$

Now, since $\alpha+\beta+\gamma=\pi$, it follows that $\frac{\alpha+\beta}{2}=\frac{\pi}{2}-\frac{\gamma}{2}$. As such,

$$\sin\left(\frac{\alpha+\beta}{2}\right)=\sin\left(\frac{\pi}{2}-\frac{\gamma}{2}\right)=\cos\left(\frac{\gamma}{2}\right) \quad \textbf{(3)}.$$

Using **(3)** in the last line of the above string of equalities then yields

$$\frac{2\sin\left(\frac{\alpha+\beta}{2}\right)\cos\left(\frac{\alpha-\beta}{2}\right)}{2\sin\left(\frac{\gamma}{2}\right)\cos\left(\frac{\gamma}{2}\right)}=\frac{\cancel{2}\cancel{\cos\left(\frac{\gamma}{2}\right)}\cos\left(\frac{\alpha-\beta}{2}\right)}{\cancel{2}\sin\left(\frac{\gamma}{2}\right)\cancel{\cos\left(\frac{\gamma}{2}\right)}}=\frac{\cos\left(\frac{\alpha-\beta}{2}\right)}{\sin\left(\frac{\gamma}{2}\right)},$$

as desired.

60. **Claim:** $\dfrac{\tan\left(\dfrac{\alpha-\beta}{2}\right)}{\tan\left(\dfrac{\alpha+\beta}{2}\right)}=\dfrac{a-b}{a+b}$.

Proof:

$$\frac{\tan\left(\frac{\alpha-\beta}{2}\right)}{\tan\left(\frac{\alpha+\beta}{2}\right)}=\frac{\sin\left(\frac{\alpha-\beta}{2}\right)}{\cos\left(\frac{\alpha-\beta}{2}\right)}\cdot\frac{\cos\left(\frac{\alpha+\beta}{2}\right)}{\sin\left(\frac{\alpha+\beta}{2}\right)}$$

$$=\frac{\sin\left(\frac{\alpha}{2}\right)\cos\left(\frac{\beta}{2}\right)-\sin\left(\frac{\beta}{2}\right)\cos\left(\frac{\alpha}{2}\right)}{\cos\left(\frac{\alpha}{2}\right)\cos\left(\frac{\beta}{2}\right)+\sin\left(\frac{\alpha}{2}\right)\sin\left(\frac{\beta}{2}\right)}\cdot\frac{\cos\left(\frac{\alpha}{2}\right)\cos\left(\frac{\beta}{2}\right)-\sin\left(\frac{\alpha}{2}\right)\sin\left(\frac{\beta}{2}\right)}{\sin\left(\frac{\alpha}{2}\right)\cos\left(\frac{\beta}{2}\right)+\sin\left(\frac{\beta}{2}\right)\cos\left(\frac{\alpha}{2}\right)}$$

Now, multiply the top two binomials together, and the bottom two binomials together, and regroup in the following manner:

$$=\frac{\sin\left(\frac{\alpha}{2}\right)\cos\left(\frac{\alpha}{2}\right)\overbrace{\left[\cos^2\left(\frac{\beta}{2}\right)+\sin^2\left(\frac{\beta}{2}\right)\right]}^{=1}-\sin\left(\frac{\beta}{2}\right)\cos\left(\frac{\beta}{2}\right)\overbrace{\left[\cos^2\left(\frac{\alpha}{2}\right)+\sin^2\left(\frac{\alpha}{2}\right)\right]}^{=1}}{\sin\left(\frac{\alpha}{2}\right)\cos\left(\frac{\alpha}{2}\right)\underbrace{\left[\cos^2\left(\frac{\beta}{2}\right)+\sin^2\left(\frac{\beta}{2}\right)\right]}_{=1}+\sin\left(\frac{\beta}{2}\right)\cos\left(\frac{\beta}{2}\right)\underbrace{\left[\cos^2\left(\frac{\alpha}{2}\right)+\sin^2\left(\frac{\alpha}{2}\right)\right]}_{=1}}$$

$$=\frac{\sin\left(\frac{\alpha}{2}\right)\cos\left(\frac{\alpha}{2}\right)-\sin\left(\frac{\beta}{2}\right)\cos\left(\frac{\beta}{2}\right)}{\sin\left(\frac{\alpha}{2}\right)\cos\left(\frac{\alpha}{2}\right)+\sin\left(\frac{\beta}{2}\right)\cos\left(\frac{\beta}{2}\right)}=\frac{\cancel{\frac{1}{2}}\left[2\sin\left(\frac{\alpha}{2}\right)\cos\left(\frac{\alpha}{2}\right)-2\sin\left(\frac{\beta}{2}\right)\cos\left(\frac{\beta}{2}\right)\right]}{\cancel{\frac{1}{2}}\left[2\sin\left(\frac{\alpha}{2}\right)\cos\left(\frac{\alpha}{2}\right)+2\sin\left(\frac{\beta}{2}\right)\cos\left(\frac{\beta}{2}\right)\right]}$$

Now, apply double angle formula, followed by the Law of Sines on the $\sin\beta$ term (i.e., use $\dfrac{\sin\alpha}{a}=\dfrac{\sin\beta}{b}\Rightarrow\sin\beta=\dfrac{b\sin\alpha}{a}$) to obtain:

$$=\frac{\sin\alpha-\sin\beta}{\sin\alpha+\sin\beta}=\frac{\sin\alpha-\frac{b\sin\alpha}{a}}{\sin\alpha+\frac{b\sin\alpha}{a}}=\frac{\cancel{\sin\alpha}\left[1-\frac{b}{a}\right]}{\cancel{\sin\alpha}\left[1+\frac{b}{a}\right]}=\frac{\frac{a-b}{\cancel{a}}}{\frac{a+b}{\cancel{a}}}=\frac{a-b}{a+b}.$$

61. Observe that since all sides have the same lengths, the Law of Sines yields

$$\frac{\sin\alpha}{a}=\frac{\sin\beta}{a}=\frac{\sin\gamma}{a},$$

from which we immediately conclude that $\alpha=\beta=\gamma$ since the angles must sum to 180° and none of them can be obtuse.

62. Consider the following diagram:

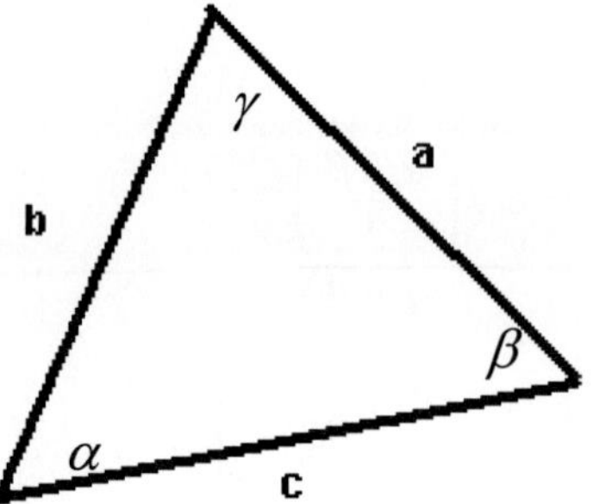

We are given that $a = \frac{1}{\sqrt{2}} b$, $c = 2$, α is an acute angle, and $\beta = 2\alpha$.

We begin by finding α. From the Law of Sines, we obtain

$$\frac{\sin\alpha}{\frac{1}{\sqrt{2}}b} = \frac{\sin\beta}{b} = \frac{\sin 2\alpha}{b}.$$

Hence, we have

$$\sin\alpha = \tfrac{1}{\sqrt{2}}\sin 2\alpha = \tfrac{2}{\sqrt{2}}\sin\alpha\cos\alpha$$
$$1 = \tfrac{2}{\sqrt{2}}\cos\alpha$$
$$\tfrac{\sqrt{2}}{2} = \cos\alpha$$
$$\alpha = \tfrac{\pi}{4}$$

Thus, we know that $\beta = 2\left(\frac{\pi}{4}\right) = \frac{\pi}{2}$ and so, $\gamma = \pi - \left(\frac{\pi}{4} + \frac{\pi}{2}\right) = \frac{\pi}{4}$.

Now, we find b. Again, using the Law of Sines yields

$$\frac{\sin\left(\frac{\pi}{4}\right)}{c} = \frac{\sin\left(\frac{\pi}{2}\right)}{b} \Rightarrow \frac{\frac{\sqrt{2}}{2}}{2} = \frac{1}{b} \Rightarrow b = 2\sqrt{2}.$$

Thus, $a = \frac{1}{\sqrt{2}}\left(2\sqrt{2}\right) = 2$. This solves the triangle.

63. The following steps constitute the program in this case:

```
Program: AYZ
:Input "SIDE A =", A
:Input "ANGLE Y =", Y
:Input "ANGLE Z =", Z
:180-Y-Z →X
:Asin(Y)/sin(X)→B
:Asin(Z)/sin(X)→C
:Disp "ANGLE X = ", X
:Disp "SIDE B = ", B
:Disp "SIDE C =", C
```

Now, in order to use this program to solve the given triangle, EXECUTE it and enter the following data at the prompts:
SIDE A = 10
ANGLE Y = 45
ANGLE Z = 65

Now, the program will display the following information that solves the triangle:
ANGLE X = 70
SIDE B = 7.524873193
SIDE C = 9.64472602

64. The following steps constitute the program in this case:

```
Program: BXY
:Input "SIDE B =", B
:Input "ANGLE X =", X
:Input "ANGLE Y =", Y
:180-X-Y →Z
:Bsin(X)/sin(Y)→A
:Bsin(Z)/sin(Y)→C
:Disp "ANGLE Z = ", Z
:Disp "SIDE A = ", A
:Disp "SIDE C =", C
```

Now, in order to use this program to solve the given triangle, EXECUTE it and enter the following data at the prompts:
SIDE B = 42.8
ANGLE X = 31.6
ANGLE Y = 82.2

Now, the program will display the following information that solves the triangle:
ANGLE Z = 66.2
SIDE A = 22.63602893
SIDE C = 39.5259744

65. The following steps constitute the program in this case:	66. The following steps constitute the program in this case:
Program: ABX :Input "SIDE A =", A :Input "SIDE B =", B :Input "ANGLE X =", X :$\sin^{-1}$(Bsin(X)/A)$\to$Y :180-Y-X $\to$Z :Asin(Z)/sin(X)$\to$C :Disp "ANGLE Y = ", Y :Disp "ANGLE Z = ", Z :Disp "SIDE C =", C	Program: BCZ :Input "SIDE B =", B :Input "SIDE C =", C :Input "ANGLE Z =", Z :$\sin^{-1}$(Bsin(Z)/C)$\to$Y :180-Y-Z $\to$X :Csin(X)/sin(Z)$\to$A :Disp "ANGLE Z = ", Z :Disp "ANGLE X = ", X :Disp "SIDE A =", A
Now, in order to use this program to solve the given triangle, EXECUTE it and enter the following data at the prompts: SIDE A = 22 SIDE B = 17 ANGLE X = 105 Now, the program will display the following information that solves the triangle: ANGLE Y = 48.27925113 ANGLE Z = 26.72074887 SIDE C = 10.24109154	Now, in order to use this program to solve the given triangle, EXECUTE it and enter the following data at the prompts: SIDE B = 16.5 SIDE C = 9.8 ANGLE Z = 79.2 Now, the program will display an error message since there is no triangle in this case.

67. The following steps constitute the program in this case:

Program: ACX
:Input "SIDE A =", A
:Input "SIDE C =", C
:Input "ANGLE X =", X
:sin^{-1}(Csin(X)/A)→Y
:180-Y-X →Z
:Asin(Z)/sin(X)→B
:Disp "ANGLE Y = ", Y
:Disp "ANGLE Z = ", Z
:Disp "SIDE B =", B

Now, in order to use this program to solve the given triangle, EXECUTE it and enter the following data at the prompts:
SIDE A = 25.7
SIDE C = 12.2
ANGLE X = 65
Now, the program will display the following information that solves the triangle:
ANGLE Y = 25.48227
ANGLE Z = 89.51773
SIDE C = 28.35581

68. The following steps constitute the program in this case:

Program: ABY
:Input "SIDE A =", A
:Input "SIDE B =", B
:Input "ANGLE Y =", Y
:sin^{-1}(Bsin(Y)/A)→X
:180-Y-X →Z
:Asin(Z)/sin(Y)→C
:Disp "ANGLE X = ", X
:Disp "ANGLE Z = ", Z
:Disp "SIDE C =", C

Now, in order to use this program to solve the given triangle, EXECUTE it and enter the following data at the prompts:
SIDE A = 54.6
SIDE B = 12.9
ANGLE Y = 23
Now, the program will display the following information that solves the triangle:
ANGLE X = 5.29684
ANGLE Z = 148.70316
SIDE C = 72.58999

69. First, note that the angle opposite the desired side a has measure 75°. By Law of Sines, we have

$$\frac{\sin 45^\circ}{20\,in.} = \frac{\sin 75^\circ}{a}$$

$$a = \frac{(20\,in.)\sin 75^\circ}{\sin 45^\circ} \approx 27\,in.$$

70. First, note that the angle opposite the side of length 200ft is $\frac{2\pi}{9}$. By Law of Sines, we have

$$\frac{\sin\frac{5\pi}{9}}{c} = \frac{\sin\frac{2\pi}{9}}{200\,ft.}$$

$$c = \frac{(200\,ft.)\sin\frac{5\pi}{9}}{\sin\frac{2\pi}{9}} \approx 306\,ft.$$

71. First, since the triangle is isosceles, the remaining two angles each has measure $\frac{3\pi}{14}$. So, by Law of Sines, we have

$$\frac{\sin\frac{4\pi}{7}}{a} = \frac{\sin\frac{3\pi}{14}}{14m}$$

$$c = \frac{(14m)\sin\frac{4\pi}{7}}{\sin\frac{3\pi}{14}} \approx 22m$$

72. First, from the Law of Sines, we have

$$\frac{\sin\beta}{30cm} = \frac{\sin 35^\circ}{45cm}$$

$$\sin\beta = \frac{(30cm)\sin 35^\circ}{45cm}$$

$$\beta \approx 22.4814495^\circ$$

Thus, $\alpha = 122.5185505^\circ$. So, by another application of Law of Sines, we have

$$\frac{\sin(22.4814495^\circ)}{a} = \frac{\sin 35^\circ}{45cm}$$

$$a = \frac{(45cm)\sin(22.4814495^\circ)}{\sin 35^\circ} \approx 66cm$$

Section 4.5 Solutions ---

1. This is SAS, so begin by using Law of Cosines:

Step 1: Find b.

$$b^2 = a^2 + c^2 - 2ac\cos\beta = (4)^2 + (3)^2 - 2(4)(3)\cos(100^\circ) = 25 - 24\cos(100^\circ)$$

$$b \approx 5.4 \approx 5$$

Step 2: Find γ.

$$\frac{\sin\beta}{b} = \frac{\sin\gamma}{c} \Rightarrow \frac{\sin 100^\circ}{5} = \frac{\sin\gamma}{3} \Rightarrow \sin\gamma = \frac{3\sin 100^\circ}{5} \Rightarrow \gamma = \sin^{-1}\left(\frac{3\sin 100^\circ}{5}\right) \approx 33^\circ$$

Step 3: Find α: $\alpha \approx 180^\circ - (33^\circ + 100^\circ) = 47^\circ$

2. This is SAS, so begin by using Law of Cosines:

Step 1: Find c.

$$c^2 = a^2 + b^2 - 2ab\cos\gamma = (6)^2 + (10)^2 - 2(6)(10)\cos(80^\circ) = 136 - 120\cos(80^\circ)$$

$$c \approx 10.731 \approx 11$$

Step 2: Find α.

$$\frac{\sin\alpha}{a} = \frac{\sin\gamma}{c} \Rightarrow \frac{\sin\alpha}{6} = \frac{\sin 80^\circ}{10.731} \Rightarrow \sin\alpha = \frac{6\sin 80^\circ}{10.731} \Rightarrow \alpha = \sin^{-1}\left(\frac{6\sin 80^\circ}{10.731}\right) \approx 33^\circ$$

Step 3: Find β.

$$\beta \approx 180^\circ - (33^\circ + 80^\circ) = 67^\circ$$

3. This is SAS, so begin by using Law of Cosines:

Step 1: Find *a*.

$$a^2 = b^2 + c^2 - 2bc\cos\alpha = (7)^2 + (2)^2 - 2(7)(2)\cos(16^\circ) = 53 - 28\cos(16^\circ)$$

$$a \approx 5.107 \approx 5$$

Step 2: Find γ.

$$\frac{\sin\alpha}{a} = \frac{\sin\gamma}{c} \Rightarrow \frac{\sin 16^\circ}{5.107} = \frac{\sin\gamma}{2} \Rightarrow \sin\gamma = \frac{2\sin 16^\circ}{5.107} \Rightarrow \gamma = \sin^{-1}\left(\frac{2\sin 16^\circ}{5.107}\right) \approx 6^\circ$$

Step 3: Find β.

$$\beta \approx 180^\circ - (6^\circ + 16^\circ) = 158^\circ$$

4. This is SAS, so begin by using Law of Cosines:

Step 1: Find *c*.

$$c^2 = a^2 + b^2 - 2ab\cos\gamma = (6)^2 + (5)^2 - 2(6)(5)\cos(170^\circ) = 61 - 60\cos(170^\circ)$$

$$c \approx 10.958 \approx 11$$

Step 2: Find α.

$$\frac{\sin\alpha}{a} = \frac{\sin\gamma}{c} \Rightarrow \frac{\sin\alpha}{6} = \frac{\sin 170^\circ}{10.958} \Rightarrow \sin\alpha = \frac{6\sin 170^\circ}{10.958} \Rightarrow \alpha = \sin^{-1}\left(\frac{6\sin 170^\circ}{10.958}\right) \approx 5^\circ$$

Step 3: Find β.

$$\beta \approx 180^\circ - (5^\circ + 170^\circ) = 5^\circ$$

5. This is SAS, so begin by using Law of Cosines:

Step 1: Find *a*.

$$a^2 = b^2 + c^2 - 2bc\cos\alpha = (5)^2 + (5)^2 - 2(5)(5)\cos(20^\circ) = 50 - 50\cos(20^\circ)$$

$$a \approx 1.736 \approx 2$$

Step 2: Find β.

$$\frac{\sin\alpha}{a} = \frac{\sin\beta}{b} \Rightarrow \frac{\sin 20^\circ}{1.736} = \frac{\sin\beta}{5} \Rightarrow \sin\beta = \frac{5\sin 20^\circ}{1.736} \Rightarrow \beta = \sin^{-1}\left(\frac{5\sin 20^\circ}{1.736}\right) \approx 80^\circ$$

Step 3: Find γ.

$$\gamma \approx 180^\circ - (80^\circ + 20^\circ) = 80^\circ$$

6. This is SAS, so begin by using Law of Cosines:
Step 1: Find *c*.

$$c^2 = a^2 + b^2 - 2ab\cos\gamma = (4.2)^2 + (7.3)^2 - 2(4.2)(7.3)\cos(25^\circ) = 70.93 - 61.32\cos(25^\circ)$$

$c \approx 3.9186 \approx 3.9$

Step 2: Find α.

$$\frac{\sin\alpha}{a} = \frac{\sin\gamma}{c} \Rightarrow \frac{\sin\alpha}{4.2} = \frac{\sin 25^\circ}{3.9186} \Rightarrow \sin\alpha = \frac{4.2\sin 25^\circ}{3.9186} \Rightarrow \alpha = \sin^{-1}\left(\frac{4.2\sin 25^\circ}{3.9186}\right) \approx 26.9^\circ$$

Step 3: Find β.

$$\beta \approx 180^\circ - (26.9^\circ + 25^\circ) = 128.1^\circ$$

7. This is SAS, so begin by using Law of Cosines:
Step 1: Find *b*.

$$b^2 = a^2 + c^2 - 2ac\cos\beta = (9)^2 + (12)^2 - 2(9)(12)\cos(23^\circ) = 225 - 216\cos(23^\circ)$$

$b \approx 5.1158 \approx 5$

Step 2: Find α.

$$\frac{\sin\beta}{b} = \frac{\sin\alpha}{a} \Rightarrow \frac{\sin 23^\circ}{5.1158} = \frac{\sin\alpha}{9} \Rightarrow \sin\alpha = \frac{9\sin 23^\circ}{5.1158} \Rightarrow \alpha = \sin^{-1}\left(\frac{9\sin 23^\circ}{5.1158}\right) \approx 43^\circ$$

Step 3: Find γ.

$$\gamma \approx 180^\circ - (43^\circ + 23^\circ) = 114^\circ$$

8. This is SAS, so begin by using Law of Cosines:
Step 1: Find *a*.

$$a^2 = b^2 + c^2 - 2bc\cos\alpha = (6)^2 + (13)^2 - 2(6)(13)\cos(16^\circ) = 205 - 156\cos(16^\circ)$$

$a \approx 7.4191 \approx 7$

Step 2: Find β.

$$\frac{\sin\alpha}{a} = \frac{\sin\beta}{b} \Rightarrow \frac{\sin 16^\circ}{7.4191} = \frac{\sin\beta}{6} \Rightarrow \sin\beta = \frac{6\sin 16^\circ}{7.4191} \Rightarrow \beta = \sin^{-1}\left(\frac{6\sin 16^\circ}{7.4191}\right) \approx 13^\circ$$

Step 3: Find γ.

$$\gamma \approx 180^\circ - (13^\circ + 16^\circ) = 151^\circ$$

9. This is SAS, so begin by using Law of Cosines:

Step 1: Find b.

$$b^2 = a^2 + c^2 - 2ac\cos\beta = (4)^2 + (8)^2 - 2(4)(8)\cos(60^\circ) = 80 - 64\cos(60^\circ)$$

$$b \approx 6.9282 \approx 7$$

Step 2: Find α.

$$\frac{\sin\beta}{b} = \frac{\sin\alpha}{a} \Rightarrow \frac{\sin 60^\circ}{6.9282} = \frac{\sin\alpha}{4} \Rightarrow \sin\alpha = \frac{4\sin 60^\circ}{6.9282} \Rightarrow \alpha = \sin^{-1}\left(\frac{4\sin 60^\circ}{6.9282}\right) \approx 30^\circ$$

Step 3: Find γ.

$$\gamma \approx 180^\circ - (30^\circ + 60^\circ) = 90^\circ$$

10. This is SAS, so begin by using Law of Cosines:

Step 1: Find a.

$$a^2 = b^2 + c^2 - 2bc\cos\alpha = (3)^2 + \left(\sqrt{18}\right)^2 - 2(3)\left(\sqrt{18}\right)\cos(45^\circ) = 27 - 6\sqrt{18}\cos(45^\circ)$$

$$a = 3$$

Step 2: Find β.

$$\frac{\sin\alpha}{a} = \frac{\sin\beta}{b} \Rightarrow \frac{\sin 45^\circ}{3} = \frac{\sin\beta}{3} \Rightarrow \sin\beta = \frac{3\sin 45^\circ}{3} \Rightarrow \beta = 45^\circ$$

Step 3: Find γ.

$$\gamma \approx 180^\circ - (45^\circ + 45^\circ) = 90^\circ$$

11. This is SSS, so begin by using Law of Cosines:

Step 1: Find the largest angle (i.e., the one opposite the longest side). Here, it is α.

$$a^2 = b^2 + c^2 - 2bc\cos\alpha \;\Rightarrow\; (8)^2 = (5)^2 + (6)^2 - 2(5)(6)\cos\alpha \;\Rightarrow\; 64 = 61 - 60\cos\alpha$$

Thus, $\cos\alpha = -\frac{3}{60}$ so that $\alpha = \cos^{-1}\left(-\frac{3}{60}\right) \approx 92.86^\circ \approx 93^\circ$.

Step 2: Find either of the remaining two angles using the Law of Sines.

$$\frac{\sin\alpha}{a} = \frac{\sin\beta}{b} \Rightarrow \frac{\sin 92.866^\circ}{8} = \frac{\sin\beta}{5} \Rightarrow \sin\beta = \frac{5\sin 92.866^\circ}{8} \Rightarrow \beta = \sin^{-1}\left(\frac{5\sin 92.866^\circ}{8}\right) \approx 39^\circ$$

Step 3: Find the third angle.

$$\gamma \approx 180^\circ - (93^\circ + 39^\circ) = 48^\circ$$

12. This is SSS, so begin by using Law of Cosines:

Step 1: Find the largest angle (i.e., the one opposite the longest side). Here, it is γ.

$$c^2 = a^2 + b^2 - 2ab\cos\gamma \Rightarrow (12)^2 = (6)^2 + (9)^2 - 2(6)(9)\cos\gamma \Rightarrow 144 = 117 - 108\cos\gamma$$

Thus, $\cos\gamma = -\frac{27}{108}$ so that $\gamma = \cos^{-1}\left(-\frac{27}{108}\right) \approx 104.477^\circ \approx 104^\circ$.

Step 2: Find either of the remaining two angles using the Law of Sines.

$$\frac{\sin\gamma}{c} = \frac{\sin\beta}{b} \Rightarrow \frac{\sin 104.477^\circ}{12} = \frac{\sin\beta}{9} \Rightarrow \sin\beta = \frac{9\sin 104.477^\circ}{12} \Rightarrow \beta = \sin^{-1}\left(\frac{9\sin 104.477^\circ}{12}\right) \approx 47^\circ$$

Step 3: Find the third angle.

$$\alpha \approx 180^\circ - (104^\circ + 47^\circ) = 29^\circ$$

13. This is SSS, so begin by using Law of Cosines:

Step 1: Find the largest angle (i.e., the one opposite the longest side). Here, it is γ.

$$c^2 = a^2 + b^2 - 2ab\cos\gamma \Rightarrow (5)^2 = (4)^2 + (4)^2 - 2(4)(4)\cos\gamma \Rightarrow 25 = 32 - 32\cos\gamma$$

Thus, $\cos\gamma = \frac{7}{32}$ so that $\gamma = \cos^{-1}\left(\frac{7}{32}\right) \approx 77^\circ$.

Step 2: Find either of the remaining two angles using the Law of Sines.

$$\frac{\sin\gamma}{c} = \frac{\sin\beta}{b} \Rightarrow \frac{\sin 77.364^\circ}{5} = \frac{\sin\beta}{4} \Rightarrow \sin\beta = \frac{4\sin 77.364^\circ}{5} \Rightarrow \beta = \sin^{-1}\left(\frac{4\sin 77.364^\circ}{5}\right) \approx 51.32^\circ$$

Step 3: Find the third angle.

$$\alpha \approx 180^\circ - (77.36^\circ + 51.32^\circ) \approx 51^\circ$$

14. This is SSS, so begin by using Law of Cosines:

Step 1: Find the largest angle (i.e., the one opposite the longest side). Here, it is γ.

$$c^2 = a^2 + b^2 - 2ab\cos\gamma \Rightarrow (33)^2 = (17)^2 + (20)^2 - 2(17)(20)\cos\gamma \Rightarrow 1089 = 689 - 680\cos\gamma$$

Thus, $\cos\gamma = -\frac{400}{680}$ so that $\gamma = \cos^{-1}\left(-\frac{400}{680}\right) \approx 126.03^\circ \approx 126^\circ$.

Step 2: Find either of the remaining two angles using the Law of Sines.

$$\frac{\sin\gamma}{c} = \frac{\sin\beta}{b} \Rightarrow \frac{\sin 126.03^\circ}{33} = \frac{\sin\beta}{20} \Rightarrow \sin\beta = \frac{20\sin 126.03^\circ}{33} \Rightarrow \beta = \sin^{-1}\left(\frac{20\sin 126.03^\circ}{33}\right) \approx 29^\circ$$

Step 3: Find the third angle.

$$\alpha \approx 180^\circ - (126^\circ + 29^\circ) = 25^\circ$$

15. This is SSS, so begin by using Law of Cosines:

Step 1: Find the largest angle (i.e., the one opposite the longest side). Here, it is α.

$a^2 = b^2 + c^2 - 2bc\cos\alpha \Rightarrow (8.2)^2 = (7.1)^2 + (6.3)^2 - 2(7.1)(6.3)\cos\alpha \Rightarrow 67.24 = 90.1 - 89.46\cos\alpha$

Thus, $\cos\alpha = \dfrac{22.86}{89.46}$ so that $\alpha = \cos^{-1}\left(\dfrac{22.86}{89.46}\right) \approx 75.19^\circ \approx 75^\circ$.

Step 2: Find either of the remaining two angles using the Law of Sines.

$$\frac{\sin\alpha}{a} = \frac{\sin\beta}{b} \Rightarrow \frac{\sin 75.19^\circ}{8.2} = \frac{\sin\beta}{7.1} \Rightarrow \sin\beta = \frac{7.1\sin 75.19^\circ}{8.2} \Rightarrow \beta = \sin^{-1}\left(\frac{7.1\sin 75.19^\circ}{8.2}\right) \approx 57^\circ$$

Step 3: Find the third angle.

$\gamma \approx 180^\circ - (75^\circ + 57^\circ) = 48^\circ$

16. This is SSS, so begin by using Law of Cosines:

Step 1: Find the largest angle (i.e., the one opposite the longest side). Here, it is β.

$$b^2 = a^2 + c^2 - 2ac\cos\beta \Rightarrow (2001)^2 = (1492)^2 + (1776)^2 - 2(1492)(1776)\cos\beta$$
$$\Rightarrow -1,376,239 = -5,299,584\cos\beta$$

Thus, $\cos\beta = \dfrac{1,376,239}{5,299,584}$ so that $\beta = \cos^{-1}\left(\dfrac{1,376,239}{5,299,584}\right) \approx 74.948^\circ \approx 75^\circ$.

Step 2: Find either of the remaining two angles using the Law of Sines.

$$\frac{\sin\alpha}{a} = \frac{\sin\beta}{b} \Rightarrow \frac{\sin\alpha}{1492} = \frac{\sin 74.948^\circ}{2001} \Rightarrow \sin\alpha = \frac{1492\sin 74.948^\circ}{2001} \Rightarrow \alpha = \sin^{-1}\left(\frac{1492\sin 74.948^\circ}{2001}\right) \approx 46^\circ$$

Step 3: Find the third angle.

$\gamma \approx 180^\circ - (75^\circ + 46^\circ) = 59^\circ$

17. Since $a + b = 9 \not> 10 = c$, there can be no triangle in this case.

18. Since $a + b = 4.0 \not> 4.2 = c$, there can be no triangle in this case.

19. This is SSS, so begin by using Law of Cosines:

Step 1: Find the largest angle (i.e., the one opposite the longest side). Here, it is γ.

$$c^2 = a^2 + b^2 - 2ab\cos\gamma \Rightarrow (13)^2 = (12)^2 + (5)^2 - 2(12)(5)\cos\gamma \Rightarrow 0 = -120\cos\gamma$$

Thus, $\cos\gamma = 0$ so that $\gamma = \cos^{-1}(0) = 90^\circ$.

Step 2: Find either of the remaining two angles using the Law of Sines.

$$\frac{\sin\gamma}{c} = \frac{\sin\beta}{b} \Rightarrow \frac{\sin 90^\circ}{13} = \frac{\sin\beta}{5} \Rightarrow \sin\beta = \frac{5\sin 90^\circ}{13} \Rightarrow \beta = \sin^{-1}\left(\frac{5\sin 90^\circ}{13}\right) \approx 23^\circ$$

Step 3: Find the third angle.

$\alpha \approx 180^\circ - (90 + 23^\circ) = 67^\circ$

20. This is SSS, so begin by using Law of Cosines:

Step 1: Find the largest angle (i.e., the one opposite the longest side). Here, it is γ.

$$c^2 = a^2 + b^2 - 2ab\cos\gamma \Rightarrow \left(\sqrt{41}\right)^2 = (4)^2 + (5)^2 - 2(4)(5)\cos\gamma \Rightarrow 0 = -40\cos\gamma$$

Thus, $\cos\gamma = 0$ so that $\gamma = \cos^{-1}(0) = 90^\circ$.

Step 2: Find either of the remaining two angles using the Law of Sines.

$$\frac{\sin\gamma}{c} = \frac{\sin\beta}{b} \Rightarrow \frac{\sin 90^\circ}{\sqrt{41}} = \frac{\sin\beta}{5} \Rightarrow \sin\beta = \frac{5\sin 90^\circ}{\sqrt{41}} \Rightarrow \beta = \sin^{-1}\left(\frac{5\sin 90^\circ}{\sqrt{41}}\right) \approx 51^\circ$$

Step 3: Find the third angle.

$$\alpha \approx 180^\circ - \left(90 + 51^\circ\right) = 39^\circ$$

21. This is AAS, so begin by using Law of Sines:

$$\gamma = 180^\circ - \left(40^\circ + 35^\circ\right) = 105^\circ$$

$$\frac{\sin\alpha}{a} = \frac{\sin\beta}{b} \Rightarrow \frac{\sin 35^\circ}{b} = \frac{\sin 40^\circ}{6} \Rightarrow b = \frac{(6)\left(\sin 35^\circ\right)}{\sin 40^\circ} \approx 5.354 \approx 5$$

$$\frac{\sin\alpha}{a} = \frac{\sin\gamma}{c} \Rightarrow \frac{\sin 40^\circ}{6} = \frac{\sin 105^\circ}{c} \Rightarrow c = \frac{(6)\left(\sin 105^\circ\right)}{\sin 40^\circ} \approx 9$$

22. This is SAS, so begin by using Law of Cosines:

Step 1: Find *c*.

$$c^2 = a^2 + b^2 - 2ab\cos\gamma = (19.0)^2 + (11.2)^2 - 2(19.0)(11.2)\cos\left(13.3^\circ\right) = 486.44 - 425.6\cos\left(13.3^\circ\right)$$

$$c \approx 8.500 \approx 8.5$$

Step 2: Find β.

$$\frac{\sin\beta}{b} = \frac{\sin\gamma}{c} \Rightarrow \frac{\sin\beta}{11.2} = \frac{\sin 13.3^\circ}{8.500} \Rightarrow \sin\beta = \frac{11.2\sin 13.3^\circ}{8.500} \Rightarrow \beta = \sin^{-1}\left(\frac{11.2\sin 13.3^\circ}{8.500}\right) \approx 17.6^\circ$$

Step 3: Find α.

$$\alpha \approx 180^\circ - \left(13.3^\circ + 17.6^\circ\right) = 149.1^\circ$$

23. This is SSA, so begin by using Law of Sines:
Step 1: Determine β.

$$\frac{\sin\alpha}{a}=\frac{\sin\beta}{b} \Rightarrow \frac{\sin 31^\circ}{12}=\frac{\sin\beta}{5} \Rightarrow \sin\beta=\frac{5\sin 31^\circ}{12} \Rightarrow \beta=\sin^{-1}\left(\frac{5\sin 31^\circ}{12}\right)\approx 12.392^\circ \approx 12^\circ$$

Note that there is only one triangle in this case since the angle in QII with the same sine as this value of α is $180^\circ-12^\circ=168^\circ$. In such case, note that $\beta+\alpha>180^\circ$, therefore preventing the formation of a triangle (since the three interior angles, two of which are β and α, must sum to 180°).

Step 2: Solve for the triangle.

$$\gamma \approx 180^\circ-\left(12^\circ+31^\circ\right)=137^\circ$$

$$\frac{\sin\gamma}{c}=\frac{\sin\alpha}{a} \Rightarrow \frac{\sin 137^\circ}{c}=\frac{\sin 31^\circ}{12} \Rightarrow c=\frac{12\sin 137^\circ}{\sin 31^\circ}\approx 16$$

24. This is SSA, so begin by using Law of Sines:
Step 1: Determine α.

$$\frac{\sin\alpha}{a}=\frac{\sin\gamma}{c} \Rightarrow \frac{\sin\alpha}{11}=\frac{\sin 60^\circ}{12} \Rightarrow \sin\alpha=\frac{11\sin 60^\circ}{12} \Rightarrow \alpha=\sin^{-1}\left(\frac{11\sin 60^\circ}{12}\right)\approx 52.547^\circ \approx 53^\circ$$

Note that there is only one triangle in this case since the angle in QII with the same sine as this value of α is $180^\circ-53^\circ=127^\circ$. In such case, note that $\gamma+\alpha>180^\circ$, therefore preventing the formation of a triangle (since the three interior angles, two of which are γ and α, must sum to 180°).

Step 2: Solve for the triangle.

$$\gamma \approx 180^\circ-\left(53^\circ+60^\circ\right)=67^\circ$$

$$\frac{\sin\gamma}{c}=\frac{\sin\beta}{b} \Rightarrow \frac{\sin 60^\circ}{12}=\frac{\sin 67^\circ}{b} \Rightarrow b=\frac{12\sin 67^\circ}{\sin 60^\circ}\approx 13$$

25. This is SSS, so begin by using Law of Cosines:
Step 1: Find the largest angle (i.e., the one opposite the longest side). Here, it is β.

$$b^2=a^2+c^2-2ac\cos\beta \Rightarrow \left(\sqrt{8}\right)^2=\left(\sqrt{7}\right)^2+\left(\sqrt{3}\right)^2-2\left(\sqrt{7}\right)\left(\sqrt{3}\right)\cos\beta$$
$$\Rightarrow -2=-2\sqrt{21}\cos\beta$$

Thus, $\cos\beta=\frac{1}{\sqrt{21}}$ so that $\beta=\cos^{-1}\left(\frac{1}{\sqrt{21}}\right)\approx 77.396^\circ \approx 77^\circ$.

Step 2: Find either of the remaining two angles using the Law of Sines.

$$\frac{\sin\alpha}{a}=\frac{\sin\beta}{b} \Rightarrow \frac{\sin\alpha}{\sqrt{7}}=\frac{\sin 77.396^\circ}{\sqrt{8}} \Rightarrow \sin\alpha=\frac{\sqrt{7}\sin 77.396^\circ}{\sqrt{8}} \Rightarrow \alpha=\sin^{-1}\left(\frac{\sqrt{7}\sin 77.396^\circ}{\sqrt{8}}\right)\approx 66^\circ$$

Step 3: Find the third angle.

$$\gamma \approx 180^\circ-\left(77^\circ+66^\circ\right)=37^\circ$$

26. This is AAS, so begin by using Law of Sines:

$\alpha = 180^\circ - \left(106^\circ + 43^\circ\right) = 31^\circ$

$$\frac{\sin\alpha}{a} = \frac{\sin\beta}{b} \Rightarrow \frac{\sin 31^\circ}{1} = \frac{\sin 106^\circ}{b} \Rightarrow b = \frac{1\left(\sin 106^\circ\right)}{\sin 31^\circ} \approx 1.866 \approx 2$$

$$\frac{\sin\alpha}{a} = \frac{\sin\gamma}{c} \Rightarrow \frac{\sin 31^\circ}{1} = \frac{\sin 43}{c} \Rightarrow c = \frac{1\left(\sin 43^\circ\right)}{\sin 31^\circ} \approx 1.32 \approx 1$$

27. This is SSA, so begin by using Law of Sines:

Step 1: Determine γ.

$$\frac{\sin\gamma}{c} = \frac{\sin\beta}{b} \Rightarrow \frac{\sin\gamma}{2} = \frac{\sin 10^\circ}{11} \Rightarrow \sin\gamma = \frac{2\sin 10^\circ}{11} \Rightarrow \gamma = \sin^{-1}\left(\frac{2\sin 10^\circ}{11}\right) \approx 1.809^\circ \approx 2^\circ$$

Note that there is only one triangle in this case since the angle in QII with the same sine as this value of γ is $180^\circ - 2^\circ = 178^\circ$. In such case, note that $\beta + \gamma > 180^\circ$, therefore preventing the formation of a triangle (since the three interior angles, two of which are β and γ, must sum to 180°).

Step 2: Solve for the triangle.

$\alpha \approx 180^\circ - \left(2^\circ + 10^\circ\right) = 168^\circ$

$$\frac{\sin\alpha}{a} = \frac{\sin\beta}{b} \Rightarrow \frac{\sin 168^\circ}{a} = \frac{\sin 10^\circ}{11} \Rightarrow a = \frac{11\sin 168^\circ}{\sin 10^\circ} \approx 13$$

28. This is SSA, so begin by using Law of Sines:

Step 1: Determine γ.

$$\frac{\sin\alpha}{a} = \frac{\sin\gamma}{c} \Rightarrow \frac{\sin 25^\circ}{6} = \frac{\sin\gamma}{9} \Rightarrow \sin\gamma = \frac{9\sin 25^\circ}{6} \Rightarrow \gamma = \sin^{-1}\left(\frac{9\sin 25^\circ}{6}\right) \approx 39^\circ$$

This is the solution in QI – label it as γ_1. The second solution in QII is given by $\gamma_2 = 180^\circ - \gamma_1 \approx 141$. Both are tenable solutions, so we need to solve for two triangles.

Step 2: Solve for both triangles.

QI triangle: $\gamma_1 \approx 39^\circ$

$\beta_1 \approx 180^\circ - \left(39^\circ + 25^\circ\right) = 116^\circ$

$$\frac{\sin\beta_1}{b_1} = \frac{\sin\gamma_1}{c} \Rightarrow \frac{\sin 116^\circ}{b_1} = \frac{\sin 39^\circ}{9} \Rightarrow b_1 = \frac{9\sin 116^\circ}{\sin 39^\circ} \approx 13$$

QII triangle: $\gamma_2 \approx 141^\circ$

$\beta_2 \approx 180^\circ - \left(141^\circ + 25^\circ\right) = 14^\circ$

$$\frac{\sin\beta_2}{b_2} = \frac{\sin\gamma_2}{c} \Rightarrow \frac{\sin 14^\circ}{b_2} = \frac{\sin 141^\circ}{9} \Rightarrow b_2 = \frac{9\sin 14^\circ}{\sin 141^\circ} \approx 3$$

29. $A=\frac{1}{2}(8)(16)\sin 60^\circ \approx \boxed{55.4}$	**30.** $A=\frac{1}{2}(6)\left(4\sqrt{3}\right)\sin 30^\circ \approx \boxed{10.4}$
31. $A=\frac{1}{2}(1)\left(\sqrt{2}\right)\sin 45^\circ \approx \boxed{0.5}$	**32.** $A=\frac{1}{2}\left(2\sqrt{2}\right)(4)\sin 45^\circ \approx \boxed{4}$
33. $A=\frac{1}{2}(6)(8)\sin 80^\circ \approx \boxed{23.6}$	**34.** $A=\frac{1}{2}(9)(10)\sin 100^\circ \approx \boxed{44.3}$
35. $A=\frac{1}{2}(4)(7)\sin 27^\circ \approx \boxed{6.4}$	**36.** $A=\frac{1}{2}(6.3)(4.8)\sin 17^\circ \approx \boxed{4.42}$
37. $A=\frac{1}{2}(100)(150)\sin 36^\circ \approx \boxed{4,408.4}$	**38.** $A=\frac{1}{2}(0.3)(0.7)\sin 145^\circ \approx \boxed{0.06}$

39. First, observe that $s=\frac{a+b+c}{2}=\frac{45}{2}=22.5$. So, the area is given by

$$A=\sqrt{s(s-a)(s-b)(s-c)}=\sqrt{(22.5)(7.5)(7.5)(7.5)}\approx \boxed{97.4}$$

40. First, observe that $s=\frac{a+b+c}{2}=\frac{3}{2}=1.5$. So, the area is given by

$$A=\sqrt{s(s-a)(s-b)(s-c)}=\sqrt{(1.5)(0.5)(0.5)(0.5)}\approx \boxed{0.4}$$

41. First, observe that $s=\frac{a+b+c}{2}=\frac{17+\sqrt{51}}{2}$. So, the area is given by

$$A=\sqrt{s(s-a)(s-b)(s-c)}$$
$$=\sqrt{\left(\frac{17+\sqrt{51}}{2}\right)\left(\frac{17+\sqrt{51}}{2}-7\right)\left(\frac{17+\sqrt{51}}{2}-\sqrt{51}\right)\left(\frac{17+\sqrt{51}}{2}-10\right)}\approx \boxed{25.0}$$

42. First, observe that $s=\frac{a+b+c}{2}=\frac{90}{2}=45$. So, the area is given by

$$A=\sqrt{s(s-a)(s-b)(s-c)}=\sqrt{(45)(36)(5)(4)}\approx \boxed{180}$$

43. First, observe that $s=\frac{a+b+c}{2}=\frac{25}{2}=12.5$. So, the area is given by

$$A=\sqrt{s(s-a)(s-b)(s-c)}=\sqrt{(12.5)(6.5)(2.5)(3.5)}\approx \boxed{26.7}$$

44. First, observe that $s=\frac{a+b+c}{2}=\frac{150}{2}=75$. So, the area is given by

$$A=\sqrt{s(s-a)(s-b)(s-c)}=\sqrt{(75)(35)(25)(15)}\approx \boxed{992.2}$$

45. First, observe that $s=\frac{a+b+c}{2}=25.05$. So, the area is given by

$$A=\sqrt{s(s-a)(s-b)(s-c)}=\sqrt{(25.05)(10.75)(9.35)(4.95)}\approx \boxed{111.64}$$

46. First, observe that $s=\dfrac{a+b+c}{2}=196.5$. So, the area is given by

$$A=\sqrt{s(s-a)(s-b)(s-c)}=\sqrt{(196.5)(50)(50)(96.5)}\approx\boxed{6,885.17}$$

47. First, observe that $s=\dfrac{a+b+c}{2}=24,600$. So, the area is given by

$$A=\sqrt{s(s-a)(s-b)(s-c)}=\sqrt{(24,600)(10,600)(8,100)(5,900)}\approx\boxed{111,632,076}$$

48. First, observe that $s=\dfrac{a+b+c}{2}=\dfrac{\sqrt{2}+\sqrt{3}+\sqrt{5}}{2}$. So, the area is given by

$$A=\sqrt{s(s-a)(s-b)(s-c)}$$

$$=\sqrt{\left(\frac{\sqrt{2}+\sqrt{3}+\sqrt{5}}{2}\right)\left(\frac{\sqrt{2}+\sqrt{3}+\sqrt{5}}{2}-\sqrt{2}\right)\left(\frac{\sqrt{2}+\sqrt{3}+\sqrt{5}}{2}-\sqrt{3}\right)\left(\frac{\sqrt{2}+\sqrt{3}+\sqrt{5}}{2}-\sqrt{5}\right)}\approx\boxed{1.2}$$

49. Observe that the sum of the lengths of the shortest two sides is 155, which does not exceed the length of the third side. Hence, there is no triangle in this case.

50. Observe that the sum of the lengths of the shortest two sides is 22, which does not exceed the length of the third side. Hence, there is no triangle in this case.

51. Consider the following diagram:

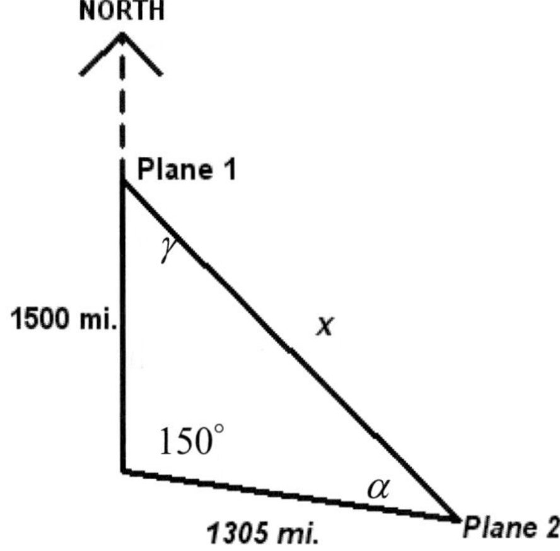

Let x = distance between Plane 1 and Plane 2 after 3 hours. In order to determine the value of x, we apply the Law of Cosines as follows:

$$x^2=1500^2+1305^2-2(1500)(1305)\cos 150^\circ$$

$$x=\sqrt{7,343,514.456}\approx 2710\text{ mi.}$$

Hence, the distance between the two planes after 3 hours is approximately $\boxed{2,710\text{ mi.}}$.

52. Consider the following diagram:

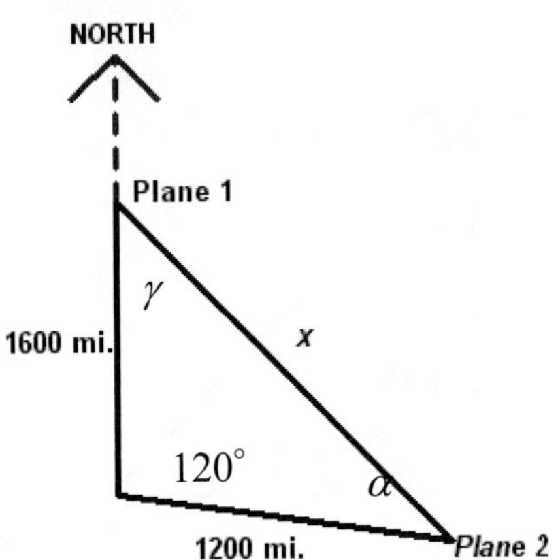

Let x = distance between Plane 1 and Plane 2 after 4 hours. In order to determine the value of x, we apply the Law of Cosines as follows:

$$x^2 = 1600^2 + 1200^2 - 2(1600)(1200)\cos 120^\circ$$

$$x = \sqrt{5{,}920{,}000} \approx 2{,}433 \text{ mi.}$$

Hence, the distance between the two planes after 4 hours is approximately $\boxed{2{,}433 \text{ mi.}}$.

53. Let P_1 be the position of plane 1, P_2 be the position of plane 2, and $\alpha = 65^\circ$ the angle between OP_1 and OP_2 (where O is the origin). Since distance = rate * time, the length of OP_1 is 1375 miles and the length of OP_2 is 875 miles. Using the Law of Cosines yields

$$\left(P_1P_2\right)^2 = 1375^2 + 875^2 - 2(1375)(875)\cos 65^\circ$$

$$P_1P_2 \approx 1280$$

So, the planes are approximately 1,280 miles apart after 2.5 hours.

54. Consider the following diagram:

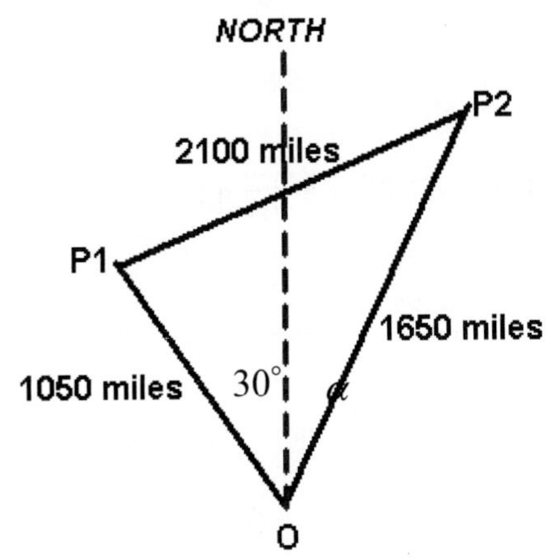

Using the Law of Cosines yields

$$2100^2 = 1050^2 + 1650^2 - 2(1050)(1650)\cos\left(30^\circ + \alpha\right)$$

$$-0.16883 = \cos\left(30^\circ + \alpha\right)$$

$$30^\circ + \alpha = \cos^{-1}\left(-0.16883\right)$$

$$\alpha \approx 50.28^\circ$$

So, the original bearing of Plane 2 is approximately 50.28° clockwise of north.

55. Consider the following diagram:

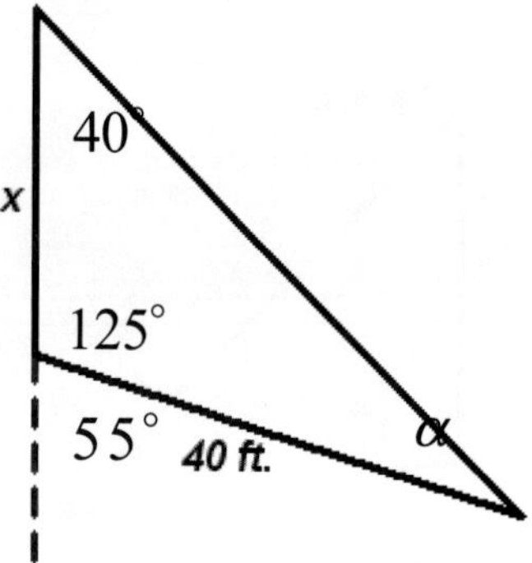

Let x = length of the window. First, observe that $\alpha \approx 180^\circ - \left(40^\circ + 125^\circ\right) = 15^\circ$.

So, $\dfrac{\sin 15^\circ}{x} = \dfrac{\sin 40^\circ}{40 \text{ ft.}} \Rightarrow x = \dfrac{40 \sin 15^\circ}{\sin 40^\circ} \approx \boxed{16 \text{ ft.}}$.

56. Consider the following diagram:

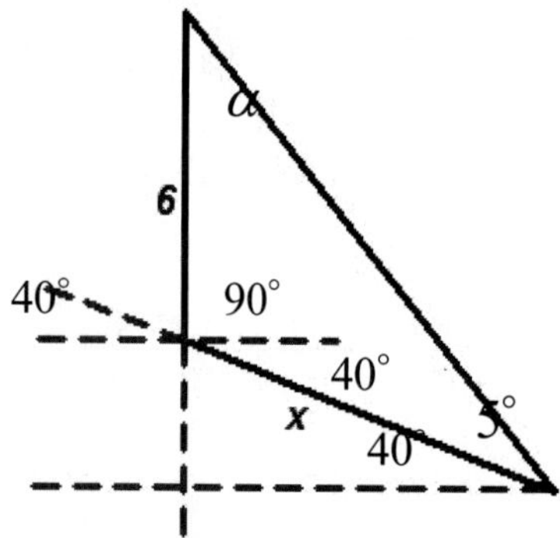

Let x = length of the slide. First, observe that $\alpha \approx 180^\circ - \left(130^\circ + 5^\circ\right) = 45^\circ$.

So, $\dfrac{\sin 45^\circ}{x} = \dfrac{\sin 5^\circ}{6} \Rightarrow x = \dfrac{6 \sin 45^\circ}{\sin 5^\circ} \approx \boxed{49 \text{ ft.}}$.

57. Consider the following diagram: 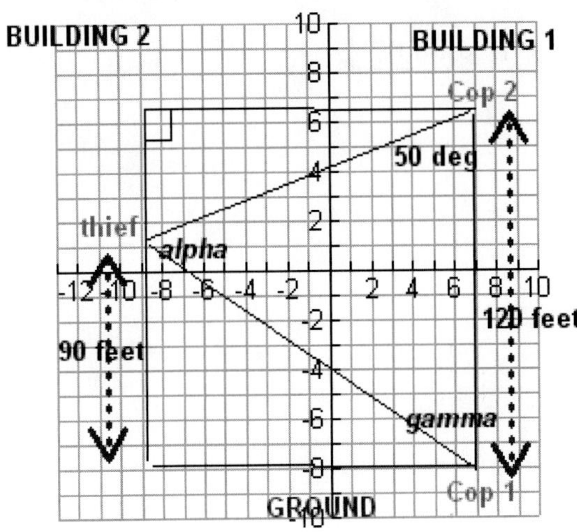	First, note that the right triangle at the top of the diagram has legs of lengths 30 feet and 40 feet. Hence, by the Pythagorean theorem, the zipline has length 50 feet. Next, find b using the Law of Cosines: $b^2 = 50^2 + 120^2 - 2(50)(120)\cos 50^\circ$ $b \approx 95.85$ Now, use the Law of Sines to find γ: $\frac{\sin\gamma}{50} = \frac{\sin 50^\circ}{95.85} \Rightarrow \gamma = \sin^{-1}\left(\frac{50\sin 50^\circ}{95.85}\right)$ $\Rightarrow \gamma \approx \boxed{21.67^\circ}$
58. Consider the following diagram: 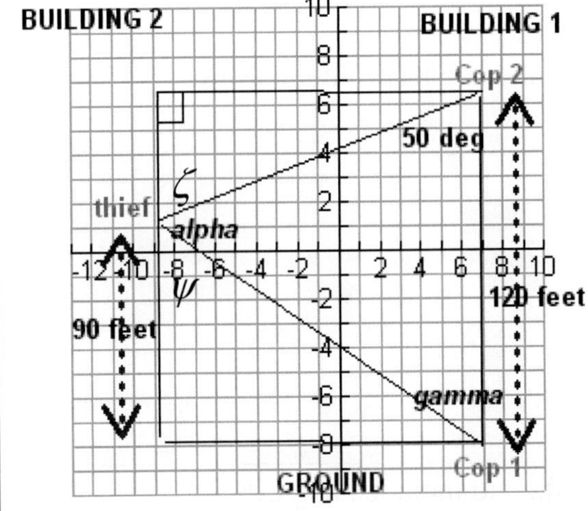	First, note that from Exercise 53, we know that $\gamma \approx 21.67^\circ$. Hence, since the buildings are parallel and γ and ψ are alternate interior angles, they must be equal. Next, since the interior angles of a triangle must sum to 180°, we see that $\alpha = 106.45^\circ$. Thus, we conclude that the desired angle ζ is found by $\zeta = 180^\circ - \left(21.67^\circ + 106.45^\circ\right) = \boxed{51.88^\circ}$.

59. Consider the following diagram:

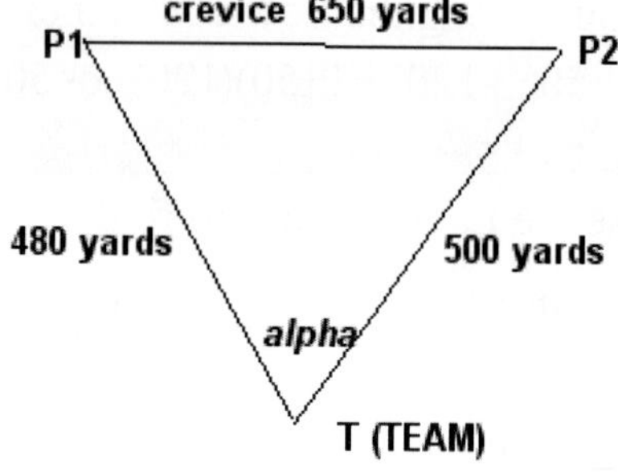

Using the Law of Cosines yields

$$650^2 = 480^2 + 500^2 - 2(480)(500)\cos\alpha$$

$$\alpha \approx \boxed{83.07^\circ}$$

60. Consider the following diagram:

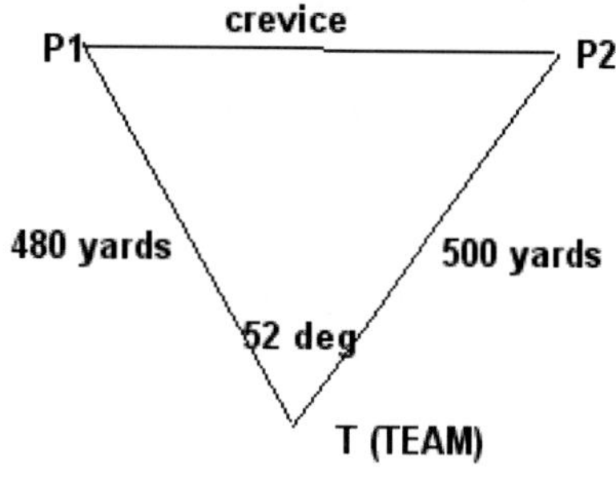

Using the Law of Cosines yields

$$(P_1P_2)^2 = 480^2 + 500^2 - 2(480)(500)\cos 52^\circ$$

$$P_1P_2 \approx 430 \text{ yards}$$

So, the crevice is about 430 yards wide.

61. Let T be a triangle with sides 200 ft., 260 ft., and included angle 65°.
The area of the parallelogram is equal to 2 × area of T.

Note that area of T = $\frac{1}{2}(200)(260)\sin 65^\circ \approx 23{,}564$ sq. ft.

So, the area of the parallelogram is approximately $\boxed{47{,}128 \text{ sq. ft.}}$

62. Let T be a triangle with sides 250 ft., 300 ft., and included angle 55°.
The area of the parallelogram is equal to 2 × area of T.

Note that area of T = $\frac{1}{2}(250)(300)\sin 545^\circ \approx 30{,}718$ sq. ft.

So, the area of the parallelogram is approximately $\boxed{61{,}436 \text{ sq. ft.}}$

63. First, note that a regular hexagon can be dissected into 6 congruent equilateral triangles – this is the case since $\theta = \frac{180^\circ(6-2)}{6} = 120^\circ$, and diagonals bisect the interior angles, thereby making two of the three angles in each of the triangles equal to 60°.
As such, area of a regular hexagon = 6 × area of one these triangles.
To find the area of one these triangles, we apply Heron's formula.
Observe that $s = \frac{3+3+3}{2} = 4.5$ and so, area $A = \sqrt{4.5(4.5-3)^3} = 3.897$ sq. ft.
Thus, the area of the regular hexagon is $6(3.897 \text{ sq. ft.}) = \boxed{23.38 \text{ sq. ft.}}$.

64. First, note that a regular decagon can be dissected into 10 congruent triangles.
As such, area of a regular decagon = 10 × area of one these triangles.
A typical triangle is illustrated below:

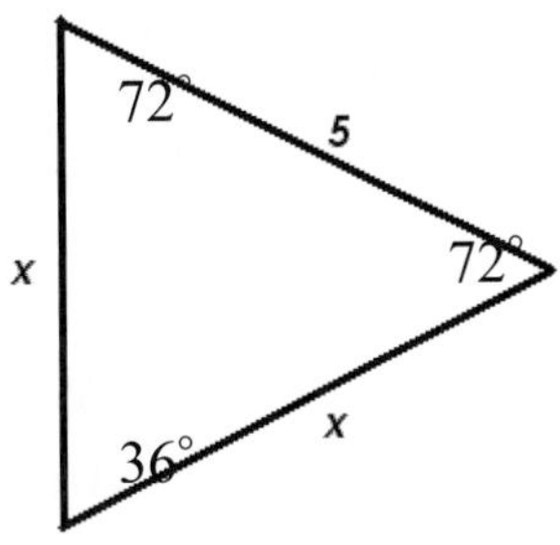

To find the area of such a triangle, we first need x. Using the Law of Sines yields

$$\frac{\sin 36^\circ}{5} = \frac{\sin 72^\circ}{x} \text{ so that } x = \frac{5\sin 72^\circ}{\sin 36^\circ} \approx 8.09017 \text{ in.}$$

Hence, the area $A \approx \frac{1}{2}(8.09017)(8.09017)\sin 36^\circ \approx 19.236$ sq. in.
Thus, the area of the regular decagon is $10(19.236 \text{ sq. in.}) = \boxed{192.36 \text{ sq. in.}}$.

65. Consider the following diagram: 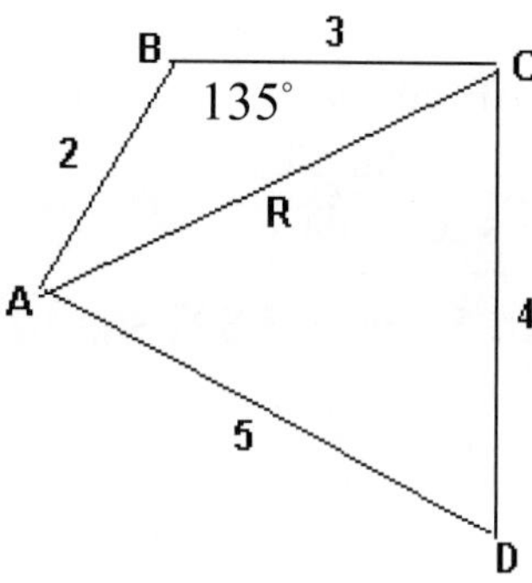	First, note that Area of $\triangle ABC = \frac{1}{2}(2)(3)\sin 135^\circ \approx 2.1213$. Next, in order to find *R*, we apply the Law of Cosines: $R^2 = 2^2 + 3^2 - 2(2)(3)\cos 135^\circ \approx 21.48528$ $R \approx 4.63522$ Now, in order to find the area of ΔACD, observe that $s = \frac{a+b+c}{2} \approx \frac{5+7+4.63522}{2} = 8.31761$. Thus, the area of ΔACD is given by $A = \sqrt{s(s-4)(s-5)(s-4.63522)} \approx \boxed{8.73}$ Hence, the area of the entire quadrilateral is approximately 10.86 square units.
66. Consider the following diagram: 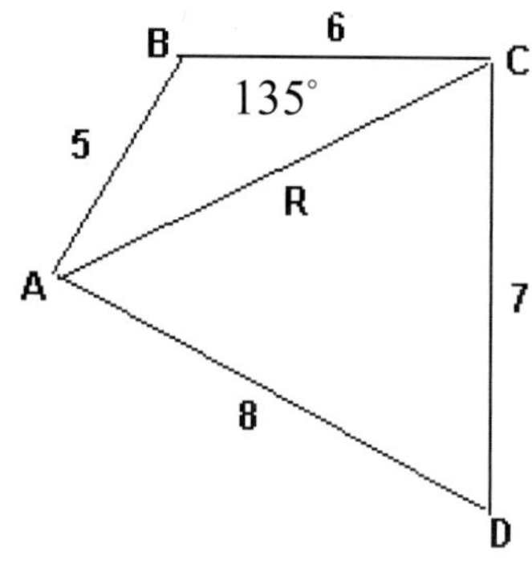	First, note that Area of $\triangle ABC = \frac{1}{2}(5)(6)\sin 135^\circ \approx 10.6066$ Next, in order to find *R*, we apply the Law of Cosines: $R^2 = 5^2 + 6^2 - 2(5)(6)\cos 135^\circ \approx 103.42687$ $R \approx 10.1699$ Now, in order to find the area of ΔACD, observe that $s = \frac{a+b+c}{2} \approx \frac{8+7+10.1699}{2} = 12.58495$. Thus, the area of ΔACD is given by $A = \sqrt{s(s-7)(s-5)8(s-10.1699)} \approx \boxed{25.0223}$ Hence, the area of the entire quadrilateral is approximately 35.629 square units
67. Should have used the smaller angle β in Step 2.	**68.** Should have used the larger angle α in Step 1.
69. False. Can use Law of Cosines to solve any such triangle.	**70.** True. There are infinitely many similar triangles to a given one for which only the angles (and no side) is known.
71. True. By the Law of Cosines, $c^2 = a^2 + b^2 - 2ab\underbrace{\cos 90^\circ}_{=0} = a^2 + b^2$ So, the Pythagorean Theorem is a special case of the Law of Cosines.	**72.** False. These only coincide when there is an angle with measure 90°.

73. True. Each of the other two angles are $\frac{180^\circ - \alpha}{2}$. If you have the length of the side adjacent to the angle α (and hence opposite one of the two congruent sides), then you have the other side as well since the triangle is isosceles. Hence, by SAS, you can apply the Law of Cosines.

74. False. If you only have three angles, you could not apply the Law of Cosines.

75. <u>Claim</u>: $\frac{\cos\alpha}{a}+\frac{\cos\beta}{b}+\frac{\cos\gamma}{c}=\frac{a^2+b^2+c^2}{2abc}$

Proof: First, observe that the left-side simplifies to:

$$\frac{\cos\alpha}{a}+\frac{\cos\beta}{b}+\frac{\cos\gamma}{c}=\frac{bc\cos\alpha+ac\cos\beta+ab\cos\gamma}{abc} \quad \textbf{(1)}.$$

Next, note that the Law of Cosines yields the following three identities:

$$bc\cos\alpha=\frac{a^2-\left(b^2+c^2\right)}{-2}, \quad ac\cos\beta=\frac{b^2-\left(a^2+c^2\right)}{-2}, \quad ab\cos\gamma=\frac{c^2-\left(a^2+b^2\right)}{-2}$$

Now, substituting these into the right-side of **(1)** yields:

$$\begin{aligned}\frac{\cos\alpha}{a}+\frac{\cos\beta}{b}+\frac{\cos\gamma}{c}&=\frac{bc\cos\alpha+ac\cos\beta+ab\cos\gamma}{abc}\\&=\frac{\left(a^2-b^2-c^2\right)+\left(b^2-a^2-c^2\right)+\left(c^2-a^2-b^2\right)}{-2abc}\\&=-\frac{a^2+b^2+c^2}{-2abc}=\frac{a^2+b^2+c^2}{2abc}\end{aligned}$$

76. <u>Claim</u>: $\alpha=c\cos\beta-b\cos\gamma$.

Proof: From the Law of Cosines, we have the following relationships:

$$a^2=b^2+c^2-2bc\cos\alpha \quad \textbf{(1)}$$

$$b^2=a^2+c^2-2ac\cos\beta \quad \textbf{(2)}$$

$$c^2=a^2+b^2-2ab\cos\gamma \quad \textbf{(3)}$$

Solving equation **(2)** for $c\cos\beta$ yields: $c\cos\beta=\frac{a^2+c^2-b^2}{2a}$ **(4)**

Solving equation **(3)** for $b\cos\gamma$ yields: $b\cos\gamma=\frac{-c^2+a^2+b^2}{2a}$ **(5)**

Now, adding equations **(4)** and **(5)** yields:

$$c\cos\beta+b\cos\gamma=\frac{a^2+c^2-b^2}{2a}+\frac{-c^2+a^2+b^2}{2a}=\frac{a^2+\cancel{c^2}\cancel{-b^2}\cancel{-c^2}+a^2\cancel{+b^2}}{2a}=\frac{2a^2}{2a}=a,$$

as desired.

77. Consider the following diagram:

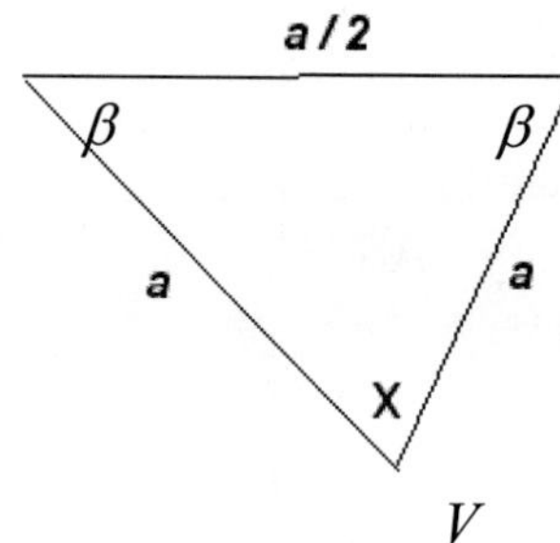

Observe that since $2\beta + X = 180^\circ$, we know that $\beta = \frac{180^\circ - X}{2}$. By the Law of Cosines, we have

$$a^2 = a^2 + \left(\tfrac{a}{2}\right)^2 - 2a\left(\tfrac{a}{2}\right)\cos\beta \quad \Rightarrow \quad -\tfrac{a^2}{4} = -a^2\cos\beta$$

$$\Rightarrow \quad \tfrac{1}{4} = \cos\beta \quad \Rightarrow \quad \beta = \cos^{-1}\left(\tfrac{1}{4}\right)$$

Hence,

$$\cos^{-1}\left(\tfrac{1}{4}\right) = \tfrac{180^\circ - X}{2} \quad \Rightarrow X = 180^\circ - 2\cos^{-1}\left(\tfrac{1}{4}\right).$$

So, we now have

$$\cos\left(\tfrac{X}{2}\right) = \sqrt{\frac{1+\cos X}{2}} = \sqrt{\frac{1+\cos\left(180^\circ - 2\cos^{-1}\left(\frac{1}{4}\right)\right)}{2}} = \boxed{\sqrt{\frac{1-\cos\left(2\cos^{-1}\left(\frac{1}{4}\right)\right)}{2}}}$$

<u>An improved alternative solution is as follows:</u>

Since the triangle is isosceles triangle, the leg obtained by connecting vertex V to the opposite base is a perpendicular bisector of the base (so that each half has length $\frac{a}{4}$), thus creating two congruent right triangles. The height of these triangles, from the Pythagorean theorem is $\frac{\sqrt{15}}{4}a$. Hence, $\cos\left(\frac{X}{2}\right) = \frac{\frac{\sqrt{15}}{4}a}{a} = \frac{\sqrt{15}}{4}$. (It is instructive to show these two answers are equivalent.)

78. From Exercise #77, we know that $X = 180^\circ - 2\cos^{-1}\left(\frac{1}{4}\right)$, so that

$$\cos X = \cos\left(180^\circ - 2\cos^{-1}\left(\tfrac{1}{4}\right)\right) = -\cos\left(2\cos^{-1}\left(\tfrac{1}{4}\right)\right).$$

Hence,

$$\tan\left(\tfrac{X}{2}\right) = \sqrt{\frac{1-\cos X}{1+\cos X}} = \boxed{\sqrt{\frac{1+\cos\left(2\cos^{-1}\left(\frac{1}{4}\right)\right)}{1-\cos\left(2\cos^{-1}\left(\frac{1}{4}\right)\right)}}}.$$

79. Consider the following diagram:

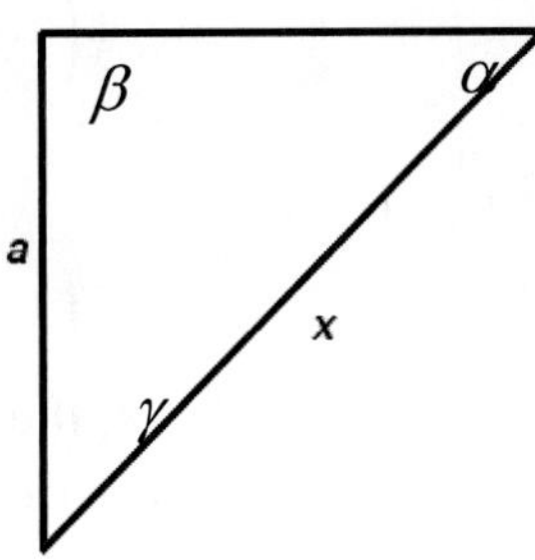

<u>Claim</u>: Area $= \dfrac{a^2 \sin\beta \sin\gamma}{2\sin\alpha}$.

Proof: First, the Law of Sines gives yields $\dfrac{\sin\alpha}{a} = \dfrac{\sin\beta}{x} \Rightarrow x = \dfrac{a\sin\beta}{\sin\alpha}$.

Hence, the area $= \dfrac{1}{2}(a)\left(\dfrac{a\sin\beta}{\sin\alpha}\right)\sin\gamma = \dfrac{a^2\sin\beta\sin\gamma}{2\sin\alpha}$.

80. Use the formula $A = \dfrac{1}{2}ab\sin\theta$, where θ is the angle between the sides measuring a and b with $a = b = s$ (for an isosceles triangle). In such case, $A = \dfrac{1}{2}s^2\sin\theta$.

81. Consider the following diagram:

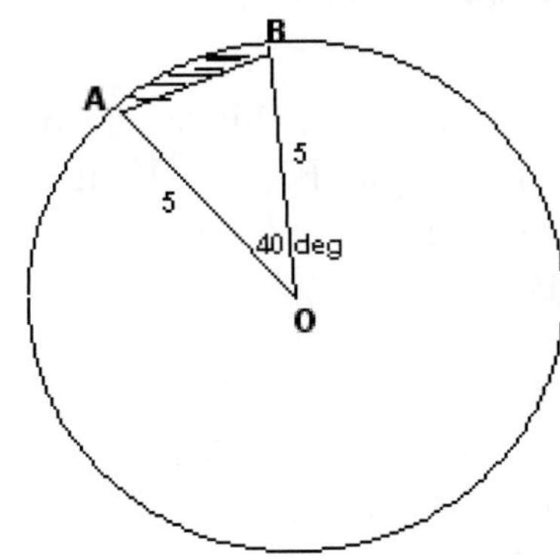

Observe that

Area of sector AOB =

$\frac{1}{2}(5)^2(40^\circ)\left(\frac{\pi}{180^\circ}\right) \approx 8.72664626$

and

Area of ΔAOB =

$\frac{1}{2}(5)(5)\sin 40^\circ \approx 8.034845$.

Thus, the area of the shaded region is

Area of sector AOB - Area of ΔAOB $\approx$ 0.69 square units.

82. Consider the following diagram:

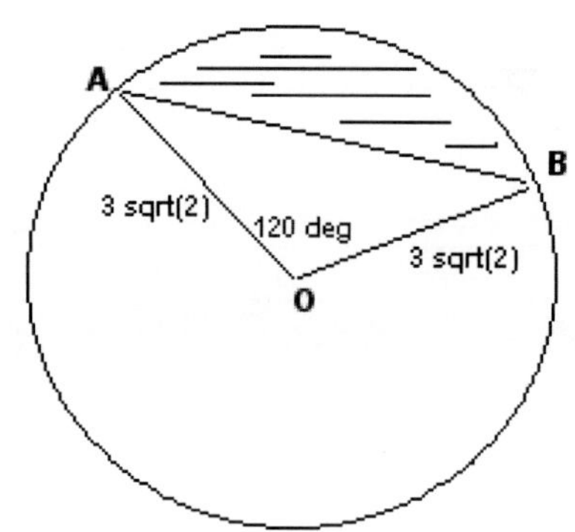

Observe that

Area of sector AOB =

$$\frac{1}{2}\left(3\sqrt{2}\right)^2\left(120^\circ\right)\left(\frac{\pi}{180^\circ}\right) \approx 18.849566$$

and

Area of ΔAOB =

$$\frac{1}{2}\left(3\sqrt{2}\right)\left(3\sqrt{2}\right)\sin 120^\circ \approx 7.79423.$$

Thus, the area of the shaded region is
Area of sector AOB - Area of ΔAOB $\approx$ 11.06 square units

83. The following steps constitute the program in this case:

```
Program: BCX
:Input "SIDE B =", B
:Input "SIDE C =", C
:Input "ANGLE X =", X
:√ (B²+C²-2BCcos(X))→A
:sin⁻¹(Bsin(X)/A)→Y
:180-Y-X →Z
:Disp "SIDE A = ", A
:Disp "ANGLE Y = ", Y
:Disp "ANGLE Z =", Z
```

Now, in order to use this program to solve the given triangle, EXECUTE it and enter the following data at the prompts:
SIDE B = 45
SIDE C = 57
ANGLE X = 43
Now, the program will display the following information that solves the triangle:
SIDE A = 39.01481143
ANGLE Y = 51.87098421
ANGLE Z = 85.12901579

84. The following steps constitute the program in this case:

```
Program: BCX
:Input "SIDE B =", B
:Input "SIDE C =", C
:Input "ANGLE X =", X
:√ (B²+C²-2BCcos(X))→A
:sin⁻¹(Bsin(X)/A)→Y
:180-Y-X →Z
:Disp "SIDE A = ", A
:Disp "ANGLE Y = ", Y
:Disp "ANGLE Z =", Z
```

Now, in order to use this program to solve the given triangle, EXECUTE it and enter the following data at the prompts:
SIDE B = 24.5
SIDE C = 31.6
ANGLE X = 81.5
Now, the program will display the following information that solves the triangle:
SIDE A = 37.0127263
ANGLE Y = 40.89415545
ANGLE Z = 57.60584455

85. The following steps constitute the program:

Program: ABC
:Input "SIDE A =", A
:Input "SIDE B =", B
:Input "SIDE C =", C
:$\cos^{-1}((B^2+C^2-A^2)/(2BC)) \to X$
: $\cos^{-1}((A^2+C^2-B^2)/(2AC)) \to Y$
:180-Y-X $\to$ Z
:Disp "ANGLE X = ", X
:Disp "ANGLE Y = ", Y
:Disp "ANGLE Z =", Z

Now, in order to use this program to solve the given triangle, EXECUTE it and enter the following data at the prompts:

SIDE A = 29.8
SIDE B = 37.6
SIDE C = 53.2

Now, the program will display the following information that solves the triangle:

ANGLE X = 32.98051035
ANGLE Y = 43.38013581
ANGLE Z = 103.6393538

86. The following steps constitute the program:

Program: ABC
:Input "SIDE A =", A
:Input "SIDE B =", B
:Input "SIDE C =", C
:$\cos^{-1}((B^2+C^2-A^2)/(2BC)) \to X$
: $\cos^{-1}((A^2+C^2-B^2)/(2AC)) \to Y$
:180-Y-X $\to$ Z
:Disp "ANGLE X = ", X
:Disp "ANGLE Y = ", Y
:Disp "ANGLE Z =", Z

Now, in order to use this program to solve the given triangle, EXECUTE it and enter the following data at the prompts:

SIDE A = 100
SIDE B = 170
SIDE C = 250

Now, the program will display the following information that solves the triangle:

ANGLE X = 16.73494404
ANGLE Y = 29.30810924
ANGLE Z = 133.9569467

87. The following steps constitute the program:

Program: ABX
:Input "SIDE A =", A
:Input "SIDE B =", B
:Input "ANGLE X =", X
:$\sqrt{\ }(A^2+B^2-2AB\cos(X)) \to C$
:$\sin^{-1}(A\sin(X)/C) \to Y$
:180-Y-X $\to$ Z
:Disp "SIDE C = ", C
:Disp "ANGLE Y = ", Y
:Disp "ANGLE Z =", Z

Now, in order to use this program to solve the given triangle, EXECUTE it and enter the following data at the prompts:

SIDE A = $\sqrt{12}$
SIDE B = $\sqrt{21}$
ANGLE X = 43

Now, the program will display the following information that solves the triangle:

SIDE C =
ANGLE Y =
ANGLE Z =

88. The following steps constitute the program:

Program: ABC
:Input "SIDE A =", A
:Input "SIDE B =", B
:Input "SIDE C =", C
:$\cos^{-1}((B^2+C^2-A^2)/(2BC)) \to X$
: $\cos^{-1}((A^2+C^2-B^2)/(2AC)) \to Y$
:180-Y-X $\to$ Z
:Disp "ANGLE X = ", X
:Disp "ANGLE Y = ", Y
:Disp "ANGLE Z =", Z

Now, in order to use this program to solve the given triangle, EXECUTE it and enter the following data at the prompts:

SIDE A = 1235
SIDE B = 987
SIDE C = 1456

Now, the program will display the following information that solves the triangle:

ANGLE X =
ANGLE Y =
ANGLE Z =

89. Let d be the distance between the two ships. Then, from the Law of Cosines, we have

$$d^2 = (160mi)^2 + (200mi)^2 - 2(160mi)(200mi)\cos 135^\circ, \text{ so that } d \approx 333mi.$$

90. Let α be the desired angle. By the Law of Cosines, we have $$80^2 = 60^2 + 70^2 - 2(60)(70)\cos\alpha \Rightarrow -2100 = -8400\cos\alpha \Rightarrow 0.25 = \cos\alpha,$$ so that $\alpha \approx 75.5^\circ$.
91. Let d be the distance between B and C. By the Law of Cosines, we have $$d^2 = (153)^2 + (200)^2 - 2(153)(200)\cos\tfrac{2\pi}{9}, \text{ so that } d \approx 129m.$$ Thus, the athlete run 153m + 129m = 282m.
92. For a regular n-gon, the central angle has measure $\frac{2\pi}{n}$. Since the segments connecting the center to the vertices all have length 10ft, using the Law of Cosines yields $$d^2 = (10)^2 + (10)^2 - 2(10)(10)\cos\tfrac{2\pi}{5}, \text{ so that } d \approx 11.75570504.$$ Thus, the perimeter is $5d = 58.8$ ft.

Chapter 4 Review Solutions ---

1. **a)** <u>complement</u>: $90^\circ - 28^\circ = \boxed{62^\circ}$ **b)** <u>supplement</u>: $180^\circ - 28^\circ = \boxed{152^\circ}$	**2.** **a)** <u>complement</u>: $90^\circ - 17^\circ = \boxed{73^\circ}$ **b)** <u>supplement</u>: $180^\circ - 17^\circ = \boxed{163^\circ}$
3. **a)** <u>complement</u>: $90^\circ - 35^\circ = \boxed{55^\circ}$ **b)** <u>supplement</u>: $180^\circ - 35^\circ = \boxed{145^\circ}$	**4.** **a)** <u>complement</u>: $90^\circ - 78^\circ = \boxed{12^\circ}$ **b)** <u>supplement</u>: $180^\circ - 78^\circ = \boxed{102^\circ}$
5. **a)** <u>complement</u>: $90^\circ - 89.01^\circ = \boxed{0.99^\circ}$ **b)** <u>supplement</u>: $180^\circ - 89.01^\circ = \boxed{90.99^\circ}$	**6.** **a)** <u>complement</u>: $90^\circ - 0.013^\circ = \boxed{89.987^\circ}$ **b)** <u>supplement</u>: $180^\circ - 0.013^\circ = \boxed{179.987^\circ}$
7. $135^\circ = 135^\circ \cdot \frac{\pi}{180^\circ} = 3(\cancel{45^\circ}) \cdot \frac{\pi}{4(\cancel{45^\circ})} = \boxed{\frac{3\pi}{4}}$	**8.** $240^\circ = 240^\circ \cdot \frac{\pi}{180^\circ} = 4(\cancel{60^\circ}) \cdot \frac{\pi}{3(\cancel{60^\circ})} = \boxed{\frac{4\pi}{3}}$
9. $330^\circ = 330^\circ \cdot \frac{\pi}{180^\circ} = 11(\cancel{30^\circ}) \cdot \frac{\pi}{6(\cancel{30^\circ})} = \boxed{\frac{11\pi}{6}}$	**10.** $180^\circ = 180^\circ \cdot \frac{\pi}{180^\circ} = \boxed{\pi}$

11. $216^\circ = 216^\circ \cdot \dfrac{\pi}{180^\circ} = 6\left(\cancel{36^\circ}\right) \cdot \dfrac{\pi}{5\left(\cancel{36^\circ}\right)} = \boxed{\dfrac{6\pi}{5}}$	**12.** $108^\circ = 108^\circ \cdot \dfrac{\pi}{180^\circ} = 3\left(\cancel{36^\circ}\right) \cdot \dfrac{\pi}{5\left(\cancel{36^\circ}\right)} = \boxed{\dfrac{3\pi}{5}}$
13. $1{,}620^\circ = 1{,}620^\circ \cdot \dfrac{\pi}{180^\circ} = 9\left(\cancel{180^\circ}\right) \cdot \dfrac{\pi}{\cancel{180^\circ}} = \boxed{9\pi}$	**14.** $900^\circ = 900^\circ \cdot \dfrac{\pi}{180^\circ} = 5\left(\cancel{180^\circ}\right) \cdot \dfrac{\pi}{\cancel{180^\circ}} = \boxed{5\pi}$
15. $\dfrac{\pi}{3} = \dfrac{\cancel{\pi}}{3} \cdot \dfrac{180^\circ}{\cancel{\pi}} = \dfrac{\cancel{3}\left(60^\circ\right)}{\cancel{3}} = \boxed{60^\circ}$	**16.** $\dfrac{11\pi}{6} = \dfrac{11\cancel{\pi}}{6} \cdot \dfrac{180^\circ}{\cancel{\pi}} = \dfrac{11\cancel{6}\left(30^\circ\right)}{\cancel{6}} = \boxed{330^\circ}$
17. $\dfrac{5\pi}{4} = \dfrac{5\cancel{\pi}}{4} \cdot \dfrac{180^\circ}{\cancel{\pi}} = \dfrac{5\cancel{4}\left(45^\circ\right)}{\cancel{4}} = \boxed{225^\circ}$	**18.** $\dfrac{2\pi}{3} = \dfrac{2\cancel{\pi}}{3} \cdot \dfrac{180^\circ}{\cancel{\pi}} = \dfrac{2\cancel{3}\left(60^\circ\right)}{\cancel{3}} = \boxed{120^\circ}$
19. $\dfrac{5\pi}{9} = \dfrac{5\cancel{\pi}}{9} \cdot \dfrac{180^\circ}{\cancel{\pi}} = \dfrac{5\cancel{9}\left(20^\circ\right)}{\cancel{9}} = \boxed{100^\circ}$	**20.** $\dfrac{17\pi}{10} = \dfrac{17\cancel{\pi}}{10} \cdot \dfrac{180^\circ}{\cancel{\pi}} = \dfrac{17\cancel{10}\left(18^\circ\right)}{\cancel{10}} = \boxed{306^\circ}$
21. $10\pi = 10\cancel{\pi} \cdot \dfrac{180^\circ}{\cancel{\pi}} = \boxed{1{,}800^\circ}$	**22.** $\dfrac{31\pi}{2} = \dfrac{31\cancel{\pi}}{2} \cdot \dfrac{180^\circ}{\cancel{\pi}} = \boxed{2{,}790^\circ}$

23. Consider the following diagram:

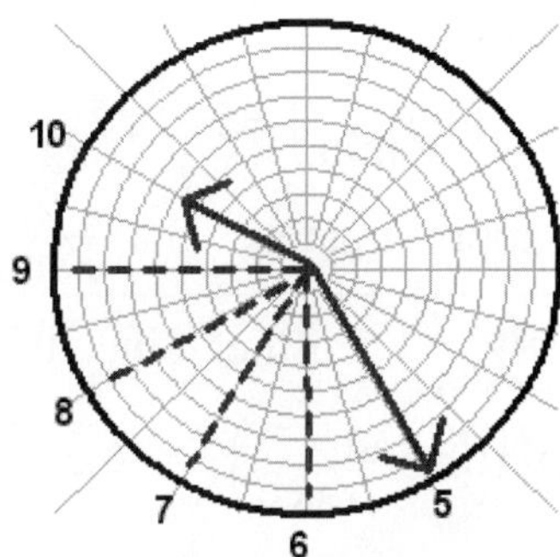

Let $\alpha =$ measure of the desired angle between the hour hand and minute hand. Since the measure of the angle formed using two rays emanating from the center of the clock out toward consecutive hours is always $\dfrac{1}{12}\left(360^\circ\right) = 30^\circ$, it immediately follows that $\alpha = 5 \cdot 30^\circ = \boxed{150^\circ}$.

24. Consider the following diagram:

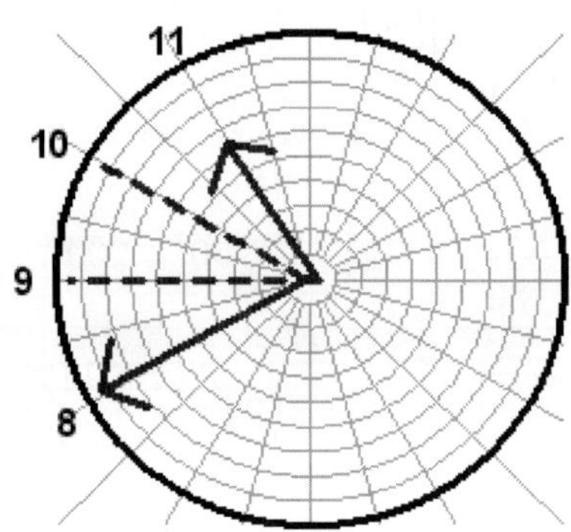

Let $\alpha =$ measure of the desired angle between the hour hand and minute hand.
Since the measure of the angle formed using two rays emanating from the center of the clock out toward consecutive hours is always $\frac{1}{12}(360^\circ) = 30^\circ$, it immediately follows that $\alpha = 3 \cdot 30^\circ = \boxed{90^\circ}$.

25. First, note that the radius $r = 2$ in. (half the diameter). Next, observe that $\omega = 60 \text{ revolutions per min.} = \frac{60(2\pi) \text{ rad.}}{1 \text{ min.}} = 120\pi \ \text{rad.}/\text{min.}$ So,

$$v = r\omega = (2 \text{ in.})\left(120\pi \ \text{rad.}/\text{min.}\right) = 240\pi \ \text{ft.}/\text{min.} \approx 754 \ \text{in.}/\text{min.}.$$

Thus, the lady bug is traveling at approximately $754 \ \text{in.}/\text{min.}$

26. We are given that $\omega = 10\pi \ \text{rad.}/\text{sec.}$. Further, since the diameter is 30 in., the radius is 15 in. Hence, $v = r\omega = (15 \text{ in.})\left(10\pi \ \text{rad.}/\text{sec.}\right) = 150\pi \ \text{in.}/\text{sec.}$.
Converting the *mph* yields:

$$v = 150\pi \ \text{in.}/\text{sec.} = \left(\frac{150\pi \text{ in.}}{1 \text{ sec.}}\right)\left(\frac{3,600 \text{ sec.}}{1 \text{ hr.}}\right)\left(\frac{1 \text{ mi.}}{5,280(12) \text{ in.}}\right) \approx \boxed{27 \text{ mph}}.$$

27. Consider the following diagram:

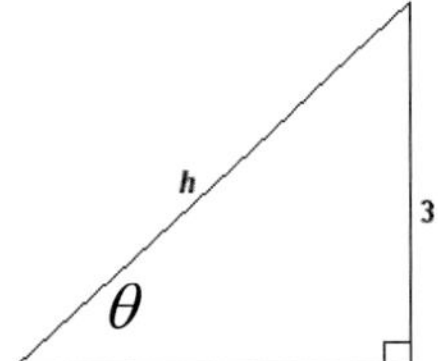

From the Pythagorean Theorem, we see that $h = \sqrt{3^2 + 2^2} = \sqrt{13}$. Thus,

$$\cos\theta = \frac{2}{\sqrt{13}} = \boxed{\frac{2\sqrt{13}}{13}}.$$

28. Using the diagram and value of h in Exercise 27, we see that $\sin\theta = \frac{3}{\sqrt{13}} = \boxed{\frac{3\sqrt{13}}{13}}$.	**29.** Using the diagram and information from Exercise 27, we see that $\sec\theta = \frac{1}{\cos\theta} = \boxed{\frac{\sqrt{13}}{2}}$.
30. Using Exercise 28, we see that $\csc\theta = \frac{1}{\sin\theta} = \boxed{\frac{\sqrt{13}}{3}}$.	**31.** First, note that using the diagram and value of h in Exercise 27, we see that $\sin\theta = \frac{3}{\sqrt{13}}$. So, we have $\tan\theta = \frac{\sin\theta}{\cos\theta} = \frac{\frac{3}{\sqrt{13}}}{\frac{2}{\sqrt{13}}} = \boxed{\frac{3}{2}}$.

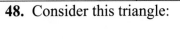

32. Using Exercise 31, we see that $\cot\theta = \frac{1}{\tan\theta} = \boxed{\frac{2}{3}}$.

33. b	**34.** a	**35.** b	**36.** a	**37.** c	**38.** c

39. 0.6691	**40.** 0.5446	**41.** 0.9548	**42.** 0.4706
43. 1.5399	**44.** 1.0446	**45.** 1.5477	**46.** 2.7837

47. Consider this triangle:

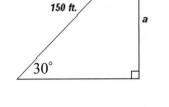

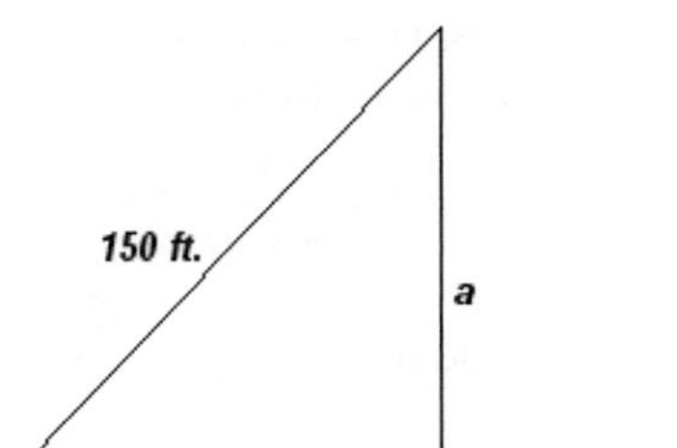

Observe that $\sin 30^\circ = \frac{a}{150 \text{ ft.}}$, so that

$$a = (\sin 30^\circ)(150 \text{ ft.}) \approx 75 \text{ ft.}.$$

So, the altitude should be about $\boxed{75 \text{ ft.}}$

48. Consider this triangle:

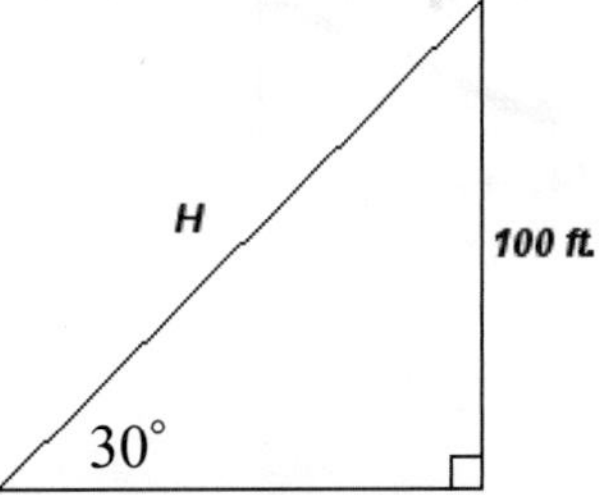

Observe that $\sin 30^\circ = \frac{100 \text{ ft.}}{H}$, so that

$$H = \frac{100 \text{ ft.}}{\sin 30^\circ} \approx 200 \text{ ft.}.$$

So, the hose should be about $\boxed{200 \text{ ft.}}$ long.

49. Consider the following diagram:

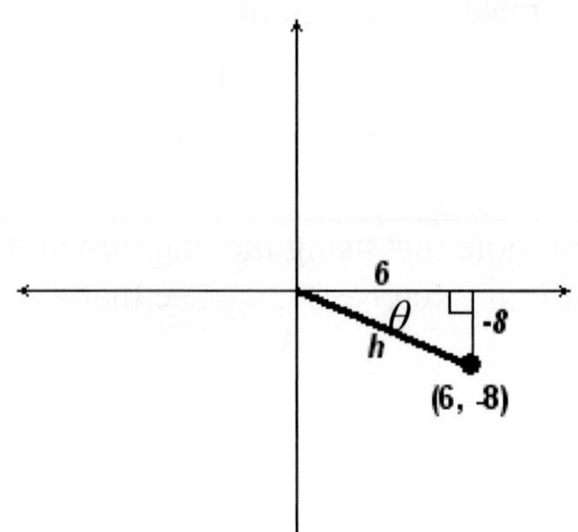

Using the Pythagorean Theorem, we obtain $h=\sqrt{6^2+(-8)^2}=10$.

$$\sin\theta=\frac{\text{opp}}{\text{hyp}}=\frac{-8}{10}=\frac{-4}{5}$$
$$\cos\theta=\frac{\text{adj}}{\text{hyp}}=\frac{6}{10}=\frac{3}{5}$$
$$\tan\theta=\frac{\text{opp}}{\text{adj}}=\frac{-8}{6}=\frac{-4}{3}$$
$$\cot\theta=\frac{1}{\tan\theta}=\frac{-3}{4}$$
$$\sec\theta=\frac{1}{\cos\theta}=\frac{5}{3}$$
$$\csc\theta=\frac{1}{\sin\theta}=\frac{-5}{4}$$

50. Consider the following diagram:

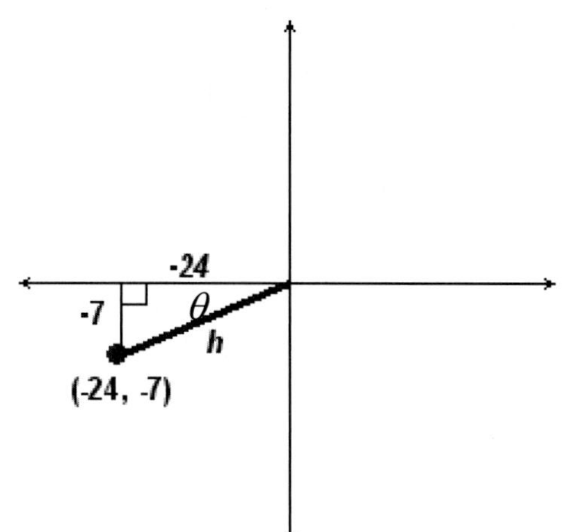

Using the Pythagorean Theorem, we obtain $h=\sqrt{(-7)^2+(-24)^2}=25$.

$$\sin\theta=\frac{\text{opp}}{\text{hyp}}=\frac{-7}{25}$$
$$\cos\theta=\frac{\text{adj}}{\text{hyp}}=\frac{-24}{25}$$
$$\tan\theta=\frac{\text{opp}}{\text{adj}}=\frac{-7}{-24}=\frac{7}{24}$$
$$\cot\theta=\frac{1}{\tan\theta}=\frac{24}{7}$$
$$\sec\theta=\frac{1}{\cos\theta}=\frac{-25}{24}$$
$$\csc\theta=\frac{1}{\sin\theta}=\frac{-25}{7}$$

51. Consider the following diagram:

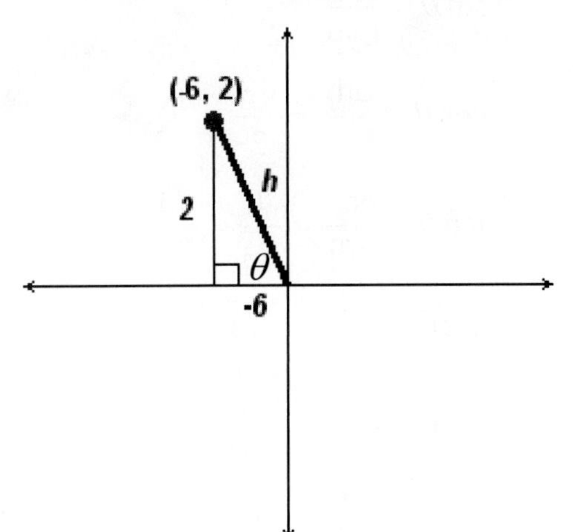

Using the Pythagorean Theorem, we obtain $h = \sqrt{(-6)^2 + 2^2} = \sqrt{40} = 2\sqrt{10}$.

$$\sin\theta = \frac{\text{opp}}{\text{hyp}} = \frac{2}{2\sqrt{10}} = \frac{\sqrt{10}}{10}$$

$$\cos\theta = \frac{\text{adj}}{\text{hyp}} = \frac{-6}{2\sqrt{10}} = \frac{-3\sqrt{10}}{10}$$

$$\tan\theta = \frac{\text{opp}}{\text{adj}} = \frac{2}{-6} = \frac{-1}{3}$$

$$\cot\theta = \frac{1}{\tan\theta} = -3$$

$$\sec\theta = \frac{1}{\cos\theta} = \frac{-\sqrt{10}}{3}$$

$$\csc\theta = \frac{1}{\sin\theta} = \sqrt{10}$$

52. Consider the following diagram:

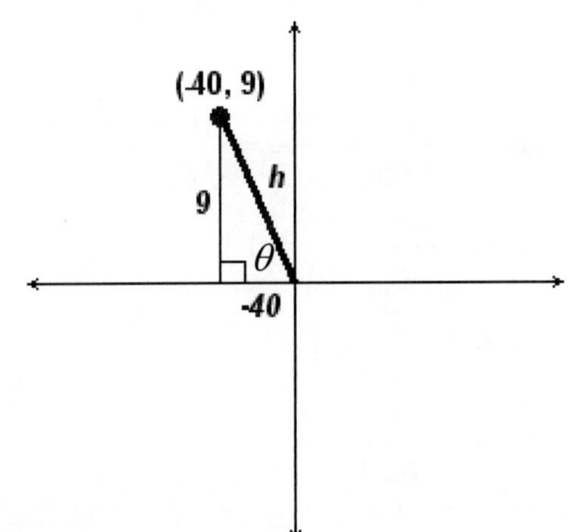

Using the Pythagorean Theorem, we obtain $h = \sqrt{(-40)^2 + 9^2} = 41$.

$$\sin\theta = \frac{\text{opp}}{\text{hyp}} = \frac{9}{41}$$

$$\cos\theta = \frac{\text{adj}}{\text{hyp}} = \frac{-40}{41}$$

$$\tan\theta = \frac{\text{opp}}{\text{adj}} = \frac{9}{-40} = \frac{-9}{40}$$

$$\cot\theta = \frac{1}{\tan\theta} = \frac{-40}{9}$$

$$\sec\theta = \frac{1}{\cos\theta} = \frac{-41}{40}$$

$$\csc\theta = \frac{1}{\sin\theta} = \frac{41}{9}$$

53. Consider the following diagram:

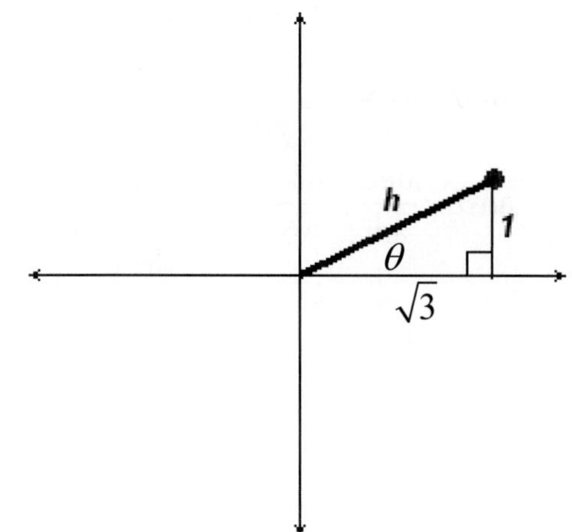

Using the Pythagorean Theorem, we obtain $h = \sqrt{\left(\sqrt{3}\right)^2 + 1^2} = 2$.

$$\sin\theta = \frac{\text{opp}}{\text{hyp}} = \frac{1}{2}$$

$$\cos\theta = \frac{\text{adj}}{\text{hyp}} = \frac{\sqrt{3}}{2}$$

$$\tan\theta = \frac{\text{opp}}{\text{adj}} = \frac{1}{\sqrt{3}} = \frac{\sqrt{3}}{3}$$

$$\cot\theta = \frac{1}{\tan\theta} = \sqrt{3}$$

$$\sec\theta = \frac{1}{\cos\theta} = \frac{2}{\sqrt{3}} = \frac{2\sqrt{3}}{3}$$

$$\csc\theta = \frac{1}{\sin\theta} = 2$$

54. Consider the following diagram:

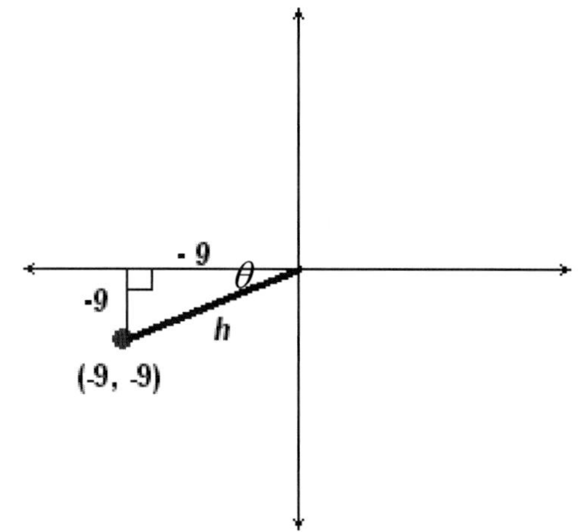

Using the Pythagorean Theorem, we obtain $h = \sqrt{(-9)^2 + (-9)^2} = \sqrt{162} = 9\sqrt{2}$.

$$\sin\theta = \frac{\text{opp}}{\text{hyp}} = \frac{-9}{9\sqrt{2}} = \frac{-1}{\sqrt{2}} = \frac{-\sqrt{2}}{2}$$

$$\cos\theta = \frac{\text{adj}}{\text{hyp}} = \frac{-9}{9\sqrt{2}} = \frac{-1}{\sqrt{2}} = \frac{-\sqrt{2}}{2}$$

$$\tan\theta = \frac{\text{opp}}{\text{adj}} = \frac{-9}{-9} = 1$$

$$\cot\theta = \frac{1}{\tan\theta} = 1$$

$$\sec\theta = \frac{1}{\cos\theta} = -\sqrt{2}$$

$$\csc\theta = \frac{1}{\sin\theta} = -\sqrt{2}$$

55. Consider the following diagram:

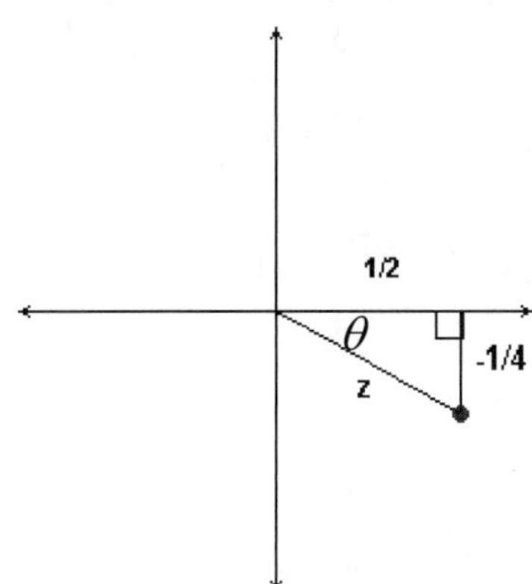

Using the Pythagorean theorem, we obtain $z=\sqrt{\left(\frac{1}{2}\right)^2+\left(-\frac{1}{4}\right)^2}=\frac{\sqrt{5}}{4}$.

$$\sin\theta=\frac{\text{opp}}{\text{hyp}}=\frac{-1}{\sqrt{5}}=\frac{-\sqrt{5}}{5}$$

$$\cos\theta=\frac{\text{adj}}{\text{hyp}}=\frac{\frac{1}{2}}{\frac{\sqrt{5}}{4}}=\frac{2\sqrt{5}}{5}$$

$$\tan\theta=\frac{\text{opp}}{\text{adj}}=\frac{-\frac{1}{4}}{\frac{1}{2}}=-\frac{1}{2}$$

$$\cot\theta=\frac{1}{\tan\theta}=-2$$

$$\sec\theta=\frac{1}{\cos\theta}=\frac{5}{2\sqrt{5}}=\frac{\sqrt{5}}{2}$$

$$\csc\theta=\frac{1}{\sin\theta}=-\sqrt{5}$$

56. Consider the following diagram:

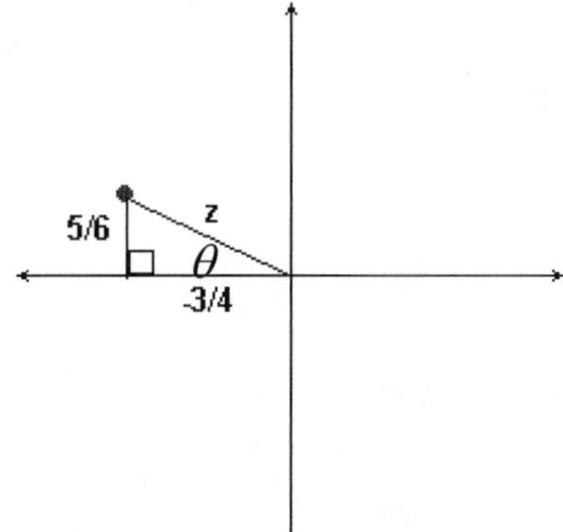

Using the Pythagorean theorem, we obtain $z=\sqrt{\left(\frac{5}{6}\right)^2+\left(-\frac{3}{4}\right)^2}=\frac{\sqrt{181}}{12}$.

$$\sin\theta=\frac{\text{opp}}{\text{hyp}}=\frac{\frac{5}{6}}{\frac{\sqrt{181}}{12}}=\frac{10\sqrt{181}}{181}$$

$$\cos\theta=\frac{\text{adj}}{\text{hyp}}=\frac{-\frac{3}{4}}{\frac{\sqrt{181}}{12}}=\frac{-9\sqrt{181}}{181}$$

$$\tan\theta=\frac{\text{opp}}{\text{adj}}=-\frac{10}{9}$$

$$\cot\theta=\frac{1}{\tan\theta}=-\frac{9}{10}$$

$$\sec\theta=\frac{1}{\cos\theta}=\frac{-\sqrt{181}}{9}$$

$$\csc\theta=\frac{1}{\sin\theta}=\frac{\sqrt{181}}{10}$$

57. Consider the following diagram:

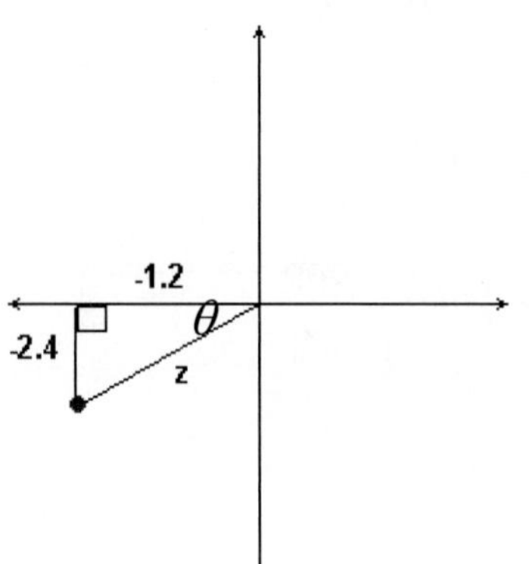

Using the Pythagorean theorem, we obtain $z = \sqrt{(-1.2)^2 + (-2.4)^2} = \sqrt{7.20}$.

$$\sin\theta = \frac{\text{opp}}{\text{hyp}} = \frac{-2.4}{\sqrt{7.2}} = \frac{-\sqrt{7.2}}{3}$$

$$\cos\theta = \frac{\text{adj}}{\text{hyp}} = \frac{-1.2}{\sqrt{7.2}} = \frac{-\sqrt{7.2}}{6}$$

$$\tan\theta = \frac{\text{opp}}{\text{adj}} = 2$$

$$\cot\theta = \frac{1}{\tan\theta} = \frac{1}{2}$$

$$\sec\theta = \frac{1}{\cos\theta} = \frac{-\sqrt{7.2}}{1.2}$$

$$\csc\theta = \frac{1}{\sin\theta} = \frac{-\sqrt{7.2}}{2.4}$$

58. Consider the following diagram:

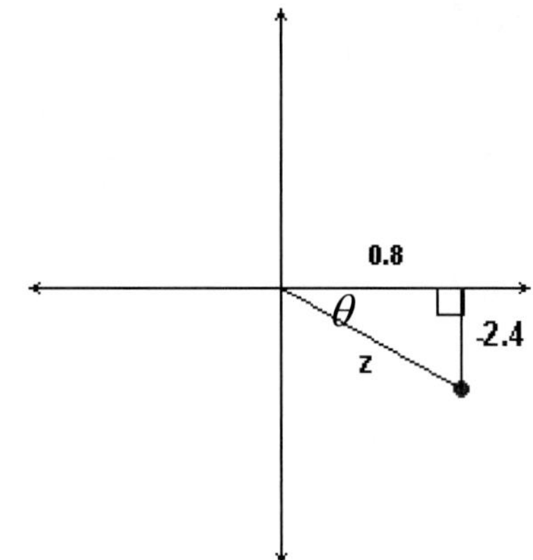

Using the Pythagorean theorem, we obtain $z = \sqrt{(0.8)^2 + (-2.4)^2} = 0.8\sqrt{10}$.

$$\sin\theta = \frac{\text{opp}}{\text{hyp}} = \frac{-2.4}{0.8\sqrt{10}} = \frac{-3\sqrt{10}}{10}$$

$$\cos\theta = \frac{\text{adj}}{\text{hyp}} = \frac{0.8}{0.8\sqrt{10}} = \frac{\sqrt{10}}{10}$$

$$\tan\theta = \frac{\text{opp}}{\text{adj}} = -3$$

$$\cot\theta = \frac{1}{\tan\theta} = -\frac{1}{3}$$

$$\sec\theta = \frac{1}{\cos\theta} = \sqrt{10}$$

$$\csc\theta = \frac{1}{\sin\theta} = \frac{-\sqrt{10}}{3}$$

59.

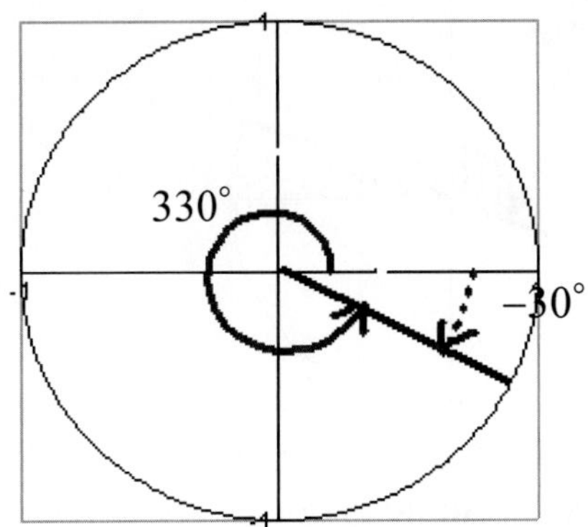

The given angle is shown in **bold**, while the reference angle is *dotted.*

$$\sin(330^\circ) = -\sin(30^\circ) = \boxed{-\frac{1}{2}}$$

60.

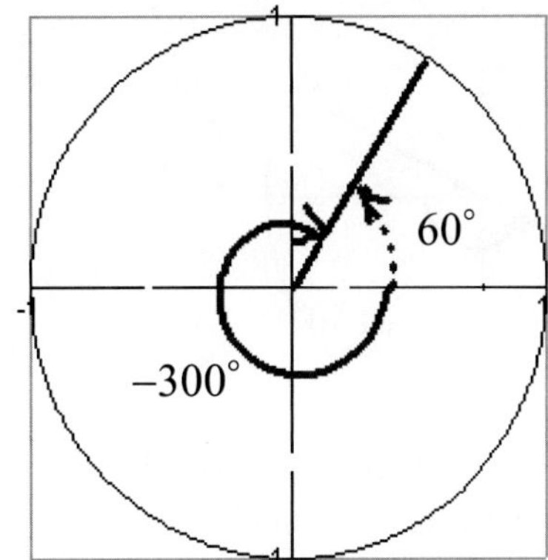

The given angle is shown in **bold**, while the reference angle is *dotted.*

$$\cos(-300^\circ) = \cos(60^\circ) = \boxed{\frac{1}{2}}$$

61.

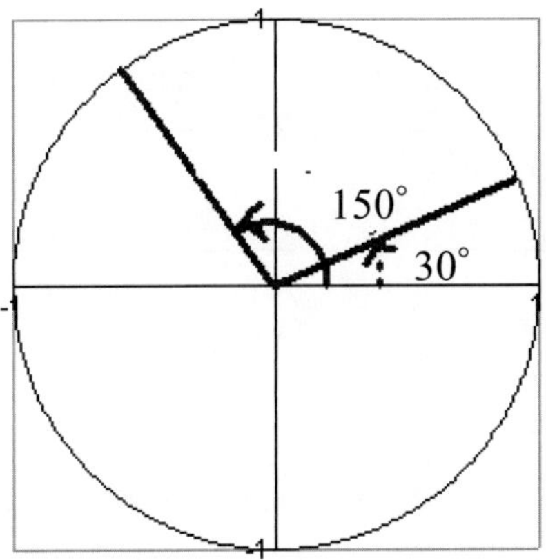

The given angle is shown in **bold**, while the reference angle is *dotted.*

$$\tan(150^\circ) = \frac{\sin(150^\circ)}{\cos(150^\circ)} = \frac{\sin 30^\circ}{-\cos 30^\circ}$$

$$= \frac{1/2}{-\sqrt{3}/2} = \boxed{-\frac{\sqrt{3}}{3}}$$

62.

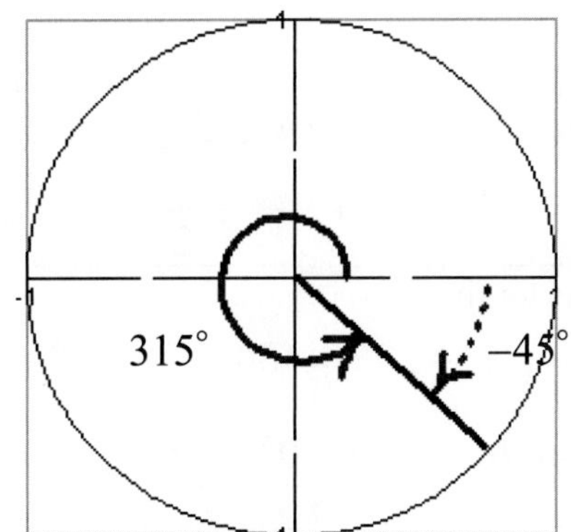

The given angle is shown in **bold**, while the reference angle is *dotted.*

$$\cot(315^\circ) = \frac{\cos(315^\circ)}{\sin(315^\circ)} = \frac{\cos 45^\circ}{-\sin 45^\circ} = \boxed{-1}$$

63.

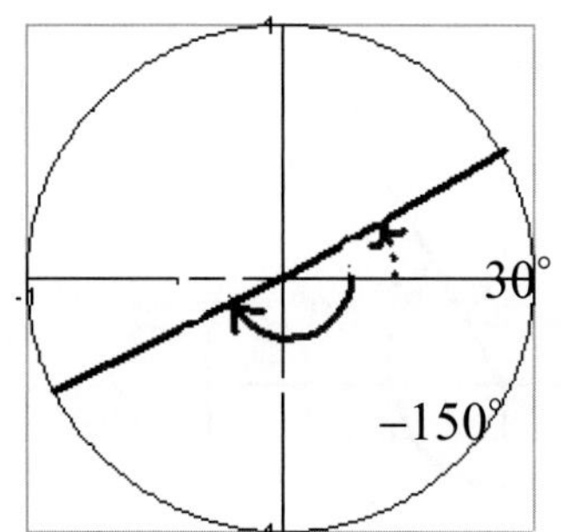

The given angle is shown in **bold**, while the reference angle is *dotted.*

$$\sec(-150^\circ) = \frac{1}{\cos(-150^\circ)} = \frac{1}{-\cos(30^\circ)} = \frac{-2}{\sqrt{3}} = \boxed{\frac{-2\sqrt{3}}{3}}$$

64.

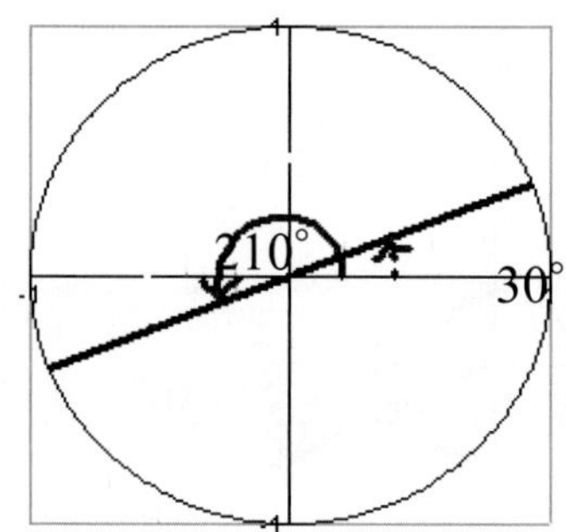

The given angle is shown in **bold**, while the reference angle is *dotted.*

$$\csc(210^\circ) = \frac{1}{\sin(210^\circ)} = \frac{-1}{\sin(30^\circ)} = \boxed{-2}$$

65. $-\frac{\sqrt{2}}{2}$	**66.** $-\frac{\sqrt{3}}{2}$	**67.** $\frac{-\frac{\sqrt{3}}{2}}{-\frac{1}{2}} = \sqrt{3}$	**68.** $\frac{-\frac{1}{2}}{-\frac{\sqrt{3}}{2}} = \frac{\sqrt{3}}{3}$
69. $\frac{1}{-\frac{\sqrt{2}}{2}} = -\sqrt{2}$	**70.** $\frac{1}{-\frac{\sqrt{3}}{2}} = -\frac{2\sqrt{3}}{3}$	**71.** $\frac{1}{-\frac{\sqrt{3}}{2}} = -\frac{2\sqrt{3}}{3}$	**72.** $\frac{\sqrt{3}}{2}$

73. Observe that:

$\gamma = 180^\circ - (10^\circ + 20^\circ) = 150^\circ$

$$\frac{\sin\alpha}{a} = \frac{\sin\beta}{b} \Rightarrow \frac{\sin 10^\circ}{4} = \frac{\sin 20^\circ}{b} \Rightarrow b = \frac{(4)(\sin 20^\circ)}{\sin 10^\circ} \approx 8$$

$$\frac{\sin\alpha}{a} = \frac{\sin\gamma}{c} \Rightarrow \frac{\sin 10^\circ}{4} = \frac{\sin 150^\circ}{c} \Rightarrow c = \frac{(4)(\sin 150^\circ)}{\sin 10^\circ} \approx 12$$

74. Observe that:

$\alpha = 180^\circ - (40^\circ + 60^\circ) = 80^\circ$

$$\frac{\sin\alpha}{a} = \frac{\sin\beta}{b} \Rightarrow \frac{\sin 80^\circ}{a} = \frac{\sin 40^\circ}{10} \Rightarrow a = \frac{(10)(\sin 80^\circ)}{\sin 40^\circ} \approx 15$$

$$\frac{\sin\beta}{b} = \frac{\sin\gamma}{c} \Rightarrow \frac{\sin 40^\circ}{10} = \frac{\sin 60^\circ}{c} \Rightarrow c = \frac{(10)(\sin 60^\circ)}{\sin 40^\circ} \approx 13$$

75. Observe that:

$$\gamma = 180^\circ - \left(45^\circ + 5^\circ\right) = 130^\circ$$

$$\frac{\sin\alpha}{a} = \frac{\sin\gamma}{c} \Rightarrow \frac{\sin 5^\circ}{a} = \frac{\sin 130^\circ}{10} \Rightarrow a = \frac{(10)\left(\sin 5^\circ\right)}{\sin 130^\circ} \approx 1.14 \approx 1$$

$$\frac{\sin\beta}{b} = \frac{\sin\gamma}{c} \Rightarrow \frac{\sin 45^\circ}{b} = \frac{\sin 130^\circ}{10} \Rightarrow b = \frac{(10)\left(\sin 45^\circ\right)}{\sin 130^\circ} \approx 9.23 \approx 9$$

76. Observe that:

$$\alpha = 180^\circ - \left(70^\circ + 60^\circ\right) = 50^\circ$$

$$\frac{\sin\alpha}{a} = \frac{\sin\beta}{b} \Rightarrow \frac{\sin 50^\circ}{20} = \frac{\sin 60^\circ}{b} \Rightarrow b = \frac{(20)\left(\sin 60^\circ\right)}{\sin 50^\circ} \approx 23$$

$$\frac{\sin\alpha}{a} = \frac{\sin\gamma}{c} \Rightarrow \frac{\sin 50^\circ}{20} = \frac{\sin 70^\circ}{c} \Rightarrow c = \frac{(20)\left(\sin 70^\circ\right)}{\sin 50^\circ} \approx 25$$

77. Observe that:

$$\beta = 180^\circ - \left(11^\circ + 11^\circ\right) = 158^\circ$$

$$\frac{\sin\alpha}{a} = \frac{\sin\gamma}{c} \Rightarrow \frac{\sin 11^\circ}{a} = \frac{\sin 11^\circ}{11} \Rightarrow a = 11$$

$$\frac{\sin\beta}{b} = \frac{\sin\gamma}{c} \Rightarrow \frac{\sin 158^\circ}{b} = \frac{\sin 11^\circ}{11} \Rightarrow b = \frac{(11)\left(\sin 158^\circ\right)}{\sin 11^\circ} \approx 22$$

78. Observe that:

$$\alpha = 180^\circ - \left(20^\circ + 50^\circ\right) = 110^\circ$$

$$\frac{\sin\alpha}{a} = \frac{\sin\beta}{b} \Rightarrow \frac{\sin 110^\circ}{a} = \frac{\sin 20^\circ}{8} \Rightarrow a = \frac{(8)\left(\sin 110^\circ\right)}{\sin 20^\circ} \approx 22$$

$$\frac{\sin\beta}{b} = \frac{\sin\gamma}{c} \Rightarrow \frac{\sin 20^\circ}{8} = \frac{\sin 50^\circ}{c} \Rightarrow c = \frac{(8)\left(\sin 50^\circ\right)}{\sin 20^\circ} \approx 18$$

79. Observe that:

$$\beta = 180^\circ - \left(45^\circ + 45^\circ\right) = 90^\circ$$

$$\frac{\sin\alpha}{a} = \frac{\sin\beta}{b} \Rightarrow \frac{\sin 45^\circ}{a} = \frac{\sin 90^\circ}{2} \Rightarrow a = \frac{(2)\left(\sin 45^\circ\right)}{\sin 90^\circ} \approx \sqrt{2}$$

$$\frac{\sin\beta}{b} = \frac{\sin\gamma}{c} \Rightarrow \frac{\sin 90^\circ}{2} = \frac{\sin 45^\circ}{c} \Rightarrow c = \frac{(2)\left(\sin 45^\circ\right)}{\sin 90^\circ} \approx \sqrt{2}$$

80. Observe that:

$\gamma = 180^\circ - \left(60^\circ + 20^\circ\right) = 100^\circ$

$$\frac{\sin\alpha}{a} = \frac{\sin\gamma}{c} \Rightarrow \frac{\sin 60^\circ}{a} = \frac{\sin 100^\circ}{17} \Rightarrow a = \frac{(17)\left(\sin 60^\circ\right)}{\sin 100^\circ} \approx 15$$

$$\frac{\sin\beta}{b} = \frac{\sin\gamma}{c} \Rightarrow \frac{\sin 20^\circ}{b} = \frac{\sin 100^\circ}{17} \Rightarrow b = \frac{(17)\left(\sin 20^\circ\right)}{\sin 100^\circ} \approx 6$$

81. Observe that:

$\beta = 180^\circ - \left(12^\circ + 22^\circ\right) = 146^\circ$

$$\frac{\sin\alpha}{a} = \frac{\sin\beta}{b} \Rightarrow \frac{\sin 12^\circ}{99} = \frac{\sin 146^\circ}{b} \Rightarrow b = \frac{(99)\left(\sin 146^\circ\right)}{\sin 12^\circ} \approx 266$$

$$\frac{\sin\alpha}{a} = \frac{\sin\gamma}{c} \Rightarrow \frac{\sin 12^\circ}{99} = \frac{\sin 22^\circ}{c} \Rightarrow c = \frac{(99)\left(\sin 22^\circ\right)}{\sin 12^\circ} \approx 178$$

82. Observe that:

$\alpha = 180^\circ - \left(102^\circ + 27^\circ\right) = 51^\circ$

$$\frac{\sin\alpha}{a} = \frac{\sin\beta}{b} \Rightarrow \frac{\sin 51^\circ}{24} = \frac{\sin 102^\circ}{b} \Rightarrow b = \frac{(24)\left(\sin 102^\circ\right)}{\sin 51^\circ} \approx 30$$

$$\frac{\sin\beta}{b} = \frac{\sin\gamma}{c} \Rightarrow \frac{\sin 51^\circ}{24} = \frac{\sin 27^\circ}{c} \Rightarrow c = \frac{(24)\left(\sin 27^\circ\right)}{\sin 51^\circ} \approx 14$$

83. Step 1: Determine β.

$$\frac{\sin\alpha}{a} = \frac{\sin\beta}{b} \Rightarrow \frac{\sin 20^\circ}{7} = \frac{\sin\beta}{9} \Rightarrow \sin\beta = \frac{9\sin 20^\circ}{7} \Rightarrow \beta = \sin^{-1}\left(\frac{9\sin 20^\circ}{7}\right) \approx 26^\circ$$

This is the solution in QI – label it as β_1. The second solution in QII is given by $\beta_2 = 180^\circ - \beta_1 \approx 154^\circ$. Both are tenable solutions, so we need to solve for two triangles.

Step 2: Solve for both triangles.

QI triangle: $\beta_1 \approx 26^\circ$

$\gamma_1 \approx 180^\circ - \left(26^\circ + 20^\circ\right) = 134^\circ$

$$\frac{\sin\beta_1}{b} = \frac{\sin\gamma_1}{c_1} \Rightarrow \frac{\sin 26^\circ}{9} = \frac{\sin 134^\circ}{c_1} \Rightarrow c_1 = \frac{9\sin 134^\circ}{\sin 26^\circ} \approx 15$$

QII triangle: $\beta_2 \approx 154^\circ$

$\gamma_2 \approx 180^\circ - \left(154^\circ + 20^\circ\right) = 6^\circ$

$$\frac{\sin\beta_2}{b} = \frac{\sin\gamma_2}{c_2} \Rightarrow \frac{\sin 154^\circ}{9} = \frac{\sin 6^\circ}{c_2} \Rightarrow c_2 = \frac{9\sin 6^\circ}{\sin 154^\circ} \approx 2$$

84. Step 1: Determine γ.

$$\frac{\sin\gamma}{c}=\frac{\sin\beta}{b} \Rightarrow \frac{\sin\gamma}{30}=\frac{\sin 16^\circ}{24} \Rightarrow \sin\gamma=\frac{30\sin 16^\circ}{24} \Rightarrow \gamma=\sin^{-1}\left(\frac{30\sin 16^\circ}{24}\right)\approx 20.154^\circ\approx 20^\circ$$

This is the solution in QI – label it as γ_1. The second solution in QII is given by $\gamma_2=180^\circ-\gamma_1\approx 160^\circ$. Both are tenable solutions, so we need to solve for two triangles.

Step 2: Solve for both triangles.

QI triangle: $\gamma_1\approx 20.154^\circ\approx 20^\circ$

$$\alpha_1\approx 180^\circ-\left(20.154^\circ+16^\circ\right)=143.84587^\circ\approx 144^\circ$$

$$\frac{\sin\alpha_1}{a_1}=\frac{\sin\gamma_1}{c} \Rightarrow \frac{\sin 143.84587^\circ}{a_1}=\frac{\sin 16^\circ}{24} \Rightarrow a_1=\frac{24\sin 143.84587^\circ}{\sin 16^\circ}\approx 51.36\approx 51$$

QII triangle: $\gamma_2\approx 160^\circ$

$$\alpha_2\approx 180^\circ-\left(160^\circ+16^\circ\right)=4^\circ$$

$$\frac{\sin\alpha_2}{a_2}=\frac{\sin\gamma_2}{c} \Rightarrow \frac{\sin 4^\circ}{a_2}=\frac{\sin 160^\circ}{30} \Rightarrow a_2=\frac{30\sin 4^\circ}{\sin 160^\circ}\approx 6$$

85. Step 1: Determine γ.

$$\frac{\sin\gamma}{c}=\frac{\sin\alpha}{a} \Rightarrow \frac{\sin\gamma}{12}=\frac{\sin 24^\circ}{10} \Rightarrow \sin\gamma=\frac{12\sin 24^\circ}{10} \Rightarrow \gamma=\sin^{-1}\left(\frac{12\sin 24^\circ}{10}\right)\approx 29^\circ$$

This is the solution in QI – label it as γ_1. The second solution in QII is given by $\gamma_2=180^\circ-\gamma_1\approx 151^\circ$. Both are tenable solutions, so we need to solve for two triangles.

Step 2: Solve for both triangles.

QI triangle: $\gamma_1\approx 29^\circ$

$$\alpha_1\approx 180^\circ-\left(29^\circ+24^\circ\right)=127^\circ$$

$$\frac{\sin\beta_1}{b_1}=\frac{\sin\alpha}{a} \Rightarrow \frac{\sin 127^\circ}{b_1}=\frac{\sin 24^\circ}{10} \Rightarrow b_1=\frac{10\sin 127^\circ}{\sin 24^\circ}\approx 20$$

QII triangle: $\gamma_2\approx 151^\circ$

$$\beta_2\approx 180^\circ-\left(151^\circ+24^\circ\right)=5^\circ$$

$$\frac{\sin\beta_2}{b_2}=\frac{\sin\alpha}{a} \Rightarrow \frac{\sin 5^\circ}{b_2}=\frac{\sin 24^\circ}{10} \Rightarrow b_2=\frac{10\sin 5^\circ}{\sin 24^\circ}\approx 2$$

86. Step 1: Determine γ.

$$\frac{\sin\gamma}{c}=\frac{\sin\beta}{b} \Rightarrow \frac{\sin\gamma}{116}=\frac{\sin 12^\circ}{100} \Rightarrow \sin\gamma=\frac{116\sin 12^\circ}{100} \Rightarrow \gamma=\sin^{-1}\left(\frac{116\sin 12^\circ}{100}\right)\approx 14^\circ$$

This is the solution in QI – label it as γ_1. The second solution in QII is given by $\gamma_2 = 180^\circ - \gamma_1 \approx 166^\circ$. Both are tenable solutions, so we need to solve for two triangles.

Step 2: Solve for both triangles.

QI triangle: $\gamma_1 \approx 14^\circ$

$$\alpha_1 \approx 180^\circ - (12^\circ + 14^\circ) = 154^\circ$$

$$\frac{\sin\alpha_1}{a_1}=\frac{\sin\beta}{b} \Rightarrow \frac{\sin 154^\circ}{a_1}=\frac{\sin 12^\circ}{100} \Rightarrow a_1=\frac{100\sin 154^\circ}{\sin 12^\circ}\approx 211$$

QII triangle: $\gamma_2 \approx 166^\circ$

$$\alpha_2 \approx 180^\circ - (166^\circ + 12^\circ) = 2^\circ$$

$$\frac{\sin\alpha_2}{a_2}=\frac{\sin\beta}{b} \Rightarrow \frac{\sin 2^\circ}{a_2}=\frac{\sin 12^\circ}{100} \Rightarrow a_2=\frac{100\sin 2^\circ}{\sin 12^\circ}\approx 17$$

87. First, determine α:

$$\frac{\sin\alpha}{a}=\frac{\sin\beta}{b} \Rightarrow \frac{\sin\alpha}{40}=\frac{\sin 150^\circ}{30} \Rightarrow \sin\alpha=\frac{40\sin 150^\circ}{30} \Rightarrow \alpha=\sin^{-1}\left(\frac{40\sin 150^\circ}{30}\right)\approx 42^\circ$$

Hence, there is no triangle in this case since $\alpha+\beta>180^\circ$.

88. Step 1: Determine β.

$$\frac{\sin\gamma}{c}=\frac{\sin\beta}{b} \Rightarrow \frac{\sin 165^\circ}{3}=\frac{\sin\beta}{2} \Rightarrow \sin\beta=\frac{2\sin 165^\circ}{3} \Rightarrow \beta=\sin^{-1}\left(\frac{2\sin 165^\circ}{3}\right)\approx 10^\circ$$

Note that there is only one triangle in this case since the angle in QII with the same sine as this value of β is $180^\circ - 10^\circ = 170^\circ$. In such case, note that $\beta+\gamma>180^\circ$, therefore preventing the formation of a triangle (since the three interior angles, two of which are β and γ, must sum to 180°).

Step 2: Solve for the triangle.

$$\alpha \approx 180^\circ - (10^\circ + 165^\circ) = 5^\circ$$

$$\frac{\sin\gamma}{c}=\frac{\sin\alpha}{a} \Rightarrow \frac{\sin 165^\circ}{3}=\frac{\sin 5^\circ}{a} \Rightarrow a=\frac{3\sin 5^\circ}{\sin 165^\circ}\approx 1$$

89. Step 1: Determine β.

$$\frac{\sin\alpha}{a}=\frac{\sin\beta}{b} \Rightarrow \frac{\sin 10^\circ}{4}=\frac{\sin\beta}{6} \Rightarrow \sin\beta=\frac{6\sin 10^\circ}{4} \Rightarrow \beta=\sin^{-1}\left(\frac{6\sin 10^\circ}{4}\right)\approx 15^\circ$$

This is the solution in QI – label it as β_1. The second solution in QII is given by $\beta_2=180^\circ-\beta_1\approx 165^\circ$. Both are tenable solutions, so we need to solve for two triangles.

Step 2: Solve for both triangles.

QI triangle: $\beta_1\approx 15^\circ$

$$\gamma_1\approx 180^\circ-\left(15^\circ+10^\circ\right)=155^\circ$$

$$\frac{\sin\beta_1}{b}=\frac{\sin\gamma_1}{c_1} \Rightarrow \frac{\sin 15^\circ}{6}=\frac{\sin 155^\circ}{c_1} \Rightarrow c_1=\frac{6\sin 155^\circ}{\sin 15^\circ}\approx 10$$

QII triangle: $\beta_2\approx 165^\circ$

$$\gamma_2\approx 180^\circ-\left(165^\circ+10^\circ\right)=5^\circ$$

$$\frac{\sin\beta_2}{b}=\frac{\sin\gamma_2}{c_2} \Rightarrow \frac{\sin 165^\circ}{6}=\frac{\sin 5^\circ}{c_2} \Rightarrow c_2=\frac{6\sin 5^\circ}{\sin 165^\circ}\approx 2$$

90. Step 1: Determine α.

$$\frac{\sin\alpha}{a}=\frac{\sin\gamma}{c} \Rightarrow \frac{\sin\alpha}{37}=\frac{\sin 4^\circ}{25} \Rightarrow \sin\alpha=\frac{37\sin 4^\circ}{25} \Rightarrow \alpha=\sin^{-1}\left(\frac{37\sin 4^\circ}{25}\right)\approx 6^\circ$$

This is the solution in QI – label it as α_1. The second solution in QII is given by $\alpha_2=180^\circ-\alpha_1\approx 174^\circ$. Both are tenable solutions, so we need to solve for two triangles.

Step 2: Solve for both triangles.

QI triangle: $\alpha_1\approx 6^\circ$

$$\beta_1\approx 180^\circ-\left(6^\circ+4^\circ\right)=170^\circ$$

$$\frac{\sin\beta_1}{b_1}=\frac{\sin\gamma}{c} \Rightarrow \frac{\sin 170^\circ}{b_1}=\frac{\sin 4^\circ}{25} \Rightarrow b_1=\frac{25\sin 170^\circ}{\sin 4^\circ}\approx 62$$

QII triangle: $\alpha_2\approx 174^\circ$

$$\beta_2\approx 180^\circ-\left(174^\circ+4^\circ\right)=2^\circ$$

$$\frac{\sin\beta_2}{b_2}=\frac{\sin\gamma}{c} \Rightarrow \frac{\sin 2^\circ}{b_2}=\frac{\sin 4^\circ}{25} \Rightarrow b_2=\frac{25\sin 2^\circ}{\sin 4^\circ}\approx 13$$

91. This is SAS, so begin by using Law of Cosines:

Step 1: Find c.

$$c^2 = a^2 + b^2 - 2ab\cos\gamma = (40)^2 + (60)^2 - 2(40)(60)\cos(50^\circ) = 5200 - 4800\cos(50^\circ)$$

$$c \approx 45.9849 \approx 46$$

Step 2: Find α.

$$\frac{\sin\alpha}{a} = \frac{\sin\gamma}{c} \Rightarrow \frac{\sin\alpha}{40} = \frac{\sin 50^\circ}{45.9849} \Rightarrow \sin\alpha = \frac{40\sin 50^\circ}{45.9849} \Rightarrow \alpha = \sin^{-1}\left(\frac{40\sin 50^\circ}{45.9849}\right) \approx 42^\circ$$

Step 3: Find β.

$$\beta \approx 180^\circ - (42^\circ + 50^\circ) = 88^\circ$$

92. This is SAS, so begin by using Law of Cosines:

Step 1: Find a.

$$a^2 = b^2 + c^2 - 2bc\cos\alpha = (15)^2 + (12)^2 - 2(15)(12)\cos(140^\circ) = 369 - 360\cos(140^\circ)$$

$$a \approx 25.392 \approx 25$$

Step 2: Find β.

$$\frac{\sin\alpha}{a} = \frac{\sin\beta}{b} \Rightarrow \frac{\sin 40^\circ}{25.392} = \frac{\sin\beta}{15} \Rightarrow \sin\beta = \frac{15\sin 140^\circ}{25.392} \Rightarrow \beta = \sin^{-1}\left(\frac{15\sin 140^\circ}{25.392}\right) \approx 22^\circ$$

Step 3: Find γ.

$$\gamma \approx 180^\circ - (22^\circ + 140^\circ) = 18^\circ$$

93. This is SSS, so begin by using Law of Cosines:

Step 1: Find the largest angle (i.e., the one opposite the longest side). Here, it is γ.

$$c^2 = a^2 + b^2 - 2ab\cos\gamma \Rightarrow (30)^2 = (24)^2 + (25)^2 - 2(24)(25)\cos\gamma$$

$$\Rightarrow 900 = 1201 - 1200\cos\gamma$$

Thus, $\cos\gamma = \frac{301}{1200}$ so that $\gamma = \cos^{-1}\left(\frac{301}{1200}\right) \approx 75.47^\circ \approx 75^\circ$.

Step 2: Find either of the remaining two angles using the Law of Sines.

$$\frac{\sin\gamma}{c} = \frac{\sin\beta}{b} \Rightarrow \frac{\sin 75.47^\circ}{30} = \frac{\sin\beta}{25} \Rightarrow \sin\beta = \frac{25\sin 75.47^\circ}{30} \Rightarrow \beta = \sin^{-1}\left(\frac{25\sin 75.47^\circ}{30}\right) \approx 54^\circ$$

Step 3: Find the third angle.

$$\alpha \approx 180^\circ - (54^\circ + 75^\circ) = 51^\circ$$

94. This is SSS, so begin by using Law of Cosines:

Step 1: Find the largest angle (i.e., the one opposite the longest side). Here, it is γ.

$$c^2 = a^2 + b^2 - 2ab\cos\gamma \Rightarrow (8)^2 = (6)^2 + (6)^2 - 2(6)(6)\cos\gamma$$
$$\Rightarrow 64 = 72 - 72\cos\gamma$$

Thus, $\cos\gamma = \frac{8}{72}$ so that $\gamma = \cos^{-1}\left(\frac{8}{72}\right) \approx 83.6206^\circ \approx 84^\circ$.

Step 2: Find either of the remaining two angles using the Law of Sines.

$$\frac{\sin\gamma}{c} = \frac{\sin\beta}{b} \Rightarrow \frac{\sin 83.6206^\circ}{8} = \frac{\sin\beta}{6} \Rightarrow \sin\beta = \frac{6\sin 83.6206^\circ}{8} \Rightarrow \beta = \sin^{-1}\left(\frac{6\sin 83.6206^\circ}{8}\right) \approx 48^\circ$$

Step 3: Find the third angle.

$$\alpha \approx 180^\circ - (48^\circ + 84^\circ) = 48^\circ$$

95. This is SSS, so begin by using Law of Cosines:

Step 1: Find the largest angle (i.e., the one opposite the longest side). Here, it is γ.

$$c^2 = a^2 + b^2 - 2ab\cos\gamma \Rightarrow (5)^2 = \left(\sqrt{11}\right)^2 + \left(\sqrt{14}\right)^2 - 2\left(\sqrt{11}\right)\left(\sqrt{14}\right)\cos\gamma$$
$$\Rightarrow 25 = 25 - 2\sqrt{11}\sqrt{14}\cos\gamma$$

Thus, $\cos\gamma = 0$ so that $\gamma = \cos^{-1}(0) = 90^\circ$.

Step 2: Find either of the remaining two angles using the Law of Sines.

$$\frac{\sin\gamma}{c} = \frac{\sin\beta}{b} \Rightarrow \frac{\sin 90^\circ}{5} = \frac{\sin\beta}{\sqrt{14}} \Rightarrow \sin\beta = \frac{\sqrt{14}\sin 90^\circ}{5} \Rightarrow \beta = \sin^{-1}\left(\frac{\sqrt{14}\sin 90^\circ}{5}\right) \approx 48^\circ$$

Step 3: Find the third angle.

$$\alpha \approx 180^\circ - (48^\circ + 90^\circ) = 42^\circ$$

96. This is SSS, so begin by using Law of Cosines:

Step 1: Find the largest angle (i.e., the one opposite the longest side). Here, it is γ.

$$c^2 = a^2 + b^2 - 2ab\cos\gamma \Rightarrow (122)^2 = (120)^2 + (22)^2 - 2(120)(22)\cos\gamma$$
$$\Rightarrow 14{,}884 = 14{,}884 - 5280\cos\gamma$$

Thus, $\cos\gamma = 0$ so that $\gamma = \cos^{-1}(0) = 90^\circ$.

Step 2: Find either of the remaining two angles using the Law of Sines.

$$\frac{\sin\gamma}{c} = \frac{\sin\beta}{b} \Rightarrow \frac{\sin 90^\circ}{122} = \frac{\sin\beta}{120} \Rightarrow \sin\beta = \frac{120\sin 90^\circ}{122} \Rightarrow \beta = \sin^{-1}\left(\frac{120\sin 90^\circ}{122}\right) \approx 80^\circ$$

Step 3: Find the third angle.

$$\alpha \approx 180^\circ - (80^\circ + 90^\circ) = 10^\circ$$

97. This is SAS, so begin by using Law of Cosines:

Step 1: Find *a*.

$$a^2 = b^2 + c^2 - 2bc\cos\alpha = (7)^2 + (10)^2 - 2(7)(10)\cos(14^\circ) = 149 - 140\cos(14^\circ)$$

$$a \approx 4$$

Step 2: Find β.

$$\frac{\sin\alpha}{a} = \frac{\sin\beta}{b} \Rightarrow \frac{\sin 14^\circ}{3.627} = \frac{\sin\beta}{7} \Rightarrow \sin\beta = \frac{7\sin 14^\circ}{3.627} \Rightarrow \beta = \sin^{-1}\left(\frac{7\sin 14^\circ}{3.627}\right) \approx 28^\circ$$

Step 3: Find γ.

$$\gamma \approx 180^\circ - (28^\circ + 14^\circ) = 138^\circ$$

98. This is SAS, so begin by using Law of Cosines:

Step 1: Find *c*.

$$c^2 = a^2 + b^2 - 2ab\cos\gamma = (6)^2 + (12)^2 - 2(6)(12)\cos(80^\circ) = 180 - 144\cos(80^\circ)$$

$$c \approx 12.449 \approx 12$$

Step 2: Find β.

$$\frac{\sin\beta}{b} = \frac{\sin\gamma}{c} \Rightarrow \frac{\sin\beta}{12} = \frac{\sin 80^\circ}{12.449} \Rightarrow \sin\beta = \frac{12\sin 80^\circ}{12.449} \Rightarrow \beta = \sin^{-1}\left(\frac{12\sin 80^\circ}{12.449}\right) \approx 72^\circ$$

Step 3: Find α.

$$\alpha \approx 180^\circ - (72^\circ + 80^\circ) = 28^\circ$$

99. This is SAS, so begin by using Law of Cosines:

Step 1: Find *a*.

$$a^2 = b^2 + c^2 - 2bc\cos\alpha = (10)^2 + (4)^2 - 2(10)(4)\cos(90^\circ) = 116 - 80\cos(90^\circ)$$

$$a \approx 10.770 \approx 11$$

Step 2: Find β.

$$\frac{\sin\alpha}{a} = \frac{\sin\beta}{b} \Rightarrow \frac{\sin 90^\circ}{10.770} = \frac{\sin\beta}{10} \Rightarrow \sin\beta = \frac{10\sin 90^\circ}{10.770} \Rightarrow \beta = \sin^{-1}\left(\frac{10\sin 90^\circ}{10.770}\right) \approx 68^\circ$$

Step 3: Find γ.

$$\gamma \approx 180^\circ - (68^\circ + 90^\circ) = 22^\circ$$

100. This is SAS, so begin by using Law of Cosines:

<u>Step 1</u>: Find c.

$$c^2 = a^2 + b^2 - 2ab\cos\gamma = (4)^2 + (5)^2 - 2(4)(5)\cos(75^\circ) = 41 - 40\cos(75^\circ)$$

$$c \approx 5.536 \approx 6$$

<u>Step 2</u>: Find β.

$$\frac{\sin\beta}{b} = \frac{\sin\gamma}{c} \Rightarrow \frac{\sin\beta}{5} = \frac{\sin 75^\circ}{5.536} \Rightarrow \sin\beta = \frac{5\sin 75^\circ}{5.536} \Rightarrow \beta = \sin^{-1}\left(\frac{5\sin 75^\circ}{5.536}\right) \approx 61^\circ$$

<u>Step 3</u>: Find α.

$$\alpha \approx 180^\circ - (61^\circ + 75^\circ) = 44^\circ$$

101. This is SSS, so begin by using Law of Cosines:

<u>Step 1</u>: Find the largest angle (i.e., the one opposite the longest side). Here, it is γ.

$$c^2 = a^2 + b^2 - 2ab\cos\gamma \;\Rightarrow\; (12)^2 = (10)^2 + (11)^2 - 2(10)(11)\cos\gamma$$

$$\Rightarrow\; 144 = 221 - 220\cos\gamma$$

Thus, $\cos\gamma = \dfrac{77}{220}$ so that $\gamma = \cos^{-1}\left(\dfrac{77}{220}\right) \approx 69.512^\circ \approx 70^\circ$.

<u>Step 2</u>: Find either of the remaining two angles using the Law of Sines.

$$\frac{\sin\gamma}{c} = \frac{\sin\beta}{b} \Rightarrow \frac{\sin 69.512^\circ}{12} = \frac{\sin\beta}{11} \Rightarrow \sin\beta = \frac{11\sin 69.512^\circ}{12} \Rightarrow \beta = \sin^{-1}\left(\frac{11\sin 69.512^\circ}{12}\right) \approx 59^\circ$$

<u>Step 3</u>: Find the third angle.

$$\alpha \approx 180^\circ - (70^\circ + 59^\circ) = 51^\circ$$

102. This is SSS, so begin by using Law of Cosines:

<u>Step 1</u>: Find the largest angle (i.e., the one opposite the longest side). Here, it is γ.

$$c^2 = a^2 + b^2 - 2ab\cos\gamma \;\Rightarrow\; (25)^2 = (22)^2 + (24)^2 - 2(22)(24)\cos\gamma$$

$$\Rightarrow\; 625 = 1060 - 1056\cos\gamma$$

Thus, $\cos\gamma = \dfrac{435}{1056}$ so that $\gamma = \cos^{-1}\left(\dfrac{435}{1056}\right) \approx 65.673^\circ \approx 66^\circ$.

<u>Step 2</u>: Find either of the remaining two angles using the Law of Sines.

$$\frac{\sin\gamma}{c} = \frac{\sin\beta}{b} \Rightarrow \frac{\sin 65.673^\circ}{25} = \frac{\sin\beta}{24} \Rightarrow \sin\beta = \frac{24\sin 65.673^\circ}{25} \Rightarrow \beta = \sin^{-1}\left(\frac{24\sin 65.673^\circ}{25}\right) \approx 61^\circ$$

<u>Step 3</u>: Find the third angle.

$$\alpha \approx 180^\circ - (61^\circ + 66^\circ) = 53^\circ$$

103. This is SAS, so begin by using Law of Cosines:

Step 1: Find *a*.

$$a^2 = b^2 + c^2 - 2bc\cos\alpha = (16)^2 + (18)^2 - 2(16)(18)\cos(100^\circ) = 580 - 576\cos(100^\circ)$$

$$a \approx 26.077 \approx 26$$

Step 2: Find β.

$$\frac{\sin\alpha}{a} = \frac{\sin\beta}{b} \Rightarrow \frac{\sin 100^\circ}{26.077} = \frac{\sin\beta}{16} \Rightarrow \sin\beta = \frac{16\sin 100^\circ}{26.077} \Rightarrow \beta = \sin^{-1}\left(\frac{16\sin 100^\circ}{26.077}\right) \approx 37^\circ$$

Step 3: Find γ.

$$\gamma \approx 180^\circ - (37^\circ + 100^\circ) = 43^\circ$$

104. This is SAS, so begin by using Law of Cosines:

Step 1: Find *b*.

$$b^2 = a^2 + c^2 - 2ac\cos\beta = (25)^2 + (25)^2 - 2(25)(25)\cos(9^\circ) = 1250 - 1250\cos(9^\circ)$$

$$b \approx 3.923 \approx 4$$

Step 2: Find α.

$$\frac{\sin\beta}{b} = \frac{\sin\gamma}{c} \Rightarrow \frac{\sin 9^\circ}{3.923} = \frac{\sin\gamma}{25} \Rightarrow \sin\gamma = \frac{25\sin 9^\circ}{3.923} \Rightarrow \gamma = \sin^{-1}\left(\frac{25\sin 9^\circ}{3.923}\right) \approx 85.5^\circ$$

Step 3: Find γ.

$$\gamma \approx 180^\circ - (9^\circ + 85.5^\circ) = 85.5^\circ$$

105. This is SAS, so begin by using Law of Cosines:

Step 1: Find *a*.

$$a^2 = b^2 + c^2 - 2bc\cos\alpha = (12)^2 + (40)^2 - 2(12)(40)\cos(10^\circ) = 1744 - 960\cos(10^\circ)$$

$$a \approx 28.259 \approx 28$$

Step 2: Find β.

$$\frac{\sin\alpha}{a} = \frac{\sin\beta}{b} \Rightarrow \frac{\sin 10^\circ}{28.259} = \frac{\sin\beta}{12} \Rightarrow \sin\beta = \frac{12\sin 10^\circ}{28.259} \Rightarrow \beta = \sin^{-1}\left(\frac{12\sin 10^\circ}{28.259}\right) \approx 4^\circ$$

Step 3: Find γ.

$$\gamma \approx 180^\circ - (4^\circ + 10^\circ) = 166^\circ$$

106. This is SSS, so begin by using Law of Cosines:

Step 1: Find the largest angle (i.e., the one opposite the longest side). Here, it is α.

$a^2 = b^2 + c^2 - 2bc\cos\alpha \Rightarrow (26)^2 = (20)^2 + (10)^2 - 2(20)(10)\cos\alpha \Rightarrow 676 = 500 - 400\cos\alpha$

Thus, $\cos\alpha = -\frac{176}{400}$ so that $\alpha = \cos^{-1}\left(-\frac{176}{400}\right) \approx 116.103^\circ \approx 116^\circ$.

Step 2: Find either of the remaining two angles using the Law of Sines.

$$\frac{\sin\alpha}{a} = \frac{\sin\beta}{b} \Rightarrow \frac{\sin 116.103^\circ}{26} = \frac{\sin\beta}{20} \Rightarrow \sin\beta = \frac{20\sin 116.103^\circ}{26} \Rightarrow \beta = \sin^{-1}\left(\frac{20\sin 116.103^\circ}{26}\right) \approx 44^\circ$$

Step 3: Find the third angle - $\gamma \approx 180^\circ - (116^\circ + 44^\circ) = 20^\circ$

107. There is no triangle in this case since $a + c = 39 \not> 40 = b$.

108. There is no triangle in this case since $a + b = 3 \not> 3 = c$.

109. This is SSA, so use Law of Sines:

Step 1: Determine β.

$$\frac{\sin\alpha}{a} = \frac{\sin\beta}{b} \Rightarrow \frac{\sin 15^\circ}{6.3} = \frac{\sin\beta}{4.2} \Rightarrow \sin\beta = \frac{4.2\sin 15^\circ}{6.3} \Rightarrow \beta = \sin^{-1}\left(\frac{4.2\sin 15^\circ}{6.3}\right) \approx 10^\circ$$

Note that there is only one triangle in this case since the angle in QII with the same sine as this value of β is $180^\circ - 10^\circ = 170^\circ$. In such case, note that $\beta + \alpha > 180^\circ$, therefore preventing the formation of a triangle (since the three interior angles, two of which are β and α, must sum to 180°).

Step 2: Solve for the triangle.

$$\gamma \approx 180^\circ - (10^\circ + 15^\circ) = 155^\circ$$

$$\frac{\sin\gamma}{c} = \frac{\sin\alpha}{a} \Rightarrow \frac{\sin 155^\circ}{c} = \frac{\sin 15^\circ}{6.3} \Rightarrow c = \frac{6.3\sin 155^\circ}{\sin 15^\circ} \approx 10.3$$

110. This is SSA, so use Law of Sines:

Step 1: Determine γ.

$$\frac{\sin\gamma}{c} = \frac{\sin\beta}{b} \Rightarrow \frac{\sin\gamma}{6} = \frac{\sin 35^\circ}{5} \Rightarrow \sin\gamma = \frac{6\sin 35^\circ}{5} \Rightarrow \gamma = \sin^{-1}\left(\frac{6\sin 35^\circ}{5}\right) \approx 43^\circ$$

Note that there is only one triangle in this case since the angle in QII with the same sine as this value of γ is $180^\circ - 43^\circ = 137^\circ$. In such case, note that $\beta + \gamma > 180^\circ$, therefore preventing the formation of a triangle (since the three interior angles, two of which are β and γ, must sum to 180°).

Step 2: Solve for the triangle.

$$\alpha \approx 180^\circ - (35^\circ + 43^\circ) = 102^\circ$$

$$\frac{\sin\alpha}{a} = \frac{\sin\beta}{b} \Rightarrow \frac{\sin 102^\circ}{a} = \frac{\sin 35^\circ}{5} \Rightarrow a = \frac{5\sin 102^\circ}{\sin 35^\circ} \approx 9$$

111. $A=\frac{1}{2}(16)(18)\sin 100^\circ \approx \boxed{141.8}$	**112.** $A=\frac{1}{2}(25)(25)\sin 9^\circ \approx \boxed{48.89}$

113. First, observe that $s=\frac{a+b+c}{2}=\frac{33}{2}=16.5$. So, the area is given by

$$A=\sqrt{s(s-a)(s-b)(s-c)}=\sqrt{(16.5)(6.5)(5.5)(4.5)}\approx \boxed{51.5}$$

114. First, observe that $s=\frac{a+b+c}{2}=\frac{71}{2}=35.5$. So, the area is given by

$$A=\sqrt{s(s-a)(s-b)(s-c)}=\sqrt{(35.5)(13.5)(11.5)(10.5)}\approx \boxed{240.6}$$

115. First, observe that $s=\frac{a+b+c}{2}=\frac{56}{2}=28$. So, the area is given by

$$A=\sqrt{s(s-a)(s-b)(s-c)}=\sqrt{(28)(2)(8)(18)}\approx \boxed{89.8}$$

116. First, observe that $s=\frac{a+b+c}{2}=\frac{96}{2}=48$. So, the area is given by

$$A=\sqrt{s(s-a)(s-b)(s-c)}=\sqrt{(48)(24)(16)(8)}\approx \boxed{384}$$

117. $A=\frac{1}{2}(12)(40)\sin 10^\circ \approx \boxed{41.7}$	**118.** $A=\frac{1}{2}(21)(75)\sin 60^\circ \approx \boxed{682.0}$

119. Use the formula $A=\frac{abc}{4r}$ with $a=b=c=9$ in. and $A=35$ in.^{2} to find r:

$$35=\frac{(9)(9)(9)}{4r} \text{ so that } 140r=729 \text{ and hence, } r=\frac{729}{140}\approx \boxed{5.2 \text{ in.}}$$

120. Use the formula $A=\frac{abc}{4r}$ with $a=9$ in., $b=12$ in., $c=15$ in. and $A=54$ in.^{2} to find r:

$$54=\frac{(9)(12)(15)}{4r} \text{ so that } 216r=1620 \text{ and hence, } r=\boxed{7.5 \text{ in.}}$$

Chapter 4 Practice Test Solutions--

1. Consider the following two diagrams:

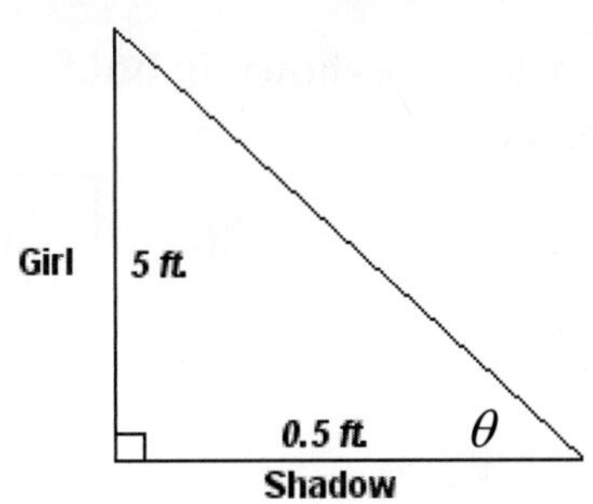

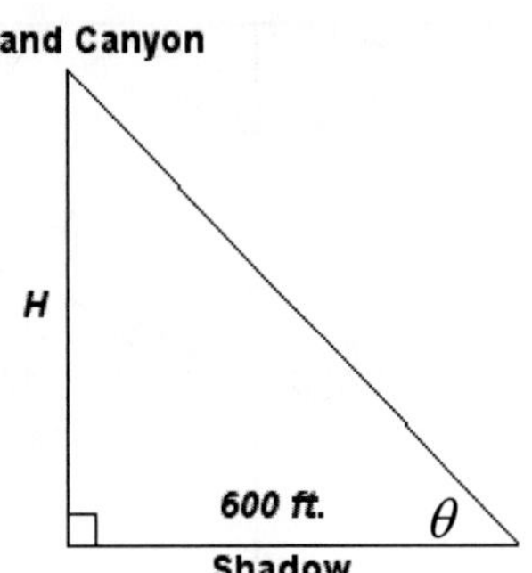

Let H = height of the Grand Canyon (in feet). Then, using similarity (which applies since sunlight rays act like parallel lines – see Text Section 6.1 Example 7), we obtain

$$\frac{5\text{ ft.}}{\frac{1}{2}\text{ ft.}} = \frac{H}{600\text{ ft.}} \quad \text{so that } H = (600\text{ ft.})\left(\frac{5\text{ ft.}}{\frac{1}{2}\text{ ft.}}\right) = 6{,}000\text{ ft.}.$$

So, the Grand Canyon is $\boxed{6{,}000\text{ ft}}$. tall.

2. In the following computations, we use

$$\tan\theta = \frac{\sin\theta}{\cos\theta},\quad \cot\theta = \frac{1}{\tan\theta},\quad \sec\theta = \frac{1}{\cos\theta},\quad \csc\theta = \frac{1}{\sin\theta}$$

θ	$\sin\theta$	$\cos\theta$	$\tan\theta$	$\cot\theta$	$\sec\theta$	$\csc\theta$
30°	$\frac{1}{2}$	$\frac{\sqrt{3}}{2}$	$\frac{\sqrt{3}}{3}$	$\sqrt{3}$	$\frac{2\sqrt{3}}{3}$	2
45°	$\frac{\sqrt{2}}{2}$	$\frac{\sqrt{2}}{2}$	1	1	$\sqrt{2}$	$\sqrt{2}$
60°	$\frac{\sqrt{3}}{2}$	$\frac{1}{2}$	$\sqrt{3}$	$\frac{\sqrt{3}}{3}$	2	$\frac{2\sqrt{3}}{3}$

3. The first value ($\cos\theta = \frac{2}{3}$) is an exact value of the cosine function, while the second ($\cos\theta = 0.66667$) would serve as an approximation to the first.

4.

	0°	**QI**	90°	**QII**	180°	**QIII**	270°	**QIV**	360°
$\sin\theta$	0	+	1	+	0	-	-1	-	0
$\cos\theta$	1	+	0	-	-1	-	0	+	1

5. Since $\sec\theta = \frac{1}{\cos\theta} > 0$, it follows that $\cos\theta > 0$. Since also we have

$\cot\theta = \frac{\cos\theta}{\sin\theta} < 0$, it follows that $\sin\theta < 0$, so that the terminal side of θ is in $\boxed{\text{QIV}}$.

6. Consider the following diagram:

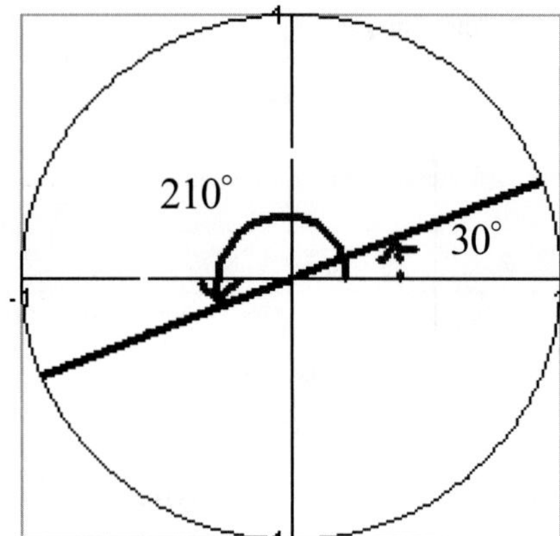

The given angle is shown in **bold**, while the reference angle is *dotted*.

$$\sin(210^\circ) = -\sin(30^\circ) = \boxed{-\frac{1}{2}}$$

7. $\frac{13\pi}{4} = \frac{13\not{\pi}}{4} \cdot \frac{180^\circ}{\not{\pi}} = \frac{13\not{4}(45^\circ)}{\not{4}} = \boxed{585^\circ}$

8. $260^\circ = 260^\circ \cdot \frac{\pi}{180^\circ} = 13\not{(20^\circ)} \cdot \frac{\pi}{9\not{(20^\circ)}} = \boxed{\frac{13\pi}{9}}$

9. Since the central angle swept out in this scenario is 150°, we see that the area of this sector is $\frac{1}{2}(3)^2(150^\circ)\left(\frac{\pi}{180^\circ}\right) = \frac{15\pi}{4}$ sq. in.

10. Consider the following diagram:

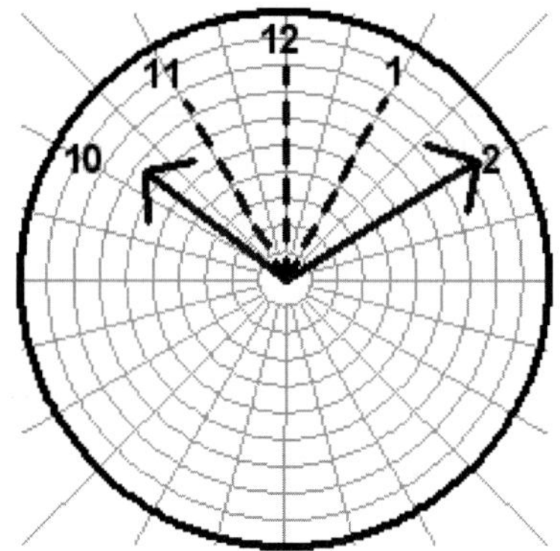

Suppose the central angle, θ, of a sector was formed by having the hands of the clock lie identically on the **10** and the **2**. Since the radian measure of *one* of the twelve congruent sectors (formed between consecutive hours) is $\frac{2\pi}{12}$, we see that

$$\theta = 4\left(\frac{2\pi}{12}\right) = \frac{2\pi}{3}.$$

However, at the time **10:10**, once the minute hand is on the **2**, the hour hand has moved $\frac{1}{6}$ the distance from the **10** to the **11** – this slight movement corresponds to a central angle of 5°, or $5^\circ\left(\frac{\pi}{180^\circ}\right) = \frac{\pi}{36}$ radians. Hence, the desired angle is $\left(\frac{2\pi}{3} - \frac{\pi}{36}\right) = \boxed{\frac{23\pi}{36}}$.

11. This is AAS, so we use Law of Sines. Observe that:

$\gamma = 180^\circ - (30^\circ + 40^\circ) = 110^\circ$

$$\frac{\sin\alpha}{a} = \frac{\sin\beta}{b} \Rightarrow \frac{\sin 30^\circ}{a} = \frac{\sin 40^\circ}{10} \Rightarrow a = \frac{(10)(\sin 30^\circ)}{\sin 40^\circ} \approx 7.8$$

$$\frac{\sin\beta}{b} = \frac{\sin\gamma}{c} \Rightarrow \frac{\sin 40^\circ}{10} = \frac{\sin 110^\circ}{c} \Rightarrow c = \frac{(10)(\sin 110^\circ)}{\sin 40^\circ} \approx 14.6$$

12. Since only three angles are given, and none of the side lengths is prescribed, there are infinitely many such triangles by similarity.

13. This is SSS, so begin by using Law of Cosines:

Step 1: Find the largest angle (i.e., the one opposite the longest side). Here, it is γ.

$$c^2 = a^2 + b^2 - 2ab\cos\gamma \Rightarrow (12)^2 = (7)^2 + (9)^2 - 2(7)(9)\cos\gamma$$
$$\Rightarrow 144 = 130 - 126\cos\gamma$$

Thus, $\cos\gamma = -\frac{14}{126}$ so that $\gamma = \cos^{-1}\left(-\frac{14}{126}\right) \approx 96.4^\circ$.

Step 2: Find either of the remaining two angles using the Law of Sines.

$$\frac{\sin\gamma}{c} = \frac{\sin\beta}{b} \Rightarrow \frac{\sin 96.4^\circ}{12} = \frac{\sin\beta}{9} \Rightarrow \sin\beta = \frac{9\sin 96.4^\circ}{12} \Rightarrow \beta = \sin^{-1}\left(\frac{9\sin 96.4^\circ}{12}\right) \approx 48.2^\circ$$

Step 3: Find the third angle - $\alpha \approx 180^\circ - (96.4^\circ + 48.2^\circ) = 35.4^\circ$

14. This is SSA, so we use Law of Sines:

Step 1: Determine β.

$$\frac{\sin\alpha}{a} = \frac{\sin\beta}{b} \Rightarrow \frac{\sin 45^\circ}{8} = \frac{\sin\beta}{10} \Rightarrow \sin\beta = \frac{10\sin 45^\circ}{8} \Rightarrow \beta = \sin^{-1}\left(\frac{10\sin 45^\circ}{8}\right) \approx 62^\circ$$

This is the solution in QI – label it as β_1. The second solution in QII is given by $\beta_2 = 180^\circ - \beta_1 \approx 118^\circ$. Both are tenable solutions, so we need to solve for two triangles.

Step 2: Solve for both triangles.

QI triangle: $\beta_1 \approx 62^\circ$

$\gamma_1 \approx 180^\circ - (62^\circ + 45^\circ) = 73^\circ$

$$\frac{\sin\alpha}{a} = \frac{\sin\gamma_1}{c_1} \Rightarrow \frac{\sin 45^\circ}{8} = \frac{\sin 73^\circ}{c_1} \Rightarrow c_1 = \frac{8\sin 73^\circ}{\sin 45^\circ} \approx 10.8$$

QII triangle: $\beta_2 \approx 118^\circ$

$\gamma_2 \approx 180^\circ - (118^\circ + 45^\circ) = 17^\circ$

$$\frac{\sin\alpha}{a} = \frac{\sin\gamma_2}{c_2} \Rightarrow \frac{\sin 45^\circ}{8} = \frac{\sin 17^\circ}{c_2} \Rightarrow c_2 = \frac{8\sin 17^\circ}{\sin 45^\circ} \approx 3.3$$

15. There is no triangle in this case since the triangle inequality is violated (since $a+b \not> c$).
16. This is SAS, so begin by using Law of Cosines: Step 1: Find the side opposite the given angle β. $b^2 = a^2 + c^2 - 2ac\cos\beta \Rightarrow b^2 = \left(\tfrac{23}{7}\right)^2 + \left(\tfrac{5}{7}\right)^2 - 2\left(\tfrac{23}{7}\right)\left(\tfrac{5}{7}\right)\cos 61.2^\circ \Rightarrow b \approx 3.01$ Step 2: Find either of the remaining two angles using the Law of Sines. $\frac{\sin\alpha}{a} = \frac{\sin\beta}{b} \Rightarrow \frac{\sin\alpha}{23/7} = \frac{\sin 61.2^\circ}{3.01} \Rightarrow \sin\alpha = \frac{\frac{23}{7}\sin 61.2^\circ}{3.01} \Rightarrow \alpha = \sin^{-1}\left(\frac{\frac{23}{7}\sin 61.2^\circ}{3.01}\right) \approx 73.21^\circ$ Step 3: Find the third angle - $\gamma \approx 180^\circ - (73.21^\circ + 61.2^\circ) = 45.59^\circ$
17. This is AAS, so we use Law of Sines. Observe that: $\gamma = 180^\circ - (110^\circ + 20^\circ) = 50^\circ$ $\frac{\sin\alpha}{a} = \frac{\sin\beta}{b} \Rightarrow \frac{\sin 110^\circ}{5} = \frac{\sin 20^\circ}{b} \Rightarrow b = \frac{(5)(\sin 20^\circ)}{\sin 110^\circ} \approx 1.82$ $\frac{\sin\alpha}{a} = \frac{\sin\gamma}{c} \Rightarrow \frac{\sin 110^\circ}{5} = \frac{\sin 50^\circ}{c} \Rightarrow c = \frac{(5)(\sin 50^\circ)}{\sin 110^\circ} \approx 4.08$
18. This is SAS, so begin by using Law of Cosines: Step 1: Find the side opposite the given angle α. $a^2 = b^2 + c^2 - 2bc\cos\alpha \Rightarrow a^2 = \left(\tfrac{\sqrt{5}}{2}\right)^2 + \left(3\sqrt{5}\right)^2 - 2\left(\tfrac{\sqrt{5}}{2}\right)\left(3\sqrt{5}\right)\cos 45^\circ \Rightarrow a \approx 5.97$ Step 2: Find either of the remaining two angles using the Law of Sines. $\frac{\sin\alpha}{a} = \frac{\sin\beta}{b} \Rightarrow \frac{\sin 45^\circ}{5.97} = \frac{\sin\beta}{\sqrt{5}/2} \Rightarrow \sin\beta = \frac{\sqrt{5}/2\,\sin 45^\circ}{5.97} \Rightarrow \beta = \sin^{-1}\left(\frac{\sqrt{5}/2\,\sin 45^\circ}{5.97}\right) \approx 7.61^\circ$ Step 3: Find the third angle - $\gamma \approx 180^\circ - (45^\circ + 7.61^\circ) = 127.39^\circ$
19. $A = \frac{1}{2}(10)(12)\sin 72^\circ \approx \boxed{57}$
20. First, observe that $s = \frac{a+b+c}{2} = \frac{30}{2} = 15$. So, the area is given by $A = \sqrt{s(s-a)(s-b)(s-c)} = \sqrt{(15)(8)(5)(3)} \approx \boxed{34.64}$

Chapter 4 Cumulative Test

1. $\dfrac{f(4)-f(2)}{4-2}=\dfrac{\frac{5}{4}-\frac{5}{2}}{2}=\boxed{-\dfrac{5}{8}}$

2. We must have $x^2-25\geq 0$, which is equivalent to $(x-5)(x+5)\geq 0$. The critical points are -5, 5:

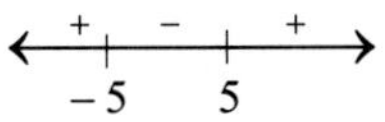

So, the domain is $\boxed{(-\infty,-5]\cup[5,\infty)}$.

3.

$$\frac{f(x+h)-f(x)}{h}=\frac{\left(5-(x+h)^2\right)-\left(5-x^2\right)}{h}$$
$$=\frac{5-x^2-2hx-h^2-5+x^2}{h}$$
$$=\frac{-h\left(2x+h\right)}{h}=\boxed{-2x-h}$$

4.

a. $f(0)=-1$

b. $f(4)=7$

c. $f(5)=0$

d. $f(-4)=16$

e. The domain and range are both all reals.

f. increasing on (0,5) and decreasing on $(-\infty,0)\cup(5,\infty)$. Never constant.

5. $f(-1)=\sqrt[3]{-1-7}=-2 \qquad g(f(-1))=g(-2)=\dfrac{5}{3-(-2)}=\boxed{1}$

6. In order to find the inverse of $f(x)=\dfrac{5x+2}{x-3}$, switch the *x* and *y*, and then solve for *y*:

$$y=\frac{5x+2}{x-3}$$
$$x=\frac{5y+2}{y-3}$$
$$x(y-3)=5y+2$$
$$xy-3x=5y+2$$
$$xy-5y=3x+2$$
$$y=\frac{3x+2}{x-5}$$

So, $\boxed{f^{-1}(x)=\dfrac{3x+2}{x-5}}$.

7. We need to identify *a*, *h*, and *k* in the equation $y=a(x-h)^2+k$. Since the vertex is (0,7), we know that $y=a(x)^2+7$. To find *a*, substitute in (2,-1):

$$-1=a(2)^2+7 \;\Rightarrow\; -8=4a \;\Rightarrow\; a=-2$$

Thus, the equation is $\boxed{y=-2x^2+7}$.

8. Observe that $f(x)=\frac{1}{7}x^5+\frac{2}{9}x^3=x^3\left(\frac{1}{7}x^2+\frac{2}{9}\right)$. The only real zero is 0, with multiplicity 3. The other factor is an irreducible quadratic factor (with only imaginary roots).

9.
Vertical asymptote: $x=2$
Horizontal asymptote: none
Slant asymptote:

$$\begin{array}{r|l} & x+2 \\ x-2 & x^2+0x+3 \\ & \underline{-(x^2-2x)} \\ & \quad 2x+3 \\ & \quad \underline{-(2x-4)} \\ & \qquad 7 \end{array}$$

So, the slant asymptote is $y=x+2$.

The graph is as follows:

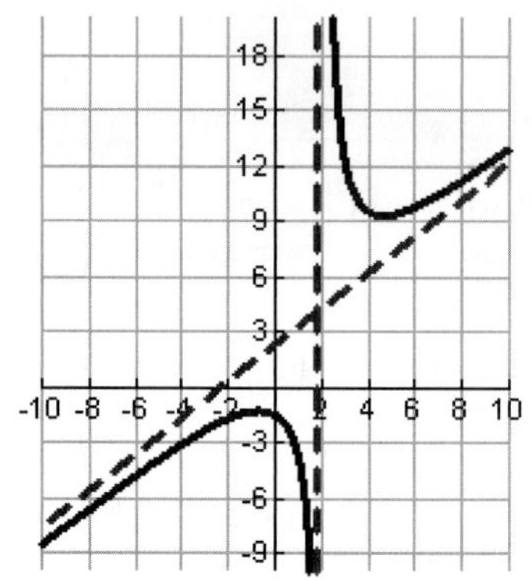

10. Factors of 5: $\pm1,\ \pm5$
Factors of 4: $\pm1,\ \pm2,\ \pm4$
Possible rational zeros: $\pm1,\ \pm5,\ \pm\frac{1}{2},\ \pm\frac{5}{2},\ \pm\frac{1}{4},\ \pm\frac{5}{4}$
Observe that

$$\begin{array}{r|rrrrr} -\frac{1}{2} & 4 & -4 & 13 & 18 & 5 \\ & & -2 & 3 & -8 & -5 \\ \hline -\frac{1}{2} & 4 & -6 & 16 & 10 & 0 \\ & & -2 & 4 & -10 & \\ \hline & 4 & -8 & 20 & 0 & \end{array}$$

So, $P(x)=\left(x+\frac{1}{2}\right)^2\left(4x^2-8x+20\right)=\boxed{(2x+1)^2\left(x^2-2x+5\right)}$.

11. Use $A=Pe^{rt}$. Here,

$$A=85{,}000,\ r=0.055,\ t=15.$$

Solving for P yields

$$85{,}000=Pe^{0.055\,(15)}$$

$$p=\frac{85{,}000}{e^{0.055\,(15)}}\approx 37{,}250$$

So, about \$37,250.

12. $\log_{4.7} 8.9 = \dfrac{\log 8.9}{\log 4.7} \approx \boxed{1.413}$	
13. Observe that $$5\left(10^{2x}\right) = 37 \;\Rightarrow\; 10^{2x} = 7.6 \;\Rightarrow\; 2x = \log_{10} 7.6 \;\Rightarrow\; \boxed{x \approx 0.440}$$	
14. $$\begin{aligned} \ln\sqrt{6-3x} - \tfrac{1}{2}\ln(x+2) &= \ln x \\ \tfrac{1}{2}\ln(6-3x) - \tfrac{1}{2}\ln(x+2) &= \ln x \\ \tfrac{1}{2}\ln\left(\frac{6-3x}{x+2}\right) &= \ln x \\ \ln\left(\frac{6-3x}{x+2}\right)^{\frac{1}{2}} &= \ln x \\ \left(\frac{6-3x}{x+2}\right)^{\frac{1}{2}} &= x \\ \frac{6-3x}{x+2} &= x^2 \\ 6-3x &= x^2(x+2) \\ x^3 + 2x^2 + 3x - 6 &= 0 \\ (x-1)\left(x^2+3x+6\right) &= 0 \\ \boxed{x=1} \end{aligned}$$	
15. Since the leg has length $x = 15$ ft. and it is a $45^\circ - 45^\circ - 90^\circ$ triangle, the hypotenuse is $\boxed{15\sqrt{2}\text{ ft.}}$	
16. First, convert all measurements to improper fractions. Then, using similar triangles yields the proportion $\dfrac{T}{46/3} = \dfrac{6}{23/10} \Rightarrow \left(23/10\right)T = 6\left(46/3\right) \Rightarrow \boxed{T = \tfrac{920}{23}\text{ ft.}}$	
17. $432^\circ = 432^\circ \cdot \dfrac{\pi}{180^\circ} \approx \boxed{7.54}$ To find the exact answer, observe that $432^\circ = 360^\circ + 72^\circ$, and $72^\circ = 72^\circ \cdot \dfrac{\pi}{180^\circ} = \dfrac{2\pi}{5}$. Thus, $432^\circ = 2\pi + \dfrac{2\pi}{5} = \dfrac{12\pi}{5}$.	
18. $\dfrac{5\pi}{9} = \dfrac{5\pi}{9} \cdot \dfrac{180^\circ}{\pi} \approx \boxed{100^\circ}$	**19.** $\dfrac{-\frac{\sqrt{3}}{2}}{-\frac{1}{2}} = \sqrt{3}$
20. $\dfrac{1}{-\frac{\sqrt{3}}{2}} = \dfrac{-2\sqrt{3}}{3}$	**21.** 1.6616

22. First, note that $\theta = 180^\circ - \left(43^\circ + 90^\circ\right) = 47^\circ$. Using the Law of Sines then yields

$$\frac{\sin 90^\circ}{12cm} = \frac{\sin 43^\circ}{b} \Rightarrow b = \frac{(12cm)\sin 43^\circ}{\sin 90^\circ} \approx 8.2cm$$

$$\frac{\sin 90^\circ}{12cm} = \frac{\sin 47^\circ}{a} \Rightarrow a = \frac{(12cm)\sin 47^\circ}{\sin 90^\circ} \approx 8.8cm$$

23. First, note that $\alpha = 180^\circ - \left(19^\circ + 23\right) = 138^\circ$. Using the Law of Sines then yields

$$\frac{\sin 19^\circ}{c} = \frac{\sin 138^\circ}{16cm} \Rightarrow c = \frac{(16cm)\sin 19^\circ}{\sin 138^\circ} \approx 8cm$$

$$\frac{\sin 23^\circ}{c} = \frac{\sin 138^\circ}{16cm} \Rightarrow c = \frac{(16cm)\sin 23^\circ}{\sin 138^\circ} \approx 9cm$$

24. First, determine γ using the Law of Cosines:

$$5^2 = 2^2 + 4^2 - 2(2)(4)\cos\gamma \Rightarrow -\tfrac{5}{16} = \cos\gamma \Rightarrow \gamma \approx 108^\circ.$$

Next, find β using the Law of Cosines:

$$\frac{\sin\beta}{4} = \frac{\sin 108^\circ}{5} \Rightarrow \sin\beta = \frac{4\sin 108^\circ}{5} \Rightarrow \beta \approx 50^\circ.$$

Finally, conclude that $\alpha = 180^\circ - \left(108^\circ + 50^\circ\right) = 22^\circ$.

CHAPTER 5

Section 5.1 Solutions --

1. $\sin\left(\frac{5\pi}{3}\right)=\boxed{-\frac{\sqrt{3}}{2}}$	**2.** $\cos\left(\frac{5\pi}{3}\right)=\boxed{\frac{1}{2}}$
3. $\cos\left(\frac{7\pi}{6}\right)=\boxed{-\frac{\sqrt{3}}{2}}$	**4.** $\sin\left(\frac{7\pi}{6}\right)=\boxed{-\frac{1}{2}}$
5. $\sin\left(\frac{3\pi}{4}\right)=\boxed{\frac{\sqrt{2}}{2}}$	**6.** $\cos\left(\frac{3\pi}{4}\right)=\boxed{-\frac{\sqrt{2}}{2}}$
7. $\tan\left(\frac{7\pi}{4}\right)=\frac{\sin\left(\frac{7\pi}{4}\right)}{\cos\left(\frac{7\pi}{4}\right)}=\frac{-\frac{\sqrt{2}}{2}}{\frac{\sqrt{2}}{2}}=\boxed{-1}$	**8.** Using Exercise 7, we see that $\cot\left(\frac{7\pi}{4}\right)=\frac{1}{\tan\left(\frac{7\pi}{4}\right)}=\frac{1}{-1}=\boxed{-1}$
9. $\sec\left(\frac{5\pi}{4}\right)=\frac{1}{\cos\left(\frac{5\pi}{4}\right)}=\frac{1}{-\frac{\sqrt{2}}{2}}=-\frac{2}{\sqrt{2}}=\boxed{-\sqrt{2}}$	**10.** $\csc\left(\frac{5\pi}{3}\right)=\frac{1}{\sin\left(\frac{5\pi}{3}\right)}=\frac{1}{-\frac{\sqrt{3}}{2}}=-\frac{2}{\sqrt{3}}=\boxed{-\frac{2\sqrt{3}}{3}}$
11. $\tan\left(\frac{4\pi}{3}\right)=\frac{\sin\left(\frac{4\pi}{3}\right)}{\cos\left(\frac{4\pi}{3}\right)}=\frac{-\frac{\sqrt{3}}{2}}{-\frac{1}{2}}=\boxed{\sqrt{3}}$	**12.** $\cot\left(\frac{11\pi}{6}\right)=\frac{\cos\left(\frac{11\pi}{6}\right)}{\sin\left(\frac{11\pi}{6}\right)}=\frac{\frac{\sqrt{3}}{2}}{-\frac{1}{2}}=\boxed{-\sqrt{3}}$
13. $\csc\left(\frac{5\pi}{6}\right)=\frac{1}{\sin\left(\frac{5\pi}{6}\right)}=\frac{1}{\frac{1}{2}}=\boxed{2}$	**14.** $\cot\left(\frac{2\pi}{3}\right)=\frac{\cos\left(\frac{2\pi}{3}\right)}{\sin\left(\frac{2\pi}{3}\right)}=\frac{-\frac{1}{2}}{\frac{\sqrt{3}}{2}}=\boxed{-\frac{\sqrt{3}}{3}}$
15. $\sin\left(-\frac{2\pi}{3}\right)=-\sin\left(\frac{2\pi}{3}\right)=\boxed{-\frac{\sqrt{3}}{2}}$	**16.** $\sin\left(-\frac{5\pi}{4}\right)=-\sin\left(\frac{5\pi}{4}\right)=-\left(-\frac{\sqrt{2}}{2}\right)=\boxed{\frac{\sqrt{2}}{2}}$
17. $\sin\left(-\frac{\pi}{3}\right)=-\sin\left(\frac{\pi}{3}\right)=\boxed{-\frac{\sqrt{3}}{2}}$	**18.** $\sin\left(-\frac{7\pi}{6}\right)=-\sin\left(\frac{7\pi}{6}\right)=-\left(-\frac{1}{2}\right)=\boxed{\frac{1}{2}}$

19. $\cos\left(-\frac{3\pi}{4}\right)=\cos\left(\frac{3\pi}{4}\right)=\boxed{-\frac{\sqrt{2}}{2}}$	**20.** $\cos\left(-\frac{5\pi}{3}\right)=\cos\left(\frac{5\pi}{3}\right)=\boxed{\frac{1}{2}}$
21. $\cos\left(-\frac{5\pi}{6}\right)=\cos\left(\frac{5\pi}{6}\right)=\boxed{-\frac{\sqrt{3}}{2}}$	**22.** $\cos\left(-\frac{7\pi}{4}\right)=\cos\left(\frac{7\pi}{4}\right)=\boxed{\frac{\sqrt{2}}{2}}$
23. $\sin\left(-\frac{5\pi}{4}\right)=-\sin\left(\frac{5\pi}{4}\right)=-\left(-\frac{\sqrt{2}}{2}\right)=\boxed{\frac{\sqrt{2}}{2}}$	**24.** $\sin(-\pi)=-\sin(\pi)=\boxed{0}$
25. $\sin\left(-\frac{3\pi}{2}\right)=-\sin\left(\frac{3\pi}{2}\right)=-(-1)=\boxed{1}$	**26.** $\sin\left(-\frac{\pi}{3}\right)=-\sin\left(\frac{\pi}{3}\right)=\boxed{-\frac{\sqrt{3}}{2}}$
27. $\cos\left(-\frac{\pi}{4}\right)=\cos\left(\frac{\pi}{4}\right)=\boxed{\frac{\sqrt{2}}{2}}$	**28.** $\cos\left(-\frac{3\pi}{4}\right)=\cos\left(\frac{3\pi}{4}\right)=\boxed{-\frac{\sqrt{2}}{2}}$
29. $\cos\left(-\frac{\pi}{2}\right)=\cos\left(\frac{\pi}{2}\right)=\boxed{0}$	**30.** $\cos\left(-\frac{7\pi}{6}\right)=\cos\left(\frac{7\pi}{6}\right)=\boxed{-\frac{\sqrt{3}}{2}}$
31. $\csc\left(-\frac{5\pi}{6}\right)=\csc\left(\frac{5\pi}{6}\right)=\frac{1}{\sin\left(\frac{5\pi}{6}\right)}=\frac{1}{-\frac{1}{2}}=\boxed{-2}$	**32.** $\sec\left(-\frac{7\pi}{4}\right)=\sec\left(\frac{\pi}{4}\right)=\frac{1}{\cos\left(\frac{\pi}{4}\right)}=\frac{1}{\frac{\sqrt{2}}{2}}=\boxed{\sqrt{2}}$
33. $\tan\left(-\frac{11\pi}{6}\right)=\tan\left(\frac{\pi}{6}\right)=\frac{\sin\left(\frac{\pi}{6}\right)}{\cos\left(\frac{\pi}{6}\right)}=\frac{\frac{1}{2}}{\frac{\sqrt{3}}{2}}=\boxed{\frac{\sqrt{3}}{3}}$	**34.** $\cot\left(-\frac{11\pi}{6}\right)=\cot\left(\frac{\pi}{6}\right)=\frac{\cos\left(\frac{\pi}{6}\right)}{\sin\left(\frac{\pi}{6}\right)}=\frac{\frac{\sqrt{3}}{2}}{\frac{1}{2}}=\boxed{\sqrt{3}}$
35. $\theta=\frac{\pi}{6},\frac{11\pi}{6}$	**36.** $\theta=\frac{5\pi}{6},\frac{7\pi}{6}$
37. $\theta=\frac{4\pi}{3},\frac{5\pi}{3}$	**38.** $\theta=\frac{\pi}{3},\frac{2\pi}{3}$
39. $\theta=0,\pi,2\pi,3\pi,4\pi$	**40.** $\theta=\frac{3\pi}{2},\frac{7\pi}{2}$
41. $\theta=\pi,3\pi$	**42.** $\theta=\frac{\pi}{2},\frac{3\pi}{2},\frac{5\pi}{2},\frac{7\pi}{2}$
43. We seek angles θ for which $\lvert\sin\theta\rvert=\lvert\cos\theta\rvert$ <u>and</u> $\sin\theta, \cos\theta$ have opposite signs. These two conditions occur when $\boxed{\theta=\frac{3\pi}{4},\frac{7\pi}{4}}$.	**44.** We seek angles θ for which $\lvert\sin\theta\rvert=\lvert\cos\theta\rvert$ <u>and</u> $\sin\theta, \cos\theta$ have the same sign. These two conditions occur when $\boxed{\theta=\frac{\pi}{4},\frac{5\pi}{4}}$.

45. We seek angles θ for which $\frac{1}{\cos\theta} = -\sqrt{2}$, which is equivalent to

$$\cos\theta = -\frac{1}{\sqrt{2}} = -\frac{\sqrt{2}}{2}.$$

This occurs when $\boxed{\theta = \frac{3\pi}{4}, \frac{5\pi}{4}}$.

46. We seek angles θ for which $\frac{1}{\sin\theta} = \sqrt{2}$, which is equivalent to

$$\sin\theta = \frac{1}{\sqrt{2}} = \frac{\sqrt{2}}{2}.$$

This occurs when $\boxed{\theta = \frac{\pi}{4}, \frac{3\pi}{4}}$.

47. We seek angles θ for which $\sin\theta = 0$ (since then $\csc\theta = \frac{1}{\sin\theta}$ is undefined). This occurs when $\boxed{\theta = 0, \pi, 2\pi}$.

48. We seek angles θ for which $\cos\theta = 0$ (since then $\sec\theta = \frac{1}{\cos\theta}$ is undefined). This occurs when $\boxed{\theta = \frac{\pi}{2}, \frac{3\pi}{2}}$.

49. We seek angles θ for which $\cos\theta = 0$ (since then $\tan\theta = \frac{\sin\theta}{\cos\theta}$ is undefined). This occurs when $\boxed{\theta = \frac{\pi}{2}, \frac{3\pi}{2}}$.

50. We seek angles θ for which $\sin\theta = 0$ (since then $\cot\theta = \frac{\cos\theta}{\sin\theta}$ is undefined). This occurs when $\boxed{\theta = 0, \pi, 2\pi}$.

51. We seek angles θ for which $\frac{1}{\sin\theta} = -2$, which is equivalent to

$$\sin\theta = -\frac{1}{2}.$$

This occurs when $\boxed{\theta = \frac{7\pi}{6}, \frac{11\pi}{6}}$.

52. We seek angles θ for which $\cot\theta = -\sqrt{3}$, which is equivalent to

$$\frac{\cos\theta}{\sin\theta} = -\frac{\frac{\sqrt{3}}{2}}{\frac{1}{2}}.$$

This occurs when $\boxed{\theta = \frac{5\pi}{6}, \frac{11\pi}{6}}$.

53. We seek angles θ for which $\frac{1}{\cos\theta} = \frac{2\sqrt{3}}{3}$, which is equivalent to

$$\cos\theta = \frac{\sqrt{3}}{2}.$$

This occurs when $\boxed{\theta = \frac{\pi}{6}, \frac{11\pi}{6}}$.

54. We seek angles θ for which $\tan\theta = \frac{\sqrt{3}}{3}$, which is equivalent to

$$\frac{\sin\theta}{\cos\theta} = \frac{\frac{1}{2}}{\frac{\sqrt{3}}{2}}.$$

This occurs when $\boxed{\theta = \frac{\pi}{6}, \frac{7\pi}{6}}$.

55. Note that February 15 corresponds to $x = 46$. Observe that $T(46) = 50 - 28\cos\left(\frac{2\pi(46-31)}{365}\right) \approx \boxed{22.9^\circ F}$	**56.** Note that August 15 corresponds to $x = 227$. Observe that $T(227) = 50 - 28\cos\left(\frac{2\pi(227-31)}{365}\right) \approx \boxed{77.2^\circ F}$
57. Note that 6 am corresponds to $x = 6$. Observe that $T(6) = 99.1 - 0.5\sin\left(6 + \frac{\pi}{12}\right) \approx \boxed{99.1^\circ F}$	**58.** Note that 9 pm corresponds to $x = 21$. Observe that $T(21) = 99.1 - 0.5\sin\left(21 + \frac{\pi}{12}\right) \approx \boxed{98.8^\circ F}$
59. Note that 3 pm corresponds to $x = 15$. Observe that $h(15) = 5 + 4.8\sin\left(\frac{\pi}{6}(15+4)\right) \approx \boxed{2.6 \text{ ft.}}$	**60.** Note that 5 am corresponds to $x = 5$. Observe that $h(15) = 5 + 4.8\sin\left(\frac{\pi}{6}(5+4)\right) \approx \boxed{0.2 \text{ ft.}}$
61. Since $x = 6$ corresponds to June, we see that $w(6) = 145 + 10\cos\left(\frac{\pi}{6}(6)\right) = 145 + 10\cos(\pi) = \boxed{135 \text{ lbs.}}$	**62.** Since $x = 12$ corresponds to December, we see that $w(12) = 145 + 10\cos\left(\frac{\pi}{6}(12)\right) = 145 + 10\cos(2\pi) = \boxed{155 \text{ lbs.}}$
63. Since $x = 2$ corresponds to February, we see that $n(2) = 30{,}000 + 20{,}000\sin\left(\frac{\pi}{2}(2+1)\right) = \boxed{10{,}000 \text{ guests}}$	**64.** Since $x = 12$ corresponds to December, we see that $n(2) = 30{,}000 + 20{,}000\sin\underbrace{\left(\frac{\pi}{2}(12+1)\right)}_{\frac{13\pi}{2}} = \boxed{50{,}000 \text{ guests}}$
65. Should have used $\cos\left(\frac{5\pi}{6}\right) = -\frac{\sqrt{3}}{2}$ and $\sin\left(\frac{5\pi}{6}\right) = \frac{1}{2}$.	**66.** Should have used $\cos\left(\frac{11\pi}{6}\right) = \frac{\sqrt{3}}{2}$.
67. True. The angles $(2n\pi + \theta)$ and θ are coterminal for any integer n. Hence, they have the same cosine and sine values.	**68.** True. The angles $(2n\pi + \theta)$ and θ are coterminal for any integer n. Hence, they have the same cosine and sine values.
69. False. For instance, $\sin\left(\frac{3\pi}{2}\right) = -1$ (which corresponds to $n = 1$).	**70.** False. For instance, $\cos(\pi) = -1$ (which corresponds to $n = 1$).

71. True, because $\tan(\theta+2n\pi)=\tan((\theta+n\pi)+n\pi)$ $=\tan(\theta+n\pi)$ $=\tan\theta$	**72.** False, because $\tan\theta=0 \Rightarrow \sin\theta=0 \Rightarrow \theta=n\pi$, where n is an integer.

73. Cosecant is an odd function because

$$\csc(-\theta)=\frac{1}{\sin(-\theta)}=\frac{1}{-\sin\theta}=-\frac{1}{\sin\theta}=-\csc(\theta).$$

74. Tangent is an odd function because

$$\tan(-\theta)=\frac{\sin(-\theta)}{\cos(-\theta)}=\frac{-\sin\theta}{\cos\theta}=-\frac{\sin\theta}{\cos\theta}=-\tan(\theta).$$

75. When considering the restriction $0\le\theta\le 2\pi$, the equation $\sin\theta=\cos\theta$ is satisfied when $\boxed{\theta=\frac{\pi}{4},\frac{5\pi}{4}}$.

76. If one removes the restriction in Exercise 63, then the equation $\sin\theta=\cos\theta$ is satisfied not only when $\theta=\frac{\pi}{4},\frac{5\pi}{4}$, but also at any angle coterminal with either of these. As such, the solutions are given by

$$\theta=\frac{\pi}{4}+2n\pi,\ \frac{5\pi}{4}+2n\pi,\ \text{where } n \text{ is an integer.}$$

This can be written more succinctly as $\theta=\frac{\pi}{4}+n\pi$, where n is an integer.

77. Observe that

$$2\sin\theta=\csc\theta \Rightarrow 2\sin\theta=\tfrac{1}{\sin\theta} \Rightarrow 2\sin^2\theta=1 \Rightarrow \sin\theta=\pm\tfrac{1}{\sqrt{2}}$$

This occurs when $\theta=\frac{\pi}{4},\frac{3\pi}{4},\frac{5\pi}{4},\frac{7\pi}{4}$.

78. Observe that

$$\cos\theta=\tfrac{1}{4}\sec\theta \Rightarrow \cos\theta=\tfrac{1}{4\cos\theta} \Rightarrow 4\cos^2\theta=1 \Rightarrow \cos\theta=\pm\tfrac{1}{2}.$$

This occurs when $\theta=\frac{\pi}{3},\frac{2\pi}{3},\frac{4\pi}{3},\frac{5\pi}{3}$.

79. Observe that

$$3\csc\theta=4\sin\theta=\csc\theta \Rightarrow \tfrac{3}{\sin\theta}=4\sin\theta \Rightarrow \sin^2\theta=\tfrac{3}{4} \Rightarrow \sin\theta=\pm\tfrac{\sqrt{3}}{2}$$

This occurs when $\theta=\frac{\pi}{3}+n\pi,\ \frac{2\pi}{3}+n\pi$, where n is an integer.

80. Observe that $$4\cos\theta = 3\sec\theta \Rightarrow 4\cos\theta = \tfrac{3}{\cos\theta} \Rightarrow \cos^2\theta = \tfrac{3}{4} \Rightarrow \cos\theta = \pm\tfrac{\sqrt{3}}{2}$$ his occurs when $\theta = \frac{\pi}{6} + n\pi, \frac{5\pi}{6} + n\pi$, where n is an integer.	
81. Observe that $$\begin{aligned} \tan\theta = \cot\theta &\Rightarrow \frac{\sin\theta}{\cos\theta} = \frac{\cos\theta}{\sin\theta} \\ &\Rightarrow \sin^2\theta = \cos^2\theta \\ &\Rightarrow \sin^2\theta - \cos^2\theta = 0 \\ &\Rightarrow (\sin\theta - \cos\theta)(\sin\theta + \cos\theta) = 0 \\ &\Rightarrow \sin\theta = \pm\cos\theta \end{aligned}$$ This occurs when $\theta = \frac{\pi}{4}, \frac{3\pi}{4}, \frac{5\pi}{4}, \frac{7\pi}{4}$.	
82. Observe that $$\sec\theta = \csc(-\theta) \Rightarrow \frac{1}{\cos\theta} = \frac{1}{\sin(-\theta)} = -\frac{1}{\sin\theta} \Rightarrow \sin\theta = -\cos\theta.$$ This occurs when $\theta = \frac{3\pi}{4}, \frac{7\pi}{4}$.	
83. $\sin(423^\circ) \approx 0.891$ $\sin(-423^\circ) = -\sin(423^\circ) \approx -0.891$	**84.** $\cos(227^\circ) \approx -0.682$ $\cos(-227^\circ) = \cos(227^\circ) \approx -0.682$
85. $\tan(81^\circ) \approx 6.314$ $\tan(-81^\circ) \approx -6.314$	**86.** $\csc(211^\circ) = \frac{1}{\sin(211^\circ)} \approx -1.942$ $\csc(-211^\circ) = \frac{1}{\sin(-211^\circ)} \approx 1.942$
87. $\cos\left(\frac{\pi}{3}\right) = \boxed{0.5}$	**88.** $\sin\left(\frac{\pi}{3}\right) \approx \boxed{0.866}$
89. $\sin\left(\frac{2\pi}{3}\right) \approx \boxed{0.866}$	**90.** $\cos\left(\frac{5\pi}{4}\right) = \boxed{-0.707}$
91. $[-\cos\pi] - [-\cos 0] = 1 + 1 = 2$	**92.** $\sin\frac{5\pi}{6} - \sin\frac{\pi}{4} = \frac{1}{2} - \frac{\sqrt{2}}{2} = \frac{1-\sqrt{2}}{2}$

93. $\tan\frac{5\pi}{4}-\tan\frac{7\pi}{6}=1-\frac{1}{\sqrt{3}}=\frac{3-\sqrt{3}}{3}$	**94.** $\left[-\csc\frac{11\pi}{6}\right]-\left[-\csc\frac{5\pi}{3}\right]=-\frac{1}{-\frac{1}{2}}+\frac{1}{-\frac{\sqrt{3}}{2}}$ $=2-\frac{2}{\sqrt{3}}=\boxed{2-\frac{2\sqrt{3}}{3}}$

Section 5.2 Solutions ---

1. c (Observe that this graph is the graph of $y=\sin x$ reflected over the x-axis.)	**2. d**
3. a	**4. j** (Observe that this graph is the graph of $y=\cos x$ reflected over the x-axis.)
5. h (Observe that the amplitude is 2, and the graph has the same intercepts and shape as $y=\sin x$.)	**6. i** (Observe that the amplitude is 2, and the graph has the same intercepts and shape as $y=\cos x$.)
7. b (Observe that the graph has the same shape as $y=\sin x$, but is horizontally stretched by a factor of 2.)	**8. f** (Observe that the graph has the same shape as $y=\cos x$, but is horizontally stretched by a factor of 2.)
9. e (Observe that this graph has the same shape as $y=\cos\left(\frac{1}{2}x\right)$, but it has an amplitude of 2, is stretched by a factor of 2, and reflected over the x-axis.)	**10. g** (Observe that this graph has the same shape as $y=\sin\left(\frac{1}{2}x\right)$, but it has an amplitude of 2, is stretched by a factor of 2, and reflected over the x-axis.)
11. Since $A=\frac{3}{2}$ and $B=3$, we conclude that the amplitude is $\frac{3}{2}$ and the period is $p=\frac{2\pi}{3}$.	**12.** Since $A=\frac{2}{3}$ and $B=4$, we conclude that the amplitude is $\frac{2}{3}$ and the period is $p=\frac{2\pi}{4}=\frac{\pi}{2}$.

13. Since $A=-1$ and $B=5$, we conclude that the amplitude is 1 and the period is $p=\frac{2\pi}{5}$.	**14.** Since $A=-1$ and $B=7$, we conclude that the amplitude is 1 and the period is $p=\frac{2\pi}{7}$.
15. Since $A=\frac{2}{3}$ and $B=\frac{3}{2}$, we conclude that the amplitude is $\frac{2}{3}$ and the period is $p=\frac{2\pi}{3/2}=\frac{4\pi}{3}$.	**16.** Since $A=\frac{3}{2}$ and $B=\frac{2}{3}$, we conclude that the amplitude is $\frac{3}{2}$ and the period is $p=\frac{2\pi}{2/3}=3\pi$.
17. Since $A=-3$ and $B=\pi$, we conclude that the amplitude is 3 and the period is $p=\frac{2\pi}{\pi}=2$.	**18.** Since $A=-2$ and $B=\pi$, we conclude that the amplitude is 2 and the period is $p=\frac{2\pi}{\pi}=2$.
19. Since $A=5$ and $B=\frac{\pi}{3}$, we conclude that the amplitude is 5 and the period is $p=\frac{2\pi}{\pi/3}=6$.	**20.** Since $A=4$ and $B=\frac{\pi}{4}$, we conclude that the amplitude is 4 and the period is $p=\frac{2\pi}{\pi/4}=8$.

21. We have the following information about the graph of $y = 8\cos x$ on $[0, 2\pi]$:

<table>
<tr><td>Amplitude: 8</td><td>Period: $\frac{2\pi}{1} = 2\pi$</td></tr>
<tr><td colspan="2">x-intercepts occur when $\cos x = 0$. This happens when $x = \frac{\pi}{2}, \frac{3\pi}{2}$. So, the points $\left(\frac{\pi}{2}, 0\right)$, $\left(\frac{3\pi}{2}, 0\right)$ are on the graph.</td></tr>
<tr><td colspan="2">The maximum value of the graph is 8 since the maximum of $\cos x$ is 1, and the amplitude is 8. This occurs when $x = 0, 2\pi$. So, the points $(0,8)$, $(2\pi,8)$ are on the graph.</td></tr>
<tr><td colspan="2">The minimum value of the graph is – 8 since the minimum of $\cos x$ is – 1, and the amplitude is 8. This occurs when $x = \pi$. So, the point $(\pi,-8)$ is on the graph.</td></tr>
<tr><td colspan="2">Since the coefficient used to compute the amplitude is nonnegative, there is no reflection over the x-axis.</td></tr>
</table>

Using this information, we find that the graph of $y = 8\cos x$ on $[0, 2\pi]$ is given by:

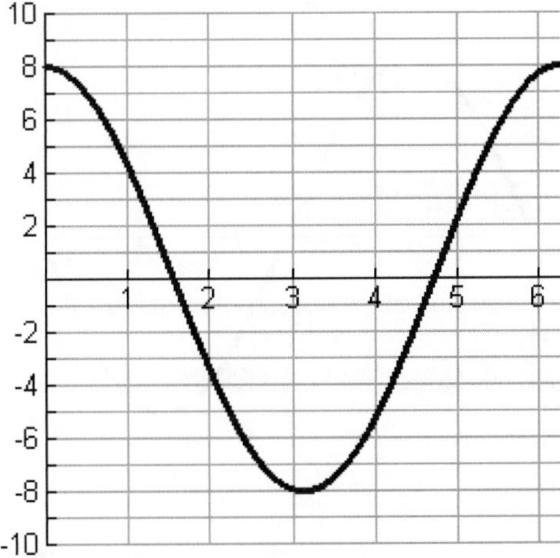

22. We have the following information about the graph of $y = 7\sin x$ on $[0, 2\pi]$:

Amplitude: 7	Period: $\frac{2\pi}{1} = 2\pi$
x-intercepts occur when $\sin x = 0$. This happens when $x = 0, \pi, 2\pi$. So, the points $(0, 0)$, $(\pi, 0)$, $(2\pi, 0)$ are on the graph.	
The maximum value of the graph is 7 since the maximum of $\sin x$ is 1, and the amplitude is 7. This occurs when $x = \frac{\pi}{2}$. So, the point $\left(\frac{\pi}{2}, 7\right)$ is on the graph.	
The minimum value of the graph is -7 since the minimum of $\sin x$ is -1, and the amplitude is 7. This occurs when $x = \frac{3\pi}{2}$. So, the point $\left(\frac{3\pi}{2}, -7\right)$ is on the graph.	
Since the coefficient used to compute the amplitude is nonnegative, there is no reflection over the x-axis.	

Using this information, we find that the graph of $y = 7\sin x$ on $[0, 2\pi]$ is given by:

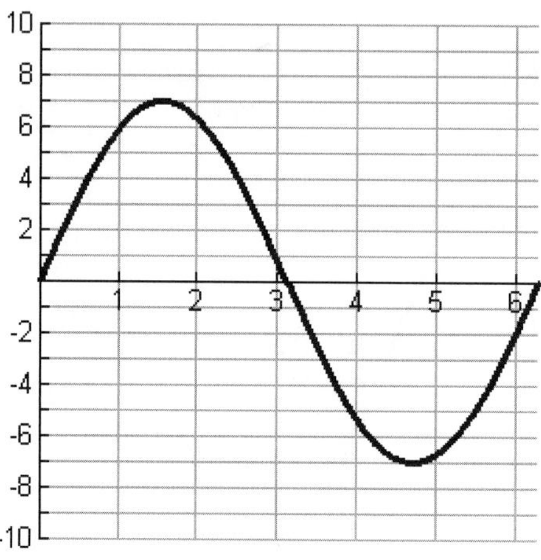

23. We have the following information about the graph of $y = \sin 4x$ on $\left[0, \frac{\pi}{2}\right]$:

Amplitude: 1	Period: $\frac{2\pi}{4} = \frac{\pi}{2}$
x-intercepts occur when $\sin 4x = 0$. This happens when $4x = n\pi$, and so $x = \frac{n\pi}{4}$, where n is an integer. So, the points $(0, 0)$, $\left(\frac{\pi}{4}, 0\right)$, $\left(\frac{\pi}{2}, 0\right)$ are on the graph.	
The maximum value of the graph is 1 since the maximum of $\sin 4x$ is 1, and the amplitude is 1. This occurs when $4x = \frac{\pi}{2}$, so that $x = \frac{\pi}{8}$. So, the point $\left(\frac{\pi}{8}, 1\right)$ is on the graph.	
The minimum value of the graph is – 1 since the minimum of $\sin 4x$ is – 1, and the amplitude is 1. This occurs when $4x = \frac{3\pi}{2}$, so that $x = \frac{3\pi}{8}$. So, the point $\left(\frac{3\pi}{8}, -1\right)$ is on the graph.	
Since the coefficient used to compute the amplitude is nonnegative, there is no reflection over the x-axis.	

Using this information, we find that the graph of $y = \sin 4x$ on $\left[0, \frac{\pi}{2}\right]$ is given by:

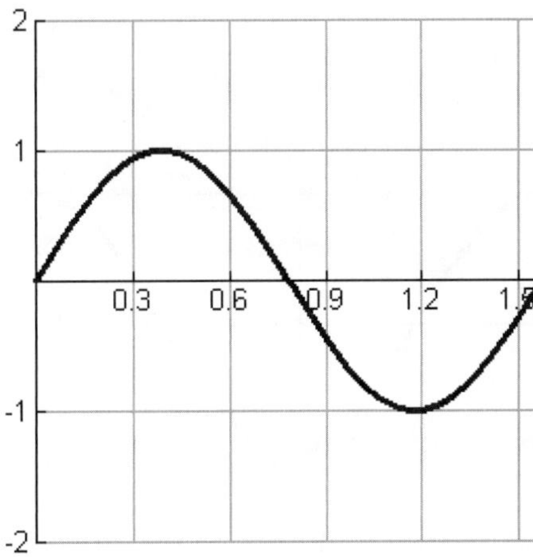

24. We have the following information about the graph of $y = \cos 3x$ on $\left[0, \frac{2\pi}{3}\right]$:

<table>
<tr><td>Amplitude: 1</td><td>Period: $\frac{2\pi}{3}$</td></tr>
<tr><td colspan="2">x-intercepts occur when $\cos 3x = 0$. This happens when $3x = \frac{(2n+1)\pi}{2}$, and so $x = \frac{(2n+1)\pi}{6}$, where n is an integer. So, the points $\left(\frac{\pi}{6}, 0\right), \left(\frac{\pi}{2}, 0\right)$ are on the graph.</td></tr>
<tr><td colspan="2">The maximum value of the graph is 1 since the maximum of $\cos 3x$ is 1, and the amplitude is 1. This occurs when $3x = 0,\ 2\pi$, so that $x = 0, \frac{2\pi}{3}$. So, the points $(0,1), \left(\frac{2\pi}{3}, 1\right)$ are on the graph.</td></tr>
<tr><td colspan="2">The minimum value of the graph is – 1 since the minimum of $\cos 3x$ is – 1, and the amplitude is 1. This occurs when $3x = \pi$, so that $x = \frac{\pi}{3}$. So, the point $\left(\frac{\pi}{3}, -1\right)$ is on the graph.</td></tr>
<tr><td colspan="2">Since the coefficient used to compute the amplitude is nonnegative, there is no reflection over the x-axis.</td></tr>
</table>

Using this information, we find that the graph of $y = \cos 3x$ on $\left[0, \frac{2\pi}{3}\right]$ is given by:

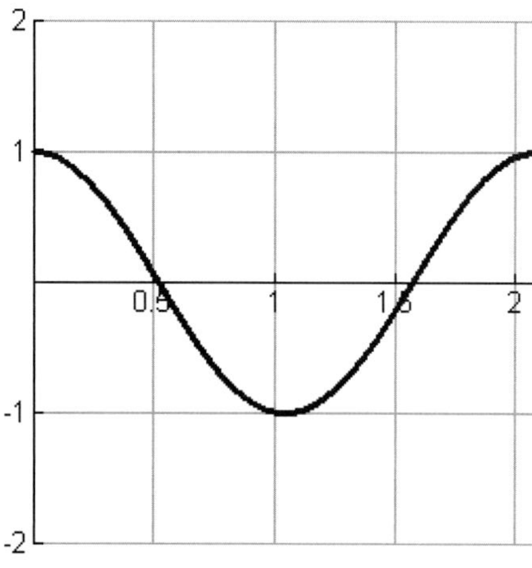

25. We have the following information about the graph of $y = -3\cos\left(\frac{1}{2}x\right)$ on $[0, 4\pi]$:

<u>Amplitude</u>: $\lvert -3 \rvert = 3$	<u>Period</u>: $\frac{2\pi}{1/2} = 4\pi$
<u>x-intercepts</u> occur when $\cos\left(\frac{1}{2}x\right) = 0$. This happens when $\frac{1}{2}x = \frac{\pi}{2}, \frac{3\pi}{2}$, and so $x = \pi, 3\pi$. So, the points $(\pi, 0)$, $(3\pi, 0)$ are on the graph.	
The <u>maximum value</u> of the graph is 3 since the maximum of $\cos\left(\frac{1}{2}x\right)$ is 1, and the amplitude is 3. This occurs when $\frac{1}{2}x = 0, 2\pi$, and so $x = 0, 4\pi$. So, the points $(0,3)$, $(4\pi,3)$ are on the graph.	
The <u>minimum value</u> of the graph is – 3 since the minimum of $\cos\left(\frac{1}{2}x\right)$ is – 1, and the amplitude is 3. This occurs when $\frac{1}{2}x = \pi$, and so $x = 2\pi$. So, the point $(2\pi, -3)$ is on the graph.	
Since the coefficient used to compute the amplitude is negative, we reflect the graph of $y = 3\cos\left(\frac{1}{2}x\right)$ over the x-axis.	

Using this information, we find that the graph of $y = -3\cos\left(\frac{1}{2}x\right)$ on $[0, 4\pi]$ is given by:

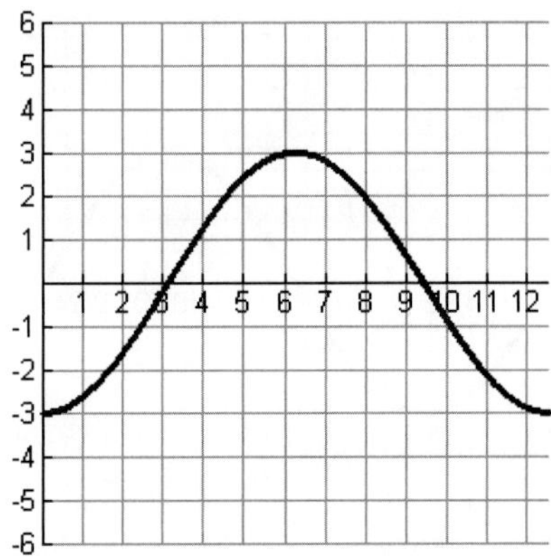

26. We have the following information about the graph of $y=-2\sin\left(\frac{1}{4}x\right)$ on $[0,8\pi]$:

Amplitude: $\lvert -2 \rvert = 2$	Period: $\frac{2\pi}{1/4} = 8\pi$
x-intercepts occur when $\sin\left(\frac{1}{4}x\right)=0$. This happens when $\frac{1}{4}x=0,\ \pi,\ 2\pi$, and so $x=0,\ 4\pi,\ 8\pi$. So, the points $(0,\ 0)$, $(4\pi,\ 0)$, $(8\pi,\ 0)$ are on the graph.	
The maximum value of the graph is 2 since the maximum of $\sin\left(\frac{1}{4}x\right)$ is 1, and the amplitude is 2. This occurs when $\frac{1}{4}x=\frac{\pi}{2}$, and so $x=2\pi$. So, the point $(2\pi,2)$ is on the graph.	
The minimum value of the graph is – 2 since the minimum of $\sin\left(\frac{1}{4}x\right)$ is – 1, and the amplitude is 2. This occurs when $\frac{1}{4}x=\frac{3\pi}{2}$, and so $x=6\pi$. So, the point $(6\pi,-2)$ is on the graph.	
Since the coefficient used to compute the amplitude is negative, we reflect the graph of $y=2\sin\left(\frac{1}{4}x\right)$ over the x-axis.	

Using this information, we find that the graph of $y=-2\sin\left(\frac{1}{4}x\right)$ on $[0,8\pi]$ is given by:

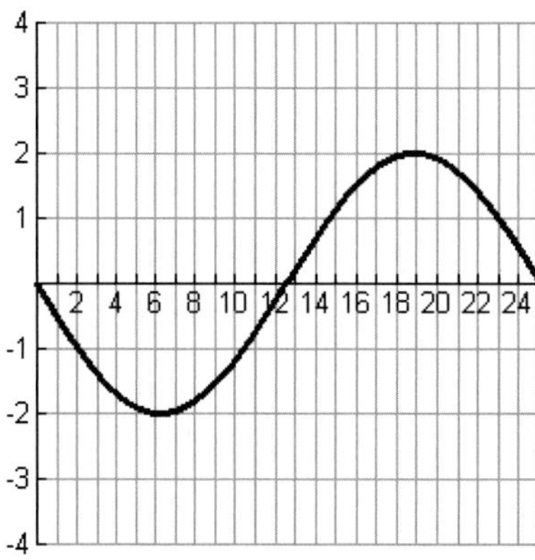

27. We have the following information about the graph of $y = -3\sin(\pi x)$ on $[0, 2]$:

Amplitude: $\lvert -3 \rvert = 3$	Period: $\frac{2\pi}{\pi} = 2$
x-intercepts occur when $\sin(\pi x) = 0$. This happens when $\pi x = 0,\ \pi,\ 2\pi$, and so $x = 0, 1, 2$. So, the points $(0, 0)$, $(1, 0)$, $(2, 0)$ are on the graph.	
The maximum value of the graph is 3 since the maximum of $\sin(\pi x)$ is 1, and the amplitude is 3. This occurs when $\pi x = \frac{\pi}{2}$, and so $x = \frac{1}{2}$. So, the point $\left(\frac{1}{2}, 3\right)$ is on the graph.	
The minimum value of the graph is – 3 since the minimum of $\sin(\pi x)$ is – 1, and the amplitude is 3. This occurs when $\pi x = \frac{3\pi}{2}$, and so $x = \frac{3}{2}$. So, the point $\left(\frac{3}{2}, -3\right)$ is on the graph.	
Since the coefficient used to compute the amplitude is negative, we reflect the graph of $y = 3\sin(\pi x)$ over the x-axis.	

Using this information, we find that the graph of $y = -3\sin(\pi x)$ on $[0, 2]$ is given by:

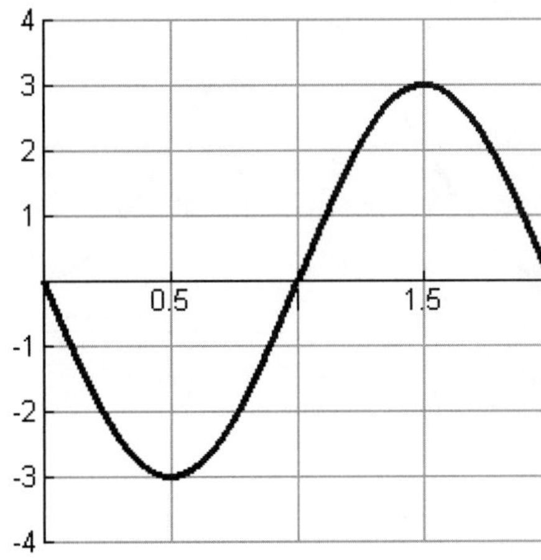

28. We have the following information about the graph of $y=-2\cos(\pi x)$ on $[0,2]$:

<table>
<tr><td>Amplitude: $|-2|=2$</td><td>Period: $\frac{2\pi}{\pi}=2$</td></tr>
<tr><td colspan="2">x-intercepts occur when $\cos(\pi x)=0$. This happens when $\pi x=\frac{\pi}{2},\frac{3\pi}{2}$, and so $x=\frac{1}{2},\frac{3}{2}$. So, the points $\left(\frac{1}{2},0\right)$, $\left(\frac{3}{2},0\right)$ are on the graph.</td></tr>
<tr><td colspan="2">The maximum value of the graph is 2 since the maximum of $\cos(\pi x)$ is 1, and the amplitude is 2. This occurs when $\pi x=0,\ 2\pi$, and so $x=0,\ 2$. So, the points $(0,2)$, $(2,2)$ are on the graph.</td></tr>
<tr><td colspan="2">The minimum value of the graph is – 2 since the minimum of $\cos(\pi x)$ is – 1, and the amplitude is 2. This occurs when $\pi x=\pi$, and so $x=1$. So, the point $(1,-2)$ is on the graph.</td></tr>
<tr><td colspan="2">Since the coefficient used to compute the amplitude is negative, we reflect the graph of $y=2\cos(\pi x)$ over the x-axis.</td></tr>
</table>

Using this information, we find that the graph of $y=-2\cos(\pi x)$ on $[0,2]$ is given by:

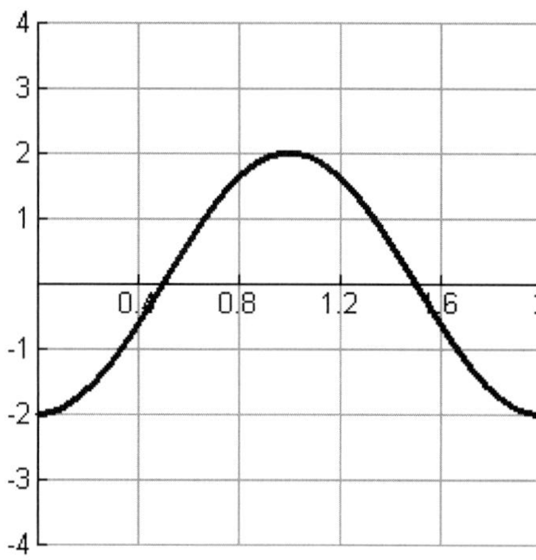

29. We have the following information about the graph of $y = 5\cos 2\pi x$ on $[0,1]$:

Amplitude: 5	Period: $\frac{2\pi}{2\pi} = 1$
x-intercepts occur when $\cos 2\pi x = 0$. This happens when $2\pi x = \frac{(2n+1)\pi}{2}$ and so, $x = \frac{(2n+1)}{4}$, where n is an integer. So, the points $\left(\frac{1}{4}, 0\right), \left(\frac{3}{4}, 0\right)$ are on the graph.	
The maximum value of the graph is 5 since the maximum of $\cos 2\pi x$ is 1, and the amplitude is 5. This occurs when $2\pi x = 0,\ 2\pi$, so that $x = 0, 1$. So, the points $(0,5), (1,5)$ are on the graph.	
The minimum value of the graph is – 5 since the minimum of $\cos 2\pi x$ is – 1, and the amplitude is 5. This occurs when $2\pi x = \pi$, so that $x = \frac{1}{2}$. So, the point $\left(\frac{1}{2}, -5\right)$ is on the graph.	
Since the coefficient used to compute the amplitude is nonnegative, there is no reflection over the x-axis.	

Using this information, we find that the graph of $y = 5\cos 2\pi x$ on $[0,1]$ is given by:

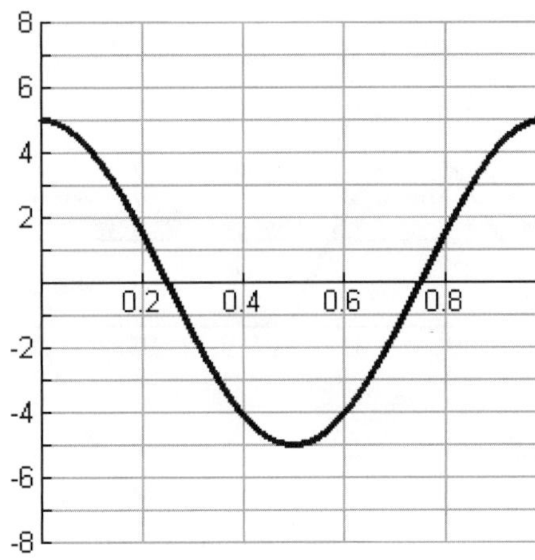

30. We have the following information about the graph of $y = 4\sin 2\pi x$ on $[0,1]$:

Amplitude: 4	Period: $\dfrac{2\pi}{2\pi} = 1$
x-intercepts occur when $\sin 2\pi x = 0$. This happens when $2\pi x = n\pi$, and so $x = \dfrac{n}{2}$, where *n* is an integer. So, the points $(0, 0)$, $\left(\dfrac{1}{2}, 0\right)$, $(1, 0)$ are on the graph.	
The maximum value of the graph is 4 since the maximum of $\sin 2\pi x$ is 1, and the amplitude is 4. This occurs when $2\pi x = \dfrac{\pi}{2}$, so that $x = \dfrac{1}{4}$. So, the point $\left(\dfrac{1}{4}, 4\right)$ is on the graph.	
The minimum value of the graph is – 4 since the minimum of $\sin 2\pi x$ is – 1, and the amplitude is 4. This occurs when $2\pi x = \dfrac{3\pi}{2}$, so that $x = \dfrac{3}{4}$. So, the point $\left(\dfrac{3}{4}, -4\right)$ is on the graph.	
Since the coefficient used to compute the amplitude is nonnegative, there is no reflection over the *x*-axis.	

Using this information, we find that the graph of $y = 4\sin 2\pi x$ on $[0,1]$ is given by:

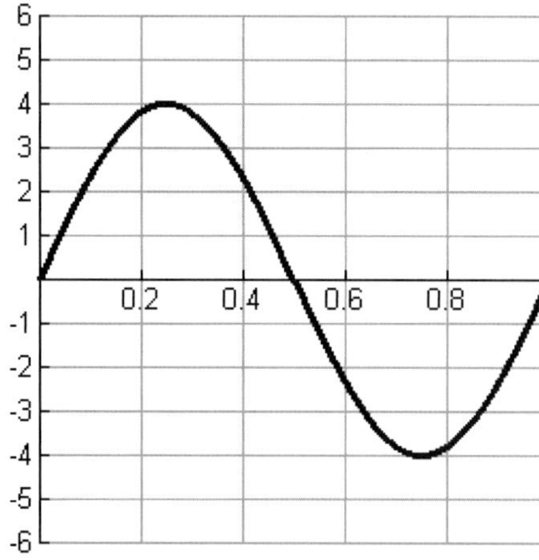

31. We have the following information about the graph of $y = -3\sin\left(\frac{\pi}{4}x\right)$ on $[0,8]$:

Amplitude: $\lvert -3 \rvert = 3$	Period: $\frac{2\pi}{\pi/4} = 8$
x-intercepts occur when $\sin\left(\frac{\pi}{4}x\right) = 0$. This happens when $\frac{\pi}{4}x = 0,\ \pi,\ 2\pi$, and so $x = 0,\ 4,\ 8$. So, the points $(0,\ 0)$, $(4,\ 0)$, $(8,\ 0)$ are on the graph.	
The maximum value of the graph is 3 since the maximum of $\sin\left(\frac{\pi}{4}x\right)$ is 1, and the amplitude is 3. This occurs when $\frac{\pi}{4}x = \frac{\pi}{2}$, and so $x = 2$. So, the point $(2,\ 3)$ is on the graph.	
The minimum value of the graph is – 3 since the minimum of $\sin\left(\frac{\pi}{4}x\right)$ is – 1, and the amplitude is 3. This occurs when $\frac{\pi}{4}x = \frac{3\pi}{2}$, and so $x = 6$. So, the point $(6,\ -3)$ is on the graph.	
Since the coefficient used to compute the amplitude is negative, we reflect the graph of $y = 3\sin\left(\frac{\pi}{4}x\right)$ over the x-axis.	

Using this information, we find that the graph of $y = -3\sin\left(\frac{\pi}{4}x\right)$ on $[0,8]$ is given by:

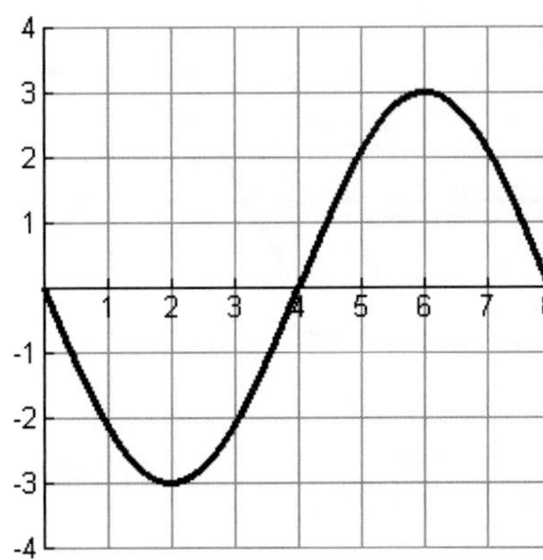

32. We have the following information about the graph of $y = -4\sin\left(\frac{\pi}{2}x\right)$ on $[0,4]$:

Amplitude: $\lvert -4 \rvert = 4$	Period: $\frac{2\pi}{\pi/2} = 4$
x-intercepts occur when $\sin\left(\frac{\pi}{2}x\right) = 0$. This happens when $\frac{\pi}{2}x = 0,\ \pi,\ 2\pi$, and so $x = 0,\ 2,\ 4$. So, the points $(0,\ 0)$, $(2,\ 0)$, $(4,\ 0)$ are on the graph.	
The maximum value of the graph is 4 since the maximum of $\sin\left(\frac{\pi}{2}x\right)$ is 1, and the amplitude is 4. This occurs when $\frac{\pi}{2}x = \frac{\pi}{2}$, and so $x = 1$. So, the point $(1,\ 4)$ is on the graph.	
The minimum value of the graph is – 4 since the minimum of $\sin\left(\frac{\pi}{2}x\right)$ is – 1, and the amplitude is 4. This occurs when $\frac{\pi}{2}x = \frac{3\pi}{2}$, and so $x = 3$. So, the point $(3,\ -4)$ is on the graph.	
Since the coefficient used to compute the amplitude is negative, we reflect the graph of $y = 4\sin\left(\frac{\pi}{2}x\right)$ over the *x*-axis.	

Using this information, we find that the graph of $y = -4\sin\left(\frac{\pi}{2}x\right)$ on $[0,4]$ is given by:

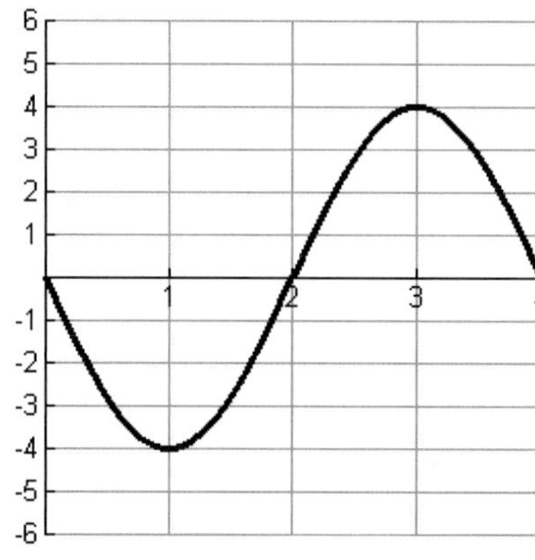

33. In order to construct the graph of $y=-4\cos\left(\frac{1}{2}x\right)$ on $[-8\pi,8\pi]$, we first gather the information essential to forming the graph for one period, namely on the interval $[0,4\pi]$. Then, we extend this picture to cover the entire given interval. Indeed, we have the following:

<table>
<tr><td>Amplitude: $|-4|=4$</td><td>Period: $\frac{2\pi}{1/2}=4\pi$</td></tr>
<tr><td colspan="2">x-intercepts occur when $\cos\left(\frac{1}{2}x\right)=0$. This happens when $\frac{1}{2}x=\frac{\pi}{2},\frac{3\pi}{2}$, and so $x=\pi,\ 3\pi$. So, the points $(\pi,\ 0)$, $(3\pi,\ 0)$ are on the graph.</td></tr>
<tr><td colspan="2">The maximum value of the graph is 4 since the maximum of $\cos\left(\frac{1}{2}x\right)$ is 1, and the amplitude is 4. This occurs when $\frac{1}{2}x=0,\ 2\pi$, and so $x=0,\ 4\pi$. So, the points $(0,4)$, $(4\pi,4)$ are on the graph.</td></tr>
<tr><td colspan="2">The minimum value of the graph is – 4 since the minimum of $\cos\left(\frac{1}{2}x\right)$ is – 1, and the amplitude is 4. This occurs when $\frac{1}{2}x=\pi$, and so $x=2\pi$. So, the point $(2\pi,-4)$ is on the graph.</td></tr>
<tr><td colspan="2">Since the coefficient used to compute the amplitude is negative, we reflect the graph of $y=4\cos\left(\frac{1}{2}x\right)$ over the x-axis.</td></tr>
</table>

Now, using this information, we can form the graph of $y=-3\cos\left(\frac{1}{2}x\right)$ on $[0,4\pi]$, and then extend it to the interval $[-8\pi,8\pi]$, as shown below:

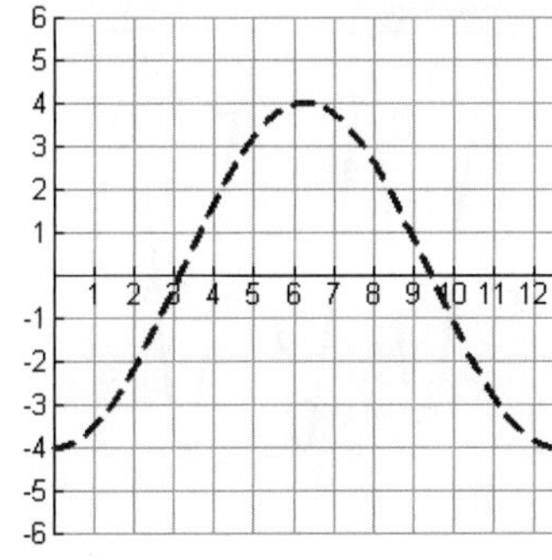

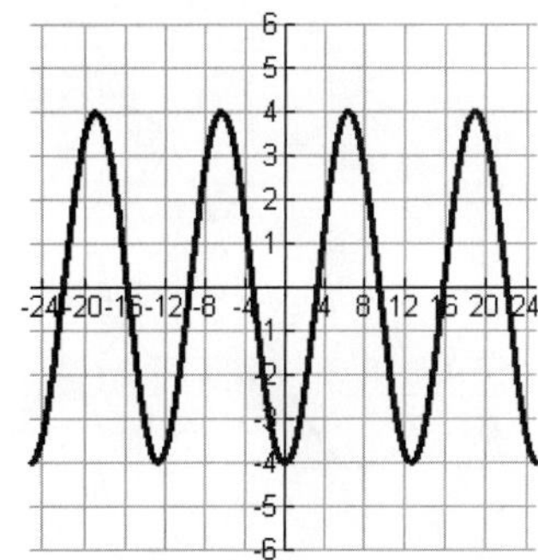

34. In order to construct the graph of $y=-5\sin\left(\frac{1}{2}x\right)$ on $[-8\pi,8\pi]$, we first gather the information essential to forming the graph for one period, namely on the interval $[0,4\pi]$. Then, we extend this picture to cover the entire given interval. Indeed, we have the following:

<u>Amplitude</u>: $\lvert -5\rvert = 5$	<u>Period</u>: $\frac{2\pi}{1/2} = 4\pi$
<u>x-intercepts</u> occur when $\sin\left(\frac{1}{2}x\right)=0$. This happens when $\frac{1}{2}x = 0,\ \pi,\ 2\pi$, and so $x = 0,\ 2\pi,\ 4\pi$. So, the points $(0,\ 0),\ (2\pi,\ 0),\ (4\pi,\ 0)$ are on the graph.	
The <u>maximum value</u> of the graph is 5 since the maximum of $\sin\left(\frac{1}{2}x\right)$ is 1, and the amplitude is 5. This occurs when $\frac{1}{2}x=\frac{\pi}{2}$, and so $x=\pi$. So, the point $(\pi,5)$ is on the graph.	
The <u>minimum value</u> of the graph is – 5 since the minimum of $\sin\left(\frac{1}{2}x\right)$ is – 1, and the amplitude is 5. This occurs when $\frac{1}{2}x=\frac{3\pi}{2}$, and so $x=3\pi$. So, the point $(3\pi,-5)$ is on the graph.	
Since the coefficient used to compute the amplitude is negative, we reflect the graph of $y=5\sin\left(\frac{1}{2}x\right)$ over the x-axis.	

Now, using this information, we can form the graph of $y=-5\sin\left(\frac{1}{2}x\right)$ on $[0,4\pi]$, and then extend it to the interval $[-8\pi,8\pi]$, as shown below:

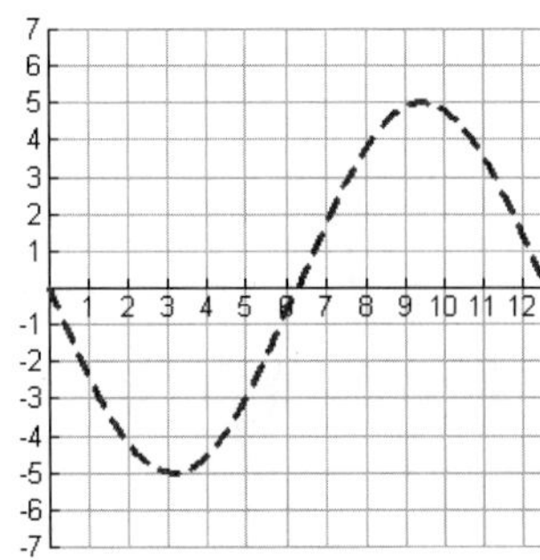

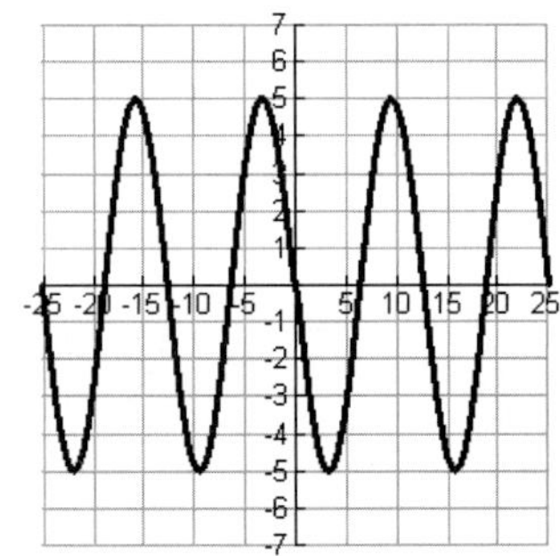

35. In order to construct the graph of $y=-\sin(6x)$ on $\left[-\frac{2\pi}{3},\frac{2\pi}{3}\right]$, we first gather the information essential to forming the graph for one period, namely on the interval $\left[0,\frac{\pi}{3}\right]$. Then, extend this picture to cover the entire interval. Indeed, we have the following:

Amplitude: $\lvert -1\rvert = 1$	Period: $\frac{2\pi}{6}=\frac{\pi}{3}$
x-intercepts occur when $\sin 6x = 0$. This happens when $6x = n\pi$, and so $x=\frac{n\pi}{6}$, where n is an integer. So, the points $(0, 0)$, $\left(\frac{\pi}{6}, 0\right)$, $\left(\frac{\pi}{3}, 0\right)$ are on the graph.	
The maximum value of the graph is 1 since the maximum of $\sin 6x$ is 1, and the amplitude is 1. This occurs when $6x=\frac{\pi}{2}$, so that $x=\frac{\pi}{12}$. So, the point $\left(\frac{\pi}{12},1\right)$ is on the graph.	
The minimum value of the graph is – 1 since the minimum of $\sin 6x$ is – 1, and the amplitude is 1. This occurs when $6x=\frac{3\pi}{2}$, so that $x=\frac{\pi}{4}$. So, the point $\left(\frac{\pi}{4},-1\right)$ is on the graph.	
Since the coefficient used to compute the amplitude is negative, we reflect the graph of $y=\sin(6x)$ over the x-axis.	

Now, using this information, we can form the graph of $y=-\sin(6x)$ on $\left[0,\frac{\pi}{3}\right]$, and then extend it to the interval $\left[-\frac{2\pi}{3},\frac{2\pi}{3}\right]$, as shown below:

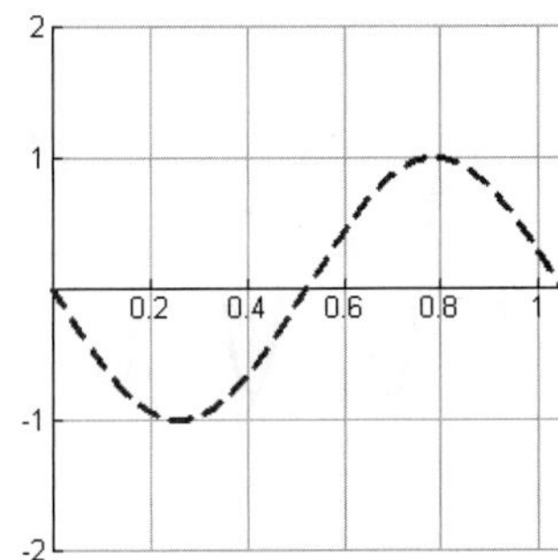

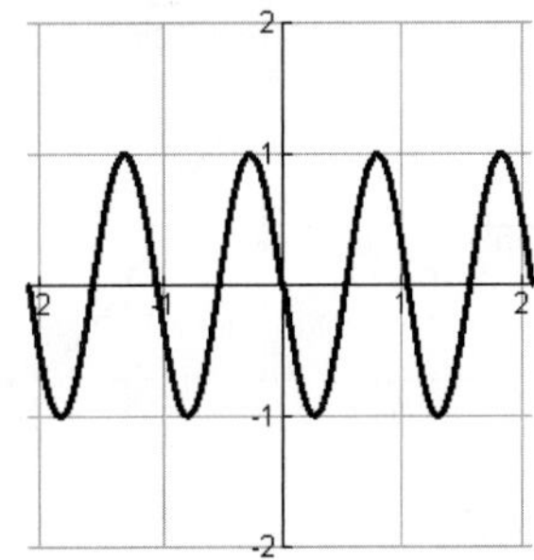

36. In order to construct the graph of $y=-\cos(4x)$ on $[-\pi,\pi]$, we first gather the information essential to forming the graph for one period, namely on the interval $\left[0,\frac{\pi}{2}\right]$. Then, extend this picture to cover the entire interval. Indeed, we have the following:

Amplitude: $\lvert -1\rvert =1$	Period: $\frac{2\pi}{4}=\frac{\pi}{2}$
x-intercepts occur when $\cos 4x=0$. This happens when $4x=\frac{(2n+1)\pi}{2}$, and so $x=\frac{(2n+1)\pi}{8}$, where *n* is an integer. So, the points $\left(\frac{\pi}{8},0\right),\left(\frac{3\pi}{8},0\right)$ are on the graph.	
The maximum value of the graph is 1 since the maximum of $\cos 4x$ is 1, and the amplitude is 1. This occurs when $4x=0,\ 2\pi$, so that $x=0,\ \frac{\pi}{2}$. So, the points $(0,1),\left(\frac{\pi}{2},1\right)$ are on the graph.	
The minimum value of the graph is – 1 since the minimum of $\cos 4x$ is – 1, and the amplitude is 1. This occurs when $4x=\pi$, so that $x=\frac{\pi}{4}$. So, the point $\left(\frac{\pi}{4},-1\right)$ is on the graph.	
Since the coefficient used to compute the amplitude is negative, we reflect the graph of $y=\cos(4x)$ over the *x*-axis.	

Now, using this information, we can form the graph of $y=-\cos 4x$ on $\left[0,\frac{\pi}{2}\right]$, and then extend it to the interval $[-\pi,\pi]$, as shown below:

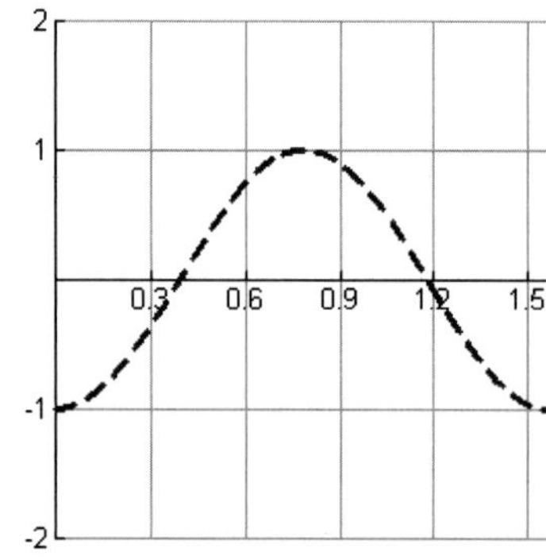

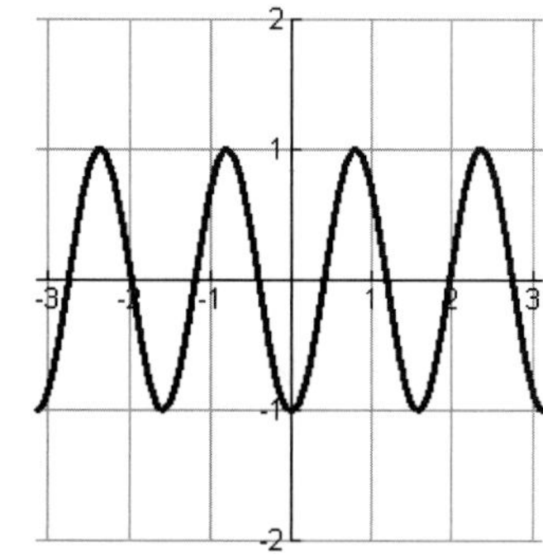

37. In order to construct the graph of $y = 3\cos\left(\frac{\pi}{4}x\right)$ on $[-16,16]$, we first gather the information essential to forming the graph for one period, namely on the interval $[0,8]$. Then, extend this picture to cover the entire interval. Indeed, we have the following:

Amplitude: 3	Period: $\frac{2\pi}{\pi/4} = 8$
x-intercepts occur when $\cos\left(\frac{\pi}{4}x\right) = 0$. This happens when $\frac{\pi}{4}x = \frac{(2n+1)\pi}{2}$ and so, $x = 2(2n+1)$, where n is an integer. So, the points $(2, 0)$, $(6, 0)$ are on the graph.	
The maximum value of the graph is 3 since the maximum of $\cos\left(\frac{\pi}{4}x\right)$ is 1, and the amplitude is 3. This occurs when $\frac{\pi}{4}x = 0,\ 2\pi$, so that $x = 0,\ 8$. So, the points $(0,3)$, $(8,3)$ are on the graph.	
The minimum value of the graph is – 3 since the minimum of $\cos\left(\frac{\pi}{4}x\right)$ is – 1, and the amplitude is 3. This occurs when $\frac{\pi}{4}x = \pi$, so that $x = 4$. So, the point $(4,-3)$ is on the graph.	
Since the coefficient used to compute the amplitude is nonnegative, there is no reflection over the x-axis.	

Now, using this information, we can form the graph of $y = 3\cos\left(\frac{\pi}{4}x\right)$ on $[0,8]$, and then extend it to the interval $[-16,16]$, as shown below:

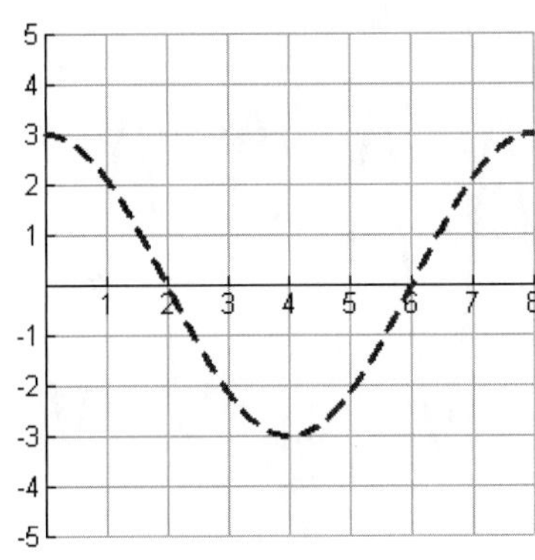

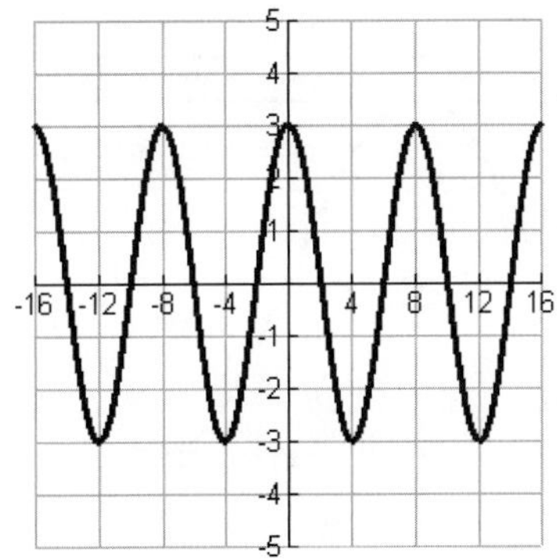

38. In order to construct the graph of $y = 4\sin\left(\frac{\pi}{4}x\right)$ on $[-16,16]$, we first gather the information essential to forming the graph for one period, namely on the interval $[0,8]$. Then, extend this picture to cover the entire interval. Indeed, we have the following:

Amplitude: 4	Period: $\frac{2\pi}{\pi/4} = 8$
x-intercepts occur when $\sin\left(\frac{\pi}{4}x\right) = 0$. This happens when $\frac{\pi}{4}x = n\pi$ and so, $x = 4n$, where *n* is an integer. So, the points $(0, 0), (4, 0), (8, 0)$ are on the graph.	
The maximum value of the graph is 4 since the maximum of $\sin\left(\frac{\pi}{4}x\right)$ is 1, and the amplitude is 4. This occurs when $\frac{\pi}{4}x = \frac{\pi}{2}$, so that $x = 2$. So, the point $(2,4)$ is on the graph.	
The minimum value of the graph is – 4 since the minimum of $\sin\left(\frac{\pi}{4}x\right)$ is – 1, and the amplitude is 4. This occurs when $\frac{\pi}{4}x = \frac{3\pi}{2}$, so that $x = 6$. So, the point $(6,-4)$ is on the graph.	
Since the coefficient used to compute the amplitude is nonnegative, there is no reflection over the *x*-axis.	

Now, using this information, we can form the graph of $y = 4\sin\left(\frac{\pi}{4}x\right)$ on $[0,8]$, and then extend it to the interval $[-16,16]$, as shown below:

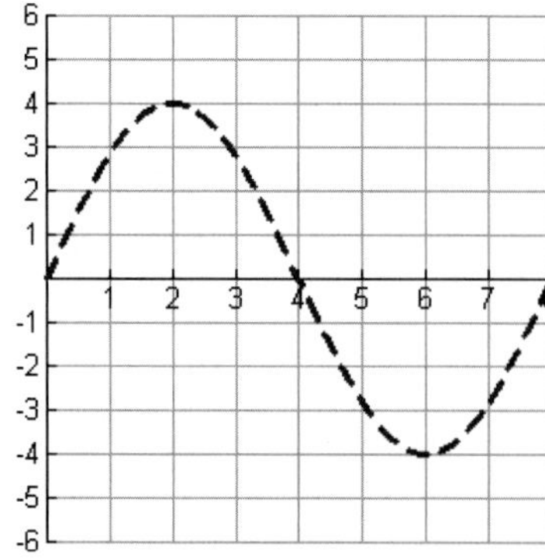

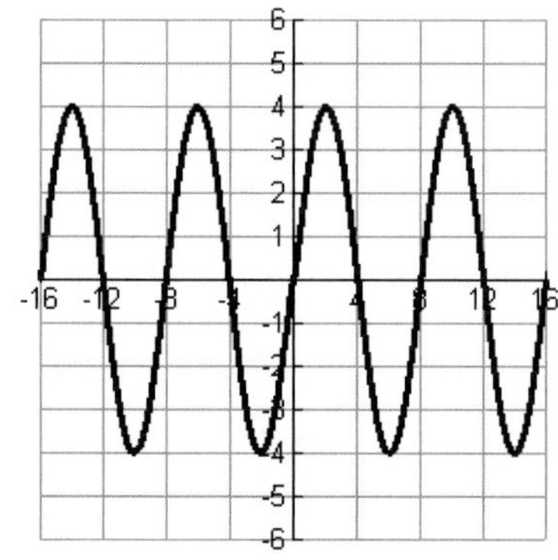

39. In order to construct the graph of $y=\sin(4\pi x)$ on $[-1,1]$, we first gather the information essential to forming the graph for one period, namely on the interval $\left[0,\frac{1}{2}\right]$. Then, extend this picture to cover the entire interval. Indeed, we have the following:

Amplitude: 1	Period: $\frac{2\pi}{4\pi}=\frac{1}{2}$
x-intercepts occur when $\sin 4\pi x=0$. This happens when $4\pi x=n\pi$, and so $x=\frac{n}{4}$, where n is an integer. So, the points $(0,\,0)$, $\left(\frac{1}{4},\,0\right)$, $\left(\frac{1}{2},\,0\right)$ are on the graph.	
The maximum value of the graph is 1 since the maximum of $\sin 4\pi x$ is 1, and the amplitude is 1. This occurs when $4\pi x=\frac{\pi}{2}$, so that $x=\frac{1}{8}$. So, the point $\left(\frac{1}{8},1\right)$ is on the graph.	
The minimum value of the graph is – 1 since the minimum of $\sin 4\pi x$ is – 1, and the amplitude is 1. This occurs when $4\pi x=\frac{3\pi}{2}$, so that $x=\frac{3}{8}$. So, the point $\left(\frac{3}{8},-1\right)$ is on the graph.	
Since the coefficient used to compute the amplitude is nonnegative, there is no reflection over the x-axis.	

Now, using this information, we can form the graph of $y=\sin(4\pi x)$ on $\left[0,\frac{1}{2}\right]$, and then extend it to the interval $[-1,1]$, as shown below:

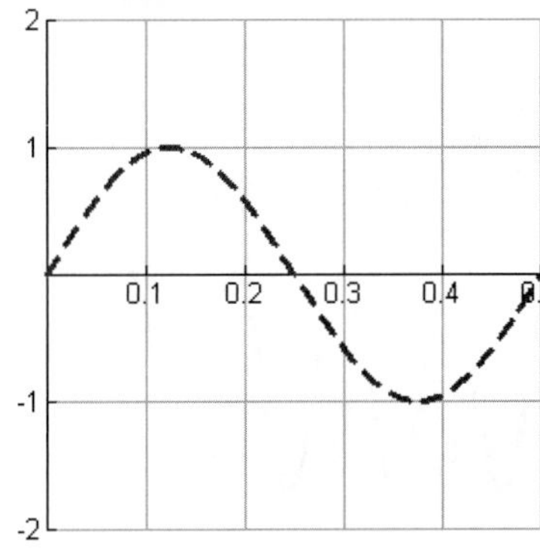

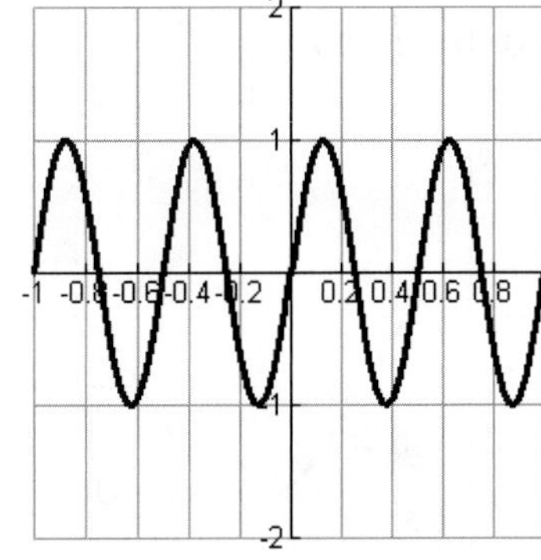

40. In order to construct the graph of $y=\cos(6\pi x)$ on $\left[-\frac{2}{3},\frac{2}{3}\right]$, we first gather the information essential to forming the graph for one period, namely on the interval $\left[0,\frac{1}{3}\right]$. Then, extend this picture to cover the entire interval. Indeed, we have the following:

Amplitude: 1	Period: $\frac{2\pi}{6\pi}=\frac{1}{3}$
x-intercepts occur when $\cos 6\pi x=0$. This happens when $6\pi x=\frac{(2n+1)\pi}{2}$ and so, $x=\frac{(2n+1)}{12}$, where n is an integer. So, the points $\left(\frac{1}{12},0\right),\left(\frac{1}{4},0\right)$ are on the graph.	
The maximum value of the graph is 1 since the maximum of $\cos 6\pi x$ is 1, and the amplitude is 1. This occurs when $6\pi x=0,\ 2\pi$, so that $x=0,\ \frac{1}{3}$. So, the points $(0,1),\left(\frac{1}{3},1\right)$ are on the graph.	
The minimum value of the graph is – 1 since the minimum of $\cos 6\pi x$ is – 1, and the amplitude is 1. This occurs when $6\pi x=\pi$, so that $x=\frac{1}{6}$. So, the point $\left(\frac{1}{6},-1\right)$ is on the graph.	
Since the coefficient used to compute the amplitude is nonnegative, there is no reflection over the x-axis.	

Now, using this information, we can form the graph of $y=\cos(6\pi x)$ on $\left[0,\frac{1}{3}\right]$, and then extend it to the interval $\left[-\frac{2}{3},\frac{2}{3}\right]$, as shown below:

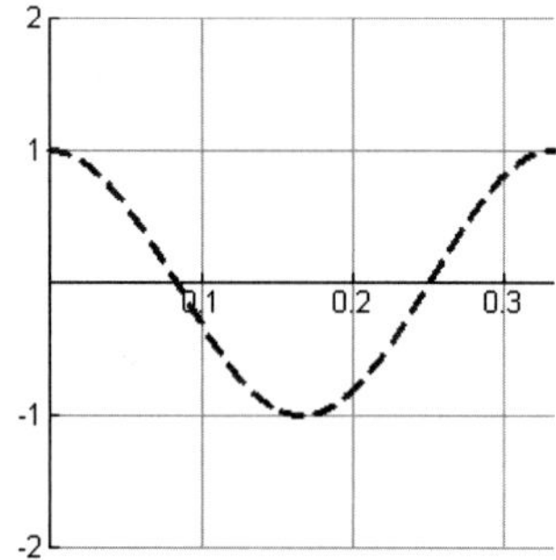

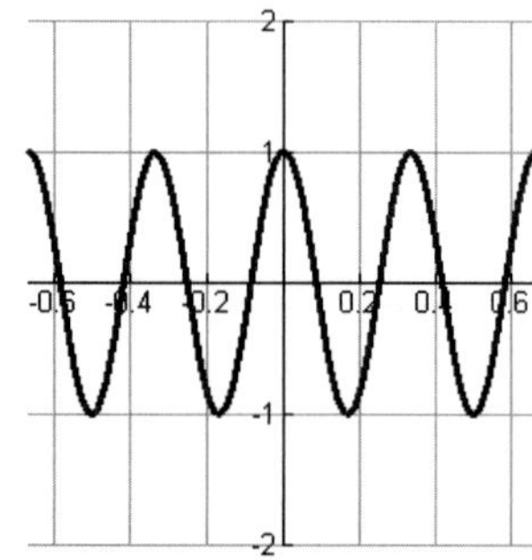

41. In order to determine the equation, we make the following key observations that enable us to determine the values of A and B in the appropriate choice of standard form (namely $y = A\sin Bx$ or $y = A\cos Bx$).

- y-intercept is (0,0).
- Since one full period occurs in the interval $[0,\pi]$, the period is π and so, we can determine the value of B by solving $\frac{2\pi}{B} = \pi$. Doing so yields $B = 2$.
- Since the maximum y-value is 1 and the minimum y-value is –1, we know that the amplitude $|A| = \frac{\text{maximum } y - \text{value} \; - \; \text{minimum } y - \text{value}}{2} = 1$.
- The original graph of $y = \sin x$ is reflected over the x-axis to result in this graph; so we know that the correct choice of A is –1.

Thus, the equation is $\boxed{y = -\sin 2x}$.

42. In order to determine the equation, we make the following key observations that enable us to determine the values of A and B in the appropriate choice of standard form (namely $y = A\sin Bx$ or $y = A\cos Bx$).

- y-intercept is (0, -2).
- Since one full period occurs in the interval $[0,2\pi]$, the period is 2π and so, we can determine the value of B by solving $\frac{2\pi}{B} = 2\pi$. Doing so yields $B = 1$.
- Since the maximum y-value is 2 and the minimum y-value is – 2, we know that the amplitude $|A| = \frac{\text{maximum } y - \text{value} \; - \; \text{minimum } y - \text{value}}{2} = 2$.
- The original graph of $y = \cos x$ is reflected over the x-axis to result in this graph; so we know that the correct choice of A is –2.

Thus, the equation is $\boxed{y = -2\cos x}$.

43. In order to determine the equation, we make the following key observations that enable us to determine the values of A and B in the appropriate choice of standard form (namely $y = A\sin Bx$ or $y = A\cos Bx$).

- y-intercept is (0, 1).
- Since one full period occurs in the interval $[0,2]$, the period is 2 and so, we can determine the value of B by solving $\frac{2\pi}{B} = 2$. Doing so yields $B = \pi$.
- Since the maximum y-value is 1 and the minimum y-value is – 1, we know that the amplitude $|A| = \frac{\text{maximum } y - \text{value } - \text{ minimum } y - \text{value}}{2} = 1$.
- The shape of the graph coincides with the graph of $y = \cos x$ and hence, it is not reflected over the x-axis; so we know that the correct choice of A is 1.

Thus, the equation is $\boxed{y = \cos \pi x}$.

44. In order to determine the equation, we make the following key observations that enable us to determine the values of A and B in the appropriate choice of standard form (namely $y = A\sin Bx$ or $y = A\cos Bx$).

- y-intercept is (0, 0).
- Since one full period occurs in the interval $[0,2]$, the period is 2 and so, we can determine the value of B by solving $\frac{2\pi}{B} = 2$. Doing so yields $B = \pi$.
- Since the maximum y-value is 1 and the minimum y-value is – 1, we know that the amplitude $|A| = \frac{\text{maximum } y - \text{value } - \text{ minimum } y - \text{value}}{2} = 1$.
- The shape of the graph coincides with the graph of $y = \sin x$ and hence, it is not reflected over the x-axis; so we know that the correct choice of A is 1.

Thus, the equation is $\boxed{y = \sin \pi x}$.

45. In order to determine the equation, we make the following key observations that enable us to determine the values of A and B in the appropriate choice of standard form (namely $y = A\sin Bx$ or $y = A\cos Bx$).

- y-intercept is (0,0).
- Since one full period occurs in the interval $[0,4]$, the period is 4 and so, we can determine the value of B by solving $\frac{2\pi}{B} = 4$. Doing so yields $B = \frac{\pi}{2}$.
- Since the maximum y-value is 2 and the minimum y-value is –2, we know that the amplitude $|A| = \frac{\text{maximum } y-\text{value} \; - \; \text{minimum } y-\text{value}}{2} = 2$.
- The original graph of $y = \sin x$ is reflected over the x-axis to result in this graph; so we know that the correct choice of A is –2.

Thus, the equation is $\boxed{y = -2\sin\left(\frac{\pi}{2}x\right)}$.

46. In order to determine the equation, we make the following key observations that enable us to determine the values of A and B in the appropriate choice of standard form (namely $y = A\sin Bx$ or $y = A\cos Bx$).

- y-intercept is (0,-3).
- Since one full period occurs in the interval $[0,4]$, the period is 4 and so, we can determine the value of B by solving $\frac{2\pi}{B} = 4$. Doing so yields $B = \frac{\pi}{2}$.
- Since the maximum y-value is 3 and the minimum y-value is –3, we know that the amplitude $|A| = \frac{\text{maximum } y-\text{value} \; - \; \text{minimum } y-\text{value}}{2} = 3$.
- The original graph of $y = \cos x$ is reflected over the x-axis to result in this graph; so we know that the correct choice of A is –3.

Thus, the equation is $\boxed{y = -3\cos\left(\frac{\pi}{2}x\right)}$.

47. In order to determine the equation, we make the following key observations that enable us to determine the values of A and B in the appropriate choice of standard form (namely $y = A\sin Bx$ or $y = A\cos Bx$).

- y-intercept is (0, 0).
- Since one full period occurs in the interval $\left[0, \frac{1}{4}\right]$, the period is $\frac{1}{4}$ and so, we can determine the value of B by solving $\frac{2\pi}{B} = \frac{1}{4}$. Doing so yields $B = 8\pi$.
- Since the maximum y-value is 1 and the minimum y-value is – 1, we know that the amplitude $|A| = \frac{\text{maximum } y - \text{value} \ - \ \text{minimum } y - \text{value}}{2} = 1$.
- The shape of the graph coincides with the graph of $y = \sin x$ and hence, it is not reflected over the x-axis; so we know that the correct choice of A is 1.

Thus, the equation is $\boxed{y = \sin 8\pi x}$.

48. In order to determine the equation, we make the following key observations that enable us to determine the values of A and B in the appropriate choice of standard form (namely $y = A\sin Bx$ or $y = A\cos Bx$).

- y-intercept is (0, 1).
- Since one full period occurs in the interval $\left[0, \frac{1}{2}\right]$, the period is $\frac{1}{2}$ and so, we can determine the value of B by solving $\frac{2\pi}{B} = \frac{1}{2}$. Doing so yields $B = 4\pi$.
- Since the maximum y-value is 1 and the minimum y-value is – 1, we know that the amplitude $|A| = \frac{\text{maximum } y - \text{value} \ - \ \text{minimum } y - \text{value}}{2} = 1$.
- The shape of the graph coincides with the graph of $y = \cos x$ and hence, it is not reflected over the x-axis; so we know that the correct choice of A is 1.

Thus, the equation is $\boxed{y = \cos 4\pi x}$.

49. Amplitude 2, Period $\frac{2\pi}{\pi} = 2$, Phase Shift $\frac{1}{\pi}$

50. Amplitude 4, Period $\frac{2\pi}{1} = 2\pi$, Phase Shift $-\frac{\pi}{1} = -\pi$

51. Amplitude 5, Period $\frac{2\pi}{3}$, Phase Shift $-\frac{2}{3}$

52. Amplitude 7, Period $\frac{2\pi}{4} = \frac{\pi}{2}$, Phase Shift $\frac{3}{4}$

53. Amplitude 6, Period $\frac{2\pi}{\lvert -\pi \rvert} = 2$, Phase Shift $\frac{2\pi}{-\pi} = -2$
54. Amplitude 3, Period $\frac{2\pi}{\lvert -\pi/2 \rvert} = 4$, Phase Shift $\frac{-\pi/2}{-\pi/2} = 1$
55. Amplitude: 3 Period: π Phase Shift: $\frac{\pi}{2}$ left
56. Amplitude: 4 Period: π Phase Shift: $-\frac{\pi}{2}$ right
57. Amplitude: $\frac{1}{4}$ Period: 8π Phase Shift: 2π right
58. Amplitude: $\frac{1}{2}$ Period: 6π Phase Shift: 3π left
59. Amplitude: 2 Period: 4 Phase Shift: - 4 right
60. Amplitude: 5 Period: 2 Phase Shift: 1 left

61. First, we have the following information for the equation $y = \frac{1}{2} + \frac{3}{2}\cos(2x+\pi)$:

Amplitude $\frac{3}{2}$, Period $\frac{2\pi}{2} = \pi$, Phase Shift $\frac{-\pi}{2}$, Vertical Shift $\frac{1}{2}$ unit up

In order to graph $y = \frac{1}{2} + \frac{3}{2}\cos(2x+\pi)$ on $\left[-\frac{3\pi}{2}, \frac{3\pi}{2}\right]$, follow these steps:

Step 1: Graph $y = \frac{3}{2}\cos(2x+\pi)$ on $\left[-\frac{\pi}{2}, -\frac{\pi}{2}+\pi\right] = \left[-\frac{\pi}{2}, \frac{\pi}{2}\right]$:

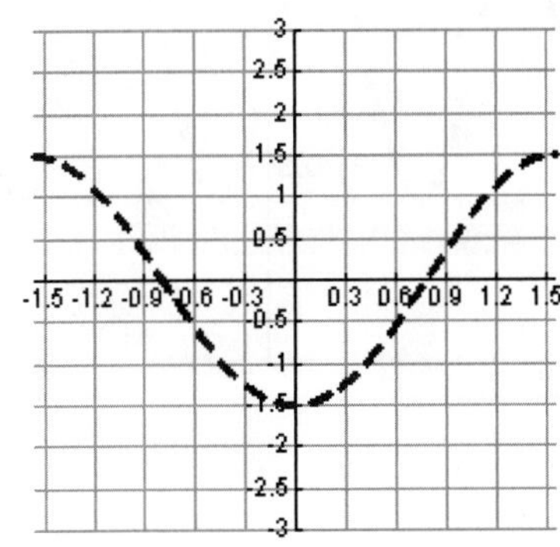

Step 2: Extend the graph in Step 1 to the interval $\left[-\frac{3\pi}{2}, \frac{3\pi}{2}\right]$ using periodicity:

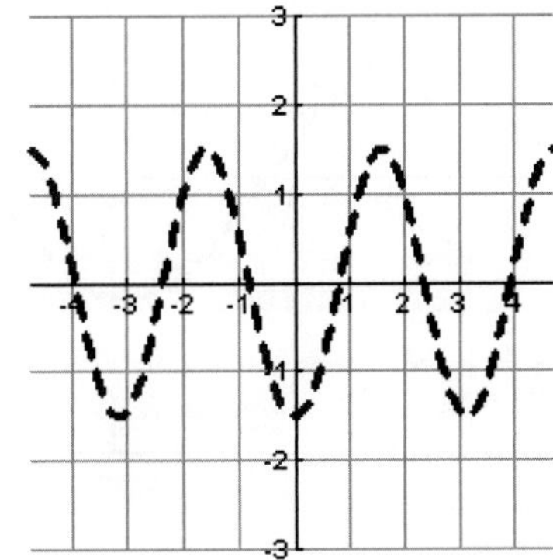

Step 3: Translate the graph in Step 2 vertically up $\frac{1}{2}$ unit to obtain the graph of $y=\frac{1}{2}+\frac{3}{2}\cos(2x+\pi)$ on $\left[-\frac{3\pi}{2},\frac{3\pi}{2}\right]$:

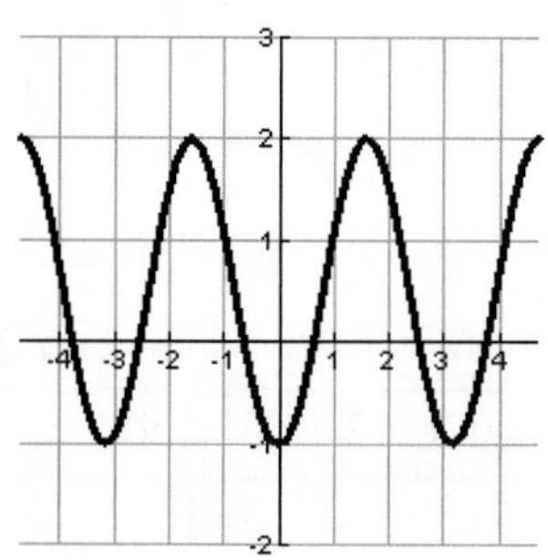

62. First, we have the following information for the equation $y=\frac{1}{3}+\frac{2}{3}\sin(2x-\pi)$:

Amplitude $\frac{2}{3}$, Period $\frac{2\pi}{2}=\pi$, Phase Shift $\frac{\pi}{2}$, Vertical Shift $\frac{1}{3}$ unit up

In order to graph $y=\frac{1}{3}+\frac{2}{3}\sin(2x-\pi)$ on $\left[-\frac{3\pi}{2},\frac{3\pi}{2}\right]$, follow these steps:

Step 1: Graph $y=\frac{2}{3}\sin(2x-\pi)$ on $\left[\frac{\pi}{2},\frac{\pi}{2}+\pi\right]=\left[\frac{\pi}{2},\frac{3\pi}{2}\right]$:

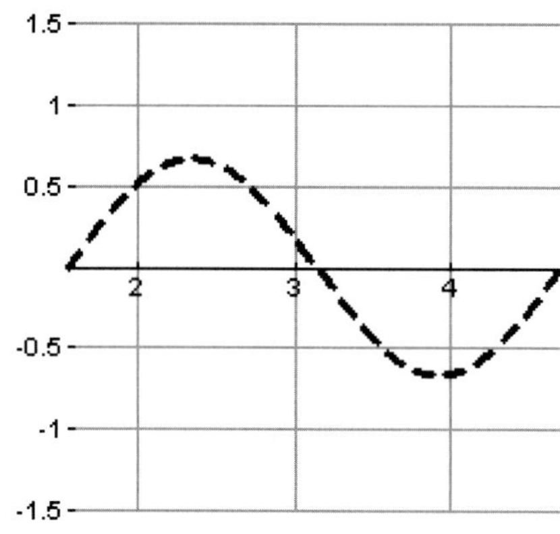

Step 2: Extend the graph in Step 1 to the interval $\left[-\frac{3\pi}{2},\frac{3\pi}{2}\right]$ using periodicity:

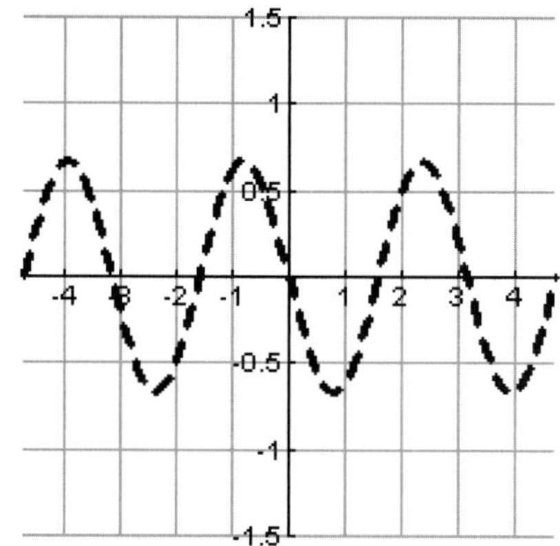

Step 3: Translate the graph in Step 2 vertically up $\frac{1}{3}$ unit to obtain the graph of $y=\frac{1}{3}+\frac{2}{3}\sin(2x-\pi)$ on $\left[-\frac{3\pi}{2},\frac{3\pi}{2}\right]$:

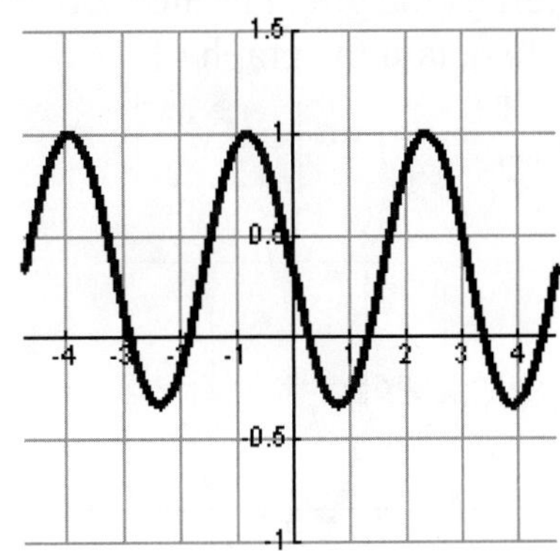

63. First, we have the following information for the equation $y=\frac{1}{2}-\frac{1}{2}\sin\left(\frac{1}{2}x-\frac{\pi}{4}\right)$:

Amplitude $\left|-\frac{1}{2}\right|=\frac{1}{2}$ (negative in front indicates reflection over x-axis),

Period $\frac{2\pi}{1/2}=4\pi$, Phase Shift $\frac{\pi/4}{1/2}=\frac{\pi}{2}$, Vertical Shift $\frac{1}{2}$ unit up

In order to graph $y=\frac{1}{2}-\frac{1}{2}\sin\left(\frac{1}{2}x-\frac{\pi}{4}\right)$ on $\left[-\frac{7\pi}{2},\frac{9\pi}{2}\right]$, follow these steps:

Step 1: Graph $y=\frac{1}{2}\sin\left(\frac{1}{2}x-\frac{\pi}{4}\right)$ on $\left[\frac{\pi}{2},\frac{\pi}{2}+4\pi\right]=\left[\frac{\pi}{2},\frac{9\pi}{2}\right]$:

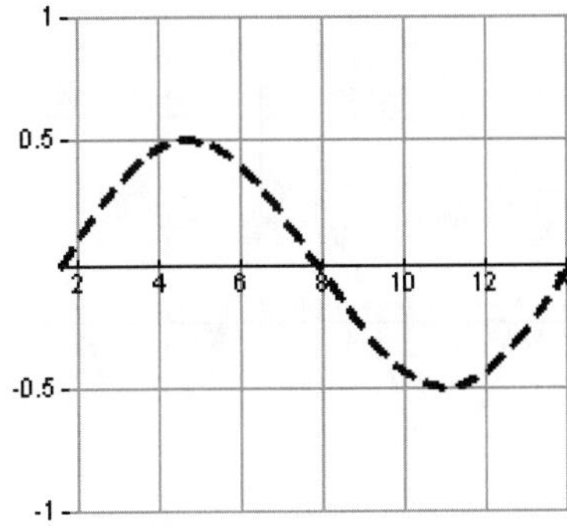

Step 2: Extend the graph in Step 1 to the interval $\left[-\frac{7\pi}{2},\frac{9\pi}{2}\right]$ using periodicity:

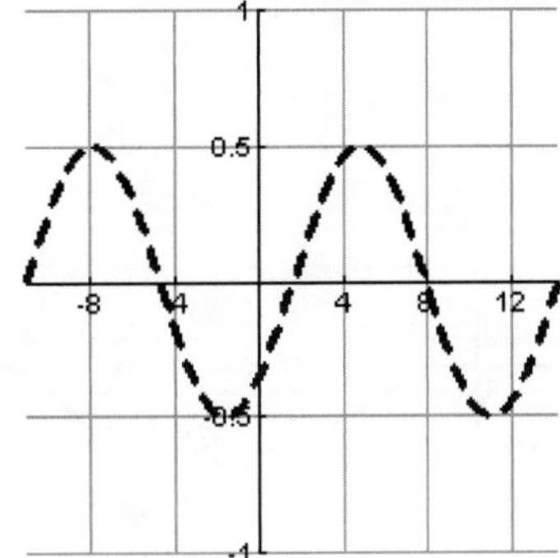

Step 3: Reflect the graph in Step 2 over the x-axis to obtain the graph of $y = -\frac{1}{2}\sin\left(\frac{1}{2}x - \frac{\pi}{4}\right)$ on $\left[-\frac{7\pi}{2}, \frac{9\pi}{2}\right]$:

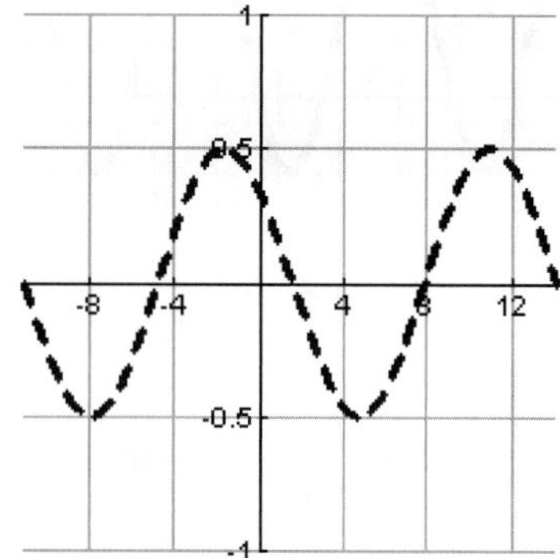

Step 4: Translate the graph in Step 3 vertically up $\frac{1}{2}$ unit to obtain the graph of $y = \frac{1}{2} - \frac{1}{2}\sin\left(\frac{1}{2}x - \frac{\pi}{4}\right)$ on $\left[-\frac{7\pi}{2}, \frac{9\pi}{2}\right]$:

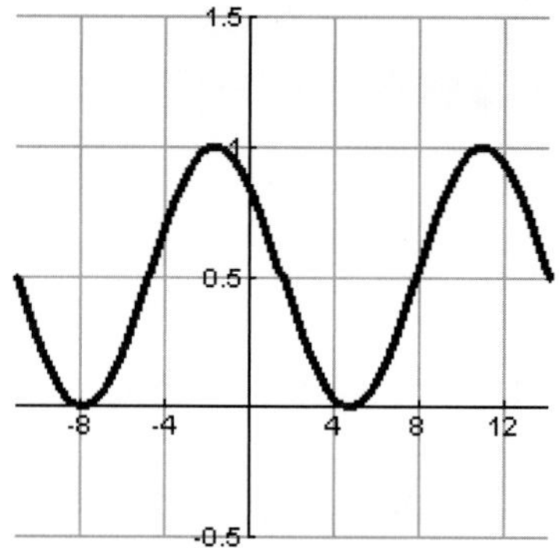

64. First, we have the following information for the equation $y = -\frac{1}{2} + \frac{1}{2}\cos\left(\frac{1}{2}x + \frac{\pi}{4}\right)$:

Amplitude $\frac{1}{2}$, Period $\frac{2\pi}{\frac{1}{2}} = 4\pi$, Phase Shift $-\frac{\frac{\pi}{4}}{\frac{1}{2}} = -\frac{\pi}{2}$, Vertical Shift $\frac{1}{2}$ unit down

In order to graph $y = -\frac{1}{2} + \frac{1}{2}\cos\left(\frac{1}{2}x + \frac{\pi}{4}\right)$ on $\left[-\frac{9\pi}{2}, \frac{7\pi}{2}\right]$, follow these steps:

Step 1: Graph $y = \frac{1}{2}\cos\left(\frac{1}{2}x + \frac{\pi}{4}\right)$ on $\left[-\frac{\pi}{2}, -\frac{\pi}{2} + 4\pi\right] = \left[-\frac{\pi}{2}, \frac{7\pi}{2}\right]$:

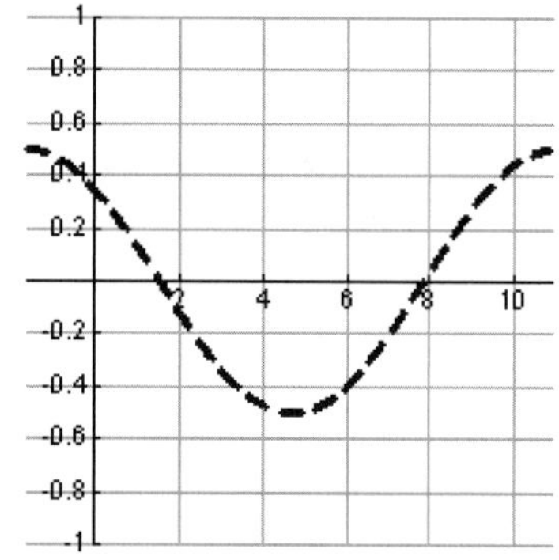

Step 2: Extend the graph in Step 1 to the interval $\left[-\frac{9\pi}{2}, \frac{7\pi}{2}\right]$ using periodicity:

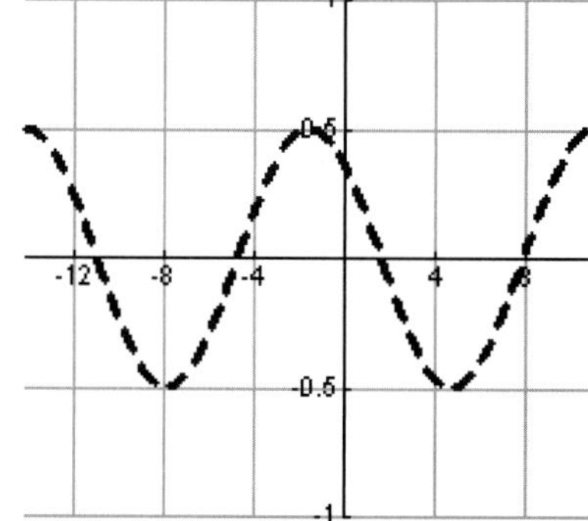

Step 3: Translate the graph in Step 2 vertically down $\frac{1}{2}$ unit to obtain the graph of $y=-\frac{1}{2}+\frac{1}{2}\cos\left(\frac{1}{2}x+\frac{\pi}{4}\right)$ on $\left[-\frac{9\pi}{2},\frac{7\pi}{2}\right]$:

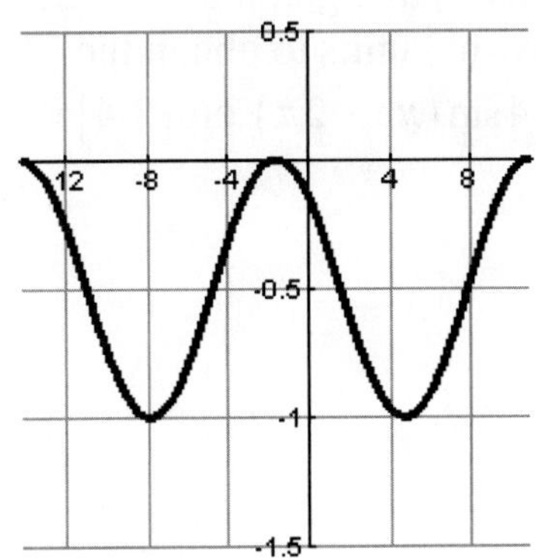

65. First, we have the following information for the equation $y=-3+4\sin(\pi x-2\pi)$:

Amplitude 4, Period $\frac{2\pi}{\pi}=2$, Phase Shift $\frac{2\pi}{\pi}=2$, Vertical Shift 3 units down

In order to graph $y=-3+4\sin(\pi x-2\pi)$ on $[0,4]$, follow these steps:

Step 1: Graph $y=4\sin(\pi x-2\pi)$ on $[2,2+2]=[2,4]$:

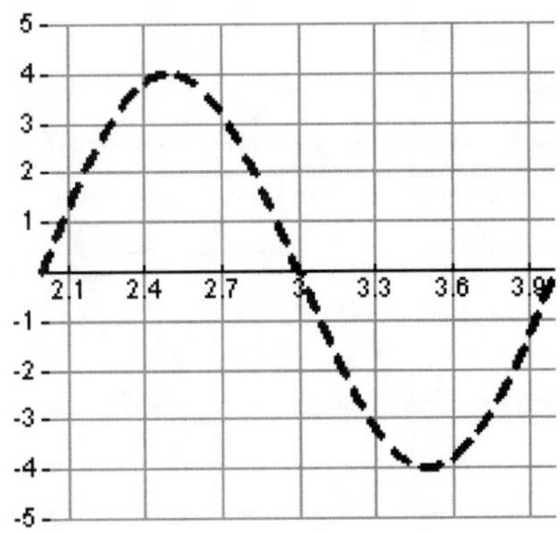

Step 2: Extend the graph in Step 1 to the interval $[0,4]$ using periodicity:

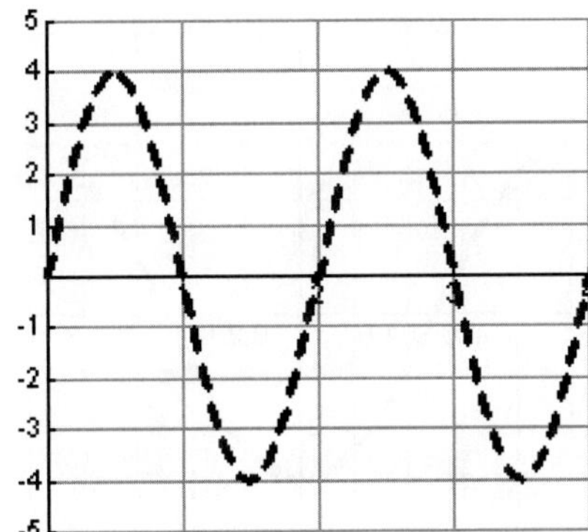

Step 3: Translate the graph in Step 2 vertically down 3 units to obtain the graph of $y=-3+4\sin(\pi x-2\pi)$ on $[0,4]$:

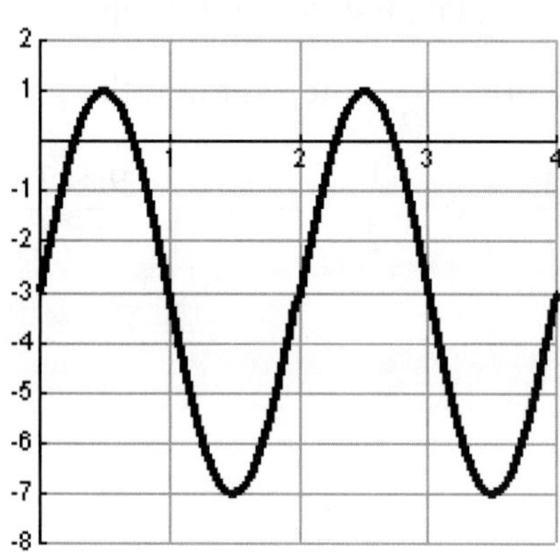

66. First, we have the following information for the equation $y=4-3\cos(\pi x+\pi)$:

Amplitude $|-3|=3$, Period $\frac{2\pi}{\pi}=2$, Phase Shift $-\frac{\pi}{\pi}=-1$, Vertical Shift 4 units up

In order to graph $y=4-3\cos(\pi x+\pi)$ on $[-1,3]$, follow these steps:

Step 1: Graph $y=3\cos(\pi x+\pi)$ on $[-1,-1+2]=[-1,1]$:

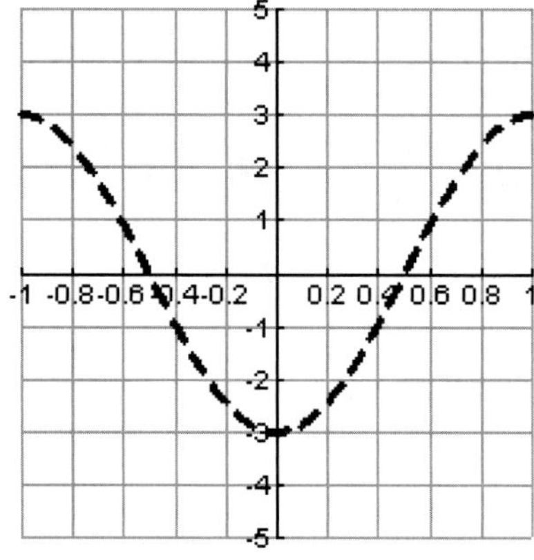

Step 2: Extend the graph in Step 1 to the interval $[-1,3]$ using periodicity:

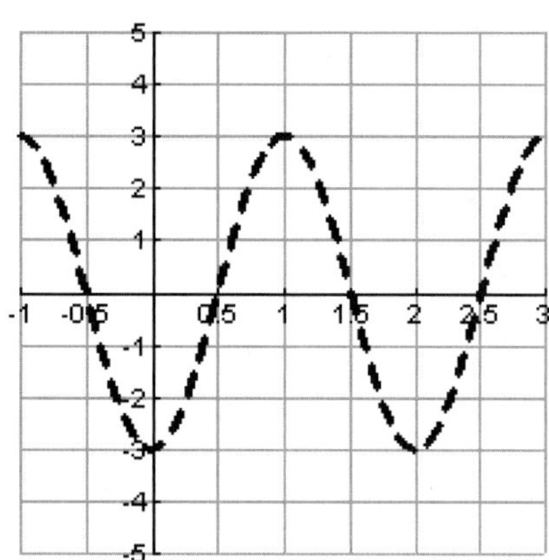

<table>
<tr>
<td>

<u>Step 3</u>: Reflect the graph in Step 2 over the x-axis to obtain the graph of $y=-3\cos(\pi x+\pi)$ on $[-1,3]$:

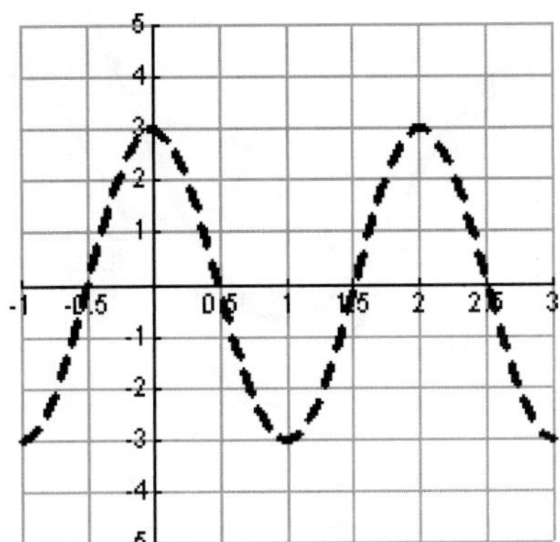

</td>
<td>

<u>Step 4</u>: Translate the graph in Step 3 vertically up 4 units to obtain the graph of $y=4-3\cos(\pi x+\pi)$ on $[-1,3]$:

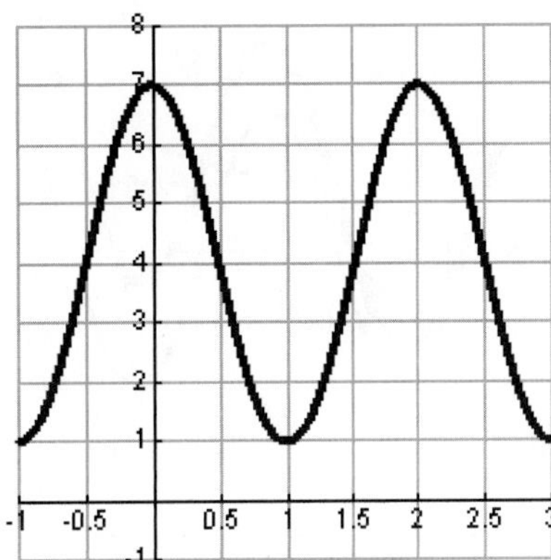

</td>
</tr>
<tr>
<td colspan="2">

Note: Regarding problems 67–94, the convention for the three graphs displayed in each case is as follows. The dotted graph corresponds to the first term of the sum (including the negative if it is present); the dashed graph corresponds to the second term (no negative, even if it is present), and the solid graph is the sum (or difference if the second term has a negative in front of it) of the two functions.

</td>
</tr>
<tr>
<td>

67.

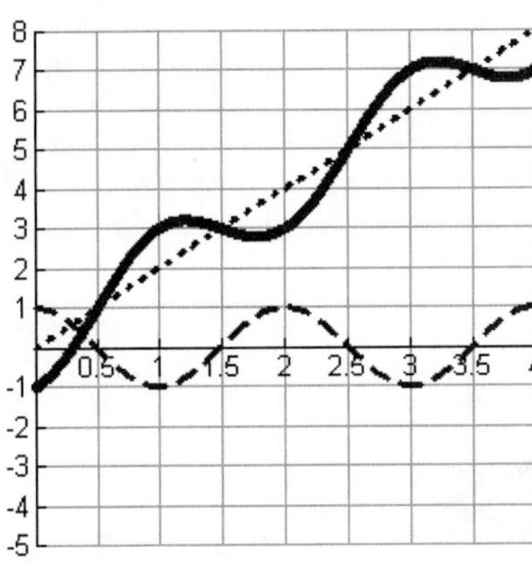

</td>
<td>

68.

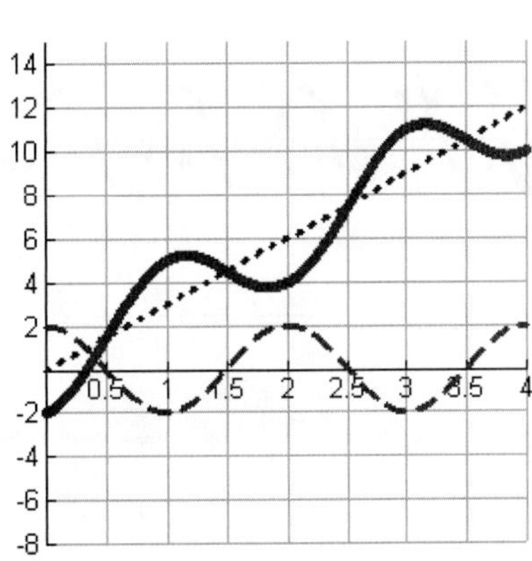

</td>
</tr>
</table>

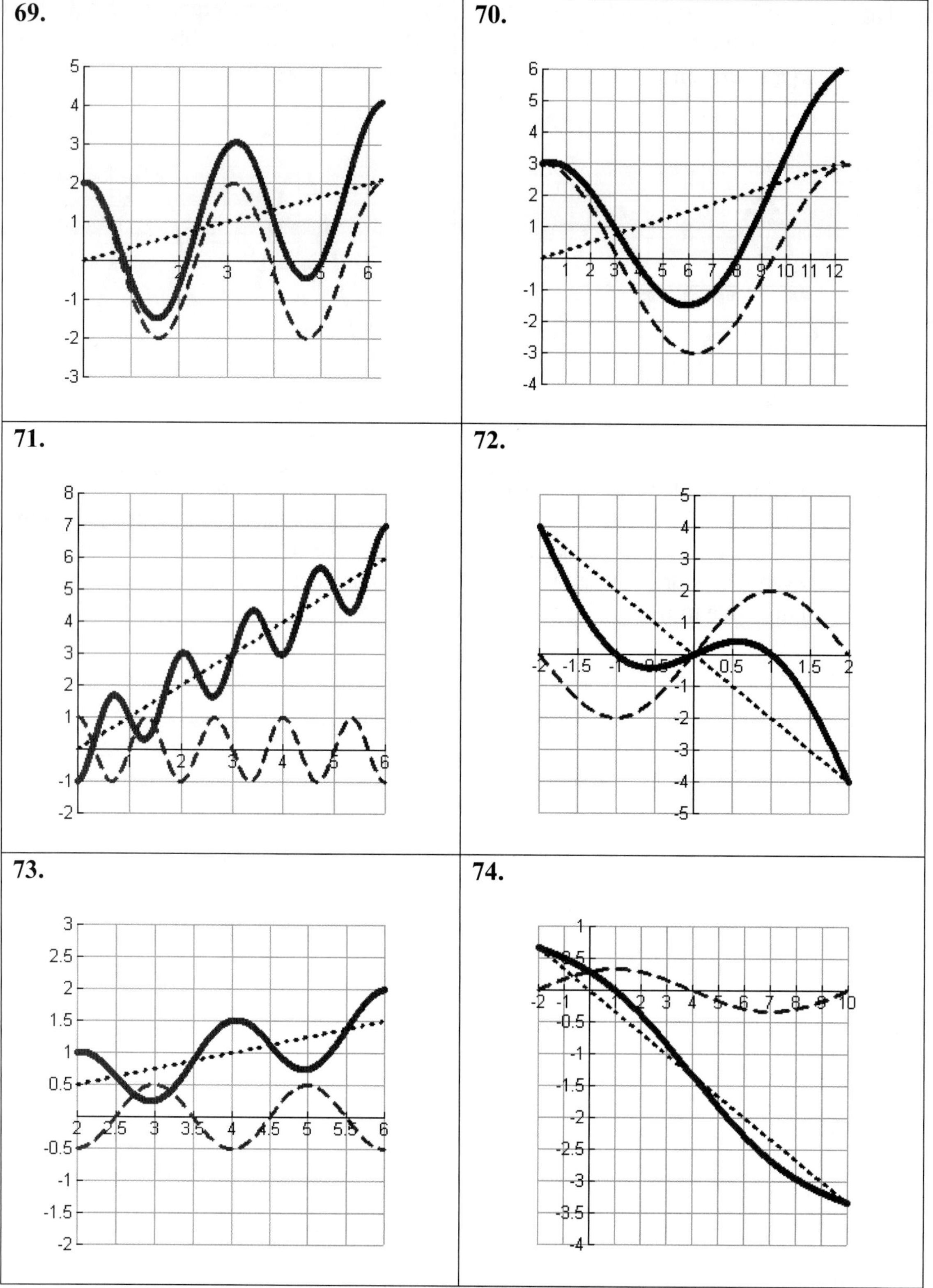
69.
70.
71.
72.
73.
74.

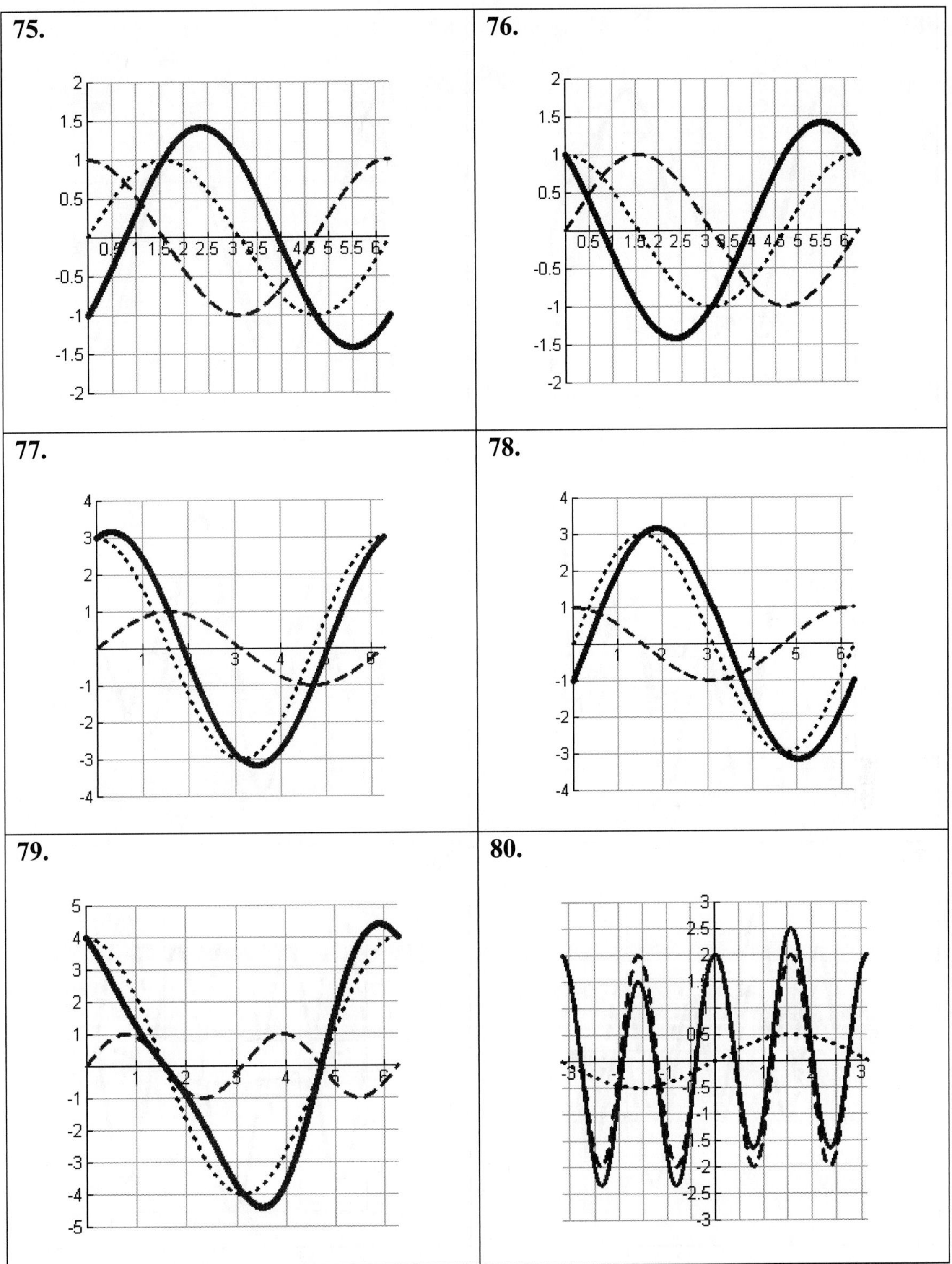
75.
76.
77.
78.
79.
80.

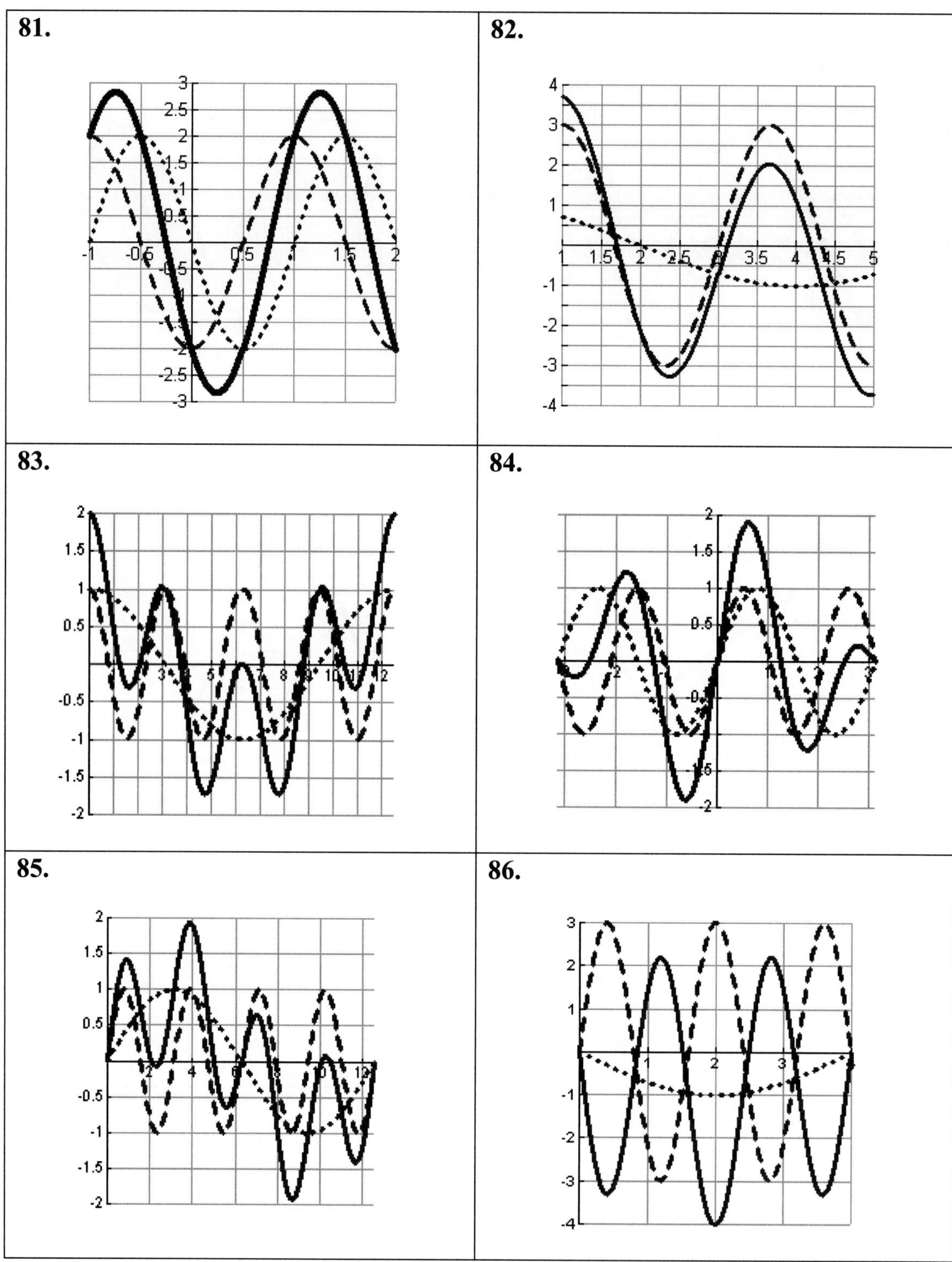

81.
82.
83.
84.
85.
86.

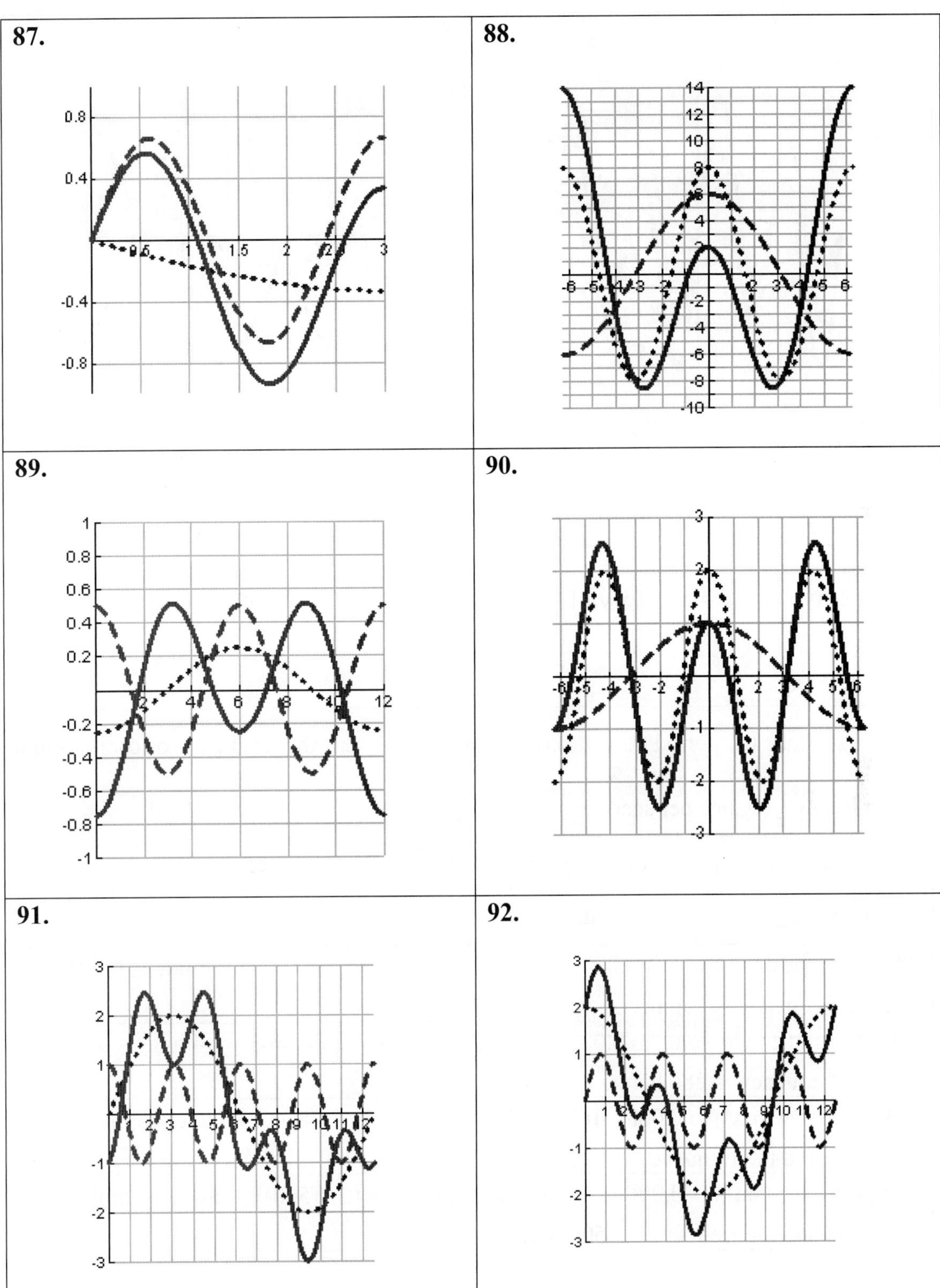
87.
88.
89.
90.
91.
92.

93.

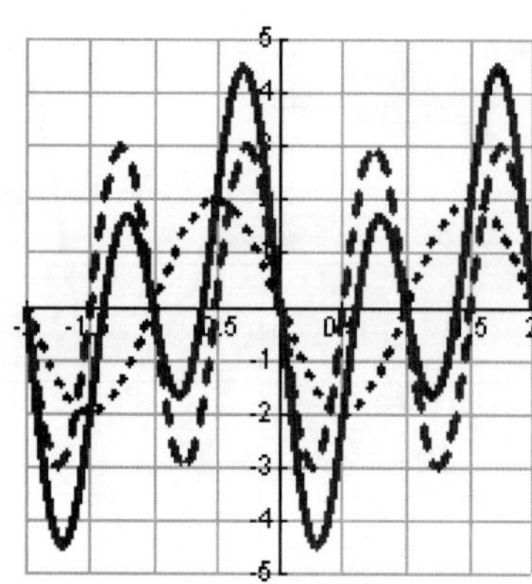

94.

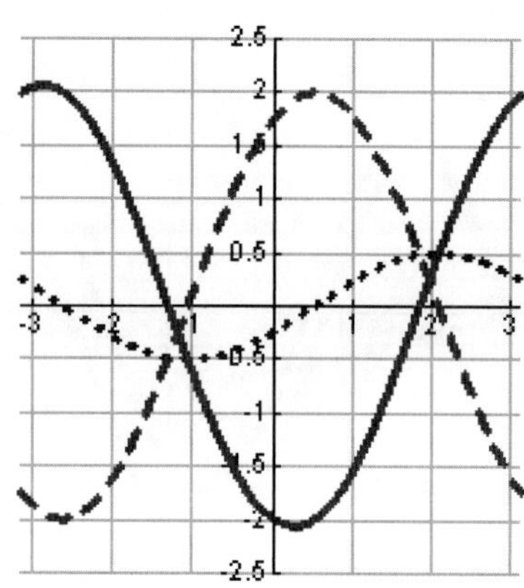

95. Observe that $y = 4\cos\left(\frac{t\sqrt{k}}{2}\right) = 4\cos\left(t\sqrt{\frac{k}{4}}\right)$. So, the amplitude is 4 cm and the mass $m = 4$ g.

96. Observe that $y = 3\cos\left(3t\sqrt{k}\right) = 3\cos\left(t\sqrt{\frac{k}{1/9}}\right)$. So, the amplitude is 3 cm and the mass $m = 1/9$ g.

97. First, since the period of $y = 3\cos\left(\frac{t}{2}\right)$ is $2\pi/\tfrac{1}{2} = 4\pi$, the frequency of oscillation is given by $1/4\pi$ cycles per second.

98. First, since the period of $y = 3.5\cos(3t)$ is $2\pi/3$, the frequency of oscillation is given by $3/2\pi$ cycles per second.

99. From the equation $y = 0.005\sin\left(2\pi(256t)\right)$ we see that the amplitude is 0.005 cm and the frequency is 256 hertz.

100. From the equation $y = 0.005\sin\left(2\pi(288t)\right)$ we see that the amplitude is 0.005 cm and the frequency is 288 hertz.

101. Observe that $y = 0.008\sin(750\pi t) = 0.008\sin\left(2\pi(375t)\right)$. So, the amplitude is 0.008 cm and the frequency is 375 hertz.

102. Observe that $y = 0.006\cos(1{,}000\pi t) = 0.006\cos\left(2\pi(500t)\right)$. So, the amplitude is 0.006 cm and the frequency is 500 hertz.

103. Using the equation $\frac{330 \ ^{m}/_{\text{sec.}}}{V} = \frac{1}{M}$ with Mach number $M = 2$, we see that the speed of the plane $V = 2\left(330 \ ^{m}/_{\text{sec.}}\right) = \boxed{660 \ ^{m}/_{\text{sec.}}}$

104. Using the equation $\frac{330 \ ^{m}/_{\text{sec.}}}{V} = \frac{1}{M}$ with speed $V = 990 \ ^{m}/_{\text{sec.}}$, we see that the Mach number M satisfies the following:

$$\frac{330 \ ^{m}/_{\text{sec.}}}{990 \ ^{m}/_{\text{sec.}}} = \frac{1}{M} \text{ so that } \boxed{M = 3}.$$

105. Use the equation $\sin\left(\frac{\theta}{2}\right) = \frac{330 \ ^{m}/_{\text{sec.}}}{V}$ with $\theta = 60^\circ$ to find V.

Observe that $\sin\left(\frac{60^\circ}{2}\right) = \sin\left(30^\circ\right) = \frac{1}{2}$. Substituting this into the above equation yields

$\frac{1}{2} = \frac{330 \ ^{m}/_{\text{sec.}}}{V}$ so that $\boxed{V = 660 \ ^{m}/_{\text{sec.}}}$.

106. Use the equation $\sin\left(\frac{\theta}{2}\right) = \frac{330 \ ^{m}/_{\text{sec.}}}{V}$ with $\theta = 30^\circ$ to find V.

Observe that $\sin\left(\frac{30^\circ}{2}\right) = \sin\left(15^\circ\right)$. Substituting this into the above equation yields

$V = \frac{330 \ ^{m}/_{\text{sec.}}}{\sin\left(15^\circ\right)} \approx \boxed{1{,}275 \ ^{m}/_{\text{sec.}}}$

107. Assume that the equation is of the form $y = K + A\sin(Bt + C)$, where t is measured in seconds. We make the following preliminary observations:

- Since the period is 4 seconds, we know that $\frac{2\pi}{B} = 4$, so that $B = \frac{\pi}{2}$.
- Since the maximum y-value is 50 and the minimum y-value is 0, we have:
  - Vertical shift $K = \frac{50+0}{2} = 25$
  - Amplitude $|A| = \frac{50-0}{2} = 25$

So, the equation thus far is $y = 25 + 25\sin\left(\frac{\pi}{2}t + C\right)$. To find C, we use the fact that the point $(0,0)$ is on the graph:

$$0 = 25 + 25\sin\left(\frac{\pi}{2}(0) + C\right)$$
$$-1 = \sin(C)$$

Though this equation has infinitely many solutions, the simplest one is given by $C = -\frac{\pi}{2}$. So, the equation is $\boxed{y = 25 + 25\sin\left(\frac{\pi}{2}t - \frac{\pi}{2}\right) = 25 + 25\sin\left(\frac{\pi}{2}(t-1)\right)}$, or equivalently, $\boxed{y = 25 - 25\cos\left(\frac{\pi t}{2}\right)}$. Also, observe that the intensity at 4 minutes is 0 candelas per m^2 because 4 minutes = 4(60) seconds, and the period of the cycle is 4 seconds.

108. The graph of $25 + 25\sin\left(\frac{\pi}{2}(t-1)\right)$ for ½ minute occurs on the interval [0, 30], since t is measured in seconds. The graph is given by:

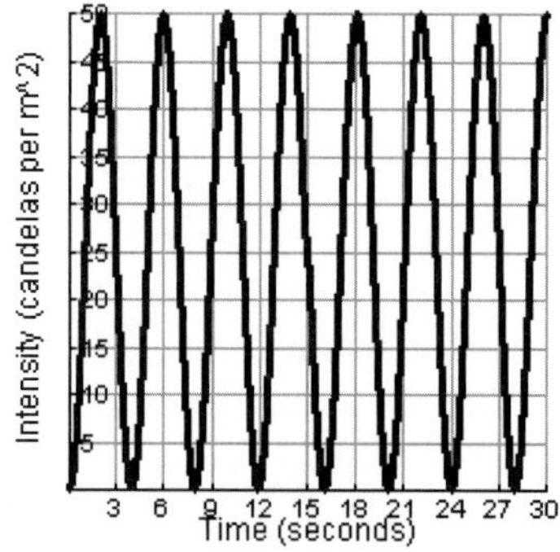

109. The correct graph is reflected over the x-axis.

110. The correct graph is reflected over the x-axis. Also, since the period is $\frac{2\pi}{2}=\pi$, one should form the table in steps of $\frac{\pi}{4}$ so that points other than the x-intercepts (such as the maximum and minimum points) are obtained.
111. True. In general, the graphs of $y=f(x)$ and $y=-f(x)$ are reflections of each other over the x-axis.
112. True. Observe that since sine is odd, $y=A\sin(-Bx)=-A\sin(Bx)$, which is a reflection of $y=A\sin(Bx)$ over the x-axis.
113. False. Observe that since cosine is even, $y=-A\cos(-Bx)=-A\cos(Bx)$, which is a reflection of $y=A\cos(Bx)$ over the x-axis.
114. True. Observe that since sine is odd, $y=-A\sin(-Bx)=-\left(-A\sin(Bx)\right)=A\sin(Bx)$.
115. Since $A\cos\left(B\cdot 0\right)=A(1)=A$, the y-intercept of $y=A\cos(Bx)$ is $(0, A)$.
116. Since $A\sin\left(B\cdot 0\right)=A(0)=0$, the y-intercept of $y=A\sin(Bx)$ is $(0, 0)$.
117. Assuming that $A\neq 0$, the x-intercepts of $y=A\sin(Bx)$ occur when $\sin(Bx)=0$, which happens when $Bx=n\pi$, where n is an integer. So, $x=\frac{n\pi}{B}$, where n is an integer.
118. Assuming that $A\neq 0$, the x-intercepts of $y=A\cos(Bx)$ occur when $\cos(Bx)=0$, which happens when $Bx=\frac{(2n+1)\pi}{2}$, where n is an integer. So, $x=\frac{(2n+1)\pi}{2B}$, where n is an integer.
119. Evaluating at $x=0$ yields $y=-A\sin\left(\frac{\pi}{6}\right)=-\frac{A}{2}$. So, the y-intercept is $\left(0,-\frac{A}{2}\right)$.
120. Evaluating at $x=0$ yields $y=A\cos(-\pi)+C=-A+C$. So, the y-intercept is $\left(0,-A+C\right)$.
121. We determine the x-intercepts by solving the following equation: $A\sin(Bx)+A=0$; $\sin Bx=-1$; $Bx=\frac{3\pi}{2}+2n\pi$; $x=\frac{3\pi}{2B}+\frac{2n\pi}{B}$ where n is an integer. So, the x-intercepts are $\left(\frac{3\pi}{2B}+\frac{2n\pi}{B},0\right)$, where n is an integer.

122. We determine the x-intercepts by solving the following equation: $$A\cos(Bx)+C=0$$ $$\cos Bx=-\tfrac{C}{A}$$ Since the range of cosine is [-1,1], we know that this equation has no solution if $-\frac{C}{A}<-1$ or $-\frac{C}{A}>1$. This is equivalent to saying $C>A$ or $C<-A$, which is further more succinctly stated as $\lvert C\rvert>A$. So, if this occurs, then there are no x-intercepts.
123. First, note that the amplitude of $y=2A\sin(Bx+C)$. The range of this function is [-2A, 2A]. Since subtracting $\frac{A}{2}$ from it shifts the graph down that many units, we do the same to the endpoints of the range of the non-translated function to obtain $\left[-\frac{5A}{2},\frac{3A}{2}\right]$.
124. No. The maximum value that $A\sin Bx$ can attain is A and the maximum value that $3A\cos\left(\frac{B}{2}x\right)$ can attain is $3A$. Hence, the maximum of the sum is $4A$.
125. No. Observe that the graphs of $y_1=\sin 5x$ and $y_2=5\sin x$ are different, as seen below: 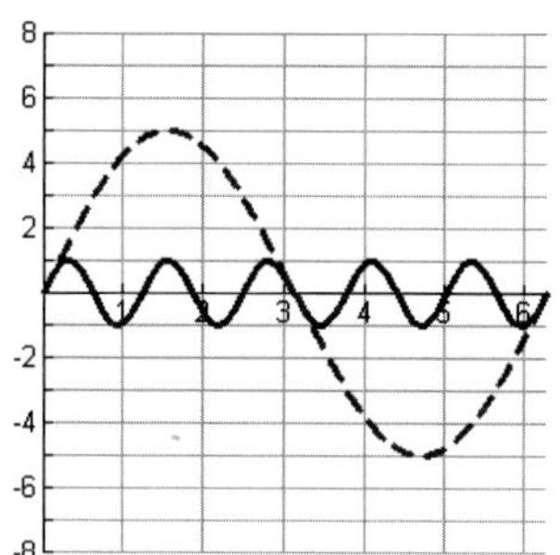
126. No. Observe that the graphs of $y_1=\cos 3x$ and $y_2=3\cos x$ are different, as seen below: 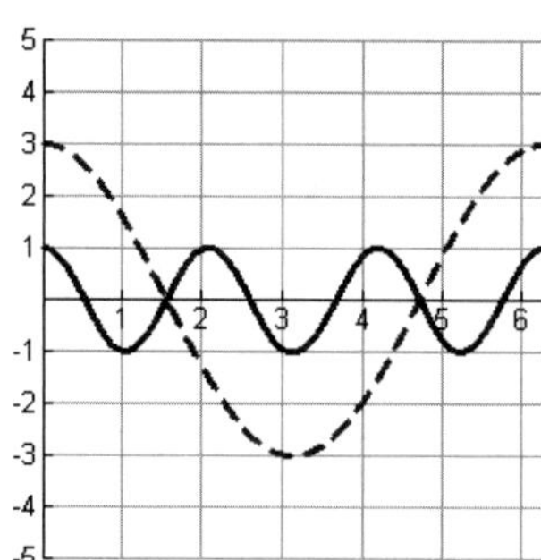

127. The graphs of $y_1 = \sin x$ and $y_2 = \cos\left(x - \frac{\pi}{2}\right)$ are the same, as seen below:

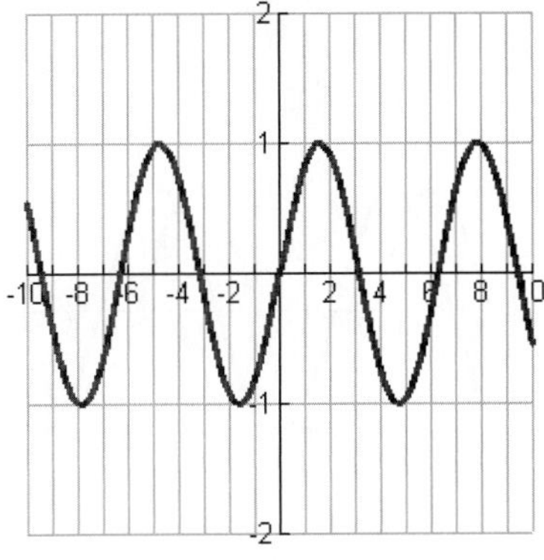

128. The graphs of $y_1 = \cos x$ and $y_2 = \sin\left(x + \frac{\pi}{2}\right)$ are the same, as seen below:

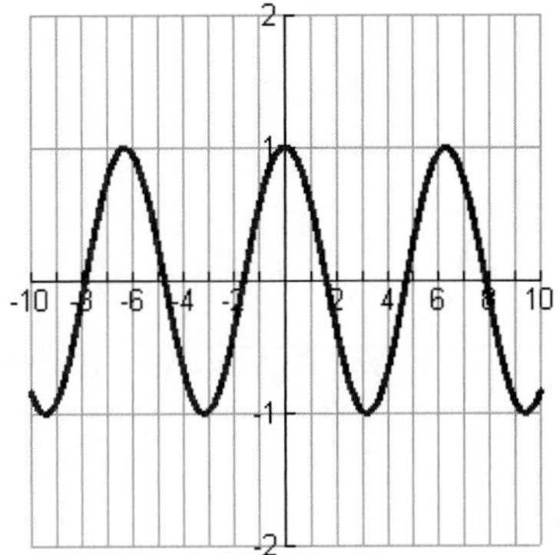

129. a. The graph of $y_2 = \cos\left(x + \frac{\pi}{3}\right)$ is the graph of $y_1 = \cos x$ (dotted graph) shifted to the <u>left</u> $\frac{\pi}{3}$ units, as seen below:

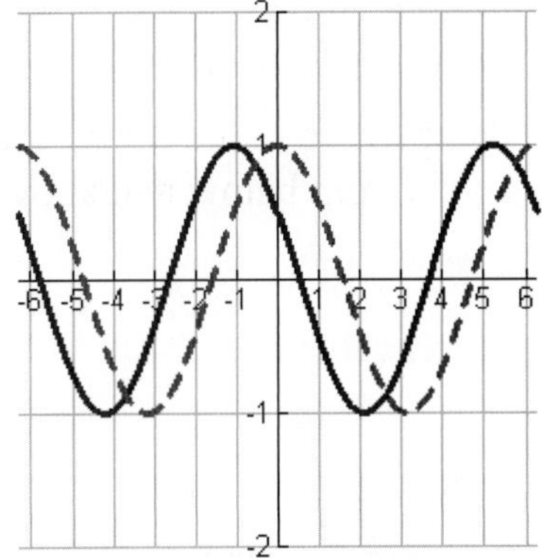

b. The graph of $y_2 = \cos\left(x - \frac{\pi}{3}\right)$ is the graph of $y_1 = \cos x$ (dotted graph) shifted to the <u>right</u> $\frac{\pi}{3}$ units, as seen below:

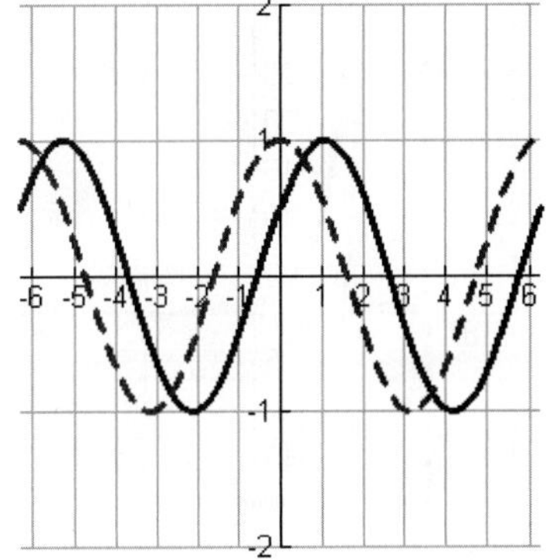

130. a. The graph of $y_1 = \sin\left(x + \frac{\pi}{3}\right)$ is the graph of $y_1 = \sin x$ (dotted graph) shifted to the <u>left</u> $\frac{\pi}{3}$ units, as seen below:

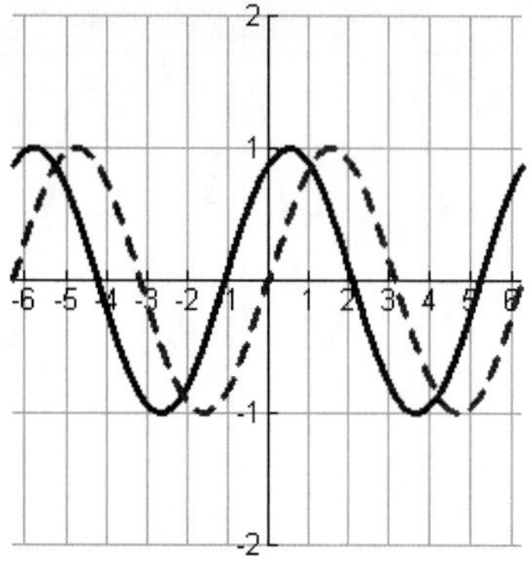

b. The graph of $y_2 = \sin\left(x - \frac{\pi}{3}\right)$ is the graph of $y_1 = \sin x$ (dotted graph) shifted to the <u>right</u> $\frac{\pi}{3}$ units, as seen below:

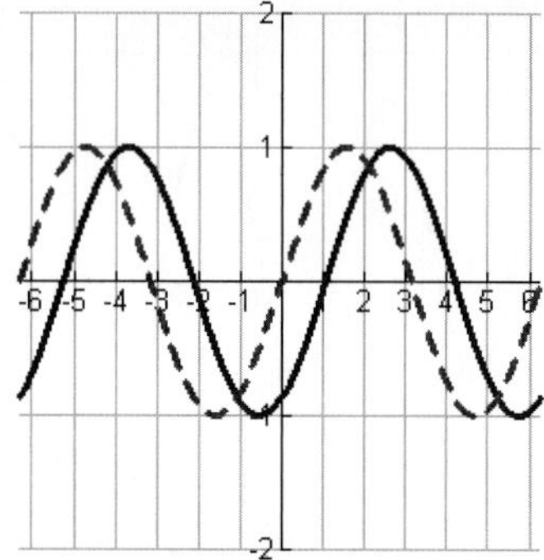

131. Consider the graphs of the following functions on the interval $[0, 2\pi]$:

$$y_1 = e^{-t},\ y_2 = \sin t,\ y_3 = e^{-t}\sin t$$

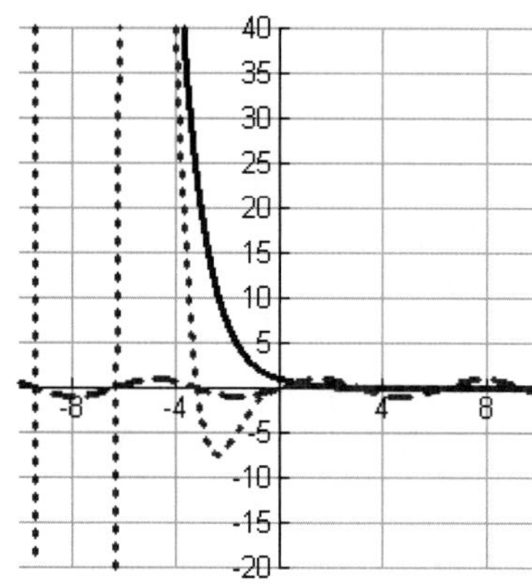

<u>Observation</u>: As t increases,

- the graph of y_1 approaches the t-axis from above,
- the graph of y_2 oscillates, keeping its same form,
- the graph of y_3 oscillates, but dampens so that it approaches the t-axis from below and above.

132. Consider the graphs of the following functions (on two different windows):

$$y_1 = e^{-t}\sin t$$
$$y_2 = e^{-2t}\sin t$$
$$y_3 = e^{-4t}\sin t$$

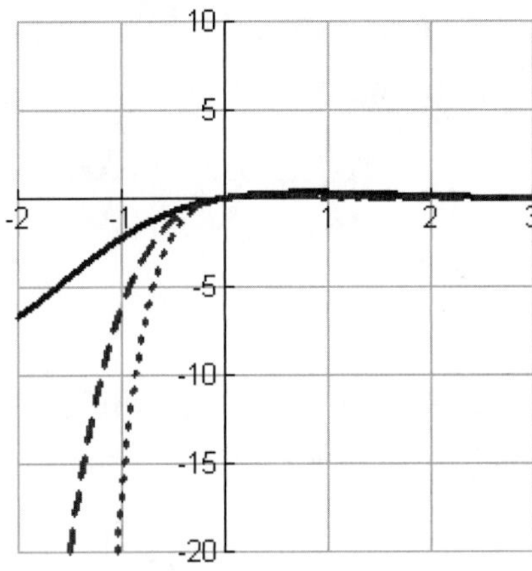

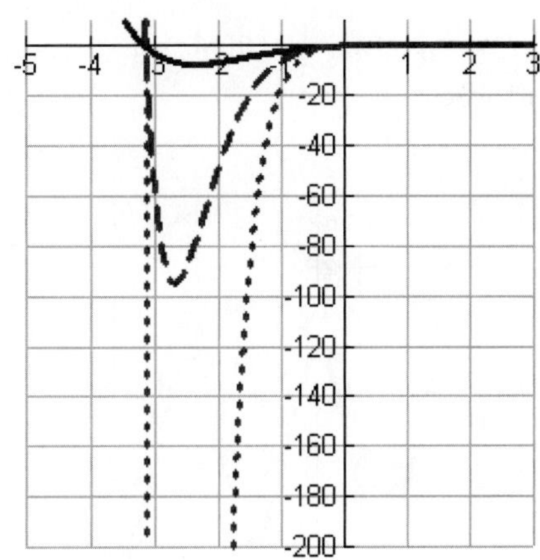

Observation: As the value of *k* increases, the effect of the damping increases, causing the graph to approach the *t*-axis more quickly.

133. a. The graphs of

$$y_1 = \sin x \quad \text{(solid)}$$
$$y_2 = \sin x + 1 \quad \text{(dashed)}$$

are as follows:

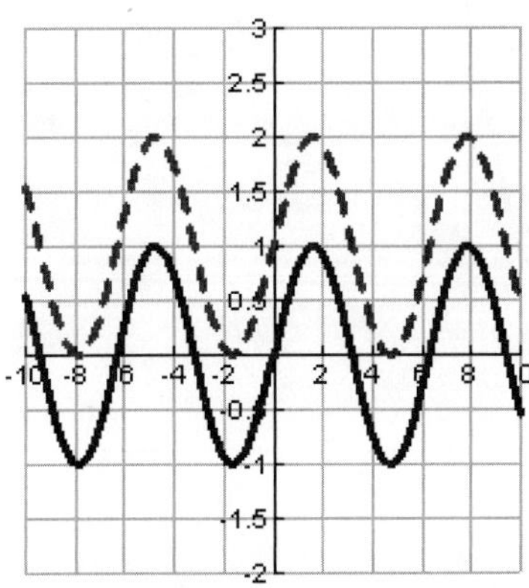

The graph of y_2 is that of y_1 shifted up 1 unit.

b. The graphs of

$$y_1 = \sin x \quad \text{(solid)}$$
$$y_2 = \sin x - 1 \quad \text{(dashed)}$$

are as follows:

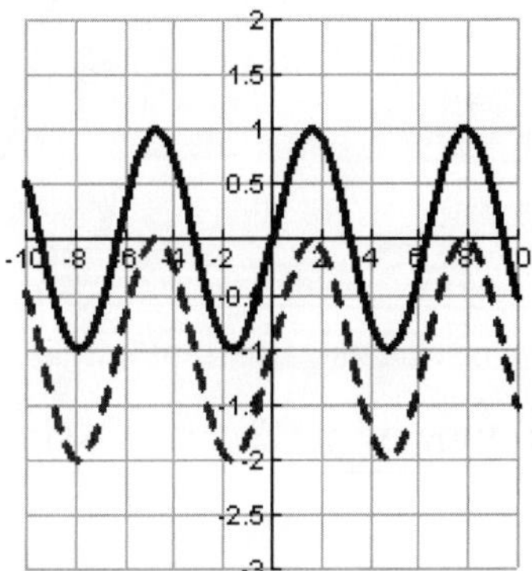

The graph of y_2 is that of y_1 shifted down 1 unit.

134. a. The graphs of

$y_1 = \cos x$ (solid)

$y_2 = \cos x + \frac{1}{2}$ (dashed)

are as follows:

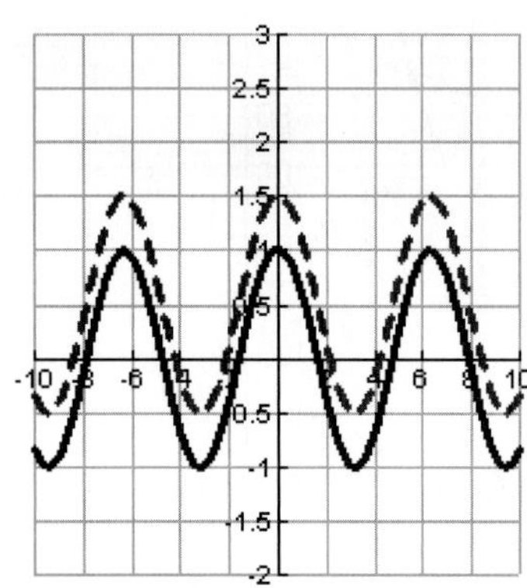

The graph of y_2 is that of y_1 shifted up $\frac{1}{2}$ unit.

b. The graphs of

$y_1 = \cos x$ (solid)

$y_2 = \cos x - \frac{1}{2}$ (dashed)

are as follows:

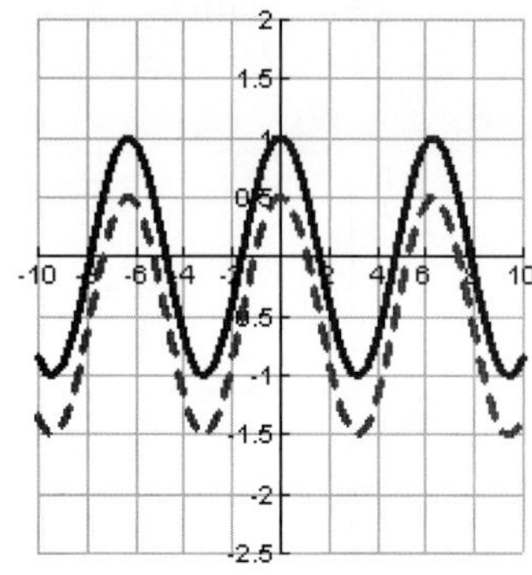

The graph of y_2 is that of y_1 shifted down $\frac{1}{2}$ unit.

135. Consider the graphs of the following functions on the interval $[-2\pi, 2\pi]$:

$y_1 = 3\cos x$ (dashed bold),
$y_2 = 4\sin x$ (dotted bold),
$y_3 = 3\cos x + 4\sin x$ (solid bold)

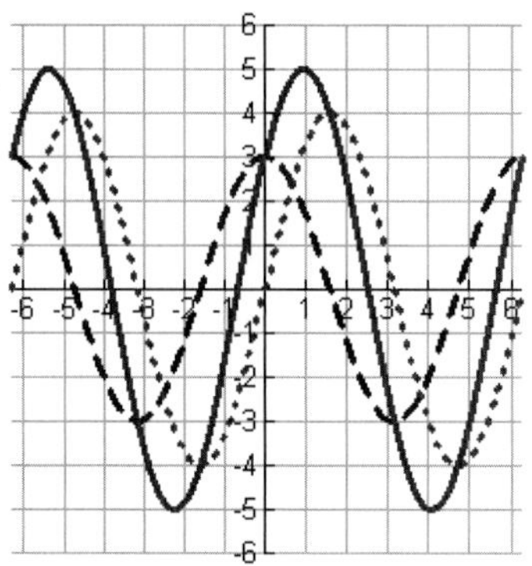

The amplitude of $y = 3\cos x + 4\sin x$ appears to be 5.

136. Consider the graphs of the following functions on the interval $[-2\pi, 2\pi]$:

$$y_1 = \sqrt{3}\cos x \quad \text{(dashed bold)},$$
$$y_2 = \sin x \quad \text{(dotted bold)},$$
$$y_3 = \sqrt{3}\cos x - \sin x \quad \text{(solid bold)}$$

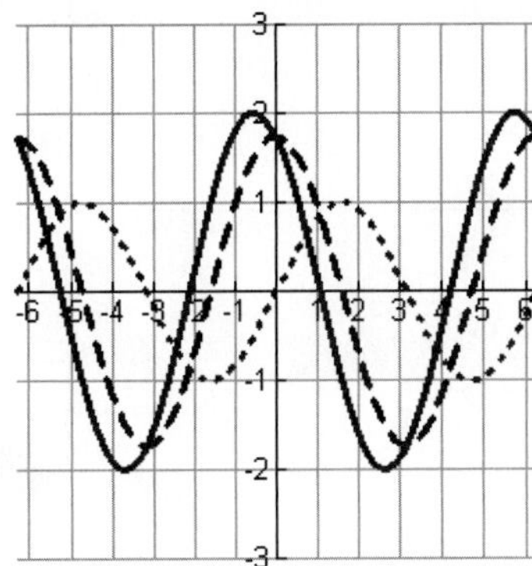

The amplitude of $y = \sqrt{3}\cos x - \sin x$ appears to be approximately 2.

137.

$$\int_0^{\pi} \sin x dx = (-\cos\pi) - (-\cos 0)$$
$$= 1 + 1 = 2$$

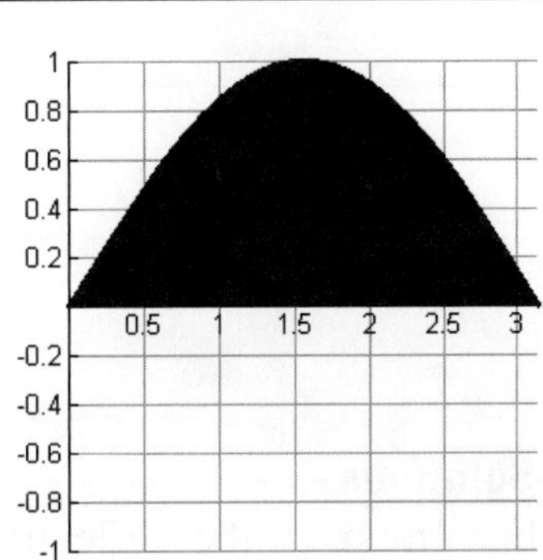

138.

$$\int_{-\frac{\pi}{2}}^{\frac{\pi}{2}} \cos x dx = \sin\tfrac{\pi}{2} - \sin\left(-\tfrac{\pi}{2}\right)$$
$$= 1 + 1 = 2$$

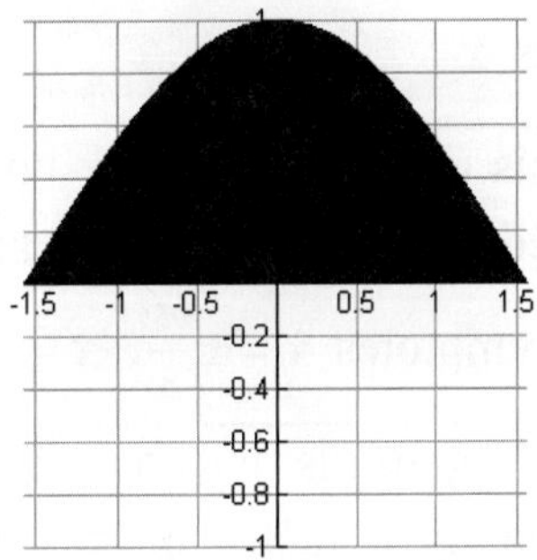

139. $\int_0^{\frac{\pi}{2}} \cos x dx = \sin\frac{\pi}{2} - \sin 0$ $= 1 - 0 = 1$	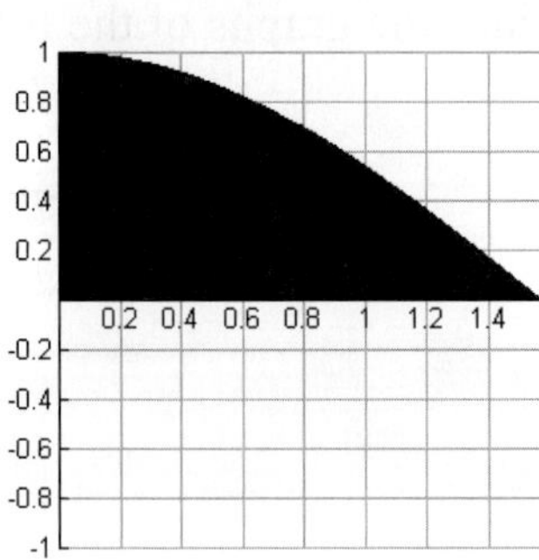
140. $\int_0^{\frac{\pi}{2}} \sin x dx = \left(-\cos\frac{\pi}{2}\right) - \left(-\cos 0\right)$ $= 0 + 1 = 1$	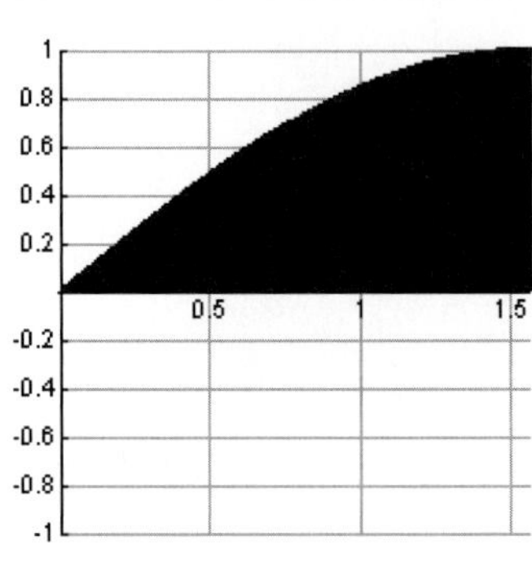

Section 5.3 Solutions ---

1. b This graph is the reflection of $y = \tan x$ over the x-axis, and has vertical asymptotes $x = \pm\frac{\pi}{2}$.	**2. f** This graph is the reflection of $y = \csc x$ over the x-axis, and has vertical asymptotes $x = 0,\ x = \pm\pi$.
3. h This graph is $y = \sec x$ horizontally compressed by a factor of 2, and it has vertical asymptotes $x = \pm\frac{\pi}{4},\ x = \pm\frac{3\pi}{4}$.	**4. a** This graph is $y = \csc x$ horizontally compressed by a factor of 2, and it has vertical asymptotes $x = 0,\ x = \pm\frac{\pi}{2},\ x = \pm\pi$.
5. c This graph is $y = \cot x$ horizontally compressed by a factor of π, and it has vertical asymptotes $x = 0,\ x = \pm 1$.	**6. e** This is the graph of Exercise 5 reflected over the x-axis.
7. d This graph has the same shape and vertical asymptotes as $y = \sec x$, but with vertices at ± 3 rather than ± 1.	**8. g** This graph has the same shape and vertical asymptotes as $y = \csc x$, but with vertices at ± 3 rather than ± 1.

9. We have the following information about the graph of $y = \tan\left(\frac{1}{2}x\right)$ on $[-2\pi, 2\pi]$:

<table>
<tr><td><u>Period</u>:
$\frac{\pi}{1/2} = 2\pi$</td><td><u>Reflection about x-axis</u>?
No</td><td><u>Phase Shift</u>:
None</td></tr>
<tr><td colspan="3"><u>x-intercepts</u> occur when $\sin\left(\frac{1}{2}x\right) = 0$. This happens when $\frac{1}{2}x = 0,\ \pm\pi$, so that $x = 0,\ \pm 2\pi$. So, the points $(0,\ 0),\ (2\pi,\ 0),\ (-2\pi,\ 0)$ are on the graph.</td></tr>
<tr><td colspan="3"><u>Vertical asymptotes</u> occur when $\cos\left(\frac{1}{2}x\right) = 0$. This happens when $\frac{1}{2}x = \pm\frac{\pi}{2}$, so that $x = \pm\pi$.</td></tr>
</table>

Using this information, we find that the graph of $y = \tan\left(\frac{1}{2}x\right)$ on $[-2\pi, 2\pi]$ is given by:

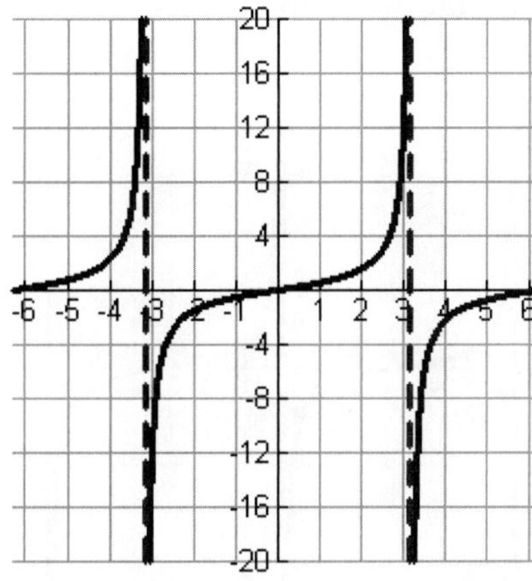

10. We have the following information about the graph of $y = \cot\left(\frac{1}{2}x\right)$ on $[-2\pi, 2\pi]$:

Period: $\dfrac{\pi}{1/2} = 2\pi$	Reflection about x-axis? No	Phase Shift: None
x-intercepts occur when $\cos\left(\frac{1}{2}x\right) = 0$. This happens when $\frac{1}{2}x = \pm\frac{\pi}{2}$, so that $x = \pm\pi$. So, the points $(-\pi, 0), (\pi, 0)$ are on the graph.		
Vertical asymptotes occur when $\sin\left(\frac{1}{2}x\right) = 0$. This happens when $\frac{1}{2}x = 0, \pm\pi$, so that $x = 0, \pm 2\pi$.		

Using this information, we find that the graph of $y = \cot\left(\frac{1}{2}x\right)$ on $[-2\pi, 2\pi]$ is given by:

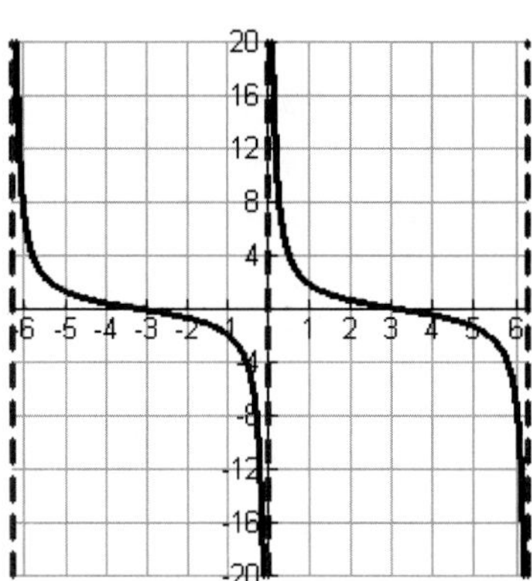

11. We have the following information about the graph of $y=-\cot(2\pi x)$ on $[-1,1]$:

Period: $\frac{\pi}{2\pi}=\frac{1}{2}$	Reflection about x-axis? Yes	Phase Shift: None
x-intercepts occur when $\cos(2\pi x)=0$. This happens when $2\pi x=\pm\frac{\pi}{2},\ \pm\frac{3\pi}{2}$, so that $x=\pm\frac{1}{4},\ \pm\frac{3}{4}$. So, the points $\left(\pm\frac{1}{4},\ 0\right),\left(\pm\frac{3}{4},\ 0\right)$ are on the graph.		
Vertical asymptotes occur when $\sin(2\pi x)=0$. This happens when $2\pi x=0,\ \pm\pi,\ \pm 2\pi$, so that $x=0,\ \pm\frac{1}{2},\ \pm 1$.		

Using this information, we construct the graph of $y=-\cot(2\pi x)$ on $[-1,1]$ as follows:

Step 1: Graph $y=\cot(2\pi x)$ on $[-1,1]$:

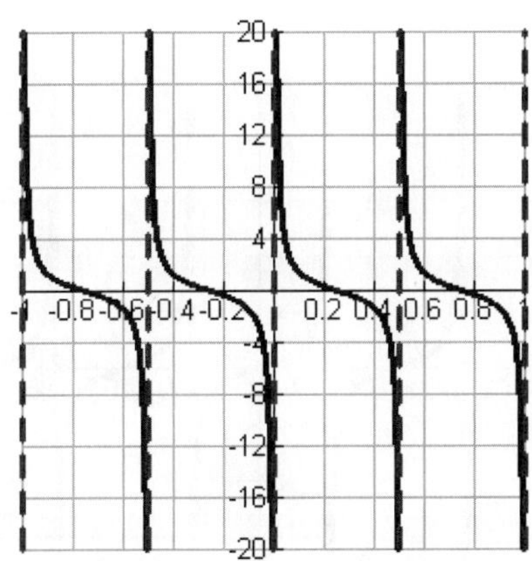

Step 2: Reflect the graph in Step 1 over the x-axis to obtain the graph of $y=-\cot(2\pi x)$ on $[-1,1]$:

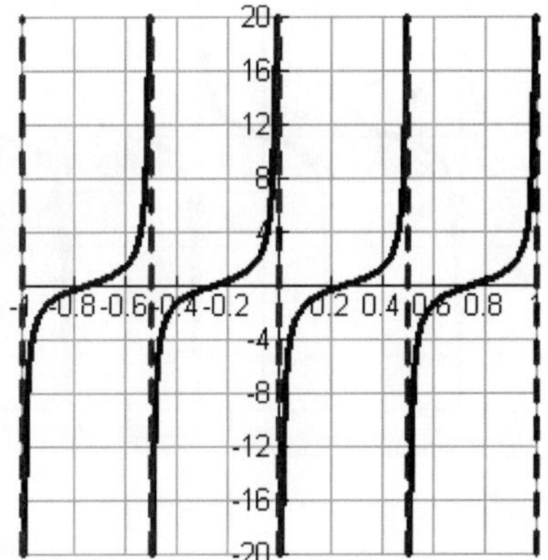

12. We have the following information about the graph of $y = -\tan(2\pi x)$ on $[-1,1]$:

<u>Period</u>: $\dfrac{\pi}{2\pi} = \dfrac{1}{2}$	<u>Reflection about *x*-axis</u>? Yes	<u>Phase Shift</u>: None
<u>*x*-intercepts</u> occur when $\sin(2\pi x) = 0$. This happens when $2\pi x = 0,\ \pm\pi,\ \pm 2\pi$, so that $x = 0,\ \pm\frac{1}{2},\ \pm 1$. So, the points $(0,0)$, $\left(\pm\frac{1}{2}, 0\right)$, $(\pm 1, 0)$ are on the graph.		
<u>Vertical asymptotes</u> occur when $\cos(2\pi x) = 0$. This happens when $2\pi x = \pm\frac{\pi}{2},\ \pm\frac{3\pi}{2}$, so that $x = \pm\frac{1}{4},\ \pm\frac{3}{4}$.		

Using this information, we construct the graph of $y = -\tan(2\pi x)$ on $[-1,1]$ as follows:

<u>Step 1</u>: Graph $y = \tan(2\pi x)$ on $[-1,1]$:

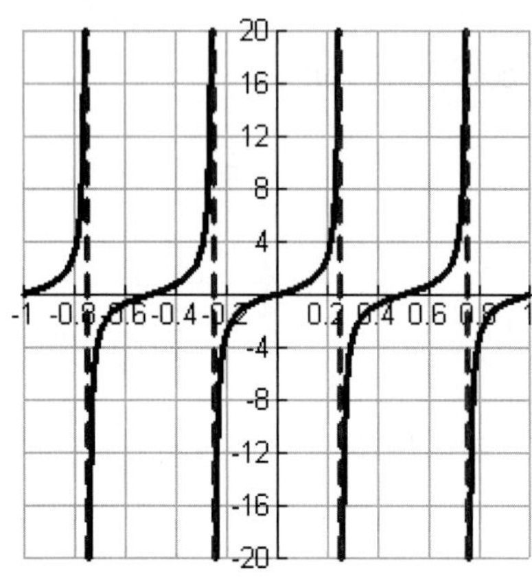

<u>Step 2</u>: Reflect the graph in Step 1 over the *x*-axis to obtain the graph of $y = -\tan(2\pi x)$ on $[-1,1]$:

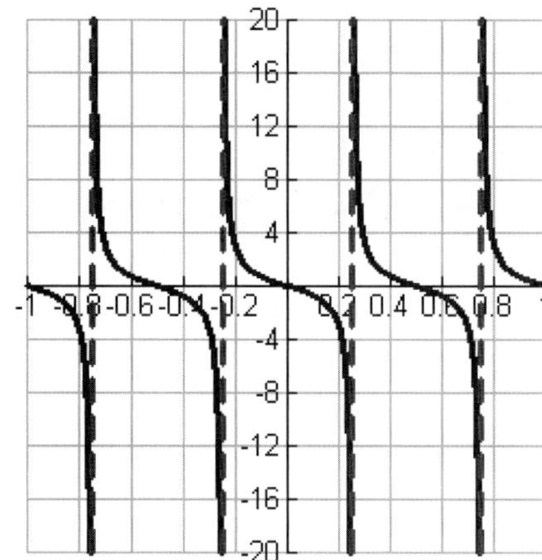

13. We have the following information about the graph of $y = 2\tan(3x)$ on $[-\pi, \pi]$:

Period:	Reflection about x-axis?	Phase Shift:
$\frac{\pi}{3}$	No	None

<u>x-intercepts</u> occur when $\sin(3x) = 0$. This happens when $3x = 0, \pm\pi, \pm 2\pi, \pm 3\pi$, so that $x = 0, \pm\frac{\pi}{3}, \pm\frac{2\pi}{3}, \pm\pi$. So, the points $(0,0), \left(\pm\frac{\pi}{3}, 0\right), \left(\pm\frac{2\pi}{3}, 0\right), (\pm\pi, 0)$ are on the graph.

<u>Vertical asymptotes</u> occur when $\cos(3x) = 0$. This happens when $3x = \pm\frac{\pi}{2}, \pm\frac{3\pi}{2}, \pm\frac{5\pi}{2}$, so that $x = \pm\frac{\pi}{6}, \pm\frac{\pi}{2}, \pm\frac{5\pi}{6}$.

<u>Stretching</u>: The coefficient 2 has the effect of making the graph of $y = 2\tan(3x)$ approach the vertical asymptotes twice as quickly as the graph of $y = \tan(3x)$.

Using this information, we find that the graph of $y = 2\tan(3x)$ on $[-\pi, \pi]$ is given by:

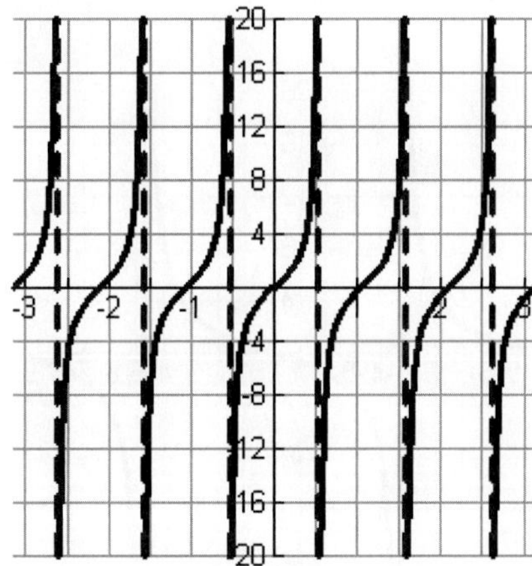

14. We have the following information about the graph of $y = 2\tan\left(\frac{1}{3}x\right)$ on $[-3\pi, 3\pi]$:

<table>
<tr><td><u>Period</u>:
$\frac{\pi}{1/3} = 3\pi$</td><td><u>Reflection about x-axis?</u>
No</td><td><u>Phase Shift</u>:
None</td></tr>
<tr><td colspan="3"><u>x-intercepts</u> occur when $\sin\left(\frac{1}{3}x\right) = 0$. This happens when $\frac{1}{3}x = 0, \pm\pi$, so that $x = 0, \pm 3\pi$. So, the points $(0,0)$, $(\pm 3\pi, 0)$ are on the graph.</td></tr>
<tr><td colspan="3"><u>Vertical asymptotes</u> occur when $\cos\left(\frac{1}{3}x\right) = 0$. This happens when $\frac{1}{3}x = \pm\frac{\pi}{2}$, so that $x = \pm\frac{3\pi}{2}$.</td></tr>
<tr><td colspan="3"><u>Stretching</u>: The coefficient 2 has the effect of making the graph of $y = 2\tan\left(\frac{1}{3}x\right)$ approach the vertical asymptotes twice as quickly as the graph of $y = \tan\left(\frac{1}{3}x\right)$.</td></tr>
</table>

Using this information, we find that the graph of $y = 2\tan\left(\frac{1}{3}x\right)$ on $[-3\pi, 3\pi]$ is given by:

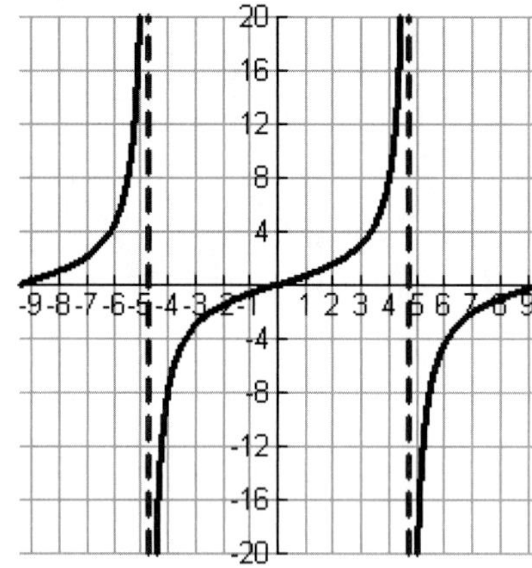

15. We have the following information about the graph of $y=-\frac{1}{4}\cot\left(\frac{1}{2}x\right)$ on $[-2\pi,2\pi]$:

<table>
<tr><td><u>Period</u>:
$\frac{\pi}{1/2}=2\pi$</td><td><u>Reflection about x-axis?</u>
Yes</td><td><u>Phase Shift</u>:
None</td></tr>
<tr><td colspan="3"><u>x-intercepts</u> occur when $\cos\left(\frac{1}{2}x\right)=0$. This happens when $\frac{1}{2}x=\pm\frac{\pi}{2}$, so that $x=\pm\pi$. So, the points $(\pm\pi,0)$ are on the graph.</td></tr>
<tr><td colspan="3"><u>Vertical asymptotes</u> occur when $\sin\left(\frac{1}{2}x\right)=0$. This happens when $\frac{1}{2}x=0,\pm\pi$, so that $x=0,\pm2\pi$. .</td></tr>
<tr><td colspan="3"><u>Stretching</u>: The coefficient $\frac{1}{4}$ has the effect of making the graph approach the vertical asymptotes more slowly than the corresponding graph without it.</td></tr>
</table>

Using this information, we find that the graph of $y=-\frac{1}{4}\cot\left(\frac{1}{2}x\right)$ on $[-2\pi,2\pi]$ is given by:

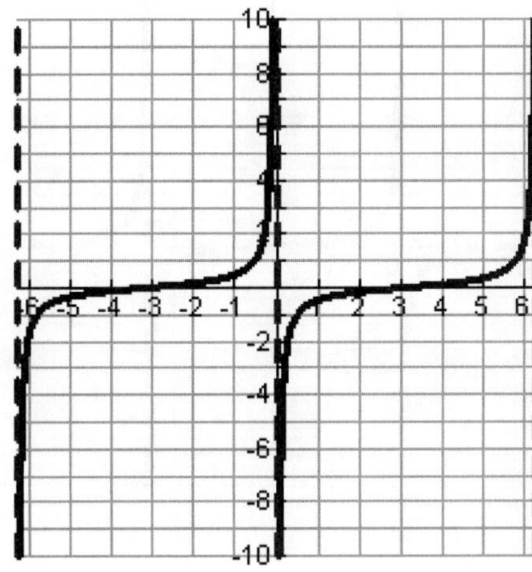

16. We have the following information about the graph of $y=-\frac{1}{2}\tan\left(\frac{1}{4}x\right)$ on $[-4\pi, 4\pi]$:

Period: $\dfrac{\pi}{1/4}=4\pi$	Reflection about x-axis? Yes	Phase Shift: None
x-intercepts occur when $\sin\left(\frac{1}{4}x\right)=0$. This happens when $\frac{1}{4}x=0, \pm\pi$, so that $x=0, \pm 4\pi$. So, the points, (0,0), $(\pm 4\pi, 0)$ are on the graph.		
Vertical asymptotes occur $\cos\left(\frac{1}{4}x\right)=0$. This happens when $\frac{1}{4}x=\pm\frac{\pi}{2}$, so that $x=\pm 2\pi$.		
Stretching: The coefficient $\frac{1}{2}$ has the effect of making the graph approach the vertical asymptotes more slowly than the corresponding graph without it.		

Using this information, we find that the graph of $y=-\frac{1}{2}\tan\left(\frac{1}{4}x\right)$ on $[-4\pi, 4\pi]$ is given by:

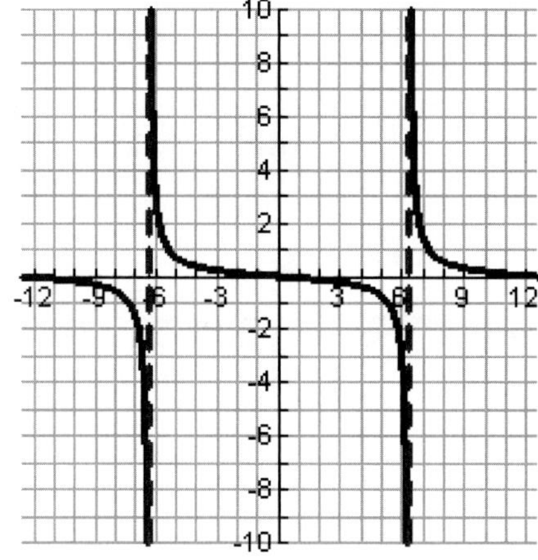

17. We have the following information about the graph of $y = -\tan\left(x - \frac{\pi}{2}\right)$ on $[-\pi, \pi]$:

<table>
<tr><td><u>Period</u>:
$\frac{\pi}{1} = \pi$</td><td><u>Reflection about x-axis?</u>
Yes</td><td><u>Phase Shift</u>:
Right $\frac{\pi}{2}$ units</td></tr>
<tr><td colspan="3"><u>x-intercepts</u> occur when $\sin\left(x - \frac{\pi}{2}\right) = 0$. This happens when $x - \frac{\pi}{2} = 0, \pm\pi$, so that $x = \pm\frac{\pi}{2}, \underbrace{\cancel{\frac{3\pi}{2}}}_{\text{outside}}$. So, the points $\left(\pm\frac{\pi}{2}, 0\right)$ are on the graph.</td></tr>
<tr><td colspan="3"><u>Vertical asymptotes</u> occur when $\cos\left(x - \frac{\pi}{2}\right) = 0$. This happens when $x - \frac{\pi}{2} = \pm\frac{\pi}{2}, \pm\frac{3\pi}{2}$, so that $x = 0, \pm\pi, \underbrace{\cancel{2\pi}}_{\text{outside}}$.</td></tr>
</table>

Using this information, we construct the graph of $y = -\tan\left(x - \frac{\pi}{2}\right)$ on $[-\pi, \pi]$ as follows:

<u>Step 1</u>: Graph $y = \tan\left(x - \frac{\pi}{2}\right)$ on $[-\pi, \pi]$:

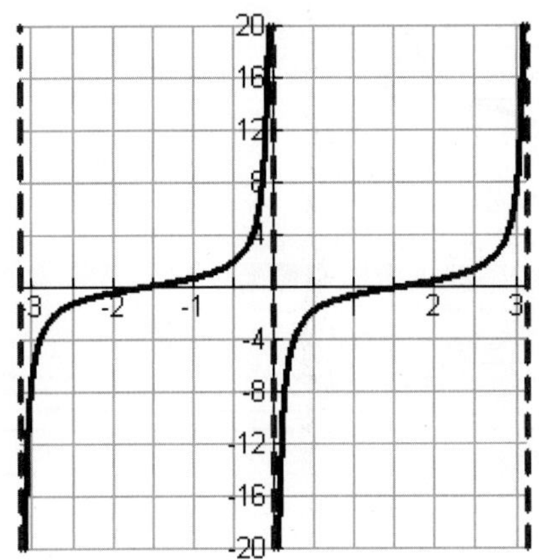

<u>Step 2</u>: Reflect the graph in Step 1 over the *x*-axis to obtain the graph of $y = -\tan\left(x - \frac{\pi}{2}\right)$ on $[-\pi, \pi]$:

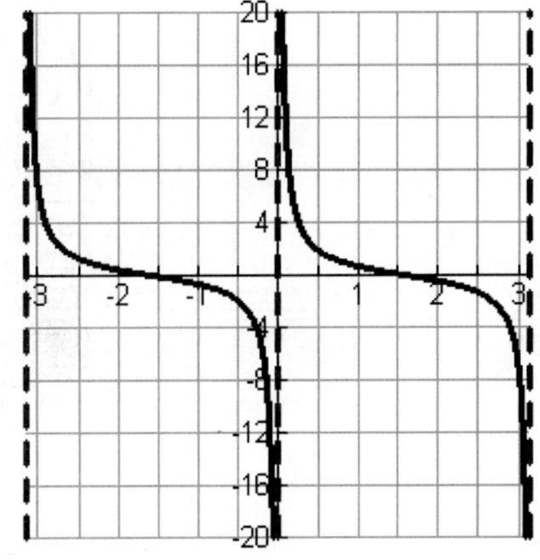

18. We have the following information about the graph of $y = \tan\left(x + \frac{\pi}{4}\right)$ on $[-\pi, \pi]$:

Period:	Reflection about x-axis?	Phase Shift:
$\frac{\pi}{1} = \pi$	No	Left $\frac{\pi}{4}$ units

x-intercepts occur when $\sin\left(x + \frac{\pi}{4}\right) = 0$. This happens when $x + \frac{\pi}{4} = 0, \pm\pi$, so that $x = -\frac{\pi}{4}, \frac{3\pi}{4}, \underbrace{-\frac{5\pi}{4}}_{\text{outside}}$. So, the points $\left(-\frac{\pi}{4}, 0\right), \left(\frac{3\pi}{4}, 0\right)$ are on the graph.

Vertical asymptotes occur when $\cos\left(x + \frac{\pi}{4}\right) = 0$. This happens when $x + \frac{\pi}{4} = \pm\frac{\pi}{2}$, so that $x = -\frac{3\pi}{4}, \frac{\pi}{4}$.

Using this information, we construct the graph of $y = \tan\left(x + \frac{\pi}{4}\right)$ on $[-\pi, \pi]$ as follows:

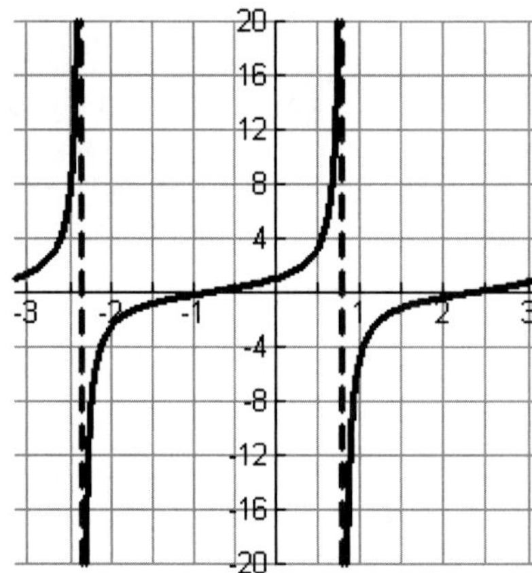

19. We have the following information about the graph of $y = 2\tan\left(x + \frac{\pi}{6}\right)$ on $[-\pi, \pi]$:

<u>Period</u>: $\frac{\pi}{1} = \pi$	<u>Reflection about *x*-axis?</u> No	<u>Phase Shift</u>: Left $\frac{\pi}{6}$ units
<u>*x*-intercepts</u> occur when $\sin\left(x + \frac{\pi}{6}\right) = 0$. This happens when $x + \frac{\pi}{6} = 0, \pm\pi$, so that $x = -\frac{\pi}{6}, \frac{5\pi}{6}, \underbrace{-\frac{7\pi}{6}}_{\text{outside}}$. So, the points $\left(-\frac{\pi}{6}, 0\right), \left(\frac{5\pi}{6}, 0\right)$ are on the graph.		
<u>Vertical asymptotes</u> occur when $\cos\left(x + \frac{\pi}{6}\right) = 0$. This happens when $x + \frac{\pi}{6} = \pm\frac{\pi}{2}$, so that $x = -\frac{2\pi}{3}, \frac{\pi}{3}$.		
<u>Stretching</u>: The coefficient 2 has the effect of making the graph approach the vertical asymptotes twice as quickly than the corresponding graph without it.		

Using this information, we construct the graph of $y = 2\tan\left(x + \frac{\pi}{6}\right)$ on $[-\pi, \pi]$ as follows:

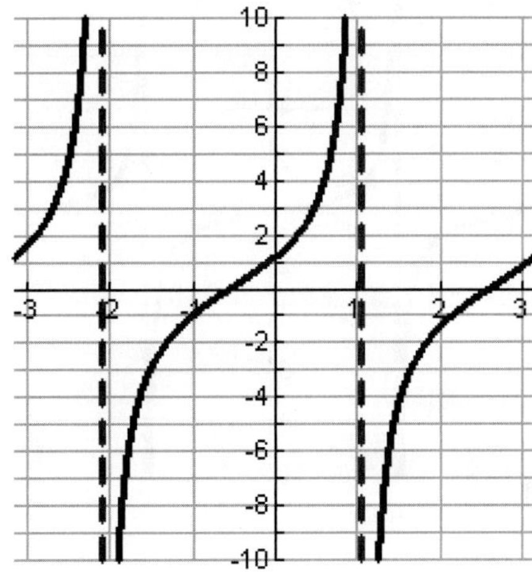

20. We have the following information about the graph of $y=-\frac{1}{2}\tan(x+\pi)$ on $[-\pi,\pi]$:

Period: $\frac{\pi}{1}=\pi$	Reflection about x-axis? Yes	Phase Shift: Left π units
x-intercepts occur when $\sin(x+\pi)=0$. This happens when $x+\pi=0, \pm\pi$, so that $x=-\pi, 0, \underbrace{\cancel{2\pi}}_{\text{outside}}$. So, the points $(-\pi, 0), (0, 0)$ are on the graph.		
Vertical asymptotes occur when $\cos(x+\pi)=0$. This happens when $x+\pi=\pm\frac{\pi}{2}$, so that $x=\cancel{-\frac{3\pi}{2}}, -\frac{\pi}{2}$.		
Stretching: The coefficient $\frac{1}{2}$ has the effect of making the graph approach the vertical asymptotes more slowly than the corresponding graph without it.		

Using this information, we construct the graph of $y=-\frac{1}{2}\tan(x+\pi)$ on $[-\pi,\pi]$ as follows:

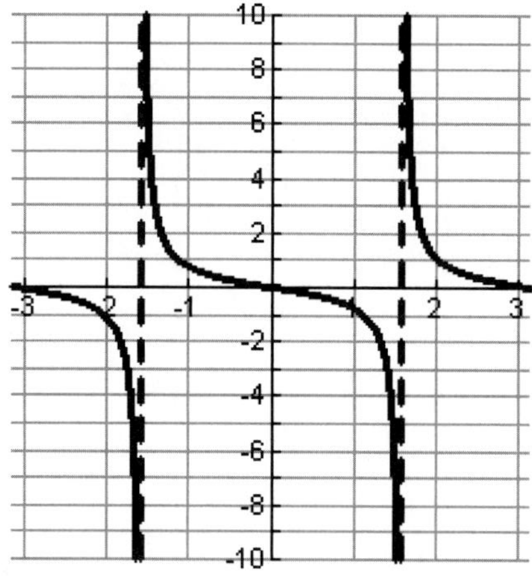

21. We have the following information about the graph of $y = \cot\left(x - \frac{\pi}{4}\right)$ on $[-\pi, \pi]$:

Period: $\frac{\pi}{1} = \pi$	Reflection about x-axis? No	Phase Shift: Right $\frac{\pi}{4}$ units
x-intercepts occur when $\cos\left(x - \frac{\pi}{4}\right) = 0$. This happens when $x - \frac{\pi}{4} = \pm\frac{\pi}{2}$, so that $x = \frac{3\pi}{4}, -\frac{\pi}{4}$. So, the points $\left(-\frac{\pi}{4}, 0\right), \left(\frac{3\pi}{4}, 0\right)$ are on the graph.		
Vertical asymptotes occur when $\sin\left(x - \frac{\pi}{4}\right) = 0$. This happens when $x - \frac{\pi}{4} = 0, \pm\pi$, so that $x = \frac{\pi}{4}, -\frac{3\pi}{4}, \underbrace{\frac{5\pi}{4}}_{\text{outside}}$ (crossed out).		

Using this information, we construct the graph of $y = \cot\left(x - \frac{\pi}{4}\right)$ on $[-\pi, \pi]$ as follows:

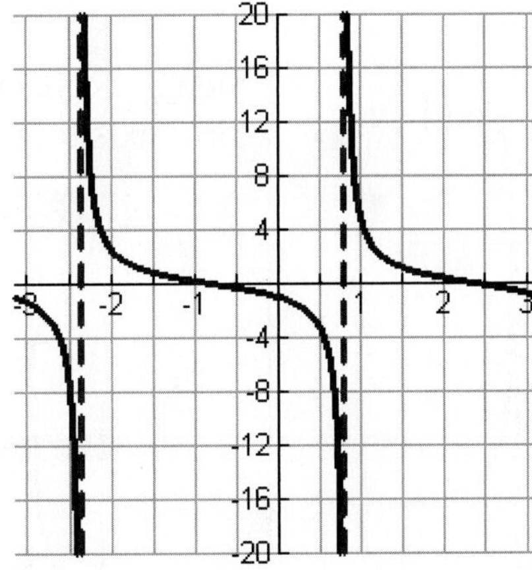

22. We have the following information about the graph of $y = -\cot\left(x + \frac{\pi}{2}\right)$ on $[-\pi, \pi]$:

Period: $\frac{\pi}{1} = \pi$	Reflection about x-axis? Yes	Phase Shift: Left $\frac{\pi}{2}$ units
x-intercepts occur when $\cos\left(x + \frac{\pi}{2}\right) = 0$. This happens when $x + \frac{\pi}{2} = \pm\frac{\pi}{2}, \pm\frac{3\pi}{2}$, so that $x = 0, \pm\pi, \underbrace{\cancel{-2\pi}}_{\text{outside}}$. So, the points $(0, 0), (\pm\pi, 0)$ are on the graph.		
Vertical asymptotes occur when $\sin\left(x + \frac{\pi}{2}\right) = 0$. This happens when $x + \frac{\pi}{2} = 0, \pm\pi$, so that $x = \pm\frac{\pi}{2}, \underbrace{\cancel{\frac{3\pi}{2}}}_{\text{outside}}$.		

Using this information, we construct the graph of $y = -\cot\left(x + \frac{\pi}{2}\right)$ on $[-\pi, \pi]$ as follows:

Step 1: Graph $y = \cot\left(x + \frac{\pi}{2}\right)$ on $[-\pi, \pi]$:

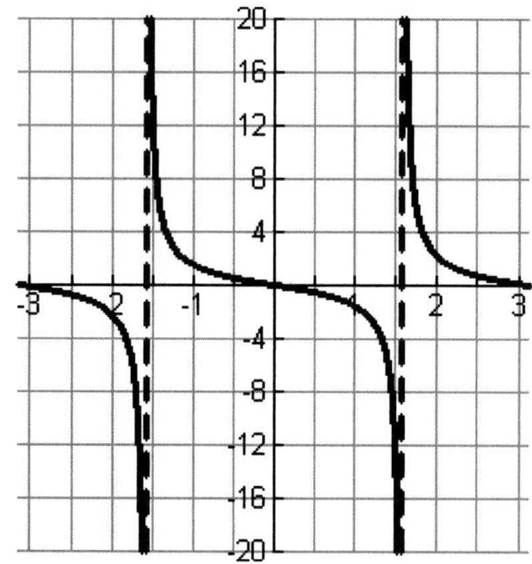

Step 2: Reflect the graph in Step 1 over the x-axis to obtain the graph of $y = -\cot\left(x + \frac{\pi}{2}\right)$ on $[-\pi, \pi]$:

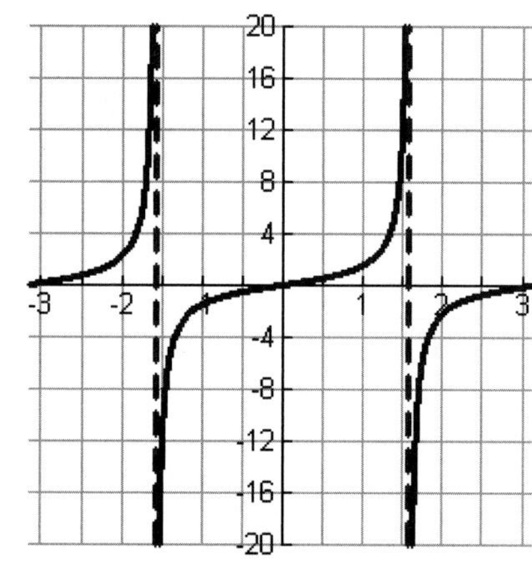

23. We have the following information about the graph of $y = -\frac{1}{2}\cot\left(x + \frac{\pi}{3}\right)$ on $[-\pi, \pi]$:

<table>
<tr><td><u>Period</u>:
$\frac{\pi}{1} = \pi$</td><td><u>Reflection about x-axis?</u>
Yes</td><td><u>Phase Shift</u>:
Left $\frac{\pi}{3}$ units</td></tr>
<tr><td colspan="3"><u>x-intercepts</u> occur when $\cos\left(x + \frac{\pi}{3}\right) = 0$. This happens when $x + \frac{\pi}{3} = \pm\frac{\pi}{2}$, so that $x = \frac{\pi}{6}, -\frac{5\pi}{6}$. So, the points $\left(\frac{\pi}{6}, 0\right), \left(-\frac{5\pi}{6}, 0\right)$ are on the graph.</td></tr>
<tr><td colspan="3"><u>Vertical asymptotes</u> occur when $\sin\left(x + \frac{\pi}{3}\right) = 0$. This happens when $x + \frac{\pi}{3} = 0, \pm\pi$, so that $x = -\frac{\pi}{3}, \frac{2\pi}{3}, \underbrace{\cancel{-\frac{4\pi}{3}}}_{\text{outside}}$.</td></tr>
<tr><td colspan="3"><u>Stretching</u>: The coefficient $\frac{1}{2}$ has the effect of making the graph approach the vertical asymptotes more slowly than the corresponding graph without it.</td></tr>
</table>

Using this information, we construct the graph of $y = -\frac{1}{2}\cot\left(x + \frac{\pi}{3}\right)$ on $[-\pi, \pi]$ as follows:

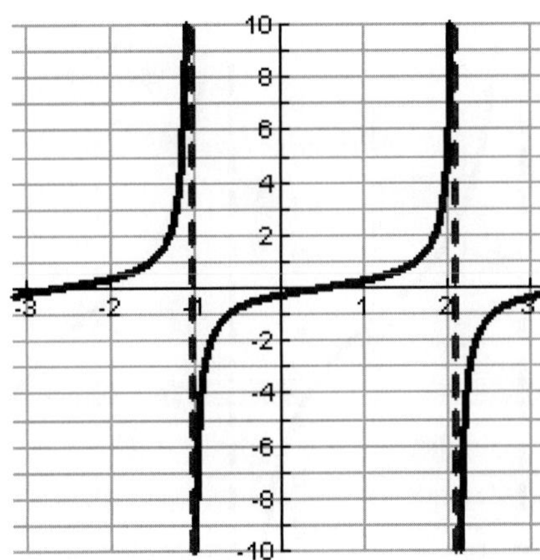

24. We have the following information about the graph of $y = 3\cot\left(x - \frac{\pi}{6}\right)$ on $[-\pi, \pi]$:

Period: $\frac{\pi}{1} = \pi$	Reflection about x-axis? No	Phase Shift: Right $\frac{\pi}{6}$ units
x-intercepts occur when $\cos\left(x - \frac{\pi}{6}\right) = 0$. This happens when $x - \frac{\pi}{6} = \pm\frac{\pi}{2}$, so that $x = \frac{\pi}{3}, -\frac{2\pi}{3}$. So, the points $\left(-\frac{2\pi}{3}, 0\right), \left(\frac{\pi}{3}, 0\right)$ are on the graph.		
Vertical asymptotes occur when $\sin\left(x - \frac{\pi}{6}\right) = 0$. This happens when $x - \frac{\pi}{6} = 0, \pm\pi$, so that $x = \frac{\pi}{6}, -\frac{5\pi}{6}, \underbrace{\cancel{\frac{7\pi}{6}}}_{\text{outside}}$.		
Stretching: The coefficient 3 has the effect of making the graph approach the vertical asymptotes more quickly than the corresponding graph without it.		

Using this information, we construct the graph of $y = 3\cot\left(x - \frac{\pi}{6}\right)$ on $[-\pi, \pi]$ as follows:

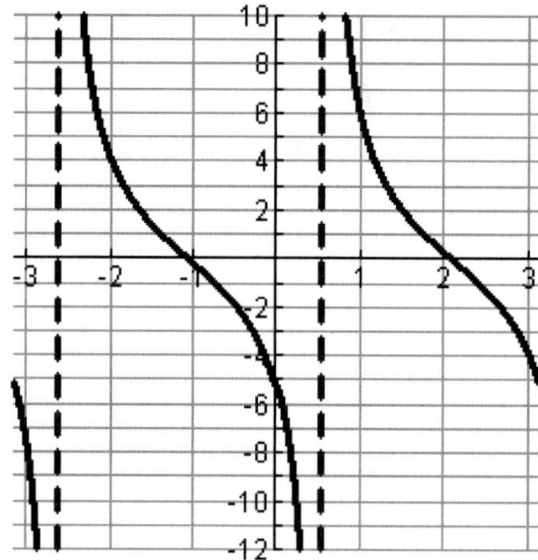

25. We have the following information about the graph of $y = \tan(2x - \pi)$ on $[-2\pi, 2\pi]$:

Period:	Reflection about x-axis?	Phase Shift:
$\frac{\pi}{2}$	No	Right $\frac{\pi}{2}$ units

Step 1: Graph $y = \tan(2x)$ on $\left[-\frac{\pi}{2}, \frac{\pi}{2}\right]$.

We have the following information about this particular graph

x-intercepts occur when $\sin 2x = 0$. This happens when $2x = 0, \pm\pi$, so that $x = 0, \pm\frac{\pi}{2}$. So, the points $(0, 0), \left(\pm\frac{\pi}{2}, 0\right)$ are on the graph.
Vertical asymptotes occur when $\cos 2x = 0$. This happens when $2x = \pm\frac{\pi}{2}$ so that $x = \pm\frac{\pi}{4}$.

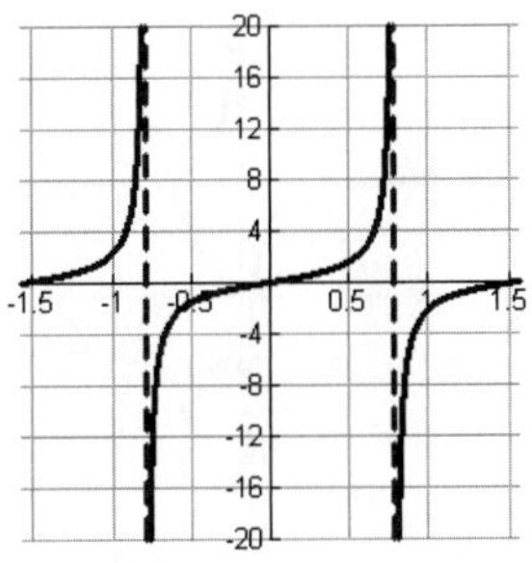

Step 2: Extend the graph in Step 1 to $[-2\pi, 2\pi]$:

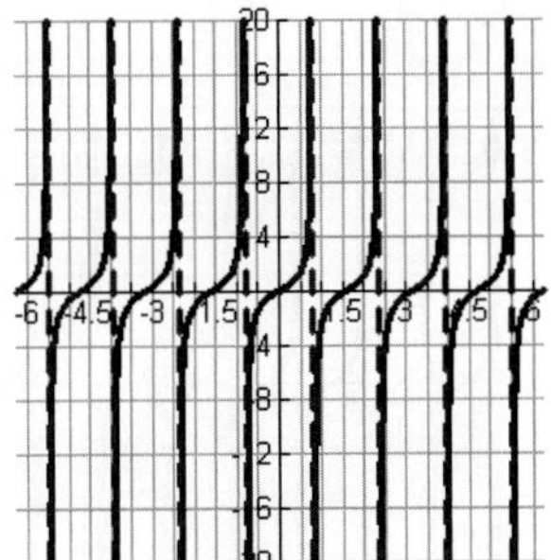

Step 3: Translate the graph in Step 2 to the right $\pi/2$ units. (No visible change since the period is $\pi/2$.)

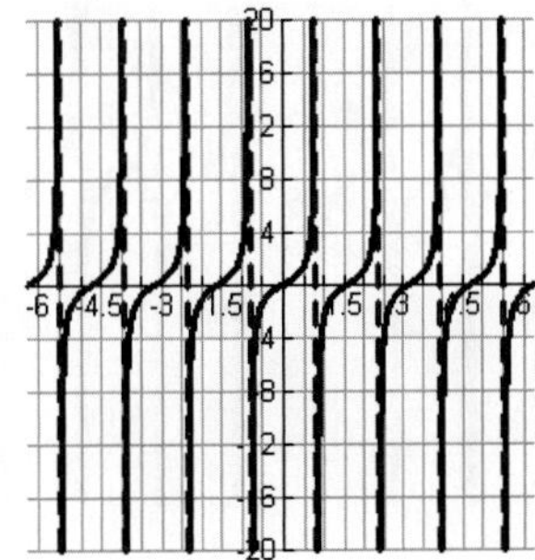

26. We have the following information about the graph of $y = \cot(2x - \pi)$ on $[-2\pi, 2\pi]$:

Period:	Reflection about x-axis?	Phase Shift:
$\frac{\pi}{2}$	No	Right $\frac{\pi}{2}$ units

Step 1: Graph $y = \cot(2x)$ on $\left[-\frac{\pi}{2}, \frac{\pi}{2}\right]$.

We have the following information about this particular graph

x-intercepts occur when $\cos 2x = 0$. This happens when $2x = \pm\frac{\pi}{2}$ so that $x = \pm\frac{\pi}{4}$. So, the points $\left(\pm\frac{\pi}{4}, 0\right)$ are on the graph.
Vertical asymptotes occur when $\sin 2x = 0$. This happens when $2x = 0, \pm\pi$, so that $x = 0, \pm\frac{\pi}{2}$.

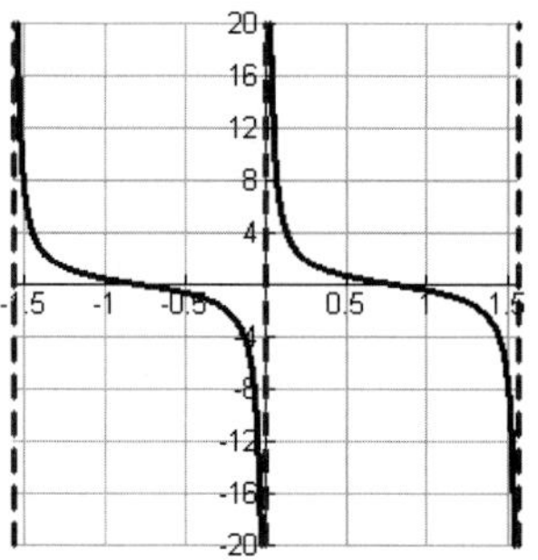

Step 2: Extend the graph in Step 1 to $[-2\pi, 2\pi]$:

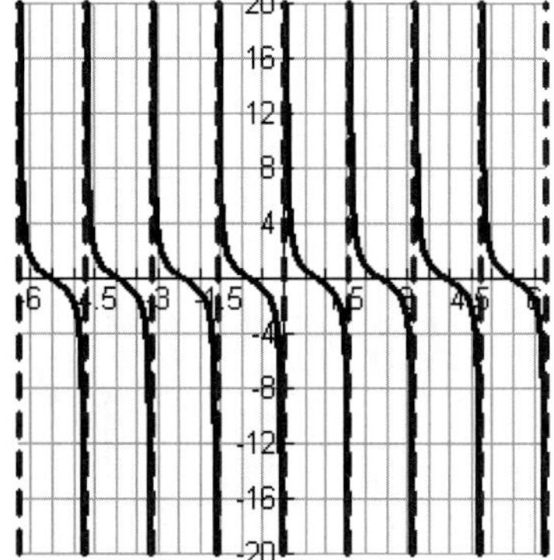

Step 3: Translate the graph in Step 2 to the right $\pi/2$ units. (No visible change since the period is $\pi/2$.)

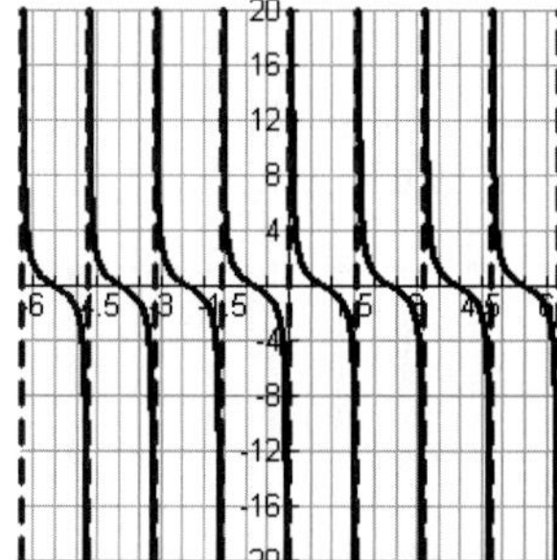

27. We have the following information about the graph of $y=\cot\left(\frac{x}{2}+\frac{\pi}{4}\right)$ on $[-\pi,\pi]$:

Period: $\dfrac{\pi}{1/2}=2\pi$	Reflection about x-axis? No	Phase Shift: Left $\dfrac{\pi/4}{1/2}=\dfrac{\pi}{2}$ units
x-intercepts occur when $\cos\left(\frac{x}{2}+\frac{\pi}{4}\right)=0$. This happens when $\frac{x}{2}+\frac{\pi}{4}=\pm\frac{\pi}{2},\pm\frac{3\pi}{2}$, so that $x=\frac{\pi}{2}$. So, the points $\left(\frac{\pi}{2},0\right)$ are on the graph.		
Vertical asymptotes occur when $\sin\left(\frac{x}{2}+\frac{\pi}{4}\right)=0$. This happens when $\frac{x}{2}+\frac{\pi}{4}=0,\pm\pi$, so that $x=-\frac{\pi}{2}$.		

Using this information, we construct the graph of $y=\cot\left(\frac{x}{2}+\frac{\pi}{4}\right)$ on $[-\pi,\pi]$ as follows:

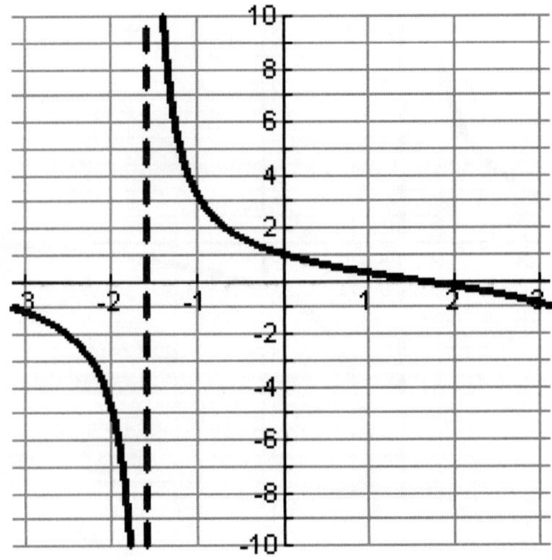

28. We have the following information about the graph of $y = \tan\left(\frac{x}{3} - \frac{\pi}{3}\right)$ on $[-\pi, \pi]$:

Period: $\frac{\pi}{1/3} = 3\pi$	Reflection about x-axis? No	Phase Shift: Right $\frac{\pi/3}{1/3} = \pi$ units
x-intercepts occur when $\sin\left(\frac{x}{3} - \frac{\pi}{3}\right) = 0$. This happens when $\frac{x}{3} - \frac{\pi}{3} = 0$, so that $x = -\pi$. So, the points $(-\pi, 0)$ are on the graph.		
Vertical asymptotes occur when $\cos\left(\frac{x}{3} - \frac{\pi}{3}\right) = 0$. This happens when $\frac{x}{3} - \frac{\pi}{3} = -\frac{\pi}{2}$, so that $x = -\frac{\pi}{2}$.		

Using this information, we construct the graph of $y = \tan\left(\frac{x}{3} - \frac{\pi}{3}\right)$ on $[-\pi, \pi]$ as follows:

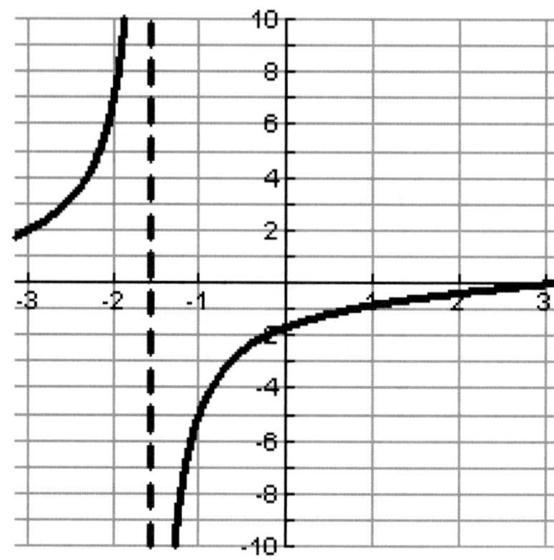

29. In order to graph $y = \sec\left(\frac{1}{2}x\right)$ on $[-2\pi, 2\pi]$, proceed as described below:

- Graph the *guide function* $y = \cos\left(\frac{1}{2}x\right)$ on $[-2\pi, 2\pi]$ using the approach discussed earlier in Chapter 6.

  x-intercepts: Occur when $\frac{1}{2}x = \pm\frac{\pi}{2}$, so that $x = \pm\pi$

- At each x-intercept of this guide function, draw a vertical asymptote.
- On the intervals where this guide function is above the x-axis, draw a $\cup$–shaped graph with vertex at the maximum of the guide function, opening upward toward the asymptotes.
- On the intervals where this guide function is below the x-axis, draw a $\cap$–shaped graph with vertex at the minimum of the guide function, opening downward toward the asymptotes.

The graph of the guide function (dotted) and the given function $y = \sec\left(\frac{1}{2}x\right)$ on $[-2\pi, 2\pi]$ is as follows:

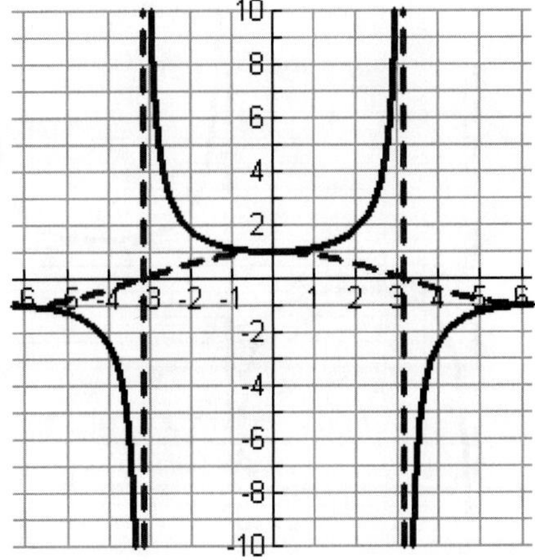

30. In order to graph $y = \csc\left(\frac{1}{2}x\right)$ on $[-2\pi, 2\pi]$, proceed as described below:

- Graph the *guide function* $y = \sin\left(\frac{1}{2}x\right)$ on $[-2\pi, 2\pi]$ using the approach discussed earlier in Chapter 6.

  x-intercepts: Occur when $\frac{1}{2}x = 0, \pm\pi$, so that $x = 0, \pm 2\pi$

  - At each *x*-intercept of this guide function, draw a vertical asymptote.
  - On the intervals where this guide function is above the *x*-axis, draw a $\cup$ – shaped graph with vertex at the maximum of the guide function, opening upward toward the asymptotes.
  - On the intervals where this guide function is below the *x*-axis, draw a $\cap$ – shaped graph with vertex at the minimum of the guide function, opening downward toward the asymptotes.

The graph of the guide function (dotted) and the given function $y = \csc\left(\frac{1}{2}x\right)$ on $[-2\pi, 2\pi]$ is as follows:

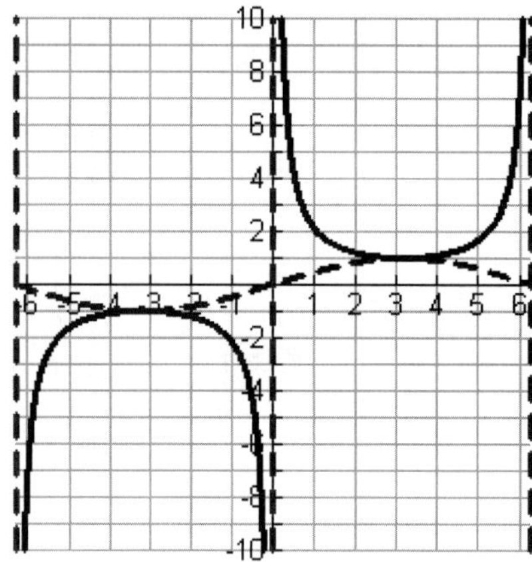

31. In order to graph $y=-\csc(2\pi x)$ on $[-1,1]$, proceed as described below:

Step 1: Graph $y=\csc(2\pi x)$ on $[-1,1]$

- Graph the *guide function* $y=\sin(2\pi x)$ on $[-1,1]$ using the approach discussed earlier in Chapter 6.

  x-intercepts: Occur when $2\pi x=0,\ \pm\pi,\ \pm 2\pi$, so that $x=0,\ \pm\frac{1}{2},\ \pm 1$

  - At each x-intercept of this guide function, draw a vertical asymptote.
  - On the intervals where this guide function is above the x-axis, draw a $\cup$-shaped graph with vertex at the maximum of the guide function, opening upward toward the asymptotes.
  - On the intervals where this guide function is below the x-axis, draw a $\cap$-shaped graph with vertex at the minimum of the guide function, opening downward toward the asymptotes.

The graph of the guide function (dotted) and the function $y=\csc(2\pi x)$ on $[-1,1]$ is as follows:

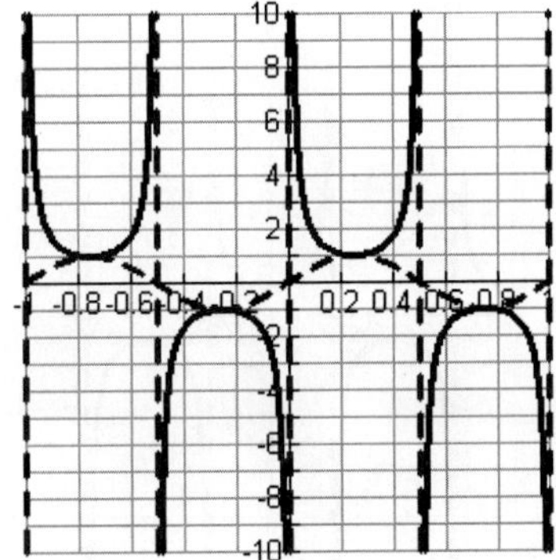

Step 2: Now, reflect the graph in Step 1 over the x-axis to obtain the graph of $y=-\csc(2\pi x)$ on $[-1,1]$, as shown below:

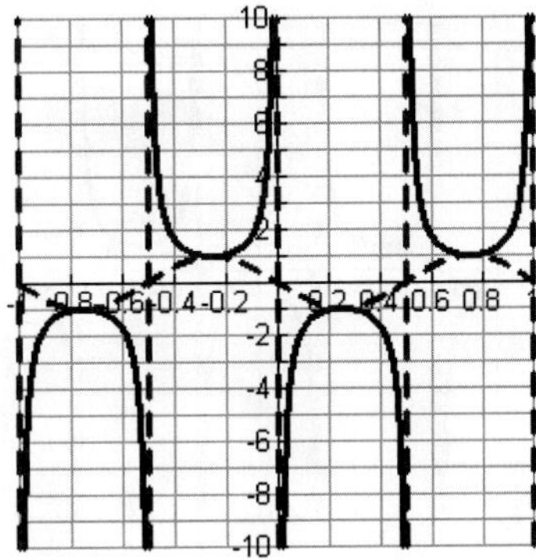

32. In order to graph $y = -\sec(2\pi x)$ on $[-1,1]$, proceed as described below:

Step 1: Graph $y = \sec(2\pi x)$ on $[-1,1]$

- Graph the *guide function* $y = \cos(2\pi x)$ on $[-1,1]$ using the approach discussed earlier in Chapter 6.

  x-intercepts: Occur when $2\pi x = \pm\frac{\pi}{2}, \pm\frac{3\pi}{2}$, so that $x = \pm\frac{1}{4}, \pm\frac{3}{4}$

  - At each x-intercept of this guide function, draw a vertical asymptote.
  - On the intervals where this guide function is above the x-axis, draw a $\cup$–shaped graph with vertex at the maximum of the guide function, opening
    upward toward the asymptotes.
  - On the intervals where this guide function is below the x-axis, draw a $\cap$–shaped graph with vertex at the minimum of the guide function, opening downward toward the asymptotes.

The graph of the guide function (dotted) and the function $y = \sec(2\pi x)$ on $[-1,1]$ is as follows:

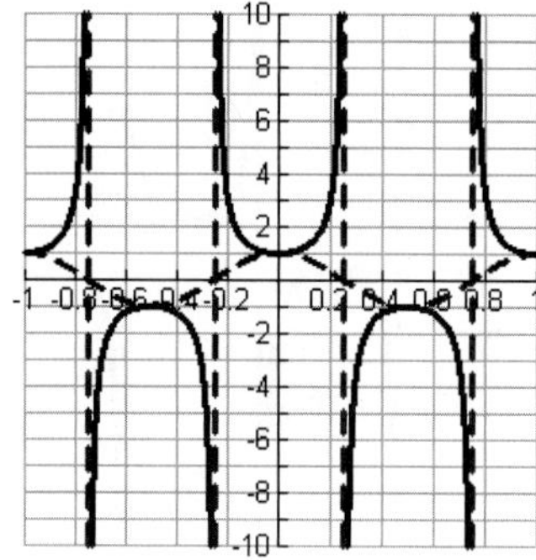

Step 2: Now, reflect the graph in Step 1 over the x-axis to obtain the graph of $y = -\sec(2\pi x)$ on $[-1,1]$, as shown below:

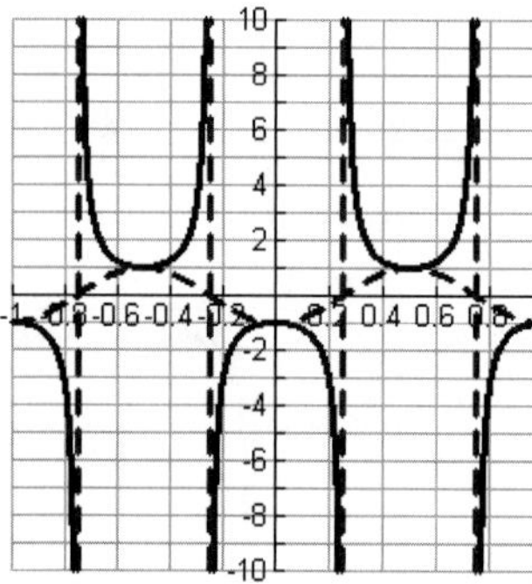

33. In order to graph $y = \frac{1}{3}\sec\left(\frac{\pi}{2}x\right)$ on $[-4,4]$, proceed as described below:

- Graph the *guide function* $y = \frac{1}{3}\cos\left(\frac{\pi}{2}x\right)$ on $[-4,4]$ using the approach discussed earlier in Chapter 6.

  x-intercepts: Occur when $\frac{\pi}{2}x = \pm\frac{\pi}{2}, \pm\frac{3\pi}{2}$, so that $x = \pm 1, \pm 3$

  - At each x-intercept of this guide function, draw a vertical asymptote.
  - On the intervals where this guide function is above the x-axis, draw a $\cup$–shaped graph with vertex at the maximum of the guide function, opening upward toward the asymptotes.
  - On the intervals where this guide function is below the x-axis, draw a $\cap$–shaped graph with vertex at the minimum of the guide function, opening downward toward the asymptotes.

The graph of the guide function (dotted) and the given function $y = \frac{1}{3}\sec\left(\frac{\pi}{2}x\right)$ on $[-4,4]$ is as follows:

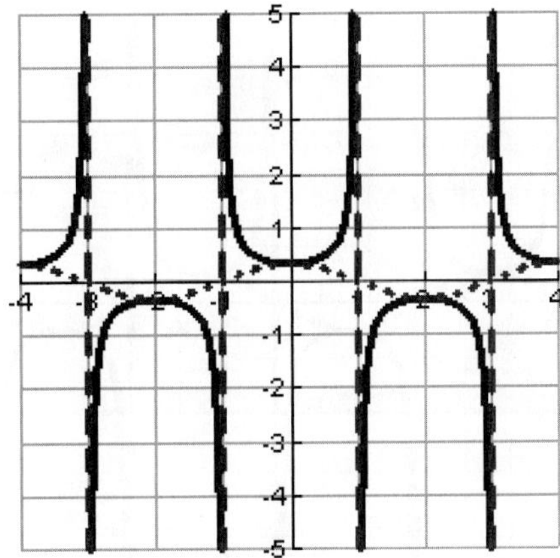

34. In order to graph $y = \frac{1}{2}\csc\left(\frac{\pi}{3}x\right)$ on $[-6,6]$, proceed as described below:

- Graph the *guide function* $y = \frac{1}{2}\sin\left(\frac{\pi}{3}x\right)$ on $[-6,6]$ using the approach discussed earlier in Chapter 6.

  x-intercepts: Occur when $\frac{\pi}{3}x = 0, \pm\pi, \pm 2\pi$, so that $x = 0, \pm 3, \pm 6$

- At each x-intercept of this guide function, draw a vertical asymptote.
- On the intervals where this guide function is above the x-axis, draw a $\cup$-shaped graph with vertex at the maximum of the guide function, opening upward toward the asymptotes.
- On the intervals where this guide function is below the x-axis, draw a $\cap$-shaped graph with vertex at the minimum of the guide function, opening downward toward the asymptotes.

The graph of the guide function (dotted) and the given function $y = \frac{1}{2}\csc\left(\frac{\pi}{3}x\right)$ on $[-6,6]$ is as follows:

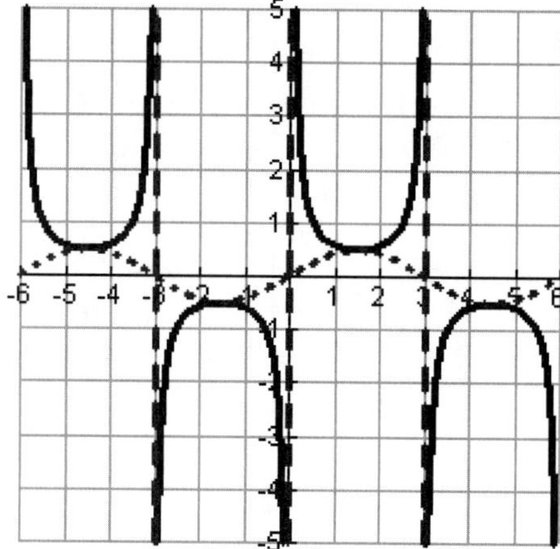

35. In order to graph $y = -3\csc\left(\frac{x}{3}\right)$ on $[-6\pi, 0]$, proceed as described below:

Step 1: Graph $y = 3\csc\left(\frac{x}{3}\right)$ on $[-6\pi, 0]$.

- Graph the *guide function* $y = 3\sin\left(\frac{x}{3}\right)$ on $[-6\pi, 0]$ using the approach discussed earlier in Chapter 6.

  x-intercepts: Occur when $\frac{x}{3} = 0, -\pi, -2\pi$, so that $x = 0, \ -3\pi, -6\pi$

  - At each x-intercept of this guide function, draw a vertical asymptote.
  - On the intervals where this guide function is above the x-axis, draw a $\cup$– shaped graph with vertex at the maximum of the guide function, opening upward toward the asymptotes.
  - On the intervals where this guide function is below the x-axis, draw a $\cap$– shaped graph with vertex at the minimum of the guide function, opening downward toward the asymptotes.

The graph of the guide function (dotted) and the function $y = 3\csc\left(\frac{x}{3}\right)$ on $[-6\pi, 0]$ is as follows:

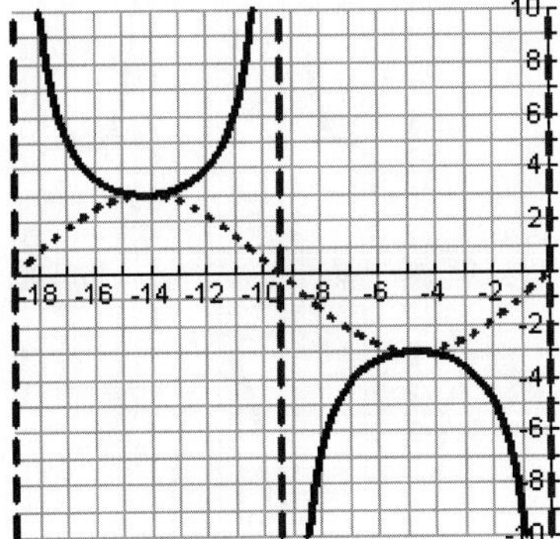

Step 2: Now, reflect the graph in Step 1 over the x-axis to obtain the graph of $y = -3\csc\left(\frac{x}{3}\right)$ on $[-6\pi, 0]$, as shown below:

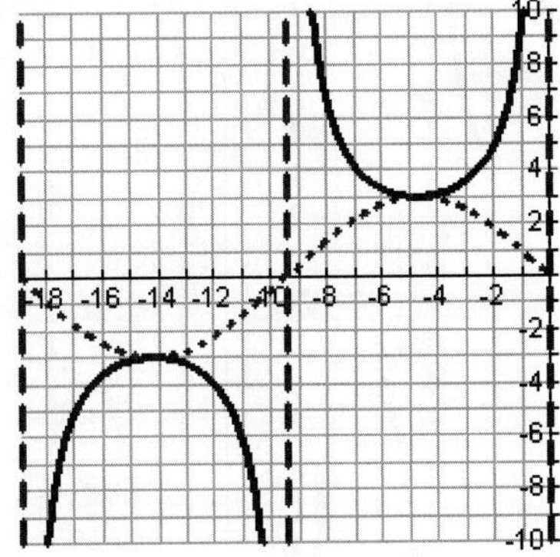

36. In order to graph $y = -4\sec\left(\frac{x}{2}\right)$ on $[-4\pi, 4\pi]$, proceed as described below:

Step 1: Graph $y = 4\sec\left(\frac{x}{2}\right)$ on $[-4\pi, 4\pi]$.

- Graph the *guide function* $y = 4\cos\left(\frac{x}{2}\right)$ on $[-4\pi, 4\pi]$ using the approach discussed earlier in Chapter 6.

  x-intercepts: Occur when $\frac{x}{2} = \pm\frac{\pi}{2},\ \pm\frac{3\pi}{2}$, so that $x = \pm\pi,\ \pm 3\pi$

  - At each x-intercept of this guide function, draw a vertical asymptote.
  - On the intervals where this guide function is above the x-axis, draw a $\cup$ – shaped graph with vertex at the maximum of the guide function, opening upward toward the asymptotes.
  - On the intervals where this guide function is below the x-axis, draw a $\cap$ – shaped graph with vertex at the minimum of the guide function, opening downward toward the asymptotes.

The graph of the guide function (dotted) and the function $y = 4\sec\left(\frac{x}{2}\right)$ on $[-4\pi, 4\pi]$ is as follows:

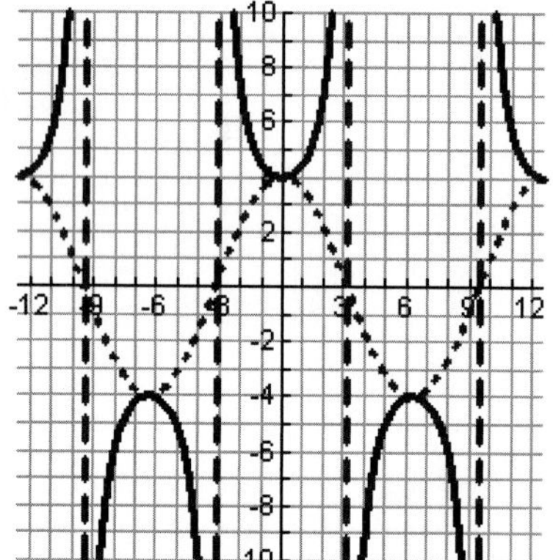

Step 2: Now, reflect the graph in Step 1 over the x-axis to obtain the graph of $y = -4\sec\left(\frac{x}{2}\right)$ on $[-4\pi, 4\pi]$, as shown below:

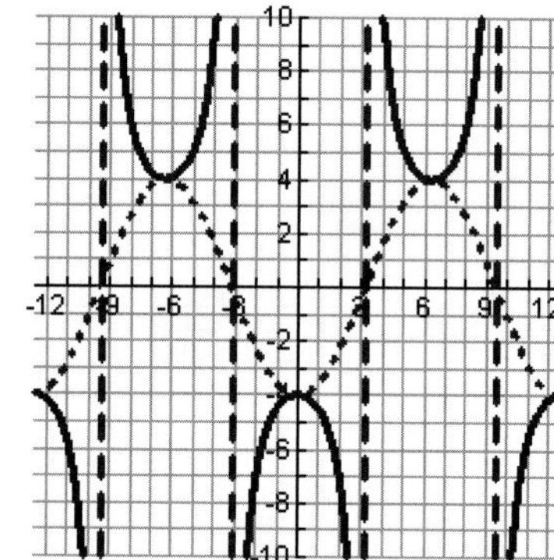

37. In order to graph $y = 2\sec(3x)$ on $[0, 2\pi]$, proceed as described below:

- Graph the *guide function* $y = 2\cos(3x)$ on $[0, 2\pi]$ using the approach discussed earlier in Chapter 6.

  x-intercepts: Occur when $3x = \frac{\pi}{2}, \frac{3\pi}{2}, \frac{5\pi}{2}, \frac{7\pi}{2}, \frac{9\pi}{2}, \frac{11\pi}{2}$, so that

  $x = \frac{\pi}{6}, \frac{\pi}{2}, \frac{5\pi}{6}, \frac{7\pi}{6}, \frac{3\pi}{2}, \frac{11\pi}{6}$

- At each x-intercept of this guide function, draw a vertical asymptote.
- On the intervals where this guide function is above the x-axis, draw a $\cup$ – shaped graph with vertex at the maximum of the guide function, opening upward toward the asymptotes.
- On the intervals where this guide function is below the x-axis, draw a $\cap$ – shaped graph with vertex at the minimum of the guide function, opening downward toward the asymptotes.

The graph of the guide function (dotted) and the given function $y = 2\sec(3x)$ on $[0, 2\pi]$ is as follows:

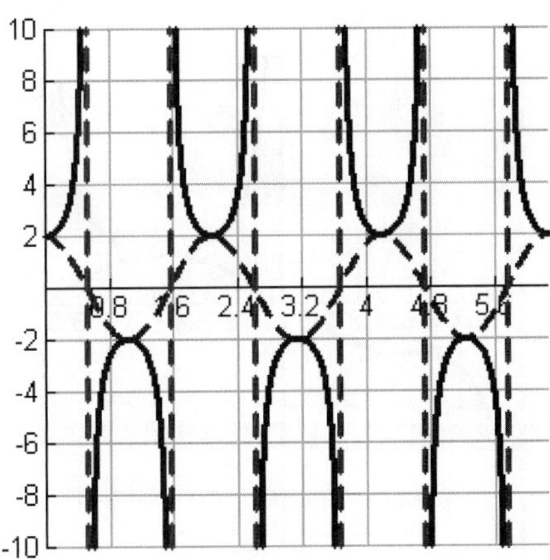

38. In order to graph $y = 2\csc\left(\frac{1}{3}x\right)$ on $[-3\pi, 3\pi]$, proceed as described below:

- Graph the *guide function* $y = 2\sin\left(\frac{1}{3}x\right)$ on $[-3\pi, 3\pi]$ using the approach discussed earlier in Chapter 6.

  x-intercepts: Occur when $\frac{1}{3}x = 0, \pm\pi$, so that $x = 0, \pm 3\pi$.

  - At each x-intercept of this guide function, draw a vertical asymptote.
  - On the intervals where this guide function is above the x-axis, draw a $\cup$ – shaped graph with vertex at the maximum of the guide function, opening upward toward the asymptotes.
  - On the intervals where this guide function is below the x-axis, draw a $\cap$ – shaped graph with vertex at the minimum of the guide function, opening downward toward the asymptotes.

The graph of the guide function (dotted) and the given function $y = 2\csc\left(\frac{1}{3}x\right)$ on $[-3\pi, 3\pi]$ is as follows:

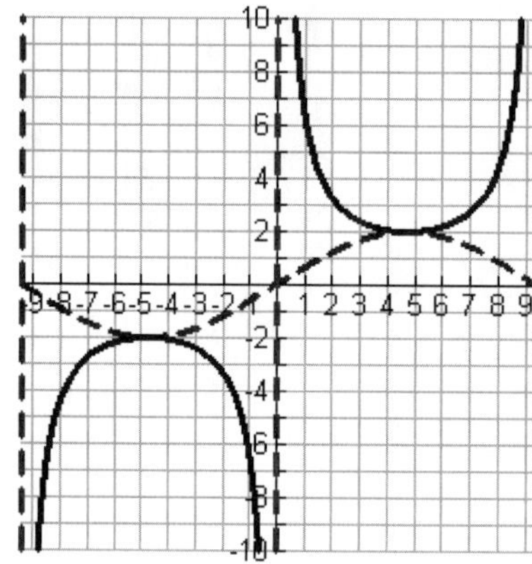

39. First, note that the period is 2π and the phase shift is $\frac{\pi}{2}$ units to the right. Hence, one period of this function would occur on the interval $\left[\frac{\pi}{2},\frac{5\pi}{2}\right]$, for instance.

In order to graph $y=-3\csc\left(x-\frac{\pi}{2}\right)$ on $\left[\frac{\pi}{2},\frac{5\pi}{2}\right]$, proceed as described below:

Step 1: Graph $y=3\csc\left(x-\frac{\pi}{2}\right)$ on $\left[\frac{\pi}{2},\frac{5\pi}{2}\right]$.

- Graph the *guide function* $y=3\sin\left(x-\frac{\pi}{2}\right)$ on $\left[\frac{\pi}{2},\frac{5\pi}{2}\right]$ using the approach discussed earlier in Chapter 6.

  x-intercepts: Occur when $x-\frac{\pi}{2}=0,\ \pi,\ 2\pi$, so that $x=\frac{\pi}{2},\frac{3\pi}{2},\frac{5\pi}{2}$

  - At each x-intercept of this guide function, draw a vertical asymptote.
  - On the intervals where this guide function is above the x-axis, draw a $\cup$–shaped graph with vertex at the maximum of the guide function, opening upward toward the asymptotes.
  - On the intervals where this guide function is below the x-axis, draw a $\cap$–shaped graph with vertex at the minimum of the guide function, opening downward toward the asymptotes.

The graph of the guide function (dotted) and the function $y=3\csc\left(x-\frac{\pi}{2}\right)$ on $\left[\frac{\pi}{2},\frac{5\pi}{2}\right]$ is as follows:

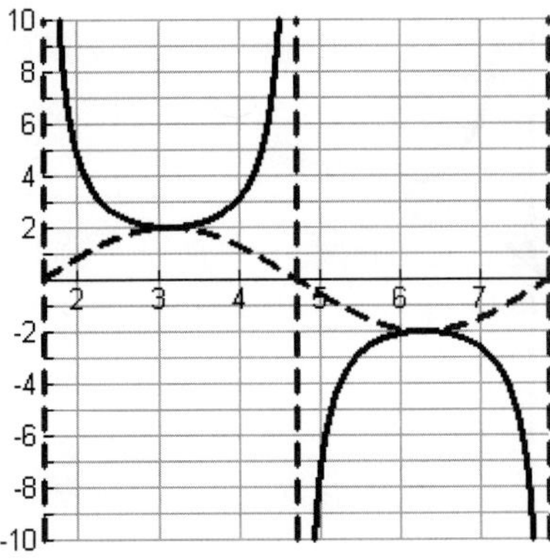

Step 2: Now, reflect the graph in Step 1 over the x-axis to obtain the graph of $y=-3\csc\left(x-\frac{\pi}{2}\right)$ on $\left[\frac{\pi}{2},\frac{5\pi}{2}\right]$, as shown below:

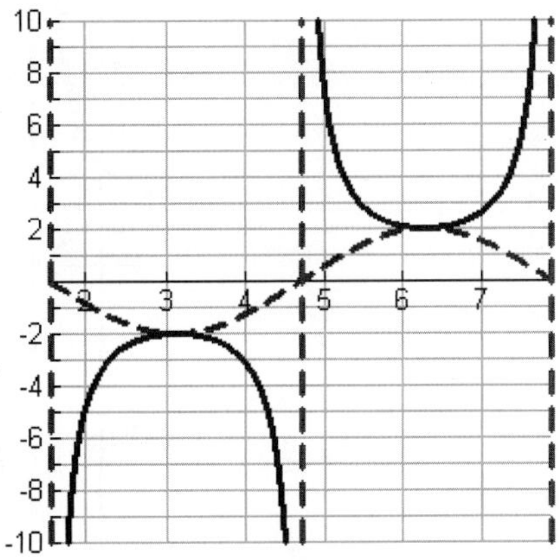

40. First, note that the period is 2π and the phase shift is $\frac{\pi}{4}$ units to the left. Hence, one period of this function would occur on the interval $\left[-\frac{\pi}{4}, \frac{7\pi}{4}\right]$, for instance.

In order to graph $y = 5\sec\left(x + \frac{\pi}{4}\right)$ on $\left[-\frac{\pi}{4}, \frac{7\pi}{4}\right]$, proceed as described below:

- Graph the *guide function* $y = 5\cos\left(x + \frac{\pi}{4}\right)$ on $\left[-\frac{\pi}{4}, \frac{7\pi}{4}\right]$ using the approach discussed earlier in Chapter 6.

  x-intercepts: Occur when $x + \frac{\pi}{4} = \frac{\pi}{2}, \frac{3\pi}{2}$, so that $x = \frac{\pi}{4}, \frac{5\pi}{4}$

  - At each x-intercept of this guide function, draw a vertical asymptote.
  - On the intervals where this guide function is above the x-axis, draw a $\cup$–shaped graph with vertex at the maximum of the guide function, opening upward toward the asymptotes.
  - On the intervals where this guide function is below the x-axis, draw a $\cap$–shaped graph with vertex at the minimum of the guide function, opening downward toward the asymptotes.

The graph of the guide function (dotted) and the function $y = 5\sec\left(x + \frac{\pi}{4}\right)$ on $\left[-\frac{\pi}{4}, \frac{7\pi}{4}\right]$ is as follows:

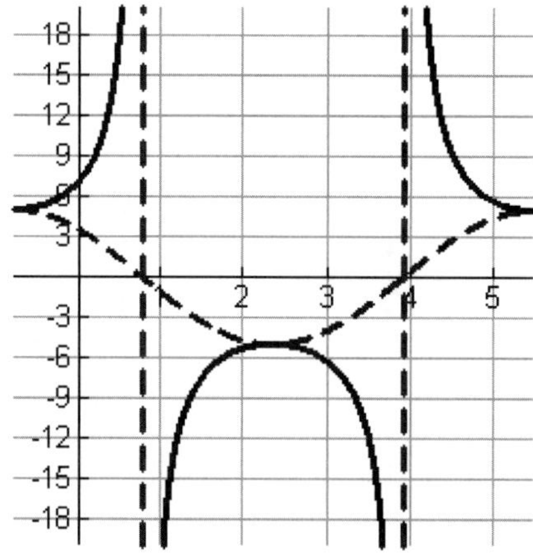

41. First, note that the period is 2π and the phase shift is π units to the right. Hence, one period of this function would occur on the interval $[\pi, 3\pi]$, for instance.

In order to graph $y = \frac{1}{2}\sec(x-\pi)$ on $[\pi, 3\pi]$, proceed as described below:

- Graph the *guide function* $y = \frac{1}{2}\cos(x-\pi)$ on $[\pi, 3\pi]$ using the approach discussed earlier in Chapter 6.

  x-intercepts: Occur when $x - \pi = \frac{\pi}{2}, \frac{3\pi}{2}$, so that $x = \frac{3\pi}{2}, \frac{5\pi}{2}$

- At each x-intercept of this guide function, draw a vertical asymptote.
- On the intervals where this guide function is above the x-axis, draw a $\cup$ – shaped graph with vertex at the maximum of the guide function, opening upward toward the asymptotes.
- On the intervals where this guide function is below the x-axis, draw a $\cap$ – shaped graph with vertex at the minimum of the guide function, opening downward toward the asymptotes.

The graph of the guide function (dotted) and the function $y = \frac{1}{2}\sec(x-\pi)$ on $[\pi, 3\pi]$ is as follows:

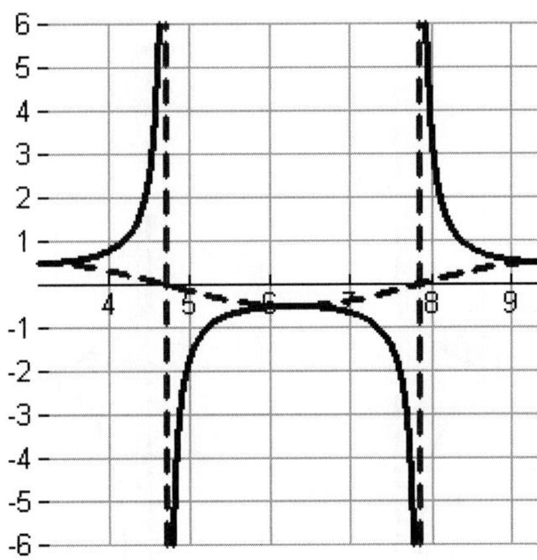

42. First, note that the period is 2π and the phase shift is π units to the left. Hence, one period of this function would occur on the interval $[-\pi,\pi]$, for instance.

In order to graph $y=-4\csc(x+\pi)$ on $[-\pi,\pi]$, proceed as described below:

Step 1: Graph $y=4\csc(x+\pi)$ on $[-\pi,\pi]$.

- Graph the *guide function* $y=4\sin(x+\pi)$ on $[-\pi,\pi]$ using the approach discussed earlier in Chapter 6.
  x-intercepts: Occur when $x+\pi=0,\ \pi,\ 2\pi$, so that $x=-\pi,\ 0,\ \pi$

  - At each x-intercept of this guide function, draw a vertical asymptote.
  - On the intervals where this guide function is above the x-axis, draw a $\cup$-shaped graph with vertex at the maximum of the guide function, opening upward toward the asymptotes.
  - On the intervals where this guide function is below the x-axis, draw a $\cap$-shaped graph with vertex at the minimum of the guide function, opening downward toward the asymptotes.

The graph of the guide function (dotted) and the function $y=4\csc(x+\pi)$ on $[-\pi,\pi]$ is as follows:

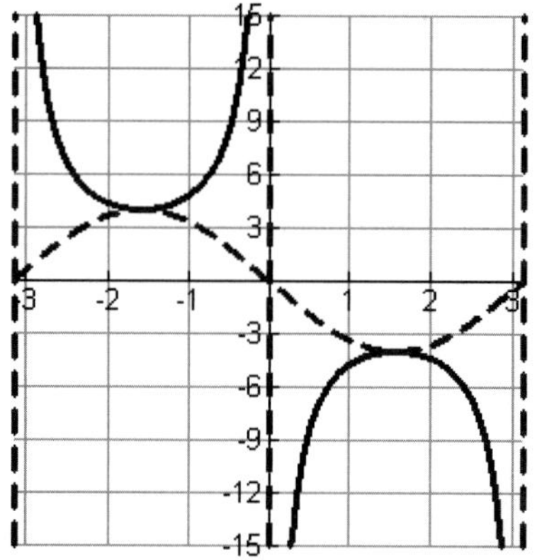

Step 2: Now, reflect the graph in Step 1 over the x-axis to obtain the graph of $y=-4\csc(x+\pi)$ on $[-\pi,\pi]$, as shown below:

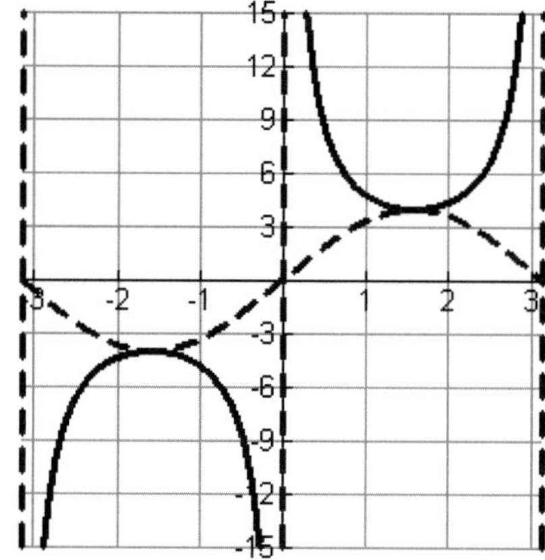

43. First, note that the period is π and the phase shift is $\frac{\pi}{2}$ units to the right. Hence, one period of this function would occur on the interval $\left[\frac{\pi}{2},\frac{3\pi}{2}\right]$, for instance.

In order to graph $y=2\sec(2x-\pi)$ on $[-2\pi,2\pi]$, proceed as described below:

Step 1: Graph $y=2\sec(2x-\pi)$ on $\left[\frac{\pi}{2},\frac{3\pi}{2}\right]$.

- Graph the *guide function* $y=2\cos(2x-\pi)$ on $\left[\frac{\pi}{2},\frac{3\pi}{2}\right]$ using the approach discussed earlier in Chapter 6.

  x-intercepts: Occur when $2x-\pi=\frac{\pi}{2},\frac{3\pi}{2}$, so that $x=\frac{3\pi}{4},\frac{5\pi}{4}$

  - At each x-intercept of this guide function, draw a vertical asymptote.
  - On the intervals where this guide function is above the x-axis, draw a $\cup$ – shaped graph with vertex at the maximum of the guide function, opening upward toward the asymptotes.
  - On the intervals where this guide function is below the x-axis, draw a $\cap$ – shaped graph with vertex at the minimum of the guide function, opening downward toward the asymptotes.

The graph of the guide function (dotted) and the function $y=2\sec(2x-\pi)$ on $\left[\frac{\pi}{2},\frac{3\pi}{2}\right]$ is as follows:

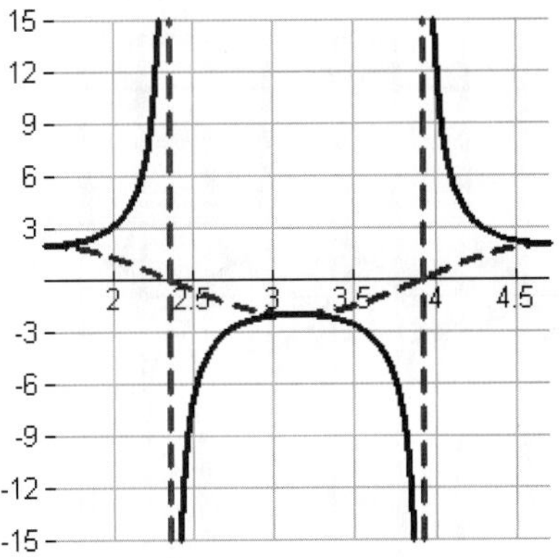

Step 2: Extend the graph using periodicity to the interval $[-2\pi,2\pi]$, as shown below:

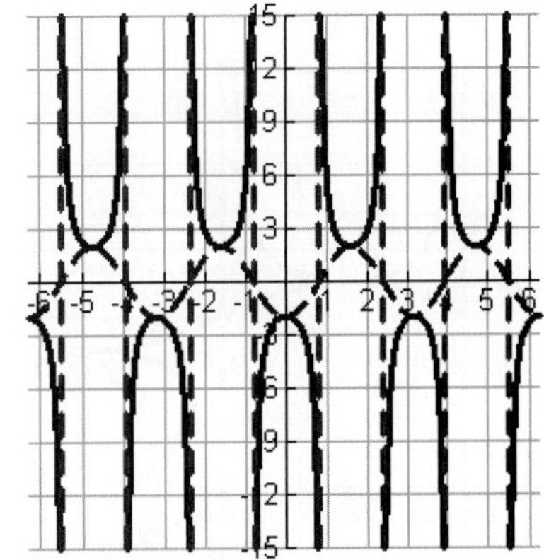

44. First, note that the period is π and the phase shift is $\frac{\pi}{2}$ units to the left. Hence, one period of this function would occur on the interval $\left[-\frac{\pi}{2},\frac{\pi}{2}\right]$, for instance.

In order to graph $y=2\csc(2x+\pi)$ on $[-2\pi,2\pi]$, proceed as described below:

Step 1: Graph $y=2\csc(2x+\pi)$ on $\left[-\frac{\pi}{2},\frac{\pi}{2}\right]$.

- Graph the *guide function* $y=2\sin(2x+\pi)$ on $\left[-\frac{\pi}{2},\frac{\pi}{2}\right]$ using the approach discussed earlier in Chapter 6.

  x-intercepts: Occur when $2x+\pi=0,\ \pi,\ 2\pi$, so that $x=-\frac{\pi}{2},\ 0,\ \frac{\pi}{2}$

  - At each x-intercept of this guide function, draw a vertical asymptote.
  - On the intervals where this guide function is above the x-axis, draw a $\cup$-shaped graph with vertex at the maximum of the guide function, opening upward toward the asymptotes.
  - On the intervals where this guide function is below the x-axis, draw a $\cap$-shaped graph with vertex at the minimum of the guide function, opening downward toward the asymptotes.

The graph of the guide function (dotted) and the function $y=2\csc(2x+\pi)$ on $\left[-\frac{\pi}{2},\frac{\pi}{2}\right]$ is as follows:

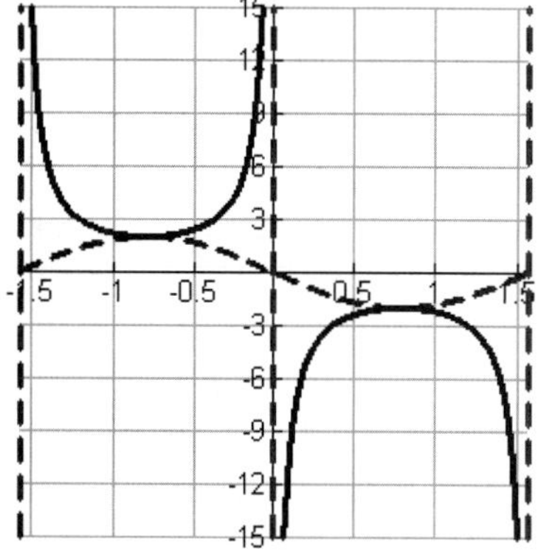

Step 2: Extend the graph using periodicity to the interval $[-2\pi,2\pi]$, as shown below:

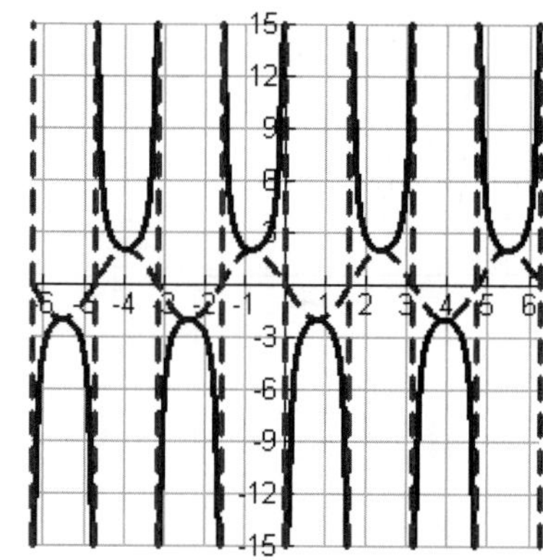

45. First, note that the period is $\frac{2\pi}{3}$ and the phase shift is $\frac{\pi}{3}$ units to the left. Hence, one period of this function would occur on the interval $\left[-\frac{\pi}{3}, \frac{\pi}{3}\right]$, for instance.

In order to graph $y = -\frac{1}{4}\sec(3x+\pi)$ on $[-\pi, \pi]$, proceed as described below:

Step 1: Graph $y = \frac{1}{4}\sec(3x+\pi)$ on $[-\pi, \pi]$.

- Graph the *guide function* $y = \frac{1}{4}\cos(3x+\pi)$ on $[-\pi, \pi]$ using the approach discussed earlier in Chapter 6.

  x-intercepts: Occur when $3x+\pi = \pm\frac{\pi}{2}, \pm\frac{3\pi}{2}$, so that $x = -\frac{\pi}{2}, \pm\frac{\pi}{6}$

  - At each x-intercept of this guide function, draw a vertical asymptote.
  - On the intervals where this guide function is above the x-axis, draw a $\cup$ – shaped graph with vertex at the maximum of the guide function, opening upward toward the asymptotes.
  - On the intervals where this guide function is below the x-axis, draw a $\cap$ – shaped graph with vertex at the minimum of the guide function, opening downward toward the asymptotes.

The graph of the guide function (dotted) and the function $y = \frac{1}{4}\sec(3x+\pi)$ on $[-\pi, \pi]$ is as follows:

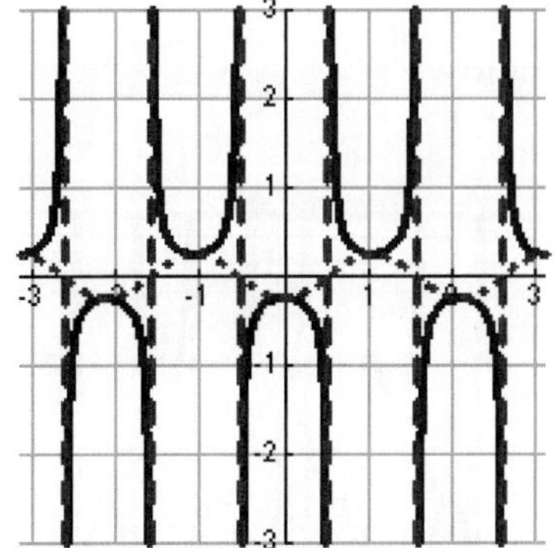

Step 2: Now, reflect the graph in Step 1 over the x-axis to obtain the graph of $y = -\frac{1}{4}\sec(3x+\pi)$ on $[-\pi, \pi]$, as shown below:

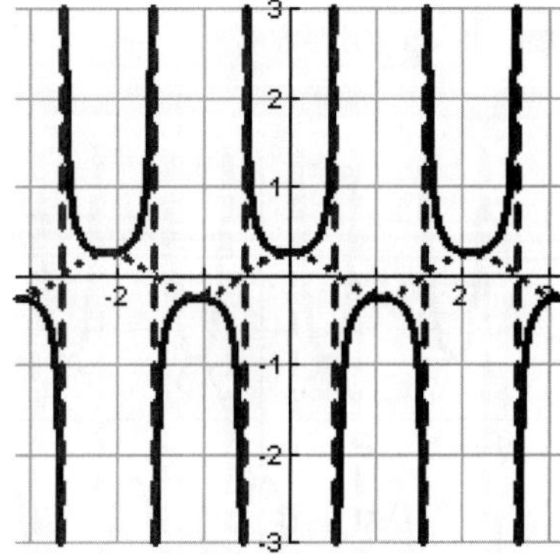

46. First, note that the period is $\frac{2\pi}{4} = \frac{\pi}{2}$ and the phase shift is $\frac{\pi}{8}$ units to the right

Hence, one period of this function would occur on the interval $\left[\frac{\pi}{8}, \frac{5\pi}{8}\right]$, for instance.

In order to graph $y = -\frac{2}{3}\csc\left(4x - \frac{\pi}{2}\right)$ on $[-\pi, \pi]$, proceed as described below:

Step 1: Graph $y = \frac{2}{3}\csc\left(4x - \frac{\pi}{2}\right)$ on $[-\pi, \pi]$.

- Graph the *guide function* $y = \frac{2}{3}\sin\left(4x - \frac{\pi}{2}\right)$ on $[-\pi, \pi]$ using the approach discussed earlier in Chapter 6.

  x-intercepts: Occur when $4x - \frac{\pi}{2} = 0, \pm\pi, \pm 2\pi, \pm 3\pi$, so that

  $x = \pm\frac{\pi}{8}, \pm\frac{3\pi}{8}, \pm\frac{5\pi}{8}, \pm\frac{7\pi}{8}$
  - At each x-intercept of this guide function, draw a vertical asymptote.
  - On the intervals where this guide function is above the x-axis, draw a $\cup$–shaped graph with vertex at the maximum of the guide function, opening upward toward the asymptotes.
  - On the intervals where this guide function is below the x-axis, draw a $\cap$–shaped graph with vertex at the minimum of the guide function, opening downward toward the asymptotes.

The graph of the guide function (dotted) and the function $y = \frac{2}{3}\csc\left(4x - \frac{\pi}{2}\right)$ on $[-\pi, \pi]$ is as follows:

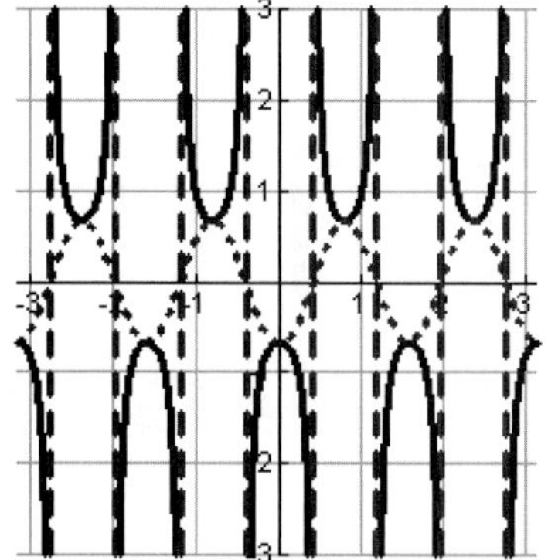

Step 2: Now, reflect the graph in Step 1 over the x-axis to obtain the graph of $y = -\frac{2}{3}\csc\left(4x - \frac{\pi}{2}\right)$ on $[-\pi, \pi]$, as shown below:

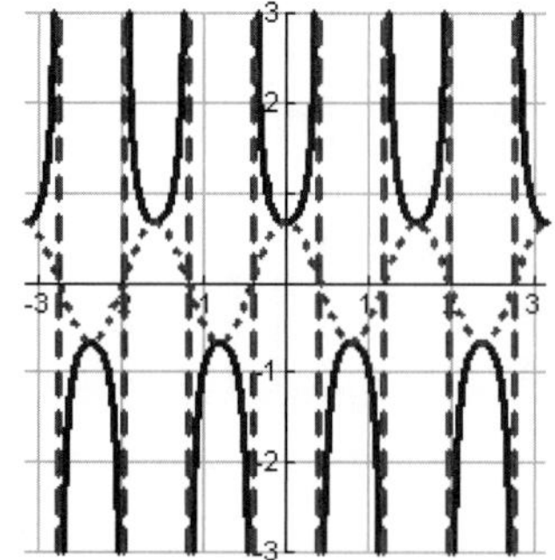

47. First, note that the period is 2π and the phase shift is $\frac{\pi}{2}$ units to the right. Hence, one period of this function would occur on the interval $\left[\frac{\pi}{2}, \frac{5\pi}{2}\right]$, for instance.

In order to graph $y = 3 - 2\sec\left(x - \frac{\pi}{2}\right)$ on $\left[\frac{\pi}{2}, \frac{5\pi}{2}\right]$, proceed as described below:

Step 1: Graph $y = -2\sec\left(x - \frac{\pi}{2}\right)$ on $\left[\frac{\pi}{2}, \frac{5\pi}{2}\right]$.

- Graph the *guide function* $y = -2\cos\left(x - \frac{\pi}{2}\right)$ on $\left[\frac{\pi}{2}, \frac{5\pi}{2}\right]$ using the approach discussed earlier in Chapter 6. (**Note**: You will need to reflect the graph of $y = 2\cos\left(x - \frac{\pi}{2}\right)$ over the x-axis.)

  x-intercepts: Occur when $x - \frac{\pi}{2} = \frac{\pi}{2}, \frac{3\pi}{2}$, so that $x = \pi,\ 2\pi$

  - At each x-intercept of this guide function, draw a vertical asymptote.
  - On the intervals where this guide function is above the x-axis, draw a $\cup$–shaped graph with vertex at the maximum of the guide function, opening upward toward the asymptotes.
  - On the intervals where this guide function is below the x-axis, draw a $\cap$–shaped graph with vertex at the minimum of the guide function, opening downward toward the asymptotes.

The graph of the guide function (dotted) and the function $y = -2\sec\left(x - \frac{\pi}{2}\right)$ on $\left[\frac{\pi}{2}, \frac{5\pi}{2}\right]$ is as follows:

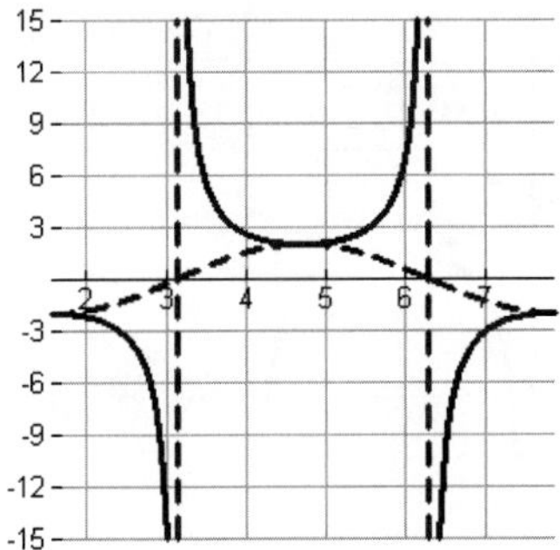

Step 2: Translate the graph is Step 1 vertically up 3 units to obtain the graph of $y = 3 - 2\sec\left(x - \frac{\pi}{2}\right)$ on $\left[\frac{\pi}{2}, \frac{5\pi}{2}\right]$, as shown below:

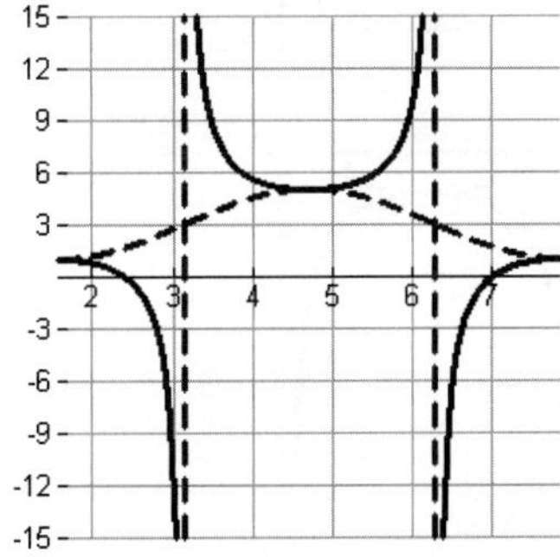

48. First, note that the period is 2π and the phase shift is $\frac{\pi}{2}$ units to the left. Hence, one period of this function would occur on the interval $\left[-\frac{\pi}{2}, \frac{3\pi}{2}\right]$, for instance.

In order to graph $y = -3 + 2\csc\left(x + \frac{\pi}{2}\right)$ on $\left[-\frac{\pi}{2}, \frac{3\pi}{2}\right]$, proceed as described below:

<u>Step 1</u>: Graph $y = 2\csc\left(x + \frac{\pi}{2}\right)$ on $\left[-\frac{\pi}{2}, \frac{3\pi}{2}\right]$.

- Graph the *guide function* $y = 2\sin\left(x + \frac{\pi}{2}\right)$ on $\left[-\frac{\pi}{2}, \frac{3\pi}{2}\right]$ using the approach discussed earlier in Chapter 6.

  <u>*x*-intercepts</u>: Occur when $x + \frac{\pi}{2} = 0,\ \pi,\ 2\pi$, so that $x = -\frac{\pi}{2}, \frac{\pi}{2}, \frac{3\pi}{2}$
  - At each *x*-intercept of this guide function, draw a vertical asymptote.
  - On the intervals where this guide function is <u>above</u> the *x*-axis, draw a $\cup$ – shaped graph with vertex at the maximum of the guide function, opening upward toward the asymptotes.
  - On the intervals where this guide function is <u>below</u> the *x*-axis, draw a $\cap$ – shaped graph with vertex at the minimum of the guide function, opening downward toward the asymptotes.

The graph of the guide function (dotted) and the function $y = 2\csc\left(x + \frac{\pi}{2}\right)$ on $\left[-\frac{\pi}{2}, \frac{3\pi}{2}\right]$ is as follows:

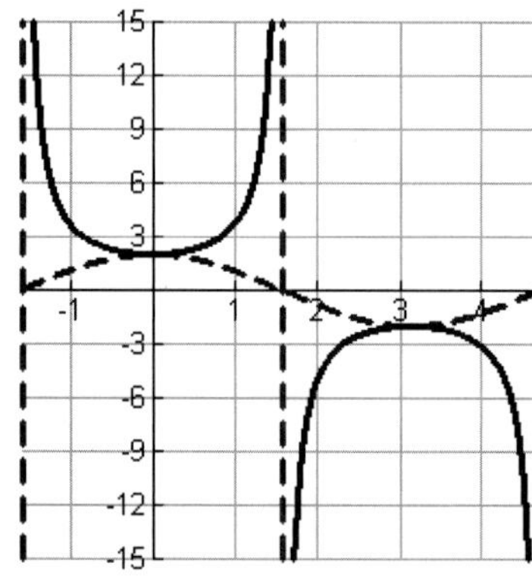

<u>Step 2</u>: Translate the graph is Step 1 vertically down 3 units to obtain the graph of $y = -3 + 2\csc\left(x + \frac{\pi}{2}\right)$ on $\left[-\frac{\pi}{2}, \frac{3\pi}{2}\right]$, as shown below:

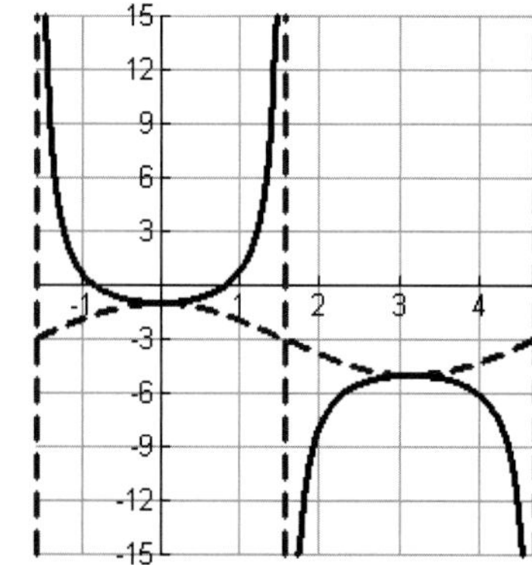

49. We have the following information about the graph of $y=\frac{1}{2}+\frac{1}{2}\tan\left(x-\frac{\pi}{2}\right)$:

Period:	Vertical Translation:	Reflection about x-axis?	Phase Shift:
$\frac{\pi}{1}=\pi$	Up $\frac{1}{2}$ unit	No	Right $\frac{\pi}{2}$ units

Step 1: Graph $y=\frac{1}{2}\tan\left(x-\frac{\pi}{2}\right)$ on $[0,\pi]$.

We have the following information about this particular graph

x-intercepts occur when $\sin\left(x-\frac{\pi}{2}\right)=0$. This happens when $x-\frac{\pi}{2}=0,\pi$, so that $x=\frac{\pi}{2},\frac{3\pi}{2}$. So, the points $\left(\frac{\pi}{2},0\right)$, $\left(\frac{3\pi}{2},0\right)$ (Outisde) are on this graph. (Note: These will change in Step 2 when we translate this graph vertically.)
Vertical asymptotes occur when $\cos\left(x-\frac{\pi}{2}\right)=0$. This happens when $x-\frac{\pi}{2}=\frac{\pi}{2}$ so that $x=\pi$. (Note: These remain unchanged by vertical translation.)

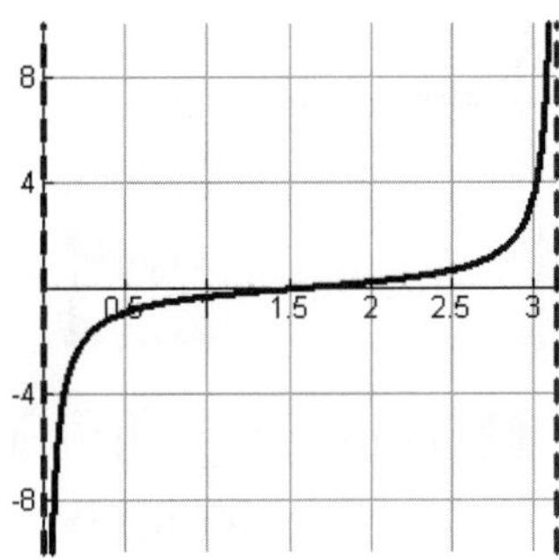

Step 2: Translate the graph in Step 1 vertically up $\frac{1}{2}$ unit.

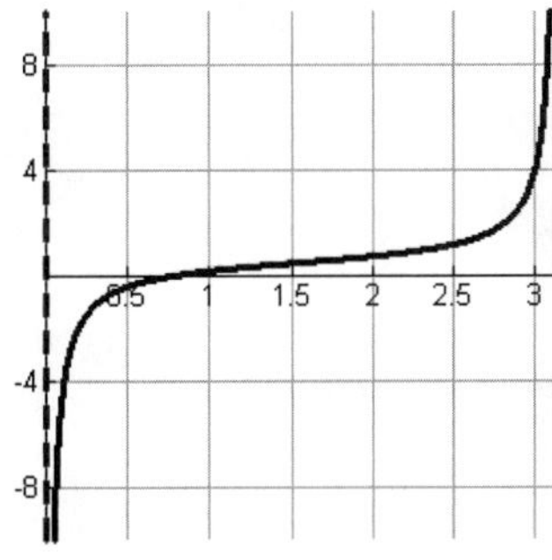

50. We have the following information about the graph of $y = \frac{3}{4} - \frac{1}{4}\cot\left(x + \frac{\pi}{2}\right)$:

Period:	Vertical Translation:	Reflection about x-axis?	Phase Shift:
$\frac{\pi}{1} = \pi$	Up $\frac{3}{4}$ unit	Yes	Left $\frac{\pi}{2}$ units

Step 1: Graph $y = \frac{1}{4}\cot\left(x + \frac{\pi}{2}\right)$ on $\left[-\frac{\pi}{2}, \frac{\pi}{2}\right]$.

We have the following information about this particular graph

x-intercepts occur when $\cos\left(x + \frac{\pi}{2}\right) = 0$. This happens when $x + \frac{\pi}{2} = \frac{\pi}{2}$ so that $x = 0$. So, the point $(0, 0)$ is on this graph. (Note: This will change in Step 2 when we translate this graph vertically.)
Vertical asymptotes occur when $\sin\left(x + \frac{\pi}{2}\right) = 0$. This happens when $x + \frac{\pi}{2} = 0, \pi$, so that $x = \pm\frac{\pi}{2}$. (Note: These remain unchanged by vertical translation.)

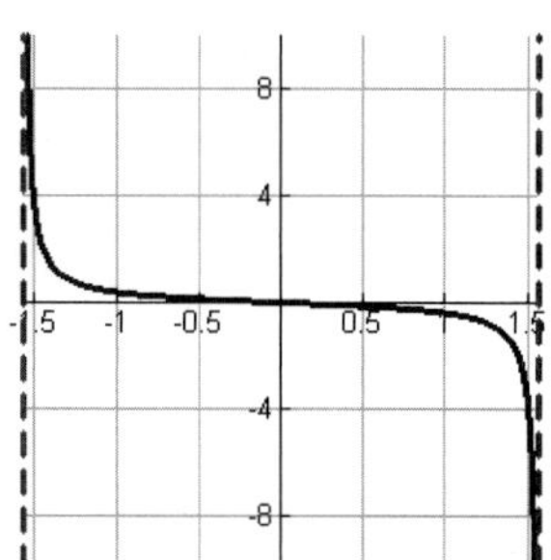

Step 2: Reflect the graph in Step 1 over the x-axis to obtain the graph of $y = -\frac{1}{4}\cot\left(x + \frac{\pi}{2}\right)$ on $\left[-\frac{\pi}{2}, \frac{\pi}{2}\right]$.

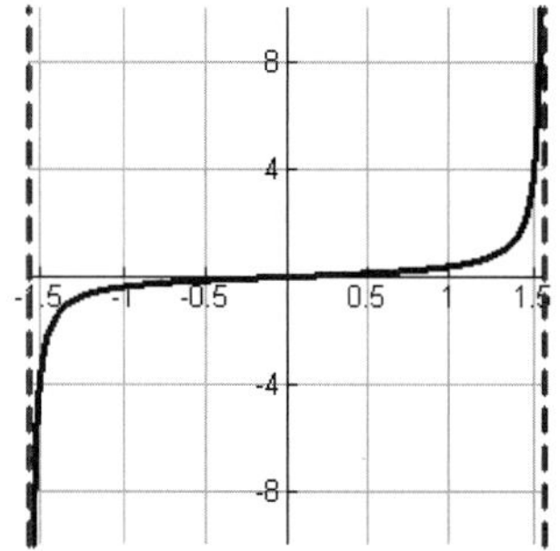

Step 3: Translate the graph in Step 2 vertically up $\frac{3}{4}$ unit to obtain the graph of $y = \frac{3}{4} - \frac{1}{4}\cot\left(x + \frac{\pi}{2}\right)$ on $\left[-\frac{\pi}{2}, \frac{\pi}{2}\right]$.

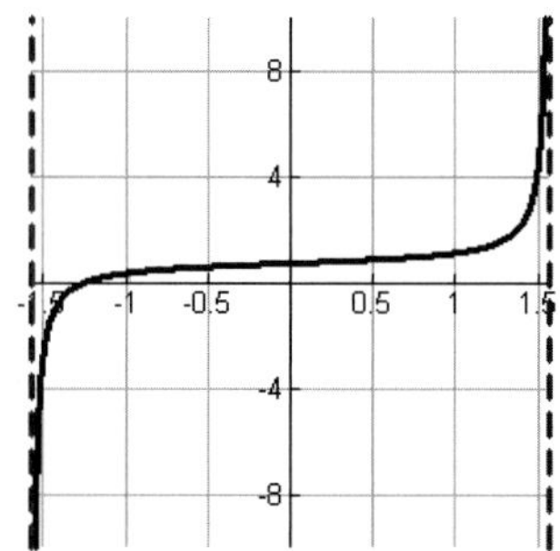

51. First, note that the period is π and the phase shift is $\frac{\pi}{2}$ units to the right. Hence, one period of this function would occur on the interval $\left[\frac{\pi}{2},\frac{3\pi}{2}\right]$, for instance.

In order to graph $y=-2+3\csc(2x-\pi)$ on $\left[\frac{\pi}{2},\frac{3\pi}{2}\right]$, proceed as described below:

Step 1: Graph $y=3\csc(2x-\pi)$ on $\left[\frac{\pi}{2},\frac{3\pi}{2}\right]$.

- Graph the *guide function* $y=3\sin(2x-\pi)$ on $\left[\frac{\pi}{2},\frac{3\pi}{2}\right]$ using the approach discussed earlier in Chapter 6.

  x-intercepts: Occur when $2x-\pi=0,\ \pi,\ 2\pi$, so that $x=\frac{\pi}{2},\ \pi,\ \frac{3\pi}{2}$

  - At each x-intercept of this guide function, draw a vertical asymptote.
  - On the intervals where this guide function is above the x-axis, draw a $\cup$-shaped graph with vertex at the maximum of the guide function, opening upward toward the asymptotes.
  - On the intervals where this guide function is below the x-axis, draw a $\cap$-shaped graph with vertex at the minimum of the guide function, opening downward toward the asymptotes.

The graph of the guide function (dotted) and the function $y=3\csc(2x-\pi)$ on $\left[\frac{\pi}{2},\frac{3\pi}{2}\right]$ is as follows:

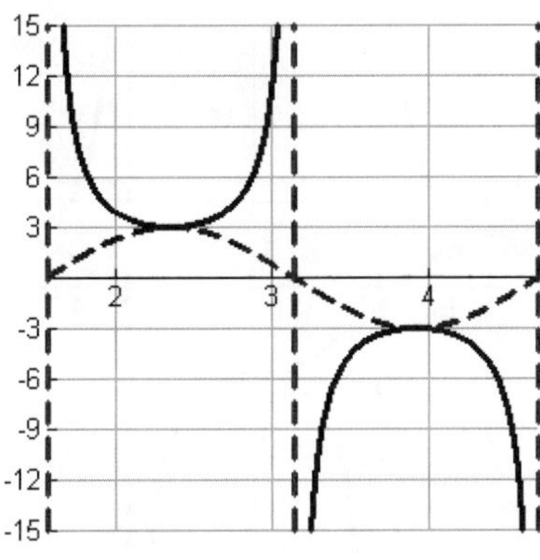

Step 2: Translate the graph is Step 1 vertically down 2 units to obtain the graph of $y=-2+3\csc(2x-\pi)$ on $\left[\frac{\pi}{2},\frac{3\pi}{2}\right]$ as shown below:

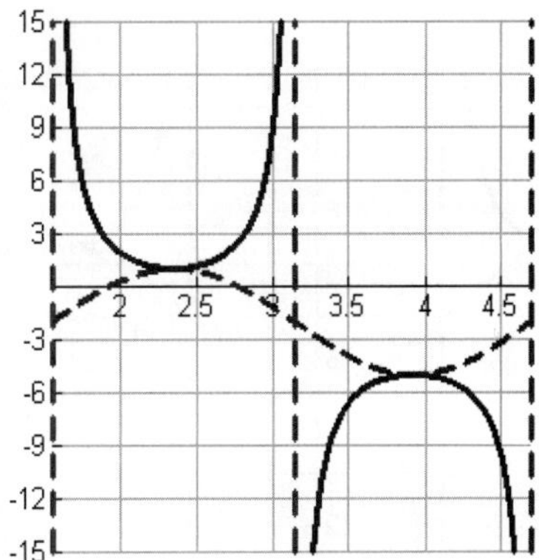

52. First, note that the period is π and the phase shift is $\frac{\pi}{2}$ units to the left. Hence, one period of this function would occur on the interval $\left[-\frac{\pi}{2},\frac{\pi}{2}\right]$, for instance.

In order to graph $y=-1+4\sec(2x+\pi)$ on $\left[-\frac{\pi}{2},\frac{\pi}{2}\right]$, proceed as described below:

Step 1: Graph $y=4\sec(2x+\pi)$ on $\left[-\frac{\pi}{2},\frac{\pi}{2}\right]$.

- Graph the *guide function* $y=4\cos(2x+\pi)$ on $\left[-\frac{\pi}{2},\frac{\pi}{2}\right]$ using the approach discussed earlier in Chapter 6.

  x-intercepts: Occur when $2x+\pi=\frac{\pi}{2},\frac{3\pi}{2}$, so that $x=-\frac{\pi}{4},\frac{\pi}{4}$

  - At each x-intercept of this guide function, draw a vertical asymptote.
  - On the intervals where this guide function is above the x-axis, draw a $\cup$ – shaped graph with vertex at the maximum of the guide function, opening upward toward the asymptotes.
  - On the intervals where this guide function is below the x-axis, draw a $\cap$ – shaped graph with vertex at the minimum of the guide function, opening downward toward the asymptotes.

The graph of the guide function (dotted) and the function $y=4\sec(2x+\pi)$ on $\left[-\frac{\pi}{2},\frac{\pi}{2}\right]$ is as follows:

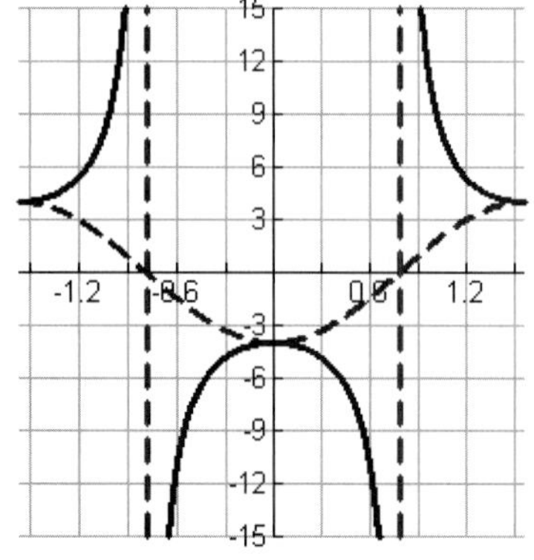

Step 2: Translate the graph is Step 1 vertically down 1 unit to obtain the graph of $y=-1+4\sec(2x+\pi)$ on $\left[-\frac{\pi}{2},\frac{\pi}{2}\right]$ as shown below:

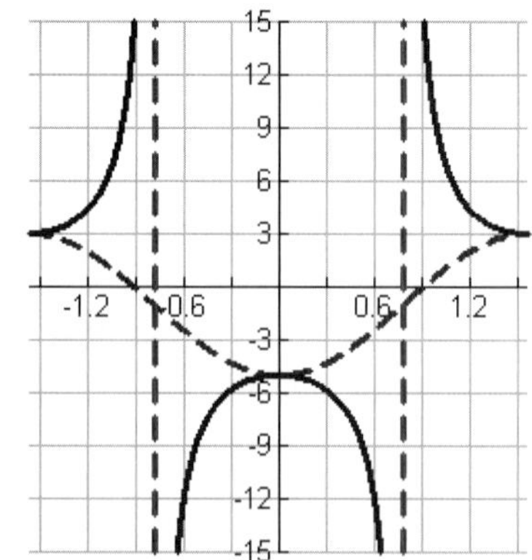

53. First, note that the period is $\frac{2\pi}{\frac{1}{2}} = 4\pi$ and the phase shift is $\frac{\pi}{2}$ units to the right.

Hence, one period of this function would occur on the interval $\left[\frac{\pi}{2}, \frac{9\pi}{2}\right]$, for instance.

In order to graph $y = -1 - \sec\left(\frac{1}{2}x - \frac{\pi}{4}\right)$ on $\left[\frac{\pi}{2}, \frac{9\pi}{2}\right]$, proceed as described below:

Step 1: Graph $y = -\sec\left(\frac{1}{2}x - \frac{\pi}{4}\right)$ on $\left[\frac{\pi}{2}, \frac{9\pi}{2}\right]$:

- Graph the *guide function* $y = -\cos\left(\frac{1}{2}x - \frac{\pi}{4}\right)$ on $\left[\frac{\pi}{2}, \frac{9\pi}{2}\right]$ using the approach discussed earlier in Chapter 6.

  x-intercepts: Occur when $\frac{1}{2}x - \frac{\pi}{4} = \pm\frac{\pi}{2}, \pm\frac{3\pi}{2}$, so that $x = \frac{3\pi}{2}, \frac{7\pi}{2}$

  - At each x-intercept of this guide function, draw a vertical asymptote.
  - On the intervals where this guide function is above the x-axis, draw a $\cup$ – shaped graph with vertex at the maximum of the guide function, opening upward toward the asymptotes.
  - On the intervals where this guide function is below the x-axis, draw a $\cap$ – shaped graph with vertex at the minimum of the guide function, opening downward toward the asymptotes.

The graph of the guide function (dotted) and the function $y = -\sec\left(\frac{1}{2}x - \frac{\pi}{4}\right)$ on $\left[\frac{\pi}{2}, \frac{9\pi}{2}\right]$ is as follows:

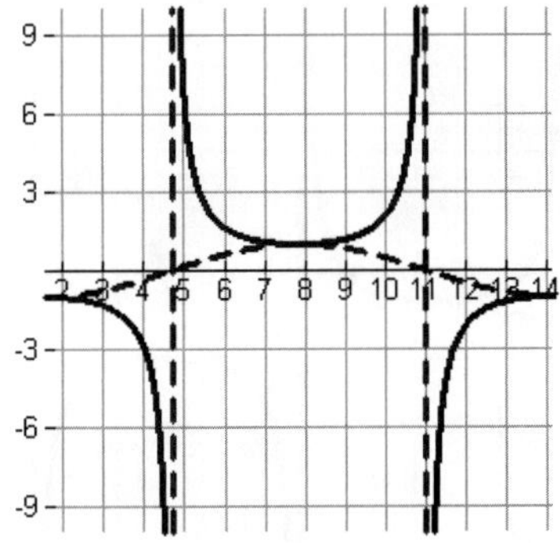

Step 2: Translate the graph is Step 1 vertically down 1 unit to obtain the graph of $y = -1 - \sec\left(\frac{1}{2}x - \frac{\pi}{4}\right)$ on $\left[\frac{\pi}{2}, \frac{9\pi}{2}\right]$ as shown below:

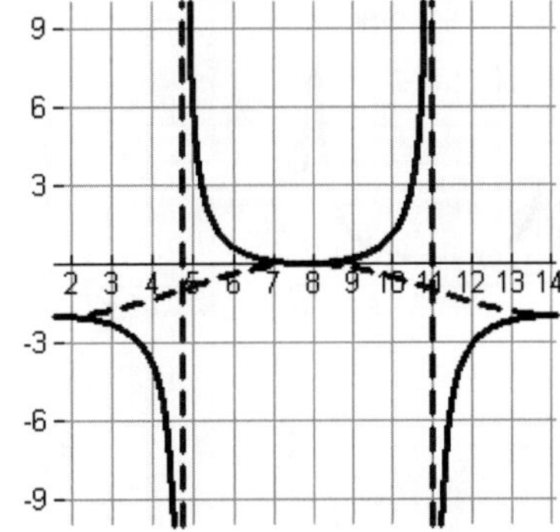

54. First, note that the period is $\frac{2\pi}{1/2} = 4\pi$ and the phase shift is $\frac{\pi}{2}$ units to the left.

Hence, one period of this function would occur on the interval $\left[-\frac{\pi}{2}, \frac{7\pi}{2}\right]$, for instance.

In order to graph $y = -2 + \csc\left(\frac{1}{2}x + \frac{\pi}{4}\right)$ on $\left[-\frac{\pi}{2}, \frac{7\pi}{2}\right]$, proceed as described below:

Step 1: Graph $y = \csc\left(\frac{1}{2}x + \frac{\pi}{4}\right)$ on $\left[-\frac{\pi}{2}, \frac{7\pi}{2}\right]$.

- Graph the *guide function* $y = \sin\left(\frac{1}{2}x + \frac{\pi}{4}\right)$ on $\left[-\frac{\pi}{2}, \frac{7\pi}{2}\right]$ using the approach discussed earlier in Chapter 6.

  x-intercepts: Occur when $\frac{1}{2}x + \frac{\pi}{4} = 0,\ \pi,\ 2\pi$, so that $x = -\frac{\pi}{2}, \frac{3\pi}{2}, \frac{7\pi}{2}$

  - At each x-intercept of this guide function, draw a vertical asymptote.
  - On the intervals where this guide function is above the x-axis, draw a $\cup$–shaped graph with vertex at the maximum of the guide function, opening upward toward the asymptotes.
  - On the intervals where this guide function is below the x-axis, draw a $\cap$–shaped graph with vertex at the minimum of the guide function, opening downward toward the asymptotes.

The graph of the guide function (dotted) and the function $y = \csc\left(\frac{1}{2}x + \frac{\pi}{4}\right)$ on $\left[-\frac{\pi}{2}, \frac{7\pi}{2}\right]$ is as follows:

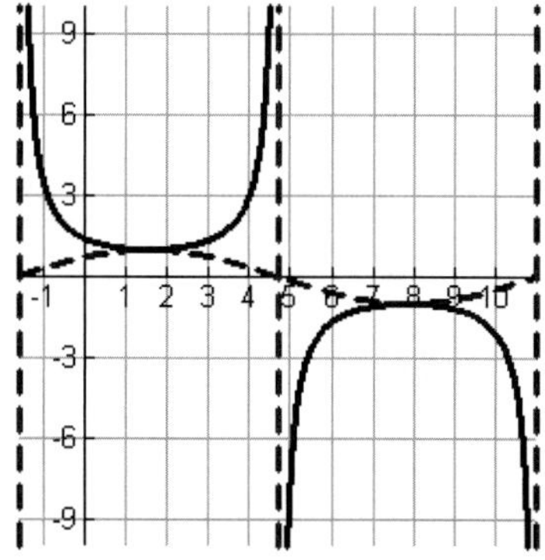

Step 2: Translate the graph is Step 1 vertically down 2 units to obtain the graph of $y = -2 + \csc\left(\frac{1}{2}x + \frac{\pi}{4}\right)$ on $\left[-\frac{\pi}{2}, \frac{7\pi}{2}\right]$ as shown below:

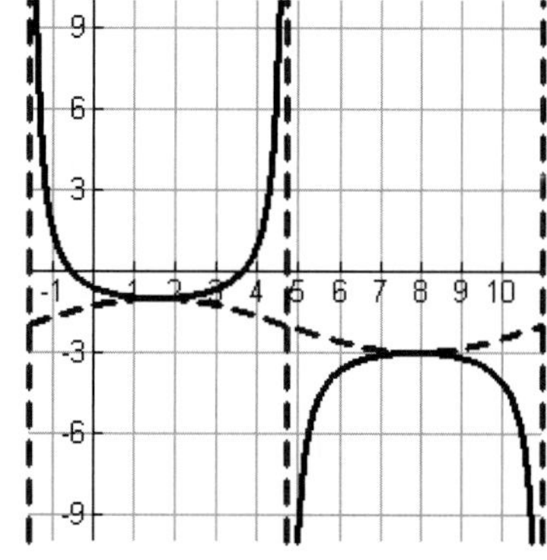

55. We have the following information about the graph of $y=-2-3\cot\left(2x-\frac{\pi}{4}\right)$:

Period:	Vertical Translation:	Reflection about x-axis?	Phase Shift:
$\frac{\pi}{2}$	Down 2 units	Yes	Right $\frac{\pi}{8}$ units

Step 1: Graph $y=3\cot\left(2x-\frac{\pi}{4}\right)$ on $[-\pi,\pi]$.

We have the following information about this particular graph

x-intercepts occur when $\cos\left(2x-\frac{\pi}{4}\right)=0$. This happens when $2x-\frac{\pi}{4}=\pm\frac{\pi}{2},\pm\frac{3\pi}{2}$ so that $x=-\frac{5\pi}{8},-\frac{\pi}{8},\frac{3\pi}{8},\frac{7\pi}{8}$. (Note: This will change in Step 2 when we translate this graph vertically.)
Vertical asymptotes occur when $\sin\left(2x-\frac{\pi}{4}\right)=0$. This happens when $2x-\frac{\pi}{4}=0,\pm\pi,-2\pi$, so that $x=-\frac{3\pi}{8},\frac{\pi}{8},\frac{5\pi}{8}$. (Note: These remain unchanged by vertical translation.)

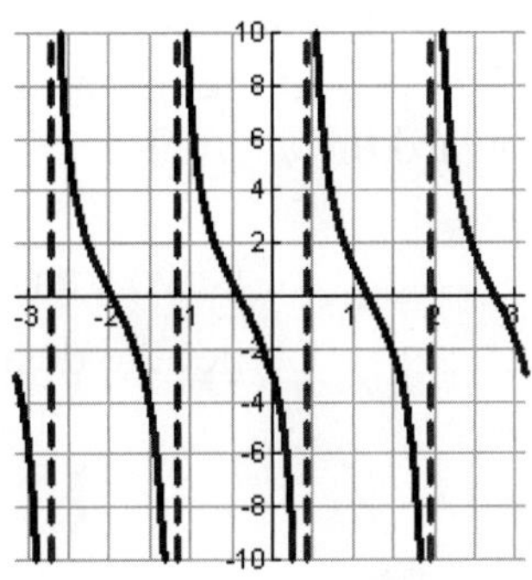

Now, reflect this graph over the x-axis to obtain:

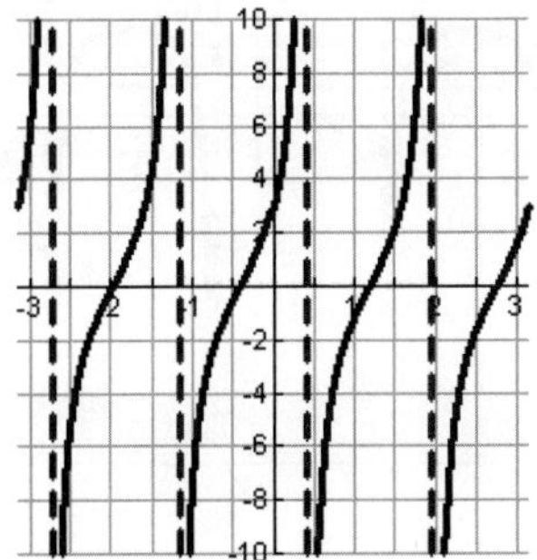

Finally, translate the graph to the left down 2 units:

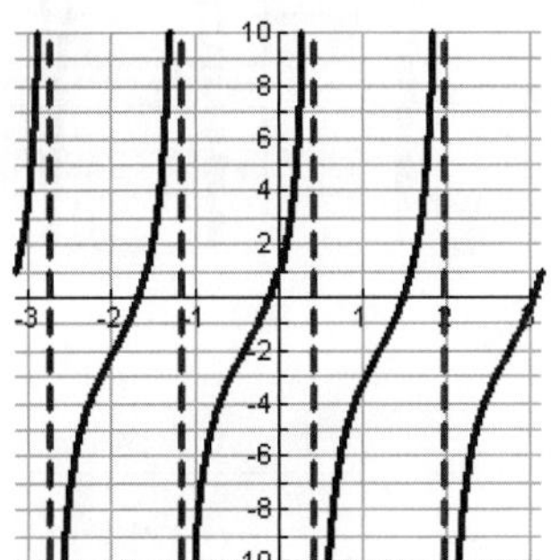

56. First, note that the period is $\frac{2\pi}{\pi}=2$ and the phase shift is $\frac{1}{2}$ units to the left.

Hence, one period of this function would occur on the interval $\left[-\frac{1}{2},\frac{3}{2}\right]$, for instance.

In order to graph $y=-\frac{1}{4}+\frac{1}{2}\sec\left(\pi x+\frac{\pi}{4}\right)$ on $[-2,2]$, proceed as described below:

<u>Step 1</u>: Graph $y=\frac{1}{2}\sec\left(\pi x+\frac{\pi}{4}\right)$ on $[-2,2]$.

- Graph the *guide function* $y=\frac{1}{2}\cos\left(\pi x+\frac{\pi}{4}\right)$ on $[-2,2]$ using the approach discussed earlier in Chapter 6.

  <u>x-intercepts</u>: Occur when $\pi x+\frac{\pi}{4}=\pm\frac{\pi}{2}, \pm\frac{3\pi}{2}$, so that $x=-\frac{7}{4}, -\frac{3}{4}, \frac{1}{4}, \frac{5}{4}$

  - At each x-intercept of this guide function, draw a vertical asymptote.
  - On the intervals where this guide function is <u>above</u> the x-axis, draw a $\cup$ – shaped graph with vertex at the maximum of the guide function, opening upward toward the asymptotes.
  - On the intervals where this guide function is <u>below</u> the x-axis, draw a $\cap$ – shaped graph with vertex at the minimum of the guide function, opening downward toward the asymptotes.

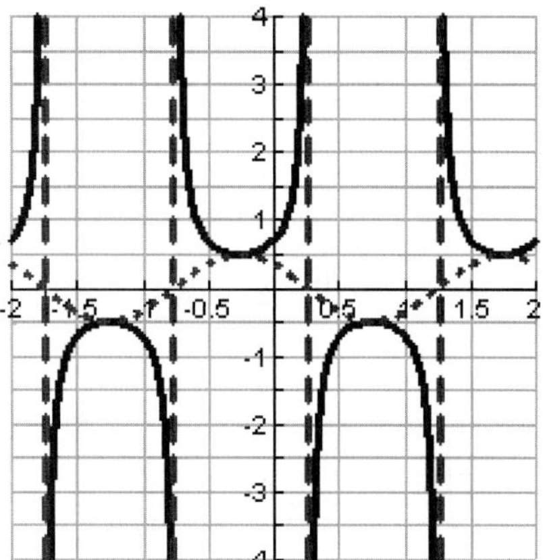

Now, translate this graph down $\frac{1}{4}$ unit to get the desired graph:

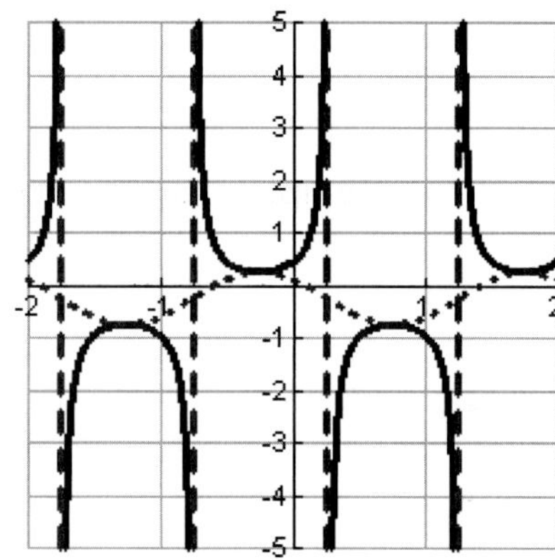

57. Consider the function $y = \tan\left(\pi x - \frac{\pi}{2}\right)$.

Domain: All real numbers x for which $\cos\left(\pi x - \frac{\pi}{2}\right) \neq 0$. Observe that

$\cos\left(\pi x - \frac{\pi}{2}\right) = 0$ precisely when

$$\pi x - \frac{\pi}{2} = (2n+1)\frac{\pi}{2}$$
$$\pi x = \frac{\pi}{2} + (2n+1)\frac{\pi}{2}$$
$$x = \frac{1}{\pi}\left[\frac{\pi}{2} + (2n+1)\frac{\pi}{2}\right] = \frac{1}{2}(2n+2) = n+1,$$

where n is an integer. Hence, the domain is the set of all real numbers x such that $x \neq n$, where n is an integer.

Range: All real numbers.

58. Consider the function $y = \cot\left(x - \frac{\pi}{2}\right)$.

Domain: All real numbers x for which $\sin\left(x - \frac{\pi}{2}\right) \neq 0$. Observe that $\sin\left(x - \frac{\pi}{2}\right) = 0$ precisely when $x - \frac{\pi}{2} = n\pi$, so that $x = \frac{\pi}{2} + n\pi = \frac{\pi}{2}(1+2n)$, where n is an integer.

Hence, the domain is the set of all real numbers x such that $x \neq \frac{2n+1}{2}\pi$, where n is an integer.

Range: All real numbers.

59. Consider the function $y = 2\sec(5x)$.

Domain: All real numbers x for which $\cos(5x) \neq 0$. Observe that $\cos(5x) = 0$ precisely when $5x = (2n+1)\frac{\pi}{2}$, so that $x = \frac{2n+1}{10}\pi$, where n is an integer. Hence, the domain is the set of all real numbers x such that $x \neq \frac{2n+1}{10}\pi$, where n is an integer.

Range: Note that the maximum and minimum y-values for the guide function $y = 2\cos(5x)$ are 2 and –2, respectively. Hence, the range is $(-\infty,-2]\cup[2,\infty)$.

60. Consider the function $y = -4\sec(3x)$.

Domain: All real numbers x for which $\cos(3x) \neq 0$. Observe that $\cos(3x) = 0$ precisely when $3x = (2n+1)\frac{\pi}{2}$, so that $x = \frac{2n+1}{6}\pi$, where n is an integer. Hence, the domain is the set of all real numbers x such that $x \neq \frac{2n+1}{6}\pi$, where n is an integer.

Range: Note that the maximum and minimum y-values for the guide function $y = 4\cos(3x)$ are 4 and –4, respectively. Hence, the range is $(-\infty,-4]\cup[4,\infty)$.

61. Consider the function $y = 2-\csc\left(\frac{1}{2}x-\pi\right)$.

Domain: All real numbers x for which $\sin\left(\frac{1}{2}x-\pi\right) \neq 0$. Observe that $\sin\left(\frac{1}{2}x-\pi\right) = 0$ precisely when $\frac{1}{2}x-\pi = n\pi$, so that $x = \underbrace{2(n+1)\pi}_{\text{Any even multiple of }\pi}$ where n is an integer. Hence, the domain is the set of all real numbers x such that $x \neq 2n\pi$, where n is an integer.

Range: Note that the maximum and minimum y-values for the guide function $y = 2-\sin\left(\frac{1}{2}x-\pi\right)$ are 3 and 1, respectively. Hence, the range is $(-\infty,1]\cup[3,\infty)$.

62. Consider the function $y = 1-2\sec\left(\frac{1}{2}x+\pi\right)$.

Domain: All real numbers x for which $\cos\left(\frac{1}{2}x+\pi\right) \neq 0$. Observe that $\cos\left(\frac{1}{2}x+\pi\right) = 0$ precisely when $\frac{1}{2}x+\pi = \frac{2n+1}{2}\pi$, so that $x = \underbrace{(2n-1)\pi}_{\text{Any odd multiple of }\pi}$ where n is an integer.

Hence, the domain is the set of all real numbers x such that $x \neq (2n-1)\pi$, where n is an integer.

Range: Note that the maximum and minimum y-values for the guide function $y = 1-2\cos\left(\frac{1}{2}x+\pi\right)$ are 3 and – 1, respectively. Hence, the range is $(-\infty,-1]\cup[3,\infty)$.

63. Consider the function $y = -3\tan\left(\frac{\pi}{4}x - \pi\right) + 1$.

Domain: All real numbers x for which $\cos\left(\frac{\pi}{4}x - \pi\right) \neq 0$. Observe that

$\cos\left(\frac{\pi}{4}x - \pi\right) = 0$ precisely when

$$\frac{\pi}{4}x - \pi = (2n+1)\frac{\pi}{2}$$
$$\frac{\pi}{4}x = \pi + (2n+1)\frac{\pi}{2}$$
$$x = \frac{4}{\pi}\left[\pi + (2n+1)\frac{\pi}{2}\right] = 4\left[1 + \tfrac{1}{2}(2n+1)\right] = 4n+6,$$

where n is an integer. Hence, the domain is the set of all real numbers x such that $x \neq 4n+6$, where n is an integer.
Range: All real numbers.

64. Consider the function $y = \frac{1}{4}\cot\left(2\pi x + \frac{\pi}{3}\right) - 3$.

Domain: All real numbers x for which $\sin\left(2\pi x + \frac{\pi}{3}\right) \neq 0$. Observe that

$\sin\left(2\pi x + \frac{\pi}{3}\right) = 0$ precisely when $2\pi x + \frac{\pi}{3} = n\pi$, so that $x = \frac{1}{2\pi}\left(n\pi - \frac{\pi}{3}\right) = \frac{3n-1}{6}$
where n is an integer. Hence, the domain is the set of all real numbers x such that $x \neq \frac{3n-1}{6}$, where n is an integer.
Range: All real numbers.

65. Consider the function $y = -2 + \frac{1}{2}\sec\left(\pi x + \frac{\pi}{2}\right)$.

Domain: All real numbers x for which $\cos\left(\pi x + \frac{\pi}{2}\right) \neq 0$. Observe that $\cos\left(\pi x + \frac{\pi}{2}\right) = 0$ precisely when $\pi x + \frac{\pi}{2} = \frac{2n+1}{2}\pi$, so that $x = n$ where n is an integer. Hence, the domain is the set of all real numbers x such that $x \neq n$, where n is an integer.
Range: Note that the maximum and minimum y-values for the guide function $y = -2 + \frac{1}{2}\cos\left(\pi x + \frac{\pi}{2}\right)$ are $-\frac{3}{2}$ and $-\frac{5}{2}$, respectively. Hence, the range is $\left(-\infty, -\frac{5}{2}\right] \cup \left[-\frac{3}{2}, \infty\right)$.

66. Consider the function $y = \frac{1}{2} - \frac{1}{3}\csc\left(3x - \frac{\pi}{2}\right)$.

Domain: All real numbers x for which $\sin\left(3x - \frac{\pi}{2}\right) \neq 0$. Observe that $\sin\left(3x - \frac{\pi}{2}\right) = 0$ precisely when $3x - \frac{\pi}{2} = n\pi$, so that $x = \frac{(2n+1)\pi}{6}$ where n is an integer. Hence, the domain is the set of all real numbers x such that $x \neq \frac{(2n+1)\pi}{6}$, where n is an integer.

Range: Note that the maximum and minimum y-values for the guide function $y = \frac{1}{2} - \frac{1}{3}\sin\left(3x - \frac{\pi}{2}\right)$ are $\frac{5}{6}$ and $\frac{1}{6}$, respectively. Hence, the range is $\left(-\infty, \frac{1}{6}\right] \cup \left[\frac{5}{6}, \infty\right)$.

67. Consider the triangle:

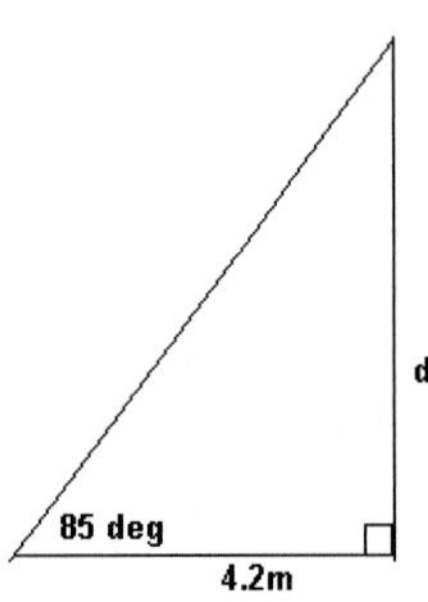

$$\tan 85^\circ = \frac{d}{4.2m}$$

$$d = \left(\tan 85^\circ\right)\left(4.2m\right) \approx 48m$$

68. Consider the diagram:

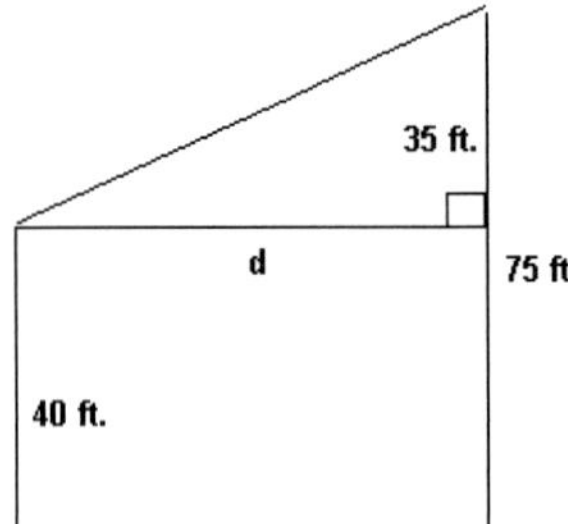

$$\tan \tfrac{\pi}{6} = \frac{35\ ft.}{d}$$

$$d = \frac{35\ ft.}{\tan \frac{\pi}{6}} \approx 60.6\ ft.$$

69. **a.)** $x\left(\frac{2}{3}\right) = 3\tan\left(\frac{2\pi}{3}\right) = 3\frac{\frac{\sqrt{3}}{2}}{-\frac{1}{2}} = -3\sqrt{3} \approx -5.2\,mi.$

b.) $x\left(\frac{3}{4}\right) = 3\tan\left(\frac{3\pi}{4}\right) = -3\,mi.$ **c.)** $x(1) = 3\tan(\pi) = 0$

d.) $x\left(\frac{5}{4}\right) = 3\tan\left(\frac{5\pi}{4}\right) = 3\,mi.$ **e.)** $x\left(\frac{4}{3}\right) = 3\tan\left(\frac{4\pi}{3}\right) = 3\frac{-\frac{\sqrt{3}}{2}}{-\frac{1}{2}} = 3\sqrt{3} \approx 5.2\,mi.$

70. .**a)** $y\left(\frac{2}{3}\right)=3\left|\sec\left(\frac{2\pi}{3}\right)\right|=\dfrac{3}{\left|\cos\left(\frac{2\pi}{3}\right)\right|}=6\ mi.$

b) $y\left(\frac{3}{4}\right)=3\left|\sec\left(\frac{3\pi}{4}\right)\right|=\dfrac{3}{\left|\cos\left(\frac{3\pi}{4}\right)\right|}=3\sqrt{2}\approx 4.2\,mi.$

c) $y(1)=3\left|\sec(\pi)\right|=\dfrac{3}{\left|\cos(\pi)\right|}=3\,mi.$

d) $y\left(\frac{5}{4}\right)=3\left|\sec\left(\frac{5\pi}{4}\right)\right|=\dfrac{3}{\left|\cos\left(\frac{5\pi}{4}\right)\right|}=3\sqrt{2}\approx 4.2\,mi.$

e) $y\left(\frac{4}{3}\right)=3\left|\sec\left(\frac{4\pi}{3}\right)\right|=\dfrac{3}{\left|\cos\left(\frac{4\pi}{3}\right)\right|}=6\ mi.$

71. Forgot that the amplitude is 3, not 1. So, the guide function should have been $y=3\sin(2x)$.

72. The vertical asymptotes for $y=\tan x$ occur at $x=\pm\dfrac{\pi}{2}$, while the x-intercept occurs at $x=0$. Hence, the quantities used in Steps 2 and 3 are incorrect. (In fact, they coincide with those for cotangent.)

73. True. This is a consequence of the fact that $\cos\left(x-\dfrac{\pi}{2}\right)=\sin x$. Indeed, observe that $\sec\left(x-\dfrac{\pi}{2}\right)=\dfrac{1}{\cos\left(x-\dfrac{\pi}{2}\right)}=\dfrac{1}{\sin x}=\csc x$.

74. False. This is a consequence of the fact that $\sin\left(x-\dfrac{\pi}{2}\right)\neq\cos x$. For instance, for $x=\pi$, $\sin\left(\pi-\dfrac{\pi}{2}\right)=\sin\left(\dfrac{\pi}{2}\right)=1$, while $\cos\pi=-1$. So, in general, $\dfrac{1}{\sin\left(x-\dfrac{\pi}{2}\right)}\neq\dfrac{1}{\cos x}$, so that $\csc\left(x-\dfrac{\pi}{2}\right)\neq\sec x$, in general.

75. Since $y=\tan x$ has period π, we know by definition that $\tan x=\tan(x-n\pi)$, where n is an integer. So, any *integer* value n gives equality.

76. Since $y=\csc x$ has period 2π, we know by definition that $\csc x=\csc(x-2k\pi)$, where k is an integer. So, if $n=2k$ (i.e., n is any even integer) the equality holds.

77. Consider the following graph of $y = \tan(2x - \pi)$:

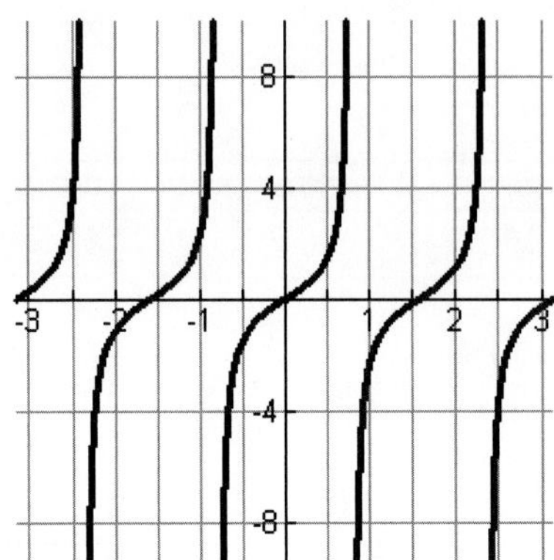

The solutions of the equation $\tan(2x - \pi) = 0$ are the x-intercepts of this graph. These values of x are $0, \pm\frac{\pi}{2}, \pm\pi$.

78. Consider the following graph of $y = \csc(2x + \pi)$:

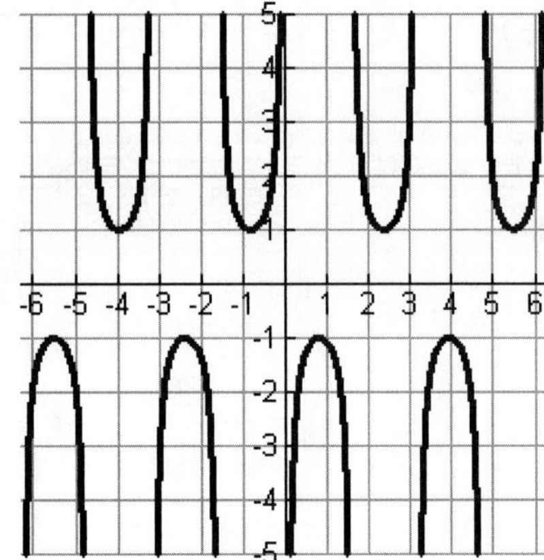

The solutions of the equation $\csc(2x + \pi) = 0$ would be the x-intercepts of this graph. However, there are no such intercepts, so that this equation has no solution.

79. The x-intercepts occur when $\sin(Bx + C) = 0$. This occurs when

$$Bx + C = n\pi$$

$$x = \frac{n\pi - C}{B},$$

where n is an integer.

80. $y = -A\sec\left(\frac{\pi}{2}x\right) > 0$ when

$$\frac{-A}{\cos\left(\frac{\pi}{2}x\right)} > 0 \Rightarrow \cos\left(\frac{\pi}{2}x\right) < 0.$$

This occurs when $\frac{\pi}{2}x$ is in QII or QIII, which is equivalent to

$$\frac{\pi}{2} + 2n\pi < \frac{\pi}{2}x < \frac{3\pi}{2} + 2n\pi$$

$$2 + 4n < x < 3 + 4n,$$

where n is an integer.

81. There are infinitely many solutions of this equation since tangent is periodic and extends to infinity in abutting intervals of length π. Since $y = x$ is increasing, its graph intersects that of $y = \tan(x)$ in every such interval, and thus infinitely often.

82. Since the range of $y = -2\csc\left(\frac{\pi}{6}x - \pi\right)$ is $(-\infty, -2] \cup [2, \infty)$, we note that if $|A| < 2$, then the graph of $y = A\sin(Bx + C)$ never intersects this graph.

83. Consider the graphs of the following functions on the interval $[-2\pi, 2\pi]$:

$y_1 = \cos x$ (dotted bold),
$y_2 = \sin x$ (dashed bold),
$y_3 = \cos x + \sin x$ (solid bold)

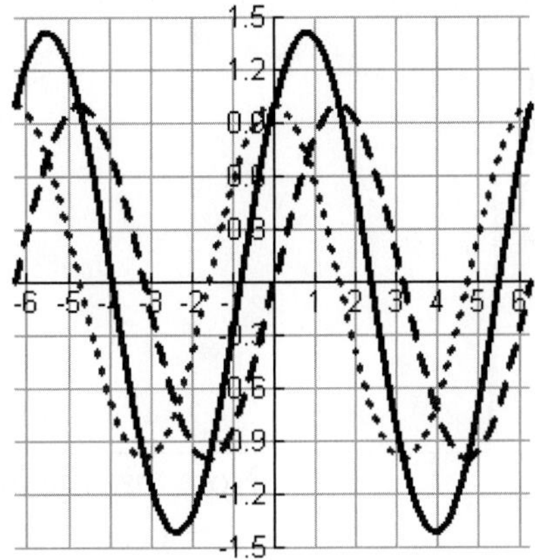

In order to determine the amplitude of y_3, we need to determine its maximum and minimum y-values. From the graph, observe that both the maximum and minimum values of y_3 occur at the x-values where y_1 and y_2 intersect. But, $\sin x = \cos x$ when $x = \frac{\pi}{4}, \frac{5\pi}{4}$. As such,

$$y_3\left(\frac{\pi}{4}\right) = \cos\left(\frac{\pi}{4}\right) + \sin\left(\frac{\pi}{4}\right) = \frac{\sqrt{2}}{2} + \frac{\sqrt{2}}{2} = \sqrt{2}$$

$$y_3\left(\frac{5\pi}{4}\right) = \cos\left(\frac{5\pi}{4}\right) + \sin\left(\frac{5\pi}{4}\right) = -\left[\frac{\sqrt{2}}{2} + \frac{\sqrt{2}}{2}\right] = -\sqrt{2}$$

So, the amplitude of y_3 is $\dfrac{\text{maximum of } y_3 - \text{minimum of } y_3}{2} = \boxed{\sqrt{2}}$.

84. Consider the graphs of the following functions on the interval $[-2\pi, 2\pi]$: $y_1 = \cos x + \sin x$ (dashed bold), $y_2 = \sec x + \csc x$ (solid bold) 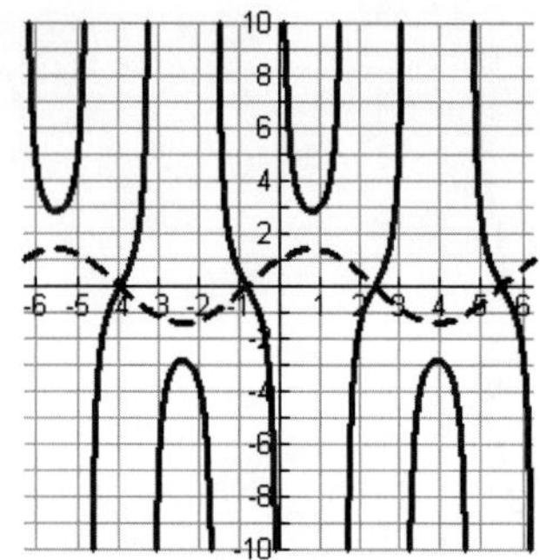	Notice that y_1 is not the guide function of y_2.
85. Below are the graphs of $y_1 = \tan x + \cot x$ $y_2 = 2\csc 2x$ 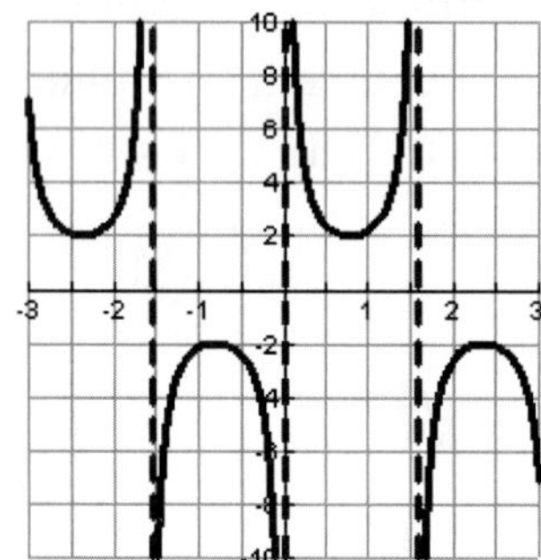	Note that the graphs coincide since $y_1 = y_2$. Observe that the period of both graphs is $\boxed{\pi}$.
86. Below are the graphs of $y_1 = \tan\left(2x + \frac{\pi}{2}\right)$ (solid) $y_2 = \tan\left(2x - \frac{\pi}{2}\right)$ $y_3 = \tan\left(-2x + \frac{\pi}{2}\right)$ (dotted) 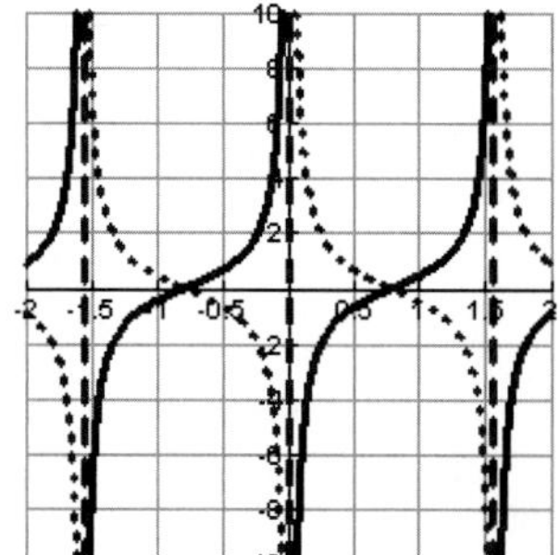	Note that the graphs of y_1 and y_2 coincide, while the graph of y_3 is the reflection of the graph of y_2 (and hence of y_1) over the x-axis.

87.

$$\int_0^{\frac{\pi}{4}} \tan x dx = \left(-\ln\left|\cos\tfrac{\pi}{4}\right|\right) - \left(-\ln\left|\cos 0\right|\right)$$
$$= \left(-\ln\tfrac{\sqrt{2}}{2}\right) + \ln 1 = -\ln\tfrac{\sqrt{2}}{2}$$

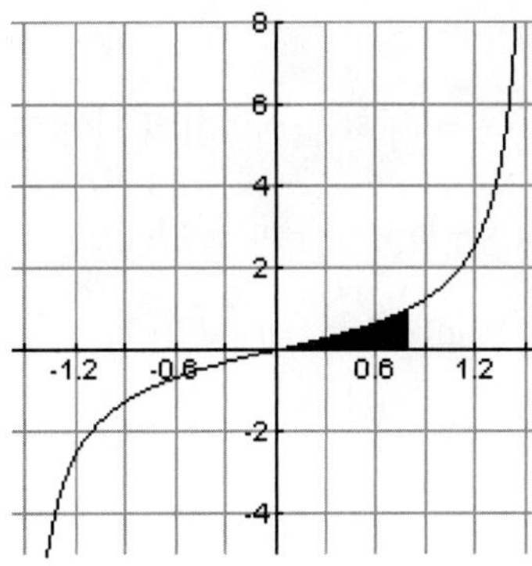

88.

$$\int_{\frac{\pi}{4}}^{\frac{\pi}{2}} \cot x dx = \left(\ln\left|\sin\tfrac{\pi}{2}\right|\right) - \left(\ln\left|\sin\tfrac{\pi}{4}\right|\right)$$
$$= \ln 1 - \ln\tfrac{\sqrt{2}}{2} = -\ln\tfrac{\sqrt{2}}{2}$$

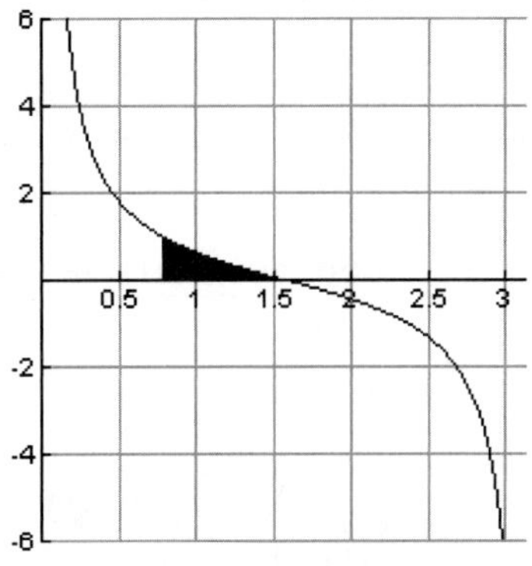

89.

$$\int_0^{\frac{\pi}{4}} \sec x dx = \ln\left|\sec\tfrac{\pi}{4} + \tan\tfrac{\pi}{4}\right| - \ln\left|\sec 0 + \tan 0\right|$$
$$= \ln\left|\frac{1}{\frac{\sqrt{2}}{2}} + 1\right| - \ln\left|1 + 0\right|$$
$$= \ln\left(\tfrac{2\sqrt{2}+2}{2}\right) = \ln\left(\sqrt{2} + 1\right)$$

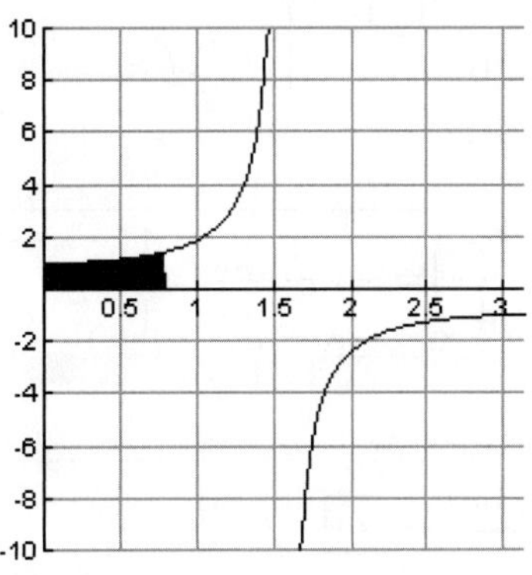

90.

$$\int_{\frac{\pi}{4}}^{\frac{\pi}{2}} \csc x\,dx = -\ln\left|\csc\tfrac{\pi}{2}+\cot\tfrac{\pi}{2}\right| - \left(-\ln\left|\csc\tfrac{\pi}{4}+\cot\tfrac{\pi}{4}\right|\right)$$

$$= -\ln|1+0| + \ln\left|\tfrac{1}{\frac{\sqrt{2}}{2}}+1\right|$$

$$= \ln\left(\tfrac{2\sqrt{2}+2}{2}\right) = \ln\left(\sqrt{2}+1\right)$$

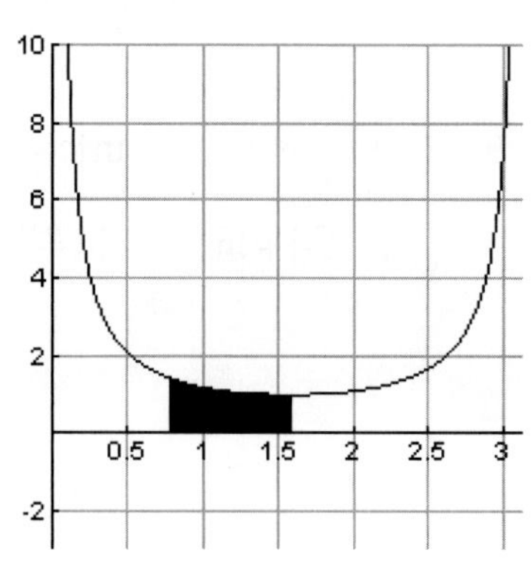

Chapter 5 Review Solutions

1. $\tan\left(\frac{5\pi}{6}\right) = \dfrac{\sin\left(\frac{5\pi}{6}\right)}{\cos\left(\frac{5\pi}{6}\right)} = \dfrac{\frac{1}{2}}{-\frac{\sqrt{3}}{2}} = \dfrac{-1}{\sqrt{3}} = \boxed{\dfrac{-\sqrt{3}}{3}}$	**2.** $\cos\left(\frac{5\pi}{6}\right) = \boxed{-\dfrac{\sqrt{3}}{2}}$
3. $\sin\left(\frac{11\pi}{6}\right) = \boxed{-\dfrac{1}{2}}$	**4.** $\sec\left(\frac{11\pi}{6}\right) = \dfrac{1}{\cos\left(\frac{11\pi}{6}\right)} = \dfrac{1}{\frac{\sqrt{3}}{2}} = \dfrac{2}{\sqrt{3}} = \dfrac{2\sqrt{3}}{3}$
5. $\cot\left(\frac{5\pi}{4}\right) = \dfrac{1}{\tan\left(\frac{5\pi}{4}\right)} = \dfrac{1}{1} = \boxed{1}$	**6.** $\csc\left(\frac{5\pi}{4}\right) = \dfrac{1}{\sin\left(\frac{5\pi}{4}\right)} = \dfrac{1}{-\frac{\sqrt{2}}{2}} = \dfrac{-2}{\sqrt{2}} = \boxed{-\sqrt{2}}$
7. $\sin\left(\frac{3\pi}{2}\right) = \boxed{-1}$	**8.** $\cos\left(\frac{3\pi}{2}\right) = \boxed{0}$
9. $\cos(\pi) = \boxed{-1}$	**10.** $\tan\left(\frac{7\pi}{4}\right) = \dfrac{\sin\left(\frac{7\pi}{4}\right)}{\cos\left(\frac{7\pi}{4}\right)} = \dfrac{-\frac{\sqrt{2}}{2}}{\frac{\sqrt{2}}{2}} = \boxed{-1}$
11. $\cos\left(\frac{\pi}{3}\right) = \boxed{\dfrac{1}{2}}$	**12.** $\sin\left(\frac{11\pi}{6}\right) = \boxed{-\dfrac{1}{2}}$

13. $\sin\left(-\frac{7\pi}{4}\right)=\boxed{\frac{\sqrt{2}}{2}}$	**14.** $\tan\left(-\frac{2\pi}{3}\right)=\frac{-\frac{\sqrt{3}}{2}}{-\frac{1}{2}}=\boxed{\sqrt{3}}$
15. $\csc\left(-\frac{3\pi}{2}\right)=\frac{1}{\sin\left(-\frac{3\pi}{2}\right)}=\boxed{1}$	**16.** $\cot\left(-\frac{5\pi}{6}\right)=\frac{1}{\tan\left(-\frac{5\pi}{6}\right)}=\boxed{\sqrt{3}}$
17. $\cos\left(-\frac{7\pi}{6}\right)=\boxed{-\frac{\sqrt{3}}{2}}$	**18.** $\sec\left(-\frac{3\pi}{4}\right)=\frac{1}{\cos\left(-\frac{3\pi}{4}\right)}=\boxed{-\sqrt{2}}$
19. $\tan\left(-\frac{13\pi}{6}\right)=\frac{-\frac{1}{2}}{\frac{\sqrt{3}}{2}}=\boxed{\frac{-\sqrt{3}}{3}}$	**20.** $\cos\left(-\frac{14\pi}{3}\right)=\cos\left(-\frac{2\pi}{3}\right)=\boxed{-\frac{1}{2}}$

21. The distance from maximum to next consecutive maximum is 2π. Hence, the period is $\boxed{2\pi}$.

22. Since

$$A=\frac{\text{maximum } y-\text{value } -\text{minimum } y-\text{value}}{2}=\frac{4-(-4)}{2}=4$$

the amplitude is 4.

23. Assume that the equation of the form $y=K+A\cos(Bx+C)$. From Exercises 1 and 2, we know that since the general shape of the curve is that of a non-reflected cosine. So,

- Since the period is 2π, we know that $\frac{2\pi}{B}=2\pi$, so that $B=1$.
- Since the maximum y-value is 4 and the minimum y-value is -4, we have:
  - Vertical shift $K=\frac{4+(-4)}{2}=0$ -- so there is no vertical shift
  - Amplitude $|A|=4$, and in fact, $A=4$ (since there is no reflection).

So, the equation thus far is $y=4\cos(x+C)$. To find C, use the fact that the point $(0,4)$ is on the graph:

$$4=4\cos(\pi(0)+C)$$
$$1=\cos(C)$$

Though this equation has infinitely many solutions, the simplest one is given by $C=0$. Doing so results in the equation $\boxed{y=4\cos x}$.

24. The distance from maximum to next consecutive maximum is π. Hence, the period is $\boxed{\pi}$.

25. Since

$$|A|=\frac{\text{maximum } y-\text{value } -\text{minimum } y-\text{value}}{2}=\frac{5-(-5)}{2}=5,$$

the amplitude is 5.

26. Assume that the equation of the form $y = K + A\sin(Bx + C)$. From Exercises 4 and 5, we know that since the general shape of the curve is that of a reflected sine. So,

- Since the period is π, we know that $\frac{2\pi}{B} = \pi$, so that $B = 2$.
- Since the maximum y-value is 5 and the minimum y-value is -5, we have:
  - Vertical shift $K = \frac{5+(-5)}{2} = 0$ -- so there is no vertical shift
  - Amplitude $|A| = 5$, and in fact, $A = -5$ (to account for the reflection).

So, the equation thus far is $y = -5\sin(2x + C)$. To find C, use the fact that the point $(0,0)$ is on the graph:

$$0 = -5\sin(2(0) + C)$$
$$0 = \sin(C)$$

Though this equation has infinitely many solutions, the simplest one is given by $C = 0$. Doing so results in the equation $\boxed{y = -5\sin 2x}$.

27. Since $A = 2$ and $B = 2\pi$, we conclude that the amplitude is 2 and the period is $p = \frac{2\pi}{2\pi} = 1$.

28. Since $A = \frac{1}{3}$ and $B = \frac{\pi}{2}$, we conclude that the amplitude is $\frac{1}{3}$ and the period is $p = \frac{2\pi}{\pi/2} = 4$.

29. Since $A = \frac{1}{5}$ and $B = 3$, we conclude that the amplitude is $\frac{1}{5}$ and the period is $p = \frac{2\pi}{3}$.

30. Since $A = \frac{7}{6}$ and $B = 6$, we conclude that the amplitude is $\frac{7}{6}$ and the period is $p = \frac{2\pi}{6} = \frac{\pi}{3}$.

31. We have the following information about the graph of $y=-2\sin\left(\frac{1}{2}x\right)$ on $[-2\pi,2\pi]$:

<table>
<tr><td><u>Amplitude</u>: $|-2|=2$</td><td><u>Period</u>: $\frac{2\pi}{1/2}=4\pi$</td></tr>
<tr><td colspan="2"><u>x-intercepts</u> occur when $\sin\left(\frac{1}{2}x\right)=0$. This happens when $\frac{1}{2}x=0,\pm\pi$, and so $x=0,\pm 2\pi$. So, the points $(0,0)$, $(-2\pi,0)$, $(2\pi,0)$ are on the graph.</td></tr>
<tr><td colspan="2">The <u>maximum value</u> of the graph is 2 since the maximum of $\sin\left(\frac{1}{2}x\right)$ is 1, and the amplitude is 2. This occurs when $\frac{1}{2}x=\frac{\pi}{2}$, and so $x=\pi$. So, the point $(\pi,2)$ is on the graph.</td></tr>
<tr><td colspan="2">The <u>minimum value</u> of the graph is – 2 since the minimum of $\sin\left(\frac{1}{2}x\right)$ is – 1, and the amplitude is 2. This occurs when $\frac{1}{2}x=-\frac{\pi}{2}$, and so $x=-\pi$. So, the point $(-\pi,-2)$ is on the graph.</td></tr>
<tr><td colspan="2">Since the coefficient used to compute the amplitude is negative, we reflect the graph of $y=2\sin\left(\frac{1}{2}x\right)$ over the x-axis.</td></tr>
</table>

Using this information, we find that the graph of $y=-2\sin\left(\frac{1}{2}x\right)$ on $[-2\pi,2\pi]$ is given by:

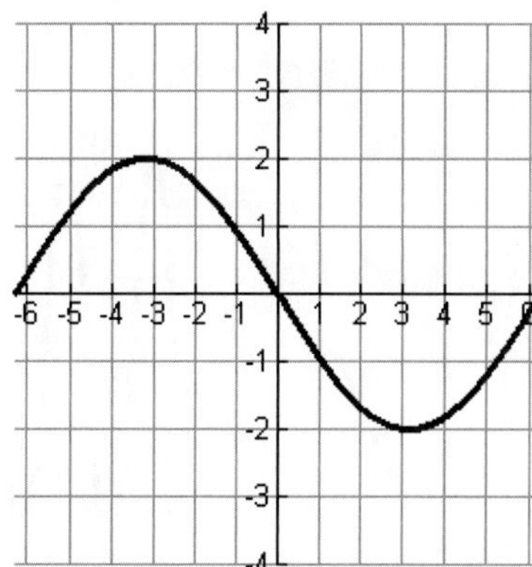

32. In order to construct the graph of $y = 3\sin(3x)$ on $[-2\pi, 2\pi]$, we first gather the information essential to forming the graph for one period, namely on the interval $\left[0, \frac{2\pi}{3}\right]$. Then, extend this picture to cover the entire interval. Indeed, we have the following:

<table>
<tr><td>Amplitude: 3</td><td>Period: $\frac{2\pi}{3}$</td></tr>
<tr><td colspan="2">x-intercepts occur when $\sin 3x = 0$. This happens when $3x = n\pi$, and so $x = \frac{n\pi}{3}$, where n is an integer. So, the points $(0, 0)$, $\left(\frac{\pi}{3}, 0\right)$, $\left(\frac{2\pi}{3}, 0\right)$ are on the graph.</td></tr>
<tr><td colspan="2">The maximum value of the graph is 3 since the maximum of $\sin 3x$ is 1, and the amplitude is 3. This occurs when $3x = \frac{\pi}{2}$, so that $x = \frac{\pi}{6}$. So, the point $\left(\frac{\pi}{6}, 3\right)$ is on the graph.</td></tr>
<tr><td colspan="2">The minimum value of the graph is – 3 since the minimum of $\sin 3x$ is – 1, and the amplitude is 3. This occurs when $3x = \frac{3\pi}{2}$, so that $x = \frac{\pi}{2}$. So, the point $\left(\frac{\pi}{2}, -3\right)$ is on the graph.</td></tr>
<tr><td colspan="2">Since the coefficient used to compute the amplitude is nonnegative, there is no reflection over the x-axis.</td></tr>
</table>

Now, using this information, we can form the graph of $y = 3\sin(3x)$ on $\left[0, \frac{2\pi}{3}\right]$, and then extend it to the interval $[-2\pi, 2\pi]$, as shown below:

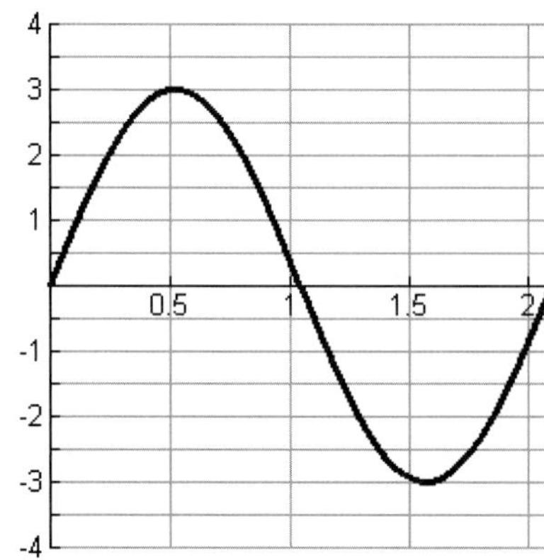

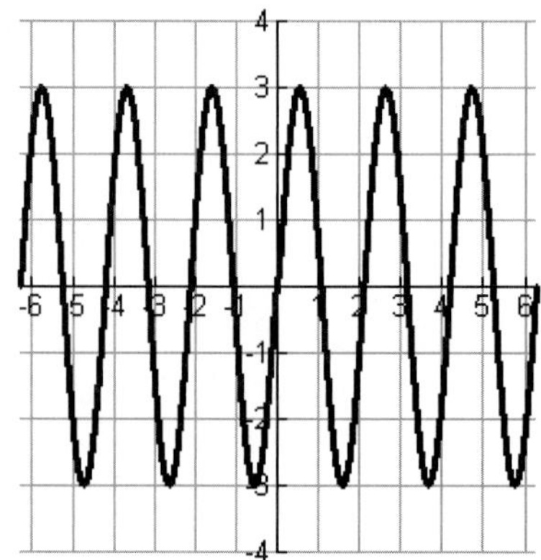

33. In order to construct the graph of $y = \frac{1}{2}\cos(2x)$ on $[-2\pi, 2\pi]$, we first gather the information essential to forming the graph for one period, namely on the interval $[0, \pi]$. Then, extend this picture to cover the entire interval. Indeed, we have the following:

Amplitude: $\frac{1}{2}$	Period: $\frac{2\pi}{2} = \pi$
x-intercepts occur when $\cos 2x = 0$. This happens when $2x = \frac{(2n+1)\pi}{2}$ and so, $x = \frac{(2n+1)\pi}{4}$, where n is an integer. So, the points $\left(\frac{\pi}{4}, 0\right), \left(\frac{3\pi}{4}, 0\right)$ are on the graph.	
The maximum value of the graph is $\frac{1}{2}$ since the maximum of $\cos 2x$ is 1, and the amplitude is $\frac{1}{2}$. This occurs when $2x = 0, 2\pi$, so that $x = 0, \pi$. So, the points $\left(0, \frac{1}{2}\right), \left(\pi, \frac{1}{2}\right)$ are on the graph.	
The minimum value of the graph is $-\frac{1}{2}$ since the minimum of $\cos 2x$ is -1, and the amplitude is $\frac{1}{2}$. This occurs when $2x = \pi$, so that $x = \frac{\pi}{2}$. So, the point $\left(\frac{\pi}{2}, -\frac{1}{2}\right)$ is on the graph.	
Since the coefficient used to compute the amplitude is nonnegative, there is no reflection over the x-axis.	

Now, using this information, we can form the graph of $y = \frac{1}{2}\cos(2x)$ on $[0, \pi]$, and then extend it to the interval $[-2\pi, 2\pi]$, as shown below:

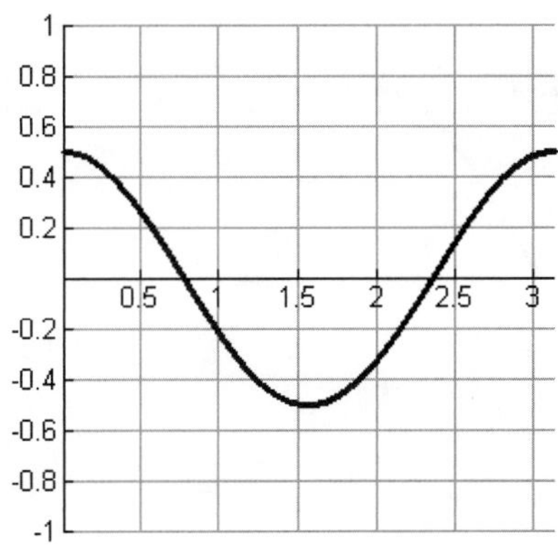

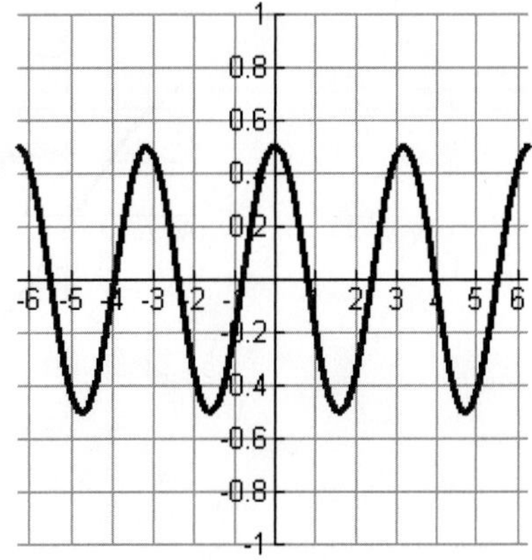

34. We have the following information about the graph of $y=-\frac{1}{4}\cos\left(\frac{1}{2}x\right)$ on $[-2\pi, 2\pi]$:

<table>
<tr><td><u>Amplitude</u>: $\left|-\frac{1}{4}\right|=\frac{1}{4}$</td><td><u>Period</u>: $\frac{2\pi}{\frac{1}{2}}=4\pi$</td></tr>
<tr><td colspan="2"><u>x-intercepts</u> occur when $\cos\left(\frac{1}{2}x\right)=0$. This happens when $\frac{1}{2}x=\pm\frac{\pi}{2}$, and so $x=\pm\pi$. So, the points $(-\pi, 0), (\pi, 0)$ are on the graph.</td></tr>
<tr><td colspan="2">The <u>maximum value</u> of the graph is $\frac{1}{4}$ since the maximum of $\cos\left(\frac{1}{2}x\right)$ is 1, and the amplitude is $\frac{1}{4}$. This occurs when $\frac{1}{2}x=0$, and so $x=0$. So, the point $\left(0, \frac{1}{4}\right)$ is on the graph.</td></tr>
<tr><td colspan="2">The <u>minimum value</u> of the graph is $-\frac{1}{4}$ since the minimum of $\cos\left(\frac{1}{2}x\right)$ is -1, and the amplitude is $\frac{1}{4}$. This occurs when $\frac{1}{2}x=\pm\pi$, and so $x=\pm2\pi$. So, the points $\left(2\pi,-\frac{1}{4}\right)$ and $\left(-2\pi,-\frac{1}{4}\right)$ on the graph.</td></tr>
<tr><td colspan="2">Since the coefficient used to compute the amplitude is negative, we reflect the graph of $y=4\cos\left(\frac{1}{2}x\right)$ over the x-axis.</td></tr>
</table>

Now, using this information, we can form the graph of $y=-\frac{1}{4}\cos\left(\frac{1}{2}x\right)$ on $[-2\pi, 2\pi]$:

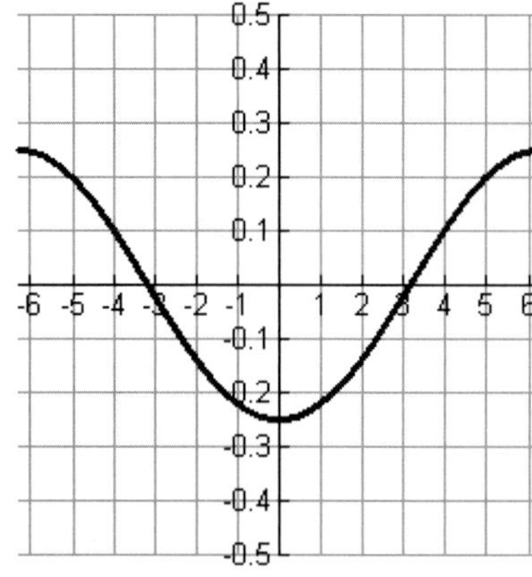

35.

Amplitude 3,

Period $\frac{2\pi}{1} = 2\pi$,

Phase Shift $\frac{\pi}{2}$

Vertical Shift 2 (Up)

36.

Amplitude $\frac{1}{2}$

Period $\frac{2\pi}{1} = 2\pi$,

Phase Shift $-\frac{\pi}{4}$

Vertical Shift 3 (Up)

37.

Amplitude 4,

Period $\frac{2\pi}{3}$,

Phase Shift $-\frac{\pi}{4}$

Vertical Shift - 2 (Down)

38.

Amplitude 2,

Period $\frac{2\pi}{2} = \pi$,

Phase Shift $\frac{\pi}{3}$

Vertical Shift - 1 (Down)

39.

Amplitude $\frac{1}{3}$,

Period 2,

Phase Shift $\frac{1}{2\pi}$

Vertical Shift $\frac{1}{2}$ (Down)

40.

Amplitude $\frac{1}{6}$,

Period 12,

Phase Shift 2

Vertical Shift $\frac{3}{4}$ (Up)

Note: Regarding problems 41-44, the convention for the three graphs displayed in each case is as follows. The dotted graph corresponds to the first term of the sum (including the negative if it is present); the dashed graph corresponds to the second term (no negative, even if it is present), and the solid graph is the sum (or difference if the second term has a negative in front of it) of the two functions.

41. **42.**

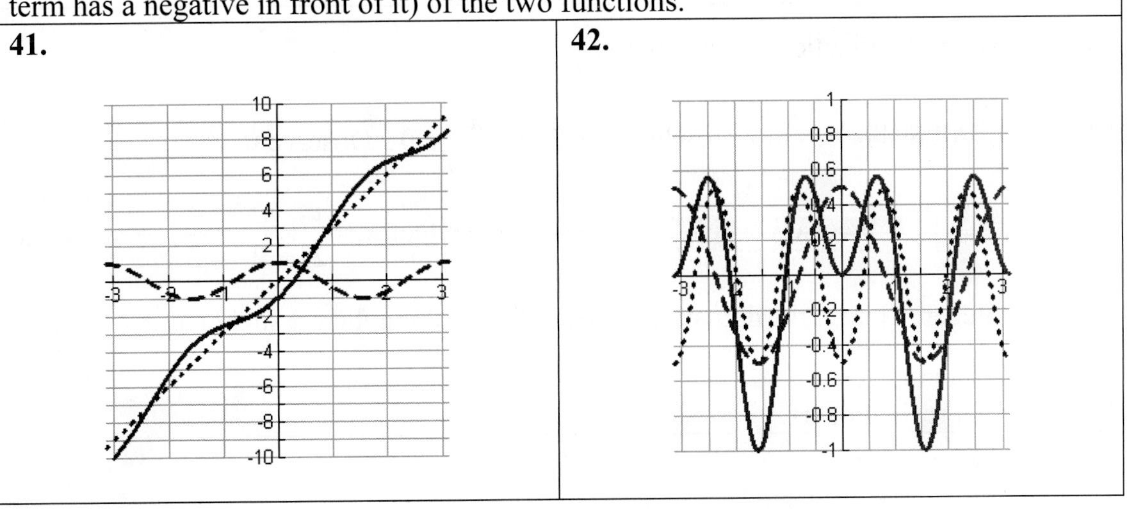

43.

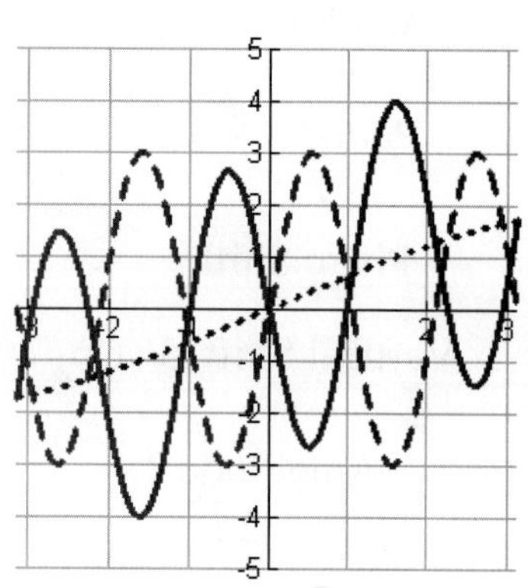

44.

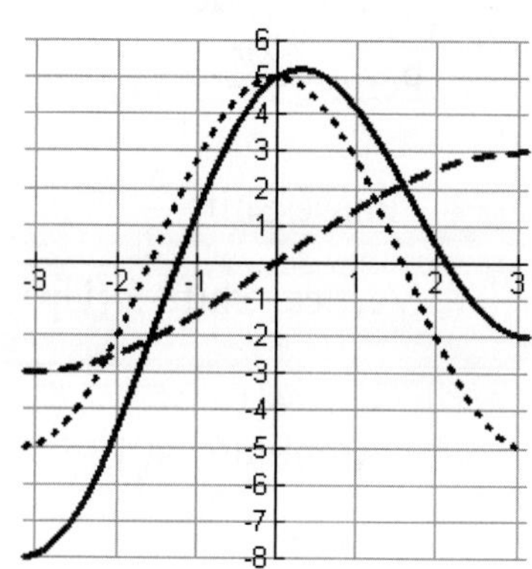

45. Consider the function $y = 4\tan\left(x+\frac{\pi}{2}\right)$.

Domain: All real numbers x for which $\cos\left(x+\frac{\pi}{2}\right) \neq 0$. Observe that $\cos\left(x+\frac{\pi}{2}\right) = 0$ precisely when

$$\begin{aligned} x+\frac{\pi}{2} &= (2n+1)\frac{\pi}{2} \\ x &= -\frac{\pi}{2}+(2n+1)\frac{\pi}{2} \\ x &= n\pi \end{aligned}$$

where n is an integer. Hence, the domain is the set of all real numbers x such that $x \neq n\pi$, where n is an integer.
Range: All real numbers.

46. Consider the function $y = \cot 2\left(x-\frac{\pi}{2}\right)$.

Domain: All real numbers x for which $\sin 2\left(x-\frac{\pi}{2}\right) \neq 0$. Observe that $\sin 2\left(x-\frac{\pi}{2}\right) = 0$ precisely when $2\left(x-\frac{\pi}{2}\right) = n\pi$, so that $x = \frac{\pi}{2}+\frac{n\pi}{2} = \frac{\pi}{2}(1+n)$, where n is an integer. Hence, the domain is the set of all real numbers x such that $x \neq \frac{n\pi}{2}$, where n is an integer.
Range: All real numbers.

47. Consider the function $y = 3\sec(2x)$.

Domain: All real numbers x for which $\cos(2x) \neq 0$. Observe that $\cos(2x) = 0$ precisely when $2x = (2n+1)\frac{\pi}{2}$, so that $x = \frac{2n+1}{4}\pi$, where n is an integer. Hence, the domain is the set of all real numbers x such that $x \neq \frac{2n+1}{4}\pi$, where n is an integer.

Range: Note that the maximum and minimum y-values for the guide function $y = 3\cos(2x)$ are 3 and –3, respectively. Hence, the range is $(-\infty,-3]\cup[3,\infty)$.

48. Consider the function $y = 1 + 2\csc x$.

Domain: All real numbers x for which $\sin x \neq 0$. Observe that $\sin x = 0$ precisely when $x = n\pi$, where n is an integer. Hence, the domain is the set of all real numbers x such that $x \neq n\pi$, where n is an integer.

Range: Note that the maximum and minimum y-values for the guide function $y = 1 + 2\sin x$ are 3 and – 1, respectively. Hence, the range is $(-\infty,-1]\cup[3,\infty)$.

49. Consider the function $y = \frac{1}{4}\sec\left(\pi x - \frac{2\pi}{3}\right) - \frac{1}{2}$.

Domain: All real numbers x for which $\cos\left(\pi x - \frac{2\pi}{3}\right) \neq 0$. Observe that this occurs precisely when $\pi x - \frac{2\pi}{3} = \frac{2n+1}{2}\pi$, so that $x = \frac{6n+7}{6}$, where n is an integer. Hence, the domain is the set of all real numbers x such that $x \neq \frac{6n+7}{6}$, where n is an integer.

Range: Note that the maximum and minimum y-values for the guide function $y = \frac{1}{4}\cos\left(\pi x - \frac{2\pi}{3}\right) - \frac{1}{2}$ are $-\frac{1}{4}$ and $-\frac{3}{4}$, respectively. Hence, the range is $\left(-\infty,-\frac{3}{4}\right]\cup\left[-\frac{1}{4},\infty\right)$.

50. Consider the function $y = 3 - \frac{1}{2}\csc(2x - \pi)$.

Domain: All real numbers x for which $\sin(2x - \pi) \neq 0$. Observe that this occurs precisely when $2x - \pi = n\pi$, so that $x = \frac{n\pi}{2}$, where n is an integer. Hence, the domain is the set of all real numbers x such that $x \neq \frac{n\pi}{2}$, where n is an integer.

Range: Note that the maximum and minimum y-values for the guide function $y = 3 - \frac{1}{2}\sin(2x - \pi)$ are $\frac{7}{2}$ and $\frac{5}{2}$, respectively. Hence, the range is $\left(-\infty,\frac{5}{2}\right]\cup\left[\frac{7}{2},\infty\right)$.

51. We have this information about the graph of $y = -\tan\left(x - \frac{\pi}{4}\right)$ on $[-2\pi, 2\pi]$:

Period: $\frac{\pi}{1} = \pi$	Reflection about x-axis? Yes	Phase Shift: Right $\frac{\pi}{4}$ units
x-intercepts occur when $\sin\left(x - \frac{\pi}{4}\right) = 0$. This happens when $x - \frac{\pi}{4} = 0, \pm\pi$, so that $x = -\frac{3\pi}{4}, \frac{\pi}{4}, \frac{5\pi}{4}$. So, the points $\left(-\frac{3\pi}{4}, 0\right), \left(\frac{\pi}{4}, 0\right), \left(\frac{5\pi}{4}, 0\right)$ are on the graph.		
Vertical asymptotes occur when $\cos\left(x - \frac{\pi}{4}\right) = 0$. This happens when $x - \frac{\pi}{4} = \pm\frac{\pi}{2}, \pm\frac{3\pi}{2}$, so that $x = -\frac{\pi}{4}, \frac{3\pi}{4}$.		

Using this information, we construct the graph of $y = -\tan\left(x - \frac{\pi}{4}\right)$ on $[-2\pi, 2\pi]$:

as follows:

Step 1: Graph $y = \tan\left(x - \frac{\pi}{4}\right)$ on $\left[\frac{\pi}{4}, \frac{\pi}{4} + \pi\right] = \left[\frac{\pi}{4}, \frac{5\pi}{4}\right]$:

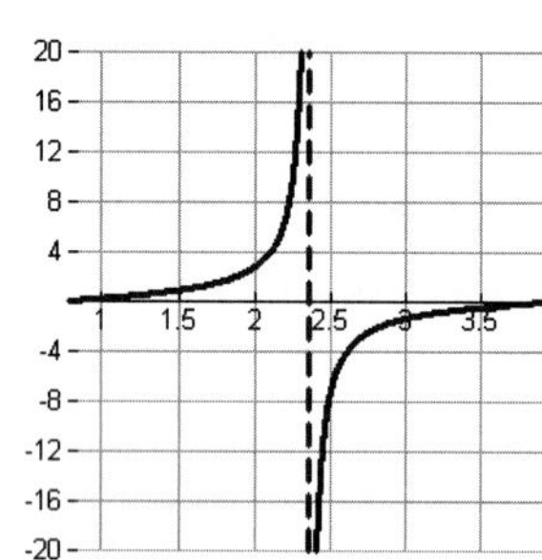

Step 2: Extend the graph in Step 1 to $[-2\pi, 2\pi]$:

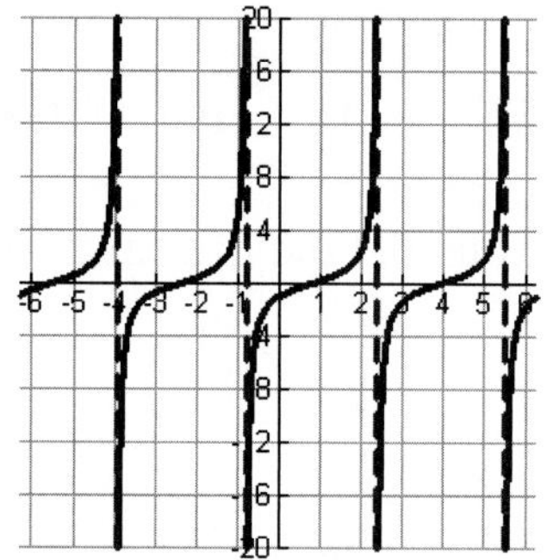

Step 3: Reflect the graph in Step 2 over the x-axis to obtain the graph of $y = -\tan\left(x - \frac{\pi}{4}\right)$ on $[-2\pi, 2\pi]$:

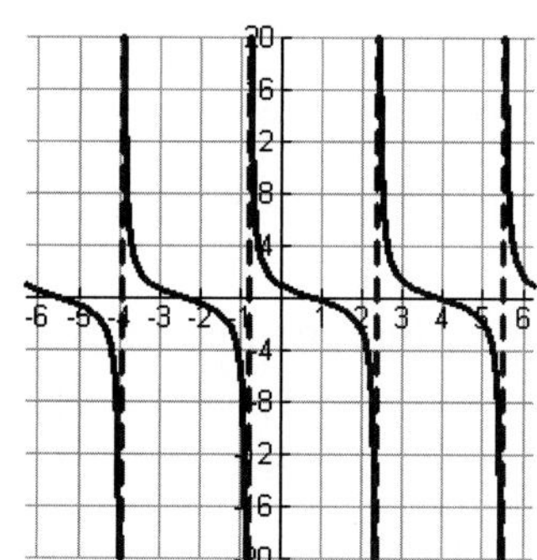

52. We have the following information about the graph of $y = 1 + \cot(2x)$:

Period:	Vertical Translation:	Reflection about *x*-axis?	Phase Shift:
$\frac{\pi}{2}$	Up 1 unit	No	None

Step 1: Graph $y = \cot(2x)$ on $\left(0, \frac{\pi}{2}\right)$.

We have the following information about this particular graph

x-intercepts occur when $\cos(2x) = 0$. This happens when $2x = \frac{\pi}{2}$ so that $x = \frac{\pi}{4}$. So, the point $\left(\frac{\pi}{4}, 0\right)$ is on this graph. (Note: This will change in Step 2 when we translate this graph vertically.)
Vertical asymptotes occur when $\sin(2x) = 0$. This happens when $2x = 0, \pi$, so that $x = 0, \frac{\pi}{2}$. (Note: These remain unchanged by vertical translation.)

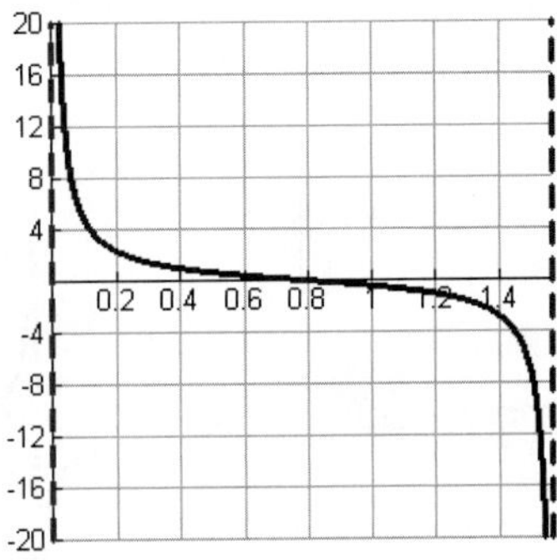

Step 2: Extend the graph in Step 1 to $[-2\pi, 2\pi]$:

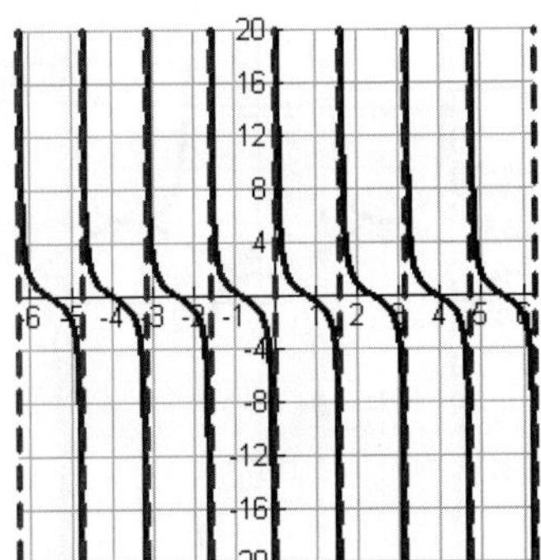

Step 3: Translate the graph in Step 2 vertically up 1 unit to obtain the graph of $y = 1 + \cot(2x)$ on $[-2\pi, 2\pi]$:

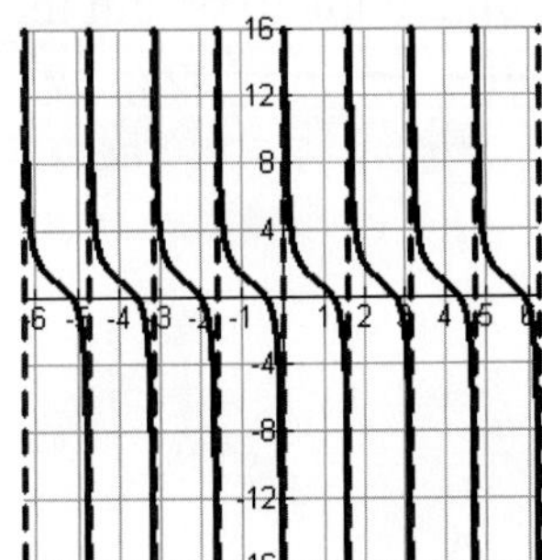

53. First, note that the period is 2π and the phase shift is π units to the right. Hence, one period of this function would occur on the interval $[\pi, 3\pi]$, for instance.

In order to graph $y = 2 + \sec(x - \pi)$ on $[-2\pi, 2\pi]$, proceed as described below:

Step 1: Graph $y = \sec(x - \pi)$ on $[\pi, 3\pi]$.

- Graph the *guide function* $y = \cos(x - \pi)$ on $[\pi, 3\pi]$ using the approach discussed earlier in Chapter 6.

  x-intercepts: Occur when $x - \pi = \dfrac{\pi}{2}, \dfrac{3\pi}{2}$, so that $x = \dfrac{3\pi}{2}, \dfrac{5\pi}{2}$

- At each x-intercept of this guide function, draw a vertical asymptote.
- On the intervals where this guide function is above the x-axis, draw a $\cup$-shaped graph with vertex at the maximum of the guide function, opening upward toward the asymptotes.
- On the intervals where this guide function is below the x-axis, draw a $\cap$-shaped graph with vertex at the minimum of the guide function, opening downward toward the asymptotes.

The graph of guide function (dotted) and the function $y = \sec(x - \pi)$ on $[\pi, 3\pi]$:

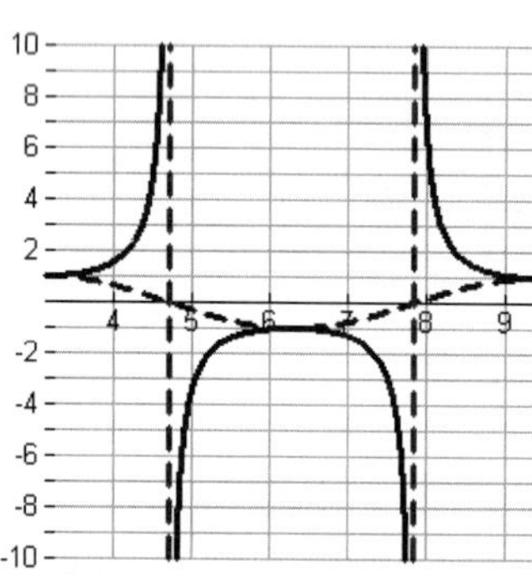

Step 2: Extend the graph of Step 1 to $[-2\pi, 2\pi]$:

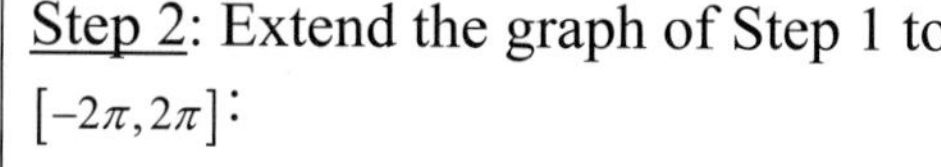

Step 3: Translate the graph is Step 1 vertically up 2 units to obtain the graph of $y = 2 + \sec(x - \pi)$ on $[-2\pi, 2\pi]$ as shown below

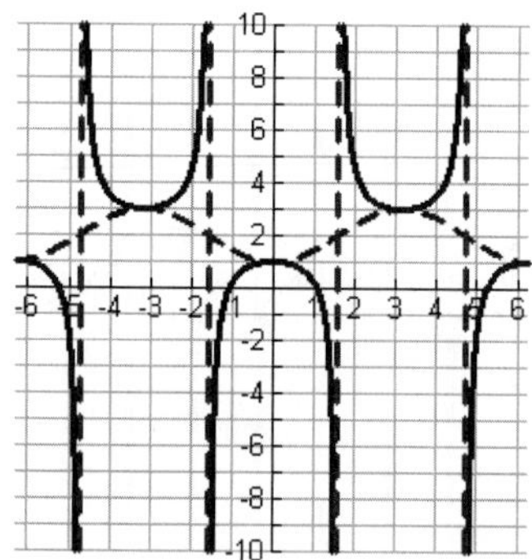

54. First, note that the period is 2π and the phase shift is $\frac{\pi}{4}$ units to the left. Hence, one period of this function would occur on the interval $\left[-\frac{\pi}{4}, \frac{7\pi}{4}\right]$, for instance.

In order to graph $y = -\csc\left(x + \frac{\pi}{4}\right)$ on $[-2\pi, 2\pi]$, proceed as described below:

Step 1: Graph $y = \csc\left(x + \frac{\pi}{4}\right)$ on $\left[-\frac{\pi}{4}, \frac{7\pi}{4}\right]$.

- Graph the *guide function* $y = \sin\left(x + \frac{\pi}{4}\right)$ on $\left[-\frac{\pi}{4}, \frac{7\pi}{4}\right]$ using the approach discussed earlier in Chapter 6.

  x-intercepts: Occur when $x + \frac{\pi}{4} = 0,\ \pi,\ 2\pi$, so that $x = -\frac{\pi}{4}, \frac{3\pi}{4}, \frac{7\pi}{4}$
- At each x-intercept of this guide function, draw a vertical asymptote.
- On the intervals where this guide function is above the x-axis, draw a $\cup$-shaped graph with vertex at the maximum of the guide function, opening upward toward the asymptotes.
- On the intervals where this guide function is below the x-axis, draw a $\cap$-shaped graph with vertex at the minimum of the guide function, opening downward toward the asymptotes.

The graph of guide function (dotted) and the function $y = \csc\left(x + \frac{\pi}{4}\right)$ on $\left[-\frac{\pi}{4}, \frac{7\pi}{4}\right]$:

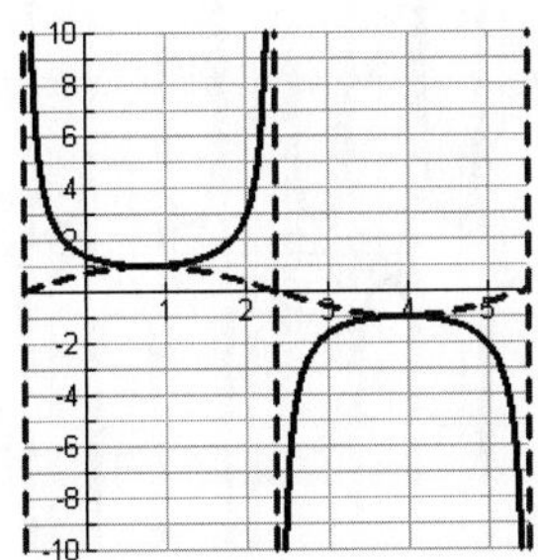

Step 2: Extend the graph using periodicity to the interval $[-2\pi, 2\pi]$:

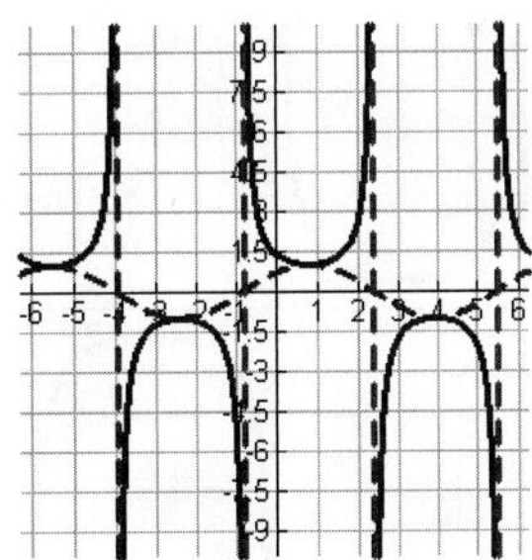

Step 3: Reflect the graph in Step 2 over the x-axis to obtain the graph of $y = -\csc\left(x + \frac{\pi}{4}\right)$ on $[-2\pi, 2\pi]$:

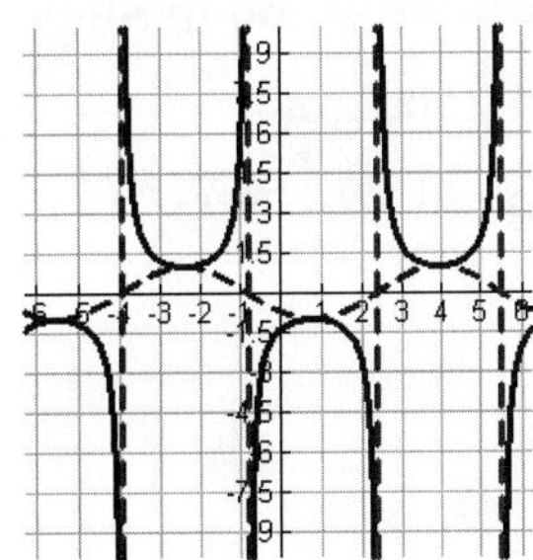

55. First, note that the period is π and the phase shift is $\frac{\pi}{4}$ units to the right. Hence, one period of this function would occur on the interval $\left[\frac{\pi}{4}, \frac{5\pi}{4}\right]$, for instance.

In order to graph $y=\frac{1}{2}+\csc\left(2x-\frac{\pi}{2}\right)$ on $[-2\pi, 2\pi]$, proceed as described below:

<u>Step 1</u>: Graph $y=\csc\left(2x-\frac{\pi}{2}\right)$ on $\left[\frac{\pi}{4}, \frac{5\pi}{4}\right]$.

- Graph the *guide function* $y=\sin\left(2x-\frac{\pi}{2}\right)$ on $\left[\frac{\pi}{4}, \frac{5\pi}{4}\right]$ using the approach discussed earlier in Chapter 6.

  <u>x-intercepts</u>: Occur when $2x-\frac{\pi}{2}=0,\ \pi$, so that $x=\frac{\pi}{4}, \frac{3\pi}{4}$
- At each x-intercept of this guide function, draw a vertical asymptote.
- On the intervals where this guide function is <u>above</u> the x-axis, draw a $\cup$−shaped graph with vertex at the maximum of the guide function, opening upward toward the asymptotes.
- On the intervals where this guide function is <u>below</u> the x-axis, draw a $\cap$−shaped graph with vertex at the minimum of the guide function, opening downward toward the asymptotes.

The graph of guide function (dotted) and the function $y=\csc\left(2x-\frac{\pi}{2}\right)$ on $\left[\frac{\pi}{4}, \frac{5\pi}{4}\right]$:

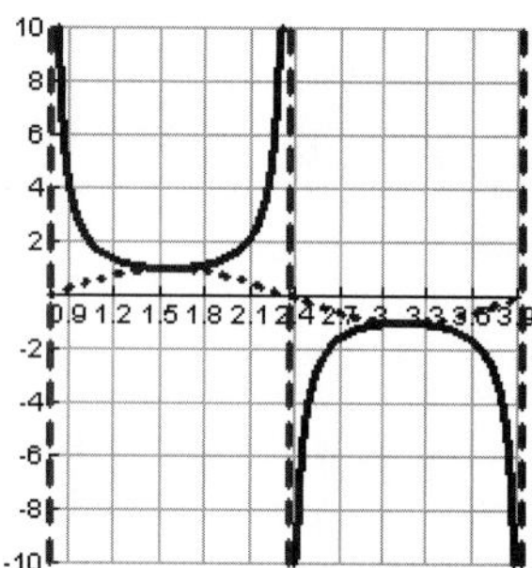

<u>Step 2</u>: Extend the graph using periodicity to the interval $[-2\pi, 2\pi]$:

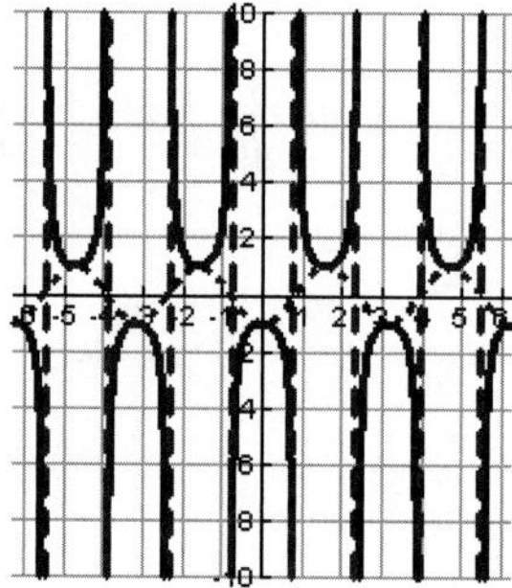

<u>Step 3</u>: Shift the graph in Step 2 up $\frac{1}{2}$ unit to obtain the graph of $y=\frac{1}{2}+\csc\left(2x-\frac{\pi}{2}\right)$ on $[-2\pi, 2\pi]$:

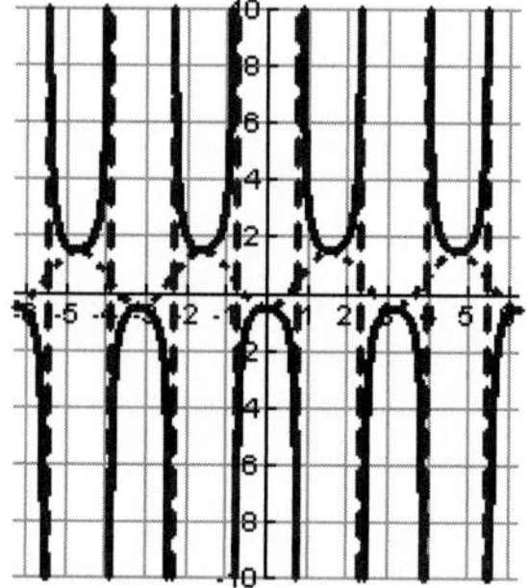

56. First, note that the period is 2 and the phase shift is $\frac{3}{4}$ units to the left. Hence, one period of this function would occur on the interval $\left[-\frac{3}{4},\frac{5}{4}\right]$, for instance.

In order to graph $y=-1-\frac{1}{2}\sec\left(\pi x-\frac{3\pi}{4}\right)$ on $[-2\pi,2\pi]$, proceed as described below:

Step 1: Graph $y=\frac{1}{2}\sec\left(\pi x-\frac{3\pi}{4}\right)$ on $\left[-\frac{3}{4},\frac{5}{4}\right]$.

- Graph the *guide function* $y=\frac{1}{2}\cos\left(\pi x-\frac{3\pi}{4}\right)$ on $\left[-\frac{3}{4},\frac{5}{4}\right]$ using the approach discussed earlier in Chapter 6.
  x-intercepts: Occur when $\pi x-\frac{3\pi}{4}=\frac{\pi}{2},\frac{3\pi}{2}$, so that $x=-\frac{1}{4},\frac{3}{4}$
- At each x-intercept of this guide function, draw a vertical asymptote.
- On the intervals where this guide function is above the x-axis, draw a $\cup$–shaped graph with vertex at the maximum of the guide function, opening upward toward the asymptotes.
- On the intervals where this guide function is below the x-axis, draw a $\cap$–shaped graph with vertex at the minimum of the guide function, opening downward toward the asymptotes.

The graph of guide function (dotted) and the function $y=\frac{1}{2}\sec\left(\pi x-\frac{3\pi}{4}\right)$ on $\left[-\frac{3}{4},\frac{5}{4}\right]$:

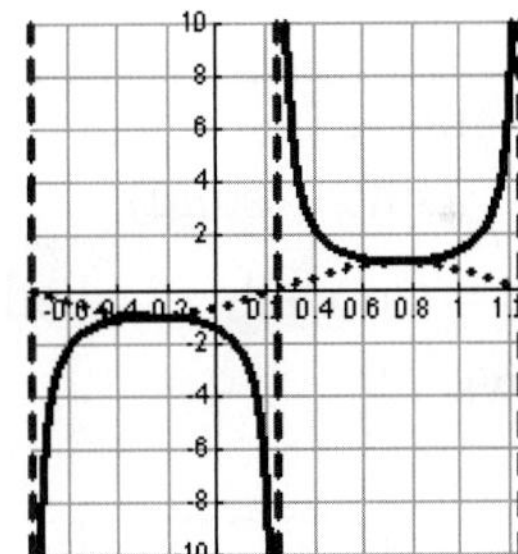

Step 2: Extend the graph using periodicity to the interval $[-2\pi,2\pi]$.

Step 3: Reflect the graph in Step 2 over the x-axis to obtain the graph of $y=-\frac{1}{2}\sec\left(\pi x-\frac{3\pi}{4}\right)$ on $[-2\pi,2\pi]$.

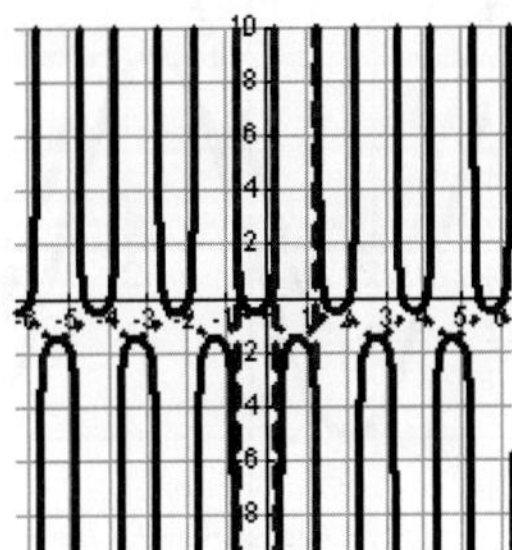

Step 4: Finally, shift the graph in Step 3 down 1 unit to obtain the graph of $y=-1-\frac{1}{2}\sec\left(\pi x-\frac{3\pi}{4}\right)$ on $[-2\pi,2\pi]$.
This results in:

Note: For graphical clarity, only the vertical asymptotes corresponding to one period are shown.

57. -0.9659	**58.** 0.5000

59. a. The graphs of

$y_1 = \cos x$ (solid)

$y_2 = \cos\left(x + \frac{\pi}{6}\right)$ (dashed)

are as follows:

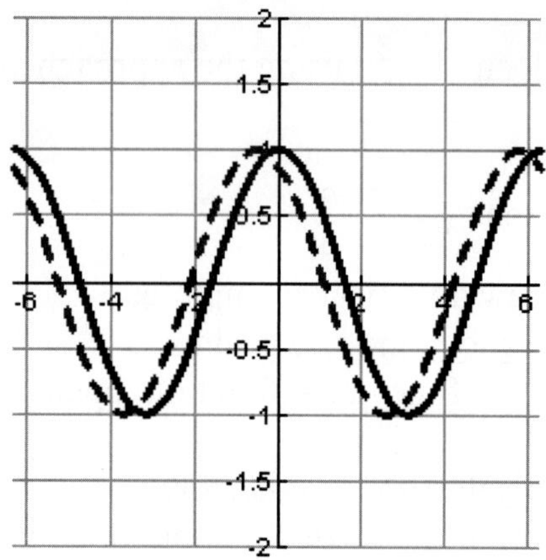

The graph of y_2 is that of y_1 shifted left $\frac{\pi}{6}$ units.

b. The graphs of

$y_1 = \cos x$ (solid)

$y_2 = \cos\left(x - \frac{\pi}{6}\right)$ (dashed)

are as follows:

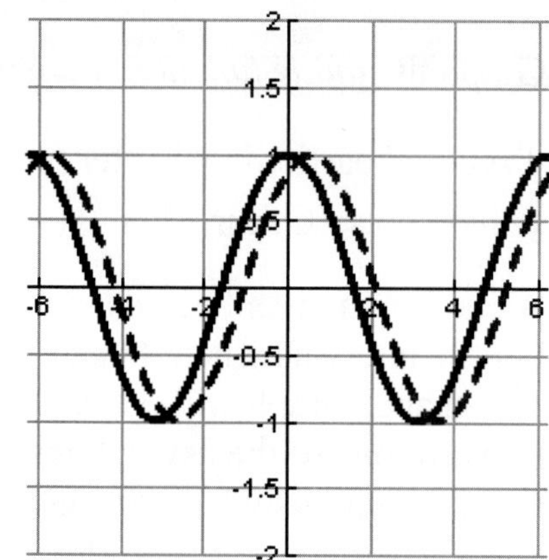

The graph of y_2 is that of y_1 shifted right $\frac{\pi}{6}$ units.

60. a. The graphs of

$y_1 = \sin x$ (solid)

$y_2 = \sin x + \frac{1}{2}$ (dashed)

are as follows:

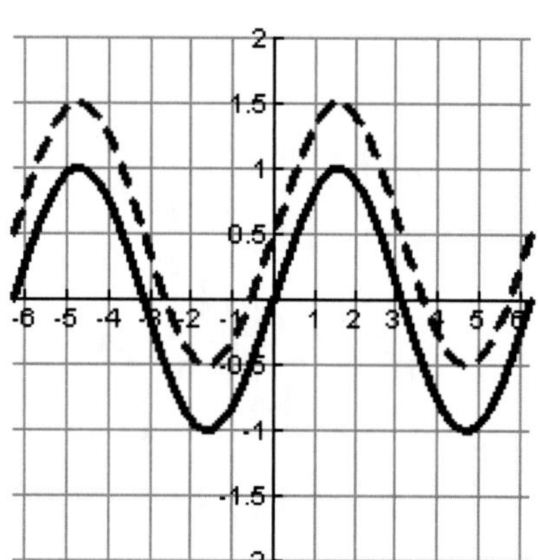

The graph of y_2 is that of y_1 shifted up $\frac{1}{2}$ unit.

b. The graphs of

$y_1 = \sin x$ (solid)

$y_2 = \sin x - \frac{1}{2}$ (dashed)

are as follows:

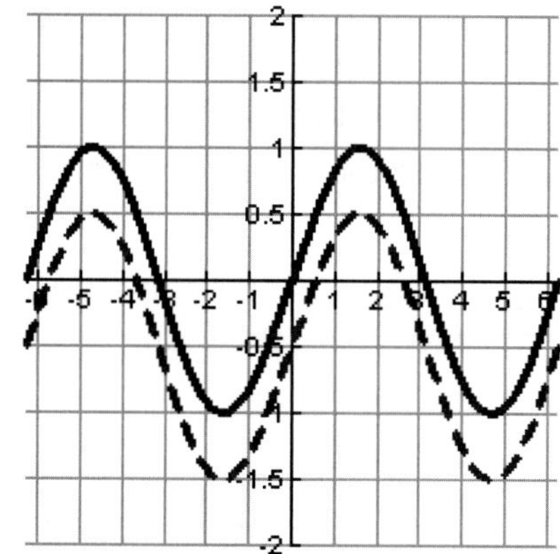

The graph of y_2 is that of y_1 shifted down $\frac{1}{2}$ unit.

61. Consider the graphs of the following functions on the interval $[-2\pi, 2\pi]$:

$$y_1 = 4\cos x \quad \text{(dashed bold),}$$
$$y_2 = 3\sin x \quad \text{(dotted bold),}$$
$$y_3 = 4\cos x - 3\sin x \quad \text{(solid bold)}$$

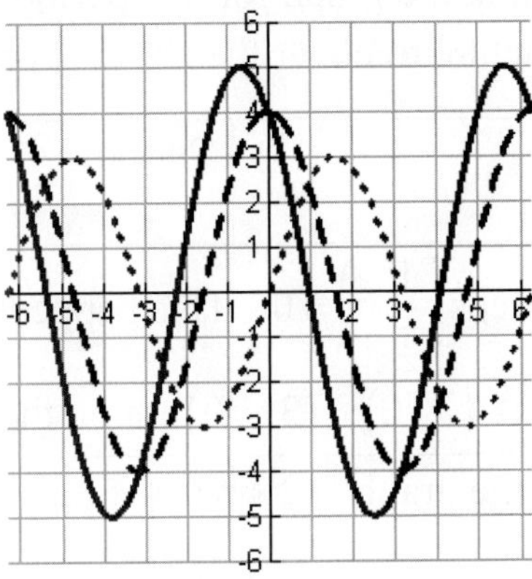

The amplitude of $y = 4\cos x - 3\sin x$ appears to be approximately 5.

62. Consider the graphs of the following functions on the interval $[-2\pi, 2\pi]$:

$$y_1 = \cos x \quad \text{(dashed bold),}$$
$$y_2 = \sqrt{3}\sin x \quad \text{(dotted bold),}$$
$$y_3 = \cos x + \sqrt{3}\sin x \quad \text{(solid bold)}$$

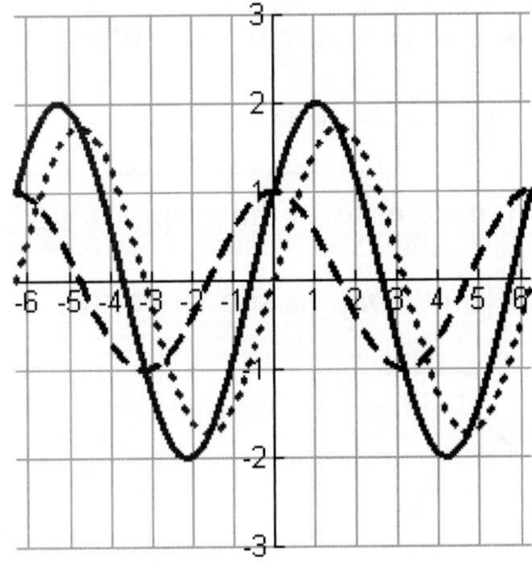

The amplitude of $y = \sqrt{3}\cos x - \sin x$ appears to be approximately 2.

Chapter 5 Practice Test Solutions --

1. Amplitude is 5 and period is $2\pi/3$

2. In order to construct the graph of $y=-2\cos\left(\frac{1}{2}x\right)$ on $[-4\pi,4\pi]$, we first gather the information essential to forming the graph for one period, namely on the interval $[0,4\pi]$. Then, extend this picture to cover the entire given interval. Indeed, we have:

Amplitude: $\lvert -2\rvert = 2$	Period: $\frac{2\pi}{1/2}=4\pi$
x-intercepts occur when $\cos\left(\frac{1}{2}x\right)=0$. This happens when $\frac{1}{2}x=\frac{\pi}{2},\frac{3\pi}{2}$, and so $x=\pi,\ 3\pi$. So, the points $(\pi,\ 0)$, $(3\pi,\ 0)$ are on the graph.	
The maximum value of the graph is 2 since the maximum of $\cos\left(\frac{1}{2}x\right)$ is 1, and the amplitude is 2. This occurs when $\frac{1}{2}x=0,\ 2\pi$, and so $x=0,\ 4\pi$. So, the points $(0,2)$, $(4\pi,2)$ are on the graph.	
The minimum value of the graph is – 2 since the minimum of $\cos\left(\frac{1}{2}x\right)$ is – 1, and the amplitude is 2. This occurs when $\frac{1}{2}x=\pi$, and so $x=2\pi$. So, the point $(2\pi,-2)$ is on the graph.	
Since the coefficient used to compute the amplitude is negative, we reflect the graph of $y=2\cos\left(\frac{1}{2}x\right)$ over the x-axis.	

Now, using this information, we can form the graph of $y=-2\cos\left(\frac{1}{2}x\right)$ on $[0,4\pi]$, and then extend it to the interval $[-4\pi,4\pi]$, as shown below:

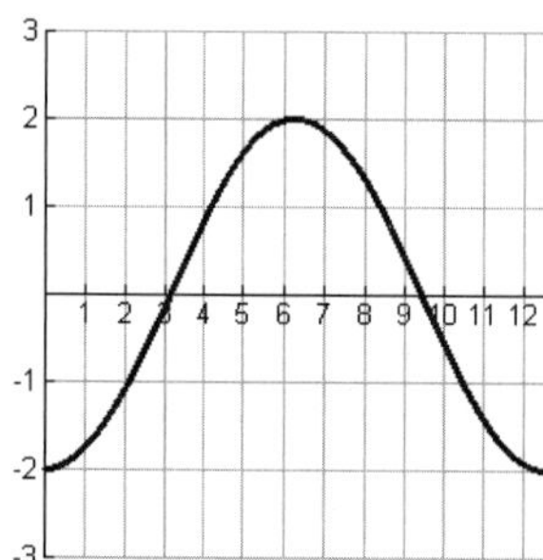

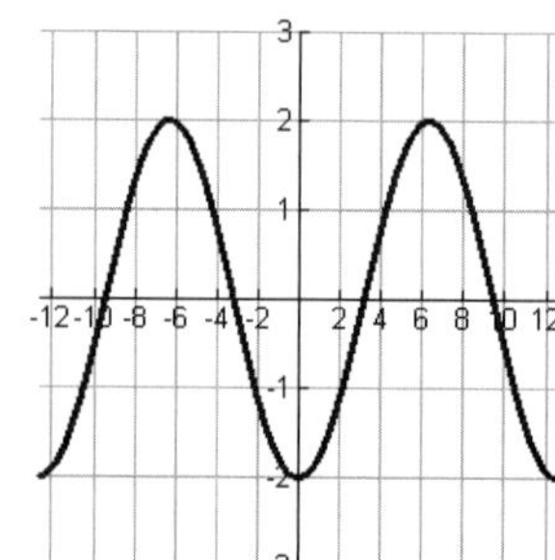

3. Amplitude: 3
Phase Shift: $-\pi$ (left)
Vertical Shift: 1 (up)

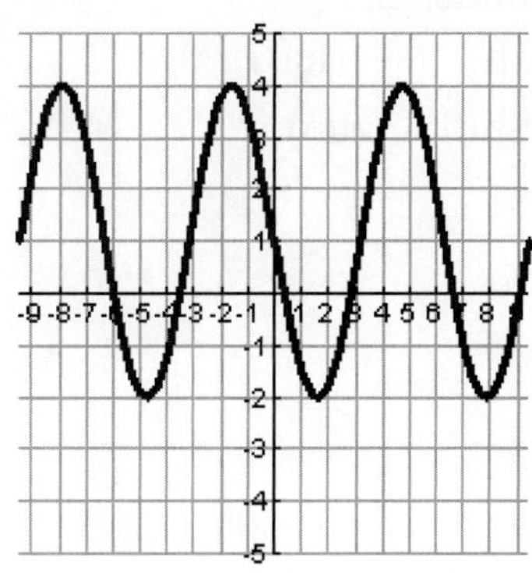

4. Amplitude: 1
Phase Shift: $\pi/2$ (right)
Vertical Shift: 4 (up)
Reflect over x-axis

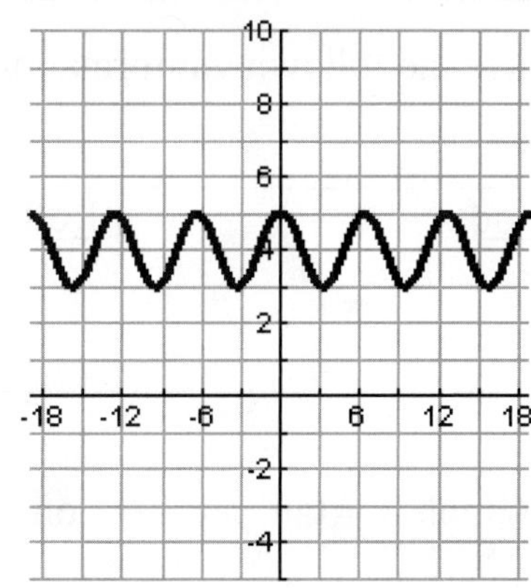

5. Amplitude: 1
Phase Shift: $-\pi/2$ (left)
Vertical Shift: 2 (down)
Reflect over x-axis

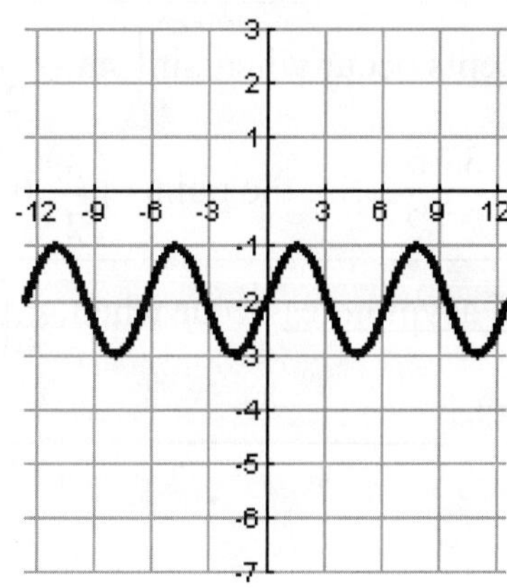

6. Amplitude: 2
Phase Shift: $-\frac{3\pi}{2}$ (left)
Vertical Shift: 3 (up)

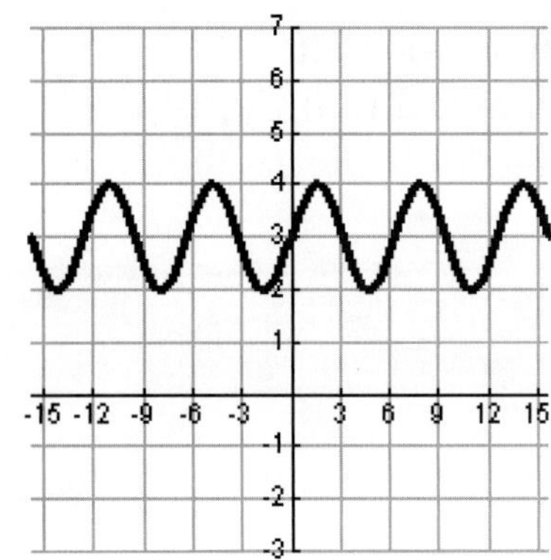

7. We have the following information about the graph of $y = \tan\left(\pi x - \frac{\pi}{2}\right)$:

Period:	Reflection about x-axis?	Phase Shift:
$\frac{\pi}{\pi} = 1$	No	Right $\frac{\pi/2}{\pi} = \frac{1}{2}$ units

Step 1: Graph $y = \tan\left(\pi x - \frac{\pi}{2}\right)$ on $\left[\frac{1}{2}, \frac{3}{2}\right]$.

We have the following information about this particular graph

x-intercepts occur when $\sin\left(\pi x - \frac{\pi}{2}\right) = 0$. This happens when $\pi x - \frac{\pi}{2} = 0, \pm\pi$, so that $x = \frac{3}{2}, \frac{5}{2}$. So, the points $\left(\frac{3}{2}, 0\right), \left(\frac{5}{2}, 0\right)$ are on the graph.
Vertical asymptotes occur when $\cos\left(\pi x - \frac{\pi}{2}\right) = 0$. This happens when $\pi x - \frac{\pi}{2} = \pm\frac{\pi}{2}$ so that $x = 1, 2$.

The graph of $y = \tan\left(\pi x - \frac{\pi}{2}\right)$ on $\left[\frac{1}{2}, \frac{3}{2}\right]$ is as follows: 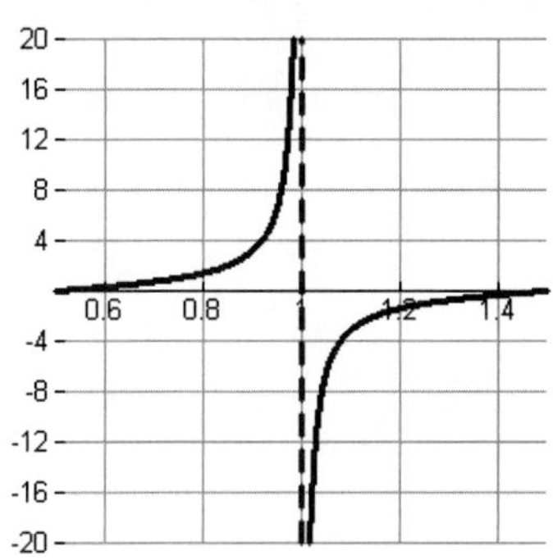	Step 2: Extend the graph in Step 1 to $\left[\frac{1}{2}, \frac{5}{2}\right]$ to show two periods: 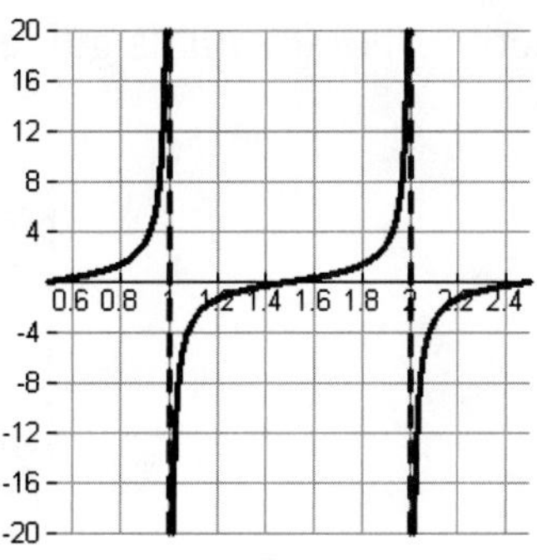

8. The vertical asymptotes of $y = 2\csc(3x - \pi)$ correspond to the x-intercepts of $y = 2\sin(3x - \pi)$.
9. The x-intercepts of $y = \tan 2x$ occur when $\sin 2x = 0$. This happens when $2x = n\pi$, or $x = \frac{n\pi}{2}$, where n is an integer.
10. Phase shift: - 3 Vertical shift: none
11. The range of $y = -3\sec\left(2x + \frac{\pi}{3}\right)$ is $(-\infty, -3] \cup [3, \infty)$. So, shifting it down one unit results in the range of $y = -1 - 3\sec\left(2x + \frac{\pi}{3}\right)$ being $(-\infty, -4] \cup [2, \infty)$.
12. The domain is the set of all real numbers x for which $\cos\left(2x - \frac{\pi}{6}\right) \neq 0$. This occurs precisely when $2x - \frac{\pi}{6} = \frac{2n+1}{2}\pi$, so that $x = (3n + 2)\pi$, where n is an integer. So, the domain is the set of all real numbers for which $x \neq (3n + 2)\pi$, where n is an integer.

13. The period is 2π and the phase shift is $\frac{\pi}{2}$ units to the left. Hence, one period of this function would occur on the interval $\left[-\frac{\pi}{2}, 2\pi - \frac{\pi}{2}\right]$, for instance. As such, the graph of this function, along with its guide function (dotted), on $\left[-\frac{\pi}{2}, 4\pi - \frac{\pi}{2}\right]$ are as follows: 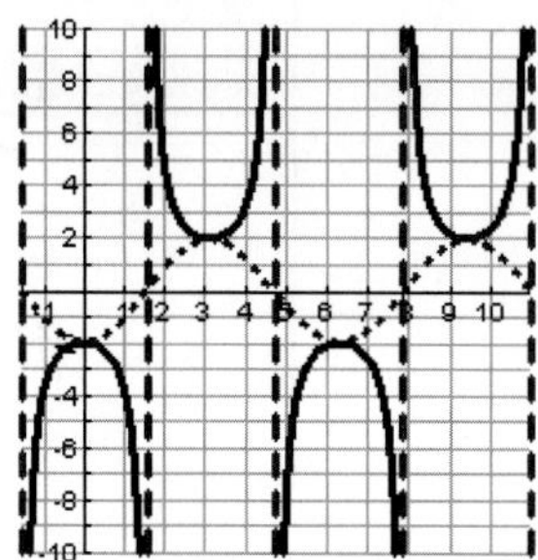
14. To find the x-intercepts of $y = \frac{6}{\sqrt{3}} - 3\sec\left(6x - \frac{5\pi}{6}\right)$, we solve the following equation: $\frac{6}{\sqrt{3}} - 3\sec\left(6x - \frac{5\pi}{6}\right) = 0$ $\sec\left(6x - \frac{5\pi}{6}\right) = \frac{2}{\sqrt{3}}$ $\cos\left(6x - \frac{5\pi}{6}\right) = \frac{\sqrt{3}}{2}$ $6x - \frac{5\pi}{6} = \frac{\pi}{6} + 2n\pi$ or $\frac{11\pi}{6} + 2n\pi$ Solving for x then yields the x-intercepts are: $x = \left(\frac{(2n+1)\pi}{6}, 0\right), \left(\frac{(3n+4)\pi}{3}, 0\right)$, where n is an integer.
15. True. $\sin\theta = 1.00005$ has no solution since the range of sine is [-1,1].
16. The graph of $y = \cos(2x)$ lies below the x-axis precisely when $\cos(2x) < 0$. This occurs when the input $2x$ is in the second or third quadrants, or equivalently, $\frac{\pi}{2} + 2n\pi < 2x < \frac{3\pi}{2} + 2n\pi$, where n is an integer. Solving for x then implies that the graph is below the x-axis precisely when $\frac{\pi}{4} + n\pi < x < \frac{3\pi}{4} + n\pi$, where n is an integer.
17. Taking into account all of the features provided yields the function $y = 4\sin\left(2\left(x + \frac{3}{2}\right)\right) - \frac{1}{2}$.
18. Taking into account all of the features provided yields the function $y = \cot x + 0.01$.

19. The convention for the three graphs displayed in each case is as follows. The dotted graph corresponds to the first term of the sum; the dashed graph corresponds to the second term (no negative), and the solid graph is the difference of the two functions.

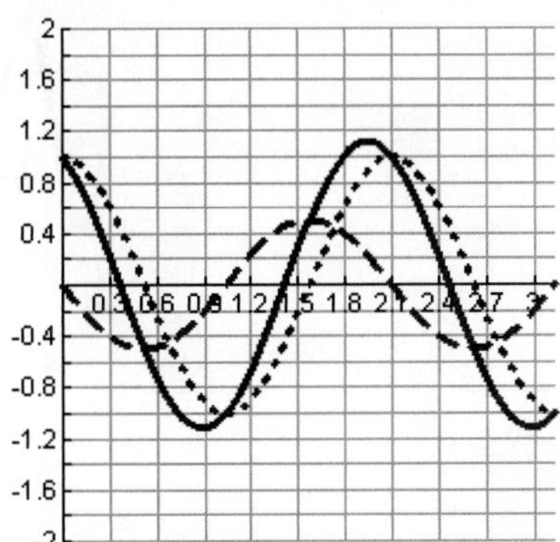

20. b.) Amplitude = $\frac{1}{5}$

c.) Period = $\dfrac{2\pi}{\frac{1}{3}} = 6\pi$

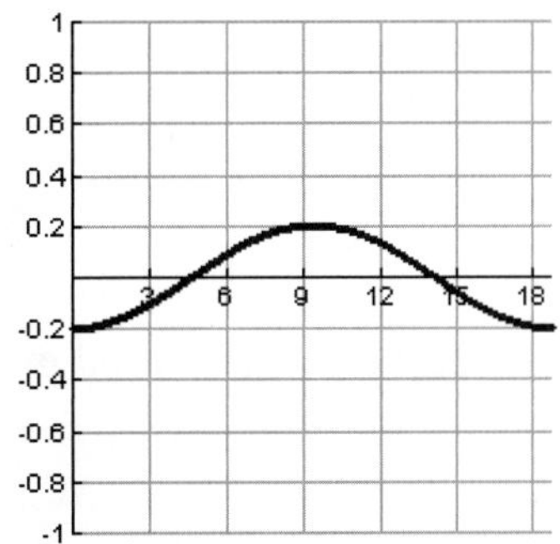

21. b.) Amplitude = 4

c.) Period = $\dfrac{2\pi}{2\pi} = 1$

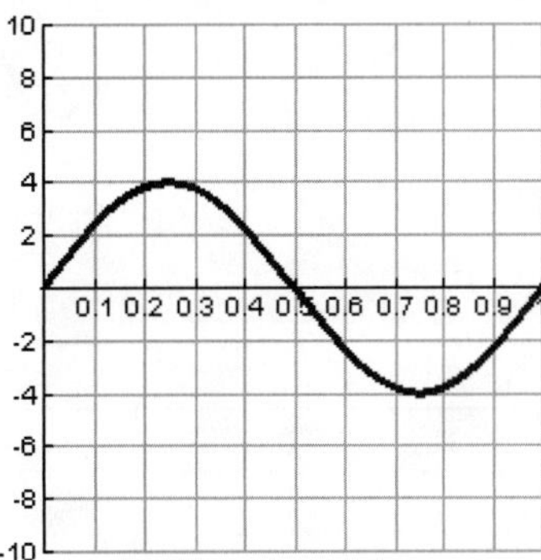

22. a.) $y=-2\sin\left(3\left(x+\frac{4\pi}{3}\right)\right)+1$ (graph to the right) **b.)** Amplitude = 2 Period = $\dfrac{2\pi}{3}$ Phase Shift = $-\dfrac{4\pi}{3}$ (left)	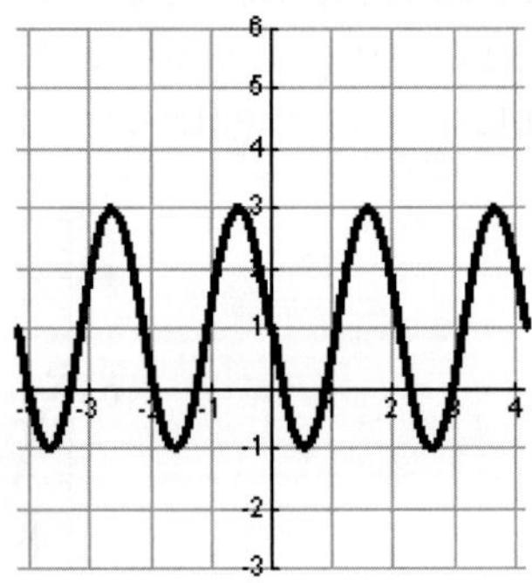
23. a.) $y=6+5\cos\left(2\left(x-\frac{\pi}{2}\right)\right)$ (graph to the right) **b.)** Amplitude = 5 Period = $\dfrac{2\pi}{2}=\pi$ Phase Shift = $\dfrac{\pi}{2}$ (right)	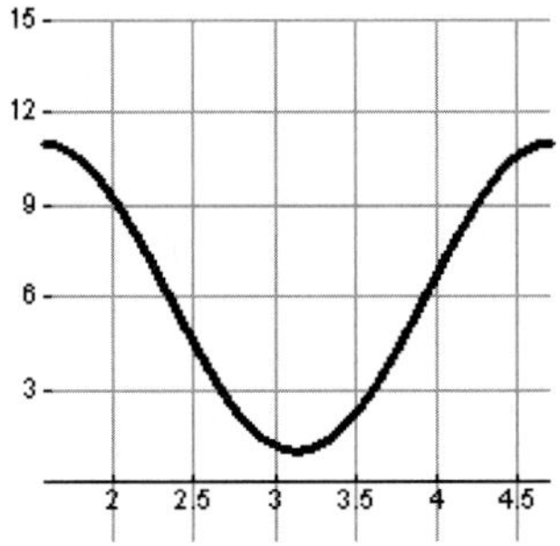
24. The solid curve is the desired graph: 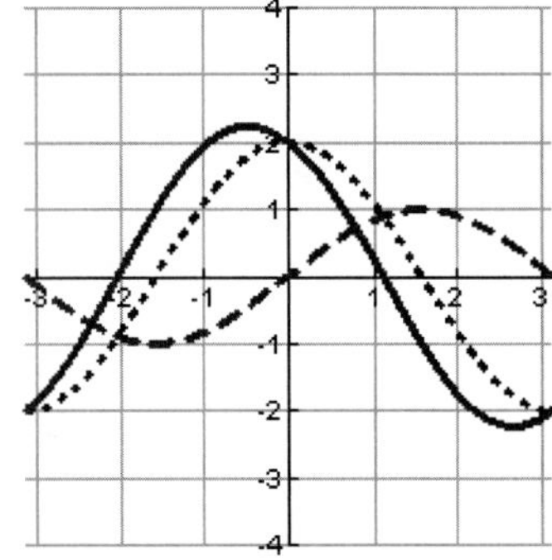	

Chapter 5 Cumulative Test ---

<table>
<tr><td>1. Must have $15+3x>0$, so that $x>-5$. So, the domain is $(-5,\infty)$.</td><td>2. Identify February as 0 and so May corresponds to 3. So, average rate of change is $\dfrac{2.39-1.94}{3-0}=\$0.15$ per month.</td></tr>
<tr><td>3. The desired function is $y=2\left|x+6\right|+4$, whose graph is given by
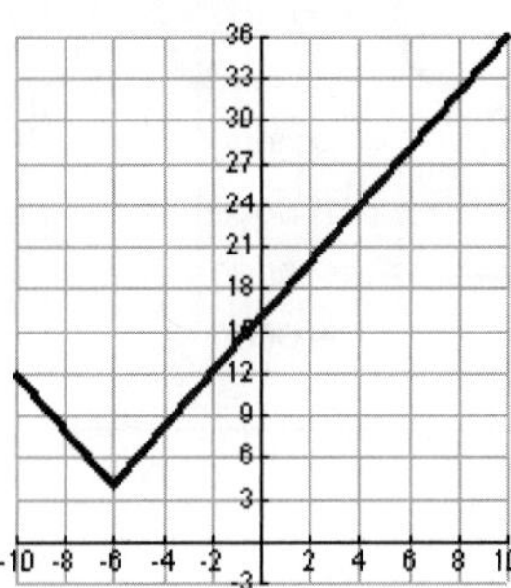</td><td>4.
$f(x)+g(x)=(2x-5)+(x^2+7)=x^2+2x+2$
$f(x)-g(x)=(2x-5)-(x^2+7)=-x^2+2x-12$
$f(x)\bullet g(x)=(2x-5)\bullet(x^2+7)=2x^3-5x^2+14x-35$
$\dfrac{f(x)}{g(x)}=\dfrac{2x-5}{x^2+7}$
$(f\circ g)(x)=2(x^2+7)-5=2x^2+9$</td></tr>
<tr><td colspan="2">5. First, note that
$$dom(f)=rng(f^{-1})=\left(-\infty,-\tfrac{5}{3}\right)\cup\left(-\tfrac{5}{3},\infty\right)$$ $$rng(f)=dom(f^{-1})=\left(-\infty,\tfrac{1}{3}\right)\cup\left(\tfrac{1}{3},\infty\right)$$ Now, the inverse is determined as follows:
$$y=\frac{x-2}{3x+5}$$ $$y(3x+5)=x-2$$ $$3yx+5y=x-2$$ $$x(3y-1)=-5y-2$$ $$x=\frac{-5y-2}{3y-1}=\frac{5y+2}{1-3y}$$ So, $f^{-1}(x)=\dfrac{5x+2}{1-3x}$.</td></tr>
</table>

6. a. 0 (multiplicity 2)
3 (multiplicity 3)
-5 (multiplicity 4)

b.

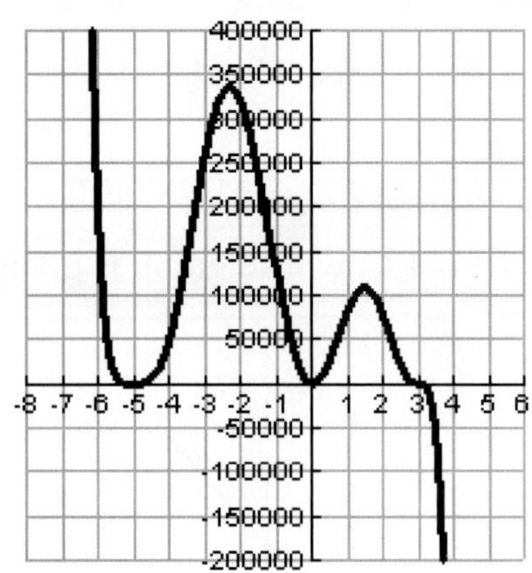

7.

$$\begin{array}{r} 3x^2 - \tfrac{5}{2}x + \tfrac{9}{2} \\ 2x^2 + 0x - 1 \overline{\big) 6x^4 - 5x^3 + 6x^2 + 7x - 4} \\ \underline{-\left(6x^4 + 0x^3 - 3x^2\right)} \\ -5x^3 + 9x^2 + 7x \\ \underline{-\left(-5x^3 + 0x^2 + \tfrac{5}{2}x\right)} \\ 9x^2 + \tfrac{9}{2}x - 4 \\ \underline{-\left(9x^2 + 0x - \tfrac{9}{2}\right)} \\ \tfrac{9}{2}x + \tfrac{1}{2} \end{array}$$

$Q(x) = 3x^2 - \frac{5}{2}x + \frac{9}{2}, \quad r(x) = \frac{9}{2}x + \frac{1}{2}$

8. a. Possible rational zeros are:

$$\pm 1, \pm 2, \pm 3, \pm 6$$

Using synthetic division yields

$$\begin{array}{r|rrrrr} -3 & 1 & 1 & -7 & -1 & 6 \\ & & -3 & 6 & 3 & -6 \\ \hline -1 & 1 & -2 & -1 & 2 & 0 \\ & & -1 & 3 & -2 & \\ \hline 1 & 1 & -3 & 2 & & \\ & & 1 & -2 & & \\ \hline & 1 & -2 & 0 & & \end{array}$$

Thus, $f(x) = (x+3)(x+1)(x-1)(x-2)$

b.

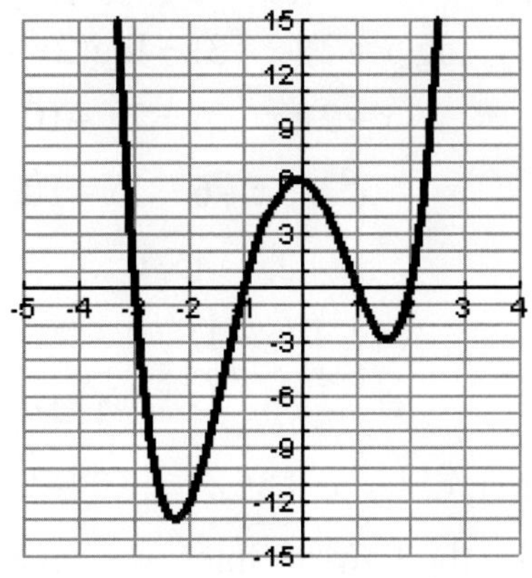

9. $P(x) = \left(x^2 - 5\right)\left(x^2 + 1\right) = (x - \sqrt{5})(x + \sqrt{5})(x - i)(x + i)$

10. a. Observe that

$$f(x) = \frac{(x-7)(x+2)}{2(x+5)(x+2)} = \frac{x-7}{2(x+5)}$$

VA: $x = -5$

HA: $y = \frac{1}{2}$

No slant asymptotes

Note: There is an open hole at $x = -2$ (not visible on the graph).

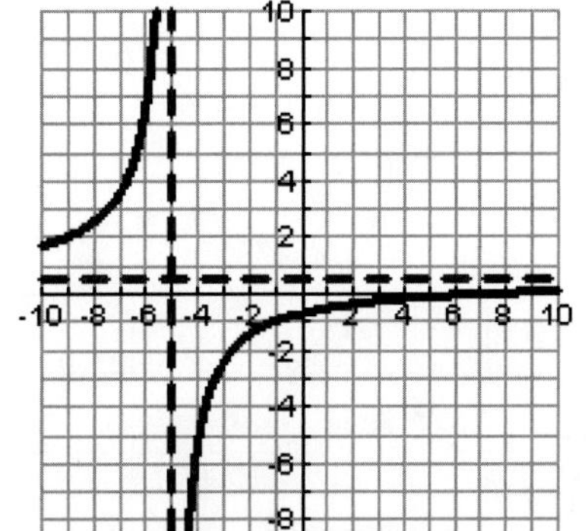

11. Use $A = P\left(1 + \frac{r}{n}\right)^{nt}$.

Here, $P = 3000, r = 0.012, n = 4, t = 10$.

So, $A = 3000\left(1 + \frac{0.012}{4}\right)^{4(10)} \approx \$3,382$.

12. $\log_2 19 = \frac{\log 19}{\log 2} \approx 4.25$

13.

$$\ln\left(a^3\right) - \ln\left(b^2 c^5\right) = \ln\left(a^3\right) - \left[\ln\left(b^2 c^5\right) - \ln\left(c^5\right)\right]$$
$$= 3\ln a - 2\ln b - 5\ln c$$

14.

$$4^{3x-2} = 18$$
$$\log_4 18 = 3x - 2$$
$$x = \tfrac{1}{3}\left(2 + \log_4 18\right) \approx 1.362$$

15.

$$\log(x+2)+\log(x+3)=\log(2x+10)$$
$$\log(x+2)(x+3)=\log(2x+10)$$
$$(x+2)(x+3)=2x+10$$
$$x^2+5x+6=2x+10$$
$$x^2+3x-4=0$$
$$(x+4)(x-1)=0$$
$$x=\not{-4},1$$

16.

$$A=\tfrac{1}{2}r^2\theta=\tfrac{1}{2}(6.5cm)^2\left(\tfrac{4\pi}{5}\right)$$
$$\approx 53\,cm^2$$

17. First, observe that $\alpha=180^\circ-\left(90^\circ+27^\circ\right)=63^\circ$.

Next, by the Law of Sines, we have

$$\frac{\sin 90^\circ}{14\,in.}=\frac{\sin 27^\circ}{b}\Rightarrow b=\frac{(14\,in.)\left(\sin 27^\circ\right)}{\sin 90^\circ}\approx 6.36\,in.$$

Finally, using the Pythagorean theorem, we see that

$$a^2+(6.36\,in.)^2=(14\,in.)^2\Rightarrow a\approx 12.47\,in.$$

18.

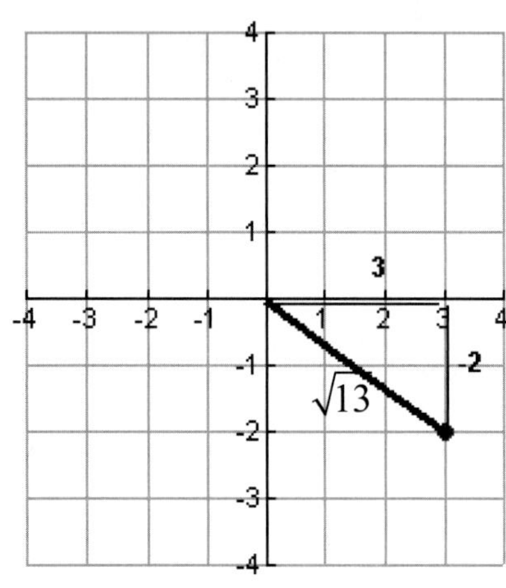

$$\sin\theta=\frac{\text{opp}}{\text{hyp}}=\frac{-2}{\sqrt{13}}=\frac{-2\sqrt{13}}{13}$$
$$\cos\theta=\frac{\text{adj}}{\text{hyp}}=\frac{3}{\sqrt{13}}=\frac{3\sqrt{13}}{13}$$
$$\tan\theta=\frac{\text{opp}}{\text{adj}}=-\frac{2}{3}$$
$$\cot\theta=\frac{1}{\tan\theta}=-\frac{3}{2}$$
$$\sec\theta=\frac{1}{\cos\theta}=\frac{\sqrt{13}}{3}$$
$$\csc\theta=\frac{1}{\sin\theta}=-\frac{\sqrt{13}}{2}$$

19. Solution 1: Using the Law of Sines, we have

$$\frac{\sin 68^\circ}{24m} = \frac{\sin\beta}{24.5m} \Rightarrow \sin\beta = \frac{(24.5m)(\sin 68^\circ)}{24m} \Rightarrow \beta \approx 71.17^\circ.$$

Thus, we see that $\gamma = 180^\circ - (68^\circ + 71.17^\circ) = 40.83^\circ$.

Finally, another application of the Law of Sines yields

$$\frac{\sin 68^\circ}{24m} = \frac{\sin 40.83^\circ}{c} \Rightarrow c = \frac{(24.5m)(\sin 40.83^\circ)}{\sin 68^\circ} \Rightarrow c \approx 16.92m$$

Solution 2: A second possibility in QII for β is $\beta = 180^\circ - 71.17^\circ = 108.83^\circ$.

Note that in such case $\gamma = 180^\circ - (108.83^\circ + 68^\circ) = 3.17^\circ$. Then, using the Law of Sines yields

$$\frac{\sin 68^\circ}{24m} = \frac{\sin 3.17^\circ}{c} \Rightarrow c = \frac{(24.5m)(\sin 3.17^\circ)}{\sin 68^\circ} \Rightarrow c \approx 1.43m$$

20. Using the Law of Cosines yields

$$7^2 = 5^2 + 6^2 - 2(5)(6)\cos\gamma$$

$$\tfrac{1}{5} = \cos\gamma$$

$$\gamma \approx 78.46^\circ$$

Next, using the Law of Sines yields

$$\frac{\sin 78.46^\circ}{7} = \frac{\sin\beta}{6}$$

$$\sin\beta = \frac{6\sin 78.46^\circ}{7}$$

$$\beta \approx 57.1^\circ$$

Thus,

$$\alpha = 180^\circ - (78.46^\circ + 57.1^\circ) = 44.4^\circ$$

21.

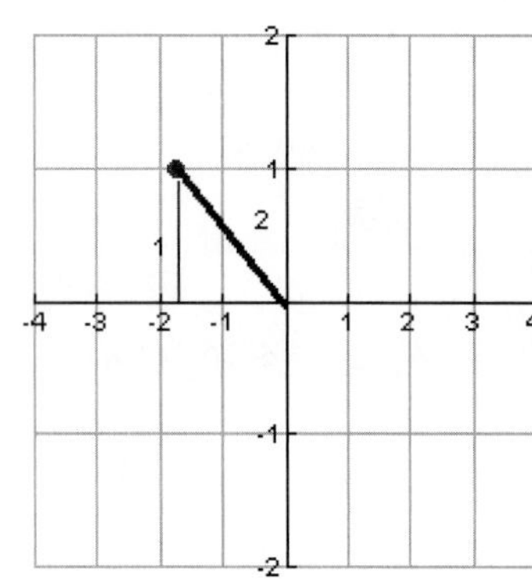

$$\cos\theta = \frac{\text{adj}}{\text{hyp}} = -\frac{\sqrt{3}}{2}$$

$$\tan\theta = \frac{\text{opp}}{\text{adj}} = -\frac{1}{\sqrt{3}} = -\frac{\sqrt{3}}{3}$$

$$\cot\theta = \frac{1}{\tan\theta} = -\sqrt{3}$$

$$\sec\theta = \frac{1}{\cos\theta} = -\frac{2}{\sqrt{3}} = -\frac{2\sqrt{3}}{3}$$

$$\csc\theta = \frac{1}{\sin\theta} = 2$$

22. Amplitude = 4

Period = $\dfrac{2\pi}{2} = \pi$

Phase Shift = $-\frac{\pi}{2}$ (left)

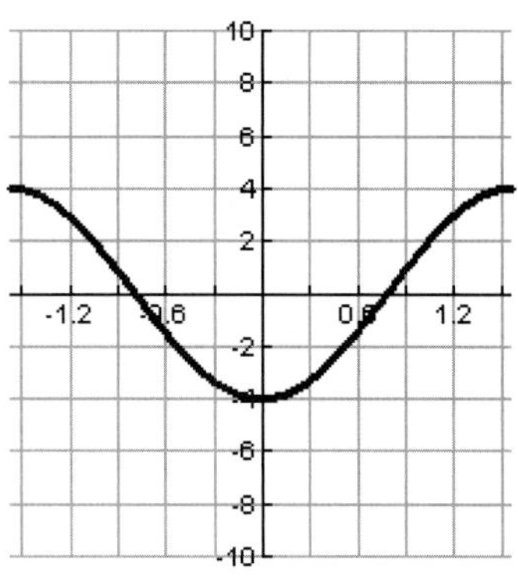

23. The period p is $\frac{2\pi}{4} = \frac{\pi}{2}$. So,

$$f = \tfrac{1}{p} = \tfrac{2}{\pi}.$$

24. The range is $(-\infty, -7] \cup [3, \infty)$

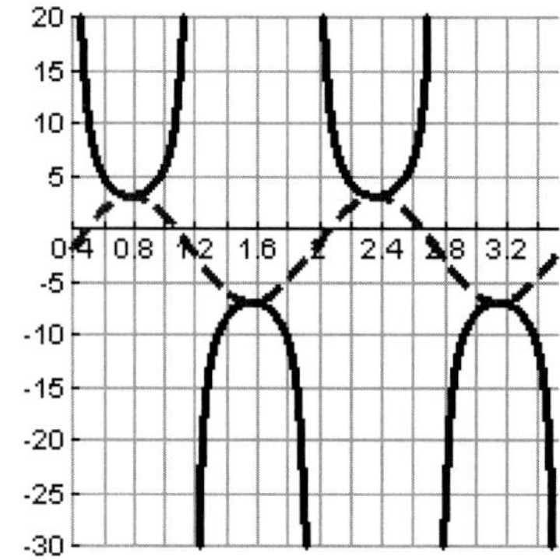

CHAPTER 6

Section 6.1 Solutions--

1. $\sin(60^\circ) = \cos(90^\circ - 60^\circ) = \cos\left(\boxed{30^\circ}\right)$

2. $\sin(45^\circ) = \cos(90^\circ - 45^\circ) = \cos\left(\boxed{45^\circ}\right)$

3. $\cos(x) = \sin\left(\boxed{90^\circ - x}\right)$

4. $\cot(A) = \tan\left(\boxed{90^\circ - A}\right)$

5. $\csc(30^\circ) = \sec(90^\circ - 30^\circ) = \sec\left(\boxed{60^\circ}\right)$

6. $\sec(B) = \csc\left(\boxed{90^\circ - B}\right)$

7.
$$\sin(x+y) = \cos(90^\circ - (x+y)) = \cos\left(\boxed{90^\circ - x - y}\right)$$

8.
$$\sin(60^\circ - x) = \cos(90^\circ - (60^\circ - x)) = \cos\left(\boxed{30^\circ + x}\right)$$

9.
$$\cos(20^\circ + A) = \sin(90^\circ - (20^\circ + A)) = \sin\left(\boxed{70^\circ - A}\right)$$

10.
$$\cos(A+B) = \sin(90^\circ - (A+B)) = \sin\left(\boxed{90^\circ - A - B}\right)$$

11.
$$\cot(45^\circ - x) = \tan(90^\circ - (45^\circ - x)) = \tan\left(\boxed{45^\circ + x}\right)$$

12.
$$\sec(30^\circ - \theta) = \csc(90^\circ - (30^\circ - \theta)) = \csc\left(\boxed{60^\circ + \theta}\right)$$

13.
$$\csc(60^\circ - \theta) = \sec(90^\circ - (60^\circ - \theta)) = \sec(30^\circ + \theta)$$

14.
$$\tan(40^\circ + \theta) = \cot(90^\circ - (40^\circ + \theta)) = \cot(50^\circ - \theta)$$

15. $\sin x \csc x = \cancel{\sin x}\left(\dfrac{1}{\cancel{\sin x}}\right) = \boxed{1}$

16. $\tan x \cot x = \left(\dfrac{\cancel{\sin x}}{\cancel{\cos x}}\right)\left(\dfrac{\cancel{\cos x}}{\cancel{\sin x}}\right) = \boxed{1}$

17. $\sec(-x)\cot x = \dfrac{1}{\cos(-x)}\left(\dfrac{\cos x}{\sin x}\right) = \dfrac{1}{\cancel{\cos(x)}}\left(\dfrac{\cancel{\cos x}}{\sin x}\right) = \dfrac{1}{\sin(x)} = \boxed{\csc x}$

18. $\tan(-x)\cos(-x) = \dfrac{\sin(-x)}{\cancel{\cos(-x)}}\cancel{\cos(-x)} = \sin(-x) = \boxed{-\sin x}$

19. $\csc(-x)\sin x = \dfrac{1}{\sin(-x)}\sin x = \dfrac{1}{-\cancel{\sin x}}\cancel{\sin x} = \boxed{-1}$

20. $\cot(-x)\tan x = -\cot x \tan x = -\dfrac{1}{\cancel{\tan x}}\cancel{\tan x} = \boxed{-1}$

21. $\sec x\cos(-x) + \tan^2 x = \dfrac{1}{\cancel{\cos x}}\cancel{\cos x} + \tan^2 x = 1 + \tan^2 x = \boxed{\sec^2 x}$

22. $\sec(-x)\tan(-x)\cos(-x) = \dfrac{1}{\cancel{\cos(-x)}}[-\tan x]\cancel{\cos(-x)} = \boxed{-\tan x}$

23. $\sin^2 x\left(\cot^2 x + 1\right) = \sin^2 x \csc^2 x = \cancel{\sin^2 x}\left(\dfrac{1}{\cancel{\sin^2 x}}\right) = \boxed{1}$

24. $\cos^2 x\left(\tan^2 x + 1\right) = \cos^2 x \sec^2 x = \cancel{\cos^2 x}\left(\dfrac{1}{\cancel{\cos^2 x}}\right) = \boxed{1}$

25. $\left(\sin x - \cos x\right)\left(\sin x + \cos x\right) = \boxed{\sin^2 x - \cos^2 x}$

26.
$\left(\sin x + \cos x\right)^2 = \sin^2 x + 2\sin x\cos x + \cos^2 x = \left(\sin^2 x + \cos^2 x\right) + 2\sin x\cos x = \boxed{1 + 2\sin x\cos x}$

27. $\dfrac{\csc x}{\cot x} = \dfrac{\dfrac{1}{\cancel{\sin x}}}{\dfrac{\cos x}{\cancel{\sin x}}} = \dfrac{1}{\cos x} = \boxed{\sec x}$

28. $\dfrac{\sec x}{\tan x} = \dfrac{\dfrac{1}{\cancel{\cos x}}}{\dfrac{\sin x}{\cancel{\cos x}}} = \dfrac{1}{\sin x} = \boxed{\csc x}$

29. $\dfrac{1 - \cot(-x)}{1 + \cot x} = \dfrac{1 + \cot x}{1 + \cot x} = \boxed{1}$

30.
$$\begin{aligned}\sec^2 x - \tan^2(-x) &= \sec^2 x - [\tan(-x)]^2 \\ &= \sec^2 x - [-\tan x]^2 \\ &= \sec^2 x - \tan^2 x \\ &= 1 + \tan^2 x - \tan^2 x = \boxed{1}\end{aligned}$$

31. $\dfrac{1 - \cos^4 x}{1 + \cos^2 x} = \dfrac{\left(1 - \cos^2 x\right)\cancel{\left(1 + \cos^2 x\right)}}{\cancel{1 + \cos^2 x}} = 1 - \cos^2 x = \boxed{\sin^2 x}$

32. $\dfrac{1 - \sin^4 x}{1 + \sin^2 x} = \dfrac{\left(1 - \sin^2 x\right)\cancel{\left(1 + \sin^2 x\right)}}{\cancel{1 + \sin^2 x}} = 1 - \sin^2 x = \boxed{\cos^2 x}$

33 $\dfrac{1 - \cot^4 x}{1 - \cot^2 x} = \dfrac{\left(1 + \cot^2 x\right)\cancel{\left(1 - \cot^2 x\right)}}{\cancel{1 - \cot^2 x}} = 1 + \cot^2 x = \boxed{\csc^2 x}$

34.

$$\frac{1-\tan^4(-x)}{1-\tan^2 x}=\frac{1-[\tan(-x)]^4}{1-\tan^2 x}=\frac{1-\tan^4 x}{1-\tan^2 x}=\frac{(1+\tan^2 x)\cancel{(1-\tan^2 x)}}{\cancel{1-\tan^2 x}}=1+\tan^2 x=\boxed{\sec^2 x}$$

35.

$$1-\frac{\sin^2 x}{1-\cos x}=1-\frac{1-\cos^2 x}{1-\cos x}=1-\frac{\cancel{(1-\cos x)}(1+\cos x)}{\cancel{1-\cos x}}=1-(1+\cos x)=1-1-\cos x=\boxed{-\cos x}$$

36.

$$1-\frac{\cos^2 x}{1+\sin x}=1-\frac{1-\sin^2 x}{1+\sin x}=1-\frac{\cancel{(1+\sin x)}(1-\sin x)}{\cancel{1+\sin x}}=1-(1-\sin x)$$

$$=1-1+\sin x=\boxed{\sin x}$$

37.

$$\frac{\tan x-\cot x}{\tan x+\cot x}+2\cos^2 x=\frac{\frac{\sin x}{\cos x}-\frac{\cos x}{\sin x}}{\frac{\sin x}{\cos x}+\frac{\cos x}{\sin x}}+2\cos^2 x=\frac{\frac{\sin^2 x-\cos^2 x}{\cancel{\sin x\cos x}}}{\frac{\sin^2 x+\cos^2 x}{\cancel{\sin x\cos x}}}+2\cos^2 x$$

$$=\frac{\sin^2 x-\cos^2 x}{\underbrace{\sin^2 x+\cos^2 x}_{=1}}+2\cos^2 x=\sin^2 x-\cos^2 x+2\cos^2 x$$

$$=\sin^2 x+\cos^2 x=\boxed{1}$$

38.

$$\frac{\tan x-\cot x}{\tan x+\cot x}+\cos^2 x=\frac{\frac{\sin x}{\cos x}-\frac{\cos x}{\sin x}}{\frac{\sin x}{\cos x}+\frac{\cos x}{\sin x}}+\cos^2 x=\frac{\frac{\sin^2 x-\cos^2 x}{\cancel{\sin x\cos x}}}{\frac{\sin^2 x+\cos^2 x}{\cancel{\sin x\cos x}}}+\cos^2 x$$

$$=\frac{\sin^2 x-\cos^2 x}{\underbrace{\sin^2 x+\cos^2 x}_{=1}}+\cos^2 x=\sin^2 x-\cos^2 x+\cos^2 x=\boxed{\sin^2 x}$$

39.

$$(\sin x+\cos x)^2+(\sin x-\cos x)^2$$

$$=\underbrace{\left(\sin^2 x+\cancel{2\sin x\cos x}+\cos^2 x\right)}_{\text{What remains after cancellation}=1}+\underbrace{\left(\sin^2 x-\cancel{2\sin x\cos x}+\cos^2 x\right)}_{\text{What remains after cancellation}=1}=2,$$

as claimed.

40. $(1-\sin x)(1+\sin x)=1-\sin^2 x=\cos^2 x$, as claimed.

41. $(\csc x+1)(\csc x-1)=\csc^2 x-1=(1+\cot^2 x)-1=\cot^2 x$, as claimed.
42. $(\sec x+1)(\sec x-1)=\sec^2 x-1=(1+\tan^2 x)-1=\tan^2 x$, as claimed.
43. $\tan x+\cot x=\frac{\sin x}{\cos x}+\frac{\cos x}{\sin x}=\frac{\sin^2 x+\cos^2 x}{\sin x\cos x}=\frac{1}{\sin x\cos x}=\left(\frac{1}{\sin x}\right)\left(\frac{1}{\cos x}\right)=\csc x\sec x$, as claimed.
44. $\csc x-\sin x=\frac{1}{\sin x}-\sin x=\frac{1-\sin^2 x}{\sin x}=\frac{\cos^2 x}{\sin x}=\left(\frac{\cos x}{\sin x}\right)(\cos x)=\cot x\cos x$, as claimed.
45. $\frac{2-\sin^2 x}{\cos x}=\frac{1+(1-\sin^2 x)}{\cos x}=\frac{1+\cos^2 x}{\cos x}=\frac{1}{\cos x}+\frac{\cos^2 x}{\cos x}=\boxed{\sec x+\cos x}$
46. $\frac{2-\cos^2 x}{\sin x}=\frac{1+(1-\cos^2 x)}{\sin x}=\frac{1+\sin^2 x}{\sin x}=\frac{1}{\sin x}+\frac{\sin^2 x}{\sin x}=\boxed{\csc x+\sin x}$
47. $[\cos(-x)-1][1+\cos x]=[\cos x-1][\cos x+1]=\cos^2 x-1=\boxed{-\sin^2 x}$
48. $\tan(-x)\cot x=-\tan x\cot x=-\cancel{\tan x}\,\frac{1}{\cancel{\tan x}}=\boxed{-1}$
49. $\frac{\sec(-x)\cot x}{\csc(-x)}=\frac{\frac{1}{\cos(-x)}\,\frac{\cos x}{\sin x}}{\frac{1}{\sin(-x)}}=\frac{\sin(-x)}{\cos(-x)}\,\frac{\cos x}{\sin x}=-\frac{\sin x}{\cos x}\,\frac{\cos x}{\sin x}=\boxed{-1}$
50. $\frac{\cot^2 x}{\csc(-x)+1}=\frac{\csc^2 x-1}{-\csc x+1}=\frac{\cancel{(\csc x-1)}(\csc x+1)}{-\cancel{(\csc x-1)}}=-(\csc x+1)=-\csc x-1=\csc(-x)-1$, as claimed.
51. $\frac{1}{\csc^2 x}+\frac{1}{\sec^2 x}=\frac{1}{\frac{1}{\sin^2 x}}+\frac{1}{\frac{1}{\cos^2 x}}=\sin^2 x+\cos^2 x=1$, as claimed.

52.

$$\frac{1}{\cot^2 x}-\frac{1}{\tan^2 x}=\frac{\tan^2 x-\cot^2 x}{\underbrace{\tan^2 x\cot^2 x}_{=1}}=\tan^2 x-\cot^2 x=\frac{\sin^2 x}{\cos^2 x}-\frac{\cos^2 x}{\sin^2 x}$$

$$=\frac{\sin^4 x-\cos^4 x}{\sin^2 x\cos^2 x}=\frac{\left(\sin^2 x-\cos^2 x\right)\overbrace{\left(\sin^2 x+\cos^2 x\right)}^{=1}}{\sin^2 x\cos^2 x}$$

$$=\frac{\cancel{\sin^2 x}}{\cancel{\sin^2 x}\cos^2 x}-\frac{\cancel{\cos^2 x}}{\sin^2 x\cancel{\cos^2 x}}=\frac{1}{\cos^2 x}-\frac{1}{\sin^2 x}=\sec^2 x-\csc^2 x,$$

as claimed.

53. $\frac{1}{1-\sin x}+\frac{1}{1+\sin x}=\frac{(1+\sin x)+(1-\sin x)}{(1+\sin x)(1-\sin x)}=\frac{2}{1-\sin^2 x}=\frac{2}{\cos^2 x}=2\sec^2 x$, as claimed.

54. $\frac{1}{1-\cos x}+\frac{1}{1+\cos x}=\frac{(1+\cos x)+(1-\cos x)}{(1+\cos x)(1-\cos x)}=\frac{2}{1-\cos^2 x}=\frac{2}{\sin^2 x}=2\csc^2 x$, as claimed.

55. $\frac{\sin^2 x}{1-\cos x}=\frac{1-\cos^2 x}{1-\cos x}=\frac{\cancel{(1-\cos x)}(1+\cos x)}{\cancel{1-\cos x}}=1+\cos x$, as claimed.

56. $\frac{\cos^2 x}{1-\sin x}=\frac{1-\sin^2 x}{1-\sin x}=\frac{\cancel{(1-\sin x)}(1+\sin x)}{\cancel{1-\sin x}}=1+\sin x$, as claimed.

57.

$$\sec x+\tan x=\frac{(\sec x+\tan x)(\sec x-\tan x)}{\sec x-\tan x}=\frac{\sec^2 x-\tan^2 x}{\sec x-\tan x}$$

$$=\frac{\left(1+\cancel{\tan^2 x}\right)-\cancel{\tan^2 x}}{\sec x-\tan x}=\frac{1}{\sec x-\tan x},$$

as claimed.

58.

$$\csc x+\cot x=\frac{(\csc x+\cot x)(\csc x-\cot x)}{\csc x-\cot x}=\frac{\csc^2 x-\cot^2 x}{\csc x-\cot x}$$

$$=\frac{\left(1+\cancel{\cot^2 x}\right)-\cancel{\cot^2 x}}{\csc x-\cot x}=\frac{1}{\csc x-\cot x},$$

as claimed.

59. $\dfrac{\csc x-\tan x}{\sec x+\cot x}=\dfrac{\dfrac{1}{\sin x}-\dfrac{\sin x}{\cos x}}{\dfrac{1}{\cos x}+\dfrac{\cos x}{\sin x}}=\dfrac{\dfrac{\cos x-\sin^2 x}{\cancel{\sin x\cos x}}}{\dfrac{\sin x+\cos^2 x}{\cancel{\sin x\cos x}}}=\dfrac{\cos x-\sin^2 x}{\sin x+\cos^2 x}$, as claimed.

60. $\dfrac{\sec x+\tan x}{\csc x+1}=\dfrac{\dfrac{1}{\cos x}+\dfrac{\sin x}{\cos x}}{\dfrac{1}{\sin x}+1}=\dfrac{\dfrac{1+\sin x}{\cos x}}{\dfrac{1+\sin x}{\sin x}}=\dfrac{\cancel{1+\sin x}}{\cos x}\cdot\dfrac{\sin x}{\cancel{1+\sin x}}=\dfrac{\sin x}{\cos x}=\tan x$, as claimed.

61.

$$\frac{\cos^2 x+1+\sin x}{\cos^2 x+3}=\frac{\left(1-\sin^2 x\right)+1+\sin x}{\left(1-\sin^2 x\right)+3}=\frac{-\left[\sin^2 x-\sin x-2\right]}{-\left[\sin^2 x-4\right]}$$

$$=\frac{-(\sin x+1)\,\cancel{(\sin x-2)}}{-\cancel{(\sin x-2)}\,(\sin x+2)}=\frac{\sin x+1}{\sin x+2}=\frac{1+\sin x}{2+\sin x},$$

as claimed.

62.

$$\frac{\sin x+1-\cos^2 x}{\cos^2 x}=\frac{\sin x+1-\left(1-\sin^2 x\right)}{1-\sin^2 x}=\frac{\sin^2 x+\sin x}{1-\sin^2 x}=\frac{\sin x\,\cancel{(\sin x+1)}}{(1-\sin x)\,\cancel{(1+\sin x)}}$$

$$=\frac{\sin x}{1-\sin x},$$

as claimed.

63.

$$\sec x\left(\tan x+\cot x\right)=\frac{1}{\cos x}\left[\frac{\sin x}{\cos x}+\frac{\cos x}{\sin x}\right]=\frac{1}{\cos x}\left[\frac{\overbrace{\sin^2 x+\cos^2 x}^{=1}}{\cos x\sin x}\right]=\frac{1}{\cos x}\left[\frac{1}{\cos x\sin x}\right]$$

$$=\left(\frac{1}{\cos^2 x}\right)\left(\frac{1}{\sin x}\right)=\left(\frac{1}{\cos^2 x}\right)\left(\csc x\right)=\frac{\csc x}{\cos^2 x},$$

as claimed.

64. $\tan x\left(\csc x-\sin x\right)=\dfrac{\sin x}{\cos x}\left(\dfrac{1}{\sin x}-\sin x\right)=\dfrac{\cancel{\sin x}}{\cos x}\left(\dfrac{1-\sin^2 x}{\cancel{\sin x}}\right)=\dfrac{\cos^2 x}{\cos x}=\cos x$, as claimed.

65. Observe that the left-side of the equation simplifies as follows:

$$\begin{aligned}\cos^2 x(\tan x-\sec x)(\tan x+\sec x) &= \cos^2 x\left(\tan^2 x-\sec^2 x\right)\\ &= \cos^2 x\left(\tan^2 x-\left(1+\tan^2 x\right)\right)\\ &= -\cos^2 x\end{aligned}$$

Since $-\cos^2 x$ is, in fact, never equal to 1 (being a non-positive quantity), the given equation is a *conditional*.

66. Observe that

$$\begin{aligned}\cos^2 x(\tan x-\sec x)(\tan x+\sec x) &= \cos^2 x\left(\tan^2 x-\sec^2 x\right)\\ &= \cos^2 x\left(\tan^2 x-\left(1+\tan^2 x\right)\right)\\ &= -\cos^2 x = -(1-\sin^2 x) = \sin^2 x - 1\end{aligned}$$

So, the given equation is an *identity*.

67. Observe that

$$\frac{\csc x\cot x}{\sec x\tan x} = \frac{\frac{1}{\sin x}\cot x}{\frac{1}{\cos x}\tan x} = \left(\frac{1}{\sin x}\cot x\right)\left(\frac{1}{\frac{1}{\cos x}\tan x}\right) = \left(\frac{1}{\sin x}\cot x\right)\left(\cos x\frac{1}{\tan x}\right) = \underbrace{\frac{\cos x}{\sin x}}_{=\cot x}\cot x\underbrace{\frac{1}{\tan x}}_{=\cot x} = \cot^3 x$$

So, the given equation is an *identity*.

68. The equation $\sin x\cos x = 0$ is a *conditional* since, for instance, it does not hold if $x=\frac{\pi}{4}$ (since the left-side reduces to $\frac{1}{2}$ in such case).

69. The equation $\sin x+\cos x=\sqrt{2}$ is a *conditional* since, for instance, it does not hold if $x=0$ (since the left-side reduces to 1).

70. The equation $\sin^2 x+\cos^2 x=1$ is a known *identity*.

71. Observe that

$$\tan^2 x-\sec^2 x=\tan^2 x-\left(1+\tan^2 x\right)=-1\neq 1.$$

So, the given equation is a *conditional* (which, incidentally, is never true).

72. Observe that

$$\sec^2 x-\tan^2 x=\left(1+\tan^2 x\right)-\tan^2 x=1.$$

So, the given equation is an *identity*.

73. The equation $\sin x=\sqrt{1-\cos^2 x}$ is a *conditional*. To see this, observe that the right-side reduces as follows: $\sqrt{1-\cos^2 x}=\sqrt{\sin^2 x}=|\sin x|$. As such, the given equation is not true for $x=\frac{3\pi}{2}$, for instance.

(**Note:** The equation IS an identity if one restricts attention to only those x-values for which $\sin x>0$.)

74. The equation $\csc x = \sqrt{1+\cot^2 x}$ is a *conditional.* To see this, observe that the right-side reduces as follows: $\sqrt{1+\cot^2 x} = \sqrt{\csc^2 x} = |\csc x|$. As such, the given equation is not true for $x = \frac{3\pi}{2}$, for instance.
(**Note:** The equation IS an identity if one restricts attention to only those x-values for which $\sin x > 0$ (and hence, $\csc x > 0$).)

75. Observe that $\sqrt{\sin^2 x + \cos^2 x} = \sqrt{1} = 1$. So, the given equation is an *identity.*

76. Observe that $\sqrt{\sin^2 x + \cos^2 x} = \sqrt{1} = 1$. So, the equation $\sqrt{\sin^2 x + \cos^2 x} = \sin x + \cos x$ is an *conditional.* Indeed, take for instance $x = \frac{\pi}{4}$. Then, the left-side = 1, while the right-side is $\sqrt{2}$.
(**Note**: A common algebraic error is to say $\sqrt{a^2+b^2} = a+b$, which is false.)

77. Conditional. Note that it is false when $x = 0$.

78. Conditional. This holds only when $x = \frac{2n+1}{2}\pi$

79. $\pi\left(\sec^2 x\right) = \pi\left(1+\tan^2 x\right) = \pi + \pi\tan^2 x$

80. $\frac{1}{2}(\cos x)(\sec x) = \frac{1}{2}(\cos x)\left(\dfrac{1}{\cos x}\right) = \frac{1}{2}$

81. $1^2 + \tan^2\theta = h^2 \Rightarrow h = \sqrt{1^2+\tan^2\theta} = \sqrt{\sec^2\theta} = |\sec\theta|$.

82. $1^2 + \cot^2\theta = h^2 \Rightarrow h = \sqrt{1^2+\cot^2\theta} = \sqrt{\csc^2\theta} = |\csc\theta|$

83. Simplified the two fractions in Step 2 incorrectly. The correct computations are:

$$\frac{\cos x}{1-\dfrac{\sin x}{\cos x}} = \frac{\cos x}{\dfrac{\cos x - \sin x}{\cos x}} = \frac{\cos^2 x}{\cos x - \sin x}$$

$$\frac{\sin x}{1-\dfrac{\cos x}{\sin x}} = \frac{\sin x}{\dfrac{\sin x - \cos x}{\sin x}} = \frac{\sin^2 x}{\sin x - \cos x}$$

84. There are two errors, as follows:

1. $\sec x = \dfrac{1}{\cos x}$, not $\dfrac{1}{\sin x}$.

2. $\dfrac{\frac{1}{\sin x}}{1-\sin x} = \dfrac{1}{\sin x(1-\sin x)} = \dfrac{1}{\sin x - \sin^2 x}$, not $\dfrac{1}{1-\sin^2 x}$.

85. Need to observe that $\tan^2 x = 1$ is a *conditional*, which does not hold for all values of x. Verifying the truth of the equation for a particular value of x (here, $x = \frac{\pi}{4}$) is not enough to prove that it is an identity.

86. Verifying the truth of the equation for particular values of x (here, odd multiples of $\frac{\pi}{2}$) is not enough to prove that it is an identity. Indeed, the given equation is actually a conditional. To see this, take $x = 0$. Then, $|\sin 0| - \cos 0 = -1 \neq 1$.

87. False. The equation $\sin^2 x = 1$ is not an identity since even though it is true for $x = \frac{\pi}{2} + n\pi$ (n is an integer), it is false for all other values of x for which $\sin x$ is defined.

88. False. The equation $\sin x = 1$ has infinitely many solutions, but it is not always true.

89. The equation $\cos\theta = \sqrt{1-\sin^2\theta}$ is true whenever $\cos\theta > 0$ (since $\sqrt{1-\sin^2\theta} = \sqrt{\cos^2\theta} = |\cos\theta|$.) This occurs for those angles θ whose terminal side is in QI or QIV.

90. The equation $-\cos\theta = \sqrt{1-\sin^2\theta}$ is true whenever $\cos\theta < 0$ (since $\sqrt{1-\sin^2\theta} = \sqrt{\cos^2\theta} = |\cos\theta|$.) This occurs for those angles θ whose terminal side is in QII or QIII.

91. $\csc\theta = -\sqrt{\csc^2\theta}$ occurs for those angles θ whose terminal side is in QIII or QIV.

92. $\sec\theta = \sqrt{1+\tan^2\theta}$ occurs for those angles θ whose terminal side is in QI or QIV.

93. The equation $\sin(A+B) = \sin A + \sin B$ is not true in general. For instance, let $A = 30°$ and $B = 60°$. Then, $\sin\left(30° + 60°\right) = \sin\left(90°\right) = 1$, whereas

$$\sin\left(30°\right) + \sin\left(60°\right) = \frac{1}{2} + \frac{\sqrt{3}}{2} = \frac{1+\sqrt{3}}{2} \neq 1.$$

94. The equation $\cos\left(\frac{1}{2}A\right) = \frac{1}{2}\cos A$ is not true in general. For instance, let $A = \frac{\pi}{2}$.

Then, $\cos\left(\frac{1}{2}\cdot\frac{\pi}{2}\right) = \cos\left(\frac{\pi}{4}\right) = \frac{\sqrt{2}}{2}$, whereas $\frac{1}{2}\cos\left(\frac{\pi}{2}\right) = 0$.

95. No. For instance, take $A = \frac{\pi}{4}$. Note that $\tan(2A)$ is not defined, whereas $2\tan\left(\frac{\pi}{4}\right) = 2$.

96. No, they are not the same, as is apparent from the difference in their graphs shown below:

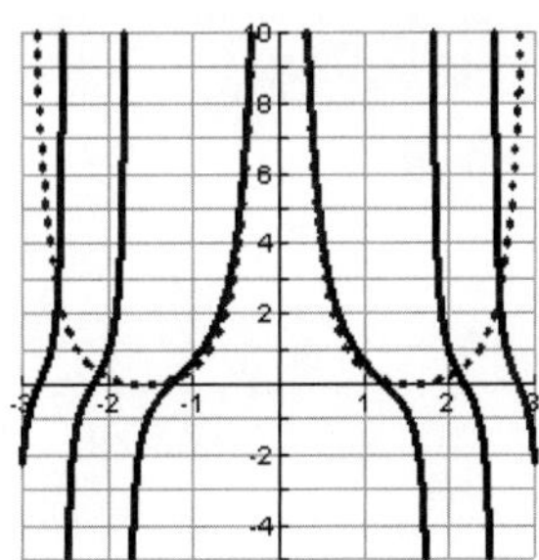

97.

$$(a\sin x+b\cos x)^2+(b\sin x-a\cos x)^2$$
$$=a^2\sin^2 x+\cancel{2ab\sin x\cos x}+b^2\cos^2 x+b^2\sin^2 x-\cancel{2ab\sin x\cos x}+a^2\cos^2 x$$
$$=(a^2+b^2)\sin^2 x+(a^2+b^2)\cos^2 x=(a^2+b^2)\underbrace{(\sin^2 x+\cos^2 x)}_{=1}=\boxed{a^2+b^2}$$

98.

$$\frac{1+\cot^3 x}{1+\cot x}+\cot x=\frac{1+\cot^3 x+\cot x(1+\cot x)}{1+\cot x}$$
$$=\frac{1+\cot^3 x+\cot x+\cot^2 x}{1+\cot x}$$
$$=\frac{(1+\cot x)+(\cot^3 x+\cot^2 x)}{1+\cot x}$$
$$=\frac{(1+\cot x)+\cot^2 x(\cot x+1)}{1+\cot x}$$
$$=\frac{\cancel{(1+\cot x)}(1+\cot^2 x)}{\cancel{1+\cot x}}$$
$$=\csc^2 x$$

99. Observe that for any integer n,

$$\csc\left(\tfrac{\pi}{2}+\theta+2n\pi\right)=\frac{1}{\sin\left(\frac{\pi}{2}+\theta+2n\pi\right)}$$
$$=\frac{1}{\sin\left(\frac{\pi}{2}+\theta\right)}$$
$$=\frac{1}{\cos\theta}$$
$$=\sec\theta$$

100. Observe that for any integer n,

$$\sec\left(-\tfrac{\pi}{2}-\theta-2n\pi\right)=\frac{1}{\cos\left(-\left(\frac{\pi}{2}+\theta+2n\pi\right)\right)}$$
$$=\frac{1}{\cos\left(\frac{\pi}{2}+\theta+2n\pi\right)}$$
$$=\frac{1}{\cos\left(\frac{\pi}{2}+\theta\right)}$$
$$=\frac{1}{\sin\theta}=\csc\theta$$

101. Observe that

$$\begin{aligned}
\csc\left(2\pi-\frac{\pi}{2}-\theta\right)\cdot\sec\left(\theta-\frac{\pi}{2}\right)\cdot\sin(-\theta) &= \frac{1}{\sin\left(-\left(\frac{\pi}{2}+\theta\right)+2\pi\right)}\cdot\frac{1}{\cos\left(-\left(\frac{\pi}{2}-\theta\right)\right)}\cdot\sin(-\theta)\\
&= \frac{1}{\sin\left(-\left(\frac{\pi}{2}+\theta\right)\right)}\cdot\frac{1}{\cos\left(-\left(\frac{\pi}{2}-\theta\right)\right)}\cdot\sin(-\theta)\\
&= -\frac{1}{\sin\left(\frac{\pi}{2}+\theta\right)}\cdot\frac{1}{\cos\left(\frac{\pi}{2}-\theta\right)}[-\sin\theta]\\
&= \frac{1}{\cos\theta}\cdot\frac{1}{\sin\theta}\cdot\sin\theta\\
&= \frac{1}{\cos\theta}\\
&= \sec\theta
\end{aligned}$$

102. $\tan\theta\cdot\cot(2\pi-\theta)=\tan\theta\cdot\cot(-\theta)=-\tan\theta\cdot\cot(\theta)=-\tan\theta\cdot\frac{1}{\tan\theta}=-1$

103. The correct identity is $\cos(A+B)=\cos A\cos B\,\boxed{-}\,\sin A\sin B$.

Consider the graphs of the following functions. Note that the graphs of y_1 (**BOLD**) and y_3 are the same. (The graph of y_2 is ***dotted***.)

$$y_1=\cos\left(x+\frac{\pi}{4}\right)$$

$$y_2=\cos(x)\cos\left(\frac{\pi}{4}\right)+\sin(x)\sin\left(\frac{\pi}{4}\right)$$

$$y_3=\cos(x)\cos\left(\frac{\pi}{4}\right)-\sin(x)\sin\left(\frac{\pi}{4}\right)$$

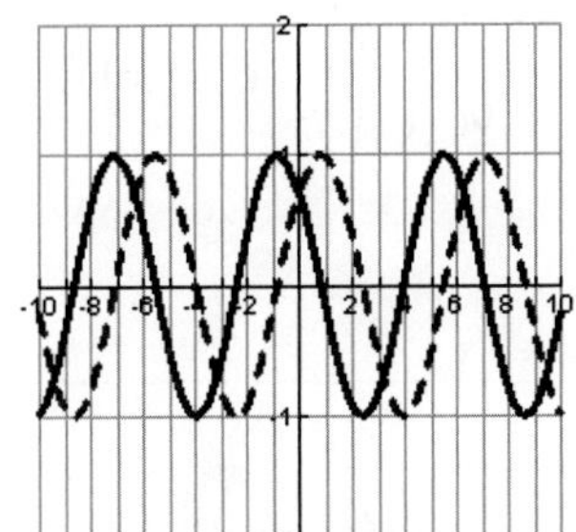

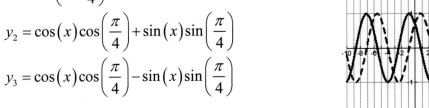

104. The correct identity is $\cos(A-B)=\cos A\cos B\,\boxed{+}\,\sin A\sin B$.

Consider the graphs of the following functions. Note that the graphs of y_1 (**BOLD**) and y_3 are the same. (The graph of y_2 is ***dotted***.)

$$y_1=\cos\left(x-\frac{\pi}{4}\right)$$

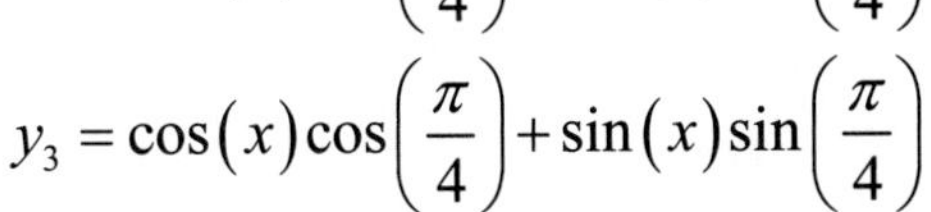

$$y_2=\cos(x)\cos\left(\frac{\pi}{4}\right)-\sin(x)\sin\left(\frac{\pi}{4}\right)$$

$$y_3=\cos(x)\cos\left(\frac{\pi}{4}\right)+\sin(x)\sin\left(\frac{\pi}{4}\right)$$

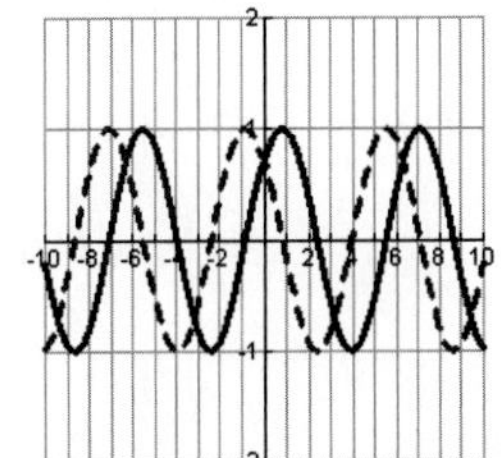

105. The correct identity is $\sin(A+B)=\sin A\cos B \boxed{+} \cos A\sin B$.
Consider the graphs of the following functions. Note that the graphs of y_1 (**BOLD**) and y_2 are the same. (The graph of y_3 is ***dotted***.)

$$y_1=\sin\left(x+\frac{\pi}{4}\right)$$

$$y_2=\sin(x)\cos\left(\frac{\pi}{4}\right)+\cos(x)\sin\left(\frac{\pi}{4}\right)$$

$$y_3=\sin(x)\cos\left(\frac{\pi}{4}\right)-\cos(x)\sin\left(\frac{\pi}{4}\right)$$

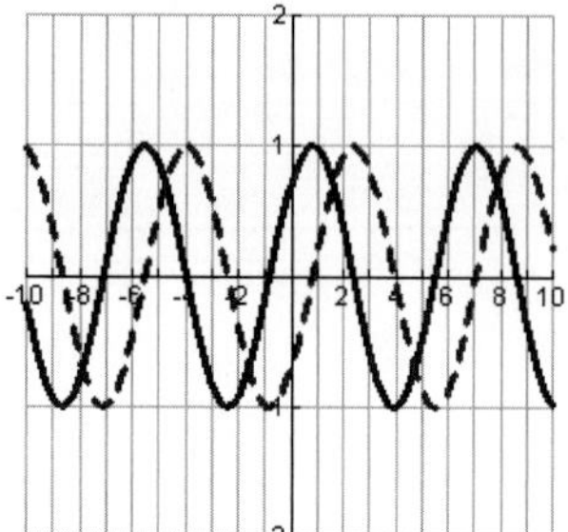

106. The correct identity is $\sin(A-B)=\sin A\cos B \boxed{-} \cos A\sin B$.
Consider the graphs of the following functions. Note that the graphs of y_1 (**BOLD**) and y_3 are the same. (The graph of y_2 is ***dotted***.)

$$y_1=\sin\left(x-\frac{\pi}{4}\right)$$

$$y_2=\sin(x)\cos\left(\frac{\pi}{4}\right)+\cos(x)\sin\left(\frac{\pi}{4}\right)$$

$$y_3=\sin(x)\cos\left(\frac{\pi}{4}\right)-\cos(x)\sin\left(\frac{\pi}{4}\right)$$

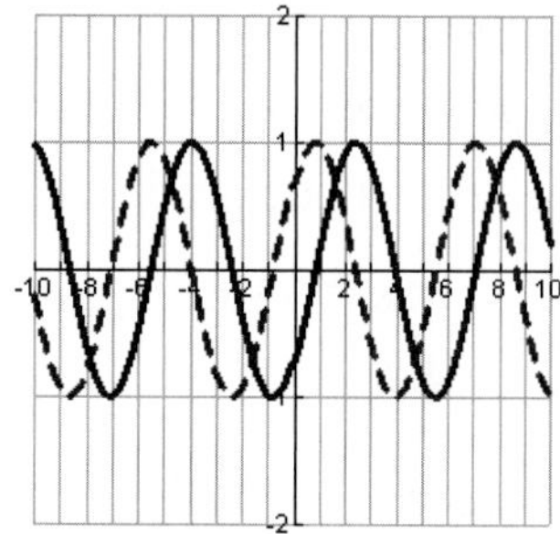

107.

$$\begin{aligned}\sqrt{a^2-(a\sin\theta)^2}&=|a|\sqrt{1-\sin^2\theta}\\&=|a|\sqrt{\cos^2\theta}\\&=|a|\cos\theta,\end{aligned}$$

since $-\frac{\pi}{2}\le\theta\le\frac{\pi}{2}$.

108.

$$\begin{aligned}\sqrt{a^2+(a\tan\theta)^2}&=|a|\sqrt{1+\tan^2\theta}\\&=|a|\sqrt{\sec^2\theta}\\&=|a|\sec\theta,\end{aligned}$$

since $-\frac{\pi}{2}\le\theta\le\frac{\pi}{2}$.

109.

$$\begin{aligned}\sqrt{(a\sec\theta)^2-a^2}&=|a|\sqrt{\sec^2\theta-1}\\&=|a|\sqrt{\tan^2\theta}\\&=|a|\tan\theta,\end{aligned}$$

since $0\le\theta<\frac{\pi}{2}$.

110.

$$\begin{aligned}\sqrt{9-(3\sin\theta)^2}&=|3|\sqrt{1-\sin^2\theta}\\&=|3|\sqrt{\cos^2\theta}\end{aligned}$$

Section 6.2 Solutions --

1. Using the addition formula $\sin(A-B)=\sin A\cos B-\cos A\sin B$ yields

$$\sin\left(\frac{\pi}{12}\right)=\sin\left(\frac{\pi}{3}-\frac{\pi}{4}\right)=\sin\left(\frac{\pi}{3}\right)\cos\left(\frac{\pi}{4}\right)-\cos\left(\frac{\pi}{3}\right)\sin\left(\frac{\pi}{4}\right)$$

$$=\left(\frac{\sqrt{3}}{2}\right)\left(\frac{\sqrt{2}}{2}\right)-\left(\frac{1}{2}\right)\left(\frac{\sqrt{2}}{2}\right)=\left(\frac{\sqrt{2}}{2}\right)\left(\frac{\sqrt{3}}{2}-\frac{1}{2}\right)=\boxed{\frac{\sqrt{6}-\sqrt{2}}{4}}$$

2. Using the addition formula $\cos(A-B)=\cos A\cos B+\sin A\sin B$ yields

$$\cos\left(\frac{\pi}{12}\right)=\cos\left(\frac{\pi}{3}-\frac{\pi}{4}\right)=\cos\left(\frac{\pi}{3}\right)\cos\left(\frac{\pi}{4}\right)+\sin\left(\frac{\pi}{3}\right)\sin\left(\frac{\pi}{4}\right)$$

$$=\left(\frac{1}{2}\right)\left(\frac{\sqrt{2}}{2}\right)+\left(\frac{\sqrt{3}}{2}\right)\left(\frac{\sqrt{2}}{2}\right)=\left(\frac{\sqrt{2}}{2}\right)\left(\frac{1}{2}+\frac{\sqrt{3}}{2}\right)=\boxed{\frac{\sqrt{2}+\sqrt{6}}{4}}$$

3. Using the even identity for cosine, along with the addition formula $\cos(A+B)=\cos A\cos B-\sin A\sin B$ yields

$$\cos\left(-\frac{5\pi}{12}\right)=\cos\left(\frac{5\pi}{12}\right)=\cos\left(\frac{\pi}{6}+\frac{\pi}{4}\right)=\cos\left(\frac{\pi}{6}\right)\cos\left(\frac{\pi}{4}\right)-\sin\left(\frac{\pi}{6}\right)\sin\left(\frac{\pi}{4}\right)$$

$$=\left(\frac{\sqrt{3}}{2}\right)\left(\frac{\sqrt{2}}{2}\right)-\left(\frac{1}{2}\right)\left(\frac{\sqrt{2}}{2}\right)=\left(\frac{\sqrt{2}}{2}\right)\left(\frac{\sqrt{3}}{2}-\frac{1}{2}\right)=\boxed{\frac{\sqrt{6}-\sqrt{2}}{4}}$$

4. Using the odd identity for sine, along with the addition formula $\sin(A+B)=\sin A\cos B+\cos A\sin B$, yields

$$\sin\left(-\frac{5\pi}{12}\right)=-\sin\left(\frac{5\pi}{12}\right)=-\sin\left(\frac{\pi}{6}+\frac{\pi}{4}\right)=-\left[\sin\left(\frac{\pi}{6}\right)\cos\left(\frac{\pi}{4}\right)+\cos\left(\frac{\pi}{6}\right)\sin\left(\frac{\pi}{4}\right)\right]$$

$$=-\left[\left(\frac{1}{2}\right)\left(\frac{\sqrt{2}}{2}\right)+\left(\frac{\sqrt{3}}{2}\right)\left(\frac{\sqrt{2}}{2}\right)\right]=-\left(\frac{\sqrt{2}}{2}\right)\left(\frac{1}{2}+\frac{\sqrt{3}}{2}\right)=\boxed{\frac{-\sqrt{2}-\sqrt{6}}{4}}$$

5. First, we need to compute both $\sin\left(\frac{\pi}{12}\right)$ and $\cos\left(\frac{\pi}{12}\right)$. Then, we use the definition of tangent to compute $\tan\left(\frac{\pi}{12}\right)$. To this end, we have the following:

Step 1: Using the odd identity for sine and Exercise 1 yields

$$\sin\left(-\frac{\pi}{12}\right)=-\sin\left(\frac{\pi}{12}\right)=-\left[\frac{\sqrt{6}-\sqrt{2}}{4}\right]$$

Step 2: Using the even identity for cosine and the addition formula $\cos(A-B)=\cos A\cos B+\sin A\sin B$ yields

$$\cos\left(-\frac{\pi}{12}\right)=\cos\left(\frac{\pi}{12}\right)=\cos\left(\frac{\pi}{3}-\frac{\pi}{4}\right)=\cos\left(\frac{\pi}{3}\right)\cos\left(\frac{\pi}{4}\right)+\sin\left(\frac{\pi}{3}\right)\sin\left(\frac{\pi}{4}\right)$$

$$=\left(\frac{1}{2}\right)\left(\frac{\sqrt{2}}{2}\right)+\left(\frac{\sqrt{3}}{2}\right)\left(\frac{\sqrt{2}}{2}\right)=\left(\frac{\sqrt{2}}{2}\right)\left(\frac{1}{2}+\frac{\sqrt{3}}{2}\right)=\frac{\sqrt{2}+\sqrt{6}}{4}$$

Step 3: Using the definition of tangent and the computations in Steps 1 and 2 yields

$$\tan\left(-\frac{\pi}{12}\right)=\frac{\sin\left(-\frac{\pi}{12}\right)}{\cos\left(-\frac{\pi}{12}\right)}=\frac{-\sin\left(\frac{\pi}{12}\right)}{\cos\left(\frac{\pi}{12}\right)}=\frac{-\left[\frac{\sqrt{6}-\sqrt{2}}{4}\right]}{\frac{\sqrt{6}+\sqrt{2}}{4}}=\frac{-\sqrt{6}+\sqrt{2}}{\sqrt{6}+\sqrt{2}}$$

$$=\frac{-\sqrt{6}+\sqrt{2}}{\sqrt{6}+\sqrt{2}}\cdot\frac{\sqrt{6}-\sqrt{2}}{\sqrt{6}-\sqrt{2}}=\frac{-6+2\sqrt{2}\sqrt{6}-2}{6-2}=\frac{-8+4\sqrt{3}}{4}=\boxed{-2+\sqrt{3}}$$

6. Using the periodicity and odd identities of tangent, followed by Exercise 5, yields

$$\tan\left(\frac{13\pi}{12}\right)=\tan\left(\frac{\pi}{12}+\pi\right)=\tan\left(\frac{\pi}{12}\right)=-\left[-\tan\left(\frac{\pi}{12}\right)\right]=-\left[\tan\left(-\frac{\pi}{12}\right)\right]$$

$$=-\left[-2+\sqrt{3}\right]=\boxed{2-\sqrt{3}}$$

7. Using the addition formula $\sin(A+B)=\sin A\cos B+\cos A\sin B$ yields

$$\sin\left(105^\circ\right)=\sin\left(60^\circ+45^\circ\right)=\sin\left(60^\circ\right)\cos\left(45^\circ\right)+\cos\left(60^\circ\right)\sin\left(45^\circ\right)=$$

$$=\left(\frac{\sqrt{3}}{2}\right)\left(\frac{\sqrt{2}}{2}\right)+\left(\frac{1}{2}\right)\left(\frac{\sqrt{2}}{2}\right)=\boxed{\frac{\sqrt{2}+\sqrt{6}}{4}}$$

8. First, applying the addition formula $\cos(A-B)=\cos A\cos B+\sin A\sin B$ yields

$$\cos\left(195^\circ\right)=\cos\left(225^\circ-30^\circ\right)=\cos\left(225^\circ\right)\cos\left(30^\circ\right)+\sin\left(225^\circ\right)\sin\left(30^\circ\right)\quad\textbf{(1)}.$$

Next, observe that the reference angle for 225° is 45°. Using this observation (and affixing appropriate signs), we further simplify **(1)** as follows:

$$=\cos\left(-45^\circ\right)\cos\left(30^\circ\right)+\sin\left(-45^\circ\right)\sin\left(30^\circ\right)$$

$$=\left(-\frac{\sqrt{2}}{2}\right)\left(\frac{\sqrt{3}}{2}\right)+\left(\frac{-\sqrt{2}}{2}\right)\left(\frac{1}{2}\right)=\left(-\frac{\sqrt{2}}{2}\right)\left(\frac{\sqrt{3}}{2}+\frac{1}{2}\right)=\boxed{\frac{-\sqrt{2}-\sqrt{6}}{4}}$$

9. First, we need to compute both $\sin\left(-105^\circ\right)$ and $\cos\left(-105^\circ\right)$. Then, we use the definition of tangent to compute $\tan\left(-105^\circ\right)$. To this end, we have the following:

Step 1: Using the odd identity for sine and Exercise 7 yields

$$\sin\left(-105^\circ\right) = -\sin\left(105^\circ\right) = -\left[\frac{\sqrt{2}+\sqrt{6}}{4}\right]$$

Step 2: Using the even identity for cosine and the addition formula $\cos(A+B) = \cos A \cos B - \sin A \sin B$ yields

$$\cos\left(-105^\circ\right) = \cos\left(105^\circ\right) = \cos\left(60^\circ + 45^\circ\right) = \cos\left(60^\circ\right)\cos\left(45^\circ\right) - \sin\left(60^\circ\right)\sin\left(45^\circ\right) =$$

$$= \left(\frac{1}{2}\right)\left(\frac{\sqrt{2}}{2}\right) - \left(\frac{\sqrt{3}}{2}\right)\left(\frac{\sqrt{2}}{2}\right) = \frac{\sqrt{2}-\sqrt{6}}{4}$$

Step 3: Using the definition of tangent and the computations in Steps 1 and 2 yields

$$\tan\left(-105^\circ\right) = \frac{\sin\left(-105^\circ\right)}{\cos\left(-105^\circ\right)} = \frac{-\left[\frac{\sqrt{2}+\sqrt{6}}{4}\right]}{\frac{\sqrt{2}-\sqrt{6}}{4}} = \frac{-\sqrt{6}-\sqrt{2}}{\sqrt{2}-\sqrt{6}}$$

$$= \frac{-\sqrt{6}-\sqrt{2}}{\sqrt{2}-\sqrt{6}} \cdot \frac{\sqrt{2}+\sqrt{6}}{\sqrt{2}+\sqrt{6}} = \frac{-6-2\sqrt{2}\sqrt{6}-2}{2-6} = \frac{-8-4\sqrt{3}}{-4} = \boxed{2+\sqrt{3}}$$

10. Using the addition formula $\tan(A+B) = \dfrac{\tan A + \tan B}{1 - \tan A \tan B}$ yields

$$\tan(165^\circ) = \tan(120^\circ + 45^\circ) = \frac{\tan 120^\circ + \tan 45^\circ}{1 - \tan 120^\circ \tan 45^\circ} \quad \textbf{(1)}$$

Now, observe that $\tan 45^\circ = 1$, and since the reference angle of 120° is 60°, we have (affixing the appropriate signs)

$$\tan 120^\circ = \frac{\sin 120^\circ}{\cos 120^\circ} = \frac{\sin 60^\circ}{-\cos 60^\circ} = \frac{\frac{\sqrt{3}}{2}}{-\frac{1}{2}} = -\sqrt{3}$$

So, continuing the computation in **(1)** yields

$$\tan(165^\circ) = \frac{-\sqrt{3}+1}{1-\left(-\sqrt{3}\right)} = \frac{-\sqrt{3}+1}{1+\sqrt{3}} = \frac{-\sqrt{3}+1}{1+\sqrt{3}} \cdot \frac{1-\sqrt{3}}{1-\sqrt{3}} = \frac{-2\sqrt{3}+3+1}{1-3}$$

$$= \frac{4-2\sqrt{3}}{-2} = \boxed{-2+\sqrt{3}}$$

11. Using the odd identity of tangent, followed by Exercise 5, yields

$$\cot\left(\frac{\pi}{12}\right)=\frac{1}{\tan\left(\frac{\pi}{12}\right)}=\frac{1}{-\left[-\tan\left(\frac{\pi}{12}\right)\right]}=\frac{1}{-\tan\left(-\frac{\pi}{12}\right)}=\frac{1}{-\left[-2+\sqrt{3}\right]}$$

$$=\frac{1}{2-\sqrt{3}}=\frac{1}{2-\sqrt{3}}\cdot\frac{2+\sqrt{3}}{2+\sqrt{3}}=\frac{2+\sqrt{3}}{4-3}=\boxed{2+\sqrt{3}}$$

12. First, we need to compute both $\sin\left(-\frac{5\pi}{12}\right)$ and $\cos\left(-\frac{5\pi}{12}\right)$. Then, we use the definition of cotangent to compute $\cot\left(-\frac{5\pi}{12}\right)$. To this end, we have the following:

Step 1: Using the odd identity for sine and Exercise 3 yields

$$\cos\left(-\frac{5\pi}{12}\right)=\left[\frac{\sqrt{6}-\sqrt{2}}{4}\right]$$

Step 2: Using the odd identity for sine, along with the addition formula $\sin(A+B)=\sin A\cos B+\cos A\sin B$, yields

$$\sin\left(-\frac{5\pi}{12}\right)=-\sin\left(\frac{5\pi}{12}\right)=-\sin\left(\frac{\pi}{6}+\frac{\pi}{4}\right)=-\left[\sin\left(\frac{\pi}{6}\right)\cos\left(\frac{\pi}{4}\right)+\cos\left(\frac{\pi}{6}\right)\sin\left(\frac{\pi}{4}\right)\right]$$

$$=-\left[\left(\frac{1}{2}\right)\left(\frac{\sqrt{2}}{2}\right)+\left(\frac{\sqrt{3}}{2}\right)\left(\frac{\sqrt{2}}{2}\right)\right]=-\left(\frac{\sqrt{2}}{2}\right)\left(\frac{1}{2}+\frac{\sqrt{3}}{2}\right)=\frac{-\sqrt{2}-\sqrt{6}}{4}$$

Step 3: Using the definition of tangent and the computations in Steps 1 and 2 yields

$$\cot\left(-\frac{5\pi}{12}\right)=\frac{\cos\left(-\frac{5\pi}{12}\right)}{\sin\left(-\frac{5\pi}{12}\right)}=\frac{\frac{\sqrt{6}-\sqrt{2}}{4}}{\frac{-\sqrt{2}-\sqrt{6}}{4}}=\frac{\sqrt{6}-\sqrt{2}}{-\sqrt{2}-\sqrt{6}}$$

$$=\frac{\sqrt{6}-\sqrt{2}}{-\left(\sqrt{2}+\sqrt{6}\right)}\cdot\frac{\sqrt{6}-\sqrt{2}}{\sqrt{6}-\sqrt{2}}=\frac{6-2\sqrt{2}\sqrt{6}+2}{-(6-2)}=\frac{8-4\sqrt{3}}{-4}=\boxed{-2+\sqrt{3}}$$

13. First, we need to compute $\cos\left(-\frac{11\pi}{12}\right)$. Then, we use the definition of secant to compute $\sec\left(-\frac{11\pi}{12}\right)$. To this end, we have the following:

Step 1: Using the even identity for cosine and the addition formula $\cos(A-B)=\cos A\cos B+\sin A\sin B$ yields

$$\begin{aligned}\cos\left(-\frac{11\pi}{12}\right)&=\cos\left(\frac{11\pi}{12}\right)=\cos\left(\pi-\frac{11\pi}{12}\right)\\&=\underbrace{\cos\pi}_{=-1}\cos\left(\frac{11\pi}{12}\right)+\underbrace{\sin\pi}_{=0}\sin\left(\frac{11\pi}{12}\right)=-\cos\left(\frac{\pi}{12}\right)\end{aligned}\qquad\textbf{(1)}$$

Using the addition formula $\cos(A-B)=\cos A\cos B+\sin A\sin B$ (again) now yields

$$\begin{aligned}\cos\left(\frac{\pi}{12}\right)&=\cos\left(\frac{\pi}{3}-\frac{\pi}{4}\right)=\cos\left(\frac{\pi}{3}\right)\cos\left(\frac{\pi}{4}\right)+\sin\left(\frac{\pi}{3}\right)\sin\left(\frac{\pi}{4}\right)\\&=\left(\frac{1}{2}\right)\left(\frac{\sqrt{2}}{2}\right)+\left(\frac{\sqrt{3}}{2}\right)\left(\frac{\sqrt{2}}{2}\right)=\left(\frac{\sqrt{2}}{2}\right)\left(\frac{1}{2}+\frac{\sqrt{3}}{2}\right)=\frac{\sqrt{2}+\sqrt{6}}{4}\end{aligned}$$

Hence, using this in **(1)** yields $\cos\left(-\frac{11\pi}{12}\right)=-\left[\frac{\sqrt{2}+\sqrt{6}}{4}\right]$.

Step 2: Using the definition of secant and the computation in Step 1 now yields

$$\begin{aligned}\sec\left(-\frac{11\pi}{12}\right)&=\frac{1}{\cos\left(-\frac{11\pi}{12}\right)}=\frac{1}{-\left[\frac{\sqrt{2}+\sqrt{6}}{4}\right]}=\frac{-4}{\sqrt{2}+\sqrt{6}}\\&=\frac{-4}{\sqrt{2}+\sqrt{6}}\cdot\frac{\sqrt{2}-\sqrt{6}}{\sqrt{2}-\sqrt{6}}=\frac{-4\left(\sqrt{2}-\sqrt{6}\right)}{2-6}=\boxed{\sqrt{2}-\sqrt{6}}\end{aligned}$$

14. First, we need to compute $\cos\left(-\frac{13\pi}{12}\right)$. Then, we use the definition of secant to compute $\sec\left(-\frac{13\pi}{12}\right)$. To this end, we have the following:

Step 1: Using the even identity for cosine and the addition formula $\cos(A+B)=\cos A\cos B-\sin A\sin B$ yields

$$\begin{aligned}\cos\left(-\frac{13\pi}{12}\right)&=\cos\left(\frac{13\pi}{12}\right)=\cos\left(\pi+\frac{11\pi}{12}\right)\\&=\underbrace{\cos\pi}_{=-1}\cos\left(\frac{11\pi}{12}\right)-\underbrace{\sin\pi}_{=0}\sin\left(\frac{11\pi}{12}\right)=-\cos\left(\frac{\pi}{12}\right)\end{aligned}\quad\textbf{(1)}$$

Using the addition formula $\cos(A-B)=\cos A\cos B+\sin A\sin B$ (again) now yields

$$\begin{aligned}\cos\left(\frac{\pi}{12}\right)&=\cos\left(\frac{\pi}{3}-\frac{\pi}{4}\right)=\cos\left(\frac{\pi}{3}\right)\cos\left(\frac{\pi}{4}\right)+\sin\left(\frac{\pi}{3}\right)\sin\left(\frac{\pi}{4}\right)\\&=\left(\frac{1}{2}\right)\left(\frac{\sqrt{2}}{2}\right)+\left(\frac{\sqrt{3}}{2}\right)\left(\frac{\sqrt{2}}{2}\right)=\left(\frac{\sqrt{2}}{2}\right)\left(\frac{1}{2}+\frac{\sqrt{3}}{2}\right)=\frac{\sqrt{2}+\sqrt{6}}{4}\end{aligned}$$

Hence, using this in **(1)** yields $\cos\left(-\frac{13\pi}{12}\right)=-\left[\frac{\sqrt{2}+\sqrt{6}}{4}\right]$.

Step 2: Using the definition of secant and the computation in Step 1 now yields

$$\begin{aligned}\sec\left(-\frac{13\pi}{12}\right)&=\frac{1}{\cos\left(-\frac{13\pi}{12}\right)}=\frac{1}{-\left[\frac{\sqrt{2}+\sqrt{6}}{4}\right]}=\frac{-4}{\sqrt{2}+\sqrt{6}}\\&=\frac{-4}{\sqrt{2}+\sqrt{6}}\cdot\frac{\sqrt{2}-\sqrt{6}}{\sqrt{2}-\sqrt{6}}=\frac{-4\left(\sqrt{2}-\sqrt{6}\right)}{2-6}=\boxed{\sqrt{2}-\sqrt{6}}\end{aligned}$$

15.
$$\begin{aligned}\csc\left(-255^\circ\right)&=\frac{1}{\sin\left(45^\circ-300^\circ\right)}=\frac{1}{\sin\left(45^\circ\right)\cos\left(300^\circ\right)-\cos\left(45^\circ\right)\sin\left(300^\circ\right)}\\&=\frac{1}{\left(\frac{\sqrt{2}}{2}\right)\left(\frac{1}{2}\right)-\left(\frac{\sqrt{2}}{2}\right)\left(-\frac{\sqrt{3}}{2}\right)}\\&=\frac{4}{\sqrt{2}\left(1+\sqrt{3}\right)}\cdot\frac{\sqrt{2}-\sqrt{6}}{\sqrt{2}-\sqrt{6}}=\boxed{\sqrt{6}-\sqrt{2}}\end{aligned}$$

16. $\csc(-15^\circ) = \dfrac{1}{\sin(45^\circ - 60^\circ)} = \dfrac{1}{\sin(45^\circ)\cos(60^\circ) - \cos(45^\circ)\sin(60^\circ)}$ $= \dfrac{1}{\left(\frac{\sqrt{2}}{2}\right)\left(\frac{1}{2}\right) - \left(\frac{\sqrt{2}}{2}\right)\left(\frac{\sqrt{3}}{2}\right)} = \boxed{\dfrac{4}{\sqrt{2}(1-\sqrt{3})}}$
17. Using the addition formula $\cos(A-B) = \cos A \cos B + \sin A \sin B$ with $A = 2x$ and $B = 3x$, followed by the even identity for cosine at the very end, yields $\sin 2x \sin 3x + \cos 2x \cos 3x = \cos(2x - 3x) = \cos(-x) = \boxed{\cos x}$.
18. Using the addition formula $\cos(A+B) = \cos A \cos B - \sin A \sin B$ with $A = x$ and $B = 2x$ yields $\sin x \sin 2x - \cos x \cos 2x = -[\cos x \cos 2x - \sin x \sin 2x] = -\cos(x + 2x) = \boxed{-\cos 3x}$
19. Using the addition formula $\sin(A-B) = \sin A \cos B - \cos A \sin B$ with $A = x$ and $B = 2x$, followed by the odd identity for sine at the very end, yields $\sin x \cos 2x - \cos x \sin 2x = \sin(x - 2x) = \sin(-x) = \boxed{-\sin x}$
20. Using the addition formula $\sin(A+B) = \sin A \cos B + \cos A \sin B$ with $A = 2x$ and $B = 3x$ yields $\sin 2x \cos 3x + \cos 2x \sin 3x = \sin(2x + 3x) = \boxed{\sin 5x}$.
21. Using the addition formula $\sin(A+B) = \sin A \cos B + \cos A \sin B$ with $A = \pi - x$ and $B = x$ yields $\sin x \cos(\pi - x) + \cos x \sin(\pi - x) = \sin(x + \pi - x) = \sin \pi = \boxed{0}$.
22. Using the addition formula $\sin(A-B) = \sin A \cos B - \cos A \sin B$ with $A = \frac{\pi}{3}x$ and $B = -\frac{\pi}{2}x$ yields $\sin\left(\frac{\pi}{3}x\right)\cos\left(-\frac{\pi}{2}x\right) - \cos\left(\frac{\pi}{3}x\right)\sin\left(-\frac{\pi}{2}x\right) = \sin\left(\frac{\pi}{3}x - \left(-\frac{\pi}{2}x\right)\right) = \boxed{\sin\left(\frac{5\pi}{6}x\right)}$.
23. Observe that $(\sin A - \sin B)^2 + (\cos A - \cos B)^2 - 2 =$ $\sin^2 A - 2\sin A \sin B + \sin^2 B + \cos^2 A - 2\cos A \cos B + \cos^2 B - 2 =$ $\underbrace{(\sin^2 A + \cos^2 A)}_{=1} + \underbrace{(\sin^2 B + \cos^2 B)}_{=1} - 2[\sin A \sin B + \cos A \cos B] - 2 =$ $-2[\sin A \sin B + \cos A \cos B] = -2[\cos A \cos B + \sin A \sin B] = \boxed{-2\cos(A-B)}$

24. Observe that

$$(\sin A+\sin B)^2+(\cos A+\cos B)^2-2=$$
$$\sin^2 A+2\sin A\sin B+\sin^2 B+\cos^2 A+2\cos A\cos B+\cos^2 B-2=$$
$$\underbrace{\left(\sin^2 A+\cos^2 A\right)}_{=1}+\underbrace{\left(\sin^2 B+\cos^2 B\right)}_{=1}+2\left[\sin A\sin B+\cos A\cos B\right]-2=$$
$$2\left[\sin A\sin B+\cos A\cos B\right]=2\left[\cos A\cos B+\sin A\sin B\right]=\boxed{2\cos(A-B)}$$

25. Observe that

$$2-(\sin A+\cos B)^2-(\cos A+\sin B)^2=$$
$$2-\left(\sin^2 A+2\sin A\cos B+\cos^2 B\right)-\left(\cos^2 A+2\cos A\sin B+\sin^2 B\right)$$
$$2-\underbrace{\left(\sin^2 A+\cos^2 A\right)}_{=1}-\underbrace{\left(\sin^2 B+\cos^2 B\right)}_{=1}-2\left[\sin A\cos B+\cos A\sin B\right]=$$
$$-2\left[\sin A\cos B+\cos A\sin B\right]=\boxed{-2\sin(A+B)}$$

26. Observe that

$$2-(\sin A-\cos B)^2-(\cos A+\sin B)^2=$$
$$2-\left(\sin^2 A-2\sin A\cos B+\cos^2 B\right)-\left(\cos^2 A+2\cos A\sin B+\sin^2 B\right)$$
$$2-\underbrace{\left(\sin^2 A+\cos^2 A\right)}_{=1}-\underbrace{\left(\sin^2 B+\cos^2 B\right)}_{=1}+2\left[\sin A\cos B-\cos A\sin B\right]=$$
$$2\left[\sin A\cos B-\cos A\sin B\right]=\boxed{2\sin(A-B)}$$

27. Using the addition formula $\tan(A-B)=\dfrac{\tan A-\tan B}{1+\tan A\tan B}$ with $A=49^\circ$ and $B=23^\circ$ yields $\dfrac{\tan 49^\circ-\tan 23^\circ}{1+\tan 49^\circ\tan 23^\circ}=\tan(49^\circ-23^\circ)=\boxed{\tan(26^\circ)}$

28. Using the addition formula $\tan(A+B)=\dfrac{\tan A+\tan B}{1-\tan A\tan B}$ with $A=49^\circ$ and $B=23^\circ$ yields $\dfrac{\tan 49^\circ+\tan 23^\circ}{1-\tan 49^\circ\tan 23^\circ}=\tan(49^\circ+23^\circ)=\boxed{\tan(72^\circ)}$

29. Observe that $\cos(\alpha+\beta)=\cos\alpha\cos\beta-\sin\alpha\sin\beta=\left(-\frac{1}{3}\right)\left(-\frac{1}{4}\right)-\sin\alpha\sin\beta$. We need to compute both $\sin\alpha$ and $\sin\beta$. We proceed as follows:

Since the terminal side of α is in QIII, we know that $\sin\alpha<0$. As such, since we are given that $\cos\alpha=-\frac{1}{3}$, the diagram is:

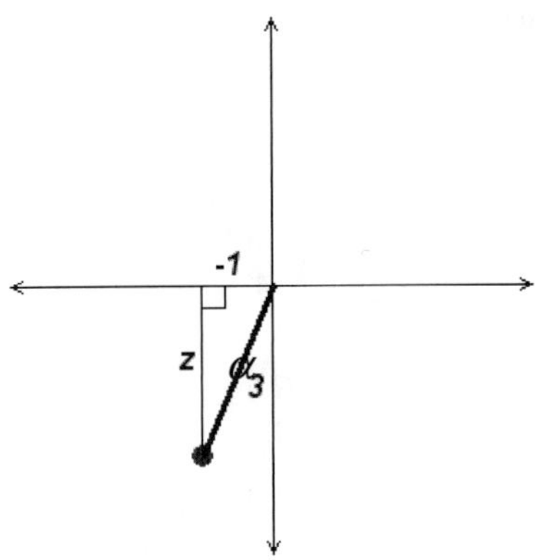

From the Pythagorean Theorem, we have $z^2+(-1)^2=3^2$, so that $z=-2\sqrt{2}$.

Thus, $\sin\alpha=-\frac{2\sqrt{2}}{3}$.

Since the terminal side of β is in QII, we know that $\sin\beta>0$. As such, since we are given that $\cos\beta=-\frac{1}{4}$, the diagram is:

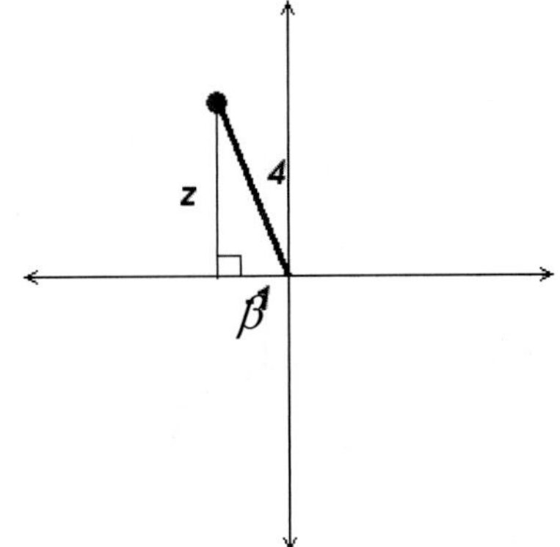

From the Pythagorean Theorem, we have $z^2+(-1)^2=4^2$, so that $z=\sqrt{15}$.

Thus, $\sin\beta=\frac{\sqrt{15}}{4}$.

Hence,

$$\cos(\alpha+\beta)=\left(-\frac{1}{3}\right)\left(-\frac{1}{4}\right)-\sin\alpha\sin\beta=\left(-\frac{1}{3}\right)\left(-\frac{1}{4}\right)-\left(-\frac{2\sqrt{2}}{3}\right)\left(\frac{\sqrt{15}}{4}\right)=\frac{1}{12}+\frac{2\sqrt{30}}{12}=\boxed{\frac{1+2\sqrt{30}}{12}}$$

30. Observe that $\cos(\alpha-\beta)=\cos\alpha\cos\beta+\sin\alpha\sin\beta=\left(\frac{1}{3}\right)\left(-\frac{1}{4}\right)+\sin\alpha\sin\beta$. We need to compute both $\sin\alpha$ and $\sin\beta$. We proceed as follows:

Since the terminal side of α is in QIV, we know that $\sin\alpha<0$. As such, since we are given that $\cos\alpha=\frac{1}{3}$, the diagram is as follows:

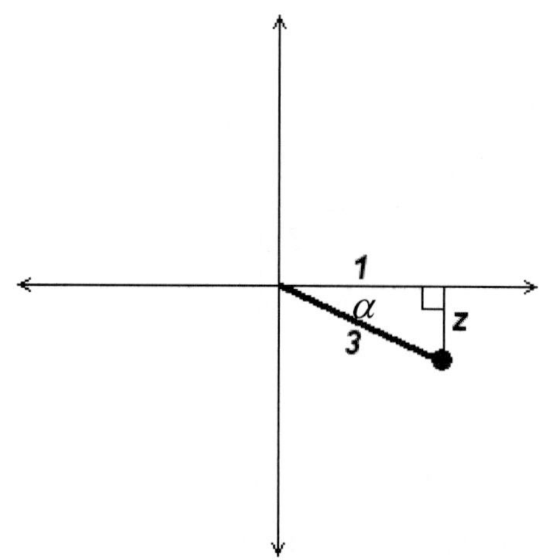

From the Pythagorean Theorem, we have $z^2+1^2=3^2$, so that $z=-2\sqrt{2}$.

Thus, $\sin\alpha=-\frac{2\sqrt{2}}{3}$.

Since the terminal side of β is in QII, we know that $\sin\beta>0$. As such, since we are given that $\cos\beta=-\frac{1}{4}$, the diagram is as follows:

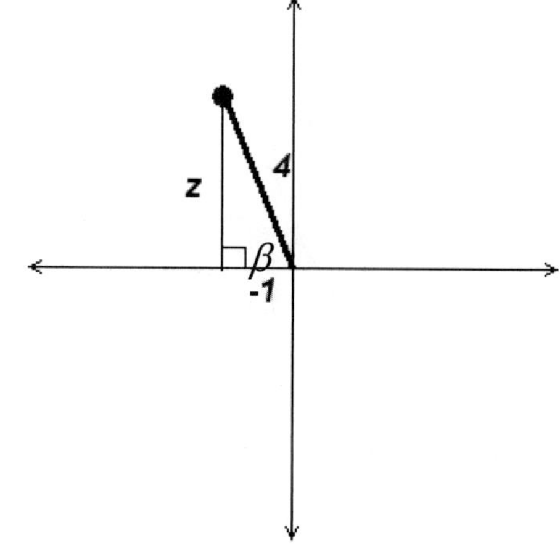

From the Pythagorean Theorem, we have $z^2+(-1)^2=4^2$, so that $z=\sqrt{15}$.

Thus, $\sin\beta=\frac{\sqrt{15}}{4}$.

Hence,

$$\cos(\alpha-\beta)=\left(\frac{1}{3}\right)\left(-\frac{1}{4}\right)+\sin\alpha\sin\beta=\left(\frac{1}{3}\right)\left(-\frac{1}{4}\right)+\left(-\frac{2\sqrt{2}}{3}\right)\left(\frac{\sqrt{15}}{4}\right)$$

$$=-\frac{1}{12}-\frac{2\sqrt{30}}{12}=\boxed{\frac{-\left(1+2\sqrt{30}\right)}{12}}$$

31. Observe that $\sin(\alpha-\beta)=\sin\alpha\cos\beta-\cos\alpha\sin\beta=\left(-\frac{3}{5}\right)\cos\beta-\cos\alpha\left(\frac{1}{5}\right)$. We need to compute both $\cos\alpha$ and $\cos\beta$. We proceed as follows:

Since the terminal side of α is in QIII, we know that $\cos\alpha<0$. As such, since we are given that $\sin\alpha=-\frac{3}{5}$, the diagram is as follows:

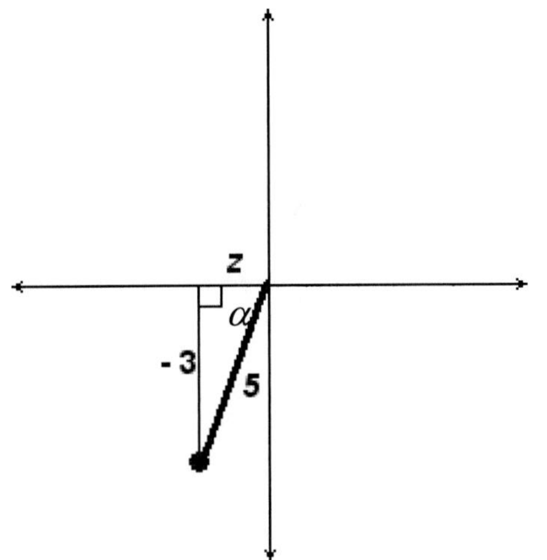

From the Pythagorean Theorem, we have $z^2+3^2=5^2$, so that $z=-4$.
Thus, $\cos\alpha=-\frac{4}{5}$.

Since the terminal side of β is in QI, we know that $\cos\beta>0$. As such, since we are given that $\sin\beta=\frac{1}{5}$, the diagram is as follows:

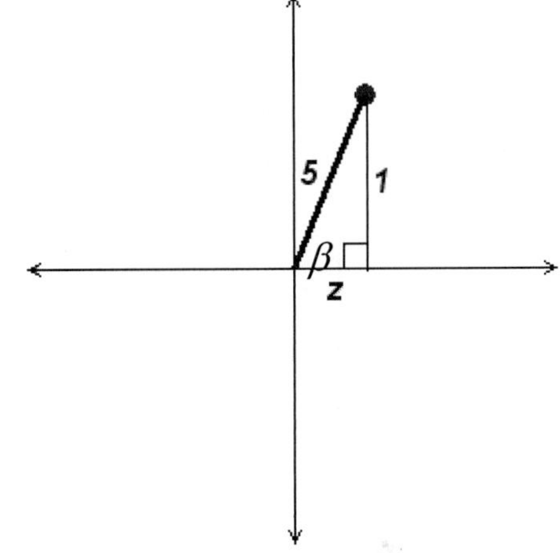

From the Pythagorean Theorem, we have $z^2+1^2=5^2$, so that $z=\sqrt{24}=2\sqrt{6}$.
Thus, $\cos\beta=\frac{2\sqrt{6}}{5}$.

Hence,

$$\sin(\alpha-\beta)=\left(-\frac{3}{5}\right)\cos\beta-\cos\alpha\left(\frac{1}{5}\right)=\left(-\frac{3}{5}\right)\left(\frac{2\sqrt{6}}{5}\right)-\left(-\frac{4}{5}\right)\left(\frac{1}{5}\right)=\frac{-6\sqrt{6}}{25}+\frac{4}{25}=\boxed{\frac{-6\sqrt{6}+4}{25}}$$

32. Observe that $\sin(\alpha+\beta)=\sin\alpha\cos\beta+\cos\alpha\sin\beta=\left(-\frac{3}{5}\right)\cos\beta+\cos\alpha\left(\frac{1}{5}\right)$. We need to compute both $\cos\alpha$ and $\cos\beta$. We proceed as follows:

Since the terminal side of α is in QIII, we know that $\cos\alpha<0$. As such, since we are given that $\sin\alpha=-\frac{3}{5}$, the diagram is as follows:

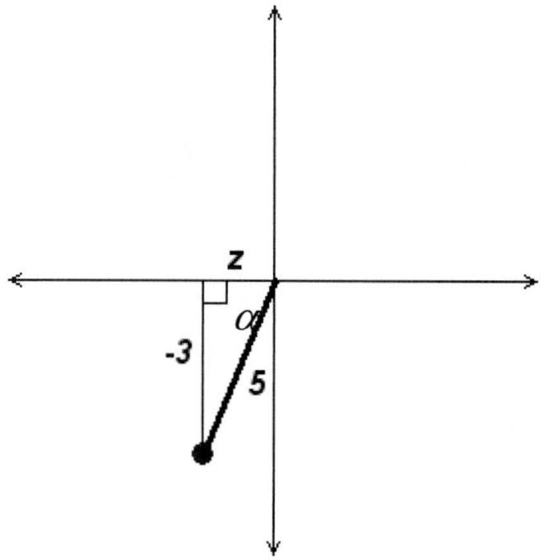

From the Pythagorean Theorem, we have $z^2+(-3)^2=5^2$, so that $z=-4$.

Thus, $\cos\alpha=-\frac{4}{5}$.

Since the terminal side of β is in QII, we know that $\cos\beta<0$. As such, since we are given that $\sin\beta=\frac{1}{5}$, the diagram is as follows:

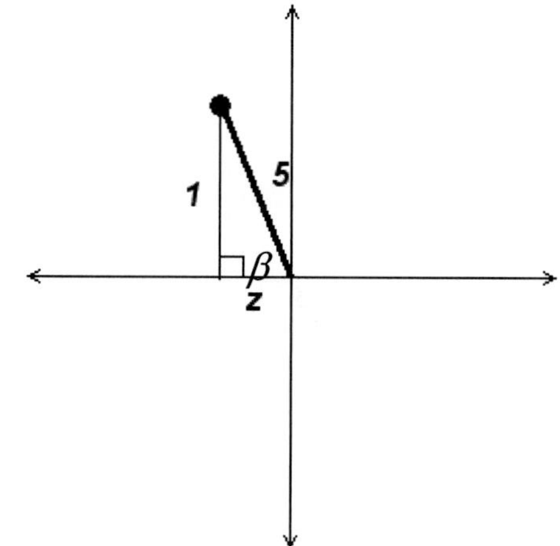

From the Pythagorean Theorem, we have $z^2+1^2=5^2$, so that $z=-\sqrt{24}=-2\sqrt{6}$.

Thus, $\cos\beta=-\frac{2\sqrt{6}}{5}$.

Hence,

$$\sin(\alpha+\beta)=\left(-\frac{3}{5}\right)\cos\beta+\cos\alpha\left(\frac{1}{5}\right)=\left(-\frac{3}{5}\right)\left(-\frac{2\sqrt{6}}{5}\right)+\left(-\frac{4}{5}\right)\left(\frac{1}{5}\right)=\frac{6\sqrt{6}}{25}-\frac{4}{25}=\boxed{\frac{6\sqrt{6}-4}{25}}$$

33. Observe that since $\tan(\alpha+\beta)=\dfrac{\tan\alpha+\tan\beta}{1-\tan\alpha\tan\beta}$, we need to determine the values of $\cos\alpha$ and $\sin\beta$ so that we can then use this information in combination with the given values of $\sin\alpha$ and $\cos\beta$ in order to compute $\tan\alpha$ and $\tan\beta$. We proceed as follows:

Since the terminal side of α is in QIII, we know that $\cos\alpha<0$. As such, since we are given that $\sin\alpha=-\dfrac{3}{5}$, the diagram is:

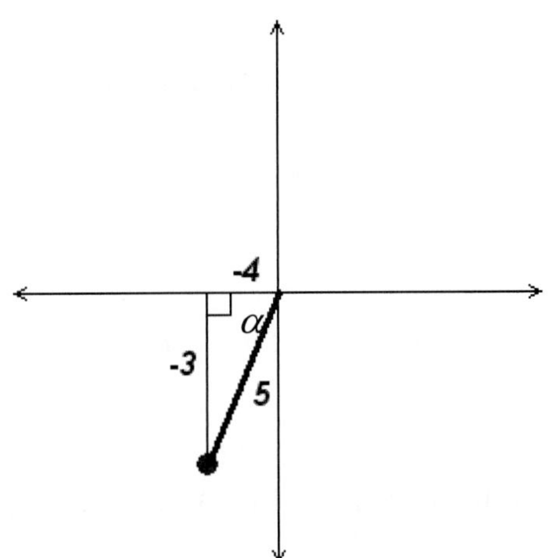

From the Pythagorean Theorem, we have $z^2+(-3)^2=5^2$, so that $z=-4$.

As such, $\cos\alpha=-\dfrac{4}{5}$, and so,

$$\tan\alpha=\frac{\sin\alpha}{\cos\alpha}=\frac{-3/5}{-4/5}=\frac{3}{4}$$

Since the terminal side of β is in QII, we know that $\sin\beta>0$. As such, since we are given that $\cos\beta=-\dfrac{1}{4}$, the diagram is:

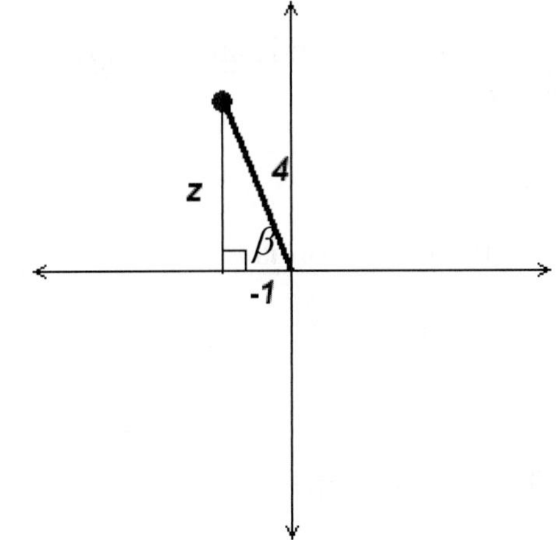

From the Pythagorean Theorem, we have $z^2+(-1)^2=4^2$, so that $z=\sqrt{15}$.

As such, $\sin\beta=\dfrac{\sqrt{15}}{4}$, and so

$$\tan\beta=\frac{\sin\beta}{\cos\beta}=\frac{\sqrt{15}/4}{-1/4}=-\sqrt{15}$$

Hence,

$$\tan(\alpha+\beta)=\frac{\tan\alpha+\tan\beta}{1-\tan\alpha\tan\beta}=\frac{\frac{3}{4}-\sqrt{15}}{1-\left(\frac{3}{4}\right)\left(-\sqrt{15}\right)}=\frac{\frac{3-4\sqrt{15}}{4}}{\frac{4+3\sqrt{15}}{4}}$$

$$=\frac{3-4\sqrt{15}}{4+3\sqrt{15}}\cdot\frac{4-3\sqrt{15}}{4-3\sqrt{15}}=\frac{12-16\sqrt{15}-9\sqrt{15}+12(15)}{16-9(15)}=\boxed{\frac{192-25\sqrt{15}}{-119}}$$

34. Using the values of $\tan\alpha$ and $\tan\beta$ obtained in Exercise 29, we obtain $$\tan(\alpha-\beta)=\frac{\tan\alpha-\tan\beta}{1+\tan\alpha\tan\beta}=\frac{\frac{3}{4}+\sqrt{15}}{1+\left(\frac{3}{4}\right)\left(-\sqrt{15}\right)}=\frac{\frac{3+4\sqrt{15}}{4}}{\frac{4-3\sqrt{15}}{4}}$$ $$=\frac{3+4\sqrt{15}}{4-3\sqrt{15}}\cdot\frac{4+3\sqrt{15}}{4+3\sqrt{15}}=\frac{12+16\sqrt{15}+9\sqrt{15}+12(15)}{16-9(15)}=\boxed{\frac{192+25\sqrt{15}}{-119}}$$
35. Observe that $$\sin(A+B)+\sin(A-B)=(\sin A\cos B+\cancel{\sin B\cos A})+(\sin A\cos B-\cancel{\sin B\cos A})$$ $$=2\sin A\cos B$$ So, the given equation is an identity.
36. Observe that $$\cos(A+B)+\cos(A-B)=(\cos A\cos B-\cancel{\sin A\sin B})+(\cos A\cos B+\cancel{\sin A\sin B})$$ $$=2\cos A\cos B$$ So, the given equation is an identity.
37. The equation $\sin\left(x-\frac{\pi}{2}\right)=\cos\left(x+\frac{\pi}{2}\right)$ is a conditional since it is false for $x=\frac{\pi}{2}$, for instance.
38. The equation $\sin\left(x+\frac{\pi}{2}\right)=\cos\left(x+\frac{\pi}{2}\right)$ is a conditional since it is false for $x=\frac{\pi}{2}$, for instance.
39. Observe that $\sin\left(x+\frac{\pi}{4}\right)=\sin x\cos\left(\frac{\pi}{4}\right)+\sin\left(\frac{\pi}{4}\right)\cos x=\frac{\sqrt{2}}{2}\left(\sin x+\cos x\right)$. So, the given equation is an identity.
40. The equation is a conditional since it is false for $x=\frac{\pi}{2}$, for instance.
41. Observe that $$\frac{1-\cos 2x}{2}=\frac{1-\left(\cos x\cos x-\sin x\sin x\right)}{2}=\frac{1-\left(\cos^2 x-\left(1-\cos^2 x\right)\right)}{2}$$ $$=\frac{2-2\cos^2 x}{2}=1-\cos^2 x=\sin^2 x$$ So, the given equation is an identity.

42. Observe that

$$\frac{1+\cos 2x}{2}=\frac{1+(\cos x\cos x-\sin x\sin x)}{2}=\frac{1+\left(\left(1-\sin^2 x\right)-\sin^2 x\right)}{2}$$
$$=\frac{2-2\sin^2 x}{2}=1-\sin^2 x=\cos^2 x$$

So, the given equation is an identity.

43. Observe that

$$\sin 2x=\sin(x+x)=\sin x\cos x+\sin x\cos x=2\sin x\cos x.$$

So, the given equation is an identity.

44. Observe that

$$\cos 2x=\cos(x+x)=\cos x\cos x-\sin x\sin x=\cos^2 x-\sin^2 x.$$

So, the given equation is an identity.

45. The equation $\sin(A+B)=\sin A+\sin B$ is a conditional since it is false for $A=B=\frac{\pi}{2}$, for instance.

46. The equation $\cos(A+B)=\cos A+\cos B$ is a conditional since it is false for $A=B=\frac{\pi}{2}$, for instance.

47. Observe that

$$\tan(\pi+\beta)=\frac{\tan\pi+\tan\beta}{1-\tan\pi\tan\beta}=\frac{0+\tan\beta}{1-(0)\tan\beta}=\tan\beta.$$

So, the given equation is an identity.

48. Observe that

$$\tan(A-\pi)=\frac{\tan A-\tan\pi}{1+\tan A\tan\pi}=\frac{\tan A-0}{1+\tan A(0)}=\tan A.$$

So, the given equation is an identity.

49. Observe that

$$\cot(3\pi+x)=\frac{1}{\tan(3\pi+x)}=\frac{1}{\tan x}.$$

So, the given equation is an identity.

50. The given equation is never true because

$$\csc 2x=\frac{1}{\sin 2x}=\frac{1}{2\sin x\cos x}=\tfrac{1}{2}\sec x\csc x.$$

So, it is a conditional (in a big way!).

51. The given equation is a conditional since it is false when $x=\frac{\pi}{4}$, for instance.

52. Observe that

$$\cot\left(x+\tfrac{\pi}{4}\right)=\frac{1}{\tan\left(x+\tfrac{\pi}{4}\right)}=\frac{1}{\dfrac{\tan x+\tan\left(\tfrac{\pi}{4}\right)}{1-\tan x\tan\left(\tfrac{\pi}{4}\right)}}=\frac{1-\tan x}{\tan x+1}=\frac{1-\tan x}{1+\tan x}.$$

So, the given equation is an identity.

53. Using the addition formula $\sin(A+B)=\sin A\cos B+\cos A\sin B$ with $A=x$ and $B=\frac{\pi}{3}$ yields $y=\cos\left(\frac{\pi}{3}\right)\sin x+\sin\left(\frac{\pi}{3}\right)\cos x=\sin\left(x+\frac{\pi}{3}\right)$.

So, the graph is the graph of $y=\sin x$ shifted to the left $\frac{\pi}{3}$ units, as seen below.

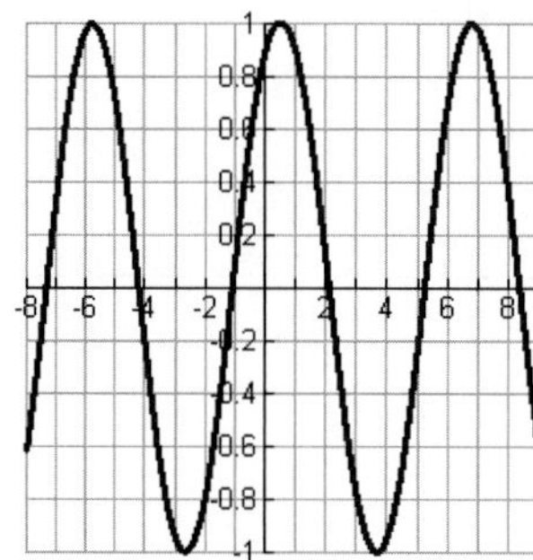

54. Using the addition formula $\sin(A-B)=\sin A\cos B-\cos A\sin B$ with $A=x$ and $B=\frac{\pi}{3}$ yields $y=\cos\left(\frac{\pi}{3}\right)\sin x-\sin\left(\frac{\pi}{3}\right)\cos x=\sin\left(x-\frac{\pi}{3}\right)$.

So, the graph is the graph of $y=\sin x$ shifted to the right $\frac{\pi}{3}$ units, as seen below.

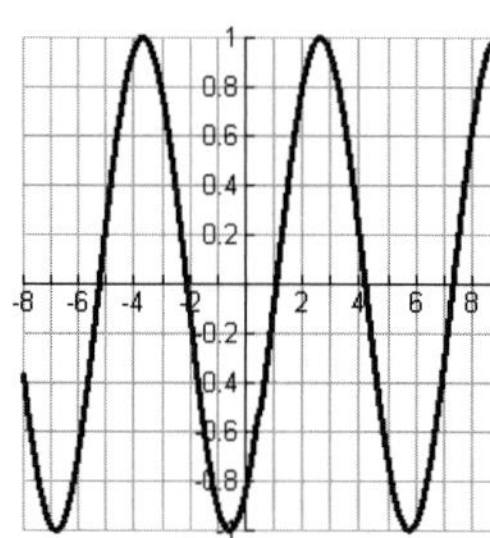

55. Using the addition formula $\cos(A-B)=\cos A\cos B+\sin A\sin B$ with $A=x$ and $B=\frac{\pi}{4}$ yields

$$y=\sin x\sin\left(\frac{\pi}{4}\right)+\cos x\cos\left(\frac{\pi}{4}\right)=\cos x\cos\left(\frac{\pi}{4}\right)+\sin x\sin\left(\frac{\pi}{4}\right)=\cos\left(x-\frac{\pi}{4}\right).$$

So, the graph is the graph of $y = \cos x$ shifted to the right $\frac{\pi}{4}$ units, as seen below.

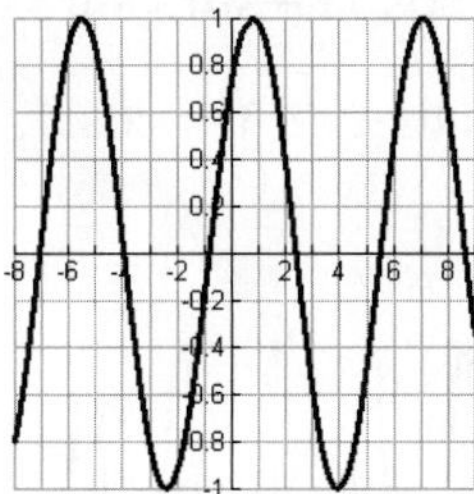

56. Using the addition formula $\cos(A+B) = \cos A \cos B - \sin A \sin B$ with $A = x$ and $B = \frac{\pi}{4}$ yields

$$y = \sin x \sin\left(\frac{\pi}{4}\right) - \cos x \cos\left(\frac{\pi}{4}\right) = -\left[\cos x \cos\left(\frac{\pi}{4}\right) - \sin x \sin\left(\frac{\pi}{4}\right)\right] = -\cos\left(x + \frac{\pi}{4}\right).$$

So, the graph is the graph of $y = \cos x$ shifted to the left $\frac{\pi}{4}$ units and then reflected over the *x*-axis, as seen below.

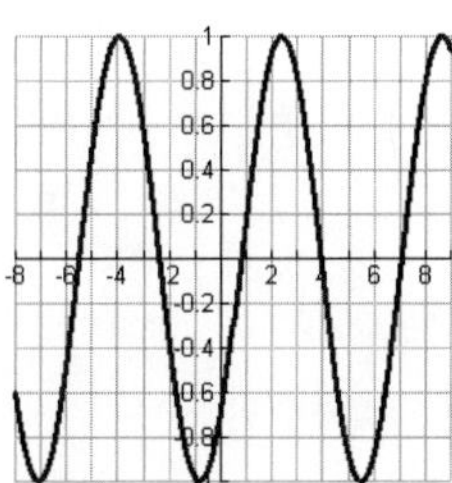

57. Using the addition formula $\sin(A+B) = \sin A \cos B + \cos A \sin B$ with $A = x$ and $B = 3x$ yields

$$y = -\sin x \cos 3x - \cos x \sin 3x = -[\sin x \cos 3x + \cos x \sin 3x] = -\sin(x+3x) = -\sin 4x$$

So, the general shape of the graph is that of $y = \sin x$, but with period is $\frac{2\pi}{4} = \frac{\pi}{2}$ and then reflected over the *x*-axis, as seen below.

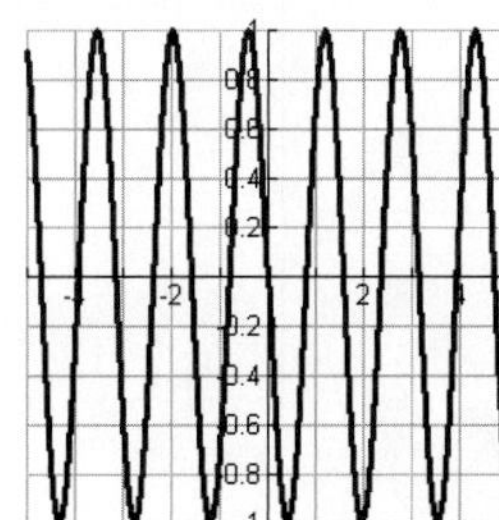

58. Using the addition formula $\cos(A-B)=\cos A\cos B+\sin A\sin B$ with $A=x$ and $B=3x$ yields

$$y=\sin x\sin 3x+\cos x\cos 3x=\cos x\cos 3x+\sin x\sin 3x=\cos\left(x-3x\right)=\cos(-2x)=\cos 2x$$

So, the general shape of the graph is that of $y=\cos x$, but the period is $\frac{2\pi}{2}=\pi$, as seen below.

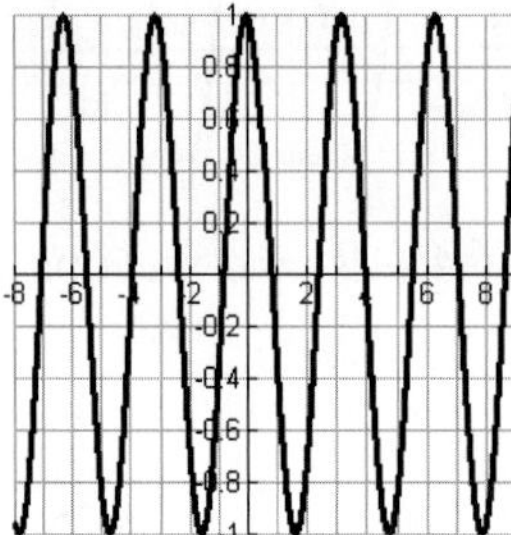

59. Observe that $\frac{1+\tan x}{1-\tan x}=\tan\left(x-\frac{\pi}{4}\right)$.

So, the graph is as follows:

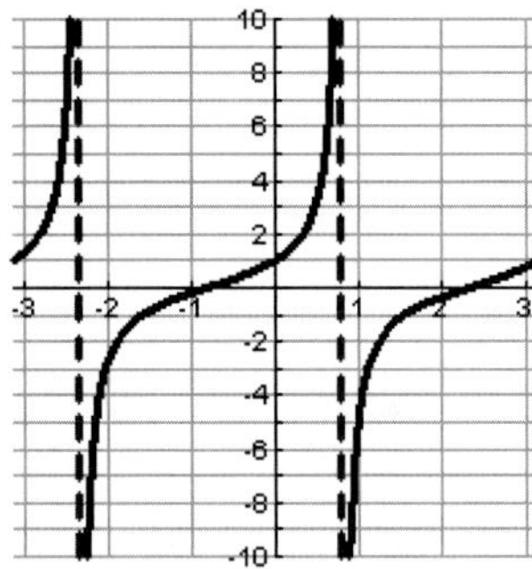

60. Observe that $\frac{\sqrt{3}-\tan x}{1+\sqrt{3}\tan x}=\tan\left(\frac{\pi}{3}-x\right)$.

So, the graph is as follows:

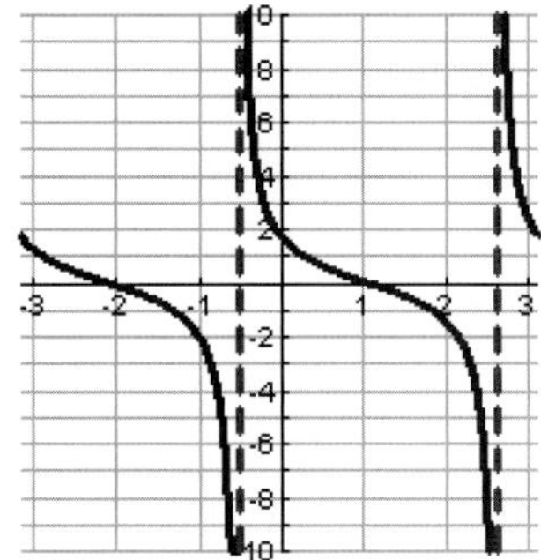

61. Observe that $\frac{1+\sqrt{3}\tan x}{\sqrt{3}-\tan x}=\tan\left(x+\frac{\pi}{6}\right)$.

So, the graph is as follows:

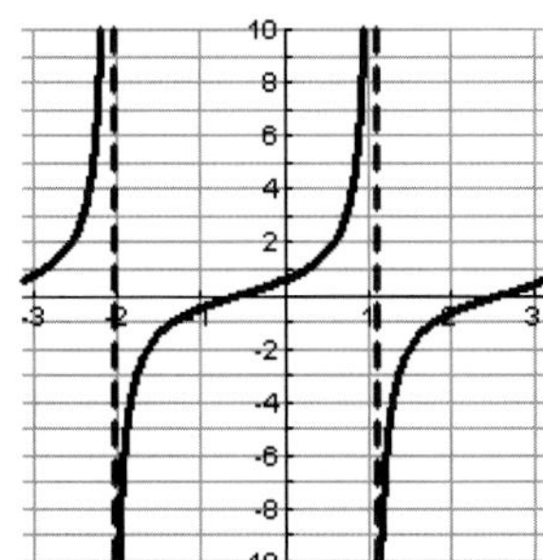

62. Observe that $\frac{1-\tan x}{1+\tan x}=\tan\left(\frac{\pi}{4}-x\right)$.

So, the graph is as follows:

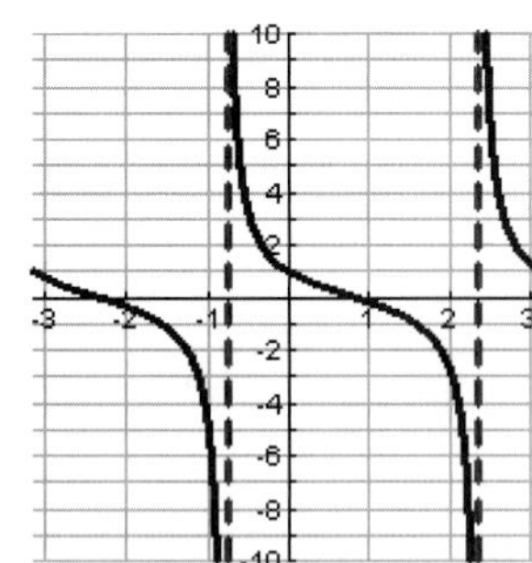

63.	**64.**
$\cos\left(x-\frac{\pi}{4}\right)=\cos x\cos\frac{\pi}{4}+\sin x\sin\frac{\pi}{4}$ $=\frac{\sqrt{2}}{2}\left(\cos x+\sin x\right)$ $=\frac{\sqrt{2}}{2}\left(1+x-\frac{x^2}{2!}-\frac{x^3}{3!}+\frac{x^4}{4!}+\frac{x^5}{5!}--++...\right)$	$\sin\left(x+\frac{3\pi}{2}\right)=\sin x\cos\frac{3\pi}{2}+\cos x\sin\frac{3\pi}{2}$ $=-\cos x$ $=-1+\frac{x^2}{2!}-\frac{x^4}{4!}+-...$

65. Observe that $\tan\left(\theta_2-\theta_1\right)=\dfrac{\tan\theta_2-\tan\theta_1}{1+\tan\theta_2\tan\theta_1}=\boxed{\dfrac{m_2-m_1}{1+m_2m_1}}$.

66. Observe that $\tan\left(\theta_1-\theta_2\right)=\dfrac{\tan\theta_1-\tan\theta_2}{1+\tan\theta_1\tan\theta_2}=\boxed{\dfrac{m_1-m_2}{1+m_1m_2}}$.

67. $E=A\cos\left(kz-ct\right)=A\left[\cos\left(kz\right)\cos\left(ct\right)+\sin\left(kz\right)\sin\left(ct\right)\right]$

If $\dfrac{z}{\lambda}=n$, an integer, then $z=n\lambda$, so that $kz=kn\lambda=kn\left(\dfrac{2\pi}{k}\right)=2n\pi$. Hence,

$\sin(kz)=0$. So, the formula for E simplifies to $E=A\cos\left(kz-ct\right)=A\cos\left(kz\right)\cos\left(ct\right)$.

68. $E=A\cos\left(kz-ct\right)=A\left[\cos\left(kz\right)\cos\left(ct\right)+\sin\left(kz\right)\sin\left(ct\right)\right]$

So, in particular, $E(0)=A\cos\left(kz\right)$.

69. $\tan(A+B)\neq\tan A+\tan B$. Should have used $\tan(A+B)=\dfrac{\tan A+\tan B}{1-\tan A\tan B}$.

70. Tangent is an odd function, not an even function. Hence, the very first step should have been: $\tan\left(-\dfrac{7\pi}{6}\right)=-\tan\left(\dfrac{7\pi}{6}\right)$. The remaining calculations are correct.

71. False. Observe that

$$\cos\left(15^\circ\right)=\cos\left(45^\circ-30^\circ\right)=\cos\left(45^\circ\right)\cos\left(30^\circ\right)+\sin\left(45^\circ\right)\sin\left(30^\circ\right)$$
$$=\left(\frac{\sqrt{2}}{2}\right)\left(\frac{\sqrt{3}}{2}\right)+\left(\frac{\sqrt{2}}{2}\right)\left(\frac{1}{2}\right)=\frac{\sqrt{6}+\sqrt{2}}{4}$$

while

$$\cos\left(45^\circ\right)-\cos\left(30^\circ\right)=\left(\frac{\sqrt{2}}{2}\right)-\left(\frac{\sqrt{3}}{2}\right)=\frac{\sqrt{2}-\sqrt{3}}{2}.$$

72. False. Since $\sin\left(\dfrac{\pi}{2}\right)=1$, $\sin\left(\dfrac{\pi}{3}\right)=\dfrac{\sqrt{3}}{2}$, and $\sin\left(\dfrac{\pi}{6}\right)=\dfrac{1}{2}$ and it is clear that $1\neq\dfrac{\sqrt{3}+1}{2}$, the given equality is not true.

73. False. Observe that $\dfrac{1+\tan x}{1-\tan x}=\tan\left(x+\frac{\pi}{4}\right)$.

74. True. Observe that $\cot\left(\frac{\pi}{4}-x\right)=\dfrac{1}{\tan\left(\frac{\pi}{4}-x\right)}=\dfrac{1+\tan x}{1-\tan x}$.

75. Approaches may vary slightly here, but they will result in equivalent answers as long as the identities are applied correctly. The final form of the simplification depends on how you initially choose to group the input quantities. For instance, the following is a legitimate simplification:

$$\begin{aligned}\sin\left(A+B+C\right)&=\sin\left(\left(A+B\right)+C\right)=\sin(A+B)\cos C+\sin C\cos(A+B)\\&=\left[\sin A\cos B+\sin B\cos A\right]\cos C+\sin C\left[\cos A\cos B-\sin A\sin B\right]\\&=\sin A\cos B\cos C+\sin B\cos A\cos C+\cos A\cos B\sin C-\sin A\sin B\sin C\end{aligned}$$

76. Approaches may vary slightly here, but they will result in equivalent answers as long as the identities are applied correctly. The final form of the simplification depends on how you initially choose to group the input quantities. For instance, the following is a legitimate simplification:

$$\begin{aligned}\cos\left(A+B+C\right)&=\cos\left(\left(A+B\right)+C\right)=\cos(A+B)\cos C-\sin C\sin(A+B)\\&=\left[\cos A\cos B-\sin B\sin A\right]\cos C-\sin C\left[\sin A\cos B+\sin B\cos A\right]\\&=\cos A\cos B\cos C-\sin B\sin A\cos C-\sin A\cos B\sin C-\sin B\cos A\sin C\end{aligned}$$

77. Observe that $\sin(A-B)=\sin A\cos B-\sin B\cos A$. We seek values of A and B such that the right-side equals $\sin A-\sin B$. Certainly, if we choose values of A and B such that $\cos B=\cos A=1$, then this occurs. And this occurs precisely when A and B are integer multiples of 2π (and they need not be the same multiple of 2π !). So, the solutions are $\boxed{B=2m\pi,\ A=2n\pi\text{, where } n \text{ and } m \text{ are integers}}$.

78. Observe that $\sin(A+B)=\sin A\cos B+\sin B\cos A$. We seek values of A and B such that the right-side equals $\sin A+\sin B$. Certainly, if we choose values of A and B such that $\cos B=\cos A=1$, then this occurs. And this occurs precisely when A and B are integer multiples of 2π (and they need not be the same multiple of 2π !). So, the solutions are $\boxed{B=2m\pi,\ A=2n\pi\text{, where } n \text{ and } m \text{ are integers}}$.

79. a. The following is the graph of $y_1 = \cos x\left(\frac{\sin 1}{1}\right) - \sin x\left(\frac{1-\cos 1}{1}\right)$, along with the graph of $y = \cos x$ (dotted graph):

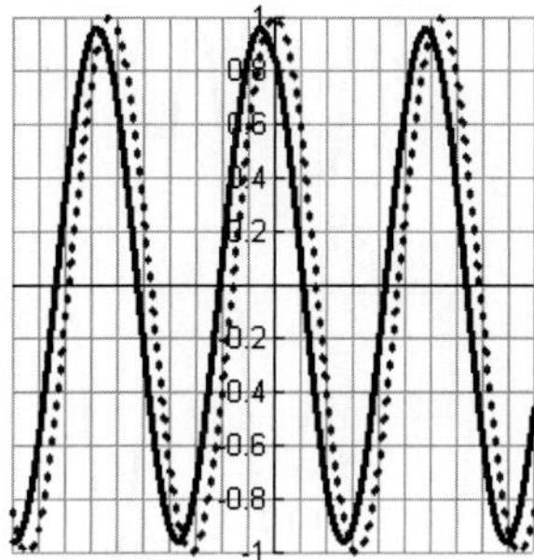

b. The following is the graph of $y_2 = \cos x\left(\frac{\sin 0.1}{0.1}\right) - \sin x\left(\frac{1-\cos 0.1}{0.1}\right)$, along with the graph of $y = \cos x$ (dotted graph):

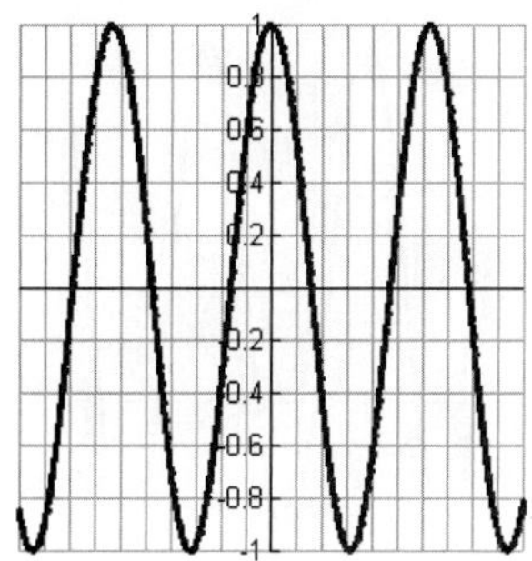

c. The following is the graph of $y_3 = \cos x\left(\frac{\sin 0.01}{0.01}\right) - \sin x\left(\frac{1-\cos 0.01}{0.01}\right)$, along with the graph of $y = \cos x$ (dotted graph):

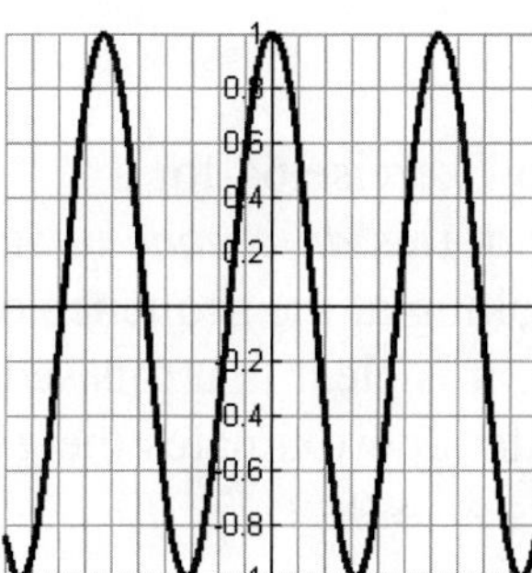

Refer back to Exercise 63 for a development of this *difference quotient* formula. Looking at the sequence of graphs in parts **a** – **c**, it is clear that this sequence of graphs better approximates the graph of $y = \cos x$ as $h \to 0$.

80. a. The following is the graph of $y_1 = -\sin x\left(\frac{\sin 1}{1}\right) - \cos x\left(\frac{1-\cos 1}{1}\right)$, along with the graph of $y = -\sin x$ (dotted graph):

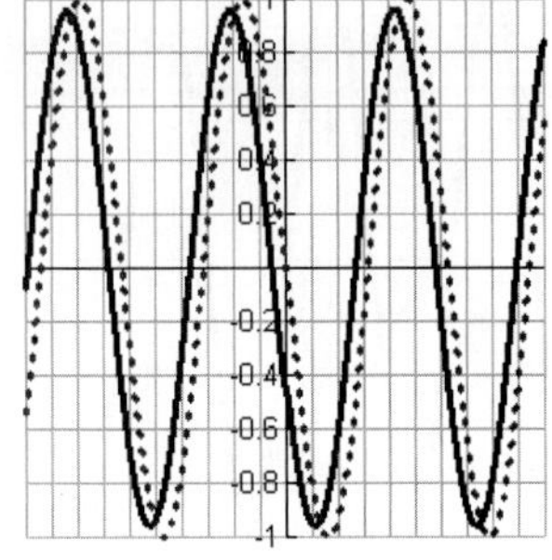

b. The following is the graph of $y_2 = -\sin x\left(\frac{\sin 0.1}{0.1}\right) - \cos x\left(\frac{1-\cos 0.1}{0.1}\right)$, along with the graph of $y = -\sin x$ (dotted graph):

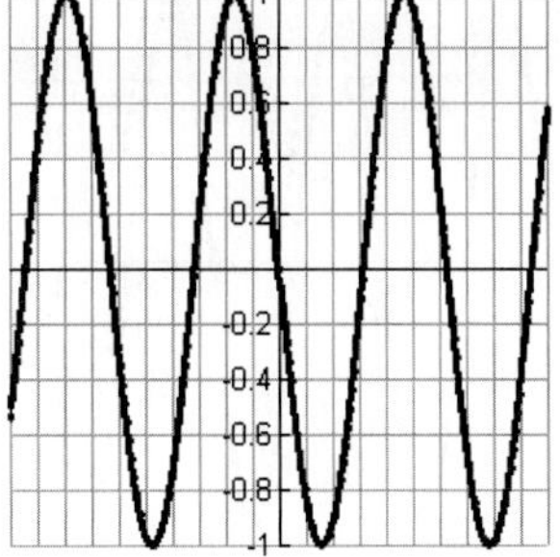

c. The following is the graph of $y_3 = -\sin x\left(\frac{\sin 0.01}{0.01}\right) - \cos x\left(\frac{1-\cos 0.01}{0.01}\right)$, along with the graph of $y = -\sin x$ (dotted graph):

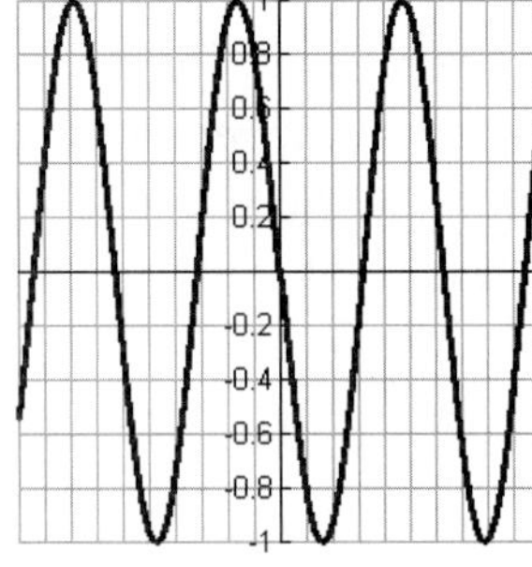

Refer back to Exercise 64 for a development of this *difference quotient* formula. Looking at the sequence of graphs in parts **a** – **c**, it is clear that this sequence of graphs better approximates the graph of $y = -\sin x$ as $h \to 0$.

81. Observe that

$$\begin{aligned}\frac{\sin(2(x+h)) - \sin 2x}{h} &= \frac{\sin 2x\cos(2h) + \sin(2h)\cos 2x - \sin 2x}{h}\\ &= \frac{\sin 2x(\cos(2h)-1) + \sin(2h)\cos 2x}{h}\\ &= \frac{\sin 2x(\cos(2h)-1)}{h} + \frac{\sin(2h)\cos 2x}{h}\\ &= \cos 2x\left(\frac{\sin(2h)}{h}\right) - \sin 2x\left(\frac{1-\cos(2h)}{h}\right)\end{aligned}$$

a. The following is the graph of
$$y_1 = \cos 2x\left(\frac{\sin(2)}{1}\right) - \sin 2x\left(\frac{1-\cos(2)}{1}\right),$$
along with the graph of $y = 2\cos 2x$ (dotted graph):

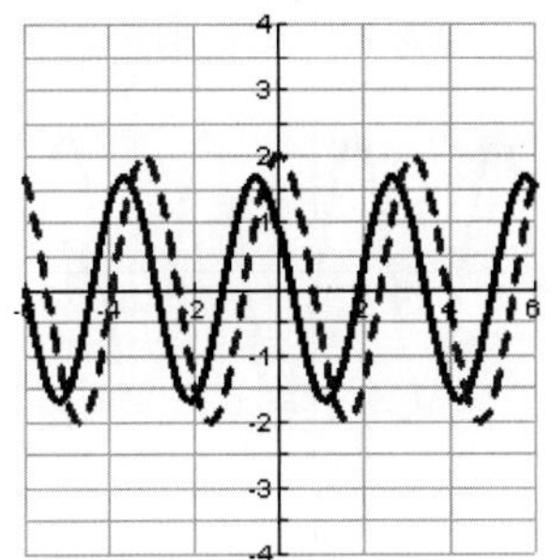

b. The following is the graph of
$$y_2 = \cos 2x\left(\frac{\sin(0.2)}{0.1}\right) - \sin 2x\left(\frac{1-\cos(0.2)}{0.1}\right),$$
along with the graph of $y = 2\cos 2x$ (dotted graph):

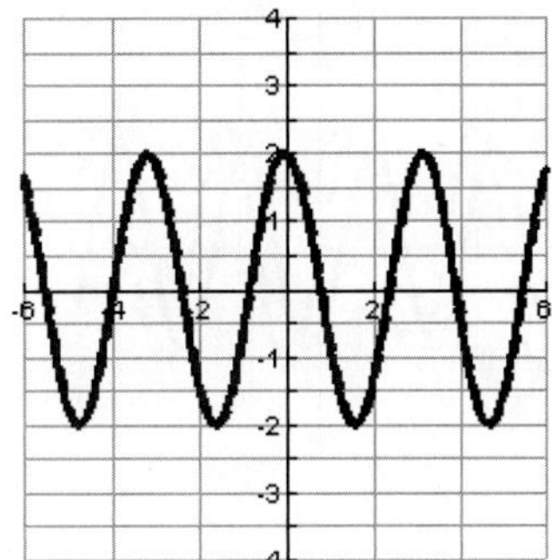

c. The following is the graph of
$$y_3 = \cos 2x\left(\frac{\sin(0.02)}{0.01}\right) - \sin 2x\left(\frac{1-\cos(0.02)}{0.01}\right),$$
along with the graph of $y = 2\cos 2x$ (dotted graph):

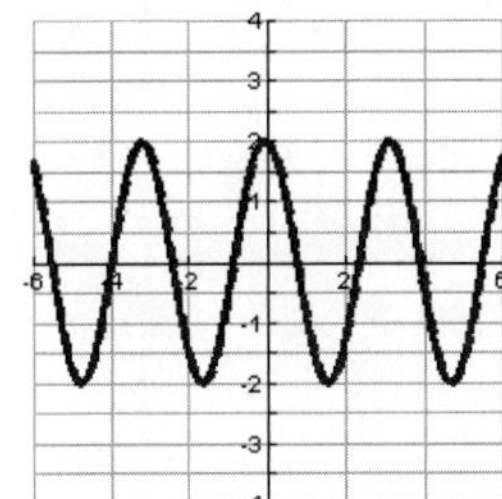

Looking at the sequence of graphs in parts **a** – **c**, it is clear that this sequence of graphs better approximates the graph of $y = 2\cos 2x$ as $h \to 0$.

82. Observe that
$$\begin{aligned}\frac{\cos(2(x+h)) - \cos 2x}{h} &= \frac{\cos 2x\cos(2h) - \sin 2x\sin(2h) - \cos 2x}{h} \\ &= \frac{\cos 2x(\cos(2h)-1) - \sin 2x\sin(2h)}{h} \\ &= \cos 2x\left(\frac{\cos(2h)-1}{h}\right) - \sin 2x\left(\frac{\sin(2h)}{h}\right) \\ &= -\cos 2x\left(\frac{1-\cos(2h)}{h}\right) - \sin 2x\left(\frac{\sin(2h)}{h}\right)\end{aligned}$$

a. The following is the graph of

$y_1 = -\cos 2x\left(\dfrac{1-\cos(2)}{1}\right) - \sin 2x\left(\dfrac{\sin(2)}{1}\right)$,

along with the graph of $y = 2\cos 2x$ (dotted graph):

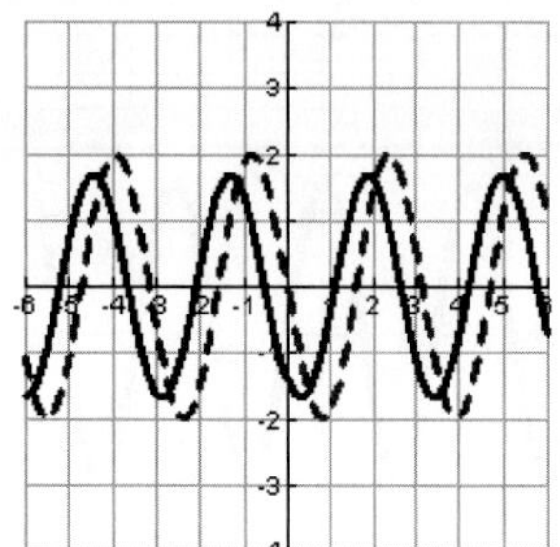

b. The following is the graph of

$y_2 = -\cos 2x\left(\dfrac{1-\cos(0.2)}{0.1}\right) - \sin 2x\left(\dfrac{\sin(0.2)}{0.1}\right)$,

along with the graph of $y = 2\cos 2x$ (dotted graph):

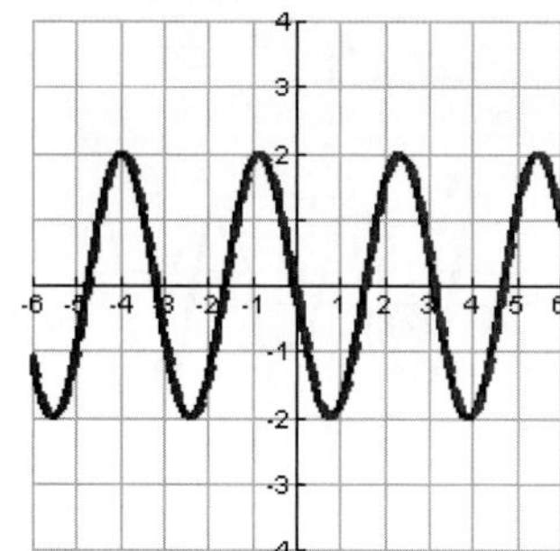

c. The following is the graph of

$y_2 = -\cos 2x\left(\dfrac{1-\cos(0.02)}{0.01}\right) - \sin 2x\left(\dfrac{\sin(0.02)}{0.01}\right)$,

along with the graph of $y = 2\cos 2x$ (dotted graph):

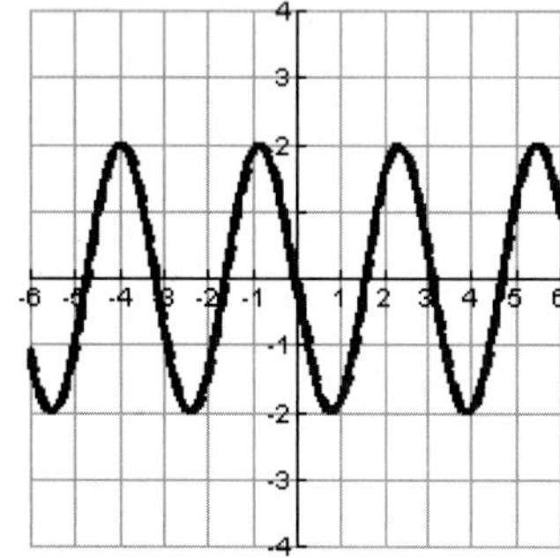

Looking at the sequence of graphs in parts **a** – **c**, it is clear that this sequence of graphs better approximates the graph of $y = -2\sin 2x$ as $h \to 0$.

83. Observe that

$$\begin{aligned}\frac{\sin(x+h)-\sin x}{h} &= \frac{\sin x\cos(h)+\sin(h)\cos x-\sin x}{h}\\ &= \frac{\left(\sin x\cos(h)-\sin x\right)+\sin(h)\cos x}{h}\\ &= \frac{\sin x\left(\cos(h)-1\right)+\sin(h)\cos x}{h}\\ &= \frac{\sin x\left(\cos(h)-1\right)}{h}+\frac{\sin(h)\cos x}{h}\\ &= \sin x\left(\frac{\cos(h)-1}{h}\right)+\cos x\left(\frac{\sin(h)}{h}\right)\\ &= \cos x\left(\frac{\sin(h)}{h}\right)-\sin x\left(\frac{1-\cos(h)}{h}\right)\end{aligned}$$

84. Observe that

$$\begin{aligned}\frac{\cos(x+h)-\cos x}{h} &= \frac{\cos x\cos(h)-\sin x\sin(h)-\cos x}{h} = \frac{\left(\cos x\cos(h)-\cos x\right)-\sin x\sin(h)}{h}\\ &= \frac{\cos x\left(\cos(h)-1\right)-\sin x\sin(h)}{h} = \cos x\left(\frac{\cos(h)-1}{h}\right)-\sin x\left(\frac{\sin(h)}{h}\right)\\ &= -\cos x\left(\frac{1-\cos(h)}{h}\right)-\sin x\left(\frac{\sin(h)}{h}\right) = -\sin x\left(\frac{\sin(h)}{h}\right)-\cos x\left(\frac{1-\cos(h)}{h}\right)\end{aligned}$$

85.

$$\begin{aligned}\sin(x+y)&=0\\ \sin x\cos y+\sin y\cos x&=0\\ \sin x\cos y&=-\sin y\cos x\\ \frac{\sin x}{\cos x}&=-\frac{\sin y}{\cos y}\\ \tan x&=-\tan y\end{aligned}$$

86.

$$\begin{aligned}\cos(x-y)&=0\\ \cos x\cos y+\sin x\sin y&=0\\ \cos x\cos y&=-\sin x\sin y\\ \frac{\cos x}{\sin x}&=-\frac{\sin y}{\cos y}\\ \cot x&=-\tan y\end{aligned}$$

87.	**88.**
$\tan(x+y)=2$ $\dfrac{\tan x+\tan y}{1-\tan x\tan y}=2$ $\tan x+\tan y=2-2\tan x\tan y$ $\tan x(1+2\tan y)=2-\tan y$ $\tan x=\dfrac{2-\tan y}{1+2\tan y}$	$\cos(x+y)=\sin y$ $\cos x\cos y-\sin x\sin y=\sin y$ $\cos x\cos y=\sin y(1+\sin x)$ $\dfrac{\cos x}{1+\sin x}=\tan y$

Section 6.3 Solutions ---

1. Since $\sin x=\frac{1}{\sqrt{5}}$ and $\cos x<0$, the terminal side of x must be in QII.
The diagram is as follows:

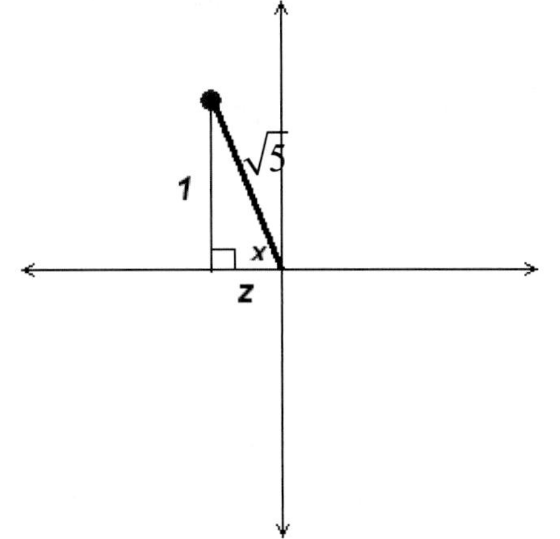

From the Pythagorean Theorem, we have $z^2+1^2=\left(\sqrt{5}\right)^2$, so that $z=-2$.

As such, $\cos x=-\frac{2}{\sqrt{5}}$. Therefore,

$$\sin 2x=2\sin x\cos x$$
$$=2\left(\frac{1}{\sqrt{5}}\right)\left(-\frac{2}{\sqrt{5}}\right)=\boxed{-\frac{4}{5}}$$

2. Since $\sin x=\frac{1}{\sqrt{5}}$ and $\cos x<0$, the terminal side of x must be in QII.
The diagram is as follows:

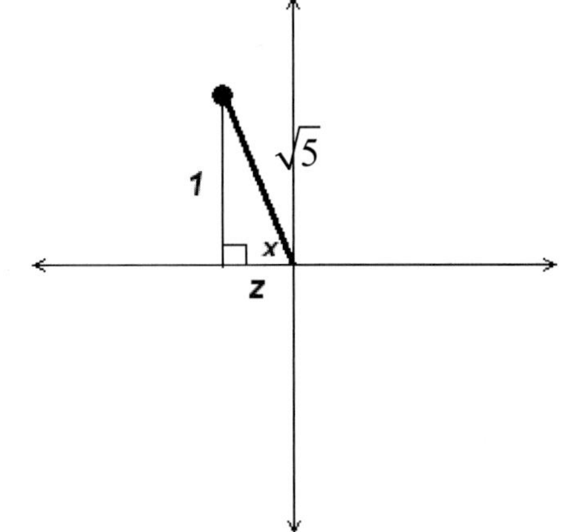

From the Pythagorean Theorem, we have $z^2+1^2=\left(\sqrt{5}\right)^2$, so that $z=-2$.

As such, $\cos x=-\frac{2}{\sqrt{5}}$. Therefore,

$$\cos 2x=\cos^2 x-\sin^2 x$$
$$=\left(-\frac{2}{\sqrt{5}}\right)^2-\left(\frac{1}{\sqrt{5}}\right)^2=\frac{4}{5}-\frac{1}{5}=\boxed{\frac{3}{5}}$$

3. Since $\cos x = \frac{5}{13}$ and $\sin x < 0$, the terminal side of x must be in QIV. The diagram is as follows:

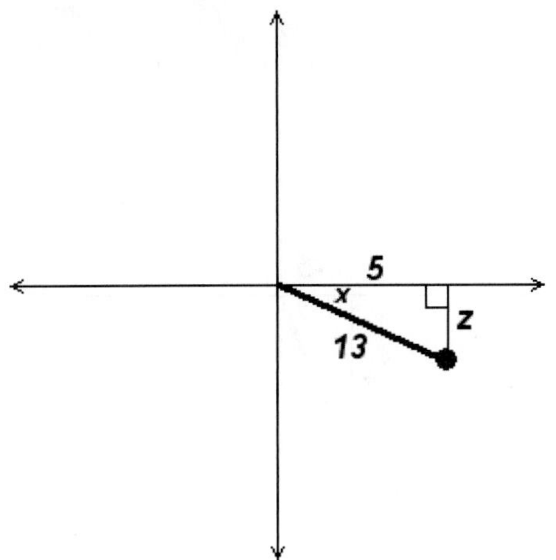

From the Pythagorean Theorem, we have $z^2 + 5^2 = 13^2$, so that $z = -12$.

As such, $\sin x = -\frac{12}{13}$, so that

$$\tan x = \frac{\sin x}{\cos x} = \frac{-\frac{12}{13}}{\frac{5}{13}} = -\frac{12}{5}.$$

Therefore,

$$\tan 2x = \frac{2\tan x}{1-\tan^2 x} = \frac{2\left(-\frac{12}{5}\right)}{1-\left(-\frac{12}{5}\right)^2} = \frac{-\frac{24}{5}}{-\frac{119}{25}} = \boxed{\frac{120}{119}}$$

4. Since $\cos x = -\frac{5}{13}$ and $\sin x < 0$, the terminal side of x must be in QIII. The diagram is as follows:

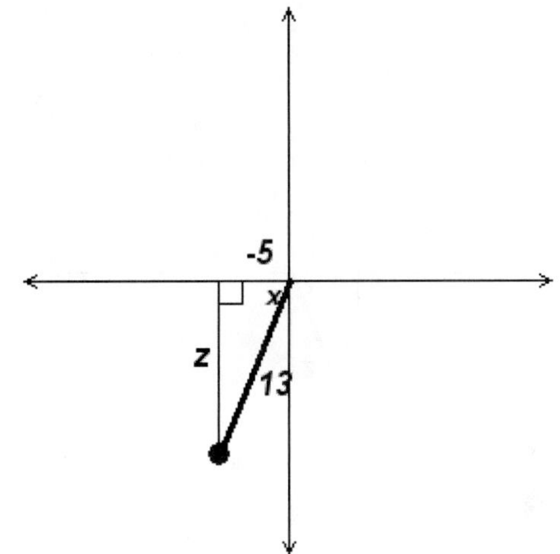

From the Pythagorean Theorem, we have $z^2 + (-5)^2 = 13^2$, so that $z = -12$.

As such, $\sin x = -\frac{12}{13}$, so that

$$\tan x = \frac{\sin x}{\cos x} = \frac{-\frac{12}{13}}{-\frac{5}{13}} = \frac{12}{5}.$$

Therefore,

$$\tan 2x = \frac{2\tan x}{1-\tan^2 x} = \frac{2\left(\frac{12}{5}\right)}{1-\left(\frac{12}{5}\right)^2} = \frac{\frac{24}{5}}{-\frac{119}{25}} = \boxed{-\frac{120}{119}}$$

5. Since $\tan x = \frac{-12}{-5} = \frac{12}{5}$ and $\pi < x < \frac{3\pi}{2}$, the diagram is as follows:

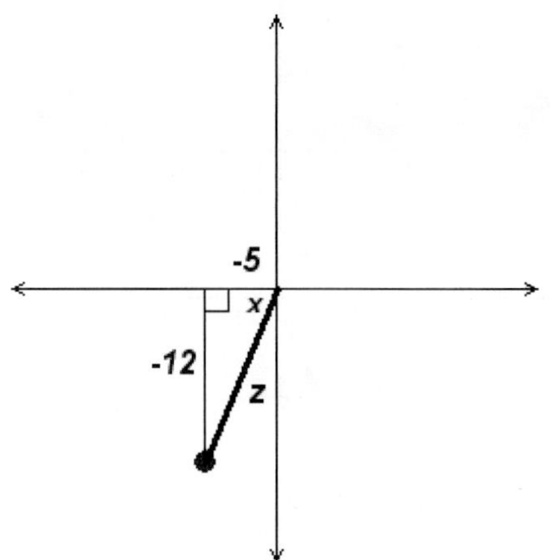

From the Pythagorean Theorem, we have $(-5)^2 + (-12)^2 = z^2$, so that $z = 13$.

As such, $\sin x = -\frac{12}{13}$ and $\cos x = -\frac{5}{13}$.

Therefore,

$$\sin 2x = 2\sin x \cos x = 2\left(-\frac{12}{13}\right)\left(-\frac{5}{13}\right)$$

$$= \boxed{\frac{120}{169}}$$

6. Since $\tan x = \frac{-12}{-5} = \frac{12}{5}$ and $\pi < x < \frac{3\pi}{2}$, the diagram is as follows:

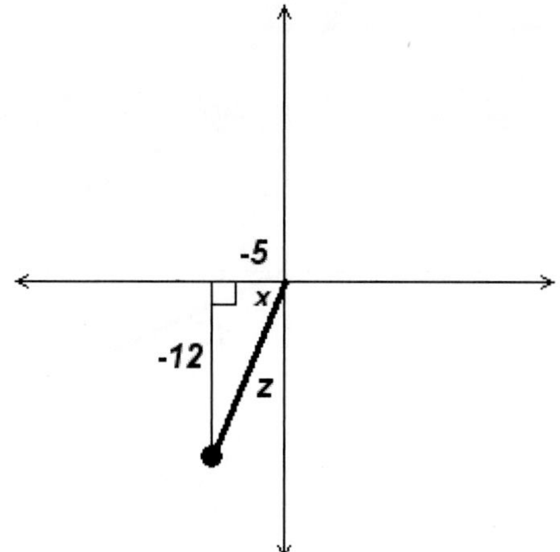

From the Pythagorean Theorem, we have $(-5)^2 + (-12)^2 = z^2$, so that $z = 13$.

As such, $\sin x = -\frac{12}{13}$ and $\cos x = -\frac{5}{13}$.

Therefore,

$$\cos 2x = \cos^2 x - \sin^2 x = \left(-\frac{5}{13}\right)^2 - \left(-\frac{12}{13}\right)^2$$

$$= \boxed{-\frac{119}{169}}$$

7. Since $\sec x = \sqrt{5}$, it follows that $\dfrac{1}{\cos x} = \sqrt{5}$ and so, $\cos x = \dfrac{1}{\sqrt{5}}$. Since also $\sin x > 0$, the terminal side of x must be in QI. The diagram is as follows:

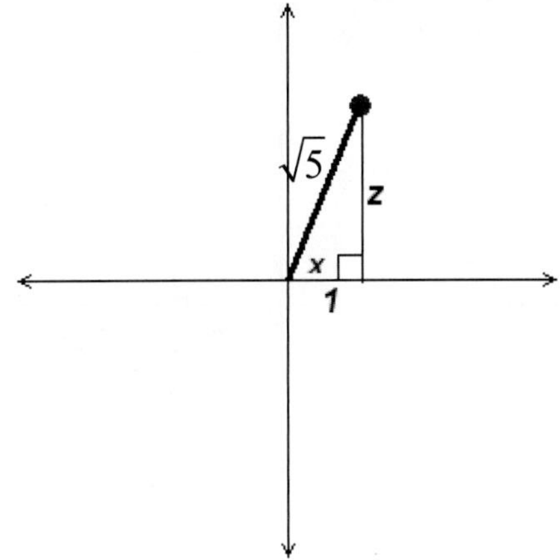

From the Pythagorean Theorem, we have $z^2 + 1^2 = \left(\sqrt{5}\right)^2$, so that $z = 2$.

As such, $\sin x = \dfrac{2}{\sqrt{5}}$, so that

$$\tan x = \frac{\sin x}{\cos x} = \frac{\frac{2}{\sqrt{5}}}{\frac{1}{\sqrt{5}}} = 2.$$

Therefore,

$$\tan 2x = \frac{2\tan x}{1-\tan^2 x} = \frac{2(2)}{1-2^2} = \frac{4}{-3} = \boxed{-\frac{4}{3}}.$$

8. Since $\sec x = \sqrt{3}$, it follows that $\dfrac{1}{\cos x} = \sqrt{3}$ and so, $\cos x = \dfrac{1}{\sqrt{3}}$. Since also $\sin x < 0$, the terminal side of x must be in QIV. The diagram is as follows:

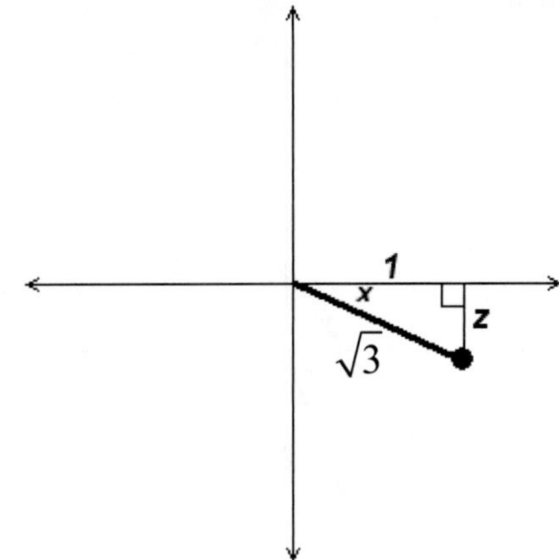

From the Pythagorean Theorem, we have $z^2 + 1^2 = \left(\sqrt{3}\right)^2$, so that $z = -\sqrt{2}$.

As such, $\sin x = -\dfrac{\sqrt{2}}{\sqrt{3}}$, so that

$$\tan x = \frac{\sin x}{\cos x} = \frac{-\frac{\sqrt{2}}{\sqrt{3}}}{\frac{1}{\sqrt{3}}} = -\sqrt{2}.$$

Therefore,

$$\tan 2x = \frac{2\tan x}{1-\tan^2 x} = \frac{2\left(-\sqrt{2}\right)}{1-\left(-\sqrt{2}\right)^2} = \frac{-2\sqrt{2}}{1-2} = \boxed{2\sqrt{2}}$$

.

9. Since $\csc x = -2\sqrt{5}$, it follows that $\frac{1}{\sin x} = -2\sqrt{5}$ and so, $\sin x = -\frac{1}{2\sqrt{5}} = -\frac{\sqrt{5}}{10}$. Since also $\cos x < 0$, the terminal side of x must be in QIII. The diagram is as follows:

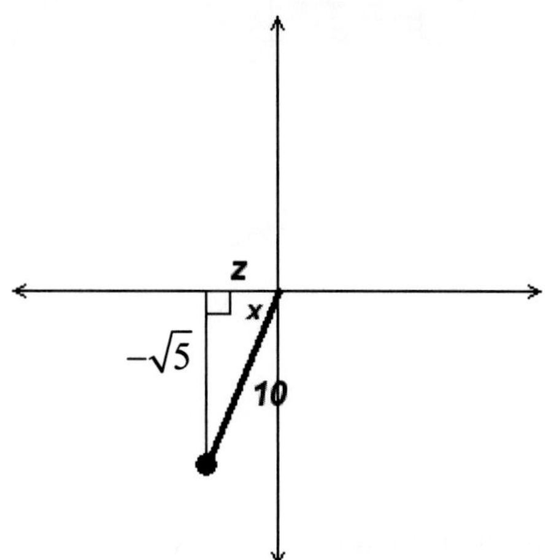

From the Pythagorean Theorem, we have $z^2 + \left(-\sqrt{5}\right)^2 = (10)^2$, so that $z = -\sqrt{95}$.

As such, $\cos x = -\frac{\sqrt{95}}{10}$. Therefore,

$$\sin 2x = 2\sin x \cos x = 2\left(-\frac{\sqrt{5}}{10}\right)\left(-\frac{\sqrt{95}}{10}\right)$$

$$= \frac{10\sqrt{19}}{100} = \boxed{\frac{\sqrt{19}}{10}}$$

10. Since $\csc x = -\sqrt{13}$, it follows that $\frac{1}{\sin x} = -\sqrt{13}$ and so, $\sin x = -\frac{1}{\sqrt{13}}$. Since also $\cos x > 0$, the terminal side of x must be in QIV. The diagram is as follows:

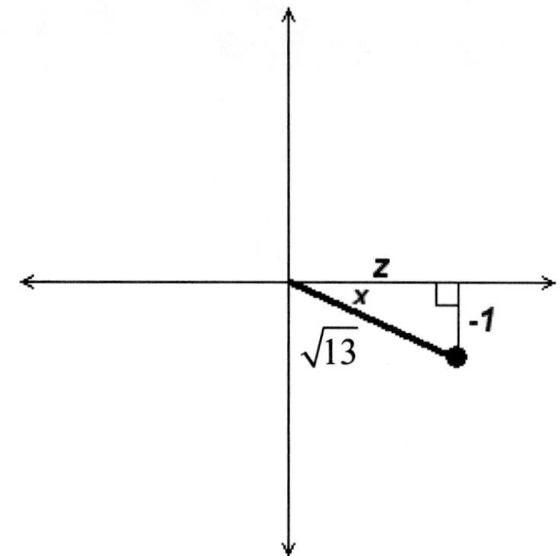

From the Pythagorean Theorem, we have $z^2 + (-1)^2 = \left(\sqrt{13}\right)^2$, so that $z = \sqrt{12}$.

As such, $\cos x = \frac{\sqrt{12}}{\sqrt{13}}$. Therefore,

$$\sin 2x = 2\sin x \cos x = 2\left(-\frac{1}{\sqrt{13}}\right)\left(\frac{\sqrt{12}}{\sqrt{13}}\right)$$

$$= -\frac{2\sqrt{12}}{13} = \boxed{-\frac{4\sqrt{3}}{13}}$$

11. Since $\csc x < 0$, it follows that $\sin x < 0$. Since also $\cos x = -\frac{12}{13}$, the terminal side of x must be in QIII. The diagram is as follows:

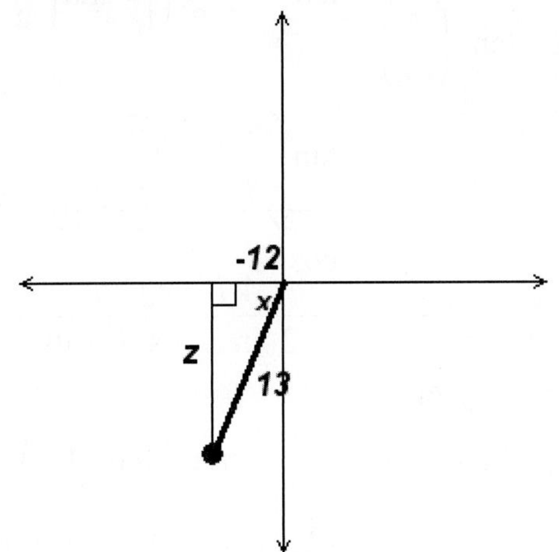

From the Pythagorean Theorem, we have $z^2 + (-12)^2 = 13^2$, so that $z = -5$.

As such, $\sin x = -\frac{5}{13}$, so that

$$\tan x = \frac{\sin x}{\cos x} = \frac{-\frac{5}{13}}{-\frac{12}{13}} = \frac{5}{12}.$$

Therefore,

$$\cot 2x = \frac{1}{\tan 2x} = \frac{1}{\frac{2\tan x}{1-\tan^2 x}} = \frac{1-\tan^2 x}{2\tan x}$$

$$= \frac{1-\left(\frac{5}{12}\right)^2}{2\left(\frac{5}{12}\right)} = \frac{119}{144}\cdot\frac{6}{5} = \boxed{\frac{119}{120}}$$

12. Since $\cot x < 0$ and $\sin x = \frac{12}{13}$, it follows that $\cos x < 0$, and so the terminal side of x must be in QII. The diagram is as follows:

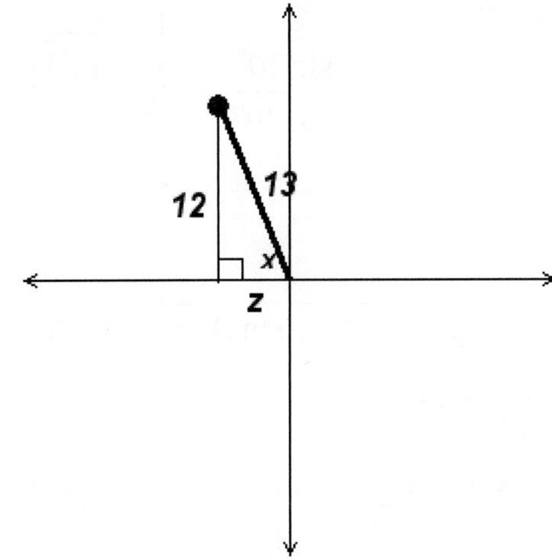

From the Pythagorean Theorem, we have $z^2 + 12^2 = 13^2$, so that $z = -5$.

As such, $\cos x = -\frac{5}{13}$. Therefore,

$$\csc 2x = \frac{1}{\sin 2x} = \frac{1}{2\sin x\cos x}$$

$$= \frac{1}{2\left(\frac{12}{13}\right)\left(-\frac{5}{13}\right)} = \frac{1}{-\frac{120}{169}} = \boxed{-\frac{169}{120}}$$

13. Using $\tan 2A = \dfrac{2\tan A}{1-\tan^2 A}$ with $A = 15^\circ$ yields

$$\frac{2\tan 15^\circ}{1-\tan^2 15^\circ} = \tan\left(2\cdot 15^\circ\right) = \tan 30^\circ$$

$$= \frac{\sin 30^\circ}{\cos 30^\circ} = \frac{\frac{1}{2}}{\frac{\sqrt{3}}{2}} = \boxed{\frac{\sqrt{3}}{3}}$$

14. Using $\tan 2A = \dfrac{2\tan A}{1-\tan^2 A}$ with $A = \dfrac{\pi}{8}$ yields

$$\frac{2\tan\left(\frac{\pi}{8}\right)}{1-\tan^2\left(\frac{\pi}{8}\right)} = \tan\left(2\cdot\left(\frac{\pi}{8}\right)\right) = \tan\left(\frac{\pi}{4}\right)$$

$$= \frac{\sin\left(\frac{\pi}{4}\right)}{\cos\left(\frac{\pi}{4}\right)} = \frac{\frac{\sqrt{2}}{2}}{\frac{\sqrt{2}}{2}} = \boxed{1}$$

15. Using $\sin 2A = 2\sin A\cos A$ with $A = \dfrac{\pi}{8}$ yields

$$\sin\left(\frac{\pi}{8}\right)\cos\left(\frac{\pi}{8}\right) = \frac{1}{2}\left[2\sin\left(\frac{\pi}{8}\right)\cos\left(\frac{\pi}{8}\right)\right]$$

$$= \frac{1}{2}\sin\left(\frac{\pi}{4}\right) = \frac{1}{2}\left(\frac{\sqrt{2}}{2}\right)$$

$$= \boxed{\frac{\sqrt{2}}{4}}$$

16. Using $\sin 2A = 2\sin A\cos A$ with $A = 15^\circ$ yields

$$\sin\left(15^\circ\right)\cos\left(15^\circ\right) = \frac{1}{2}\left[2\sin\left(15^\circ\right)\cos\left(15^\circ\right)\right]$$

$$= \frac{1}{2}\sin\left(2\cdot 15^\circ\right) = \frac{1}{2}\sin\left(30^\circ\right)$$

$$= \frac{1}{2}\left(\frac{1}{2}\right) = \boxed{\frac{1}{4}}$$

17. Using $\cos 2A = \cos^2 A - \sin^2 A$ with $A = 2x$ yields

$$\cos^2 2x - \sin^2 2x = \cos(2\cdot 2x) = \boxed{\cos 4x}$$

18. Using $\cos 2A = \cos^2 A - \sin^2 A$ with $A = x+2$ yields

$$\cos^2(x+2) - \sin^2(x+2) = \cos(2\cdot(x+2))$$

$$= \boxed{\cos(2x+4)}$$

19.

$$\frac{2\tan\left(\frac{5\pi}{12}\right)}{1-\tan^2\left(\frac{5\pi}{12}\right)} = \tan\left(2\cdot\tfrac{5\pi}{12}\right) = \tan\left(\tfrac{5\pi}{6}\right)$$

$$= \frac{\sin\left(\frac{5\pi}{6}\right)}{\cos\left(\frac{5\pi}{6}\right)} = \frac{1/2}{-\sqrt{3}/2} = \boxed{-\frac{\sqrt{3}}{3}}$$

20. $\dfrac{2\tan\left(\frac{x}{2}\right)}{1-\tan^2\left(\frac{x}{2}\right)} = \tan\left(2\cdot\tfrac{x}{2}\right) = \boxed{\tan x}$

21.

$$1-2\sin^2\left(\tfrac{7\pi}{12}\right) = \cos\left(2\cdot\tfrac{7\pi}{12}\right)$$

$$= \cos\left(\tfrac{7\pi}{6}\right) = \boxed{-\tfrac{\sqrt{3}}{2}}$$

22.

$$2\sin^2\left(-\tfrac{5\pi}{8}\right) - 1 = -\cos\left(2\cdot\tfrac{5\pi}{8}\right)$$

$$= -\cos\left(\tfrac{5\pi}{4}\right) = \boxed{\tfrac{\sqrt{2}}{2}}$$

23.	24.
$2\cos^2\left(-\frac{7\pi}{12}\right)-1=\cos\left(-2\cdot\frac{7\pi}{12}\right)$ $=\cos\left(-\frac{7\pi}{6}\right)=\cos\left(\frac{7\pi}{6}\right)=-\frac{\sqrt{3}}{2}$	$1-2\cos^2\left(-\frac{\pi}{8}\right)=-\cos\left(-2\cdot\frac{\pi}{8}\right)$ $=-\cos\left(\frac{\pi}{4}\right)=-\frac{\sqrt{2}}{2}$

25. Observe that

$$\csc 2A=\frac{1}{\sin 2A}=\frac{1}{2\sin A\cos A}=\frac{1}{2}\left(\frac{1}{\sin A}\right)\left(\frac{1}{\cos A}\right)=\frac{1}{2}\csc A\sec A.$$

26. Observe that

$$\cot 2A=\frac{1}{\tan 2A}=\frac{1}{\frac{2\tan A}{1-\tan^2 A}}=\frac{1-\tan^2 A}{2\tan A}=\frac{1}{2}\left[\frac{1}{\tan A}-\frac{\tan^2 A}{\tan A}\right]=\frac{1}{2}\left[\cot A-\tan A\right].$$

27. Observe that

$$(\sin x-\cos x)(\cos x+\sin x)=\sin^2 x-\cos^2 x=-\left[\cos^2 x-\sin^2 x\right]=-\cos 2x.$$

28. Observe that

$$(\sin x+\cos x)^2=\sin^2 x+2\sin x\cos x+\cos^2 x=\underbrace{\left(\sin^2 x+\cos^2 x\right)}_{=1}+\underbrace{2\sin x\cos x}_{=\sin 2x}$$
$$=1+\sin 2x$$

29. Note that starting with the right-side is easier this time. Indeed, observe that

$$\frac{1+\cos 2x}{2}=\frac{1+\left(\cos^2 x-\sin^2 x\right)}{2}$$
$$=\frac{\left(1-\sin^2 x\right)+\cos^2 x}{2}$$
$$=\frac{\cos^2 x+\cos^2 x}{2}$$
$$=\frac{2\cos^2 x}{2}=\cos^2 x$$

30. Note that starting with the right-side is easier this time. Indeed, observe that

$$\frac{1-\cos 2x}{2}=\frac{1-\left(\cos^2 x-\sin^2 x\right)}{2}$$
$$=\frac{\left(1-\cos^2 x\right)+\sin^2 x}{2}$$
$$=\frac{\sin^2 x+\sin^2 x}{2}$$
$$=\frac{2\sin^2 x}{2}=\sin^2 x$$

31. Observe that

$$\cos^4 x - \sin^4 x = \left(\cos^2 x - \sin^2 x\right)\underbrace{\left(\cos^2 x + \sin^2 x\right)}_{=1} = \cos 2x$$

32. Note that starting with the right-side is easier this time. Indeed, observe that

$$\begin{aligned} 1 - \frac{1}{2}\sin^2(2x) &= 1 - \frac{1}{2}[2\sin x \cos x]^2 \\ &= 1 - 2\sin^2 x \cos^2 x \\ &= \left(\sin^2 x + \cos^2 x\right)^2 - 2\sin^2 x \cos^2 x \\ &= \sin^4 x + \cancel{2\sin^2 x \cos^2 x} + \cos^4 x - \cancel{2\sin^2 x \cos^2 x} \\ &= \sin^4 x + \cos^4 x \end{aligned}$$

33. Observe that

$$\begin{aligned} 8\sin^2 x \cos^2 x &= 2 \cdot 4\sin^2 x \cos^2 x \\ &= 2[2\sin x \cos x]^2 \\ &= 2[\sin 2x]^2 \\ &= 2\left[1 - \cos^2 2x\right] \\ &= 2 - 2\cos^2 2x \\ &= 1 + 1 - 2\cos^2 2x \\ &= \left(\cos^2 2x + \sin^2 2x\right) + 1 - 2\cos^2 2x \\ &= 1 - \cos^2 2x + \sin^2 2x \\ &= 1 - \left[\cos^2 2x - \sin^2 2x\right] \\ &= 1 - \cos(2 \cdot 2x) = 1 - \cos 4x \end{aligned}$$

34. Observe that

$$\left(\cos 2x - \sin 2x\right)\left(\sin 2x + \cos 2x\right) = \cos^2 2x - \sin^2 2x = \cos\left(2 \cdot 2x\right) = \cos 4x\,.$$

35. Note that starting with the right-side is easier this time. Indeed, observe that

$$-2\sin^2 x \csc^2 2x = \frac{-2\sin^2 x}{\sin^2 2x} = \frac{-2\sin^2 x}{[2\sin x \cos x]^2} = \frac{-2\cancel{\sin^2 x}}{4\cancel{\sin^2 x}\cos^2 x} = -\frac{1}{2\cos^2 x} = -\frac{1}{2}\sec^2 x$$

36. Note that starting with the right-side is easier this time. Indeed, observe that

$$\frac{\sec x \csc x}{\cos 2x} = \frac{1}{\cos x \sin x \cos 2x}$$
$$= \frac{1}{\frac{1}{2}[2\cos x \sin x]\cos 2x}$$
$$= \frac{1}{\frac{1}{2}\sin 2x \cos 2x}$$
$$= \frac{1}{\frac{1}{2}\cdot\frac{1}{2}\cdot[2\sin 2x \cos 2x]}$$
$$= \frac{1}{\frac{1}{4}\sin 4x} = \frac{4}{\sin 4x} = 4\csc 4x$$

37. Note that starting with the right-side is easier this time. Indeed, observe that

$$\sin x\left(4\cos^2 x - 1\right) = \sin x\left(3\cos^2 x + \cos^2 x - 1\right)$$
$$= \sin x\left(3\cos^2 x - \left(1 - \cos^2 x\right)\right)$$
$$= \sin x\left(3\cos^2 x - \sin^2 x\right)$$
$$= 3\sin x \cos^2 x - \sin^3 x$$
$$= \sin x\left(\cos^2 x - \sin^2 x\right) + 2\sin x \cos^2 x$$
$$= \sin x \cos 2x + (2\sin x \cos x)\cos x$$
$$= \sin x \cos 2x + \sin 2x \cos x$$
$$= \sin(x + 2x) = \sin 3x$$

38. Observe that

$$\tan 3x = \tan(x+2x) = \frac{\tan x + \tan 2x}{1-\tan x \tan 2x}$$

$$= \frac{\tan x + \dfrac{2\tan x}{1-\tan^2 x}}{1-\tan x\left[\dfrac{2\tan x}{1-\tan^2 x}\right]}$$

$$= \frac{\dfrac{\tan x\left(1-\tan^2 x\right)+2\tan x}{\cancel{1-\tan^2 x}}}{\dfrac{\left(1-\tan^2 x\right)-2\tan^2 x}{\cancel{1-\tan^2 x}}}$$

$$= \frac{3\tan x - \tan^3 x}{1-3\tan^2 x}$$

$$= \frac{\tan x\left(3-\tan^2 x\right)}{1-3\tan^2 x}$$

39. Note that starting with the right-side is easier this time. Indeed, observe that

$$2\sin x\cos x - 4\sin^3 x\cos x = 2\sin x\cos x\left(1-2\sin^2 x\right)$$

$$= 2\sin x\cos x\left(\underbrace{1-\sin^2 x}_{=\cos^2 x}-\sin^2 x\right)$$

$$= \sin 2x\left(\cos^2 x - \sin^2 x\right)$$

$$= \sin 2x\cos 2x$$

$$= \frac{1}{2}\left(2\sin 2x\cos 2x\right)$$

$$= \frac{1}{2}\sin 4x$$

40. Note that starting with the right-side is easier this time. Indeed, observe that

$$\left(\cos 2x - \sin 2x\right)\left(\cos 2x + \sin 2x\right) = \left(\cos^2 2x - \sin^2 2x\right) = \cos\left(2\cdot 2x\right) = \cos 4x$$

41. Before graphing, observe that

$$y=\frac{\sin 2x}{1-\cos 2x}=\frac{2\sin x\cos x}{1-\left(\cos^2 x-\sin^2 x\right)}=\frac{2\sin x\cos x}{\left(\cos^2 x+\sin^2 x\right)-\left(\cos^2 x-\sin^2 x\right)}$$

$$=\frac{2\sin x\cos x}{2\sin^2 x}=\frac{\cos x}{\sin x}=\cot x$$

So, a snapshot of the graph is as follows:

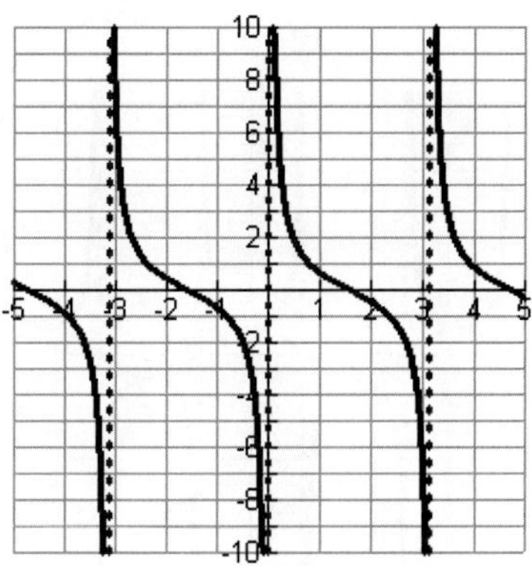

42. Before graphing, observe that

$$y=\frac{2\tan x}{2-\sec^2 x}=\frac{2\tan x}{2-\left(1+\tan^2 x\right)}=\frac{2\tan x}{1-\tan^2 x}=\tan 2x$$

So, the period of this function is $\frac{\pi}{2}$ and the vertical asymptotes occur when $\cos 2x=0$, namely when $2x=\frac{(2n+1)\pi}{2}$, and so $x=\frac{(2n+1)\pi}{4}$. The graph is as follows:

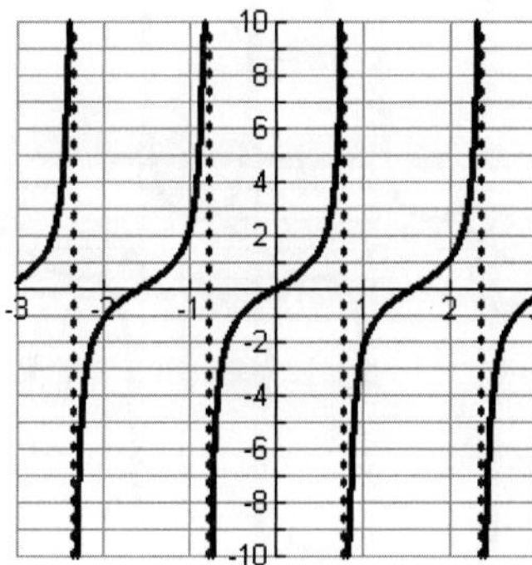

43. Before graphing, observe that

$$y=\frac{\cot x+\tan x}{\cot x-\tan x}=\frac{\frac{\cos x}{\sin x}+\frac{\sin x}{\cos x}}{\frac{\cos x}{\sin x}-\frac{\sin x}{\cos x}}=\frac{\frac{\cos^2 x+\sin^2 x}{\sin x\cos x}}{\frac{\cos^2 x-\sin^2 x}{\sin x\cos x}}=\frac{\overbrace{\cos^2 x+\sin^2 x}^{=1}}{\underbrace{\cos^2 x-\sin^2 x}_{=\cos 2x}}=\frac{1}{\cos 2x}=\sec 2x$$

So, the period of this function is $\frac{2\pi}{2}=\pi$. The graph is as follows:

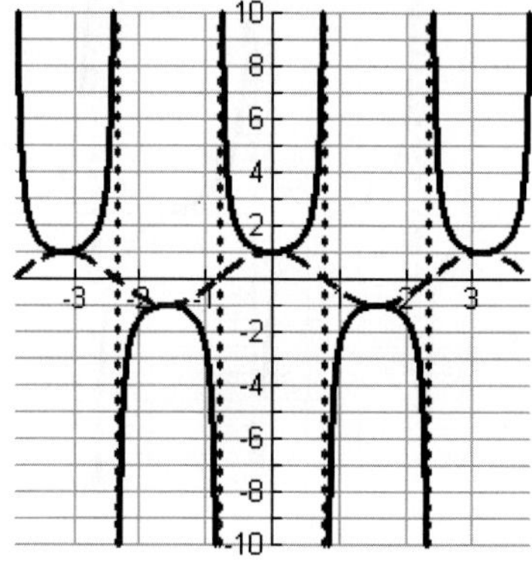

44. Before graphing, observe that

$$y=\frac{1}{2}\underbrace{\tan x\cot x}_{=1}\sec x\csc x=\frac{1}{2}\cdot\frac{1}{\cos x}\cdot\frac{1}{\sin x}=\frac{1}{\sin 2x}=\csc 2x$$

So, the period of this function is $\frac{2\pi}{2}=\pi$. The graph is as follows:

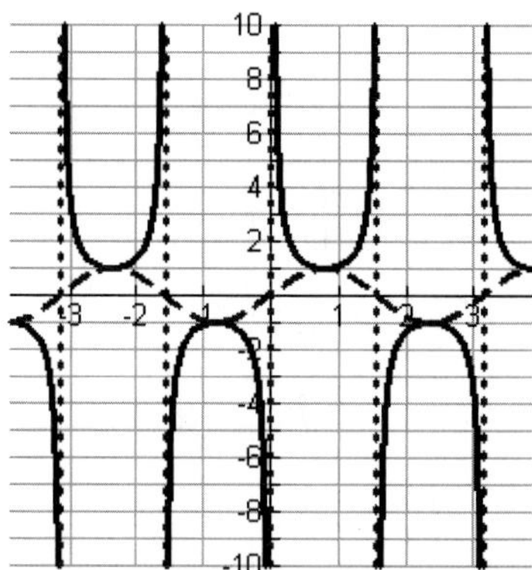

45. Before graphing, observe that $\sin(2x)\cos(2x)=\frac{1}{2}\sin(4x)$. The graph is as follows:

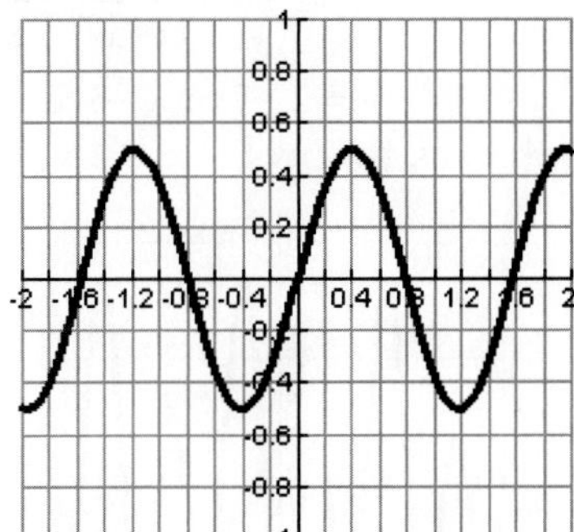

46. Before graphing, observe that

$$3\sin(3x)\cos(-3x)=\tfrac{3}{2}\big(2\sin(3x)\cos(3x)\big)$$
$$=\tfrac{3}{2}\sin(6x)$$

The graph is as follows:

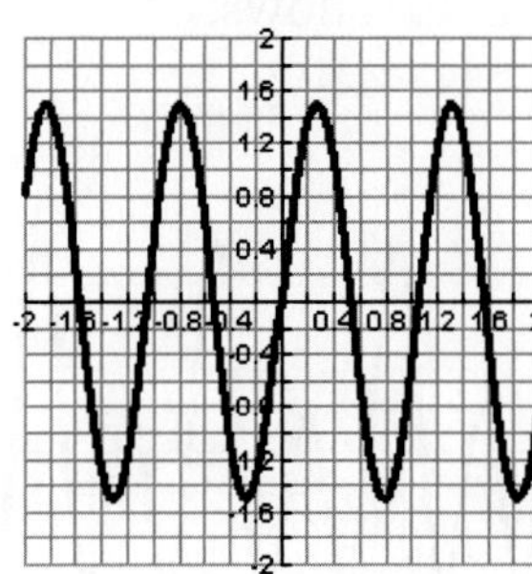

47. Before graphing, observe that

$$1-\frac{\tan x\cot x}{\sec x\csc x}=1-\frac{1}{\dfrac{1}{\cos x}\cdot\dfrac{1}{\sin x}}$$
$$=1-\sin x\cos x$$
$$=1-\tfrac{1}{2}\sin(2x)$$

The graph is as follows:

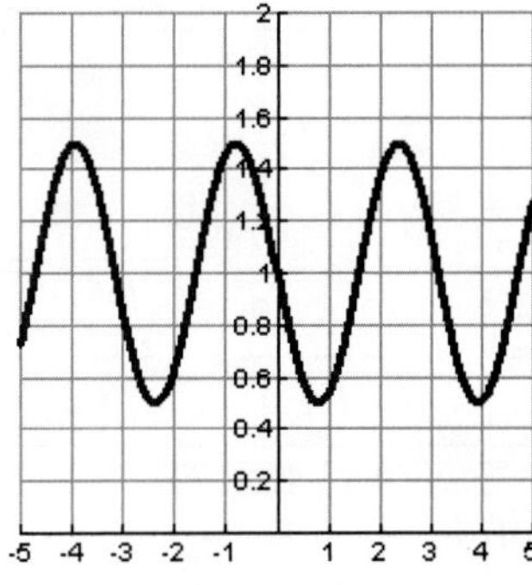

48. Before graphing, observe that

$$3-2\frac{\sec 2x}{\csc 2x}=3-2\cdot\frac{\frac{1}{\cos 2x}}{\frac{1}{\sin 2x}}=2-3\tan 2x\,.$$

The graph is as follows:

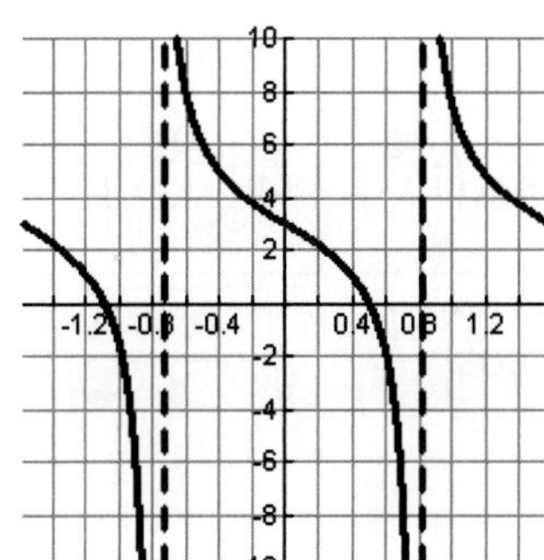

49. Before graphing, observe that

$$\frac{\sin 2x}{\cos x} - 3\cos 2x = \frac{2\sin x \cancel{\cos x}}{\cancel{\cos x}} - 3\cos 2x$$
$$= 2\sin x - 3\cos 2x$$

The graph is as follows:

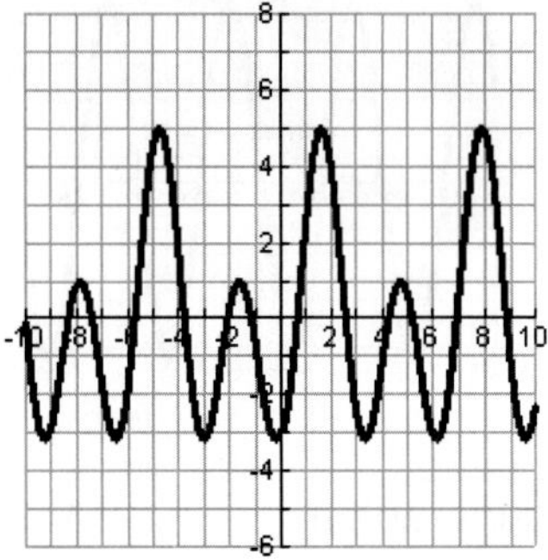

50. Before graphing, observe that

$$2 + \frac{\sin 2x}{\cos x} - 3\cos 2x = 2 + \frac{\sin 2x}{\cos x} - 3\cos 2x$$
$$= 2 + \frac{2\sin x \cancel{\cos x}}{\cancel{\cos x}} - 3\cos 2x$$
$$= 2 + 2\sin x - 3\cos 2x$$

The graph is as follows:

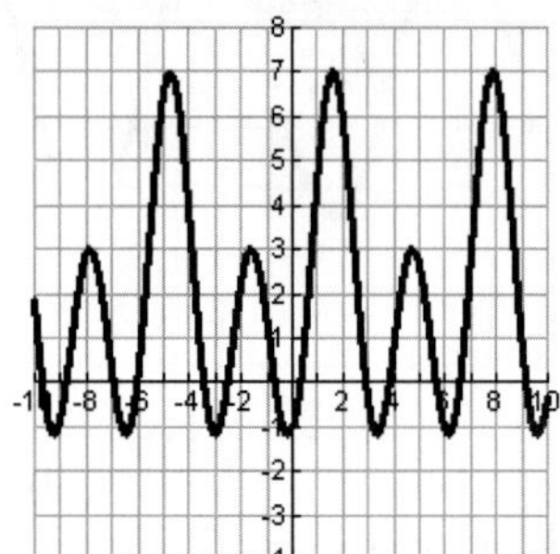

51. Use $\sin\left(\frac{A}{2}\right) = \sqrt{\frac{1-\cos A}{2}}$ with $A = 30^\circ$ to obtain

$$\sin\left(15^\circ\right) = \sin\left(\frac{30^\circ}{2}\right) = \sqrt{\frac{1-\cos 30^\circ}{2}} = \sqrt{\frac{1-\frac{\sqrt{3}}{2}}{2}} = \sqrt{\frac{2-\sqrt{3}}{4}} = \boxed{\frac{\sqrt{2-\sqrt{3}}}{2}}.$$

52. Use $\cos\left(\frac{A}{2}\right) = \sqrt{\frac{1+\cos A}{2}}$ with $A = 45^\circ$ to obtain

$$\cos\left(22.5^\circ\right) = \cos\left(\frac{45^\circ}{2}\right) = \sqrt{\frac{1+\cos 45^\circ}{2}} = \sqrt{\frac{1+\frac{\sqrt{2}}{2}}{2}} = \sqrt{\frac{2+\sqrt{2}}{4}} = \boxed{\frac{\sqrt{2+\sqrt{2}}}{2}}.$$

53. Use $\cos\left(\frac{A}{2}\right) = -\sqrt{\frac{1+\cos A}{2}}$ with $A = \frac{11\pi}{6}$ to obtain

$$\cos\left(\frac{11\pi}{12}\right) = \cos\left(\frac{\frac{11\pi}{6}}{2}\right) = -\sqrt{\frac{1+\cos\frac{11\pi}{6}}{2}} = -\sqrt{\frac{1+\frac{\sqrt{3}}{2}}{2}} = -\sqrt{\frac{2+\sqrt{3}}{4}} = \boxed{-\frac{\sqrt{2+\sqrt{3}}}{2}}.$$

54. Use $\sin\left(\frac{A}{2}\right)=\sqrt{\frac{1-\cos A}{2}}$ with $A=\frac{\pi}{4}$ to obtain

$$\sin\left(\frac{\pi}{8}\right)=\sin\left(\frac{\frac{\pi}{4}}{2}\right)=\sqrt{\frac{1-\cos\frac{\pi}{4}}{2}}=\sqrt{\frac{1-\frac{\sqrt{2}}{2}}{2}}=\sqrt{\frac{2-\sqrt{2}}{4}}=\boxed{\frac{\sqrt{2-\sqrt{2}}}{2}}.$$

55. Use $\cos\left(\frac{A}{2}\right)=\sqrt{\frac{1+\cos A}{2}}$ with $A=150^\circ$ to obtain

$$\cos\left(75^\circ\right)=\cos\left(\frac{150^\circ}{2}\right)=\sqrt{\frac{1+\cos 150^\circ}{2}}=\sqrt{\frac{1-\frac{\sqrt{3}}{2}}{2}}=\sqrt{\frac{2-\sqrt{3}}{4}}=\boxed{\frac{\sqrt{2-\sqrt{3}}}{2}}.$$

56. Use $\sin\left(\frac{A}{2}\right)=\sqrt{\frac{1-\cos A}{2}}$ with $A=150^\circ$ to obtain

$$\sin\left(75^\circ\right)=\sin\left(\frac{150^\circ}{2}\right)=\sqrt{\frac{1-\cos 150^\circ}{2}}=\sqrt{\frac{1+\frac{\sqrt{3}}{2}}{2}}=\sqrt{\frac{2+\sqrt{3}}{4}}=\boxed{\frac{\sqrt{2+\sqrt{3}}}{2}}.$$

Alternatively, you could use Exercise 5 together with the identity $\sin A=\sqrt{1-\cos^2 A}$ with $A=75^\circ$ to obtain

$$\sin 75^\circ=\sqrt{1-\cos^2 75^\circ}=\sqrt{1-\left(\frac{\sqrt{2-\sqrt{3}}}{2}\right)^2}=\sqrt{1-\frac{2-\sqrt{3}}{4}}=\sqrt{\frac{4-2+\sqrt{3}}{4}}=\frac{\sqrt{2+\sqrt{3}}}{2}$$

57. Use $\tan\left(\frac{A}{2}\right)=\sqrt{\frac{1-\cos A}{1+\cos A}}$ with $A=135^\circ$ to obtain

$$\tan\left(67.5^\circ\right)=\tan\left(\frac{135^\circ}{2}\right)=\sqrt{\frac{1-\cos 135^\circ}{1+\cos 135^\circ}}.$$

Since $\cos 135^\circ=-\cos 45^\circ=-\frac{\sqrt{2}}{2}$, the above further simplifies to

$$\tan\left(67.5^\circ\right)=\sqrt{\frac{1-\left(-\frac{\sqrt{2}}{2}\right)}{1+\left(-\frac{\sqrt{2}}{2}\right)}}=\sqrt{\frac{\frac{2+\sqrt{2}}{\cancel{2}}}{\frac{2-\sqrt{2}}{\cancel{2}}}}=\sqrt{\frac{2+\sqrt{2}}{2-\sqrt{2}}\cdot\frac{2+\sqrt{2}}{2+\sqrt{2}}}=\sqrt{\frac{4+4\sqrt{2}+2}{4-2}}=\boxed{\sqrt{3+2\sqrt{2}}}$$

58. Use $\tan\left(\frac{A}{2}\right)=-\sqrt{\frac{1-\cos A}{1+\cos A}}$ with $A=202.5^\circ$ to obtain

$$\tan\left(202.5^\circ\right)=\tan\left(\frac{405^\circ}{2}\right)=-\sqrt{\frac{1-\cos 405^\circ}{1+\cos 405^\circ}}.$$

Since $\cos 405^\circ=\cos\left(360^\circ+45^\circ\right)=\cos\left(45^\circ\right)=\frac{\sqrt{2}}{2}$, the above further simplifies to

$$\tan\left(202.5^\circ\right)=-\sqrt{\frac{1-\frac{\sqrt{2}}{2}}{1+\frac{\sqrt{2}}{2}}}=-\sqrt{\frac{\frac{2-\sqrt{2}}{\cancel{2}}}{\frac{2+\sqrt{2}}{\cancel{2}}}}=-\sqrt{\frac{2-\sqrt{2}}{2+\sqrt{2}}\cdot\frac{2-\sqrt{2}}{2-\sqrt{2}}}=-\sqrt{\frac{4-4\sqrt{2}+2}{4-2}}=\boxed{-\sqrt{3-2\sqrt{2}}}$$

59. Observe that

$$\sec\left(-\frac{9\pi}{8}\right)=\sec\left(\frac{9\pi}{8}\right)=\frac{1}{\cos\left(\frac{9\pi}{8}\right)}=\frac{1}{\cos\left(\frac{9\pi/4}{2}\right)}\ \textbf{(1)}.$$

Also, note that $\cos\left(\frac{9\pi}{4}\right)=\cos\left(2\pi+\frac{\pi}{4}\right)=\cos\left(\frac{\pi}{4}\right)=\frac{\sqrt{2}}{2}$. Hence, using the formula

$\cos\left(\frac{A}{2}\right)=-\sqrt{\frac{1+\cos A}{2}}$ with $A=\frac{9\pi}{4}$ gives $\cos\left(\frac{9\pi/4}{2}\right)=-\sqrt{\frac{1+\cos\left(\frac{9\pi}{4}\right)}{2}}=-\sqrt{\frac{1+\frac{\sqrt{2}}{2}}{2}}=-\frac{\sqrt{2+\sqrt{2}}}{2}$.

Using this in **(1)** then yields: $\sec\left(-\frac{9\pi}{8}\right)=\frac{1}{-\frac{\sqrt{2+\sqrt{2}}}{2}}=\boxed{-\frac{2}{\sqrt{2+\sqrt{2}}}}$

60. First, note that using the formula $\cos\left(\frac{A}{2}\right)=-\sqrt{\frac{1+\cos A}{2}}$ with $A=\frac{9\pi}{4}$ yields

$$\cos\left(\frac{9\pi}{8}\right)=\cos\left(\frac{9\pi/4}{2}\right)=-\sqrt{\frac{1+\cos\left(\frac{9\pi}{4}\right)}{2}}=-\sqrt{\frac{1+\frac{\sqrt{2}}{2}}{2}}=-\frac{\sqrt{2+\sqrt{2}}}{2}.$$

Now, using $\sin A=-\sqrt{1-\cos^2 A}$ with $A=\frac{9\pi}{8}$ subsequently yields

$$\csc\left(\frac{9\pi}{8}\right)=\frac{1}{\sin\left(\frac{9\pi}{8}\right)}=\frac{1}{-\sqrt{1-\cos^2\left(\frac{9\pi}{8}\right)}}=\frac{-1}{\sqrt{1-\left[-\frac{\sqrt{2+\sqrt{2}}}{2}\right]^2}}=\frac{-1}{\sqrt{1-\frac{2+\sqrt{2}}{4}}}=\frac{-1}{\sqrt{\frac{2-\sqrt{2}}{4}}}=\boxed{\frac{-2}{\sqrt{2-\sqrt{2}}}}$$

61. First, note that using $\tan\left(\frac{A}{2}\right)=\sqrt{\frac{1-\cos A}{1+\cos A}}$ with $A=\frac{13\pi}{4}$, we obtain

$$\cot\left(\frac{13\pi}{8}\right)=\frac{1}{\tan\left(\frac{13\pi}{8}\right)}=\frac{1}{\tan\left(\frac{13\pi/4}{2}\right)}=\frac{1}{\sqrt{\frac{1-\cos\left(13\pi/4\right)}{1+\cos\left(13\pi/4\right)}}} \quad \textbf{(1)}$$

Next, observe that $\cos\left(13\pi/4\right)=\cos\left(2\pi+5\pi/4\right)=\cos\left(5\pi/4\right)=-\frac{\sqrt{2}}{2}$. Using this in **(1)** subsequently yields:

$$\cot\left(\frac{13\pi}{8}\right)=\frac{1}{\sqrt{\frac{1-\left(-\frac{\sqrt{2}}{2}\right)}{1+\left(-\frac{\sqrt{2}}{2}\right)}}}=\frac{1}{\sqrt{\frac{\frac{2+\sqrt{2}}{\cancel{2}}}{\frac{2-\sqrt{2}}{\cancel{2}}}}}=\frac{1}{\sqrt{\frac{2+\sqrt{2}}{2-\sqrt{2}}}}=\frac{1}{\sqrt{\frac{2+\sqrt{2}}{2-\sqrt{2}}\cdot\frac{2+\sqrt{2}}{2+\sqrt{2}}}}=\frac{1}{\sqrt{\frac{4+4\sqrt{2}+2}{4-2}}}=\boxed{\frac{-1}{\sqrt{3+2\sqrt{2}}}}$$

Note: Recall that there are two other formulae equivalent to $\tan\left(\frac{A}{2}\right)=\sqrt{\frac{1-\cos A}{1+\cos A}}$, and either one could be used in place of this one to obtain an equivalent (but perhaps different looking) result. For instance, using $\tan\left(\frac{A}{2}\right)=\frac{\sin A}{1+\cos A}$ with $A=\frac{13\pi}{4}$ yields

$$\cot\left(\frac{13\pi}{8}\right)=\frac{1}{\tan\left(\frac{13\pi}{8}\right)}=\frac{1}{\tan\left(\frac{13\pi/4}{2}\right)}=\frac{1}{\frac{\sin\left(13\pi/4\right)}{1+\cos\left(13\pi/4\right)}} \quad \textbf{(2)}$$

Since

$$\cos\left(13\pi/4\right)=\cos\left(2\pi+5\pi/4\right)=\cos\left(5\pi/4\right)=-\frac{\sqrt{2}}{2}$$

$$\sin\left(13\pi/4\right)=\sin\left(2\pi+5\pi/4\right)=\sin\left(5\pi/4\right)=-\frac{\sqrt{2}}{2}$$

we can use these in **(2)** to obtain

$$\cot\left(\frac{13\pi}{8}\right)=\frac{1}{\frac{-\frac{\sqrt{2}}{2}}{1-\frac{\sqrt{2}}{2}}}=\frac{2-\sqrt{2}}{-\sqrt{2}}=\frac{2-\sqrt{2}}{-\sqrt{2}}\cdot\frac{\sqrt{2}}{\sqrt{2}}=\frac{2\sqrt{2}-2}{-2}=\boxed{1-\sqrt{2}}.$$

Even though the forms of the two answers are different, they are equivalent.

62. First, note that using $\tan\left(\frac{A}{2}\right) = -\sqrt{\frac{1-\cos A}{1+\cos A}}$ with $A = \frac{7\pi}{4}$, we obtain

$$\cot\left(\frac{7\pi}{8}\right) = \frac{1}{\tan\left(\frac{7\pi}{8}\right)} = \frac{1}{\tan\left(\frac{7\pi/4}{2}\right)} = \frac{1}{-\sqrt{\frac{1-\cos\left(7\pi/4\right)}{1+\cos\left(7\pi/4\right)}}} \quad \textbf{(1)}$$

Next, observe that $\cos\left(7\pi/4\right) = \cos\left(\pi/4\right) = \frac{\sqrt{2}}{2}$. Using this in **(1)** subsequently yields:

$$\cot\left(\frac{7\pi}{8}\right) = \frac{1}{-\sqrt{\frac{1-\left(\frac{\sqrt{2}}{2}\right)}{1+\left(\frac{\sqrt{2}}{2}\right)}}} = \frac{1}{-\sqrt{\frac{\frac{2-\sqrt{2}}{\not{2}}}{\frac{2+\sqrt{2}}{\not{2}}}}} = \frac{1}{-\sqrt{\frac{2-\sqrt{2}}{2+\sqrt{2}}}} = \frac{1}{-\sqrt{\frac{2-\sqrt{2}}{2+\sqrt{2}} \cdot \frac{2-\sqrt{2}}{2-\sqrt{2}}}}$$

$$= \frac{1}{-\sqrt{\frac{4-4\sqrt{2}+2}{4-2}}} = \boxed{\frac{-1}{\sqrt{3-2\sqrt{2}}}}$$

Note: Recall that there are two other formulae equivalent to $\tan\left(\frac{A}{2}\right) = -\sqrt{\frac{1-\cos A}{1+\cos A}}$, and either one could be used in place of this one to obtain an equivalent (but perhaps different looking) result. For instance, using $\tan\left(\frac{A}{2}\right) = \frac{\sin A}{1+\cos A}$ with $A = \frac{7\pi}{4}$ yields

$$\cot\left(\frac{7\pi}{8}\right) = \frac{1}{\tan\left(\frac{7\pi}{8}\right)} = \frac{1}{\tan\left(\frac{7\pi/4}{2}\right)} = \frac{1}{\frac{\sin\left(7\pi/4\right)}{1+\cos\left(7\pi/4\right)}} \quad \textbf{(2)}$$

Since

$$\cos\left(7\pi/4\right) = \cos\left(\pi/4\right) = \frac{\sqrt{2}}{2} \qquad \sin\left(7\pi/4\right) = -\sin\left(\pi/4\right) = -\frac{\sqrt{2}}{2}$$

we can use these in **(2)** to obtain

$$\cot\left(\frac{7\pi}{8}\right) = \frac{1}{\frac{-\frac{\sqrt{2}}{2}}{1+\frac{\sqrt{2}}{2}}} = \frac{2+\sqrt{2}}{-\sqrt{2}} = \frac{2+\sqrt{2}}{-\sqrt{2}} \cdot \frac{\sqrt{2}}{\sqrt{2}} = \frac{2\sqrt{2}+2}{-2} = \boxed{-\left(1+\sqrt{2}\right)}.$$

Even though the forms of the two answers are different, they are equivalent.

63. First, note that using the formula $\cos\left(\frac{A}{2}\right)=-\sqrt{\frac{1+\cos A}{2}}$ with $A=\frac{5\pi}{4}$ yields

$$\cos\left(\frac{5\pi}{8}\right)=\cos\left(\frac{5\pi/4}{2}\right)=-\sqrt{\frac{1+\cos\left(\frac{5\pi}{4}\right)}{2}}=-\sqrt{\frac{1-\frac{\sqrt{2}}{2}}{2}}=-\frac{\sqrt{2-\sqrt{2}}}{2}.$$

Then, we have

$$\sec\left(\frac{5\pi}{8}\right)=\frac{1}{\cos\left(\frac{5\pi}{8}\right)}=\frac{1}{-\frac{\sqrt{2-\sqrt{2}}}{2}}=\boxed{\frac{-2}{\sqrt{2-\sqrt{2}}}}.$$

64. First, note that using the formula $\sin\left(\frac{A}{2}\right)=\sqrt{\frac{1-\cos A}{2}}$ with $A=\frac{5\pi}{4}$ yields

$$\sin\left(-\frac{5\pi}{8}\right)=-\sin\left(\frac{5\pi/4}{2}\right)=\sqrt{\frac{1-\cos\left(\frac{5\pi}{4}\right)}{2}}=\sqrt{\frac{1+\frac{\sqrt{2}}{2}}{2}}=\frac{\sqrt{2+\sqrt{2}}}{2}.$$

Then, we have

$$\csc\left(-\frac{5\pi}{8}\right)=\frac{1}{\sin\left(-\frac{5\pi}{8}\right)}=\frac{1}{\frac{\sqrt{2+\sqrt{2}}}{2}}=\boxed{\frac{2}{\sqrt{2+\sqrt{2}}}}.$$

65. Observe that

$$\cot\left(-135^\circ\right)=\cot\left(-\tfrac{270^\circ}{2}\right)=\frac{1}{\tan\left(-\frac{270^\circ}{2}\right)}=-\frac{1}{\tan\left(\frac{270^\circ}{2}\right)}=\frac{1}{\sqrt{\frac{1-\cos\left(270^\circ\right)}{1+\cos\left(270^\circ\right)}}}=\frac{1}{\sqrt{\frac{1-0}{1+0}}}=\boxed{1}.$$

66. Observe that

$$\cot\left(105^\circ\right)=\cot\left(\tfrac{210^\circ}{2}\right)=\frac{1}{\tan\left(\frac{210^\circ}{2}\right)}=-\frac{1}{\sqrt{\frac{1-\cos\left(210^\circ\right)}{1+\cos\left(210^\circ\right)}}}=\frac{-1}{\sqrt{\frac{1+\frac{\sqrt{3}}{2}}{1-\frac{\sqrt{3}}{2}}}}=\frac{-1}{\sqrt{\frac{2+\sqrt{3}}{2-\sqrt{3}}}}=\boxed{-\sqrt{\frac{2-\sqrt{3}}{2+\sqrt{3}}}}.$$

67. Since $\cos x=\frac{5}{13}$ and $\sin x<0$, the terminal side of x lies in QIV. Hence, the terminal side of $\frac{x}{2}$ lies in QII, and so $\sin\left(\frac{x}{2}\right)>0$. Therefore, we have

$$\sin\left(\frac{x}{2}\right)=\sqrt{\frac{1-\cos x}{2}}=\sqrt{\frac{1-\frac{5}{13}}{2}}=\sqrt{\frac{13-5}{26}}=\sqrt{\frac{8}{26}}=\frac{2}{\sqrt{13}}=\boxed{\frac{2\sqrt{13}}{13}}.$$

68. Since $\cos x = -\frac{5}{13}$ and $\sin x < 0$, the terminal side of x lies in QIII. Hence, the terminal side of $\frac{x}{2}$ lies in QII, and so $\cos\left(\frac{x}{2}\right) < 0$. Therefore, we have

$$\cos\left(\frac{x}{2}\right) = -\sqrt{\frac{1+\cos x}{2}} = -\sqrt{\frac{1+\left(-\frac{5}{13}\right)}{2}} = -\sqrt{\frac{13-5}{26}} = -\sqrt{\frac{8}{26}} = -\frac{2}{\sqrt{13}} = \boxed{-\frac{2\sqrt{13}}{13}}.$$

69. Since $\tan x = \frac{12}{5}$ and $\pi < x < \frac{3\pi}{2}$, we know that $\sin x < 0$ and $\cos x < 0$. Also, note that since $\frac{\pi}{2} < \frac{x}{2} < \frac{3\pi}{4}$, $\sin\left(\frac{x}{2}\right) > 0$.

Consider the following diagram:

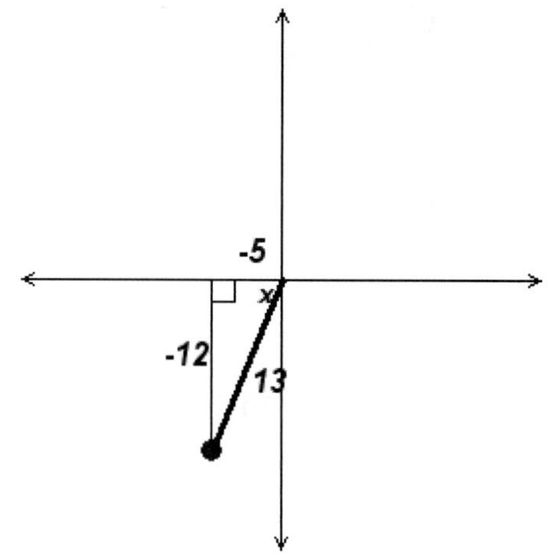

We see from the diagram that $\cos x = -\frac{5}{13}$. So, we have

$$\sin\left(\frac{x}{2}\right) = \sqrt{\frac{1-\cos x}{2}} = \sqrt{\frac{1-\left(-\frac{5}{13}\right)}{2}}$$

$$= \frac{3}{\sqrt{13}} = \boxed{\frac{3\sqrt{13}}{13}}$$

70. Since $\tan x = \frac{12}{5}$ and $\pi < x < \frac{3\pi}{2}$, we know that $\sin x < 0$ and $\cos x < 0$. Also, note that since $\frac{\pi}{2} < \frac{x}{2} < \frac{3\pi}{4}$, $\cos\left(\frac{x}{2}\right) < 0$.

Consider the following diagram:

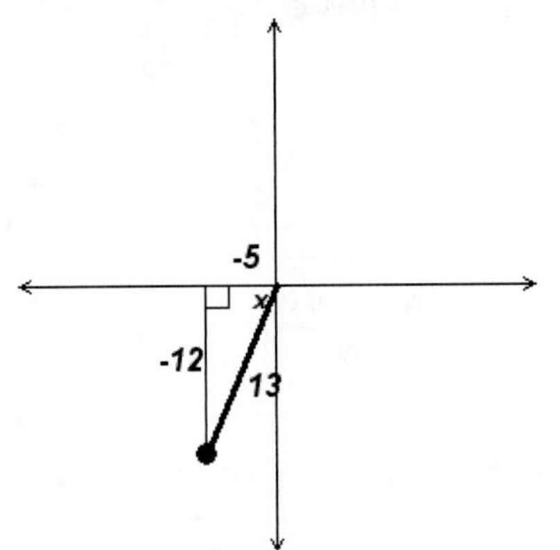

We see from the diagram that $\cos x = -\frac{5}{13}$. So, we have

$$\cos\left(\frac{x}{2}\right) = -\sqrt{\frac{1+\cos x}{2}} = -\sqrt{\frac{1+\left(-\frac{5}{13}\right)}{2}}$$

$$= -\frac{2}{\sqrt{13}} = \boxed{-\frac{2\sqrt{13}}{13}}$$

71. Since $\sec x = \sqrt{5}$, we know that $\cos x = \frac{1}{\sqrt{5}}$. This, together with $\sin x > 0$, implies that the terminal side of x is in QI. Hence, the terminal side of $\frac{x}{2}$ is also in QI. Thus, $\tan\left(\frac{x}{2}\right) > 0$.

Consider the following diagram:

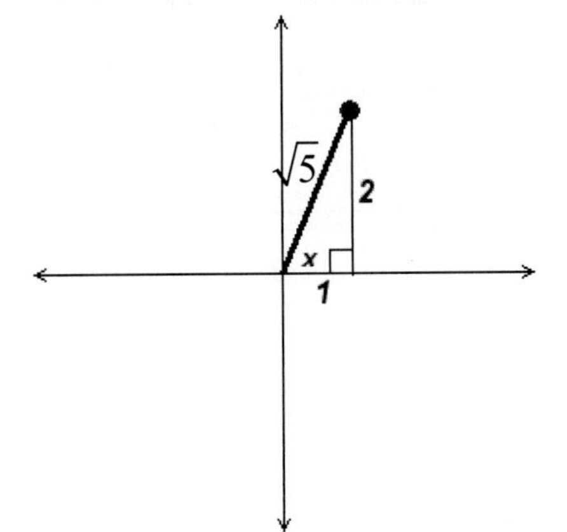

We see from the diagram that $\sin x = \frac{2}{\sqrt{5}}$. So, we have

$$\tan\left(\frac{x}{2}\right) = \frac{\sin x}{1+\cos x} = \frac{\frac{2}{\sqrt{5}}}{1+\frac{1}{\sqrt{5}}} = \frac{2}{\sqrt{5}+1}$$

$$= \frac{2\left(\sqrt{5}-1\right)}{5-1} = \boxed{\frac{\sqrt{5}-1}{2}}$$

Alternatively, we could have used one of the other two equivalent formulae for $\tan\left(\frac{x}{2}\right)$. Specifically, the following computation is also valid:

$$\tan\left(\frac{x}{2}\right) = \sqrt{\frac{1-\cos x}{1+\cos x}} = \sqrt{\frac{1-\frac{1}{\sqrt{5}}}{1+\frac{1}{\sqrt{5}}}} = \sqrt{\frac{\sqrt{5}-1}{\sqrt{5}+1}}$$

$$= \sqrt{\frac{\sqrt{5}-1}{\sqrt{5}+1}\cdot\frac{\sqrt{5}-1}{\sqrt{5}-1}} = \sqrt{\frac{6-2\sqrt{5}}{4}}$$

$$= \boxed{\sqrt{\frac{3-\sqrt{5}}{2}}}$$

72. Since $\sec x = \sqrt{3}$, we know that $\cos x = \frac{1}{\sqrt{3}}$. This, together with $\sin x < 0$, implies that the terminal side of x is in QIV. Hence, the terminal side of $\frac{x}{2}$ is in QII. Thus, $\tan\left(\frac{x}{2}\right) < 0$.

Consider the following diagram:

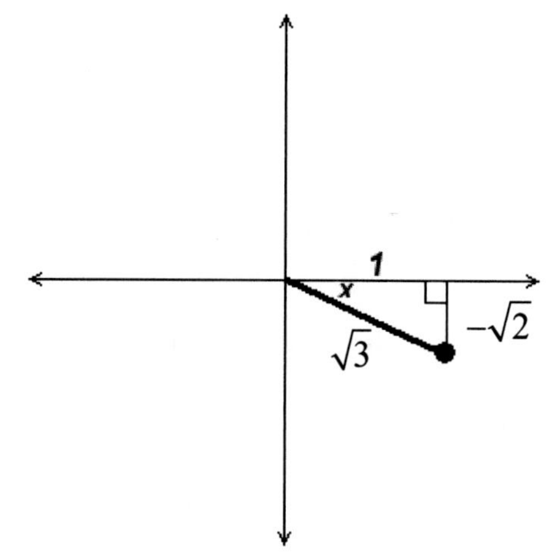

We see from the diagram that $\sin x = -\frac{\sqrt{2}}{\sqrt{3}}$. So, we have

$$\tan\left(\frac{x}{2}\right) = \frac{\sin x}{1+\cos x} = \frac{-\frac{\sqrt{2}}{\sqrt{3}}}{1+\frac{1}{\sqrt{3}}} = \frac{-\sqrt{2}}{\sqrt{3}+1}$$

$$= \frac{-\sqrt{2}}{\sqrt{3}+1} \cdot \frac{\sqrt{3}-1}{\sqrt{3}-1} = \frac{-\sqrt{2}\left(\sqrt{3}-1\right)}{3-1}$$

$$= \boxed{\frac{\sqrt{2}-\sqrt{6}}{2}}$$

Alternatively, we could have used one of the other two equivalent formulae for $\tan\left(\frac{x}{2}\right)$. Specifically, the following computation is also valid:

$$\tan\left(\frac{x}{2}\right) = -\sqrt{\frac{1-\cos x}{1+\cos x}} = -\sqrt{\frac{1-\frac{1}{\sqrt{3}}}{1+\frac{1}{\sqrt{3}}}} = -\sqrt{\frac{\sqrt{3}-1}{\sqrt{3}+1}}$$

$$= -\sqrt{\frac{\sqrt{3}-1}{\sqrt{3}+1} \cdot \frac{\sqrt{3}-1}{\sqrt{3}-1}} = -\sqrt{\frac{4-2\sqrt{3}}{2}}$$

$$= \boxed{-\sqrt{2-\sqrt{3}}}$$

73. Since $\csc x = 3$, we know that $\sin x = \frac{1}{3}$. This, together with $\cos x < 0$, implies that the terminal side of x is in QII. Hence, the terminal side of $\frac{x}{2}$ is in QI. So, $\sin\left(\frac{x}{2}\right) > 0$.

Consider the following diagram:

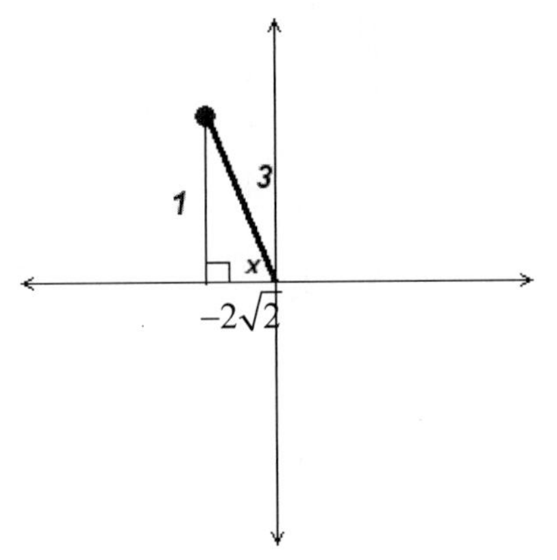

We see from the diagram that $\cos x = -\frac{2\sqrt{2}}{3}$. So, we have

$$\sin\left(\frac{x}{2}\right) = \sqrt{\frac{1-\cos x}{2}} = \sqrt{\frac{1-\left(-\frac{2\sqrt{2}}{3}\right)}{2}}$$

$$= \boxed{\sqrt{\frac{3+2\sqrt{2}}{6}}}$$

74. Since $\csc x = -3$, we know that $\sin x = -\frac{1}{3}$. This, together with $\cos x > 0$, implies that the terminal side of x is in QIV. Hence, the terminal side of $\frac{x}{2}$ is in QII. Thus, $\cos\left(\frac{x}{2}\right) < 0$.

Consider the following diagram:

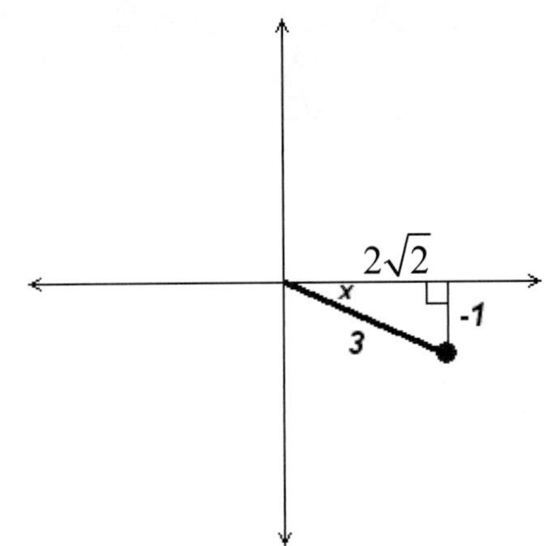

We see from the diagram that $\cos x = \frac{2\sqrt{2}}{3}$. So, we have

$$\cos\left(\frac{x}{2}\right) = -\sqrt{\frac{1+\cos x}{2}} = -\sqrt{\frac{1+\left(\frac{2\sqrt{2}}{3}\right)}{2}}$$

$$= -\sqrt{\frac{3+2\sqrt{2}}{6}} = \boxed{-\sqrt{\frac{1}{2}+\frac{\sqrt{2}}{3}}}$$

75. Since $\csc x < 0$, it follows that $\sin x < 0$. This, together with $\cos x = -\frac{1}{4}$, implies that the terminal side of x is in QIII. Hence, the terminal side of $\frac{x}{2}$ is in QII. Thus, $\cot\left(\frac{x}{2}\right) < 0$.

Consider the following diagram:

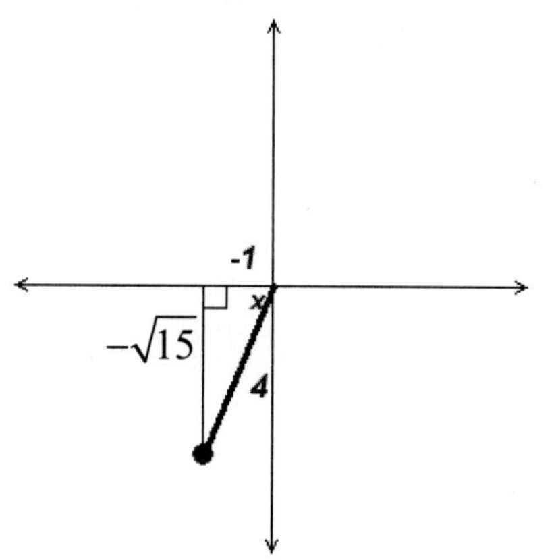

We see from the diagram that $\sin x = -\frac{\sqrt{15}}{4}$. So, we have

$$\cot\left(\frac{x}{2}\right) = \frac{1}{\tan\left(\frac{x}{2}\right)} = \frac{1}{-\sqrt{\frac{1-\cos x}{1+\cos x}}} = \frac{1}{-\sqrt{\frac{1-\left(-\frac{1}{4}\right)}{1+\left(-\frac{1}{4}\right)}}} = \frac{-1}{\sqrt{\frac{5/4}{3/4}}} = -\sqrt{\frac{3}{5}} = \boxed{-\frac{\sqrt{15}}{5}}$$

76. Since $\cot x < 0$ and $\cos x = \frac{1}{4}$, it follows that $\sin x < 0$. So, the terminal side of x is in QIV, and the terminal side of $\frac{x}{2}$ is in QII. So, $\sin\left(\frac{x}{2}\right) > 0$ and $\csc\left(\frac{x}{2}\right) > 0$.

Consider the following diagram:

We see from the diagram that $\sin x = -\frac{\sqrt{15}}{4}$. So, we have

$$\csc\left(\frac{x}{2}\right) = \frac{1}{\sin\left(\frac{x}{2}\right)} = \frac{1}{\sqrt{\frac{1-\cos x}{2}}} = \frac{1}{\sqrt{\frac{1-\frac{1}{4}}{2}}} = \frac{1}{\sqrt{\frac{3}{8}}} = \frac{1}{\frac{\sqrt{3}}{2\sqrt{2}}} = \frac{4}{\sqrt{6}} = \frac{4\sqrt{6}}{6} = \boxed{\frac{2\sqrt{6}}{3}}$$

77. Since $\cot x = -\frac{24}{5}$ and $\frac{\pi}{2} < x < \pi$, we know that $\sin x > 0$ and $\cos x < 0$. Also, note that since $0 < \frac{x}{2} < \frac{\pi}{2}$, $\cos\left(\frac{x}{2}\right) > 0$.

Consider the following diagram:

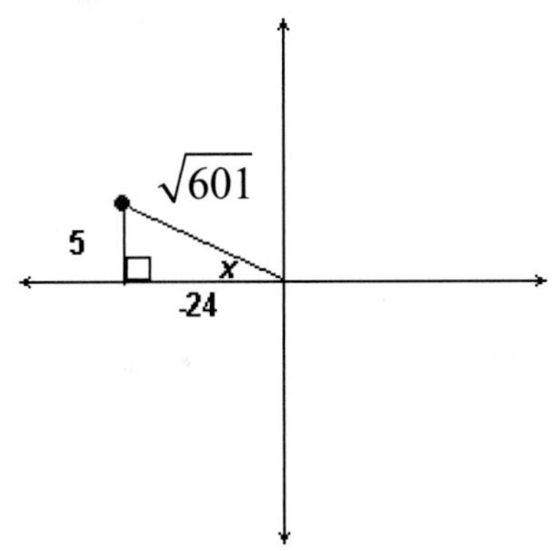

We see from the diagram that $\cos x = -\frac{24}{\sqrt{601}}$. So, we have

$$\cos\left(\frac{x}{2}\right) = \boxed{\sqrt{\frac{1-\frac{24}{\sqrt{601}}}{2}}}$$

78. Since $\cot x = -\frac{24}{5}$ and $\frac{\pi}{2} < x < \pi$, we know that $\sin x > 0$ and $\cos x < 0$. Also, note that since $0 < \frac{x}{2} < \frac{\pi}{2}$, $\sin\left(\frac{x}{2}\right) > 0$.

Consider the following diagram:

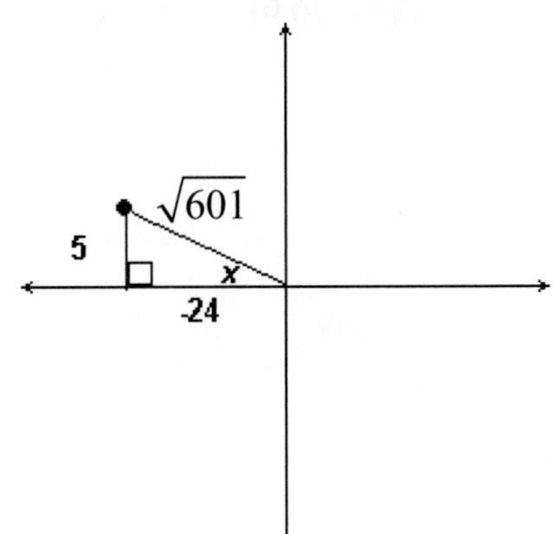

We see from the diagram that $\cos x = -\frac{24}{\sqrt{601}}$. So, we have

$$\sin\left(\frac{x}{2}\right) = \boxed{\sqrt{\frac{1+\frac{24}{\sqrt{601}}}{2}}}$$

79. Since $\sec x > 0$, it follows that $\cos x > 0$. This, together with $\sin x = -0.3$, implies that the terminal side of x is in QIV. Hence, the terminal side of $\frac{x}{2}$ is in QII. Thus, $\tan\left(\frac{x}{2}\right) < 0$.

Consider the following diagram:

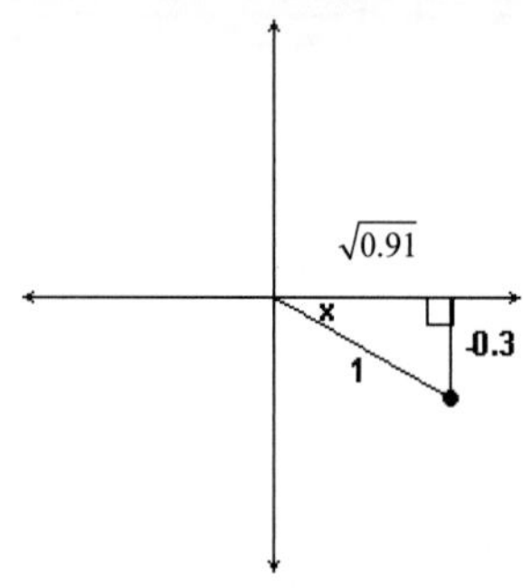

We see from the diagram that $\cos x = \sqrt{0.91}$. So, we have

$$\tan\left(\frac{x}{2}\right) = -\sqrt{\frac{1-\cos x}{1+\cos x}} = \boxed{-\sqrt{\frac{1-\sqrt{0.91}}{1+\sqrt{0.91}}}}.$$

80. Since $\sec x < 0$, it follows that $\cos x < 0$. This, together with $\sin x = -0.3$, implies that the terminal side of x is in QIII. Hence, the terminal side of $\frac{x}{2}$ is in QII. Thus, $\cot\left(\frac{x}{2}\right) < 0$.

Consider the following diagram:

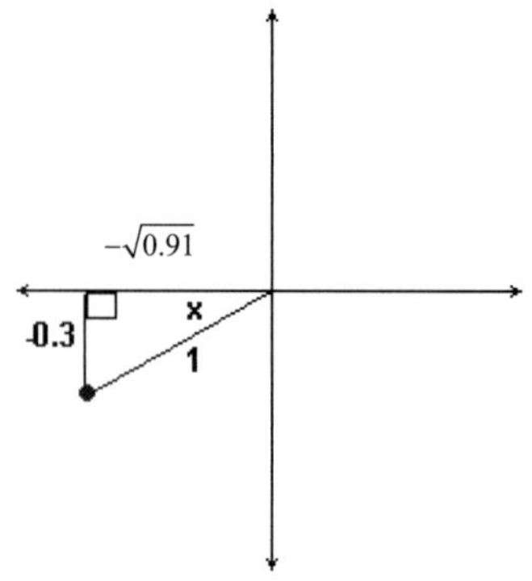

We see from the diagram that $\cos x = -\sqrt{0.91}$. So, we have

$$\cot\left(\frac{x}{2}\right) = \frac{1}{\tan\left(\frac{x}{2}\right)} = \frac{1}{-\sqrt{\frac{1-\cos x}{1+\cos x}}}.$$

$$= \frac{-1}{\sqrt{\frac{1+\sqrt{0.91}}{1-\sqrt{0.91}}}} = \boxed{-\sqrt{\frac{1-\sqrt{0.91}}{1+\sqrt{0.91}}}}$$

81. Since $\sec x = 2.5$, we know that $\cos x = \frac{1}{2.5} = 0.4$. This, together with $\tan x > 0$, implies that the terminal side of x is in QI. Hence, the terminal side of $\frac{x}{2}$ is in QI. Thus, $\cot\left(\frac{x}{2}\right) > 0$.

Consider the following diagram:

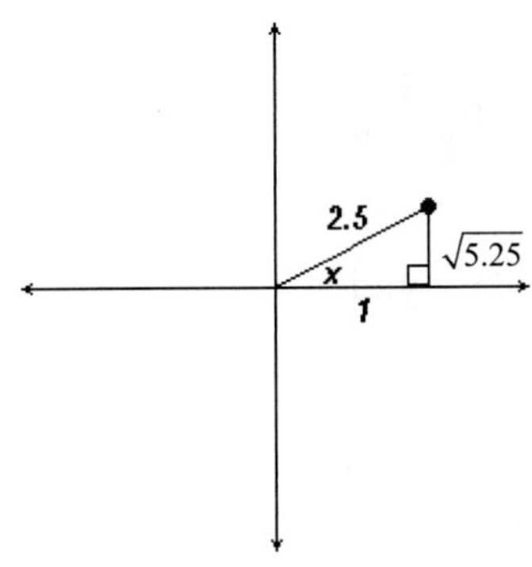

We have

$$\cot\left(\frac{x}{2}\right) = \frac{1}{\tan\left(\frac{x}{2}\right)} = \frac{1}{\sqrt{\frac{1-0.4}{1+0.4}}} = \boxed{\sqrt{\frac{7}{3}}}$$

82. Since $\sec x = -3$, we know that $\cos x = -\frac{1}{3}$. This, together with $\cot x < 0$, implies that the terminal side of x is in QII. Hence, the terminal side of $\frac{x}{2}$ is in QI. Thus, $\tan\left(\frac{x}{2}\right) > 0$.

Consider the following diagram:

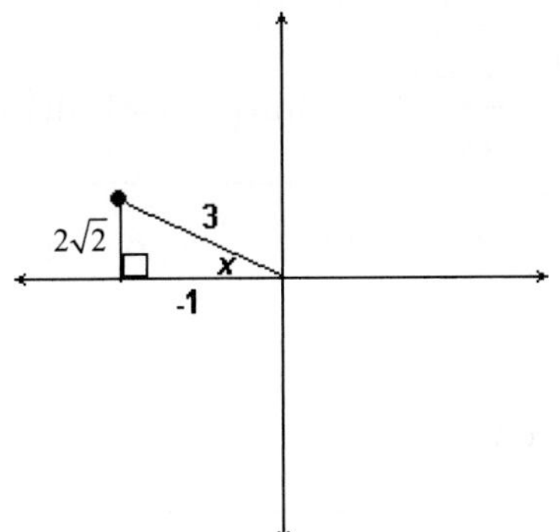

We have

$$\tan\left(\frac{x}{2}\right) = \sqrt{\frac{1-\cos x}{1+\cos x}} = \sqrt{\frac{1-\frac{1}{3}}{1+\frac{1}{3}}} = \sqrt{\frac{\frac{4}{3}}{\frac{2}{3}}} = \boxed{\sqrt{2}}$$

83. Using the formula $\cos\left(\frac{A}{2}\right)=\sqrt{\frac{1+\cos A}{2}}$ with $A=\frac{5\pi}{6}$ yields

$$\sqrt{\frac{1+\cos\left(\frac{5\pi}{6}\right)}{2}}=\cos\left(\frac{5\pi/6}{2}\right)=\boxed{\cos\left(\frac{5\pi}{12}\right)}.$$

84. Using the formula $\sin\left(\frac{A}{2}\right)=\sqrt{\frac{1-\cos A}{2}}$ with $A=\frac{\pi}{4}$ yields

$$\sqrt{\frac{1-\cos\left(\frac{\pi}{4}\right)}{2}}=\sin\left(\frac{\pi/4}{2}\right)=\boxed{\sin\left(\frac{\pi}{8}\right)}.$$

85. Using the formula $\tan\left(\frac{A}{2}\right)=\frac{\sin A}{1+\cos A}$ with $A=150^\circ$ yields

$$\frac{\sin 150^\circ}{1+\cos 150^\circ}=\tan\left(\frac{150^\circ}{2}\right)=\boxed{\tan\left(75^\circ\right)}.$$

86. Observe that

$$\frac{1-\cos 150^\circ}{\sin 150^\circ}=\frac{1-\cos 150^\circ}{\sin 150^\circ}\cdot\frac{1+\cos 150^\circ}{1+\cos 150^\circ}=\frac{1-\cos^2 150^\circ}{\left(\sin 150^\circ\right)\left(1+\cos 150^\circ\right)}$$

$$=\frac{\sin^2 150^\circ}{\left(\sin 150^\circ\right)\left(1+\cos 150^\circ\right)}=\frac{\sin 150^\circ}{1+\cos 150^\circ}$$

Now, using the formula $\tan\left(\frac{A}{2}\right)=\frac{\sin A}{1+\cos A}$ with $A=150^\circ$ yields

$$\frac{\sin 150^\circ}{1+\cos 150^\circ}=\tan\left(\frac{150^\circ}{2}\right)=\boxed{\tan\left(75^\circ\right)}.$$

87. $\sqrt{\frac{1-\cos\frac{5\pi}{4}}{1+\cos\frac{5\pi}{4}}}=-\tan\left(\frac{\frac{5\pi}{4}}{2}\right)=\boxed{-\tan\left(\frac{5\pi}{8}\right)}$

88. $\sqrt{\frac{1-\cos 15^\circ}{1+\cos 15^\circ}}=\tan\left(\frac{15^\circ}{2}\right)=\boxed{\tan\left(7.5^\circ\right)}$

89. Use the known Pythagorean identity $\sin^2\theta+\cos^2\theta=1$ with $\theta=\frac{x}{2}$ to conclude that the given equation is in fact an identity.

90. Observe that $\cos^2\left(\frac{x}{2}\right)-\sin^2\left(\frac{x}{2}\right)=\cos\left(2\cdot\frac{x}{2}\right)=\cos x$.

91. Observe that

$$-2\sin\left(\tfrac{x}{2}\right)\cos\left(\tfrac{x}{2}\right)=-\sin\left(2\cdot\tfrac{x}{2}\right)=-\sin x=\sin(-x).$$

92. Observe that $$2\cos^2\left(\tfrac{x}{4}\right)=2\left[\frac{1+\cos\left(2\cdot\tfrac{x}{4}\right)}{2}\right]=1+\cos\left(\tfrac{x}{2}\right).$$
93. Observe that $\tan^2\left(\frac{x}{2}\right)=\left[\tan\left(\frac{x}{2}\right)\right]^2=\left[\pm\sqrt{\frac{1-\cos x}{1+\cos x}}\right]^2=\frac{1-\cos x}{1+\cos x}.$
94. Note that starting with the right-side is easier. Indeed, observe that $$(\csc x-\cot x)^2=\left(\frac{1}{\sin x}-\frac{\cos x}{\sin x}\right)^2=\frac{(1-\cos x)^2}{\sin^2 x}=\frac{(1-\cos x)(1-\cos x)}{1-\cos^2 x}$$ $$=\frac{\cancel{(1-\cos x)}(1-\cos x)}{\cancel{(1-\cos x)}(1+\cos x)}=\frac{1-\cos x}{1+\cos x}=\left(\tan\left(\frac{x}{2}\right)\right)^2=\tan^2\left(\frac{x}{2}\right)$$
95. Observe that $$\tan\left(\frac{A}{2}\right)+\cot\left(\frac{A}{2}\right)=\frac{\sin A}{1+\cos A}+\frac{1}{\frac{\sin A}{1+\cos A}}=\frac{\sin A}{1+\cos A}+\frac{1+\cos A}{\sin A}$$ $$=\frac{\sin^2 A+(1+\cos A)^2}{(\sin A)(1+\cos A)}=\frac{(1-\cos^2 A)+(1+\cos A)^2}{(\sin A)(1+\cos A)}$$ $$=\frac{(1-\cos A)(1+\cos A)+(1+\cos A)^2}{(\sin A)(1+\cos A)}$$ $$=\frac{\cancel{(1+\cos A)}\left[(1-\cos A)+(1+\cos A)\right]}{(\sin A)\cancel{(1+\cos A)}}=\frac{2}{\sin A}=2\csc A$$
96. Observe that $$\cot\left(\frac{A}{2}\right)-\tan\left(\frac{A}{2}\right)=\frac{1}{\frac{\sin A}{1+\cos A}}-\frac{\sin A}{1+\cos A}=\frac{1+\cos A}{\sin A}-\frac{\sin A}{1+\cos A}$$ $$=\frac{(1+\cos A)^2-\sin^2 A}{(\sin A)(1+\cos A)}=\frac{(1+\cos A)^2-(1-\cos^2 A)}{(\sin A)(1+\cos A)}$$ $$=\frac{(1+\cos A)^2-(1-\cos A)(1+\cos A)}{(\sin A)(1+\cos A)}$$ $$=\frac{\cancel{(1+\cos A)}\left[(1+\cos A)-(1-\cos A)\right]}{(\sin A)\cancel{(1+\cos A)}}=\frac{2\cos A}{\sin A}=2\cot A$$

97. Observe that

$$\csc^2\left(\frac{A}{2}\right)=\frac{1}{\sin^2\left(\frac{A}{2}\right)}=\frac{1}{\left[\sqrt{\frac{1-\cos A}{2}}\right]^2}=\frac{2}{1-\cos A}$$

$$=\frac{2}{1-\cos A}\cdot\frac{1+\cos A}{1+\cos A}=\frac{2(1+\cos A)}{1-\cos^2 A}=\frac{2(1+\cos A)}{\sin^2 A}$$

98. Observe that

$$\sec^2\left(\frac{A}{2}\right)=\frac{1}{\cos^2\left(\frac{A}{2}\right)}=\frac{1}{\left[\sqrt{\frac{1+\cos A}{2}}\right]^2}=\frac{2}{1+\cos A}$$

$$=\frac{2}{1+\cos A}\cdot\frac{1-\cos A}{1-\cos A}=\frac{2(1-\cos A)}{1-\cos^2 A}=\frac{2(1-\cos A)}{\sin^2 A}$$

99. $\csc\left(\frac{A}{2}\right)=\pm\sqrt{\frac{2(1+\cos A)}{\sin^2 A}}=\pm\sqrt{\csc^2 A(2+2\cos A)}=\pm|\csc A|\sqrt{2+2\cos A}$.

100. $\sec\left(\frac{A}{2}\right)=\pm\sqrt{\frac{2(1-\cos A)}{\sin^2 A}}=\pm\sqrt{\csc^2 A(2-2\cos A)}=\pm|\csc A|\sqrt{2-2\cos A}$

101. Before graphing, observe that

$$y=4\cos^2\left(\frac{x}{2}\right)=4\left[\sqrt{\frac{1+\cos x}{2}}\right]^2=4\left(\frac{1+\cos x}{2}\right)=2+2\cos x.$$

Observe that the period is 2π, the amplitude is 2, there is no phase shift, and the vertical shift is up 2 units. So, following the approach in Chapter 6, we see that the graph is as follows:

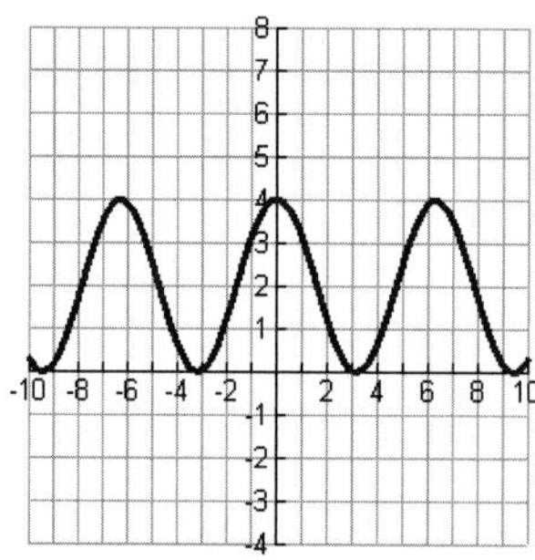

102. Before graphing, observe that

$$y=-6\sin^2\left(\frac{x}{2}\right)=-6\left[\sqrt{\frac{1-\cos x}{2}}\right]^2=-6\left(\frac{1-\cos x}{2}\right)=-3+3\cos x.$$

Observe that the period is 2π, the amplitude is 3, there is no phase shift, and the vertical shift is down 3 units. So, following the approach in Chapter 6, we see that the graph is as follows:

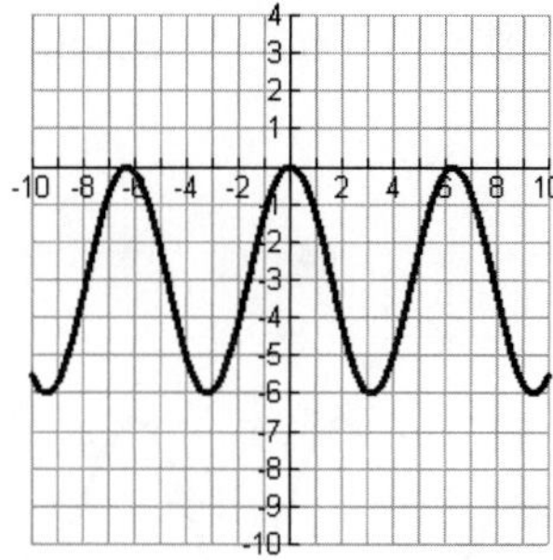

103. Before graphing, observe that

$$y=\frac{1-\tan^2\left(\frac{x}{2}\right)}{1+\tan^2\left(\frac{x}{2}\right)}=\frac{1-\left[\frac{1-\cos x}{1+\cos x}\right]}{1+\left[\frac{1-\cos x}{1+\cos x}\right]}=\frac{\frac{(1+\cos x)-(1-\cos x)}{\cancel{(1+\cos x)}}}{\frac{(1+\cos x)+(1-\cos x)}{\cancel{(1+\cos x)}}}=\frac{2\cos x}{2}=\cos x.$$

So, the graph is as follows:

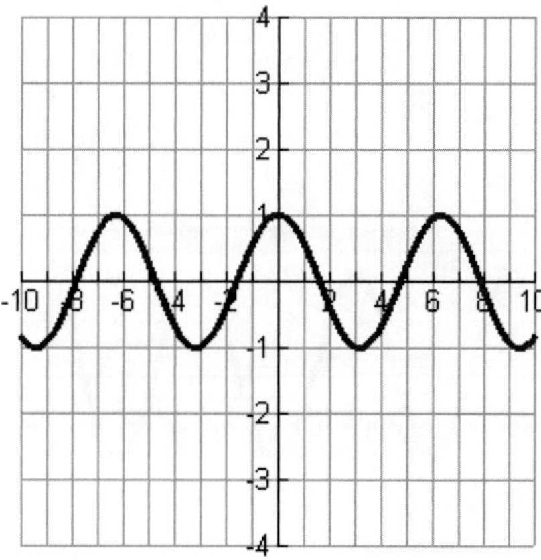

104. Before graphing, observe that

$$y = 1 - \left(\sin\left(\frac{x}{2}\right) + \cos\left(\frac{x}{2}\right)\right)^2 = 1 - \left[\sin^2\left(\frac{x}{2}\right) + 2\sin\left(\frac{x}{2}\right)\cos\left(\frac{x}{2}\right) + \cos^2\left(\frac{x}{2}\right)\right]$$

$$= 1 - \left[\underbrace{\sin^2\left(\frac{x}{2}\right) + \cos^2\left(\frac{x}{2}\right)}_{=1} + \underbrace{2\sin\left(\frac{x}{2}\right)\cos\left(\frac{x}{2}\right)}_{=\sin\left(2\cdot\frac{x}{2}\right)}\right] = -\sin x$$

So, the graph is that of $y = \sin x$ reflected over the x-axis, as shown below:

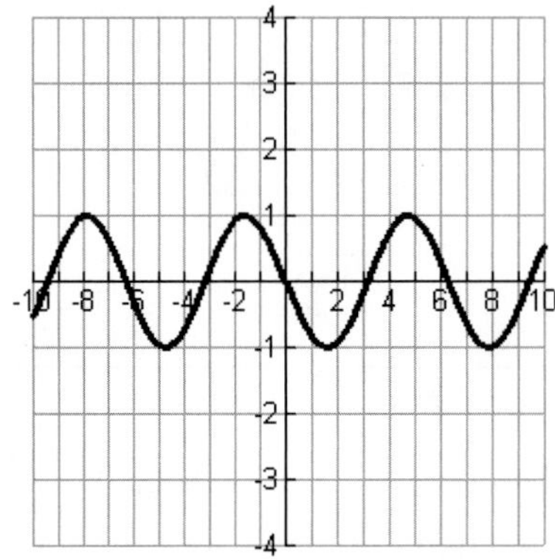

105. Before graphing, observe that

$$y = 4\sin^2\left(\tfrac{x}{2}\right) - 1 = 4\left[\frac{1 - \cos\left(2 \cdot \tfrac{x}{2}\right)}{2}\right] - 1 = 2 - 2\cos x - 1 = 1 - 2\cos x.$$

The graph is as shown below:

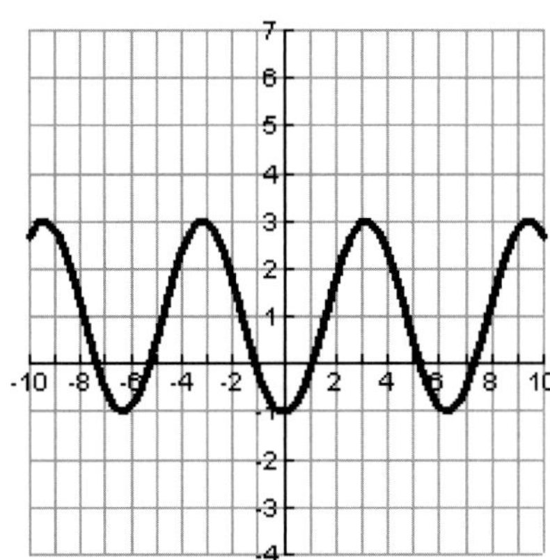

106. Before graphing, observe that

$$y=-\tfrac{1}{6}\cos^2\left(\tfrac{x}{2}\right)+2=-\tfrac{1}{6}\left[\frac{1+\cos\left(2\cdot\frac{x}{2}\right)}{2}\right]+2=-\tfrac{1}{12}-\tfrac{1}{12}\cos x+2=\tfrac{23}{12}-\tfrac{1}{12}\cos x.$$

The graph is as shown below:

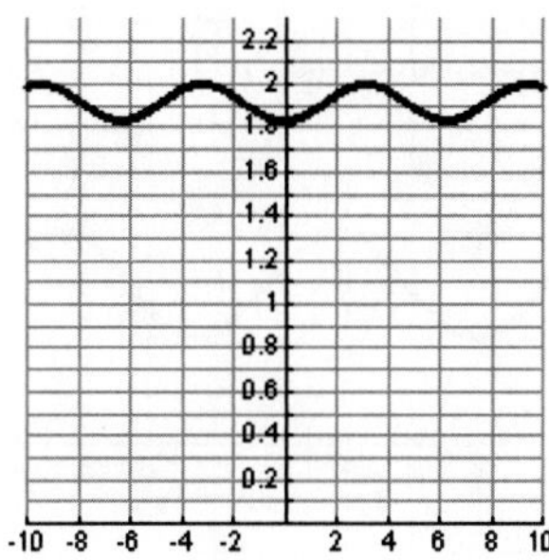

107. Before graphing, observe that

$$y=\sqrt{\frac{1-\cos 2x}{1+\cos 2x}}=\tan\left(\tfrac{2x}{2}\right)=\left|\tan x\right|.$$

The graph is as shown below:

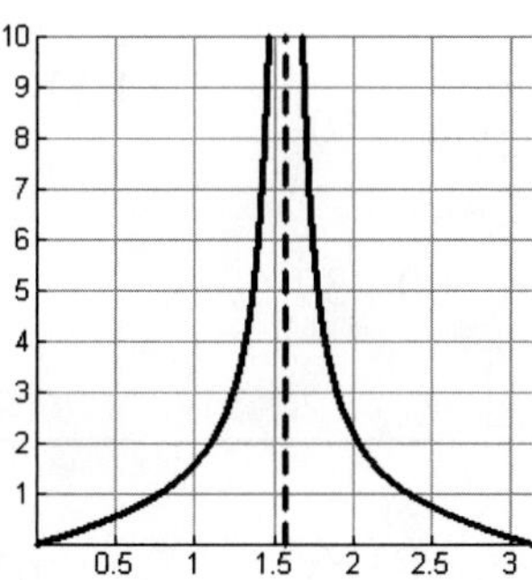

108. Before graphing, observe that

$$y = \sqrt{\frac{1+\cos 3x}{2}} + 3 = \cos\left(\tfrac{3x}{2}\right) + 3.$$

The graph is as shown below:

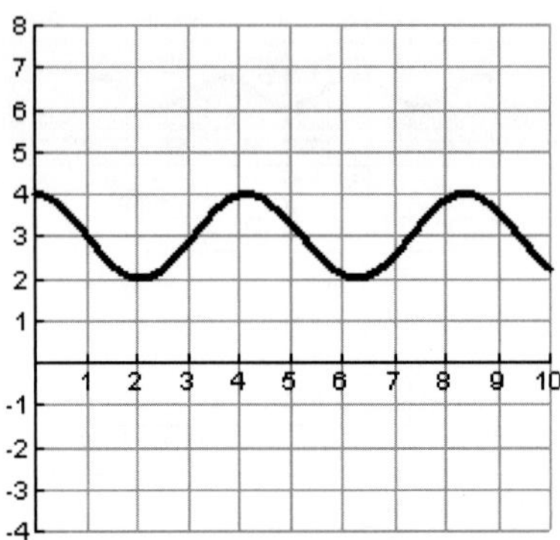

109. Observe that

$$F = 500\sin 60^\circ + \frac{1}{2}(200)^2\left[750\left(1-\cos 120^\circ\right) + 3.75\sin 120^\circ\right]$$

$$= 500\left(\frac{\sqrt{3}}{2}\right) + \frac{40,000}{2}\left[750\left(1-\left(-\frac{1}{2}\right)\right) + 3.75\left(\frac{\sqrt{3}}{2}\right)\right]$$

$$= 250\sqrt{3} + 22,500,000 + 37,500\sqrt{3}$$

$$\approx \boxed{22,565,385 \text{ lbs.}}$$

110. Observe that

$$F = 500\sin 75^\circ + \frac{1}{2}(200)^2\left[750\left(1-\cos 150^\circ\right) + 3.75\sin 150^\circ\right].$$

You can immediately enter this into the calculator to get the estimate indicated below. However, we simplify the various quantities using addition formulas first in order to provide the <u>exact</u> (not just <u>approximate</u>) answer. To this end, observe that

$$\sin 75^\circ = \sin\left(30^\circ + 45^\circ\right) = \sin 30^\circ \cos 45^\circ + \cos 30^\circ \sin 45^\circ$$

$$= \left(\frac{1}{2}\right)\left(\frac{\sqrt{2}}{2}\right) + \left(\frac{\sqrt{3}}{2}\right)\left(\frac{\sqrt{2}}{2}\right) = \frac{\sqrt{2}+\sqrt{6}}{4}$$

$$\cos 75^\circ = \cos\left(30^\circ + 45^\circ\right) = \cos 30^\circ \cos 45^\circ - \sin 30^\circ \sin 45^\circ$$

$$= \left(\frac{\sqrt{3}}{2}\right)\left(\frac{\sqrt{2}}{2}\right) - \left(\frac{1}{2}\right)\left(\frac{\sqrt{2}}{2}\right) = \frac{\sqrt{6}-\sqrt{2}}{4}$$

As such, we use this information to now compute the following:

$$\cos 150^\circ = \cos\left(2\cdot 75^\circ\right) = \cos^2 75^\circ - \sin^2 75^\circ$$
$$= \left(\frac{\sqrt{6}-\sqrt{2}}{4}\right)^2 - \left(\frac{\sqrt{6}+\sqrt{2}}{4}\right)^2$$
$$= \frac{\left(6-2\sqrt{12}+2\right)-\left(6+2\sqrt{12}+2\right)}{16} = \frac{-4\sqrt{12}}{16} = \frac{-\sqrt{3}}{2}$$

$$\sin 150^\circ = \sin\left(2\cdot 75^\circ\right) = 2\sin 75^\circ \cos 75^\circ$$
$$= 2\left(\frac{\sqrt{6}+\sqrt{2}}{4}\right)\left(\frac{\sqrt{6}-\sqrt{2}}{4}\right) = \frac{2(6-4)}{16} = \frac{1}{4}$$

So, we have

$$F = 500\left(\frac{\sqrt{2}+\sqrt{6}}{4}\right) + \frac{1}{2}(200)^2\left[750\left(1+\frac{\sqrt{3}}{2}\right)+3.75\left(\frac{1}{4}\right)\right]$$
$$= 125\sqrt{2}+125\sqrt{6}+7,500,000\sqrt{3}+15,018,750 \approx \boxed{28,009,614 \text{ lbs.}}$$

111. Observe that

$$A = bh$$
$$= \left(a\sin\frac{\theta}{2}\right)\left(a\cos\frac{\theta}{2}\right)$$
$$= a^2\cdot\frac{1}{2}\cdot\left(2\sin\frac{\theta}{2}\cos\frac{\theta}{2}\right)$$
$$= \frac{a^2}{2}\sin\theta$$

112. Consider the following triangle:

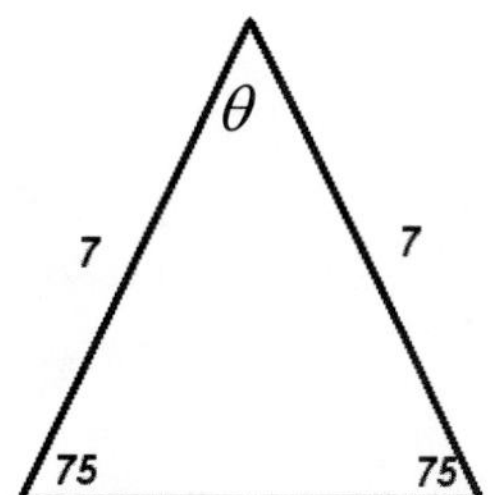

Since the sum of the interior angles in a triangle must be 180°, we see that $\theta = 30^\circ$. So, using Exercise 41, the area of this triangle is:

$$A = \frac{(7 \text{ in.})^2}{2}\sin 30^\circ = \frac{49}{4} \text{ in.}^2.$$

113. Consider the following diagram:

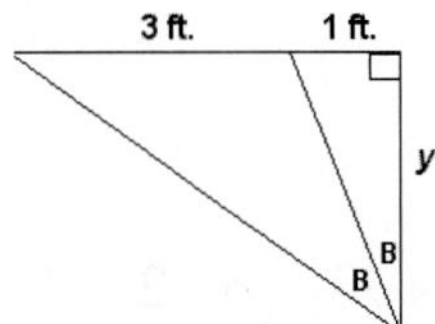

Observe that

$$\tan 2B = \frac{2\tan B}{1-\tan^2 B} = \frac{4}{y} \quad \textbf{(1)} \qquad \tan B = \frac{1}{y} \quad \textbf{(2)}$$

Substituting **(2)** into **(1)** yields the following equation that we solve for y:

$$\frac{2\left(\frac{1}{y}\right)}{1-\left(\frac{1}{y}\right)^2} = \frac{4}{y} \quad \Rightarrow \quad \frac{\frac{2}{y}}{1-\frac{1}{y^2}} = \frac{4}{y} \quad \Rightarrow \quad \frac{\frac{2}{y}}{\frac{y^2-1}{y^2}} = \frac{4}{y} \quad \Rightarrow \quad \frac{2y^2}{y\left(y^2-1\right)} = \frac{4}{y}$$

$$\Rightarrow \quad 2y^2 = 4y^2 - 4 \quad \Rightarrow \quad 4 = 2y^2 \quad \Rightarrow \quad y^2 = 2 \quad \Rightarrow \quad \boxed{y = \sqrt{2}\text{ ft.}}$$

114. Consider the following diagram:

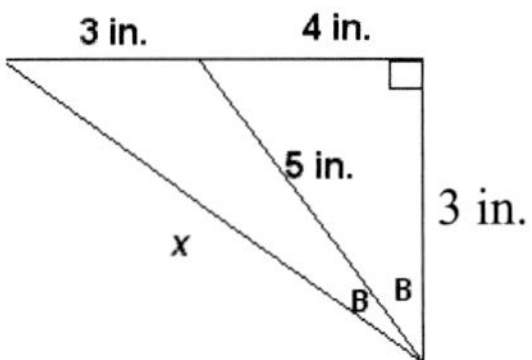

Observe that

$$\sin 2B = \frac{7}{x} \quad \Rightarrow \quad 2\sin B\cos B = \frac{7}{x}.$$

Since $\sin B = \frac{4}{5}$ and $\cos B = \frac{3}{5}$, we have

$$2\left(\tfrac{4}{5}\right)\left(\tfrac{3}{5}\right) = \tfrac{7}{x} \Rightarrow \tfrac{24}{25} = \tfrac{7}{x} \Rightarrow 24x = 175 \Rightarrow \boxed{x = \tfrac{175}{24}\text{ in.}}$$

115. Should use $\sin x = -\frac{2\sqrt{2}}{3}$ since we are assuming that $\sin x < 0$.

116. Note that $\tan 2x = \frac{\sin 2x}{\cos 2x}$, not $\frac{\sin 2x}{\cos x}$.

117. If $\pi < x < \frac{3\pi}{2}$, then $\frac{\pi}{2} < \frac{x}{2} < \frac{3\pi}{4}$, so that $\sin\left(\frac{x}{2}\right)$ is positive, not negative.

118. Note that $\tan^2\left(\frac{x}{2}\right)=\frac{\sin^2\left(\frac{x}{2}\right)}{\cos^2\left(\frac{x}{2}\right)}$, not $\frac{\sin^2\left(\frac{x}{2}\right)}{\cos^2 x}$.

119. False. Let $A=\frac{\pi}{4}$. Observe that

$$\sin\left(2\cdot\frac{\pi}{4}\right)=\sin\left(\frac{\pi}{2}\right)=1$$

$$\sin\left(4\cdot\frac{\pi}{4}\right)=\sin(\pi)=0$$

Since $1+1\neq 0$, the statement is false.

120. False. Let $A=\pi$. Observe that

$$\cos(2\cdot\pi)=1$$

$$\cos(4\cdot\pi)=1$$

Since $1-1\neq 1$, the statement is false.

121. False. Note that $\tan 2x=\frac{2\tan x}{1-\tan^2 x}$.

Take any $x\in\left(\frac{\pi}{4},\frac{\pi}{2}\right)$. For such *x*,

$\tan x>1$, so that $1-\tan^2 x<0$. So, for these values of *x*, $\tan 2x<0$.

122. False. Note that $\sin 2x=2\sin x\cos x$. For instance, there exist *x* values for which

$\sin x>0$ and $\cos x<0$ (namely $\frac{\pi}{2}<x<\pi$).

Thus, for these *x* values, $\sin 2x<0$.

123. False. Use $A=\pi$, for instance. Indeed, observe that $\sin\left(\frac{\pi}{2}\right)+\sin\left(\frac{\pi}{2}\right)=2$,

while $\sin\pi=0$.

124. False. Use $A=\pi$, for instance. Indeed, observe that $\cos\left(\frac{\pi}{2}\right)+\cos\left(\frac{\pi}{2}\right)=0$,

while $\cos\pi=-1$.

125. False. Use $x=\frac{5\pi}{4}$, for instance. Observe that $\tan\left(\frac{5\pi}{4}\right)=1>0$, while

$\tan\left(\frac{5\pi}{8}\right)<0$ since the terminal side of $\frac{5\pi}{8}$ is in QII.

126. False. Use $x=450^\circ$, for instance. Observe that

$\sin(450^\circ)=\sin(360^\circ+90^\circ)=\sin(90^\circ)=1>0$, while $\sin\left(\frac{450^\circ}{2}\right)=\sin(225^\circ)<0$ since

the terminal side of 225° is in QIII.

127. Observe that

$$\tan\left(\frac{A}{2}\right)=\pm\sqrt{\frac{1-\cos A}{1+\cos A}}=\pm\sqrt{\frac{1-\cos A}{1+\cos A}\cdot\frac{1-\cos A}{1-\cos A}}=\pm\sqrt{\frac{(1-\cos A)^2}{1-\cos^2 A}}$$

$$=\pm\sqrt{\frac{(1-\cos A)^2}{\sin^2 A}}=\pm\left|\frac{1-\cos A}{\sin A}\right|=\pm\left|\frac{1-\cos A}{\sin A}\cdot\frac{1+\cos A}{1+\cos A}\right|$$

$$=\pm\left|\frac{1-\cos^2 A}{(\sin A)(1+\cos A)}\right|=\pm\left|\frac{\sin^2 A}{(\sin A)(1+\cos A)}\right|=\pm\left|\frac{\sin A}{1+\cos A}\right|$$

Case 1: The terminal side of $\frac{A}{2}$ is in QI or QIII.

Here, we use $\tan\left(\frac{A}{2}\right)=\left|\frac{\sin A}{1+\cos A}\right|$. Further, for such angles *A*, both $\sin A>0$ and $1+\cos A\geq 0$. Hence, $\frac{\sin A}{1+\cos A}>0$, so that we have $\tan\left(\frac{A}{2}\right)=\left|\frac{\sin A}{1+\cos A}\right|=\frac{\sin A}{1+\cos A}$.

Case 2: The terminal side of $\frac{A}{2}$ is in QII or QIV.

Here, we use $\tan\left(\frac{A}{2}\right)=-\left|\frac{\sin A}{1+\cos A}\right|$. Further, for such angles *A*, both $\sin A<0$ and $1+\cos A\geq 0$. Hence, $\frac{\sin A}{1+\cos A}<0$, so that we have

$$\tan\left(\frac{A}{2}\right)=-\left|\frac{\sin A}{1+\cos A}\right|=-\left(-\frac{\sin A}{1+\cos A}\right)=\frac{\sin A}{1+\cos A}.$$

So, in either case, we conclude that $\tan\left(\frac{A}{2}\right)=\frac{\sin A}{1+\cos A}$.

One cannot evaluate the identity at $A=\pi$ since the denominator on the right-side would be 0 (hence, the fraction is not well-defined) and $\frac{\pi}{2}$ is not in the domain of tangent.

128. Observe that

$$\tan\left(\frac{A}{2}\right)=\pm\sqrt{\frac{1-\cos A}{1+\cos A}}=\pm\sqrt{\frac{1-\cos A}{1+\cos A}\cdot\frac{1-\cos A}{1-\cos A}}=\pm\sqrt{\frac{(1-\cos A)^2}{1-\cos^2 A}}$$

$$=\pm\sqrt{\frac{(1-\cos A)^2}{\sin^2 A}}=\pm\left|\frac{1-\cos A}{\sin A}\right|$$

Case 1: The terminal side of $\frac{A}{2}$ is in QI or QIII.

Here, we use $\tan\left(\frac{A}{2}\right)=\left|\frac{1-\cos A}{\sin A}\right|$. Further, for such angles *A*, both $\sin A>0$ and $1-\cos A\geq 0$. Hence, $\frac{1-\cos A}{\sin A}>0$, so that we have $\tan\left(\frac{A}{2}\right)=\left|\frac{1-\cos A}{\sin A}\right|=\frac{1-\cos A}{\sin A}$.

Case 2: The terminal side of $\frac{A}{2}$ is in QII or QIV.

Here, we use $\tan\left(\frac{A}{2}\right)=-\left|\frac{1-\cos A}{\sin A}\right|$. Further, for such angles *A*, both $\sin A<0$ and $1-\cos A\geq 0$. Hence, $\frac{1-\cos A}{\sin A}<0$, so that we have

$$\tan\left(\frac{A}{2}\right)=-\left|\frac{1-\cos A}{\sin A}\right|=-\left(-\frac{1-\cos A}{\sin A}\right)=\frac{1-\cos A}{\sin A}.$$

So, in either case, we conclude that $\tan\left(\frac{A}{2}\right)=\frac{1-\cos A}{\sin A}$.

One cannot evaluate the identity at $A=\pi$ since the denominator on the right-side would be 0 (hence, the fraction is not well-defined) and $\frac{\pi}{2}$ is not in the domain of tangent.

129. One cannot verify the identity for this value of *x* since it does not belong to the domains of the functions involved.

130. One cannot verify the identity for this value of *x* since it does not belong to the domains of the functions involved.

131. Observe that

$$\cot\left(\tfrac{A}{4}\right)=\frac{1}{\tan\left(\tfrac{A}{4}\right)}=\frac{1}{\tan\left(\tfrac{A/2}{4}\right)}=\frac{1}{\pm\sqrt{\frac{1-\cos\left(\frac{A}{2}\right)}{1+\cos\left(\frac{A}{2}\right)}}}=\pm\sqrt{\frac{1+\cos\left(\frac{A}{2}\right)}{1-\cos\left(\frac{A}{2}\right)}}.$$

132. Observe that

$$\cot\left(-\tfrac{A}{2}\right)\sec\left(\tfrac{A}{2}\right)\csc\left(-\tfrac{A}{2}\right)\tan\left(\tfrac{A}{2}\right)=\cancel{\cot\left(\tfrac{A}{2}\right)}\sec\left(\tfrac{A}{2}\right)\csc\left(\tfrac{A}{2}\right)\cancel{\tan\left(\tfrac{A}{2}\right)}$$

$$=\frac{1}{\cos\left(\frac{A}{2}\right)}\cdot\frac{1}{\sin\left(\frac{A}{2}\right)}=\frac{1}{\frac{1}{2}\sin\left(2\cdot\frac{A}{2}\right)}=2\csc A$$

133. Observe that $\tan\left(\frac{x}{2}\right) > 0$ when $0 < \frac{x}{2} < \frac{\pi}{2}$ or $\pi < \frac{x}{2} < \frac{3\pi}{2}$. So, the desired values of x in $[0, 2\pi]$ for which is true are $\boxed{0 < x < \pi}$.

134. Observe that $\cot\left(\frac{x}{2}\right) \le 0$ when $\frac{\pi}{2} \le \frac{x}{2} < \pi$. So, the desired values of x in $[0, 2\pi]$ for which is true are $\boxed{\pi \le x < 2\pi}$.

135. Consider the following graphs:

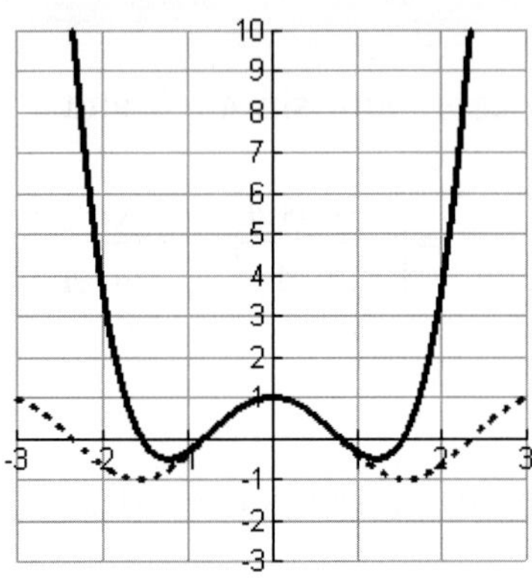

Observe that y_1 (dotted graph) is a good approximation of y_2 on $[-1,1]$.

136. Consider the following graphs:

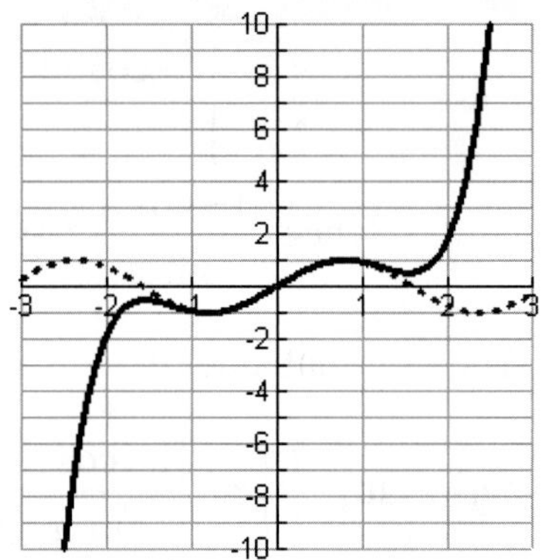

Observe that y_1 (dotted graph) is a good approximation of y_2 on $[-1,1]$.

137. Consider the graphs of the following functions:

$$y_1 = \left(\frac{x}{2}\right) - \frac{\left(\frac{x}{2}\right)^3}{3!} + \frac{\left(\frac{x}{2}\right)^5}{5!} \qquad y_2 = \sin\left(\frac{x}{2}\right)$$

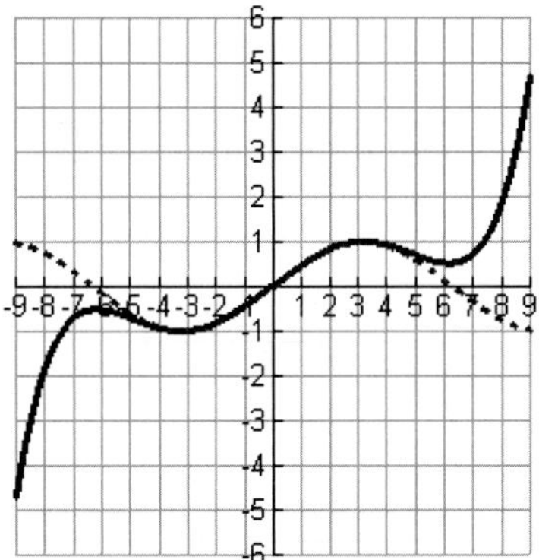

Observe that y_1 (dotted graph) is a good approximation of y_2 on $[-1,1]$.

138. Consider the graphs of the following functions:

$$y_1 = 1 - \frac{\left(\frac{x}{2}\right)^2}{2!} + \frac{\left(\frac{x}{2}\right)^4}{4!} \qquad y_2 = \cos\left(\frac{x}{2}\right)$$

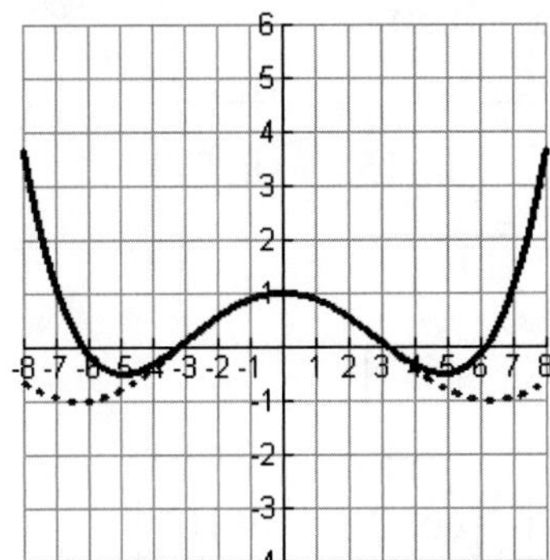

Observe that y_1 (dotted graph) is a good approximation of y_2 on $[-1,1]$.

139. $\dfrac{\frac{2\sin x}{\cos x}}{\frac{\cos^2 x - \sin^2 x}{\cos^2 x}} = \frac{2\sin x}{\cos x} \cdot \frac{\cos^2 x}{\cos^2 x - \sin^2 x} = \frac{2\sin x\cos x}{\cos 2x} = \frac{\sin 2x}{\cos 2x} = \tan 2x$

140.

$$\begin{aligned}\cos^4 x - 6\sin^2 x\cos^2 x + \sin^4 x &= \left[\cos^4 x - 2\sin^2 x\cos^2 x + \sin^4 x\right] - 4\sin^2 x\cos^2 x \\ &= \left(\cos^2 x - \sin^2 x\right)^2 - \left(2\sin x\cos x\right)^2 \\ &= \cos^2 2x - \sin^2 2x \\ &= \cos 4x\end{aligned}$$

141.

$$\begin{aligned}3\sin x\cos^2 x - \sin^3 x &= \left[\sin x\cos^2 x - \sin^3 x\right] + 2\sin x\cos^2 x \\ &= \sin x\left(\cos^2 x - \sin^2 x\right) + \left(2\sin x\cos x\right)\cos x \\ &= \sin x\cos 2x + \sin 2x\cos x \\ &= \sin(x + 2x) \\ &= \sin 3x\end{aligned}$$

142. $\sqrt{\dfrac{1-\sqrt{\frac{1+\cos x}{2}}}{1+\sqrt{\frac{1+\cos x}{2}}}} = \sqrt{\dfrac{1-\cos\left(\frac{x}{2}\right)}{1+\cos\left(\frac{x}{2}\right)}} = \tan\left(\dfrac{\frac{x}{2}}{2}\right) = \tan\left(\frac{x}{4}\right)$

Section 6.4 Solutions

1.

$$\sin 2x \cos x = \frac{1}{2}\left[\sin(2x+x)+\sin(2x-x)\right] = \boxed{\frac{1}{2}\left[\sin 3x + \sin x\right]}$$

2.

$$\cos 10x \sin 5x = \frac{1}{2}\left[\sin(5x+10x)+\sin(5x-10x)\right] =$$

$$= \frac{1}{2}\left[\sin 15x + \sin(-5x)\right] = \boxed{\frac{1}{2}\left[\sin 15x - \sin 5x\right]}$$

3.

$$5\sin 4x \sin 6x = 5\cdot\frac{1}{2}\left[\cos(4x-6x)-\cos(4x+6x)\right]$$

$$= \frac{5}{2}\left[\cos(-2x)-\cos 10x\right] = \boxed{\frac{5}{2}\left[\cos 2x - \cos 10x\right]}$$

4.

$$-3\sin 2x \sin 4x = -3\cdot\frac{1}{2}\left[\cos(2x-4x)-\cos(2x+4x)\right]$$

$$= -\frac{3}{2}\left[\cos(-2x)-\cos 6x\right] = \boxed{-\frac{3}{2}\left[\cos 2x - \cos 6x\right]}$$

5.

$$4\cos(-x)\cos 2x = 4\cdot\frac{1}{2}\left[\cos(-x+2x)+\cos(-x-2x)\right]$$

$$= 2\left[\cos x + \cos(-3x)\right] = \boxed{2\left[\cos x + \cos 3x\right]}$$

6.

$$-8\cos 3x \cos 5x = -8\cdot\frac{1}{2}\left[\cos(3x+5x)+\cos(3x-5x)\right]$$

$$= -4\left[\cos 8x + \cos(-2x)\right] = \boxed{-4\left[\cos 8x + \cos 2x\right]}$$

7.

$$\sin\left(\frac{3x}{2}\right)\sin\left(\frac{5x}{2}\right) = \frac{1}{2}\left[\cos\left(\frac{3x}{2}-\frac{5x}{2}\right)-\cos\left(\frac{3x}{2}+\frac{5x}{2}\right)\right]$$

$$= \frac{1}{2}\left[\cos(-x)-\cos 4x\right] = \boxed{\frac{1}{2}\left[\cos x - \cos 4x\right]}$$

8. $$\sin\left(\frac{\pi x}{2}\right)\sin\left(\frac{5\pi x}{2}\right)=\frac{1}{2}\left[\cos\left(\frac{\pi x}{2}-\frac{5\pi x}{2}\right)-\cos\left(\frac{\pi x}{2}+\frac{5\pi x}{2}\right)\right]$$ $$=\frac{1}{2}\left[\cos(-2\pi x)-\cos 3\pi x\right]=\boxed{\frac{1}{2}\left[\cos 2\pi x-\cos 3\pi x\right]}$$
9. $$\cos\left(\frac{2}{3}x\right)\cos\left(\frac{4}{3}x\right)=\frac{1}{2}\left[\cos\left(\frac{2}{3}x+\frac{4}{3}x\right)+\cos\left(\frac{2}{3}x-\frac{4}{3}x\right)\right]$$ $$=\frac{1}{2}\left[\cos 2x+\cos\left(-\frac{2}{3}x\right)\right]=\boxed{\frac{1}{2}\left[\cos 2x+\cos\left(\frac{2x}{3}\right)\right]}$$
10. $$\sin\left(-\frac{\pi}{4}x\right)\cos\left(-\frac{\pi}{2}x\right)=\frac{1}{2}\left[\sin\left(-\frac{\pi}{4}x-\frac{\pi}{2}x\right)+\sin\left(-\frac{\pi}{4}x-\left(-\frac{\pi}{2}x\right)\right)\right]$$ $$=\frac{1}{2}\left[\sin\left(-\frac{3\pi}{4}x\right)+\sin\left(\frac{\pi}{4}x\right)\right]=\boxed{\frac{1}{2}\left[\sin\left(\frac{\pi}{4}x\right)-\sin\left(\frac{3\pi}{4}x\right)\right]}$$
11. $-3\cos(0.4x)\cos(1.5x)=-\frac{3}{2}\left[\cos(1.9x)+\cos(-1.1x)\right]=\boxed{-\frac{3}{2}\left[\cos(1.9x)+\cos(1.1x)\right]}$
12. $2\sin(2.1x)\sin(3.4x)=\cos(-1.3x)-\cos(5.5x)=\boxed{\cos(1.3x)-\cos(5.5x)}$
13. $$4\sin\left(-\sqrt{3}x\right)\cos\left(3\sqrt{3}x\right)=2\left[\sin\left(2\sqrt{3}x\right)+\sin\left(-4\sqrt{3}x\right)\right]=\boxed{2\left[\sin\left(2\sqrt{3}x\right)-\sin\left(4\sqrt{3}x\right)\right]}$$
14. $$-5\cos\left(-\frac{\sqrt{2}}{3}x\right)\sin\left(\frac{5\sqrt{2}}{3}x\right)=-\frac{5}{2}\left[\sin\left(\frac{4\sqrt{2}}{3}x\right)+\sin\left(-\frac{6\sqrt{2}}{3}x\right)\right]=\boxed{-\frac{5}{2}\left[\sin\left(\frac{4\sqrt{2}}{3}x\right)-\sin\left(2\sqrt{2}x\right)\right]}$$
15. $$\cos 5x+\cos 3x=2\cos\left(\frac{5x+3x}{2}\right)\cos\left(\frac{5x-3x}{2}\right)=\boxed{2\cos 4x\cos x}$$
16. $$\cos 2x-\cos 4x=-2\sin\left(\frac{2x+4x}{2}\right)\sin\left(\frac{2x-4x}{2}\right)=-2\sin(3x)\sin(-x)=\boxed{2\sin 3x\sin x}$$
17. $$\sin 3x-\sin x=2\sin\left(\frac{3x-x}{2}\right)\cos\left(\frac{3x+x}{2}\right)=\boxed{2\sin x\cos 2x}$$

18.

$$\sin 10x+\sin 5x=2\sin\left(\frac{10x+5x}{2}\right)\cos\left(\frac{10x-5x}{2}\right)=\boxed{2\sin\left(\frac{15}{2}x\right)\cos\left(\frac{5}{2}x\right)}$$

19.

$$\sin\left(\frac{x}{2}\right)-\sin\left(\frac{5x}{2}\right)=2\sin\left(\frac{\frac{x}{2}-\frac{5x}{2}}{2}\right)\cos\left(\frac{\frac{x}{2}+\frac{5x}{2}}{2}\right)=2\sin(-x)\cos\left(\frac{3x}{2}\right)=\boxed{-2\sin x\cos\left(\frac{3x}{2}\right)}$$

20.

$$\cos\left(\frac{x}{2}\right)-\cos\left(\frac{5x}{2}\right)=-2\sin\left(\frac{\frac{x}{2}+\frac{5x}{2}}{2}\right)\sin\left(\frac{\frac{x}{2}-\frac{5x}{2}}{2}\right)=-2\sin\left(\frac{3x}{2}\right)\sin(-x)=\boxed{2\sin\left(\frac{3x}{2}\right)\sin x}$$

21.

$$\cos\left(\frac{2}{3}x\right)+\cos\left(\frac{7}{3}x\right)=2\cos\left(\frac{\frac{2}{3}x+\frac{7}{3}x}{2}\right)\cos\left(\frac{\frac{2}{3}x-\frac{7}{3}x}{2}\right)$$

$$=2\cos\left(\frac{3}{2}x\right)\cos\left(-\frac{5}{6}x\right)=\boxed{2\cos\left(\frac{3}{2}x\right)\cos\left(\frac{5}{6}x\right)}$$

22.

$$\sin\left(\frac{2}{3}x\right)+\sin\left(\frac{7}{3}x\right)=2\sin\left(\frac{\frac{2}{3}x+\frac{7}{3}x}{2}\right)\cos\left(\frac{\frac{2}{3}x-\frac{7}{3}x}{2}\right)$$

$$=2\sin\left(\frac{3}{2}x\right)\cos\left(-\frac{5}{6}x\right)=\boxed{2\sin\left(\frac{3}{2}x\right)\cos\left(\frac{5}{6}x\right)}$$

23.

$$\sin(0.4x)+\sin(0.6x)=2\sin\left(\frac{0.4x+0.6x}{2}\right)\cos\left(\frac{0.4x-0.6x}{2}\right)$$

$$=2\sin(0.5x)\cos(-0.1x)=\boxed{2\sin(0.5x)\cos(0.1x)}$$

24.

$$\cos(0.3x)-\cos(0.5x)=-2\sin\left(\frac{0.3x+0.5x}{2}\right)\sin\left(\frac{0.3x-0.5x}{2}\right)$$

$$=-2\sin(0.4x)\sin(-0.1x)=\boxed{2\sin(0.4x)\sin(0.1x)}$$

25.

$$\sin\left(\sqrt{5}x\right)-\sin\left(3\sqrt{5}x\right)=2\sin\left(\frac{\sqrt{5}x-3\sqrt{5}x}{2}\right)\cos\left(\frac{\sqrt{5}x+3\sqrt{5}x}{2}\right)$$

$$=2\sin\left(-\sqrt{5}x\right)\cos\left(2\sqrt{5}x\right)=\boxed{-2\sin\left(\sqrt{5}x\right)\cos\left(2\sqrt{5}x\right)}$$

26.

$$\cos\left(-3\sqrt{7}x\right)-\cos\left(2\sqrt{7}x\right)=-2\sin\left(\frac{-3\sqrt{7}x+2\sqrt{7}x}{2}\right)\sin\left(\frac{-3\sqrt{7}x-2\sqrt{7}x}{2}\right)$$

$$=-2\sin\left(-\frac{\sqrt{7}x}{2}\right)\sin\left(-\frac{5\sqrt{7}x}{2}\right)=\boxed{-2\sin\left(\frac{\sqrt{7}x}{2}\right)\sin\left(\frac{5\sqrt{7}x}{2}\right)}$$

27.

$$\cos\left(-\frac{\pi}{4}x\right)+\cos\left(\frac{\pi}{6}x\right)=2\cos\left(\frac{-\frac{\pi}{4}x+\frac{\pi}{6}x}{2}\right)\cos\left(\frac{-\frac{\pi}{4}x-\frac{\pi}{6}x}{2}\right)$$

$$=2\cos\left(-\frac{\pi}{24}x\right)\cos\left(-\frac{5\pi}{24}x\right)=\boxed{2\cos\left(\frac{\pi}{24}x\right)\cos\left(\frac{5\pi}{24}x\right)}$$

28.

$$\sin\left(\frac{3\pi}{4}x\right)+\sin\left(\frac{5\pi}{4}x\right)=2\sin\left(\frac{\frac{3\pi}{4}x+\frac{5\pi}{4}x}{2}\right)\cos\left(\frac{\frac{3\pi}{4}x-\frac{5\pi}{4}x}{2}\right)$$

$$=2\sin(\pi x)\cos\left(-\frac{\pi}{4}x\right)=\boxed{2\sin(\pi x)\cos\left(\frac{\pi}{4}x\right)}$$

29.

$$\frac{\cos 3x-\cos x}{\sin 3x+\sin x}=\frac{-\cancel{2}\cancel{\sin\left(\frac{3x+x}{2}\right)}\sin\left(\frac{3x-x}{2}\right)}{\cancel{2}\cancel{\sin\left(\frac{3x+x}{2}\right)}\cos\left(\frac{3x-x}{2}\right)}=-\tan\left(\frac{3x-x}{2}\right)=\boxed{-\tan x}$$

30.

$$\frac{\sin 4x+\sin 2x}{\cos 4x-\cos 2x}=\frac{\cancel{2}\cancel{\sin\left(\frac{4x+2x}{2}\right)}\cos\left(\frac{4x-2x}{2}\right)}{-\cancel{2}\cancel{\sin\left(\frac{4x+2x}{2}\right)}\sin\left(\frac{4x-2x}{2}\right)}=-\cot\left(\frac{4x-2x}{2}\right)=\boxed{-\cot x}$$

31.

$$\frac{\cos x - \cos 3x}{\sin 3x - \sin x} = \frac{-\cancel{2}\sin\left(\frac{x+3x}{2}\right)\sin\left(\frac{x-3x}{2}\right)}{\cancel{2}\sin\left(\frac{3x-x}{2}\right)\cos\left(\frac{3x+x}{2}\right)} = \frac{-\sin 2x \sin(-x)}{\sin x \cos 2x} = \frac{\sin 2x \cancel{\sin x}}{\cancel{\sin x} \cos 2x} = \boxed{\tan 2x}$$

32.

$$\frac{\sin 4x + \sin 2x}{\cos 4x + \cos 2x} = \frac{\cancel{2}\sin\left(\frac{4x+2x}{2}\right)\cancel{\cos\left(\frac{4x-2x}{2}\right)}}{\cancel{2}\cos\left(\frac{4x+2x}{2}\right)\cancel{\cos\left(\frac{4x-2x}{2}\right)}} = \tan\left(\frac{4x+2x}{2}\right) = \boxed{\tan 3x}$$

33.

$$\frac{\cos 5x + \cos 2x}{\sin 5x - \sin 2x} = \frac{\cancel{2}\cos\left(\frac{5x-2x}{2}\right)\cancel{\cos\left(\frac{5x+2x}{2}\right)}}{\cancel{2}\sin\left(\frac{5x-2x}{2}\right)\cancel{\cos\left(\frac{5x+2x}{2}\right)}} = \cot\left(\frac{5x-2x}{2}\right) = \boxed{\cot\left(\frac{3x}{2}\right)}$$

34.

$$\frac{\sin 7x - \sin 2x}{\cos 7x - \cos 2x} = \frac{\cancel{2}\cos\left(\frac{7x+2x}{2}\right)\cancel{\sin\left(\frac{7x-2x}{2}\right)}}{-\cancel{2}\sin\left(\frac{7x+2x}{2}\right)\cancel{\sin\left(\frac{7x-2x}{2}\right)}} = -\cot\left(\frac{7x+2x}{2}\right) = \boxed{-\cot\left(\frac{9x}{2}\right)}$$

35.

$$\frac{\sin A + \sin B}{\cos A + \cos B} = \frac{\cancel{2}\sin\left(\frac{A+B}{2}\right)\cancel{\cos\left(\frac{A-B}{2}\right)}}{\cancel{2}\cos\left(\frac{A+B}{2}\right)\cancel{\cos\left(\frac{A-B}{2}\right)}} = \boxed{\tan\left(\frac{A+B}{2}\right)}$$

36.

$$\frac{\sin A - \sin B}{\cos A + \cos B} = \frac{\cancel{2}\sin\left(\frac{A-B}{2}\right)\cancel{\cos\left(\frac{A+B}{2}\right)}}{\cancel{2}\cancel{\cos\left(\frac{A+B}{2}\right)}\cos\left(\frac{A-B}{2}\right)} = \boxed{\tan\left(\frac{A-B}{2}\right)}$$

37.

$$\frac{\cos A-\cos B}{\sin A+\sin B}=\frac{-\cancel{2}\sin\cancel{\left(\frac{A+B}{2}\right)}\sin\left(\frac{A-B}{2}\right)}{\cancel{2}\sin\cancel{\left(\frac{A+B}{2}\right)}\cos\left(\frac{A-B}{2}\right)}=\boxed{-\tan\left(\frac{A-B}{2}\right)}$$

38.

$$\frac{\cos A-\cos B}{\sin A-\sin B}=\frac{-\cancel{2}\sin\left(\frac{A+B}{2}\right)\sin\cancel{\left(\frac{A-B}{2}\right)}}{\cancel{2}\sin\cancel{\left(\frac{A-B}{2}\right)}\cos\left(\frac{A+B}{2}\right)}=\boxed{-\tan\left(\frac{A+B}{2}\right)}$$

39.

$$\frac{\sin A+\sin B}{\sin A-\sin B}=\frac{\cancel{2}\sin\left(\frac{A+B}{2}\right)\cos\left(\frac{A-B}{2}\right)}{\cancel{2}\sin\left(\frac{A-B}{2}\right)\cos\left(\frac{A+B}{2}\right)}=\frac{\sin\left(\frac{A+B}{2}\right)}{\cos\left(\frac{A+B}{2}\right)}\cdot\frac{\cos\left(\frac{A-B}{2}\right)}{\sin\left(\frac{A-B}{2}\right)}$$

$$=\boxed{\tan\left(\frac{A+B}{2}\right)\cot\left(\frac{A-B}{2}\right)}$$

40.

$$\frac{\cos A-\cos B}{\cos A+\cos B}=\frac{-\cancel{2}\sin\left(\frac{A+B}{2}\right)\sin\left(\frac{A-B}{2}\right)}{\cancel{2}\cos\left(\frac{A+B}{2}\right)\cos\left(\frac{A-B}{2}\right)}=-\frac{\sin\left(\frac{A+B}{2}\right)}{\cos\left(\frac{A+B}{2}\right)}\cdot\frac{\sin\left(\frac{A-B}{2}\right)}{\cos\left(\frac{A-B}{2}\right)}$$

$$=\boxed{-\tan\left(\frac{A+B}{2}\right)\tan\left(\frac{A-B}{2}\right)}$$

41.

$$\frac{\cos(A+B)+\cos(A-B)}{\sin(A+B)+\sin(A-B)}=\frac{\cancel{2}\cos\left(\frac{A+B+A-B}{2}\right)\cancel{\cos\left(\frac{A+B-A+B}{2}\right)}}{\cancel{2}\sin\left(\frac{A+B+A-B}{2}\right)\cancel{\cos\left(\frac{A+B-A+B}{2}\right)}}=\frac{\cos A}{\sin A}=\cot A$$

42.

$$\frac{\cos(A-B)-\cos(A+B)}{\sin(A+B)+\sin(A-B)} = \frac{-2\sin\left(\frac{A-B+A+B}{2}\right)\sin\left(\frac{A-B-A-B}{2}\right)}{2\sin\left(\frac{A+B+A-B}{2}\right)\cos\left(\frac{A+B-A+B}{2}\right)}$$

$$= -\frac{\sin(-B)}{\cos B} = \frac{\sin B}{\cos B} = \tan B$$

43. Description of G note: $\cos\left(2\pi(392)t\right)$ Description of B note: $\cos\left(2\pi(494)t\right)$

Combining the two notes: $\cos(784\pi t)+\cos(988\pi t)$.

Using the sum-to-product identity then yields:

$$\cos(784\pi t)+\cos(988\pi t) = 2\cos\left(\frac{784\pi t+988\pi t}{2}\right)\cos\left(\frac{784\pi t-988\pi t}{2}\right)$$

$$= 2\cos(886\pi t)\cos(-102\pi t)$$

$$= 2\cos(886\pi t)\cos(102\pi t) \quad \text{(since cosine is even)}$$

The beat frequency is 494 – 392 = 102 Hz.

The average frequency is $\frac{494+392}{2} = 443$ Hz.

44. Description of F note: $\cos\left(2\pi(349)t\right)$ Description of A note: $\cos\left(2\pi(440)t\right)$

Combining the two notes: $\cos(698\pi t)+\cos(880\pi t)$.

Using the sum-to-product identity then yields:

$$\cos(698\pi t)+\cos(880\pi t) = 2\cos\left(\frac{698\pi t+880\pi t}{2}\right)\cos\left(\frac{698\pi t-880\pi t}{2}\right)$$

$$= 2\cos(789\pi t)\cos(-91\pi t)$$

$$= 2\cos(789\pi t)\cos(91\pi t) \quad \text{(since cosine is even)}$$

The beat frequency is 440 – 349 = 91 Hz.

The average frequency is $\frac{440+349}{2} = 394.5$ Hz.

45. The resulting signal is

$$\sin\left(\frac{2\pi tc}{1.55\times10^{-6}}\right)+\sin\left(\frac{2\pi tc}{0.63\times10^{-6}}\right)$$

$$=2\sin\left(2\pi tc\frac{\left(\frac{1}{1.55\times10^{-6}}+\frac{1}{0.63\times10^{-6}}\right)}{2}\right)\cos\left(2\pi tc\frac{\left(\frac{1}{1.55\times10^{-6}}-\frac{1}{0.63\times10^{-6}}\right)}{2}\right)$$

$$=2\sin\left(2\pi tc\frac{\left(\frac{10^6}{1.55}+\frac{10^6}{0.63}\right)}{2}\right)\cos\left(2\pi tc\frac{\left(\frac{10^6}{1.55}-\frac{10^6}{0.63}\right)}{2}\right)$$

$$=2\sin\left(\frac{2\pi tc}{2}\left(\frac{1}{1.55}+\frac{1}{0.63}\right)10^6\right)\cos\left(\frac{2\pi tc}{2}\left(\frac{1}{1.55}-\frac{1}{0.63}\right)10^6\right)$$

46. Using the same model as in Exercise 33, we find that:

The beat frequency is $\frac{c}{1.55\times10^{-6}}-\frac{c}{0.63\times10^{-6}}=c\left[\frac{1}{1.55}-\frac{1}{0.63}\right]10^6$ Hz

The average frequency is $\frac{\frac{c}{1.55\times10^{-6}}+\frac{c}{0.63\times10^{-6}}}{2}=\frac{c}{2}\left[\frac{1}{1.55}+\frac{1}{0.63}\right]10^6$ Hz

47.

$$\sin(2\pi(770)t)+\sin(2\pi(1209)t)=2\sin\left(\frac{1540\pi t+2418\pi t}{2}\right)\cos\left(\frac{1540\pi t-2418\pi t}{2}\right)$$
$$=2\sin(1979\pi t)\cos(-439\pi t)=2\sin(1979\pi t)\cos(439\pi t)$$

48.

$$\sin(2\pi(697)t)+\sin(2\pi(1477)t)=2\sin\left(\frac{1394\pi t+2954\pi t}{2}\right)\cos\left(\frac{1394\pi t-2954\pi t}{2}\right)$$
$$=2\sin(2174\pi t)\cos(-780\pi t)=2\sin(2174\pi t)\cos(780\pi t)$$

49. Note that $A+52.5^\circ+7.5^\circ=180^\circ$, so that $A=120^\circ$. So, the area is

$$\frac{(10\text{ ft.})^2\sin(52.5^\circ)\sin(7.5^\circ)}{2\sin(120^\circ)}=\frac{100\cdot\frac{1}{2}\left[\cos(52.5^\circ-7.5^\circ)-\cos(52.5^\circ+7.5^\circ)\right]}{2\sin(120^\circ)}\text{ ft.}^2$$

$$=\frac{50\left[\cos(45^\circ)-\cos(60^\circ)\right]}{2\sin(120^\circ)}\text{ ft.}^2$$

$$=\frac{25\left[\frac{\sqrt{2}}{2}-\frac{1}{2}\right]}{\frac{\sqrt{3}}{2}}\text{ ft.}^2=\frac{25\left(\sqrt{2}-1\right)}{\sqrt{3}}\text{ ft.}^2=\frac{25\left(\sqrt{2}-1\right)\sqrt{3}}{3}\text{ ft.}^2$$

$$=\frac{25\left(\sqrt{6}-\sqrt{3}\right)}{3}\text{ ft.}^2\approx\boxed{5.98\text{ ft.}^2}$$

50. Note that $A+75^\circ+45^\circ=180^\circ$, so that $A=60^\circ$. So, the area is

$$\frac{(12\text{ in.})^2\sin(75^\circ)\sin(45^\circ)}{2\sin(60^\circ)}=\frac{144\cdot\frac{1}{2}\left[\cos(75^\circ-45^\circ)-\cos(75^\circ+45^\circ)\right]}{2\sin(60^\circ)}\text{ in.}^2$$

$$=\frac{36\left[\cos(30^\circ)-\cos(120^\circ)\right]}{\sin(60^\circ)}\text{ in.}^2$$

$$=\frac{36\left[\frac{\sqrt{3}}{2}+\frac{1}{2}\right]}{\frac{\sqrt{3}}{2}}\text{ in.}^2=\frac{36\left(\sqrt{3}+1\right)}{\sqrt{3}}\text{ in.}^2=\frac{36\left(\sqrt{3}+1\right)\sqrt{3}}{3}\text{ in.}^2$$

$$=\left(36+12\sqrt{3}\right)\text{ in.}^2\approx\boxed{56.78\text{ in.}^2}$$

51. In the final step of the computation, note that $\cos A\cos B\neq\cos AB$ and $\sin A\sin B\neq\sin AB$. Should have used the product-to-sum identities.

52. In general, $\sin A\cos B-\sin B\cos A\neq 0$. Should have used the product-to-sum identity to simplify this.

53. False. From the product-to-sum identities, we have

$$\cos A\cos B=\frac{1}{2}\left[\cos(A+B)+\cos(A-B)\right],$$

and the right-side is not, in general, expressible as the cosine of a product.

54. False. From the sum-to-product identities, we have

$$\sin A\sin B=\frac{1}{2}\left[\cos(A-B)-\cos(A+B)\right],$$

and the right-side is not, in general, expressible as the sine of a product.

55. True. From the product-to-sum identities, we have

$$\cos A\cos B=\frac{1}{2}\left[\cos(A+B)+\cos(A-B)\right].$$

56. True. From the product-to-sum identities, we have

$$\sin A\sin B=\frac{1}{2}\left[\cos(A-B)-\cos(A+B)\right].$$

57. Observe that

$$\begin{aligned}\sin A\sin B\sin C&=\left[\sin A\sin B\right]\sin C\\&=\left[\frac{1}{2}\left(\cos\left(A-B\right)-\cos(A+B)\right)\right]\sin C\\&=\frac{1}{2}\left[\cos\left(A-B\right)\sin C-\cos(A+B)\sin C\right]\\&=\frac{1}{2}\left\{\frac{1}{2}\left[\sin\left(C+A-B\right)+\sin\left(C-\left(A-B\right)\right)\right]\right.\\&\qquad\left.-\frac{1}{2}\left[\sin\left(C+A+B\right)+\sin\left(C-\left(A+B\right)\right)\right]\right\}\\&=\frac{1}{4}\left[\sin(A-B+C)+\sin(C-A+B)-\sin(A+B+C)-\sin(A+B-C)\right]\end{aligned}$$

At this point, depending on which terms you decide to apply the odd identity for sine, the answer can take on a different form.

58. Observe that

$$\begin{aligned}\cos A\cos B\cos C&=\left[\cos A\cos B\right]\cos C\\&=\left[\frac{1}{2}\left(\cos\left(A+B\right)+\cos(A-B)\right)\right]\cos C\\&=\frac{1}{2}\left[\cos\left(A+B\right)\cos C+\cos(A-B)\cos C\right]\\&=\frac{1}{2}\left\{\frac{1}{2}\left[\cos\left(A+B+C\right)+\cos\left(A+B-C\right)\right]\right.\\&\qquad\left.+\frac{1}{2}\left[\cos\left(A-B+C\right)+\cos\left(A-B-C\right)\right]\right\}\\&=\frac{1}{4}\left[\cos\left(A+B+C\right)+\cos\left(A+B-C\right)+\cos\left(A-B+C\right)+\cos\left(A-B-C\right)\right]\end{aligned}$$

At this point, depending on which terms you decide to apply the odd identity for sine, the answer can take on a different form.

59.

$$\begin{aligned}\cos A\cos B-\sin A\sin B&=\tfrac{1}{2}\left[\cos(A+B)+\cos(A-B)\right]-\tfrac{1}{2}\left[\cos(A-B)+\cos(A+B)\right]\\&=\cos(A+B)\end{aligned}$$

60.

$$\sin A\cos B-\sin B\cos A=\tfrac{1}{2}[\sin(A+B)+\sin(A-B)]-\tfrac{1}{2}[\sin(B+A)+\sin(B-A)]$$
$$=\tfrac{1}{2}\left[\sin(A-B)-\sin\left(-(A-B)\right)\right]$$
$$=\tfrac{1}{2}\left[\sin(A-B)+\sin\left(A-B\right)\right]$$
$$=\sin(A-B)$$

61. Observe that

$$y=1-3\sin(\pi x)\sin\left(-\tfrac{\pi}{6}x\right)=1-3\left[\cos\left(\pi x+\tfrac{\pi}{6}x\right)-\cos\left(\pi x-\tfrac{\pi}{6}x\right)\right]$$
$$=1-\tfrac{3}{2}\left[\cos\left(\tfrac{7\pi}{6}x\right)-\cos\left(\tfrac{5\pi}{6}x\right)\right]$$

The graph is as follows:

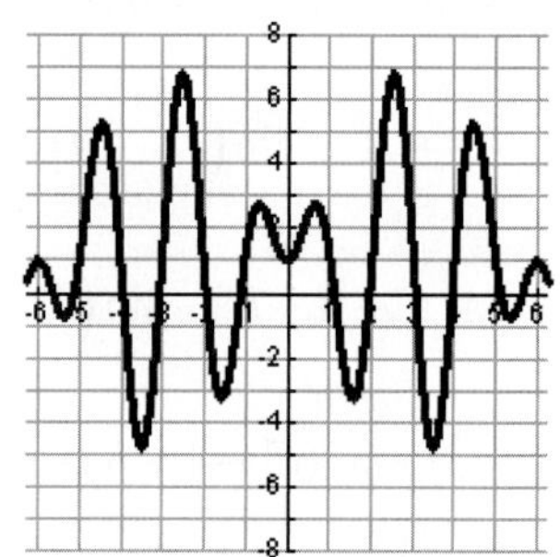

62. Observe that

$$y=4\sin(2x-1)\cos(2-x)$$
$$=2\left[\sin(2x-1+2-x)+\sin(2x-1-2+x)\right]$$
$$=2\left[\sin(x+1)+\sin(3x-3)\right]$$

The graph is as follows:

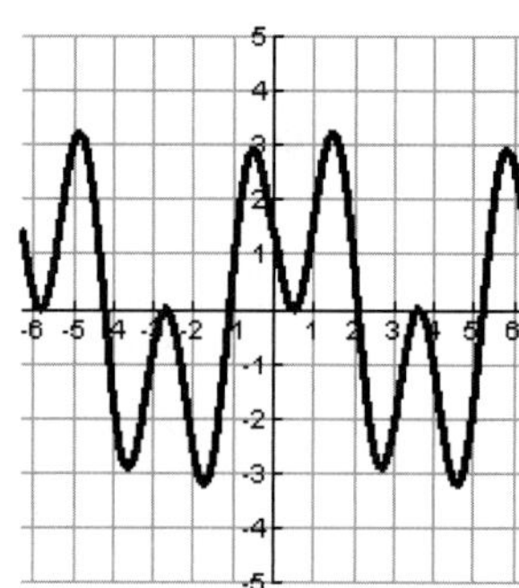

63. Observe that

$$\begin{aligned} y &= -\cos\left(\tfrac{2\pi}{3}x\right)\cos\left(\tfrac{5\pi}{6}x\right) \\ &= -\tfrac{1}{2}\left[\cos\left(\tfrac{2\pi}{3}x+\tfrac{5\pi}{6}x\right)+\cos\left(\tfrac{2\pi}{3}x-\tfrac{5\pi}{6}x\right)\right] \\ &= -\tfrac{1}{2}\left[\cos\left(\tfrac{3\pi}{2}x\right)+\cos\left(-\tfrac{\pi}{6}x\right)\right] = -\tfrac{1}{2}\left[\cos\left(\tfrac{3\pi}{2}x\right)+\cos\left(\tfrac{\pi}{6}x\right)\right] \end{aligned}$$

The graph is as follows:

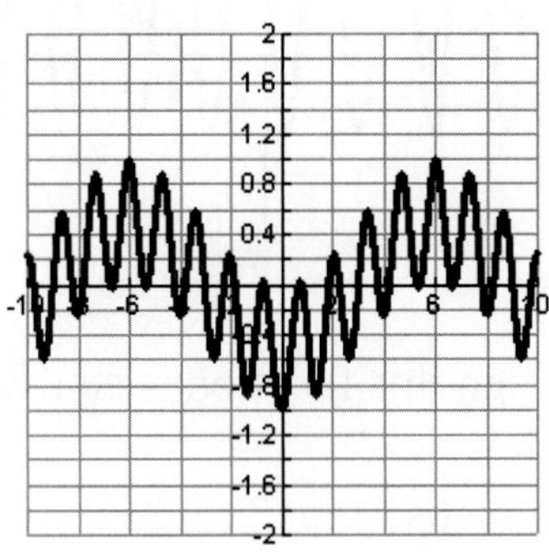

64. Observe that

$$\begin{aligned} y &= x-\cos(2x)\sin(3x) \\ &= x-\tfrac{1}{2}\left[\sin(2x+3x)+\sin(2x-3x)\right] \\ &= x-\tfrac{1}{2}\left[\sin(5x)+\sin(-x)\right] = x-\tfrac{1}{2}\left[\sin(5x)-\sin x\right] \end{aligned}$$

The graph is as follows:

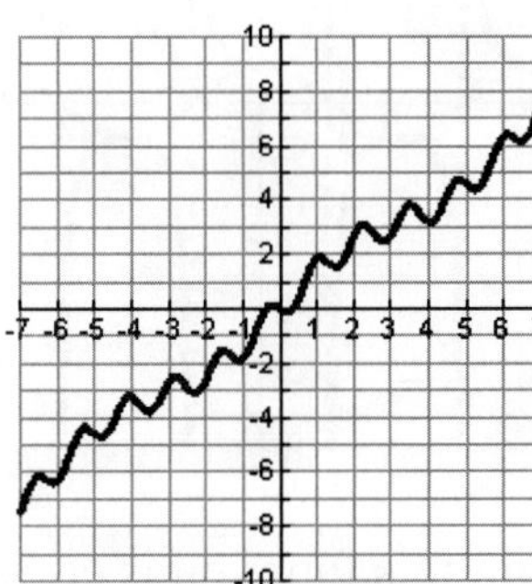

65. Consider the graph of $y = 4\sin x\cos x\cos 2x$, as seen below:

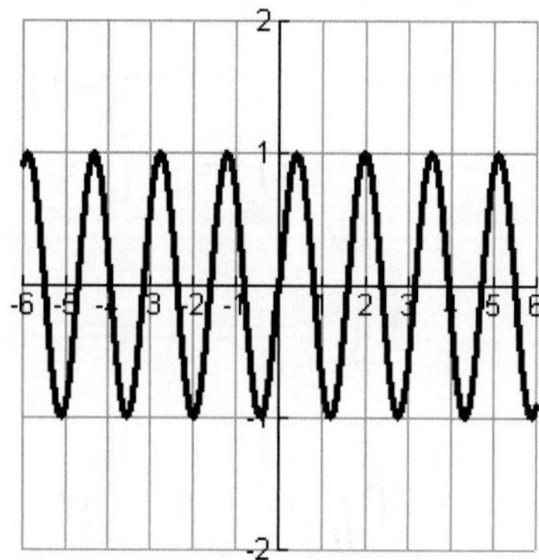

From the graph, it seems as though this function is equivalent to $\sin 4x$. We prove this identity below:

$$4\sin x\cos x\cos 2x = 2\underbrace{(2\sin x\cos x)}_{=\sin 2x}\cos 2x = 2\sin 2x\cos 2x = \sin(2\cdot 2x) = \sin 4x$$

66. Consider the graph of $y = 1+\tan x\tan 2x$, as seen below:

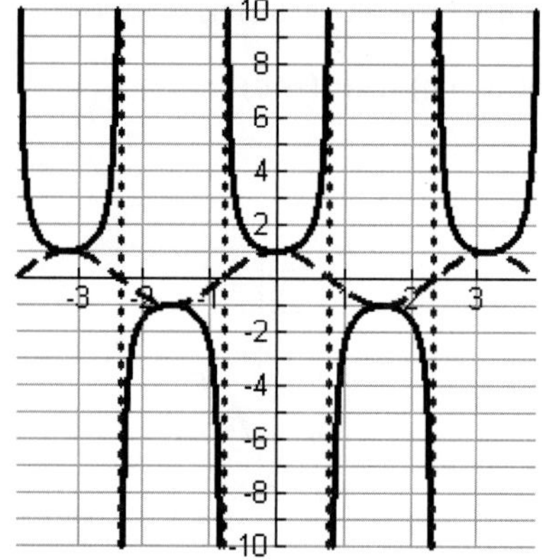

From the graph, it seems as though this function is equivalent to $\sec 2x$. We prove this identity below:

$$1+\tan x\tan 2x = 1+\frac{\sin x}{\cos x}\cdot\frac{\sin 2x}{\cos 2x} = 1+\frac{\sin x(2\sin x\cancel{\cos x})}{\cancel{\cos x}\cos 2x}$$

$$= 1+\frac{2\sin^2 x}{\cos 2x} = \frac{\cos 2x+2\sin^2 x}{\cos 2x} = \frac{(\cos^2 x-\sin^2 x)+2\sin^2 x}{\cos 2x}$$

$$= \frac{\cos^2 x+\sin^2 x}{\cos 2x} = \frac{1}{\cos 2x} = \sec 2x$$

67. To the right are the graphs of the following functions:

$$y_1 = \sin 4x \sin 2x \text{ (solid)}$$

$$y_2 = \sin 6x \text{ (dashed)}$$

$$y_3 = \tfrac{1}{2}\left[\cos 2x - \cos 6x\right]$$

Note that the graphs of y_1 and y_3 are the same.

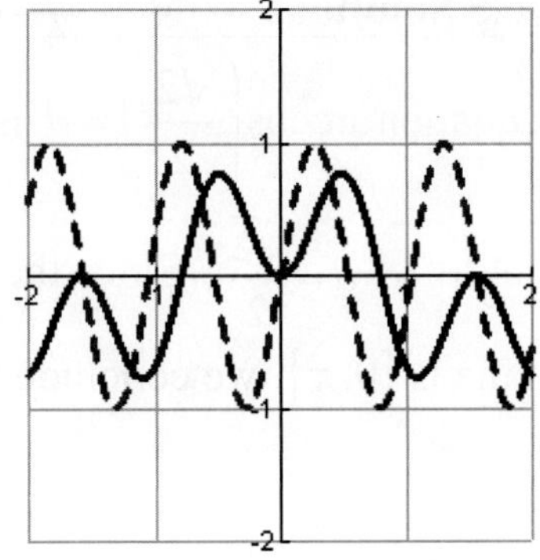

68. To the right are the graphs of the following functions:

$$y_1 = \cos 4x \cos 2x \text{ (solid)}$$

$$y_2 = \cos 6x \text{ (dashed)}$$

$$y_3 = \tfrac{1}{2}\left[\cos 6x + \cos 2x\right]$$

Note that the graphs of y_1 and y_3 are the same.

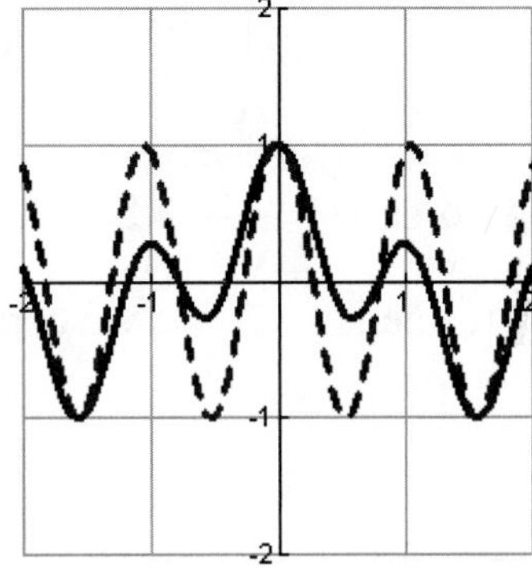

69.

$$\sin\left(\tfrac{x+y}{2}\right)\sin\left(\tfrac{x-y}{2}\right) = \tfrac{1}{5}$$

$$-\tfrac{1}{2}[\cos x - \cos y] = \tfrac{1}{5}$$

$$\cos x - \cos y = -\tfrac{2}{5}$$

$$\cos x = \cos y - \tfrac{2}{5}$$

70.

$$\tfrac{1}{2} = \sin\left(\tfrac{x+y}{2}\right)\cos\left(\tfrac{x-y}{2}\right)$$

$$1 = 2\sin\left(\tfrac{x+y}{2}\right)\cos\left(\tfrac{x-y}{2}\right)$$

$$1 = \sin x + \sin y$$

$$\sin x = 1 - \sin y$$

71.

$$\sin(x+y) = 1 + \sin(x-y)$$

$$\sin x \cos y + \sin y \cos x = 1 + \sin x \cos y - \sin y \cos x$$

$$2\sin y \cos x = 1$$

$$2\sin y = \sec x$$

72.

$$2 + \cos(x+y) = \cos(x-y)$$

$$2 + \cos x \cos y - \sin x \sin y = \cos x \cos y + \sin x \sin y$$

$$2 - \sin x \sin y = \sin x \sin y$$

$$2 = 2\sin x \sin y$$

$$1 = \sin x \sin y$$

$$\csc y = \sin x \quad \text{OR} \quad \csc x = \sin y$$

Section 6.5 Solutions ---

1. The equation $\arccos\left(\frac{\sqrt{2}}{2}\right)=\theta$ is equivalent to $\cos\theta=\frac{\sqrt{2}}{2}$. Since the range of arccosine is $[0,\pi]$, we conclude that $\boxed{\theta=\frac{\pi}{4}}$.	**2.** The equation $\arccos\left(-\frac{\sqrt{2}}{2}\right)=\theta$ is equivalent to $\cos\theta=-\frac{\sqrt{2}}{2}$. Since the range of arccosine is $[0,\pi]$, we conclude that $\boxed{\theta=\frac{3\pi}{4}}$.
3. The equation $\arcsin\left(-\frac{\sqrt{3}}{2}\right)=\theta$ is equivalent to $\sin\theta=-\frac{\sqrt{3}}{2}$. Since the range of arcsine is $\left[-\frac{\pi}{2},\frac{\pi}{2}\right]$, we conclude $\boxed{\theta=-\frac{\pi}{3}}$.	**4.** The equation $\arcsin\left(\frac{1}{2}\right)=\theta$ is equivalent to $\sin\theta=\frac{1}{2}$. Since the range of arcsine is $\left[-\frac{\pi}{2},\frac{\pi}{2}\right]$, we conclude that $\boxed{\theta=\frac{\pi}{6}}$.
5. The equation $\cot^{-1}(-1)=\theta$ is equivalent to $\cot\theta=-1=-\frac{\sqrt{2}/2}{\sqrt{2}/2}=\frac{\cos\theta}{\sin\theta}$. Since the range of inverse cotangent is $(0,\pi)$, we conclude that $\boxed{\theta=\frac{3\pi}{4}}$.	**6.** The equation $\tan^{-1}\left(\frac{\sqrt{3}}{3}\right)=\theta$ is equivalent to $\tan\theta=\frac{\sqrt{3}}{3}=\frac{1/2}{\sqrt{3}/2}=\frac{\sin\theta}{\cos\theta}$. Since the range of inverse tangent is $\left(-\frac{\pi}{2},\frac{\pi}{2}\right)$, we conclude that $\boxed{\theta=\frac{\pi}{6}}$.
7. The equation $\operatorname{arcsec}\left(\frac{2\sqrt{3}}{3}\right)=\theta$ is equivalent to $\sec\theta=\frac{2\sqrt{3}}{3}$, which is further the same as $\cos\theta=\frac{3}{2\sqrt{3}}=\frac{\sqrt{3}}{2}$. Since the range of inverse secant is $\left[0,\frac{\pi}{2}\right)\cup\left(\frac{\pi}{2},\pi\right]$, we conclude that $\boxed{\theta=\frac{\pi}{6}}$.	**8.** The equation $\operatorname{arccsc}(-1)=\theta$ is equivalent to $\csc\theta=-1$, which is further the same as $\sin\theta=-1$. Since the range of inverse cosecant is $\left[-\frac{\pi}{2},0\right)\cup\left(0,\frac{\pi}{2}\right]$, we conclude that $\boxed{\theta=-\frac{\pi}{2}}$.

9. The equation $\csc^{-1}(2)=\theta$ is equivalent to $\csc\theta = 2$, which is further the same as $\sin\theta = \frac{1}{2}$. Since the range of inverse cosecant is $\left[-\frac{\pi}{2},0\right)\cup\left(0,\frac{\pi}{2}\right]$, we conclude that $\boxed{\theta=\frac{\pi}{6}}$.	**10.** The equation $\sec^{-1}(-2)=\theta$ is equivalent to $\sec\theta=-2$, which is further the same as $\cos\theta=-\frac{1}{2}$. Since the range of inverse secant is $\left[0,\frac{\pi}{2}\right)\cup\left(\frac{\pi}{2},\pi\right]$, we conclude that $\boxed{\theta=\frac{2\pi}{3}}$.
11. The equation $\arctan\left(-\sqrt{3}\right)=\theta$ is equivalent to $\tan\theta=-\sqrt{3}=-\frac{\sqrt{3}/2}{1/2}=\frac{\sin\theta}{\cos\theta}$. Since the range of inverse tangent is $\left(-\frac{\pi}{2},\frac{\pi}{2}\right)$, we conclude that $\boxed{\theta=-\frac{\pi}{3}}$.	**12.** The equation $\operatorname{arccot}\left(\sqrt{3}\right)=\theta$ is equivalent to $\cot\theta=\sqrt{3}=\frac{\sqrt{3}/2}{1/2}=\frac{\cos\theta}{\sin\theta}$. Since the range of inverse cotangent is $(0,\pi)$, we conclude that $\boxed{\theta=\frac{\pi}{6}}$.
13. The equation $\arcsin(0)=\theta$ is equivalent to $\sin\theta=0$. Since the range of arcsine is $\left[-\frac{\pi}{2},\frac{\pi}{2}\right]$, we conclude $\boxed{\theta=0}$.	**14.** The equation $\arctan(1)=\theta$ is equivalent to $\tan\theta=1=\frac{\sqrt{2}/2}{\sqrt{2}/2}=\frac{\sin\theta}{\cos\theta}$. Since the range of inverse tangent is $\left(-\frac{\pi}{2},\frac{\pi}{2}\right)$, we conclude that $\boxed{\theta=\frac{\pi}{4}}$.
15. The equation $\sec^{-1}(-1)=\theta$ is equivalent to $\sec\theta=-1$, which is further the same as $\cos\theta=-1$. Since the range of inverse secant is $\left[0,\frac{\pi}{2}\right)\cup\left(\frac{\pi}{2},\pi\right]$, we conclude that $\boxed{\theta=\pi}$.	**16.** The equation $\operatorname{arccot}(0)=\theta$ is equivalent to $\cot\theta=0=\frac{\cos\theta}{\sin\theta}$, which implies $\cos\theta=0$. Since the range of inverse cotangent is $(0,\pi)$, we conclude that that $\boxed{\theta=\frac{\pi}{2}}$.

17. The equation $\cos^{-1}\left(\frac{1}{2}\right)=\theta$ is equivalent to $\cos\theta=\frac{1}{2}$. Since the range of arccosine is $[0,\pi]$, we conclude that $\theta=\frac{\pi}{3}$, which corresponds to $\boxed{\theta=60^\circ}$.	**18.** The equation $\cos^{-1}\left(-\frac{\sqrt{3}}{2}\right)=\theta$ is equivalent to $\cos\theta=-\frac{\sqrt{3}}{2}$. Since the range of arccosine is $[0,\pi]$, we conclude that $\theta=\frac{5\pi}{6}$, which corresponds to $\boxed{\theta=150^\circ}$.
19. The equation $\sin^{-1}\left(\frac{\sqrt{2}}{2}\right)=\theta$ is equivalent to $\sin\theta=\frac{\sqrt{2}}{2}$. Since the range of arcsine is $\left[-\frac{\pi}{2},\frac{\pi}{2}\right]$, we conclude that $\theta=\frac{\pi}{4}$, which corresponds to $\boxed{\theta=45^\circ}$.	**20.** The equation $\sin^{-1}(0)=\theta$ is equivalent to $\sin\theta=0$. Since the range of arcsine is $\left[-\frac{\pi}{2},\frac{\pi}{2}\right]$, we conclude that $\theta=0$, which corresponds to $\boxed{\theta=0^\circ}$.
21. The equation $\cot^{-1}\left(-\frac{\sqrt{3}}{3}\right)=\theta$ is equivalent to $\cot\theta=-\frac{\sqrt{3}}{3}=-\frac{1/2}{\sqrt{3}/2}=\frac{\cos\theta}{\sin\theta}$. Since the range of inverse cotangent is $(0,\pi)$, we conclude that $\theta=\frac{2\pi}{3}$, which corresponds to $\boxed{\theta=120^\circ}$.	**22.** The equation $\tan^{-1}\left(-\sqrt{3}\right)=\theta$ is equivalent to $\tan\theta=-\sqrt{3}=-\frac{\sqrt{3}/2}{1/2}=\frac{\sin\theta}{\cos\theta}$. Since the range of inverse tangent is $\left(-\frac{\pi}{2},\frac{\pi}{2}\right)$, we conclude that $\theta=-\frac{\pi}{3}$, which corresponds to $\boxed{\theta=-60^\circ}$.

23. The equation $\operatorname{arc}\tan\left(\frac{\sqrt{3}}{3}\right)=\theta$ is equivalent to $\tan\theta=\frac{\sqrt{3}}{3}=\frac{1/2}{\sqrt{3}/2}=\frac{\sin\theta}{\cos\theta}$. Since the range of inverse tangent is $\left(-\frac{\pi}{2},\frac{\pi}{2}\right)$, we conclude that $\theta=\frac{\pi}{6}$, which corresponds to $\boxed{\theta=30^\circ}$.	**24.** The equation $\operatorname{arc}\cot(1)=\theta$ is equivalent to $\cot\theta=1=\frac{\sqrt{2}/2}{\sqrt{2}/2}=\frac{\cos\theta}{\sin\theta}$. Since the range of inverse cotangent is $(0,\pi)$, we conclude that $\theta=\frac{\pi}{4}$, which corresponds to $\boxed{\theta=45^\circ}$.
25. The equation $\operatorname{arc}\csc(-2)=\theta$ is equivalent to $\csc\theta=-2$, which is further the same as $\sin\theta=-\frac{1}{2}$. Since the range of inverse cosecant is $\left[-\frac{\pi}{2},0\right)\cup\left(0,\frac{\pi}{2}\right]$, we conclude that $\theta=-\frac{\pi}{6}$, which corresponds to $\boxed{\theta=-30^\circ}$.	**26.** The equation $\csc^{-1}\left(-\frac{2\sqrt{3}}{3}\right)=\theta$ is equivalent to $\csc\theta=-\frac{2\sqrt{3}}{3}$, which is further the same as $\sin\theta=-\frac{3}{2\sqrt{3}}=-\frac{\sqrt{3}}{2}$. Since the range of inverse cosecant is $\left[-\frac{\pi}{2},0\right)\cup\left(0,\frac{\pi}{2}\right]$, we conclude that $\theta=-\frac{\pi}{3}$, which corresponds to $\boxed{\theta=-60^\circ}$.
27. The equation $\operatorname{arc}\sec\left(-\sqrt{2}\right)=\theta$ is equivalent to $\sec\theta=-\sqrt{2}$, which is further the same as $\cos\theta=-\frac{1}{\sqrt{2}}$. Since the range of inverse secant is $\left[0,\frac{\pi}{2}\right)\cup\left(\frac{\pi}{2},\pi\right]$, we conclude that $\theta=\frac{3\pi}{4}$, which corresponds to $\boxed{\theta=135^\circ}$.	**28.** The equation $\operatorname{arc}\csc\left(-\sqrt{2}\right)=\theta$ is equivalent to $\csc\theta=-\sqrt{2}$, which is further the same as $\sin\theta=-\frac{1}{\sqrt{2}}$. Since the range of inverse cosecant is $\left[-\frac{\pi}{2},0\right)\cup\left(0,\frac{\pi}{2}\right]$, we conclude that $\theta=-\frac{\pi}{4}$ which corresponds to $\boxed{\theta=-45^\circ}$.

29. The equation $\sin^{-1}(-1)=\theta$ is equivalent to $\sin\theta=-1$. Since the range of arcsine is $\left[-\frac{\pi}{2},\frac{\pi}{2}\right]$, we conclude that $\theta=-\frac{\pi}{2}$, which corresponds to $\boxed{\theta=-90^\circ}$.	**30.** The equation $\arctan(-1)=\theta$ is equivalent to $\tan\theta=-1=-\frac{\sqrt{2}/2}{\sqrt{2}/2}=\frac{\sin\theta}{\cos\theta}$. Since the range of inverse tangent is $\left(-\frac{\pi}{2},\frac{\pi}{2}\right)$, we conclude that $\theta=-\frac{\pi}{4}$, which corresponds to $\boxed{\theta=-45^\circ}$.
31. The equation $\text{arc}\cot(0)=\theta$ is equivalent to $\cot\theta=0=\frac{\cos\theta}{\sin\theta}$, which implies $\cos\theta=0$. Since the range of inverse cotangent is $(0,\pi)$, we conclude that $\theta=\frac{\pi}{2}$, which corresponds to $\boxed{\theta=90^\circ}$.	**32.** The equation $\sec^{-1}(-1)=\theta$ is equivalent to $\sec\theta=-1$, which is further the same as $\cos\theta=-1$. Since the range of inverse secant is $\left[0,\frac{\pi}{2}\right)\cup\left(\frac{\pi}{2},\pi\right]$, we conclude that $\theta=\pi$, which corresponds to $\boxed{\theta=180^\circ}$.
33. $\cos^{-1}(0.5432)\approx\boxed{57.10^\circ}$	**34.** $\sin^{-1}(0.7821)\approx\boxed{51.45^\circ}$
35. $\tan^{-1}(1.895)\approx\boxed{62.18^\circ}$	**36.** $\tan^{-1}(3.2678)\approx\boxed{72.99^\circ}$
37. $\sec^{-1}(1.4973)=\cos^{-1}\left(\frac{1}{1.4973}\right)\approx\boxed{48.10^\circ}$	**38.** $\sec^{-1}(2.7864)=\cos^{-1}\left(\frac{1}{2.7864}\right)\approx\boxed{68.97^\circ}$
39. $\csc^{-1}(-3.7893)=\sin^{-1}\left(\frac{1}{-3.7893}\right)\approx\boxed{-15.30^\circ}$	**40.** $\csc^{-1}(-6.1324)=\sin^{-1}\left(\frac{1}{-6.1324}\right)\approx\boxed{-9.39^\circ}$
41. $\cot^{-1}(-4.2319)=\pi+\tan^{-1}\left(\frac{1}{-4.2319}\right)$ $\approx180^\circ-13.30^\circ=\boxed{166.70^\circ}$	**42.** $\cot^{-1}(-0.8977)=\pi+\tan^{-1}\left(\frac{1}{-0.8977}\right)$ $\approx180^\circ-48.09^\circ=\boxed{131.91^\circ}$
43. $\sin^{-1}(-0.5878)\approx\boxed{-0.63}$	**44.** $\sin^{-1}(0.8660)\approx\boxed{1.05}$
45. $\cos^{-1}(0.1423)\approx\boxed{1.43}$	**46.** $\tan^{-1}(-0.9279)\approx\boxed{-0.75}$
47. $\tan^{-1}(1.3242)\approx\boxed{0.92}$	**48.** $\cot^{-1}(2.4142)=\tan^{-1}\left(\frac{1}{2.4142}\right)\approx\boxed{0.39}$

49. $\cot^{-1}(-0.5774) = \pi + \tan^{-1}\left(\frac{1}{-0.5774}\right) \approx \boxed{2.09}$	**50.** $\sec^{-1}(-1.0422) = \cos^{-1}\left(\frac{1}{-1.0422}\right) \approx \boxed{2.86}$
51. $\csc^{-1}(3.2361) = \sin^{-1}\left(\frac{1}{3.2361}\right) \approx \boxed{0.31}$	**52.** $\csc^{-1}(-2.9238) = \sin^{-1}\left(\frac{1}{-2.9238}\right) \approx \boxed{-0.35}$

53. $\sin^{-1}\left(\sin\left(\frac{5\pi}{12}\right)\right) = \boxed{\frac{5\pi}{12}}$ since $-\frac{\pi}{2} \le \frac{5\pi}{12} \le \frac{\pi}{2}$.

54. $\sin^{-1}\left(\sin\left(-\frac{5\pi}{12}\right)\right) = \boxed{-\frac{5\pi}{12}}$ since $-\frac{\pi}{2} \le -\frac{5\pi}{12} \le \frac{\pi}{2}$.

55. $\sin\left(\sin^{-1}(1.03)\right)$ is undefined since 1.03 is not in the domain of inverse sine.

56. $\sin\left(\sin^{-1}(1.1)\right)$ is undefined since 1.1 is not in the domain of inverse sine.

57. Note that we need to use the angle θ in $\left[-\frac{\pi}{2}, \frac{\pi}{2}\right]$ such that $\sin\theta = \sin\left(-\frac{7\pi}{6}\right)$. To this end, observe that $\sin^{-1}\left(\sin\left(-\frac{7\pi}{6}\right)\right) = \sin^{-1}\left(\sin\left(\frac{\pi}{6}\right)\right) = \boxed{\frac{\pi}{6}}$.

58. Note that we need to use the angle θ in $\left[-\frac{\pi}{2}, \frac{\pi}{2}\right]$ such that $\sin\theta = \sin\left(\frac{7\pi}{6}\right)$. To this end, observe that $\sin^{-1}\left(\sin\left(\frac{7\pi}{6}\right)\right) = \sin^{-1}\left(\sin\left(-\frac{\pi}{6}\right)\right) = \boxed{-\frac{\pi}{6}}$.

59. Note that we need to use the angle θ in $[0,\pi]$ such that $\cos\theta = \cos\left(\frac{4\pi}{3}\right)$. To this end, observe that $\cos^{-1}\left(\cos\left(\frac{4\pi}{3}\right)\right) = \cos^{-1}\left(\cos\left(\frac{2\pi}{3}\right)\right) = \boxed{\frac{2\pi}{3}}$.

60. Note that we need to use the angle θ in $[0,\pi]$ such that $\cos\theta = \cos\left(-\frac{5\pi}{3}\right)$. To this end, observe that $\cos^{-1}\left(\cos\left(-\frac{5\pi}{3}\right)\right) = \cos^{-1}\left(\cos\left(\frac{\pi}{3}\right)\right) = \boxed{\frac{\pi}{3}}$.

61. Since $\cot\left(\cot^{-1} x\right) = x$ for all $-\infty < x < \infty$, we see that $\cot\left(\cot^{-1}\sqrt{3}\right) = \boxed{\sqrt{3}}$.

62. Note that we need to use the angle θ in $[0,\pi]$ such that $\cot\theta = \cot\left(\frac{5\pi}{4}\right)$. To this end, observe that $\cot^{-1}\left(\cot\left(\frac{5\pi}{4}\right)\right) = \cot^{-1}\left(\cot\left(\frac{\pi}{4}\right)\right) = \boxed{\frac{\pi}{4}}$.
63. Note that we need to use the angle θ in $\left[0,\frac{\pi}{2}\right)\cup\left(\frac{\pi}{2},\pi\right]$ such that $\sec\theta = \sec\left(-\frac{\pi}{3}\right)$. To this end, observe that $\sec^{-1}\left(\sec\left(-\frac{\pi}{3}\right)\right) = \sec^{-1}\left(\sec\left(\frac{\pi}{3}\right)\right) = \boxed{\frac{\pi}{3}}$.
64. $\sec\left(\sec^{-1}\left(\frac{1}{2}\right)\right)$ is undefined since $\frac{1}{2}$ is not in the domain of inverse secant.
65. $\csc\left(\csc^{-1}\left(\frac{1}{2}\right)\right)$ is undefined since $\frac{1}{2}$ is not in the domain of inverse cosecant.
66. Note that we need to use the angle θ in $\left[-\frac{\pi}{2},0\right)\cup\left(0,\frac{\pi}{2}\right]$ such that $\csc\theta = \csc\left(\frac{7\pi}{6}\right)$. To this end, observe that $\csc^{-1}\left(\csc\left(\frac{7\pi}{6}\right)\right) = \csc^{-1}\left(\csc\left(-\frac{\pi}{6}\right)\right) = \boxed{-\frac{\pi}{6}}$.
67. Since $\cot\left(\cot^{-1}x\right) = x$ for all $-\infty < x < \infty$, we see that $\cot\left(\cot^{-1}0\right) = \boxed{0}$.
68. Note that we need to use the angle θ in $[0,\pi]$ such that $\cot\theta = \cot\left(-\frac{\pi}{4}\right)$. To this end, observe that $\cot^{-1}\left(\cot\left(-\frac{\pi}{4}\right)\right) = \cot^{-1}\left(\cot\left(\frac{3\pi}{4}\right)\right) = \boxed{\frac{3\pi}{4}}$.
69. Since $\tan^{-1}(\tan x) = x$ for all $-\frac{\pi}{2} < x < \frac{\pi}{2}$, we see that $\tan^{-1}\left(\tan\left(-\frac{\pi}{4}\right)\right) = \boxed{-\frac{\pi}{4}}$.
70. Since $\tan^{-1}(\tan x) = x$ for all $-\frac{\pi}{2} < x < \frac{\pi}{2}$, we see that $\tan^{-1}\left(\tan\left(\frac{\pi}{4}\right)\right) = \boxed{\frac{\pi}{4}}$.
71. Not possible
72. Not possible
73. $\cot^{-1}\left(\cot\left(\frac{8\pi}{3}\right)\right) = \cot^{-1}\left(-\frac{1}{\sqrt{3}}\right) = \boxed{\frac{2\pi}{3}}$
74. $\tan^{-1}(\tan 8\pi) = \tan^{-1}(0) = \boxed{0}$
75. $\csc^{-1}\left(\csc\frac{15\pi}{4}\right) = \csc^{-1}\left(-\frac{\sqrt{2}}{2}\right) = \boxed{-\frac{\pi}{4}}$
76. Not possible

77. Let $\theta = \sin^{-1}\left(\frac{3}{4}\right)$. Then, $\sin\theta = \frac{3}{4}$, as shown in the diagram:

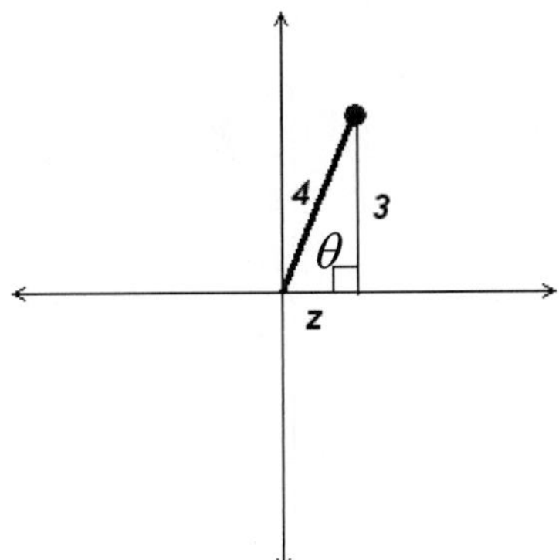

Using the Pythagorean Theorem, we see that $z^2 + 3^2 = 4^2$, so that $z = \sqrt{7}$.

Hence, $\cos\left(\sin^{-1}\left(\frac{3}{4}\right)\right) = \cos\theta = \boxed{\frac{\sqrt{7}}{4}}$.

78. Let $\theta = \cos^{-1}\left(\frac{2}{3}\right)$. Then, $\cos\theta = \frac{2}{3}$, as shown in the diagram:

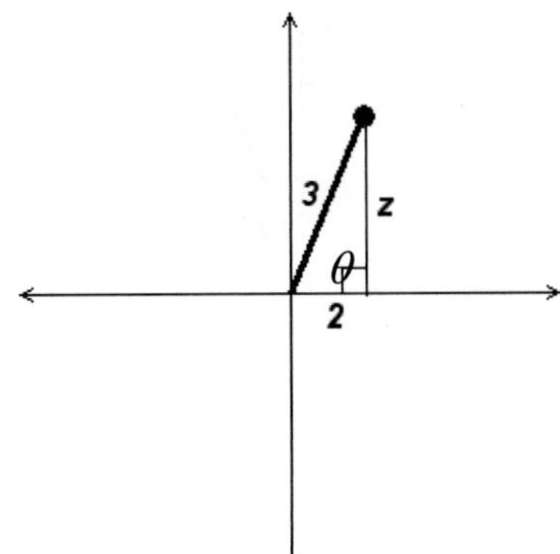

Using the Pythagorean Theorem, we see that $z^2 + 2^2 = 3^2$, so that $z = \sqrt{5}$.

Hence, $\sin\left(\cos^{-1}\left(\frac{2}{3}\right)\right) = \sin\theta = \boxed{\frac{\sqrt{5}}{3}}$.

79. Let $\theta = \tan^{-1}\left(\frac{12}{5}\right)$. Then, $\tan\theta = \frac{12}{5}$, as shown in the diagram:

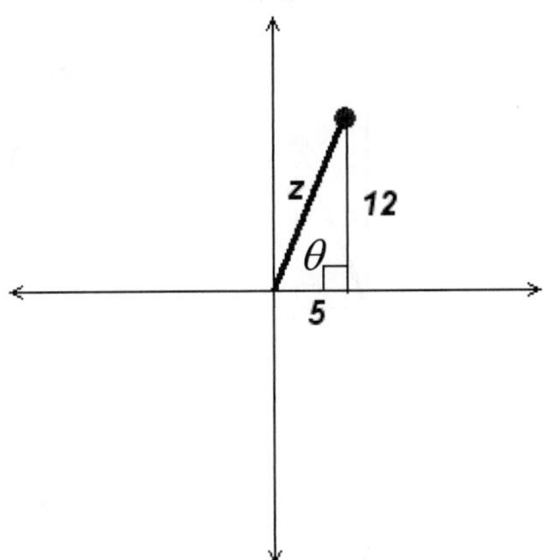

Using the Pythagorean Theorem, we see that $12^2 + 5^2 = z^2$, so that $z = 13$.

Hence, $\sin\left(\tan^{-1}\left(\frac{12}{5}\right)\right) = \sin\theta = \boxed{\frac{12}{13}}$.

80. Let $\theta = \tan^{-1}\left(\frac{7}{24}\right)$. Then, $\tan\theta = \frac{7}{24}$, as shown in the diagram:

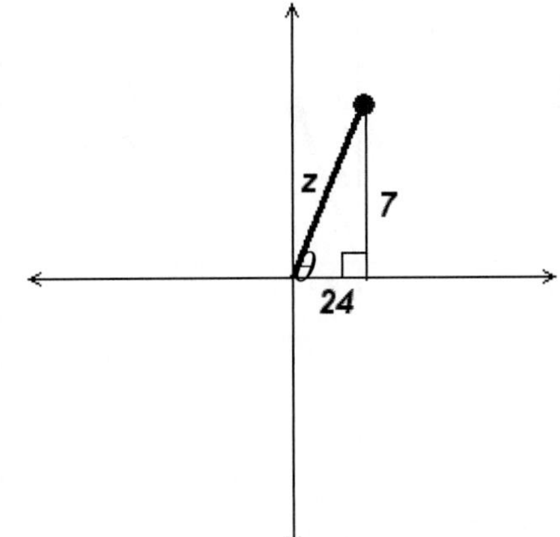

Using the Pythagorean Theorem, we see that $7^2 + 24^2 = z^2$, so that $z = 25$.

Hence, $\cos\left(\tan^{-1}\left(\frac{7}{24}\right)\right) = \cos\theta = \boxed{\frac{24}{25}}$.

81. Let $\theta = \sin^{-1}\left(\frac{3}{5}\right)$. Then, $\sin\theta = \frac{3}{5}$, as shown in the diagram:

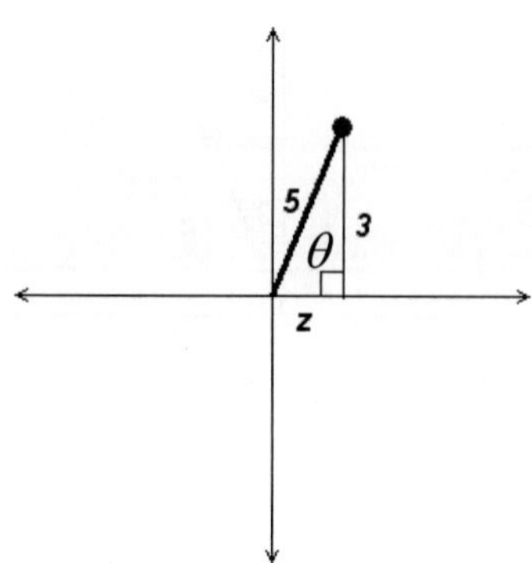

Using the Pythagorean Theorem, we see that $z^2 + 3^2 = 5^2$, so that $z = 4$.

Hence, $\tan\left(\sin^{-1}\left(\frac{3}{5}\right)\right) = \tan\theta = \boxed{\frac{3}{4}}$.

82. Let $\theta = \cos^{-1}\left(\frac{2}{5}\right)$. Then, $\cos\theta = \frac{2}{5}$, as shown in the diagram:

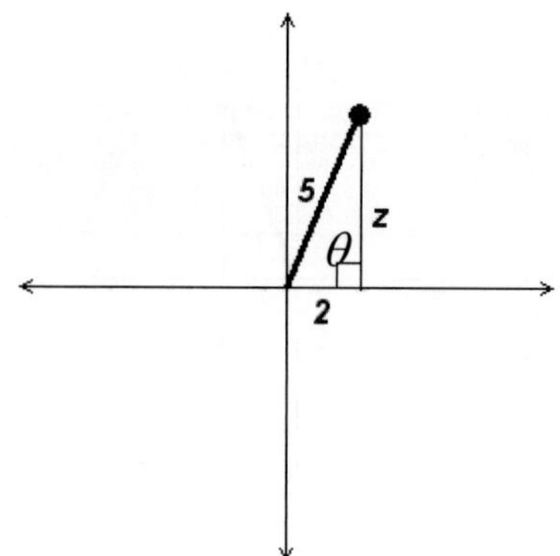

Using the Pythagorean Theorem, we see that $z^2 + 2^2 = 5^2$, so that $z = \sqrt{21}$.

Hence, $\tan\left(\cos^{-1}\left(\frac{2}{5}\right)\right) = \tan\theta = \boxed{\frac{\sqrt{21}}{2}}$.

83. Let $\theta = \sin^{-1}\left(\frac{\sqrt{2}}{5}\right)$. Then, $\sin\theta = \frac{\sqrt{2}}{5}$, as shown in the diagram:

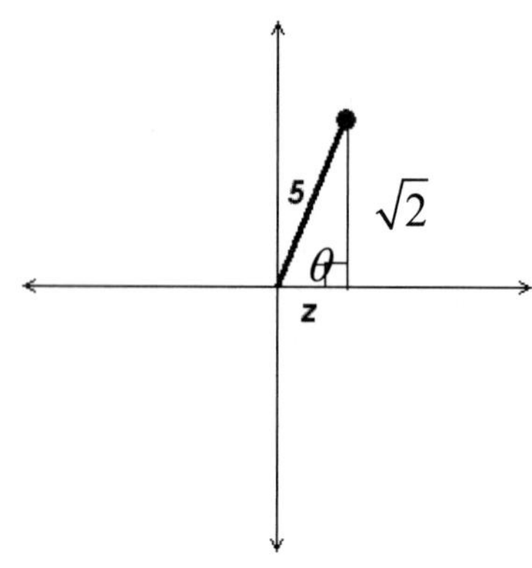

84. Let $\theta = \cos^{-1}\left(\frac{\sqrt{7}}{4}\right)$. Then, $\cos\theta = \frac{\sqrt{7}}{4}$, as shown in the diagram:

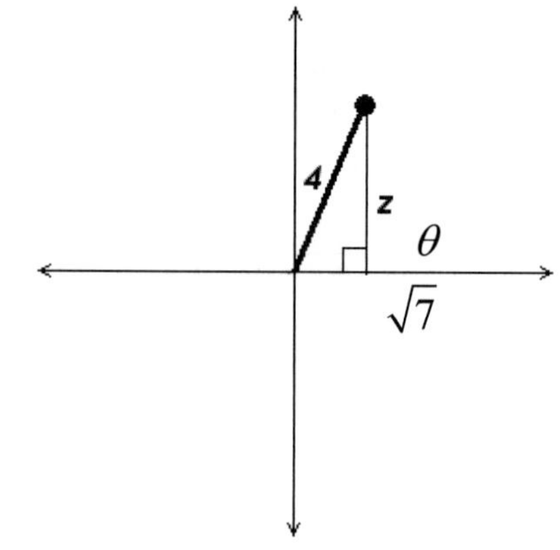

Using the Pythagorean Theorem, we see that $z^2+\left(\sqrt{2}\right)^2=5^2$, so that $z=\sqrt{23}$. So,

$$\sec\left(\sin^{-1}\left(\frac{3}{4}\right)\right)=\sec\theta=\frac{1}{\cos\theta}=\frac{5}{\sqrt{23}}=\boxed{\frac{5\sqrt{23}}{23}}$$

.

Using the Pythagorean Theorem, we see that $z^2+\left(\sqrt{7}\right)^2=4^2$, so that $z=3$. So,

$$\sec\left(\cos^{-1}\left(\frac{\sqrt{7}}{4}\right)\right)=\sec\theta=\frac{1}{\cos\theta}=\frac{4}{\sqrt{7}}=\boxed{\frac{4\sqrt{7}}{7}}.$$

Alternatively, you can use the formula $\sec^{-1}x=\cos^{-1}\left(\frac{1}{x}\right)$ here. Indeed,

$$\sec\left(\cos^{-1}\left(\frac{\sqrt{7}}{4}\right)\right)=\sec\left(\cos^{-1}\left(\frac{1}{4/\sqrt{7}}\right)\right)$$

$$=\sec\left(\sec^{-1}\left(\frac{4}{\sqrt{7}}\right)\right)=\frac{4}{\sqrt{7}}=\boxed{\frac{4\sqrt{7}}{7}}$$

85. Let $\theta=\cos^{-1}\left(\frac{1}{4}\right)$. Then, $\cos\theta=\frac{1}{4}$, as shown in the diagram:

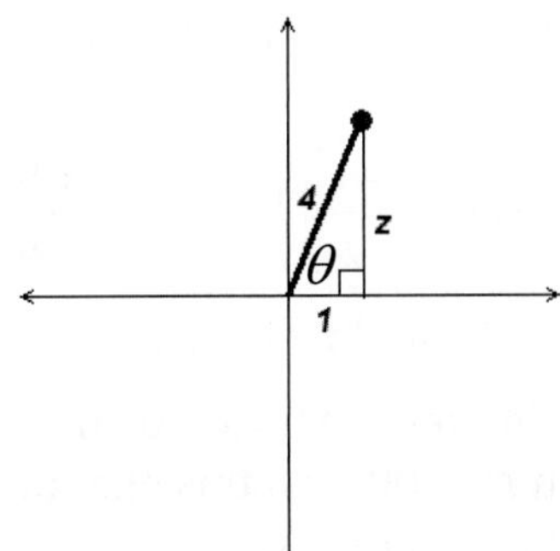

Using the Pythagorean Theorem, we see that $z^2+1^2=4^2$, so that $z=\sqrt{15}$. Hence,

$$\csc\left(\cos^{-1}\left(\frac{1}{4}\right)\right)=\csc\theta=\frac{1}{\sin\theta}=\frac{4}{\sqrt{15}}=\boxed{\frac{4\sqrt{15}}{15}}.$$

86. Let $\theta=\sin^{-1}\left(\frac{1}{4}\right)$. Then, $\sin\theta=\frac{1}{4}$, as shown in the diagram:

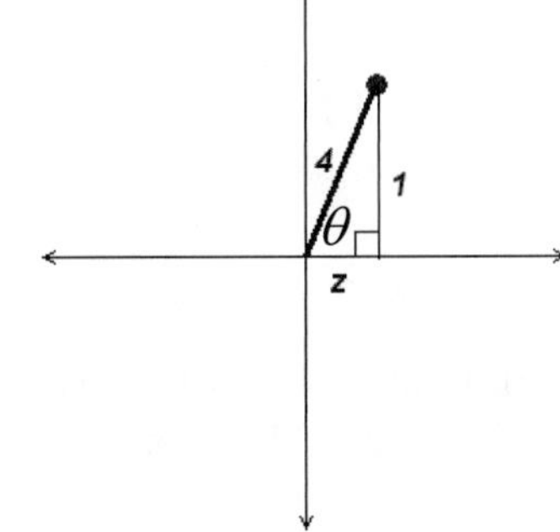

Using the Pythagorean Theorem, we see that $z^2+1^2=4^2$, so that $z=\sqrt{15}$. Hence,

$$\csc\left(\sin^{-1}\left(\frac{1}{4}\right)\right)=\csc\theta=\frac{1}{\sin\theta}=\boxed{4}.$$

Alternatively, you can use the formula $\csc^{-1}x=\sin^{-1}\left(\frac{1}{x}\right)$ here. Indeed, observe that

$$\csc\left(\sin^{-1}\left(\frac{1}{4}\right)\right)=\csc\left(\sin^{-1}\left(\frac{1}{1/4}\right)\right)$$

$$=\csc\left(\csc^{-1}(4)\right)=\boxed{4}$$

87. Let $\theta = \sin^{-1}\left(\frac{60}{61}\right)$. Then, $\sin\theta = \frac{60}{61}$, as shown in the diagram:

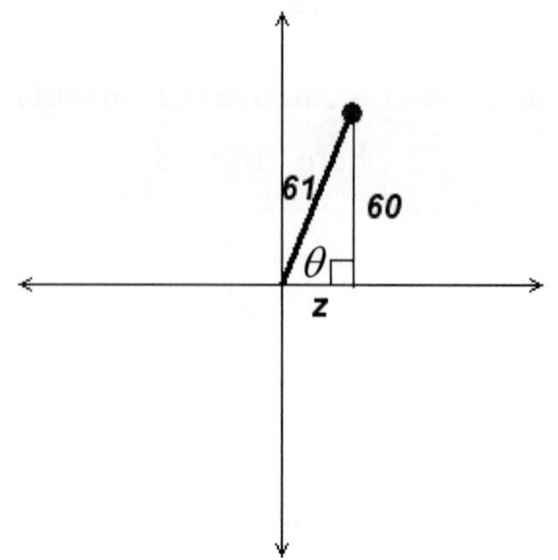

Using the Pythagorean Theorem, we see that $z^2 + 60^2 = 61^2$, so that $z = 11$. Hence,

$$\cot\left(\sin^{-1}\left(\frac{60}{61}\right)\right) = \cot\theta = \boxed{\frac{11}{60}}.$$

88. Let $\theta = \sec^{-1}\left(\frac{41}{9}\right)$. Then, $\sec\theta = \frac{41}{9}$, so that $\cos\theta = \frac{9}{41}$, as shown in the diagram:

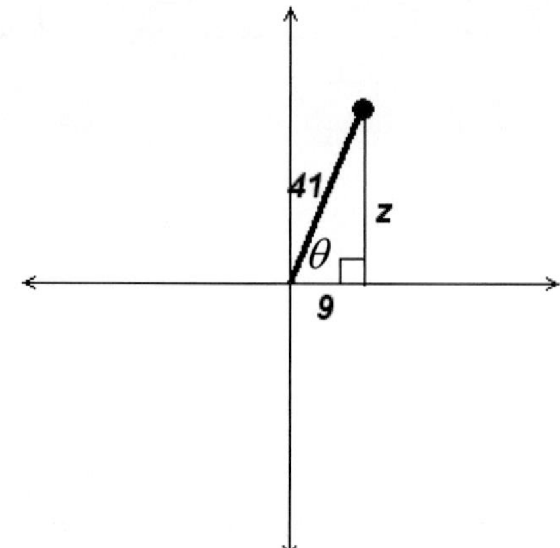

Using the Pythagorean Theorem, we see that $z^2 + 9^2 = 41^2$, so that $z = 40$. Hence,

$$\cot\left(\sec^{-1}\left(\frac{41}{9}\right)\right) = \cot\theta = \boxed{\frac{9}{40}}.$$

89. Use $i = I\sin(2\pi f t)$ with $f = 5$ and $I = 115$. Find the smallest positive value of t for which $i = 85$. To this end, observe

$$115\sin(2\pi \cdot 5t) = 85$$

$$\sin(10\pi t) = \frac{85}{115}$$

$$t = \frac{\sin^{-1}\left(\frac{85}{115}\right)}{10\pi} \approx 0.026476$$

So, $t \approx \boxed{0.026476 \text{ sec.} = 26 \text{ ms}}$.

90. Use $i = I\sin(2\pi f t)$ with $f = 100$ and $I = 240$. Find the smallest positive value of t for which $i = 100$. To this end, observe

$$240\sin(2\pi \cdot 100t) = 100$$

$$\sin(200\pi t) = \frac{100}{240}$$

$$t = \frac{\sin^{-1}\left(\frac{100}{240}\right)}{200\pi} \approx 0.000684$$

So, $t \approx \boxed{0.000684 \text{ sec.} = 0.68 \text{ ms}}$.

91. Given that $H(t) = 12 + 2.4\sin(0.017t - 1.377)$, we must find the value of t for which $H(t) = 14.4$. To this end, we have

$$12 + 2.4\sin(0.017t - 1.377) = 14.4$$
$$\sin(0.017t - 1.377) = 1$$
$$0.017t - 1.377 = \sin^{-1}(1) = \pi/2$$
$$t = \frac{\pi/2 + 1.377}{0.017} \approx 173.4$$

Now, note that $173.4 - 151 = 22.4$. As such, this corresponds to June 22-23.

92. Given that $H(t) = 12 + 2.4\sin(0.017t - 1.377)$, we must find the value of t for which $H(t) = 9.6$. To this end, we have

$$12 + 2.4\sin(0.017t - 1.377) = 9.6$$
$$\sin(0.017t - 1.377) = -1$$
$$0.017t - 1.377 = \sin^{-1}(-1) = -\pi/2$$
$$t = \frac{-\pi/2 + 1.377}{0.017} \approx -11.4$$

So, counting backward into December 12 days implies this corresponds to Dec.19.

93. We need to find the smallest value of t for which

$$12.5\cos(0.157t) + 2.5 = 0,$$

and the graph of the left-side is decreasing prior to this value. We solve this graphically. The solid graph corresponds to the left-side of the equation. We have:

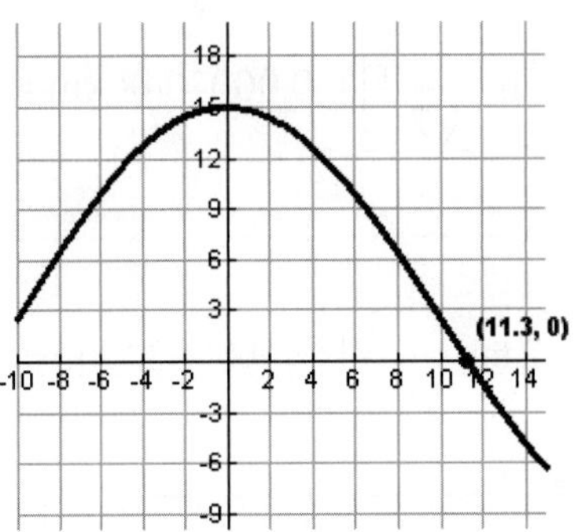

Notice that the solution is approximately $t = 11.3$, or about 11 years.

94. We need to find the smallest value of t larger than 11.3 for which

$$12.5\cos(0.157t) + 2.5 = 15.$$

We approach this graphically, where the solid graph corresponds to the left-side of the equation. We have:

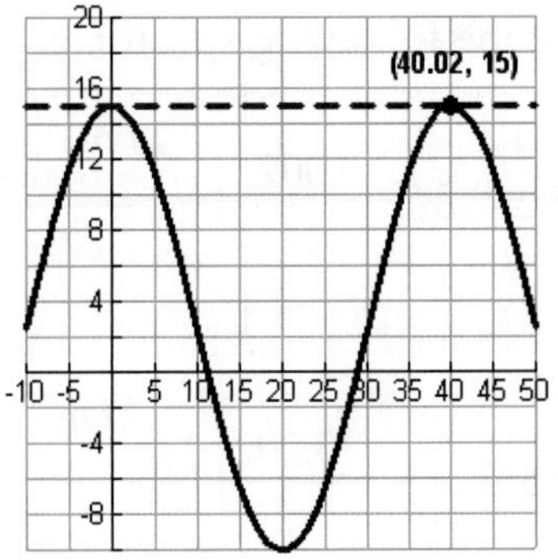

This occurs at approximately 40 years.

95. Consider the following diagram:

7 ft.
β
x
α
1 ft.

Let $\theta = \alpha + \beta$. Then,

$$\tan\theta = \tan(\alpha+\beta) = \frac{\tan\alpha + \tan\beta}{1-\tan\alpha\tan\beta}$$

$$= \frac{\frac{1}{x}+\frac{7}{x}}{1-\left(\frac{1}{x}\right)\left(\frac{7}{x}\right)} = \frac{\frac{8}{x}}{\frac{x^2-7}{x^2}} = \boxed{\frac{8x}{x^2-7}}$$

96. Using Exercise 95, we see that $\tan\theta = \frac{8x}{x^2-7}$, so that $\theta = \tan^{-1}\left(\frac{8x}{x^2-7}\right)$. Thus, we have the following specific calculations:

$$\underline{x=10}:\quad \theta = \tan^{-1}\left(\frac{80}{93}\right) \approx 0.71 \text{ radians} \approx 41^\circ$$

$$\underline{x=20}:\quad \theta = \tan^{-1}\left(\frac{160}{393}\right) \approx 0.39 \text{ radians} \approx 22^\circ$$

97. Use the formula $M = \frac{f}{2}\left[1-\frac{2\tan^{-1}\left(\frac{k}{d}\right)}{\pi}\right]$ with $f = 2\text{m} = 0.002$ km and $d = 4$ km.

We have the following specific calculation:

$$\underline{k=2\text{ km}}:\quad M = \frac{0.002}{2}\left[1-\frac{2\tan^{-1}\left(\frac{2}{4}\right)}{\pi}\right] = \frac{0.001}{\pi}\left[\pi - 2\tan^{-1}\left(\frac{1}{2}\right)\right] \approx 0.0007048 \text{ km} \approx 0.70\text{m}$$

$$\underline{k=10\text{ km}}:\quad M = \frac{0.002}{2}\left[1-\frac{2\tan^{-1}\left(\frac{10}{4}\right)}{\pi}\right] = \frac{0.001}{\pi}\left[\pi - 2\tan^{-1}\left(\frac{5}{2}\right)\right] \approx 0.00024 \text{ km} \approx 0.24\text{m}$$

98. Use the formula $M=\frac{f}{2}\left[1-\frac{2\tan^{-1}\left(\frac{k}{d}\right)}{\pi}\right]$ with $f=3\text{ m}=0.003\text{ km}$ and $d=2.5\text{ km}$.

We have the following specific calculation:

$\underline{k=5\text{ km}}$: $$M=\frac{0.003}{2}\left[1-\frac{2\tan^{-1}\left(\frac{5}{2.5}\right)}{\pi}\right]=\frac{0.0015}{\pi}\left[\pi-2\tan^{-1}(2)\right]\approx 0.00044\text{ km}\approx 0.44\text{m}$$

$\underline{k=10\text{ km}}$: $$M=\frac{0.003}{2}\left[1-\frac{2\tan^{-1}\left(\frac{10}{2.5}\right)}{\pi}\right]=\frac{0.0015}{\pi}\left[\pi-2\tan^{-1}(4)\right]\approx 0.00023\text{ km}\approx 0.23\text{ m}$$

99. Consider the following diagram:

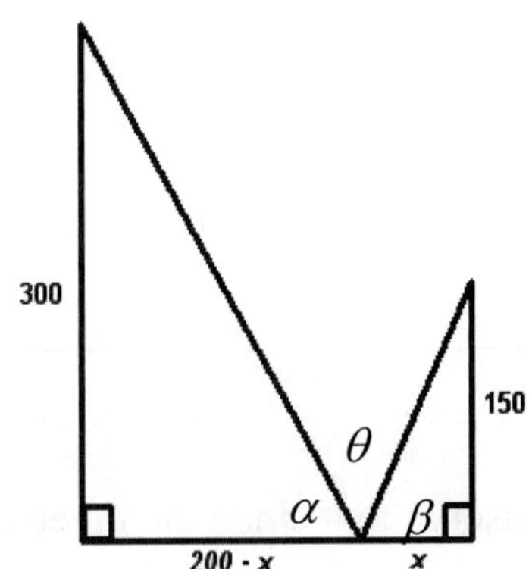

Since $\tan\alpha=\frac{300}{200-x}$, we have
$$\alpha=\tan^{-1}\left(\frac{300}{200-x}\right).$$
Also, since $\tan\beta=\frac{150}{x}$, we have
$$\beta=\tan^{-1}\left(\frac{150}{x}\right).$$
Therefore, since $\alpha+\beta+\theta=\pi$, we see that
$$\boxed{\theta=\pi-\tan^{-1}\left(\frac{300}{200-x}\right)-\tan^{-1}\left(\frac{150}{x}\right)}.$$

100. Consider the following diagram:

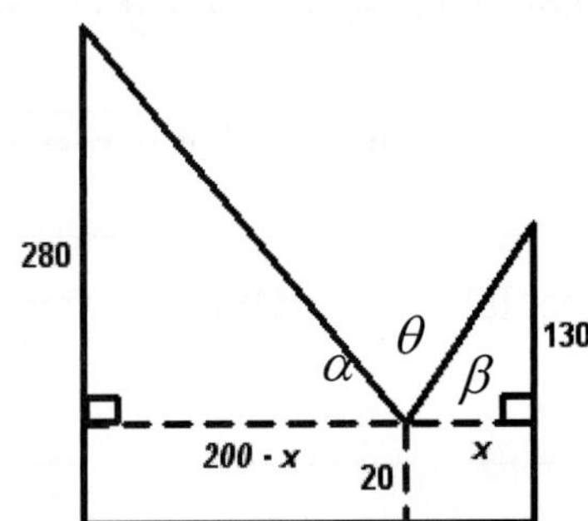

Since $\tan\alpha=\frac{280}{200-x}$, we have
$$\alpha=\tan^{-1}\left(\frac{280}{200-x}\right).$$
Also, since $\tan\beta=\frac{130}{x}$, we have
$$\beta=\tan^{-1}\left(\frac{130}{x}\right).$$
Therefore, since $\alpha+\beta+\theta=\pi$, we see that
$$\boxed{\theta=\pi-\tan^{-1}\left(\frac{280}{200-x}\right)-\tan^{-1}\left(\frac{130}{x}\right)}.$$

101. The identity $\sin^{-1}(\sin x) = x$ is valid only for x in the interval $\left[-\frac{\pi}{2}, \frac{\pi}{2}\right]$, not $[0, \pi]$.
102. The identity $\cos^{-1}(\cos x) = x$ is valid only for x in the interval $[0, \pi]$, not $\left[-\frac{\pi}{2}, \frac{\pi}{2}\right]$.
103. In general, $\cot^{-1} x \neq \frac{1}{\tan^{-1} x}$.
104. In general, $\csc^{-1} x \neq \frac{1}{\sin^{-1} x}$.
105. False. Upon inspection of the graphs, the portion to the right of the y-axis, when reflected over the y-axis does not match up identically with the left portion, as seen below: More precisely, note that for instance $\sec^{-1}(1) = \cos^{-1}\left(\frac{1}{1}\right) = 0$, while $\sec^{-1}(-1) = \cos^{-1}\left(-\frac{1}{1}\right) = \pi$. As such, $\sec^{-1}(x) \neq \sec^{-1}(-x)$, for all x in the domain of inverse secant.
106. True. First, judging from the graph of $y = \csc^{-1} x = \sin^{-1}\left(\frac{1}{x}\right)$, it *seems* as though the inverse cosecant function is odd. To prove this, observe that since the inverse sine function is odd, we obtain $$\csc^{-1}(-x) = \sin^{-1}\left(\frac{1}{-x}\right) = -\sin^{-1}\left(\frac{1}{x}\right) = -\csc^{-1}(x).$$ From the definition of *odd*, we are done. (**Note:** In general, if f and g are both odd functions, then the composition $f \circ g$ is odd on its domain.)
107. False. This holds only on a subset of the domain to which cosecant is restricted in order to define its inverse.
108. False. In general, $\csc^{-1}\theta \neq \frac{1}{\sin^{-1}\theta}$. For instance, let $x = \frac{1}{2}$. Observe that $\csc^{-1}\left(2 \cdot \frac{1}{2}\right) = 0$, but $\sin^{-1}(1) \cdot \csc^{-1}(1) = 0 \neq 1$.
109. $\sec^{-1}\left(\frac{1}{2}\right)$ does not exist since $\frac{1}{2}$ is not in the domain of the inverse secant function (which coincides with the range of the secant function).

110. $\csc^{-1}\left(\frac{1}{2}\right)$ does not exist since $\frac{1}{2}$ is not in the domain of the inverse cosecant function (which coincides with the range of the cosecant function).

111. In order to compute $\sin\left[\cos^{-1}\left(\frac{\sqrt{2}}{2}\right)+\sin^{-1}\left(-\frac{1}{2}\right)\right]$, we first simplify both $\cos^{-1}\left(\frac{\sqrt{2}}{2}\right)$ and $\sin^{-1}\left(-\frac{1}{2}\right)$:

$$\text{If } \cos^{-1}\left(\frac{\sqrt{2}}{2}\right)=\theta, \text{ then } \cos\theta=\frac{\sqrt{2}}{2}. \text{ So, } \theta=\frac{\pi}{4}.$$

$$\text{If } \sin^{-1}\left(-\frac{1}{2}\right)=\beta, \text{ then } \sin\beta=-\frac{1}{2}. \text{ So, } \beta=-\frac{\pi}{6}.$$

Hence,

$$\sin\left[\cos^{-1}\left(\frac{\sqrt{2}}{2}\right)+\sin^{-1}\left(-\frac{1}{2}\right)\right]=\sin\left[\frac{\pi}{4}-\frac{\pi}{6}\right]=\sin\left(\frac{\pi}{4}\right)\cos\left(\frac{\pi}{6}\right)-\sin\left(\frac{\pi}{6}\right)\cos\left(\frac{\pi}{4}\right)$$

$$=\left(\frac{\sqrt{2}}{2}\right)\left(\frac{\sqrt{3}}{2}\right)-\left(\frac{1}{2}\right)\left(\frac{\sqrt{2}}{2}\right)=\boxed{\frac{\sqrt{6}-\sqrt{2}}{4}}$$

112. Observe that $\sin^{-1}\left(2\sin\left(\frac{3x}{2}\right)\cos\left(\frac{3x}{2}\right)\right)=\sin^{-1}\left(\sin(3x)\right)=3x$ provided that $-\frac{\pi}{2}\le 3x\le\frac{\pi}{2}$, so that $\boxed{-\frac{\pi}{6}\le x\le\frac{\pi}{6}}$.

113. In order to compute $\sin\left(2\sin^{-1}(1)\right)$, first observe that

$$\text{If } \theta=\sin^{-1}(1), \text{ then } \sin\theta=1. \text{ So, } \theta=\frac{\pi}{2}.$$

Hence, $\sin\left(2\sin^{-1}(1)\right)=\sin\left(2\cdot\frac{\pi}{2}\right)=\sin\pi=\boxed{0}$.

114. Consider the function $f(x) = 2 - 4\sin\left(x - \frac{\pi}{2}\right)$.

a. Note that this function has a phase shift of $\pi/2$ units to the right. So, we take the interval used to define $\sin^{-1} x$, namely $\left[-\frac{\pi}{2}, \frac{\pi}{2}\right]$ and add $\frac{\pi}{2}$ to both endpoints to get the interval $[0, \pi]$. Note that *f* is, in fact, one-to-one on this interval.

b. Now, we determine a formula for f^{-1}, along with its domain:

$$y = 2 - 4\sin\left(x - \frac{\pi}{2}\right)$$

$$\frac{y-2}{-4} = \sin\left(x - \frac{\pi}{2}\right)$$

$$\sin^{-1}\left(\frac{y-2}{-4}\right) = x - \frac{\pi}{2}$$

$$\frac{\pi}{2} + \sin^{-1}\left(\frac{2-y}{4}\right) = x$$

So, the equation of the inverse of *f* is given by: $f^{-1}(x) = \frac{\pi}{2} + \sin^{-1}\left(\frac{2-x}{4}\right)$

The domain of f^{-1} is equal to the range of *f*. Since the amplitude of *f* is 4 and there is a vertical shift up of 2 units, we see that the range of *f* is $[-2, 6]$. Hence, the domain of f^{-1} is $[-2, 6]$.

115. Consider the function $f(x) = 3 + \cos\left(x - \frac{\pi}{4}\right)$.

a. Note that this function has a phase shift of $\pi/4$ units to the right. So, we take the interval used to define $\cos^{-1} x$, namely $[0, \pi]$ and add $\frac{\pi}{4}$ to both endpoints to get the interval $\left[\frac{\pi}{4}, \frac{5\pi}{4}\right]$. Note that *f* is, in fact, one-to-one on this interval.

b. Now, we determine a formula for f^{-1}, along with its domain:

$$y = 3 + \cos\left(x - \frac{\pi}{4}\right)$$

$$y - 3 = \cos\left(x - \frac{\pi}{4}\right)$$

$$\cos^{-1}(y-3) = x - \frac{\pi}{4}$$

$$\frac{\pi}{4} + \cos^{-1}(y-3) = x$$

So, the equation of the inverse of *f* is given by: $f^{-1}(x) = \frac{\pi}{4} + \cos^{-1}(x-3)$

The domain of f^{-1} is equal to the range of *f*. Since the amplitude of *f* is 1 and there is a vertical shift up of 3 units, we see that the range of *f* is $[2,4]$. Hence, the domain of f^{-1} is $[2,4]$.

116. Consider the function $f(x) = 1 - \tan\left(x + \frac{\pi}{3}\right)$.

(a) We know that $y = \tan x$ is 1-1 on $\left(-\frac{\pi}{2}, \frac{\pi}{2}\right)$. As such, since $y = \tan\left(x + \frac{\pi}{3}\right)$ is simply a horizontal shift of $y = \tan x$ to the left $\frac{\pi}{3}$ units, we conclude that it is 1-1 on the interval $\left(-\frac{\pi}{2} - \frac{\pi}{3}, \frac{\pi}{2} - \frac{\pi}{3}\right) = \left(-\frac{5\pi}{6}, \frac{\pi}{6}\right)$. Since reflecting over the *x*-axis and shifting it vertically do not affect whether or not it is 1-1, we conclude that *f* is 1-1 on this interval as well.

(b) Restricting our attention to *x* values in $\left(-\frac{5\pi}{6}, \frac{\pi}{6}\right)$, we determine the inverse as follows:

$$y = 1 - \tan\left(x + \frac{\pi}{3}\right)$$

$$x = 1 - \tan\left(y + \frac{\pi}{3}\right)$$

$$\tan\left(y + \frac{\pi}{3}\right) = 1 - x$$

$$y + \frac{\pi}{3} = \tan^{-1}(1-x)$$

$$y = -\frac{\pi}{3} + \tan^{-1}(1-x)$$

Hence, the inverse is $f^{-1}(x) = -\frac{\pi}{3} + \tan^{-1}(1-x)$ with domain $\mathbb{R}$.

117. Consider the function $f(x) = 2 + \frac{1}{4}\cot\left(2x - \frac{\pi}{6}\right)$.

(a) We know that $y = \cot 2x$ is 1-1 on $\left(0, \frac{\pi}{2}\right)$. As such, since $y = \cot\left(2x - \frac{\pi}{6}\right)$ is simply a horizontal shift of $y = \cot x$ to the right $\frac{\pi}{12}$ units, we conclude that it is 1-1 on the interval $\left(0 - \frac{\pi}{12}, \frac{\pi}{2} - \frac{\pi}{12}\right) = \left(-\frac{\pi}{12}, \frac{5\pi}{12}\right)$. Since multiplying by a constant and shifting it vertically do not affect whether or not it is 1-1, we conclude that *f* is 1-1 on this interval as well.

(b) Restricting our attention to x values in $\left(-\frac{\pi}{12}, \frac{5\pi}{12}\right)$, we determine the inverse as follows:

$$\begin{aligned}
y &= 2+\tfrac{1}{4}\cot\left(2x-\tfrac{\pi}{6}\right)\\
x &= 2+\tfrac{1}{4}\cot\left(2y-\tfrac{\pi}{6}\right)\\
x-2 &= \tfrac{1}{4}\cot\left(2y-\tfrac{\pi}{6}\right)\\
4(x-2) &= \cot\left(2y-\tfrac{\pi}{6}\right)\\
\cot^{-1}\left(4(x-2)\right) &= 2y-\tfrac{\pi}{6}\\
\tfrac{\pi}{6}+\cot^{-1}\left(4x-8\right) &= 2y\\
\tfrac{\pi}{12}+\tfrac{1}{2}\cot^{-1}\left(4x-8\right) &= y
\end{aligned}$$

Hence, the inverse is $f^{-1}(x)=\frac{\pi}{12}+\frac{1}{2}\cot^{-1}\left(4x-8\right)$ with domain $\mathbb{R}$.

118. Consider the function $f(x)=-\csc\left(\frac{\pi}{4}x-1\right)$.

(a) We know that $y=\csc x$ is 1-1 on $\left[-\frac{\pi}{2},0\right)\cup\left(0,\frac{\pi}{2}\right]$. As such, $y=\csc\left(\frac{\pi}{4}x-1\right)$ is 1-1 on the following interval:

$$\begin{aligned}
-\tfrac{\pi}{2}\le\tfrac{\pi}{4}x-1<0 \quad &\text{or} \quad 0<\tfrac{\pi}{4}x-1\le\tfrac{\pi}{2}\\
\tfrac{4}{\pi}\left(1-\tfrac{\pi}{2}\right)\le x<\tfrac{4}{\pi}(1) \quad &\text{or} \quad \tfrac{4}{\pi}(1)<x\le\tfrac{4}{\pi}\left(1+\tfrac{\pi}{2}\right)\\
\tfrac{4}{\pi}-2\le x<\tfrac{4}{\pi} \quad &\text{or} \quad \tfrac{4}{\pi}<x\le\tfrac{4}{\pi}+2
\end{aligned}$$

Since multiplying by a constant does not affect whether or not it is 1-1, we conclude that f is 1-1 on this set as well.

(b) Restricting our attention to x values listed above, we determine the inverse as follows:

$$\begin{aligned}
y &= -\csc\left(\tfrac{\pi}{4}x-1\right)\\
x &= -\csc\left(\tfrac{\pi}{4}y-1\right)\\
-x &= \csc\left(\tfrac{\pi}{4}y-1\right)\\
\csc^{-1}(-x) &= \tfrac{\pi}{4}y-1\\
\tfrac{4}{\pi}\left[1+\csc^{-1}(-x)\right] &= y
\end{aligned}$$

Hence, the inverse is $f^{-1}(x)=\frac{4}{\pi}\left[1+\csc^{-1}(-x)\right]$ with domain $(-\infty,-1]\cup[1,\infty)$.

119. The graphs of the following two functions on the interval $[-3,3]$ is below:

$$Y_1 = \sin\left(\sin^{-1} x\right), \quad Y_2 = x$$

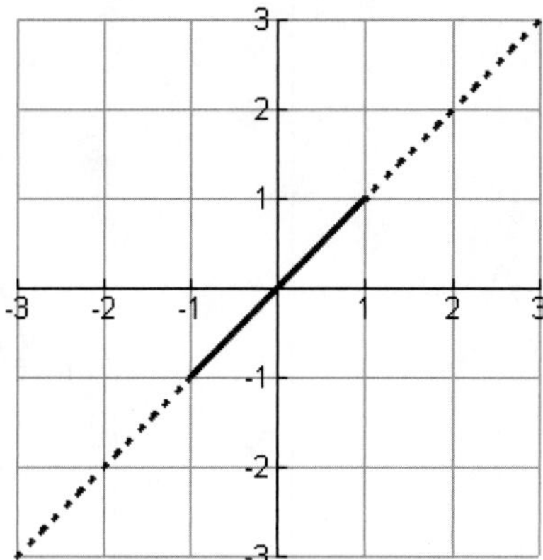

The graphs are different outside the interval $[-1,1]$ because the identity $\sin\left(\sin^{-1} x\right) = x$ only holds for $-1 \le x \le 1$.

120. The graphs of the following two functions on the interval $[-3,1]$ is below:

$$Y_1 = \cos\left(\cos^{-1} x\right), \quad Y_2 = x$$

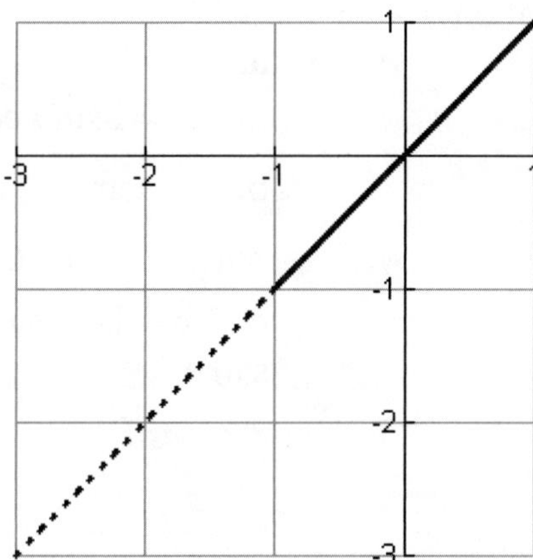

The results are different outside the interval $[-1,1]$ because the identity $\cos\left(\cos^{-1} x\right) = x$ only holds for $-1 \le x \le 1$.

121. The graphs of the following two functions on the interval $\left[-\frac{\pi}{2}, \frac{\pi}{2}\right]$ is below: $Y_1 = \csc^{-1}(\csc x)$, $Y_2 = x$

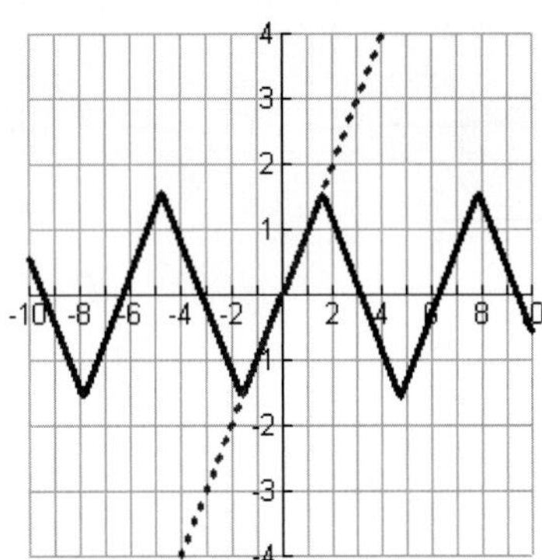

Observe that the graphs do indeed coincide on this interval. This occurs since $\csc^{-1}(\csc x) = x$ holds when $x \in \left[-\frac{\pi}{2}, 0\right) \cup \left(0, \frac{\pi}{2}\right]$.

122. The graphs of the following two functions on the interval $[0, \pi]$ is below: $Y_1 = \sec^{-1}(\sec x)$, $Y_2 = x$

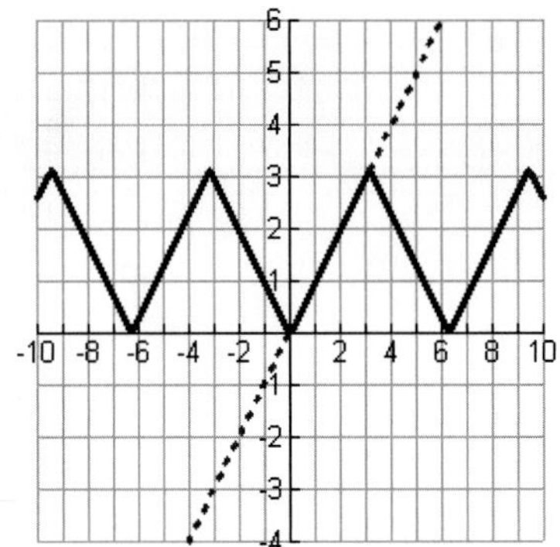

Observe that the graphs do indeed coincide on this interval. This occurs since $\sec^{-1}(\sec x) = x$ holds when $0 \le x \le \pi$.

123. From the given information, we have the following diagram:

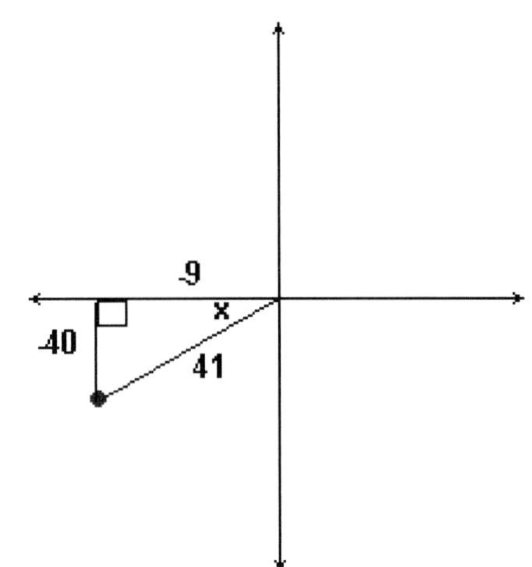

a.
$\sin 2x = 2\sin x \cos x = 2\left(-\frac{40}{41}\right)\left(-\frac{9}{41}\right) = \frac{720}{1681}$

b. $x = \tan^{-1}\left(\frac{40}{9}\right) \approx 1.34948$. So,
$\sin 2x = \sin(2.69896) = 0.42832$

c. Yes, the results in parts **a.** and **b.** are the same.

124. From the given information, we have the following diagram:

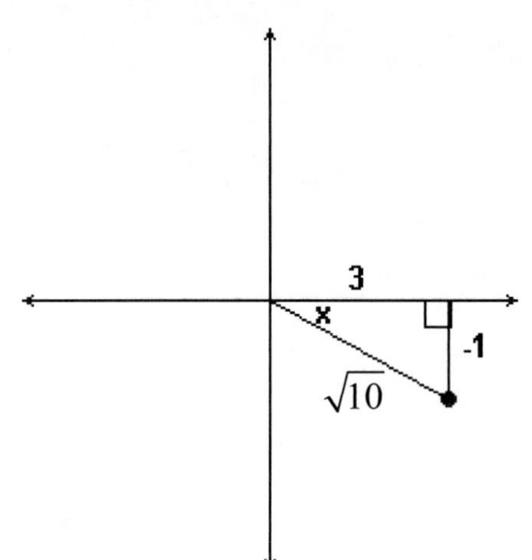

a.

$$\tan 2x = \frac{2\tan x}{1-\tan^2 x} = \frac{2\left(-\frac{1}{3}\right)}{1-\left(-\frac{1}{3}\right)^2} = \frac{-\frac{2}{3}}{\frac{8}{9}} = -\frac{3}{4}$$

b. $x = \sin^{-1}\left(-\frac{1}{\sqrt{10}}\right) \approx -0.32175$. So,

$$\tan 2x = \tan\left(-0.64350\right) = -0.7500$$

c. Yes, the results in parts **a.** and **b.** are the same.

125. Note that $y = \tan^{-1} x \Rightarrow \tan y = x$. This yields the following triangle:

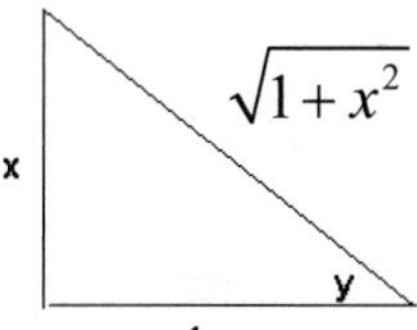

So,

$$\sec y = \frac{1}{\cos y} = \frac{1}{\frac{1}{\sqrt{1+x^2}}} = \sqrt{1+x^2}$$

and thus,

$$\sec^2 y = 1 + x^2.$$

126. Note that $y = \cos^{-1} x \Rightarrow \cos y = x$. This yields the following triangle:

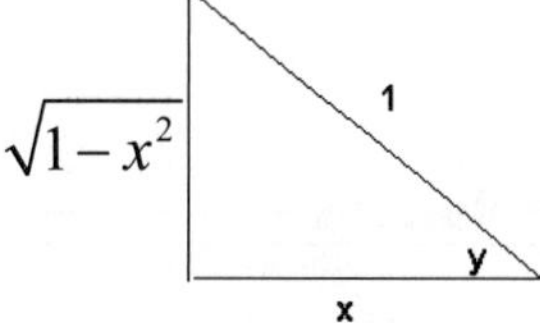

So,

$$\sin y = \sqrt{1-x^2}.$$

127. Note that $y = \sec^{-1} x \Rightarrow \sec y = x$. This yields the following triangle: 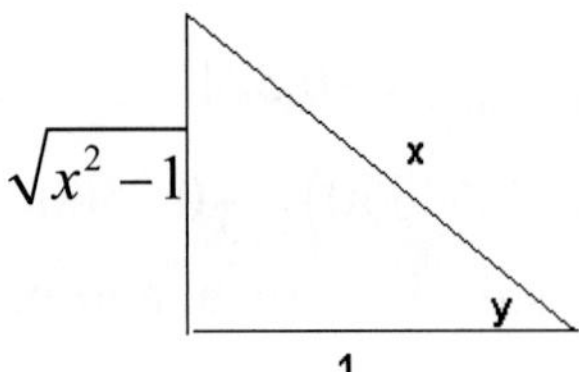 So, $\tan y = \sqrt{x^2-1}$ and thus, $$\sec y \tan y = x\sqrt{x^2-1}.$$	**128.** Note that $y = \cot^{-1} x \Rightarrow \cot y = x$. This yields the following triangle: 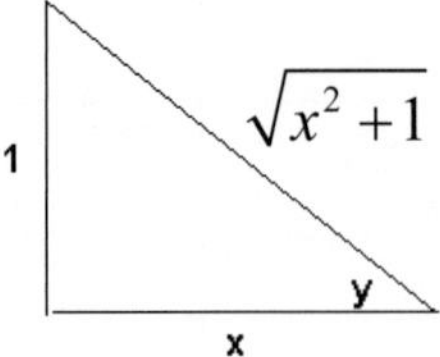 So, $$\csc y = \frac{1}{\sin y} = \frac{1}{\frac{1}{\sqrt{x^2+1}}} = \sqrt{x^2+1}$$ and thus, $$\csc^2 y = 1 + x^2.$$

Section 6.6 Solutions ---

1. The values of θ in $[0, 2\pi]$ that satisfy $\cos\theta = -\frac{\sqrt{2}}{2}$ are $\theta = \boxed{\frac{3\pi}{4}, \frac{5\pi}{4}}$.
2. The values of θ in $[0, 2\pi]$ that satisfy $\sin\theta = -\frac{\sqrt{2}}{2}$ are $\theta = \boxed{\frac{5\pi}{4}, \frac{7\pi}{4}}$.
3. First, observe that $\csc\theta = -2$ is equivalent to $\sin\theta = -\frac{1}{2}$. The values of θ in $[0, 4\pi]$ that satisfy $\sin\theta = -\frac{1}{2}$ are $$\theta = \frac{7\pi}{6}, \frac{7\pi}{6} + 2\pi, \frac{11\pi}{6}, \frac{11\pi}{6} + 2\pi = \boxed{\frac{7\pi}{6}, \frac{11\pi}{6}, \frac{19\pi}{6}, \frac{23\pi}{6}}.$$

4. First, observe that $\sec\theta = -2$ is equivalent to $\cos\theta = -\frac{1}{2}$. The values of θ in $[0, 4\pi]$ that satisfy $\cos\theta = -\frac{1}{2}$ are

$$\theta = \frac{2\pi}{3},\ \frac{2\pi}{3} + 2\pi,\ \frac{4\pi}{3},\ \frac{4\pi}{3} + 2\pi = \boxed{\frac{2\pi}{3}, \frac{4\pi}{3}, \frac{8\pi}{3}, \frac{10\pi}{3}}.$$

5. The only way $\tan\theta = 0$ is for $\sin\theta = 0$. The values of θ in $\mathbb{R}$ that satisfy $\sin\theta = 0$ (and hence, the original equation) are $\theta = \boxed{n\pi, \text{ where } n \text{ is an integer}}$.

6. The only way $\cot\theta = 0$ is for $\cos\theta = 0$. The values of θ in $\mathbb{R}$ that satisfy $\cos\theta = 0$ (and hence, the original equation) are $\theta = \boxed{\frac{(2n+1)\pi}{2}, \text{ where } n \text{ is an integer}}$.

7. The values of θ in $[0, 2\pi]$ that satisfy $\sin 2\theta = -\frac{1}{2}$ must satisfy

$$2\theta = \frac{7\pi}{6},\ \frac{7\pi}{6} + 2\pi,\ \frac{11\pi}{6},\ \frac{11\pi}{6} + 2\pi = \frac{7\pi}{6}, \frac{11\pi}{6}, \frac{19\pi}{6}, \frac{23\pi}{6}.$$

So, dividing all values by 2 yields the following values of θ which satisfy the original equation:

$$\theta = \boxed{\frac{7\pi}{12}, \frac{11\pi}{12}, \frac{19\pi}{12}, \frac{23\pi}{12}}$$

8. The values of θ in $[0, 2\pi]$ that satisfy $\cos 2\theta = \frac{\sqrt{3}}{2}$ must satisfy

$$2\theta = \frac{\pi}{6},\ \frac{\pi}{6} + 2\pi,\ \frac{11\pi}{6},\ \frac{11\pi}{6} + 2\pi = \frac{\pi}{6}, \frac{11\pi}{6}, \frac{13\pi}{6}, \frac{23\pi}{6}.$$

So, dividing all values by 2 yields the following values of θ which satisfy the original equation:

$$\theta = \boxed{\frac{\pi}{12}, \frac{11\pi}{12}, \frac{13\pi}{12}, \frac{23\pi}{12}}$$

9. The values of θ in $\mathbb{R}$ that satisfy $\sin\left(\frac{\theta}{2}\right)=-\frac{1}{2}$ must satisfy

$$\frac{\theta}{2}=\frac{7\pi}{6}+2n\pi,\ \frac{11\pi}{6}+2n\pi \text{, where } n \text{ is an integer.}$$

So, multiplying all values by 2 yields the following values of θ which satisfy the original equation:

$$\theta=2\left(\frac{7\pi}{6}+2n\pi\right),\ 2\left(\frac{11\pi}{6}+2n\pi\right)=\frac{7\pi}{3}+4n\pi,\ \frac{11\pi}{3}+4n\pi$$

$$=\boxed{\frac{7\pi}{3}+4n\pi,\ \frac{11\pi}{3}+4n\pi, \text{where } n \text{ is an integer}}.$$

10. The values of θ in $\mathbb{R}$ that satisfy $\cos\left(\frac{\theta}{2}\right)=-1$ must satisfy

$$\frac{\theta}{2}=(2n+1)\pi \text{, where } n \text{ is an integer.}$$

So, multiplying by 2 yields the following values of θ which satisfy the original equation:

$$\boxed{\theta=2(2n+1)\pi, \text{ where } n \text{ is an integer}}.$$

11. The values of θ in $[-2\pi,2\pi]$ that satisfy $\tan 2\theta=\sqrt{3}$ must satisfy

$$2\theta=\frac{\pi}{3},\ \frac{\pi}{3}+2\pi,\ \frac{4\pi}{3},\ \frac{4\pi}{3}+2\pi,\ -\frac{2\pi}{3},\ -\frac{2\pi}{3}-2\pi,\ -\frac{5\pi}{3},\ -\frac{5\pi}{3}-2\pi$$

$$=\frac{\pi}{3},\frac{4\pi}{3},\frac{7\pi}{3},\frac{10\pi}{3},-\frac{2\pi}{3},-\frac{5\pi}{3},-\frac{8\pi}{3},-\frac{11\pi}{3}$$

So, dividing all values by 2 yields the following values of θ which satisfy the original equation:

$$\theta=\frac{\pi}{6},\frac{4\pi}{6},\frac{7\pi}{6},\frac{10\pi}{6},-\frac{2\pi}{6},-\frac{5\pi}{6},-\frac{8\pi}{6},-\frac{11\pi}{6}$$

$$=\boxed{\frac{\pi}{6},\frac{2\pi}{3},\frac{7\pi}{6},\frac{5\pi}{3},-\frac{\pi}{3},-\frac{5\pi}{6},-\frac{4\pi}{3},-\frac{11\pi}{6}}$$

12. The values of θ in $\mathbb{R}$ that satisfy $\tan 2\theta = -\sqrt{3}$ must satisfy

$$2\theta = \frac{2\pi}{3} + 2n\pi,\ \frac{5\pi}{3} + 2n\pi \text{, where } n \text{ is an integer.}$$

This reduces to

$$2\theta = \frac{2\pi}{3} + n\pi \text{, where } n \text{ is an integer.}$$

So, dividing all values by 2 yields the following values of θ which satisfy the original equation:

$$\theta = \frac{\pi}{3} + \frac{n\pi}{2} = \boxed{\frac{\pi(2+3n)}{6}, \text{ where } n \text{ is an integer}}.$$

13. First, observe that $\sec\theta = -2$ is equivalent to $\cos\theta = -\frac{1}{2}$. The values of θ in $[-2\pi, 0]$ that satisfy $\cos\theta = -\frac{1}{2}$ are $\theta = \boxed{-\frac{2\pi}{3},\ -\frac{4\pi}{3}}$.

14. First, observe that $\csc\theta = \frac{2\sqrt{3}}{3}$ is equivalent to $\sin\theta = \frac{3}{2\sqrt{3}} = \frac{\sqrt{3}}{2}$. The values of θ in $[-\pi, \pi]$ that satisfy $\sin\theta = \frac{\sqrt{3}}{2}$ are $\theta = \boxed{\frac{\pi}{3}, \frac{2\pi}{3}}$.

15. The values of θ in $\mathbb{R}$ that satisfy $\cot 4\theta = -\frac{\sqrt{3}}{3} = -\frac{\frac{1}{2}}{\frac{\sqrt{3}}{2}}$ must satisfy

$$4\theta = \frac{2\pi}{3} + 2n\pi,\ \frac{5\pi}{3} + 2n\pi \text{, where } n \text{ is an integer.}$$

This reduces to

$$4\theta = \frac{2\pi}{3} + n\pi \text{, where } n \text{ is an integer.}$$

So, dividing all values by 4 yields the following values of θ which satisfy the original equation:

$$\theta = \frac{\pi}{6} + \frac{n\pi}{4} = \boxed{\frac{\pi(2+3n)}{12}, \text{ where } n \text{ is an integer}}.$$

16. The values of θ in $\mathbb{R}$ that satisfy $\tan 5\theta = 1$ must satisfy

$$5\theta = \frac{\pi}{4} + 2n\pi,\ \frac{5\pi}{4} + 2n\pi \text{, where } n \text{ is an integer.}$$

This reduces to

$$5\theta = \frac{\pi}{4} + n\pi \text{, where } n \text{ is an integer.}$$

So, dividing all values by 5 yields the following values of θ which satisfy the original equation:

$$\theta = \frac{\pi}{20} + \frac{n\pi}{5} = \boxed{\frac{\pi(1+4n)}{20}\text{, where } n \text{ is an integer}}.$$

17. First, observe that $\sec 3\theta = -1$ is equivalent to $\cos 3\theta = -1$. The values of θ in $\mathbb{R}$ for which this is true must satisfy

$$3\theta = \pi + 2n\pi = (2n+1)\pi \text{, where } n \text{ is an integer.}$$

So, dividing all values by 5 yields the following values of θ which satisfy the original equation:

$$\theta = \boxed{\frac{(2n+1)\pi}{3}\text{, where } n \text{ is an integer}}.$$

18. First, observe that $\sec 4\theta = \sqrt{2}$ is equivalent to $\cos 4\theta = \frac{1}{\sqrt{2}}$. The values of θ in $\mathbb{R}$ for which this is true must satisfy

$$4\theta = \frac{\pi}{4} + 2n\pi,\ \frac{7\pi}{4} + 2n\pi \text{, where } n \text{ is an integer.}$$

So, dividing all values by 5 yields the following values of θ in $[0,\pi]$ which satisfy the original equation:

$$\boxed{\theta = \frac{\pi}{16}, \frac{7\pi}{16}, \frac{9\pi}{16}, \frac{15\pi}{16}}$$

19. First, observe that $\csc 3\theta = 1$ is equivalent to $\sin 3\theta = 1$. The values of θ in $\mathbb{R}$ for which this is true must satisfy

$$3\theta = \frac{\pi}{2} + 2n\pi = \frac{(4n+1)\pi}{2} \text{, where } n \text{ is an integer.}$$

So, dividing all values by 5 yields the following values of θ in $[-2\pi, 0]$ which satisfy the original equation:

$$\boxed{\theta = -\frac{\pi}{2}, -\frac{7\pi}{6}, -\frac{11\pi}{6}}$$

20. First, observe that $\csc 6\theta = -\frac{2\sqrt{3}}{3}$ is equivalent to $\sin 6\theta = -\frac{\sqrt{3}}{2}$. The values of θ in $\mathbb{R}$ for which this is true must satisfy

$$6\theta = \frac{4\pi}{3} + 2n\pi,\ \frac{5\pi}{3} + 2n\pi = \frac{(4+6n)\pi}{3},\ \frac{(5+6n)\pi}{3},\text{ where } n \text{ is an integer.}$$

So, dividing all values by 6 yields the following values of θ in $[0,\pi]$ which satisfy the original equation:

$$\boxed{\theta = \frac{2\pi}{9}, \frac{5\pi}{18}}$$

21. First, observe that $2\sin 2\theta = \sqrt{3}$ is equivalent to $\sin 2\theta = \frac{\sqrt{3}}{2}$. The values of θ in $[0,2\pi]$ that satisfy $\sin 2\theta = \frac{\sqrt{3}}{2}$ must satisfy

$$2\theta = \frac{\pi}{3},\ \frac{\pi}{3} + 2\pi,\ \frac{2\pi}{3},\ \frac{2\pi}{3} + 2\pi = \frac{\pi}{3}, \frac{2\pi}{3}, \frac{7\pi}{3}, \frac{8\pi}{3}.$$

So, dividing all values by 2 yields the following values of θ which satisfy the original equation:

$$\theta = \frac{\pi}{6}, \frac{2\pi}{6}, \frac{7\pi}{6}, \frac{8\pi}{6} = \boxed{\frac{\pi}{6}, \frac{\pi}{3}, \frac{7\pi}{6}, \frac{4\pi}{3}}.$$

22. First, observe that $2\cos\left(\frac{\theta}{2}\right) = -\sqrt{2}$ is equivalent to $\cos\left(\frac{\theta}{2}\right) = \frac{-\sqrt{2}}{2}$. The values of θ in $[0,2\pi]$ that satisfy $\cos\left(\frac{\theta}{2}\right) = \frac{-\sqrt{2}}{2}$ must satisfy $\frac{\theta}{2} = \frac{3\pi}{4}$, so that $\theta = \boxed{\frac{3\pi}{2}}$.

23. First, observe that $3\tan 2\theta - \sqrt{3} = 0$ is equivalent to $\tan 2\theta = \frac{\sqrt{3}}{3}$. The values of θ in $[0,2\pi]$ that satisfy $\tan 2\theta = \frac{\sqrt{3}}{3}$ must satisfy

$$2\theta = \frac{\pi}{6},\ \frac{\pi}{6} + 2\pi,\ \frac{7\pi}{6},\ \frac{7\pi}{6} + 2\pi = \frac{\pi}{6}, \frac{7\pi}{6}, \frac{13\pi}{6}, \frac{19\pi}{6}.$$

So, dividing all values by 2 yields the following values of θ which satisfy the original equation:

$$\theta = \boxed{\frac{\pi}{12}, \frac{7\pi}{12}, \frac{13\pi}{12}, \frac{19\pi}{12}}.$$

24. First, observe that $4\tan\left(\frac{\theta}{2}\right)-4=0$ is equivalent to $\tan\left(\frac{\theta}{2}\right)=1$. The values of θ in $[0,2\pi]$ that satisfy $\tan\left(\frac{\theta}{2}\right)=1$ must satisfy $\frac{\theta}{2}=\frac{\pi}{4}$, so that $\theta=\boxed{\frac{\pi}{2}}$.

25. First, observe that $2\cos(2\theta)+1=0$ is equivalent to $\cos(2\theta)=-\frac{1}{2}$. The values of θ in $[0,2\pi]$ that satisfy $\cos(2\theta)=-\frac{1}{2}$ must satisfy

$$2\theta=\frac{2\pi}{3},\ \frac{2\pi}{3}+2\pi,\ \frac{4\pi}{3},\ \frac{4\pi}{3}+2\pi=\frac{2\pi}{3},\frac{4\pi}{3},\frac{8\pi}{3},\frac{10\pi}{3}.$$

So, dividing all values by 2 yields the following values of θ which satisfy the original equation:

$$\theta=\frac{2\pi}{6},\frac{4\pi}{6},\frac{8\pi}{6},\frac{10\pi}{6}=\boxed{\frac{\pi}{3},\frac{2\pi}{3},\frac{4\pi}{3},\frac{5\pi}{3}}$$

26. First, observe that $4\csc(2\theta)+8=0$ is equivalent to $\csc(2\theta)=-2$, which can be further simplified to $\sin(2\theta)=-\frac{1}{2}$. The values of θ in $[0,2\pi]$ that satisfy $\sin(2\theta)=-\frac{1}{2}$ must satisfy

$$2\theta=\frac{7\pi}{6},\ \frac{7\pi}{6}+2\pi,\ \frac{11\pi}{6},\ \frac{11\pi}{6}+2\pi=\frac{7\pi}{6},\frac{11\pi}{6},\frac{19\pi}{6},\frac{23\pi}{6}.$$

So, dividing all values by 2 yields the following values of θ which satisfy the original equation:

$$\theta=\boxed{\frac{7\pi}{12},\frac{11\pi}{12},\frac{19\pi}{12},\frac{23\pi}{12}}$$

27. First, observe that $\sqrt{3}\cot\left(\frac{\theta}{2}\right)-3=0$ is equivalent to

$$\cot\left(\frac{\theta}{2}\right)=\frac{3}{\sqrt{3}}=\frac{1/2}{\sqrt{3}/2}=\frac{\cos\left(\frac{\theta}{2}\right)}{\sin\left(\frac{\theta}{2}\right)}.$$

The value of θ in $[0,2\pi]$ that satisfy this equation must satisfy $\frac{\theta}{2}=\frac{\pi}{3}$, so that $\theta=\boxed{\frac{2\pi}{3}}$.

28. First, observe that $\sqrt{3}\sec(2\theta)+2=0$ is equivalent to $\sec(2\theta)=-\frac{2}{\sqrt{3}}$, which can be further simplified to $\cos(2\theta)=-\frac{\sqrt{3}}{2}$. The values of θ in $[0,2\pi]$ that satisfy $\cos(2\theta)=-\frac{\sqrt{3}}{2}$ must satisfy

$$2\theta=\frac{5\pi}{6},\ \frac{5\pi}{6}+2\pi,\ \frac{7\pi}{6},\ \frac{7\pi}{6}+2\pi=\frac{5\pi}{6},\frac{7\pi}{6},\frac{17\pi}{6},\frac{19\pi}{6}.$$

So, dividing all values by 2 yields the following values of θ which satisfy the original equation:

$$\theta=\boxed{\frac{5\pi}{12},\frac{7\pi}{12},\frac{17\pi}{12},\frac{19\pi}{12}}$$

29. Factoring the left-side of $\tan^2\theta-1=0$ yields the equivalent equation $(\tan\theta-1)(\tan\theta+1)=0$ which is satisfied when either $\tan\theta-1=0$ or $\tan\theta+1=0$. The values of θ in $[0,2\pi]$ that satisfy $\tan\theta=1$ are $\theta=\frac{\pi}{4},\frac{5\pi}{4}$, and those which satisfy $\tan\theta=-1$ are $\theta=\frac{3\pi}{4},\frac{7\pi}{4}$. Thus, the solutions to the original equation are

$$\theta=\boxed{\frac{\pi}{4},\frac{3\pi}{4},\frac{5\pi}{4},\frac{7\pi}{4}}.$$

30. Factoring the left-side of $\sin^2\theta+2\sin\theta+1=0$ yields the equivalent equation $(\sin\theta+1)^2=0$ which is satisfied when $\sin\theta+1=0$. The value of θ in $[0,2\pi]$ that satisfies $\sin\theta=-1$ is $\theta=\boxed{\frac{3\pi}{2}}$.

31. Factoring the left-side of $2\cos^2\theta-\cos\theta=0$ yields the equivalent equation $\cos\theta(2\cos\theta-1)=0$ which is satisfied when either $\cos\theta=0$ or $2\cos\theta-1=0$. The values of θ in $[0,2\pi]$ that satisfy $\cos\theta=0$ are $\theta=\frac{\pi}{2},\frac{3\pi}{2}$, and those which satisfy $\cos\theta=\frac{1}{2}$ are $\theta=\frac{\pi}{3},\frac{5\pi}{3}$. Thus, the solutions to the original equation are

$$\theta=\boxed{\frac{\pi}{2},\frac{3\pi}{2},\frac{\pi}{3},\frac{5\pi}{3}}.$$

32. Factoring the left-side of $\tan^2\theta - \sqrt{3}\tan\theta = 0$ yields the equivalent equation $\tan\theta\left(\tan\theta - \sqrt{3}\right) = 0$ which is satisfied when either $\tan\theta = 0$ or $\tan\theta - \sqrt{3} = 0$. The values of θ in $[0, 2\pi]$ that satisfy $\tan\theta = 0$ are $\theta = 0, \pi, 2\pi$, and those which satisfy $\tan\theta = \sqrt{3} = \dfrac{\sqrt{3}/2}{1/2}$ are $\theta = \dfrac{\pi}{3}, \dfrac{4\pi}{3}$. Thus, the solutions to the original equation are

$$\theta = \boxed{0,\ \pi,\ 2\pi,\ \frac{\pi}{3},\ \frac{4\pi}{3}}.$$

33. Factoring the left-side of $\csc^2\theta + 3\csc\theta + 2 = 0$ yields the equivalent equation $(\csc\theta + 2)(\csc\theta + 1) = 0$ which is satisfied when either $\csc\theta + 2 = 0$ or $\csc\theta + 1 = 0$. The values of θ in $[0, 2\pi]$ that satisfy $\csc\theta = -2$ (or equivalently $\sin\theta = -\frac{1}{2}$) are $\theta = \frac{7\pi}{6}, \frac{11\pi}{6}$, and those which satisfy $\csc\theta = -1$ (or equivalently $\sin\theta = -1$) are $\theta = \frac{3\pi}{2}$. Thus, the solutions to the original equation are $\theta = \boxed{\frac{7\pi}{6}, \frac{11\pi}{6}, \frac{3\pi}{2}}$.

34. Observe that $\cot^2\theta = 1$ is equivalent to $\cot^2\theta - 1 = 0$. Factoring the left-side of this equation yields the equivalent equation $(\cot\theta - 1)(\cot\theta + 1) = 0$ which is satisfied when either $\cot\theta - 1 = 0$ or $\cot\theta + 1 = 0$. The values of θ in $[0, 2\pi]$ that satisfy $\cot\theta = 1$ are $\theta = \frac{\pi}{4}, \frac{5\pi}{4}$ and those which satisfy $\cot\theta = -1$ are $\theta = \frac{3\pi}{4}, \frac{7\pi}{4}$. Thus, the solutions to the original equation are $\theta = \boxed{\frac{\pi}{4}, \frac{3\pi}{4}, \frac{5\pi}{4}, \frac{7\pi}{4}}$.

35. Factoring the left-side of $\sin^2\theta + 2\sin\theta - 3 = 0$ yields the equivalent equation $(\sin\theta + 3)(\sin\theta - 1) = 0$ which is satisfied when either $\sin\theta + 3 = 0$ or $\sin\theta - 1 = 0$. Note that the equation $\sin\theta + 3 = 0$ has no solution since –3 is not in the range of sine. The value of θ in $[0, 2\pi]$ that satisfies $\sin\theta = 1$ is $\theta = \boxed{\frac{\pi}{2}}$.

36. Factoring the left-side of $2\sec^2\theta + \sec\theta - 1 = 0$ yields the equivalent equation $(2\sec\theta - 1)(\sec\theta + 1) = 0$ which is satisfied when either $2\sec\theta - 1 = 0$ or $\sec\theta + 1 = 0$. Note that the equation $\sec\theta = \frac{1}{2}$ (or equivalently $\cos\theta = 2$) has no solution since 2 is not in the range of cosine. The value of θ in $[0, 2\pi]$ that satisfies $\sec\theta = -1$ is $\theta = \boxed{\pi}$.

37. Factoring the left-side of $\sec^2\theta - 1 = 0$ yields the equivalent equation $(\sec\theta - 1)(\sec\theta + 1) = 0$, which is satisfied whenever $\sec\theta = \pm 1$, or equivalently $\cos\theta = \pm 1$. The values of θ in $(0, 2\pi)$ for which this is true are $\theta = \boxed{0, \pi}$.
38. Factoring the left-side of $\csc^2\theta - 1 = 0$ yields the equivalent equation $(\csc\theta - 1)(\csc\theta + 1) = 0$, which is satisfied whenever $\csc\theta = \pm 1$, or equivalently $\sin\theta = \pm 1$. The values of θ in $[0, 2\pi]$ for which this is true are $\theta = \boxed{\frac{\pi}{2}, \frac{3\pi}{2}}$.
39. Factoring the left-side of $\sec^2(2\theta) - \frac{4}{3} = 0$ yields the equivalent equation $\left(\sec 2\theta - \frac{2}{\sqrt{3}}\right)\left(\sec 2\theta + \frac{2}{\sqrt{3}}\right) = 0$, which is satisfied whenever $\sec 2\theta = \pm\frac{2}{\sqrt{3}}$, or equivalently $\cos 2\theta = \pm\frac{\sqrt{3}}{2}$. The values of θ for which this is true must satisfy $$2\theta = \frac{\pi}{6}, \frac{5\pi}{6}, \frac{7\pi}{6}, \frac{11\pi}{6}, \frac{13\pi}{6}, \frac{17\pi}{6}, \frac{19\pi}{6}, \frac{23\pi}{6}$$ So, dividing by 2 then yields the values of θ in $[0, 2\pi]$ for which this is true: $$\boxed{\theta = \frac{\pi}{12}, \frac{5\pi}{12}, \frac{7\pi}{12}, \frac{11\pi}{12}, \frac{13\pi}{12}, \frac{17\pi}{12}, \frac{19\pi}{12}, \frac{23\pi}{12}}$$
40. Factoring the left-side of $\csc^2(2\theta) - 4 = 0$ yields the equivalent equation $(\csc 2\theta - 2)(\csc 2\theta + 2) = 0$, which is satisfied whenever $\csc 2\theta = \pm 2$, or equivalently $\sin 2\theta = \pm\frac{1}{2}$. The values of θ for which this is true must satisfy $$2\theta = \frac{\pi}{6}, \frac{5\pi}{6}, \frac{7\pi}{6}, \frac{11\pi}{6}, \frac{13\pi}{6}, \frac{17\pi}{6}, \frac{19\pi}{6}, \frac{23\pi}{6}$$ So, dividing by 2 then yields the values of θ in $[0, 2\pi]$ for which this is true: $$\boxed{\theta = \frac{\pi}{12}, \frac{5\pi}{12}, \frac{7\pi}{12}, \frac{11\pi}{12}, \frac{13\pi}{12}, \frac{17\pi}{12}, \frac{19\pi}{12}, \frac{23\pi}{12}}$$

41. In order to find all of the values of θ in $\left[0^\circ, 360^\circ\right]$ that satisfy $\sin 2\theta = -0.7843$, we proceed as follows:

Step 1: Find the values of 2θ whose sine is -0.7843.

Indeed, observe that one solution is $\sin^{-1}(-0.7843) \approx -51.655^\circ$, which is in QIV. Since the angles we seek have positive measure, we use the representative $360^\circ - 51.655^\circ \approx 308.345^\circ$. A second solution occurs in QIII, and has value $180^\circ + 51.655^\circ = 231.655^\circ$.

Step 2: Use periodicity to find all values of θ that satisfy the original equation.

Using periodicity with the solutions obtained in Step 1, we see that

$$2\theta = 308.345^\circ,\ 308.345^\circ + 360^\circ,\ 231.655^\circ,\ 231.655^\circ + 360^\circ$$

$$= 308.345^\circ,\ 668.345^\circ,\ 231.655^\circ,\ 591.655^\circ$$

and so, the solutions to the original equation are approximately:

$$\theta \approx \boxed{115.83^\circ,\ 295.83^\circ,\ 154.17^\circ,\ 334.17^\circ}$$

42. In order to find all of the values of θ in $\left[0^\circ, 360^\circ\right]$ that satisfy $\cos 2\theta = 0.5136$, we proceed as follows:

Step 1: Find the values of 2θ whose cosine is 0.5136.

Indeed, observe that one solution is $\cos^{-1}(0.5136) \approx 59.096^\circ$, which is in QI.

A second solution occurs in QIV, and has value $360^\circ - 59.096^\circ = 300.904^\circ$.

Step 2: Use periodicity to find all values of θ that satisfy the original equation.

Using periodicity with the solutions obtained in Step 1, we see that

$$2\theta = 59.096^\circ,\ 59.096^\circ + 360^\circ,\ 300.904^\circ, 300.904^\circ + 360^\circ$$

$$= 59.096^\circ,\ 419.096^\circ,\ 300.904^\circ,\ 660.904^\circ$$

and so, the solutions to the original equation are approximately:

$$\theta \approx \boxed{29.55^\circ,\ 209.55^\circ,\ 150.45^\circ,\ 330.45^\circ}$$

43. In order to find all of the values of θ in $\left[0^\circ, 360^\circ\right]$ that satisfy $\tan\left(\frac{\theta}{2}\right) = -0.2343$, we proceed as follows:

Step 1: Find the values of $\frac{\theta}{2}$ whose tangent is -0.2343.

Indeed, observe that one solution is $\tan^{-1}(-0.2343) \approx -13.187^\circ$, which is in QIV. Since the angles we seek have positive measure, we use the representative $360^\circ - 13.187^\circ \approx 346.813^\circ$. A second solution occurs in QII, and has value $346.813^\circ - 180^\circ = 166.813^\circ$.

Step 2: Use Step 1 to find all values of θ that satisfy the original equation, and exclude any value of θ that satisfies the equation, but lies outside the interval $\left[0^\circ, 360^\circ\right]$.

The solutions obtained in Step 1 are

$$\frac{\theta}{2} = 166.813^\circ,\ 346.813^\circ.$$

When multiplied by 2, the solution corresponding to 346.813° will no longer be in the interval. So, the solution to the original equation in $\left[0^\circ, 360^\circ\right]$ is approximately

$\theta \approx \boxed{333.63^\circ}$.

44. In order to find all of the values of θ in $\left[0^\circ, 360^\circ\right]$ that satisfy $\sec\left(\frac{\theta}{2}\right) = 1.4275$, we proceed as follows:

Step 1: Find the values of $\frac{\theta}{2}$ whose secant is 1.4275.

Indeed, observe that one solution is $\sec^{-1}(1.4275) = \cos^{-1}\left(\frac{1}{1.4275}\right) \approx 45.531^\circ$, which is in QI. A second solution occurs in QIV, and has value $360^\circ - 45.531^\circ \approx 314.469^\circ$.

Step 2: Use Step 1 to find all values of θ that satisfy the original equation, and exclude any value of θ that satisfies the equation, but lies outside the interval $\left[0^\circ, 360^\circ\right]$.

The solutions obtained in Step 1 are

$$\frac{\theta}{2} = 45.531^\circ,\ 314.469^\circ.$$

When multiplied by 2, the solution corresponding to 314.469° will no longer be in the interval. So, the solution to the original equation in $\left[0^\circ, 360^\circ\right]$ is approximately

$\theta \approx \boxed{91.06^\circ}$.

45. In order to find all of the values of θ in $\left[0^\circ, 360^\circ\right]$ that satisfy $5\cot\theta - 9 = 0$, we proceed as follows:

Step 1: First, observe that the equation is equivalent to $\cot\theta = \frac{9}{5}$.

Step 2: Find the values of θ whose cotangent is $\frac{9}{5}$.

Indeed, observe that one solution is $\cot^{-1}\left(\frac{9}{5}\right) = \tan^{-1}\left(\frac{1}{9/5}\right) = \tan^{-1}\left(\frac{5}{9}\right) \approx 29.0546^\circ$,

which is in QI. A second solution occurs in QIII, and has value $29.0546^\circ + 180^\circ = 209.0546^\circ$. Since the input of cotangent is simply θ, and not some multiple thereof, we conclude that the solutions to the original equation are approximately $\theta \approx \boxed{29.05^\circ,\ 209.05^\circ}$.

46. In order to find all of the values of θ in $\left[0^\circ, 360^\circ\right]$ that satisfy $5\sec\theta + 6 = 0$, we proceed as follows:

Step 1: First, observe that the equation is equivalent to $\sec\theta = -\frac{6}{5}$.

Step 2: Find the values of θ whose secant is $-\frac{6}{5}$.

Indeed, note that one solution is

$$\sec^{-1}\left(-\frac{6}{5}\right) = \cos^{-1}\left(\frac{1}{-6/5}\right) = \cos^{-1}\left(-\frac{5}{6}\right) \approx 146.4427^\circ,$$

which is in QII. A second solution occurs in QIII, and has value $360^\circ - 146.4427^\circ = 213.5573^\circ$. Since the input of secant is simply θ, and not some multiple thereof, we conclude that the solutions to the original equation are approximately $\theta \approx \boxed{146.44^\circ,\ 213.56^\circ}$.

47. In order to find all of the values of θ in $\left[0^\circ, 360^\circ\right]$ that satisfy $4\sin\theta + \sqrt{2} = 0$, we proceed as follows:

Step 1: First, observe that the equation is equivalent to $\sin\theta = -\frac{\sqrt{2}}{4}$.

Step 2: Find the values of θ whose sine is $-\frac{\sqrt{2}}{4}$.

Indeed, observe that one solution is $\sin^{-1}\left(-\frac{\sqrt{2}}{4}\right) \approx -20.7048^\circ$, which is in QIV.

Since the angles we seek have positive measure, we use the representative $360^\circ - 20.7048^\circ \approx 339.30^\circ$. A second solution occurs in QIII, and has value $20.7048^\circ + 180^\circ = 200.70^\circ$. Since the input of sine is simply θ, and not some multiple thereof, we conclude that the solutions to the original equation are approximately $\theta \approx \boxed{200.70^\circ,\ 339.30^\circ}$.

48. In order to find all of the values of θ in $\left[0^\circ, 360^\circ\right]$ that satisfy $3\cos\theta - \sqrt{5} = 0$, we proceed as follows:

Step 1: First, observe that the equation is equivalent to $\cos\theta = \frac{\sqrt{5}}{3}$.

Step 2: Find the values of θ whose cosine is $\frac{\sqrt{5}}{3}$.

Indeed, observe that one solution is $\cos^{-1}\left(\frac{\sqrt{5}}{3}\right) \approx 41.8103^\circ$, which is in QI. A second solution occurs in QIV, and has value $360^\circ - 41.8103^\circ = 318.19^\circ$. Since the input of cosine is simply θ, and not some multiple thereof, we conclude that the solutions to the original equation are approximately $\theta \approx \boxed{41.81^\circ,\ 318.19^\circ}$.

49. In order to find all of the values of θ in $\left[0^\circ, 360^\circ\right]$ that satisfy $4\cos^2\theta + 5\cos\theta - 6 = 0$, we proceed as follows:

Step 1: Simplify the equation algebraically.
Factoring the left-side of the equation yields:

$$(4\cos\theta - 3)(\cos\theta + 2) = 0$$

$$4\cos\theta - 3 = 0 \quad \text{or} \quad \cos\theta + 2 = 0$$

$$\cos\theta = \frac{3}{4} \quad \text{or} \quad \underbrace{\cos\theta = -2}_{\text{No solution}}$$

Step 2: Find the values of θ whose cosine is $\frac{3}{4}$.

Indeed, observe that one solution is $\cos^{-1}\left(\frac{3}{4}\right) \approx 41.4096^\circ$, which is in QI. A second solution occurs in QIV, and has value $360^\circ - 41.4096^\circ = 318.59^\circ$. Since the input of cosine is simply θ, and not some multiple thereof, we conclude that the solutions to the original equation are approximately $\theta \approx \boxed{41.41^\circ,\ 318.59^\circ}$.

50. In order to find all of the values of θ in $\left[0^\circ, 360^\circ\right]$ that satisfy $6\sin^2\theta - 13\sin\theta - 5 = 0$, we proceed as follows:

Step 1: Simplify the equation algebraically.
Factoring the left-side of the equation yields:

$$(3\sin\theta + 1)(2\sin\theta - 5) = 0$$

$$3\sin\theta + 1 = 0 \quad \text{or} \quad 2\sin\theta - 5 = 0$$

$$\sin\theta = -\frac{1}{3} \quad \text{or} \quad \underbrace{\sin\theta = \frac{5}{2}}_{\text{No solution}}$$

Step 2: Find the values of θ whose sine is $-\frac{1}{3}$.

Indeed, observe that one solution is $\sin^{-1}\left(-\frac{1}{3}\right) \approx -19.4712^\circ$, which is in QIV. Since the angles we seek have positive measure, we use the representative $360^\circ - 19.4712^\circ \approx 340.53^\circ$. A second solution occurs in QIII, and has value $180^\circ + 19.47^\circ = 199.47^\circ$. Since the input of cosine is simply θ, and not some multiple thereof, we conclude that the solutions to the original equation are approximately $\theta \approx \boxed{199.47^\circ,\ 340.53^\circ}$.

51. In order to find all of the values of θ in $\left[0^\circ, 360^\circ\right]$ that satisfy $6\tan^2\theta - \tan\theta - 12 = 0$, we proceed as follows:

Step 1: Simplify the equation algebraically.
Factoring the left-side of the equation yields:

$$(3\tan\theta + 4)(2\tan\theta - 3) = 0$$

$$3\tan\theta + 4 = 0 \quad \text{or} \quad 2\tan\theta - 3 = 0$$

$$\tan\theta = -\frac{4}{3} \quad \text{or} \quad \tan\theta = \frac{3}{2}$$

Step 2: Solve $\tan\theta = -\frac{4}{3}$.

To do so, we must find the values of θ whose tangent is $-\frac{4}{3}$.

Indeed, observe that one solution is $\tan^{-1}\left(-\frac{4}{3}\right) \approx -53.13^\circ$, which is in QIV. Since the angles we seek have positive measure, we use the representative $360^\circ - 53.13^\circ \approx 306.87^\circ$. A second solution occurs in QII, and has value $180^\circ - 53.13^\circ = 126.87^\circ$. Since the input of tangent is simply θ, and not some multiple thereof, we conclude that the solutions to this equation are approximately $\theta \approx 126.87^\circ, 306.87^\circ$.

Step 3: Solve $\tan\theta = \frac{3}{2}$

To do so, we must find the values of θ whose tangent is $\frac{3}{2}$.

Indeed, observe that one solution is $\tan^{-1}\left(\frac{3}{2}\right) \approx 56.31^\circ$, which is in QI. A second solution occurs in QIII, and has value $180^\circ + 56.31^\circ = 236.31^\circ$. Since the input of tangent is simply θ, and not some multiple thereof, we conclude that the solutions to this equation are approximately $\theta \approx 56.31^\circ, 236.31^\circ$.

Step 4: Conclude that the solutions to the original equation are

$$\theta \approx \boxed{56.31^\circ, 126.87^\circ, 236.31^\circ, 306.87^\circ}.$$

52. In order to find all of the values of θ in $\left[0^\circ, 360^\circ\right]$ that satisfy $6\sec^2\theta - 7\sec\theta - 20 = 0$, we proceed as follows:

Step 1: Simplify the equation algebraically.
Factoring the left-side of the equation yields:

$$(2\sec\theta - 5)(3\sec\theta + 4) = 0$$

$$2\sec\theta - 5 = 0 \quad \text{or} \quad 3\sec\theta + 4 = 0$$

$$\sec\theta = \frac{5}{2} \quad \text{or} \quad \sec\theta = -\frac{4}{3}$$

Step 2: Solve $\sec\theta = \frac{5}{2}$, or equivalently $\cos\theta = \frac{2}{5}$.

To do so, we must find the values of θ whose cosine is $\frac{2}{5}$.

Indeed, observe that one solution is $\cos^{-1}\left(\frac{2}{5}\right) \approx 66.42^\circ$, which is in QI. A second solution occurs in QIV, and has value $360^\circ - 66.42^\circ = 293.58^\circ$. Since the input of cosine is simply θ, and not some multiple thereof, we conclude that the solutions to this equation are approximately $\theta \approx 66.42^\circ, 293.58^\circ$.

Step 3: Solve $\sec\theta = -\frac{4}{3}$, or equivalently, $\cos\theta = -\frac{3}{4}$.

To do so, we must find the values of θ whose cosine is $-\frac{3}{4}$.

Indeed, observe that one solution is $\cos^{-1}\left(-\frac{3}{4}\right) \approx 138.59^\circ$, which is in QII. A second

solution occurs in QIII, and has value $360^\circ - 138.59^\circ = 221.41^\circ$. Since the input of cosine is simply θ, and not some multiple thereof, we conclude that the solutions to this equation are approximately $\theta \approx 138.59^\circ, 221.41^\circ$.

Step 4: Conclude that the solutions to the original equation are

$$\theta \approx \boxed{66.42^\circ, 138.59^\circ, 221.41^\circ, 293.58^\circ}.$$

53. In order to find all of the values of θ in $\left[0^\circ, 360^\circ\right]$ that satisfy $15\sin^2 2\theta + \sin 2\theta - 2 = 0$, we proceed as follows:

Step 1: Simplify the equation algebraically. Factoring the left-side of the equation yields:

$$(5\sin 2\theta + 2)(3\sin 2\theta - 1) = 0$$

$$5\sin 2\theta + 2 = 0 \quad \text{or} \quad 3\sin 2\theta - 1 = 0$$

$$\sin 2\theta = -\frac{2}{5} \quad \text{or} \quad \sin 2\theta = \frac{1}{3}$$

Step 2: Solve $\sin 2\theta = -\frac{2}{5}$.

Step a: Find the values of 2θ whose sine is $-\frac{2}{5}$.

Indeed, observe that one solution is $\sin^{-1}\left(-\frac{2}{5}\right) \approx -23.578^\circ$, which is in QIV. Since the angles we seek have positive measure, we use the representative $360^\circ - 23.578^\circ \approx 336.42^\circ$. A second solution occurs in QIII, namely $180^\circ + 23.578^\circ = 203.578^\circ$.

Step b: Use periodicity to find all values of θ that satisfy $\sin 2\theta = -\frac{2}{5}$.

Using periodicity with the solutions obtained in Step a, we see that

$$2\theta = 203.578^\circ,\ 203.578^\circ + 360^\circ,\ 336.42^\circ,\ 336.42^\circ + 360^\circ$$

$$= 203.578^\circ,\ 563.578^\circ,\ 336.42^\circ,\ 696.42^\circ$$

and so, the solutions are approximately: $\theta \approx 101.79^\circ,\ 281.79^\circ,\ 168.21^\circ,\ 348.21^\circ$

Step 3: Solve $\sin 2\theta = \frac{1}{3}$.

Step a: Find the values of 2θ whose sine is $\frac{1}{3}$.

Indeed, observe that one solution is $\sin^{-1}\left(\frac{1}{3}\right) \approx 19.4712^\circ$, which is in QI.

A second solution occurs in QII, and has value $180^\circ - 19.4712^\circ = 160.528^\circ$.

Step b: Use periodicity to find all values of θ that satisfy $\sin 2\theta = \frac{1}{3}$.

Using periodicity with the solutions obtained in Step a, we see that

$$2\theta = 19.4712^\circ,\ 19.4712^\circ + 360^\circ,\ 160.528^\circ,\ 160.528^\circ + 360^\circ$$

$$= 19.4712^\circ,\ 160.528^\circ,\ 379.4712^\circ,\ 520.528^\circ$$

and so, the solutions are approximately: $\theta \approx 9.74^\circ,\ 189.74^\circ,\ 80.26^\circ,\ 260.26^\circ$

Step 4: Conclude that the solutions to the original equation are

$$\theta \approx \boxed{101.79^\circ,\ 281.79^\circ,\ 168.21^\circ,\ 348.21^\circ,\ 9.74^\circ,\ 189.74^\circ,\ 80.26^\circ,\ 260.26^\circ}.$$

54. In order to find all of the values of θ in $\left[0°,360°\right]$ that satisfy $12\cos^2\left(\frac{\theta}{2}\right)-13\cos\left(\frac{\theta}{2}\right)+3=0$, we proceed as follows:

Step 1: Simplify the equation algebraically.

Factoring the left-side of the equation yields:

$$\left(3\cos\left(\frac{\theta}{2}\right)-1\right)\left(4\cos\left(\frac{\theta}{2}\right)-3\right)=0$$

$$3\cos\left(\frac{\theta}{2}\right)-1=0 \text{ or } 4\cos\left(\frac{\theta}{2}\right)-3=0 \text{ or equivalently } \cos\left(\frac{\theta}{2}\right)=\frac{1}{3} \text{ or } \cos\left(\frac{\theta}{2}\right)=\frac{3}{4}$$

Step 2: Solve $\cos\left(\theta/2\right)=\frac{1}{3}$.

Step a: Find the values of $\frac{\theta}{2}$ whose cosine is $\frac{1}{3}$.

Indeed, observe that one solution is $\cos^{-1}\left(\frac{1}{3}\right)\approx 70.5288°$, which is in QI. A second solution occurs in QIV, and has value $360°-70.5288°=289.47°$.

Step b: Use Step a to find all values of θ that satisfy the original equation, and exclude any value of θ that satisfies the equation, but lies outside $\left[0°,360°\right]$.

The solutions obtained is Step a are

$$\theta/2=70.5288°,\ 289.47°.$$

When multiplied by 2, the solution corresponding to $289.47°$ will no longer be in the interval. So, the solution to the equation in $\left[0°,360°\right]$ is approximately $\theta\approx 141.06°$.

Step 3: Solve $\cos\left(\theta/2\right)=\frac{3}{4}$.

Step a: Find the values of $\theta/2$ whose cosine is $\frac{3}{4}$.

Indeed, observe that one solution is $\cos^{-1}\left(\frac{3}{4}\right)\approx 41.4096°$, which is in QI. A second solution occurs in QIV, and has value $360°-41.4096°=318.59°$.

Step b: Use Step a to find all values of θ that satisfy the original equation, and exclude any value of θ that satisfies the equation, but lies outside $\left[0°,360°\right]$.

The solutions obtained is Step a are $\theta/2=41.4096°,\ 318.59°$.

When multiplied by 2, the solution corresponding to $318.59°$ will no longer be in the interval. So, the solution to the original equation in $\left[0°,360°\right]$ is approximately

$\boxed{\theta\approx 82.82°}$.

Step 4: Conclude that the solutions to the original equation are $\theta\approx\boxed{82.82°,\ 141.06°}$.

55. In order to find all of the values of θ in $\left[0°,360°\right]$ that satisfy $\cos^2\theta - 6\cos\theta + 1 = 0$, we proceed as follows:

Step 1: Simplify the equation algebraically.

Since the left-side does not factor nicely, we apply the quadratic formula (treating $\cos\theta$ as the variable): $\cos\theta = \dfrac{-(-6)\pm\sqrt{(-6)^2-4(1)(1)}}{2(1)} = \dfrac{6\pm4\sqrt{2}}{2} = 3\pm2\sqrt{2}$

So, θ is a solution to the original equation if $\underbrace{\cos\theta = 3+2\sqrt{2}}_{\text{No solution since } 3+2\sqrt{2}>1}$ or $\cos\theta = 3-2\sqrt{2}$.

Step 2: Find the values of θ whose cosine is $3-2\sqrt{2}$.

Indeed, observe that one solution is $\cos^{-1}\left(3-2\sqrt{2}\right) \approx 80.1207°$, which is in QI. A second solution occurs in QIV, and has value $360° - 80.1207° = 279.88°$. Since the input of cosine is simply θ, and not some multiple thereof, we conclude that the solutions to this equation are approximately $\theta \approx 80.12°,\ 279.88°$.

Step 3: Conclude that the solutions to the original equation are $\theta \approx \boxed{80.12°,\ 279.88°}$.

56. In order to find all of the values of θ in $\left[0°,360°\right]$ that satisfy $\sin^2\theta + 3\sin\theta - 3 = 0$, we proceed as follows:

Step 1: Simplify the equation algebraically.

Since the left-side does not factor nicely, we apply the quadratic formula (treating $\cos\theta$ as the variable): $\sin\theta = \dfrac{-3\pm\sqrt{3^2-4(1)(-3)}}{2(1)} = \dfrac{-3\pm\sqrt{21}}{2}$

So, θ is a solution to the original equation if

$\underbrace{\sin\theta = \dfrac{-3-\sqrt{21}}{2}}_{\text{No solution since } \frac{-3-\sqrt{21}}{2}<-1}$ or $\sin\theta = \dfrac{-3+\sqrt{21}}{2}$.

Step 2: Find the values of θ whose sine is $\frac{-3+\sqrt{21}}{2}$.

Indeed, observe that one solution is $\sin^{-1}\left(\frac{-3+\sqrt{21}}{2}\right) \approx 52.306^\circ$, which is in QI. A second solution occurs in QII, and has value $180^\circ - 52.306^\circ = 127.69^\circ$. Since the input of sine is simply θ, and not some multiple thereof, we conclude that the solutions to this equation are approximately $\theta \approx 52.306^\circ, 127.69^\circ$.

Step 3: Conclude that the solutions to the original equation are $\theta \approx \boxed{52.306^\circ, 127.69^\circ}$.

57. In order to find all of the values of θ in $\left[0^\circ, 360^\circ\right]$ that satisfy $2\tan^2\theta - \tan\theta - 7 = 0$, we proceed as follows:

Step 1: Simplify the equation algebraically.

Since the left-side does not factor nicely, we apply the quadratic formula (treating $\cos\theta$ as the variable):

$$\tan\theta = \frac{-(-1) \pm \sqrt{(-1)^2 - 4(2)(-7)}}{2(2)} = \frac{1 \pm \sqrt{57}}{4}$$

So, θ is a solution to the original equation if either

$$\tan\theta = \frac{1-\sqrt{57}}{4} \quad \text{or} \quad \tan\theta = \frac{1+\sqrt{57}}{4}.$$

Step 2: Solve $\tan\theta = \frac{1-\sqrt{57}}{4}$.

To do so, we must find the values of θ whose tangent is $\frac{1-\sqrt{57}}{4}$.

Indeed, observe that one solution is $\tan^{-1}\left(\frac{1-\sqrt{57}}{4}\right) \approx -58.587^\circ$, which is in QIV.

Since the angles we seek have positive measure, we use the representative $360^\circ - 58.587^\circ \approx 301.41^\circ$. A second solution occurs in QII, and has value $180^\circ - 58.587^\circ = 121.41^\circ$. Since the input of tangent is simply θ, and not some multiple thereof, we conclude that the solutions to this equation are approximately $\theta \approx 121.41^\circ, 301.41^\circ$.

Step 3: Solve $\tan\theta = \frac{1+\sqrt{57}}{4}$

To do so, we must find the values of θ whose tangent is $\frac{1+\sqrt{57}}{4}$.

Indeed, observe that one solution is $\tan^{-1}\left(\frac{1+\sqrt{57}}{4}\right) \approx 64.93^\circ$, which is in QI. A second solution occurs in QIII, and has value $180^\circ + 64.93^\circ = 244.93^\circ$. Since the input of tangent is simply θ, and not some multiple thereof, we conclude that the solutions to this equation are approximately $\theta \approx 64.93^\circ,\ 244.93^\circ$.

Step 4: Conclude that the solutions to the original equation are

$$\theta \approx \boxed{64.93^\circ,\ 121.41^\circ,\ 244.93^\circ,\ 301.41^\circ}.$$

58. In order to find all of the values of θ in $\left[0^\circ, 360^\circ\right]$ that satisfy $3\cot^2\theta + 2\cot\theta - 4 = 0$, we proceed as follows:

Step 1: Simplify the equation algebraically.
Since the left-side does not factor nicely, we apply the quadratic formula (treating $\cos\theta$ as the variable):

$$\cot\theta = \frac{-2 \pm \sqrt{2^2 - 4(3)(-4)}}{2(3)} = \frac{-2 \pm \sqrt{52}}{6} = \frac{-1 \pm \sqrt{13}}{3}$$

So, θ is a solution to the original equation if either

$$\cot\theta = \frac{-1-\sqrt{13}}{3} \quad \text{or} \quad \cot\theta = \frac{-1+\sqrt{13}}{3}.$$

Step 2: Solve $\cot\theta = \frac{-1-\sqrt{13}}{3}$.

To do so, we must find the values of θ whose cotangent is $\frac{-1-\sqrt{13}}{3}$.

Indeed, observe that one solution is

$$\cot^{-1}\left(\frac{-1-\sqrt{13}}{3}\right) = 180^\circ + \tan^{-1}\left(\frac{3}{-1-\sqrt{13}}\right) \approx 146.92^\circ, \text{ which is in QII.}$$

A second solution occurs in QIV, and has value $180^\circ + 146.92^\circ = 326.92^\circ$. Since the input of cotangent is simply θ, and not some multiple thereof, we conclude that the solutions to this equation are approximately $\theta \approx 146.92^\circ,\ 326.92^\circ$.

Step 3: Solve $\cot\theta = \dfrac{-1+\sqrt{13}}{3}$

To do so, we must find the values of θ whose cotangent is $\dfrac{-1+\sqrt{13}}{3}$.

Indeed, observe that one solution is $\cot^{-1}\left(\dfrac{-1+\sqrt{13}}{3}\right) = \tan^{-1}\left(\dfrac{3}{-1+\sqrt{13}}\right) \approx 49.025^\circ$, which is in QI. A second solution occurs in QIII, and has value $180^\circ + 49.025^\circ = 229.025^\circ$. Since the input of cotangent is simply θ, and not some multiple thereof, we conclude that the solutions to this equation are approximately $\theta \approx 49.03^\circ, 229.03^\circ$.

Step 4: Conclude that the solutions to the original equation are

$$\theta \approx \boxed{49.03^\circ, 146.92^\circ, 229.03^\circ, 326.92^\circ}.$$

59. In order to find all of the values of θ in $\left[0^\circ, 360^\circ\right]$ that satisfy $\csc^2(3\theta) - 2 = 0$, we proceed as follows:

Step 1: Simplify the equation algebraically.
Factoring the left-side of $\csc^2(3\theta) - 2 = 0$ yields the equivalent equation $\left(\csc 3\theta - \sqrt{2}\right)\left(\csc 3\theta + \sqrt{2}\right) = 0$, which is satisfied whenever $\csc 3\theta = \pm\sqrt{2}$, or equivalently $\sin 3\theta = \pm\dfrac{\sqrt{2}}{2}$. The values of θ for which this is true must satisfy

$$3\theta = \frac{\pi}{4} + 2n\pi,\ \frac{3\pi}{4} + 2n\pi,\ \frac{5\pi}{4} + 2n\pi,\ \frac{7\pi}{4} + 2n\pi, \text{ where } n \text{ is an integer,}$$

which reduces to

$$3\theta = \frac{\pi}{4} + \frac{n\pi}{2}, \text{ where } n \text{ is an integer.}$$

So, dividing by 3 then yields the values of θ in $\left[0, 2\pi\right]$ for which this is true.

Step 2: Convert the values obtained in Step 1 to degrees.

$$\boxed{\theta = 15^\circ, 45^\circ, 75^\circ, 105^\circ, 135^\circ, 165^\circ, 195^\circ, 225^\circ, 255^\circ, 285^\circ, 315^\circ, 345^\circ}$$

60. In order to find all of the values of θ in $\left[0^\circ, 360^\circ\right]$ that satisfy $\sec^2\left(\frac{\theta}{2}\right)-2=0$, we proceed as follows:

Step 1: Simplify the equation algebraically.
Factoring the left-side of $\sec^2\left(\frac{\theta}{2}\right)-2=0$ yields the equivalent equation $\left(\sec\left(\frac{\theta}{2}\right)-\sqrt{2}\right)\left(\sec\left(\frac{\theta}{2}\right)+\sqrt{2}\right)=0$, which is satisfied whenever $\sec\left(\frac{\theta}{2}\right)=\pm\sqrt{2}$, or equivalently $\cos\left(\frac{\theta}{2}\right)=\pm\frac{\sqrt{2}}{2}$. The values of θ for which this is true must satisfy

$$\frac{\theta}{2}=\frac{\pi}{4}+2n\pi,\ \frac{3\pi}{4}+2n\pi,\ \frac{5\pi}{4}+2n\pi,\ \frac{7\pi}{4}+2n\pi, \text{ where } n \text{ is an integer,}$$

which reduces to

$$\theta=\frac{\pi}{2}+4n\pi,\ \frac{3\pi}{2}+4n\pi, \text{ where } n \text{ is an integer.}$$

Step 2: Convert the appropriate values obtained in Step 1 to degrees.

$$\boxed{\theta=90^\circ,\ 270^\circ}$$

61. By inspection, the values of x in $\left[0, 2\pi\right]$ that satisfy the equation $\sin x=\cos x$ are $x=\boxed{\frac{\pi}{4}, \frac{5\pi}{4}}$.

62. By inspection, the values of x in $\left[0, 2\pi\right]$ that satisfy the equation $\sin x=-\cos x$ are $x=\boxed{\frac{3\pi}{4}, \frac{7\pi}{4}}$.

63. Observe that

$$\sec x+\cos x=-2$$
$$\frac{1}{\cos x}+\cos x=-2$$
$$1+\cos^2 x=-2\cos x$$
$$\cos^2 x+2\cos x+1=0$$
$$(\cos x+1)^2=0$$
$$\cos x+1=0$$
$$\cos x=-1$$

The value of x in $\left[0, 2\pi\right]$ that satisfies the equation $\cos x=-1$ is $x=\boxed{\pi}$. Substituting this value into the original equation shows that it is, in fact, a solution to the original equation.

64. Observe that

$$\sin x+\csc x=2$$
$$\sin x+\frac{1}{\sin x}=2$$
$$\sin^2 x+1=2\sin x$$
$$\sin^2 x-2\sin x+1=0$$
$$(\sin x-1)^2=0$$
$$\sin x-1=0$$
$$\sin x=1$$

The value of x in $\left[0, 2\pi\right]$ that satisfies the equation $\sin x=1$ is $x=\boxed{\pi/2}$. Substituting this value into the original equation shows that it is, in fact, a solution to the original equation.

65. Observe that

$$\sec x - \tan x = \frac{\sqrt{3}}{3}$$

$$\frac{1}{\cos x} - \frac{\sin x}{\cos x} = \frac{\sqrt{3}}{3}$$

$$\frac{1-\sin x}{\cos x} = \frac{\sqrt{3}}{3}$$

$$\frac{(1-\sin x)^2}{\cos^2 x} = \frac{1}{3}$$

$$\frac{(1-\sin x)(1-\sin x)}{1-\sin^2 x} = \frac{1}{3}$$

$$\frac{\cancel{(1-\sin x)}(1-\sin x)}{\cancel{(1-\sin x)}(1+\sin x)} = \frac{1}{3}$$

$$3(1-\sin x) = 1+\sin x$$

$$4\sin x = 2$$

$$\sin x = \frac{1}{2}$$

The values of x in $[0,2\pi]$ that satisfy the equation $\sin x = \frac{1}{2}$ are $x = \boxed{\frac{\pi}{6}, \frac{5\pi}{6}}$.

Substituting these values into the original equation shows that while $\frac{\pi}{6}$ is a solution, $\frac{5\pi}{6}$ is extraneous. Indeed, note that

$$\sec\left(\frac{5\pi}{6}\right) - \tan\left(\frac{5\pi}{6}\right) = \frac{1}{-\sqrt{3}/2} - \frac{1/2}{-\sqrt{3}/2}$$

$$= -\frac{\sqrt{3}}{3} \neq \frac{\sqrt{3}}{3}$$

So, the only solution is $\boxed{\frac{\pi}{6}}$.

66. Observe that

$$\sec x + \tan x = 1$$

$$\frac{1}{\cos x} + \frac{\sin x}{\cos x} = 1$$

$$\frac{1+\sin x}{\cos x} = 1$$

$$\frac{(1+\sin x)^2}{\cos^2 x} = 1$$

$$\frac{(1+\sin x)(1+\sin x)}{1-\sin^2 x} = 1$$

$$\frac{\cancel{(1+\sin x)}(1+\sin x)}{\cancel{(1+\sin x)}(1-\sin x)} = 1$$

$$1-\sin x = 1+\sin x$$

$$2\sin x = 0$$

$$\sin x = 0$$

The values of x in $[0,2\pi]$ that satisfy the equation $\sin x = 0$ are $x = \boxed{0, \pi, 2\pi}$.

Substituting these values into the original equation shows that while $0, 2\pi$ are solutions, π is extraneous. Indeed, note that

$$\sec(\pi) + \tan(\pi) = \frac{1}{-1} + \frac{0}{-1} = -1 \neq 1$$

So, the only solutions are $\boxed{0, 2\pi}$.

67. Observe that

$$\csc x+\cot x=\sqrt{3}$$
$$\frac{1}{\sin x}+\frac{\cos x}{\sin x}=\sqrt{3}$$
$$\frac{1+\cos x}{\sin x}=\sqrt{3}$$
$$\frac{(1+\cos x)^2}{\sin^2 x}=3$$
$$\frac{(1+\cos x)(1+\cos x)}{1-\cos^2 x}=3$$
$$\frac{\cancel{(1+\cos x)}\,(1+\cos x)}{\cancel{(1+\cos x)}\,(1-\cos x)}=3$$
$$3(1-\cos x)=1+\cos x$$
$$3-3\cos x=1+\cos x$$
$$4\cos x=2$$
$$\cos x=\frac{1}{2}$$

The values of x in $[0,2\pi]$ that satisfy the equation $\cos x=\frac{1}{2}$ are $x=\frac{\pi}{3},\frac{5\pi}{3}$. Substituting this value into the original equation shows that while $\frac{\pi}{3}$ is a solution, $\frac{5\pi}{3}$ is extraneous. Indeed, note that

$$\csc\left(\frac{5\pi}{3}\right)+\cot\left(\frac{5\pi}{3}\right)=\frac{1}{-\sqrt{3}/2}+\frac{1/2}{-\sqrt{3}/2}$$
$$=-\frac{3}{\sqrt{3}}=-\sqrt{3}\neq\sqrt{3}$$

So, the only solution is $\boxed{\frac{\pi}{3}}$.

68. Observe that

$$\csc x-\cot x=\frac{\sqrt{3}}{3}$$
$$\frac{1}{\sin x}-\frac{\cos x}{\sin x}=\frac{\sqrt{3}}{3}$$
$$\frac{1-\cos x}{\sin x}=\frac{\sqrt{3}}{3}$$
$$\frac{(1-\cos x)^2}{\sin^2 x}=\frac{1}{3}$$
$$\frac{(1-\cos x)(1-\cos x)}{1-\cos^2 x}=\frac{1}{3}$$
$$\frac{\cancel{(1-\cos x)}\,(1-\cos x)}{\cancel{(1-\cos x)}\,(1+\cos x)}=\frac{1}{3}$$
$$3(1-\cos x)=1+\cos x$$
$$3-3\cos x=1+\cos x$$
$$4\cos x=2$$
$$\cos x=\frac{1}{2}$$

The values of x in $[0,2\pi]$ that satisfy the equation $\cos x=\frac{1}{2}$ are $x=\frac{\pi}{3},\frac{5\pi}{3}$. Substituting this value into the original equation shows that while $\frac{5\pi}{3}$ is a solution, $\frac{\pi}{3}$ is extraneous. Indeed, note that

$$\csc\left(\frac{\pi}{3}\right)-\cot\left(\frac{\pi}{3}\right)=\frac{1}{\sqrt{3}/2}-\frac{1/2}{\sqrt{3}/2}$$
$$=-\frac{\sqrt{3}}{3}\neq\frac{\sqrt{3}}{3}$$

So, the only solution is $\boxed{\frac{5\pi}{3}}$.

69. Observe that

$$2\sin x - \csc x = 0$$
$$2\sin x - \frac{1}{\sin x} = 0$$
$$\frac{2\sin^2 x - 1}{\sin x} = 0$$
$$2\sin^2 x - 1 = 0$$
$$\sin^2 x = \frac{1}{2}$$
$$\sin x = -\frac{1}{\sqrt{2}} \text{ or } \sin x = \frac{1}{\sqrt{2}}$$

The solutions to these equations in $[0, 2\pi]$ are $x = \boxed{\frac{\pi}{4}, \frac{3\pi}{4}, \frac{5\pi}{4}, \frac{7\pi}{4}}$.
Substituting these into the original equation shows that they are all, in fact, solutions to the original equation.

70. Observe that

$$2\sin x + \csc x = 3$$
$$2\sin x + \frac{1}{\sin x} = 3$$
$$\frac{2\sin^2 x + 1}{\sin x} = 3$$
$$2\sin^2 x + 1 = 3\sin x$$
$$2\sin^2 x - 3\sin x + 1 = 0$$
$$(2\sin x - 1)(\sin x - 1) = 0$$
$$2\sin x - 1 = 0 \text{ or } \sin x - 1 = 0$$
$$\sin x = \frac{1}{2} \text{ or } \sin x = 1$$

The solutions to these equations in $[0, 2\pi]$ are $x = \boxed{\frac{\pi}{6}, \frac{5\pi}{6}, \frac{\pi}{2}}$. Substituting these into the original equation shows that they are all, in fact, solutions to the original equation.

71. Observe that

$$\sin 2x = 4\cos x$$
$$2\sin x\cos x = 4\cos x$$
$$2\cos x(\sin x - 2) = 0$$
$$\cos x = 0 \text{ or } \sin x - 2 = 0$$
$$\cos x = 0 \text{ or } \underbrace{\sin x = 2}_{\text{No solution}}$$

The solutions to these equations in $[0, 2\pi]$ are $x = \boxed{\frac{\pi}{2}, \frac{3\pi}{2}}$. Substituting these into the original equation shows that they are all, in fact, solutions to the original equation.

72. Observe that

$$\sin 2x = \sqrt{3}\sin x$$
$$2\sin x\cos x = \sqrt{3}\sin x$$
$$\sin x\left(2\cos x - \sqrt{3}\right) = 0$$
$$\sin x = 0 \text{ or } 2\cos x - \sqrt{3} = 0$$
$$\sin x = 0 \text{ or } \cos x = \frac{\sqrt{3}}{2}$$

The solutions to these equations in $[0, 2\pi]$ are $x = \boxed{0, \pi, 2\pi, \frac{\pi}{6}, \frac{11\pi}{6}}$. Substituting these into the original equation shows that they are all, in fact, solutions to the original equation.

73. Observe that

$$\sqrt{2}\sin x = \tan x$$
$$\sqrt{2}\sin x = \frac{\sin x}{\cos x}$$
$$\sqrt{2}\sin x\cos x = \sin x$$
$$\sin x\left(\sqrt{2}\cos x - 1\right) = 0$$
$$\sin x = 0 \quad \text{or} \quad \sqrt{2}\cos x - 1 = 0$$
$$\sin x = 0 \quad \text{or} \quad \cos x = \frac{1}{\sqrt{2}}$$

The solutions to these equations in $[0, 2\pi]$ are $x = \boxed{0,\ \pi,\ 2\pi,\ \frac{\pi}{4},\ \frac{7\pi}{4}}$.

Substituting these into the original equation shows that they are all, in fact, solutions to the original equation.

74. Observe that

$$\cos 2x = \sin x$$
$$\cos^2 x - \sin^2 x = \sin x$$
$$\left(1 - \sin^2 x\right) - \sin^2 x = \sin x$$
$$2\sin^2 x + \sin x - 1 = 0$$
$$(2\sin x - 1)(\sin x + 1) = 0$$
$$2\sin x - 1 = 0 \quad \text{or} \quad \sin x + 1 = 0$$
$$\sin x = \frac{1}{2} \quad \text{or} \quad \sin x = -1$$

The solutions to these equations in $[0, 2\pi]$ are $x = \boxed{\frac{\pi}{6},\ \frac{5\pi}{6},\ \frac{3\pi}{2}}$. Substituting these into the original equation shows that they are all, in fact, solutions to the original equation.

75. Observe that

$$\tan 2x = \cot x$$
$$\frac{\sin 2x}{\cos 2x} = \frac{\cos x}{\sin x}$$
$$\cos x(\cos 2x) - (\sin 2x)\sin x = 0$$
$$\cos(x + 2x) = 0$$
$$\cos 3x = 0$$

Note that the solutions of $\cos 3x = 0$ are

$$3x = \frac{\pi}{2},\ \frac{\pi}{2} + 2\pi,\ \frac{\pi}{2} + 4\pi,\ \frac{3\pi}{2},\ \frac{3\pi}{2} + 2\pi,\ \frac{3\pi}{2} + 4\pi$$

so that

$$x = \boxed{\frac{\pi}{6},\ \frac{5\pi}{6},\ \frac{3\pi}{2},\ \frac{\pi}{2},\ \frac{7\pi}{6},\ \frac{11\pi}{6}}.$$

Substituting these into the original equation shows that they are all, in fact, solutions to the original equation.

76. Observe that

$$3\cot 2x = \cot x$$

$$\frac{3\cos 2x}{\sin 2x} = \frac{\cos x}{\sin x}$$

$$3\cos 2x \sin x = \cos x \sin 2x$$

$$3\cos 2x \sin x - \cos x \sin 2x = 0$$

$$2\cos 2x \sin x + \left(\cos 2x \sin x - \cos x \sin 2x\right) = 0$$

$$2\cos 2x \sin x + \sin(x - 2x) = 0$$

$$2\cos 2x \sin x + \sin(-x) = 0$$

$$2\cos 2x \sin x - \sin x = 0$$

$$\sin x\left(2\cos 2x - 1\right) = 0$$

$$\sin x = 0 \quad \text{or} \quad 2\cos 2x - 1 = 0$$

$$\sin x = 0 \quad \text{or} \quad \cos 2x = \frac{1}{2}$$

Note that the solutions of $\sin x = 0$ in $\left[0, 2\pi\right]$ are $x = 0,\ \pi,\ 2\pi$. However, substituting these values into the <u>original</u> equation show that NONE of them are solutions since the right-side is undefined at each of these values.

Next, the solutions of $\cos 2x = \frac{1}{2}$ are

$$2x = \frac{\pi}{3},\ \frac{\pi}{3} + 2\pi,\ \frac{5\pi}{3},\ \frac{5\pi}{3} + 2\pi$$

so that

$$x = \boxed{\frac{\pi}{6},\ \frac{5\pi}{6},\ \frac{7\pi}{6},\ \frac{11\pi}{6}}.$$

Substituting all of <u>these</u> into the original equation shows that they are all, in fact, solutions to the original equation.

77. Observe that

$$\sqrt{3}\sec x = 4\sin x$$
$$\frac{\sqrt{3}}{\cos x} = 4\sin x$$
$$\sqrt{3} = 4\sin x\cos x$$
$$\sqrt{3} = 2(2\sin x\cos x)$$
$$\sqrt{3} = 2\sin 2x$$
$$\frac{\sqrt{3}}{2} = \sin 2x$$

Next, the solutions of $\sin 2x = \frac{\sqrt{3}}{2}$ are

$$2x = \frac{\pi}{3}, \frac{\pi}{3}+2\pi, \frac{2\pi}{3}, \frac{2\pi}{3}+2\pi$$

so that

$$x = \boxed{\frac{\pi}{6}, \frac{\pi}{3}, \frac{7\pi}{6}, \frac{4\pi}{3}}.$$

Substituting all of these into the original equation shows that they are all, in fact, solutions to the original equation.

78. Observe that

$$\sqrt{3}\tan x = 2\sin x$$
$$\frac{\sqrt{3}\sin x}{\cos x} = 2\sin x$$
$$\sqrt{3}\sin x = 2\sin x\cos x$$
$$\sqrt{3}\sin x - 2\sin x\cos x = 0$$
$$\sin x\left(\sqrt{3} - 2\cos x\right) = 0$$
$$\sin x = 0 \quad \text{or} \quad \sqrt{3} - 2\cos x = 0$$
$$\sin x = 0 \quad \text{or} \quad \cos x = \frac{\sqrt{3}}{2}$$

The solutions to these equations in $[0, 2\pi]$ are $x = \boxed{0, \pi, 2\pi, \frac{\pi}{6}, \frac{11\pi}{6}}$. Substituting these into the original equation shows that they are all, in fact, solutions to the original equation.

79. Observe that

$$\sin^2 x - \cos 2x = -\frac{1}{4}$$
$$\sin^2 x - \left(\cos^2 x - \sin^2 x\right) = -\frac{1}{4}$$
$$\sin^2 x - \left(1 - \sin^2 x - \sin^2 x\right) = -\frac{1}{4}$$
$$3\sin^2 x - 1 = -\frac{1}{4}$$
$$\sin^2 x = \frac{1}{4}$$
$$\sin x = -\frac{1}{2} \quad \text{or} \quad \sin x = \frac{1}{2}$$

The solutions to these equations in $[0, 2\pi]$ are $x = \boxed{\frac{\pi}{6}, \frac{5\pi}{6}, \frac{7\pi}{6}, \frac{11\pi}{6}}$.

Substituting these into the original equation shows that they are all solutions to original equation.

80. Observe that

$$\sin^2 x - 2\sin x = 0$$
$$\sin x(\sin x - 2) = 0$$
$$\sin x = 0 \quad \text{or} \quad \sin x - 2 = 0$$
$$\sin x = 0 \quad \text{or} \quad \underbrace{\sin x = 2}_{\text{No solution}}$$

The solutions to these equations in $[0, 2\pi]$ are $x = \boxed{0, \pi, 2\pi}$. Substituting these into the original equation shows that they are all, in fact, solutions to the original equation.

81. Observe that

$$\cos^2 x + 2\sin x + 2 = 0$$
$$\left(1-\sin^2 x\right) + 2\sin x + 2 = 0$$
$$\sin^2 x - 2\sin x - 3 = 0$$
$$(\sin x + 1)(\sin x - 3) = 0$$
$$\sin x + 1 = 0 \quad \text{or} \quad \sin x - 3 = 0$$
$$\sin x = -1 \quad \text{or} \quad \underbrace{\sin x = 3}_{\text{No solution}}$$

The solution to these equations in $[0, 2\pi]$ is $x = \boxed{\frac{3\pi}{2}}$. Substituting this into the original equation shows that it is, in fact, a solution to the original equation.

82. Observe that

$$2\cos^2 x = \sin x + 1$$
$$2\left(1-\sin^2 x\right) = \sin x + 1$$
$$2 - 2\sin^2 x = \sin x + 1$$
$$2\sin^2 x + \sin x - 1 = 0$$
$$(\sin x + 1)(2\sin x - 1) = 0$$
$$\sin x + 1 = 0 \quad \text{or} \quad 2\sin x - 1 = 0$$
$$\sin x = -1 \quad \text{or} \quad \sin x = \frac{1}{2}$$

The solutions to these equations in $[0, 2\pi]$ are $x = \boxed{\frac{\pi}{6}, \frac{5\pi}{6}, \frac{3\pi}{2}}$. Substituting these into the original equation shows that they are, in fact, solutions to the original equation.

83. Observe that

$$2\sin^2 x + 3\cos x = 0$$
$$2\left(1-\cos^2 x\right) + 3\cos x = 0$$
$$2 - 2\cos^2 x + 3\cos x = 0$$
$$2\cos^2 x - 3\cos x - 2 = 0$$
$$(2\cos x + 1)(\cos x - 2) = 0$$
$$2\cos x + 1 = 0 \quad \text{or} \quad \cos x - 2 = 0$$
$$\cos x = -\frac{1}{2} \quad \text{or} \quad \underbrace{\cos x = 2}_{\text{No solution}}$$

The solutions to these equations in $[0, 2\pi]$ are $x = \boxed{\frac{2\pi}{3}, \frac{4\pi}{3}}$. Substituting these into the original equation shows that they are, in fact, solutions to the original equation.

84. Observe that

$$4\cos^2 x - 4\sin x = 5$$
$$4\left(1-\sin^2 x\right) - 4\sin x = 5$$
$$4 - 4\sin^2 x - 4\sin x = 5$$
$$4\sin^2 x + 4\sin x + 1 = 0$$
$$(2\sin x + 1)^2 = 0$$
$$2\sin x + 1 = 0$$
$$\sin x = -\frac{1}{2}$$

The solutions to this equation in $[0, 2\pi]$ are $x = \boxed{\frac{7\pi}{6}, \frac{11\pi}{6}}$. Substituting these into the original equation shows that they are, in fact, solutions to the original equation.

85. Observe that

$$\cos 2x + \cos x = 0$$
$$\left(\cos^2 x - \sin^2 x\right) + \cos x = 0$$
$$\left(\cos^2 x - \left(1 - \cos^2 x\right)\right) + \cos x = 0$$
$$2\cos^2 x + \cos x - 1 = 0$$
$$(2\cos x - 1)(\cos x + 1) = 0$$
$$2\cos x - 1 = 0 \quad \text{or} \quad \cos x + 1 = 0$$
$$\cos x = \frac{1}{2} \quad \text{or} \quad \cos x = -1$$

The solutions to these equations in $[0, 2\pi]$ are $x = \boxed{\frac{\pi}{3}, \frac{5\pi}{3}, \pi}$. Substituting these into the original equation shows that they are, in fact, solutions to the original equation.

86. Observe that

$$2\cot x = \csc x$$
$$\frac{2\cos x}{\sin x} = \frac{1}{\sin x}$$
$$2\cos x \sin x = \sin x$$
$$\sin x(2\cos x - 1) = 0$$
$$\sin x = 0 \quad \text{or} \quad 2\cos x - 1 = 0$$
$$\sin x = 0 \quad \text{or} \quad \cos x = \frac{1}{2}$$

Note that the solutions of $\sin x = 0$ in $[0, 2\pi]$ are $x = 0, \pi, 2\pi$. However, substituting these values into the original equation show that NONE of them are solutions since both the left- and right-sides are undefined at each of these values.

Next, the solutions of $\cos x = \frac{1}{2}$ are

$$x = \boxed{\frac{\pi}{3}, \frac{5\pi}{3}}$$

Substituting all of these into the original equation shows that they are all, in fact, solutions to the original equation.

87. Observe that

$$\tfrac{1}{4}\sec 2x = \sin 2x$$
$$\frac{1}{4\cos 2x} = \sin 2x$$
$$1 = 4\sin 2x \cos 2x$$
$$1 = 2\sin 4x$$
$$\tfrac{1}{2} = \sin 4x$$

The solutions to this equation must satisfy

$$4x = \frac{\pi}{6}, \frac{5\pi}{6}, \frac{13\pi}{6}, \frac{17\pi}{6}, \frac{25\pi}{6}, \frac{29\pi}{6}, \frac{37\pi}{6}, \frac{41\pi}{6}$$

and so, the solutions are

$$\boxed{x = \frac{\pi}{24}, \frac{5\pi}{24}, \frac{13\pi}{24}, \frac{17\pi}{24}, \frac{25\pi}{24}, \frac{29\pi}{24}, \frac{37\pi}{24}, \frac{41\pi}{24}}$$

88. Observe that

$$-\tfrac{1}{4}\csc\left(\tfrac{x}{2}\right) = \cos\left(\tfrac{x}{2}\right)$$
$$-\frac{1}{4\sin\left(\frac{x}{2}\right)} = \cos\left(\tfrac{x}{2}\right)$$
$$-1 = 4\sin\left(\tfrac{x}{2}\right)\cos\left(\tfrac{x}{2}\right)$$
$$-1 = 2\sin\left(2\cdot\tfrac{x}{2}\right)$$
$$-\tfrac{1}{2} = \sin x$$

The solutions of this equation are

$$\boxed{x = \frac{7\pi}{6}, \frac{11\pi}{6}}.$$

89. In order to find all of the values of x in $\left[0^\circ, 360^\circ\right]$ that satisfy $\cos(2x)+\frac{1}{2}\sin x=0$, we proceed as follows:

<u>Step 1</u>: Simplify the equation algebraically.

$$\begin{aligned}\cos(2x)+\frac{1}{2}\sin x=0 \;&\Rightarrow\; \cos^2 x-\sin^2 x+\frac{1}{2}\sin x=0\\ &\Rightarrow\; \left(1-\sin^2 x\right)-\sin^2 x+\frac{1}{2}\sin x=0\\ &\Rightarrow\; 4\sin^2 x-\sin x-2=0\\ &\Rightarrow\; \sin x=\frac{-(-1)\pm\sqrt{(-1)^2-4(4)(-2)}}{2(4)}=\frac{1\pm\sqrt{33}}{8}\end{aligned}$$

<u>Step 2</u>: Solve $\sin x=\frac{1+\sqrt{33}}{8}$.

To do so, we must find the values of x whose sine is $\frac{1+\sqrt{33}}{8}$.

Indeed, observe that one solution is $\sin^{-1}\left(\frac{1+\sqrt{33}}{8}\right)\approx 57.47^\circ$, which is in QI. A second solution occurs in QII, and has value $180^\circ-57.47^\circ=122.53^\circ$. Since the input of sine is simply x, and not some multiple thereof, we conclude that the solutions to this equation are approximately $x\approx 57.47^\circ, 122.53^\circ$.

<u>Step 3</u>: Solve $\sin x=\frac{1-\sqrt{33}}{8}$.

To do so, we must find the values of x whose sine is $\frac{1-\sqrt{33}}{8}$.

Indeed, observe that one solution is $\sin^{-1}\left(\frac{1-\sqrt{33}}{8}\right)\approx -36.38^\circ$, which is in QIV.

Since the angles we seek have positive measure, we use the representative $360^\circ-36.38^\circ=323.62^\circ$. A second solution occurs in QIII, and has value $180^\circ+36.38^\circ=216.38^\circ$. Since the input of sine is simply x, and not some multiple thereof, we conclude that the solutions to this equation are approximately $x\approx 216.38^\circ, 323.62^\circ$.

<u>Step 4</u>: Conclude that the solutions to the original equation are

$$x\approx \boxed{57.47^\circ, 122.53^\circ, 216.38^\circ, 323.62^\circ}.$$

Substituting these into the original equation shows that they are, in fact, solutions to the original equation.

90. Observe that

$$\sec^2 x = \tan x + 1$$
$$1 + \tan^2 x = \tan x + 1$$
$$\tan x(\tan x - 1) = 0$$
$$\tan x = 0 \quad \text{or} \quad \tan x - 1 = 0 \quad \text{so that} \quad \tan x = 0 \quad \text{or} \quad \tan x = 1$$

The solutions to these equations in $[0^\circ, 360^\circ]$ are $x = \boxed{0^\circ, 180^\circ, 360^\circ, 45^\circ, 225^\circ}$.
Substituting these into the original equation shows that they are all solutions.

91. In order to find all of the values of *x* in $[0^\circ, 360^\circ]$ that satisfy $6\cos^2 x + \sin x = 5$, we proceed as follows:

Step 1: Simplify the equation algebraically.

$$6\cos^2 x + \sin x = 5$$
$$6(1 - \sin^2 x) + \sin x = 5$$
$$6 - 6\sin^2 x + \sin x = 5$$
$$(3\sin x + 1)(2\sin x - 1) = 0$$
$$3\sin x + 1 = 0 \quad \text{or} \quad 2\sin x - 1 = 0 \quad \text{so that} \quad \sin x = -\frac{1}{3} \quad \text{or} \quad \sin x = \frac{1}{2}$$

Step 2: Solve $\sin x = -\frac{1}{3}$.

To do so, we must find the values of *x* whose sine is $-\frac{1}{3}$.

Indeed, observe that one solution is $\sin^{-1}\left(-\frac{1}{3}\right) \approx -19.47^\circ$, which is in QIV.

Since the angles we seek have positive measure, we use the representative $360^\circ - 19.47^\circ = 340.53^\circ$. A second solution occurs in QIII, and has value $180^\circ + 19.47^\circ = 199.47^\circ$. Since the input of sine is simply *x*, and not some multiple thereof, we conclude that the solutions to this equation are $x \approx 199.47^\circ, 340.53^\circ$.

Step 3: Solve $\sin x = \frac{1}{2}$.

Indeed, observe that one solution is $\sin^{-1}\left(\frac{1}{2}\right) = 30^\circ$, which is in QI. A second solution occurs in QII, and has value $180^\circ - 30^\circ = 150^\circ$. Since the input of sine is simply *x*, and not some multiple thereof, we see that the solutions are $x = 30^\circ, 150^\circ$.

Step 4: Conclude that the solutions to the original equation are

$$x \approx \boxed{30^\circ, 150^\circ, 199.47^\circ, 340.53^\circ}.$$

Substituting these into the original equation shows that they are all solutions.

92. In order to find all of the values of x in $\left[0°,360°\right]$ that satisfy $\sec^2 x = 2\tan x + 4$, we proceed as follows:

Step 1: Simplify the equation algebraically.

$$\sec^2 x = 2\tan x + 4$$
$$1 + \tan^2 x = 2\tan x + 4$$
$$\tan^2 x - 2\tan x - 3 = 0$$
$$(\tan x - 3)(\tan x + 1) = 0$$
$$\tan x - 3 = 0 \quad \text{or} \quad \tan x + 1 = 0$$
$$\tan x = 3 \quad \text{or} \quad \tan x = -1$$

Step 2: Solve $\tan x = 3$.

To do so, we must find the values of x whose tangent is 3.
Indeed, observe that one solution is $\tan^{-1}(3) \approx 71.57°$, which is in QI.

A second solution occurs in QIII, and has value $180° + 71.57° = 251.57°$. Since the input of tangent is simply x, and not some multiple thereof, we conclude that the solutions to this equation are approximately $x \approx 71.57°, 251.57°$.

Step 3: Solve $\tan x = -1$.

To do so, we must find the values of x whose tangent is -1.
Indeed, observe that one solution is $\tan^{-1}(-1) = 135°$, which is in QII.

A second solution occurs in QIV, and has value $180° + 135° = 315°$. Since the input of tangent is simply x, and not some multiple thereof, we conclude that the solutions to this equation are approximately $x = 135°, 315°$.

Step 4: Conclude that the solutions to the original equation are

$$x \approx \boxed{135°, 315°, 71.57°, 251.57°}.$$

Substituting these into the original equation shows that they are, in fact, solutions to the original equation.

93. In order to find all of the values of x in $[0°, 360°]$ that satisfy $\cot^2 x - 3\csc x - 3 = 0$, we proceed as follows:

Step 1: Simplify the equation algebraically.

$$\cot^2 x - 3\csc x - 3 = 0$$
$$\left(\csc^2 x - 1\right) - 3\csc x - 3 = 0$$
$$\csc^2 x - 3\csc x - 4 = 0$$
$$\left(\csc x + 1\right)\left(\csc x - 4\right) = 0$$
$$\csc x + 1 = 0 \quad \text{or} \quad \csc x - 4 = 0$$
$$\csc x = -1 \quad \text{or} \quad \csc x = 4$$
$$\sin x = -1 \quad \text{or} \quad \sin x = \frac{1}{4}$$

Step 2: Solve $\sin x = -1$.
To do so, we must find the values of x whose sine is -1.
Indeed, the only solution is $\sin^{-1}(-1) = 270°$, which is in QIII.

Step 3: Solve $\sin x = \frac{1}{4}$.

Indeed, observe that one solution is $\sin^{-1}\left(\frac{1}{4}\right) \approx 14.48°$, which is in QI. A second solution occurs in QII, and has value $180° - 14.48° = 165.52°$. Since the input of sine is simply x, and not some multiple thereof, we conclude that the solutions to this equation are $x \approx 14.48°, 165.52°$.

Step 4: Conclude that the solutions to the original equation are

$$x \approx \boxed{14.48°, 165.52°, 270°}.$$

Substituting these into the original equation shows that they are, in fact, solutions to the original equation.

94. In order to find all of the values of x in $\left[0^\circ, 360^\circ\right]$ that satisfy $\csc^2 x + \cot x = 7$, we proceed as follows:

Step 1: Simplify the equation algebraically:

$$\csc^2 x + \cot x = 7$$
$$\left(1+\cot^2 x\right) + \cot x = 7$$
$$\cot^2 x + \cot x - 6 = 0$$
$$\left(\cot x + 3\right)\left(\cot x - 2\right) = 0$$
$$\cot x + 3 = 0 \quad \text{or} \quad \cot x - 2 = 0$$
$$\cot x = -3 \quad \text{or} \quad \cot x = 2$$

Step 2: Solve $\cot x = -3$.

To do so, we must find the values of x whose cotangent is -3.
Indeed, observe that one solution is
$\cot^{-1}(-3) = 180^\circ + \tan^{-1}\left(-\frac{1}{3}\right) \approx 161.57^\circ$, which is in QII. A second solution occurs in QIV, and has value $180^\circ + 161.57^\circ = 341.57^\circ$. Since the input of cotangent is simply x, and not some multiple thereof, we conclude that the solutions to this equation are approximately $x \approx 161.57^\circ, 341.57^\circ$.

Step 3: Solve $\cot x = 2$

To do so, we must find the values of x whose cotangent is 2.
Indeed, observe that one solution is $\cot^{-1}(2) = \tan^{-1}\left(\frac{1}{2}\right) \approx 26.57^\circ$, which is in QI. A second solution occurs in QIII, and has value $180^\circ + 26.57^\circ = 206.57^\circ$. Since the input of cotangent is simply x, and not some multiple thereof, we conclude that the solutions to this equation are approximately $x \approx 26.57^\circ, 206.57^\circ$.

Step 4: Conclude that the solutions to the original equation are

$$x \approx \boxed{26.57^\circ,\ 206.57^\circ,\ 161.57^\circ,\ 341.57^\circ}.$$

95. In order to find all of the values of x in $[0°, 360°]$ that satisfy
$2\sin^2 x + 2\cos x - 1 = 0$,
we proceed as follows:

Step 1: Simplify the equation algebraically.

$$2\sin^2 x + 2\cos x - 1 = 0$$
$$2\left(1-\cos^2 x\right) + 2\cos x - 1 = 0$$
$$2 - 2\cos^2 x + 2\cos x - 1 = 0$$
$$2\cos^2 x - 2\cos x - 1 = 0$$

$$\cos x = \frac{-(-2) \pm \sqrt{(-2)^2 - 4(2)(-1)}}{2(2)} = \frac{2 \pm 2\sqrt{3}}{4} = \frac{1 \pm \sqrt{3}}{2}$$

$$\cos x = \frac{1-\sqrt{3}}{2} \quad \text{or} \quad \underbrace{\cos x = \frac{1+\sqrt{3}}{2}}_{\text{No solution since } \frac{1+\sqrt{3}}{2} > 1}.$$

Step 2: Solve $\cos x = \dfrac{1-\sqrt{3}}{2}$.

To do so, we must find the values of x whose cosine is $\dfrac{1-\sqrt{3}}{2}$.

Indeed, the only solution is $\cos^{-1}\left(\dfrac{1-\sqrt{3}}{2}\right) \approx 111.47°$, which is in QII. A second solution occurs in QIII and has value $360° - 111.47° = 248.53°$. Since the input of cosine is simply x, and not some multiple thereof, we conclude that the solutions to this equation are approximately $x \approx 111.47°, 248.53°$.

Step 3: Conclude that the solutions to the original equation are

$$x \approx \boxed{111.47°,\ 248.53°}.$$

Substituting these into the original equation shows that they are, in fact, solutions to the original equation.

96. In order to find all of the values of x in $\left[0^\circ, 360^\circ\right]$ that satisfy $\sec^2 x + \tan x - 2 = 0$, we proceed as follows:

Step 1: Simplify the equation algebraically.

$$\sec^2 x + \tan x - 2 = 0$$
$$\left(1 + \tan^2 x\right) + \tan x - 2 = 0$$
$$\tan^2 x + \tan x - 1 = 0$$
$$\tan x = \frac{-1 \pm \sqrt{1^2 - 4(1)(-1)}}{2(1)} = \frac{-1 \pm \sqrt{5}}{2}$$

Step 2: Solve $\tan x = \frac{-1-\sqrt{5}}{2}$.

To do so, we must find the values of x whose tangent is $\frac{-1-\sqrt{5}}{2}$.

Indeed, observe that one solution is $\tan^{-1}\left(\frac{-1-\sqrt{5}}{2}\right) \approx -58.28^\circ$, which is in QIV.

Since the angles we seek have positive measure, we use the representative $360^\circ - 58.28^\circ = 301.72^\circ$ A second solution occurs in QII, and has value $301.72^\circ - 180^\circ = 121.72^\circ$. Since the input of tangent is simply x, and not some multiple thereof, we conclude that the solutions to this equation are approximately $x \approx 121.72^\circ, 301.72^\circ$.

Step 3: Solve $\tan x = \frac{-1+\sqrt{5}}{2}$.

To do so, we must find the values of x whose tangent is $\frac{-1+\sqrt{5}}{2}$.

Indeed, observe that one solution is $\tan^{-1}\left(\frac{-1+\sqrt{5}}{2}\right) \approx 31.72^\circ$, which is in QI.

A second solution occurs in QIII, and has value $180^\circ + 31.72^\circ = 211.72^\circ$. Since the input of tangent is simply x, and not some multiple thereof, we conclude that the solutions to this equation are approximately $x \approx 31.72^\circ, 211.72^\circ$.

Step 4: Conclude that the solutions to the original equation are

$$x \approx \boxed{31.72^\circ, 211.72^\circ, 121.72^\circ, 301.72^\circ}.$$

Substituting these into the original equation shows that they are, in fact, solutions to the original equation.

97. Observe that

$$\begin{aligned}
\tfrac{1}{16}\csc^2\left(\tfrac{x}{4}\right)-\cos^2\left(\tfrac{x}{4}\right)&=0\\
\left(\tfrac{1}{4}\csc\left(\tfrac{x}{4}\right)-\cos\left(\tfrac{x}{4}\right)\right)\left(\tfrac{1}{4}\csc\left(\tfrac{x}{4}\right)+\cos\left(\tfrac{x}{4}\right)\right)&=0\\
\tfrac{1}{4}\csc\left(\tfrac{x}{4}\right)&=\pm\cos\left(\tfrac{x}{4}\right)\\
\tfrac{1}{4}&=\pm\sin\left(\tfrac{x}{4}\right)\cos\left(\tfrac{x}{4}\right)\\
\tfrac{1}{4}&=\pm\tfrac{1}{2}\sin\left(2\cdot\tfrac{x}{4}\right)\\
\pm\tfrac{1}{2}&=\sin\left(\tfrac{x}{2}\right)
\end{aligned}$$

The values of x that satisfy these equation must satisfy $\frac{x}{2}=\frac{\pi}{6},\frac{5\pi}{6}$. So, the solutions are

$$x=\boxed{\frac{\pi}{3},\frac{5\pi}{3}}.$$

98. Observe that

$$\begin{aligned}
-\tfrac{1}{4}\sec^2\left(\tfrac{x}{8}\right)+\sin^2\left(\tfrac{x}{8}\right)&=0\\
\left(\sin\left(\tfrac{x}{8}\right)-\tfrac{1}{2}\sec\left(\tfrac{x}{8}\right)\right)\left(\sin\left(\tfrac{x}{8}\right)+\tfrac{1}{2}\sec\left(\tfrac{x}{8}\right)\right)&=0\\
\sin\left(\tfrac{x}{8}\right)&=\pm\tfrac{1}{2}\sec\left(\tfrac{x}{8}\right)\\
\tfrac{1}{2}&=\pm\sin\left(\tfrac{x}{8}\right)\cos\left(\tfrac{x}{8}\right)\\
\tfrac{1}{2}&=\pm\tfrac{1}{2}\sin\left(2\cdot\tfrac{x}{8}\right)\\
\pm1&=\sin\left(\tfrac{x}{4}\right)
\end{aligned}$$

The values of x that satisfy these equation must satisfy $\frac{x}{4}=\frac{\pi}{2},\frac{3\pi}{2}$. So, the desired solution is $x=\boxed{2\pi}$.

99. Solving for x yields:

$$\begin{aligned}
2,400&=400\sin\left(\frac{\pi}{6}x\right)+2,000\\
400&=400\sin\left(\frac{\pi}{6}x\right)\\
1&=\sin\left(\frac{\pi}{6}x\right)
\end{aligned}$$

This equation is satisfied when $\frac{\pi}{6}x=\frac{\pi}{2}$, so that $x=\frac{\pi}{2}\cdot\frac{6}{\pi}=3$. So, the sales reach 2,400 in $\boxed{\text{March}}$.

100. Solving for x yields:

$$1,800 = 400\sin\left(\frac{\pi}{6}x\right) + 2,000$$
$$-200 = 400\sin\left(\frac{\pi}{6}x\right)$$
$$-\frac{1}{2} = \sin\left(\frac{\pi}{6}x\right)$$

This equation is satisfied when $\frac{\pi}{6}x = \frac{7\pi}{6}, \frac{11\pi}{6}$, so that $x = 7, 11$. So, the sales reach 1,800 during in $\boxed{\text{July and November}}$.

101. Consider the following diagram:

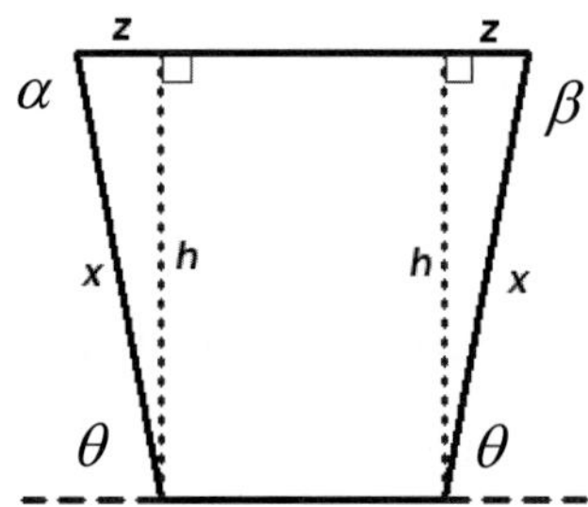

Let h = height of the trapezoid, and x = length of one base and two edges of the trapezoid, as labeled above.
Note that $\alpha = \beta = \theta$ since they are alternate interior angles. As such,

$$\sin\theta = \frac{h}{x}, \text{ so that } h = x\sin\theta.$$

Furthermore, using the Pythagorean Theorem enables us to find z:

$$z^2 + h^2 = x^2, \text{ so that } z = \sqrt{x^2 - h^2} = \sqrt{x^2 - (x\sin\theta)^2} = \sqrt{x^2\underbrace{\left(1-\sin^2\theta\right)}_{=\cos^2\theta}} = x\left|\cos\theta\right|.$$

Since $0 \le \theta \le \pi$, we conclude that $z = x\cos\theta$. Hence,

$$b_1 = x \text{ and } b_2 = x + 2z = x + 2x\cos\theta = x\left(1 + 2\cos\theta\right).$$

Thus, the area A of the cross-section of the rain gutter is

$$A = \frac{1}{2}h\left(b_1 + b_2\right) = \frac{1}{2}(x\sin\theta)\underbrace{\left[x + x\left(1+2\cos\theta\right)\right]}_{=2x(1+\cos\theta)}$$
$$= (x\sin\theta)\left[x(1+\cos\theta)\right]$$
$$= x^2\left[\sin\theta + \sin\theta\cos\theta\right]$$
$$= x^2\left[\sin\theta + \frac{1}{2}(2\sin\theta\cos\theta)\right]$$
$$= x^2\left[\sin\theta + \frac{\sin 2\theta}{2}\right]$$

102. Observe that

$$\cos^2\theta - \sin^2\theta + \cos\theta = 0$$
$$\cos^2\theta - \left(1-\cos^2\theta\right) + \cos\theta = 0$$
$$2\cos^2\theta + \cos\theta - 1 = 0$$
$$(2\cos\theta - 1)(\cos\theta + 1) = 0$$

So, θ satisfies the original equation if either $2\cos\theta - 1 = 0$ or $\cos\theta + 1 = 0$.

Observe that $2\cos\theta - 1 = 0$ is equivalent to $\cos\theta = \frac{1}{2}$, which is satisfied when $\theta = \frac{\pi}{3}, \frac{5\pi}{3}$.

Also, $\cos\theta + 1 = 0$ is equivalent to $\cos\theta = -1$, which is satisfied when $\theta = \pi$.

So, we conclude that the solutions to the original equation are $\theta = \boxed{\frac{\pi}{3}, \pi, \frac{5\pi}{3}}$.

103. Solving the equation $200 + 100\sin\left(\frac{\pi}{2}x\right) = 300$, for $x > 2{,}000$ yields:

$$200 + 100\sin\left(\frac{\pi}{2}x\right) = 300$$
$$100\sin\left(\frac{\pi}{2}x\right) = 100$$
$$\sin\left(\frac{\pi}{2}x\right) = 1$$

Observe that this equation is satisfied when $\frac{\pi}{2}x = \frac{\pi}{2} + 2n\pi$, where n is an integer, so that $x = 1 + 2n\pi\left(\frac{2}{\pi}\right) = 1 + 4n$, where n is an integer. So, the first value of n for which $1 + 4n > 2{,}000$ is $n = 500$. The resulting year is $x = 1 + 4(500) = \boxed{2001}$.

104. Solving the equation $200 + 100\sin\left(\frac{\pi}{6}x\right) = 150$ yields:

$$200 + 100\sin\left(\frac{\pi}{6}x\right) = 150$$
$$100\sin\left(\frac{\pi}{6}x\right) = -50$$
$$\sin\left(\frac{\pi}{6}x\right) = -\frac{1}{2}$$

Observe that this equation is satisfied when $\frac{\pi}{6}x = \frac{7\pi}{6}, \frac{11\pi}{6}$, so that

$x = \frac{6}{\pi}\left[\frac{7\pi}{6}\right], \frac{6}{\pi}\left[\frac{11\pi}{6}\right] = 7,$ ~~11~~. So, the first year after 2000 in which this occurs is $\boxed{2007}$.

105. Use $n_i \sin(\theta_i) = n_r \sin(\theta_r)$ with the given information to obtain

$$1.00\sin(75^\circ) = 2.417\sin(\theta_r)$$

so that

$$\frac{1.00\sin(75^\circ)}{2.417} = \sin(\theta_r)$$

$$\boxed{24^\circ} \approx \sin^{-1}\left(\frac{1.00\sin(75^\circ)}{2.417}\right) = \theta_r$$

106. Use $n_i \sin(\theta_i) = n_r \sin(\theta_r)$ with the given information to obtain

$$2.417\sin(15^\circ) = 1.00\sin(\theta_r)$$

so that

$$\frac{2.417\sin(15^\circ)}{1.00} = \sin(\theta_r)$$

$$\boxed{39^\circ} \approx \sin^{-1}\left(\frac{2.417\sin(15^\circ)}{1.00}\right) = \theta_r$$

107. Observe that using the identity $\sin 2A = 2\sin A\cos A$ with $A = \frac{\pi}{3}x$ yields

$$2\sin\left(\frac{\pi}{3}x\right)\cos\left(\frac{\pi}{3}x\right) + 3 = 4$$

$$\sin\left(2\cdot\frac{\pi}{3}x\right) = 1$$

This equation is satisfied when $\frac{2\pi}{3}x = \frac{\pi}{2}$, so that $x = \frac{3}{4}$. So, it takes $\boxed{\frac{3}{4}\text{ sec.}}$ for the volume of air to equal 4 liters.

108. Observe that using the identity $\sin 2A = 2\sin A\cos A$ with $A = \frac{\pi}{3}x$ yields

$$2\sin\left(\frac{\pi}{3}x\right)\cos\left(\frac{\pi}{3}x\right) + 3 = 2$$

$$\sin\left(2\cdot\frac{\pi}{3}x\right) = -1$$

This equation is satisfied when $\frac{2\pi}{3}x = \frac{3\pi}{2}$, so that $x = \frac{9}{4}$. So, it takes $\boxed{\frac{9}{4}\text{ sec.}}$ for the volume of air to equal 2 liters.

109. First, we solve $-2\sin x + 2\sin 2x = 0$ on $[0, 2\pi]$:

$$-2\sin x + 2\sin 2x = 0$$
$$-2\sin x + 2[2\sin x\cos x] = 0$$
$$2\sin x(-1 + 2\cos x) = 0$$

so that

$$2\sin x = 0 \quad\text{or}\quad -1 + 2\cos x = 0$$
$$2\sin x = 0 \quad\text{or}\quad \cos x = \frac{1}{2}$$

These equations are satisfied when $x = 0,\ \pi,\ 2\pi,\ \frac{\pi}{3},\ \frac{5\pi}{3}$. We now need to determine the corresponding y-coordinates.

x	$y(x) = 2\cos x - \cos 2x$	Point
0	$y(0) = 2\cos(0) - \cos(2\cdot 0) = 2 - 1 = 1$	$(0,1)$
$\pi/3$	$y(\pi/3) = 2\cos(\pi/3) - \cos(2\cdot\pi/3) = 2\left(\frac{1}{2}\right) - \left(-\frac{1}{2}\right) = \frac{3}{2}$	$\left(\frac{\pi}{3}, \frac{3}{2}\right)$

$5\pi/3$	$y\left(5\pi/3\right)=2\cos\left(5\pi/3\right)-\cos\left(2\cdot 5\pi/3\right)=2\left(\frac{1}{2}\right)-\left(-\frac{1}{2}\right)=\frac{3}{2}$	$\left(\frac{5\pi}{3},\frac{3}{2}\right)$
π	$y(\pi)=2\cos(\pi)-\cos(2\cdot\pi)=2(-1)-1=-3$	$(\pi,-3)$
2π	$y(2\pi)=2\cos(2\pi)-\cos(2\cdot 2\pi)=2(1)-1=1$	$(2\pi,1)$

So, the turning points are $\boxed{(0,1),\ \left(\frac{\pi}{3},\frac{3}{2}\right),\ \left(\frac{5\pi}{3},\frac{3}{2}\right),\ (\pi,-3),\text{ and }(2\pi,1).}$

110. First, we solve $-2\sin x+2\sin 2x=0$ on $[-2\pi,0]$:

$$\begin{aligned}-2\sin x+2\sin 2x&=0\\ -2\sin x+2[2\sin x\cos x]&=0\\ 2\sin x(-1+2\cos x)&=0\end{aligned}$$

so that

$$\begin{aligned}2\sin x=0 \quad&\text{or}\quad -1+2\cos x=0\\ 2\sin x=0 \quad&\text{or}\quad \cos x=\frac{1}{2}\end{aligned}$$

These equations are satisfied when $x=0,\ -\pi,\ -2\pi,-\frac{\pi}{3},\ -\frac{5\pi}{3}$. We now need to determine the corresponding y-coordinates.

x	$y(x)=2\cos x-\cos 2x$	Point
0	$y(0)=2\cos(0)-\cos(2\cdot 0)=2-1=1$	$(0,1)$
$-\pi/3$	$y\left(-\pi/3\right)=2\cos\left(-\pi/3\right)-\cos\left(-2\cdot\pi/3\right)=2\left(\frac{1}{2}\right)-\left(-\frac{1}{2}\right)=\frac{3}{2}$	$\left(-\frac{\pi}{3},\frac{3}{2}\right)$
$-5\pi/3$	$y\left(-5\pi/3\right)=2\cos\left(-5\pi/3\right)-\cos\left(-2\cdot 5\pi/3\right)=2\left(\frac{1}{2}\right)-\left(-\frac{1}{2}\right)=\frac{3}{2}$	$\left(-\frac{5\pi}{3},\frac{3}{2}\right)$
$-\pi$	$y(-\pi)=2\cos(-\pi)-\cos(-2\cdot\pi)=2(-1)-1=-3$	$(-\pi,-3)$
-2π	$y(-2\pi)=2\cos(-2\pi)-\cos(-2\cdot 2\pi)=2(1)-1=1$	$(-2\pi,1)$

So, the turning points are $\boxed{(0,1),\ \left(-\frac{\pi}{3},\frac{3}{2}\right),\ \left(-\frac{5\pi}{3},\frac{3}{2}\right),\ (-\pi,-3),\text{ and }(-2\pi,1).}$

111. The value $\theta=3\pi/2$ does not satisfy the original equation. Indeed, observe that $\sqrt{2+\sin\left(\frac{3\pi}{2}\right)}=\sqrt{2-1}=1$, while $\sin\left(\frac{3\pi}{2}\right)=-1$. So, this value of θ is an extraneous solution.

112. The value $\theta = \pi/2$ does not satisfy the original equation. Indeed, observe that $\sqrt{3\sin\left(\frac{\pi}{2}\right)-2} = \sqrt{3-2} = 1$, while $-\sin\left(\frac{\pi}{2}\right) = -1$. So, this value of θ is an extraneous solution.

113. Cannot divide by $\cos x$ since it could be zero. Rather, should factor as follows:

$$6\sin x\cos x = 2\cos x$$
$$6\sin x\cos x - 2\cos x = 0$$
$$2\cos x(3\sin x - 1) = 0$$

Now, proceed…

114. Forgot to check for extraneous solutions. Note that for $x = \pi$, we have $\sqrt{1+\sin\pi} = \sqrt{1} = 1$, while $\cos\pi = -1$. Hence, $x = \pi$ is not a solution to the equation. The remaining values ARE solutions.

115. False. For instance, $\sin\theta = \frac{\sqrt{3}}{2}$ has two solutions on $[0, 2\pi]$, namely $\theta = \frac{\pi}{3}, \frac{2\pi}{3}$.

116. False. For instance, $\sin^2\theta = \frac{3}{2}$ has four solutions on $[0, 2\pi]$, namely $\theta = \frac{\pi}{3}, \frac{2\pi}{3}, \frac{4\pi}{3}, \frac{5\pi}{3}$.

117. True. This follows by definition of an identity.

118. False. This is not sufficient. For instance, the equation $\sin x = 1$ has infinitely many solutions, but there are values of x in the domain for which it is not true (for example, $x = 0$).

119. Solving the equation $16\sin^4\theta - 8\sin^2\theta + 1 = 0$ on $[0, 2\pi]$ yields

$$16\sin^4\theta - 8\sin^2\theta + 1 = 0$$
$$\left(4\sin^2\theta - 1\right)^2 = 0$$
$$(2\sin\theta - 1)(2\sin\theta + 1) = 0$$

So, θ satisfies the original equation if either $2\sin\theta - 1 = 0$ or $2\sin\theta + 1 = 0$.

Observe that $2\sin\theta - 1 = 0$ is equivalent to $\sin\theta = \frac{1}{2}$, which is satisfied when $\theta = \frac{\pi}{6}, \frac{5\pi}{6}$.

Also, $2\sin\theta + 1 = 0$ is equivalent to $\sin\theta = -\frac{1}{2}$, which is satisfied when $\theta = \frac{7\pi}{6}, \frac{11\pi}{6}$.

So, we conclude that the solutions to the original equation are $\theta = \boxed{\frac{\pi}{6}, \frac{5\pi}{6}, \frac{7\pi}{6}, \frac{11\pi}{6}}$.

120. Solving the equation $\left|\cos\left(\theta+\frac{\pi}{4}\right)\right|=\frac{\sqrt{3}}{2}$ on $\mathbb{R}$ is equivalent to

$$\cos\left(\theta+\frac{\pi}{4}\right)=\frac{\sqrt{3}}{2} \text{ or } \cos\left(\theta+\frac{\pi}{4}\right)=-\frac{\sqrt{3}}{2}.$$

Observe that $\cos\left(\theta+\frac{\pi}{4}\right)=\frac{\sqrt{3}}{2}$ is satisfied when $\theta+\frac{\pi}{4}=\frac{\pi}{6}+2n\pi, \frac{11\pi}{6}+2n\pi$, so that $\theta=\frac{\pi}{6}-\frac{\pi}{4}+2n\pi, \frac{11\pi}{6}-\frac{\pi}{4}+2n\pi=-\frac{\pi}{12}+2n\pi, \frac{19\pi}{12}+2n\pi$, where n is an integer.

Also, $\cos\left(\theta+\frac{\pi}{4}\right)=-\frac{\sqrt{3}}{2}$ is satisfied when $\theta+\frac{\pi}{4}=\frac{5\pi}{6}+2n\pi, \frac{7\pi}{6}+2n\pi$, so that $\theta=\frac{5\pi}{6}-\frac{\pi}{4}+2n\pi, \frac{7\pi}{6}-\frac{\pi}{4}+2n\pi=\frac{7\pi}{12}+2n\pi, \frac{11\pi}{6}+2n\pi$, where n is an integer.

So, we conclude that the solutions to the original equation are

$$\theta=-\frac{\pi}{12}+2n\pi, \frac{19\pi}{12}+2n\pi, \frac{7\pi}{12}+2n\pi, \frac{11\pi}{6}+2n\pi, \text{ where } n \text{ is an integer,}$$

which can be further simplified as:

$$\boxed{\theta=-\frac{\pi}{12}+n\pi, \frac{7\pi}{12}+2n\pi, \text{ where } n \text{ is an integer.}}$$

121. First, observe that using the addition formulae for $\sin(A \pm B)$ yields

$$\sin\left(x+\frac{\pi}{4}\right)+\sin\left(x-\frac{\pi}{4}\right)=\left[\sin x\cos\left(\frac{\pi}{4}\right)+\cancel{\cos x\sin\left(\frac{\pi}{4}\right)}\right]+\left[\sin x\cos\left(\frac{\pi}{4}\right)-\cancel{\cos x\sin\left(\frac{\pi}{4}\right)}\right]$$

$$=\frac{\sqrt{2}}{2}(2\sin x)$$

$$=\sqrt{2}\sin x$$

We substitute this for the left-side of the given equation to obtain:

$$\sin\left(x+\frac{\pi}{4}\right)+\sin\left(x-\frac{\pi}{4}\right)=\frac{\sqrt{2}}{2}$$

$$\sqrt{2}\sin x=\frac{\sqrt{2}}{2}$$

$$\sin x=\frac{1}{2}$$

The smallest positive value of x for which this is true is $x=\boxed{\frac{\pi}{6} \text{ or } 30^\circ}$.

122. First, observe that using the addition formula $\cos(A-B)=\cos A\cos B+\sin A\sin B$ yields

$$\cos x\cos\left(15^\circ\right)+\sin x\sin\left(15^\circ\right)=0.7$$

$$\cos\left(x-15^\circ\right)=0.7$$

We now find the solutions to this equation in $\left[0^\circ,360^\circ\right]$:

One solution is $x=15^\circ+\cos^{-1}(0.7)\approx 15^\circ+45.57^\circ=60.57^\circ$, which is in QI. The second solution is in QIV and has value $15^\circ+314.43^\circ=329.43^\circ$. Observe that the smallest positive solution is approximately $x\approx\boxed{60.57^\circ}$.

123. Observe that using a half-angle formula, we see that $\dfrac{1-\cos\left(\frac{x}{3}\right)}{1+\cos\left(\frac{x}{3}\right)}=\tan^2\left(\frac{x}{6}\right)$, thereby resulting in the equivalent equation $\tan^2\left(\frac{x}{6}\right)=-1$, which has $\boxed{\text{no solution}}$.

124. Factoring the left-side of $\sec^4\left(\frac{\theta}{3}\right)-1=0$ yields the equivalent equation $\left(\sec^2\left(\frac{\theta}{3}\right)-1\right)\left(\sec^2\left(\frac{\theta}{3}\right)+1\right)=0$, which is satisfied whenever either $\sec^2\left(\frac{\theta}{3}\right)-1=0$ or $\sec^2\left(\frac{\theta}{3}\right)+1=0$. The second equation has no solution since the left-side is always greater than or equal to 1. The first equation is equivalent to $\cos^2\left(\frac{\theta}{3}\right)=1$, which holds whenever $\cos\left(\frac{\theta}{3}\right)=\pm 1$. The values of θ for which this is true must satisfy

$$\frac{\theta}{3}=n\pi \text{, where } n \text{ is an integer.}$$

So, dividing by 3 then yields the values of θ in $\mathbb{R}$ for which this is true:

$$\boxed{\theta = 3n\pi, \text{ where } n \text{ is an integer}}.$$

125. Factoring the left-side of $\csc^4\left(\frac{\pi}{4}\theta-\pi\right)-4=0$ yields the equivalent equation $\left(\csc^2\left(\frac{\pi}{4}\theta-\pi\right)-2\right)\left(\csc^2\left(\frac{\pi}{4}\theta-\pi\right)+2\right)=0$, which is satisfied whenever either $\csc^2\left(\frac{\pi}{4}\theta-\pi\right)-2=0$ or $\csc^2\left(\frac{\pi}{4}\theta-\pi\right)+2=0$. The second equation has no solution since the left-side is always greater than or equal to 2. The first equation is equivalent to $\sin^2\left(\frac{\pi}{4}\theta-\pi\right)=\frac{1}{2}$, which holds whenever $\sin\left(\frac{\pi}{4}\theta-\pi\right)=\pm\frac{\sqrt{2}}{2}$. The values of θ for which this is true must satisfy

$$\frac{\pi}{4}\theta-\pi=\frac{\pi}{4}+\frac{n\pi}{2}\text{, where } n \text{ is an integer.}$$

Solving this equation for θ yields the values of θ in $\mathbb{R}$ for which this is true:

$$\boxed{\theta = 5+2n, \text{ where } n \text{ is an integer}}.$$

126. Observe that

$$2\tan 3x=\sqrt{3}-\sqrt{3}\tan^2(3x)$$
$$\frac{2\tan 3x}{1-\tan^2(3x)}=\sqrt{3}$$
$$\tan 6x=\sqrt{3}=\frac{\frac{\sqrt{3}}{2}}{\frac{1}{2}}$$

The values of x for which this equation holds are

$$6x=\frac{\pi}{3}+n\pi\text{, where } n \text{ is an integer,}$$

which is equivalent to

$$x=\frac{\pi}{18}+\frac{n\pi}{6}=\boxed{\frac{(1+3n)\pi}{18}}\text{, where } n \text{ is an integer.}$$

127. Consider the graphs below of
$y_1 = \sin\theta,\ y_2 = \cos 2\theta$.

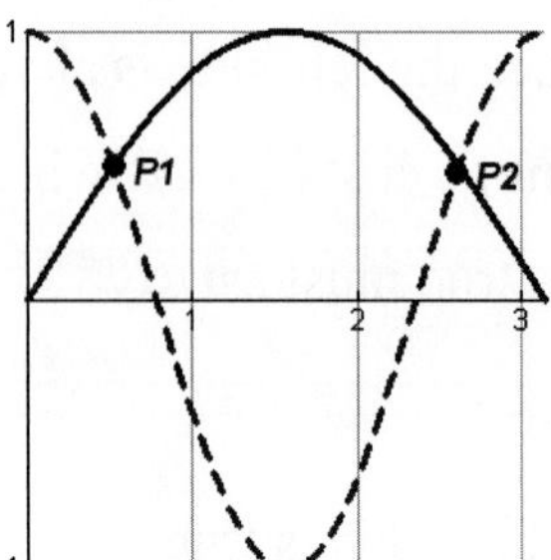

Observe that the solutions of the equation $\sin\theta = \cos 2\theta$ on $[0,\pi]$ are approximately
$P1(0.524, 0.5),\ P2(2.618, 0.5)$.

The exact solutions are $\frac{\pi}{6}, \frac{5\pi}{6}$.

128. Consider the graphs below of
$y_1 = \csc\theta,\ y_2 = \sec\theta$.

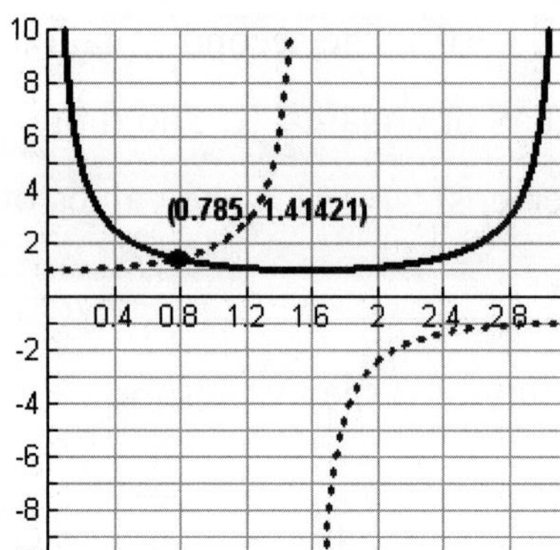

Observe that the approximate solution to this equation is 0.785. The exact solution is $\frac{\pi}{4}$.

129. Consider the graphs below of
$y_1 = \sin\theta,\ y_2 = \sec\theta$.

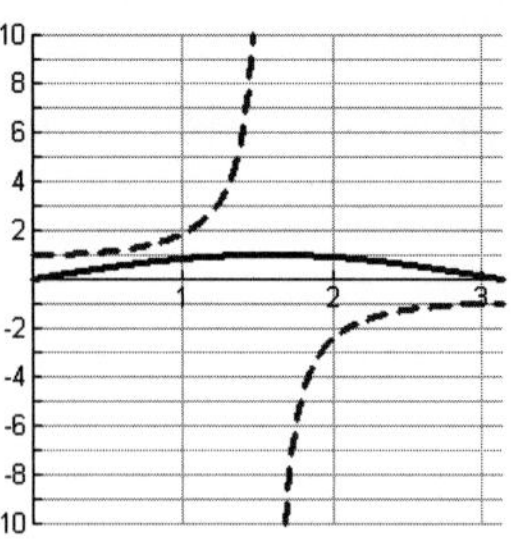

Since the curves never intersect, there are no solutions of $\sin\theta = \sec\theta$ on $[0,\pi]$.

130. Consider the graphs below of
$y_1 = \cos\theta,\ y_2 = \csc\theta$.

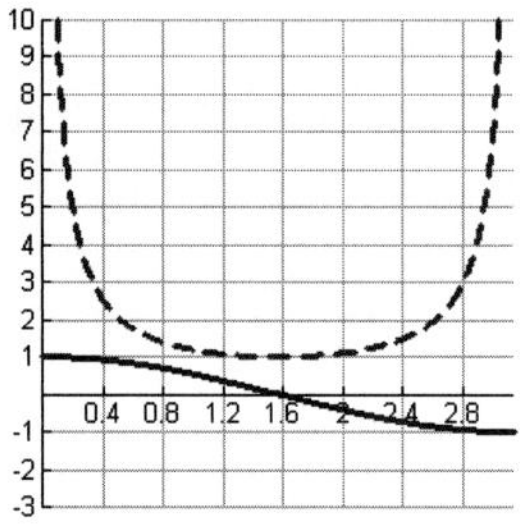

Since the curves never intersect, there are no solutions of $\sin\theta = \sec\theta$ on $[0,\pi]$.

131. Consider the graphs below of

$$y_1 = \sin\theta,\ y_2 = e^{\theta}.$$

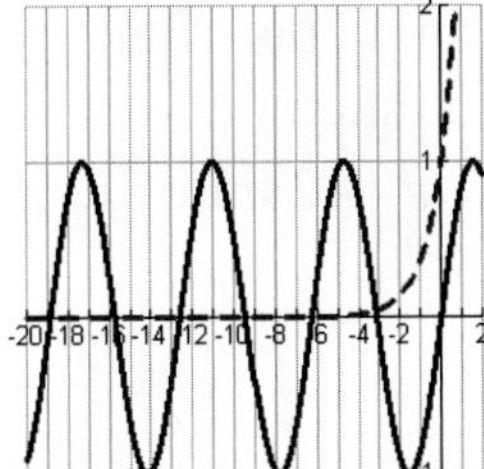

First, while there are no positive solutions of the equation $\sin\theta = e^{\theta}$, there are infinitely many negative solutions (at least one between each consecutive pair of x-intercepts). They are all irrational, and there is no apparent closed-form formula to generate them.

132. Consider the graphs below of

$$y_1 = \cos\theta,\ y_2 = e^{\theta}.$$

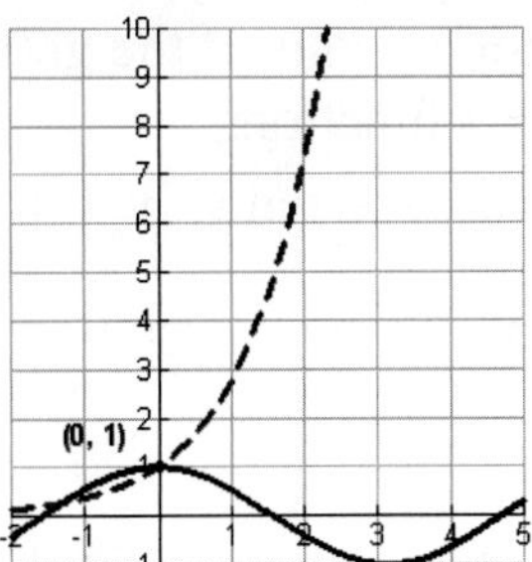

First, note that while there are no *positive* solutions of the equation $\cos\theta = e^{\theta}$, the two curves do intersect at $\theta = 0$.

133. To determine the smallest positive solution (approximately) of the equation $\sec 3x + \csc 2x = 5$ graphically, we search for the intersection points of the graphs of the following two functions:

$$y_1 = \sec 3x + \csc 2x,\ y_2 = 5$$

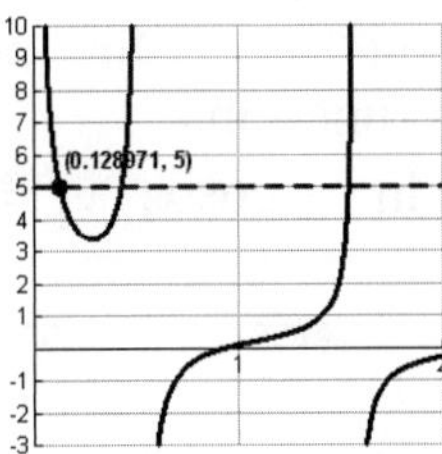

The x-coordinate of the intersection point is in radians. Observe that the smallest positive solution, *in degrees*, is approximately $\boxed{7.39^{\circ}}$.

134. To determine the smallest positive solution (approximately) of the equation $\cot 5x + \tan 2x = -3$ graphically, we search for the intersection points of the graphs of the following two functions:

$$y_1 = \cot 5x + \tan 2x,\ y_2 = -3$$

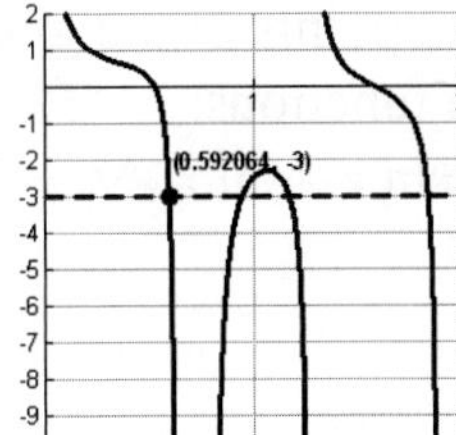

The x-coordinate of the intersection point is in radians. Observe that the smallest positive solution, *in degrees*, is approximately $\boxed{33.92^{\circ}}$.

135. To determine the smallest positive solution (approximately) of the equation $e^x - \tan x = 0$ graphically, we search for the intersection points of the graphs of the following two functions: $$y_1 = e^x - \tan x,\ y_2 = 0$$ 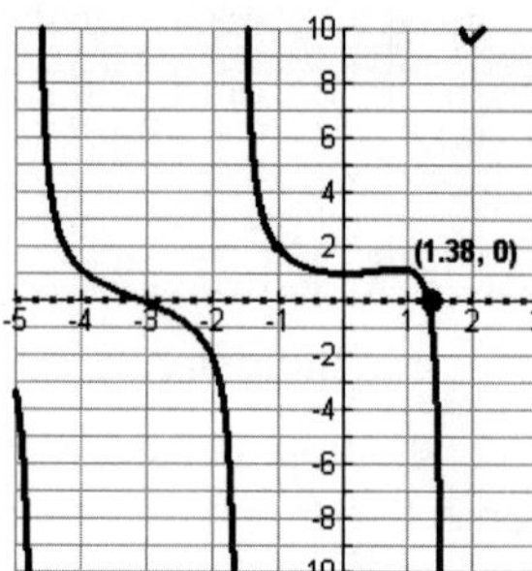 Observe that the graphs seem to intersect at approximately 1.3 radians, which is about $\boxed{79.07^\circ}$	**136.** To determine the smallest positive solution (approximately) of the equation $e^x + 2\sin x = 1$ graphically, we search for the intersection points of the graphs of the following two functions: $$y_1 = e^x + 2\sin x,\ y_2 = 1$$ 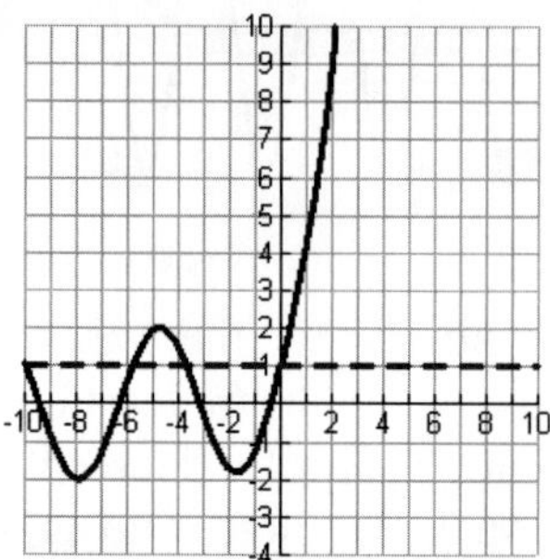 Observe that the two graphs intersect at 0, but never intersect at any *positive* x-value.
137. To determine the smallest positive solution (approximately) of the equation $\ln x - \sin x = 0$ graphically, we search for the intersection points of the graphs of the following two functions: $$y_1 = \ln x - \sin x,\ y_2 = 0$$ 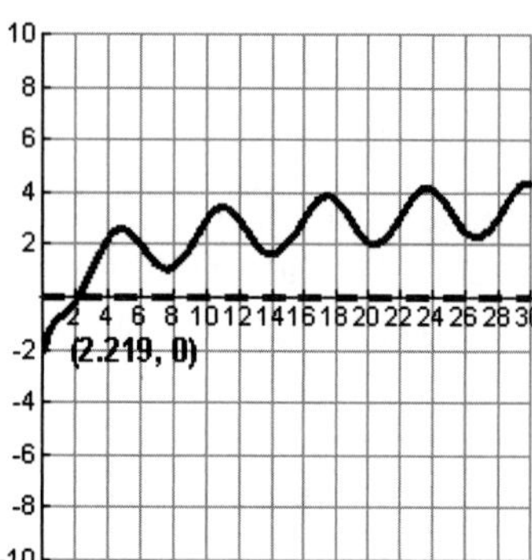 Observe that the graphs seems to intersect at approximately 2.219 radians, which is about $\boxed{127.1^\circ}$.	**138.** To determine the smallest positive solution (approximately) of the equation $\ln x - \cos x = 0$ graphically, we search for the intersection points of the graphs of the following two functions: $$y_1 = \ln x - \cos x,\ y_2 = 0$$ 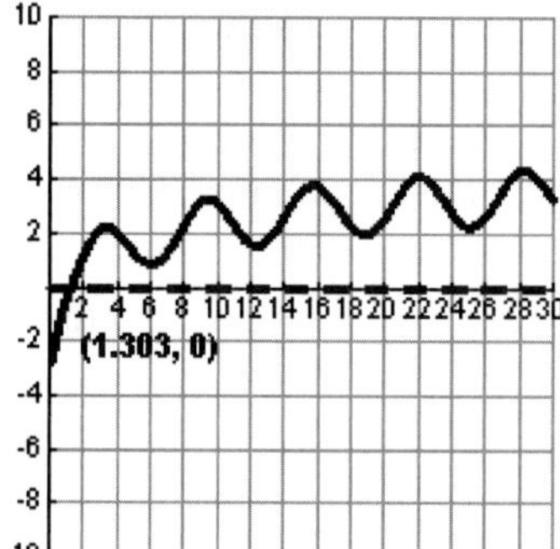 Observe that the graphs seems to intersect at approximately 1.303 radians, which is about $\boxed{74.66^\circ}$.

139. $$\begin{aligned}\cos x &= 2-\cos x\\ 2\cos x &= 2\\ \cos x &= 1\\ x &= 0, 2\pi\end{aligned}$$	**140.** $$\begin{aligned}\cos 2x &= \sin x\\ 1-2\sin^2 x &= \sin x\\ 2\sin^2 x+\sin x-1 &= 0\\ (2\sin x-1)(\sin x+1) &= 0\\ \sin x = \tfrac{1}{2} \text{ OR } \sin x &= -1\\ x &= \tfrac{\pi}{6}, -\tfrac{\pi}{2}\end{aligned}$$
141. $$\begin{aligned}2\sin x &= \frac{\sin x}{\cos x}\\ 2\sin x\cos x &= \sin x\\ \sin x(2\cos x-1) &= 0\\ \sin x = 0 \ OR\ 2\cos x-1 &= 0\\ x &= 0, \tfrac{\pi}{3}, -\tfrac{\pi}{3}\end{aligned}$$	**142.** $$\begin{aligned}\sin 2x-\cos 2x &= 0\\ (\sin 2x-\cos 2x)^2 &= 0\\ \sin^2 2x-2\sin 2x\cos 2x+\cos^2 2x &= 0\\ 1-\sin 4x &= 0\\ \sin 4x &= 1\\ 4x &= \tfrac{\pi}{2}, \tfrac{5\pi}{2}\\ x &= \tfrac{\pi}{8}, \tfrac{5\pi}{8}\end{aligned}$$

Chapter 6 Review Solutions --

1. $\sin(30^\circ) = \cos(90^\circ - 30^\circ) = \cos\left(\boxed{60^\circ}\right)$	**2.** $\cos(A) = \sin\left(\boxed{90^\circ - A}\right)$
3. $\tan(45^\circ) = \cot(90^\circ - 45^\circ) = \cot\left(\boxed{45^\circ}\right)$	**4.** $\csc(60^\circ) = \sec(90^\circ - 60^\circ) = \sec\left(\boxed{30^\circ}\right)$
5. $\sec(30^\circ) = \csc\left(\boxed{60^\circ}\right)$	**6.** $\cot(60^\circ) = \tan\left(\boxed{30^\circ}\right)$
7. $\tan x(\cot x+\tan x) = \tan x\cot x+\tan^2 x = 1+\tan^2 x = \sec^2 x$	
8. $(\sec x+1)(\sec x-1) = \sec^2 x-1 = (1+\tan^2 x)-1 = \tan^2 x$	
9. $\dfrac{\tan^4 x-1}{\tan^2 x-1} = \dfrac{\cancel{(\tan^2 x-1)}(\tan^2 x+1)}{\cancel{(\tan^2 x-1)}} = \tan^2 x+1 = \sec^2 x$	

10.	$\sec^2 x\left(\cot^2 x-\cos^2 x\right)=\frac{1}{\cos^2 x}\left(\frac{\cos^2 x}{\sin^2 x}-\cos^2 x\right)=\frac{1}{\sin^2 x}-1$ $=\frac{1-\sin^2 x}{\sin^2 x}=\frac{\cos^2 x}{\sin^2 x}=\cot^2 x$
11.	$\cos x\left(\cos(-x)-\tan(-x)\right)-\sin x=\cos x(\cos x+\tan x)-\sin x$ $=\cos^2 x+\cancel{\cos x}\left(\frac{\sin x}{\cancel{\cos x}}\right)-\sin x$ $=\cos^2 x+\sin x-\sin x=\cos^2 x$
12.	$\frac{\tan^2 x+1}{2\sec^2 x}=\frac{\cancel{\sec^2 x}}{2\cancel{\sec^2 x}}=\frac{1}{2}$
13.	$\frac{\csc^3(-x)+8}{\csc x-2}=\frac{8-\csc^3 x}{-(2-\csc x)}=\frac{\cancel{(2-\csc x)}\left(4+2\csc x+\csc^2 x\right)}{-\cancel{(2-\csc x)}}$ $=-\left(4+2\csc x+\csc^2 x\right)$
14.	$\frac{\csc^2 x-1}{\cot x}=\frac{\cot^2 x}{\cot x}=\cot x$
15.	$\left(\tan x+\cot x\right)^2-2=\tan^2 x+2\underbrace{\tan x\cot x}_{=1}+\cot^2 x-2$ $=\tan^2 x+2+\cot^2 x-2=\tan^2 x+\cot^2 x$
16.	$\csc^2 x-\cot^2 x=\frac{1}{\sin^2 x}-\frac{\cos^2 x}{\sin^2 x}=\frac{1-\cos^2 x}{\sin^2 x}=\frac{\sin^2 x}{\sin^2 x}=1$
17.	$\frac{1}{\sin^2 x}-\frac{1}{\tan^2 x}=\frac{1}{\sin^2 x}-\frac{\cos^2 x}{\sin^2 x}=\frac{1-\cos^2 x}{\sin^2 x}=\frac{\sin^2 x}{\sin^2 x}=1$
18.	$\frac{1}{\csc x+1}+\frac{1}{\csc x-1}=\frac{\left(\csc x-1\right)+\left(\csc x+1\right)}{\left(\csc x-1\right)\left(\csc x+1\right)}=\frac{2\csc x}{\csc^2 x-1}$ $=\frac{2\csc x}{\left(1+\cot^2 x\right)-1}=\frac{2\csc x}{\cot^2 x}=\frac{2\cdot\frac{1}{\sin x}}{\frac{\cos^2 x}{\sin^2 x}}=\frac{2}{\sin x}\cdot\frac{\sin^2 x}{\cos^2 x}$ $=\frac{2\sin x}{\cos x\cos x}=2\cdot\frac{\sin x}{\cos x}\cdot\frac{1}{\cos x}=\frac{2\tan x}{\cos x}$

19.

$$\frac{\tan^2 x - 1}{\sec^2 x + 3\tan x + 1} = \frac{\tan^2 x - 1}{\left(1+\tan^2 x\right) + 3\tan x + 1} = \frac{\tan^2 x - 1}{\tan^2 x + 3\tan x + 2}$$

$$= \frac{(\tan x - 1)\cancel{(\tan x + 1)}}{(\tan x + 2)\cancel{(\tan x + 1)}} = \frac{(\tan x - 1)}{(\tan x + 2)}$$

20.

$$\cot x(\sec x - \cos x) = \frac{\cos x}{\sin x}\left(\frac{1}{\cos x} - \cos x\right)$$

$$= \frac{\cos x}{\sin x}\left(\frac{1-\cos^2 x}{\cos x}\right) = \frac{\cancel{\cos x}}{\sin x}\left(\frac{\sin^2 x}{\cancel{\cos x}}\right) = \sin x$$

21. Observe that

$$2\tan^2 x + 1 = 2\cdot\frac{\sin^2 x}{\cos^2 x} + 1 = \frac{2\sin^2 x + \cos^2 x}{\cos^2 x}$$

$$= \frac{\sin^2 x + \left(\sin^2 x + \cos^2 x\right)}{\cos^2 x} = \frac{\sin^2 x + 1}{\cos^2 x} = \frac{1+\sin^2 x}{\cos^2 x}$$

So, the given equation is an *identity*.

22. The equation $\sin x - \cos x = 0$ is a *conditional* since, for instance, it does not hold if $x = \pi$ (since the left-side reduces to 1 in such case).

23. The equation $\cot^2 x - 1 = \tan^2 x$ is a *conditional* since, for instance, it does not hold if $x = \frac{\pi}{4}$ (since the left-side reduces to 0, while the right-side equals 1 in such case).

24. Observe that $\cos^2 x\left(1+\cot^2 x\right) = \cos^2 x \csc^2 x = \cos^2 x \cdot \dfrac{1}{\sin^2 x} = \cot^2 x$.

So, the given equation is an *identity*.

25. Since $\cot x = \dfrac{1}{\tan x}$, the given equation is an identity.

26. Observe that

$$\csc x + \sec x = \frac{1}{\sin x} + \frac{1}{\cos x} = \frac{\cos x + \sin x}{\sin x \cos x},$$

which is not equal to $\dfrac{1}{\sin x + \cos x}$ for all x. As such, the equation is a conditional.

27. Using the addition formula $\cos(A+B) = \cos A\cos B - \sin A\sin B$ yields

$$\cos\left(\frac{7\pi}{12}\right) = \cos\left(\frac{\pi}{3} + \frac{\pi}{4}\right) = \cos\left(\frac{\pi}{3}\right)\cos\left(\frac{\pi}{4}\right) - \sin\left(\frac{\pi}{3}\right)\sin\left(\frac{\pi}{4}\right)$$

$$= \left(\frac{1}{2}\right)\left(\frac{\sqrt{2}}{2}\right) - \left(\frac{\sqrt{3}}{2}\right)\left(\frac{\sqrt{2}}{2}\right) = \left(\frac{\sqrt{2}}{2}\right)\left(\frac{1}{2} - \frac{\sqrt{3}}{2}\right) = \boxed{\frac{\sqrt{2}-\sqrt{6}}{4}}$$

28. Using the addition formula $\sin(A-B)=\sin A\cos B-\cos A\sin B$ yields

$$\sin\left(\frac{\pi}{12}\right)=\sin\left(\frac{\pi}{3}-\frac{\pi}{4}\right)=\sin\left(\frac{\pi}{3}\right)\cos\left(\frac{\pi}{4}\right)-\cos\left(\frac{\pi}{3}\right)\sin\left(\frac{\pi}{4}\right)$$

$$=\left(\frac{\sqrt{3}}{2}\right)\left(\frac{\sqrt{2}}{2}\right)-\left(\frac{1}{2}\right)\left(\frac{\sqrt{2}}{2}\right)=\left(\frac{\sqrt{2}}{2}\right)\left(\frac{\sqrt{3}}{2}-\frac{1}{2}\right)=\boxed{\frac{\sqrt{6}-\sqrt{2}}{4}}$$

29. Using the addition formula $\tan(A-B)=\dfrac{\tan A-\tan B}{1+\tan A\tan B}$ yields

$$\tan(-15^\circ)=\tan(30^\circ-45^\circ)=\frac{\tan 30^\circ-\tan 45^\circ}{1+\tan 30^\circ\tan 45^\circ}\quad\textbf{(1)}$$

Now, observe that $\tan 30^\circ=\dfrac{1/2}{\sqrt{3}/2}=\dfrac{1}{\sqrt{3}}=\dfrac{\sqrt{3}}{3}$ and $\tan 45^\circ=1$. So, continuing the computation in **(1)** yields

$$\tan(-15^\circ)=\frac{\frac{\sqrt{3}}{3}-1}{1+\left(\frac{\sqrt{3}}{3}\right)}=\frac{\frac{\sqrt{3}-3}{\cancel{3}}}{\frac{3+\sqrt{3}}{\cancel{3}}}=\frac{\sqrt{3}-3}{3+\sqrt{3}}\cdot\frac{3-\sqrt{3}}{3-\sqrt{3}}=\frac{3\sqrt{3}-9-3+3\sqrt{3}}{9-3}=\frac{6\sqrt{3}-12}{6}=\boxed{\sqrt{3}-2}$$

30. First, we use the addition formula $\tan(A+B)=\dfrac{\tan A+\tan B}{1-\tan A\tan B}$ to compute $\tan 105^\circ$:

$$\tan(105^\circ)=\tan(60^\circ+45^\circ)=\frac{\tan 60^\circ+\tan 45^\circ}{1-\tan 60^\circ\tan 45^\circ}\quad\textbf{(1)}$$

Now, observe that $\tan 45^\circ=1$, and $\tan 60^\circ=\dfrac{\sin 60^\circ}{\cos 60^\circ}=\dfrac{\frac{\sqrt{3}}{2}}{\frac{1}{2}}=\sqrt{3}$. So, continuing the computation in **(1)** yields

$$\tan(105^\circ)=\frac{\sqrt{3}+1}{1-\sqrt{3}}=\frac{\sqrt{3}+1}{1-\sqrt{3}}\cdot\frac{1+\sqrt{3}}{1+\sqrt{3}}=\frac{1+2\sqrt{3}+3}{1-3}=\frac{4+2\sqrt{3}}{-2}=-2-\sqrt{3}.$$

Thus, $\cot 105^\circ=\dfrac{1}{\tan 105^\circ}=\dfrac{1}{-2-\sqrt{3}}=\dfrac{1}{-2-\sqrt{3}}\cdot\dfrac{-2+\sqrt{3}}{-2+\sqrt{3}}=\dfrac{-2+\sqrt{3}}{4-3}=\boxed{-2+\sqrt{3}}$

31. Using the addition formula $\sin(A-B)=\sin A\cos B-\cos A\sin B$ with $A=4x$ and $B=3x$ yields $\sin 4x\cos 3x-\cos 4x\sin 3x=\sin(4x-3x)=\boxed{\sin x}$.

32. Using the even/odd identities for cosine and sine, followed by the addition formula $\cos(A-B)=\cos A\cos B+\sin A\sin B$ with $A=x$ and $B=2x$, and finally one more application of the even identity of cosine, yields

$$\sin(-x)\sin(-2x)+\cos(-x)\cos(-2x)=\sin x\sin 2x+\cos x\cos 2x$$

$$=\cos x\cos 2x+\sin x\sin 2x=\cos(x-2x)=\cos(-x)=\boxed{\cos x}.$$

33. Using the addition formula $\tan(A-B)=\dfrac{\tan A-\tan B}{1+\tan A\tan B}$ with $A=5x$ and $B=4x$ yields $\dfrac{\tan 5x-\tan 4x}{1+\tan 5x\tan 4x}=\tan(5x-4x)=\boxed{\tan x}$.

34. Using the addition formula $\tan(A+B)=\dfrac{\tan A+\tan B}{1-\tan A\tan B}$ with $A=\dfrac{\pi}{4}$ and $B=\dfrac{\pi}{3}$ yields $\dfrac{\tan\frac{\pi}{4}+\tan\frac{\pi}{3}}{1-\tan\frac{\pi}{4}\tan\frac{\pi}{3}}=\tan\left(\dfrac{\pi}{4}+\dfrac{\pi}{3}\right)=\boxed{\tan\left(\dfrac{7\pi}{12}\right)}$.

35. Observe that since $\tan(\alpha-\beta)=\dfrac{\tan\alpha-\tan\beta}{1+\tan\alpha\tan\beta}$, we need to determine the values of $\cos\alpha$ and $\sin\beta$ so that we can then use this information in combination with the given values of $\sin\alpha$ and $\cos\beta$ in order to compute $\tan\alpha$ and $\tan\beta$. We proceed as follows:

Since the terminal side of α is in QIV, we know that $\cos\alpha>0$. As such, since we are given that $\sin\alpha=-\dfrac{3}{5}$, the diagram is as follows:

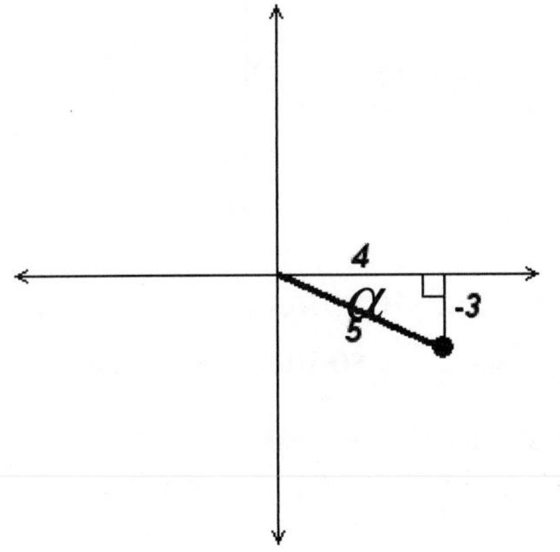

From the Pythagorean Theorem, we have $z^2+(-3)^2=5^2$, so that $z=4$. As such, $\cos\alpha=\dfrac{4}{5}$, and so,

$$\tan\alpha=\frac{\sin\alpha}{\cos\alpha}=\frac{-3/5}{4/5}=-\frac{3}{4}$$

Since the terminal side of β is in QIII, we know that $\sin\beta<0$. As such, since we are given that $\sin\beta=-\dfrac{24}{25}$, the diagram is as follows:

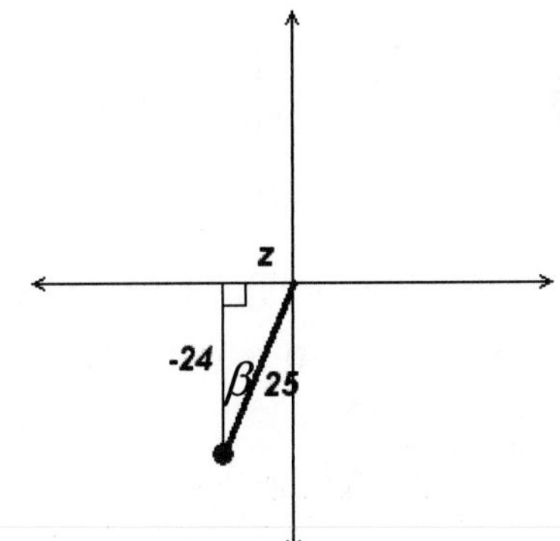

From the Pythagorean Theorem, we have $z^2+(-24)^2=25^2$, so that $z=-7$. As such, $\cos\beta=-\dfrac{7}{25}$, so

$$\tan\beta=\frac{\sin\beta}{\cos\beta}=\frac{-24/25}{-7/25}=\frac{24}{7}.$$

Hence,

$$\tan(\alpha-\beta)=\frac{\tan\alpha-\tan\beta}{1+\tan\alpha\tan\beta}=\frac{-\frac{3}{4}-\frac{24}{7}}{1+\left(-\frac{3}{4}\right)\left(\frac{24}{7}\right)}=\frac{\frac{-21-96}{28}}{\frac{28-72}{28}}=\boxed{\frac{117}{44}}.$$

36. Observe that $\cos(\alpha+\beta)=\cos\alpha\cos\beta-\sin\alpha\sin\beta=\left(-\frac{5}{13}\right)\cos\beta-\sin\alpha\left(\frac{7}{25}\right)$.

We need to compute both $\sin\alpha$ and $\sin\beta$. We proceed as follows:

Since the terminal side of α is in QII, we know that $\sin\alpha>0$. As such, since we are given that $\cos\alpha=-\frac{5}{13}$, the diagram is as follows:

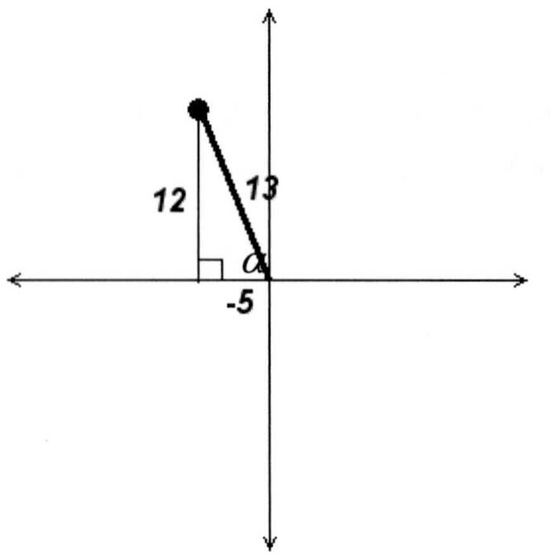

From the Pythagorean Theorem, we have $z^2+(-5)^2=13^2$, so that $z=12$.

Thus, $\sin\alpha=\frac{12}{13}$.

Since the terminal side of β is in QII, we know that $\cos\beta<0$. As such, since we are given that $\sin\beta=\frac{7}{25}$, the diagram is as follows:

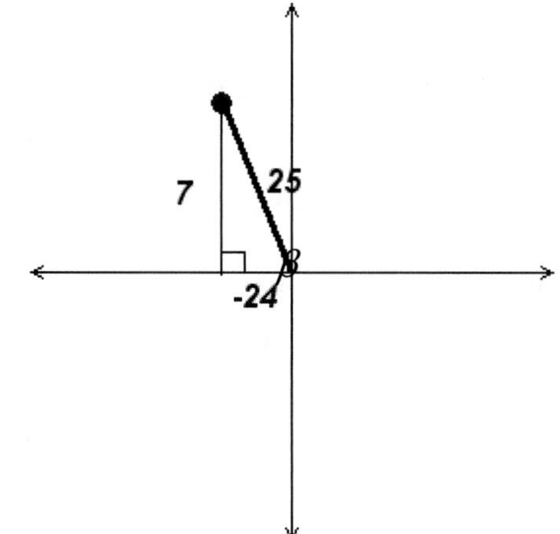

From the Pythagorean Theorem, we have $z^2+7^2=25^2$, so that $z=-24$.

Thus, $\cos\beta=-\frac{24}{25}$.

Hence,

$$\cos(\alpha+\beta)=\cos\alpha\cos\beta-\sin\alpha\sin\beta=\left(-\frac{5}{13}\right)\left(-\frac{24}{25}\right)-\left(\frac{12}{13}\right)\left(\frac{7}{25}\right)=\boxed{\frac{36}{325}}.$$

37. Observe that $\cos(\alpha-\beta)=\cos\alpha\cos\beta+\sin\alpha\sin\beta=\left(\frac{9}{41}\right)\left(\frac{7}{25}\right)+\sin\alpha\sin\beta$. We need to compute both $\sin\alpha$ and $\sin\beta$. We proceed as follows:

Since the terminal side of α is in QIV, we know that $\sin\alpha<0$. As such, since we are given that $\cos\alpha=\frac{9}{41}$, the diagram is as follows:

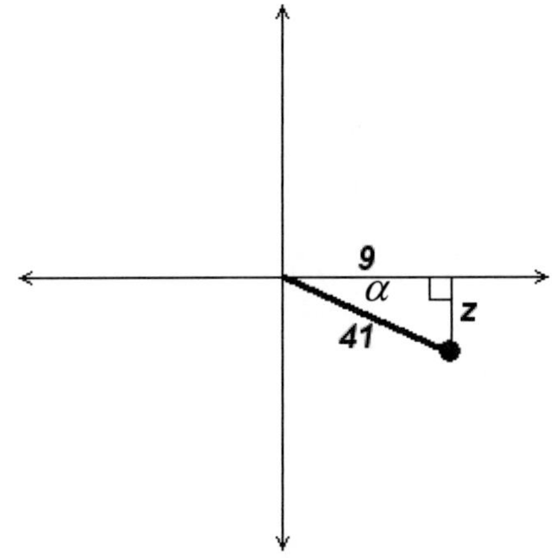

From the Pythagorean Theorem, we have $z^2+9^2=41^2$, so that $z=-40$.
Thus, $\sin\alpha=-\frac{40}{41}$.

Since the terminal side of β is in QI, we know that $\sin\beta>0$. As such, since we are given that $\cos\beta=\frac{7}{25}$, the diagram is as follows:

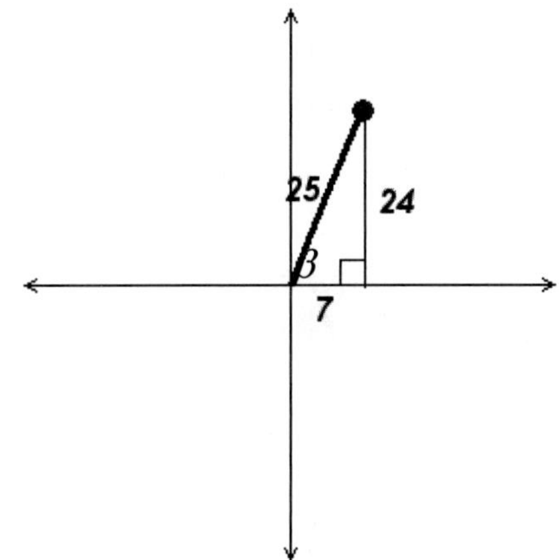

From the Pythagorean Theorem, we have $z^2+7^2=25^2$, so that $z=24$.
Thus, $\sin\beta=\frac{24}{25}$.

Hence,

$$\cos(\alpha-\beta)=\cos\alpha\cos\beta+\sin\alpha\sin\beta=\left(\frac{9}{41}\right)\left(\frac{7}{25}\right)+\left(-\frac{40}{41}\right)\left(\frac{24}{25}\right)=\boxed{-\frac{897}{1{,}025}}.$$

38. Observe that $\sin(\alpha-\beta)=\sin\alpha\cos\beta-\cos\alpha\sin\beta=\left(-\frac{5}{13}\right)\left(-\frac{4}{5}\right)-\cos\alpha\sin\beta$.

We need to compute both $\cos\alpha$ and $\sin\beta$. We proceed as follows:

Since the terminal side of α is in QIII, we know that $\cos\alpha<0$. As such, since we are given that $\sin\alpha=-\frac{5}{13}$, the diagram is as follows:

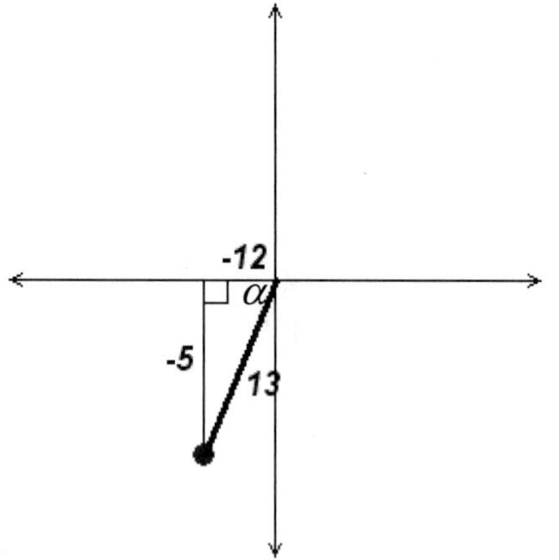

From the Pythagorean Theorem, we have $z^2+(-5)^2=13^2$, so that $z=-12$.

Thus, $\cos\alpha=-\frac{12}{13}$.

Since the terminal side of β is in QII, we know that $\sin\beta>0$. As such, since we are given that $\cos\beta=-\frac{4}{5}$, the diagram is as follows:

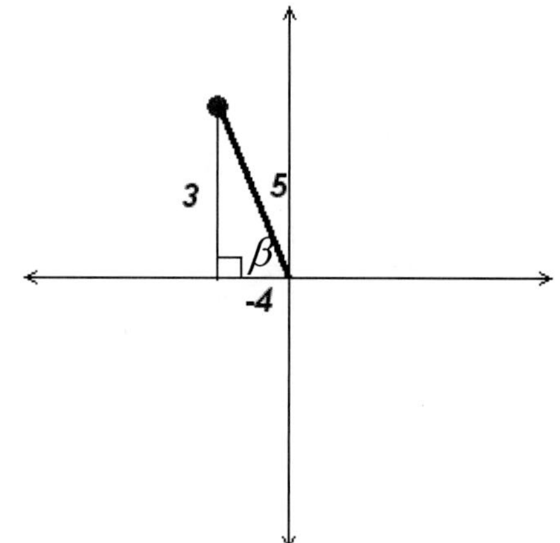

From the Pythagorean Theorem, we have $z^2+(-4)^2=5^2$, so that $z=3$.

Thus, $\sin\beta=\frac{3}{5}$.

Hence,

$$\sin(\alpha-\beta)=\sin\alpha\cos\beta-\cos\alpha\sin\beta=\left(-\frac{5}{13}\right)\left(-\frac{4}{5}\right)-\left(-\frac{12}{13}\right)\left(\frac{3}{5}\right)=\boxed{\frac{56}{65}}.$$

39. Observe that

$$\begin{aligned}\cos(A+B)+\cos(A-B)&=(\cos A\cos B-\cancel{\sin A\sin B})+(\cos A\cos B+\cancel{\sin A\sin B})\\&=2\cos A\cos B\end{aligned}$$

So, the given equation is an *identity*.

40. Observe that

$$\begin{aligned}\cos(A-B)-\cos(A+B)&=(\cancel{\cos A\cos B}+\sin A\sin B)-(\cancel{\cos A\cos B}-\sin A\sin B)\\&=2\sin A\sin B\end{aligned}$$

So, the given equation is an *identity*.

41. Before graphing, observe that

$$y = \underbrace{\cos\left(\frac{\pi}{2}\right)}_{=0}\cos x - \underbrace{\sin\left(\frac{\pi}{2}\right)}_{=1}\sin x = -\sin x.$$

So, the graph is simply the graph of $y = \sin x$ reflected over the x-axis, as seen below:

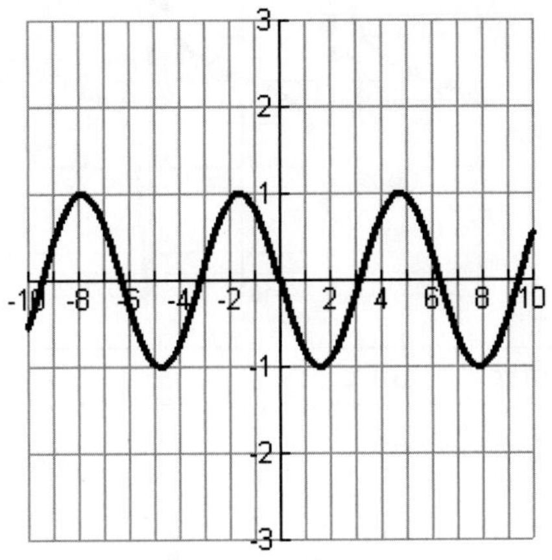

42. Using the addition formula $\sin(A+B) = \sin A\cos B + \cos A\sin B$ with $A = x$ and $B = \frac{2\pi}{3}$ yields $y = \cos x \sin\left(\frac{2\pi}{3}\right) + \sin x \cos\left(\frac{2\pi}{3}\right) = \sin\left(x + \frac{2\pi}{3}\right)$.

So, the graph is the graph of $y = \sin x$ shifted to the left $\frac{2\pi}{3}$ units, as seen below.

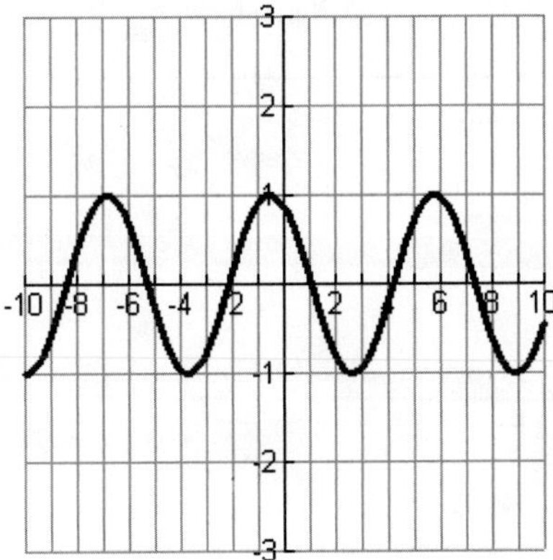

43. Observe that $y = \dfrac{2\tan\left(\frac{x}{3}\right)}{1-\tan^2\left(\frac{x}{3}\right)} = \tan\left(\frac{2x}{3}\right)$. The graph is as follows:

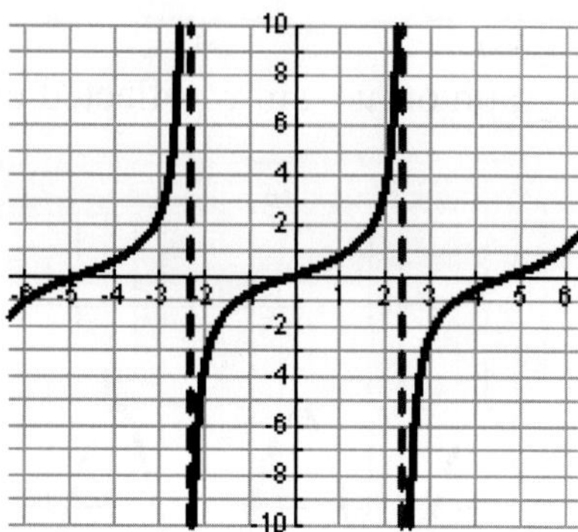

44. Observe that $y = \dfrac{\tan(\pi x)-\tan x}{1+\tan(\pi x)\tan x} = \tan\left((\pi-1)x\right)$. The graph is as follows:

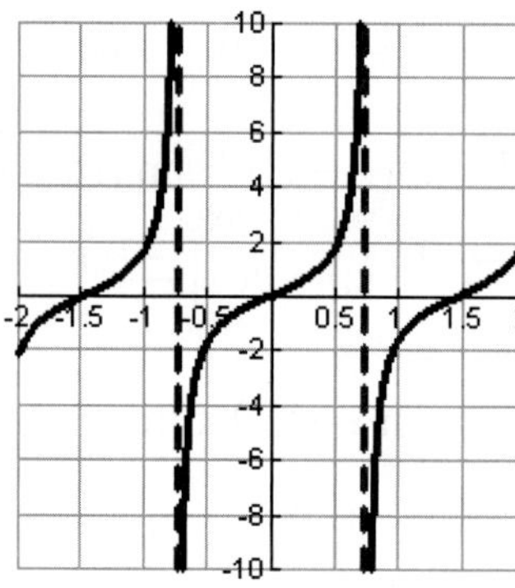

45. Since the terminal side of x is in QII, $\sin x = \frac{3}{5}$ and $\cos x < 0$. The diagram is as follows:

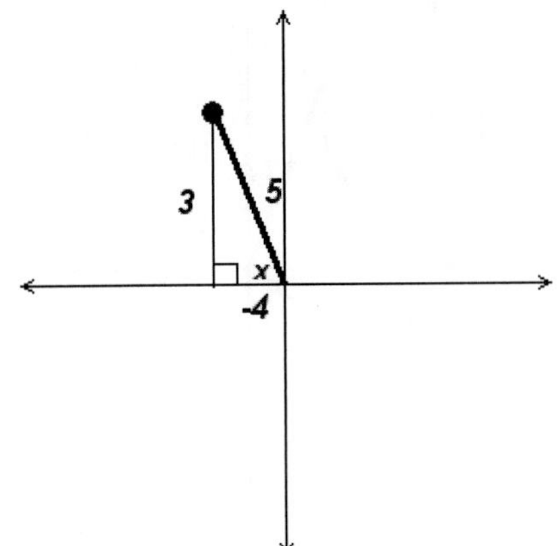

From the Pythagorean Theorem, we have $z^2 + 3^2 = 5^2$, so that $z = -4$.

As such, $\cos x = -\frac{4}{5}$. Therefore,

$$\cos 2x = \cos^2 x - \sin^2 x$$
$$= \left(-\frac{4}{5}\right)^2 - \left(\frac{3}{5}\right)^2 = \frac{16}{25} - \frac{9}{25} = \boxed{\frac{7}{25}}$$

46. Since the terminal side of x is in QIV, $\cos x = \frac{7}{25}$ and $\sin x < 0$. The diagram is as follows:

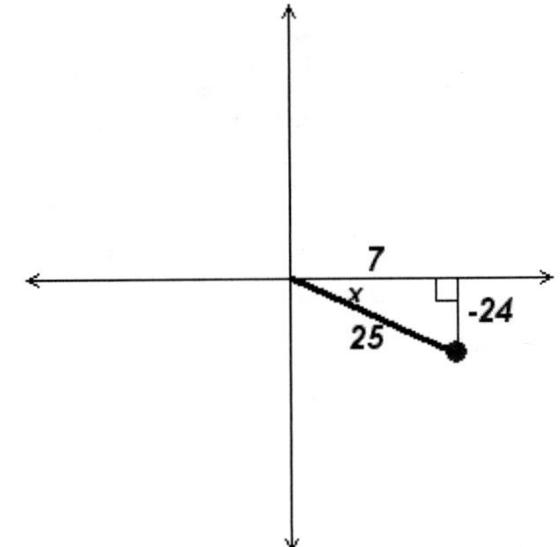

From the Pythagorean Theorem, we have $z^2 + 7^2 = 25^2$, so that $z = -24$.

As such, $\sin x = -\frac{24}{25}$. Therefore,

$$\sin 2x = 2 \sin x \cos x$$
$$= 2\left(-\frac{24}{25}\right)\left(\frac{7}{25}\right) = \boxed{-\frac{336}{625}}$$

47. Observe that since $\cot x = -\frac{11}{61}$, it follows that $\tan x = -\frac{61}{11}$. So,

$$\tan 2x = \frac{2\tan x}{1-\tan^2 x} = \frac{2\left(-\frac{61}{11}\right)}{1-\left(-\frac{61}{11}\right)^2}$$

$$= \frac{-\frac{122}{11}}{-\frac{3{,}600}{121}} = \boxed{\frac{671}{1{,}800}}$$

48. Since $\tan x = \frac{12}{-5}$ and $\frac{\pi}{2} < x < \pi$, the diagram is as follows:

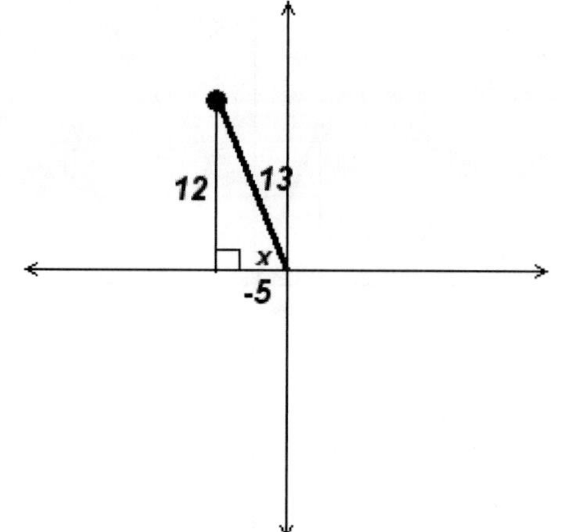

From the Pythagorean Theorem, we have $(-5)^2 + 12^2 = z^2$, so that $z = 13$.

As such, $\sin x = \frac{12}{13}$ and $\cos x = -\frac{5}{13}$.

Therefore,

$$\cos 2x = \cos^2 x - \sin^2 x = \left(-\frac{5}{13}\right)^2 - \left(\frac{12}{13}\right)^2$$

$$= \boxed{-\frac{119}{169}}$$

49. Since $\sec x = \frac{25}{24}$, it follows that $\frac{1}{\cos x} = \frac{25}{24}$ and so, $\cos x = \frac{24}{25}$. Since the terminal side of x must be in QI, we see $\sin x > 0$. The diagram is as follows:

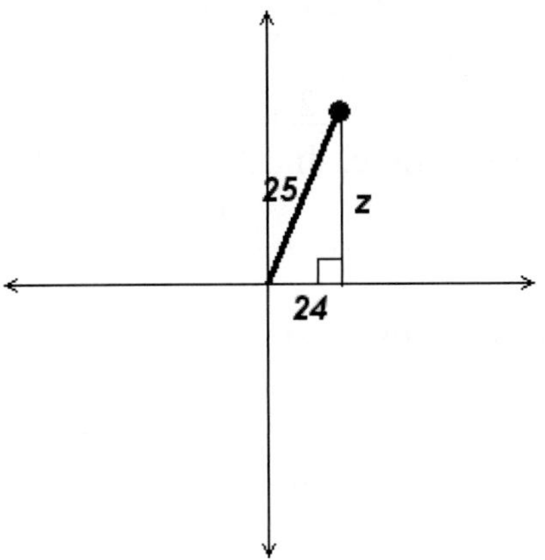

From the Pythagorean Theorem, we have $z^2 + 24^2 = 25^2$, so that $z = 7$.

As such, $\sin x = \frac{7}{25}$. Therefore,

$$\sin 2x = 2\sin x \cos x = 2\left(\frac{7}{25}\right)\left(\frac{24}{25}\right) = \boxed{\frac{336}{625}}$$

50. Since $\csc x = \frac{5}{4}$, it follows that $\frac{1}{\sin x} = \frac{5}{4}$ and so, $\sin x = \frac{4}{5}$. Since the terminal side of x must be in QII, we see $\cos x < 0$, The diagram is as follows:

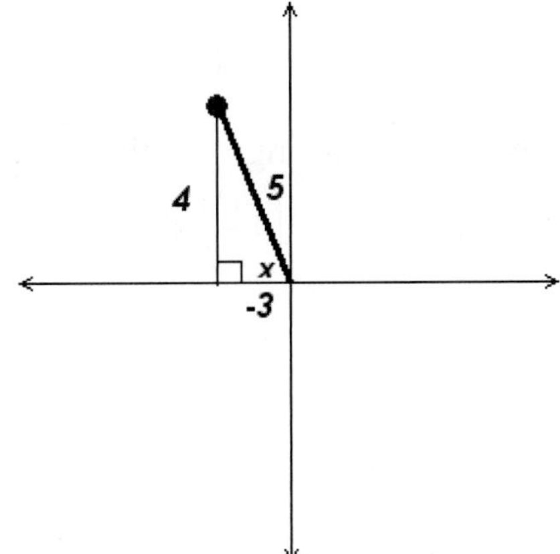

From the Pythagorean Theorem, we have $z^2 + 4^2 = 5^2$, so that $z = -3$.

As such, $\tan x = -\frac{4}{3}$. Therefore,

$$\tan 2x = \frac{2\tan x}{1-\tan^2 x} = \frac{2\left(-\frac{4}{3}\right)}{1-\left(-\frac{4}{3}\right)^2} = \boxed{\frac{24}{7}}.$$

51. Using $\cos 2A = \cos^2 A - \sin^2 A$ with $A = 15^\circ$ yields

$$\cos^2 15^\circ - \sin^2 15^\circ = \cos(2 \cdot 15^\circ) = \cos\left(30^\circ\right) = \boxed{\frac{\sqrt{3}}{2}}.$$

52. Using $\tan 2A = \frac{2\tan A}{1-\tan^2 A}$ with $A = -\frac{\pi}{12}$ yields

$$\frac{2\tan\left(-\frac{\pi}{12}\right)}{1-\tan^2\left(-\frac{\pi}{12}\right)} = \tan\left(2\cdot\left(-\frac{\pi}{12}\right)\right) = \tan\left(-\frac{\pi}{6}\right) = \frac{\sin\left(-\frac{\pi}{6}\right)}{\cos\left(-\frac{\pi}{6}\right)} = \frac{-\sin\left(\frac{\pi}{6}\right)}{\cos\left(\frac{\pi}{6}\right)} = -\frac{\frac{1}{2}}{\frac{\sqrt{3}}{2}} = \boxed{-\frac{\sqrt{3}}{3}}$$

53. Using $\sin 2A = 2\sin A\cos A$ with $A = \frac{\pi}{12}$ yields

$$6\sin\left(\frac{\pi}{12}\right)\cos\left(\frac{\pi}{12}\right) = 3\left[2\sin\left(\frac{\pi}{12}\right)\cos\left(\frac{\pi}{12}\right)\right] = 3\sin\left(2\cdot\frac{\pi}{12}\right) = 3\sin\left(\frac{\pi}{6}\right) = 3\left(\frac{1}{2}\right) = \boxed{\frac{3}{2}}.$$

54. Using $\cos 2A = 1 - 2\sin^2 A$ with $A = \frac{\pi}{8}$ yields

$$1 - 2\sin^2\left(\frac{\pi}{8}\right) = \cos\left(2\cdot\frac{\pi}{8}\right) = \cos\left(\frac{\pi}{4}\right) = \boxed{\frac{\sqrt{2}}{2}}.$$

55. Note that starting with the right-side is easier this time. Indeed, observe that

$$\begin{aligned}(\sin A - \cos A)\left(1 + \frac{1}{2}\sin 2A\right) &= (\sin A - \cos A)\left(1 + \frac{1}{2}2\sin A\cos A\right)\\ &= \sin A + \sin^2 A\cos A - \cos A - \sin A\cos^2 A\\ &= \sin A + \left(1 - \cos^2 A\right)\cos A - \cos A - \sin A\left(1 - \sin^2 A\right)\\ &= \cancel{\sin A} + \cancel{\cos A} - \cos^3 A - \cancel{\cos A} - \cancel{\sin A} + \sin^3 A\\ &= \sin^3 A - \cos^3 A\end{aligned}$$

56. Observe that

$$2\sin A\cos^3 A - 2\sin^3 A\cos A = (2\sin A\cos A)\left(\cos^2 A - \sin^2 A\right) = (\sin 2A)(\cos 2A).$$

57. Observe that

$$\frac{\sin 2A}{1 + \cos 2A} = \tan\left(2\cdot\frac{A}{2}\right) = \tan A.$$

58. Observe that

$$\frac{1 - \cos 2A}{\sin 2A} = \tan\left(2\cdot\frac{A}{2}\right) = \tan A.$$

59. Observe that

$$V = \frac{2\cos 2\theta}{1 + \cos 2\theta} = \frac{2\left[\cos^2\theta - \sin^2\theta\right]}{1 + \left(\cos^2\theta - \sin^2\theta\right)} = \frac{2\left[\cos^2\theta - \sin^2\theta\right]}{1 + \cos^2\theta - \left(1 - \cos^2\theta\right)} = \frac{\cancel{2}\left[\cos^2\theta - \sin^2\theta\right]}{\cancel{2}\cos^2\theta}$$

$$= \frac{\cos^2\theta}{\cos^2\theta} - \frac{\sin^2\theta}{\cos^2\theta} = 1 - \tan^2\theta$$

60. Using Exercise 59, observe that

$$V\left(\frac{\pi}{6}\right) = 1 - \tan^2\left(\frac{\pi}{6}\right) = 1 - \left[\frac{\sin\left(\frac{\pi}{6}\right)}{\cos\left(\frac{\pi}{6}\right)}\right]^2 = 1 - \left[\frac{1/2}{\sqrt{3}/2}\right]^2 = 1 - \frac{1}{3} = \boxed{\frac{2}{3}}.$$

61. Use $\sin\left(\frac{A}{2}\right) = -\sqrt{\frac{1 - \cos A}{2}}$ with $A = -45^\circ$ to obtain

$$\sin\left(-22.5^\circ\right) = \sin\left(\frac{-45^\circ}{2}\right) = -\sqrt{\frac{1 - \cos\left(-45^\circ\right)}{2}} = -\sqrt{\frac{1 - \frac{\sqrt{2}}{2}}{2}} = -\sqrt{\frac{2 - \sqrt{2}}{4}} = \boxed{-\frac{\sqrt{2 - \sqrt{2}}}{2}}.$$

62. Use $\cos\left(\frac{A}{2}\right)=\sqrt{\frac{1+\cos A}{2}}$ with $A=135^\circ$ to obtain

$$\cos\left(67.5^\circ\right)=\cos\left(\frac{135^\circ}{2}\right)=\sqrt{\frac{1+\cos 135^\circ}{2}}=\sqrt{\frac{1+\left(-\frac{\sqrt{2}}{2}\right)}{2}}=\sqrt{\frac{2-\sqrt{2}}{4}}=\boxed{\frac{\sqrt{2-\sqrt{2}}}{2}}.$$

63. First, note that using $\tan\left(\frac{A}{2}\right)=\sqrt{\frac{1-\cos A}{1+\cos A}}$ with $A=\frac{3\pi}{4}$, we obtain

$$\cot\left(\frac{3\pi}{8}\right)=\frac{1}{\tan\left(\frac{3\pi}{8}\right)}=\frac{1}{\tan\left(\frac{3\pi/4}{2}\right)}=\frac{1}{\sqrt{\frac{1-\cos\left(3\pi/4\right)}{1+\cos\left(3\pi/4\right)}}} \quad \textbf{(1)}$$

Next, using $\cos\left(3\pi/4\right)=-\frac{\sqrt{2}}{2}$ in **(1)** subsequently yields:

$$\cot\left(\frac{3\pi}{8}\right)=\frac{1}{\sqrt{\frac{1-\left(-\frac{\sqrt{2}}{2}\right)}{1+\left(-\frac{\sqrt{2}}{2}\right)}}}=\frac{1}{\sqrt{\frac{\frac{2+\sqrt{2}}{\not 2}}{\frac{2-\sqrt{2}}{\not 2}}}}=\frac{1}{\sqrt{\frac{2+\sqrt{2}}{2-\sqrt{2}}}}=\frac{1}{\sqrt{\frac{2+\sqrt{2}}{2-\sqrt{2}}\cdot\frac{2+\sqrt{2}}{2+\sqrt{2}}}}=\frac{1}{\sqrt{\frac{4+4\sqrt{2}+2}{4-2}}}=\boxed{\frac{1}{\sqrt{3+2\sqrt{2}}}}$$

<u>Note</u>: Recall that there are two other formulae equivalent to $\tan\left(\frac{A}{2}\right)=\sqrt{\frac{1-\cos A}{1+\cos A}}$, and either one could be used in place of this one to obtain an equivalent (but perhaps different looking) result. For instance, using $\tan\left(\frac{A}{2}\right)=\frac{\sin A}{1+\cos A}$ with $A=\frac{3\pi}{4}$ yields

$$\cot\left(\frac{3\pi}{8}\right)=\frac{1}{\tan\left(\frac{3\pi}{8}\right)}=\frac{1}{\tan\left(\frac{3\pi/4}{2}\right)}=\frac{1}{\frac{\sin\left(3\pi/4\right)}{1+\cos\left(3\pi/4\right)}} \quad \textbf{(2)}$$

Since $\cos\left(3\pi/4\right)=-\frac{\sqrt{2}}{2}$ and $\sin\left(3\pi/4\right)=\frac{\sqrt{2}}{2}$, we can use these in **(2)** to obtain

$$\cot\left(\frac{13\pi}{8}\right)=\frac{1}{\frac{\frac{\sqrt{2}}{2}}{1-\frac{\sqrt{2}}{2}}}=\frac{2-\sqrt{2}}{\sqrt{2}}=\frac{2-\sqrt{2}}{\sqrt{2}}\cdot\frac{\sqrt{2}}{\sqrt{2}}=\frac{2\sqrt{2}-2}{2}=\boxed{\sqrt{2}-1}.$$

Even though the forms of the two answers are different, they are equivalent.

64. First, note that using the formula $\cos\left(\frac{A}{2}\right)=-\sqrt{\frac{1+\cos A}{2}}$ with $A=\frac{7\pi}{4}$ yields

$$\cos\left(\frac{7\pi}{8}\right)=\cos\left(\frac{7\pi/4}{2}\right)=-\sqrt{\frac{1+\cos\left(\frac{7\pi}{4}\right)}{2}}=-\sqrt{\frac{1+\frac{\sqrt{2}}{2}}{2}}=-\frac{\sqrt{2+\sqrt{2}}}{2}.$$

Now, using $\sin A=-\sqrt{1-\cos^2 A}$ with $A=\frac{7\pi}{8}$ subsequently yields

$$\csc\left(\frac{7\pi}{8}\right)=\frac{1}{\sin\left(\frac{7\pi}{8}\right)}=\frac{1}{-\sqrt{1-\cos^2\left(\frac{7\pi}{8}\right)}}=\frac{-1}{\sqrt{1-\left[-\frac{\sqrt{2+\sqrt{2}}}{2}\right]^2}}=\frac{-1}{\sqrt{1-\frac{2+\sqrt{2}}{4}}}=\frac{-1}{\sqrt{\frac{2-\sqrt{2}}{4}}}=\boxed{\frac{-2}{\sqrt{2-\sqrt{2}}}}$$

So, $\csc\left(-\frac{7\pi}{8}\right)=-\csc\left(\frac{7\pi}{8}\right)=\frac{2}{\sqrt{2-\sqrt{2}}}$.

65. First, note that $\sec\left(-165^\circ\right)=\sec\left(-\frac{330^\circ}{2}\right)=\frac{1}{\cos\left(-\frac{330^\circ}{2}\right)}=\frac{1}{\cos\left(\frac{30^\circ}{2}\right)}$. Now, use

$\cos\left(\frac{A}{2}\right)=\sqrt{\frac{1+\cos A}{2}}$ with $A=30^\circ$ to obtain

$$\cos\left(\frac{30^\circ}{2}\right)=\sqrt{\frac{1+\cos\left(30^\circ\right)}{2}}=\sqrt{\frac{1+\frac{\sqrt{3}}{2}}{2}}.$$

Hence,

$$\sec\left(-165^\circ\right)=\frac{1}{\cos\left(\frac{30^\circ}{2}\right)}=\frac{1}{\sqrt{\frac{1+\frac{\sqrt{3}}{2}}{2}}}=\boxed{\frac{2}{\sqrt{2+\sqrt{3}}}}.$$

66. Using $\tan\left(\frac{A}{2}\right)=-\sqrt{\frac{1-\cos A}{1+\cos A}}$ with $A=-150^\circ$ yields

$$\tan\left(-75^\circ\right)=\tan\left(\frac{-150^\circ}{2}\right)=-\sqrt{\frac{1-\cos\left(-150^\circ\right)}{1+\cos\left(-150^\circ\right)}}=-\sqrt{\frac{1+\frac{\sqrt{3}}{2}}{1-\frac{\sqrt{3}}{2}}}=\boxed{-\sqrt{\frac{2+\sqrt{3}}{2-\sqrt{3}}}}.$$

67. Since the terminal side of x lies in QIII, $\sin x = -\frac{7}{25}$ and $\cos x < 0$, The diagram is as follows:

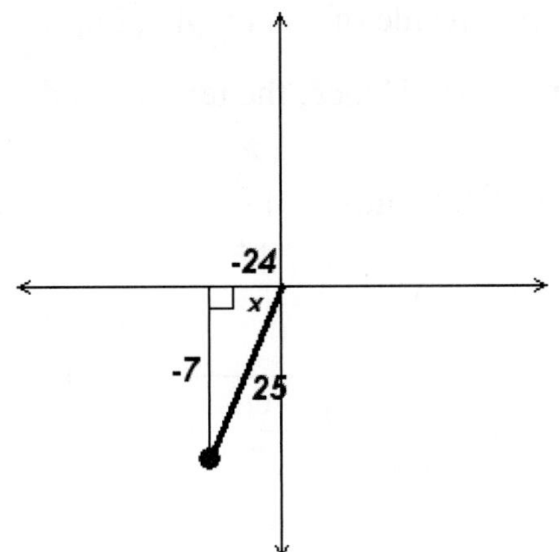

Hence, the terminal side of $\frac{x}{2}$ lies in QII, and so $\sin\left(\frac{x}{2}\right) > 0$. Therefore, we have

$$\sin\left(\frac{x}{2}\right) = \sqrt{\frac{1-\cos x}{2}} = \sqrt{\frac{1-\left(-\frac{24}{25}\right)}{2}}$$

$$= \sqrt{\frac{49}{50}} = \frac{7}{5\sqrt{2}} = \boxed{\frac{7\sqrt{2}}{10}}.$$

68. Since the terminal side of x lies in QII $\cos x = -\frac{4}{5}$ and $\sin x > 0$. Hence, the terminal side of $\frac{x}{2}$ lies in QI, and so $\cos\left(\frac{x}{2}\right) > 0$. Therefore, we have

$$\cos\left(\frac{x}{2}\right) = \sqrt{\frac{1+\cos x}{2}} = \sqrt{\frac{1+\left(-\frac{4}{5}\right)}{2}}.$$

$$= \sqrt{\frac{1}{10}} = \boxed{\frac{\sqrt{10}}{10}}.$$

69. The diagram is as follows:

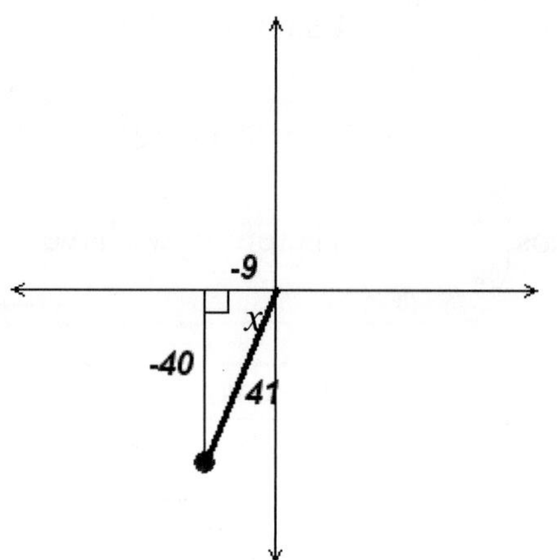

Note that $\cos x = -\frac{9}{41}$.

Since the terminal side of x is in QIII, the terminal side of $\frac{x}{2}$ lies in QII, and so $\tan\left(\frac{x}{2}\right) < 0$. Therefore, we have

$$\tan\left(\frac{x}{2}\right) = -\sqrt{\frac{1-\cos x}{1+\cos x}} = -\sqrt{\frac{1-\left(-\frac{9}{41}\right)}{1+\left(-\frac{9}{41}\right)}} =$$

$$= -\sqrt{\frac{50}{32}} = -\sqrt{\frac{25}{16}} = \boxed{-\frac{5}{4}}$$

70. Since $\sec x = \frac{17}{15}$, we know that $\cos x = \frac{15}{17}$. This, together with the terminal side of x is in QIV, implies that $\sin x < 0$. Hence, the terminal side of $\frac{x}{2}$ is in QII. Thus, $\sin\left(\frac{x}{2}\right) > 0$.

So, we have

$$\sin\left(\frac{x}{2}\right) = \sqrt{\frac{1-\cos x}{2}} = \sqrt{\frac{1-\left(\frac{15}{17}\right)}{2}}$$

$$= \sqrt{\frac{2}{2(17)}} = \boxed{\frac{\sqrt{17}}{17}}$$

71. Using the formula $\sin\left(\frac{A}{2}\right) = \sqrt{\frac{1-\cos A}{2}}$ with $A = \frac{\pi}{6}$ yields

$$\sqrt{\frac{1-\cos\left(\frac{\pi}{6}\right)}{2}} = \sin\left(\frac{\pi/6}{2}\right) = \boxed{\sin\left(\frac{\pi}{12}\right)}.$$

72. Using the formula $\tan\left(\frac{A}{2}\right)=\sqrt{\frac{1-\cos A}{1+\cos A}}$ with $A=\frac{11\pi}{6}$ yields

$$\sqrt{\frac{1-\cos\frac{11\pi}{6}}{1+\cos\frac{11\pi}{6}}}=\tan\left(\frac{\frac{11\pi}{6}}{2}\right)=\boxed{\tan\left(\frac{11\pi}{12}\right)}.$$

73.

$$\left(\sin\frac{A}{2}+\cos\frac{A}{2}\right)^2=\sin^2\frac{A}{2}+2\sin\frac{A}{2}\cos\frac{A}{2}+\cos^2\frac{A}{2}$$

$$=\left(\sin^2\frac{A}{2}+\cos^2\frac{A}{2}\right)+\left(2\sin\frac{A}{2}\cos\frac{A}{2}\right)=1+\sin\left(2\cdot\frac{A}{2}\right)=1+\sin A$$

74.

$$\sec^2\frac{A}{2}+\tan^2\frac{A}{2}=\frac{1}{\cos^2\frac{A}{2}}+\tan^2\frac{A}{2}=\frac{1}{\frac{1+\cos A}{2}}+\frac{1-\cos A}{1+\cos A}$$

$$=\frac{2}{1+\cos A}+\frac{1-\cos A}{1+\cos A}=\frac{3-\cos A}{1+\cos A}$$

75. Since $\cot^2\frac{A}{2}=\frac{1}{\tan^2\frac{A}{2}}=\frac{1}{\frac{1-\cos A}{1+\cos A}}$, we see that

$$\csc^2\frac{A}{2}+\cot^2\frac{A}{2}=\frac{1}{\sin^2\frac{A}{2}}+\cot^2\frac{A}{2}=\frac{1}{\frac{1-\cos A}{2}}+\frac{1+\cos A}{1-\cos A}$$

$$=\frac{2}{1-\cos A}+\frac{1+\cos A}{1-\cos A}=\frac{3+\cos A}{1-\cos A}$$

76. Use the known identity $1+\tan^2\theta=\sec^2\theta$ with $\theta=\frac{A}{2}$ to immediately conclude that $1+\tan^2\frac{A}{2}=\sec^2\frac{A}{2}$.

77. Before graphing, observe that

$$y=\sqrt{\frac{1-\cos\left(\frac{\pi}{12}x\right)}{2}}=\left|\sin\left(\frac{\frac{\pi}{12}x}{2}\right)\right|=\left|\sin\left(\frac{\pi}{24}x\right)\right|.$$

This graph has the same shape as the graph of $y=|\sin x|$ with period

$\frac{\pi}{\pi/24}=24$, as seen below:

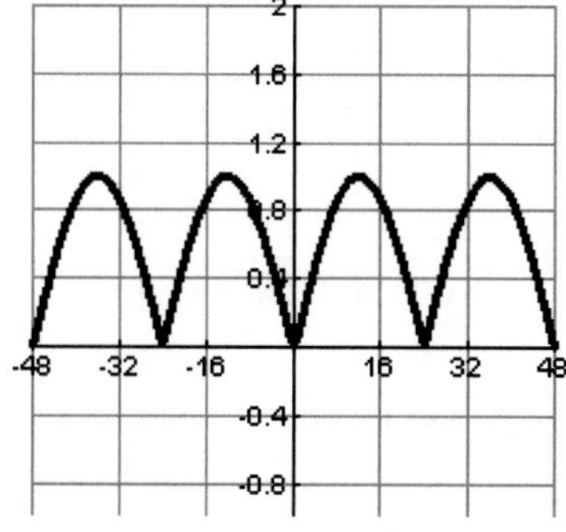

78. Before graphing, observe that

$$y=\cos^2\frac{x}{2}-\sin^2\frac{x}{2}=\cos\left(2\cdot\frac{x}{2}\right)=\cos x$$

So, this graph is simply the graph of $y=\cos x$, as seen below:

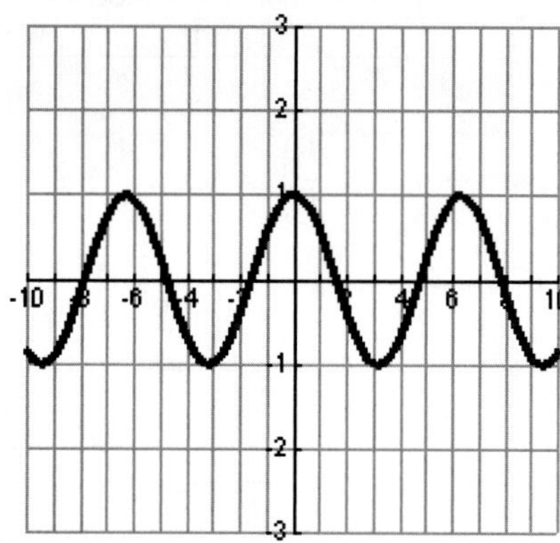

79. Observe that

$$y=-\sqrt{\frac{1-\cos x}{1+\cos x}}=-\tan\left(\frac{x}{2}\right).$$

The graph is as follows:

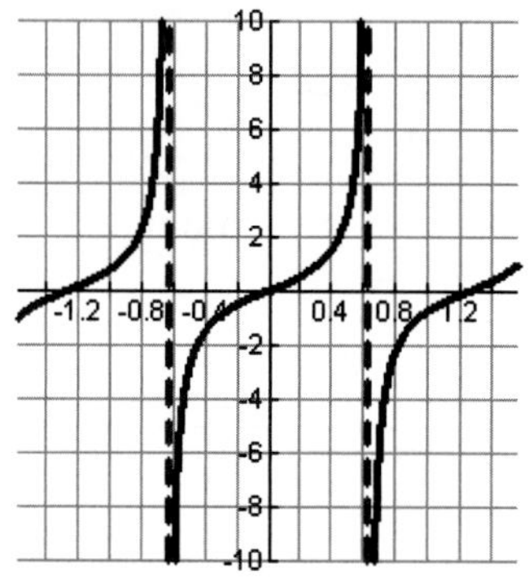

80. Observe that

$$y=\sqrt{\frac{1+\cos(3x-1)}{2}}=\cos\left(\frac{3x-1}{2}\right).$$

The graph is as follows:

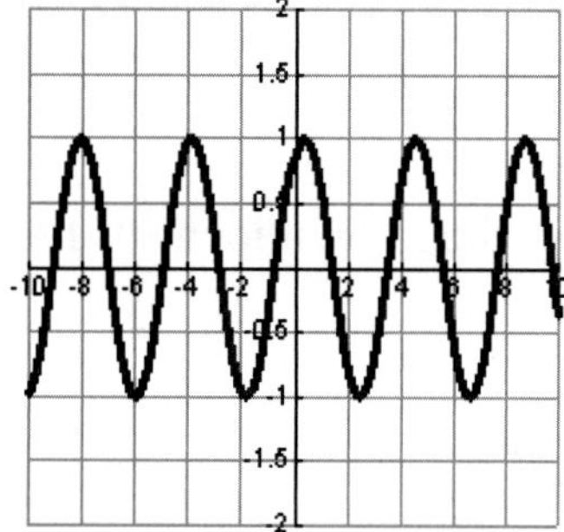

81. $6\sin 5x\cos 2x=6\cdot\frac{1}{2}[\sin(5x+2x)+\sin(5x-2x)]=\boxed{3[\sin 7x+\sin 3x]}$

82. $3\sin 4x\sin 2x=3\cdot\frac{1}{2}[\cos(4x-2x)-\cos(4x+2x)]=\boxed{\frac{3}{2}[\cos 2x-\cos 6x]}$

83. $$\cos 5x - \cos 3x = -2\sin\left(\frac{5x+3x}{2}\right)\sin\left(\frac{5x-3x}{2}\right) = \boxed{-2\sin(4x)\sin(x)}$$

84. $$\sin\left(\frac{5}{2}x\right)+\sin\left(\frac{3}{2}x\right) = 2\sin\left(\frac{\frac{5}{2}x+\frac{3}{2}x}{2}\right)\cos\left(\frac{\frac{5}{2}x-\frac{3}{2}x}{2}\right) = \boxed{2\sin(2x)\cos\left(\frac{x}{2}\right)}$$

85. $$\sin\left(\frac{4x}{3}\right)-\sin\left(\frac{2x}{3}\right) = 2\sin\left(\frac{\frac{4x}{3}-\frac{2x}{3}}{2}\right)\cos\left(\frac{\frac{4x}{3}+\frac{2x}{3}}{2}\right) = \boxed{2\sin\left(\frac{x}{3}\right)\cos x}$$

86. $$\cos 7x + \cos x = 2\cos\left(\frac{7x+x}{2}\right)\cos\left(\frac{7x-x}{2}\right) = \boxed{2\cos(4x)\cos(3x)}$$

87. $$\frac{\cos 8x+\cos 2x}{\sin 8x-\sin 2x} = \frac{\cancel{2}\cos\cancel{\left(\frac{8x+2x}{2}\right)}\cos\left(\frac{8x-2x}{2}\right)}{\cancel{2}\sin\left(\frac{8x-2x}{2}\right)\cancel{\cos\left(\frac{8x+2x}{2}\right)}} = \cot\left(\frac{8x-2x}{2}\right) = \boxed{\cot 3x}$$

88. $$\frac{\sin 5x+\sin 3x}{\cos 5x+\cos 3x} = \frac{\cancel{2}\sin\left(\frac{5x+3x}{2}\right)\cancel{\cos\left(\frac{5x-3x}{2}\right)}}{\cancel{2}\cos\left(\frac{5x+3x}{2}\right)\cancel{\cos\left(\frac{5x-3x}{2}\right)}} = \tan\left(\frac{5x+3x}{2}\right) = \boxed{\tan 4x}$$

89. $$\frac{\sin A+\sin B}{\cos A-\cos B} = \frac{\cancel{2}\cancel{\sin\left(\frac{A+B}{2}\right)}\cos\left(\frac{A-B}{2}\right)}{-\cancel{2}\cancel{\sin\left(\frac{A+B}{2}\right)}\sin\left(\frac{A-B}{2}\right)} = \boxed{-\cot\left(\frac{A-B}{2}\right)}$$

90. $$\frac{\sin A-\sin B}{\cos A-\cos B} = \frac{\cancel{2}\cancel{\sin\left(\frac{A-B}{2}\right)}\cos\left(\frac{A+B}{2}\right)}{-\cancel{2}\sin\left(\frac{A+B}{2}\right)\cancel{\sin\left(\frac{A-B}{2}\right)}} = \boxed{-\cot\left(\frac{A+B}{2}\right)}$$

91.

$$\csc\left(\frac{A-B}{2}\right) = \frac{1}{\sin\left(\frac{A-B}{2}\right)} = \frac{-2\sin\left(\frac{A+B}{2}\right)}{-2\sin\left(\frac{A+B}{2}\right)\sin\left(\frac{A-B}{2}\right)} = \frac{2\sin\left(\frac{A+B}{2}\right)}{\cos B-\cos A}$$

92.	
$\sec\left(\frac{A+B}{2}\right)=\frac{1}{\cos\left(\frac{A+B}{2}\right)}=\frac{2\sin\left(\frac{A-B}{2}\right)}{2\sin\left(\frac{A-B}{2}\right)\cos\left(\frac{A+B}{2}\right)}=\frac{2\sin\left(\frac{A-B}{2}\right)}{\sin A-\sin B}$	
93. The equation $\arctan(1)=\theta$ is equivalent to $\tan\theta=1=\frac{\sin\theta}{\cos\theta}$. Since the range of inverse tangent is $\left(-\frac{\pi}{2},\frac{\pi}{2}\right)$, we conclude that $\boxed{\theta=\frac{\pi}{4}}$.	**94.** The equation $\operatorname{arc}\csc(-2)=\theta$ is equivalent to $\csc\theta=-2$, which is further the same as $\sin\theta=-\frac{1}{2}$. Since the range of inverse cosecant is $\left[-\frac{\pi}{2},0\right)\cup\left(0,\frac{\pi}{2}\right]$, we conclude that $\boxed{\theta=-\frac{\pi}{6}}$.
95. The equation $\cos^{-1}(0)=\theta$ is equivalent to $\cos\theta=0$. Since the range of arccosine is $[0,\pi]$, we conclude that $\boxed{\theta=\frac{\pi}{2}}$.	**96.** The equation $\sin^{-1}(-1)=\theta$ is equivalent to $\sin\theta=-1$. Since the range of arcsine is $\left[-\frac{\pi}{2},\frac{\pi}{2}\right]$, we conclude that $\theta=\boxed{-\frac{\pi}{2}}$.
97. The equation $\sec^{-1}\left(\frac{2}{\sqrt{3}}\right)=\theta$ is equivalent to $\sec\theta=\frac{2}{\sqrt{3}}$, which is further the same as $\cos\theta=\frac{\sqrt{3}}{2}$. So, we conclude that $\boxed{\theta=\frac{\pi}{6}}$	**98.** The equation $\cot^{-1}\left(-\sqrt{3}\right)=\theta$ is equivalent to $\cot\theta=-\sqrt{3}=-\frac{\frac{\sqrt{3}}{2}}{\frac{1}{2}}=\frac{\cos\theta}{\sin\theta}$. Since the range of inverse cotangent is $(0,\pi)$, we conclude that $\boxed{\theta=\frac{5\pi}{6}}$.
99. The equation $\csc^{-1}(-1)=\theta$ is equivalent to $\csc\theta=-1$, which is further the same as $\sin\theta=-1$. Since the range of inverse cosecant is $\left[-\frac{\pi}{2},0\right)\cup\left(0,\frac{\pi}{2}\right]$, we conclude that $\theta=-\frac{\pi}{2}$, which corresponds to $\boxed{\theta=-90^\circ}$.	**100.** The equation $\arctan(-1)=\theta$ is equivalent to $\tan\theta=-1=\frac{\sin\theta}{\cos\theta}$. Since the range of inverse tangent is $\left(-\frac{\pi}{2},\frac{\pi}{2}\right)$, we conclude that $\theta=-\frac{\pi}{4}$, which corresponds to $\boxed{\theta=-45^\circ}$.

101. The equation $\operatorname{arc\,cot}\left(\frac{\sqrt{3}}{3}\right)=\theta$ is equivalent to $\cot\theta=\frac{\sqrt{3}}{3}=\frac{1}{\sqrt{3}}=\frac{1/2}{\sqrt{3}/2}=\frac{\cos\theta}{\sin\theta}$. Since the range of inverse cotangent is $(0,\pi)$, we conclude that $\theta=\frac{\pi}{3}$, which corresponds to $\boxed{\theta=60^\circ}$.	**102.** The equation $\cos^{-1}\left(\frac{\sqrt{2}}{2}\right)=\theta$ is equivalent to $\cos\theta=\frac{\sqrt{2}}{2}$. Since the range of arccosine is $[0,\pi]$, we conclude that $\theta=\frac{\pi}{4}$, which corresponds to $\boxed{\theta=45^\circ}$.
103. The equation $\sin^{-1}\left(-\frac{\sqrt{3}}{2}\right)=\theta$ is equivalent to $\sin\theta=-\frac{\sqrt{3}}{2}$. Since the range of arcsine is $\left[-\frac{\pi}{2},\frac{\pi}{2}\right]$, we conclude that $\theta=\boxed{-60^\circ}$.	**104.** The equation $\sec^{-1}(1)=\theta$ is equivalent to $\sec\theta=1$, which is further the same as $\cos\theta=1$. Hence, we conclude that $\boxed{\theta=0^\circ}$.
105. $\sin^{-1}(-0.6088)\approx\boxed{-37.50^\circ}$	**106.** $\tan^{-1}(1.1918)\approx\boxed{50.00^\circ}$
107. $\sec^{-1}(1.0824)=\cos^{-1}\left(\frac{1}{1.0824}\right)\approx\boxed{22.50^\circ}$	**108.** $\cot^{-1}(-3.7321)=180^\circ+\tan^{-1}\left(\frac{1}{-3.7321}\right)\approx\boxed{165.00^\circ}$
109. $\cos^{-1}(-0.1736)\approx\boxed{1.75}$	**110.** $\tan^{-1}(0.1584)\approx\boxed{0.16}$
111. $\csc^{-1}(-10.0167)=\sin^{-1}\left(\frac{1}{-10.0167}\right)\approx\boxed{-0.10}$	**112.** $\sec^{-1}(-1.1223)=\cos^{-1}\left(\frac{1}{-1.1223}\right)\approx\boxed{2.67}$

113. $\sin^{-1}\left(\sin\left(-\frac{\pi}{4}\right)\right)=\boxed{-\frac{\pi}{4}}$ since $-1\le-\frac{\pi}{4}\le1$.

114. $\cos\left(\cos^{-1}\left(-\frac{\sqrt{2}}{2}\right)\right)=\boxed{-\frac{\sqrt{2}}{2}}$ since $-1\le-\frac{\sqrt{2}}{2}\le1$.

115. Since $\tan\left(\tan^{-1}x\right)=x$ for all $-\infty<x<\infty$, we see that $\tan\left(\tan^{-1}\left(-\sqrt{3}\right)\right)=\boxed{-\sqrt{3}}$.

116. Note that we need to use the angle θ in $[0,\pi]$ such that $\cot\theta = \cot\left(\frac{11\pi}{6}\right)$. To this end, observe that $\cot^{-1}\left(\cot\left(\frac{11\pi}{6}\right)\right) = \cot^{-1}\left(\cot\left(\frac{5\pi}{6}\right)\right) = \boxed{\frac{5\pi}{6}}$.

117. Note that we need to use the angle θ in $\left[-\frac{\pi}{2},0\right)\cup\left(0,\frac{\pi}{2}\right]$ such that $\csc\theta = \csc\left(\frac{2\pi}{3}\right)$. To this end, observe that $\csc^{-1}\left(\csc\left(\frac{2\pi}{3}\right)\right) = \csc^{-1}\left(\csc\left(\frac{\pi}{3}\right)\right) = \boxed{\frac{\pi}{3}}$.

118. Since $\sec\left(\sec^{-1}\theta\right) = \theta$ for all $\theta \in (-\infty,-1]\cup[1,\infty)$, we see that $\sec\left(\sec^{-1}\left(-\frac{2\sqrt{3}}{3}\right)\right) = \boxed{-\frac{2\sqrt{3}}{3}}$.

119. Let $\theta = \cos^{-1}\left(\frac{11}{61}\right)$. Then, $\cos\theta = \frac{11}{61}$, as shown in the diagram:

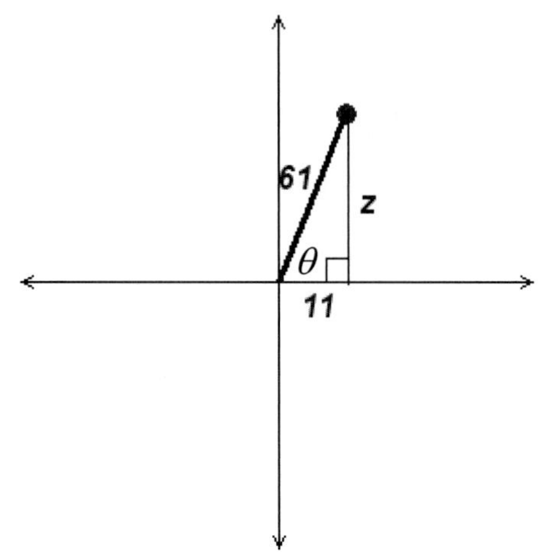

Using the Pythagorean Theorem, we see that $z^2 + 11^2 = 61^2$, so that $z = 60$.

Hence, $\sin\left(\cos^{-1}\left(\frac{11}{61}\right)\right) = \sin\theta = \boxed{\frac{60}{61}}$.

120. Let $\theta = \tan^{-1}\left(\frac{40}{9}\right)$. Then, $\tan\theta = \frac{40}{9}$, as shown in the diagram:

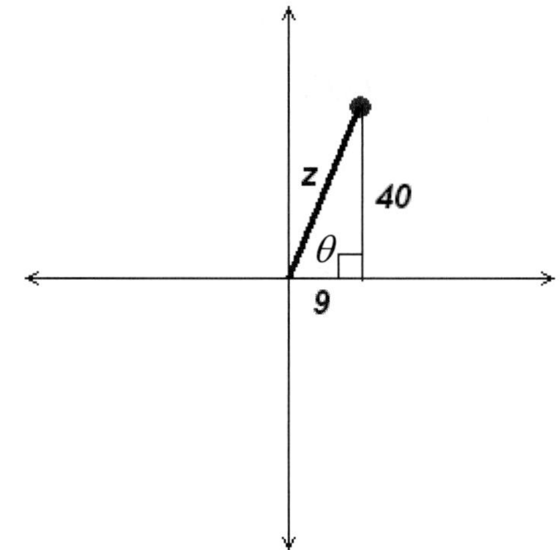

Using the Pythagorean Theorem, we see that $9^2 + 40^2 = z^2$, so that $z = 41$.

Hence, $\cos\left(\tan^{-1}\left(\frac{40}{9}\right)\right) = \cos\theta = \boxed{\frac{9}{41}}$.

121. Let $\theta = \cot^{-1}\left(\frac{6}{7}\right)$. Then, $\cot\theta = \frac{6}{7}$. So, $\tan\left(\cot^{-1}\left(\frac{6}{7}\right)\right) = \tan\theta = \frac{1}{\cot\theta} = \boxed{\frac{7}{6}}$.

122. Let $\theta = \sec^{-1}\left(\frac{25}{7}\right)$. Then, $\sec\theta = \frac{25}{7}$, so that $\cos\theta = \frac{7}{25}$, as shown in the diagram:

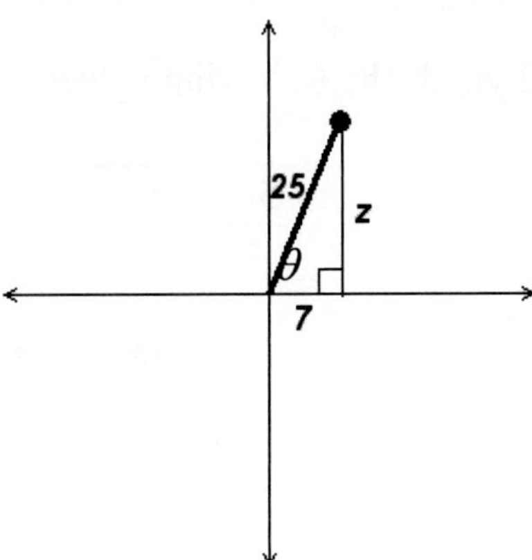

Using the Pythagorean Theorem, we see that $z^2 + 7^2 = 25^2$, so that $z = 24$.
Hence, $\cot\left(\sec^{-1}\left(\frac{25}{7}\right)\right) = \cot\theta = \boxed{\frac{7}{24}}$.

123. Let $\theta = \sin^{-1}\left(\frac{1}{6}\right)$. Then, $\sin\theta = \frac{1}{6}$, as shown in the diagram:

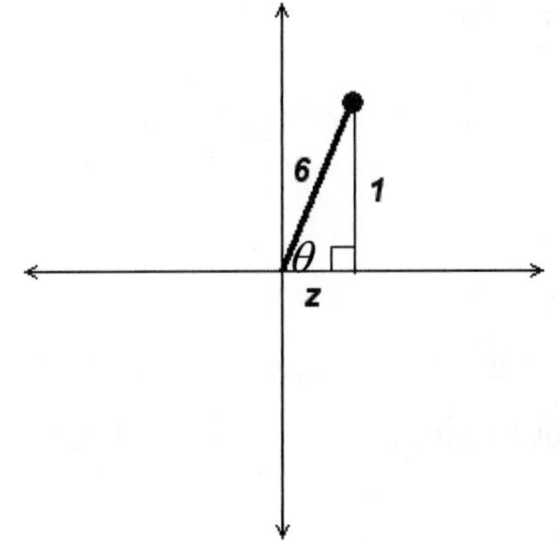

Using the Pythagorean Theorem, we see that $z^2 + 1^2 = 6^2$, so that $z = \sqrt{35}$. So,
$\sec\left(\sin^{-1}\left(\frac{1}{6}\right)\right) = \sec\theta = \frac{1}{\cos\theta}$
$= \frac{6}{\sqrt{35}} = \boxed{\frac{6\sqrt{35}}{35}}$

124. Let $\theta = \cot^{-1}\left(\frac{5}{12}\right)$. Then, $\cot\theta = \frac{5}{12}$, as shown in the diagram:

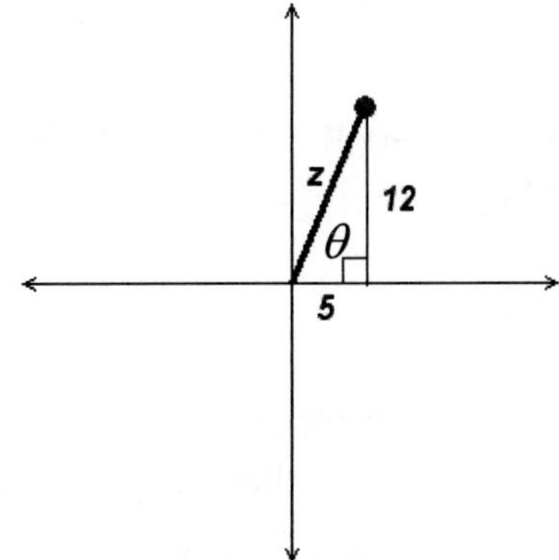

Using the Pythagorean Theorem, we see that $5^2 + 12^2 = z^2$, so that $z = 13$. So,
$\csc\left(\cot^{-1}\left(\frac{5}{12}\right)\right) = \csc\theta$
$= \frac{1}{\sin\theta} = \boxed{\frac{31}{12}}$

125. The values of θ in $[0,2\pi]$ that satisfy $\sin 2\theta = -\frac{\sqrt{3}}{2}$ must satisfy

$$2\theta = \frac{4\pi}{3},\ \frac{4\pi}{3}+2\pi,\ \frac{5\pi}{3},\ \frac{5\pi}{3}+2\pi = \frac{4\pi}{3},\ \frac{10\pi}{3},\ \frac{5\pi}{3},\ \frac{11\pi}{3}.$$

So, dividing all values by 2 yields the following values of θ which satisfy the original equation:

$$\theta = \boxed{\frac{2\pi}{3},\ \frac{5\pi}{3},\ \frac{5\pi}{6},\ \frac{11\pi}{6}}$$

126. First, observe that $\sec\left(\frac{\theta}{2}\right) = 2$ is equivalent to $\cos\left(\frac{\theta}{2}\right) = \frac{1}{2}$. The values of θ in $[-2\pi,2\pi]$ that satisfy $\cos\left(\frac{\theta}{2}\right) = \frac{1}{2}$ must satisfy $\frac{\theta}{2} = \pm\frac{\pi}{3},\ \pm\frac{5\pi}{3}$.

So, multiplying all values by 2 yields the following values of θ which satisfy the original equation:

$$\theta = \boxed{\pm\frac{2\pi}{3}},\ \underbrace{\cancel{\pm\frac{10\pi}{3}}}_{\text{outside interval}}$$

127. The values of θ in $[-2\pi,2\pi]$ that satisfy $\sin\left(\frac{\theta}{2}\right) = -\frac{\sqrt{2}}{2}$ must satisfy

$$\frac{\theta}{2} = -\frac{\pi}{4},\ -\frac{3\pi}{4},\ \frac{5\pi}{4},\ \frac{7\pi}{4}.$$

So, multiplying all values by 2 yields the following values of θ which satisfy the original equation:

$$\theta = \boxed{-\frac{\pi}{2},\ -\frac{3\pi}{2}},\ \underbrace{\cancel{\frac{5\pi}{2},\ \frac{7\pi}{2}}}_{\text{Outside interval}}.$$

128. First, observe that $\csc(2\theta) = 2$ is equivalent to $\sin(2\theta) = \frac{1}{2}$. The values of θ in $[0,2\pi]$ that satisfy $\sin(2\theta) = \frac{1}{2}$ must satisfy

$$2\theta = \frac{\pi}{6},\ \frac{\pi}{6}+2\pi,\ \frac{5\pi}{6},\ \frac{5\pi}{6}+2\pi = \frac{\pi}{6},\ \frac{13\pi}{6},\ \frac{5\pi}{6},\ \frac{17\pi}{6}.$$

So, dividing all values by 2 yields the following values of θ which satisfy the original equation:

$$\theta = \boxed{\frac{\pi}{12},\ \frac{13\pi}{12},\ \frac{5\pi}{12},\ \frac{17\pi}{12}}$$

129. Observe that $\tan\left(\frac{1}{3}\theta\right)=-1$ holds when $\frac{1}{3}\theta=\frac{3\pi}{4},\ \frac{7\pi}{4}$, which is equivalent to $\boxed{\theta=\frac{9\pi}{4},\ \frac{21\pi}{4}}$.

130. First, observe that $\cot(4\theta)=-\sqrt{3}=-\frac{\frac{\sqrt{3}}{2}}{\frac{1}{2}}$. The values of θ that satisfy this must satisfy

$$4\theta=\frac{5\pi}{6}+2n\pi,\ \frac{11\pi}{6}+2n\pi$$

$$\theta=\frac{5\pi}{24}+\frac{n\pi}{2},\ \frac{11\pi}{24}+\frac{n\pi}{2},\ \text{where } n \text{ is an integer}$$

As such, the values of θ in $[0,2\pi]$ are $\boxed{\theta=-\frac{7\pi}{24},\ -\frac{\pi}{24},\ \frac{5\pi}{24},\ \frac{11\pi}{24}}$.

131. First, observe that $4\cos(2\theta)+2=0$ is equivalent to $\cos(2\theta)=-\frac{1}{2}$. The values of θ in $[0,2\pi]$ that satisfy $\cos(2\theta)=-\frac{1}{2}$ must satisfy

$$2\theta=\frac{2\pi}{3},\ \frac{2\pi}{3}+2\pi,\ \frac{4\pi}{3},\ \frac{4\pi}{3}+2\pi=\frac{2\pi}{3},\frac{4\pi}{3},\frac{8\pi}{3},\frac{10\pi}{3}.$$

So, dividing all values by 2 yields the following values of θ which satisfy the original equation: $\theta=\frac{2\pi}{6},\frac{4\pi}{6},\frac{8\pi}{6},\frac{10\pi}{6}=\boxed{\frac{\pi}{3},\frac{2\pi}{3},\frac{4\pi}{3},\frac{5\pi}{3}}$

132. First, observe that $\sqrt{3}\tan\left(\frac{\theta}{2}\right)-1=0$ is equivalent to

$$\tan\left(\frac{\theta}{2}\right)=\frac{1}{\sqrt{3}}=\frac{1/2}{\sqrt{3}/2}=\frac{\sin\left(\frac{\theta}{2}\right)}{\cos\left(\frac{\theta}{2}\right)}.$$

The values of θ in $[0,2\pi]$ that satisfy this must satisfy $\frac{\theta}{2}=\frac{\pi}{6},\frac{7\pi}{6}$, so that

$$\theta=\boxed{\frac{\pi}{3}},\ \underbrace{\cancel{\frac{7\pi}{3}}}_{\text{Outside interval}}.$$

133. First, observe that $2\tan 2\theta+2=0$ is equivalent to $\tan 2\theta=-1$. The values of θ in $[0,2\pi]$ that satisfy $\tan 2\theta=-1$ must satisfy

$$2\theta=\frac{3\pi}{4},\ \frac{3\pi}{4}+2\pi,\ \frac{7\pi}{4},\ \frac{7\pi}{4}+2\pi=\frac{3\pi}{4},\frac{11\pi}{4},\frac{7\pi}{4},\frac{15\pi}{4}.$$

So, dividing all values by 2 yields the following values of θ which satisfy the original equation:

$$\theta=\boxed{\frac{3\pi}{8},\frac{11\pi}{8},\frac{7\pi}{8},\frac{15\pi}{8}}.$$

134. Factoring the left-side of $2\sin^2\theta+\sin\theta-1=0$ yields

$$(\sin\theta+1)(2\sin\theta-1)=0$$
$$\sin\theta+1=0 \quad \text{or} \quad 2\sin\theta-1=0$$
$$\sin\theta=-1 \quad \text{or} \quad \sin\theta=\frac{1}{2}$$

The values of θ in $[0,2\pi]$ that satisfy these equations are $\theta=\boxed{\frac{\pi}{6},\frac{5\pi}{6},\frac{3\pi}{2}}$.

135. Factoring the left-side of $\tan^2\theta+\tan\theta=0$ yields the equivalent equation $\tan\theta(\tan\theta+1)=0$ which is satisfied when either $\tan\theta=0$ or $\tan\theta+1=0$. The values of θ in $[0,2\pi]$ that satisfy $\tan\theta=0$ are $\theta=0,\pi,2\pi$, and those which satisfy $\tan\theta=-1$ are $\theta=\frac{3\pi}{4},\frac{7\pi}{4}$. Thus, the solutions to the original equation are

$$\theta=\boxed{0,\ \pi,\ 2\pi,\ \frac{3\pi}{4},\frac{7\pi}{4}}.$$

136. Factoring the left-side of $\sec^2\theta-3\sec\theta+2=0$ yields the equivalent equation $(\sec\theta-2)(\sec\theta-1)=0$ which is satisfied when either $\sec\theta-2=0$ or $\sec\theta-1=0$. Note that the values of θ in $[0,2\pi]$ that satisfy the equation $\sec\theta=2$ (or equivalently $\cos\theta=\frac{1}{2}$) are $\theta=\frac{\pi}{3},\frac{5\pi}{3}$. Also, the values of θ in $[0,2\pi]$ that satisfy $\sec\theta=1$ are $\theta=0,\ 2\pi$. Thus, the solutions to the original equation are

$$\theta=\boxed{\frac{\pi}{3},\frac{5\pi}{3},\ 0,\ 2\pi}$$

137. In order to find all of the values of θ in $\left[0^\circ, 360^\circ\right]$ that satisfy $\tan(2\theta) = -0.3459$, we proceed as follows:

Step 1: Find the values of 2θ whose tangent is -0.3459.

Indeed, observe that one solution is $\tan^{-1}(-0.3459) \approx -19.081^\circ$, which is in QIV. Since the angles we seek have positive measure, we use the representative $360^\circ - 19.081^\circ \approx 340.92^\circ$. A second solution occurs in QII, and has value $340.92^\circ - 180^\circ = 160.92^\circ$.

Step 2: Use Step 1 to find all values of θ that satisfy the original equation.

The solutions obtained in Step 1 are

$$2\theta = 160.92^\circ,\ 160.92^\circ + 360^\circ,\ 340.92^\circ,\ 340.92^\circ + 360^\circ.$$

When divided by 2, we see that the solutions to the original equation in $\left[0^\circ, 360^\circ\right]$ are approximately $\theta \approx \boxed{80.46^\circ,\ 260.46^\circ,\ 170.46^\circ,\ 350.46^\circ}$.

138. In order to find all of the values of θ in $\left[0^\circ, 360^\circ\right]$ that satisfy $6\sin\theta - 5 = 0$, we proceed as follows:

Step 1: First, observe that the equation is equivalent to $\sin\theta = \frac{5}{6}$.

Step 2: Find the values of θ whose sine is $\frac{5}{6}$.

Indeed, observe that one solution is $\sin^{-1}\left(\frac{5}{6}\right) \approx 56.44^\circ$, which is in QI.

A second solution occurs in QII, and has value $180^\circ - 56.443^\circ = 123.56^\circ$. Since the input of sine is simply θ, and not some multiple thereof, we conclude that the solutions to the original equation are approximately $\theta \approx \boxed{56.44^\circ,\ 123.56^\circ}$.

139. In order to find all of the values of θ in $\left[0^\circ, 360^\circ\right]$ that satisfy $4\cos^2\theta + 3\cos\theta = 0$, we proceed as follows:

Step 1: Simplify the equation algebraically.

Factoring the left-side of the equation yields: $\cos\theta\left(4\cos\theta + 3\right) = 0$

$$\cos\theta = 0 \quad \text{or} \quad 4\cos\theta + 3 = 0 \;\text{ so that }\; \cos\theta = 0 \quad \text{or} \quad \cos\theta = -\frac{3}{4}$$

Step 2: Observe that the values of θ that satisfy $\cos\theta = 0$ are $\theta = 90^\circ, 270^\circ$.

Step 3: Find the values of θ whose cosine is $-\frac{3}{4}$.

Indeed, observe that one solution is $\cos^{-1}\left(-\frac{3}{4}\right) \approx 138.59^\circ$, which is in QII. A second solution occurs in QIII, and has value $360^\circ - 138.59^\circ = 221.41^\circ$. Since the input of cosine is simply θ, and not some multiple thereof, we conclude that the solutions to the original equation are approximately $\theta \approx 138.59^\circ, 221.41^\circ$.

Step 4: Conclude the solutions in $\left[0^\circ, 360^\circ\right]$ are $\theta \approx \boxed{90^\circ, 270^\circ, 138.59^\circ, 221.41^\circ}$.

140. In order to find all of the values of θ in $\left[0^\circ, 360^\circ\right]$ that satisfy $12\cos^2\theta - 7\cos\theta + 1 = 0$, we proceed as follows:

Step 1: Simplify the equation algebraically.

Factoring the left-side of the equation yields: $\left(4\cos\theta - 1\right)\left(3\cos\theta - 1\right) = 0$

$$4\cos\theta - 1 = 0 \quad \text{or} \quad 3\cos\theta - 1 = 0 \;\text{ so that }\; \cos\theta = \frac{1}{4} \quad \text{or} \quad \cos\theta = \frac{1}{3}$$

Step 2: Find the values of θ whose cosine is $\frac{1}{4}$.

Indeed, observe that one solution is $\cos^{-1}\left(\frac{1}{4}\right) \approx 75.52^\circ$, which is in QI. A second solution occurs in QIV, and has value $360^\circ - 75.52^\circ = 284.48^\circ$. Since the input of cosine is simply θ, and not some multiple thereof, we conclude that the solutions to the original equation are approximately $\theta \approx 75.52^\circ, 284.48^\circ$.

Step 3: Find the values of θ whose cosine is $\frac{1}{3}$.

Indeed, observe that one solution is $\cos^{-1}\left(\frac{1}{3}\right) \approx 70.53^\circ$, which is in QI. A second solution occurs in QIV, and has value $360^\circ - 70.53^\circ = 289.47^\circ$. Since the input of cosine is simply θ, and not some multiple thereof, we conclude that the solutions to the original equation are approximately $\theta \approx 70.53^\circ, 289.47^\circ$.

141. In order to find all of the values of θ in $\left[0°,360°\right]$ that satisfy $\csc^2\theta - 3\csc\theta - 1 = 0$, we proceed as follows:

Step 1: Simplify the equation algebraically.

Since the left-side does not factor nicely, we apply the quadratic formula (treating $\csc\theta$ as the variable):

$$\csc\theta = \frac{-(-3) \pm \sqrt{(-3)^2 - 4(1)(-1)}}{2(1)} = \frac{3 \pm \sqrt{13}}{2}$$

So, θ is a solution to the original equation if either

$$\csc\theta = \frac{3+\sqrt{13}}{2} \quad \text{or} \quad \csc\theta = \frac{3-\sqrt{13}}{2},$$

which is equivalent to

$$\sin\theta = \frac{2}{3+\sqrt{13}} \quad \text{or} \quad \underbrace{\sin\theta = \frac{2}{3-\sqrt{13}}}_{\text{No solution since } \frac{2}{3-\sqrt{13}} < -1}$$

Step 2: Solve $\sin\theta = \dfrac{2}{3+\sqrt{13}}$.

To do so, we must find the values of θ whose sine is $\dfrac{2}{3+\sqrt{13}}$.

Indeed, observe that one solution is $\sin^{-1}\left(\dfrac{2}{3+\sqrt{13}}\right) \approx 17.62°$, which is in QI.

A second solution occurs in QII, and has value $180° - 17.62° = 162.38°$. Since the input of sine is simply θ, and not some multiple thereof, we conclude that the solutions to this equation are approximately $\theta \approx 17.62°, 162.38°$.

Step 3: Conclude that the solutions to the original equation are

$$\theta \approx \boxed{17.62°,\ 162.38°}.$$

142. In order to find all of the values of θ in $\left[0^\circ, 360^\circ\right]$ that satisfy $2\cot^2\theta + 5\cot\theta - 4 = 0$, we proceed as follows:

Step 1: Simplify the equation algebraically.

Since the left-side does not factor nicely, we apply the quadratic formula (treating $\cot\theta$ as the variable):

$$\cot\theta = \frac{-5 \pm \sqrt{5^2 - 4(2)(-4)}}{2(2)} = \frac{-5 \pm \sqrt{57}}{4}$$

So, θ is a solution to the original equation if either

$$\cot\theta = \frac{-5-\sqrt{57}}{4} \quad \text{or} \quad \cot\theta = \frac{-5+\sqrt{57}}{4}.$$

Step 2: Solve $\cot\theta = \frac{-5-\sqrt{57}}{4}$.

To do so, we must find the values of θ whose cotangent is $\frac{-5-\sqrt{57}}{4}$.

Indeed, observe that one solution is

$\cot^{-1}\left(\frac{-5-\sqrt{57}}{4}\right) = 180^\circ + \tan^{-1}\left(\frac{4}{-5-\sqrt{57}}\right) \approx 162.32^\circ$, which is in QII.

A second solution occurs in QIV, and has value $180^\circ + 162.32^\circ = 342.32^\circ$. Since the input of cotangent is simply θ, and not some multiple thereof, we conclude that the solutions to this equation are approximately $\theta \approx 162.32^\circ, 342.32^\circ$.

Step 3: Solve $\cot\theta = \frac{-5+\sqrt{57}}{4}$.

To do so, we must find the values of θ whose cotangent is $\frac{-5+\sqrt{57}}{4}$.

Indeed, observe that one solution is $\cot^{-1}\left(\frac{-5+\sqrt{57}}{4}\right) = \tan^{-1}\left(\frac{4}{-5+\sqrt{57}}\right) \approx 57.48^\circ$, which is in QI. A second solution occurs in QIII, and has value $180^\circ + 57.48^\circ = 237.48^\circ$. Since the input of cotangent is simply θ, and not some multiple thereof, we conclude that the solutions to this equation are approximately $\theta \approx 57.48^\circ, 237.48^\circ$.

Step 4: Conclude that the solutions to the original equation are

$$\theta \approx \boxed{57.48^\circ, 237.48^\circ, 162.32^\circ, 342.32^\circ}.$$

143. Observe that

$$\sec x = 2\sin x$$
$$\frac{1}{\cos x} = 2\sin x$$
$$1 = 2\sin x\cos x$$
$$1 = \sin 2x$$

Note that the solutions of $\sin 2x = 1$ are $2x = \frac{\pi}{2}, \frac{\pi}{2} + 2\pi$, so that $x = \boxed{\frac{\pi}{4}, \frac{5\pi}{4}}$.

Substituting these into the original equation shows that they are all, in fact, solutions to the original equation.

144. Observe that

$$3\tan x + \cot x = 2\sqrt{3}$$
$$3\frac{\sin x}{\cos x} + \frac{\cos x}{\sin x} = 2\sqrt{3}$$
$$\frac{3\sin^2 x + \cos^2 x}{\sin x\cos x} = 2\sqrt{3}$$
$$3\sin^2 x + \cos^2 x = 2\sqrt{3}\sin x\cos x$$
$$\left(3\sin^2 x + \cos^2 x\right)^2 = \left(2\sqrt{3}\sin x\cos x\right)^2$$
$$9\sin^4 x + 6\sin^2 x\cos^2 x + \cos^4 x = 12\sin^2 x\cos^2 x$$
$$9\sin^4 x - 6\sin^2 x\cos^2 x + \cos^4 x = 0$$
$$\left(3\sin^2 x - \cos^2 x\right)^2 = 0$$
$$3\sin^2 x - \cos^2 x = 0$$
$$3\sin^2 x - \left(1 - \sin^2 x\right) = 0$$
$$4\sin^2 x = 1$$
$$\sin x = \pm\frac{1}{2}$$

The values of x in $\left[0, 2\pi\right]$ that satisfy these equations are $x = \frac{\pi}{6}, \frac{5\pi}{6}, \frac{7\pi}{6}, \frac{11\pi}{6}$.

Substituting these into the original equation shows that while $\frac{\pi}{6}, \frac{7\pi}{6}$ are solutions, $\frac{5\pi}{6}, \frac{11\pi}{6}$ is extraneous. So, the only solutions to the original equation are $\boxed{\frac{\pi}{6}, \frac{7\pi}{6}}$.

145. Observe that

$$\sqrt{3}\tan x - \sec x = 1$$
$$\sqrt{3}\left(\frac{\sin x}{\cos x}\right) - \frac{1}{\cos x} = 1$$
$$\sqrt{3}\sin x - 1 = \cos x$$
$$\left(\sqrt{3}\sin x - 1\right)^2 = \cos^2 x$$
$$3\sin^2 x - 2\sqrt{3}\sin x + 1 = 1 - \sin^2 x$$
$$4\sin^2 x - 2\sqrt{3}\sin x = 0$$
$$2\sin x\left(2\sin x - \sqrt{3}\right) = 0$$
$$\sin x = 0 \quad \text{or} \quad 2\sin x - \sqrt{3} = 0$$
$$\sin x = 0 \quad \text{or} \quad \sin x = \frac{\sqrt{3}}{2}$$

The values of x in $[0, 2\pi]$ that satisfy these equations are $x = 0,\ \pi,\ 2\pi,\ \frac{\pi}{3},\ \frac{2\pi}{3}$.

Substituting these into the original equation shows that while $\pi,\ \frac{\pi}{3}$ are solutions, $0,\ \frac{2\pi}{3},\ 2\pi$ is extraneous. So, the only solutions to the original equation are $\boxed{\pi,\ \frac{\pi}{3}}$.

146. Observe that

$$2\sin 2x = \cot x$$
$$2(2\sin x\cos x) = \frac{\cos x}{\sin x}$$
$$4\sin^2 x\cos x = \cos x$$
$$4\sin^2 x\cos x - \cos x = 0$$
$$\cos x\left(4\sin^2 x - 1\right) = 0$$
$$\cos x = 0 \quad \text{or} \quad 4\sin^2 x - 1 = 0$$
$$\cos x = 0 \quad \text{or} \quad \sin^2 x = \frac{1}{4}$$
$$\cos x = 0 \quad \text{or} \quad \sin x = \pm\frac{1}{2}$$

Note that the solutions of these equations are $x = \boxed{\frac{\pi}{2},\ \frac{3\pi}{2},\ \frac{\pi}{6},\ \frac{5\pi}{6},\ \frac{7\pi}{6},\ \frac{11\pi}{6}}$.

Substituting these into the original equation shows that they are all, in fact, solutions to the original equation.

147. Observe that

$$\sqrt{3}\tan x = 2\sin x$$
$$\sqrt{3}\frac{\sin x}{\cos x} = 2\sin x$$
$$\sqrt{3}\sin x = 2\sin x\cos x$$
$$\sqrt{3}\sin x - 2\sin x\cos x = 0$$
$$\sin x\left(\sqrt{3} - 2\cos x\right) = 0$$
$$\sin x = 0 \quad \text{or} \quad \sqrt{3} - 2\cos x = 0$$
$$\sin x = 0 \quad \text{or} \quad \cos x = \frac{\sqrt{3}}{2}$$

Note that the solutions of these equations are $x = \boxed{0,\ \pi,\ 2\pi,\ \frac{\pi}{6},\ \frac{11\pi}{6}}$.

Substituting these into the original equation shows that they are all, in fact, solutions to the original equation.

148. Observe that

$$3\cot x = 2\sin x$$
$$3\frac{\cos x}{\sin x} = 2\sin x$$
$$3\cos x = 2\sin^2 x$$
$$3\cos x = 2\left(1 - \cos^2 x\right)$$
$$3\cos x = 2 - 2\cos^2 x$$
$$2\cos^2 x + 3\cos x - 2 = 0$$
$$(2\cos x - 1)(\cos x + 2) = 0$$
$$2\cos x - 1 = 0 \quad \text{or} \quad \cos x + 2 = 0$$
$$\cos x = \frac{1}{2} \quad \text{or} \quad \underbrace{\cos x = -2}_{\text{No solution}}$$

Note that the solutions of these equations are $x = \boxed{\frac{\pi}{3},\ \frac{5\pi}{3}}$. Substituting these into the original equation shows that they are all, in fact, solutions to the original equation.

149. Observe that

$$\cos^2 x + \sin x + 1 = 0$$
$$\left(1 - \sin^2 x\right) + (\sin x + 1) = 0$$
$$(1 - \sin x)(1 + \sin x) + (\sin x + 1) = 0$$
$$(1 + \sin x)[1 - \sin x + 1] = 0$$
$$(1 + \sin x)(2 - \sin x) = 0$$
$$1 + \sin x = 0 \quad \text{or} \quad 2 - \sin x = 0$$
$$\sin x = -1 \quad \text{or} \quad \underbrace{\sin x = 2}_{\text{No solution}}$$

Note that the solution of these equations is $x = \boxed{\frac{3\pi}{2}}$. Substituting into the original equation shows it is, in fact, a solution.

150. Observe that

$$2\cos^2 x - \sqrt{3}\cos x = 0$$
$$\cos x\left(2\cos x - \sqrt{3}\right) = 0$$
$$\cos x = 0 \quad \text{or} \quad 2\cos x - \sqrt{3} = 0$$
$$\cos x = 0 \quad \text{or} \quad \cos x = \frac{\sqrt{3}}{2}$$

Note that the solutions of these equations are $x = \boxed{\frac{\pi}{2},\ \frac{3\pi}{2},\ \frac{\pi}{6},\ \frac{11\pi}{6}}$. Substituting these into the original equation shows that they are all, in fact, solutions to the original equation.

151. Observe that

$$\cos 2x + 4\cos x + 3 = 0$$
$$\left(2\cos^2 x - 1\right) + 4\cos x + 3 = 0$$
$$2\cos^2 x + 4\cos x + 2 = 0$$
$$2\left(\cos x + 1\right)^2 = 0$$
$$\cos x + 1 = 0$$
$$\cos x = -1$$

The solution to this equation is $x = \boxed{\pi}$. Substituting it back into the original equation shows that it is, in fact, a solution of the original equation.

152. Observe that

$$\sin 2x + \sin x = 0$$
$$2\sin x\cos x + \sin x = 0$$
$$\sin x\left(2\cos x + 1\right) = 0$$
$$\sin x = 0 \text{ or } 2\cos x + 1 = 0$$
$$\sin x = 0 \text{ or } \cos x = -\frac{1}{2}$$

Note that the solutions of these equations are $x = \boxed{0,\ \pi,\ 2\pi,\ \frac{2\pi}{3},\ \frac{4\pi}{3}}$. Substituting these into the original equation shows that they are all solutions to the original equation.

153. Observe that

$$\tan^2\left(\tfrac{x}{2}\right) - 1 = 0$$
$$\left(\tan\left(\tfrac{x}{2}\right) - 1\right)\left(\tan\left(\tfrac{x}{2}\right) + 1\right) = 0$$
$$\tan\left(\tfrac{x}{2}\right) = \pm 1$$

The values of x that satisfy this are such that $\frac{x}{2} = \frac{\pi}{4}, \frac{3\pi}{4}$, so that $\boxed{x = \frac{\pi}{2}, \frac{3\pi}{2}}$.

154. Observe that

$$\cot^2\left(\tfrac{x}{3}\right) - 1 = 0$$
$$\left(\cot\left(\tfrac{x}{3}\right) - 1\right)\left(\cot\left(\tfrac{x}{3}\right) + 1\right) = 0$$
$$\cot\left(\tfrac{x}{3}\right) = \pm 1$$

The values of x that satisfy this are such that $\frac{x}{3} = \frac{\pi}{4}, \frac{3\pi}{4}$, so that $\boxed{x = \frac{3\pi}{4}}$.

155. In order to find all of the values of x in $\left[0°, 360°\right]$ that satisfy $\csc^2 x + \cot x = 1$, we proceed as follows:

Step 1: Simplify the equation algebraically:

$$\csc^2 x + \cot x = 1$$
$$\left(1 + \cot^2 x\right) + \cot x = 1$$
$$\cot^2 x + \cot x = 0$$
$$\cot x\left(\cot x + 1\right) = 0$$
$$\cot x = 0 \quad \text{or} \quad \cot x = -1$$

Step 2: Observe that the solutions to these equations are given by $x = \boxed{90°, 270°, 135°, 315°}$. Substituting them back into the original equation shows that they are all, in fact, solutions of the original equation.

156. In order to find all of the values of x in $\left[0^\circ, 360^\circ\right]$ that satisfy $8\cos^2 x + 6\sin x = 9$, we proceed as follows:

Step 1: Simplify the equation algebraically.

$$8\cos^2 x + 6\sin x = 9$$
$$8\left(1-\sin^2 x\right) + 6\sin x = 9$$
$$8 - 8\sin^2 x + 6\sin x = 9$$
$$8\sin^2 x - 6\sin x + 1 = 0$$
$$(4\sin x - 1)(2\sin x - 1) = 0$$
$$4\sin x - 1 = 0 \quad \text{or} \quad 2\sin x - 1 = 0$$
$$\sin x = \frac{1}{4} \quad \text{or} \quad \sin x = \frac{1}{2}$$

Step 2: Solve $\sin x = \frac{1}{4}$.

To do so, we must find the values of x whose sine is $\frac{1}{4}$.

Indeed, observe that one solution is $\sin^{-1}\left(\frac{1}{4}\right) \approx 14.48^\circ$, which is in QI.

A second solution occurs in QII, and has value $180^\circ - 14.48^\circ = 165.52^\circ$. Since the input of sine is simply x, and not some multiple thereof, we conclude that the solutions to this equation are approximately $x \approx 14.48^\circ, 165.52^\circ$.

Step 3: Solve $\sin x = \frac{1}{2}$.

Indeed, observe that one solution is $\sin^{-1}\left(\frac{1}{2}\right) = 30^\circ$, which is in QI. A second solution occurs in QII, and has value $180^\circ - 30^\circ = 150^\circ$. Since the input of sine is simply x, and not some multiple thereof, we conclude that the solutions to this equation are $x = 30^\circ, 150^\circ$.

Step 4: Conclude that the solutions to the original equation are

$$x \approx \boxed{30^\circ, 150^\circ, 14.48^\circ, 165.52^\circ}.$$

Substituting these into the original equation shows that they are, in fact, solutions to the original equation.

157. Observe that

$$\sin^2 x + 2 = 2\cos x$$
$$1 - \cos^2 x + 2 = 2\cos x$$
$$\cos^2 x + 2\cos x - 3 = 0$$
$$(\cos x - 1)(\cos x + 3) = 0$$
$$\cos x - 1 = 0 \text{ or } \cos x + 3 = 0$$
$$\cos x = 1 \text{ or } \underbrace{\cos x = -3}_{\text{No solution}}$$

Note that the solutions of these equations are $x = \boxed{0°, 360°}$. Substituting these into the original equation shows that they are all, in fact, solutions to the original equation.

158. Observe that

$$\cos 2x = 3\sin x - 1$$
$$1 - 2\sin^2 x = 3\sin x - 1$$
$$2\sin^2 x + 3\sin x - 2 = 0$$
$$(2\sin x - 1)(\sin x + 2) = 0$$
$$2\sin x - 1 = 0 \text{ or } \sin x + 2 = 0$$
$$\sin x = \frac{1}{2} \text{ or } \underbrace{\sin x = -2}_{\text{No solution}}$$

Note that the solutions of these equations are $x = \boxed{30°, 150°}$. Substituting these into the original equation shows that they are all, in fact, solutions to the original equation.

159. Observe that

$$\cos x - 1 = \cos 2x$$
$$\cos x - 1 = 2\cos^2 x - 1$$
$$2\cos^2 x - \cos x = 0$$
$$\cos x(2\cos x - 1) = 0$$
$$\cos x = 0 \text{ or } 2\cos x - 1 = 0$$
$$\cos x = 0 \text{ or } \cos x = \frac{1}{2}$$

Note that the solutions of these equations are $x = \boxed{90°, 270°, 60°, 300°}$. Substituting these into the original equation shows that they are all, in fact, solutions to the original equation.

160. In order to find all of the values of x in $[0°, 360°]$ that satisfy $12\cos^2 x + 4\sin x = 11$, we proceed as follows:

Step 1: Simplify the equation algebraically.

$$12\cos^2 x + 4\sin x = 11$$
$$12(1-\sin^2 x) + 4\sin x = 11$$
$$12 - 12\sin^2 x + 4\sin x - 11 = 0$$
$$12\sin^2 x - 4\sin x - 1 = 0$$
$$(6\sin x + 1)(2\sin x - 1) = 0$$
$$6\sin x + 1 = 0 \text{ or } 2\sin x - 1 = 0$$
$$\sin x = -\frac{1}{6} \quad \text{or} \quad \sin x = \frac{1}{2}$$

Step 2: Solve $\sin x = -\frac{1}{6}$.

To do so, we must find the values of x whose sine is $-\frac{1}{6}$.

Indeed, observe that one solution is $\sin^{-1}\left(-\frac{1}{6}\right) \approx -9.59°$, which is in QIV. Since the angles we seek have positive measure, we use the representative $360° - 9.59° \approx 350.41°$. A second solution occurs in QIII, and has value $180° + 9.59° = 189.59°$. Since the input of sine is simply x, and not some multiple thereof, we conclude that the solutions to this equation are approximately $x \approx 189.59°, 350.41°$.

Step 3: Solve $\sin x = \frac{1}{2}$.

Indeed, observe that one solution is $\sin^{-1}\left(\frac{1}{2}\right) = 30°$, which is in QI. A second solution occurs in QII, and has value $180° - 30° = 150°$. Since the input of sine is simply x, and not some multiple thereof, we conclude that the solutions to this equation are $x = 30°, 150°$.

Step 4: Conclude that the solutions to the original equation are

$$x \approx \boxed{30°, 150°, 189.59°, 350.41°}.$$

Substituting these into the original equation shows that they are, in fact, solutions to the original equation.

161. **a.** $\cos(73^\circ) \approx 0.2924$ **b.** $1-\sin(73^\circ) \approx 0.0437$ **c.** $\sqrt{1-\sin^2(73^\circ)} \approx 0.2924$

The values in parts **a.** and **c.** are the same.

162. **a.** $\cot(28^\circ) \approx 1.8807$ **b.** $1+\cot(28^\circ) \approx 2.8807$ **c.** $\sqrt{1+\cot^2(28^\circ)} \approx 2.1301$

None of the values are the same.

163. Observe that

$$\begin{aligned}\frac{\sin(3(x+h))-\sin 3x}{h} &= \frac{\sin 3x\cos(3h)+\sin(3h)\cos 3x-\sin 3x}{h}\\ &= \frac{\sin 3x(\cos(3h)-1)+\sin(3h)\cos 3x}{h}\\ &= \frac{\sin 3x(\cos(3h)-1)}{h}+\frac{\sin(3h)\cos 3x}{h}\\ &= \cos 3x\left(\frac{\sin(3h)}{h}\right)-\sin 3x\left(\frac{1-\cos(3h)}{h}\right)\end{aligned}$$

a. The following is the graph of
$y_1 = \cos 3x\left(\frac{\sin(3)}{1}\right)-\sin 3x\left(\frac{1-\cos(3)}{1}\right)$,
along with the graph of $y = 3\cos 3x$ (dotted graph):

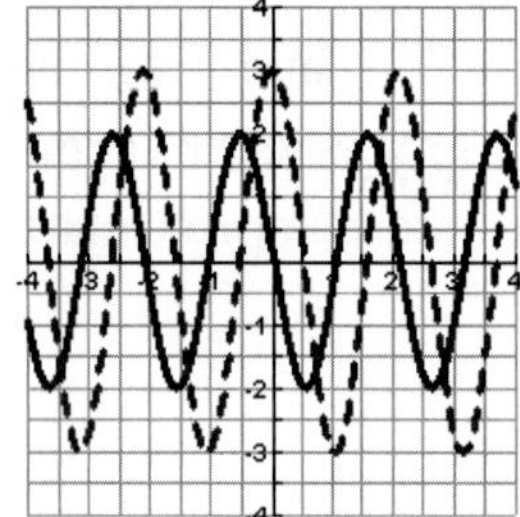

b. The following is the graph of
$y_2 = \cos 3x\left(\frac{\sin(0.3)}{0.1}\right)-\sin 3x\left(\frac{1-\cos(0.3)}{0.1}\right)$
, along with the graph of $y = 3\cos 3x$ (dotted graph):

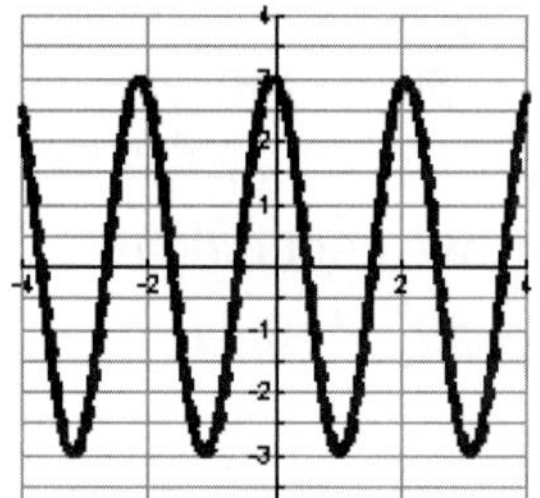

c. The following is the graph of
$y_3 = \cos 3x\left(\frac{\sin(0.03)}{0.01}\right)-\sin 3x\left(\frac{1-\cos(0.03)}{0.01}\right)$,
along with the graph of $y = 3\cos 3x$ (dotted graph):

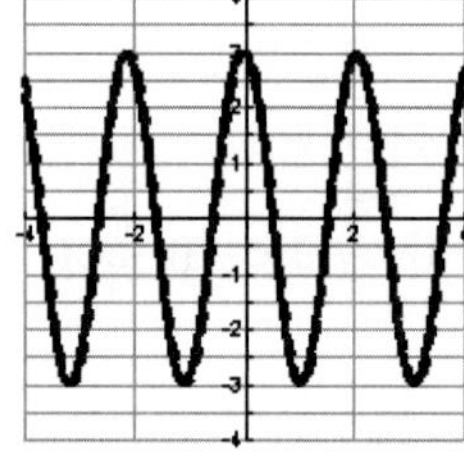

Looking at the sequence of graphs in parts **a** – **c**, it is clear that this sequence of graphs better approximates the graph of $y = 3\cos 3x$ as $h \to 0$.

164. Observe that

$$\begin{aligned}\frac{\cos\left(3(x+h)\right)-\cos 3x}{h} &= \frac{\cos 3x\cos(3h)-\sin 3x\sin(3h)-\cos 3x}{h} \\ &= \frac{\cos 3x\left(\cos(3h)-1\right)-\sin 3x\sin(3h)}{h} \\ &= \cos 3x\left(\frac{\cos(3h)-1}{h}\right)-\sin 3x\left(\frac{\sin(3h)}{h}\right) \\ &= -\cos 3x\left(\frac{1-\cos(3h)}{h}\right)-\sin 3x\left(\frac{\sin(3h)}{h}\right)\end{aligned}$$

a. The following is the graph of

$y_1 = -\cos 3x\left(\frac{1-\cos(3)}{1}\right)-\sin 3x\left(\frac{\sin(3)}{1}\right)$,

along with the graph of $y=-3\sin 3x$ (dotted graph):

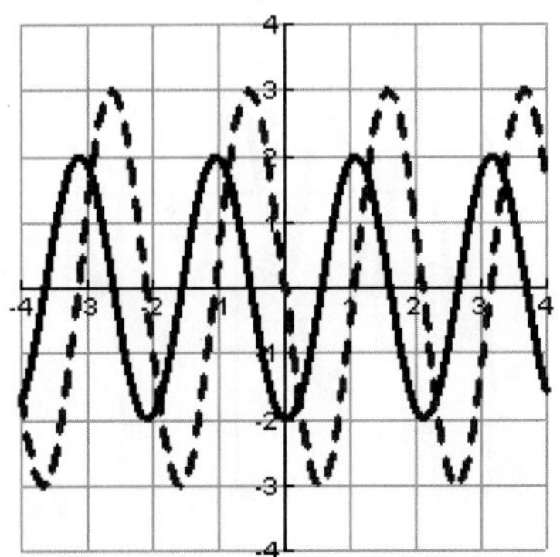

b. The following is the graph of

$y_2 = -\cos 3x\left(\frac{1-\cos(0.3)}{0.1}\right)-\sin 3x\left(\frac{\sin(0.3)}{0.1}\right)$,

along with the graph of $y=-3\sin 3x$ (dotted graph):

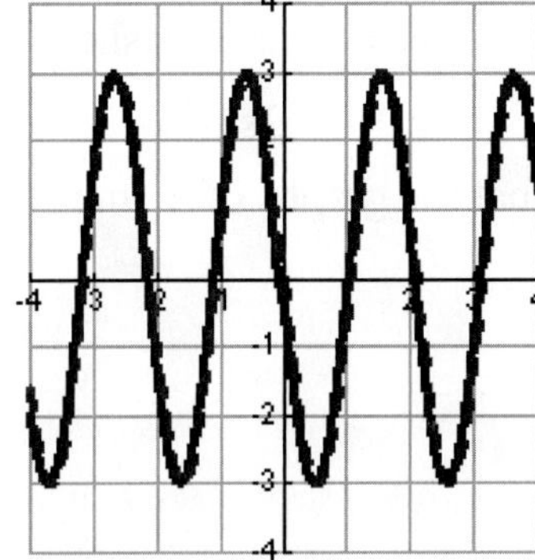

c. The following is the graph of

$y_2 = -\cos 3x\left(\frac{1-\cos(0.03)}{0.01}\right)-\sin 3x\left(\frac{\sin(0.03)}{0.01}\right)$,

along with the graph of $y=-3\sin 3x$ (dotted graph):

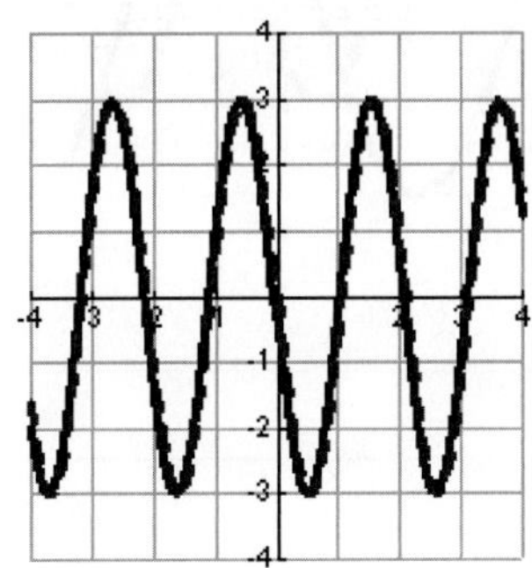

Looking at the sequence of graphs in parts **a** – **c**, it is clear that this sequence of graphs better approximates the graph of $y=-3\sin 3x$ as $h \to 0$.

165. To the right are the graphs of the following functions:

$$y_1 = \tan 2x \text{ (solid)}$$

$$y_2 = 2\tan x \text{ (dashed)}$$

$$y_3 = \frac{2\tan x}{1-\tan^2 x}$$

Note that the graphs of y_1 and y_3 are the same.

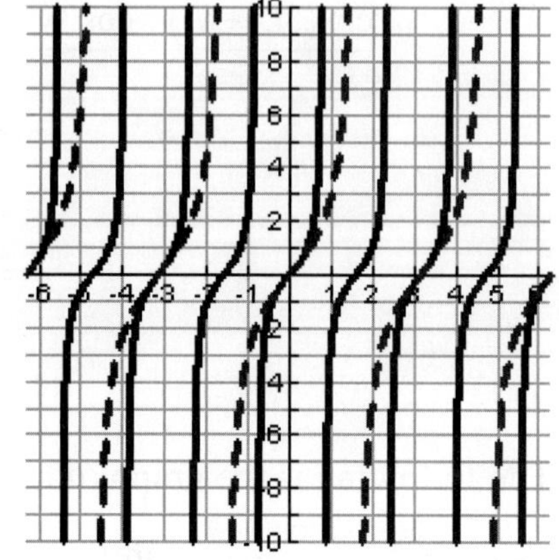

166. To the right are the graphs of the following functions:

$$y_1 = \cos 2x \text{ (solid)}$$

$$y_2 = 2\cos x \text{ (dashed)}$$

$$y_3 = 1-2\sin^2 x$$

Note that the graphs of y_1 and y_3 are the same.

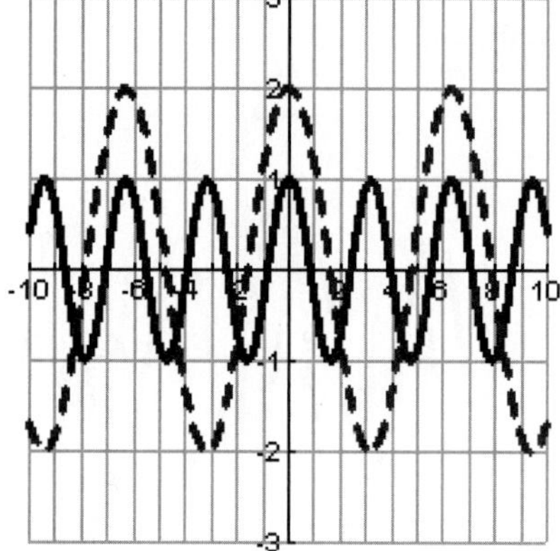

167. To the right are the graphs of the following functions:

$$y_1 = \cos\left(\tfrac{x}{2}\right) \text{ (solid)}$$

$$y_2 = \tfrac{1}{2}\cos x \text{ (dashed)}$$

$$y_3 = \sqrt{\frac{1+\cos x}{2}}$$

Note that the graphs of y_1 and y_3 are the same.

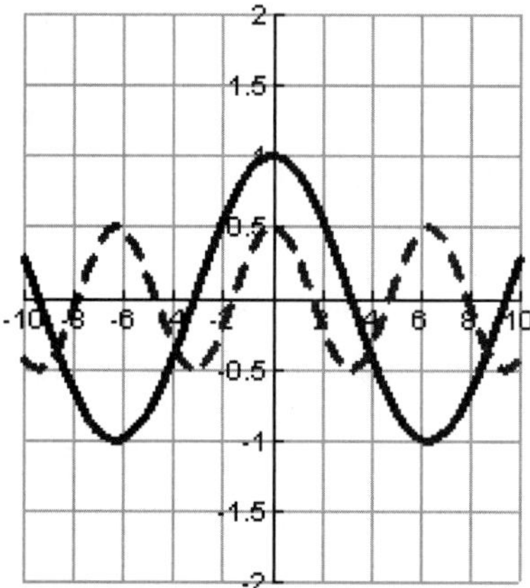

168. To the right are the graphs of the following functions:

$y_1 = \sin\left(\frac{x}{2}\right)$ (solid)

$y_2 = \frac{1}{2}\sin x$ (dashed)

$y_3 = \sqrt{\frac{1-\cos x}{2}}$

Note that the graphs of y_1 and y_3 are the same.

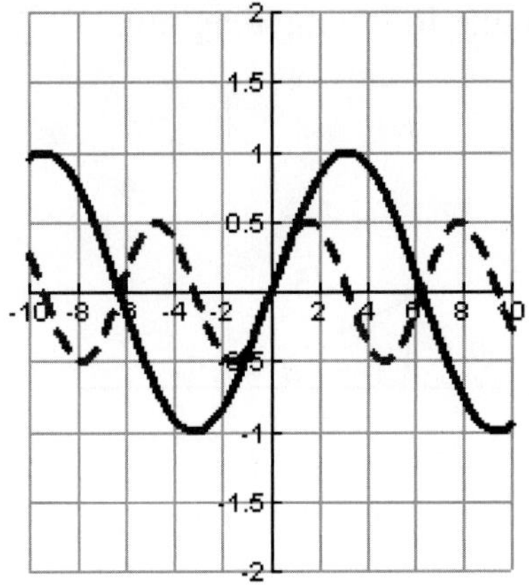

169. To the right are the graphs of the following functions:

$y_1 = \sin 5x \cos 3x$ (solid)

$y_2 = \sin 4x$ (dashed)

$y_3 = \frac{1}{2}[\sin 8x + \sin 2x]$

Note that the graphs of y_1 and y_3 are the same.

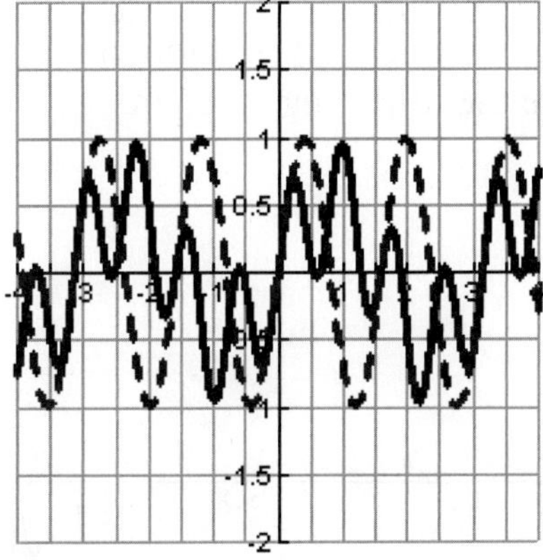

170. To the right are the graphs of the following functions:

$y_1 = \sin 3x \cos 5x$ (solid)

$y_2 = \cos 4x$ (dashed)

$y_3 = \frac{1}{2}[\sin 8x - \sin 2x]$

Note that the graphs of y_1 and y_3 are the same.

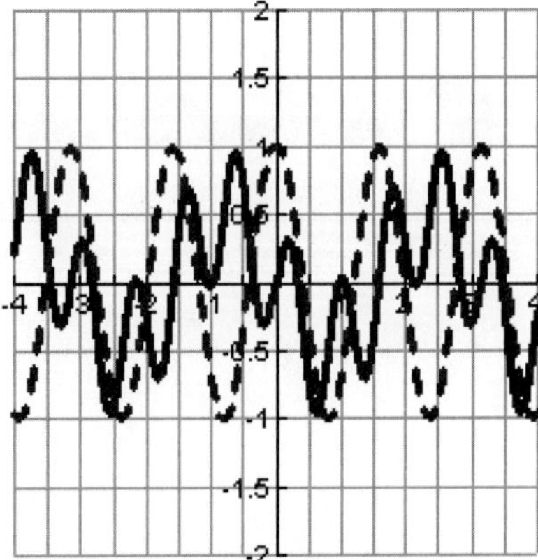

171. From the given information, we have the following diagram:

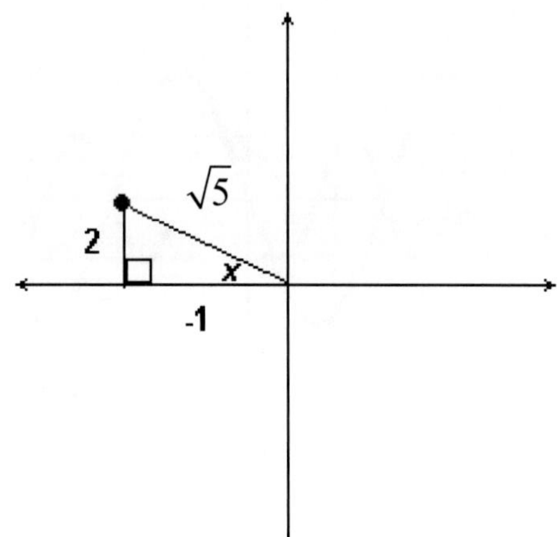

a.

$$\cos 2x = \cos^2 x - \sin^2 x$$
$$= \left(-\tfrac{1}{\sqrt{5}}\right)^2 - \left(\tfrac{2}{\sqrt{5}}\right)^2 = -\tfrac{3}{5}$$

b. $x = \cos^{-1}\left(-\tfrac{1}{\sqrt{5}}\right) \approx 2.0344$. So,

$\cos 2x = \cos(4.0689) \approx -0.6$

c. Yes, the results in parts **a.** and **b.** are the same.

172. From the given information, we have the following diagram:

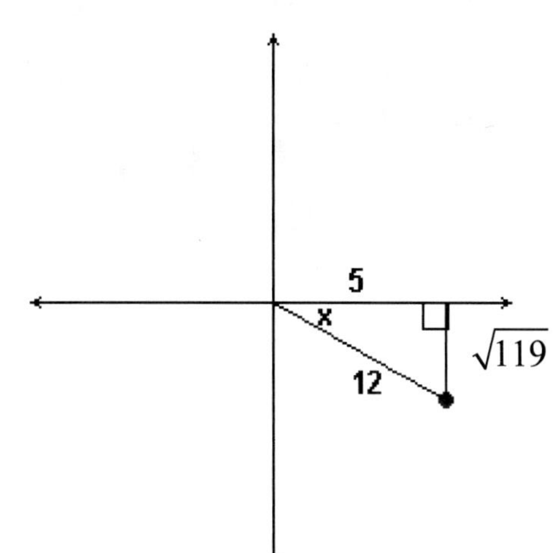

a. Since x is in QIV, $\tfrac{x}{2}$ is in QII. Hence,

$$\cos\left(\tfrac{x}{2}\right) = -\sqrt{\frac{1+\cos x}{2}} = -\sqrt{\frac{1+\frac{5}{12}}{2}} = -\sqrt{\frac{17}{24}}$$

b. $x = \cos^{-1}\left(\tfrac{5}{12}\right) \approx 1.1410$. So,

$\cos\left(\tfrac{x}{2}\right) = \cos(0.5705) \approx 0.8416$

c. Yes, the results in parts **a.** and **b.** are the same.

173. To determine the smallest positive solution (approximately) of the equation $\ln x + \sin x = 0$ graphically, we search for the intersection points of the graphs of the following two functions:

$$y_1 = \ln x + \sin x,\ \ y_2 = 0$$

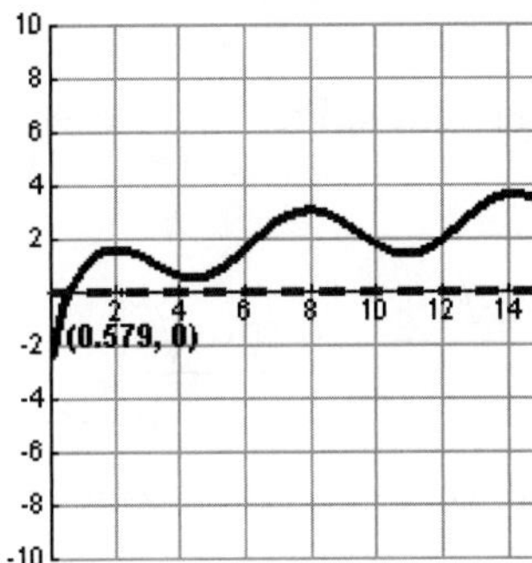

Observe that the graphs seems to intersect at approximately 0.579 radians, which is about $\boxed{33.17^\circ}$.

174. To determine the smallest positive solution (approximately) of the equation $\ln x + \cos x = 0$ graphically, we search for the intersection points of the graphs of the following two functions:

$$y_1 = \ln x + \cos x,\ \ y_2 = 0$$

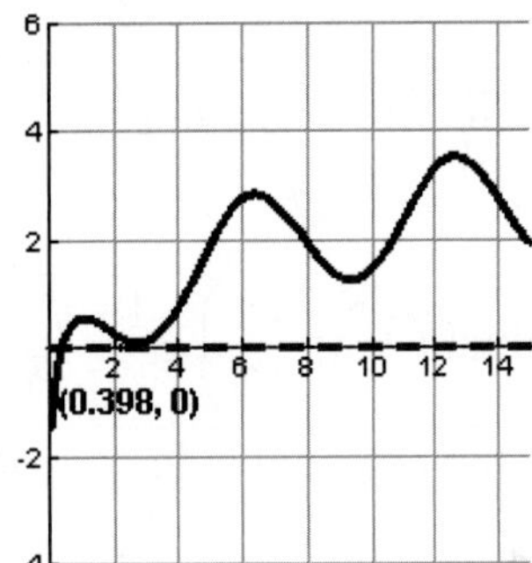

Observe that the graphs seems to intersect at approximately 0.398 radians, which is about $\boxed{22.80^\circ}$.

Chapter 6 Practice Test Solutions --

1. The identity $\tan x = \frac{\sin x}{\cos x}$ does not hold for $x = \frac{(2n+1)\pi}{2}$, where n is an integer.

2. Observe that $\sqrt{\sin^2 x + \cos^2 x} = \sqrt{1} = 1$. So, the equation $\sqrt{\sin^2 x + \cos^2 x} = \sin x + \cos x$ is a *conditional*. Indeed, take for instance $x = \frac{\pi}{4}$. Then, the left-side = 1, while the right-side is $\sqrt{2}$. (**Note**: A common algebraic error is to say $\sqrt{a^2 + b^2} = a + b$, which is false.)

3. Use the odd identity for sine, followed by $\sin\left(\frac{A}{2}\right) = \sqrt{\frac{1-\cos A}{2}}$ with $A = \frac{\pi}{4}$ to obtain

$$\sin\left(-\frac{\pi}{8}\right) = -\sin\left(\frac{\pi}{8}\right) = -\sin\left(\frac{\frac{\pi}{4}}{2}\right) = -\sqrt{\frac{1-\cos\frac{\pi}{4}}{2}} = -\sqrt{\frac{1-\frac{\sqrt{2}}{2}}{2}} = -\sqrt{\frac{2-\sqrt{2}}{4}} = \boxed{-\frac{\sqrt{2-\sqrt{2}}}{2}}.$$

4. Using the formula $\tan(A+B)=\dfrac{\tan A+\tan B}{1-\tan A\tan B}$ with $A=\dfrac{\pi}{4},\ B=\dfrac{\pi}{3}$ yields

$$\tan\left(\frac{7\pi}{12}\right)=\tan\left(\frac{\pi}{4}+\frac{\pi}{3}\right)=\frac{\tan\frac{\pi}{4}+\tan\frac{\pi}{3}}{1-\tan\frac{\pi}{4}\tan\frac{\pi}{3}}.\ \textbf{(1)}$$

Since $\tan\dfrac{\pi}{4}=1$ and $\tan\dfrac{\pi}{3}=\dfrac{\sin\frac{\pi}{3}}{\cos\frac{\pi}{3}}=\dfrac{\sqrt{3}/2}{1/2}=\sqrt{3}$, we can substitute these into **(1)** to further obtain $\tan\left(\dfrac{7\pi}{12}\right)=\dfrac{1+\sqrt{3}}{1-\sqrt{3}}=\dfrac{1+\sqrt{3}}{1-\sqrt{3}}\cdot\dfrac{1+\sqrt{3}}{1+\sqrt{3}}=\dfrac{1+2\sqrt{3}+3}{1-3}=\boxed{-2-\sqrt{3}}$.

5. Since the terminal side of x lies in QIV, $\cos x=\dfrac{2}{5}$ and $\sin x<0$. Hence, the terminal side of $\dfrac{x}{2}$ lies in QII, and so $\sin\left(\dfrac{x}{2}\right)>0$. Therefore, we have

$$\sin\left(\frac{x}{2}\right)=\sqrt{\frac{1-\cos x}{2}}=\sqrt{\frac{1-\frac{2}{5}}{2}}$$
$$=\sqrt{\frac{3}{10}}=\boxed{\frac{\sqrt{30}}{10}}.$$

6. Since $\sin x=-\dfrac{1}{5}$ and $\cos x<0$, the terminal side of x must be in QIII. The diagram is as follows:

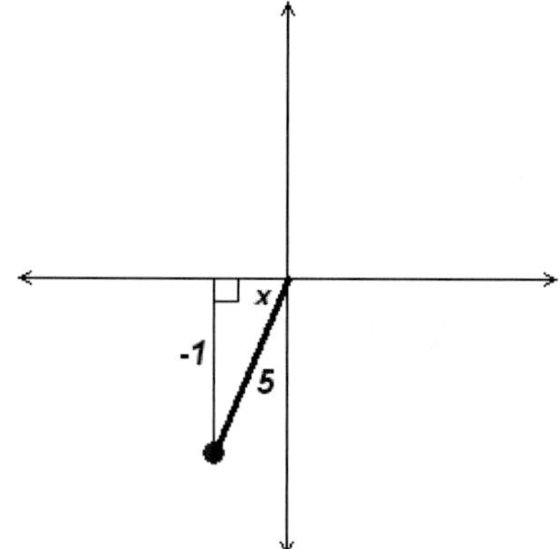

From the Pythagorean Theorem, we have $z^2+(-1)^2=5^2$, so that $z=-2\sqrt{6}$.

As such, $\cos x=-\dfrac{2\sqrt{6}}{5}$. Therefore,

$$\cos 2x=\cos^2 x-\sin^2 x$$
$$=\left(-\frac{2\sqrt{6}}{5}\right)^2-\left(-\frac{1}{5}\right)^2$$
$$=\frac{24-1}{25}=\boxed{\frac{23}{25}}$$

7. Using $\cos(A+B)=\cos A\cos B-\sin A\sin B$ with $A=3x,\ B=7x$ yields

$$\cos 3x\cos 7x-\sin 3x\sin 7x=\cos(3x+7x)=\boxed{\cos(10x)}.$$

8. Observe that $\dfrac{-2\tan x}{1-\tan^2 x}=-\left[\dfrac{2\tan x}{1-\tan^2 x}\right]=\boxed{-\tan 2x}$.

9. If the terminal side of $a+b$ is in QII, then $\sqrt{\dfrac{1+\cos(a+b)}{2}}=\cos\left(\dfrac{a+b}{2}\right)$, since the terminal side of $\dfrac{a+b}{2}$ is in QI.

10. Observe that

$$2\sin\left(\frac{x+3}{2}\right)\cos\left(\frac{x-3}{2}\right)=2\cdot\frac{1}{2}\left[\sin\left(\frac{x+3}{2}+\frac{x-3}{2}\right)+\sin\left(\frac{x+3}{2}-\frac{x-3}{2}\right)\right]=\boxed{\sin x+\sin 3}$$

11. Observe that

$$\begin{aligned}10\cos(3-x)+10\cos(x+3)&=10\left[\cos(3-x)+10\cos(x+3)\right]\\&=10\left[2\cos\left(\frac{(3-x)+(x+3)}{2}\right)\cos\left(\frac{(3-x)-(x+3)}{2}\right)\right]\\&=20\cos 3\cos(-x)=\boxed{20\cos x\cos 3}\end{aligned}$$

12. Let $u=3\sin x$, where $-\frac{\pi}{2}\leq x\leq\frac{\pi}{2}$. Then,

$$\sqrt{9-u^2}=\sqrt{9-(3\sin x)^2}=\sqrt{9-9\sin^2 x}=\sqrt{9(1-\sin^2 x)}=\sqrt{9\cos^2 x}=3|\cos x|.$$

13. Observe that $2\sin\theta=-\sqrt{3}$ is equivalent to $\sin\theta=-\frac{\sqrt{3}}{2}$. The values of θ in $\mathbb{R}$ that satisfy this equation are $\boxed{\theta=\frac{4\pi}{3}+2n\pi,\ \frac{5\pi}{3}+2n\pi\text{, where } n \text{ is an integer}}$.

14. In order to find all of the values of θ in $[0,2\pi]$ that satisfy $2\cos^2\theta+\cos\theta-1=0$, we proceed as follows:

Step 1: Simplify the equation algebraically.

Factoring the left-side yields $(\cos\theta+1)(2\cos\theta-1)=0$

So, θ is a solution to the original equation if either

$$\cos\theta=-1 \quad\text{or}\quad \cos\theta=\frac{1}{2}.$$

Step 2: Conclude that the solutions to the original equation are $\theta=\boxed{\pi,\ \frac{\pi}{3},\ \frac{5\pi}{3}}$.

15. In order to find all of the values of θ in $\left[0°, 360°\right]$ that satisfy $\sin 2\theta = \frac{1}{2}\cos\theta$, we proceed as follows:

Step 1: Simplify the equation algebraically.

$$\sin 2\theta = \frac{1}{2}\cos\theta$$
$$2\sin\theta\cos\theta = \frac{1}{2}\cos\theta$$
$$4\sin\theta\cos\theta = \cos\theta$$
$$\cos\theta(4\sin\theta - 1) = 0$$
$$\cos\theta = 0 \text{ or } 4\sin\theta - 1 = 0$$
$$\cos\theta = 0 \text{ or } \sin\theta = \frac{1}{4}$$

Step 2: Observe that the solutions of $\cos\theta = 0$ are $\theta = 90°, 270°$.

Step 3: Solve $\sin\theta = \frac{1}{4}$.

To do so, we must find the values of θ whose sine is $\frac{1}{4}$.

Indeed, observe that one solution is $\sin^{-1}\left(\frac{1}{4}\right) \approx 14.48°$, which is in QI.

A second solution occurs in QIII, and has value $180° - 14.48° = 165.52°$. Since the input of sine is simply θ, and not some multiple thereof, we conclude that the solutions to this equation are approximately $14.48°, 165.52°$.

Step 4: We conclude that the solutions of the original equation in $\left[0°, 360°\right]$ are approximately $\theta \approx \boxed{14.48°, 165.52°, 90°, 270°}$.

16. The equation $\sqrt{\sin x + \cos x} = -1$ has no solution since the left-side is always non-negative (for any value of x), while the right-side is negative.

17. $(1 + \cot x)^2 = \csc^2 x$ is a conditional since it is false when $x = \frac{3\pi}{4}$, for instance.

18. Observe that

$$\csc\left(-\frac{\pi}{12}\right) = \frac{1}{\sin\left(-\frac{\pi}{12}\right)} = \frac{1}{\sin\left(\frac{\pi}{4} - \frac{\pi}{3}\right)} = \frac{1}{\sin\left(\frac{\pi}{4}\right)\cos\left(\frac{\pi}{3}\right) - \cos\left(\frac{\pi}{4}\right)\sin\left(\frac{\pi}{3}\right)}$$
$$= \frac{1}{\frac{\sqrt{2}}{2}\left(\frac{1}{2} - \frac{\sqrt{3}}{2}\right)} = \boxed{\frac{4}{\sqrt{2}\left(1 - \sqrt{3}\right)}}$$

19. Since $\sin x = -\dfrac{5}{13}$ and the terminal side of x is in QIII, we know that $\cos x = -\dfrac{12}{13}$. Moreover, the terminal side of $\frac{x}{2}$ is in QII, so that $\cos\left(\dfrac{x}{2}\right) < 0$. Therefore, we have

$$\cos\left(\frac{x}{2}\right) = -\sqrt{\frac{1+\cos x}{2}} = -\sqrt{\frac{1-\frac{12}{13}}{2}} = \boxed{-\frac{\sqrt{26}}{26}}.$$

20. Since $\cos x = -0.26$ and the terminal side of x is in QII, we can use the Pythagorean theorem in the following triangle to conclude that $\sin x \approx \sqrt{0.9324}$.

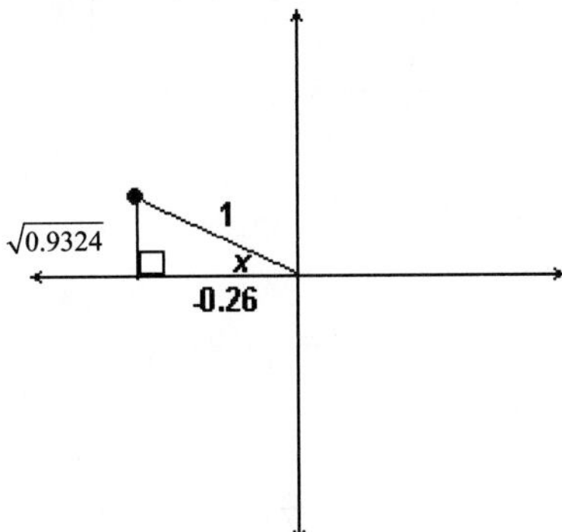

Thus, we have

$$\sin 2x = 2\sin x\cos x = 2\sqrt{0.9324}\,(-0.26) \approx \boxed{0.251}.$$

21. Observe that

$$y = \sqrt{\frac{1+\frac{\sqrt{2}}{2}\left(\cos\left(\frac{\pi}{3}x\right)+\sin\left(\frac{\pi}{3}x\right)\right)}{1-\frac{\sqrt{2}}{2}\left(\cos\left(\frac{\pi}{3}x\right)+\sin\left(\frac{\pi}{3}x\right)\right)}} = \sqrt{\frac{1+\left(\cos\left(\frac{\pi}{3}x\right)\cos\left(\frac{\pi}{4}\right)+\sin\left(\frac{\pi}{3}x\right)\sin\left(\frac{\pi}{4}\right)\right)}{1-\left(\cos\left(\frac{\pi}{3}x\right)\cos\left(\frac{\pi}{4}\right)+\sin\left(\frac{\pi}{3}x\right)\sin\left(\frac{\pi}{4}\right)\right)}}$$

$$= \sqrt{\frac{1+\cos\left(\frac{\pi}{3}x+\frac{\pi}{4}\right)}{1-\cos\left(\frac{\pi}{3}x+\frac{\pi}{4}\right)}} = \frac{1}{\sqrt{\dfrac{1-\cos\left(\frac{\pi}{3}x+\frac{\pi}{4}\right)}{1+\cos\left(\frac{\pi}{3}x+\frac{\pi}{4}\right)}}} = \frac{1}{\tan\left(\dfrac{\frac{\pi}{3}x+\frac{\pi}{4}}{2}\right)} = \boxed{\cot\left(\frac{\pi}{6}x+\frac{\pi}{8}\right)}$$

22. $\csc\left(\csc^{-1}\left(\sqrt{2}\right)\right) = \csc\left(\frac{\pi}{4}\right) = \boxed{\sqrt{2}}$

23. First, note that $y=\csc x$ is 1-1 on the set $\left[-\frac{\pi}{2},0\right)\cup\left(0,\frac{\pi}{2}\right]$ and so, $y=\csc\pi x$ is 1-1 on the set $\left[-\frac{1}{2},0\right)\cup\left(0,\frac{1}{2}\right]$. Hence, translating *c* units to the right, we see that $y=\csc(\pi x+c)$ is 1-1 on the set $\left[c-\frac{1}{2},c\right)\cup\left(c,c+\frac{1}{2}\right]$. Since multiplying by a constant and subsequently adding a constant do not change the set on which a function is 1-1, we conclude that *f* is 1-1 on $\left[c-\frac{1}{2},c\right)\cup\left(c,c+\frac{1}{2}\right]$.

Next, we determine the inverse of *f*, on this set, as follows:

$$y=a+b\csc(\pi x+c)$$

$$x=a+b\csc(\pi y+c)$$

$$\frac{x-a}{b}=\csc(\pi y+c)$$

$$\csc^{-1}\left(\frac{x-a}{b}\right)=\pi y+c$$

$$y=\frac{1}{\pi}\csc^{-1}\left(\frac{x-a}{b}\right)-\frac{c}{\pi}$$

Thus, $f^{-1}(x)=\frac{1}{\pi}\csc^{-1}\left(\frac{x-a}{b}\right)-\frac{c}{\pi}$.

24. The range of $y=\arctan(2x-3)$ is $\left(-\frac{\pi}{2},\frac{\pi}{2}\right)$. Hence, shifting this down by $\frac{\pi}{4}$ units, we conclude that the range of $y=-\frac{\pi}{4}+\arctan(2x-3)$ is $\boxed{\left(-\frac{3\pi}{4},\frac{\pi}{4}\right)}$.

25. Observe that the values of θ in $\mathbb{R}$ for which $\cos\left(\frac{\pi}{4}\theta\right)=-\frac{1}{2}$ must satisfy

$$\frac{\pi}{4}\theta=\frac{2\pi}{3}+2n\pi,\ \frac{4\pi}{3}+2n\pi, \text{ where } n \text{ is an integer.}$$

Thus, solving for θ yields

$$\theta=\frac{8}{3}+8n,\ \frac{16}{3}+8n, \text{ where } n \text{ is an integer.}$$

26. First, note that $\sqrt{\frac{1-\cos(2\pi x)}{1+\cos(2\pi x)}}=\tan(\pi x)$. Hence, we solve the equivalent equation $\tan(\pi x)=-\frac{1}{\sqrt{3}}=-\frac{\frac{1}{2}}{\frac{\sqrt{3}}{2}}$. The *x*-values for which this equation holds must satisfy:

$$\pi x=\frac{5\pi}{6}+2n\pi,\ \frac{11\pi}{6}+2n\pi, \text{ where } n \text{ is an integer.}$$

Solving for *x* yields

$$x=\frac{5+12n}{6},\ \frac{11+12n}{6} \text{ where } n \text{ is an integer.}$$

27. Observe that

$$\frac{\sqrt{3}}{\csc\left(\frac{x}{3}\right)} = \cos\left(\frac{x}{3}\right)$$

$$\sqrt{3}\sin\left(\frac{x}{3}\right) = \cos\left(\frac{x}{3}\right)$$

$$\sqrt{3} = \frac{\cos\left(\frac{x}{3}\right)}{\sin\left(\frac{x}{3}\right)} = \cot\left(\frac{x}{3}\right)$$

The x-values for which this equation holds must satisfy:

$$\frac{x}{3} = \frac{\pi}{6} + 2n\pi,\ \frac{7\pi}{6} + 2n\pi \text{, where } n \text{ is an integer.}$$

Solving for x yields

$$x = \frac{\pi}{2} + 6n\pi,\ \frac{7\pi}{2} + 6n\pi, \text{ where } n \text{ is an integer.}$$

28. Observe that

$$\begin{aligned}
\frac{\cos\left(\frac{1}{2}(x+h)\right) - \cos\frac{1}{2}x}{h} &= \frac{\cos\frac{1}{2}x\cos(\frac{1}{2}h) - \sin\frac{1}{2}x\sin(\frac{1}{2}h) - \cos\frac{1}{2}x}{h} \\
&= \frac{\cos\frac{1}{2}x\left(\cos(\frac{1}{2}h) - 1\right) - \sin\frac{1}{2}x\sin(\frac{1}{2}h)}{h} \\
&= \cos\frac{1}{2}x\left(\frac{\cos(\frac{1}{2}h) - 1}{h}\right) - \sin\frac{1}{2}x\left(\frac{\sin(\frac{1}{2}h)}{h}\right) \\
&= -\cos\frac{1}{2}x\left(\frac{1 - \cos(\frac{1}{2}h)}{h}\right) - \sin\frac{1}{2}x\left(\frac{\sin(\frac{1}{2}h)}{h}\right)
\end{aligned}$$

a. The following is the graph of

$$y_1 = -\cos\tfrac{1}{2}x\left(\frac{1-\cos(\frac{1}{2})}{1}\right) - \sin\tfrac{1}{2}x\left(\frac{\sin(\frac{1}{2})}{1}\right),$$

along with the graph of $y = -\frac{1}{2}\sin\frac{1}{2}x$ (dotted graph):

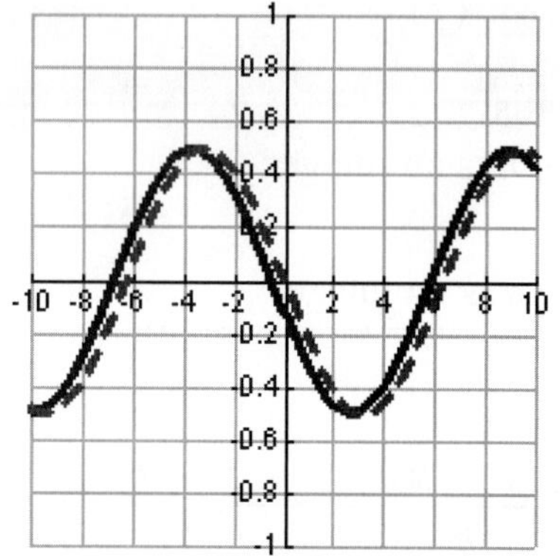

b. The following is the graph of

$$y_2 = -\cos\tfrac{1}{2}x\left(\frac{1-\cos(0.05)}{0.1}\right) - \sin\tfrac{1}{2}x\left(\frac{\sin(0.05)}{0.1}\right)$$

, along with the graph of $y = -\frac{1}{2}\sin\frac{1}{2}x$ (dotted graph):

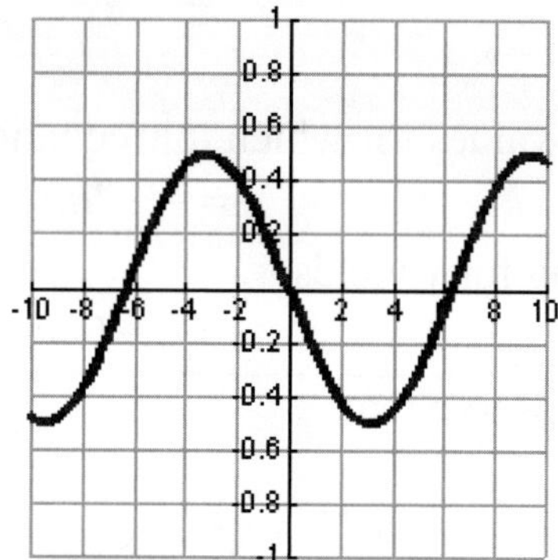

c. The following is the graph of

$$y_3 = -\cos\tfrac{1}{2}x\left(\frac{1-\cos(0.005)}{0.01}\right) - \sin\tfrac{1}{2}x\left(\frac{\sin(0.005)}{0.01}\right),$$

along with the graph of $y = -\frac{1}{2}\sin\frac{1}{2}x$ (dotted graph):

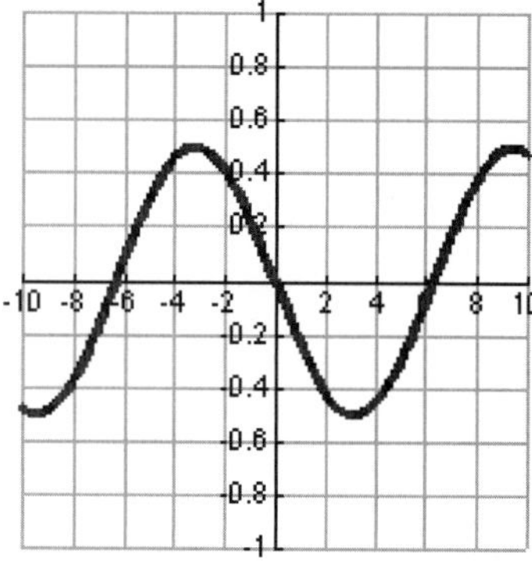

Looking at the sequence of graphs in parts **a** – **c**, it is clear that this sequence of graphs better approximates the graph of $y = -\frac{1}{2}\sin\frac{1}{2}x$ as $h \to 0$.

29. From the given information, we have the following diagram: 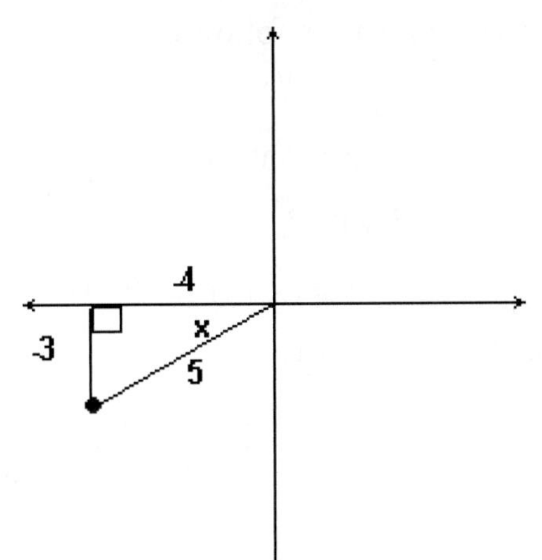	**a.** Since x is in QIII, $\frac{x}{2}$ is in QII. Hence, $\sin\left(\frac{x}{2}\right)=\sqrt{\frac{1-\cos x}{2}}=\sqrt{\frac{1+\frac{4}{5}}{2}}=\sqrt{\frac{9}{10}}=\frac{3\sqrt{10}}{10}$ **b.** $x=\tan^{-1}\left(\frac{3}{4}\right)+\pi\approx 3.785094$. So, $\sin\left(\frac{x}{2}\right)=\sin\left(\frac{3.785094}{2}\right)\approx 0.94868$ **c.** Yes, the results in parts **a.** and **b.** are the same.

Chapter 6 Cumulative Test--

1. a.) $\frac{\sqrt{3}}{2}$ **b.)** $\sqrt{3}$ **c.)** $\dfrac{1}{-\frac{1}{2}}=-2$	**2. a.)** $\dfrac{1}{-\frac{\sqrt{3}}{2}}=-\dfrac{2\sqrt{3}}{3}$ **b.)** $-\dfrac{\sqrt{2}}{2}$ **c.)** $\sqrt{3}$
3. a.) $\dfrac{2\pi}{3}$ **b.)** $-\dfrac{\pi}{6}$ **c.)** $\dfrac{5\pi}{6}$	**4.** The relation $x^2-y^2=25$ is not a function since for $x=6$, for instance, we have both $\left(6,\sqrt{11}\right)$ and $\left(6,-\sqrt{11}\right)$ on the graph. So, it doesn't pass the vertical line test and hence, cannot be a function.
5. Observe that $g(-x)=\sqrt{2-(-x)^2}=\sqrt{2-x^2}=g(x)$. Thus, g is even.	**6.** In order to graph $y=5(x-4)^2$, first translate the graph of $y=x^2$ to the right 4 units. Then, vertically stretch it by a factor of 5.

7. Observe that

$$(f \circ g)(x) = f(g(x)) = f\left(\tfrac{1}{x}\right) = \tfrac{1}{x^3} - 1.$$

The domain is $(-\infty, 0) \cup (0, \infty)$.

8. To find the inverse of $f(x) = \sqrt[3]{x} - 1,\ x \geq 0$, switch the x and y and solve for y, as follows:

$$\begin{aligned} y &= \sqrt[3]{x} - 1 \\ x &= \sqrt[3]{y} - 1 \\ x + 1 &= \sqrt[3]{y} \\ y &= (x+1)^3,\ x \geq -1 \end{aligned}$$

So, $f^{-1}(x) = (x+1)^3$.

9. Completing the square yields

$$\begin{aligned} f(x) &= \tfrac{1}{4}x^2 + \tfrac{3}{5}x - \tfrac{6}{25} \\ &= \tfrac{1}{4}\left(x^2 + \tfrac{12}{5}x\right) - \tfrac{6}{25} \\ &= \tfrac{1}{4}\left(x^2 + \tfrac{12}{5}x + \tfrac{36}{25}\right) - \tfrac{6}{25} - \tfrac{1}{4}\left(\tfrac{36}{25}\right) \\ &= \tfrac{1}{4}\left(x + \tfrac{6}{5}\right)^2 - \tfrac{3}{5} \end{aligned}$$

So, the vertex is $\boxed{\left(-\tfrac{6}{5}, -\tfrac{3}{5}\right)}$.

10. $P(x) = cx^3\left(x + \sqrt{7}\right)^2\left(x - \sqrt{7}\right)^2$, for any constant c.

11. Long division yields

$$\begin{array}{r} 5x - 4 \\ x^2 + 0x + 1 \overline{\big) \; 5x^3 - 4x^2 + 0x + 3} \\ \underline{-\left(5x^3 + 0x^2 + 5x\right)} \\ -4x^2 - 5x + 3 \\ \underline{-\left(-4x^2 + 0x - 4\right)} \\ -5x + 7 \end{array}$$

So,

$$\boxed{Q(x) = 5x - 4, \quad r(x) = -5x + 7}.$$

12. Since $x = 4i$ is a zero of $P(x)$, so is $x = -4i$. Hence,

$$(x - 4i)(x + 4i) = x^2 + 16$$

must divide $P(x)$ evenly. Long division yields

$$\begin{array}{r} x^2 + 2x - 15 \\ x^2 + 0x + 16 \overline{\big) \; x^4 + 2x^3 + x^2 + 32x - 240} \\ \underline{-\left(x^4 + 0x^3 + 16x^2\right)} \\ 2x^3 - 15x^2 + 32x \\ \underline{-\left(2x^3 + 0x^2 + 32x\right)} \\ -15x^2 + 0x - 240 \\ \underline{-\left(-15x^2 + 0x - 240\right)} \\ 0 \end{array}$$

So,

$$\begin{aligned} P(x) &= \left(x^2 + 16\right)\left(x^2 + 16\right) \\ &= \boxed{\left(x^2 + 16\right)(x - 3)(x + 5)} \end{aligned}$$

13. Observe that

$$f(x)=\frac{0.7x^2-5x+11}{x^2-x-6}=\frac{0.7x^2-5x+11}{(x-3)(x+2)}$$

So, the horizontal asymptote is $y=0.7$ and the vertical asymptotes are $x=-2, x=3$.

14. Use $A=Pe^{rt}$. Here

$$P=5400,\ r=0.0225,\ t=4.$$

So,

$$A=5400e^{0.0225(4)}\approx\boxed{5{,}908.54}$$

15. Must have $x+3>0$. So, the domain is $\boxed{(-3,\infty)}$.

16. Observe that

$$\log_\pi 1=\frac{\ln 1}{\ln \pi}=\boxed{0}$$

17. Observe that

$$\log_5(x+2)+\log_5(6-x)=\log_5(3x)$$
$$\log_5(x+2)(6-x)=\log_5(3x)$$
$$(x+2)(6-x)=3x$$
$$x^2-x-12=0$$
$$(x-4)(x+3)=0$$
$$x=\cancel{-3},\ \boxed{4}$$

18. Use $A=Pe^{0.04t}$. We must find *t* such that $A=3P$. Observe that

$$3P=Pe^{0.04t}$$
$$3=e^{0.04t}$$
$$\ln 3=0.04t$$
$$t=\tfrac{1}{0.04}\ln 3\approx\boxed{27.47\text{ years}}$$

19. $\cos\left(62^\circ\right)\approx\boxed{0.4695}$

20. From the information, we know that

$$\sin\left(63^\circ\right)=\frac{4200\text{ ft.}}{z}$$
$$z=\frac{4200\text{ ft.}}{\sin\left(63^\circ\right)}\approx\boxed{4{,}713.77\text{ ft.}}$$

21. $-105^\circ=-105^\circ\left(\frac{\pi}{180^\circ}\right)=\boxed{-\frac{7\pi}{12}}$

22. Note that $\tan\theta=1$ when $\sin\theta=\cos\theta$. This occurs when $\boxed{\theta=\frac{\pi}{4},\frac{5\pi}{4}}$.

23. This is a conditional statement since it does not hold for $x=n\pi$, where *n* is an integer.

24. Observe that

$$\frac{2\tan\left(-\frac{\pi}{8}\right)}{1-\tan^2\left(-\frac{\pi}{8}\right)} = \tan\left(2\cdot\left(-\frac{\pi}{8}\right)\right)$$
$$= \tan\left(-\frac{\pi}{4}\right) = \boxed{-1}$$

25. Let $\theta = \sin^{-1}\left(\frac{5}{13}\right)$. Then, $\sin\theta = \frac{5}{13}$, as shown below:

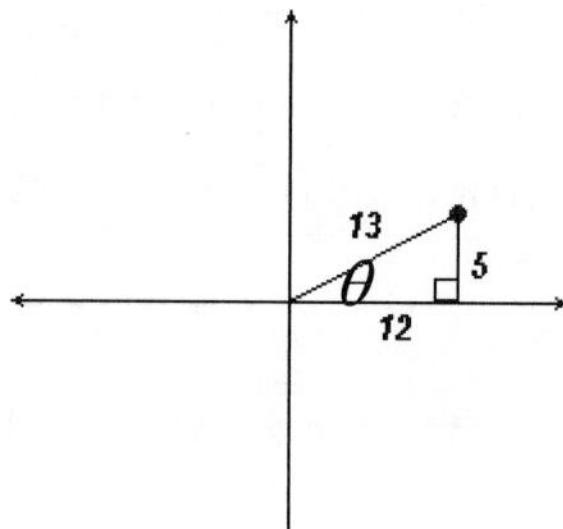

So, $\tan\left(\sin^{-1}\left(\frac{5}{13}\right)\right) = \boxed{\frac{5}{12}}$.

26. To the right are the graphs of the following functions:

$y_1 = \sin x \cos 3x$ (solid)

$y_2 = \cos 4x$ (dashed)

$y_3 = \frac{1}{2}[\sin 4x - \sin 2x]$

Note that the graphs of y_1 and y_3 are the same.

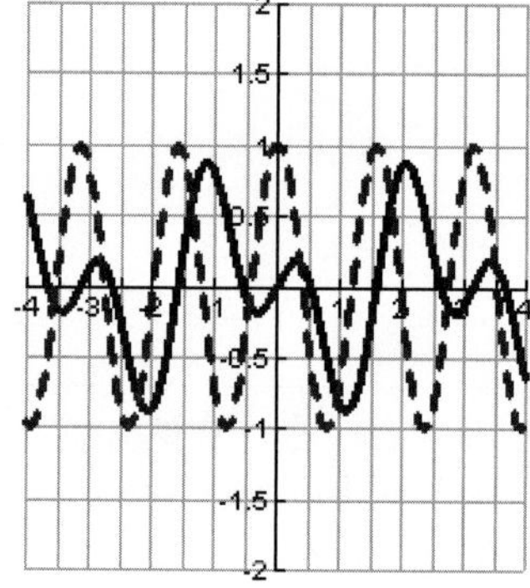

27. To determine the smallest positive solution (approximately) of the equation $\ln x + \sin x = 0$ graphically, we search for the intersection points of the graphs of the following two functions:

$$y_1 = \ln x - \sin 2x, \;\; y_2 = 0$$

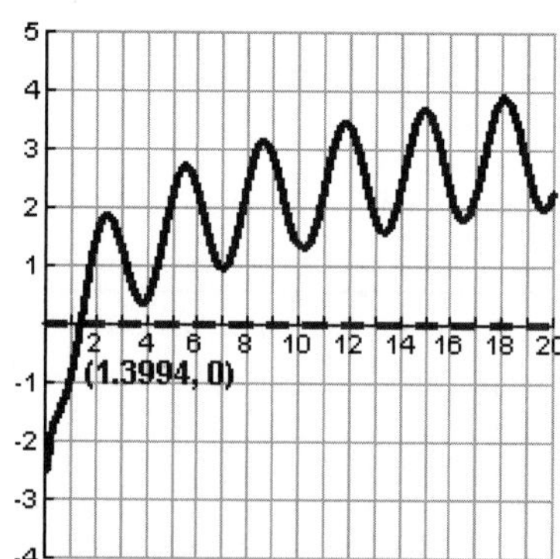

Observe that the graphs seems to intersect at approximately $\boxed{1.3994}$ radians.

CHAPTER 7

Section 7.1 Solutions --

1. $\overrightarrow{AB}=\langle 5-2,9-7\rangle=\langle 3,2\rangle$ $\lvert\overrightarrow{AB}\rvert=\sqrt{(3)^2+(2)^2}=\boxed{\sqrt{13}}$	**2.** $\overrightarrow{AB}=\langle 3-(-2),-4-3\rangle=\langle 5,-7\rangle$ $\lvert\overrightarrow{AB}\rvert=\sqrt{(5)^2+(-7)^2}=\boxed{\sqrt{74}}$
3. $\overrightarrow{AB}=\langle -3-4,0-1\rangle=\langle -7,-1\rangle$ $\lvert\overrightarrow{AB}\rvert=\sqrt{(-7)^2+(-1)^2}=\sqrt{50}=\boxed{5\sqrt{2}}$	**4.** $\overrightarrow{AB}=\langle 2-(-1),-5-(-1)\rangle=\langle 3,-4\rangle$ $\lvert\overrightarrow{AB}\rvert=\sqrt{(3)^2+(-4)^2}=\boxed{5}$
5. $\overrightarrow{AB}=\langle -24-0,0-7\rangle=\langle -24,-7\rangle$ $\lvert\overrightarrow{AB}\rvert=\sqrt{(-24)^2+(-7)^2}=\boxed{25}$	**6.** $\overrightarrow{AB}=\langle 4-(-2),9-1\rangle=\langle 6,8\rangle$ $\lvert\overrightarrow{AB}\rvert=\sqrt{(6)^2+(8)^2}=\boxed{10}$
7. Given that $\vec{u}=\langle 3,8\rangle$, we have $\lvert\vec{u}\rvert=\sqrt{(3)^2+(8)^2}=\sqrt{73}$ $\tan\theta=\frac{8}{3}$ so that $\theta=\tan^{-1}\left(\frac{8}{3}\right)\approx 69.4^\circ$	**8.** Given that $\vec{u}=\langle 4,7\rangle$, we have $\lvert\vec{u}\rvert=\sqrt{(4)^2+(7)^2}=\sqrt{65}$ $\tan\theta=\frac{7}{4}$ so that $\theta=\tan^{-1}\left(\frac{7}{4}\right)\approx 60.3^\circ$
9. Given that $\vec{u}=\langle 5,-1\rangle$, we have $\lvert\vec{u}\rvert=\sqrt{(5)^2+(-1)^2}=\sqrt{26}$ $\tan\theta=\frac{-1}{5}$ so that $\theta=\tan^{-1}\left(\frac{-1}{5}\right)\approx 348.7^\circ$	**10.** Given that $\vec{u}=\langle -6,-2\rangle$, we have $\lvert\vec{u}\rvert=\sqrt{(-6)^2+(-2)^2}=\sqrt{40}=2\sqrt{10}$ $\tan\theta=\frac{2}{6}$ so that since the head is in QIII, $\theta=180^\circ+\tan^{-1}\left(\frac{2}{6}\right)\approx 198.4^\circ$
11. Given that $\vec{u}=\langle -4,1\rangle$, we have $\lvert\vec{u}\rvert=\sqrt{(-4)^2+(1)^2}=\sqrt{17}$ $\tan\theta=\frac{1}{-4}$ so that since the head is in QII, $\theta=180^\circ+\tan^{-1}\left(\frac{1}{-4}\right)\approx 166.0^\circ$	**12.** Given that $\vec{u}=\langle -6,3\rangle$, we have $\lvert\vec{u}\rvert=\sqrt{(-6)^2+(3)^2}=\sqrt{45}=3\sqrt{5}$ $\tan\theta=\frac{3}{-6}$ so that since the head is in QII, $\theta=180^\circ+\tan^{-1}\left(\frac{3}{-6}\right)\approx 153.4^\circ$

13. Given that $\vec{u}=\langle -8,0\rangle$, we have $\lvert \vec{u}\rvert=\sqrt{(-8)^2+(0)^2}=8$ Since the vector is on the x-axis pointing in the negative direction, $\theta=180^\circ$.	**14.** Given that $\vec{u}=\langle 0,7\rangle$, we have $\lvert \vec{u}\rvert=\sqrt{(0)^2+(7)^2}=7$ Since the vector is on the y-axis pointing in the positive direction, $\theta=90^\circ$.
15. Given that $\vec{u}=\langle \sqrt{3},3\rangle$, we have $\lvert \vec{u}\rvert=\sqrt{\left(\sqrt{3}\right)^2+(3)^2}=2\sqrt{3}$ $\tan\theta=\frac{3}{\sqrt{3}}$ so that $\theta=\tan^{-1}\left(\frac{3}{\sqrt{3}}\right)\approx 60^\circ$	**16.** Given that $\vec{u}=\langle -5,-5\rangle$, we have $\lvert \vec{u}\rvert=\sqrt{(-5)^2+(-5)^2}=\sqrt{50}=5\sqrt{2}$ $\tan\theta=\frac{-5}{-5}=1$ so that since the head is in QIII, $\theta=180^\circ+\tan^{-1}(1)\approx 225^\circ$

17. $\vec{u}+\vec{v}=\langle -4,3\rangle+\langle 2,-5\rangle=\langle -4+2,3-5\rangle=\boxed{\langle -2,-2\rangle}$.

18. $\vec{u}-\vec{v}=\langle -4,3\rangle-\langle 2,-5\rangle=\langle -4-2,3-(-5)\rangle=\boxed{\langle -6,8\rangle}$

19. $3\vec{u}=3\langle -4,3\rangle=\boxed{\langle -12,9\rangle}$

20. $-2\vec{u}=-2\langle -4,3\rangle=\boxed{\langle 8,-6\rangle}$

21. $2\vec{u}+4\vec{v}=2\langle -4,3\rangle+4\langle 2,-5\rangle=\langle -8,6\rangle+\langle 8,-20\rangle=\boxed{\langle 0,-14\rangle}$

22. $5(\vec{u}+\vec{v})=5\vec{u}+5\vec{v}=5\langle -4,3\rangle+5\langle 2,-5\rangle=\langle -20,15\rangle+\langle 10,-25\rangle=\boxed{\langle -10,-10\rangle}$

23. $6(\vec{u}-\vec{v})=6\vec{u}-6\vec{v}=6\langle -4,3\rangle-6\langle 2,-5\rangle=\langle -24,18\rangle+\langle -12,30\rangle=\boxed{\langle -36,48\rangle}$

24. $2\vec{u}-3\vec{v}+4\vec{u}=6\vec{u}-3\vec{v}=6\langle -4,3\rangle-3\langle 2,-5\rangle=\langle -24,18\rangle+\langle -6,15\rangle=\boxed{\langle -30,33\rangle}$

25. $\vec{u}=\langle \lvert\vec{u}\rvert\cos\theta,\ \lvert\vec{u}\rvert\sin\theta\rangle=\langle 7\cos 25^\circ,\ 7\sin 25^\circ\rangle\approx\boxed{\langle 6.3,\ 3.0\rangle}$

26. $\vec{u}=\langle \lvert\vec{u}\rvert\cos\theta,\ \lvert\vec{u}\rvert\sin\theta\rangle=\langle 5\cos 75^\circ,\ 5\sin 75^\circ\rangle\approx\boxed{\langle 1.3,\ 4.8\rangle}$

27. $\vec{u}=\langle \lvert\vec{u}\rvert\cos\theta,\ \lvert\vec{u}\rvert\sin\theta\rangle=\langle 16\cos 100^\circ,\ 16\sin 100^\circ\rangle\approx\boxed{\langle -2.8,\ 15.8\rangle}$

28. $\vec{u}=\langle \lvert\vec{u}\rvert\cos\theta,\ \lvert\vec{u}\rvert\sin\theta\rangle=\langle 8\cos 200^\circ,\ 8\sin 200^\circ\rangle\approx\boxed{\langle -7.5,\ -2.7\rangle}$

29. $\vec{u}=\langle \lvert\vec{u}\rvert\cos\theta,\ \lvert\vec{u}\rvert\sin\theta\rangle=\langle 4\cos 310^\circ,\ 4\sin 310^\circ\rangle\approx\boxed{\langle 2.6,\ -3.1\rangle}$

30. $\vec{u}=\langle \lvert\vec{u}\rvert\cos\theta,\ \lvert\vec{u}\rvert\sin\theta\rangle=\langle 8\cos 225^\circ,\ 8\sin 225^\circ\rangle\approx\boxed{\langle -5.7,\ -5.7\rangle}$

31. $\vec{u}=\langle \lvert\vec{u}\rvert\cos\theta,\ \lvert\vec{u}\rvert\sin\theta\rangle=\langle 9\cos(335^\circ),\ 9\sin(335^\circ)\rangle\approx\boxed{\langle 8.2,\ -3.8\rangle}$

32. $\vec{u}=\langle \lvert\vec{u}\rvert\cos\theta,\ \lvert\vec{u}\rvert\sin\theta\rangle=\langle 3\cos(315^\circ),\ 3\sin(315^\circ)\rangle\approx\boxed{\langle 2.1,\ -2.1\rangle}$

33. $\vec{u}=\left\langle \lvert\vec{u}\rvert\cos\theta,\ \lvert\vec{u}\rvert\sin\theta \right\rangle=\left\langle 2\cos 120^\circ,\ 2\sin 120^\circ\right\rangle \approx \boxed{\langle -1,\ 1.7\rangle}$	
34. $\vec{u}=\left\langle \lvert\vec{u}\rvert\cos\theta,\ \lvert\vec{u}\rvert\sin\theta \right\rangle=\left\langle 6\cos 330^\circ,\ 6\sin 330^\circ\right\rangle \approx \boxed{\langle 5.2,\ -3\rangle}$	
35. $\frac{\vec{v}}{\lvert\vec{v}\rvert}=\frac{\langle -5,-12\rangle}{\sqrt{(-5)^2+(-12)^2}}=\frac{\langle -5,-12\rangle}{13}=\boxed{\left\langle \frac{-5}{13},\frac{-12}{13}\right\rangle}$	
36. $\frac{\vec{v}}{\lvert\vec{v}\rvert}=\frac{\langle 3,4\rangle}{\sqrt{(3)^2+(4)^2}}=\frac{\langle 3,4\rangle}{5}=\boxed{\left\langle \frac{3}{5},\frac{4}{5}\right\rangle}$	
37. $\frac{\vec{v}}{\lvert\vec{v}\rvert}=\frac{\langle 60,11\rangle}{\sqrt{(60)^2+(11)^2}}=\frac{\langle 60,11\rangle}{61}=\boxed{\left\langle \frac{60}{61},\frac{11}{61}\right\rangle}$	
38. $\frac{\vec{v}}{\lvert\vec{v}\rvert}=\frac{\langle -7,24\rangle}{\sqrt{(-7)^2+(24)^2}}=\frac{\langle -7,24\rangle}{25}=\boxed{\left\langle \frac{-7}{25},\frac{24}{25}\right\rangle}$	
39. $\frac{\vec{v}}{\lvert\vec{v}\rvert}=\frac{\langle 24,-7\rangle}{\sqrt{(24)^2+(-7)^2}}=\frac{\langle 24,-7\rangle}{25}=\boxed{\left\langle \frac{24}{25},\frac{-7}{25}\right\rangle}$	
40. $\frac{\vec{v}}{\lvert\vec{v}\rvert}=\frac{\langle -10,24\rangle}{\sqrt{(-10)^2+(24)^2}}=\frac{\langle -10,24\rangle}{26}=\left\langle \frac{-10}{26},\frac{24}{26}\right\rangle=\boxed{\left\langle \frac{-5}{13},\frac{12}{13}\right\rangle}$	
41. $\frac{\vec{v}}{\lvert\vec{v}\rvert}=\frac{\langle -9,-12\rangle}{\sqrt{(-9)^2+(-12)^2}}=\frac{\langle -9,-12\rangle}{15}=\left\langle \frac{-9}{15},\frac{-12}{15}\right\rangle=\boxed{\left\langle \frac{-3}{5},\frac{-4}{5}\right\rangle}$	
42. $\frac{\vec{v}}{\lvert\vec{v}\rvert}=\frac{\langle 40,-9\rangle}{\sqrt{(40)^2+(-9)^2}}=\frac{\langle 40,-9\rangle}{41}=\boxed{\left\langle \frac{40}{41},\frac{-9}{41}\right\rangle}$	
43. $\frac{\vec{v}}{\lvert\vec{v}\rvert}=\frac{\langle \sqrt{2},3\sqrt{2}\rangle}{\sqrt{(\sqrt{2})^2+(3\sqrt{2})^2}}=\frac{\langle \sqrt{2},3\sqrt{2}\rangle}{2\sqrt{5}}=\left\langle \frac{\sqrt{2}}{2\sqrt{5}},\frac{3\sqrt{2}}{2\sqrt{5}}\right\rangle=\boxed{\left\langle \frac{\sqrt{10}}{10},\frac{3\sqrt{10}}{10}\right\rangle}$	
44. $\frac{\vec{v}}{\lvert\vec{v}\rvert}=\frac{\langle -4\sqrt{3},-2\sqrt{3}\rangle}{\sqrt{(-4\sqrt{3})^2+(-2\sqrt{3})^2}}=\frac{\langle -4\sqrt{3},-2\sqrt{3}\rangle}{2\sqrt{15}}=\left\langle \frac{-4\sqrt{3}}{2\sqrt{15}},\frac{-2\sqrt{3}}{2\sqrt{15}}\right\rangle=\boxed{\left\langle \frac{-2\sqrt{5}}{5},\frac{-\sqrt{5}}{5}\right\rangle}$	
45. $7\vec{i}+3\vec{j}$	**46.** $-2\vec{i}+4\vec{j}$
47. $5\vec{i}-3\vec{j}$	**48.** $-6\vec{i}-2\vec{j}$
49. $-\vec{i}+0\vec{j}$	**50.** $0\vec{i}+2\vec{j}$
51. $2\vec{i}+0\vec{j}$	**52.** $7\vec{i}-7\vec{j}$

53. $-5\vec{i}+5\vec{j}$	**54.** $3\vec{i}-4\vec{j}$
55. $7\vec{i}+0\vec{j}$	**56.** $0\vec{i}-3\vec{j}$

57. The velocity vector is given by $\vec{v}=\langle 2200\cos 30^\circ,\ 2200\sin 30^\circ\rangle \approx \langle 1905,\ 1100\rangle$.

So, the horizontal component is approximately 1,905 $^{\text{ft.}}/_{\text{sec.}}$ and the vertical component is approximately 1,100 $^{\text{ft.}}/_{\text{sec.}}$.

58. The force vector is given by $\langle 50\cos 40^\circ,\ 50\sin 40^\circ\rangle \approx \langle 38.30,\ 32.14\rangle$.

So, the component perpendicular to the bench is approximately 38.30 lbs., and the component parallel to the bench is approximately 32.14 lbs.

59. $\text{weight}=\left(\dfrac{630}{\sin 13^\circ}\right)\text{ lbs.}\approx 2800.609\text{ lbs.}\approx \boxed{2801\text{ lbs.}}$

60. $\text{weight}=\left(\dfrac{500}{\sin 16^\circ}\right)\text{ lbs.}\approx 1813.97\text{ lbs.}\approx \boxed{1814\text{ lbs.}}$

61. Consider the following diagram:

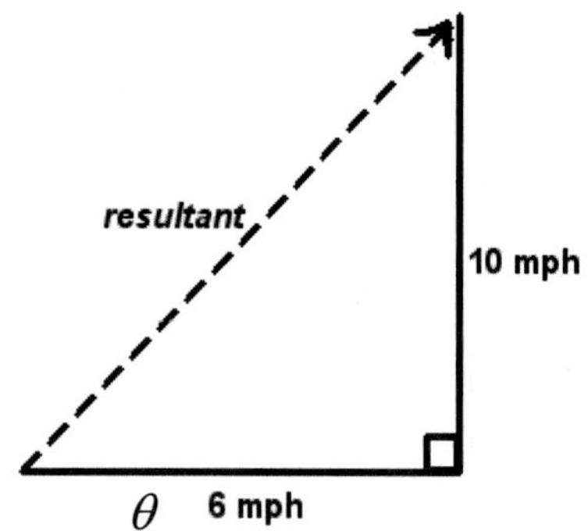

By the Pythagorean Theorem, the actual velocity of the ship is

$$\sqrt{10^2+6^2}\approx 11.6619\approx \boxed{11.7\text{ mph}}.$$

In order to determine the direction, we use the Law of Sines:

$$\frac{\sin\theta}{10}=\frac{\sin 90^\circ}{11.6619}$$

$$\theta=\sin^{-1}\left(\frac{10\sin 90^\circ}{11.6619}\right)\approx 59.036^\circ$$

So the direction is approximately $\boxed{31^\circ \text{ west of due north}}$.

62. Consider the following diagram:

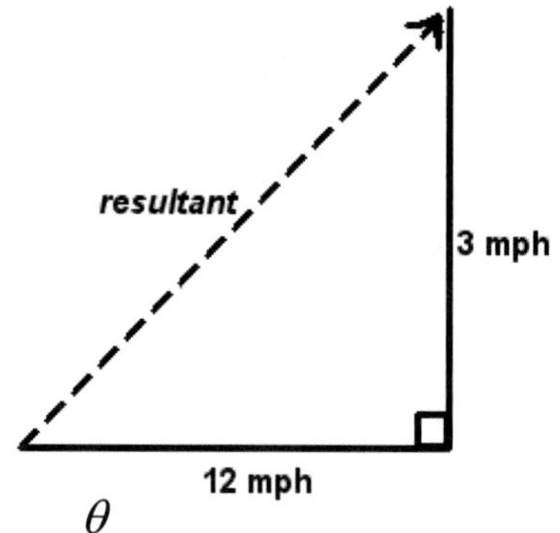

By the Pythagorean Theorem, the actual velocity of the ship is

$$\sqrt{12^2+3^2}\approx 12.369\approx \boxed{12.4\text{ mph}}.$$

In order to determine the direction, we use the Law of Sines:

$$\frac{\sin\theta}{3}=\frac{\sin 90^\circ}{12.369}$$

$$\theta=\sin^{-1}\left(\frac{3\sin 90^\circ}{12.369}\right)\approx 14.036^\circ$$

So the direction is approximately $\boxed{76^\circ \text{ west of due north}}$.

63. Consider the following diagram:

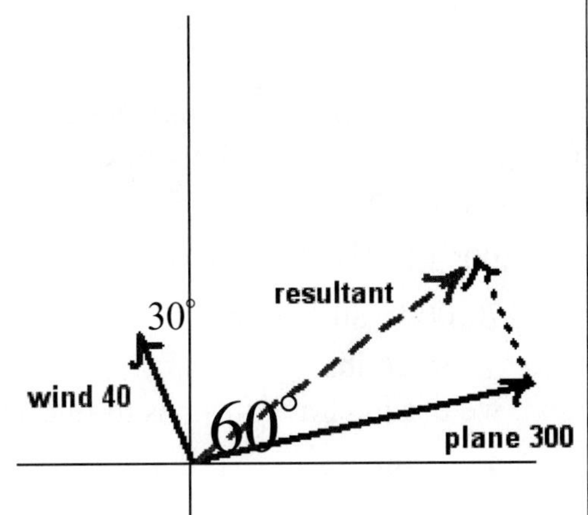

Observe that

$$\text{Plane} = \langle 300\cos 30^\circ,\ 300\sin 30^\circ \rangle$$

$$\text{Wind} = \langle 40\cos 120^\circ, 40\sin 120^\circ \rangle$$

Thus, the

$$\text{Resultant} = \text{Plane} + \text{Wind} \approx \langle 239.81,\ 184.64 \rangle$$

Hence,

Airspeed = |Resultant| $\approx$ 303 mph

Heading $= 90^\circ - \tan^{-1}\left(\frac{184.64}{239.81}\right) \approx 52.41^\circ$

(East of due North)

64. Consider the following diagram:

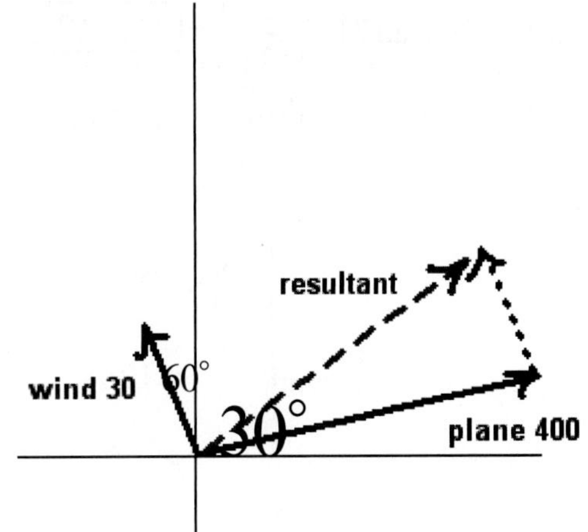

Observe that

$$\text{Plane} = \langle 400\cos 60^\circ,\ 400\sin 60^\circ \rangle$$

$$\text{Wind} = \langle 30\cos 150^\circ, 30\sin 150^\circ \rangle$$

Thus, the

$$\text{Resultant} = \text{Plane} + \text{Wind} \approx \langle 174.02, 361.41 \rangle$$

Hence,

Airspeed = |Resultant| $\approx$ 401 mph

Heading $= 90^\circ - \tan^{-1}\left(\frac{361.41}{174.02}\right) \approx 25.71^\circ$

(East of due North)

65. The force required to hold the box in place is $500\sin 30^\circ = \boxed{250 \text{ lbs.}}$

66. The force required to hold the box in place is $500\sin 10^\circ = \boxed{87 \text{ lbs.}}$

67. vertical component of velocity $= 80\sin 40^\circ = \boxed{51.4 \text{ ft./sec.}}$

horizontal component of velocity $= 80\cos 40^\circ = \boxed{61.3 \text{ ft./sec.}}$

68. vertical component of velocity $= 100\sin 5^\circ = \boxed{8.7 \text{ ft./sec.}}$

horizontal component of velocity $= 100\cos 5^\circ = \boxed{99.6 \text{ ft./sec.}}$

69. We need the component form for each of the three vectors involved. Indeed, they are:

$$\vec{A} = \langle 0,4 \rangle,\ \vec{B} = \langle 12,0 \rangle,\ \vec{C} = \langle 20\cos 330^\circ,\ 20\sin 330^\circ \rangle \approx \langle 17.32,\ -10 \rangle$$

As such, $\vec{A}+\vec{B}+\vec{C} = \langle 29.32,\ -6 \rangle$. So, $\left|\vec{A}+\vec{B}+\vec{C}\right| = \sqrt{29.32^2 + (-6)^2} \approx \boxed{29.93 \text{ yards}}$.

70. Using Exercise 69, we know that $\vec{A}+\vec{B}+\vec{C}=\langle 29.32, -6\rangle$. So, the direction angle is given by $\tan^{-1}\left(\dfrac{-6}{29.32}\right)\approx\boxed{-11.57^\circ}$.

71. Consider the following diagram:

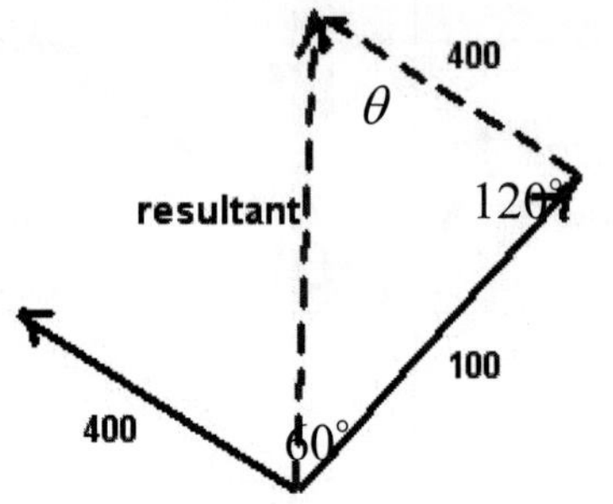

The given information yields a SAS triangle. So, we use the Law of Cosines:

$$c^2 = a^2 + b^2 - 2ab\cos\gamma$$
$$= 400^2 + 100^2 - 2(400)(100)\cos 120^\circ$$
$$= 170{,}000 - 80{,}000\cos 120^\circ$$

So, $c \approx 458.26$ lbs.

Next, we use the Law of Sines to find θ:

$$\frac{\sin\theta}{100} = \frac{\sin 120^\circ}{458.26}$$
$$\theta = \sin^{-1}\left(\frac{100\sin 120^\circ}{458.26}\right)\approx\boxed{10.9^\circ}$$

72. From Exercise 71, we see that $\boxed{c \approx 458 \text{ lbs.}}$

73. Consider the following diagram:

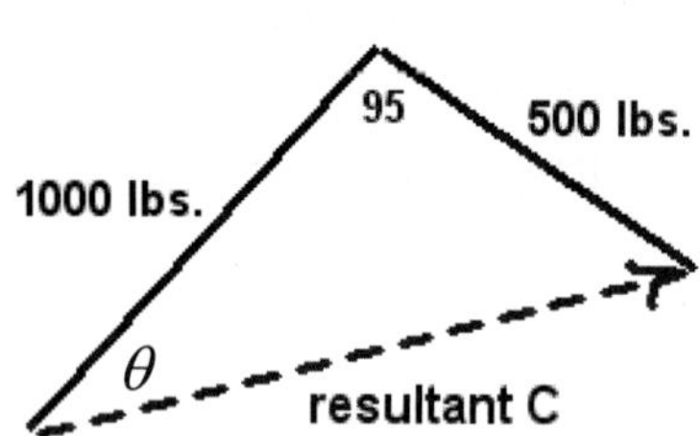

The given information yields a SAS triangle. So, we use the Law of Cosines:

$$c^2 = a^2 + b^2 - 2ab\cos\gamma$$
$$= 1000^2 + 500^2 - 2(1000)(500)\cos 95^\circ$$
$$= 1{,}250{,}000 - 1{,}000{,}000\cos 95^\circ$$
$$\approx 1156.3545 \approx \boxed{1156 \text{ lbs.}}$$

74. Next, we use the Law of Sines to find θ:

$$\frac{\sin\theta}{500} = \frac{\sin 95^\circ}{1156} \quad \text{so that } \theta = \sin^{-1}\left(\frac{500\sin 95^\circ}{1156.3545}\right)\approx\boxed{25.5^\circ}$$

75. The two forces acting on the hook are given in vector form by

$$\vec{u} = \langle 200\cos 45^\circ, 200\sin 45^\circ \rangle = \langle 100\sqrt{2}, 100\sqrt{2} \rangle \text{ and } \vec{v} = \langle 180, 0 \rangle.$$

The resultant force is thus $\vec{u}+\vec{v} = \langle 100\sqrt{2}+180, 100\sqrt{2} \rangle$. The magnitude of this force is

$$|\vec{u}+\vec{v}| = \sqrt{\left(100\sqrt{2}+180\right)^2 + \left(100\sqrt{2}\right)^2} \approx 351.16$$

and the direction is

$$\tan\theta = \frac{100\sqrt{2}}{100\sqrt{2}+180} \Rightarrow \theta = \tan^{-1}\left(\frac{100\sqrt{2}}{100\sqrt{2}+180}\right) \approx 23.75^\circ.$$

76. The two forces acting on the hook are given in vector form by

$$\vec{u} = \langle 100\cos 30^\circ, 100\sin 30^\circ \rangle = \langle \tfrac{100\sqrt{3}}{2}, 50 \rangle \text{ and } \vec{v} = \langle 50, 0 \rangle.$$

The resultant force is therefore $\vec{u}+\vec{v} = \langle 50\sqrt{3}+50, 50 \rangle$. The magnitude of this force is

$$|\vec{u}+\vec{v}| = \sqrt{\left(50\sqrt{3}+50\right)^2 + (50)^2} \approx 145.47$$

and the direction is

$$\tan\theta = \frac{50}{50\sqrt{3}+50} \Rightarrow \theta = \tan^{-1}\left(\frac{50}{50\sqrt{3}+50}\right) \approx 20.10^\circ.$$

77. Consider the following diagram:

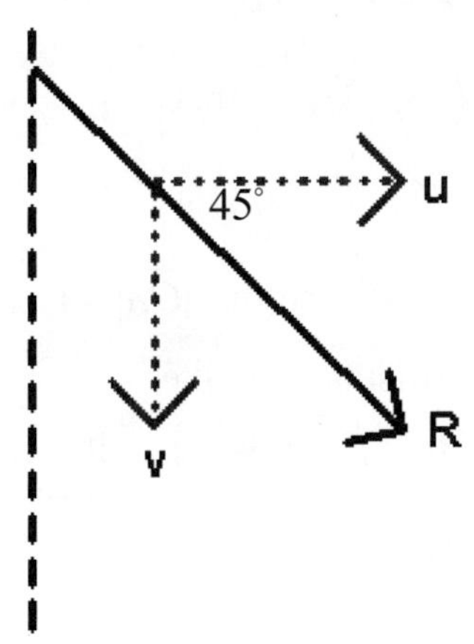

Using the information provided in the problem, we have

$$\vec{u} = \langle |\vec{u}|, 0 \rangle$$

$$\vec{R} = \langle |\vec{u}|, -5 \rangle = \langle |\vec{R}|\cos 45^\circ, -|\vec{R}|\sin 45^\circ \rangle$$

As such, we have

$$\begin{cases} |\vec{R}|\cos 45^\circ = |\vec{u}| \\ -|\vec{R}|\sin 45^\circ = -5 \end{cases}$$

From the second equation, we see that

$$|\vec{R}| = \tfrac{5}{\sin 45^\circ} = 5\sqrt{2}$$

Using this in the first equation yields

$$|\vec{u}| = \left(5\sqrt{2}\right)\cos 45^\circ = 5$$

78. Consider the following diagram: 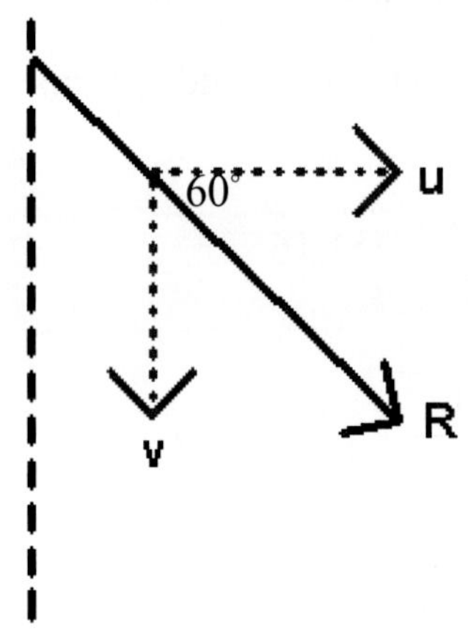	Using the information provided in the problem, we have $\vec{u} = \langle \lvert\vec{u}\rvert, 0 \rangle$ $\vec{R} = \langle \lvert\vec{u}\rvert, -8 \rangle = \langle \lvert\vec{R}\rvert \cos 60^\circ, -\lvert\vec{R}\rvert \sin 60^\circ \rangle$ As such, we have $\begin{cases} \lvert\vec{R}\rvert \cos 60^\circ = \lvert\vec{u}\rvert \\ -\lvert\vec{R}\rvert \sin 60^\circ = -8 \end{cases}$ From the second equation, we see that $\lvert\vec{R}\rvert = \frac{8}{\sin 60^\circ} = \frac{16\sqrt{3}}{3}$ Using this in the first equation yields $\lvert\vec{u}\rvert = \left(\frac{16\sqrt{3}}{3}\right) \cos 60^\circ = \frac{8\sqrt{3}}{3}$
79. Consider the following diagram: 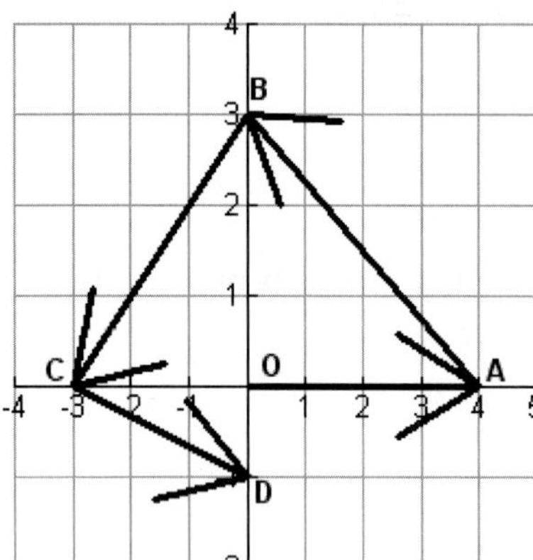	Note that the resultant vector is the zero vector since the person starts and ends at the same point. To determine how far he or she walked, we need to find the lengths of all of the vectors pictured to the left, and sum them: $\overrightarrow{OA} = \langle 4, 0 \rangle$, so that $\lvert\overrightarrow{OA}\rvert = 4$ $\overrightarrow{OA} = \langle -4, 3 \rangle$, so that $\lvert\overrightarrow{OA}\rvert = 5$ $\overrightarrow{OA} = \langle -3, -3 \rangle$, so that $\lvert\overrightarrow{OA}\rvert = 3\sqrt{2}$ $\overrightarrow{OA} = \langle -3, -1 \rangle$, so that $\lvert\overrightarrow{OA}\rvert = \sqrt{10}$ $\overrightarrow{OA} = \langle 0, -1 \rangle$, so that $\lvert\overrightarrow{OA}\rvert = 1$ So, the total distance walked is $\left(10 + \sqrt{10} + 3\sqrt{2}\right)$ units.

80. Consider the following diagram:

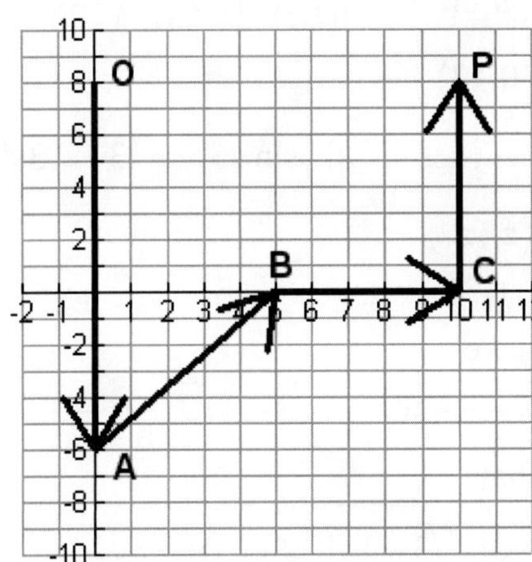

Note that the resultant vector is the vector $\overrightarrow{OP}$ since the person starts at O and ends at P. To determine how far he or she walked, we need to find the lengths of all of the vectors pictured to the left, and sum them:

$$\overrightarrow{AB} = \langle 0, -14 \rangle, \text{ so that } \left|\overrightarrow{AB}\right| = 14$$

$$\overrightarrow{BC} = \langle 5, 6 \rangle, \text{ so that } \left|\overrightarrow{BC}\right| = \sqrt{61}$$

$$\overrightarrow{CD} = \langle 5, 0 \rangle, \text{ so that } \left|\overrightarrow{CD}\right| = 5$$

$$\overrightarrow{DP} = \langle -10, 8 \rangle, \text{ so that } \left|\overrightarrow{DP}\right| = \sqrt{164}$$

So, the total distance walked is $\left(19 + \sqrt{164} + \sqrt{61}\right)$ units.

81. The torque is given by $\tau = 45(0.2)\sin 85^\circ \approx \boxed{8.97\text{Nm}}$.

82. The torque is given by $\tau = 0.85(40)\sin 90^\circ \approx \boxed{34\text{Nm}}$.

83. The torque is given by $\tau = 0.85(40)\sin 110^\circ \approx \boxed{31.95\text{Nm}}$.

84. We applied the force parallel to the door, so it is sliding straight down it rather than pushing it forward at any angle. As such, the angle is 0°.

85. Consider the following diagram:

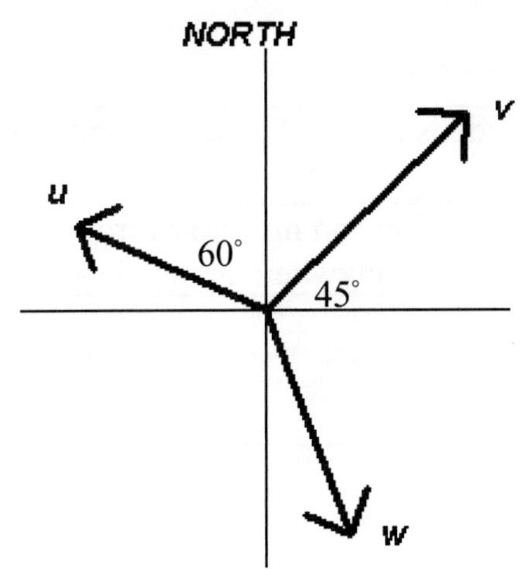

Observe that

$$\vec{u} = \left\langle |\vec{u}|\cos 150^\circ, |\vec{u}|\sin 150^\circ \right\rangle = \left\langle -4\sqrt{3}, 4 \right\rangle$$

$$\vec{v} = \left\langle |\vec{v}|\cos 45^\circ, |\vec{v}|\sin 45^\circ \right\rangle = \left\langle 3\sqrt{2}, 3\sqrt{2} \right\rangle$$

$$\vec{w} = \left\langle w_1, w_2 \right\rangle$$

We want $\vec{u} + \vec{v} + \vec{w} = \vec{0}$. As such, we must have $\vec{w} = -\vec{u} - \vec{v} = \left\langle 4\sqrt{3} - 3\sqrt{2}, -4 - 3\sqrt{2} \right\rangle$.

Note that

$$|\vec{w}| = \sqrt{\left(4\sqrt{3} - 3\sqrt{2}\right)2 + \left(-4 - 3\sqrt{2}\right)^2}$$
$$\approx 8.67$$

$$\tan\theta = \frac{-4 - 3\sqrt{2}}{4\sqrt{3} - 3\sqrt{2}} \quad \Rightarrow \quad \theta = \tan^{-1}\left(\frac{-4 - 3\sqrt{2}}{4\sqrt{3} - 3\sqrt{2}}\right)$$
$$\approx -71.95^\circ$$

So, the direction of $\vec{w}$ is 18.05° counterclockwise of SOUTH.

86. Consider the following diagram:

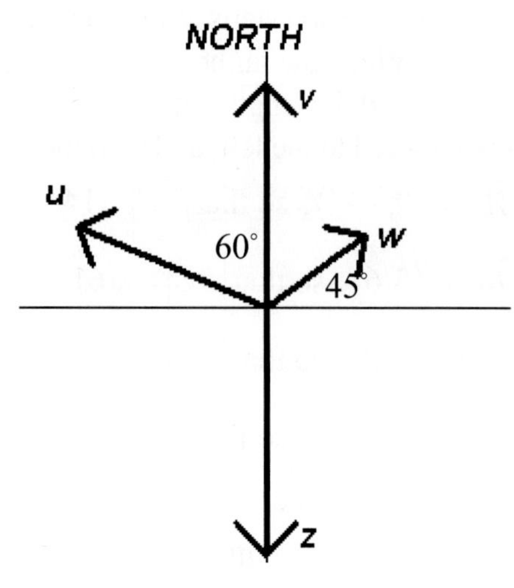

Observe that

$$\vec{u} = \left\langle |\vec{u}|\cos 150^\circ, |\vec{u}|\sin 150^\circ \right\rangle = \left\langle -4\sqrt{3}, 4 \right\rangle$$

$$\vec{v} = \langle 0, 12 \rangle$$

$$\vec{w} = \left\langle |\vec{w}|\cos 45^\circ, |\vec{w}|\sin 45^\circ \right\rangle = \left\langle 3\sqrt{2}, 3\sqrt{2} \right\rangle$$

$$\vec{z} = \langle z_1, z_2 \rangle$$

We want $\vec{u}+\vec{v}+\vec{w}+\vec{z}=\vec{0}$. As such, we must have

$$\vec{z} = -\vec{u}-\vec{v}-\vec{w} = \left\langle 4\sqrt{3}-3\sqrt{2}, -16-3\sqrt{2} \right\rangle.$$

Note that

$$|\vec{z}| = \sqrt{\left(4\sqrt{3}-3\sqrt{2}\right)2 + \left(-16-3\sqrt{2}\right)^2}$$
$$\approx 20.42$$

$$\tan\theta = \frac{-16-3\sqrt{2}}{4\sqrt{3}-3\sqrt{2}} \;\Rightarrow\; \theta = \tan^{-1}\left(\frac{-16-3\sqrt{2}}{4\sqrt{3}-3\sqrt{2}}\right)$$
$$\approx -82.44^\circ$$

So, the direction of $\vec{w}$ is 7.56° counterclockwise of SOUTH.

87. Magnitude cannot be negative. Observe that

$$|\langle -2,-8 \rangle| = \sqrt{(-2)^2+(-8)^2} = \sqrt{68} = 2\sqrt{17}.$$

88. Inverse tangent gives an angle whose terminal side is in QI. As such, we must add 180° to obtain the QIII solution.

89. False. $|\vec{i}\,| = |\langle 1,0 \rangle| = 1$.

90. False. You can move the arrow representative anywhere in the plane.

91. True. Let $\vec{u} = \langle u_1, u_2 \rangle$. Observe that

$$|\vec{u}| = |\langle u_1, u_2 \rangle| = \sqrt{(u_1)^2+(u_2)^2} \geq \sqrt{(u_1)^2} = |u_1| = u_1.$$

92. True. Let $\vec{u} = \langle u_1, u_2 \rangle$. Observe that

$$|\vec{u}| = |\langle u_1, u_2 \rangle| = \sqrt{(u_1)^2+(u_2)^2} \geq \sqrt{(u_2)^2} = |u_2| = u_2.$$

93. Vector (since need two pieces of information to precisely define this quantity).

94. Scalar (since only need one piece of information to precisely define the quantity).

95. $\left\|\langle -a,b\rangle\right\| = \sqrt{(-a)^2 + b^2} = \sqrt{a^2+b^2}$	**96.** Since $\langle -a,b\rangle$ is in QII, the direction angle must be $\underbrace{\tan^{-1}\left(-\frac{b}{a}\right)}_{\text{in QIV}} + 180°$
97. Assume that $\vec{u} = c\vec{v}$, where $c > 0$, and that $\|\vec{u}\| = 1$. Then, we have $c\|\vec{v}\| = 1$, so that $c = \frac{1}{\|\vec{v}\|}$. Thus, $\vec{u} = \left(\frac{1}{\|\vec{v}\|}\right)\vec{v}$, as needed.	**98.** Since $\vec{u} = a\vec{i} + b\vec{j}$ is a unit vector, we know that $\|\vec{u}\| = \sqrt{a^2+b^2} = 1$, so that $a^2 + b^2 = 1$. But, this is precisely what it means for a point (a, b) to lie on the graph of $x^2 + y^2 = 1$, the equation of the unit circle.
99. $\langle -6,4\rangle = \frac{1}{2}\langle -8,4\rangle - 2\langle 1,-1\rangle$	**100.** $\langle -\frac{2}{9}a, \frac{8}{9}b\rangle = -\frac{1}{3}\langle a,3b\rangle - \frac{1}{9}\langle -a,-b\rangle$
101. Observe that $\vec{u} + 3(2\vec{v} - \vec{u}) = \langle u_1,u_2\rangle + 3(2\langle v_1,v_2\rangle - \langle u_1,u_2\rangle)$ $= \langle u_1,u_2\rangle + 3(\langle 2v_1,2v_2\rangle - \langle u_1,u_2\rangle)$ $= \langle u_1,u_2\rangle + 3\langle 2v_1 - u_1, 2v_2 - u_2\rangle$ $= \langle u_1,u_2\rangle + \langle 6v_1 - 3u_1, 6v_2 - 3u_2\rangle$ $= \langle 6v_1 - 2u_1, 6v_2 - 2u_2\rangle$ $= 6\vec{v} - 2\vec{u}$	**102.** Let $\vec{u} = \langle 2a,a\rangle$, $\vec{v} = \langle -a,-2a\rangle$. Observe that $\left\|\frac{2\vec{u}}{\|\vec{v}\|} - \frac{3\vec{v}}{\|\vec{u}\|}\right\| = \left\|\frac{\langle 4a,2a\rangle}{\sqrt{a^2+4a^2}} + \frac{\langle 3a,6a\rangle}{\sqrt{a^2+4a^2}}\right\|$ $= \frac{1}{a\sqrt{5}}\left\|\langle 7a,8a\rangle\right\|$ $= \frac{a\sqrt{113}}{a\sqrt{5}} = \frac{\sqrt{565}}{5}$
103. Enter the following to compute the given quantity: [[8] [-5]] + 3[[-7] [11]] The output is: [[-13] [28]].	**104.** Enter the following to compute the given quantity: -9([[8] [-5]] -2[[-7] [11]]) The output is: [[-198] [243]].
105. First, compute the length of the vector, as follows: √ $(10^2 + (-24)^2)$ (output 26) Then, to obtain the unit vector, divide both components of the given vector by this length (26) – to list as a FRACTION, enter the following: 1/26 [[10] [-24]] ▲Frac Output is: [[5/13] [-12/13]]	**106.** First, compute the length of the vector, as follows: √ $((-9)^2 + (-40)^2)$ (output 41) Then, to obtain the unit vector, divide both components of the given vector by this length (41) – to list as a FRACTION, enter the following: 1/41 [[-9] [-40]] ▲Frac Output is: [[-9/41] [-40/41]]

107.	**108.**
Magnitude = $\sqrt{\ }$ Sum $((-33)^\wedge 2, (180)^\wedge 2)$ Direction angle = $\tan^{-1} 180 \div (-33)$ The output will be: Magnitude = 183 Direction Angle = -79.61114	Magnitude = $\sqrt{\ }$ Sum $((-20)^\wedge 2, (30\sqrt{5})^\wedge 2)$ Direction angle = $\tan^{-1} 30\sqrt{5} \div (-20)$ The output will be: Magnitude = 23.45208 Direction Angle = -73.39845
109. $\left\|\vec{v}(t)\right\| = \sqrt{\cos^2 t + \sin^2 t} = 1$	**110.** $\tan(\theta(t)) = \frac{-4t}{-3t} = \frac{4}{3}$ $\theta = \tan^{-1}\left(\frac{4}{3}\right) \approx 53.1^\circ$ Since the head of the vectors occur in QIII, we add 180° to this to conclude that the direction of $\vec{v}(t)$ is 233.1°.

111.

$$\frac{\vec{v}(t+h)-\vec{v}(t)}{h} = \frac{\langle t+h,(t+h)^2\rangle - \langle t,t^2\rangle}{h} = \left\langle \frac{t+h-t}{h}, \frac{(t+h)^2-t^2}{h} \right\rangle = \left\langle 1, \frac{2ht+h^2}{h} \right\rangle = \langle 1, 2t+h \rangle$$

112.

$$\frac{\vec{v}(t+h)-\vec{v}(t)}{h} = \frac{\langle (t+h)^2+1,(t+h)^3\rangle - \langle t^2+1,t^3\rangle}{h} = \left\langle \frac{(t+h)^2+1-(t^2+1)}{h}, \frac{(t+h)^3-t^3}{h} \right\rangle$$

$$= \left\langle \frac{2ht+h^2}{h}, \frac{3h^2t+3ht^2+h^3}{h} \right\rangle = \langle 2t+h, 3ht+3t^2+h^2 \rangle$$

Section 7.2 Solutions ---

1. $\langle 4,-2\rangle \cdot \langle 3,5\rangle = (4)(3)+(-2)(5) = \boxed{2}$	**2.** $\langle 7,8\rangle \cdot \langle 2,-1\rangle = (7)(2)+(8)(-1) = \boxed{6}$
3. $\langle -5,6\rangle \cdot \langle 3,2\rangle = (-5)(3)+(6)(2) = \boxed{-3}$	**4.** $\langle 6,-3\rangle \cdot \langle 2,1\rangle = (6)(2)+(-3)(1) = \boxed{9}$
5. $\langle -7,-4\rangle \cdot \langle -2,-7\rangle = (-7)(-2)+(-4)(-7)$ $= \boxed{42}$	**6.** $\langle 5,-2\rangle \cdot \langle -1,-1\rangle = (5)(-1)+(-2)(-1)$ $= \boxed{-3}$

7. $\langle\sqrt{3},-2\rangle\cdot\langle 3\sqrt{3},-1\rangle=(\sqrt{3})(3\sqrt{3})+(-2)(-1)$ $=\boxed{11}$	**8.** $\langle 4\sqrt{2},\sqrt{7}\rangle\cdot\langle-\sqrt{2},-\sqrt{7}\rangle$ $=(4\sqrt{2})(-\sqrt{2})+(\sqrt{7})(-\sqrt{7})=\boxed{-15}$
9. $\langle 5,a\rangle\cdot\langle-3a,2\rangle=(5)(-3a)+(a)(2)$ $=\boxed{-13a}$	**10.** $\langle 4x,3y\rangle\cdot\langle 2y,-5x\rangle=(4x)(2y)+(3y)(-5x)$ $=\boxed{-7xy}$
11. $\langle 0.8,-0.5\rangle\cdot\langle 2,6\rangle=(0.8)(2)+(-0.5)(6)$ $=\boxed{-1.4}$	**12.** $\langle-18,3\rangle\cdot\langle 10,-300\rangle=(-18)(10)+(3)(-300)$ $=\boxed{-1080}$

13. Use the formula $\theta=\cos^{-1}\left(\dfrac{\vec{u}\cdot\vec{v}}{|\vec{u}||\vec{v}|}\right)$ with the following computations:

$$\vec{u}\cdot\vec{v}=\langle-4,3\rangle\cdot\langle-5,-9\rangle=-7,\quad |\vec{u}|=\sqrt{(-4)^2+(3)^2}=5,\quad |\vec{v}|=\sqrt{(-5)^2+(-9)^2}=\sqrt{106}$$

So, $\theta=\cos^{-1}\left(\dfrac{-7}{(5)(\sqrt{106})}\right)\approx\boxed{98^\circ}$.

14. Use the formula $\theta=\cos^{-1}\left(\dfrac{\vec{u}\cdot\vec{v}}{|\vec{u}||\vec{v}|}\right)$ with the following computations:

$$\vec{u}\cdot\vec{v}=\langle 2,-4\rangle\cdot\langle 4,-1\rangle=12,\quad |\vec{u}|=\sqrt{(2)^2+(-4)^2}=2\sqrt{5},\quad |\vec{v}|=\sqrt{(4)^2+(-1)^2}=\sqrt{17}$$

So, $\theta=\cos^{-1}\left(\dfrac{12}{(2\sqrt{5})(\sqrt{17})}\right)\approx\boxed{49^\circ}$.

15. Use the formula $\theta=\cos^{-1}\left(\dfrac{\vec{u}\cdot\vec{v}}{|\vec{u}||\vec{v}|}\right)$ with the following computations:

$$\vec{u}\cdot\vec{v}=\langle-2,-3\rangle\cdot\langle-3,4\rangle=-6,\quad |\vec{u}|=\sqrt{(-2)^2+(-3)^2}=\sqrt{13},\quad |\vec{v}|=\sqrt{(-3)^2+(4)^2}=5$$

So, $\theta=\cos^{-1}\left(\dfrac{-6}{(5)(\sqrt{13})}\right)\approx\boxed{109^\circ}$.

16. Use the formula $\theta = \cos^{-1}\left(\dfrac{\vec{u}\cdot\vec{v}}{|\vec{u}||\vec{v}|}\right)$ with the following computations:

$$\vec{u}\cdot\vec{v} = \langle 6,5\rangle\cdot\langle 3,-2\rangle = 8,\quad |\vec{u}| = \sqrt{(6)^2+(5)^2} = \sqrt{61},\quad |\vec{v}| = \sqrt{(3)^2+(-2)^2} = \sqrt{13}$$

So, $\theta = \cos^{-1}\left(\dfrac{8}{\left(\sqrt{61}\right)\left(\sqrt{13}\right)}\right) \approx \boxed{73^\circ}$.

17. Use the formula $\theta = \cos^{-1}\left(\dfrac{\vec{u}\cdot\vec{v}}{|\vec{u}||\vec{v}|}\right)$ with the following computations:

$$\vec{u}\cdot\vec{v} = \langle -4,6\rangle\cdot\langle -6,8\rangle = 72,\quad |\vec{u}| = \sqrt{(-4)^2+(6)^2} = 2\sqrt{13},\quad |\vec{v}| = \sqrt{(-6)^2+(8)^2} = 10$$

So, $\theta = \cos^{-1}\left(\dfrac{72}{\left(2\sqrt{13}\right)(10)}\right) \approx \boxed{3^\circ}$.

18. Use the formula $\theta = \cos^{-1}\left(\dfrac{\vec{u}\cdot\vec{v}}{|\vec{u}||\vec{v}|}\right)$ with the following computations:

$$\vec{u}\cdot\vec{v} = \langle 1,5\rangle\cdot\langle -3,-2\rangle = -13,\quad |\vec{u}| = \sqrt{(1)^2+(5)^2} = \sqrt{26},\quad |\vec{v}| = \sqrt{(-3)^2+(-2)^2} = \sqrt{13}$$

So, $\theta = \cos^{-1}\left(\dfrac{-13}{\left(\sqrt{26}\right)\left(\sqrt{13}\right)}\right) = \boxed{135^\circ}$.

19. Use the formula $\theta = \cos^{-1}\left(\dfrac{\vec{u}\cdot\vec{v}}{|\vec{u}||\vec{v}|}\right)$ with the following computations:

$$\vec{u}\cdot\vec{v} = \left\langle -2,2\sqrt{3}\right\rangle\cdot\left\langle -\sqrt{3},1\right\rangle = 4\sqrt{3},\quad |\vec{u}| = \sqrt{(-2)^2+\left(2\sqrt{3}\right)^2} = 4,\quad |\vec{v}| = \sqrt{\left(-\sqrt{3}\right)^2+(1)^2} = 2$$

So, $\theta = \cos^{-1}\left(\dfrac{4\sqrt{3}}{(4)(2)}\right) = \boxed{30^\circ}$.

20. Use the formula $\theta = \cos^{-1}\left(\dfrac{\vec{u}\cdot\vec{v}}{|\vec{u}||\vec{v}|}\right)$ with the following computations:

$$\vec{u}\cdot\vec{v} = \left\langle -3\sqrt{3},-3\right\rangle\cdot\left\langle -2\sqrt{3},2\right\rangle = 12,\quad |\vec{u}| = \sqrt{\left(-3\sqrt{3}\right)^2+(-3)^2} = 6,\quad |\vec{v}| = \sqrt{\left(-2\sqrt{3}\right)^2+(2)^2} = 4$$

So, $\theta = \cos^{-1}\left(\dfrac{12}{(6)(4)}\right) = \boxed{60^\circ}$.

21. Use the formula $\theta=\cos^{-1}\left(\frac{\vec{u}\cdot\vec{v}}{|\vec{u}||\vec{v}|}\right)$ with the following computations:

$$\vec{u}\cdot\vec{v}=\left\langle -5\sqrt{3},-5\right\rangle\cdot\left\langle \sqrt{2},-\sqrt{2}\right\rangle=5\left(\sqrt{2}-\sqrt{6}\right),$$

$$|\vec{u}|=\sqrt{\left(-5\sqrt{3}\right)^2+(-5)^2}=10,\ |\vec{v}|=\sqrt{\left(\sqrt{2}\right)^2+\left(-\sqrt{2}\right)^2}=2$$

So, $\theta=\cos^{-1}\left(\frac{5\left(\sqrt{2}-\sqrt{6}\right)}{(10)(2)}\right)=\boxed{105^\circ}$.

22. Use the formula $\theta=\cos^{-1}\left(\frac{\vec{u}\cdot\vec{v}}{|\vec{u}||\vec{v}|}\right)$ with the following computations:

$$\vec{u}\cdot\vec{v}=\left\langle -5,-5\sqrt{3}\right\rangle\cdot\left\langle 2,-\sqrt{2}\right\rangle=-10+5\sqrt{6},$$

$$|\vec{u}|=\sqrt{(-5)^2+\left(-5\sqrt{3}\right)^2}=10,\ |\vec{v}|=\sqrt{(2)^2+\left(-\sqrt{2}\right)^2}=\sqrt{6}$$

So, $\theta=\cos^{-1}\left(\frac{-10+5\sqrt{6}}{(10)\left(\sqrt{6}\right)}\right)=\boxed{85^\circ}$.

23. Use the formula $\theta=\cos^{-1}\left(\frac{\vec{u}\cdot\vec{v}}{|\vec{u}||\vec{v}|}\right)$ with the following computations:

$$\vec{u}\cdot\vec{v}=\langle 4,6\rangle\cdot\langle -6,-9\rangle=-78,\quad |\vec{u}|=\sqrt{(4)^2+(6)^2}=2\sqrt{13},\ \ |\vec{v}|=\sqrt{(-6)^2+(-9)^2}=3\sqrt{13}$$

So, $\theta=\cos^{-1}\left(\frac{-78}{\left(2\sqrt{13}\right)\left(3\sqrt{13}\right)}\right)=\boxed{180^\circ}$.

24. Use the formula $\theta=\cos^{-1}\left(\frac{\vec{u}\cdot\vec{v}}{|\vec{u}||\vec{v}|}\right)$ with the following computations:

$$\vec{u}\cdot\vec{v}=\langle 2,8\rangle\cdot\langle -12,3\rangle=0$$

So, $\theta=\cos^{-1}(0)=\boxed{90^\circ}$.

25. Since $\langle -6,8\rangle\cdot\langle -8,6\rangle=96\neq 0$, these two vectors are $\boxed{\text{not orthogonal}}$.

26. Since $\langle 5,-2\rangle\cdot\langle -5,2\rangle=-29\neq 0$, these two vectors are $\boxed{\text{not orthogonal}}$.

27. Since $\langle 6,-4\rangle\cdot\langle -6,-9\rangle=0$, these two vectors are $\boxed{\text{orthogonal}}$.

28. Since $\langle 8,3\rangle\cdot\langle -6,16\rangle=0$, these two vectors are $\boxed{\text{orthogonal}}$.

29. Since $\langle 0.8,4\rangle \cdot \langle 3,-6\rangle = -21.6 \neq 0$, these two vectors are $\boxed{\text{not orthogonal}}$.
30. Since $\langle -7,3\rangle \cdot \left\langle \frac{1}{7},-\frac{1}{3}\right\rangle = -2 \neq 0$, these two vectors are $\boxed{\text{not orthogonal}}$.
31. Since $\langle 5,-0.4\rangle \cdot \langle 1.6,20\rangle = 0$, these two vectors are $\boxed{\text{orthogonal}}$.
32. Since $\langle 12,9\rangle \cdot \langle 3,-4\rangle = 0$, these two vectors are $\boxed{\text{orthogonal}}$.
33. Since $\left\langle \sqrt{3},\sqrt{6}\right\rangle \cdot \left\langle -\sqrt{2},1\right\rangle = 0$, these two vectors are $\boxed{\text{orthogonal}}$.
34. Since $\left\langle \sqrt{7},-\sqrt{3}\right\rangle \cdot \langle 3,7\rangle = 3\sqrt{7}-7\sqrt{3} \neq 0$, these two vectors are $\boxed{\text{not orthogonal}}$.
35. Since $\left\langle \frac{4}{3},\frac{8}{15}\right\rangle \cdot \left\langle -\frac{1}{12},\frac{5}{24}\right\rangle = 0$, these two vectors are $\boxed{\text{orthogonal}}$.
36. Since $\left\langle \frac{5}{6},\frac{6}{7}\right\rangle \cdot \left\langle \frac{36}{25},-\frac{49}{36}\right\rangle = \frac{6}{5}-\frac{7}{6} \neq 0$, these two vectors are $\boxed{\text{not orthogonal}}$.
37. $W = \left\lvert \vec{F} \right\rvert d = (100 \text{ lbs.})(4 \text{ ft.}) = \boxed{400 \text{ ft. lbs.}}$.
38. $W = \left\lvert \vec{F} \right\rvert d = (150 \text{ lbs.})(3.5 \text{ ft.}) = \boxed{525 \text{ ft. lbs.}}$
39. Note that 1 ton = 2000 lbs. So, $W = \left\lvert \vec{F} \right\rvert d = (4000 \text{ lbs.})(20 \text{ ft.}) = \boxed{80{,}000 \text{ ft. lbs.}}$
40. Note that 1 ton = 2000 lbs. So, $W = \left\lvert \vec{F} \right\rvert d = (5000 \text{ lbs.})(25 \text{ ft.}) = \boxed{125{,}000 \text{ ft. lbs.}}$
41. Here, $\left\lvert \vec{F} \right\rvert = (50 \text{ lbs.})\cos 30^\circ \approx 43.30$ lbs. So, $$W = \left\lvert \vec{F} \right\rvert d = (43.30 \text{ lbs.})(30 \text{ ft.}) = \boxed{1299 \text{ ft. lbs.}}$$
42. Here, $\left\lvert \vec{F} \right\rvert = (800 \text{ lbs.})\cos 20^\circ \approx 751.75$ lbs. So, $$W = \left\lvert \vec{F} \right\rvert d = (751.75 \text{ lbs.})(50 \text{ ft.}) = \boxed{37{,}588 \text{ ft. lbs.}}$$
43. Here, $\left\lvert \vec{F} \right\rvert = (35 \text{ lbs.})\cos 45^\circ \approx 24.748$ lbs. So, $$W = \left\lvert \vec{F} \right\rvert d = (24.748 \text{ lbs.})(6 \text{ ft.}) = \boxed{148 \text{ ft. lbs.}}$$
44. Here, $\left\lvert \vec{F} \right\rvert = (45 \text{ lbs.})\cos 55^\circ \approx 25.810$ lbs. So, $$W = \left\lvert \vec{F} \right\rvert d = (25.810 \text{ lbs.})(6 \text{ ft.}) = \boxed{155 \text{ ft. lbs.}}$$
45. Weight of the car = 2500 lbs. Since we don't want the car to roll down the hill, we want to apply the correct amount of force that would keep the vertical position of the car the same. That component is $(2500 \text{ lbs.})\sin 40^\circ \approx 1607$ lbs. Hence, the force should be $\boxed{1607 \text{ lbs.}}$ to keep the car from rolling down the hill.

46. Using the same setting as in Exercise 45, we now seek the horizontal component of the weight of the car. This is given by $(2500 \text{ lbs.})\cos 40^\circ \approx 1915 \text{ lbs.}$

Hence, $W = \left|\vec{F}\right| d \approx (1915 \text{ lbs.})(120 \text{ ft.}) = \boxed{229{,}813 \text{ ft. lbs.}}$

47. $W = \left|\vec{F}\right| d = \left[(40{,}000 \text{ lbs.})\cos 80^\circ\right](100 \text{ ft.}) \approx \boxed{694{,}593 \text{ ft. lbs.}}$

48. We seek the vertical component of the weight of the car. This is given by

$$(40{,}000 \text{ lbs.})\sin 10^\circ \approx \boxed{6946 \text{ lbs.}}$$

49. Observe that $\vec{u}\cdot\vec{v} = (2000)(8.4) + (5000)(6.50) = 49{,}300$. So, the total cost of both types is $49,300.

50. Observe that $\vec{u}\cdot\vec{v} = 120(7.2) + 80(5.3) = 1{,}288$. So, the total number of minutes spent registering is 1,288.

51. Consider a rhombus whose adjacent sides are formed using vectors $\vec{v}$ and $\vec{w}$, as shown below:

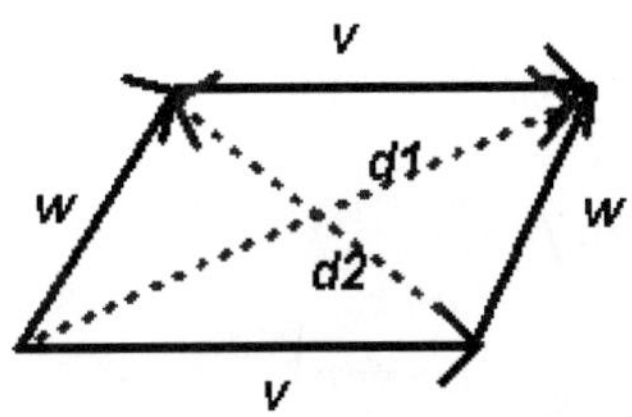

Observe that

$$\vec{d_1} = \vec{v} + \vec{w} \text{ and } \vec{d_2} = \vec{w} - \vec{v}.$$

Hence,

$$\begin{aligned}\vec{d_1}\cdot\vec{d_2} &= (\vec{v}+\vec{w})\cdot(\vec{w}-\vec{v}) \\ &= \vec{v}\cdot\vec{w} - \vec{w}\cdot\vec{v} \\ &= 0\end{aligned}$$

Hence, the diagonals are perpendicular to each other.

52. Consider the following diagram:

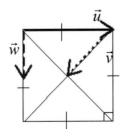

First, note that since

$$|\vec{u}| = 1,\ |\vec{w}| = \tfrac{1}{2},\ \theta = 90^\circ$$

we conclude that

$$\vec{u}\cdot\vec{w} = |\vec{u}||\vec{w}|\cos\theta = 0.$$

Next, since the diagonals each have length $\sqrt{2}$ (from the Pythagorean theorem) and they bisect each other, we see that

$$\vec{u}\cdot\vec{v} = |\vec{u}||\vec{v}|\cos 45^\circ = \tfrac{1}{2}$$

53. Consider the following diagram:

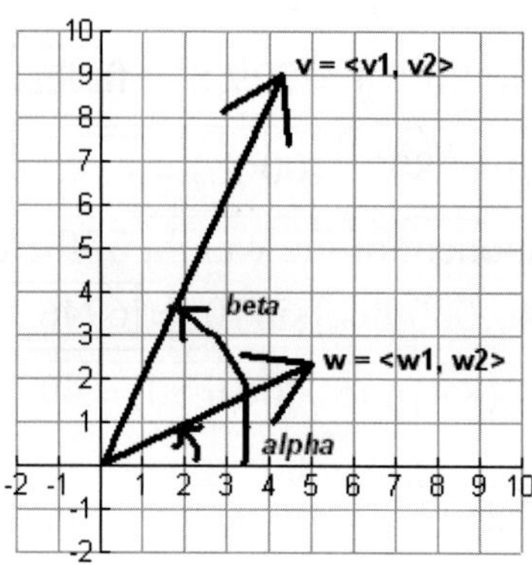

(i) Observe that from the triangle definitions of the trigonometric functions, we have

$$\cos\beta = \frac{v_1}{\sqrt{v_1^2+v_2^2}}, \quad \sin\beta = \frac{v_2}{\sqrt{v_1^2+v_2^2}}$$
$$\cos\alpha = \frac{w_1}{\sqrt{w_1^2+w_2^2}}, \quad \sin\alpha = \frac{w_2}{\sqrt{w_1^2+w_2^2}}.$$

(ii) Now, using (i) and the addition formula for cosine, we see that

$$\begin{aligned}\cos(\alpha-\beta) &= \cos\alpha\cos\beta + \sin\alpha\sin\beta \\ &= \left(\frac{w_1}{\sqrt{w_1^2+w_2^2}}\right)\left(\frac{v_1}{\sqrt{v_1^2+v_2^2}}\right) + \left(\frac{w_2}{\sqrt{w_1^2+w_2^2}}\right)\left(\frac{v_2}{\sqrt{v_1^2+v_2^2}}\right) \\ &= \frac{w_1v_1 + w_2v_2}{\sqrt{v_1^2+v_2^2}\,\sqrt{w_1^2+w_2^2}} \\ &= \frac{\vec{v}\cdot\vec{w}}{\sqrt{\vec{v}\cdot\vec{v}}\,\sqrt{\vec{w}\cdot\vec{w}}}\end{aligned}$$

54. Consider the following diagram:

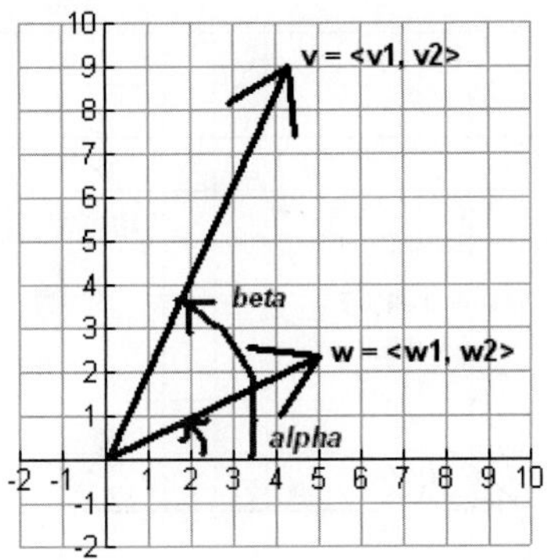

(i) Observe that from the triangle definitions of the trigonometric functions, we have

$$\cos\beta = \frac{v_1}{\sqrt{v_1^2+v_2^2}}, \quad \sin\beta = \frac{v_2}{\sqrt{v_1^2+v_2^2}}$$
$$\cos\alpha = \frac{w_1}{\sqrt{w_1^2+w_2^2}}, \quad \sin\alpha = \frac{w_2}{\sqrt{w_1^2+w_2^2}}.$$

(ii) Now, using (i) and the addition formula for cosine, we see that

$$\begin{aligned}\cos(\alpha+\beta) &= \cos\alpha\cos\beta - \sin\alpha\sin\beta \\ &= \left(\frac{w_1}{\sqrt{w_1^2+w_2^2}}\right)\left(\frac{v_1}{\sqrt{v_1^2+v_2^2}}\right) - \left(\frac{w_2}{\sqrt{w_1^2+w_2^2}}\right)\left(\frac{v_2}{\sqrt{v_1^2+v_2^2}}\right) \\ &= \frac{w_1v_1 - w_2v_2}{\sqrt{v_1^2+v_2^2}\,\sqrt{w_1^2+w_2^2}} \\ &= \frac{\vec{v}\cdot\langle w_1, -w_2\rangle}{\sqrt{\vec{v}\cdot\vec{v}}\,\sqrt{\vec{w}\cdot\vec{w}}}\end{aligned}$$

55. Consider the following diagram:

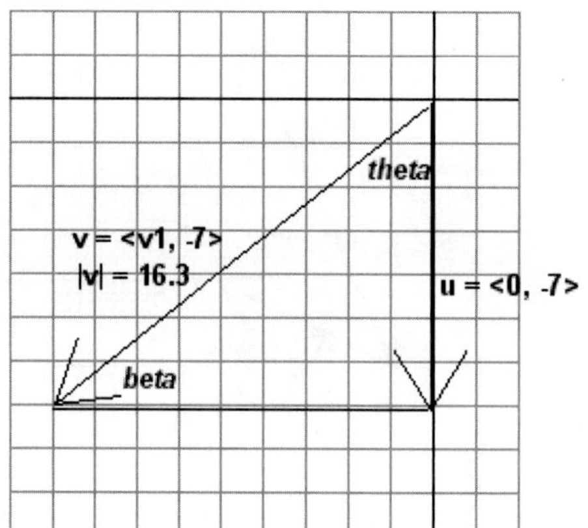

First, note that $|\vec{v}| = \sqrt{v_1^2 + (-7)^2} = \sqrt{v_1^2 + 49} = 16.3$ so that $v_1^2 + 49 = 16.3^2$. Hence, $v_1 = \pm\sqrt{16.3^2 - 49} \approx \pm 14.72$. We choose $v_1 = -14.72$ since the head of the vector with this x-component is in QIII.

Next, observe that

$$\langle 0,-7\rangle \cdot \langle -14.72,-7\rangle = 7(16.3)\cos\theta$$
$$49 = 114.1\cos\theta$$
$$\theta = \cos^{-1}\left(\tfrac{49}{114.1}\right) \approx 64.57^\circ$$

56. From Exercise 55, $\beta = 90^\circ - 64.57^\circ = 25.43^\circ$.

57. Let $\vec{u} = \langle a,b\rangle$ be a given vector and suppose that the head of $\vec{n} = \langle n_1, n_2\rangle$ lies on the circle $x^2 + y^2 = r^2$. We want to find the vector $\vec{n}$ such that $\vec{u}\cdot\vec{n}$ is as big as possible. Observe that

$$\vec{u}\cdot\vec{n} = |\vec{u}||\vec{n}|\cos\theta = |\vec{u}|r\cos\theta.$$

This quantity is largest when $\cos\theta = 1$, which occurs when $\theta = 0^\circ$. Hence, the vector that works is $\vec{n} = \langle r,0\rangle$ and in such case, $\vec{u}\cdot\vec{n} = |\vec{u}|r$.

58. Let $\vec{u} = \langle a,b\rangle$ be a given vector and suppose that the head of $\vec{n} = \langle n_1, n_2\rangle$ lies on the circle $x^2 + y^2 = r^2$. We want to find the vector $\vec{n}$ such that $\vec{u}\cdot\vec{n}$ is as small as possible. Observe that

$$\vec{u}\cdot\vec{n} = |\vec{u}||\vec{n}|\cos\theta = |\vec{u}|r\cos\theta.$$

This quantity is largest when $\cos\theta = -1$, which occurs when $\theta = 180^\circ$. Hence, the vector that works is $\vec{n} = \langle -r,0\rangle$ and in such case, $\vec{u}\cdot\vec{n} = -|\vec{u}|r$.

59. Consider the following diagram:

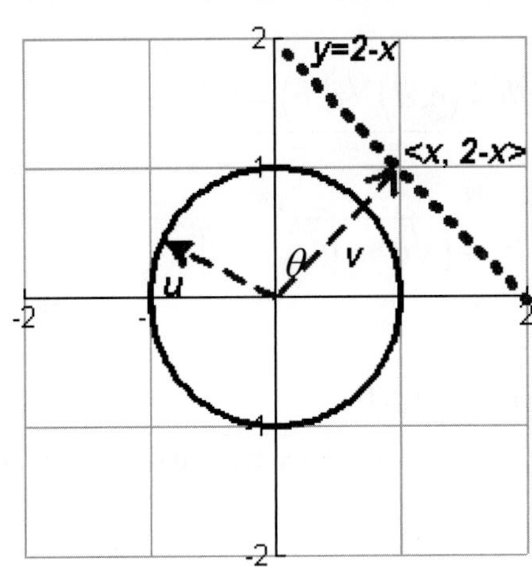

Observe that

$$\begin{aligned}\vec{u}\cdot\vec{v} &= |\vec{u}||\vec{v}|\cos\theta \\ &= 1\left[x^2+(2-x)^2\right]^{1/2}\cos\theta \\ &= \left[2(x-1)^2+1\right]^{1/2}\cos\theta\end{aligned}$$

Use $x = 2$ and $\theta = \pi$ to obtain the minimal result of $\vec{u}\cdot\vec{v} = \boxed{-2}$.

60. Consider the following diagram:

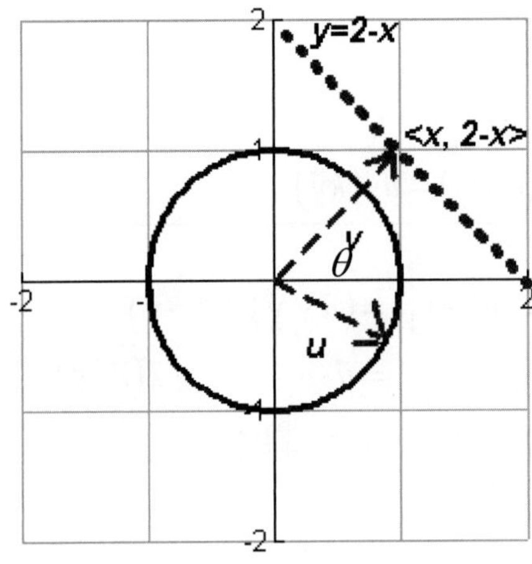

Observe that

$$\begin{aligned}\vec{u}\cdot\vec{v} &= |\vec{u}||\vec{v}|\cos\theta \\ &= 1\left[x^2+(2-x)^2\right]^{1/2}\cos\theta \\ &= \left[2(x-1)^2+1\right]^{1/2}\cos\theta\end{aligned}$$

Use $x = 2$ and $\theta = 0$ to obtain the maximal result of $\vec{u}\cdot\vec{v} = \boxed{2}$.

61. The dot product of two vectors is a scalar, not a vector. Should have summed:
$(-3)(2)+(2)(5)=4$

62. The dot product is the sum of the products of corresponding components (first × first, second × second).

63. False. By definition, the dot product of two vectors is a scalar.

64. True. By definition.

65. True. Since $\vec{u}\cdot\vec{v} = |\vec{u}||\vec{v}|\cos\theta$, if $\theta = 90^\circ$, then $\cos 90^\circ = 0$, so that $\vec{u}\cdot\vec{v} = 0$.

66. True. If $\vec{u}\cdot\vec{v} = |\vec{u}||\vec{v}|\cos\theta = 0$, then assuming that $\vec{u},\vec{v}$ are not the zero vector, then $\cos\theta = 0$, so that $\theta = 90^\circ$.

67. $\langle 3,7,-5\rangle\cdot\langle -2,4,1\rangle = (3)(-2)+(7)(4)+(-5)(1) = \boxed{17}$

68. $\langle 1,0,-2,3\rangle\cdot\langle 5,2,3,1\rangle=(1)(5)+(0)(2)+(-2)(3)+(3)(1)=\boxed{2}$

69. Let $\vec{u}=\langle a,b\rangle$, $\vec{v}=\langle c,d\rangle$. Observe that

$$\vec{u}\cdot\vec{v}=\langle a,b\rangle\cdot\langle c,d\rangle=ac+bd=ca+db=\langle c,d\rangle\cdot\langle a,b\rangle=\vec{v}\cdot\vec{u}\,.$$

70. Let $\vec{u}=\langle a,b\rangle$. Observe that

$$\vec{u}\cdot\vec{u}=\langle a,b\rangle\cdot\langle a,b\rangle=aa+bb=a^2+b^2=\left(\sqrt{a^2+b^2}\right)^2=|\vec{u}|^2$$

71. Let $\vec{u}=\langle a,b\rangle$. Observe that

$$\vec{0}\cdot\vec{u}=\langle 0,0\rangle\cdot\langle a,b\rangle=0(a)+0(b)=0$$

72. We prove the two equalities separately. Then, we conclude by transitivity that they must all be equal.

Claim 1: $k(\vec{u}\cdot\vec{v})=(k\vec{u})\cdot\vec{v}$

$$\begin{aligned}k(\vec{u}\cdot\vec{v})&=k(\langle a,b\rangle\cdot\langle c,d\rangle)=k(ac+bd)=kac+kbd=(ka)c+(kb)d\\&=\langle ka,kb\rangle\cdot\langle c,d\rangle=(k\langle a,b\rangle)\cdot\langle c,d\rangle=(k\vec{u})\cdot\vec{v}\end{aligned}$$

Claim 2: $(k\vec{u})\cdot\vec{v}=\vec{u}\cdot(k\vec{v})$

$$\begin{aligned}(k\vec{u})\cdot\vec{v}&=(k\langle a,b\rangle)\cdot\langle c,d\rangle=\langle ka,kb\rangle\cdot\langle c,d\rangle=(ka)c+(kb)d\\&=a(kc)+b(kd)=\langle a,b\rangle\cdot(k\langle c,d\rangle)=\vec{u}\cdot(k\vec{v})\end{aligned}$$

73. Observe that

$$\begin{aligned}\vec{u}\cdot(\vec{v}+\vec{w})&=\langle u_1,u_2\rangle\cdot(\langle v_1,v_2\rangle+\langle w_1,w_2\rangle)\\&=\langle u_1,u_2\rangle\cdot\langle v_1+w_1,v_2+w_2\rangle\\&=u_1(v_1+w_1)+u_2(v_2+w_2)\\&=u_1v_1+u_1w_1+u_2v_2+u_2w_2\\&=(u_1v_1+u_2v_2)+(u_1w_1+u_2w_2)\\&=\vec{u}\cdot\vec{v}+\vec{u}\cdot\vec{w}\end{aligned}$$

74.

$$\begin{aligned}|\vec{u}-\vec{v}|^2&=(\vec{u}-\vec{v})\cdot(\vec{u}-\vec{v})\\&=\vec{u}\cdot\vec{u}+\vec{v}\cdot\vec{v}-2(\vec{u}\cdot\vec{v})\\&=|\vec{u}|^2+|\vec{v}|^2-2(\vec{u}\cdot\vec{v})\end{aligned}$$

75. Using $\text{proj}_{\vec{u}}\vec{v}=\left(\dfrac{\vec{u}\cdot\vec{v}}{|\vec{u}|^2}\right)\vec{u}$, observe that

(i) $\text{proj}_{-\vec{u}}\,2\vec{u}=-\left(\dfrac{2\vec{u}\cdot\vec{u}}{|\vec{u}|^2}\right)\vec{u}=-2\vec{u}$.

(ii) More generally, for any $c>0$, $\text{proj}_{-\vec{u}}\,c\vec{u}=-\left(\dfrac{c\vec{u}\cdot\vec{u}}{|\vec{u}|^2}\right)\vec{u}=-c\vec{u}$

76. Using $\text{proj}_{\vec{u}}\vec{v} = \left(\dfrac{\vec{u}\cdot\vec{v}}{|\vec{u}|^2}\right)\vec{u}$, observe that

(i) $\text{proj}_{\vec{u}}\, 2\vec{u} = \left(\dfrac{2\vec{u}\cdot\vec{u}}{|\vec{u}|^2}\right)\vec{u} = 2\vec{u}$.

(ii) More generally, for any $c > 0$, $\text{proj}_{\vec{u}}\, c\vec{u} = \left(\dfrac{c\vec{u}\cdot\vec{u}}{|\vec{u}|^2}\right)\vec{u} = c\vec{u}$.

77. Since $\text{proj}_{\vec{u}}\vec{v} = \left(\dfrac{\vec{u}\cdot\vec{v}}{|\vec{u}|^2}\right)\vec{u}$, we need $\vec{u}\cdot\vec{v} = 0$. In such case, $\vec{u}$ is perpendicular to $\vec{v}$, so that $\theta = 90^\circ$.

78. True. This holds because $\vec{u}\cdot(\vec{v}+\vec{w}) = (\vec{u}\cdot\vec{v}) + (\vec{u}\cdot\vec{w})$.

79. Observe that

$$(-2\vec{u})\cdot(3\vec{v}) = -6(\vec{u}\cdot\vec{v}) = -6(|\vec{u}||\vec{v}|\cos\theta) = -6\cos\theta .$$

So, the maximum value is 6 and occurs when $\theta = 0$, and the minimum value is -6 and occurs when $\theta = \pi$.

80. Observe that

$$\begin{aligned}\frac{(\vec{u}\cdot\vec{v})\vec{u}}{|\vec{v}|} - \frac{(\vec{v}\cdot\vec{u})\vec{v}}{|\vec{u}|} &= \frac{(\vec{u}\cdot\vec{v})|\vec{u}|\vec{u} - (\vec{v}\cdot\vec{u})|\vec{v}|\vec{v}}{|\vec{v}||\vec{u}|} \\ &= \frac{(\vec{u}\cdot\vec{v})\left[|\vec{u}|\vec{u} - |\vec{v}|\vec{v}\right]}{|\vec{v}||\vec{u}|} \\ &= \frac{(\vec{u}\cdot\vec{v})}{|\vec{v}||\vec{u}|}\left[|\vec{u}|\vec{u} - |\vec{v}|\vec{v}\right] \\ &= \cos\left(\tfrac{\pi}{3}\right)\left[|\vec{u}|\vec{u} - |\vec{v}|\vec{v}\right] \\ &= \frac{|\vec{u}|\vec{u} - |\vec{v}|\vec{v}}{2}\end{aligned}$$

81. Enter the following two 2×1 matrices:

$$[A] = \begin{bmatrix} -11 \\ 34 \end{bmatrix}, \quad [B] = \begin{bmatrix} 15 \\ -27 \end{bmatrix}$$

Then, compute $A^T B$ (using the keystrokes given in the exercise) to obtain the dot product: -1083

82. Enter the following two 2×1 matrices:

$$[A] = \begin{bmatrix} 23 \\ -350 \end{bmatrix}, \quad [B] = \begin{bmatrix} 45 \\ 202 \end{bmatrix}$$

Then, compute $A^T B$ (using the keystrokes given in the exercise) to obtain the dot product: -69665

83. Let $\vec{a} = \langle 18,0 \rangle$ and $\vec{b} = \langle 18,11 \rangle$. Note that $$\theta = \cos^{-1}\left(\frac{\vec{a}\cdot\vec{b}}{\lVert\vec{a}\rVert\lVert\vec{b}\rVert}\right) = \cos^{-1}\left(\frac{(18)(18)+(0)(11)}{\sqrt{18^2}\sqrt{18^2+11^2}}\right) \approx \boxed{31.4^\circ}$$	
84. Let $\vec{a} = \langle 0,7,0 \rangle$ and $\vec{b} = \langle 12,7,9 \rangle$. Note that $$\theta = \cos^{-1}\left(\frac{\vec{a}\cdot\vec{b}}{\lVert\vec{a}\rVert\lVert\vec{b}\rVert}\right) = \cos^{-1}\left(\frac{(0)(12)+(7)(7)+(0)(9)}{\sqrt{7^2}\sqrt{12^2+7^2+9^2}}\right) \approx \boxed{65.0^\circ}$$	
85. 1220	**86.** 135
87. $\vec{u}(t)\cdot\vec{v}(t) = (2t)(t)+(t^2)(-3t) = 2t^2-3t^3$	**88.** $\vec{u}(t)\cdot\vec{v}(t) = \cos^2 t - \sin^2 t$
89. $$\theta = \cos^{-1}\left(\frac{\vec{u}(t)\cdot\vec{v}(t)}{\lvert\vec{u}(t)\rvert\lvert\vec{v}(t)\rvert}\right) = \cos^{-1}\left(\frac{(\sin t)(\csc t)-\cos^2 t}{\sqrt{\sin^2 t+\cos^2 t}\sqrt{\csc^2 t+\cos^2 t}}\right) = \cos^{-1}\left(\frac{1-\cos^2 t}{\sqrt{\frac{1+\sin^2 t\cos^2 t}{\sin^2 t}}}\right)$$ $$= \cos^{-1}\left(\frac{\sin^2 t(\sin t)}{\sqrt{1+\sin^2 t\cos^2 t}}\right) = \cos^{-1}\left(\frac{\sin^3 t}{\sqrt{1+\sin^2 t\cos^2 t}}\right)$$ So, at $t=\frac{\pi}{6}$, the angle is 1.518 radians.	
90. Find t such that $\vec{u}(t)\cdot\vec{v}(t) = 0$: $$\sin t\cos t - \sin^2 t = \sin t(\cos t - \sin t) = 0 \Rightarrow \sin t = 0 \text{ or } \cos t = \sin t$$ So, the values of t that work are $t = n\pi,\ \frac{\pi}{4}+n\pi$, where n is an integer.	

Section 7.3 Solutions ---

1. – 8. The points listed in Exercises 1 – 8 are all plotted on the axes below:

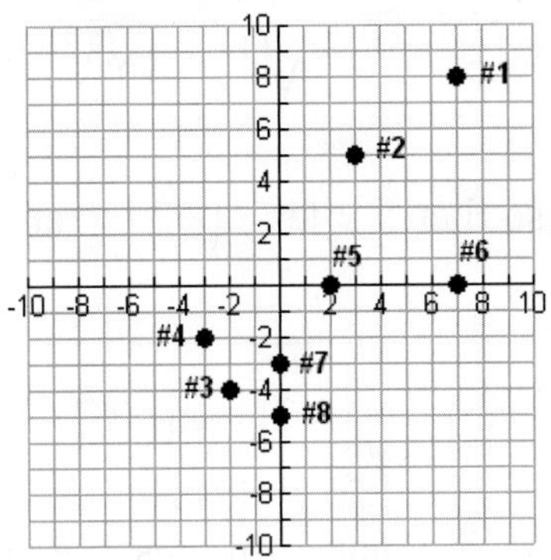

9. In order to express $1-i$ in polar form, observe that $x=1,\ y=-1$, so that the point is in QIV. Now,

$$r=\sqrt{x^2+y^2}=\sqrt{(1)^2+(-1)^2}=\sqrt{2}$$

$$\tan\theta=\frac{y}{x}=\frac{-1}{1}=-1\text{, so that }\theta=\tan^{-1}(-1)=-45^\circ\text{ or }-\frac{\pi}{4}.$$

Since $0\le\theta<2\pi$, we must use the reference angle 315° or $\frac{7\pi}{4}$.

Hence, $\boxed{1-i=\sqrt{2}\left[\cos\left(\frac{7\pi}{4}\right)+i\sin\left(\frac{7\pi}{4}\right)\right]=\sqrt{2}\left[\cos\left(315^\circ\right)+i\sin\left(315^\circ\right)\right]}$.

10. In order to express $2+2i$ in polar form, observe that $x=2,\ y=2$, so that the point is in QI. Now,

$$r=\sqrt{x^2+y^2}=\sqrt{(2)^2+(2)^2}=2\sqrt{2}$$

$$\tan\theta=\frac{y}{x}=\frac{2}{2}=1\text{, so that }\theta=\tan^{-1}(1)=45^\circ\text{ or }\frac{\pi}{4}.$$

Hence, $\boxed{2+2i=2\sqrt{2}\left[\cos\left(\frac{\pi}{4}\right)+i\sin\left(\frac{\pi}{4}\right)\right]=2\sqrt{2}\left[\cos\left(45^\circ\right)+i\sin\left(45^\circ\right)\right]}$.

11. In order to express $1+\sqrt{3}\,i$ in polar form, observe that $x=1,\ y=\sqrt{3}$, so that the point is in QI. Now,

$$r=\sqrt{x^2+y^2}=\sqrt{(1)^2+\left(\sqrt{3}\right)^2}=2$$

$$\tan\theta=\frac{y}{x}=\frac{\sqrt{3}}{1}=\sqrt{3}\text{, so that }\theta=\tan^{-1}\left(\sqrt{3}\right)=\tan^{-1}\left(\frac{\sqrt{3}/2}{1/2}\right)=60^\circ\text{ or }\frac{\pi}{3}.$$

Hence, $\boxed{1+\sqrt{3}\,i=2\left[\cos\left(\frac{\pi}{3}\right)+i\sin\left(\frac{\pi}{3}\right)\right]=2\left[\cos\left(60^\circ\right)+i\sin\left(60^\circ\right)\right]}$.

12. In order to express $-3-\sqrt{3}\,i$ in polar form, observe that $x=-3,\ y=-\sqrt{3}$, so that the point is in QIII. Now,

$$r=\sqrt{x^2+y^2}=\sqrt{(-3)^2+\left(-\sqrt{3}\right)^2}=2\sqrt{3}$$

$$\tan\theta=\frac{y}{x}=\frac{-\sqrt{3}}{-3}=\frac{1}{\sqrt{3}}\text{, so that }\theta=\tan^{-1}\left(\frac{1}{\sqrt{3}}\right)=\pi+\frac{\pi}{6}=\frac{7\pi}{6}\text{ or }210^\circ.$$

(Remember to add π to $\tan^{-1}\left(\frac{1}{\sqrt{3}}\right)$ since the angle has terminal side in QIII.)

Hence, $\boxed{-3-\sqrt{3}\,i=2\sqrt{3}\left[\cos\left(\frac{7\pi}{6}\right)+i\sin\left(\frac{7\pi}{6}\right)\right]=2\sqrt{3}\left[\cos\left(210^\circ\right)+i\sin\left(210^\circ\right)\right]}$.

13. In order to express $-4+4i$ in polar form, observe that $x=-4,\ y=4$, so that the point is in QII. Now,

$$r=\sqrt{x^2+y^2}=\sqrt{(-4)^2+(4)^2}=4\sqrt{2}$$

$$\tan\theta=\frac{y}{x}=\frac{4}{-4}=-1\text{, so that }\theta=\pi+\tan^{-1}(-1)=\pi-\frac{\pi}{4}=\frac{3\pi}{4}\text{ or }135^\circ.$$

(Remember to add π to $\tan^{-1}(-1)$ since the angle has terminal side in QII.)

Hence, $\boxed{-4+4i=4\sqrt{2}\left[\cos\left(\frac{3\pi}{4}\right)+i\sin\left(\frac{3\pi}{4}\right)\right]=4\sqrt{2}\left[\cos\left(135^\circ\right)+i\sin\left(135^\circ\right)\right]}$.

14. In order to express $\sqrt{5}-\sqrt{5}i$ in polar form, observe that $x=\sqrt{5},\ y=-\sqrt{5}$, so that the point is in QIV. Now,

$$r=\sqrt{x^2+y^2}=\sqrt{\left(\sqrt{5}\right)^2+\left(-\sqrt{5}\right)^2}=\sqrt{10}$$

$$\tan\theta=\frac{y}{x}=\frac{-\sqrt{5}}{\sqrt{5}}=-1\text{, so that }\theta=\tan^{-1}(-1)=-45^\circ\text{ or }-\frac{\pi}{4}.$$

Since $0\le\theta<2\pi$, we must use the reference angle 315° or $\frac{7\pi}{4}$.

Hence, $\boxed{\sqrt{5}-\sqrt{5}i=\sqrt{10}\left[\cos\left(\frac{7\pi}{4}\right)+i\sin\left(\frac{7\pi}{4}\right)\right]=\sqrt{10}\left[\cos(315^\circ)+i\sin(315^\circ)\right]}$.

15. In order to express $\sqrt{3}-3i$ in polar form, observe that $x=\sqrt{3},\ y=-3$, so that the point is in QIV. Now,

$$r=\sqrt{x^2+y^2}=\sqrt{\left(\sqrt{3}\right)^2+(-3)^2}=2\sqrt{3}$$

$$\tan\theta=\frac{y}{x}=\frac{-3}{\sqrt{3}}=-\sqrt{3}\text{, so that }\theta=\tan^{-1}\left(-\sqrt{3}\right)=-60^\circ\text{ or }-\frac{\pi}{3}.$$

Since $0\le\theta<2\pi$, we must use the reference angle 300° or $\frac{5\pi}{3}$.

Hence, $\boxed{\sqrt{3}-3i=2\sqrt{3}\left[\cos\left(\frac{5\pi}{3}\right)+i\sin\left(\frac{5\pi}{3}\right)\right]=2\sqrt{3}\left[\cos(300^\circ)+i\sin(300^\circ)\right]}$.

16. In order to express $-\sqrt{3}+i$ in polar form, observe that $x=-\sqrt{3},\ y=1$, so that the point is in QII. Now,

$$r=\sqrt{x^2+y^2}=\sqrt{\left(-\sqrt{3}\right)^2+(1)^2}=2$$

$$\tan\theta=\frac{y}{x}=\frac{-3}{\sqrt{3}}=-\frac{1}{\sqrt{3}}\text{, so that }\theta=\pi+\tan^{-1}\left(-\frac{1}{\sqrt{3}}\right)=\pi-\frac{\pi}{6}=\frac{5\pi}{6}\text{ or }150^\circ.$$

(Remember to add π to $\tan^{-1}\left(-\frac{1}{\sqrt{3}}\right)$ since the angle has terminal side in QII.)

Hence, $\boxed{-\sqrt{3}+i=2\left[\cos\left(\frac{5\pi}{6}\right)+i\sin\left(\frac{5\pi}{6}\right)\right]=2\left[\cos(150^\circ)+i\sin(150^\circ)\right]}$.

17. In order to express $3+0i$ in polar form, observe that $x=3,\ y=0$, so that the point is on the positive x-axis. Now,

$$r=\sqrt{x^2+y^2}=\sqrt{(3)^2+(0)^2}=3 \text{ and } \theta=0 \text{ or } 0^\circ.$$

Hence, $\boxed{3+0i=3\left[\cos(0)+i\sin(0)\right]=3\left[\cos(0^\circ)+i\sin(0^\circ)\right]}$.

18. In order to express $-2+0i$ in polar form, observe that $x=-2,\ y=0$, so that the point is on the negative x-axis. Now,

$$r=\sqrt{x^2+y^2}=\sqrt{(-2)^2+(0)^2}=2 \text{ and } \theta=\pi \text{ or } 180^\circ.$$

Hence, $\boxed{-2+0i=2\left[\cos(\pi)+i\sin(\pi)\right]=2\left[\cos(180^\circ)+i\sin(180^\circ)\right]}$.

19. In order to express $-\frac{1}{2}-\frac{1}{2}i$ in polar form, observe that $x=-\frac{1}{2},\ y=-\frac{1}{2}$, so that the point is in QIII. Now,

$$r=\sqrt{x^2+y^2}=\sqrt{\left(-\tfrac{1}{2}\right)^2+\left(-\tfrac{1}{2}\right)^2}=\frac{\sqrt{2}}{2}$$

$$\tan\theta=\frac{y}{x}=\frac{-\frac{1}{2}}{-\frac{1}{2}}=1, \text{ so that } \theta=\pi+\tan^{-1}(1)=\pi+\frac{\pi}{4}=\frac{5\pi}{4} \text{ or } 225^\circ.$$

(Remember to add π to $\tan^{-1}(1)$ since the angle has terminal side in QIII.)

Hence, $\boxed{-\frac{1}{2}-\frac{1}{2}i=\frac{\sqrt{2}}{2}\left[\cos\left(\frac{5\pi}{4}\right)+i\sin\left(\frac{5\pi}{4}\right)\right]=\frac{\sqrt{2}}{2}\left[\cos(225^\circ)+i\sin(225^\circ)\right]}$.

20. In order to express $\frac{1}{6}-\frac{1}{6}i$ in polar form, observe that $x=\frac{1}{6},\ y=-\frac{1}{6}$, so that the point is in QIV. Now,

$$r=\sqrt{x^2+y^2}=\sqrt{\left(\tfrac{1}{6}\right)^2+\left(-\tfrac{1}{6}\right)^2}=\frac{\sqrt{3}}{3}$$

$$\tan\theta=\frac{y}{x}=\frac{-\frac{1}{6}}{\frac{1}{6}}=-1, \text{ so that } \theta=\tan^{-1}(-1)=315^\circ \text{ or } \frac{7\pi}{4}.$$

Since $0\le\theta<2\pi$, we must use the reference angle 315° or $\frac{7\pi}{4}$.

Hence, $\boxed{\frac{1}{6}-\frac{1}{6}i=\frac{\sqrt{3}}{3}\left[\cos\left(\frac{7\pi}{4}\right)+i\sin\left(\frac{7\pi}{4}\right)\right]=\frac{\sqrt{3}}{3}\left[\cos(315^\circ)+i\sin(315^\circ)\right]}$.

21. In order to express $-\sqrt{6}-\sqrt{6}i$ in polar form, observe that $x=-\sqrt{6},\ y=-\sqrt{6}$, so that the point is in QIII. Now,

$$r=\sqrt{x^2+y^2}=\sqrt{\left(-\sqrt{6}\right)^2+\left(-\sqrt{6}\right)^2}=\sqrt{12}=2\sqrt{3}$$

$$\tan\theta=\frac{y}{x}=\frac{-\sqrt{6}}{-\sqrt{6}}=1\text{, so that }\theta=\pi+\tan^{-1}(1)=\pi+\frac{\pi}{4}=\frac{5\pi}{4}\text{ or }225^\circ.$$

(Remember to add π to $\tan^{-1}(1)$ since the angle has terminal side in QIII.)

Hence, $\boxed{-\sqrt{6}-\sqrt{6}i=2\sqrt{3}\left[\cos\left(\frac{5\pi}{4}\right)+i\sin\left(\frac{5\pi}{4}\right)\right]=2\sqrt{3}\left[\cos\left(225^\circ\right)+i\sin\left(225^\circ\right)\right]}$.

22. In order to express $\frac{1}{3}-\frac{1}{3}i$ in polar form, observe that $x=\frac{1}{3},\ y=-\frac{1}{3}$, so that the point is in QIV. Now,

$$r=\sqrt{x^2+y^2}=\sqrt{\left(\tfrac{1}{3}\right)^2+\left(-\tfrac{1}{3}\right)^2}=\frac{\sqrt{2}}{3}$$

$$\tan\theta=\frac{y}{x}=\frac{-\frac{1}{3}}{\frac{1}{3}}=-1\text{, so that }\theta=\tan^{-1}(-1)=315^\circ\text{ or }\frac{7\pi}{4}.$$

Since $0\le\theta<2\pi$, we must use the reference angle 315° or $\frac{7\pi}{4}$.

Hence, $\boxed{\frac{1}{3}-\frac{1}{3}i=\frac{\sqrt{2}}{3}\left[\cos\left(\frac{7\pi}{4}\right)+i\sin\left(\frac{7\pi}{4}\right)\right]=\frac{\sqrt{2}}{3}\left[\cos\left(315^\circ\right)+i\sin\left(315^\circ\right)\right]}$

23. In order to express $-5+5i$ in polar form, observe that $x=-5,\ y=5$, so that the point is in QII. Now,

$$r=\sqrt{x^2+y^2}=\sqrt{(-5)^2+(5)^2}=5\sqrt{2}$$

$$\tan\theta=\frac{y}{x}=\frac{5}{-5}=-1\text{, so that }\theta=\pi+\tan^{-1}(-1)=\pi-\frac{\pi}{4}=\frac{3\pi}{4}\text{ or }135^\circ.$$

(Remember to add π to $\tan^{-1}(-1)$ since the angle has terminal side in QII.)

Hence, $\boxed{-5+5i=5\sqrt{2}\left[\cos\left(\frac{3\pi}{4}\right)+i\sin\left(\frac{3\pi}{4}\right)\right]=5\sqrt{2}\left[\cos\left(135^\circ\right)+i\sin\left(135^\circ\right)\right]}$

24. In order to express $3+3i$ in polar form, observe that $x=3,\ y=3$, so that the point is in QI. Now,

$$r=\sqrt{x^2+y^2}=\sqrt{(3)^2+(3)^2}=3\sqrt{2}$$

$$\tan\theta=\frac{y}{x}=\frac{3}{3}=1,\text{ so that }\theta=\tan^{-1}(1)=45^\circ\text{ or }\frac{\pi}{4}.$$

Hence, $\boxed{3+3i=3\sqrt{2}\left[\cos\left(\frac{\pi}{4}\right)+i\sin\left(\frac{\pi}{4}\right)\right]=3\sqrt{2}\left[\cos\left(45^\circ\right)+i\sin\left(45^\circ\right)\right]}$

25. In order to express $3-7i$ in polar form, observe that $x=3,\ y=-7$, so that the point is in QIV. Now,

$$r=\sqrt{x^2+y^2}=\sqrt{(3)^2+(-7)^2}=\sqrt{58}$$

$$\tan\theta=\frac{y}{x}=\frac{-7}{3},\text{ so that }\theta=\tan^{-1}\left(-\frac{7}{3}\right)\approx-66.80^\circ.$$

Since $0\le\theta<360^\circ$, we must use the reference angle 293.2°.

Hence, $\boxed{3-7i\approx\sqrt{58}\left[\cos\left(293.2^\circ\right)+i\sin\left(293.2^\circ\right)\right]}$.

26. In order to express $2+3i$ in polar form, observe that $x=2,\ y=3$, so that the point is in QI. Now,

$$r=\sqrt{x^2+y^2}=\sqrt{(2)^2+(3)^2}=\sqrt{13}$$

$$\tan\theta=\frac{y}{x}=\frac{3}{2},\text{ so that }\theta=\tan^{-1}\left(\frac{3}{2}\right)\approx56.3^\circ.$$

Hence, $\boxed{2+3i\approx\sqrt{13}\left[\cos\left(56.3^\circ\right)+i\sin\left(56.3^\circ\right)\right]}$.

27. In order to express $-6+5i$ in polar form, observe that $x=-6,\ y=5$, so that the point is in QII. Now,

$$r=\sqrt{x^2+y^2}=\sqrt{(-6)^2+(5)^2}=\sqrt{61}$$

$$\tan\theta=\frac{y}{x}=\frac{5}{-6},\text{ so that }\theta=180^\circ+\tan^{-1}\left(-\frac{5}{6}\right)\approx140.2^\circ.$$

(Remember to add 180° to $\tan^{-1}\left(-\frac{5}{6}\right)$ since the angle has terminal side in QII.)

Hence, $\boxed{-6+5i\approx\sqrt{61}\left[\cos\left(140.2^\circ\right)+i\sin\left(140.2^\circ\right)\right]}$.

28. In order to express $-4-3i$ in polar form, observe that $x=-4,\ y=-3$, so that the point is in QIII. Now,

$$r=\sqrt{x^2+y^2}=\sqrt{(-4)^2+(-3)^2}=5$$

$$\tan\theta=\frac{y}{x}=\frac{3}{4}\text{, so that ..}$$

(Remember to add 180° to $\tan^{-1}\left(\frac{3}{4}\right)$ since the angle has terminal side in QIII.)

Hence, $\boxed{-4-3i\approx 5\left[\cos\left(216.9^\circ\right)+i\sin\left(216.9^\circ\right)\right]}$.

29. In order to express $-5+12i$ in polar form, observe that $x=-5,\ y=12$, so that the point is in QII. Now,

$$r=\sqrt{x^2+y^2}=\sqrt{(-5)^2+(12)^2}=13$$

$$\tan\theta=\frac{y}{x}=\frac{-12}{5}\text{, so that }\theta=180^\circ+\tan^{-1}\left(-\frac{12}{5}\right)\approx 112.6^\circ.$$

(Remember to add 180° to $\tan^{-1}\left(-\frac{12}{5}\right)$ since the angle has terminal side in QII.)

Hence, $\boxed{-5+12i\approx 13\left[\cos\left(112.6^\circ\right)+i\sin\left(112.6^\circ\right)\right]}$.

30. In order to express $24+7i$ in polar form, observe that $x=24,\ y=7$, so that the point is in QI. Now,

$$r=\sqrt{x^2+y^2}=\sqrt{(24)^2+(7)^2}=25$$

$$\tan\theta=\frac{y}{x}=\frac{7}{24}\text{, so that }\theta=\tan^{-1}\left(\frac{7}{24}\right)\approx 16.3^\circ.$$

Hence, $\boxed{24+7i\approx 25\left[\cos\left(16.3^\circ\right)+i\sin\left(16.3^\circ\right)\right]}$.

31. In order to express $8-6i$ in polar form, observe that $x=8,\ y=-6$, so that the point is in QIV. Now,

$$r=\sqrt{x^2+y^2}=\sqrt{(8)^2+(-6)^2}=10$$

$$\tan\theta=\frac{y}{x}=\frac{-6}{8}\text{, so that }\theta=\tan^{-1}\left(-\frac{6}{8}\right)\approx -36.87^\circ.$$

Since $0\le\theta<360^\circ$, we must use the reference angle 323.1°.

Hence, $\boxed{8-6i\approx 10\left[\cos\left(323.1^\circ\right)+i\sin\left(323.1^\circ\right)\right]}$.

32. In order to express $-3+4i$ in polar form, observe that $x=-3,\ y=4$, so that the point is in QII. Now,

$$r=\sqrt{x^2+y^2}=\sqrt{(-3)^2+(4)^2}=5$$

$$\tan\theta=\frac{y}{x}=\frac{4}{-3},\text{ so that }\theta=180^\circ+\tan^{-1}\left(-\frac{4}{3}\right)\approx 126.9^\circ.$$

(Remember to add 180° to $\tan^{-1}\left(-\frac{4}{3}\right)$ since the angle has terminal side in QII.)

Hence, $\boxed{-3+4i\approx 5\left[\cos(126.9^\circ)+i\sin(126.9^\circ)\right]}$.

33. In order to express $-\frac{1}{2}+\frac{3}{4}i$ in polar form, observe that $x=-\frac{1}{2},\ y=\frac{3}{4}$, so that the point is in QII. Now,

$$r=\sqrt{x^2+y^2}=\sqrt{\left(-\tfrac{1}{2}\right)^2+\left(\tfrac{3}{4}\right)^2}=\frac{\sqrt{13}}{4}$$

$$\tan\theta=\frac{y}{x}=\frac{3/4}{-1/2}=-\frac{3}{2},\text{ so that }\theta=180^\circ+\tan^{-1}\left(-\frac{3}{2}\right)\approx 123.7^\circ.$$

(Remember to add 180° to $\tan^{-1}\left(-\frac{3}{2}\right)$ since the angle has terminal side in QII.)

Hence, $\boxed{-\frac{1}{2}+\frac{3}{4}i=\frac{\sqrt{13}}{4}\left[\cos(123.7^\circ)+i\sin(123.7^\circ)\right]}$.

34. In order to express $-\frac{5}{8}-\frac{11}{4}i$ in polar form, observe that $x=-\frac{5}{8},\ y=-\frac{11}{4}$, so that the point is in QIII. Now,

$$r=\sqrt{x^2+y^2}=\sqrt{\left(-\tfrac{5}{8}\right)^2+\left(-\tfrac{11}{4}\right)^2}=\frac{\sqrt{509}}{8}$$

$$\tan\theta=\frac{y}{x}=\frac{-\frac{11}{4}}{-\frac{5}{8}},\text{ so that }\theta=180^\circ+\tan^{-1}\left(\frac{22}{5}\right)\approx 257.20^\circ.$$

(Remember to add 180° to $\tan^{-1}\left(\frac{22}{5}\right)$ since the angle has terminal side in QIII.)

Hence, $\boxed{-\frac{5}{8}-\frac{11}{4}i\approx\frac{\sqrt{509}}{8}\left[\cos(257.20^\circ)+i\sin(257.20^\circ)\right]}$.

35. In order to express $5.1+2.3i$ in polar form, observe that $x=5.1,\ y=2.3$, so that the point is in QI. Now,

$$r=\sqrt{x^2+y^2}=\sqrt{(5.1)^2+(2.3)^2}\approx 5.59$$

$$\tan\theta=\frac{y}{x}=\frac{2.3}{5.1},\text{ so that }\theta=\tan^{-1}\left(\frac{2.3}{5.1}\right)\approx 24.27^\circ.$$

Hence, $\boxed{5.1+2.3i\approx 5.59\left[\cos\left(24.27^\circ\right)+i\sin\left(24.27^\circ\right)\right]}$.

36. In order to express $1.8-0.9i$ in polar form, observe that $x=1.8,\ y=-0.9$, so that the point is in QIV. Now,

$$r=\sqrt{x^2+y^2}=\sqrt{(1.8)^2+(-0.9)^2}\approx 2.01$$

$$\tan\theta=\frac{y}{x}=\frac{-0.9}{1.8},\text{ so that }\theta=\tan^{-1}\left(\frac{-0.9}{1.8}\right)\approx -26.57^\circ.$$

Since $0\le\theta<360^\circ$, we must use the reference angle 333.43°.

Hence, $\boxed{1.8-0.9i\approx 2.01\left[\cos\left(333.43^\circ\right)+i\sin\left(333.43^\circ\right)\right]}$.

37. In order to express $-2\sqrt{3}-\sqrt{5}\,i$ in polar form, observe that $x=-2\sqrt{3},\ y=-\sqrt{5}$, so that the point is in QIII. Now,

$$r=\sqrt{x^2+y^2}=\sqrt{\left(-2\sqrt{3}\right)^2+\left(-\sqrt{5}\right)^2}=\sqrt{17}$$

$$\tan\theta=\frac{y}{x}=\frac{-\sqrt{5}}{-2\sqrt{3}},\text{ so that }\theta=180^\circ+\tan^{-1}\left(\frac{-\sqrt{5}}{-2\sqrt{3}}\right)\approx 212.84^\circ.$$

(Remember to add 180° to $\tan^{-1}\left(\frac{-\sqrt{5}}{-2\sqrt{3}}\right)$ since the angle has terminal side in QIII.)

Hence, $\boxed{-2\sqrt{3}-\sqrt{5}\,i\approx\sqrt{17}\left[\cos\left(212.84^\circ\right)+i\sin\left(212.84^\circ\right)\right]}$.

38. In order to express $-\frac{4\sqrt{5}}{3}+\frac{\sqrt{5}}{2}i$ in polar form, observe that $x=-\frac{4\sqrt{5}}{3}$, $y=\frac{\sqrt{5}}{2}$, so that the point is in QII. Now,

$$r=\sqrt{x^2+y^2}=\sqrt{\left(-\tfrac{4\sqrt{5}}{3}\right)^2+\left(\tfrac{\sqrt{5}}{2}\right)^2}=\frac{\sqrt{85}}{2}$$

$$\tan\theta=\frac{y}{x}=\frac{\sqrt{5}/2}{-4\sqrt{5}/3},\text{ so that }\theta=180^\circ+\tan^{-1}\left(\frac{\sqrt{5}/2}{-4\sqrt{5}/3}\right)\approx 159.44^\circ.$$

(Remember to add 180° to $\tan^{-1}\left(\frac{\sqrt{5}/2}{-4\sqrt{5}/3}\right)$ since the angle has terminal side in QII.)

Hence, $\boxed{-\frac{4\sqrt{5}}{3}+\frac{\sqrt{5}}{2}i=\frac{\sqrt{85}}{2}\left[\cos\left(159.44^\circ\right)+i\sin\left(159.44^\circ\right)\right]}$.

39. In order to express $4.02-2.11i$ in polar form, observe that $x=4.2,\ y=-2.11$, so that the point is in QIV. Now,

$$r=\sqrt{x^2+y^2}=\sqrt{(4.2)^2+(-2.11)^2}\approx 4.54$$

$$\tan\theta=\frac{y}{x}=\frac{-2.11}{4.02},\text{ so that }\theta=\tan^{-1}\left(\frac{-2.11}{4.02}\right)\approx -27.69^\circ.$$

Since $0\le\theta<360^\circ$, we must use the reference angle 332.31°.

Hence, $\boxed{4.02-2.11i\approx 4.54\left[\cos\left(332.31^\circ\right)+i\sin\left(332.31^\circ\right)\right]}$.

40. In order to express $1.78-0.12i$ in polar form, observe that $x=1.78,\ y=-0.12$, so that the point is in QIV. Now,

$$r=\sqrt{x^2+y^2}=\sqrt{(1.78)^2+(-0.12)^2}\approx 1.78$$

$$\tan\theta=\frac{y}{x}=\frac{-0.12}{1.78},\text{ so that }\theta=\tan^{-1}\left(\frac{-0.12}{1.78}\right)\approx -3.86^\circ.$$

Since $0\le\theta<360^\circ$, we must use the reference angle 356.14°.

Hence, $\boxed{1.78-0.12i\approx 1.78\left[\cos\left(356.14^\circ\right)+i\sin\left(356.14^\circ\right)\right]}$.

41. $5\left[\cos\left(180^\circ\right)+i\sin\left(180^\circ\right)\right]=5[-1+0i]=\boxed{-5}$

42. $2\left[\cos\left(135^\circ\right)+i\sin\left(135^\circ\right)\right]=2\left[-\frac{\sqrt{2}}{2}+\frac{\sqrt{2}}{2}i\right]=\boxed{-\sqrt{2}+\sqrt{2}i}$

43. $2\left[\cos\left(315^\circ\right)+i\sin\left(315^\circ\right)\right]=2\left[\frac{\sqrt{2}}{2}-\frac{\sqrt{2}}{2}i\right]=\boxed{\sqrt{2}-\sqrt{2}i}$

44. $3\left[\cos\left(270^\circ\right)+i\sin\left(270^\circ\right)\right]=3[0-i]=\boxed{-3i}$

45. $-4\left[\cos(60^\circ)+i\sin(60^\circ)\right]=-4\left[\frac{1}{2}+\frac{\sqrt{3}}{2}i\right]=\boxed{-2-2\sqrt{3}i}$
46. $-4\left[\cos(210^\circ)+i\sin(210^\circ)\right]=-4\left[-\frac{\sqrt{3}}{2}-\frac{1}{2}i\right]=\boxed{2\sqrt{3}+2i}$
47. $\sqrt{3}\left[\cos(150^\circ)+i\sin(150^\circ)\right]=\sqrt{3}\left[-\frac{\sqrt{3}}{2}+\frac{1}{2}i\right]=\boxed{-\frac{3}{2}+\frac{\sqrt{3}}{2}i}$
48. $\sqrt{3}\left[\cos(330^\circ)+i\sin(330^\circ)\right]=\sqrt{3}\left[\frac{\sqrt{3}}{2}-\frac{1}{2}i\right]=\boxed{\frac{3}{2}-\frac{\sqrt{3}}{2}i}$
49. $\sqrt{2}\left[\cos\left(\frac{\pi}{4}\right)+i\sin\left(\frac{\pi}{4}\right)\right]=\sqrt{2}\left[\frac{\sqrt{2}}{2}+\frac{\sqrt{2}}{2}i\right]=\boxed{1+i}$
50. $2\left[\cos\left(\frac{5\pi}{6}\right)+i\sin\left(\frac{5\pi}{6}\right)\right]=2\left[-\frac{\sqrt{3}}{2}+\frac{1}{2}i\right]=\boxed{-\sqrt{3}+i}$
51. $6\left[\cos\left(\frac{3\pi}{4}\right)+i\sin\left(\frac{3\pi}{4}\right)\right]=6\left[-\frac{\sqrt{2}}{2}+\frac{\sqrt{2}}{2}i\right]=\boxed{-3\sqrt{2}+3\sqrt{2}i}$
52. $4\left[\cos\left(\frac{11\pi}{6}\right)+i\sin\left(\frac{11\pi}{6}\right)\right]=4\left[\frac{\sqrt{3}}{2}-\frac{1}{2}i\right]=\boxed{2\sqrt{3}-2i}$
53. $5\left[\cos(295^\circ)+i\sin(295^\circ)\right]\approx\boxed{2.1131-4.5315i}$
54. $4\left[\cos(35^\circ)+i\sin(35^\circ)\right]\approx\boxed{3.2766+2.2943i}$
55. $3\left[\cos(100^\circ)+i\sin(100^\circ)\right]\approx\boxed{-0.5209+2.9544i}$
56. $6\left[\cos(250^\circ)+i\sin(250^\circ)\right]\approx\boxed{-2.0521-5.6382i}$
57. $-7\left[\cos(140^\circ)+i\sin(140^\circ)\right]\approx\boxed{5.3623-4.4995i}$
58. $-5\left[\cos(320^\circ)+i\sin(320^\circ)\right]\approx\boxed{-3.8302+3.2139i}$
59. $3\left[\cos\left(\frac{11\pi}{12}\right)+i\sin\left(\frac{11\pi}{12}\right)\right]\approx\boxed{-2.8978+0.7765i}$
60. $2\left[\cos\left(\frac{4\pi}{7}\right)+i\sin\left(\frac{4\pi}{7}\right)\right]=\boxed{-0.4450+1.9499i}$
61. $-2\left[\cos\left(\frac{3\pi}{5}\right)+i\sin\left(\frac{3\pi}{5}\right)\right]\approx\boxed{0.6180-1.9021i}$

62. $-4\left[\cos\left(\frac{15\pi}{11}\right)+i\sin\left(\frac{15\pi}{11}\right)\right]\approx\boxed{1.6617+3.6385i}$
63. $-5\left[\cos\left(\frac{4\pi}{9}\right)+i\sin\left(\frac{4\pi}{9}\right)\right]\approx\boxed{-0.87-4.92i}$
64. $6\left[\cos\left(\frac{13\pi}{8}\right)+i\sin\left(\frac{13\pi}{8}\right)\right]\approx\boxed{2.30-5.54i}$

65. First, note that $m(\angle ABC)=180^\circ-28^\circ=152^\circ$.

a) Using the Law of Cosines, we see that

$$\overline{AC}^2=46^2+15^2-2(46)(15)\cos 152^\circ=3559.467678 \Rightarrow \overline{AC}=59.7\,mi.$$

b) Consider the following diagram:

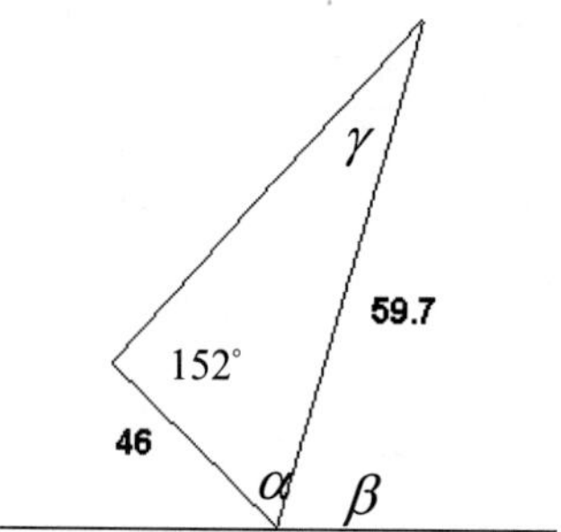

We must find α and β. First, we use the Law of Sines to find γ:

$$\frac{\sin 152^\circ}{59.7}=\frac{\sin\gamma}{46} \Rightarrow \sin\gamma=\frac{46\sin 152^\circ}{59.7} \Rightarrow \gamma=\sin^{-1}\left(\frac{46\sin 152^\circ}{59.7}\right)\approx 21.2^\circ$$

Thus, $\alpha=180^\circ-152^\circ-21.2^\circ=6.8^\circ$. Also, $\beta=90^\circ-6.8^\circ=83.2^\circ$. As such, the polar form of AC : $59.7\left[\cos 68.8^\circ+i\sin 68.8^\circ\right]$.

c.) $m(\angle BAC)=6.8^\circ$ (from part (b)).

66. First, note that $m(\angle ABC) = 124^\circ$.

a) Using the Law of Cosines, we see that

$$\overline{AC}^2 = 14.3^2 + 25^2 - 2(14.3)(25)\cos 124^\circ = 1229.312926 \Rightarrow \overline{AC} = 35mi.$$

b) Consider the following diagram:

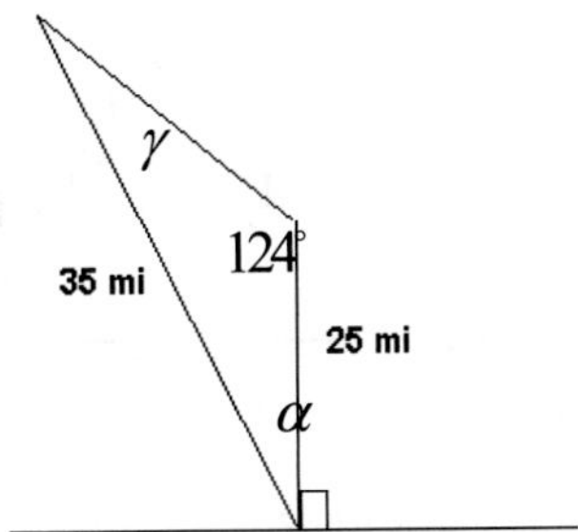

We must find α. First, we find γ using the Law of Sines:

$$\frac{\sin 124^\circ}{35} = \frac{\sin \gamma}{25} \Rightarrow \sin \gamma = \frac{25\sin 124^\circ}{35} \Rightarrow \gamma = \sin^{-1}\left(\frac{25\sin 124^\circ}{35}\right) \approx 36.3^\circ$$

Thus, $\alpha = 180^\circ - 36.2^\circ - 124^\circ = 19.8^\circ$. Hence, the direction angle for $\overline{AC}$ is $90^\circ + 19.8^\circ = 109.8^\circ$. As such, the polar form of AC: $35\left[\cos 126.3^\circ + i\sin 126.3^\circ\right]$.

c) $m(\angle BAC) = 19.8^\circ$ (from part (b)).

67. First, note that

Point	Polar Form
A	0 + 0i
B	0 – 2i
C	3 – 5i
D	3 – 6i

a)

$$\overrightarrow{AB} = B - A = -2i$$

$$\overrightarrow{BC} = C - B = 3 - 3i$$

$$\overrightarrow{CD} = D - C = -i$$

b) Observe that $\overrightarrow{AD} = D - A = 3 - 6i$.
In polar form:

$$r = \sqrt{3^2 + 6^2} = \sqrt{45} = 3\sqrt{5}$$

$$\tan\theta = -\tfrac{6}{3} = -2 \Rightarrow \theta = \tan^{-1}(-2) \approx -63.4^\circ$$

which corresponds to $360^\circ - 63.4^\circ \approx 297^\circ$.
Hence, the polar form of *AD* is:

$$3\sqrt{5}\left[\cos 297^\circ + i\sin 297^\circ\right]$$

68. First, note that

Point	Polar Form
A	0 + 0i
B	0 + 2i
C	3 + 7i
D	6 + 7i

a)

$$\overrightarrow{AB} = B - A = 2i$$
$$\overrightarrow{BC} = C - B = 3 + 5i$$
$$\overrightarrow{CD} = D - C = 3 + 0i$$

b) Observe that $\overrightarrow{AD} = D - A = 6 + 7i$.
In polar form:

$$r = \sqrt{6^2 + 7^2} = \sqrt{85}$$
$$\tan\theta = \tfrac{7}{6} \Rightarrow \theta = \tan^{-1}(\tfrac{7}{6}) \approx 49^\circ$$

Hence, the polar form of AD is:

$$\sqrt{85}\left[\cos 49^\circ + i\sin 49^\circ\right]$$

69. Use $z = |z|[\cos x + i\sin x]$, where $|z| = \sqrt{R^2 + I^2}$, $x = \arctan\left(\frac{I}{R}\right)$, with $R = 4, I = 6$ to obtain $|z| = \sqrt{4^2 + 6^2} = \sqrt{52} = 2\sqrt{13}$ and $x = \arctan\left(\frac{6}{4}\right) \approx 56.31^\circ$. Thus,

$$z = 2\sqrt{13}\left[\cos(56.31^\circ) + i\sin(56.31^\circ)\right].$$

70. Use $z = |z|[\cos x + i\sin x]$, where $|z| = \sqrt{R^2 + I^2}$, $x = \arctan\left(\frac{I}{R}\right)$, with $R = 7, I = -5$ to obtain $|z| = \sqrt{7^2 + (-5)^2} = \sqrt{74}$ and $x = \arctan\left(-\frac{5}{7}\right) \approx 324.46^\circ$. Thus,

$$z = \sqrt{74}\left[\cos(324.46^\circ) + i\sin(324.46^\circ)\right].$$

71. The point is in QIII, not QI. As such, you should add 180° to $\tan^{-1}\left(\frac{8}{3}\right)$.

72. The point is in QII, not QIV. As such, you should add 180° to $\tan^{-1}\left(-\frac{8}{3}\right)$.

73. True. All points that lie along the *x*-axis are of the form $a + 0i$. So, they must be real numbers.

74. True. All points that lie along the *y*-axis are of the form $0 + bi$. So, they must be (purely) imaginary numbers.

75. True. Let $z = a + bi$. Then, $\overline{z} = a - bi$. Observe that

$$|z| = \sqrt{a^2 + b^2} \text{ and } |\overline{z}| = \sqrt{a^2 + (-b)^2} = \sqrt{a^2 + b^2}.$$

76. False. Consider, for example, $z = i$. Then, $\overline{z} = -i$. Note that the argument of z is 90°, while the argument of $\overline{z}$ is 270°.

77. The argument of $z = a$, where a is a positive real number, is 0° since it lies on the positive *x*-axis.

78. The argument of $z = bi$, where b is a positive real number, is 90° since it lies on the positive *y*-axis.

79. The modulus of $z = bi$ is $\sqrt{b^2} = |b|$.

80. The modulus of $z = a$ is $\sqrt{a^2} = |a|$.

81. In order to express $a - 2ai,\ a > 0,$ in polar form, observe that $x = a,\ y = -2a$, so that the point is in QIV. Now,

$$r = \sqrt{x^2 + y^2} = \sqrt{(a)^2 + (-2a)^2} = a\sqrt{5}$$

$$\tan\theta = \frac{y}{x} = \frac{-2a}{a} = -2\text{, so that } \theta = \tan^{-1}(-2) \approx -63.43^\circ.$$

Since $0 \le \theta < 360^\circ$, we must use the reference angle 296.6°.

Hence, $\boxed{a - 2ai \approx a\sqrt{5}\left[\cos(296.6^\circ) + i\sin(296.6^\circ)\right]}$.

82. In order to express $-3a - 4ai,\ a > 0,$ in polar form, observe that $x = -3a,\ y = -4a$, so that the point is in QIII. Now,

$$r = \sqrt{x^2 + y^2} = \sqrt{(-3a)^2 + (-4a)^2} = 5a$$

$$\tan\theta = \frac{y}{x} = \frac{-4a}{-3a} = \frac{4}{3}\text{, so that } \theta = 180^\circ + \tan^{-1}\left(\frac{4}{3}\right) \approx 233.1^\circ.$$

(Remember to add 180° to $\tan^{-1}\left(\frac{4}{3}\right)$ since the angle has terminal side in QIII.)

Hence, $\boxed{-3a - 4ai \approx 5a\left[\cos(233.1^\circ) + i\sin(233.1^\circ)\right]}$.

83. First, we know that $|\vec{z}| = \pi$. Also,

$$\cos\left(\tfrac{\theta}{2}\right) = \tfrac{1}{2} \Rightarrow \tfrac{\theta}{2} = \tfrac{\pi}{3}, \tfrac{5\pi}{3} \Rightarrow \theta = \cancel{\tfrac{2\pi}{3}}, \tfrac{10\pi}{3}.$$

So, $\vec{z} = \pi\left(\cos\left(\tfrac{10\pi}{3}\right) + i\sin\left(\tfrac{10\pi}{3}\right)\right) = \boxed{-\tfrac{\pi}{2} - \tfrac{\pi\sqrt{3}}{2}i}$.

(Note: Although the complex numbers are the same, this doesn't satisfy the original conditions.)

84. First, we know that $|\vec{z}| = 2\sqrt{2}$. Also,

$$\sin\left(\tfrac{\theta}{2}\right) = -\tfrac{\sqrt{3}}{2} \Rightarrow \tfrac{\theta}{2} = \tfrac{4\pi}{3}, \tfrac{5\pi}{3} \Rightarrow \theta = \cancel{\tfrac{5\pi}{3}}, \tfrac{8\pi}{3}.$$

So, $\vec{z} = 2\sqrt{2}\left(\cos\left(\tfrac{8\pi}{3}\right) + i\sin\left(\tfrac{8\pi}{3}\right)\right) = \boxed{-\sqrt{2} + i\sqrt{6}}$.

(Note: Although the complex numbers are the same, this doesn't satisfy the original conditions.)

85. Consider the following diagram:

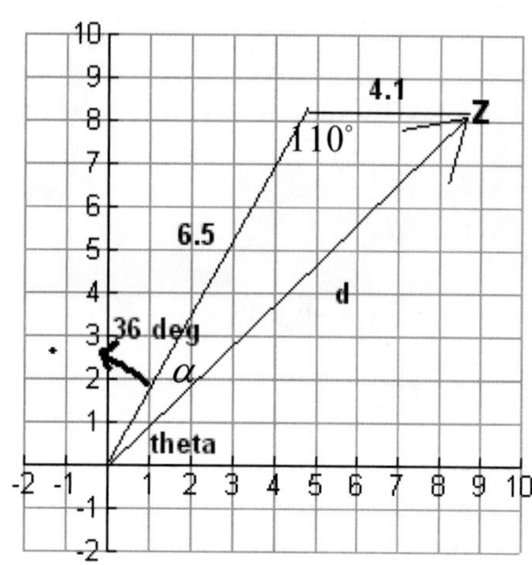

We need to find d and θ. Using the Law of Cosines, we obtain

$$d^2 = 6.5^2 + 4.1^2 - 2(6.5)(4.1)\cos 110^\circ$$
$$d \approx 8.79$$

Next, we find α using the Law of Sines:

$$\frac{\sin\alpha}{4.1} = \frac{\sin 110^\circ}{8.79} \Rightarrow \sin\alpha = \frac{4.1\sin 110^\circ}{8.79} \approx 26^\circ$$

Hence, $\theta \approx 90^\circ - (36^\circ + 26^\circ) = 28^\circ$.

Thus,

$$\vec{z} = 8.79(\cos 28^\circ + i\sin 28^\circ).$$

86. Consider the following diagram:

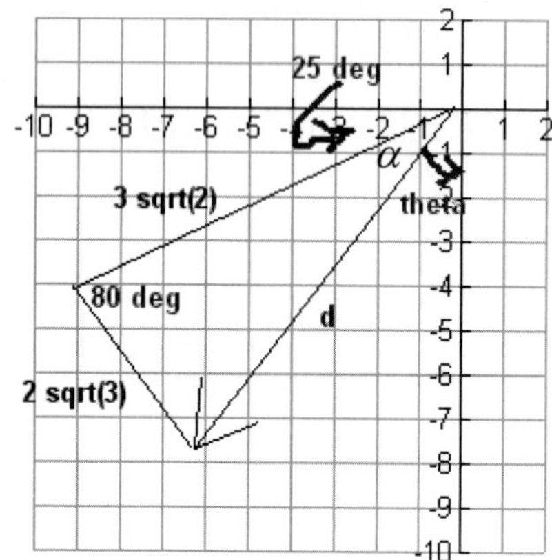

We need to find d and θ. Using the Law of Cosines, we obtain

$$d^2 = \left(2\sqrt{3}\right)^2 + \left(3\sqrt{2}\right)^2 - 2\left(2\sqrt{3}\right)\left(3\sqrt{2}\right)\cos 80^\circ$$
$$d \approx 4.99$$

Next, we find α using the Law of Sines:

$$\frac{\sin\alpha}{2\sqrt{3}} = \frac{\sin 80^\circ}{4.99} \Rightarrow \sin\alpha = \frac{2\sqrt{3}\sin 80^\circ}{4.99} \approx 43.13^\circ$$

Hence, $\theta \approx 180^\circ + 25^\circ + 43.13^\circ = 248.13^\circ$.
Thus,

$$\vec{z} = 4.99(\cos 248.13^\circ + i\sin 248.13^\circ).$$

87. Let $z = r[\cos\theta + i\sin\theta]$. Then, $-z$ is the reflection of z about the origin. So, the angle increases by π and r remains the same. So,

$$-z = r\left[\cos(\theta+\pi) + i\sin(\theta+\pi)\right].$$

88. Let $z = r[\cos\theta + i\sin\theta]$. Then, $\overline{z}$ is the reflection of z over the real axis. So, the angle becomes $2\pi - \theta$ and r remains the same. So,

$$\overline{z} = r\left[\cos(2\pi-\theta) + i\sin(2\pi-\theta)\right].$$

89. From the calculator, we obtain:

$$abs(1+i) \approx 1.41421 \text{ (actually} = \sqrt{2}\text{)} \quad angle(1+i) = 45^\circ$$

So, $1+i \approx 1.4142(\cos 45^\circ + i\sin 45^\circ)$.

90. From the calculator, we obtain:

$$abs(1-i) \approx 1.41421 \text{ (actually} = \sqrt{2}\text{)} \quad angle(1-i) = 315^\circ$$

So, $1-i \approx 1.4142(\cos 315^\circ + i\sin 315^\circ)$.

91. In order to express $2+i$ in polar form, observe that $x=2,\ y=1$, so that the point is in QI. Now,

$$r = \sqrt{x^2+y^2} = \sqrt{(2)^2+(1)^2} = \sqrt{5}$$

$$\tan\theta = \frac{y}{x} = \frac{1}{2}, \text{ so that } \theta = \tan^{-1}\left(\frac{1}{2}\right) \approx 26.57^\circ.$$

Hence, $\boxed{2+i \approx \sqrt{5}\left[\cos(26.57^\circ)+i\sin(26.57^\circ)\right]}$.

92. $3(\cos 45^\circ + i\sin 45^\circ) = 3\left(\frac{\sqrt{2}}{2}+\frac{\sqrt{2}}{2}i\right) = \boxed{\frac{3\sqrt{2}}{2}+\frac{3\sqrt{2}}{2}i}$

93. From the calculator, we obtain:

$$|28-21i| = 35$$

$$angle(28-21i) = -37^\circ \text{, so we use } 323^\circ$$

So, $28-21i \approx 35(\cos 323^\circ + i\sin 323^\circ)$.

94. From the calculator, we obtain:

$$\left|-\sqrt{21}+10i\right| = 11$$

$$angle\left(-\sqrt{21}+10i\right) = -65^\circ \text{, so we use } 295^\circ$$

So, $-\sqrt{21}+10i \approx 11(\cos 295^\circ + i\sin 295^\circ)$.

95. $x^2+y^2=25 \Rightarrow r^2=25 \Rightarrow r=5$

96.

$$x^2+y^2=4x \Rightarrow r^2=4r\cos\theta$$
$$\Rightarrow r=4\cos\theta$$

97.

$$y^2-2y=-x^2 \Rightarrow x^2+y^2=2y$$
$$\Rightarrow r^2=2r\sin\theta$$
$$\Rightarrow r=2\sin\theta$$

98.

$$\left(x^2+y^2\right)^2-16\left(x^2-y^2\right)=0$$
$$\left(r^2\right)^2-16\left(r^2\cos^2\theta-r^2\sin^2\theta\right)=0$$
$$r^2\left[r^2-16\cos 2\theta\right]=0$$
$$r^2=16\cos 2\theta$$

assuming that $r \neq 0$.

Section 7.4 Solutions

1. Using the formula $z_1z_2 = r_1r_2\left[\cos(\theta_1+\theta_2)+i\sin(\theta_1+\theta_2)\right]$ yields

$$z_1z_2 = (4)(3)\left[\cos(40^\circ+80^\circ)+i\sin(40^\circ+80^\circ)\right]$$
$$= 12\left[\cos(120^\circ)+i\sin(120^\circ)\right] = 12\left[-\frac{1}{2}+\frac{\sqrt{3}}{2}i\right] = \boxed{-6+6\sqrt{3}i}$$

2. Using the formula $z_1z_2 = r_1r_2\left[\cos(\theta_1+\theta_2)+i\sin(\theta_1+\theta_2)\right]$ yields

$$z_1z_2 = (2)(5)\left[\cos(100^\circ+50^\circ)+i\sin(100^\circ+50^\circ)\right]$$
$$= 10\left[\cos(150^\circ)+i\sin(150^\circ)\right] = 10\left[-\frac{\sqrt{3}}{2}+\frac{1}{2}i\right] = \boxed{-5\sqrt{3}+5i}$$

3. Using the formula $z_1z_2 = r_1r_2\left[\cos(\theta_1+\theta_2)+i\sin(\theta_1+\theta_2)\right]$ yields

$$z_1z_2 = (4)(2)\left[\cos(80^\circ+145^\circ)+i\sin(80^\circ+145^\circ)\right]$$
$$= 8\left[\cos(225^\circ)+i\sin(225^\circ)\right] = 8\left[-\frac{\sqrt{2}}{2}-\frac{\sqrt{2}}{2}i\right] = \boxed{-4\sqrt{2}-4\sqrt{2}i}$$

4. Using the formula $z_1z_2 = r_1r_2\left[\cos(\theta_1+\theta_2)+i\sin(\theta_1+\theta_2)\right]$ yields

$$z_1z_2 = (3)(4)\left[\cos(130^\circ+170^\circ)+i\sin(130^\circ+170^\circ)\right]$$
$$= 12\left[\cos(300^\circ)+i\sin(300^\circ)\right] = 12\left[\frac{1}{2}-\frac{\sqrt{3}}{2}i\right] = \boxed{6-6\sqrt{3}i}$$

5. Using the formula $z_1z_2 = r_1r_2\left[\cos(\theta_1+\theta_2)+i\sin(\theta_1+\theta_2)\right]$ yields

$$z_1z_2 = (2)(4)\left[\cos(10^\circ+80^\circ)+i\sin(10^\circ+80^\circ)\right]$$
$$= 8\left[\cos(90^\circ)+i\sin(90^\circ)\right] = 8[0+i] = \boxed{0+8i}$$

6. Using the formula $z_1z_2 = r_1r_2\left[\cos(\theta_1+\theta_2)+i\sin(\theta_1+\theta_2)\right]$ yields

$$z_1z_2 = (3)(5)\left[\cos(190^\circ+80^\circ)+i\sin(190^\circ+80^\circ)\right]$$
$$= 15\left[\cos(270^\circ)+i\sin(270^\circ)\right] = 15[0-i] = \boxed{0-15i}$$

7. Using the formula $z_1z_2 = r_1r_2\left[\cos(\theta_1+\theta_2)+i\sin(\theta_1+\theta_2)\right]$ yields $$z_1z_2 = \left(\sqrt{3}\right)\left(\sqrt{27}\right)\left[\cos\left(\frac{\pi}{12}+\frac{\pi}{6}\right)+i\sin\left(\frac{\pi}{12}+\frac{\pi}{6}\right)\right]$$ $$=9\left[\cos\left(\frac{\pi}{4}\right)+i\sin\left(\frac{\pi}{4}\right)\right]=9\left[\frac{\sqrt{2}}{2}+\frac{\sqrt{2}}{2}i\right]=\boxed{\frac{9\sqrt{2}}{2}+\frac{9\sqrt{2}}{2}i}$$
8. Using the formula $z_1z_2 = r_1r_2\left[\cos(\theta_1+\theta_2)+i\sin(\theta_1+\theta_2)\right]$ yields $$z_1z_2 = \left(\sqrt{5}\right)\left(\sqrt{5}\right)\left[\cos\left(\frac{\pi}{15}+\frac{4\pi}{15}\right)+i\sin\left(\frac{\pi}{15}+\frac{4\pi}{15}\right)\right]$$ $$=5\left[\cos\left(\frac{\pi}{3}\right)+i\sin\left(\frac{\pi}{3}\right)\right]=5\left[\frac{1}{2}+\frac{\sqrt{3}}{2}i\right]=\boxed{\frac{5}{2}+\frac{5\sqrt{3}}{2}i}$$
9. Using the formula $z_1z_2 = r_1r_2\left[\cos(\theta_1+\theta_2)+i\sin(\theta_1+\theta_2)\right]$ yields $$z_1z_2 = (4)(3)\left[\cos\left(\frac{3\pi}{8}+\frac{\pi}{8}\right)+i\sin\left(\frac{3\pi}{8}+\frac{\pi}{8}\right)\right]$$ $$=12\left[\cos\left(\frac{\pi}{2}\right)+i\sin\left(\frac{\pi}{2}\right)\right]=12[0+i]=\boxed{0+12i}$$
10. Using the formula $z_1z_2 = r_1r_2\left[\cos(\theta_1+\theta_2)+i\sin(\theta_1+\theta_2)\right]$ yields $$z_1z_2 = (6)(5)\left[\cos\left(\frac{2\pi}{9}+\frac{\pi}{9}\right)+i\sin\left(\frac{2\pi}{9}+\frac{\pi}{9}\right)\right]$$ $$=30\left[\cos\left(\frac{\pi}{3}\right)+i\sin\left(\frac{\pi}{3}\right)\right]=30\left[\frac{1}{2}+\frac{\sqrt{3}}{2}i\right]=\boxed{15+15\sqrt{3}i}$$
11. Using the formula $\frac{z_1}{z_2}=\frac{r_1}{r_2}\left[\cos(\theta_1-\theta_2)+i\sin(\theta_1-\theta_2)\right]$ yields $$\frac{z_1}{z_2}=\frac{6}{2}\left[\cos\left(100^\circ-40^\circ\right)+i\sin\left(100^\circ-40^\circ\right)\right]$$ $$=3\left[\cos\left(60^\circ\right)+i\sin\left(60^\circ\right)\right]=3\left[\frac{1}{2}+\frac{\sqrt{3}}{2}i\right]=\boxed{\frac{3}{2}+\frac{3\sqrt{3}}{2}i}$$

12. Using the formula $\frac{z_1}{z_2}=\frac{r_1}{r_2}\left[\cos(\theta_1-\theta_2)+i\sin(\theta_1-\theta_2)\right]$ yields $\frac{z_1}{z_2}=\frac{8}{2}\left[\cos\left(80^\circ-35^\circ\right)+i\sin\left(80^\circ-35^\circ\right)\right]$ $=4\left[\cos\left(45^\circ\right)+i\sin\left(45^\circ\right)\right]=4\left[\frac{\sqrt{2}}{2}+\frac{\sqrt{2}}{2}i\right]=\boxed{2\sqrt{2}+2\sqrt{2}i}$
13. Using the formula $\frac{z_1}{z_2}=\frac{r_1}{r_2}\left[\cos(\theta_1-\theta_2)+i\sin(\theta_1-\theta_2)\right]$ yields $\frac{z_1}{z_2}=\frac{10}{5}\left[\cos\left(200^\circ-65^\circ\right)+i\sin\left(200^\circ-65^\circ\right)\right]$ $=2\left[\cos\left(135^\circ\right)+i\sin\left(135^\circ\right)\right]=2\left[-\frac{\sqrt{2}}{2}+\frac{\sqrt{2}}{2}i\right]=\boxed{-\sqrt{2}+\sqrt{2}i}$
14. Using the formula $\frac{z_1}{z_2}=\frac{r_1}{r_2}\left[\cos(\theta_1-\theta_2)+i\sin(\theta_1-\theta_2)\right]$ yields $\frac{z_1}{z_2}=\frac{4}{4}\left[\cos\left(280^\circ-55^\circ\right)+i\sin\left(280^\circ-55^\circ\right)\right]$ $=\left[\cos\left(225^\circ\right)+i\sin\left(225^\circ\right)\right]=\boxed{-\frac{\sqrt{2}}{2}-\frac{\sqrt{2}}{2}i}$
15. Using the formula $\frac{z_1}{z_2}=\frac{r_1}{r_2}\left[\cos(\theta_1-\theta_2)+i\sin(\theta_1-\theta_2)\right]$ yields $\frac{z_1}{z_2}=\frac{\sqrt{12}}{\sqrt{3}}\left[\cos\left(350^\circ-80^\circ\right)+i\sin\left(350^\circ-80^\circ\right)\right]$ $=2\left[\cos\left(270^\circ\right)+i\sin\left(270^\circ\right)\right]=\boxed{0-2i}$
16. Using the formula $\frac{z_1}{z_2}=\frac{r_1}{r_2}\left[\cos(\theta_1-\theta_2)+i\sin(\theta_1-\theta_2)\right]$ yields $\frac{z_1}{z_2}=\frac{\sqrt{40}}{\sqrt{10}}\left[\cos\left(110^\circ-20^\circ\right)+i\sin\left(110^\circ-20^\circ\right)\right]$ $=2\left[\cos\left(90^\circ\right)+i\sin\left(90^\circ\right)\right]=\boxed{0+2i}$

17. Using the formula $\frac{z_1}{z_2}=\frac{r_1}{r_2}\left[\cos(\theta_1-\theta_2)+i\sin(\theta_1-\theta_2)\right]$ yields

$$\frac{z_1}{z_2}=\frac{9}{3}\left[\cos\left(\frac{5\pi}{12}-\frac{\pi}{12}\right)+i\sin\left(\frac{5\pi}{12}-\frac{\pi}{12}\right)\right]$$

$$=3\left[\cos\left(\frac{\pi}{3}\right)+i\sin\left(\frac{\pi}{3}\right)\right]=3\left[\frac{1}{2}+\frac{\sqrt{3}}{2}i\right]=\boxed{\frac{3}{2}+\frac{3\sqrt{3}}{2}i}$$

18. Using the formula $\frac{z_1}{z_2}=\frac{r_1}{r_2}\left[\cos(\theta_1-\theta_2)+i\sin(\theta_1-\theta_2)\right]$ yields

$$\frac{z_1}{z_2}=\frac{8}{4}\left[\cos\left(\frac{5\pi}{8}-\frac{3\pi}{8}\right)+i\sin\left(\frac{5\pi}{8}-\frac{3\pi}{8}\right)\right]$$

$$=2\left[\cos\left(\frac{\pi}{4}\right)+i\sin\left(\frac{\pi}{4}\right)\right]=2\left[\frac{\sqrt{2}}{2}+\frac{\sqrt{2}}{2}i\right]=\boxed{\sqrt{2}+\sqrt{2}i}$$

19. Using the formula $\frac{z_1}{z_2}=\frac{r_1}{r_2}\left[\cos(\theta_1-\theta_2)+i\sin(\theta_1-\theta_2)\right]$ yields

$$\frac{z_1}{z_2}=\frac{45}{9}\left[\cos\left(\frac{22\pi}{15}-\frac{2\pi}{15}\right)+i\sin\left(\frac{22\pi}{15}-\frac{2\pi}{15}\right)\right]$$

$$=5\left[\cos\left(\frac{4\pi}{3}\right)+i\sin\left(\frac{4\pi}{3}\right)\right]=5\left[-\frac{1}{2}-\frac{\sqrt{3}}{2}i\right]=\boxed{-\frac{5}{2}-\frac{5\sqrt{3}}{2}i}$$

20. Using the formula $\frac{z_1}{z_2}=\frac{r_1}{r_2}\left[\cos(\theta_1-\theta_2)+i\sin(\theta_1-\theta_2)\right]$ yields

$$\frac{z_1}{z_2}=\frac{22}{11}\left[\cos\left(\frac{11\pi}{18}-\frac{5\pi}{18}\right)+i\sin\left(\frac{11\pi}{18}-\frac{5\pi}{18}\right)\right]=2\left[\cos\left(\frac{\pi}{3}\right)+i\sin\left(\frac{\pi}{3}\right)\right]=2\left[\frac{1}{2}+\frac{\sqrt{3}}{2}i\right]=\boxed{1+\sqrt{3}i}$$

21. In order to express $(-1+i)^5$ in rectangular form, follow these steps:

Step 1: Write $-1+i$ in polar form.
Since $x=-1,\ y=1$, the point is in QII. So,

$$r=\sqrt{x^2+y^2}=\sqrt{(-1)^2+(1)^2}=\sqrt{2}$$

$$\tan\theta=\frac{y}{x}=\frac{1}{-1}=-1,\text{ so that } \theta=\pi+\tan^{-1}(-1)=\pi-\frac{\pi}{4}=\frac{3\pi}{4}.$$

(Remember to add π to $\tan^{-1}(-1)$ since the angle has terminal side in QII.)

Hence, $-1+i=\sqrt{2}\left[\cos\left(\frac{3\pi}{4}\right)+i\sin\left(\frac{3\pi}{4}\right)\right]$.

Step 2: Apply DeMoivre's theorem $z^n=r^n\left[\cos(n\theta)+i\sin(n\theta)\right]$.

$$(-1+i)^5=\left(\sqrt{2}\right)^5\left[\cos\left(5\cdot\frac{3\pi}{4}\right)+i\sin\left(5\cdot\frac{3\pi}{4}\right)\right]=4\sqrt{2}\left[\frac{\sqrt{2}}{2}-\frac{\sqrt{2}}{2}i\right]=\boxed{4-4i}$$

22. In order to express $(1-i)^4$ in rectangular form, follow these steps:

Step 1: Write $1-i$ in polar form.
Since $x=1,\ y=-1$, the point is in QIV. So,

$$r=\sqrt{x^2+y^2}=\sqrt{(1)^2+(-1)^2}=\sqrt{2}$$

$$\tan\theta=\frac{y}{x}=\frac{-1}{1}=-1,\text{ so that } \theta=\tan^{-1}(-1)=-\frac{\pi}{4}.$$

Since $0\le\theta<2\pi$, we must use the reference angle $\frac{7\pi}{4}$.

Hence, $1-i=\sqrt{2}\left[\cos\left(\frac{7\pi}{4}\right)+i\sin\left(\frac{7\pi}{4}\right)\right]$.

Step 2: Apply DeMoivre's theorem $z^n=r^n\left[\cos(n\theta)+i\sin(n\theta)\right]$.

$$(1-i)^4=\left(\sqrt{2}\right)^4\left[\cos\left(4\cdot\frac{7\pi}{4}\right)+i\sin\left(4\cdot\frac{7\pi}{4}\right)\right]=4\left[\cos(7\pi)+i\sin(7\pi)\right]$$

$$=4\left[-1+0i\right]=\boxed{-4+0i}$$

23. In order to express $\left(-\sqrt{3}+i\right)^6$ in rectangular form, follow these steps:

Step 1: Write $-\sqrt{3}+i$ in polar form.

Since $x=-\sqrt{3},\ y=1$, the point is in QII. So, $r=\sqrt{x^2+y^2}=\sqrt{\left(-\sqrt{3}\right)^2+(1)^2}=2$

$$\tan\theta=\frac{y}{x}=\frac{-3}{\sqrt{3}}=-\frac{1}{\sqrt{3}},\ \text{so that}\ \theta=\pi+\tan^{-1}\left(-\frac{1}{\sqrt{3}}\right)=\pi-\frac{\pi}{6}=\frac{5\pi}{6}.$$

(Remember to add π to $\tan^{-1}\left(-\frac{1}{\sqrt{3}}\right)$ since the angle has terminal side in QII.)

Hence, $-\sqrt{3}+i=2\left[\cos\left(\frac{5\pi}{6}\right)+i\sin\left(\frac{5\pi}{6}\right)\right]$.

Step 2: Apply DeMoivre's theorem $z^n=r^n\left[\cos(n\theta)+i\sin(n\theta)\right]$.

$\left(-\sqrt{3}+i\right)^6=(2)^6\left[\cos\left(6\cdot\frac{5\pi}{6}\right)+i\sin\left(6\cdot\frac{5\pi}{6}\right)\right]=64\left[\cos(5\pi)+i\sin(5\pi)\right]=64\left[-1+0i\right]=\boxed{-64+0i}$

24. In order to express $\left(\sqrt{3}-i\right)^8$ in rectangular form, follow these steps:

Step 1: Write $\sqrt{3}-i$ in polar form.

Since $x=\sqrt{3},\ y=-1$, the point is in QIV. So, $r=\sqrt{x^2+y^2}=\sqrt{\left(\sqrt{3}\right)^2+(-1)^2}=2$

$$\tan\theta=\frac{y}{x}=\frac{-1}{\sqrt{3}},\ \text{so that}\ \theta=\tan^{-1}\left(-\frac{1}{\sqrt{3}}\right)=-\frac{\pi}{6}.$$

Since $0\le\theta<2\pi$, we must use the reference angle $\frac{11\pi}{6}$.

Hence, $\sqrt{3}-i=2\left[\cos\left(\frac{11\pi}{6}\right)+i\sin\left(\frac{11\pi}{6}\right)\right]$.

Step 2: Apply DeMoivre's theorem $z^n=r^n\left[\cos(n\theta)+i\sin(n\theta)\right]$.

$$\left(\sqrt{3}-i\right)^8=2^8\left[\cos\left(8\cdot\frac{11\pi}{6}\right)+i\sin\left(8\cdot\frac{11\pi}{6}\right)\right]$$

In order to simplify this further, observe that $\frac{88\pi}{6}=\frac{44\pi}{3}=14\pi+\frac{2}{3}\pi$ and so,

$\cos\left(\frac{88\pi}{6}\right)=\cos\left(14\pi+\frac{2}{3}\pi\right)=\cos\left(\frac{2}{3}\pi\right)=-\frac{1}{2}\qquad \sin\left(\frac{88\pi}{6}\right)=\sin\left(14\pi+\frac{2}{3}\pi\right)=\sin\left(\frac{2}{3}\pi\right)=\frac{\sqrt{3}}{2}$

Thus, $\left(\sqrt{3}-i\right)^8=2^8\left[-\frac{1}{2}+\frac{\sqrt{3}}{2}i\right]=\boxed{-128+128\sqrt{3}i}$.

25. In order to express $\left(1-\sqrt{3}i\right)^4$ in rectangular form, follow these steps:

<u>Step 1</u>: Write $1-\sqrt{3}i$ in polar form.
Since $x=1,\ y=-\sqrt{3}$, the point is in QIV. So,

$$r=\sqrt{x^2+y^2}=\sqrt{(1)^2+\left(-\sqrt{3}\right)^2}=2$$

$$\tan\theta=\frac{y}{x}=\frac{-\sqrt{3}}{1},\ \text{so that}\ \theta=\tan^{-1}\left(-\frac{\sqrt{3}}{1}\right)=-\frac{\pi}{3}.$$

Since $0\le\theta<2\pi$, we must use the reference angle $\frac{5\pi}{3}$.

Hence, $1-\sqrt{3}i=2\left[\cos\left(\frac{5\pi}{3}\right)+i\sin\left(\frac{5\pi}{3}\right)\right]$.

<u>Step 2</u>: Apply DeMoivre's theorem $z^n=r^n\left[\cos(n\theta)+i\sin(n\theta)\right]$.

$$\left(1-\sqrt{3}i\right)^4=2^4\left[\cos\left(4\cdot\frac{5\pi}{3}\right)+i\sin\left(4\cdot\frac{5\pi}{3}\right)\right]$$

In order to simplify this further, observe that $\frac{20\pi}{3}=6\pi+\frac{2}{3}\pi$ and so,

$$\cos\left(\frac{20\pi}{3}\right)=\cos\left(6\pi+\frac{2}{3}\pi\right)=\cos\left(\frac{2}{3}\pi\right)=-\frac{1}{2}$$

$$\sin\left(\frac{20\pi}{3}\right)=\sin\left(6\pi+\frac{2}{3}\pi\right)=\sin\left(\frac{2}{3}\pi\right)=\frac{\sqrt{3}}{2}$$

Thus, $\left(1-\sqrt{3}i\right)^4=2^4\left[-\frac{1}{2}+\frac{\sqrt{3}}{2}i\right]=\boxed{-8+8\sqrt{3}i}$.

26. In order to express $\left(-1+\sqrt{3}i\right)^5$ in rectangular form, follow these steps:

Step 1: Write $-1+\sqrt{3}i$ in polar form.
Since $x=-1,\ y=\sqrt{3}$, the point is in QII. So,

$$r=\sqrt{x^2+y^2}=\sqrt{(-1)^2+\left(\sqrt{3}\right)^2}=2$$

$$\tan\theta=\frac{y}{x}=\frac{-\sqrt{3}}{1}, \text{ so that } \theta=\pi+\tan^{-1}\left(-\sqrt{3}\right)=\pi-\frac{\pi}{3}=\frac{2\pi}{3}.$$

(Remember to add π to $\tan^{-1}\left(-\sqrt{3}\right)$ since the angle has terminal side in QII.)

Hence, $-1+\sqrt{3}i=2\left[\cos\left(\frac{2\pi}{3}\right)+i\sin\left(\frac{2\pi}{3}\right)\right]$.

Step 2: Apply DeMoivre's theorem $z^n=r^n\left[\cos(n\theta)+i\sin(n\theta)\right]$.

$$\left(-1+\sqrt{3}i\right)^5=(2)^5\left[\cos\left(5\cdot\frac{2\pi}{3}\right)+i\sin\left(5\cdot\frac{2\pi}{3}\right)\right]=32\left[\cos\left(\frac{10\pi}{3}\right)+i\sin\left(\frac{10\pi}{3}\right)\right]$$

$$=32\left[-\frac{1}{2}-\frac{\sqrt{3}}{2}i\right]=\boxed{-16-16\sqrt{3}i}$$

27. In order to express $(4-4i)^8$ in rectangular form, follow these steps:

Step 1: Write $4-4i$ in polar form.
Since $x=4,\ y=-4$, the point is in QIV. So,

$$r=\sqrt{x^2+y^2}=\sqrt{(4)^2+(-4)^2}=4\sqrt{2}$$

$$\tan\theta=\frac{y}{x}=\frac{-4}{4}=-1, \text{ so that } \theta=\tan^{-1}(-1)=-\frac{\pi}{4}.$$

Since $0\le\theta<2\pi$, we must use the reference angle $\frac{7\pi}{4}$.

Hence, $4-4i=4\sqrt{2}\left[\cos\left(\frac{7\pi}{4}\right)+i\sin\left(\frac{7\pi}{4}\right)\right]$.

Step 2: Apply DeMoivre's theorem $z^n=r^n\left[\cos(n\theta)+i\sin(n\theta)\right]$.

$$(4-4i)^8=\left(4\sqrt{2}\right)^8\left[\cos\left(8\cdot\frac{7\pi}{4}\right)+i\sin\left(8\cdot\frac{7\pi}{4}\right)\right]=1{,}048{,}576\left[\cos(14\pi)+i\sin(14\pi)\right]$$

In order to simplify this further, observe that $14\pi=7(2\pi)$ and so,

$$\cos(14\pi)=1 \quad \sin(14\pi)=0$$

Thus, $(4-4i)^8=1{,}048{,}576[1+0i]=\boxed{1{,}048{,}576+0i}$.

28. In order to express $(-3+3i)^{10}$ in rectangular form, follow these steps:

Step 1: Write $-3+3i$ in polar form.
Since $x=-3,\ y=3$, the point is in QII. So,

$$r=\sqrt{x^2+y^2}=\sqrt{(-3)^2+(3)^2}=3\sqrt{2}$$

$$\tan\theta=\frac{y}{x}=\frac{3}{-3}=-1,\text{ so that }\theta=\pi+\tan^{-1}(-1)=\pi-\frac{\pi}{4}=\frac{3\pi}{4}.$$

(Remember to add π to $\tan^{-1}(-1)$ since the angle has terminal side in QII.)

Hence, $-3+3i=3\sqrt{2}\left[\cos\left(\frac{3\pi}{4}\right)+i\sin\left(\frac{3\pi}{4}\right)\right]$.

Step 2: Apply DeMoivre's theorem $z^n=r^n\left[\cos(n\theta)+i\sin(n\theta)\right]$.

$$(-3+3i)^{10}=\left(3\sqrt{2}\right)^{10}\left[\cos\left(10\cdot\frac{3\pi}{4}\right)+i\sin\left(10\cdot\frac{3\pi}{4}\right)\right]=1,889,568\left[\cos\left(\frac{15\pi}{2}\right)+i\sin\left(\frac{15\pi}{2}\right)\right]$$

In order to simplify this further, observe that $\frac{15\pi}{2}=6\pi+\frac{3}{2}\pi$ and so,

$$\cos\left(\frac{15\pi}{2}\right)=\cos\left(6\pi+\frac{3}{2}\pi\right)=\cos\left(\frac{3}{2}\pi\right)=0$$

$$\sin\left(\frac{15\pi}{2}\right)=\sin\left(6\pi+\frac{3}{2}\pi\right)=\sin\left(\frac{3}{2}\pi\right)=-1$$

Thus, $(-3+3i)^{10}=1,889,568[0-i]=\boxed{0-1,889,568i}$

29. In order to express $\left(4\sqrt{3}+4i\right)^7$ in rectangular form, follow these steps:

Step 1: Write $4\sqrt{3}+4i$ in polar form.
Since $x=4\sqrt{3},\ y=4$, the point is in QII. So,

$$r=\sqrt{x^2+y^2}=\sqrt{\left(4\sqrt{3}\right)^2+(4)^2}=8$$

$$\tan\theta=\frac{y}{x}=\frac{4}{4\sqrt{3}}=\frac{1}{\sqrt{3}},\text{ so that }\theta=\tan^{-1}\left(\frac{1}{\sqrt{3}}\right)=\frac{\pi}{6}.$$

Hence, $4\sqrt{3}+4i=8\left[\cos\left(\frac{\pi}{6}\right)+i\sin\left(\frac{\pi}{6}\right)\right]$.

Step 2: Apply DeMoivre's theorem $z^n=r^n\left[\cos(n\theta)+i\sin(n\theta)\right]$.

$$\left(4\sqrt{3}+4i\right)^7=8^7\left[\cos\left(7\cdot\frac{\pi}{6}\right)+i\sin\left(7\cdot\frac{\pi}{6}\right)\right]=2,097,152\left[-\frac{\sqrt{3}}{2}-\frac{1}{2}i\right]$$

$$=\boxed{-1,048,576\sqrt{3}-1,048,576i}$$

30. In order to express $\left(-5+5\sqrt{3}i\right)^7$ in rectangular form, follow these steps:

Step 1: Write $-5+5\sqrt{3}i$ in polar form.
Since $x=-5,\ y=5\sqrt{3}$, the point is in QII. So,

$$r=\sqrt{x^2+y^2}=\sqrt{(-5)^2+\left(5\sqrt{3}\right)^2}=10$$

$$\tan\theta=\frac{y}{x}=\frac{5\sqrt{3}}{-5}=-\sqrt{3}\text{, so that }\theta=\pi+\tan^{-1}\left(-\sqrt{3}\right)=\pi-\frac{\pi}{3}=\frac{2\pi}{3}.$$

(Remember to add π to $\tan^{-1}\left(-\sqrt{3}\right)$ since the angle has terminal side in QII.)

Hence, $-5+5\sqrt{3}i=10\left[\cos\left(\frac{2\pi}{3}\right)+i\sin\left(\frac{2\pi}{3}\right)\right]$.

Step 2: Apply DeMoivre's theorem $z^n=r^n\left[\cos(n\theta)+i\sin(n\theta)\right]$.

$$\left(-5+5\sqrt{3}i\right)^7=10^7\left[\cos\left(7\cdot\frac{2\pi}{3}\right)+i\sin\left(7\cdot\frac{2\pi}{3}\right)\right]$$

In order to simplify this further, observe that $\frac{14\pi}{3}=4\pi+\frac{2}{3}\pi$ and so,

$$\cos\left(\frac{14\pi}{3}\right)=\cos\left(4\pi+\frac{2}{3}\pi\right)=\cos\left(\frac{2}{3}\pi\right)=-\frac{1}{2}$$

$$\sin\left(\frac{14\pi}{3}\right)=\sin\left(4\pi+\frac{2}{3}\pi\right)=\sin\left(\frac{2}{3}\pi\right)=\frac{\sqrt{3}}{2}$$

Thus, $\left(-5+5\sqrt{3}i\right)^7=10^7\left[-\frac{1}{2}+\frac{\sqrt{3}}{2}i\right]=\boxed{-5,000,000+5,000,000\sqrt{3}i}$

31. Given $z = 2 - 2\sqrt{3}i$, in order to compute $z^{1/2}$, follow these steps:

Step 1: Write $2 - 2\sqrt{3}i$ in polar form.
Since $x = 2,\ y = -2\sqrt{3}$, the point is in QIV. So,

$$r = \sqrt{x^2 + y^2} = \sqrt{(2)^2 + \left(-2\sqrt{3}\right)^2} = 4$$

$$\tan\theta = \frac{y}{x} = \frac{-2\sqrt{3}}{2} = -\sqrt{3}\text{, so that } \theta = \tan^{-1}\left(-\sqrt{3}\right) = -60^\circ.$$

Since $0 \le \theta < 360^\circ$, we must use the reference angle 300°.
Hence, $z = 2 - 2\sqrt{3}i = 4\left[\cos\left(300^\circ\right) + i\sin\left(300^\circ\right)\right]$.

Step 2: Now, apply $z^{1/n} = r^{1/n}\left[\cos\left(\frac{\theta}{n} + \frac{360^\circ k}{n}\right) + i\sin\left(\frac{\theta}{n} + \frac{360^\circ k}{n}\right)\right],\quad k = 0, 1, \ldots, n-1$:

$$z^{1/2} = \sqrt{4}\left[\cos\left(\frac{300^\circ}{2} + \frac{360^\circ k}{2}\right) + i\sin\left(\frac{300^\circ}{2} + \frac{360^\circ k}{2}\right)\right],\quad k = 0, 1$$

$$= \boxed{2\left[\cos\left(150^\circ\right) + i\sin\left(150^\circ\right)\right],\ 2\left[\cos\left(330^\circ\right) + i\sin\left(330^\circ\right)\right]}$$

Step 3: Plot the roots in the complex plane:

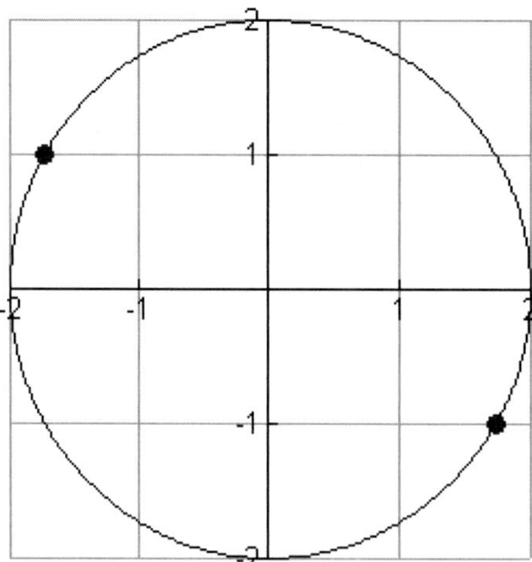

32. Given $z = 2 + 2\sqrt{3}i$, in order to compute $z^{1/2}$, follow these steps:

Step 1: Write $2 + 2\sqrt{3}i$ in polar form.

Since $x = 2,\ y = 2\sqrt{3}$, the point is in QI. So,

$$r = \sqrt{x^2 + y^2} = \sqrt{(2)^2 + \left(2\sqrt{3}\right)^2} = 4$$

$$\tan\theta = \frac{y}{x} = \frac{2\sqrt{3}}{2} = \sqrt{3}\text{, so that } \theta = \tan^{-1}\left(\sqrt{3}\right) = 60^\circ.$$

Hence, $z = 2 + 2\sqrt{3}i = 4\left[\cos\left(60^\circ\right) + i\sin\left(60^\circ\right)\right]$.

Step 2: Now, apply $z^{1/n} = r^{1/n}\left[\cos\left(\frac{\theta}{n} + \frac{360^\circ k}{n}\right) + i\sin\left(\frac{\theta}{n} + \frac{360^\circ k}{n}\right)\right],\quad k = 0, 1, \ldots, n-1$:

$$z^{1/2} = \sqrt{4}\left[\cos\left(\frac{60^\circ}{2} + \frac{360^\circ k}{2}\right) + i\sin\left(\frac{60^\circ}{2} + \frac{360^\circ k}{2}\right)\right],\quad k = 0, 1$$

$$= \boxed{2\left[\cos\left(30^\circ\right) + i\sin\left(30^\circ\right)\right],\ 2\left[\cos\left(210^\circ\right) + i\sin\left(210^\circ\right)\right]}$$

Step 3: Plot the roots in the complex plane:

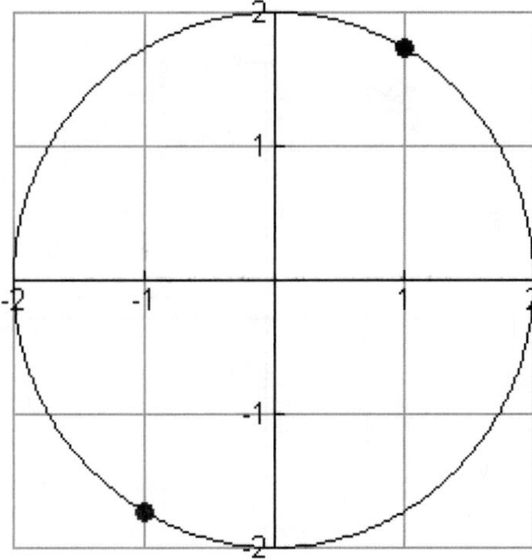

33. Given $z=\sqrt{18}-\sqrt{18}i$, in order to compute $z^{1/2}$, follow these steps:

Step 1: Write $z=\sqrt{18}-\sqrt{18}i$ in polar form.
Since $x=\sqrt{18},\ y=-\sqrt{18}$, the point is in QIV. So,

$$r=\sqrt{x^2+y^2}=\sqrt{\left(\sqrt{18}\right)^2+\left(-\sqrt{18}\right)^2}=6$$

$$\tan\theta=\frac{y}{x}=\frac{-\sqrt{18}}{\sqrt{18}}=-1,\text{ so that }\theta=\tan^{-1}(-1)=-45^\circ.$$

Since $0\le\theta<360^\circ$, we must use the reference angle 315°.
Hence, $z=\sqrt{18}-\sqrt{18}i=6\left[\cos\left(315^\circ\right)+i\sin\left(315^\circ\right)\right]$.

Step 2: Now, apply $z^{1/n}=r^{1/n}\left[\cos\left(\frac{\theta}{n}+\frac{360^\circ k}{n}\right)+i\sin\left(\frac{\theta}{n}+\frac{360^\circ k}{n}\right)\right],\quad k=0,1,\ldots,n-1$:

$$z^{1/2}=\sqrt{6}\left[\cos\left(\frac{315^\circ}{2}+\frac{360^\circ k}{2}\right)+i\sin\left(\frac{315^\circ}{2}+\frac{360^\circ k}{2}\right)\right],\quad k=0,1$$

$$=\boxed{\sqrt{6}\left[\cos\left(157.5^\circ\right)+i\sin\left(157.5^\circ\right)\right],\ \sqrt{6}\left[\cos\left(337.5^\circ\right)+i\sin\left(337.5^\circ\right)\right]}$$

Step 3: Plot the roots in the complex plane:

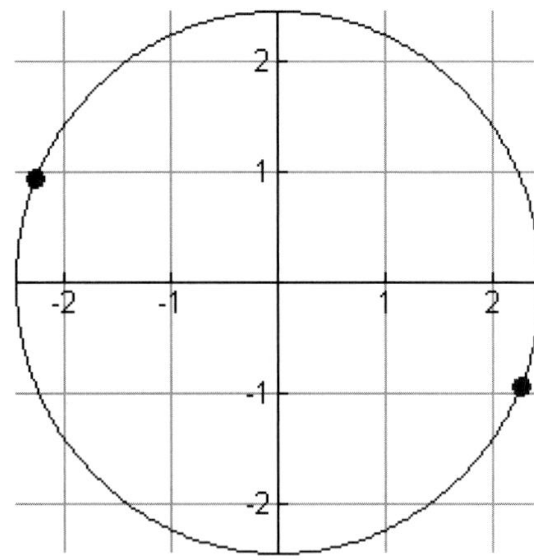

34. Given $z=-\sqrt{2}+\sqrt{2}i$, in order to compute $z^{1/2}$, follow these steps:

Step 1: Write $z=-\sqrt{2}+\sqrt{2}i$ in polar form.
Since $x=-\sqrt{2},\ y=\sqrt{2}$, the point is in QII. So,

$$r=\sqrt{x^2+y^2}=\sqrt{\left(-\sqrt{2}\right)^2+\left(\sqrt{2}\right)^2}=2$$

$$\tan\theta=\frac{y}{x}=\frac{\sqrt{2}}{-\sqrt{2}}=-1\text{, so that }\theta=180^\circ+\tan^{-1}(-1)=135^\circ.$$

(Remember to add 180° to $\tan^{-1}(-1)$ since the angle has terminal side in QII.)
Hence, $z=-\sqrt{2}+\sqrt{2}i=2\left[\cos\left(135^\circ\right)+i\sin\left(135^\circ\right)\right]$.

Step 2: Now, apply $z^{1/n}=r^{1/n}\left[\cos\left(\frac{\theta}{n}+\frac{360^\circ k}{n}\right)+i\sin\left(\frac{\theta}{n}+\frac{360^\circ k}{n}\right)\right],\quad k=0,1,\ldots,n-1$:

$$z^{1/2}=\sqrt{2}\left[\cos\left(\frac{135^\circ}{2}+\frac{360^\circ k}{2}\right)+i\sin\left(\frac{135^\circ}{2}+\frac{360^\circ k}{2}\right)\right],\quad k=0,1$$

$$=\boxed{\sqrt{2}\left[\cos\left(67.5^\circ\right)+i\sin\left(67.5^\circ\right)\right],\ \sqrt{2}\left[\cos\left(247.5^\circ\right)+i\sin\left(247.5^\circ\right)\right]}$$

Step 3: Plot the roots in the complex plane:

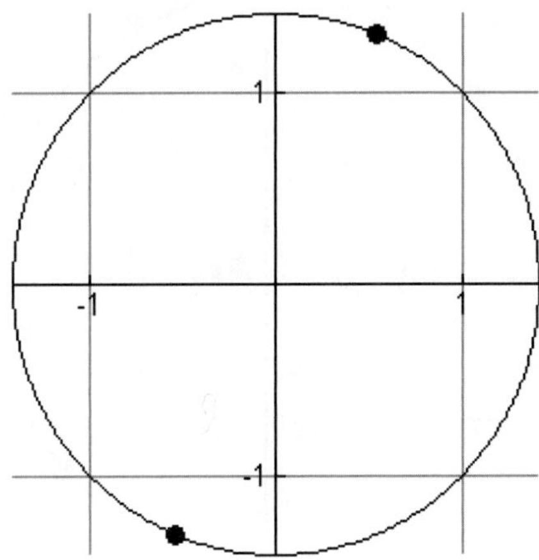

35. Given $z = 4 + 4\sqrt{3}i$, in order to compute $z^{1/3}$, follow these steps:

Step 1: Write $4 + 4\sqrt{3}i$ in polar form.
Since $x = 4,\ y = 4\sqrt{3}$, the point is in QI. So,

$$r = \sqrt{x^2 + y^2} = \sqrt{(4)^2 + \left(4\sqrt{3}\right)^2} = 8$$

$$\tan\theta = \frac{y}{x} = \frac{4\sqrt{3}}{4} = \sqrt{3}\text{, so that } \theta = \tan^{-1}\left(\sqrt{3}\right) = 60^\circ.$$

Hence, $z = 4 + 4\sqrt{3}i = 8\left[\cos\left(60^\circ\right) + i\sin\left(60^\circ\right)\right]$.

Step 2: Now, apply $z^{1/n} = r^{1/n}\left[\cos\left(\frac{\theta}{n} + \frac{360^\circ k}{n}\right) + i\sin\left(\frac{\theta}{n} + \frac{360^\circ k}{n}\right)\right],\quad k = 0, 1, \ldots, n-1$:

$$z^{1/3} = \sqrt[3]{8}\left[\cos\left(\frac{60^\circ}{3} + \frac{360^\circ k}{3}\right) + i\sin\left(\frac{60^\circ}{3} + \frac{360^\circ k}{3}\right)\right],\quad k = 0, 1, 2$$

$$= \boxed{2\left[\cos\left(20^\circ\right) + i\sin\left(20^\circ\right)\right],\ 2\left[\cos\left(140^\circ\right) + i\sin\left(140^\circ\right)\right],\ 2\left[\cos\left(260^\circ\right) + i\sin\left(260^\circ\right)\right]}$$

Step 3: Plot the roots in the complex plane:

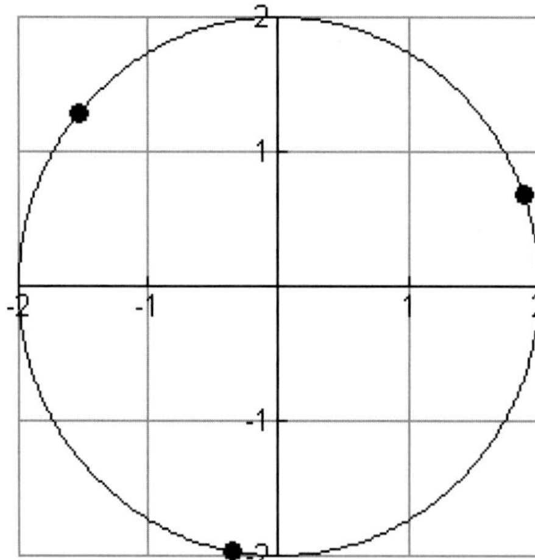

36. Given $z=-\frac{27}{2}+\frac{27\sqrt{3}}{2}i$, in order to compute $z^{1/3}$, follow these steps:

Step 1: Write $z=-\frac{27}{2}+\frac{27\sqrt{3}}{2}i$ in polar form.

Since $x=-\frac{27}{2},\ y=\frac{27\sqrt{3}}{2}$, the point is in QII. So,

$$r=\sqrt{x^2+y^2}=\sqrt{\left(-\frac{27}{2}\right)^2+\left(\frac{27\sqrt{3}}{2}\right)^2}=27$$

$$\tan\theta=\frac{y}{x}=\frac{27\sqrt{3}/2}{-27/2}=-\sqrt{3}\text{, so that }\theta=180^\circ+\tan^{-1}\left(-\sqrt{3}\right)=120^\circ.$$

(Remember to add 180° to $\tan^{-1}\left(-\sqrt{3}\right)$ since the angle has terminal side in QII.)

Hence, $z=-\frac{27}{2}+\frac{27\sqrt{3}}{2}i=27\left[\cos\left(120^\circ\right)+i\sin\left(120^\circ\right)\right]$.

Step 2: Now, apply $z^{1/n}=r^{1/n}\left[\cos\left(\frac{\theta}{n}+\frac{360^\circ k}{n}\right)+i\sin\left(\frac{\theta}{n}+\frac{360^\circ k}{n}\right)\right],\quad k=0,1,\ldots,n-1$:

$$z^{1/3}=\sqrt[3]{27}\left[\cos\left(\frac{120^\circ}{3}+\frac{360^\circ k}{3}\right)+i\sin\left(\frac{120^\circ}{3}+\frac{360^\circ k}{3}\right)\right],\quad k=0,1,2$$

$$=\boxed{3\left[\cos\left(40^\circ\right)+i\sin\left(40^\circ\right)\right],\ 3\left[\cos\left(160^\circ\right)+i\sin\left(160^\circ\right)\right],\ 3\left[\cos\left(280^\circ\right)+i\sin\left(280^\circ\right)\right]}$$

Step 3: Plot the roots in the complex plane:

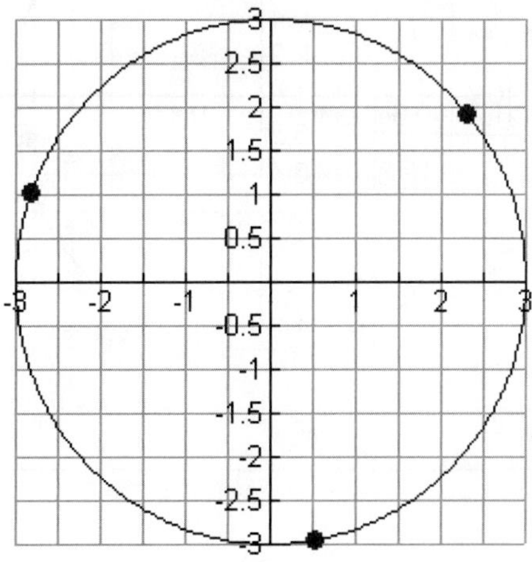

37. Given $z = \sqrt{3} - i$, in order to compute $z^{1/3}$, follow these steps:

Step 1: Write $z = \sqrt{3} - i$ in polar form.
Since $x = \sqrt{3},\ y = -1$, the point is in QIV. So,

$$r = \sqrt{x^2 + y^2} = \sqrt{\left(\sqrt{3}\right)^2 + (-1)^2} = 2$$

$$\tan\theta = \frac{y}{x} = \frac{-1}{\sqrt{3}},\ \text{so that } \theta = \tan^{-1}\left(\frac{-1}{\sqrt{3}}\right) = -30^\circ.$$

Since $0 \le \theta < 360^\circ$, we must use the reference angle 330°.
Hence, $z = \sqrt{3} - i = 2\left[\cos\left(330^\circ\right) + i\sin\left(330^\circ\right)\right]$.

Step 2: Now, apply $z^{1/n} = r^{1/n}\left[\cos\left(\frac{\theta}{n} + \frac{360^\circ k}{n}\right) + i\sin\left(\frac{\theta}{n} + \frac{360^\circ k}{n}\right)\right],\quad k = 0, 1, \ldots, n-1$:

$$z^{1/3} = \sqrt[3]{2}\left[\cos\left(\frac{330^\circ}{3} + \frac{360^\circ k}{3}\right) + i\sin\left(\frac{330^\circ}{3} + \frac{360^\circ k}{3}\right)\right],\quad k = 0, 1, 2$$

$$= \boxed{\begin{array}{l}\sqrt[3]{2}\left[\cos\left(110^\circ\right) + i\sin\left(110^\circ\right)\right],\ \sqrt[3]{2}\left[\cos\left(230^\circ\right) + i\sin\left(230^\circ\right)\right], \\ \sqrt[3]{2}\left[\cos\left(350^\circ\right) + i\sin\left(350^\circ\right)\right]\end{array}}$$

Step 3: Plot the roots in the complex plane:

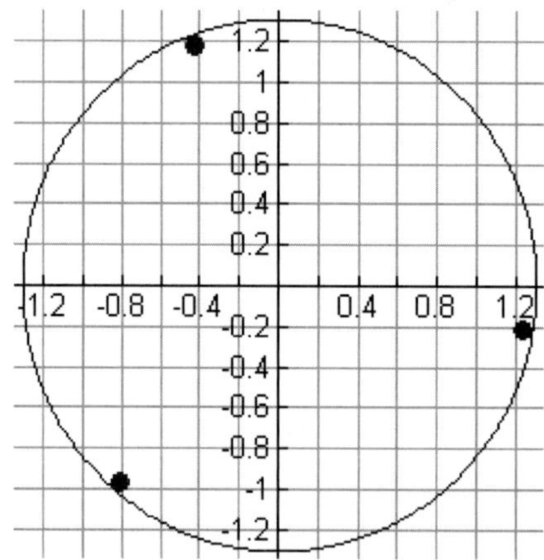

38. Given $z = 4\sqrt{2} + 4\sqrt{2}i$, in order to compute $z^{1/3}$, follow these steps:

Step 1: Write $4\sqrt{2} + 4\sqrt{2}i$ in polar form.
Since $x = 4\sqrt{2},\ y = 4\sqrt{2}$, the point is in QI. So,

$$r = \sqrt{x^2 + y^2} = \sqrt{\left(4\sqrt{2}\right)^2 + \left(4\sqrt{2}\right)^2} = 8$$

$$\tan\theta = \frac{y}{x} = \frac{4\sqrt{2}}{4\sqrt{2}} = 1, \text{ so that } \theta = \tan^{-1}(1) = 45^\circ.$$

Hence, $z = 4\sqrt{2} + 4\sqrt{2}i = 8\left[\cos(45^\circ) + i\sin(45^\circ)\right]$.

Step 2: Now, apply $z^{1/n} = r^{1/n}\left[\cos\left(\frac{\theta}{n} + \frac{360^\circ k}{n}\right) + i\sin\left(\frac{\theta}{n} + \frac{360^\circ k}{n}\right)\right],\quad k = 0, 1, \ldots, n-1$:

$$z^{1/3} = \sqrt[3]{8}\left[\cos\left(\frac{45^\circ}{3} + \frac{360^\circ k}{3}\right) + i\sin\left(\frac{45^\circ}{3} + \frac{360^\circ k}{3}\right)\right],\quad k = 0, 1, 2$$

$$= \boxed{2\left[\cos(15^\circ) + i\sin(15^\circ)\right],\ 2\left[\cos(135^\circ) + i\sin(135^\circ)\right],\ 2\left[\cos(255^\circ) + i\sin(255^\circ)\right]}$$

Step 3: Plot the roots in the complex plane:

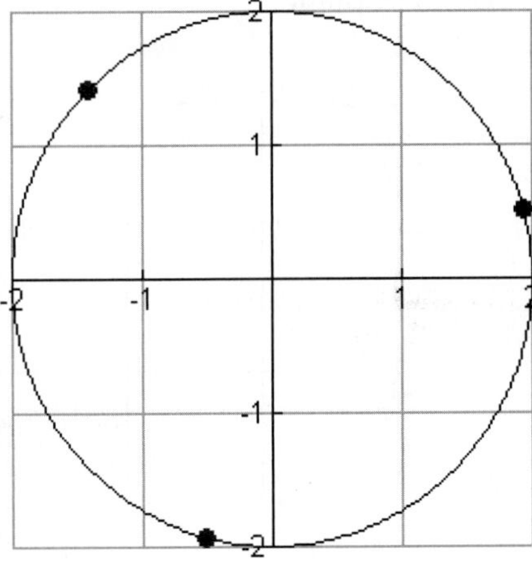

39. Given $z = 8\sqrt{2} - 8\sqrt{2}i$, in order to compute $z^{1/4}$, follow these steps:

Step 1: Write $z = 8\sqrt{2} - 8\sqrt{2}i$ in polar form.
Since $x = 8\sqrt{2},\ y = -8\sqrt{2}$, the point is in QIV. So,

$$r = \sqrt{x^2 + y^2} = \sqrt{\left(8\sqrt{2}\right)^2 + \left(-8\sqrt{2}\right)^2} = 16$$

$$\tan\theta = \frac{y}{x} = \frac{-8\sqrt{2}}{8\sqrt{2}} = -1, \text{ so that } \theta = \tan^{-1}(-1) = -45^\circ.$$

Since $0 \le \theta < 360^\circ$, we must use the reference angle 315°.
Hence, $z = 8\sqrt{2} - 8\sqrt{2}i = 16\left[\cos(315^\circ) + i\sin(315^\circ)\right]$.

Step 2: Now, apply $z^{1/n} = r^{1/n}\left[\cos\left(\frac{\theta}{n} + \frac{360^\circ k}{n}\right) + i\sin\left(\frac{\theta}{n} + \frac{360^\circ k}{n}\right)\right],\quad k = 0, 1, \ldots, n-1$:

$$z^{1/4} = \sqrt[4]{16}\left[\cos\left(\frac{315^\circ}{4} + \frac{360^\circ k}{4}\right) + i\sin\left(\frac{315^\circ}{4} + \frac{360^\circ k}{4}\right)\right],\quad k = 0, 1, 2, 3$$

$$= \boxed{\begin{array}{l} 2\left[\cos(78.75^\circ) + i\sin(78.75^\circ)\right],\ 2\left[\cos(168.75^\circ) + i\sin(168.75^\circ)\right], \\ 2\left[\cos(258.75^\circ) + i\sin(258.75^\circ)\right],\ 2\left[\cos(348.75^\circ) + i\sin(348.75^\circ)\right] \end{array}}$$

Step 3: Plot the roots in the complex plane:

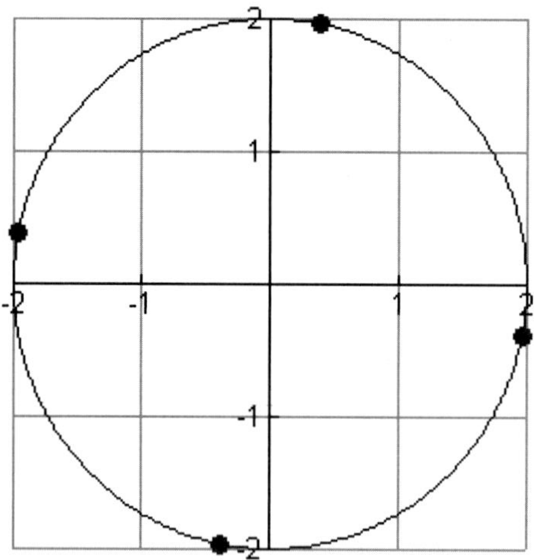

40. Given $z=-\sqrt{128}+\sqrt{128}i$, in order to compute $z^{1/4}$, follow these steps:

Step 1: Write $z=-\sqrt{128}+\sqrt{128}i$ in polar form.
Since $x=-\sqrt{128},\ y=\sqrt{128}$, the point is in QII. So,

$$r=\sqrt{x^2+y^2}=\sqrt{\left(-\sqrt{128}\right)^2+\left(\sqrt{128}\right)^2}=16$$

$$\tan\theta=\frac{y}{x}=\frac{\sqrt{128}}{-\sqrt{128}}=-1\text{, so that }\theta=180^\circ+\tan^{-1}(-1)=135^\circ.$$

(Remember to add 180° to $\tan^{-1}(-1)$ since the angle has terminal side in QII.)
Hence, $z=-\sqrt{128}+\sqrt{128}i=16\left[\cos(135^\circ)+i\sin(135^\circ)\right]$.

Step 2: Now, apply $z^{1/n}=r^{1/n}\left[\cos\left(\frac{\theta}{n}+\frac{360^\circ k}{n}\right)+i\sin\left(\frac{\theta}{n}+\frac{360^\circ k}{n}\right)\right],\quad k=0,1,\ldots,n-1$:

$$z^{1/4}=\sqrt[4]{16}\left[\cos\left(\frac{135^\circ}{4}+\frac{360^\circ k}{4}\right)+i\sin\left(\frac{135^\circ}{4}+\frac{360^\circ k}{4}\right)\right],\quad k=0,1,2,3$$

$$=\boxed{\begin{array}{l}2\left[\cos(33.75^\circ)+i\sin(33.75^\circ)\right],\ 2\left[\cos(123.75^\circ)+i\sin(123.75^\circ)\right],\\ 2\left[\cos(213.75^\circ)+i\sin(213.75^\circ)\right],\ 2\left[\cos(303.75^\circ)+i\sin(303.75^\circ)\right]\end{array}}$$

Step 3: Plot the roots in the complex plane:

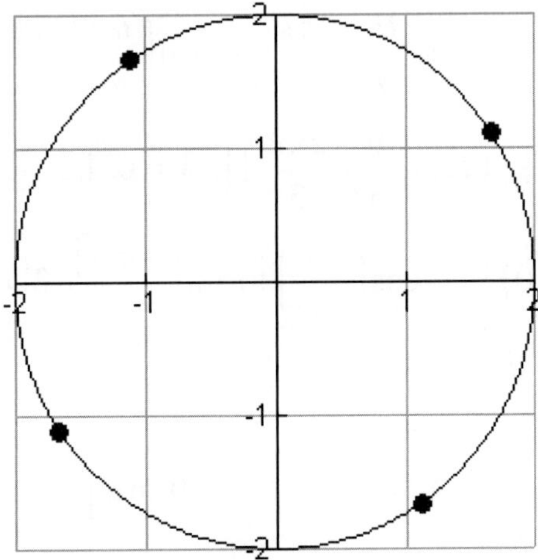

41. We seek all complex numbers x such that $x^4-16=0$. Observe that this equation can be written equivalently as:

$$\left(x^2-4\right)\left(x^2+4\right)=0$$
$$(x-2)(x+2)\left(x^2+4\right)=0$$

The solutions are therefore $x=\boxed{\pm 2,\ \pm 2i}$.

(Note: We could have alternatively used the formula for complex roots, namely

$$z^{1/n}=r^{1/n}\left[\cos\left(\frac{\theta}{n}+\frac{360^\circ k}{n}\right)+i\sin\left(\frac{\theta}{n}+\frac{360^\circ k}{n}\right)\right],\quad k=0,\,1,\,\ldots,\,n-1.$$

However, the left-side readily factored in a way that is easy to see what *all* of the solutions were. As such, factoring was the more expedient route for this problem.)

42. We seek all complex numbers x such that $x^3-8=0$, or equivalently $x^3=8$. While it is clear that $x=2$ is a solution by inspection, the remaining two complex solutions aren't immediately discernible. As such, we shall apply the approach for computing complex roots involving the formula

$$z^{1/n}=r^{1/n}\left[\cos\left(\frac{\theta}{n}+\frac{2\pi k}{n}\right)+i\sin\left(\frac{\theta}{n}+\frac{2\pi k}{n}\right)\right],\quad k=0,\,1,\,\ldots,\,n-1.$$

To this end, we follow these steps.

Step 1: Write $z=8$ in polar form.
Since $x=8,\ y=0$, the point is on the positive x-axis. So, $r=8,\ \theta=0$.
Hence, $8=8\left[\cos(0)+i\sin(0)\right]$.

Step 2: Now, apply $z^{1/n}=r^{1/n}\left[\cos\left(\frac{\theta}{n}+\frac{2\pi k}{n}\right)+i\sin\left(\frac{\theta}{n}+\frac{2\pi k}{n}\right)\right],\quad k=0,\,1,\,\ldots,\,n-1$:

$$8^{1/3}=\sqrt[3]{8}\left[\cos\left(\frac{0}{3}+\frac{2\pi k}{3}\right)+i\sin\left(\frac{0}{3}+\frac{2\pi k}{3}\right)\right],\quad k=0,\,1,\,2$$
$$=2\left[\cos(0)+i\sin(0)\right],\ 2\left[\cos\left(\frac{2\pi}{3}\right)+i\sin\left(\frac{2\pi}{3}\right)\right],\ 2\left[\cos\left(\frac{4\pi}{3}\right)+i\sin\left(\frac{4\pi}{3}\right)\right]$$
$$=2,\ -1+\sqrt{3}i,\ -1-\sqrt{3}i$$

Step 3: Hence, the complex solutions of $x^3-8=0$ are $\boxed{2,\ -1+\sqrt{3}i,\ -1-\sqrt{3}i}$.

43. We seek all complex numbers x such that $x^3+8=0$, or equivalently $x^3=-8$. While it is clear that $x=-2$ is a solution by inspection, the remaining two complex solutions aren't immediately discernible. As such, we shall apply the approach for computing complex roots involving the formula

$$z^{1/n}=r^{1/n}\left[\cos\left(\frac{\theta}{n}+\frac{2\pi k}{n}\right)+i\sin\left(\frac{\theta}{n}+\frac{2\pi k}{n}\right)\right],\quad k=0,1,\ldots,n-1.$$

To this end, we follow these steps.

Step 1: Write $z=-8$ in polar form.

Since $x=-8,\ y=0$, the point is on the negative x-axis. So, $r=8,\ \theta=\pi$.

Hence, $-8=8\left[\cos(\pi)+i\sin(\pi)\right]$.

Step 2: Now, apply $z^{1/n}=r^{1/n}\left[\cos\left(\frac{\theta}{n}+\frac{2\pi k}{n}\right)+i\sin\left(\frac{\theta}{n}+\frac{2\pi k}{n}\right)\right],\quad k=0,1,\ldots,n-1$:

$$(-8)^{1/3}=\sqrt[3]{8}\left[\cos\left(\frac{\pi}{3}+\frac{2\pi k}{3}\right)+i\sin\left(\frac{\pi}{3}+\frac{2\pi k}{3}\right)\right],\quad k=0,1,2$$

$$=2\left[\cos\left(\frac{\pi}{3}\right)+i\sin\left(\frac{\pi}{3}\right)\right],\ 2\left[\cos(\pi)+i\sin(\pi)\right],\ 2\left[\cos\left(\frac{5\pi}{3}\right)+i\sin\left(\frac{5\pi}{3}\right)\right]=-2,\ 1-\sqrt{3}i,\ 1+\sqrt{3}i$$

Step 3: Hence, the complex solutions of $x^3+8=0$ are $\boxed{-2,\ 1-\sqrt{3}i,\ 1+\sqrt{3}i}$.

44. We seek all complex numbers x such that $x^3+1=0$, or equivalently $x^3=-1$. While it is clear that $x=-1$ is a solution by inspection, the remaining two complex solutions aren't immediately discernible. As such, we shall apply the approach for computing complex roots involving the formula

$$z^{1/n}=r^{1/n}\left[\cos\left(\frac{\theta}{n}+\frac{2\pi k}{n}\right)+i\sin\left(\frac{\theta}{n}+\frac{2\pi k}{n}\right)\right],\quad k=0,1,\ldots,n-1.$$

To this end, we follow these steps.

Step 1: Write $z=-1$ in polar form.

Since $x=-1,\ y=0$, the point is on the negative x-axis. So, $r=1,\ \theta=\pi$.

Hence, $-1=1\left[\cos(\pi)+i\sin(\pi)\right]$.

Step 2: Now, apply $z^{1/n}=r^{1/n}\left[\cos\left(\frac{\theta}{n}+\frac{2\pi k}{n}\right)+i\sin\left(\frac{\theta}{n}+\frac{2\pi k}{n}\right)\right],\quad k=0,1,\ldots,n-1$:

$$(-1)^{1/3}=\sqrt[3]{1}\left[\cos\left(\frac{\pi}{3}+\frac{2\pi k}{3}\right)+i\sin\left(\frac{\pi}{3}+\frac{2\pi k}{3}\right)\right],\quad k=0,1,2$$

$$=1\left[\cos\left(\frac{\pi}{3}\right)+i\sin\left(\frac{\pi}{3}\right)\right],\ 1\left[\cos(\pi)+i\sin(\pi)\right],\ 1\left[\cos\left(\frac{5\pi}{3}\right)+i\sin\left(\frac{5\pi}{3}\right)\right]=-1,\ \frac{1}{2}-\frac{\sqrt{3}}{2}i,\ \frac{1}{2}+\frac{\sqrt{3}}{2}i$$

Step 3: Hence, the complex solutions of $x^3+1=0$ are $\boxed{-1,\ \frac{1}{2}-\frac{\sqrt{3}}{2}i,\ \frac{1}{2}+\frac{\sqrt{3}}{2}i}$.

45. We seek all complex numbers x such that $x^4+16=0$, or equivalently $x^4=-16$. We shall apply the approach for computing complex roots involving the formula

$$z^{1/n}=r^{1/n}\left[\cos\left(\frac{\theta}{n}+\frac{2\pi k}{n}\right)+i\sin\left(\frac{\theta}{n}+\frac{2\pi k}{n}\right)\right],\ k=0,1,\ldots,n-1.$$

To this end, we follow these steps.

Step 1: Write $z=-16$ in polar form.
Since $x=-16,\ y=0$, the point is on the negative x-axis. So, $r=16,\ \theta=\pi$.
Hence, $-16=16[\cos(\pi)+i\sin(\pi)]$.

Step 2: Now, apply $z^{1/n}=r^{1/n}\left[\cos\left(\frac{\theta}{n}+\frac{2\pi k}{n}\right)+i\sin\left(\frac{\theta}{n}+\frac{2\pi k}{n}\right)\right],\ k=0,1,\ldots,n-1$:

$$\begin{aligned}(-16)^{1/4}&=\sqrt[4]{16}\left[\cos\left(\frac{\pi}{4}+\frac{2\pi k}{4}\right)+i\sin\left(\frac{\pi}{4}+\frac{2\pi k}{4}\right)\right],\ k=0,1,2,3\\&=2\left[\cos\left(\frac{\pi}{4}\right)+i\sin\left(\frac{\pi}{4}\right)\right],\ 2\left[\cos\left(\frac{3\pi}{4}\right)+i\sin\left(\frac{3\pi}{4}\right)\right],\\&\quad 2\left[\cos\left(\frac{5\pi}{4}\right)+i\sin\left(\frac{5\pi}{4}\right)\right],\ 2\left[\cos\left(\frac{7\pi}{4}\right)+i\sin\left(\frac{7\pi}{4}\right)\right]\\&=\sqrt{2}+\sqrt{2}i,\ -\sqrt{2}+\sqrt{2}i,\ -\sqrt{2}-\sqrt{2}i,\ \sqrt{2}-\sqrt{2}i\end{aligned}$$

Step 3: Hence, the complex solutions of $x^4+16=0$ are

$$\boxed{\sqrt{2}+\sqrt{2}i,\ -\sqrt{2}+\sqrt{2}i,\ -\sqrt{2}-\sqrt{2}i,\ \sqrt{2}-\sqrt{2}i}.$$

46. We seek all complex numbers x such that $x^6+1=0$, or equivalently $x^6=-1$. We shall apply the approach for computing complex roots involving the formula

$$z^{1/n}=r^{1/n}\left[\cos\left(\frac{\theta}{n}+\frac{2\pi k}{n}\right)+i\sin\left(\frac{\theta}{n}+\frac{2\pi k}{n}\right)\right],\ k=0,1,\ldots,n-1.$$

To this end, we follow these steps.

Step 1: Write $z=-1$ in polar form.
Since $x=-1,\ y=0$, the point is on the negative x-axis. So, $r=1,\ \theta=\pi$.
Hence, $-1=1[\cos(\pi)+i\sin(\pi)]$.

Step 2: Now, apply $z^{1/n} = r^{1/n}\left[\cos\left(\frac{\theta}{n}+\frac{2\pi k}{n}\right)+i\sin\left(\frac{\theta}{n}+\frac{2\pi k}{n}\right)\right]$, $k = 0, 1, \ldots, n-1$:

$$(-1)^{1/6} = \sqrt[6]{1}\left[\cos\left(\frac{\pi}{6}+\frac{2\pi k}{6}\right)+i\sin\left(\frac{\pi}{6}+\frac{2\pi k}{6}\right)\right],\ k = 0, 1, 2, 3, 4, 5$$

$$= 1\left[\cos\left(\frac{\pi}{6}\right)+i\sin\left(\frac{\pi}{6}\right)\right],\ 1\left[\cos\left(\frac{\pi}{2}\right)+i\sin\left(\frac{\pi}{2}\right)\right],\ 1\left[\cos\left(\frac{5\pi}{6}\right)+i\sin\left(\frac{5\pi}{6}\right)\right]$$

$$1\left[\cos\left(\frac{7\pi}{6}\right)+i\sin\left(\frac{7\pi}{6}\right)\right],\ 1\left[\cos\left(\frac{3\pi}{2}\right)+i\sin\left(\frac{3\pi}{2}\right)\right],\ 1\left[\cos\left(\frac{11\pi}{6}\right)+i\sin\left(\frac{11\pi}{6}\right)\right]$$

$$= \frac{\sqrt{3}}{2}+\frac{1}{2}i,\ i,\ -\frac{\sqrt{3}}{2}+\frac{1}{2}i,\ -\frac{\sqrt{3}}{2}-\frac{1}{2}i,\ -i,\ \frac{\sqrt{3}}{2}-\frac{1}{2}i$$

Step 3: Hence, the complex solutions of $x^6+1=0$ are

$$\boxed{\frac{\sqrt{3}}{2}+\frac{1}{2}i,\ i,\ -\frac{\sqrt{3}}{2}+\frac{1}{2}i,\ -\frac{\sqrt{3}}{2}-\frac{1}{2}i,\ -i,\ \frac{\sqrt{3}}{2}-\frac{1}{2}i}.$$

47. We seek all complex numbers x such that $x^6-1=0$, or equivalently $x^6=1$. While it is clear that $x = 1$ is a solution by inspection, the remaining complex solutions aren't immediately discernible. As such, we shall apply the approach for computing complex roots involving the formula

$$z^{1/n} = r^{1/n}\left[\cos\left(\frac{\theta}{n}+\frac{2\pi k}{n}\right)+i\sin\left(\frac{\theta}{n}+\frac{2\pi k}{n}\right)\right],\ k = 0, 1, \ldots, n-1.$$

To this end, we follow these steps.

Step 1: Write $z=1$ in polar form.
Since $x=1,\ y=0$, the point is on the positive x-axis. So, $r=1,\ \theta=0$.
Hence, $1=1\left[\cos(0)+i\sin(0)\right]$.

Step 2: Now, apply $z^{1/n} = r^{1/n}\left[\cos\left(\frac{\theta}{n}+\frac{2\pi k}{n}\right)+i\sin\left(\frac{\theta}{n}+\frac{2\pi k}{n}\right)\right],\ k=0,1,\ldots,n-1$:

$$(1)^{1/6} = \sqrt[6]{1}\left[\cos\left(0+\frac{2\pi k}{6}\right)+i\sin\left(0+\frac{2\pi k}{6}\right)\right],\ k=0,1,2,3,4,5$$

$$=1\left[\cos(0)+i\sin(0)\right],\ 1\left[\cos\left(\frac{\pi}{3}\right)+i\sin\left(\frac{\pi}{3}\right)\right],\ 1\left[\cos\left(\frac{2\pi}{3}\right)+i\sin\left(\frac{2\pi}{3}\right)\right]$$

$$1\left[\cos(\pi)+i\sin(\pi)\right],\ 1\left[\cos\left(\frac{4\pi}{3}\right)+i\sin\left(\frac{4\pi}{3}\right)\right],\ 1\left[\cos\left(\frac{5\pi}{3}\right)+i\sin\left(\frac{5\pi}{3}\right)\right]$$

$$=1,\ \frac{1}{2}+\frac{\sqrt{3}}{2}i,\ -\frac{1}{2}+\frac{\sqrt{3}}{2}i,\ -1,\ -\frac{1}{2}-\frac{\sqrt{3}}{2}i,\ \frac{1}{2}-\frac{\sqrt{3}}{2}i$$

Step 3: Hence, the complex solutions of $x^6-1=0$ are

$$\boxed{1,\ \frac{1}{2}+\frac{\sqrt{3}}{2}i,\ -\frac{1}{2}+\frac{\sqrt{3}}{2}i,\ -1,\ -\frac{1}{2}-\frac{\sqrt{3}}{2}i,\ \frac{1}{2}-\frac{\sqrt{3}}{2}i}.$$

48. We seek all complex numbers x such that $4x^2+1=0$. Observe that this equation can be written equivalently as:

$$x^2 = -\frac{1}{4} \quad \text{so that} \quad x = \pm\sqrt{-\frac{1}{4}} = \pm\frac{1}{2}i$$

The solutions are therefore $x = \boxed{-\frac{1}{2}i,\ \frac{1}{2}i}$.

(Note: We could have alternatively used the formula for complex roots, namely

$$z^{1/n} = r^{1/n}\left[\cos\left(\frac{\theta}{n}+\frac{360^\circ k}{n}\right)+i\sin\left(\frac{\theta}{n}+\frac{360^\circ k}{n}\right)\right],\ k=0,1,\ldots,n-1.$$

However, the left-side readily factored in a way that is easy to see what *all* of the solutions were. As such, factoring was the more expedient route for this problem.)

49. We seek all complex numbers x such that $x^2+i=0$. Observe that this equation can be written equivalently as:

$$x^2=-i \text{ so that } x=\pm\sqrt{-i}$$

Now, in order to express these solutions in rectangular form, we proceed as follows:
Step 1: Write $z=-i$ in polar form.
Since $x=0,\ y=-1$, the point is on the negative y-axis. So, $r=1,\ \theta=\frac{3\pi}{2}$.

Hence, $-i=1\left[\cos\left(\frac{3\pi}{2}\right)+i\sin\left(\frac{3\pi}{2}\right)\right]$.

Step 2: Now, apply $z^{1/n}=r^{1/n}\left[\cos\left(\frac{\theta}{n}+\frac{2\pi k}{n}\right)+i\sin\left(\frac{\theta}{n}+\frac{2\pi k}{n}\right)\right],\quad k=0,1,\ldots,n-1$:

$$(-i)^{1/2}=\sqrt{1}\left[\cos\left(\frac{3\pi/2}{2}+\frac{2\pi k}{2}\right)+i\sin\left(\frac{3\pi/2}{2}+\frac{2\pi k}{2}\right)\right],\quad k=0,1$$

$$=1\left[\cos\left(\frac{3\pi}{4}\right)+i\sin\left(\frac{3\pi}{4}\right)\right],\ 1\left[\cos\left(\frac{7\pi}{4}\right)+i\sin\left(\frac{7\pi}{4}\right)\right]=-\frac{\sqrt{2}}{2}+\frac{\sqrt{2}}{2}i,\ \frac{\sqrt{2}}{2}-\frac{\sqrt{2}}{2}i$$

The solutions are therefore $x=\boxed{-\frac{\sqrt{2}}{2}+\frac{\sqrt{2}}{2}i,\ \frac{\sqrt{2}}{2}-\frac{\sqrt{2}}{2}i}$.

50. We seek all complex numbers x such that $x^2-i=0$. Observe that this equation can be written equivalently as:

$$x^2=i \text{ so that } x=\pm\sqrt{i}$$

Now, in order to express these solutions in rectangular form, we proceed as follows:
Step 1: Write $z=i$ in polar form.
Since $x=0,\ y=1$, the point is on the positive y-axis. So, $r=1,\ \theta=\frac{\pi}{2}$.

Hence, $i=1\left[\cos\left(\frac{\pi}{2}\right)+i\sin\left(\frac{\pi}{2}\right)\right]$.

Step 2: Now, apply $z^{1/n} = r^{1/n}\left[\cos\left(\frac{\theta}{n}+\frac{2\pi k}{n}\right)+i\sin\left(\frac{\theta}{n}+\frac{2\pi k}{n}\right)\right]$, $k = 0, 1, \ldots, n-1$:

$$(i)^{1/2} = \sqrt{1}\left[\cos\left(\frac{\pi/2}{2}+\frac{2\pi k}{2}\right)+i\sin\left(\frac{\pi/2}{2}+\frac{2\pi k}{2}\right)\right],\quad k=0,1$$

$$=1\left[\cos\left(\frac{\pi}{4}\right)+i\sin\left(\frac{\pi}{4}\right)\right],\ 1\left[\cos\left(\frac{5\pi}{4}\right)+i\sin\left(\frac{5\pi}{4}\right)\right]=\frac{\sqrt{2}}{2}+\frac{\sqrt{2}}{2}i,\ -\frac{\sqrt{2}}{2}-\frac{\sqrt{2}}{2}i$$

The solutions are therefore $x = \boxed{\frac{\sqrt{2}}{2}+\frac{\sqrt{2}}{2}i,\ -\frac{\sqrt{2}}{2}-\frac{\sqrt{2}}{2}i}$.

51. We seek all complex numbers x such that $x^4 - 2i = 0$. Observe that this equation can be written equivalently as:

$$x^4 = 2i \text{ so that } x = (2i)^{1/4}$$

Now, in order to express these solutions in rectangular form, we proceed as follows:
Step 1: Write $z = 2i$ in polar form.

Since $x = 0,\ y = 2$, the point is on the positive y-axis. So, $r = 2,\ \theta = \frac{\pi}{2}$.

Hence, $2i = 2\left[\cos\left(\frac{\pi}{2}\right)+i\sin\left(\frac{\pi}{2}\right)\right]$.

Step 2: Now, apply $z^{1/n} = r^{1/n}\left[\cos\left(\frac{\theta}{n}+\frac{2\pi k}{n}\right)+i\sin\left(\frac{\theta}{n}+\frac{2\pi k}{n}\right)\right]$, $k = 0, 1, \ldots, n-1$:

$$(2i)^{1/4} = \sqrt[4]{2}\left[\cos\left(\frac{\pi/2}{4}+\frac{2\pi k}{4}\right)+i\sin\left(\frac{\pi/2}{4}+\frac{2\pi k}{4}\right)\right],\quad k=0,1,2,3$$

$$=\boxed{\begin{array}{l}\sqrt[4]{2}\left[\cos\left(\frac{\pi}{8}\right)+i\sin\left(\frac{\pi}{8}\right)\right],\ \sqrt[4]{2}\left[\cos\left(\frac{5\pi}{8}\right)+i\sin\left(\frac{5\pi}{8}\right)\right],\\ \sqrt[4]{2}\left[\cos\left(\frac{9\pi}{8}\right)+i\sin\left(\frac{9\pi}{8}\right)\right],\sqrt[4]{2}\left[\cos\left(\frac{13\pi}{8}\right)+i\sin\left(\frac{13\pi}{8}\right)\right]\end{array}}$$

52. We seek all complex numbers x such that $x^4 + 2i = 0$. Observe that this equation can be written equivalently as:

$$x^4 = -2i \text{ so that } x = (-2i)^{1/4}$$

Now, in order to express these solutions in rectangular form, we proceed as follows:
Step 1: Write $z = -2i$ in polar form.
Since $x = 0,\ y = -2$, the point is on the negative y-axis. So, $r = 2,\ \theta = \frac{3\pi}{2}$.

Hence, $-2i = 2\left[\cos\left(\frac{3\pi}{2}\right) + i\sin\left(\frac{3\pi}{2}\right)\right]$.

Step 2: Now, apply $z^{1/n} = r^{1/n}\left[\cos\left(\frac{\theta}{n} + \frac{2\pi k}{n}\right) + i\sin\left(\frac{\theta}{n} + \frac{2\pi k}{n}\right)\right],\ k = 0, 1, \ldots, n-1$:

$$(-2i)^{1/4} = \sqrt[4]{2}\left[\cos\left(\frac{3\pi/2}{4} + \frac{2\pi k}{4}\right) + i\sin\left(\frac{3\pi/2}{4} + \frac{2\pi k}{4}\right)\right],\ k = 0,1,2,3$$

$$= \boxed{\begin{array}{l} \sqrt[4]{2}\left[\cos\left(\frac{3\pi}{8}\right) + i\sin\left(\frac{3\pi}{8}\right)\right],\ \sqrt[4]{2}\left[\cos\left(\frac{7\pi}{8}\right) + i\sin\left(\frac{7\pi}{8}\right)\right], \\ \sqrt[4]{2}\left[\cos\left(\frac{11\pi}{8}\right) + i\sin\left(\frac{11\pi}{8}\right)\right], \sqrt[4]{2}\left[\cos\left(\frac{15\pi}{8}\right) + i\sin\left(\frac{15\pi}{8}\right)\right] \end{array}}$$

53. We seek all complex numbers x such that $x^5 + 32 = 0$. Observe that this equation can be written equivalently as:

$$x^5 = -32 \text{ so that } x = (-32)^{1/5}$$

Now, in order to express these solutions in rectangular form, we proceed as follows:
Step 1: Write $z = -32$ in polar form.
Since $x = -32,\ y = 0$, the point is on the negative x-axis. So, $r = 32,\ \theta = \pi$.
Hence, $-32 = 32\left[\cos(\pi) + i\sin(\pi)\right]$.

Step 2: Now, apply $z^{1/n} = r^{1/n}\left[\cos\left(\frac{\theta}{n} + \frac{2\pi k}{n}\right) + i\sin\left(\frac{\theta}{n} + \frac{2\pi k}{n}\right)\right],\ k = 0, 1, \ldots, n-1$:

$$(-32)^{1/5} = \sqrt[5]{32}\left[\cos\left(\frac{\pi}{5} + \frac{2\pi k}{5}\right) + i\sin\left(\frac{\pi}{5} + \frac{2\pi k}{5}\right)\right],\ k = 0,1,2,3,4$$

$$= \boxed{\begin{array}{l} 2\left[\cos\left(\frac{\pi}{5}\right) + i\sin\left(\frac{\pi}{5}\right)\right],\ 2\left[\cos\left(\frac{3\pi}{5}\right) + i\sin\left(\frac{3\pi}{5}\right)\right], 2\left[\cos(\pi) + i\sin(\pi)\right] \\ 2\left[\cos\left(\frac{7\pi}{5}\right) + i\sin\left(\frac{7\pi}{5}\right)\right], 2\left[\cos\left(\frac{9\pi}{5}\right) + i\sin\left(\frac{9\pi}{5}\right)\right] \end{array}}$$

54. We seek all complex numbers x such that $x^5 - 32 = 0$. Observe that this equation can be written equivalently as:

$$x^5 = 32 \text{ so that } x = (32)^{1/5}$$

Now, in order to express these solutions in rectangular form, we proceed as follows:

Step 1: Write $z = 32$ in polar form.
Since $x = -32,\ y = 0$, the point is on the positive x-axis. So, $r = 32,\ \theta = 0$.
Hence, $32 = 32\left[\cos(0) + i\sin(0)\right]$.

Step 2: Now, apply $z^{1/n} = r^{1/n}\left[\cos\left(\frac{\theta}{n} + \frac{2\pi k}{n}\right) + i\sin\left(\frac{\theta}{n} + \frac{2\pi k}{n}\right)\right],\quad k = 0, 1, \ldots, n-1$:

$$(32)^{1/5} = \sqrt[5]{32}\left[\cos\left(0 + \frac{2\pi k}{5}\right) + i\sin\left(0 + \frac{2\pi k}{5}\right)\right],\ k = 0,1,2,3,4$$

$$= \boxed{\begin{array}{l} 2\left[\cos(0) + i\sin(0)\right],\ 2\left[\cos\left(\frac{2\pi}{5}\right) + i\sin\left(\frac{2\pi}{5}\right)\right], 2\left[\cos\left(\frac{4\pi}{5}\right) + i\sin\left(\frac{4\pi}{5}\right)\right] \\ 2\left[\cos\left(\frac{6\pi}{5}\right) + i\sin\left(\frac{6\pi}{5}\right)\right], 2\left[\cos\left(\frac{8\pi}{5}\right) + i\sin\left(\frac{8\pi}{5}\right)\right] \end{array}}$$

55. We seek all complex numbers x such that $x^7 - \pi^{14} i = 0$. Observe that this equation can be written equivalently as:

$$x^7 = \pi^{14} i \text{ so that } x = \left(\pi^{14} i\right)^{1/7}$$

Now, in order to express these solutions in rectangular form, we proceed as follows:

Step 1: Write $z = \pi^{14} i$ in polar form.
Since $x = 0,\ y = \pi^{14}$, the point is on the negative x-axis. So, $r = \pi^{14},\ \theta = \frac{\pi}{2}$.
Hence, $\pi^{14} i = \pi^{14}\left[\cos\left(\frac{\pi}{2}\right) + i\sin\left(\frac{\pi}{2}\right)\right]$.

Step 2: Now, apply $z^{1/n} = r^{1/n}\left[\cos\left(\frac{\theta}{n} + \frac{2\pi k}{n}\right) + i\sin\left(\frac{\theta}{n} + \frac{2\pi k}{n}\right)\right],\quad k = 0, 1, \ldots, n-1$:

$$\left(\pi^{14} i\right)^{1/7} = \sqrt[7]{\pi^{14}}\left[\cos\left(\frac{\pi/2}{7} + \frac{2\pi k}{7}\right) + i\sin\left(\frac{\pi/2}{7} + \frac{2\pi k}{7}\right)\right],\ k = 0,1,2,3,4,5,6$$

$$= \boxed{\begin{array}{l} \pi^2\left[\cos\left(\frac{\pi}{14}\right) + i\sin\left(\frac{\pi}{14}\right)\right],\ \pi^2\left[\cos\left(\frac{5\pi}{14}\right) + i\sin\left(\frac{5\pi}{14}\right)\right],\ \pi^2\left[\cos\left(\frac{9\pi}{14}\right) + i\sin\left(\frac{9\pi}{14}\right)\right] \\ \pi^2\left[\cos\left(\frac{13\pi}{14}\right) + i\sin\left(\frac{13\pi}{14}\right)\right],\ \pi^2\left[\cos\left(\frac{17\pi}{14}\right) + i\sin\left(\frac{17\pi}{14}\right)\right] \pi^2\left[\cos\left(\frac{21\pi}{14}\right) + i\sin\left(\frac{21\pi}{14}\right)\right] \\ \pi^2\left[\cos\left(\frac{25\pi}{14}\right) + i\sin\left(\frac{25\pi}{14}\right)\right] \end{array}}$$

56. We seek all complex numbers x such that $x^7 + \pi^{14} i = 0$. Observe that this equation can be written equivalently as:

$$x^7 = -\pi^{14}\text{i} \text{ so that } x = \left(-\pi^{14} i\right)^{1/7}$$

Now, in order to express these solutions in rectangular form, we proceed as follows:

Step 1: Write $z = -\pi^{14} i$ in polar form.
Since $x = 0,\ y = -\pi^{14}$, the point is on the negative x-axis. So, $r = \pi^{14},\ \theta = \frac{3\pi}{2}$.
Hence, $-\pi^{14} i = \pi^{14}\left[\cos\left(\frac{3\pi}{2}\right) + i\sin\left(\frac{3\pi}{2}\right)\right]$.

Step 2: Now, apply $z^{1/n} = r^{1/n}\left[\cos\left(\frac{\theta}{n} + \frac{2\pi k}{n}\right) + i\sin\left(\frac{\theta}{n} + \frac{2\pi k}{n}\right)\right],\ k = 0, 1, \ldots, n-1$:

$$\left(-\pi^{14} i\right)^{1/7} = \sqrt[7]{\pi^{14}}\left[\cos\left(\frac{3\pi}{14} + \frac{2\pi k}{7}\right) + i\sin\left(\frac{3\pi}{2} + \frac{2\pi k}{7}\right)\right],\ k = 0,1,2,3,4,5,6$$

57. Given $z = -\frac{\sqrt{2}}{2} - \frac{\sqrt{2}}{2} i$, in order to compute $z^{1/5}$, follow these steps:

Step 1: Write $z = -\frac{\sqrt{2}}{2} - \frac{\sqrt{2}}{2} i$ in polar form.

Since $x = -\frac{\sqrt{2}}{2},\ y = -\frac{\sqrt{2}}{2}$, the point is in QIII. So,

$$r = \sqrt{x^2 + y^2} = \sqrt{\left(-\frac{\sqrt{2}}{2}\right)^2 + \left(-\frac{\sqrt{2}}{2}\right)^2} = 1$$

$$\tan\theta = \frac{y}{x} = \frac{-\sqrt{2}/2}{-\sqrt{2}/2} = 1, \text{ so that } \theta = 180^\circ + \tan^{-1}(1) = 225^\circ.$$

(Remember to add 180° to $\tan^{-1}(1)$ since the angle has terminal side in QIII.)

Hence, $z = -\frac{\sqrt{2}}{2} - \frac{\sqrt{2}}{2} i = 1\left[\cos\left(225^\circ\right) + i\sin\left(225^\circ\right)\right]$.

Step 2: Now, apply $z^{1/n} = r^{1/n}\left[\cos\left(\frac{\theta}{n}+\frac{360^\circ k}{n}\right)+i\sin\left(\frac{\theta}{n}+\frac{360^\circ k}{n}\right)\right],\quad k=0, 1, \ldots, n-1$:

$$z^{1/5} = \sqrt[5]{1}\left[\cos\left(\frac{225^\circ}{5}+\frac{360^\circ k}{5}\right)+i\sin\left(\frac{225^\circ}{5}+\frac{360^\circ k}{5}\right)\right],\quad k=0, 1, 2, 3, 4$$

$$= \boxed{\begin{array}{l}\left[\cos(45^\circ)+i\sin(45^\circ)\right], \left[\cos(117^\circ)+i\sin(117^\circ)\right], \left[\cos(189^\circ)+i\sin(189^\circ)\right],\\ \left[\cos(261^\circ)+i\sin(261^\circ)\right], \left[\cos(333^\circ)+i\sin(333^\circ)\right]\end{array}}$$

Step 3: Plot the roots in the complex plane and connect consecutive roots to form a regular pentagon:

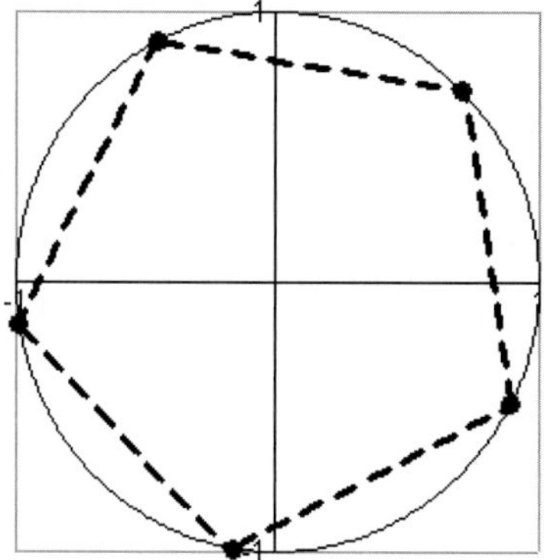

58. Given $z = 16i$, in order to compute $z^{1/4}$, follow these steps:

Step 1: Write $z = 16i$ in polar form.
Since $x = 0,\ y = 16$, the point is on the positive y-axis. So,

$$r = 16,\ \theta = 90^\circ$$

Hence, $z = 16i = 16\left[\cos(90^\circ)+i\sin(90^\circ)\right]$.

Step 2: Now, apply $z^{1/n} = r^{1/n}\left[\cos\left(\frac{\theta}{n}+\frac{360^\circ k}{n}\right)+i\sin\left(\frac{\theta}{n}+\frac{360^\circ k}{n}\right)\right]$, $k = 0, 1, \ldots, n-1$:

$$z^{1/4} = \sqrt[4]{16}\left[\cos\left(\frac{90^\circ}{4}+\frac{360^\circ k}{4}\right)+i\sin\left(\frac{90^\circ}{4}+\frac{360^\circ k}{4}\right)\right], \quad k = 0, 1, 2, 3$$

$$= \boxed{\begin{array}{l} 2\left[\cos(22.5^\circ)+i\sin(22.5^\circ)\right],\ 2\left[\cos(112.5^\circ)+i\sin(112.5^\circ)\right], \\ 2\left[\cos(202.5^\circ)+i\sin(202.5)\right],\ 2\left[\cos(292.5^\circ)+i\sin(292.5^\circ)\right] \end{array}}$$

Step 3: Plot the roots in the complex plane and connect consecutive roots to form a square:

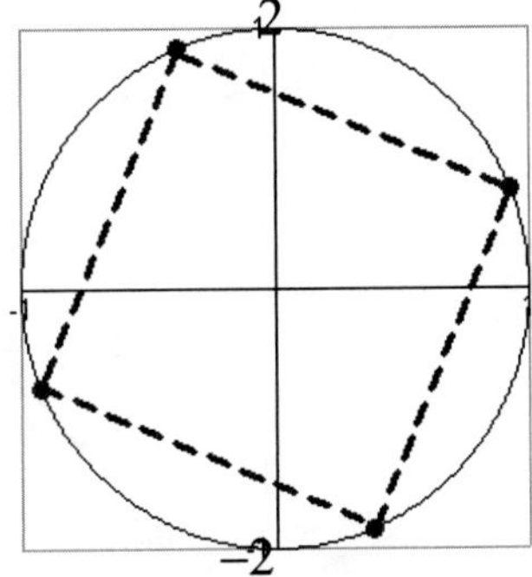

59. Given $z = \frac{1}{2}-\frac{\sqrt{3}}{2}i$, in order to compute $z^{1/6}$, follow these steps:

Step 1: Write $z = \frac{1}{2}-\frac{\sqrt{3}}{2}i$ in polar form.

Since $x = \frac{1}{2}$, $y = -\frac{\sqrt{3}}{2}$, the point is in QIV. So,

$$r = \sqrt{x^2+y^2} = \sqrt{\left(\frac{1}{2}\right)^2+\left(-\frac{\sqrt{3}}{2}\right)^2} = 1$$

$$\tan\theta = \frac{y}{x} = \frac{-\sqrt{3}/2}{1/2} = -\sqrt{3}\text{, so that } \theta = -60^\circ\text{, so we use } 300^\circ.$$

Hence, $z = \frac{1}{2}-\frac{\sqrt{3}}{2}i = 1\left[\cos(300^\circ)+i\sin(300^\circ)\right]$.

Step 2: Now, apply $z^{1/n} = r^{1/n}\left[\cos\left(\frac{\theta}{n}+\frac{360^\circ k}{n}\right)+i\sin\left(\frac{\theta}{n}+\frac{360^\circ k}{n}\right)\right], \quad k=0, 1, \ldots, n-1$:

$$z^{1/6} = \sqrt[6]{1}\left[\cos\left(\frac{300^\circ}{6}+\frac{360^\circ k}{6}\right)+i\sin\left(\frac{300^\circ}{6}+\frac{360^\circ k}{6}\right)\right], \quad k=0, 1, 2, 3, 4, 5$$

$$=\left[\cos\left(50^\circ+60^\circ k\right)+i\sin\left(50^\circ+60^\circ k\right)\right], \quad k=0, 1, 2, 3, 4, 5$$

Step 3: Plot the roots in the complex plane and connect consecutive roots to form a regular hexagon:

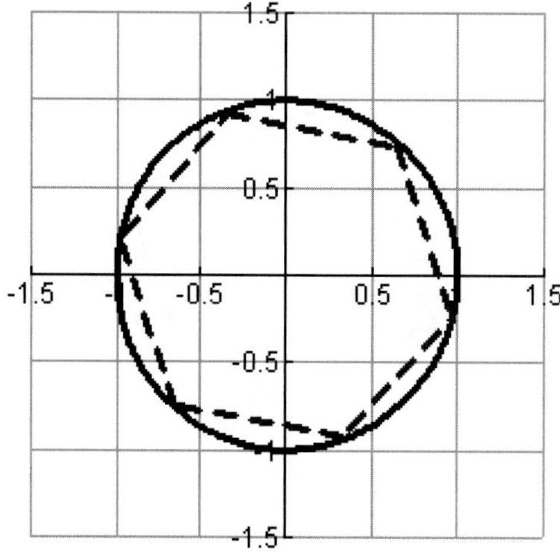

60. Given $z = 2i$, in order to compute $z^{1/8}$, follow these steps:
Step 1: Write $z = 2i$ in polar form.
Since $x = 0, y = 2$, the point is on the positive y-axis. So, $r = 2, \theta = 90^\circ$.
Hence, $z = 2i = 2\left[\cos\left(90^\circ\right)+i\sin\left(90^\circ\right)\right]$.

Step 2: Now, apply $z^{1/n} = r^{1/n}\left[\cos\left(\frac{\theta}{n}+\frac{360°k}{n}\right)+i\sin\left(\frac{\theta}{n}+\frac{360°k}{n}\right)\right]$, $k = 0, 1, \ldots, n-1$:

$$z^{1/8} = \sqrt[8]{2}\left[\cos\left(\frac{90°}{8}+\frac{360°k}{8}\right)+i\sin\left(\frac{90°}{8}+\frac{360°k}{8}\right)\right], \quad k = 0, 1, 2, 3, 4, 5, 6, 7$$

$$= \sqrt[8]{2}\left[\cos\left(\frac{90°}{8}+45°k\right)+i\sin\left(\frac{90°}{8}+45°k\right)\right], \quad k = 0, 1, 2, 3, 4, 5, 6, 7$$

Step 3: Plot the roots in the complex plane and connect consecutive roots to form a regular octagon:

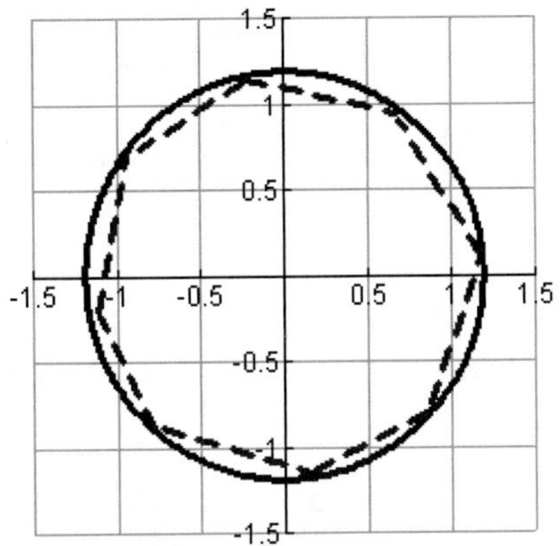

61. Reversed order of angles being subtracted.	**62.** Should use the product formula, namely $z_1z_2 = r_1r_2\left[\cos(\theta_1+\theta_2)+i\sin(\theta_1+\theta_2)\right]$, to multiply.
63. Should use DeMoivre's formula. In general, $(a+b)^6 \neq a^6+b^6$.	**64.** Only found one of the five solutions.

65. True. From the product formula, we see that

$$z_1z_2 = r_1r_2\left[\cos(\theta_1+\theta_2)+i\sin(\theta_1+\theta_2)\right] = r_1r_2\cos(\theta_1+\theta_2)+\left[r_1r_2\sin(\theta_1+\theta_2)\right]i,$$

which is of the form of a complex number.

66. True. From the quotient formula, we see that $\frac{z_1}{z_2} = \frac{r_1}{r_2}\left[\cos(\theta_1-\theta_2)+i\sin(\theta_1-\theta_2)\right]$.

67. False. For instance, the only real solutions of $x^4-1=0$ are -1 and 1.

68. True.

69. True.

70. False. All complex numbers have a complex square root.

71. True. A regular n-gon is formed and so, all edges have the same length.

72. False. Rotate an equilateral triangle so that one of the vertices is at (0,2) and the remaining two are appropriately positioned. Then, the vertices would represent the complex 3rd roots of the number.

73. Observe that

$$z_1 z_2 = \left(r_1 e^{i\theta_1}\right)\left(r_2 e^{i\theta_2}\right) = r_1 r_2 e^{i(\theta_1+\theta_2)} = r_1 r_2\left[\cos\left(\theta_1+\theta_2\right)+i\sin\left(\theta_1+\theta_2\right)\right]$$

74. Observe that

$$\frac{z_1}{z_2} = \frac{r_1 e^{i\theta_1}}{r_2 e^{i\theta_2}} = \frac{r_1}{r_2} e^{i(\theta_1-\theta_2)} = \frac{r_1}{r_2}\left[\cos\left(\theta_1-\theta_2\right)+i\sin\left(\theta_1-\theta_2\right)\right]$$

75. Observe that $z^n = \left(re^{i\theta}\right)^n = r^n e^{in\theta} = r^n\left[\cos(n\theta)+i\sin(n\theta)\right]$

76. Observe that $z^{1/n} = \left(re^{i\theta}\right)^{1/n} = r^{1/n} e^{i(\theta/n)} = r^{1/n}\left[\cos\left(\frac{\theta}{n}\right)+i\sin\left(\frac{\theta}{n}\right)\right]$. Hence, by periodicity (period $= 2\pi/n$), we have $w_k = r^{1/n}\left[\cos\left(\frac{\theta}{n}+\frac{2k\pi}{n}\right)+i\sin\left(\frac{\theta}{n}+\frac{2k\pi}{n}\right)\right]$, where $k = 0, \dots, n-1$

77. DeMoivre's theorem with $n = 2$ states

$$\left(\cos x + i\sin x\right)^2 = \cos 2x + i\sin 2x. \quad \textbf{(1)}$$

Multiply out the left to see that

$$\left(\cos x + i\sin x\right)^2 = \cos^2 x + 2i\sin x\cos x - \sin^2 x. \quad \textbf{(2)}$$

Substituting **(2)** in for the left-side of **(1)**, and simplifying using the double-angle formula for sine yields

$$\cos^2 x + \cancel{2i\sin x\cos x} - \sin^2 x = \cos 2x + \cancel{i\sin 2x}$$

$$\cos^2 x - \sin^2 x = \cos 2x$$

78. Observe that using DeMoivre's theorem yields

$$\cos 3x = \cos^3 x + 3\cos x\left(\cos^2 x - 1\right) = 4\cos^3 x - 3\cos x.$$

79. Observe that using DeMoivre's theorem yields

$$\sin 3x = 3\cos^2 x\sin x - \sin^3 x = 3\sin x - 4\sin^3 x.$$

80. $$\frac{\left(\frac{1}{2}+\frac{\sqrt{3}}{2}i\right)^{14}}{\left(\frac{1}{2}-\frac{\sqrt{3}}{2}i\right)^{20}} = \frac{\left(e^{i\frac{\pi}{3}}\right)^{14}}{\left(e^{-i\frac{\pi}{3}}\right)^{20}} = \frac{e^{i\frac{14\pi}{3}}}{e^{-i\frac{20\pi}{3}}} = e^{i\frac{34\pi}{3}} = e^{i\left(10+\frac{4}{3}\right)\pi} = e^{i\frac{4}{3}\pi} = \boxed{-\frac{1}{2}-\frac{\sqrt{3}}{2}i}$$

81. $$(1-i)^n\cdot(1+i)^m = \left(\sqrt{2}\,e^{i\left(-\frac{\pi}{4}\right)}\right)^n\left(\sqrt{2}\,e^{i\left(\frac{\pi}{4}\right)}\right)^m = \boxed{2^{\frac{n+m}{2}}\,e^{\frac{\pi}{4}(m-n)i}}$$

82. $$\frac{(1+i)^n}{(1-i)^m} = \frac{\left(\sqrt{2}\,e^{i\left(\frac{\pi}{4}\right)}\right)^n}{\left(\sqrt{2}\,e^{i\left(-\frac{\pi}{4}\right)}\right)^m} = \boxed{2^{\frac{n-m}{2}}\,e^{\frac{\pi}{4}(m+n)i}}$$

83. Given $z = \frac{\sqrt{3}}{2} - \frac{1}{2}i$, in order to compute $z^{1/5}$, follow these steps:

Step 1: Write $z = \frac{\sqrt{3}}{2} - \frac{1}{2}i$ in polar form.

Since $x = \frac{\sqrt{3}}{2},\ y = -\frac{1}{2}$, the point is in QIV. Also, we have $r = 1$ and

$$\tan\theta = \frac{y}{x} \Rightarrow \sin\theta = -\frac{1}{2} \text{ and } \cos\theta = \frac{\sqrt{3}}{2}, \text{ so that } \theta = \frac{11\pi}{6}.$$

Hence, $z = \left[\cos\left(\frac{11\pi}{6}\right) + i\sin\left(\frac{11\pi}{6}\right)\right]$.

Step 2: Now, apply $z^{1/n} = r^{1/n}\left[\cos\left(\frac{\theta}{n} + \frac{2k\pi}{n}\right) + i\sin\left(\frac{\theta}{n} + \frac{2k\pi}{n}\right)\right],\ k = 0, 1, \ldots, n-1$:

$$z^{1/5} = \left[\cos\left(\frac{11\pi}{6(5)} + \frac{2k\pi}{5}\right) + i\sin\left(\frac{11\pi}{6(5)} + \frac{2k\pi}{5}\right)\right], \quad k = 0, 1, 2, 3, 4$$

Step 3: Plot the roots in the complex plane:

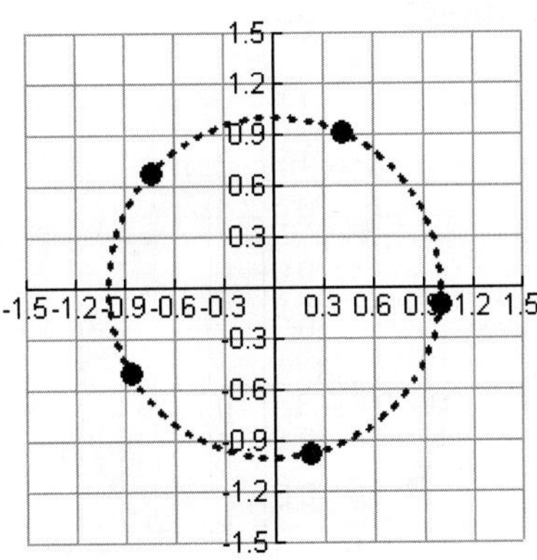

84. Given $z=-\frac{\sqrt{2}}{2}+\frac{\sqrt{2}}{2}i$, in order to compute $z^{1/4}$, follow these steps:

Step 1: Write $z=-\frac{\sqrt{2}}{2}+\frac{\sqrt{2}}{2}i$ in polar form.

Since $x=-\frac{\sqrt{2}}{2},\ y=\frac{\sqrt{2}}{2}$, the point is in QII. Also, we have $r=1$ and

$$\tan\theta=\frac{y}{x}=-1\text{ , so that }\theta=\frac{3\pi}{4}.$$

Hence, $z=\left[\cos\left(\frac{3\pi}{4}\right)+i\sin\left(\frac{3\pi}{4}\right)\right]$.

Step 2: Now, apply $z^{1/n}=r^{1/n}\left[\cos\left(\frac{\theta}{n}+\frac{2k\pi}{n}\right)+i\sin\left(\frac{\theta}{n}+\frac{2k\pi}{n}\right)\right],\quad k=0,1,\ldots,n-1$:

$$z^{1/4}=\left[\cos\left(\frac{3\pi}{4(4)}+\frac{2k\pi}{4}\right)+i\sin\left(\frac{3\pi}{4(4)}+\frac{2k\pi}{4}\right)\right],\quad k=0,1,2,3$$

Step 3: Plot the roots in the complex plane:

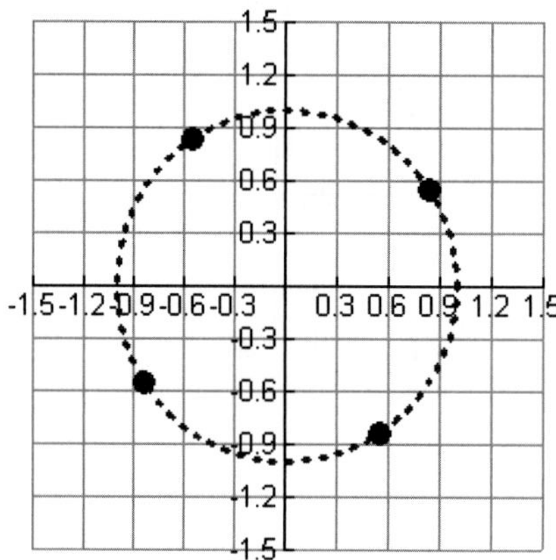

85. Given $z=-\frac{1}{2}-\frac{\sqrt{3}}{2}i$, in order to compute $z^{1/4}$, follow these steps:

Step 1: Write $z=-\frac{1}{2}-\frac{\sqrt{3}}{2}i$ in polar form.

Since $x=-\frac{1}{2},\ y=-\frac{\sqrt{3}}{2}$, the point is in QIII. Also, we have $r=1$ and

$$\tan\theta=\frac{y}{x}\Rightarrow \sin\theta=-\frac{\sqrt{3}}{2} \text{ and } \cos\theta=-\frac{1}{2}, \text{ so that } \theta=\frac{4\pi}{3}.$$

Hence, $z=\left[\cos\left(\frac{4\pi}{3}\right)+i\sin\left(\frac{4\pi}{3}\right)\right]$.

Step 2: Now, apply $z^{1/n}=r^{1/n}\left[\cos\left(\frac{\theta}{n}+\frac{2k\pi}{n}\right)+i\sin\left(\frac{\theta}{n}+\frac{2k\pi}{n}\right)\right],\quad k=0,1,\ldots,n-1$:

$$z^{1/6}=\left[\cos\left(\frac{4\pi}{3(6)}+\frac{2k\pi}{6}\right)+i\sin\left(\frac{4\pi}{3(6)}+\frac{2k\pi}{6}\right)\right],\quad k=0,1,2,3,4,5$$

Step 3: Plot the roots in the complex plane:

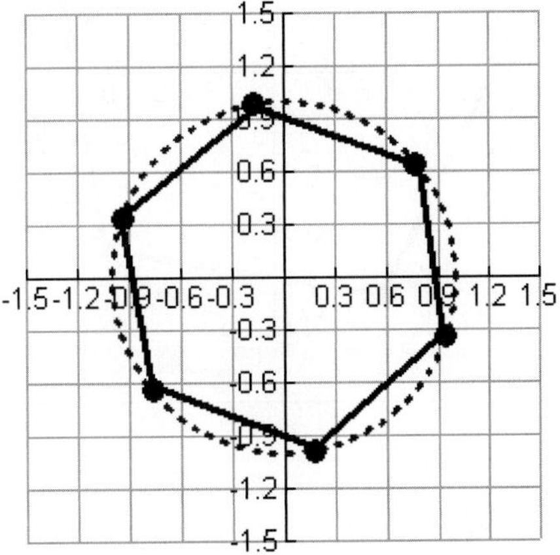

86. Given $z=-4+4i$, in order to compute $z^{1/5}$, follow these steps:

Step 1: Write $z=-4+4i$ in polar form.
Since $x=-4,\ y=4$, the point is in QII. Also, we have

$$r=\sqrt{(-4)^2+4^2}=4\sqrt{2}\text{ and }\tan\theta=\frac{y}{x}=-1\text{, so that }\theta=\frac{3\pi}{4}.$$

Hence, $z=4\sqrt{2}\left[\cos\left(\frac{3\pi}{4}\right)+i\sin\left(\frac{3\pi}{4}\right)\right]$.

Step 2: Now, apply $z^{1/n}=r^{1/n}\left[\cos\left(\frac{\theta}{n}+\frac{2k\pi}{n}\right)+i\sin\left(\frac{\theta}{n}+\frac{2k\pi}{n}\right)\right],\quad k=0,1,\ldots,n-1$:

$$z^{1/5}=\left(4\sqrt{2}\right)^{1/5}\left[\cos\left(\frac{3\pi}{4(5)}+\frac{2k\pi}{5}\right)+i\sin\left(\frac{3\pi}{4(5)}+\frac{2k\pi}{5}\right)\right],\quad k=0,1,2,3,4$$

Step 3: Plot the roots in the complex plane:

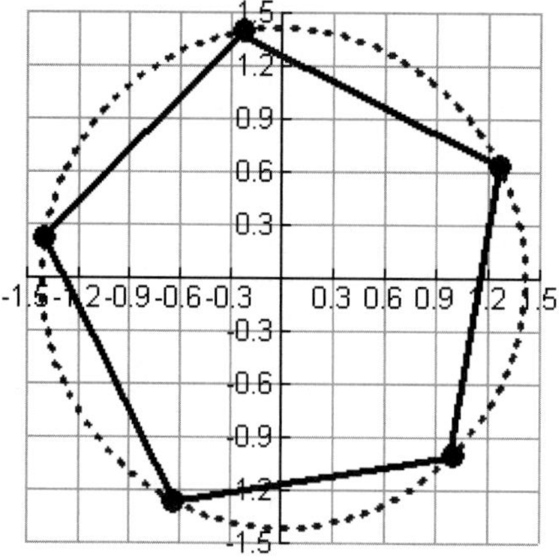

87. Given $z = \frac{27\sqrt{2}}{2} + \frac{27\sqrt{2}}{2}i$, in order to compute $z^{1/3}$, follow these steps:

Step 1: Write $z = \frac{27\sqrt{2}}{2} + \frac{27\sqrt{2}}{2}i$ in polar form.

Since $x = \frac{27\sqrt{2}}{2}$, $y = \frac{27\sqrt{2}}{2}$, the point is in QI. Using the calculator, we have

$$r = 27 \text{ and } \tan\theta = \frac{y}{x} = 1, \text{ so that } \theta = \frac{\pi}{4}.$$

Hence, $z = 27\left[\cos\left(\frac{\pi}{4}\right) + i\sin\left(\frac{\pi}{4}\right)\right]$.

Step 2: Now, apply $z^{1/n} = r^{1/n}\left[\cos\left(\frac{\theta}{n} + \frac{2k\pi}{n}\right) + i\sin\left(\frac{\theta}{n} + \frac{2k\pi}{n}\right)\right]$, $k = 0, 1, \ldots, n-1$:

$$z^{1/3} = (27)^{1/3}\left[\cos\left(\frac{\pi/4}{3} + \frac{2k\pi}{3}\right) + i\sin\left(\frac{\pi/4}{3} + \frac{2k\pi}{3}\right)\right], \quad k = 0, 1, 2$$

Step 3: Plot the roots in the complex plane:

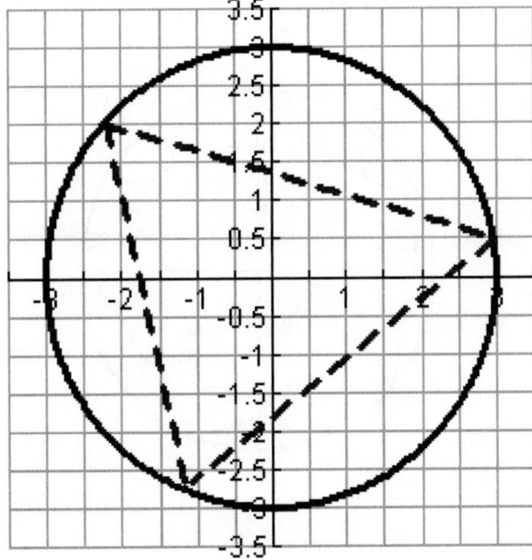

88. Given $z = 8\sqrt{2}\left(\sqrt{3}-1\right) + 8\sqrt{2}\left(\sqrt{3}-1\right)i$, in order to compute $z^{1/5}$, follow these steps:

Step 1: Write $z = 8\sqrt{2}\left(\sqrt{3}-1\right) + 8\sqrt{2}\left(\sqrt{3}-1\right)i$ in polar form.

Since $x = 8\sqrt{2}\left(\sqrt{3}-1\right)$, $y = 8\sqrt{2}\left(\sqrt{3}-1\right)$, the point is in QI. Using the calculator, we have

$$r \approx 11.712 \text{ and } \tan\theta = \frac{y}{x} = 1, \text{ so that } \theta = \frac{\pi}{4}.$$

Hence, $z = 11.712\left[\cos\left(\frac{\pi}{4}\right) + i\sin\left(\frac{\pi}{4}\right)\right]$.

Step 2: Now, apply $z^{1/n} = r^{1/n}\left[\cos\left(\frac{\theta}{n} + \frac{2k\pi}{n}\right) + i\sin\left(\frac{\theta}{n} + \frac{2k\pi}{n}\right)\right]$, $k = 0, 1, \ldots, n-1$:

$$z^{1/5} = (11.712)^{1/5}\left[\cos\left(\frac{\pi}{4(5)} + \frac{2k\pi}{5}\right) + i\sin\left(\frac{\pi}{4(5)} + \frac{2k\pi}{5}\right)\right], \quad k = 0, 1, 2, 3, 4$$

Step 3: Plot the roots in the complex plane:

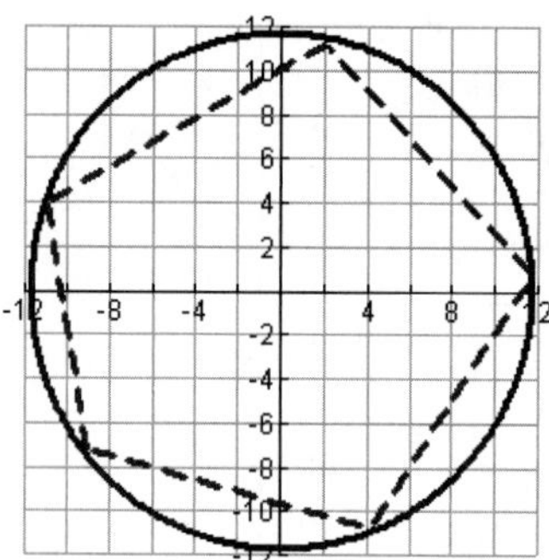

89. First, note that

$$\begin{aligned}\cos 2\theta + i\sin 2\theta &= \left(\cos\theta + i\sin\theta\right)^2 \\ &= \cos^2\theta + 2i\sin\theta\cos\theta + i^2\sin^2\theta \\ &= \left(\cos^2\theta - \sin^2\theta\right) + (2\sin\theta\cos\theta)i.\end{aligned}$$

Hence, by equating real parts, we conclude that $\cos 2\theta = \cos^2\theta - \sin^2\theta$.

90. First, note that

$$\begin{aligned}\cos 2\theta + i\sin 2\theta &= (\cos\theta + i\sin\theta)^2 \\ &= \cos^2\theta + 2i\sin\theta\cos\theta + i^2\sin^2\theta \\ &= (\cos^2\theta - \sin^2\theta) + (2\sin\theta\cos\theta)i.\end{aligned}$$

Hence, by equating imaginary parts, we conclude that $\sin 2\theta = 2\cos\theta\sin\theta$.

91. First, note that

$$\begin{aligned}\cos 3\theta + i\sin 3\theta &= (\cos\theta + i\sin\theta)^3 \\ &= \cos^3\theta + 3i\cos^2\theta\sin\theta + 3i^2\cos\theta\sin^2\theta + i^3\sin^3\theta \\ &= (\cos^3\theta - 3\cos\theta\sin^2\theta) + (3\cos^2\theta\sin\theta - \sin^3\theta)i \\ &= (\cos^3\theta - 3\cos\theta(1-\cos^2\theta)) + (3(1-\sin^2\theta)\sin\theta - \sin^3\theta)i \\ &= (4\cos^3\theta - 3\cos\theta) + (3\sin\theta - 4\sin^3\theta)i.\end{aligned}$$

Hence, by equating real parts, we conclude that $\cos 3\theta = 4\cos^3\theta - 3\cos\theta$.

92. First, note that

$$\begin{aligned}\cos 3\theta + i\sin 3\theta &= (\cos\theta + i\sin\theta)^3 \\ &= \cos^3\theta + 3i\cos^2\theta\sin\theta + 3i^2\cos\theta\sin^2\theta + i^3\sin^3\theta \\ &= (\cos^3\theta - 3\cos\theta\sin^2\theta) + (3\cos^2\theta\sin\theta - \sin^3\theta)i \\ &= (\cos^3\theta - 3\cos\theta(1-\cos^2\theta)) + (3(1-\sin^2\theta)\sin\theta - \sin^3\theta)i \\ &= (4\cos^3\theta - 3\cos\theta) + (3\sin\theta - 4\sin^3\theta)i.\end{aligned}$$

Hence, by equating imaginary parts, we conclude that $\sin 3\theta = 3\sin\theta - 4\sin^3\theta$.

Section 7.5 Solutions

1. – 10. The following graph shows the points in Exercises 1 – 10:

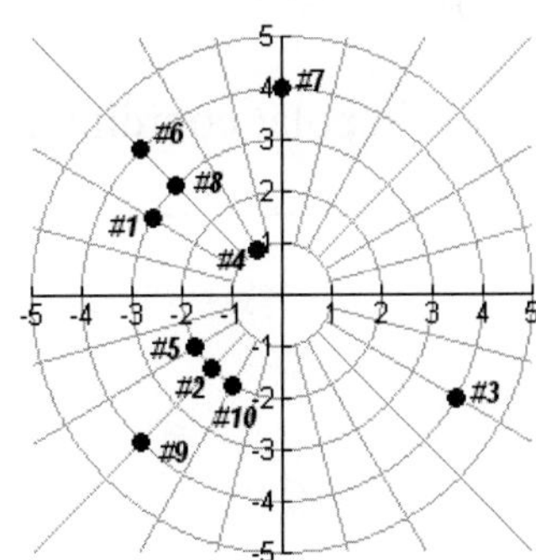

11. In order to convert the point $\left(2, 2\sqrt{3}\right)$ (expressed in rectangular coordinates) to polar coordinates, first observe that $x = 2,\ y = 2\sqrt{3}$, so that the point is in QI. Hence,

$$r = \sqrt{(2)^2 + \left(2\sqrt{3}\right)^2} = 4$$

$$\tan\theta = \frac{y}{x} = \frac{2\sqrt{3}}{2} = \sqrt{3} \text{ so that } \theta = \tan^{-1}\left(\sqrt{3}\right) = \frac{\pi}{3}.$$

So, the given point can be expressed in polar coordinates as $\boxed{\left(4, \frac{\pi}{3}\right)}$.

12. In order to convert the point $(3, -3)$ (expressed in rectangular coordinates) to polar coordinates, first observe that $x = 3,\ y = -3$, so that the point is in QIV. Hence,

$$r = \sqrt{(3)^2 + (-3)^2} = 3\sqrt{2}$$

$$\tan\theta = \frac{y}{x} = \frac{-3}{3} = -1 \text{ so that } \theta = \tan^{-1}(-1) = -\frac{\pi}{4}.$$

Since $0 \le \theta < 2\pi$, we use the reference angle $\theta = \frac{7\pi}{4}$.

So, the given point can be expressed in polar coordinates as $\boxed{\left(3\sqrt{2}, \frac{7\pi}{4}\right)}$.

13. In order to convert the point $\left(-1, -\sqrt{3}\right)$ (expressed in rectangular coordinates) to polar coordinates, first observe that $x = -1,\ y = -\sqrt{3}$, so that the point is in QIII. Hence,

$$r = \sqrt{(-1)^2 + \left(-\sqrt{3}\right)^2} = 2$$

$$\tan\theta = \frac{y}{x} = \frac{-\sqrt{3}}{-1} = \sqrt{3} \text{ so that } \theta = \pi + \tan^{-1}\left(\sqrt{3}\right) = \frac{4\pi}{3}.$$

(Remember to add π to $\tan^{-1}\left(\sqrt{3}\right)$ since the point is in QIII.)

So, the given point can be expressed in polar coordinates as $\boxed{\left(2, \frac{4\pi}{3}\right)}$.

14. In order to convert the point $\left(6,6\sqrt{3}\right)$ (expressed in rectangular coordinates) to polar coordinates, first observe that $x=6,\ y=6\sqrt{3}$, so that the point is in QI. Hence,

$$r=\sqrt{(6)^2+\left(6\sqrt{3}\right)^2}=12$$

$$\tan\theta=\frac{y}{x}=\frac{6\sqrt{3}}{6}=\sqrt{3} \text{ so that } \theta=\tan^{-1}\left(\sqrt{3}\right)=\frac{\pi}{3}.$$

So, the given point can be expressed in polar coordinates as $\boxed{\left(12,\frac{\pi}{3}\right)}$.

15. In order to convert the point $(-4,4)$ (expressed in rectangular coordinates) to polar coordinates, first observe that $x=-4,\ y=4$, so that the point is in QII. Hence,

$$r=\sqrt{(-4)^2+(4)^2}=4\sqrt{2}$$

$$\tan\theta=\frac{y}{x}=\frac{-4}{4}=-1 \text{ so that } \theta=\pi+\tan^{-1}(-1)=\frac{3\pi}{4}.$$

(Remember to add π to $\tan^{-1}(-1)$ since the point is in QII.)

So, the given point can be expressed in polar coordinates as $\boxed{\left(4\sqrt{2},\frac{3\pi}{4}\right)}$.

16. In order to convert the point $\left(0,\sqrt{2}\right)$ (expressed in rectangular coordinates) to polar coordinates, first observe that $x=0,\ y=\sqrt{2}$, so that the point is on the positive *y*-axis. Hence,

$$r=\sqrt{(0)^2+\left(\sqrt{2}\right)^2}=\sqrt{2},\quad \theta=\frac{\pi}{2}$$

So, the given point can be expressed in polar coordinates as $\boxed{\left(\sqrt{2},\frac{\pi}{2}\right)}$.

17. In order to convert the point $(3,0)$ (expressed in rectangular coordinates) to polar coordinates, first observe that $x=3,\ y=0$, so that the point is on the positive *x*-axis. Hence,

$$r=\sqrt{(3)^2+(0)^2}=3,\quad \theta=0$$

So, the given point can be expressed in polar coordinates as $\boxed{(3,0)}$.

18. In order to convert the point $(-7,-7)$ (expressed in rectangular coordinates) to polar coordinates, first observe that $x=-7,\ y=-7$, so that the point is in QIII. Hence,

$$r=\sqrt{(-7)^2+(-7)^2}=7\sqrt{2}$$

$$\tan\theta=\frac{y}{x}=\frac{-7}{-7}=1 \text{ so that } \theta=\pi+\tan^{-1}(1)=\frac{5\pi}{4}.$$

(Remember to add π to $\tan^{-1}(1)$ since the point is in QIII.)

So, the given point can be expressed in polar coordinates as $\boxed{\left(7\sqrt{2},\frac{5\pi}{4}\right)}$.

19. In order to convert the point $\left(-\sqrt{3},-1\right)$ (expressed in rectangular coordinates) to polar coordinates, first observe that $x=-\sqrt{3},\ y=-1$, so that the point is in QIII. Hence,

$$r=\sqrt{\left(-\sqrt{3}\right)^2+(-1)^2}=2$$

$$\tan\theta=\frac{y}{x}=\frac{-1}{-\sqrt{3}}=\frac{1}{\sqrt{3}} \text{ so that } \theta=\pi+\tan^{-1}\left(\frac{1}{\sqrt{3}}\right)=\frac{7\pi}{6}.$$

(Remember to add π to $\tan^{-1}\left(\frac{1}{\sqrt{3}}\right)$ since the point is in QIII.)

So, the given point can be expressed in polar coordinates as $\boxed{\left(2,\frac{7\pi}{6}\right)}$.

20. In order to convert the point $\left(2\sqrt{3},-2\right)$ (expressed in rectangular coordinates) to polar coordinates, first observe that $x=2\sqrt{3},\ y=-2$, so that the point is in QIV. Hence,

$$r=\sqrt{\left(2\sqrt{3}\right)^2+(-2)^2}=4$$

$$\tan\theta=\frac{y}{x}=\frac{-2}{2\sqrt{3}}=-\frac{1}{\sqrt{3}} \text{ so that } \theta=\tan^{-1}\left(-\frac{1}{\sqrt{3}}\right)=-\frac{\pi}{6}.$$

Since $0\le\theta<2\pi$, we use the reference angle $\theta=\frac{11\pi}{6}$.

So, the given point can be expressed in polar coordinates as $\boxed{\left(4,\frac{11\pi}{6}\right)}$.

21. In order to convert the point $\left(4, \frac{5\pi}{3}\right)$ (expressed in polar coordinates) to rectangular coordinates, first observe that $r = 4,\ \theta = \frac{5\pi}{3}$. Hence,

$$x = r\cos\theta = 4\cos\left(\frac{5\pi}{3}\right) = 4\left(\frac{1}{2}\right) = 2$$

$$y = r\sin\theta = 4\sin\left(\frac{5\pi}{3}\right) = 4\left(-\frac{\sqrt{3}}{2}\right) = -2\sqrt{3}$$

So, the given point can be expressed in rectangular coordinates as $\boxed{\left(2,\ -2\sqrt{3}\right)}$.

22. In order to convert the point $\left(2, \frac{3\pi}{4}\right)$ (expressed in polar coordinates) to rectangular coordinates, first observe that $r = 2,\ \theta = \frac{3\pi}{4}$. Hence,

$$x = r\cos\theta = 2\cos\left(\frac{3\pi}{4}\right) = 2\left(-\frac{\sqrt{2}}{2}\right) = -\sqrt{2}$$

$$y = r\sin\theta = 2\sin\left(\frac{3\pi}{4}\right) = 2\left(\frac{\sqrt{2}}{2}\right) = \sqrt{2}$$

So, the given point can be expressed in rectangular coordinates as $\boxed{\left(-\sqrt{2},\ \sqrt{2}\right)}$.

23. In order to convert the point $\left(-1, \frac{5\pi}{6}\right)$ (expressed in polar coordinates) to rectangular coordinates, first observe that $r = -1,\ \theta = \frac{5\pi}{6}$. Hence,

$$x = r\cos\theta = -\cos\left(\frac{5\pi}{6}\right) = -\left(-\frac{\sqrt{3}}{2}\right) = \frac{\sqrt{3}}{2}$$

$$y = r\sin\theta = -\sin\left(\frac{5\pi}{6}\right) = -\left(\frac{1}{2}\right) = -\frac{1}{2}$$

So, the given point can be expressed in rectangular coordinates as $\boxed{\left(\frac{\sqrt{3}}{2}, -\frac{1}{2}\right)}$.

24. In order to convert the point $\left(-2, \frac{7\pi}{4}\right)$ (expressed in polar coordinates) to rectangular coordinates, first observe that $r = -2,\ \theta = \frac{7\pi}{4}$. Hence,

$$x = r\cos\theta = -2\cos\left(\frac{7\pi}{4}\right) = -2\left(\frac{\sqrt{2}}{2}\right) = -\sqrt{2}$$

$$y = r\sin\theta = -2\sin\left(\frac{7\pi}{4}\right) = -2\left(-\frac{\sqrt{2}}{2}\right) = \sqrt{2}$$

So, the given point can be expressed in rectangular coordinates as $\boxed{\left(-\sqrt{2}, \sqrt{2}\right)}$.

25. In order to convert the point $\left(0, \frac{11\pi}{6}\right)$ (expressed in polar coordinates) to rectangular coordinates, first observe that $r = 0,\ \theta = \frac{11\pi}{6}$. Hence,

$$x = r\cos\theta = 0\cos\left(\frac{11\pi}{6}\right) = 0$$

$$y = r\sin\theta = 0\sin\left(\frac{11\pi}{6}\right) = 0$$

So, the given point can be expressed in rectangular coordinates as $\boxed{(0,0)}$.

26. In order to convert the point $(6,0)$ (expressed in polar coordinates) to rectangular coordinates, first observe that $r = 6,\ \theta = 0$. Hence,

$$x = r\cos\theta = 6\cos(0) = 6(1) = 6$$

$$y = r\sin\theta = 6\sin(0) = 6(0) = 0$$

So, the given point can be expressed in rectangular coordinates as $\boxed{(6,0)}$.

27. In order to convert the point $(2,\ 240^\circ)$ (expressed in polar coordinates) to rectangular coordinates, first observe that $r = 2,\ \theta = 240^\circ$. Hence,

$$x = r\cos\theta = 2\cos\left(240^\circ\right) = 2\left(-\frac{1}{2}\right) = -1$$

$$y = r\sin\theta = 2\sin\left(240^\circ\right) = 2\left(-\frac{\sqrt{3}}{2}\right) = -\sqrt{3}$$

So, the given point can be expressed in rectangular coordinates as $\boxed{\left(-1, -\sqrt{3}\right)}$.

28. In order to convert the point $(-3, 150^\circ)$ (expressed in polar coordinates) to rectangular coordinates, first observe that $r = -3,\ \theta = 150^\circ$. Hence,

$$x = r\cos\theta = -3\cos(150^\circ) = -3\left(-\frac{\sqrt{3}}{2}\right) = \frac{3\sqrt{3}}{2}$$

$$y = r\sin\theta = -3\sin(150^\circ) = -3\left(\frac{1}{2}\right) = -\frac{3}{2}$$

So, the given point can be expressed in rectangular coordinates as $\boxed{\left(\frac{3\sqrt{3}}{2}, -\frac{3}{2}\right)}$.

29. In order to convert the point $(-1, 135^\circ)$ (expressed in polar coordinates) to rectangular coordinates, first observe that $r = -1,\ \theta = 135^\circ$. Hence,

$$x = r\cos\theta = -\cos(135^\circ) = -\left(-\frac{\sqrt{2}}{2}\right) = \frac{\sqrt{2}}{2}$$

$$y = r\sin\theta = -\sin(135^\circ) = -\left(\frac{\sqrt{2}}{2}\right) = -\frac{\sqrt{2}}{2}$$

So, the given point can be expressed in rectangular coordinates as $\boxed{\left(\frac{\sqrt{2}}{2}, -\frac{\sqrt{2}}{2}\right)}$.

30. In order to convert the point $(5, 315^\circ)$ (expressed in polar coordinates) to rectangular coordinates, first observe that $r = 5,\ \theta = 315^\circ$. Hence,

$$x = r\cos\theta = 5\cos(315^\circ) = 5\left(\frac{\sqrt{2}}{2}\right) = \frac{5\sqrt{2}}{2} \qquad y = r\sin\theta = 5\sin(315^\circ) = 5\left(-\frac{\sqrt{2}}{2}\right) = -\frac{5\sqrt{2}}{2}$$

So, the given point can be expressed in rectangular coordinates as $\boxed{\left(\frac{5\sqrt{2}}{2}, -\frac{5\sqrt{2}}{2}\right)}$.

31. d	**32. b**	**33. a**	**34. c**

35. The graph of $r = 5$ is a circle centered at the origin with radius 5 since for any given angle, the same distance from the origin is used to plot the point.

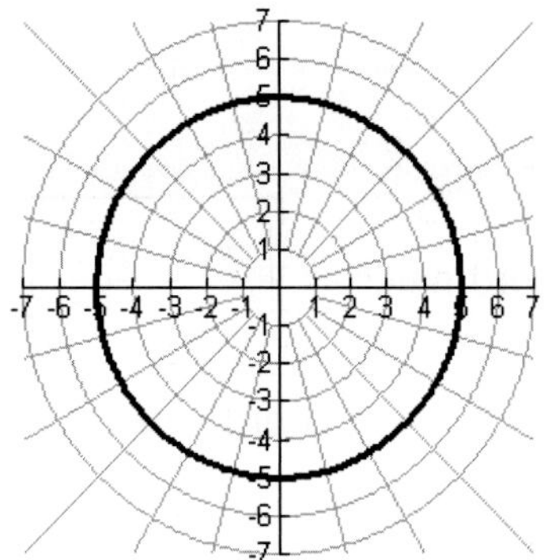

36. The graph $\theta = -\pi/3$ is a line through the origin which makes an angle of $-\pi/3$ with the positive x-axis.

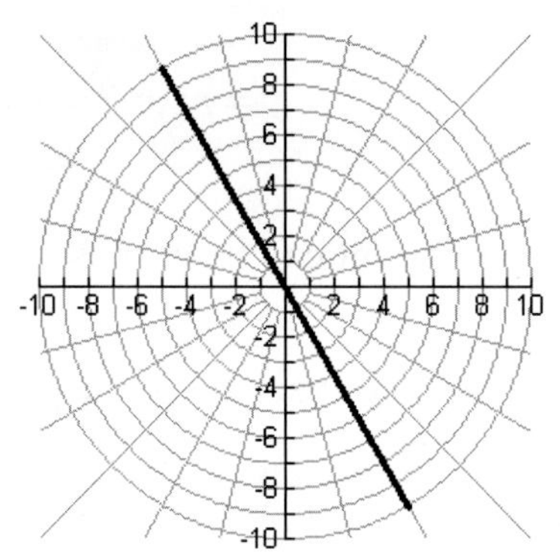

37. In order to graph $r = 2\cos\theta$, consider the following table of points:

θ	$r = 2\cos\theta$	(r, θ)
0	2	$(2, 0)$
$\pi/2$	0	$(0, \pi/2)$
π	-2	$(-2, \pi)$
$3\pi/2$	0	$(0, 3\pi/2)$
2π	2	$(2, 2\pi)$

The graph is as follows:

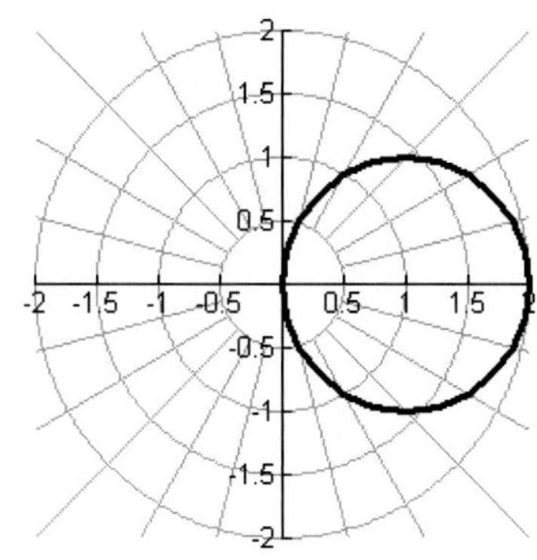

38. In order to graph $r = 3\sin\theta$, consider the following table of points:

θ	$r = 3\sin\theta$	(r, θ)
0	0	$(0, 0)$
$\pi/2$	3	$(3, \pi/2)$
π	0	$(0, \pi)$
$3\pi/2$	-3	$(-3, 3\pi/2)$
2π	0	$(0, 2\pi)$

The graph is as follows:

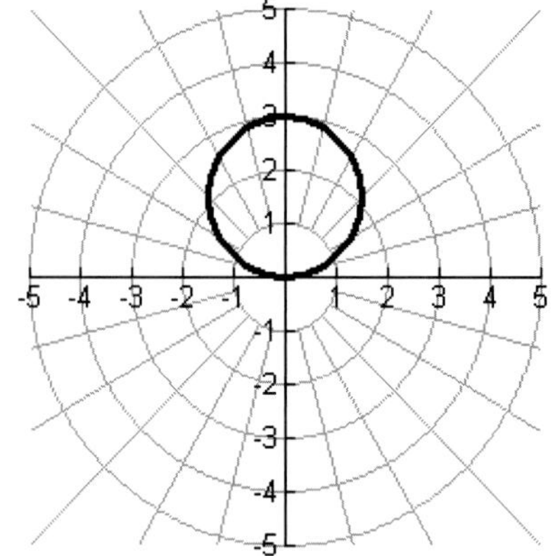

39. In order to graph $r = 4\sin 2\theta$, consider the following table of points:

θ	$r = 4\sin 2\theta$	(r,θ)
0	0	$(0,0)$
$\pi/4$	4	$(4,\pi/4)$
$\pi/2$	0	$(0,\pi/2)$
$3\pi/4$	-4	$(-4,3\pi/4)$
π	0	$(0,\pi)$
$5\pi/4$	4	$(4,5\pi/4)$
$3\pi/2$	0	$(0,3\pi/2)$
$7\pi/4$	-4	$(-4,7\pi/4)$
2π	0	$(0,2\pi)$

The graph is as follows:

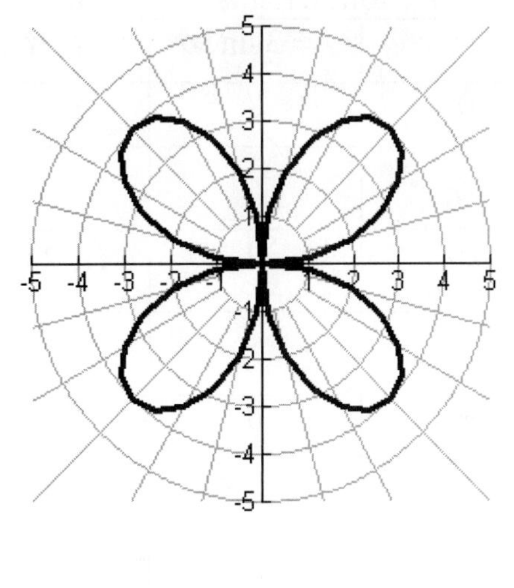

40. In order to graph $r = 5\cos 2\theta$, consider the following table of points:

θ	$r = 5\cos 2\theta$	(r,θ)
0	5	$(5,0)$
$\pi/4$	0	$(0,\pi/4)$
$\pi/2$	-5	$(-5,\pi/2)$
$3\pi/4$	0	$(0,3\pi/4)$
π	5	$(5,\pi)$
$5\pi/4$	0	$(0,5\pi/4)$
$3\pi/2$	-5	$(-5,3\pi/2)$
$7\pi/4$	0	$(0,7\pi/4)$
2π	5	$(5,2\pi)$

The graph is as follows:

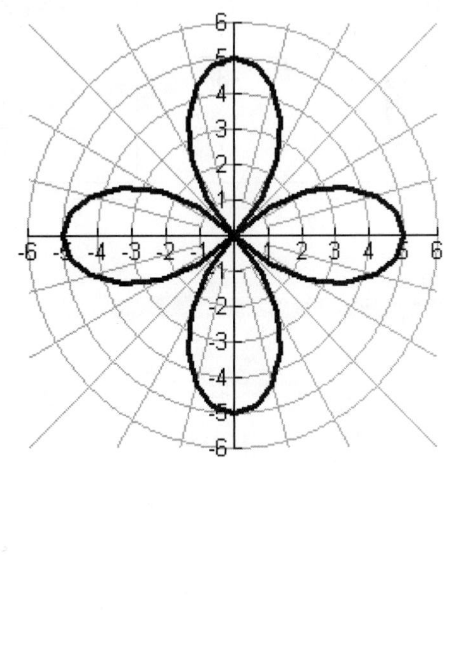

41. In order to graph $r = 3\sin 3\theta$, consider the following table of points:

θ	$r = 3\sin 3\theta$	(r,θ)
0	0	$(0,0)$
$\pi/6$	3	$(3,\pi/6)$
$2\pi/6$	0	$(0,2\pi/6)$
$\pi/2$	-3	$(-3,\pi/2)$
$4\pi/6$	0	$(0,4\pi/6)$
$5\pi/6$	3	$(3,5\pi/6)$
π	0	$(0,\pi)$
$7\pi/6$	-3	$(-3,7\pi/6)$
$8\pi/6$	0	$(0,8\pi/6)$
$3\pi/2$	3	$(3,3\pi/2)$
$10\pi/6$	0	$(0,10\pi/6)$
$11\pi/6$	-3	$(-3,11\pi/6)$
2π	0	$(0,2\pi)$

The graph is as follows:

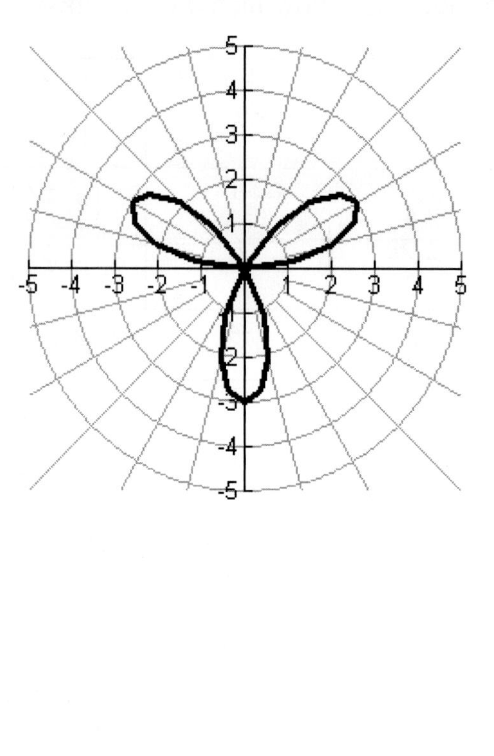

42. In order to graph $r = 4\cos 3\theta$, consider the following table of points:

θ	$r = 4\cos 3\theta$	(r, θ)
0	4	$(4, 0)$
$\pi/6$	0	$(0, \pi/6)$
$2\pi/6$	-4	$(-4, 2\pi/6)$
$\pi/2$	0	$(0, \pi/2)$
$4\pi/6$	4	$(4, 4\pi/6)$
$5\pi/6$	0	$(0, 5\pi/6)$
π	-4	$(-4, \pi)$
$7\pi/6$	0	$(0, 7\pi/6)$
$8\pi/6$	4	$(4, 8\pi/6)$
$3\pi/2$	0	$(0, 3\pi/2)$
$10\pi/6$	-4	$(-4, 10\pi/6)$
$11\pi/6$	0	$(0, 11\pi/6)$
2π	4	$(4, 2\pi)$

The graph is as follows:

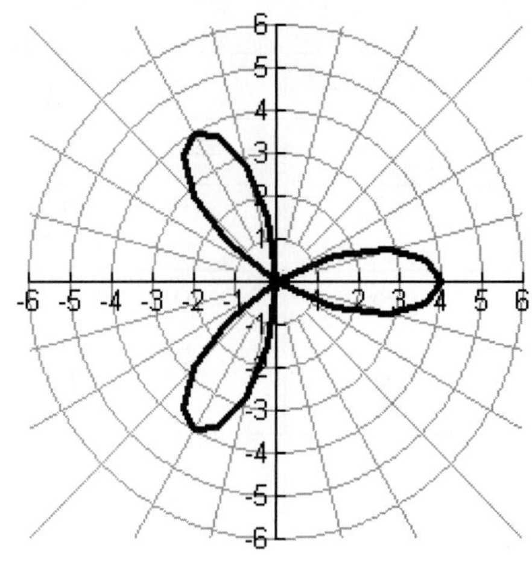

43. In order to graph $r^2 = 9\cos 2\theta$, consider the following table of points:

θ	$r^2 = 9\cos 2\theta$	r	(r,θ)
0	9	±3	$(\pm 3, 0)$
$\pi/4$	0	0	$(0, \pi/4)$
$\pi/2$	-9	No points	
$3\pi/4$	0	0	$(0, 3\pi/4)$
π	9	±3	$(\pm 3, \pi)$
$5\pi/4$	0	0	$(0, 5\pi/4)$
$3\pi/2$	-9	No points	
$7\pi/4$	0	0	$(0, 7\pi/4)$
2π	9	±3	$(\pm 3, 2\pi)$

The graph is as follows:

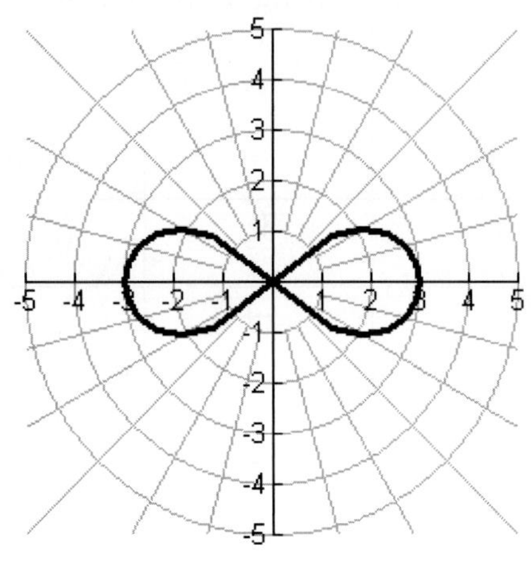

44. In order to graph $r^2 = 16\sin 2\theta$, consider the following table of points:

θ	$r^2 = 16\sin 2\theta$	r	(r,θ)
0	0	0	$(0,0)$
$\pi/4$	16	±4	$(\pm 4, \pi/4)$
$\pi/2$	0	0	$(0, \pi/2)$
$3\pi/4$	-16	No points	
π	0	0	$(0,\pi)$
$5\pi/4$	16	±4	$(\pm 4, 5\pi/4)$
$3\pi/2$	0	0	$(0, 3\pi/2)$
$7\pi/4$	-16	No points	
2π	0	0	$(0, 2\pi)$

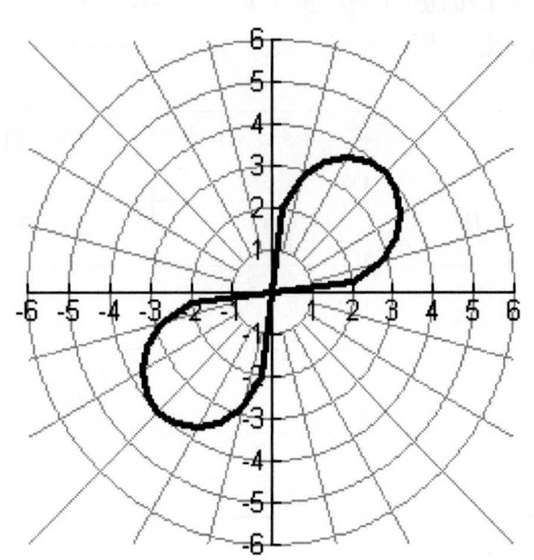

45. In order to graph $r = -2\cos\theta$, consider the following table of points:

θ	$r = -2\cos\theta$	(r,θ)
0	-2	$(-2,0)$
$\pi/2$	0	$(0, \pi/2)$
π	2	$(2,\pi)$
$3\pi/2$	0	$(0, 3\pi/2)$
2π	-2	$(-2, 2\pi)$

The graph is as follows:

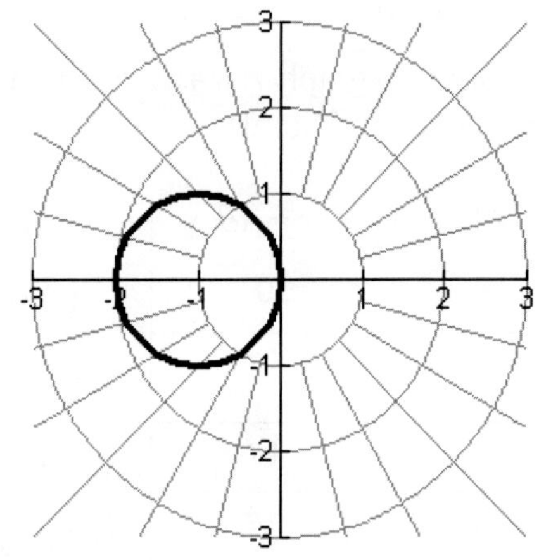

46. In order to graph $r = -3\sin 3\theta$, consider the following table of points:

θ	$r = -3\sin 3\theta$	(r,θ)
0	0	$(0,0)$
$\pi/6$	-3	$(-3,\pi/6)$
$2\pi/6$	0	$(0,2\pi/6)$
$\pi/2$	3	$(3,\pi/2)$
$4\pi/6$	0	$(0,4\pi/6)$
$5\pi/6$	-3	$(-3,5\pi/6)$
π	0	$(0,\pi)$
$7\pi/6$	3	$(3,7\pi/6)$
$8\pi/6$	0	$(0,8\pi/6)$
$3\pi/2$	-3	$(-3,3\pi/2)$
$10\pi/6$	0	$(0,10\pi/6)$
$11\pi/6$	3	$(3,11\pi/6)$
2π	0	$(0,2\pi)$

The graph is as follows:

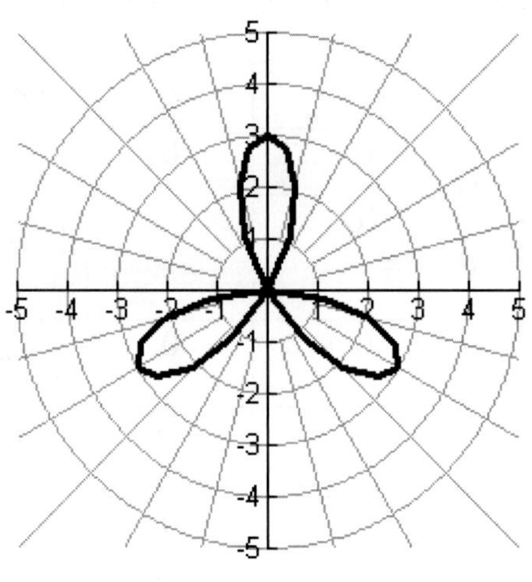

47. In order to graph $r = 4\theta$, consider the following table of points:

θ	$r = 4\theta$	(r,θ)
0	0	$(0,0)$
$\pi/2$	2π	$(2\pi,\pi/2)$
π	4π	$(4\pi,\pi)$
$3\pi/2$	6π	$(6\pi,3\pi/2)$
2π	8π	$(8\pi,2\pi)$

The graph is as follows:

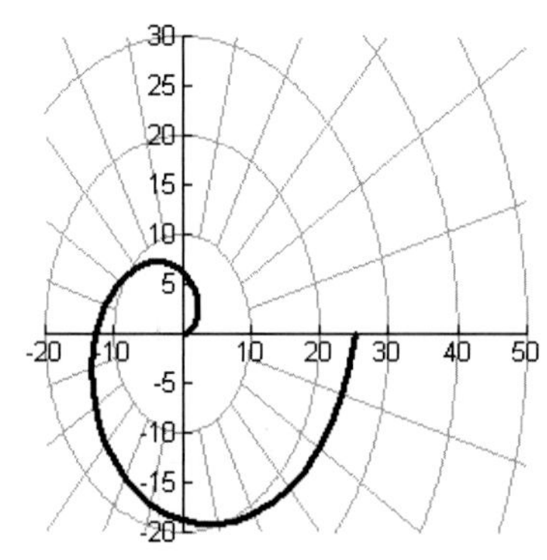

48. In order to graph $r=-2\theta$, consider the following table of points:

θ	$r=-2\theta$	(r,θ)
0	0	$(0,0)$
$\pi/2$	$-\pi$	$(-\pi,\pi/2)$
π	-2π	$(-2\pi,\pi)$
$3\pi/2$	-3π	$(-3\pi,3\pi/2)$
2π	-4π	$(-4\pi,2\pi)$

The graph is as follows:

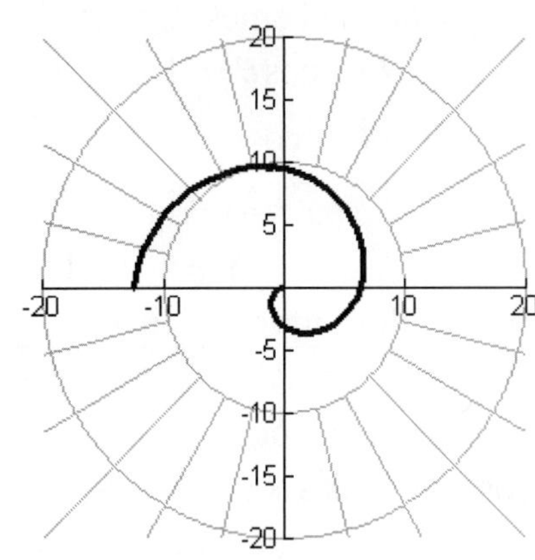

49. In order to graph $r=-3+2\cos\theta$, consider the following table of points:

θ	$r=-3+2\cos\theta$	(r,θ)
0	-1	$(-1,0)$
$\pi/2$	-3	$(-3,\pi/2)$
π	-5	$(-5,\pi)$
$3\pi/2$	-3	$(-3,3\pi/2)$
2π	-1	$(-1,2\pi)$

The graph is as follows:

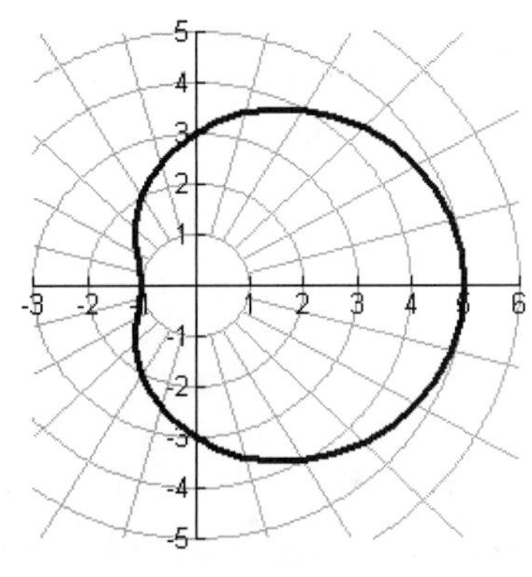

50. In order to graph $r=2+3\sin\theta$, consider the following table of points:

θ	$r=2+3\sin\theta$	(r,θ)
0	2	$(2,0)$
$\pi/2$	5	$(5,\pi/2)$
π	2	$(2,\pi)$
$3\pi/2$	-1	$(-1,3\pi/2)$
2π	2	$(2,2\pi)$

The graph is as follows:

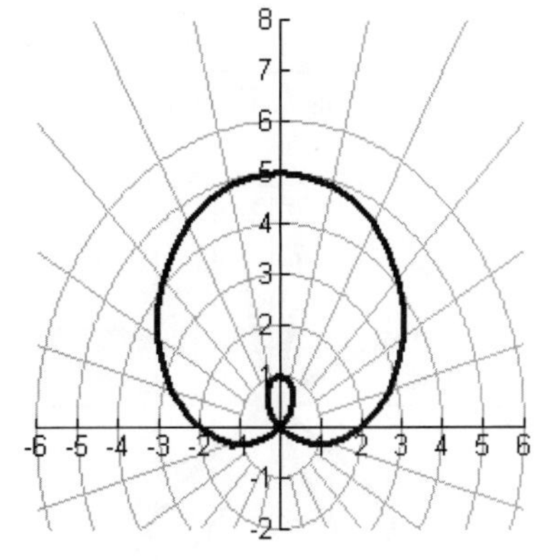

51. Using $x = r\cos\theta,\ y = r\sin\theta$, we see that the equation becomes

$$r(\sin\theta + 2\cos\theta) = 1$$
$$r\sin\theta + 2r\cos\theta = 1$$
$$y + 2x = 1$$
$$y = -2x + 1$$

The graph of this curve is a line.

52. Using $x = r\cos\theta,\ y = r\sin\theta$, we see that the equation becomes

$$r(\sin\theta - 3\cos\theta) = 2$$
$$r\sin\theta - 3r\cos\theta = 2$$
$$y - 3x = 2$$
$$y = 3x + 2$$

The graph of this curve is a line.

53. Using $x = r\cos\theta,\ y = r\sin\theta$, we see that the equation becomes

$$r^2\cos^2\theta - 2r\cos\theta + r^2\sin^2\theta = 8$$
$$x^2 - 2x + y^2 = 8$$
$$x^2 - 2x + 1 + y^2 = 8 + 1$$
$$(x-1)^2 + y^2 = 9$$

The graph of this curve is a circle.

54. Using $x = r\cos\theta,\ y = r\sin\theta$, we see that the equation becomes

$$r^2\cos^2\theta - r\sin\theta = -2$$
$$x^2 - y = -2$$
$$y = x^2 + 2$$

The graph of this curve is a parabola.

55. The graph is as follows:

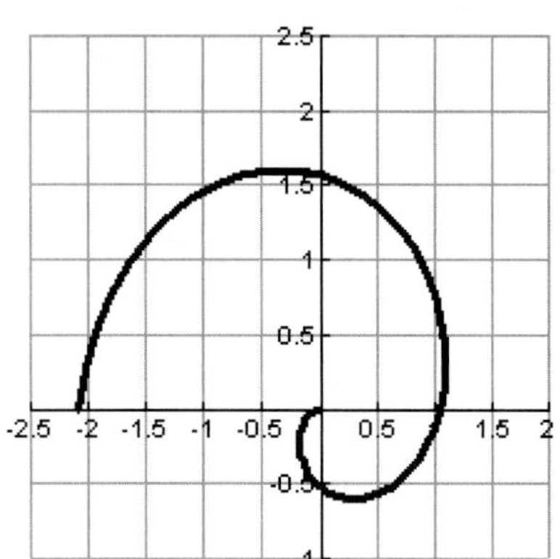

56. The graph is as follows:

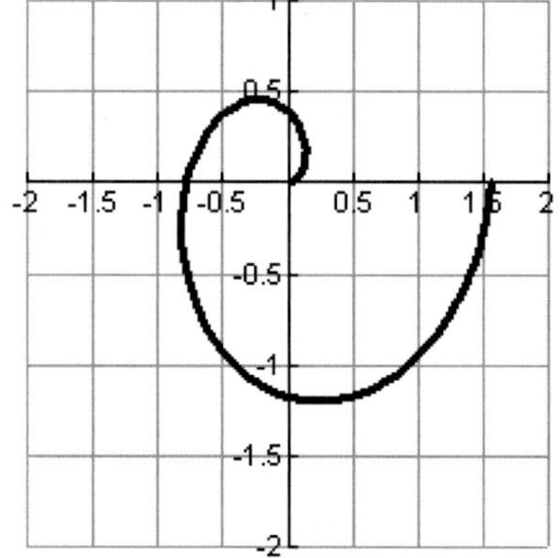

57. The graph is as follows:

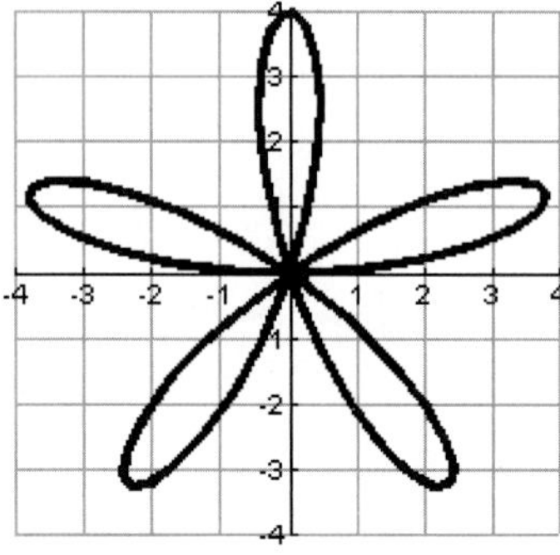

58. The graph is as follows:

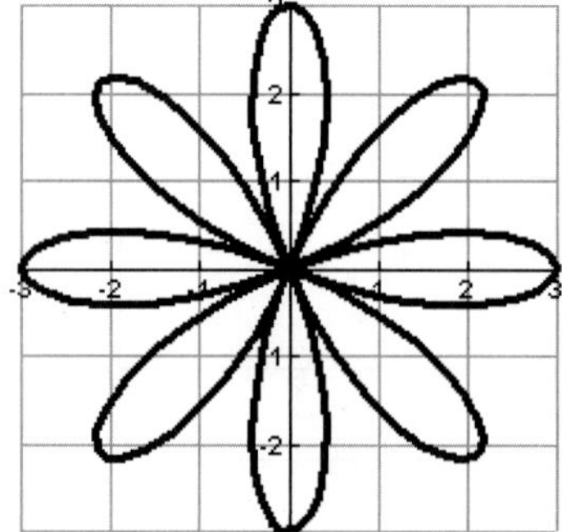

59. The graph is as follows:

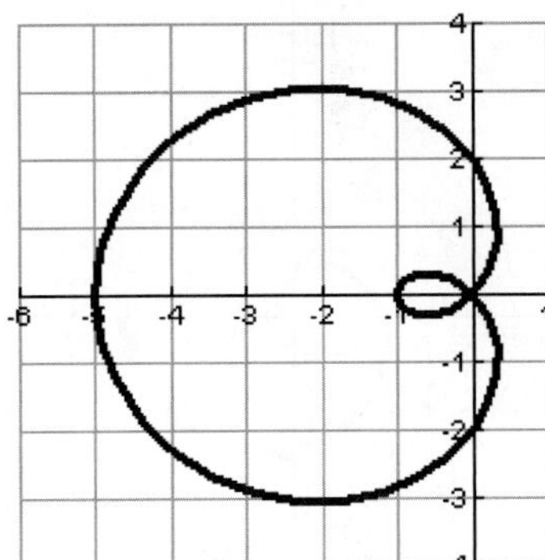

60. The graph is as follows:

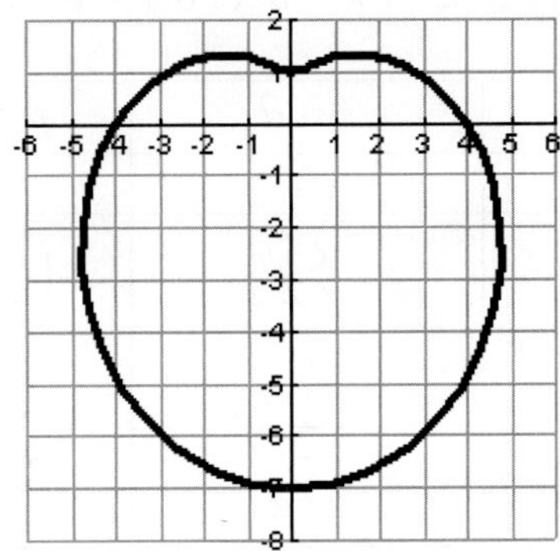

61. The graph of $r = \dfrac{0.587(1+0.967)}{1-0.967\cos\theta}$ is as follows:

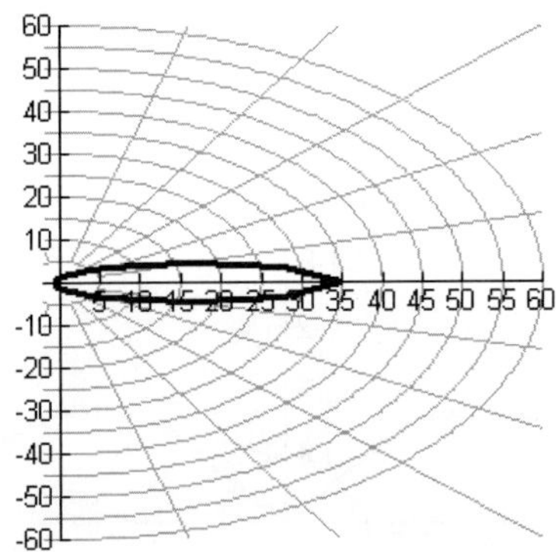

62. The graph of $r = \dfrac{29.62(1+0.249)}{1-0.249\cos\theta}$ is as follows:

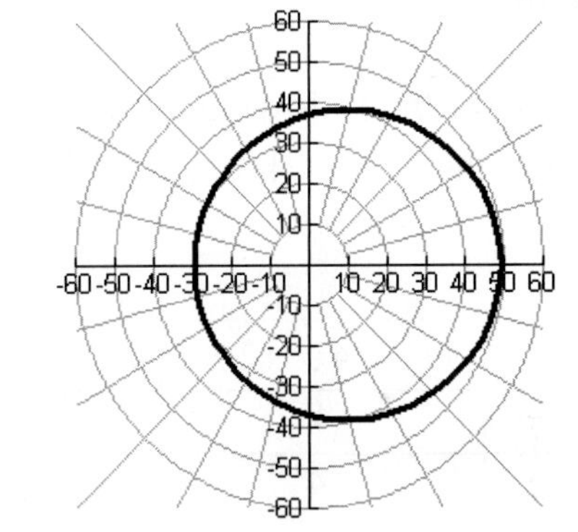

63. Consider the following table of values for the graphs of $r=\theta$ and $r=\sqrt{\theta}$:

θ	$r=\theta$	$r=\sqrt{\theta}$
0	0	0
$\pi/2$	$\pi/2$	$\sqrt{\pi/2}$
π	π	$\sqrt{\pi}$
$3\pi/2$	$3\pi/2$	$\sqrt{3\pi/2}$
2π	2π	$\sqrt{2\pi}$

Notice that the graph of $r=\sqrt{\theta}$ (dotted) is more tightly wound than is the graph of $r=\theta$ (solid).

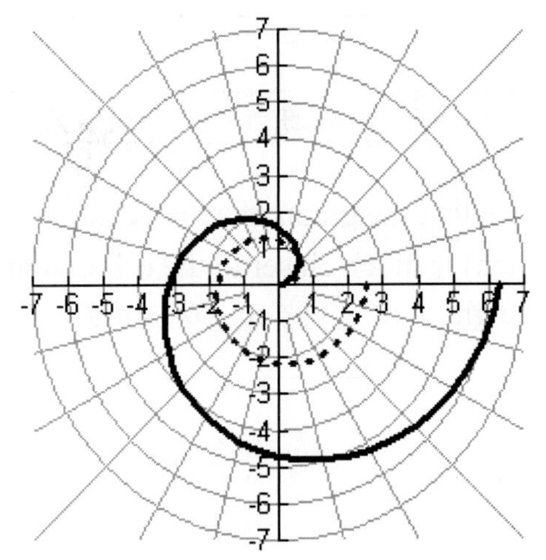

64. Consider the following table of values for the graphs of $r=\theta$ and $r=\theta^{4/3}$:

θ	$r=\theta$	$r=\theta^{4/3}$
0	0	0
$\pi/2$	$\pi/2$	$(\pi/2)^{4/3}$
π	π	$(\pi)^{4/3}$
$3\pi/2$	$3\pi/2$	$(3\pi/2)^{4/3}$
2π	2π	$(2\pi)^{4/3}$

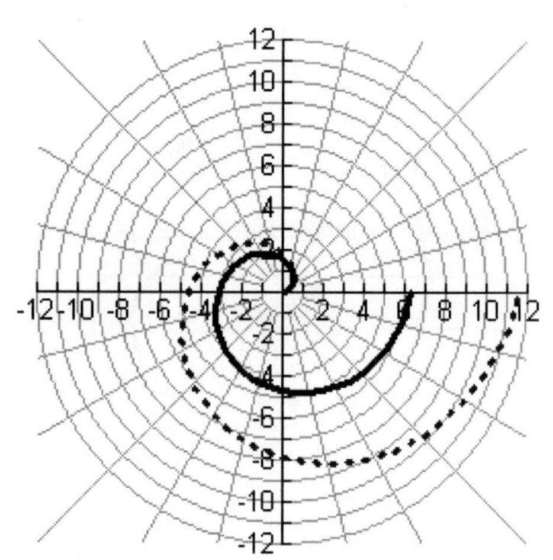

Notice that the graph of $r=\theta^{4/3}$ (dotted) is more loosely wound than is the graph of $r=\theta$ (solid).

65. Consider the following table of values for the graphs of $r^2=4\cos 2\theta$ and $r^2=\frac{1}{4}\cos 2\theta$:

θ	$r^2=4\cos 2\theta$	$r^2=\frac{1}{4}\cos 2\theta$
0	4	$1/4$
$\pi/2$	-4	$-1/4$
π	4	$1/4$
$3\pi/2$	-4	$-1/4$
2π	4	$1/4$

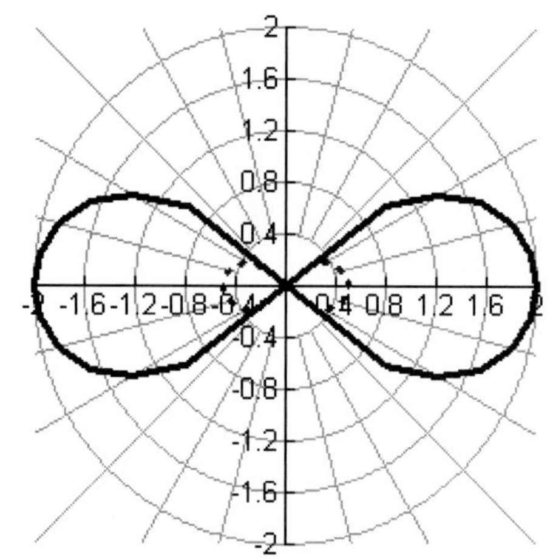

Notice that the graph of $r^2=\frac{1}{4}\cos 2\theta$ (dotted) is much closer to the origin than is the graph of $r^2=4\cos 2\theta$ (solid).

66. The graphs of $r^2 = 4\cos 2\theta$ and $r^2 = 4\cos(2\theta + 2)$ are given to the right. Notice that the graph of $r^2 = 4\cos(2\theta + 2)$ (dotted) is simply a rotation of the graph of $r^2 = 4\cos 2\theta$ (solid). The same set of points is generated, but starting/ending at different positions on the graph.

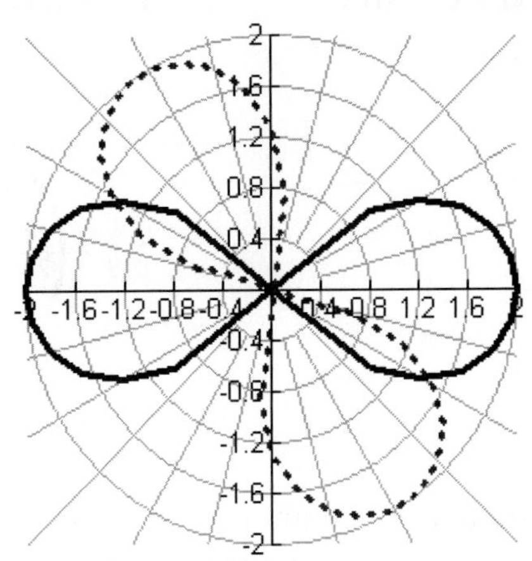

67. In order to graph $r = 2 + 2\sin\theta$, consider the following table of points:

θ	$r=2+2\sin\theta$	(r,θ)
0	2	$(2,0)$
$\pi/2$	4	$(4,\pi/2)$
π	2	$(2,\pi)$
$3\pi/2$	0	$(0,3\pi/2)$
2π	2	$(2,2\pi)$

The graph is as follows:

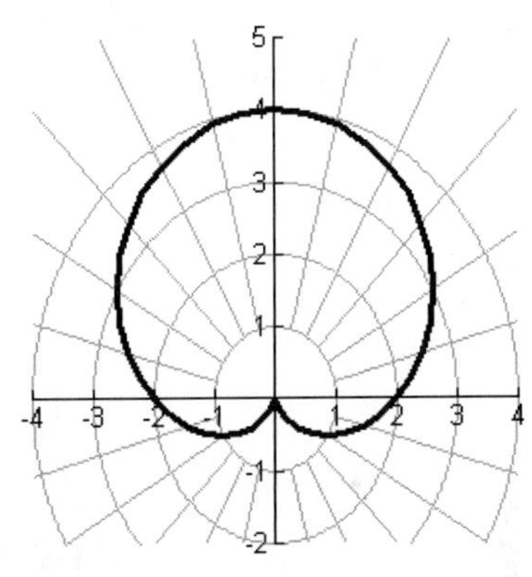

68. In order to graph $r = -4 - 4\sin\theta$, consider the following table of points:

θ	$r=-4-4\sin\theta$	(r,θ)
0	-4	$(-4,0)$
$\pi/2$	-8	$(-8,\pi/2)$
π	-4	$(-4,\pi)$
$3\pi/2$	0	$(0,3\pi/2)$
2π	-4	$(-4,2\pi)$

The graph is as follows:

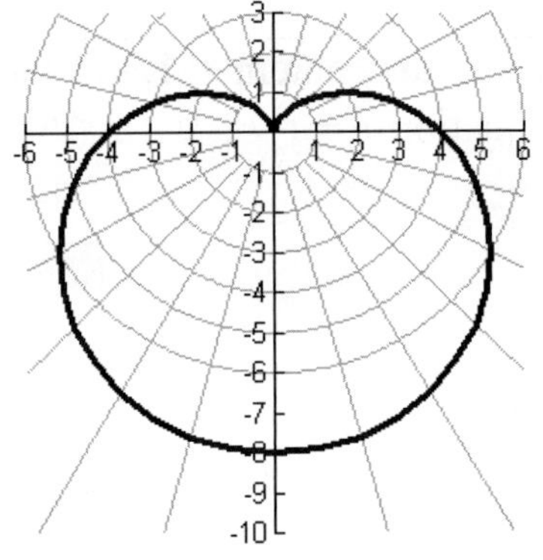

69. a) – c) All three graphs generate the same set of points, as seen below:

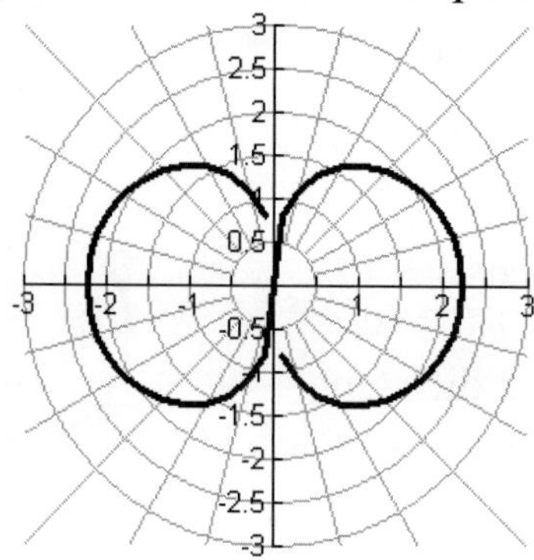

Note that all three graphs are figure eights. Extending the domain in **b** results in twice as fast movement, while doing so in **c** results in movement that is four times as fast.

70. We must choose $L = 12$, and choose a such that $r^2 = \cos(a\theta)$ makes 8 complete cycles in the given interval. Using the observations in Exercise 69, we see that a must be 8, so that the equation is $\boxed{r^2 = 12\cos(8\theta)}$.

71. You need to start at 0 and hit $x = 5$ at $\theta = 2\pi$, and then $x = 10$ at $\theta = 4\pi$, and so on until $x = 30$ is hit at $\theta = 12\pi$. Thus, we use the function $r = \frac{5}{2\pi}\theta,\ 0 \le \theta \le 12\pi$, which is pictured below:

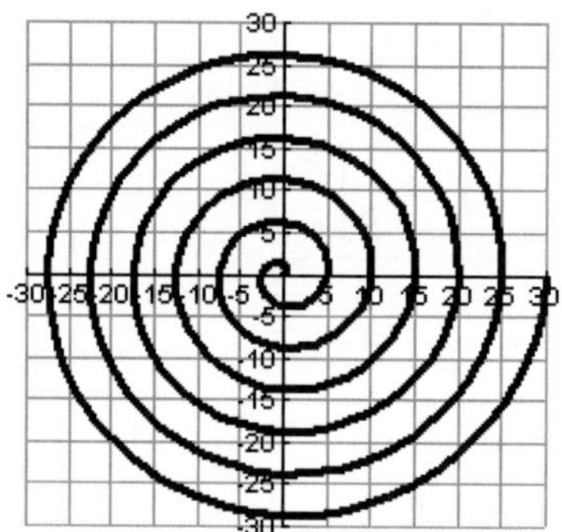

72. Similar to problem #71, the walkway can now be 7 feet wide. The path of the spiral equation $r = \frac{10}{2\pi}\theta,\ 0 \le \theta \le 6\pi$ is as follows:

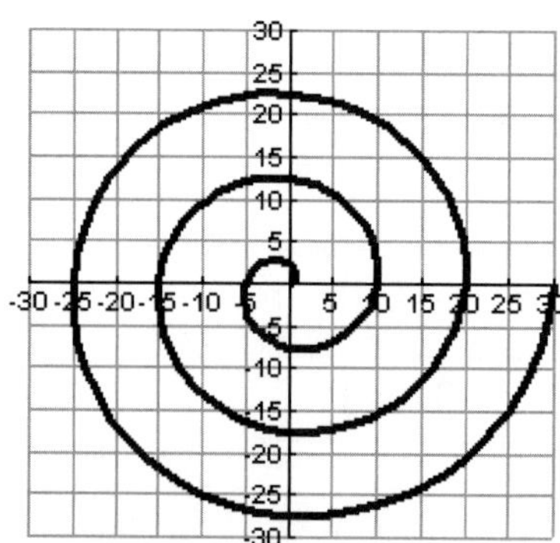

73. **(i)** The equation of the polar curve that describes this scenario is $r = 8\sin 3\theta,\ 0 \le \theta \le 2\pi$.

(ii) It retraces the path 50 times in the interval $[0,100\pi]$ since each complete trace occurs in an interval of length 2π.

74. In order to create an 8-petaled rose of the sort described in #73, we can use the polar equation $r = 8\cos 4\theta,\ 0 \le \theta \le 2\pi$, whose graph is below:

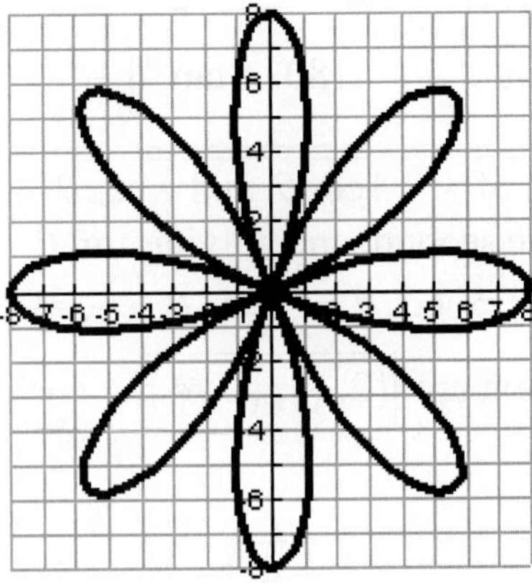

75. The point is in QIII, so needed to add π to the angle.	**76.** The point is in QII, so needed to add π to the angle.

77. True. By definition.

78. False. The equation of a cardioid is either $r = a + a\cos\theta$ or $r = a + a\sin\theta$, while the more general limacon has equation either $r = a + b\cos\theta$ or $r = a + b\sin\theta$. So, simply choose $a \ne b$ to get a limacon that is not a cardioid.

79. Observe that if $x = a$, then $r\cos\theta = a$, so that $r = \dfrac{a}{\cos\theta}$, which represents a vertical line.	**80.** Observe that if $y = b$, then $r\sin\theta = b$, so that $r = \dfrac{b}{\sin\theta}$, which represents a horizontal line.

81. The point (a,θ) can also be represented as $\left(-a,\theta \pm 180^\circ\right)$. This is because the point $(-a,\theta)$ is diametrically opposed (through the origin) to the given point. As such, we need to rotate from it 180° clockwise or counterclockwise to arrive at the given point.

82. In order to convert the point $(-a,b)$ (expressed in rectangular coordinates) to polar coordinates, first observe that $x=-a,\ y=b$, so that the point is in QII. Hence,

$$r=\sqrt{(-a)^2+(b)^2}=\sqrt{a^2+b^2}$$

$$\tan\theta=\frac{y}{x}=\frac{b}{-a}=-\frac{b}{a} \text{ so that } \theta=180^\circ+\tan^{-1}\left(-\frac{b}{a}\right).$$

So, the given point can be expressed in polar coordinates as

$$\boxed{\left(\sqrt{a^2+b^2},180^\circ+\tan^{-1}\left(-\frac{b}{a}\right)\right)}.$$

83. The points of intersection of $r=4\cos\theta$ and $r\cos\theta=1$ are found by substituting the first equation in for r in the second, and solving for θ:

$$4\cos^2\theta=1 \Rightarrow \cos\theta=\tfrac{1}{2} \Rightarrow \theta=\tfrac{\pi}{3}, \tfrac{5\pi}{3} \text{ (since } r>0)$$

Hence, the points of intersection are $\boxed{\left(2,\tfrac{\pi}{3}\right), \left(2,\tfrac{5\pi}{3}\right)}$. The graphs of these two curves are displayed below:

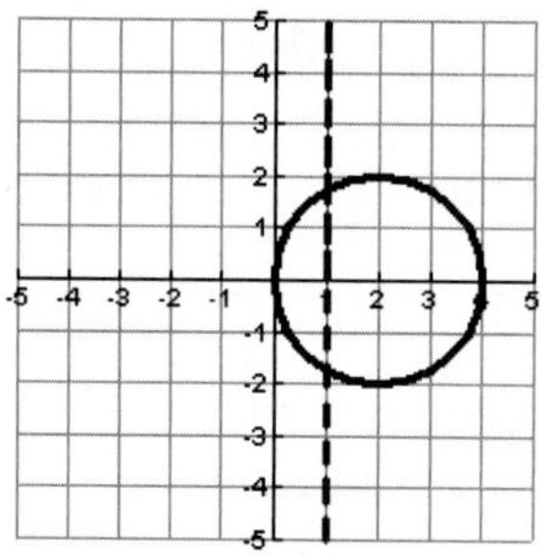

84. We convert $r=a\sin\theta+b\cos\theta$ to a Cartesian equation as follows:

$$r=a\sin\theta+b\cos\theta$$

$$r^2=ar\sin\theta+br\cos\theta$$

$$x^2+y^2=ay+bx$$

$$x^2-bx+y^2-ay=0$$

$$\left(x-\tfrac{b}{2}\right)^2+\left(y-\tfrac{a}{2}\right)^2=\frac{a^2+b^2}{4} \quad \text{(upon completing the square)}$$

This is the equation of a circle.

85. We convert $r=\dfrac{a\sin 2\theta}{\cos^3\theta-\sin^3\theta}$ to a Cartesian equation as follows: $$r=\frac{a\sin 2\theta}{\cos^3\theta-\sin^3\theta}$$ $$r\left(\cos^3\theta-\sin^3\theta\right)=a\sin 2\theta$$ $$r^3\left(\cos^3\theta-\sin^3\theta\right)=ar^2\sin 2\theta$$ $$\left(r\cos\theta\right)^3-\left(r\sin\theta\right)^3=ar^2\left(2\sin\theta\cos\theta\right)$$ $$\left(r\cos\theta\right)^3-\left(r\sin\theta\right)^3=2a\left(r\sin\theta\right)\left(r\cos\theta\right)$$ $$x^3-y^3-2axy=0$$
86. The equation for this graph would be $r=10\sin(9\theta),\ 0\le\theta\le 2\pi$.
87. Graphing the equation $r=2a\cos(\theta-b)$ for different values of a and b always yields a circle with radius a centered at (a, b).
88. Note that the period of $\sin b\theta$ (and of $a\sin b\theta$) is $\frac{2\pi}{b}$. As such, the smallest value of M for which the graph starts to repeat is $\frac{2\pi}{b}$.

89. The graph of $r=\cos\left(\dfrac{\theta}{2}\right)$ is as follows:

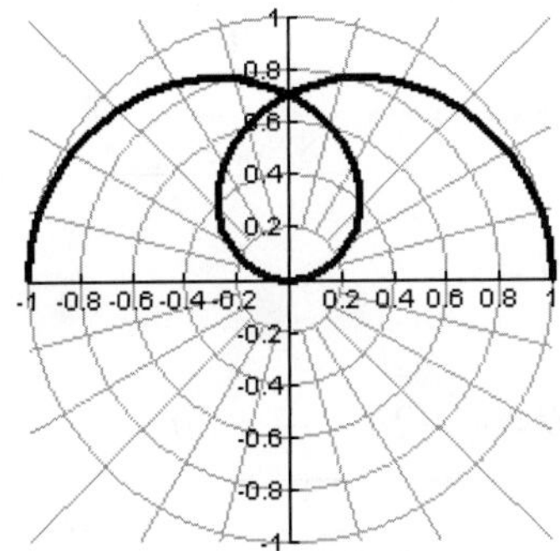

The inner loop is generated beginning with $\theta=\dfrac{\pi}{2}$ and ending with $\dfrac{3\pi}{2}$.

90. The graph of $r=2\cos\left(\dfrac{3\theta}{2}\right)$ is as follows:

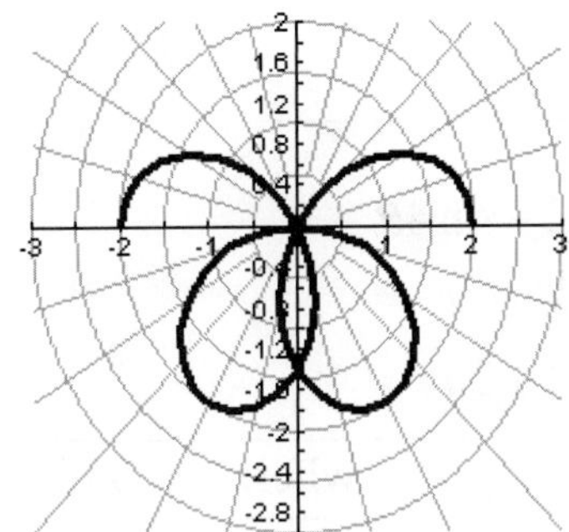

The petal in the 1st quadrant is generated beginning with $\theta=0$ and ending with $\frac{\pi}{3}$.

91. The graph of $r = 1 + 3\cos\theta$ is as follows:

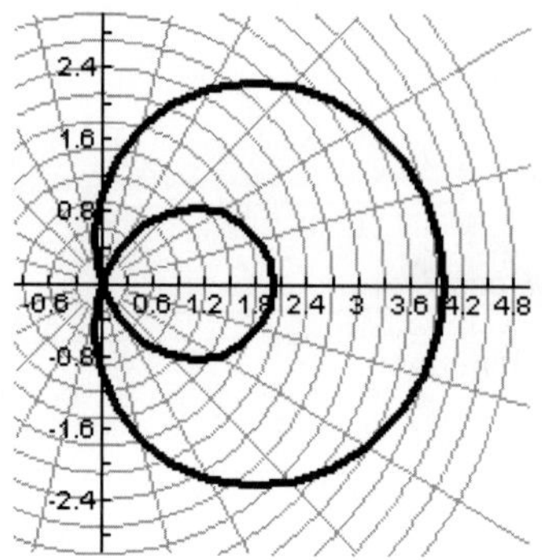

The very tip of the inner loop begins with $\theta = \frac{\pi}{2}$, then it crosses the origin (the first time) at $\theta = \cos^{-1}\left(-\frac{1}{3}\right)$, winds around, and eventually ends with $\theta = \frac{3\pi}{2}$.

92. The graphs of $r = 1 + \sin(2\theta)$ (solid) and $r = 1 - \cos(2\theta)$ (dashed) are as follows:

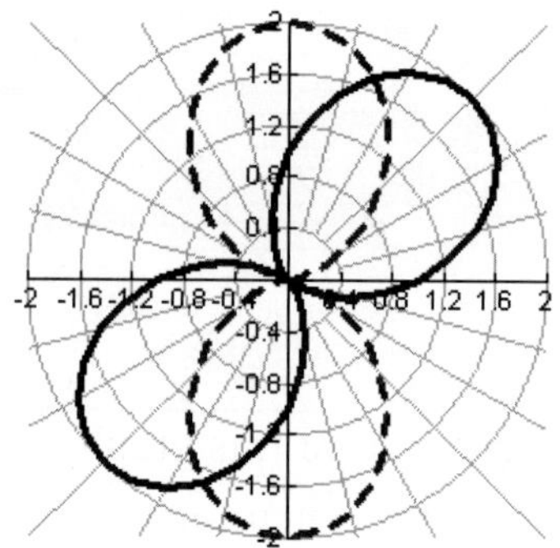

Solving $1 + \sin(2\theta) = 1 - \cos(2\theta)$,
we see that $\sin(2\theta) + \cos(2\theta) = 0$, which is satisfied when 2θ is an integer multiple of $\frac{\pi}{4}$ that has terminal side in QII or QIV. As such, we see that $2\theta = \frac{3\pi}{4}, \frac{7\pi}{4}, \frac{11\pi}{4}, \frac{15\pi}{4}$ and hence, $\theta = \frac{3\pi}{8}, \frac{7\pi}{8}, \frac{11\pi}{8}, \frac{15\pi}{8}$.

93.

$$1 = 2 + \sin 4\theta$$
$$1 = \sin 4\theta$$
$$4\theta = \frac{\pi}{2} + 2n\pi$$
$$\boxed{\theta = \frac{\pi}{8} + \frac{n\pi}{2}, \text{ where } n \text{ is an integer}}$$

94.

$$2 - \cos 3\theta = 1.5$$
$$\cos 3\theta = 0.5$$
$$3\theta = \frac{\pi}{3} + 2n\pi,\ \frac{5\pi}{3} + 2n\pi$$
$$\boxed{\theta = \frac{\pi}{9} + \frac{2n\pi}{3},\ \frac{5\pi}{9} + \frac{2n\pi}{3},\ n \text{ an integer}}$$

95.

$$4\sin\theta = 4\cos\theta$$
$$\sin\theta = \cos\theta$$
$$\theta = \frac{\pi}{4}, \frac{5\pi}{4}$$

The corresponding r-values are

$$4\sin\left(\frac{\pi}{4}\right) = 4\left(\frac{\sqrt{2}}{2}\right) = 2\sqrt{2}$$
$$4\sin\left(\frac{5\pi}{4}\right) = 4\left(-\frac{\sqrt{2}}{2}\right) = -2\sqrt{2}$$

So, the points of intersection are

$$\left(2\sqrt{2}, \frac{\pi}{4}\right), \left(-2\sqrt{2}, \frac{5\pi}{4}\right).$$

96.

$$\cos\theta = 2 + 3\cos\theta$$
$$-1 = \cos\theta$$
$$\theta = \pi$$

The corresponding r-value is -1. So, the point of intersection is $(-1, \pi)$.

97.	**98.**
$1-\sin\theta = 1+\cos\theta$ $-\sin\theta = \cos\theta$ $\theta = \frac{3\pi}{4}, \frac{7\pi}{4}$ The corresponding r-values are: $1-\sin\left(\frac{3\pi}{4}\right) = 1-\frac{\sqrt{2}}{2} = \frac{2-\sqrt{2}}{2}$ $1-\sin\left(\frac{7\pi}{4}\right) = 1+\frac{\sqrt{2}}{2} = \frac{2+\sqrt{2}}{2}$ So, the points of intersection are $\left(\frac{2-\sqrt{2}}{2}, \frac{3\pi}{4}\right), \left(\frac{2+\sqrt{2}}{2}, \frac{7\pi}{4}\right)$	$1+\sin\theta = 1+\cos\theta$ $\sin\theta = \cos\theta$ $\theta = \frac{\pi}{4}, \frac{5\pi}{4}$ The corresponding r-values are: $1+\sin\left(\frac{\pi}{4}\right) = 1+\frac{\sqrt{2}}{2} = \frac{2+\sqrt{2}}{2}$ $1+\sin\left(\frac{5\pi}{4}\right) = 1-\frac{\sqrt{2}}{2} = \frac{2-\sqrt{2}}{2}$ So, the points of intersection are $\left(\frac{2+\sqrt{2}}{2}, \frac{\pi}{4}\right), \left(\frac{2-\sqrt{2}}{2}, \frac{5\pi}{4}\right)$

Chapter 7 Review Solutions--

1.	**2.**
$\overrightarrow{AB} = \langle -8-4, 2-(-3)\rangle = \langle -12, 5\rangle$ $\left\|\overrightarrow{AB}\right\| = \sqrt{(-12)^2+(5)^2} = \boxed{13}$	$\overrightarrow{AB} = \langle 2-(-2), 8-11\rangle = \langle 4, -3\rangle$ $\left\|\overrightarrow{AB}\right\| = \sqrt{(4)^2+(-3)^2} = \boxed{5}$
3. $\overrightarrow{AB} = \langle 5-0, 9-(-3)\rangle = \langle 5, 12\rangle$ $\left\|\overrightarrow{AB}\right\| = \sqrt{(5)^2+(12)^2} = \boxed{13}$	**4.** $\overrightarrow{AB} = \langle 9-3, -3-(-11)\rangle = \langle 6, 8\rangle$ $\left\|\overrightarrow{AB}\right\| = \sqrt{(6)^2+(8)^2} = \boxed{10}$
5. Given that $\vec{u} = \langle -10, 24\rangle$, we have $\|\vec{u}\| = \sqrt{(-10)^2+(24)^2} = 26$ $\tan\theta = \frac{24}{-10}$ so that since the head is in QII, $\theta = 180^\circ + \tan^{-1}\left(\frac{24}{-10}\right) \approx 112.6^\circ$	**6.** Given that $\vec{u} = \langle -5, -12\rangle$, we have $\|\vec{u}\| = \sqrt{(-5)^2+(-12)^2} = 13$ $\tan\theta = \frac{12}{5}$ so that $\theta = \tan^{-1}\left(\frac{12}{5}\right) \approx 67.4^\circ$
7. Given that $\vec{u} = \langle 16, -12\rangle$, we have $\|\vec{u}\| = \sqrt{(16)^2+(-12)^2} = 20$ $\tan\theta = \frac{-12}{16}$ so that $\theta = \tan^{-1}\left(\frac{-12}{16}\right) \approx 323.1^\circ$	**8.** Given that $\vec{u} = \langle 0, 3\rangle$, we have $\|\vec{u}\| = \sqrt{(0)^2+(3)^2} = 3$ Since the vector is on the y-axis pointing in the positive direction, $\theta = 90^\circ$.

9. $2\vec{u}+3\vec{v}=2\langle 7,-2\rangle+3\langle -4,5\rangle=\langle 14,-4\rangle+\langle -12,15\rangle=\langle 14-12,-4+15\rangle=\boxed{\langle 2,11\rangle}$	
10. $\vec{u}-\vec{v}=\langle 7,-2\rangle-\langle -4,5\rangle=\langle 7-(-4),-2-5\rangle=\boxed{\langle 11,-7\rangle}$	
11. $6\vec{u}+\vec{v}=6\langle 7,-2\rangle+\langle -4,5\rangle=\langle 42,-12\rangle+\langle -4,5\rangle=\langle 42-4,-12+5\rangle=\boxed{\langle 38,-7\rangle}$	
12. $-3(\vec{u}+2\vec{v})=-3(\langle 7,-2\rangle+2\langle -4,5\rangle)=-3(\langle 7,-2\rangle+\langle -8,10\rangle)=-3\langle -1,8\rangle=\boxed{\langle 3,-24\rangle}$	
13. $\vec{u}=\langle \lvert\vec{u}\rvert\cos\theta,\ \lvert\vec{u}\rvert\sin\theta\rangle=\langle 10\cos 75^\circ,\ 10\sin 75^\circ\rangle\approx\boxed{\langle 2.6,\ 9.7\rangle}$	
14. $\vec{u}=\langle \lvert\vec{u}\rvert\cos\theta,\ \lvert\vec{u}\rvert\sin\theta\rangle=\langle 8\cos 225^\circ,\ 8\sin 225^\circ\rangle\approx\boxed{\langle -5.7,\ -5.7\rangle}$	
15. $\vec{u}=\langle \lvert\vec{u}\rvert\cos\theta,\ \lvert\vec{u}\rvert\sin\theta\rangle=\langle 12\cos 105^\circ,\ 12\sin 105^\circ\rangle\approx\boxed{\langle -3.1,\ 11.6\rangle}$	
16. $\vec{u}=\langle \lvert\vec{u}\rvert\cos\theta,\ \lvert\vec{u}\rvert\sin\theta\rangle=\langle 20\cos 15^\circ,\ 20\sin 15^\circ\rangle\approx\boxed{\langle 19.3,\ 5.2\rangle}$	
17. $\dfrac{\vec{v}}{\lvert\vec{v}\rvert}=\dfrac{\langle\sqrt{6},-\sqrt{6}\rangle}{\sqrt{(\sqrt{6})^2+(-\sqrt{6})^2}}=\dfrac{\langle\sqrt{6},-\sqrt{6}\rangle}{2\sqrt{3}}=\boxed{\left\langle\dfrac{\sqrt{2}}{2},-\dfrac{\sqrt{2}}{2}\right\rangle}$	
18. $\dfrac{\vec{v}}{\lvert\vec{v}\rvert}=\dfrac{\langle -11,60\rangle}{\sqrt{(-11)^2+(60)^2}}=\dfrac{\langle -11,60\rangle}{61}=\boxed{\left\langle\dfrac{-11}{61},\dfrac{60}{61}\right\rangle}$	
19. $5\vec{i}+\vec{j}$	**20.** $-15\vec{i}+2\vec{j}$
21. $\langle 6,-3\rangle\cdot\langle 1,4\rangle=(6)(1)+(-3)(4)=\boxed{-6}$	**22.** $\langle -6,5\rangle\cdot\langle -4,2\rangle=(-6)(-4)+(5)(2)=\boxed{34}$
23. $\langle 3,3\rangle\cdot\langle 3,-6\rangle=(3)(3)+(3)(-6)=\boxed{-9}$	**24.** $\langle -2,-8\rangle\cdot\langle -1,1\rangle=(-2)(-1)+(-8)(1)=\boxed{-6}$
25. $\langle 0,8\rangle\cdot\langle 1,2\rangle=(0)(1)+(8)(2)=\boxed{16}$	**26.** $\langle 4,-3\rangle\cdot\langle -1,0\rangle=(4)(-1)+(-3)(0)=\boxed{-4}$

27. Use the formula $\theta=\cos^{-1}\left(\dfrac{\vec{u}\cdot\vec{v}}{\lvert\vec{u}\rvert\lvert\vec{v}\rvert}\right)$ with the following computations:

$$\vec{u}\cdot\vec{v}=\langle 3,4\rangle\cdot\langle -5,12\rangle=33,\quad \lvert\vec{u}\rvert=\sqrt{(3)^2+(4)^2}=5,\quad \lvert\vec{v}\rvert=\sqrt{(-5)^2+(12)^2}=13$$

So, $\theta=\cos^{-1}\left(\dfrac{33}{(5)(13)}\right)\approx\boxed{59^\circ}$.

28. Use the formula $\theta = \cos^{-1}\left(\dfrac{\vec{u}\cdot\vec{v}}{\lvert\vec{u}\rvert\lvert\vec{v}\rvert}\right)$ with the following computations: $$\vec{u}\cdot\vec{v} = \langle -4,5\rangle\cdot\langle 5,-4\rangle = -40,\quad \lvert\vec{u}\rvert = \sqrt{(-4)^2+(5)^2} = \sqrt{41},\quad \lvert\vec{v}\rvert = \sqrt{(5)^2+(-4)^2} = \sqrt{41}$$ So, $\theta = \cos^{-1}\left(\dfrac{-40}{(\sqrt{41})(\sqrt{41})}\right) \approx \boxed{167^\circ}$.
29. Use the formula $\theta = \cos^{-1}\left(\dfrac{\vec{u}\cdot\vec{v}}{\lvert\vec{u}\rvert\lvert\vec{v}\rvert}\right)$ with the following computations: $$\vec{u}\cdot\vec{v} = \langle 1,\sqrt{2}\rangle\cdot\langle -1,3\sqrt{2}\rangle = 5,\quad \lvert\vec{u}\rvert = \sqrt{(1)^2+(\sqrt{2})^2} = \sqrt{3},\quad \lvert\vec{v}\rvert = \sqrt{(-1)^2+(3\sqrt{2})^2} = \sqrt{19}$$ So, $\theta = \cos^{-1}\left(\dfrac{5}{(\sqrt{3})(\sqrt{19})}\right) \approx \boxed{49^\circ}$.
30. Use the formula $\theta = \cos^{-1}\left(\dfrac{\vec{u}\cdot\vec{v}}{\lvert\vec{u}\rvert\lvert\vec{v}\rvert}\right)$ with the following computations: $$\vec{u}\cdot\vec{v} = \langle 7,-24\rangle\cdot\langle -6,8\rangle = -234,\quad \lvert\vec{u}\rvert = \sqrt{(7)^2+(-24)^2} = 25,\quad \lvert\vec{v}\rvert = \sqrt{(-6)^2+(8)^2} = 10$$ So, $\theta = \cos^{-1}\left(\dfrac{-234}{(25)(10)}\right) \approx \boxed{159^\circ}$.
31. Use the formula $\theta = \cos^{-1}\left(\dfrac{\vec{u}\cdot\vec{v}}{\lvert\vec{u}\rvert\lvert\vec{v}\rvert}\right)$ with the following computations: $$\vec{u}\cdot\vec{v} = \langle 3,5\rangle\cdot\langle -4,-4\rangle = -32,\quad \lvert\vec{u}\rvert = \sqrt{(3)^2+(5)^2} = \sqrt{34},\quad \lvert\vec{v}\rvert = \sqrt{(-4)^2+(-4)^2} = \sqrt{32}$$ So, $\theta = \cos^{-1}\left(\dfrac{-32}{(\sqrt{34})(\sqrt{32})}\right) \approx \boxed{166^\circ}$.
32. Use the formula $\theta = \cos^{-1}\left(\dfrac{\vec{u}\cdot\vec{v}}{\lvert\vec{u}\rvert\lvert\vec{v}\rvert}\right)$ with the following computations: $$\vec{u}\cdot\vec{v} = \langle -1,6\rangle\cdot\langle 2,-2\rangle = -14,\quad \lvert\vec{u}\rvert = \sqrt{(-1)^2+(6)^2} = \sqrt{37},\quad \lvert\vec{v}\rvert = \sqrt{(2)^2+(-2)^2} = \sqrt{8}$$ So, $\theta = \cos^{-1}\left(\dfrac{-14}{(\sqrt{37})(\sqrt{8})}\right) \approx \boxed{144^\circ}$.

33. Since $\langle 8,3\rangle\cdot\langle -3,12\rangle = 12 \neq 0$, these two vectors are $\boxed{\text{not orthogonal}}$.
34. Since $\langle -6,2\rangle\cdot\langle 4,12\rangle = 0$, these two vectors are $\boxed{\text{orthogonal}}$.
35. Since $\langle 5,-6\rangle\cdot\langle -12,-10\rangle = 0$, these two vectors are $\boxed{\text{orthogonal}}$.
36. Since $\langle 1,1\rangle\cdot\langle -4,4\rangle = 0$, these two vectors are $\boxed{\text{orthogonal}}$.
37. Since $\langle 0,4\rangle\cdot\langle 0,-4\rangle = -16 \neq 0$, these two vectors are $\boxed{\text{not orthogonal}}$.
38. Since $\langle -7,2\rangle\cdot\left\langle \frac{1}{7},-\frac{1}{2}\right\rangle = -2 \neq 0$, these two vectors are $\boxed{\text{not orthogonal}}$.
39. Since $\langle 6z, a-b\rangle\cdot\langle a+b,-6z\rangle = 6za+6zb-6za+6bz = 12zb \neq 0$ in general, these two vectors are $\boxed{\text{not orthogonal}}$.
40. Since $\langle a-b,-1\rangle\cdot\langle a+b,a^2-b^2\rangle = (a-b)(a+b)-\left(a^2-b^2\right) = 0$, these two vectors are $\boxed{\text{orthogonal}}$.

41. & 42. The complex numbers –6 + 2*i* and 5*i* are plotted below:

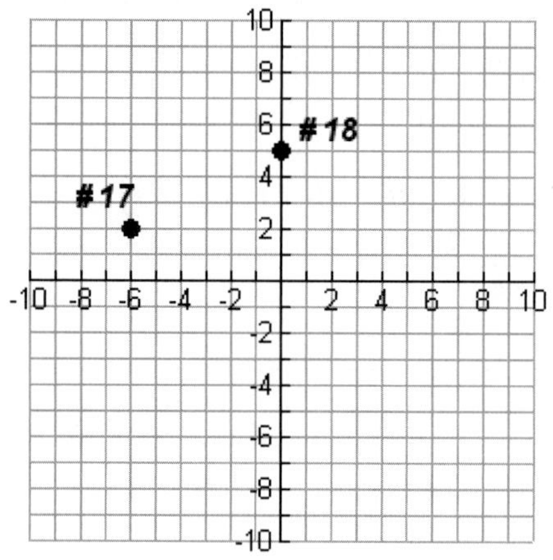

43. In order to express $\sqrt{2}-\sqrt{2}i$ in polar form, observe that $x=\sqrt{2},\ y=-\sqrt{2}$, so that the point is in QIV. Now,

$$r=\sqrt{x^2+y^2}=\sqrt{\left(\sqrt{2}\right)^2+\left(-\sqrt{2}\right)^2}=2$$

$$\tan\theta=\frac{y}{x}=\frac{-\sqrt{2}}{\sqrt{2}}=-1\text{, so that }\theta=\tan^{-1}(-1)=-45^\circ\text{ or }-\frac{\pi}{4}.$$

Since $0\le\theta<2\pi$, we must use the reference angle 315° or $\frac{7\pi}{4}$.

Hence, $\boxed{\sqrt{2}-\sqrt{2}i=2\left[\cos\left(\frac{7\pi}{4}\right)+i\sin\left(\frac{7\pi}{4}\right)\right]=2\left[\cos\left(315^\circ\right)+i\sin\left(315^\circ\right)\right]}$.

44. In order to express $\sqrt{3}+i$ in polar form, observe that $x=\sqrt{3},\ y=1$, so that the point is in QI. Now,

$$r=\sqrt{x^2+y^2}=\sqrt{\left(\sqrt{3}\right)^2+(1)^2}=2$$

$$\tan\theta=\frac{y}{x}=\frac{3}{\sqrt{3}}=\frac{1}{\sqrt{3}},\text{ so that }\theta=\tan^{-1}\left(\frac{1}{\sqrt{3}}\right)=\frac{\pi}{6}\text{ or }30^\circ.$$

Hence, $\boxed{\sqrt{3}+i=2\left[\cos\left(\frac{\pi}{6}\right)+i\sin\left(\frac{\pi}{6}\right)\right]=2\left[\cos\left(30^\circ\right)+i\sin\left(30^\circ\right)\right]}$.

45. In order to express $0-8i$ in polar form, observe that $x=0,\ y=-8$, so that the point is on the negative y-axis. Now,

$$r=\sqrt{x^2+y^2}=\sqrt{(0)^2+(-8)^2}=8\ \text{ and }\ \theta=270^\circ.$$

Hence, $\boxed{0-8i=8\left[\cos\left(270^\circ\right)+i\sin\left(270^\circ\right)\right]}$.

46. In order to express $-8-8i$ in polar form, observe that $x=-8,\ y=-8$, so that the point is in QIII. Now,

$$r=\sqrt{x^2+y^2}=\sqrt{(-8)^2+(-8)^2}=8\sqrt{2}$$

$$\tan\theta=\frac{y}{x}=\frac{-8}{-8}=1,\text{ so that }\theta=180^\circ+\tan^{-1}(1)=225^\circ.$$

(Remember to add 180° to $\tan^{-1}(1)$ since the point is in QIII.)

Hence, $\boxed{-8-8i=8\sqrt{2}\left[\cos\left(225^\circ\right)+i\sin\left(225^\circ\right)\right]}$.

47. In order to express $-60+11i$ in polar form, observe that $x=-60,\ y=11$, so that the point is in QII. Now,

$$r=\sqrt{x^2+y^2}=\sqrt{(-60)^2+(11)^2}=61$$

$$\tan\theta=\frac{y}{x}=\frac{-11}{60},\text{ so that }\theta=180^\circ+\tan^{-1}\left(\frac{-11}{60}\right)\approx169.6^\circ.$$

(Remember to add 180° to $\tan^{-1}\left(\frac{-11}{60}\right)$ since the angle has terminal side in QII.)

Hence, $\boxed{-60+11i\approx61\left[\cos\left(169.6^\circ\right)+i\sin\left(169.6^\circ\right)\right]}$.

48. In order to express $9-40i$ in polar form, observe that $x=9,\ y=-40$, so that the point is in QIV. Now,

$$r=\sqrt{x^2+y^2}=\sqrt{(9)^2+(-40)^2}=41$$

$$\tan\theta=\frac{y}{x}=\frac{-40}{9},\text{ so that }\theta=\tan^{-1}\left(\frac{-40}{9}\right)\approx-77.3196^\circ.$$

Since $0\le\theta<360^\circ$, we must use the reference angle 282.7°.

Hence, $\boxed{9-40i\approx41\left[\cos\left(282.7^\circ\right)+i\sin\left(282.7^\circ\right)\right]}$.

49. In order to express $15+8i$ in polar form, observe that $x=15,\ y=8$, so that the point is in QI. Now,

$$r=\sqrt{x^2+y^2}=\sqrt{(15)^2+(8)^2}=17$$

$$\tan\theta=\frac{y}{x}=\frac{8}{15},\text{ so that }\theta=\tan^{-1}\left(\frac{8}{15}\right)\approx28.1^\circ.$$

Hence, $\boxed{15+8i\approx17\left[\cos\left(28.1^\circ\right)+i\sin\left(28.1^\circ\right)\right]}$.

50. In order to express $-10-24i$ in polar form, observe that $x=-10,\ y=-24$, so that the point is in QIII. Now,

$$r=\sqrt{x^2+y^2}=\sqrt{(-10)^2+(-24)^2}=26$$

$$\tan\theta=\frac{y}{x}=\frac{24}{10},\text{ so that }\theta=180^\circ+\tan^{-1}\left(\frac{24}{10}\right)\approx247.4^\circ.$$

(Remember to add 180° to $\tan^{-1}\left(\frac{24}{10}\right)$ since the angle has terminal side in QIII.)

Hence, $\boxed{-10-24i\approx26\left[\cos\left(247.4^\circ\right)+i\sin\left(247.4^\circ\right)\right]}$.

51. $6\left[\cos\left(300^\circ\right)+i\sin\left(300^\circ\right)\right]=6\left[\frac{1}{2}-\frac{\sqrt{3}}{2}i\right]=\boxed{3-3\sqrt{3}i}$

52. $4\left[\cos\left(210^\circ\right)+i\sin\left(210^\circ\right)\right]=4\left[-\frac{\sqrt{3}}{2}-\frac{1}{2}i\right]=\boxed{-2\sqrt{3}-2i}$

53. $\sqrt{2}\left[\cos\left(135^\circ\right)+i\sin\left(135^\circ\right)\right]=\sqrt{2}\left[-\frac{\sqrt{2}}{2}+\frac{\sqrt{2}}{2}i\right]=\boxed{-1+i}$

54. $4\left[\cos\left(150^\circ\right)+i\sin\left(150^\circ\right)\right]=4\left[-\frac{\sqrt{3}}{2}+\frac{1}{2}i\right]=\boxed{-2\sqrt{3}+2i}$

55. $4\left[\cos\left(200^\circ\right)+i\sin\left(200^\circ\right)\right]\approx\boxed{-3.7588-1.3681i}$

56. $3\left[\cos\left(350^\circ\right)+i\sin\left(350^\circ\right)\right]\approx\boxed{2.9544-0.5209i}$

57. Using the formula $z_1z_2=r_1r_2\left[\cos\left(\theta_1+\theta_2\right)+i\sin\left(\theta_1+\theta_2\right)\right]$ yields

$$\begin{aligned}z_1z_2&=(3)(4)\left[\cos\left(200^\circ+70^\circ\right)+i\sin\left(200^\circ+70^\circ\right)\right]\\&=12\left[\cos\left(270^\circ\right)+i\sin\left(270^\circ\right)\right]=12[0-i]=\boxed{-12i}\end{aligned}$$

58. Using the formula $z_1z_2=r_1r_2\left[\cos\left(\theta_1+\theta_2\right)+i\sin\left(\theta_1+\theta_2\right)\right]$ yields

$$\begin{aligned}z_1z_2&=(3)(4)\left[\cos\left(20^\circ+220^\circ\right)+i\sin\left(20^\circ+220^\circ\right)\right]\\&=12\left[\cos\left(240^\circ\right)+i\sin\left(240^\circ\right)\right]=12\left[-\frac{1}{2}-\frac{\sqrt{3}}{2}i\right]=\boxed{-6-6\sqrt{3}i}\end{aligned}$$

59. Using the formula $z_1z_2=r_1r_2\left[\cos\left(\theta_1+\theta_2\right)+i\sin\left(\theta_1+\theta_2\right)\right]$ yields

$$\begin{aligned}z_1z_2&=(7)(3)\left[\cos\left(100^\circ+140^\circ\right)+i\sin\left(100^\circ+140^\circ\right)\right]\\&=21\left[\cos\left(240^\circ\right)+i\sin\left(240^\circ\right)\right]=21\left[-\frac{1}{2}-\frac{\sqrt{3}}{2}i\right]=\boxed{-\frac{21}{2}-\frac{21\sqrt{3}}{2}i}\end{aligned}$$

60. Using the formula $z_1z_2=r_1r_2\left[\cos\left(\theta_1+\theta_2\right)+i\sin\left(\theta_1+\theta_2\right)\right]$ yields

$$\begin{aligned}z_1z_2&=(1)(4)\left[\cos\left(290^\circ+40^\circ\right)+i\sin\left(290^\circ+40^\circ\right)\right]\\&=4\left[\cos\left(330^\circ\right)+i\sin\left(330^\circ\right)\right]=4\left[\frac{\sqrt{3}}{2}-\frac{1}{2}i\right]=\boxed{2\sqrt{3}-2i}\end{aligned}$$

61. Using the formula $\frac{z_1}{z_2}=\frac{r_1}{r_2}\left[\cos\left(\theta_1-\theta_2\right)+i\sin\left(\theta_1-\theta_2\right)\right]$ yields

$$\begin{aligned}\frac{z_1}{z_2}&=\frac{\sqrt{6}}{\sqrt{6}}\left[\cos\left(200^\circ-50^\circ\right)+i\sin\left(200^\circ-50^\circ\right)\right]\\&=\left[\cos\left(150^\circ\right)+i\sin\left(150^\circ\right)\right]=\boxed{-\frac{\sqrt{3}}{2}+\frac{1}{2}i}\end{aligned}$$

62. Using the formula $\frac{z_1}{z_2}=\frac{r_1}{r_2}\left[\cos\left(\theta_1-\theta_2\right)+i\sin\left(\theta_1-\theta_2\right)\right]$ yields

$$\begin{aligned}\frac{z_1}{z_2}&=\frac{18}{2}\left[\cos\left(190^\circ-100^\circ\right)+i\sin\left(190^\circ-100^\circ\right)\right]\\&=9\left[\cos\left(90^\circ\right)+i\sin\left(90^\circ\right)\right]=\boxed{9i}\end{aligned}$$

63. Using the formula $\frac{z_1}{z_2}=\frac{r_1}{r_2}\left[\cos(\theta_1-\theta_2)+i\sin(\theta_1-\theta_2)\right]$ yields

$$\frac{z_1}{z_2}=\frac{24}{4}\left[\cos\left(290^\circ-110^\circ\right)+i\sin\left(290^\circ-110^\circ\right)\right]$$
$$=6\left[\cos\left(180^\circ\right)+i\sin\left(180^\circ\right)\right]=\boxed{-6}$$

64. Using the formula $\frac{z_1}{z_2}=\frac{r_1}{r_2}\left[\cos(\theta_1-\theta_2)+i\sin(\theta_1-\theta_2)\right]$ yields

$$\frac{z_1}{z_2}=\frac{\sqrt{200}}{\sqrt{2}}\left[\cos\left(93^\circ-48^\circ\right)+i\sin\left(93^\circ-48^\circ\right)\right]$$
$$=10\left[\cos\left(45^\circ\right)+i\sin\left(45^\circ\right)\right]=10\left[\frac{\sqrt{2}}{2}+\frac{\sqrt{2}}{2}i\right]=\boxed{5\sqrt{2}+5\sqrt{2}i}$$

65. In order to express $(3+3i)^4$ in rectangular form, follow these steps:

Step 1: Write $3+3i$ in polar form.
Since $x=3,\ y=3$, the point is in QI. So,

$$r=\sqrt{x^2+y^2}=\sqrt{(3)^2+(3)^2}=3\sqrt{2}$$
$$\tan\theta=\frac{y}{x}=\frac{3}{3}=1\text{, so that } \theta=\tan^{-1}(1)=45^\circ.$$

Hence, $3+3i=3\sqrt{2}\left[\cos\left(45^\circ\right)+i\sin\left(45^\circ\right)\right]$.

Step 2: Apply DeMoivre's theorem $z^n=r^n\left[\cos(n\theta)+i\sin(n\theta)\right]$.

$$(3+3i)^4=\left(3\sqrt{2}\right)^4\left[\cos\left(4\cdot45^\circ\right)+i\sin\left(4\cdot45^\circ\right)\right]=324\left[\cos\left(180^\circ\right)+i\sin\left(180^\circ\right)\right]$$
$$=324\left[-1+0i\right]=\boxed{-324}$$

66. In order to express $\left(3+\sqrt{3}i\right)^4$ in rectangular form, follow these steps:

Step 1: Write $3+\sqrt{3}i$ in polar form.
Since $x=3,\ y=\sqrt{3}$, the point is in QI. So,

$$r=\sqrt{x^2+y^2}=\sqrt{(3)^2+\left(\sqrt{3}\right)^2}=2\sqrt{3}$$

$$\tan\theta=\frac{y}{x}=\frac{\sqrt{3}}{3}=\frac{1}{\sqrt{3}},\ \text{so that } \theta=\tan^{-1}\left(\frac{1}{\sqrt{3}}\right)=30^\circ.$$

Hence, $3+\sqrt{3}i=2\sqrt{3}\left[\cos\left(30^\circ\right)+i\sin\left(30^\circ\right)\right]$.

Step 2: Apply DeMoivre's theorem $z^n=r^n\left[\cos(n\theta)+i\sin(n\theta)\right]$.

$$\left(3+\sqrt{3}i\right)^4=\left(2\sqrt{3}\right)^4\left[\cos\left(4\cdot30^\circ\right)+i\sin\left(4\cdot30^\circ\right)\right]=144\left[\cos\left(120^\circ\right)+i\sin\left(120^\circ\right)\right]$$

$$=144\left[-\frac{1}{2}+\frac{\sqrt{3}}{2}i\right]=\boxed{-72+72\sqrt{3}i}$$

67. In order to express $\left(1+\sqrt{3}i\right)^5$ in rectangular form, follow these steps:

Step 1: Write $1+\sqrt{3}i$ in polar form.
Since $x=1,\ y=\sqrt{3}$, the point is in QI. So,

$$r=\sqrt{x^2+y^2}=\sqrt{(1)^2+\left(\sqrt{3}\right)^2}=2$$

$$\tan\theta=\frac{y}{x}=\frac{\sqrt{3}}{1},\ \text{so that } \theta=\tan^{-1}\left(\sqrt{3}\right)=60^\circ.$$

Hence, $1+\sqrt{3}i=2\left[\cos\left(60^\circ\right)+i\sin\left(60^\circ\right)\right]$.

Step 2: Apply DeMoivre's theorem $z^n=r^n\left[\cos(n\theta)+i\sin(n\theta)\right]$.

$$\left(1+\sqrt{3}i\right)^5=(2)^5\left[\cos\left(5\cdot60^\circ\right)+i\sin\left(5\cdot60^\circ\right)\right]=32\left[\cos\left(300^\circ\right)+i\sin\left(300^\circ\right)\right]$$

$$=32\left[\frac{1}{2}-\frac{\sqrt{3}}{2}i\right]=\boxed{16-16\sqrt{3}i}$$

68. In order to express $(-2-2i)^7$ in rectangular form, follow these steps:

Step 1: Write $-2-2i$ in polar form.
Since $x=-2,\ y=-2$, the point is in QIII. So,

$$r=\sqrt{x^2+y^2}=\sqrt{(-2)^2+(-2)^2}=2\sqrt{2}$$

$$\tan\theta=\frac{y}{x}=\frac{-2}{-2}=1,\text{ so that }\theta=180^\circ+\tan^{-1}(1)=225^\circ.$$

(Remember to add 180° to $\tan^{-1}(1)$ since the angle has terminal side in QII.)
Hence, $-2-2i=2\sqrt{2}\left[\cos\left(225^\circ\right)+i\sin\left(225^\circ\right)\right]$.

Step 2: Apply DeMoivre's theorem $z^n=r^n\left[\cos(n\theta)+i\sin(n\theta)\right]$.

$$(-2-2i)^7=\left(2\sqrt{2}\right)^7\left[\cos\left(7\cdot225^\circ\right)+i\sin\left(7\cdot225^\circ\right)\right]=1024\sqrt{2}\left[-\frac{\sqrt{2}}{2}+\frac{\sqrt{2}}{2}i\right]$$

$$=\boxed{-1024+1024i}$$

69. Given $z=2+2\sqrt{3}i$, in order to compute $z^{1/2}$, follow these steps:

Step 1: Write $2+2\sqrt{3}i$ in polar form.
Since $x=2,\ y=2\sqrt{3}$, the point is in QI. So,

$$r=\sqrt{x^2+y^2}=\sqrt{(2)^2+\left(2\sqrt{3}\right)^2}=4$$

$$\tan\theta=\frac{y}{x}=\frac{2\sqrt{3}}{2}=\sqrt{3},\text{ so that }\theta=\tan^{-1}\left(\sqrt{3}\right)=60^\circ.$$

Hence, $z=2+2\sqrt{3}i=4\left[\cos\left(60^\circ\right)+i\sin\left(60^\circ\right)\right]$.

Step 2: Now, apply $z^{1/n}=r^{1/n}\left[\cos\left(\frac{\theta}{n}+\frac{360^\circ k}{n}\right)+i\sin\left(\frac{\theta}{n}+\frac{360^\circ k}{n}\right)\right],\quad k=0,1,\ldots,n-1$:

$$z^{1/2}=\sqrt{4}\left[\cos\left(\frac{60^\circ}{2}+\frac{360^\circ k}{2}\right)+i\sin\left(\frac{60^\circ}{2}+\frac{360^\circ k}{2}\right)\right],\quad k=0,1$$

$$=\boxed{2\left[\cos\left(30^\circ\right)+i\sin\left(30^\circ\right)\right],\ 2\left[\cos\left(210^\circ\right)+i\sin\left(210^\circ\right)\right]}$$

Step 3: Plot the roots in the complex plane:

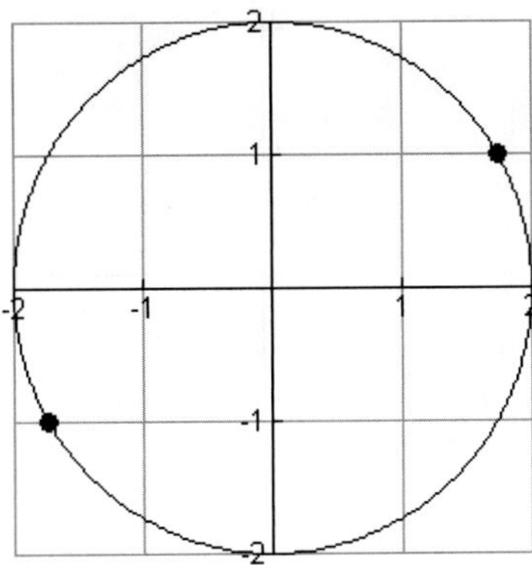

70. Given $z=-8+8\sqrt{3}i$, in order to compute $z^{1/4}$, follow these steps:

Step 1: Write $z=-8+8\sqrt{3}i$ in polar form.
Since $x=-8,\ y=8\sqrt{3}$, the point is in QII. So,

$$r=\sqrt{x^2+y^2}=\sqrt{(-8)^2+\left(8\sqrt{3}\right)^2}=16$$

$$\tan\theta=\frac{y}{x}=\frac{8\sqrt{3}}{-8}=-\sqrt{3}\text{, so that }\theta=180^\circ+\tan^{-1}\left(-\sqrt{3}\right)=120^\circ.$$

(Remember to add 180° to $\tan^{-1}\left(-\sqrt{3}\right)$ since the angle has terminal side in QII.)
Hence, $z=-8+8\sqrt{3}i=16\left[\cos(120^\circ)+i\sin(120^\circ)\right]$.

Step 2: Now, apply $z^{1/n}=r^{1/n}\left[\cos\left(\frac{\theta}{n}+\frac{360^\circ k}{n}\right)+i\sin\left(\frac{\theta}{n}+\frac{360^\circ k}{n}\right)\right],\quad k=0,1,\ldots,n-1$:

$$z^{1/4}=\sqrt[4]{16}\left[\cos\left(\frac{120^\circ}{4}+\frac{360^\circ k}{4}\right)+i\sin\left(\frac{120^\circ}{4}+\frac{360^\circ k}{4}\right)\right],\quad k=0,1,2,3$$

$$=\boxed{\begin{array}{l}2\left[\cos(30^\circ)+i\sin(30^\circ)\right],\ 2\left[\cos(120^\circ)+i\sin(120^\circ)\right],\\ 2\left[\cos(210^\circ)+i\sin(210^\circ)\right],\ 2\left[\cos(300^\circ)+i\sin(300^\circ)\right]\end{array}}$$

Step 3: Plot the roots in the complex plane:

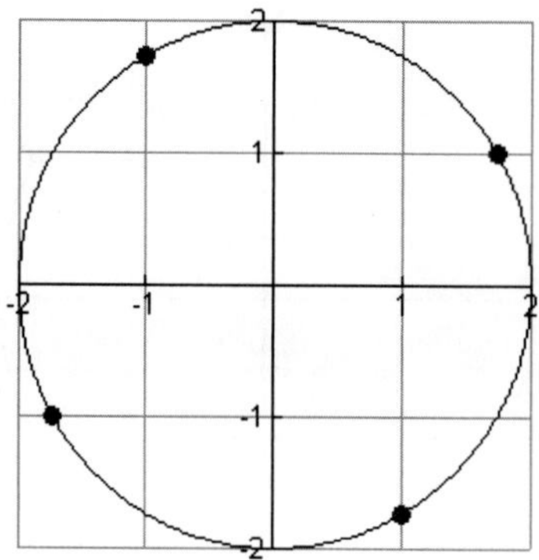

71. Given $z=-256+0i$, in order to compute $z^{1/4}$, follow these steps:

Step 1: Write $z=-256+0i$ in polar form.
Since $x=-256,\ y=0$, the point is on the negative x-axis. So,

$$r=256, \quad \theta=180^\circ$$

Hence, $z=-256+0i=256\left[\cos\left(180^\circ\right)+i\sin\left(180^\circ\right)\right]$.

Step 2: Now, apply $z^{1/n}=r^{1/n}\left[\cos\left(\frac{\theta}{n}+\frac{360^\circ k}{n}\right)+i\sin\left(\frac{\theta}{n}+\frac{360^\circ k}{n}\right)\right], \quad k=0, 1, \ldots, n-1$:

$$z^{1/4}=\sqrt[4]{256}\left[\cos\left(\frac{180^\circ}{4}+\frac{360^\circ k}{4}\right)+i\sin\left(\frac{180^\circ}{4}+\frac{360^\circ k}{4}\right)\right], \quad k=0, 1, 2, 3$$

$$=\boxed{\begin{array}{l} 4\left[\cos\left(45^\circ\right)+i\sin\left(45^\circ\right)\right],\ 4\left[\cos\left(135^\circ\right)+i\sin\left(135^\circ\right)\right], \\ 4\left[\cos\left(225^\circ\right)+i\sin\left(225^\circ\right)\right],\ 4\left[\cos\left(315^\circ\right)+i\sin\left(315^\circ\right)\right] \end{array}}$$

Step 3: Plot the roots in the complex plane:

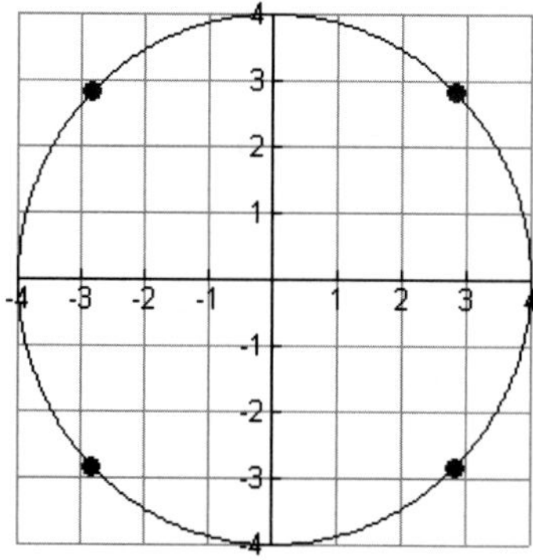

72. Given $z = 0 - 18i$, in order to compute $z^{1/2}$, follow these steps:

Step 1: Write $z = 0 - 18i$ in polar form.
Since $x = 0,\ y = -18$, the point is on the negative y-axis. So,

$$r = 18, \quad \theta = 270^\circ$$

Hence, $z = 0 - 18i = 18\left[\cos(270^\circ) + i\sin(270^\circ)\right]$.

Step 2: Now, apply $z^{1/n} = r^{1/n}\left[\cos\left(\frac{\theta}{n} + \frac{360^\circ k}{n}\right) + i\sin\left(\frac{\theta}{n} + \frac{360^\circ k}{n}\right)\right]$, $k = 0, 1, \ldots, n-1$:

$$z^{1/2} = \sqrt{18}\left[\cos\left(\frac{270^\circ}{2} + \frac{360^\circ k}{2}\right) + i\sin\left(\frac{270^\circ}{2} + \frac{360^\circ k}{2}\right)\right], \quad k = 0, 1$$

$$= \boxed{3\sqrt{2}\left[\cos(135^\circ) + i\sin(135^\circ)\right],\ 3\sqrt{2}\left[\cos(315^\circ) + i\sin(315^\circ)\right]}$$

Step 3: Plot the roots in the complex plane:

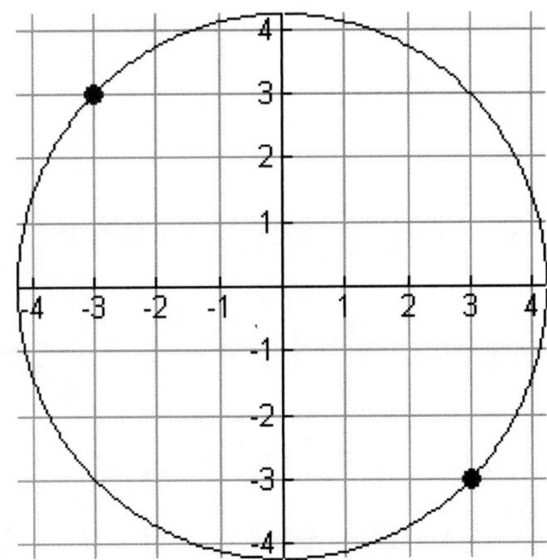

73. We seek all complex numbers x such that $x^3+216=0$, or equivalently $x^3=-216$. While it is clear that $x=-6$ is a solution by inspection, the remaining two complex solutions aren't immediately discernible. As such, we shall apply the approach for computing complex roots involving the formula

$$z^{1/n}=r^{1/n}\left[\cos\left(\frac{\theta}{n}+\frac{2\pi k}{n}\right)+i\sin\left(\frac{\theta}{n}+\frac{2\pi k}{n}\right)\right],\quad k=0,1,\dots,n-1.$$

To this end, we follow these steps.

Step 1: Write $z=-216$ in polar form.
Since $x=-216,\ y=0$, the point is on the negative x-axis. So, $r=216,\ \theta=\pi$.
Hence, $-216=216\left[\cos(\pi)+i\sin(\pi)\right]$.

Step 2: Now, apply $z^{1/n}=r^{1/n}\left[\cos\left(\frac{\theta}{n}+\frac{2\pi k}{n}\right)+i\sin\left(\frac{\theta}{n}+\frac{2\pi k}{n}\right)\right],\quad k=0,1,\dots,n-1$:

$$\begin{aligned}(-216)^{1/3}&=\sqrt[3]{216}\left[\cos\left(\frac{\pi}{3}+\frac{2\pi k}{3}\right)+i\sin\left(\frac{\pi}{3}+\frac{2\pi k}{3}\right)\right],\quad k=0,1,2\\&=6\left[\cos\left(\frac{\pi}{3}\right)+i\sin\left(\frac{\pi}{3}\right)\right],\ 6\left[\cos(\pi)+i\sin(\pi)\right],\ 6\left[\cos\left(\frac{5\pi}{3}\right)+i\sin\left(\frac{5\pi}{3}\right)\right]\\&=3+3\sqrt{3}i,\ -6,\ 3-3\sqrt{3}i\end{aligned}$$

Step 3: Hence, the complex solutions of $x^3+216=0$ are $\boxed{3+3\sqrt{3}i,\ -6,\ 3-3\sqrt{3}i}$.

74. We seek all complex numbers x such that $x^4-1=0$. Observe that this equation can be written equivalently as:

$$\begin{aligned}(x^2-1)(x^2+1)&=0\\(x-1)(x+1)(x^2+1)&=0\end{aligned}$$

The solutions are therefore $x=\boxed{\pm1,\ \pm i}$.

(Note: We could have alternatively used the formula for complex roots, namely

$$z^{1/n}=r^{1/n}\left[\cos\left(\frac{\theta}{n}+\frac{360^\circ k}{n}\right)+i\sin\left(\frac{\theta}{n}+\frac{360^\circ k}{n}\right)\right],\quad k=0,1,\dots,n-1.$$

However, the left-side readily factored in a way that is easy to see what *all* of the solutions were. As such, factoring was the more expedient route for this problem.)

75. We seek all complex numbers x such that $x^4+1=0$, or equivalently $x^4=-1$. We shall apply the approach for computing complex roots involving the formula

$$z^{1/n}=r^{1/n}\left[\cos\left(\frac{\theta}{n}+\frac{2\pi k}{n}\right)+i\sin\left(\frac{\theta}{n}+\frac{2\pi k}{n}\right)\right],\quad k=0,1,\ldots,n-1.$$

To this end, we follow these steps.

Step 1: Write $z=-1$ in polar form.
Since $x=-1,\ y=0$, the point is on the negative x-axis. So, $r=1,\ \theta=\pi$.
Hence, $-1=1\left[\cos(\pi)+i\sin(\pi)\right]$.

Step 2: Now, apply $z^{1/n}=r^{1/n}\left[\cos\left(\frac{\theta}{n}+\frac{2\pi k}{n}\right)+i\sin\left(\frac{\theta}{n}+\frac{2\pi k}{n}\right)\right],\quad k=0,1,\ldots,n-1$:

$$(-1)^{1/4}=\sqrt[4]{1}\left[\cos\left(\frac{\pi}{4}+\frac{2\pi k}{4}\right)+i\sin\left(\frac{\pi}{4}+\frac{2\pi k}{4}\right)\right],\quad k=0,1,2,3$$

$$=\left[\cos\left(\frac{\pi}{4}\right)+i\sin\left(\frac{\pi}{4}\right)\right],\ \left[\cos\left(\frac{3\pi}{4}\right)+i\sin\left(\frac{3\pi}{4}\right)\right],$$

$$\left[\cos\left(\frac{5\pi}{4}\right)+i\sin\left(\frac{5\pi}{4}\right)\right],\ \left[\cos\left(\frac{7\pi}{4}\right)+i\sin\left(\frac{7\pi}{4}\right)\right]$$

$$=\frac{\sqrt{2}}{2}+\frac{\sqrt{2}}{2}i,\ -\frac{\sqrt{2}}{2}+\frac{\sqrt{2}}{2}i,\ -\frac{\sqrt{2}}{2}-\frac{\sqrt{2}}{2}i,\ \frac{\sqrt{2}}{2}-\frac{\sqrt{2}}{2}i$$

Step 3: Hence, the complex solutions of $x^4+1=0$ are

$$\boxed{\frac{\sqrt{2}}{2}+\frac{\sqrt{2}}{2}i,\ -\frac{\sqrt{2}}{2}+\frac{\sqrt{2}}{2}i,\ -\frac{\sqrt{2}}{2}-\frac{\sqrt{2}}{2}i,\ \frac{\sqrt{2}}{2}-\frac{\sqrt{2}}{2}i}.$$

76. We seek all complex numbers x such that $x^3-125=0$, or equivalently $x^3=125$. While it is clear that $x = 5$ is a solution by inspection, the remaining two complex solutions aren't immediately discernible. As such, we shall apply the approach for computing complex roots involving the formula

$$z^{1/n}=r^{1/n}\left[\cos\left(\frac{\theta}{n}+\frac{2\pi k}{n}\right)+i\sin\left(\frac{\theta}{n}+\frac{2\pi k}{n}\right)\right],\quad k=0,1,\ldots,n-1.$$

To this end, we follow these steps.

Step 1: Write $z=125$ in polar form.

Since $x=125,\ y=0$, the point is on the positive x-axis. So, $r=125,\ \theta=0$.

Hence, $125=125\left[\cos(0)+i\sin(0)\right]$.

Step 2: Now, apply $z^{1/n}=r^{1/n}\left[\cos\left(\frac{\theta}{n}+\frac{2\pi k}{n}\right)+i\sin\left(\frac{\theta}{n}+\frac{2\pi k}{n}\right)\right],\ k=0,1,\ldots,n-1$:

$$125^{1/3}=\sqrt[3]{125}\left[\cos\left(\frac{0}{3}+\frac{2\pi k}{3}\right)+i\sin\left(\frac{0}{3}+\frac{2\pi k}{3}\right)\right],\quad k=0,1,2$$

$$=5\left[\cos(0)+i\sin(0)\right],\ 5\left[\cos\left(\frac{2\pi}{3}\right)+i\sin\left(\frac{2\pi}{3}\right)\right],\ 5\left[\cos\left(\frac{4\pi}{3}\right)+i\sin\left(\frac{4\pi}{3}\right)\right]=5,\ -\frac{5}{2}+\frac{5\sqrt{3}}{2}i,\ -\frac{5}{2}-\frac{5\sqrt{3}}{2}i$$

Step 3: Hence, the complex solutions of $x^3-125=0$ are $\boxed{5,\ -\frac{5}{2}+\frac{5\sqrt{3}}{2}i,\ -\frac{5}{2}-\frac{5\sqrt{3}}{2}i}$.

77. In order to convert the point $(-2,2)$ (expressed in rectangular coordinates) to polar coordinates, first observe that $x=-2,\ y=2$, so that the point is in QII. Hence,

$$r=\sqrt{(-2)^2+(2)^2}=2\sqrt{2}$$

$$\tan\theta=\frac{y}{x}=\frac{-2}{2}=-1 \text{ so that } \theta=\pi+\tan^{-1}(-1)=\frac{3\pi}{4}.$$

(Remember to add π to $\tan^{-1}(-1)$ since the point is in QII.)

So, the point can be expressed in polar coordinates as $\boxed{\left(2\sqrt{2},\frac{3\pi}{4}\right)}$, plotted below:

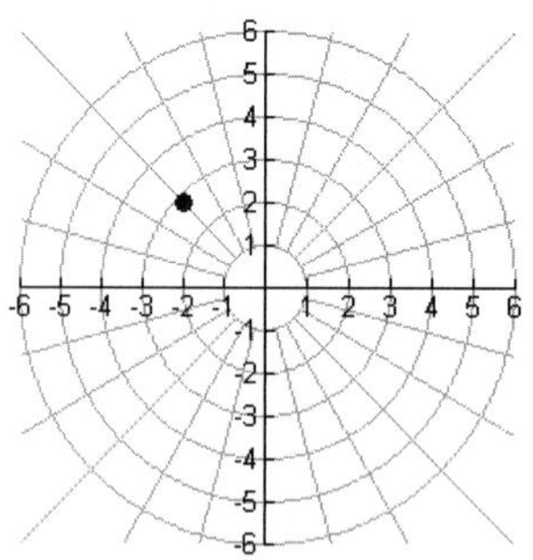

78. In order to convert the point $\left(4,-4\sqrt{3}\right)$ (expressed in rectangular coordinates) to polar coordinates, first observe that $x=4,\ y=-4\sqrt{3}$, so that the point is in QIV. Hence,

$$r=\sqrt{(4)^2+\left(-4\sqrt{3}\right)^2}=8$$

$$\tan\theta=\frac{y}{x}=\frac{-4\sqrt{3}}{4}=-\sqrt{3} \text{ so that } \theta=\tan^{-1}\left(-\sqrt{3}\right)=-\frac{\pi}{3}.$$

Since $0\le\theta<2\pi$, we use the reference angle $\theta=\dfrac{5\pi}{3}$.

So, the point can be expressed in polar coordinates as $\boxed{\left(8,\frac{5\pi}{3}\right)}$, plotted below:

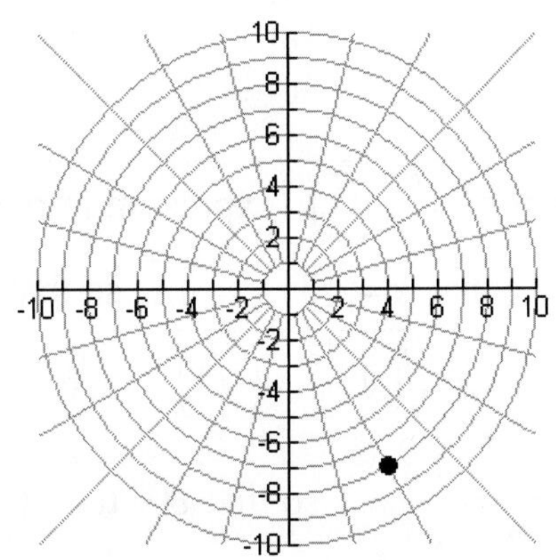

79. In order to convert the point $\left(-5\sqrt{3},-5\right)$ (expressed in rectangular coordinates) to polar coordinates, first observe that $x=-5\sqrt{3},\ y=-5$, so that the point is in QIII. Hence,

$$r=\sqrt{\left(-5\sqrt{3}\right)^2+(-5)^2}=10$$

$$\tan\theta=\frac{y}{x}=\frac{-5}{-5\sqrt{3}}=\frac{1}{\sqrt{3}} \text{ so that } \theta=\pi+\tan^{-1}\left(\frac{1}{\sqrt{3}}\right)=\frac{7\pi}{6}.$$

(Remember to add π to $\tan^{-1}\left(\dfrac{1}{\sqrt{3}}\right)$ since the point is in QIII.)

So, the point can be expressed in polar coordinates as $\boxed{\left(10, \frac{7\pi}{6}\right)}$, which is plotted below:

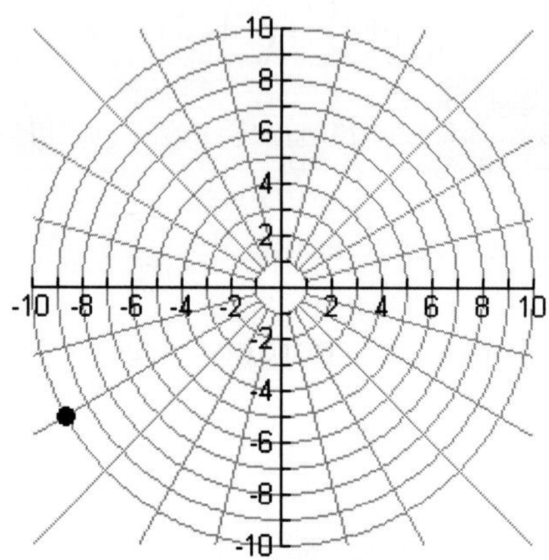

80. In order to convert the point $\left(\sqrt{3}, \sqrt{3}\right)$ (expressed in rectangular coordinates) to polar coordinates, first observe that $x = \sqrt{3},\ y = \sqrt{3}$, so that the point is in QI. Hence,

$$r = \sqrt{\left(\sqrt{3}\right)^2 + \left(\sqrt{3}\right)^2} = \sqrt{6}$$

$$\tan\theta = \frac{y}{x} = \frac{\sqrt{3}}{\sqrt{3}} = 1 \text{ so that } \theta = \tan^{-1}(1) = \frac{\pi}{4}.$$

So, the given point can be expressed in polar coordinates as $\boxed{\left(\sqrt{6}, \frac{\pi}{4}\right)}$, which is plotted below:

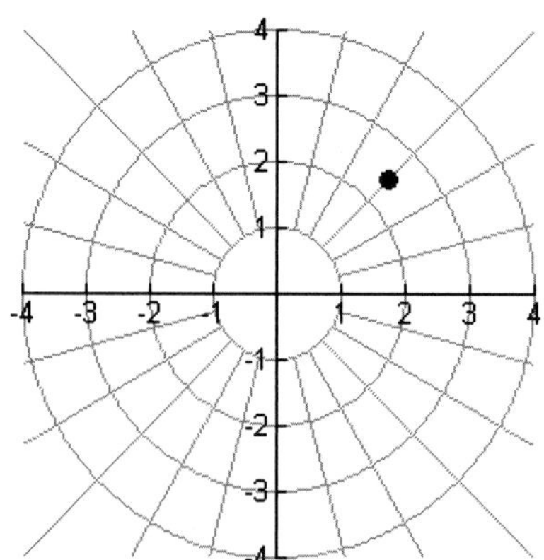

81. In order to convert the point $(0,-2)$ (expressed in rectangular coordinates) to polar coordinates, first observe that $x=0,\ y=-2$, so that the point is on the negative y-axis. Hence,

$$r=\sqrt{(0)^2+(-2)^2}=2, \quad \theta=\frac{3\pi}{2}$$

So, the given point can be expressed in polar coordinates as $\boxed{\left(2,\frac{3\pi}{2}\right)}$, which is plotted below:

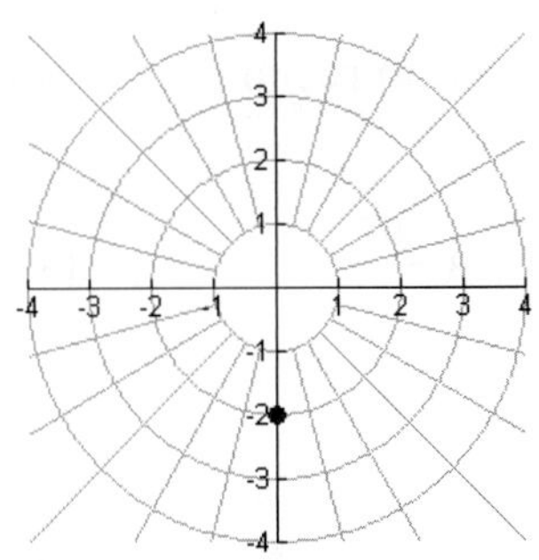

82. In order to convert the point $(11,0)$ (expressed in rectangular coordinates) to polar coordinates, first observe that $x=11,\ y=0$, so that the point is on the positive x-axis. Hence,

$$r=\sqrt{(11)^2+(0)^2}=11, \quad \theta=0$$

So, the given point can be expressed in polar coordinates as $\boxed{(11,0)}$, which is plotted below:

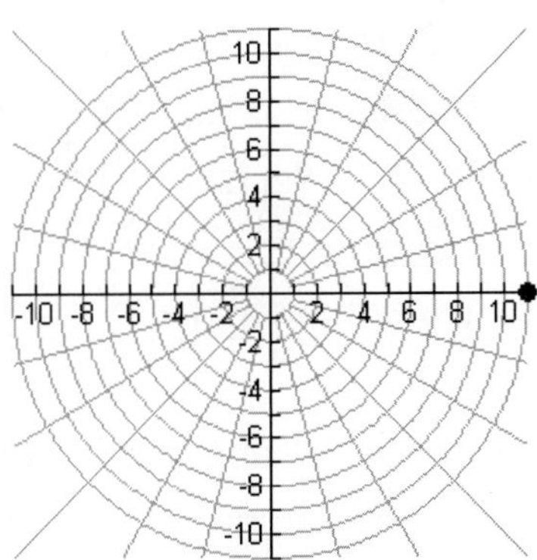

83. In order to convert the point $\left(-3, \frac{5\pi}{3}\right)$ (expressed in polar coordinates) to rectangular coordinates, first observe that $r = -3,\ \theta = \frac{5\pi}{3}$. Hence,

$$x = r\cos\theta = -3\cos\left(\frac{5\pi}{3}\right) = -3\left(\frac{1}{2}\right) = -\frac{3}{2}$$

$$y = r\sin\theta = -3\sin\left(\frac{5\pi}{3}\right) = -3\left(-\frac{\sqrt{3}}{2}\right) = \frac{3\sqrt{3}}{2}$$

So, the given point can be expressed in rectangular coordinates as $\boxed{\left(-\frac{3}{2}, \frac{3\sqrt{3}}{2}\right)}$.

84. In order to convert the point $\left(4, \frac{5\pi}{4}\right)$ (expressed in polar coordinates) to rectangular coordinates, first observe that $r = 4,\ \theta = \frac{5\pi}{4}$. Hence,

$$x = r\cos\theta = 4\cos\left(\frac{5\pi}{4}\right) = 4\left(-\frac{\sqrt{2}}{2}\right) = -2\sqrt{2}$$

$$y = r\sin\theta = 4\sin\left(\frac{5\pi}{4}\right) = 4\left(-\frac{\sqrt{2}}{2}\right) = -2\sqrt{2}$$

So, the given point can be expressed in rectangular coordinates as $\boxed{\left(-2\sqrt{2}, -2\sqrt{2}\right)}$.

85. In order to convert the point $\left(2, \frac{\pi}{3}\right)$ (expressed in polar coordinates) to rectangular coordinates, first observe that $r = 2,\ \theta = \frac{\pi}{3}$. Hence,

$$x = r\cos\theta = 2\cos\left(\frac{\pi}{3}\right) = 2\left(\frac{1}{2}\right) = 1$$

$$y = r\sin\theta = 2\sin\left(\frac{\pi}{3}\right) = 2\left(\frac{\sqrt{3}}{2}\right) = \sqrt{3}$$

So, the given point can be expressed in rectangular coordinates as $\boxed{\left(1, \sqrt{3}\right)}$.

86. In order to convert the point $\left(6, \frac{7\pi}{6}\right)$ (expressed in polar coordinates) to rectangular coordinates, first observe that $r = 6,\ \theta = \frac{7\pi}{6}$. Hence,

$$x = r\cos\theta = 6\cos\left(\frac{7\pi}{6}\right) = 6\left(-\frac{\sqrt{3}}{2}\right) = -3\sqrt{3}$$

$$y = r\sin\theta = 6\sin\left(\frac{7\pi}{6}\right) = 6\left(-\frac{1}{2}\right) = -3$$

So, the given point can be expressed in rectangular coordinates as $\boxed{\left(-3\sqrt{3}, -3\right)}$.

87. In order to convert the point $\left(1, \frac{4\pi}{3}\right)$ (expressed in polar coordinates) to rectangular coordinates, first observe that $r = 1,\ \theta = \frac{4\pi}{3}$. Hence,

$$x = r\cos\theta = 1\cos\left(\frac{4\pi}{3}\right) = -\frac{1}{2}$$

$$y = r\sin\theta = 1\sin\left(\frac{4\pi}{3}\right) = -\frac{\sqrt{3}}{2}$$

So, the given point can be expressed in rectangular coordinates as $\boxed{\left(-\frac{1}{2}, -\frac{\sqrt{3}}{2}\right)}$.

88. In order to convert the point $\left(-3, \frac{7\pi}{4}\right)$ (expressed in polar coordinates) to rectangular coordinates, first observe that $r = -3,\ \theta = \frac{7\pi}{4}$. Hence,

$$x = r\cos\theta = -3\cos\left(\frac{7\pi}{4}\right) = -3\left(\frac{\sqrt{2}}{2}\right) = -\frac{3\sqrt{2}}{2}$$

$$y = r\sin\theta = -3\sin\left(\frac{7\pi}{4}\right) = -3\left(-\frac{\sqrt{2}}{2}\right) = \frac{3\sqrt{2}}{2}$$

So, the given point can be expressed in rectangular coordinates as $\boxed{\left(-\frac{3\sqrt{2}}{2}, \frac{3\sqrt{2}}{2}\right)}$.

89. In order to graph $r = 4\cos 2\theta$, consider the following table of points:

θ	$r = 4\cos 2\theta$	(r,θ)
0	4	$(4,0)$
$\pi/4$	0	$(0,\pi/4)$
$\pi/2$	-4	$(-4,\pi/2)$
$3\pi/4$	0	$(0,3\pi/4)$
π	4	$(4,\pi)$
$5\pi/4$	0	$(0,5\pi/4)$
$3\pi/2$	-4	$(-4,3\pi/2)$
$7\pi/4$	0	$(0,7\pi/4)$
2π	4	$(4,2\pi)$

The graph is as follows:

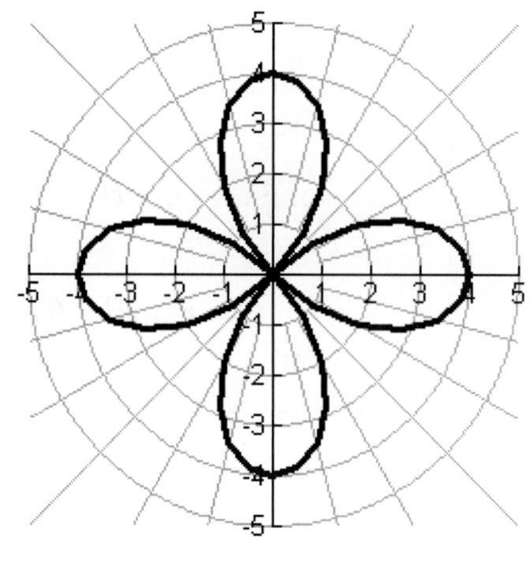

90. In order to graph $r = \sin 3\theta$, consider the following table of points:

θ	$r = 3\sin 3\theta$	(r,θ)
0	0	$(0,0)$
$\pi/6$	1	$\left(1,\pi/6\right)$
$2\pi/6$	0	$\left(0,2\pi/6\right)$
$\pi/2$	-1	$\left(-1,\pi/2\right)$
$4\pi/6$	0	$\left(0,4\pi/6\right)$
$5\pi/6$	1	$\left(1,5\pi/6\right)$
π	0	$(0,\pi)$
$7\pi/6$	-1	$\left(-1,7\pi/6\right)$
$8\pi/6$	0	$\left(0,8\pi/6\right)$
$3\pi/2$	1	$\left(1,3\pi/2\right)$
$10\pi/6$	0	$\left(0,10\pi/6\right)$
$11\pi/6$	-1	$\left(-1,11\pi/6\right)$
2π	0	$(0,2\pi)$

The graph is as follows:

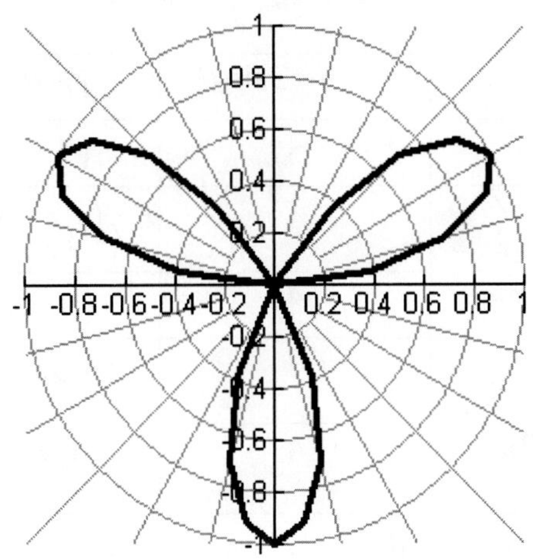

91. In order to graph $r = -\theta$, consider the following table of points:

θ	$r = -2\theta$	(r,θ)
0	0	$(0,0)$
$\pi/2$	$-\pi/2$	$\left(-\pi/2,-\pi/2\right)$
π	$-\pi$	$(-\pi,-\pi)$
$3\pi/2$	$-3\pi/2$	$\left(-3\pi/2,-3\pi/2\right)$
2π	-2π	$(-2\pi,-2\pi)$

The graph is as follows:

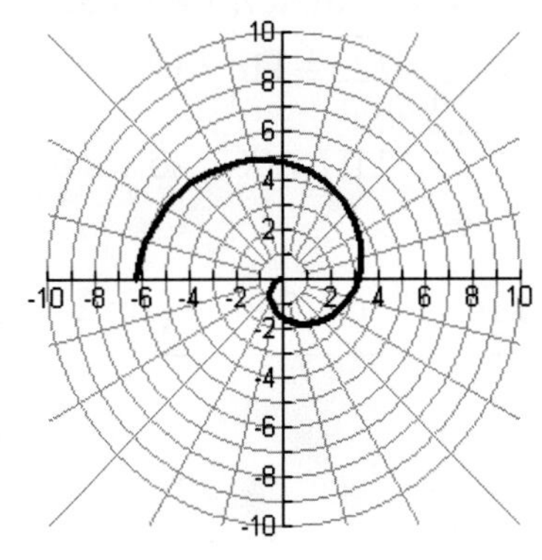

92. In order to graph $r = 4 - 3\sin\theta$, consider the following table of points:

θ	$r = 4 - 3\sin\theta$	(r,θ)
0	4	$(4,0)$
$\pi/2$	1	$(1,\pi/2)$
π	4	$(4,\pi)$
$3\pi/2$	7	$(7,3\pi/2)$
2π	4	$(4,2\pi)$

The graph is as follows:

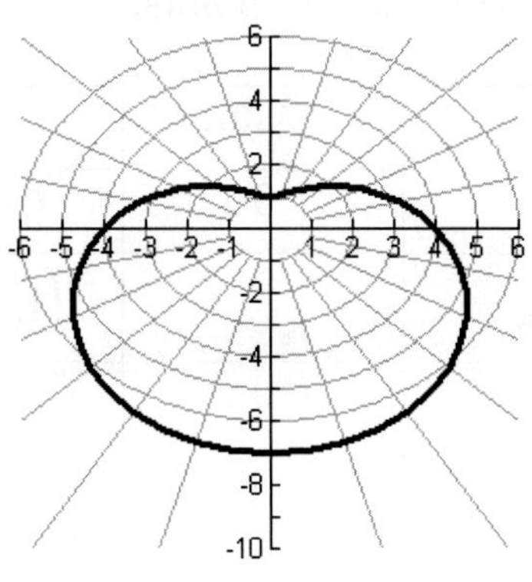

93. Magnitude = $\sqrt{\ }$ Sum ({25,−60}^2)
Direction angle = $\tan^{-1}$ (−60) ÷ (25)

The output will be:
Magnitude = 65
Direction Angle = -67.38014

94. Magnitude =
$\sqrt{\ }$ Sum ({−70,10∗$\sqrt{\ }$15}^2)
Direction angle =
$\tan^{-1}$ (10∗$\sqrt{\ }$15) ÷ (−70)

The output will be:
Magnitude = 80
Direction Angle = -28.95502

95.
Sum({14,37}*{9,−26})/
(Sum({14,37}^2)*Sum({9,−26}^2) =
0.76807
$\cos^{-1}(Ans) = 39.81911$
So, the angle is approximately 40°.

96.
Sum({−23,−8}*{18,−32})/
(Sum({−23,−8}^2)*Sum({18,−32}^2) =
−0.17672
$\cos^{-1}(Ans) = 100.17877$
So, the angle is approximately 100°.

97. From the calculator, we obtain:

$$\left|-\sqrt{23} - 11i\right| = 12$$

$angle\left(-\sqrt{23} - 11i\right) \approx 66^\circ$, so we use 246°

So,
$-\sqrt{23} - 11i \approx 12\left(\cos 246^\circ + i\sin 246^\circ\right)$.

98. From the calculator, we obtain:

$$\left|11 + \sqrt{23}i\right| = 12$$

$$angle\left(11 + \sqrt{23}i\right) \approx 78^\circ$$

So, $11 + \sqrt{23}i \approx 12\left(\cos 78^\circ + i\sin 78^\circ\right)$.

99. Given $z = -8 + 8\sqrt{3}\,i$, in order to compute $z^{1/4}$, follow these steps:

Step 1: Write $z = -8 + 8\sqrt{3}\,i$ in polar form.
Since $x = -8, y = 8\sqrt{3}$, the point is in QII. So, using the calculator yields

$$r = 16, \ \ \theta = 120^\circ$$

Hence, $z = -8 + 8\sqrt{3}\,i = 16\left[\cos\left(120^\circ\right) + i\sin\left(120^\circ\right)\right]$.

Step 2: Now, apply $z^{1/n} = r^{1/n}\left[\cos\left(\frac{\theta}{n} + \frac{360^\circ k}{n}\right) + i\sin\left(\frac{\theta}{n} + \frac{360^\circ k}{n}\right)\right], \quad k = 0, 1, \ldots, n-1$:

$$z^{1/4} = \sqrt[4]{16}\left[\cos\left(\frac{120^\circ}{4} + \frac{360^\circ k}{4}\right) + i\sin\left(\frac{120^\circ}{4} + \frac{360^\circ k}{4}\right)\right], \quad k = 0, 1, 2, 3$$

Step 3: Plot the roots in the complex plane and connect consecutive roots to form a square:

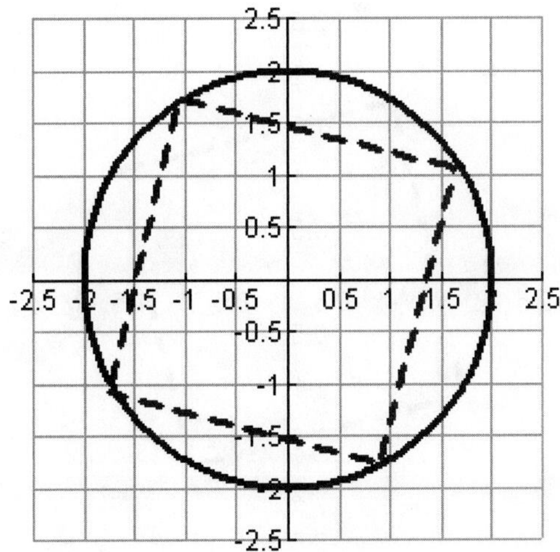

100. Given $z = 8\sqrt{3} + 8i$, in order to compute $z^{1/4}$, follow these steps:

Step 1: Write $z = 8\sqrt{3} + 8i$ in polar form.

Since $x = 8\sqrt{3}, y = 8$, the point is in QI. So, using the calculator yields

$$r = 16,\ \theta = 30^\circ$$

Hence, $z = -8 + 8\sqrt{3}\,i = 16\left[\cos\left(30^\circ\right) + i\sin\left(30^\circ\right)\right]$.

Step 2: Now, apply $z^{1/n} = r^{1/n}\left[\cos\left(\frac{\theta}{n} + \frac{360^\circ k}{n}\right) + i\sin\left(\frac{\theta}{n} + \frac{360^\circ k}{n}\right)\right]$, $k = 0, 1, \ldots, n-1$:

$$z^{1/4} = \sqrt[4]{16}\left[\cos\left(\frac{30^\circ}{4} + \frac{360^\circ k}{4}\right) + i\sin\left(\frac{30^\circ}{4} + \frac{360^\circ k}{4}\right)\right], \quad k = 0, 1, 2, 3$$

Step 3: Plot the roots in the complex plane and connect consecutive roots to form a square:

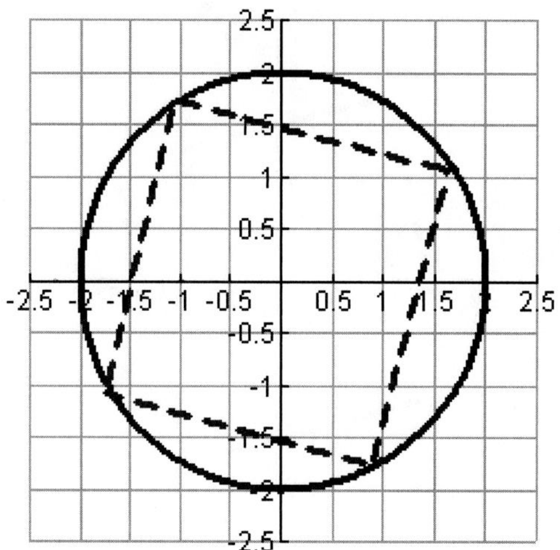

101. Consider the following graph of $r = 1 - 2\sin 3\theta$:

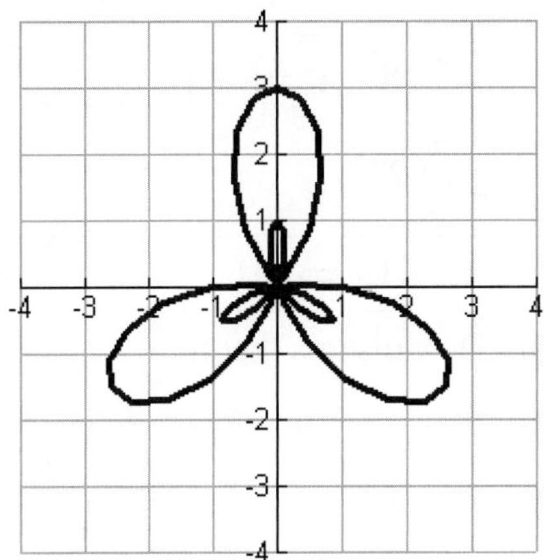

Note that the curve self-intersects at the origin, multiple times. This occurs when $r = 0$. To find these angles for which this is true, we solve the following equation:

$$0 = 1 - 2\sin 3\theta$$
$$\sin 3\theta = \tfrac{1}{2}$$
$$3\theta = \tfrac{\pi}{6}, \tfrac{5\pi}{6}, \tfrac{13\pi}{6}, \tfrac{17\pi}{6}, \tfrac{25\pi}{6}, \tfrac{29\pi}{6}$$
$$\theta = \boxed{\tfrac{\pi}{18}, \tfrac{5\pi}{18}, \tfrac{13\pi}{18}, \tfrac{17\pi}{18}, \tfrac{25\pi}{18}, \tfrac{29\pi}{18}}$$

102. Consider the following graph of $r = 1 + 2\cos 3\theta$:

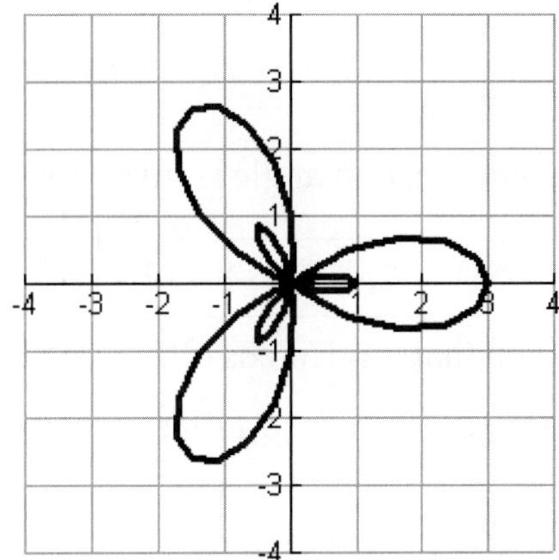

Note that the curve self-intersects at the origin, multiple times. This occurs when $r = 0$. To find these angles for which this is true, we solve the following equation:

$$0 = 1 + 2\cos 3\theta$$
$$\cos 3\theta = \tfrac{1}{2}$$
$$3\theta = \tfrac{\pi}{3}, \tfrac{5\pi}{3}, \tfrac{7\pi}{3}, \tfrac{11\pi}{3}, \tfrac{13\pi}{3}, \tfrac{17\pi}{3}$$
$$\theta = \boxed{\tfrac{\pi}{9}, \tfrac{5\pi}{9}, \tfrac{7\pi}{9}, \tfrac{11\pi}{9}, \tfrac{13\pi}{9}, \tfrac{17\pi}{9}}$$

Chapter 7 Practice Test Solutions

1. Given that $\vec{u} = \langle -5, 12 \rangle$, we have

$$|\vec{u}| = \sqrt{(-5)^2 + (12)^2} = 13$$

$\tan\theta = \dfrac{12}{-5}$ so that since the head is in QII, $\theta = 180^\circ + \tan^{-1}\left(\dfrac{12}{-5}\right) \approx 112.6^\circ$

2.

$$\frac{\vec{v}}{|\vec{v}|} = \frac{\langle -3, -4 \rangle}{\sqrt{(-3)^2 + (-4)^2}} = \frac{\langle -3, -4 \rangle}{5} = \boxed{\left\langle \frac{-3}{5}, \frac{-4}{5} \right\rangle}$$

3. a. $2\langle -1, 4 \rangle - 3\langle 4, 1 \rangle = \langle -2, 8 \rangle - \langle 12, 3 \rangle = \langle -2 - 12, 8 - 3 \rangle = \boxed{\langle -14, 5 \rangle}$

b. $\langle -7, -1 \rangle \cdot \langle 2, 2 \rangle = (-7)(2) + (-1)(2) = \boxed{-16}$

4. We need the component form for each of the three vectors involved. Indeed, they are:

$$\vec{A}=\langle 0,3\rangle,\ \ \vec{B}=\langle 12,0\rangle,\ \ \vec{C}=\left\langle 18\cos 330^\circ,\ 18\sin 330^\circ\right\rangle \approx \langle 15.59,\ -9\rangle$$

As such, $\vec{A}+\vec{B}+\vec{C}=\langle 27.59,\ -6\rangle$. So, $\left|\vec{A}+\vec{B}+\vec{C}\right|=\sqrt{27.59^2+(-6)^2}\approx \boxed{28.23\text{ yards}}$.

Also, the direction angle is given by $\tan^{-1}\left(\dfrac{-6}{27.59}\right)\approx \boxed{-12.27^\circ}$.

5. $\langle 4,-51\rangle\cdot\langle -2,-\frac{1}{3}\rangle=-8+17=\boxed{9}$

6. $4a-10a=18 \Rightarrow -6a=18 \Rightarrow \boxed{a=-3}$

7. Given that $z=16\left[\cos 120^\circ+i\sin 120^\circ\right]$, we apply DeMoivre's theorem $z^n=r^n\left[\cos(n\theta)+i\sin(n\theta)\right]$ to compute z^4. Indeed, observe that

$$z^4=(16)^4\left[\cos(4\cdot 120^\circ)+i\sin(4\cdot 120^\circ)\right]=65{,}536\left[-\frac{1}{2}+\frac{\sqrt{3}}{2}i\right]=\boxed{32{,}768\left[-1+\sqrt{3}i\right]}$$

8. Given that $z=16\left[\cos 120^\circ+i\sin 120^\circ\right]$, we apply

$$z^{1/n}=r^{1/n}\left[\cos\left(\frac{\theta}{n}+\frac{360^\circ k}{n}\right)+i\sin\left(\frac{\theta}{n}+\frac{360^\circ k}{n}\right)\right],\quad k=0,1,\ldots,n-1,$$

to compute $z^{1/4}$. Indeed, observe that

$$\begin{aligned} z^{1/4}&=\sqrt[4]{16}\left[\cos\left(\frac{120^\circ}{4}+\frac{360^\circ k}{4}\right)+i\sin\left(\frac{120^\circ}{4}+\frac{360^\circ k}{4}\right)\right],\quad k=0,1,2,3\\ &=2\left[\cos(30^\circ)+i\sin(30^\circ)\right],\ 2\left[\cos(120^\circ)+i\sin(120^\circ)\right],\\ &\quad 2\left[\cos(210^\circ)+i\sin(210^\circ)\right],\ 2\left[\cos(300^\circ)+i\sin(300^\circ)\right],\\ &=\boxed{\sqrt{3}+i,\ -1+\sqrt{3}i,\ -\sqrt{3}-i,\ 1-\sqrt{3}i} \end{aligned}$$

9. In order to convert the point $(3,\ 210^\circ)$ (expressed in polar coordinates) to rectangular coordinates, first observe that $r=3,\ \theta=210^\circ$. Hence,

$$x=r\cos\theta=3\cos(210^\circ)=3\left(-\frac{\sqrt{3}}{2}\right)=-\frac{3\sqrt{3}}{2}$$

$$y=r\sin\theta=3\sin(210^\circ)=3\left(-\frac{1}{2}\right)=-\frac{3}{2}$$

So, the given point can be expressed in rectangular coordinates as $\boxed{-\dfrac{3\sqrt{3}}{2}-\dfrac{3}{2}i}$.

10. Since $r = 4,\ \theta = \frac{5\pi}{4}$. Then,

$$x + iy = 4\left[\cos\tfrac{5\pi}{4} + i\sin\tfrac{5\pi}{4}\right] = 4\left[-\tfrac{\sqrt{2}}{2} - i\tfrac{\sqrt{2}}{2}\right] = -2\sqrt{2} - \left(2\sqrt{2}\right)i.$$

So, the rectangular coordinates are $\left(-2\sqrt{2}, -2\sqrt{2}\right)$.

11. Since $x = 30,\ y = -15$, we see that

$$r = \sqrt{30^2 + (-15)^2} = \sqrt{1125} = 15\sqrt{5}$$

$$\tan\theta = -\tfrac{15}{30} \Rightarrow \theta = \tan^{-1}\left(-\tfrac{1}{2}\right) = -26.6^\circ$$

In QIV, the angle is equivalent to 333.4°. So, the polar coordinates are $\left(15\sqrt{5}, 333.4^\circ\right)$.

12. In order to graph $r = 6\sin 2\theta$, consider the following table of points:

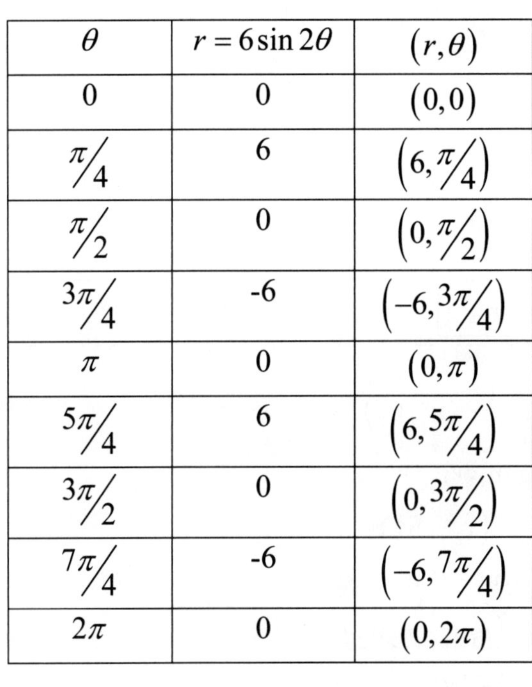

θ	$r = 6\sin 2\theta$	(r,θ)
0	0	$(0,0)$
$\pi/4$	6	$(6,\pi/4)$
$\pi/2$	0	$(0,\pi/2)$
$3\pi/4$	-6	$(-6,3\pi/4)$
π	0	$(0,\pi)$
$5\pi/4$	6	$(6,5\pi/4)$
$3\pi/2$	0	$(0,3\pi/2)$
$7\pi/4$	-6	$(-6,7\pi/4)$
2π	0	$(0,2\pi)$

The graph is as follows:

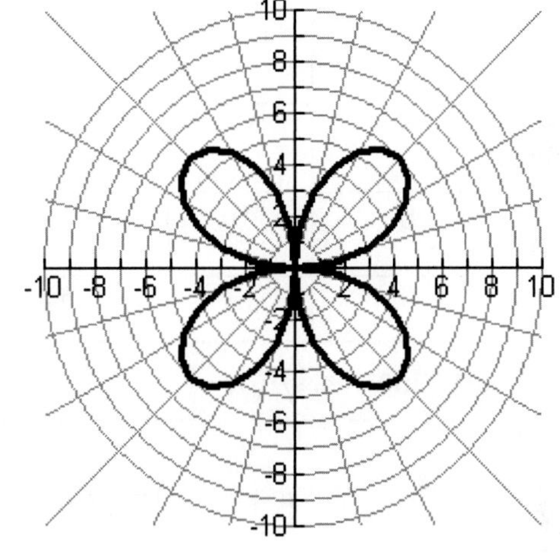

13. In order to graph $r^2 = 9\cos 2\theta$, consider the following table of points:

θ	$r^2 = 9\cos 2\theta$	r	(r,θ)
0	9	± 3	$(\pm 3, 0)$
$\pi/4$	0	0	$(0, \pi/4)$
$\pi/2$	-9	No points	
$3\pi/4$	0	0	$(0, 3\pi/4)$
π	9	± 3	$(\pm 3, \pi)$
$5\pi/4$	0	0	$(0, 5\pi/4)$
$3\pi/2$	-9	No points	
$7\pi/4$	0	0	$(0, 7\pi/4)$
2π	9	± 3	$(\pm 3, 2\pi)$

The graph is as follows:

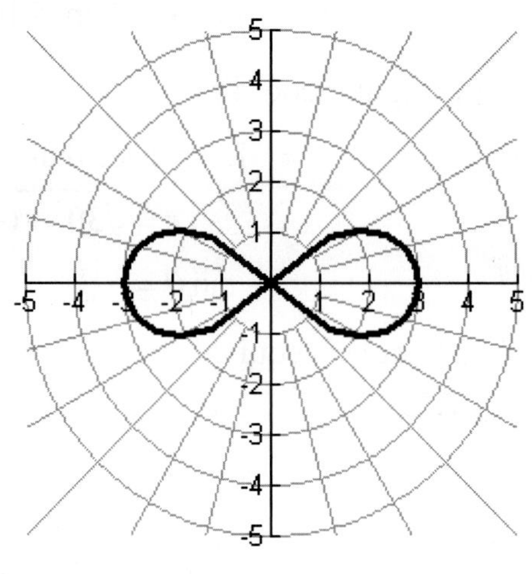

14. Observe that $\langle x,1\rangle \cdot \langle 3,-4\rangle = 3x-4=0 \Rightarrow \boxed{x=\frac{4}{3}}$

15. Observe that

$$\begin{aligned}\vec{u}\cdot(\vec{v}-\vec{w}) &= \langle u_1,u_2\rangle \cdot (\langle v_1,v_2\rangle - \langle w_1,w_2\rangle) \\ &= \langle u_1,u_2\rangle \cdot (\langle v_1-w_1, v_2-w_2\rangle) \\ &= u_1(v_1-w_1)+u_2(v_2-w_2) \\ &= u_1v_1 - u_1w_1 + u_2v_2 - u_2w_2 \\ &= (u_1v_1+u_2v_2)-(u_1w_1+u_2w_2) \\ &= \langle u_1,u_2\rangle\cdot\langle v_1,v_2\rangle - \langle u_1,u_2\rangle\cdot\langle w_1,w_2\rangle \\ &= \vec{u}\cdot\vec{v}-\vec{u}\cdot\vec{w}\end{aligned}$$

16. The desired vector is

$$-\left(\frac{1}{\left|\langle 3,5\rangle\right|}\right)\langle 3,5\rangle = -\frac{1}{\sqrt{34}}\langle 3,5\rangle = \left\langle -\frac{3}{\sqrt{34}}, -\frac{5}{\sqrt{34}}\right\rangle = \boxed{\left\langle -\frac{3\sqrt{34}}{34}, -\frac{5\sqrt{34}}{34}\right\rangle}$$

17. $\vec{u}\cdot\vec{v} = 4(10)\cos\left(\frac{2\pi}{3}\right) = 40\left(-\frac{1}{2}\right) = \boxed{-20}$

18. Since $\vec{u}\cdot\vec{v} = (\sin\theta)(-\cos\theta)+(\cos\theta)(\sin\theta)=0$, we conclude that $\vec{u}$ and $\vec{v}$ are perpendicular.

19. $|\vec{v}| = \sqrt{(-1)^2 + (-1)^2} = \boxed{\sqrt{2}}$

20. Consider the following diagram:

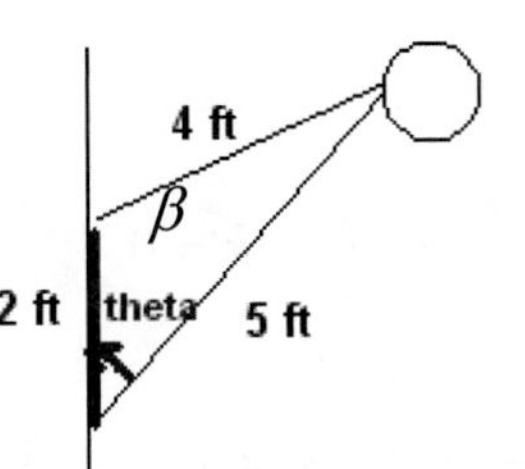

We first determine β using the Law of Cosines:

$$5^2 = 4^2 + 2^2 - 2(2)(4)\cos\beta$$
$$-\tfrac{5}{16} = \cos\beta$$
$$\beta \approx -18.209^\circ$$

So, we use the reference angle $\beta = 161.791^\circ$.

Now, by the Law of Sines, we obtain

$$\frac{\sin\theta}{4} = \frac{\sin 161.791^\circ}{5} \Rightarrow \sin\theta = \frac{4\sin 161.791^\circ}{5}$$
$$\Rightarrow \theta = \sin^{-1}\left(\frac{4\sin 161.791^\circ}{5}\right)$$
$$\approx \boxed{14.48^\circ}$$

21. We first use the Law of Cosines to find the resultant force F:

$$F^2 = 27{,}000^2 + 25{,}000^2 - 2(27{,}000)(25{,}000)\cos 155^\circ$$
$$F = 50{,}769\, lbs.$$

Next, we use the Law of Sines to find α:

$$\frac{\sin\alpha}{25{,}000} = \frac{\sin 155^\circ}{50{,}769} \Rightarrow \alpha = \sin^{-1}\left(\frac{25{,}000\sin 155^\circ}{50{,}769}\right) \approx 12^\circ.$$

22. True. Observe that

$$(\vec{u}+\vec{v})\cdot(\vec{u}-\vec{v}) = |\vec{u}|^2 - |\vec{v}|^2 + \underbrace{\vec{u}\cdot\vec{v} - \vec{v}\cdot\vec{u}}_{=0} = 0 \Rightarrow |\vec{u}|^2 = |\vec{v}|^2 \Rightarrow |\vec{u}| = |\vec{v}|.$$

23. Solving the equation $z^4+256i=0$ is equivalent to computing $z^{1/4}$, where $z=0-256i$, Follow these steps:

Step 1: Write $z=0-256i$ in polar form.

Since $x=0,\ y=-256$, the point is on the negative y-axis. So,

$$r=256,\quad \theta=270^\circ$$

Hence, $z=0-256i=256\left[\cos\left(270^\circ\right)+i\sin\left(270^\circ\right)\right]$.

Step 2: Now, apply $z^{1/n}=r^{1/n}\left[\cos\left(\frac{\theta}{n}+\frac{360^\circ k}{n}\right)+i\sin\left(\frac{\theta}{n}+\frac{360^\circ k}{n}\right)\right],\quad k=0,1,\ldots,n-1$:

$$z^{1/4}=\sqrt[4]{256}\left[\cos\left(\frac{270^\circ}{4}+\frac{360^\circ k}{4}\right)+i\sin\left(\frac{270^\circ}{4}+\frac{360^\circ k}{4}\right)\right],\quad k=0,1,2,3$$

$$=\boxed{4\left[\cos\left(67.5^\circ+90^\circ k\right)+i\sin\left(67.5^\circ+90^\circ k\right)\right],\quad k=0,1,2,3}$$

24. Observe that

$$r^2=\tan\theta$$

$$r^2=\frac{\sin\theta}{\cos\theta}=\frac{r\sin\theta}{r\cos\theta}$$

$$\boxed{x^2+y^2=\frac{y}{x}}$$

25.

$$\text{Sum}(\{-8,-11\}*\{-16,26\})/$$

$$(\text{Sum}(\{-8,-11\}\wedge 2)*\text{Sum}(\{-16,26\}\wedge 2)=$$

$$-0.38051$$

$$\cos^{-1}(Ans)=112.36528$$

So, the angle is approximately 112°.

26. Given $z=-8\sqrt{3}-8i$, in order to compute $z^{1/4}$, follow these steps:

Step 1: Write $z=-8\sqrt{3}-8i$ in polar form.
Since $x=-8\sqrt{3}, y=-8$, the point is in QIII. So, using the calculator yields

$$r=16,\ \theta=210^\circ$$

Hence, $z=-8-8\sqrt{3}i=16\left[\cos(210^\circ)+i\sin(210^\circ)\right]$.

Step 2: Now, apply $z^{1/n}=r^{1/n}\left[\cos\left(\frac{\theta}{n}+\frac{360^\circ k}{n}\right)+i\sin\left(\frac{\theta}{n}+\frac{360^\circ k}{n}\right)\right],\quad k=0, 1, \ldots, n-1$:

$$z^{1/4}=\sqrt[4]{16}\left[\cos\left(\frac{210^\circ}{4}+\frac{360^\circ k}{4}\right)+i\sin\left(\frac{210^\circ}{4}+\frac{360^\circ k}{4}\right)\right],\quad k=0, 1, 2, 3$$

Step 3: Plot the roots in the complex plane and connect consecutive roots to form a square:

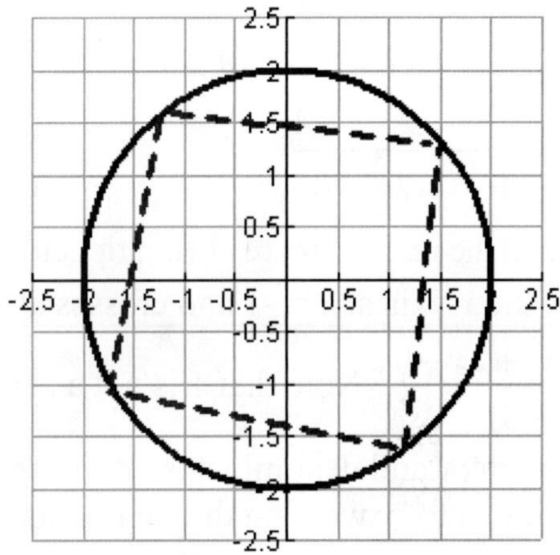

Chapter 7 Cumulative Test

1. $(f\circ g)(x)=\left(\frac{1}{\sqrt{3x+5}}\right)^2-4=\frac{1}{3x+5}-4=\frac{-12x-19}{3x+5}$

$dom(f)=(-\infty,\infty),\quad dom(g)=\left(-\frac{5}{3},\infty\right)=dom(f\circ g)$

2. $f(-x)=\left|(-x)^3\right|=\left|-x^3\right|=\left|-1\right|\left|x^3\right|=\left|x^3\right|=f(x)$. So, f is even.

3. Since the vertex is $(-1,2)$, the function has the form $y=a(x+1)^2+2$. To find a, use the fact that the point $(2,-1)$ is on the graph:

$$-1=a(2+1)^2+2$$
$$-1=9a+2$$
$$-\tfrac{1}{3}=a$$

So, the function is $y=-\tfrac{1}{3}(x+1)^2+2=-\tfrac{1}{3}x^2-\tfrac{2}{3}x+\tfrac{5}{3}$.

4. a.) Possible rational zeros are: $\pm1, \pm2, \pm4, \pm8$. Using synthetic division yields

$$\begin{array}{r|rrrrrr}
-1 & 1 & -4 & 1 & 10 & -4 & -8 \\
 & & -1 & 5 & -6 & -4 & 8 \\ \hline
-1 & 1 & -5 & 6 & 4 & -8 & 0 \\
 & & -1 & 6 & -12 & 8 & \\ \hline
2 & 1 & -6 & 12 & -8 & 0 & \\
 & & 2 & -8 & 8 & & \\ \hline
2 & 1 & -4 & 4 & 0 & & \\
 & & 2 & -4 & & & \\ \hline
 & 1 & -2 & 0 & & &
\end{array}$$

So, $f(x)=(x+1)^2(x-2)^3$ and the real zeros are -1 (multiplicity 2) and 2 (multiplicity 3).

b.) The graph of f touches the x-axis at x = -1 and crosses at x = 2.

5. Consider $f(x)=\dfrac{x^3-3x^2+2x-1}{(x-1)^2}$. Note that 1 is not a zero of the numerator and so, no factors cancel in the numerator and denominator. As such, $x=1$ is a VA and there is no HA. As for the slant asymptote, we long divide to obtain:

$$\begin{array}{r}
x-1 \\
x^2-2x+1\overline{\smash{\big)}\,x^3-3x^2+2x-1} \\
-\left(x^3-2x^2+x\right) \\ \hline
-x^2+x-1 \\
-\left(-x^2+10x-5\right) \\ \hline
-9x+4
\end{array}$$

So, the slant asymptote is $y=x-1$.

6. Use $P=P_0e^{rt}$ with $P(8)=5000$, $t=0.014$, $t=8$ to find P_0:

$$5000=P_0e^{0.014(8)} \Rightarrow P_0=\tfrac{5000}{e^{0.112}}\approx\boxed{\$4,470}$$

7. $\log_{625} 5 = \frac{1}{4}$	**8.** $305^\circ\left(\dfrac{\pi}{180^\circ}\right) = \dfrac{61\pi}{36}$

9. Consider the following diagram:

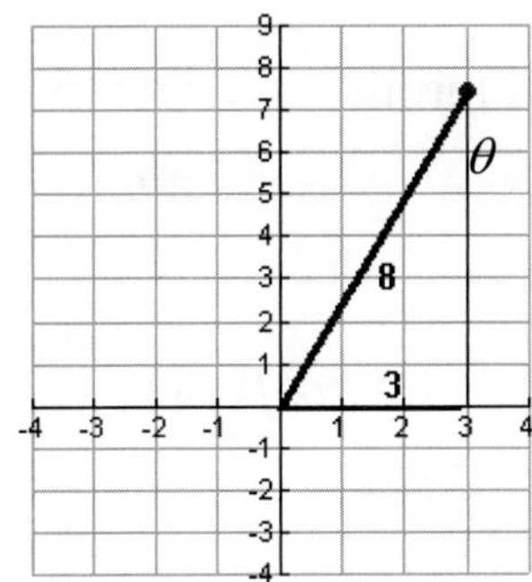

$$\sin\theta = \frac{\text{opp}}{\text{hyp}} = \frac{3}{8}$$
$$\cos\theta = \frac{\text{adj}}{\text{hyp}} = \frac{\sqrt{55}}{8}$$
$$\tan\theta = \frac{\text{opp}}{\text{adj}} = \frac{3\sqrt{55}}{55}$$
$$\cot\theta = \frac{1}{\tan\theta} = \frac{\sqrt{55}}{3}$$
$$\sec\theta = \frac{1}{\cos\theta} = \frac{8\sqrt{55}}{55}$$
$$\csc\theta = \frac{1}{\sin\theta} = \frac{8}{3}$$

10. First, observe that $\gamma = 180^\circ - (30^\circ + 30^\circ) = 120^\circ$.

Note that $a = b$ (since the triangle is isosceles). Using the Law of Sines yields

$$\frac{\sin 30^\circ}{a} = \frac{\sin 120^\circ}{4\,in.} \Rightarrow a = \frac{(4\,in.)\sin 30^\circ}{\sin 120^\circ} = \frac{4\sqrt{3}}{3}\,in.$$

Thus, $a = b = \frac{4\sqrt{3}}{3}\,in.$ and $\gamma = 120^\circ$.

11. Using the Law of Cosines yields

$$c^2 = 14.2^2 + 16.5^2 - 2(14.2)(16.5)\cos 50^\circ = 172.679m^2 \Rightarrow c \approx 13.1m.$$

Next, using the Law of Sines yields

$$\frac{\sin\alpha}{14.2} = \frac{\sin 50^\circ}{13.1} \Rightarrow \sin\alpha = \frac{14.2\sin 50^\circ}{13.1} \Rightarrow \alpha = \sin^{-1}\left(\frac{14.2\sin 50^\circ}{13.1}\right) \approx 56.1^\circ.$$

Thus, $\beta = 180^\circ - (56.1^\circ + 50^\circ) = 73.9^\circ$.

12. **a.)** -1 **b.)** 1
c.) 1 **d.)** $-\sqrt{3}$
e.) - 2 **f.)** 2

13. Amplitude = $\left|-\frac{1}{3}\right| = \frac{1}{3}$
Period = $\frac{2\pi}{4} = \frac{\pi}{2}$
Phase Shift = $\frac{\pi}{4}$ (right)
Vertical Shift = 4 (up)

14. $\cos^2 x + \sin^2 x = 1 \Rightarrow \cos x = -\sqrt{1-\sin^2 x} = -\sqrt{1-\left(\frac{1}{\sqrt{3}}\right)^2} = -\sqrt{\frac{2}{3}}$. So,

$$\cos 2x = \cos^2 x - \sin^2 x = \tfrac{2}{3} - \tfrac{1}{3} = \tfrac{1}{3}.$$

15. Let $\theta = \cos^{-1}\left(-\frac{3}{5}\right)$, $\beta = \sin^{-1}\left(\frac{1}{2}\right)$. We wish to compute $\tan(\alpha+\beta) = \dfrac{\tan\alpha + \tan\beta}{1-\tan\alpha\tan\beta}$. To do so, we first compute $\tan\alpha$ and $\tan\beta$. To this end, note that

$\theta = \cos^{-1}\left(-\frac{3}{5}\right) \Rightarrow \cos\theta = -\frac{3}{5}$. So, $\tan\theta = \frac{4}{-3}$ (obtained forming the right triangle governed by the given value of $\cos\theta$). Similarly, $\beta = \sin^{-1}\left(\frac{1}{2}\right) \Rightarrow \sin\beta = \frac{1}{2}$. So, $\tan\beta = \frac{\sqrt{3}}{3}$. Thus,

$$\tan(\alpha+\beta) = \frac{-\frac{4}{3}+\frac{\sqrt{3}}{3}}{1-\left(-\frac{4}{3}\right)\left(\frac{\sqrt{3}}{3}\right)} = \frac{-12+3\sqrt{3}}{9+4\sqrt{3}} = \frac{\left(-12+3\sqrt{3}\right)\left(9-4\sqrt{3}\right)}{\left(9+4\sqrt{3}\right)\left(9+4\sqrt{3}\right)} = \frac{75\sqrt{3}-144}{33} = \frac{25\sqrt{3}-48}{11}$$

16. In order to express $\left(1+\sqrt{3}i\right)^8$ in rectangular form, follow these steps:

<u>Step 1</u>: Write $1+\sqrt{3}i$ in polar form.

Since $x=1$, $y=\sqrt{3}$, the point is in QI. So,

$$r = \sqrt{x^2+y^2} = \sqrt{(1)^2+\left(\sqrt{3}\right)^2} = 2, \qquad \tan\theta = \frac{y}{x} = \frac{\sqrt{3}}{1} = \sqrt{3}, \text{ so that } \theta = \tan^{-1}\left(\sqrt{3}\right) = \frac{\pi}{3}.$$

Hence, $1+\sqrt{3}i = 2\left[\cos\left(\frac{\pi}{3}\right) + i\sin\left(\frac{\pi}{3}\right)\right]$.

<u>Step 2</u>: Apply DeMoivre's theorem $z^n = r^n\left[\cos(n\theta) + i\sin(n\theta)\right]$.

$$\left(1+\sqrt{3}i\right)^8 = 2^8\left[\cos\left(8\cdot\frac{\pi}{3}\right) + i\sin\left(8\cdot\frac{\pi}{3}\right)\right] = 256\left[-\frac{1}{2}+\frac{\sqrt{3}}{2}i\right] = \boxed{-128+128\sqrt{3}i}$$

CHAPTER 8

Section 8.1 Solutions --

<table>
<tr>
<td>

1.

Solve the system: $\begin{cases} x+y=7 & \textbf{(1)} \\ x-y=9 & \textbf{(2)} \end{cases}$

Solve **(1)** for y: $y=7-x$ **(3)**

Substitute **(3)** into **(2)** and solve for x:

$$x-(7-x)=9$$
$$x-7+x=9$$
$$x=8$$

Substitute $x=8$ into **(1)** to find y: $y=-1$.

So, the solution is $\boxed{(8,-1)}$.

</td>
<td>

2.

Solve the system: $\begin{cases} x-y=-10 & \textbf{(1)} \\ x+y=4 & \textbf{(2)} \end{cases}$

Solve **(1)** for x: $x=y-10$ **(3)**

Substitute **(3)** into **(2)** and solve for y:

$$(y-10)+y=4$$
$$2y=14$$
$$y=7$$

Substitute $y=7$ into **(1)** to find x: $x=-3$.

So, the solution is $\boxed{(-3,7)}$.

</td>
</tr>
<tr>
<td>

3.

Solve the system: $\begin{cases} 2x-y=3 & \textbf{(1)} \\ x-3y=4 & \textbf{(2)} \end{cases}$

Solve **(2)** for x: $x=3y+4$ **(3)**

Substitute **(3)** into **(1)** and solve for y:

$$2(3y+4)-y=3$$
$$5y=-5$$
$$y=-1$$

Substitute $y=-1$ into **(1)** to find x:

$$2x-(-1)=3 \Rightarrow 2x=2 \Rightarrow x=1.$$

So, the solution is $\boxed{(1,-1)}$.

</td>
<td>

4.

Solve the system: $\begin{cases} 4x+3y=3 & \textbf{(1)} \\ 2x+y=1 & \textbf{(2)} \end{cases}$

Solve **(2)** for y: $y=1-2x$ **(3)**

Substitute **(3)** into **(1)** and solve for x:

$$4x+3(1-2x)=3$$
$$4x+3-6x=3$$
$$-2x=0$$
$$x=0$$

Substitute $x=0$ into **(3)** to find that $y=1$

So, the solution is $\boxed{(0,1)}$.

</td>
</tr>
</table>

5.

Solve the system: $\begin{cases} 3x+y=5 & \textbf{(1)} \\ 2x-5y=-8 & \textbf{(2)} \end{cases}$

Solve **(1)** for y: $y=5-3x$ **(3)**

Substitute **(3)** into **(2)** and solve for x:

$$2x-5(5-3x)=-8$$
$$2x-25+15x=-8$$
$$17x=17$$
$$x=1$$

Substitute $x=1$ into **(3)** to find y: $y=2$

So, the solution is $\boxed{(1,2)}$.

6.

Solve the system: $\begin{cases} 6x-y=-15 & \textbf{(1)} \\ 2x-4y=-16 & \textbf{(2)} \end{cases}$

Solve **(1)** for y: $y=6x+15$ **(3)**

Substitute **(3)** into **(2)** and solve for x:

$$2x-4(6x+15)=-16$$
$$2x-24x-60=-16$$
$$-22x=44$$
$$x=-2$$

Substitute $x=-2$ into **(3)** to find y: $y=3$

So, the solution is $\boxed{(-2,3)}$.

7.

Solve the system: $\begin{cases} 2u+5v=7 & \textbf{(1)} \\ 3u-v=5 & \textbf{(2)} \end{cases}$

Solve **(2)** for v: $v=3u-5$ **(3)**

Substitute **(3)** into **(1)** and solve for u:

$$2u+5(3u-5)=7$$
$$17u=32$$
$$u=\tfrac{32}{17}$$

Substitute $u=\tfrac{32}{17}$ into **(2)** to find v:

$$3(\tfrac{32}{17})-v=5$$
$$v=\tfrac{96-85}{17}=\tfrac{11}{17}.$$

So, the solution is $\boxed{u=\tfrac{32}{17},\ v=\tfrac{11}{17}}$.

8.

Solve the system: $\begin{cases} m-2n=4 & \textbf{(1)} \\ 3m+2n=1 & \textbf{(2)} \end{cases}$

Solve **(1)** for m: $m=2n+4$ **(3)**

Substitute **(3)** into **(2)** and solve for n:

$$3(2n+4)+2n=1$$
$$8n=-11$$
$$n=-\tfrac{11}{8}$$

Substitute $n=-\tfrac{11}{8}$ into **(1)** to find m:

$$m-2(-\tfrac{11}{8})=4$$
$$m=4-\tfrac{11}{4}=\tfrac{5}{4}.$$

So, the solution is $\boxed{m=\tfrac{5}{4},\ n=-\tfrac{11}{8}}$.

9.

Solve the system: $\begin{cases} 2x+y=7 & \textbf{(1)} \\ -2x-y=5 & \textbf{(2)} \end{cases}$

Solve **(1)** for y: $y=7-2x$ **(3)**

Substitute **(3)** into **(2)** and solve for x:

$$-2x-(7-2x)=5$$
$$-7=5$$

So, the system is inconsistent. Thus, there is $\boxed{\text{no solution}}$.

10.

Solve the system: $\begin{cases} 3x-y=2 & \textbf{(1)} \\ 3x-y=4 & \textbf{(2)} \end{cases}$

Solve **(1)** for y: $y=3x-2$ **(3)**

Substitute **(3)** into **(2)** and solve for x:

$$3x-(3x-2)=4$$
$$2=4$$

So, the system is inconsistent. Thus, there is $\boxed{\text{no solution}}$.

11.

Solve the system: $\begin{cases} 4r - s = 1 & \textbf{(1)} \\ 8r - 2s = 2 & \textbf{(2)} \end{cases}$

Solve **(1)** for s: $s = 4r - 1$ **(3)**

Substitute **(3)** into **(2)** and solve for r:

$$8r - 2(4r - 1) = 2$$
$$2 = 2$$

So, the system is consistent. There are $\boxed{\text{infinitely many solutions}}$.

12.

Solve the system: $\begin{cases} -3p + q = -4 & \textbf{(1)} \\ 6p - 2q = 8 & \textbf{(2)} \end{cases}$

Solve **(1)** for q: $q = 3p - 4$ **(3)**

Substitute **(3)** into **(2)** and solve for p:

$$6p - 2(3p - 4) = 8$$
$$8 = 8$$

So, the system is consistent. There are $\boxed{\text{infinitely many solutions}}$.

13.

Solve the system: $\begin{cases} 5r - 3s = 15 & \textbf{(1)} \\ -10r + 6s = -30 & \textbf{(2)} \end{cases}$

Solve **(1)** for r: $r = \frac{1}{5}(3s + 15)$ **(3)**

Substitute **(3)** into **(2)** and solve for s:

$$-10\left[\tfrac{1}{5}(3s + 15)\right] + 6s = -30$$
$$-6s - 30 + 6s = -30$$
$$-30 = -30$$

So, the system is consistent. There are $\boxed{\text{infinitely many solutions}}$.

14.

Solve the system: $\begin{cases} -5p - 3q = -1 & \textbf{(1)} \\ 10p + 6q = 2 & \textbf{(2)} \end{cases}$

Solve **(1)** for p: $p = \frac{1}{5}(-3q + 1)$ **(3)**

Substitute **(3)** into **(2)** and solve for q:

$$10\left[\tfrac{1}{5}(-3q + 1)\right] + 6q = 2$$
$$-6q + 2 + 6q = 2$$
$$2 = 2$$

So, the system is consistent. There are $\boxed{\text{infinitely many solutions}}$.

15. Solve the system:

$\begin{cases} 2x - 3y = -7 & \textbf{(1)} \\ 3x + 7y = 24 & \textbf{(2)} \end{cases}$

Solve **(1)** for x: $x = \frac{1}{2}(3y - 7)$ **(3)**

Substitute **(3)** into **(2)** and solve for y:

$$3 \cdot \tfrac{1}{2}(3y - 7) + 7y = 24$$
$$\tfrac{9}{2}y - \tfrac{21}{2} + 7y = 24$$
$$\tfrac{23}{2}y = \tfrac{69}{2}$$
$$\boxed{y = 3}$$

Substitute this value back into **(3)** to obtain that $\boxed{x = 1}$. So, the solution is $(1,3)$.

16. Solve the system:

$\begin{cases} 4x - 5y = -7 & \textbf{(1)} \\ 3x + 8y = 30 & \textbf{(2)} \end{cases}$

Solve **(1)** for x: $x = \frac{1}{4}(5y - 7)$ **(3)**

Substitute **(3)** into **(2)** and solve for y:

$$3 \cdot \tfrac{1}{4}(5y - 7) + 8y = 30$$
$$\tfrac{15}{4}y - \tfrac{21}{4} + 8y = 30$$
$$\tfrac{47}{4}y = \tfrac{141}{4}$$
$$\boxed{y = 3}$$

Substitute this value back into **(3)** to obtain that $\boxed{x = 2}$. So, the solution is $(2,3)$.

17. <u>Solve the system</u>: $\begin{cases} \frac{1}{3}x-\frac{1}{4}y=0 & \textbf{(1)} \\ -\frac{2}{3}x+\frac{3}{4}y=2 & \textbf{(2)} \end{cases}$ First, clear the fractions by multiplying both equations by 12 to obtain the equivalent system: $\begin{cases} 4x-3y=0 & \textbf{(3)} \\ -8x+9y=24 & \textbf{(4)} \end{cases}$ Solve **(3)** for *x*: $x=\frac{3}{4}y$ **(5)** Substitute **(5)** into **(4)** and solve for *y*: $-8\left(\frac{3}{4}y\right)+9y=24$ $3y=24$ $\boxed{y=8}$ Substitute this back into **(5)** to obtain that $\boxed{x=6}$. So, the solution is $(6,8)$.	**18.** <u>Solve the system</u>: $\begin{cases} \frac{1}{5}x+\frac{2}{3}y=10 & \textbf{(1)} \\ -\frac{1}{2}x-\frac{1}{6}y=-7 & \textbf{(2)} \end{cases}$ First, clear the fractions by multiplying both equations by their respective LCDs to obtain the equivalent system: $\begin{cases} 3x+10y=150 & \textbf{(3)} \\ 3x+y=42 & \textbf{(4)} \end{cases}$ Solve **(4)** for *y*: $y=42-3x$ **(5)** Substitute **(5)** into **(3)** and solve for *x*: $3x+10(42-3x)=150$ $-27x+420=150$ $\boxed{x=10}$ Substitute this back into **(5)** to obtain that $\boxed{y=12}$. So, the solution is $(10,12)$.
19. <u>Solve the system</u>: $\begin{cases} -3.9x+4.2y=15.3 & \textbf{(1)} \\ -5.4x+7.9y=16.7 & \textbf{(2)} \end{cases}$ Solve **(1)** for *y*: $y=\frac{1}{4.2}(15.3+3.9x)$ **(3)** Substitute **(3)** into **(2)** and solve for *x*: $-5.4x+\frac{7.9}{4.2}(15.3+3.9x)=16.7$ $120.87+8.13x=70.14$ $\boxed{x=-6.24}$ Substitute this back into **(3)** to obtain that $\boxed{y=-2.15}$. So, solution is $(-6.24,-2.15)$.	**20.** <u>Solve the system</u>: $\begin{cases} 6.3x-7.4y=18.6 & \textbf{(1)} \\ 2.4x+3.5y=10.2 & \textbf{(2)} \end{cases}$ Solve **(1)** for *y*: $y=\frac{1}{7.4}(6.3x-18.6)$ **(3)** Substitute **(3)** into **(2)** and solve for *x*: $2.4x+\frac{3.5}{7.4}(6.3x-18.6)=10.2$ $39.81x-65.1=75.48$ $\boxed{x=3.53}$ Substitute this back into **(3)** to obtain that $\boxed{y=0.49}$. So, solution is $(3.53,0.49)$.

21.	22.
Solve the system: $\begin{cases} x-y=-3 & \textbf{(1)} \\ x+y=7 & \textbf{(2)} \end{cases}$ Add **(1)** and **(2)** to eliminate *y*: $2x=4$ $x=2 \quad \textbf{(3)}$ Substitute **(3)** into **(2)** and solve for *y*: $2+y=7$ $y=5$ So, the solution is $\boxed{(2,5)}$.	Solve the system: $\begin{cases} x-y=-10 & \textbf{(1)} \\ x+y=8 & \textbf{(2)} \end{cases}$ Add **(1)** and **(2)** to eliminate *y*: $2x=-2$ $x=-1 \quad \textbf{(3)}$ Substitute **(3)** into **(2)** and solve for *y*: $-1+y=8$ $y=9$ So, the solution is $\boxed{(-1,9)}$.
23.	**24.**
Solve the system: $\begin{cases} 5x+3y=-3 & \textbf{(1)} \\ 3x-3y=-21 & \textbf{(2)} \end{cases}$ Add **(1)** and **(2)** to eliminate *y*: $8x=-24$ $x=-3 \quad \textbf{(3)}$ Substitute **(3)** into **(1)** and solve for *y*: $5(-3)+3y=-3$ $3y=12$ $y=4$ So, the solution is $\boxed{(-3,4)}$.	Solve the system: $\begin{cases} -2x+3y=1 & \textbf{(1)} \\ 2x-y=7 & \textbf{(2)} \end{cases}$ Add **(1)** and **(2)** to eliminate *x*: $2y=8$ $y=4 \quad \textbf{(3)}$ Substitute **(3)** into **(2)** and solve for *x*: $2x-4=7$ $2x=11$ $x=\frac{11}{2}$ So, the solution is $\boxed{\left(\frac{11}{2},4\right)}$.
25.	**26.**
Solve the system: $\begin{cases} 2x-7y=4 & \textbf{(1)} \\ 5x+7y=3 & \textbf{(2)} \end{cases}$ Add **(1)** and **(2)** to eliminate *y*: $7x=7$ $x=1 \quad \textbf{(3)}$ Substitute **(3)** into **(2)** and solve for *y*: $5(1)+7y=3$ $7y=-2$ $y=-\frac{2}{7}$ So, the solution is $\boxed{\left(1,-\frac{2}{7}\right)}$.	Solve the system: $\begin{cases} 3x+2y=6 & \textbf{(1)} \\ -3x+6y=18 & \textbf{(2)} \end{cases}$ Add **(1)** and **(2)** to eliminate *x*: $8y=24$ $y=3 \quad \textbf{(3)}$ Substitute **(3)** into **(1)** and solve for *x*: $3x+2(3)=6$ $3x=0$ $x=0$ So, the solution is $\boxed{(0,3)}$.

27.

Solve the system: $\begin{cases} 2x + 5y = 7 & \textbf{(1)} \\ 3x - 10y = 5 & \textbf{(2)} \end{cases}$

Multiply **(1)** by 2: $4x + 10y = 14$ **(3)**

Add **(2)** and **(3)** to eliminate *y*:

$$7x = 19$$

$$x = \tfrac{19}{7} \quad \textbf{(4)}$$

Substitute **(4)** into **(1)** and solve for *y*:

$$2(\tfrac{19}{7}) + 5y = 7$$

$$y = \tfrac{11}{35}$$

So, the solution is $\boxed{\left(\tfrac{19}{7}, \tfrac{11}{35}\right)}$.

28.

Solve the system: $\begin{cases} 6x - 2y = 3 & \textbf{(1)} \\ -3x + 2y = -2 & \textbf{(2)} \end{cases}$

Add **(1)** and **(2)** to eliminate *y*:

$$3x = 1$$

$$x = \tfrac{1}{3} \quad \textbf{(3)}$$

Substitute **(3)** into **(1)** and solve for *y*:

$$6(\tfrac{1}{3}) - 2y = 3$$

$$2 - 2y = 3$$

$$y = -\tfrac{1}{2}$$

So, the solution is $\boxed{\left(\tfrac{1}{3}, -\tfrac{1}{2}\right)}$.

29.

Solve the system: $\begin{cases} 2x + 5y = 5 & \textbf{(1)} \\ -4x - 10y = -10 & \textbf{(2)} \end{cases}$

Multiply **(1)** by 2: $4x + 10y = 10$ **(3)**

Add **(2)** and **(3)** to eliminate *y*:

$$0 = 0 \quad \textbf{(4)}$$

So, the system is consistent. Thus, there are $\boxed{\text{infinitely many solutions}}$.

30.

Solve the system: $\begin{cases} 11x + 3y = 3 & \textbf{(1)} \\ 22x + 6y = 6 & \textbf{(2)} \end{cases}$

Multiply **(1)** by -2: $-22x - 6y = -6$ **(3)**

Add **(2)** and **(3)** to eliminate *y*:

$$0 = 0 \quad \textbf{(4)}$$

So, the system is consistent. Thus, there are $\boxed{\text{infinitely many solutions}}$.

31.

Solve the system: $\begin{cases} 3x - 2y = 12 & \textbf{(1)} \\ 4x + 3y = 16 & \textbf{(2)} \end{cases}$

Multiply **(1)** by 4: $12x - 8y = 48$ **(3)**

Multiply **(2)** by -3: $-12x - 9y = -48$ **(4)**

Add **(3)** and **(4)** to eliminate *x*:

$$-17y = 0$$

$$y = 0 \quad \textbf{(5)}$$

Substitute **(5)** into **(1)** and solve for *x*:

$$3x - 2(0) = 12$$

$$x = 4$$

So, the solution is $\boxed{(4, 0)}$.

32.

Solve the system: $\begin{cases} 5x - 2y = 7 & \textbf{(1)} \\ 3x + 5y = 29 & \textbf{(2)} \end{cases}$

Multiply **(1)** by 5: $25x - 10y = 35$ **(3)**

Multiply **(2)** by 2: $6x + 10y = 58$ **(4)**

Add **(3)** and **(4)** to eliminate *y*:

$$31x = 93$$

$$x = 3 \quad \textbf{(5)}$$

Substitute **(5)** into **(1)** and solve for *x*:

$$5(3) - 2y = 7$$

$$y = 4$$

So, the solution is $\boxed{(3, 4)}$.

33.

Solve the system: $\begin{cases} 6x-3y=-15 & \textbf{(1)} \\ 7x+2y=-12 & \textbf{(2)} \end{cases}$

Multiply **(1)** by 2: $12x-6y=-30$ **(3)**

Multiply **(2)** by 3: $21x+6y=-36$ **(4)**

Add **(3)** and **(4)** to eliminate y:

$$33x=-66 \Rightarrow x=-2 \quad \textbf{(5)}$$

Substitute **(5)** into **(1)** and solve for y:

$$6(-2)-3y=-15$$

$$-3y=-3$$

$$y=1$$

So, the solution is $\boxed{(-2,1)}$.

34.

Solve the system: $\begin{cases} 7x-4y=-1 & \textbf{(1)} \\ 3x-5y=16 & \textbf{(2)} \end{cases}$

Multiply **(1)** by 5: $35x-20y=-5$ **(3)**

Multiply **(2)** by -4: $-12x+20y=-64$ **(4)**

Add **(3)** and **(4)** to eliminate y:

$$23x=-69$$

$$x=-3 \quad \textbf{(5)}$$

Substitute **(5)** into **(1)** and solve for y:

$$7(-3)-4(y)=-1$$

$$y=-5$$

So, the solution is $\boxed{(-3,-5)}$.

35.

Solve the system: $\begin{cases} 4x-5y=22 & \textbf{(1)} \\ 3x+4y=1 & \textbf{(2)} \end{cases}$

Multiply **(1)** by 4: $16x-20y=88$ **(3)**

Multiply **(2)** by 5: $15x+20y=5$ **(4)**

Add **(3)** and **(4)** to eliminate y:

$$31x=93 \Rightarrow x=3 \quad \textbf{(5)}$$

Substitute **(5)** into **(1)** and solve for y:

$$4(3)-5y=22$$

$$5y=-10$$

$$y=-2$$

So, the solution is $\boxed{(3,-2)}$.

36.

Solve the system: $\begin{cases} 6x-5y=32 & \textbf{(1)} \\ 2x-6y=2 & \textbf{(2)} \end{cases}$

Multiply **(2)** by -3: $-6x+18y=-6$ **(3)**

Add **(1)** and **(3)** to eliminate x

$$13y=26$$

$$y=2 \quad \textbf{(4)}$$

Substitute **(4)** into **(2)** and solve for x:

$$2x-6(2)=2$$

$$x=7$$

So, the solution is $\boxed{(7,2)}$.

37. Solve the system: $\begin{cases} \frac{1}{3}x+\frac{1}{2}y=1 & \textbf{(1)} \\ \frac{1}{5}x+\frac{7}{2}y=2 & \textbf{(2)} \end{cases}$ Multiply **(1)** by -7: $-\frac{7}{3}x-\frac{7}{2}y=-7$ **(3)** Add **(2)** and **(3)** to eliminate y: $-\frac{32}{15}x=-5$ $x=\frac{75}{32}$ **(4)** Substitute **(4)** into **(1)** and solve for y: $\frac{1}{3}(\frac{75}{32})+\frac{1}{2}y=1$ $\frac{1}{2}y=\frac{21}{96}$ $y=\frac{7}{16}$ So, the solution is $\boxed{\left(\frac{75}{32},\frac{7}{16}\right)}$.	**38.** Solve the system: $\begin{cases} \frac{1}{2}x-\frac{1}{3}y=0 & \textbf{(1)} \\ \frac{3}{2}x+\frac{1}{2}y=\frac{3}{4} & \textbf{(2)} \end{cases}$ Multiply **(1)** by -3: $-\frac{3}{2}x+y=0$ **(3)** Add **(2)** and **(3)** to eliminate x: $\frac{3}{2}y=\frac{3}{4}$ $y=\frac{1}{2}$ **(4)** Substitute **(4)** into **(1)** and solve for x: $\frac{1}{2}x-\frac{1}{3}(\frac{1}{2})=0$ $\frac{1}{2}x=\frac{1}{6}$ $x=\frac{1}{3}$ So, the solution is $\boxed{\left(\frac{1}{3},\frac{1}{2}\right)}$.
39. Solve the system: $\begin{cases} 3.4x+1.7y=8.33 & \textbf{(1)} \\ -2.7x-7.8y=15.96 & \textbf{(2)} \end{cases}$ Multiply **(1)** by 2.7: $9.18x+4.59y=22.491$ **(3)** Multiply **(2)** by 3.4: $-9.18x-26.52y=54.264$ **(4)** Add **(3)** and **(4)** to eliminate x: $-21.93y=76.755$ $y=-3.5$ **(5)** Substitute **(5)** into **(1)** and solve for x: $3.4x+1.7(-3.5)=8.33$ $x=4.2$ So, the solution is $\boxed{(4.2,-3.5)}$.	**40.** Solve the system: $\begin{cases} -0.04x+1.12y=9.815 & \textbf{(1)} \\ 2.79x+1.19y=-0.165 & \textbf{(2)} \end{cases}$ Multiply **(1)** by 2.79: $-0.1116x+3.1248y=27.38385$ **(3)** Multiply **(2)** by 0.04: $0.1116x+0.0476y=-0.0066$ **(4)** Add **(3)** and **(4)** to eliminate x: $3.1724y=27.37725$ $y=8.63$ **(5)** Substitute **(5)** into **(1)** and solve for x: $-0.04x+1.12(8.63)=9.815$ $x=-3.74$ So, the solution is $\boxed{(-3.74,8.63)}$.
41. c Adding the equations yields $6x=6$, so that $x=1$. Thus, $3(1)-y=1$, so that $y=2$.	**42. b** Multiply first equation by 2 and add to the second. This yields $y=1$, and upon substitution, $x=3$.
43. d Subtracting the equations yields $0=-4$. Thus, the system is inconsistent and so, the lines should be parallel.	**44. a** Multiply first equation by 2 and add to the second. This yields $0=0$. Thus, the system is consistent, and the lines should be the same.

45.

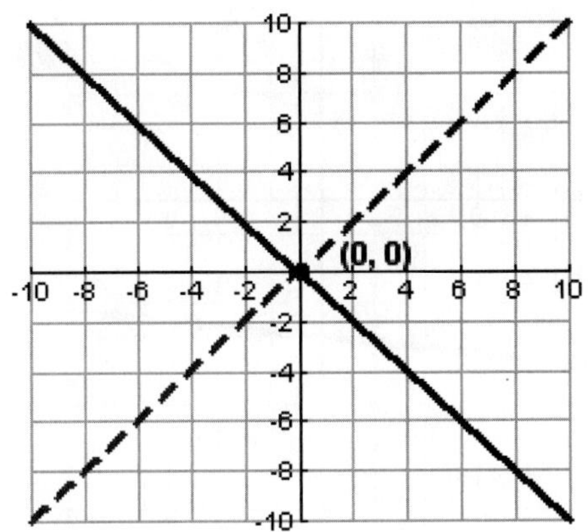

Notes on the graph:
Solid curve is $y=-x$
Dashed curve is $y=x$
So, the solution is (0,0).

46.

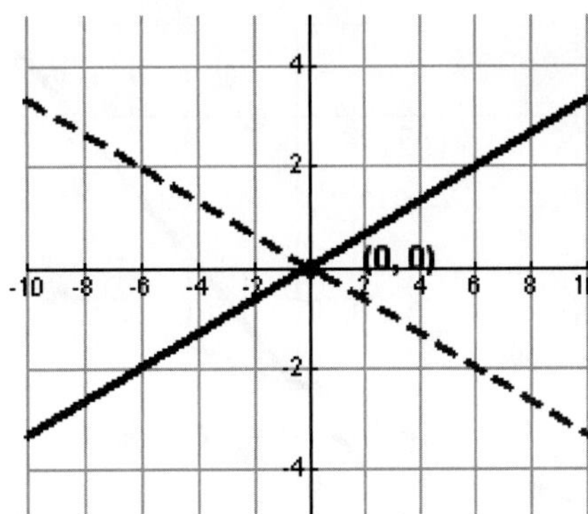

Notes on the graph:
Solid curve is $x-3y=0$
Dashed curve is $x+3y=0$
So, the solution is (0,0).

47.

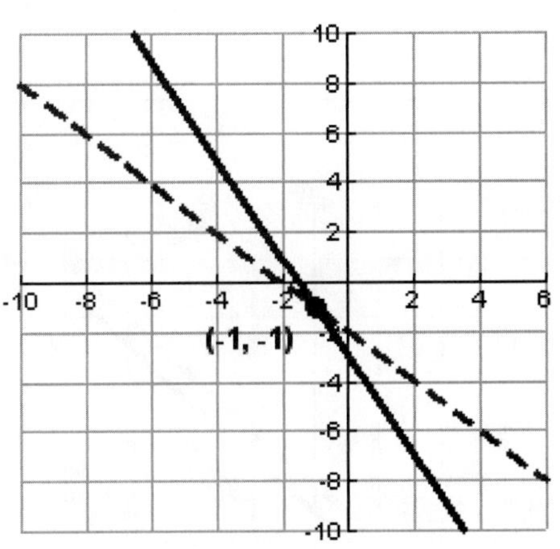

Notes on the graph:
Solid curve is $2x+y=-3$
Dashed curve is $x+y=-2$
So, the solution is $(-1,-1)$.

48.

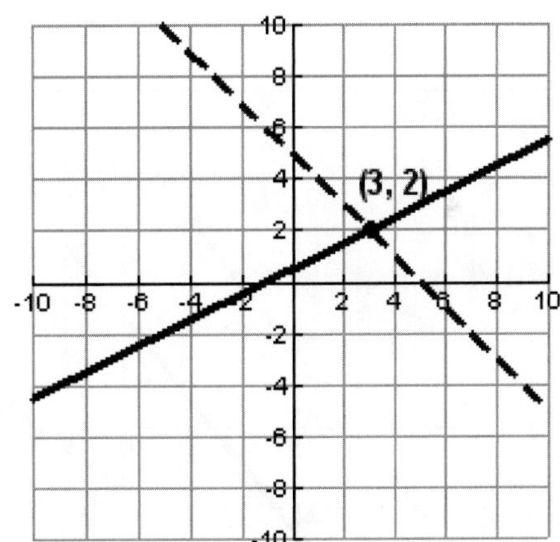

Notes on the graph:
Solid curve is $x-2y=-1$
Dashed curve is $-x-y=-5$
So, the solution is $(3,2)$.

49.

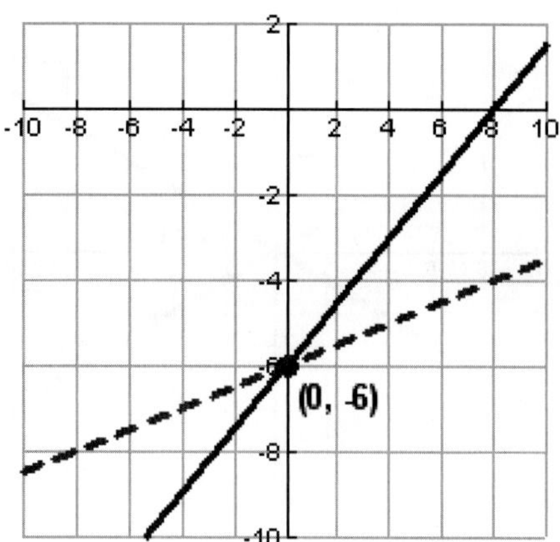

Notes on the graph:
Solid curve is $\frac{1}{2}x - \frac{2}{3}y = 4$
Dashed curve is $\frac{1}{4}x - y = 6$
So, the solution is $(0, -6)$.

50.

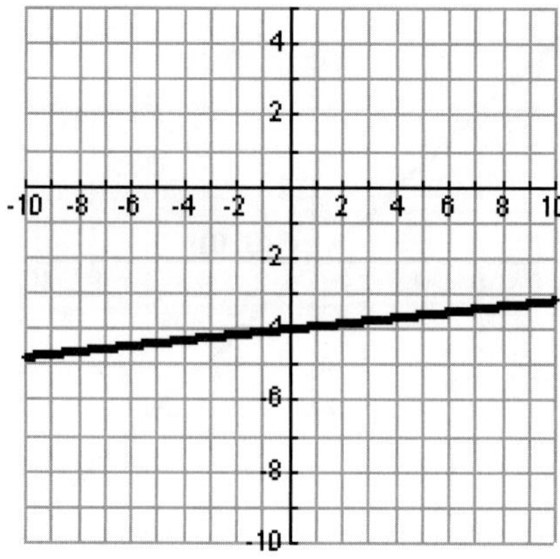

Notes on the graph:
The curves $\frac{1}{5}x - \frac{5}{2}y = 10$ and $\frac{1}{15}x - \frac{5}{6}y = \frac{10}{3}$ have the same graph pictured above. So, there are infinitely many solutions.

51.

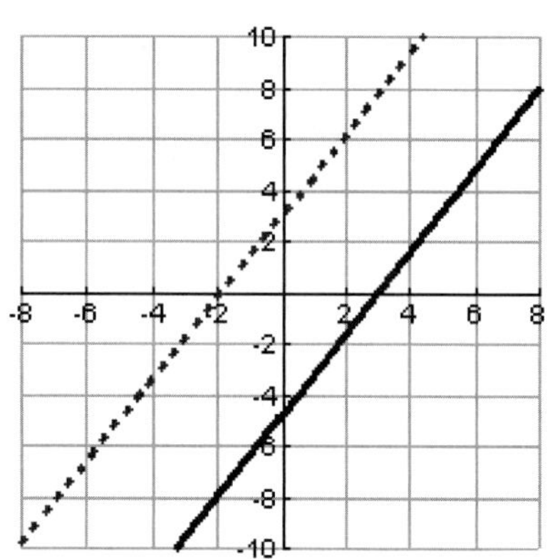

Notes on the graph:
Solid curve is $1.6x - y = 4.8$
Dashed curve is $-0.8x + 0.5y = 1.5$
So, there is no solution.

52.

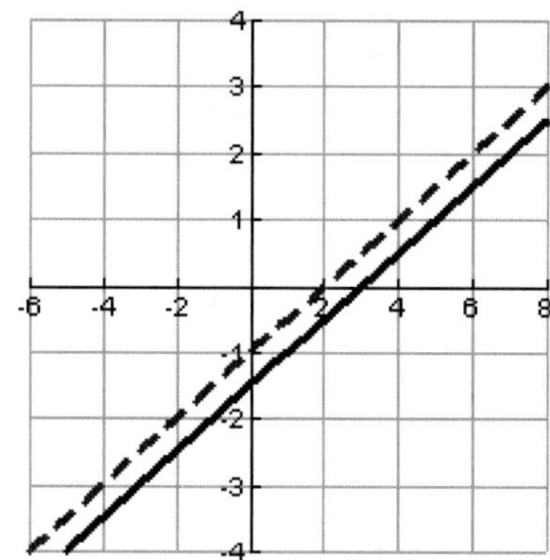

Notes on the graph:
Solid curve is $1.1x - 2.2y = 3.3$
Dashed curve is $-3.3x + 6.6y = -6.6$
So, there is no solution.

53. Let x = number of ml of 8% HCl.
y = number of ml of 15% HCl.

Must solve the system:

$$\begin{cases} 0.08x + 0.15y = (0.12)(37) & \textbf{(1)} \\ x + y = 37 & \textbf{(2)} \end{cases}$$

Multiply **(2)** by -0.08:

$$-0.08x - 0.08y = -2.96 \quad \textbf{(3)}$$

Add **(1)** and **(3)** to eliminate x:

$$0.07y = 1.48 \Rightarrow y \cong 21.14 \quad \textbf{(4)}$$

Substitute **(4)** into **(2)** and solve for x:

$$x \cong 15.86$$

So, should use approximately 15.86 ml of 8% HCl and 21.14 ml of 15% HCl.

54. Let x = # gallons of gas in 2.5% mix.
y = # gallons of oil in 2.5% mix.

Must solve the system:

$$\begin{cases} 0.025(x+y) = y & \textbf{(1)} \\ 0.04(350 - (x+y)) = 10 - y & \textbf{(2)} \end{cases}$$

(1) is equivalent to:

$$0.975y - 0.025x = 0 \quad \textbf{(3)}$$

(2) is equivalent to:

$$0.04x - 0.96y = 4 \quad \textbf{(4)}$$

<u>Solve the equivalent system **(3) & (4)**</u>:

Solve **(3)** for y: $y = \frac{0.025}{0.975}x$ **(5)**

Substitute **(5)** into **(4)** and simplify:

$$x = 260$$

Substitute this value of x into **(5)** to get

$$y = \tfrac{20}{3}.$$

So, the 2.5% mixture contains 260 gallons of gas and $\frac{20}{3}$ gallons of oil. The 4% mixture contains 80 gallons of gas and $\frac{10}{3}$ gallons of oil.

Thus, the 2.5% mixture contains $\frac{800}{3}$ gallons and the 4% mixture contains $\frac{250}{3}$ gallons.

55. Let x = total annual sales.

<u>Salary at Autocount</u>: $15{,}000 + 0.10x$

<u>Salary at Polk</u>: $30{,}000 + 0.05x$

We need to find x such that

$$15{,}000 + 0.10x > 30{,}000 + 0.05x.$$

Solving this yields:

$$0.05x > 15{,}000 \Rightarrow x > 300{,}000$$

So, he needs to make at least $300,000 of sales in order to make more money at Autocount.

56. Let x = total made from homes sold.

<u>Salary for agents in resale</u>: $0.06x$

<u>Salary for agents in new homes</u>:

$$15{,}000 + 0.015x$$

We need to find x such that

$$0.06x > 15{,}000 + 0.015x.$$

Solving this yields:

$$0.045x > 15{,}000 \Rightarrow x > 333{,}333.3\overline{3}$$

So, an agent would need to sell at least $333,333.33 worth of homes to make more money in resale.

57. Let x= gallons used for highway miles
y = gallons used for city miles
Must solve the system:

$$\begin{cases} 26x+19y=349.5 & \textbf{(1)} \\ x + y = 16 & \textbf{(2)} \end{cases}$$

Multiply **(2)** by -19:

$$-19x-19y=-304 \quad \textbf{(3)}$$

Add **(1)** and **(3)** to eliminate y:

$$7x=45.5 \Rightarrow x=6.5 \quad \textbf{(4)}$$

Substitute **(4)** into **(2)** and solve for y:

$$y=9.5$$

Thus, there are 169 highway miles and 180.5 city miles.

58. Let x = number of minutes
Assume that $600 < x < 800$.
<u>Cost of 800-minute plan</u>: 79
<u>Cost of 600-minute plan</u>: $59+0.13x$
We need to find x such that

$$79 < 59+0.13x.$$

Solving this yields:

$$20 < 0.13x$$

$$153.85 < x$$

So, you would need to talk for at least 753.85 minutes in order for the 800-minute plan to be the better deal.

59. Let x = speed of the plane and y = wind speed.

	Rate (mph)	**Time (hours)**	**Distance**
Atlanta to Paris	$x+y$	8	4000
Paris to Atlanta	$x-y$	10	4000

So, using Distance = Rate × Time, we see that we must solve the system:

$$\begin{cases} 8(x+y)=4000 \\ 10(x-y)=4000 \end{cases}, \text{ which is equivalent to } \begin{cases} x+y=500 & \textbf{(1)} \\ x-y=400 & \textbf{(2)} \end{cases}$$

Adding **(1)** and **(2)** yields: $2x=900$, so that $x=450$. Substituting this into **(1)** then yields $y=50$. So, the average ground speed of the plane is 450 mph, and the average wind speed is 50 mph.

60. Let x = speed of the plane and y = wind speed.

	Rate (mph)	**Time (hours)**	**Distance**
Trip	$x+y$	3	500
Return Trip	$x-y$	4	500

So, using Distance = Rate × Time, we see that we must solve the system:

$$\begin{cases} 3(x+y)=500 \\ 4(x-y)=500 \end{cases}, \text{ which is equivalent to } \begin{cases} x+y=\frac{500}{3} & \textbf{(1)} \\ x-y=125 & \textbf{(2)} \end{cases}$$

Adding **(1)** and **(2)** yields: $2x=\frac{875}{3}$, so that $x=\frac{875}{6}$. Substituting this into **(1)** then yields $y=\frac{125}{6}$. So, the average ground speed of the plane is approximately145.83 mph, and the average wind speed is approximately 20.83 mph.

61. Let x = amount invested in 10% stock
y = amount invested in 14% stock

We must solve the system:

$$\begin{cases} 0.10x + 0.14y = 1260 & \textbf{(1)} \\ x + y = 10{,}000 & \textbf{(2)} \end{cases}$$

Multiply **(2)** by -0.10:

$$-0.10x - 0.10y = -1000 \quad \textbf{(3)}$$

Add **(1)** and **(3)** to eliminate x:

$$0.04y = 260$$

$$y = 6500 \quad \textbf{(4)}$$

Substitute **(4)** into **(2)** and solve for y:

$$x = 3500$$

So, should invest \$3500 in the 10% stock, and \$6500 in the 14% stock.

62. Let x = amount invested in 5% stock
y = amount invested in 7% stock

We must solve the system:

$$\begin{cases} 0.05x + 0.07y = 665 & \textbf{(1)} \\ y = 2x & \textbf{(2)} \end{cases}$$

Substitute **(2)** into **(1)** and solve for x:

$$0.05x + 0.07(2x) = 665$$

$$0.19x = 665$$

$$x = 3500 \quad \textbf{(3)}$$

Substitute **(3)** into **(2)** to find y:

$$y = 7000.$$

So, should invest \$3500 in the 5% stock, and \$7000 in the 7% stock.

63. Let x = # CD players sold
Cost equation: $y = 15x + 120$ **(1)**
Revenue equation: $y = 30x$ **(2)**
We want the intersection of these two equations to find the break even point. To this end, substitute **(2)** into **(1)** to find that $x = 8$. So, must sell 8 CD players to break even.

64. Let x = # of glasses of lemonade sold
Cost equation: $y = 0.10x + 15$ **(1)**
Revenue equation: $y = 0.25x$ **(2)**
We want the intersection of these two equations to find the break even point. To this end, substitute **(2)** into **(1)** to find that $x = 100$. So, must sell 100 glasses of lemonade to break even.

65. Let C_x = cost of a Type I meal
C_y = cost of a Type II meal

Solve the system:

$$\begin{cases} 75C_x + 75C_y = 1275 \\ 90C_x + 60C_y = 1260 \end{cases}$$

which is equivalent to

$$\begin{cases} C_x + C_y = 17 & \textbf{(1)} \\ 3C_x + 2C_y = 42 & \textbf{(2)} \end{cases}$$

Multiply **(1)** by -2 and add to **(2)**: $C_x = 8$

Substitute this into **(1)** to see that $C_y = 9$.

66. Let x = cost of a meal plan 1 (the more expensive one), and y = cost of meal plan 2

Solve the system:

$$\begin{cases} x + y = 1.2 & \textbf{(1)} \\ y = x + 0.05x = 1.05x & \textbf{(2)} \end{cases}$$

Substitute **(2)** into **(1)**:

$$x + 1.05x = 1.2$$

$$x = 0.585366$$

Substitute this into **(2)**: $y = 0.614634$.

So, meal plan 1 costs \$585,366 and meal plan 2 costs \$614,634.

67. Let x = number of females y = number of males. Solve the system: $\begin{cases} x+y=18,328,340 & \textbf{(1)} \\ x=y+329,910 & \textbf{(2)} \end{cases}$ Substitute **(2)** into **(1)**: $y+329,910+y=18,328,340$ $y=8,999,215$ Substitute this into **(2)**: $x=9,329,125$. Thus, there are 8,999,215 males and 9,329,125 females.	**68.** Let x = number of senior citizens in 2000 and y = number of non-senior citizens in 2000. Solve the system: $\begin{cases} x+y=281,420,906 & \textbf{(1)} \\ 1.30x+1.20y=341,250,007 & \textbf{(2)} \end{cases}$ Multiply **(1)** by -1.30 and add to **(2)**: $-0.10y=-24,597,170.8$ $y=245,971,708$ Substitute this into **(1)**: $x=35,449,198$. Thus, there are 35,449,198 senior citizens and 245,971,708 non-senior citizens in 2000.
69. Every term in the first equation is not multiplied by -1 correctly. The equation should be $-2x-y=3$, and the resulting solution should be $x=11,\ y=-25$.	**70.** 0 = 12 is an inconsistent statement. Hence, there is no solution to the system.
71. Did not distribute -1 correctly. In Step 3, the calculation should be $-(-3y-4)=3y+4$ and the resulting answer should be (2, -2).	**72.** Actually, the lines are coincident, so that there are infinitely many solutions to the system.
73. False. If the lines are coincident, then there are infinitely many solutions.	**74.** True. The intersection of such lines is guaranteed.
75. False. The lines could be coincident.	**76.** False. $\begin{cases} Ax-By=1 \\ -Ax+By=-1 \end{cases}$ Adding the two equations results in 0 = 0, which means there are infinitely many solutions.

77. Substitute the pair $(2,-3)$ into both equations to get the system:

$$\begin{cases} 2A-3B=-29 & \textbf{(1)} \\ 2A+3B=13 & \textbf{(2)} \end{cases}$$

Add **(1)** and **(2)** to eliminate B:

$$4A=-16$$

$$A=-4 \quad \textbf{(3)}$$

Substitute **(3)** into **(1)** and solve for B:

$$2(-4)-3B=-29$$

$$-3B=-21$$

$$B=7$$

Hence, the solution is $A=-4,\ B=7$.

78. The slopes are very close together, but are not equal. It may be difficult to distinguish on a graphing calculator unless just the right window is chosen.

79. Let $x=$ # cups of pineapple juice for 2% drink, and $y=$ # cups of pomegranate juice for 2% drink.

Must solve the system:

$$\begin{cases} \frac{y}{x+y}=0.02 & \textbf{(1)} \\ \frac{4-y}{(100-x)+(4-y)}=0.04 & \textbf{(2)} \end{cases}$$

This system, after simplification, is equivalent to:

$$\begin{cases} 0.02x-0.98y=0 & \textbf{(3)} \\ 0.04x-0.96y=0.16 & \textbf{(4)} \end{cases}$$

Substitute **(3)** into **(4)**: $y=\frac{0.02}{0.98}x$ **(5)**

Substitute **(5)** into **(4)** and simplify:

$$x=7.84$$

Substitute this value of x back into **(5)**:

$$y=0.16$$

So, she can make 8 cups of the 2% drink and 96 cups of the 4% drink.

80. Let $x=$ # tablespoons of red for light purple solution, and $y=$ # tablespoons of blue for light purple solution.

Must solve the system:

$$\begin{cases} \frac{y}{x+y}=0.02 & \textbf{(1)} \\ \frac{2-y}{(30-x)+(2-y)}=0.10 & \textbf{(2)} \end{cases}$$

This system, after simplification, is equivalent to:

$$\begin{cases} 0.02x-0.98y=0 & \textbf{(3)} \\ 0.10x-0.90y=1.2 & \textbf{(4)} \end{cases}$$

Substitute **(3)** into **(4)**: $y=\frac{0.02}{0.98}x$ **(5)**

Substitute **(5)** into **(4)** and simplify:

$$x=14.7$$

Substitute this value of x back into **(5)**:

$$y=0.3$$

So, they can make 15 tablespoons of light purple and 17 tablespoons of deep purple.

81. Solve the system:

$$\begin{cases} 4 = m(-2) + b & \textbf{(1)} \\ -2 = m(4) + b & \textbf{(2)} \end{cases}$$

Subtract **(1) – (2):** $6 = -6m \Rightarrow m = -1$

Substitute into **(1)**: $b = 2$.

82. Solve the system:

$$\begin{cases} 7 = 2^2 + b(2) + c \\ 7 = (-6)^2 + b(-6) + c \end{cases}$$

which is equivalent to

$$\begin{cases} 3 = 2b + c & \textbf{(1)} \\ -29 = -6b + c & \textbf{(2)} \end{cases}$$

Subtract **(1) – (2):** $b = 4$

Substitute this into **(1)**: $c = -5$.

83. Solve the system:

$$\begin{cases} 46 = b(4)^2 + b(4) + c \\ 10 = b(-2)^2 + b(-2) + c \end{cases}$$

which is equivalent to

$$\begin{cases} 46 = 20b + c & \textbf{(1)} \\ 10 = 2b + c & \textbf{(2)} \end{cases}$$

Subtract **(1) – (2):** $b = 2$

Substitute this into **(1)**: $c = 6$.

84. Solve the system:

$$\begin{cases} u + v = 4 & \textbf{(1)} \\ u - v = 2 & \textbf{(2)} \end{cases}$$

Add **(1)** and **(2)**: $u = 3$

Substitute this into **(1)**: $v = 1$

As such, we have

$$u = x^2 \Rightarrow 3 = x^2 \Rightarrow x = \pm\sqrt{3}$$

$$v = y^2 \Rightarrow 1 = y^2 \Rightarrow y = \pm 1$$

Thus, there are four solutions:

$$\left(\sqrt{3}, -1\right), \left(\sqrt{3}, 1\right), \left(-\sqrt{3}, -1\right), \left(-\sqrt{3}, 1\right)$$

85. Solve the system:

$$\begin{cases} u + 2v = 11 & \textbf{(1)} \\ 4u + v = 16 & \textbf{(2)} \end{cases}$$

Multiply **(2)** by -2 and then add to **(1)**:

$$-7u = -21 \Rightarrow u = 3$$

Substitute into **(2)**: $v = 4$

As such, we have

$$u = x^2 \Rightarrow 3 = x^2 \Rightarrow x = \pm\sqrt{3}$$

$$v = y^2 \Rightarrow 4 = y^2 \Rightarrow y = \pm 2$$

Thus, there are four solutions:

$$\left(\sqrt{3}, -2\right), \left(\sqrt{3}, 2\right), \left(-\sqrt{3}, -2\right), \left(-\sqrt{3}, 2\right)$$

86. Solve the system:

$$\begin{cases} a = b(-2)^2 - 2(-2) - a \\ b - 2 = b(-1)^2 - 2(-1) - a \end{cases}$$

which is equivalent to

$$\begin{cases} 2a - 4b = 4 & \textbf{(1)} \\ a = 4 & \textbf{(2)} \end{cases}$$

Substitute **(2)** into **(1)**: $b = 1$

So, the solution is (4, 1).

87.

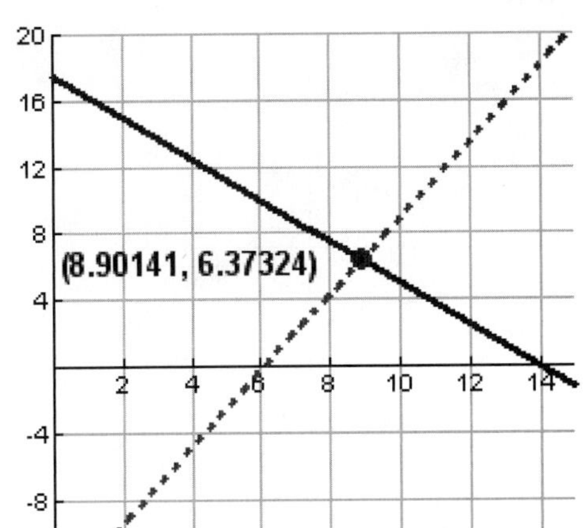

Notes on the graph:
Solid curve is $y = -1.25x + 17.5$
Dashed curve is $y = 2.3x - 14.1$
The approximate solution is (8.9, 6.4).

88.

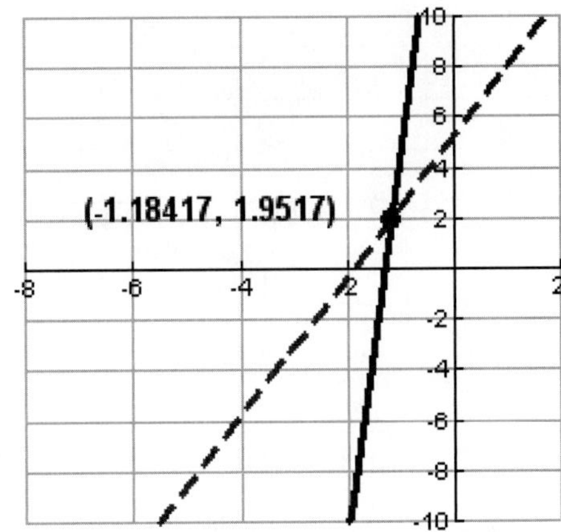

Notes on the graph:
Solid curve is $y = 14.76x + 19.43$
Dashed curve is $y = 2.76x + 5.22$
The approximate solution is $(-1.2, 2.0)$.

89.

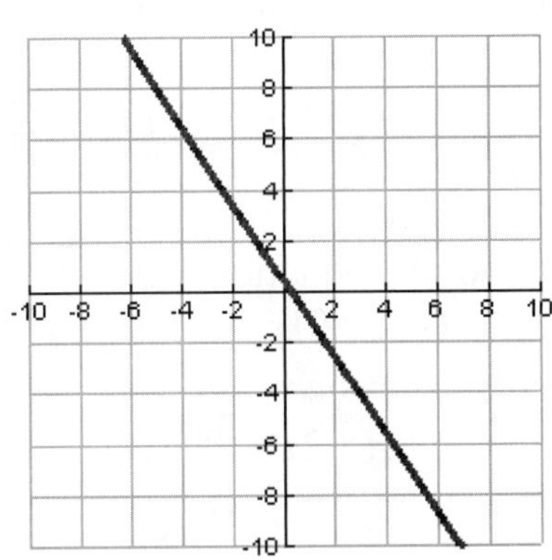

Notes on the graph:
Solid curve is $23x + 15y = 7$
Dashed curve is $46x + 30y = 14$
Note that the graphs are the same line, so there are infinitely many solutions to this system.

90.

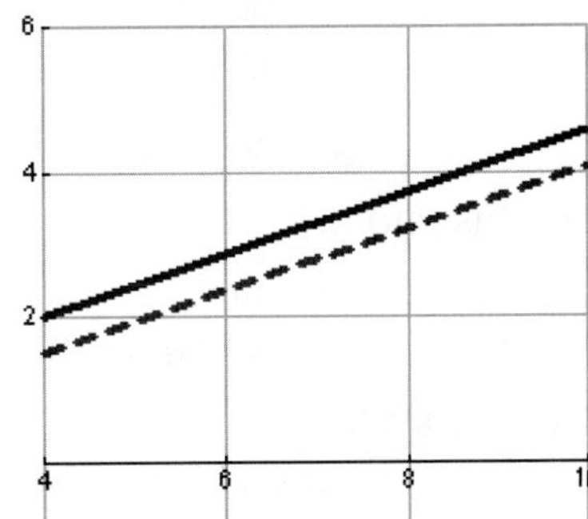

Notes on the graph:
Solid curve is $-3x + 7y = 2$
Dashed curve is $6x - 14y = 3$
Since the lines are parallel, there is no solution.

91. The graph of the system of equations is as follows:

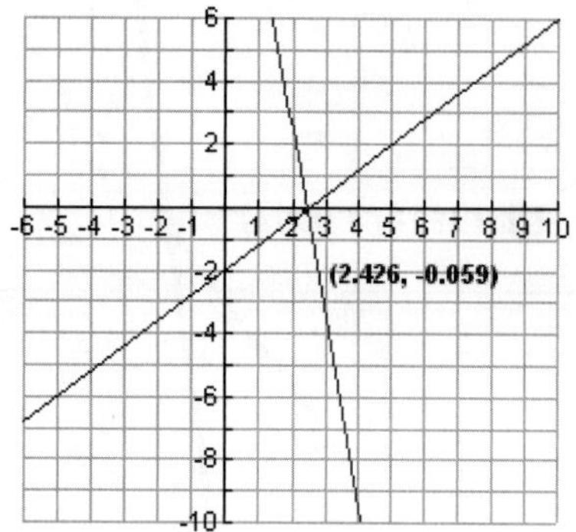

The solution is approximately (2.426, -0.059).

92. The graph of the system of equations is as follows:

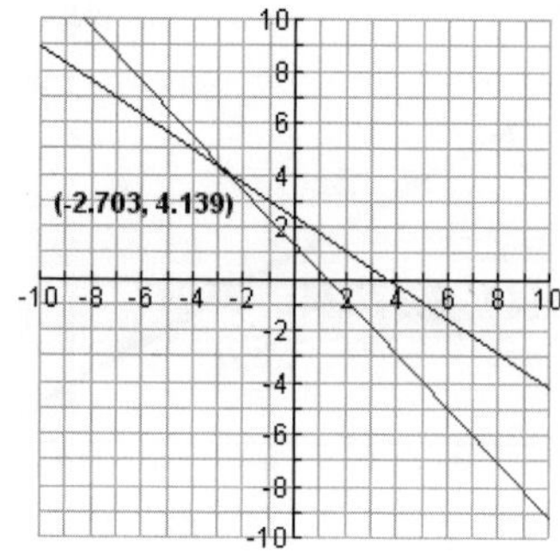

The solution is approximately (-2.703, 4.139).

93. Consider the equation:

$$x+5=(A+B)x+(2A-4B)$$

This yields the system:

$$\begin{cases} A+B=1 & \textbf{(1)} \\ 2A-4B=5 & \textbf{(2)} \end{cases}$$

Solve **(1)** for A: $A=1-B$ **(3)**

Substitute **(3)** into **(2)** and solve for B:

$$2(1-B)-4B=5$$

$$B=-\tfrac{1}{2}$$

Substitute this into **(3)**: $A=\tfrac{3}{2}$

94. Consider the equation:

$$6x=(A+B)x+(A-2B)$$

This yields the system:

$$\begin{cases} A+B=6 & \textbf{(1)} \\ A-2B=0 & \textbf{(2)} \end{cases}$$

Subtract **(1) – (2):** $B=2$

Substitute this into **(1)**: $A=4$

95. Consider the equation:

$$x+1=(A+B)x+(2A-3B)$$

This yields the system:

$$\begin{cases} A+B=1 & \textbf{(1)} \\ 2A-3B=1 & \textbf{(2)} \end{cases}$$

Solve **(1)** for A: $A=1-B$ **(3)**

Substitute **(3)** into **(2)** and solve for B:

$$2(1-B)-3B=1$$

$$B=\tfrac{1}{5}$$

Substitute this into **(3)**: $A=\tfrac{4}{5}$

96. Consider the equation:

$$5=(A+2B)x+(-2A+B)$$

This yields the system:

$$\begin{cases} A+2B=0 & \textbf{(1)} \\ -2A+B=5 & \textbf{(2)} \end{cases}$$

Multiply **(1)** by 2 and then add to **(2)**: $B = 1$

Substitute this into **(1)**: $A = -2$.

Section 8.2 Solutions ---

1. Solve the system:

$$\begin{cases} x - y + z = 6 & \textbf{(1)} \\ -x + y + z = 3 & \textbf{(2)} \\ -x - y - z = 0 & \textbf{(3)} \end{cases}$$

Step 1: Obtain a system of 2 equations in 2 unknowns by eliminating the same variable in two pairs of equations.

Add **(1)** and **(2)**:

$$2z = 9 \Rightarrow \boxed{z = \tfrac{9}{2}} \quad \textbf{(4)}$$

Add **(2)** and **(3)**:

$$-2x = 3 \Rightarrow \boxed{x = -\tfrac{3}{2}} \quad \textbf{(5)}$$

Step 2: Solve for the remaining variable using **(4)** and **(5)**.

Substitute **(4)** and **(5)** into **(1)**:

$$-\tfrac{3}{2} - y + \tfrac{9}{2} = 6 \Rightarrow \boxed{y = -3}$$

2. Solve the system:

$$\begin{cases} -x - y + z = -1 & \textbf{(1)} \\ -x + y - z = 3 & \textbf{(2)} \\ x - y - z = 5 & \textbf{(3)} \end{cases}$$

Step 1: Obtain a system of 2 equations in 2 unknowns by eliminating the same variable in two pairs of equations.

Add **(1)** and **(2)**:

$$-2x = 2 \Rightarrow \boxed{x = -1} \quad \textbf{(4)}$$

Add **(2)** and **(3)**:

$$-2z = 8 \Rightarrow \boxed{z = -4} \quad \textbf{(5)}$$

Step 2: Solve for the remaining variable using **(4)** and **(5)**.

Substitute **(4)** and **(5)** into **(1)**:

$$-(-1) - y + (-4) = -1 \Rightarrow \boxed{y = -2}$$

3. Solve the system:

$$\begin{cases} x + y - z = 2 & \textbf{(1)} \\ -x - y - z = -3 & \textbf{(2)} \\ -x + y - z = 6 & \textbf{(3)} \end{cases}$$

Step 1: Obtain a system of 2 equations in 2 unknowns by eliminating the same variable in two pairs of equations.

Add **(1)** and **(2)**: $-2z = -1 \Rightarrow \boxed{z = \tfrac{1}{2}}$ **(4)**

Add **(2)** and **(3)**, and sub. in **(4)**:

$$-2x - 2z = 3 \Rightarrow -2x - 2\left(\tfrac{1}{2}\right) = 3$$

$$\Rightarrow -2x = 4 \Rightarrow \boxed{x = -2} \quad \textbf{(5)}$$

Step 2: Solve for the remaining variable using **(4)** and **(5)**.

Substitute **(4)** and **(5)** into **(1)**:

$$-2 + y - \tfrac{1}{2} = 2 \Rightarrow \boxed{y = \tfrac{9}{2}}$$

4. Solve the system:

$$\begin{cases} x + y + z = -1 & \textbf{(1)} \\ -x + y - z = 3 & \textbf{(2)} \\ -x - y + z = 8 & \textbf{(3)} \end{cases}$$

Step 1: Obtain a system of 2 equations in 2 unknowns by eliminating the same variable in two pairs of equations.

Add **(1)** and **(3)**: $2z = 7 \Rightarrow \boxed{z = \tfrac{7}{2}}$ **(4)**

Add **(1)** and **(2)**:

$$2y = 2 \Rightarrow \boxed{y = 1} \quad \textbf{(5)}$$

Step 2: Solve for the remaining variable using **(4)** and **(5)**.

Substitute **(4)** and **(5)** into **(1)**:

$$x + 1 + \tfrac{7}{2} = -1 \Rightarrow \boxed{x = -\tfrac{11}{2}}$$

5. Solve the system:

$$\begin{cases} -x + y - z = -1 & \textbf{(1)} \\ x - y - z = 3 & \textbf{(2)} \\ x + y - z = 9 & \textbf{(3)} \end{cases}$$

Step 1: Obtain a system of 2 equations in 2 unknowns by eliminating the same variable in two pairs of equations.

Add **(1)** and **(2)**: $-2z = 2 \Rightarrow \boxed{z = -1}$ **(4)**

Add **(2)** and **(3)**, and sub. in **(4)**:

$2x - 2z = 12 \Rightarrow 2x - 2(-1) = 12$

$\Rightarrow 2x = 10 \Rightarrow \boxed{x = 5}$ **(5)**

Step 2: Solve for the remaining variable using **(4)** and **(5)**.

Substitute **(4)** and **(5)** into **(1)**:

$-5 + y - (-1) = -1 \Rightarrow \boxed{y = 3}$

6. Solve the system:

$$\begin{cases} x - y - z = 2 & \textbf{(1)} \\ -x - y + z = 4 & \textbf{(2)} \\ -x + y - z = 6 & \textbf{(3)} \end{cases}$$

Step 1: Obtain a system of 2 equations in 2 unknowns by eliminating the same variable in two pairs of equations.

Add **(1)** and **(2)**:

$-2y = 6 \Rightarrow \boxed{y = -3}$ **(4)**

Add **(2)** and **(3)**:

$-2x = 10 \Rightarrow \boxed{x = -5}$ **(5)**

Step 2: Solve for the remaining variable using **(4)** and **(5)**.

Substitute **(4)** and **(5)** into **(1)**:

$-5 - (-3) - z = 2 \Rightarrow \boxed{z = -4}$

7. Solve the system:

$$\begin{cases} 2x - 3y + 4z = -3 & \textbf{(1)} \\ -x + y + 2z = 1 & \textbf{(2)} \\ 5x - 2y - 3z = 7 & \textbf{(3)} \end{cases}$$

Step 1: Obtain a system of 2 equations in 2 unknowns by eliminating the same variable in two pairs of equations.

Multiply **(2)** by 3:

$-3x + 3y + 6z = 3$ **(4)**

Add **(4)** and **(1)** to eliminate y:

$-x + 10z = 0$ **(5)**

Next, multiply **(2)** by 2:

$-2x + 2y + 4z = 2$ **(6)**

Add **(6)** and **(3)** to eliminate y:

$3x + z = 9$ **(7)**

These steps yield the following system:

$$(*) \begin{cases} -x + 10z = 0 & \textbf{(5)} \\ 3x + z = 9 & \textbf{(7)} \end{cases}$$

Step 2: Solve system $(*)$ from Step 1.

Multiply **(5)** by 3:

$-3x + 30z = 0$ **(8)**

Add **(8)** and **(7)** to eliminate x and solve for z:

$31z = 9$ so that $z = \frac{9}{31}$ **(9)**

Substitute **(9)** into **(7)** to find x:

$-x + 10(\frac{9}{31}) = 0$

$x = \frac{90}{31}$ **(10)**

Step 3: Use the solution of the system in Step 2 to find the value of the third variable in the original system.

Substitute **(9)** and **(10)** into **(2)** to find y:

$-(\frac{90}{31}) + y + 2(\frac{9}{31}) = 1$

$y - \frac{72}{31} = 1$ so that $y = \frac{103}{31}$

Thus, the solution is:

$\boxed{x = \frac{90}{31},\ y = \frac{103}{31},\ z = \frac{9}{31}}$.

8. Solve the system:

$$\begin{cases} x - 2y + z = 0 & \textbf{(1)} \\ -2x + y - z = -5 & \textbf{(2)} \\ 13x + 7y + 5z = 6 & \textbf{(3)} \end{cases}$$

Step 1: Obtain a system of 2 equations in 2 unknowns by eliminating the same variable in two pairs of equations.

Multiply **(1)** by 2:

$$2x - 4y + 2z = 0 \quad \textbf{(4)}$$

Add **(4)** and **(2)** to eliminate x:

$$-3y + z = -5 \quad \textbf{(5)}$$

Multiply **(1)** by -13 and then add to **(3)**:

$$33y - 8z = 6 \quad \textbf{(6)}$$

This yields the following 2×2 system:

$$(*)\begin{cases} -3y + z = -5 & \textbf{(5)} \\ 33y - 8z = 6 & \textbf{(6)} \end{cases}$$

Step 2: Solve system **(*)** from Step 1.

Multiply **(5)** by 8 and then add to **(6)**::

$$9y = -34$$

$$y = -\tfrac{34}{9} \quad \textbf{(7)}$$

Substitute **(7)** into **(5)** to solve for z:

$$-3(-\tfrac{34}{9}) + z = -5$$

$$z = -\tfrac{49}{3} \quad \textbf{(8)}$$

Substitute **(7)** and **(8)** into **(1)** to find x:

$$x - 2(-\tfrac{34}{9}) + (-\tfrac{49}{3}) = 0$$

$$x = \tfrac{79}{9} \quad \textbf{(9)}$$

Hence, the solution is:

$$\boxed{x = \tfrac{79}{9},\ y = -\tfrac{34}{9},\ z = -\tfrac{49}{3}}.$$

9. Solve the system:

$$\begin{cases} -4x + 3y + 5z = 2 & \textbf{(1)} \\ 2x - 3y - 2z = -3 & \textbf{(2)} \\ -2x + 4y + 3z = 1 & \textbf{(3)} \end{cases}$$

Step 1: Obtain a system of 2 equations in 2 unknowns by eliminating the same variable in two pairs of equations.

Add **(2)** and **(3)** to eliminate x:

$$y + z = -2 \quad \textbf{(4)}$$

Next, multiply **(2)** by 2:

$$4x - 6y - 4z = -6 \quad \textbf{(5)}$$

Add **(5)** and **(1)** to eliminate x:

$$-3y + z = -4 \quad \textbf{(6)}$$

These steps yield the following system:

$$(*)\begin{cases} y + z = -2 & \textbf{(4)} \\ -3y + z = -4 & \textbf{(6)} \end{cases}$$

Step 2: Solve system **(*)** from Step 1.

Multiply **(6)** by -1: $3y - z = 4$ **(7)**

Add **(4)** and **(7)** to eliminate z and solve for y:

$$4y = 2 \Rightarrow y = \tfrac{1}{2} \quad \textbf{(8)}$$

Substitute **(8)** into **(4)** to find z:

$$\tfrac{1}{2} + z = -2$$

$$z = -\tfrac{5}{2} \quad \textbf{(9)}$$

Step 3: Use the solution of the system in Step 2 to find the value of the third variable in the original system.

Substitute **(8)** and **(9)** into **(2)** to find y:

$$2x - 3(\tfrac{1}{2}) - 2(-\tfrac{5}{2}) = -3$$

$$2x = -\tfrac{13}{2}$$

$$x = -\tfrac{13}{4} \quad \textbf{(10)}$$

Thus, the solution is:

$$\boxed{x = -\tfrac{13}{4},\ y = \tfrac{1}{2},\ z = -\tfrac{5}{2}}.$$

10. Solve the system:

$$\begin{cases} -x+2y+z=5 & \textbf{(1)} \\ 2x-2y+3z=0 & \textbf{(2)} \\ -4x+y-2z=3 & \textbf{(3)} \end{cases}$$

Step 1: Obtain a system of 2 equations in 2 unknowns by eliminating the same variable in two pairs of equations.

Add **(1)** and **(2)** to eliminate y:

$$x+4z=5 \quad \textbf{(4)}$$

Next, multiply **(3)** by 2:

$$-8x+2y-4z=6 \quad \textbf{(5)}$$

Add **(2)** and **(5)** to eliminate y:

$$-6x-z=6 \quad \textbf{(6)}$$

These steps yield the following system:

$$(*)\begin{cases} x+4z=5 & \textbf{(4)} \\ -6x-z=6 & \textbf{(6)} \end{cases}$$

Step 2: Solve system **(*)** from Step 1.

Multiply **(6)** by 4:

$$-24x-4z=24 \quad \textbf{(7)}$$

Add **(4)** and **(7)** to eliminate z and solve for x:

$$-23x=29$$

$$x=-\tfrac{29}{23} \quad \textbf{(8)}$$

Substitute **(8)** into **(4)** to find z:

$$-\tfrac{29}{23}+4z=5$$

$$4z=\tfrac{144}{23}$$

$$z=\tfrac{36}{23} \quad \textbf{(9)}$$

Step 3: Use the solution of the system in Step 2 to find the value of the third variable in the original system.

Substitute **(8)** and **(9)** into **(1)** to find y:

$$\tfrac{29}{23}+2y+\tfrac{36}{23}=5$$

$$2y=\tfrac{50}{23}$$

$$y=\tfrac{25}{23} \quad \textbf{(10)}$$

Thus, the solution is:

$$\boxed{x=-\tfrac{29}{23},\ y=\tfrac{25}{23},\ z=\tfrac{36}{23}}.$$

11. Solve the system:

$$\begin{cases} x-y+z=-1 & \textbf{(1)} \\ y-z=-1 & \textbf{(2)} \\ -x+y+z=1 & \textbf{(3)} \end{cases}$$

Add **(1)** and **(3)** to eliminate both x and y:

$$2z=0$$

$$z=0 \quad \textbf{(4)}$$

Substitute **(4)** into **(2)** to find y:

$$y-0=-1 \quad \textbf{(5)}$$

Substitute **(4)** and **(5)** into **(1)** to find x:

$$x-(-1)+0=-1$$

$$x=-2 \ \textbf{(6)}$$

Thus, the solution is:

$$\boxed{x=-2,\ y=-1,\ z=0}.$$

12. Solve the system:

$$\begin{cases} -y+z=1 & \textbf{(1)} \\ x-y+z=-1 & \textbf{(2)} \\ x-y-z=-1 & \textbf{(3)} \end{cases}$$

Multiply **(2)** by -1:

$$-x+y-z=1 \quad \textbf{(4)}$$

Add **(3)** and **(4)** to eliminate both x and y:

$$-2z=0$$

$$z=0 \quad \textbf{(5)}$$

Substitute **(5)** into **(1)** to find y:

$$-y+0=1$$

$$y=-1 \ \textbf{(6)}$$

Substitute **(5)** and **(6)** into **(2)** to find x:

$$x-(-1)+0=-1$$

$$x=-2 \ \textbf{(6)}$$

Thus, the solution is:

$$\boxed{x=-2,\ y=-1,\ z=0}.$$

13. Solve the system:

$$\begin{cases} 3x - 2y - 3z = -1 & \textbf{(1)} \\ x - y + z = -4 & \textbf{(2)} \\ 2x + 3y + 5z = 14 & \textbf{(3)} \end{cases}$$

Step 1: Obtain a system of 2 equations in 2 unknowns by eliminating the same variable in two pairs of equations.

Multiply **(2)** by 3:

$$3x - 3y + 3z = -12 \quad \textbf{(4)}$$

Add **(4)** and **(1)** to eliminate z:

$$6x - 5y = -13 \quad \textbf{(5)}$$

Next, multiply **(2)** by -5:

$$-5x + 5y - 5z = 20 \quad \textbf{(6)}$$

Add **(6)** and **(3)** to eliminate z:

$$-3x + 8y = 34 \quad \textbf{(7)}$$

These steps yield the following system:

$$\textbf{(*)} \begin{cases} 6x - 5y = -13 & \textbf{(5)} \\ -3x + 8y = 34 & \textbf{(7)} \end{cases}$$

Step 2: Solve system **(*)** from Step 1.

Multiply **(7)** by 2:

$$-6x + 16y = 68 \quad \textbf{(8)}$$

Add **(5)** and **(8)** to eliminate x and solve for y:

$$11y = 55$$

$$y = 5 \quad \textbf{(9)}$$

Substitute **(9)** into **(5)** to find x:

$$6x - 5(5) = -13$$

$$6x = 12$$

$$x = 2 \quad \textbf{(10)}$$

Step 3: Use the solution of the system in Step 2 to find the value of the third variable in the original system.

Substitute **(9)** and **(10)** into **(2)** to find z:

$$2 - 5 + z = -4$$

$$z = -1 \quad \textbf{(11)}$$

Thus, the solution is:

$$\boxed{x = 2,\ \ y = 5,\ \ z = -1}.$$

14. Solve the system:

$$\begin{cases} 3x - y + z = 2 & \textbf{(1)} \\ x - 2y + 3z = 1 & \textbf{(2)} \\ 2x + y - 3z = -1 & \textbf{(3)} \end{cases}$$

Step 1: Obtain a system of 2 equations in 2 unknowns by eliminating the same variable in two pairs of equations.

Add **(1)** and **(3)** to eliminate y:

$$5x - 2z = 1 \quad \textbf{(4)}$$

Next, multiply **(3)** by 2:

$$4x + 2y - 6z = -2 \quad \textbf{(5)}$$

Add **(2)** and **(5)** to eliminate y:

$$5x - 3z = -1 \quad \textbf{(6)}$$

These steps yield the following system:

$$\textbf{(*)} \begin{cases} 5x - 2z = 1 & \textbf{(4)} \\ 5x - 3z = -1 & \textbf{(6)} \end{cases}$$

Step 2: Solve system **(*)** from Step 1.

Multiply **(4)** by -1:

$$-5x + 2z = -1 \quad \textbf{(7)}$$

Add **(6)** and **(7)** to eliminate x and solve for z:

$$-z = -2$$

$$z = 2 \quad \textbf{(8)}$$

Substitute **(8)** into **(4)** to find x:

$$5x - 2(2) = 1$$

$$x = 1 \quad \textbf{(9)}$$

Step 3: Use the solution of the system in Step 2 to find the value of the third variable in the original system.

Substitute **(8)** and **(9)** into **(2)** to find y:

$$1 - 2y + 3(2) = 1$$

$$y = 3 \quad \textbf{(10)}$$

Thus, the solution is:

$$\boxed{x = 1,\ \ y = 3,\ \ z = 2}.$$

15. Solve the system:

$$\begin{cases} -3x - y - z = 2 & \textbf{(1)} \\ x + 2y - 3z = 4 & \textbf{(2)} \\ 2x - y + 4z = 6 & \textbf{(3)} \end{cases}$$

Step 1: Obtain a system of 2 equations in 2 unknowns by eliminating the same variable in two pairs of equations.

Multiply **(2)** by 3:

$$3x+6y-9z=12 \quad \textbf{(4)}$$

Add **(1)** and **(4)** to eliminate x:

$$5y-10z=14 \quad \textbf{(5)}$$

Next, multiply **(2)** by -2:

$$-2x-4y+6z=-8 \quad \textbf{(6)}$$

Add **(3)** and **(6)** to eliminate x:

$$-5y+10z=-2 \quad \textbf{(7)}$$

These steps yield the following system:

$$(*) \begin{cases} 5y-10z=14 & \textbf{(5)} \\ -5y+10z=-2 & \textbf{(7)} \end{cases}$$

Step 2: Solve system **(*)** from Step 1.

Adding **(5)** and **(7)** yields the false statement $0 = 12$. Hence, this system has no solution.

16. Solve the system:

$$\begin{cases} 2x - 3y + z = 1 & \textbf{(1)} \\ x + 4y - 2z = 2 & \textbf{(2)} \\ 3x - y + 4z = -3 & \textbf{(3)} \end{cases}$$

Step 1: Obtain a system of 2 equations in 2 unknowns by eliminating the same variable in two pairs of equations.

Multiply **(2)** by -2:

$$-2x-8y+4z=-4 \quad \textbf{(4)}$$

Add **(1)** and **(4)** to eliminate x:

$$-11y+5z=-3 \quad \textbf{(5)}$$

Next, multiply **(2)** by -3:

$$-3x-12y+6z=-6 \quad \textbf{(6)}$$

Add **(3)** and **(6)** to eliminate x:

$$-13y+10z=-9 \quad \textbf{(7)}$$

These steps yield the following system:

$$(*) \begin{cases} -11y+5z=-3 & \textbf{(5)} \\ -13y+10z=-9 & \textbf{(7)} \end{cases}$$

Step 2: Solve system **(*)** from Step 1.

Multiply **(5)** by -2:

$$22y-10z=6 \quad \textbf{(8)}$$

Add **(7)** and **(8)** to eliminate z and solve for y:

$$9y=-3 \Rightarrow y=-\tfrac{1}{3} \quad \textbf{(9)}$$

Substitute **(9)** into **(5)** to find x:

$$-11\left(-\tfrac{1}{3}\right)+5z=-3 \Rightarrow z=-\tfrac{4}{3} \quad \textbf{(10)}$$

Step 3: Use the solution of the system in Step 2 to find the value of the third variable in the original system.

Substitute **(9)** and **(10)** into **(1)** to find x:

$$2x-3\left(-\tfrac{1}{3}\right)-\tfrac{4}{3}=1 \Rightarrow x=\tfrac{2}{3}$$

Thus, the solution is:

$$\boxed{x=\tfrac{2}{3},\ y=-\tfrac{1}{3},\ z=-\tfrac{4}{3}}.$$

17. Solve the system:

$$\begin{cases} 3x+2y+z=4 & \textbf{(1)} \\ -4x-3y-z=-15 & \textbf{(2)} \\ x-2y+3z=12 & \textbf{(3)} \end{cases}$$

Step 1: Obtain a system of 2 equations in 2 unknowns by eliminating the same variable in two pairs of equations.

Multiply **(1)** by -3:

$$-9x-6y-3z=-12 \quad \textbf{(4)}$$

Add **(3)** and **(4)** to eliminate z:

$$-8x-8y=0 \Rightarrow x+y=0 \ \textbf{(5)}$$

Next, multiply **(2)** by 3:

$$-12x-9y-3z=-45 \quad \textbf{(6)}$$

Add **(3)** and **(6)** to eliminate z:

$$-11x-11y=-33 \Rightarrow x+y=3 \quad \textbf{(7)}$$

These steps yield the following system:

$$(*) \begin{cases} x+y=0 & \textbf{(5)} \\ x+y=3 & \textbf{(7)} \end{cases}$$

Step 2: Solve system **(*)** from Step 1. Subtracting **(5)** and **(7)** yields the false statement $0=3$. Hence, this system has no solution.

18. Solve the system:

$$\begin{cases} 3x-y+4z=13 & \textbf{(1)} \\ -4x-3y-z=-15 & \textbf{(2)} \\ x-2y+3z=12 & \textbf{(3)} \end{cases}$$

Step 1: Obtain a system of 2 equations in 2 unknowns by eliminating the same variable in two pairs of equations.

Multiply **(3)** by -3:

$$-3x+6y-9z=-36 \quad \textbf{(4)}$$

Add **(1)** and **(4)** to eliminate z:

$$5y-5z=-23 \Rightarrow y-z=-\tfrac{23}{5} \ \textbf{(5)}$$

Next, multiply **(3)** by 4:

$$4x-8y+12z=48 \quad \textbf{(6)}$$

Add **(2)** and **(6)** to eliminate z:

$$-11y+11z=33 \Rightarrow -y+z=3 \quad \textbf{(7)}$$

These steps yield the following system:

$$(*) \begin{cases} y-z=-\tfrac{23}{5} & \textbf{(5)} \\ -y+z=3 & \textbf{(7)} \end{cases}$$

Step 2: Solve system **(*)** from Step 1. Adding **(5)** and **(7)** yields the false statement $0=-\frac{8}{5}$. Hence, this system has no solution.

19. Solve the system:

$$\begin{cases} -x + 2y + z = -2 & \textbf{(1)} \\ 3x - 2y + z = 4 & \textbf{(2)} \\ 2x - 4y - 2z = 4 & \textbf{(3)} \end{cases}$$

Step 1: Obtain a system of 2 equations in 2 unknowns by eliminating the same variable in two pairs of equations.

Multiply **(1)** by 3:

$$-3x + 6y + 3z = -6 \quad \textbf{(4)}$$

Add **(4)** and **(2)** to eliminate x:

$$4y + 4z = -2 \quad \textbf{(5)}$$

Next, multiply **(1)** by 2:

$$-2x + 4y + 2z = -4 \quad \textbf{(6)}$$

Add **(6)** and **(3)** to eliminate x:

$$0 = 0 \quad \textbf{(7)}$$

Hence, we know that the system has infinitely many solutions.

Let $z = a$. Then, substituting this value into **(5)**, we can find the value of y:

$$2y = -2a - 1$$
$$y = -(a + \tfrac{1}{2})$$

Now, substitute the values of z and y into **(1)** to find x:

$$-x - 2(a + \tfrac{1}{2}) + a = -2$$
$$-x - 2a - 1 + a = -2$$
$$x = 1 - a$$

Thus, the solution is

$\boxed{x = 1 - a,\ \ y = -(a + \tfrac{1}{2}),\ \ z = a}$, where a is any real number.

20. Solve the system:

$$\begin{cases} 2x - y = 1 & \textbf{(1)} \\ -x + z = -2 & \textbf{(2)} \\ -2x + y = -1 & \textbf{(3)} \end{cases}$$

First, we solve the 2×2 system consisting of equations **(1)** and **(3):**
Add **(1)** and **(3)**: $0 = 0$

So, there are infinitely many solutions to this system.

To determine them, let $y = a$. Substitute this into **(1)** to find x:

$$2x - a = 1$$
$$x = \tfrac{1+a}{2}$$

Now, substitute the values of x and y into **(2)** to find z:

$$-(\tfrac{1+a}{2}) + z = -2$$
$$z = \tfrac{1+a}{2} - 2$$
$$z = \tfrac{a-3}{2}$$

Thus, the solution is:

$\boxed{x = \tfrac{1+a}{2},\ \ y = a,\ \ z = \tfrac{a-3}{2}}$.

21. Solve the system:

$$\begin{cases} x - y - z = 10 & \textbf{(1)} \\ 2x - 3y + z = -11 & \textbf{(2)} \\ -x + y + z = -10 & \textbf{(3)} \end{cases}$$

Step 1: Obtain a system of 2 equations in 2 unknowns by eliminating the same variable in two pairs of equations.

Add **(1)** and **(3)**:

$$0 = 0 \quad \textbf{(4)}$$

Add **(1)** and **(2)**:

$$3x - 4y = -1 \quad \textbf{(5)}$$

Hence, we know that the system has infinitely many solutions.

To determine them, let $x = a$. Substitute this into **(5)** to find y:

$$3a - 4y = -1$$

$$y = \tfrac{1}{4}(3a+1)$$

Now, substitute the values of x and y into **(1)** to find z:

$$a - \tfrac{1}{4}(3a+1) - z = 10$$

$$z = \tfrac{4a-3a-1-40}{4}$$

$$z = \tfrac{a-41}{4}$$

Thus, the solution is:

$$x = a,\ y = \tfrac{1}{4}(3a+1),\ z = \tfrac{a-41}{4}.$$

Equivalently,

$$\boxed{x = 41 + 4a,\ y = 31 + 3a,\ z = a}.$$

22. Solve the system:

$$\begin{cases} 2x + y + z = -3 & \textbf{(1)} \\ x + 2y - z = 0 & \textbf{(2)} \\ x + y + 2z = 5 & \textbf{(3)} \end{cases}$$

Step 1: Obtain a system of 2 equations in 2 unknowns by eliminating the same variable in two pairs of equations.

Add **(1)** and **(2)** to eliminate z:

$$3x + 3y = -3 \quad \text{or}$$

$$x + y = -1 \quad \textbf{(4)}$$

Multiply **(2)** by 2:

$$2x + 4y - 2z = 0 \quad \textbf{(5)}$$

Add **(5)** and **(3)** to eliminate z:

$$3x + 5y = 5 \quad \textbf{(6)}$$

These steps yield the following system:

$$\textbf{(*)} \begin{cases} x + y = -1 & \textbf{(4)} \\ 3x + 5y = 5 & \textbf{(6)} \end{cases}$$

Step 2: Solve system **(*)** from Step 1.

Multiply **(4)** by -3:

$$-3x - 3y = 3 \quad \textbf{(7)}$$

Add **(6)** and **(7)** to eliminate x and solve for y:

$$2y = 8$$

$$y = 4 \quad \textbf{(8)}$$

Substitute **(8)** into **(4)** to find x:

$$x + 4 = -1$$

$$x = -5 \quad \textbf{(9)}$$

Step 3: Use the solution of the system in Step 2 to find the value of the third variable in the original system.

Substitute **(8)** and **(9)** into **(2)** to find z:

$$-5 + 2(4) - z = 0$$

$$z = 3 \ \textbf{(10)}$$

Thus, the solution is:

$$\boxed{x = -5,\ y = 4,\ z = 3}.$$

23. Solve the system:

$$\begin{cases} 3x_1 + x_2 - x_3 = 1 & \textbf{(1)} \\ x_1 - x_2 + x_3 = -3 & \textbf{(2)} \\ 2x_1 + x_2 + x_3 = 0 & \textbf{(3)} \end{cases}$$

Step 1: Obtain a system of 2 equations in 2 unknowns by eliminating the same variable in two pairs of equations.

Add **(1)** and **(2)** to eliminate x_2 and x_3:

$$4x_1 = -2$$

$$x_1 = -\tfrac{1}{2} \quad \textbf{(4)}$$

Add **(2)** and **(3)** to eliminate x_2:

$$3x_1 + 2x_3 = -3 \quad \textbf{(5)}$$

These steps yield the following system:

$$(*) \begin{cases} x_1 = -\tfrac{1}{2} & \textbf{(4)} \\ 3x_1 + 2x_3 = -3 & \textbf{(5)} \end{cases}$$

Step 2: Solve system **(*)** from Step 1.

Substitute **(4)** into **(5)**:

$$3(-\tfrac{1}{2}) + 2x_3 = -3$$

$$2x_3 = -\tfrac{3}{2}$$

$$x_3 = -\tfrac{3}{4} \quad \textbf{(6)}$$

Step 3: Use the solution of the system in Step 2 to find the value of the third variable in the original system.

Substitute **(4)** and **(6)** into **(1)** to find x_2:

$$3(-\tfrac{1}{2}) + x_2 - (-\tfrac{3}{4}) = 1$$

$$x_2 = \tfrac{7}{4} \quad \textbf{(7)}$$

Thus, the solution is:

$$\boxed{x_1 = -\tfrac{1}{2},\ x_2 = \tfrac{7}{4},\ x_3 = -\tfrac{3}{4}}.$$

24. Solve the system:

$$\begin{cases} 2x_1 + x_2 + x_3 = -1 & \textbf{(1)} \\ x_1 + x_2 - x_3 = 5 & \textbf{(2)} \\ 3x_1 - x_2 - x_3 = 1 & \textbf{(3)} \end{cases}$$

Step 1: Obtain a system of 2 equations in 2 unknowns by eliminating the same variable in two pairs of equations.

Add **(1)** and **(3)** to eliminate x_2 and x_3:

$$5x_1 = 0$$

$$x_1 = 0 \quad \textbf{(4)}$$

Add **(1)** and **(2)** to eliminate x_3:

$$3x_1 + 2x_2 = 4 \quad \textbf{(5)}$$

These steps yield the following system:

$$(*) \begin{cases} x_1 = 0 & \textbf{(4)} \\ 3x_1 + 2x_2 = 4 & \textbf{(5)} \end{cases}$$

Step 2: Solve system **(*)** from Step 1.

Substitute **(4)** into **(5)**:

$$3(0) + 2x_2 = 4$$

$$2x_2 = 4$$

$$x_2 = 2 \quad \textbf{(6)}$$

Step 3: Use the solution of the system in Step 2 to find the value of the third variable in the original system.

Substitute **(4)** and **(6)** into **(2)** to find x_3:

$$0 + 2 - x_3 = 5$$

$$x_3 = -3 \quad \textbf{(7)}$$

Thus, the solution is:

$$\boxed{x_1 = 0,\ x_2 = 2,\ x_3 = -3}.$$

25. Solve the system:

$$\begin{cases} 2x + 5y \qquad = 9 & \textbf{(1)} \\ x + 2y - z = 3 & \textbf{(2)} \\ -3x - 4y + 7z = 1 & \textbf{(3)} \end{cases}$$

Step 1: Obtain a system of 2 equations in 2 unknowns by eliminating the same variable in two pairs of equations.

Multiply **(2)** by 7:

$$7x + 14y - 7z = 21 \quad \textbf{(4)}$$

Add **(4)** and **(3)** to eliminate z:

$$4x + 10y = 22 \quad \textbf{(5)}$$

These steps yield the following system:

$$\textbf{(*)} \begin{cases} 2x + 5y = 9 & \textbf{(1)} \\ 4x + 10y = 22 & \textbf{(5)} \end{cases}$$

Step 2: Solve system **(*)** from Step 1.

Multiply **(1)** by -2:

$$-4x - 10y = -18 \quad \textbf{(6)}$$

Add **(5)** and **(6)** to eliminate x and solve for y:

$$0 = 4 \quad \textbf{(7)}$$

Hence, we conclude from **(7)** that the system has $\boxed{\text{no solution}}$.

26. Solve the system:

$$\begin{cases} x - 2y + 3z = 1 & \textbf{(1)} \\ -2x + 7y - 9z = 4 & \textbf{(2)} \\ x \qquad + z = 9 & \textbf{(3)} \end{cases}$$

Step 1: Obtain a system of 2 equations in 2 unknowns by eliminating the same variable in two pairs of equations.

Solve **(3)** for z:

$$z = 9 - x \quad \textbf{(4)}$$

Substitute **(4)** into **(1)** to eliminate z:

$$x - 2y + 3(9 - x) = 1$$

$$-2x - 2y = -26$$

$$x + y = 13 \quad \textbf{(5)}$$

Substitute **(4)** into **(2)** to eliminate z:

$$-2x + 7y - 9(9 - x) = 4$$

$$7x + 7y = 85 \quad \textbf{(6)}$$

These steps yield the following system:

$$\textbf{(*)} \begin{cases} x + y = 13 & \textbf{(5)} \\ 7x + 7y = 85 & \textbf{(6)} \end{cases}$$

Step 2: Solve system **(*)** from Step 1.

Multiply **(5)** by -7:

$$-7x - 7y = -91 \quad \textbf{(7)}$$

Add **(6)** and **(7)** to eliminate x and solve for y:

$$0 = -6 \quad \textbf{(8)}$$

Hence, we conclude from **(8)** that the system has $\boxed{\text{no solution}}$.

27. Solve the system:

$$\begin{cases} 2x_1 - x_2 + x_3 = 3 & \textbf{(1)} \\ x_1 - x_2 + x_3 = 2 & \textbf{(2)} \\ -2x_1 + 2x_2 - 2x_3 = -4 & \textbf{(3)} \end{cases}$$

Step 1: Obtain a system of 2 equations in 2 unknowns by eliminating the same variable in two pairs of equations.

Multiply **(1)** by -1:

$$-2x_1 + x_2 - x_3 = -3 \quad \textbf{(4)}$$

Add **(2)** and **(4)** to eliminate x_2 and x_3:

$$-x_1 = -1$$

$$x_1 = 1 \quad \textbf{(5)}$$

Substitute **(5)** into **(2)**:

$$1 - x_2 + x_3 = 2$$

$$-x_2 + x_3 = 1 \quad \textbf{(6)}$$

Substitute **(5)** into **(3)**:

$$-2 + 2x_2 - 2x_3 = -4$$

$$x_2 - x_3 = -1 \quad \textbf{(7)}$$

These steps yield the following system:

$$(*) \begin{cases} -x_2 + x_3 = 1 & \textbf{(6)} \\ x_2 - x_3 = -1 & \textbf{(7)} \end{cases}$$

Step 2: Solve system **(*)** from Step 1.

Add **(6)** and **(7)**: $0 = 0$ **(8)**

Hence, we conclude from **(8)** that there are infinitely many solutions. To determine them, let $x_3 = a$. Substitute this value into **(7)** to see that $x_2 = a - 1$.

Thus, the solution is:

$$\boxed{x_1 = 1,\ x_2 = -1 + a,\ x_3 = a}.$$

28. Solve the system:

$$\begin{cases} x_1 - x_2 - 2x_3 = 0 & \textbf{(1)} \\ -2x_1 + 5x_2 + 10x_3 = -3 & \textbf{(2)} \\ 3x_1 + x_2 = 0 & \textbf{(3)} \end{cases}$$

Step 1: Obtain a system of 2 equations in 2 unknowns by eliminating the same variable in two pairs of equations.

Solve **(3)** for x_2: $x_2 = -3x_1$ **(4)**

Substitute **(4)** into **(1)**:

$$x_1 - (-3x_1) - 2x_3 = 0$$

$$4x_1 - 2x_3 = 0 \quad \textbf{(5)}$$

Substitute **(4)** into **(2)**:

$$-2x_1 + 5(-3x_1) + 10x_3 = -3$$

$$-17x_1 + 10x_3 = -3 \quad \textbf{(6)}$$

These steps yield the following system:

$$(*) \begin{cases} 4x_1 - 2x_3 = 0 & \textbf{(5)} \\ -17x_1 + 10x_3 = -3 & \textbf{(6)} \end{cases}$$

Step 2: Solve system **(*)** from Step 1.

Multiply **(5)** by 5:

$$20x_1 - 10x_3 = 0 \quad \textbf{(7)}$$

Add **(6)** and **(7)**:

$$3x_1 = -3$$

$$x_1 = -1 \quad \textbf{(8)}$$

Substitute **(8)** into **(5)** to find x_3:

$$4(-1) - 2x_3 = 0$$

$$-2 = x_3 \quad \textbf{(9)}$$

Step 3: Use the solution of the system in Step 2 to find the value of the third variable in the original system.

Substitute **(8)** into **(4)** to find x_2:

$$x_2 = -3(-1) = 3$$

Thus, the solution is:

$$\boxed{x_1 = -1,\ x_2 = -3,\ x_3 = -2}.$$

29. Solve the system:

$$\begin{cases} 2x + y - z = 2 & \textbf{(1)} \\ x - y - z = 6 & \textbf{(2)} \end{cases}$$

Since there are two equations and three unknowns, we know there are infinitely many solutions (as long as neither statement is inconsistent). To this end, let $z = a$.

Add **(1)** and **(2)**: $3x - 2z = 8$ **(3)**

Substitute $z = a$ into **(3)** to find x:

$$x = \tfrac{2}{3}a + \tfrac{8}{3} \quad \textbf{(4)}$$

Finally, substitute $z = a$ and **(4)** into **(2)** to find y: $y = -\tfrac{1}{3}a - \tfrac{10}{3}$

Thus, the solution is

$$\boxed{x = \tfrac{2}{3}a + \tfrac{8}{3},\ y = -\tfrac{1}{3}a - \tfrac{10}{3},\ z = a}.$$

30. Solve the system:

$$\begin{cases} 3x + y - z = 0 & \textbf{(1)} \\ x + y + 7z = 4 & \textbf{(2)} \end{cases}$$

Since there are two equations and three unknowns, we know there are infinitely many solutions (as long as neither statement is inconsistent). To this end, let $x = a$. Multiply **(2)** by -1:

$$-x - y - 7z = -4 \quad \textbf{(3)}$$

Add **(1)** and **(3)**:

$$2x - 8z = -4 \quad \text{or}$$

$$x - 4z = -2 \quad \textbf{(4)}$$

Substitute $x = a$ into **(4)** to find z:

$$a - 4z = -2 \Rightarrow z = \tfrac{1}{4}(a+2) \quad \textbf{(5)}$$

Finally, substitute $x = a$ and **(5)** into **(1)** to find y:

$$3a + y - \tfrac{1}{4}(a+2) = 0 \Rightarrow y = -\tfrac{11}{4}a + \tfrac{1}{2}$$

Thus, the solution is

$$\boxed{x = a,\ y = -\tfrac{11}{4}a + \tfrac{1}{2},\ z = \tfrac{1}{4}(a+2)}.$$

31. Solve the system:

$$\begin{cases} 4x + 3y - 3z = 5 & \textbf{(1)} \\ 6x \qquad + 2z = 10 & \textbf{(2)} \end{cases}$$

Since there are two equations and three unknowns, we know there are infinitely many solutions (as long as neither statement is inconsistent). To this end, let $x = a$.

Then **(2)** implies: $z = \frac{10-6a}{2} = 5 - 3a$.

Substitute both into **(1)** to find y:

$$4a + 3y - 3(5 - 3a) = 5$$

$$3y = 20 - 13a$$

$$y = \tfrac{20}{3} - \tfrac{13}{3}a$$

Thus, the solution is

$$\boxed{x = a,\ y = \tfrac{20}{3} - \tfrac{13}{3}a,\ z = 5 - 3a}.$$

32. Solve the system:

$$\begin{cases} x + 2y + 4z = 12 & \textbf{(1)} \\ -3x - 4y + 7z = 21 & \textbf{(2)} \end{cases}$$

Since there are two equations and three unknowns, we know there are infinitely many solutions (as long as neither statement is inconsistent). To this end, let $x = a$. Substituting this into both **(1)** and **(2)** yields the system:

$$\begin{cases} 2y + 4z = 12 - a & \textbf{(3)} \\ -4y + 7z = 21 + 3a & \textbf{(4)} \end{cases}$$

Multiply **(3)** by 2 and add to **(4):**

$$15z = 45 + a \Rightarrow z = 3 + \tfrac{a}{15}$$

Now, substitute this and $x = a$ into **(1)** to find y:

$$a + 2y + 4\left(3 + \tfrac{a}{15}\right) = 12 \Rightarrow y = -\tfrac{19}{30}a.$$

Thus, the solution is

$$\boxed{x = a,\ y = -\tfrac{19}{30}a,\ z = 3 + \tfrac{a}{15}}.$$

33. Let $x = \#$ Mediterranean chicken sand.
$y = \#$ Six-inch tuna sandwiches
$z = \#$ Six-inch roast beef sandwiches

We must solve the system:

$$(*)\begin{cases} x+y+z=14 & \textbf{(1)} \\ 350x+430y+290z=4840 & \textbf{(2)} \\ 18x+19y+5z=190 & \textbf{(3)} \end{cases}$$

Step 1: Obtain a system of 2 equations in 2 unknowns by eliminating the same variable in two pairs of equations.

Solve **(1)** for x: $x=14-y-z$ **(4)**

Substitute **(4)** into **(2)** and simplify:

$4y-3z=-3$ **(5)**

Substitute **(4)** into **(3)** and simplify:

$y-13z=-62$ **(6)**

These steps yield the following system:

$$(*)\begin{cases} 4y-3z=-3 & \textbf{(5)} \\ y-13z=-62 & \textbf{(6)} \end{cases}$$

Step 2: Solve system **(*)** from Step 1.

Solve **(6)** for y:

$y=13z-62$ **(7)**

Substitute **(7)** into **(5)** and simplify:

$z=5$ **(8)**

Substitute **(8)** into **(7)** and simplify:

$y=3$ **(8)**

Step 3: Use the solution of the system in Step 2 to find the value of the third variable in the original system.

Substitute **(8)** and **(9)** into **(1)** to find x:

$x=6$ **(10)**

Thus, there are:
6 Mediterranean chicken sandwiches,
3 Six-inch tuna sandwiches,
5 Six-inch roast beef sandwiches.

34. Let $x = \#$ Mediterranean chicken sand.
$y = \#$ Six-inch tuna sandwiches
$z = \#$ Six-inch roast beef sandwiches

We must solve the system:

$$(*)\begin{cases} x+y+z=14 & \textbf{(1)} \\ 350x+430y+290z=4380 & \textbf{(2)} \\ 18x+19y+5z=123 & \textbf{(3)} \end{cases}$$

Step 1: Obtain a system of 2 equations in 2 unknowns by eliminating the same variable in two pairs of equations.

Solve **(1)** for x: $x=14-y-z$ **(4)**

Substitute **(4)** into **(2)** and simplify:

$4y-3z=-26$ **(5)**

Substitute **(4)** into **(3)** and simplify:

$y-13z=-129$ **(6)**

These steps yield the following system:

$$(*)\begin{cases} 4y-3z=-26 & \textbf{(5)} \\ y-13z=-129 & \textbf{(6)} \end{cases}$$

Step 2: Solve system **(*)** from Step 1.

Solve **(6)** for y:

$y=13z-129$ **(7)**

Substitute **(7)** into **(5)** and simplify:

$z=10$ **(8)**

Substitute **(8)** into **(7)** and simplify:

$y=1$ **(8)**

Step 3: Use the solution of the system in Step 2 to find the value of the third variable in the original system.

Substitute **(8)** and **(9)** into **(1)** to find x:

$x=3$ **(10)**

Thus, there are:
3 Mediterranean chicken sandwiches,
1 Six-inch tuna sandwiches,
10 Six-inch roast beef sandwiches.

35. The system that must be solved is:

$$\begin{cases} 36 = \frac{1}{2}a(1)^2 + v_0(1) + h_0 \\ 40 = \frac{1}{2}a(2)^2 + v_0(2) + h_0 \\ 12 = \frac{1}{2}a(3)^2 + v_0(3) + h_0 \end{cases}$$

which is equivalent to

$$\begin{cases} 72 = a + 2v_0 + 2h_0 & \textbf{(1)} \\ 40 = 2a + 2v_0 + h_0 & \textbf{(2)} \\ 24 = 9a + 6v_0 + 2h_0 & \textbf{(3)} \end{cases}$$

Step 1: Obtain a 2×2 system:

Multiply **(1)** by -2, and add to **(2)**:

$$-2v_0 - 3h_0 = -104 \quad \textbf{(4)}$$

Multiply **(1)** by -9, and add to **(3)**:

$$-12v_0 - 16h_0 = -624 \quad \textbf{(5)}$$

These steps yield the following 2×2 system:

$$\begin{cases} -2v_0 - 3h_0 = -104 & \textbf{(4)} \\ -12v_0 - 16h_0 = -624 & \textbf{(5)} \end{cases}$$

Step 2: Solve the system in Step 1.

Multiply **(4)** by -6 and add to **(5)**:

$$2h_0 = 0$$

$$\boxed{h_0 = 0} \quad \textbf{(6)}$$

Substitute **(6)** into **(4)**:

$$-2v_0 - 3(0) = -104$$

$$-2v_0 = -104$$

$$\boxed{v_0 = 52} \quad \textbf{(7)}$$

Step 3: Find values of remaining variables.

Substitute **(6)** and **(7)** into **(1)**:

$$72 = a + 2(52) + 2(0)$$

$$\boxed{-32 = a}$$

Thus, the polynomial has the equation $h = -16t^2 + 52t$.

36. The system that must be solved is:

$$\begin{cases} 84 = \frac{1}{2}a(1)^2 + v_0(1) + h_0 \\ 136 = \frac{1}{2}a(2)^2 + v_0(2) + h_0 \\ 156 = \frac{1}{2}a(3)^2 + v_0(3) + h_0 \end{cases}$$

which is equivalent to

$$\begin{cases} 168 = a + 2v_0 + 2h_0 & \textbf{(1)} \\ 136 = 2a + 2v_0 + h_0 & \textbf{(2)} \\ 312 = 9a + 6v_0 + 2h_0 & \textbf{(3)} \end{cases}$$

Step 1: Obtain a 2×2 system:

Multiply **(1)** by -2, and add to **(2)**:

$$-2v_0 - 3h_0 = -200 \quad \textbf{(4)}$$

Multiply **(1)** by -9, and add to **(3)**:

$$-12v_0 - 16h_0 = -1200 \quad \textbf{(5)}$$

These steps yield the following 2×2 system:

$$\begin{cases} -2v_0 - 3h_0 = -200 & \textbf{(4)} \\ -12v_0 - 16h_0 = -1200 & \textbf{(5)} \end{cases}$$

Step 2: Solve the system in Step 1.

Multiply **(4)** by -6 and add to **(5)**:

$$2h_0 = 0$$

$$\boxed{h_0 = 0} \quad \textbf{(6)}$$

Substitute **(6)** into **(4)**:

$$-2v_0 - 3(0) = -200$$

$$-2v_0 = -200$$

$$\boxed{v_0 = 100} \quad \textbf{(7)}$$

Step 3: Find values of remaining variables.

Substitute **(6)** and **(7)** into **(1)**:

$$168 = a + 2(100) + 2(0)$$

$$\boxed{-32 = a}$$

Thus, the polynomial has the equation

$$h = \underbrace{\tfrac{1}{2}(-32)}_{-16}t^2 + 100t.$$

37. Since we are given that the points $(20,30)$, $(40,60)$, and $(60,40)$, the system that must be solved is:

$$\begin{cases} 30 = a(20)^2 + b(20) + c \\ 60 = a(40)^2 + b(40) + c \\ 40 = a(60)^2 + b(60) + c \end{cases}$$

which is equivalent to

$$\begin{cases} 30 = 400a + 20b + c & \textbf{(1)} \\ 60 = 1600a + 40b + c & \textbf{(2)} \\ 40 = 3600a + 60b + c & \textbf{(3)} \end{cases}$$

Step 1: Obtain a 2×2 system:

Multiply **(1)** by -4, and add to **(2)**:

$$-40b - 3c = -60 \quad \textbf{(4)}$$

Multiply **(1)** by -9, and add to **(3)**:

$$-120b - 8c = -230 \quad \textbf{(5)}$$

These steps yield the following 2×2 system:

$$\begin{cases} -40b - 3c = -60 & \textbf{(4)} \\ -120b - 8c = -230 & \textbf{(5)} \end{cases}$$

Step 2: Solve the system in Step 1.

Multiply **(4)** by -3 and add to **(5)**:

$$c = -50 \quad \textbf{(6)}$$

Substitute **(6)** into **(4)**:

$$-40b - 3(-50) = -60$$

$$-40b = -210$$

$$b = \tfrac{21}{4} = 5.25 \quad \textbf{(7)}$$

Step 3: Find values of remaining variables.

Substitute **(6)** and **(7)** into **(1)**:

$$30 = 400a + 20(5.25) + (-50)$$

$$-25 = 400a$$

$$0.0625 = a$$

Thus, the polynomial has the equation $\boxed{y = -0.0625x^2 + 5.25x - 50}$.

38. Since we are given that the points $(0,18.6)$, $(20,20.2)$, and $(72,25.3)$, the system that must be solved is:

$$\begin{cases} 18.6 = a(0)^2 + b(0) + c \\ 20.2 = a(20)^2 + b(20) + c \\ 25.3 = a(72)^2 + b(72) + c \end{cases}$$

which is equivalent to

$$\begin{cases} 18.6 = c & \textbf{(1)} \\ 20.2 = 400a + 20b + c & \textbf{(2)} \\ 25.3 = 5184a + 72b + c & \textbf{(3)} \end{cases}$$

Step 1: Obtain a 2×2 system:

Substitute **(1)** into both **(2)** and **(3)**:

$$20.2 = 400a + 20b + 18.6$$

$$25.3 = 5184a + 72b + 18.6$$

This yields the following 2×2 system:

$$\begin{cases} 1.6 = 400a + 20b & \textbf{(4)} \\ 6.7 = 5184a + 72b & \textbf{(5)} \end{cases}$$

Step 2: Solve the system in Step 1.

Multiply **(4)** by -72 , **(5)** by 20, and then add them to find a:

$$74880a = 18.8$$

$$a = 0.000251 \quad \textbf{(6)}$$

Substitute **(6)** into **(4)** to find b:

$$1.6 = 400(0.000251) + 20b$$

$$0.07498 = b \quad \textbf{(7)}$$

Thus, the polynomial has the equation $\boxed{y = 0.000251x^2 + 0.07498x + 18.6}$.

39. Let x = amount in money market
y = amount in mutual fund
z = amount in stock

The system we must solve is:

$$\begin{cases} x = y + 6000 & \textbf{(1)} \\ x + y + z = 20{,}000 & \textbf{(2)} \\ 0.03x + 0.07y + 0.10z = 1180 & \textbf{(3)} \end{cases}$$

Step 1: Obtain a 2×2 system:
Substitute **(1)** into both **(2)** and **(3)**:

$$(y+6000) + y + z = 20{,}000$$

$$0.03(y+6000) + 0.07y + 0.10z = 1180$$

This yields the following system:

$$\begin{cases} 2y + z = 14{,}000 & \textbf{(4)} \\ 0.10y + 0.10z = 1000 & \textbf{(5)} \end{cases}$$

Step 2: Solve the system in Step 1.
Multiply **(4)** by -0.10 and add to **(5)** to find y:

$$-0.10y = -400$$

$$y = 4000 \quad \textbf{(6)}$$

Substitute **(6)** into **(1)** to find x:

$$x = 6000 + 4000 = 10{,}000 \quad \textbf{(7)}$$

Substitute **(6)** and **(7)** into **(2)** to find z:

$$10{,}000 + 4000 + z = 20{,}000$$

$$z = 6000$$

Thus, the following allocation of funds should be made:

\$10,000 in money market
\$4000 in mutual fund
\$6000 in stock

40. Let x = amount in money market
y = amount in mutual fund
z = amount in stock

The system we must solve is:

$$\begin{cases} x + y + z = 20{,}000 & \textbf{(1)} \\ z = y + 6000 & \textbf{(2)} \\ 0.03x + 0.07y + 0.10z = 1680 & \textbf{(3)} \end{cases}$$

Step 1: Obtain a 2×2 system:
Substitute **(2)** into both **(1)** and **(3)**:

$$x + y + (y+6000) = 20{,}000$$

$$0.03x + 0.07y + 0.10(y+6000) = 1680$$

This yields the following system:

$$\begin{cases} x + 2y = 14{,}000 & \textbf{(4)} \\ 0.03x + 0.17y = 1080 & \textbf{(5)} \end{cases}$$

Step 2: Solve the system in Step 1.
Solve **(4)** for x:

$$x = 14{,}000 - 2y \quad \textbf{(6)}$$

Substitute **(6)** into **(5)** and solve for y:

$$0.03(14{,}000 - 2y) + 0.17y = 1080$$

$$0.11y = 660$$

$$y = 6000 \quad \textbf{(7)}$$

Substitute **(7)** into **(6)** to find x:

$$x = 14{,}000 - 2(6000) = 2000 \quad \textbf{(8)}$$

Finally, substitute **(7)** into **(2)** to find z:

$$z = 6000 + 6000 = 12{,}000$$

Thus, the following allocation of funds should be made:

\$2,000 in money market
\$6,000 in mutual fund
\$12,000 in stock

41. Let $x = \#$ regular models skis
$y = \#$ trick skis
$z = \#$ slalom skis
Solve the system:

$$\begin{cases} x+y+z=110 & \textbf{(1)} \\ 2x+3y+3z=297 & \textbf{(2)} \\ x+2y+5z=202 & \textbf{(3)} \end{cases}$$

Step 1: Obtain a 2×2 system:
Solve **(1)** for x: $x=110-y-z$ **(4)**
Substitute **(4)** into **(2)**, and simplify:

$$y+z=77 \quad \textbf{(5)}$$

Substitute **(4)** into **(3)**, and simplify:

$$y+4z=92 \quad \textbf{(6)}$$

This yields the following system:

$$(*)\begin{cases} y+z=77 & \textbf{(5)} \\ y+4z=92 & \textbf{(6)} \end{cases}$$

Step 2: Solve system (*) from Step 1.
Subtract **(6) – (5)**:

$$3z=15 \Rightarrow \boxed{z=5} \quad \textbf{(7)}$$

Substitute **(7)** into **(5)**:

$$\boxed{y=72} \quad \textbf{(8)}$$

Substitute **(7)** and **(8)** into **(1):**

$$x+72+5=110 \Rightarrow \boxed{x=33}$$

So, need to sell 33 regular model skis, 72 trick skis, and 5 slalom skis.

42. Let $x = \#$ compact cars
$y = \#$ intermediate cars
$z = \#$ luxury models
Solve the system:

$$\begin{cases} x+y+z=500 & \textbf{(1)} \\ 200x+300y+250z=128,750 & \textbf{(2)} \\ 30x+20y+45z=15,625 & \textbf{(3)} \end{cases}$$

Step 1: Obtain a 2×2 system:
Solve **(1)** for x: $x=500-y-z$ **(4)**
Substitute **(4)** into **(2)**, and simplify:

$$2y+z=575 \quad \textbf{(5)}$$

Substitute **(4)** into **(3)**, and simplify:

$$-2y+3z=125 \quad \textbf{(6)}$$

This yields the following system:

$$(*)\begin{cases} 2y+z=575 & \textbf{(5)} \\ -2y+3z=125 & \textbf{(6)} \end{cases}$$

Step 2: Solve system (*) from Step 1.
Add **(6) + (5)**:

$$4z=700 \Rightarrow \boxed{z=175} \quad \textbf{(7)}$$

Substitute **(7)** into **(5)**:

$$\boxed{y=200} \quad \textbf{(8)}$$

Substitute **(7)** and **(8)** into **(1):**

$$x+200+175=500 \Rightarrow \boxed{x=125}$$

So, 125 compact cars, 200 intermediate cars, and 175 luxury models.

43. Let x = # points scored in game 1
y = # points scored in game 2
z = # points scored in game 3
Solve the system:

$$\begin{cases} x+y+z=2{,}591 & \textbf{(1)} \\ x-y=62 & \textbf{(2)} \\ x-z=2 & \textbf{(3)} \end{cases}$$

Step 1: Obtain a 2×2 system:
Solve **(1)** for x: $x=2591-y-z$ **(4)**
Substitute **(4)** into **(2)**, and simplify:

$$2y+z=2529 \quad \textbf{(5)}$$

Substitute **(4)** into **(3)**, and simplify:

$$y+2z=2589 \quad \textbf{(6)}$$

This yields the following system:

$$(*)\begin{cases} 2y+z=2529 & \textbf{(5)} \\ y+2z=2589 & \textbf{(6)} \end{cases}$$

Step 2: Solve system (*) from Step 1.
Multiply **(6)** by -2:

$$-2y-4z=-5178 \quad \textbf{(7)}$$

Add **(7)** + **(5)**:

$$-3z=-2649 \;\Rightarrow\; \boxed{z=883} \quad \textbf{(8)}$$

Substitute **(8)** into **(6)**:

$$\boxed{y=823} \quad \textbf{(9)}$$

Substitute **(9)** and **(8)** into **(1):**

$$x+823+883=2591 \;\Rightarrow\; \boxed{x=885}$$

So, 885 points scored in game 1, 823 points scored in game 2, and 883 points scored in game 3.

44. Let x = speed of IBM's Blue Gene/L
y = speed of IBM's BGW
z = speed of IBM's ASC Purple
Solve the system:

$$\begin{cases} x=y+245 & \textbf{(1)} \\ y=z+22 & \textbf{(2)} \\ x+y+z=568 & \textbf{(3)} \end{cases}$$

Step 1: Obtain a 2×2 system:
Substitute **(1)** into **(3)**: $2y+z=323$ **(4)**

Solve the system:

$$(*)\begin{cases} y=z+22 & \textbf{(2)} \\ 2y+z=323 & \textbf{(4)} \end{cases}$$

Step 2: Solve system (*) from Step 1.
Substitute **(2)** into **(4)**, and simplify:

$$\boxed{z=93} \quad \textbf{(5)}$$

Substitute **(5)** into **(2)**, and simplify:

$$\boxed{y=115} \quad \textbf{(6)}$$

Substitute **(5)** and **(6)** into **(3):**

$$x+115+93=568 \;\Rightarrow\; \boxed{x=360}$$

So, the Blue Gene/L runs at 360 teraflops, the BGW at 115 teraflops, and the ASC Purple at 93 teraflops.

45. Let x = number of eagle balls, y = number of birdie balls, and z = number of bogey balls.
<u>Solve the system</u>:

$$\begin{cases} x+y+z=10{,}000 & \textbf{(1)} \\ x+y=z & \textbf{(2)} \\ y=3x & \textbf{(3)} \end{cases}$$

Substitute **(3)** into **(1)** and **(2)** to obtain the system:

$$\begin{cases} 4x+z=10{,}000 & \textbf{(4)} \\ 4x=z & \textbf{(5)} \end{cases}$$

Substitute **(5)** into **(4)**: $x = 1250$
Substitute this into **(5)**: $z = 5000$
Finally, substitute both into **(1):** $y = 3750$.
So, there are 1250 eagle balls, 3750 birdie balls, and 5000 bogey balls.

46. Let x = cost per part of cheese A, y = cost per part of cheese B, and z = cost per part of cheese C.
<u>Solve the system</u>:

$$\begin{cases} 2x+2y+2z=2.40 & \textbf{(1)} \\ 2x+y+2z=2.20 & \textbf{(2)} \\ 2x+2y+3z=2.70 & \textbf{(3)} \end{cases}$$

Subtract **(1) – (3):** $z = 0.30$ **(4)**
Substitute **(4)** into both **(1)** and **(2)** to obtain the system:

$$\begin{cases} 2x+2y+2(0.30)=2.40 & \textbf{(5)} \\ 2x+y+2(0.30)=2.20 & \textbf{(6)} \end{cases}$$

Subtract **(5) – (6)**: $y = 0.20$
Substitute both values into **(1)**: $x =$ 0.70.

47. Let x = number of 10 sec. commercials, y = number of 20 sec. commercials, and z = number of 40 sec. commercials.
<u>Solve the system</u>:

$$\begin{cases} 100x+180y+320z=1060 & \textbf{(1)} \\ x+y+z=6 & \textbf{(2)} \\ x=2z & \textbf{(3)} \end{cases}$$

Substitute **(3)** into both **(1)** and **(2)** to obtain the system:

$$\begin{cases} 180y+520z=1060 & \textbf{(4)} \\ y+3z=6 & \textbf{(5)} \end{cases}$$

Solve **(5)** for y: $y=6-3z$ **(6)**
Substitute **(6)** into **(4)**: $z = 1$.
Now, substitute this back into **(6)**: $y = 3$.
Finally, substitute both values into **(1)**: $x =$ 2.
So, there are 2 10 sec. commercials, 3 20 sec. commercials, and 1 40 sec. commercial.

48. Let x = number of first class seats, y = number of business seats, and z = number of coach seats.
<u>Solve the system</u>:

$$\begin{cases} x+y+z=270 & \textbf{(1)} \\ x=\tfrac{1}{3}y \Rightarrow y=3x & \textbf{(2)} \\ z=x+250 & \textbf{(3)} \end{cases}$$

Substitute both **(2)** and **(3)** into **(1)**:

$$x+3x+x+250=270$$
$$x=4$$

Now, substitute this value back into **(2)** and **(3)** to obtain $y=12$, $z=254$.
So, there are 4 first class seats, 12 business seats, and 254 coach seats.

49. Equation (2) and Equation (3) must be added correctly – should be $2x-y+z=2$. Also, should begin by eliminating one variable from Equation (1).

50. From the system

$$\begin{cases} y+3z=5 \\ y+3z=9 \end{cases}$$

one should (upon subtracting the equations) conclude that $0=-4$. So, the system is inconsistent, and hence, has no solution.

51. True

52 False. Consider the system
$$\begin{cases} x+y=1 \\ x+y=-1 \end{cases}$$

53. False. It could represent an empty graph. For instance, take $A=B=0$ and $C\neq 0$.

54. True, just divide all equations in the second system by 2.

55. Substitute the given points into the equation to obtain the following system:
$$\begin{cases} (-2)^2+4^2+a(-2)+b(4)+c=0 \\ 1^2+1^2+a(1)+b(1)+c=0 \\ (-2)^2+(-2)^2+a(-2)+b(-2)+c=0 \end{cases}$$
which is equivalent to (after simplification)
$$\begin{cases} -2a+4b+c=-20 & \textbf{(1)} \\ a+b+c=-2 & \textbf{(2)} \\ -2a-2b+c=-8 & \textbf{(3)} \end{cases}$$
Multiply **(2)** by 2 and then, add to **(3)**:
$$3c=-12 \Rightarrow c=-4 \quad \textbf{(4)}$$
Multiply **(2)** by 2 and then, add to **(1)**:
$$6b+3c=-24 \quad \textbf{(5)}$$
Substitute **(4)** into **(5)** to find b:
$$6b+3(-4)=-24$$
$$6b=-12$$
$$b=-2 \quad \textbf{(6)}$$
Finally, substitute **(4)** and **(6)** into **(2)** to find a:
$$a+(-2)+(-4)=-2$$
$$a=4$$
Thus, the equation is
$$\boxed{x^2+y^2+4x-2y-4=0}.$$

56. Substitute the given points into the equation to obtain the following system:
$$\begin{cases} (0)^2+7^2+a(0)+b(7)+c=0 \\ 6^2+1^2+a(6)+b(1)+c=0 \\ (5)^2+(4)^2+a(5)+b(4)+c=0 \end{cases}$$
which is equivalent to (after simplification)
$$\begin{cases} 7b+c=-49 & \textbf{(1)} \\ 6a+b+c=-37 & \textbf{(2)} \\ 5a+4b+c=-41 & \textbf{(3)} \end{cases}$$
Solve **(1)** for c: $c=-7b-49$ **(4)**

Substitute **(4)** into both **(2)** and **(3)**:
$$6a+b+(-7b-49)=-37$$
$$5a+4b+(-7b-49)=-41$$
which is equivalent to
$$\begin{cases} 6a-6b=12 & \textbf{(5)} \\ 5a-3b=8 & \textbf{(6)} \end{cases}$$
Multiply **(6)** by -2 and then, add to **(5)**:
$$-4a=-4$$
$$a=1 \quad \textbf{(7)}$$
Substitute **(7)** into **(6)** to find b:
$$5(1)-3b=8$$
$$b=-1 \quad \textbf{(8)}$$
Finally, substitute **(8)** into **(4)** to find c:
$$c=-7(-1)-49=-42$$
Thus, the equation is approximately
$$\boxed{x^2+y^2+x-y-42=0}.$$

57. We deduce from the diagram that the following points are on the curve:

$(-2,46)$, $(-1,51)$, $(0,44)$, $(1,51)$, $(2,43)$

Substituting these points into the given equation gives rise to the system:

$$\begin{cases} 46 = a(-2)^4 + b(-2)^3 + c(-2)^2 + d(-2) + e \\ 51 = a(-1)^4 + b(-1)^3 + c(-1)^2 + d(-1) + e \\ 44 = a(0)^4 + b(0)^3 + c(0)^2 + d(0) + e \\ 51 = a(1)^4 + b(1)^3 + c(1)^2 + d(1) + e \\ 43 = a(2)^4 + b(2)^3 + c(2)^2 + d(2) + e \end{cases}$$

Observe that the third equation above simplifies to

$$e = 44.$$

Substitute this value into the remaining four equations to obtain the following system:

$$\begin{cases} 16a - 8b + 4c - 2d = 2 & \textbf{(1)} \\ a - b + c - d = 7 & \textbf{(2)} \\ a + b + c + d = 7 & \textbf{(3)} \\ 16a + 8b + 4c + 2d = -1 & \textbf{(4)} \end{cases}$$

Now, add **(2)** and **(3)**: $2a + 2c = 14$ **(5)**

Add **(1)** and **(4)**: $32a + 8c = 1$ **(6)**

Solve the system: $\begin{cases} 2a + 2c = 14 & \textbf{(5)} \\ 32a + 8c = 1 & \textbf{(6)} \end{cases}$

Multiply **(5)** by -4, and then add to **(6)**:

$$24a = -55$$

$$\boxed{a = -\tfrac{55}{24}} \quad \textbf{(7)}$$

Substitute **(7)** into **(5)** to find c:

$2(-\tfrac{55}{24}) + 2c = 14$

$$\boxed{c = 7 + \tfrac{55}{24} = \tfrac{223}{24}} \quad \textbf{(8)}$$

At this point, we have values for three of the five unknowns. We now use **(7)** and **(8)** to obtain a 2×2 system involving these two remaining unknowns:

Substitute **(7)** and **(8)** into **(1):**

$$2 = 16(-\tfrac{55}{24}) - 8b + 4(\tfrac{223}{24}) - 2d$$

$$36 = -192b - 48d \quad \textbf{(9)}$$

Substitute **(7)** and **(8)** into **(2):**

$$7 = -\tfrac{55}{24} - b + \tfrac{223}{24} - d$$

$$0 = b + d \quad \textbf{(10)}$$

Solve the system:

$$\begin{cases} 36 = -192b - 48d & \textbf{(9)} \\ 0 = b + d & \textbf{(10)} \end{cases}$$

Solve **(10)** for b: $b = -d$ **(11)**

Substitute **(11)** into **(9)** to find d:

$$36 = -192(-d) - 48d$$

$$36 = 144d$$

$$\boxed{\tfrac{1}{4} = d} \quad \textbf{(12)}$$

Substitute **(12)** into **(10)** to find b:

$$\boxed{b = -\tfrac{1}{4}}$$

Hence, the equation of the polynomial is

$y = -\tfrac{55}{24}x^4 - \tfrac{1}{4}x^3 + \tfrac{223}{24}x^2 + \tfrac{1}{4}x + 44$.

58. Let x = number of nickels
y = number of dimes
z = number of quarters

The system that must be solved is:

$$\begin{cases} x+y+z=30 & \textbf{(1)} \\ 0.05x+0.10y+0.25z=4.60 & \textbf{(2)} \\ z=x+4 & \textbf{(3)} \end{cases}$$

Substitute **(3)** into both **(1)** and **(2)**:

$$x+y+(x+4)=30$$

$$0.05x+0.10y+0.25(x+4)=4.60$$

This simplifies to the following system:

$$\begin{cases} 2x+y=26 & \textbf{(4)} \\ 0.30x+0.10y=3.60 & \textbf{(5)} \end{cases}$$

Multiply **(4)** by -0.10, and then add to **(5)**:

$$0.10x=1.0$$

$$x=10 \quad \textbf{(6)}$$

Substitute **(6)** into **(4)** to find y:

$$2(10)+y=26$$

$$y=6 \quad \textbf{(7)}$$

Finally, substitute **(6)** and **(7)** into **(1)** to find z:

$$10+6+z=30$$

$$z=14$$

Thus, there are 10 nickels, 6 dimes, and 14 quarters.

59. Solve the system

$$\begin{cases} 2y+z=3 & \textbf{(1)} \\ 4x-z=-3 & \textbf{(2)} \\ 7x-3y-3z=2 & \textbf{(3)} \\ x-y-z=-2 & \textbf{(4)} \end{cases}$$

Step 1: Obtain a 3×3 system:

Solve **(2)** for z: $z=4x+3$ **(5)**

Substitute **(5)** into each of **(1)**, **(3)**, and **(4)**.

Simplifying yields the following system:

$$\begin{cases} 2y+4x=0 & \textbf{(6)} \\ -3y-5x=11 & \textbf{(7)} \\ -y-3x=1 & \textbf{(8)} \end{cases}$$

Step 2: Solve the system in Step 1.

Solve **(6)** for y: $y=-2x$ **(9)**

Substitute **(9)** into both **(7)** and **(8)** to obtain the following two equations:

$-3(-2x)-5x=11$ so that $x=11$ **(10)**

$-(-2x)-3x=1$ so that $x=-1$ **(11)**

Observe that **(10)** and **(11)** yield different values of x. As such, there can be $\boxed{\text{no solution}}$ to this system.

60. Solve the system:

$$\begin{cases} -2x-y+2z=3 & \textbf{(1)} \\ 3x-4z=2 & \textbf{(2)} \\ 2x+y=-1 & \textbf{(3)} \\ -x+y-z=-8 & \textbf{(4)} \end{cases}$$

Solve **(3)** for y: $y=-2x-1$ **(5)**

Substitute **(5)** into **(1)**:

$$-2x-(-2x-1)+2z=3$$

$$z=1 \quad \textbf{(6)}$$

Substitute **(5)** into **(4)**:

$$-x+(-2x-1)-z=-8$$

$$-3x-z=-7 \quad \textbf{(7)}$$

Substitute **(6)** into **(7)** to find x:

$$-3x-1=-7$$

$$x=2 \quad \textbf{(8)}$$

Observe that substituting **(6)** and **(8)** into both **(2)** and **(7)** yields true statements. As such, it remains to find y. To do so, substitute **(8)** into **(5)**:

$$y=-2(2)-1=-5$$

All equations are satisfied by the solution: $\boxed{x=2,\ y=-5,\ z=1}$.

61. Solve the system

$$\begin{cases} 3x_1 - 2x_2 + x_3 + 2x_4 = -2 & \textbf{(1)} \\ -x_1 + 3x_2 + 4x_3 + 3x_4 = 4 & \textbf{(2)} \\ x_1 + x_2 + x_3 + x_4 = 0 & \textbf{(3)} \\ 5x_1 + 3x_2 + x_3 + 2x_4 = -1 & \textbf{(4)} \end{cases}$$

Step 1: Obtain a 3×3 system:

Add **(2)** and **(3)**:

$$4x_2 + 5x_3 + 4x_4 = 4 \quad \textbf{(5)}$$

Multiply **(2)** by 3, and then add to **(1)**:

$$7x_2 + 13x_3 + 11x_4 = 10 \quad \textbf{(6)}$$

Multiply **(2)** by 5, and then add to **(4)**:

$$18x_2 + 21x_3 + 17x_4 = 19 \quad \textbf{(7)}$$

These steps yield the following system:

$$\begin{cases} 4x_2 + 5x_3 + 4x_4 = 4 & \textbf{(5)} \\ 7x_2 + 13x_3 + 11x_4 = 10 & \textbf{(6)} \\ 18x_2 + 21x_3 + 17x_4 = 19 & \textbf{(7)} \end{cases}$$

Step 2: Solve the system in Step 1.

Multiply **(5)** by -7, **(6)** by 4, and then add them: $17x_3 + 16x_4 = 12$ **(8)**

Multiply **(6)** by 18, **(7)** by -7, and then add them: $87x_3 + 79x_4 = 47$ **(9)**

These steps yield the following system:

$$\begin{cases} 17x_3 + 16x_4 = 12 & \textbf{(8)} \\ 87x_3 + 79x_4 = 47 & \textbf{(9)} \end{cases}$$

Solve this system:

Multiply **(8)** by -87, **(9)** by 17, and then add :

$$-49x_4 = -245$$

$$x_4 = 5 \quad \textbf{(10)}$$

Substitute **(10)** into **(8)**:

$$17x_3 + 16(5) = 12$$

$$x_3 = -4 \quad \textbf{(11)}$$

Now, substitute **(10)** and **(11)** into **(5)**:

$$4x_2 + 5(-4) + 4(5) = 4$$

$$x_2 = 1 \quad \textbf{(12)}$$

Finally, substitute **(10)** – **(12)** into **(3)**:

$$x_1 + 1 - 4 + 5 = 0$$

$$x_1 = -2$$

Thus, the solution is:

$$\boxed{x_1 = -2,\ x_2 = 1, x_3 = -4,\ x_4 = 5}.$$

62. Solve the system

$$\begin{cases} 5x_1 + 3x_2 + 8x_3 + x_4 = 1 & \textbf{(1)} \\ x_1 + 2x_2 + 5x_3 + 2x_4 = 3 & \textbf{(2)} \\ 4x_1 + x_3 - 2x_4 = -3 & \textbf{(3)} \\ x_2 + x_3 + x_4 = 0 & \textbf{(4)} \end{cases}$$

Step 1: Obtain a 3×3 system:

Multiply **(2)** by -5, and then add to **(1)**:

$$-7x_2 - 17x_3 - 9x_4 = -14 \quad \textbf{(5)}$$

Multiply **(2)** by -4, and then add to **(3)**:

$$-8x_2 - 19x_3 - 10x_4 = -15 \quad \textbf{(6)}$$

These steps yield the following system:

$$\begin{cases} x_2 + x_3 + x_4 = 0 & \textbf{(4)} \\ -7x_2 - 17x_3 - 9x_4 = -14 & \textbf{(5)} \\ -8x_2 - 19x_3 - 10x_4 = -15 & \textbf{(6)} \end{cases}$$

Step 2: Solve the system in Step 1.

Multiply **(4)** by 7, and then add to **(5)**:

$$-10x_3 - 2x_4 = -14 \quad \textbf{(7)}$$

Multiply **(4)** by 8, and then add to **(6)**:

$$-11x_3 - 2x_4 = -15 \quad \textbf{(8)}$$

These steps lead to the following system:

$$\begin{cases} -10x_3 - 2x_4 = -14 & \textbf{(7)} \\ -11x_3 - 2x_4 = -15 & \textbf{(8)} \end{cases}$$

Solve this system:

Multiply **(7)** by -1, and then add to **(8)**:

$$-x_3 = -1$$

$$x_3 = 1 \quad \textbf{(9)}$$

Substitute **(9)** into **(7)**:

$$-10 - 2x_4 = -14$$

$$x_4 = 2 \quad \textbf{(10)}$$

Now, substitute **(9)** and **(10)** into **(4)**:

$$x_2 + 1 + 2 = 0$$

$$x_2 = -3 \quad \textbf{(11)}$$

Finally, substitute **(9)** and **(10)** into **(3)**:

$$4x_1 + 1 - 2(2) = -3$$

$$x_1 = 0$$

Thus, the solution is:

$$\boxed{x_1 = 0,\ x_2 = -3,\ x_3 = 1,\ x_4 = 2}.$$

63. Consider the equation:

$$x^3 + x^2 + 2x + 3 = (A + C)x^3 + (B + D)x^2 + (3A + 2C)x + (3B + 2D)$$

This yields the system:

$$\begin{cases} A + C = 1 & \textbf{(1)} \\ B + D = 1 & \textbf{(2)} \\ 3A + 2C = 2 & \textbf{(3)} \\ 3B + 2D = 3 & \textbf{(4)} \end{cases}$$

Solve the system comprised of **(1) & (3)**:

Solve **(1)** for A: $A = 1 - C$ **(5)**

Substitute **(5)** into **(3)**: $C = 1$. Now, substitute this into **(5)** to get $A = 0$.

Solve the system comprised of **(1) & (4)**:

Solve **(2)** for B: $B = 1 - D$ **(6)**

Substitute **(6)** into **(4)**: $D = 0$. Now, substitute this into **(6)** to get $B = 1$.

Thus, $A = 0$, $B = 1$, $C = 1$, and $D = 0$.

64. Consider the equation:

$$Ax^4 + Ax^3 + Bx^3 + Bx^2 + Cx^2 + Cx + Dx + D + Ex^4 = 4x^4 + x + 1$$

This yields the system:

$$\begin{cases} A+E=4 & \textbf{(1)} \\ A+B=0 & \textbf{(2)} \\ B+C=0 & \textbf{(3)} \\ C+D=1 & \textbf{(4)} \\ D=1 & \textbf{(5)} \end{cases}$$

Substitute **(5)** into **(4)**: $C = 0$
Substitute this into **(3)**: $B = 0$
Substitute this into **(2)**: $A = 0$
Substitute this into **(1)**: $E = 4$.

Thus, $A = 0$, $B = 0$, $C = 0$, $D = 1$, and $E = 4$.

65. See the solution to #21.

66. See the solution to #22.

67. Write the system in the form:

$$\begin{cases} z = x - y - 10 \\ z = -2x + 3y - 11 \\ z = x - y - 10 \end{cases}$$

A graphical solution is given to the right: Notice that there are only two planes since the first and third equations of the system are the same. Hence, the line of intersection of these two planes constitutes the infinitely many solutions of the system. This is precisely what was found to be the case in Exercises 21 and 65.

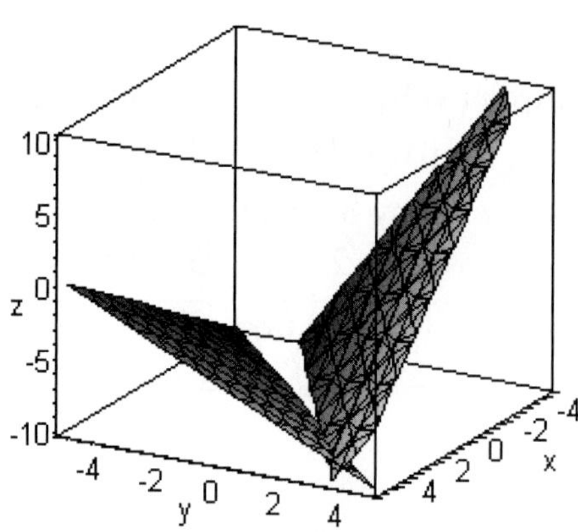

68. Write the system in the form:

$$\begin{cases} z=-2x-y-3 \\ z=x+2y \\ z=\frac{1}{2}(-x-y+5) \end{cases}$$

A graphical solution is given to the right:

In this case, there are three planes which intersect in the point (-5, 4, 3). It is difficult to see in the picture, but they all meet at this point.

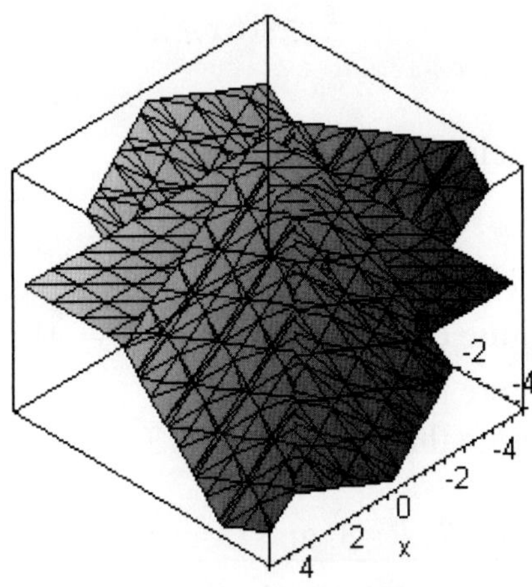

69. If your calculator has 3-dimensional graphing capabilities, then generate the graph of the system to obtain:

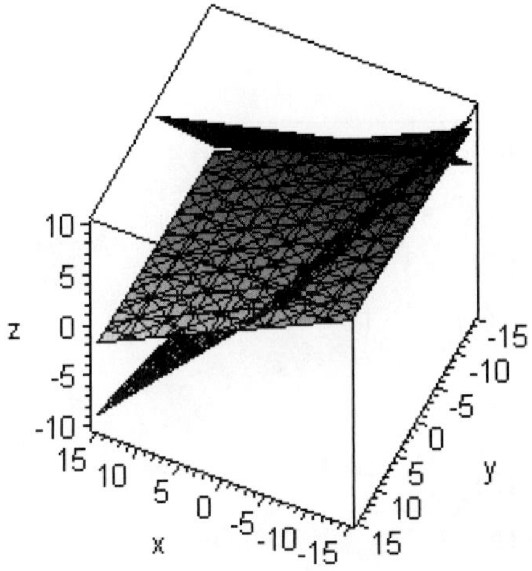

It's a bit difficult to tell from the graph, but the three planes intersect in a common point with coordinates $\left(-\frac{80}{7},-\frac{80}{7},\frac{48}{7}\right)$, which can be easily verified using your calculator.

70. If your calculator has 3-dimensional graphing capabilities, the generate the graph of the system to obtain:

The two planes intersect in a line, meaning there are infinitely many solutions to the system. This line is described by $x=a,\ y=\frac{30a-24}{5},\ z=\frac{3a-2}{6}$.

71. Consider the equation:

$$5x^2+6x+2=(A+B)x^2+(2A+2B+C)x+(5A+2C)$$

This yields the system: $\begin{cases} A+B=5 & \textbf{(1)} \\ 2A+2B+C=6 & \textbf{(2)} \\ 5A+2C=2 & \textbf{(3)} \end{cases}$

Solve **(1)** for B: $B=5-A$ **(4)**

Substitute **(4)** into **(2)**: $2A+2(5-A)+C=6 \Rightarrow 10+C=6 \Rightarrow C=-4$

Substitute this into **(3)**: $A=2$

Substitute this into **(4)**: $B=3$

Thus, $A=2$, $B=3$, and $C=-4$.

72. Consider the equation: $2x^2-3x+2=(A+B)x^2+Cx+A$

This yields the system: $\begin{cases} A+B=2 & \textbf{(1)} \\ C=-3 & \textbf{(2)} \\ A=2 & \textbf{(3)} \end{cases}$

Substitute **(3)** into **(1)**: $B=0$.

Thus, $A=2$, $B=0$, and $C=-3$.

73. Consider the equation: $3x+8=(A+B+C)x^2+(5A+3B+2C)x+6A$

This yields the system: $\begin{cases} A+B+C=0 & \textbf{(1)} \\ 5A+3B+2C=3 & \textbf{(2)} \\ 6A=8 & \textbf{(3)} \end{cases}$

Solve **(3)** for A: $A=\frac{4}{3}$ **(4)**

Substitute **(4)** into **(1)** and **(2)** to obtain the system:

$$\begin{cases} B+C=-\frac{4}{3} & \textbf{(5)} \\ 3B+2C=-\frac{11}{3} & \textbf{(6)} \end{cases}$$

Solve **(5)** for B: $B=-C-\frac{4}{3}$ **(7)**

Substitute **(7)** into **(6)**: $C=-\frac{1}{3}$

Substitute this into **(7)**: $B=-1$.

Thus, $A=\frac{4}{3}$, $B=-1$, and $C=-\frac{1}{3}$.

74. Consider the equation:

$$x^2+x+1=(A+B+C)x^2+(5A+4B+3C)x+(6A+3B+2C)$$

This yields the system:

$$\begin{cases} A+B+C=1 & \textbf{(1)} \\ 5A+4B+3C=1 & \textbf{(2)} \\ 6A+3B+2C=1 & \textbf{(3)} \end{cases}$$

Multiply **(1)** by -5 and add to **(2)**: $-B-2C=-4$ **(4)**

Multiply **(1)** by -6 and add to **(3)**: $-3B-4C=-5$ **(5)**

<u>Solve the system comprised of **(4) & (5)**</u>:

Multiply **(4)** by -3 and add to **(5)**: $C=\frac{7}{2}$

Substitute this into **(4)**: $B=-3$

Substitute this into **(1)**: $A=\frac{1}{2}$

Thus, $A=\frac{1}{2}$, $B=-3$, and $C=\frac{7}{2}$.

Section 8.3 Solutions --

1. 2×3	**2.** 3×2
3. 1×4	**4.** 4×1
5. 1×1	**6.** 4×4
7. $\left[\begin{array}{cc\|c} 3 & -2 & 7 \\ -4 & 6 & -3 \end{array}\right]$	**8.** $\left[\begin{array}{cc\|c} -1 & 1 & 2 \\ 1 & -1 & -4 \end{array}\right]$
9. $\left[\begin{array}{ccc\|c} 2 & -3 & 4 & -3 \\ -1 & 1 & 2 & 1 \\ 5 & -2 & -3 & 7 \end{array}\right]$	**10.** $\left[\begin{array}{ccc\|c} 1 & -2 & 1 & 0 \\ -2 & 1 & -1 & -5 \\ 13 & 7 & 5 & 6 \end{array}\right]$
11. $\left[\begin{array}{ccc\|c} 1 & 1 & 0 & 3 \\ 1 & 0 & -1 & 2 \\ 0 & 1 & 1 & 5 \end{array}\right]$	**12.** $\left[\begin{array}{ccc\|c} 1 & -1 & 0 & -4 \\ 0 & 1 & 1 & 3 \end{array}\right]$
13. $\left[\begin{array}{ccc\|c} -4 & 3 & 5 & 2 \\ 2 & -3 & -2 & -3 \\ -2 & 4 & 3 & 1 \end{array}\right]$	**14.** $\left[\begin{array}{ccc\|c} -1 & 2 & 1 & 5 \\ 2 & -2 & 3 & 0 \\ -4 & 1 & -2 & 3 \end{array}\right]$

15. $\begin{cases} -3x+7y=2 \\ x+5y=8 \end{cases}$	**16.** $\begin{cases} -x+2y+4z=4 \\ 7x+9y+3z=-3 \\ 4x+6y-5z=8 \end{cases}$
17. $\begin{cases} -x=4 \\ 7x+9y+3z=-3 \\ 4x+6y-5z=8 \end{cases}$	**18.** $\begin{cases} 2x+3y-4z=6 \\ 7x-y+5z=9 \end{cases}$
19. $\begin{cases} x=a \\ y=b \end{cases}$	**20.** $\begin{cases} 3x+5z=1 \\ -4y+7z=-3 \\ 2x-y=8 \end{cases}$
21. Not in row echelon form since Condition 3 is violated.	**22.** Not in row echelon form since Condition 3 is violated.
23. In reduced form.	**24.** In reduced form.
25. Not in row echelon form since Condition 4 is violated.	**26.** In reduced form.
27. In reduced form.	**28.** Not in row echelon form since Condition 2 is violated.
29. In reduced form.	**30.** In reduced form.
31. $\left[\begin{array}{cc\|c} 1 & -2 & -3 \\ 0 & 7 & 5 \end{array}\right]$	**32.** $\left[\begin{array}{cc\|c} 1 & 2 & 5 \\ 2 & -3 & -4 \end{array}\right]$
33. $\left[\begin{array}{ccc\|c} 1 & -2 & -1 & 3 \\ 0 & 5 & -1 & 0 \\ 3 & -2 & 5 & 8 \end{array}\right]$	**34.** $\left[\begin{array}{ccc\|c} 1 & -2 & 1 & 3 \\ 0 & 1 & -2 & 6 \\ 0 & -6 & 2 & 4 \end{array}\right]$
35. $\left[\begin{array}{cccc\|c} 1 & -2 & 5 & -1 & 2 \\ 0 & 1 & 1 & -3 & 3 \\ 0 & -2 & 1 & -2 & 5 \\ 0 & 0 & 1 & -1 & -6 \end{array}\right]$	**36.** $\left[\begin{array}{cccc\|c} 1 & 0 & 5 & -10 & 15 \\ 0 & 1 & 2 & -3 & 4 \\ 0 & 0 & \frac{7}{2} & -3 & \frac{9}{2} \\ 0 & 0 & 1 & -1 & -3 \end{array}\right]$
37. $\left[\begin{array}{cccc\|c} 1 & 0 & 5 & -10 & -5 \\ 0 & 1 & 2 & -3 & -2 \\ 0 & 0 & -7 & 6 & 3 \\ 0 & 0 & 8 & -10 & -9 \end{array}\right]$	**38.** $\left[\begin{array}{cccc\|c} 1 & 0 & 0 & 0 & 1 \\ 0 & 1 & 0 & 0 & -2 \\ 0 & 0 & 1 & 0 & 0 \\ 0 & 0 & 0 & 1 & -3 \end{array}\right]$

39.
$$\left[\begin{array}{cccc|c} 1 & 0 & 4 & 0 & 27 \\ 0 & 1 & 2 & 0 & -11 \\ 0 & 0 & 1 & 0 & 21 \\ 0 & 0 & 0 & 1 & -3 \end{array}\right]$$

40.
$$\left[\begin{array}{cccc|c} 1 & 0 & -1 & 0 & -3 \\ 0 & 1 & 2 & 0 & -8 \\ 0 & 0 & 1 & 0 & 4 \\ 0 & 0 & 0 & 1 & 1 \end{array}\right]$$

41.
$$\left[\begin{array}{cc|c} 1 & 2 & 4 \\ 2 & 3 & 2 \end{array}\right] \xrightarrow{R_1 \leftrightarrow R_2} \left[\begin{array}{cc|c} 2 & 3 & 2 \\ 1 & 2 & 4 \end{array}\right]$$
$$\xrightarrow{R_1 - 2R_2 \to R_2} \left[\begin{array}{cc|c} 2 & 3 & 2 \\ 0 & -1 & -6 \end{array}\right]$$
$$\xrightarrow{-R_2 \to R_2} \left[\begin{array}{cc|c} 2 & 3 & 2 \\ 0 & 1 & 6 \end{array}\right]$$
$$\xrightarrow{\frac{1}{2}R_1 \to R_1} \left[\begin{array}{cc|c} 1 & \frac{3}{2} & 1 \\ 0 & 1 & 6 \end{array}\right]$$
$$\xrightarrow{R_1 - \frac{3}{2}R_2 \to R_1} \left[\begin{array}{cc|c} 1 & 0 & -8 \\ 0 & 1 & 6 \end{array}\right]$$

42.
$$\left[\begin{array}{cc|c} 1 & -1 & 3 \\ -3 & 2 & 2 \end{array}\right] \xrightarrow{R_1 \leftrightarrow R_2} \left[\begin{array}{cc|c} -3 & 2 & 2 \\ 1 & -1 & 3 \end{array}\right]$$
$$\xrightarrow{R_1 + 3R_2 \to R_2} \left[\begin{array}{cc|c} -3 & 2 & 2 \\ 0 & -1 & 11 \end{array}\right]$$
$$\xrightarrow{-R_2 \to R_2} \left[\begin{array}{cc|c} -3 & 2 & 2 \\ 0 & 1 & -11 \end{array}\right]$$
$$\xrightarrow{-\frac{1}{3}R_1 \to R_1} \left[\begin{array}{cc|c} 1 & -\frac{2}{3} & -\frac{2}{3} \\ 0 & 1 & -11 \end{array}\right]$$
$$\xrightarrow{R_1 + \frac{2}{3}R_2 \to R_1} \left[\begin{array}{cc|c} 1 & 0 & -8 \\ 0 & 1 & -11 \end{array}\right]$$

43.
$$\left[\begin{array}{ccc|c} 1 & -1 & 1 & -1 \\ 0 & 1 & -1 & -1 \\ -1 & 1 & 1 & 1 \end{array}\right] \xrightarrow{R_1 + R_3 \to R_3} \left[\begin{array}{ccc|c} 1 & -1 & 1 & -1 \\ 0 & 1 & -1 & -1 \\ 0 & 0 & 2 & 0 \end{array}\right]$$
$$\xrightarrow{\frac{1}{2}R_3 \to R_3} \left[\begin{array}{ccc|c} 1 & -1 & 1 & -1 \\ 0 & 1 & -1 & -1 \\ 0 & 0 & 1 & 0 \end{array}\right] \xrightarrow[R_1 + R_2 \to R_1]{R_2 + R_3 \to R_2} \left[\begin{array}{ccc|c} 1 & 0 & 0 & -2 \\ 0 & 1 & 0 & -1 \\ 0 & 0 & 1 & 0 \end{array}\right]$$

44.
$$\left[\begin{array}{ccc|c} 0 & -1 & 1 & 1 \\ 1 & -1 & 1 & -1 \\ 1 & -1 & -1 & -1 \end{array}\right] \xrightarrow{R_2 - R_3 \to R_3} \left[\begin{array}{ccc|c} 0 & -1 & 1 & 1 \\ 1 & -1 & 1 & -1 \\ 0 & 0 & 2 & 0 \end{array}\right] \xrightarrow{R_1 \leftrightarrow R_2} \left[\begin{array}{ccc|c} 1 & -1 & 1 & -1 \\ 0 & -1 & 1 & -1 \\ 0 & 0 & 2 & 0 \end{array}\right]$$
$$\xrightarrow[\frac{1}{2}R_3 \to R_3]{-R_2 \to R_2} \left[\begin{array}{ccc|c} 1 & -1 & 1 & -1 \\ 0 & 1 & -1 & 1 \\ 0 & 0 & 1 & 0 \end{array}\right] \xrightarrow[R_2 + R_3 \to R_2]{R_1 + R_2 + R_3 \to R_1} \left[\begin{array}{ccc|c} 1 & 0 & 0 & 0 \\ 0 & 1 & 0 & 1 \\ 0 & 0 & 1 & 0 \end{array}\right]$$

45.

$$\left[\begin{array}{ccc|c} 3 & -2 & -3 & -1 \\ 1 & -1 & 1 & -4 \\ 2 & 3 & 5 & 14 \end{array}\right] \xrightarrow[R_3-2R_2\to R_3]{R_1-3R_2\to R_1} \left[\begin{array}{ccc|c} 0 & 1 & -6 & 11 \\ 1 & -1 & 1 & -4 \\ 0 & 5 & 3 & 22 \end{array}\right] \xrightarrow{R_1\leftrightarrow R_2} \left[\begin{array}{ccc|c} 1 & -1 & 1 & -4 \\ 0 & 1 & -6 & 11 \\ 0 & 5 & 3 & 22 \end{array}\right]$$

$$\xrightarrow{R_3-5R_2\to R_3} \left[\begin{array}{ccc|c} 1 & -1 & 1 & -4 \\ 0 & 1 & -6 & 11 \\ 0 & 0 & 33 & -33 \end{array}\right] \xrightarrow{\frac{1}{33}R_3\to R_3} \left[\begin{array}{ccc|c} 1 & -1 & 1 & -4 \\ 0 & 1 & -6 & 11 \\ 0 & 0 & 1 & -1 \end{array}\right]$$

$$\xrightarrow{R_2+6R_3\to R_2} \left[\begin{array}{ccc|c} 1 & -1 & 1 & -4 \\ 0 & 1 & 0 & 5 \\ 0 & 0 & 1 & -1 \end{array}\right] \xrightarrow{R_1+R_2-R_3\to R_1} \left[\begin{array}{ccc|c} 1 & 0 & 0 & 2 \\ 0 & 1 & 0 & 5 \\ 0 & 0 & 1 & -1 \end{array}\right]$$

46.

$$\left[\begin{array}{ccc|c} 3 & -1 & 1 & 2 \\ 1 & -2 & 3 & 1 \\ 2 & 1 & -3 & -1 \end{array}\right] \xrightarrow{R_1-3R_2\to R_1} \left[\begin{array}{ccc|c} 0 & 5 & -8 & -1 \\ 1 & -2 & 3 & 1 \\ 2 & 1 & -3 & -1 \end{array}\right] \xrightarrow{R_3-2R_2\to R_3} \left[\begin{array}{ccc|c} 0 & 5 & -8 & -1 \\ 1 & -2 & 3 & 1 \\ 0 & 5 & -9 & -3 \end{array}\right]$$

$$\xrightarrow{R_1\leftrightarrow R_2} \left[\begin{array}{ccc|c} 1 & -2 & 3 & 1 \\ 0 & 5 & -8 & -1 \\ 0 & 5 & -9 & -3 \end{array}\right] \xrightarrow{R_3-R_2\to R_3} \left[\begin{array}{ccc|c} 1 & -2 & 3 & 1 \\ 0 & 5 & -8 & -1 \\ 0 & 0 & -1 & -2 \end{array}\right]$$

$$\xrightarrow[-R_3\to R_3]{\frac{1}{5}R_2\to R_2} \left[\begin{array}{ccc|c} 1 & -2 & 3 & 1 \\ 0 & 1 & -\frac{8}{5} & -\frac{1}{5} \\ 0 & 0 & 1 & 2 \end{array}\right] \xrightarrow{R_2+\frac{8}{5}R_3\to R_2} \left[\begin{array}{ccc|c} 1 & -2 & 3 & 1 \\ 0 & 1 & 0 & 3 \\ 0 & 0 & 1 & 2 \end{array}\right]$$

$$\xrightarrow{R_1+2R_2-3R_3\to R_1} \left[\begin{array}{ccc|c} 1 & 0 & 0 & 1 \\ 0 & 1 & 0 & 3 \\ 0 & 0 & 1 & 2 \end{array}\right]$$

47.

$$\left[\begin{array}{ccc|c} 2 & 1 & -6 & 4 \\ 1 & -2 & 2 & -3 \end{array}\right] \xrightarrow{R_1-2R_2\to R_2} \left[\begin{array}{ccc|c} 2 & 1 & -6 & 4 \\ 0 & 5 & -10 & 10 \end{array}\right]$$

$$\xrightarrow[\frac{1}{5}R_2\to R_2]{\frac{1}{2}R_1\to R_1} \left[\begin{array}{ccc|c} 1 & \frac{1}{2} & -3 & 2 \\ 0 & 1 & -2 & 2 \end{array}\right] \xrightarrow{R_1-\frac{1}{2}R_2\to R_1} \left[\begin{array}{ccc|c} 1 & 0 & -2 & 1 \\ 0 & 1 & -2 & 2 \end{array}\right]$$

48.

$$\begin{bmatrix} -3 & -1 & 2 & | & -1 \\ -1 & -2 & 1 & | & -3 \end{bmatrix} \xrightarrow{R_1-3R_2\to R_1} \begin{bmatrix} 0 & 5 & -1 & | & 8 \\ -1 & -2 & 1 & | & -3 \end{bmatrix} \xrightarrow[\frac{1}{5}R_1\to R_1]{-R_2\to R_2} \begin{bmatrix} 0 & 1 & -\frac{1}{5} & | & \frac{8}{5} \\ 1 & 2 & -1 & | & 3 \end{bmatrix}$$

$$\xrightarrow{R_1\leftrightarrow R_2} \begin{bmatrix} 1 & 2 & -1 & | & 3 \\ 0 & 1 & -\frac{1}{5} & | & \frac{8}{5} \end{bmatrix} \xrightarrow{R_1-2R_2\to R_1} \begin{bmatrix} 1 & 0 & -\frac{3}{5} & | & -\frac{1}{5} \\ 0 & 1 & -\frac{1}{5} & | & \frac{8}{5} \end{bmatrix}$$

49.

$$\begin{bmatrix} -1 & 2 & 1 & | & -2 \\ 3 & -2 & 1 & | & 4 \\ 2 & -4 & -2 & | & 4 \end{bmatrix} \xrightarrow{R_2+3R_1\to R_2} \begin{bmatrix} -1 & 2 & 1 & | & -2 \\ 0 & 4 & 4 & | & -2 \\ 2 & -4 & -2 & | & 4 \end{bmatrix} \xrightarrow{R_3+2R_1\to R_3} \begin{bmatrix} -1 & 2 & 1 & | & -2 \\ 0 & 4 & 4 & | & -2 \\ 0 & 0 & 0 & | & 0 \end{bmatrix}$$

$$\xrightarrow[\frac{1}{4}R_2\to R_2]{-R_1\to R_1} \begin{bmatrix} 1 & -2 & -1 & | & 2 \\ 0 & 1 & 1 & | & -\frac{1}{2} \\ 0 & 0 & 0 & | & 0 \end{bmatrix} \xrightarrow{R_1+2R_2\to R_1} \begin{bmatrix} 1 & 0 & 1 & | & 1 \\ 0 & 1 & 1 & | & -\frac{1}{2} \\ 0 & 0 & 0 & | & 0 \end{bmatrix}$$

50.

$$\begin{bmatrix} 2 & -1 & 0 & | & 1 \\ -1 & 0 & 1 & | & -2 \\ -2 & 1 & 0 & | & -1 \end{bmatrix} \xrightarrow{R_1+2R_2\to R_1} \begin{bmatrix} 0 & -1 & 2 & | & -3 \\ -1 & 0 & 1 & | & -2 \\ -2 & 1 & 0 & | & -1 \end{bmatrix} \xrightarrow{R_3-2R_2\to R_3} \begin{bmatrix} 0 & -1 & 2 & | & -3 \\ -1 & 0 & 1 & | & -2 \\ 0 & 1 & -2 & | & 3 \end{bmatrix}$$

$$\xrightarrow{R_1+R_3\to R_3} \begin{bmatrix} 0 & -1 & 2 & | & -3 \\ -1 & 0 & 1 & | & -2 \\ 0 & 0 & 0 & | & 0 \end{bmatrix} \xrightarrow{R_1\leftrightarrow R_2} \begin{bmatrix} -1 & 0 & 1 & | & -2 \\ 0 & -1 & 2 & | & -3 \\ 0 & 0 & 0 & | & 0 \end{bmatrix}$$

$$\xrightarrow[-R_2\to R_2]{-R_1\to R_1} \begin{bmatrix} 1 & 0 & -1 & | & 2 \\ 0 & 1 & -2 & | & 3 \\ 0 & 0 & 0 & | & 0 \end{bmatrix}$$

51. First, reduce the augmented matrix down to row echelon form:

$$\begin{bmatrix} 2 & 3 & | & 1 \\ 1 & 1 & | & -2 \end{bmatrix} \xrightarrow{R_1-2R_2\to R_2} \begin{bmatrix} 2 & 3 & | & 1 \\ 0 & 1 & | & 5 \end{bmatrix} \xrightarrow{\frac{1}{2}R_1\to R_1} \begin{bmatrix} 1 & \frac{3}{2} & | & \frac{1}{2} \\ 0 & 1 & | & 5 \end{bmatrix}$$

Now, from this we see that $y=5$, and then substituting this value into the equation obtained from the first row yields:

$$x+\tfrac{3}{2}(5)=\tfrac{1}{2}$$
$$x=-7$$

Hence, the solution is $\boxed{x=-7,\ y=5}$.

52. First, reduce the augmented matrix down to row echelon form:

$$\left[\begin{array}{cc|c} 3 & 2 & 11 \\ 1 & -1 & 12 \end{array}\right] \xrightarrow{R_1-3R_2\to R_2} \left[\begin{array}{cc|c} 3 & 2 & 11 \\ 0 & 5 & -25 \end{array}\right] \xrightarrow[\frac{1}{5}R_2\to R_2]{\frac{1}{3}R_1\to R_1} \left[\begin{array}{cc|c} 1 & \frac{2}{3} & \frac{11}{3} \\ 0 & 1 & -5 \end{array}\right]$$

Now, from this we see that $y=-5$, and then substituting this value into the equation obtained from the first row yields:

$$x+\tfrac{2}{3}(-5)=\tfrac{11}{3}$$
$$x=7$$

Hence, the solution is $\boxed{x=7,\ y=-5}$.

53. First, reduce the augmented matrix down to row echelon form:

$$\left[\begin{array}{cc|c} -1 & 2 & 3 \\ 2 & -4 & -6 \end{array}\right] \xrightarrow{2R_1+R_2\to R_2} \left[\begin{array}{cc|c} -1 & 2 & 3 \\ 0 & 0 & 0 \end{array}\right]$$

From the bottom row, we conclude that there are infinitely many solutions. To get the precise form of these solutions, let $y=a$. Then, substituting this value into the equation obtained from the first row yields:

$$-x+2a=3$$
$$x=2a-3$$

Hence, the solution is $\boxed{x=2a-3,\ y=a}$.

54. First, reduce the augmented matrix down to row echelon form:

$$\left[\begin{array}{cc|c} 3 & -1 & -1 \\ 6 & 2 & 2 \end{array}\right] \xrightarrow{2R_1-R_2\to R_2} \left[\begin{array}{cc|c} 3 & -1 & -1 \\ 0 & -4 & -4 \end{array}\right] \xrightarrow[-\frac{1}{4}R_2\to R_2]{\frac{1}{3}R_1\to R_1} \left[\begin{array}{cc|c} 1 & -\frac{1}{3} & -\frac{1}{3} \\ 0 & 1 & 1 \end{array}\right]$$

Now, from this we see that $y=1$, and then substituting this value into the equation obtained from the first row yields:

$$x-\tfrac{1}{3}(1)=-\tfrac{1}{3}$$
$$x=0$$

Hence, the solution is $\boxed{x=0,\ y=1}$.

55. First, reduce the augmented matrix down to row echelon form:

$$\left[\begin{array}{cc|c} \frac{2}{3} & \frac{1}{3} & \frac{8}{9} \\ \frac{1}{2} & \frac{1}{4} & \frac{3}{4} \end{array}\right] \xrightarrow[2R_2\to R_2]{3R_1\to R_1} \left[\begin{array}{cc|c} 2 & 1 & \frac{8}{3} \\ 1 & \frac{1}{2} & \frac{3}{2} \end{array}\right] \xrightarrow{R_1-2R_2\to R_2} \left[\begin{array}{cc|c} 2 & 1 & \frac{8}{3} \\ 0 & 0 & -\frac{1}{3} \end{array}\right]$$

From the second row, we see that $0=-\frac{1}{3}$, which means that the system has $\boxed{\text{no solution}}$.

56. First, reduce the augmented matrix down to row echelon form:

$$\left[\begin{array}{cc|c} 0.4 & -0.5 & 2.08 \\ -0.3 & 0.7 & 1.88 \end{array}\right] \xrightarrow[10R_2 \to R_2]{10R_1 \to R_1} \left[\begin{array}{cc|c} 4 & -5 & 20.8 \\ -3 & 7 & 18.8 \end{array}\right] \xrightarrow{\frac{1}{4}R_1 \to R_1} \left[\begin{array}{cc|c} 1 & -1.25 & 5.2 \\ -3 & 7 & 18.8 \end{array}\right]$$

$$\xrightarrow{3R_1 + R_2 \to R_2} \left[\begin{array}{cc|c} 1 & -1.25 & 5.2 \\ 0 & 3.25 & 34.4 \end{array}\right] \xrightarrow{\frac{1}{3.25}R_2 \to R_2} \left[\begin{array}{cc|c} 1 & -1.25 & 5.2 \\ 0 & 1 & 10.58 \end{array}\right]$$

Now, from this we see that $y \cong 10.58$, and then substituting this value into the equation obtained from the first row yields:

$$x - 1.25(10.58) = 5.2$$

$$x \cong 18.425$$

Hence, the solution is $\boxed{x = 18.425,\ y = 10.58}$.

57. First, rewrite the system as

$$\begin{cases} x - y - z = 10 \\ 2x - 3y + z = -11 \\ -x + y + z = -10 \end{cases}$$

Now, reduce the augmented matrix down to row echelon form:

$$\left[\begin{array}{ccc|c} 1 & -1 & -1 & 10 \\ 2 & -3 & 1 & -11 \\ -1 & 1 & 1 & -10 \end{array}\right] \xrightarrow[R_2 - 2R_1 \to R_2]{R_1 + R_3 \to R_3} \left[\begin{array}{ccc|c} 1 & -1 & -1 & 10 \\ 0 & -1 & 3 & -31 \\ 0 & 0 & 0 & 0 \end{array}\right] \xrightarrow{-R_2 \to R_2} \left[\begin{array}{ccc|c} 1 & -1 & -1 & 10 \\ 0 & 1 & -3 & 31 \\ 0 & 0 & 0 & 0 \end{array}\right]$$

From the last row, we conclude that the system has infinitely many solutions. To find them, let $z = a$. Then, the second row implies that $y - 3a = 31$ so that $y = 31 + 3a$. So, substituting these values of y and z into the first row yields $x - (3a + 31) - a = 10$, so that $x = 4a + 41$. Hence, the solutions are $\boxed{x = 4a + 41,\ y = 31 + 3a,\ z = a}$.

58. First, rewrite the system as

$$\begin{cases} 2x+y+z=-3 \\ x+2y-z=0 \\ x+y+2z=5 \end{cases}$$

Now, reduce the augmented matrix down to row echelon form:

$$\left[\begin{array}{ccc|c} 2 & 1 & 1 & -3 \\ 1 & 2 & -1 & 0 \\ 1 & 1 & 2 & 5 \end{array}\right] \xrightarrow[R_1-2R_2\to R_2]{R_2-R_3\to R_3} \left[\begin{array}{ccc|c} 2 & 1 & 1 & -3 \\ 0 & -3 & 3 & -3 \\ 0 & 1 & -3 & -5 \end{array}\right] \xrightarrow{\frac{1}{3}R_2\to R_2} \left[\begin{array}{ccc|c} 2 & 1 & 1 & -3 \\ 0 & -1 & 1 & -1 \\ 0 & 1 & -3 & -5 \end{array}\right]$$

$$\xrightarrow{R_2+R_3\to R_3} \left[\begin{array}{ccc|c} 2 & 1 & 1 & -3 \\ 0 & -1 & 1 & -1 \\ 0 & 0 & -2 & -6 \end{array}\right] \xrightarrow[-\frac{1}{2}R_3\to R_3]{\frac{1}{2}R_1\to R_1,\ -R_2\to R_2} \left[\begin{array}{ccc|c} 1 & \frac{1}{2} & \frac{1}{2} & \frac{-3}{2} \\ 0 & 1 & -1 & 1 \\ 0 & 0 & 1 & 3 \end{array}\right]$$

From the last row, we conclude that $z=3$. Then, the second row implies that $y-3=1$ so that $y=4$. So, substituting these values of y and z into the first row yields $x+\frac{1}{2}(4)+\frac{1}{2}(3)=-\frac{3}{2}$, so that $x=-5$. Hence, the solution is $\boxed{x=-5,\ y=4,\ z=3}$.

59. Reduce the augmented matrix down to row echelon form:

$$\left[\begin{array}{ccc|c} 3 & 1 & -1 & 1 \\ 1 & -1 & 1 & -3 \\ 2 & 1 & 1 & 0 \end{array}\right] \xrightarrow[R_1-3R_2\to R_2]{R_3-2R_2\to R_3} \left[\begin{array}{ccc|c} 3 & 1 & -1 & 1 \\ 0 & 4 & -4 & 10 \\ 0 & 3 & -1 & 6 \end{array}\right] \xrightarrow[\frac{1}{3}R_3\to R_3]{\frac{1}{3}R_1\to R_1,\ \frac{1}{4}R_2\to R_2} \left[\begin{array}{ccc|c} 1 & \frac{1}{3} & -\frac{1}{3} & \frac{1}{3} \\ 0 & 1 & -1 & \frac{5}{2} \\ 0 & 1 & -\frac{1}{3} & 2 \end{array}\right]$$

$$\xrightarrow{R_2-R_3\to R_3} \left[\begin{array}{ccc|c} 1 & \frac{1}{3} & -\frac{1}{3} & \frac{1}{3} \\ 0 & 1 & -1 & \frac{5}{2} \\ 0 & 0 & -\frac{2}{3} & \frac{1}{2} \end{array}\right] \xrightarrow{-\frac{3}{2}R_3\to R_3} \left[\begin{array}{ccc|c} 1 & \frac{1}{3} & -\frac{1}{3} & \frac{1}{3} \\ 0 & 1 & -1 & \frac{5}{2} \\ 0 & 0 & 1 & -\frac{3}{4} \end{array}\right]$$

From the last row, we conclude that $x_3=-\frac{3}{4}$. Then, the second row implies that $x_2-(-\frac{3}{4})=\frac{5}{2}$ so that $x_2=\frac{7}{4}$. So, substituting these values of x_2 and x_3 into the first row yields $x_1+\frac{1}{3}(\frac{7}{4})-\frac{1}{3}(-\frac{3}{4})=\frac{1}{3}$, so that $x_1=-\frac{1}{2}$. Hence, the solution is

$\boxed{x_1=-\frac{1}{2},\ x_2=\frac{7}{4},\ x_3=-\frac{3}{4}}$.

60. Reduce the augmented matrix down to row echelon form:

$$\left[\begin{array}{ccc|c} 2 & 1 & 1 & -1 \\ 1 & 1 & -1 & 5 \\ 3 & -1 & -1 & 1 \end{array}\right] \xrightarrow[R_1-2R_2\to R_2]{R_3-3R_2\to R_3} \left[\begin{array}{ccc|c} 2 & 1 & 1 & -1 \\ 0 & -1 & 3 & -11 \\ 0 & -4 & 2 & -14 \end{array}\right] \xrightarrow{R_3-4R_2\to R_3} \left[\begin{array}{ccc|c} 2 & 1 & 1 & -1 \\ 0 & -1 & 3 & -11 \\ 0 & 0 & -10 & 30 \end{array}\right]$$

$$\xrightarrow[\substack{-R_2\to R_2 \\ -\frac{1}{10}R_3\to R_3}]{\frac{1}{2}R_1\to R_1} \left[\begin{array}{ccc|c} 1 & \frac{1}{2} & \frac{1}{2} & -\frac{1}{2} \\ 0 & 1 & -3 & 11 \\ 0 & 0 & 1 & -3 \end{array}\right]$$

From the last row, we conclude that $z=-3$. Then, the second row implies that $y-3(-3)=11$ so that $y=2$. So, substituting these values of y and z into the first row yields $x+\frac{1}{2}(2)+\frac{1}{2}(-3)=-\frac{1}{2}$, so that $x=0$. Hence, the solution is $\boxed{x=0,\ y=2,\ z=-3}$.

61. Reduce the augmented matrix down to row echelon form:

$$\left[\begin{array}{ccc|c} 2 & 5 & 0 & 9 \\ 1 & 2 & -1 & 3 \\ -3 & -4 & 7 & 1 \end{array}\right] \xrightarrow[R_1-2R_2\to R_2]{R_3+3R_2\to R_3} \left[\begin{array}{ccc|c} 2 & 5 & 0 & 9 \\ 0 & 1 & 2 & 3 \\ 0 & 2 & 4 & 10 \end{array}\right] \xrightarrow{R_3-2R_2\to R_3} \left[\begin{array}{ccc|c} 2 & 5 & 0 & 9 \\ 0 & 1 & 2 & 3 \\ 0 & 0 & 0 & 4 \end{array}\right]$$

$$\xrightarrow{\frac{1}{2}R_1\to R_1} \left[\begin{array}{ccc|c} 1 & \frac{5}{2} & 0 & \frac{9}{2} \\ 0 & 1 & 2 & 3 \\ 0 & 0 & 0 & 4 \end{array}\right]$$

From the last row, we have the false statement 0 = 4, so that we can conclude the system has $\boxed{\text{no solution}}$.

62. Reduce the augmented matrix down to row echelon form:

$$\left[\begin{array}{ccc|c} 1 & -2 & 3 & 1 \\ -2 & 7 & -9 & 4 \\ 1 & 0 & 1 & 9 \end{array}\right] \xrightarrow[R_1-R_3\to R_3]{R_2+2R_3\to R_2} \left[\begin{array}{ccc|c} 1 & -2 & 3 & 1 \\ 0 & 7 & -7 & 22 \\ 0 & -2 & 2 & -8 \end{array}\right] \xrightarrow{\frac{1}{7}R_2\to R_2} \left[\begin{array}{ccc|c} 1 & -2 & 3 & 1 \\ 0 & 1 & -1 & \frac{22}{7} \\ 0 & -2 & 2 & -8 \end{array}\right]$$

$$\xrightarrow{R_3+2R_2\to R_3} \left[\begin{array}{ccc|c} 1 & -2 & 3 & 1 \\ 0 & 1 & -1 & \frac{22}{7} \\ 0 & 0 & 0 & -\frac{12}{7} \end{array}\right]$$

From the last row, we have the false statement $0=-\frac{12}{7}$, so that we can conclude the system has $\boxed{\text{no solution}}$.

63. Reduce the augmented matrix down to row echelon form:

$$\left[\begin{array}{ccc|c} 2 & -1 & 1 & 3 \\ 1 & -1 & 1 & 2 \\ -2 & 2 & -2 & -4 \end{array}\right] \xrightarrow[2R_2+R_3\to R_3]{\frac{1}{2}R_1\to R_1} \left[\begin{array}{ccc|c} 1 & -\frac{1}{2} & \frac{1}{2} & \frac{3}{2} \\ 0 & 1 & -1 & -1 \\ 0 & 0 & 0 & 0 \end{array}\right]$$

From the last row, we conclude that the system has infinitely many solutions. To find them, let $x_3 = a$. Then, the second row implies that $x_2 - a = -1$ so that $x_2 = a-1$. So, substituting these values of x_2 and x_3 into the first row yields $x_1 - \frac{1}{2}(a-1) + \frac{1}{2}a = \frac{3}{2}$, so that $x_1 = 1$. Hence, the solutions are $\boxed{x_1 = 1,\ x_2 = a-1,\ x_3 = a}$.

64. Reduce the augmented matrix down to row echelon form:

$$\left[\begin{array}{ccc|c} 1 & -1 & -2 & 0 \\ -2 & 5 & 10 & -3 \\ 3 & 1 & 0 & 0 \end{array}\right] \xrightarrow[R_2+2R_1\to R_2]{R_3-3R_1\to R_3} \left[\begin{array}{ccc|c} 1 & -1 & -2 & 0 \\ 0 & 3 & 6 & -3 \\ 0 & 4 & 6 & 0 \end{array}\right] \xrightarrow[\frac{1}{4}R_3\to R_3]{\frac{1}{3}R_2\to R_2} \left[\begin{array}{ccc|c} 1 & -1 & -2 & 0 \\ 0 & 1 & 2 & -1 \\ 0 & 1 & \frac{3}{2} & 0 \end{array}\right]$$

$$\xrightarrow{R_2-R_3\to R_3} \left[\begin{array}{ccc|c} 1 & -1 & -2 & 0 \\ 0 & 1 & 2 & -1 \\ 0 & 0 & \frac{1}{2} & -1 \end{array}\right] \xrightarrow{2R_3\to R_3} \left[\begin{array}{ccc|c} 1 & -1 & -2 & 0 \\ 0 & 1 & 2 & -1 \\ 0 & 0 & 1 & -2 \end{array}\right]$$

From the last row, we conclude that $x_3 = -2$. Then, the second row implies that $x_2 + 2(-2) = -1$ so that $x_2 = 3$. So, substituting these values of x_2 and x_3 into the first row yields $x_1 - 3 - 2(-2) = 0$, so that $x_1 = -1$. Hence, the solution is $\boxed{x_1 = -1,\ x_2 = 3,\ x_3 = -2}$.

65. Reduce the augmented matrix down to row echelon form:

$$\left[\begin{array}{ccc|c} 2 & 1 & -1 & 2 \\ 1 & -1 & -1 & 6 \end{array}\right] \xrightarrow{R_1-2R_2\to R_2} \left[\begin{array}{ccc|c} 2 & 1 & -1 & 2 \\ 0 & 3 & 1 & -10 \end{array}\right] \xrightarrow[\frac{1}{3}R_2\to R_2]{\frac{1}{2}R_1\to R_1} \left[\begin{array}{ccc|c} 1 & \frac{1}{2} & -\frac{1}{2} & 1 \\ 0 & 1 & \frac{1}{3} & -\frac{10}{3} \end{array}\right]$$

Since there are two rows and three unknowns, we know from the above calculation that there are infinitely many solutions to this system. (Note: The only other possibility in such case would be that there was no solution, which would occur if one of the rows yielded a false statement.) To find the solutions, let $z = a$. Then, from row 2, we observe that $y + \frac{1}{3}a = -\frac{10}{3}$ so that $y = -\frac{1}{3}(a+10)$. Then, substituting these values of y and z into the equation obtained from row 1, we see that

$$\begin{aligned} x + \tfrac{1}{2}\left(-\tfrac{1}{3}(a+10)\right) - \tfrac{1}{2}a &= 1 \\ x - \tfrac{2}{3}a - \tfrac{5}{3} &= 1 \\ x &= \tfrac{2}{3}a + \tfrac{8}{3} \end{aligned}$$

Hence, the solutions are $\boxed{x = \frac{2}{3}(a+4),\ y = -\frac{1}{3}(a+10),\ z = a}$.

66. Reduce the augmented matrix down to row echelon form:

$$\left[\begin{array}{ccc|c} 3 & 1 & -1 & 0 \\ 1 & 1 & 7 & 4 \end{array}\right] \xrightarrow{R_1-3R_2\to R_2} \left[\begin{array}{ccc|c} 3 & 1 & -1 & 0 \\ 0 & -2 & -22 & -12 \end{array}\right] \xrightarrow[-\frac{1}{2}R_2\to R_2]{\frac{1}{3}R_1\to R_1} \left[\begin{array}{ccc|c} 1 & \frac{1}{3} & -\frac{1}{3} & 0 \\ 0 & 1 & 11 & 6 \end{array}\right]$$

Since there are two rows and three unknowns, we know from the above calculation that there are infinitely many solutions to this system. (Note: The only other possibility in such case would be that there was no solution, which would occur if one of the rows yielded a false statement.) To find the solutions, let $z = a$. Then, from row 2, we observe that $y + 11a = 6$ so that $y = -11a + 6$. Then, substituting these values of y and z into the equation obtained from row 1, we see that

$$x + \tfrac{1}{3}(-11a + 6) - \tfrac{1}{3}a = 0$$
$$x - \tfrac{11}{3}a + 2 - \tfrac{1}{3}a = 0$$
$$x = 4a - 2$$

Hence, the solutions are $\boxed{x = 4a - 2,\ y = -11a + 6,\ z = a}$.

67. Reduce the augmented matrix down to row echelon form:

$$\left[\begin{array}{ccc|c} 0 & 2 & 1 & 3 \\ 4 & 0 & -1 & -3 \\ 7 & -3 & -3 & 2 \\ 1 & -1 & -1 & -2 \end{array}\right] \xrightarrow{R_1\leftrightarrow R_2} \left[\begin{array}{ccc|c} 4 & 0 & -1 & -3 \\ 0 & 2 & 1 & 3 \\ 7 & -3 & -3 & 2 \\ 1 & -1 & -1 & -2 \end{array}\right] \xrightarrow[R_1-4R_4\to R_4]{R_3-7R_4\to R_3} \left[\begin{array}{ccc|c} 4 & 0 & -1 & -3 \\ 0 & 2 & 1 & 3 \\ 0 & 4 & 4 & 16 \\ 0 & 4 & 3 & 5 \end{array}\right]$$

$$\xrightarrow[R_4-2R_2\to R_4]{R_3-2R_2\to R_3} \left[\begin{array}{ccc|c} 4 & 0 & -1 & -3 \\ 0 & 2 & 1 & 3 \\ 0 & 0 & 2 & 10 \\ 0 & 0 & 1 & -1 \end{array}\right] \xrightarrow[\frac{1}{2}R_3\to R_3]{\frac{1}{4}R_1\to R_1,\ \frac{1}{2}R_2\to R_2} \left[\begin{array}{ccc|c} 1 & 0 & -\frac{1}{4} & -\frac{3}{4} \\ 0 & 1 & \frac{1}{2} & \frac{3}{2} \\ 0 & 0 & 1 & 5 \\ 0 & 0 & 1 & -1 \end{array}\right]$$

The last two rows require that $z = 5$ and $z = -1$ simultaneously. Since this is not possible, we conclude that the system has $\boxed{\text{no solution}}$.

68. Reduce the augmented matrix down to row echelon form:

$$\left[\begin{array}{ccc|c} -2 & -1 & 2 & 3 \\ 3 & 0 & -4 & 2 \\ 2 & 1 & 0 & -1 \\ -1 & 1 & -1 & -8 \end{array}\right] \xrightarrow[\substack{R_3+2R_4 \to R_3}]{\substack{R_1-2R_4 \to R_4 \\ R_2+3R_4 \to R_2}} \left[\begin{array}{ccc|c} -2 & -1 & 2 & 3 \\ 0 & 3 & -7 & -22 \\ 0 & 3 & -2 & -17 \\ 0 & -3 & 4 & 19 \end{array}\right]$$

$$\xrightarrow{\substack{R_2-R_3 \to R_3 \\ R_3+R_4 \to R_4}} \left[\begin{array}{ccc|c} -2 & -1 & 2 & 3 \\ 0 & 3 & -1 & -22 \\ 0 & 0 & -5 & -5 \\ 0 & 0 & 2 & 2 \end{array}\right] \xrightarrow{\substack{-\frac{1}{2}R_1 \to R_1 \\ \frac{1}{3}R_2 \to R_2 \\ -\frac{1}{5}R_3 \to R_3 \\ \frac{1}{2}R_4 \to R_4}} \left[\begin{array}{ccc|c} 1 & \frac{1}{2} & -1 & -\frac{3}{2} \\ 0 & 1 & -\frac{1}{3} & -\frac{22}{3} \\ 0 & 0 & 1 & 1 \\ 0 & 0 & 1 & 1 \end{array}\right]$$

Thus, Row 3now implies that $z = 1$.

Then, Row 2implies that $y - \frac{1}{3}(1) = -\frac{22}{3}$ so that $y = -7$.

Then, Row 1implies that $x + \frac{1}{2}(-7) - (1) = -\frac{3}{2}$ so that $x = 3$.

Hence, the solution is $\boxed{x = 3,\ y = -7,\ z = 1}$.

69. Reduce the augmented matrix down to row echelon form:

$$\left[\begin{array}{cccc|c} 3 & -2 & 1 & 2 & -2 \\ -1 & 3 & 4 & 3 & 4 \\ 1 & 1 & 1 & 1 & 0 \\ 5 & 3 & 1 & 2 & -1 \end{array}\right] \xrightarrow{\substack{R_1+3R_2 \to R_2 \\ R_1-3R_3 \to R_3 \\ R_4-5R_3 \to R_4}} \left[\begin{array}{cccc|c} 0 & -5 & -2 & -1 & -2 \\ 0 & 4 & 5 & 4 & 4 \\ 1 & 1 & 1 & 1 & 0 \\ 0 & -2 & -4 & -3 & -1 \end{array}\right] \xrightarrow{R_1 \leftrightarrow R_3} \left[\begin{array}{cccc|c} 1 & 1 & 1 & 1 & 0 \\ 0 & 4 & 5 & 4 & 4 \\ 0 & -5 & -2 & -1 & -2 \\ 0 & -2 & -4 & -3 & -1 \end{array}\right]$$

$$\xrightarrow{\substack{\frac{1}{4}R_2 \to R_2 \\ -\frac{1}{5}R_3 \to R_3 \\ -\frac{1}{2}R_4 \to R_4}} \left[\begin{array}{cccc|c} 1 & 1 & 1 & 1 & 0 \\ 0 & 1 & \frac{5}{4} & 1 & 1 \\ 0 & 1 & \frac{2}{5} & \frac{1}{5} & \frac{2}{5} \\ 0 & 1 & 2 & \frac{3}{2} & \frac{1}{2} \end{array}\right] \xrightarrow{\substack{R_2-R_3 \to R_2 \\ R_4-R_3 \to R_4}} \left[\begin{array}{cccc|c} 1 & 1 & 1 & 1 & 0 \\ 0 & 0 & \frac{17}{20} & \frac{4}{5} & \frac{3}{5} \\ 0 & 1 & \frac{2}{5} & \frac{1}{5} & \frac{2}{5} \\ 0 & 1 & \frac{8}{5} & \frac{13}{10} & \frac{1}{10} \end{array}\right] \xrightarrow{R_2 \leftrightarrow R_3} \left[\begin{array}{cccc|c} 1 & 1 & 1 & 1 & 0 \\ 0 & 1 & \frac{2}{5} & \frac{1}{5} & \frac{2}{5} \\ 0 & 0 & \frac{17}{20} & \frac{4}{5} & \frac{3}{5} \\ 0 & 1 & \frac{8}{5} & \frac{13}{10} & \frac{1}{10} \end{array}\right]$$

$$\xrightarrow{\substack{\frac{20}{17}R_3 \to R_3 \\ \frac{5}{8}R_4 \to R_4}} \left[\begin{array}{cccc|c} 1 & 1 & 1 & 1 & 0 \\ 0 & 1 & \frac{2}{5} & \frac{1}{5} & \frac{2}{5} \\ 0 & 0 & 1 & \frac{16}{17} & \frac{12}{17} \\ 0 & 0 & 1 & \frac{13}{16} & \frac{1}{16} \end{array}\right] \xrightarrow{R_3-R_4 \to R_4} \left[\begin{array}{cccc|c} 1 & 1 & 1 & 1 & 0 \\ 0 & 1 & \frac{2}{5} & \frac{1}{5} & \frac{2}{5} \\ 0 & 0 & 1 & \frac{16}{17} & \frac{12}{17} \\ 0 & 0 & 0 & \frac{35}{272} & \frac{175}{272} \end{array}\right] \xrightarrow{\frac{272}{35}R_4 \to R_4} \left[\begin{array}{cccc|c} 1 & 1 & 1 & 1 & 0 \\ 0 & 1 & \frac{2}{5} & \frac{1}{5} & \frac{2}{5} \\ 0 & 0 & 1 & \frac{16}{17} & \frac{12}{17} \\ 0 & 0 & 0 & 1 & 5 \end{array}\right]$$

Thus, Row 4 now implies that $x_4 = 5$.

Then, Row 3 implies that $x_3 + \frac{16}{17}(5) = \frac{12}{17}$ so that $x_3 = -4$.

Then, Row 2 implies that $x_2 + \frac{2}{5}(-4) + \frac{1}{5}(5) = \frac{2}{5}$ so that $x_2 = 1$.

Finally, Row 1 implies that $x_1 + 1 - 4 + 5 = 0$ so that $x_1 = -2$.

Hence, the solution is $\boxed{x_1 = -2,\ x_2 = 1,\ x_3 = -4,\ x_4 = 5}$.

70. Reduce the augmented matrix down to row echelon form:

$$\left[\begin{array}{cccc|c} 5 & 3 & 8 & 1 & 1 \\ 1 & 2 & 5 & 2 & 3 \\ 4 & 0 & 1 & -2 & -3 \\ 0 & 1 & 1 & 1 & 0 \end{array}\right] \xrightarrow[R_3-4R_2\to R_3]{R_1-5R_2\to R_2} \left[\begin{array}{cccc|c} 5 & 3 & 8 & 1 & 1 \\ 0 & -7 & -17 & -9 & -14 \\ 0 & -8 & -19 & -10 & -15 \\ 0 & 1 & 1 & 1 & 0 \end{array}\right]$$

$$\xrightarrow[R_3+8R_4\to R_3]{R_2+7R_4\to R_2} \left[\begin{array}{cccc|c} 5 & 3 & 8 & 1 & 1 \\ 0 & 0 & -10 & -2 & -14 \\ 0 & 0 & -11 & -2 & -15 \\ 0 & 1 & 1 & 1 & 0 \end{array}\right] \xrightarrow{R_4\leftrightarrow R_2\leftrightarrow R_3} \left[\begin{array}{cccc|c} 5 & 3 & 8 & 1 & 1 \\ 0 & 1 & 1 & 1 & 0 \\ 0 & 0 & -10 & -2 & -14 \\ 0 & 0 & -11 & -2 & -15 \end{array}\right]$$

$$\xrightarrow[\substack{-\frac{1}{10}R_3\to R_3 \\ -\frac{1}{11}R_4\to R_4}]{\frac{1}{5}R_1\to R_1} \left[\begin{array}{cccc|c} 1 & \frac{3}{5} & \frac{8}{5} & \frac{1}{5} & \frac{1}{5} \\ 0 & 1 & 1 & 1 & 0 \\ 0 & 0 & 1 & \frac{1}{5} & \frac{7}{5} \\ 0 & 0 & 1 & \frac{2}{11} & \frac{15}{11} \end{array}\right] \xrightarrow{R_3-R_4\to R_4} \left[\begin{array}{cccc|c} 1 & \frac{3}{5} & \frac{8}{5} & \frac{1}{5} & \frac{1}{5} \\ 0 & 1 & 1 & 1 & 0 \\ 0 & 0 & 1 & \frac{1}{5} & \frac{7}{5} \\ 0 & 0 & 0 & \frac{1}{55} & \frac{2}{55} \end{array}\right] \xrightarrow{55R_4\to R_4} \left[\begin{array}{cccc|c} 1 & \frac{3}{5} & \frac{8}{5} & \frac{1}{5} & \frac{1}{5} \\ 0 & 1 & 1 & 1 & 0 \\ 0 & 0 & 1 & \frac{1}{5} & \frac{7}{5} \\ 0 & 0 & 0 & 1 & 2 \end{array}\right]$$

Thus, Row 4 now implies that $x_4 = 2$.
Then, Row 3 implies that $x_3 + \frac{1}{5}(2) = \frac{7}{5}$ so that $x_3 = 1$.
Then, Row 2 implies that $x_2 + 1 + 2 = 0$ so that $x_2 = -3$.
Finally, Row 1 implies that $x_1 + \frac{3}{5}(-3) + \frac{8}{5}(1) + \frac{1}{5}(2) = \frac{1}{5}$ so that $x_1 = 0$.

71. Reduce the matrix down to reduced row echelon form:

$$\left[\begin{array}{cc|c} 1 & 3 & -5 \\ -2 & -1 & 0 \end{array}\right] \xrightarrow{R_2+2R_1\to R_2} \left[\begin{array}{cc|c} 1 & 3 & -5 \\ 0 & 5 & -10 \end{array}\right] \xrightarrow{\frac{1}{5}R_2\to R_2} \left[\begin{array}{cc|c} 1 & 3 & -5 \\ 0 & 1 & -2 \end{array}\right]$$

$$\xrightarrow{R_1-3R_2\to R_1} \left[\begin{array}{cc|c} 1 & 0 & 1 \\ 0 & 1 & -2 \end{array}\right]$$

Hence, the solution is $\boxed{(1,-2)}$.

72. Reduce the matrix down to reduced row echelon form:

$$\left[\begin{array}{cc|c} 5 & -4 & 31 \\ 3 & 7 & -19 \end{array}\right] \xrightarrow{R_2-\frac{3}{5}R_1\to R_2} \left[\begin{array}{cc|c} 5 & -4 & 31 \\ 0 & \frac{47}{5} & -\frac{188}{5} \end{array}\right] \xrightarrow{\frac{5}{47}R_2\to R_2} \left[\begin{array}{cc|c} 5 & -4 & 31 \\ 0 & 1 & -4 \end{array}\right]$$

$$\xrightarrow{R_1+4R_2\to R_1} \left[\begin{array}{cc|c} 5 & 0 & 15 \\ 0 & 1 & -4 \end{array}\right] \xrightarrow{\frac{1}{5}R_1\to R_1} \left[\begin{array}{cc|c} 1 & 0 & 3 \\ 0 & 1 & -4 \end{array}\right]$$

Hence, the solution is $\boxed{(3,-4)}$.

73. Reduce the matrix down to reduced row echelon form:

$$\left[\begin{array}{cc|c} 1 & 1 & 4 \\ -3 & -3 & 10 \end{array}\right] \xrightarrow{3R_1+R_2\to R_1} \left[\begin{array}{cc|c} 0 & 0 & 22 \\ -3 & -3 & 10 \end{array}\right] \xrightarrow{-\frac{1}{3}R_2\to R_2} \left[\begin{array}{cc|c} 0 & 0 & 22 \\ 1 & 1 & -\frac{10}{3} \end{array}\right]$$

Since the first row translates to $0 = 22$, which is false, the solution has $\boxed{\text{no solution}}$.

74. Reduce the matrix down to reduced row echelon form:

$$\left[\begin{array}{cc|c} 3 & -4 & 12 \\ -6 & 8 & -24 \end{array}\right] \xrightarrow{2R_1+R_2\to R_1} \left[\begin{array}{cc|c} 0 & 0 & 0 \\ -6 & 8 & -24 \end{array}\right] \xrightarrow{-\frac{1}{6}R_2\to R_2} \left[\begin{array}{cc|c} 0 & 0 & 0 \\ 1 & -\frac{4}{3} & 4 \end{array}\right]$$

Since the first row holds for all values of x and y, there are infinitely many solutions to the system. Let $x = a$, so that from the second row, $a-\frac{4}{3}y=4 \Rightarrow y=\frac{3}{4}a-3$. So, the solution is $\boxed{x=a,\ y=\frac{3}{4}a-3}$.

75. Reduce the matrix down to reduced row echelon form:

$$\left[\begin{array}{ccc|c} 1 & -2 & 3 & 5 \\ 3 & 6 & -4 & -12 \\ -1 & -4 & 6 & 16 \end{array}\right] \xrightarrow[-3R_1+R_2\to R_2]{R_1+R_3\to R_3} \left[\begin{array}{ccc|c} 1 & -2 & 3 & 5 \\ 0 & 12 & -13 & -27 \\ 0 & -6 & 9 & 21 \end{array}\right] \xrightarrow{R_2+2R_3\to R_3} \left[\begin{array}{ccc|c} 1 & -2 & 3 & 5 \\ 0 & 12 & -13 & -27 \\ 0 & 0 & 5 & 15 \end{array}\right]$$

$$\xrightarrow{\frac{1}{5}R_3\to R_3} \left[\begin{array}{ccc|c} 1 & -2 & 3 & 5 \\ 0 & 12 & -13 & -27 \\ 0 & 0 & 1 & 3 \end{array}\right] \xrightarrow{13R_3+R_2\to R_2} \left[\begin{array}{ccc|c} 1 & -2 & 3 & 5 \\ 0 & 12 & 0 & 12 \\ 0 & 0 & 1 & 3 \end{array}\right]$$

$$\xrightarrow{\frac{1}{12}R_2\to R_2} \left[\begin{array}{ccc|c} 1 & -2 & 3 & 5 \\ 0 & 1 & 0 & 1 \\ 0 & 0 & 1 & 3 \end{array}\right] \xrightarrow{R_1+2R_2-3R_3\to R_1} \left[\begin{array}{ccc|c} 1 & 0 & 0 & -2 \\ 0 & 1 & 0 & 1 \\ 0 & 0 & 1 & 3 \end{array}\right]$$

Hence, the solution is $\boxed{(-2,1,3)}$.

76. Reduce the matrix down to reduced row echelon form:

$$\left[\begin{array}{ccc|c} 1 & 2 & -1 & 6 \\ 2 & -1 & 3 & -13 \\ 3 & -2 & 3 & -16 \end{array}\right] \xrightarrow[R_3-3R_1\to R_3]{R_2-2R_1\to R_2} \left[\begin{array}{ccc|c} 1 & 2 & -1 & 6 \\ 0 & -5 & 5 & -25 \\ 0 & -8 & 6 & -34 \end{array}\right] \xrightarrow[\frac{1}{2}R_3\to R_3]{\frac{1}{5}R_2\to R_2} \left[\begin{array}{ccc|c} 1 & 2 & -1 & 6 \\ 0 & -1 & 1 & -5 \\ 0 & -4 & 3 & -17 \end{array}\right]$$

$$\xrightarrow{R_3-4R_2\to R_3} \left[\begin{array}{ccc|c} 1 & 2 & -1 & 6 \\ 0 & -1 & 1 & -5 \\ 0 & 0 & -1 & 3 \end{array}\right] \xrightarrow[-R_3\to R_3]{R_2+R_3\to R_2} \left[\begin{array}{ccc|c} 1 & 2 & -1 & 6 \\ 0 & -1 & 0 & -2 \\ 0 & 0 & 1 & -3 \end{array}\right]$$

$$\xrightarrow[-R_2\to R_2]{R_1+2R_2+R_3\to R_1} \left[\begin{array}{ccc|c} 1 & 0 & 0 & -1 \\ 0 & 1 & 0 & 2 \\ 0 & 0 & 1 & -3 \end{array}\right]$$

Hence, the solution is $\boxed{(-1,2,-3)}$.

77. Reduce the matrix down to reduced row echelon form:

$$\left[\begin{array}{ccc|c} 1 & 1 & 1 & 3 \\ 1 & 0 & -1 & 1 \\ 0 & 1 & -1 & -4 \end{array}\right] \xrightarrow{R_1-R_2\to R_2} \left[\begin{array}{ccc|c} 1 & 1 & 1 & 3 \\ 0 & 1 & 2 & 2 \\ 0 & 1 & -1 & -4 \end{array}\right] \xrightarrow{R_2-R_3\to R_3} \left[\begin{array}{ccc|c} 1 & 1 & 1 & 3 \\ 0 & 1 & 2 & 2 \\ 0 & 0 & 3 & 6 \end{array}\right]$$

$$\xrightarrow{\frac{1}{3}R_3\to R_3} \left[\begin{array}{ccc|c} 1 & 1 & 1 & 3 \\ 0 & 1 & 2 & 2 \\ 0 & 0 & 1 & 2 \end{array}\right] \xrightarrow{R_2-2R_3\to R_2} \left[\begin{array}{ccc|c} 1 & 1 & 1 & 3 \\ 0 & 1 & 0 & -2 \\ 0 & 0 & 1 & 2 \end{array}\right]$$

$$\xrightarrow{R_1-R_2-R_3\to R_1} \left[\begin{array}{ccc|c} 1 & 0 & 0 & 3 \\ 0 & 1 & 0 & -2 \\ 0 & 0 & 1 & 2 \end{array}\right]$$

Hence, the solution is $\boxed{(3,-2,2)}$.

78. Reduce the matrix down to reduced row echelon form:

$$\left[\begin{array}{ccc|c} 1 & -2 & 4 & 2 \\ 2 & -3 & -2 & -3 \\ \frac{1}{2} & \frac{1}{4} & 1 & -2 \end{array}\right] \xrightarrow{4R_3 \to R_3} \left[\begin{array}{ccc|c} 1 & -2 & 4 & 2 \\ 2 & -3 & -2 & -3 \\ 2 & 1 & 4 & -8 \end{array}\right] \xrightarrow[R_3-2R_1 \to R_3]{R_2-2R_1 \to R_2} \left[\begin{array}{ccc|c} 1 & -2 & 4 & 2 \\ 0 & 1 & -10 & -7 \\ 0 & 5 & -4 & -12 \end{array}\right]$$

$$\xrightarrow{R_3-5R_2 \to R_3} \left[\begin{array}{ccc|c} 1 & -2 & 4 & 2 \\ 0 & 1 & -10 & -7 \\ 0 & 0 & 46 & 23 \end{array}\right] \xrightarrow{\frac{1}{46}R_3 \to R_3} \left[\begin{array}{ccc|c} 1 & -2 & 4 & 2 \\ 0 & 1 & -10 & -7 \\ 0 & 0 & 1 & \frac{1}{2} \end{array}\right]$$

$$\xrightarrow{R_2+10R_3 \to R_2} \left[\begin{array}{ccc|c} 1 & -2 & 4 & 2 \\ 0 & 1 & 0 & -2 \\ 0 & 0 & 1 & \frac{1}{2} \end{array}\right] \xrightarrow{R_1+2R_2-4R_3 \to R_1} \left[\begin{array}{ccc|c} 1 & 0 & 0 & -4 \\ 0 & 1 & 0 & -2 \\ 0 & 0 & 1 & \frac{1}{2} \end{array}\right]$$

Hence, the solution is $\boxed{\left(-4,-2,\frac{1}{2}\right)}$.

79. Reduce the matrix down to reduced row echelon form:

$$\left[\begin{array}{ccc|c} 1 & 2 & 1 & 3 \\ 2 & -1 & 3 & 7 \\ 3 & 1 & 4 & 5 \end{array}\right] \xrightarrow[R_3-3R_1 \to R_3]{R_2-2R_1 \to R_2} \left[\begin{array}{ccc|c} 1 & 2 & 1 & 3 \\ 0 & -5 & 1 & 1 \\ 0 & -5 & 1 & -4 \end{array}\right] \xrightarrow{R_2-R_3 \to R_3} \left[\begin{array}{ccc|c} 1 & 2 & 1 & 3 \\ 0 & -5 & 1 & 1 \\ 0 & 0 & 0 & 5 \end{array}\right]$$

The last row is equivalent to the statement $0 = 5$, which is false. Hence, the system has $\boxed{\text{no solution}}$.

80. Reduce the matrix down to reduced row echelon form:

$$\left[\begin{array}{ccc|c} 1 & 2 & 1 & 3 \\ 2 & -1 & 3 & 7 \\ 3 & 1 & 4 & 10 \end{array}\right] \xrightarrow[R_3-3R_1 \to R_3]{R_2-2R_1 \to R_2} \left[\begin{array}{ccc|c} 1 & 2 & 1 & 3 \\ 0 & -5 & 1 & 1 \\ 0 & -5 & 1 & 1 \end{array}\right] \xrightarrow{R_2-R_3 \to R_3} \left[\begin{array}{ccc|c} 1 & 2 & 1 & 3 \\ 0 & -5 & 1 & 1 \\ 0 & 0 & 0 & 0 \end{array}\right]$$

$$\xrightarrow{-\frac{1}{5}R_2 \to R_2} \left[\begin{array}{ccc|c} 1 & 2 & 1 & 3 \\ 0 & 1 & -\frac{1}{5} & -\frac{1}{5} \\ 0 & 0 & 0 & 0 \end{array}\right] \xrightarrow{R_1-2R_2 \to R_1} \left[\begin{array}{ccc|c} 1 & 0 & \frac{7}{5} & \frac{17}{5} \\ 0 & 1 & -\frac{1}{5} & -\frac{1}{5} \\ 0 & 0 & 0 & 0 \end{array}\right]$$

Now, we must solve the dependent system:

$$\begin{cases} x+\frac{7}{5}z=\frac{17}{5} \\ y-\frac{1}{5}z=-\frac{1}{5} \end{cases} \text{ which is equivalent to } \begin{cases} 5x+7z=17 & \textbf{(1)} \\ 5y-z=-1 & \textbf{(2)} \end{cases}$$

Multiply **(2)** by 7 and add to **(1)** to obtain $5x+35y=10$, which is equivalent to $x+7y=2$. Let $x=a$. Then, $y=\frac{2-a}{7}$ and $z=\frac{17-5a}{7}$.

81. Reduce the matrix down to reduced row echelon form:

$$\left[\begin{array}{ccc|c} 3 & -1 & 1 & 8 \\ 1 & 1 & -2 & 4 \end{array}\right] \xrightarrow{R_1 - 3R_2 \to R_2} \left[\begin{array}{ccc|c} 3 & -1 & 1 & 8 \\ 0 & -4 & 7 & -4 \end{array}\right] \xrightarrow[-\frac{1}{4}R_2 \to R_2]{R_1 - \frac{1}{4}R_2 \to R_1} \left[\begin{array}{ccc|c} 3 & 0 & -\frac{3}{4} & 9 \\ 0 & 1 & -\frac{7}{4} & 1 \end{array}\right]$$

$$\xrightarrow{\frac{1}{3}R_1 \to R_1} \left[\begin{array}{ccc|c} 1 & 0 & -\frac{1}{4} & 3 \\ 0 & 1 & -\frac{7}{4} & 1 \end{array}\right]$$

Now, must solve the dependent system:

$$\begin{cases} x - \frac{1}{4}z = 3 & \textbf{(1)} \\ y - \frac{7}{4}z = 1 & \textbf{(2)} \end{cases}$$

Let $z = a$. Then, $x = \frac{a}{4} + 3$ and $y = \frac{7a}{4} + 1$.

82. Reduce the matrix down to reduced row echelon form:

$$\left[\begin{array}{ccc|c} 1 & -2 & 3 & 10 \\ -3 & 0 & 1 & 9 \end{array}\right] \xrightarrow{3R_1 + R_2 \to R_2} \left[\begin{array}{ccc|c} 1 & -2 & 3 & 10 \\ 0 & -6 & 10 & 39 \end{array}\right] \xrightarrow{-3R_1 + R_2 \to R_1} \left[\begin{array}{ccc|c} -3 & 0 & 1 & 9 \\ 0 & -6 & 10 & 39 \end{array}\right]$$

Now, must solve the dependent system:

$$\begin{cases} -3x + z = 9 & \textbf{(1)} \\ -6y + 10z = 39 & \textbf{(2)} \end{cases}$$

Multiply **(1)** by -10 and add to **(2)** to obtain $30x - 6y = -51$. Let $x = a$. Then, $y = \frac{30a+51}{6}$ and $z = 3a + 9$.

83. Reduce the matrix down to reduced row echelon form:

$$\left[\begin{array}{ccc|c} 4 & -2 & 5 & 20 \\ 1 & 3 & -2 & 6 \end{array}\right] \xrightarrow{R_1 - 4R_2 \to R_2} \left[\begin{array}{ccc|c} 4 & -2 & 5 & 20 \\ 0 & -14 & 13 & -4 \end{array}\right] \xrightarrow{-7R_1 + R_2 \to R_1} \left[\begin{array}{ccc|c} -28 & 0 & -22 & -144 \\ 0 & -14 & 13 & -4 \end{array}\right]$$

$$\xrightarrow{-\frac{1}{2}R_1 \to R_1} \left[\begin{array}{ccc|c} 14 & 0 & 11 & 72 \\ 0 & -14 & 13 & -4 \end{array}\right]$$

Now, must solve the dependent system:

$$\begin{cases} 14x + 11y = 72 & \textbf{(1)} \\ -14y + 13z = -4 & \textbf{(2)} \end{cases}$$

Multiply **(1)** by 14 and **(2)** by 11 and add to obtain $196x + 143z = 964$. Let $z = a$. Then, $x = \frac{72-11a}{14}$ and $y = \frac{13a+4}{14}$.

84. Reduce the matrix down to reduced row echelon form:

$$\left[\begin{array}{ccc|c} 0 & 1 & 1 & 4 \\ 1 & 1 & 0 & 8 \end{array}\right] \xrightarrow{R_1 \leftrightarrow R_2} \left[\begin{array}{ccc|c} 1 & 1 & 0 & 8 \\ 0 & 1 & 1 & 4 \end{array}\right] \xrightarrow{R_1 - R_2 \to R_1} \left[\begin{array}{ccc|c} 1 & 0 & -1 & 4 \\ 0 & 1 & 1 & 4 \end{array}\right]$$

Now, must solve the dependent system:

$$\begin{cases} x - z = 4 & \textbf{(1)} \\ y + z = 4 & \textbf{(2)} \end{cases}$$

Add **(1)** and **(2)** to obtain $x + y = 4$. Let $x = a$. Then, $y = 4 - a$ and $z = a - 4$.

85. Reduce the matrix down to reduced row echelon form:

$$\left[\begin{array}{cccc|c} 1 & -1 & -1 & -1 & 1 \\ 2 & 1 & 1 & 2 & 3 \\ 1 & -2 & -2 & -3 & 0 \\ 3 & -4 & 1 & 5 & -3 \end{array}\right] \xrightarrow[\substack{R_2 - 2R_3 \to R_3 \\ R_4 - 3R_3 \to R_4}]{R_2 - 2R_1 \to R_2} \left[\begin{array}{cccc|c} 1 & -1 & -1 & -1 & 1 \\ 0 & 3 & 3 & 4 & 1 \\ 0 & 5 & 5 & 8 & 3 \\ 0 & 2 & 7 & 14 & -3 \end{array}\right] \xrightarrow{\frac{1}{5}R_3 \to R_3} \left[\begin{array}{cccc|c} 1 & -1 & -1 & -1 & 1 \\ 0 & 3 & 3 & 4 & 1 \\ 0 & 1 & 1 & \frac{8}{5} & \frac{3}{5} \\ 0 & 2 & 7 & 14 & -3 \end{array}\right]$$

$$\xrightarrow[R_4 - 2R_3 \to R_4]{R_2 - 3R_3 \to R_2} \left[\begin{array}{cccc|c} 1 & -1 & -1 & -1 & 1 \\ 0 & 0 & 0 & -\frac{4}{5} & -\frac{4}{5} \\ 0 & 1 & 1 & \frac{8}{5} & \frac{3}{5} \\ 0 & 0 & 5 & \frac{54}{5} & -\frac{21}{5} \end{array}\right] \xrightarrow[R_3 \leftrightarrow R_2]{R_1 \leftrightarrow R_3} \left[\begin{array}{cccc|c} 1 & -1 & -1 & -1 & 1 \\ 0 & 1 & 1 & \frac{8}{5} & \frac{3}{5} \\ 0 & 0 & 5 & \frac{54}{5} & -\frac{21}{5} \\ 0 & 0 & 0 & -\frac{4}{5} & -\frac{4}{5} \end{array}\right]$$

$$\xrightarrow[-\frac{5}{4}R_4 \to R_4]{\frac{1}{5}R_3 \to R_3} \left[\begin{array}{cccc|c} 1 & -1 & -1 & -1 & 1 \\ 0 & 1 & 1 & \frac{8}{5} & \frac{3}{5} \\ 0 & 0 & 1 & \frac{54}{25} & -\frac{21}{25} \\ 0 & 0 & 0 & 1 & 1 \end{array}\right] \xrightarrow{R_3 - \frac{54}{25}R_4 \to R_3} \left[\begin{array}{cccc|c} 1 & -1 & -1 & -1 & 1 \\ 0 & 1 & 1 & \frac{8}{5} & \frac{3}{5} \\ 0 & 0 & 1 & 0 & -3 \\ 0 & 0 & 0 & 1 & 1 \end{array}\right]$$

$$\xrightarrow{R_2 - R_3 - \frac{8}{5}R_4 \to R_2} \left[\begin{array}{cccc|c} 1 & -1 & -1 & -1 & 1 \\ 0 & 1 & 0 & 0 & 2 \\ 0 & 0 & 1 & 0 & -3 \\ 0 & 0 & 0 & 1 & 1 \end{array}\right] \xrightarrow{R_1 + R_2 + R_3 + R_4 \to R_1} \left[\begin{array}{cccc|c} 1 & 0 & 0 & 0 & 1 \\ 0 & 1 & 0 & 0 & 2 \\ 0 & 0 & 1 & 0 & -3 \\ 0 & 0 & 0 & 1 & 1 \end{array}\right]$$

Hence, the solution is $\boxed{w = 1, z = -3, y = 2, x = 1}$

86. Reduce the matrix down to reduced row echelon form:

$$\left[\begin{array}{cccc|c} 1 & -3 & 3 & -2 & 4 \\ 1 & 2 & -1 & 0 & -3 \\ 1 & 0 & 3 & 2 & 3 \\ 0 & 1 & 1 & 5 & 6 \end{array}\right] \xrightarrow[R_1-R_3 \to R_3]{R_1-R_2 \to R_2} \left[\begin{array}{cccc|c} 1 & -3 & 3 & -2 & 4 \\ 0 & -5 & 4 & -2 & 7 \\ 0 & -3 & 0 & -4 & 1 \\ 0 & 1 & 1 & 5 & 6 \end{array}\right] \xrightarrow[R_3+3R_4 \to R_3]{R_2+5R_4 \to R_2} \left[\begin{array}{cccc|c} 1 & -3 & 3 & -2 & 4 \\ 0 & 0 & 9 & 23 & 37 \\ 0 & 0 & 3 & 11 & 19 \\ 0 & 1 & 1 & 5 & 6 \end{array}\right]$$

$$\xrightarrow{R_2-3R_3 \to R_2} \left[\begin{array}{cccc|c} 1 & -3 & 3 & -2 & 4 \\ 0 & 0 & 0 & -10 & -20 \\ 0 & 0 & 3 & 11 & 19 \\ 0 & 1 & 1 & 5 & 6 \end{array}\right] \xrightarrow{R_2 \leftrightarrow R_4} \left[\begin{array}{cccc|c} 1 & -3 & 3 & -2 & 4 \\ 0 & 1 & 1 & 5 & -20 \\ 0 & 0 & 3 & 11 & 31 \\ 0 & 0 & 0 & -10 & 6 \end{array}\right]$$

$$\xrightarrow[\frac{1}{3}R_3 \to R_3]{-\frac{1}{10}R_4 \to R_4} \left[\begin{array}{cccc|c} 1 & -3 & 3 & -2 & 4 \\ 0 & 1 & 1 & 5 & 6 \\ 0 & 0 & 1 & \frac{11}{3} & \frac{31}{3} \\ 0 & 0 & 0 & 1 & 2 \end{array}\right] \xrightarrow{R_3-\frac{11}{3}R_4 \to R_3} \left[\begin{array}{cccc|c} 1 & -3 & 3 & -2 & 4 \\ 0 & 1 & 1 & 5 & 6 \\ 0 & 0 & 1 & 0 & 3 \\ 0 & 0 & 0 & 1 & 2 \end{array}\right]$$

$$\xrightarrow{R_2-R_3-5R_4 \to R_2} \left[\begin{array}{cccc|c} 1 & -3 & 3 & -2 & 4 \\ 0 & 1 & 0 & 0 & -7 \\ 0 & 0 & 1 & 0 & 3 \\ 0 & 0 & 0 & 1 & 2 \end{array}\right] \xrightarrow{R_1+3R_2-3R_3+2R_4 \to R_1} \left[\begin{array}{cccc|c} 1 & 0 & 0 & 0 & -22 \\ 0 & 1 & 0 & 0 & -7 \\ 0 & 0 & 1 & 0 & 3 \\ 0 & 0 & 0 & 1 & 2 \end{array}\right]$$

Hence, the solution is $\boxed{w=2, x=-22, y=-7, z=3}$.

87. Let x = number of red dwarfs, y = number of yellow stars, and z = number of blue stars (all measured in millions).

Solve the system:

$$\begin{cases} x = 120z & \textbf{(1)} \\ y = 3000x & \textbf{(2)} \\ x+y+z = 2.880968 & \textbf{(3)} \end{cases}$$

Solve **(1)** for z: $z = \frac{1}{120}x$ **(4)**

Substitute **(2)** and **(4)** into **(3)**: $x + 3000x + \frac{1}{120}x = 2.880968 \Rightarrow x = 0.00960$

Substitute this into **(4)** and **(2)** to find z and y: $z = 0.000008,\ y = 2{,}880{,}000$

So, there are about 960 red dwarfs, 8 blue stars, and 2,880,000 yellow stars.

88. Let x = amount of Hamlin oranges, y = amount of Valencia oranges, and z = amount of Navel oranges (all measured in gallons)
Solve the system:

$$\begin{cases} 2.50x+3.40y+2.80z=3.00 & \textbf{(1)} \\ y=z & \textbf{(2)} \\ x+y+z=1 & \textbf{(3)} \end{cases}$$

Substitute **(2)** into **(1)** and **(3)** to obtain the system:

$$\begin{cases} 2.5x+6.2y=3.0 & \textbf{(4)} \\ x+2y=1 & \textbf{(5)} \end{cases}$$

To solve this system, multiply **(5)** by -2.5 and add to **(4)**: $1.2y=0.5 \Rightarrow y=\frac{5}{12}$
Substitute this into **(5)**: $x=\frac{1}{6}$
Substitute both of these into **(3)** to obtain: $z=\frac{5}{12}$
So, there is $\frac{1}{6}$ gallon of Hamlin oranges, $\frac{5}{12}$ gallon of Valencia oranges, and $\frac{5}{12}$ gallon of Navel oranges.

89. Let w = number of Mediterranean chicken sandwiches
x = number of 6-Inch Tuna sandwiches
y = number of 6-Inch Roast Beef sandwiches
z = number of Turkey Bacon wraps

We must solve the system:

$$\begin{cases} w+x+y+z=14 \\ 17w+46x+45y+20z=526 \\ 18w+19x+5y+27z=168 \\ 36w+20x+19y+34z=332 \end{cases}$$

Now, write down the augmented matrix, and reduce it down to row echelon form:

$$\left[\begin{array}{cccc|c} 1 & 1 & 1 & 1 & 14 \\ 17 & 46 & 45 & 20 & 526 \\ 18 & 19 & 5 & 27 & 168 \\ 36 & 20 & 19 & 34 & 332 \end{array}\right] \xrightarrow[R_2-17R_1\to R_2]{\substack{R_4-36R_1\to R_4 \\ R_3-18R_1\to R_3}} \left[\begin{array}{cccc|c} 1 & 1 & 1 & 1 & 14 \\ 0 & 29 & 28 & 3 & 288 \\ 0 & 1 & -13 & 9 & -84 \\ 0 & -16 & -17 & -2 & -172 \end{array}\right]$$

$$\xrightarrow[R_4+16R_3\to R_4]{R_2-29R_3\to R_2} \left[\begin{array}{cccc|c} 1 & 1 & 1 & 1 & 14 \\ 0 & 0 & 405 & -258 & 2724 \\ 0 & 1 & -13 & 9 & -84 \\ 0 & 0 & -225 & 142 & -1516 \end{array}\right] \xrightarrow{R_2\leftrightarrow R_3} \left[\begin{array}{cccc|c} 1 & 1 & 1 & 1 & 14 \\ 0 & 1 & -13 & 9 & -84 \\ 0 & 0 & 405 & -258 & 2724 \\ 0 & 0 & -225 & 142 & -1516 \end{array}\right]$$

$$\xrightarrow{\frac{1}{405}R_3\to R_3} \left[\begin{array}{cccc|c} 1 & 1 & 1 & 1 & 14 \\ 0 & 1 & -13 & 9 & -84 \\ 0 & 0 & 1 & -\frac{258}{405} & \frac{2724}{405} \\ 0 & 0 & -225 & 142 & -1516 \end{array}\right] \xrightarrow{R_4+225R_3\to R_4} \left[\begin{array}{cccc|c} 1 & 1 & 1 & 1 & 14 \\ 0 & 1 & -13 & 9 & -84 \\ 0 & 0 & 1 & -\frac{258}{405} & \frac{2724}{405} \\ 0 & 0 & 0 & -\frac{540}{405} & -\frac{1080}{405} \end{array}\right]$$

$$\xrightarrow{-\frac{405}{540}R_4\to R_4} \left[\begin{array}{cccc|c} 1 & 1 & 1 & 1 & 14 \\ 0 & 1 & -13 & 9 & -84 \\ 0 & 0 & 1 & -\frac{258}{405} & \frac{2724}{405} \\ 0 & 0 & 0 & 1 & 2 \end{array}\right]$$

Thus, Row 4 now implies that $z=2$.
Then, Row 3 implies that $y-\frac{258}{405}(2)=\frac{2724}{405}$ so that $y=8$.
Then, Row 2 implies that $x-13(8)+9(2)=-84$ so that $x=2$.
Then, Row 1 implies that $w+2+8+2=14$ so that $w=2$.
So, there were 2 Mediterranean chicken sandwiches, 2 6-Inch Tuna sandwich, 8 6-Inch Roast Beef sandwiches, and 2 Turkey Bacon wraps.

90. Let w = number of Mediterranean chicken sandwiches
x = number of 6-Inch Tuna sandwiches
y = number of 6-Inch Roast Beef sandwiches
z = number of Turkey Bacon wraps

We must solve the system:

$$\begin{cases} w+x+y+z=14 \\ 350w+430x+290y+430z=5180 \\ 17w+46x+45y+20z=335 \\ 18+19x+5y+27z=263 \end{cases}$$

Now, write down the augmented matrix, and reduce it down to row echelon form:

$$\left[\begin{array}{cccc|c} 1 & 1 & 1 & 1 & 14 \\ 350 & 430 & 290 & 430 & 5180 \\ 17 & 46 & 45 & 20 & 335 \\ 18 & 19 & 5 & 27 & 263 \end{array}\right] \xrightarrow[\substack{R_3-17R_1\to R_3 \\ R_4-18R_1\to R_4}]{R_2-350R_1\to R_2} \left[\begin{array}{cccc|c} 1 & 1 & 1 & 1 & 14 \\ 0 & 80 & -60 & 80 & 280 \\ 0 & 29 & 28 & 3 & 97 \\ 0 & 1 & -13 & 9 & 11 \end{array}\right]$$

$$\xrightarrow{R_2\leftrightarrow R_4} \left[\begin{array}{cccc|c} 1 & 1 & 1 & 1 & 14 \\ 0 & 1 & -13 & 9 & 11 \\ 0 & 29 & 28 & 3 & 97 \\ 0 & 80 & -60 & 80 & 280 \end{array}\right] \xrightarrow[R_4-80R_2\to R_4]{R_3-29R_2\to R_3} \left[\begin{array}{cccc|c} 1 & 1 & 1 & 1 & 14 \\ 0 & 1 & -13 & 9 & 11 \\ 0 & 0 & 405 & -258 & -222 \\ 0 & 0 & 980 & -640 & -600 \end{array}\right]$$

$$\xrightarrow{\frac{1}{405}R_3\to R_3} \left[\begin{array}{cccc|c} 1 & 1 & 1 & 1 & 14 \\ 0 & 1 & -13 & 9 & 11 \\ 0 & 0 & 1 & -\frac{258}{405} & -\frac{222}{405} \\ 0 & 0 & 980 & -640 & -600 \end{array}\right] \xrightarrow{R_4-980R_3\to R_3} \left[\begin{array}{cccc|c} 1 & 1 & 1 & 1 & 14 \\ 0 & 1 & -13 & 9 & 11 \\ 0 & 0 & 1 & -\frac{258}{405} & -\frac{222}{405} \\ 0 & 0 & 0 & -\frac{6360}{405} & -\frac{25,440}{405} \end{array}\right]$$

$$\xrightarrow{-\frac{405}{6360}R_4\to R_4} \left[\begin{array}{cccc|c} 1 & 1 & 1 & 1 & 14 \\ 0 & 1 & -13 & 9 & 11 \\ 0 & 0 & 1 & -\frac{258}{405} & -\frac{222}{405} \\ 0 & 0 & 0 & 1 & 4 \end{array}\right]$$

Thus, Row 4 now implies that $z=4$.
Then, Row 3 implies that $y-\frac{258}{405}(4)=-\frac{222}{405}$ so that $y=2$.
Then, Row 2 implies that $x-13(2)+9(4)=11$ so that $x=1$.
Then, Row 1 implies that $w+1+2+4=14$ so that $w=7$.
So, there were 7 Mediterranean chicken sandwiches, 1 6-Inch Tuna sandwich, 2 6-Inch Roast Beef sandwiches, and 4 Turkey Bacon wraps.

91. From the given information, we must solve the following system:

$$\begin{cases} 34 = \frac{1}{2}a(1)^2 + v_0(1) + h_0 \\ 36 = \frac{1}{2}a(2)^2 + v_0(2) + h_0 \\ 6 = \frac{1}{2}a(3)^2 + v_0(3) + h_0 \end{cases} \text{ is equivalent to } \begin{cases} 68 = a + 2v_0 + 2h_0 \\ 72 = 4a + 4v_0 + 2h_0 \\ 12 = 9a + 6v_0 + 2h_0 \end{cases}$$

Now, write down the augmented matrix, and reduce it down to row echelon form:

$$\left[\begin{array}{ccc|c} 1 & 2 & 2 & 68 \\ 4 & 4 & 2 & 72 \\ 9 & 6 & 2 & 12 \end{array}\right] \xrightarrow[R_3-9R_1 \to R_3]{R_2-4R_1 \to R_2} \left[\begin{array}{ccc|c} 1 & 2 & 2 & 68 \\ 0 & -4 & -6 & -200 \\ 0 & -12 & -16 & -600 \end{array}\right] \xrightarrow{R_3-3R_2 \to R_3} \left[\begin{array}{ccc|c} 1 & 2 & 2 & 68 \\ 0 & -4 & -6 & -200 \\ 0 & 0 & 2 & 0 \end{array}\right]$$

Thus, Row 3 now implies that $h_0 = 0$ ft. = initial height .

Then, Row 2 implies that $-4v_0 - 6(0) = -200$ so that $v_0 = 50$ ft./sec. = initial velocity .

Then, Row 1 implies that $a + 2(50) + 2(0) = 68$ so that $a = -32$ ft./sec^2=acceleration. .

Thus, the equation of the curve is $y = -\frac{1}{2}(32)t^2 + 50t + 0 = -16t^2 + 50t$.

92. From the given information, we must solve the following system:

$$\begin{cases} 54 = \frac{1}{2}a(1)^2 + v_0(1) + h_0 \\ 66 = \frac{1}{2}a(2)^2 + v_0(2) + h_0 \\ 46 = \frac{1}{2}a(3)^2 + v_0(3) + h_0 \end{cases} \text{ is equivalent to } \begin{cases} 108 = a + 2v_0 + 2h_0 \\ 132 = 4a + 4v_0 + 2h_0 \\ 92 = 9a + 6v_0 + 2h_0 \end{cases}$$

Now, write down the augmented matrix, and reduce it down to row echelon form:

$$\left[\begin{array}{ccc|c} 1 & 2 & 2 & 108 \\ 4 & 4 & 2 & 132 \\ 9 & 6 & 2 & 92 \end{array}\right] \xrightarrow[R_3-9R_1 \to R_3]{R_2-4R_1 \to R_2} \left[\begin{array}{ccc|c} 1 & 2 & 2 & 108 \\ 0 & -4 & -6 & -300 \\ 0 & -12 & -16 & -880 \end{array}\right] \xrightarrow{R_3-3R_2 \to R_3} \left[\begin{array}{ccc|c} 1 & 2 & 2 & 108 \\ 0 & -4 & -6 & -300 \\ 0 & 0 & 2 & 20 \end{array}\right]$$

Thus, Row 3 now implies that $2h_0 = 20$ so that $h_0 = 10$ ft. = initial height .

Then, Row 2 implies that $-4v_0 - 6(10) = -300$ so that $v_0 = 60$ ft./sec. = initial velocity .

Then, Row 1 implies that $a + 2(60) + 2(10) = 108$ so that $a = -32$ ft./sec.^{2} = acceleration .

Thus, the equation of the curve is $y = -\frac{1}{2}(32)t^2 + 60t + 10 = -16t^2 + 60t + 10$.

93. From the given information, we must solve the following system:

$$\begin{cases} 25 = a(16)^2 + b(16) + c \\ 64 = a(40)^2 + b(40) + c \\ 40 = a(65)^2 + b(65) + c \end{cases} \text{ is equivalent to } \begin{cases} 25 = 256a + 16b + c \\ 64 = 1600a + 40b + c \\ 40 = 4225a + 65b + c \end{cases}$$

Now, write down the augmented matrix, and reduce it down to row echelon form:

$$\left[\begin{array}{ccc|c} 256 & 16 & 1 & 25 \\ 1600 & 40 & 1 & 64 \\ 4225 & 65 & 1 & 40 \end{array}\right] \xrightarrow{\frac{1}{256}R_1 \to R_1} \left[\begin{array}{ccc|c} 1 & \frac{1}{16} & \frac{1}{256} & \frac{25}{256} \\ 1600 & 40 & 1 & 64 \\ 4225 & 65 & 1 & 40 \end{array}\right]$$

$$\xrightarrow[R_3 - 4225R_1 \to R_3]{R_2 - 1600R_1 \to R_1} \left[\begin{array}{ccc|c} 1 & \frac{1}{16} & \frac{1}{256} & \frac{25}{256} \\ 0 & -60 & -\frac{1344}{256} & -\frac{23616}{256} \\ 0 & -\frac{3185}{16} & -\frac{3969}{256} & -\frac{95385}{256} \end{array}\right] \xrightarrow[-\frac{16}{3185}R_3 \to R_3]{-\frac{1}{60}R_2 \to R_2} \left[\begin{array}{ccc|c} 1 & \frac{1}{16} & \frac{1}{256} & \frac{25}{256} \\ 0 & 1 & \frac{1344}{15360} & \frac{23616}{15360} \\ 0 & 1 & \frac{63504}{815360} & \frac{1526160}{815360} \end{array}\right]$$

$$\xrightarrow{R_3 - R_2 \to R_3} \left[\begin{array}{ccc|c} 1 & \frac{1}{16} & \frac{1}{256} & \frac{25}{256} \\ 0 & 1 & \frac{1344}{15360} & \frac{23616}{15360} \\ 0 & 0 & \frac{63504}{815360} - \frac{1344}{15360} & \frac{1526160}{815360} - \frac{23616}{15360} \end{array}\right]$$

$$\xrightarrow{\frac{1}{\frac{1344}{15360} - \frac{63504}{815360}}R_3 \to R_3} \left[\begin{array}{ccc|c} 1 & \frac{1}{16} & \frac{1}{256} & \frac{25}{256} \\ 0 & 1 & \frac{1344}{15360} & \frac{23616}{15360} \\ 0 & 0 & 1 & -34.76326 \end{array}\right] \cong \left[\begin{array}{ccc|c} 1 & 0.0625 & 0.0039 & 0.0977 \\ 0 & 1 & 0.0875 & 1.5375 \\ 0 & 0 & 1 & -34.76326 \end{array}\right]$$

Thus, Row 3 now implies that $c \cong -34.76326$.
Then, Row 2 implies that $b + 0.0875(-34.76326) = 1.5375$ so that $b \cong 4.57928$.
Then, Row 1 implies that
$a + 0.0625(4.57928) + 0.0039(-34.76326) = 0.0977$ so that $a \cong -0.052755$.

Thus, the approximate equation of the curve is $\boxed{y = -0.053x^2 + 4.58x - 34.76}$.

94. As instructed in the problem, make the following identifications:

Year 1920 corresponds to $x = 0$
Year 1960 corresponds to $x = 40$
Year 2002 corresponds to $x = 82$

From the given information, we know that the points (0, 18.4), (40, 20.3), and (82, 25.3) lie on the graph. Hence, we must solve the following system:

$$\begin{cases} 18.4 = a(0)^2 + b(0) + c \\ 20.3 = a(40)^2 + b(40) + c \\ 25.3 = a(82)^2 + b(82) + c \end{cases} \text{ is equivalent to } \begin{cases} c = 18.4 \quad \textbf{(1)} \\ 1600a + 40b + c = 20.3 \quad \textbf{(2)} \\ 6724a + 82b + c = 25.3 \quad \textbf{(3)} \end{cases}$$

Given the fact that we know the value of c (given in **(1)**), it makes sense to first substitute **(1)** into both **(2)** and **(3)** to obtain a 2×2 system:

$$\begin{cases} 1600a + 40b + 18.4 = 20.3 \quad \textbf{(2)} \\ 6724a + 82b + 18.4 = 25.3 \quad \textbf{(3)} \end{cases} \text{ which is equivalent to } \begin{cases} 1600a + 40b = 1.9 \quad \textbf{(2)} \\ 6724a + 82b = 6.9 \quad \textbf{(3)} \end{cases}$$

Now, write down the augmented matrix, and reduce it down to row echelon form:

$$\left[\begin{array}{cc|c} 1600 & 40 & 1.9 \\ 6724 & 82 & 6.9 \end{array}\right] \xrightarrow{\frac{1}{1600}R_1 \to R_1} \left[\begin{array}{cc|c} 1 & \frac{1}{40} & \frac{1.9}{1600} \\ 6724 & 82 & 6.9 \end{array}\right]$$

$$\xrightarrow{R_2 - 6724R_1 \to R_2} \left[\begin{array}{cc|c} 1 & \frac{1}{40} & \frac{1.9}{1600} \\ 0 & 82 - \frac{6724}{40} & 6.9 - \frac{(1.9)(6724)}{1600} \end{array}\right] \cong \left[\begin{array}{cc|c} 1 & 0.025 & 0.0011875 \\ 0 & -86.1 & -1.08475 \end{array}\right]$$

$$\xrightarrow{\frac{1}{-86.1}R_2 \to R_2} \left[\begin{array}{cc|c} 1 & 0.025 & 0.0011875 \\ 0 & 1 & 0.0126 \end{array}\right]$$

So, Row 2 implies that $b \cong 0.0126$.
Then, Row 1 implies that $a + 0.025(0.0126) = 0.0011875$ so that $a \cong 0.0008725$.

Thus, the approximate equation of the curve is $\boxed{y = 0.0008725x^2 + 0.0126x + 18.4}$.

Finally, using this line to predict ages in future years, we see that in the year 2010 (which corresponds to $x = 90$), the average age would be $\boxed{y(90) \cong 26.60 \text{ years}}$.

95. Let x = amount in money market (3%)
y = amount in mutual fund (7%)
z = amount in stock (10%)

We must solve the system:

$$\begin{cases} x+y+z=10{,}000 \\ x=y+3000 \\ 0.03x+0.07y+0.10z=540 \end{cases} \text{ which is equivalent to } \begin{cases} x+y+z=10{,}000 \\ x-y=3000 \\ 3x+7y+10z=54{,}000 \end{cases}$$

Write down the augmented matrix, and reduce it down to row echelon form:

$$\left[\begin{array}{ccc|c} 1 & 1 & 1 & 10000 \\ 1 & -1 & 0 & 3000 \\ 3 & 7 & 10 & 54000 \end{array}\right] \xrightarrow[R_2-R_1\to R_2]{R_3-3R_1\to R_3} \left[\begin{array}{ccc|c} 1 & 1 & 1 & 10000 \\ 0 & -2 & -1 & -7000 \\ 0 & 4 & 7 & 24000 \end{array}\right] \xrightarrow{R_3+2R_2\to R_3} \left[\begin{array}{ccc|c} 1 & 1 & 1 & 10000 \\ 0 & -2 & -1 & -7000 \\ 0 & 0 & 5 & 10000 \end{array}\right]$$

$$\xrightarrow[-\frac{1}{2}R_2\to R_2]{\frac{1}{5}R_3\to R_3} \left[\begin{array}{ccc|c} 1 & 1 & 1 & 10000 \\ 0 & 1 & \frac{1}{2} & 3500 \\ 0 & 0 & 1 & 2000 \end{array}\right] \xrightarrow{R_2-\frac{1}{2}R_3\to R_2} \left[\begin{array}{ccc|c} 1 & 1 & 1 & 10000 \\ 0 & 1 & 0 & 2500 \\ 0 & 0 & 1 & 2000 \end{array}\right] \xrightarrow{R_1-R_2-R_3\to R_1} \left[\begin{array}{ccc|c} 1 & 0 & 0 & 5500 \\ 0 & 1 & 0 & 2500 \\ 0 & 0 & 1 & 2000 \end{array}\right]$$

So, he should invest \$5500 in money market, \$2500 in mutual fund, and \$2000 in stock.

96. Let x = amount in money market (3%)
y = amount in mutual fund (7%)
z = amount in stock (10%)

We must solve the system:

$$\begin{cases} x+y+z=10{,}000 \\ z=y+3000 \\ 0.03x+0.07y+0.10z=840 \end{cases} \text{ which is equivalent to } \begin{cases} x+y+z=10{,}000 \\ -y+z=3000 \\ 3x+7y+10z=84{,}000 \end{cases}$$

Write down the augmented matrix, and reduce it down to row echelon form:

$$\left[\begin{array}{ccc|c} 1 & 1 & 1 & 10000 \\ 0 & -1 & 1 & 3000 \\ 3 & 7 & 10 & 84000 \end{array}\right] \xrightarrow{R_3-3R_1\to R_3} \left[\begin{array}{ccc|c} 1 & 1 & 1 & 10000 \\ 0 & -1 & 1 & 3000 \\ 0 & 4 & 7 & 54000 \end{array}\right] \xrightarrow{4R_2+R_3\to R_3} \left[\begin{array}{ccc|c} 1 & 1 & 1 & 10000 \\ 0 & -1 & 1 & 3000 \\ 0 & 0 & 11 & 66000 \end{array}\right]$$

$$\xrightarrow[\frac{1}{11}R_3\to R_3]{-R_2\to R_2} \left[\begin{array}{ccc|c} 1 & 1 & 1 & 10000 \\ 0 & 1 & -1 & -3000 \\ 0 & 0 & 1 & 6000 \end{array}\right]$$

Then, Row 3 implies that $z=6000$.
Then, Row 2 implies that $y-6000=-3000$ so that $y=3000$.
Then, Row 1 implies that $x+3000+6000=10000$ so that $x=1000$.
So, he should invest \$1000 in money market, \$3000 in mutual fund, and \$6000 in stock.

97. Let a = units of product x, b = units of product y, and c = units of product z. We employ matrix methods to solve the system:

$$\begin{cases} 20a+25b+150c=2400 \\ 2a+5b+10c=310 \\ a+0b+\frac{1}{2}c=28 \end{cases}$$

$$\left[\begin{array}{ccc|c} 20 & 25 & 150 & 2400 \\ 2 & 5 & 10 & 310 \\ 1 & 0 & \frac{1}{2} & 28 \end{array}\right] \xrightarrow[R_2-2R_3\to R_3]{\frac{1}{5}R_1\to R_1} \left[\begin{array}{ccc|c} 4 & 5 & 30 & 480 \\ 2 & 5 & 10 & 310 \\ 0 & 5 & 9 & 254 \end{array}\right] \xrightarrow{R_1-2R_2\to R_2} \left[\begin{array}{ccc|c} 4 & 5 & 30 & 480 \\ 0 & -5 & 10 & -140 \\ 0 & 5 & 9 & 254 \end{array}\right]$$

$$\xrightarrow{R_2+R_3\to R_3} \left[\begin{array}{ccc|c} 4 & 5 & 30 & 480 \\ 0 & -5 & 10 & -140 \\ 0 & 0 & 19 & 114 \end{array}\right] \xrightarrow[\frac{1}{19}R_3\to R_3]{-\frac{1}{5}R_2\to R_2} \left[\begin{array}{ccc|c} 4 & 5 & 30 & 480 \\ 0 & 1 & -2 & 28 \\ 0 & 0 & 1 & 6 \end{array}\right] \xrightarrow[R_2+2R_3\to R_2]{\frac{1}{4}R_1\to R_1}$$

$$\left[\begin{array}{ccc|c} 1 & \frac{5}{4} & \frac{15}{2} & 120 \\ 0 & 1 & 0 & 40 \\ 0 & 0 & 1 & 6 \end{array}\right] \xrightarrow{R_1-\frac{5}{4}R_2-\frac{15}{2}R_3\to R_1} \left[\begin{array}{ccc|c} 1 & 0 & 0 & 25 \\ 0 & 1 & 0 & 40 \\ 0 & 0 & 1 & 6 \end{array}\right]$$

Hence, there are 25 units of product x, 40 units of product y, and 6 units of product z.

98. Using the given three points generates the following system that must be solved in order to determine the coefficients of the quadratic function on which they lie:

$$\begin{cases} a+b+c=5 \\ 4a-2b+c=-10 \\ c=4 \end{cases}$$

We solve using matrix methods:

$$\left[\begin{array}{ccc|c} 1 & 1 & 1 & 5 \\ 4 & -2 & 1 & -10 \\ 0 & 0 & 1 & 4 \end{array}\right] \xrightarrow{R_2-4R_1\to R_2} \left[\begin{array}{ccc|c} 1 & 1 & 1 & 5 \\ 0 & -6 & -3 & -25 \\ 0 & 0 & 1 & 4 \end{array}\right] \xrightarrow{R_2+3R_3\to R_2} \left[\begin{array}{ccc|c} 1 & 1 & 1 & 5 \\ 0 & -6 & 0 & -13 \\ 0 & 0 & 1 & 4 \end{array}\right]$$

$$\xrightarrow{-\frac{1}{6}R_2\to R_2} \left[\begin{array}{ccc|c} 1 & 1 & 1 & 5 \\ 0 & 1 & 0 & \frac{13}{6} \\ 0 & 0 & 1 & 4 \end{array}\right] \xrightarrow{R_1-R_2-R_3\to R_1} \left[\begin{array}{ccc|c} 1 & 0 & 0 & -\frac{7}{6} \\ 0 & 1 & 0 & \frac{13}{6} \\ 0 & 0 & 1 & 4 \end{array}\right]$$

Hence, the equation is $\boxed{y=-\frac{7}{6}x^2+\frac{13}{6}x+4}$.

99. Let x = # of general admission tickets, y = # of reserved tickets, and z = # of end zone tickets.

We solve the following system using matrix methods:

$$\begin{cases} y = x+5 \\ z = x+20 \\ 20x+40y+15z = 2375 \end{cases}$$

$$\left[\begin{array}{ccc|c} 1 & -1 & 0 & -5 \\ 1 & 0 & -1 & -20 \\ 20 & 40 & 15 & 2375 \end{array}\right] \xrightarrow[\frac{1}{5}R_3 \to R_3]{R_1 - R_2 \to R_2} \left[\begin{array}{ccc|c} 1 & -1 & 0 & -5 \\ 0 & -1 & 1 & 15 \\ 4 & 8 & 3 & 475 \end{array}\right] \xrightarrow{R_3 - 4R_1 \to R_3} \left[\begin{array}{ccc|c} 1 & -1 & 0 & -5 \\ 0 & -1 & 1 & 15 \\ 0 & 12 & 3 & 495 \end{array}\right]$$

$$\xrightarrow[-R_2 \to R_2]{R_3 + 12R_2 \to R_3} \left[\begin{array}{ccc|c} 1 & -1 & 0 & -5 \\ 0 & 1 & -1 & -15 \\ 0 & 0 & 15 & 675 \end{array}\right] \xrightarrow{\frac{1}{15}R_3 \to R_3} \left[\begin{array}{ccc|c} 1 & -1 & 0 & -5 \\ 0 & 1 & -1 & -15 \\ 0 & 0 & 1 & 45 \end{array}\right]$$

$$\xrightarrow{R_2 + R_3 \to R_2} \left[\begin{array}{ccc|c} 1 & -1 & 0 & -5 \\ 0 & 1 & 0 & 30 \\ 0 & 0 & 1 & 45 \end{array}\right] \xrightarrow{R_1 + R_2 \to R_1} \left[\begin{array}{ccc|c} 1 & 0 & 0 & 25 \\ 0 & 1 & 0 & 30 \\ 0 & 0 & 1 & 45 \end{array}\right]$$

Hence, there are 25 general admission tickets, 30 reserved tickets, and 45 end zone tickets.

100. Let $x = \#$ of minutes walking, $y = \#$ of minutes step-up exercise, and $z = \#$ of minutes weight training.

a. # calories per minute for each exercise:

$$\text{walking: } \frac{85 \text{ cal}}{15 \text{ min}} = \frac{17}{3} \text{ cal/min}$$

$$\text{step-up: } \frac{45 \text{ cal}}{10 \text{ min}} = \frac{9}{2} \text{ cal/min}$$

$$\text{weight training: } \frac{137 \text{ cal}}{20 \text{ min}} = \frac{137}{20} \text{ cal/min}$$

b. Solve the system using matrix methods:

$$\begin{cases} z = 2x \\ x + y + z = 60 \\ \frac{17}{3}x + \frac{9}{2}y + \frac{137}{20}z = 358 \end{cases}$$

$$\left[\begin{array}{ccc|c} 2 & 0 & -1 & 0 \\ 1 & 1 & 1 & 60 \\ \frac{17}{3} & \frac{9}{2} & \frac{137}{20} & 358 \end{array}\right] \xrightarrow[60R_3 \to R_3]{R_1 - 2R_2 \to R_1} \left[\begin{array}{ccc|c} 0 & -2 & -3 & -120 \\ 1 & 1 & 1 & 60 \\ 340 & 270 & 411 & 21{,}480 \end{array}\right]$$

$$\xrightarrow{R_3 - 340R_2 \to R_3} \left[\begin{array}{ccc|c} 0 & -2 & -3 & -120 \\ 1 & 1 & 1 & 60 \\ 0 & -70 & 71 & 1080 \end{array}\right] \xrightarrow{R_1 \leftrightarrow R_2} \left[\begin{array}{ccc|c} 1 & 1 & 1 & 60 \\ 0 & -2 & -3 & -120 \\ 0 & -70 & 71 & 1080 \end{array}\right]$$

$$\xrightarrow{R_3 - 35R_2 \to R_3} \left[\begin{array}{ccc|c} 1 & 1 & 1 & 60 \\ 0 & -2 & -3 & -120 \\ 0 & 0 & 176 & 5280 \end{array}\right] \xrightarrow{\frac{1}{176}R_3 \to R_3} \left[\begin{array}{ccc|c} 1 & 1 & 1 & 60 \\ 0 & -2 & -3 & -120 \\ 0 & 0 & 1 & 30 \end{array}\right]$$

$$\xrightarrow{R_2 + 3R_3 \to R_2} \left[\begin{array}{ccc|c} 1 & 1 & 1 & 60 \\ 0 & -2 & 0 & -30 \\ 0 & 0 & 1 & 30 \end{array}\right] \xrightarrow{-\frac{1}{2}R_2 \to R_2} \left[\begin{array}{ccc|c} 1 & 1 & 1 & 60 \\ 0 & 1 & 0 & 15 \\ 0 & 0 & 1 & 30 \end{array}\right] \xrightarrow{R_1 - R_2 - R_3 \to R_1} \left[\begin{array}{ccc|c} 1 & 0 & 0 & 15 \\ 0 & 1 & 0 & 15 \\ 0 & 0 & 1 & 30 \end{array}\right]$$

Hence, 15 minutes walking, 15 minutes step-up exercise, and 30 minutes weight training.

101. Since the points $(4,4)$, $(-3,-1)$, and $(1,-3)$ are assumed to lie on the circle, they must each satisfy the equation. This generates the following system:

$$\begin{cases} (4)^2+(4)^2+a(4)+b(4)+c=0 \\ (-3)^2+(-1)^2+a(-3)+b(-1)+c=0 \\ (1)^2+(-3)^2+a(1)+b(-3)+c=0 \end{cases} \text{ which is equivalent to } \begin{cases} 4a+4b+c=-32 \\ -3a-b+c=-10 \\ a-3b+c=-10 \end{cases}$$

Write down the augmented matrix, and reduce it down to row echelon form:

$$\left[\begin{array}{ccc|c} 4 & 4 & 1 & -32 \\ -3 & -1 & 1 & -10 \\ 1 & -3 & 1 & -10 \end{array}\right] \xrightarrow[R_2+3R_3\to R_2]{R_1-4R_3\to R_3} \left[\begin{array}{ccc|c} 4 & 4 & 1 & -32 \\ 0 & -10 & 4 & -40 \\ 0 & 16 & -3 & 8 \end{array}\right] \xrightarrow[-\frac{1}{10}R_2\to R_2]{\frac{1}{4}R_1\to R_1} \left[\begin{array}{ccc|c} 1 & 1 & \frac{1}{4} & -8 \\ 0 & 1 & -\frac{2}{5} & 4 \\ 0 & 16 & -3 & 8 \end{array}\right]$$

$$\xrightarrow{R_3-16R_2\to R_3} \left[\begin{array}{ccc|c} 1 & 1 & \frac{1}{4} & -8 \\ 0 & 1 & -\frac{2}{5} & 4 \\ 0 & 0 & \frac{17}{5} & -56 \end{array}\right] \xrightarrow{\frac{5}{17}R_3\to R_3} \left[\begin{array}{ccc|c} 1 & 1 & \frac{1}{4} & -8 \\ 0 & 1 & -\frac{2}{5} & 4 \\ 0 & 0 & 1 & -\frac{280}{17} \end{array}\right]$$

Row 3 implies that $\boxed{c=-\frac{280}{17}}$.

Then, Row 2 implies that $b-\frac{2}{5}(-\frac{280}{17})=4$ so that $\boxed{b=-\frac{44}{17}}$.

Then, Row 1 implies that $a+(-\frac{44}{17})+\frac{1}{4}(-\frac{280}{17})=-8$ so that $\boxed{a=-\frac{22}{17}}$.

Hence, the equation of the circle is $x^2+y^2-\frac{22}{7}x-\frac{44}{17}y-\frac{280}{17}=0$.

102. Since the points $(0,7)$, $(6,1)$, and $(5,4)$ are assumed to lie on the circle, they must each satisfy the equation. This generates the following system:

$$\begin{cases} (0)^2+(7)^2+a(0)+b(7)+c=0 \\ (6)^2+(1)^2+a(6)+b(1)+c=0 \\ (5)^2+(4)^2+a(5)+b(4)+c=0 \end{cases} \text{ which is equivalent to } \begin{cases} 7b+c=-49 \\ 6a+b+c=-37 \\ 5a+4b+c=-41 \end{cases}$$

Write down the augmented matrix, and reduce it down to row echelon form:

$$\left[\begin{array}{ccc|c} 0 & 7 & 1 & -49 \\ 6 & 1 & 1 & -37 \\ 5 & 4 & 1 & -41 \end{array}\right] \xrightarrow{R_2 \leftrightarrow R_1} \left[\begin{array}{ccc|c} 6 & 1 & 1 & -37 \\ 0 & 7 & 1 & -49 \\ 5 & 4 & 1 & -41 \end{array}\right] \xrightarrow{\frac{1}{6}R_1 \to R_1} \left[\begin{array}{ccc|c} 1 & \frac{1}{6} & \frac{1}{6} & -\frac{37}{6} \\ 0 & 7 & 1 & -49 \\ 5 & 4 & 1 & -41 \end{array}\right]$$

$$\xrightarrow{R_3-5R_1 \to R_3} \left[\begin{array}{ccc|c} 1 & \frac{1}{6} & \frac{1}{6} & -\frac{37}{6} \\ 0 & 7 & 1 & -49 \\ 0 & \frac{19}{6} & \frac{1}{6} & -\frac{61}{6} \end{array}\right] \xrightarrow{\frac{1}{7}R_2 \to R_2} \left[\begin{array}{ccc|c} 1 & \frac{1}{6} & \frac{1}{6} & -\frac{37}{6} \\ 0 & 1 & \frac{1}{7} & -7 \\ 0 & \frac{19}{6} & \frac{1}{6} & -\frac{61}{6} \end{array}\right]$$

$$\xrightarrow{R_3-\frac{19}{6}R_2 \to R_3} \left[\begin{array}{ccc|c} 1 & \frac{1}{6} & \frac{1}{6} & -\frac{37}{6} \\ 0 & 1 & \frac{1}{7} & -7 \\ 0 & 0 & -\frac{2}{7} & 12 \end{array}\right] \xrightarrow{-\frac{7}{2}R_3 \to R_3} \left[\begin{array}{ccc|c} 1 & \frac{1}{6} & \frac{1}{6} & -\frac{37}{6} \\ 0 & 1 & \frac{1}{7} & -7 \\ 0 & 0 & 1 & -42 \end{array}\right]$$

Row 3 implies that $\boxed{c=-42}$.

Then, Row 2 implies that $b+\frac{1}{7}(-42)=-7$ so that $\boxed{b=-1}$.

Then, Row 1 implies that $a+\frac{1}{6}(-1)+\frac{1}{6}(-42)=-\frac{37}{6}$ so that $\boxed{a=1}$.

Hence, the equation of the circle is $x^2+y^2+x-y-42=0$.

103. Need to line up a single variable in a given column before forming the augmented matrix. Specifically, write the system as follows:

$$\begin{cases} -x+y+\ \ z=2 \\ \ \ x+y-2z=-3 \\ \ \ x+y+\ \ z=6 \end{cases}$$

So, the correct matrix is $\left[\begin{array}{ccc|c} -1 & 1 & 1 & 2 \\ 1 & 1 & -2 & -3 \\ 1 & 1 & 1 & 6 \end{array}\right]$, and after reducing, $\left[\begin{array}{ccc|c} 1 & 0 & 0 & 2 \\ 0 & 1 & 0 & 1 \\ 0 & 0 & 1 & 3 \end{array}\right]$.

104. **a.** Need to subtract all entries of $2R_1$ from the respective entries of R_2. So, the second row should be $0 \quad -1 \quad -1 \mid 0$.

b. Same error as in part **a**. The third row should be $0 \quad 4 \quad -1 \mid -12$.

105. Row 3 implies that $1z=0$, which is valid. Also, Rows 2 and 1 imply $y=2,\ x=1$, respectively. So, the solution is $x=1,\ y=2,\ z=0$.

<table>
<tr><td colspan="4">106. Row 3 implies that the system is inconsistent since it requires $0z = 4$.</td></tr>
<tr><td>107. False</td><td>108. False (See #47 for example.)</td><td>109. True</td><td>110. True</td></tr>
<tr><td colspan="4">111. False. For instance, the system $\begin{bmatrix} 1 & 1 \\ 0 & 0 \end{bmatrix}\begin{bmatrix} x \\ y \end{bmatrix} = \begin{bmatrix} 2 \\ 0 \end{bmatrix}$ has infinitely many solutions.</td></tr>
<tr><td colspan="4">112. True</td></tr>
<tr><td colspan="4">113. False. The system $\begin{bmatrix} 1 & 1 \\ 0 & 0 \end{bmatrix}\begin{bmatrix} x \\ y \end{bmatrix} = \begin{bmatrix} 2 \\ 2 \end{bmatrix}$ has no solution.</td></tr>
<tr><td colspan="4">114. False.</td></tr>
</table>

115. Assume year 1999 corresponds to 0, 2000 to 1, 2001 to 2, 2002 to 3, and 2003 to 4. Then, from the given data, we can infer that the following four points are on the graph of $f(x) = ax^4 + bx^3 + cx^2 + dx + e$:

$$(0,44),\ (1,51),\ (2,46),\ (3,51),\ (4,44)$$

Hence, they must all satisfy the equation. This leads to the following 5×5 system:

$$\begin{cases} 44 = a(0)^4 + b(0)^3 + c(0)^2 + d(0) + e \\ 51 = a(1)^4 + b(1)^3 + c(1)^2 + d(1) + e \\ 46 = a(2)^4 + b(2)^3 + c(2)^2 + d(2) + e \\ 51 = a(3)^4 + b(3)^3 + c(3)^2 + d(3) + e \\ 44 = a(4)^4 + b(4)^3 + c(4)^2 + d(4) + e \end{cases}$$

which is equivalent to:

$$\begin{cases} 44 = e \\ 51 = a + b + c + d + e \\ 46 = 16a + 8b + 4c + 2d + e \\ 51 = 81a + 27b + 9c + 3d + e \\ 44 = 256a + 64b + 16c + 4d + e \end{cases}$$

The corresponding augmented matrix is:

$$\left[\begin{array}{ccccc|c} 0 & 0 & 0 & 0 & 1 & 44 \\ 1 & 1 & 1 & 1 & 1 & 51 \\ 16 & 8 & 4 & 2 & 1 & 46 \\ 81 & 27 & 9 & 3 & 1 & 51 \\ 256 & 64 & 16 & 4 & 1 & 44 \end{array}\right]$$

One can perform Gaussian elimination as in the previous problems, but this would be a fine example as to when using technology can be useful. Indeed, using a calculator with matrix capabilities can be used to verify the following choices for *a*, *b*, *c*, *d*, and *e*: $a = -\frac{11}{6}$, $b = \frac{44}{3}$, $c = -\frac{223}{6}$, $d = \frac{94}{3}$, $e = 44$. Hence, the equation of the polynomial is

$$\boxed{y = -\tfrac{11}{6}x^4 + \tfrac{44}{3}x^3 - \tfrac{223}{6}x^2 + \tfrac{94}{3}x + 44}.$$

116. Let x = number of nickels, y = number of dimes, z = number of quarters.
We must solve the system:

$$\begin{cases} x + y + z = 30 \\ 0.05x + 0.10y + 0.25z = 4.60 \end{cases}$$

The corresponding augmented matrix is:

$$\left[\begin{array}{ccc|c} 1 & 1 & 1 & 30 \\ 0.05 & 0.10 & 0.25 & 4.60 \end{array}\right]$$

We solve this system by reducing the augmented matrix to row echelon form:

$$\left[\begin{array}{ccc|c} 1 & 1 & 1 & 30 \\ 0.05 & 0.10 & 0.25 & 4.60 \end{array}\right] \xrightarrow{100R_2 \to R_2} \left[\begin{array}{ccc|c} 1 & 1 & 1 & 30 \\ 5 & 10 & 25 & 460 \end{array}\right]$$

$$\xrightarrow{R_2 - 5R_1 \to R_2} \left[\begin{array}{ccc|c} 1 & 1 & 1 & 30 \\ 0 & 5 & 20 & 310 \end{array}\right] \xrightarrow{\frac{1}{5}R_2 \to R_2} \left[\begin{array}{ccc|c} 1 & 1 & 1 & 30 \\ 0 & 1 & 4 & 62 \end{array}\right]$$

Let $z = t$. Then, substituting this value into the equation obtained from the second row yields $y = 62 - 4t$, and then substituting these values of z and y into the equation obtained form the first row, we conclude that $x = 3t - 32$.

Now, the restrictions on t in this problem are that **(i)** $0 \le 62 - 4t \le 30$, **(ii)** $0 \le 3t - 32 \le 30$, and **(iii)** t is an integer. Solving these inequalities yields

$$\begin{array}{rcccl} 0 \le & 62 - 4t & \le 30 \\ -62 \le & -4t & \le -32 \\ 15.5 \ge & t & \ge 8 \end{array} \quad \text{and} \quad \begin{array}{rcccl} 0 \le & 3t - 32 & \le 30 \\ 32 \le & 3t & \le 62 \\ 10.\overline{6} \le & t & \le 20.\overline{6} \end{array}$$

Since both of these inequalities must hold simultaneously and t must be an integer, we conclude that the possible values of t are 11, 12, 13, 14, and 15. We verify that each of these is a solution below:

t	$x = 3t - 32$	$y = 62 - 4t$	$z = t$
11	1	18	11
12	4	14	12
13	7	10	13
14	10	6	14
15	13	2	15

117. Let x = rate of the boat in still water, y = rate of current
Using Distance = Rate x Time yields the following two equations:
City A to City B: $(x+y)(5)=d$, which is equivalent to $x+y=\frac{d}{5}$ **(1)**.

City B to City A: $(x-y)(7)=d$, which is equivalent to $x-y=\frac{d}{7}$ **(2)**.

Adding **(1)** and **(2)** yields: $x=\frac{6d}{35}$.

Substituting this into **(2)** then yields: $y=\frac{d}{35}$

Hence, the time it takes to make the trip from City A to City B if the rate is $\frac{d}{35}$ mph is

$$\tfrac{d}{35}t=d \Rightarrow t=35 \text{ hours}.$$

118. Solve the system:

$$\begin{cases} \frac{3}{x}-\frac{4}{y}+\frac{6}{z}=1 & \textbf{(1)} \\ \frac{9}{x}+\frac{8}{y}-\frac{12}{z}=3 & \textbf{(2)} \\ \frac{9}{x}-\frac{4}{y}+\frac{12}{z}=4 & \textbf{(3)} \end{cases}$$

Let

$$u=\tfrac{1}{x},\ v=\tfrac{1}{y},\ w=\tfrac{1}{z} \quad \textbf{(*)}$$

in this system to obtain the equivalent system in simpler form:

$$\begin{cases} 3u-4v+6w=1 & \textbf{(4)} \\ 9u+8v-12w=3 & \textbf{(5)} \\ 9u-4v+12w=4 & \textbf{(6)} \end{cases}$$

Eliminate *u*:
Multiply **(4)** by -3 and add to **(5)**: $20v-30w=0$ **(7)**
Multiply **(4)** by -3 and add to **(6)**: $8v-6w=1$ **(8)**
Now, solve the system comprised of equations **(7)** and **(8)**:
Multiply **(8)** by -5 and add to **(7)**: $v=\frac{1}{4}$
Substitute this into **(7)**: $w=\frac{1}{6}$
Finally, substitute both of these values into **(4)**: $3u-4\left(\frac{1}{4}\right)+6\left(\frac{1}{6}\right)=1 \Rightarrow u=\frac{1}{3}$.

Now, find the corresponding values of the original variables x, y, and z by substituting these values into **(*)**: $x=3, y=4, z=6$

119. We must find pairwise intersections of these lines. This requires that we solve three 2x2 systems.

Solve $\begin{cases} x - y = -3 & \textbf{(1)} \\ 3x + 4y = 5 & \textbf{(2)} \end{cases}$

Solve **(1)** for y: $y = x + 3$

Substitute this into **(2)**: $x = -1$

Substitute this back into **(1)**: $y = 2$.

So, one vertex is (-1, 2).

Solve $\begin{cases} x - y = -3 & \textbf{(3)} \\ 6x + y = 17 & \textbf{(4)} \end{cases}$

Add **(3)** and **(4)**: $x = 2$

Substitute this into **(3)**: $y = -1$

So, one vertex is (2, -1).

Solve $\begin{cases} 3x + 4y = 5 & \textbf{(5)} \\ 6x + y = 17 & \textbf{(6)} \end{cases}$

Multiply **(6)** by -4 and add to **(5)**: $x = 3$

Substitute this into **(5)**: $y = -1$

So, one vertex is (3, -1).

Thus, the vertices are (-1, 2), (2, -1), and (3, -1).

120. Solve the following system obtained from analyzing each of the three wires:

$$\begin{cases} \frac{1}{6}A + \frac{1}{5}B + \frac{3}{19}C = 17 \\ \frac{1}{3}A + \frac{1}{3}B + \frac{7}{19}C = 35 \\ \frac{1}{2}A + \frac{7}{15}B + \frac{9}{19}C = 47 \end{cases}$$

Using Gaussian elimination, we proceed as follows:

$$\left[\begin{array}{ccc|c} \frac{1}{6} & \frac{1}{5} & \frac{3}{19} & 17 \\ \frac{1}{3} & \frac{1}{3} & \frac{7}{19} & 35 \\ \frac{1}{2} & \frac{7}{15} & \frac{9}{19} & 47 \end{array}\right] \xrightarrow{6R_1 \to R_1} \left[\begin{array}{ccc|c} 1 & \frac{6}{5} & \frac{18}{19} & 102 \\ \frac{1}{3} & \frac{1}{3} & \frac{7}{19} & 35 \\ \frac{1}{2} & \frac{7}{15} & \frac{9}{19} & 47 \end{array}\right] \longrightarrow \left[\begin{array}{ccc|c} 1 & \frac{6}{5} & \frac{18}{19} & 102 \\ 0 & -\frac{1}{15} & \frac{1}{19} & 1 \\ 0 & -\frac{2}{15} & 0 & -4 \end{array}\right]$$

$$\xrightarrow[-15R_3 \to R_3]{-15R_2 \to R_2} \left[\begin{array}{ccc|c} 1 & \frac{6}{5} & \frac{18}{19} & 102 \\ 0 & 1 & -\frac{15}{19} & -15 \\ 0 & 1 & 0 & 30 \end{array}\right] \xrightarrow{R_2 \leftrightarrow R_3} \left[\begin{array}{ccc|c} 1 & \frac{6}{5} & \frac{18}{19} & 102 \\ 0 & 1 & 0 & 30 \\ 0 & 1 & -\frac{15}{19} & -15 \end{array}\right] \xrightarrow{R_3 - R_2 \to R_3} \left[\begin{array}{ccc|c} 1 & \frac{6}{5} & \frac{18}{19} & 102 \\ 0 & 1 & 0 & 30 \\ 0 & 0 & -\frac{15}{19} & -45 \end{array}\right]$$

$$\xrightarrow{-\frac{19}{15}R_3 \to R_3} \left[\begin{array}{ccc|c} 1 & \frac{6}{5} & \frac{18}{19} & 102 \\ 0 & 1 & 0 & 30 \\ 0 & 0 & 1 & 57 \end{array}\right] \xrightarrow{R_1 - \frac{18}{19}R_3 \to R_1} \left[\begin{array}{ccc|c} 1 & \frac{6}{5} & 0 & 48 \\ 0 & 1 & 0 & 30 \\ 0 & 0 & 1 & 57 \end{array}\right] \xrightarrow{R_1 - \frac{6}{5}R_2 \to R_1} \left[\begin{array}{ccc|c} 1 & 0 & 0 & 12 \\ 0 & 1 & 0 & 30 \\ 0 & 0 & 1 & 57 \end{array}\right]$$

So, there are 12 liters of A, 30 liters of B, and 57 liters of C.

121. The answers coincide.

$rref([A])$

$$\begin{matrix} [[1 & 0 & -4 & 41] \\ [0 & 1 & -3 & 31] \\ [0 & 0 & 0 & 0]] \end{matrix}$$

122. The answers coincide.

123.

a.

```
rref([A])
[[1 0 0 -.23653…
 [0 1 0 .928846…
 [0 0 1 6.08846…
```

$y = -0.24x^2 + 0.93x + 6.09$

b.

```
QuadReg
 y=ax²+bx+c
 a=-.2365384615
 b=.9288461538
 c=6.088461538
```

$y = -0.24x^2 + 0.93x + 6.09$

124.

a.

```
rref([A])
[[1 0 0 .361616…
 [0 1 0 -.92323…
 [0 0 1 -17.6   …
```

$y = 0.36x^2 - 0.92x - 17.6$

b.

```
QuadReg
 y=ax²+bx+c
 a=.3616161616
 b=-.9232323232
 c=-17.6
```

$y = 0.36x^2 - 0.92x - 17.6$

125. Solve the system:

$$\begin{cases} c_1 + c_2 = 0 & \textbf{(1)} \\ c_1 + 5c_2 = -3 & \textbf{(2)} \end{cases}$$

Solve **(1)** for c_1: $c_1 = -c_2$ **(3)**

Substitute **(3)** into **(2)**: $c_2 = -\frac{3}{4}$

Substitute this back into **(3)**: $c_1 = \frac{3}{4}$.

Thus, $c_1 = \frac{3}{4}$, $c_2 = -\frac{3}{4}$.

126. Solve the system:

$$\begin{cases} 3c_1 + 3c_2 = 0 & \textbf{(1)} \\ 2c_1 + 3c_2 = 0 & \textbf{(2)} \end{cases}$$

Solve **(1)** for c_1: $c_2 = -c_1$ **(3)**

Substitute **(3)** into **(2)**: $c_1 = 0$

Substitute this back into **(3)**: $c_2 = 0$.

Thus, $c_1 = 0$, $c_2 = 0$.

127. Solve the system: $\begin{cases} 2c_1+2c_2+2c_3=0 & \textbf{(1)} \\ 2c_1-2c_3=2 & \textbf{(2)} \\ c_1-c_2+c_3=6 & \textbf{(3)} \end{cases}$ Solve **(2)** for c_3: $c_3=c_1-1$ **(4)** Substitute **(4)** into **(1)** and **(3)** to obtain the system: $\begin{cases} 4c_1+2c_2=2 & \textbf{(5)} \\ 2c_1-c_2=7 & \textbf{(6)} \end{cases}$	To solve this system, multiply **(6)** by 2 and add to **(5)**: $c_1=2$ Substitute this into **(6)**: $c_2=-3$ Substitute these both into **(1)**: $c_3=1$. Thus, $c_1=2$, $c_2=-3$, and $c_3=1$.
128. Solve the system: $\begin{cases} c_1+c_4=1 & \textbf{(1)} \\ c_3=1 & \textbf{(2)} \\ c_2+3c_4=1 & \textbf{(3)} \\ c_1-2c_3=1 & \textbf{(4)} \end{cases}$	Substitute **(2)** into **(4)**: $c_1=3$ Substitute this into **(1)**: $c_4=-2$ Substitute this into **(3)**: $c_2=7$. Thus, $c_1=3$, $c_2=7$, $c_3=1$, and $c_4=-2$.

Section 8.4 Solutions --

1. 2×3	**2.** 3×2
3. 2×2	**4.** 1×4
5. 3×3	**6.** 4×1
7. 4×4	**8.** 2×4
9. Since corresponding entries of equal matrices must themselves be equal, we have $x=-5, y=1$.	**10.** Since corresponding entries of equal matrices must themselves be equal, we have $x=10, y=12$
11. Since corresponding entries of equal matrices must themselves be equal, we have $\begin{cases} x+y=-5 & \textbf{(1)} \\ x-y=-1 & \textbf{(2)} \\ z=3 & \end{cases}$ <u>Solve the system</u>: Add **(1)** and **(2)**: $2x=-6 \Rightarrow x=-3$ Substitute this into **(1)** to obtain $y=-2$.	**12.** Since corresponding entries of equal matrices must themselves be equal, we have $\begin{cases} x=2+y & \textbf{(1)} \\ y=5 & \textbf{(2)} \end{cases}$ <u>Solve the system</u>: Substitute **(2)** into **(1)** to obtain $x=7$.

13. Since corresponding entries of equal matrices must themselves be equal, we have $\begin{cases} x-y=3 & \textbf{(1)} \\ x+2y=12 & \textbf{(2)} \end{cases}$ Solve the system: Multiply **(1)** by -1: $-x+y=-3$ **(3)** Add **(2)** and **(3)**: $3y=9 \Rightarrow y=3$ Substitute this into **(1)** to obtain $x=6$.	**14.** Since corresponding entries of equal matrices must themselves be equal, we have $\begin{cases} a^2=9 & \textbf{(1)} \\ 2b+1=9 & \textbf{(2)} \\ b^2=16 & \textbf{(3)} \\ 2a+1=-5 & \textbf{(4)} \end{cases}$ Note that **(2)** implies $b=4$, and this satisfies **(3)**. Also, **(4)** implies $a=-3$, and this satisfies **(1)**.
15. $\begin{bmatrix} -1 & 3 & 0 \\ 2 & 4 & 1 \end{bmatrix}+\begin{bmatrix} 0 & 2 & 1 \\ 3 & -2 & 4 \end{bmatrix}=\begin{bmatrix} -1 & 5 & 1 \\ 5 & 2 & 5 \end{bmatrix}$	**16.** $\begin{bmatrix} 2 & -2 \\ 2 & 0 \\ 7 & -1 \end{bmatrix}$
17. $\begin{bmatrix} -2 & 4 \\ 2 & -2 \\ -1 & 3 \end{bmatrix}$	**18.** $\begin{bmatrix} -1 & 3 & 0 \\ 2 & 4 & 1 \end{bmatrix}-\begin{bmatrix} 0 & 2 & 1 \\ 3 & -2 & 4 \end{bmatrix}=\begin{bmatrix} -1 & 1 & -1 \\ -1 & 6 & -3 \end{bmatrix}$
19. Not defined since *B* is 2×3 and *C* is 3×2	**20.** Not defined since *A* is 2×3 and *D* is 3×2
21. This is not defined since the matrices have different orders.	**22.** This is not defined since the matrices have different orders.

23.

$$2\begin{bmatrix} -1 & 3 & 0 \\ 2 & 4 & 1 \end{bmatrix}+3\begin{bmatrix} 0 & 2 & 1 \\ 3 & -2 & 4 \end{bmatrix}=\begin{bmatrix} -2 & 6 & 0 \\ 4 & 8 & 2 \end{bmatrix}+\begin{bmatrix} 0 & 6 & 3 \\ 9 & -6 & 12 \end{bmatrix}=\begin{bmatrix} -2 & 12 & 3 \\ 13 & 2 & 14 \end{bmatrix}$$

24.

$$2\begin{bmatrix} 0 & 2 & 1 \\ 3 & -2 & 4 \end{bmatrix}-3\begin{bmatrix} -1 & 3 & 0 \\ 2 & 4 & 1 \end{bmatrix}=\begin{bmatrix} 0 & 4 & 2 \\ 6 & -4 & 8 \end{bmatrix}+\begin{bmatrix} 3 & -9 & 0 \\ -6 & -12 & -3 \end{bmatrix}=\begin{bmatrix} 3 & -5 & 2 \\ 0 & -16 & 5 \end{bmatrix}$$

25. $\begin{bmatrix} 8 & 3 \\ 11 & 5 \end{bmatrix}$	**26.** $[5]$
27. $\begin{bmatrix} -3 & 21 & 6 \\ -4 & 7 & 1 \\ 13 & 14 & 9 \end{bmatrix}$	**28.** $\begin{bmatrix} 2 & -2 \\ 20 & 21 \\ 13 & 14 \end{bmatrix}$

29. $\begin{bmatrix} 3 & 6 \\ -2 & -2 \\ 17 & 24 \end{bmatrix}$	**30.** $\begin{bmatrix} -1 & 4 & 1 \\ 6 & 9 & 14 \\ 1 & 3 & 11 \end{bmatrix}$
31. Not possible you cannot multiply a 2×2 matrix by a 3×2 matrix.	**32.** Not possible since ED is 3×2 while C is 2×3. Matrices must have the same orders to be added.
33. $-4\begin{bmatrix} 0 & -15 \end{bmatrix} = \begin{bmatrix} 0 & 60 \end{bmatrix}$	**34.** $-3\begin{bmatrix} -1 & 5 \\ 15 & 19 \\ 2 & 24 \end{bmatrix} = \begin{bmatrix} 3 & -15 \\ -45 & -57 \\ -6 & -72 \end{bmatrix}$
35. $\begin{bmatrix} 2 & 0 & -3 \end{bmatrix}\begin{bmatrix} 0 & 2 & 0 \\ 2 & 4 & 5 \\ 2 & 1 & 3 \end{bmatrix} = \begin{bmatrix} -6 & 1 & -9 \end{bmatrix}$	**36.** $\begin{bmatrix} 5 & 7 & 4 \\ 9 & 21 & 10 \end{bmatrix} + \begin{bmatrix} -5 & 35 & 10 \\ 15 & 0 & 5 \end{bmatrix} = \begin{bmatrix} 0 & 42 & 14 \\ 24 & 21 & 15 \end{bmatrix}$
37. $\begin{bmatrix} 2 & 0 & -3 \\ 0 & 0 & 0 \\ -2 & 0 & 3 \end{bmatrix} + \begin{bmatrix} 5 & 10 & -5 \\ 0 & 15 & 5 \\ 25 & 0 & -10 \end{bmatrix} = \begin{bmatrix} 7 & 10 & -8 \\ 0 & 15 & 5 \\ 23 & 0 & -7 \end{bmatrix}$	**38.** $AA = \begin{bmatrix} -4 & 8 & 3 \\ 5 & 9 & 1 \\ -5 & 10 & -1 \end{bmatrix}$
39 $\begin{bmatrix} 7 & 10 \\ 15 & 22 \end{bmatrix} + \begin{bmatrix} 5 & 10 \\ 15 & 20 \end{bmatrix} = \begin{bmatrix} 12 & 20 \\ 30 & 42 \end{bmatrix}$	**40** $\begin{bmatrix} -1 & 7 & 2 \\ 3 & 0 & 1 \end{bmatrix}\begin{bmatrix} -2 & 0 & 2 \\ 4 & 2 & 8 \\ -6 & 2 & 10 \end{bmatrix} = \begin{bmatrix} 18 & 18 & 74 \\ -12 & 2 & 16 \end{bmatrix}$
41 $\begin{bmatrix} -2 & 0 & 2 \\ 4 & 2 & 8 \\ -6 & 2 & 10 \end{bmatrix}\begin{bmatrix} 1 \\ 0 \\ -1 \end{bmatrix} = \begin{bmatrix} -4 \\ -4 \\ -16 \end{bmatrix}$	**42** $\begin{bmatrix} 9 & 19 & 4 \\ 8 & 6 & -5 \end{bmatrix} + \begin{bmatrix} -5 & 35 & 10 \\ 15 & 0 & 5 \end{bmatrix} = \begin{bmatrix} 4 & 54 & 14 \\ 23 & 6 & 0 \end{bmatrix}$
43. Not possible since the inner dimensions are different.	**44.** $\begin{bmatrix} 6 & 1 & 4 \\ 3 & 4 & 17 \\ 1 & -2 & -5 \end{bmatrix}$

45. Since $AB = \begin{bmatrix} 8 & -11 \\ -5 & 7 \end{bmatrix} \cdot \begin{bmatrix} 7 & 11 \\ 5 & 8 \end{bmatrix} = \begin{bmatrix} 1 & 0 \\ 0 & 1 \end{bmatrix} = I$, B must be the inverse of A.

46. Since $AB = \begin{bmatrix} 7 & -9 \\ -3 & 4 \end{bmatrix} \cdot \begin{bmatrix} 4 & 9 \\ 3 & 7 \end{bmatrix} = \begin{bmatrix} 1 & 0 \\ 0 & 1 \end{bmatrix} = I$, B must be the inverse of A.

47. Since $AB = \begin{bmatrix} 3 & 1 \\ 1 & -2 \end{bmatrix} \cdot \begin{bmatrix} \frac{2}{7} & \frac{1}{7} \\ \frac{1}{7} & -\frac{3}{7} \end{bmatrix} = \begin{bmatrix} 1 & 0 \\ 0 & 1 \end{bmatrix} = I$, B must be the inverse of A.

48. Since $AB = \begin{bmatrix} 2 & 3 \\ 1 & -1 \end{bmatrix} \cdot \begin{bmatrix} \frac{1}{5} & \frac{3}{5} \\ \frac{1}{5} & -\frac{2}{5} \end{bmatrix} = \begin{bmatrix} 1 & 0 \\ 0 & 1 \end{bmatrix} = I$, B is the inverse of A.

49. Since $AB = \begin{bmatrix} 1 & -1 & 1 \\ 1 & 0 & -1 \\ 0 & 1 & -1 \end{bmatrix} \cdot \begin{bmatrix} 1 & 0 & 1 \\ 1 & -1 & 2 \\ 1 & -1 & 1 \end{bmatrix} = \begin{bmatrix} 1 & 0 & 0 \\ 0 & 1 & 0 \\ 0 & 0 & 1 \end{bmatrix} = I$, B must be the inverse of A.

50. Since $AB = \begin{bmatrix} -1 & 0 & -1 \\ -1 & 1 & -2 \\ -1 & 1 & -1 \end{bmatrix} \cdot \begin{bmatrix} -1 & 1 & -1 \\ -1 & 0 & 1 \\ 0 & -1 & 1 \end{bmatrix} = \begin{bmatrix} 1 & 0 & 0 \\ 0 & 1 & 0 \\ 0 & 0 & 1 \end{bmatrix} = I$, B must be the inverse of A.

51. $A^{-1} = \frac{1}{1}\begin{bmatrix} 0 & -1 \\ 1 & 2 \end{bmatrix} = \begin{bmatrix} 0 & -1 \\ 1 & 2 \end{bmatrix}$.

52. $A^{-1} = \frac{1}{1}\begin{bmatrix} 1 & -1 \\ -2 & 3 \end{bmatrix} = \begin{bmatrix} 1 & -1 \\ -2 & 3 \end{bmatrix}$.

53. $A^{-1} = \frac{1}{-\frac{39}{4}}\begin{bmatrix} \frac{3}{4} & -2 \\ -5 & \frac{1}{3} \end{bmatrix} = \begin{bmatrix} -\frac{4}{39}(\frac{3}{4}) & -\frac{4}{39}(-2) \\ -\frac{4}{39}(-5) & -\frac{4}{39}(\frac{1}{3}) \end{bmatrix} = \begin{bmatrix} -\frac{1}{13} & \frac{8}{39} \\ \frac{20}{39} & -\frac{4}{117} \end{bmatrix}$.

54. $A^{-1} = \frac{1}{-\frac{1}{2}}\begin{bmatrix} \frac{2}{3} & -2 \\ -\frac{1}{3} & \frac{1}{4} \end{bmatrix} = \begin{bmatrix} -2(\frac{2}{3}) & -2(-2) \\ -2(-\frac{1}{3}) & -2(\frac{1}{4}) \end{bmatrix} = \begin{bmatrix} -\frac{4}{3} & 4 \\ \frac{2}{3} & -\frac{1}{2} \end{bmatrix}$.

55. Using Gaussian elimination, we obtain the following:

$$\left[\begin{array}{ccc|ccc} 1 & 1 & 1 & 1 & 0 & 0 \\ 1 & -1 & -1 & 0 & 1 & 0 \\ -1 & 1 & -1 & 0 & 0 & 1 \end{array}\right] \xrightarrow{R_2+R_3 \to R_3} \left[\begin{array}{ccc|ccc} 1 & 1 & 1 & 1 & 0 & 0 \\ 1 & -1 & -1 & 0 & 1 & 0 \\ 0 & 0 & -2 & 0 & 1 & 1 \end{array}\right]$$

$$\xrightarrow{R_1+R_2 \to R_1} \left[\begin{array}{ccc|ccc} 2 & 0 & 0 & 1 & 1 & 0 \\ 1 & -1 & -1 & 0 & 1 & 0 \\ 0 & 0 & -2 & 0 & 1 & 1 \end{array}\right] \xrightarrow[-\frac{1}{2}R_3 \to R_3]{\frac{1}{2}R_1 \to R_1} \left[\begin{array}{ccc|ccc} 1 & 0 & 0 & \frac{1}{2} & \frac{1}{2} & 0 \\ 1 & -1 & -1 & 0 & 1 & 0 \\ 0 & 0 & 1 & 0 & -\frac{1}{2} & -\frac{1}{2} \end{array}\right]$$

$$\xrightarrow{R_1-R_2 \to R_2} \left[\begin{array}{ccc|ccc} 1 & 0 & 0 & \frac{1}{2} & \frac{1}{2} & 0 \\ 0 & 1 & 1 & \frac{1}{2} & -\frac{1}{2} & 0 \\ 0 & 0 & 1 & 0 & -\frac{1}{2} & -\frac{1}{2} \end{array}\right] \xrightarrow{R_2-R_3 \to R_2} \left[\begin{array}{ccc|ccc} 1 & 0 & 0 & \frac{1}{2} & \frac{1}{2} & 0 \\ 0 & 1 & 0 & \frac{1}{2} & 0 & \frac{1}{2} \\ 0 & 0 & 1 & 0 & -\frac{1}{2} & -\frac{1}{2} \end{array}\right]$$

So, $A^{-1} = \begin{bmatrix} \frac{1}{2} & \frac{1}{2} & 0 \\ \frac{1}{2} & 0 & \frac{1}{2} \\ 0 & -\frac{1}{2} & -\frac{1}{2} \end{bmatrix}$.

56. Using Gaussian elimination, we obtain the following:

$$\left[\begin{array}{ccc|ccc} 1 & -1 & 1 & 1 & 0 & 0 \\ 1 & 1 & 1 & 0 & 1 & 0 \\ -1 & 2 & -3 & 0 & 0 & 1 \end{array}\right] \xrightarrow{R_1-R_2\to R_2} \left[\begin{array}{ccc|ccc} 1 & -1 & 1 & 1 & 0 & 0 \\ 0 & -2 & 0 & 1 & -1 & 0 \\ -1 & 2 & -3 & 0 & 0 & 1 \end{array}\right]$$

$$\xrightarrow{R_1+R_3\to R_3} \left[\begin{array}{ccc|ccc} 1 & -1 & 1 & 1 & 0 & 0 \\ 0 & -2 & 0 & 1 & -1 & 0 \\ 0 & 1 & -2 & 1 & 0 & 1 \end{array}\right] \xrightarrow{2R_3+R_2\to R_3} \left[\begin{array}{ccc|ccc} 1 & -1 & 1 & 1 & 0 & 0 \\ 0 & -2 & 0 & 1 & -1 & 0 \\ 0 & 0 & -4 & 3 & -1 & 2 \end{array}\right]$$

$$\xrightarrow[-\frac{1}{4}R_3\to R_3]{-\frac{1}{2}R_2\to R_2} \left[\begin{array}{ccc|ccc} 1 & -1 & 1 & 1 & 0 & 0 \\ 0 & 1 & 0 & -\frac{1}{2} & \frac{1}{2} & 0 \\ 0 & 0 & 1 & -\frac{3}{4} & \frac{1}{4} & -\frac{1}{2} \end{array}\right] \xrightarrow{R_1+R_2\to R_1} \left[\begin{array}{ccc|ccc} 1 & 0 & 1 & \frac{1}{2} & \frac{1}{2} & 0 \\ 0 & 1 & 0 & -\frac{1}{2} & \frac{1}{2} & 0 \\ 0 & 0 & 1 & -\frac{3}{4} & \frac{1}{4} & -\frac{1}{2} \end{array}\right]$$

$$\xrightarrow{R_1-R_3\to R_1} \left[\begin{array}{ccc|ccc} 1 & 0 & 0 & \frac{5}{4} & \frac{1}{4} & \frac{1}{2} \\ 0 & 1 & 0 & -\frac{1}{2} & \frac{1}{2} & 0 \\ 0 & 0 & 1 & -\frac{3}{4} & \frac{1}{4} & -\frac{1}{2} \end{array}\right] \quad \text{So, } A^{-1} = \begin{bmatrix} \frac{5}{4} & \frac{1}{4} & \frac{1}{2} \\ -\frac{1}{2} & \frac{1}{2} & 0 \\ -\frac{3}{4} & \frac{1}{4} & -\frac{1}{2} \end{bmatrix}.$$

57. Using Gaussian elimination, we obtain the following:

$$\left[\begin{array}{ccc|ccc} 1 & 0 & 1 & 1 & 0 & 0 \\ 0 & 1 & 1 & 0 & 1 & 0 \\ 1 & -1 & 0 & 0 & 0 & 1 \end{array}\right] \xrightarrow{R_2+R_3\to R_3} \left[\begin{array}{ccc|ccc} 1 & 0 & 1 & 1 & 0 & 0 \\ 0 & 1 & 1 & 0 & 1 & 0 \\ 1 & 0 & 1 & 0 & 1 & 1 \end{array}\right] \xrightarrow{R_1-R_3\to R_3} \left[\begin{array}{ccc|ccc} 1 & 0 & 1 & 1 & 0 & 0 \\ 0 & 1 & 1 & 0 & 1 & 1 \\ 0 & 0 & 0 & 1 & -1 & -1 \end{array}\right]$$

From the bottom row, we deduce that A is not invertible.

58. Using Gaussian elimination, we obtain the following:

$$\left[\begin{array}{ccc|ccc} 1 & 2 & -3 & 1 & 0 & 0 \\ 1 & -1 & -1 & 0 & 1 & 0 \\ 1 & 0 & -4 & 0 & 0 & 1 \end{array}\right] \xrightarrow[R_1-R_3\to R_3]{R_1-R_2\to R_2} \left[\begin{array}{ccc|ccc} 1 & 2 & -3 & 1 & 0 & 0 \\ 0 & 3 & -2 & 1 & -1 & 0 \\ 0 & 2 & 1 & 1 & 0 & -1 \end{array}\right]$$

$$\xrightarrow{\frac{1}{3}R_2\to R_2} \left[\begin{array}{ccc|ccc} 1 & 2 & -3 & 1 & 0 & 0 \\ 0 & 1 & -\frac{2}{3} & \frac{1}{3} & -\frac{1}{3} & 0 \\ 0 & 2 & 1 & 1 & 0 & -1 \end{array}\right] \xrightarrow{R_3-2R_2\to R_3} \left[\begin{array}{ccc|ccc} 1 & 2 & -3 & 1 & 0 & 0 \\ 0 & 1 & -\frac{2}{3} & \frac{1}{3} & -\frac{1}{3} & 0 \\ 0 & 0 & \frac{7}{3} & \frac{1}{3} & \frac{2}{3} & -1 \end{array}\right]$$

$$\xrightarrow{\frac{3}{7}R_3\to R_3} \left[\begin{array}{ccc|ccc} 1 & 2 & -3 & 1 & 0 & 0 \\ 0 & 1 & -\frac{2}{3} & \frac{1}{3} & -\frac{1}{3} & 0 \\ 0 & 0 & 1 & \frac{1}{7} & \frac{2}{7} & -\frac{3}{7} \end{array}\right] \xrightarrow{R_2+\frac{2}{3}R_3\to R_2} \left[\begin{array}{ccc|ccc} 1 & 2 & -3 & 1 & 0 & 0 \\ 0 & 1 & 0 & \frac{3}{7} & -\frac{1}{7} & -\frac{2}{7} \\ 0 & 0 & 1 & \frac{1}{7} & \frac{2}{7} & -\frac{3}{7} \end{array}\right]$$

$$\xrightarrow{R_1-2R_2+3R_3\to R_1} \left[\begin{array}{ccc|ccc} 1 & 0 & 0 & \frac{4}{7} & \frac{8}{7} & -\frac{5}{7} \\ 0 & 1 & 0 & \frac{3}{7} & -\frac{1}{7} & -\frac{2}{7} \\ 0 & 0 & 1 & \frac{1}{7} & \frac{2}{7} & -\frac{3}{7} \end{array}\right]. \quad \text{So, } A^{-1} = \begin{bmatrix} \frac{4}{7} & \frac{8}{7} & -\frac{5}{7} \\ \frac{3}{7} & -\frac{1}{7} & -\frac{2}{7} \\ \frac{1}{7} & \frac{2}{7} & -\frac{3}{7} \end{bmatrix}$$

59. Using Gaussian elimination, we obtain the following:

$$\left[\begin{array}{ccc|ccc} 2 & 4 & 1 & 1 & 0 & 0 \\ 1 & 1 & -1 & 0 & 1 & 0 \\ 1 & 1 & 0 & 0 & 0 & 1 \end{array}\right] \xrightarrow[R_2 \leftrightarrow R_3]{\frac{1}{2}R_1 \to R_1} \left[\begin{array}{ccc|ccc} 1 & 2 & \frac{1}{2} & \frac{1}{2} & 0 & 0 \\ 1 & 1 & 0 & 0 & 0 & 1 \\ 1 & 1 & -1 & 0 & 1 & 0 \end{array}\right]$$

$$\xrightarrow[R_1 - R_3 \to R_3]{R_2 - R_3 \to R_2} \left[\begin{array}{ccc|ccc} 1 & 2 & \frac{1}{2} & \frac{1}{2} & 0 & 0 \\ 0 & 0 & 1 & 0 & -1 & 1 \\ 0 & 1 & \frac{3}{2} & \frac{1}{2} & -1 & 0 \end{array}\right] \xrightarrow{R_2 \leftrightarrow R_3} \left[\begin{array}{ccc|ccc} 1 & 2 & \frac{1}{2} & \frac{1}{2} & 0 & 0 \\ 0 & 1 & \frac{3}{2} & \frac{1}{2} & -1 & 0 \\ 0 & 0 & 1 & 0 & -1 & 1 \end{array}\right]$$

$$\xrightarrow{R_2 - \frac{3}{2}R_3 \to R_2} \left[\begin{array}{ccc|ccc} 1 & 2 & \frac{1}{2} & \frac{1}{2} & 0 & 0 \\ 0 & 1 & 0 & \frac{1}{2} & \frac{1}{2} & -\frac{3}{2} \\ 0 & 0 & 1 & 0 & -1 & 1 \end{array}\right] \xrightarrow{R_1 - 2R_2 - \frac{1}{2}R_3 \to R_1} \left[\begin{array}{ccc|ccc} 1 & 0 & 0 & -\frac{1}{2} & -\frac{1}{2} & \frac{5}{2} \\ 0 & 1 & 0 & \frac{1}{2} & \frac{1}{2} & -\frac{3}{2} \\ 0 & 0 & 1 & 0 & -1 & 1 \end{array}\right]$$

So, $A^{-1} = \begin{bmatrix} -\frac{1}{2} & -\frac{1}{2} & \frac{5}{2} \\ \frac{1}{2} & \frac{1}{2} & -\frac{3}{2} \\ 0 & -1 & 1 \end{bmatrix}$.

60. Using Gaussian elimination, we obtain the following:

$$\left[\begin{array}{ccc|ccc} 1 & 0 & 1 & 1 & 0 & 0 \\ 1 & 1 & -1 & 0 & 1 & 0 \\ 2 & 1 & -1 & 0 & 0 & 1 \end{array}\right] \xrightarrow[R_2 - R_1 \to R_2]{R_3 - 2R_1 \to R_3} \left[\begin{array}{ccc|ccc} 1 & 0 & 1 & 1 & 0 & 0 \\ 0 & 1 & -2 & -1 & 1 & 0 \\ 0 & 1 & -3 & -2 & 0 & 1 \end{array}\right]$$

$$\xrightarrow{R_3 - R_2 \to R_3} \left[\begin{array}{ccc|ccc} 1 & 0 & 1 & 1 & 0 & 0 \\ 0 & 1 & -2 & -1 & 1 & 1 \\ 0 & 0 & -1 & -1 & -1 & 1 \end{array}\right] \xrightarrow[-R_3 \to R_3]{R_2 - 2R_3 \to R_2} \left[\begin{array}{ccc|ccc} 1 & 0 & 1 & 1 & 0 & 0 \\ 0 & 1 & 0 & 1 & 3 & -2 \\ 0 & 0 & 1 & 1 & 1 & -1 \end{array}\right]$$

$$\xrightarrow{R_1 - R_3 \to R_1} \left[\begin{array}{ccc|ccc} 1 & 0 & 0 & 0 & -1 & 1 \\ 0 & 1 & 0 & 1 & 3 & -2 \\ 0 & 0 & 1 & 1 & 1 & -1 \end{array}\right]$$

So, $A^{-1} = \begin{bmatrix} 0 & -1 & 1 \\ 1 & 3 & -2 \\ 1 & 1 & -1 \end{bmatrix}$.

61. Using Gaussian elimination, we obtain the following:

$$\left[\begin{array}{ccc|ccc} 1 & 1 & -1 & 1 & 0 & 0 \\ 1 & -1 & 1 & 0 & 1 & 0 \\ 2 & -1 & -1 & 0 & 0 & 1 \end{array}\right] \xrightarrow[R_2 - R_1 \to R_2]{R_3 - 2R_1 \to R_3} \left[\begin{array}{ccc|ccc} 1 & 1 & -1 & 1 & 0 & 0 \\ 0 & -2 & 2 & -1 & 1 & 0 \\ 0 & -3 & 1 & -2 & 0 & 1 \end{array}\right]$$

$$\xrightarrow[-\frac{1}{3}R_3 \to R_3]{-\frac{1}{2}R_2 \to R_2} \left[\begin{array}{ccc|ccc} 1 & 1 & -1 & 1 & 0 & 0 \\ 0 & 1 & -1 & \frac{1}{2} & -\frac{1}{2} & 0 \\ 0 & 1 & -\frac{1}{3} & \frac{2}{3} & 0 & -\frac{1}{3} \end{array}\right] \xrightarrow{R_3 - R_2 \to R_3} \left[\begin{array}{ccc|ccc} 1 & 1 & -1 & 1 & 0 & 0 \\ 0 & 1 & -1 & \frac{1}{2} & -\frac{1}{2} & 0 \\ 0 & 0 & \frac{2}{3} & \frac{1}{6} & \frac{1}{2} & -\frac{1}{3} \end{array}\right]$$

$$\xrightarrow{\frac{3}{2}R_3 \to R_3} \left[\begin{array}{ccc|ccc} 1 & 1 & -1 & 1 & 0 & 0 \\ 0 & 1 & -1 & \frac{1}{2} & -\frac{1}{2} & 0 \\ 0 & 0 & 1 & \frac{1}{4} & \frac{3}{4} & -\frac{1}{2} \end{array}\right] \xrightarrow{R_2 + R_3 \to R_2} \left[\begin{array}{ccc|ccc} 1 & 1 & -1 & 1 & 0 & 0 \\ 0 & 1 & 0 & \frac{3}{4} & \frac{1}{4} & -\frac{1}{2} \\ 0 & 0 & 1 & \frac{1}{4} & \frac{3}{4} & -\frac{1}{2} \end{array}\right]$$

$$\xrightarrow{R_1 - R_2 + R_3 \to R_1} \left[\begin{array}{ccc|ccc} 1 & 0 & 0 & \frac{1}{2} & \frac{1}{2} & 0 \\ 0 & 1 & 0 & \frac{3}{4} & \frac{1}{4} & -\frac{1}{2} \\ 0 & 0 & 1 & \frac{1}{4} & \frac{3}{4} & -\frac{1}{2} \end{array}\right]$$

So, $A^{-1} = \begin{bmatrix} \frac{1}{2} & \frac{1}{2} & 0 \\ \frac{3}{4} & \frac{1}{4} & -\frac{1}{2} \\ \frac{1}{4} & \frac{3}{4} & -\frac{1}{2} \end{bmatrix}$.

62. Using Gaussian elimination, we obtain the following:

$$\left[\begin{array}{ccc|ccc} 1 & -1 & -1 & 1 & 0 & 0 \\ 1 & 1 & -3 & 0 & 1 & 0 \\ 3 & -5 & 1 & 0 & 0 & 1 \end{array}\right] \xrightarrow{R_3 \leftrightarrow R_1} \left[\begin{array}{ccc|ccc} 3 & -5 & 1 & 0 & 0 & 1 \\ 1 & 1 & -3 & 0 & 1 & 0 \\ 1 & -1 & -1 & 1 & 0 & 0 \end{array}\right]$$

$$\xrightarrow[R_1-3R_3\to R_3]{R_1-3R_2\to R_2} \left[\begin{array}{ccc|ccc} 3 & -5 & 1 & 0 & 0 & 1 \\ 0 & -8 & 10 & 0 & -3 & 1 \\ 0 & -2 & 4 & -3 & 0 & 1 \end{array}\right] \xrightarrow{R_2-4R_3\to R_3} \left[\begin{array}{ccc|ccc} 3 & -5 & 1 & 0 & 0 & 1 \\ 0 & -8 & 10 & 0 & -3 & 1 \\ 0 & 0 & -6 & 12 & -3 & -3 \end{array}\right]$$

$$\xrightarrow[-\frac{1}{8}R_2\to R_2]{-\frac{1}{6}R_3\to R_3} \left[\begin{array}{ccc|ccc} 3 & -5 & 1 & 0 & 0 & 1 \\ 0 & 1 & -\frac{5}{4} & 0 & \frac{3}{8} & -\frac{1}{8} \\ 0 & 0 & 1 & -2 & \frac{1}{2} & \frac{1}{2} \end{array}\right] \xrightarrow{R_2+\frac{5}{4}R_3\to R_2} \left[\begin{array}{ccc|ccc} 3 & -5 & 1 & 0 & 0 & 1 \\ 0 & 1 & 0 & -\frac{5}{2} & 1 & \frac{1}{2} \\ 0 & 0 & 1 & -2 & \frac{1}{2} & \frac{1}{2} \end{array}\right]$$

$$\xrightarrow{R_1+5R_2-R_3\to R_1} \left[\begin{array}{ccc|ccc} 3 & 0 & 0 & -\frac{21}{2} & \frac{9}{2} & 3 \\ 0 & 1 & 0 & -\frac{5}{2} & 1 & \frac{1}{2} \\ 0 & 0 & 1 & -2 & \frac{1}{2} & \frac{1}{2} \end{array}\right] \xrightarrow{\frac{1}{3}R_1\to R_1} \left[\begin{array}{ccc|ccc} 1 & 0 & 0 & -\frac{7}{2} & \frac{3}{2} & 1 \\ 0 & 1 & 0 & -\frac{5}{2} & 1 & \frac{1}{2} \\ 0 & 0 & 1 & -2 & \frac{1}{2} & \frac{1}{2} \end{array}\right]$$

So, $A^{-1} = \begin{bmatrix} -\frac{7}{2} & \frac{3}{2} & 1 \\ -\frac{5}{2} & 1 & \frac{1}{2} \\ -2 & \frac{1}{2} & \frac{1}{2} \end{bmatrix}$.

63. First, write the system in the form $AX = B$:

$$\begin{bmatrix} 2 & -1 \\ 1 & 1 \end{bmatrix} \begin{bmatrix} x \\ y \end{bmatrix} = \begin{bmatrix} 5 \\ 1 \end{bmatrix}$$

The solution is $\begin{bmatrix} x \\ y \end{bmatrix} = \begin{bmatrix} 2 & -1 \\ 1 & 1 \end{bmatrix}^{-1} \begin{bmatrix} 5 \\ 1 \end{bmatrix}$. Since $\begin{bmatrix} 2 & -1 \\ 1 & 1 \end{bmatrix}^{-1} = \frac{1}{3}\begin{bmatrix} 1 & 1 \\ -1 & 2 \end{bmatrix}$, the solution is $\begin{bmatrix} x \\ y \end{bmatrix} = \frac{1}{3}\begin{bmatrix} 1 & 1 \\ -1 & 2 \end{bmatrix}\begin{bmatrix} 5 \\ 1 \end{bmatrix} = \begin{bmatrix} 2 \\ -1 \end{bmatrix}$. So, $\boxed{x = 2\,,\, y = -1}$.

64. First, write the system in the form $AX = B$:

$$\begin{bmatrix} 2 & -3 \\ 1 & 1 \end{bmatrix}\begin{bmatrix} x \\ y \end{bmatrix} = \begin{bmatrix} 12 \\ 1 \end{bmatrix}$$

The solution is $\begin{bmatrix} x \\ y \end{bmatrix} = \begin{bmatrix} 2 & -3 \\ 1 & 1 \end{bmatrix}^{-1}\begin{bmatrix} 12 \\ 1 \end{bmatrix}$. Since $\begin{bmatrix} 2 & -3 \\ 1 & 1 \end{bmatrix}^{-1} = \frac{1}{5}\begin{bmatrix} 1 & 3 \\ -1 & 2 \end{bmatrix}$, the solution is $\begin{bmatrix} x \\ y \end{bmatrix} = \frac{1}{5}\begin{bmatrix} 1 & 3 \\ -1 & 2 \end{bmatrix}\begin{bmatrix} 12 \\ 1 \end{bmatrix} = \begin{bmatrix} 3 \\ -2 \end{bmatrix}$. So, $\boxed{x = 3, y = -2}$

65. First, write the system in the form $AX = B$:

$$\begin{bmatrix} 4 & -9 \\ 7 & -3 \end{bmatrix}\begin{bmatrix} x \\ y \end{bmatrix} = \begin{bmatrix} -1 \\ \frac{5}{2} \end{bmatrix}$$

The solution is $\begin{bmatrix} x \\ y \end{bmatrix} = \begin{bmatrix} 4 & -9 \\ 7 & -3 \end{bmatrix}^{-1}\begin{bmatrix} -1 \\ \frac{5}{2} \end{bmatrix}$. Since $\begin{bmatrix} 4 & -9 \\ 7 & -3 \end{bmatrix}^{-1} = \frac{1}{51}\begin{bmatrix} -3 & 9 \\ -7 & 4 \end{bmatrix}$, the solution is $\begin{bmatrix} x \\ y \end{bmatrix} = \frac{1}{51}\begin{bmatrix} -3 & 9 \\ -7 & 4 \end{bmatrix}\begin{bmatrix} -1 \\ \frac{5}{2} \end{bmatrix} = \begin{bmatrix} \frac{1}{2} \\ \frac{1}{3} \end{bmatrix}$. So, $\boxed{x = \frac{1}{2}, y = \frac{1}{3}}$.

66. First, write the system in the form $AX = B$:

$$\begin{bmatrix} 7 & -3 \\ 4 & -5 \end{bmatrix}\begin{bmatrix} x \\ y \end{bmatrix} = \begin{bmatrix} 1 \\ -\frac{7}{5} \end{bmatrix}$$

The solution is $\begin{bmatrix} x \\ y \end{bmatrix} = \begin{bmatrix} 7 & -3 \\ 4 & -5 \end{bmatrix}^{-1}\begin{bmatrix} 1 \\ -\frac{7}{5} \end{bmatrix}$. Since $\begin{bmatrix} 7 & -3 \\ 4 & -5 \end{bmatrix}^{-1} = -\frac{1}{23}\begin{bmatrix} -5 & 3 \\ -4 & 7 \end{bmatrix}$, the solution is $\begin{bmatrix} x \\ y \end{bmatrix} = -\frac{1}{23}\begin{bmatrix} -5 & 3 \\ -4 & 7 \end{bmatrix}\begin{bmatrix} 1 \\ -\frac{7}{5} \end{bmatrix} = \begin{bmatrix} \frac{2}{5} \\ \frac{3}{5} \end{bmatrix}$. So, $\boxed{x = \frac{2}{5}, y = \frac{3}{5}}$.

67. From #55, we know that $A^{-1} = \begin{bmatrix} \frac{1}{2} & \frac{1}{2} & 0 \\ \frac{1}{2} & 0 & \frac{1}{2} \\ 0 & -\frac{1}{2} & -\frac{1}{2} \end{bmatrix}$. Further, in the current problem, we identify $B = \begin{bmatrix} 1 \\ -1 \\ -1 \end{bmatrix}$. Since $A\begin{bmatrix} x \\ y \\ z \end{bmatrix} = B$ is equivalent to $\begin{bmatrix} x \\ y \\ z \end{bmatrix} = A^{-1}B = \begin{bmatrix} 0 \\ 0 \\ 1 \end{bmatrix}$, we conclude that the solution of this system is $\boxed{x = 0,\ y = 0,\ z = 1}$.

68. From #56, we know that $A^{-1} = \begin{bmatrix} \frac{5}{4} & \frac{1}{4} & \frac{1}{2} \\ -\frac{1}{2} & \frac{1}{2} & 0 \\ -\frac{3}{4} & \frac{1}{4} & -\frac{1}{2} \end{bmatrix}$. Further, in the current problem, we identify $B = \begin{bmatrix} 0 \\ 2 \\ 1 \end{bmatrix}$. Since $A\begin{bmatrix} x \\ y \\ z \end{bmatrix} = B$ is equivalent to $\begin{bmatrix} x \\ y \\ z \end{bmatrix} = A^{-1}B = \begin{bmatrix} 1 \\ 1 \\ 0 \end{bmatrix}$, we conclude that the solution of this system is $\boxed{x = 1,\ y = 1,\ z = 0}$.

69. From #57, we know that A^{-1} does not exist.

70. From #58, we know that $A^{-1} = \begin{bmatrix} \frac{4}{7} & \frac{8}{7} & -\frac{5}{7} \\ \frac{3}{7} & -\frac{1}{7} & -\frac{2}{7} \\ \frac{1}{7} & \frac{2}{7} & -\frac{3}{7} \end{bmatrix}$. Further, in the current problem, we identify $B = \begin{bmatrix} 1 \\ 3 \\ 0 \end{bmatrix}$. Since $A\begin{bmatrix} x \\ y \\ z \end{bmatrix} = B$ is equivalent to $\begin{bmatrix} x \\ y \\ z \end{bmatrix} = A^{-1}B = \begin{bmatrix} 4 \\ 0 \\ 1 \end{bmatrix}$, we conclude that the solution of this system is $\boxed{x = 4,\ y = 0,\ z = 1}$.

71. From #59, we know that $A^{-1} = \begin{bmatrix} -\frac{1}{2} & -\frac{1}{2} & \frac{5}{2} \\ \frac{1}{2} & \frac{1}{2} & -\frac{3}{2} \\ 0 & -1 & 1 \end{bmatrix}$. Further, in the current problem, we identify $B = \begin{bmatrix} -5 \\ 7 \\ 0 \end{bmatrix}$. Since $A\begin{bmatrix} x \\ y \\ z \end{bmatrix} = B$ is equivalent to $\begin{bmatrix} x \\ y \\ z \end{bmatrix} = A^{-1}B = \begin{bmatrix} -1 \\ 1 \\ -7 \end{bmatrix}$, we conclude that the solution of this system is $\boxed{x = -1,\ y = 1,\ z = -7}$.

72. From #60, we know that $A^{-1} = \begin{bmatrix} 0 & -1 & 1 \\ 1 & 3 & -2 \\ 1 & 1 & -1 \end{bmatrix}$. Further, in the current problem, we identify $B = \begin{bmatrix} 3 \\ -3 \\ -5 \end{bmatrix}$. Since $A\begin{bmatrix} x \\ y \\ z \end{bmatrix} = B$ is equivalent to $\begin{bmatrix} x \\ y \\ z \end{bmatrix} = A^{-1}B = \begin{bmatrix} -2 \\ -21 \\ 5 \end{bmatrix}$, we conclude that the solution of this system is $\boxed{x = -2,\ y = 4,\ z = 5}$.

73. From #61, we know that $A^{-1} = \begin{bmatrix} \frac{1}{2} & \frac{1}{2} & 0 \\ \frac{3}{4} & \frac{1}{4} & -\frac{1}{2} \\ \frac{1}{4} & \frac{3}{4} & -\frac{1}{2} \end{bmatrix}$. Further, in the current problem, we identify $B = \begin{bmatrix} 4 \\ 2 \\ -3 \end{bmatrix}$. Since $A\begin{bmatrix} x \\ y \\ z \end{bmatrix} = B$ is equivalent to $\begin{bmatrix} x \\ y \\ z \end{bmatrix} = A^{-1}B = \begin{bmatrix} 1 \\ \frac{19}{2} \\ \frac{17}{2} \end{bmatrix}$, we conclude that the solution of this system is $\boxed{x = 3,\ y = 5,\ z = 4}$.

74. From #62, we know that $A^{-1} = \begin{bmatrix} -\frac{7}{2} & \frac{3}{2} & 1 \\ -\frac{5}{2} & 1 & \frac{1}{2} \\ -2 & \frac{1}{2} & \frac{1}{2} \end{bmatrix}$. Further, in the current problem, we identify $B = \begin{bmatrix} 0 \\ 2 \\ 4 \end{bmatrix}$. Since $A\begin{bmatrix} x \\ y \\ z \end{bmatrix} = B$ is equivalent to $\begin{bmatrix} x \\ y \\ z \end{bmatrix} = A^{-1}B = \begin{bmatrix} \frac{23}{3} \\ \frac{16}{3} \\ \frac{11}{3} \end{bmatrix}$, we conclude that the solution of this system is $\boxed{x = 7,\ y = 4,\ z = 3}$.

75. $A = \begin{bmatrix} \textit{Yes}\text{ response} \\ \textit{No}\text{ response} \end{bmatrix} = \begin{bmatrix} 0.70 \\ 0.30 \end{bmatrix}$ $\quad B = \begin{bmatrix} \textit{Yes}\text{ to "increase lung cancer"} \\ \textit{Yes}\text{ to "would shorten lives"} \end{bmatrix} = \begin{bmatrix} 0.89 \\ 0.84 \end{bmatrix}$

a. $46A = 46\begin{bmatrix} 0.70 \\ 0.30 \end{bmatrix} = \begin{bmatrix} 32.2 \\ 13.8 \end{bmatrix}$. This matrix tells us that out of 46 million people, 32.2 million said that they had tried to quit smoking, while 13.8 million said that they had not.

b. $46B = 46\begin{bmatrix} 0.89 \\ 0.84 \end{bmatrix} = \begin{bmatrix} 40.94 \\ 38.64 \end{bmatrix}$. This matrix tells us that out of 46 million people, 40.94 million believed that smoking would increase the chance of getting lung cancer, and that 38.64 million believed that smoking would shorten their lives.

76. $A=\begin{bmatrix}0.24\\0.23\end{bmatrix}$ (1981) $B=\begin{bmatrix}0.32\\0.21\end{bmatrix}$ (1991) $C=\begin{bmatrix}0.38\\0.30\end{bmatrix}$ (2001)

$$C-B=\begin{bmatrix}0.38\\0.30\end{bmatrix}-\begin{bmatrix}0.32\\0.21\end{bmatrix}=\begin{bmatrix}0.06\\0.09\end{bmatrix}$$

This matrix tells us the percent increase in the ten year span from 1991 to 2001 in the number of female graduate students in mathematics and computer science. In particular, there was a 6% increase in the number of female graduate students in mathematics from 1991 to 2001, and a 9% increase in the number of female graduate students in computer science in the same time span.

$$B-A=\begin{bmatrix}0.32\\0.21\end{bmatrix}-\begin{bmatrix}0.24\\0.23\end{bmatrix}=\begin{bmatrix}0.08\\-0.02\end{bmatrix}$$

This matrix tells us the percent increase in the ten year span from 1981 to 1991 in the number of female graduate students in mathematics and computer science. In particular, there was an 8% increase in the number of female graduate students in mathematics from 1981 to 1991, and a 2% decrease in the number of female graduate students in computer science in the same time span.

77. Since $A=\begin{bmatrix}0.589 & 0.628\\0.414 & 0.430\end{bmatrix}$ and $B=\begin{bmatrix}100M\\110M\end{bmatrix}$, it follows that $AB=\begin{bmatrix}127.98M\\88.7M\end{bmatrix}$.

This matrix tells us the number of registered voters and the number of actual voters. In particular, there are 127.98 million registered voters, and of those 88.7 million actually vote.

78. Here, $BA=\begin{bmatrix}8 & 7 & 5\\6 & 8 & 8\end{bmatrix}\cdot\begin{bmatrix}0.5 & 0.6\\0.3 & 0.1\\0.2 & 0.3\end{bmatrix}=\begin{bmatrix}7.1 & 7\\7 & 6.8\end{bmatrix}$. (The order is 2×2.)

Each entry of this matrix gives the respective applicant's total score according to each of the two rubrics. Specifically,
Applicant 1's score according to Rubric 1 is 7.1,
Applicant 1's score according to Rubric 2 is 7,
Applicant 2's score according to Rubric 1 is 7,
Applicant 2's score according to Rubric 2 is 6.8.

79.

$A=\begin{bmatrix}0.45 & 0.50 & 1.00\end{bmatrix}$ $B=\begin{bmatrix}7,523\\2,700\\15,200\end{bmatrix}$

$AB=\begin{bmatrix}19,935.35\end{bmatrix}$

80.

$A=\begin{bmatrix}85 & 75 & 100\end{bmatrix}$ $B=\begin{bmatrix}0.25\\0.20\\0.15\end{bmatrix}$

$AB=\begin{bmatrix}51.25\end{bmatrix}$

81.

$$A=\begin{bmatrix}0.06 & 0\\ 0.02 & 0.1\\ 0 & 0.3\end{bmatrix}\quad B=\begin{bmatrix}3.80\\ 0.05\end{bmatrix}\quad AB=\begin{bmatrix}0.228\\ 0.081\\ 0.015\end{bmatrix}$$

The entries in AB represent the total cost to run each type of automobile per mile.

82.

$$12{,}000AB=\begin{bmatrix}2{,}736\\ 972\\ 180\end{bmatrix}$$

Cost to run SUV: \$2,736
Cost to run Hybrid: \$972
Cost to run Electric Car: \$180

83. For Problems 83–88, we need to compute K^{-1}:

$$\left[\begin{array}{ccc|ccc}1 & 1 & 0 & 1 & 0 & 0\\ -1 & 0 & 1 & 0 & 1 & 0\\ 2 & 0 & -1 & 0 & 0 & 1\end{array}\right]\xrightarrow[R_3-2R_1\to R_3]{R_1+R_2\to R_2}\left[\begin{array}{ccc|ccc}1 & 1 & 0 & 1 & 0 & 0\\ 0 & 1 & 1 & 1 & 1 & 0\\ 0 & -2 & -1 & -2 & 0 & 1\end{array}\right]\xrightarrow{R_3+2R_2\to R_3}$$

$$\left[\begin{array}{ccc|ccc}1 & 1 & 0 & 1 & 0 & 0\\ 0 & 1 & 1 & 1 & 1 & 0\\ 0 & 0 & 1 & 0 & 2 & 1\end{array}\right]\xrightarrow{R_2-R_3\to R_2}\left[\begin{array}{ccc|ccc}1 & 1 & 0 & 1 & 0 & 0\\ 0 & 1 & 0 & 1 & -1 & -1\\ 0 & 0 & 1 & 0 & 2 & 1\end{array}\right]\xrightarrow{R_1-R_2\to R_1}$$

$$\left[\begin{array}{ccc|ccc}1 & 0 & 0 & 0 & 1 & 1\\ 0 & 1 & 0 & 1 & -1 & -1\\ 0 & 0 & 1 & 0 & 2 & 1\end{array}\right].\ \text{So, } K^{-1}=\begin{bmatrix}0 & 1 & 1\\ 1 & -1 & -1\\ 0 & 2 & 1\end{bmatrix}.$$

As such, $\begin{bmatrix}55 & 10 & -22\end{bmatrix}\begin{bmatrix}0 & 1 & 1\\ 1 & -1 & -1\\ 0 & 2 & 1\end{bmatrix}=\begin{bmatrix}10 & 1 & 23\end{bmatrix}$, which corresponds to JAW.

84. Since $K^{-1}=\begin{bmatrix}0 & 1 & 1\\ 1 & -1 & -1\\ 0 & 2 & 1\end{bmatrix}$ (see Problem #83), we see that

$\begin{bmatrix}31 & 8 & -7\end{bmatrix}\begin{bmatrix}0 & 1 & 1\\ 1 & -1 & -1\\ 0 & 2 & 1\end{bmatrix}=\begin{bmatrix}8 & 9 & 16\end{bmatrix}$, which corresponds to HIP.

85. Since $K^{-1} = \begin{bmatrix} 0 & 1 & 1 \\ 1 & -1 & -1 \\ 0 & 2 & 1 \end{bmatrix}$ (see Problem #83), we see that

$\begin{bmatrix} 21 & 12 & -2 \end{bmatrix} \begin{bmatrix} 0 & 1 & 1 \\ 1 & -1 & -1 \\ 0 & 2 & 1 \end{bmatrix} = \begin{bmatrix} 12 & 5 & 7 \end{bmatrix}$, which corresponds to LEG.

86. Since $K^{-1} = \begin{bmatrix} 0 & 1 & 1 \\ 1 & -1 & -1 \\ 0 & 2 & 1 \end{bmatrix}$ (see Problem #83), we see that

$\begin{bmatrix} 9 & 1 & 5 \end{bmatrix} \begin{bmatrix} 0 & 1 & 1 \\ 1 & -1 & -1 \\ 0 & 2 & 1 \end{bmatrix} = \begin{bmatrix} 1 & 18 & 13 \end{bmatrix}$, which corresponds to ARM.

87. Since $K^{-1} = \begin{bmatrix} 0 & 1 & 1 \\ 1 & -1 & -1 \\ 0 & 2 & 1 \end{bmatrix}$ (see Problem #83), we see that

$\begin{bmatrix} -10 & 5 & 20 \end{bmatrix} \begin{bmatrix} 0 & 1 & 1 \\ 1 & -1 & -1 \\ 0 & 2 & 1 \end{bmatrix} = \begin{bmatrix} 5 & 25 & 5 \end{bmatrix}$, which corresponds to EYE.

88. Since $K^{-1} = \begin{bmatrix} 0 & 1 & 1 \\ 1 & -1 & -1 \\ 0 & 2 & 1 \end{bmatrix}$ (see Problem #83), we see that

$\begin{bmatrix} 40 & 5 & -17 \end{bmatrix} \begin{bmatrix} 0 & 1 & 1 \\ 1 & -1 & -1 \\ 0 & 2 & 1 \end{bmatrix} = \begin{bmatrix} 5 & 1 & 18 \end{bmatrix}$, which corresponds to EAR.

89. Not multiplying correctly. It should be:

$$\begin{bmatrix} 3 & 2 \\ 1 & 4 \end{bmatrix} \cdot \begin{bmatrix} -1 & 3 \\ -2 & 5 \end{bmatrix} = \begin{bmatrix} 3(-1)+2(-2) & 3(3)+2(5) \\ 1(-1)+4(-2) & 1(3)+4(5) \end{bmatrix} = \begin{bmatrix} -7 & 19 \\ -9 & 23 \end{bmatrix}$$

90. Not multiplying correctly. It should be:

$$\begin{bmatrix} 3 & 2 \\ 1 & 4 \end{bmatrix} \cdot \begin{bmatrix} -1 & 3 \\ -2 & 5 \end{bmatrix} = \begin{bmatrix} 3(-1)+2(-2) & 3(3)+2(5) \\ 1(-1)+4(-2) & 1(3)+4(5) \end{bmatrix} = \begin{bmatrix} -7 & 19 \\ -9 & 23 \end{bmatrix}$$

91. The third row of A is comprised of all zeros. Hence, $\det(A)=0$, which implies A is not invertible. Also, note that the identity matrix doesn't occur on the left as it should once you have computed the inverse.

92. This is not how the inverse is defined. Rather,

$$A^{-1}=\frac{1}{2(10)-5(3)}\begin{bmatrix}10 & -5\\ -3 & 2\end{bmatrix}=\frac{1}{5}\begin{bmatrix}10 & -5\\ -3 & 2\end{bmatrix}=\begin{bmatrix}2 & -1\\ -\frac{3}{5} & \frac{2}{5}\end{bmatrix}.$$

93. False. In general, $AB=\begin{bmatrix}a_{11} & a_{12}\\ a_{21} & a_{22}\end{bmatrix}\cdot\begin{bmatrix}b_{11} & b_{12}\\ b_{21} & b_{22}\end{bmatrix}=\begin{bmatrix}a_{11}b_{11}+a_{12}b_{21} & a_{11}b_{12}+a_{12}b_{22}\\ a_{21}b_{11}+a_{22}b_{21} & a_{21}b_{12}+a_{22}b_{22}\end{bmatrix}$

94. False.

Let $A=\begin{bmatrix}2 & 0\\ 1 & 4\\ -1 & 3\end{bmatrix}$ and $B=\begin{bmatrix}4\\ 1\end{bmatrix}$.

AB is defined, but BA is not.
So, $AB\neq BA$.

95. True

96. True

97. False. In general, $A^{-1}=\frac{1}{a_{22}a_{11}-a_{21}a_{12}}\begin{bmatrix}a_{22} & -a_{12}\\ -a_{21} & a_{11}\end{bmatrix}$, provided that $a_{22}a_{11}-a_{21}a_{12}\neq 0$

98. False. $\begin{bmatrix}0 & 0\\ 0 & 0\end{bmatrix}$ is a square matrix with determinant = 0, so that it has no inverse.

99. $A^2=AA=\begin{bmatrix}a_{11}^2+a_{12}a_{21} & a_{11}a_{12}+a_{12}a_{22}\\ a_{21}a_{11}+a_{22}a_{21} & a_{22}^2+a_{21}a_{12}\end{bmatrix}$

100. $m=n$ since then the number of columns of A is equal to the number of rows of the matrix it is to be multiplied by, namely A.

101. By definition of inverse, the value of x for which A^{-1} does not exist would be the one for which $2x$ - 18 is 0, namely $\boxed{x=9}$, since we need to divide by the quantity 2(6) – 3(4) in order to form the inverse.

102. In this case, realize that if any of a, b, or c were 0, then the matrix A would have a row consisting of all zeros. In such case, the determinant of A would be 0, and hence, A would not be invertible. Assuming that none of them is zero, we obtain:

$$\left[\begin{array}{ccc|ccc}a & 0 & 0 & 1 & 0 & 0\\ 0 & b & 0 & 0 & 1 & 0\\ 0 & 0 & c & 0 & 0 & 1\end{array}\right]\xrightarrow[\frac{1}{c}R_3\to R_3]{\substack{\frac{1}{a}R_1\to R_1\\ \frac{1}{b}R_2\to R_2}}\left[\begin{array}{ccc|ccc}1 & 0 & 0 & \frac{1}{a} & 0 & 0\\ 0 & 1 & 0 & 0 & \frac{1}{b} & 0\\ 0 & 0 & 1 & 0 & 0 & \frac{1}{c}\end{array}\right].\ \text{So, } A^{-1}=\begin{bmatrix}\frac{1}{a} & 0 & 0\\ 0 & \frac{1}{b} & 0\\ 0 & 0 & \frac{1}{c}\end{bmatrix}.$$

103. Observe that

$$A = \begin{bmatrix} 1 & 1 \\ 1 & 1 \end{bmatrix},$$

$$A^2 = AA = \begin{bmatrix} 2 & 2 \\ 2 & 2 \end{bmatrix},$$

$$A^3 = A^2 A = \begin{bmatrix} 4 & 4 \\ 4 & 4 \end{bmatrix} = \begin{bmatrix} 2^2 & 2^2 \\ 2^2 & 2^2 \end{bmatrix} = 2^2 \begin{bmatrix} 1 & 1 \\ 1 & 1 \end{bmatrix} = 2^2 A$$

$$\vdots$$

$$A^n = \begin{bmatrix} 2^{n-1} & 2^{n-1} \\ 2^{n-1} & 2^{n-1} \end{bmatrix} = 2^{n-1} \begin{bmatrix} 1 & 1 \\ 1 & 1 \end{bmatrix} = 2^{n-1} A, \quad n \geq 1$$

104. Observe that

$$A = \begin{bmatrix} 1 & 0 \\ 0 & 1 \end{bmatrix}$$

$$A^2 = AA = \begin{bmatrix} 1 & 0 \\ 0 & 1 \end{bmatrix} = A$$

$$\vdots$$

$$A^n = \begin{bmatrix} 1 & 0 \\ 0 & 1 \end{bmatrix} = A, \quad n \geq 1$$

105. $A_{m\times n} B_{n\times p}$ is a matrix of order $m \times p$. Since a product CD is defined if and only if the number of columns of C equals the number of rows of D, if $C = D$, then C has the same number of columns and rows. Applying this to the given product, we see that in order for $\left(A_{m\times n} B_{n\times p}\right)^2$ to be defined, we need $m = p$.

106. AC is an $m \times m$ matrix, and CB is an $n \times n$ matrix. Since $m \neq n$, $AC \neq CB$ (since they have different orders).

107. Observe that

$$A \cdot A^{-1} = \begin{bmatrix} a & b \\ c & d \end{bmatrix} \cdot \left(\frac{1}{ad-bc} \begin{bmatrix} d & -b \\ -c & a \end{bmatrix} \right) = \frac{1}{ad-bc} \left(\begin{bmatrix} a & b \\ c & d \end{bmatrix} \cdot \begin{bmatrix} d & -b \\ -c & a \end{bmatrix} \right)$$

$$= \frac{1}{ad-bc} \begin{bmatrix} ad-bc & 0 \\ 0 & ad-bc \end{bmatrix} = \begin{bmatrix} \frac{ad-bc}{ad-bc} & 0 \\ 0 & \frac{ad-bc}{ad-bc} \end{bmatrix} = \begin{bmatrix} 1 & 0 \\ 0 & 1 \end{bmatrix} = I$$

108. Assume that $a \neq 0$. Applying Gaussian elimination to obtain the inverse reveals:

$$\left[\begin{array}{cc|cc} a & b & 1 & 0 \\ c & d & 0 & 1 \end{array}\right] \xrightarrow{R_2-\frac{c}{a}R_1 \to R_2} \left[\begin{array}{cc|cc} a & b & 1 & 0 \\ 0 & d-\frac{bc}{a} & -\frac{c}{a} & 1 \end{array}\right]$$

$$\xrightarrow{\frac{1}{a}R_1 \to R_1} \left[\begin{array}{cc|cc} 1 & \frac{b}{a} & \frac{1}{a} & 0 \\ 0 & d-\frac{bc}{a} & -\frac{c}{a} & 1 \end{array}\right]$$

$$\xrightarrow{\frac{a}{da-bc}R_2 \to R_2} \left[\begin{array}{cc|cc} 1 & \frac{b}{a} & \frac{1}{a} & 0 \\ 0 & 1 & -\frac{c}{a}\left(\frac{a}{da-bc}\right) & \frac{a}{da-bc} \end{array}\right]$$

$$\xrightarrow{R_1-\frac{b}{a}R_2 \to R_1} \left[\begin{array}{cc|cc} 1 & 0 & \frac{1}{a}+\frac{b}{a}\left(\frac{c}{\cancel{a}}\left(\frac{\cancel{a}}{da-bc}\right)\right) & -\frac{b}{\cancel{a}}\left(\frac{\cancel{a}}{da-bc}\right) \\ 0 & 1 & -\frac{c}{\cancel{a}}\left(\frac{\cancel{a}}{da-bc}\right) & \frac{a}{da-bc} \end{array}\right]$$

Now, simplifying the result above yields:

$$\left[\begin{array}{cc|cc} 1 & 0 & \frac{1}{a}+\frac{b}{a}\left(\frac{c}{\cancel{a}}\left(\frac{\cancel{a}}{da-bc}\right)\right) & -\frac{b}{\cancel{a}}\left(\frac{\cancel{a}}{da-bc}\right) \\ 0 & 1 & -\frac{c}{\cancel{a}}\left(\frac{\cancel{a}}{da-bc}\right) & \frac{a}{da-bc} \end{array}\right] = \left[\begin{array}{cc|cc} 1 & 0 & \frac{da-\cancel{bc}+\cancel{bc}}{a(da-bc)} & -\frac{b}{da-bc} \\ 0 & 1 & -\frac{c}{da-bc} & \frac{a}{da-bc} \end{array}\right]$$

$$= \left[\begin{array}{cc|cc} 1 & 0 & \frac{d}{da-bc} & -\frac{b}{da-bc} \\ 0 & 1 & -\frac{c}{da-bc} & \frac{a}{da-bc} \end{array}\right]$$

Hence, $A^{-1} = \begin{bmatrix} \frac{d}{da-bc} & -\frac{b}{da-bc} \\ -\frac{c}{da-bc} & \frac{a}{da-bc} \end{bmatrix} = \frac{1}{da-bc}\begin{bmatrix} d & -b \\ -c & a \end{bmatrix}$.

109. Since $da-bc=0$ the inverse does not exist.

110. In the attempt to calculate the inverse, applying the row operation $R_2-2R_1 \to R_1$

yields $\left[\begin{array}{ccc|ccc} 0 & 0 & 0 & -2 & 1 & 0 \\ 2 & 4 & -2 & 0 & 1 & 0 \\ 0 & 1 & 3 & 0 & 0 & 1 \end{array}\right]$. As such, the 3×3 matrix to the left of the vertical bar can never be transformed into the identity matrix, and so, the inverse doesn't exist.

111. $AB = \begin{bmatrix} 33 & 35 \\ -96 & -82 \\ 31 & 19 \\ 146 & 138 \end{bmatrix}$

112. *BA* is not defined.

113. *BB* is not defined.

114. $AA = \begin{bmatrix} -2 & 8 & 124 & 126 \\ 114 & 148 & -131 & 14 \\ 36 & 28 & 29 & 48 \\ -6 & 11 & 189 & 87 \end{bmatrix}$

115. $\begin{bmatrix} 5 & -4 & 4 \\ 2 & -15 & -3 \\ 26 & 4 & -8 \end{bmatrix}$

116. $\begin{bmatrix} 74 & 121 & 233 \\ -503 & -560 & 312 \\ 1072 & -1006 & -462 \end{bmatrix}$

117. $A^{-1} = \begin{bmatrix} -\frac{115}{6008} & \frac{431}{6008} & \frac{-1067}{6008} & \frac{103}{751} \\ \frac{411}{6008} & \frac{-391}{6008} & \frac{731}{6008} & \frac{-22}{751} \\ \frac{57}{751} & \frac{28}{751} & \frac{-85}{751} & \frac{3}{751} \\ \frac{-429}{6008} & \frac{145}{6008} & \frac{1035}{6008} & \frac{12}{751} \end{bmatrix}$

118. $AA^{-1} = \begin{bmatrix} 1 & 0 & 0 & 0 \\ 0 & 1 & 0 & 0 \\ 0 & 0 & 1 & 0 \\ 0 & 0 & 0 & 1 \end{bmatrix} = I$

119. $A^{-1} = \dfrac{1}{(2x)(-2y)-(2y)(2x)} \begin{bmatrix} -2y & -2y \\ -2x & 2x \end{bmatrix} = -\dfrac{1}{8xy} \begin{bmatrix} -2y & -2y \\ -2x & 2x \end{bmatrix} = \begin{bmatrix} \frac{1}{4x} & \frac{1}{4x} \\ \frac{1}{4y} & -\frac{1}{4y} \end{bmatrix}$

120. $A^{-1} = \dfrac{1}{4x-4y} \begin{bmatrix} 4x & -1 \\ -4y & 1 \end{bmatrix} = \begin{bmatrix} \frac{x}{x-y} & \frac{1}{4(y-x)} \\ \frac{y}{y-x} & \frac{1}{4(x-y)} \end{bmatrix}$

121. $A^{-1} = \dfrac{1}{\cos^2\theta + \sin^2\theta} \begin{bmatrix} \cos\theta & -\sin\theta \\ \sin\theta & \cos\theta \end{bmatrix} = \begin{bmatrix} \cos\theta & -\sin\theta \\ \sin\theta & \cos\theta \end{bmatrix}$

122.

$$\left[\begin{array}{ccc|ccc} \cos\theta & -r\sin\theta & 0 & 1 & 0 & 0 \\ \sin\theta & r\cos\theta & 0 & 0 & 1 & 0 \\ 0 & 0 & 1 & 0 & 0 & 1 \end{array}\right] \xrightarrow[\cos\theta R_2 \to R_2]{-\sin\theta R_1 \to R_1} \left[\begin{array}{ccc|ccc} -\sin\theta\cos\theta & r\sin^2\theta & 0 & -\sin\theta & 0 & 0 \\ \sin\theta\cos\theta & r\cos^2\theta & 0 & 0 & \cos\theta & 0 \\ 0 & 0 & 1 & 0 & 0 & 1 \end{array}\right]$$

$$\xrightarrow{R_1 + R_2 \to R_2} \left[\begin{array}{ccc|ccc} -\sin\theta\cos\theta & r\sin^2\theta & 0 & -\sin\theta & 0 & 0 \\ 0 & r & 0 & -\sin\theta & \cos\theta & 0 \\ 0 & 0 & 1 & 0 & 0 & 1 \end{array}\right]$$

$$\xrightarrow{\frac{1}{r}R_2 \to R_2} \left[\begin{array}{ccc|ccc} -\sin\theta\cos\theta & r\sin^2\theta & 0 & -\sin\theta & 0 & 0 \\ 0 & 1 & 0 & -\frac{\sin\theta}{r} & \frac{\cos\theta}{r} & 0 \\ 0 & 0 & 1 & 0 & 0 & 1 \end{array}\right]$$

$$\xrightarrow{R_1 - r\sin^2\theta R_2 \to R_1} \left[\begin{array}{ccc|ccc} -\sin\theta\cos\theta & 0 & 0 & -\sin\theta + \sin^3\theta & -\sin^2\theta\cos\theta & 0 \\ 0 & 1 & 0 & -\frac{\sin\theta}{r} & \frac{\cos\theta}{r} & 0 \\ 0 & 0 & 1 & 0 & 0 & 1 \end{array}\right]$$

$$\xrightarrow{-\frac{1}{\sin\theta\cos\theta}R_1 \to R_1} \left[\begin{array}{ccc|ccc} 1 & 0 & 0 & \frac{1}{\cos\theta} - \frac{\sin^2\theta}{\cos\theta} & \sin\theta & 0 \\ 0 & 1 & 0 & -\frac{\sin\theta}{r} & \frac{\cos\theta}{r} & 0 \\ 0 & 0 & 1 & 0 & 0 & 1 \end{array}\right]$$

So, $A^{-1} = \begin{bmatrix} \frac{1-\sin^2\theta}{\cos\theta} & \sin\theta & 0 \\ -\frac{\sin\theta}{r} & \frac{\cos\theta}{r} & 0 \\ 0 & 0 & 1 \end{bmatrix} = \begin{bmatrix} \frac{\cos^2\theta}{\cos\theta} & \sin\theta & 0 \\ -\frac{\sin\theta}{r} & \frac{\cos\theta}{r} & 0 \\ 0 & 0 & 1 \end{bmatrix} = \begin{bmatrix} \cos\theta & \sin\theta & 0 \\ -\frac{\sin\theta}{r} & \frac{\cos\theta}{r} & 0 \\ 0 & 0 & 1 \end{bmatrix}$.

Section 8.5 Solutions ---

1. $(1)(4)-(3)(2)=-2$	**2.** $(1)(-4)-(-3)(-2)=-10$
3. $(7)(-2)-(-5)(9)=31$	**4.** $(-3)(15)-(7)(-11)=32$
5. $(0)(-1)-(4)7=-28$	**6.** $(0)(0)-(1)(0)=0$
7. $(-1.2)(1.5)-(-0.5)(2.4)=-0.6$	**8.** $(-1.0)(-2.8)-(1.5)(1.4)=0.7$
9. $\left(\tfrac{3}{4}\right)\left(\tfrac{8}{9}\right)-(2)\left(\tfrac{1}{3}\right)=0$	**10.** $\left(-\tfrac{1}{2}\right)\left(-\tfrac{8}{9}\right)-\left(\tfrac{2}{3}\right)\left(\tfrac{1}{4}\right)=\tfrac{5}{18}$
11. $D=\begin{vmatrix}1 & 1\\ 1 & -1\end{vmatrix}=(1)(-1)-(1)(1)=-2$ $D_x=\begin{vmatrix}-1 & 1\\ 11 & -1\end{vmatrix}=(-1)(-1)-(11)(1)=-10$ $D_y=\begin{vmatrix}1 & -1\\ 1 & 11\end{vmatrix}=(1)(11)-(1)(-1)=12$	So, from Cramer's Rule, we have: $x=\dfrac{D_x}{D}=\dfrac{-10}{-2}=5$ $y=\dfrac{D_y}{D}=\dfrac{12}{-2}=-6$ Thus, the solution is $x=5,\ y=-6$.
12. $D=\begin{vmatrix}1 & 1\\ 1 & -1\end{vmatrix}=(1)(-1)-(1)(1)=-2$ $D_x=\begin{vmatrix}-1 & 1\\ -9 & -1\end{vmatrix}=(-1)(-1)-(-9)(1)=10$ $D_y=\begin{vmatrix}1 & -1\\ 1 & -9\end{vmatrix}=(1)(-9)-(1)(-1)=-8$	So, from Cramer's Rule, we have: $x=\dfrac{D_x}{D}=\dfrac{10}{-2}=-5$ $y=\dfrac{D_y}{D}=\dfrac{-8}{-2}=4$ Thus, the solution is $x=-5,\ y=4$.
13. $D=\begin{vmatrix}3 & 2\\ -2 & 1\end{vmatrix}=(3)(1)-(-2)(2)=7$ $D_x=\begin{vmatrix}-4 & 2\\ 5 & 1\end{vmatrix}=(-4)(1)-(5)(2)=-14$ $D_y=\begin{vmatrix}3 & -4\\ -2 & 5\end{vmatrix}=(3)(5)-(-2)(-4)=7$	So, from Cramer's Rule, we have: $x=\dfrac{D_x}{D}=\dfrac{-14}{7}=-2$ $y=\dfrac{D_y}{D}=\dfrac{7}{7}=1$ Thus, the solution is $x=-2,\ y=1$.

14.

$$D=\begin{vmatrix}5 & 3\\ 4 & -7\end{vmatrix}=(5)(-7)-(4)(3)=-47$$

$$D_x=\begin{vmatrix}1 & 3\\ -18 & -7\end{vmatrix}=(1)(-7)-(-18)(3)=47$$

$$D_y=\begin{vmatrix}5 & 1\\ 4 & -18\end{vmatrix}=(5)(-18)-(4)(1)=-94$$

So, from Cramer's Rule, we have:

$$x=\frac{D_x}{D}=\frac{47}{-47}=-1$$

$$y=\frac{D_y}{D}=\frac{-94}{-47}=2$$

Thus, the solution is $x=-1,\ y=2$.

15.

$$D=\begin{vmatrix}3 & -2\\ 5 & 4\end{vmatrix}=(3)(4)-(5)(-2)=22$$

$$D_x=\begin{vmatrix}-1 & -2\\ -31 & 4\end{vmatrix}=(-1)(4)-(-31)(-2)=-66$$

$$D_y=\begin{vmatrix}3 & -1\\ 5 & -31\end{vmatrix}=(3)(-31)-(5)(-1)=-88$$

So, from Cramer's Rule, we have:

$$x=\frac{D_x}{D}=\frac{-66}{22}=-3$$

$$y=\frac{D_y}{D}=\frac{-88}{22}=-4$$

Thus, the solution is $x=-3,\ y=-4$.

16.

$$D=\begin{vmatrix}1 & -4\\ 3 & 8\end{vmatrix}=(1)(8)-(3)(-4)=20$$

$$D_x=\begin{vmatrix}-7 & -4\\ 19 & 8\end{vmatrix}=(-7)(8)-(19)(-4)=20$$

$$D_y=\begin{vmatrix}1 & -7\\ 3 & 19\end{vmatrix}=(1)(19)-(3)(-7)=40$$

So, from Cramer's Rule, we have:

$$x=\frac{D_x}{D}=\frac{20}{20}=1$$

$$y=\frac{D_y}{D}=\frac{40}{20}=2$$

Thus, the solution is $x=1,\ y=2$.

17.

$$D=\begin{vmatrix}7 & -3\\ 5 & 2\end{vmatrix}=(7)(2)-(5)(-3)=29$$

$$D_x=\begin{vmatrix}-29 & -3\\ 0 & 2\end{vmatrix}=(-29)(2)-(0)(-3)=-58$$

$$D_y=\begin{vmatrix}7 & -29\\ 5 & 0\end{vmatrix}=(7)(0)-(5)(-29)=145$$

So, from Cramer's Rule, we have:

$$x=\frac{D_x}{D}=\frac{-58}{29}=-2$$

$$y=\frac{D_y}{D}=\frac{145}{29}=5$$

Thus, the solution is $x=-2,\ y=5$.

18.

$D=\begin{vmatrix} 6 & -2 \\ 4 & 7 \end{vmatrix}=(6)(7)-(4)(-2)=50$

$D_x=\begin{vmatrix} 24 & -2 \\ 41 & 7 \end{vmatrix}=(24)(7)-(41)(-2)=250$

$D_y=\begin{vmatrix} 6 & 24 \\ 4 & 41 \end{vmatrix}=(6)(41)-(4)(24)=150$

So, from Cramer's Rule, we have:

$$x=\frac{D_x}{D}=\frac{250}{50}=5$$
$$y=\frac{D_y}{D}=\frac{150}{50}=3$$

Thus, the solution is $x=5$, $y=3$.

19.

$D=\begin{vmatrix} 3 & 5 \\ -1 & 1 \end{vmatrix}=(3)(1)-(-1)(5)=8$

$D_x=\begin{vmatrix} 16 & 5 \\ 0 & 1 \end{vmatrix}=(16)(1)-(0)(5)=16$

$D_y=\begin{vmatrix} 3 & 16 \\ -1 & 0 \end{vmatrix}=(3)(0)-(-1)(16)=16$

So, from Cramer's Rule, we have:

$$x=\frac{D_x}{D}=\frac{16}{8}=2$$
$$y=\frac{D_y}{D}=\frac{16}{8}=2$$

Thus, the solution is $x=2$, $y=2$.

20.

$D=\begin{vmatrix} -2 & -3 \\ 4 & 7 \end{vmatrix}=(-2)(7)-(4)(-3)=-2$

$D_x=\begin{vmatrix} 15 & -3 \\ -33 & 7 \end{vmatrix}=(15)(7)-(-33)(-3)=6$

$D_y=\begin{vmatrix} -2 & 15 \\ 4 & -33 \end{vmatrix}=(-2)(-33)-(4)(15)=6$

So, from Cramer's Rule, we have:

$$x=\frac{D_x}{D}=\frac{6}{-2}=-3$$
$$y=\frac{D_y}{D}=\frac{6}{-2}=-3$$

Thus, the solution is $x=-3$, $y=-3$.

21.

$D=\begin{vmatrix} 3 & -5 \\ -6 & 10 \end{vmatrix}=(3)(10)-(-6)(-5)=0$

$D_x=\begin{vmatrix} 7 & -5 \\ -21 & 10 \end{vmatrix}=(7)(10)-(-21)(-5)$

$=-35$

From this information alone, we can conclude that the system is inconsistent or dependent.

22.

$D=\begin{vmatrix} 3 & -5 \\ 6 & -10 \end{vmatrix}=(3)(-10)-(6)(-5)=0$

$D_x=\begin{vmatrix} 7 & -5 \\ 14 & -10 \end{vmatrix}=(7)(-10)-(14)(-5)=0$

$D_y=\begin{vmatrix} 3 & 7 \\ 6 & 14 \end{vmatrix}=(3)(14)-(6)(7)=0$

Hence, the system is inconsistent or dependent.

23.

$$D=\begin{vmatrix} 2 & -3 \\ -10 & 15 \end{vmatrix}=(2)(15)-(-10)(-3)=0$$

$$D_x=\begin{vmatrix} 4 & -3 \\ -20 & 15 \end{vmatrix}=(4)(15)-(-20)(-3)=0$$

$$D_y=\begin{vmatrix} 2 & 4 \\ -10 & -20 \end{vmatrix}=(2)(-20)-(-10)(4)=0$$

Hence, the system is inconsistent or dependent.

24.

$$D=\begin{vmatrix} 2 & -3 \\ 10 & -15 \end{vmatrix}=(2)(-15)-(10)(-3)=0$$

$$D_x=\begin{vmatrix} 2 & -3 \\ 20 & -15 \end{vmatrix}=(2)(-15)-(20)(-3)=30$$

From this information alone, we can conclude that the system is inconsistent or dependent.

25. First, rewrite the system as:

$$\begin{cases} 6x+y=2 \\ 12x+y=5 \end{cases}.$$

Now, compute the determinants from Cramer's Rule:

$$D=\begin{vmatrix} 6 & 1 \\ 12 & 1 \end{vmatrix}=(6)(1)-(12)(1)=-6$$

$$D_x=\begin{vmatrix} 2 & 1 \\ 5 & 1 \end{vmatrix}=(2)(1)-(5)(1)=-3$$

$$D_y=\begin{vmatrix} 6 & 2 \\ 12 & 5 \end{vmatrix}=(6)(5)-(12)(2)=6$$

So, from Cramer's Rule, we have:

$$x=\frac{D_x}{D}=\frac{-3}{-6}=\frac{1}{2}$$

$$y=\frac{D_y}{D}=\frac{6}{-6}=-1$$

Thus, the solution is $x=\frac{1}{2},\ y=-1$.

26. First, rewrite the system as:

$$\begin{cases} 12x+18y=9 \\ 4x+3y=1 \end{cases}.$$

Now, compute the determinants from Cramer's Rule:

$$D=\begin{vmatrix} 12 & 18 \\ 4 & 3 \end{vmatrix}=(12)(3)-(4)(18)=-36$$

$$D_x=\begin{vmatrix} 9 & 18 \\ 1 & 3 \end{vmatrix}=(9)(3)-(1)(18)=9$$

$$D_y=\begin{vmatrix} 12 & 9 \\ 4 & 1 \end{vmatrix}=(12)(1)-(4)(9)=-24$$

So, from Cramer's Rule, we have:

$$x=\frac{D_x}{D}=\frac{9}{-36}=-\frac{1}{4}$$

$$y=\frac{D_y}{D}=\frac{-24}{-36}=\frac{2}{3}$$

Thus, the solution is $x=-\frac{1}{4},\ y=\frac{2}{3}$.

27.

$D=\begin{vmatrix}0.3 & -0.5\\0.2 & 1\end{vmatrix}=(0.3)(1)-(0.2)(-0.5)$
$=0.4$

$D_x=\begin{vmatrix}-0.6 & -0.5\\2.4 & 1\end{vmatrix}=(-0.6)(1)-(2.4)(-0.5)$
$=-0.6$

$D_y=\begin{vmatrix}0.3 & -0.6\\0.2 & 2.4\end{vmatrix}=(0.3)(2.4)-(0.2)(-0.6)$
$=0.84$

So, from Cramer's Rule, we have:

$$x=\frac{D_x}{D}=\frac{0.6}{0.4}=1.5$$
$$y=\frac{D_y}{D}=\frac{0.84}{0.4}=2.1$$

Thus, the solution is $x=1.5,\ y=2.1$.

28.

$D=\begin{vmatrix}0.5 & -0.4\\10 & 3.6\end{vmatrix}$
$=(0.5)(3.6)-(10)(-0.4)=5.8$

$D_x=\begin{vmatrix}-3.6 & -0.4\\-14 & 3.6\end{vmatrix}$
$=(-3.6)(3.6)-(-14)(-0.4)=-18.56$

$D_y=\begin{vmatrix}0.5 & -3.6\\10 & -14\end{vmatrix}$
$=(0.5)(-14)-(10)(-3.6)=29$

So, from Cramer's Rule, we have:

$$x=\frac{D_x}{D}=\frac{-18.56}{5.8}\cong -3.2$$
$$y=\frac{D_y}{D}=\frac{29}{5.8}\cong 5$$

Thus, the solution is $x\cong -3.2,\ y\cong 5$.

29. First, rewrite the system as:

$$\begin{cases}17x-y=-7\\-15x-y=-7\end{cases}.$$

Now, compute the determinants from Cramer's Rule:

$D=\begin{vmatrix}17 & -1\\-15 & -1\end{vmatrix}=(17)(-1)-(-15)(-1)$
$=-32$

$D_x=\begin{vmatrix}-7 & -1\\-7 & -1\end{vmatrix}=(-7)(-1)-(-7)(-1)$
$=0$

$D_y=\begin{vmatrix}17 & -7\\-15 & -7\end{vmatrix}=(17)(-7)-(-15)(-7)$
$=-224$

So, from Cramer's Rule, we have:

$$x=\frac{D_x}{D}=\frac{0}{-32}=0$$
$$y=\frac{D_y}{D}=\frac{-224}{-32}=7$$

Thus, the solution is $x=0,\ y=7$.

30. First, rewrite the system as:

$$\begin{cases} 9x+2y=-45 \\ 4x+3y=-20 \end{cases}.$$

Now, compute the determinants from Cramer's Rule:

$$D=\begin{vmatrix} 9 & 2 \\ 4 & 3 \end{vmatrix}=(9)(3)-(4)(2)=19$$

$$D_x=\begin{vmatrix} -45 & 2 \\ -20 & 3 \end{vmatrix}=(-45)(3)-(-20)(2)=-95$$

$$D_y=\begin{vmatrix} 9 & -45 \\ 4 & -20 \end{vmatrix}=(9)(-20)-(4)(-45)=0$$

So, from Cramer's Rule, we have:

$$x=\frac{D_x}{D}=\frac{-95}{19}=-5$$

$$y=\frac{D_y}{D}=\frac{0}{19}=0$$

Thus, the solution is $x=-5,\ y=0$.

31.

$$\begin{vmatrix} 3 & 1 & 0 \\ 2 & 0 & -1 \\ -4 & 1 & 0 \end{vmatrix} = 3\begin{vmatrix} 0 & -1 \\ 1 & 0 \end{vmatrix} - 1\begin{vmatrix} 2 & -1 \\ -4 & 0 \end{vmatrix} + 0\begin{vmatrix} 2 & 0 \\ -4 & 1 \end{vmatrix}$$

$$= 3\underbrace{\left[(0)(0)-(1)(-1)\right]}_{=1} - 1\underbrace{\left[(2)(0)-(-4)(-1)\right]}_{=-4} + 0\underbrace{\left[(2)(1)-(-4)(0)\right]}_{=2} = \boxed{7}$$

32.

$$\begin{vmatrix} 1 & 1 & 0 \\ 0 & 2 & -1 \\ 0 & -3 & 5 \end{vmatrix} = \begin{vmatrix} 2 & -1 \\ -3 & 5 \end{vmatrix} - \begin{vmatrix} 0 & -1 \\ 0 & 5 \end{vmatrix} + 0\begin{vmatrix} 0 & 2 \\ 0 & -3 \end{vmatrix}$$

$$= \underbrace{\left[(2)(5)-(-3)(-1)\right]}_{=7} - \underbrace{\left[(0)(5)-(0)(-1)\right]}_{=0} + 0\underbrace{\left[(0)(-3)-(0)(2)\right]}_{=0} = \boxed{7}$$

33.

$$\begin{vmatrix} 2 & 1 & -5 \\ 3 & 0 & -1 \\ 4 & 0 & 7 \end{vmatrix} = 2\begin{vmatrix} 0 & -1 \\ 0 & 7 \end{vmatrix} - \begin{vmatrix} 3 & -1 \\ 4 & 7 \end{vmatrix} + (-5)\begin{vmatrix} 3 & 0 \\ 4 & 0 \end{vmatrix}$$

$$= 2\underbrace{\left[(0)(7)-(0)(-1)\right]}_{=0} - \underbrace{\left[(3)(7)-(4)(-1)\right]}_{=25} - 5\underbrace{\left[(3)(0)-(4)(0)\right]}_{=0} = \boxed{-25}$$

34.

$$\begin{vmatrix} 2 & 1 & -5 \\ 3 & -7 & 0 \\ 4 & -6 & 0 \end{vmatrix} = 2\begin{vmatrix} -7 & 0 \\ -6 & 0 \end{vmatrix} - \begin{vmatrix} 3 & 0 \\ 4 & 0 \end{vmatrix} + (-5)\begin{vmatrix} 3 & -7 \\ 4 & -6 \end{vmatrix}$$

$$= 2\underbrace{\left[(-7)(0)-(-6)(0)\right]}_{=0} - \underbrace{\left[(3)(0)-(4)(0)\right]}_{=0} - 5\underbrace{\left[(3)(-6)-(4)(-7)\right]}_{=10} = \boxed{-50}$$

35.

$$\begin{vmatrix} 1 & 1 & -5 \\ 3 & -7 & -4 \\ 4 & -6 & 9 \end{vmatrix} = \begin{vmatrix} -7 & -4 \\ -6 & 9 \end{vmatrix} - \begin{vmatrix} 3 & -4 \\ 4 & 9 \end{vmatrix} + (-5)\begin{vmatrix} 3 & -7 \\ 4 & -6 \end{vmatrix}$$

$$= \underbrace{\left[(-7)(9)-(-6)(-4)\right]}_{=-87} - \underbrace{\left[(3)(9)-(4)(-4)\right]}_{=43} - 5\underbrace{\left[(3)(-6)-(4)(-7)\right]}_{=10}$$

$$= \boxed{-180}$$

36.

$$\begin{vmatrix} -3 & 2 & -5 \\ 1 & 8 & 2 \\ 4 & -6 & 9 \end{vmatrix} = -3\begin{vmatrix} 8 & 2 \\ -6 & 9 \end{vmatrix} - 2\begin{vmatrix} 1 & 2 \\ 4 & 9 \end{vmatrix} + (-5)\begin{vmatrix} 1 & 8 \\ 4 & -6 \end{vmatrix}$$

$$= \underbrace{-3\left[(8)(9)-(-6)(2)\right]}_{=84} - 2\underbrace{\left[(1)(9)-(4)(2)\right]}_{=1} - 5\underbrace{\left[(1)(-6)-(4)(8)\right]}_{=-38}$$

$$= \boxed{-64}$$

37.

$$\begin{vmatrix} 1 & 3 & 4 \\ 2 & -1 & 1 \\ 3 & -2 & 1 \end{vmatrix} = \begin{vmatrix} -1 & 1 \\ -2 & 1 \end{vmatrix} - 3\begin{vmatrix} 2 & 1 \\ 3 & 1 \end{vmatrix} + 4\begin{vmatrix} 2 & -1 \\ 3 & -2 \end{vmatrix}$$

$$= \underbrace{\left[(-1)(1)-(-2)(1)\right]}_{=1} - 3\underbrace{\left[(2)(1)-(3)(1)\right]}_{=-1} + 4\underbrace{\left[(2)(-2)-(3)(-1)\right]}_{=-1}$$

$$= \boxed{0}$$

38.

$$\begin{vmatrix} -7 & 2 & 5 \\ \frac{7}{8} & 3 & 4 \\ -1 & 4 & 6 \end{vmatrix} = -7\begin{vmatrix} 3 & 4 \\ 4 & 6 \end{vmatrix} - 2\begin{vmatrix} \frac{7}{8} & 4 \\ -1 & 6 \end{vmatrix} + 5\begin{vmatrix} \frac{7}{8} & 3 \\ -1 & 4 \end{vmatrix}$$

$$= -7\underbrace{\left[(3)(6)-(4)(4)\right]}_{=2} - 2\underbrace{\left[\left(\tfrac{7}{8}\right)(6)-(-1)(4)\right]}_{=9.25} + 5\underbrace{\left[\left(\tfrac{7}{8}\right)(4)-(-1)(3)\right]}_{=6.5}$$

$$= \boxed{0}$$

39.

$$\begin{vmatrix} -3 & 1 & 5 \\ 2 & 0 & 6 \\ 4 & 7 & -9 \end{vmatrix} = -3\begin{vmatrix} 0 & 6 \\ 7 & -9 \end{vmatrix} - 1\begin{vmatrix} 2 & 6 \\ 4 & -9 \end{vmatrix} + 5\begin{vmatrix} 2 & 0 \\ 4 & 7 \end{vmatrix}$$

$$= -3\underbrace{\left[(0)(-9)-(6)(7)\right]}_{-42} - \underbrace{\left[(2)(-9)-(6)(4)\right]}_{-42} + 5\underbrace{\left[(2)(7)-(4)(0)\right]}_{14}$$

$$= \boxed{238}$$

40.

$$\begin{vmatrix} 1 & -1 & 5 \\ 3 & -3 & 6 \\ 4 & 9 & 0 \end{vmatrix} = 1\begin{vmatrix} -3 & 6 \\ 9 & 0 \end{vmatrix} - 1\begin{vmatrix} 3 & 6 \\ 4 & 0 \end{vmatrix} + (-5)\begin{vmatrix} 3 & -3 \\ 4 & 9 \end{vmatrix}$$

$$= \underbrace{\left[(-3)(0)-(9)(6)\right]}_{-54} - \underbrace{\left[(3)(0)-(4)(6)\right]}_{-24} - 5\underbrace{\left[(3)(9)-(4)(-3)\right]}_{39}$$

$$= \boxed{-225}$$

41.

$$\begin{vmatrix} -2 & 1 & -7 \\ 4 & -2 & 14 \\ 0 & 1 & 8 \end{vmatrix} = -2\begin{vmatrix} -2 & 14 \\ 1 & 8 \end{vmatrix} - 1\begin{vmatrix} 4 & 14 \\ 0 & 8 \end{vmatrix} + (-7)\begin{vmatrix} 4 & -2 \\ 0 & 1 \end{vmatrix}$$

$$= -2\underbrace{\left[(-2)(8)-(1)(14)\right]}_{-30} - 1\underbrace{\left[(4)(8)-(0)(14)\right]}_{32} - 7\underbrace{\left[(4)(1)-(0)(-2)\right]}_{4}$$

$$= \boxed{0}$$

42.

$$\begin{vmatrix} 5 & -2 & -1 \\ 4 & -9 & -3 \\ 2 & 8 & -6 \end{vmatrix} = 5\begin{vmatrix} -9 & -3 \\ 8 & -6 \end{vmatrix} - (-2)\begin{vmatrix} 4 & -3 \\ 2 & -6 \end{vmatrix} - 1\begin{vmatrix} 4 & -9 \\ 2 & 8 \end{vmatrix}$$

$$= 5\underbrace{\left[(-9)(-6)-(8)(-3)\right]}_{78} + 2\underbrace{\left[(4)(-6)-(2)(-3)\right]}_{-18} - \underbrace{\left[(4)(8)-(2)(-9)\right]}_{50}$$

$$= \boxed{304}$$

43.

$$D=\begin{vmatrix}1&1&-1\\1&-1&1\\1&1&1\end{vmatrix}=1\begin{vmatrix}-1&1\\1&1\end{vmatrix}-1\begin{vmatrix}1&1\\1&1\end{vmatrix}+(-1)\begin{vmatrix}1&-1\\1&1\end{vmatrix}$$

$$=1\underbrace{\left[(-1)(1)-(1)(1)\right]}_{=-2}-1\underbrace{\left[(1)(1)-(1)(1)\right]}_{=0}+(-1)\underbrace{\left[(1)(1)-(1)(-1)\right]}_{=2}=-4$$

$$D_x=\begin{vmatrix}0&1&-1\\4&-1&1\\10&1&1\end{vmatrix}=0\begin{vmatrix}-1&1\\1&1\end{vmatrix}-1\begin{vmatrix}4&1\\10&1\end{vmatrix}+(-1)\begin{vmatrix}4&-1\\10&1\end{vmatrix}$$

$$=0\underbrace{\left[(-1)(1)-(1)(1)\right]}_{=-2}-1\underbrace{\left[(4)(1)-(10)(1)\right]}_{=-6}+(-1)\underbrace{\left[(4)(1)-(10)(-1)\right]}_{=14}=-8$$

$$D_y=\begin{vmatrix}1&0&-1\\1&4&1\\1&10&1\end{vmatrix}=1\begin{vmatrix}4&1\\10&1\end{vmatrix}-0\begin{vmatrix}1&1\\1&1\end{vmatrix}+(-1)\begin{vmatrix}1&4\\1&10\end{vmatrix}$$

$$=1\underbrace{\left[(4)(1)-(10)(1)\right]}_{=-6}-0\underbrace{\left[(1)(1)-(1)(1)\right]}_{=0}+(-1)\underbrace{\left[(1)(10)-(1)(4)\right]}_{=6}=-12$$

$$D_z=\begin{vmatrix}1&1&0\\1&-1&4\\1&1&10\end{vmatrix}=1\begin{vmatrix}-1&4\\1&10\end{vmatrix}-1\begin{vmatrix}1&4\\1&10\end{vmatrix}+0\begin{vmatrix}1&-1\\1&1\end{vmatrix}$$

$$=1\underbrace{\left[(-1)(10)-(1)(4)\right]}_{=-14}-1\underbrace{\left[(1)(10)-(1)(4)\right]}_{=6}+0\underbrace{\left[(1)(1)-(1)(-1)\right]}_{=2}=-20$$

So, by Cramer's Rule,

$$\boxed{x=\frac{D_x}{D}=\frac{-8}{-4}=2\quad y=\frac{D_y}{D}=\frac{-12}{-4}=3\quad z=\frac{D_z}{D}=\frac{-20}{-4}=5}$$

44.

$$D=\begin{vmatrix}-1 & 1 & 1\\ 1 & 1 & -1\\ 1 & 1 & 1\end{vmatrix}=(-1)\begin{vmatrix}1 & -1\\ 1 & 1\end{vmatrix}-1\begin{vmatrix}1 & -1\\ 1 & 1\end{vmatrix}+1\begin{vmatrix}1 & 1\\ 1 & 1\end{vmatrix}$$

$$=(-1)\underbrace{\left[(1)(1)-(1)(-1)\right]}_{=2}-1\underbrace{\left[(1)(1)-(1)(-1)\right]}_{=2}+1\underbrace{\left[(1)(1)-(1)(1)\right]}_{=0}=-4$$

$$D_x=\begin{vmatrix}-4 & 1 & 1\\ 0 & 1 & -1\\ 2 & 1 & 1\end{vmatrix}=-4\begin{vmatrix}1 & -1\\ 1 & 1\end{vmatrix}-1\begin{vmatrix}0 & -1\\ 2 & 1\end{vmatrix}+1\begin{vmatrix}0 & 1\\ 2 & 1\end{vmatrix}$$

$$=\underbrace{-4\left[(1)(1)-(1)(-1)\right]}_{=2}-1\underbrace{\left[(0)(1)-(2)(-1)\right]}_{=2}+1\underbrace{\left[(0)(1)-(2)(1)\right]}_{=-2}=-12$$

$$D_y=\begin{vmatrix}-1 & -4 & 1\\ 1 & 0 & -1\\ 1 & 2 & 1\end{vmatrix}=(-1)\begin{vmatrix}0 & -1\\ 2 & 1\end{vmatrix}-(-4)\begin{vmatrix}1 & -1\\ 1 & 1\end{vmatrix}+1\begin{vmatrix}1 & 0\\ 1 & 2\end{vmatrix}$$

$$=(-1)\underbrace{\left[(0)(1)-(2)(-1)\right]}_{=2}+4\underbrace{\left[(1)(1)-(1)(-1)\right]}_{=2}+1\underbrace{\left[(1)(2)-(1)(0)\right]}_{=2}=8$$

$$D_z=\begin{vmatrix}-1 & 1 & -4\\ 1 & 1 & 0\\ 1 & 1 & 2\end{vmatrix}=(-1)\begin{vmatrix}1 & 0\\ 1 & 2\end{vmatrix}-1\begin{vmatrix}1 & 0\\ 1 & 2\end{vmatrix}+(-4)\begin{vmatrix}1 & 1\\ 1 & 1\end{vmatrix}$$

$$=(-1)\underbrace{\left[(1)(2)-(1)(0)\right]}_{=2}-1\underbrace{\left[(1)(2)-(1)(0)\right]}_{=2}+(-4)\underbrace{\left[(1)(1)-(1)(1)\right]}_{=0}=-4$$

So, by Cramer's Rule,

$$\boxed{x=\frac{D_x}{D}=\frac{-12}{-4}=3 \quad y=\frac{D_y}{D}=\frac{8}{-4}=-2 \quad z=\frac{D_z}{D}=\frac{-4}{-4}=1}$$

45.

$$D=\begin{vmatrix}3&8&2\\-2&5&3\\4&9&2\end{vmatrix}=3\begin{vmatrix}5&3\\9&2\end{vmatrix}-8\begin{vmatrix}-2&3\\4&2\end{vmatrix}+2\begin{vmatrix}-2&5\\4&9\end{vmatrix}$$

$$=3\underbrace{\left[(5)(2)-(9)(3)\right]}_{=-17}-8\underbrace{\left[(-2)(2)-(4)(3)\right]}_{=-16}+2\underbrace{\left[(-2)(9)-(4)(5)\right]}_{=-38}=1$$

$$D_x=\begin{vmatrix}28&8&2\\34&5&3\\29&9&2\end{vmatrix}=28\begin{vmatrix}5&3\\9&2\end{vmatrix}-8\begin{vmatrix}34&3\\29&2\end{vmatrix}+2\begin{vmatrix}34&5\\29&9\end{vmatrix}$$

$$=28\underbrace{\left[(5)(2)-(9)(3)\right]}_{=-17}-8\underbrace{\left[(34)(2)-(29)(3)\right]}_{=-19}+2\underbrace{\left[(34)(9)-(29)(5)\right]}_{=161}=-2$$

$$D_y=\begin{vmatrix}3&28&2\\-2&34&3\\4&29&2\end{vmatrix}=3\begin{vmatrix}34&3\\29&2\end{vmatrix}-28\begin{vmatrix}-2&3\\4&2\end{vmatrix}+2\begin{vmatrix}-2&34\\4&29\end{vmatrix}$$

$$=3\underbrace{\left[(34)(2)-(29)(3)\right]}_{=-19}-28\underbrace{\left[(-2)(2)-(4)(3)\right]}_{=-16}+2\underbrace{\left[(-2)(29)-(4)(34)\right]}_{=-194}=3$$

$$D_z=\begin{vmatrix}3&8&28\\-2&5&34\\4&9&29\end{vmatrix}=3\begin{vmatrix}5&34\\9&29\end{vmatrix}-8\begin{vmatrix}-2&34\\4&29\end{vmatrix}+28\begin{vmatrix}-2&5\\4&9\end{vmatrix}$$

$$=3\underbrace{\left[(5)(29)-(9)(34)\right]}_{=-161}-8\underbrace{\left[(-2)(29)-(4)(34)\right]}_{=-194}+28\underbrace{\left[(-2)(9)-(4)(5)\right]}_{=-38}=5$$

So, by Cramer's Rule,

$$\boxed{x=\frac{D_x}{D}=-2 \qquad y=\frac{D_y}{D}=3 \qquad z=\frac{D_z}{D}=5}$$

46.

$$D=\begin{vmatrix}7 & 2 & -1\\ 6 & 5 & 1\\ -5 & -4 & 3\end{vmatrix}=7\begin{vmatrix}5 & 1\\ -4 & 3\end{vmatrix}-2\begin{vmatrix}6 & 1\\ -5 & 3\end{vmatrix}+(-1)\begin{vmatrix}6 & 5\\ -5 & -4\end{vmatrix}$$

$$=7\underbrace{\left[(5)(3)-(-4)(1)\right]}_{=19}-2\underbrace{\left[(6)(3)-(-5)(1)\right]}_{=23}+(-1)\underbrace{\left[(6)(-4)-(-5)(5)\right]}_{=1}=86$$

$$D_x=\begin{vmatrix}-1 & 2 & -1\\ 16 & 5 & 1\\ -5 & -4 & 3\end{vmatrix}=(-1)\begin{vmatrix}5 & 1\\ -4 & 3\end{vmatrix}-2\begin{vmatrix}16 & 1\\ -5 & 3\end{vmatrix}+(-1)\begin{vmatrix}16 & 5\\ -5 & -4\end{vmatrix}$$

$$=(-1)\underbrace{\left[(5)(3)-(-4)(1)\right]}_{=19}-2\underbrace{\left[(16)(3)-(-5)(1)\right]}_{=53}+(-1)\underbrace{\left[(16)(-4)-(-5)(5)\right]}_{=-39}=-86$$

$$D_y=\begin{vmatrix}7 & -1 & -1\\ 6 & 16 & 1\\ -5 & -5 & 3\end{vmatrix}=7\begin{vmatrix}16 & 1\\ -5 & 3\end{vmatrix}-(-1)\begin{vmatrix}6 & 1\\ -5 & 3\end{vmatrix}+(-1)\begin{vmatrix}6 & 16\\ -5 & -5\end{vmatrix}$$

$$=7\underbrace{\left[(16)(3)-(-5)(1)\right]}_{=53}-(-1)\underbrace{\left[(6)(3)-(-5)(1)\right]}_{=23}+(-1)\underbrace{\left[(6)(-5)-(-5)(16)\right]}_{=50}=344$$

$$D_z=\begin{vmatrix}7 & 2 & -1\\ 6 & 5 & 16\\ -5 & -4 & -5\end{vmatrix}=7\begin{vmatrix}5 & 16\\ -4 & -5\end{vmatrix}-2\begin{vmatrix}6 & 16\\ -5 & -5\end{vmatrix}+(-1)\begin{vmatrix}6 & 5\\ -5 & -4\end{vmatrix}$$

$$=7\underbrace{\left[(5)(-5)-(-4)(16)\right]}_{=39}-2\underbrace{\left[(6)(-5)-(-5)(16)\right]}_{=50}+(-1)\underbrace{\left[(6)(-4)-(-5)(5)\right]}_{=1}=172$$

So, by Cramer's Rule,

$$x=\frac{D_x}{D}=\frac{-86}{86}=-1 \qquad y=\frac{D_y}{D}=\frac{344}{86}=4 \qquad z=\frac{D_z}{D}=\frac{172}{86}=2$$

47.

$$D=\begin{vmatrix}3&0&5\\0&4&3\\2&-1&0\end{vmatrix}=3\begin{vmatrix}4&3\\-1&0\end{vmatrix}-0\begin{vmatrix}0&3\\2&0\end{vmatrix}+5\begin{vmatrix}0&4\\2&-1\end{vmatrix}$$

$$=3\underbrace{\left[(4)(0)-(-1)(3)\right]}_{=3}-0\underbrace{\left[(0)(0)-(2)(3)\right]}_{=-6}+5\underbrace{\left[(0)(-1)-(2)(4)\right]}_{=-8}=-31$$

$$D_x=\begin{vmatrix}11&0&5\\-9&4&3\\7&-1&0\end{vmatrix}=11\begin{vmatrix}4&3\\-1&0\end{vmatrix}-0\begin{vmatrix}-9&3\\7&0\end{vmatrix}+5\begin{vmatrix}-9&4\\7&-1\end{vmatrix}$$

$$=11\underbrace{\left[(4)(0)-(-1)(3)\right]}_{=3}-0\underbrace{\left[(-9)(0)-(7)(3)\right]}_{=-21}+5\underbrace{\left[(-9)(-1)-(7)(4)\right]}_{=-19}=-62$$

$$D_y=\begin{vmatrix}3&11&5\\0&-9&3\\2&7&0\end{vmatrix}=3\begin{vmatrix}-9&3\\7&0\end{vmatrix}-11\begin{vmatrix}0&3\\2&0\end{vmatrix}+5\begin{vmatrix}0&-9\\2&7\end{vmatrix}$$

$$=3\underbrace{\left[(-9)(0)-(7)(3)\right]}_{=-21}-11\underbrace{\left[(0)(0)-(2)(3)\right]}_{=-6}+5\underbrace{\left[(0)(7)-(2)(-9)\right]}_{=18}=93$$

$$D_z=\begin{vmatrix}3&0&11\\0&4&-9\\2&-1&7\end{vmatrix}=3\begin{vmatrix}4&-9\\-1&7\end{vmatrix}-0\begin{vmatrix}0&-9\\2&7\end{vmatrix}+11\begin{vmatrix}0&4\\2&-1\end{vmatrix}$$

$$=3\underbrace{\left[(4)(7)-(-1)(-9)\right]}_{=19}-0\underbrace{\left[(0)(7)-(2)(-9)\right]}_{=18}+11\underbrace{\left[(0)(-1)-(2)(4)\right]}_{=-8}=-31$$

So, by Cramer's Rule,

$$\boxed{x=\frac{D_x}{D}=\frac{-62}{-31}=2 \qquad y=\frac{D_y}{D}=\frac{93}{-31}=-3 \qquad z=\frac{D_z}{D}=\frac{-31}{-31}=1}$$

48.

$$D=\begin{vmatrix}3 & 0 & -2\\ 4 & 0 & 1\\ 6 & -2 & 0\end{vmatrix}=3\begin{vmatrix}0 & 1\\ -2 & 0\end{vmatrix}-0\begin{vmatrix}4 & 1\\ 6 & 0\end{vmatrix}+(-2)\begin{vmatrix}4 & 0\\ 6 & -2\end{vmatrix}$$

$$=3\underbrace{\left[(0)(0)-(-2)(1)\right]}_{=2}-0\underbrace{\left[(4)(0)-(6)(1)\right]}_{=-6}+(-2)\underbrace{\left[(4)(-2)-(6)(0)\right]}_{=-8}=22$$

$$D_x=\begin{vmatrix}7 & 0 & -2\\ 24 & 0 & 1\\ 10 & -2 & 0\end{vmatrix}=7\begin{vmatrix}0 & 1\\ -2 & 0\end{vmatrix}-0\begin{vmatrix}24 & 1\\ 10 & 0\end{vmatrix}+(-2)\begin{vmatrix}24 & 0\\ 10 & -2\end{vmatrix}$$

$$=7\underbrace{\left[(0)(0)-(-2)(1)\right]}_{=2}-0\underbrace{\left[(24)(0)-(10)(1)\right]}_{=-10}+(-2)\underbrace{\left[(24)(-2)-(10)(0)\right]}_{=-48}=110$$

$$D_y=\begin{vmatrix}3 & 7 & -2\\ 4 & 24 & 1\\ 6 & 10 & 0\end{vmatrix}=3\begin{vmatrix}24 & 1\\ 10 & 0\end{vmatrix}-7\begin{vmatrix}4 & 1\\ 6 & 0\end{vmatrix}+(-2)\begin{vmatrix}4 & 24\\ 6 & 10\end{vmatrix}$$

$$=3\underbrace{\left[(24)(0)-(10)(1)\right]}_{=-10}-7\underbrace{\left[(4)(0)-(6)(1)\right]}_{=-6}+(-2)\underbrace{\left[(4)(10)-(6)(24)\right]}_{=-104}=220$$

$$D_z=\begin{vmatrix}3 & 0 & 7\\ 4 & 0 & 24\\ 6 & -2 & 10\end{vmatrix}=3\begin{vmatrix}0 & 24\\ -2 & 10\end{vmatrix}-0\begin{vmatrix}4 & 24\\ 6 & 10\end{vmatrix}+7\begin{vmatrix}4 & 0\\ 6 & -2\end{vmatrix}$$

$$=3\underbrace{\left[(0)(10)-(-2)(24)\right]}_{=48}-0\underbrace{\left[(4)(10)-(6)(24)\right]}_{=-104}+7\underbrace{\left[(4)(-2)-(6)(0)\right]}_{=-8}=88$$

So, by Cramer's Rule,

$$\boxed{x=\frac{D_x}{D}=\frac{110}{22}=5 \qquad y=\frac{D_y}{D}=\frac{220}{22}=10 \qquad z=\frac{D_z}{D}=\frac{88}{22}=4}$$

49.

$$D=\begin{vmatrix}1 & 1 & -1\\ 1 & -1 & 1\\ -2 & -2 & 2\end{vmatrix}=1\begin{vmatrix}-1 & 1\\ -2 & 2\end{vmatrix}-1\begin{vmatrix}1 & 1\\ -2 & 2\end{vmatrix}+(-1)\begin{vmatrix}1 & -1\\ -2 & -2\end{vmatrix}$$

$$=1\underbrace{\left[(-1)(2)-(-2)(1)\right]}_{=0}-1\underbrace{\left[(1)(2)-(-2)(1)\right]}_{=4}+(-1)\underbrace{\left[(1)(-2)-(-2)(-1)\right]}_{=-4}=0$$

So, Cramer's Rule implies that the system is either inconsistent or dependent.

50.

$$D=\begin{vmatrix}1 & 1 & -1\\ 1 & -1 & 1\\ -2 & -2 & 2\end{vmatrix}=1\begin{vmatrix}-1 & 1\\ -2 & 2\end{vmatrix}-1\begin{vmatrix}1 & 1\\ -2 & 2\end{vmatrix}+(-1)\begin{vmatrix}1 & -1\\ -2 & -2\end{vmatrix}$$

$$=1\underbrace{\left[(-1)(2)-(-2)(1)\right]}_{=0}-1\underbrace{\left[(1)(2)-(-2)(1)\right]}_{=4}+(-1)\underbrace{\left[(1)(-2)-(-2)(-1)\right]}_{=-4}=0$$

So, Cramer's Rule implies that the system is either inconsistent or dependent.

51.

$$D=\begin{vmatrix}1 & 1 & 1\\ 1 & -1 & 1\\ -1 & 1 & -1\end{vmatrix}=1\begin{vmatrix}-1 & 1\\ 1 & -1\end{vmatrix}-1\begin{vmatrix}1 & 1\\ -1 & -1\end{vmatrix}+1\begin{vmatrix}1 & -1\\ -1 & 1\end{vmatrix}$$

$$=1\underbrace{\left[(-1)(-1)-(1)(1)\right]}_{=0}-1\underbrace{\left[(1)(-1)-(-1)(1)\right]}_{=0}+1\underbrace{\left[(1)(1)-(-1)(-1)\right]}_{=0}=0$$

So, Cramer's Rule implies that the system is either inconsistent or dependent.

52.

$$D=\begin{vmatrix}1 & 1 & 1\\ 1 & -1 & -1\\ -1 & 1 & 1\end{vmatrix}=1\begin{vmatrix}-1 & -1\\ 1 & 1\end{vmatrix}-1\begin{vmatrix}1 & -1\\ -1 & 1\end{vmatrix}+1\begin{vmatrix}1 & -1\\ -1 & 1\end{vmatrix}$$

$$=1\underbrace{\left[(-1)(1)-(1)(-1)\right]}_{=0}-1\underbrace{\left[(1)(1)-(-1)(-1)\right]}_{=0}+1\underbrace{\left[(1)(1)-(-1)(-1)\right]}_{=0}=0$$

So, Cramer's Rule implies that the system is either inconsistent or dependent.

53. First, write the system in matrix form as:

$$\begin{bmatrix} 1 & 2 & 3 \\ -2 & 3 & 5 \\ 4 & -1 & 8 \end{bmatrix} \begin{bmatrix} x \\ y \\ z \end{bmatrix} = \begin{bmatrix} 11 \\ 29 \\ 19 \end{bmatrix}$$

Now, apply Cramer's rule to solve the system:

$$D = \begin{vmatrix} 1 & 2 & 3 \\ -2 & 3 & 5 \\ 4 & -1 & 8 \end{vmatrix} = 1\begin{vmatrix} 3 & 5 \\ -1 & 8 \end{vmatrix} - 2\begin{vmatrix} -2 & 5 \\ 4 & 8 \end{vmatrix} + 3\begin{vmatrix} -2 & 3 \\ 4 & -1 \end{vmatrix} = 71$$

$$D_x = \begin{vmatrix} 11 & 2 & 3 \\ 29 & 3 & 5 \\ 19 & -1 & 8 \end{vmatrix} = 11\begin{vmatrix} 3 & 5 \\ -1 & 8 \end{vmatrix} - 2\begin{vmatrix} 29 & 5 \\ 19 & 8 \end{vmatrix} + 3\begin{vmatrix} 29 & 3 \\ 19 & -1 \end{vmatrix} = -213$$

$$D_y = \begin{vmatrix} 1 & 11 & 3 \\ -2 & 29 & 5 \\ 4 & 19 & 8 \end{vmatrix} = 1\begin{vmatrix} 29 & 5 \\ 19 & 8 \end{vmatrix} - 11\begin{vmatrix} -2 & 5 \\ 4 & 8 \end{vmatrix} + 3\begin{vmatrix} -2 & 29 \\ 4 & 19 \end{vmatrix} = 71$$

$$D_z = \begin{vmatrix} 1 & 2 & 11 \\ -2 & 3 & 29 \\ 4 & -1 & 19 \end{vmatrix} = 1\begin{vmatrix} 3 & 29 \\ -1 & 19 \end{vmatrix} - 2\begin{vmatrix} -2 & 29 \\ 4 & 19 \end{vmatrix} + 11\begin{vmatrix} -2 & 3 \\ 4 & -1 \end{vmatrix} = 284$$

So, by Cramer's rule:

$$\boxed{x = \frac{D_x}{D} = \frac{-213}{71} = -3 \quad y = \frac{D_y}{D} = \frac{71}{71} = 1 \quad z = \frac{D_z}{D} = \frac{284}{71} = 4}$$

54. First, write the system in matrix form as:

$$\begin{bmatrix} 8 & -2 & 5 \\ 3 & 1 & -1 \\ 2 & -6 & 4 \end{bmatrix} \begin{bmatrix} x \\ y \\ z \end{bmatrix} = \begin{bmatrix} 36 \\ 17 \\ -2 \end{bmatrix}$$

Now, apply Cramer's rule to solve the system:

$$D = \begin{vmatrix} 8 & -2 & 5 \\ 3 & 1 & -1 \\ 2 & -6 & 4 \end{vmatrix} = 8\begin{vmatrix} 1 & -1 \\ -6 & 4 \end{vmatrix} + 2\begin{vmatrix} 3 & -1 \\ 2 & 4 \end{vmatrix} + 5\begin{vmatrix} 3 & 1 \\ 2 & -6 \end{vmatrix} = -88$$

$$D_x = \begin{vmatrix} 36 & -2 & 5 \\ 17 & 1 & -1 \\ -2 & 6 & 4 \end{vmatrix} = 36\begin{vmatrix} 1 & -1 \\ 6 & 4 \end{vmatrix} + 2\begin{vmatrix} 17 & -1 \\ -2 & 4 \end{vmatrix} + 5\begin{vmatrix} 17 & 1 \\ -2 & 6 \end{vmatrix} = 1{,}012$$

$$D_y = \begin{vmatrix} 8 & 36 & 5 \\ 3 & 17 & -91 \\ 2 & -2 & 4 \end{vmatrix} = 8\begin{vmatrix} 17 & -1 \\ -2 & 4 \end{vmatrix} - 36\begin{vmatrix} 3 & -1 \\ 2 & 4 \end{vmatrix} + 5\begin{vmatrix} 3 & 17 \\ 2 & -2 \end{vmatrix} = -176$$

$$D_z = \begin{vmatrix} 8 & -2 & 36 \\ 3 & 1 & 17 \\ 2 & -6 & -2 \end{vmatrix} = 8\begin{vmatrix} 1 & 17 \\ 6 & -2 \end{vmatrix} + 2\begin{vmatrix} 3 & 17 \\ 2 & -2 \end{vmatrix} + 36\begin{vmatrix} 3 & 1 \\ 2 & -6 \end{vmatrix} = -1{,}632$$

So, by Cramer's rule:

$$\boxed{x = \frac{D_x}{D} = -\frac{23}{2} \quad y = \frac{D_y}{D} = 2 \quad z = \frac{D_z}{D} = \frac{204}{11}}$$

55. First, write the system in matrix form as:

$$\begin{bmatrix} 1 & -4 & 7 \\ -3 & 2 & -1 \\ 5 & 8 & -2 \end{bmatrix}\begin{bmatrix} x \\ y \\ z \end{bmatrix} = \begin{bmatrix} 49 \\ -17 \\ -24 \end{bmatrix}$$

Now, apply Cramer's rule to solve the system:

$$D = \begin{vmatrix} 1 & -4 & 7 \\ -3 & 2 & -1 \\ 5 & 8 & -2 \end{vmatrix} = 1\begin{vmatrix} 2 & -1 \\ 8 & -2 \end{vmatrix} + 4\begin{vmatrix} -3 & -1 \\ 5 & -2 \end{vmatrix} + 7\begin{vmatrix} -3 & 2 \\ 5 & 8 \end{vmatrix} = -190$$

$$D_x = \begin{vmatrix} 49 & -4 & 7 \\ -17 & 2 & -1 \\ -24 & 8 & -2 \end{vmatrix} = 49\begin{vmatrix} 2 & -1 \\ 8 & -2 \end{vmatrix} + 4\begin{vmatrix} -17 & -1 \\ -24 & -2 \end{vmatrix} + 7\begin{vmatrix} -17 & 2 \\ -24 & 8 \end{vmatrix} = -380$$

$$D_y = \begin{vmatrix} 1 & 49 & 7 \\ -3 & -17 & -1 \\ 5 & -24 & -2 \end{vmatrix} = 1\begin{vmatrix} -17 & -1 \\ -24 & -2 \end{vmatrix} - 49\begin{vmatrix} -3 & -1 \\ 5 & -2 \end{vmatrix} + 7\begin{vmatrix} -3 & -17 \\ 5 & -24 \end{vmatrix} = 570$$

$$D_z = \begin{vmatrix} 1 & -4 & 49 \\ -3 & 2 & -17 \\ 5 & 8 & -24 \end{vmatrix} = 1\begin{vmatrix} 2 & -17 \\ 8 & -24 \end{vmatrix} + 4\begin{vmatrix} -3 & -17 \\ 5 & -24 \end{vmatrix} + 49\begin{vmatrix} -3 & 2 \\ 5 & 8 \end{vmatrix} = -950$$

So, by Cramer's rule:

$$x = \frac{D_x}{D} = \frac{-380}{-190} = 2 \qquad y = \frac{D_y}{D} = \frac{570}{-190} = -3 \qquad z = \frac{D_z}{D} = \frac{-950}{-190} = 5$$

56. First, write the system in matrix form as:

$$\begin{bmatrix} \frac{1}{2} & -2 & 7 \\ 1 & \frac{1}{4} & -4 \\ -4 & 5 & 0 \end{bmatrix}\begin{bmatrix} x \\ y \\ z \end{bmatrix}=\begin{bmatrix} 25 \\ -2 \\ -56 \end{bmatrix}$$

Now, apply Cramer's rule to solve the system:

$$D=\begin{vmatrix} \frac{1}{2} & -2 & 7 \\ 1 & \frac{1}{4} & -4 \\ -4 & 5 & 0 \end{vmatrix}=\frac{1}{2}\begin{vmatrix} \frac{1}{4} & -4 \\ 5 & 0 \end{vmatrix}+2\begin{vmatrix} 1 & -4 \\ -4 & 0 \end{vmatrix}+7\begin{vmatrix} 1 & \frac{1}{4} \\ -4 & 5 \end{vmatrix}=20$$

$$D_x=\begin{vmatrix} 25 & -2 & 7 \\ -2 & \frac{1}{4} & -4 \\ -56 & 5 & 0 \end{vmatrix}=25\begin{vmatrix} \frac{1}{4} & -4 \\ 5 & 0 \end{vmatrix}+2\begin{vmatrix} -2 & -4 \\ -56 & 0 \end{vmatrix}+7\begin{vmatrix} -2 & \frac{1}{4} \\ -56 & 5 \end{vmatrix}=80$$

$$D_y=\begin{vmatrix} \frac{1}{2} & 25 & 7 \\ 1 & -2 & -4 \\ -4 & -56 & 0 \end{vmatrix}=\frac{1}{2}\begin{vmatrix} -2 & -4 \\ -56 & 0 \end{vmatrix}-25\begin{vmatrix} 1 & -4 \\ -4 & 0 \end{vmatrix}+7\begin{vmatrix} 1 & -2 \\ -4 & -56 \end{vmatrix}=-160$$

$$D_z=\begin{vmatrix} \frac{1}{2} & -2 & 25 \\ 1 & \frac{1}{4} & -2 \\ -4 & 5 & -56 \end{vmatrix}=\frac{1}{2}\begin{vmatrix} \frac{1}{4} & -2 \\ 5 & -56 \end{vmatrix}+2\begin{vmatrix} 1 & -2 \\ -4 & -56 \end{vmatrix}+25\begin{vmatrix} 1 & \frac{1}{4} \\ -4 & 5 \end{vmatrix}=20$$

So, by Cramer's rule:

$$x=\frac{D_x}{D}=\frac{80}{20}=4 \quad y=\frac{D_y}{D}=\frac{-160}{20}=-8 \quad z=\frac{D_z}{D}=\frac{20}{20}=1$$

57.

$$D = \begin{vmatrix} 2 & 7 & -4 \\ -1 & -4 & -5 \\ 4 & -2 & -9 \end{vmatrix}$$

$$= 2\underbrace{\left[(-4)(-9)-(-2)(-5)\right]}_{=26} - 7\underbrace{\left[(-1)(-9)-(4)(-5)\right]}_{=29} + (-4)\underbrace{\left[(-1)(-2)-(4)(-4)\right]}_{=18} = -223$$

$$D_x = \begin{vmatrix} -5.5 & 7 & -4 \\ -19 & -4 & -5 \\ -38 & -2 & -9 \end{vmatrix}$$

$$= -5.5\underbrace{\left[(-4)(-9)-(-2)(-5)\right]}_{=26} - 7\underbrace{\left[(-19)(-9)-(-38)(-5)\right]}_{=-19} + (-4)\underbrace{\left[(-19)(-2)-(-38)(-4)\right]}_{=-114} = 446$$

$$D_y = \begin{vmatrix} 2 & -5.5 & -4 \\ -1 & -19 & -5 \\ 4 & -38 & -9 \end{vmatrix}$$

$$= 2\underbrace{\left[(-19)(-9)-(-38)(-5)\right]}_{=-19} - (-5.5)\underbrace{\left[(-1)(-9)-(4)(-5)\right]}_{=29} + (-4)\underbrace{\left[(-1)(-38)-(4)(-19)\right]}_{=114} = -334.5$$

$$D_z = \begin{vmatrix} 2 & 7 & -5.5 \\ -1 & -4 & -19 \\ 4 & -2 & -38 \end{vmatrix}$$

$$= 2\underbrace{\left[(-4)(-38)-(-2)(-19)\right]}_{=114} - 7\underbrace{\left[(-1)(-38)-(4)(-19)\right]}_{=114} + (-5.5)\underbrace{\left[(-1)(-2)-(4)(-4)\right]}_{=18} = -669$$

So, by Cramer's Rule,

$$\boxed{x = \frac{D_x}{D} = -2 \qquad y = \frac{D_y}{D} = \frac{3}{2} \qquad z = \frac{D_z}{D} = 3}$$

58.

$$D = \begin{vmatrix} 4 & -2 & 1 \\ 3 & 1 & -2 \\ -6 & 1 & 5 \end{vmatrix}$$

$$= 4\underbrace{\left[(1)(5)-(1)(-2)\right]}_{=7} - (-2)\underbrace{\left[(3)(5)-(-6)(-2)\right]}_{=3} + (1)\underbrace{\left[(3)(1)-(-6)(1)\right]}_{=9} = 43$$

$$D_x = \begin{vmatrix} -15 & -2 & 1 \\ -20 & 1 & -2 \\ 51 & 1 & 5 \end{vmatrix}$$

$$= -15\underbrace{\left[(1)(5)-(1)(-2)\right]}_{=7} - (-2)\underbrace{\left[(-20)(5)-(51)(-2)\right]}_{=2} + (1)\underbrace{\left[(-20)(1)-(51)(1)\right]}_{=-71} = -172$$

$$D_y = \begin{vmatrix} 4 & -15 & 1 \\ 3 & -20 & -2 \\ -6 & 51 & 5 \end{vmatrix}$$

$$= 4\underbrace{\left[(-20)(5)-(51)(-2)\right]}_{=2} - (-15)\underbrace{\left[(3)(5)-(-6)(-2)\right]}_{=3} + (1)\underbrace{\left[(3)(5)-(-6)(-20)\right]}_{=33} = 86$$

$$D_z = \begin{vmatrix} 4 & -2 & -15 \\ 3 & 1 & -20 \\ -6 & 1 & 51 \end{vmatrix}$$

$$= 4\underbrace{\left[(1)(51)-(1)(-20)\right]}_{=71} - (-2)\underbrace{\left[(3)(5)-(-6)(-20)\right]}_{=33} + (-15)\underbrace{\left[(3)(1)-(-6)(1)\right]}_{=9} = 215$$

So, by Cramer's Rule,

$$\boxed{x = \frac{D_x}{D} = -4 \qquad y = \frac{D_y}{D} = 2 \qquad z = \frac{D_z}{D} = 5}$$

59.

$$D = \begin{vmatrix} -2 & -1 & 1 \\ 1 & 5 & 1 \\ 3 & 9 & 1 \end{vmatrix} = -2\underbrace{\left[(5)(1)-(1)(9)\right]}_{=-4} - (-1)\underbrace{\left[(1)(1)-(3)(1)\right]}_{=-2} + (1)\underbrace{\left[(1)(9)-(3)(5)\right]}_{=-6} = 0$$

So, yes, collinear.

60.

$$D = \begin{vmatrix} 2 & -6 & 1 \\ -7 & 30 & 1 \\ 5 & -18 & 1 \end{vmatrix} = 2\underbrace{\left[(30)(1)-(-18)(1)\right]}_{=48} - (-6)\underbrace{\left[(-7)(1)-(5)(1)\right]}_{=-12} + (1)\underbrace{\left[(-7)(-18)-(5)(30)\right]}_{=-24} = 0$$

So, yes, collinear.

61. Let $x_1 = 3,\ x_2 = 5,\ x_3 = 3,\ y_1 = 2,\ y_2 = 2,\ y_3 = -4$. Then, we have

$$\text{Area} = \pm\tfrac{1}{2}\begin{vmatrix} 3 & 2 & 1 \\ 5 & 2 & 1 \\ 3 & -4 & 1 \end{vmatrix} = \pm\tfrac{1}{2}\left[3\begin{vmatrix} 2 & 1 \\ -4 & 1 \end{vmatrix} - 2\begin{vmatrix} 5 & 1 \\ 3 & 1 \end{vmatrix} + 1\begin{vmatrix} 5 & 2 \\ 3 & -4 \end{vmatrix}\right]$$

$$= \pm\tfrac{1}{2}\left(3\underbrace{\left[(2)(1)-(-4)(1)\right]}_{=6} - 2\underbrace{\left[(5)(1)-(3)(1)\right]}_{=2} + 1\underbrace{\left[(5)(-4)-(3)(2)\right]}_{=-26}\right)$$

$$= \pm\tfrac{1}{2}(-12)$$

Hence, choosing the positive value above, we conclude that the area is 6 units^2. Alternatively, we could compute the area by identifying the height and base of the triangle. Indeed, consider the following diagram:

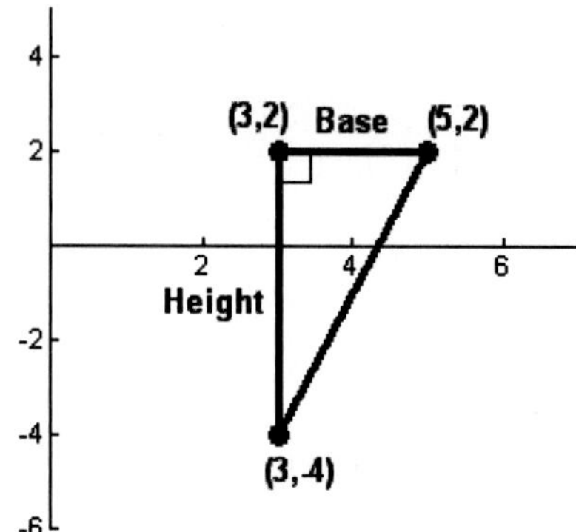

Base = 2
Height = 6

Hence, Area =
$\frac{1}{2}(\text{Base})(\text{Height}) = \frac{1}{2}(2)(6) = 6$ units^2.

62. Let $x_1 = 2,\ x_2 = 7,\ x_3 = 7,\ y_1 = 3,\ y_2 = 3,\ y_3 = 7$. Then, we have

$$\text{Area} = \pm\tfrac{1}{2}\begin{vmatrix} 2 & 3 & 1 \\ 7 & 3 & 1 \\ 7 & 7 & 1 \end{vmatrix} = \pm\tfrac{1}{2}\left[2\begin{vmatrix} 3 & 1 \\ 7 & 1 \end{vmatrix} - 3\begin{vmatrix} 7 & 1 \\ 7 & 1 \end{vmatrix} + 1\begin{vmatrix} 7 & 3 \\ 7 & 7 \end{vmatrix}\right]$$

$$= \pm\tfrac{1}{2}\left(2\underbrace{\left[(3)(1)-(7)(1)\right]}_{=-4} - 3\underbrace{\left[(7)(1)-(7)(1)\right]}_{=0} + 1\underbrace{\left[(7)(7)-(7)(3)\right]}_{=28} \right)$$

$$= \pm\tfrac{1}{2}(20)$$

Hence, choosing the positive value above, we conclude that the area is 10 units^2. Alternatively, we could compute the area by identifying the height and base of the triangle. Indeed, consider the following diagram:

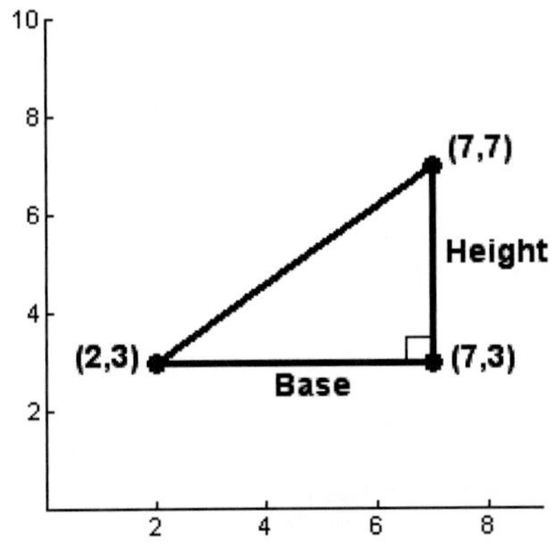

Base = 5
Height = 4

Hence, we see that

$$\text{Area} = \tfrac{1}{2}(\text{Base})(\text{Height})$$
$$= \tfrac{1}{2}(5)(4) = 10 \text{ units}^2$$

63. Let $x_1 = 1,\ x_2 = 3,\ x_3 = -2,\ y_1 = 2,\ y_2 = 4,\ y_3 = 5$. Then, we have

$$\text{Area} = \pm\tfrac{1}{2}\begin{vmatrix} 1 & 2 & 1 \\ 3 & 4 & 1 \\ -2 & 5 & 1 \end{vmatrix} = \pm\tfrac{1}{2}\left[1\begin{vmatrix} 4 & 1 \\ 5 & 1 \end{vmatrix} - 2\begin{vmatrix} 3 & 1 \\ -2 & 1 \end{vmatrix} + 1\begin{vmatrix} 3 & 4 \\ -2 & 5 \end{vmatrix}\right]$$

$$= \pm\tfrac{1}{2}\left(1\underbrace{\left[(4)(1)-(5)(1)\right]}_{=-1} - 2\underbrace{\left[(3)(1)-(-2)(1)\right]}_{=5} + 1\underbrace{\left[(3)(5)-(-2)(4)\right]}_{=23} \right) = \pm\tfrac{1}{2}(12)$$

Hence, choosing the positive value above, we conclude that the area is 6 units^2.

64. Let $x_1 = -1,\ x_2 = 3,\ x_3 = 2,\ y_1 = -2,\ y_2 = 4,\ y_3 = 1$. Then, we have

$$\text{Area} = \pm\tfrac{1}{2}\begin{vmatrix} -1 & -2 & 1 \\ 3 & 4 & 1 \\ 2 & 1 & 1 \end{vmatrix} = \pm\tfrac{1}{2}\left[(-1)\begin{vmatrix} 4 & 1 \\ 1 & 1 \end{vmatrix} - (-2)\begin{vmatrix} 3 & 1 \\ 2 & 1 \end{vmatrix} + 1\begin{vmatrix} 3 & 4 \\ 2 & 1 \end{vmatrix}\right]$$

$$= \pm\tfrac{1}{2}\left((-1)\underbrace{\left[(4)(1) - (1)(1)\right]}_{=3} - (-2)\underbrace{\left[(3)(1) - (2)(1)\right]}_{=1} + 1\underbrace{\left[(3)(1) - (2)(4)\right]}_{=-5} \right) = \pm\tfrac{1}{2}(-6)$$

Hence, choosing the positive value above, we conclude that the area is 3 units^2.

65. Let $x_1 = 1,\ x_2 = 2,\ y_1 = 2,\ y_2 = 4$. The equation of the line through the two points (x_1, y_1) and (x_2, y_2) is:

$$0 = \begin{vmatrix} x & y & 1 \\ 1 & 2 & 1 \\ 2 & 4 & 1 \end{vmatrix} \text{ which simplifies to}$$

$$0 = x\begin{vmatrix} 2 & 1 \\ 4 & 1 \end{vmatrix} - y\begin{vmatrix} 1 & 1 \\ 2 & 1 \end{vmatrix} + 1\begin{vmatrix} 1 & 2 \\ 2 & 4 \end{vmatrix}$$

$$= x\underbrace{\left[(2)(1) - (4)(1)\right]}_{=-2} - y\underbrace{\left[(1)(1) - (2)(1)\right]}_{=-1} + 1\underbrace{\left[(1)(4) - (2)(2)\right]}_{=0} = -2x + y + 0$$

Hence, the equation of this line is $\boxed{y = 2x}$.

66. Let $x_1 = 0,\ x_2 = 2,\ x_3 = 1,\ y_1 = 5,\ y_2 = 0,\ y_3 = 2$. Observe that

$$\begin{vmatrix} 0 & 5 & 1 \\ 2 & 0 & 1 \\ 1 & 2 & 1 \end{vmatrix} = 0\begin{vmatrix} 0 & 1 \\ 2 & 1 \end{vmatrix} - 5\begin{vmatrix} 2 & 1 \\ 1 & 1 \end{vmatrix} + 1\begin{vmatrix} 2 & 0 \\ 1 & 2 \end{vmatrix}$$

$$= 0\underbrace{\left[(0)(1) - (2)(1)\right]}_{=-2} - 5\underbrace{\left[(2)(1) - (1)(1)\right]}_{=1} + 1\underbrace{\left[(2)(2) - (1)(0)\right]}_{=4}$$

$$= -1$$

Since this determinant is not 0, we conclude that the points (x_1, y_1), (x_2, y_2), and (x_3, y_3) are $\boxed{\text{not collinear}}$.

67. The system we must solve is

$$\begin{cases} I_1 - I_2 - I_3 = 0 \\ 4I_1 + 0I_2 + 2I_3 = 16 \\ 4I_1 + 4I_2 + 0I_3 = 24 \end{cases}, \text{ which is equivalent to } \begin{bmatrix} 1 & -1 & -1 \\ 4 & 0 & 2 \\ 4 & 4 & 0 \end{bmatrix}\begin{bmatrix} I_1 \\ I_2 \\ I_3 \end{bmatrix} = \begin{bmatrix} 0 \\ 16 \\ 24 \end{bmatrix}.$$

The solution is

$$\begin{bmatrix} I_1 \\ I_2 \\ I_3 \end{bmatrix} = \begin{bmatrix} 1 & -1 & -1 \\ 4 & 0 & 2 \\ 4 & 4 & 0 \end{bmatrix}^{-1}\begin{bmatrix} 0 \\ 16 \\ 24 \end{bmatrix}.$$

We now compute the inverse:

$$\left[\begin{array}{ccc|ccc} & -1 & -1 & 1 & 0 & 0 \\ 4 & 0 & 2 & 0 & 1 & 0 \\ 4 & 4 & 0 & 0 & 0 & 1 \end{array}\right] \xrightarrow[R_2-4R_1\to R_2]{R_3-4R_1\to R_3} \left[\begin{array}{ccc|ccc} 1 & -1 & -1 & -1 & 0 & 0 \\ 0 & 4 & 6 & -4 & 1 & 0 \\ 0 & 8 & 4 & -4 & 0 & 1 \end{array}\right] \xrightarrow{R_3-2R_2\to R_3}$$

$$\left[\begin{array}{ccc|ccc} 1 & -1 & -1 & 1 & 0 & 0 \\ 0 & 4 & 6 & -4 & 1 & 0 \\ 0 & 0 & -8 & 4 & -2 & 1 \end{array}\right] \xrightarrow{-\frac{1}{8}R_3\to R_3} \left[\begin{array}{ccc|ccc} 1 & -1 & -1 & 1 & 0 & 0 \\ 0 & 4 & 6 & -4 & 1 & 0 \\ 0 & 0 & 1 & -\frac{1}{2} & \frac{1}{4} & -\frac{1}{8} \end{array}\right] \xrightarrow{R_2-6R_3\to R_2}$$

$$\left[\begin{array}{ccc|ccc} 1 & -1 & -1 & 1 & 0 & 0 \\ 0 & 4 & 0 & -1 & -\frac{1}{2} & \frac{3}{4} \\ 0 & 0 & 1 & -\frac{1}{2} & \frac{1}{4} & -\frac{1}{8} \end{array}\right] \xrightarrow{\frac{1}{4}R_2\to R_2} \left[\begin{array}{ccc|ccc} 1 & -1 & -1 & 1 & 0 & 0 \\ 0 & 1 & 0 & -\frac{1}{4} & -\frac{1}{8} & \frac{3}{16} \\ 0 & 0 & 1 & -\frac{1}{2} & \frac{1}{4} & -\frac{1}{8} \end{array}\right] \xrightarrow{R_1+R_2+R_3\to R_1}$$

$$\left[\begin{array}{ccc|ccc} 1 & 0 & 0 & \frac{1}{4} & \frac{1}{8} & \frac{1}{16} \\ 0 & 1 & 0 & -\frac{1}{4} & -\frac{1}{8} & \frac{3}{16} \\ 0 & 0 & 1 & -\frac{1}{2} & \frac{1}{4} & -\frac{1}{8} \end{array}\right]$$

So, the solution is

$$\begin{bmatrix} I_1 \\ I_2 \\ I_3 \end{bmatrix} = \begin{bmatrix} \frac{1}{4} & \frac{1}{8} & \frac{1}{16} \\ -\frac{1}{4} & -\frac{1}{8} & \frac{3}{16} \\ -\frac{1}{2} & \frac{1}{4} & -\frac{1}{8} \end{bmatrix}\begin{bmatrix} 0 \\ 16 \\ 24 \end{bmatrix} = \begin{bmatrix} \frac{7}{2} \\ \frac{5}{2} \\ 1 \end{bmatrix}.$$

So, $I_1 = \frac{7}{2}$, $I_2 = \frac{5}{2}$, $I_3 = 1$.

68. The system we must solve is

$$\begin{cases} I_1 - I_2 - I_3 = 0 \\ 6I_1 + 0I_2 + 3I_3 = 24 \\ 6I_1 + 6I_2 + 0I_3 = 36 \end{cases}, \text{ which is equivalent to } \begin{bmatrix} 1 & -1 & -1 \\ 2 & 0 & 1 \\ 1 & 1 & 0 \end{bmatrix} \begin{bmatrix} I_1 \\ I_2 \\ I_3 \end{bmatrix} = \begin{bmatrix} 0 \\ 8 \\ 6 \end{bmatrix}.$$

The solution is

$$\begin{bmatrix} I_1 \\ I_2 \\ I_3 \end{bmatrix} = \begin{bmatrix} 1 & -1 & -1 \\ 2 & 0 & 1 \\ 1 & 1 & 0 \end{bmatrix}^{-1} \begin{bmatrix} 0 \\ 8 \\ 6 \end{bmatrix}.$$

We now compute the inverse:

$$\left[\begin{array}{ccc|ccc} 1 & -1 & -1 & 1 & 0 & 0 \\ 2 & 0 & 1 & 0 & 1 & 0 \\ 1 & 1 & 0 & 0 & 0 & 1 \end{array}\right] \xrightarrow[R_2 - 2R_3 \to R_3]{R_2 - 2R_1 \to R_1} \left[\begin{array}{ccc|ccc} 0 & 2 & 3 & -2 & 1 & 0 \\ 2 & 0 & 1 & 0 & 1 & 0 \\ 0 & -2 & 1 & 0 & 1 & -2 \end{array}\right] \xrightarrow{R_1 + R_3 \to R_3}$$

$$\left[\begin{array}{ccc|ccc} 0 & 2 & 3 & -2 & 1 & 0 \\ 2 & 0 & 1 & 0 & 1 & 0 \\ 0 & 0 & 4 & -2 & 2 & -2 \end{array}\right] \xrightarrow{\frac{1}{4}R_3 \to R_3} \left[\begin{array}{ccc|ccc} 0 & 2 & 3 & -2 & 1 & 0 \\ 2 & 0 & 1 & 0 & 1 & 0 \\ 0 & 0 & 1 & -\frac{1}{2} & \frac{1}{2} & -\frac{1}{2} \end{array}\right] \xrightarrow{R_1 - 3R_3 \to R_1}$$

$$\left[\begin{array}{ccc|ccc} 0 & 2 & 0 & -\frac{1}{2} & -\frac{1}{2} & \frac{3}{2} \\ 2 & 0 & 1 & 0 & 1 & 0 \\ 0 & 0 & 1 & -\frac{1}{2} & \frac{1}{2} & -\frac{1}{2} \end{array}\right] \xrightarrow{\frac{1}{2}R_1 \to R_1} \left[\begin{array}{ccc|ccc} 0 & 1 & 0 & -\frac{1}{4} & -\frac{1}{4} & \frac{3}{4} \\ 2 & 0 & 1 & 0 & 1 & 0 \\ 0 & 0 & 1 & -\frac{1}{2} & \frac{1}{2} & -\frac{1}{2} \end{array}\right] \xrightarrow{R_2 - R_3 \to R_2}$$

$$\left[\begin{array}{ccc|ccc} 0 & 1 & 0 & -\frac{1}{4} & -\frac{1}{4} & \frac{3}{4} \\ 2 & 0 & 0 & \frac{1}{2} & \frac{1}{2} & \frac{1}{2} \\ 0 & 0 & 1 & -\frac{1}{2} & \frac{1}{2} & -\frac{1}{2} \end{array}\right] \xrightarrow[R_2 \leftrightarrow R_1]{\frac{1}{2}R_2 \to R_2} \left[\begin{array}{ccc|ccc} 1 & 0 & 0 & \frac{1}{4} & \frac{1}{4} & \frac{1}{4} \\ 0 & 1 & 0 & -\frac{1}{4} & -\frac{1}{4} & \frac{3}{4} \\ 0 & 0 & 1 & -\frac{1}{2} & \frac{1}{2} & -\frac{1}{2} \end{array}\right]$$

So, the solution is:

$$\begin{bmatrix} I_1 \\ I_2 \\ I_3 \end{bmatrix} = \begin{bmatrix} \frac{1}{4} & \frac{1}{4} & \frac{1}{4} \\ -\frac{1}{4} & -\frac{1}{4} & \frac{3}{4} \\ -\frac{1}{2} & \frac{1}{2} & -\frac{1}{2} \end{bmatrix} \begin{bmatrix} 0 \\ 8 \\ 6 \end{bmatrix} = \begin{bmatrix} \frac{7}{2} \\ \frac{5}{2} \\ 1 \end{bmatrix}$$

So, $I_1 = \frac{7}{2}$, $I_2 = \frac{5}{2}$, $I_3 = 1$.

69. The second determinant should be subtracted; that is, it should be $-1\begin{vmatrix} -3 & 2 \\ 1 & -1 \end{vmatrix}$.

70. The third determinant should be $3\begin{vmatrix} -3 & 0 \\ 1 & 4 \end{vmatrix}$.

71. In D_x and D_y, the column $\begin{bmatrix} 6 \\ -3 \end{bmatrix}$ should replace the column corresponding to the variable that is being solved for in each case. Precisely,

D_x should be $\begin{vmatrix} 6 & 3 \\ -3 & -1 \end{vmatrix}$ and

D_y should be $\begin{vmatrix} 2 & 6 \\ -1 & -3 \end{vmatrix}$.

72. $x = \dfrac{D_x}{D}$, not $\dfrac{D}{D_x}$.

73 True

74. True

75. False. Observe that

$$\begin{vmatrix} 2 & 6 & 4 \\ 0 & 2 & 8 \\ 4 & 0 & 10 \end{vmatrix} = 2\begin{vmatrix} 2 & 8 \\ 0 & 10 \end{vmatrix} - 6\begin{vmatrix} 0 & 8 \\ 4 & 10 \end{vmatrix} + 4\begin{vmatrix} 0 & 2 \\ 4 & 0 \end{vmatrix}$$

$$= 2\underbrace{\left[(2)(10)-(0)(8)\right]}_{=20} - 6\underbrace{\left[(0)(10)-(4)(8)\right]}_{=-32} + 4\underbrace{\left[(0)(0)-(4)(2)\right]}_{=-8} = 200$$

whereas

$$2\begin{vmatrix} 1 & 3 & 2 \\ 0 & 1 & 4 \\ 2 & 0 & 5 \end{vmatrix} = 2\left(1\begin{vmatrix} 1 & 4 \\ 0 & 5 \end{vmatrix} - 3\begin{vmatrix} 0 & 4 \\ 2 & 5 \end{vmatrix} + 2\begin{vmatrix} 0 & 1 \\ 2 & 0 \end{vmatrix} \right)$$

$$= 2\left(1\underbrace{\left[(1)(5)-(0)(4)\right]}_{=5} - 3\underbrace{\left[(0)(5)-(2)(4)\right]}_{=-8} + 2\underbrace{\left[(0)(0)-(2)(1)\right]}_{=-2} \right) = 2(25) = 50$$

76. True. (Note also that two rows are identical – the determinant will always be 0 in such case.)

$$\begin{vmatrix} 3 & 1 & 2 \\ 0 & 2 & 8 \\ 3 & 1 & 2 \end{vmatrix} = 3\begin{vmatrix} 2 & 8 \\ 1 & 2 \end{vmatrix} - 1\begin{vmatrix} 0 & 8 \\ 3 & 2 \end{vmatrix} + 2\begin{vmatrix} 0 & 2 \\ 3 & 1 \end{vmatrix}$$

$$= 3\underbrace{\left[(2)(2) - (1)(8)\right]}_{=-4} - 1\underbrace{\left[(0)(2) - (3)(8)\right]}_{=-24} + 2\underbrace{\left[(0)(1) - (3)(2)\right]}_{=-6} = 0$$

77.

$$\begin{vmatrix} a & 0 & 0 \\ 0 & b & 0 \\ 0 & 0 & c \end{vmatrix} = a\begin{vmatrix} b & 0 \\ 0 & c \end{vmatrix} - 0\begin{vmatrix} 0 & 0 \\ 0 & c \end{vmatrix} + 0\begin{vmatrix} 0 & b \\ 0 & 0 \end{vmatrix}$$

$$= a\underbrace{\left[(b)(c) - (0)(0)\right]}_{=bc} - 0\underbrace{\left[(0)(c) - (0)(0)\right]}_{=0} + 0\underbrace{\left[(0)(0) - (0)(b)\right]}_{=0} = \boxed{abc}$$

78.

$$\begin{vmatrix} a_1 & b_1 & c_1 \\ 0 & b_2 & c_2 \\ 0 & 0 & c_3 \end{vmatrix} = a_1\begin{vmatrix} b_2 & c_2 \\ 0 & c_3 \end{vmatrix} - b_1\begin{vmatrix} 0 & c_2 \\ 0 & c_3 \end{vmatrix} + c_1\begin{vmatrix} 0 & b_2 \\ 0 & 0 \end{vmatrix}$$

$$= a_1\underbrace{\left[(b_2)(c_3) - (0)(c_2)\right]}_{=b_2c_3} - b_1\underbrace{\left[(0)(c_3) - (0)(c_2)\right]}_{=0} + c_1\underbrace{\left[(0)(0) - (0)(b_2)\right]}_{=0} = \boxed{a_1b_2c_3}$$

79.

$$\begin{vmatrix} 1 & -2 & -1 & 3 \\ 4 & 0 & 1 & 2 \\ 0 & 3 & 2 & 4 \\ 1 & -3 & 5 & -4 \end{vmatrix} = 1\begin{vmatrix} 0 & 1 & 2 \\ 3 & 2 & 4 \\ -3 & 5 & -4 \end{vmatrix} - (-2)\begin{vmatrix} 4 & 1 & 2 \\ 0 & 2 & 4 \\ 1 & 5 & -4 \end{vmatrix} - 1\begin{vmatrix} 4 & 0 & 2 \\ 0 & 3 & 4 \\ 1 & -3 & -4 \end{vmatrix} - 3\begin{vmatrix} 4 & 0 & 1 \\ 0 & 3 & 2 \\ 1 & -3 & 5 \end{vmatrix} \quad \textbf{(1)}$$

where

$$\begin{vmatrix} 0 & 1 & 2 \\ 3 & 2 & 4 \\ -3 & 5 & -4 \end{vmatrix} = 0\begin{vmatrix} 2 & 4 \\ 5 & -4 \end{vmatrix} - 1\begin{vmatrix} 3 & 4 \\ -3 & -4 \end{vmatrix} + 2\begin{vmatrix} 3 & 2 \\ -3 & 5 \end{vmatrix}$$

$$= 0\underbrace{\left[(2)(-4)-(5)(4)\right]}_{=-28} - 1\underbrace{\left[(3)(-4)-(-3)(4)\right]}_{=0} + 2\underbrace{\left[(3)(5)-(-3)(2)\right]}_{=21} = 42$$

$$\begin{vmatrix} 4 & 1 & 2 \\ 0 & 2 & 4 \\ 1 & 5 & -4 \end{vmatrix} = 4\begin{vmatrix} 2 & 4 \\ 5 & -4 \end{vmatrix} - 1\begin{vmatrix} 0 & 4 \\ 1 & -4 \end{vmatrix} + 2\begin{vmatrix} 0 & 2 \\ 1 & 5 \end{vmatrix}$$

$$= 4\underbrace{\left[(2)(-4)-(5)(4)\right]}_{=-28} - 1\underbrace{\left[(0)(-4)-(1)(4)\right]}_{=-4} + 2\underbrace{\left[(0)(5)-(1)(2)\right]}_{=-2} = -112$$

$$\begin{vmatrix} 4 & 0 & 2 \\ 0 & 3 & 4 \\ 1 & -3 & -4 \end{vmatrix} = 4\begin{vmatrix} 3 & 4 \\ -3 & -4 \end{vmatrix} - 0\begin{vmatrix} 0 & 4 \\ 1 & -4 \end{vmatrix} + 2\begin{vmatrix} 0 & 3 \\ 1 & -3 \end{vmatrix}$$

$$= 4\underbrace{\left[(3)(-4)-(-3)(4)\right]}_{=0} - 0\underbrace{\left[(0)(-4)-(1)(4)\right]}_{=-4} + 2\underbrace{\left[(0)(-3)-(1)(3)\right]}_{=-3} = -6$$

$$\begin{vmatrix} 4 & 0 & 1 \\ 0 & 3 & 2 \\ 1 & -3 & 5 \end{vmatrix} = 4\begin{vmatrix} 3 & 2 \\ -3 & 5 \end{vmatrix} - 0\begin{vmatrix} 0 & 2 \\ 1 & 5 \end{vmatrix} + 1\begin{vmatrix} 0 & 3 \\ 1 & -3 \end{vmatrix}$$

$$= 4\underbrace{\left[(3)(5)-(-3)(2)\right]}_{=21} - 0\underbrace{\left[(0)(5)-(1)(2)\right]}_{=-2} + 1\underbrace{\left[(0)(-3)-(1)(3)\right]}_{=-3} = 81$$

So, substituting the values of the determinants of the four individual 3×3 matrices, we conclude that the given determinant is $1(42)+2(-112)-(-6)-3(81)=\boxed{-419}$

80. Observe that $D = \begin{vmatrix} 3 & 2 \\ a & -4 \end{vmatrix} = (3)(-4)-(a)(2) = -12-2a$. So, if a is chosen such that $D=0$, then there wouldn't be a unique solution to the system. In this case, choose $\boxed{a=-6}$.

81.

$$\begin{vmatrix} a_1 & b_1 & c_1 \\ a_2 & b_2 & c_2 \\ a_3 & b_3 & c_3 \end{vmatrix} = -b_1 \begin{vmatrix} a_2 & c_2 \\ a_3 & c_3 \end{vmatrix} + b_2 \begin{vmatrix} a_1 & c_1 \\ a_3 & c_3 \end{vmatrix} - b_3 \begin{vmatrix} a_1 & c_1 \\ a_2 & c_2 \end{vmatrix}$$

$$= -b_1\left[(a_2)(c_3)-(a_3)(c_2)\right] + b_2\left[(a_1)(c_3)-(a_3)(c_1)\right] - b_3\left[(a_1)(c_2)-(a_2)(c_1)\right]$$

$$= -a_2b_1c_3 + a_3b_1c_2 + a_1b_2c_3 - a_3b_2c_1 - a_1b_3c_2 + a_2b_3c_1$$

The last line above is equal to the given right-side since addition and multiplication of real numbers is commutative.

82.

$$\begin{vmatrix} a_1 & b_1 & c_1 \\ a_2 & b_2 & c_2 \\ a_3 & b_3 & c_3 \end{vmatrix} = a_3 \begin{vmatrix} b_1 & c_1 \\ b_2 & c_2 \end{vmatrix} - b_3 \begin{vmatrix} a_1 & c_1 \\ a_2 & c_2 \end{vmatrix} + c_3 \begin{vmatrix} a_1 & b_1 \\ a_2 & b_2 \end{vmatrix}$$

$$= a_3\left[(b_1)(c_2)-(b_2)(c_1)\right] - b_3\left[(a_1)(c_2)-(a_2)(c_1)\right] + c_3\left[(a_1)(b_2)-(a_2)(b_1)\right]$$

$$= a_3b_1c_2 - a_3b_2c_1 - a_1b_3c_2 + a_2b_3c_1 + a_1b_2c_3 - a_2b_1c_3$$

$$= a_1b_2c_3 + a_2b_3c_1 + a_3b_1c_2 - a_1b_3c_2 - a_2b_1c_3 - a_3b_2c_1$$

The last line above is equal to the given right-side since addition and multiplication of real numbers is commutative. (Note also that this is equivalent to the answer found in Problem 81.)

83.

$$D = \begin{vmatrix} a^2 & a & 1 \\ b^2 & b & 1 \\ c^2 & c & 1 \end{vmatrix} = a^2 \underbrace{\left[(b)(1)-(c)(1)\right]}_{=b-c} - a \underbrace{\left[(b^2)(1)-(c^2)(1)\right]}_{=b^2-c^2} + (1) \underbrace{\left[(b^2)(c)-(b)(c^2)\right]}_{=b^2c-bc^2}$$

$$= a^2b - a^2c - ab^2 + ac^2 + b^2c - bc^2$$

$$= \left(a^2 - ab - ac + bc\right)(b-c)$$

$$= (a-b)(a-c)(b-c)$$

84. We must find a value of a such that $D = \begin{vmatrix} 1 & 3 & 2 \\ 1 & a & 4 \\ 0 & 2 & a \end{vmatrix} = 0$. Computing the determinant yields: $1\left(a^2-8\right) - 3(a) + 2(2) = a^2 - 3a - 4 = (a-4)(a+1) = 0 \Rightarrow a = -1, 4$.

85. –180	**86.** –64	**87.** –1019	**88.** –2287

89. First, write the system in matrix form as:

$$\begin{bmatrix} 3.1 & 1.6 & -4.8 \\ 5.2 & -3.4 & 0.5 \\ 0.5 & -6.4 & 11.4 \end{bmatrix}\begin{bmatrix} x \\ y \\ z \end{bmatrix}=\begin{bmatrix} -33.76 \\ -36.68 \\ 25.96 \end{bmatrix}$$

Now, apply Cramer's rule to solve the system:

$$D=\begin{vmatrix} 3.1 & 1.6 & -4.8 \\ 5.2 & -3.4 & 0.5 \\ 0.5 & -6.4 & 11.4 \end{vmatrix},\quad D_x=\begin{vmatrix} -33.76 & 1.6 & -4.8 \\ -36.68 & -3.4 & 0.5 \\ 25.96 & -6.4 & 11.4 \end{vmatrix},$$

$$D_y=\begin{vmatrix} 3.1 & -33.76 & -4.8 \\ 5.2 & -36.68 & 0.5 \\ 0.5 & 25.96 & 11.4 \end{vmatrix},\ D_z=\begin{vmatrix} 3.1 & 1.6 & -33.76 \\ 5.2 & -3.4 & -36.68 \\ 0.5 & -6.4 & 25.96 \end{vmatrix}$$

Using a calculator to compute the determinants, we obtain from Cramer's rule that:

$$\boxed{x=\frac{D_x}{D}=-6.4 \quad y=\frac{D_y}{D}=1.5 \quad z=\frac{D_z}{D}=3.4}$$

90. First, write the system in matrix form as:

$$\begin{bmatrix} -9.2 & 2.7 & 5.1 \\ 4.3 & -6.9 & -7.6 \\ 2.8 & -3.9 & -3.5 \end{bmatrix}\begin{bmatrix} x \\ y \\ z \end{bmatrix}=\begin{bmatrix} -89.2 \\ 38.89 \\ 34.08 \end{bmatrix}$$

Now, apply Cramer's rule to solve the system:

$$D=\begin{vmatrix} -9.2 & 2.7 & 5.1 \\ 4.3 & -6.9 & -7.6 \\ 2.8 & -3.9 & -3.5 \end{vmatrix},\quad D_x=\begin{vmatrix} -89.2 & 2.7 & 5.1 \\ 38.89 & -6.9 & -7.6 \\ 34.08 & -3.9 & -3.5 \end{vmatrix},$$

$$D_y=\begin{vmatrix} -9.2 & -89.2 & 5.1 \\ 4.3 & 38.89 & -7.6 \\ 2.8 & 34.08 & -3.5 \end{vmatrix},\ D_z=\begin{vmatrix} -9.2 & 2.7 & -89.2 \\ 4.3 & -6.9 & 38.89 \\ 2.8 & -3.9 & 34.08 \end{vmatrix}$$

Using a calculator to compute the determinants, we obtain from Cramer's rule that:

$$\boxed{x=\frac{D_x}{D}=12.5 \quad y=\frac{D_y}{D}=-8.2 \quad z=\frac{D_z}{D}=9.4}$$

91.	$\begin{vmatrix} \cos\theta & -r\sin\theta \\ \sin\theta & r\cos\theta \end{vmatrix} = r\cos^2\theta + r\sin^2\theta = r\left(\cos^2\theta + \sin^2\theta\right) = r$
92.	$\begin{vmatrix} 2x & 2y \\ 2x & 2y-2 \end{vmatrix} = 2x(2y-2) - 2x(2y) = -4x$

93.

$$\begin{vmatrix} \sin\varphi\cos\theta & -\rho\sin\varphi\sin\theta & \rho\cos\varphi\cos\theta \\ \sin\varphi\sin\theta & \rho\sin\varphi\cos\theta & \rho\cos\varphi\sin\theta \\ \cos\varphi & 0 & -\rho\sin\varphi \end{vmatrix} =$$

$$\sin\varphi\cos\theta\left[\rho\sin\varphi\cos\theta(-\rho\sin\varphi) - 0\right]$$
$$-(-\rho\sin\varphi\sin\theta)\left[\sin\varphi\sin\theta(-\rho\sin\varphi) - \cos\varphi(\rho\cos\varphi\sin\theta)\right]$$
$$+\rho\cos\varphi\cos\theta\left[0 - \cos\varphi(\rho\sin\varphi\cos\theta)\right] =$$

$$-\rho^2\sin^3\varphi\cos^2\theta - \rho^2\sin^3\varphi\sin^2\theta - \rho^2\sin\varphi\cos^2\varphi\sin^2\theta - \rho^2\cos^2\varphi\cos^2\theta\sin\varphi =$$
$$-\rho^2\left[\sin^3\varphi\cos^2\theta + \sin^3\varphi\sin^2\theta + \sin\varphi\cos^2\varphi\sin^2\theta + \cos^2\varphi\cos^2\theta\sin\varphi\right] =$$
$$-\rho^2\left[\sin\varphi\cos^2\theta\left(\sin^2\varphi + \cos^2\varphi\right) + \sin\varphi\sin^2\theta\left(\sin^2\varphi + \cos^2\varphi\right)\right] =$$
$$-\rho^2\left[\sin\varphi\cos^2\theta + \sin\varphi\sin^2\theta\right] =$$
$$-\rho^2\sin\varphi\left(\cos^2\theta + \sin^2\theta\right) = -\rho^2\sin\varphi$$

94.

$$\begin{vmatrix} \cos\theta & -r\sin\theta & 0 \\ \sin\theta & r\cos\theta & 0 \\ 0 & 0 & 1 \end{vmatrix} = 1\left[\cos\theta(r\cos\theta) - (-r\sin\theta)\sin\theta\right] = r\cos^2\theta + r\sin^2\theta = r$$

Section 8.6 Solutions --

1. d $x\left(x^2-25\right) = x(x-5)(x+5)$ has three distinct linear factors	**2. c** $x\left(x^2+25\right)$ has one linear factor and one irreducible quadratic factor.
3. a $x^2\left(x^2+25\right)$ has one repeated linear factor and one irreducible quadratic factor.	**4. f** $x^2(x-5)(x+5)$ has one repeated linear factor, and two other distinct linear factors.

5. b $x\left(x^2+25\right)^2$ has one linear factor and one repeated irreducible quadratic factor.	**6. e** $x^2\left(x^2+25\right)^2$ has one repeated linear factor and one repeated irreducible quadratic factor.
7. $\dfrac{9}{x^2-x-20}=\dfrac{9}{(x-5)(x+4)}=\dfrac{A}{x-5}+\dfrac{B}{x+4}$	**8.** $\dfrac{8}{x^2-3x-10}=\dfrac{8}{(x-5)(x+2)}=\dfrac{A}{x-5}+\dfrac{B}{x+2}$
9. $\dfrac{2x+5}{x^3-4x^2}=\dfrac{2x+5}{x^2(x-4)}=\dfrac{A}{x-4}+\dfrac{B}{x}+\dfrac{C}{x^2}$	**10.** $\dfrac{x^2+2x-1}{x^4-9x^2}=\dfrac{x^2+2x-1}{x^2(x-3)(x+3)}=\dfrac{A}{x}+\dfrac{B}{x^2}+\dfrac{C}{x-3}+\dfrac{D}{x+3}$
11. Must long divide in this case. $x^2+x+5\,\overline{)\,2x^3-4x^2+7x+3}$ quotient $2x-6$; $-(2x^3+2x^2+10x)$; $-6x^2-3x+3$; $-(-6x^2-6x-30)$; $3x+33$. So, $\dfrac{2x^3-4x^2+7x+3}{x^2+x+5}=2x-6+\dfrac{3x+33}{x^2+x+5}$	**12.** Must long divide in this case. $x^2-3x+7\,\overline{)\,2x^3+5x^2+0x+6}$ quotient $2x+11$; $-(2x^3-6x^2+14x)$; $11x^2-14x+6$; $-(11x^2-33x+77)$; $19x-71$. So, $\dfrac{2x^3+5x^2+6}{x^2-3x+7}=2x+11+\dfrac{19x-71}{x^2-3x+7}$
13. $\dfrac{3x^3-x+9}{\left(x^2+10\right)^2}=\dfrac{Ax+B}{x^2+10}+\dfrac{Cx+D}{\left(x^2+10\right)^2}$	**14.** $\dfrac{5x^3+2x^2+4}{\left(x^2+13\right)^2}=\dfrac{Ax+B}{x^2+13}+\dfrac{Cx+D}{\left(x^2+13\right)^2}$

15. The partial fraction decomposition has the form:

$$\frac{1}{x(x+1)} = \frac{A}{x} + \frac{B}{x+1} \quad \textbf{(1)}$$

To find the coefficients, multiply both sides of **(1)** by $x(x+1)$, and gather like terms:

$$1 = A(x+1) + Bx$$

$$1 = (A+B)x + A \quad \textbf{(2)}$$

Equate corresponding coefficients in **(2)** to obtain the following system:

$$(*) \begin{cases} A+B=0 & \textbf{(3)} \\ A=1 & \textbf{(4)} \end{cases}$$

Now, solve system $(*)$:

Substitute **(4)** into **(3)** to see that $B=-1$.

Thus, the partial fraction decomposition **(1)** becomes:

$$\boxed{\frac{1}{x(x+1)} = \frac{1}{x} - \frac{1}{x+1}}$$

16. The partial fraction decomposition has the form:

$$\frac{1}{x(x-1)} = \frac{A}{x} + \frac{B}{x-1} \quad \textbf{(1)}$$

To find the coefficients, multiply both sides of **(1)** by $x(x-1)$, and gather like terms:

$$1 = A(x-1) + Bx$$

$$1 = (A+B)x - A \quad \textbf{(2)}$$

Equate corresponding coefficients in **(2)** to obtain the following system:

$$(*) \begin{cases} A+B=0 & \textbf{(3)} \\ -A=1 & \textbf{(4)} \end{cases}$$

Now, solve system $(*)$:

From **(4)**, we see that $A=-1$.

Substitute this value into **(3)** to see that $B=1$.

Thus, the partial fraction decomposition **(1)** becomes:

$$\boxed{\frac{1}{x(x-1)} = \frac{1}{x-1} - \frac{1}{x}}$$

17. Observe that simplifying the expression first yields

$$\frac{\cancel{x}}{\cancel{x}(x-1)} = \frac{1}{x-1}.$$

This IS the partial fraction decomposition.

18. Observe that simplifying the expression first yields

$$\frac{\cancel{x}}{\cancel{x}(x+1)} = \frac{1}{x+1}.$$

This IS the partial fraction decomposition.

19. The partial fraction decomposition has the form:

$$\frac{9x-11}{(x-3)(x+5)}=\frac{A}{x-3}+\frac{B}{x+5} \quad \textbf{(1)}$$

To find the coefficients, multiply both sides of **(1)** by $(x-3)(x+5)$, and gather like terms:

$$9x-11=A(x+5)+B(x-3)$$

$$9x-11=(A+B)x+(5A-3B) \quad \textbf{(2)}$$

Equate corresponding coefficients in **(2)** to obtain the following system:

$$(*)\begin{cases} A+B=9 & \textbf{(3)} \\ 5A-3B=-11 & \textbf{(4)} \end{cases}$$

Now, solve system **(*)**:

Multiply **(3)** by -5:

$$-5A-5B=-45 \quad \textbf{(5)}$$

Add **(5)** and **(3)** to solve for B:

$$-8B=-56$$

$$B=7 \quad \textbf{(6)}$$

Substitute **(6)** into **(3)** to find A:

$$A=2$$

Thus, the partial fraction decomposition **(1)** becomes:

$$\boxed{\frac{9x-11}{(x-3)(x+5)}=\frac{2}{x-3}+\frac{7}{x+5}}$$

20. The partial fraction decomposition has the form:

$$\frac{8x-13}{(x-2)(x+1)}=\frac{A}{x-2}+\frac{B}{x+1} \quad \textbf{(1)}$$

To find the coefficients, multiply both sides of **(1)** by $(x-2)(x+1)$, and gather like terms:

$$8x-13=A(x+1)+B(x-2)$$

$$8x-13=(A+B)x+(A-2B) \quad \textbf{(2)}$$

Equate corresponding coefficients in **(2)** to obtain the following system:

$$(*)\begin{cases} A+B=8 & \textbf{(3)} \\ A-2B=-13 & \textbf{(4)} \end{cases}$$

Now, solve system **(*)**:

Multiply **(3)** by 2:

$$2A+2B=16 \quad \textbf{(5)}$$

Add **(5)** and **(4)** to solve for A:

$$3A=3$$

$$A=1 \quad \textbf{(6)}$$

Substitute **(6)** into **(3)** to find B:

$$1+B=8$$

$$B=7$$

Thus, the partial fraction decomposition **(1)** becomes:

$$\boxed{\frac{8x-13}{(x-2)(x+1)}=\frac{1}{x-2}+\frac{7}{x+1}}$$

21. The partial fraction decomposition has the form:

$$\frac{3x+1}{(x-1)^2}=\frac{A}{x-1}+\frac{B}{(x-1)^2} \quad \textbf{(1)}$$

To find the coefficients, multiply both sides of **(1)** by $(x-1)^2$, and gather like terms:

$$3x+1=A(x-1)+B$$
$$3x+1=Ax+(B-A) \quad \textbf{(2)}$$

Equate corresponding coefficients in **(2)** to obtain the following system:

$$(*)\begin{cases} A=3 & \textbf{(3)} \\ B-A=1 & \textbf{(4)} \end{cases}$$

Substitute **(3)** into **(4)** to find B:

$$B-3=1 \Rightarrow B=4$$

Thus, the partial fraction decomposition **(1)** becomes:

$$\boxed{\frac{3x+1}{(x-1)^2}=\frac{3}{x-1}+\frac{4}{(x-1)^2}}$$

22. The partial fraction decomposition has the form:

$$\frac{9y-2}{(y-1)^2}=\frac{A}{y-1}+\frac{B}{(y-1)^2} \quad \textbf{(1)}$$

To find the coefficients, multiply both sides of **(1)** by $(y-1)^2$, and gather like terms:

$$9y-2=A(y-1)+B$$
$$9y-2=Ay+(B-A) \quad \textbf{(2)}$$

Equate corresponding coefficients in **(2)** to obtain the following system:

$$(*)\begin{cases} A=9 & \textbf{(3)} \\ B-A=-2 & \textbf{(4)} \end{cases}$$

Substitute **(3)** into **(4)** to find B:

$$B-9=-2 \Rightarrow B=7$$

Thus, the partial fraction decomposition **(1)** becomes:

$$\boxed{\frac{9y-2}{(y-1)^2}=\frac{9}{y-1}+\frac{7}{(y-1)^2}}$$

23. The partial fraction decomposition has the form: $\frac{4x-3}{(x+3)^2}=\frac{A}{x+3}+\frac{B}{(x+3)^2}$ **(1)**

To find the coefficients, multiply both sides of **(1)** by $(x+3)^2$, gather like terms:

$$4x-3=A(x+3)+B$$
$$4x-3=Ax+(3A+B) \quad \textbf{(2)}$$

Equate corresponding coefficients in **(2)** to obtain the following system:

$$(*)\begin{cases} A=4 & \textbf{(3)} \\ 3A+B=-3 & \textbf{(4)} \end{cases}$$

Substitute **(3)** into **(4)** to find B:

$$3(4)+B=-3 \Rightarrow B=-15$$

Thus, the partial fraction decomposition **(1)** becomes:

$$\boxed{\frac{4x-3}{(x+3)^2}=\frac{4}{x+3}+\frac{-15}{(x+3)^2}}$$

24. The partial fraction decomposition has the form: $\frac{3x+1}{(x+2)^2}=\frac{A}{x+2}+\frac{B}{(x+2)^2}$ **(1)**

To find the coefficients, multiply both sides of **(1)** by $(x+2)^2$, and gather like terms:

$$3x+1=A(x+2)+B$$
$$3x+1=Ax+(2A+B) \quad \textbf{(2)}$$

Equate corresponding coefficients in **(2)** to obtain the following system:

$$(*)\begin{cases} A=3 & \textbf{(3)} \\ 2A+B=1 & \textbf{(4)} \end{cases}$$

Substitute **(3)** into **(4)** to find B:

$$2(3)+B=1 \Rightarrow B=-5$$

Thus, the partial fraction decomposition **(1)** becomes: $\boxed{\frac{3x+1}{(x+2)^2}=\frac{3}{x+2}+\frac{-5}{(x+2)^2}}$

25. The partial fraction decomposition has the form:

$$\frac{4x^2-32x+72}{(x+1)(x-5)^2}=\frac{A}{x+1}+\frac{B}{x-5}+\frac{C}{(x-5)^2} \quad \textbf{(1)}$$

To find the coefficients, multiply both sides of **(1)** by $(x+1)(x-5)^2$, and gather like terms:

$$\begin{aligned} 4x^2-32x+72 &= A(x-5)^2+B(x-5)(x+1)+C(x+1) \\ &= A(x^2-10x+25)+B(x^2-4x-5)+C(x+1) \\ &= (A+B)x^2+(-10A-4B+C)x+(25A-5B+C) \quad \textbf{(2)} \end{aligned}$$

Equate corresponding coefficients in **(2)** to obtain the following system:

$$(*)\begin{cases} A+B=4 & \textbf{(3)} \\ -10A-4B+C=-32 & \textbf{(4)} \\ 25A-5B+C=72 & \textbf{(5)} \end{cases}$$

Solve **(3)** for *B*: $B=4-A$ **(6)**

Substitute **(6)** into **(4)**:

$-10A-4(4-A)+C=-32$ which is equivalent to $-6A+C=-16$ **(7)**

Substitute **(6)** into **(5)**:

$25A-5(4-A)+C=72$ which is equivalent to $30A+C=92$ **(8)**

Now, solve the 2×2 system:

$$\begin{cases} -6A+C=-16 & \textbf{(7)} \\ 30A+C=92 & \textbf{(8)} \end{cases}$$

Multiply **(7)** by -1, and add to **(8)** to find *A*: $36A=108$ so that $A=3$.
Substitute this value for *A* into **(7)** to find *C*: $-6(3)+C=-16$ so that $C=2$.
Finally, substitute the value of *A* into **(3)** to find *B*: $3+B=4$ so that $B=1$.

Thus, the partial fraction decomposition **(1)** becomes:

$$\boxed{\frac{4x^2-32x+72}{(x+1)(x-5)^2}=\frac{3}{x+1}+\frac{1}{x-5}+\frac{2}{(x-5)^2}}$$

26. The partial fraction decomposition has the form:

$$\frac{4x^2-7x-3}{(x+2)(x-1)^2}=\frac{A}{x+2}+\frac{B}{x-1}+\frac{C}{(x-1)^2} \quad \textbf{(1)}$$

To find the coefficients, multiply both sides of **(1)** by $(x+2)(x-1)^2$, and gather like terms:

$$\begin{aligned} 4x^2-7x-3 &= A(x-1)^2+B(x-1)(x+2)+C(x+2) \\ &= A(x^2-2x+1)+B(x^2+x-2)+C(x+2) \\ &= (A+B)x^2+(-2A+B+C)x+(A-2B+2C) \quad \textbf{(2)} \end{aligned}$$

Equate corresponding coefficients in **(2)** to obtain the following system:

$$(*)\begin{cases} A+B=4 & \textbf{(3)} \\ -2A+B+C=-7 & \textbf{(4)} \\ A-2B+2C=-3 & \textbf{(5)} \end{cases}$$

Solve **(3)** for B: $B=4-A$ **(6)**

Substitute **(6)** into **(4)**:

$-2A+(4-A)+C=-7$ which is equivalent to $-3A+C=-11$ **(7)**

Substitute **(6)** into **(5)**:

$A-2(4-A)+2C=-3$ which is equivalent to $3A+2C=5$ **(8)**

Now, solve the 2×2 system:

$$\begin{cases} -3A+C=-11 & \textbf{(7)} \\ 3A+2C=5 & \textbf{(8)} \end{cases}$$

Add **(7)** and **(8)** to find C: $3C=-6$ so that $C=-2$.
Substitute this value for C into **(7)** to find A: $-3A-2=-11$ so that $A=3$.
Finally, substitute the value of A into **(3)** to find B: $3+B=4$ so that $B=1$.

Thus, the partial fraction decomposition **(1)** becomes:

$$\boxed{\frac{4x^2-7x-3}{(x+2)(x-1)^2}=\frac{3}{x+2}+\frac{1}{x-1}-\frac{2}{(x-1)^2}}$$

27. The partial fraction decomposition has the form:

$$\frac{5x^2+28x-6}{(x+4)(x^2+3)}=\frac{A}{x+4}+\frac{Bx+C}{x^2+3} \quad \textbf{(1)}$$

To find the coefficients, multiply both sides of **(1)** by $(x+4)(x^2+3)$, and gather like terms:

$$\begin{aligned}5x^2+28x-6 &= A(x^2+3)+(Bx+C)(x+4)\\ &= Ax^2+3A+Bx^2+Cx+4Bx+4C\\ &=(A+B)x^2+(4B+C)x+(3A+4C) \quad \textbf{(2)}\end{aligned}$$

Equate corresponding coefficients in **(2)** to obtain the following system:

$$(*)\begin{cases} A+B=5 & \textbf{(3)}\\ 4B+C=28 & \textbf{(4)}\\ 3A+4C=-6 & \textbf{(5)}\end{cases}$$

Solve **(3)** for B: $B=5-A$ **(6)**
Substitute **(6)** into **(4)**:

$4(5-A)+C=28$ which is equivalent to $-4A+C=8$ **(7)**

Now, solve the 2×2 system:

$$\begin{cases} 3A+4C=-6 & \textbf{(5)}\\ -4A+C=8 & \textbf{(7)}\end{cases}$$

Multiply **(7)** by –4 and then add to **(5)** to find A: $19A=-38$ so that $A=-2$.
Substitute this value for A into **(7)** to find C: $-4(-2)+C=8$ so that $C=0$.
Finally, substitute the value of A into **(3)** to find B: $-2+B=5$ so that $B=7$.

Thus, the partial fraction decomposition **(1)** becomes:

$$\boxed{\frac{5x^2+28x-6}{(x+4)(x^2+3)}=\frac{-2}{x+4}+\frac{7x}{x^2+3}}$$

28. The partial fraction decomposition has the form:

$$\frac{x^2+5x+4}{(x-2)(x^2+2)}=\frac{A}{x-2}+\frac{Bx+C}{x^2+2} \quad \textbf{(1)}$$

To find the coefficients, multiply both sides of **(1)** by $(x-2)(x^2+2)$, and gather like terms:

$$\begin{aligned} x^2+5x+4 &= A(x^2+2)+(Bx+C)(x-2) \\ &= Ax^2+2A+Bx^2+Cx-2Bx-2C \\ &= (A+B)x^2+(C-2B)x+(2A-2C) \quad \textbf{(2)} \end{aligned}$$

Equate corresponding coefficients in **(2)** to obtain the following system:

$$(*) \begin{cases} A+B=1 & \textbf{(3)} \\ -2B+C=5 & \textbf{(4)} \\ 2A-2C=4 & \textbf{(5)} \end{cases}$$

Solve **(3)** for *B*: $B=1-A$ **(6)**
Substitute **(6)** into **(4)**:

$$-2(1-A)+C=5 \text{ which is equivalent to } 2A+C=7 \quad \textbf{(7)}$$

Now, solve the 2×2 system:

$$\begin{cases} 2A-2C=4 & \textbf{(5)} \\ 2A+C=7 & \textbf{(7)} \end{cases}$$

Multiply **(7)** by -1 and then add to **(5)** to find *C*: $-3C=-3$ so that $C=1$.
Substitute this value for *C* into **(7)** to find *A*: $2A+1=7$ so that $A=3$.
Finally, substitute the value of *A* into **(3)** to find *B*: $3+B=1$ so that $B=-2$.

Thus, the partial fraction decomposition **(1)** becomes:

$$\boxed{\frac{x^2+5x+4}{(x-2)(x^2+2)}=\frac{3}{x-2}+\frac{-2x+1}{x^2+2}}$$

29. The partial fraction decomposition has the form:

$$\frac{-2x^2-17x+11}{(x-7)\underbrace{\left(3x^2-7x+5\right)}_{\text{Irreducible Quadratic Term}}}=\frac{A}{x-7}+\frac{Bx+C}{3x^2-7x+5} \quad \textbf{(1)}$$

To find the coefficients, multiply both sides of **(1)** by $(x-7)\left(3x^2-7x+5\right)$, and gather like terms:

$$\begin{aligned}-2x^2-17x+11&=A\left(3x^2-7x+5\right)+(Bx+C)(x-7)\\&=3Ax^2-7Ax+5A+Bx^2+Cx-7Bx-7C\\&=(3A+B)x^2+(-7A-7B+C)x+(5A-7C) \quad \textbf{(2)}\end{aligned}$$

Equate corresponding coefficients in **(2)** to obtain the following system:

$$(*)\begin{cases}3A+B=-2 & \textbf{(3)}\\-7A-7B+C=-17 & \textbf{(4)}\\5A-7C=11 & \textbf{(5)}\end{cases}$$

Solve **(3)** for B: $B=-2-3A$ **(6)**
Substitute **(6)** into **(4)**:

$-7A-7(-2-3A)+C=-17$ which is equivalent to $14A+C=-31$ **(7)**

Now, solve the 2×2 system:

$$\begin{cases}5A-7C=11 & \textbf{(5)}\\14A+C=-31 & \textbf{(7)}\end{cases}$$

Multiply **(7)** by 7 and then add to **(5)** to find A: $103A=-206$ so that $A=-2$.
Substitute this value for A into **(5)** to find C: $5(-2)-7C=11$ so that $C=-3$.
Finally, substitute the value of A into **(3)** to find B: $3(-2)+B=-2$ so that $B=4$.

Thus, the partial fraction decomposition **(1)** becomes:

$$\boxed{\frac{-2x^2-17x+11}{(x-7)\left(3x^2-7x+5\right)}=\frac{-2}{x-7}+\frac{4x-3}{3x^2-7x+5}}$$

30. The partial fraction decomposition has the form:

$$\frac{14x^2+8x+40}{(x+5)\underbrace{\left(2x^2-3x+5\right)}_{\text{Irreducible Quadratic Term}}}=\frac{A}{x+5}+\frac{Bx+C}{2x^2-3x+5} \quad \textbf{(1)}$$

To find the coefficients, multiply both sides of **(1)** by $(x+5)\left(2x^2-3x+5\right)$, and gather like terms:

$$\begin{aligned}14x^2+8x+40&=A\left(2x^2-3x+5\right)+(Bx+C)(x+5)\\&=2Ax^2-3Ax+5A+Bx^2+Cx+5Bx+5C\\&=(2A+B)x^2+(-3A+5B+C)x+(5A+5C) \quad \textbf{(2)}\end{aligned}$$

Equate corresponding coefficients in **(2)** to obtain the following system:

$$(*)\begin{cases}2A+B=14 & \textbf{(3)}\\-3A+5B+C=8 & \textbf{(4)}\\5A+5C=40 & \textbf{(5)}\end{cases}$$

Solve **(3)** for B: $B=14-2A$ **(6)**
Substitute **(6)** into **(4)**:

$-3A+5(14-2A)+C=8$ which is equivalent to $-13A+C=-62$ **(7)**

Now, solve the 2×2 system:

$$\begin{cases}5A+5C=40 & \textbf{(5)}\\-13A+C=-62 & \textbf{(7)}\end{cases}$$

Multiply **(7)** by -5 and then add to **(5)** to find A: $70A=350$ so that $A=5$.
Substitute this value for A into **(5)** to find C: $5(5)+5C=40$ so that $C=3$.
Finally, substitute the value of A into **(3)** to find B: $2(5)+B=14$ so that $B=4$.

Thus, the partial fraction decomposition **(1)** becomes:

$$\boxed{\frac{14x^2+8x+40}{(x+5)\left(2x^2-3x+5\right)}=\frac{5}{x+5}+\frac{4x+3}{2x^2-3x+5}}$$

31. The partial fraction decomposition has the form:

$$\frac{x^3}{\left(x^2+9\right)^2}=\frac{Ax+B}{x^2+9}+\frac{Cx+D}{\left(x^2+9\right)^2} \quad \textbf{(1)}$$

To find the coefficients, multiply both sides of **(1)** by $\left(x^2+9\right)^2$, and gather like terms:

$$\begin{aligned} x^3 &= (Ax+B)\left(x^2+9\right)+(Cx+D) \\ &= Ax^3+Bx^2+9Ax+9B+Cx+D \\ &= Ax^3+Bx^2+(9A+C)x+(9B+D) \quad \textbf{(2)} \end{aligned}$$

Equate corresponding coefficients in **(2)** to obtain the following system:

$$(*)\begin{cases} A=1 & \textbf{(3)} \\ B=0 & \textbf{(4)} \\ 9A+C=0 & \textbf{(5)} \\ 9B+D=0 & \textbf{(6)} \end{cases}$$

Substitute **(3)** into **(5)** to find C: $9(1)+C=0$ so that $C=-9$
Substitute **(4)** into **(6)** to find D: $9(0)+D=0$ so that $D=0$

Thus, the partial fraction decomposition **(1)** becomes:

$$\boxed{\frac{x^3}{\left(x^2+9\right)^2}=\frac{x}{x^2+9}-\frac{9x}{\left(x^2+9\right)^2}}$$

32. The partial fraction decomposition has the form:

$$\frac{x^2}{\left(x^2+9\right)^2}=\frac{Ax+B}{x^2+9}+\frac{Cx+D}{\left(x^2+9\right)^2} \quad \textbf{(1)}$$

To find the coefficients, multiply both sides of **(1)** by $\left(x^2+9\right)^2$, and gather like terms:

$$\begin{aligned} x^2 &= (Ax+B)\left(x^2+9\right)+(Cx+D) \\ &= Ax^3+Bx^2+9Ax+9B+Cx+D \\ &= Ax^3+Bx^2+(9A+C)x+(9B+D) \quad \textbf{(2)} \end{aligned}$$

Equate corresponding coefficients in **(2)** to obtain the following system:

$$(*)\begin{cases} A=0 & \textbf{(3)} \\ B=1 & \textbf{(4)} \\ 9A+C=0 & \textbf{(5)} \\ 9B+D=0 & \textbf{(6)} \end{cases}$$

Substitute **(3)** into **(5)** to find *C*: $9(0)+C=0$ so that $C=0$
Substitute **(4)** into **(6)** to find *D*: $9(1)+D=0$ so that $D=-9$

Thus, the partial fraction decomposition **(1)** becomes:

$$\boxed{\frac{x^3}{\left(x^2+9\right)^2}=\frac{1}{x^2+9}-\frac{9}{\left(x^2+9\right)^2}}$$

33. The partial fraction decomposition has the form:

$$\frac{2x^3-3x^2+7x-2}{\left(x^2+1\right)^2}=\frac{Ax+B}{x^2+1}+\frac{Cx+D}{\left(x^2+1\right)^2} \quad \textbf{(1)}$$

To find the coefficients, multiply both sides of **(1)** by $\left(x^2+1\right)^2$, and gather like terms:

$$\begin{aligned} 2x^3-3x^2+7x-2 &= (Ax+B)\left(x^2+1\right)+(Cx+D) \\ &= Ax^3+Bx^2+Ax+B+Cx+D \\ &= Ax^3+Bx^2+(A+C)x+(B+D) \quad \textbf{(2)} \end{aligned}$$

Equate corresponding coefficients in **(2)** to obtain the following system:

$$\textbf{(*)} \begin{cases} A=2 & \textbf{(3)} \\ B=-3 & \textbf{(4)} \\ A+C=7 & \textbf{(5)} \\ B+D=-2 & \textbf{(6)} \end{cases}$$

Substitute **(3)** into **(5)** to find *C*: $2+C=7$ so that $C=5$
Substitute **(4)** into **(6)** to find *D*: $-3+D=-2$ so that $D=1$

Thus, the partial fraction decomposition **(1)** becomes:

$$\boxed{\frac{2x^3-3x^2+7x-2}{\left(x^2+1\right)^2}=\frac{2x-3}{x^2+1}+\frac{5x+1}{\left(x^2+1\right)^2}}$$

34. The partial fraction decomposition has the form:

$$\frac{-x^3+2x^2-3x+15}{\left(x^2+8\right)^2}=\frac{Ax+B}{x^2+8}+\frac{Cx+D}{\left(x^2+8\right)^2} \quad \textbf{(1)}$$

To find the coefficients, multiply both sides of **(1)** by $\left(x^2+8\right)^2$, and gather like terms:

$$\begin{aligned}-x^3+2x^2-3x+15&=(Ax+B)\left(x^2+8\right)+(Cx+D)\\&=Ax^3+Bx^2+8Ax+8B+Cx+D\\&=Ax^3+Bx^2+(8A+C)x+(8B+D) \quad \textbf{(2)}\end{aligned}$$

Equate corresponding coefficients in **(2)** to obtain the following system:

$$(*)\begin{cases}A=-1 & \textbf{(3)}\\ B=2 & \textbf{(4)}\\ 8A+C=-3 & \textbf{(5)}\\ 8B+D=15 & \textbf{(6)}\end{cases}$$

Substitute **(3)** into **(5)** to find C: $8(-1)+C=-3$ so that $C=5$
Substitute **(4)** into **(6)** to find D: $8(2)+D=15$ so that $D=-1$

Thus, the partial fraction decomposition **(1)** becomes:

$$\boxed{\frac{-x^3+2x^2-3x+15}{\left(x^2+8\right)^2}=\frac{-x+2}{x^2+8}+\frac{5x-1}{\left(x^2+8\right)^2}}$$

35. The partial fraction decomposition has the form:

$$\frac{3x+1}{x^4-1}=\frac{3x+1}{\left(x^2-1\right)\left(x^2+1\right)}=\frac{3x+1}{(x-1)(x+1)\left(x^2+1\right)}=\frac{A}{x-1}+\frac{B}{x+1}+\frac{Cx+D}{x^2+1} \quad \textbf{(1)}$$

To find the coefficients, multiply both sides of **(1)** by $(x-1)(x+1)\left(x^2+1\right)$, and gather like terms:

$$\begin{aligned}3x+1&=A(x+1)\left(x^2+1\right)+B(x-1)\left(x^2+1\right)+\left(Cx+D\right)(x-1)(x+1)\\&=Ax^3+Ax^2+Ax+A+Bx^3-Bx^2+Bx-B+Cx^3+Dx^2-Cx-D\\&=(A+B+C)x^3+(A-B+D)x^2+(A+B-C)x+(A-B-D) \quad \textbf{(2)}\end{aligned}$$

Equate corresponding coefficients in **(2)** to obtain the following system:

$$\textbf{(*)}\begin{cases}A+B+C=0 & \textbf{(3)}\\ A-B+D=0 & \textbf{(4)}\\ A+B-C=3 & \textbf{(5)}\\ A-B-D=1 & \textbf{(6)}\end{cases}$$

To solve this system, first obtain a 3×3 system:

Add **(4)** and **(6)**: $2A-2B=1$ **(7)**

This enables us to consider the following 3×3 system:

$$\begin{cases}A+B+C=0 & \textbf{(3)}\\ A+B-C=3 & \textbf{(5)}\\ 2A-2B=1 & \textbf{(7)}\end{cases}$$

Now, to solve this system, obtain a 2×2 system:

Add **(3)** and **(5)**: $2A+2B=3$ **(8)**

This enables us to consider the following 2×2 system:

$$\begin{cases}2A-2B=1 & \textbf{(7)}\\ 2A+2B=3 & \textbf{(8)}\end{cases}$$

Add **(7)** and **(8)** to find A: $4A=4$ so that $A=1$

Substitute this value of A into **(7)** to find B**:** $2(1)-2B=1$ so that $B=\frac{1}{2}$**.**

Now, substitute these values of A and B into **(6)** to find D:

$$1-\tfrac{1}{2}-D=1 \text{ so that } D=-\tfrac{1}{2}$$

Finally, substitute these values of A and B into **(3)** to find C:

$$1+\tfrac{1}{2}+C=0 \text{ so that } C=-\tfrac{3}{2}.$$

Thus, the partial fraction decomposition **(1)** becomes:

$$\boxed{\frac{3x+1}{x^4-1}=\frac{1}{x-1}+\frac{1}{2(x+1)}+\frac{-3x-1}{2\left(x^2+1\right)}}$$

36. The partial fraction decomposition has the form:

$$\frac{2-x}{x^4-81}=\frac{2-x}{\left(x^2-9\right)\left(x^2+9\right)}=\frac{2-x}{(x-3)(x+3)\left(x^2+9\right)}=\frac{A}{x-3}+\frac{B}{x+3}+\frac{Cx+D}{x^2+9} \quad \textbf{(1)}$$

To find the coefficients, multiply both sides of **(1)** by $(x-3)(x+3)\left(x^2+9\right)$, and gather like terms:

$$\begin{aligned} 2-x &= A(x+3)\left(x^2+9\right)+B(x-3)\left(x^2+9\right)+\left(Cx+D\right)(x-3)(x+3) \\ &= Ax^3+3Ax^2+9Ax+27A+Bx^3-3Bx^2+9Bx-27B+Cx^3+Dx^2-9Cx-9D \\ &= (A+B+C)x^3+(3A-3B+D)x^2+(9A+9B-9C)x+(27A-27B-9D) \quad \textbf{(2)} \end{aligned}$$

Equate corresponding coefficients in **(2)** to obtain the following system:

$$\textbf{(*)} \begin{cases} A+B+C=0 & \textbf{(3)} \\ 3A-3B+D=0 & \textbf{(4)} \\ 9A+9B-9C=-1 & \textbf{(5)} \\ 27A-27B-9D=2 & \textbf{(6)} \end{cases}$$

To solve this system, first obtain a 3×3 system:

Multiply **(4)** by 9, and then add to **(6)**: $54A-54B=2$ **(7)**

This enables us to consider the following 3×3 system:

$$\begin{cases} A+B+C=0 & \textbf{(3)} \\ 9A+9B-9C=-1 & \textbf{(5)} \\ 54A-54B=2 & \textbf{(7)} \end{cases}$$

Now, to solve this system, obtain a 2×2 system:

Multiply **(3)** by 9, and then add to **(5)**: $18A+18B=-1$ **(8)**

This enables us to consider the following 2×2 system:

$$\begin{cases} 54A-54B=2 & \textbf{(7)} \\ 18A+18B=-1 & \textbf{(8)} \end{cases}$$

Multiply **(8)** by 3, and then add to **(7)** to find A: $108A=-1$ so that $A=-\frac{1}{108}$

Substitute this value of A into **(8)** to find B**:** $18(-\frac{1}{108})+18B=-1$ so that $B=-\frac{5}{108}$.

Now, substitute these values of A and B into **(3)** to find C:

$$-\tfrac{1}{108}-\tfrac{5}{108}+C=0 \text{ so that } C=\tfrac{1}{18}$$

Finally, substitute these values of A and B into **(4)** to find D:

$$3(-\tfrac{1}{108})-3(-\tfrac{5}{108})+D=0 \text{ so that } D=-\tfrac{1}{9}.$$

Thus, the partial fraction decomposition **(1)** becomes:

$$\boxed{\frac{2-x}{x^4-81}=\frac{-\frac{1}{108}}{x-3}+\frac{-\frac{5}{108}}{x+3}+\frac{\frac{1}{18}x-\frac{1}{9}}{x^2+9}}$$

37. The partial fraction decomposition has the form:

$$\frac{5x^2+9x-8}{(x-1)\underbrace{\left(x^2+2x-1\right)}_{\text{Irreducible Quadratic Term}}}=\frac{A}{x-1}+\frac{Bx+C}{x^2+2x-1}\quad \textbf{(1)}$$

To find the coefficients, multiply both sides of **(1)** by $(x+5)\left(2x^2-3x+5\right)$, and gather like terms:

$$\begin{aligned}5x^2+9x-8&=A\left(x^2+2x-1\right)+(Bx+C)(x-1)\\&=Ax^2+2Ax-A+Bx^2+Cx-Bx-C\\&=(A+B)x^2+(2A-B+C)x+(-A-C)\quad \textbf{(2)}\end{aligned}$$

Equate corresponding coefficients in **(2)** to obtain the following system:

$$\textbf{(*)}\ \begin{cases}A+B=5 & \textbf{(3)}\\ 2A-B+C=9 & \textbf{(4)}\\ -A-C=-8 & \textbf{(5)}\end{cases}$$

Solve **(3)** for *B*: $B=5-A$ **(6)**
Substitute **(6)** into **(4)**:

$$3A+C=14\quad \textbf{(7)}$$

Now, solve the 2×2 system:

$$\begin{cases}-A-C=-8 & \textbf{(5)}\\ 3A+C=14 & \textbf{(7)}\end{cases}$$

Add **(5)** and **(7)** to find *A*: $2A=6$ so that $A=3$.
Substitute this value for *A* into **(7)** to find *C*: $3(3)+C=14$ so that $C=5$.
Finally, substitute the value of *A* into **(3)** to find *B*: $B=2$.

Thus, the partial fraction decomposition **(1)** becomes:

$$\boxed{\frac{5x^2+9x-8}{(x-1)\left(x^2+2x-1\right)}=\frac{3}{x-1}+\frac{2x+5}{x^2+2x-1}}$$

38. The partial fraction decomposition has the form:

$$\frac{10x^2-5x+29}{(x-3)\underbrace{\left(x^2+4x+5\right)}_{\text{Irreducible Quadratic Term}}}=\frac{A}{x-3}+\frac{Bx+C}{x^2+4x+5} \quad \textbf{(1)}$$

To find the coefficients, multiply both sides of **(1)** by $(x-3)\left(x^2+4x+5\right)$, and gather like terms:

$$\begin{aligned}10x^2-5x+29 &= A\left(x^2+4x+5\right)+(Bx+C)(x-3)\\ &= Ax^2+4Ax+5A+Bx^2+Cx-3Bx-3C\\ &=(A+B)x^2+(4A-3B+C)x+(5A-3C) \quad \textbf{(2)}\end{aligned}$$

Equate corresponding coefficients in **(2)** to obtain the following system:

$$(*)\begin{cases} A+B=10 & \textbf{(3)}\\ 4A-3B+C=-5 & \textbf{(4)}\\ 5A-3C=29 & \textbf{(5)}\end{cases}$$

Solve **(3)** for *B*: $B=10-A$ **(6)**
Substitute **(6)** into **(4)**:

$$7A+C=25 \quad \textbf{(7)}$$

Now, solve the 2×2 system:

$$\begin{cases} 5A-3C=29 & \textbf{(5)}\\ 7A+C=25 & \textbf{(7)}\end{cases}$$

Multiply **(7)** by 3: $21A+3C=75$ **(8)**
Add **(5)** and **(8)** to find *A*: $26A=104$ so that $A=4$.
Substitute this value for *A* into **(5)** to find *C*: $5(4)-3C=29$ so that $C=-3$.
Finally, substitute the value of *A* into **(6)** to find *B*: $B=6$.

Thus, the partial fraction decomposition **(1)** becomes:

$$\boxed{\frac{10x^2-5x+29}{(x-3)\left(x^2+4x+5\right)}=\frac{4}{x-3}+\frac{6x-3}{x^2+4x+5}}$$

39. The partial fraction decomposition has the form:

$$\frac{3x}{x^3-1}=\frac{3x}{(x-1)\left(x^2+x+1\right)}=\frac{A}{x-1}+\frac{Bx+C}{x^2+x+1} \quad \textbf{(1)}$$

To find the coefficients, multiply both sides of **(1)** by $(x-1)\left(x^2+x+1\right)$, and gather like terms:

$$\begin{aligned} 3x &= A\left(x^2+x+1\right)+\left(Bx+C\right)(x-1) \\ &= Ax^2+Ax+A+Bx^2+Cx-Bx-C \\ &= (A+B)x^2+\left(A-B+C\right)x+\left(A-C\right) \quad \textbf{(2)} \end{aligned}$$

Equate corresponding coefficients in **(2)** to obtain the following system:

$$(*)\begin{cases} A+B=0 & \textbf{(3)} \\ A-B+C=3 & \textbf{(4)} \\ A-C=0 & \textbf{(5)} \end{cases}$$

Solve **(3)** for B: $B=-A$ **(6)**
Solve **(5)** for C: $C=A$ **(7)**
Substitute **(6)** and **(7)** into **(4)**: $A-(-A)+(A)=3$ so that $A=1$.
Substitute this value of A into **(6)** to find B: $B=-1$
Finally, substitute this value of A into **(7)** to find C: $C=1$

Thus, the partial fraction decomposition **(1)** becomes:

$$\boxed{\frac{3x}{x^3-1}=\frac{1}{x-1}+\frac{1-x}{x^2+x+1}}$$

40. The partial fraction decomposition has the form:

$$\frac{5x+2}{x^3-8}=\frac{5x+2}{(x-2)\left(x^2+2x+4\right)}=\frac{A}{x-2}+\frac{Bx+C}{x^2+2x+4}\quad \textbf{(1)}$$

To find the coefficients, multiply both sides of **(1)** by $(x-2)\left(x^2+2x+4\right)$, and gather like terms:

$$\begin{aligned}5x+2&=A\left(x^2+2x+4\right)+\left(Bx+C\right)(x-2)\\&=Ax^2+2Ax+4A+Bx^2+Cx-2Bx-2C\\&=(A+B)x^2+\left(2A-2B+C\right)x+\left(4A-2C\right)\quad \textbf{(2)}\end{aligned}$$

Equate corresponding coefficients in **(2)** to obtain the following system:

$$\textbf{(*)}\begin{cases}A+B=0 & \textbf{(3)}\\2A-2B+C=5 & \textbf{(4)}\\4A-2C=2 & \textbf{(5)}\end{cases}$$

Solve **(3)** for *B*: $B=-A$ **(6)**
Solve **(5)** for *C*: $C=\frac{1}{2}(4A-2)=2A-1$ **(7)**
Substitute **(6)** and **(7)** into **(4)**: $2A-2(-A)+2A-1=5$ so that $A=1$.
Substitute this value of *A* into **(6)** to find *B*: $B=-1$
Finally, substitute this value of *A* into **(7)** to find *C*: $C=2(1)-1=1$

Thus, the partial fraction decomposition **(1)** becomes: $\boxed{\frac{5x+2}{x^3-8}=\frac{1}{x-2}+\frac{-x+1}{x^2+2x+4}}$

41. The partial fraction decomposition has the form:

$$\frac{f\left(d_i+d_0\right)}{d_i d_0}=\frac{A}{d_i}+\frac{B}{d_0} \quad \textbf{(1)}$$

To find the coefficients, multiply both sides of **(1)** by $d_i d_0$, and gather like terms:

$$f\left(d_i+d_0\right)=Ad_0+Bd_i$$
$$fd_i+fd_0=Ad_0+Bd_i$$

Equate corresponding coefficients in **(2)** to obtain the following system:

$$\textbf{(*)}\ \begin{cases} A=f & \textbf{(3)} \\ B=f & \textbf{(4)} \end{cases}$$

Thus, the partial fraction decomposition **(1)** becomes

$$\frac{f\left(d_i+d_0\right)}{d_i d_0}=\frac{f}{d_i}+\frac{f}{d_0} \quad \textbf{(5)}$$

Hence, using **(5)** enables us to write the lens law as $\frac{f}{d_i}+\frac{f}{d_0}=1$, or as $\frac{1}{d_0}+\frac{1}{d_i}=\frac{1}{f}$.

42. The partial fraction decomposition has the form:

$$\frac{1}{n(n+1)}=\frac{A}{n}+\frac{B}{n+1} \quad \textbf{(1)}$$

To find the coefficients, multiply both sides of **(1)** by $n(n+1)$, and gather like terms:

$$1=A(n+1)+Bn$$
$$=(A+B)n+A \quad \textbf{(2)}$$

Equate corresponding coefficients in **(2)** to obtain the following system:

$$\textbf{(*)}\ \begin{cases} A+B=0 & \textbf{(3)} \\ A=1 & \textbf{(4)} \end{cases}$$

Substitute **(4)** into **(3)** to find B: $1+B=0$ so that $B=-1$

Thus, the partial fraction decomposition **(1)** becomes

$$\boxed{\frac{1}{n(n+1)}=\frac{1}{n}-\frac{1}{n+1}}$$

Thus, we can use this to compute the following sum:

$$\frac{1}{1\cdot 2}+\frac{1}{2\cdot 3}+\ldots+\frac{1}{998\cdot 999}+\frac{1}{999\cdot 1000}=$$

$$\left[\frac{1}{1}-\frac{1}{2}\right]+\left[\frac{1}{2}-\frac{1}{3}\right]+\ldots+\left[\frac{1}{998}-\frac{1}{999}\right]+\left[\frac{1}{999}-\frac{1}{1000}\right]=1-\frac{1}{1000}=\boxed{\frac{999}{1000}}$$

43. The partial fraction decomposition has the form:

$$\frac{9+s}{(2-s)(2+s)} = \frac{A}{2-s} + \frac{B}{2+s} \quad \textbf{(1)}$$

To find the coefficients, multiply both sides of **(1)** by $(2-s)(2+s)$, and gather like terms:

$$9+s = A(2+s)+B(2-s)$$
$$9+s = (2A+2B)+(A-B)s \quad \textbf{(2)}$$

Equate corresponding coefficients in **(2)** to obtain the following system:

$$\textbf{(*)} \begin{cases} 2A+2B=9 & \textbf{(3)} \\ A-B=1 & \textbf{(4)} \end{cases}$$

Now, solve system **(*)**:

Multiply **(2)** by 2 and add to **(3)** to solve for A:

$$4A = 11$$
$$A = \tfrac{11}{4} \quad \textbf{(5)}$$

Substitute **(5)** into **(4)** to find B:

$$\tfrac{11}{4} - B = 1$$
$$B = \tfrac{7}{4}$$

Thus, the partial fraction decomposition **(1)** becomes: $\dfrac{9+s}{4-s^2} = \dfrac{-\frac{11}{4}}{s-2} + \dfrac{\frac{7}{4}}{s+2}$

As such, we have

$$\begin{aligned}
\mathfrak{L}^{-1}\left\{\frac{9+s}{4-s^2}\right\} &= \mathfrak{L}^{-1}\left\{\frac{-\frac{11}{4}}{s-2}\right\} + \mathfrak{L}^{-1}\left\{\frac{\frac{7}{4}}{s+2}\right\} \\
&= -\tfrac{11}{4}\mathfrak{L}^{-1}\left\{\frac{1}{s-2}\right\} + \tfrac{7}{4}\mathfrak{L}^{-1}\left\{\frac{1}{s+2}\right\} \\
&= -\tfrac{11}{4}e^{2t} + \tfrac{7}{4}e^{-2t}
\end{aligned}$$

44. The partial fraction decomposition has the form:

$$\frac{2s^2+3s-2}{s(s+1)(s-2)}=\frac{A}{s}+\frac{B}{s+1}+\frac{C}{s-2} \quad \textbf{(1)}$$

To find the coefficients, multiply both sides of **(1)** by $s(s+1)(s-2)$, and gather like terms:

$$\begin{aligned}2s^2+3s-2 &= A(s+1)(s-2)+Bs(s-2)+Cs(s+1)\\ &=(A+B+C)s^2+(-A-2B+C)s+(-2A) \quad \textbf{(2)}\end{aligned}$$

Equate corresponding coefficients in **(2)** to obtain the following system:

$$(*)\begin{cases} A+B+C=2 & \textbf{(3)}\\ -A-2B+C=3 & \textbf{(4)}\\ -2A=-2 & \textbf{(5)}\end{cases}$$

Solve **(3)** for A: $A=1$ **(6)**
Substitute **(6)** into **(3)** and **(4)** and solve the resulting 2×2 system:

$$\begin{cases} B+C=1 & \textbf{(7)}\\ -2B+C=4 & \textbf{(8)}\end{cases}$$

Subtract **(7)** – **(8)** to find B: $3B=-3$ so that $B=-1$.
Substitute this value for B into **(7)** to find C: $C=2$.

Thus, the partial fraction decomposition **(1)** becomes:

$$\frac{2s^2+3s-2}{s(s+1)(s-2)}=\frac{1}{s}+\frac{-1}{s+1}+\frac{2}{s-2}$$

Thus, $\mathcal{L}^{-1}\left\{\frac{2s^2+3s-2}{s(s+1)(s-2)}\right\}=\mathcal{L}^{-1}\left\{\frac{1}{s}\right\}-\mathcal{L}^{-1}\left\{\frac{1}{s+1}\right\}+2\mathcal{L}^{-1}\left\{\frac{1}{s-2}\right\}=1-e^{-t}+2e^{2t}$

45. The form of the decomposition is incorrect. It should be $\frac{A}{x}+\frac{Bx+C}{x^2+1}$. Once this correction is made, the correct decomposition is $\frac{1}{x}+\frac{2x+3}{x^2+1}$.

46. Should long divide first since the degree of the numerator is larger than the degree of the denominator.

47. False. The degree of the numerator must be less than or equal to the degree of the denominator in order to apply the partial fraction decomposition procedure.

48. True.	**49.** True	**50.** True

51. False. $\frac{1}{(x-2)^3}$ is its own partial fraction decomposition, but the coefficients $A_1=A_2=0$.

52. False. Factor the denominator as $(x+1)(x^2-x+1)$. Then, find A and B such that

$$\frac{1}{x^3+1}=\frac{A}{x+1}+\frac{B}{x^2-x+1}$$

53. The first step in forming the partial fraction decomposition of $\dfrac{x^2+4x-8}{x^3-x^2-4x+4}$ is to factor the denominator. To do so, we begin by applying the Rational Root Test:

Factors of 4: $\pm 1, \pm 2, \pm 4$

Factors of 1: ± 1

Possible Rational Zeros: $\pm 1, \pm 2, \pm 4$

Applying synthetic division to the zeros, one can see that 1 is a rational zero:

$$\begin{array}{r|rrrr} 1 & 1 & -1 & -4 & 4 \\ & & 1 & 0 & -4 \\ \hline & 1 & 0 & -4 & 0 \end{array}$$

So, $x^3-x^2-4x+4=(x-1)(x^2-4)=(x-1)(x-2)(x+2)$.

Now, the partial fraction decomposition has the form:

$$\frac{x^2+4x-8}{x^3-x^2-4x+4}=\frac{x^2+4x-8}{(x-1)(x-2)(x+2)}=\frac{A}{x-1}+\frac{B}{x-2}+\frac{C}{x+2} \quad \textbf{(1)}$$

To find the coefficients, multiply both sides of **(1)** by $(x-1)(x-2)(x+2)$, and gather like terms:

$$\begin{aligned} x^2+4x-8 &= A(x-2)(x+2)+B(x-1)(x+2)+C(x-1)(x-2) \\ &= A(x^2-4)+B(x^2+x-2)+C(x^2-3x+2) \\ &= (A+B+C)x^2+(B-3C)x+(-4A-2B+2C) \quad \textbf{(2)} \end{aligned}$$

Equate corresponding coefficients in **(2)** to obtain the following system:

$$(*) \begin{cases} A+B+C=1 & \textbf{(3)} \\ B-3C=4 & \textbf{(4)} \\ -4A-2B+2C=-8 & \textbf{(5)} \end{cases}$$

Now, solve system **(*)**:

Multiply **(3)** by 4 and then, add to **(5)**: $2B+6C=-4$ **(6)**

This leads to the following 2×2 system:

$$\begin{cases} B-3C=4 & \textbf{(4)} \\ 2B+6C=-4 & \textbf{(6)} \end{cases}$$

Multiply **(4)** by -2 and then add to **(6)** to find C: $12C=-12$ so that $C=-1$

Substitute this value of C into **(4)** to find B: $B-3(-1)=4$ so that $B=1$.

Finally, substitute these values of B and C into **(3)** to find A: $A+1-1=1$ so that $A=1$.

Thus, the partial fraction decomposition **(1)** becomes:

$$\boxed{\frac{x^2+4x-8}{x^3-x^2-4x+4}=\frac{1}{x-1}-\frac{1}{x+2}+\frac{1}{x-2}}$$

54. Case 1: Assume $c \neq 0$. Then, the partial fraction decomposition has the form:

$$\frac{ax+b}{x^2-c^2} = \frac{ax+b}{(x-c)(x+c)} = \frac{A}{x-c} + \frac{B}{x+c} \quad \textbf{(1)}$$

To find the coefficients, multiply both sides of **(1)** by $(x-c)(x+c)$, and gather like terms:

$$\begin{aligned} ax+b &= A(x+c)+B(x-c) \\ &= (A+B)x+(Ac-Bc) \quad \textbf{(2)} \end{aligned}$$

Equate corresponding coefficients in **(2)** to obtain the following system:

$$(*)\begin{cases} A+B=a \\ Ac-Bc=b \end{cases} \quad \text{which is equivalent to } (*)\begin{cases} A+B=a & \textbf{(3)} \\ A-B=\frac{b}{c} & \textbf{(4)} \end{cases}$$

Now, solve system $(*)$:

Add **(3)** and **(4)** to find A: $2A = a+\frac{b}{c}$ so that $A = \frac{1}{2}\left(a+\frac{b}{c}\right) = \frac{ac+b}{2c}$.
Substitute this value of A into **(3)** to find B:

$$\begin{aligned} &\tfrac{ac+b}{2c} + B = a \\ &B = a - \tfrac{1}{2}\left(a+\tfrac{b}{c}\right) = \tfrac{a}{2} - \tfrac{b}{2c} = \tfrac{ac-b}{2c} \end{aligned}$$

Thus, the partial fraction decomposition **(1)** in this case becomes:

$$\boxed{\frac{ax+b}{x^2-c^2} = \frac{\frac{ac+b}{2c}}{x-c} + \frac{\frac{ac-b}{2c}}{x+c} = \frac{1}{2c}\left[\frac{ac+b}{x-c} + \frac{ac-b}{x+c}\right]}$$

Case 2: Assume $c = 0$. Then, $\dfrac{ax+b}{x^2-c^2} = \dfrac{ax+b}{x^2} = \dfrac{a}{x} + \dfrac{b}{x^2}$.

55. The partial fraction decomposition has the form:

$$\frac{2x^3+x^2-x-1}{x^4+x^3}=\frac{2x^3+x^2-x-1}{x^3(x+1)}=\frac{A}{x}+\frac{B}{x^2}+\frac{C}{x^3}+\frac{D}{x+1} \quad \textbf{(1)}$$

To find the coefficients, multiply both sides of **(1)** by $x^3(x+1)$, and gather like terms:

$$\begin{aligned} 2x^3+x^2-x-1 &= Ax^2(x+1)+Bx(x+1)+C(x+1)+Dx^3 \\ &= Ax^3+Ax^2+Bx^2+Bx+Cx+C+Dx^3 \\ &= (A+D)x^3+(A+B)x^2+(B+C)x+C \quad \textbf{(2)} \end{aligned}$$

Equate corresponding coefficients in **(2)** to obtain the following system:

$$\textbf{(*)} \begin{cases} A+D=2 & \textbf{(3)} \\ A+B=1 & \textbf{(4)} \\ B+C=-1 & \textbf{(5)} \\ C=-1 & \textbf{(6)} \end{cases}$$

Substitute **(6)** into **(5)** to find *B*: $B-1=-1$ so that $B=0$.
Substitute this value of *B* into **(4)** to find *A*: $A+0=1$ so that $A=1$.
Substitute this value of *A* into **(3)** to find *D*: $1+D=2$ so that $D=1$.

Thus, the partial fraction decomposition **(1)** becomes:

$$\boxed{\frac{2x^3+x^2-x-1}{x^4+x^3}=\frac{1}{x}+\frac{1}{x+1}-\frac{1}{x^3}}$$

56. The partial fraction decomposition has the form:

$$\frac{-x^3+2x-2}{x^5-x^4}=\frac{-x^3+2x-2}{x^4(x-1)}=\frac{A}{x}+\frac{B}{x^2}+\frac{C}{x^3}+\frac{D}{x^4}+\frac{E}{x-1} \quad \textbf{(1)}$$

To find the coefficients, multiply both sides of **(1)** by $x^4(x-1)$, and gather like terms:

$$\begin{aligned}-x^3+2x-2 &= Ax^3(x-1)+Bx^2(x-1)+Cx(x-1)+D(x-1)+Ex^4\\ &= Ax^4-Ax^3+Bx^3-Bx^2+Cx^2-Cx+Dx-D+Ex^4\\ &=(A+E)x^4+(-A+B)x^3+(-B+C)x^2+(-C+D)x-D \quad \textbf{(2)}\end{aligned}$$

Equate corresponding coefficients in **(2)** to obtain the following system:

$$(*)\begin{cases} A+E=0 & \textbf{(3)}\\ -A+B=-1 & \textbf{(4)}\\ -B+C=0 & \textbf{(5)}\\ -C+D=2 & \textbf{(6)}\\ -D=-2 & \textbf{(7)}\end{cases}$$

From **(7)**, we know that $D=2$.
Substitute this value of D into **(6)** to find C: $-C+2=2$ so that $C=0$.
Substitute this value of C into **(5)** to find B: $-B+0=0$ so that $B=0$.
Substitute this value of B into **(4)** to find A: $-A+0=-1$ so that $A=1$.
Substitute this value of A into **(3)** to find E: $1+E=0$ so that $E=-1$.

Thus, the partial fraction decomposition **(1)** becomes:

$$\boxed{\frac{-x^3+2x-2}{x^5-x^4}=\frac{1}{x}+\frac{2}{x^4}-\frac{1}{x-1}}$$

57. The partial fraction decomposition has the form:

$$\frac{x^5+2}{\left(x^2+1\right)^3}=\frac{Ax+B}{x^2+1}+\frac{Cx+D}{\left(x^2+1\right)^2}+\frac{Ex+F}{\left(x^2+1\right)^3} \quad \textbf{(1)}$$

To find the coefficients, multiply both sides of **(1)** by $\left(x^2+1\right)^3$, and gather like terms:

$$\begin{aligned}x^5+2&=(Ax+B)\underbrace{\left(x^2+1\right)^2}_{x^4+2x^2+1}+(Cx+D)\left(x^2+1\right)+(Ex+F)\\&=Ax^5+Bx^4+2Ax^3+2Bx^2+Ax+B+Cx^3+Dx^2+Cx+D+Ex+F\\&=Ax^5+Bx^4+(2A+C)x^3+(2B+D)x^2+(A+C+E)x+(B+D+F) \quad \textbf{(2)}\end{aligned}$$

Equate corresponding coefficients in **(2)** to obtain the following system:

$$\textbf{(*)}\begin{cases}A=1 & \textbf{(3)}\\ B=0 & \textbf{(4)}\\ 2A+C=0 & \textbf{(5)}\\ 2B+D=0 & \textbf{(6)}\\ A+C+E=0 & \textbf{(7)}\\ B+D+F=2 & \textbf{(8)}\end{cases}$$

Substitute **(3)** into **(5)** to find *C*: $2+C=0$ so that $C=-2$.
Substitute **(4)** into **(6)** to find *D*: $0+D=0$ so that $D=0$.
Substitute the values of *A* and *C* into **(7)** to find *E*: $1+(-2)+E=0$ so that $E=1$.
Substitute the values of *B* and *D* into **(8)** to find *F*: $0+0+F=2$ so that $F=2$.

Thus, the partial fraction decomposition **(1)** becomes:

$$\boxed{\frac{x^5+2}{\left(x^2+1\right)^3}=\frac{x}{x^2+1}-\frac{2x}{\left(x^2+1\right)^2}+\frac{x+2}{\left(x^2+1\right)^3}}$$

58. The partial fraction decomposition has the form:

$$\frac{x^2-4}{\left(x^2+1\right)^3}=\frac{Ax+B}{x^2+1}+\frac{Cx+D}{\left(x^2+1\right)^2}+\frac{Ex+F}{\left(x^2+1\right)^3} \quad \textbf{(1)}$$

To find the coefficients, multiply both sides of **(1)** by $\left(x^2+1\right)^3$, and gather like terms:

$$x^2-4=(Ax+B)\underbrace{\left(x^2+1\right)^2}_{x^4+2x^2+1}+(Cx+D)\left(x^2+1\right)+(Ex+F)$$

$$=Ax^5+Bx^4+2Ax^3+2Bx^2+Ax+B+Cx^3+Dx^2+Cx+D+Ex+F$$

$$=Ax^5+Bx^4+(2A+C)x^3+(2B+D)x^2+(A+C+E)x+(B+D+F) \quad \textbf{(2)}$$

Equate corresponding coefficients in **(2)** to obtain the following system:

$$(*)\begin{cases} A=0 & \textbf{(3)} \\ B=0 & \textbf{(4)} \\ 2A+C=0 & \textbf{(5)} \\ 2B+D=1 & \textbf{(6)} \\ A+C+E=0 & \textbf{(7)} \\ B+D+F=-4 & \textbf{(8)} \end{cases}$$

Substitute **(3)** into **(5)** to find C: $0+C=0$ so that $C=0$.
Substitute **(4)** into **(6)** to find D: $2(0)+D=1$ so that $D=1$.
Substitute the values of A and C into **(7)** to find E: $0+0+E=0$ so that $E=0$.
Substitute the values of B and D into **(8)** to find F: $0+1+F=-4$ so that $F=-5$.

Thus, the partial fraction decomposition **(1)** becomes:

$$\boxed{\frac{x^2-4}{\left(x^2+1\right)^3}=\frac{1}{\left(x^2+1\right)^2}-\frac{5}{\left(x^2+1\right)^3}}$$

59.

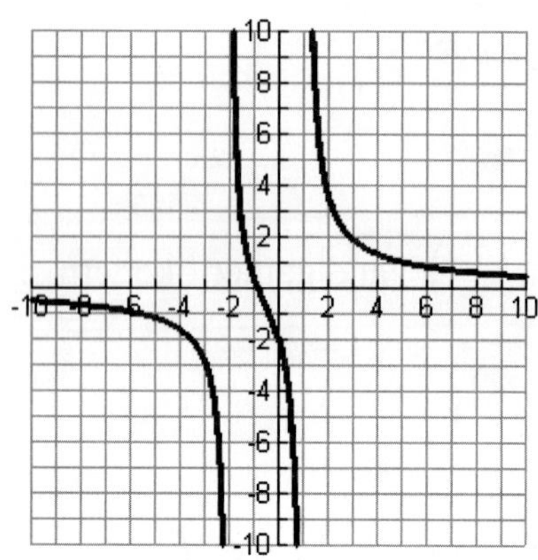

Notes on the graph:
Solid curve: Graph of y_1,
Dashed curve: Graph of y_2
In this case, since the graphs coincide, we know y_2 is, in fact, the partial fraction decomposition of y_1.

60.

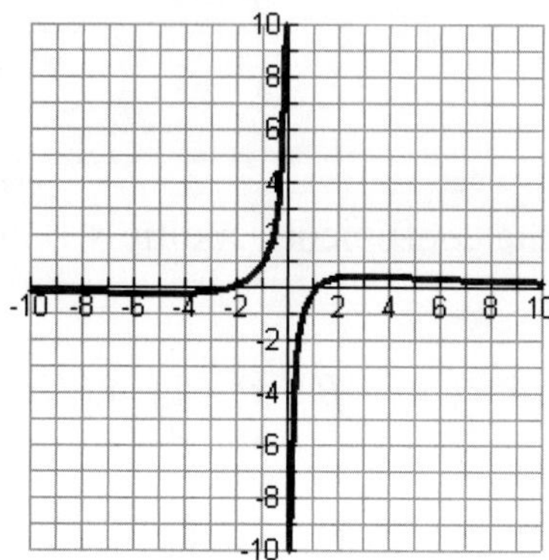

Notes on the graph:
Solid curve: Graph of y_1,
Dashed curve: Graph of y_2
In this case, since the graphs coincide, we know y_2 is, in fact, the partial fraction decomposition of y_1.

61.

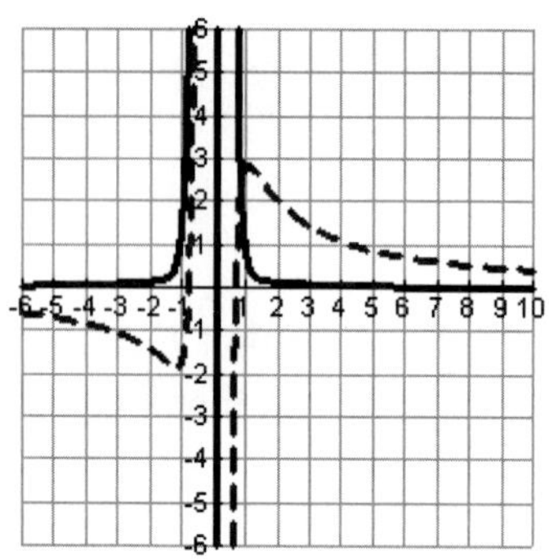

Notes on the graph:
Solid curve: Graph of y_1,
Dashed curve: Graph of y_2
In this case, since the graphs do not coincide, we know y_2 is not the partial fraction decomposition of y_1.

62.

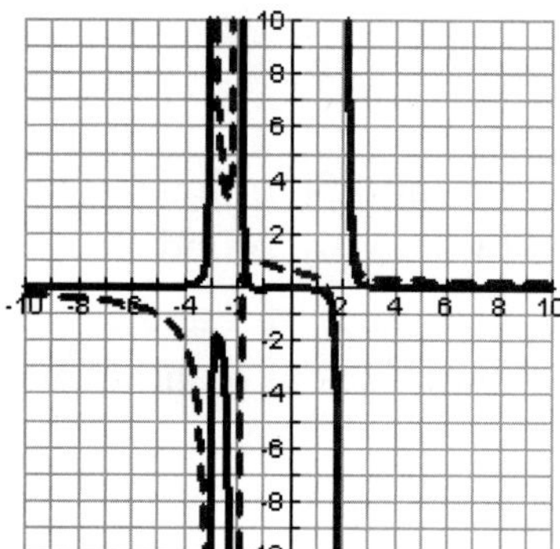

Notes on the graph:
Solid curve: Graph of y_1,
Dashed curve: Graph of y_2
In this case, since the graphs do not coincide, we know y_2 is not the partial fraction decomposition of y_1.

63 Since the graphs coincide, as seen below, y2 is the partial fraction decomposition of y1.

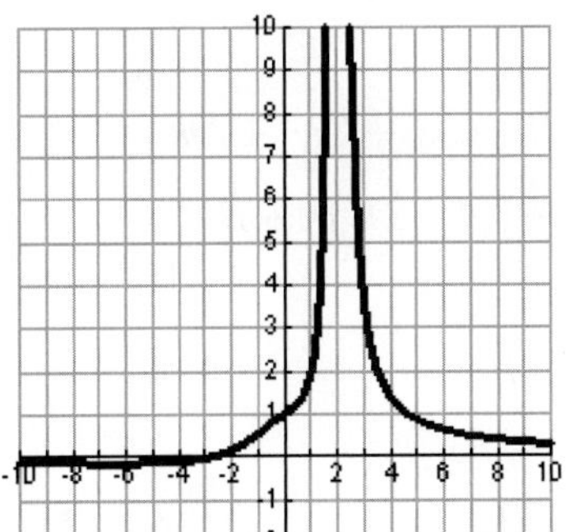

64. Since the graphs do not coincide, as seen below, y2 is not the partial fraction decomposition of y1.

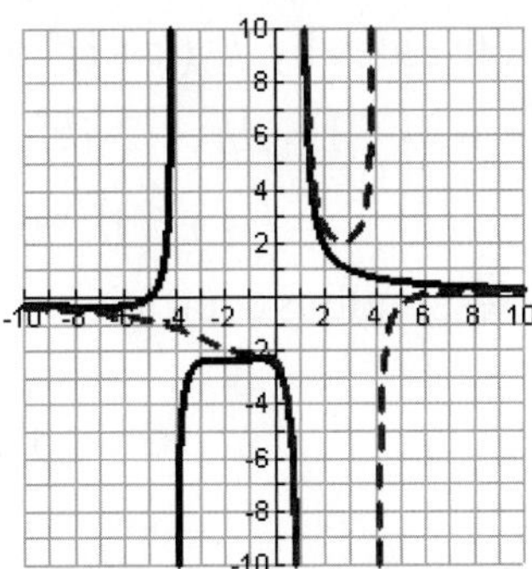

65. First, we find the partial fraction decomposition:

$$\frac{9}{k(k+3)} = \frac{A}{k} + \frac{B}{k+3}.$$

Multiply both sides by $k(k+3)$, simplify, and equate corresponding coefficients to obtain the system:

$$\begin{cases} 3A = 9 & \textbf{(1)} \\ A + B = 0 & \textbf{(2)} \end{cases}$$

Solve **(1)** for A: $A = 3$
Substitute this into **(2)** to find B: $B = -3$

Hence, $\frac{9}{k(k+3)} = \frac{3}{k} - \frac{3}{k+3}$.

Thus, $\sum \frac{9}{k(k+3)} = \sum \left(\frac{3}{k} - \frac{3}{k+3} \right)$.

66. First, we find the partial fraction decomposition:

$$\frac{1}{k(k+1)} = \frac{A}{k} + \frac{B}{k+1}.$$

Multiply both sides by $k(k+1)$, simplify, and equate corresponding coefficients to obtain the system:

$$\begin{cases} A = 1 & \textbf{(1)} \\ A + B = 0 & \textbf{(2)} \end{cases}$$

Solve **(1)** for A: $A = 1$
Substitute this into **(2)** to find B: $B = -1$

Hence, $\frac{1}{k(k+1)} = \frac{1}{k} - \frac{1}{k+1}$.

Thus, $\sum \frac{1}{k(k+1)} = \sum \left(\frac{1}{k} - \frac{1}{k+3} \right)$.

67. First, we find the partial fraction decomposition:

$$\frac{2k+1}{k^2(k+1)^2} = \frac{A}{k} + \frac{B}{k^2} + \frac{C}{k+1} + \frac{D}{(k+1)^2}.$$

Multiply both sides by $k^2(k+1)^2$ and simplify:

$$\begin{aligned} 2k+1 &= Ak^3 + 2Ak^2 + Ak + Bk^2 + 2Bk + B + Ck^3 + Ck^2 + Dk^2 \\ &= (A+C)k^3 + (2A+B+C+D)k^2 + (A+2B)k + B \end{aligned}$$

Equate corresponding coefficients to obtain the system:

$$\begin{cases} A+C=0 & \textbf{(1)} \\ 2A+B+C+D=0 & \textbf{(2)} \\ A+2B=2 & \textbf{(3)} \\ B=1 & \textbf{(4)} \end{cases}$$

Substitute **(4)** into **(3)** to find A: $A=0$
Substitute this into **(1)** to find C: $C=0$
Substitute these into **(2)** to find D: $D=-1$.

Thus, $\dfrac{2k+1}{k^2(k+1)^2} = \dfrac{1}{k^2} - \dfrac{1}{(k+1)^2}$.

68. First, we find the partial fraction decomposition:

$$\frac{4}{k(k+1)(k+2)} = \frac{A}{k} + \frac{B}{k+1} + \frac{C}{k+2}.$$

Multiply both sides by $k(k+1)(k+2)$ and simplify:

$$\begin{aligned} 4 &= Ak^2 + 3Ak + 2A + Bk^2 + 2Bk + Ck^2 + Ck \\ &= (A+B+C)k^2 + (3A+2B+C)k + (2A) \end{aligned}$$

Equate corresponding coefficients to obtain the system:

$$\begin{cases} A+B+C=0 & \textbf{(1)} \\ 3A+2B+C=0 & \textbf{(2)} \\ 2A=4 & \textbf{(3)} \end{cases}$$

Solve **(3)**: $A=2$
Substitute this into **(1)** and **(2)** to obtain the 2x2 system:

$$\begin{cases} B+C=-2 & \textbf{(4)} \\ 2B+C=-6 & \textbf{(5)} \end{cases}$$

Subtract **(4)** – **(5)**: $B=-4$
Substitute this into **(4)**: $C=2$

Thus, $\dfrac{4}{k(k+1)(k+2)} = \dfrac{2}{k} - \dfrac{4}{k+1} + \dfrac{2}{k+2}$.

Section 8.7 Solutions---

1. d Above the line $y = x$, and do not include the line itself.	**2. c** Above the line $y = x$, and do include the line itself.
3. b Below the line $y = x$, and do not include the line itself.	**4. a** Below the line $y = x$, and do include the line itself.
5. 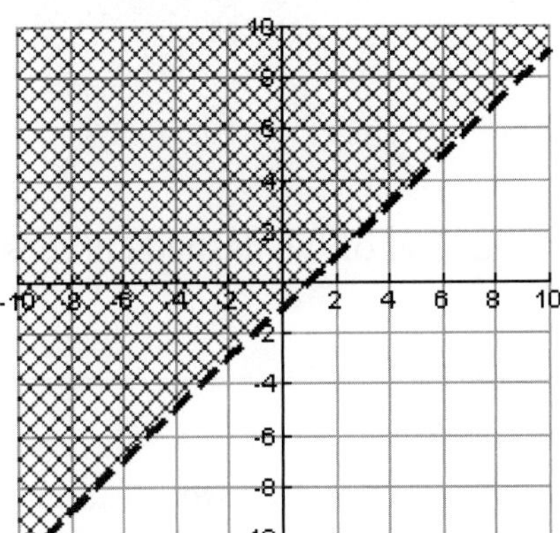 The equation of dashed curve is $y = x - 1$.	**6.** 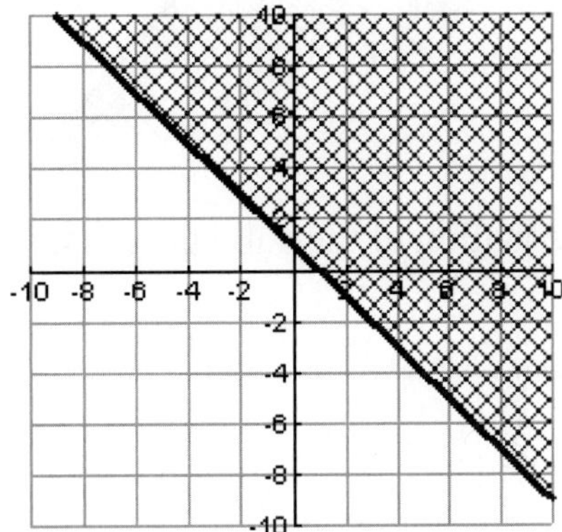 The equation of solid curve is $y = -x + 1$.
7. 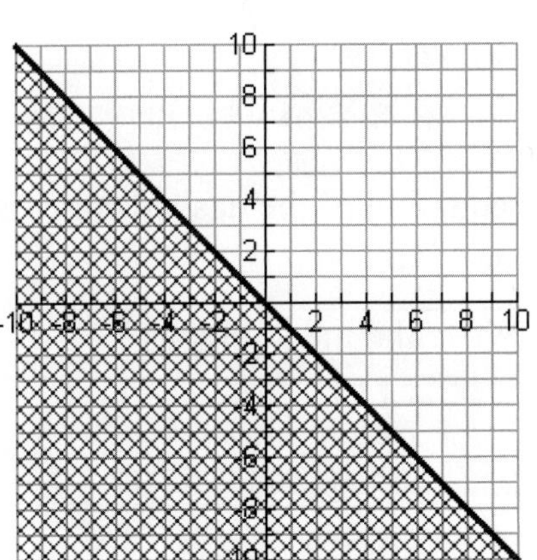 The equation of the solid curve is $y = -x$.	**8.** 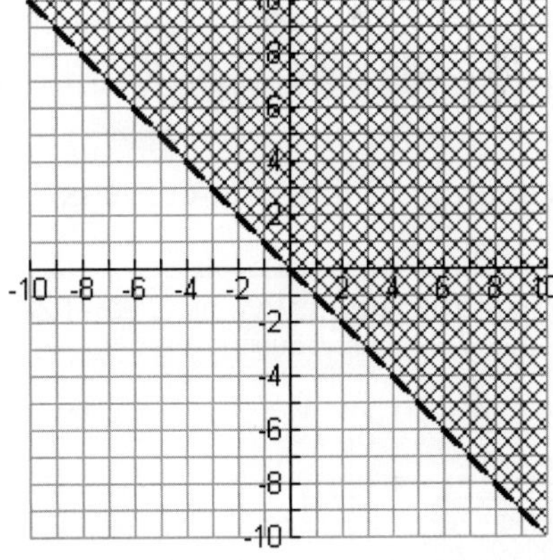 The equation of the dashed curve is $y = -x$.

9. 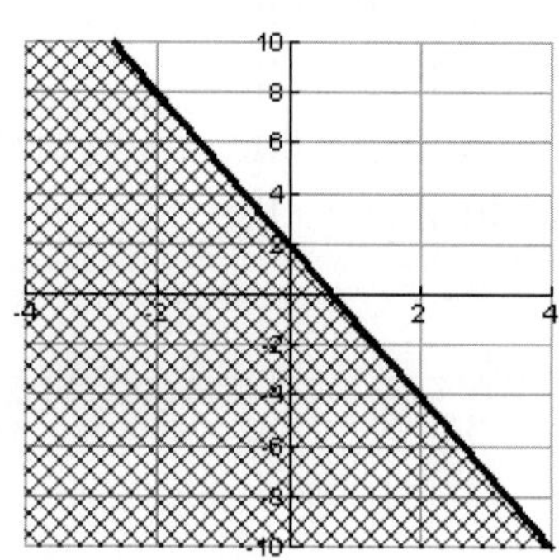 The equation of the solid curve is $y = -3x + 2$.	**10.** 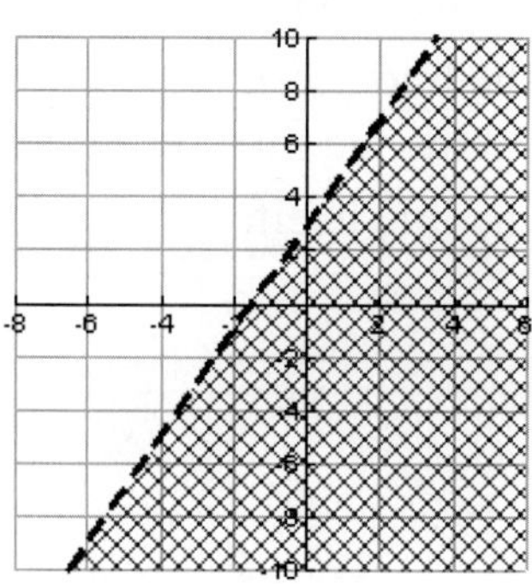 The equation of the dashed curve is $y = 2x + 3$.
11. 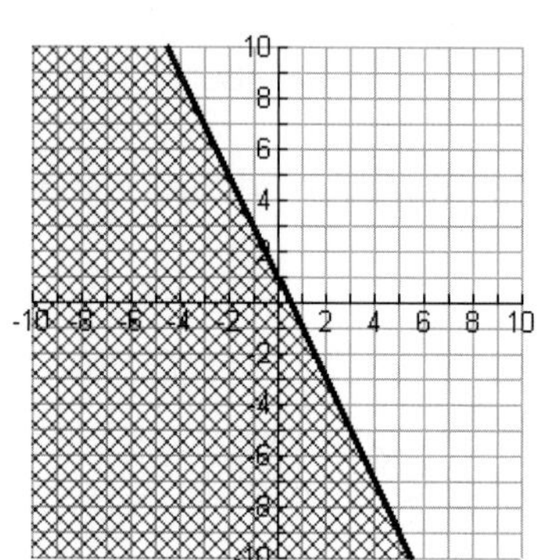 The equation of the solid curve is $y = -2x + 1$	**12.** 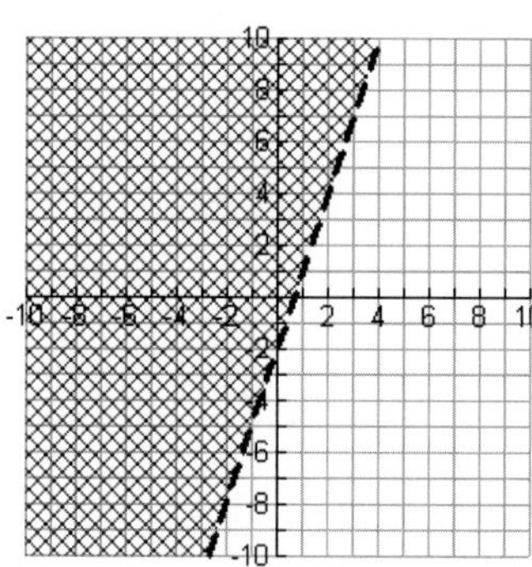 The equation of the dashed curve is $y = 3x - 2$.
13. Write the inequality as $y < \frac{1}{4}(2 - 3x)$. 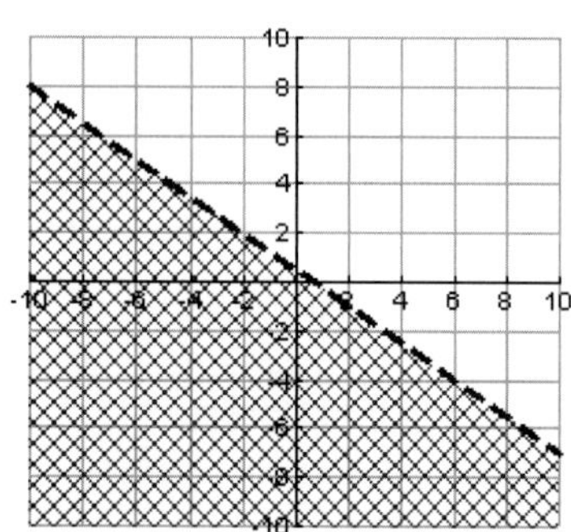 The equation of the dashed curve is $y = \frac{1}{4}(2 - 3x)$.	**14.** Write the inequality as $y > -\frac{1}{3}(2x + 6)$. 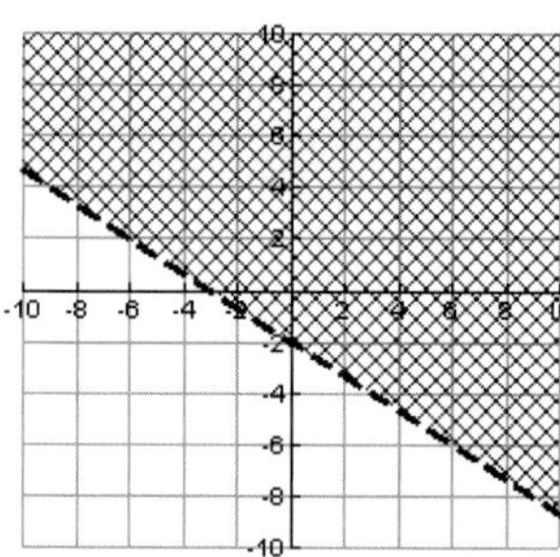 The equation of the dashed curve is $y = -\frac{1}{3}(2x + 6)$.

15.

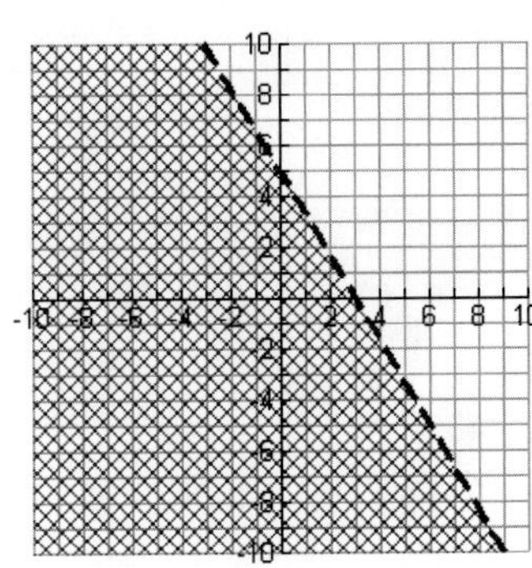

The equation of the dashed curve is $y = -\frac{5}{3}x + 5$.

16.

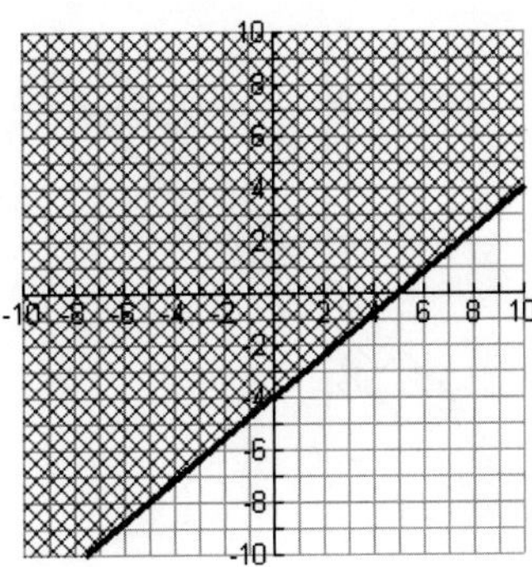

The equation of the solid curve is $y = \frac{4}{5}x - 4$.

17. Write the inequality as $y \le 2x - 3$.

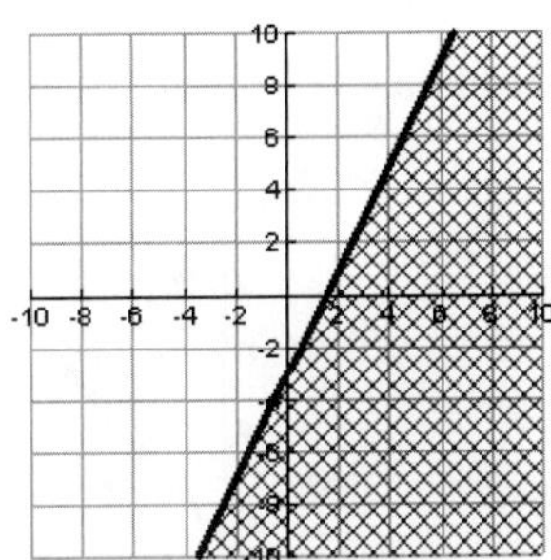

The equation of the solid curve is $y = 2x - 3$.

18. Write the inequality as $y \le 2x - 3$.

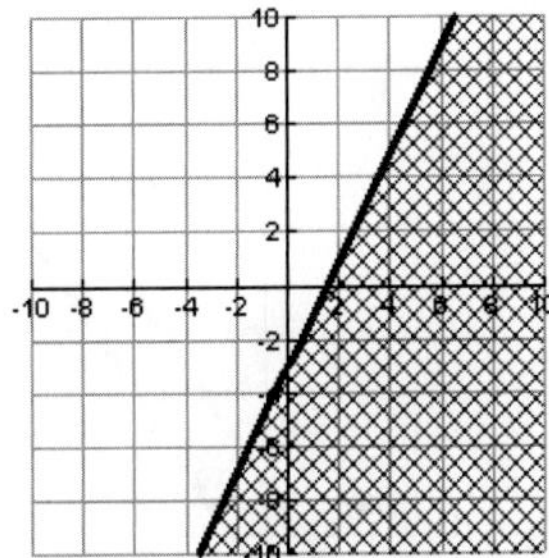

The equation of the solid curve is $y = 2x - 3$.

19.

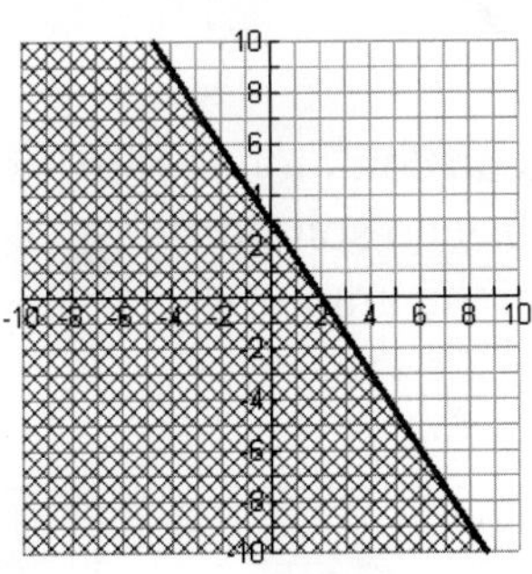

The equation of the solid curve is $y = -\frac{3}{2}x + 3$.

20.

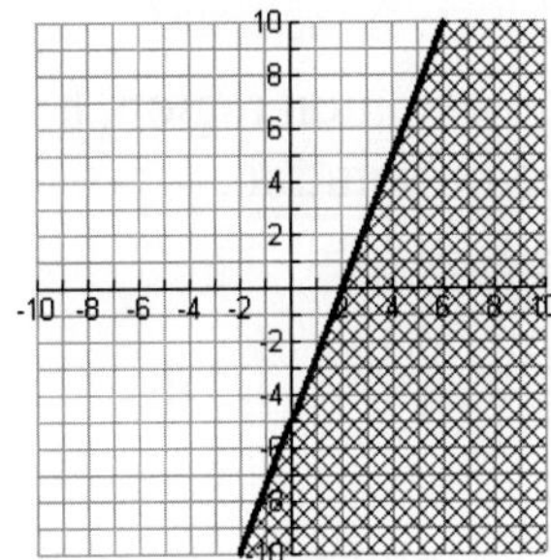

The equation of the solid curve is $y = \frac{5}{2}x - 5$.

21.

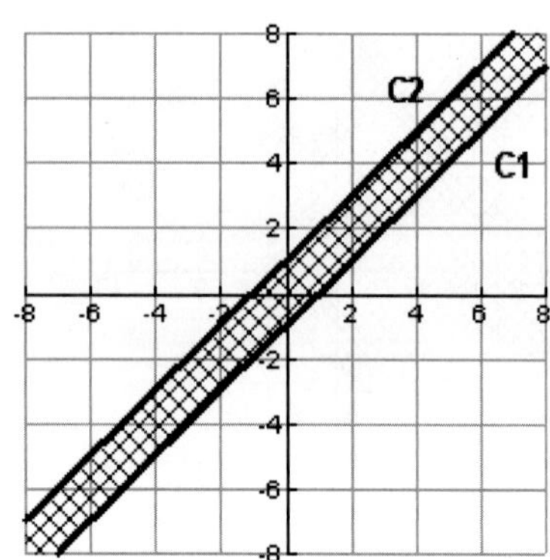

Notes on the graph:

C1: $y = x - 1$

C2: $y = x + 1$

22.

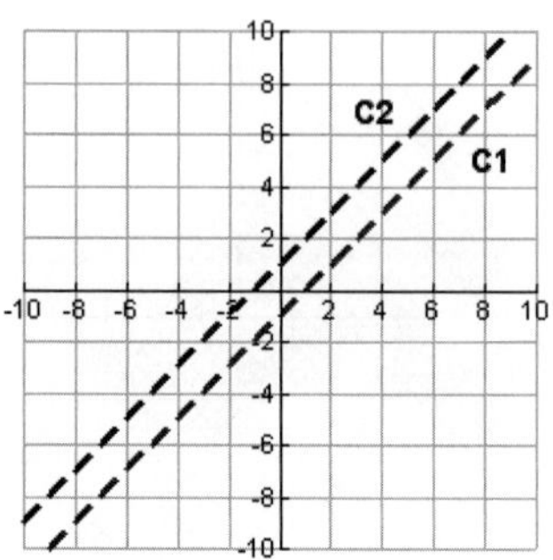

There is no common region, hence the system has no solution.

Notes on the graph:

C1: $y = x + 1$

C2: $y = x - 1$

23.

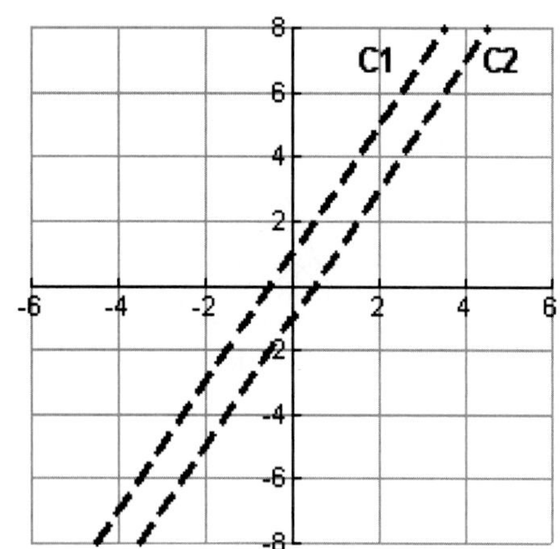

There is no common region, hence the system has no solution.

Notes on the graph:

C1: $y = 2x + 1$

C2: $y = 2x - 1$

24.

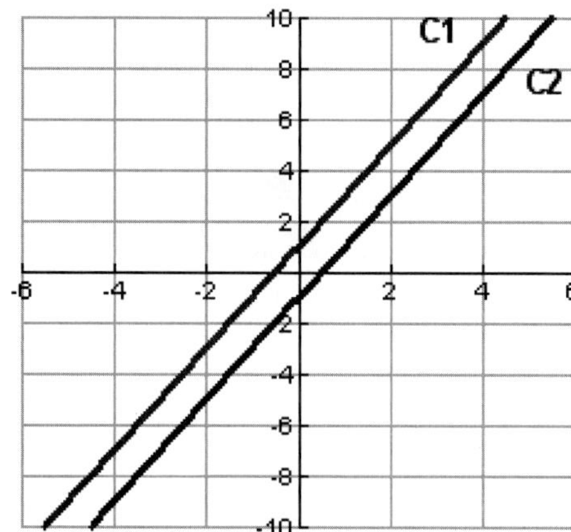

There is no common region, hence the system has no solution.

Notes on the graph:

C1: $y = 2x + 1$

C2: $y = 2x - 1$

25.

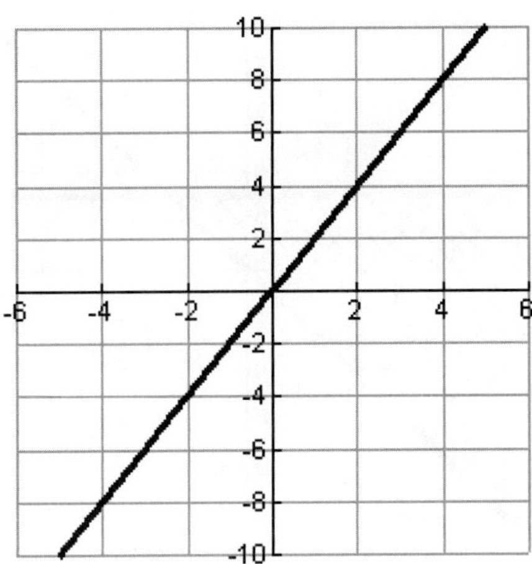

In this case, the common region is the line itself.

<u>Notes on the graph</u>:

C1 and **C2:** $y = 2x$

26.

There is no solution in this case since there are no points that lie both strictly above and strictly below the given line.

27.

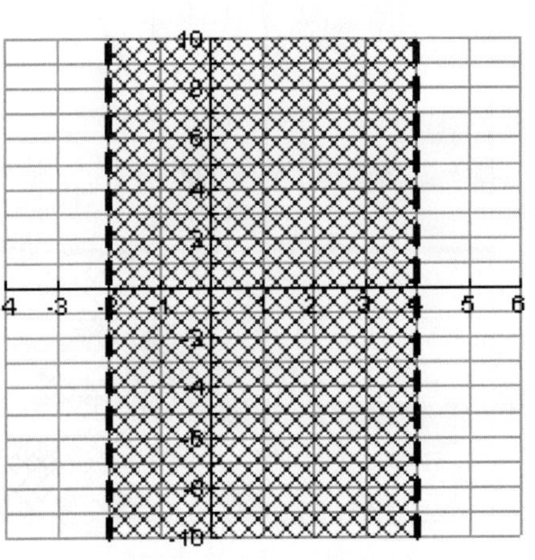

28.

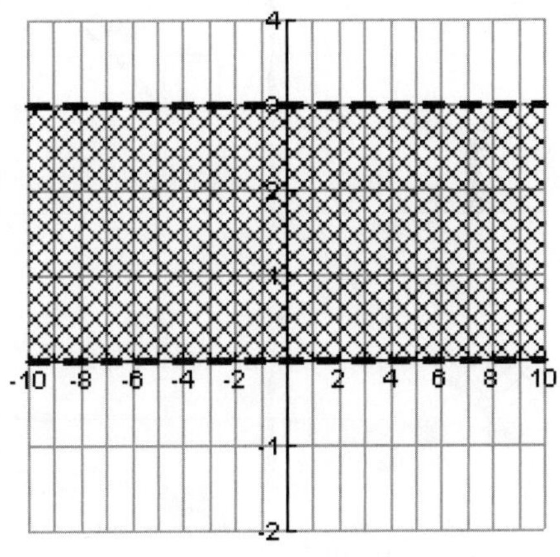

29.

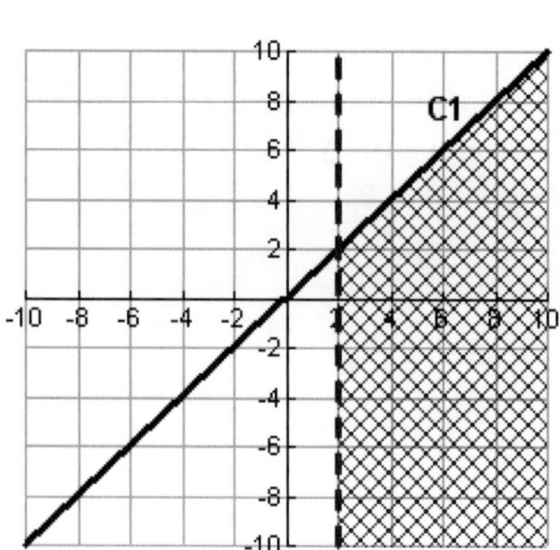

Notes on the graph:
C1: $y = x$

30.

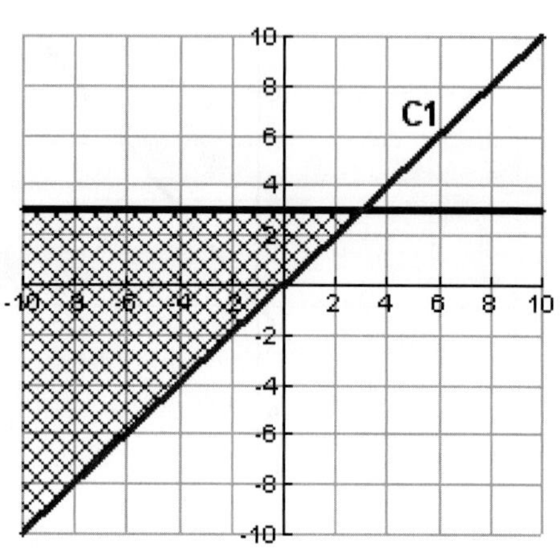

Notes on the graph:
C1: $y = x$

31.

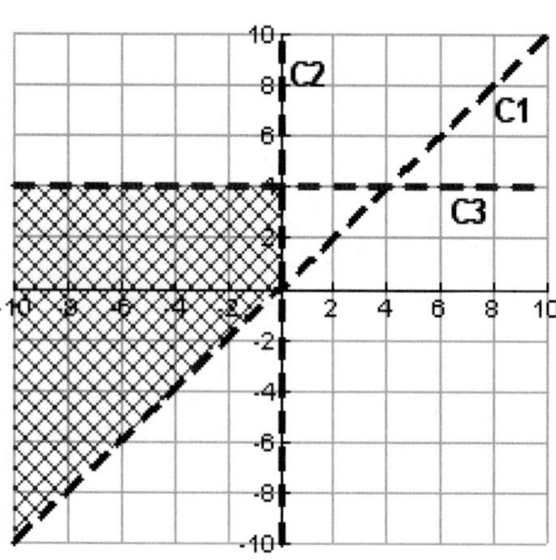

Notes on the graph:
C1: $y = x$
C2: $x = 0$
C3: $y = 4$

32.

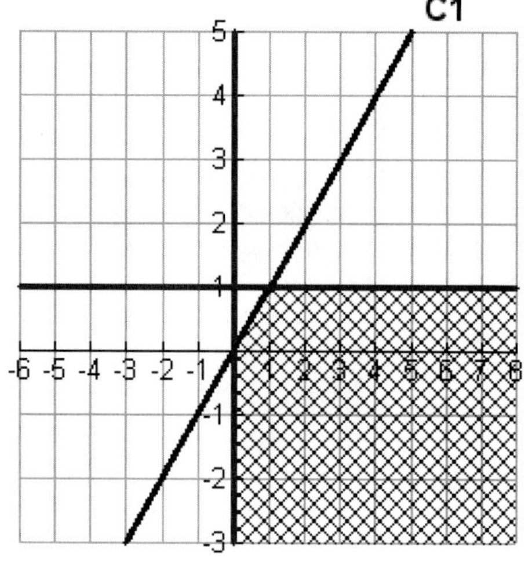

Notes on the graph:
C1: $y = x$

33.

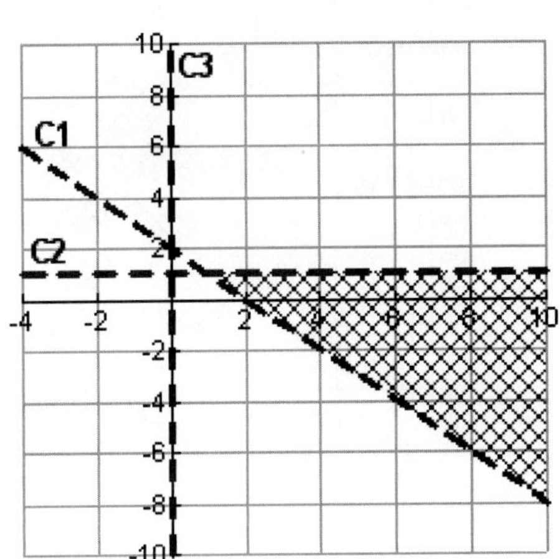

Notes on the graph:
C1: $y = -x + 2$
C2: $y = 1$
C3: $x = 0$

34.

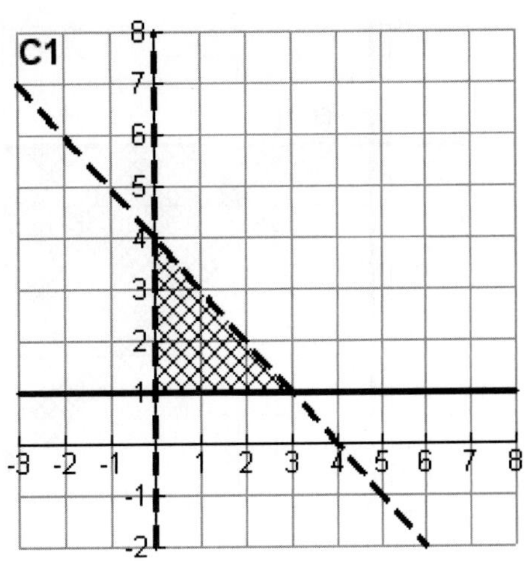

Notes on the graph:
C1: $y = -x + 4$

35.

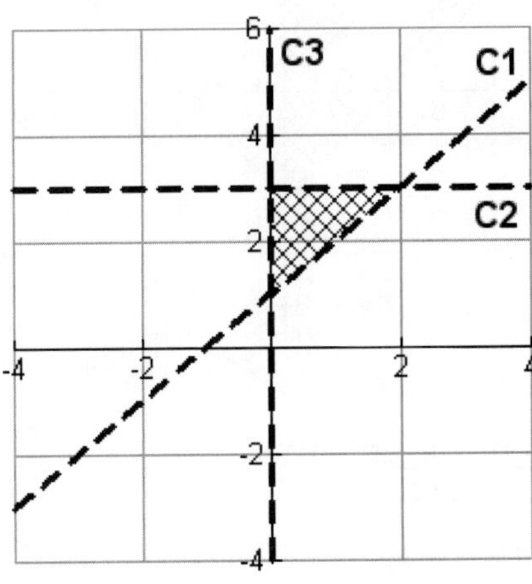

Notes on the graph:
C1: $y = x + 1$
C2: $y = 3$
C3: $x = 0$

36.

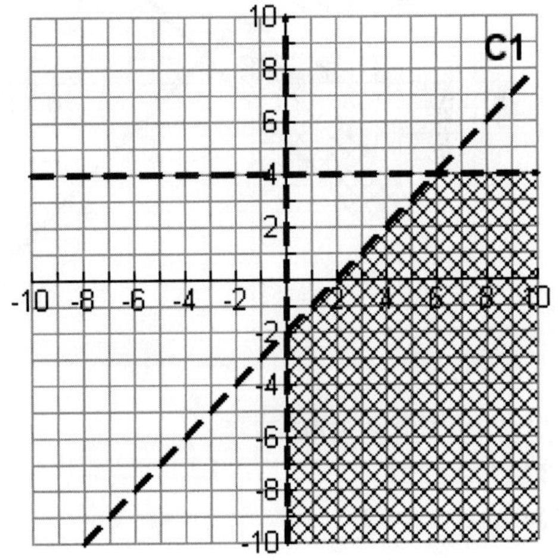

Notes on the graph:
C1: $y = x - 2$

37. 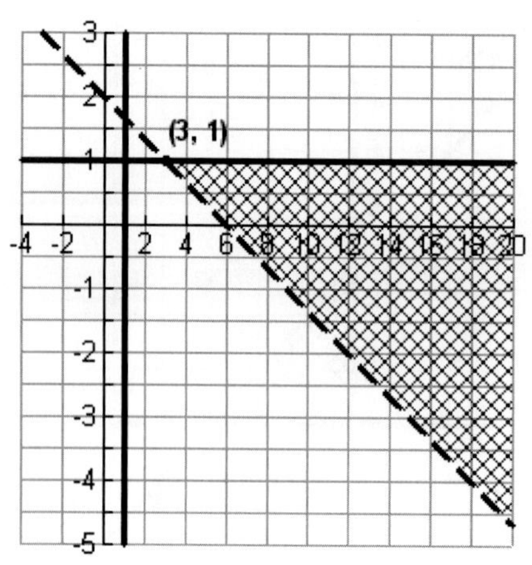 	**38.** 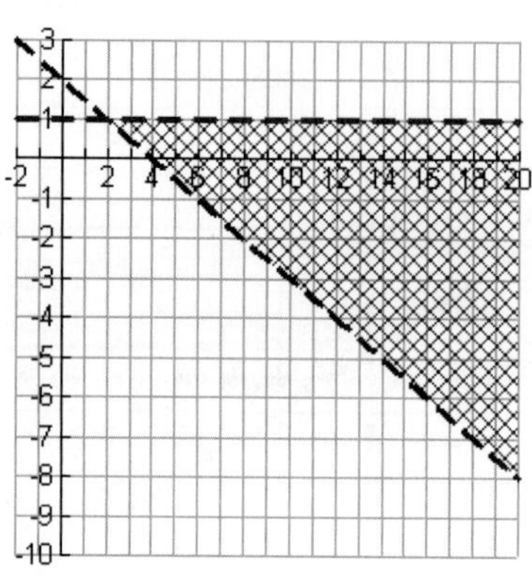
39. 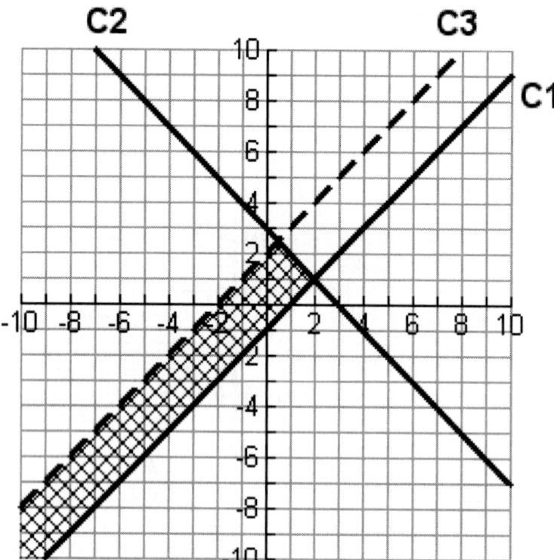 <u>Notes on the graph</u>: **C1:** $y = x - 1$ **C2:** $y = -x + 3$ **C3:** $y = x + 2$	**40.** 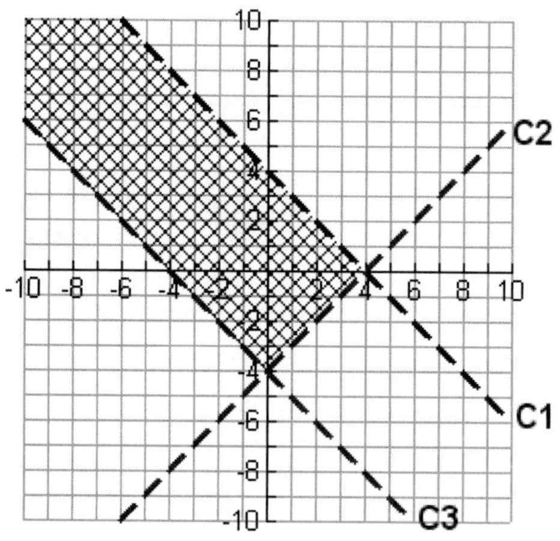 <u>Notes on the graph</u>: **C1:** $y = -x + 4$ **C2:** $y = x - 4$ **C3:** $y = -x - 4$

41.

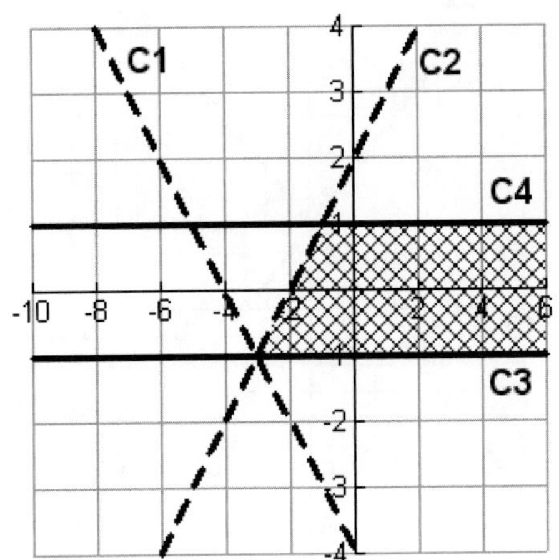

Notes on the graph:
C1: $y = -x - 4$
C2: $y = x + 2$
C3: $y = -1$
C4: $y = 1$

42.

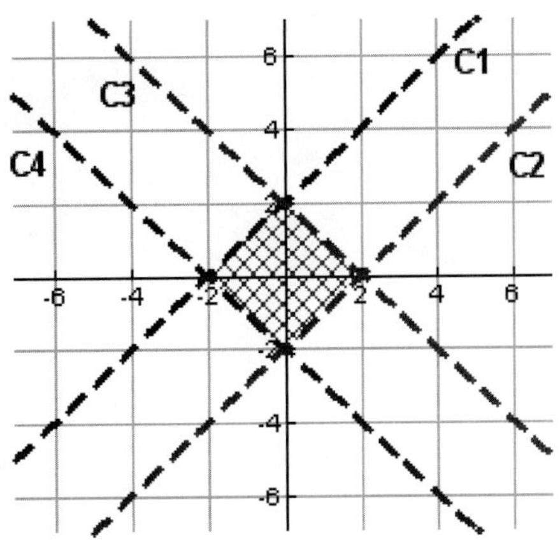

Notes on the graph:
C1: $y = x + 2$
C2: $y = x - 2$
C3: $y = -x + 2$
C4: $y = -x - 2$

43.

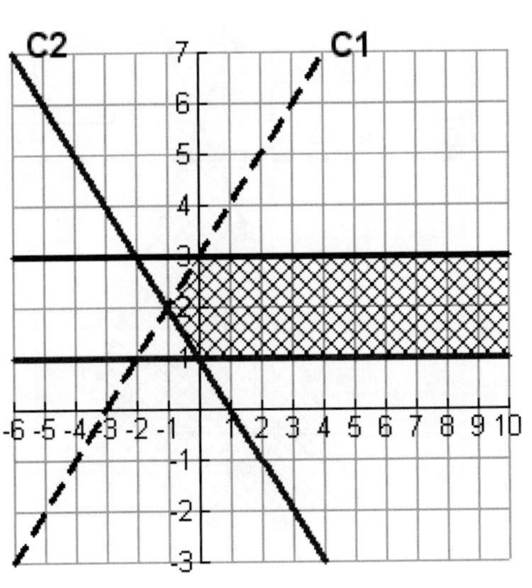

Notes on the graph:
C1: $y = x + 3$
C2: $y = -x + 1$

44.

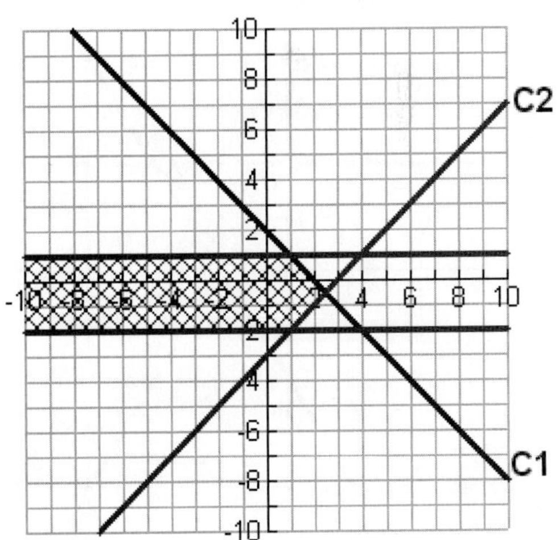

Notes on the graph:
C1: $y = -x + 2$
C2: $y = x - 3$

45. There is no solution since the regions do not overlap.

46.

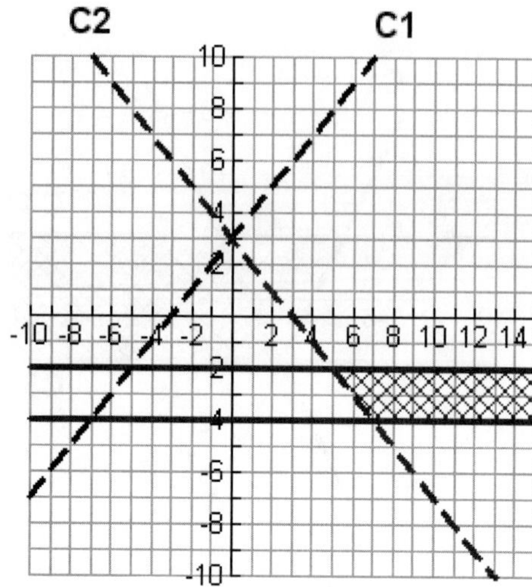

Notes on the graph:
C1: $y = -x + 2$
C2: $y = -x + 4$

47.

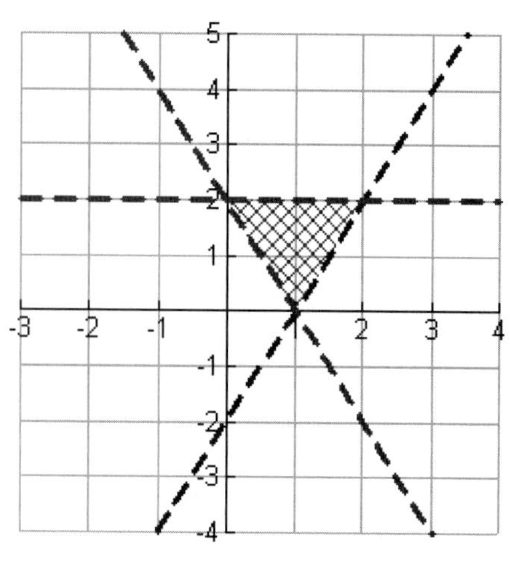

48.

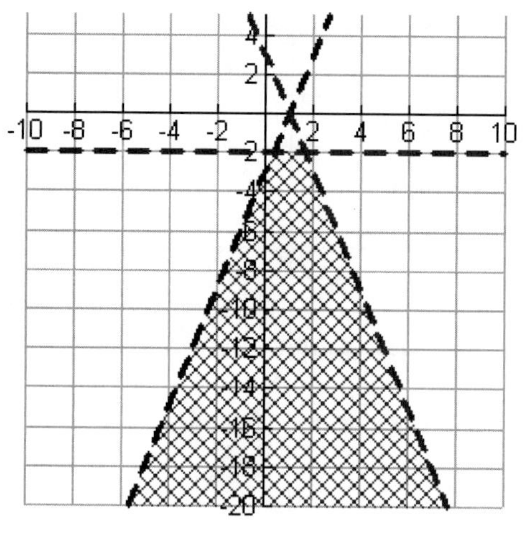

49.

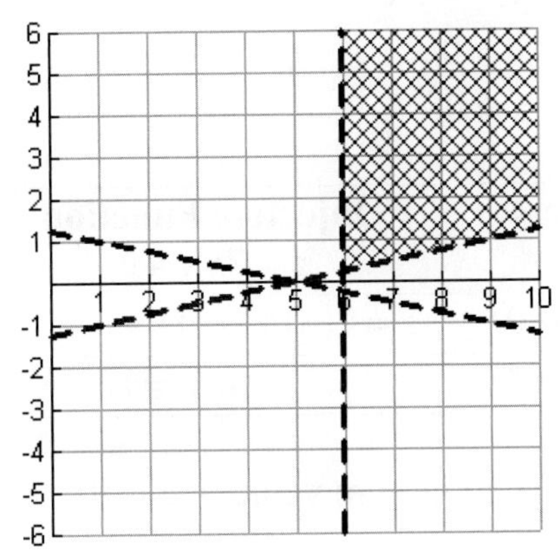

50.

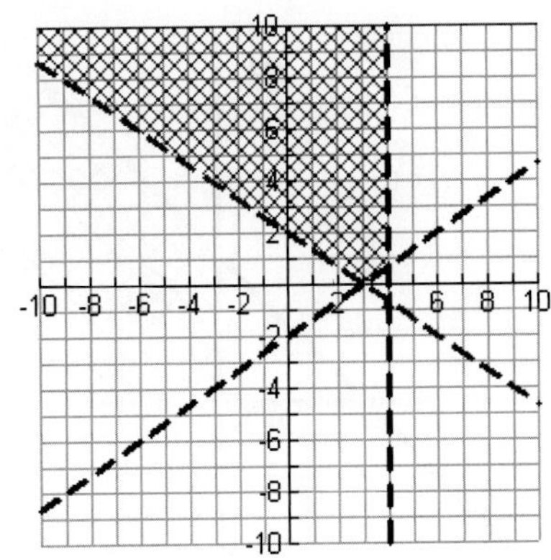

51.

Vertex	Objective Function $z = 2x + 3y$
$(-1,4)$	$z = 2(-1) + 3(4) = 10$
$(2,4)$	$z = 2(2) + 3(4) = 16$
$(-2,-1)$	$z = 2(-2) + 3(-1) = -7$
$(1,-1)$	$z = 2(1) + 3(-1) = -1$

So, the maximum value of z is 16, and the minimum value of z is -7.

52.

Vertex	Objective Function $z = 3x + 2y$
$(-1,4)$	$z = 3(-1) + 2(4) = 5$
$(2,4)$	$z = 3(2) + 2(4) = 14$
$(-2,-1)$	$z = 3(-2) + 2(-1) = -8$
$(1,-1)$	$z = 3(1) + 3(-1) = 0$

So, the maximum value of z is 14, and the minimum value of z is -8.

53.

Vertex	Objective Function $z = 1.5x + 4.5y$
$(-1,4)$	$z = 1.5(-1) + 4.5(4) = 16.5$
$(2,4)$	$z = 1.5(2) + 4.5(4) = 21$
$(-2,-1)$	$z = 1.5(-2) + 4.5(-1) = -7.5$
$(1,-1)$	$z = 1.5(1) + 4.5(-1) = -3$

So, the maximum value of z is 21, and the minimum value of z is -7.5.

54.

Vertex	Objective Function $z = \frac{2}{3}x + \frac{3}{5}y$
$(-1,4)$	$z = \frac{2}{3}(-1) + \frac{3}{5}(4) = \frac{26}{15}$
$(2,4)$	$z = \frac{2}{3}(2) + \frac{3}{5}(4) = \frac{56}{15}$
$(-2,-1)$	$z = \frac{2}{3}(-2) + \frac{3}{5}(-1) = -\frac{29}{15}$
$(1,-1)$	$z = \frac{2}{3}(1) + \frac{3}{5}(-1) = \frac{1}{15}$

So, the maximum value of z is $\frac{56}{15}$, and the minimum value of z is $-\frac{29}{15}$.

55. The region in this case is:

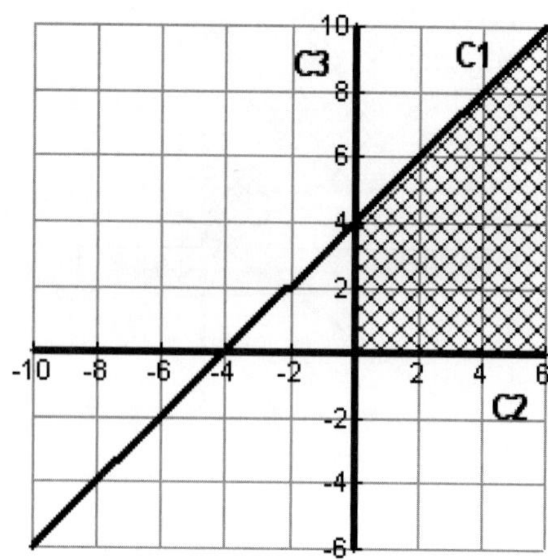

Notes on the graph:
C1: $y = 4 + x$
C2: $y = 0$
C3: $x = 0$

Vertex	Objective Function $z = 7x + 4y$
$(0,0)$	$z = 7(0) + 4(0) = 0$
$(0,4)$	$z = 7(0) + 4(4) = 16$

So, the minimum value of z is 0.

56. The region in this case is:

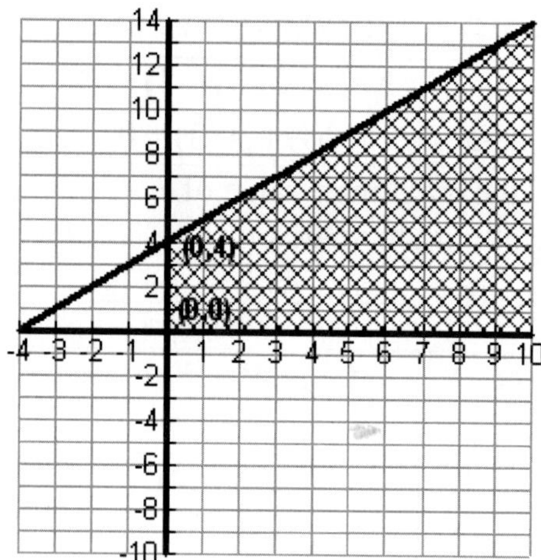

Vertex	Objective Function $z = 3x + 5y$
$(0,0)$	0
$(0,4)$	20
(5,2)	**25**

Since the region is unbounded and the maximum must occur at a vertex, we conclude that the objective function does not have a maximum in this case.

57. The region in this case is:

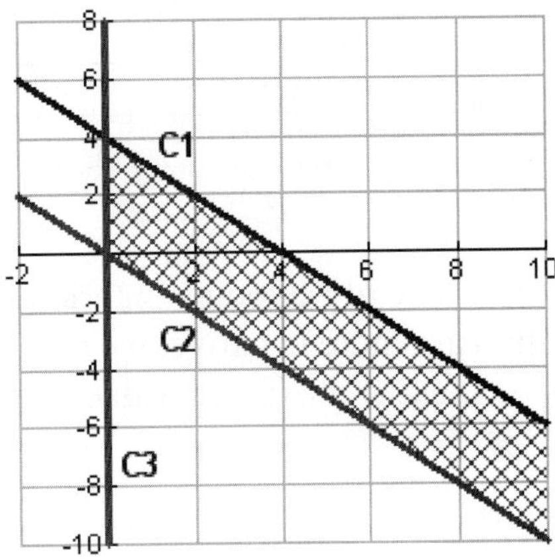

Notes on the graph:

C1: $y = 4 - x$ **C2:** $y = -x$ **C3:** $x = 0$

Vertex	Objective Function $z = 4x + 3y$
$(0,0)$	$z = 4(0) + 3(0) = 0$
$(0,4)$	$z = 4(0) + 3(4) = 12$
Additional point $(4,0)$	$z = 4(4) + 3(0) = 16$

Since the region is unbounded and the maximum must occur at a vertex, we conclude that the objective function does not have a maximum in this case.

58. The region in this case is:

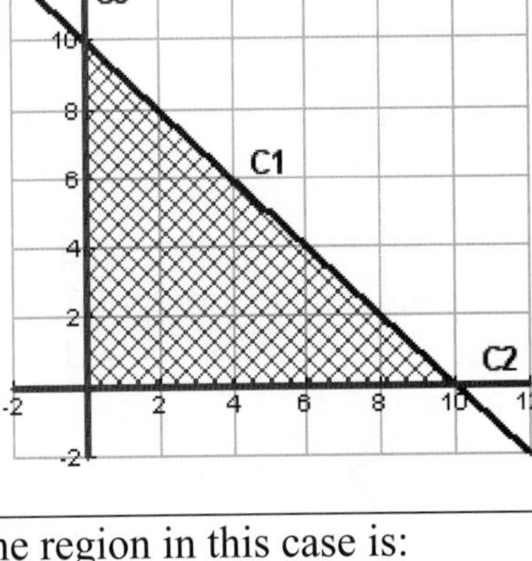

Notes on the graph:

C1: $y = 10 - x$ **C2:** $y = 0$ **C3:** $x = 0$

Vertex	Objective Function $z = 4x + 3y$
$(0,0)$	$z = 4(0) + 3(0) = 0$
$(10,0)$	$z = 4(10) + 3(0) = 40$
$(0,10)$	$z = 4(0) + 3(10) = 30$

So, the minimum value of z is 0.

59. The region in this case is:

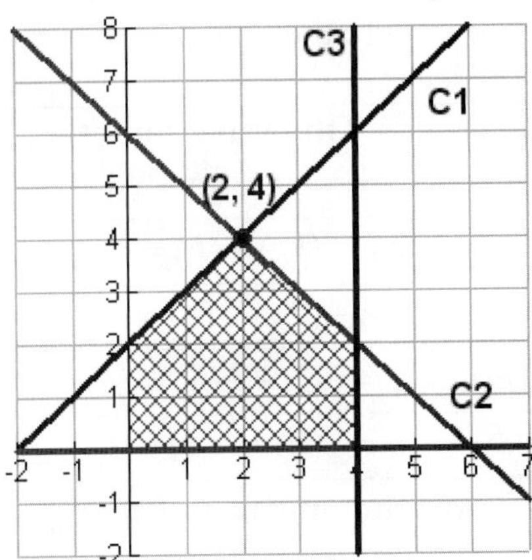

Notes on the graph:

C1: $y = x + 2$ **C2:** $y = -x + 6$ **C3:** $x = 4$

Vertex	Objective Function $z = 2.5x + 3.1y$
$(2,4)$	$z = 2.5(2) + 3.1(4) = 17.4$
$(0,0)$	$z = 2.5(0) + 3.1(0) = 0$
$(4,2)$	$z = 2.5(4) + 3.1(2) = 16.2$
$(0,2)$	$z = 2.5(0) + 3.1(2) = 6.2$

So, the minimum value of z is 0.

60. The region in this case is:

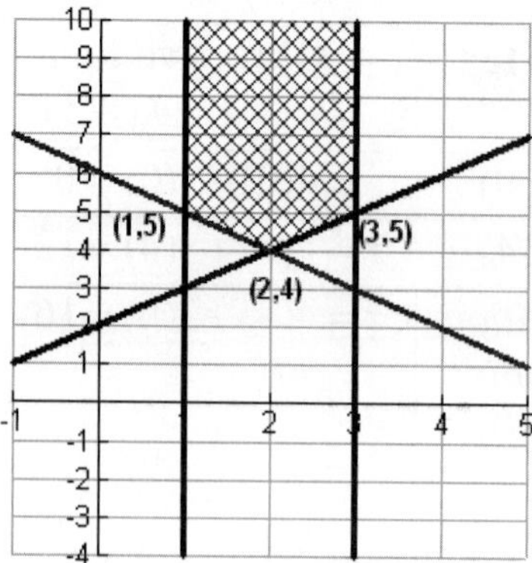

Vertex	Objective Function $z = 2.5x - 3.1y$
(1,5)	-13
(3,5)	-8
(2,4)	-7.4

Since the value of z at points interior to the region are more negative, we can conclude that the maximum value of z is -7.4.

61. The region in this case is:

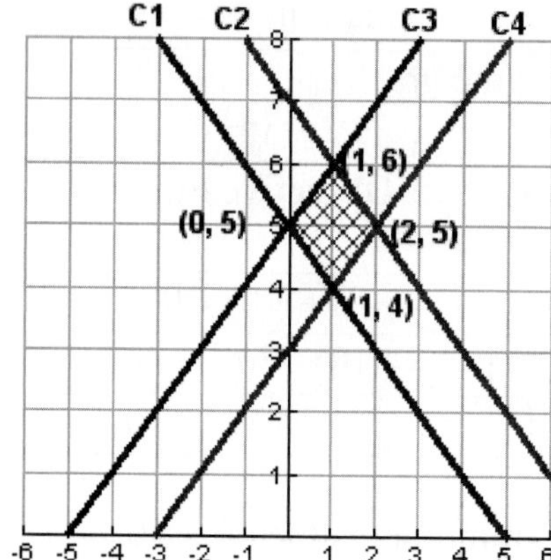

Notes on the graph:
C1: $y = -x + 5$ **C2:** $y = -x + 7$
C3: $y = x + 5$ **C4:** $y = x + 3$

We need all of the intersection points since they constitute the vertices:

Intersection of $y = x + 5$ and $y = -x + 7$:

$$x + 5 = -x + 7$$
$$2x = 2$$
$$x = 1$$

So, the intersection point is (1,6).

Intersection of $y = x + 3$ and $y = -x + 7$:

$$x + 3 = -x + 7$$
$$2x = 4 \text{ so that } x = 2$$

So, the intersection point is (2,5).

Intersection of $y = x + 3$ and $y = -x + 5$:

$$x + 3 = -x + 5$$
$$2x = 2 \text{ so that } x = 1$$

So, the intersection point is (1,4).

Intersection of $y = x + 5$ and $y = -x + 5$:

$$x + 5 = -x + 5$$
$$2x = 0 \text{ so that } x = 0$$

So, the intersection point is (0,5).

Now, compute the objective function at the vertices:

Vertex	Objective Function $z = \frac{1}{4}x + \frac{2}{5}y$
(1,6)	$z = \frac{1}{4}(1) + \frac{2}{5}(6) = \frac{53}{20}$
(2,5)	$z = \frac{1}{4}(2) + \frac{2}{5}(5) = \frac{5}{2}$
(1,4)	$z = \frac{1}{4}(1) + \frac{2}{5}(4) = \frac{37}{20}$
(0,5)	$z = \frac{1}{4}(0) + \frac{2}{5}(5) = 2$

So, the maximum value of z is $\frac{53}{20} = 2.65$.

62. The region in this case is:

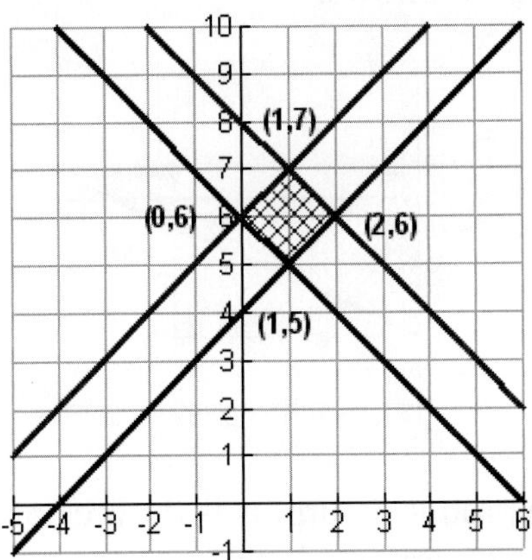

Now, compute the objective function at the vertices:

Vertex	**Objective Function** $z = \frac{1}{3}x - \frac{2}{5}y$
(0,6)	$-\frac{12}{5}$
(1,5)	$-\frac{5}{3}$
(2,6)	$-\frac{26}{15}$
(1,7)	$-\frac{37}{15}$

So, the minimum value of z is $-\frac{37}{15}$.

63. First, find the point of intersection of

$$\begin{cases} P = 80 - 0.01x \\ P = 20 + 0.02x \end{cases}$$

Equating these and solving for x yields $x = 2000$. Then, substituting this into the first equation yields $P = 60$. Hence, the system for consumer surplus is:

$$\begin{cases} P \le 80 - 0.01x \\ P \ge 60 \\ x \ge 0 \end{cases}$$

64. First, find the point of intersection of

$$\begin{cases} P = 80 - 0.01x \\ P = 20 + 0.02x \end{cases}$$

Equating these and solving for x yields $x = 2000$. Then, substituting this into the first equation yields $P = 60$. Hence, the system for producer surplus is:

$$\begin{cases} P \ge 20 + 0.02x \\ P \le 60 \\ x \ge 0 \end{cases}$$

65. The graph of the system in Exercise 63 is as follows:

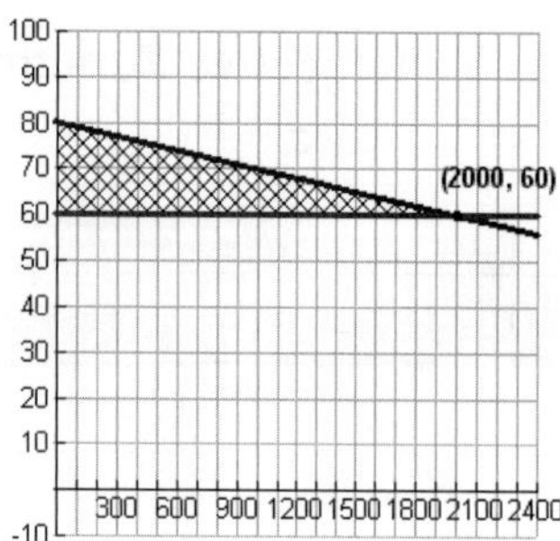

The consumer surplus is the area of the shaded region, which is

$$\tfrac{1}{2}(20)(2000) = 20{,}000 \text{ units}^2.$$

66. The graph of the system in Exercise 64 is as follows:

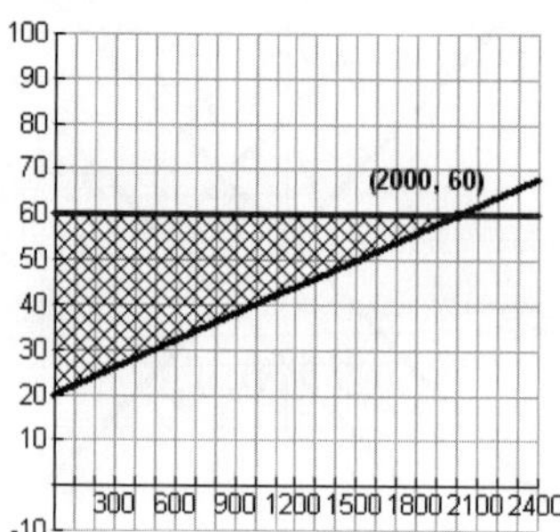

The producer surplus is the area of the shaded region, which is

$$\tfrac{1}{2}(40)(2000) = 40{,}000 \text{ units}^2.$$

67. Let $x =$ number of cases of water

$y =$ number of generators

Certainly, $x \geq 0,\ y \geq 0$. Then, we also have the following restrictions:

Cubic feet restriction:

$$x + 20y \leq 2400$$

Weight restriction:

$$25x + 150y \leq 6000$$

So, we obtain the following system of inequalities:

$$\begin{cases} x \geq 0,\ y \geq 0 \\ x + 20y \leq 2400 \\ 25x + 150y \leq 6000 \end{cases}$$

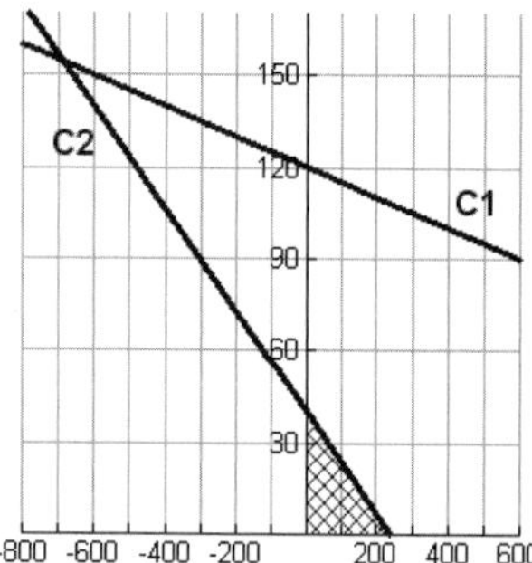

Notes on the graph:

C1: $x + 20y = 2400$ **C2:** $25x + 150y = 6000$

68. Let $x =$ number of cases of plywood

$y =$ number of cases of tarps

Certainly, $x \geq 0,\ y \geq 0$. Then, we also have the following restrictions:

Cubic feet restriction:

$$60x + 10y \leq 1500$$

Weight restriction:

$$500x + 50y \leq 2000$$

So, we obtain the following system of inequalities:

$$\begin{cases} x \geq 0,\ y \geq 0 \\ 60x + 10y \leq 1500 \\ 500x + 50y \leq 2000 \end{cases}$$

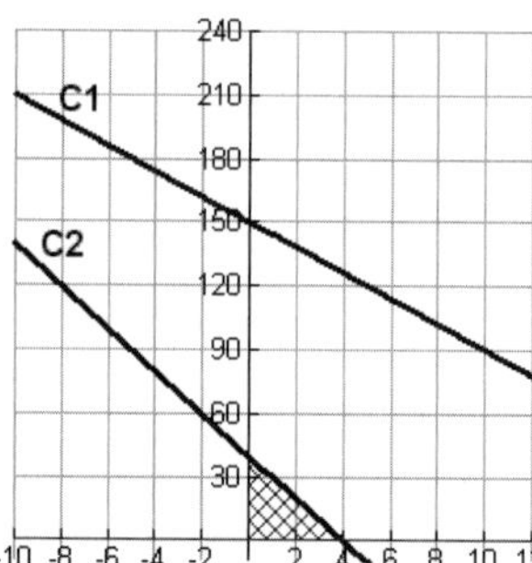

Notes on the graph:

C1: $60x + 10y = 1500$ **C2:** $500x + 50y = 2000$

69. Let $x =$ number of Charley T-shirts
$y =$ number of Francis T-shirts

Profit from Charley T-shirts:
Revenue – cost $= 13x - 7x = 6x$
Profit from Francis T-shirts:
Revenue – cost $= 10y - 5y = 5y$

So, to maximize profit, we must maximize the objective function $z = 6x + 5y$. We have the following constraints:

$$\begin{cases} 7x + 5y \leq 1000 \\ x + y \leq 180 \\ x \geq 0,\ y \geq 0 \end{cases}$$

The region in this case is:

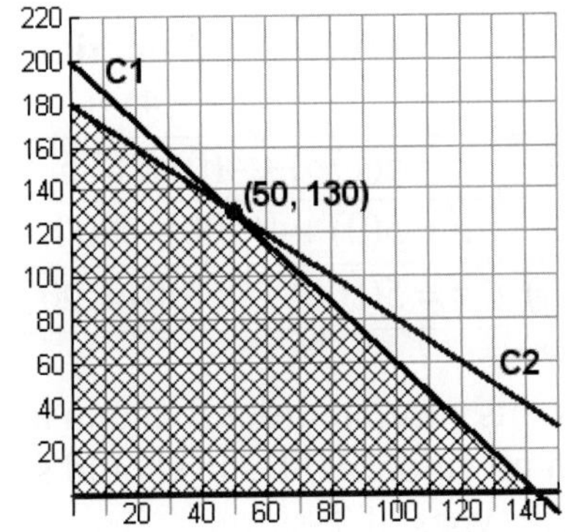

Notes on the graph:
C1: $y = -\frac{7}{5}x + 200$
C2: $y = -x + 180$
C3: $x = 0$

We need all of the intersection points since they constitute the vertices:
Intersection of $7x + 5y = 1000$ and $x + y = 180$:

$$-x + 180 = -\tfrac{7}{5}x + 200$$
$$\tfrac{2}{5}x = 20$$
$$x = 50$$

So, the intersection point is (50, 130).

Now, compute the objective function at the vertices:

Vertex	**Objective Function** $z = 6x + 5y$
(0,180)	$z = 6(0) + 5(180) = 900$
$(\frac{1000}{7}, 0)$	$z = 6(\frac{1000}{7}) + 5(0) \cong 857.14$
(50, 130)	$z = 6(50) + 5(130) = 950$

So, to attain a maximum profit of \$950, she should sell 130 Francis T-shirts, and 50 Charley T-shirts.

70. Let $x = \#$ of "Got Plywood" T-shirts
$y = \#$ of "Got Gas" T-shirts

<u>Profit from "Got Plywood" T-shirts:</u>
Revenue – cost = $13x - 8x = 5x$
<u>Profit from "Got Gas" T-shirts:</u>
Revenue – cost = $10y - 6y = 4y$

So, to maximize profit, we must maximize the objective function $z = 5x + 4y$. We have the following constraints:

$$\begin{cases} 8x + 6y \le 1400 \\ x + y \le 200 \\ x \ge 0,\ y \ge 0 \end{cases}$$

The region in this case is:

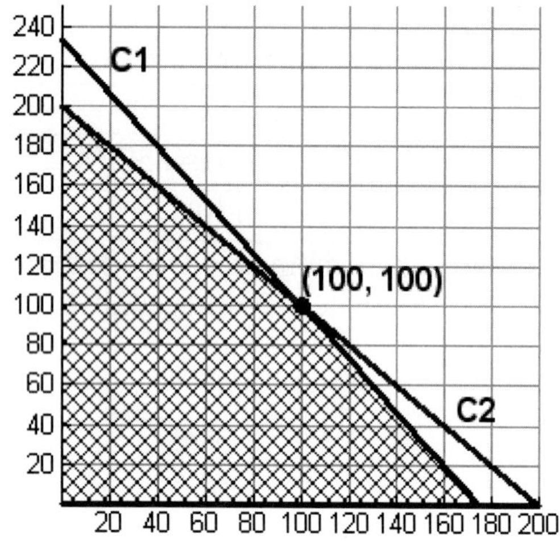

<u>Notes on the graph:</u>
C1: $y = \frac{700}{3} - \frac{4}{3}x$
C2: $y = 200 - x$
C3: $x = 0$

<u>We need all of the intersection points since they constitute the vertices:</u>
Intersection of $8x + 6y = 1400$ and $x + y = 200$:

$$\frac{1400-8x}{6} = 200 - x$$
$$1400 - 8x = 1200 - 6x$$
$$100 = x$$

So, the intersection point is (100,100).

Now, compute the objective function at the vertices:

Vertex	**Objective Function** $z = 5x + 4y$
(175,0)	$z = 5(175) + 4(0) = 875$
(0, 200)	$z = 5(0) + 4(200) = 800$
(100,100)	$z = 5(100) + 4(100) = 900$

Hence, in this case, in order to attain a maximum profit of \$900, she should sell 100 of both types of T-shirts.

71. Let $x = \#$ of desktops
$y = \#$ of laptops

We must maximize the objective function $z = 500x + 300y$ subject to the following constraints:

$$\begin{cases} 5x + 3y \le 90 \\ 700x + 400y \le 10{,}000 \\ y \ge 3x \\ x \ge 0,\ y \ge 0 \end{cases}$$

The region in this case is:

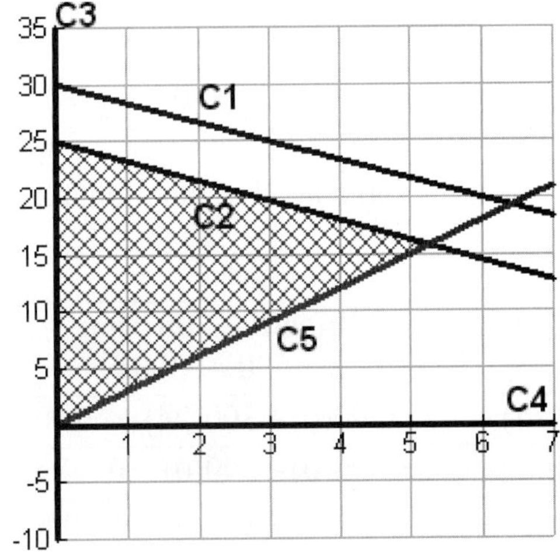

Notes on the graph:
C1: $y = 30 - \frac{5}{3}x$ **C2:** $y = 25 - \frac{7}{4}x$
C3: $x = 0$ **C4:** $y = 0$ **C5:** $y = 3x$

We need all of the intersection points since they constitute the vertices:

Intersection of $5x + 3y = 90$ and $700x + 400y = 10{,}000$:

$$30 - \tfrac{5}{3}x = 25 - \tfrac{7}{4}x$$
$$360 - 20x = 300 - 21x$$
$$x = -60$$

So, the intersection point is $(-60, 130)$. Not a vertex, however, since x must be non-negative (see the region).

Intersection of **C5** and **C2**:

$$3x = 25 - \tfrac{7}{4}x$$
$$\tfrac{19}{4}x = 25 \text{ so that } x = \tfrac{100}{19}$$

So, the intersection point is $\left(\frac{100}{19}, \frac{300}{19}\right)$

Now, compute the objective function at the vertices:

Vertex	Objective Function $z = 500x + 300y$
(0,25)	$z = 500(0) + 300(25) = 7500$
(0,0)	$z = 500(0) + 300(0) = 0$
$\left(\frac{100}{19}, \frac{300}{19}\right)$	$z = 500(\frac{100}{19}) + 300(\frac{300}{19}) = \frac{140{,}000}{19}$

In order to attain a maximum profit of \$7500, he must sell 25 laptops and 0 desktops.

72. Let $x = \#$ of desktops
$y = \#$ of laptops

We must maximize the objective function $z = 500x + 300y$ subject to the following constraints:

$$\begin{cases} 5x + 3y \le 120 \\ 700x + 400y \le 30{,}000 \\ y \ge 3x \\ x \ge 0,\ y \ge 0 \end{cases}$$

The region in this case is:

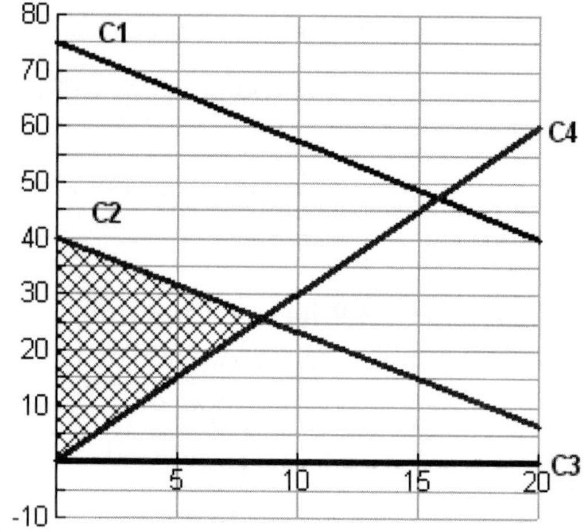

Notes on the graph:
C1: $y = 75 - \frac{7}{4}x$ **C2:** $y = 40 - \frac{5}{3}x$
C3: $x = 0$ **C4:** $y = 3x$

We need all of the intersection points since they constitute the vertices:

Intersection of $5x + 3y = 120$ and $700x + 400y = 30{,}000$:

$$75 - \tfrac{7}{4}x = 40 - \tfrac{5}{3}x$$
$$900 - 21x = 480 - 20x$$
$$420 = x$$

So, the intersection point is $(420, -660)$. Not a vertex, however, since y must be non-negative (see the region).

Intersection of **C4** and **C2**:

$$3x = 40 - \tfrac{5}{3}x$$
$$\tfrac{14}{3}x = 40 \text{ so that } x = \tfrac{60}{7}$$

So, the intersection point is $\left(\frac{60}{7}, \frac{180}{7}\right)$.

Now, compute the objective function at the vertices:

Vertex	Objective Function $z = 500x + 300y$
(0,24)	$z = 500(0) + 300(24) = 7200$
(0, 0)	$z = 500(0) + 300(0) = 0$
$\left(\frac{60}{7}, \frac{180}{7}\right)$	$z = 500(\frac{60}{7}) + 300(\frac{180}{7}) = 12{,}000$

So, he should sell 30 laptops and 6 desktops in order to obtain maximum profit of \$12,000.

73. Let $x = \#$ first class cars
$y = \#$ second class cars

Let p denote the profit for each second class car. Then, the profit for each first class car is $2p$. Hence, we seek to maximize the objective function

$$z = py + 2px = p(y + 2x).$$

We have the following constraints:

$$\begin{cases} x + y = 30 \\ 2 \le x \le 4 \\ y \ge 8x \end{cases}$$

The region in this case is:

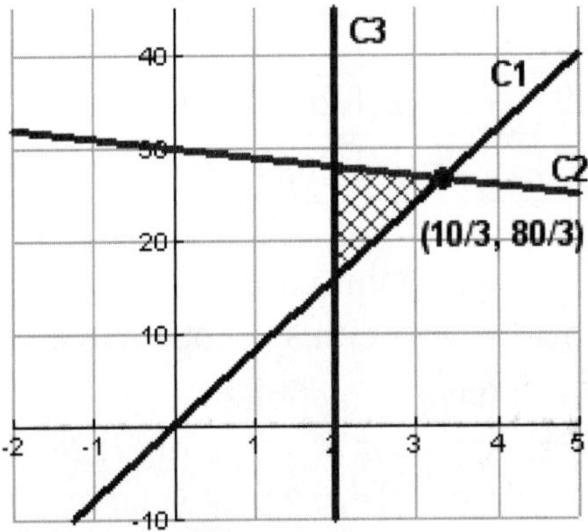

Notes on the graph:

C1: $y = 8x$

C2: $y = -x + 30$

C3: $x = 2$

We need all of the intersection points since they constitute the vertices:

Intersection of $x + y = 30$ and $y = 8x$:

$$-x + 30 = 8x$$

$$30 = 9x \text{ so that } \tfrac{10}{3} = x$$

So, the intersection point is $(\tfrac{10}{3}, \tfrac{80}{3})$.

Now, compute the objective function at the vertices:

Vertex	**Objective Function** $z = p(2x + y)$
(2, 16)	$z = p(2(2) + 16) = 20p$
(2, 28)	$z = p(2(2) + 28) = 32p$
$(\tfrac{10}{3}, \tfrac{80}{3})$	$z = p(2(\tfrac{10}{3}) + \tfrac{80}{3}) = \tfrac{100}{3}p$
(3, 27)	$z = p(2(3) + 27) = 33p$

The maximum in this case would occur at $(\tfrac{10}{3}, \tfrac{80}{3})$. However, this is not tenable since one cannot have a fraction of a car. However, very near at the vertex (3, 27) the profit is very near to this one, as are the number of cars. Hence, to maximize profit, they should use:

3 first class cars and 27 second class cars.

74. Let $x = \#$ first class cars
$y = \#$ second class cars

Let p denote the profit for each second class car. Then, the profit for each first class car is $2p$. Hence, we seek to maximize the objective function

$$z = py + 1.2px = p(y + 1.2x).$$

We have the following constraints:

$$\begin{cases} x + y = 30 \\ 1 \le x \le 4 \\ y \ge 10x \end{cases}$$

The region in this case is:

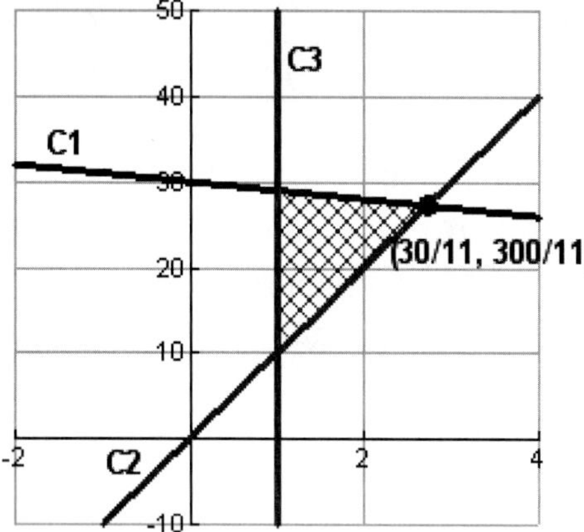

Notes on the graph:
C1: $y = -x + 30$
C2: $y = 10x$
C3: $x = 1$

We need all of the intersection points since they constitute the vertices:

Intersection of $x + y = 30$ and $y = 10x$:

$$-x + 30 = 10x$$
$$30 = 11x$$
$$\tfrac{30}{11} = x$$

So, the intersection point is $\left(\frac{30}{11}, \frac{300}{11}\right)$.

Now, compute the objective function at the vertices:

Vertex	Objective Function $z = p(1.2x + y)$
(1, 10)	$z = p(1.2(1) + 10) = 11.2p$
(1, 29)	$z = p(1.2(1) + 29) = 30.2p$
$\left(\frac{30}{11}, \frac{300}{11}\right)$	$z = p(1.2(\frac{30}{11}) + \frac{300}{11}) = 30.54p$

The maximum in this case would occur at $\left(\frac{30}{11}, \frac{300}{11}\right)$. However, this is not tenable since one cannot have a fraction of a car. However, very near at the point (2, 28), the profit would be $30.4p$, which is larger than the profit at vertex (1, 29). Even though this is not a vertex, it is still in the region, and the restriction that the values of x and y requires one to consider pairs in the region close to the vertex at which the maximum would have occurred, and compare these values to the other vertices. In this case, to maximize profit, they should use:
2 first class cars and 28 second class cars.

75. Let $x =$ # of regular skis
$y =$ # of slalom skis

We must maximize the objective function $z = 25x + 50y$ subject to the following constraints:

$$\begin{cases} x + y \le 400 \\ x \ge 200, \ \ y \ge 80 \\ x \ge 0, \ y \ge 0 \end{cases}$$

The region in this case is:

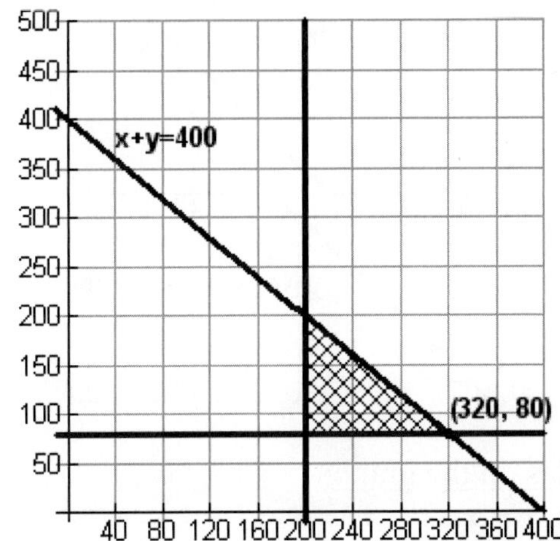

Now, compute the objective function at the vertices:

Vertex	Objective Function $z = 25x + 50y$
(200,80)	$z = 25(200) + 50(80) = 9000$
(200,200)	$z = 25(200) + 50(200) = 15,000$
(320,80)	$z = 25(320) + 50(80) = 12,000$

So, he should sell 200 of each type of ski.

76. Let $x =$ # dozen of crème-filled donuts
$y =$ # dozen of jelly-filled donuts

We must maximize the objective function $z = 1.20x + 1.80y$ subject to the following constraints:

$$\begin{cases} 10 \le x + y \le 30 \\ 3x + 2y \le 120 \\ x \ge 0, \ y \ge 0 \end{cases}$$

The region in this case is:

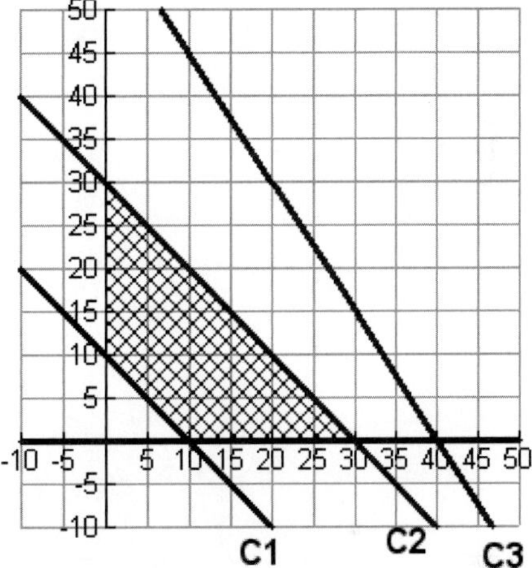

Notes on the graph:
C1: $y = -x + 10$ **C2:** $y = -x + 30$
C3: $y = -\frac{3}{2}x + 60$

Now, compute the objective function at the vertices:

Vertex	Objective Function $z = 1.20x + 1.80y$
(0,10)	18
(0,30)	54
(10,0)	12
(30,0)	36

So, he should sell 0 crème-filled donuts and 30 jelly-filled donuts.

77. The shading should be <u>above</u> the line.	**78.** The shading should not include the actual line (the line should be dashed).
79. True. The line cuts the plane into two half planes, and one must either shade above or below the line.	**80.** True.
81. False A dashed curve is used.	**82.** False. The following linear system has no solution, as is seen graphically below: $\begin{cases} y > x+1 \\ y < x-1 \end{cases}$ No common region. Notes on the graph: **C1:** $y = x+1$ **C2:** $y = x-1$
83. False. The region could be the entire plane (i.e., unconstrained).	
84. True	

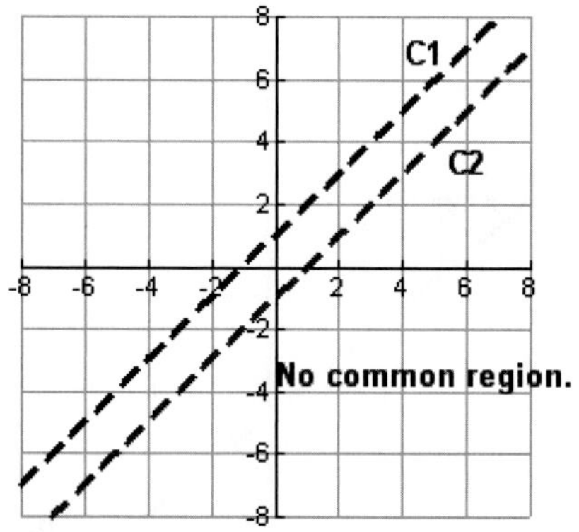

85. Given that $a < b$ and $c < d$, the solution region is a shaded rectangle which includes the upper and left sides – shown below:

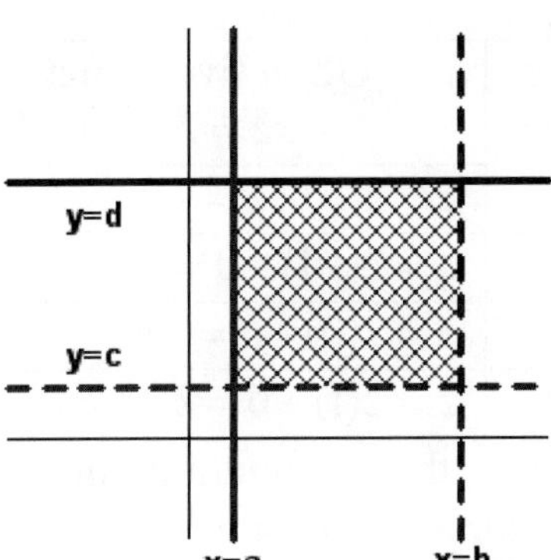

86. In the case when $a > b$ and $c > d$, there is no solution since there is no common region.

87. For any value of *b*, the following system has a solution:

$$\begin{cases} y \le ax + b \\ y \ge -ax + b \end{cases}$$

The solution region occurs in the first and fourth quadrants, and is shown graphically below:

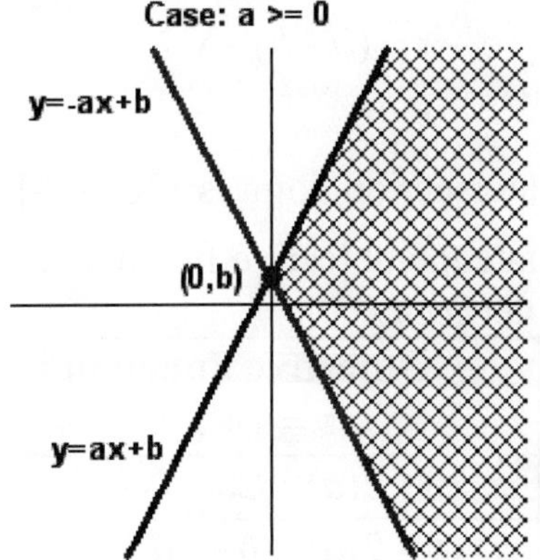

If $a = 0$, then the system becomes:

$$\begin{cases} y \le b \\ y \ge b \end{cases}$$

The solution to this system is the horizontal line $y = b$.

88. For any value of *b*, the following system has a solution:

$$\begin{cases} y \le ax + b \\ y \ge -ax + b \end{cases}$$

The solution region occurs in the second and third quadrants, and is shown graphically below:

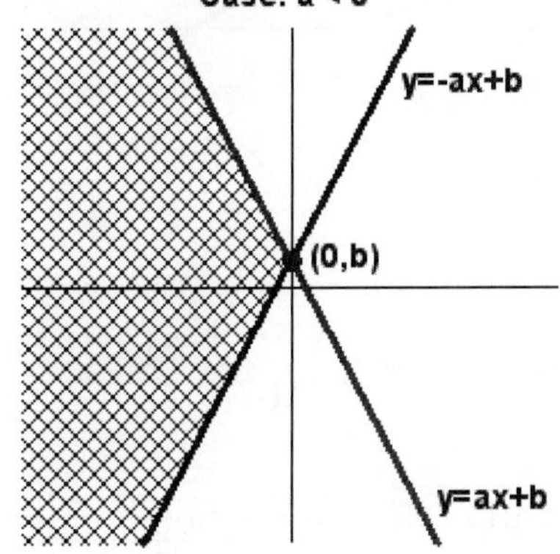

89. Assume that $a > 2$. Then, the region looks like:

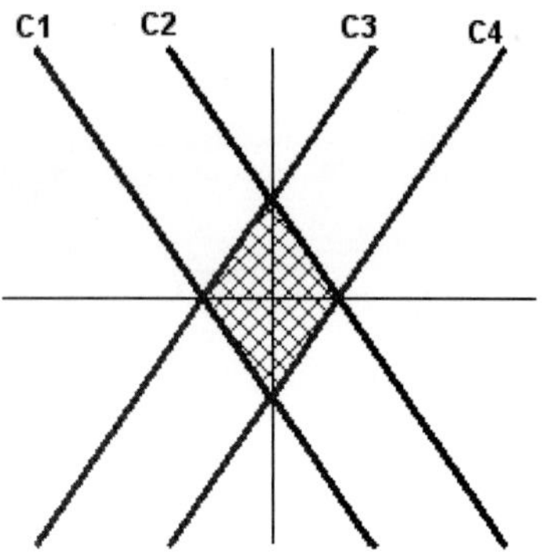

Notes on the graph:

C1: $y = -ax - a$ **C3:** $y = ax + a$

C2: $y = -ax + a$ **C4:** $y = ax - a$

Now, compute the objective function at the vertices:

Vertex	Objective Function $z = 2x + y$
$(0, a)$	$z = 2(0) + a = a$
$(-1, 0)$	$z = 2(-1) + 0 = -2$
$(0, -a)$	$z = 2(0) - a = -a$
$(1, 0)$	$z = 2(1) + 0 = 2$

Since $a > 2$, the maximum in this case would occur at $(0, a)$ and is a.

90. Assume that $a > b > 0$. Then, the region looks like:

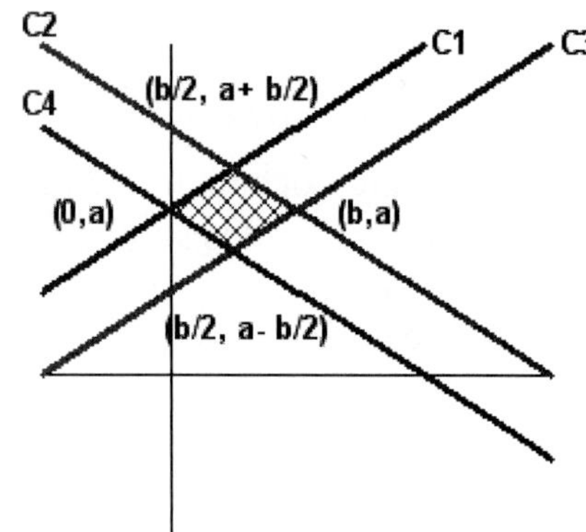

Notes on the graph:

C1: $y = a + x$ **C3:** $y = (a - b) + x$

C2: $y = (a + b) - x$ **C4:** $y = a - x$

We need all of the intersection points since they constitute the vertices:

Intersection of $y = (a + b) - x$ and $y = (a - b) + x$:

$$(a - b) + x = (a + b) - x$$

$$x = b$$

So, the intersection point is (b, a).

Intersection of $y = a - x$ and $y = (a - b) + x$:

$$(a - b) + x = a - x$$

$$x = \tfrac{b}{2}$$

So, the intersection point is $\left(\tfrac{b}{2}, a - \tfrac{b}{2}\right)$.

Intersection of $y = (a + b) - x$ and $y = a + x$:

$$a + x = (a + b) - x$$

$$x = \tfrac{b}{2}$$

So, the intersection point is $\left(\tfrac{b}{2}, a + \tfrac{b}{2}\right)$

Now, evaluate the objective function at each of the vertices:

Vertex	Objective Function $z = x + 2y$
$(0, a)$	$z = 0 + 2(a) = 2a$
(b, a)	$z = b + 2(a) = b + 2a$
$\left(\tfrac{b}{2}, a + \tfrac{b}{2}\right)$	$z = \tfrac{b}{2} + 2\left(a + \tfrac{b}{2}\right) = 2a + \tfrac{3b}{2}$
$\left(\tfrac{b}{2}, a - \tfrac{b}{2}\right)$	$z = \tfrac{b}{2} + 2\left(a - \tfrac{b}{2}\right) = 2a - \tfrac{b}{2}$

Since $a > b > 0$, the maximum in this case would occur at $\left(\tfrac{b}{2}, a + \tfrac{b}{2}\right)$ and is $2a + \tfrac{3b}{2}$.

91.

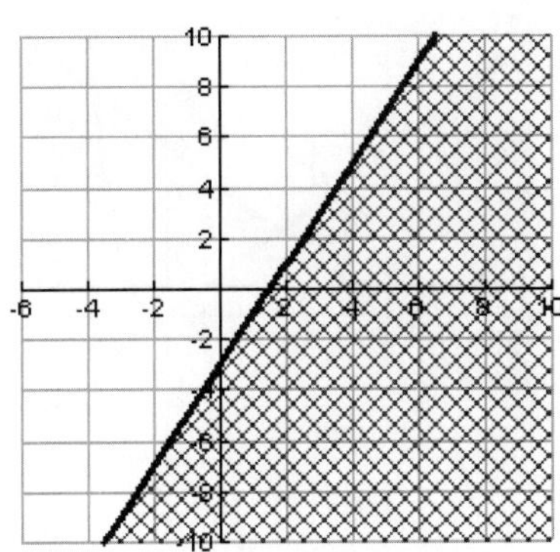

The solid curve is the graph of $y = 2x - 3$.

92.

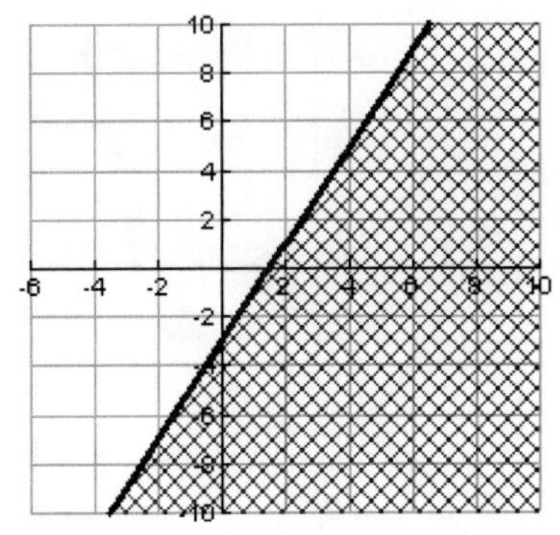

The solid curve is the graph of $y = 2x - 3$.

93.

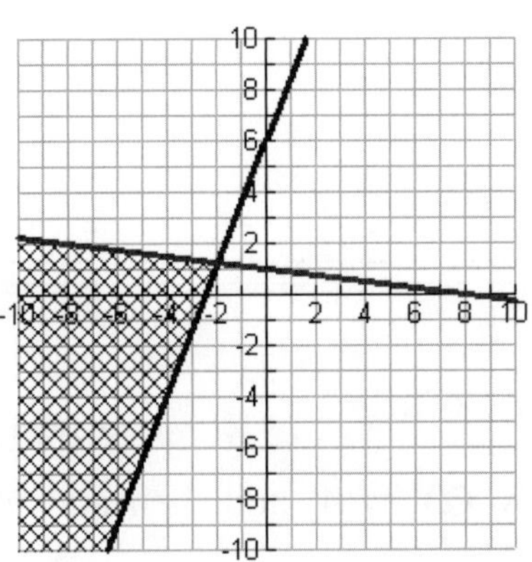

94.

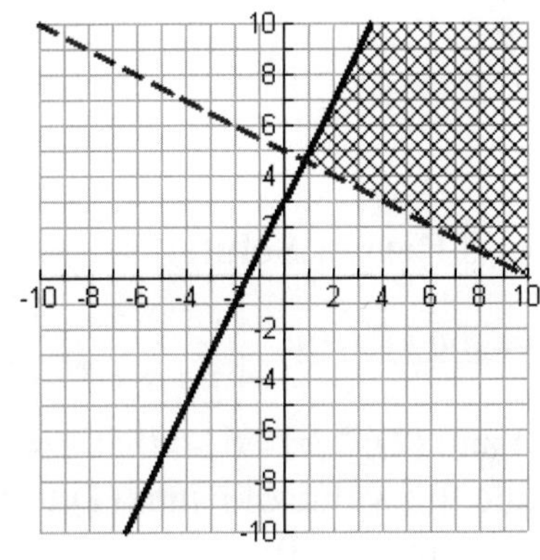

95.

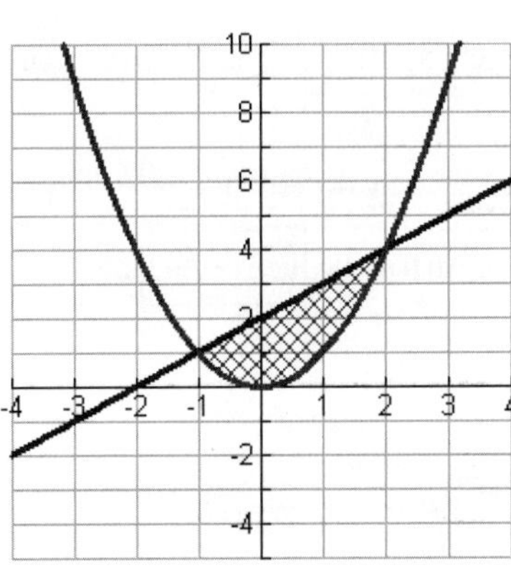

96.

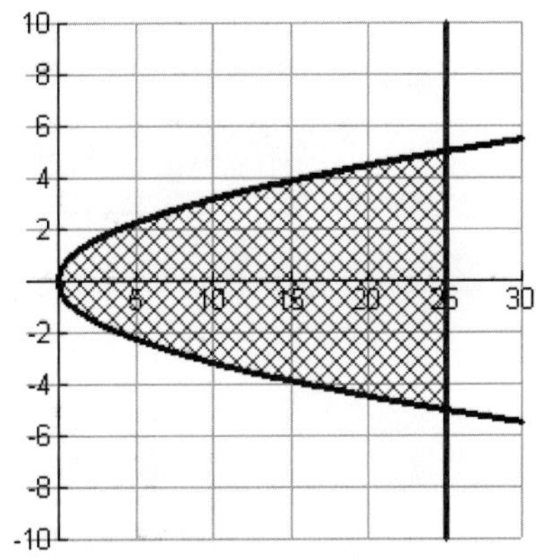

97.

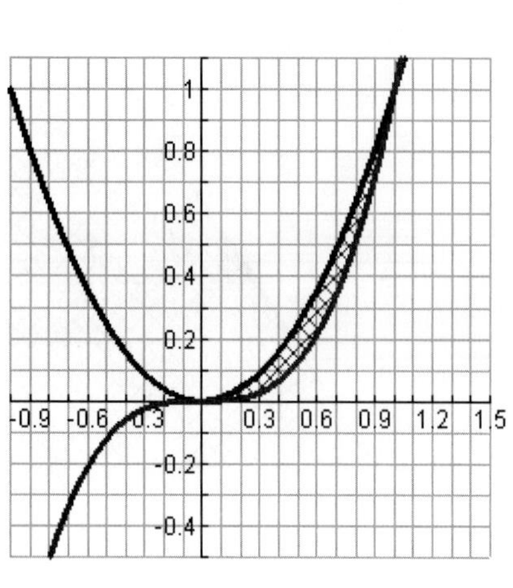

98.

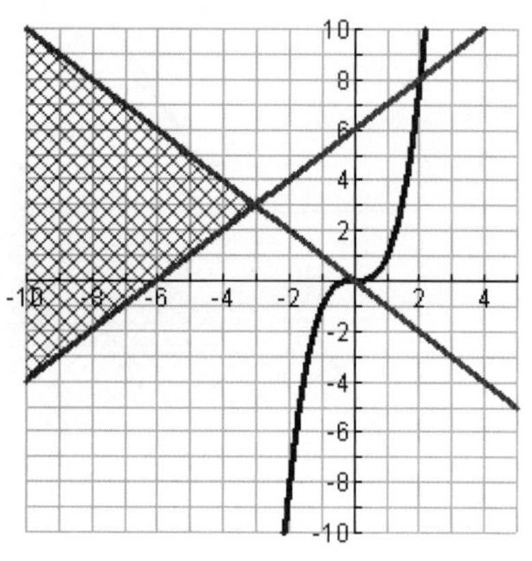

Chapter 8 Review Solutions--

1. Solve the system: $\begin{cases} r-s=3 & \textbf{(1)} \\ r+s=3 & \textbf{(2)} \end{cases}$

Add **(1)** and **(2)**: $2r=6$ so that $r=3$

Substitute this value of r into **(1)** to find s: $3-s=3$ so that $s=0$.

So, the solution is $\boxed{(3,0)}$.

2.

Solve the system: $\begin{cases} 3x+4y=2 & \textbf{(1)} \\ x-y=6 & \textbf{(2)} \end{cases}$

Multiply **(2)** by 4, and then add to **(1)** to eliminate y:

$$7x=26$$

$$x=\tfrac{26}{7} \quad \textbf{(3)}$$

Substitute **(3)** into **(2)** and solve for y:

$$\tfrac{26}{7}-y=6$$

$$y=\tfrac{26}{7}-6=-\tfrac{16}{7}$$

So, the solution is $\boxed{\left(\tfrac{26}{7},-\tfrac{16}{7}\right)}$.

3. Solve the system:

$$\begin{cases} -4x+2y=3 & \textbf{(1)} \\ 4x-y=5 & \textbf{(2)} \end{cases}$$

Add **(1)** and **(2)** to eliminate x:

$$y=8 \quad \textbf{(3)}$$

Substitute **(3)** into **(2)** and solve for x:

$$4x-8=5$$
$$4x=13$$
$$x=\tfrac{13}{4}$$

So, the solution is $\boxed{\left(\tfrac{13}{4},8\right)}$.

4. Solve the system:

$$\begin{cases} 0.25x-0.5y=0.6 & \textbf{(1)} \\ 0.50x+0.25y=0.8 & \textbf{(2)} \end{cases}$$

Multiply **(1)** by 0.5 and then add to **(2)** to eliminate y:

$$0.625x=1.1$$
$$x=1.76 \quad \textbf{(3)}$$

Substitute **(3)** into **(2)** to find y:

$$0.5(1.76)+0.25y=0.8$$
$$0.25y=-0.08$$
$$y=-0.32$$

So, the solution is $\boxed{(1.76,\,-0.32)}$.

5. Solve the system: $\begin{cases} x+y=3 & \textbf{(1)} \\ x-y=1 & \textbf{(2)} \end{cases}$

Solve **(1)** for y: $y=3-x$ **(3)**

Substitute **(3)** into **(2)** and solve for x:

$$x-(3-x)=1$$
$$x-3+x=1$$
$$2x=4 \text{ so that } x=2$$

Substitute this value of x into **(3)** to find y:

$$y=1.$$

So, the solution is $\boxed{(2,\,1)}$.

6. Solve the system: $\begin{cases} 3x+y=4 & \textbf{(1)} \\ 2x+y=1 & \textbf{(2)} \end{cases}$

Solve **(1)** for y: $y=4-3x$ **(3)**

Substitute **(3)** into **(2)** and solve for x:

$$2x+(4-3x)=1$$
$$4-x=1$$
$$x=3$$

Substitute this value of x into **(3)** to find y:

$$y=4-3(3)=-5.$$

So, the solution is $\boxed{(3,\,-5)}$.

7. Solve the system: $\begin{cases} 4c-4d=3 & \textbf{(1)} \\ c+d=4 & \textbf{(2)} \end{cases}$

Solve **(2)** for c: $c=4-d$ **(3)**

Substitute **(3)** into **(1)** and solve for d:

$$4(4-d)-4d=3$$
$$16-8d=3$$
$$8d=13$$
$$d=\tfrac{13}{8}$$

Now, substitute this value of d into **(3)** to find c: $c=4-\tfrac{13}{8}=\tfrac{19}{8}$

So, the solution is $\boxed{\left(\tfrac{19}{8},\tfrac{13}{8}\right)}$.

8. Solve the system: $\begin{cases} 5r+2s=1 & \textbf{(1)} \\ r-s=-3 & \textbf{(2)} \end{cases}$

Solve **(2)** for r: $r=s-3$ **(3)**

Substitute **(3)** into **(1)** and solve for s:

$$5(s-3)+2s=1$$
$$7s-15=1$$
$$7s=16$$
$$s=\tfrac{16}{7}$$

Now, substitute this value of s into **(3)** to find r: $r=\tfrac{16}{7}-3=-\tfrac{5}{7}$

So, the solution is $\boxed{\left(-\tfrac{5}{7},\tfrac{16}{7}\right)}$.

9.

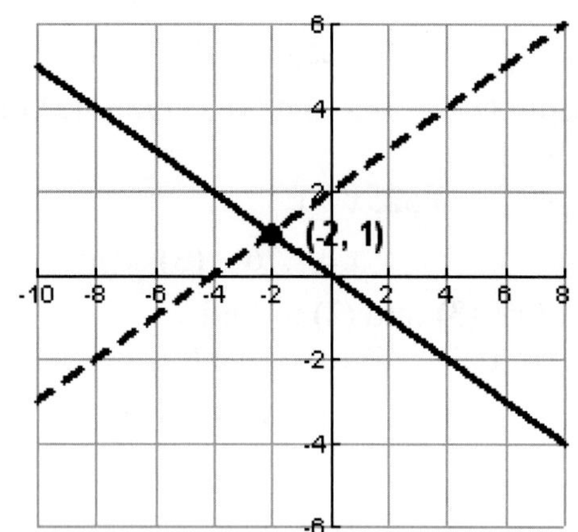

Notes on the graph:
Solid curve: $y=-\frac{1}{2}x$
Dashed curve: $y=\frac{1}{2}x+2$
So, the solution is $(-2,1)$.

10.

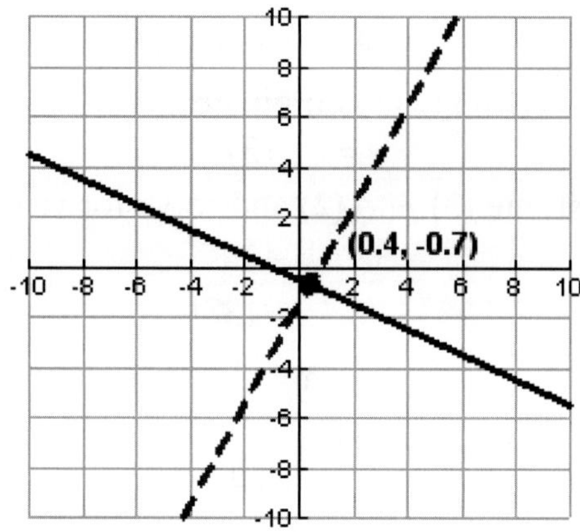

Notes on the graph:
Solid curve: $2x+4y=-2$
Dashed curve: $4x-2y=3$
So, the solution is $(0.4,-0.7)$.

11.

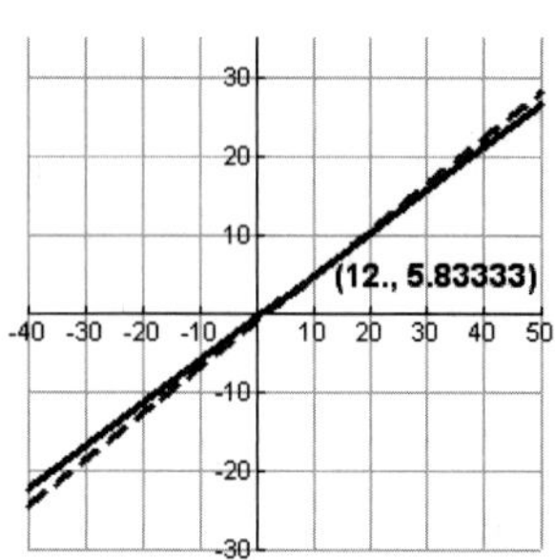

Notes on the graph: (Careful! The curves are very close together in a vicinity of the point of intersection.)
Solid curve: $1.3x-2.4y=1.6$
Dashed curve: $0.7x-1.2y=1.4$
So, the solution is $(12,5.8\overline{3})$.

12.

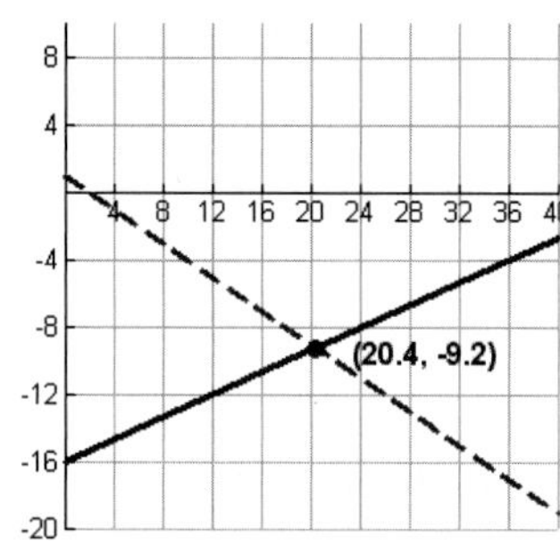

Notes on the graph:
Solid curve: $\frac{1}{4}x-\frac{3}{4}y=12$
Dashed curve: $\frac{1}{2}y+\frac{1}{4}x=\frac{1}{2}$
So, the solution is $(20.4,-9.2)$.

13. Solve the system: $\begin{cases} 5x-3y=21 & \textbf{(1)} \\ -2x+7y=-20 & \textbf{(2)} \end{cases}$ To eliminate x, multiply **(1)** by 2: $10x-6y=42$ **(3)** Multiply **(2)** by 5: $-10x+35y=-100$ **(4)** Add **(3)** and **(4)**: $29y=-58 \Rightarrow y=-2$ Substitute this into **(1)** to find x: $5x=15 \Rightarrow x=3$ So, the solution is $\boxed{(3,-2)}$.	**14.** Solve the system: $\begin{cases} 6x-2y=-2 & \textbf{(1)} \\ 4x+3y=16 & \textbf{(2)} \end{cases}$ To eliminate y, multiply **(1)** by 3: $18x-6y=-6$ **(3)** Multiply **(2)** by 2: $8x+6y=32$ **(4)** Add **(3)** and **(4)**: $26x=26 \Rightarrow x=1$ Substitute this into **(1)** to find x: $6(1)-2y=-2 \Rightarrow y=4$ So, the solution is $\boxed{(1,4)}$.
15. Solve the system: $\begin{cases} 10x-7y=-24 & \textbf{(1)} \\ 7x+4y=1 & \textbf{(2)} \end{cases}$ To eliminate y, multiply **(1)** by 4: $40x-28y=-96$ **(3)** Multiply **(2)** by 7: $49x+28y=7$ **(4)** Add **(3)** and **(4)**: $89x=-89 \Rightarrow x=-1$ Substitute this into **(1)** to find y: $-7y=-14 \Rightarrow y=2$ So, the solution is $\boxed{(-1,2)}$.	**16.** Solve the system: $\begin{cases} \frac{1}{3}x-\frac{2}{9}y=\frac{2}{9} & \text{(1)} \\ \frac{4}{5}x+\frac{3}{4}y=-\frac{3}{4} & \text{(2)} \end{cases}$ First, clear the fractions by multiplying **(1)** by 9 and **(2)** by 20: $3x-2y=2$ **(3)** $16x+15y=-15$ **(4)** To eliminate y, multiply **(3)** by 15: $45x-30y=30$ **(5)** Multiply **(4)** by 2: $32x+30y=-30$ **(6)** Add **(5)** and **(6)**: $77x=0 \Rightarrow x=0$ Substitute this into **(3)** to find y: $-2y=2 \Rightarrow y=-1$ So, the solution is $\boxed{(0,-1)}$.
17. c The intersection point is $(\frac{25}{11},\frac{2}{11})$	**18. b** Subtracting the two equations yields the false statement 0 = 4, so the lines must be parallel.
19. d Multiplying the first equation by 2 reveals that the two equations are equivalent. Hence, the graphs are the same line.	**20. a** The intersection point is $(1,-1)$

21. Let x = number of ml of 6% NaCl.
y = number of ml of 18% NaCl.
Must solve the system:

$$\begin{cases} 0.06x + 0.18y = (0.15)(42) & \textbf{(1)} \\ x + y = 42 & \textbf{(2)} \end{cases}$$

First, for convenience, simplify **(1)** to get the equivalent system:

$$\begin{cases} x + 3y = 105 & \textbf{(3)} \\ x + y = 42 & \textbf{(4)} \end{cases}$$

Multiply **(1)** by -1, and then add to **(2)**:

$$-2y = -63$$

$$y = \tfrac{63}{2} = 31.5 \quad \textbf{(6)}$$

Substitute **(6)** into **(4)** to find x:

$$x + 31.5 = 42$$

$$x = 10.5$$

So, should use approximately 10.5 ml of 6% NaCl and 31.5 ml of 18% NaCl.

22. Let x = gallons used for highway miles
y = gallons used for city miles
Must solve the system:

$$\begin{cases} 32x + 18y = 265 & \textbf{(1)} \\ x + y = 12 & \textbf{(2)} \end{cases}$$

Solve **(2)** for y: $y = 12 - x$ **(3)**

Substitute **(3)** into **(1)** and solve for x:

$$32x + 18(12 - x) = 265$$

$$14x + 216 = 265$$

$$14x = 49$$

$$x = 3.5$$

Now, substitute this value into **(3)** to find y:

$$y = 12 - 3.5 = 8.5$$

Thus, there are
$32(3.5) = 112$ highway miles
$18(8.5) = 153$ city miles.

23. Solve the system:

$$\begin{cases} x + y + z = 1 & \textbf{(1)} \\ x - y - z = -3 & \textbf{(2)} \\ -x + y + z = 3 & \textbf{(3)} \end{cases}$$

Add **(2)** and **(3)**: $0 = 0$ **(4)**

Hence, we know that the system has infinitely many solutions.
Add **(1)** and **(2)**:

$$2x = -2 \text{ so that } x = -1 \quad \textbf{(5)}$$

Substitute this value of x into **(1)**:

$$-1 + y + z = 1$$

$$y + z = 2 \quad \textbf{(6)}$$

Let $z = a$. Substitute this into **(6)** to find z:

$$y = -a + 2$$

Thus, the solutions are:

$$\boxed{x = -1,\ y = -a + 2,\ z = a}.$$

24. Solve the system:

$$\begin{cases} x - 2y + z = 3 & \textbf{(1)} \\ 2x - y + z = -4 & \textbf{(2)} \\ 3x - 3y - 5z = 2 & \textbf{(3)} \end{cases}$$

Step 1: Obtain a system of 2 equations in 2 unknowns by eliminating the same variable in two pairs of equations.

Multiply **(1)** by -1, and then add to **(2)** to eliminate z:

$$x + y = -7 \quad \textbf{(4)}$$

Next, multiply **(2)** by 5 and add to **(3)** to eliminate z:

$$13x - 8y = -18 \quad \textbf{(5)}$$

These steps yield the following system:

$$(*) \begin{cases} x + y = -7 & \textbf{(4)} \\ 13x - 8y = -18 & \textbf{(5)} \end{cases}$$

Step 2: Solve system **(*)** from Step 1.

Solve **(4)** for y: $y = -x - 7$ **(6)**

Substitute **(6)** into **(5)** and solve for x:

$$13x + 8(x + 7) = -18$$

$$21x + 56 = -18$$

$$x \cong -3.524$$

Substitute this value of x back into **(6)** to find y: $y = -(-3.524) - 7 \cong -3.476$

Step 3: Use the solution of the system in Step 2 to find the value of the third variable in the original system.

Substitute the above values of x and y into **(1)** to find z:

$$-3.524 - 2(-3.476) + z = 3$$

$$z \cong -0.429$$

Thus, the approximate solution is:

$$\boxed{x = -3.524, \; y = -3.476, \; z = -0.429}.$$

25. Solve the system:

$$\begin{cases} x + y + z = 7 & \textbf{(1)} \\ x - y - z = 17 & \textbf{(2)} \\ y + z = 5 & \textbf{(3)} \end{cases}$$

Step 1: Obtain a system of 2 equations in 2 unknowns by eliminating the same variable in two pairs of equations.

Subtract **(1) - (2)** and simplify:

$$y + z = -5 \quad \textbf{(4)}$$

Solve the system:

$$(*) \begin{cases} y + z = 5 & \textbf{(3)} \\ y + z = -5 & \textbf{(4)} \end{cases}$$

Note that equating **(3)** and **(4)** yields the false statement 5 = -5. Hence, the system has $\boxed{\text{no solution}}$.

26. Solve the system:

$$\begin{cases} x + z = 3 & \textbf{(1)} \\ -x + y - z = -1 & \textbf{(2)} \\ x + y + z = 5 & \textbf{(3)} \end{cases}$$

Step 1: Obtain a system of 2 equations in 2 unknowns by eliminating the same variable in two pairs of equations.

Add **(2)** and **(3)**:

$$2y = 4 \text{ so that } y = 2 \quad \textbf{(4)}$$

Now, substitute **(4)** into **(2)** and consider **(1)** and **(2)** together. That is, consider the

$$2 \times 2 \text{ system: } \begin{cases} x + z = 3 & \textbf{(5)} \\ -x - z = -3 & \textbf{(6)} \end{cases}$$

Add **(5)** and **(6)**: $0 = 0$

Hence, we know that the system has infinitely many solutions. To determine them, let $x = a$. Substitute this into **(5)** to find z: $a + z = 3 \Rightarrow z = 3 - a$

Thus, the solutions are:

$$\boxed{x = a, \; y = 2, \; z = 3 - a}.$$

27. Since we are given that the points $(16,2)$, $(40,6)$, and $(65,4)$, the system that must be solved is:

$$\begin{cases} 2 = a(16)^2 + b(16) + c \\ 6 = a(40)^2 + b(40) + c \\ 4 = a(65)^2 + b(65) + c \end{cases}$$

which is equivalent to

$$\begin{cases} 2 = \;\; 256a + 16b + c & \textbf{(1)} \\ 6 = 1600a + 40b + c & \textbf{(2)} \\ 4 = 4225a + 65b + c & \textbf{(3)} \end{cases}$$

Step 1: Obtain a 2×2 system:

Multiply **(1)** by -1, and add to **(2)**:

$$1344a + 24b = 4 \quad \textbf{(4)}$$

Multiply **(1)** by -1, and add to **(3)**:

$$3969a + 49b = 2 \quad \textbf{(5)}$$

These steps yield the following 2×2 system:

Step 2: Solve the system in Step 1.

$$\begin{cases} 1344a + 24b = 4 & \textbf{(4)} \\ 3969a + 49b = 2 & \textbf{(5)} \end{cases}$$

Multiply **(4)** by -49 and **(5)** by 24. Then, add them together:

$$29,400a = -148$$

$$a \cong -0.0050 \quad \textbf{(6)}$$

Substitute **(6)** into **(4)**:

$$1344(-0.005) + 24b = 4$$

$$b \cong 0.4486 \quad \textbf{(7)}$$

Step 3: Find values of remaining variables.

Substitute **(6)** and **(7)** into **(1)**:

$$256(-0.005) + 16(0.447) + c = 2$$

$$c \cong -3.8884$$

Thus, the polynomial has the approximate equation $\boxed{y = -0.0050x^2 + 0.4486x - 3.8884}$.

28. Let x = amount in IRA
y = amount in mutual fund
z = amount in stock

The system we must solve is:

$$\begin{cases} x = y + 4000 \\ x + y + z = 20,000 \\ 0.045x + 0.08y + 0.12z = 1525 \end{cases}$$

We first simplify the third equation by multiplying 100 to get the following equivalent system:

$$\begin{cases} x = y + 4000 & \textbf{(1)} \\ x + y + z = 20,000 & \textbf{(2)} \\ 45x + 80y + 120z = 1,525,000 & \textbf{(3)} \end{cases}$$

Step 1: Obtain a 2×2 system:

Substitute **(1)** into both **(2)** and **(3)** to obtain the system

$$\begin{cases} 45(y + 4000) + 80y + 120z = 1,525,000 \\ (y + 4000) + y + z = 20,000 \end{cases}$$

which is equivalent to

$$\begin{cases} 125y + 120z = 1,345,000 & \textbf{(4)} \\ 2y + z = 16,000 & \textbf{(5)} \end{cases}$$

Step 2: Solve the system in Step 1.

Multiply **(5)** by -120 and add to **(4)** to find y:

$$-115y = -575,000$$

$$y = 5000 \quad \textbf{(6)}$$

Substitute **(6)** into **(1)** to find x:

$$x = 5000 + 4000 = 9000 \quad \textbf{(7)}$$

Substitute **(6)** and **(7)** into **(2)** to find z:

$$5000 + 9000 + z = 20,000$$

$$z = 6000$$

Thus, the following approximate allocation of funds should be made:

\$9000 in IRA
\$5000 in mutual fund
\$6000 in stock

29. $\left[\begin{array}{cc\|c} 5 & 7 & 2 \\ 3 & -4 & -2 \end{array}\right]$	**30.** $\left[\begin{array}{cc\|c} 2.3 & -4.5 & 6.8 \\ -0.4 & 2.1 & -9.1 \end{array}\right]$
31. $\left[\begin{array}{ccc\|c} 2 & 0 & -1 & 3 \\ 0 & 1 & -3 & -2 \\ 1 & 0 & 4 & -3 \end{array}\right]$	**32.** $\left[\begin{array}{ccc\|c} -1 & 2 & 3 & 1 \\ 3 & -2 & 4 & -2 \\ 1 & -1 & -4 & 0 \end{array}\right]$
33. No, it is not reduced row echelon form.	**34.** Yes
35. No, condition 2 is not met.	**36.** No, condition 3 is not met.
37. $\left[\begin{array}{cc\|c} 1 & -2 & 1 \\ 0 & 1 & -1 \end{array}\right]$	**38.** $\left[\begin{array}{cc\|c} 1 & 4 & 1 \\ 0 & -10 & 1 \end{array}\right]$
39. $\left[\begin{array}{ccc\|c} 1 & -4 & 3 & -1 \\ 0 & -2 & 3 & -2 \\ 0 & 1 & -4 & 8 \end{array}\right]$	**40.** $\left[\begin{array}{cccc\|c} -2 & 0 & -4 & -9 & -2 \\ 0 & 2 & 2 & 3 & -2 \\ 0 & 0 & 1 & -2 & 4 \\ 0 & -1 & 4 & -5 & 7 \end{array}\right]$

41.

$$\left[\begin{array}{cc|c} 1 & 3 & 0 \\ 3 & 4 & 1 \end{array}\right] \xrightarrow{R_1 \leftrightarrow R_2} \left[\begin{array}{cc|c} 3 & 4 & 1 \\ 1 & 3 & 0 \end{array}\right] \xrightarrow{R_1 - 3R_2 \to R_2} \left[\begin{array}{cc|c} 3 & 4 & 1 \\ 0 & -5 & 1 \end{array}\right]$$

$$\xrightarrow[-\frac{1}{5}R_2 \to R_2]{\frac{1}{3}R_1 \to R_1} \left[\begin{array}{cc|c} 1 & \frac{4}{3} & \frac{1}{3} \\ 0 & 1 & -\frac{1}{5} \end{array}\right] \xrightarrow{R_1 - \frac{4}{3}R_2 \to R_1} \left[\begin{array}{cc|c} 1 & 0 & \frac{3}{5} \\ 0 & 1 & -\frac{1}{5} \end{array}\right]$$

42.

$$\left[\begin{array}{ccc|c} 1 & 2 & -1 & 0 \\ 0 & 1 & 1 & -1 \\ -2 & 0 & 1 & -2 \end{array}\right] \xrightarrow{R_2 + 2R_1 \to R_2} \left[\begin{array}{ccc|c} 1 & 2 & -1 & 0 \\ 0 & 1 & 1 & -1 \\ 0 & 4 & -1 & -2 \end{array}\right] \xrightarrow{R_3 - 4R_2 \to R_3} \left[\begin{array}{ccc|c} 1 & 2 & -1 & 0 \\ 0 & 1 & 1 & -1 \\ 0 & 0 & -5 & 2 \end{array}\right]$$

$$\xrightarrow{-\frac{1}{5}R_3 \to R_3} \left[\begin{array}{ccc|c} 1 & 2 & -1 & 0 \\ 0 & 1 & 1 & -1 \\ 0 & 0 & 1 & -\frac{2}{5} \end{array}\right] \xrightarrow[R_1 - 2R_2 + R_3 \to R_1]{R_2 - R_3 \to R_2} \left[\begin{array}{ccc|c} 1 & 0 & 0 & \frac{4}{5} \\ 0 & 1 & 0 & -\frac{3}{5} \\ 0 & 0 & 1 & -\frac{2}{5} \end{array}\right]$$

43.

$$\left[\begin{array}{ccc|c} 4 & 1 & -2 & 0 \\ 1 & 0 & -1 & 0 \\ -2 & 1 & 1 & 12 \end{array}\right] \xrightarrow[R_1+2R_3\to R_3]{R_1-4R_2\to R_2} \left[\begin{array}{ccc|c} 4 & 1 & -2 & 0 \\ 0 & 1 & 2 & 0 \\ 0 & 3 & 0 & 24 \end{array}\right] \xrightarrow{R_3-3R_2\to R_2} \left[\begin{array}{ccc|c} 4 & 1 & -2 & 0 \\ 0 & 0 & -6 & 24 \\ 0 & 3 & 0 & 24 \end{array}\right]$$

$$\xrightarrow{R_2\leftrightarrow R_3} \left[\begin{array}{ccc|c} 4 & 1 & -2 & 0 \\ 0 & 3 & 0 & 24 \\ 0 & 0 & -6 & 24 \end{array}\right] \xrightarrow[-\frac{1}{6}R_3\to R_3]{\frac{1}{4}R_1\to R_1,\ \frac{1}{3}R_2\to R_2} \left[\begin{array}{ccc|c} 1 & \frac{1}{4} & -\frac{1}{2} & 0 \\ 0 & 1 & 0 & 8 \\ 0 & 0 & 1 & -4 \end{array}\right] \xrightarrow{R_1-\frac{1}{4}R_2+\frac{1}{2}R_3\to R_1} \left[\begin{array}{ccc|c} 1 & 0 & 0 & -4 \\ 0 & 1 & 0 & 8 \\ 0 & 0 & 1 & -4 \end{array}\right]$$

44.

$$\left[\begin{array}{ccc|c} 2 & 3 & 2 & 1 \\ 0 & -1 & 1 & -2 \\ 1 & 1 & -1 & 6 \end{array}\right] \xrightarrow{R_1-2R_3\to R_3} \left[\begin{array}{ccc|c} 2 & 3 & 2 & 1 \\ 0 & -1 & 1 & -2 \\ 0 & 1 & 4 & -11 \end{array}\right] \xrightarrow{R_2+R_3\to R_3} \left[\begin{array}{ccc|c} 2 & 3 & 2 & 1 \\ 0 & 1 & -1 & -2 \\ 0 & 0 & 5 & -13 \end{array}\right]$$

$$\xrightarrow[\frac{1}{5}R_3\to R_3]{\frac{1}{2}R_1\to R_1,\ -R_2\to R_2} \left[\begin{array}{ccc|c} 1 & \frac{3}{2} & 1 & \frac{1}{2} \\ 0 & 1 & -1 & 2 \\ 0 & 0 & 1 & -\frac{13}{5} \end{array}\right] \xrightarrow[R_1-\frac{3}{2}R_2+R_3\to R_1]{R_2+R_3\to R_2} \left[\begin{array}{ccc|c} 1 & 0 & 0 & 4 \\ 0 & 1 & 0 & -\frac{3}{5} \\ 0 & 0 & 1 & -\frac{13}{5} \end{array}\right]$$

45. First, reduce the augmented matrix down to row echelon form:

$$\left[\begin{array}{cc|c} 3 & -2 & 2 \\ -2 & 4 & 1 \end{array}\right] \xrightarrow{-\frac{1}{2}R_2\to R_2} \left[\begin{array}{cc|c} 3 & -2 & 2 \\ 1 & -2 & -\frac{1}{2} \end{array}\right] \xrightarrow{R_1-3R_2\to R_2} \left[\begin{array}{cc|c} 3 & -2 & 2 \\ 0 & 4 & \frac{7}{2} \end{array}\right]$$

$$\xrightarrow[\frac{1}{4}R_2\to R_2]{\frac{1}{3}R_1\to R_1} \left[\begin{array}{cc|c} 1 & -\frac{2}{3} & \frac{2}{3} \\ 0 & 1 & \frac{7}{8} \end{array}\right]$$

Now, from this we see that $y=\frac{7}{8}$, and then substituting this value into the equation obtained from the first row yields:

$$x-\tfrac{2}{3}\left(\tfrac{7}{8}\right)=\tfrac{2}{3}$$
$$x=\tfrac{5}{4}$$

Hence, the solution is $\boxed{x=\frac{5}{4},\ y=\frac{7}{8}}$.

46. First, reduce the augmented matrix down to reduced row echelon form:

$$\left[\begin{array}{cc|c} 2 & -7 & 22 \\ 1 & 5 & -23 \end{array}\right] \xrightarrow{R_2-2R_1\to R_2} \left[\begin{array}{cc|c} 2 & -7 & 22 \\ 0 & -17 & 68 \end{array}\right] \xrightarrow[-\frac{1}{17}R_2\to R_2]{\frac{1}{2}R_1\to R_1} \left[\begin{array}{cc|c} 1 & -\frac{7}{2} & 11 \\ 0 & 1 & -4 \end{array}\right]$$

$$\xrightarrow{R_1+\frac{7}{2}R_2\to R_1} \left[\begin{array}{cc|c} 1 & 0 & -3 \\ 0 & 1 & -4 \end{array}\right]$$

Hence, the solution is $\boxed{x=-3,\ y=-4}$.

47. First, reduce the augmented matrix down to reduced row echelon form:

$$\left[\begin{array}{cc|c} 5 & -1 & 9 \\ 1 & 4 & 6 \end{array}\right] \xrightarrow[\frac{1}{5}R_1 \to R_1]{R_1 - 5R_2 \to R_2} \left[\begin{array}{cc|c} 1 & -\frac{1}{5} & \frac{9}{5} \\ 0 & -21 & -21 \end{array}\right] \xrightarrow{-\frac{1}{21}R_2 \to R_2} \left[\begin{array}{cc|c} 1 & -\frac{1}{5} & \frac{9}{5} \\ 0 & 1 & 1 \end{array}\right]$$

$$\xrightarrow{R_1 + \frac{1}{5}R_2 \to R_1} \left[\begin{array}{cc|c} 1 & 0 & 2 \\ 0 & 1 & 1 \end{array}\right]$$

Hence, the solution is $\boxed{x = 2, y = 1}$.

48. First, reduce the augmented matrix down to reduced row echelon form:

$$\left[\begin{array}{cc|c} 8 & 7 & 10 \\ -3 & 5 & 42 \end{array}\right] \xrightarrow[\frac{1}{8}R_1 \to R_1]{\frac{1}{3}R_2 \to R_2} \left[\begin{array}{cc|c} 1 & \frac{7}{8} & \frac{5}{4} \\ -1 & \frac{5}{3} & 14 \end{array}\right] \xrightarrow{R_1 + R_2 \to R_2} \left[\begin{array}{cc|c} 1 & \frac{7}{8} & \frac{5}{4} \\ 0 & \frac{61}{24} & \frac{61}{4} \end{array}\right]$$

$$\xrightarrow{\frac{24}{61}R_2 \to R_2} \left[\begin{array}{cc|c} 1 & \frac{7}{8} & \frac{5}{4} \\ 0 & 1 & 6 \end{array}\right] \xrightarrow{R_1 - \frac{7}{8}R_2 \to R_1} \left[\begin{array}{cc|c} 1 & 0 & -4 \\ 0 & 1 & 6 \end{array}\right]$$

Hence, the solution is $\boxed{x = -4, y = 6}$.

49. First, reduce the augmented matrix down to row echelon form:

$$\left[\begin{array}{ccc|c} 1 & -2 & 1 & 3 \\ 2 & -1 & 1 & -4 \\ 3 & -3 & -5 & 2 \end{array}\right] \xrightarrow[R_2 - 2R_1 \to R_2]{R_3 - 3R_1 \to R_3} \left[\begin{array}{ccc|c} 1 & -2 & 1 & 3 \\ 0 & 3 & -1 & -10 \\ 0 & 3 & -8 & -7 \end{array}\right] \xrightarrow{R_2 - R_3 \to R_3} \left[\begin{array}{ccc|c} 1 & -2 & 1 & 3 \\ 0 & 3 & -1 & -10 \\ 0 & 0 & 7 & -3 \end{array}\right]$$

$$\xrightarrow[\frac{1}{3}R_2 \to R_2]{\frac{1}{7}R_3 \to R_3} \left[\begin{array}{ccc|c} 1 & -2 & 1 & 3 \\ 0 & 1 & -\frac{1}{3} & -\frac{10}{3} \\ 0 & 0 & 1 & -\frac{3}{7} \end{array}\right]$$

Now, from this we see that $z = -\frac{3}{7}$. Then, substituting this value into the equation obtained from the second row yields:

$$y - \tfrac{1}{3}(-\tfrac{3}{7}) = -\tfrac{10}{3} \text{ so that } y = -\tfrac{73}{21}$$

Finally, substituting these values of y and z into the equation obtained from the first row subsequently yields:

$$x - 2(-\tfrac{73}{21}) + (-\tfrac{3}{7}) = 3 \text{ so that } x = -\tfrac{74}{21}.$$

Hence, the solution is $\boxed{x = -\frac{74}{21},\ y = -\frac{73}{21},\ z = -\frac{3}{7}}$.

50. We reduce the matrix down to reduced row echelon form.

$$\left[\begin{array}{ccc|c} 3 & -1 & 4 & 18 \\ 5 & 2 & -1 & -20 \\ 1 & 7 & -6 & -38 \end{array}\right] \xrightarrow[R_2-5R_3\to R_2]{R_1-3R_3\to R_1} \left[\begin{array}{ccc|c} 0 & -22 & 22 & 132 \\ 0 & -33 & 29 & 170 \\ 1 & 7 & -6 & -38 \end{array}\right] \xrightarrow{-\frac{1}{22}R_1\to R_1} \left[\begin{array}{ccc|c} 0 & 1 & -1 & -6 \\ 0 & -33 & 29 & 170 \\ 1 & 7 & -6 & -38 \end{array}\right]$$

$$\xrightarrow{R_2+33R_1\to R_2} \left[\begin{array}{ccc|c} 0 & 1 & -1 & -6 \\ 0 & 0 & -4 & -28 \\ 1 & 7 & -6 & -38 \end{array}\right] \xrightarrow[-\frac{1}{4}R_2\to R_2]{R_1\leftrightarrow R_3} \left[\begin{array}{ccc|c} 1 & 7 & -6 & -38 \\ 0 & 0 & 1 & 7 \\ 0 & 1 & -1 & 6 \end{array}\right]$$

$$\xrightarrow{R_2+R_3\to R_3} \left[\begin{array}{ccc|c} 1 & 7 & -6 & -38 \\ 0 & 0 & 1 & 7 \\ 0 & 1 & 0 & 1 \end{array}\right] \xrightarrow[R_2\leftrightarrow R_3]{R_1+6R_2-7R_3\to R_1} \left[\begin{array}{ccc|c} 1 & 0 & 0 & -3 \\ 0 & 1 & 0 & 1 \\ 0 & 0 & 1 & 7 \end{array}\right]$$

Hence, the solution is $\boxed{x=-3, y=1, z=7}$.

51. We reduce the matrix down to reduced row echelon form:

$$\left[\begin{array}{ccc|c} 1 & -4 & 10 & -61 \\ 3 & -5 & 8 & -52 \\ -5 & 1 & -2 & 8 \end{array}\right] \xrightarrow[R_3+5R_1\to R_3]{R_2-3R_1\to R_2} \left[\begin{array}{ccc|c} 1 & -4 & 10 & -61 \\ 0 & 7 & -22 & 131 \\ 0 & -19 & 48 & -297 \end{array}\right] \xrightarrow[-\frac{1}{19}R_3\to R_3]{\frac{1}{7}R_2\to R_2} \left[\begin{array}{ccc|c} 1 & -4 & 10 & -61 \\ 0 & 1 & -\frac{22}{7} & \frac{131}{7} \\ 0 & 1 & -\frac{48}{19} & \frac{297}{19} \end{array}\right]$$

$$\xrightarrow{R_2-R_3\to R_3} \left[\begin{array}{ccc|c} 1 & -4 & 10 & -61 \\ 0 & 1 & -\frac{22}{7} & \frac{131}{7} \\ 0 & 0 & -\frac{82}{19(7)} & \frac{410}{19(7)} \end{array}\right] \xrightarrow{-\frac{19(7)}{82}R_3\to R_3} \left[\begin{array}{ccc|c} 1 & -4 & 10 & -61 \\ 0 & 1 & -\frac{22}{7} & \frac{131}{7} \\ 0 & 0 & 1 & -5 \end{array}\right] \xrightarrow{R_2+\frac{22}{7}R_3\to R_2}$$

$$\left[\begin{array}{ccc|c} 1 & -4 & 10 & -61 \\ 0 & 1 & 0 & 3 \\ 0 & 0 & 1 & -5 \end{array}\right] \xrightarrow{R_1+4R_2-10R_3\to R_1} \left[\begin{array}{ccc|c} 1 & 0 & 0 & 1 \\ 0 & 1 & 0 & 3 \\ 0 & 0 & 1 & -5 \end{array}\right]$$

Hence, the solution is $\boxed{x=1, y=3, z=-5}$.

52. We reduce the matrix down to reduced row echelon form:

$$\left[\begin{array}{ccc|c} 4 & -2 & 5 & 17 \\ 1 & 6 & -3 & -\frac{17}{2} \\ -2 & 5 & 1 & 2 \end{array}\right] \xrightarrow[R_3+2R_2\to R_3]{R_1-4R_2\to R_1} \left[\begin{array}{ccc|c} 0 & -26 & 17 & 51 \\ 1 & 6 & -3 & -\frac{17}{2} \\ 0 & 17 & -5 & -15 \end{array}\right] \xrightarrow{R_1\leftrightarrow R_2} \left[\begin{array}{ccc|c} 1 & 6 & -3 & -\frac{17}{2} \\ 0 & -26 & 17 & 51 \\ 0 & 17 & -5 & -15 \end{array}\right]$$

$$\xrightarrow[\frac{1}{17}R_3\to R_3]{-\frac{1}{26}R_2\to R_2} \left[\begin{array}{ccc|c} 1 & 6 & -3 & -\frac{17}{2} \\ 0 & 1 & -\frac{17}{26} & -\frac{51}{26} \\ 0 & 1 & -\frac{5}{17} & -\frac{15}{17} \end{array}\right] \xrightarrow{R_2-R_3\to R_3} \left[\begin{array}{ccc|c} 1 & 6 & -3 & -\frac{17}{2} \\ 0 & 1 & -\frac{17}{26} & -\frac{51}{26} \\ 0 & 0 & -\frac{159}{17(26)} & -\frac{477}{17(26)} \end{array}\right] \xrightarrow{-\frac{26(17)}{159}R_3\to R_3}$$

$$\left[\begin{array}{ccc|c} 1 & 6 & -3 & -\frac{17}{2} \\ 0 & 1 & -\frac{17}{26} & -\frac{51}{26} \\ 0 & 0 & 1 & 3 \end{array}\right] \xrightarrow{R_2+\frac{17}{26}R_3\to R_2} \left[\begin{array}{ccc|c} 1 & 6 & -3 & -\frac{17}{2} \\ 0 & 1 & 0 & 0 \\ 0 & 0 & 1 & 3 \end{array}\right] \xrightarrow{R_1-6R_2+3R_3\to R_1} \left[\begin{array}{ccc|c} 1 & 0 & 0 & \frac{1}{2} \\ 0 & 1 & 0 & 0 \\ 0 & 0 & 1 & 3 \end{array}\right]$$

Hence, the solution is $\boxed{x=\frac{1}{2}, y=0, z=3}$.

53. First, reduce the augmented matrix down to row echelon form:

$$\left[\begin{array}{ccc|c} 3 & 1 & 1 & -4 \\ 1 & -2 & 1 & -6 \end{array}\right] \xrightarrow{R_1-3R_2\to R_2} \left[\begin{array}{ccc|c} 3 & 1 & 1 & -4 \\ 0 & 7 & -2 & 14 \end{array}\right] \xrightarrow[\frac{1}{7}R_2\to R_2]{\frac{1}{3}R_1\to R_1} \left[\begin{array}{ccc|c} 1 & \frac{1}{3} & \frac{1}{3} & -\frac{4}{3} \\ 0 & 1 & -\frac{2}{7} & 2 \end{array}\right]$$

Since there are two rows and three unknowns, we know from the above calculation that there are infinitely many solutions to this system. (Note: The only other possibility in such case would be that there was no solution, which would occur if one of the rows yielded a false statement.) To find the solutions, let $z=a$. Then, from row 2, we observe that $y-\frac{2}{7}a=2$ so that $y=\frac{2}{7}a+2$. Then, substituting these values of y and z into the equation obtained from row 1, we see that

$$x+\tfrac{1}{3}\left(\tfrac{2}{7}a+2\right)+\tfrac{1}{3}a=-\tfrac{4}{3}$$
$$x=-\tfrac{3}{7}a-2$$

Hence, the solutions are $\boxed{x=-\frac{3}{7}a-2,\ y=\frac{2}{7}a+2,\ z=a}$.

54. First, reduce the augmented matrix down to row echelon form:

$$\left[\begin{array}{ccc|c} 2 & -1 & 3 & 6 \\ 3 & 2 & -1 & 12 \end{array}\right] \xrightarrow[\frac{1}{3}R_2 \to R_2]{\frac{1}{2}R_1 \to R_1} \left[\begin{array}{ccc|c} 1 & -\frac{1}{2} & \frac{3}{2} & 3 \\ 1 & \frac{2}{3} & -\frac{1}{3} & 4 \end{array}\right] \xrightarrow{R_1 - R_2 \to R_2} \left[\begin{array}{ccc|c} 1 & -\frac{1}{2} & \frac{3}{2} & 3 \\ 0 & -\frac{7}{6} & \frac{11}{6} & -1 \end{array}\right]$$

Since there are two rows and three unknowns, we know from the above calculation that there are infinitely many solutions to this system. (Note: The only other possibility in such case would be that there was no solution, which would occur if one of the rows yielded a false statement.) To find the solutions, let $y = a$. Then, from row 2, $-\frac{7}{6}a + \frac{11}{6}z = -1 \Rightarrow z = \frac{7}{11}a - \frac{6}{11}$. Substituting these values into the equation obtained from row 1, we see that

$$x - \tfrac{1}{2}a + \tfrac{3}{2}\left(\tfrac{7}{11}a - \tfrac{6}{11}\right) = 3$$

$$x + \tfrac{10}{22}a = \tfrac{42}{11}$$

$$x = -\tfrac{5}{11}a + \tfrac{42}{11}$$

Hence, the solutions are $\boxed{x = -\frac{5}{11}a + \frac{42}{11}, y = a, z = \frac{7}{11}a - \frac{6}{11}}$.

55. From the given information, we must solve the following system:

$$\begin{cases} 2 = a(16)^2 + b(16) + c \\ 6 = a(40)^2 + b(40) + c \\ 4 = a(65)^2 + b(65) + c \end{cases} \text{ is equivalent to } \begin{cases} 2 = 256a + 16b + c \\ 6 = 1600a + 40b + c \\ 4 = 4225a + 65b + c \end{cases}$$

Identify $A = \begin{bmatrix} 256 & 16 & 1 \\ 1600 & 40 & 1 \\ 4225 & 65 & 1 \end{bmatrix}$ and $B = \begin{bmatrix} 2 \\ 6 \\ 4 \end{bmatrix}$, and rewrite this system in matrix form:

$$A\begin{bmatrix} a \\ b \\ c \end{bmatrix} = B$$

The solution of this system is $\begin{bmatrix} a \\ b \\ c \end{bmatrix} = A^{-1}B$. In this case, rather than using Gaussian elimination to compute the inverse, using technology would be beneficial. Indeed, in this case, the above simplifies to:

$$\begin{bmatrix} a \\ b \\ c \end{bmatrix} = A^{-1}B = \begin{bmatrix} -0.00503 \\ 0.44857 \\ -3.888 \end{bmatrix}$$

Thus, the equation of the curve is $\boxed{y = -0.005x^2 + 0.45x - 3.89}$.

56. Let x = amount invested in IRA account (4.5%)
y = amount invested in mutual fund (8%)
z = amount invested in stock (12%)

Solve the system:

$$\begin{cases} x+y+z=20,000 & \textbf{(1)} \\ y=x+3000 & \textbf{(2)} \\ 0.045x+0.08y+0.12z=1877.50 & \textbf{(3)} \end{cases}$$

First, observe that substituting **(2)** into **(1)** and **(3)** to obtain the following 2×2 system:

$$\begin{cases} x+(x+3000)+z=20,000 \\ 0.045x+0.08(x+3000)+0.12z=1877.50 \end{cases}$$ which is equivalent to

$$\begin{cases} 2x+z=17,000 & \textbf{(4)} \\ 0.125x+0.12z=1637.50 & \textbf{(5)} \end{cases}$$

$$\left[\begin{array}{cc|c} 2 & 1 & 17,000 \\ 0.125 & 0.12 & 1637.50 \end{array}\right] \xrightarrow{\frac{1}{0.125}R_2 \to R_2} \left[\begin{array}{cc|c} 2 & 1 & 17,000 \\ 1 & 0.96 & 13,100 \end{array}\right]$$

$$\xrightarrow{R_1-2R_2\to R_2} \left[\begin{array}{cc|c} 2 & 1 & 17,000 \\ 0 & -0.92 & -9200 \end{array}\right] \xrightarrow[-\frac{1}{0.92}R_2\to R_2]{\frac{1}{2}R_1\to R_1} \left[\begin{array}{cc|c} 1 & 0.5 & 8500 \\ 0 & 1 & 10,000 \end{array}\right]$$

Now, from this we see that $z=10,000$. Then, substituting this value into the equation obtained from the second row yields:

$$x+0.5(10,000)=8500 \text{ so that } x=3500$$

Finally, substituting this value of x into **(2)** yields:

$$y=3500+3000=6500.$$

Therefore, he should make the following investments:
\$3500 in IRA, \$6500 in mutual fund, and \$10,000 in stock.

57. Not defined since the matrices have different orders.

58. Not defined since the matrices have different orders.

59. $\begin{bmatrix} 3 & 5 & 2 \\ 7 & 8 & 1 \end{bmatrix}$

60. $\begin{bmatrix} 7 & -1 \\ 9 & 8 \end{bmatrix}$

61. $\begin{bmatrix} 4 & -6 \\ 0 & 2 \end{bmatrix} + \begin{bmatrix} 5 & 2 \\ 9 & 7 \end{bmatrix} = \begin{bmatrix} 9 & -4 \\ 9 & 9 \end{bmatrix}$

62. $\begin{bmatrix} 6 & 0 & 9 \\ 12 & 3 & -3 \end{bmatrix} + \begin{bmatrix} 1 & 5 & -1 \\ 3 & 7 & 2 \end{bmatrix} = \begin{bmatrix} 7 & 5 & 8 \\ 15 & 10 & -1 \end{bmatrix}$

63. $\begin{bmatrix} 10 & 4 \\ 18 & 14 \end{bmatrix} - \begin{bmatrix} 6 & -9 \\ 0 & 3 \end{bmatrix} = \begin{bmatrix} 4 & 13 \\ 18 & 11 \end{bmatrix}$

64. $\begin{bmatrix} 3 & 15 & -3 \\ 9 & 21 & 6 \end{bmatrix} - \begin{bmatrix} 8 & 0 & 12 \\ 16 & 4 & -4 \end{bmatrix} = \begin{bmatrix} -5 & 15 & -15 \\ -7 & 17 & 10 \end{bmatrix}$

65. $\begin{bmatrix} 10 & -15 \\ 0 & 5 \end{bmatrix} - \begin{bmatrix} 10 & 4 \\ 18 & 14 \end{bmatrix} = \begin{bmatrix} 0 & -19 \\ -18 & -9 \end{bmatrix}$	**66.** $\begin{bmatrix} 5 & 25 & -5 \\ 15 & 35 & 10 \end{bmatrix} - \begin{bmatrix} 8 & 0 & 12 \\ 16 & 4 & -4 \end{bmatrix} = \begin{bmatrix} -3 & 25 & -17 \\ -1 & 31 & 14 \end{bmatrix}$
67. $\begin{bmatrix} -7 & -11 & -8 \\ 3 & 7 & 2 \end{bmatrix}$	**68.** $\begin{bmatrix} 15 & -8 & 15 \\ 29 & -1 & 43 \end{bmatrix}$
69. $\begin{bmatrix} 10 & -13 \\ 18 & -20 \end{bmatrix}$	**70.** $\begin{bmatrix} -17 & -17 \\ 9 & 7 \end{bmatrix}$
71. $\begin{bmatrix} 15 & -8 & 15 \\ 29 & -1 & 43 \end{bmatrix} + \begin{bmatrix} 2 & 0 & 3 \\ 4 & 1 & -1 \end{bmatrix} = \begin{bmatrix} 17 & -8 & 18 \\ 33 & 0 & 42 \end{bmatrix}$	**72.** $\begin{bmatrix} 11 & 39 & -1 \\ 30 & 94 & 5 \end{bmatrix}$
73. $\begin{bmatrix} 10 & 9 & 20 \\ 22 & -4 & 2 \end{bmatrix}$	**74.** Not defined

75. Since $AB = \begin{bmatrix} 6 & 4 \\ 4 & 2 \end{bmatrix} \cdot \begin{bmatrix} -0.5 & 1 \\ 1 & -1.5 \end{bmatrix} = \begin{bmatrix} 1 & 0 \\ 0 & 1 \end{bmatrix} = I$, B must be the inverse of A.

76. Since $AB = \begin{bmatrix} 1 & -2 \\ 2 & -4 \end{bmatrix} \cdot \begin{bmatrix} 1 & 2 \\ 2 & -2 \end{bmatrix} = \begin{bmatrix} -3 & 6 \\ -6 & 12 \end{bmatrix} \neq I$, B cannot be the inverse of A.

77. Since $AB = \begin{bmatrix} 1 & -2 & 6 \\ 2 & 3 & -2 \\ 0 & -1 & 1 \end{bmatrix} \cdot \begin{bmatrix} -\frac{1}{7} & \frac{4}{7} & 2 \\ \frac{2}{7} & -\frac{1}{7} & -2 \\ \frac{2}{7} & -\frac{1}{7} & -1 \end{bmatrix} = \begin{bmatrix} 1 & 0 & 0 \\ 0 & 1 & 0 \\ 0 & 0 & 1 \end{bmatrix} = I$, B must be the inverse of A.

78. Since $AB = \begin{bmatrix} 0 & 7 & 6 \\ 1 & 0 & -4 \\ -2 & 1 & 0 \end{bmatrix} \cdot \begin{bmatrix} 1 & 1 & 1 \\ -2 & -2 & -2 \\ 2 & 0 & 6 \end{bmatrix} = \begin{bmatrix} -2 & -14 & 22 \\ -7 & 1 & -23 \\ -4 & -4 & -4 \end{bmatrix} \neq I$, B cannot be the inverse of A.

79. $A^{-1} = \frac{1}{4(1)-(-3)(2)} \begin{bmatrix} 4 & -2 \\ 3 & 1 \end{bmatrix} = \frac{1}{10} \begin{bmatrix} 4 & -2 \\ 3 & 1 \end{bmatrix} = \begin{bmatrix} 0.4 & -0.2 \\ 0.3 & 0.1 \end{bmatrix}$

80 $A^{-1} = \frac{1}{(-2)(6)-(-4)(7)} \begin{bmatrix} 6 & -7 \\ 4 & -2 \end{bmatrix} = \frac{1}{16} \begin{bmatrix} 6 & -7 \\ 4 & -2 \end{bmatrix} = \begin{bmatrix} \frac{3}{8} & -\frac{7}{16} \\ \frac{1}{4} & -\frac{1}{8} \end{bmatrix}$

81. $A^{-1} = \frac{1}{(0)(0)-(-2)(1)} \begin{bmatrix} 0 & -1 \\ 2 & 0 \end{bmatrix} = \frac{1}{2} \begin{bmatrix} 0 & -1 \\ 2 & 0 \end{bmatrix} = \begin{bmatrix} 0 & -\frac{1}{2} \\ 1 & 0 \end{bmatrix}$

82. $A^{-1} = \frac{1}{(3)(2)-(-1)(-2)} \begin{bmatrix} 2 & 1 \\ 2 & 3 \end{bmatrix} = \frac{1}{4} \begin{bmatrix} 2 & 1 \\ 2 & 3 \end{bmatrix} = \begin{bmatrix} \frac{1}{2} & \frac{1}{4} \\ \frac{1}{2} & \frac{3}{4} \end{bmatrix}$

83. Using Gaussian elimination, we obtain the following:

$$\left[\begin{array}{ccc|ccc} 1 & 3 & -2 & 1 & 0 & 0 \\ 2 & 1 & -1 & 0 & 1 & 0 \\ 0 & 1 & -3 & 0 & 0 & 1 \end{array}\right] \xrightarrow{R_2 - 2R_1 \to R_2} \left[\begin{array}{ccc|ccc} 1 & 3 & -2 & 1 & 0 & 0 \\ 0 & -5 & 3 & -2 & 1 & 0 \\ 0 & 1 & -3 & 0 & 0 & 1 \end{array}\right]$$

$$\xrightarrow{R_2 + 5R_3 \to R_3} \left[\begin{array}{ccc|ccc} 1 & 3 & -2 & 1 & 0 & 0 \\ 0 & -5 & 3 & -2 & 1 & 0 \\ 0 & 0 & -12 & -2 & 1 & 5 \end{array}\right] \xrightarrow[-\frac{1}{12}R_3 \to R_3]{-\frac{1}{5}R_2 \to R_2} \left[\begin{array}{ccc|ccc} 1 & 3 & -2 & 1 & 0 & 0 \\ 0 & 1 & -\frac{3}{5} & \frac{2}{5} & -\frac{1}{5} & 0 \\ 0 & 0 & 1 & \frac{1}{6} & -\frac{1}{12} & -\frac{5}{12} \end{array}\right]$$

$$\xrightarrow{R_2 + \frac{3}{5}R_3 \to R_2} \left[\begin{array}{ccc|ccc} 1 & 3 & -2 & 1 & 0 & 0 \\ 0 & 1 & 0 & \frac{1}{2} & -\frac{1}{4} & -\frac{1}{4} \\ 0 & 0 & 1 & \frac{1}{6} & -\frac{1}{12} & -\frac{5}{12} \end{array}\right] \xrightarrow{R_1 - 3R_2 + 2R_3 \to R_1} \left[\begin{array}{ccc|ccc} 1 & 0 & 0 & -\frac{1}{6} & \frac{7}{12} & -\frac{1}{12} \\ 0 & 1 & 0 & \frac{1}{2} & -\frac{1}{4} & -\frac{1}{4} \\ 0 & 0 & 1 & \frac{1}{6} & -\frac{1}{12} & -\frac{5}{12} \end{array}\right]$$

So, $A^{-1} = \begin{bmatrix} -\frac{1}{6} & \frac{7}{12} & -\frac{1}{12} \\ \frac{1}{2} & -\frac{1}{4} & -\frac{1}{4} \\ \frac{1}{6} & -\frac{1}{12} & -\frac{5}{12} \end{bmatrix}$.

84. Using Gaussian elimination, we obtain the following:

$$\left[\begin{array}{ccc|ccc} 0 & 1 & 0 & 1 & 0 & 0 \\ 4 & 1 & 2 & 0 & 1 & 0 \\ -3 & -2 & 1 & 0 & 0 & 1 \end{array}\right] \xrightarrow[R_1 \leftrightarrow R_2]{-\frac{1}{3}R_3 \to R_3} \left[\begin{array}{ccc|ccc} 4 & 1 & 2 & 0 & 1 & 0 \\ 0 & 1 & 0 & 1 & 0 & 0 \\ 1 & \frac{2}{3} & -\frac{1}{3} & 0 & 0 & -\frac{1}{3} \end{array}\right]$$

$$\xrightarrow{R_1 - 4R_3 \to R_3} \left[\begin{array}{ccc|ccc} 4 & 1 & 2 & 0 & 1 & 0 \\ 0 & 1 & 0 & 1 & 0 & 0 \\ 0 & -\frac{5}{3} & \frac{10}{3} & 0 & 1 & \frac{4}{3} \end{array}\right] \xrightarrow{R_3 + \frac{5}{3}R_2 \to R_3} \left[\begin{array}{ccc|ccc} 4 & 1 & 2 & 0 & 1 & 0 \\ 0 & 1 & 0 & 1 & 0 & 0 \\ 0 & 0 & \frac{10}{3} & \frac{5}{3} & 1 & \frac{4}{3} \end{array}\right]$$

$$\xrightarrow[\frac{1}{4}R_1 \to R_1]{\frac{3}{10}R_3 \to R_3} \left[\begin{array}{ccc|ccc} 1 & \frac{1}{4} & \frac{1}{2} & 0 & \frac{1}{4} & 0 \\ 0 & 1 & 0 & 1 & 0 & 0 \\ 0 & 0 & 1 & \frac{1}{2} & \frac{3}{10} & \frac{2}{5} \end{array}\right] \xrightarrow{R_1 - \frac{1}{4}R_2 - \frac{1}{2}R_3 \to R_1} \left[\begin{array}{ccc|ccc} 1 & 0 & 0 & -\frac{1}{2} & \frac{1}{10} & -\frac{1}{5} \\ 0 & 1 & 0 & 1 & 0 & 0 \\ 0 & 0 & 1 & \frac{1}{2} & \frac{3}{10} & \frac{2}{5} \end{array}\right]$$

So, $A^{-1} = \begin{bmatrix} -\frac{1}{2} & \frac{1}{10} & -\frac{1}{5} \\ 1 & 0 & 0 \\ \frac{1}{2} & \frac{3}{10} & \frac{2}{5} \end{bmatrix}$.

85. Using Gaussian elimination, we obtain the following:

$$\left[\begin{array}{ccc|ccc} -1 & 1 & 0 & 1 & 0 & 0 \\ -2 & 1 & 2 & 0 & 1 & 0 \\ 1 & 2 & 4 & 0 & 0 & 1 \end{array}\right] \xrightarrow[R_1+R_3 \to R_3]{R_2-2R_1 \to R_2} \left[\begin{array}{ccc|ccc} -1 & 1 & 0 & 1 & 0 & 0 \\ 0 & -1 & 2 & -2 & 1 & 0 \\ 0 & 3 & 4 & 1 & 0 & 1 \end{array}\right]$$

$$\xrightarrow[-R_1 \to R_1]{R_3+3R_2 \to R_3} \left[\begin{array}{ccc|ccc} 1 & -1 & 0 & -1 & 0 & 0 \\ 0 & -1 & 2 & -2 & 1 & 0 \\ 0 & 0 & 10 & -5 & 3 & 1 \end{array}\right] \xrightarrow[\frac{1}{10}R_3 \to R_3]{R_3-5R_2 \to R_2} \left[\begin{array}{ccc|ccc} 1 & -1 & 0 & -1 & 0 & 0 \\ 0 & 5 & 0 & 5 & -2 & 1 \\ 0 & 0 & 1 & -\frac{1}{2} & \frac{3}{10} & \frac{1}{10} \end{array}\right]$$

$$\xrightarrow{\frac{1}{5}R_2 \to R_2} \left[\begin{array}{ccc|ccc} 1 & -1 & 0 & -1 & 0 & 0 \\ 0 & 1 & 0 & 1 & -\frac{2}{5} & \frac{1}{5} \\ 0 & 0 & 1 & -\frac{1}{2} & \frac{3}{10} & \frac{1}{10} \end{array}\right] \xrightarrow{R_1+R_2 \to R_1} \left[\begin{array}{ccc|ccc} 1 & 0 & 0 & 0 & -\frac{2}{5} & \frac{1}{5} \\ 0 & 1 & 0 & 1 & -\frac{2}{5} & \frac{1}{5} \\ 0 & 0 & 1 & -\frac{1}{2} & \frac{3}{10} & \frac{1}{10} \end{array}\right]$$

So, $A^{-1} = \begin{bmatrix} 0 & -\frac{2}{5} & \frac{1}{5} \\ 1 & -\frac{2}{5} & \frac{1}{5} \\ -\frac{1}{2} & \frac{3}{10} & \frac{1}{10} \end{bmatrix}$.

86. Using Gaussian elimination, we obtain the following:

$$\left[\begin{array}{ccc|ccc} -4 & 4 & 3 & 1 & 0 & 0 \\ 1 & 2 & 2 & 0 & 1 & 0 \\ 3 & -1 & 6 & 0 & 0 & 1 \end{array}\right] \xrightarrow[R_1+4R_2 \to R_2]{R_3-3R_2 \to R_3} \left[\begin{array}{ccc|ccc} -4 & 4 & 3 & 1 & 0 & 0 \\ 0 & 12 & 11 & 1 & 4 & 0 \\ 0 & -7 & 0 & 0 & -3 & 1 \end{array}\right]$$

$$\xrightarrow{R_2 \leftrightarrow R_3} \left[\begin{array}{ccc|ccc} -4 & 4 & 3 & 1 & 0 & 0 \\ 0 & -7 & 0 & 0 & -3 & 1 \\ 0 & 12 & 11 & 1 & 4 & 0 \end{array}\right] \xrightarrow[\frac{1}{12}R_3 \to R_3]{-\frac{1}{4}R_1 \to R_1,\ -\frac{1}{7}R_2 \to R_2} \left[\begin{array}{ccc|ccc} 1 & -1 & -\frac{3}{4} & -\frac{1}{4} & 0 & 0 \\ 0 & 1 & 0 & 0 & \frac{3}{7} & -\frac{1}{7} \\ 0 & 1 & \frac{11}{12} & \frac{1}{12} & \frac{1}{3} & 0 \end{array}\right]$$

$$\xrightarrow{R_3-R_2 \to R_3} \left[\begin{array}{ccc|ccc} 1 & -1 & -\frac{3}{4} & -\frac{1}{4} & 0 & 0 \\ 0 & 1 & 0 & 0 & \frac{3}{7} & -\frac{1}{7} \\ 0 & 0 & \frac{11}{12} & \frac{1}{12} & -\frac{2}{21} & \frac{1}{7} \end{array}\right] \xrightarrow{\frac{12}{11}R_3 \to R_3} \left[\begin{array}{ccc|ccc} 1 & -1 & -\frac{3}{4} & -\frac{1}{4} & 0 & 0 \\ 0 & 1 & 0 & 0 & \frac{3}{7} & -\frac{1}{7} \\ 0 & 0 & 1 & \frac{1}{11} & -\frac{8}{77} & \frac{12}{77} \end{array}\right]$$

$$\xrightarrow{R_1+R_2+\frac{3}{4}R_3 \to R_1} \left[\begin{array}{ccc|ccc} 1 & 0 & 0 & -\frac{2}{11} & \frac{27}{77} & -\frac{2}{77} \\ 0 & 1 & 0 & 0 & \frac{3}{7} & -\frac{1}{7} \\ 0 & 0 & 1 & \frac{1}{11} & -\frac{8}{77} & \frac{12}{77} \end{array}\right] \quad \text{So, } A^{-1} = \begin{bmatrix} -\frac{2}{11} & \frac{27}{77} & -\frac{2}{77} \\ 0 & \frac{3}{7} & -\frac{1}{7} \\ \frac{1}{11} & -\frac{8}{77} & \frac{12}{77} \end{bmatrix}.$$

87. The system in matrix form is:

$$\begin{bmatrix} 3 & -1 \\ 5 & 2 \end{bmatrix} \begin{bmatrix} x \\ y \end{bmatrix} = \begin{bmatrix} 11 \\ 33 \end{bmatrix}.$$

The solution is

$$\begin{bmatrix} x \\ y \end{bmatrix} = \begin{bmatrix} 3 & -1 \\ 5 & 2 \end{bmatrix}^{-1} \begin{bmatrix} 11 \\ 33 \end{bmatrix} = \frac{1}{11} \begin{bmatrix} 2 & 1 \\ -5 & 3 \end{bmatrix} \begin{bmatrix} 11 \\ 33 \end{bmatrix} = \begin{bmatrix} 5 \\ 4 \end{bmatrix}.$$

Hence, $\boxed{x = 5, y = 4}$.

88. The system in matrix form is:

$$\begin{bmatrix} 6 & 4 \\ -3 & -2 \end{bmatrix}\begin{bmatrix} x \\ y \end{bmatrix}=\begin{bmatrix} 15 \\ -1 \end{bmatrix}$$

The solution would be of the form

$$\begin{bmatrix} x \\ y \end{bmatrix}=\begin{bmatrix} 6 & 4 \\ -3 & -2 \end{bmatrix}^{-1}\begin{bmatrix} 15 \\ -1 \end{bmatrix},$$

but since $\begin{bmatrix} 6 & 4 \\ -3 & -2 \end{bmatrix}^{-1}$ does not exist, there is $\boxed{\text{no solution}}$.

89. The system in matrix form is:

$$\begin{bmatrix} \frac{5}{8} & -\frac{2}{3} \\ \frac{3}{4} & \frac{5}{6} \end{bmatrix}\begin{bmatrix} x \\ y \end{bmatrix}=\begin{bmatrix} -3 \\ 16 \end{bmatrix}$$

The solution is:

$$\begin{bmatrix} x \\ y \end{bmatrix}=\begin{bmatrix} \frac{5}{8} & -\frac{2}{3} \\ \frac{3}{4} & \frac{5}{6} \end{bmatrix}^{-1}\begin{bmatrix} -3 \\ 16 \end{bmatrix}=\frac{48}{49}\begin{bmatrix} \frac{5}{6} & \frac{2}{3} \\ -\frac{3}{4} & \frac{5}{8} \end{bmatrix}\begin{bmatrix} -3 \\ 16 \end{bmatrix}=\begin{bmatrix} 8 \\ 12 \end{bmatrix}.$$

Hence, $\boxed{x=8, y=12}$.

90. The system in matrix form is:

$$\begin{bmatrix} 1 & 1 & -1 \\ 2 & -1 & 3 \\ 3 & -2 & 1 \end{bmatrix}\begin{bmatrix} x \\ y \\ z \end{bmatrix} = \begin{bmatrix} 0 \\ 18 \\ 17 \end{bmatrix}$$

The solution is:

$$\begin{bmatrix} x \\ y \\ z \end{bmatrix} = \begin{bmatrix} 1 & 1 & -1 \\ 2 & -1 & 3 \\ 3 & -2 & 1 \end{bmatrix}^{-1}\begin{bmatrix} 0 \\ 18 \\ 17 \end{bmatrix}.$$

We calculate the inverse below:

$$\left[\begin{array}{ccc|ccc} 1 & 1 & -1 & 1 & 0 & 0 \\ 2 & -1 & 3 & 0 & 1 & 0 \\ 3 & -2 & 1 & 0 & 0 & 1 \end{array}\right] \xrightarrow[R_3-3R_1\to R_3]{R_2-2R_1\to R_2} \left[\begin{array}{ccc|ccc} 1 & 1 & -1 & 1 & 0 & 0 \\ 0 & -3 & 5 & -2 & 1 & 0 \\ 0 & -5 & 4 & -3 & 0 & 1 \end{array}\right] \xrightarrow[-\frac{1}{5}R_3\to R_3]{-\frac{1}{3}R_2\to R_2} \left[\begin{array}{ccc|ccc} 1 & 1 & -1 & 1 & 0 & 0 \\ 0 & 1 & -\frac{5}{3} & \frac{2}{3} & -\frac{1}{3} & 0 \\ 0 & 1 & -\frac{4}{5} & \frac{3}{5} & 0 & -\frac{1}{5} \end{array}\right]$$

$$\xrightarrow{R_2-R_3\to R_3} \left[\begin{array}{ccc|ccc} 1 & 1 & -1 & 1 & 0 & 0 \\ 0 & 1 & -\frac{5}{3} & \frac{2}{3} & -\frac{1}{3} & 0 \\ 0 & 0 & -\frac{13}{15} & \frac{1}{15} & -\frac{1}{3} & \frac{1}{5} \end{array}\right] \xrightarrow{-\frac{15}{13}R_3\to R_3} \left[\begin{array}{ccc|ccc} 1 & 1 & -1 & 1 & 0 & 0 \\ 0 & 1 & -\frac{5}{3} & \frac{2}{3} & -\frac{1}{3} & 0 \\ 0 & 0 & 1 & -\frac{1}{13} & \frac{5}{13} & -\frac{3}{13} \end{array}\right] \xrightarrow{R_2+\frac{5}{3}R_3\to R_2}$$

$$\left[\begin{array}{ccc|ccc} 1 & 1 & -1 & 1 & 0 & 0 \\ 0 & 1 & 0 & \frac{7}{13} & -\frac{2}{39} & -\frac{5}{13} \\ 0 & 0 & 1 & -\frac{1}{13} & \frac{5}{13} & -\frac{3}{13} \end{array}\right] \xrightarrow{R_1-R_2+R_3\to R_1} \left[\begin{array}{ccc|ccc} 1 & 0 & 0 & \frac{5}{13} & \frac{1}{3} & \frac{2}{13} \\ 0 & 1 & 0 & \frac{7}{13} & -\frac{2}{39} & -\frac{5}{13} \\ 0 & 0 & 1 & -\frac{1}{13} & \frac{5}{13} & -\frac{3}{13} \end{array}\right]$$

Hence, the solution of the system is

$$\begin{bmatrix} x \\ y \\ z \end{bmatrix} = \begin{bmatrix} \frac{5}{13} & \frac{1}{3} & \frac{2}{13} \\ \frac{7}{13} & -\frac{2}{39} & -\frac{5}{13} \\ -\frac{1}{13} & \frac{5}{13} & -\frac{3}{13} \end{bmatrix}\begin{bmatrix} 0 \\ 18 \\ 17 \end{bmatrix} = \begin{bmatrix} \frac{112}{13} \\ -\frac{97}{13} \\ 3 \end{bmatrix}$$

Thus, the solution is $\boxed{x = \frac{112}{13},\ y = -\frac{97}{13},\ z = 3}$.

91. We reduce down to reduced row echelon form:

$$\left[\begin{array}{ccc|c} 3 & -2 & 4 & 11 \\ 6 & 3 & -2 & 6 \\ 1 & -1 & 7 & 20 \end{array}\right] \xrightarrow[R_1-3R_3\to R_3]{R_2-6R_3\to R_2} \left[\begin{array}{ccc|c} 3 & -2 & 4 & 11 \\ 0 & 9 & -44 & -114 \\ 0 & 1 & -17 & -49 \end{array}\right] \xrightarrow{R_2-9R_3\to R_3} \left[\begin{array}{ccc|c} 3 & -2 & 4 & 11 \\ 0 & 9 & -44 & -114 \\ 0 & 0 & 109 & 327 \end{array}\right]$$

$$\xrightarrow[\frac{1}{109}R_3\to R_3]{\frac{1}{9}R_2\to R_2} \left[\begin{array}{ccc|c} 3 & -2 & 4 & 11 \\ 0 & 1 & -\frac{44}{9} & -114 \\ 0 & 0 & 109 & 327 \end{array}\right] \xrightarrow[R_2+\frac{44}{9}R_3\to R_2]{\frac{1}{3}R_1\to R_1} \left[\begin{array}{ccc|c} 1 & -\frac{2}{3} & \frac{4}{3} & \frac{11}{3} \\ 0 & 1 & 0 & 2 \\ 0 & 0 & 1 & 3 \end{array}\right] \xrightarrow{R_1+\frac{2}{3}R_2-\frac{4}{3}R_3\to R_1} \left[\begin{array}{ccc|c} 1 & 0 & 0 & 1 \\ 0 & 1 & 0 & 2 \\ 0 & 0 & 1 & 3 \end{array}\right]$$

Hence, the solution is $\boxed{x = 1, y = 2, z = 3}$.

92. We reduce down to reduced row echelon form:

$$\left[\begin{array}{ccc|c} 2 & 6 & -4 & 11 \\ -1 & -3 & 2 & -\frac{11}{2} \\ 4 & 5 & 6 & 20 \end{array}\right] \xrightarrow[R_3+4R_2\to R_3]{R_1+2R_2\to R_2} \left[\begin{array}{ccc|c} 2 & 6 & -4 & 11 \\ 0 & 0 & 0 & 0 \\ 0 & -7 & 14 & -2 \end{array}\right] \xrightarrow[-\frac{1}{7}R_3\to R_3]{\frac{1}{2}R_1\to R_1} \left[\begin{array}{ccc|c} 1 & 3 & -2 & \frac{11}{2} \\ 0 & 0 & 0 & 0 \\ 0 & 1 & -2 & \frac{2}{7} \end{array}\right]$$

Let $x = a$. Then, substituting into the equation corresponding to row 3 yields

$$a - 2y = \tfrac{2}{7} \Rightarrow y = \tfrac{1}{2}a - \tfrac{1}{7}.$$

Then, substituting these two values into the equation corresponding to row 1 yields

$$a + 3\left(\tfrac{1}{2}a - \tfrac{1}{7}\right) - 2z = \tfrac{11}{2}$$
$$2z = \tfrac{5}{2}a - \tfrac{3}{7} - \tfrac{11}{2}$$
$$z = \tfrac{5}{4}a - \tfrac{83}{28}$$

Thus, the solution is $\boxed{x = a,\ y = \tfrac{1}{2}a - \tfrac{1}{7},\ z = \tfrac{5}{4}a - \tfrac{83}{28}}$.

93. $\begin{vmatrix} 2 & 4 \\ 3 & 2 \end{vmatrix} = (2)(2) - (3)(4) = \boxed{-8}$

94. $\begin{vmatrix} -2 & -4 \\ -3 & 2 \end{vmatrix} = (-2)(2) - (-3)(-4) = \boxed{-16}$

95.
$\begin{vmatrix} 2.4 & -2.3 \\ 3.6 & -1.2 \end{vmatrix} = (2.4)(-1.2) - (3.6)(-2.3) = \boxed{5.4}$

96.
$D = \begin{vmatrix} -\frac{1}{4} & 4 \\ \frac{3}{4} & -4 \end{vmatrix} = \left(-\tfrac{1}{4}\right)(-4) - \left(\tfrac{3}{4}\right)(4) = \boxed{-2}$

97.

$$D = \begin{vmatrix} 1 & -1 \\ 1 & 1 \end{vmatrix} = (1)(1) - (1)(-1) = 2$$

$$D_x = \begin{vmatrix} 2 & -1 \\ 4 & 1 \end{vmatrix} = (2)(1) - (4)(-1) = 6$$

$$D_y = \begin{vmatrix} 1 & 2 \\ 1 & 4 \end{vmatrix} = (1)(4) - (1)(2) = 2$$

So, from Cramer's Rule, we have:

$$x = \frac{D_x}{D} = \frac{6}{2} = 3$$
$$y = \frac{D_y}{D} = \frac{2}{2} = 1$$

Thus, the solution is $\boxed{x = 3,\ y = 1}$.

98. First, write the system as

$$\begin{bmatrix} 3 & -1 \\ -1 & 5 \end{bmatrix}\begin{bmatrix} x \\ y \end{bmatrix} = \begin{bmatrix} -17 \\ 43 \end{bmatrix}.$$

$$D = \begin{vmatrix} 3 & -1 \\ -1 & 5 \end{vmatrix} = (3)(5) - (-1)(-1) = 14$$

$$D_x = \begin{vmatrix} -17 & -1 \\ 43 & 5 \end{vmatrix} = (-17)(5) - (43)(-1) = -42$$

$$D_y = \begin{vmatrix} 3 & -17 \\ -1 & 43 \end{vmatrix} = (3)(43) - (-1)(-17) = 112$$

So, from Cramer's Rule, we have:

$$x = \frac{D_x}{D} = \frac{-42}{14} = -3$$
$$y = \frac{D_y}{D} = \frac{112}{14} = 8$$

Thus, the solution is $\boxed{x = -3, y = 8}$

99. Divide the first equation by 2. Then, we have:

$$D=\begin{vmatrix}1 & 2\\ 1 & -2\end{vmatrix}=(1)(-2)-(1)(2)=-4$$

$$D_x=\begin{vmatrix}6 & 2\\ 6 & -2\end{vmatrix}=(6)(-2)-(6)(2)=-24$$

$$D_y=\begin{vmatrix}1 & 6\\ 1 & 6\end{vmatrix}=(1)(6)-(1)(6)=0$$

So, from Cramer's Rule, we have:

$$x=\frac{D_x}{D}=\frac{-24}{-4}=6$$

$$y=\frac{D_y}{D}=\frac{0}{-4}=0$$

Thus, the solution is $\boxed{x=6,\ y=0}$.

100.

$$D=\begin{vmatrix}-1 & 1\\ 2 & -6\end{vmatrix}=(-1)(-6)-(2)(1)=4$$

$$D_x=\begin{vmatrix}4 & 1\\ -5 & -6\end{vmatrix}=(4)(-6)-(-5)(1)=-19$$

$$D_y=\begin{vmatrix}-1 & 4\\ 2 & -5\end{vmatrix}=(-1)(-5)-(2)(4)=-3$$

So, from Cramer's Rule, we have:

$$x=\frac{D_x}{D}=\frac{-19}{4}$$

$$y=\frac{D_y}{D}=\frac{-3}{4}$$

Thus, the solution is $\boxed{x=-\frac{19}{4},\ y=-\frac{3}{4}}$.

101. First, write the system as:

$$\begin{cases}-3x+2y=40\\ 2x-y=25\end{cases}$$

$$D=\begin{vmatrix}-3 & 2\\ 2 & -1\end{vmatrix}=(-3)(-1)-(2)(2)=-1$$

$$D_x=\begin{vmatrix}40 & 2\\ 25 & -1\end{vmatrix}=(40)(-1)-(25)(2)=-90$$

$$D_y=\begin{vmatrix}-3 & 40\\ 2 & 25\end{vmatrix}=(-3)(25)-(2)(40)=-155$$

So, from Cramer's Rule, we have:

$$x=\frac{D_x}{D}=\frac{-90}{-1}=90$$

$$y=\frac{D_y}{D}=\frac{-155}{-1}=155$$

Thus, the solution is $\boxed{x=90,\ y=155}$.

102. First, write the system as

$$\begin{bmatrix}3 & -4\\ -1 & 1\end{bmatrix}\begin{bmatrix}x\\ y\end{bmatrix}=\begin{bmatrix}20\\ -6\end{bmatrix}$$

$$D=\begin{vmatrix}3 & -4\\ -1 & 1\end{vmatrix}=(3)(1)-(-1)(-4)=-1$$

$$D_x=\begin{vmatrix}20 & -4\\ -6 & 1\end{vmatrix}=(20)(1)-(-6)(-4)=-4$$

$$D_y=\begin{vmatrix}3 & 20\\ -1 & -6\end{vmatrix}=(3)(-6)-(-1)(20)=2$$

So, from Cramer's Rule, we have:

$$x=\frac{D_x}{D}=\frac{-4}{-1}=4$$

$$y=\frac{D_y}{D}=\frac{2}{-1}=-2$$

Thus, the solution is $\boxed{x=4, y=-2}$.

103.

$$\begin{vmatrix} 1 & 2 & 2 \\ 0 & 1 & 3 \\ 2 & -1 & 0 \end{vmatrix} = 1\begin{vmatrix} 1 & 3 \\ -1 & 0 \end{vmatrix} - 2\begin{vmatrix} 0 & 3 \\ 2 & 0 \end{vmatrix} + 2\begin{vmatrix} 0 & 1 \\ 2 & -1 \end{vmatrix}$$

$$= 1\underbrace{\left[(1)(0)-(-1)(3)\right]}_{=3} - 2\underbrace{\left[(0)(0)-(2)(3)\right]}_{=-6} + 2\underbrace{\left[(0)(-1)-(2)(1)\right]}_{=-2}$$

$$= \boxed{11}$$

104.

$$\begin{vmatrix} 0 & -2 & 1 \\ 0 & -3 & 7 \\ 1 & -10 & -3 \end{vmatrix} = 0\begin{vmatrix} -3 & 7 \\ -10 & -3 \end{vmatrix} - (-2)\begin{vmatrix} 0 & 7 \\ 1 & -3 \end{vmatrix} + 1\begin{vmatrix} 0 & -3 \\ 1 & -10 \end{vmatrix}$$

$$= 0\underbrace{\left[(-3)(-3)-(-10)(7)\right]}_{=79} - (-2)\underbrace{\left[(0)(-3)-(1)(7)\right]}_{=-7} + 1\underbrace{\left[(0)(-10)-(1)(-3)\right]}_{=3}$$

$$= \boxed{-11}$$

105.

$$\begin{vmatrix} a & 0 & -b \\ -a & b & c \\ 0 & 0 & -d \end{vmatrix} = a\begin{vmatrix} b & c \\ 0 & -d \end{vmatrix} - 0\begin{vmatrix} -a & c \\ 0 & -d \end{vmatrix} + (-b)\begin{vmatrix} -a & b \\ 0 & 0 \end{vmatrix}$$

$$= a\underbrace{\left[(b)(-d)-(0)(c)\right]}_{=-bd} - 0\underbrace{\left[(-a)(-d)-(0)(c)\right]}_{=ad} + (-b)\underbrace{\left[(-a)(0)-(0)(b)\right]}_{=0}$$

$$= \boxed{-abd}$$

106.

$$\begin{vmatrix} -2 & -4 & 6 \\ 2 & 0 & 3 \\ -1 & 2 & \frac{3}{4} \end{vmatrix} = (-2)\begin{vmatrix} 0 & 3 \\ 2 & \frac{3}{4} \end{vmatrix} - (-4)\begin{vmatrix} 2 & 3 \\ -1 & \frac{3}{4} \end{vmatrix} + 6\begin{vmatrix} 2 & 0 \\ -1 & 2 \end{vmatrix}$$

$$= (-2)\underbrace{\left[(0)(\tfrac{3}{4})-(2)(3)\right]}_{=-6} - (-4)\underbrace{\left[(2)(\tfrac{3}{4})-(-1)(3)\right]}_{=\frac{9}{2}} + 6\underbrace{\left[(2)(2)-(-1)(0)\right]}_{=4}$$

$$= \boxed{54}$$

107.

$$D=\begin{vmatrix}1&1&-2\\2&-1&1\\1&1&1\end{vmatrix}=1\begin{vmatrix}-1&1\\1&1\end{vmatrix}-1\begin{vmatrix}2&1\\1&1\end{vmatrix}+(-2)\begin{vmatrix}2&-1\\1&1\end{vmatrix}$$

$$=1\underbrace{\left[(-1)(1)-(1)(1)\right]}_{=-2}-1\underbrace{\left[(2)(1)-(1)(1)\right]}_{=1}+(-2)\underbrace{\left[(2)(1)-(1)(-1)\right]}_{=3}=-9$$

$$D_x=\begin{vmatrix}-2&1&-2\\3&-1&1\\4&1&1\end{vmatrix}=(-2)\begin{vmatrix}-1&1\\1&1\end{vmatrix}-1\begin{vmatrix}3&1\\4&1\end{vmatrix}+(-2)\begin{vmatrix}3&-1\\4&1\end{vmatrix}$$

$$=(-2)\underbrace{\left[(-1)(1)-(1)(1)\right]}_{=-2}-1\underbrace{\left[(3)(1)-(4)(1)\right]}_{=-1}+(-2)\underbrace{\left[(3)(1)-(4)(-1)\right]}_{=7}=-9$$

$$D_y=\begin{vmatrix}1&-2&-2\\2&3&1\\1&4&1\end{vmatrix}=1\begin{vmatrix}3&1\\4&1\end{vmatrix}-(-2)\begin{vmatrix}2&1\\1&1\end{vmatrix}+(-2)\begin{vmatrix}2&3\\1&4\end{vmatrix}$$

$$=1\underbrace{\left[(3)(1)-(4)(1)\right]}_{=-1}-(-2)\underbrace{\left[(2)(1)-(1)(1)\right]}_{=1}+(-2)\underbrace{\left[(2)(4)-(1)(3)\right]}_{=5}=-9$$

$$D_z=\begin{vmatrix}1&1&-2\\2&-1&3\\1&1&4\end{vmatrix}=1\begin{vmatrix}-1&3\\1&4\end{vmatrix}-1\begin{vmatrix}2&3\\1&4\end{vmatrix}+(-2)\begin{vmatrix}2&-1\\1&1\end{vmatrix}$$

$$=1\underbrace{\left[(-1)(4)-(1)(3)\right]}_{=-7}-1\underbrace{\left[(2)(4)-(1)(3)\right]}_{=5}+(-2)\underbrace{\left[(2)(1)-(1)(-1)\right]}_{=3}=-18$$

So, by Cramer's Rule,

$$\boxed{x=\frac{D_x}{D}=\frac{-9}{-9}=1\quad y=\frac{D_y}{D}=\frac{-9}{-9}=1\quad z=\frac{D_z}{D}=\frac{-18}{-9}=2}$$

108.

$$D=\begin{vmatrix}-1 & -1 & 1\\ 1 & 2 & -2\\ 2 & 1 & 4\end{vmatrix}=(-1)\begin{vmatrix}2 & -2\\ 1 & 4\end{vmatrix}-(-1)\begin{vmatrix}1 & -2\\ 2 & 4\end{vmatrix}+1\begin{vmatrix}1 & 2\\ 2 & 1\end{vmatrix}$$

$$=(-1)\underbrace{\left[(2)(4)-(1)(-2)\right]}_{=10}-(-1)\underbrace{\left[(1)(4)-(2)(-2)\right]}_{=8}+1\underbrace{\left[(1)(1)-(2)(2)\right]}_{=-3}=-5$$

$$D_x=\begin{vmatrix}3 & -1 & 1\\ 8 & 2 & -2\\ -4 & 1 & 4\end{vmatrix}=3\begin{vmatrix}2 & -2\\ 1 & 4\end{vmatrix}-(-1)\begin{vmatrix}8 & -2\\ -4 & 4\end{vmatrix}+1\begin{vmatrix}8 & 2\\ -4 & 1\end{vmatrix}$$

$$=3\underbrace{\left[(2)(4)-(1)(-2)\right]}_{=10}-(-1)\underbrace{\left[(8)(4)-(-4)(-2)\right]}_{=24}+1\underbrace{\left[(8)(1)-(-4)(2)\right]}_{=16}=70$$

$$D_y=\begin{vmatrix}-1 & 3 & 1\\ 1 & 8 & -2\\ 2 & -4 & 4\end{vmatrix}=(-1)\begin{vmatrix}8 & -2\\ -4 & 4\end{vmatrix}-3\begin{vmatrix}1 & -2\\ 2 & 4\end{vmatrix}+1\begin{vmatrix}1 & 8\\ 2 & -4\end{vmatrix}$$

$$=(-1)\underbrace{\left[(8)(4)-(-4)(-2)\right]}_{=24}-3\underbrace{\left[(1)(4)-(2)(-2)\right]}_{=8}+1\underbrace{\left[(1)(-4)-(2)(8)\right]}_{=-20}=-68$$

$$D_z=\begin{vmatrix}-1 & -1 & 3\\ 1 & 2 & 8\\ 2 & 1 & -4\end{vmatrix}=(-1)\begin{vmatrix}2 & 8\\ 1 & -4\end{vmatrix}-(-1)\begin{vmatrix}1 & 8\\ 2 & -4\end{vmatrix}+3\begin{vmatrix}1 & 2\\ 2 & 1\end{vmatrix}$$

$$=(-1)\underbrace{\left[(2)(-4)-(1)(8)\right]}_{=-16}-(-1)\underbrace{\left[(1)(-4)-(2)(8)\right]}_{=-20}+3\underbrace{\left[(1)(1)-(2)(2)\right]}_{=-3}=-13$$

So, by Cramer's Rule,

$$\boxed{x=\frac{D_x}{D}=-14 \qquad y=\frac{D_y}{D}=13.6 \qquad z=\frac{D_z}{D}=2.6}$$

109.

$$D = \begin{vmatrix} 3 & 0 & 4 \\ 1 & 1 & 2 \\ 0 & 1 & -4 \end{vmatrix} = 3\begin{vmatrix} 1 & 2 \\ 1 & -4 \end{vmatrix} - 0\begin{vmatrix} 1 & 2 \\ 0 & -4 \end{vmatrix} + 4\begin{vmatrix} 1 & 1 \\ 0 & 1 \end{vmatrix}$$

$$= 3\underbrace{\left[(1)(-4)-(1)(2)\right]}_{=-6} - 0\underbrace{\left[(1)(-4)-(0)(2)\right]}_{=-4} + 4\underbrace{\left[(1)(1)-(0)(1)\right]}_{=1} = -14$$

$$D_x = \begin{vmatrix} -1 & 0 & 4 \\ -3 & 1 & 2 \\ -9 & 1 & -4 \end{vmatrix} = (-1)\begin{vmatrix} 1 & 2 \\ 1 & -4 \end{vmatrix} - 0\begin{vmatrix} -3 & 2 \\ -9 & -4 \end{vmatrix} + 4\begin{vmatrix} -3 & 1 \\ -9 & 1 \end{vmatrix}$$

$$= (-1)\underbrace{\left[(1)(-4)-(1)(2)\right]}_{=-6} - 0\underbrace{\left[(-3)(-4)-(-9)(2)\right]}_{=30} + 4\underbrace{\left[(-3)(1)-(-9)(1)\right]}_{=6} = 30$$

$$D_y = \begin{vmatrix} 3 & -1 & 4 \\ 1 & -3 & 2 \\ 0 & -9 & -4 \end{vmatrix} = 3\begin{vmatrix} -3 & 2 \\ -9 & -4 \end{vmatrix} - (-1)\begin{vmatrix} 1 & 2 \\ 0 & -4 \end{vmatrix} + 4\begin{vmatrix} 1 & -3 \\ 0 & -9 \end{vmatrix}$$

$$= 3\underbrace{\left[(-3)(-4)-(-9)(2)\right]}_{=30} - (-1)\underbrace{\left[(1)(-4)-(0)(2)\right]}_{=-4} + 4\underbrace{\left[(1)(-9)-(0)(-3)\right]}_{=-9} = 50$$

$$D_z = \begin{vmatrix} 3 & 0 & -1 \\ 1 & 1 & -3 \\ 0 & 1 & -9 \end{vmatrix} = 3\begin{vmatrix} 1 & -3 \\ 1 & -9 \end{vmatrix} - 0\begin{vmatrix} 1 & -3 \\ 0 & -9 \end{vmatrix} + (-1)\begin{vmatrix} 1 & 1 \\ 0 & 1 \end{vmatrix}$$

$$= 3\underbrace{\left[(1)(-9)-(1)(-3)\right]}_{=-6} - 0\underbrace{\left[(1)(-9)-(0)(-3)\right]}_{=-9} + (-1)\underbrace{\left[(1)(1)-(0)(1)\right]}_{=1} = -19$$

So, by Cramer's Rule,

$$\boxed{x = \frac{D_x}{D} = \frac{30}{-14} = -\frac{15}{7} \quad y = \frac{D_y}{D} = \frac{50}{-14} = -\frac{25}{7} \quad z = \frac{D_z}{D} = \frac{-19}{-14} = \frac{19}{14}}$$

110.

$$D=\begin{vmatrix}1 & 1 & 1\\ -1 & -3 & 5\\ 2 & 1 & -3\end{vmatrix}=1\begin{vmatrix}-3 & 5\\ 1 & -3\end{vmatrix}-1\begin{vmatrix}-1 & 5\\ 2 & -3\end{vmatrix}+1\begin{vmatrix}-1 & -3\\ 2 & 1\end{vmatrix}$$

$$=1\underbrace{\left[(-3)(-3)-(1)(5)\right]}_{=4}-1\underbrace{\left[(-1)(-3)-(2)(5)\right]}_{=-7}+1\underbrace{\left[(-1)(1)-(2)(-3)\right]}_{=5}=16$$

$$D_x=\begin{vmatrix}0 & 1 & 1\\ -2 & -3 & 5\\ -4 & 1 & -3\end{vmatrix}=0\begin{vmatrix}-3 & 5\\ 1 & -3\end{vmatrix}-1\begin{vmatrix}-2 & 5\\ -4 & -3\end{vmatrix}+1\begin{vmatrix}-2 & -3\\ -4 & 1\end{vmatrix}$$

$$=0\underbrace{\left[(-3)(-3)-(1)(5)\right]}_{=4}-1\underbrace{\left[(-2)(-3)-(-4)(5)\right]}_{=26}+1\underbrace{\left[(-2)(1)-(-4)(-3)\right]}_{=-14}=-40$$

$$D_y=\begin{vmatrix}1 & 0 & 1\\ -1 & -2 & 5\\ 2 & -4 & -3\end{vmatrix}=1\begin{vmatrix}-2 & 5\\ -4 & -3\end{vmatrix}-0\begin{vmatrix}-1 & 5\\ 2 & -3\end{vmatrix}+1\begin{vmatrix}-1 & -2\\ 2 & -4\end{vmatrix}$$

$$=1\underbrace{\left[(-2)(-3)-(-4)(5)\right]}_{=26}-0\underbrace{\left[(-1)(-3)-(2)(5)\right]}_{=-7}+1\underbrace{\left[(-1)(-4)-(2)(-2)\right]}_{=8}=34$$

$$D_z=\begin{vmatrix}1 & 1 & 0\\ -1 & -3 & -2\\ 2 & 1 & -4\end{vmatrix}=1\begin{vmatrix}-3 & -2\\ 1 & -4\end{vmatrix}-1\begin{vmatrix}-1 & -2\\ 2 & -4\end{vmatrix}+0\begin{vmatrix}-1 & -3\\ 2 & 1\end{vmatrix}$$

$$=1\underbrace{\left[(-3)(-4)-(1)(-2)\right]}_{=14}-1\underbrace{\left[(-1)(-4)-(2)(-2)\right]}_{=8}+0\underbrace{\left[(-1)(1)-(2)(-3)\right]}_{=5}=6$$

So, by Cramer's Rule,

$$\boxed{x=\frac{D_x}{D}=\frac{-40}{16}=-\frac{5}{2}\qquad y=\frac{D_y}{D}=\frac{34}{16}=\frac{17}{8}\qquad z=\frac{D_z}{D}=\frac{6}{16}=\frac{3}{8}}$$

111.

$$\frac{4}{(x-1)^2(x+3)(x-5)}=\frac{A}{x-1}+\frac{B}{(x-1)^2}+\frac{C}{x+3}+\frac{D}{x-5}$$

112.

$$\frac{7}{(x-9)(3x+5)^2(x+4)}=\frac{A}{x-9}+\frac{B}{3x+5}+\frac{C}{(3x+5)^2}+\frac{D}{x+4}$$

113.

$$\frac{12}{x(4x+5)(2x+1)^2}=\frac{A}{x}+\frac{B}{4x+5}+\frac{C}{2x+1}+\frac{D}{(2x+1)^2}$$

114.

$$\frac{2}{(x+1)(x-5)(x-9)^2}=\frac{A}{x+1}+\frac{B}{x-5}+\frac{C}{x-9}+\frac{D}{(x-9)^2}$$

115.	**116.**
$\frac{3}{x^2+x-12}=\frac{3}{(x-3)(x+4)}$ $=\frac{A}{x-3}+\frac{B}{x+4}$	$\frac{x^2+3x-2}{x^3+6x^2}=\frac{x^2+3x-2}{x^2(x+6)}$ $=\frac{A}{x}+\frac{B}{x^2}+\frac{C}{x+6}$
117. $\frac{3x^3+4x^2+56x+62}{(x^2+17)^2}=\frac{Ax+B}{x^2+17}+\frac{Cx+D}{(x^2+17)^2}$	**118.** $\frac{x^3+7x^2+10}{(x^2+13)^2}=\frac{Ax+B}{x^2+13}+\frac{Cx+D}{(x^2+13)^2}$

119. The partial fraction decomposition has the form:

$$\frac{9x+23}{(x-1)(x+7)}=\frac{A}{x-1}+\frac{B}{x+7} \quad \textbf{(1)}$$

To find the coefficients, multiply both sides of **(1)** by $(x-1)(x+7)$, and gather like terms:

$$9x+23=A(x+7)+B(x-1)=(A+B)x+(7A-B) \quad \textbf{(2)}$$

Equate corresponding coefficients in **(2)** to obtain the following system:

$$(*)\begin{cases} A+B=9 & \textbf{(3)} \\ 7A-B=23 & \textbf{(4)} \end{cases}$$

Now, solve system **(*)**:

Add **(3)** and **(4)**: $8A=32 \Rightarrow A=4$

Substitute this value of A into **(3)** to see that $B=5$.

Thus, the partial fraction decomposition **(1)** becomes:

$$\boxed{\frac{9x+23}{(x-1)(x+7)}=\frac{4}{x-1}+\frac{5}{x+7}}$$

120. The partial fraction decomposition has the form:

$$\frac{12x+1}{(3x+2)(2x-1)}=\frac{A}{3x+2}+\frac{B}{2x-1} \quad \textbf{(1)}$$

To find the coefficients, multiply both sides of **(1)** by $(3x+2)(2x-1)$, and gather like terms:

$$12x+1=A(2x-1)+B(3x+2)=(2A+3B)x+(-A+2B) \quad \textbf{(2)}$$

Equate corresponding coefficients in **(2)** to obtain the following system:

$$\textbf{(*)} \begin{cases} 2A+3B=12 & \textbf{(3)} \\ -A+2B=1 & \textbf{(4)} \end{cases}$$

Now, solve system **(*)**:

Multiply **(4)** by 2: $-2A+4B=2$ **(5)**
Add **(3)** and **(5)**: $7B=14 \Rightarrow B=2$
Substitute this value of B into **(4)** to see that $A=3$.

Thus, the partial fraction decomposition **(1)** becomes:

$$\boxed{\frac{12x+1}{(3x+2)(2x-1)}=\frac{3}{3x+2}+\frac{2}{2x-1}}$$

121. The partial fraction decomposition has the form:

$$\frac{13x^2+90x-25}{2x^3-50x}=\frac{13x^2+90x-25}{2x(x-5)(x+5)}=\frac{A}{2x}+\frac{B}{x-5}+\frac{C}{x+5} \quad \textbf{(1)}$$

To find the coefficients, multiply both sides of **(1)** by $2x(x-5)(x+5)$, and gather like terms:

$$\begin{aligned}13x^2+90x-25 &= A(x^2-25)+B(2x)(x+5)+C(2x)(x-5)\\ &=(A+2B+2C)x^2+(10B-10C)x-25A\end{aligned} \quad \textbf{(2)}$$

Equate corresponding coefficients in **(2)** to obtain the following system:

$$(*)\begin{cases} A+2B+2C=13 & \textbf{(3)}\\ 10B-10C=90 & \textbf{(4)}\\ -25A=-25 & \textbf{(5)}\end{cases}$$

Now, solve system $(*)$:

Solve **(5)**: $A=1$
Substitute this value of A into **(3)**: $B+C=6$ **(6)**
Note that **(4)** is equivalent to $B-C=9$ **(7)**. So, solve the system:

$$\begin{cases} B+C=6 & \textbf{(6)}\\ B-C=9 & \textbf{(7)}\end{cases}$$

Add **(6)** and **(7)**: $2B=15 \Rightarrow B=\frac{15}{2}$
Substitute this value of B into **(6)**: $C=-\frac{3}{2}$

Thus, the partial fraction decomposition **(1)** becomes:

$$\boxed{\frac{13x^2+90x-25}{2x^3-50x}=\frac{1}{2x}+\frac{15}{2(x-5)}-\frac{3}{2(x+5)}}$$

122. The partial fraction decomposition has the form:

$$\frac{5x^2+x+24}{x^3+8x}=\frac{5x^2+x+24}{x\left(x^2+8\right)}=\frac{A}{x}+\frac{Bx+C}{x^2+8} \quad \textbf{(1)}$$

To find the coefficients, multiply both sides of **(1)** by $x\left(x^2+8\right)$, and gather like terms:

$$5x^2+x+24=A(x^2+8)+x(Bx+C)=(A+B)x^2+Cx+8A \quad \textbf{(2)}$$

Equate corresponding coefficients in **(2)** to obtain the following system:

$$\textbf{(*)}\begin{cases} A+B=5 & \textbf{(3)} \\ C=1 & \textbf{(4)} \\ 8A=24 & \textbf{(5)} \end{cases}$$

Solve **(6)**: $A=3$

Substitute this value of A into **(4)**: $3+B=5 \Rightarrow B=2$

Thus, the partial fraction decomposition **(1)** becomes:

$$\boxed{\frac{5x^2+x+24}{x^3+8x}=\frac{3}{x}+\frac{2x+1}{x^2+8}}$$

123.
The partial fraction decomposition has the form:

$$\frac{2}{x(x+1)} = \frac{A}{x} + \frac{B}{x+1} \quad \textbf{(1)}$$

To find the coefficients, multiply both sides of **(1)** by $x(x+1)$, and gather like terms:

$$2 = A(x+1) + Bx$$
$$2 = (A+B)x + A \quad \textbf{(2)}$$

Equate corresponding coefficients in **(2)** to obtain the following system:

$$\textbf{(*)} \begin{cases} A+B=0 & \textbf{(3)} \\ A=2 & \textbf{(4)} \end{cases}$$

Now, solve system **(*)**:

Substitute **(4)** into **(3)** to see that $B=-2$.

Thus, the partial fraction decomposition **(1)** becomes:

$$\boxed{\frac{2}{x(x+1)} = \frac{-2}{x+1} + \frac{2}{x}}$$

124.
Observe that simplifying the expression first yields

$$\frac{\cancel{x}}{\cancel{x}(x+3)} = \frac{1}{x+3}.$$

This IS the partial fraction decomposition.

125. The partial fraction decomposition has the form:

$$\frac{5x-17}{(x+2)^2} = \frac{A}{x+2} + \frac{B}{(x+2)^2} \quad \textbf{(1)}$$

To find the coefficients, multiply both sides of **(1)** by $(x+2)^2$, and gather like terms:

$$5x-17 = A(x+2) + B$$
$$5x-17 = Ax + (2A+B) \quad \textbf{(2)}$$

Equate corresponding coefficients in **(2)** to obtain the following system:

$$\textbf{(*)} \begin{cases} A=5 & \textbf{(3)} \\ 2A+B=-17 & \textbf{(4)} \end{cases}$$

Substitute **(3)** into **(4)** to find B:

$$2(5)+B=-17 \text{ so that } B=-27$$

Thus, the partial fraction decomposition **(1)** becomes:

$$\boxed{\frac{5x-17}{(x+2)^2} = \frac{5}{x+2} - \frac{27}{(x+2)^2}}$$

126. The partial fraction decomposition has the form:

$$\frac{x^3}{\left(x^2+64\right)^2}=\frac{Ax+B}{x^2+64}+\frac{Cx+D}{\left(x^2+64\right)^2} \quad \textbf{(1)}$$

To find the coefficients, multiply both sides of **(1)** by $\left(x^2+64\right)^2$, and gather like terms:

$$\begin{aligned} x^3 &= (Ax+B)\left(x^2+64\right)+(Cx+D) \\ &= Ax^3+Bx^2+64Ax+64B+Cx+D \\ &= Ax^3+Bx^2+(64A+C)x+(64B+D) \quad \textbf{(2)} \end{aligned}$$

Equate corresponding coefficients in **(2)** to obtain the following system:

$$(*)\begin{cases} A=1 & \textbf{(3)} \\ B=0 & \textbf{(4)} \\ 64A+C=0 & \textbf{(5)} \\ 64B+D=0 & \textbf{(6)} \end{cases}$$

Substitute **(3)** into **(5)** to find C: $64(1)+C=0$ so that $C=-64$

Substitute **(4)** into **(6)** to find D: $64(0)+D=0$ so that $D=0$

Thus, the partial fraction decomposition **(1)** becomes:

$$\boxed{\frac{x^3}{\left(x^2+64\right)^2}=\frac{x}{x^2+64}-\frac{64x}{\left(x^2+64\right)^2}}$$

127.

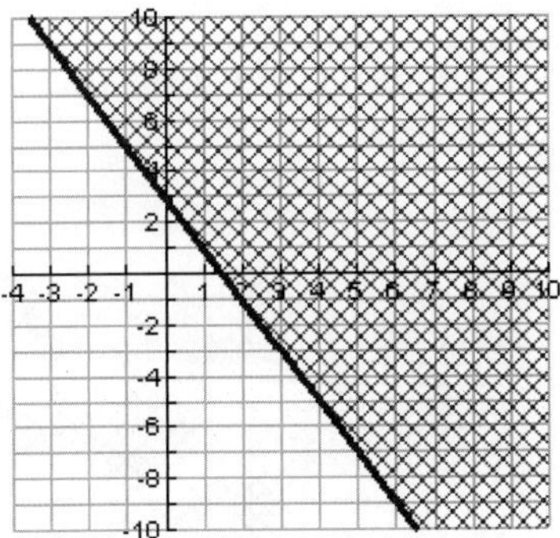

The solid curve is the graph of $y=-2x+3$.

128.

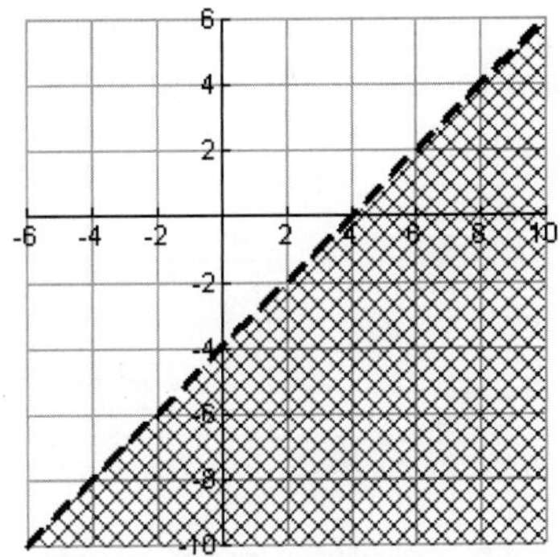

The dashed curve is the graph of $y=x-4$.

129. Shade above the dashed line $y = \frac{1}{4}(5 - 2x)$.

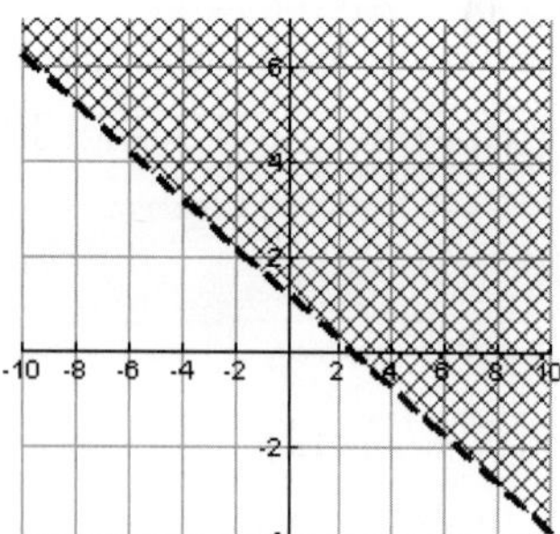

130. Shade below the solid line $y = \frac{1}{2}(4 - 5x)$.

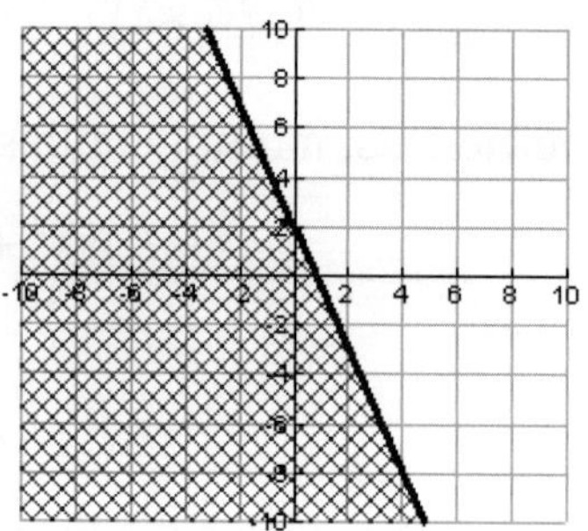

131. Shade above the solid line $y = -3x + 2$.

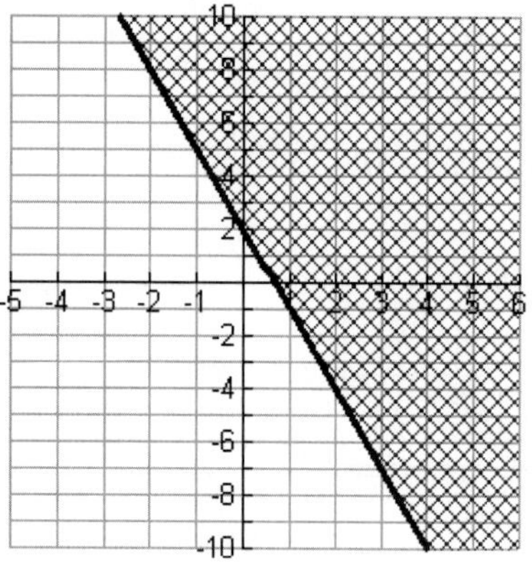

132. Shade below the dashed line $y = x - 2$.

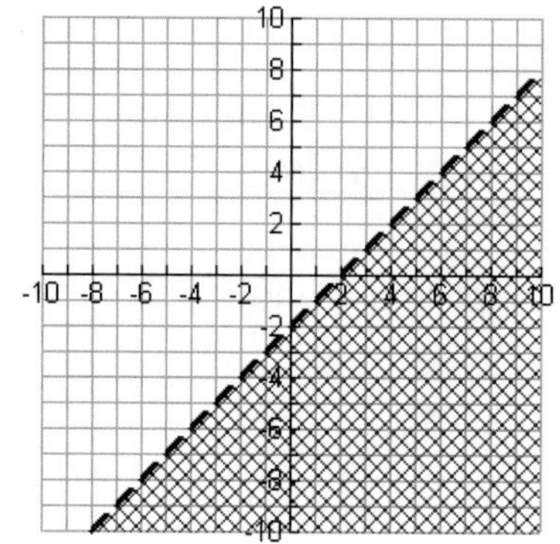

133. Shade below the solid line $y = -\frac{3}{8}x + 2$.

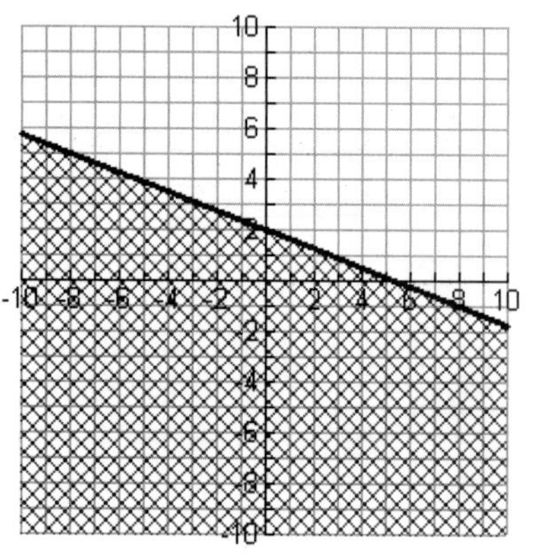

134. Shade above the solid line $y = \frac{2}{9}x - 2$

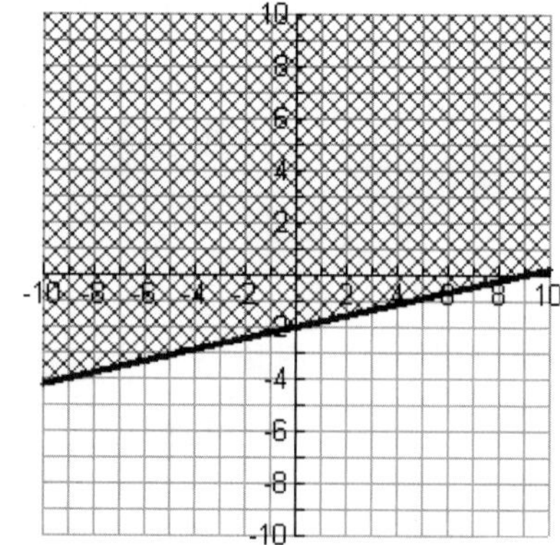

135.

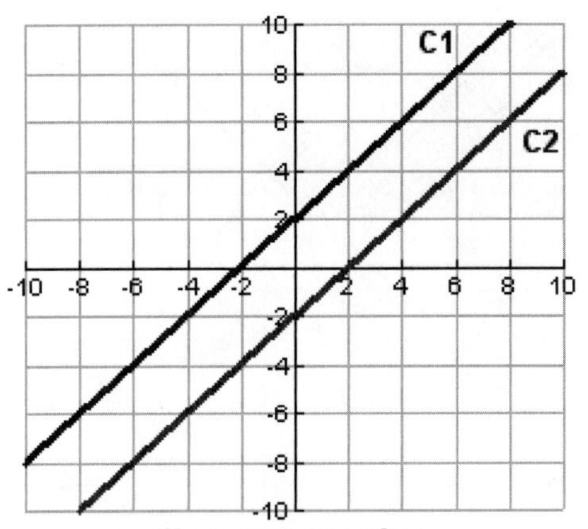

Notes on the graph:

C1: $y = x + 2$

C2: $y = x - 2$

Since there is no region in common with both inequalities, the system has no solution.

136.

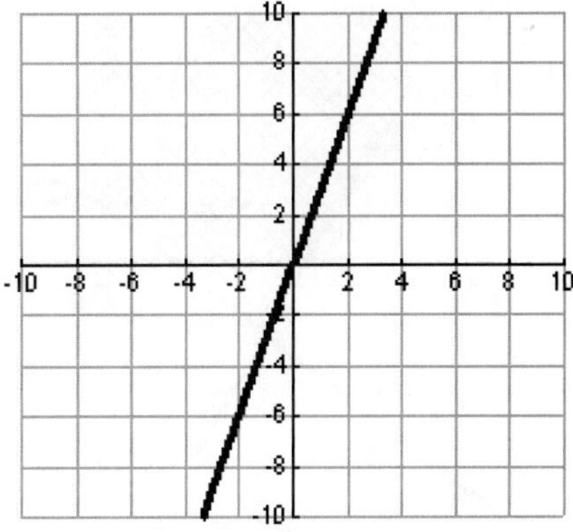

The common region this time is the line $y = 3x$ itself.

137.

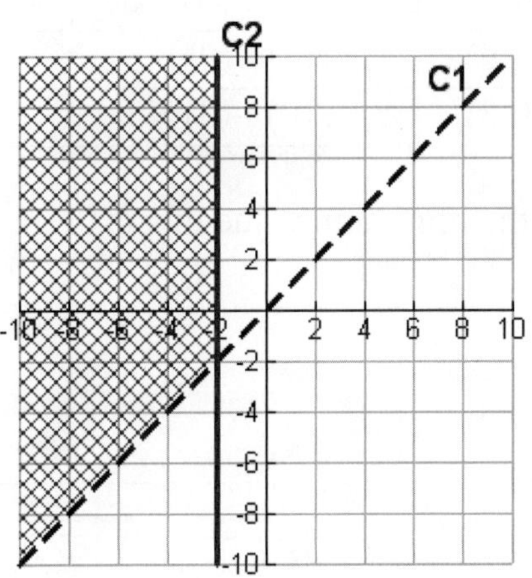

Notes on the graph:

C1: $y = x$

C2: $x = -2$

138.

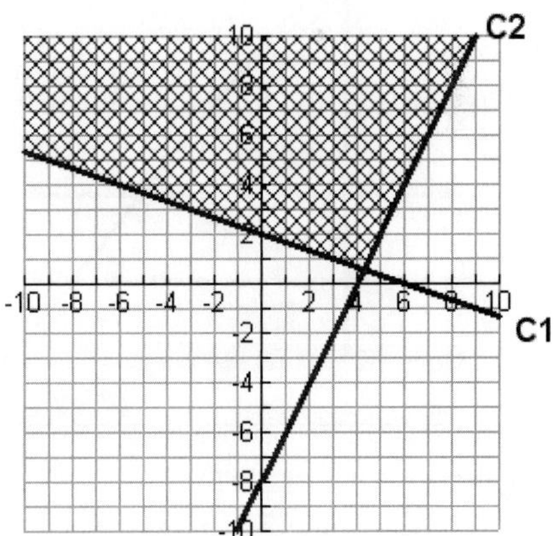

Notes on the graph:

C1: $y = -\frac{1}{3}x + 2$

C2: $y = 2x - 8$

139.

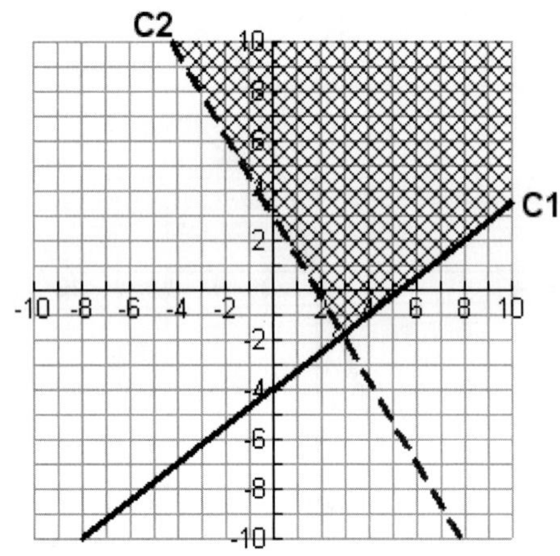

Notes on the graph:

C1: $y = \frac{3}{4}x - 4$

C2: $y = 3 - \frac{5}{3}x$

140.

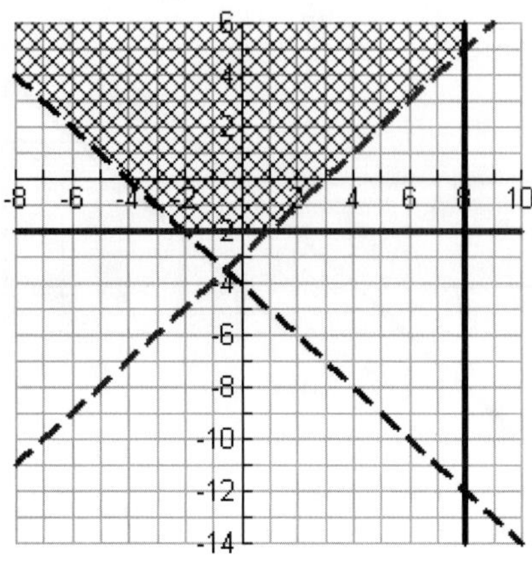

Notes on the graph:

C1: $y = -x - 4$ **C3:** $y = -2$

C2: $y = x - 3$ **C4:** $x = 8$

141. The region in this case is:

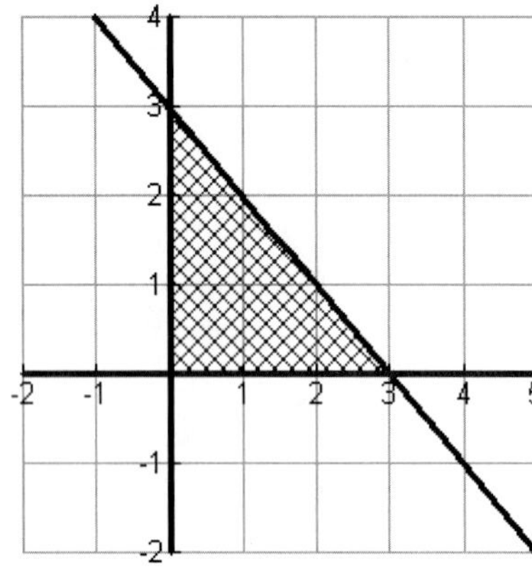

Vertex	Objective Function $z = 2x + y$
$(0,0)$	$z = 2(0) + (0) = 0$
$(3,0)$	$z = 2(3) + (0) = 6$
$(0,3)$	$z = 2(0) + (3) = 3$

So, the minimum value of z is 0.

142. The region in this case is:

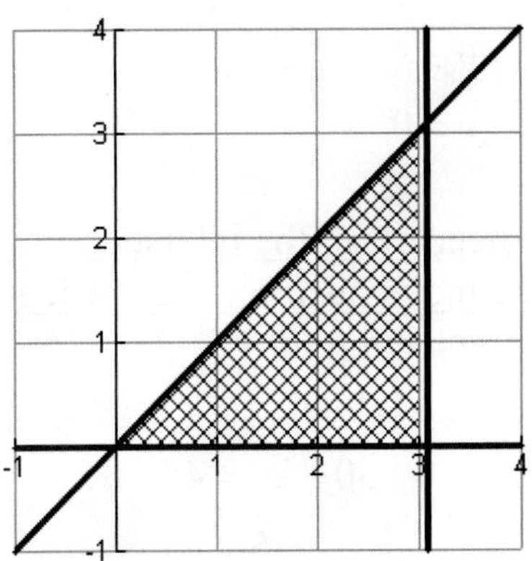

Vertex	Objective Function $z = 2x + 3y$
$(0,0)$	$z = 2(0) + 3(0) = 0$
$(3,0)$	$z = 2(3) + 3(0) = 6$
$(3,3)$	$z = 2(3) + 3(3) = 15$

So, the maximum value of z is 15.

143. The region in this case is:

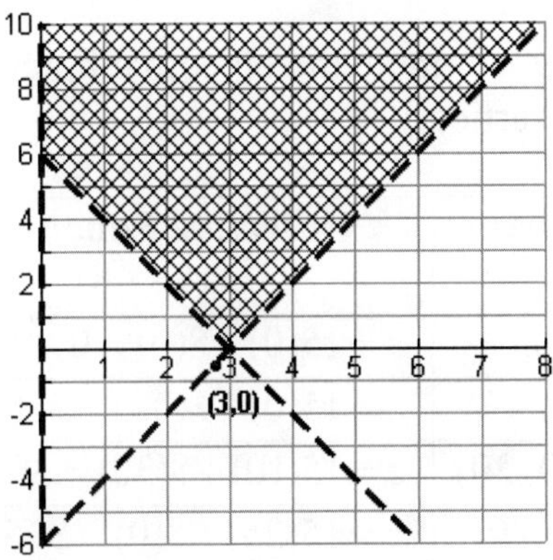

Vertex	Objective Function $z = 3x - 5y$
$(3,0)$	9
$(0,6)$	-30

So, the minimum value of z is -30 and occurs at (0,6).

144. The region in this case is:

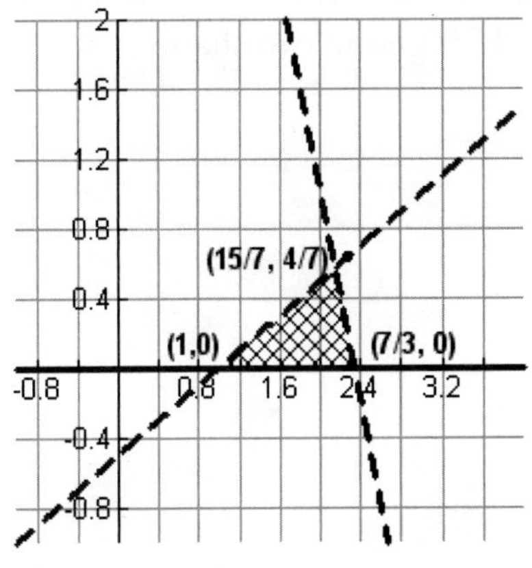

Vertex	Objective Function $z = -2x + 7y$
$\left(\frac{7}{3},0\right)$	$-\frac{14}{3}$
$\left(\frac{15}{7},\frac{4}{7}\right)$	$-\frac{2}{7}$
$(1,0)$	-2

So, the maximum value of z is $-\frac{2}{7}$.

145. Let x = number of ocean watercolor coaster sets
y = number of geometric shape coaster sets

Profit from ocean watercolor sets:
Revenue – cost = $15x$
Profit from geometric shape sets:
Revenue – cost = $8y$

So, to maximize profit, we must maximize the objective function $z = 15x + 8y$. We have the following constraints:

$$\begin{cases} 4x + 2y \le 100 \\ 3x + 2y \le 90 \\ x \ge 0,\ y \ge 0 \end{cases}$$

The region in this case is:

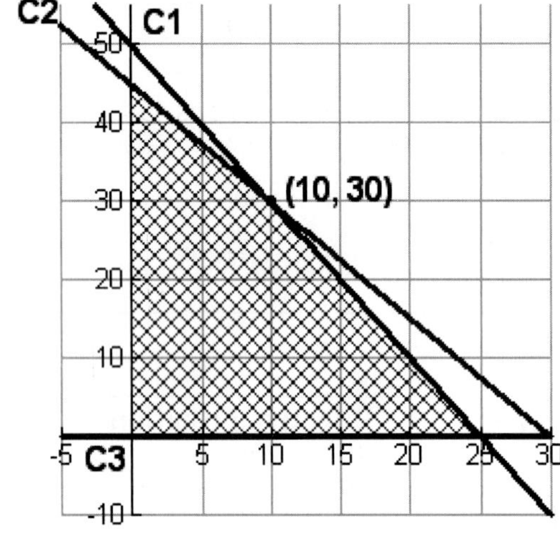

Notes on the graph:
C1: $y = 50 - 2x$
C2: $y = 45 - \frac{3}{2}x$
C3: $y = 0$

We need all of the intersection points since they constitute the vertices:

Intersection of $4x + 2y = 100$ and $3x + 2y = 90$:

$$50 - 2x = 45 - \tfrac{3}{2}x$$
$$5 = \tfrac{1}{2}x$$
$$x = 10$$

So, the intersection point is (10, 30).

Now, compute the objective function at the vertices:

Vertex	Objective Function $z = 15x + 8y$
(0,0)	$z = 15(0) + 8(0) = 0$
(0, 45)	$z = 15(0) + 8(45) = 360$
(10, 30)	$z = 15(10) + 8(30) = 390$
(25,0)	$z = 15(25) + 8(0) = 375$

So, to attain maximum profit, she should sell 10 ocean watercolor coaster sets and 30 geometric shape coaster sets.

146. Let x = number of ocean watercolor coaster sets
y = number of geometric shape coaster sets

Profit from ocean watercolor sets:
Revenue – cost = $15x$
Profit from geometric shape sets:
Revenue – cost = $8y$

So, to maximize profit, we must maximize the objective function $z = 15x + 8y$. We have the following constraints:

$$\begin{cases} 4x + 2y \le 300 \\ 3x + 2y \le 90 \\ x \ge 0,\ y \ge 0 \end{cases}$$

The region in this case is:

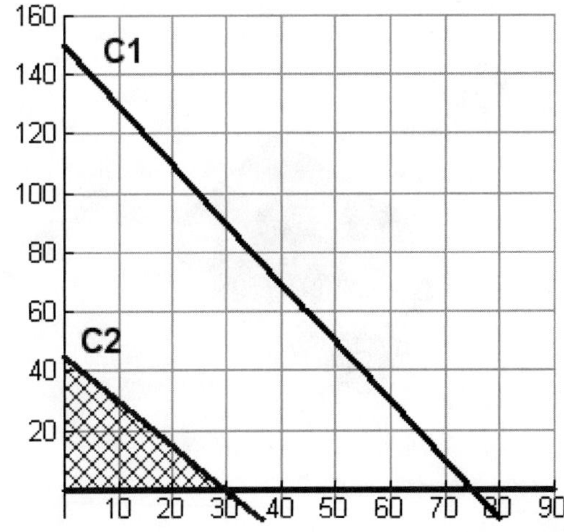

Notes on the graph:
C1: $y = 150 - 2x$
C2: $y = 45 - \frac{3}{2}x$
C3: $y = 0$

We need all of the intersection points since they constitute the vertices:
Intersection of $4x + 2y = 300$ and $3x + 2y = 90$:

$$150 - 2x = 45 - \tfrac{3}{2}x$$
$$105 = \tfrac{1}{2}x$$
$$x = 210$$

So, the intersection point is (210, -270), which is not used to form the region since both x and y need to be nonnegative.

Now, compute the objective function at the vertices:

Vertex	Objective Function $z = 15x + 8y$
(0, 45)	$z = 15(0) + 8(45) = 360$
(30, 0)	$z = 15(30) + 8(0) = 450$
(0,0)	$z = 15(0) + 8(0) = 0$

So, to attain a maximum profit of $450, she should sell 30 ocean watercolor coaster sets and 0 geometric shape coaster sets.

147. The graph of this system of equation is as follows: 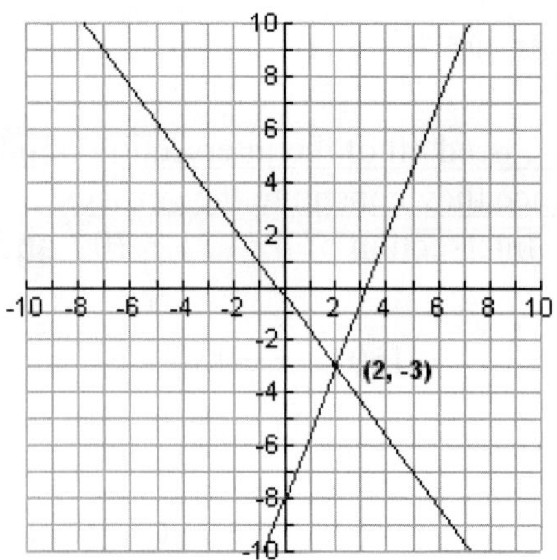 The solution is (2, -3).	**148.** The graph of this system of equation is as follows: 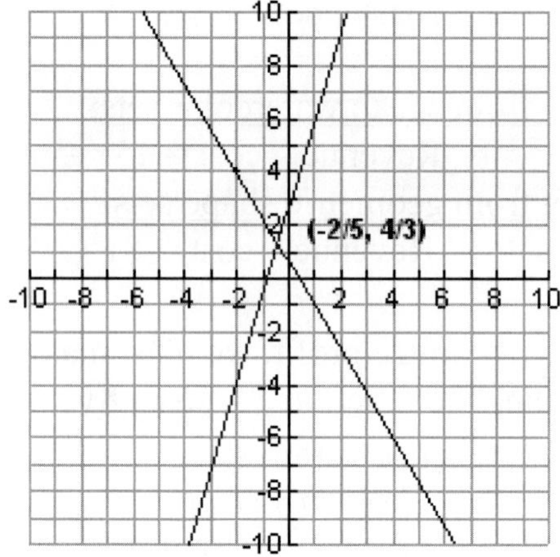 The solution is $\left(-\frac{2}{5},\frac{4}{3}\right)$.
149. If your calculator has 3-dimensional graphing capabilities, graph the system to obtain the following: 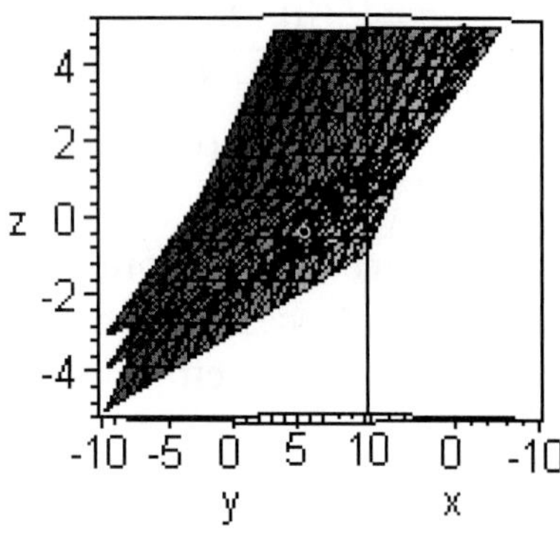 It is difficult to see from this graph, but zooming in will enable you to see that the solution is (3.6, 3, 0.8).	**150.** If your calculator has 3-dimensional graphing capabilities, graph the system to obtain the following:  Although this may seem strange, observe that if you divide the first equation by 4 on both sides, you obtain the second equation. Hence, every point on this plane is a solution. These points are expressed as $(0.75a-1.5b-13, a.b)$.

151.

a.

```
rref([A])
[[1 0 0 .161066…
 [0 1 0 -.04946…
 [0 0 1 -4.1012…
```

$y = 0.16x^2 - 0.05x - 4.10$

b.

```
QuadReg
 y=ax²+bx+c
 a=.1610661269
 b=-.0494601889
 c=-4.101214575
```

$y = 0.16x^2 - 0.05x - 4.10$

152.

a.

```
QuadReg
 y=ax²+bx+c
 a=.2862337662
 b=-2.570649351
 c=-4.862337662
```

$y = 0.29x^2 - 2.57x - 4.86$

b.

```
QuadReg
 y=ax²+bx+c
 a=.3616161616
 b=-.9232323232
 c=-17.6
```

$y = 0.29x^2 - 2.57x - 4.86$

153. $\begin{bmatrix} -238 & 206 & 50 \\ -113 & 159 & 135 \\ 40 & -30 & 0 \end{bmatrix}$

154. $\begin{bmatrix} -143 & -41 \\ -82 & 64 \end{bmatrix}$

155. The augmented matrix to enter into the calculator is

$$\left[\begin{array}{cc|c} 6.1 & -14.2 & 75.495 \\ -2.3 & 7.2 & -36.495 \end{array}\right].$$

Solving using the calculator then yields the solution of the system as (2.25, -4.35).

156. The augmented matrix to enter into the calculator is

$$\left[\begin{array}{ccc|c} 7.2 & 3.2 & -1.7 & 5.53 \\ -1.3 & 4.1 & 2.8 & -23.949 \\ 5.2 & -1.8 & 6.2 & 48.596 \end{array}\right]$$

Solving using the calculator then yields the solution of the system as (4.15, -6.26, 2.54).

157. First, write the system in matrix form as:

$$\begin{bmatrix} 4.5 & -8.7 \\ -1.4 & 5.3 \end{bmatrix}\begin{bmatrix} x \\ y \end{bmatrix} = \begin{bmatrix} -72.33 \\ 31.32 \end{bmatrix}$$

Now, apply Cramer's rule to solve the system:

$$D = \begin{vmatrix} 4.5 & -8.7 \\ -1.4 & 5.3 \end{vmatrix}, \quad D_x = \begin{vmatrix} -72.33 & -8.7 \\ 31.32 & 5.3 \end{vmatrix}, \quad D_y = \begin{vmatrix} 4.5 & -72.33 \\ -1.4 & 31.32 \end{vmatrix}$$

Using a calculator to compute the determinants, we obtain from Cramer's rule that:

$$\boxed{x = \frac{D_x}{D} = -9.5 \quad y = \frac{D_y}{D} = 3.4}$$

158. First, write the system in matrix form as:

$$\begin{bmatrix} 1.4 & 3.6 & 7.5 \\ 2.1 & -5.7 & -4.2 \\ 1.8 & -2.8 & -6.2 \end{bmatrix}\begin{bmatrix} x \\ y \\ z \end{bmatrix} = \begin{bmatrix} 42.08 \\ 5.37 \\ -9.86 \end{bmatrix}$$

Now, apply Cramer's rule to solve the system:

$$D = \begin{vmatrix} 1.4 & 3.6 & 7.5 \\ 2.1 & -5.7 & -4.2 \\ 1.8 & -2.8 & -6.2 \end{vmatrix}, \quad D_x = \begin{vmatrix} 42.08 & 3.6 & 7.5 \\ 5.37 & -5.7 & -4.2 \\ -9.86 & -2.8 & -6.2 \end{vmatrix},$$

$$D_y = \begin{vmatrix} 1.4 & 42.08 & 7.5 \\ 2.1 & 5.37 & -4.2 \\ 1.8 & -9.86 & -6.2 \end{vmatrix}, \quad D_z = \begin{vmatrix} 1.4 & 3.6 & 42.08 \\ 2.1 & -5.7 & 5.37 \\ 1.8 & -2.8 & -9.86 \end{vmatrix}$$

Using a calculator to compute the determinants, we obtain from Cramer's rule that:

$$\boxed{x = \frac{D_x}{D} = 8.5 \quad y = \frac{D_y}{D} = -1.2 \quad z = \frac{D_z}{D} = 4.6}$$

159. Since the graphs coincide, y2 is the partial fraction decomposition of y1.

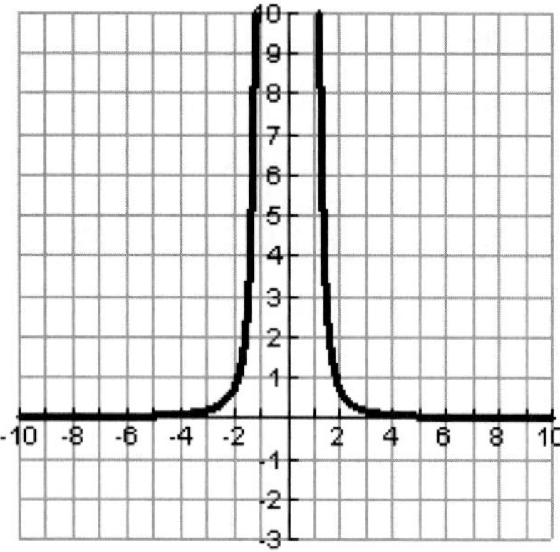

160. Since the graphs do not coincide, y2 is not the partial fraction decomposition of y1.

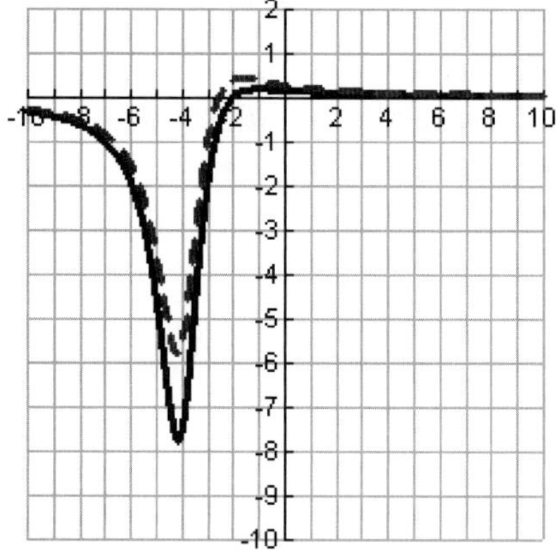

161. The region is as follows.

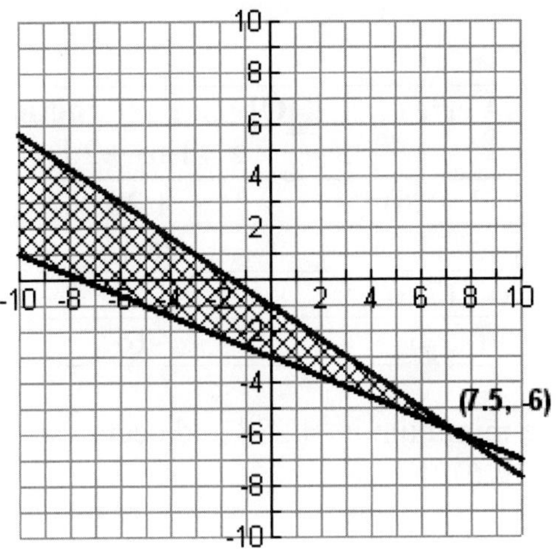

162. The region is as follows.

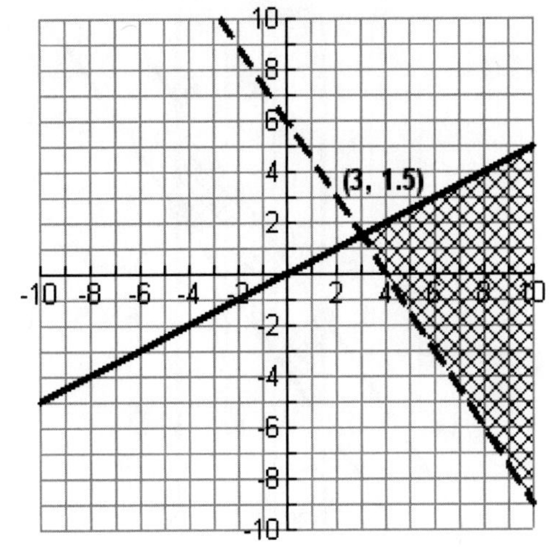

163.

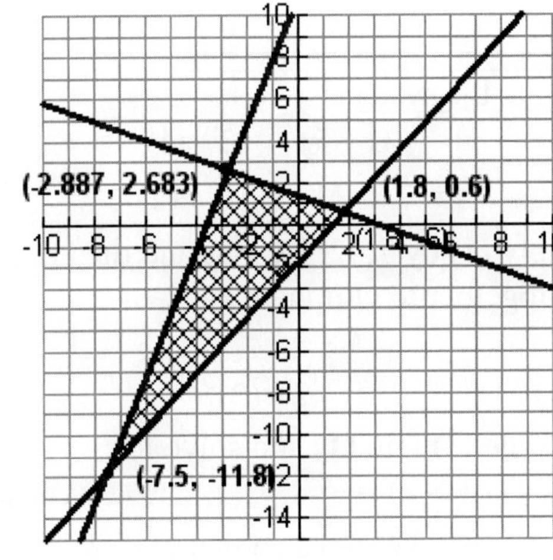

Vertex	Objective Function $z = 6.2x + 1.5y$
(-2.7, 2.6)	-12.84
(-7.5, -11.8)	-64.2
(1.8, 0.6)	12.06

The maximum is 12.06 and occurs at (1.8, 0.6).

164.

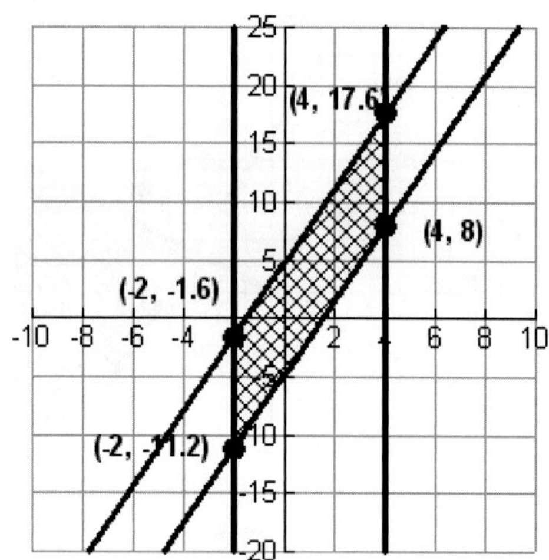

Vertex	Objective Function $z = 1.6x - 2.8y$
(-2, -1.6)	1.28
(-2, -11.2)	28.16
(4, 17.6)	-42.88
(4, 8)	-16

The minimum is -42.88 and occurs at (4, 17.6).

Chapter 8 Practice Test Solutions

1. Solve the system:

$$\begin{cases} x - 2y = 1 & \textbf{(1)} \\ -x + 3y = 2 & \textbf{(2)} \end{cases}$$

Add **(1)** and **(2)**: $y = 3$

Substitute $y = 3$ into **(1)** to find x:

$$x - 2(3) = 1$$

$$x = 7$$

So, the solution is $\boxed{(7,3)}$.

2. Solve the system:

$$\begin{cases} 3x + 5y = -2 & \textbf{(1)} \\ 7x + 11y = -6 & \textbf{(2)} \end{cases}$$

Multiply **(1)** by 7, **(2)** by -3, and then add them together to eliminate x:

$$2y = 4$$

$$y = 2 \quad \textbf{(3)}$$

Substitute **(3)** into **(1)** and solve for x:

$$3x + 5(2) = -2$$

$$3x = -12 \text{ so that } x = -4$$

So, the solution is $\boxed{(-4,2)}$.

3. Solve the system:

$$\begin{cases} x - y = 2 & \textbf{(1)} \\ -2x + 2y = -4 & \textbf{(2)} \end{cases}$$

Multiply **(1)** by 2, and then add to **(2)**:

$$0 = 0$$

So, the system is consistent. There are infinitely many solutions of the form $\boxed{x = a,\ y = a - 2}$.

4. Solve the system: $\begin{cases} 3x - 2y = 5 & \textbf{(1)} \\ 6x - 4y = 0 & \textbf{(2)} \end{cases}$

Multiply **(1)** by -2, and then add to **(2)**:

$$0 = -10$$

Since this is a false statement, we conclude that the system has $\boxed{\text{no solution}}$.

5. Solve the system:

$$\begin{cases} x+y+z=-1 & \textbf{(1)} \\ 2x+y+z=0 & \textbf{(2)} \\ -x+y+2z=0 & \textbf{(3)} \end{cases}$$

Step 1: Obtain a system of 2 equations in 2 unknowns by eliminating the same variable in two pairs of equations.

Multiply **(1)** by -2, and then add to **(2)** to eliminate x:

$$-y-z=2 \quad \textbf{(4)}$$

Add **(1)** and **(3)** to eliminate x:

$$2y+3z=-1 \quad \textbf{(5)}$$

These steps yield system:

$$\textbf{(*)} \begin{cases} -y-z=2 & \textbf{(4)} \\ 2y+3z=-1 & \textbf{(5)} \end{cases}$$

Step 2: Solve system **(*)** from Step 1.

Multiply **(4)** by 2, and then add to **(5)**:

$$z=3 \quad \textbf{(6)}$$

Substitute **(6)** into **(4)** to find y:

$$-y-3=2 \text{ so that } y=-5 \quad \textbf{(7)}$$

Step 3: Use the solution of the system in Step 2 to find the value of the third variable in the original system.

Substitute **(6)** and **(7)** into **(1)** to find x:

$$x-5+3=-1$$

$$x=1$$

Thus, the solution is:

$\boxed{x=1,\ y=-5,\ z=3}$.

6. Solve the system:

$$\begin{cases} 6x+9y+z=5 & \textbf{(1)} \\ 2x-3y+z=3 & \textbf{(2)} \\ 10x+12y+2z=9 & \textbf{(3)} \end{cases}$$

Step 1: Obtain a system of 2 equations in 2 unknowns by eliminating the same variable in two pairs of equations.

Multiply **(2)** by -3, and then add to **(1)** to eliminate x:

$$18y-2z=-4 \quad \textbf{(4)}$$

Multiply **(2)** by -5, and then add to **(3)** to eliminate x:

$$27y-3z=-6 \quad \textbf{(5)}$$

These steps lead to the system:

$$\begin{cases} 18y-2z=-4 & \textbf{(4)} \\ 27y-3z=-6 & \textbf{(5)} \end{cases}$$

which is equivalent to

$$\begin{cases} 9y-z=-2 \\ 9y-z=-2 \end{cases} \quad \textbf{(6)}$$

Hence, we know that the system has infinitely many solutions.

To determine them, let $z=a$. Substitute this into **(6)** to find y:

$$y=\tfrac{1}{9}(a-2)$$

Now, substitute the values of y and z into **(2)** to find x:

$$2x-3(\tfrac{1}{9}(a-2))+a=3$$

$$2x=\tfrac{1}{3}(a-2)-a+3$$

$$2x=-\tfrac{2}{3}a+\tfrac{7}{3}$$

$$x=-\tfrac{1}{3}a+\tfrac{7}{6}$$

Thus, the solution is:

$\boxed{x=-\tfrac{1}{3}a+\tfrac{7}{6},\ y=\tfrac{1}{9}(a-2),\ z=a}$.

7. Augmented matrix is $\left[\begin{array}{ccc|c} 6 & 9 & 1 & 5 \\ 2 & -3 & 1 & 3 \\ 10 & 12 & 2 & 9 \end{array}\right]$ Matrix equation is $\begin{bmatrix} 6 & 9 & 1 \\ 2 & -3 & 1 \\ 10 & 12 & 2 \end{bmatrix}\begin{bmatrix} x \\ y \\ z \end{bmatrix}=\begin{bmatrix} 5 \\ 3 \\ 9 \end{bmatrix}$

8. Augmented matrix is $\left[\begin{array}{ccc|c} 3 & 2 & -10 & 2 \\ 1 & 1 & -1 & 5 \end{array}\right]$ You cannot solve by finding the inverse since there are fewer equations than variables. .

9.

$$\begin{bmatrix} 1 & 3 & 5 \\ 2 & 7 & -1 \\ -3 & -2 & 0 \end{bmatrix} \xrightarrow{R_2-2R_1\to R_2} \begin{bmatrix} 1 & 3 & 5 \\ 0 & 1 & -11 \\ -3 & -2 & 0 \end{bmatrix} \xrightarrow{R_3+3R_1\to R_3} \begin{bmatrix} 1 & 3 & 5 \\ 0 & 1 & -11 \\ 0 & 7 & 15 \end{bmatrix}$$

10.

$$\left[\begin{array}{ccc|c} 2 & -1 & 1 & 3 \\ 1 & 1 & -1 & 0 \\ 3 & 2 & -2 & 1 \end{array}\right] \xrightarrow[R_3-3R_2\to R_3]{R_1-2R_2\to R_1} \left[\begin{array}{ccc|c} 0 & -3 & 3 & 3 \\ 1 & 1 & -1 & 0 \\ 0 & -1 & 1 & 1 \end{array}\right] \xrightarrow{R_1\leftrightarrow R_2} \left[\begin{array}{ccc|c} 1 & 1 & -1 & 0 \\ 0 & -3 & 3 & 3 \\ 0 & -1 & 1 & 1 \end{array}\right]$$

$$\xrightarrow{R_2-3R_3\to R_3} \left[\begin{array}{ccc|c} 1 & 1 & -1 & 0 \\ 0 & -3 & 3 & 3 \\ 0 & 0 & 0 & 0 \end{array}\right] \xrightarrow{-\frac{1}{3}R_2\to R_2} \left[\begin{array}{ccc|c} 1 & 1 & -1 & 0 \\ 0 & 1 & -1 & -1 \\ 0 & 0 & 0 & 0 \end{array}\right] \xrightarrow{R_1-R_2\to R_1} \left[\begin{array}{ccc|c} 1 & 0 & 0 & 1 \\ 0 & 1 & -1 & -1 \\ 0 & 0 & 0 & 0 \end{array}\right]$$

11. Reduce the augmented matrix down to row echelon form:

$$\left[\begin{array}{ccc|c} 6 & 9 & 1 & 5 \\ 2 & -3 & 1 & 3 \\ 10 & 12 & 2 & 9 \end{array}\right] \xrightarrow[R_3-5R_2\to R_3]{R_1-3R_2\to R_2} \left[\begin{array}{ccc|c} 6 & 9 & 1 & 5 \\ 0 & 18 & -2 & -4 \\ 0 & 27 & -3 & -6 \end{array}\right] \xrightarrow[\frac{1}{6}R_1\to R_1]{\frac{1}{18}R_2\to R_2,\ \frac{1}{27}R_3\to R_3} \left[\begin{array}{ccc|c} 1 & \frac{3}{2} & \frac{1}{6} & \frac{5}{6} \\ 0 & 1 & -\frac{1}{9} & -\frac{2}{9} \\ 0 & 1 & -\frac{1}{9} & -\frac{2}{9} \end{array}\right]$$

$$\xrightarrow{R_2-R_3\to R_3} \left[\begin{array}{ccc|c} 1 & \frac{3}{2} & \frac{1}{6} & \frac{5}{6} \\ 0 & 1 & -\frac{1}{9} & -\frac{2}{9} \\ 0 & 0 & 0 & 0 \end{array}\right] \xrightarrow{R_1-\frac{3}{2}R_2\to R_1} \left[\begin{array}{ccc|c} 1 & 0 & \frac{1}{3} & \frac{7}{6} \\ 0 & 1 & -\frac{1}{9} & -\frac{2}{9} \\ 0 & 0 & 0 & 0 \end{array}\right]$$

From the last row, we conclude that the system has infinitely many solutions. To find them, let $z=a$. Then, the second row implies that $y-\frac{1}{9}a=-\frac{2}{9}$ so that $y=\frac{1}{9}a-\frac{2}{9}$. So, substituting these values of y and z into the first row yields $x+\frac{3}{2}(\frac{1}{9}a-\frac{2}{9})+\frac{1}{6}a=\frac{5}{6}$, so that $x=-\frac{1}{3}a+\frac{7}{6}$. Hence, the solutions are $\boxed{x=-\frac{1}{3}a+\frac{7}{6},\ y=\frac{1}{9}a-\frac{2}{9},\ z=a}$.

12. Reduce the augmented matrix down to row echelon form:

$$\left[\begin{array}{ccc|c} 3 & 2 & -10 & 2 \\ 1 & 1 & -1 & 5 \end{array}\right] \xrightarrow[\frac{1}{3}R_1\to R_1]{-3R_2+R_1\to R_2} \left[\begin{array}{ccc|c} 1 & \frac{2}{3} & -\frac{10}{3} & \frac{2}{3} \\ 0 & -1 & -7 & -13 \end{array}\right] \xrightarrow{-R_2\to R_2} \left[\begin{array}{ccc|c} 1 & \frac{2}{3} & -\frac{10}{3} & \frac{2}{3} \\ 0 & 1 & 7 & 13 \end{array}\right]$$

From the last row, we conclude that the system has infinitely many solutions. To find them, let $z=t$. Then, the second row implies that $y+7t=13$ so that $y=-7t+13$. So, substituting these values of y and z into the first row yields $x+\frac{2}{3}(-7t+13)-\frac{10}{3}(t)=\frac{2}{3}$, so that $x=8t-8$. Hence, the solutions are $\boxed{x=8t-8,\ y=-7t+13,\ z=t}$.

13. $\begin{bmatrix} 1 & -2 & 5 \\ 0 & -1 & 3 \end{bmatrix} \cdot \begin{bmatrix} 0 & 4 \\ 3 & -5 \\ -1 & 1 \end{bmatrix} = \begin{bmatrix} -11 & 19 \\ -6 & 8 \end{bmatrix}$	**14.** Not possible since the matrices have different dimensions.

15. $\begin{bmatrix} 4 & 3 \\ 5 & -1 \end{bmatrix}^{-1} = \frac{1}{4(-1)-5(3)} \begin{bmatrix} -1 & -3 \\ -5 & 4 \end{bmatrix} = \frac{1}{-19} \begin{bmatrix} -1 & -3 \\ -5 & 4 \end{bmatrix} = \begin{bmatrix} \frac{1}{19} & \frac{3}{19} \\ \frac{5}{19} & -\frac{4}{19} \end{bmatrix}$

16. Using Gaussian elimination, we obtain the following:

$$\left[\begin{array}{ccc|ccc} 1 & -3 & 2 & 1 & 0 & 0 \\ 4 & 2 & 0 & 0 & 1 & 0 \\ -1 & 2 & 5 & 0 & 0 & 1 \end{array}\right] \xrightarrow[R_3+R_1 \to R_3]{R_2-4R_1 \to R_2} \left[\begin{array}{ccc|ccc} 1 & -3 & 2 & 1 & 0 & 0 \\ 0 & 14 & -8 & -4 & 1 & 0 \\ 0 & -1 & 7 & 1 & 0 & 1 \end{array}\right]$$

$$\xrightarrow{R_2+14R_3 \to R_3} \left[\begin{array}{ccc|ccc} 1 & -3 & 2 & 1 & 0 & 0 \\ 0 & 14 & -8 & -4 & 1 & 0 \\ 0 & 0 & 90 & 10 & 1 & 14 \end{array}\right] \xrightarrow{\frac{1}{90}R_3 \to R_3} \left[\begin{array}{ccc|ccc} 1 & -3 & 2 & 1 & 0 & 0 \\ 0 & 14 & -8 & -4 & 1 & 0 \\ 0 & 0 & 1 & \frac{1}{9} & \frac{1}{90} & \frac{14}{90} \end{array}\right]$$

$$\xrightarrow{R_2+8R_3 \to R_2} \left[\begin{array}{ccc|ccc} 1 & -3 & 2 & 1 & 0 & 0 \\ 0 & 14 & 0 & -\frac{28}{9} & \frac{98}{90} & \frac{112}{90} \\ 0 & 0 & 1 & \frac{1}{9} & \frac{1}{90} & \frac{14}{90} \end{array}\right] \xrightarrow{\frac{1}{14}R_2 \to R_2} \left[\begin{array}{ccc|ccc} 1 & -3 & 2 & 1 & 0 & 0 \\ 0 & 1 & 0 & -\frac{28}{9(14)} & \frac{98}{90(14)} & \frac{112}{90(14)} \\ 0 & 0 & 1 & \frac{1}{9} & \frac{1}{90} & \frac{7}{45} \end{array}\right]$$

$$= \left[\begin{array}{ccc|ccc} 1 & -3 & 2 & 1 & 0 & 0 \\ 0 & 1 & 0 & -\frac{2}{9} & \frac{7}{90} & \frac{4}{45} \\ 0 & 0 & 1 & \frac{1}{9} & \frac{1}{90} & \frac{7}{45} \end{array}\right] \xrightarrow{R_1+3R_2-2R_3 \to R_1} \left[\begin{array}{ccc|ccc} 1 & 0 & 0 & \frac{1}{9} & \frac{19}{90} & -\frac{2}{45} \\ 0 & 1 & 0 & -\frac{2}{9} & \frac{7}{90} & \frac{4}{45} \\ 0 & 0 & 1 & \frac{1}{9} & \frac{1}{90} & \frac{7}{45} \end{array}\right]$$

So, $A^{-1} = \begin{bmatrix} \frac{1}{9} & \frac{19}{90} & -\frac{2}{45} \\ -\frac{2}{9} & \frac{7}{90} & \frac{4}{45} \\ \frac{1}{9} & \frac{1}{90} & \frac{7}{45} \end{bmatrix}$.

17. The system in matrix form is:

$$\begin{bmatrix} 3 & -1 & 4 \\ 1 & 2 & 3 \\ -4 & 6 & -1 \end{bmatrix}\begin{bmatrix} x \\ y \\ z \end{bmatrix} = \begin{bmatrix} 18 \\ 20 \\ 11 \end{bmatrix}.$$

The solution is:

$$\begin{bmatrix} x \\ y \\ z \end{bmatrix} = \begin{bmatrix} 3 & -1 & 4 \\ 1 & 2 & 3 \\ -4 & 6 & -1 \end{bmatrix}^{-1}\begin{bmatrix} 18 \\ 20 \\ 11 \end{bmatrix}$$

We calculate the inverse below:

$$\left[\begin{array}{ccc|ccc} 3 & -1 & 4 & 1 & 0 & 0 \\ 1 & 2 & 3 & 0 & 1 & 0 \\ -4 & 6 & -1 & 0 & 0 & 1 \end{array}\right] \xrightarrow{R_1 \leftrightarrow R_2} \left[\begin{array}{ccc|ccc} 1 & 2 & 3 & 0 & 1 & 0 \\ 3 & -1 & 4 & 1 & 0 & 0 \\ -4 & 6 & -1 & 0 & 0 & 1 \end{array}\right] \xrightarrow[R_3+4R_1 \to R_3]{R_2-3R_1 \to R_2} \left[\begin{array}{ccc|ccc} 1 & 2 & 3 & 0 & 1 & 0 \\ 0 & -7 & -5 & 1 & -3 & 0 \\ 0 & 14 & 11 & 0 & 4 & 1 \end{array}\right]$$

$$\xrightarrow{R_3+2R_2 \to R_3} \left[\begin{array}{ccc|ccc} 1 & 2 & 3 & 0 & 1 & 0 \\ 0 & -7 & -5 & 1 & -3 & 0 \\ 0 & 0 & 1 & 2 & -2 & 1 \end{array}\right] \xrightarrow{-\frac{1}{7}R_2 \to R_2} \left[\begin{array}{ccc|ccc} 1 & 2 & 3 & 0 & 1 & 0 \\ 0 & 1 & \frac{5}{7} & -\frac{1}{7} & \frac{3}{7} & 0 \\ 0 & 0 & 1 & 2 & -2 & 1 \end{array}\right]$$

$$\xrightarrow{R_2-\frac{5}{7}R_3 \to R_2} \left[\begin{array}{ccc|ccc} 1 & 2 & 3 & 0 & 1 & 0 \\ 0 & 1 & 0 & -\frac{11}{7} & \frac{13}{7} & -\frac{5}{7} \\ 0 & 0 & 1 & 2 & -2 & 1 \end{array}\right] \xrightarrow{R_1-2R_2-3R_3 \to R_1} \left[\begin{array}{ccc|ccc} 1 & 0 & 0 & -\frac{20}{7} & \frac{23}{7} & -\frac{11}{7} \\ 0 & 1 & 0 & -\frac{11}{7} & \frac{13}{7} & -\frac{5}{7} \\ 0 & 0 & 1 & 2 & -2 & 1 \end{array}\right]$$

Hence, the solution is

$$\begin{bmatrix} x \\ y \\ z \end{bmatrix} = \begin{bmatrix} -\frac{20}{7} & \frac{23}{7} & -\frac{11}{7} \\ -\frac{11}{7} & \frac{13}{7} & -\frac{5}{7} \\ 2 & -2 & 1 \end{bmatrix}\begin{bmatrix} 18 \\ 20 \\ 11 \end{bmatrix} = \begin{bmatrix} -3 \\ 1 \\ 7 \end{bmatrix}.$$

Thus, $\boxed{x = -3, y = 1, z = 7}$.

18. $\begin{vmatrix} 7 & -5 \\ 2 & -1 \end{vmatrix} = (7)(-1) - (2)(-5) = \boxed{3}$

19.

$$\begin{vmatrix} 1 & -2 & -1 \\ 3 & -5 & 2 \\ 4 & -1 & 0 \end{vmatrix} = 1\begin{vmatrix} -5 & 2 \\ -1 & 0 \end{vmatrix} - (-2)\begin{vmatrix} 3 & 2 \\ 4 & 0 \end{vmatrix} + (-1)\begin{vmatrix} 3 & -5 \\ 4 & -1 \end{vmatrix}$$

$$= 1\underbrace{\left[(-5)(0) - (-1)(2)\right]}_{=2} - (-2)\underbrace{\left[(3)(0) - (4)(2)\right]}_{=-8} + (-1)\underbrace{\left[(3)(-1) - (4)(-5)\right]}_{=17} = \boxed{-31}$$

20.

$$D=\begin{vmatrix}1 & -2\\ -1 & 3\end{vmatrix}=(1)(3)-(-1)(-2)=1$$

$$D_x=\begin{vmatrix}1 & -2\\ 2 & 3\end{vmatrix}=(1)(3)-(2)(-2)=7$$

$$D_y=\begin{vmatrix}1 & 1\\ -1 & 2\end{vmatrix}=(1)(2)-(-1)(1)=3$$

So, from Cramer's Rule, we have:

$$x=\frac{D_x}{D}=7$$

$$y=\frac{D_y}{D}=3$$

Thus, the solution is $\boxed{x=7,\ y=3}$.

21.

$$D=\begin{vmatrix}3 & 5 & -2\\ 7 & 11 & 3\\ 1 & -1 & 1\end{vmatrix}=3\begin{vmatrix}11 & 3\\ -1 & 1\end{vmatrix}-5\begin{vmatrix}7 & 3\\ 1 & 1\end{vmatrix}-2\begin{vmatrix}7 & 11\\ 1 & -1\end{vmatrix}$$

$$=3\underbrace{\left[(11)(1)-(-1)(3)\right]}_{=14}-5\underbrace{\left[(7)(1)-(1)(3)\right]}_{=4}-2\underbrace{\left[(7)(-1)-(1)(11)\right]}_{=-18}=58$$

$$D_x=\begin{vmatrix}-6 & 5 & -2\\ 2 & 11 & 3\\ 4 & -1 & 1\end{vmatrix}=-6\begin{vmatrix}11 & 3\\ -1 & 1\end{vmatrix}-5\begin{vmatrix}2 & 3\\ 4 & 1\end{vmatrix}-2\begin{vmatrix}2 & 11\\ 4 & -1\end{vmatrix}$$

$$=-6\underbrace{\left[(11)(1)-(-1)(3)\right]}_{=14}-5\underbrace{\left[(2)(1)-(4)(3)\right]}_{=-10}-2\underbrace{\left[(2)(-1)-(4)(11)\right]}_{=-46}=58$$

$$D_y=\begin{vmatrix}3 & -6 & -2\\ 7 & 2 & 3\\ 1 & 4 & 1\end{vmatrix}=3\begin{vmatrix}2 & 3\\ 4 & 1\end{vmatrix}+6\begin{vmatrix}7 & 3\\ 1 & 1\end{vmatrix}-2\begin{vmatrix}7 & 2\\ 1 & 4\end{vmatrix}$$

$$=3\underbrace{\left[(2)(1)-(4)(3)\right]}_{=-10}+6\underbrace{\left[(7)(1)-(1)(3)\right]}_{=4}-2\underbrace{\left[(7)(4)-(1)(2)\right]}_{=26}=-58$$

$$D_z=\begin{vmatrix}3 & 5 & -6\\ 7 & 11 & 2\\ 1 & -1 & 4\end{vmatrix}=3\begin{vmatrix}11 & 2\\ -1 & 4\end{vmatrix}-5\begin{vmatrix}7 & 2\\ 1 & 4\end{vmatrix}-6\begin{vmatrix}7 & 11\\ 1 & -1\end{vmatrix}$$

$$=3\underbrace{\left[(11)(4)-(-1)(2)\right]}_{=46}-5\underbrace{\left[(7)(4)-(1)(2)\right]}_{=26}-6\underbrace{\left[(7)(-1)-(1)(11)\right]}_{=-18}=116$$

So, by Cramer's Rule,

$$\boxed{x=\frac{D_x}{D}=1\quad y=\frac{D_y}{D}=-1\quad z=\frac{D_z}{D}=2}$$

22. Here, $BA=\begin{bmatrix}4&7&3\\6&5&4\end{bmatrix}\cdot\begin{bmatrix}0.4&0.6\\0.5&0.1\\0.1&0.3\end{bmatrix}=\begin{bmatrix}5.4&4\\5.3&5.3\end{bmatrix}$. (The order is 2×2.)

Each entry of this matrix gives the respective applicant's total score according to each of the two rubrics. Specifically,
Applicant 1's score according to Rubric 1 is 5.4,
Applicant 1's score according to Rubric 2 is 4,
Applicant 2's score according to Rubric 1 is 5.3,
Applicant 2's score according to Rubric 2 is 5.3.

23. The partial fraction decomposition has the form:

$$\frac{2x+5}{x(x+1)}=\frac{A}{x}+\frac{B}{x+1} \quad \textbf{(1)}$$

To find the coefficients, multiply both sides of **(1)** by $x(x+1)$, and gather like terms:

$$2x+5=A(x+1)+Bx$$
$$2x+5=(A+B)x+A \quad \textbf{(2)}$$

Equate corresponding coefficients in **(2)** to obtain the following system:

$$\textbf{(*)}\ \begin{cases} A+B=2 & \textbf{(3)}\\ A=5 & \textbf{(4)}\end{cases}$$

Now, solve system **(*)**:

Substitute **(4)** into **(3)** to see that $B=-3$.

Thus, the partial fraction decomposition **(1)** becomes:

$$\boxed{\frac{2x+5}{x(x+1)}=\frac{5}{x}-\frac{3}{x+1}}$$

24. The partial fraction decomposition has the form:

$$\frac{3x-13}{(x-5)^2}=\frac{A}{x-5}+\frac{B}{(x-5)^2} \quad \textbf{(1)}$$

To find the coefficients, multiply both sides of **(1)** by $(x-5)^2$, and gather like terms:

$$3x-13=A(x-5)+B=Ax+(-5A+B) \quad \textbf{(2)}$$

Equate corresponding coefficients in **(2)** to obtain the following system:

$$\textbf{(*)}\ \begin{cases} A=3 & \textbf{(3)}\\ -5A+B=-13 & \textbf{(4)}\end{cases}$$

Now, solve system **(*)**:

Substitute **(3)** into **(4)** to see that $B=2$.

Thus, the partial fraction decomposition **(1)** becomes:

$$\boxed{\frac{3x-13}{(x-5)^2}=\frac{3}{x-5}+\frac{2}{(x-5)^2}}$$

25. The partial fraction decomposition has the form:

$$\frac{5x-3}{x(x-3)(x+3)}=\frac{A}{x}+\frac{B}{x-3}+\frac{C}{x+3} \quad \textbf{(1)}$$

To find the coefficients, multiply both sides of **(1)** by $x(x-3)(x+3)$, and gather like terms:

$$\begin{aligned}5x-3&=A(x-3)(x+3)+Bx(x+3)+Cx(x-3)\\&=A(x^2-9)+B(x^2+3x)+C(x^2-3x)\\&=(A+B+C)x^2+(3B-3C)x+(-9A) \quad \textbf{(2)}\end{aligned}$$

Equate corresponding coefficients in **(2)** to obtain the following system:

$$(*)\begin{cases}A+B+C=0 & \textbf{(3)}\\ 3B-3C=5 & \textbf{(4)}\\ -9A=-3 & \textbf{(5)}\end{cases}$$

Solve **(5)** for A: $A=\frac{1}{3}$ **(6)**
Substitute **(6)** into **(1)** to eliminate A: $B+C=-\frac{1}{3}$ **(7)**
Now, solve the 2×2 system:

$$\begin{cases}3B-3C=5 & \textbf{(4)}\\ B+C=-\frac{1}{3} & \textbf{(7)}\end{cases}$$

Multiply **(7)** by 3, and then add to **(4)** to find B: $6B=4$ so that $B=\frac{2}{3}$.
Substitute this value for B into **(7)** to find C: $\frac{2}{3}+C=-\frac{1}{3}$ so that $C=-1$.
Finally, substitute the values of B and C into **(1)** to find A: $A+\frac{2}{3}-1=0$ so that $A=\frac{1}{3}$.

Thus, the partial fraction decomposition **(1)** becomes:

$$\boxed{\frac{5x-3}{x(x-3)(x+3)}=\frac{1}{3x}+\frac{2}{3(x-3)}-\frac{1}{x+3}}$$

26. The partial fraction decomposition has the form:

$$\frac{1}{2x^2+5x-3}=\frac{1}{(2x-1)(x+3)}=\frac{A}{2x-1}+\frac{B}{x+3} \quad \textbf{(1)}$$

To find the coefficients, multiply both sides of **(1)** by $(2x-1)(x+3)$, and gather like terms:

$$1=A(x+3)+B(2x-1)=(A+2B)x+(3A-B) \quad \textbf{(2)}$$

Equate corresponding coefficients in **(2)** to obtain the following system:

$$\textbf{(*)} \begin{cases} A+2B=0 & \textbf{(3)} \\ 3A-B=1 & \textbf{(4)} \end{cases}$$

Now, solve system **(*)**:

Multiply **(4)** by 2, and then add to **(3)**: $7A=2$ so that $A=\frac{2}{7}$

Substitute this value of A into **(3)** to find B:

$$\tfrac{2}{7}+2B=0$$
$$B=-\tfrac{1}{7}$$

Thus, the partial fraction decomposition **(1)** becomes:

$$\boxed{\frac{1}{2x^2+5x-3}=\frac{\frac{2}{7}}{2x-1}-\frac{\frac{1}{7}}{x+3}}$$

27.

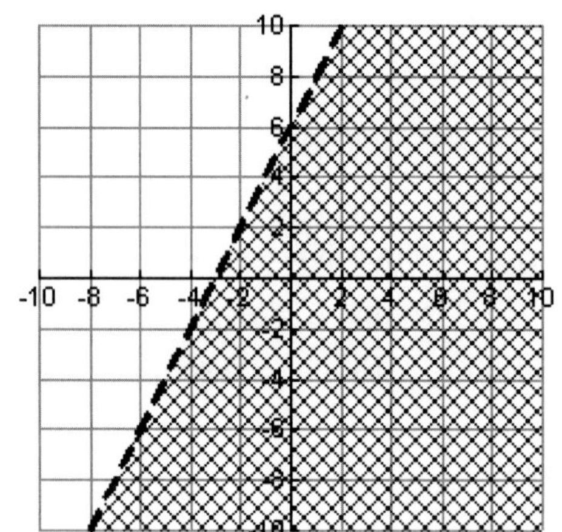

The dashed curve is the graph of $y=2x+6$.

28.

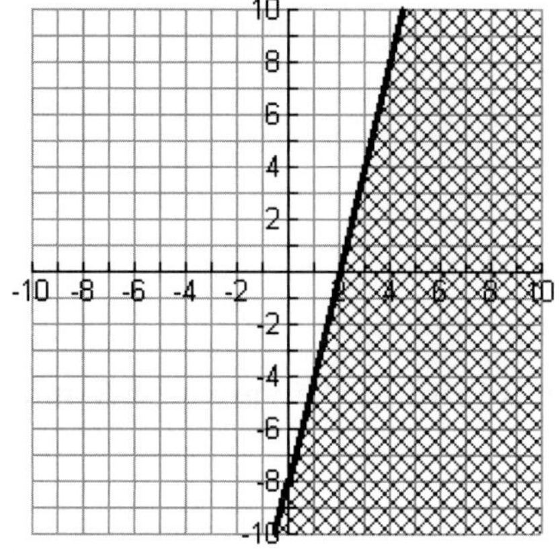

The solid curve is the graph of $y=4x-8$.

29.

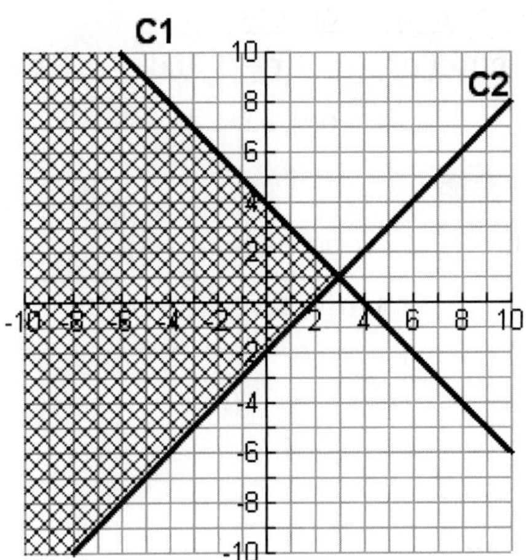

Notes on the graph:

C1: $y = 4 - x$

C2: $y = x - 2$

30.

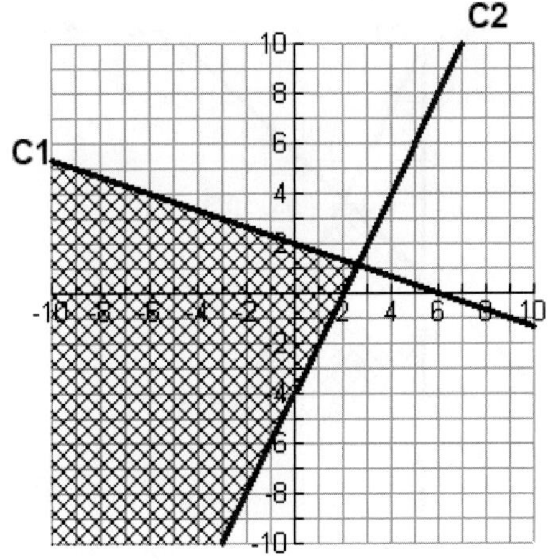

Notes on the graph:

C1: $y = -\frac{1}{3}x + 2$

C2: $y = 2x - 4$

31. The region in this case is:

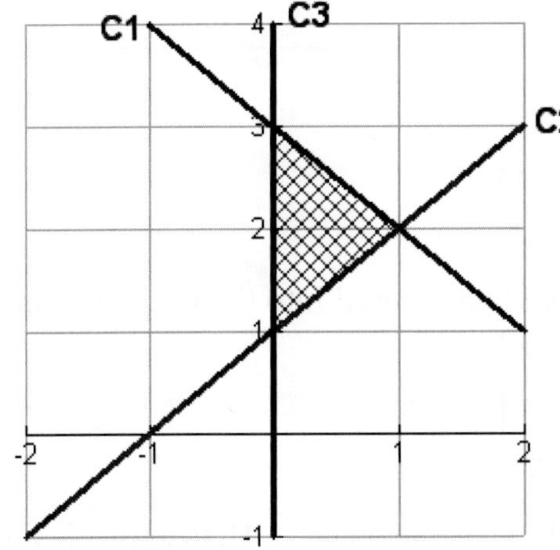

Notes on the graph:

C1: $y = -x + 3$

C2: $y = x + 1$

C3: $x = 0$

We need all of the intersection points since they constitute the vertices:

Intersection of $y = -x + 3$ and $y = x + 1$:

$$-x + 3 = x + 1$$

$$2x = 2$$

$$x = 1$$

So, the intersection point is (1,2).

Vertex	**Objective Function** $z = 5x + 7y$
$(0,3)$	$z = 5(0) + 7(3) = 21$
$(0,1)$	$z = 5(0) + 7(1) = 7$
$(1,2)$	$z = 5(1) + 7(2) = 19$

So, the minimum value of z is 7.

32. The region in this case is:

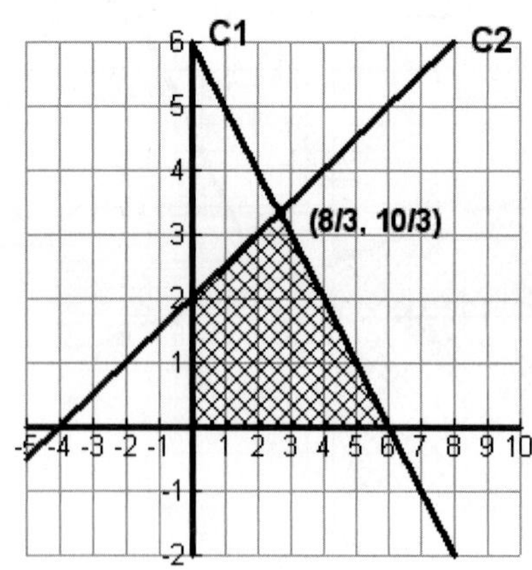

Notes on the graph:

C1: $y=-x+6$

C2: $y=\frac{1}{2}x+2$

We need all of the intersection points since they constitute the vertices:

Intersection of $y=-x+6$ and $y=\frac{1}{2}x+2$:

$$-x+6=\frac{1}{2}x+2$$

$$x=\frac{8}{3}$$

So, the intersection point is $\left(\frac{8}{3},\frac{10}{3}\right)$.

Vertex	**Objective Function** $z=3x+6y$
$(0,0)$	0
$(0,2)$	12
$(6,0)$	18
$\left(\frac{8}{3},\frac{10}{3}\right)$	28 MAX

So, the maximum value of z is 28.

33. The augmented matrix that should be entered into the calculator is

$$\left[\begin{array}{cc|c} 5.6 & -2.7 & 87.28 \\ -4.2 & 8.4 & -106.26 \end{array}\right]$$

Solving yields the solution (12.5, -6.4).

34. a. The system we must solve is:

$$\begin{cases} 9a-3b+c=6 \\ a+b+c=12 \\ 25a+5b+c=7 \end{cases}$$

b. The corresponding augmented matrix that should be entered into the calculator is

$$\left[\begin{array}{ccc|c} 9 & -3 & 1 & 6 \\ 1 & 1 & 1 & 12 \\ 25 & 5 & 1 & 7 \end{array}\right]$$

Solving this system yields $a=-0.34375$, $b=0.8125$, $c=11.53125$. Hence, the desired quadratic function is $y=-0.34375x^2+0.8125x+11.53125$.

Chapter 8 Cumulative Test---

1. $g(f(-1)) = \sqrt{9} = \boxed{3}$

2. We must have $1-5x>0$, which is equivalent to $x<\frac{1}{5}$. In interval notation, this can be written as $\left(-\infty, \frac{1}{5}\right)$.

3.

$$\begin{aligned}&\frac{f(x+h)-f(x)}{h}\\&=\frac{(x+h)^2-3(x+h)+2-\left(x^2-3x+2\right)}{h}\\&=\frac{x^2+2hx+h^2-3x-3h+2-x^2+3x-2}{h}\\&=\frac{h\left(2x+h-3\right)}{h}=2x+h-3\end{aligned}$$

4. The real zeros (with multiplicity) are:

0 (multiplicity 1)
7 (multiplicity 2)
-13 (multiplicity 3)

5. Observe that

$$\begin{aligned}f(x)&=-0.04x^2+1.2x-3\\&=-0.04\left(x^2-30x\right)-3\\&=-0.04\left(x^2-30x+225\right)-3+9\\&=-0.04(x-15)^2+6\end{aligned}$$

So, the vertex is (15,6).

6. $P(x)=\left(x^2+9\right)\left(x^2-1\right)=(x+3i)(x-3i)(x-1)(x+1)$

7. Vertical asymptote: $x=3$ Horizontal asymptote: $y=-5$

8. 23.14

9. $\log_5 0.2 = \log_5\left(\frac{1}{5}\right) = \log_5\left(5^{-1}\right) = \boxed{-1}$

10.

$$\begin{aligned}5^{2x-1}&=11\\2x-1&=\log_5 11\\2x&=1+\log_5 11\\x&=\tfrac{1+\log_5 11}{2}\end{aligned}$$

11. $\log_2 6 = \frac{\ln 6}{\ln 2} \approx 2.585$

12.

$$\begin{aligned}\ln(5x-6)&=2\\5x-6&=e^2\\x&=\frac{6+e^2}{5}\approx 2.678\end{aligned}$$

13. $\frac{\sqrt{3}}{2}$

14. Use $A = P\left(1+\frac{r}{n}\right)^{nt}$.

Here, $A = 65,000$, $r = 0.047$, $n = 52$, $t = 17$.

So, we have

$$65,000 = P\left(1+\frac{0.047}{52}\right)^{52(17)}$$

$$29,246.21 \approx \frac{65,000}{\left(1+\frac{0.047}{52}\right)^{52(17)}} = P$$

So, about $29,246.21.

15. Consider the following diagram:

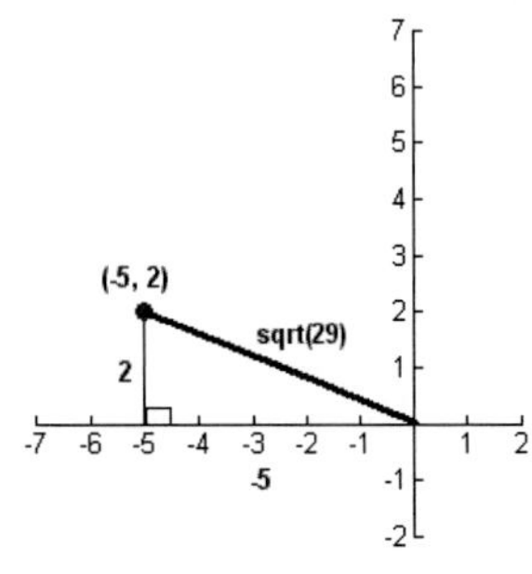

$$\sin\theta = \frac{2}{\sqrt{29}} = \frac{2\sqrt{29}}{29}$$

$$\cos\theta = \frac{-5}{\sqrt{29}} = \frac{-5\sqrt{29}}{29}$$

$$\tan\theta = \frac{-2}{5}$$

$$\cot\theta = -\frac{5}{2}$$

$$\sec\theta = -\frac{\sqrt{29}}{5}$$

$$\csc\theta = \frac{\sqrt{29}}{2}$$

16. 150°, 210°

17.

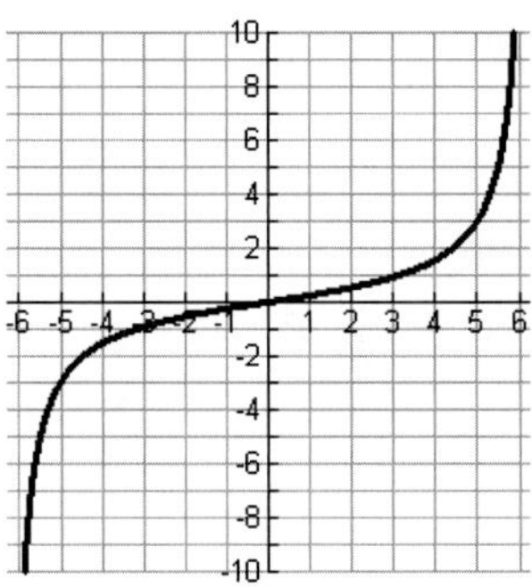

18.

$$\begin{aligned}
\cos x\left(1-4\sin^2 x\right) &= \cos x\left[\left(1-2\sin^2 x\right)-2\sin^2 x\right] \\
&= \cos x\left[\left(\cos 2x\right)-\left(2\sin x\sin x\right)\right] \\
&= \cos x\cos 2x - \sin x\left(2\sin x\cos x\right) \\
&= \cos x\cos 2x - \sin x\sin 2x \\
&= \cos(x+2x) \\
&= \cos 3x
\end{aligned}$$

19. <u>Domain</u>: $x-\frac{\pi}{2}\neq\frac{(2n+1)\pi}{2}$, which is equivalent to $x\neq\frac{(2n+1)\pi}{2}+\frac{\pi}{2}=\frac{(2n+2)\pi}{2}=(n+1)\pi$ for any integer n. Said differently, this means that the domain consists of all real numbers except integer multiples of π. <u>Range</u>: All reals.
20. $\frac{\sec^4 x-1}{\sec^2 x+1}=\frac{\cancel{(\sec^2 x+1)}(\sec^2 x-1)}{\cancel{\sec^2 x+1}}=\sec^2 x-1=\tan^2 x$
21. $\frac{2\tan\left(-\frac{3\pi}{8}\right)}{1-\tan^2\left(-\frac{3\pi}{8}\right)}=\tan\left(2\cdot\left(-\frac{3\pi}{8}\right)\right)=\tan\left(-\frac{3\pi}{4}\right)=\boxed{1}$
22. $7\sin(-2x)\sin(5x)=7\left[\frac{1}{2}\left(\cos(-2x-5x)-\cos(-2x+5x)\right)\right]$ $=\frac{7}{2}\left[\cos(-7x)-\cos(3x)\right]=\frac{7}{2}\left[\cos(7x)-\cos(3x)\right]$
23. Observe that: $\alpha=180^\circ-\left(106.3^\circ+37.4^\circ\right)=36.3^\circ$ $\frac{\sin 106.3^\circ}{b}=\frac{\sin 36.3^\circ}{76.1}\Rightarrow b=\frac{(76.1)\left(\sin 106.3^\circ\right)}{\sin 36.3^\circ}\approx 123.4m$ $\frac{\sin 37.4^\circ}{c}=\frac{\sin 36.3^\circ}{76.1}\Rightarrow c=\frac{(76.1)\left(\sin 37.4^\circ\right)}{\sin 36.3^\circ}\approx 78.1m$
24. Use the formula $\theta=\cos^{-1}\left(\frac{\vec{u}\cdot\vec{v}}{\lvert\vec{u}\rvert\lvert\vec{v}\rvert}\right)$ with the following computations: $\vec{u}\cdot\vec{v}=\langle 2,3\rangle\cdot\langle -4,-5\rangle=-23,\ \lvert\vec{u}\rvert=\sqrt{2^2+3^2}=\sqrt{13},\ \lvert\vec{v}\rvert=\sqrt{(-4)^2+(-5)^2}=\sqrt{41}$ So, $\theta=\cos^{-1}\left(\frac{-23}{\sqrt{13}\sqrt{41}}\right)\approx\boxed{175^\circ}$.

25. We seek all complex numbers x such that $x^3 - 27 = 0$, or equivalently $x^3 = 27$. While it is clear that $x = 3$ is a solution by inspection, the remaining two complex solutions aren't immediately discernible. As such, we shall apply the approach for computing complex roots involving the formula

$$z^{1/n} = r^{1/n}\left[\cos\left(\frac{\theta}{n}+\frac{2\pi k}{n}\right)+i\sin\left(\frac{\theta}{n}+\frac{2\pi k}{n}\right)\right],\quad k = 0, 1, \ldots, n-1.$$

To this end, we follow these steps.

Step 1: Write $z = 27$ in polar form.

Since $x = 27,\ y = 0$, the point is on the positive x-axis. So, $r = 27,\ \theta = 0$.

Hence, $27 = 27\left[\cos(0)+i\sin(0)\right]$.

Step 2: Now, apply $z^{1/n} = r^{1/n}\left[\cos\left(\frac{\theta}{n}+\frac{2\pi k}{n}\right)+i\sin\left(\frac{\theta}{n}+\frac{2\pi k}{n}\right)\right],\ k = 0, 1, \ldots, n-1$:

$$(27)^{1/3} = \sqrt[3]{27}\left[\cos\left(\frac{2\pi k}{3}\right)+i\sin\left(\frac{2\pi k}{3}\right)\right],\ k = 0, 1, 2$$

$$= 3\left[\cos(0)+i\sin(0)\right],\ 3\left[\cos\left(\frac{2\pi}{3}\right)+i\sin\left(\frac{2\pi}{3}\right)\right],\ 3\left[\cos\left(\frac{4\pi}{3}\right)+i\sin\left(\frac{4\pi}{3}\right)\right]$$ Step 3:

$$= 3,\ -\tfrac{3\sqrt{3}}{2}+\tfrac{3}{2}i,\ -\tfrac{3\sqrt{3}}{2}-\tfrac{3}{2}i$$

Hence, the complex solutions of $x^3 + 8 = 0$ are $\boxed{-2,\ 1-\sqrt{3}i,\ 1+\sqrt{3}i}$.

26.

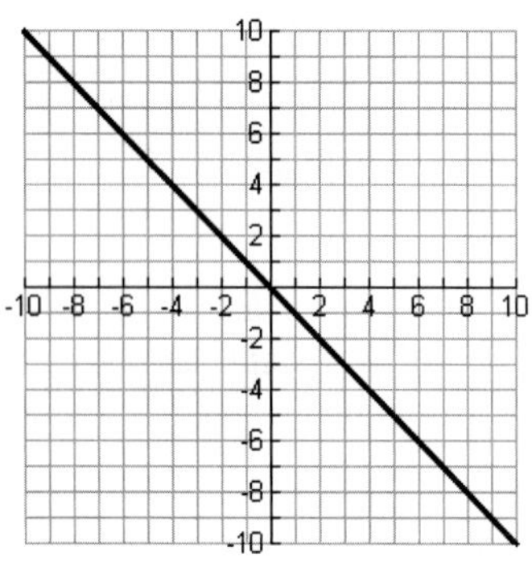

27.

$$\begin{bmatrix} 72 & -18 & 54 \\ 26 & -2 & 4 \end{bmatrix}$$

28. Observe that

$$\left[\begin{array}{ccc|c} 1 & -2 & 3 & 11 \\ 4 & 5 & -1 & -8 \\ 3 & 1 & -2 & 1 \end{array}\right] \xrightarrow[R_3-3R_1\to R_3]{R_2-4R_1\to R_2} \left[\begin{array}{ccc|c} 1 & -2 & 3 & 11 \\ 0 & 13 & -13 & -52 \\ 0 & 7 & -11 & -32 \end{array}\right] \xrightarrow[\frac{1}{13}R_2\to R_2]{\frac{1}{7}R_3\to R_3} \left[\begin{array}{ccc|c} 1 & -2 & 3 & 11 \\ 0 & 1 & -1 & -4 \\ 0 & 1 & -\frac{11}{7} & -\frac{32}{7} \end{array}\right]$$

$$\xrightarrow{R_2-R_3\to R_3} \left[\begin{array}{ccc|c} 1 & -2 & 3 & 11 \\ 0 & 1 & -1 & -4 \\ 0 & 0 & \frac{4}{7} & \frac{4}{7} \end{array}\right] \xrightarrow{\frac{7}{4}R_3\to R_3} \left[\begin{array}{ccc|c} 1 & -2 & 3 & 11 \\ 0 & 1 & -1 & -4 \\ 0 & 0 & 1 & 1 \end{array}\right] \xrightarrow{R_2+R_3\to R_2} \left[\begin{array}{ccc|c} 1 & -2 & 3 & 11 \\ 0 & 1 & 0 & -3 \\ 0 & 0 & 1 & 1 \end{array}\right]$$

$$\xrightarrow{R_1+2R_2-3R_3\to R_1} \left[\begin{array}{ccc|c} 1 & 0 & 0 & 2 \\ 0 & 1 & 0 & -3 \\ 0 & 0 & 1 & 1 \end{array}\right]$$

So, the solution is $\boxed{x=2, y=-3, z=1}$.

29. First, write the system as

$$\begin{bmatrix} 7 & 5 \\ -1 & 4 \end{bmatrix}\begin{bmatrix} x \\ y \end{bmatrix}=\begin{bmatrix} 1 \\ -1 \end{bmatrix}.$$

$$D=\begin{vmatrix} 7 & 5 \\ -1 & 4 \end{vmatrix}=(7)(4)-(-1)(5)=33$$

$$D_x=\begin{vmatrix} 1 & 5 \\ -1 & 4 \end{vmatrix}=(1)(4)-(-1)(5)=9$$

$$D_y=\begin{vmatrix} 7 & 1 \\ -1 & -1 \end{vmatrix}=(7)(-1)-(-1)(1)=-6$$

So, by Cramer's Rule, we have

$$x=\frac{D_x}{D}=\frac{3}{11}$$

$$y=\frac{D_y}{D}=-\frac{2}{11}$$

30. First, write the system as

$$\begin{bmatrix} 2 & 5 \\ -1 & 4 \end{bmatrix}\begin{bmatrix} x \\ y \end{bmatrix}=\begin{bmatrix} -1 \\ 7 \end{bmatrix}.$$

The solution is given by

$$\begin{bmatrix} x \\ y \end{bmatrix}=\begin{bmatrix} 2 & 5 \\ -1 & 4 \end{bmatrix}^{-1}\begin{bmatrix} -1 \\ 7 \end{bmatrix}=\frac{1}{13}\begin{bmatrix} 4 & -5 \\ 1 & 2 \end{bmatrix}\begin{bmatrix} -1 \\ 7 \end{bmatrix}=\begin{bmatrix} -3 \\ 1 \end{bmatrix}$$

So, the solution is $\boxed{x=-3, y=1}$

31. The region is as follows:

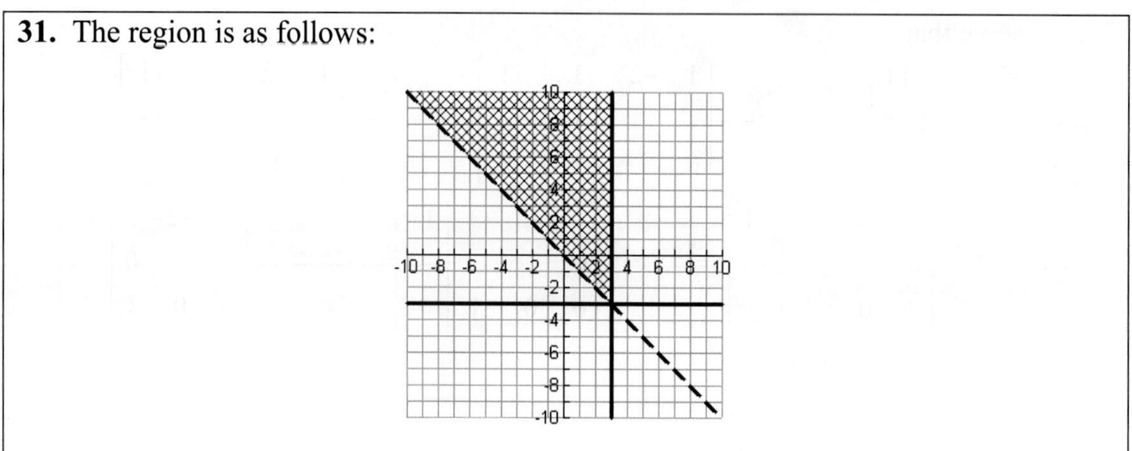

CHAPTER 9

Section 9.1 Solutions

1. Since $A=1,\ B=1,\ C=-1$, $B^2-4AC=1+4=5>0$. Hence, this is the equation of a $\boxed{\text{hyperbola}}$.	**2.** Since $A=1,\ B=1,\ C=1$, $B^2-4AC=1-4=-3<0$. Hence, this is the equation of a $\boxed{\text{ellipse}}$.
3. Since $A=2,\ B=0,\ C=2$, $B^2-4AC=0-4(2)(2)=-16<0$. Hence, this is the equation of a $\boxed{\text{circle}}$.	**4.** Since $A=1,\ B=0,\ C=1$, $B^2-4AC=0-4(1)(1)<0$. Hence, this is the equation of a $\boxed{\text{circle}}$.
5. Hyperbola	**6.** Hyperbola
7. Ellipse	**8.** Ellipse
9. Parabola	**10.** Parabola
11. Circle	**12.** Circle

Section 9.2 Solutions

1. c Opens to the right	**2. a** Opens to the left
3. d Opens down	**4. b** Opens up
5. c Vertex is $(1,1)$, opens to the right	**6. d** Vertex is $(1,-1)$, opens to the left
7. a Vertex is $(-1,-1)$, opens down	**8. b** Vertex is $(1,1)$, opens up
9. Vertex is $(0,0)$ and focus is $(0,3)$. So, the parabola opens up. As such, the general form is $x^2=4py$. In this case, $p=3$, so the equation is $\boxed{x^2=12y}$.	**10.** Vertex is $(0,0)$ and focus is $(2,0)$. So, the parabola opens to the right. As such, the general form is $y^2=4px$. In this case, $p=2$, so the equation is $\boxed{y^2=8x}$.
11. Vertex is $(0,0)$ and focus is $(-5,0)$. So, the parabola opens to the left. As such, the general form is $y^2=4px$. In this case, $p=-5$, so the equation is $\boxed{y^2=-20x}$.	**12.** Vertex is $(0,0)$ and focus is $(0,-4)$. So, the parabola opens down. As such, the general form is $x^2=4py$. In this case, $p=-4$, so the equation is $\boxed{x^2=-16y}$.

13. Vertex is $(3,5)$ and focus is $(3,7)$. So, the parabola opens up. As such, the general form is $(x-3)^2=4p(y-5)$. In this case, $p=2$, so the equation is $\boxed{(x-3)^2=8(y-5)}$.	**14.** Vertex is $(3,5)$ and focus is $(7,5)$. So, the parabola opens to the right. As such, the general form is $(y-5)^2=4p(x-3)$. In this case, $p=4$, so the equation is $\boxed{(y-5)^2=16(x-3)}$.
15. Vertex is $(2,4)$ and focus is $(0,4)$. So, the parabola opens to the left. As such, the general form is $(y-4)^2=4p(x-2)$. In this case, $p=-2$, so the equation is $\boxed{(y-4)^2=-8(x-2)}$.	**16.** Vertex is $(2,4)$ and focus is $(2,-1)$. So, the parabola opens down. As such, the general form is $(x-2)^2=4p(y-4)$. In this case, $p=-5$, so the equation is $\boxed{(x-2)^2=-20(y-4)}$.
17. Focus is $(2,4)$ and the directrix is $y=-2$. So, the parabola opens up. Since the distance between the focus and directrix is 6 and the vertex must occur halfway between them, we know $p=3$ and the vertex is $(2,1)$. Hence, the equation is $\boxed{(x-2)^2=4(3)(y-1)=12(y-1)}$.	**18.** Focus is $(2,-2)$ and the directrix is $y=4$. So, the parabola opens down. Since the distance between the focus and directrix is 6 and the vertex must occur halfway between them, we know $p=-3$ and the vertex is $(2,1)$. Hence, the equation is $\boxed{(x-2)^2=4(-3)(y-1)=-12(y-1)}$.
19. Focus is $(3,-1)$ and the directrix is $x=1$. So, the parabola opens to the right. Since the distance between the focus and directrix is 2 and the vertex must occur halfway between them, we know $p=1$ and the vertex is $(2,-1)$. Hence, the equation is $\boxed{(y+1)^2=4(1)(x-2)=4(x-2)}$.	**20.** Focus is $(-1,5)$ and the directrix is $x=5$. So, the parabola opens to the left. Since the distance between the focus and directrix is 6 and the vertex must occur halfway between them, we know $p=-3$ and the vertex is $(2,5)$. Hence, the equation is $\boxed{(y-5)^2=4(-3)(x-2)=-12(x-2)}$.
21. Since the vertex is $(-1,2)$ and the parabola opens to the right, the general form of its equation is $(y-2)^2=4p(x+1)$, for some $p>0$. The fact that $(1,6)$ is on the graph implies: $$(6-2)^2=4p(2)$$ $$16=8p$$ $$2=p$$ Hence, the equation is $\boxed{(y-2)^2=8(x+1)}$.	**22.** Since the vertex is $(2,-1)$ and the parabola opens up, the general form of its equation is $(x-2)^2=4p(y+1)$, for some $p>0$. The fact that $(6,1)$ is on the graph implies: $$(6-2)^2=4p(2)$$ $$16=8p$$ $$2=p$$ Hence, the equation is $\boxed{(x-2)^2=8(y+1)}$.

23. Since the vertex is $(2,-1)$ and the parabola opens down, the general form of its equation is $(x-2)^2 = 4p(y+1)$, for some $p<0$. The fact that $(6,-3)$ is on the graph implies: $$(6-2)^2 = 4p(-3+1)$$ $$16 = -8p$$ $$-2 = p$$ Hence, the equation is $\boxed{(x-2)^2 = -8(y+1)}$.	**24.** Since the vertex is $(-1,2)$ and the parabola opens to the left, the general form of its equation is $(y-2)^2 = 4p(x+1)$, for some $p<0$. The fact that $(-3,6)$ is on the graph implies: $$(6-2)^2 = 4p(-3+1)$$ $$16 = -8p$$ $$-2 = p$$ Hence, the equation is $\boxed{(y-2)^2 = -8(x+1)}$.

25.

Equation: $x^2 = 8y = 4(2)y$ **(1)**

So, $p = 2$ and opens up.

Vertex: $(0,0)$

Focus: $(0,2)$

Directrix: $y = -2$

Latus Rectum: Connects x-values corresponding to $y = 2$. Substituting this into **(1)** yields:

$$x^2 = 16 \text{ so that } x = \pm 4$$

So, the length of the latus rectum is 8.

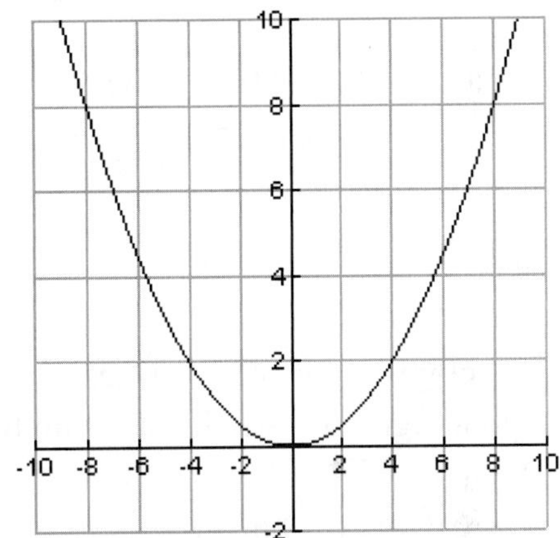

26.

Equation: $x^2 = -12y = 4(-3)y$ **(1)**

So, $p = -3$ and opens down.

Vertex: $(0,0)$

Focus: $(0,-3)$

Directrix: $y = 3$

Latus Rectum: Connects x-values corresponding to $y = -3$. Substituting this into **(1)** yields:

$$x^2 = 36 \text{ so that } x = \pm 6$$

So, the length of the latus rectum is 12.

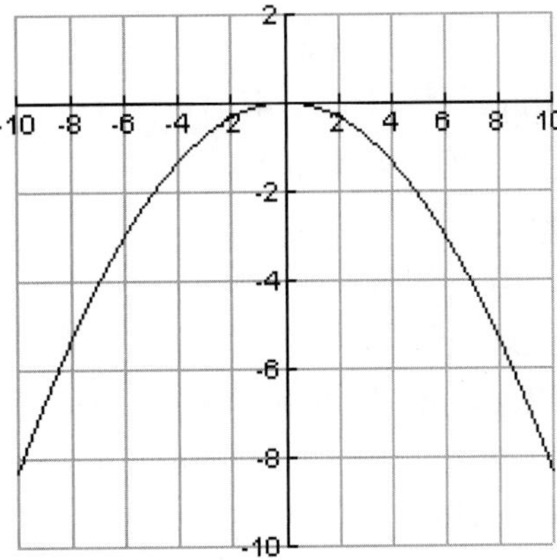

27.

Equation: $y^2 = -2x = 4(-\frac{1}{2})x$ **(1)**

So, $p = -\frac{1}{2}$ and opens to the left.

Vertex: $(0,0)$

Focus: $(-\frac{1}{2},0)$

Directrix: $x = \frac{1}{2}$

Latus Rectum: Connects y-values corresponding to $x = -\frac{1}{2}$. Substituting this into **(1)** yields:

$$y^2 = 1 \text{ so that } y = \pm 1$$

So, the length of the latus rectum is 2.

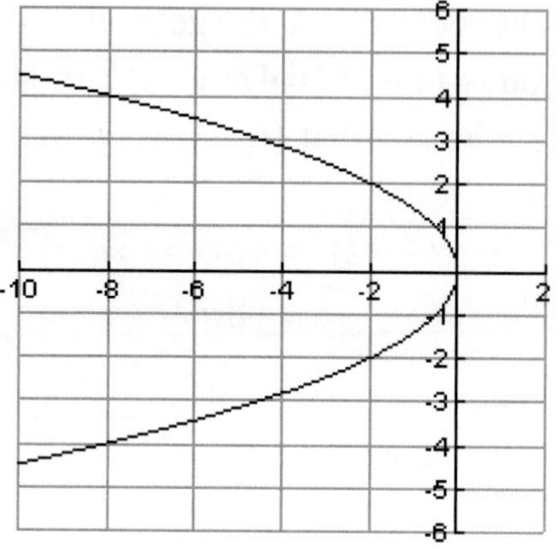

28.

Equation: $y^2 = 6x = 4(\frac{3}{2})x$ **(1)**

So, $p = \frac{3}{2}$ and opens to the right.

Vertex: $(0,0)$

Focus: $(\frac{3}{2},0)$

Directrix: $x = -\frac{3}{2}$

Latus Rectum: Connects y-values corresponding to $x = \frac{3}{2}$. Substituting this into **(1)** yields:

$$y^2 = 9 \text{ so that } y = \pm 3$$

So, the length of the latus rectum is 6.

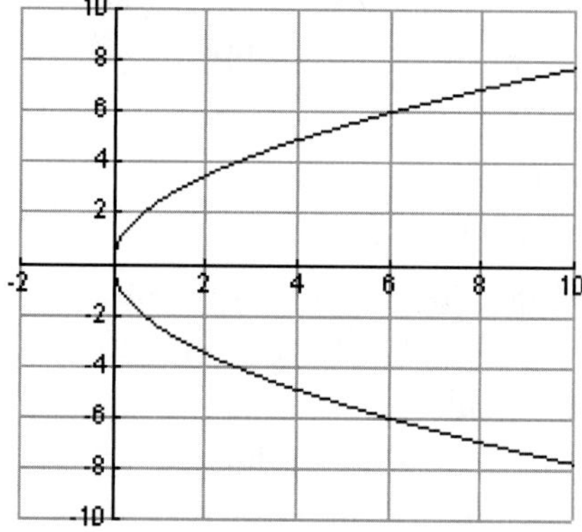

29.

Equation: $x^2 = 16y = 4(4)y$ **(1)**

So, $p = 4$ and opens up.

Vertex: $(0,0)$

Focus: $(0,4)$

Directrix: $y = -4$

Latus Rectum: Connects x-values corresponding to $y = 4$. Substituting this into **(1)** yields:

$$x^2 = 64 \text{ so that } x = \pm 8$$

So, the length of the latus rectum is 16.

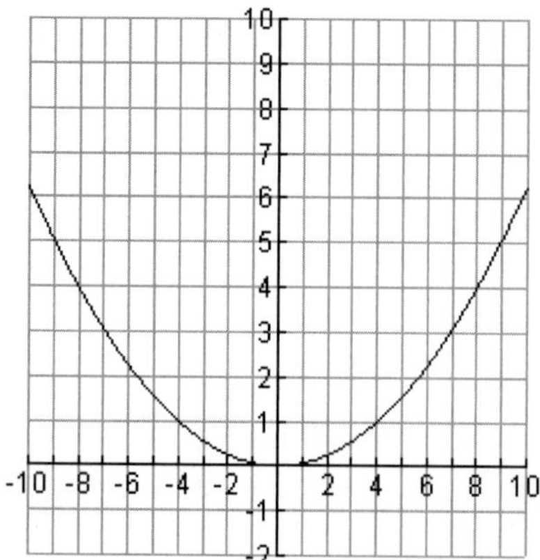

30.

Equation: $x^2 = -8y = 4(-2)y$ **(1)**

So, $p = -2$ and opens down.

Vertex: $(0,0)$

Focus: $(0,-2)$

Directrix: $y = 2$

Latus Rectum: Connects x-values corresponding to $y = -2$. Substituting this into **(1)** yields:

$$x^2 = 16 \text{ so that } x = \pm 4$$

So, the length of the latus rectum is 8.

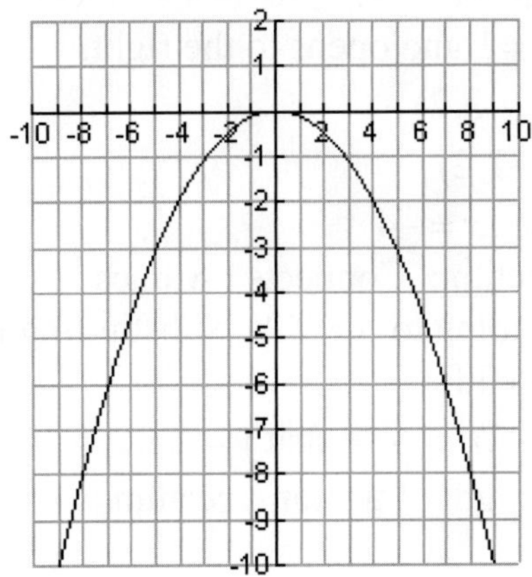

31.

Equation: $y^2 = 4x = 4(1)x$ **(1)**

So, $p = 1$ and opens to the right.

Vertex: $(0,0)$

Focus: $(1,0)$

Directrix: $x = -1$

Latus Rectum: Connects y-values corresponding to $x = 1$. Substituting this into **(1)** yields:

$$y^2 = 4 \text{ so that } y = \pm 2$$

So, the length of the latus rectum is 4.

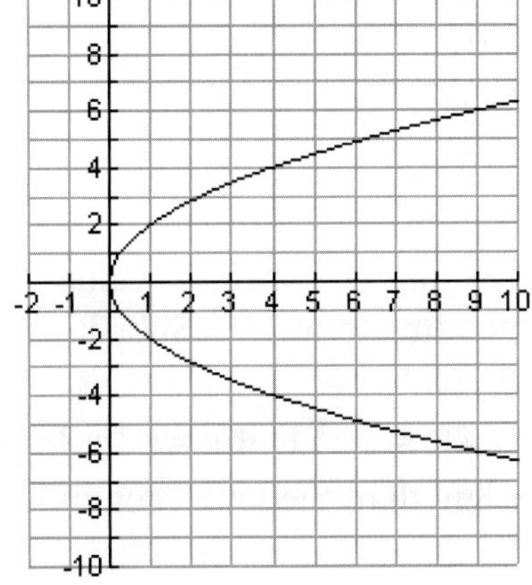

32.

Equation: $y^2 = -16x = 4(-4)x$ **(1)**

So, $p = -4$ and opens to the left.

Vertex: $(0,0)$

Focus: $(-4,0)$

Directrix: $x = 4$

Latus Rectum: Connects y-values corresponding to $x = -4$. Substituting this into **(1)** yields:

$$y^2 = 64 \text{ so that } y = \pm 8$$

So, the length of the latus rectum is 16.

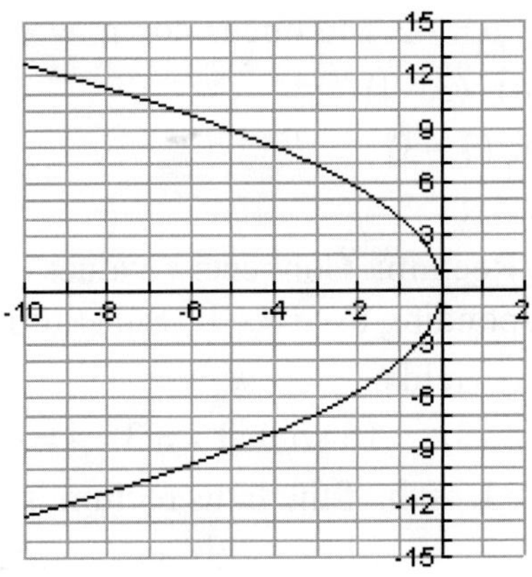

33.

Equation: $(y-2)^2 = 4(1)(x+3)$ **(1)**

So, $p = 1$ and opens to the right.

Vertex: $(-3, 2)$

Focus: $(-2, 2)$

Directrix: $x = -4$

Latus Rectum: Connects y-values corresponding to $x = -2$. Substituting this into **(1)** yields:

$$(y-2)^2 = 4 \text{ so that } y = 2 \pm 2 = 4,\ 0$$

So, the length of the latus rectum is 4.

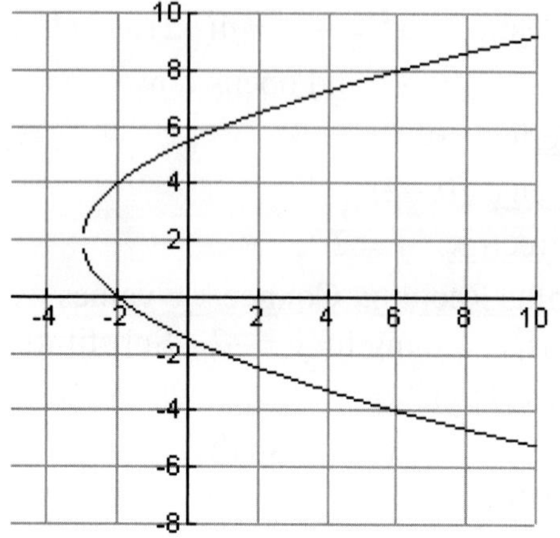

34.

Equation: $(y+2)^2 = 4(-1)(x-1)$ **(1)**

So, $p = -1$ and opens to the left.

Vertex: $(1, -2)$

Focus: $(0, -2)$

Directrix: $x = 2$

Latus Rectum: Connects y-values corresponding to $x = 0$. Substituting this into **(1)** yields:

$$(y+2)^2 = 4 \text{ so that } y = -2 \pm 2 = -4,\ 0$$

So, the length of the latus rectum is 4.

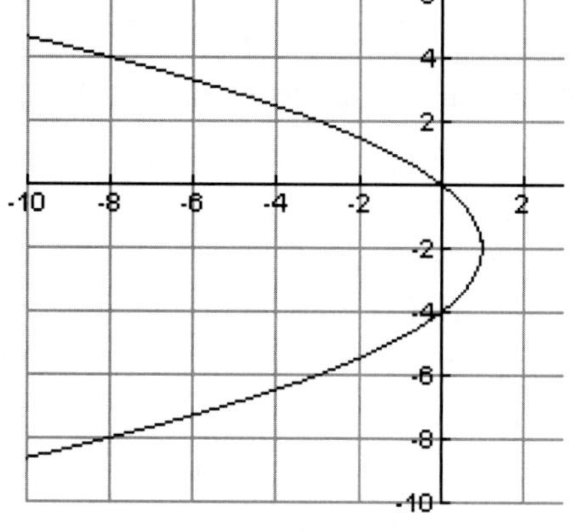

35.

Equation: $(x-3)^2 = 4(-2)(y+1)$ **(1)**

So, $p = -2$ and opens down.

Vertex: $(3, -1)$

Focus: $(3, -3)$

Directrix: $y = 1$

Latus Rectum: Connects x-values corresponding to $y = -3$. Substituting this into **(1)** yields:

$$(x-3)^2 = 16 \text{ so that } x = 3 \pm 4 = -1,\ 7$$

So, the length of the latus rectum is 8.

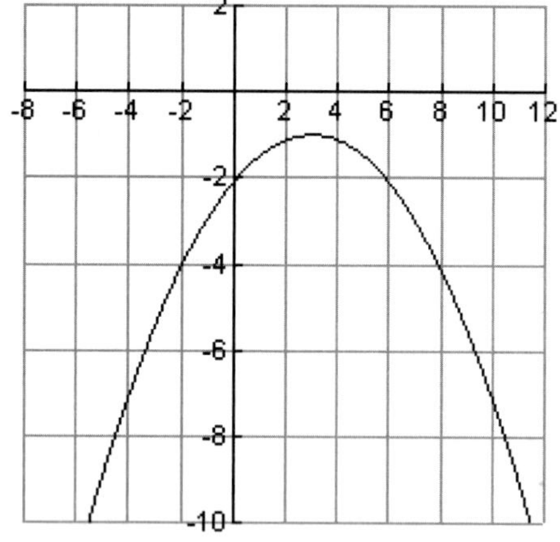

36.

Equation: $(x+3)^2 = 4(-2)(y-2)$ **(1)**

So, $p = -2$ and opens down.

Vertex: $(-3,2)$

Focus: $(-3,0)$

Directrix: $y = 4$

Latus Rectum: Connects x-values corresponding to $y = 0$. Substituting this into **(1)** yields:

$(x+3)^2 = 16$ so that $x = -3 \pm 4 = 1, -7$

So, the length of the latus rectum is 8.

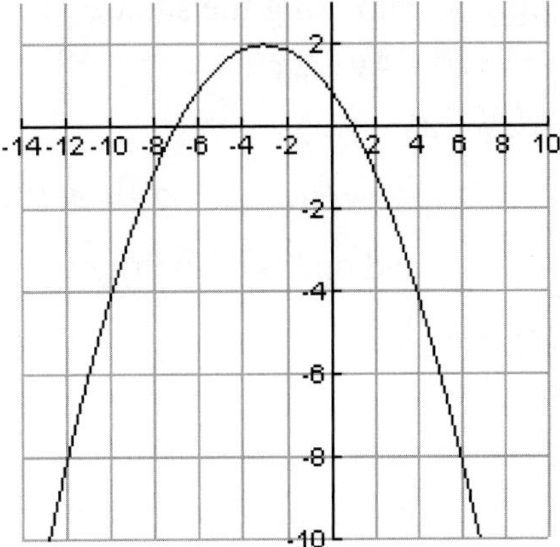

37.

Equation: $(x+5)^2 = 4(-\frac{1}{2})(y-0)$ **(1)**

So, $p = -\frac{1}{2}$ and opens down.

Vertex: $(-5,0)$

Focus: $(-5,-\frac{1}{2})$

Directrix: $y = \frac{1}{2}$

Latus Rectum: Connects x-values corresponding to $y = -\frac{1}{2}$. Substituting this into **(1)** yields:

$(x+5)^2 = 1$ so that $x = -5 \pm 1 = -6, -4$

So, the length of the latus rectum is 2.

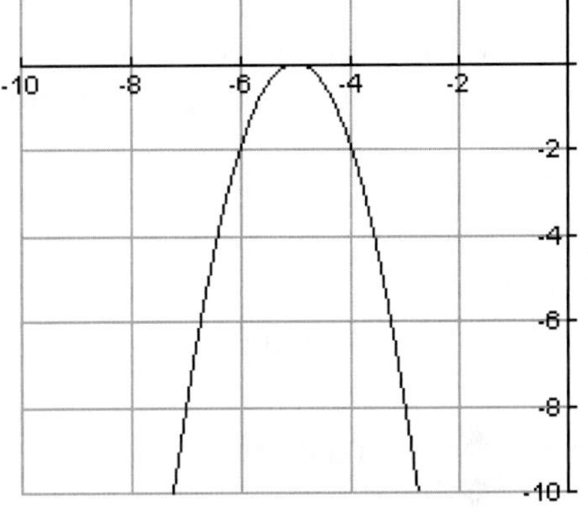

38.

Equation: $(y-0)^2 = 4(-4)(x+1)$ **(1)**

So, $p = -4$ and opens to the left.

Vertex: $(-1,0)$

Focus: $(-5,0)$

Directrix: $x = 3$

Latus Rectum: Connects y-values corresponding to $x = -5$. Substituting this into **(1)** yields:

$y^2 = 64$ so that $y = \pm 8$

So, the length of the latus rectum is 16.

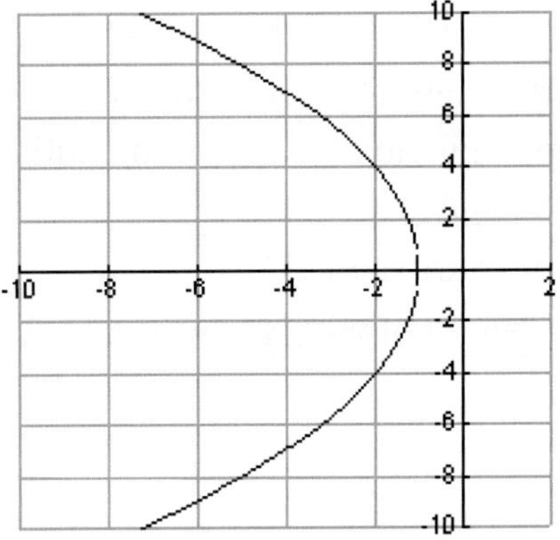

39.

Equation: Completing the square yields:

$$y^2 - 4y - 2x + 4 = 0$$

$$(y^2 - 4y + 4) - 4 - 2x + 4 = 0$$

$$(y-2)^2 = 2x = 4(\tfrac{1}{2})x \quad \textbf{(1)}$$

So, $p = \frac{1}{2}$ and opens to the right.

Vertex: $(0,2)$

Focus: $(\frac{1}{2},2)$

Directrix: $x = -\frac{1}{2}$

Latus Rectum: Connects y-values corresponding to $x = \frac{1}{2}$. Substituting this into **(1)** yields:

$$(y-2)^2 = 1 \text{ so that } y = 2 \pm 1 = 1,\ 3$$

So, the length of the latus rectum is 2.

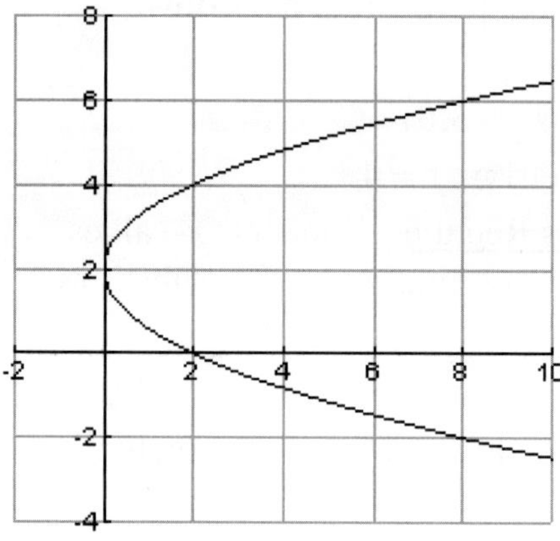

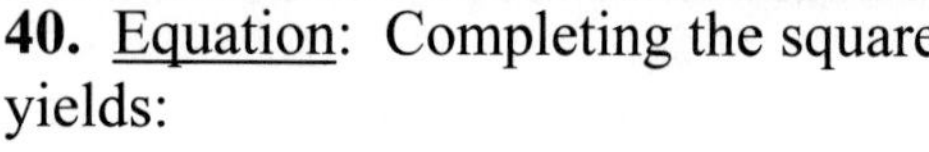

40. Equation: Completing the square yields:

$$x^2 - 6x + 2y + 9 = 0$$

$$(x^2 - 6x + 9) - 9 + 2y + 9 = 0$$

$$(x-3)^2 = -2y = 4(-\tfrac{1}{2})y \quad \textbf{(1)}$$

So, $p = -\frac{1}{2}$ and opens down.

Vertex: $(3,0)$

Focus: $(3,-\frac{1}{2})$

Directrix: $y = \frac{1}{2}$

Latus Rectum: Connects x-values corresponding to $y = -\frac{1}{2}$. Substituting this into **(1)** yields:

$(x-3)^2 = 1$ so that $x = 3 \pm 1 = 2,\ 4$

So, the length of the latus rectum is 2.

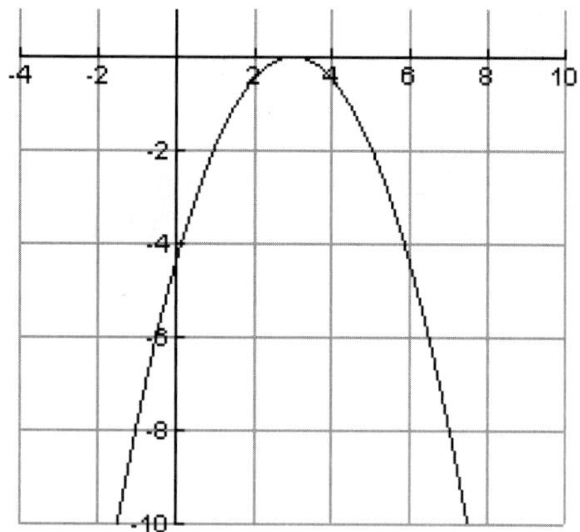

41.

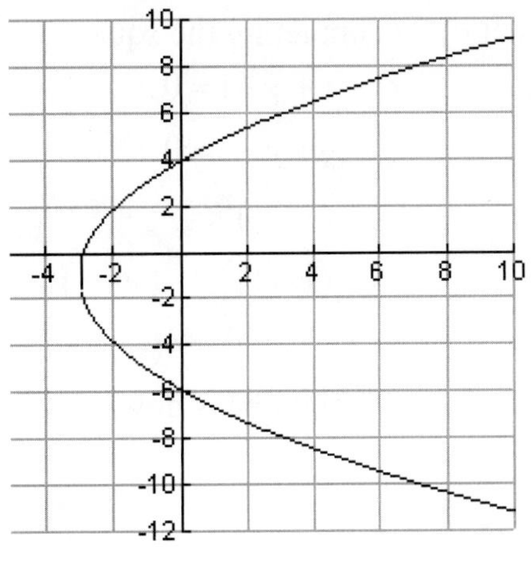

Equation: Completing the square yields:

$$y^2+2y-8x-23=0$$
$$(y^2+2y+1)-1-8x-23=0$$
$$(y+1)^2=8x+24$$
$$(y+1)^2=8(x+3)$$
$$(y+1)^2=4(2)(x+3) \quad \textbf{(1)}$$

So, $p=2$ and opens to the right.

Vertex: $(-3,-1)$

Focus: $(-1,-1)$

Directrix: $x=-5$

Latus Rectum: Connects y-values corresponding to $x=-1$. Substituting this into **(1)** yields:

$(y+1)^2=16$ so that $y=-1\pm 4=3,-5$

So, the length of the latus rectum is 8.

42.

Equation: Completing the square yields:

$$x^2-6x-4y+10=0$$
$$(x^2-6x+9)-9-4y+10=0$$
$$(x-3)^2=4y-1$$
$$(x-3)^2=4(y-\tfrac{1}{4}) \quad \textbf{(1)}$$

So, $p=1$ and opens up.

Vertex: $(3,\tfrac{1}{4})$

Focus: $(3,\tfrac{5}{4})$

Directrix: $y=-\tfrac{3}{4}$

Latus Rectum: Connects x-values corresponding to $y=\tfrac{5}{4}$. Substituting this into **(1)** yields:

$(x-3)^2=4$ so that $x=3\pm 2=1,\,5$

So, the length of the latus rectum is 4.

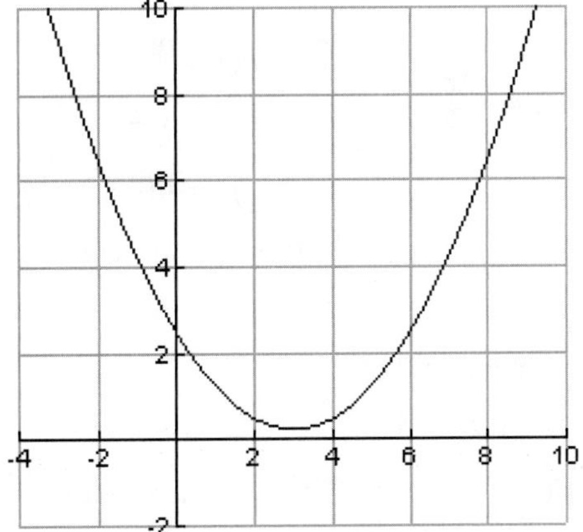

43.

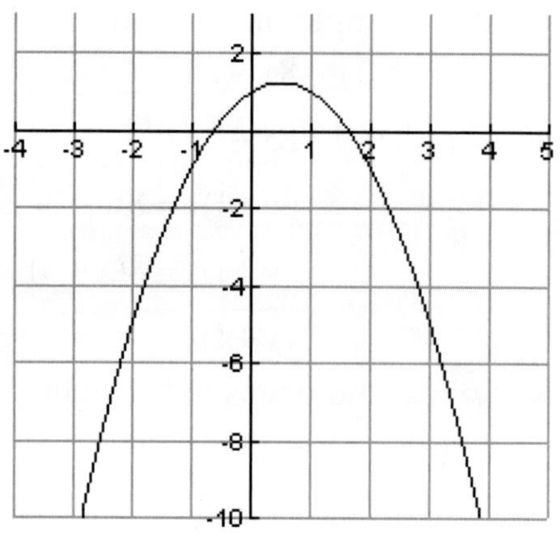

Equation: Completing the square yields:

$$x^2 - x + y - 1 = 0$$
$$(x^2 - x + \tfrac{1}{4}) - \tfrac{1}{4} + y - 1 = 0$$
$$(x - \tfrac{1}{2})^2 = -y + \tfrac{5}{4}$$
$$(x - \tfrac{1}{2})^2 = -(y - \tfrac{5}{4})$$
$$(x - \tfrac{1}{2})^2 = 4(-\tfrac{1}{4})(y - \tfrac{5}{4}) \quad \textbf{(1)}$$

So, $p = -\frac{1}{4}$ and opens down.

Vertex: $(\frac{1}{2}, \frac{5}{4})$

Focus: $(\frac{1}{2}, 1)$

Directrix: $y = \frac{3}{2}$

Latus Rectum: Connects x-values corresponding to $y = 1$. Substituting this into **(1)** yields:

$$(x - \tfrac{1}{2})^2 = \tfrac{1}{4} \text{ so that } x = \tfrac{1}{2} \pm \tfrac{1}{2} = 1, 0$$

So, the length of the latus rectum is 1.

44.

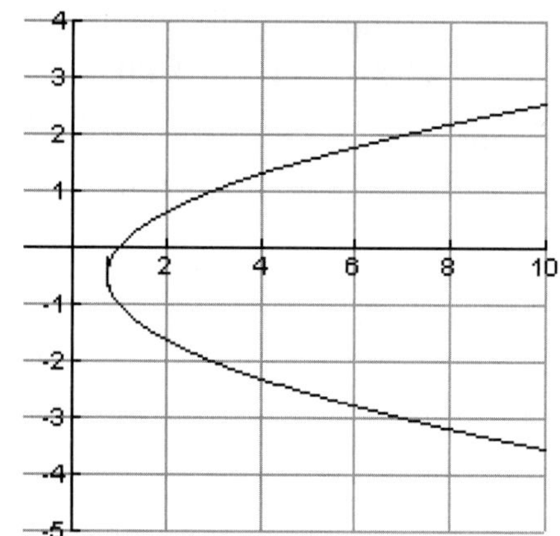

Equation: Completing the square yields:

$$y^2 + y - x + 1 = 0$$
$$(y^2 + y + \tfrac{1}{4}) - \tfrac{1}{4} - x + 1 = 0$$
$$(y + \tfrac{1}{2})^2 = x - \tfrac{3}{4}$$
$$(y + \tfrac{1}{2})^2 = 4(\tfrac{1}{4})(x - \tfrac{3}{4}) \quad \textbf{(1)}$$

So, $p = \frac{1}{4}$ and opens to the right.

Vertex: $(\frac{3}{4}, -\frac{1}{2})$

Focus: $(1, -\frac{1}{2})$

Directrix: $x = \frac{1}{2}$

Latus Rectum: Connects y-values corresponding to $x = 1$. Substituting this into **(1)** yields:

$$(y + \tfrac{1}{2})^2 = \tfrac{1}{4} \text{ so that } y = -\tfrac{1}{2} \pm \tfrac{1}{2} = 0, -1$$

So, the length of the latus rectum is 1.

45. Assume the vertex is at (0,0). The distance from (0,0) to the opening of the dish is 2 feet. Identifying the opening as the latus rectum, we know that the focus will be at (0,2). So, the receiver should be placed 2 feet from the vertex.

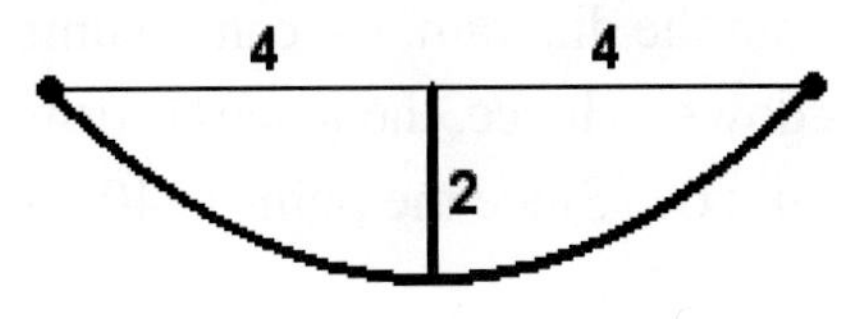

46. Assume the vertex is at (0,0). The distance from (0,0) to the opening of the dish is 5 feet. Identifying the opening as the latus rectum, we know that the focus will be at (0,5). So, the receiver should be placed 5 feet from the vertex.

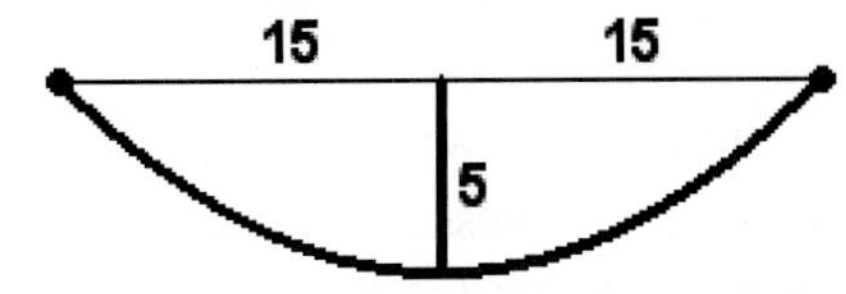

47. Assume the vertex is at the origin and that the parabola opens up. The general form of the equation, therefore, is $x^2 = 4py$. Since the focus is at (0,2), $p = 2$. Hence, the equation governing the shape of the lens is $x^2 = 8y$, or equivalently, $y = \frac{1}{8}x^2$. Since the latus rectum has length 5, the values of x are restricted to the interval $[-2.5, 2.5]$ for this model. Thus, the equation (with the restricted domain) is:

$$y = \tfrac{1}{8}x^2, \text{ for any } x \text{ in } [-2.5, 2.5].$$

Alternatively, assuming the dish opens to the right, we obtain a parabola whose equation is the same as above, but with the roles of x and y switched, namely,

$$x = \tfrac{1}{8}y^2, \text{ for any } y \text{ in } [-2.5, 2.5]$$

48. Assume the vertex is at the origin and that the parabola opens up. Since the focus is at (0,3), $p = 3$. So, the equation is $x^2 = 4(3)y = 12y$, or equivalently, $y = \frac{1}{12}x^2$. We need the x-values corresponding to $y = \frac{1}{2}$: $x^2 = 12(\frac{1}{2}) = 6$ so that $x = \pm\sqrt{6}$. So, the lens is $2\sqrt{6}$ cm wide, and is 0.5 cm from the vertex.

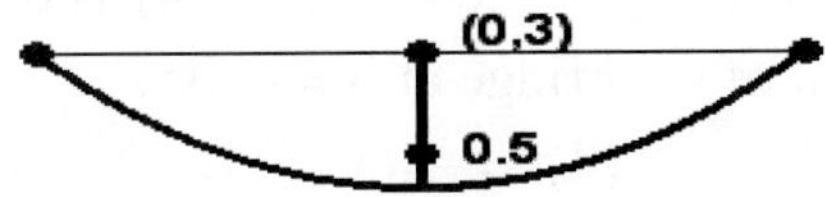

49. Assume the vertex is at (0,0) and that the parabola opens up. Since the rays are focused at (0, 40), we know that $p = 40$. Hence, the equation is $x^2 = 4(40)y = 160y$.

50. Assume the vertex is at (0,0) and that the parabola opens up. Since sunlight is focused at (0, 25), we know that $p = 25$. Hence, the equation is $x^2 = 4(25)y = 100y$.

51. From the diagram, we can assume that the vertex is at (0, 20) and that the parabola opens down. Hence, the general form of the equation is $(x-0)^2 = 4p(y-20)$ **(1)**, for some $p<0$. Since the point $(-40,0)$ is on the graph, we plug it into **(1)** to find p:

$$(-40-0)^2 = 4p(0-20)$$
$$1600 = -80p$$
$$-20 = p$$

So, the equation is $(x-0)^2 = 4(-20)(y-20)$ which simplifies to $x^2 = -80(y-20)$. Now, if the boat passes under the bridge 10 feet from the center, we need to compare the height of the bridge at $x=10$ (or $=-10$) to the height of the boat (which is 17 feet). Indeed, from the equation, the height of the bridge at $x=10$ is:

$$(10)^2 = -80(y-20)$$
$$100 = -80y+1600$$
$$-1500 = -80$$
$$18.75 = y$$

So, yes, the boat will pass under the bridge without scraping the mast.

52. From the diagram, we can assume that the vertex is at (0, 25) and that the parabola opens down. Hence, the general form of the equation is $(x-0)^2 = 4p(y-25)$ **(1)**, for some $p<0$. Since the point $(10,0)$ is on the graph, we plug it into **(1)** to find p:

$$(10-0)^2 = 4p(0-25)$$
$$100 = -100p$$
$$-1 = p$$

So, the equation is $(x-0)^2 = 4(-1)(y-25)$ which simplifies to $x^2 = -4(y-25)$. Now, if the RV straddles the center line, it will be able to pass under the bridge unhindered if the height of the bridge at $x=4$ (and -4) is bigger than 10 feet. Indeed, from the equation, the height of the bridge at $x=4$ is:

$$(4)^2 = -4(y-25)$$
$$16 = -4y+100$$
$$-84 = -4y$$
$$21 = y$$

Since the RV is only 10 feet high, it will make it under the bridge.

53. The focal length is 374.25 feet. Since the standard form of the equation is $x^2 = 4py$, we see that $p = 374.25$. So, the equation is $x^2 = 1497y$.

54. From the given information, we infer that the vertex is at (0, 10) and opens up. So, the general form of the equation is $(y-10)=4p(x-0)^2$ **(1)**. Since the point $(-150,60)$ lies on the graph, substitute it into **(1)** to find p:

$$(60-10)=4p(-150-0)^2$$
$$50=4p(22500)$$
$$p=0.000555$$

Thus, the equation is $\boxed{(y-10)=0.0022x^2}$.

55. We must find the vertex of $p(t)=0.18t^2-5.4t+95.5$. Completing the square yields

$$\begin{aligned}p(t)&=0.18\left(t^2-30t\right)+95.5\\&=0.18\left(t^2-30t+225\right)+95.5-40.5\\&=0.18(t-15)^2+55\end{aligned}$$

So, the minimum is 55.

56. We must find the vertex of $p(t)=-1.1t^2+22t+80$. Completing the square yields

$$\begin{aligned}p(t)&=-1.1t^2+22t+80\\&=-1.1\left(t^2-20t+100\right)+80+110\\&=-1.1(t-10)^2+190\end{aligned}$$

So, the max pulse rate is 90 pulses per minute.

57. If the vertex is at the origin and the focus is at (3,0), then the parabola must open to the right. So, the general equation is $y^2=4px$, for some $p>0$.

58. If the vertex is at (3,2) and the focus is at (5,2), then the parabola must open to the right. So, the general equation is $(y-2)^2=4p(x-3)$, for some $p>0$.

59. True

60. False

61. False. It is the same distance that the focus is from the vertex, but on the opposite side.

62. True. The endpoints of the latus rectum correspond to points on the parabola determined by the segment parallel to the directrix which passes through the focus.

63. Observe that

$$(y-k)^2=4(x-h)=4(1)(x-h).$$

So, the directrix is $x=-1+h=h-1$.

64. Since $p>1$, it opens to the right.

65. Directrix: $y = 4 = -p + k$

Axis of Symmetry: $x = 6 = h$

Focus: $(6,9) = (h, p+k) \Rightarrow p+k = 9$

So, we solve the system to find k:

$$\begin{cases} -p+k=4 \\ p+k=9 \end{cases}$$

Add the two equations: k = 13/2, and so [= 5/2. Hence, the vertex is (6, 13/2).

66. The equation of the parabola is

$$(x-6)^2 = 4\left(\tfrac{5}{2}\right)\left(y-\tfrac{13}{2}\right)$$

$$(x-6)^2 = 5(2y-3)$$

67. Derive the equation for $x^2 = 4py$, where $p > 0$.

Let (x, y) be any point on this parabola. Consult the diagram to the right. Using the distance formula, we find that

$$d_1 = \sqrt{(x-0)^2 + (y-p)^2}$$

$$d_2 = \sqrt{(x-x)^2 + (y-(-p))^2} = |y+p|$$

Equate d_1 and d_2 and simplify:

$$\sqrt{(x-0)^2 + (y-p)^2} = |y+p|$$

$$x^2 + (y-p)^2 = (y+p)^2$$

$$x^2 + y^2 - 2py + p^2 = y^2 + 2py + p^2$$

$$x^2 = 4py,$$

as desired.

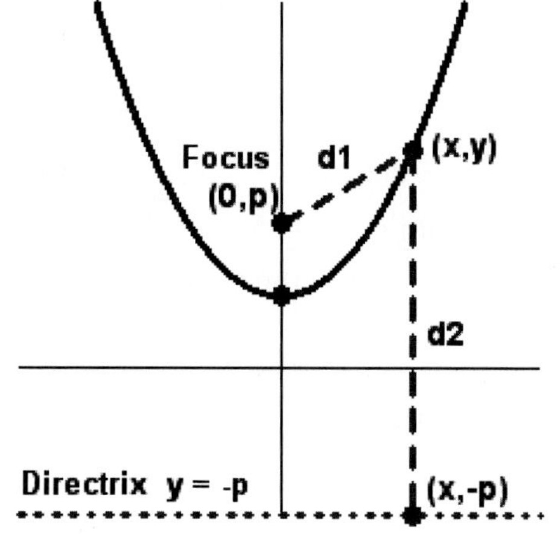

68. Derive the equation for $y^2 = 4px$, where $p > 0$.

Let (x, y) be any point on this parabola. Consult the diagram to the right. Using the distance formula, we find that

$$d_1 = \sqrt{(x-p)^2 + (y-0)^2}$$

$$d_2 = \sqrt{(x-(-p))^2 + (y-y)^2} = |x+p|$$

Equate d_1 and d_2 and simplify:

$$\sqrt{(x-p)^2 + y^2} = |x+p|$$

$$(x-p)^2 + y^2 = (x+p)^2$$

$$x^2 - 2px + p^2 + y^2 = x^2 + 2px + p^2$$

$$y^2 = 4px,$$

as desired.

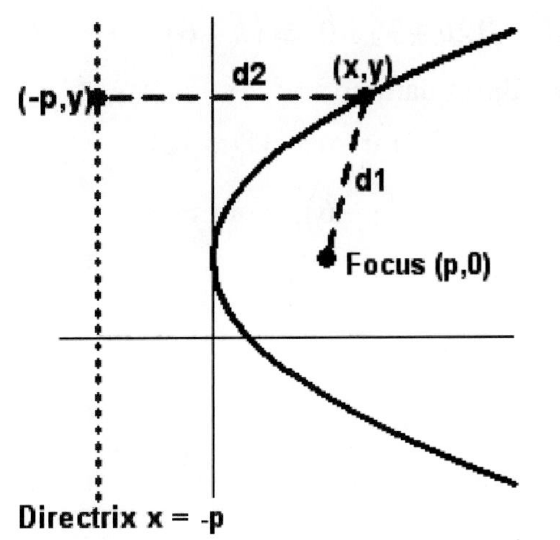

69. Parabola 1: Vertex $(h, 6)$

Directrix $x = 0 \Rightarrow 0 = -p + h \Rightarrow h = p$

So, its equation is $(y-6)^2 = 4h(x-h)$.

Parabola 2: Vertex $(\bar{h}, 6)$.

Directrix $x = 8 \Rightarrow 8 = -\bar{p} + \bar{h} \Rightarrow \bar{p} = \bar{h} - 8$

So, its equation is $(y-6)^2 = 4(\bar{h}-8)(x-\bar{h})$.

Using the fact that (4,2) is on both graphs yields the two equations:

$$16 = 4h(4-h) \quad \textbf{(1)}$$

$$16 = 4(\bar{h}-8)(4-\bar{h}) \quad \textbf{(2)}$$

Solve **(1)**:

$$-4h^2 + 16h - 16 = 0 \Rightarrow (h-2)^2 = 0 \Rightarrow h = 2$$

70. Parabola 1: Vertex $(9, k)$

Directrix $y = -11 \Rightarrow -11 = -p + k \Rightarrow p = k + 11$

So, its equation is $(x-9)^2 = 4(k+11)(y-k)$.

Parabola 2: Vertex $(9, \bar{k})$.

Directrix $y = 1 \Rightarrow 1 = -\bar{p} + \bar{k} \Rightarrow \bar{p} = \bar{k} - 1$

So, its equation is $(x-9)^2 = 4(\bar{k}-1)(y-\bar{k})$.

Using the fact that (6,-5) is on both graphs yields the two equations:

$$9 = 4(k+11)(-5-k) \quad \textbf{(1)}$$

$$9 = 4(\bar{k}-1)(-5-\bar{k}) \quad \textbf{(2)}$$

Solve **(2)**:

$$\bar{h}^2 - 12\bar{h} + 36 = 0 \Rightarrow \left(\bar{h} - 6\right)^2 = 0 \Rightarrow \bar{h} = 6$$

So, the equations are:

$$\left(y-6\right)^2 = 8(x-2)$$

$$\left(y-6\right)^2 = -4(x-6)$$

Simplify both equations and solve them using the quadratic formula to get that

$$k = -8 \pm \tfrac{3}{2}\sqrt{3},\quad \bar{k} = -8 \pm \tfrac{3}{2}\sqrt{3}$$

So, there are two possible equations in each case:

Parabola 1:

$$(x-9)^2 = 4(-8-\tfrac{3}{2}\sqrt{3}+11)\left(y-\left(-8-\tfrac{3}{2}\sqrt{3}\right)\right)$$

$$(x-9)^2 = 4(-8+\tfrac{3}{2}\sqrt{3}+11)\left(y-\left(-8+\tfrac{3}{2}\sqrt{3}\right)\right)$$

Parabola 2:

$$(x-9)^2 = 4\left(-8-\tfrac{3}{2}\sqrt{3}-1\right)\left(y-\left(-8-\tfrac{3}{2}\sqrt{3}\right)\right)$$

$$(x-9)^2 = 4\left(-8+\tfrac{3}{2}\sqrt{3}-1\right)\left(y-\left(-8+\tfrac{3}{2}\sqrt{3}\right)\right)$$

71. Parabola 1:

Focus $(0, \tfrac{3}{2}) \Rightarrow (h, p+k) = (0, \tfrac{3}{2})$

Directrix $y = \tfrac{1}{2} \Rightarrow -p+k = \tfrac{1}{2}$

So, h =0. We solve the system for p and k:

$$\begin{cases} p+k = \tfrac{3}{2} \\ -p+k = \tfrac{1}{2} \end{cases}$$

Add the equations: $k = 1$

Substitute into either equation to get $p = ½$

So, the equation is

$$(x-0)^2 = 4(\tfrac{1}{2})(y-1) \Rightarrow x^2 = 2(y-1)$$

Parabola 2:

Focus $(0, -\tfrac{3}{4}) \Rightarrow (h, p+k) = (0, -\tfrac{3}{4})$

Directrix $y = -\tfrac{5}{4} \Rightarrow -p+k = -\tfrac{5}{4}$

So, h =0. We solve the system for p and k:

$$\begin{cases} p+k = -\tfrac{3}{4} \\ -p+k = -\tfrac{5}{4} \end{cases}$$

Add the equations: k = -1

Substitute into either equation to get $p = ¼$

So, the equation is

$$(x-0)^2 = 4(\tfrac{1}{4})(y+1) \Rightarrow x^2 = y+1$$

Now, equate the two equations to find the points of intersection:

$$2(y-1) = y+1 \;\Rightarrow\; y = 3$$

Thus, $x^2 = 4 \Rightarrow x = \pm 2$.

So, the points of intersection are (2,3) and (-2,3).

72. Both parabolas either open up or down based on the respective locations of their foci and vertices. As such, two such parabolas cannot intersect.

73.

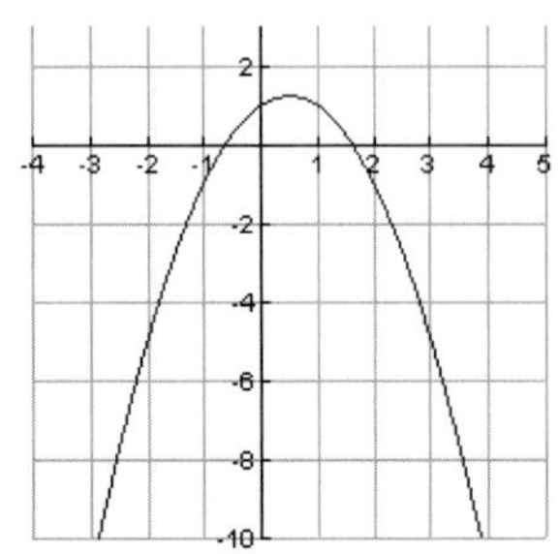

74.

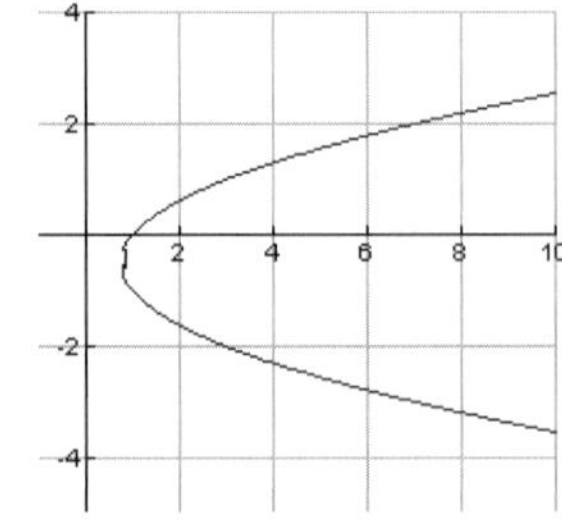

75. The vertex is $(2.5, -3.5)$. The parabola opens to the right since the square term is y^2 and $p > 0$. The graph of $(y+3.5)^2 = 10(x-2.5)$ is:

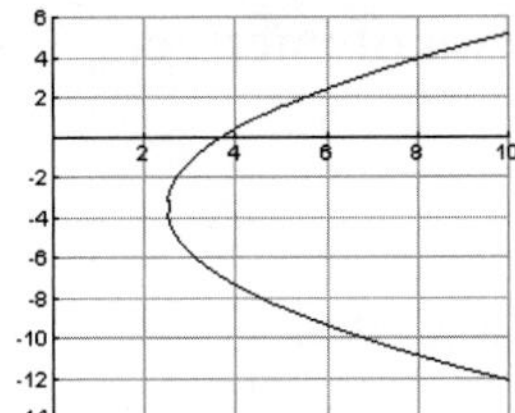

76. The vertex is $(-1.4, -1.7)$. The parabola opens down since the square term is x^2 and $p < 0$. The graph of $(x+1.4)^2 = -5(y+1.7)$ is:

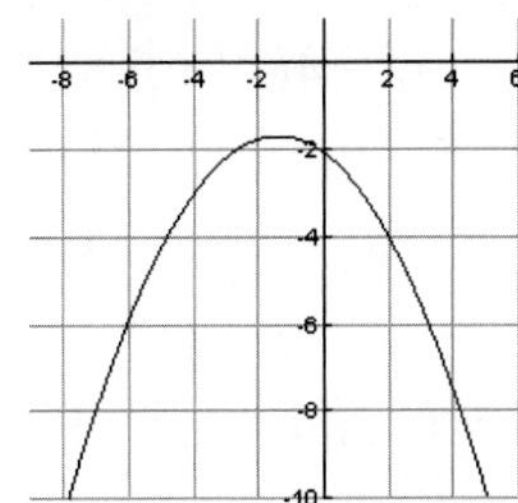

77. The vertex is at (1.8, 1.5) and it should open to the left. Graphing this parabola on the calculator confirms this:

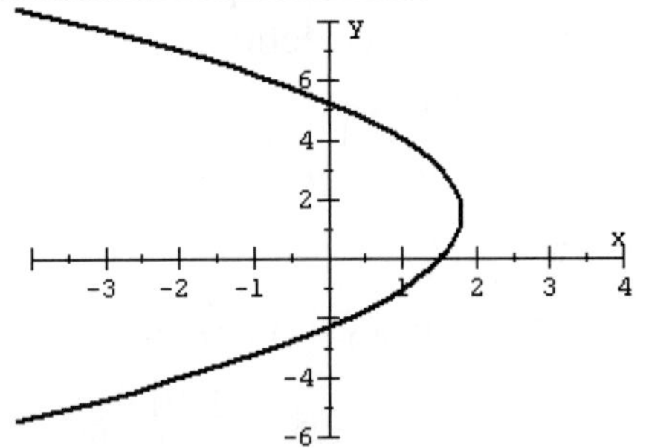

78. The vertex is at (-2.4, 3.2) and it should open up. Graphing this parabola on the calculator confirms this:

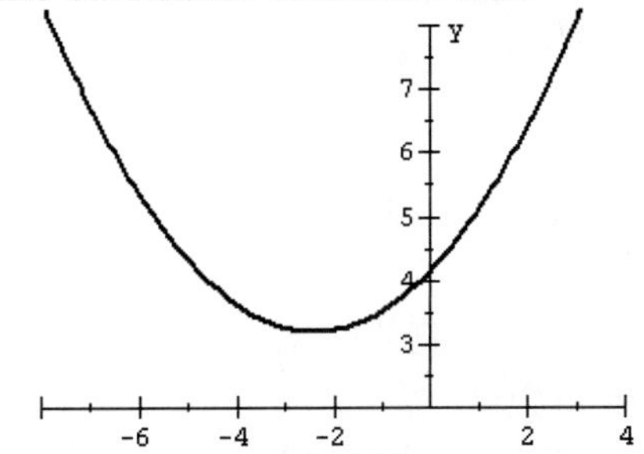

79. Parabola 1:

Vertex (0, −1) $\Rightarrow (h,k) = (0,-1)$

Directrix $y = -\frac{5}{4} \Rightarrow -p + k = -\frac{5}{4}$

So, h =0, k = -1, and so, p = ¼ .

So, the equation is

$$(x-0)^2 = 4\left(\tfrac{1}{4}\right)(y+1) \Rightarrow x^2 = y+1$$

Parabola 2:

Vertex (0, 7) $\Rightarrow (h,k) = (0,7)$

Directrix $y = \frac{29}{4} \Rightarrow -p + k = \frac{29}{4}$

So, h =0, k = 7, and so, p =-¼ .

So, the equation is

$$(x-0)^2 = 4\left(-\tfrac{1}{4}\right)(y-7) \Rightarrow x^2 = -(y-7)$$

Now, equate the two equations to find the points of intersection:

$$y+1 = -(y-7) \Rightarrow y = 3$$

Thus, $x^2 = 4 \Rightarrow x = \pm 2$.

So, the points of intersection are (2,3) and (-2,3).

80. Parabola 1:

Vertex (0, 0)$\Rightarrow (h,k)=(0,0)$

Focus (0,1) $\Rightarrow p+k=1$

So, h =0, k = 0, and so, p = 1.

So, the equation is

$$(x-0)^2=4(y-0)\Rightarrow x^2=4y$$

Parabola 2:

Vertex (1, 0)$\Rightarrow (h,k)=(1,0)$

Focus (1,1) $\Rightarrow p+k=1$

So, h =1, k = 0, and so, p = 1.

So, the equation is

$$(x-1)^2=4(y-0)\Rightarrow (x-1)^2=4y$$

Now, equate the two equations to find the points of intersection:

$$x^2=(x-1)^2 \Rightarrow 0=-2x+1 \Rightarrow x=\tfrac{1}{2}$$

Thus, $4y=\left(\tfrac{1}{2}\right)^2 \Rightarrow y=\tfrac{1}{16}$.

So, the point of intersection is $\left(\tfrac{1}{2},\tfrac{1}{16}\right)$

81. Parabola 1:

Vertex $(5,\tfrac{5}{3})\Rightarrow (h,k)=(5,\tfrac{5}{3})$

Focus $(5,\tfrac{29}{12}) \Rightarrow p+k=\tfrac{29}{12}$

So, h =5, k = 5/3, and so, p = 3/4.

So, the equation is

$$(x-5)^2=4\left(\tfrac{3}{4}\right)\left(y-\tfrac{5}{3}\right)\Rightarrow 3y=5+(x-5)^2$$

Parabola 2:

Vertex $(\tfrac{13}{2},\tfrac{289}{24})=(h,k)$

Focus $(\tfrac{13}{2},\tfrac{253}{24}) \Rightarrow p+k=\tfrac{253}{24}$

So, h =13/2, k = 289/24, and so, p = 253/24.

So, the equation is

$$\left(x-\tfrac{13}{2}\right)^2=4\left(-\tfrac{3}{2}\right)\left(y-\tfrac{289}{24}\right)\Rightarrow 3y=-\tfrac{289}{8}-\tfrac{1}{2}(x-\tfrac{13}{2})^2$$

Now, equate the two equations to find the points of intersection:

$$5+(x-5)^2=-\tfrac{289}{8}-\tfrac{1}{2}(x-\tfrac{13}{2})^2$$

$$40+8x^2-80x+200=0$$

$$12x^2-132x+120=0$$

$$12(x-10)(x-1)=0$$

$$x=1,10$$

Substitute these into either equation to find y:

$$x=1 \Rightarrow y=7$$

$$x=10 \Rightarrow y=10$$

So, the points of intersection are (1,7) and (10,10).

82. Parabola 1:

Directrix $y=-\tfrac{37}{4} \Rightarrow =-p+k=-\tfrac{37}{4}$

Focus $(-2,-\tfrac{35}{4}) \Rightarrow p+k=-\tfrac{35}{4}$

Solving the following system

$$\begin{cases} -p+k=-\tfrac{37}{4} \\ p+k=-\tfrac{35}{4} \end{cases}$$

yields $p=\tfrac{1}{4}$, $k=-9$. So, the equation is

$$(x+2)^2=4\left(\tfrac{1}{4}\right)(y+9)\Rightarrow y=-9+(x+2)^2$$

Now, equate the two equations to find the points of intersection:

$$-9+(x+2)^2=-25-(x-2)^2$$

$$2x^2+24=0$$

which has no solution.

Thus, these two parabolas do not intersect.

Parabola 2:
Directrix $y = -\frac{99}{4} \Rightarrow = -p + k = -\frac{99}{4}$
Focus $(-2, -\frac{101}{4}) \Rightarrow p + k = -\frac{101}{4}$
Solving the following system

$$\begin{cases} -p + k = -\frac{99}{4} \\ p + k = -\frac{101}{4} \end{cases}$$

yields $p = -\frac{1}{4}$, $k = -25$. So, the equation is

$$(x-2)^2 = 4\left(-\tfrac{1}{4}\right)(y+25) \Rightarrow y = -25 - (x-2)^2$$

Section 9.3 Solutions --

1. d x-axis is the major axis. The intercepts are $(6,0)$, $(-6,0)$, $(0,4)$, $(0,-4)$.	**2. b** y-axis is the major axis. The intercepts are $(4,0)$, $(-4,0)$, $(0,6)$, $(0,-6)$.
3. a y-axis is the major axis. The intercepts are $(2\sqrt{2},0)$, $(-2\sqrt{2},0)$, $(0,6\sqrt{2})$, $(0,-6\sqrt{2})$.	**4. c** x-axis is the major axis. The intercepts are $(\frac{1}{2},0)$, $(-\frac{1}{2},0)$, $(0,1)$, $(0,-1)$.
5. 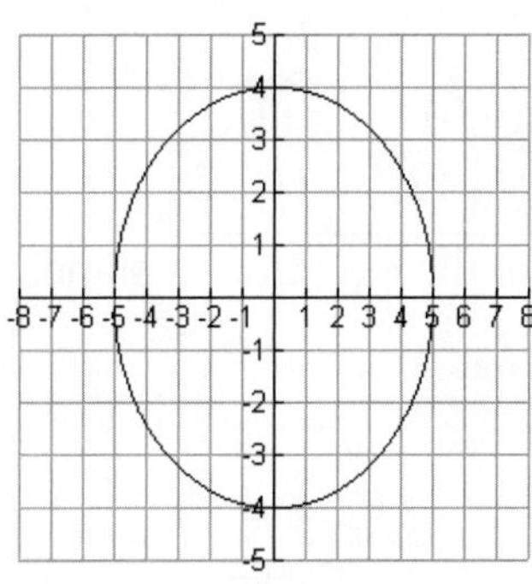 Center: (0,0) Vertices: $(\pm5,0)$, $(0,\pm4)$	**6.** 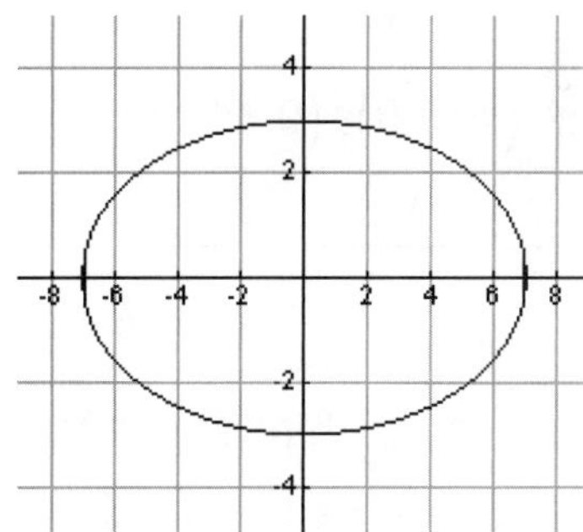 Center: (0,0) Vertices: $(\pm7,0)$, $(0,\pm3)$

7.

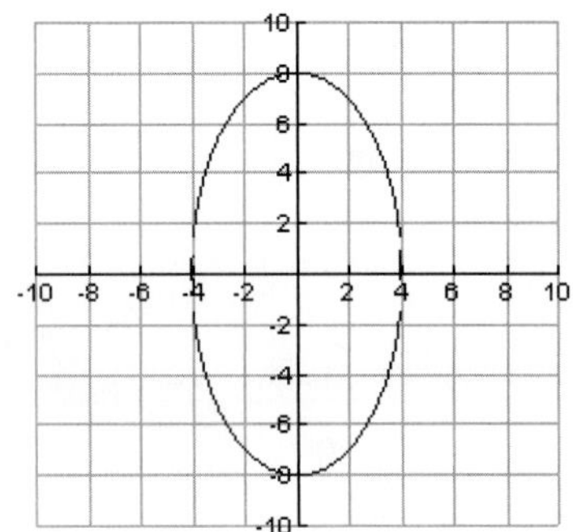

Center: (0,0)
Vertices: $(\pm 4, 0), (0, \pm 8)$

8.

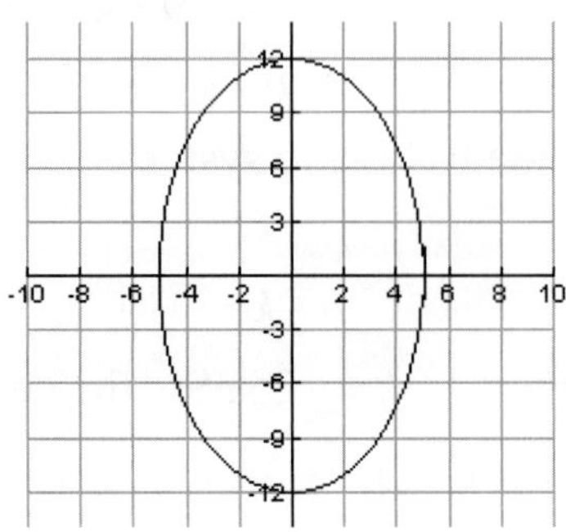

Center: (0,0)
Vertices: $(\pm 5, 0), (0, \pm 12)$

9.

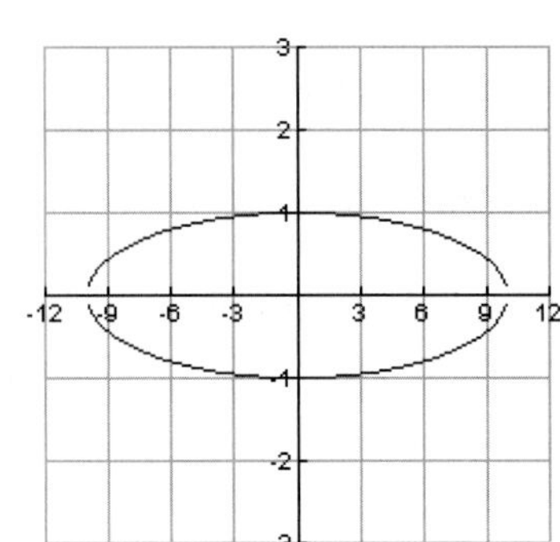

Center: (0,0)
Vertices: $(\pm 10, 0), (0, \pm 1)$

10. Write the equation in standard form as $\frac{x^2}{4} + \frac{y^2}{9} = 1$.

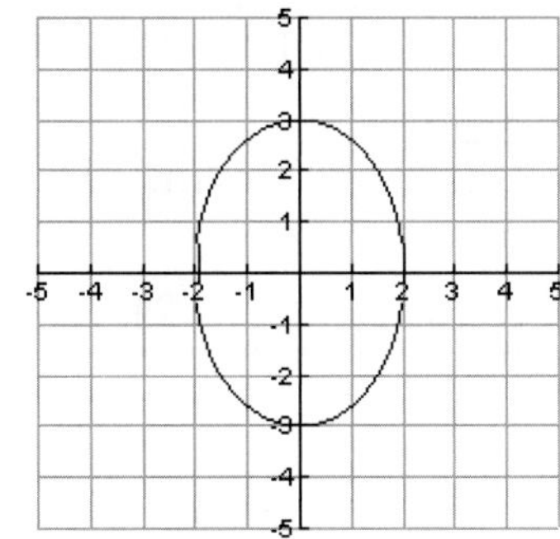

Center: (0,0)
Vertices: $(\pm 2, 0), (0, \pm 3)$

11.

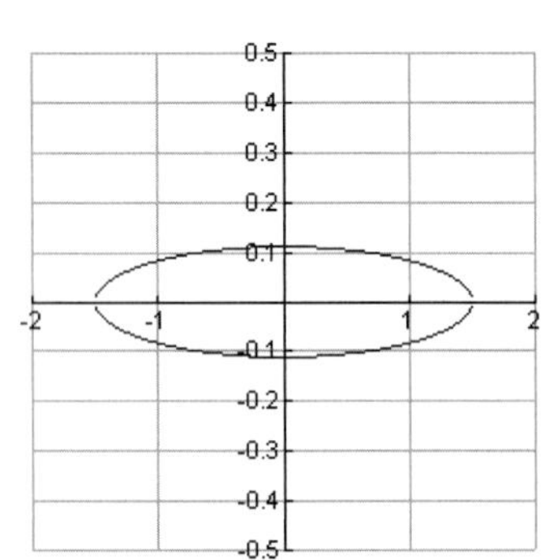

Center: (0,0)
Vertices: $(\pm \frac{3}{2}, 0), (0, \pm \frac{1}{9})$

12. Write the equation in standard form as $\frac{x^2}{\frac{25}{4}} + \frac{y^2}{\frac{9}{100}} = 1$.

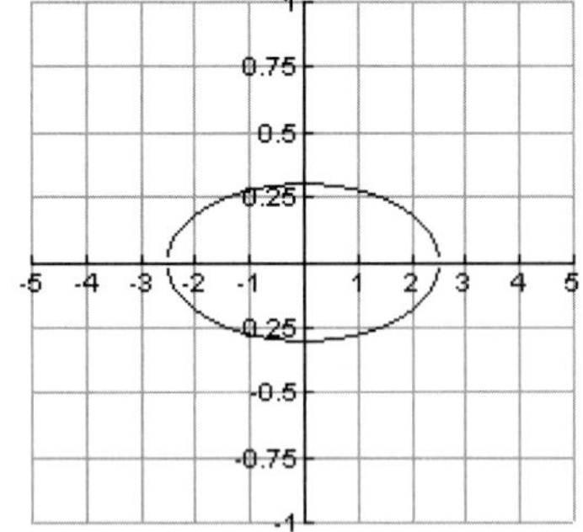

Center: (0,0)
Vertices: $(\pm \frac{5}{2}, 0), (0, \pm \frac{3}{10})$

13. Write the equation in standard form as $\frac{x^2}{4}+\frac{y^2}{16}=1$.

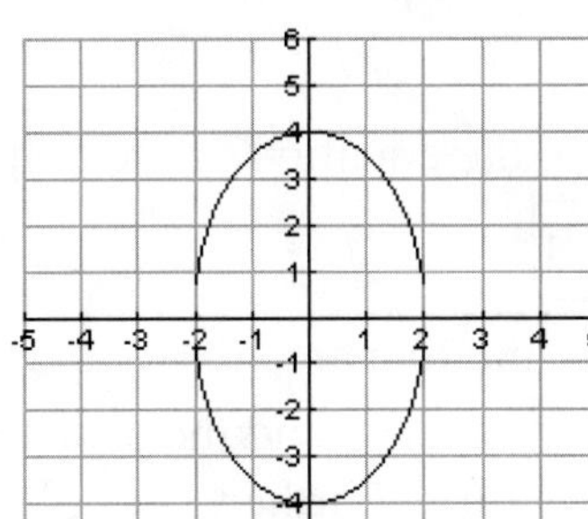

Center: (0,0)
Vertices: $(\pm 2,0)$, $(0,\pm 4)$

14.

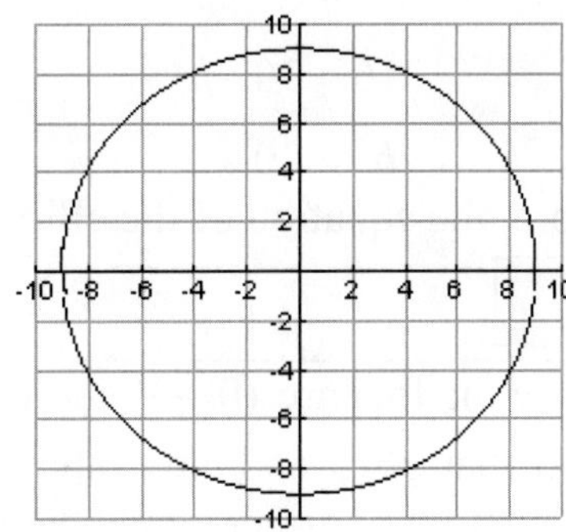

Center: (0,0)
Vertices: $(\pm 9,0)$, $(0,\pm 9)$

15. Write the equation in standard form as $\frac{x^2}{4}+\frac{y^2}{2}=1$.

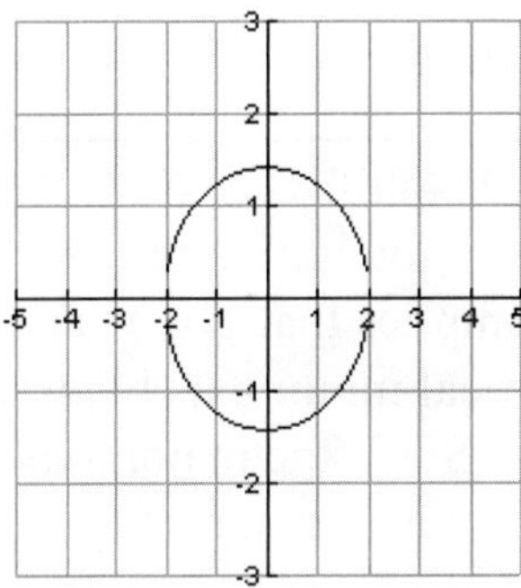

Center: (0,0)
Vertices: $(\pm 2,0)$, $\left(0,\pm\sqrt{2}\right)$

16. Write the equation in standard form as $\frac{x^2}{5}+\frac{y^2}{2}=1$.

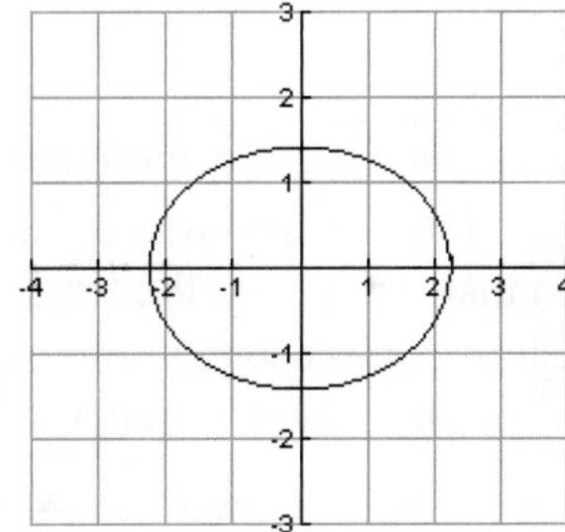

Center: (0,0)
Vertices: $\left(\pm\sqrt{5},0\right)$, $\left(0,\pm\sqrt{2}\right)$

17. Since the foci are $(-4,0)$ and $(4,0)$, we know that $c=4$ and the center of the ellipse is $(0,0)$. Further, since the vertices are $(-6,0)$ and $(6,0)$ and they lie on the *x*-axis, the ellipse is horizontal and $a=6$.

18. Since the foci are $(-1,0)$ and $(1,0)$, we know that $c=1$ and the center of the ellipse is $(0,0)$. Further, since the vertices are $(-3,0)$ and $(3,0)$ they lie on the *x*-axis, the ellipse is horizontal and so, $a=3$.

Hence, $$c^2 = a^2 - b^2$$ $$16 = 36 - b^2$$ $$b^2 = 20$$ Therefore, the equation of the ellipse is $\boxed{\frac{x^2}{36} + \frac{y^2}{20} = 1}$.	Hence, $$c^2 = a^2 - b^2$$ $$1 = 9 - b^2$$ $$b^2 = 8$$ Therefore, the equation of the ellipse is $\boxed{\frac{x^2}{9} + \frac{y^2}{8} = 1}$.
19. Since the foci are $(0,-3)$ and $(0,3)$, we know that $c = 3$ and the center of the ellipse is $(0,0)$. Further, since the vertices are $(0,-4)$ and $(0,4)$ and they lie on the *y*-axis, the ellipse is vertical and $a = 4$. Hence, $$c^2 = a^2 - b^2$$ $$9 = 16 - b^2$$ $$b^2 = 7$$ Therefore, the equation of the ellipse is $\boxed{\frac{x^2}{7} + \frac{y^2}{16} = 1}$.	**20.** Since the foci are $(0,-1)$ and $(0,1)$, we know that $c = 1$ and the center of the ellipse is $(0,0)$. Further, since the vertices are $(0,-2)$ and $(0,2)$ and they lie on the *y*-axis, the ellipse is vertical and $a = 2$. Hence, $$c^2 = a^2 - b^2$$ $$1 = 4 - b^2$$ $$b^2 = 3$$ Therefore, the equation of the ellipse is $\boxed{\frac{x^2}{3} + \frac{y^2}{4} = 1}$.
21. Since the ellipse is centered at $(0,0)$, the length of the vertical major axis being 8 implies that $a = 4$, and the length of the horizontal minor axis being 4 implies that $b = 2$. Since the major axis is vertical, the equation of the ellipse must be $\boxed{\frac{x^2}{4} + \frac{y^2}{16} = 1}$.	**22.** Since the ellipse is centered at $(0,0)$, the length of the horizontal major axis being 10 implies that $a = 5$, and the length of the vertical minor axis being 2 implies that $b = 1$. Since the major axis is horizontal, the equation of the ellipse is $\boxed{\frac{x^2}{25} + y^2 = 1}$.
23. Since the vertices are $(0,-7), (0,7)$, we know that the ellipse is centered at $(0,0)$ (since it is halfway between the vertices), the major axis is vertical, and $a = 7$. Further, since the endpoints of the minor axis are $(-3,0), (3,0)$, we know that $b = 3$. Thus, the equation of the ellipse is $\boxed{\frac{x^2}{9} + \frac{y^2}{49} = 1}$.	**24.** Since the vertices are $(-9,0), (9,0)$, we know that the ellipse is centered at $(0,0)$ (since it is halfway between the vertices), the major axis is horizontal, and $a = 9$. Further, since the endpoints of the minor axis are $(0,-4), (0,4)$, we know that $b = 4$. Thus, the equation of the ellipse is $\boxed{\frac{x^2}{81} + \frac{y^2}{16} = 1}$.
25. c Center is $(3,-2)$ and the major axis is vertical.	**26. d** Center is $(-3,2)$ and the major axis is vertical.

27. b Center is $(3,-2)$ and the major axis is horizontal.	**28. a** Center is $(-3,2)$ and the major axis is horizontal.
29. 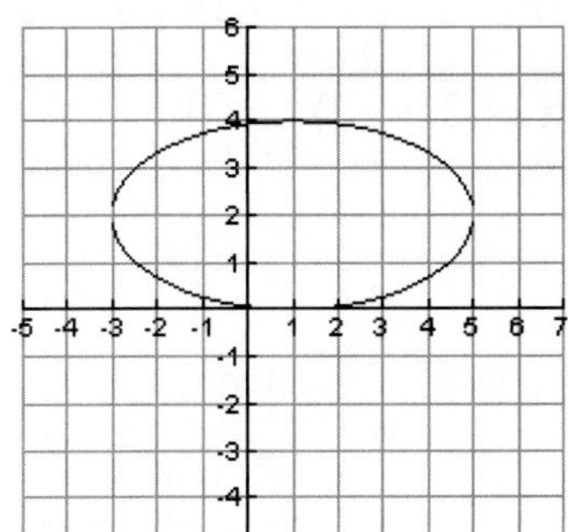 Center: (1,2) Vertices: (-3,2), (5,2), (1,0), (1,4)	**30.** 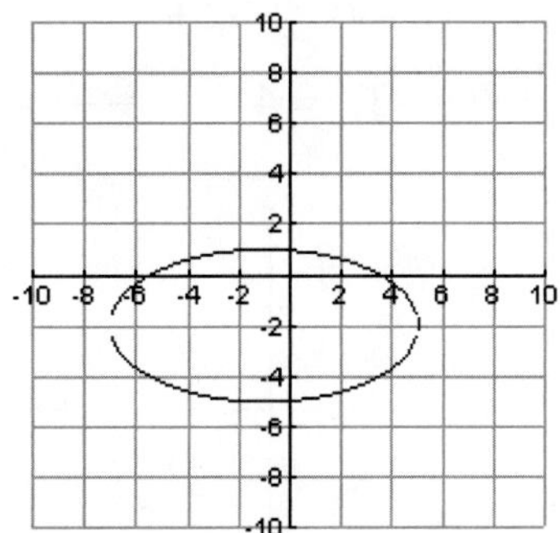 Center: (-1,-2) Vertices: (-7, -2), (5, -2), (-1,1), (-1,-5)
31. First, write the equation in standard form: $\frac{(x+3)^2}{8}+\frac{(y-4)^2}{80}=1$ 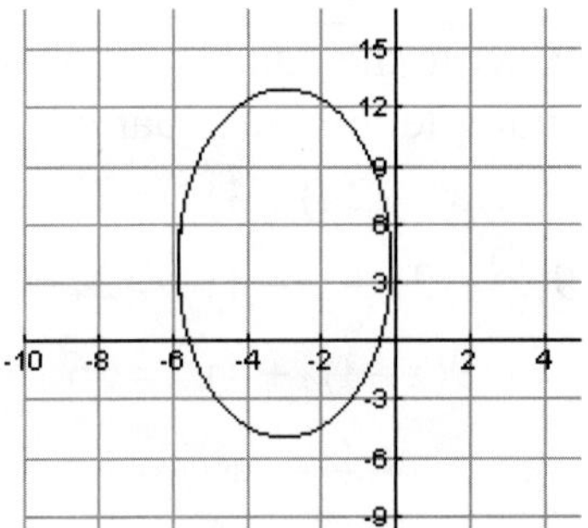 Center: (-3,4) Vertices: $\left(-2\sqrt{2}-3,4\right),\left(2\sqrt{2}-3,4\right),$ $\left(-3,4+4\sqrt{5}\right),\left(-3,4-4\sqrt{5}\right)$	**32.** First, write the equation in standard form: $\frac{(x+3)^2}{12}+\frac{(y-4)^2}{3}=1$ 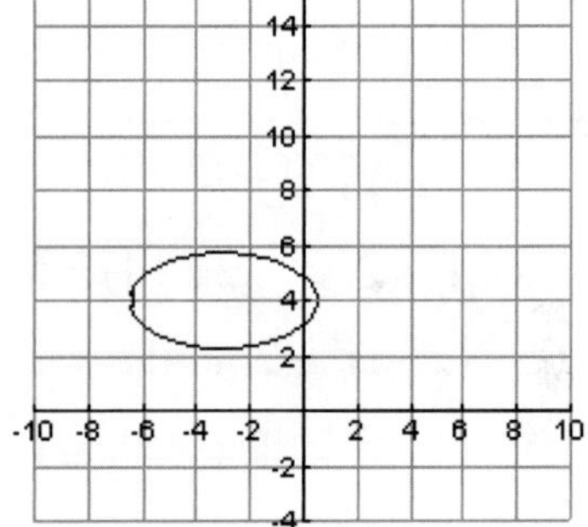 Center: (-3,4) Vertices: $\left(-2\sqrt{3}-3,4\right),\left(2\sqrt{3}-3,4\right),$ $\left(-3,4-\sqrt{3}\right),\left(-3,4+\sqrt{3}\right)$

33. First, write the equation in standard form by completing the square:

$$x^2+4(y^2-6y)=-32$$

$$x^2+4(y^2-6y+9)=-32+36$$

$$x^2+4(y-3)^2=4 \Rightarrow \tfrac{x^2}{4}+(y-3)^2=1$$

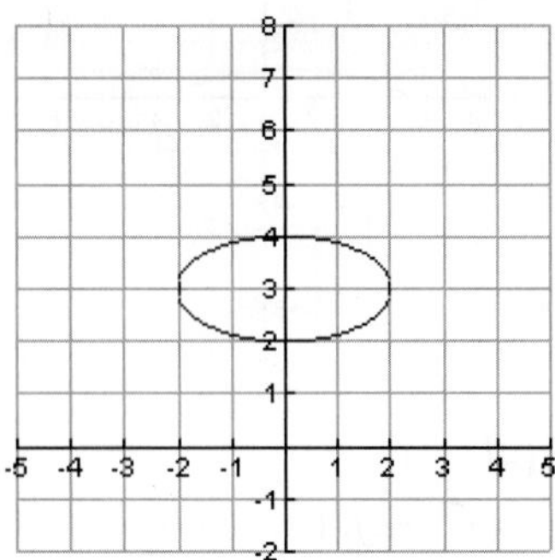

Center: (0,3)
Vertices: (-2,3), (2,3), (0,3), (0,4)

34. First, write the equation in standard form by completing the square:

$$25x^2+2(y^2-2y)=48$$

$$25x^2+2(y^2-2y+1)=48+2$$

$$25x^2+2(y-1)^2=50 \Rightarrow \tfrac{x^2}{2}+\tfrac{(y-1)^2}{25}=1$$

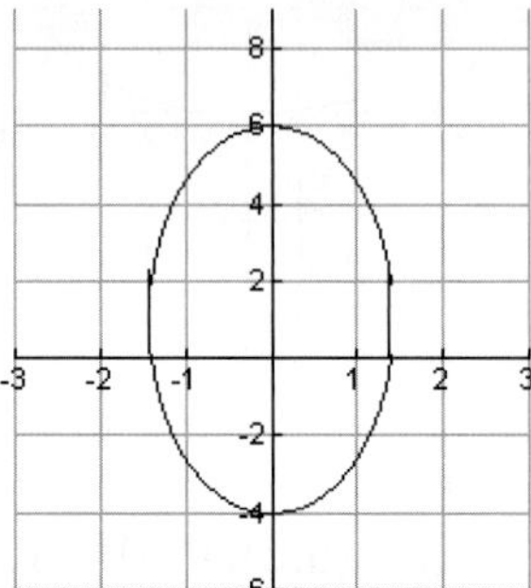

Center: (0,1)
Vertices: $\left(\pm\sqrt{2},1\right)$, $(0,-4)$, $(0,6)$

35. First, write the equation in standard form by completing the square:

$$(x^2-2x)+2(y^2-2y)=5$$

$$(x^2-2x+1)+2(y^2-2y+1)=5+1+2$$

$$(x-1)^2+2(y-1)^2=8$$

$$\tfrac{(x-1)^2}{8}+\tfrac{(y-1)^2}{4}=1$$

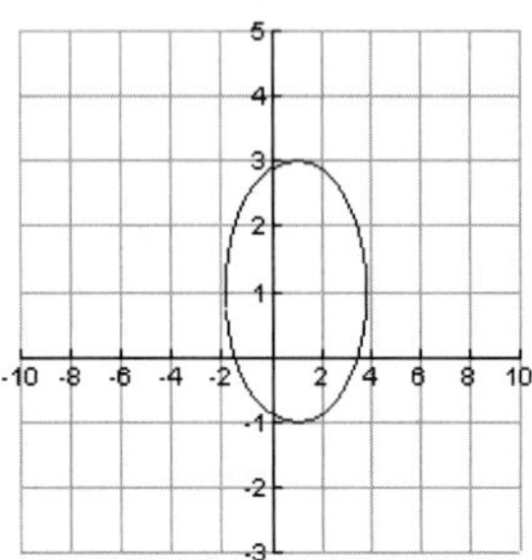

Center: (1,1)
Vertices: $\left(1\pm2\sqrt{2},1\right)$, $(1,3)$, $(1,-1)$

36. First, write the equation in standard form by completing the square:

$$9(x^2-2x)+4y^2=27$$

$$9(x^2-2x+1)+4y^2=27+9$$

$$9(x-1)^2+4y^2=36$$

$$\tfrac{(x-1)^2}{4}+\tfrac{y^2}{9}=1$$

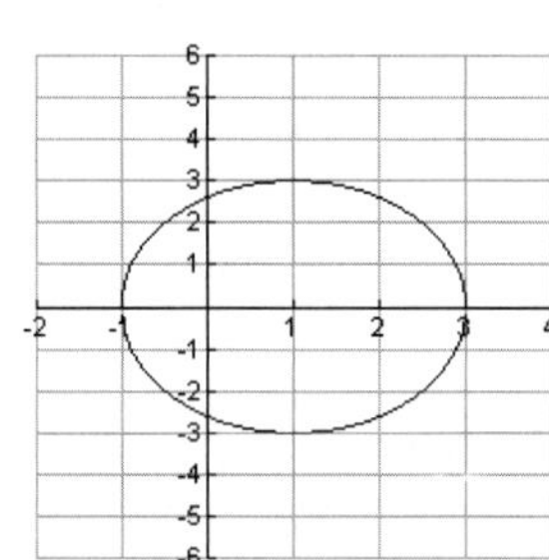

Center: (1,0)
Vertices: (-1,0), (3,0), (1,3), (1,-3)

37. First, write the equation in standard form by completing the square:

$$5(x^2+4x)+(y^2+6y)=21$$
$$5(x^2+4x+4)+(y^2+6y+9)=21+20+9$$
$$5(x+2)^2+(y+3)^2=50$$
$$\frac{(x+2)^2}{10}+\frac{(y+3)^2}{50}=1$$

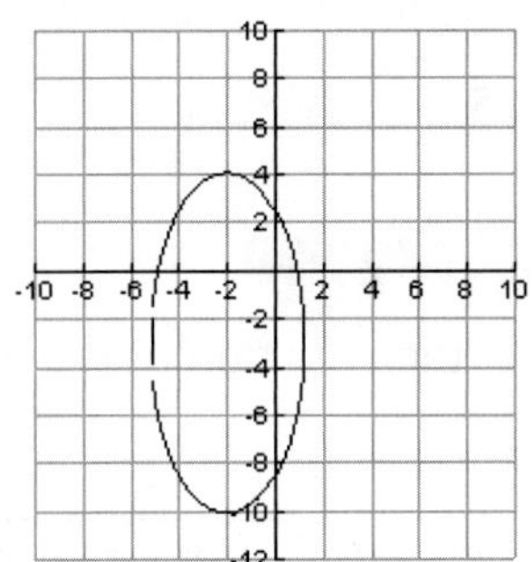

Center: (-2,-3)

Vertices: $\left(-2\pm\sqrt{10},-3\right),\left(-2,-3\pm5\sqrt{2}\right)$

38. First, write the equation in standard form by completing the square:

$$9(x^2+4x)+(y^2+2y)=-36$$
$$9(x^2+4x+4)+(y^2+2y+1)=-36+36+1$$
$$9(x+2)^2+(y+1)^2=1$$
$$\frac{(x+2)^2}{\frac{1}{9}}+(y+1)^2=1$$

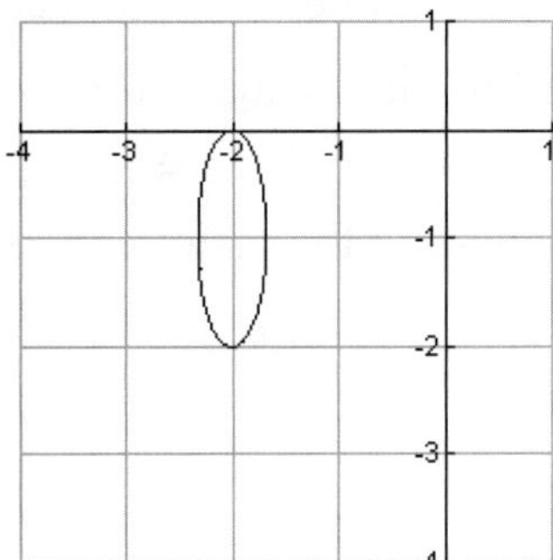

Center: (-2,-1)

Vertices: $\left(-\frac{7}{3},-1\right),\left(-\frac{5}{3},-1\right)$, $(-2,0)$, $(-2,-2)$

39. Since the foci are $(-2,5)$, $(6,5)$ and the vertices are $(-3,5)$, $(7,5)$, we know:

i) the ellipse is horizontal with center at $\left(\frac{-3+7}{2},5\right)=(2,5)$,

ii) $2-c=-2$ so that $c=4$,

iii) the length of the major axis is $7-(-3)=10$. Hence, $a=5$.

Now, since $c^2=a^2-b^2$, $16=25-b^2$ so that $b^2=9$. Hence, the equation of the ellipse is $\boxed{\frac{(x-2)^2}{25}+\frac{(y-5)^2}{9}=1}$.

40. Since the foci are $(2,-2)$, $(4,-2)$ and the vertices are $(0,-2)$, $(6,-2)$, we know:

i) the ellipse is horizontal with center at $\left(\frac{0+6}{2},-2\right)=(3,-2)$,

ii) $3-c=2$ so that $c=1$,

iii) the length of the major axis is $6-0=6$. Hence, $a=3$.

Now, since $c^2=a^2-b^2$, $1=9-b^2$ so that $b^2=8$. Hence, the equation of the ellipse is $\boxed{\frac{(x-3)^2}{9}+\frac{(y+2)^2}{8}=1}$.

41. Since the foci are $(4,-7),\ (4,-1)$ and the vertices are $(4,-8),\ (4,0)$, we know:

i) the ellipse is vertical with center at $(4,\frac{-8+0}{2})=(4,-4)$,

ii) $-4-c=-7$ so that $c=3$,

iii) the length of the major axis is $0-(-8)=8$. Hence, $a=4$.

Now, since $c^2=a^2-b^2$, $9=16-b^2$ so that $b^2=7$. Hence, the equation of the ellipse is $\boxed{\frac{(x-4)^2}{7}+\frac{(y+4)^2}{16}=1}$.

42. Since the foci are $(2,-6),\ (2,-4)$ and the vertices are $(2,-7),\ (2,-3)$, we know:

i) the ellipse is vertical with center at $(2,\frac{-7+(-3)}{2})=(2,-5)$,

ii) $-5-c=-6$ so that $c=1$,

iii) the length of the major axis is $-3-(-7)=4$. Hence, $a=2$.

Now, since $c^2=a^2-b^2$, $1=4-b^2$ so that $b^2=3$. Hence, the equation of the ellipse is $\boxed{\frac{(x-2)^2}{3}+\frac{(y+5)^2}{4}=1}$.

43. Since the length of the major axis is 8, we know that $a=4$; since the length of the minor axis is 4, we know that $b=2$. As such, since the center is $(3,2)$ and the major axis is vertical, the equation of the ellipse must be $\boxed{\frac{(x-3)^2}{4}+\frac{(y-2)^2}{16}=1}$.

44. Since the length of the major axis is 10, we know that $a=5$; since the length of the minor axis is 2, we know that $b=1$. As such, since the center is $(-4,3)$ and the major axis is horizontal, the equation of the ellipse must be $\boxed{\frac{(x+4)^2}{25}+(y-3)^2=1}$.

45. Since the vertices are $(-1,-9),\ (-1,1)$, we know that:

i) the ellipse is vertical with center at $(-1,\frac{-9+1}{2})=(-1,-4)$,

ii) $a=\frac{1-(-9)}{2}=5$.

Now, since the endpoints of the minor axis are $(-4,-4),\ (2,-4)$, we know that $b=\frac{2-(-4)}{2}=3$. Hence, the equation of the ellipse is $\boxed{\frac{(x+1)^2}{9}+\frac{(y+4)^2}{25}=1}$.

46. Since the vertices are $(-2,3),\ (6,3)$, we know that:

i) the ellipse is horizontal with center at $(\frac{-2+6}{2},3)=(2,3)$,

ii) $a=\frac{6-(-2)}{2}=4$.

Now, since the endpoints of the minor axis are $(2,1),\ (2,5)$, we know that $b=\frac{5-1}{2}=2$.

Hence, the equation of the ellipse is $\boxed{\frac{(x-2)^2}{16}+\frac{(y-3)^2}{4}=1}$.

47. Since the major axis has length 150 feet, we know that $a=75$; since the minor axis has length 30 feet, we know that $b=15$. From the diagram we know that the y-axis is the major axis. So, since the ellipse is centered at the origin, its equation must be $\frac{x^2}{225}+\frac{y^2}{5625}=1$	**48.** Since both the major and minor axes have length 180 feet, we know that $a=b=90$. So, since the ellipse is centered at the origin, its equation must be $\frac{x^2}{8100}+\frac{y^2}{8100}=1$, which simplifies to $\boxed{x^2+y^2=8100}$ (a circle).

49. a. Since the major axis has length 150 yards, we know that $a=75$; since the minor axis has length 40 yards, we know that $b=20$. So, since the ellipse is centered at the origin, its equation must be $\frac{x^2}{5625}+\frac{y^2}{400}=1$.

b. Since the football field is 120 yards long and it is completely surrounded by an elliptical track, we can compute the equation in part **a.** at $x=60$ to find the corresponding y-values, and then subtract them to find the width at that particular position in the field:

$$y^2=400(1-\tfrac{3600}{5625})=\tfrac{400(2025)}{5625}=144 \text{ so that } y=\pm 12.$$

Hence, the width of the track at the end of the football field is 24 yards. Since the field itself is 30 yards wide, the track will NOT encompass the field.

50. Here we need to find the value of b so that when the equation of the track, namely $\frac{x^2}{5625}+\frac{y^2}{b^2}=1$, is evaluated at $x=60$, the difference in the corresponding y-values is at least 30. To find the smallest such value of b, we must solve:

$$\frac{60^2}{5625}+\frac{30^2}{b^2}=1$$

$$\frac{900}{b^2}=\frac{2025}{5625}$$

$$b^2=\frac{5625(900)}{2025}=2500$$

$$b=50$$

Hence, the minor axis should be at least 100 yards wide in order to enclose the field.

51. Assume that the sun is at the origin. Then, the vertices of Pluto's horizontal elliptical trajectory are: $(-4.447\times10^8,0)$, $(7.38\times10^8,0)$

From this, we know that:

i) The length of the major axis is $7.38\times10^8-(-4.447\times10^8)=11.827\times10^8$, and so the value of a is half of this, namely 5.9135×10^8;

ii) The center of the ellipse is $\left(\frac{7.38\times10^8+(-4.447\times10^8)}{2},0\right)=(1.4665\times10^8,0)$;

iii) The value of c is 1.4665×10^{8}.

Now, since $c^2=a^2-b^2$, we have $\left(1.4665\times10^{8}\right)^2=\left(5.9135\times10^{8}\right)^2-b^2$, so that $b^2=3.2818\times10^{17}$. Hence, the equation of the ellipse must be $\boxed{\frac{\left(x-(1.4665\times10^{8})\right)^2}{3.4969\times10^{17}}+\frac{(y-0)^2}{3.2818\times10^{17}}=1}$.

(Note: There is more than one correct way to set up this problem, and it begins with choosing the center of the ellipse. If, alternatively, you choose the center to be at (0,0), then the actual coordinates of the vertices and foci will change, and will result in the following slightly different equation for the trajectory: $\frac{x^2}{5{,}914{,}000{,}000^2}+\frac{y^2}{5{,}729{,}000{,}000^2}=1$

Both are equally correct!.)

52. Assume that the sun is at the origin. Then, the vertices of Earth's horizontal elliptical trajectory are: $(-1.471\times10^{8},0)$, $(1.526\times10^{8},0)$

From this, we know that:

i) The length of the major axis is $1.526\times10^{8}-(-1.471\times10^{8})=2.997\times10^{8}$, and so the value of a is half of this, namely 1.4985×10^{8};

ii) The center of the ellipse is $\left(\frac{1.526\times10^{8}+(-1.471\times10^{8})}{2},0\right)=(0.0275\times10^{8},0)=(2.75\times10^{6},0)$;

iii) The value of c is 2.75×10^{6}.

Now, since $c^2=a^2-b^2$, we have $\left(2.75\times10^{6}\right)^2=\left(1.4985\times10^{8}\right)^2-b^2$, so that $b^2=2.245\times10^{16}$. Hence, the equation of the ellipse must be $\boxed{\frac{\left(x-(2.75\times10^{6})\right)^2}{2.246\times10^{16}}+\frac{(y-0)^2}{2.245\times10^{16}}=1}$.

53. The length of the semi-major axis $=a=150\times10^{6}$, so that $a^2=2.25\times10^{16}$.

Observe that the formula for eccentricity can be rewritten as follows:

$$e^2=1-\frac{b^2}{a^2}$$
$$\frac{b^2}{a^2}=1-e^2$$
$$b^2=a^2(1-e^2)$$
$$b=a\sqrt{1-e}$$

Since $e=0.223$ and (from above) $a=150{,}000{,}000$, we find that $b=146{,}000{,}000$.

Hence, the equation of the ellipse is $\boxed{\frac{x^2}{150{,}000{,}000^2}+\frac{y^2}{146{,}000{,}000^2}=1}$.

54. The length of the semi-major axis = $a = 350\times10^6$, so that $a^2 = 1.225\times10^{17}$. Observe that

$$0.634 = \sqrt{1-\tfrac{b^2}{1.225\times10^{17}}}$$
$$0.40196 = \tfrac{1.225\times10^{17}-b^2}{1.225\times10^{17}}$$
$$4.9239\times10^{16} = 1.225\times10^{17} - b^2$$
$$7.326\times10^{16} = b^2$$

Hence, the equation of the ellipse is $\boxed{\tfrac{x^2}{1.225\times10^{17}} + \tfrac{y^2}{7.326\times10^{16}} = 1}$.

55. It approaches the graph of a straight line. Indeed, note that

$$1 = \sqrt{1-\frac{b^2}{a^2}} \Rightarrow 1-\frac{b^2}{a^2} = 1 \Rightarrow \frac{b^2}{a^2} = 0 \Rightarrow b = 0.$$

Hence, $a^2x^2 + b^2y^2 = a^2b^2$ becomes $a^2x^2 = 0$, and since $a \neq 0$, we conclude that $x = 0$.

56. The length of the semi-major axis is a = 17.8 AU, so that $a^2 = 316.84$. Since we also know that $e \approx 0.967$, we can use $e^2 = 1-\frac{b^2}{a^2}$ to find b:

$$(0.967)^2 = 1-\tfrac{b^2}{316.84} \Rightarrow b \approx \sqrt{20.56640} \approx 4.53475 \text{ AU}.$$

So, the equation of the comet (using AU units) is $\dfrac{x^2}{316.84} + \dfrac{y^2}{20.5664} = 1$.

57. Observe that

$$\left(t^2-6t+9\right)+9\left(c^2-2c+1\right)+9-9-9=0$$
$$\left(t-3\right)^2+9\left(c-1\right)^2=9$$
$$\tfrac{(t-3)^2}{9}+(c-1)^2=1$$

The vertices are (-3, -6), (-3, 0), (1,2), and (1,0). Using the vertex (1,2) gives the maximum amount that the ellipse extends in the c-direction. Thus, the maximum concentration is 2 mg/cm^3.

58. Observe that $2a = 8$ and $2b = 6$ implies that $a = 4$ and $b = 3$. So, the area of the elliptical cross-section is $\pi ab = 12\pi$ ft^2. So, the volume of the tank is 1,131 ft^2.

59. It should be $a^2 = 6,\ b^2 = 4$, so that $a = \pm\sqrt{6},\ b = \pm 2$.

60. It should be $c^2 = a^2 - b^2$ in place of $c^2 = a^2 + b^2$.

61. False. Since you could not deduce the coordinates of the endpoints of the minor axis from this information alone. Hence, you couldn't determine the value of b.

62. True. The information provided enable you to determine the values of b and c. Then, using this information, the center and value of a can be found.

63. True. The equation would have the general form $\frac{x^2}{a^2}+\frac{y^2}{b^2}=1$, so that substituting in $-x$ for *x* and $-y$ for *y* yields the exact same equation.	**64.** False. All circles are ellipses (with $a=b$), but not all ellipses are circles.

65. Assume that $(h, k+c) = (-2,0)$. Also, since (-2,2) is on the ellipse, it must be a vertex of the ellipse that lies along the same vertical line as the focus. So, $(-2,2) = (h, k+a)$ (since the vertex is above the focus).
The other two scenarios are for which the vertices in alignment with this focus occur on the horizontal line through them. Then, (-2,2) is simply a point on the ellipse. Whether the vertex corresponding to this focus is on the ellipse or not, having this point on the ellipse ensures that there is only one such ellipse for each positioning. Thus, there are three ellipses.

66. Since the vertices are (3,0) and (-3,0), the center must be (0,0). So, we know that *a* $= \pm 3$. As such, the ellipse is of the form $\frac{x^2}{9}+\frac{y^2}{b^2}=1$. We can choose *b* to be any positive real number and still have an ellipse with these vertices. So, there are infinitely many.

67. Two ellipses can intersect at a vertex. So, they could intersect in as few as one point.	**68.** The maximum number of points that two nonoverlapping ellipses can intersect in is 4.

69. Pluto: $e=\frac{1.4665\times10^8}{5.9135\times10^8}\cong 0.25$ (from Problem 51)

Earth: $e=\frac{2.75\times10^6}{1.4985\times10^8}\cong 0.02$ (from Problem 52)

70.
a. If *e* is close to 1, then *c* is close to *a*. As such, using this fact, together with $c^2=a^2-b^2$, we see that $b^2=a^2-c^2$ would be close to 0. So, the ellipse would be very narrow in the *y*-direction.
b. If *e* is close to 0, then *c* is close to 0. So, since $a^2-b^2=c^2$, this would imply that *a* is close to *b*. So, the ellipse is nearly circular.
c. If $e=\frac{1}{2}$, then $\frac{c}{a}=\frac{1}{2}$, so that $a=2c$. Then, since $c^2=a^2-b^2$,

$$c^2=4c^2-b^2$$
$$b^2=3c^2 \text{ so that } b=\sqrt{3}\,c$$

71. Substitute the given points into the general equation $\frac{x^2}{a^2}+\frac{y^2}{b^2}=1$ to obtain a system that can be used to solve for *a* and *b*:

$$\begin{cases}\frac{1}{a^2}+\frac{9}{b^2}=1 & \textbf{(1)}\\ \frac{16}{a^2}+\frac{4}{b^2}=1 & \textbf{(2)}\end{cases}$$

Let $u = \frac{1}{a^2}$, $v = \frac{1}{b^2}$ in this system to obtain an equivalent (easier-to-deal-with) system:

$$\begin{cases} u + 9v = 1 & \textbf{(3)} \\ 16u + 4v = 1 & \textbf{(4)} \end{cases}$$

To solve this latter system, multiply **(3)** by -16 and add to **(4)** to find that $v = \frac{3}{28}$.
Substitute this back into either equation to find that $u = \frac{1}{28}$.
Going back to the substitution, we see that $a^2 = 28$, $b^2 = \frac{28}{3}$. Hence, the equation of the ellipse is $\frac{x^2}{28} + \frac{3y^2}{28} = 1 \Rightarrow x^2 + 3y^2 = 28$.

72. Substitute the given points into the general equation $\frac{x^2}{a^2} + \frac{y^2}{b^2} = 1$ to obtain a system that can be used to solve for a and b:

$$\begin{cases} \frac{1}{a^2} + \frac{36/5}{b^2} = 1 & \textbf{(1)} \\ \frac{25/9}{a^2} + \frac{4}{b^2} = 1 & \textbf{(2)} \end{cases}$$

Let $u = \frac{1}{a^2}$, $v = \frac{1}{b^2}$ in this system to obtain an equivalent (easier-to-deal-with) system:

$$\begin{cases} 5u + 36v = 5 & \textbf{(3)} \\ 25u + 36v = 9 & \textbf{(4)} \end{cases}$$

To solve this latter system, Subtract **(3)** - **(4)** to find that $u = \frac{1}{5}$.
Substitute this back into either equation to find that $v = \frac{1}{9}$.
Going back to the substitution, we see that $a^2 = 5$, $b^2 = 9$. Hence, the equation of the ellipse is $\frac{x^2}{5} + \frac{y^2}{9} = 1 \Rightarrow 9x^2 + 5y^2 = 45$.

73. Substitute the given points into the general equation $\frac{(x-2)^2}{a^2} + \frac{(y+3)^2}{b^2} = 1$ to obtain a system that can be used to solve for a and b:

$$\begin{cases} \frac{1}{a^2} + \frac{64/9}{b^2} = 1 & \textbf{(1)} \\ \frac{9}{a^2} + 0 = 1 & \textbf{(2)} \end{cases}$$

To solve this system, note that **(2)** implies that $a^2 = 9$.
Substitute this back into **(1)** to find that $b^2 = 8$.
Hence, the equation of the ellipse is $\frac{(x-2)^2}{9} + \frac{(y+3)^2}{8} = 1 \Rightarrow 8x^2 + 9y^2 - 32x + 54y + 41 = 0$.

74. Substitute the given points into the general equation $\frac{(x-1)^2}{a^2}+\frac{(y+2)^2}{b^2}=1$ to obtain a system that can be used to solve for a and b:

$$\begin{cases} 0+\frac{4}{b^2}=1 & \textbf{(1)} \\ \frac{1}{a^2}+0=1 & \textbf{(2)} \end{cases}$$

Solving **(2)** yields $a^2=1$ and solving **(1)** yields $b^2=4$.

Hence, the equation of the ellipse is $(x-1)^2+\frac{(y+2)^2}{4}=1 \Rightarrow 4x^2+y^2-8x+4y+4=0$.

75. Note that from the graphs, we see that as c increases, the graph of the ellipse described by $x^2+cy^2=1$ narrows in the y-direction and becomes more elongated.

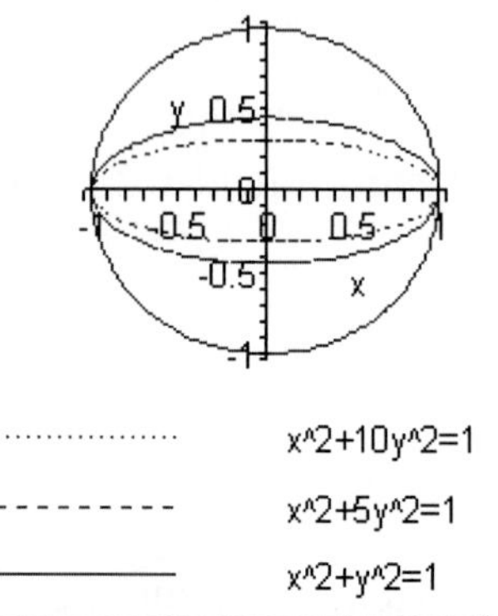

76. Note that from the graphs, we see that as c increases, the graph of the ellipse described by $cx^2+y^2=1$ narrows in the x-direction and becomes more elongated.

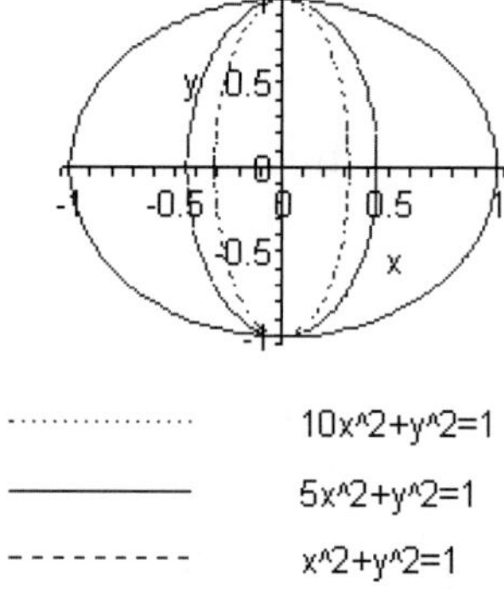

77. Note that from the graphs, we see that as c increases, the graph of the ellipse described by $cx^2+cy^2=1$, which is equivalent to $x^2+y^2=\frac{1}{c}$ (a circle) shrinks towards the origin. (This makes sense since the radius is $\frac{1}{c}$, and as c gets larger, the value of $\frac{1}{c}$ goes to 0.)

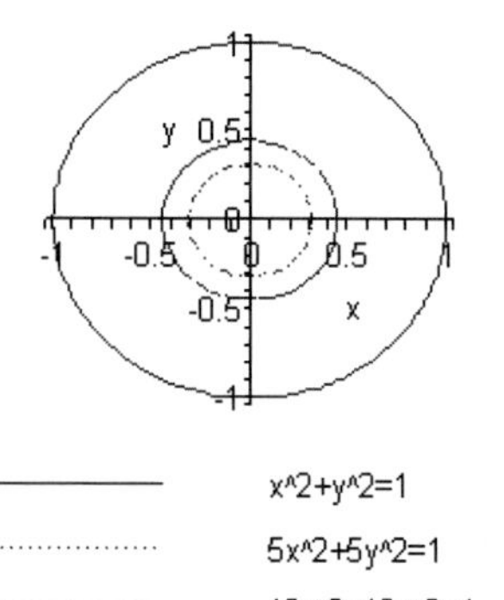

78. This is not the graph of an ellipse. Rather, it is the graph of a *hyperbola* – see Section 8.4.

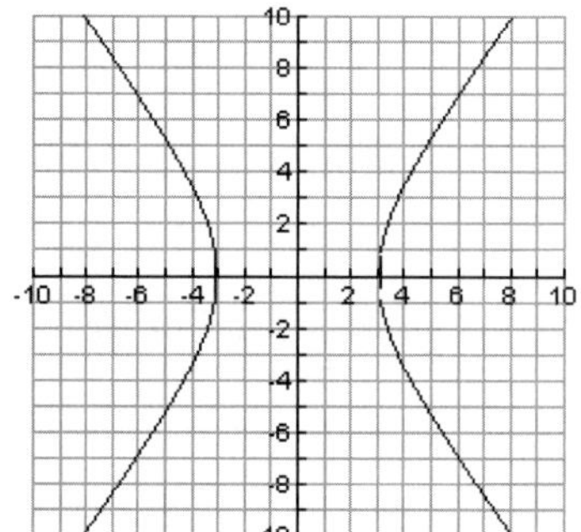

79. Note that from the graphs, we see that as c decreases, the major axis (along the x-axis) of the ellipse described by $cx^2 + y^2 = 1$ becomes longer.

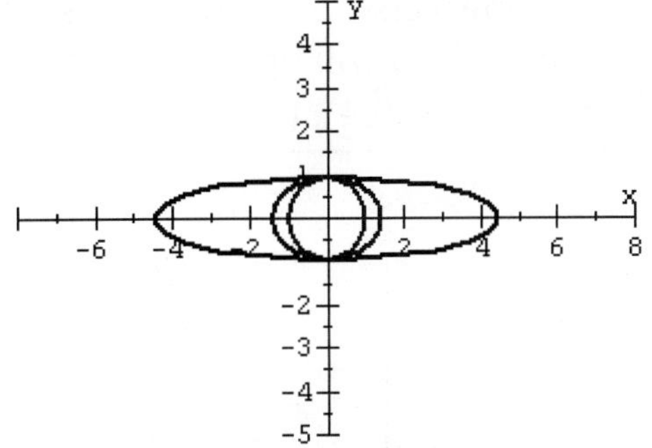

80. Note that from the graphs, we see that as c decreases, the major axis (along the y-axis) of the ellipse described by $x^2 + cy^2 = 1$ becomes longer.

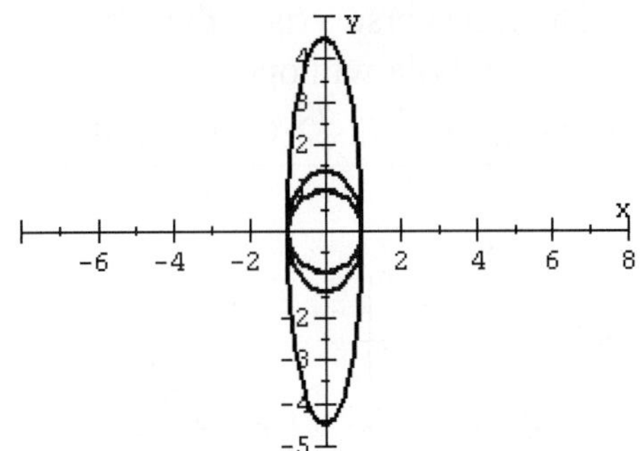

81. Complete the square and find the vertices along vertical axis:

$$4\left(x^2-6x\right)+\left(y^2+10y\right)=-57$$

$$4\left(x^2-6x+9\right)+\left(y^2+10y+25\right)=-57+36+25$$

$$4(x-3)^2+(y+5)^2=4$$

$$(x-3)^2+\tfrac{(y+5)^2}{4}=1$$

The vertices are $-5\pm2=-7,-3$. So, the maximum is -3 and the minimum is -7.

82. Complete the square and find the vertices along vertical axis:

$$9\left(x^2+8x\right)+4\left(y^2+4y\right)=-124$$

$$9\left(x^2+8x+16\right)+4\left(y^2+4y+4\right)=-124+144+16$$

$$9(x+4)^2+4(y+2)^2=36$$

$$\tfrac{(x+4)^2}{4}+\tfrac{(y+2)^2}{9}=1$$

The vertices are $-2\pm3=-5,1$. So, the maximum is 1 and the minimum is -5.

83. Complete the square and find the vertices along vertical axis:

$$81\left(x^2-12x\right)+100\left(y^2+16y\right)=-1216$$

$$81\left(x^2-12x+36\right)+100\left(y^2+16y+64\right)=-1216+2916+6400$$

$$81(x-6)^2+100(y+8)^2=8100$$

$$\tfrac{(x-6)^2}{100}+\tfrac{(y+8)^2}{81}=1$$

The vertices are $-8\pm9=-17,1$. So, the maximum is 1 and the minimum is -17.

84. Complete the square and find the vertices along vertical axis:

$$25\left(x^2+8x\right)+16\left(y^2+16y\right)=176$$

$$25\left(x^2+8x+16\right)+16\left(y^2+16y+64\right)=176+400+1024$$

$$25(x+4)^2+16(y+8)^2=1600$$

$$\tfrac{(x+4)^2}{64}+\tfrac{(y+8)^2}{100}=1$$

The vertices are $-8\pm10=-18,2$. So, the maximum is 2 and the minimum is -18.

Section 9.4 Solutions ---

1. b The transverse axis is the x-axis, so that the hyperbola will open left/right. The vertices are $(-6,0), (6,0)$.

2. a The transverse axis is the y-axis, so that the hyperbola will open up/down. The vertices are $(0,-6), (0,6)$.

3. d The transverse axis is the x-axis, so that the hyperbola will open left/right. The vertices are $(-2\sqrt{2},0), (2\sqrt{2},0)$.

4. c The transverse axis is the y-axis, so that the hyperbola will open up/down. The vertices are $(0,\frac{1}{2}), (0,-\frac{1}{2})$.

5.

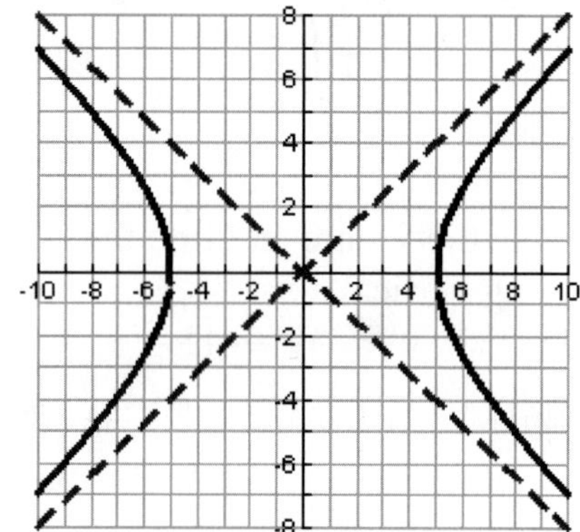

Notes on the Graph:
Equations of the asymptotes are $y=\pm\frac{4}{5}x$.

6.

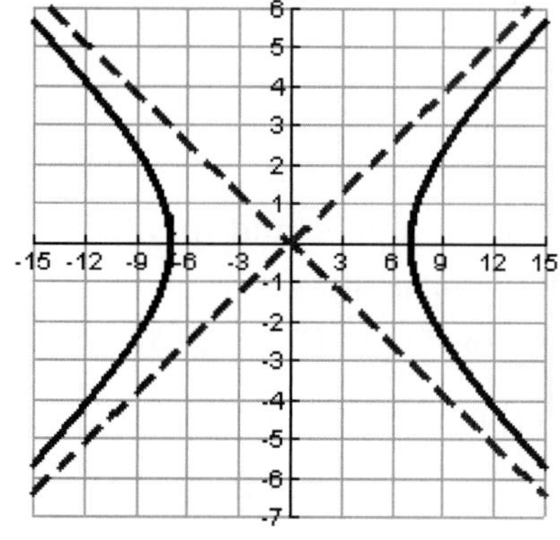

Notes on the Graph:
Equations of the asymptotes are $y=\pm\frac{3}{7}x$.

7.

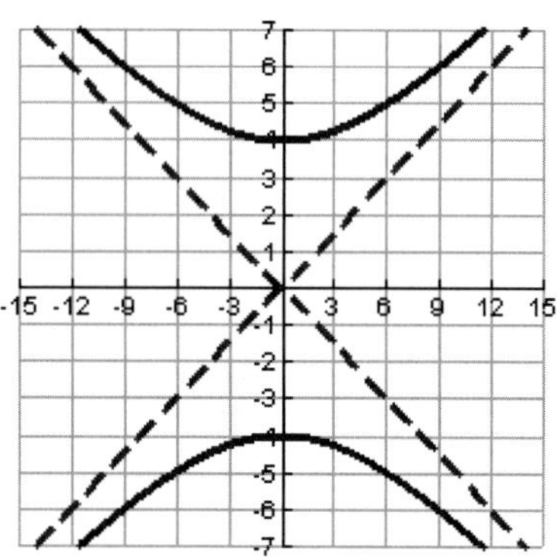

Notes on the Graph:
Equations of the asymptotes are $y=\pm\frac{1}{2}x$.

8.

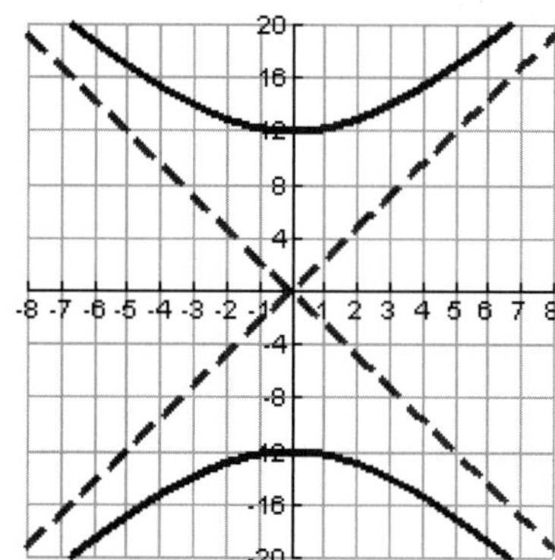

Notes on the Graph:
Equations of the asymptotes are $y=\pm\frac{12}{5}x$.

9.

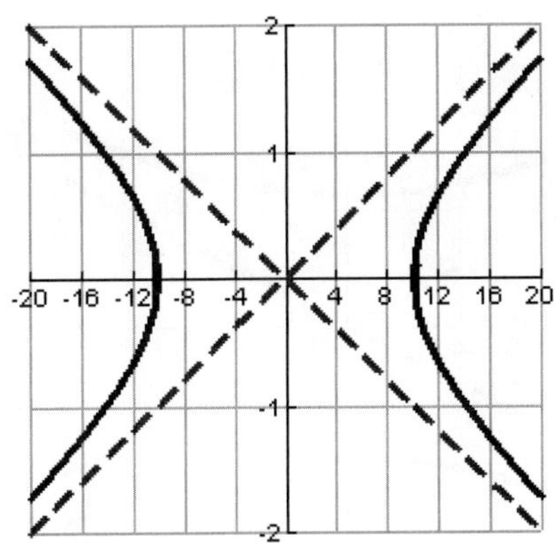

Notes on the Graph:
Equations of the asymptotes are $y = \pm \frac{1}{10} x$.

10.

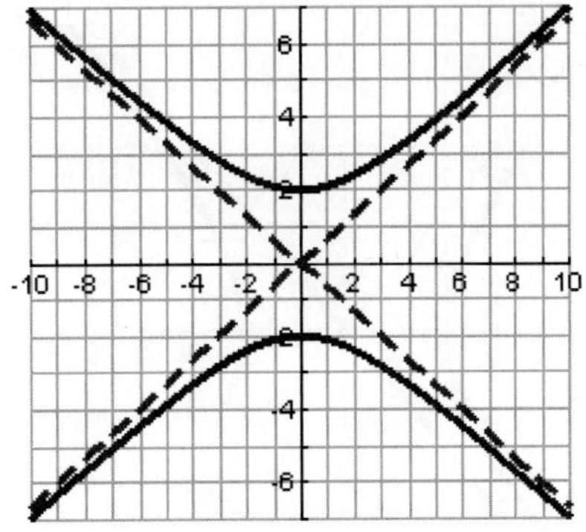

Notes on the Graph:
First, write the equation in standard form as $\frac{y^2}{4} - \frac{x^2}{9} = 1$. The equations of the asymptotes are $y = \pm \frac{2}{3} x$.

11.

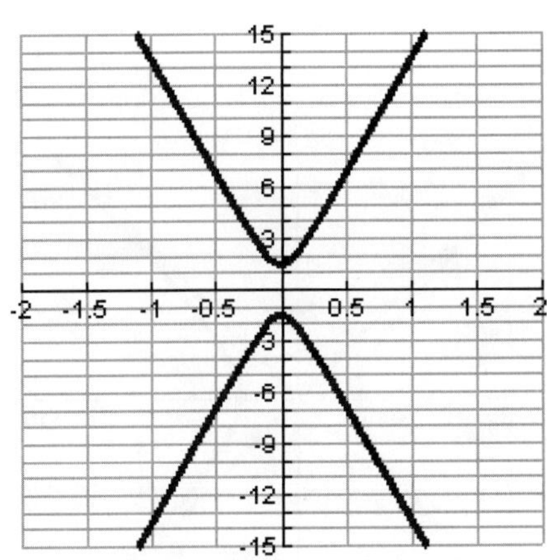

Notes on the Graph:
First, write the equation in standard form as $\frac{y^2}{\frac{9}{4}} - \frac{x^2}{\frac{1}{81}} = 1$. The equations of the asymptotes are $y = \pm \frac{27}{2} x$.

12.

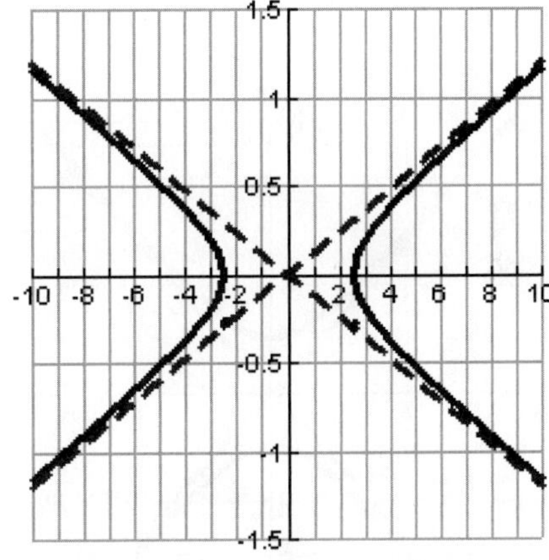

Notes on the Graph:
First, write the equation in standard form as $\frac{x^2}{\frac{25}{4}} - \frac{y^2}{\frac{9}{100}} = 1$. The equations of the asymptotes are $y = \pm \frac{3}{25} x$.

13.

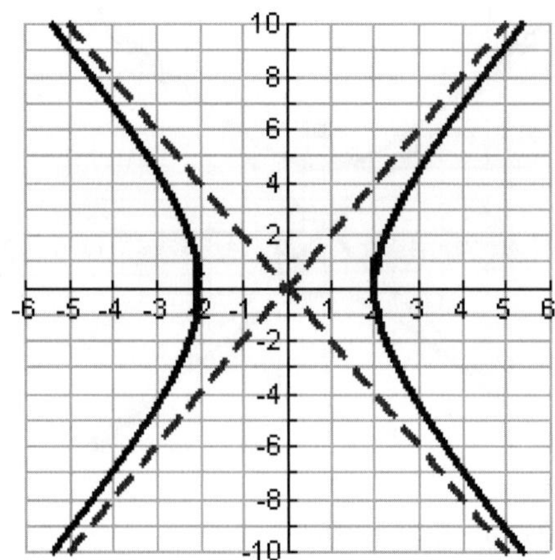

Notes on the Graph:
First, write the equation in standard form as $\frac{x^2}{4}-\frac{y^2}{16}=1$. The equations of the asymptotes are $y=\pm 2x$.

14.

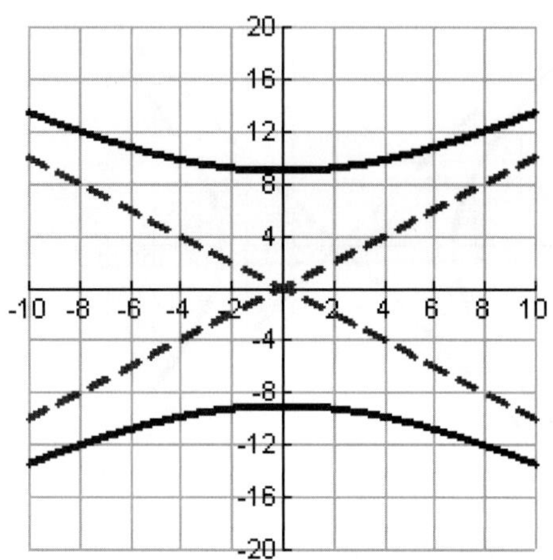

Notes on the Graph:
First, write the equation in standard form as $\frac{y^2}{81}-\frac{x^2}{81}=1$. The equations of the asymptotes are $y=\pm x$.

15.

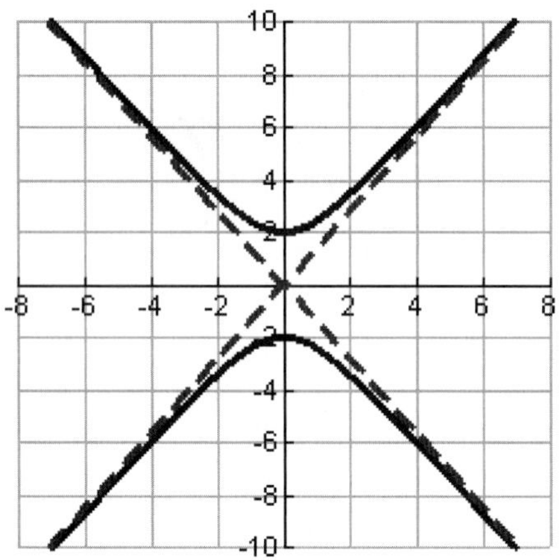

Notes on the Graph:
First, write the equation in standard form as $\frac{y^2}{4}-\frac{x^2}{2}=1$. The equations of the asymptotes are $y=\pm\frac{2}{\sqrt{2}}x$.

16.

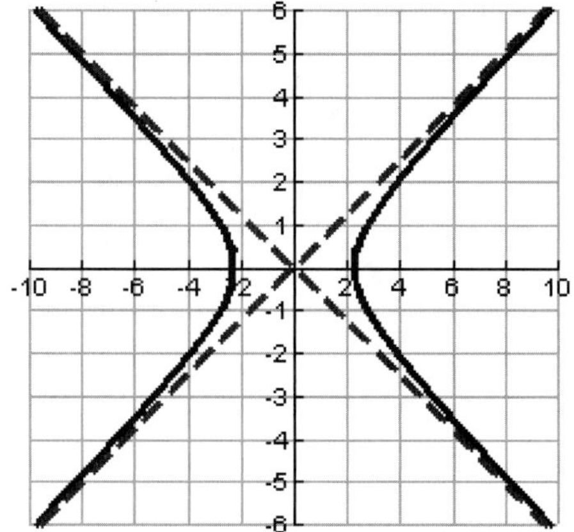

Notes on the Graph:
First, write the equation in standard form as $\frac{x^2}{5}-\frac{y^2}{2}=1$. The equations of the asymptotes are $y=\pm\sqrt{\frac{2}{5}}x=\pm\frac{\sqrt{10}}{5}x$.

17. Since the foci are $(-6,0)$, $(6,0)$ and the vertices are $(-4,0)$, $(4,0)$, we know that:
i) the hyperbola opens right/left with center at $(0,0)$,
ii) $a=4$, $c=6$,
Now, since $c^2=a^2+b^2$, $36=16+b^2$ so that $b^2=20$. Hence, the equation of the hyperbola is $\boxed{\frac{x^2}{16}-\frac{y^2}{20}=1}$.

18. Since the foci are $(-3,0)$, $(3,0)$ and the vertices are $(-1,0)$, $(1,0)$, we know that:
i) the hyperbola opens right/left with center at $(0,0)$,
ii) $a=1$, $c=3$,
Now, since $c^2=a^2+b^2$, $9=1+b^2$ so that $b^2=8$. Hence, the equation of the hyperbola is $\boxed{x^2-\frac{y^2}{8}=1}$.

19. Since the foci are $(0,-4)$, $(0,4)$ and the vertices are $(0,-3)$, $(0,3)$, we know that:
i) the hyperbola opens up/down with center at $(0,0)$,
ii) $a=3$, $c=4$,
Now, since $c^2=a^2+b^2$, $16=9+b^2$ so that $b^2=7$. Hence, the equation of the hyperbola is $\boxed{\frac{y^2}{9}-\frac{x^2}{7}=1}$.

20. Since the foci are $(0,-2)$, $(0,2)$ and the vertices are $(0,-1)$, $(0,1)$, we know that:
i) the hyperbola opens up/down with center at $(0,0)$,
ii) $a=1$, $c=2$,
Now, since $c^2=a^2+b^2$, $4=1+b^2$ so that $b^2=3$. Hence, the equation of the hyperbola is $\boxed{y^2-\frac{x^2}{3}=1}$.

21. Since the center is $(0,0)$ and the transverse axis is the *x*-axis, the general form of the equation of this hyperbola is $\frac{x^2}{a^2}-\frac{y^2}{b^2}=1$ with asymptotes $y=\pm\frac{b}{a}x$. In this case, $\frac{b}{a}=1$ so that $a=b$. Thus, the equation simplifies to $\frac{x^2}{a^2}-\frac{y^2}{a^2}=1$, or equivalently $\boxed{x^2-y^2=a^2}$.

22. Since the center is $(0,0)$ and the transverse axis is the *y*-axis, the general form of the equation of this hyperbola is $\frac{y^2}{a^2}-\frac{x^2}{b^2}=1$ with asymptotes $y=\pm\frac{a}{b}x$. In this case, $\frac{a}{b}=1$ so that $a=b$. Thus, the equation simplifies to $\frac{y^2}{a^2}-\frac{x^2}{a^2}=1$, or equivalently $\boxed{y^2-x^2=a^2}$.

23. Since the center is $(0,0)$ and the transverse axis is the *y*-axis, the general form of the equation of this hyperbola is $\frac{y^2}{a^2}-\frac{x^2}{b^2}=1$ with asymptotes $y=\pm\frac{a}{b}x$. In this case, $\frac{a}{b}=2$ so that $a=2b$. Thus, the equation simplifies to $\frac{y^2}{(2b)^2}-\frac{x^2}{b^2}=1$, or equivalently $\boxed{\frac{y^2}{4}-x^2=b^2}$.

24. Since the center is $(0,0)$ and the transverse axis is the x-axis, the general form of the equation of this hyperbola is $\frac{x^2}{a^2}-\frac{y^2}{b^2}=1$ with asymptotes $y=\pm\frac{b}{a}x$. In this case, $\frac{b}{a}=2$ so that $2a=b$. Thus, the equation simplifies to $\frac{x^2}{a^2}-\frac{y^2}{(2a)^2}=1$, or equivalently $\boxed{x^2-\frac{y^2}{4}=a^2}$.

25. c The transverse axis is parallel to the x-axis, so that the hyperbola will open left/right. The vertices are $(1,-2)$, $(5,-2)$ and the center is $(3,-2)$.

26. d The transverse axis is parallel to the x-axis, so that the hyperbola will open left/right. The vertices are $(-5,2)$, $(-1,2)$ and the center is $(-3,2)$.

27. b The transverse axis is parallel to the y-axis, so that the hyperbola will open up/down. The vertices are $(-2,8)$, $(-2,-2)$ and the center is $(-2,3)$.

28. a The transverse axis is parallel to the y-axis, so that the hyperbola will open up/down. The vertices are $(2,-8)$, $(2,2)$ and the center is $(2,-3)$.

29.

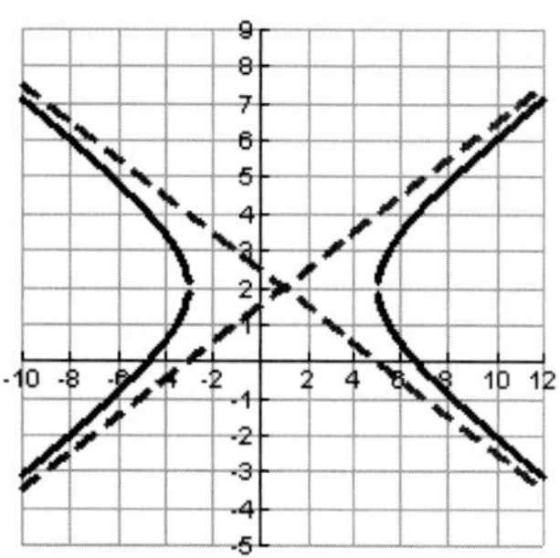

Notes on the Graph:
The equations of the asymptotes are $y=\pm\frac{1}{2}(x-1)+2$.

30.

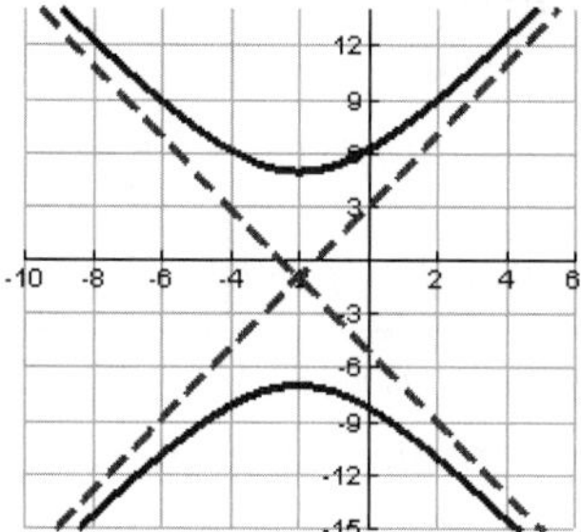

Notes on the Graph:
The equations of the asymptotes are $y=\pm 2(x+2)-1$.

31.

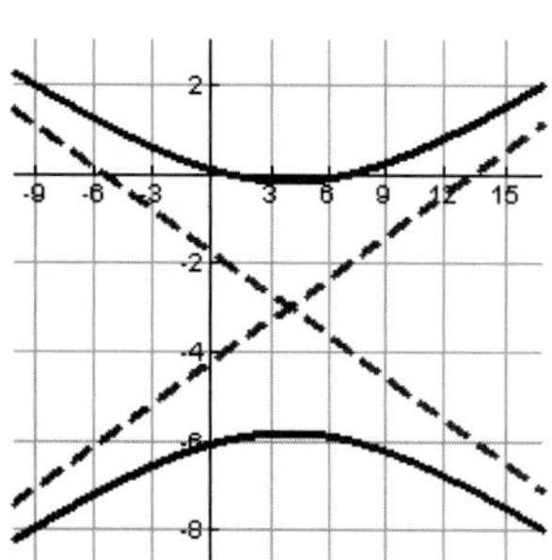

Notes on the Graph:
The equations of the asymptotes are $y=\pm\sqrt{\frac{1}{10}}(x-4)-3=\pm\frac{\sqrt{10}}{10}(x-4)-3$.

32.

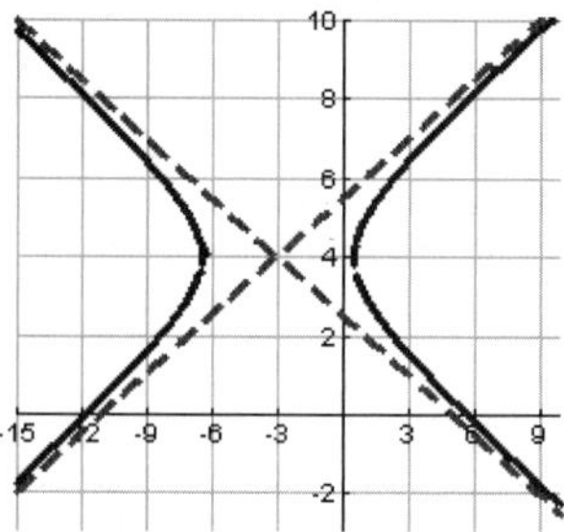

Notes on the Graph:
The equations of the asymptotes are $y=\pm\frac{1}{2}(x+3)+4$.

33.

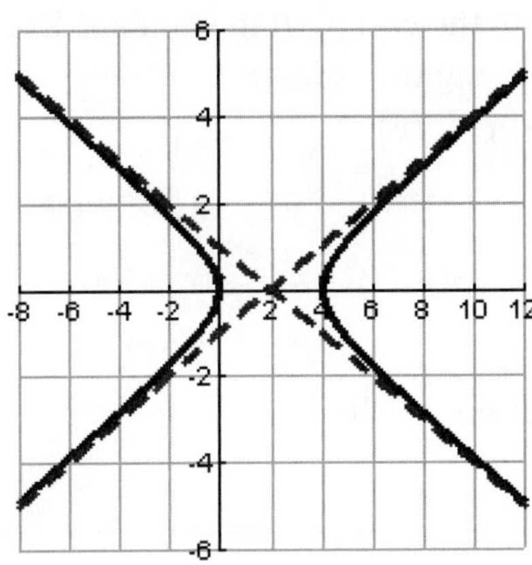

Notes on the Graph:
First, write the equation in standard form by completing the square:

$$x^2 - 4x - 4y^2 = 0$$

$$\left(x^2 - 4x + 4\right) - 4y^2 = 0 + 4$$

$$(x-2)^2 - 4y^2 = 4$$

$$\frac{(x-2)^2}{4} - y^2 = 1$$

The equations of the asymptotes are $y = \pm\frac{1}{2}(x-2)$.

34.

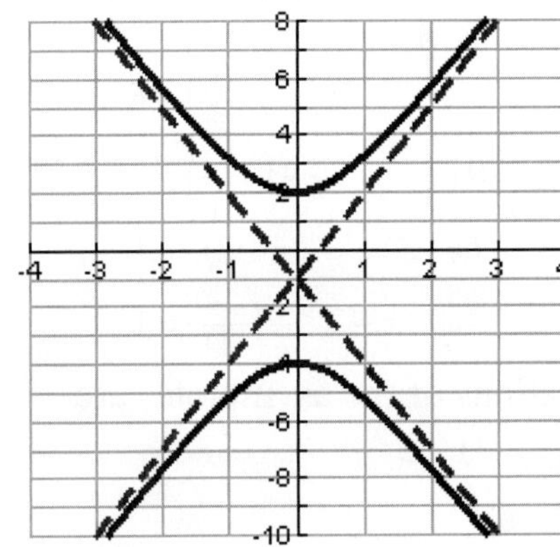

Notes on the Graph:
First, write the equation in standard form by completing the square:

$$-9x^2 + y^2 + 2y = 8$$

$$-9x^2 + \left(y^2 + 2y + 1\right) = 8 + 1$$

$$-9x^2 + \left(y+1\right)^2 = 9$$

$$-x^2 + \frac{(y+1)^2}{9} = 1$$

$$\frac{(y+1)^2}{9} - x^2 = 1$$

The equations of the asymptotes are $y = \pm 3x - 1$.

35.

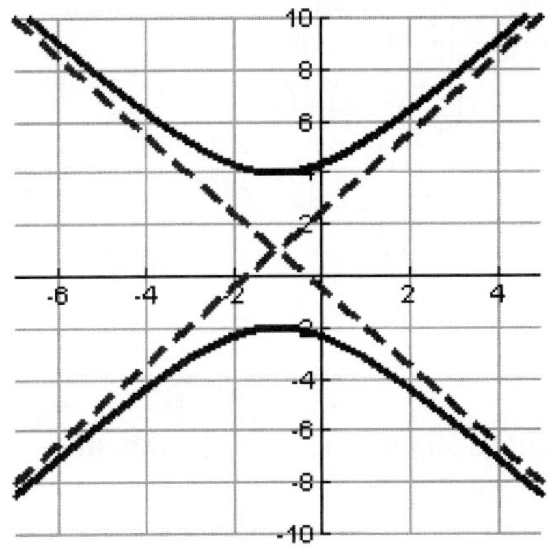

Notes on the Graph:
First, write the equation in standard form by completing the square:

$$-9\left(x^2 + 2x\right) + 4\left(y^2 - 2y\right) = 41$$

$$-9\left(x^2 + 2x + 1\right) + 4\left(y^2 - 2y + 1\right) = 41 - 9 + 4$$

$$-9(x+1)^2 + 4(y-1)^2 = 36$$

$$-\frac{(x+1)^2}{4} + \frac{(y-1)^2}{9} = 1$$

$$\frac{(y-1)^2}{9} - \frac{(x+1)^2}{4} = 1$$

The equations of the asymptotes are $y = \pm\frac{3}{2}(x+1) + 1$.

36. 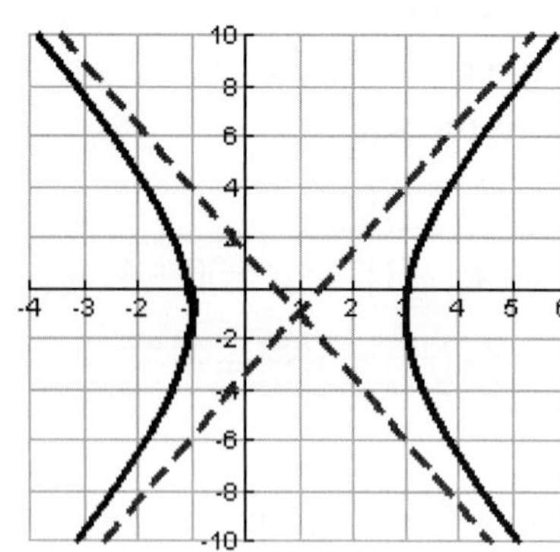	Notes on the Graph: First, write the equation in standard form by completing the square: $25(x^2-2x)-4(y^2+2y)=79$ $25(x^2-2x+1)-4(y^2+2y+1)=79+25-4$ $25(x-1)^2-4(y+1)^2=100$ $\frac{(x-1)^2}{4}-\frac{(y+1)^2}{25}=1$ The equations of the asymptotes are $y=\pm\frac{5}{2}(x-1)-1$.
37. 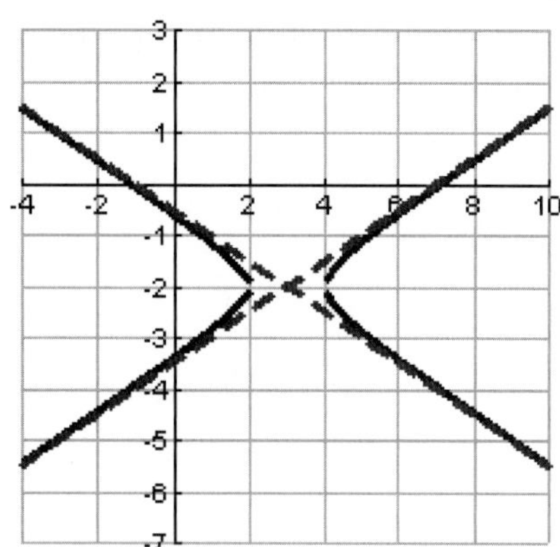	Notes on the Graph: First, write the equation in standard form by completing the square: $(x^2-6x)-4(y^2+4y)=8$ $(x^2-6x+9)-4(y^2+4y+4)=8+9-16$ $(x-3)^2-4(y+2)^2=1$ $(x-3)^2-\frac{(y+2)^2}{\frac{1}{4}}=1$ The equations of the asymptotes are $y=\pm\frac{1}{2}(x-3)-2$.
38. 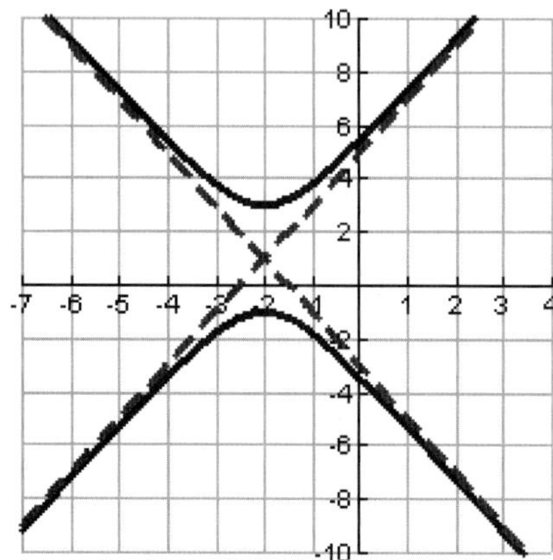	Notes on the Graph: First, write the equation in standard form by completing the square: $-4(x^2+4x)+(y^2-2y)=19$ $-4(x^2+4x+4)+(y^2-2y+1)=19-16+1$ $-4(x+2)^2+(y-1)^2=4$ $-(x+2)^2+\frac{(y-1)^2}{4}=1$ $\frac{(y-1)^2}{4}-(x+2)^2=1$ The equations of the asymptotes are $y=\pm2(x+2)+1$.

39. Since the vertices are $(-2,5)$, $(6,5)$, we know that the center is $(2,5)$ and the transverse axis is parallel to the *x*-axis. Hence, $2-a=-2$ so that $a=4$. Also, since the foci are $(-3,5)$, $(7,5)$, we know that $2+c=-3$ so that $c=-5$. Now, to find *b*, we substitute the values of *c* and *a* obtained above into $c^2=a^2+b^2$ to see that $25=16+b^2$ so that $b^2=9$. Hence, the equation of the hyperbola is $\boxed{\frac{(x-2)^2}{16}-\frac{(y-5)^2}{9}=1}$.

40. Since the vertices are $(1,-2)$, $(3,-2)$, we know that the center is $(2,-2)$ and the transverse axis is parallel to the *x*-axis. Hence, $2-a=1$ so that $a=1$. Also, since the foci are $(0,-2)$, $(4,-2)$, we know that $2+c=0$ so that $c=-2$. Now, to find *b*, we substitute the values of *c* and *a* obtained above into $c^2=a^2+b^2$ to see that $4=1+b^2$ so that $b^2=3$. Hence, the equation of the hyperbola is $\boxed{(x-2)^2-\frac{(y+2)^2}{3}=1}$.

41. Since the vertices are $(4,-7)$, $(4,-1)$, we know that the center is $(4,-4)$ and the transverse axis is parallel to the *y*-axis. Hence, $-4-a=-7$ so that $a=3$. Also, since the foci are $(4,-8)$, $(4,0)$, we know that $-4-c=-8$ so that $c=4$. Now, to find *b*, we substitute the values of *c* and *a* obtained above into $c^2=a^2+b^2$ to see that $16=9+b^2$ so that $b^2=7$. Hence, the equation of the hyperbola is $\boxed{\frac{(y+4)^2}{9}-\frac{(x-4)^2}{7}=1}$.

42. Since the vertices are $(2,-6)$, $(2,-4)$, we know that the center is $(2,-5)$ and the transverse axis is parallel to the *y*-axis. Hence, $-5-a=-6$ so that $a=1$. Also, since the foci are $(2,-7)$, $(2,-3)$, we know that $-5-c=-7$ so that $c=2$. Now, to find *b*, we substitute the values of *c* and *a* obtained above into $c^2=a^2+b^2$ to see that $4=1+b^2$ so that $b^2=3$. Hence, the equation of the hyperbola is $\boxed{(y+5)^2-\frac{(x-2)^2}{3}=1}$.

43. Assume that the stations coincide with the foci and are located at $(-75,0)$, $(75,0)$. The difference in distance between the ship and each of the two stations must remain constantly $2a$, where $(a,0)$ is the vertex. Assume that the radio signal speed is 186,000 $^{mi}/_{sec}$ and the time difference is 0.0005 sec. Then, using distance = rate × time, we obtain:

$$2a=(186{,}000)(0.0005)=93 \text{ so that } a=46.5$$

So, the ship will come ashore between the two stations 28.5 miles from one and 121.5 miles from the other.

44. Assume that the stations coincide with the foci and are located at $(-150,0), (150,0)$. The difference in distance between the ship and each of the two stations must remain constantly $2a$, where $(a,0)$ is the vertex. Assume that the radio signal speed is 186,000 $^{mi}/_{sec}$ and the time difference is 0.0007 sec. Then, using distance = rate × time, we obtain:

$$2a = (186,000)(0.0007) = 130.2 \text{ so that } a = 65.1$$

So, the ship will come ashore between the two stations 84.9 miles from one and 215.1 miles from the other.

45. Here, we want $a = 45$. So, $2a = 90$. Using the same radio speed, observe that $90 = 186,000(t)$, so that $t = 0.000484$ sec.

46. Here, we want $a = 100$. So, $2a = 200$. Using the same radio speed, observe that $200 = 186,000(t)$, so that $t = 0.00107$ sec.

47. Assume that the vertices are $(0,a), (0,-a)$. Then, $2a = 2$ so that $a = 1$, and the center is at (0,0). Further, the transverse axis is parallel to the y-axis. Since the foci are $(0,1.5), (0,-1.5)$, we know that $c = 1.5$. Thus, to find b, we substitute these values of a and c into $c^2 = a^2 + b^2$ to see that $(1.5)^2 = 1^2 + b^2$, and so $b^2 = 1.25$. Hence, the equation of the hyperbola is $y^2 - \frac{x^2}{1.25} = 1$, which is equivalent to $\boxed{y^2 - \frac{4}{5}x^2 = 1}$.

48. Consider the following diagram:

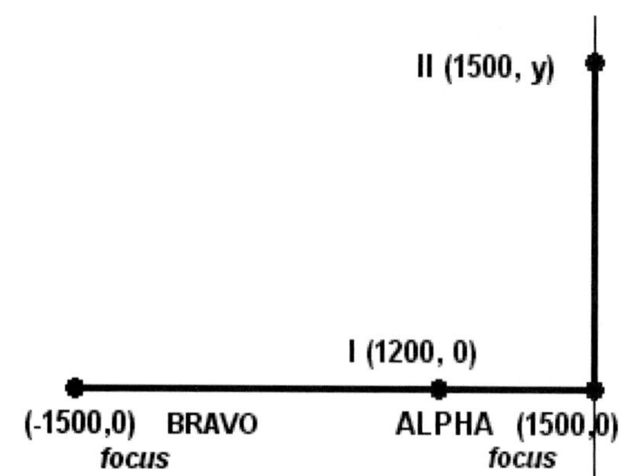

Since distance = rate × time, we have:

Time heard at Bravo:

$$t_B^I = \frac{\text{distance from } (1200,0) \text{ to } B}{r} = \frac{2700}{r}$$

Time heard at Alpha:

$$t_A^I = \frac{\text{distance from } (1200,0) \text{ to } A}{r} = \frac{300}{r}$$

We must find the value of y such that if we were to place the explosive at the point $(0,y)$, it would be heard at Alpha and Bravo at the same times as such an explosion would be heard if placed at (1200,0). To do so, we see that

Time heard at Bravo:

$$t_B^{II} = \frac{\text{distance from } (0,y) \text{ to } B}{r} = \frac{\sqrt{3000^2 + y^2}}{r}$$

Time heard at Alpha:

$$t_A^{II} = \frac{\text{distance from } (0,y) \text{ to } A}{r} = \frac{y}{r}$$

Hence, we solve $t_A^I = t_A^{II}$:

$$\frac{300}{r} = \frac{y}{r} \text{ so that } y = 300.$$

Thus, the explosive should be placed at (0,300).

49. Consider the following diagram:

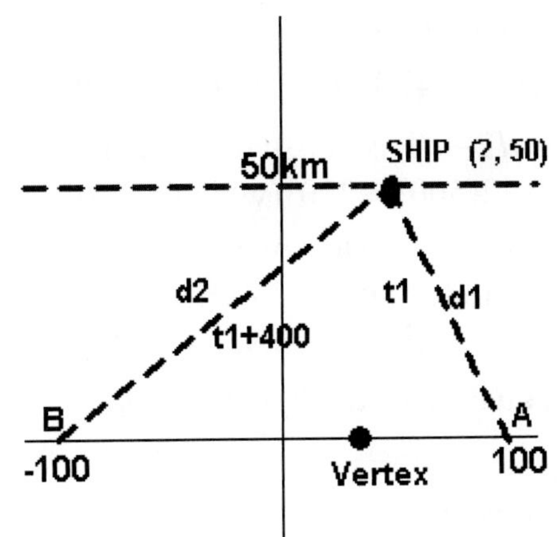

Using d = rt, we have

$$r = 320\ m/\mu s = 0.32\ km/\mu s.$$

Observe that

$$d1 = 0.32t1,\ \ d2 = 0.32(t1+400).$$

Since $d2 - d1$ is constant, we conclude that $0.32(400) = 128$ (which is $2a$).

Thus, $a = 64$. Moreover, from the foci, we conclude that $c = 100$. As such,

$$c^2 = a^2 + b^2 \Rightarrow b^2 = 5904.$$

The equation of the hyperbola is

$$\tfrac{x^2}{4096} - \tfrac{y^2}{5904} = 1.$$

Along the path $y = 50$, we have

$$\tfrac{x^2}{4096} - \tfrac{50^2}{5904} = 1 \Rightarrow x^2 = 5830 \Rightarrow x \approx 76.$$

So, the ship is located at point (76, 50).

50. Consider the following diagram:

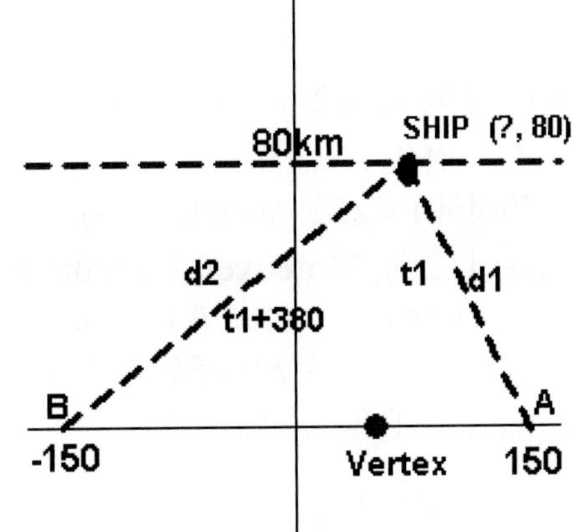

Using d = rt, we have

$$r = 350\ m/\mu s = 0.35\ km/\mu s.$$

Observe that

$$d1 = 0.35t1,\ \ d2 = 0.35(t1+380).$$

Since $d2 - d1$ is constant, we conclude that $0.35(380) = 133$ (which is $2a$).

Thus, $a = 66.5$. Moreover, from the foci, we conclude that $c = 150$. As such,

$$c^2 = a^2 + b^2 \Rightarrow b^2 = 18077.75.$$

The equation of the hyperbola is

$$\tfrac{x^2}{4422.25} - \tfrac{y^2}{18077.75} = 1.$$

Along the path $y = 80$, we have

$$\tfrac{x^2}{4422.25} - \tfrac{80^2}{18077.75} = 1 \Rightarrow x^2 = 5987.843 \Rightarrow x \approx 77.4$$

So, the ship is located at point (77.4, 80).

51. Consider the following diagram: 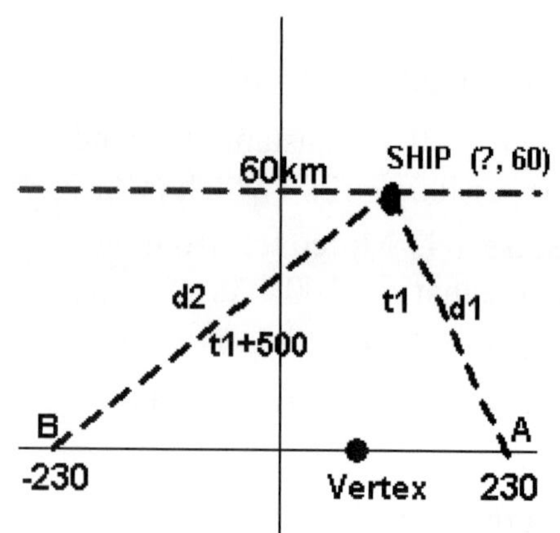 	Using d = rt, we have $r = 420\ m/\mu s = 0.42\ km/\mu s$. Observe that $d1 = 0.42t1,\ d2 = 0.42(t1+500)$. Since $d2-d1$ is constant, we conclude that $0.42(500) = 210$ (which is $2a$). Thus, $a = 105$. Moreover, from the foci, we conclude that $c = 230$. As such, $c^2 = a^2 + b^2 \Rightarrow b^2 = 41{,}875$. The equation of the hyperbola is $\frac{x^2}{11025} - \frac{y^2}{41875} = 1$. Along the path $y = 60$, we have $\frac{x^2}{11025} - \frac{60^2}{41875} = 1 \Rightarrow x^2 = 11972.82 \Rightarrow x \approx 109.4$ So, the ship is located at point (109.4, 60).
52. Consider the following diagram: 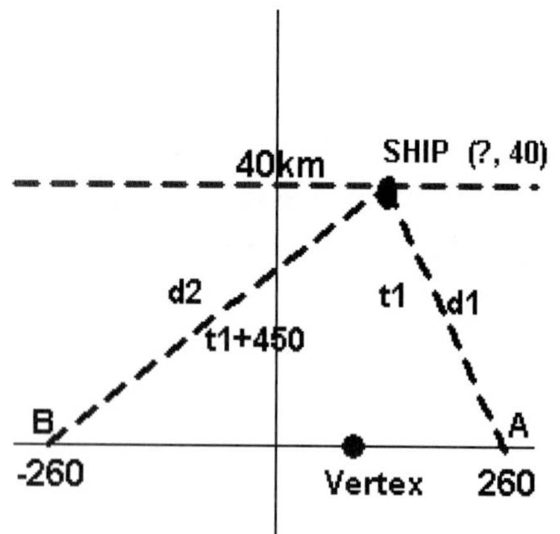 	Using d = rt, we have $r = 500\ m/\mu s = 0.50\ km/\mu s$. Observe that $d1 = 0.50t1,\ d2 = 0.50(t1+450)$. Since $d2-d1$ is constant, we conclude that $0.50(450) = 225$ (which is $2a$). Thus, $a = 112.5$. Moreover, from the foci, we conclude that $c = 260$. As such, $c^2 = a^2 + b^2 \Rightarrow b^2 = 54943.75$. The equation of the hyperbola is $\frac{x^2}{12656.25} - \frac{y^2}{54943.75} = 1$. Along the path $y = 40$, we have $\frac{x^2}{12656.25} - \frac{40^2}{54943.75} = 1 \Rightarrow x^2 = 13024.81$ $\Rightarrow x \approx 114.13$ So, the ship is located at point (114.13,40).
53. The transverse axis should be vertical. The points are $(3,0)$, $(-3,0)$ and the vertices are $(0,2)$, $(0,-2)$.	**54.** Here, $a=1$, $b=2$. So, the vertices are $(1,0)$, $(-1,0)$ and the points are $(0,2)$, $(0,-2)$.
55. False. You won't be able to find the value of b in such case without more information.	**56.** True. Since the foci and vertices yield the values of a and c, you can find b using $c^2 = a^2 + b^2$.

57. True. Since the general forms of the equations are $\frac{x^2}{a^2}-\frac{y^2}{b^2}=1$ and $\frac{y^2}{a^2}-\frac{x^2}{b^2}=1$, substituting $-x$ for x and $-y$ for y doesn't change the equation – hence, the symmetry follows.	**58.** False.
59. Using symmetry, if (p,q) is on this hyperbola, then (p, -q), (-p,q) and (-p,-q) are also on it.	**60.** The asymptotes are $y=\pm\frac{b}{a}x$. We want $\left(\frac{b}{a}\right)\left(-\frac{b}{a}\right)=-1$, which is equivalent to $\frac{b^2}{a^2}=1$, or $a^2=b^2$. Since $a^2=4$, choose $b^2=4$, so that b=2.
61. We know that (p,q) is on $\frac{x^2}{9}-\frac{y^2}{4}=1$ and (p,r) is on $\frac{x^2}{9}-\frac{y^2}{16}=1$. Substitute them into their respective equation to obtain the system: $\begin{cases}\frac{p^2}{9}-\frac{q^2}{4}=1\\ \frac{p^2}{9}-\frac{r^2}{16}=1\end{cases}$ Solve both equations for $\frac{p^2}{9}$ and equate to obtain $1+\frac{q^2}{4}=1+\frac{r^2}{16}\Rightarrow 4q^2=r^2$. From this, it follows that $r>q$ since both r and q are assumed to be positive.	**62.** If the line intersects the hyperbola, then the following system has a solution: $\begin{cases}\frac{x^2}{a^2}-\frac{y^2}{b^2}=1 & \textbf{(1)}\\ y=\frac{2b}{a}x & \textbf{(2)}\end{cases}$ Substitute **(2)** into **(1)** and simplify to obtain $-\frac{3}{a^2}x^2=1$, which has no solution. Hence, the line cannot intersect the hyperbola in this case.

63. Assume the center of the hyperbola is (0,0). Assume that the two asymptotes are perpendicular. We separate our discussion into two cases, depending on which axis is the transverse axis.

Case 1: Here, the equations of the asymptotes are $y=\pm\frac{a}{b}x$. Hence, in order for them to be perpendicular, the products of their slopes must be -1. So, we have $\left(\frac{a}{b}\right)\left(-\frac{a}{b}\right)=-1$, which simplifies to $-\frac{a^2}{b^2}=-1$ so that $a^2=b^2$. Hence, in such case, the equation is $\frac{y^2}{a^2}-\frac{x^2}{a^2}=1$, which is equivalent to $y^2-x^2=a^2$.

Case 2: Here, the equations of the asymptotes are $y=\pm\frac{b}{a}x$. Hence, in order for them to be perpendicular, the products of their slopes must be -1. So, we have $\left(\frac{b}{a}\right)\left(-\frac{b}{a}\right)=-1$, which simplifies to $-\frac{b^2}{a^2}=-1$ so that $a^2=b^2$. Hence, in such case, the equation is $\frac{x^2}{a^2}-\frac{y^2}{b^2}=1$, which is equivalent to $x^2-y^2=a^2$.

64. Since the vertices are $(3,-2), (-1,-2)$, and the center is the midpoint of the segment connecting them, the center must be $(1,-2)$. Moreover, since the form of the vertices is $(h\pm a,k)$ (which in this case is $(1\pm a,-2)$) we see that $1+a=3$ so that $a=2$. It remains to find *b*. At this point, we need to use the fact that we are given that the equations of the asymptotes are $y=2x-4,\ y=-2x$. Since the transverse axis is parallel to the *x*-axis, we know that the slopes of the asymptotes for such a hyperbola are $\pm\frac{b}{a}$. Thus, $\pm\frac{b}{a}=\pm 2$. Since $a=2$, this implies that $b=\pm 4$. Hence, the equation of the hyperbola must be $\boxed{\frac{(x-1)^2}{4}-\frac{(y+2)^2}{16}=1}$.
65. Completing the square yields $$9\left(y^2-4y\right)-16\left(x^2+2x\right)=124$$ $$9\left(y^2-4y+4\right)-16\left(x^2+2x+1\right)=124+36-16$$ $$9(y-2)^2-16(x+1)^2=144$$ $$\frac{(y-2)^2}{16}-\frac{(x+1)^2}{9}=1$$ So, $a=4,\ b=3,\ h=-1,\ k=2$. Thus, the asymptotes are $y=\pm\frac{4}{3}(x+1)+2$, which yields $$y=-\frac{4}{3}x+\frac{2}{3},\ \ y=\frac{4}{3}x+\frac{10}{3}.$$
66. Completing the square yields $$5\left(x^2+4x\right)-4\left(y^2-2y\right)=4$$ $$5\left(x^2+4x+4\right)-4\left(y^2-2y+1\right)=4+20-4$$ $$5(x+2)^2-4(y-1)^2=20$$ $$\frac{(x+2)^2}{4}-\frac{(y-1)^2}{5}=1$$ So, $a=2,\ b=\sqrt{5},\ h=-2,\ k=1$. Thus, the asymptotes are $y=\pm\frac{\sqrt{5}}{2}(x+2)+1$, which yields $y=\frac{\sqrt{5}}{2}x+\sqrt{5}+1,\ \ y=-\frac{\sqrt{5}}{2}x-\sqrt{5}+1$.
67. The slope of 3*x*+5*y*-7=0 is -3/5 and the asymptotes of $\frac{x^2}{a^2}-\frac{y^2}{b^2}=1$ are $y=\pm\frac{b}{a}x$. So, if the line is perpendicular to one of the asymptotes, then either $-\frac{3}{5}\left(-\frac{b}{a}\right)=-1$ or $-\frac{3}{5}\left(\frac{b}{a}\right)=-1$. Solving these yields $3b=\pm 5a$, or equivalently $b=\pm\frac{5a}{3}$. Now, use the fact that the vertices are $\left(\pm 3,0\right)$, so that $a=\pm 3$. Then, $b=\pm 5$. Since $c^2=a^2+b^2=34$, the foci are $\left(\pm\sqrt{34},0\right)$.

68. The slope of 2x-y+9=0 is 2 and the asymptotes of $\frac{y^2}{a^2}-\frac{x^2}{b^2}=1$ are $y=\pm\frac{a}{b}x$. So, if the line is perpendicular to one of the asymptotes, then either $2\left(-\frac{a}{b}\right)=-1$ or $2\left(\frac{a}{b}\right)=-1$. Solving these yields $2a=\pm b$.

Now, use the fact that the vertices are $(0,\pm1)$, so that $a=\pm1$. Then, $b=\pm2$. Since $c^2=a^2+b^2=5$, the foci are $\left(0,\pm\sqrt{5}\right)$.

69. Observe from the graphs that as c increases, the graphs of the hyperbolas described by $x^2-cy^2=1$ become more and more squeezed down towards the x-axis.

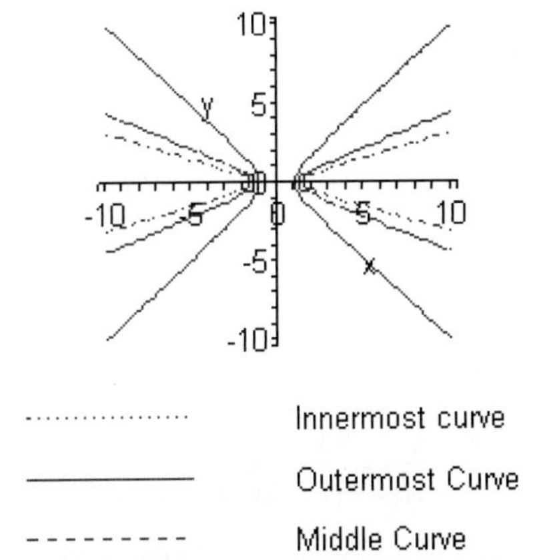

70. Observe from the graphs that as c increases, the vertices of the graphs of the hyperbolas described by $cx^2-y^2=1$ get closer to the origin, thereby causing the hyperbolas to rise more and more steeply.

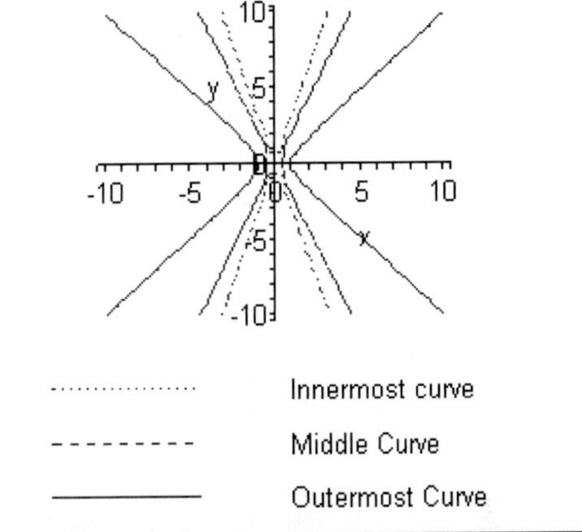

71. Note from the graphs that as c decreases, the vertices of the hyperbola whose equation is given by $cx^2-y^2=1$ are located at $\left(\pm\frac{1}{c},0\right)$, and are moving away from the origin:

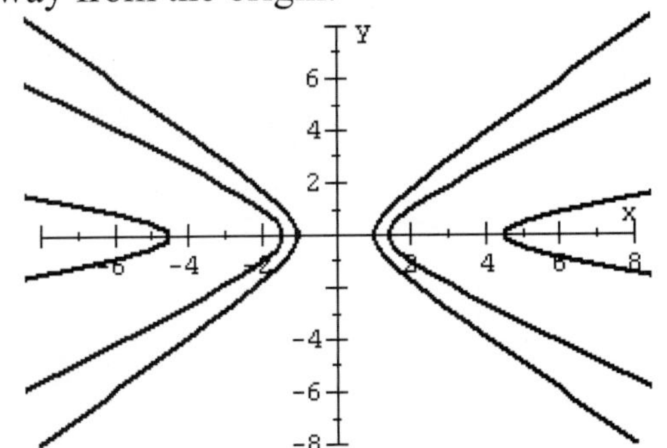

72. Note from the graphs that as c decreases, the vertices of the hyperbola whose equation is given by $x^2-cy^2=1$ remain at $(\pm1,0)$, but the graphs open outward more narrowly:

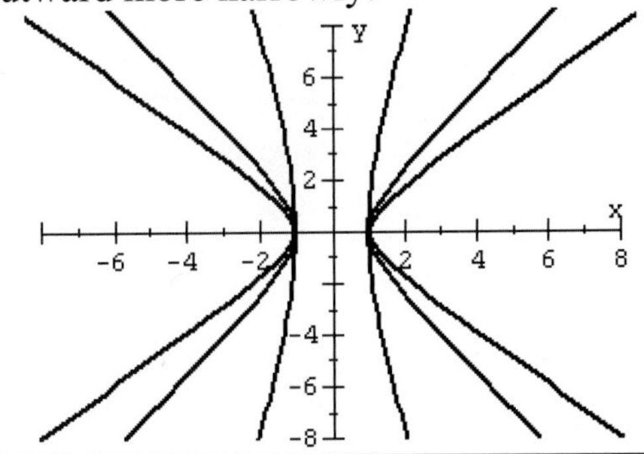

73. Observe that

$$\left(\frac{e^u+e^{-u}}{2}\right)^2-\left(\frac{e^u-e^{-u}}{2}\right)^2=$$

$$\frac{e^{2u}+2+e^{-2u}}{4}-\frac{e^{2u}-2+e^{-2u}}{4}=1$$

74. Observe that

$$\left(\frac{e^u+e^{-u}}{e^u-e^{-u}}\right)^2-\left(\frac{2}{e^u-e^{-u}}\right)^2=$$

$$\frac{e^{2u}+2+e^{-2u}-4}{e^{2u}-2+e^{-2u}}=$$

$$\frac{e^{2u}-2+e^{-2u}}{e^{2u}-2+e^{-2u}}=1$$

75. Observe that

$$y^2=1+x^2 \Rightarrow y=-\sqrt{1+x^2}$$

since $y<0$. So,

$$\frac{f(x+h)-f(x)}{h}=\frac{-\sqrt{1+(x+h)^2}+\sqrt{1+x^2}}{h}=\frac{-\sqrt{1+(x+h)^2}+\sqrt{1+x^2}}{h}\cdot\frac{-\sqrt{1+(x+h)^2}-\sqrt{1+x^2}}{-\sqrt{1+(x+h)^2}-\sqrt{1+x^2}}$$

$$=\frac{\left[1+(x+h)^2\right]-\left[1+x^2\right]}{-h\left[\sqrt{1+(x+h)^2}+\sqrt{1+x^2}\right]}=\frac{2hx+h^2}{-h\left[\sqrt{1+(x+h)^2}+\sqrt{1+x^2}\right]}=\frac{-(2x+h)}{\sqrt{1+(x+h)^2}+\sqrt{1+x^2}}$$

76. Observe that

$$y^2=1-4x^2 \Rightarrow y=\sqrt{1-4x^2}$$

since $y>0$. So,

$$\frac{f(x+h)-f(x)}{h}=\frac{\sqrt{1-4(x+h)^2}-\sqrt{1-4x^2}}{h}=\frac{\sqrt{1-4(x+h)^2}-\sqrt{1-4x^2}}{h}\cdot\frac{\sqrt{1-4(x+h)^2}+\sqrt{1-4x^2}}{\sqrt{1-4(x+h)^2}+\sqrt{1-4x^2}}$$

$$=\frac{\left(1-4(x+h)^2\right)-\left(1-4x^2\right)}{h\left[\sqrt{1-4(x+h)^2}+\sqrt{1-4x^2}\right]}=\frac{-8hx-4h^2}{h\left[\sqrt{1-4(x+h)^2}+\sqrt{1-4x^2}\right]}=\frac{-4(2x+h)}{\sqrt{1-4(x+h)^2}+\sqrt{1-4x^2}}$$

Section 9.5 Solutions --

1. Solve the system $\begin{cases} x^2 - y = -2 & \textbf{(1)} \\ -x + y = 4 & \textbf{(2)} \end{cases}$

Add **(1)** and **(2)** to get an equation in terms of x:

$$x^2 - x - 2 = 0$$
$$(x-2)(x+1) = 0$$
$$x = 2,\ -1$$

Now, substitute each value of x back into **(2)** to find the corresponding values of y:

$$x = 2: \quad -2 + y = 4 \text{ so that } y = 6$$
$$x = -1: \quad -(-1) + y = 4 \text{ so that } y = 3$$

So, the solutions are $\boxed{(2,6) \text{ and } (-1,3)}$.

2. Solve the system $\begin{cases} x^2 + y = 2 & \textbf{(1)} \\ 2x + y = -1 & \textbf{(2)} \end{cases}$

Multiply **(2)** by -1, and then add to **(1)** to get an equation in terms of x:

$$x^2 - 2x - 3 = 0$$
$$(x-3)(x+1) = 0$$
$$x = 3,\ -1$$

Now, substitute each value of x back into **(2)** to find the corresponding values of y:

$$x = 3: \quad 2(3) + y = -1 \text{ so that } y = -7$$
$$x = -1: \quad 2(-1) + y = -1 \text{ so that } y = 1$$

So, the solutions are $\boxed{(3,-7) \text{ and } (-1,1)}$.

3. Solve the system $\begin{cases} x^2 + y = 1 & \textbf{(1)} \\ 2x + y = 2 & \textbf{(2)} \end{cases}$

Multiply **(2)** by -1, and then add to **(1)** to get an equation in terms of x:

$$x^2 - 2x + 1 = 0$$
$$(x-1)^2 = 0$$
$$x = 1$$

Now, substitute this value of x back into **(1)** to find the corresponding value of y:

$$x = 1: \quad 1^2 + y = 1 \text{ so that } y = 0$$

So, the solution is $\boxed{(1,0)}$.

4. Solve the system $\begin{cases} x^2 - y = 2 & \textbf{(1)} \\ -2x + y = -3 & \textbf{(2)} \end{cases}$

Add **(1)** and **(2)** to get an equation in terms of x:

$$x^2 - 2x = -1$$
$$x^2 - 2x + 1 = 0$$
$$(x-1)^2 = 0$$
$$x = 1$$

Now, substitute this value of x back into **(1)** to find the corresponding value of y:

$$x = 1: \quad 1^2 - y = 2 \text{ so that } y = -1$$

So, the solution is $\boxed{(1,-1)}$.

5. Solve the system $\begin{cases} x^2 + y = -5 & \textbf{(1)} \\ -x + y = 3 & \textbf{(2)} \end{cases}$

Multiply **(2)** by -1 and then add to **(1)** to get an equation in terms of x:

$$x^2 + x = -8$$
$$x^2 + x + 8 = 0$$
$$x = \frac{-1 \pm \sqrt{1-4(8)}}{2}$$

Since the values of x are not real numbers, there is $\boxed{\text{no solution}}$ to this system.

6. Solve the system $\begin{cases} x^2 - y = -7 & \textbf{(1)} \\ x + y = -2 & \textbf{(2)} \end{cases}$

Add **(1)** and **(2)** to get an equation in terms of x:

$$x^2 + x = -9 \Rightarrow x^2 + x + 9 = 0 \Rightarrow x = \frac{-1 \pm \sqrt{1-4(9)}}{2}$$

Since the values of x are not real numbers, there is $\boxed{\text{no solution}}$ to this system.

7. Solve the system $\begin{cases} x^2 + y^2 = 1 & \textbf{(1)} \\ x^2 - y = -1 & \textbf{(2)} \end{cases}$

Multiply **(2)** by -1, and then add to **(1)** to get an equation in terms of y:

$$y^2 + y = 2$$
$$y^2 + y - 2 = 0$$
$$(y+2)(y-1) = 0$$
$$y = -2, 1$$

Now, substitute each value of y back into **(2)** to find the corresponding values of x:

$y=-2:\quad x^2-(-2)=-1$ so that $x^2=-3$ (No real solutions)

$y=1:\quad x^2-1=-1$

$x^2=0$ so that $x=0$

So, the solution is $\boxed{(0,1)}$.

8. Solve the system $\begin{cases} x^2+y^2=1 & \textbf{(1)} \\ x^2+y=-1 & \textbf{(2)} \end{cases}$

Multiply **(2)** by -1, and then add to **(1)** to get an equation in terms of y:

$$y^2-y=2$$
$$y^2-y-2=0$$
$$(y-2)(y+1)=0$$
$$y=2,\ -1$$

Now, substitute each value of y back into **(2)** to find the corresponding values of x:

$y=2:\quad x^2+2=-1$ so that $x^2=-3$ (No real solutions)

$y=-1:\quad x^2-1=-1$

$x^2=0$ so that $x=0$

So, the solution is $\boxed{(0,-1)}$.

9. Solve the system $\begin{cases} x^2+y^2=3 & \textbf{(1)} \\ 4x^2+y=0 & \textbf{(2)} \end{cases}$

Multiply **(1)** by -4, and then add to **(2)** to get an equation in terms of y:

$$-4y^2+y=-12$$
$$4y^2-y-12=0$$
$$y=\frac{1\pm\sqrt{1-4(4)(-12)}}{2(4)}=\frac{1\pm\sqrt{193}}{8}$$
$$\cong 1.862,\ -1.612$$

Now, substitute each value of y back into **(2)** to find the corresponding values of x:

$y=1.862:\quad 4x^2+1.862=0$ so that $x^2=-0.466$ (No real solutions)

$y=-1.612:\quad 4x^2-1.612=0$

$x^2=0.403$ so that $x\cong\pm0.635$

So, the solutions are $\boxed{(0.63,-1.61) \text{ and } (-0.63,-1.61)}$.

10. Solve the system $\begin{cases} x^2 + y^2 = 6 & \textbf{(1)} \\ -7x^2 + y = 0 & \textbf{(2)} \end{cases}$

Multiply **(1)** by 7, and then add to **(2)** to get an equation in terms of y:

$$7y^2 + y = 42$$

$$7y^2 + y - 42 = 0$$

$$y = \tfrac{-1 \pm \sqrt{1-4(7)(-42)}}{2(7)} \cong \tfrac{-1 \pm 34.307}{14}$$

$$\cong 2.379, \; -2.52$$

Now, substitute each value of y back into **(2)** to find the corresponding values of x:

$$y = 2.379: \quad -7x^2 + 2.379 = 0$$

$$x^2 = 0.340 \text{ so that } x = \pm 0.583$$

$$y = -2.52: \quad -7x^2 - 2.52 = 0$$

$$x^2 = -0.36 \text{ (No real solutions)}$$

So, the solutions are $\boxed{(0.583, 2.379) \text{ and } (-0.583, 2.379)}$.

11. Solve the system $\begin{cases} x^2 + y^2 = -6 & \textbf{(1)} \\ -2x^2 + y = 7 & \textbf{(2)} \end{cases}$

Multiply **(1)** by 2, and then add to **(2)** to get an equation in terms of y:

$$2y^2 + y = -5$$

$$2y^2 + y + 5 = 0$$

$$y = \tfrac{-1 \pm \sqrt{1-4(2)(5)}}{2(2)}$$

Since the values of y are not real numbers, there is $\boxed{\text{no solution}}$ to this system.

12. Solve the system $\begin{cases} x^2+y^2=5 & \textbf{(1)} \\ 3x^2+y=9 & \textbf{(2)} \end{cases}$

Multiply **(1)** by -3, and then add to **(2)** to get an equation in terms of y:

$$-3y^2+y=-6$$

$$3y^2-y-6=0$$

$$y=\frac{1\pm\sqrt{1-4(3)(-6)}}{2(3)}=\frac{1\pm\sqrt{73}}{6}$$

$$\cong 1.591,\ -1.257$$

Now, substitute each value of y back into **(2)** to find the corresponding values of x:

$$y=1.591:\quad 3x^2+1.591=9$$

$$x^2=2.470$$

$$x\cong\pm1.572$$

$$y=-1.257:\quad 3x^2-1.257=9$$

$$x^2=3.419$$

$$x\cong\pm1.849$$

So, the solutions are $\boxed{(1.572,1.591),\ (-1.572,1.591),\ (1.849,-1.257),\ (-1.849,-1.257)}$.

13. Solve the system $\begin{cases} x+y=2 & \textbf{(1)} \\ x^2+y^2=2 & \textbf{(2)} \end{cases}$

Solve **(1)** for y: $y=2-x$ **(3)**

Substitute **(3)** into **(2)** to get an equation in terms of x:

$$x^2+4-4x+x^2=2$$

$$2x^2-4x+2=0$$

$$2(x-1)^2=0$$

$$x=1$$

Now, substitute this value of x back into **(3)** to find the corresponding value of y:

$$y=2-(1)=1$$

So, the solution is $\boxed{(1,1)}$.

14. Solve the system $\begin{cases} x-y=-2 & \textbf{(1)} \\ x^2+y^2=2 & \textbf{(2)} \end{cases}$

Solve **(1)** for x: $x=y-2$ **(3)**

Substitute **(3)** into **(2)** to get an equation in terms of y:

$$\begin{aligned} (y-2)^2+y^2&=2 \\ y^2-4y+4+y^2&=2 \\ 2y^2-4y+2&=0 \\ 2(y-1)^2&=0 \\ y&=1 \end{aligned}$$

Now, substitute this value of y back into **(3)** to find the corresponding value of x:

$$x=1-(2)=-1$$

So, the solution is $\boxed{(-1,1)}$.

15. Solve the system $\begin{cases} xy=4 & \textbf{(1)} \\ x^2+y^2=10 & \textbf{(2)} \end{cases}$

Solve **(1)** for y: $y=\frac{4}{x}$ **(3)**

Substitute **(3)** into **(2)** to get an equation in terms of x:

$$\begin{aligned} x^2+\left(\tfrac{4}{x}\right)^2&=10 \\ x^2+\tfrac{16}{x^2}&=10 \\ \frac{x^4+16}{x^2}&=10 \\ \frac{x^4+16}{x^2}-\frac{10x^2}{x^2}&=0 \\ \frac{x^4-10x^2+16}{x^2}&=0 \\ \frac{\left(x^2-8\right)\left(x^2-2\right)}{x^2}&=0 \end{aligned}$$

$$x^2-8=0 \text{ or } x^2-2=0$$

Hence, $x=\pm2\sqrt{2}$ or $x=\pm\sqrt{2}$.

Now, substitute each of these values of x back into **(3)** to find the corresponding value of y:

$$\begin{aligned} x=2\sqrt{2}&: \quad y=\tfrac{4}{2\sqrt{2}}=\tfrac{2}{\sqrt{2}} \\ x=-2\sqrt{2}&: \quad y=\tfrac{4}{-2\sqrt{2}}=-\tfrac{2}{\sqrt{2}} \\ x=\sqrt{2}&: \quad y=\tfrac{4}{\sqrt{2}}=\tfrac{4}{\sqrt{2}} \\ x=-\sqrt{2}&: \quad y=\tfrac{4}{-\sqrt{2}}=-\tfrac{4}{\sqrt{2}} \end{aligned}$$

So, the solutions are (after rationalizing)

$$\boxed{\begin{array}{l} \left(2\sqrt{2},\sqrt{2}\right),\left(-2\sqrt{2},-\sqrt{2}\right), \\ \left(\sqrt{2},2\sqrt{2}\right), \text{ and } \left(-\sqrt{2},-2\sqrt{2}\right) \end{array}}.$$

16. Solve the system $\begin{cases} xy = -3 & \textbf{(1)} \\ x^2 + y^2 = 12 & \textbf{(2)} \end{cases}$

Solve **(1)** for y: $y = -\frac{3}{x}$ **(3)**

Substitute **(3)** into **(2)** to get an equation in terms of x:

$$x^2 + \left(-\tfrac{3}{x}\right)^2 = 12$$

$$x^2 + \tfrac{9}{x^2} = 12$$

$$\frac{x^4+9}{x^2} = 12$$

$$\frac{x^4+9}{x^2} - \frac{12x^2}{x^2} = 0 \quad \textbf{(4)}$$

$$\frac{x^4 - 12x^2 + 9}{x^2} = 0$$

Let $u = x^2$ **(5)**.

Use **(5)** to rewrite **(4)** as the equivalent equation: $u^2 - 12u + 9 = 0$

Solve this equation using the quadratic formula:

$$u = \frac{12 \pm \sqrt{144 - 4(9)}}{2} = \frac{12 \pm \sqrt{108}}{2}$$

$$= \frac{12 \pm 6\sqrt{3}}{2} = 6 \pm 3\sqrt{3}$$

Now, substituting these back into **(5)** yields the following two equations for x:

$x^2 = 6 + 3\sqrt{3}$, $x \cong \pm 3.346$ and $x^2 = 6 - 3\sqrt{3}$, $x \cong \pm 0.896$

Finally, substitute each of these values of x back into **(3)** to find the corresponding value of y:

$$x = -3.346: \quad y = -\tfrac{3}{-3.346} \cong 0.897$$

$$x = 3.346: \quad y = -\tfrac{3}{3.346} \cong -0.897$$

$$x = -0.896: \quad y = -\tfrac{3}{-0.896} \cong 3.348$$

$$x = 0.896: \quad y = -\tfrac{3}{0.896} \cong 3.348$$

So, the solutions are $\boxed{(-3.346, 0.897),\ (3.346, -0.897),\ (-0.896, 3.348), \text{ and } (0.896, -3.348)}$.

17. Solve the system $\begin{cases} y = x^2 - 3 & \textbf{(1)} \\ y = -4x + 9 & \textbf{(2)} \end{cases}$

Substitute **(1)** into **(2)** to get an equation in terms of x:

$$x^2 - 3 = -4x + 9$$

$$x^2 + 4x - 12 = 0$$

$$(x+6)(x-2) = 0$$

$$x = -6,\ 2$$

Now, substitute this value of y back into **(1)** to find the corresponding value of y:

$$x = -6: \ y = (-6)^2 - 3 = 33$$

$$x = 2: \ y = (2)^2 - 3 = 1$$

So, the solutions are $\boxed{(2,1) \text{ and } (-6,33)}$.

18. Solve the system $\begin{cases} y = -x^2 + 5 & \textbf{(1)} \\ y = 3x - 4 & \textbf{(2)} \end{cases}$

Substitute **(1)** into **(2)** to get an equation in terms of x:

$$-x^2 + 5 = 3x - 4$$

$$x^2 + 3x - 9 = 0$$

$$x = \frac{-3 \pm \sqrt{9 - 4(-9)}}{2} = \frac{-3 \pm 3\sqrt{5}}{2} \cong 1.854, \ -4.854$$

Now, substitute these values of x back into **(2)** to find the corresponding value of y:

$$x = 1.854: \quad y = 3(1.854) - 4 = 1.562$$

$$x = -4.854: \quad y = 3(-4.854) - 4 = -18.562$$

So, the solutions are $\boxed{(1.854, 1.562) \text{ and } (-4.854, -18.562)}$.

19. Solve the system $\begin{cases} x^2 + xy - y^2 = 5 & \textbf{(1)} \\ x - y = -1 & \textbf{(2)} \end{cases}$

Solve **(2)** for x: $x = y - 1$ **(3)**

Substitute **(3)** into **(1)** to get an equation in terms of y:

$$(y-1)^2 + (y-1)y - y^2 = 5$$

$$y^2 - 2y + 1 + y^2 - y - y^2 = 5$$

$$y^2 - 3y - 4 = 0$$

$$(y-4)(y+1) = 0$$

$$y = 4, \ -1$$

Now, substitute these values of y back into **(3)** to find the corresponding values of x:

$$y = 4: \quad x = 4 - 1 = 3$$

$$y = -1: \quad x = -1 - 1 = -2$$

So, the solutions are $\boxed{(3, 4) \text{ and } (-2, -1)}$.

20. Solve the system $\begin{cases} x^2 + xy + y^2 = 13 & \textbf{(1)} \\ x + y = -1 & \textbf{(2)} \end{cases}$

Solve **(2)** for x: $x = -y - 1$ **(3)**

Substitute **(3)** into **(1)** to get an equation in terms of y:

$$(-y-1)^2 + (-y-1)y + y^2 = 13$$
$$y^2 + 2y + 1 - y^2 - y + y^2 = 13$$
$$y^2 + y - 12 = 0$$
$$(y+4)(y-3) = 0$$
$$y = -4,\ 3$$

Now, substitute these values of y back into **(3)** to find the corresponding values of x:

$$y = -4:\ x = -(-4) - 1 = 3$$
$$y = 3:\quad x = -3 - 1 = -4$$

So, the solutions are $\boxed{(3,-4) \text{ and } (-4,3)}$.

21. Solve the system $\begin{cases} 2x - y = 3 & \textbf{(1)} \\ x^2 + y^2 - 2x + 6y = -9 & \textbf{(2)} \end{cases}$

Solve **(1)** for y: $y = 2x - 3$ **(3)**

Substitute **(3)** into **(2)** to get an equation in terms of x:

$$x^2 + (2x-3)^2 - 2x + 6(2x-3) = -9$$
$$x^2 + 4x^2 - 12x + 9 - 2x + 12x - 18 + 9 = 0$$
$$5x^2 - 2x = 0$$
$$x(5x-2) = 0$$
$$x = 0,\ \tfrac{2}{5}$$

Now, substitute these values of x back into **(3)** to find the corresponding values of y:

$$x = 0:\ y = 2(0) - 3 = -3$$
$$x = \tfrac{2}{5}:\ y = 2(\tfrac{2}{5}) - 3 = -\tfrac{11}{5}$$

So, the solutions are $\boxed{(0,-3) \text{ and } (\tfrac{2}{5}, -\tfrac{11}{5})}$.

22. Solve the system $\begin{cases} x^2+y^2-2x-4y=0 & \textbf{(1)} \\ -2x+y=-3 & \textbf{(2)} \end{cases}$

Solve **(2)** for *y*: $y=2x-3$ **(3)**

Substitute **(3)** into **(1)** to get an equation in terms of *x*:

$$x^2+(2x-3)^2-2x-4(2x-3)=0$$
$$x^2+4x^2-12x+9-2x-8x+12=0$$
$$5x^2-22x+21=0$$
$$(x-3)(5x-7)=0$$
$$x=3,\ \tfrac{7}{5}$$

Now, substitute these values of *x* back into **(3)** to find the corresponding values of *y*:

$$x=3:\ \ y=2(3)-3=3$$
$$x=\tfrac{7}{5}:\ \ y=2(\tfrac{7}{5})-3=-\tfrac{1}{5}$$

So, the solutions are $\boxed{(3,3) \text{ and } (\tfrac{7}{5},-\tfrac{1}{5})}$.

23. Solve the system $\begin{cases} 4x^2+12xy+9y^2=25 & \textbf{(1)} \\ -2x+y=1 & \textbf{(2)} \end{cases}$

Solve **(2)** for *y*: $y=2x+1$ **(3)**

Substitute **(3)** into **(1)** to get an equation in terms of *x*:

$$4x^2+12x(2x+1)+9(2x+1)^2=25$$
$$4x^2+24x^2+12x+\underbrace{9(4x^2+4x+1)}_{36x^2+36x+9}-25=0$$
$$64x^2+48x-16=0$$
$$4x^2+3x-1=0$$
$$(x+1)(4x-1)=0$$
$$x=-1,\ \tfrac{1}{4}$$

Now, substitute these values of *x* back into **(3)** to find the corresponding values of *y*:

$$x=-1:\ \ y=2(-1)+1=-1$$
$$x=\tfrac{1}{4}:\ \ y=2(\tfrac{1}{4})+1=\tfrac{3}{2}$$

So, the solutions are $\boxed{(-1,-1) \text{ and } (\tfrac{1}{4},\tfrac{3}{2})}$.

24. Solve the system $\begin{cases} -4xy+4y^2=8 & \textbf{(1)} \\ 3x+y=2 & \textbf{(2)} \end{cases}$

Solve **(2)** for y: $y=2-3x$ **(3)**

Substitute **(3)** into **(1)** to get an equation in terms of x:

$$-4x(2-3x)+4(2-3x)^2=8$$

$$-8x+12x^2+\underbrace{4(4-12x+9x^2)}_{16-48x+36x^2}-8=0$$

$$48x^2-56x+8=0$$

$$6x^2-7x+1=0$$

$$(x-1)(6x-1)=0$$

$$x=1,\ \tfrac{1}{6}$$

Now, substitute these values of x back into **(3)** to find the corresponding values of y:

$$x=1:\ \ y=2-3(1)=-1$$

$$x=\tfrac{1}{6}:\ \ y=2-3(\tfrac{1}{6})=\tfrac{3}{2}$$

So, the solutions are $\boxed{(1,-1) \text{ and } (\tfrac{1}{6},\tfrac{3}{2})}$.

25. Solve the system $\begin{cases} x^3-y^3=63 & \textbf{(1)} \\ x-y=3 & \textbf{(2)} \end{cases}$

Solve **(2)** for x: $x=y+3$ **(3)**

Substitute **(3)** into **(1)** to get an equation in terms of x:

$$\underbrace{(y+3)^3}_{(y+3)(y^2+6y+9)}-y^3=63$$

$$(y^3+6y^2+9y+3y^2+18y+27)-y^3-63=0$$

$$9y^2+27y-36=0$$

$$y^2+3y-4=0$$

$$(y+4)(y-1)=0$$

$$y=-4,\ 1$$

Now, substitute these values of y back into **(3)** to find the corresponding values of x:

$$y=-4:\ x=-4+3=-1$$

$$y=1:\ \ x=1+3=4$$

So, the solutions are $\boxed{(4,1) \text{ and } (-1,-4)}$.

26. Solve the system $\begin{cases} x^3 + y^3 = -26 & \textbf{(1)} \\ x + y = -2 & \textbf{(2)} \end{cases}$

Solve **(2)** for y: $y = -x - 2$ **(3)**

Substitute **(3)** into **(1)** to get an equation in terms of x:

$$x^3 + \underbrace{(-x-2)^3}_{[-(x+2)]^3 = (-1)^3(x+2)^3 = -(x+2)^3} = -26$$

$$x^3 - (x+2)^3 = -26$$

$$x^3 - (x+2)(x^2 + 4x + 4) = -26$$

$$x^3 - (x^3 + 4x^2 + 4x + 2x^2 + 8x + 8) + 26 = 0$$

$$-6x^2 - 12x + 18 = 0$$

$$x^2 + 2x - 3 = 0$$

$$(x+3)(x-1) = 0$$

$$x = -3, 1$$

Now, substitute these values of x back into **(3)** to find the corresponding values of y:

$$x = -3: \quad y = -(-3) - 2 = 1$$

$$x = 1: \quad y = -(1) - 2 = -3$$

So, the solutions are $\boxed{(-3,1) \text{ and } (1,-3)}$.

27. Solve the system $\begin{cases} 4x^2 - 3xy = -5 & \textbf{(1)} \\ -x^2 + 3xy = 8 & \textbf{(2)} \end{cases}$

Add **(1)** and **(2)** to get an equation in terms of x:

$$3x^2 = 3$$

$$x^2 = 1$$

$$x = -1, 1$$

Now, substitute these values of x back into **(2)** to find the corresponding values of y:

$$x = 1: \; -(1)^2 + 3(1)(y) = 8$$

$$3y = 9 \text{ so that } y = 3$$

$$x = -1: \quad -(-1)^2 + 3(-1)(y) = 8$$

$$-3y = 9 \text{ so that } y = -3$$

So, the solutions are $\boxed{(1,3) \text{ and } (-1,-3)}$.

28. Solve the system $\begin{cases} 2x^2+5xy=2 & \textbf{(1)} \\ x^2-xy=1 & \textbf{(2)} \end{cases}$

Multiply **(2)** by 5, and then add to **(1)** to get an equation in terms of x:

$$\begin{aligned} 7x^2 &= 7 \\ x^2 &= 1 \\ x &= -1,\ 1 \end{aligned}$$

Now, substitute these values of x back into **(2)** to find the corresponding values of y:

$$\begin{aligned} x=1: \quad & (1)^2-(1)(y)=1 \\ & y=0 \\ x=-1: \quad & (-1)^2-(-1)(y)=1 \\ & y=0 \end{aligned}$$

So, the solutions are $\boxed{(1,0) \text{ and } (-1,0)}$.

29. Solve the system $\begin{cases} 2x^2-xy=28 & \textbf{(1)} \\ 4x^2-9xy=28 & \textbf{(2)} \end{cases}$

Multiply **(1)** by -9, and then add to **(2)** to get an equation in terms of x:

$$\begin{aligned} -14x^2 &= -224 \\ x^2 &= 16 \\ x &= -4, 4 \end{aligned}$$

Now, substitute these values of x back into **(1)** to find the corresponding values of y:

$$\begin{aligned} x=4: \quad & 2(4)^2-(4)(y)=28 \\ & y=1 \\ x=-4: \quad & (-4)^2-(-4)(y)=28 \\ & y=-1 \end{aligned}$$

So, the solutions are $\boxed{(-4,-1) \text{ and } (4,1)}$.

30. Solve the system $\begin{cases} -7xy+2y^2=-3 & \textbf{(1)} \\ -3xy+y^2=0 & \textbf{(2)} \end{cases}$

Multiply **(1)** by -3, and then add to **(2)** to get an equation in terms of y:

$$y^2=9$$

$$y=-3,3$$

Now, substitute these values of y back into **(1)** to find the corresponding values of x:

$$y=-3: \quad -7x(-3)+2(-3)^2=-3$$

$$x=-1$$

$$y=3: \quad -7x(3)+2(3)^2=-3$$

$$x=1$$

So, the solutions are $\boxed{(-1,-3) \text{ and } (1,3)}$.

31. Solve the system $\begin{cases} 4x^2+10y^2=26 & \textbf{(1)} \\ -2x^2+2y^2=-6 & \textbf{(2)} \end{cases}$

Multiply **(2)** by 2, and then add to **(1)** to get an equation in terms of y:

$$14y^2=14$$

$$y^2=1$$

$$y=-1,1$$

Now, substitute these values of y back into **(1)** to find the corresponding values of x:

$$y=1: \quad 4x^2+10=26$$

$$x=\pm 2$$

$$y=-1: \quad 4x^2+10=26$$

$$x=\pm 2$$

So, the solutions are $\boxed{(-2,-1),\ (-2,1),\ (2,-1), \text{ and } (2,1)}$.

32. Solve the system $\begin{cases} x^3+y^3=19 & \textbf{(1)} \\ x^3-y^3=-35 & \textbf{(2)} \end{cases}$

Add **(1)** and **(2)** to get an equation in terms of x:

$$2x^3=-16$$

$$x=-2$$

Now, substitute this value of y back into **(1)** to find the corresponding value of y:

$$(-2)^3+y^3=19$$

$$y=3$$

So, the solution is $\boxed{(-2,3)}$.

33. Solve the system $\begin{cases} \log_x(2y) = 3 & \textbf{(1)} \\ \log_x(y) = 2 & \textbf{(2)} \end{cases}$

This system is equivalent to: $\begin{cases} x^3 = 2y & \textbf{(3)} \\ x^2 = y & \textbf{(4)} \end{cases}$

Substitute **(4)** into **(3)** to get an equation in terms of x:

$$x^3 = 2(x^2)$$
$$x^3 - 2x^2 = 0$$
$$x^2(x-2) = 0$$
$$x = \not{0},\ 2$$

Recall that 0 is not an allowable base for a logarithm. So, although $x = 0$ would give rise to a corresponding y (namely $y = 0$) such that the pair would solve system **(3) – (4)**, the equations **(1) – (2)** would not be defined for such x and y values. So, we only substitute the value $x = 2$ back into **(4)** to find the corresponding value of y that will yield a solution to the original system:

$$x = 2:\ y = 2(2) = 4$$

So, the solution is $\boxed{(2,4)}$.

34. Solve the system $\begin{cases} \log_x(y) = 1 & \textbf{(1)} \\ \log_x(2y) = \frac{1}{2} & \textbf{(2)} \end{cases}$

This system is equivalent to: $\begin{cases} x = y & \textbf{(3)} \\ x^{1/2} = 2y & \textbf{(4)} \end{cases}$

Substitute **(3)** into **(4)** to get an equation in terms of x:

$$x^{1/2} = 2x \text{ (Square both sides)}$$
$$x = 4x^2$$
$$4x^2 - x = 0$$
$$x(4x-1) = 0$$
$$x = \not{0},\ \tfrac{1}{4}$$

Recall that 0 is not an allowable base for a logarithm. So, although $x = 0$ would give rise to a corresponding y (namely $y = 0$) such that the pair would solve system **(3) – (4)**, the equations **(1) – (2)** would not be defined for such x and y values. So, we only substitute the value $x = \frac{1}{4}$ back into **(3)** to find the corresponding value of y that will yield a solution to the original system: $x = \frac{1}{4}:\ y = \frac{1}{4}$

So, the solution is $\boxed{(\frac{1}{4}, \frac{1}{4})}$.

35. Solve the system $\begin{cases} \dfrac{1}{x^3}+\dfrac{1}{y^2}=17 & \textbf{(1)} \\ \dfrac{1}{x^3}-\dfrac{1}{y^2}=-1 & \textbf{(2)} \end{cases}$

Add **(1)** and **(2)** to get an equation in terms of x:

$$\frac{2}{x^3}=16$$

$$2=16x^3$$

$$x^3=\tfrac{1}{8} \text{ so that the only real solution is } x=\tfrac{1}{2}$$

Now, substitute this value of x back into **(2)** to find the corresponding value of y:

$$\frac{1}{\left(\frac{1}{2}\right)^3}-\frac{1}{y^2}=-1$$

$$8-\frac{1}{y^2}=-1$$

$$9=\frac{1}{y^2} \text{ so that } y^2=\tfrac{1}{9}. \text{ Hence, } y=\pm\tfrac{1}{3}.$$

So, the solutions are $\boxed{\left(\tfrac{1}{2},\tfrac{1}{3}\right) \text{ and } \left(\tfrac{1}{2},-\tfrac{1}{3}\right)}$.

36. Solve the system $\begin{cases} \dfrac{2}{x^2}+\dfrac{3}{y^2}=\dfrac{5}{6} & \textbf{(1)} \\ \dfrac{4}{x^2}-\dfrac{9}{y^2}=0 & \textbf{(2)} \end{cases}$

Multiply **(1)** by 3, and then add to **(2)** to get an equation in terms of x:

$$\frac{10}{x^2}=\frac{5}{2}$$

$$5x^2=20$$

$$x^2=4 \text{ so that } x=\pm 2.$$

Now, substitute these values of x back into **(2)** to find the corresponding value of y:

$$x=2:\ \frac{4}{(2)^2}-\frac{9}{y^2}=0$$

$$1=\frac{9}{y^2} \text{ so that } y^2=9. \text{ Hence, } y=\pm 3.$$

$$x=-2:\ \frac{4}{(-2)^2}-\frac{9}{y^2}=0$$

$$1=\frac{9}{y^2} \text{ so that } y^2=9. \text{ Hence, } y=\pm 3.$$

So, the solutions are $\boxed{(2,3),\ (2,-3),\ (-2,3), \text{ and } (-2,-3)}$.

37. Solve the system $\begin{cases} 2x^2+4y^4=-2 & \textbf{(1)} \\ 6x^2+3y^4=-1 & \textbf{(2)} \end{cases}$

Note that the left side of **(1)** (and **(2)**) is always ≥ 0, while the right side is negative. Hence, there are no real values of x and y that can satisfy either equation, not to mention both simultaneously. Hence, there is $\boxed{\text{no solution}}$ of this system.

38. Solve the system $\begin{cases} x^2+y^2=-2 & \textbf{(1)} \\ x^2+y^2=-1 & \textbf{(2)} \end{cases}$

Note that the left side of **(1)** (and **(2)**) is always ≥ 0, while the right side is negative. Hence, there are no real values of x and y that can satisfy either equation, not to mention both simultaneously. Hence, there is $\boxed{\text{no solution}}$ of this system.

39. Solve the system $\begin{cases} 2x^2-5y^2=-8 & \textbf{(1)} \\ x^2-7y^2=-4 & \textbf{(2)} \end{cases}$

Multiply **(2)** by -2, and then add to **(1)** to get an equation in terms of y:

$$9y^2=0$$

$$y=0$$

Now, substitute this value of y back into **(1)** to find the corresponding value of x:

$$2x^2+8=0$$

which has no real solution. Hence, this system has no solution.

40. Solve the system $\begin{cases} x^2 + y^2 = 4x + 6y - 12 & \textbf{(1)} \\ 9x^2 + 4y^2 = 36x + 24y - 36 & \textbf{(2)} \end{cases}$

Completing the square in each equation separately yields the following equivalent system:

$$\begin{cases} (x-2)^2 + (y-3)^2 = 1 & \textbf{(3)} \\ \frac{(x-2)^2}{4} + \frac{(y-3)^2}{9} = 1 & \textbf{(4)} \end{cases}$$

Solve **(3)** for $(x-2)^2$: $(x-2)^2 = 1-(y-3)^2$ **(5)**

Substitute **(5)** into **(4)**:

$$\frac{1-(y-3)^2}{4} + \frac{(y-3)^2}{9} = 1 \Rightarrow 9\left(1-(y-3)^2\right) + 4(y-3)^2 = 36 \Rightarrow (y-3)^2 = -5$$

which has no solution. Thus, the system has no solution.

41.

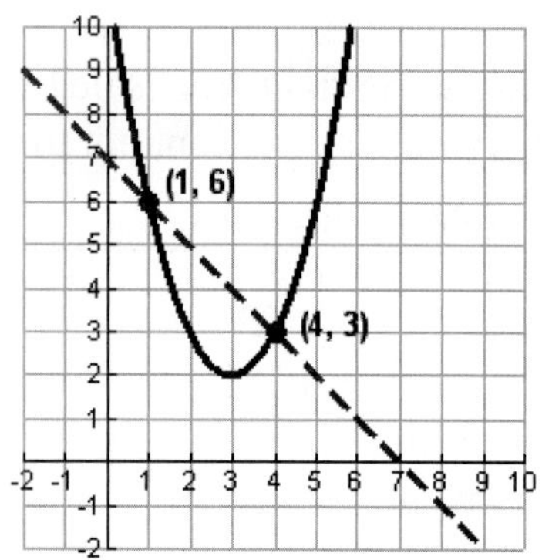

Notes on the graph:

Solid curve: $y = x^2 - 6x + 11$

Dashed curve: $y = -x + 7$

The solutions of the system are (1,6) and (4,3).

42.

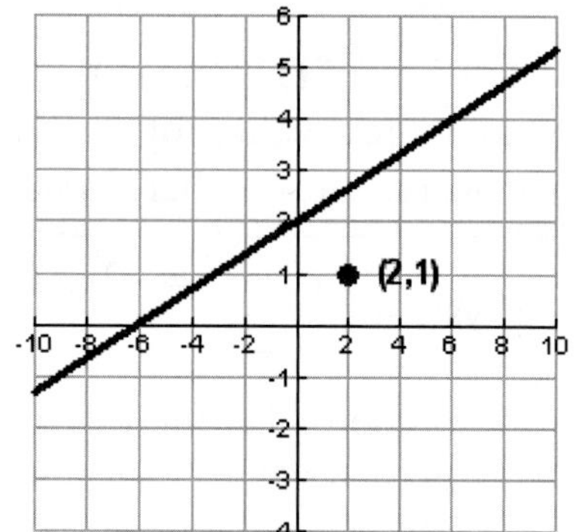

Note that in this case the graph of the equation $(x-2)^2 + (y-1)^2 = 0$ consists of the single point (2,1), which does not satisfy the second equation (the graph of which is the solid curve). Hence, this system has **no solution**.

43. Let x and y be two numbers.
The system that must be solved is:

$$\begin{cases} x+y=10 & \textbf{(1)} \\ x^2-y^2=40 & \textbf{(2)} \end{cases}$$

Solve **(1)** for y: $y=10-x$ **(3)**

Substitute **(3)** into **(2)** to get an equation in x:

$$x^2-(10-x)^2=40$$
$$\cancel{x^2}-100+20x-\cancel{x^2}=40$$
$$20x=140$$
$$x=7$$

Substitute this value of x into **(3)** to find the corresponding value of y:

$y=10-7=3$

So, the two numbers are $\boxed{3 \text{ and } 7}$.

44. Let x and y be two numbers.
The system that must be solved is:

$$\begin{cases} x-y=3 & \textbf{(1)} \\ x^2-y^2=51 & \textbf{(2)} \end{cases}$$

Solve **(1)** for x: $x=y+3$ **(3)**

Substitute **(3)** into **(2)** to get an equation in y:

$$(y+3)^2-y^2=51$$
$$y^2+6y+9-y^2=51$$
$$6y=42$$
$$y=7$$

Substitute this value of y into **(3)** to find the corresponding value of x:

$x=7+3=10$

So, the two numbers are $\boxed{10 \text{ and } 7}$.

45. Let x and y be two numbers.
The system that must be solved is:

$$\begin{cases} xy=\dfrac{1}{\frac{1}{x}-\frac{1}{y}} & \textbf{(1)} \\ xy=72 & \textbf{(2)} \end{cases}$$

First, simplify **(1)**:

$$xy=\frac{1}{\frac{1}{x}-\frac{1}{y}}=\frac{1}{\frac{y-x}{xy}}=\frac{xy}{y-x}$$
$$1=y-x \quad \textbf{(3)}$$

So, we now solve the following simplified system: $\begin{cases} xy=72 & \textbf{(2)} \\ y-x=1 & \textbf{(3)} \end{cases}$

To do so, solve **(3)** for y: $y=x+1$ **(4)**

Substitute **(4)** into **(2)** to get an equation in x:

$$x(x+1)=72$$
$$x^2+x-72=0$$
$$(x+9)(x-8)=0$$
$$x=-9,\ 8$$

Substitute these values of x into **(4)** to find the corresponding values of y:

$$x=-9: \quad y=(-9)+1=-8$$
$$x=8: \quad y=(8)+1=9$$

So, there are two pairs of numbers that will work, namely the pair 8 and 9, and the pair -8 and -9.

46. Let x and y be two numbers.

The system that must be solved is: $\begin{cases} \dfrac{x+y}{x-y} = 9 & \textbf{(1)} \\ xy = 80 & \textbf{(2)} \end{cases}$

First, simplify **(1)**:

$$x + y = 9(x - y) = 9x - 9y$$
$$8x - 10y = 0 \quad \textbf{(3)}$$

So, we now solve the following simplified system: $\begin{cases} xy = 80 & \textbf{(2)} \\ 8x - 10y = 0 & \textbf{(3)} \end{cases}$

To do so, solve **(3)** for y: $y = \frac{4}{5}x$ **(4)**

Substitute **(4)** into **(2)** to get an equation in x:

$$x(\tfrac{4}{5}x) = 80$$
$$\tfrac{4}{5}x^2 = 80$$
$$x^2 = \tfrac{5}{4}(80) = 100$$
$$x = \pm 10$$

Substitute these values of x into **(4)** to find the corresponding values of y:

$$x = -10: \quad y = \tfrac{4}{5}(-10) = -8$$
$$x = 10: \quad y = \tfrac{4}{5}(10) = 8$$

So, there are two pairs of numbers that will work, namely the pair 8 and 10, and the pair -10 and -8.

47. Let x = width of rectangular pen (in cm)
y = length of rectangular pen.

The system that must be solved is:

$$\begin{cases} 2x + 2y = 36 & \textbf{(1)} \text{ (perimeter)} \\ xy = 80 & \textbf{(2)} \text{ (area)} \end{cases}$$

Solve **(1)** for y: $y = 18 - x$ **(3)**

Substitute **(3)** into **(2)** to get an equation in x:

$$x(18 - x) = 80$$
$$18x - x^2 = 80$$
$$x^2 - 18x + 80 = 0$$
$$(x - 10)(x - 8) = 0 \text{ so that } x = 10,\ 8$$

Substitute these values of x into **(3)** to find the corresponding values of y:

$$x = 10: \quad y = 18 - 10 = 8$$
$$x = 8: \quad y = 18 - 8 = 10$$

Hence, in terms of the context of the problem, both x values result in the same solution, namely that the dimensions of the rectangular pen should be $\boxed{8\text{ cm} \times 10\text{ cm}}$.

48. Assume that circles C_1 and C_2 are centered at the same point, say the origin.
Let x = radius of circle C_1
y = radius of circle C_2

The system that must be solved is:

$$\begin{cases} 2\pi x + 2\pi y = 16\pi & \textbf{(1)} \text{ (perimeter)} \\ \pi x^2 + \pi y^2 = 34\pi & \textbf{(2)} \text{ (area)} \end{cases}$$

First, simplify equations **(1)** and **(2)** to get the following equivalent system:

$$\begin{cases} x + y = 8 & \textbf{(3)} \\ x^2 + y^2 = 34 & \textbf{(4)} \end{cases}$$

Solve **(3)** for y: $y = 8 - x$ **(5)**

Substitute **(3)** into **(4)** to get an equation in x:

$$x^2 + (8-x)^2 = 34$$
$$x^2 + 64 - 16x + x^2 = 34$$
$$2x^2 - 16x + 30 = 0$$
$$x^2 - 8x + 15 = 0$$
$$(x-5)(x-3) = 0$$
$$x = 5,\ 3$$

Substitute these values of x into **(5)** to find the corresponding values of y:

$$x = 5: \quad y = 8 - 5 = 3$$
$$x = 3: \quad y = 8 - 3 = 5$$

Hence, in terms of the context of the problem, both x values result in the same solution, namely that the radii of the two concentric circles should be 5 units and 3 units.

49. Let x = width of one of the two congruent rectangular pens (in ft)
y = length of one of the two congruent rectangular pens.

The system that must be solved is:

$$\begin{cases} 3x + 4y = 2200 & \textbf{(1)} \text{ (total amount of fence needed)} \\ x \cdot 2y = 200{,}000 & \textbf{(2)} \text{ (combined area of the two pens)} \end{cases}$$

Solve **(1)** for y: $y = \frac{1}{4}(2200 - 3x)$ **(3)**

Substitute **(3)** into **(2)** to get an equation in x:

$$x \cdot 2\left[\tfrac{1}{4}(2200 - 3x)\right] = 200{,}000$$
$$2200x - 3x^2 = 400{,}000$$
$$3x^2 - 2200x + 400{,}000 = 0$$

$$x = \frac{2200 \pm \sqrt{4{,}840{,}000 - 4{,}800{,}000}}{2(3)} = \frac{2200 \pm \sqrt{40{,}000}}{6}$$
$$= \frac{2200 \pm 200}{6} = 400,\ \frac{1000}{3}$$

Substitute these values of x into **(3)** to find the corresponding values of y:

$$x = 400: \ \tfrac{1}{4}(2200 - 3(400)) = 250$$

$$x = \tfrac{1000}{3}: \ \tfrac{1}{4}(2200 - 3(\tfrac{1000}{3})) = 300$$

Hence, in terms of the context of the problem, there are two distinct solutions.

Solution 1: Each of the two congruent rectangular pens should be $400 \text{ ft} \times 250 \text{ ft}$, so that the two combined have dimensions $\boxed{400 \text{ ft} \times 500 \text{ ft}}$.

Solution 2: Each of the two congruent rectangular pens should be $\tfrac{1000}{3} \text{ ft} \times 300 \text{ ft}$, so that the two combined have dimensions $\boxed{\tfrac{1000}{3} \text{ ft} \times 600 \text{ ft}}$.

50. Let x = width of rectangular pen (in ft)
y = length of rectangular pen.

The system that must be solved is:

$$\begin{cases} xy = 11{,}250 & \textbf{(1)} \\ y = 2x & \textbf{(2)} \end{cases}$$

Substitute **(2)** into **(1)** to get an equation in x:

$$x(2x) = 11{,}250$$
$$x^2 = 5625$$
$$x = 75, \ \cancel{-75}$$

Note that we omit $x = -75$ as a possibility since width must be nonnegative. So, we substitute only the value $x = 75$ into **(2)** to find the corresponding value of y:

$$y = 2(75) = 150$$

Hence, in terms of the context of the problem, the dimensions of the rectangular pen should be $\boxed{75 \text{ ft} \times 150 \text{ ft}}$.

51. Let r = speed of the professor (in $\frac{m}{\min}$)
t = number of minutes it took Jeremy to complete the 400m race

We have the following information:

	Distance (in m)	**Rate (in $\frac{m}{\min}$)**	**Time (in min)**
Professor	400	r	$t + \underbrace{1}_{\text{Head start}} + \underbrace{\tfrac{5}{3}}_{\text{Finished 1 min 40 sec after Jerermy}} = t + \tfrac{8}{3}$
Jeremy	400	$5r$	t

Since Distance = Rate × Time, the following system must be solved:

$$\begin{cases} 400 - r\left(t + \tfrac{8}{3}\right) & \textbf{(1)} \\ 400 = 5rt & \textbf{(2)} \end{cases}$$

Solve **(2)** for t: $t=\frac{80}{r}$ **(3)**

Substitute **(3)** into **(1)** to get an equation in x:

$$400=r\left(\tfrac{80}{r}+\tfrac{8}{3}\right)$$
$$400=\not{r}\left(\tfrac{240+8x}{3\not{r}}\right)$$
$$1200=240+8r$$
$$960=8r \text{ so that } 120=r$$

Hence, the professor's average speed was 120 $\frac{m}{min}=\boxed{2\ \frac{m}{sec}}$, while Jeremy's average speed was $600\ \frac{m}{min}=\boxed{10\ \frac{m}{sec}}$. (<u>Note</u>: Though the value of t was not needed to answer the question posed in the problem, one could substitute this value of r into **(3)** to find that the corresponding value of t is $\frac{2}{3}$, meaning that it took Jeremy 40 sec to complete the race).

52. Let x = your speed (in $\frac{m}{min}$)

y = number of minutes it took Jeremy to complete the 800m race

We have the following information:

	Distance (in m)	**Rate (in $\frac{m}{min}$)**	**Time (in min)**
You	800	x	$t+\underbrace{1}_{\text{Head start}}+\underbrace{\tfrac{1}{3}}_{\text{Finished 20 sec after Jerermy}}=t+\frac{4}{3}$
Jeremy	800	$2x$	t

Since Distance = Rate × Time, the following system must be solved:

$$\begin{cases}800=x\left(t+\frac{4}{3}\right) & \textbf{(1)}\\ 800=2xt & \textbf{(2)}\end{cases}$$

Solve **(2)** for t: $t=\frac{400}{x}$ **(3)**

Substitute **(3)** into **(1)** to get an equation in x:

$$800=x\left(\tfrac{400}{x}+\tfrac{4}{3}\right)$$
$$800=\not{x}\left(\tfrac{1200+4x}{3\not{x}}\right)$$
$$2400=1200+4x$$
$$1200=4x \text{ so that } 300=x$$

Hence, your average speed was 300 $\frac{m}{min}$, while Jeremy's average speed was 600 $\frac{m}{min}$. (<u>Note</u>: Though the value of t was not needed to answer the question posed in the problem, one could substitute this value of x into **(3)** to find that the corresponding value of y is $\frac{4}{3}$, meaning that it took Jeremy 1 min 20 sec to complete the race).

53. Let y =time Car 3 reaches Car 2 (in hours) and x = speed of Car 3.
Then, $y + \frac{3}{2}$ = time Car 3 reaches Car 1

We have the following information:

		Rate (in mph)	**Time traveled at time of 1st intersection**
1st intersection	Car 2 Car 3	40 x	$y+\frac{1}{2}$ y
2nd intersection	Car 1 Car 3	50 x	$y+2$ $y+\frac{3}{2}$

Since Distance = Rate × Time, the following system must be solved:

$$\begin{cases} 40\left(y+\frac{1}{2}\right)=xy & \textbf{(1)} \\ 50(y+2)=x\left(y+\frac{3}{2}\right) & \textbf{(2)} \end{cases}$$

This is equivalent to:

$$\begin{cases} 20+40y=xy & \textbf{(3)} \\ 100+50y=xy+\frac{3}{2}x & \textbf{(4)} \end{cases}$$

Solve **(3)** for y: $y=\frac{20}{x-40}$ **(5)**
Substitute **(5)** into **(4)**:

$$100+50\left(\frac{20}{x-40}\right)=x\left(\frac{20}{x-40}\right)+\frac{3}{2}x \text{ (Simplify...)}$$

$$3x^2-280x+6000=0$$

$$(x-60)(3x-10)=0$$

$$x=60, \frac{10}{3}$$

So, Car 3 travels at 60 mph.

54. Let d_1 = depth of box one and d_2 = depth of box two. Then,

BOX 1: width of box one = $d_1 + 16$ and length of box one = $5d_1$. So, the volume of box one = $5d_1(d_1+16)d_1$.

BOX 2: width of box two = $(d_1 + 16) - 4$ and length of box two = $5d_1 - 4$. So, the volume of box two $=(5d_1-4)(d_1+12)(\frac{5}{4}d_1)$.

Now, we must solve the equation

$$5d_1(d_1+16)d_1=(5d_1-4)(d_1+12)(\tfrac{5}{4}d_1)$$
$$d_1^2-8d_1-48=0$$
$$(d_1-12)(d_1+4)=0$$
$$d_1=12$$

So, box 2 has measurements $15\times24\times56$.

55. Let *wxyz* be a four digit number. We know the following information:

$$\begin{cases} w^2+z^2=13 & \textbf{(1)} \\ x^2+y^2=85 & \textbf{(2)} \\ x=y+1 & \textbf{(3)} \\ w=z+1 & \textbf{(4)} \end{cases}$$

Substitute **(3)** into **(2)**:

$$(y+1)^2+y^2=85$$
$$y^2+y-42=0$$
$$(y+7)(y-6)=0$$
$$y=\cancel{-7},6$$

Substitute **(4)** into **(1)**:

$$(z+1)^2+z^2=13$$
$$z^2+z-12=0$$
$$(z+4)(z-3)=0$$
$$z=\cancel{-4},3$$

Substitute these into **(3)** and **(4)**, respectively, to get $x = 7$ and $w = 4$. So, the number described is 4763.

56. Let *xyz* be a three digit number. We know the following information:

$$\begin{cases} x^3 + z^3 = 9 & \textbf{(1)} \\ y = 1 + 2x & \textbf{(2)} \\ x = 1 + z & \textbf{(3)} \end{cases}$$

Substitute **(3)** into **(1)**:

$$(1+z)^3 + z^3 = 9$$
$$2z^3 + 3z^2 + 3z - 8 = 0$$
$$(z-1)(2z^2 + 5z + 8) = 0$$
$$z = 1$$

Substitute this into **(3)** to get $x = 2$, and then substitute this into **(2)** to get $y = 5$. So, the number described is 251.

57. In general, $y^2 - y \neq 0$. Must solve this system using substitution.

58. Once you have the *x*-values, you should substitute them into the second equation to obtain:

$$x = -1:\ y = -2$$
$$x = 1:\ \ y = 2$$

So, the solutions are $(1,2)$ and $(-1,-2)$.

59. False.
There are at most two intersection points. Solving a system of the form

$$\begin{cases} y = ax^2 + bx + c \\ y = dx + e \end{cases}$$

amounts to solving the quadratic equation $ax^2 + bx + c = dx + e$. Such an equation can have at most two real solutions. So, the graphs can intersect in at most two points.

60. True.
There are at most three intersection points. Solving a system of the form

$$\begin{cases} y = ax^3 + bx^2 + cx + d \\ y = ex + f \end{cases}$$

amounts to solving the cubic equation $ax^3 + bx^2 + cx + d = ex + f$. Such an equation can have at most three real solutions. So, the graphs can intersect in at most three points.

61. False.
For example, the system

$$\begin{cases} x - y = 1 \\ x^2 + y^2 = 5 \end{cases}$$

cannot be solved using elimination since there is no common term in both equations.

62. False.
For example, consider the system

$$\begin{cases} y^4 - 5xy^3 + x^3 = -3 \\ y^7 + 5y + x^3 - 3x = 4 \end{cases}$$

63. Let $y = a_n x^n + a_{n-1}x^{n-1} + \cdots + a_1 x + a_0$ be a given *n*th degree polynomial. Then, consider a system of the form: $$\begin{cases} y = a_n x^n + a_{n-1}x^{n-1} + \cdots + a_1 x + a_0 & \textbf{(1)} \\ (x-h)^2 + (y-k)^2 = r^2 & \textbf{(2)} \end{cases}$$ Substitute **(1)** into **(2)** to obtain: $$(x-h)^2 + \underbrace{(a_n x^n + a_{n-1}x^{n-1} + \cdots + a_1 x + a_0 - k)^2}_{\text{Has degree } 2n} = r^2 \quad \textbf{(3)}$$ The degree of the left side of **(3)** is $2n$, so there can be at most $2n$ real solutions. Consequently, there are at most $2n$ intersection points of the graphs of the equations of **(1)** and **(2)**.
64. Let $y = a_n x^n + a_{n-1}x^{n-1} + \cdots + a_1 x + a_0$ be a given *n*th degree polynomial. Then, consider a system of the form: $$\begin{cases} y = a_n x^n + a_{n-1}x^{n-1} + \cdots + a_1 x + a_0 & \textbf{(1)} \\ y = mx + b & \textbf{(2)} \end{cases}$$ Substitute **(2)** into **(1)** to obtain: $$mx + b = a_n x^n + a_{n-1}x^{n-1} + \cdots + a_1 x + a_0$$ $$0 = a_n x^n + a_{n-1}x^{n-1} + \cdots + (a_1 - m)x + (a_0 - b) \quad \textbf{(3)}$$ The degree of the right side of **(3)** is *n*, so there can be at most *n* real solutions. Consequently, there are at most *n* intersection points of the graphs of the equations of **(1)** and **(2)**.

65. Consider $\begin{cases} y = x^2 + 1 \\ y = 1 \end{cases}$. Any system in which the linear equation is the tangent line to the parabola at its vertex will have only one solution.	**66.** Consider $\begin{cases} x^2 + y^2 = 1 \\ y = -x^2 - 1 \end{cases}$. Any system in which the quadratic equation is the tangent to the circle at a point on the circle will have only one solution.

67. Factoring the equations of the given system yields the equivalent system:

$$\begin{cases} \left(x^2+y^2\right)^2 = 25 \\ \left(x^2-y^2\right)^2 = 9 \end{cases}$$

Taking the square root in both equations subsequently yields four possible systems (since the right-sides become $\pm 5, \pm 3$ respectively), only two of which have solutions – namely when the right-side of the first equation equals 5.

Case 1: $\begin{cases} x^2+y^2=5 \\ x^2-y^2=3 \end{cases}$

Adding the equations yields $x = \pm 2$. Substituting each of these values back into either equation subsequently yields $y = \pm 1$. So, the solutions of this system are:

(2,-1), (2,1), (-2,-1), and (-2,1)

Case 2: $\begin{cases} x^2+y^2=5 \\ x^2-y^2=-3 \end{cases}$

Adding the equations yields $x = \pm 1$. Substituting each of these values back into either equation subsequently yields $y = \pm 2$. So, the solutions of this system are:

(1,2), (-1,2), (1,-2), and (-1,-2).

These are all solutions of the original solution.

68. Factoring the equations of the given system yields the equivalent system:

$$\begin{cases} \left(x^2+y^2\right)^2 = 169 \\ \left(x^2-y^2\right)^2 = 25 \end{cases}$$

Taking the square root in both equations subsequently yields four possible systems (since the right-sides become $\pm 13, \pm 5$ respectively), only two of which have solutions – namely when the right-side of the first equation equals 13.

Case 1: $\begin{cases} x^2+y^2=13 \\ x^2-y^2=5 \end{cases}$

Adding the equations yields $x = \pm 3$. Substituting each of these values back into either equation subsequently yields $y = \pm 2$. So, the solutions of this system are:

(3,2), (3,-2), (-3,2), and (-3,-2)

Case 2: $\begin{cases} x^2+y^2=13 \\ x^2-y^2=-5 \end{cases}$

Adding the equations yields $x = \pm 2$. Substituting each of these values back into either equation subsequently yields $y = \pm 3$. So, the solutions of this system are:

(-2,-3), (-2,3), (2,-3), and (2,3).

These are all solutions of the original solution.

69. Factoring the equations of the given system yields the equivalent system:

$$\begin{cases}\left(x^2+y^2\right)^2=-25\\ \left(x^2-y^2\right)^2=-9\end{cases}$$

Neither equation is satisfied by any ordered pair (x,y) because the left-side of both equations is always non-negative, while the right-sides are negative. Hence, the system has no solution.

70. Factoring the equations of the given system yields the equivalent system:

$$\begin{cases}\left(x^2+y^2\right)^2=-169\\ \left(x^2-y^2\right)^2=-25\end{cases}$$

Neither equation is satisfied by any ordered pair (x,y) because the left-side of both equations is always non-negative, while the right-sides are negative. Hence, the system has no solution.

71.

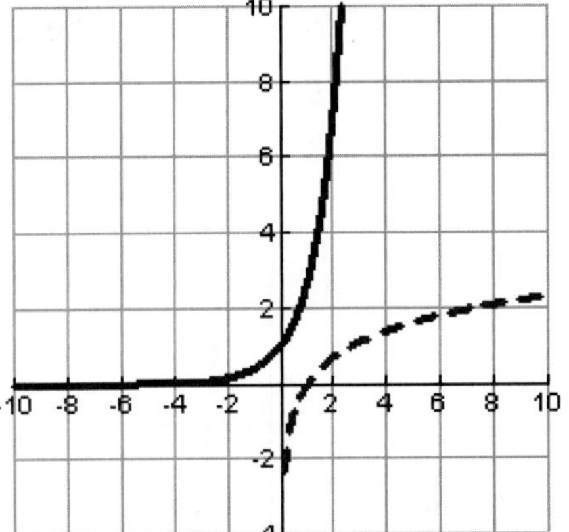

Notes on the graph:
Solid curve: $y=e^x$
Dashed curve: $y=\ln x$

Since the two graphs never intersect, the system has no solution.

72.

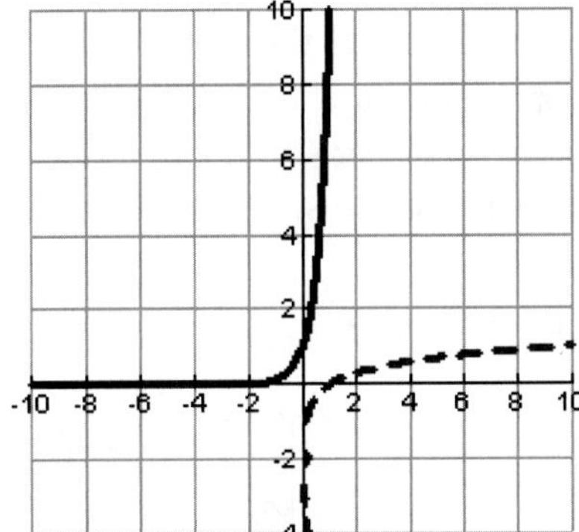

Notes on the graph:
Solid curve: $y=10^x$
Dashed curve: $y=\log x$

Since the two graphs never intersect, the system has no solution.

73.

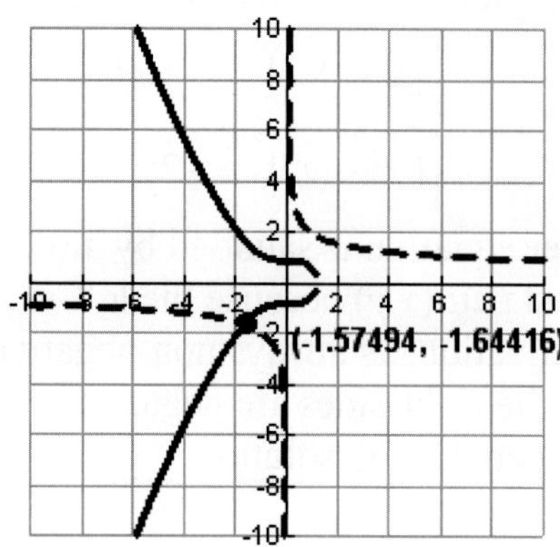

Notes on the graph:

Solid curve: $2x^3 + 4y^2 = 3$

Dashed curve: $xy^3 = 7$

The approximate solution of the system is $(-1.57, -1.64)$.

74.

Here is the graph of the first curve. Notice the scale positions it within the boundary of the second curve - you can hardly see that curve since the scale is so small.

Graph of $3x^4 - 2xy + 5y^2 = 19$:

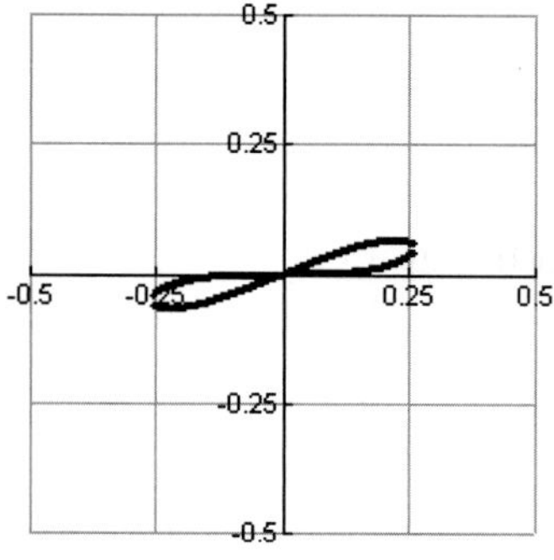

Graph of the pair of equations (the dashed curve is the graph of $x^4 y = 5$).

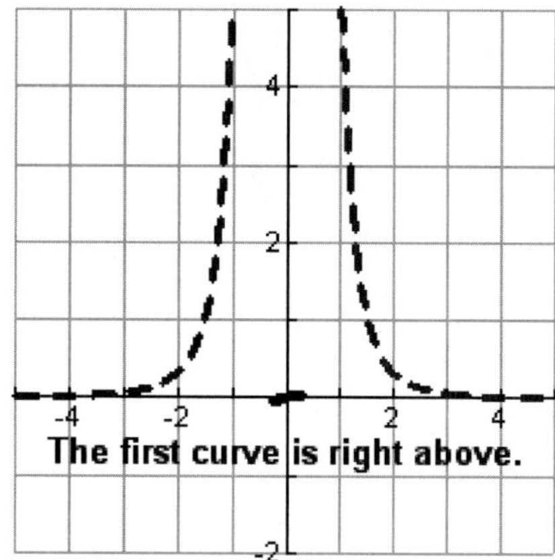

75. The graph of the nonlinear system is given by:

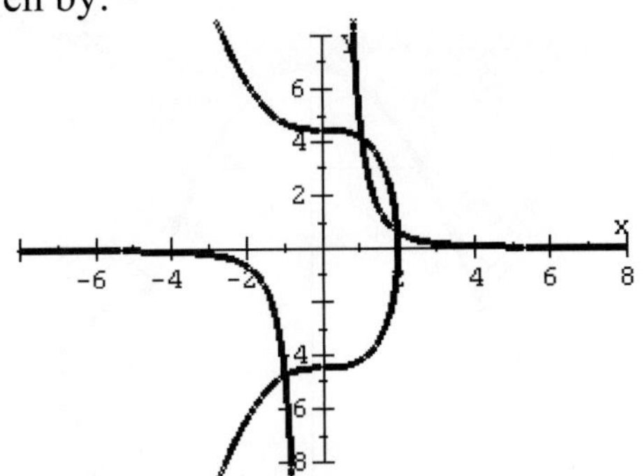

The points of intersection are approximately (1.067, 4.119), (1.986, 0.638), and (–1.017, –4.757).

76. The graph of the nonlinear system is given by:

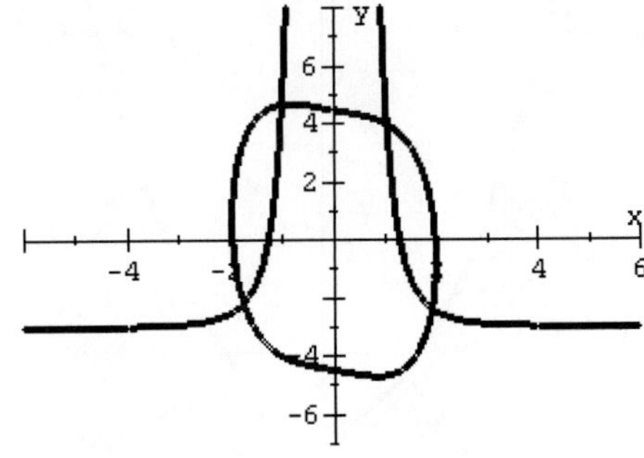

The points of intersection are approximately (1.035, 3.966), (1.899, –2.385), (–1.011, 4.664), and (–1.759, –2.165).

77. Observe that

$$4y^2 = 8 - x^2$$
$$y^2 = \tfrac{8-x^2}{4}$$
$$y = \sqrt{\tfrac{8-x^2}{4}}$$

78. Using the quadratic formula yields

$$y = \frac{-2x \pm \sqrt{(2x)^2 - 4(4)}}{2} = -x \pm \sqrt{x^2 - 4}.$$

Since $y > 0$, we use $y = -x + \sqrt{x^2 - 4}$.

79. Observe that

$$x^3 y^3 = 9y$$
$$y^2 = \tfrac{9}{x^3}$$
$$y = \sqrt{\tfrac{9}{x^3}} = \tfrac{3}{x^{3/2}}$$

80. Observe that

$$3xy + x^3 y^2 = 0$$
$$y\left(x^3 y + 3x\right) = 0$$
$$y = 0 \text{ or } y = -\tfrac{3}{x^2}$$

Since $y < 0$, we use $y = -\tfrac{3}{x^2}$.

Section 9.6 Solutions --

1. b	**2. i**
3. j	**4. g**
5. h	**6. f**
7. c	**8. l**
9. d	**10. a**
11. k	**12. e**

13.

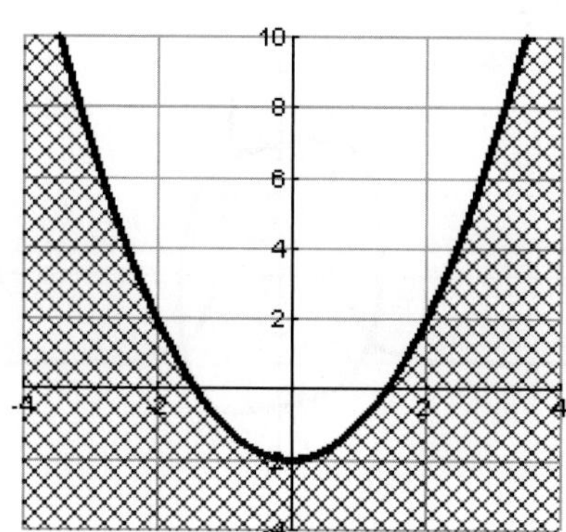

The equation of the solid curve is $y = x^2 - 2$.

14.

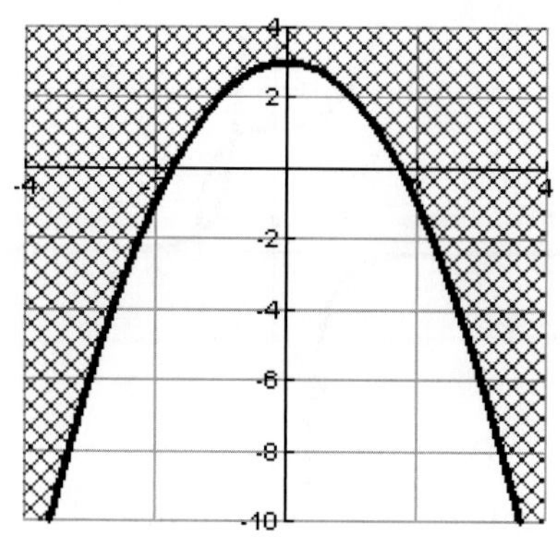

The equation of the solid curve is $y = -x^2 + 3$.

15.

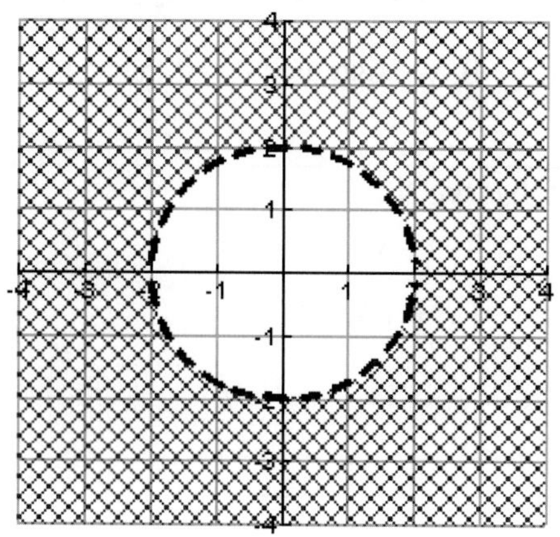

The equation of the dashed curve is $x^2 + y^2 = 4$.

16.

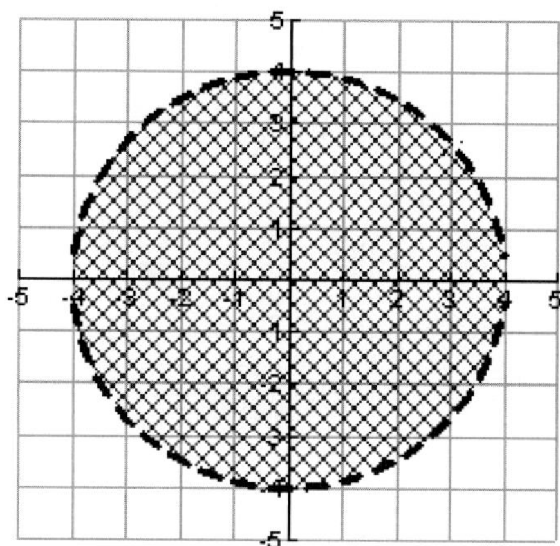

The equation of the dashed curve is $x^2 + y^2 = 16$.

17. Begin by writing the inequality in standard form (so the radius and center are easily identified):

$$\left(x^2-2x+\quad\right)+\left(y^2+4y+\quad\right)\geq -4$$

$$\left(x^2-2x+1\right)+\left(y^2+4y+4\right)\geq -4+1+4$$

$$\left(x-1\right)^2+\left(y+2\right)^2\geq 1$$

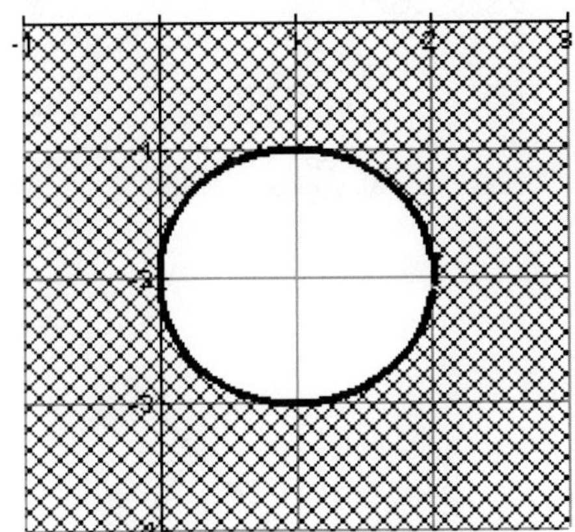

The equation of the solid curve is $\left(x-1\right)^2+\left(y+2\right)^2=1$.

18. Begin by writing the inequality in standard form (so the radius and center are easily identified):

$$\left(x^2+2x+\quad\right)+\left(y^2-2y+\quad\right)\leq 2$$

$$\left(x^2+2x+1\right)+\left(y^2-2y+1\right)\leq 2+1+1$$

$$\left(x+1\right)^2+\left(y-1\right)^2\leq 4$$

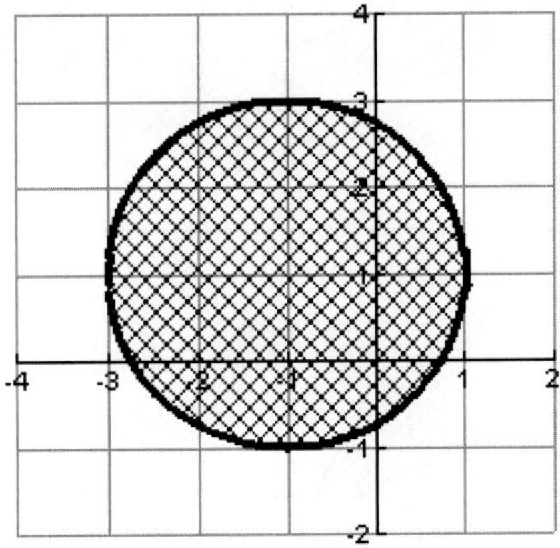

The equation of the solid curve is $\left(x+1\right)^2+\left(y-1\right)^2=4$.

19.

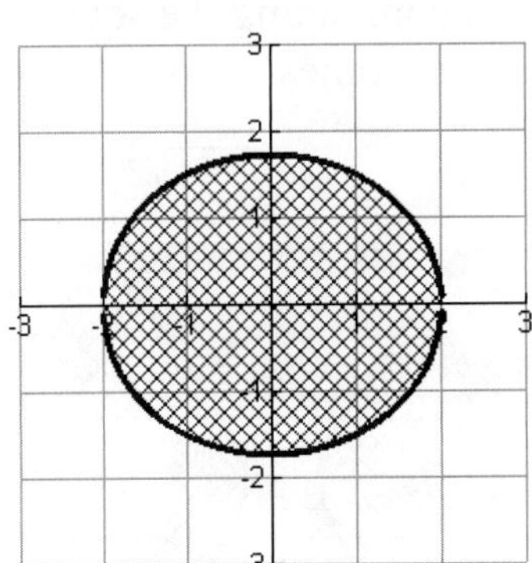

The equation of the solid curve is $\dfrac{x^2}{4}+\dfrac{y^2}{3}=1$.

20.

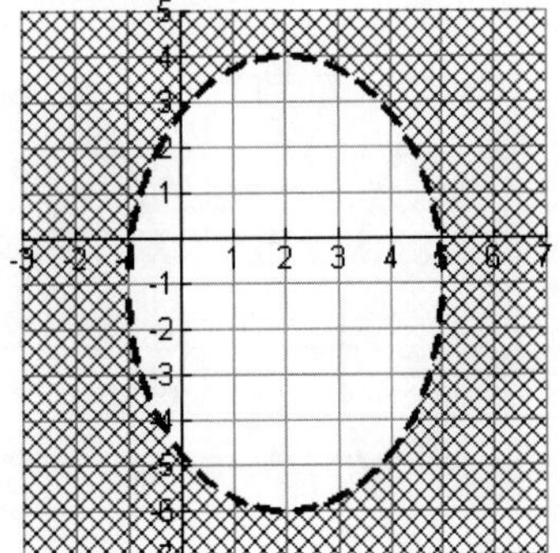

21. Upon completing the square, the inequality becomes

$$9(x-1)^2+16(y+3)^2>144$$

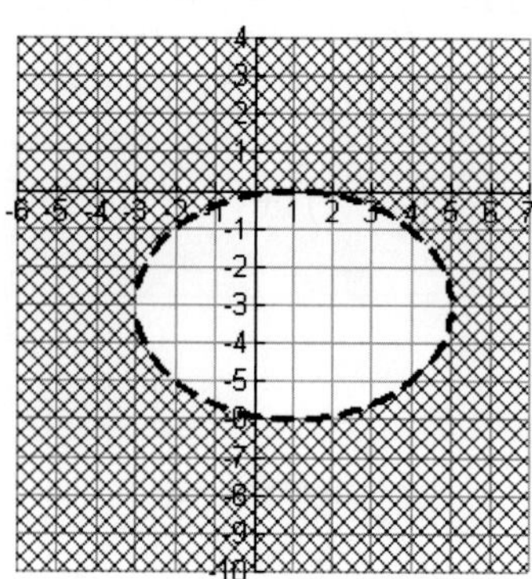

22.

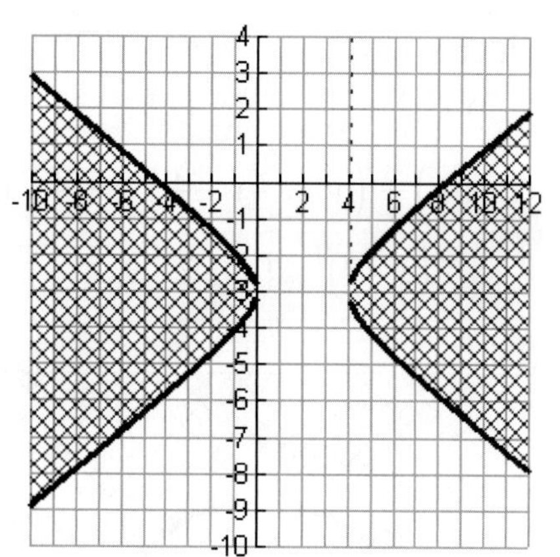

23.

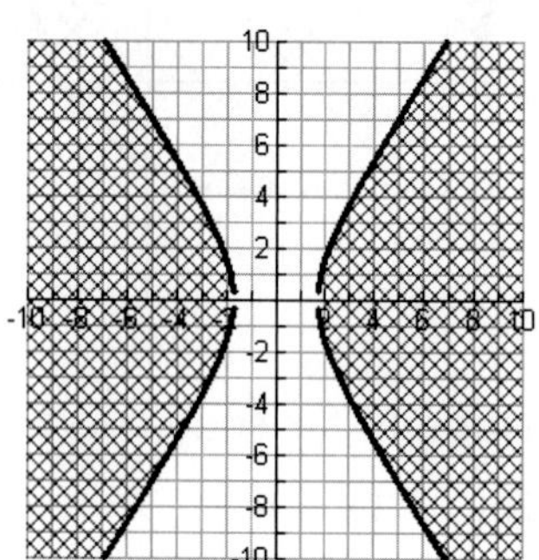

24.

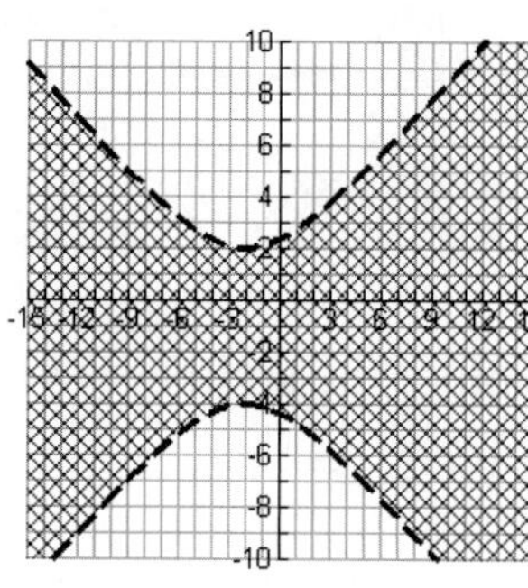

25.

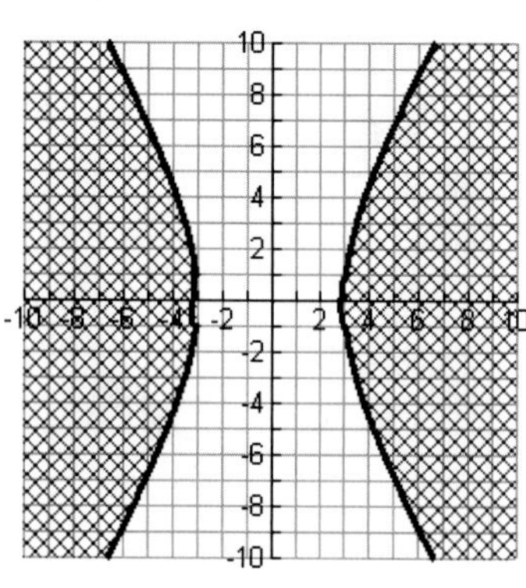

26. Upon completing the square, the inequality becomes

$$\frac{(x+4)^2}{36}-\frac{(y-2)^2}{25}\geq 1$$

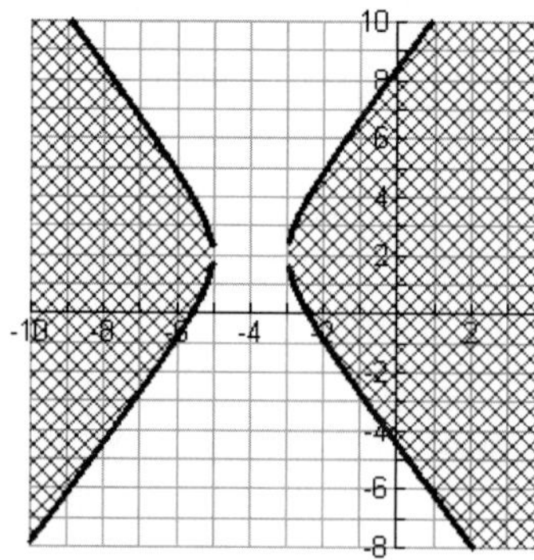

27.

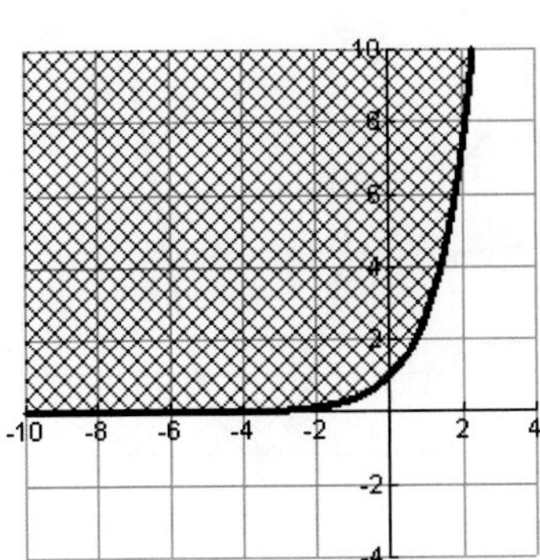

The equation of the solid curve is $y = e^x$.

28.

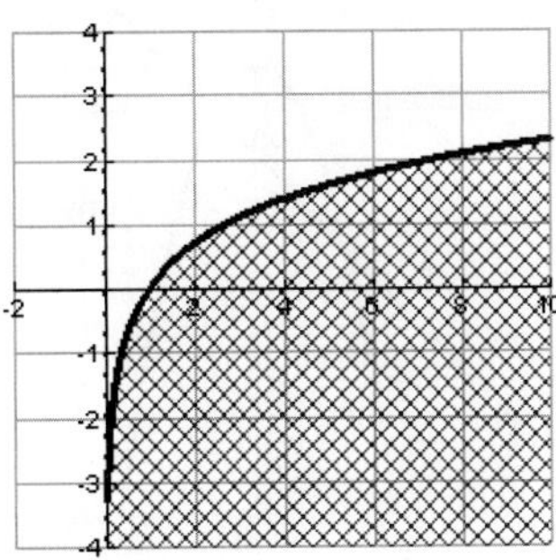

The equation of the solid curve is $y = \ln x$.

29.

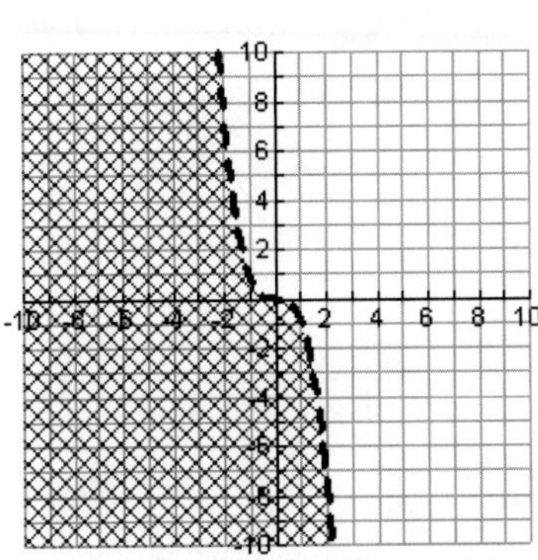

The equation of the dashed curve is $y = -x^3$.

30.

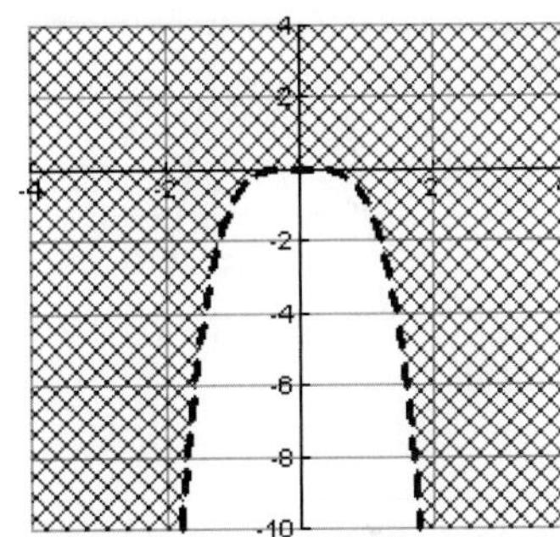

The equation of the dashed curve is $y = -x^4$.

31.

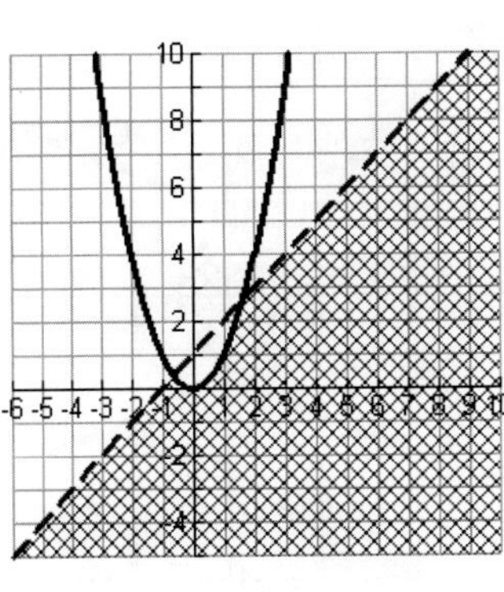

32.

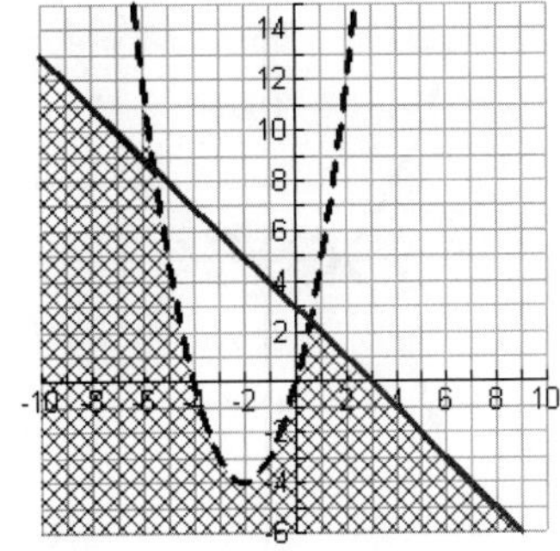

33.

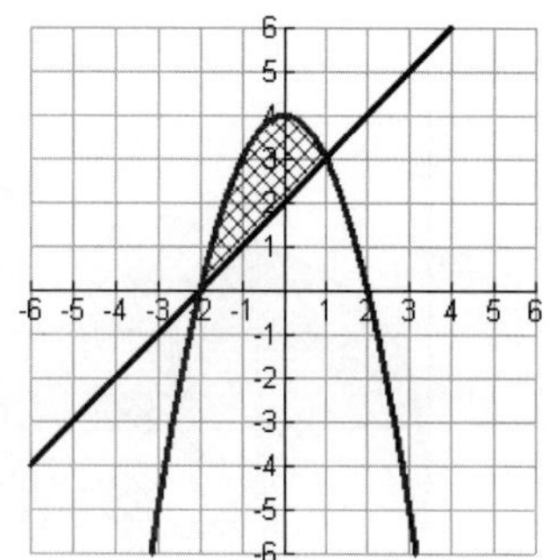

34.

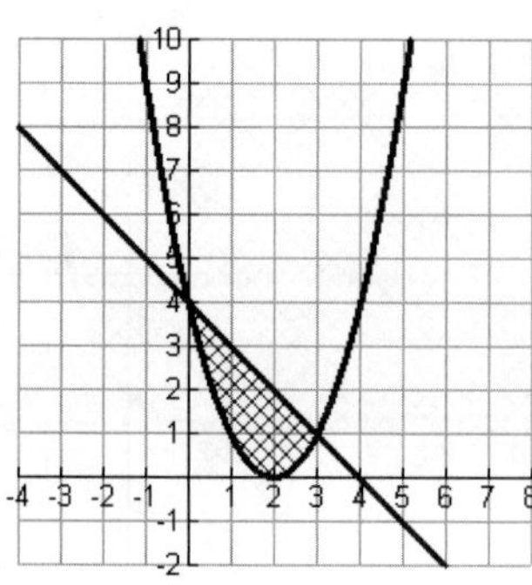

35.

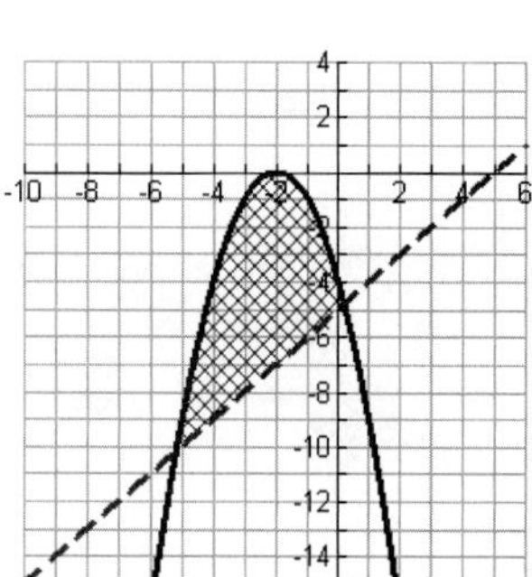

36.

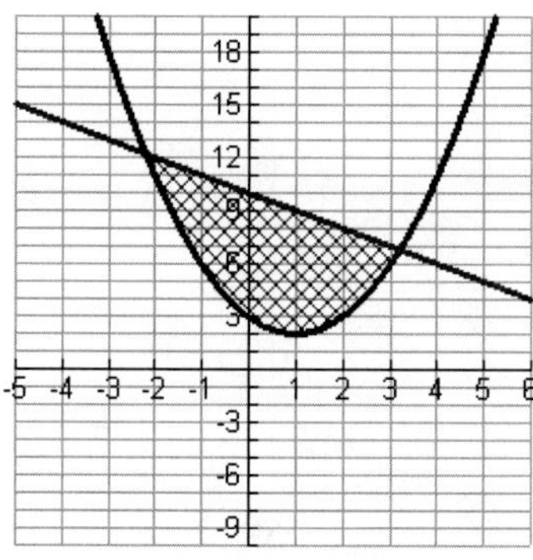

37.

First write the inequalities as:

$$y > x^2 - 1$$

$$y < 1 - x^2$$

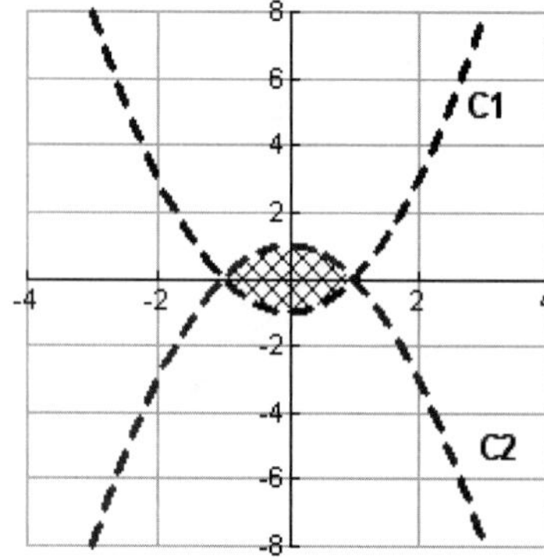

Notes on the graph:

C1: $y = x^2 - 1$ **C2:** $y = -x^2 + 1$

38.

First write the inequalities as:

$$y^2 < 1 - x$$

$$y^2 < 1 + x$$

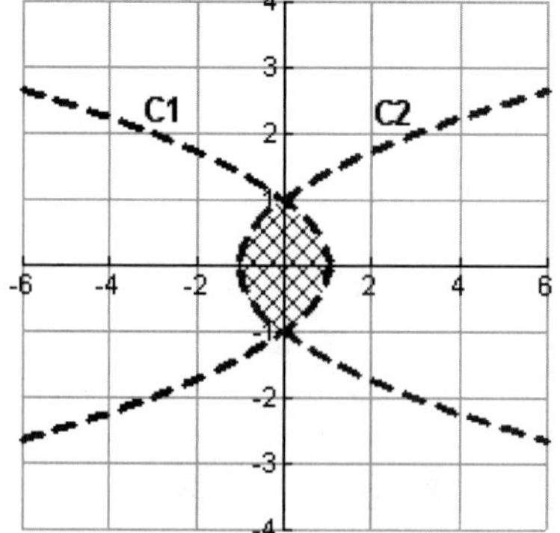

Notes on the graph:

C1: $x = -y^2 + 1$ **C2:** $x = y^2 - 1$

39.
First write the inequalities as:

$$y \geq x^2$$
$$x \geq y^2$$

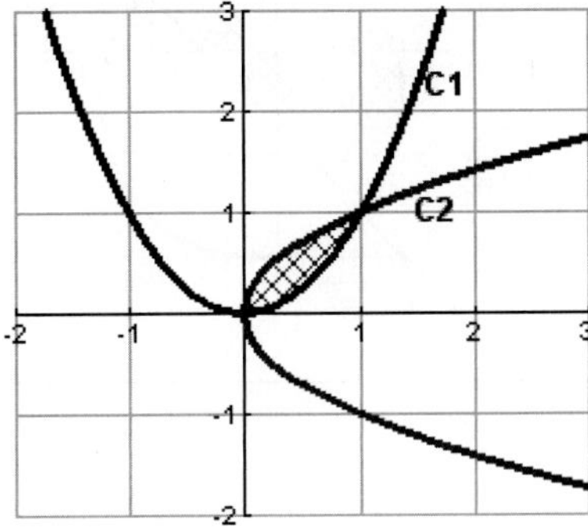

Notes on the graph:
C1: $y = x^2$ **C2:** $x = y^2$

40.

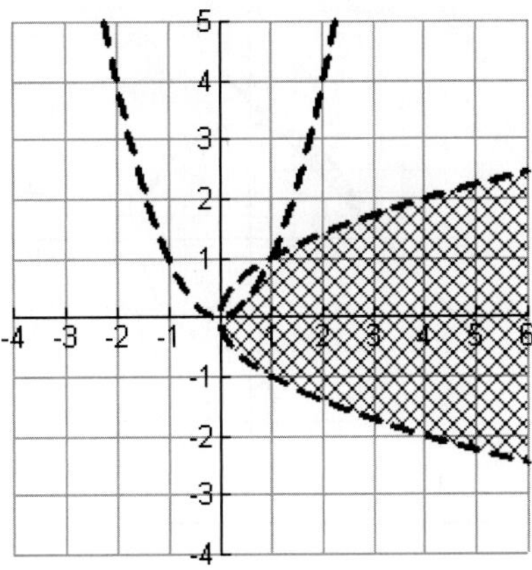

41.

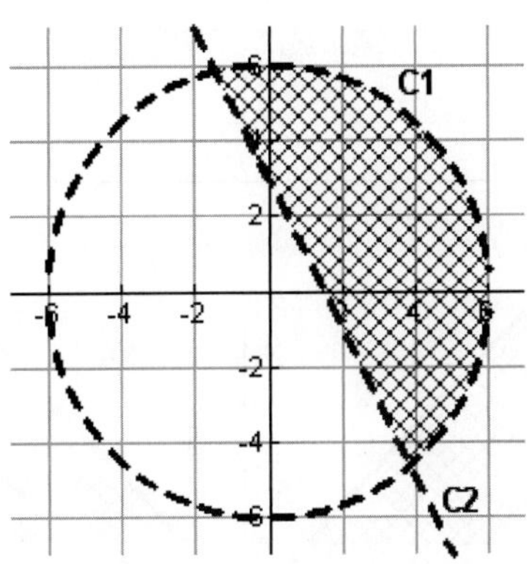

Notes on the graph:
C1: $x^2 + y^2 = 36$ **C2:** $2x + y = 3$

42.
Since the line $y = 6$ is tangent to the circle at (0,6), and neither boundary is included, there are no points (x, y) that satisfy both inequalities simultaneously. That is, there are no points strictly within the circle and strictly above the given line. Hence, the system has no solution.

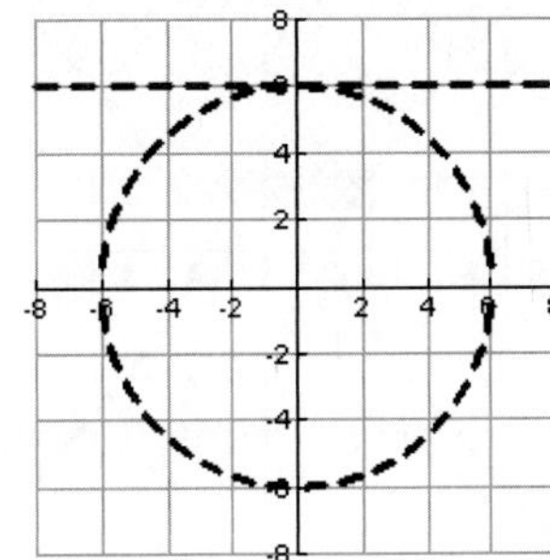

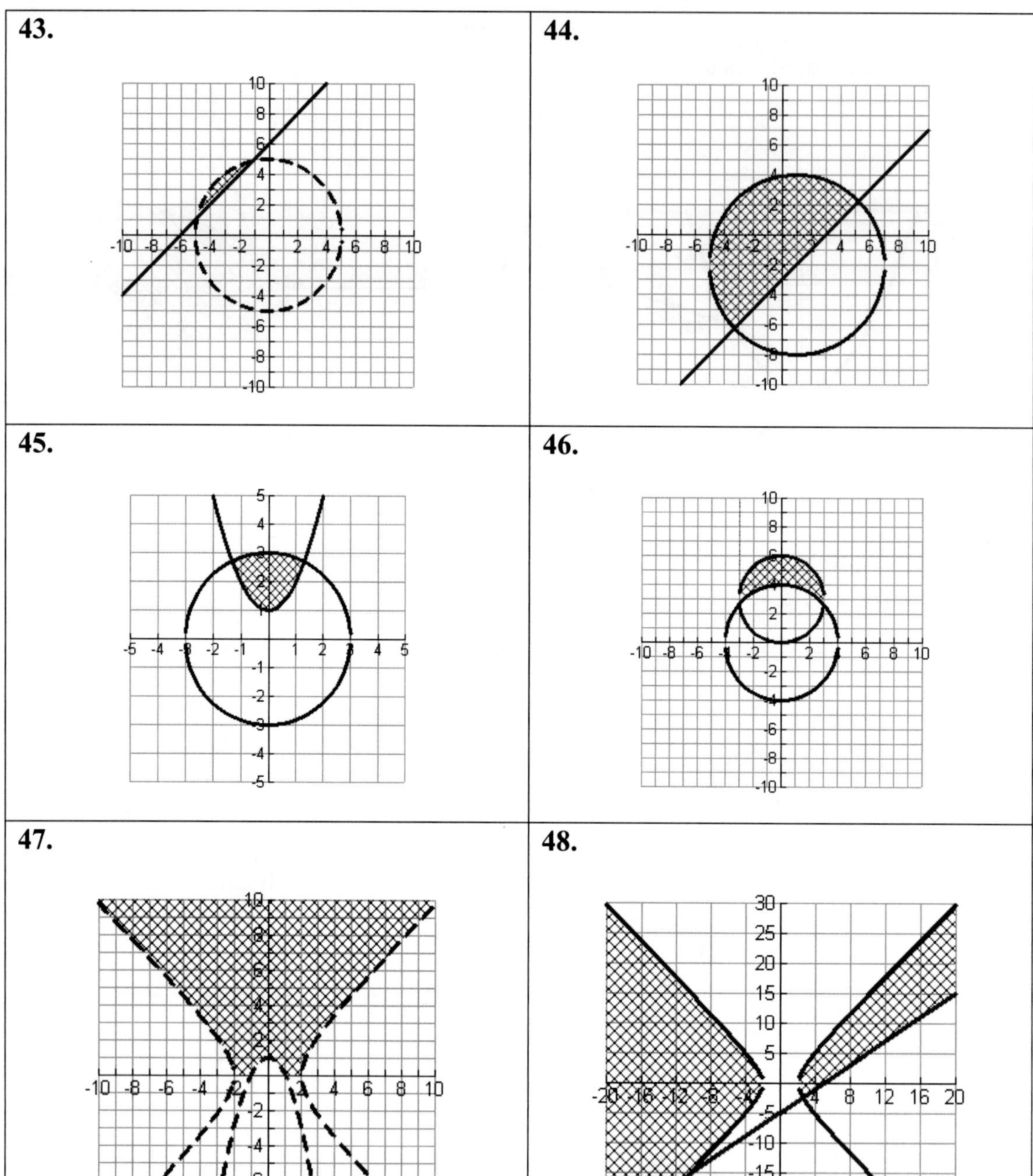
43.
44.
45.
46.
47.
48.

49.

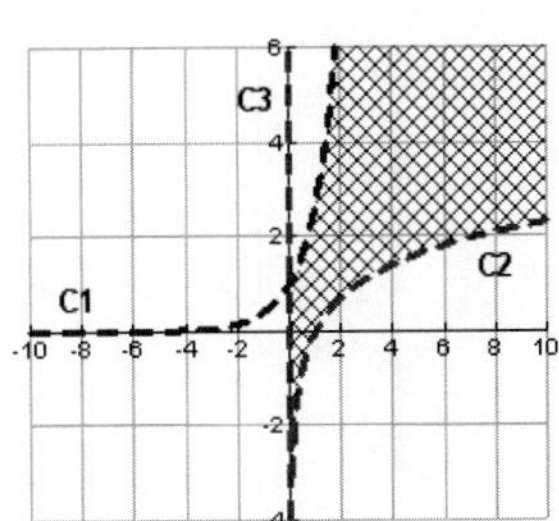

Notes on the graph:
C1: $y = \exp(x)$ **C2:** $y = \ln(x)$ **C3:** $x = 0$

50.

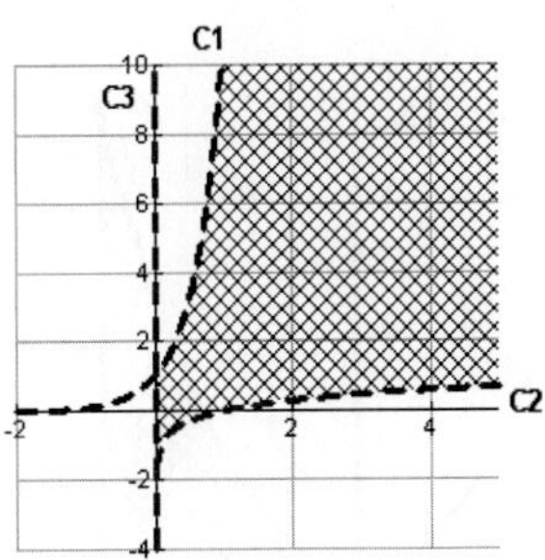

Notes on the graph:
C1: $y = 10^x$ **C2:** $y = \log(x)$ **C3:** $x = 0$

51. The shaded region is a semi-circle of radius 3, seen below. Hence, the area is:

$$\text{Area} = \tfrac{1}{2}\left(\pi(3)^2\right) = \tfrac{9}{2}\pi \text{ units}^2$$

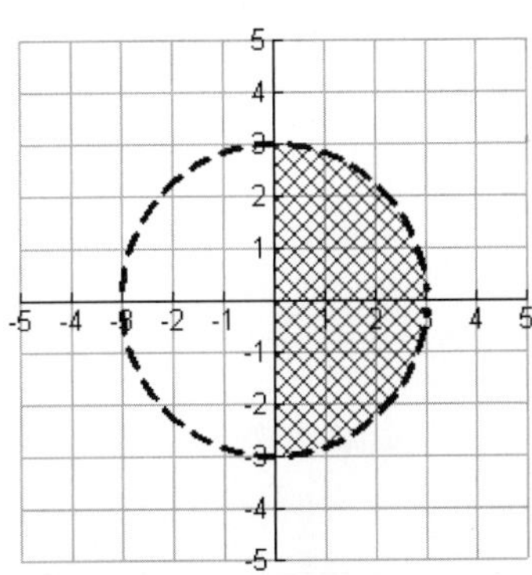

52. The shaded region is a quarter circle of radius $\sqrt{5}$ in the second quadrant, seen below. Hence, the area is:

$$\text{Area} = \tfrac{1}{4}\left(\pi\left(\sqrt{5}\right)^2\right) = \tfrac{5}{4}\pi \text{ units}^2$$

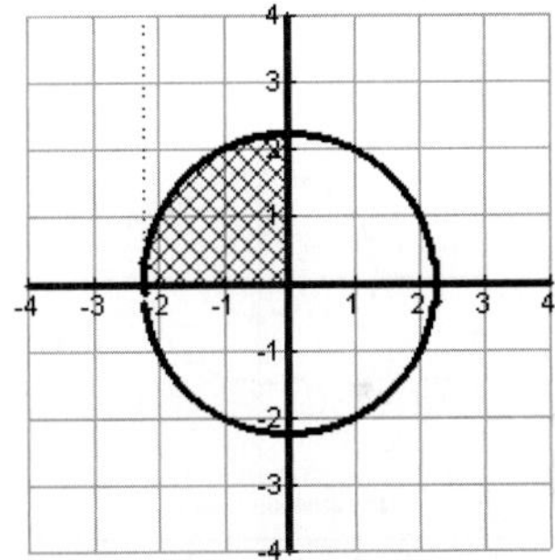

53. Consider the shaded region:

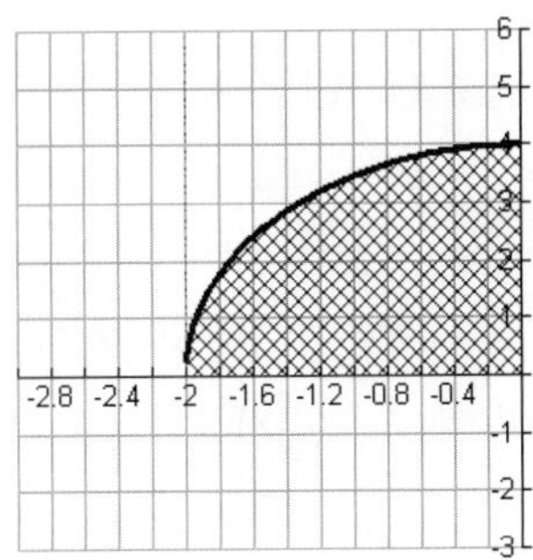

The area is $\frac{1}{4}\pi(2)(4) = 2\pi$ units^2.

54. Consider the shaded region:

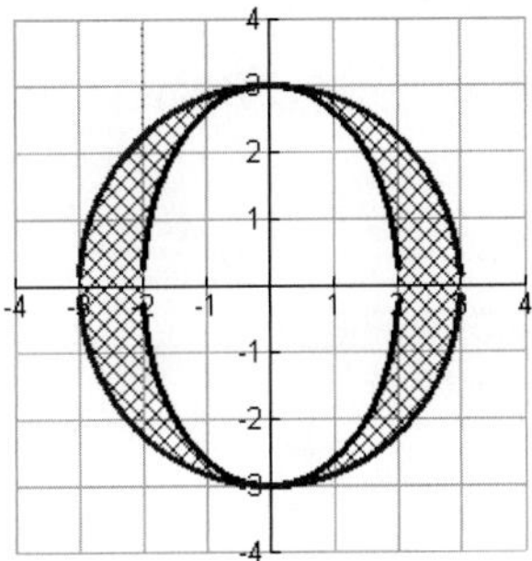

Area of shaded region =
Area of circle – area of ellipse =
$\pi(3)^2 - \pi(2)(3) = 3\pi$ units^2.

55. Consider the shaded region: 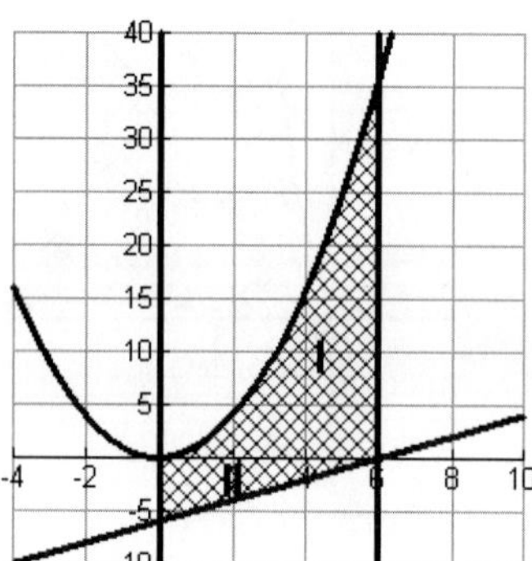 Area of shaded region = Area of I + area of II = $\frac{1}{3}(6)^3 + \frac{1}{2}(6)(6) = 90 \text{ units}^2$.	**56.** Consider the shaded region: 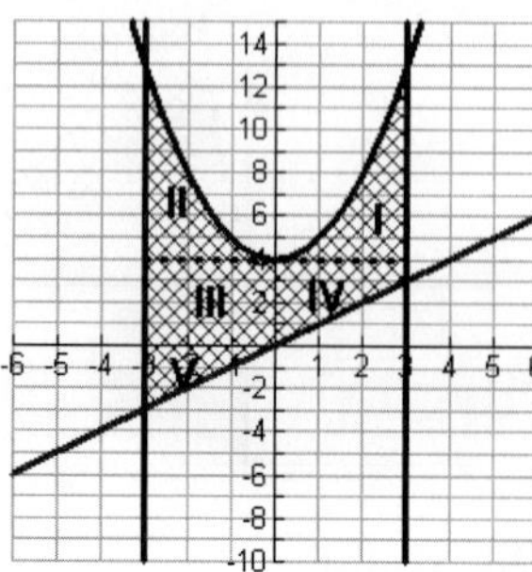 Area of I = Area of II = $\frac{1}{3}(3)^3$ Area of III = 3(4) = 12 Area of IV = Area of V = $\frac{1}{2}(4)(3) = 6$ The area of the shaded region is the sum of all of these, namely 42 units^2
57. There is no common region here – it is empty, as is seen in the graph below: 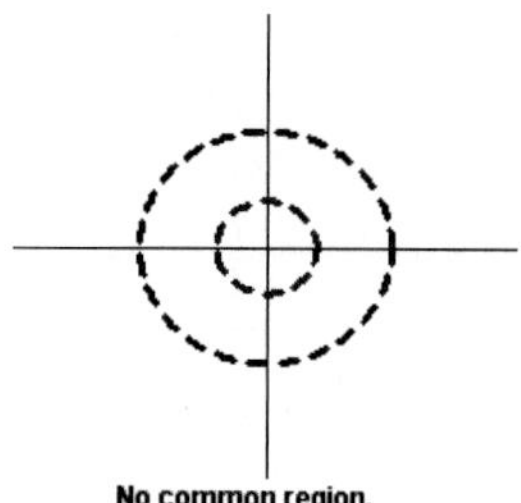 	**58.** The shading is wrong.
59. False Shade below the parabola, and include the parabola itself in the region; the inequality is $y \le x^2$	**60.** False. The following system has no solution, as is seen graphically below: $\begin{cases} y > x+1 \\ y < x-1 \end{cases}$ 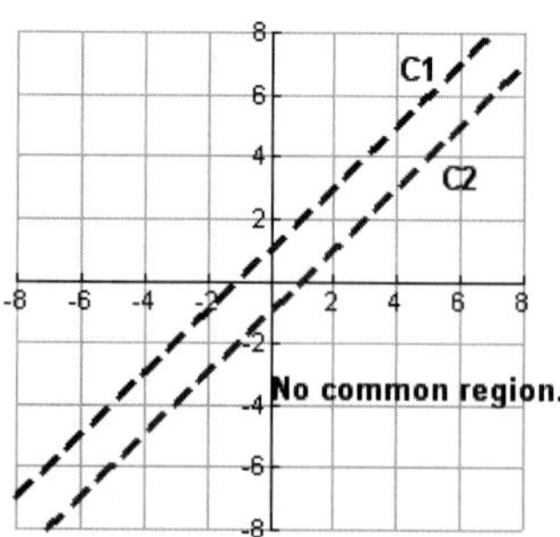 <u>Notes on the graph</u>: **C1:** $y = x+1$ **C2:** $y = x-1$

61. True. Since both conic sections are symmetric about the y-axis.	**62.** False. Consider, for instance, 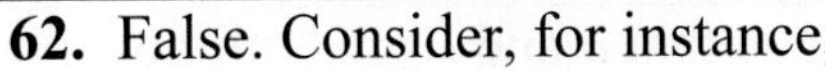 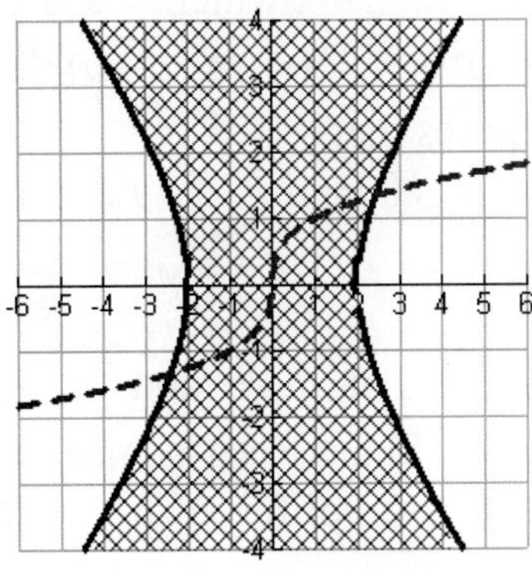
63. True. Consider the following diagram: 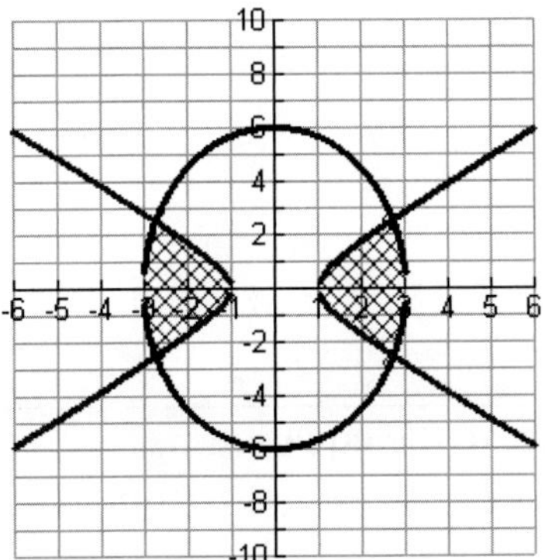	**64.** False. Consider the following diagram – the region is unbounded. 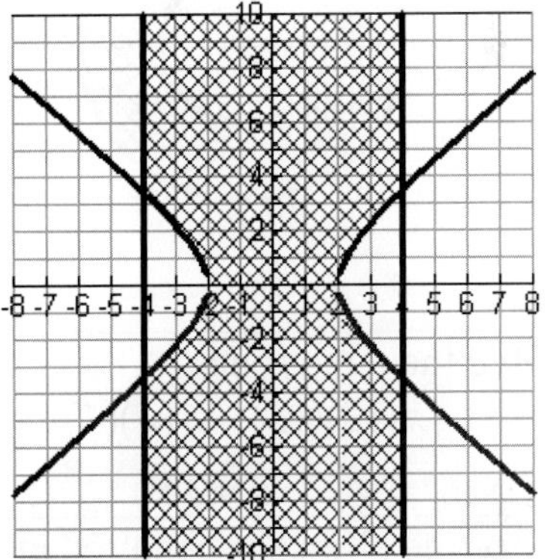
65. False. The region is rotated by 90°.	**66.** False. Consider the following diagram – the region is bounded. 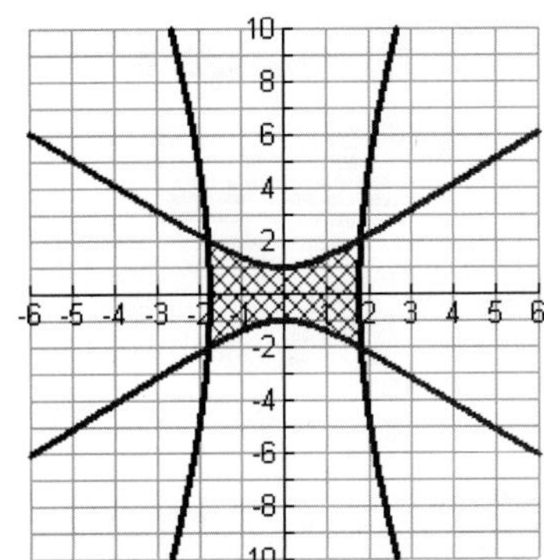

67. Assume that a and b are both positive. Must have $a^2 \le b^2$, which is equivalent to: $$a^2 - b^2 = (a-b)(a+b) \le 0$$ This requires that $$a - b \le 0 \text{ and } \underbrace{a+b \ge 0}_{\substack{\text{Automatically holds since} \\ a \text{ and } b \text{ are assumed to be} \\ \text{positive.}}}.$$ This implies that $a \le b$ and $a \ge -b$, and so $-b \le a \le b$. Since a is assumed to be positive, this simplifies to $0 \le a \le b$. 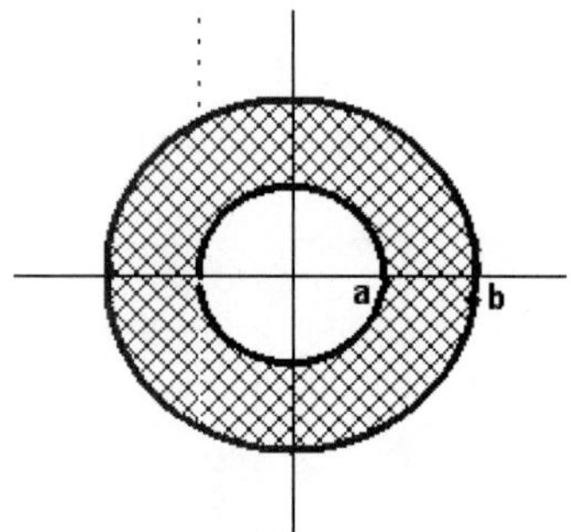	**68.** No, since $x^2 + y^2 \ge 0$, for all values of x and y, there can never be a point (x, y) that satisfies the equation if x and y are assumed to be real numbers. (If x and y can be imaginary, then there would be solutions.)
69. Solve the equation $$\pi(12)(12) = \pi(a)(4)$$ $$a = 36$$	**70.** Solve the equation $$\pi(1)(1) = \pi ab$$ $$ab = 1$$
71. Since the ellipse is centered at (0,0), the only way the region can be symmetric with respect to the y-axis is for $h = 0$.	**72.** Consider the diagram: 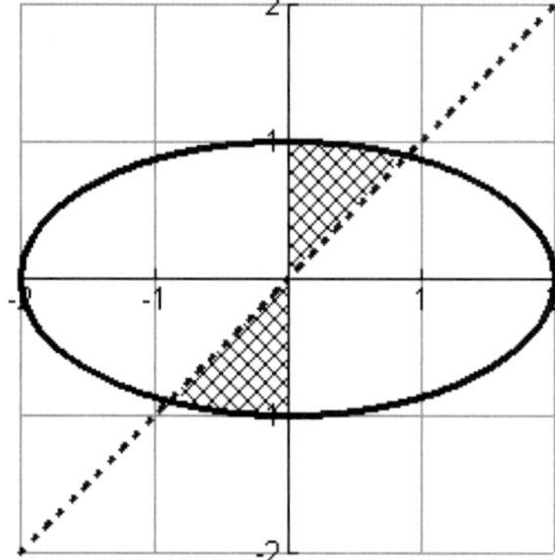 By symmetry, the points must be reflections of each other over the origin. So, $a = b$.

73. Begin by writing the inequality in standard form (so the radius and center are easily identified):

$$\left(x^2-2x+\quad\right)+\left(y^2+4y+\quad\right)\geq -4$$
$$\left(x^2-2x+1\right)+\left(y^2+4y+4\right)\geq -4+1+4$$
$$\left(x-1\right)^2+\left(y+2\right)^2\geq 1$$

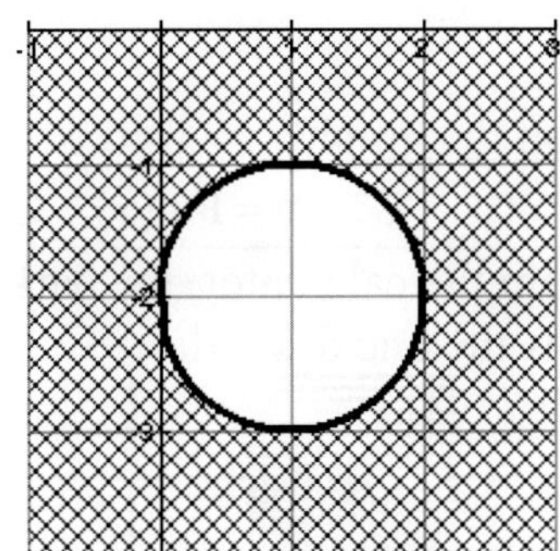

The solid curve is the graph of $\left(x-1\right)^2+\left(y+2\right)^2=1$.

74. Begin by writing the inequality in standard form (so the radius and center are easily identified):

$$\left(x^2+2x+\quad\right)+\left(y^2-2y+\quad\right)\leq 2$$
$$\left(x^2+2x+1\right)+\left(y^2-2y+1\right)\leq 2+1+1$$
$$\left(x+1\right)^2+\left(y-1\right)^2\leq 4$$

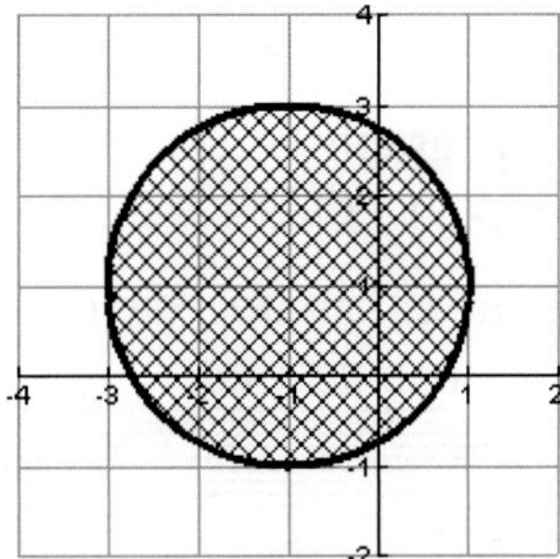

The solid curve is the graph of $\left(x+1\right)^2+\left(y-1\right)^2=4$.

75.

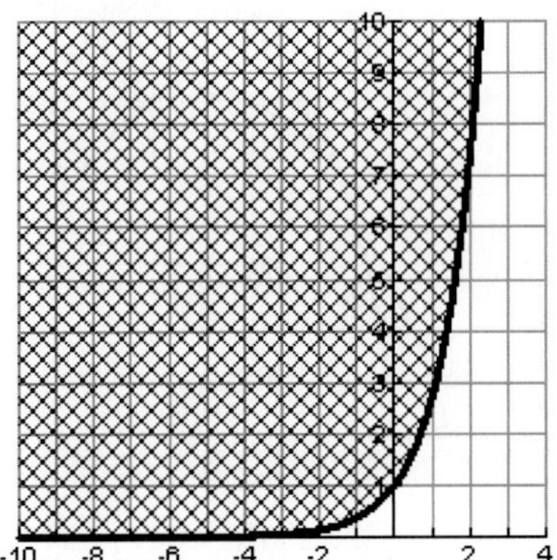

The solid curve is the graph of $y=e^x$.

76.

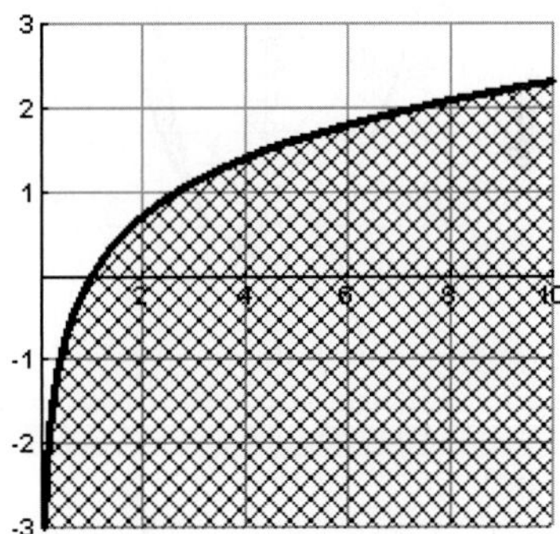

The solid curve is the graph of $y=\ln x$.

77.

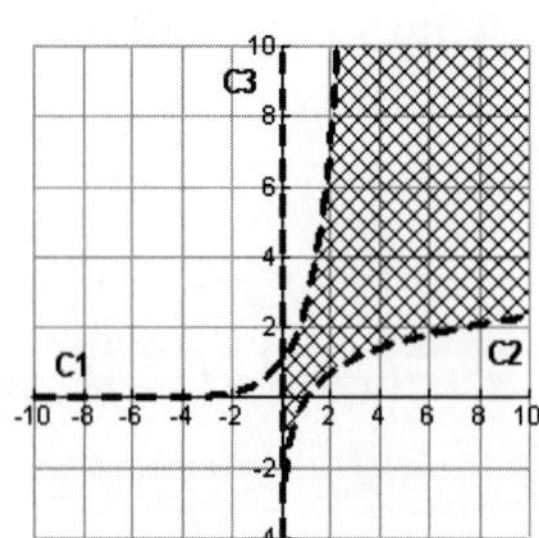

Notes on the graph:
C1: $y = \exp(x)$ **C2:** $y = \ln(x)$ **C3:** $x = 0$

78.

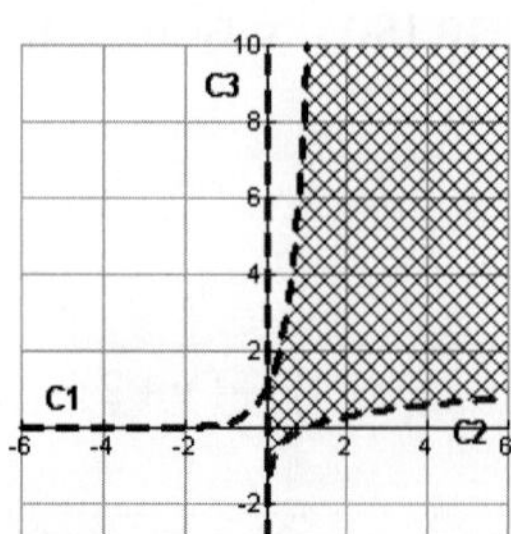

Notes on the graph:
C1: $y = 10^x$ **C2:** $y = \log(x)$ **C3:** $x = 0$

79. From the calculator, the dark shaded region is the one desired:

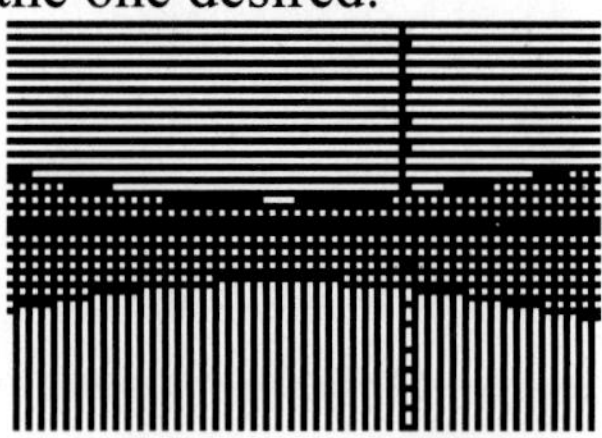

80. From the calculator, the dark shaded region is the one desired:

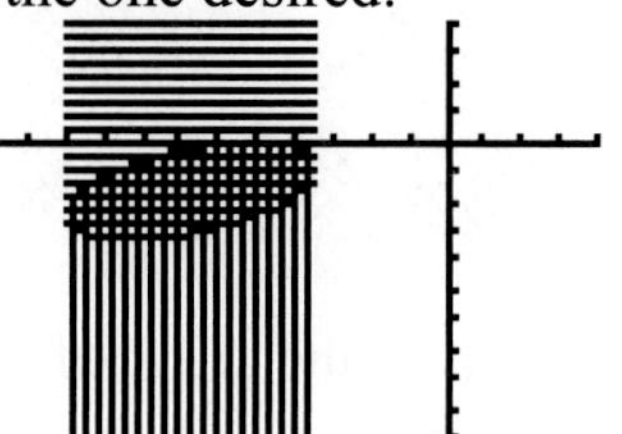

81.

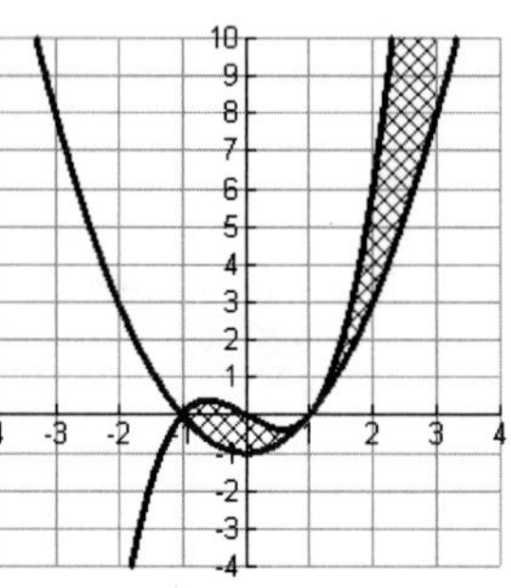

82.

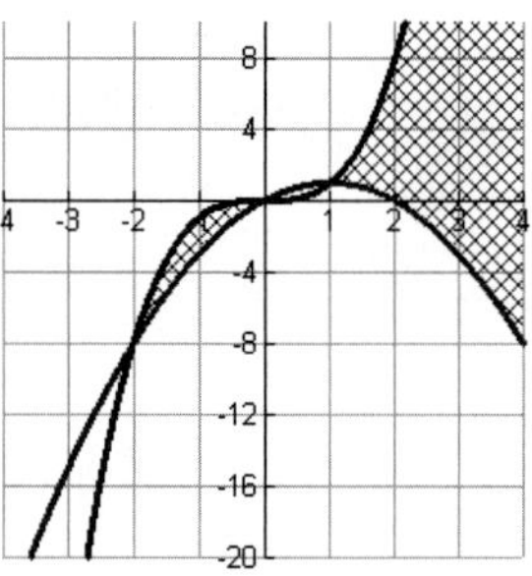

83.

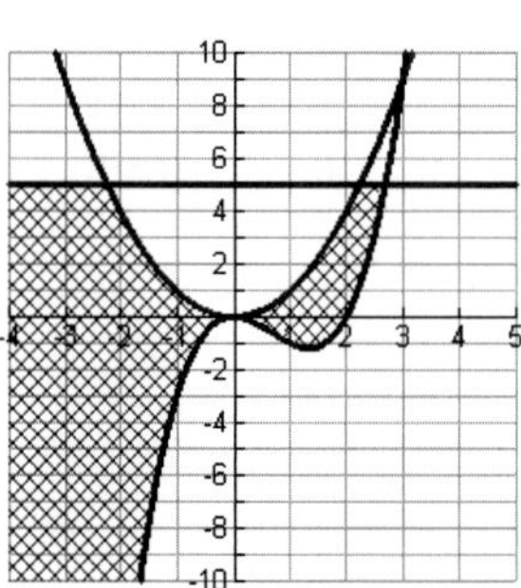

84.

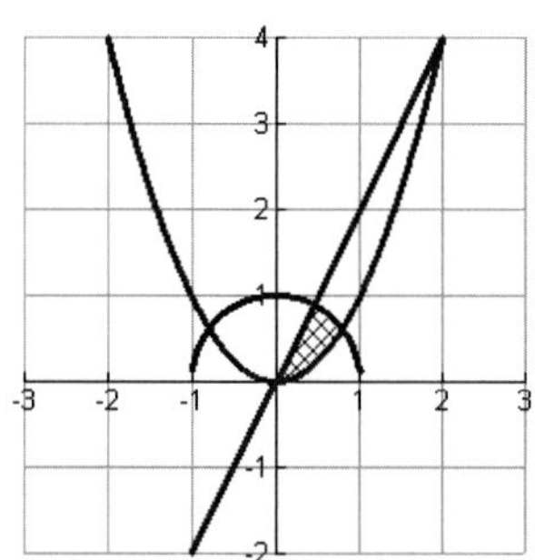

Section 9.7 Solutions --

1. The coordinates of the point obtained by rotating $(2,4)$ by an angle of 45° counterclockwise are:

$$X = x\cos\theta + y\sin\theta = 2\cos 45^\circ + 4\sin 45^\circ = 2\left(\frac{\sqrt{2}}{2}\right) + 4\left(\frac{\sqrt{2}}{2}\right) = 3\sqrt{2}$$

$$Y = -x\sin\theta + y\cos\theta = -2\sin 45^\circ + 4\cos 45^\circ = -2\left(\frac{\sqrt{2}}{2}\right) + 4\left(\frac{\sqrt{2}}{2}\right) = \sqrt{2}$$

So, the corresponding coordinates of this point in the *XY*-system are $\left(3\sqrt{2}, \sqrt{2}\right)$.

2. The coordinates of the point obtained by rotating $(5,1)$ by an angle of 60° counterclockwise are:

$$X = x\cos\theta + y\sin\theta = 5\cos 60^\circ + 1\sin 60^\circ = 5\left(\frac{1}{2}\right) + 1\left(\frac{\sqrt{3}}{2}\right) = \frac{5}{2} + \frac{\sqrt{3}}{2} = \frac{5+\sqrt{3}}{2}$$

$$Y = -x\sin\theta + y\cos\theta = -5\sin 60^\circ + 1\cos 60^\circ = -5\left(\frac{\sqrt{3}}{2}\right) + 1\left(\frac{1}{2}\right) = -\frac{5\sqrt{3}}{2} + \frac{1}{2} = \frac{1-5\sqrt{3}}{2}$$

So, the corresponding coordinates of this point in the *XY*-system are $\left(\frac{5+\sqrt{3}}{2}, \frac{1-5\sqrt{3}}{2}\right)$.

3. The coordinates of the point obtained by rotating $(-3,2)$ by an angle of 30° counterclockwise are:

$$X = x\cos\theta + y\sin\theta = -3\cos 30^\circ + 2\sin 30^\circ = -3\left(\frac{\sqrt{3}}{2}\right) + 2\left(\frac{1}{2}\right) = -\frac{3\sqrt{3}}{2} + 1$$

$$Y = -x\sin\theta + y\cos\theta = 3\sin 30^\circ + 2\cos 30^\circ = 3\left(\frac{1}{2}\right) + 2\left(\frac{\sqrt{3}}{2}\right) = \frac{3}{2} + \sqrt{3}$$

So, the corresponding coordinates of this point in the *XY*-system are $\left(-\frac{3\sqrt{3}}{2} + 1, \frac{3}{2} + \sqrt{3}\right)$.

4. The coordinates of the point obtained by rotating $(-4,6)$ by an angle of 45° counterclockwise are:

$$X = x\cos\theta + y\sin\theta \qquad Y = -x\sin\theta + y\cos\theta$$
$$= -4\cos 45^\circ + 6\sin 45^\circ \qquad = 4\sin 45^\circ + 6\cos 45^\circ$$
$$= -4\left(\frac{\sqrt{2}}{2}\right) + 6\left(\frac{\sqrt{2}}{2}\right) \qquad = 4\left(\frac{\sqrt{2}}{2}\right) + 6\left(\frac{\sqrt{2}}{2}\right)$$
$$= \sqrt{2} \qquad = 5\sqrt{2}$$

So, the corresponding coordinates of this point in the *XY*-system are $\left(\sqrt{2}, 5\sqrt{2}\right)$.

5. The coordinates of the point obtained by rotating $(-1,-3)$ by an angle of 60° counterclockwise are:

$$X = x\cos\theta + y\sin\theta \qquad Y = -x\sin\theta + y\cos\theta$$
$$= -1\cos 60^\circ - 3\sin 60^\circ \qquad = 1\sin 60^\circ - 3\cos 60^\circ$$
$$= -\left(\frac{1}{2}\right) - 3\left(\frac{\sqrt{3}}{2}\right) \qquad = 1\left(\frac{\sqrt{3}}{2}\right) - 3\left(\frac{1}{2}\right)$$
$$= -\frac{1}{2} - \frac{3\sqrt{3}}{2} = -\frac{1+3\sqrt{3}}{2} \qquad = \frac{\sqrt{3}}{2} - \frac{3}{2} = \frac{\sqrt{3}-3}{2}$$

So, the corresponding coordinates of this point in the *XY*-system are $\left(-\frac{1+3\sqrt{3}}{2}, \frac{\sqrt{3}-3}{2}\right)$.

6. The coordinates of the point obtained by rotating $(4,-4)$ by an angle of 45° counterclockwise are:

$$X = x\cos\theta + y\sin\theta \qquad Y = -x\sin\theta + y\cos\theta$$
$$= 4\cos 45^\circ - 4\sin 45^\circ \qquad = -4\sin 45^\circ - 4\cos 45^\circ$$
$$= 4\left(\frac{\sqrt{2}}{2}\right) - 4\left(\frac{\sqrt{2}}{2}\right) \qquad = -4\left(\frac{\sqrt{2}}{2}\right) - 4\left(\frac{\sqrt{2}}{2}\right)$$
$$= 0 \qquad = -4\sqrt{2}$$

So, the corresponding coordinates of this point in the *XY*-system are $\left(0, -4\sqrt{2}\right)$.

7. The coordinates of the point obtained by rotating $(0,3)$ by an angle of 60° counterclockwise are:

$$X = x\cos\theta + y\sin\theta = 0\cos 60^\circ + 3\sin 60^\circ = 0\left(\frac{1}{2}\right) + 3\left(\frac{\sqrt{3}}{2}\right) = \frac{3\sqrt{3}}{2}$$

$$Y = -x\sin\theta + y\cos\theta = 0\sin 60^\circ + 3\cos 60^\circ = 0\left(\frac{\sqrt{3}}{2}\right) + 3\left(\frac{1}{2}\right) = \frac{3}{2}$$

So, the corresponding coordinates of this point in the *XY*-system are $\left(\frac{3\sqrt{3}}{2}, \frac{3}{2}\right)$.

8. The coordinates of the point obtained by rotating $(-2,0)$ by an angle of 30° counterclockwise are:

$$X = x\cos\theta + y\sin\theta = -2\cos 30^\circ + 0\sin 30^\circ = -2\left(\frac{\sqrt{3}}{2}\right) + 0\left(\frac{1}{2}\right) = -\sqrt{3}$$

$$Y = -x\sin\theta + y\cos\theta = 2\sin 30^\circ + 0\cos 30^\circ = 2\left(\frac{1}{2}\right) + 0\left(\frac{\sqrt{3}}{2}\right) = 1$$

So, the corresponding coordinates of this point in the *XY*-system are $\left(-\sqrt{3}, 1\right)$.

9. Consider the equation $xy-1=0$.

(a) Here, $A=0,\ B=1,\ C=0$, so that $B^2-4AC=1>0$. Thus, the curve is a *hyperbola.*

(b) In order to obtain the equation of this curve in the *XY*-system, we make the following substitution into the original equation with $\theta=45^\circ$:

$$\begin{cases} x = X\cos\theta - Y\sin\theta = (X-Y)\left(\dfrac{\sqrt{2}}{2}\right) \\ y = X\sin\theta + Y\cos\theta = (X+Y)\left(\dfrac{\sqrt{2}}{2}\right) \end{cases}$$

Doing so yields the following, after collecting like terms and simplifying:

$$(X-Y)\left(\frac{\sqrt{2}}{2}\right)(X+Y)\left(\frac{\sqrt{2}}{2}\right)-1=0$$

$$(X^2-Y^2)\underbrace{\left(\frac{\sqrt{2}}{2}\right)^2}_{=\frac{1}{2}}-1=0$$

$$(X^2-Y^2)\left(\frac{1}{2}\right)-1=0$$

$$\frac{X^2}{2}-\frac{Y^2}{2}=1$$

(c) The graph of $xy-1=0$, along with the new axes $y=(\tan 45^\circ)x=x,\ y=-x$ is as follows:

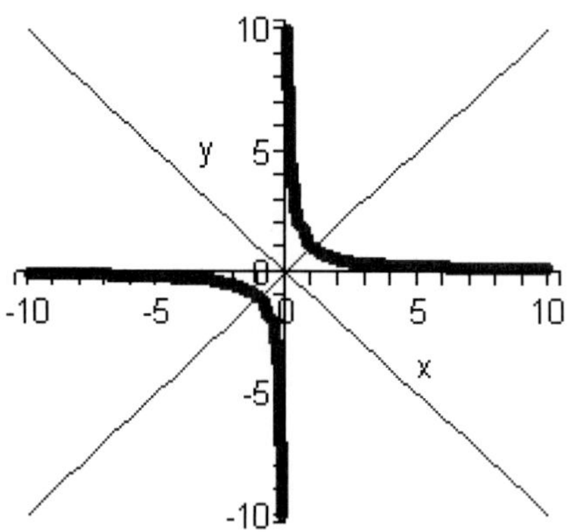

10. Consider the equation $xy - 4 = 0$.

(a) Here, $A = 0,\ B = 1,\ C = 0$, so that $B^2 - 4AC = 1 > 0$. Thus, the curve is a *hyperbola.*

(b) In order to obtain the equation of this curve in the *XY*-system, we make the following substitution into the original equation with $\theta = 45^\circ$:

$$\begin{cases} x = X\cos\theta - Y\sin\theta = (X - Y)\left(\dfrac{\sqrt{2}}{2}\right) \\ y = X\sin\theta + Y\cos\theta = (X + Y)\left(\dfrac{\sqrt{2}}{2}\right) \end{cases}$$

Doing so yields the following, after collecting like terms and simplifying:

$$(X - Y)\left(\frac{\sqrt{2}}{2}\right)(X + Y)\left(\frac{\sqrt{2}}{2}\right) - 4 = 0$$

$$(X^2 - Y^2)\underbrace{\left(\frac{\sqrt{2}}{2}\right)^2}_{=\frac{1}{2}} - 4 = 0$$

$$(X^2 - Y^2)\left(\frac{1}{2}\right) - 4 = 0$$

$$\frac{X^2}{8} - \frac{Y^2}{8} = 1$$

(c) The graph of $xy - 4 = 0$, along with the new axes $y = (\tan 45^\circ)x = x,\ y = -x$ is as follows:

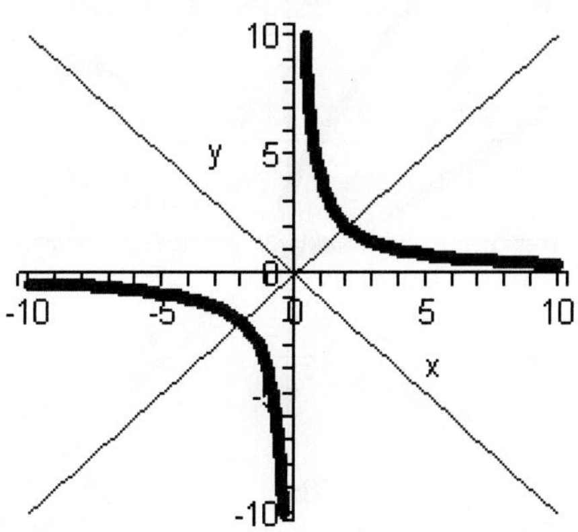

11. Consider the equation $x^2+2xy+y^2+\sqrt{2}x-\sqrt{2}y-1=0$.

(a) Here, $A=1,\ B=2,\ C=1$, so that $B^2-4AC=0$. Thus, the curve is a *parabola.*

(b) In order to obtain the equation of this curve in the *XY*-system, we make the following substitution into the original equation with $\theta=45^\circ$:

$$\begin{cases} x=X\cos\theta-Y\sin\theta=(X-Y)\left(\frac{\sqrt{2}}{2}\right) \\ y=X\sin\theta+Y\cos\theta=(X+Y)\left(\frac{\sqrt{2}}{2}\right) \end{cases}$$

Doing so yields the following, after collecting like terms and simplifying:

$$(X-Y)^2\left(\frac{\sqrt{2}}{2}\right)^2+2(X-Y)(X+Y)\left(\frac{\sqrt{2}}{2}\right)^2+(X+Y)^2\left(\frac{\sqrt{2}}{2}\right)^2$$

$$+\sqrt{2}(X-Y)\left(\frac{\sqrt{2}}{2}\right)-\sqrt{2}(X+Y)\left(\frac{\sqrt{2}}{2}\right)-1=0$$

$$\frac{1}{2}\left(X^2-2XY+Y^2+2X^2-2Y^2+X^2+2XY+Y^2\right)+X-Y-(X+Y)-1=0$$

$$\frac{1}{2}\left(4X^2\right)-2Y-1=0$$

$$2X^2-2Y-1=0$$

(c) The graph of $x^2+2xy+y^2+\sqrt{2}x-\sqrt{2}y-1=0$, along with the new axes $y=\left(\tan 45^\circ\right)x=x,\ y=-x$ is as follows:

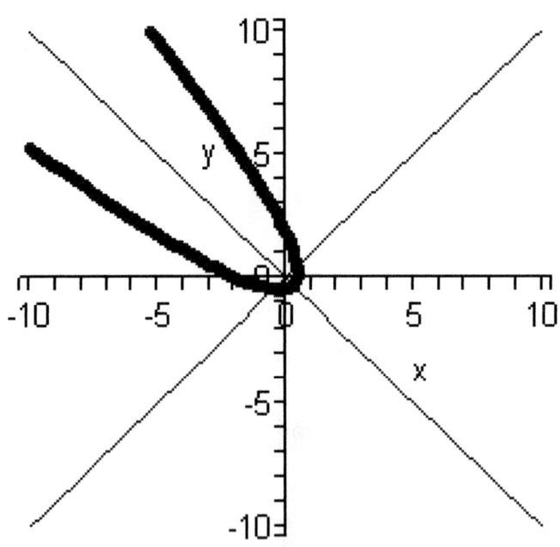

12. Consider the equation $2x^2-4xy+2y^2-\sqrt{2}x+1=0$.

(a) Here, $A=2,\ B=-4,\ C=2$, so that $B^2-4AC=0$. Thus, the curve is a *parabola.*

(b) In order to obtain the equation of this curve in the *XY*-system, we make the following substitution into the original equation with $\theta=45^\circ$:

$$\begin{cases} x = X\cos\theta - Y\sin\theta = (X-Y)\left(\frac{\sqrt{2}}{2}\right) \\ y = X\sin\theta + Y\cos\theta = (X+Y)\left(\frac{\sqrt{2}}{2}\right) \end{cases}$$

Doing so yields the following, after collecting like terms and simplifying:

$$2(X-Y)^2\left(\frac{\sqrt{2}}{2}\right)^2-4(X-Y)(X+Y)\left(\frac{\sqrt{2}}{2}\right)^2+2(X+Y)^2\left(\frac{\sqrt{2}}{2}\right)^2-\sqrt{2}(X-Y)\left(\frac{\sqrt{2}}{2}\right)+1=0$$

$$\left(X^2-2XY+Y^2-2X^2+2Y^2+X^2+2XY+Y^2\right)-X+Y+1=0$$

$$4Y^2+Y-X+1=0$$

(c) The graph of $2x^2-4xy+2y^2-\sqrt{2}x+1=0$, along with the new axes $y=\left(\tan 45^\circ\right)x=x,\ y=-x$ is as follows:

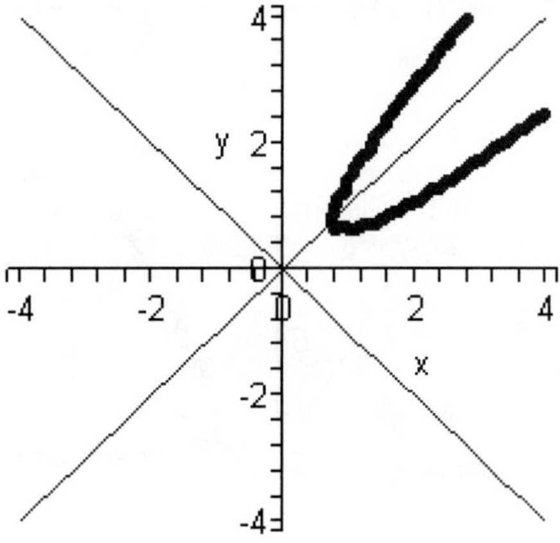

13. Consider the equation $y^2 - \sqrt{3}xy + 3 = 0$.

(a) Here, $A = 1,\ B = -\sqrt{3},\ C = 0$, so that $B^2 - 4AC = 3 > 0$. Thus, the curve is a *hyperbola.*

(b) In order to obtain the equation of this curve in the *XY*-system, we make the following substitution into the original equation with $\theta = 30^\circ$:

$$\begin{cases} x = X\cos\theta - Y\sin\theta = \left(\sqrt{3}X - Y\right)\left(\dfrac{1}{2}\right) \\ y = X\sin\theta + Y\cos\theta = \left(X + \sqrt{3}Y\right)\left(\dfrac{1}{2}\right) \end{cases}$$

Doing so yields the following, after collecting like terms and simplifying:

$$\left(X+\sqrt{3}Y\right)^2\left(\frac{1}{2}\right)^2 - \sqrt{3}\left(X+\sqrt{3}Y\right)\left(\sqrt{3}X - Y\right)\left(\frac{1}{2}\right)^2 + 3 = 0$$

$$\frac{1}{4}\left(X^2 + 2\sqrt{3}XY + 3Y^2\right) - \frac{\sqrt{3}}{4}\left(\sqrt{3}X^2 - XY + 3XY - \sqrt{3}Y^2\right) + 3 = 0$$

$$-2X^2 + 6Y^2 + 12 = 0$$

$$\frac{X^2}{6} - \frac{Y^2}{2} = 1$$

(c) The graph of $y^2 - \sqrt{3}xy + 3 = 0$, along with the new axes $y = \left(\tan 30^\circ\right)x = \left(\dfrac{1}{\sqrt{3}}\right)x$, $y = -\sqrt{3}x$ is as follows:

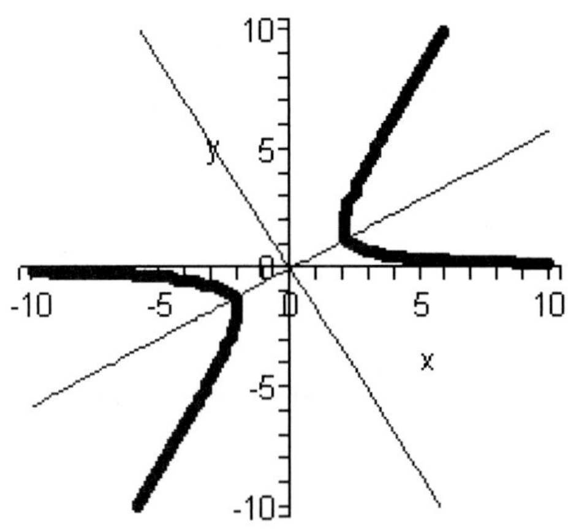

14. Consider the equation $x^2 - \sqrt{3}xy - 3 = 0$.

(a) Here, $A = 1,\ B = -\sqrt{3},\ C = 0$, so that $B^2 - 4AC = 3 > 0$. Thus, the curve is a *hyperbola*.

(b) In order to obtain the equation of this curve in the *XY*-system, we make the following substitution into the original equation with $\theta = 60^\circ$:

$$\begin{cases} x = X\cos\theta - Y\sin\theta = \left(X - \sqrt{3}Y\right)\left(\frac{1}{2}\right) \\ y = X\sin\theta + Y\cos\theta = \left(\sqrt{3}X + Y\right)\left(\frac{1}{2}\right) \end{cases}$$

Doing so yields the following, after collecting like terms and simplifying:

$$\left(X - \sqrt{3}Y\right)^2\left(\frac{1}{2}\right)^2 - \sqrt{3}\left(X - \sqrt{3}Y\right)\left(\sqrt{3}X + Y\right)\left(\frac{1}{2}\right)^2 - 3 = 0$$

$$\left(X^2 - 2\sqrt{3}XY + 3Y^2\right) - \sqrt{3}\left(\sqrt{3}X^2 - 3XY + XY - \sqrt{3}Y^2\right) - 12 = 0$$

$$-2X^2 + 6Y^2 - 12 = 0$$

$$\frac{Y^2}{2} - \frac{X^2}{6} = 1$$

(c) The graph of $x^2 - \sqrt{3}xy - 3 = 0$, along with the new axes $y = \left(\tan 60^\circ\right)x = \sqrt{3}x$, $y = -\frac{1}{\sqrt{3}}x$ is as follows:

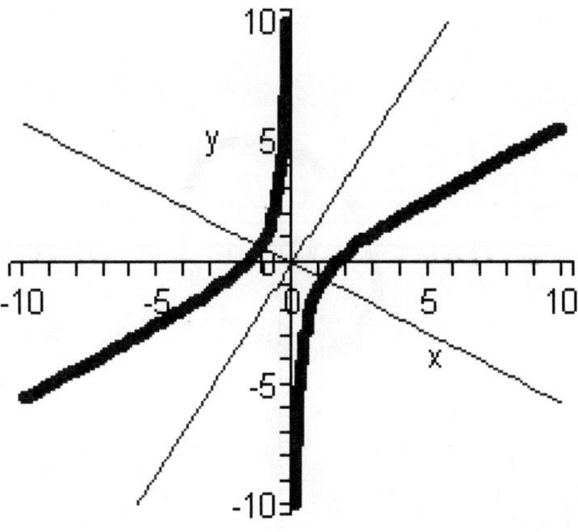

15. Consider the equation $7x^2 - 2\sqrt{3}xy + 5y^2 - 8 = 0$.

(a) Here, $A = 7,\ B = -2\sqrt{3},\ C = 5$, so that $B^2 - 4AC = -128 < 0$. Thus, the curve is an *ellipse.*

(b) In order to obtain the equation of this curve in the *XY*-system, we make the following substitution into the original equation with $\theta = 60^\circ$:

$$\begin{cases} x = X\cos\theta - Y\sin\theta = \left(X - \sqrt{3}Y\right)\left(\frac{1}{2}\right) \\ y = X\sin\theta + Y\cos\theta = \left(\sqrt{3}X + Y\right)\left(\frac{1}{2}\right) \end{cases}$$

Doing so yields the following, after collecting like terms and simplifying:

$$\begin{aligned} 7\left(X - \sqrt{3}Y\right)^2\left(\frac{1}{2}\right)^2 - 2\sqrt{3}\left(X - \sqrt{3}Y\right)\left(\sqrt{3}X + Y\right)\left(\frac{1}{2}\right)^2 + 5\left(\sqrt{3}X + Y\right)^2\left(\frac{1}{2}\right)^2 - 8 &= 0 \\ 7\left(X^2 - 2\sqrt{3}XY + 3Y^2\right) - 2\sqrt{3}\left(\sqrt{3}X^2 - 3XY + XY - \sqrt{3}Y^2\right) + 5\left(3X^2 + 2\sqrt{3}XY + Y^2\right) - 32 &= 0 \\ 16X^2 + 32Y^2 - 32 &= 0 \\ \frac{X^2}{2} + \frac{Y^2}{1} &= 1 \end{aligned}$$

(c) The graph of $7x^2 - 2\sqrt{3}xy + 5y^2 - 8 = 0$, along with the new axes $y = \left(\tan 60^\circ\right)x = \sqrt{3}x$, $y = -\frac{1}{\sqrt{3}}x$ is as follows:

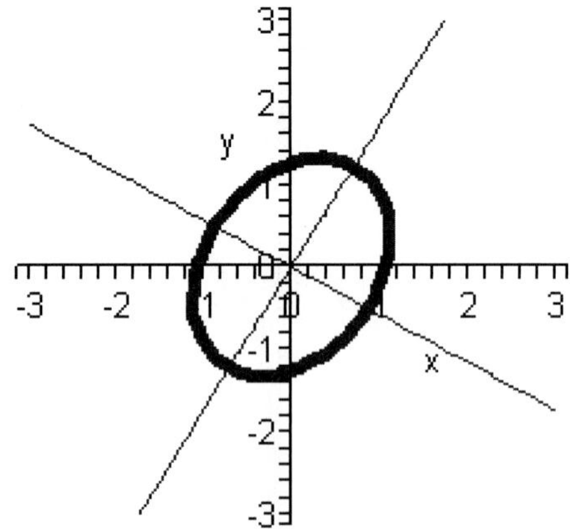

16. Consider the equation $4x^2+\sqrt{3}xy+3y^2-45=0$.

(a) Here, $A=4,\ B=\sqrt{3},\ C=3$, so that $B^2-4AC=-45<0$. Thus, the curve is an *ellipse*.

(b) In order to obtain the equation of this curve in the *XY*-system, we make the following substitution into the original equation with $\theta=30^\circ$:

$$\begin{cases} x = X\cos\theta - Y\sin\theta = \left(\sqrt{3}X-Y\right)\left(\frac{1}{2}\right) \\ y = X\sin\theta + Y\cos\theta = \left(X+\sqrt{3}Y\right)\left(\frac{1}{2}\right) \end{cases}$$

Doing so yields the following, after collecting like terms and simplifying:

$$4\left(\sqrt{3}X-Y\right)^2\left(\frac{1}{2}\right)^2+\sqrt{3}\left(X+\sqrt{3}Y\right)\left(\sqrt{3}X-Y\right)\left(\frac{1}{2}\right)^2+3\left(X+\sqrt{3}Y\right)^2\left(\frac{1}{2}\right)^2-45=0$$

$$4\left(3X^2-2\sqrt{3}XY+Y^2\right)+\sqrt{3}\left(\sqrt{3}X^2+3XY-XY-\sqrt{3}Y^2\right)+3\left(X^2+2\sqrt{3}XY+3Y^2\right)-180=0$$

$$18X^2+10Y^2-180=0$$

$$\frac{X^2}{10}+\frac{Y^2}{18}=1$$

(c) The graph of $4x^2+\sqrt{3}xy+3y^2-45=0$, along with the new axes $y=\left(\tan 30^\circ\right)x=\left(\frac{1}{\sqrt{3}}\right)x,\ y=-\sqrt{3}x$ is as follows:

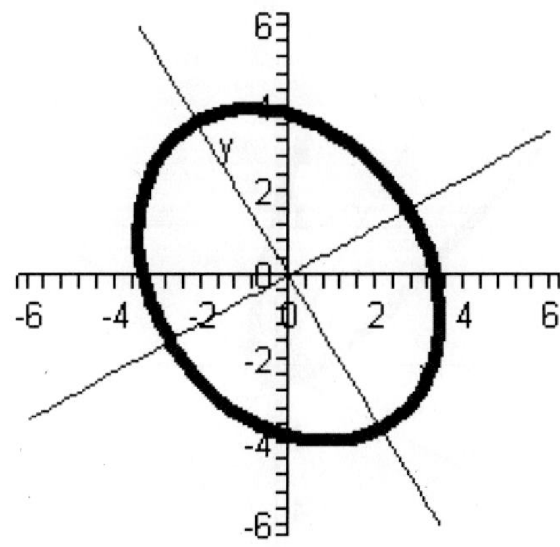

17. Consider the equation $3x^2+2\sqrt{3}xy+y^2+2x-2\sqrt{3}y-2=0$.

(a) Here, $A=3,\ B=2\sqrt{3},\ C=1$, so that $B^2-4AC=0$. Thus, the curve is a *parabola.*

(b) In order to obtain the equation of this curve in the *XY*-system, we make the following substitution into the original equation with $\theta=30^\circ$:

$$\begin{cases} x = X\cos\theta - Y\sin\theta = \left(\sqrt{3}X-Y\right)\left(\frac{1}{2}\right) \\ y = X\sin\theta + Y\cos\theta = \left(X+\sqrt{3}Y\right)\left(\frac{1}{2}\right) \end{cases}$$

Doing so yields the following, after collecting like terms and simplifying:

$$3\left(\sqrt{3}X-Y\right)^2\left(\frac{1}{2}\right)^2+2\sqrt{3}\left(X+\sqrt{3}Y\right)\left(\sqrt{3}X-Y\right)\left(\frac{1}{2}\right)^2+\left(X+\sqrt{3}Y\right)^2\left(\frac{1}{2}\right)^2+$$
$$2\left(\frac{1}{2}\right)\left(\sqrt{3}X-Y\right)-2\sqrt{3}\left(\frac{1}{2}\right)\left(X+\sqrt{3}Y\right)-2=0$$
$$3\left(3X^2-2\sqrt{3}XY+Y^2\right)+2\sqrt{3}\left(\sqrt{3}X^2+3XY-XY-\sqrt{3}Y^2\right)+\left(X^2+2\sqrt{3}XY+3Y^2\right)$$
$$4\left(\sqrt{3}X-Y\right)-4\sqrt{3}\left(X+\sqrt{3}Y\right)-8=0$$
$$16X^2-16Y-8=0$$
$$2X^2-2Y-1=0$$

(c) The graph of $3x^2+2\sqrt{3}xy+y^2+2x-2\sqrt{3}y-2=0$, along with the new axes $y=\left(\tan 30^\circ\right)x=\left(\frac{1}{\sqrt{3}}\right)x,\ y=-\sqrt{3}x$ is as follows:

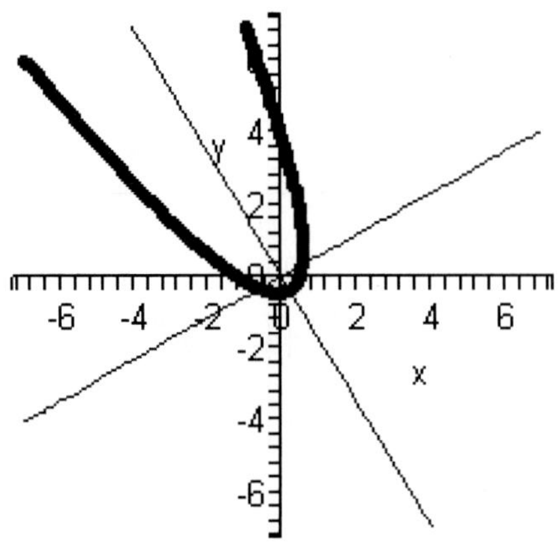

18. Consider the equation $x^2 + 2\sqrt{3}xy + 3y^2 - 2\sqrt{3}x + 2y - 4 = 0$.

(a) Here, $A = 1,\ B = 2\sqrt{3},\ C = 3$, so that $B^2 - 4AC = 0$. Thus, the curve is a *parabola.*

(b) In order to obtain the equation of this curve in the *XY*-system, we make the following substitution into the original equation with $\theta = 60^\circ$:

$$\begin{cases} x = X\cos\theta - Y\sin\theta = \left(X - \sqrt{3}Y\right)\left(\frac{1}{2}\right) \\ y = X\sin\theta + Y\cos\theta = \left(\sqrt{3}X + Y\right)\left(\frac{1}{2}\right) \end{cases}$$

Doing so yields the following, after collecting like terms and simplifying:

$$\left(X - \sqrt{3}Y\right)^2\left(\frac{1}{2}\right)^2 + 2\sqrt{3}\left(X - \sqrt{3}Y\right)\left(\sqrt{3}X + Y\right)\left(\frac{1}{2}\right)^2 + 3\left(\sqrt{3}X + Y\right)^2\left(\frac{1}{2}\right)^2$$
$$- 2\sqrt{3}\left(\frac{1}{2}\right)\left(X - \sqrt{3}Y\right) + 2\left(\frac{1}{2}\right)\left(\sqrt{3}X + Y\right) - 4 = 0$$
$$\left(X^2 - 2\sqrt{3}XY + 3Y^2\right) + 2\sqrt{3}\left(\sqrt{3}X^2 - 3XY + XY - \sqrt{3}Y^2\right) + 3\left(3X^2 + 2\sqrt{3}XY + Y^2\right)$$
$$- 4\sqrt{3}\left(X - \sqrt{3}Y\right) + 4\left(\sqrt{3}X + Y\right) - 16 = 0$$
$$16X^2 + 16Y - 16 = 0$$
$$X^2 + Y - 1 = 0$$

(c) The graph of $x^2 + 2\sqrt{3}xy + 3y^2 - 2\sqrt{3}x + 2y - 4 = 0$, along with the new axes $y = \left(\tan 60^\circ\right)x = \sqrt{3}x$, $y = -\frac{1}{\sqrt{3}}x$ is as follows:

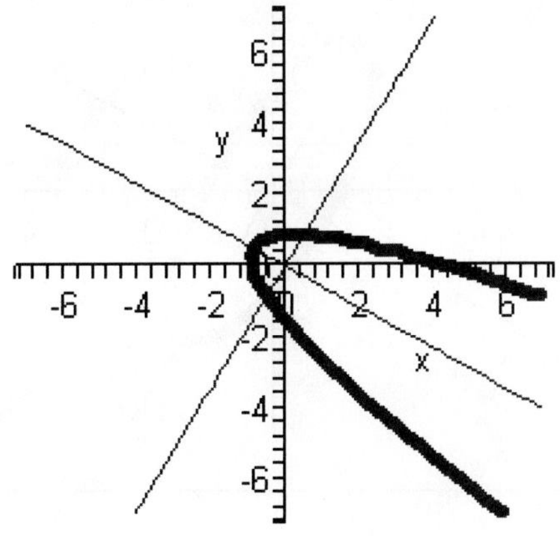

19. Consider the equation $7x^2+4\sqrt{3}xy+3y^2-9=0$.

(a) Here, $A=7,\ B=4\sqrt{3},\ C=3$, so that $B^2-4AC=-36<0$. Thus, the curve is an *ellipse.*

(b) In order to obtain the equation of this curve in the *XY*-system, we make the following substitution into the original equation with $\theta=\pi/6$:

$$\begin{cases} x = X\cos\theta - Y\sin\theta = \left(\sqrt{3}X-Y\right)\left(\frac{1}{2}\right) \\ y = X\sin\theta + Y\cos\theta = \left(X+\sqrt{3}Y\right)\left(\frac{1}{2}\right) \end{cases}$$

Doing so yields the following, after collecting like terms and simplifying:

$$7\left(\sqrt{3}X-Y\right)^2\left(\frac{1}{2}\right)^2+4\sqrt{3}\left(\sqrt{3}X-Y\right)\left(X+\sqrt{3}Y\right)\left(\frac{1}{2}\right)^2+3\left(X+\sqrt{3}Y\right)^2\left(\frac{1}{2}\right)^2-9=0$$

$$7\left(3X^2-2\sqrt{3}XY+Y^2\right)+4\sqrt{3}\left(\sqrt{3}X^2+3XY-XY-\sqrt{3}Y^2\right)+3\left(X^2+2\sqrt{3}XY+3Y^2\right)-36=0$$

$$36X^2+4Y^2-36=0$$

$$\frac{X^2}{1}+\frac{Y^2}{9}=1$$

(c) The graph of $7x^2+4\sqrt{3}xy+3y^2-9=0$, along with the new axes $y=\left(\tan \pi/6\right)x=\left(\frac{1}{\sqrt{3}}\right)x,\ y=-\sqrt{3}x$ is as follows:

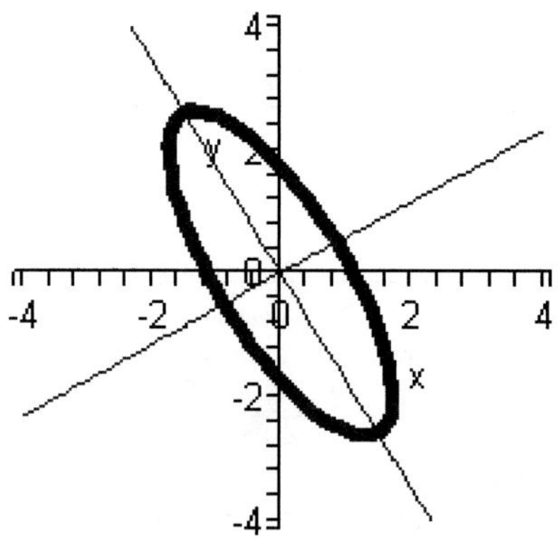

20. Consider the equation $37x^2 + 42\sqrt{3}xy + 79y^2 - 400 = 0$.

(a) Here, $A = 37,\ B = 42\sqrt{3},\ C = 79$, so that $B^2 - 4AC = -6400 < 0$. Thus, the curve is an *ellipse*.

(b) In order to obtain the equation of this curve in the *XY*-system, we make the following substitution into the original equation with $\theta = \pi/3$:

$$\begin{cases} x = X\cos\theta - Y\sin\theta = \left(X - \sqrt{3}Y\right)\left(\frac{1}{2}\right) \\ y = X\sin\theta + Y\cos\theta = \left(\sqrt{3}X + Y\right)\left(\frac{1}{2}\right) \end{cases}$$

Doing so yields the following, after collecting like terms and simplifying:

$$37\left(X - \sqrt{3}Y\right)^2\left(\frac{1}{2}\right)^2 + 42\sqrt{3}\left(X - \sqrt{3}Y\right)\left(\sqrt{3}X + Y\right)\left(\frac{1}{2}\right)^2 + 79\left(\sqrt{3}X + Y\right)^2\left(\frac{1}{2}\right)^2 - 400 = 0$$

$$37\left(X^2 - 2\sqrt{3}XY + 3Y^2\right) + 42\sqrt{3}\left(\sqrt{3}X^2 - 3XY + XY - \sqrt{3}Y^2\right) + 79\left(3X^2 + 2\sqrt{3}XY + Y^2\right) - 1600 = 0$$

$$400X^2 + 64Y^2 - 1600 = 0$$

$$\frac{X^2}{4} + \frac{Y^2}{25} = 1$$

(c) The graph of $37x^2 + 42\sqrt{3}xy + 79y^2 - 400 = 0$, along with the new axes $y = \left(\tan \pi/3\right)x = \sqrt{3}x$, $y = -\frac{1}{\sqrt{3}}x$ is as follows:

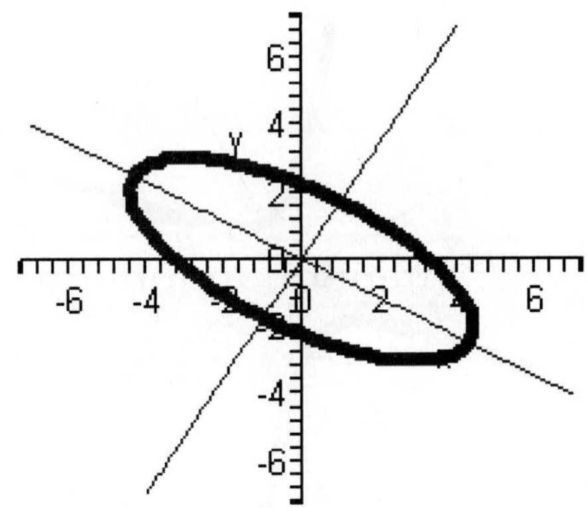

21. Consider the equation $7x^2 - 10\sqrt{3}xy - 3y^2 + 24 = 0$.

(a) Here, $A = 7,\ B = -10\sqrt{3},\ C = -3$, so that $B^2 - 4AC = 384 > 0$. Thus, the curve is a *hyperbola*.

(b) In order to obtain the equation of this curve in the *XY*-system, we make the following substitution into the original equation with $\theta = \pi/3$:

$$\begin{cases} x = X\cos\theta - Y\sin\theta = \left(X - \sqrt{3}Y\right)\left(\frac{1}{2}\right) \\ y = X\sin\theta + Y\cos\theta = \left(\sqrt{3}X + Y\right)\left(\frac{1}{2}\right) \end{cases}$$

Doing so yields the following, after collecting like terms and simplifying:

$$7\left(X - \sqrt{3}Y\right)^2\left(\frac{1}{2}\right)^2 - 10\sqrt{3}\left(X - \sqrt{3}Y\right)\left(\sqrt{3}X + Y\right)\left(\frac{1}{2}\right)^2 - 3\left(\sqrt{3}X + Y\right)^2\left(\frac{1}{2}\right)^2 + 24 = 0$$

$$7\left(X^2 - 2\sqrt{3}XY + 3Y^2\right) - 10\sqrt{3}\left(\sqrt{3}X^2 - 3XY + XY - \sqrt{3}Y^2\right) - 3\left(3X^2 + 2\sqrt{3}XY + Y^2\right) + 96 = 0$$

$$-32X^2 + 48Y^2 + 96 = 0$$

$$\frac{X^2}{3} - \frac{Y^2}{2} = 1$$

(c) The graph of $7x^2 - 10\sqrt{3}xy - 3y^2 + 24 = 0$, along with the new axes $y = \left(\tan \pi/3\right)x = \sqrt{3}x$, $y = -\frac{1}{\sqrt{3}}x$ is as follows:

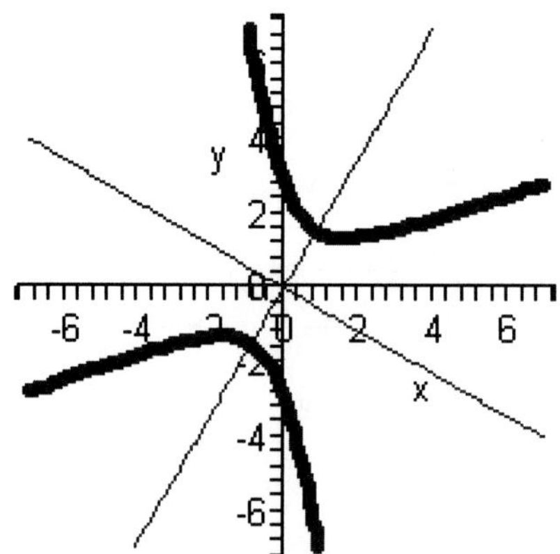

22. Consider the equation $9x^2 + 14\sqrt{3}xy - 5y^2 + 48 = 0$.

(a) Here, $A = 9,\ B = 14\sqrt{3},\ C = -5$, so that $B^2 - 4AC = 768 > 0$. Thus, the curve is a *hyperbola.*

(b) In order to obtain the equation of this curve in the *XY*-system, we make the following substitution into the original equation with $\theta = \pi/6$:

$$\begin{cases} x = X\cos\theta - Y\sin\theta = \left(\sqrt{3}X - Y\right)\left(\frac{1}{2}\right) \\ y = X\sin\theta + Y\cos\theta = \left(X + \sqrt{3}Y\right)\left(\frac{1}{2}\right) \end{cases}$$

Doing so yields the following, after collecting like terms and simplifying:

$$9\left(\sqrt{3}X - Y\right)^2\left(\frac{1}{2}\right)^2 + 14\sqrt{3}\left(\sqrt{3}X - Y\right)\left(X + \sqrt{3}Y\right)\left(\frac{1}{2}\right)^2 - 5\left(X + \sqrt{3}Y\right)^2\left(\frac{1}{2}\right)^2 + 48 = 0$$

$$9\left(3X^2 - 2\sqrt{3}XY + Y^2\right) + 14\sqrt{3}\left(\sqrt{3}X^2 + 3XY - XY - \sqrt{3}Y^2\right) - 5\left(X^2 + 2\sqrt{3}XY + 3Y^2\right) + 192 = 0$$

$$64X^2 - 48Y^2 + 192 = 0$$

$$\frac{Y^2}{4} - \frac{X^2}{3} = 1$$

(c) The graph of $9x^2 + 14\sqrt{3}xy - 5y^2 + 48 = 0$, along with the new axes $y = \left(\tan \pi/6\right)x = \left(\frac{1}{\sqrt{3}}\right)x,\ y = -\sqrt{3}x$ is as follows:

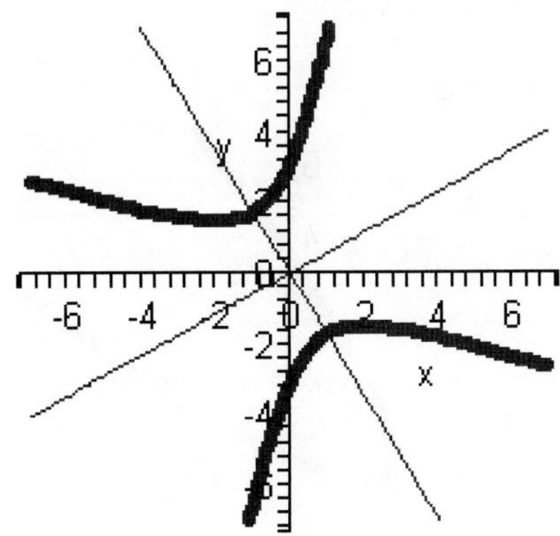

23. Consider the equation $x^2 - 2xy + y^2 - \sqrt{2}x - \sqrt{2}y - 8 = 0$.

(a) Here, $A = 1,\ B = -2,\ C = 1$, so that $B^2 - 4AC = 0$. Thus, the curve is a *parabola.*

(b) In order to obtain the equation of this curve in the *XY*-system, we make the following substitution into the original equation with $\theta = \pi/4$:

$$\begin{cases} x = X\cos\theta - Y\sin\theta = (X - Y)\left(\frac{\sqrt{2}}{2}\right) \\ y = X\sin\theta + Y\cos\theta = (X + Y)\left(\frac{\sqrt{2}}{2}\right) \end{cases}$$

Doing so yields the following, after collecting like terms and simplifying:

$$(X-Y)^2\left(\frac{\sqrt{2}}{2}\right)^2 - 2(X-Y)(X+Y)\left(\frac{\sqrt{2}}{2}\right)^2 + (X+Y)^2\left(\frac{\sqrt{2}}{2}\right)^2 - \sqrt{2}(X-Y)\left(\frac{\sqrt{2}}{2}\right) - \sqrt{2}(X+Y)\left(\frac{\sqrt{2}}{2}\right) - 8 = 0$$

$$\frac{1}{2}\left(X^2 - 2XY + Y^2\right) - \left(X^2 - Y^2\right) + \frac{1}{2}\left(X^2 + 2XY + Y^2\right) - (X - Y) - (X + Y) - 8 = 0$$

$$2Y^2 - 2X - 8 = 0$$

$$Y^2 - X - 4 = 0$$

(c) The graph of $x^2 - 2xy + y^2 - \sqrt{2}x - \sqrt{2}y - 8 = 0$, along with the new axes $y = \left(\tan \pi/4\right)x = x,\ y = -x$ is as follows:

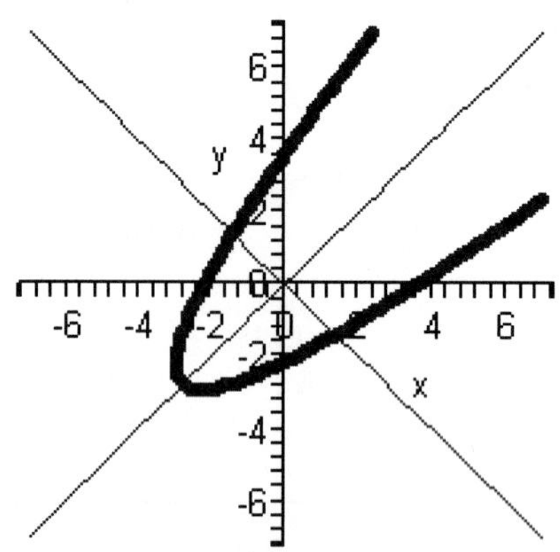

24. Consider the equation $x^2+2xy+y^2+3\sqrt{2}x+\sqrt{2}y=0$.

(a) Here, $A=1,\ B=2,\ C=1$, so that $B^2-4AC=0$. Thus, the curve is a *parabola*.

(b) In order to obtain the equation of this curve in the *XY*-system, we make the following substitution into the original equation with $\theta=\pi/4$:

$$\begin{cases} x=X\cos\theta-Y\sin\theta=(X-Y)\left(\frac{\sqrt{2}}{2}\right) \\ y=X\sin\theta+Y\cos\theta=(X+Y)\left(\frac{\sqrt{2}}{2}\right) \end{cases}$$

Doing so yields the following, after collecting like terms and simplifying:

$$(X-Y)^2\left(\frac{\sqrt{2}}{2}\right)^2+2(X-Y)(X+Y)\left(\frac{\sqrt{2}}{2}\right)^2+(X+Y)^2\left(\frac{\sqrt{2}}{2}\right)^2$$

$$+3\sqrt{2}(X-Y)\left(\frac{\sqrt{2}}{2}\right)+\sqrt{2}(X+Y)\left(\frac{\sqrt{2}}{2}\right)=0$$

$$\frac{1}{2}(X^2-2XY+Y^2)+(X^2-Y^2)+\frac{1}{2}(X^2+2XY+Y^2)+3(X-Y)+(X+Y)=0$$

$$2X^2+4X-2Y=0$$

$$X^2+2X-Y=0$$

(c) The graph of $x^2+2xy+y^2+3\sqrt{2}x+\sqrt{2}y=0$, along with the new axes $y=\left(\tan\frac{\pi}{4}\right)x=x,\ y=-x$ is as follows:

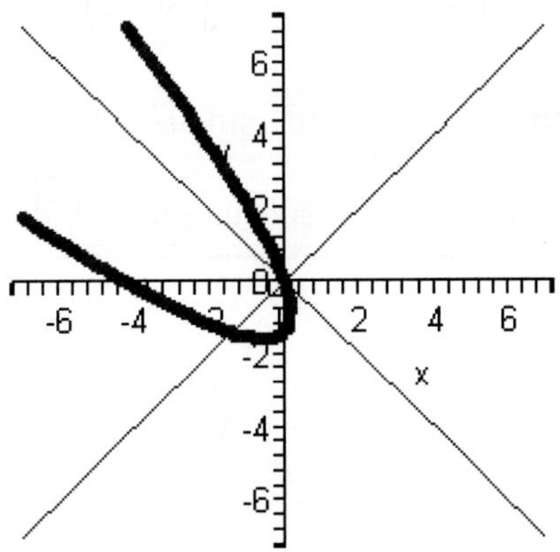

25. To find the amount of rotation needed to transform the given equation into one in the *XY*-system with no *XY* term, we seek the solution θ of the equation $\cot 2\theta = \frac{A-C}{B}$ in $[0,\pi]$. In this case, $A=1,\ B=4,\ C=1$, so that solving this equation yields:

$$\cot 2\theta = 0 \ \Rightarrow\ \cos 2\theta = 0 \ \Rightarrow\ 2\theta = 90^\circ \ \Rightarrow\ \boxed{\theta = 45^\circ}.$$

26. To find the amount of rotation needed to transform the given equation into one in the *XY*-system with no *XY* term, we seek the solution θ of the equation $\cot 2\theta = \frac{A-C}{B}$ in $[0,\pi]$. In this case, $A=3,\ B=5,\ C=3$, so that solving this equation yields:

$$\cot 2\theta = 0 \ \Rightarrow\ \cos 2\theta = 0 \ \Rightarrow\ 2\theta = 90^\circ \ \Rightarrow\ \boxed{\theta = 45^\circ}.$$

27. To find the amount of rotation needed to transform the given equation into one in the *XY*-system with no *XY* term, we seek the solution θ of the equation $\cot 2\theta = \frac{A-C}{B}$ in $[0,\pi]$. In this case, $A=2,\ B=\sqrt{3},\ C=3$, so that this equation becomes $\cot 2\theta = -\frac{1}{\sqrt{3}}$. Hence, we need to find θ in $[0,\pi]$ such that

$$\cos 2\theta = -\tfrac{1}{2} \text{ and } \sin 2\theta = \tfrac{\sqrt{3}}{2}.$$

But, this occurs precisely when $2\theta = 120^\circ$, so that $\boxed{\theta = 60^\circ}$.

28. To find the amount of rotation needed to transform the given equation into one in the *XY*-system with no *XY* term, we seek the solution θ of the equation $\cot 2\theta = \frac{A-C}{B}$ in $[0,\pi]$. In this case, $A=4,\ B=\sqrt{3},\ C=3$, so that this equation becomes $\cot 2\theta = \frac{1}{\sqrt{3}}$. Hence, we need to find θ in $[0,\pi]$ such that

$$\cos 2\theta = \tfrac{1}{2} \text{ and } \sin 2\theta = \tfrac{\sqrt{3}}{2}.$$

But, this occurs precisely when $2\theta = 60^\circ$, so that $\boxed{\theta = 30^\circ}$.

29. To find the amount of rotation needed to transform the given equation into one in the *XY*-system with no *XY* term, we seek the solution θ of the equation $\cot 2\theta = \frac{A-C}{B}$ in $[0,\pi]$. In this case, $A=2,\ B=\sqrt{3},\ C=1$, so that this equation becomes $\cot 2\theta = \frac{1}{\sqrt{3}}$. Hence, we need to find θ in $[0,\pi]$ such that $$\cos 2\theta = \tfrac{1}{2} \text{ and } \sin 2\theta = \tfrac{\sqrt{3}}{2}.$$ But, this occurs precisely when $2\theta = 60^\circ$, so that $\boxed{\theta = 30^\circ}$.
30. To find the amount of rotation needed to transform the given equation into one in the *XY*-system with no *XY* term, we seek the solution θ of the equation $\cot 2\theta = \frac{A-C}{B}$ in $[0,\pi]$. In this case, $A=2\sqrt{3},\ B=1,\ C=3\sqrt{3}$, so that this equation becomes $\cot 2\theta = -\sqrt{3}$. Hence, we need to find θ in $[0,\pi]$ such that $$\cos 2\theta = -\tfrac{\sqrt{3}}{2} \text{ and } \sin 2\theta = \tfrac{1}{2}.$$ But, this occurs precisely when $2\theta = 150^\circ$, so that $\boxed{\theta = 75^\circ}$.
31. To find the amount of rotation needed to transform the given equation into one in the *XY*-system with no *XY* term, we seek the solution θ of the equation $\cot 2\theta = \frac{A-C}{B}$ in $[0,\pi]$. In this case, $A=\sqrt{2},\ B=1,\ C=\sqrt{2}$, so that solving this equation yields: $$\cot 2\theta = 0 \Rightarrow \cos 2\theta = 0 \Rightarrow 2\theta = 90^\circ \Rightarrow \boxed{\theta = 45^\circ}.$$
32. To find the amount of rotation needed to transform the given equation into one in the *XY*-system with no *XY* term, we seek the solution θ of the equation $\cot 2\theta = \frac{A-C}{B}$ in $[0,\pi]$. In this case, $A=1,\ B=10,\ C=1$, so that solving this equation yields: $$\cot 2\theta = 0 \Rightarrow \cos 2\theta = 0 \Rightarrow 2\theta = 90^\circ \Rightarrow \boxed{\theta = 45^\circ}.$$

33. To find the amount of rotation needed to transform the given equation into one in the *XY*-system with no *XY* term, we seek the solution θ of the equation $\cot 2\theta = \frac{A-C}{B}$ in $[0,\pi]$. In this case, $A = 12\sqrt{3},\ B = 4,\ C = 8\sqrt{3}$, so that this equation becomes $\cot 2\theta = \sqrt{3}$. Hence, we need to find θ in $[0,\pi]$ such that

$$\cos 2\theta = \sqrt{3}/2 \text{ and } \sin 2\theta = 1/2.$$

But, this occurs precisely when $2\theta = 30^\circ$, so that $\boxed{\theta = 15^\circ}$.

34. To find the amount of rotation needed to transform the given equation into one in the *XY*-system with no *XY* term, we seek the solution θ of the equation $\cot 2\theta = \frac{A-C}{B}$ in $[0,\pi]$. In this case, $A = 4,\ B = 2,\ C = 2$, so that this equation becomes $\cot 2\theta = 1$. Hence, we need to find θ in $[0,\pi]$ such that

$$\cos 2\theta = \sqrt{2}/2 \text{ and } \sin 2\theta = \sqrt{2}/2.$$

But, this occurs precisely when $2\theta = 45^\circ$, so that $\boxed{\theta = 22.5^\circ}$.

35. To find the amount of rotation needed to transform the given equation into one in the *XY*-system with no *XY* term, we seek the solution θ of the equation $\cot 2\theta = \frac{A-C}{B}$ in $[0,\pi]$. In this case, $A = 5,\ B = 6,\ C = 4$, so that this equation becomes $\cot 2\theta = 1/6$. This is not satisfied by a standard angle, so we shall need to settle for an approximation:

$$2\theta = \cot^{-1}\left(1/6\right) = \tan^{-1}(6) \approx 80.537678^\circ, \text{ so that } \boxed{\theta \approx 40.3^\circ}.$$

36. To find the amount of rotation needed to transform the given equation into one in the *XY*-system with no *XY* term, we seek the solution θ of the equation $\cot 2\theta = \frac{A-C}{B}$ in $[0,\pi]$. In this case, $A = 1,\ B = 2,\ C = 12$, so that this equation becomes $\cot 2\theta = -11/2$. This is not satisfied by a standard angle, so we shall need to settle for an approximation:

$$2\theta = \cot^{-1}\left(-11/2\right) = 180^\circ + \tan^{-1}\left(-2/11\right) \approx 169.695154^\circ, \text{ so that } \boxed{\theta \approx 84.8^\circ}.$$

37. To find the amount of rotation needed to transform the given equation into one in the *XY*-system with no *XY* term, we seek the solution θ of the equation $\cot 2\theta = \frac{A-C}{B}$ in $[0,\pi]$. In this case, $A=3,\ B=10,\ C=5$, so that this equation becomes $\cot 2\theta = -\,{}^{1}\!/_{5}$. This is not satisfied by a standard angle, so we shall need to settle for an approximation:

$$2\theta = \cot^{-1}\left(-\,{}^{1}\!/_{5}\right) = 180^\circ + \tan^{-1}(-5) \approx 101.309933^\circ \text{, so that } \boxed{\theta \approx 50.7^\circ}.$$

38. To find the amount of rotation needed to transform the given equation into one in the *XY*-system with no *XY* term, we seek the solution θ of the equation $\cot 2\theta = \frac{A-C}{B}$ in $[0,\pi]$. In this case, $A=10,\ B=3,\ C=2$, so that this equation becomes $\cot 2\theta = {}^{8}\!/_{3}$. This is not satisfied by a standard angle, so we shall need to settle for an approximation:

$$2\theta = \cot^{-1}\left({}^{8}\!/_{3}\right) = \tan^{-1}\left({}^{3}\!/_{8}\right) \approx 20.556045^\circ \text{, so that } \boxed{\theta \approx 10.3^\circ}.$$

39. Consider the equation $21x^2+10\sqrt{3}xy+31y^2-144=0$.

Here, $A=21,\ B=10\sqrt{3},\ C=31$, so that $B^2-4AC=-2304<0$. Thus, the curve is an *ellipse*. In order to obtain its graph, follow these steps:

Step 1: Find the angle of counterclockwise rotation necessary to transform the equation into the *XY*-system with no *XY* term.

We seek the solution θ of the equation $\cot 2\theta=\frac{A-C}{B}=-\frac{1}{\sqrt{3}}$ in $[0,\pi]$. Hence, we need to find θ in $[0,\pi]$ such that $\cos 2\theta=-\frac{1}{2}$ and $\sin 2\theta=\frac{\sqrt{3}}{2}$.

But, this occurs precisely when $2\theta=120^\circ$, so that $\theta=60^\circ$.

Step 2: Now, determine the equation of this curve in the *XY*-system making use of the following substitution into the original equation with $\theta=60^\circ$:

$$\begin{cases} x=X\cos\theta-Y\sin\theta=\left(X-\sqrt{3}Y\right)\left(\frac{1}{2}\right) \\ y=X\sin\theta+Y\cos\theta=\left(\sqrt{3}X+Y\right)\left(\frac{1}{2}\right) \end{cases}$$

Doing so yields the following, after collecting like terms and simplifying:

$$21\left(X-\sqrt{3}Y\right)^2\left(\frac{1}{2}\right)^2+10\sqrt{3}\left(X-\sqrt{3}Y\right)\left(\sqrt{3}X+Y\right)\left(\frac{1}{2}\right)^2+31\left(\sqrt{3}X+Y\right)^2\left(\frac{1}{2}\right)^2-144=0$$

$$21\left(X^2-2\sqrt{3}XY+3Y^2\right)+10\sqrt{3}\left(\sqrt{3}X^2-3XY+XY-\sqrt{3}Y^2\right)+31\left(3X^2+2\sqrt{3}XY+Y^2\right)-576=0$$

$$144X^2+64Y^2-576=0$$

$$\frac{X^2}{4}+\frac{Y^2}{9}=1$$

Step 3: The graph of $21x^2+10\sqrt{3}xy+31y^2-144=0$, along with the new axes $y=\left(\tan 60^\circ\right)x=\sqrt{3}x,\ y=-\left(\frac{1}{\sqrt{3}}\right)x$ is as follows:

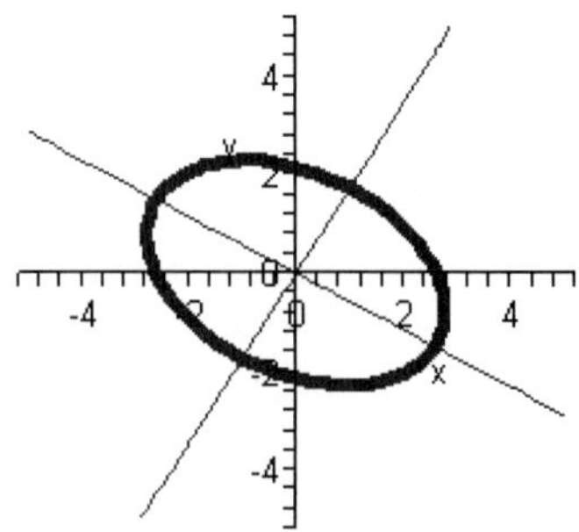

40. Consider the equation $5x^2+6xy+5y^2-8=0$.
Here, $A=5,\ B=6,\ C=5$, so that $B^2-4AC=-64<0$. Thus, the curve is an *ellipse*.
In order to obtain its graph, follow these steps:

Step 1: Find the angle of counterclockwise rotation necessary to transform the equation into the *XY*-system with no *XY* term.
We seek the solution θ of the equation $\cot 2\theta=\dfrac{A-C}{B}=0$ in $[0,\pi]$. Hence, we need to find θ in $[0,\pi]$ such that $\cos 2\theta=0$ and $\sin 2\theta=1$.
But, this occurs precisely when $2\theta=90^\circ$, so that $\theta=45^\circ$.

Step 2: Now, determine the equation of this curve in the *XY*-system making use of the following substitution into the original equation with $\theta=45^\circ$:

$$\begin{cases} x=X\cos\theta-Y\sin\theta=(X-Y)\left(\dfrac{\sqrt{2}}{2}\right) \\ y=X\sin\theta+Y\cos\theta=(X+Y)\left(\dfrac{\sqrt{2}}{2}\right) \end{cases}$$

Doing so yields the following, after collecting like terms and simplifying:

$$5(X-Y)^2\left(\frac{\sqrt{2}}{2}\right)^2+6(X-Y)(X+Y)\left(\frac{\sqrt{2}}{2}\right)^2+5(X+Y)^2\left(\frac{\sqrt{2}}{2}\right)^2-8=0$$
$$5\left(X^2-2XY+Y^2\right)+6\left(X^2-Y^2\right)+5\left(X^2+2XY+Y^2\right)-16=0$$
$$16X^2+4Y^2-16=0$$
$$\frac{X^2}{1}+\frac{Y^2}{4}=1$$

Step 3: The graph of $5x^2+6xy+5y^2-8=0$, along with the new axes $y=\left(\tan 45^\circ\right)x=x,\ y=-x$ is as follows:

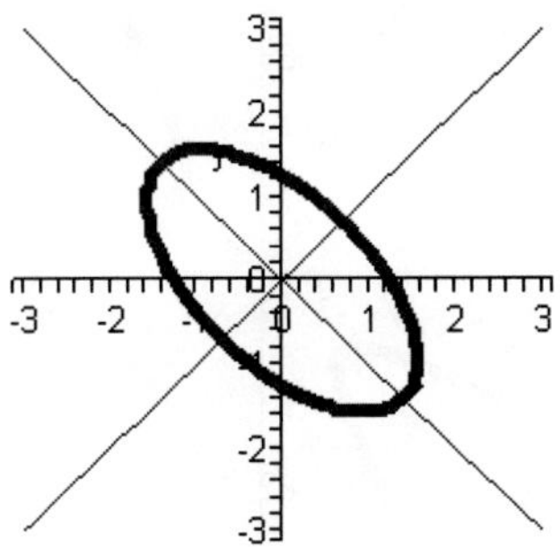

41. Consider the equation $8x^2 - 20xy + 8y^2 + 18 = 0$.
Here, $A = 8,\ B = -20,\ C = 8$, so that $B^2 - 4AC = 144 > 0$. Thus, the curve is an *hyperbola*. In order to obtain its graph, follow these steps:

Step 1: Find the angle of counterclockwise rotation necessary to transform the equation into the *XY*-system with no *XY* term.
We seek the solution θ of the equation $\cot 2\theta = \frac{A-C}{B} = 0$ in $[0, \pi]$. Hence, we need to find θ in $[0, \pi]$ such that $\cos 2\theta = 0$ and $\sin 2\theta = 1$.
But, this occurs precisely when $2\theta = 90^\circ$, so that $\theta = 45^\circ$.

Step 2: Now, determine the equation of this curve in the *XY*-system making use of the following substitution into the original equation with $\theta = 45^\circ$:

$$\begin{cases} x = X\cos\theta - Y\sin\theta = (X - Y)\left(\frac{\sqrt{2}}{2}\right) \\ y = X\sin\theta + Y\cos\theta = (X + Y)\left(\frac{\sqrt{2}}{2}\right) \end{cases}$$

Doing so yields the following, after collecting like terms and simplifying:

$$8(X-Y)^2\left(\frac{\sqrt{2}}{2}\right)^2 - 20(X-Y)(X+Y)\left(\frac{\sqrt{2}}{2}\right)^2 + 8(X+Y)^2\left(\frac{\sqrt{2}}{2}\right)^2 + 18 = 0$$

$$8(X^2 - 2XY + Y^2) - 20(X^2 - Y^2) + 8(X^2 + 2XY + Y^2) + 36 = 0$$

$$-4X^2 + 36Y^2 + 36 = 0$$

$$\frac{X^2}{9} - \frac{Y^2}{1} = 1$$

Step 3: The graph of $8x^2 - 20xy + 8y^2 + 18 = 0$, along with the new axes $y = (\tan 45^\circ)x = x,\ y = -x$ is as follows:

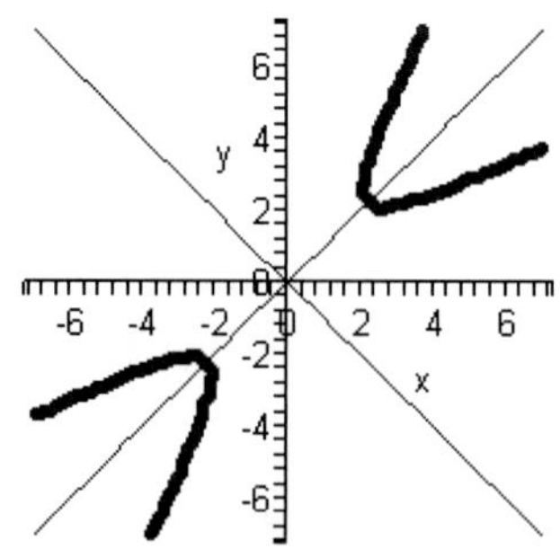

42. Consider the equation $-23x^2 - 26\sqrt{3}xy + 3y^2 - 144 = 0$.

Here, $A = -23,\ B = -26\sqrt{3},\ C = 3$, so that $B^2 - 4AC = 2304 > 0$. Thus, the curve is a *hyperbola*. In order to obtain its graph, follow these steps:

Step 1: Find the angle of counterclockwise rotation necessary to transform the equation into the *XY*-system with no *XY* term.

We seek the solution θ of the equation $\cot 2\theta = \dfrac{A-C}{B} = \dfrac{1}{\sqrt{3}}$ in $[0,\pi]$. Hence, we need to find θ in $[0,\pi]$ such that $\cos 2\theta = \frac{1}{2}$ and $\sin 2\theta = \frac{\sqrt{3}}{2}$.

But, this occurs precisely when $2\theta = 60^\circ$, so that $\theta = 30^\circ$.

Step 2: Now, determine the equation of this curve in the *XY*-system making use of the following substitution into the original equation with $\theta = 30^\circ$:

$$\begin{cases} x = X\cos\theta - Y\sin\theta = \left(\sqrt{3}X - Y\right)\left(\frac{1}{2}\right) \\ y = X\sin\theta + Y\cos\theta = \left(X + \sqrt{3}Y\right)\left(\frac{1}{2}\right) \end{cases}$$

Doing so yields the following, after collecting like terms and simplifying:

$$-23\left(\sqrt{3}X - Y\right)^2\left(\frac{1}{2}\right)^2 - 26\sqrt{3}\left(\sqrt{3}X - Y\right)\left(X + \sqrt{3}Y\right)\left(\frac{1}{2}\right)^2 + 3\left(X + \sqrt{3}Y\right)^2\left(\frac{1}{2}\right)^2 - 144 = 0$$

$$-23\left(3X^2 - 2\sqrt{3}XY + Y^2\right) - 26\sqrt{3}\left(\sqrt{3}X^2 + 3XY - XY - \sqrt{3}Y^2\right) + 3\left(X^2 + 2\sqrt{3}XY + 3Y^2\right) - 576 = 0$$

$$-144X^2 + 64Y^2 - 576 = 0$$

$$\frac{Y^2}{9} - \frac{X^2}{4} = 1$$

Step 3: The graph of $-23x^2 - 26\sqrt{3}xy + 3y^2 - 144 = 0$, along with the new axes $y = (\tan 30^\circ)x = \left(\frac{1}{\sqrt{3}}\right)x,\ y = -\sqrt{3}x$ is as follows:

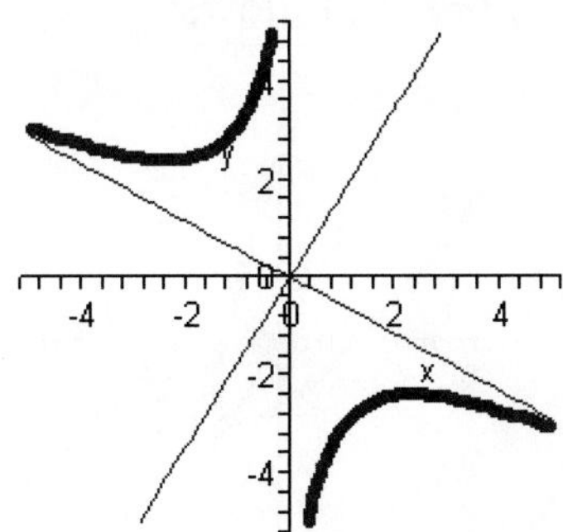

43. Consider the equation $3x^2+2\sqrt{3}xy+y^2+2x-2\sqrt{3}y-12=0$.

Here, $A=3,\ B=2\sqrt{3},\ C=1$, so that $B^2-4AC=0$. Thus, the curve is a *parabola*. In order to obtain its graph, follow these steps:

Step 1: Find the angle of counterclockwise rotation necessary to transform the equation into the *XY*-system with no *XY* term.

We seek the solution θ of the equation $\cot 2\theta=\dfrac{A-C}{B}=\dfrac{1}{\sqrt{3}}$ in $[0,\pi]$. Hence, we need to find θ in $[0,\pi]$ such that $\cos 2\theta=\frac{1}{2}$ and $\sin 2\theta=\frac{\sqrt{3}}{2}$.

But, this occurs precisely when $2\theta=60°$, so that $\theta=30°$.

Step 2: Now, determine the equation of this curve in the *XY*-system making use of the following substitution into the original equation with $\theta=30°$:

$$\begin{cases} x=X\cos\theta-Y\sin\theta=\left(\sqrt{3}X-Y\right)\left(\dfrac{1}{2}\right) \\ y=X\sin\theta+Y\cos\theta=\left(X+\sqrt{3}Y\right)\left(\dfrac{1}{2}\right) \end{cases}$$

Doing so yields the following, after collecting like terms and simplifying:

$$3\left(\sqrt{3}X-Y\right)^2\left(\frac{1}{2}\right)^2+2\sqrt{3}\left(\sqrt{3}X-Y\right)\left(X+\sqrt{3}Y\right)\left(\frac{1}{2}\right)^2+\left(X+\sqrt{3}Y\right)^2\left(\frac{1}{2}\right)^2$$
$$+2\left(\sqrt{3}X-Y\right)\left(\frac{1}{2}\right)-2\sqrt{3}\left(X+\sqrt{3}Y\right)\left(\frac{1}{2}\right)-12=0$$
$$3\left(3X^2-2\sqrt{3}XY+Y^2\right)+2\sqrt{3}\left(\sqrt{3}X^2+3XY-XY-\sqrt{3}Y^2\right)+\left(X^2+2\sqrt{3}XY+3Y^2\right)$$
$$+4\left(\sqrt{3}X-Y\right)-4\sqrt{3}\left(X+\sqrt{3}Y\right)-48=0$$
$$16X^2-16Y-48=0$$
$$X^2-Y-3=0$$

Step 3: The graph of $3x^2+2\sqrt{3}xy+y^2+2x-2\sqrt{3}y-12=0$, along with the new axes $y=\left(\tan 30°\right)x=\left(\dfrac{1}{\sqrt{3}}\right)x,\ y=-\sqrt{3}x$ is as follows:

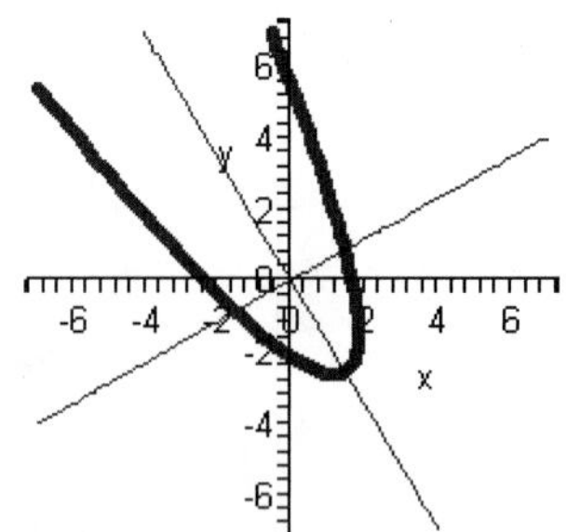

44. Consider the equation $3x^2 - 2\sqrt{3}xy + y^2 - 2x - 2\sqrt{3}y - 4 = 0$.

Here, $A = 3,\ B = -2\sqrt{3},\ C = 1$, so that $B^2 - 4AC = 0$. Thus, the curve is a *parabola.* In order to obtain its graph, follow these steps:

Step 1: Find the angle of counterclockwise rotation necessary to transform the equation into the *XY*-system with no *XY* term.

We seek the solution θ of the equation $\cot 2\theta = \dfrac{A-C}{B} = -\dfrac{1}{\sqrt{3}}$ in $[0,\pi]$. Hence, we need to find θ in $[0,\pi]$ such that $\cos 2\theta = -\frac{1}{2}$ and $\sin 2\theta = \frac{\sqrt{3}}{2}$.

But, this occurs precisely when $2\theta = 120^\circ$, so that $\theta = 60^\circ$.

Step 2: Now, determine the equation of this curve in the *XY*-system making use of the following substitution into the original equation with $\theta = 60^\circ$:

$$\begin{cases} x = X\cos\theta - Y\sin\theta = \left(X - \sqrt{3}Y\right)\left(\dfrac{1}{2}\right) \\ y = X\sin\theta + Y\cos\theta = \left(\sqrt{3}X + Y\right)\left(\dfrac{1}{2}\right) \end{cases}$$

Doing so yields the following, after collecting like terms and simplifying:

$$3\left(X - \sqrt{3}Y\right)^2\left(\frac{1}{2}\right)^2 - 2\sqrt{3}\left(X - \sqrt{3}Y\right)\left(\sqrt{3}X + Y\right)\left(\frac{1}{2}\right)^2 + \left(\sqrt{3}X + Y\right)^2\left(\frac{1}{2}\right)^2$$
$$-2\left(X - \sqrt{3}Y\right)\left(\frac{1}{2}\right) - 2\sqrt{3}\left(\sqrt{3}X + Y\right)\left(\frac{1}{2}\right) - 4 = 0$$
$$3\left(X^2 - 2\sqrt{3}XY + 3Y^2\right) - 2\sqrt{3}\left(\sqrt{3}X^2 - 3XY + XY - \sqrt{3}Y^2\right) + \left(3X^2 + 2\sqrt{3}XY + Y^2\right)$$
$$-4\left(X - \sqrt{3}Y\right) - 4\sqrt{3}\left(\sqrt{3}X + Y\right) - 16 = 0$$
$$16Y^2 - 16X - 16 = 0$$
$$Y^2 - X - 1 = 0$$

Step 3: The graph of $3x^2 - 2\sqrt{3}xy + y^2 - 2x - 2\sqrt{3}y - 4 = 0$, along with the new axes $y = \left(\tan 60^\circ\right)x = \sqrt{3}x,\ y = -\left(\dfrac{1}{\sqrt{3}}\right)x$ is as follows:

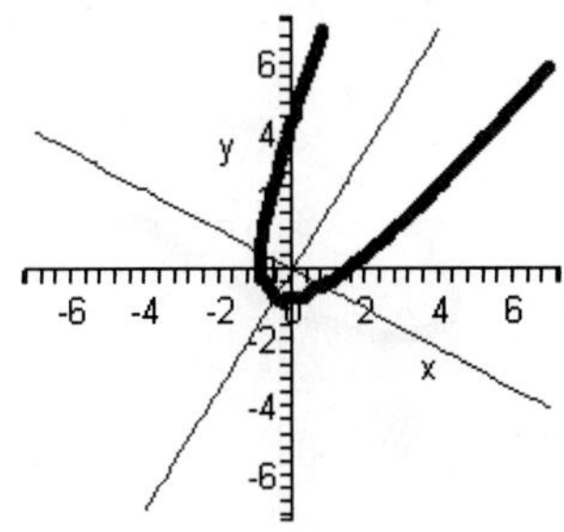

45. Consider the equation $37x^2 - 42\sqrt{3}xy + 79y^2 - 400 = 0$.

Here, $A = 37,\ B = -42\sqrt{3},\ C = 79$, so that $B^2 - 4AC = -6400 < 0$. Thus, the curve is an *ellipse*. In order to obtain its graph, follow these steps:

Step 1: Find the angle of counterclockwise rotation necessary to transform the equation into the XY-system with no XY term.

We seek the solution θ of the equation $\cot 2\theta = \frac{A-C}{B} = \frac{1}{\sqrt{3}}$ in $[0,\pi]$. Hence, we need to find θ in $[0,\pi]$ such that $\cos 2\theta = \frac{1}{2}$ and $\sin 2\theta = \frac{\sqrt{3}}{2}$.

But, this occurs precisely when $2\theta = 60^\circ$, so that $\theta = 30^\circ$.

Step 2: Now, determine the equation of this curve in the XY-system making use of the following substitution into the original equation with $\theta = 30^\circ$:

$$\begin{cases} x = X\cos\theta - Y\sin\theta = \left(\sqrt{3}X - Y\right)\left(\frac{1}{2}\right) \\ y = X\sin\theta + Y\cos\theta = \left(X + \sqrt{3}Y\right)\left(\frac{1}{2}\right) \end{cases}$$

Doing so yields the following, after collecting like terms and simplifying:

$$37\left(\sqrt{3}X - Y\right)^2\left(\frac{1}{2}\right)^2 - 42\sqrt{3}\left(\sqrt{3}X - Y\right)\left(X + \sqrt{3}Y\right)\left(\frac{1}{2}\right)^2 + 79\left(X + \sqrt{3}Y\right)^2\left(\frac{1}{2}\right)^2 - 400 = 0$$

$$37\left(3X^2 - 2\sqrt{3}XY + Y^2\right) - 42\sqrt{3}\left(\sqrt{3}X^2 + 3XY - XY - \sqrt{3}Y^2\right) + 79\left(X^2 + 2\sqrt{3}XY + 3Y^2\right) - 1600 = 0$$

$$64X^2 + 400Y^2 - 1600 = 0$$

$$\frac{X^2}{25} + \frac{Y^2}{4} = 1$$

Step 3: The graph of $37x^2 - 42\sqrt{3}xy + 79y^2 - 400 = 0$, along with the new axes $y = \left(\tan 30^\circ\right)x = \left(\frac{1}{\sqrt{3}}\right)x,\ y = -\sqrt{3}x$ is as follows:

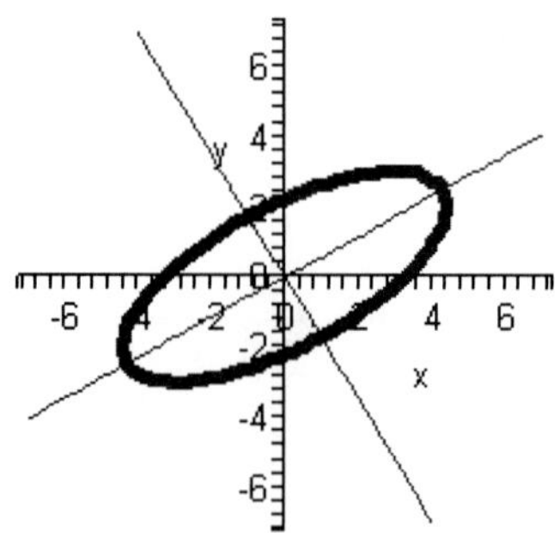

46. Consider the equation $71x^2 - 58\sqrt{3}xy + 13y^2 + 400 = 0$.

Here, $A = 71,\ B = -58\sqrt{3},\ C = 13$, so that $B^2 - 4AC = 6400 > 0$. Thus, the curve is a *hyperbola*. In order to obtain its graph, follow these steps:

Step 1: Find the angle of counterclockwise rotation necessary to transform the equation into the XY-system with no XY term.

We seek the solution θ of the equation $\cot 2\theta = \dfrac{A-C}{B} = -\dfrac{1}{\sqrt{3}}$ in $[0,\pi]$. Hence, we need to find θ in $[0,\pi]$ such that $\cos 2\theta = -\frac{1}{2}$ and $\sin 2\theta = \frac{\sqrt{3}}{2}$.

But, this occurs precisely when $2\theta = 120^\circ$, so that $\theta = 60^\circ$.

Step 2: Now, determine the equation of this curve in the XY-system making use of the following substitution into the original equation with $\theta = 60^\circ$:

$$\begin{cases} x = X\cos\theta - Y\sin\theta = \left(X - \sqrt{3}Y\right)\left(\dfrac{1}{2}\right) \\ y = X\sin\theta + Y\cos\theta = \left(\sqrt{3}X + Y\right)\left(\dfrac{1}{2}\right) \end{cases}$$

Doing so yields the following, after collecting like terms and simplifying:

$$71\left(X - \sqrt{3}Y\right)^2\left(\frac{1}{2}\right)^2 - 58\sqrt{3}\left(X - \sqrt{3}Y\right)\left(\sqrt{3}X + Y\right)\left(\frac{1}{2}\right)^2 + 13\left(\sqrt{3}X + Y\right)^2\left(\frac{1}{2}\right)^2 + 400 = 0$$

$$71\left(X^2 - 2\sqrt{3}XY + 3Y^2\right) - 58\sqrt{3}\left(\sqrt{3}X^2 - 3XY + XY - \sqrt{3}Y^2\right) + 13\left(3X^2 + 2\sqrt{3}XY + Y^2\right) + 1600 = 0$$

$$-64X^2 + 400Y^2 + 1600 = 0$$

$$\frac{X^2}{25} - \frac{Y^2}{4} = 1$$

Step 3: The graph of $71x^2 - 58\sqrt{3}xy + 13y^2 + 400 = 0$, along with the new axes $y = \left(\tan 60^\circ\right)x = \sqrt{3}x,\ y = -\left(\dfrac{1}{\sqrt{3}}\right)x$ is as follows:

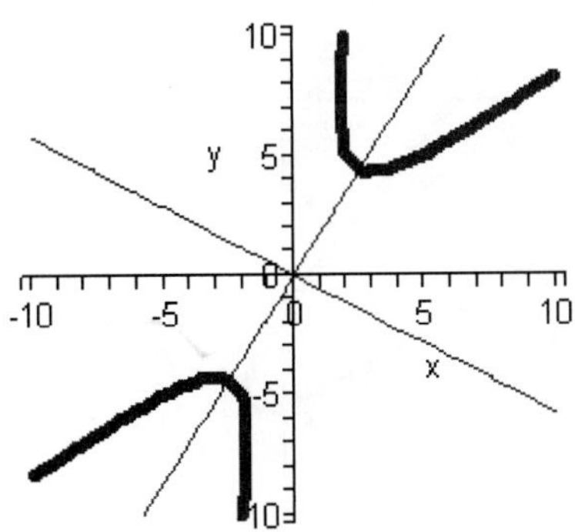

47. Consider the equation $x^2+2xy+y^2+5\sqrt{2}x+3\sqrt{2}y=0$.
Here, $A=1,\ B=2,\ C=1$, so that $B^2-4AC=0$. Thus, the curve is a *parabola*. In order to obtain its graph, follow these steps:

Step 1: Find the angle of counterclockwise rotation necessary to transform the equation into the *XY*-system with no *XY* term.
We seek the solution θ of the equation $\cot 2\theta=\frac{A-C}{B}=0$ in $[0,\pi]$. Hence, we need to find θ in $[0,\pi]$ such that $\cos 2\theta=0$ and $\sin 2\theta=1$.
But, this occurs precisely when $2\theta=90^\circ$, so that $\theta=45^\circ$.

Step 2: Now, determine the equation of this curve in the *XY*-system making use of the following substitution into the original equation with $\theta=45^\circ$:

$$\begin{cases} x=X\cos\theta-Y\sin\theta=(X-Y)\left(\frac{\sqrt{2}}{2}\right) \\ y=X\sin\theta+Y\cos\theta=(X+Y)\left(\frac{\sqrt{2}}{2}\right) \end{cases}$$

Doing so yields the following, after collecting like terms and simplifying:

$$(X-Y)^2\left(\frac{\sqrt{2}}{2}\right)^2+2(X-Y)(X+Y)\left(\frac{\sqrt{2}}{2}\right)^2+(X+Y)^2\left(\frac{\sqrt{2}}{2}\right)^2$$
$$+5\sqrt{2}(X-Y)\left(\frac{\sqrt{2}}{2}\right)+3\sqrt{2}(X+Y)\left(\frac{\sqrt{2}}{2}\right)=0$$
$$\left(X^2-2XY+Y^2\right)+2\left(X^2-Y^2\right)+\left(X^2+2XY+Y^2\right)+10(X-Y)+6(X+Y)=0$$
$$4X^2+16X-4Y=0$$
$$X^2+4X-Y=0$$

Step 3: The graph of $x^2+2xy+y^2+5\sqrt{2}x+3\sqrt{2}y=0$, along with the new axes $y=(\tan 45^\circ)x=x,\ y=-x$ is as follows:

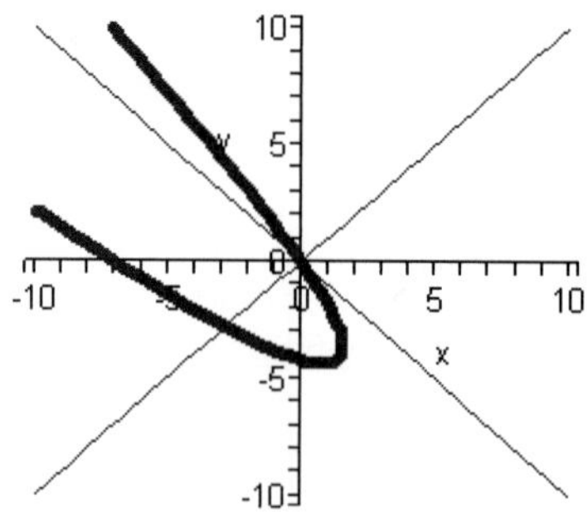

48. Consider the equation $7x^2 - 4\sqrt{3}xy + 3y^2 - 9 = 0$.

Here, $A = 7,\ B = -4\sqrt{3},\ C = 3$, so that $B^2 - 4AC = -36 < 0$. Thus, the curve is an *ellipse*. In order to obtain its graph, follow these steps:

Step 1: Find the angle of counterclockwise rotation necessary to transform the equation into the *XY*-system with no *XY* term.

We seek the solution θ of the equation $\cot 2\theta = \dfrac{A-C}{B} = -\dfrac{1}{\sqrt{3}}$ in $[0,\pi]$. Hence, we need to find θ in $[0,\pi]$ such that $\cos 2\theta = -\frac{1}{2}$ and $\sin 2\theta = \frac{\sqrt{3}}{2}$.

But, this occurs precisely when $2\theta = 120^\circ$, so that $\theta = 60^\circ$.

Step 2: Now, determine the equation of this curve in the *XY*-system making use of the following substitution into the original equation with $\theta = 60^\circ$:

$$\begin{cases} x = X\cos\theta - Y\sin\theta = \left(X - \sqrt{3}Y\right)\left(\frac{1}{2}\right) \\ y = X\sin\theta + Y\cos\theta = \left(\sqrt{3}X + Y\right)\left(\frac{1}{2}\right) \end{cases}$$

Doing so yields the following, after collecting like terms and simplifying:

$$7\left(X - \sqrt{3}Y\right)^2\left(\frac{1}{2}\right)^2 - 4\sqrt{3}\left(X - \sqrt{3}Y\right)\left(\sqrt{3}X + Y\right)\left(\frac{1}{2}\right)^2 + 3\left(\sqrt{3}X + Y\right)^2\left(\frac{1}{2}\right)^2 - 9 = 0$$

$$7\left(X^2 - 2\sqrt{3}XY + 3Y^2\right) - 4\sqrt{3}\left(\sqrt{3}X^2 - 3XY + XY - \sqrt{3}Y^2\right) + 3\left(3X^2 + 2\sqrt{3}XY + Y^2\right) - 36 = 0$$

$$4X^2 + 36Y^2 - 36 = 0$$

$$\frac{X^2}{9} + \frac{Y^2}{1} = 1$$

Step 3: The graph of $7x^2 - 4\sqrt{3}xy + 3y^2 - 9 = 0$, along with the new axes $y = \left(\tan 60^\circ\right)x = \sqrt{3}x,\ y = -\left(\dfrac{1}{\sqrt{3}}\right)x$ is as follows:

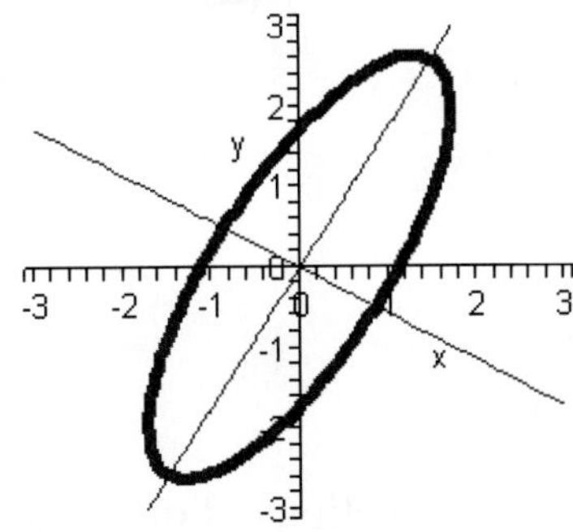

49. True. Here $A=1,\ B=k,\ C=9$. So, $B^2-4AC=k^2-4(1)(9)=k^2-36$. Now, if $0<k<6$, then $k^2-36<0$, and so the graph of the resulting curve would be an ellipse.

50. False. Here $A=1,\ B=k,\ C=9$. So, $B^2-4AC=k^2-4(1)(9)=k^2-36$. Now, if $k>6$, then $k^2-36>0$, and so the graph of the resulting curve would be a hyperbola, not a parabola.

51. True. The equation of the reciprocal function is $y=\frac{1}{x}$, which is equivalent to $xy=1$. Observe that $A=0,\ B=1,\ C=0$, and so $B^2-4AC=1^2-4(0)(0)=1>0$. Thus, the graph is a hyperbola.

52. False. Observe that the graph of $\sqrt{x}+\sqrt{y}=3$ is given below.

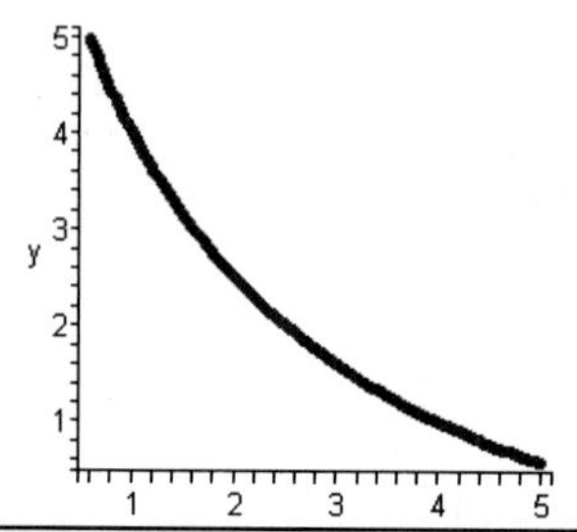

As such, it is not a conic section. In particular, no amount of rotation about the origin can produce a graph of a circle of radius 3.

53. a. $\frac{x^2}{b^2}+\frac{y^2}{a^2}=1$ **b.** The original equation

54. a. $\frac{x^2}{b^2}-\frac{y^2}{a^2}=1$ **b.** The original equation

55. If $a<0$, you get a hyperbola; if $a=0$, you get a parabola; if $a>0$, but not equal to 1, you get an ellipse; and if $a=1$, you get a circle.

56. If $a<0$, but not equal to -1, you get an ellipse; if $a=-1$, you get a circle; if $a=0$, you get a parabola; and if $a>0$, you get a hyperbola.

57. Consider the equation $3x^2+2\sqrt{3}xy+y^2+Dx+Ey+F=0$. Depending on the choices of *D*, *E*, and *F*, different graphs (if they exist) arise, as seen below:

(i)

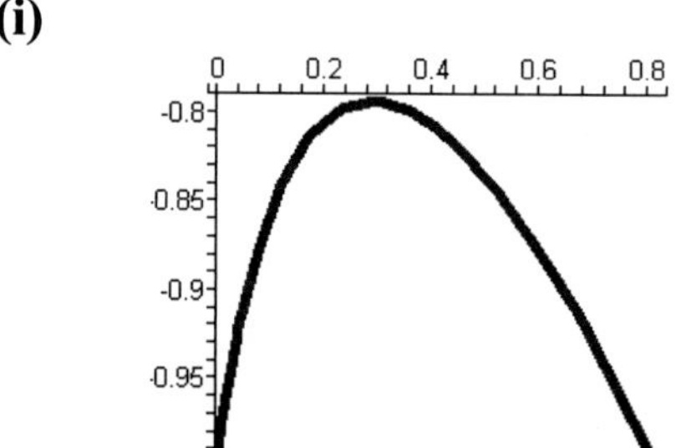

(ii)

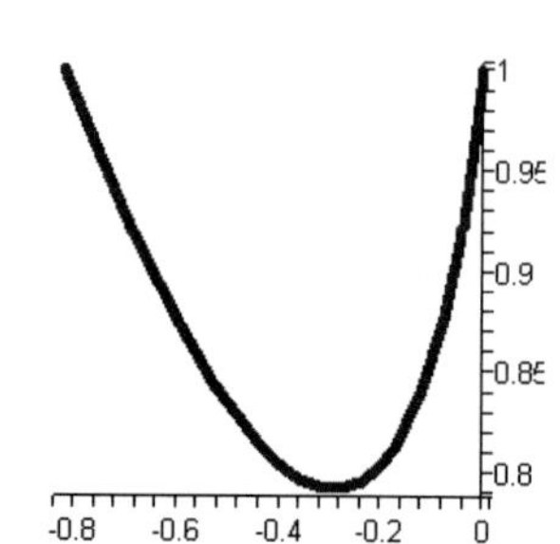

Observe that the graphs appear to be rotations of one another.

58. Consider the equation $x^2 + 3xy + 3y^2 + Dx + Ey + F = 0$. Depending on the choices of *D*, *E*, and *F*, different graphs (if they exist) arise, as seen below:

(i)

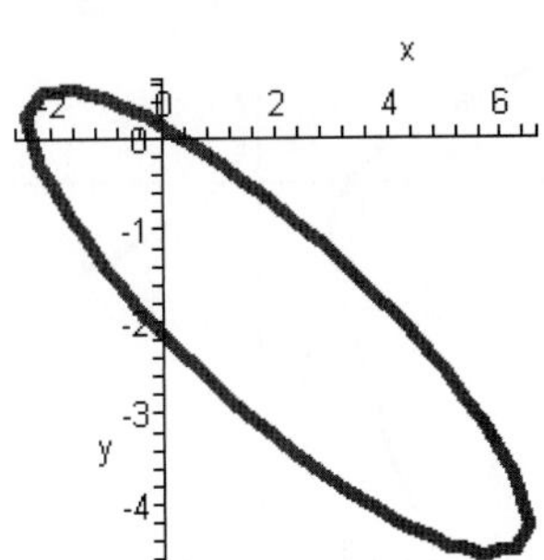

(ii)

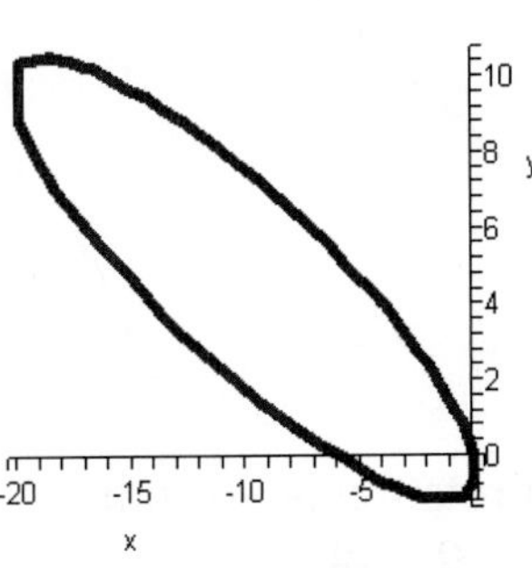

Observe that the graphs appear to be rotations of one another.

59. Consider the equation $2x^2 + 3xy + y^2 + Dx + Ey + F = 0$. Depending on the choices of *D*, *E*, and *F*, different graphs (if they exist) arise, as seen below:

(i)

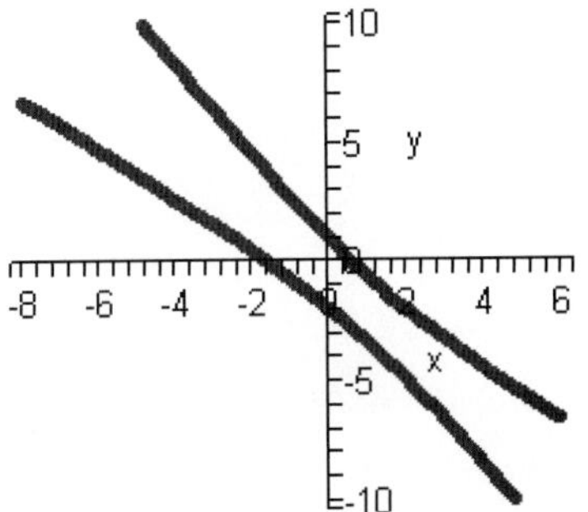

(ii)

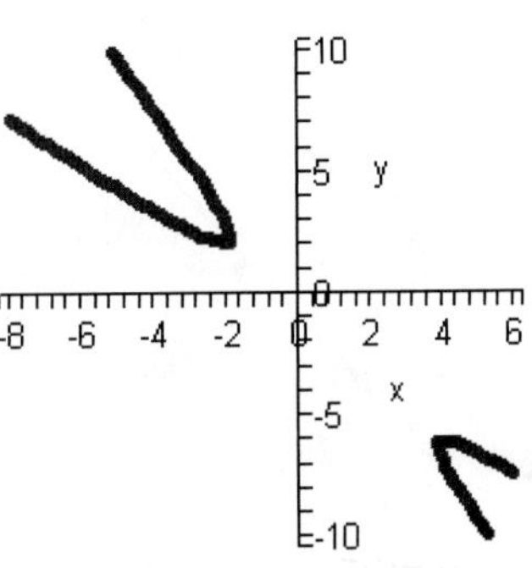

The vertices of the hyperbolae are separating as we make these changes to the coefficients.

60. Consider the equation $2\sqrt{3}x^2 + xy + \sqrt{3}y^2 + Dx + Ey + F = 0$. Depending on the choices of *D*, *E*, and *F*, different graphs (if they exist) arise, as seen below:

(i)

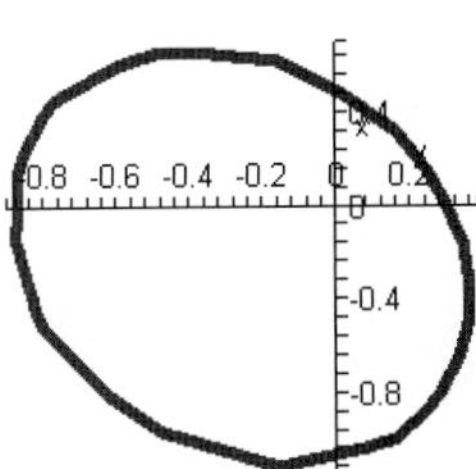

(ii)

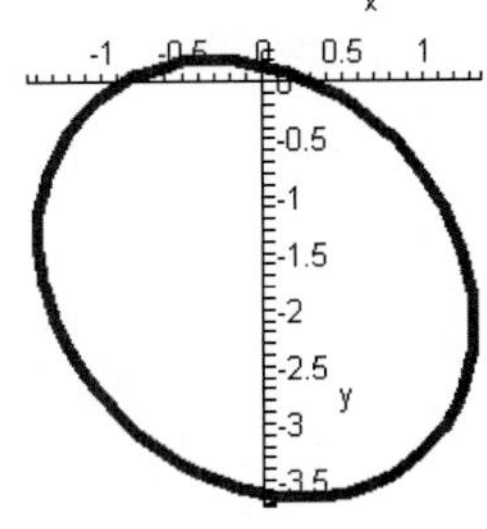

Changing the values of *D*, *E*, and *F* result in a rotation.

61. (i)

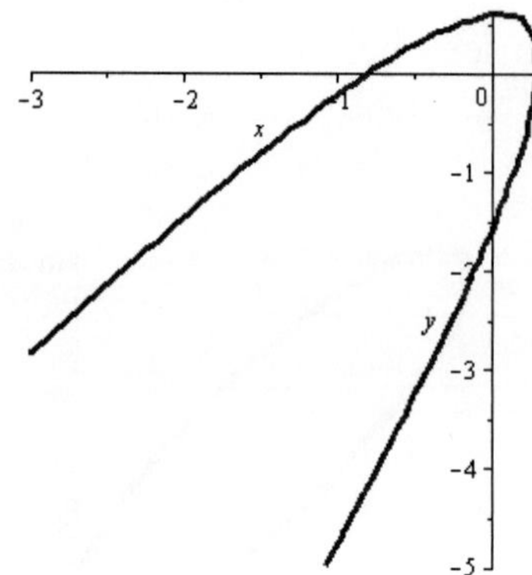

Angle of rotation is given by:

$$\cot 2\theta = \tfrac{4-1}{-4} = -\tfrac{3}{4}$$

$$\theta = \tfrac{1}{2}\cot^{-1}\left(-\tfrac{3}{4}\right) \approx -26.57^\circ$$

(ii)

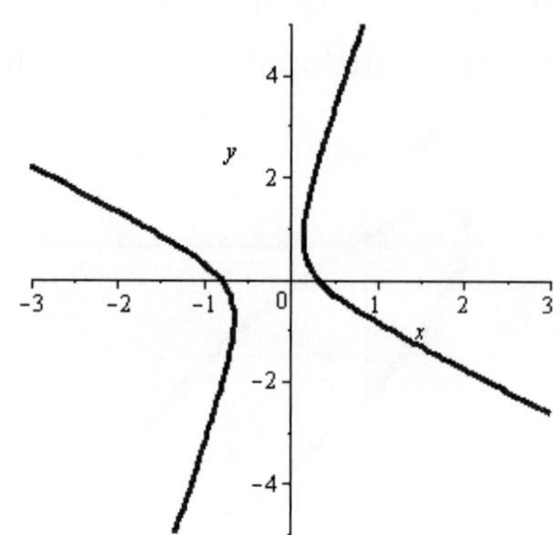

Angle of rotation is given by:

$$\cot 2\theta = \tfrac{4-(-1)}{4} = \tfrac{5}{4}$$

$$\theta = \tfrac{1}{2}\cot^{-1}\left(\tfrac{5}{4}\right) \approx 19.33^\circ$$

(iii)

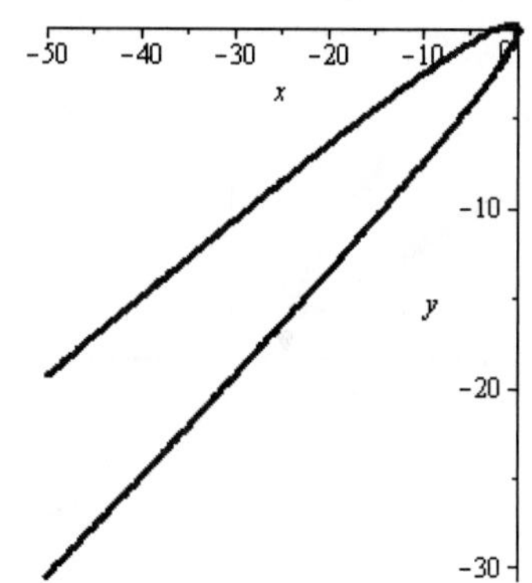

Angle of rotation is given by:

$$\cot 2\theta = \tfrac{1-4}{-4} = \tfrac{3}{4}$$

$$\theta = \tfrac{1}{2}\cot^{-1}\left(\tfrac{3}{4}\right) \approx 26.57^\circ$$

62. (i)

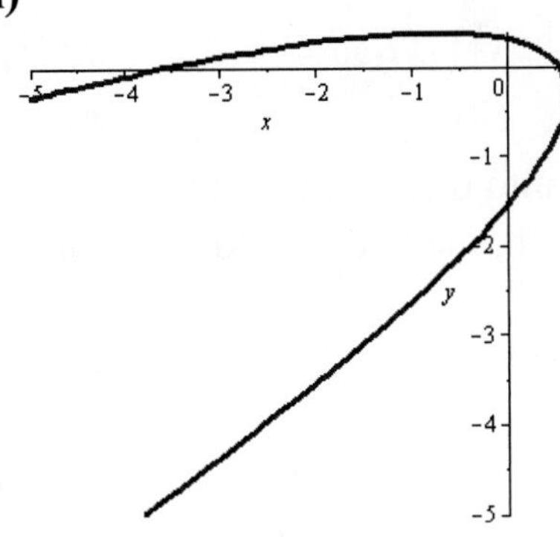

Angle of rotation is given by:

$$\cot 2\theta = \frac{1-4}{-4} = \frac{3}{4}$$

$$\theta = \frac{1}{2}\cot^{-1}\left(\frac{3}{4}\right) \approx 26.57^\circ$$

(ii)

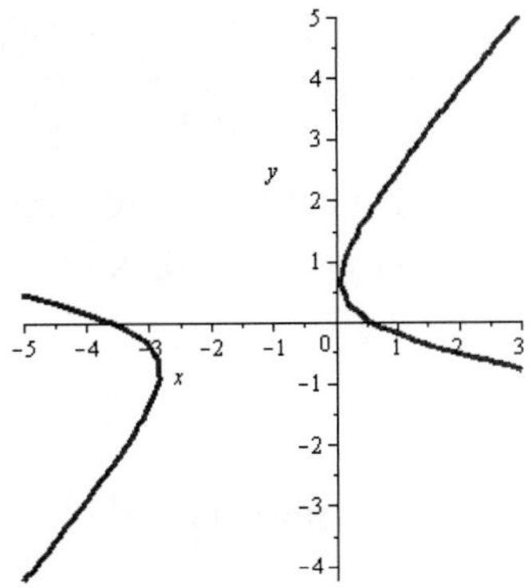

Angle of rotation is given by:

$$\cot 2\theta = \frac{1-(-4)}{4} = \frac{5}{4}$$

$$\theta = \frac{1}{2}\cot^{-1}\left(\frac{5}{4}\right) \approx 19.33^\circ$$

63. Solve the system:

$$\begin{cases} x^2 + 2xy = 10 & \textbf{(1)} \\ 3x^2 - xy = 2 & \textbf{(2)} \end{cases}$$

Solve **(2)** for y: $y = \frac{3x^2-2}{x}$ **(3)**

Substitute **(3)** into **(1)**:

$$x^2 + 2x\left(\frac{3x^2-2}{x}\right) = 10$$

$$7x^2 - 14 = 0$$

$$x = \pm\sqrt{2}$$

Substitute each of these values into **(1)** to find the corresponding y-values:

$$x = -\sqrt{2}: \ y = -2\sqrt{2}$$

$$x = \sqrt{2}: \ y = 2\sqrt{2}$$

So, the solutions of the system are

$$\left(-\sqrt{2}, -2\sqrt{2}\right), \left(\sqrt{2}, 2\sqrt{2}\right)$$

64. Solve the system:

$$\begin{cases} x^2 - 3xy + 2y^2 = 0 & \textbf{(1)} \\ x^2 + xy = 6 & \textbf{(2)} \end{cases}$$

Solve **(2)** for y: $y = \frac{6-x^2}{x}$ **(3)**

Substitute **(3)** into **(1)**:

$$x^2 - 3x\left(\frac{6-x^2}{x}\right) + 2\left(\frac{6-x^2}{x}\right)^2 = 0$$

$$6x^4 - 42x^2 + 72 = 0$$

$$\left(x^2 - 4\right)\left(x^2 - 3\right) = 0$$

$$x = \pm 2, \pm\sqrt{3}$$

Substitute each of these values into **(1)** to find the corresponding y-values:

$$x = -\sqrt{3}: \ y = -\sqrt{3}$$

$$x = \sqrt{3}: \ y = \sqrt{3}$$

$$x = -2: \ y = -1$$

$$x = 2: y = 1$$

So, the solutions of the system are

$$\left(-\sqrt{3}, -\sqrt{3}\right), \left(\sqrt{3}, \sqrt{3}\right), (-2,-1), (2,1)$$

65. Solve the system:

$$\begin{cases} 2x^2 - 7xy + 2y^2 = -1 & \textbf{(1)} \\ x^2 - 3xy + y^2 = 1 & \textbf{(2)} \end{cases}$$

Need to perform a rotation to transform this system into one that is easier to handle. In both equations, $\cot 2\theta = 0$, so that $\theta = \frac{\pi}{4}$. So, for both, make the following substitution:

$$\begin{cases} x = \frac{\sqrt{2}}{2}X - \frac{\sqrt{2}}{2}Y \\ y = \frac{\sqrt{2}}{2}X + \frac{\sqrt{2}}{2}Y \end{cases} \quad (*)$$

Substituting these into **(1)** and **(2)** and simplifying yields the following system in *X* and *Y*:

$$\begin{cases} -3X^2 + 11Y^2 = -2 & \textbf{(3)} \\ -X^2 + 5Y^2 = 2 & \textbf{(4)} \end{cases}$$

To solve this system, multiply **(4)** by -3 and add to **(3)**:

$$-4Y^2 = -8 \Rightarrow Y^2 = 2 \quad \Rightarrow \quad Y = \pm\sqrt{2}$$

Substitute these values into **(4)** to find that $X = \pm 2\sqrt{2}$.
Now, plug these values of *X* and *Y* into **(*)** to find that the corresponding values of *x* that work are: -1, 1, -3, 3. We must determine the corresponding values of *y* that then yield the solutions of the system. To this end, observe that

$$x = -1:\ y = 0, \cancel{-3}$$
$$x = 1:\ y = 0, \cancel{3}$$
$$x = -3:\ y = -1, \cancel{-8}$$
$$x = 3:\ y = 1, \cancel{8}$$

Thus, the solutions of the system are (-1,0), (1,0), (3,1), and (-3, -1).

66. Solve the system:

$$\begin{cases} 4x^2 + xy + 4y^2 = 34 & \textbf{(1)} \\ -3x^2 + 2xy - 3y^2 = -11 & \textbf{(2)} \end{cases}$$

Need to perform a rotation to transform this system into one that is easier to handle. In both equations, $\cot 2\theta = 0$, so that $\theta = \frac{\pi}{4}$. So, for both, make the following substitution:

$$\begin{cases} x = \frac{\sqrt{2}}{2}X - \frac{\sqrt{2}}{2}Y \\ y = \frac{\sqrt{2}}{2}X + \frac{\sqrt{2}}{2}Y \end{cases} \quad (*)$$

Substituting these into **(1)** and **(2)** and simplifying yields the following system in *X* and *Y*:

$$\begin{cases} 9X^2 + 7Y^2 = 68 & \textbf{(3)} \\ 2X^2 + 4Y^2 = 11 & \textbf{(4)} \end{cases}$$

Multiply **(3)** by -2 and **(4)** by 9 – then add: $22Y^2 = -37$, which has no solution. Hence, the system has no solution.

Section 9.8 Solutions --

1. The directrix is horizontal and 5 units below the pole, so the equation is $r=\dfrac{\left(\frac{1}{2}\right)(5)}{1-\left(\frac{1}{2}\right)\sin\theta}$, which simplifies to $\boxed{r=\dfrac{5}{2-\sin\theta}}$.

2. The directrix is horizontal and 3 units above the pole, so the equation is $r=\dfrac{\left(\frac{1}{3}\right)(3)}{1+\left(\frac{1}{3}\right)\sin\theta}$, which simplifies to $\boxed{r=\dfrac{3}{3+\sin\theta}}$.

3. The directrix is horizontal and 4 units above the pole, so the equation is $r=\dfrac{(2)(4)}{1+(2)\sin\theta}$, which simplifies to $\boxed{r=\dfrac{8}{1+2\sin\theta}}$.

4. The directrix is horizontal and 2 units below the pole, so the equation is $r=\dfrac{(3)(2)}{1-(3)\sin\theta}$, which simplifies to $\boxed{r=\dfrac{6}{1-3\sin\theta}}$.

5. The directrix is vertical and 1 unit to the right of the pole, so the equation is $r=\dfrac{(1)(1)}{1+(1)\cos\theta}$, which simplifies to $\boxed{r=\dfrac{1}{1+\cos\theta}}$.

6. The directrix is vertical and 1 unit to the left of the pole, so the equation is $r=\dfrac{(1)(1)}{1-(1)\cos\theta}$, which simplifies to $\boxed{r=\dfrac{1}{1-\cos\theta}}$.

7. The directrix is vertical and 2 units to the right of the pole, so the equation is $r=\dfrac{\left(\frac{3}{4}\right)(2)}{1+\left(\frac{3}{4}\right)\cos\theta}$, which simplifies to $\boxed{r=\dfrac{6}{4+3\cos\theta}}$.

8. The directrix is vertical and 4 units to the left of the pole, so the equation is $r=\dfrac{\left(\frac{2}{3}\right)(4)}{1-\left(\frac{2}{3}\right)\cos\theta}$, which simplifies to $\boxed{r=\dfrac{8}{3-2\cos\theta}}$.

9. The directrix is vertical and 3 units to the left of the pole, so the equation is $r=\dfrac{\left(\frac{4}{3}\right)(3)}{1-\left(\frac{4}{3}\right)\cos\theta}$, which simplifies to $\boxed{r=\dfrac{12}{3-4\cos\theta}}$.

10. The directrix is vertical and 5 units to the right of the pole, so the equation is $r=\dfrac{\left(\frac{3}{2}\right)(5)}{1+\left(\frac{3}{2}\right)\cos\theta}$, which simplifies to $\boxed{r=\dfrac{15}{2+3\cos\theta}}$.

11. The directrix is horizontal and 3 units below the pole, so the equation is $r = \dfrac{(1)(3)}{1-(1)\sin\theta}$, which simplifies to $\boxed{r = \dfrac{3}{1-\sin\theta}}$.

12. The directrix is horizontal and 4 units above the pole, so the equation is $r = \dfrac{(1)(4)}{1+(1)\sin\theta}$, which simplifies to $\boxed{r = \dfrac{4}{1+\sin\theta}}$.

13. The directrix is horizontal and 6 units above the pole, so the equation is $r = \dfrac{(\frac{3}{5})(6)}{1+(\frac{3}{5})\sin\theta}$, which simplifies to $\boxed{r = \dfrac{18}{5+3\sin\theta}}$.

14. The directrix is horizontal and 5 units above the pole, so the equation is $r = \dfrac{(\frac{8}{5})(5)}{1+(\frac{8}{5})\sin\theta}$, which simplifies to $\boxed{r = \dfrac{40}{5+8\sin\theta}}$.

15. Observe that $r = \dfrac{4}{1+\cos\theta} = \dfrac{(1)(4)}{1+(1)\cos\theta}$, so that $e=1$. Hence, this conic is a *parabola.*

16. Observe that $r = \dfrac{3}{2-3\sin\theta} = \dfrac{3}{2(1-\frac{3}{2}\sin\theta)} = \dfrac{\frac{3}{2}}{1-\frac{3}{2}\sin\theta}$, so that $e = \frac{3}{2} > 1$. Hence, this conic is a *hyperbola.*

17. Observe that $r = \dfrac{2}{3+2\sin\theta} = \dfrac{2}{3(1+\frac{2}{3}\sin\theta)} = \dfrac{\frac{2}{3}}{1+\frac{2}{3}\sin\theta}$, so that $e = \frac{2}{3} < 1$. Hence, this conic is an *ellipse.*

18. Observe that $r = \dfrac{3}{2-2\cos\theta} = \dfrac{3}{2(1-\cos\theta)} = \dfrac{(\frac{3}{2})(1)}{(1-\cos\theta)}$, so that $e=1$. Hence, this conic is a *parabola.*

19. Observe that $r = \dfrac{2}{4+8\cos\theta} = \dfrac{2}{4(1+2\cos\theta)} = \dfrac{\frac{1}{2}}{1+2\cos\theta}$, so that $e = 2 > 1$. Hence, this conic is a *hyperbola.*

20. Observe that $r = \dfrac{1}{4-\cos\theta} = \dfrac{1}{4(1-\frac{1}{4}\cos\theta)} = \dfrac{\frac{1}{4}}{1-\frac{1}{4}\cos\theta}$, so that $e = \frac{1}{4} < 1$. Hence, this conic is an *ellipse.*

21. Observe that $r = \dfrac{7}{3+\cos\theta} = \dfrac{7}{3(1+\frac{1}{3}\cos\theta)} = \dfrac{(7)(\frac{1}{3})}{1+\frac{1}{3}\cos\theta}$, so that $e = \frac{1}{3} < 1$. Hence, this conic is an *ellipse.*

22. Observe that $r = \dfrac{4}{5+6\sin\theta} = \dfrac{4}{5\left(1+\frac{6}{5}\sin\theta\right)} = \dfrac{\frac{4}{5}}{1+\frac{6}{5}\sin\theta}$, so that $e = \frac{6}{5} > 1$. Hence, this conic is a *hyperbola*.

23. Observe that $r = \dfrac{40}{5+5\sin\theta} = \dfrac{8}{1+\sin\theta}$, so that $e = 1$. Hence, this conic is a *parabola*.

24. Observe that $r = \dfrac{5}{5-4\sin\theta} = \dfrac{5}{5\left(1-\frac{4}{5}\sin\theta\right)} = \dfrac{1}{1-\frac{4}{5}\sin\theta}$, so that $e = \frac{4}{5} < 1$. Hence, this conic is an *ellipse*.

25. Observe that $r = \dfrac{1}{1-6\cos\theta}$, so that $e = 6 > 1$. Hence, this conic is a *hyperbola*.

26. Observe that $r = \dfrac{5}{3-3\sin\theta} = \dfrac{5}{3(1-\sin\theta)} = \dfrac{\frac{5}{3}}{1-\sin\theta}$, so that $e = 1$. Hence, this conic is a *parabola*.

27. Consider the equation $r = \dfrac{2}{1+\sin\theta}$.

(a) Observe that $r = \dfrac{2}{1+\sin\theta} = \dfrac{(2)(1)}{1+(1)\sin\theta}$, so that $e = 1$. So, this conic is a *parabola*.

(b) Inspecting the equation in (a), we see that $e = 1$ and $p = 2$.
From the general form of the equation, we know the directrix is horizontal. As such, the vertex must lie on the y-axis. Evaluating the equation at $\theta = \frac{\pi}{2}$ yields $r = 1$. So, the vertex is rectangular coordinates is $(0,1)$.

(c) Consider the following table of points:

θ	$r = \dfrac{2}{1+\sin\theta}$	(r,θ)
0	2	$(2,0)$
$\frac{\pi}{2}$	1	$\left(1,\frac{\pi}{2}\right)$
π	2	$(2,\pi)$
$\frac{3\pi}{2}$	undefined	---
2π	2	$(2,2\pi)$

The graph is as follows:

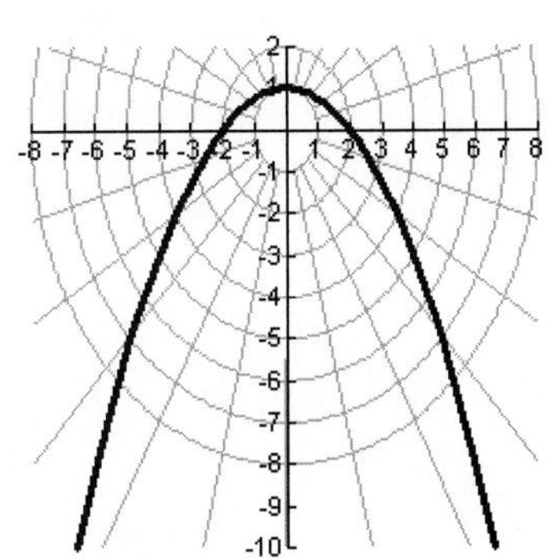

28. Consider the equation $r = \dfrac{4}{1-\cos\theta}$.

(a) Observe that $r = \dfrac{4}{1-\cos\theta} = \dfrac{(4)(1)}{1-(1)\cos\theta}$, so that $e = 1$. So, this conic is a *parabola.*

(b) Inspecting the equation in (a), we see that $e = 1$ and $p = 4$.

From the general form of the equation, we know the directrix is vertical and is to the left of the origin. As such, since the parabola opens to the right, the vertex must lie on the x-axis. Further, since the right-side of the equation is undefined at $\theta = 0$, the vertex must occur at $\theta = \pi$. Evaluating the equation at $\theta = \pi$ yields $r = 2$. So, the vertex is rectangular coordinates is $(-2,0)$.

(c) Consider the following table of points:

θ	$r = \dfrac{4}{1-\cos\theta}$	(r,θ)
0	undefined	---
$\frac{\pi}{2}$	4	$\left(4,\frac{\pi}{2}\right)$
π	2	$(2,\pi)$
$\frac{3\pi}{2}$	4	$\left(4,\frac{3\pi}{2}\right)$
2π	undefined	---

The graph is as follows:

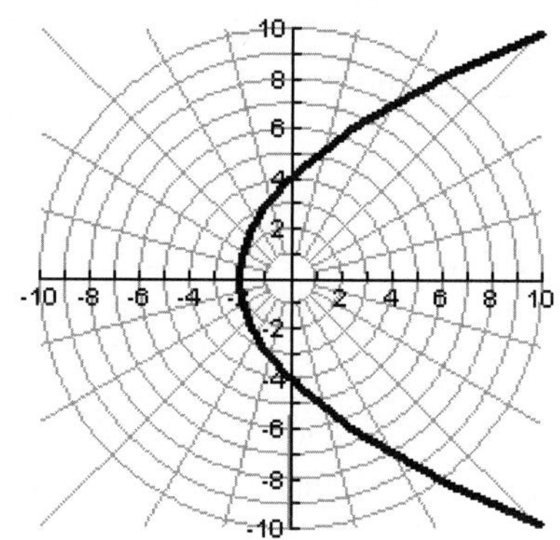

29. Consider the equation $r = \dfrac{4}{1-2\sin\theta}$.

(a) Observe that $r = \dfrac{4}{1-2\sin\theta} = \dfrac{(2)(2)}{1-(2)\sin\theta}$,so that $e = 2$. So, this conic is a *hyperbola.*

(b) Inspecting the equation in (a), we see that $e = 2$ and $p = 2$.
From the general form of the equation, we know the directrix is horizontal and is two units below the origin. Since the directrix is perpendicular to the transverse axis (the axis on which the vertices lie), we know that the transverse axis is vertical and, in fact, is the y-axis since the pole lies on it. Hence, the vertices occur when $\theta = \frac{\pi}{2}, \frac{3\pi}{2}$.
Evaluating the right-side of the equation at these angles yields: $r\left(\frac{\pi}{2}\right) = -4$ and $r\left(\frac{3\pi}{2}\right) = \frac{4}{3}$
So, in rectangular coordinates, the vertices are $(0,-4)$ and $\left(0,-\frac{4}{3}\right)$.

(c) In order to obtain the graph, we need to find the center and equations of the asymptotes.
Center = midpoint of vertices = $\left(0, \dfrac{-4-\frac{4}{3}}{2}\right) = \left(0, -\dfrac{8}{3}\right)$

The distance from the center to the focus is $c = \dfrac{8}{3}$. Now, using $e = \dfrac{c}{a}$ with $e = 2$, $c = \dfrac{8}{3}$, we see that $a = \dfrac{c}{e} = \dfrac{4}{3}$. Since $b^2 = c^2 - a^2$, we further find that $b^2 = \dfrac{48}{9} = \dfrac{16}{3}$, so that $b = \dfrac{4\sqrt{3}}{3}$. Thus, the asymptotes are $y = \pm\dfrac{a}{b}x = \pm\dfrac{4/3}{4\sqrt{3}/3}x = \pm\dfrac{1}{\sqrt{3}}x$.

The graph is as follows:

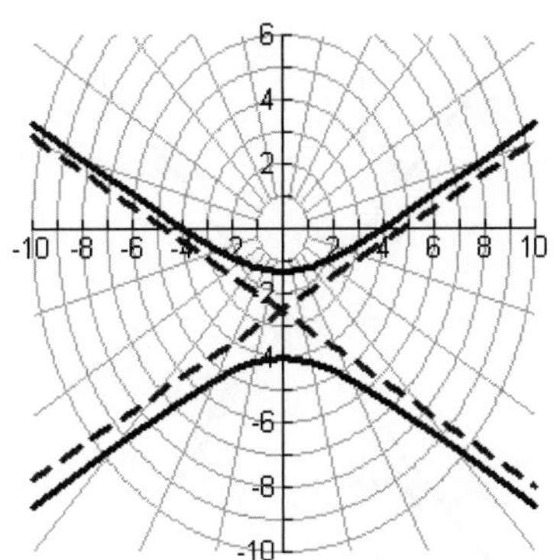

30. Consider the equation $r = \frac{3}{3+8\cos\theta}$.

(a) Observe that $r = \frac{3}{3+8\cos\theta} = \frac{3}{3\left(1+\frac{8}{3}\cos\theta\right)} = \frac{1}{1+\frac{8}{3}\cos\theta} = \frac{\left(\frac{8}{3}\right)\left(\frac{3}{8}\right)}{1+\frac{8}{3}\cos\theta}$,so that $e = \frac{8}{3} > 1$. So, this conic is a *hyperbola*.

(b) Inspecting the equation in (a), we see that $e = \frac{8}{3}$ and $p = \frac{3}{8}$.

From the general form of the equation, we know the directrix is vertical and is $\frac{3}{8}$ unit to the right of the origin. Since the directrix is perpendicular to the transverse axis (the axis on which the vertices lie), we know that the transverse axis is horizontal and, in fact, is the *x*-axis since the pole lies on it. Hence, the vertices occur when $\theta = 0, \pi$. Evaluating the right-side of the equation at these angles yields: $r(0) = \frac{3}{11}$ and $r(\pi) = -\frac{3}{5}$

So, in rectangular coordinates, the vertices are $\left(\frac{3}{11}, 0\right)$ and $\left(-\frac{3}{5}, 0\right)$.

(c) In order to obtain the graph, we need to find the center and equations of the asymptotes.

Center = midpoint of vertices = $\left(0, \frac{\frac{3}{11}+\frac{3}{5}}{2}\right) = \left(0, \frac{24}{55}\right)$

The distance from the center to the focus is $c = \frac{24}{55}$. Now, using $e = \frac{c}{a}$ with $e = \frac{8}{3}$, $c = \frac{24}{55}$, we see that $a = \frac{c}{e} = \frac{24/55}{8/3} = \frac{9}{55}$. Since $b^2 = c^2 - a^2$, we further find that $b^2 = \frac{495}{55^2}$, so that $b = \frac{3\sqrt{55}}{55}$. Thus, the asymptotes are

$$y = \pm\frac{b}{a}x = \pm\frac{\frac{3\sqrt{55}}{55}}{9/55}x = \pm\frac{\sqrt{55}}{3}x.$$

The graph is as follows:

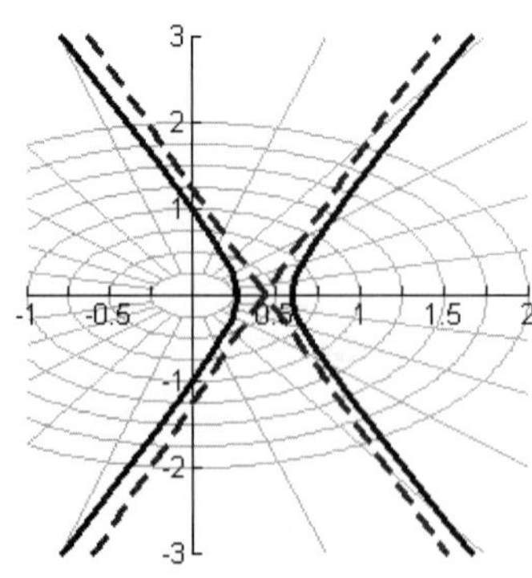

31. Consider the equation $r = \dfrac{2}{2+\sin\theta}$.

(a) Observe that $r = \dfrac{2}{2+\sin\theta} = \dfrac{2}{2\left(1+\frac{1}{2}\sin\theta\right)} = \dfrac{1}{1+\frac{1}{2}\sin\theta} = \dfrac{\left(\frac{1}{2}\right)(2)}{1+\frac{1}{2}\sin\theta}$,so that $e = \frac{1}{2} < 1$.

So, this conic is an *ellipse*.

(b) Inspecting the equation in (a), we see that $e = \frac{1}{2}$ and $p = 2$.

From the general form of the equation, we know the directrix is horizontal and is 2 units above the origin. Since the directrix is perpendicular to the major axis, we know that the major axis lies along the y-axis. Hence, the vertices (which lie on the major axis) occur when $\theta = \frac{\pi}{2}, \frac{3\pi}{2}$. Evaluating the right-side of the equation at these angles yields:

$r\left(\frac{\pi}{2}\right) = \frac{2}{3}$ and $r\left(\frac{3\pi}{2}\right) = 2$

So, in rectangular coordinates, the vertices are $\left(0, \frac{2}{3}\right)$ and $(0, -2)$.

(c) In order to obtain the graph, we need to find the center and x-intercepts.

Center = midpoint of vertices = $\left(0, \dfrac{\frac{2}{3}-2}{2}\right) = \left(0, -\dfrac{2}{3}\right)$

The length of the major axis = $2a = \frac{2}{3} - (-2) = \frac{8}{3}$, so that $a = \frac{4}{3}$. Now, using $e = \dfrac{c}{a}$, we see that $c = ae = \left(\frac{4}{3}\right)\left(\frac{1}{2}\right) = \frac{2}{3}$. Since $b^2 = a^2 - c^2$, we further find that $b^2 = \dfrac{4}{3}$, so that $b = \dfrac{2\sqrt{3}}{3}$. Hence, the length of the minor axis (which is perpendicular to the major axis through the center) = $2b$. So, the vertices of the minor axis are $\left(\pm\frac{2\sqrt{3}}{3}, -\frac{2}{3}\right)$.

The graph is as follows:

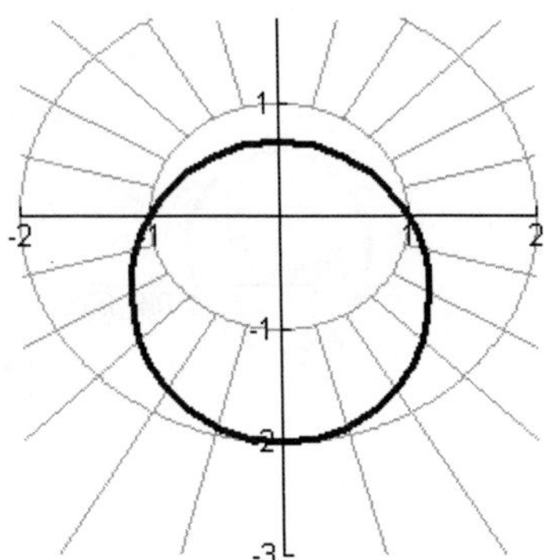

32. Consider the equation $r = \dfrac{1}{3-\sin\theta}$.

(a) Observe that $r = \dfrac{1}{3-\sin\theta} = \dfrac{1}{3\left(1-\frac{1}{3}\sin\theta\right)} = \dfrac{\left(\frac{1}{3}\right)(1)}{1-\frac{1}{3}\sin\theta}$,so that $e=\frac{1}{3}<1$. So, this conic is an *ellipse*.

(b) Inspecting the equation in (a), we see that $e=\frac{1}{3}$ and $p=1$.
From the general form of the equation, we know the directrix is horizontal and is 1 unit below the origin. Since the directrix is perpendicular to the major axis, we know that the major axis lies along the *y*-axis. Hence, the vertices (which lie on the major axis) occur when $\theta = \frac{\pi}{2}, \frac{3\pi}{2}$. Evaluating the right-side of the equation at these angles yields:
$r\left(\frac{\pi}{2}\right)=\frac{1}{2}$ and $r\left(\frac{3\pi}{2}\right)=\frac{1}{4}$
So, in rectangular coordinates, the vertices are $\left(0,\frac{1}{2}\right)$ and $\left(0,-\frac{1}{4}\right)$.

(c) In order to obtain the graph, we need to find the center and *x*-intercepts.
Center = midpoint of vertices = $\left(0, \dfrac{\frac{1}{2}-\frac{1}{4}}{2}\right) = \left(0,\dfrac{1}{8}\right)$

The length of the major axis = $2a = \frac{1}{2}-\left(-\frac{1}{4}\right)=\frac{3}{4}$, so that $a=\frac{3}{8}$. Now, using $e=\dfrac{c}{a}$, we see that $c = ae = \left(\frac{3}{8}\right)\left(\frac{1}{3}\right)=\frac{1}{8}$. Since $b^2=a^2-c^2$, we further find that $b^2 = \dfrac{1}{8}$, so that $b = \dfrac{\sqrt{2}}{4}$. Hence, the length of the minor axis (which is perpendicular to the major axis through the center) = $2b$. So, the vertices of the minor axis are $\left(\pm\frac{\sqrt{2}}{4},\frac{1}{8}\right)$.
The graph is as follows:

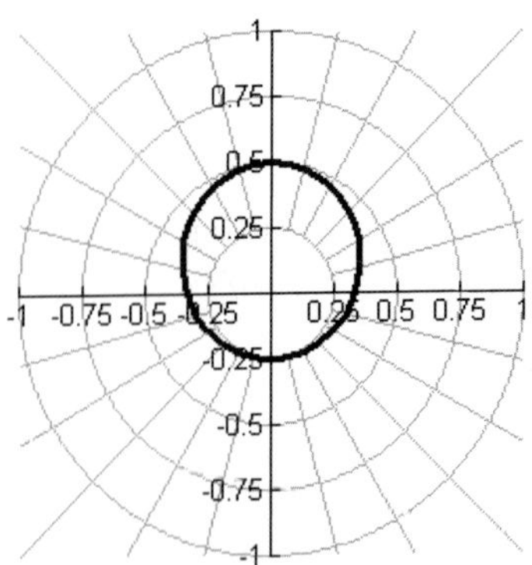

35. Consider the equation $r = \dfrac{4}{3+\cos\theta}$.

(a) Observe that $r = \dfrac{4}{3+\cos\theta} = \dfrac{4}{3\left(1+\frac{1}{3}\cos\theta\right)} = \dfrac{\frac{4}{3}}{1+\frac{1}{3}\cos\theta} = \dfrac{4\left(\frac{1}{3}\right)}{1+\frac{1}{3}\cos\theta}$,so that $e = \frac{1}{3} < 1$.

So, this conic is an *ellipse*.

(b) Inspecting the equation in (a), we see that $e = \frac{1}{3}$ and $p = 4$.
From the general form of the equation, we know the directrix is vertical and is 4 units to the right of the origin. Since the directrix is perpendicular to the major axis, we know that the major axis lies along the *x*-axis. Hence, the vertices (which lie on the major axis) occur when $\theta = 0, \pi$. Evaluating the right-side of the equation at these angles yields: $r(0) = 1$ and $r(\pi) = 2$

So, in rectangular coordinates, the vertices are $(1,0)$ and $(-2,0)$.

(c) In order to obtain the graph, we need to find the center and *y*-intercepts.

Center = midpoint of vertices = $\left(\dfrac{1-2}{2}, 0\right) = \left(-\dfrac{1}{2}, 0\right)$

The length of the major axis = $2a = 1-(-2) = 3$, so that $a = \frac{3}{2}$. Now, using $e = \dfrac{c}{a}$, we see that $c = ae = \left(\frac{3}{2}\right)\left(\frac{1}{3}\right) = \frac{1}{2}$. Since $b^2 = a^2 - c^2$, we further find that $b^2 = 2$, so that $b = \sqrt{2}$. Hence, the length of the minor axis (which is perpendicular to the major axis through the center) = $2b$. So, the vertices of the minor axis are $\left(-\frac{1}{2}, \pm\sqrt{2}\right)$.

The graph is as follows:

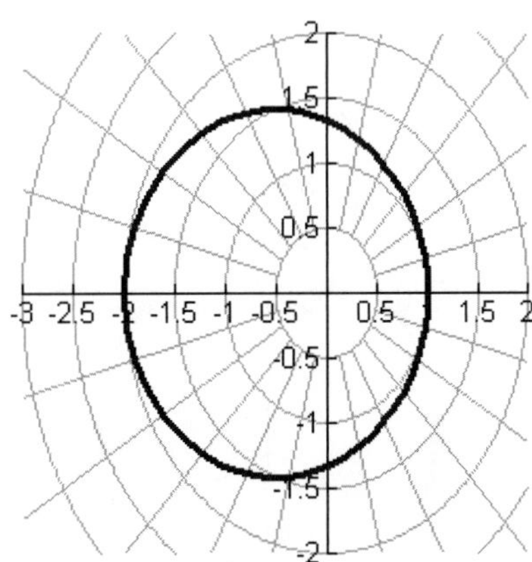

36. Consider the equation $r = \dfrac{2}{5+4\sin\theta}$.

(a) Observe that $r = \dfrac{2}{5+4\sin\theta} = \dfrac{2}{5\left(1+\frac{4}{5}\sin\theta\right)} = \dfrac{\frac{2}{5}}{1+\frac{4}{5}\sin\theta} = \dfrac{\left(\frac{4}{5}\right)\left(\frac{1}{2}\right)}{1+\frac{4}{5}\sin\theta}$,so that $e = \frac{4}{5} < 1$.

So, this conic is an *ellipse.*

(b) Inspecting the equation in (a), we see that $e = \frac{4}{5}$ and $p = \frac{1}{2}$.

From the general form of the equation, we know the directrix is horizontal and is ½ unit above the origin. Since the directrix is perpendicular to the major axis, we know that the major axis lies along the *y*-axis. Hence, the vertices (which lie on the major axis) occur when $\theta = \frac{\pi}{2}, \frac{3\pi}{2}$. Evaluating the right-side of the equation at these angles yields: $r\left(\frac{\pi}{2}\right) = \frac{2}{9}$ and $r\left(\frac{3\pi}{2}\right) = 2$

So, in rectangular coordinates, the vertices are $\left(0, \frac{2}{9}\right)$ and $\left(0, -2\right)$.

(c) In order to obtain the graph, we need to find the center and *x*-intercepts.

Center = midpoint of vertices = $\left(0, \dfrac{\frac{2}{9}-2}{2}\right) = \left(0, -\dfrac{8}{9}\right)$

The length of the major axis = $2a = \frac{2}{9} - (-2) = \frac{20}{9}$, so that $a = \frac{10}{9}$. Now, using $e = \dfrac{c}{a}$, we see that $c = ae = \left(\frac{10}{9}\right)\left(\frac{4}{5}\right) = \frac{8}{9}$. Since $b^2 = a^2 - c^2$, we further find that $b^2 = \dfrac{36}{81}$, so that $b = \dfrac{2}{3}$. Hence, the length of the minor axis (which is perpendicular to the major axis through the center) = $2b$. So, the *x*-intercepts are $\left(\pm\frac{2}{3}, -\frac{8}{9}\right)$.

The graph is as follows:

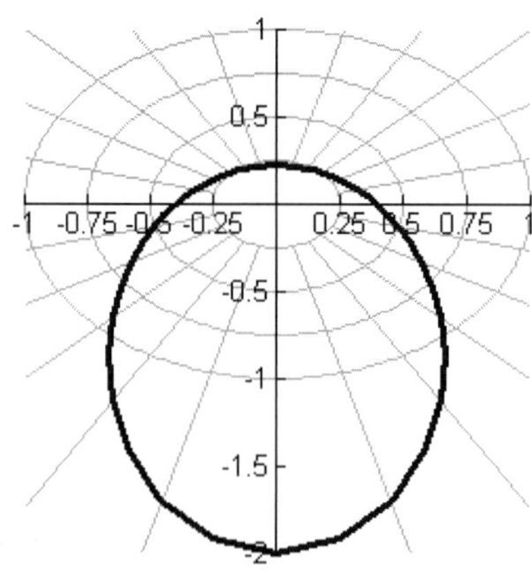

37. Consider the equation $r = \dfrac{6}{2+3\sin\theta}$.

(a) Observe that $r = \dfrac{6}{2+3\sin\theta} = \dfrac{3}{1+\left(\frac{3}{2}\right)\sin\theta} = \dfrac{\left(\frac{3}{2}\right)(2)}{1+\left(\frac{3}{2}\right)\sin\theta}$,so that $e=\frac{3}{2}$. So, this conic is a *hyperbola.*

(b) Inspecting the equation in (a), we see that $e=\frac{3}{2}$ and $p=2$.
From the general form of the equation, we know the directrix is horizontal and is two units above the origin. Since the directrix is perpendicular to the transverse axis (the axis on which the vertices lie), we know that the transverse axis is vertical and, in fact, is the y-axis since the pole lies on it. Hence, the vertices occur when $\theta = \frac{\pi}{2}, \frac{3\pi}{2}$. Evaluating the right-side of the equation at these angles yields: $r\left(\frac{\pi}{2}\right) = \frac{6}{5}$ and $r\left(\frac{3\pi}{2}\right) = -6$
So, in rectangular coordinates, the vertices are $\left(0,\frac{6}{5}\right)$ and $(0,6)$.

(c) In order to obtain the graph, we need to find the center and equations of the asymptotes.

Center = midpoint of vertices = $\left(0,\dfrac{\frac{6}{5}+6}{2}\right) = \left(0,\dfrac{18}{5}\right)$

The distance from the center to the focus is $c = \dfrac{18}{5}$. Now, using $e=\dfrac{c}{a}$ with

$e=\dfrac{3}{2}$, $c=\dfrac{18}{5}$, we see that $a = \dfrac{c}{e} = \dfrac{18/5}{3/2} = \dfrac{12}{5}$. Since $b^2 = c^2 - a^2$, we further find that

$b^2 = \dfrac{36}{5}$, so that $b = \dfrac{6\sqrt{5}}{5}$. Thus, the asymptotes are $y = \pm\dfrac{a}{b}x = \pm\dfrac{12/5}{6\sqrt{5}/5}x = \pm\dfrac{2\sqrt{5}}{5}x$.

The graph is as follows:

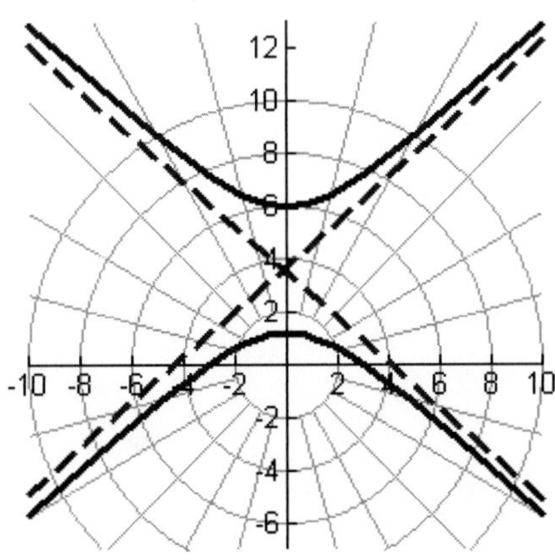

38. Consider the equation $r = \frac{6}{1+\cos\theta}$.

(a) From the equation, we see that $e = 1$. So, this conic is a *parabola.*

(b) Inspecting the equation in (a), we see that $e = 1$ and $p = 6$.

From the general form of the equation, we know the directrix is vertical and is to the right of the origin. As such, since the parabola opens to the right, the vertex must lie on the *x*-axis. Further, since the right-side of the equation is undefined at $\theta = \pi$, the vertex must occur at $\theta = 0$. Evaluating the equation at $\theta = 0$ yields $r = 3$. So, the vertex is rectangular coordinates is $(0,3)$.

(c) Consider the following table of points: The graph is as follows:

θ	$r = \frac{6}{1+\cos\theta}$	(r,θ)
0	3	$(3,0)$
$\frac{\pi}{2}$	6	$(6,\frac{\pi}{2})$
π	Undefined	---
$\frac{3\pi}{2}$	6	$(6,\frac{3\pi}{2})$
2π	3	$(3,2\pi)$

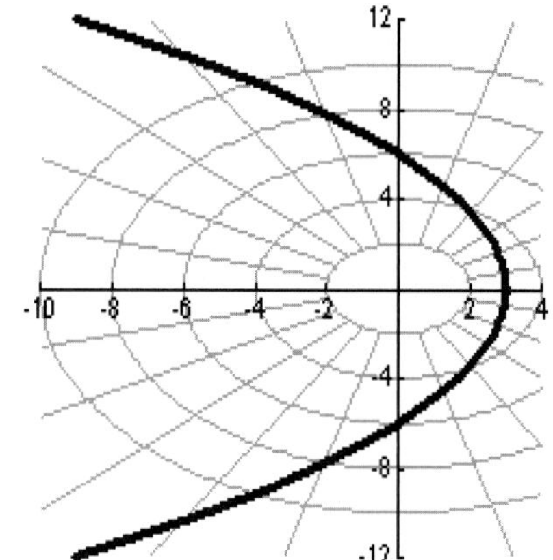

39. Consider the equation $r = \dfrac{2}{5+5\cos\theta}$.

(a) Observe that $r = \dfrac{2}{5+5\cos\theta} = \dfrac{2}{5(1+\cos\theta)} = \dfrac{(2/5)(1)}{1+\cos\theta}$ that $e=1$. So, this conic is a *parabola.*

(b) Inspecting the equation in (a), we see that $e=1$ and $p=\frac{2}{5}$.

From the general form of the equation, we know the directrix is vertical and is to the right of the origin. As such, since the parabola opens to the right, the vertex must lie on the *x*-axis. Further, since the right-side of the equation is undefined at $\theta=\pi$, the vertex must occur at $\theta=0$. Evaluating the equation at $\theta=0$ yields $r=\frac{1}{5}$. So, the vertex is rectangular coordinates is $\left(0,\frac{1}{5}\right)$.

(c) Consider the following table of points:

θ	$r=\dfrac{2}{5+5\cos\theta}$	(r,θ)
0	$1/5$	$(1/5, 0)$
$\frac{\pi}{2}$	$2/5$	$(2/5, \frac{\pi}{2})$
π	undefined	---
$\frac{3\pi}{2}$	$2/5$	$(2/5, \frac{3\pi}{2})$
2π	$1/5$	$(1/5, 2\pi)$

The graph is as follows:

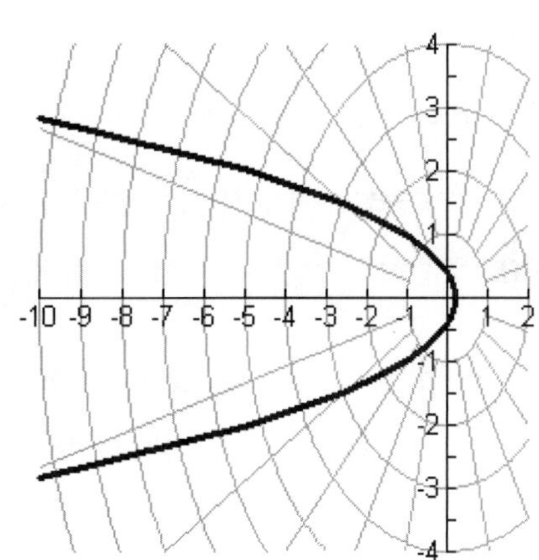

40. Consider the equation $r = \dfrac{10}{6-3\cos\theta}$.

(a) Observe that $r = \dfrac{10}{6-3\cos\theta} = \dfrac{10}{6\left(1-\frac{1}{2}\cos\theta\right)} = \dfrac{\frac{5}{3}}{1-\frac{1}{2}\sin\theta} = \dfrac{\left(\frac{10}{3}\right)\left(\frac{1}{2}\right)}{1-\frac{1}{2}\sin\theta}$, so that $e=\frac{1}{2}<1$. So, this conic is an *ellipse.*

(b) Inspecting the equation in (a), we see that $e=\frac{1}{2}$ and $p=\frac{10}{3}$.
From the general form of the equation, we know the directrix is vertical and is $\frac{10}{3}$ units to the left of the origin. Since the directrix is perpendicular to the major axis, we know that the major axis lies along the *x*-axis. Hence, the vertices (which lie on the major axis) occur when $\theta = 0, \pi$. Evaluating the right-side of the equation at these angles yields: $r(0)=\frac{10}{3}$ and $r(\pi)=\frac{10}{9}$
So, in rectangular coordinates, the vertices are $\left(\frac{10}{3},0\right)$ and $\left(-\frac{10}{9},0\right)$.

(c) In order to obtain the graph, we need to find the center and *x*-intercepts.
Center = midpoint of vertices = $\left(\dfrac{\frac{10}{3}-\frac{10}{9}}{2},0\right) = \left(\dfrac{10}{9},0\right)$

The length of the major axis = $2a = \frac{10}{3}-\left(-\frac{10}{9}\right)=\frac{40}{9}$, so that $a=\frac{20}{9}$. Now, using $e=\dfrac{c}{a}$, we see that $c=ae=\left(\frac{20}{9}\right)\left(\frac{1}{2}\right)=\frac{10}{9}$. Since $b^2=a^2-c^2$, we further find that $b^2=\dfrac{300}{81}=\dfrac{100}{27}$, so that $b=\dfrac{10\sqrt{3}}{9}$. Hence, the length of the minor axis (which is perpendicular to the major axis through the center) = $2b$. So, the vertices of the minor axis are $\left(\frac{10}{9},\pm\frac{10\sqrt{3}}{9}\right)$. The graph is as follows:

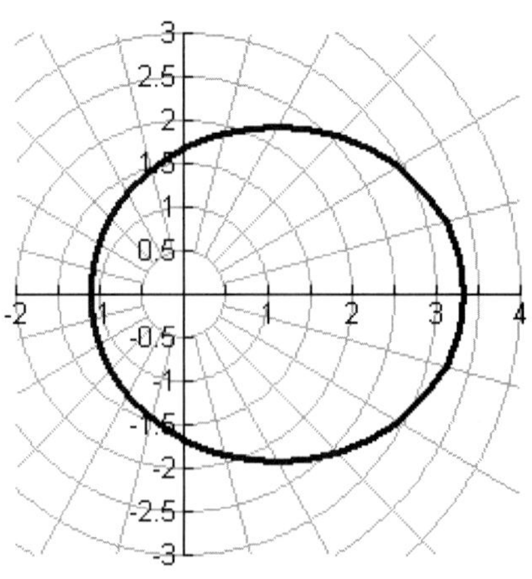

41. Consider the equation $r = \dfrac{6}{3\cos\theta + 1} = \dfrac{2(3)}{3\cos\theta + 1}$.

(a) $e = 3 > 1$, so the conic is a *hyperbola.*

(b) From the general form of the equation, we know the directrix is vertical. The vertices occur when $\theta = 0, \pi$. Evaluating the right-side of the equation at these angles yields: $r(0) = \frac{3}{2}$, $r(\pi) = -3$. So, in rectangular coordinates, the vertices are $\left(\frac{3}{2}, 0\right)$ and $\left(-3, \pi\right)$. The graph is as follows:

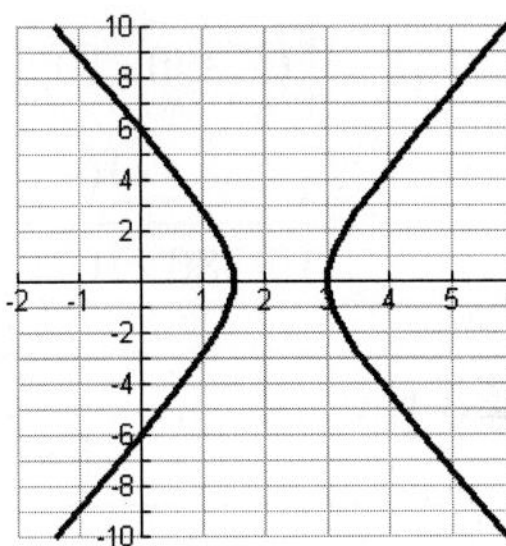

42. Consider the equation $r = \dfrac{15}{3\sin\theta + 5} = \dfrac{15}{5\left(\frac{3}{5}\sin\theta + 1\right)} = \dfrac{\frac{3}{5}(5)}{\frac{3}{5}\sin\theta + 1}$.

(a) $e = \frac{3}{5} < 1$, so the conic is a *ellipse.*

(b) From the general form of the equation, we know the directrix is horizontal. The major axis is vertical so, vertices occur when $\theta = \frac{\pi}{2}, \frac{3\pi}{2}$. Evaluating the right-side of the equation at these angles yields: $r(\frac{\pi}{2}) = \frac{15}{8}$, $r(\frac{3\pi}{2}) = \frac{15}{2}$. So, in rectangular coordinates, the vertices are $\left(\frac{15}{8}, \frac{\pi}{2}\right)$ and $\left(\frac{15}{2}, \frac{3\pi}{2}\right)$. The graph is as follows:

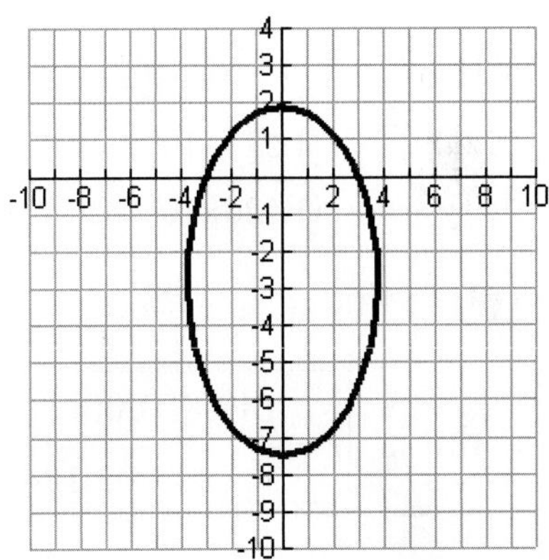

43. The general equation of a planet's orbit is $r = \frac{a(1-e^2)}{1-e\cos\theta}$. Using the information provided regarding Pluto's orbit, we have:

Perihelion = 4.447×10^9 km $= a(1-e)$ **(1)** Aphelion = 7.380×10^9 km $= a(1+e)$ **(2)**

We need to solve this system of equations. We proceed as follows:

Solve **(1)** for a: $a = \frac{4.447\times10^9}{1-e}$ **(3)** (Note: $1-e\neq 0$ since orbit is elliptical, so $e<1$.)

Substitute **(3)** into **(2)** and solve for e:

$$7.380\times10^9 = \left(\frac{4.447\times10^9}{1-e}\right)(1+e)$$

$$(7.380\times10^9)-(4.447\times10^9) = \left[(7.380\times10^9)+(4.447\times10^9)\right]e$$

$$e = \frac{(7.380\times10^9)-(4.447\times10^9)}{(7.380\times10^9)+(4.447\times10^9)} \approx 0.248$$

Substitute this value of e into **(3)** to find a: $a \approx \frac{4.447\times10^9}{1-0.248} \simeq 5{,}913{,}500$

Hence, the equation of Pluto's orbit is $\boxed{r = \frac{5{,}913{,}500{,}000(1-0.248^2)}{1-0.248\cos\theta}}$.

44. The general equation of a planet's orbit is $r = \frac{a(1-e^2)}{1-e\cos\theta}$. Using the information provided regarding Earth's orbit, we have:

Perihelion = 1.471×10^8 km $= a(1-e)$ **(1)** Aphelion = 1.526×10^8 km $= a(1+e)$ **(2)**

We need to solve this system of equations. We proceed as follows:

Solve **(1)** for a: $a = \frac{1.471\times10^8}{1-e}$ **(3)** (Note: $1-e\neq 0$ since orbit is elliptical, so $e<1$.)

Substitute **(3)** into **(2)** and solve for e:

$$1.526\times10^8 = \left(\frac{1.471\times10^8}{1-e}\right)(1+e)$$

$$(1.526\times10^8)-(1.471\times10^8) = \left[(1.526\times10^8)+(1.471\times10^8)\right]e$$

$$e = \frac{(1.526\times10^8)-(1.471\times10^8)}{(1.526\times10^8)+(1.471\times10^8)} \approx 0.0184$$

Substitute this value of e into **(3)** to find a: $a \approx \frac{1.471\times10^8}{1-0.0184} \simeq 149{,}850{,}000$

Hence, the equation of Pluto's orbit is $\boxed{r = \frac{149{,}850{,}000(1-0.0184^2)}{1-0.0184\cos\theta}}$.

45. We know that 433 Eros' orbit is elliptical, $a = 150{,}000{,}000$ km, and $e = 0.223$.
Hence, the equation of its orbit is $r = \dfrac{a(1-e^2)}{1-e\cos\theta} = \dfrac{150{,}000{,}000(1-0.223^2)}{1-0.223\cos\theta}$.

46. We know that Toutatis' orbit is elliptical, $a = 350{,}000{,}000$ km, and $e = 0.634$.
Hence, the equation of its orbit is $r = \dfrac{a(1-e^2)}{1-e\cos\theta} = \dfrac{350{,}000{,}000(1-0.634^2)}{1-0.634\cos\theta}$.

47. Observe that $r = \dfrac{1}{1+0.0167\cos\theta} = \dfrac{0.0167\left(\frac{1}{0.0167}\right)}{1+0.0167\cos\theta}$

The eccentricity $e = 0.0167 < 1$. So, the conic is an *ellipse.*
The directrix is vertical, so that the major axis is horizontal. So, we use $\theta = 0, \pi$ for the vertices. Observe that $r(0) = \frac{1}{1.0167}, r(\pi) = \frac{1}{0.9833}$. So, the vertices are (in rectangular coordinates) $(0.98357, 0)$ and $(-1.01698, 0)$.

a) The center in rectangular coordinates is the midpoint between the two vertices, namely (-0.0167,0).

b) In polar coordinates, this corresponds to $(0.0167, \pi)$.

48. Observe that $r = \dfrac{1}{1+0.0461\cos\theta} = \dfrac{0.0461\left(\frac{1}{0.0461}\right)}{1+0.0461\cos\theta}$

The eccentricity $e = 0.0461 < 1$. So, the conic is an *ellipse.*
The directrix is vertical, so that the major axis is horizontal. So, we use $\theta = 0, \pi$ for the vertices. Observe that $r(0) = \frac{1}{1.0461}, r(\pi) = \frac{1}{0.9539}$. So, the vertices are (in rectangular coordinates) $(0.95593, 0)$ and $(-1.0483, 0)$.

a) The center in rectangular coordinates is the midpoint between the two vertices, namely (-0.0462,0).

b) In polar coordinates, this corresponds to $(0.0462, \pi)$.

49. Observe that $r = \dfrac{1}{1+0.967\sin\theta}$

The eccentricity $e = 0.967 < 1$. So, the conic is an *ellipse.*
The directrix is horizontal, so that the major axis is vertical. So, we use $\theta = \frac{\pi}{2}, \frac{3\pi}{2}$ for the vertices. Observe that $r(\frac{\pi}{2}) = \frac{1}{1.967} = 0.50839$, $r(\pi) = \frac{1}{0.033} = 30.3030$. So, the vertices are (in rectangular coordinates) $(0, 0.50839)$ and $(0, -30.3030)$.

Thus, the center in rectangular coordinates is the midpoint between the two vertices, namely (0, -15.406).

50. Observe that $r = \dfrac{1}{1+0.995\sin\theta}$

The eccentricity $e = 0.995 < 1$. So, the conic is an *ellipse*.
The directrix is horizontal, so that the major axis is vertical. So, we use $\theta = \frac{\pi}{2}, \frac{3\pi}{2}$ for the vertices. Observe that $r(\frac{\pi}{2}) = \frac{1}{1.995} = 0.50125,\ r(\pi) = \frac{1}{0.005} = 200$. So, the vertices are (in rectangular coordinates) $(0, 0.50125)$ and $(0, -200)$.
Thus, the center in rectangular coordinates is the midpoint between the two vertices, namely (0, -100.251).

51. Recall that $e = \frac{c}{a}$, where c = distance between the center and the focus, and $2a$ = length of the major axis. Note that $e \to 0$ only if $c \to 0$. If this occurs, the distance between the center of the ellipse and the foci gets smaller, causing it to appear more circular. On the other hand, $e \to 1$ only if c and a get arbitrarily close, thereby causing the curve to become more elliptical.

52. Assume that the directrix is horizontal (hence parallel to the polar axis) and is p units below the pole. Consider the following diagram:

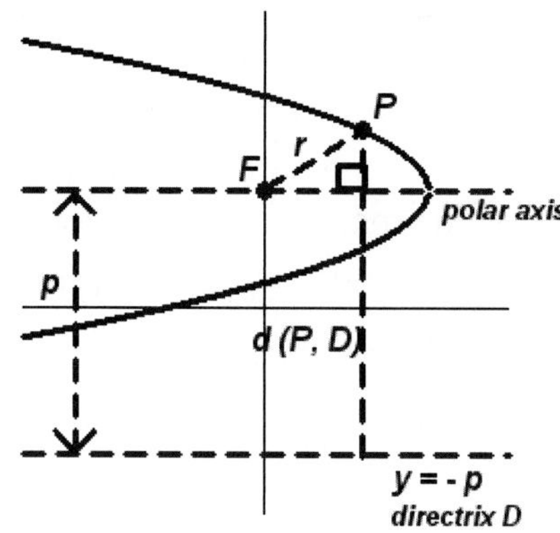

Observe that

$$d(P,F) = r \quad \textbf{(1)} \qquad d(P,D) = \text{ distance from } P \text{ to the closest point on } D. \quad \textbf{(2)}$$

Substituting **(1)** and **(2)** into $\dfrac{d(P,F)}{d(P,D)} = e$ yields $\dfrac{r}{p + r\sin\theta} = e$. Now, Solving for r gives:

$$r = e(p + r\sin\theta)$$
$$r = ep + er\sin\theta$$
$$r(1 - e\sin\theta) = ep$$
$$r = \frac{ep}{1 - e\sin\theta},$$

as claimed.

53. Let $x = r\cos\theta,\ y = r\sin\theta$. Then, we have

$$\frac{x^2}{a^2} - \frac{y^2}{b^2} = 1$$

$$\frac{(r\cos\theta)^2}{a^2} - \frac{(r\sin\theta)^2}{b^2} = 1$$

$$r^2\left[\frac{\cos^2\theta}{a^2} - \frac{\sin^2\theta}{b^2}\right] = 1$$

$$r^2\left[\frac{b^2\cos^2\theta - a^2\sin^2\theta}{a^2b^2}\right] = 1$$

$$r^2 = \frac{a^2b^2}{\left(c^2 - a^2\right)\cos^2\theta - a^2\sin^2\theta} \quad (\text{since } b^2 = c^2 - a^2)$$

$$r^2 = \frac{a^2b^2}{c^2\cos^2\theta - a^2\underbrace{\left(\cos^2\theta + \sin^2\theta\right)}_{=1}}$$

$$r^2 = \frac{a^2b^2}{c^2\cos^2\theta - a^2}$$

$$r^2 = \frac{a^2b^2}{(ae)^2\cos^2\theta - a^2} \quad (\text{since } a = \tfrac{c}{e} \Rightarrow c = ae)$$

$$r^2 = \frac{\cancel{a^2}b^2}{\cancel{a^2}\left[e^2\cos^2\theta - 1\right]}$$

$$r^2 = -\frac{b^2}{1 - e^2\cos^2\theta},$$

as claimed.

54. Let $x = r\cos\theta,\ y = r\sin\theta$. Then, we have

$$\frac{x^2}{a^2}+\frac{y^2}{b^2}=1$$

$$\frac{(r\cos\theta)^2}{a^2}+\frac{(r\sin\theta)^2}{b^2}=1$$

$$r^2\left[\frac{\cos^2\theta}{a^2}+\frac{\sin^2\theta}{b^2}\right]=1$$

$$r^2\left[\frac{b^2\cos^2\theta+a^2\sin^2\theta}{a^2b^2}\right]=1$$

$$r^2=\frac{a^2b^2}{(a^2-c^2)\cos^2\theta+a^2\sin^2\theta} \quad (\text{since } b^2=a^2-c^2)$$

$$r^2=\frac{a^2b^2}{a^2\underbrace{(\cos^2\theta+\sin^2\theta)}_{=1}-c^2\cos^2\theta}$$

$$r^2=\frac{a^2b^2}{a^2-c^2\cos^2\theta}$$

$$r^2=\frac{a^2b^2}{a^2-(ae)^2\cos^2\theta} \quad (\text{since } a=\tfrac{c}{e}\Rightarrow c=ae)$$

$$r^2=\frac{\cancel{a^2}b^2}{\cancel{a^2}\left[1-e^2\cos^2\theta\right]}$$

$$r^2=\frac{b^2}{1-e^2\cos^2\theta},$$

as claimed.

55. The major diameter of this ellipse is $\frac{2ep}{1-e^2}$.

56. The minor diameter of this ellipse is $\frac{2ep\sqrt{1-e^2}}{1-e^2}$.

57. The center of this ellipse is the midpoint between the vertices, which occur when $\theta = 0,\ \pi$. Indeed, these vertices are $\left(\frac{ep}{1+e},0\right)$ and $\left(\frac{ep}{1-e},\pi\right)$. In rectangular coordinates, these points are $\left(\frac{ep}{1+e},0\right)$ and $\left(-\frac{ep}{1-e},0\right)$. The center, therefore, is the midpoint of the segment with these endpoints, namely $\left(\frac{\frac{ep}{1+e}-\frac{ep}{1-e}}{2},0\right)$. In polar coordinates, this is equivalent to $\left(-\frac{\frac{ep}{1+e}-\frac{ep}{1-e}}{2},\pi\right)$.

58. Assume the focus is at the origin, for simplicity. We need the points of the parabola that correspond to $x = 0$. So, we use $\theta = \pm\frac{\pi}{2}$. But, r at each of these points is p. Hence, the latus rectum has length $2p$.

59. Assume that $e = 1$ (so that the curves are parabolas), and consider the following graphs:

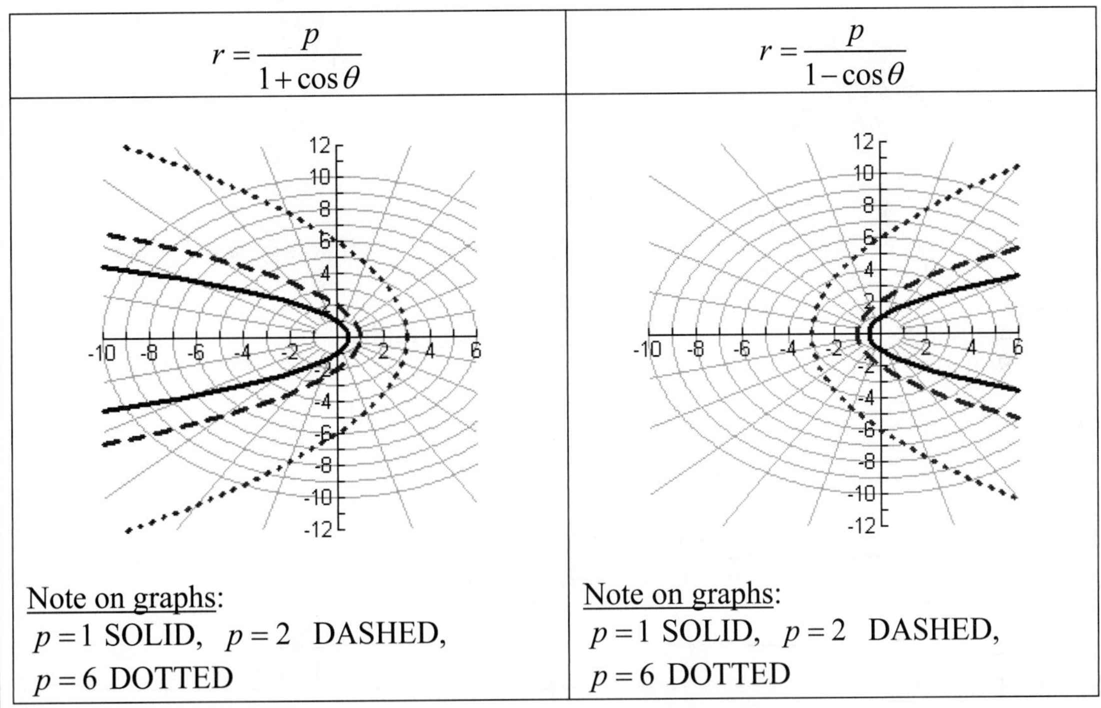

Observe that as $p \to \infty$, the graphs of both $r = \frac{p}{1+\cos\theta}$ and $r = \frac{p}{1-\cos\theta}$ get wider. They are different in that they open in opposite directions.

60. Assume that $e=1$ (so that the curves are parabolas), and consider the following graphs:

$r=\dfrac{p}{1+\sin\theta}$	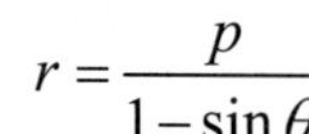 $r=\dfrac{p}{1-\sin\theta}$
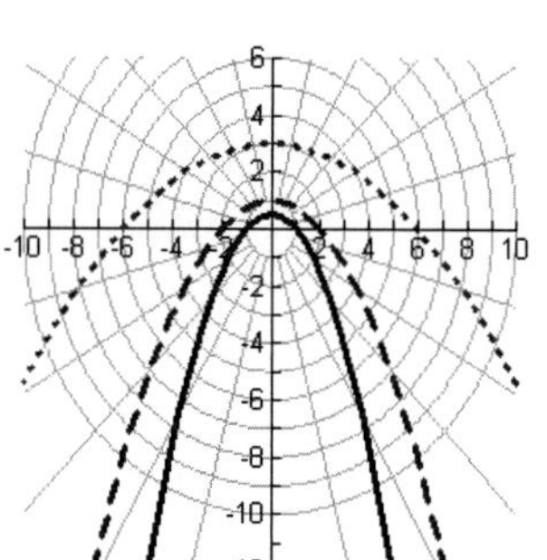	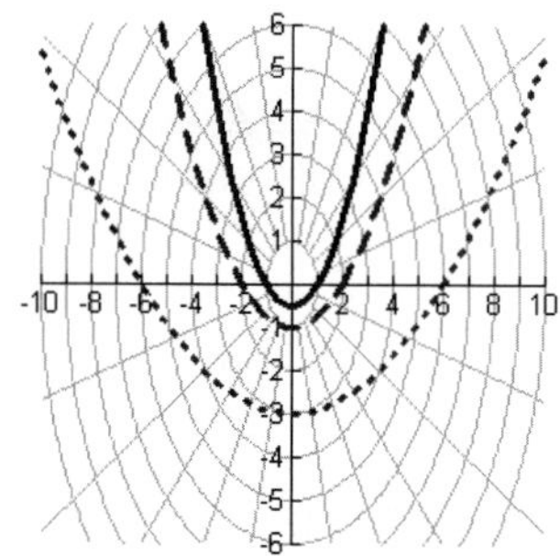
Note on graphs: $p=1$ SOLID, $p=2$ DASHED, $p=6$ DOTTED	Note on graphs: $p=1$ SOLID, $p=2$ DASHED, $p=6$ DOTTED

Observe that as $p\to\infty$, the graphs of both $r=\dfrac{p}{1+\sin\theta}$ and $r=\dfrac{p}{1-\sin\theta}$ get wider. They are different in that they open in opposite directions.

61. Assume that $p=1$ and consider the following graphs:

$r=\dfrac{e}{1+e\cos\theta}$	$r=\dfrac{e}{1-e\cos\theta}$
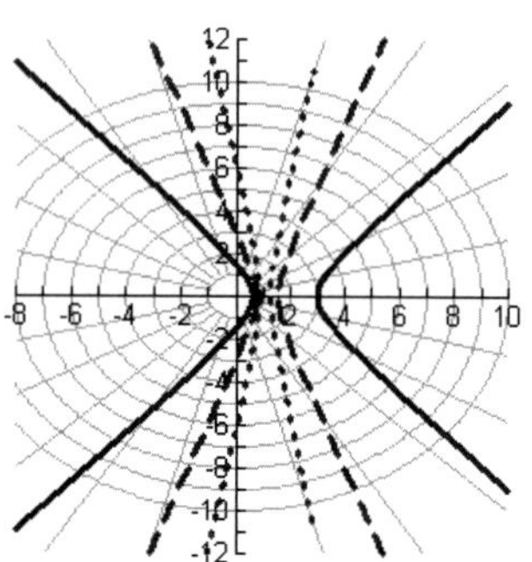	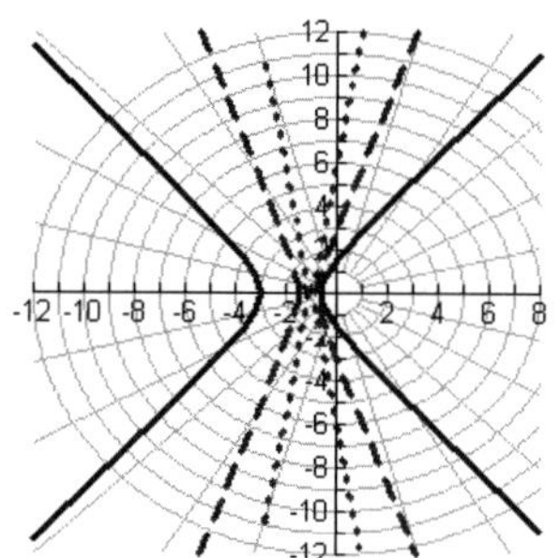
Note on graphs: $e=1.5$ SOLID, $e=3$ DASHED, $e=6$ DOTTED	Note on graphs: $e=1.5$ SOLID, $e=3$ DASHED, $e=6$ DOTTED

Observe that as $e\to\infty$, the graphs of both $r=\dfrac{e}{1+e\cos\theta}$ and $r=\dfrac{e}{1-e\cos\theta}$ get wider. They are different in that the center for the latter family of curves is shifted to the left, with respect to the first family.

62. Assume that $p=1$ and consider the following graphs:

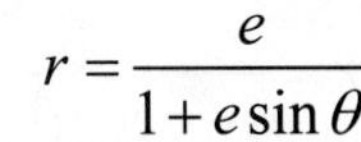 $r=\dfrac{e}{1+e\sin\theta}$	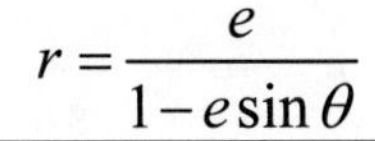 $r=\dfrac{e}{1-e\sin\theta}$

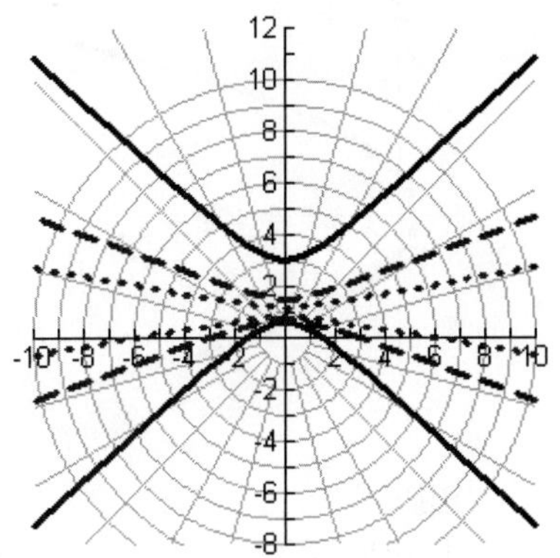

Note on graphs:
$e=1.5$ SOLID, $e=3$ DASHED,
$e=6$ DOTTED

8
6
4
2
-2
-4
-6
-8
-10
-8
-6
-4
-2
4
6
8
10

Note on graphs:
$e=1.5$ SOLID, $e=3$ DASHED,
$e=6$ DOTTED

Observe that as $e\to\infty$, the graphs of both $r=\dfrac{e}{1+e\sin\theta}$ and $r=\dfrac{e}{1-e\sin\theta}$ get wider. They are different in that the center for the latter family of curves is shifted to the left, with respect to the first family.

63. Assume that $p = 1$ and consider the following graphs:

<table>
<tr><td>$r = \dfrac{e}{1 + e\cos\theta}$</td><td>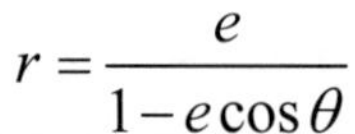
$r = \dfrac{e}{1 - e\cos\theta}$</td></tr>
<tr><td>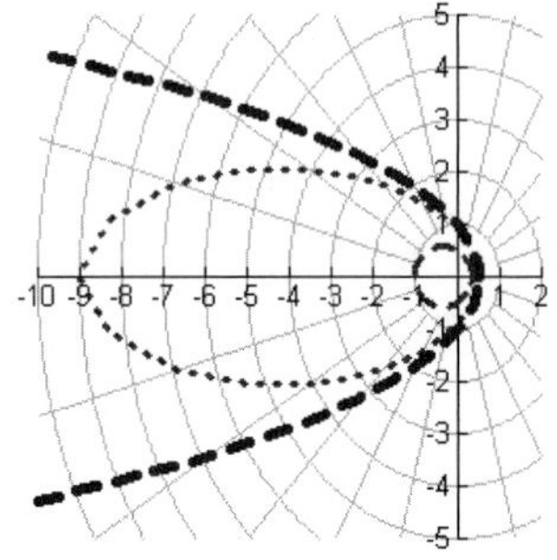

Note on graphs:
$e = 0.001$ (unobservable on window)
$e = 0.5$ DASHED,
$e = 0.9$ DOTTED
$e = 0.99$ THICK DASHED (largest one)
Also, the graph corresponding to $e = 0.99$ is very large. Any window that shows its entire graph renders the other graphs practically invisible. As such, we provide the graph on the above window, with the comment that the outermost curve is, in fact, an ellipse.</td><td>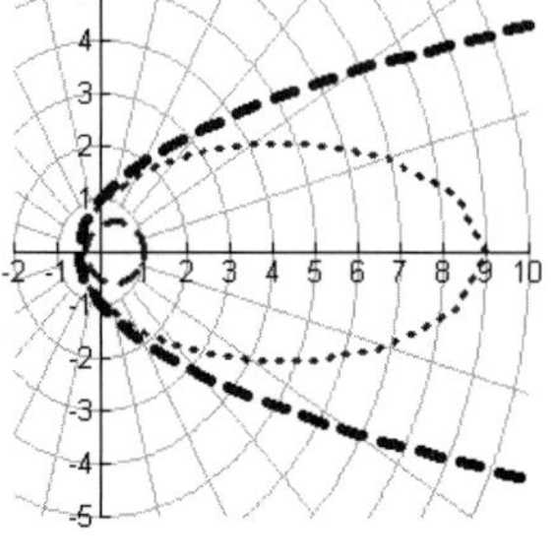

Note on graphs:
$e = 0.001$ (unobservable on window)
$e = 0.5$ DASHED, $e = 0.9$ DOTTED
$e = 0.99$ THICK DASHED (largest one)
Also, the graph corresponding to $e = 0.99$ is very large. Any window that shows its entire graph renders the other graphs practically invisible. As such, we provide the graph on the above window, with the comment that the outermost curve is, in fact, an ellipse.</td></tr>
</table>

Observe that as $e \to 1$, the graphs of both $r = \dfrac{e}{1 + e\cos\theta}$ and $r = \dfrac{e}{1 - e\cos\theta}$ get much larger, and their centers move accordingly. They are different in that the centers for the latter family of curves move to the right as $e \to 1$, while the centers for the first family move, at the same rate, but to the left.

64. Assume that $p=1$ and consider the following graphs:

$r=\dfrac{e}{1+e\sin\theta}$	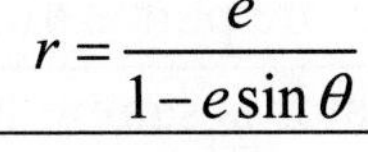 $r=\dfrac{e}{1-e\sin\theta}$
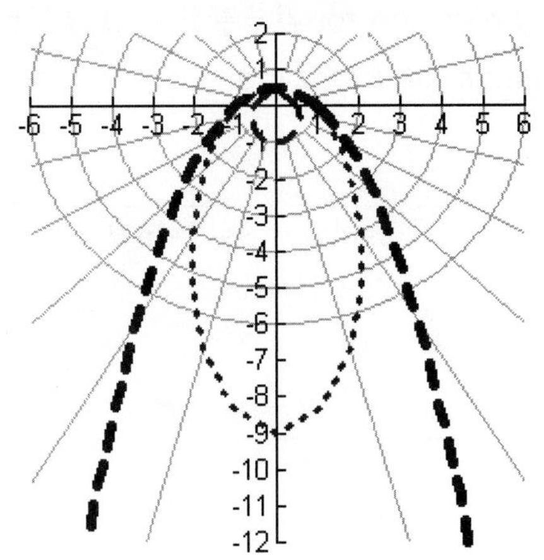	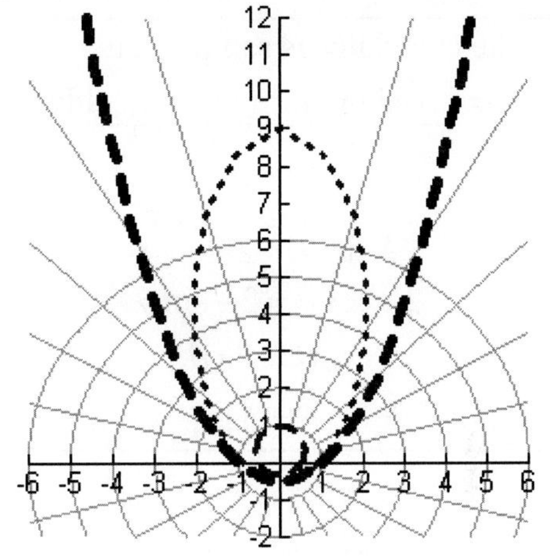
Note on graphs: $e=0.001$ (unobservable on window) $e=0.5$ DASHED, $e=0.9$ DOTTED $e=0.99$ THICK DASHED (largest one) Also, the graph corresponding to $e=0.99$ is very large. Any window that shows its entire graph renders the other graphs practically invisible. As such, we provide the graph on the above window, with the comment that the outermost curve is, in fact, an ellipse.	Note on graphs: $e=0.001$ (unobservable on window) $e=0.5$ DASHED, $e=0.9$ DOTTED $e=0.99$ THICK DASHED (largest one) Also, the graph corresponding to $e=0.99$ is very large. Any window that shows its entire graph renders the other graphs practically invisible. As such, we provide the graph on the above window, with the comment that the outermost curve is, in fact, an ellipse.

Observe that as $e\to 1$, the graphs of both $r=\dfrac{e}{1+e\sin\theta}$ and $r=\dfrac{e}{1-e\sin\theta}$ get much larger, and their centers move accordingly . They are different in that the centers for the latter family of curves move to down as $e\to 1$, while the centers for the first family move, at the same rate, but up.

65. The calculator is sampling using a more refined partition for $\theta=\frac{\pi}{3}$ than it is for $\theta=\frac{\pi}{2}$. Using more points yields a better approximation to the graph of the curve.

66. The calculator is sampling using a more refined partition for $\theta = \frac{\pi}{3}$ than it is for $\theta = \frac{\pi}{2}$. Using more points yields a better approximation to the graph of the curve.

67. The calculator is sampling using a more refined partition for $\theta = 0.4\pi$ than it is for $\theta = \frac{\pi}{2}$. Using more points yields a better approximation to the graph of the curve.

68. The calculator is sampling using a more refined partition for $\theta = \frac{\pi}{3}$ than it is for $\theta = 0.8\pi$. Using more points yields a better approximation to the graph of the curve.

69.

$$\frac{2}{2+\sin\theta} = \frac{2}{2+\cos\theta}$$
$$2+\cos\theta = 2+\sin\theta$$
$$\cos\theta = \sin\theta$$
$$\theta = \frac{\pi}{4}, \frac{5\pi}{4}$$

70.

$$\frac{1}{3+2\sin\theta} = \frac{1}{3-2\sin\theta}$$
$$3-2\sin\theta = 3+2\sin\theta$$
$$4\sin\theta = 0$$
$$\theta = 0, \pi 2\pi$$

71.

$$\frac{1}{4-3\sin\theta} = \frac{1}{-1+7\sin\theta}$$
$$-1+7\sin\theta = 4-3\sin\theta$$
$$\sin\theta = \frac{1}{2}$$
$$\theta = \frac{\pi}{6}, \frac{5\pi}{6}$$

72.

$$\frac{1}{5+2\cos\theta} = \frac{1}{10-8\cos\theta}$$
$$10-8\cos\theta = 5+2\cos\theta$$
$$\cos\theta = \frac{1}{2}$$
$$\theta = \frac{\pi}{3}, \frac{5\pi}{3}$$

Section 9.9 Solutions --

1. In order to graph the curve defined parametrically by

$$x = t+1, \; y = \sqrt{t}, \; t \geq 0,$$

it is easiest to eliminate the parameter t and write the equation in rectangular form since the result is a known graph. Indeed, observe that since $t = x-1$, substituting this into the expression for y yields

$$y = \sqrt{x-1}, \; x \geq 1.$$

The graph is as follows:

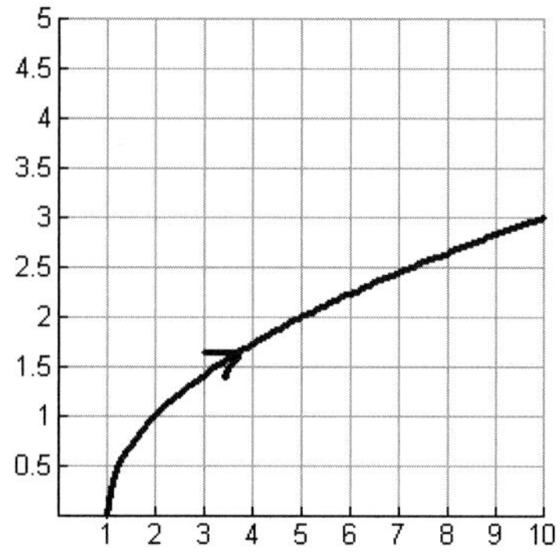

2. In order to graph the curve defined parametrically by $x = 3t,\ y = t^2 - 1,\ t \text{ in } [0,4],$ it is easiest to eliminate the parameter t and write the equation in rectangular form since the result is a known graph. Indeed, observe that since $t = x/3$, substituting this into the expression for y yields $y = \left(\frac{x}{3}\right)^2 - 1 = \frac{1}{9}x^2 - 1,\ x \text{ in } [0,12].$	The graph is as follows: 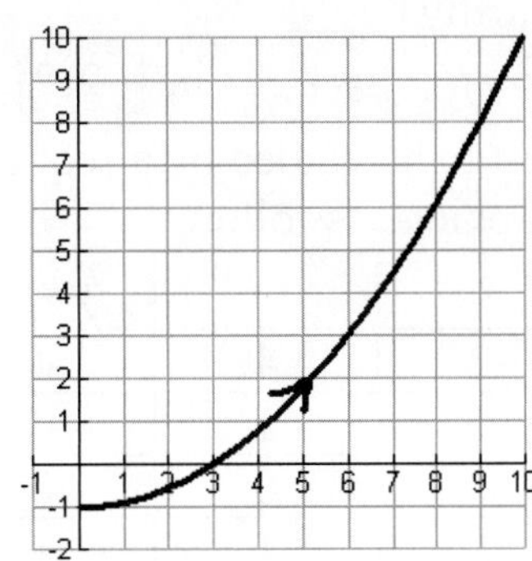
3. In order to graph the curve defined parametrically by $x = -3t,\ y = t^2 + 1,\ t \text{ in } [0,4],$ it is easiest to eliminate the parameter t and write the equation in rectangular form since the result is a known graph. Indeed, observe that since $t = -x/3$, substituting this into the expression for y yields $y = \left(-\frac{x}{3}\right)^2 + 1 = \frac{1}{9}x^2 + 1,\ x \text{ in } [-12,0].$	The graph is as follows: 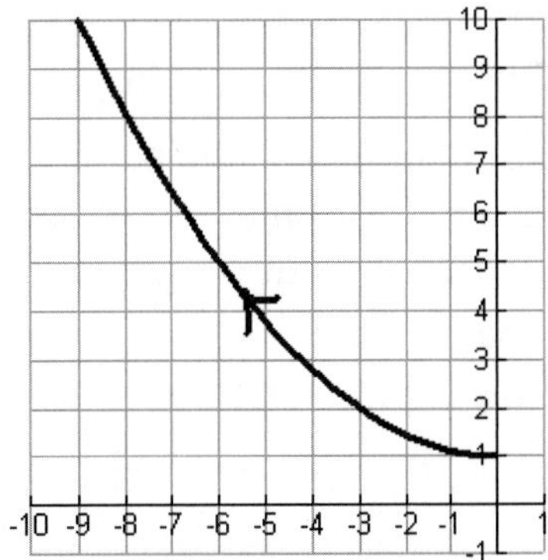
4. In order to graph the curve defined parametrically by $x = t^2 - 1,\ y = t^2 + 1,\ t \text{ in } [-3,3],$ it is easiest to eliminate the parameter t and write the equation in rectangular form since the result is a known graph. Indeed, observe that since $t^2 = x + 1$, substituting this into the expression for y yields $y = (x+1) + 1 = x + 2,\ x \text{ in } [-1,8].$	The graph is as follows: 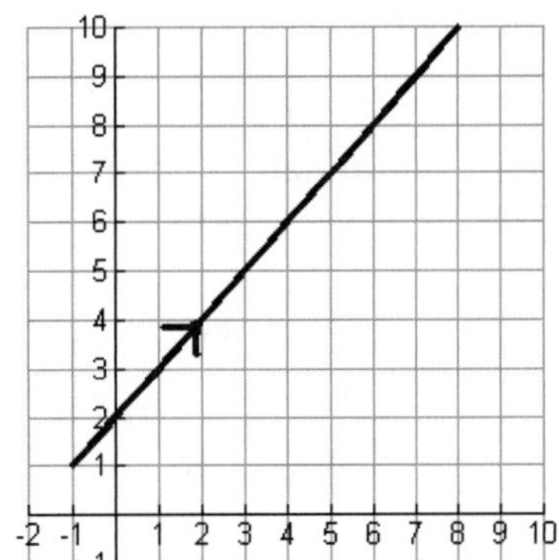

5. In order to graph the curve defined parametrically by

$$x = t^2,\ y = t^3,\ t \text{ in } [-2,2],$$

it is easiest to be guided by a sequence of tabulated points, as follows:

t	$x = t^2$	$y = t^3$
-2	4	-8
-1	1	-1
0	0	0
1	1	1
2	4	8

The graph is as follows:

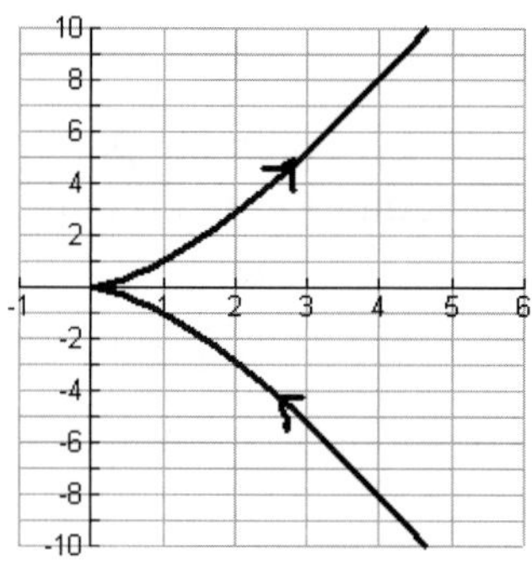

6. In order to graph the curve defined parametrically by

$$x = t^3 + 1,\ y = t^3 - 1,\ t \text{ in } [-2,2],$$

it is easiest to eliminate the parameter t and write the equation in rectangular form since the result is a known graph. Indeed, observe that since $t^3 = x - 1$, substituting this into the expression for y yields

$$y = (x-1) - 1 = x - 2,\ x \text{ in } [-7,9].$$

The graph is as follows:

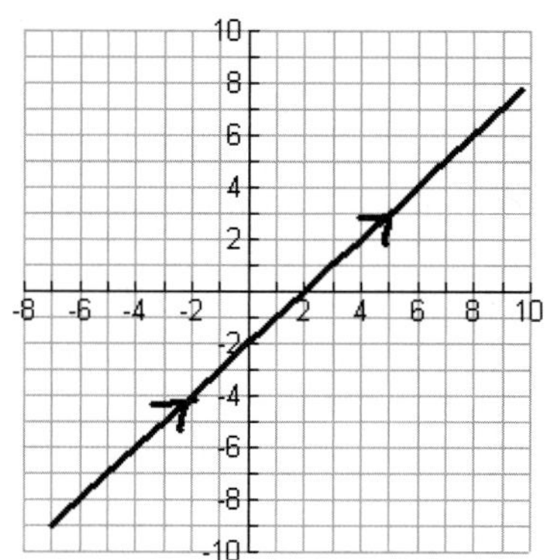

7. In order to graph the curve defined parametrically by

$$x = \sqrt{t},\ y = t,\ t \text{ in } [0,10],$$

it is easiest to eliminate the parameter t and write the equation in rectangular form since the result is a known graph. Indeed, substituting $y = t$ into the expression for x yields

$$x = \sqrt{y},\ y \text{ in } [0,10].$$

The graph is as follows:

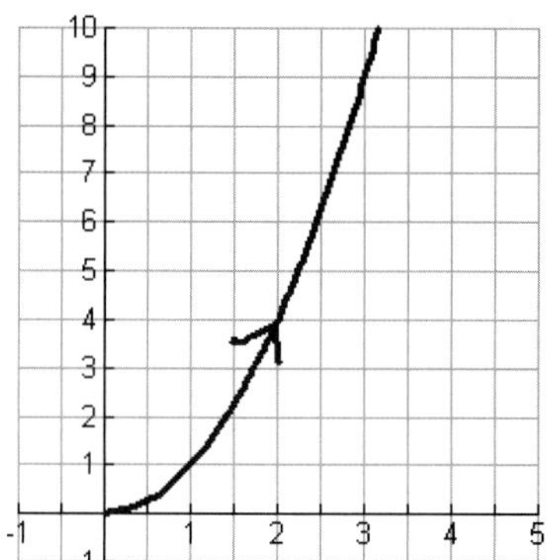

8. In order to graph the curve defined parametrically by

$$x = t,\ y = \sqrt{t^2+1},\ t \text{ in } [0,10],$$

it is easiest to eliminate the parameter t and write the equation in rectangular form since the result is a known graph. Indeed, substituting $x = t$ into the expression for y yields

$$y = \sqrt{x^2+1},\ x \text{ in } [0,10].$$

The graph is as follows:

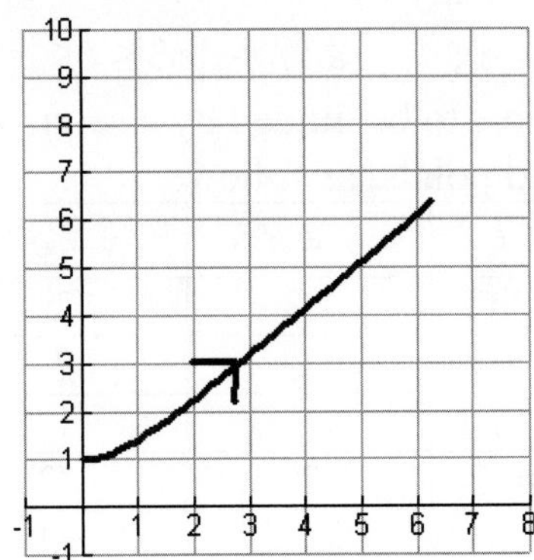

9. In order to graph the curve defined parametrically by

$$x = (t+1)^2,\ y = (t+2)^3,\ t \text{ in } [0,1],$$

it is easiest to be guided by a sequence of tabulated points, as follows:

t	$x = (t+1)^2$	$y = (t+2)^3$
0	1	8
0.25	1.5625	11.391
0.50	2.25	15.63
0.75	3.0625	20.80
1.00	4	27

The graph is as follows:

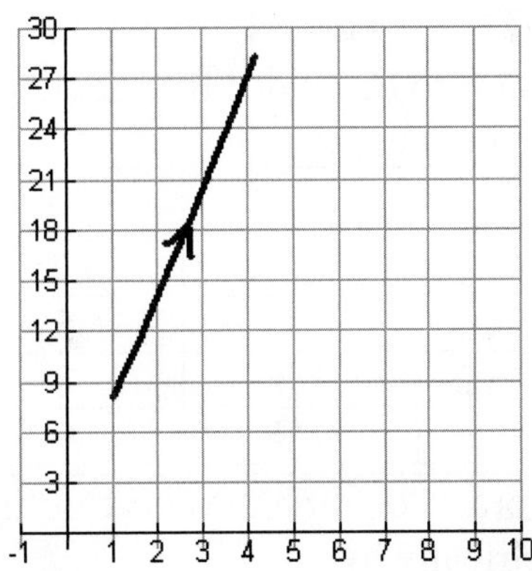

10. In order to graph the curve defined parametrically by

$$x = (t-1)^3,\ y = (t-2)^2,\ t \text{ in } [0,4],$$

it is easiest to be guided by a sequence of tabulated points, as follows:

t	$x = (t-1)^3$	$y = (t-2)^2$
0	-1	4
1	0	1
2	1	0
3	8	1
4	27	4

The graph is as follows:

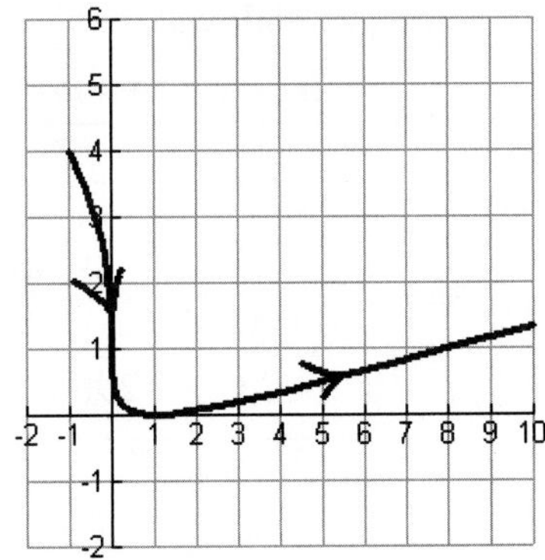

11. In order to graph the curve defined parametrically by

$$x = e^{t}, \quad y = e^{-t}, \quad -\ln 3 \le t \le \ln 3$$

it is easiest to be guided by a sequence of tabulated points, as follows:

t	$x = e^{t}$	$y = e^{-t}$
-ln3	1/3	3
-ln2	1/2	2
0	1	1
ln 2	2	1/2
ln 3	3	1/3

The graph is as follows:

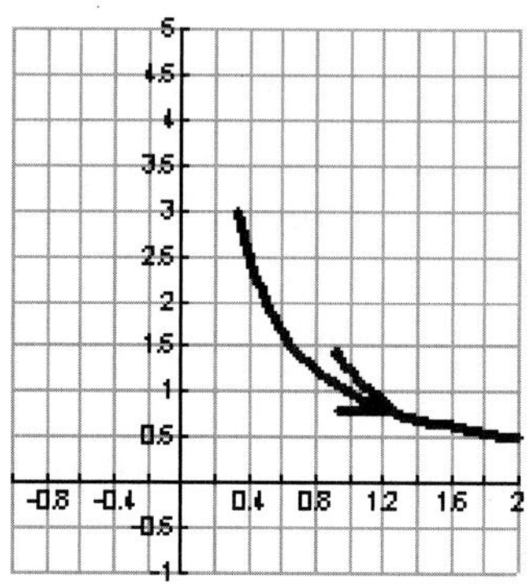

12. In order to graph the curve defined parametrically by

$$x = e^{-2t}, \quad y = e^{2t} + 4, \quad -\ln 2 \le t \le \ln 3$$

it is easiest to be guided by a sequence of tabulated points, as follows:

t	$x = e^{-2t}$	$y = e^{2t} + 4$
-ln2	4	4.25
0	1	5
ln 2	1/4	8
ln 3	1/9	9

The graph is as follows:

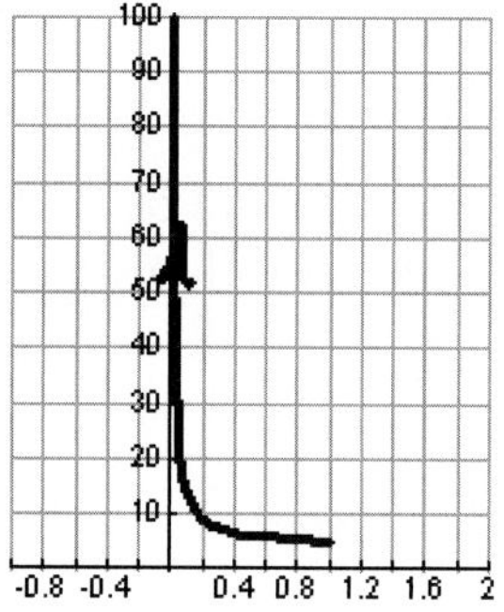

13. In order to graph the curve defined parametrically by

$$x = 2t^{4} - 1, \quad y = t^{8} + 1, \quad 0 \le t \le 4$$

it is easiest to be guided by a sequence of tabulated points, as follows:

t	$x = 2t^{4} - 1$	$y = t^{8} + 1$
0	1	5
1	0.1353	11.39
2	0.0183	58.60
3	0.0025	407.43
4	0.00003	2984.96

The graph is as follows:

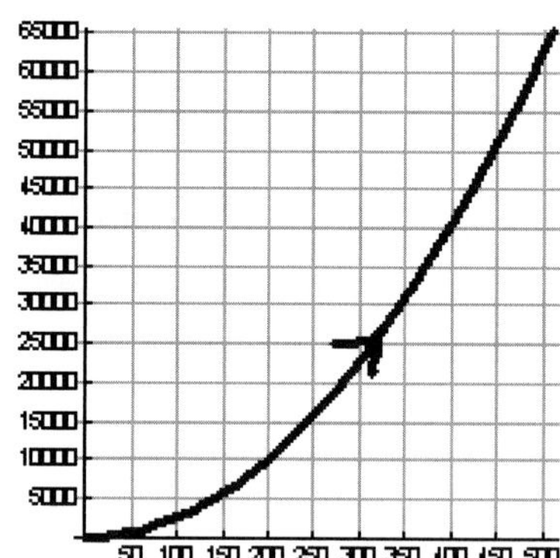

14. In order to graph the curve defined parametrically by

$$x = 3t^6 - 1, \quad y = 2t^3, \quad -1 \le t \le 1$$

it is easiest to be guided by a sequence of tabulated points, as follows:

t	$x = 3t^6 - 1$	$y = 2t^3$
-1	2	-2
0	-1	0
1	2	2

The graph is as follows:

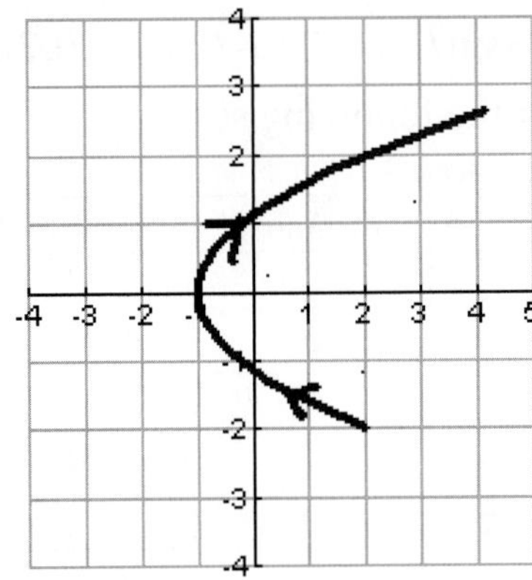

15. In order to graph the curve defined parametrically by

$$x = t(t-2)^3, \quad y = t(t-2)^3, \quad 0 \le t \le 4$$

it is easiest to be guided by a sequence of tabulated points, as follows:

t	$x = t(t-2)^3$	$y = t(t-2)^3$
0	0	0
1	-1	-1
2	0	0
3	3	3
4	32	32

The graph is as follows:

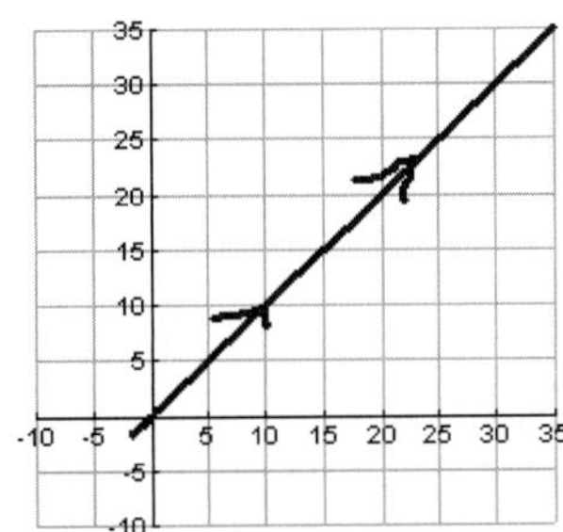

16. In order to graph the curve defined parametrically by

$$x = -t\sqrt[3]{t}, \quad y = -5t^8 - 2, \quad -3 \le t \le 3$$

it is easiest to be guided by a sequence of tabulated points, as follows:

t	$x = -t\sqrt[3]{t}$	$y = -5t^8 - 2$
-3	-4.3267	-32,807
-1	-1	-7
0	0	-2
1	-1	-7
3	-4.3267	-32.807

The graph is as follows:

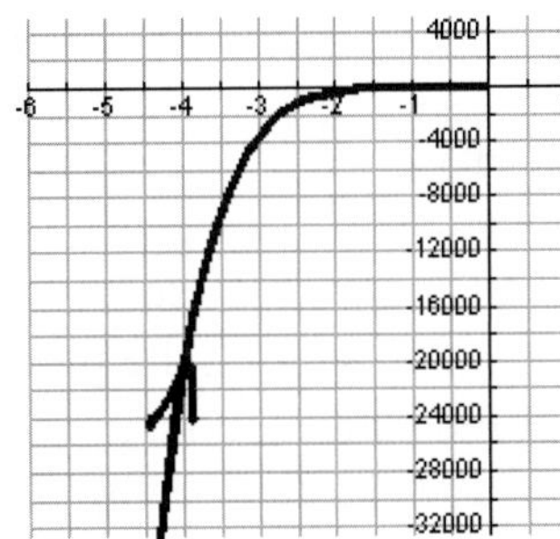

17. In order to graph the curve defined parametrically by

$$x = 3\sin t, \ \ y = 2\cos t, \ \ t \text{ in } [0, 2\pi],$$

consider the following sequence of tabulated points:

t	$x = 3\sin t$	$y = 2\cos t$
0	0	2
$\pi/2$	3	0
π	0	-2
$3\pi/2$	-3	0
2π	0	2

Eliminating the parameter reveals that the graph is an ellipse. Indeed, observe that

$$x^2 = 9\sin^2 t, \quad y^2 = 4\cos^2 t$$

so that $\frac{x^2}{9} = \sin^2 t, \quad \frac{y^2}{4} = \cos^2 t$.

Summing then yields $\frac{x^2}{9} + \frac{y^2}{4} = 1$.

The graph is as follows:

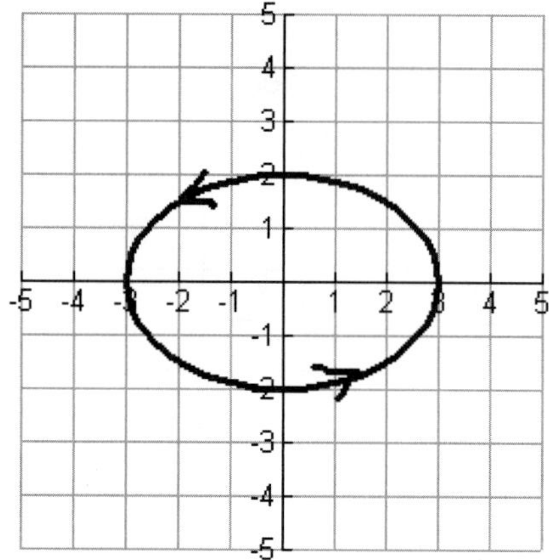

18. In order to graph the curve defined parametrically by

$$x = \cos 2t, \ \ y = \sin t, \ \ t \text{ in } [0, 2\pi],$$

consider the following sequence of tabulated points:

t	$x = \cos 2t$	$y = \sin t$
0	1	0
$\pi/2$	-1	1
π	1	0
$3\pi/2$	-1	-1
2π	1	0

Notice that the curve retraces itself within this interval. Eliminating the parameter reveals that the path that the graph takes is a portion of a parabola opening to the right. Indeed, observe that

$$x = \cos 2t = 2\sin^2 t - 1 = 2y^2 - 1.$$

The graph is as follows:

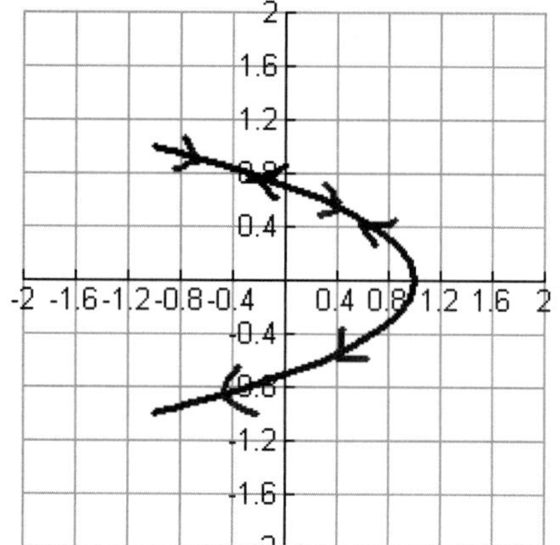

19. In order to graph the curve defined parametrically by

$$x = \sin t + 1,\ y = \cos t - 2,\ t \text{ in } [0, 2\pi],$$

consider the following sequence of tabulated points:

t	$x = \sin t + 1$	$y = \cos t - 2$
0	1	-1
$\pi/2$	2	-2
π	1	-3
$3\pi/2$	0	-2
2π	1	-1

Eliminating the parameter reveals that the graph is a circle with radius 1 centered at (1,-2). Indeed, observe that

$$x - 1 = \sin t, \qquad y + 2 = \cos t$$

$$(x-1)^2 = \sin^2 t, \quad (y+2)^2 = \cos^2 t$$

so that

$$(x-1)^2 + (y+2)^2 = 1.$$

The graph is as follows:

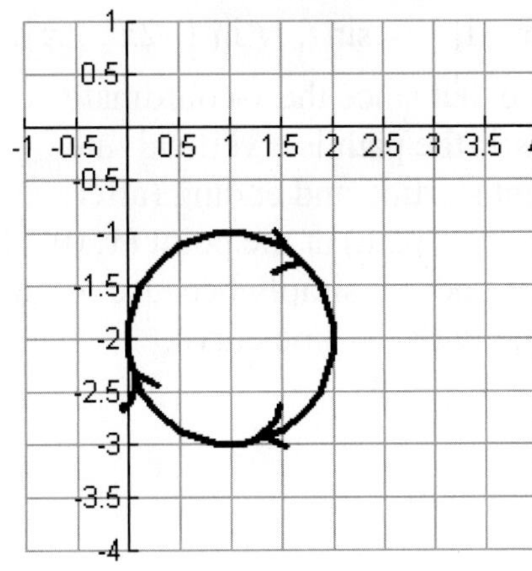

20. In order to graph the curve defined parametrically by

$$x = \tan t,\ y = 1,\ t \text{ in } \left[-\frac{\pi}{4}, \frac{\pi}{4}\right],$$

observe that since the *y*-coordinate is always 1, the path is a horizontal line segment starting at the point (-1, 1) and ending at the point (1, 1).

The graph is as follows:

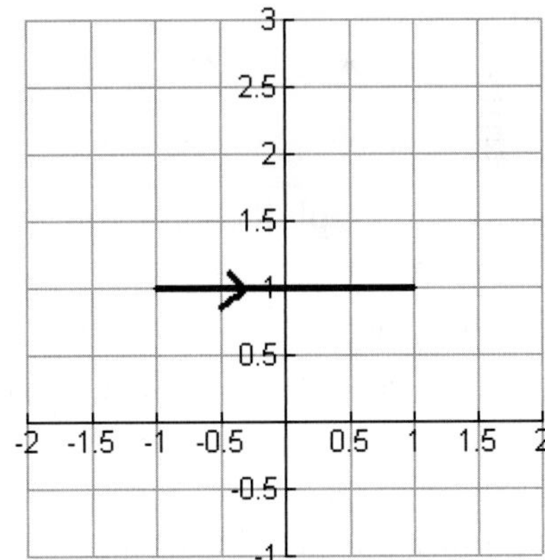

21. In order to graph the curve defined parametrically by

$$x = 1, \ y = \sin t, \ t \text{ in } [-2\pi, 2\pi],$$

observe that since the x-coordinate is always 1, the path is a vertical line segment starting and ending (after retracing its path) at the point (1, 0). The retracing occurs simply because of the periodicity of the sine curve.

The graph is as follows:

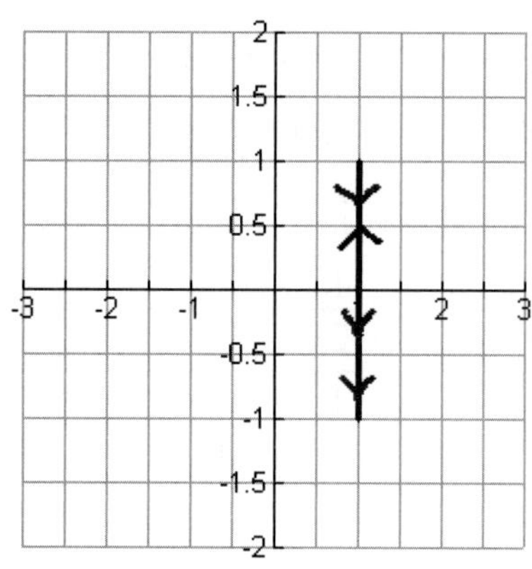

22. In order to graph the curve defined parametrically by

$$x = \sin t, \ y = 2, \ t \text{ in } [0, 2\pi],$$

observe that since the y-coordinate is always 2, the path is a horizontal line segment starting at the point (0, 2), goes to (1,2) and comes back through (0, 2) and visits (-1, 2), and then finally ends at (0,2).

The graph is as follows:

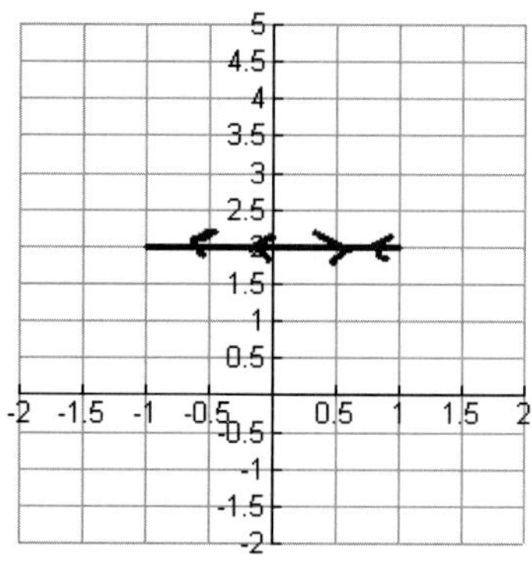

23. In order to graph the curve defined parametrically by

$$x = \sin^2 t, \ y = \cos^2 t, \ t \text{ in } [0, 2\pi],$$

it is easiest to eliminate the parameter t and write the equation in rectangular form since the result is a known graph. Indeed, simply adding the two equations yields

$$x + y = 1, \ 0 \le x \le 1, \ 0 \le y \le 1,$$

which is equivalent to

$$y = 1 - x, \ 0 \le x \le 1, \ 0 \le y \le 1.$$

The graph is as follows:

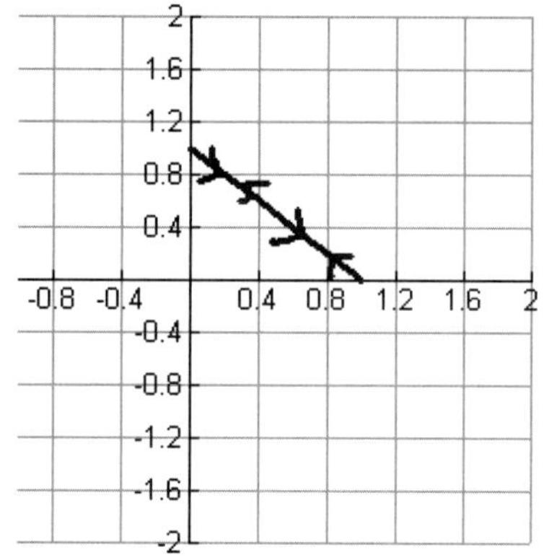

24. In order to graph the curve defined parametrically by $x = 2\sin^2 t,\ y = 2\cos^2 t,\ t \text{ in } [0, 2\pi]$, it is easiest to eliminate the parameter t and write the equation in rectangular form since the result is a known graph. Indeed, observe that $\frac{x}{2} = \sin^2 t,\ \frac{y}{2} = \cos^2 t,\ t \text{ in } [0, 2\pi]$ so that adding these two equations yields $\frac{x}{2} + \frac{y}{2} = 1,\ 0 \le x \le 2,\ 0 \le y \le 2$, which is equivalent to $y = 2 - x,\ 0 \le x \le 2,\ 0 \le y \le 2$.	The graph is as follows: 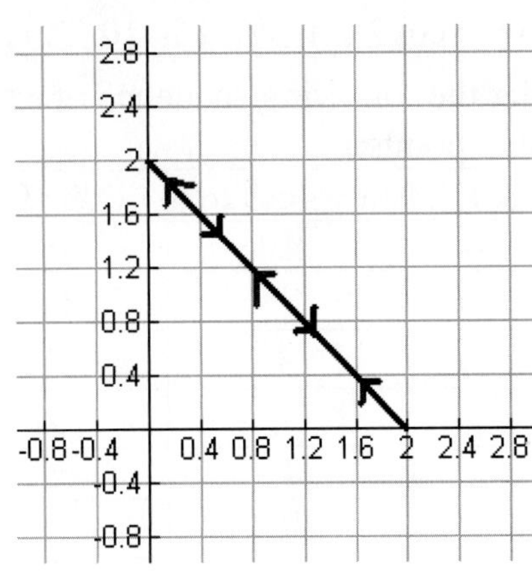
25. In order to graph the curve defined parametrically by $x = 2\sin 3t,\ y = 3\cos 2t,\ t \text{ in } [0, 2\pi]$, consider the following sequence of tabulated points:	The graph is as follows: 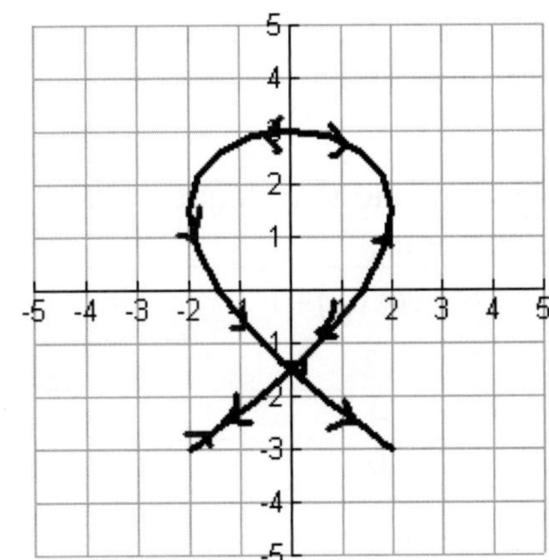

t	$x = 2\sin 3t$	$y = 3\cos 2t$
0	0	3
$\pi/6$	2	$3/2$
$\pi/4$	$\sqrt{2}$	0
$\pi/3$	0	$-3/2$
$\pi/2$	-2	-3
$2\pi/3$	0	$-3/2$
$3\pi/4$	$\sqrt{2}$	0
$5\pi/6$	2	$3/2$
π	0	3

The pattern replicates for t in $[\pi, 2\pi]$.

26. In order to graph the curve defined parametrically by

$$x = 4\cos 2t,\ \ y = t,\ \ t \text{ in } [0, 2\pi],$$

consider the following sequence of tabulated points:

t	$x = 4\cos 2t$	$y = t$
0	4	0
$\pi/4$	0	$\pi/4$
$\pi/2$	-4	$\pi/2$
$3\pi/4$	0	$3\pi/4$
π	4	π
$5\pi/4$	0	$5\pi/4$
$3\pi/2$	-4	$3\pi/2$
$7\pi/4$	0	$7\pi/4$
2π	4	2π

The graph is as follows:

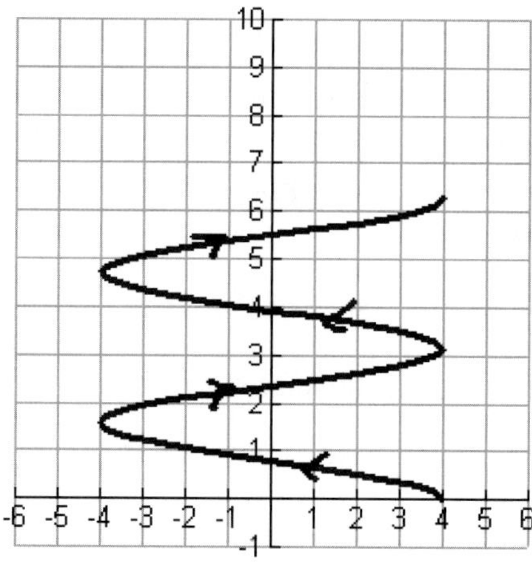

27. In order to graph the curve defined parametrically by

$$x = \cos\left(\tfrac{t}{2}\right) - 1,\ \ y = \sin\left(\tfrac{t}{2}\right) + 1,\ \ -2\pi \le t \le 2\pi$$

consider the following sequence of tabulated points:

t	$x = \cos\left(\tfrac{t}{2}\right) - 1$	$y = \sin\left(\tfrac{t}{2}\right) + 1$
-2π	-2	1
$-\pi$	-1	0
0	0	1
π	-1	2
2π	-2	1

The graph is as follows:

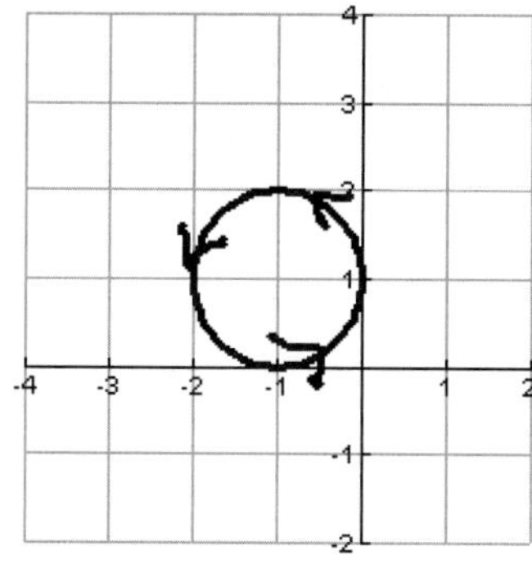

28. In order to graph the curve defined parametrically by

$$x = \sin\left(\tfrac{t}{3}\right) + 3,\ \ y = \cos\left(\tfrac{t}{3}\right) - 1,\ \ 0 \le t \le 6\pi$$

consider the following sequence of tabulated points:

t	$x = \cos\left(\tfrac{t}{2}\right) - 1$	$y = \sin\left(\tfrac{t}{2}\right) + 1$
-6π	-2	1
-3π	-1	2
0	0	1
3π	-1	0
6π	-2	1

The graph is as follows:

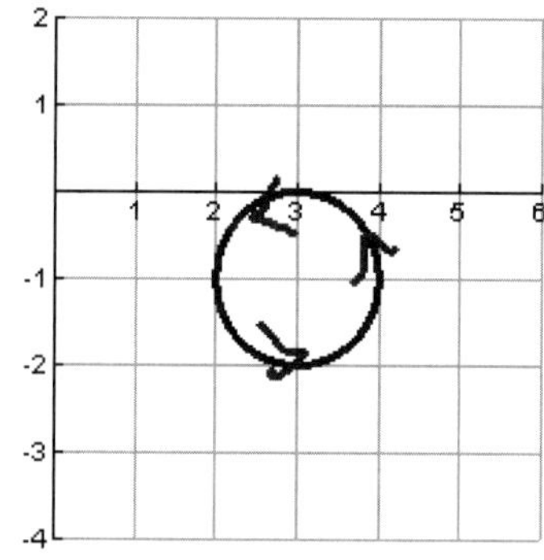

29. In order to graph the curve defined parametrically by

$x = 2\sin\left(t+\frac{\pi}{4}\right),\ y = -2\cos\left(t+\frac{\pi}{4}\right),\ -\frac{\pi}{4} \le t \le \frac{7\pi}{4}$

consider the following sequence of tabulated points:

t	$x = 2\sin\left(t+\frac{\pi}{4}\right)$	$y = -2\cos\left(t+\frac{\pi}{4}\right)$
$-\frac{\pi}{4}$	0	-2
$\frac{\pi}{4}$	2	0
$\frac{3\pi}{4}$	0	2
$\frac{5\pi}{4}$	-2	0
$\frac{7\pi}{4}$	0	-2

The graph is as follows:

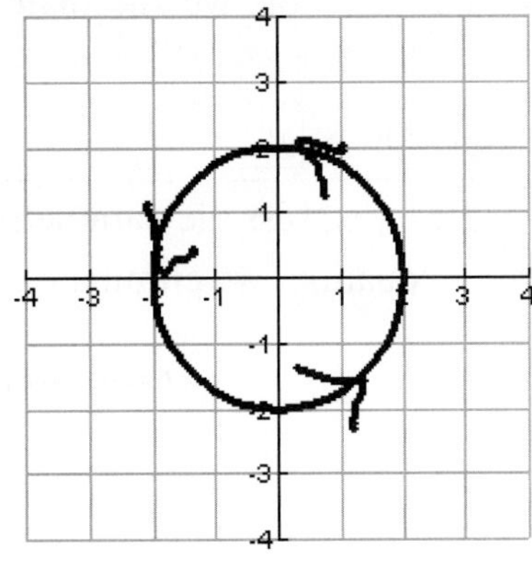

30. In order to graph the curve defined parametrically by

$x = -3\cos^2(3t),\ y = 2\cos(3t),\ -\frac{\pi}{3} \le t \le \frac{\pi}{3}$

consider the following sequence of tabulated points:

t	$x = -3\cos^2(3t)$	$y = 2\cos(3t)$
$-\frac{\pi}{3}$	-3	-2
0	-3	2
$\frac{\pi}{3}$	-3	-2

It oscillates.

The graph is as follows:

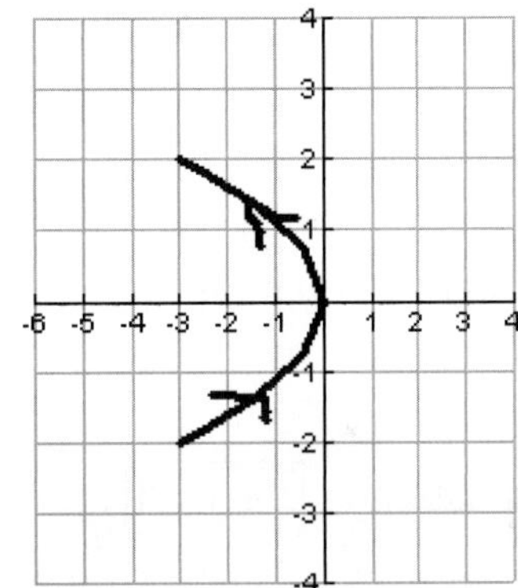

31. In order to express the parametrically-defined curve given by $x = \dfrac{1}{t},\ y = t^2$ as a rectangular equation, we eliminate the parameter t in the following sequence of steps:

$$t = \frac{1}{x} \text{ so that } y = \left(\frac{1}{x}\right)^2 = \frac{1}{x^2}.$$

So, the equation is $\boxed{y = \dfrac{1}{x^2}}$.

32. In order to express the parametrically-defined curve given by $x = t^2 - 1,\ y = t^2 + 1$ as a rectangular equation, we eliminate the parameter t in the following sequence of steps:

$$t^2 = x + 1 \text{ so that } y = (x+1) + 1 = x + 2.$$

So, the equation is $\boxed{y = x + 2}$.

33. In order to express the parametrically-defined curve given by $x = t^3 + 1,\ y = t^3 - 1$ as a rectangular equation, we eliminate the parameter t in the following sequence of steps: $$t^3 = x - 1 \text{ so that } y = (x-1) - 1 = x - 2.$$ So, the equation is $\boxed{y = x - 2}$.
34. In order to express the parametrically-defined curve given by $x = 3t,\ y = t^2 - 1$ as a rectangular equation, we eliminate the parameter t in the following sequence of steps: $$t = \frac{x}{3} \text{ so that } y = \left(\frac{x}{3}\right)^2 - 1 = \frac{x^2}{9} - 1.$$ So, the equation is $\boxed{y = \frac{1}{9}x^2 - 1}$.
35. In order to express the parametrically-defined curve given by $x = t,\ y = \sqrt{t^2 + 1}$ as a rectangular equation, we eliminate the parameter t in the following sequence of steps: $$t = x \text{ so that } y = \sqrt{x^2 + 1}.$$ So, the equation is $\boxed{y = \sqrt{x^2 + 1}}$.
36. In order to express the parametrically-defined curve given by $x = \sin^2 t,\ y = \cos^2 t$ as a rectangular equation, we eliminate the parameter t in the following sequence of steps: $$x + y = \sin^2 t + \cos^2 t = 1$$ So, the equation is $\boxed{x + y = 1}$.
37. In order to express the parametrically-defined curve given by $x = 2\sin^2 t,\ y = 2\cos^2 t$ as a rectangular equation, we eliminate the parameter t in the following sequence of steps: $$\frac{x}{2} + \frac{y}{2} = \sin^2 t + \cos^2 t = 1 \text{ so that } x + y = 2$$ So, the equation is $\boxed{x + y = 2}$.
38. In order to express the parametrically-defined curve given by $x = \sec^2 t,\ y = \tan^2 t$ as a rectangular equation, we eliminate the parameter t in the following sequence of steps: $$y = \tan^2 t = \frac{\sin^2 t}{\cos^2 t} = \frac{1 - \cos^2 t}{\cos^2 t} = \frac{1}{\cos^2 t} - 1 = \sec^2 t - 1 = x - 1$$ So, the equation is $\boxed{y = x - 1}$.

39. In order to express the parametrically-defined curve given by $x = 4(t^2 + 1),\ y = 1 - t^2$ as a rectangular equation, we eliminate the parameter t in the following sequence of steps:

Note that $\frac{x}{4} = t^2 + 1$ so that $t^2 = \frac{x}{4} - 1$. Hence, $y = 1 - \left(\frac{x}{4} - 1\right) = 2 - \frac{x}{4}$.

Simplifying, we see that the equation is $\boxed{x + 4y = 8}$.

40. In order to express the parametrically-defined curve given by $x = \sqrt{t-1},\ y = \sqrt{t}$ as a rectangular equation, we eliminate the parameter t in the following sequence of steps:

Note that $x^2 = t - 1$ so that $t = x^2 + 1$. Hence, $y = \sqrt{x^2 + 1}$.

Squaring both sides and simplifying, we see that the equation is $\boxed{y^2 - x^2 = 1}$.

41. Assuming that the initial height $h = 0$, we need to find t such that

$$y = -16t^2 + \left(400 \sin 45^\circ\right)t + 0 = 0.$$

We solve this equation as follows:

$$\begin{aligned} -16t^2 + \left(400 \sin 45^\circ\right)t + 0 &= 0 \\ -16t^2 + 200\sqrt{2}\,t &= 0 \\ t\left(-16t + 200\sqrt{2}\right) &= 0 \\ t = 0,\ \frac{200\sqrt{2}}{16} &\approx 17.7 \text{ sec.} \end{aligned}$$

So, it hits the ground approximately $\boxed{17.7 \text{ seconds}}$ later.

42. We know from Exercise 31 that the time traveled is 17.7 seconds. So, the horizontal distance traveled is $x(17.7) = \left(400 \cos 45^\circ\right)(17.7) \approx \boxed{5{,}000 \text{ ft.}}$

Also, the maximum altitude must occur at $t \approx \frac{17.7}{2} = 8.85$ sec. The height at this time is given by $y(8.85) = -16(8.85)^2 + \left(400 \sin 45^\circ\right)(8.85) \approx \boxed{1250 \text{ ft.}}$

43. First, we convert all measurements to *feet per second*:

$$\left(\frac{105 \text{ mi.}}{1 \text{ hr.}}\right)\left(\frac{1 \text{ hr.}}{3600 \text{ sec.}}\right)\left(\frac{5280 \text{ ft.}}{1 \text{ mi.}}\right) = 154 \text{ ft./sec.}$$

Now, from the given information, we know that the parametric equations are:

$$x = \left(154\cos 20^\circ\right)t$$
$$y = -16t^2 + \left(154\sin 20^\circ\right)t + 3$$

We need to determine the time t_0 corresponding to $x = 420$ ft. Then, we will compare $y(t_0)$ to 15; if $y(t_0) < 15$, then it doesn't clear the fence.

Observe that

$$420 = \left(154\cos 20^\circ\right)t_0 \text{ so that } t_0 = \frac{420}{154\cos 20^\circ} \approx 2.9023 \text{ sec.}$$

Then, we have $y(2.9023) \approx 21.0938$, so it $\boxed{\text{does clear the fence}}$.

44. We use the same information as in Exercise 33.

This time, we need to determine t_0 such that $y(t_0) = 0$. Then, we will compute

$$x(t_0) = \text{ total horizontal distance traveled}$$
$$y\left(\frac{t_0}{2}\right) = \text{ maximum height}$$

Observe that solving $-16t^2 + \left(154\sin 20^\circ\right)t + 3 = 0$ using the quadratic formula yields

$$t = \frac{-154\sin 20^\circ \pm \sqrt{\left(154\sin 20^\circ\right)^2 - 4(-16)(3)}}{2(-16)}$$
$$\approx \frac{-52.6711 \pm 54.4632}{-32} \approx \cancel{-0.056} \text{ or } 3.3479$$

Now, we have

$$\boxed{\begin{aligned} x(3.3479) &\approx 484.5 \text{ ft.} = \text{ total horizontal distance traveled} \\ y(1.67395) &\approx 46.3 \text{ ft.} = \text{ maximum height} \end{aligned}}$$

45. From the given information, we know that the parametric equations are:

$$x = \left(700\cos 60^\circ\right)t$$
$$y = -16t^2 + \left(700\sin 60^\circ\right)t + 0$$

We need to determine t_0 such that $y(t_0) = 0$. Then, we will compute

$$x(t_0) = \text{ total horizontal distance traveled}$$
$$y\left(\frac{t_0}{2}\right) = \text{ maximum height}$$

Observe that solving $-16t^2 + \left(700\sin 60^\circ\right)t = 0$ yields

$$-16t^2 + \left(700\sin 60^\circ\right)t + 0 = 0$$
$$t\left(-16t + 700\sin 60^\circ\right) = 0$$
$$t = 0,\ \frac{700\sin 60^\circ}{16} \approx 37.89 \text{ sec.}$$

Now, we have

$$\boxed{\begin{aligned} x(37.89) &\approx 13{,}261 \text{ ft.} = \text{ total horizontal distance traveled} \\ y(18.94) &\approx 5742 \text{ ft.} = \text{ maximum height} \end{aligned}}$$

46. From the given information, we know that the parametric equations are:

$$x = \left(2000\cos 60^\circ\right)t$$
$$y = -16t^2 + \left(2000\sin 60^\circ\right)t + 0$$

We need to determine t_0 such that $y(t_0) = 0$. Then, we will compute

$$x(t_0) = \text{ total horizontal distance traveled}$$
$$y\left(\frac{t_0}{2}\right) = \text{ maximum height}$$

Observe that solving $-16t^2 + \left(2000\sin 60^\circ\right)t = 0$ yields

$$-16t^2 + \left(2000\sin 60^\circ\right)t + 0 = 0$$
$$t\left(-16t + 2000\sin 60^\circ\right) = 0$$
$$t = 0,\ \frac{2000\sin 60^\circ}{16} \approx 108.25 \text{ sec.}$$

Now, we have

$$\boxed{\begin{aligned} x(108.25) &\approx 108{,}253 \text{ ft.} = \text{ total horizontal distance traveled} \\ y(54.127) &= 46{,}875 \text{ ft.} = \text{ maximum height} \end{aligned}}$$

47. From the given information, we know that the parametric equations are:

$$x = \left(4000\cos 30^\circ\right)t$$
$$y = -16t^2 + \left(4000\sin 30^\circ\right)t + 20$$

We need to determine t_0 such that $y\left(t_0\right) = 0$.

Observe that solving $-16t^2 + \left(4000\sin 30^\circ\right)t + 20 = 0$ using the quadratic formula yields

$$t = \frac{-4000\sin 30^\circ \pm \sqrt{\left(4000\sin 30^\circ\right)^2 - 4(-16)(20)}}{2(-16)}$$
$$\approx \frac{-2000 \pm \sqrt{2000^2 + 1280}}{-32} \approx \cancel{-0.001} \text{ or } 125$$

So, it stays in flight for $\boxed{\text{125 seconds}}$.

48. From the given information, we know that the parametric equations are:

$$x = \left(5000\cos 40^\circ\right)t$$
$$y = -16t^2 + \left(5000\sin 40^\circ\right)t + 20$$

We need to determine t_0 such that $y\left(t_0\right) = 0$. Then, we calculate $x\left(t_0\right)$. If this is larger than 2(5,280) = 10,560 ft., then it *can* hit the target.

Observe that solving $-16t^2 + \left(5000\sin 40^\circ\right)t + 20 = 0$ using the quadratic formula yields

$$t = \frac{-5000\sin 40^\circ \pm \sqrt{\left(5000\sin 40^\circ\right)^2 - 4(-16)(20)}}{2(-16)}$$
$$\approx \cancel{-0.006} \text{ or } 200.877 \text{ sec.}$$

So, $x(200.877) = \left(5000\cos 40^\circ\right)(200.877) \approx 769{,}405$ ft. Since this is significantly larger than 10,560 ft., $\boxed{\text{it can definitely hit a target 2 miles away}}$.

49. The parametric equations are

$$x = (100\cos 35^\circ)t$$
$$y = -16t^2 + (100\sin 35^\circ)t$$

The graph is as follows:

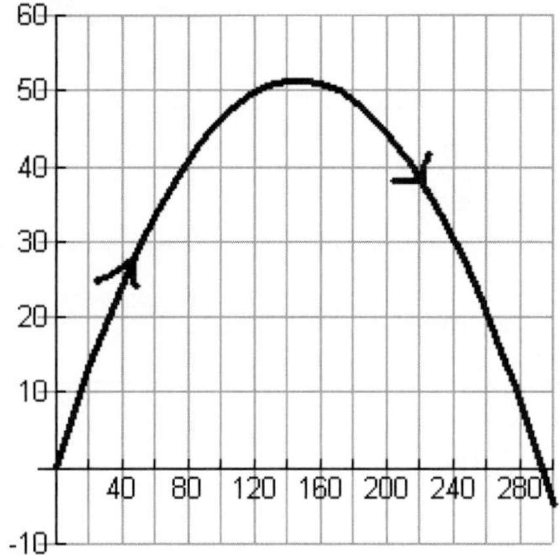

50. The parametric equations are

$$x = (150\cos 55^\circ)t$$
$$y = -16t^2 + (150\sin 55^\circ)t$$

The graph is as follows:

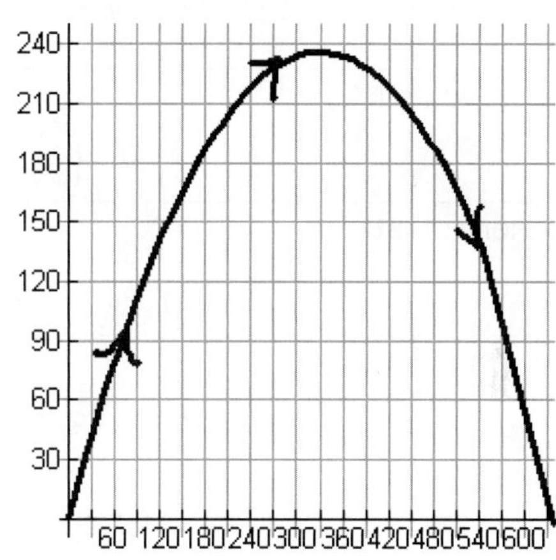

51. The points are as follows:

t	$x(t)$	$y(t)$
0	A+B	0
$\pi/2$	0	A+B
π	-A-B	0
$3\pi/2$	0	-A-B
2π	A+B	0

52. In general, multiply the arguments of the trigonometric functions by coefficients larger than 1 in order to increase speed. Note that A+B represents the length of the arm when fully extended.

53. The original domain must be $t \geq 0$. Therefore, only the portion of the parabola where $y \geq 0$ is part of the plane curve.

54. The original domain must be $t \geq 0$. Therefore, only the portion of the parabola where $x \geq 0$ is part of the plane curve.

55. False. They simply constitute a set of points in the plane.

56. True. By definition of parametric equation (viewed as how points are generated by action of a third parameter).

57. Given the parametric equations $$x=\sqrt{t},\ y=\sqrt{1-t},\ t\ge 0,$$ observe that since $t=x^2$, we have $y=\sqrt{1-x^2}$. Now, since $x=\sqrt{t}$ is always ≥ 0 and $y=\sqrt{1-x^2}$ is also always ≥ 0, this generates a $\boxed{\text{quarter circle in QI}}$.	**58.** Given the parametric equations $$x=\ln t,\ y=t,\ t>0,$$ we see that $x=\ln y$, so that $\boxed{y=e^x}$.

59. Observe that

$$\left(\tfrac{x}{a}\right)^2-\left(\tfrac{y}{b}\right)^2=\tan^2 t-\sec^2 t=1.$$

60. Observe that

$$\left(\tfrac{x}{a}\right)^2-\left(\tfrac{y}{b}\right)^2=\csc^2\left(\tfrac{t}{2}\right)-\cot^2\left(\tfrac{t}{2}\right)=1$$

61. Observe that

$$x+y=2a\sin^2 t,\quad y-x=2b\cos^2 t.$$

Hence,

$$\frac{x+y}{2a}=\sin^2 t,\quad \frac{y-x}{2b}=\cos^2 t,$$

and so,

$$\frac{x+y}{2a}+\frac{y-x}{2b}=1.$$

Solving for y then yields

$$y=\frac{a-b}{a+b}x+\frac{2ab}{a+b},$$

which is the equation of a line.

62. Observe that

$$x+y=2a\cos t,\quad x-y=2a\sin t.$$

Hence,

$$\left(\frac{x+y}{2a}\right)^2=\sin^2 t,\quad \left(\frac{x-y}{2a}\right)^2=\cos^2 t,$$

and so,

$$\left(\frac{x+y}{2a}\right)^2+\left(\frac{x-y}{2a}\right)^2=1.$$

Foiling these squared terms and then simplifying yields

$$x^2+y^2=2a^2,$$

which is the equation of a circle centered at the origin of radius $|a|\sqrt{2}$.

63. Observe that

$$y = be^t \;\Rightarrow\; \ln\left(\tfrac{y}{b}\right) = t\,.$$

Hence,

$$x = e^{at} = e^{a\ln\left(\frac{y}{b}\right)} = e^{\ln\left(\frac{y}{b}\right)^a} = \left(\tfrac{y}{b}\right)^a,$$

and so, solving for y, we obtain

$$y = bx^{1/a}.$$

If a is even, the graph looks essentially like that of $y = \sqrt{x}$, while if a is odd, the graph looks essentially like that of $y = \sqrt[3]{x}$.

64. Observe that

$$x = a\ln t \;\Rightarrow\; e^{\frac{x}{a}} = t\,.$$

Hence,

$$y = \ln\left(b\,e^{\frac{x}{a}}\right) = \ln b + \ln\left(e^{\frac{x}{a}}\right) = \ln b + \frac{x}{a}.$$

This is the equation of a line with slope $\frac{1}{a}$ and y-intercept $\ln b$.

65. Consider the curve defined parametrically by:

$$\{\, x = a\sin t - \sin at, \quad y = a\cos t + \cos at$$

We have the following graphs:

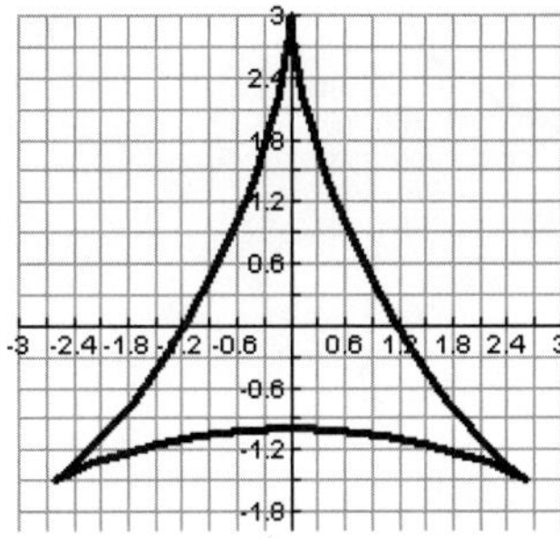

Graph for $a = 2$

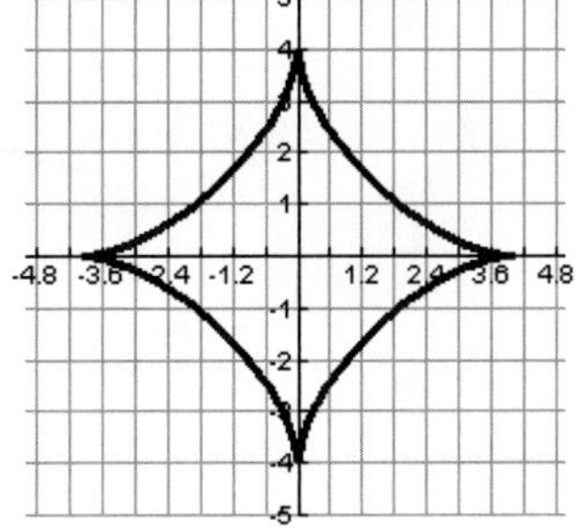

Graph for $a = 3$

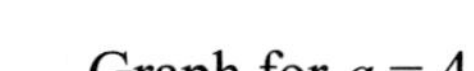

Graph for $a = 4$

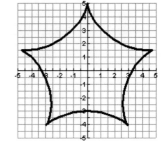

Notice that as the value of a increases, the distance between the vertices and the origin gets larger, and the number of distinct vertices (and corresponding congruent arcs) increases (and $= a$, when a is a positive integer).

66. Consider the curve defined by: $\left\{ x = a\cos t - b\cos at, \quad y = a\sin t + \sin at \right.$

We have the following graphs:

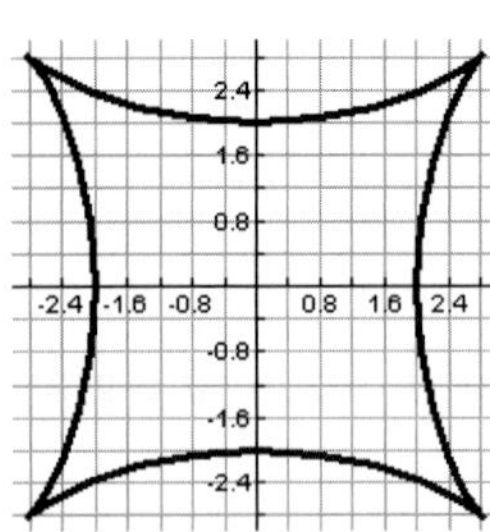

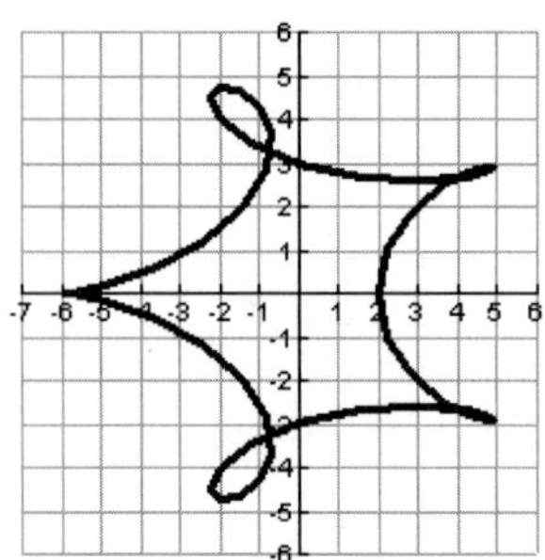

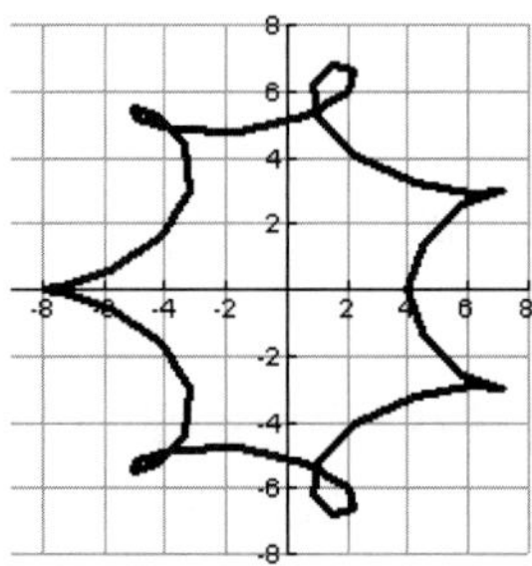

Graph for $a = 3$, $b = 1$ Graph for $a = 4$, $b = 2$ Graph for $a = 6$, $b = 2$

In each case, one cycle of the curve is completed in 2π units of time.

67. Consider the curve defined by: $\left\{ x = \cos at, \quad y = \sin bt \right.$

We have the following graphs:

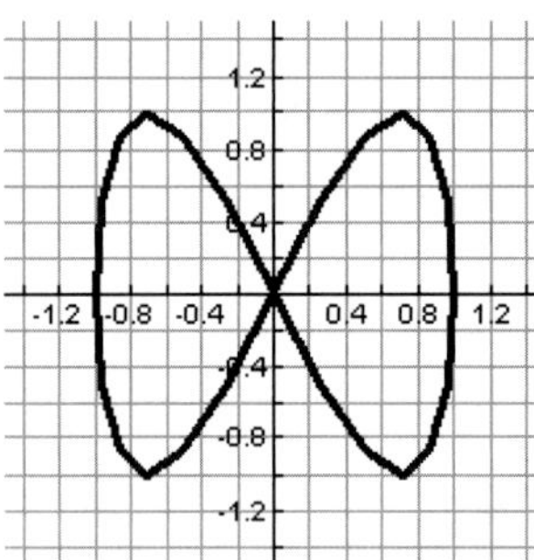

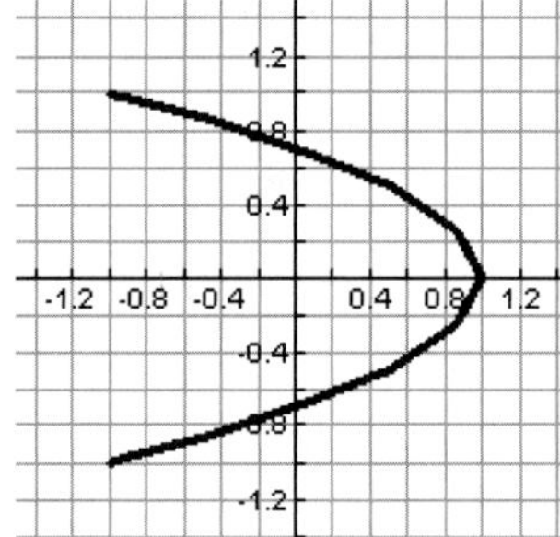

Graph for $a = 2$, $b = 4$

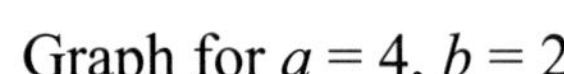

Graph for $a = 4$, $b = 2$

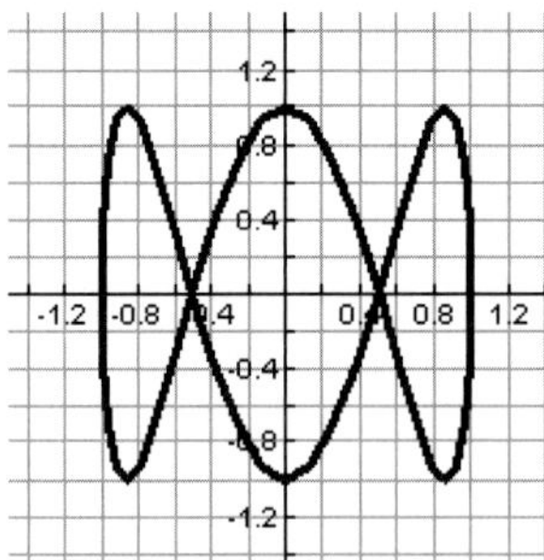

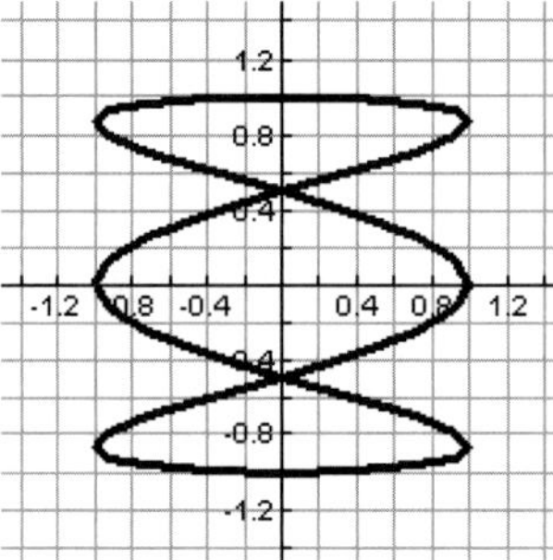

Graph for $a = 1$, $b = 3$ Graph for $a = 3$, $b = 1$

In each case, one cycle of the curve is traced out in 2π units of time.

68. First, we consider the curve defined by the following parametric equations:

$$\{\, x = a\sin at - \sin t, \quad y = a\cos at - \cos t$$

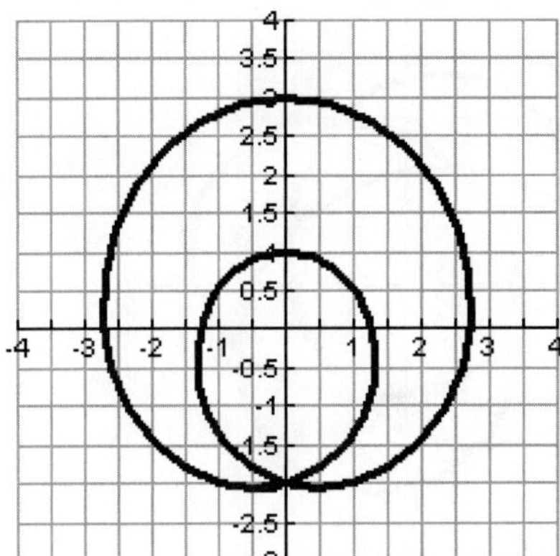

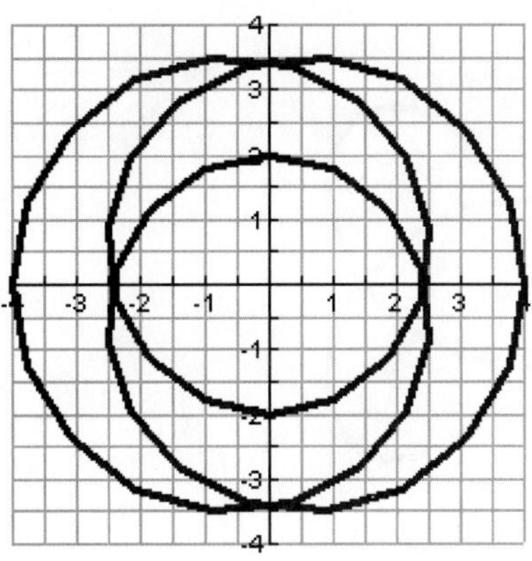

Graph for $a = 2$ Graph for $a = 3$

Now, in comparison, we consider the curve defined by the following parametric equations:

$$\{\, x = a\sin at - \sin t, \quad y = a\cos at + \cos t$$

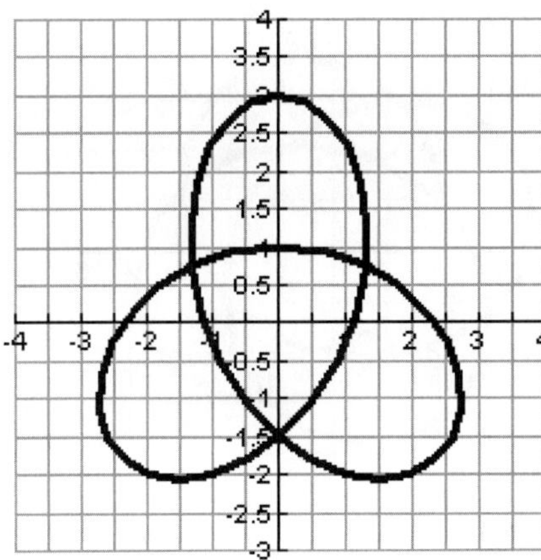

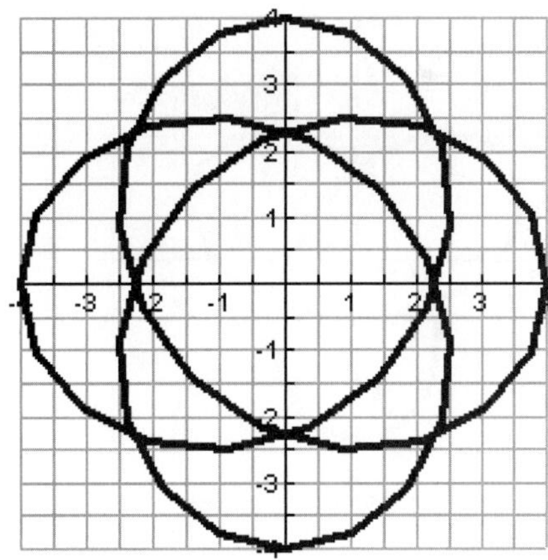

Graph for $a = 2$ Graph for $a = 3$

All curves are generated in 2π units of time.

69. Consider the curve defined by the following parametric equations:

$$x = a\cos at - \sin t, \quad y = a\sin at - \cos t$$

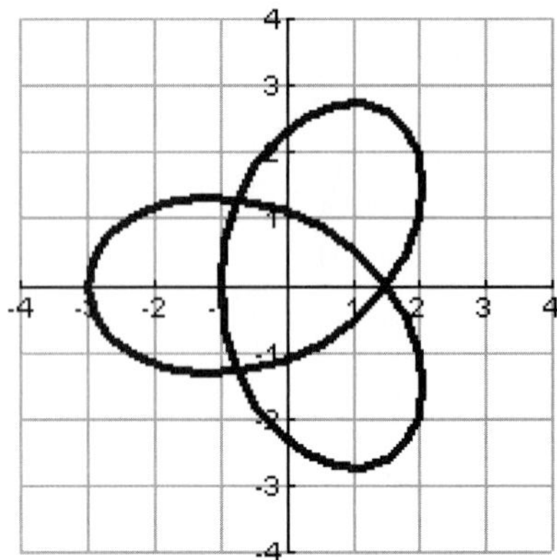

Graph for $a = 2$

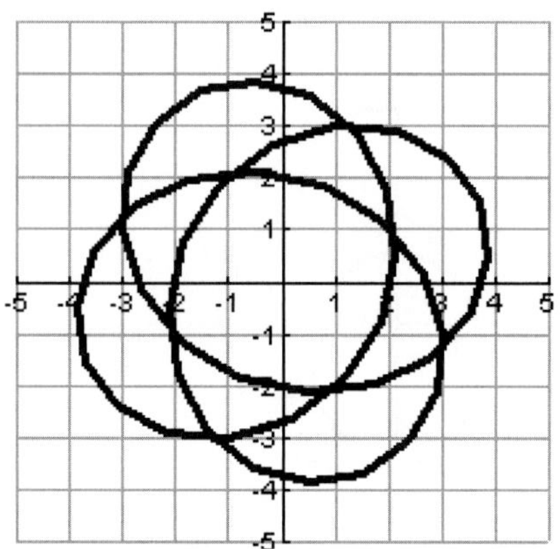

Graph for $a = 3$

All curves are generated in 2π units of time.

70. First, we consider the curve defined by the following parametric equations:

$$x = a\sin at - \cos t, \quad y = a\cos at - \sin t$$

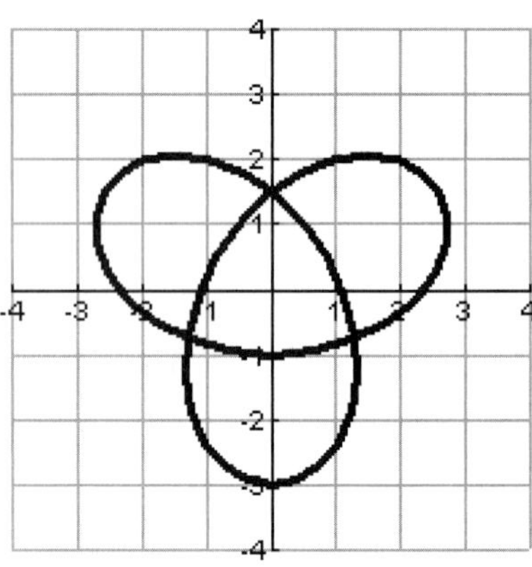

Graph for $a = 2$

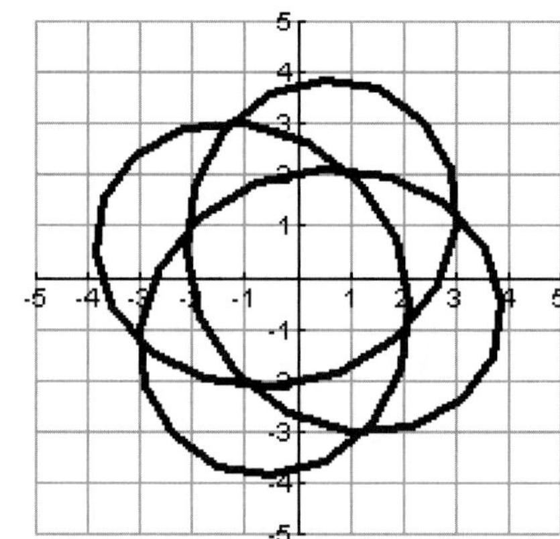

Graph for $a = 3$

All curves are generated in 2π units of time. (In comparison to #69, the graphs start at a different point corresponding to $t = 0$, but generate the same graph for an interval that constitutes one full period.)

71. Eliminate the parameter in each of the two curves to obtain equations in x and y. Then, consider them simultaneously to obtain a system that can be solved to find their points of intersection:

$$\begin{cases} y = x^2 - 1 & \textbf{(1)} \\ y = 4 - (x-1) = 5 - x & \textbf{(2)} \end{cases}$$

Substitute **(2)** into **(1)** to obtain an equation in x:

$$5 - x = x^2 - 1 \Rightarrow x^2 + x - 6 = 0 \Rightarrow (x+3)(x-2) = 0 \Rightarrow x = -3, 2$$

Now, plug these into **(2)** to find the corresponding y-values:

$$x = -3 : y = 8$$
$$x = 2 : y = 3$$

So, the points of intersection are (-3,8) and (2,3).

72. Eliminate the parameter in each of the two curves to obtain equations in x and y. Then, consider them simultaneously to obtain a system that can be solved to find their points of intersection:

$$\begin{cases} x = y^2 + 3 & \textbf{(1)} \\ x = 3 - y & \textbf{(2)} \end{cases}$$

Substitute **(1)** into **(2)** to obtain an equation in y:

$$3 - y = y^2 + 3 \Rightarrow y^2 + y = 0 \Rightarrow y(y+1) = 0 \Rightarrow y = 0, -1$$

Now, plug these into **(2)** to find the corresponding x-values:

$$y = -1 : x = 4$$
$$y = 0 : x = 3$$

So, the points of intersection are (3,0) and (4,-1).

73. Eliminate the parameter in each of the two curves to obtain equations in x and y. Then, consider them simultaneously to obtain a system that can be solved to find their points of intersection:

$$\begin{cases} y = \frac{4}{5}x - \frac{16x^2}{10000} & \textbf{(1)} \\ y = -\frac{(100-x)^2}{2500} + \frac{18(100-x)}{25} - 224 & \textbf{(2)} \end{cases}$$

Equate **(1)** and **(2)** and solve for x:

$$\frac{4}{5}x - \frac{16x^2}{10000} = -\frac{(100-x)^2}{2500} + \frac{18(100-x)}{25} - 224$$

$$-12x^2 + 14,400x + 1,560,000 = 0$$

$$x^2 - 1200x - 130,000 = 0$$

$$x = \frac{1200 \pm \sqrt{1,440,000 + 4(130,000)}}{2} = -100, 1300$$

Now, plug these into **(1)** to find the corresponding y-values:

$$x = -100: \ y = -96$$
$$x = 1300: \ y = -1664$$

So, the points of intersection are (-100,-96) and (1300,-1664).

74. Eliminate the parameter in each of the two curves to obtain equations in x and y. Then, consider them simultaneously to obtain a system that can be solved to find their points of intersection:

$$\begin{cases} x = (y-1)^2 & \textbf{(1)} \\ x = 3 - y & \textbf{(2)} \end{cases}$$

Substitute **(2)** into **(1)** to obtain an equation in y:

$$3 - y = (y-1)^2 \Rightarrow y^2 - y - 2 = 0 \Rightarrow (y-2)(y+1) = 0 \Rightarrow y = -1, 2$$

Now, plug these into **(2)** to find the corresponding x-values:

$$y = -1: x = 4$$
$$y = 2: x = 1$$

So, the points of intersection are (1,2) and (4,-1).

Chapter 9 Review Solutions --

1. False. The focus is always in the region "inside" the parabola.	**2.** False. It opens to the right (in the direction of positive *x*).
3. True.	**4.** False. The center is (-1,3).
5. Vertex is $(0,0)$ and focus is $(3,0)$. So, the parabola opens to the right. As such, the general form is $y^2 = 4px$. In this case, $p = 3$, so the equation is $\boxed{y^2 = 12x}$.	**6.** Vertex is $(0,0)$ and focus is $(0,2)$. So, the parabola opens up. As such, the general form is $x^2 = 4py$. In this case, $p = 2$, so the equation is $\boxed{x^2 = 8y}$.
7. Vertex is $(0,0)$ and the directrix is $x = 5$. So, the parabola opens to the left. Since the distance between the vertex and the directrix is 5, and the focus must be equidistant from the vertex, we know $p = -5$. Hence, the equation is $\boxed{(y-0)^2 = 4(-5)(x-0) = -20x \Rightarrow y^2 = -20x}$.	**8.** Vertex is $(0,0)$ and the directrix is $y = 4$. So, the parabola opens down. Since the distance between the vertex and the directrix is 4 and the focus must be equidistant from the vertex, we know $p = -4$. Hence, the equation is $\boxed{(x-0)^2 = 4(-4)(y-0) = -16y \Rightarrow x^2 = -16y}$.
9. Vertex is $(2,3)$ and focus is $(2,5)$. So, the parabola opens up. As such, the general form is $(x-2)^2 = 4p(y-3)$. In this case, $p = 2$, so the equation is $\boxed{(x-2)^2 = 8(y-3)}$.	**10.** Vertex is $(-1,-2)$ and focus is $(1,-2)$. So, the parabola opens to the right. As such, the general form is $(y+2)^2 = 4p(x+1)$. In this case, $p = 2$, so the equation is $\boxed{(y+2)^2 = 8(x+1)}$.
11. Focus is $(1,5)$ and the directrix is $y = 7$. So, the parabola opens down. Since the distance between the focus and directrix is 2 and the vertex must occur halfway between them, we know $p = -1$ and the vertex is $(1,6)$. Hence, the equation is $\boxed{(x-1)^2 = 4(-1)(y-6) = -4(y-6)}$.	**12.** Focus is $(2,2)$ and the directrix is $x = 0$. So, the parabola opens to the right. Since the distance between the focus and directrix is 2 and the vertex must occur halfway between them, we know $p = 1$ and the vertex is $(1,2)$. Hence, the equation is $\boxed{(y-2)^2 = 4(1)(x-1) = 4(x-1)}$.

13. Equation: $x^2 = -12y = 4(-3)y$ **(1)**

So, $p = -3$ and opens down.

Vertex: $(0,0)$

Focus: $(0,-3)$

Directrix: $y = 3$

Latus Rectum: Connects x-values corresponding to $y = -3$. Substituting this into **(1)** yields:

$$x^2 = 36 \text{ so that } x = \pm 6$$

So, the length of the latus rectum is 12.

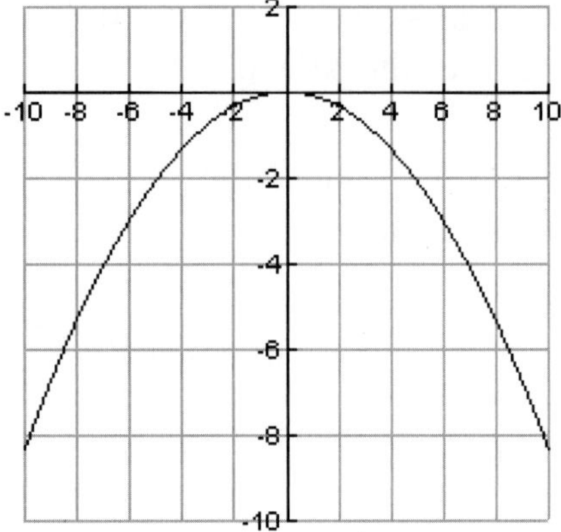

14. Equation: $x^2 = 8y = 4(2)y$ **(1)**

So, $p = 2$ and opens up.

Vertex: $(0,0)$

Focus: $(0,2)$

Directrix: $y = -2$

Latus Rectum: Connects x-values corresponding to $y = 2$. Substituting this into **(1)** yields:

$$x^2 = 16 \text{ so that } x = \pm 4$$

So, the length of the latus rectum is 8.

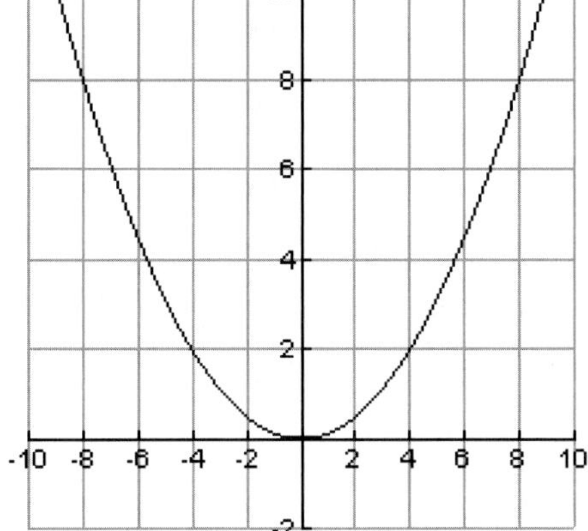

15. Equation: $y^2 = x = 4(\frac{1}{4})x$ **(1)**

So, $p = \frac{1}{4}$ and opens to the right.

Vertex: $(0,0)$

Focus: $(\frac{1}{4},0)$

Directrix: $x = -\frac{1}{4}$

Latus Rectum: Connects y-values corresponding to $x = \frac{1}{4}$. Substituting this into **(1)** yields:

$$y^2 = \frac{1}{4} \text{ so that } y = \pm\frac{1}{2}$$

So, the length of the latus rectum is 1.

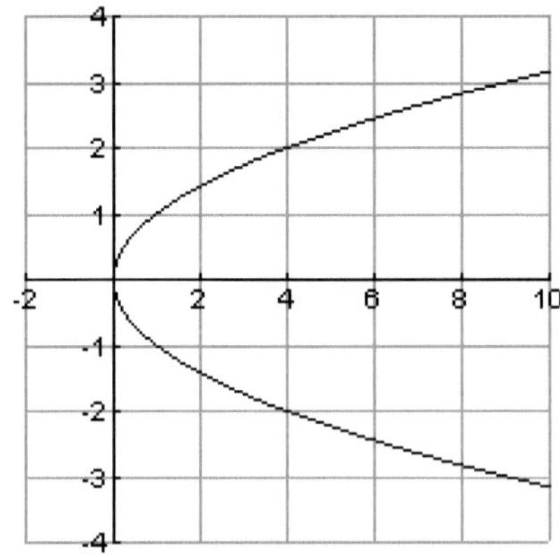

16. Equation: $y^2 = -6x = 4(-\frac{3}{2})x$ **(1)**

So, $p = -\frac{3}{2}$ and opens to the left.

Vertex: $(0,0)$

Focus: $(-\frac{3}{2},0)$

Directrix: $x = \frac{3}{2}$

Latus Rectum: Connects y-values corresponding to $x = -\frac{3}{2}$. Substituting this into **(1)** yields:

$$y^2 = 9 \text{ so that } y = \pm 3$$

So, the length of the latus rectum is 6.

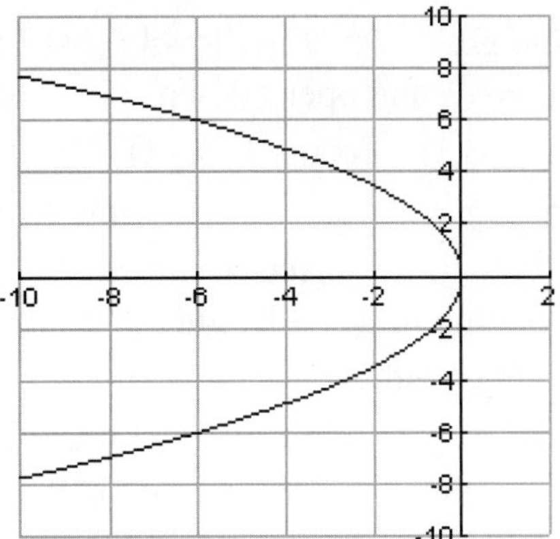

17. Equation: $(y+2)^2 = 4(1)(x-2)$ **(1)**

So, $p = 1$ and opens to the right.

Vertex: $(2,-2)$

Focus: $(3,-2)$

Directrix: $x = 1$

Latus Rectum: Connects y-values corresponding to $x = 3$. Substituting this into **(1)** yields:

$$(y+2)^2 = 4 \text{ so that } y = -2 \pm 2 = -4,\ 0$$

So, the length of the latus rectum is 4.

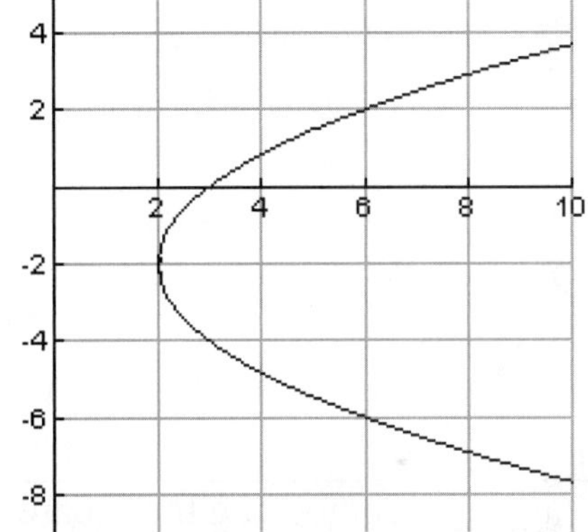

18. Equation: $(y-2)^2 = 4(-1)(x+1)$ **(1)**

So, $p = -1$ and opens to the left.

Vertex: $(-1,2)$

Focus: $(-2,2)$

Directrix: $x = 0$

Latus Rectum: Connects y-values corresponding to $x = -2$. Substituting this into **(1)** yields:

$$(y-2)^2 = 4 \text{ so that } y = 2 \pm 2 = 4,\ 0$$

So, the length of the latus rectum is 4.

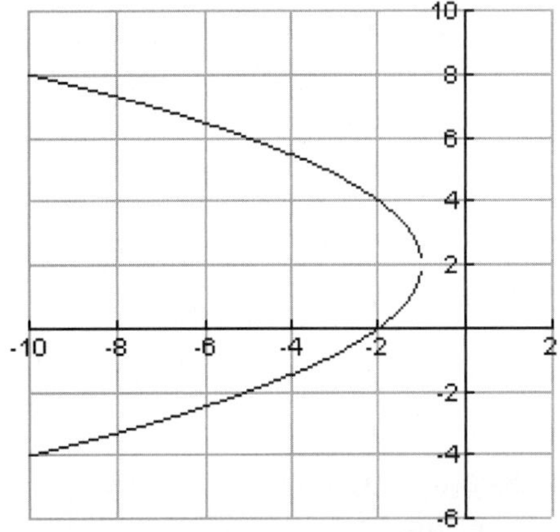

19. Equation: $(x+3)^2 = 4(-2)(y-1)$ **(1)**

So, $p = -2$ and opens down.

Vertex: $(-3,1)$ Focus: $(-3,-1)$

Directrix: $y = 3$

Latus Rectum: Connects x-values corresponding to $y = -1$. Substituting this into **(1)** yields:

$(x+3)^2 = 16$ so that $x = -3 \pm 4 = 1, -7$

So, the length of the latus rectum is 8.

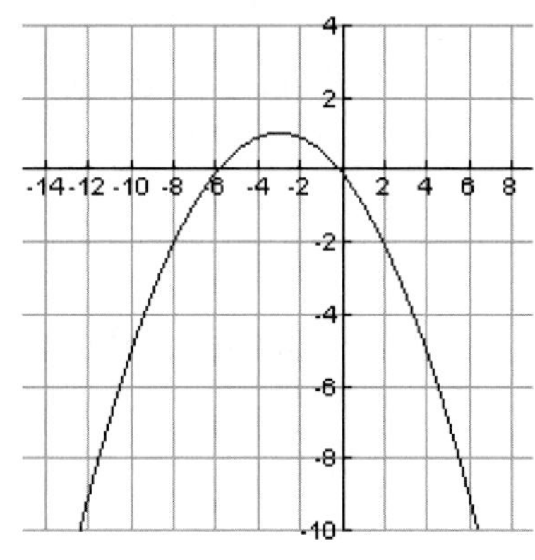

20. Equation: $(x-3)^2 = 4(-2)(y+2)$ **(1)**

So, $p = -2$ and opens down.

Vertex: $(3,-2)$ Focus: $(3,-4)$

Directrix: $y = 0$

Latus Rectum: Connects x-values corresponding to $y = -4$. Substituting this into **(1)** yields:

$(x-3)^2 = 16$ so that $x = 3 \pm 4 = -1, 7$

So, the length of the latus rectum is 8.

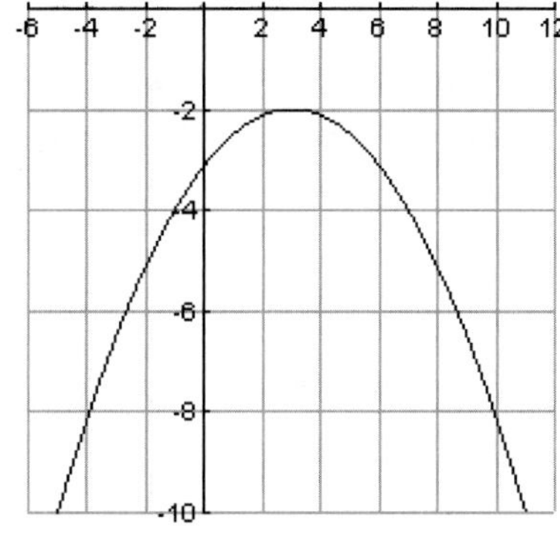

21. Equation: Completing the square yields:

$$x^2 + 5x + 2y + 25 = 0$$

$$(x^2 + 5x + \tfrac{25}{4}) = -2y - 25 + \tfrac{25}{4}$$

$$(x + \tfrac{5}{2})^2 = -2y - \tfrac{75}{4}$$

$$(x + \tfrac{5}{2})^2 = -2(y + \tfrac{75}{8})$$

$$(x + \tfrac{5}{2})^2 = 4(-\tfrac{1}{2})(y + \tfrac{75}{8}) \quad \textbf{(1)}$$

So, $p = -\frac{1}{2}$ and opens down.

Vertex: $(-\frac{5}{2}, -\frac{75}{8})$ Focus: $(-\frac{5}{2}, -\frac{79}{8})$

Directrix: $y = -\frac{71}{8}$

Latus Rectum: Connects x-values corresponding to $y = -\frac{79}{8}$. Substituting this into **(1)** yields:

$(x + \frac{5}{2})^2 = 1$ so that $x = -\frac{5}{2} \pm 1 = -\frac{3}{2}, -\frac{7}{2}$

So, the length of the latus rectum is 2.

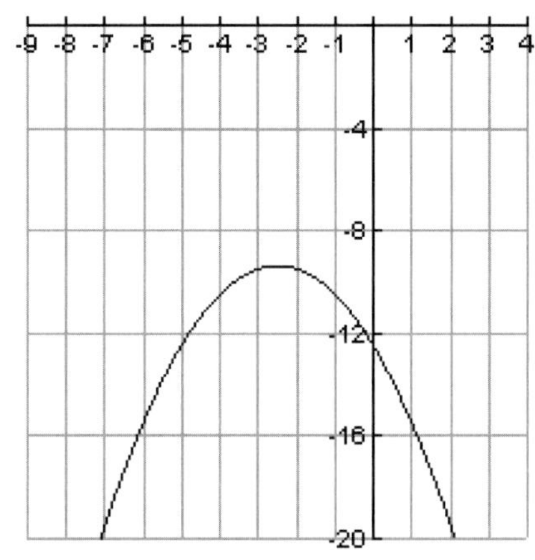

22. Equation: Completing the square yields:

$$y^2+2y-16x+1=0$$
$$(y^2+2y+1)-1=16x$$
$$(y+1)^2=16x$$
$$(y+1)^2=4(4)x \quad \textbf{(1)}$$

So, $p=4$ and opens to the right.

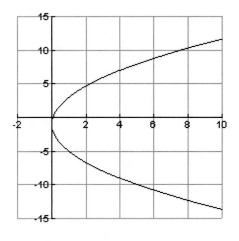

Vertex: $(0,-1)$ Focus: $(4,-1)$

Directrix: $x=-4$

Latus Rectum: Connects y-values corresponding to $x=4$. Substituting this into **(1)** yields:

$$(y+1)^2=64 \text{ so that } y=-1\pm 8=7,-9$$

So, the length of the latus rectum is 16.

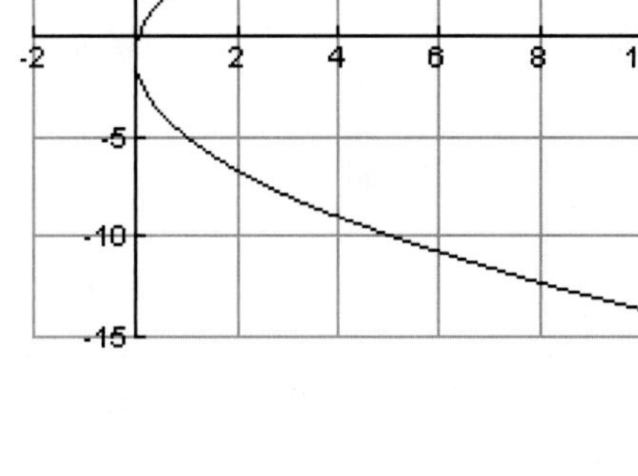

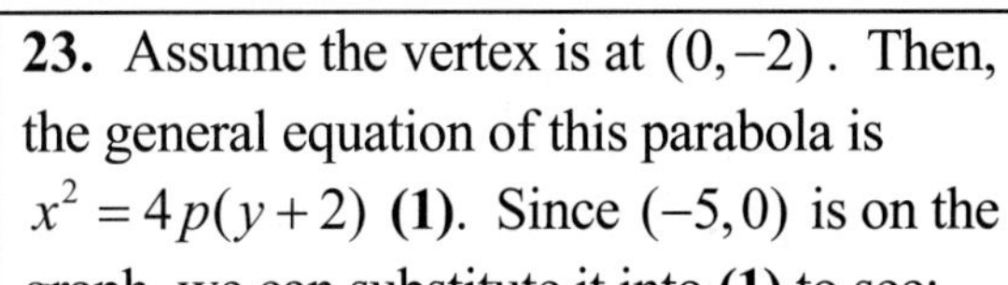

23. Assume the vertex is at $(0,-2)$. Then, the general equation of this parabola is $x^2=4p(y+2)$ **(1)**. Since $(-5,0)$ is on the graph, we can substitute it into **(1)** to see:

$$25=4p(2)=8p \text{ so that } p=\tfrac{25}{8}$$

So, the focus is at $(0,\frac{9}{8})$, and the receiver should be placed $\frac{25}{8}\approx 3.125$ feet from the center.

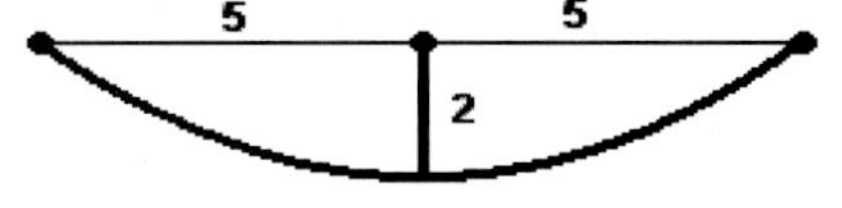

24. Assume that the parabola has vertex at (0,40) and opens down. Then, the form of the equation is $(x-0)^2=4p(y-40)$ **(1)**. Since the point (15,0) is on the graph, we can substitute it into **(1)** to find p: $225=-160p$ so that $p=-1.406$. So, **(1)** becomes $x^2=-5.625(y-40)$. We need to determine the y-value when $x=4$. If it is larger than 14, then the RV will pass under the bridge without a problem. Observe that

$$16=-5.625(y-40)=-5.625y+225$$
$$37.16=y$$

Hence, the RV will pass under the bridge without a problem.

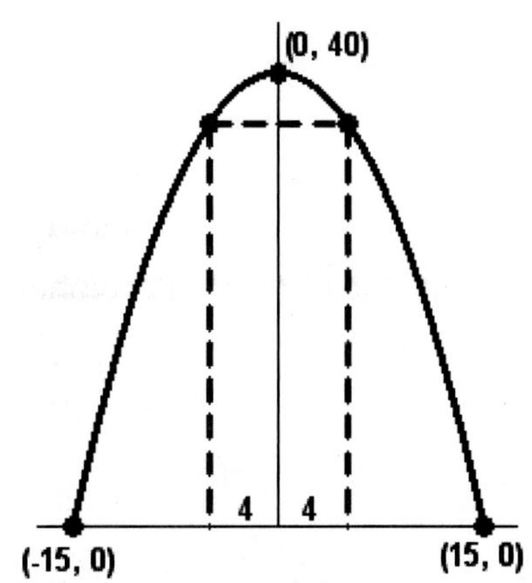

25.

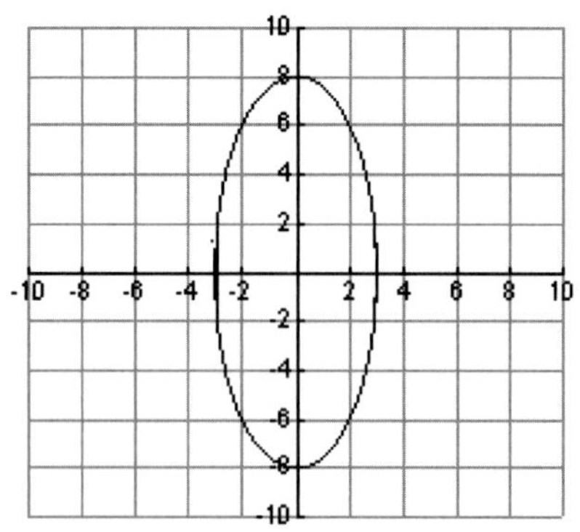

26.

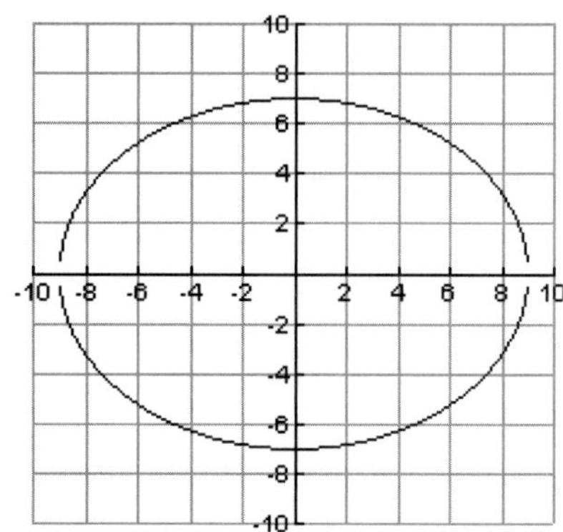

27. Write the equation in standard form as $x^2 + \frac{y^2}{25} = 1$.

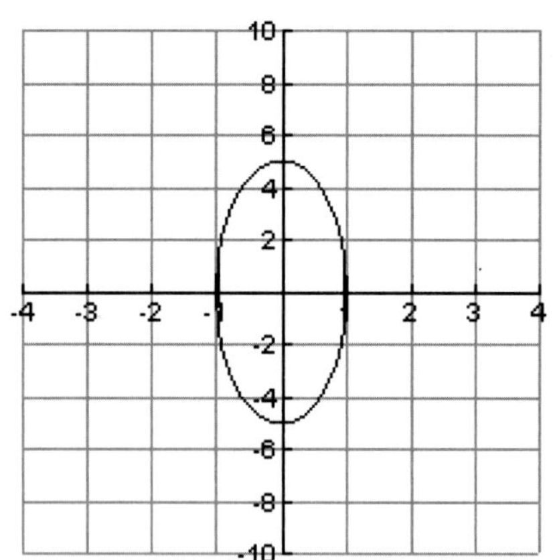

28. Write the equation in standard form as $\frac{x^2}{16} + \frac{y^2}{8} = 1$.

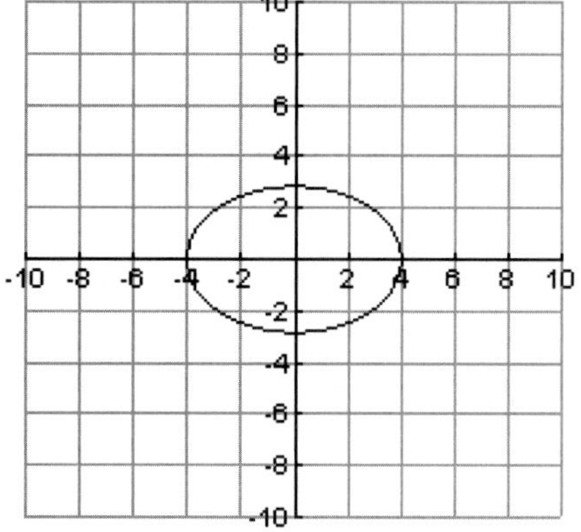

29. Since the foci are $(-3,0)$ and $(3,0)$, we know that $c = 3$ and the center of the ellipse is $(0,0)$. Further, since the vertices are $(-5,0)$ and $(5,0)$ and they lie on the x-axis, the ellipse is horizontal and $a = 5$. Hence,

$$c^2 = a^2 - b^2$$

$$9 = 25 - b^2$$

$$b^2 = 16$$

Therefore, the equation of the ellipse is $\boxed{\frac{x^2}{25} + \frac{y^2}{16} = 1}$.

30. Since the foci are $(0,-2)$ and $(0,2)$, we know that $c = 2$ and the center of the ellipse is $(0,0)$. Further, since the vertices are $(0,-3)$ and $(0,3)$ and they lie on the y-axis, the ellipse is vertical and $a = 3$. Hence,

$$c^2 = a^2 - b^2$$

$$4 = 9 - b^2$$

$$b^2 = 5$$

Therefore, the equation of the ellipse is $\boxed{\frac{x^2}{5} + \frac{y^2}{9} = 1}$.

31. Since the ellipse is centered at $(0,0)$, the length of the horizontal major axis being 16 implies that $a=8$, and the length of the horizontal minor axis being 6 implies that $b=3$. Since the major axis is vertical, the equation of the ellipse is $\boxed{\frac{x^2}{9}+\frac{y^2}{64}=1}$.

32. Since the ellipse is centered at $(0,0)$, the length of the horizontal major axis being 30 implies that $a=15$, and the length of the vertical minor axis being 20 implies that $b=10$. Since the major axis is horizontal, the equation of the ellipse is $\boxed{\frac{x^2}{225}+\frac{y^2}{100}=1}$.

33.

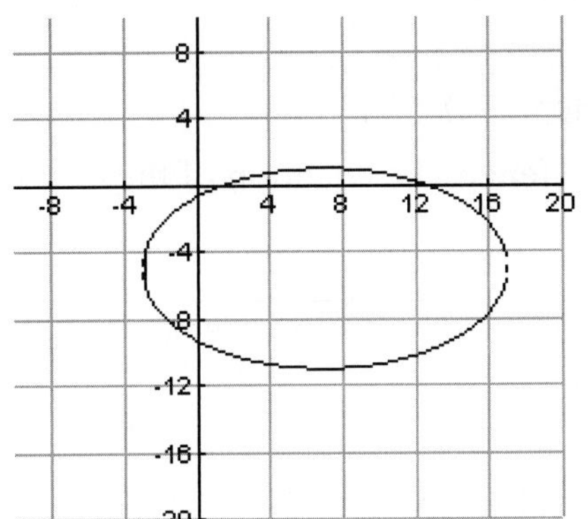

34. Write the equation in standard form as $\frac{(x+3)^2}{6}+\frac{(y-4)^2}{120}=1$.

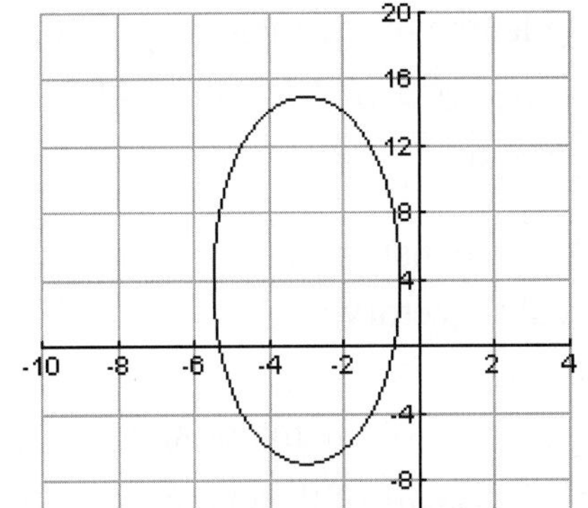

35. First, write the equation in standard form by completing the square:

$$4(x^2-4x)+12(y^2+6y)=-123$$
$$4(x^2-4x+4)+12(y^2+6y+9)=-123+16+108$$
$$4(x-2)^2+12(y+3)^2=1$$
$$\frac{(x-2)^2}{\frac{1}{4}}+\frac{(y+3)^2}{\frac{1}{12}}=1$$

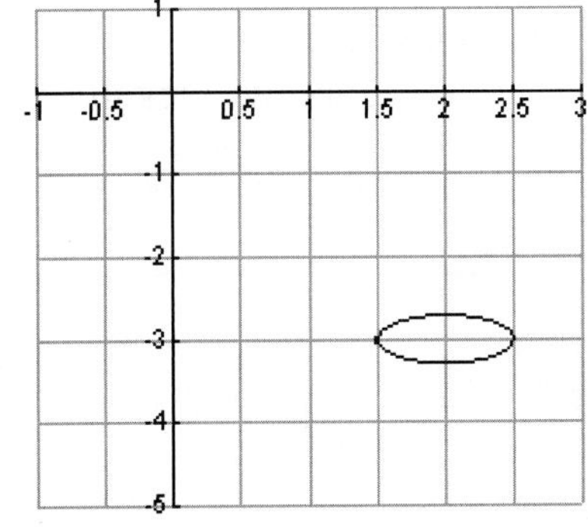

36. First, write the equation in standard form by completing the square:

$$4(x^2-2x)+9(y^2-8y)=-147$$
$$4(x^2-2x+1)+9(y^2-8y+16)=-147+4+144$$
$$4(x-1)^2+9(y-4)^2=1$$
$$\frac{(x-1)^2}{\frac{1}{4}}+\frac{(y-4)^2}{\frac{1}{9}}=1$$

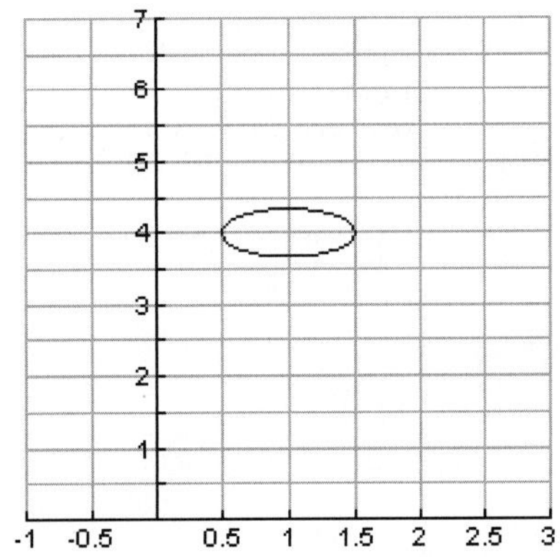

37. Since the foci are $(-1,3), (7,3)$ and the vertices are $(-2,3), (8,3)$, we know:
i) the ellipse is horizontal with center at $(\frac{-2+8}{2},3)=(3,3)$,
ii) $3-c=-1$ so that $c=4$,
iii) the length of the major axis is $8-(-2)=10$. Hence, $a=5$.
Now, since $c^2=a^2-b^2$, $16=25-b^2$ so that $b^2=9$. Hence, the equation of the ellipse is $\boxed{\frac{(x-3)^2}{25}+\frac{(y-3)^2}{9}=1}$.

38. Since the foci are $(1,-3), (1,-1)$ and the vertices are $(1,-4), (1,0)$, we know:
i) the ellipse is vertical with center at $(1,\frac{-4+0}{2})=(1,-2)$,
ii) $-2-c=-3$ so that $c=1$,
iii) the length of the major axis is $0-(-4)=4$. Hence, $a=2$.
Now, since $c^2=a^2-b^2$, $1=4-b^2$ so that $b^2=3$. Hence, the equation of the ellipse is $\boxed{\frac{(x-1)^2}{3}+\frac{(y+2)^2}{4}=1}$.

39. Assume that the sun is at the origin. Then, the vertices of Jupiter's horizontal elliptical trajectory are: $(-7.409\times10^8,0), (8.157\times10^8,0)$
From this, we know that:
i) The length of the major axis is $8.157\times10^8-(-7.409\times10^8)=1.5566\times10^9$, and so the value of a is half of this, namely 7.783×10^8;
ii) The center of the ellipse is $\left(\frac{-7.409\times10^8+(8.157\times10^8)}{2},0\right)=(3.74\times10^7,0)$;
iii) The value of c is 3.74×10^7.
Now, since $c^2=a^2-b^2$, we have $\left(3.74\times10^7\right)^2=\left(7.783\times10^8\right)^2-b^2$, so that $b^2=6.044\times10^{17}$. Hence, the equation of the ellipse must be
$$\boxed{\frac{\left(x-(3.74\times10^7)\right)^2}{6.058\times10^{17}}+\frac{(y-0)^2}{6.044\times10^{17}}=1}.$$
(Note: There is more than one correct way to set up this problem, and it begins with choosing the center of the ellipse. If, alternatively, you choose the center to be at (0,0), then the actual coordinates of the vertices and foci will change, and will result in the following slightly different equation for the trajectory: $\frac{x^2}{778{,}300{,}000^2}+\frac{y^2}{777{,}400{,}000^2}=1$
Both are equally correct!.)

40. Assume that the sun is at the origin. Then, the vertices of Mars' horizontal elliptical trajectory are: $(-2.07\times10^8,0)$, $(2.49\times10^8,0)$

From this, we know that:

i) The length of the major axis is $2.49\times10^8-(-2.077\times10^8)=4.56\times10^8$, and so the value of a is half of this, namely 2.28×10^8;

ii) The center of the ellipse is $\left(\frac{2.49\times10^8+(-2.07\times10^8)}{2},0\right)=(2.1\times10^7,0)$;

iii) The value of c is 2.1×10^7.

Now, since $c^2=a^2-b^2$, we have $\left(2.1\times10^7\right)^2=\left(2.28\times10^8\right)^2-b^2$, so that $b^2=5.154\times10^{16}$. Hence, the equation of the ellipse must be

$$\boxed{\frac{\left(x-(2.1\times10^7)\right)^2}{5.1984\times10^{16}}+\frac{(y-0)^2}{5.154\times10^{16}}=1}.$$

41.

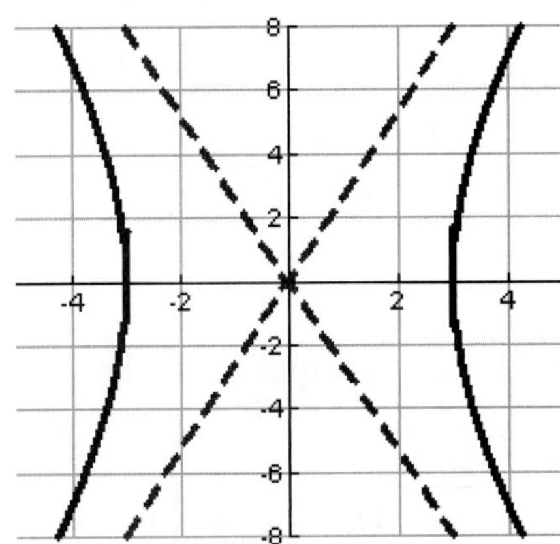

Notes on the Graph:
The equations of the asymptotes are $y=\pm\frac{8}{3}x$.

42.

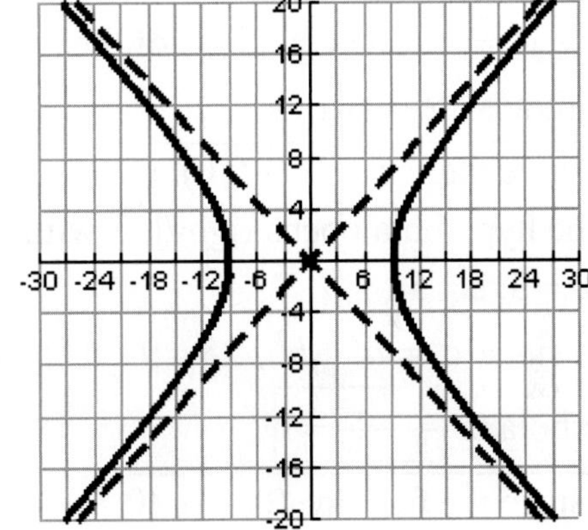

Notes on the Graph:
The equations of the asymptotes are $y=\pm\frac{7}{9}x$.

43.

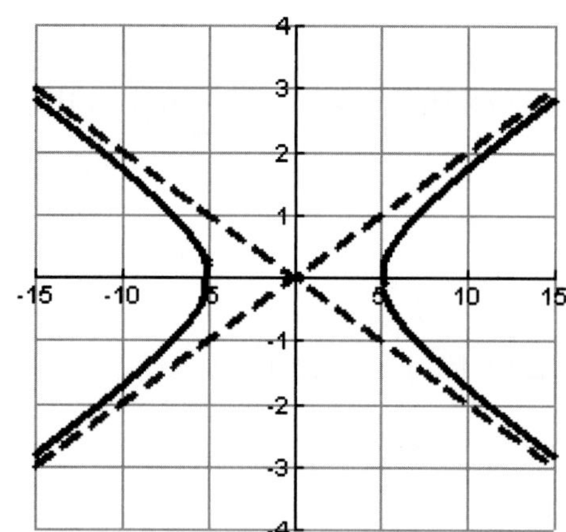

Notes on the Graph:
First, write the equation in standard form as $\frac{x^2}{25} - y^2 = 1$. The equations of the asymptotes are $y = \pm\frac{1}{5}x$.

44.

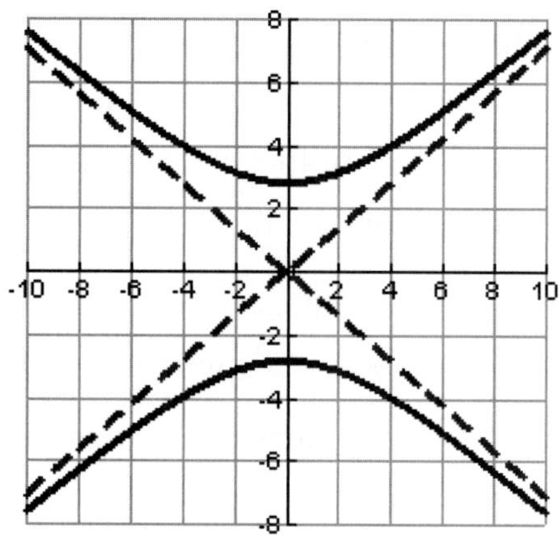

Notes on the Graph:
First, write the equation in standard form as $\frac{y^2}{8} - \frac{x^2}{16} = 1$. The equations of the asymptotes are $y = \pm\frac{\sqrt{2}}{2}x$.

45. Since the foci are $(-5,0)$, $(5,0)$ and the vertices are $(-3,0)$, $(3,0)$, we know that:
i) the hyperbola opens right/left with center at $(0,0)$,
ii) $a = 3$, $c = 5$,
Now, since $c^2 = a^2 + b^2$, $25 = 9 + b^2$ so that $b^2 = 16$. Hence, the equation of the hyperbola is $\boxed{\frac{x^2}{9} - \frac{y^2}{16} = 1}$.

46. Since the foci are $(0,-3)$, $(0,3)$ and the vertices are $(0,-1)$, $(0,1)$, we know that:
i) the hyperbola opens up/down with center at $(0,0)$,
ii) $a = 1$, $c = 3$,
Now, since $c^2 = a^2 + b^2$, $9 = 1 + b^2$ so that $b^2 = 8$. Hence, the equation of the hyperbola is $\boxed{y^2 - \frac{x^2}{8} = 1}$.

47. Since the center is $(0,0)$ and the transverse axis is the y-axis, the general form of the equation of this hyperbola is $\frac{y^2}{a^2} - \frac{x^2}{b^2} = 1$ with asymptotes $y = \pm\frac{a}{b}x$. In this case, $\frac{a}{b} = 3$ so that $a = 3b$. Thus, the equation simplifies to $\frac{y^2}{(3b)^2} - \frac{x^2}{b^2} = 1$, or equivalently $\frac{y^2}{9} - x^2 = b^2$. If we took $b = 1$, then this simplifies to $\boxed{\frac{y^2}{9} - x^2 = 1}$.

48. Since the center is $(0,0)$ and the transverse axis is the y-axis, the general form of the equation of this hyperbola is $\frac{y^2}{a^2}-\frac{x^2}{b^2}=1$ with asymptotes $y=\pm\frac{a}{b}x$. In this case, $\frac{a}{b}=\frac{1}{2}$ so that $2a=b$. Thus, the equation simplifies to $\frac{y^2}{a^2}-\frac{x^2}{(2a)^2}=1$, or equivalently $\boxed{y^2-\frac{x^2}{4}=a^2}$.

49.

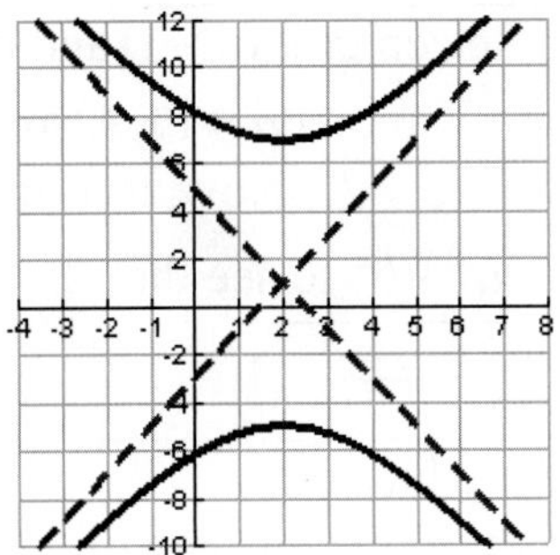

Notes on the Graph:
The equations of the asymptotes are $y=\pm 2(x-2)+1$.

50.

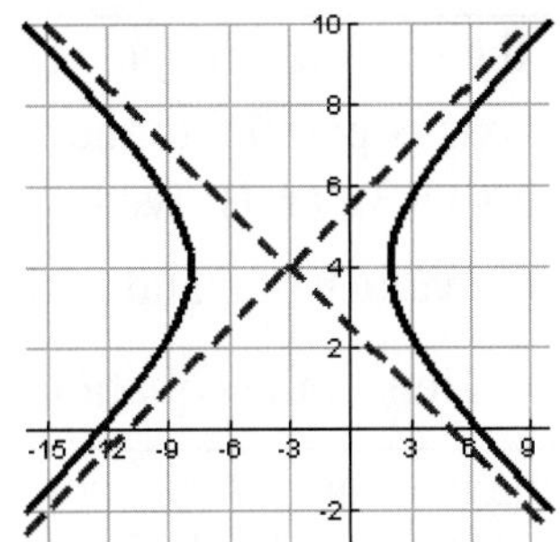

Notes on the Graph:
The equations of the asymptotes are $y=\pm\frac{1}{2}(x+3)+4$.

51.

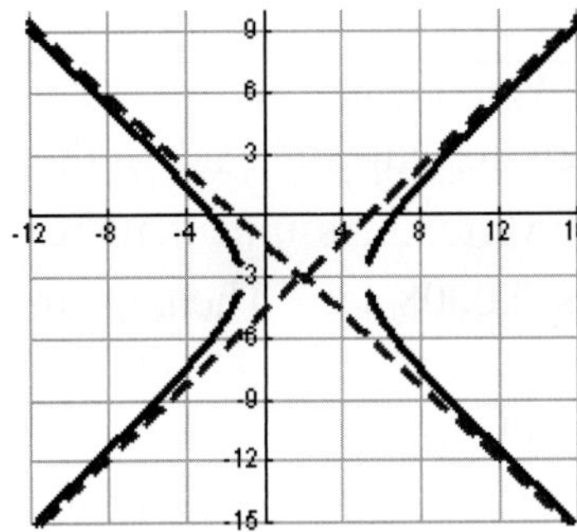

Notes on the Graph:
First, write the equation in standard form by completing the square:

$$8(x^2-4x)-10(y^2+6y)=138$$
$$8(x^2-4x+4)-10(y^2+6y+9)=138+32-90$$
$$8(x-2)^2-10(y+3)^2=80$$
$$\frac{(x-2)^2}{10}-\frac{(y+3)^2}{8}=1$$

The equations of the asymptotes are $y=\pm\frac{2}{\sqrt{5}}(x-2)-3$.

52.

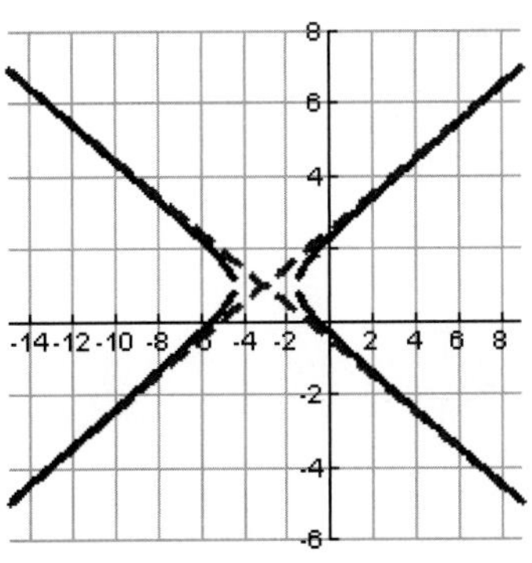

Notes on the Graph:
First, write the equation in standard form by completing the square:

$$2(x^2+6x)-8(y^2-2y)=-6$$
$$2(x^2+6x+9)-8(y^2-2y+1)=-6+18-8$$
$$2(x+3)^2-8(y-1)^2=4$$
$$\frac{(x+3)^2}{2}-\frac{(y-1)^2}{\frac{1}{2}}=1$$

The equations of the asymptotes are $y=\pm\frac{1}{2}(x+3)+1$.

53. Since the vertices are $(0,3),\ (8,3)$, we know that the center is $(4,3)$ and the transverse axis is parallel to the *x*-axis. Hence, $4-a=0$ so that $a=4$. Also, since the foci are $(-1,3),\ (9,3)$, we know that $4+c=-1$ so that $c=-5$. Now, to find *b*, we substitute the values of *c* and *a* obtained above into $c^2=a^2+b^2$ to see that $25=16+b^2$ so that $b^2=9$. Hence, the equation of the hyperbola is $\boxed{\frac{(x-4)^2}{16}-\frac{(y-3)^2}{9}=1}$.
54. Since the vertices are $(4,-2),\ (4,0)$, we know that the center is $(4,-1)$ and the transverse axis is parallel to the *y*-axis. Hence, $-1-a=-2$ so that $a=1$. Also, since the foci are $(4,-3),\ (4,1)$, we know that $-1-c=-3$ so that $c=2$. Now, to find *b*, we substitute the values of *c* and *a* obtained above into $c^2=a^2+b^2$ to see that $4=1+b^2$ so that $b^2=3$. Hence, the equation of the hyperbola is $\boxed{\frac{(y-1)^2}{1}-\frac{(x-4)^2}{3}=1}$.
55. Assume that the stations coincide with the foci and are located at $(-110,0),\ (110,0)$. The difference in distance between the ship and each of the two stations must remain constantly 2*a*, where $(a,0)$ is the vertex. Assume that the radio signal speed is 186,000 $^{mi}/_{sec}$ and the time difference is 0.00048 sec. Then, using distance = rate × time, we obtain: $$2a=(186{,}000)(0.00048)=89.28 \text{ so that } a=44.64$$ So, the ship will come ashore between the two stations 65.36 miles from one and 154.64 miles from the other.
56. Assume that the stations coincide with the foci and are located at $(-200,0),\ (200,0)$. The difference in distance between the ship and each of the two stations must remain constantly 2*a*, where $(a,0)$ is the vertex. Assume that the radio signal speed is 186,000 $^{mi}/_{sec}$ and the time difference is 0.0008 sec. Then, using distance = rate × time, we obtain: $$2a=(186{,}000)(0.0008)=148.8 \text{ so that } a=74.4$$ So, the ship will come ashore between the two stations 125.6 miles from one and 274.4 miles from the other.

57. Solve the system $\begin{cases} x^2 + y = -3 & \textbf{(1)} \\ x - y = 5 & \textbf{(2)} \end{cases}$

Add **(1)** and **(2)** to get an equation in terms of x:

$$x^2 + x - 2 = 0$$
$$(x+2)(x-1) = 0$$
$$x = -2, 1$$

Now, substitute each value of x back into **(2)** to find the corresponding values of y:

$$x = -2: \quad -2 - y = 5 \text{ so that } y = -7$$
$$x = 1: \quad 1 - y = 5 \text{ so that } y = -4$$

So, the solutions are $\boxed{(-2,-7) \text{ and } (1,-4)}$.

58. Solve the system $\begin{cases} x^2 + y = 2 & \textbf{(1)} \\ x^2 + y^2 = 4 & \textbf{(2)} \end{cases}$

Subtract **(2)** – **(1)**: $y^2 - y = 2$ **(3)**

Solve **(3)** for y: $y^2 - y - 2 = 0 \Rightarrow (y-2)(y+1) = 0 \Rightarrow y = 2, -1$

Now, substitute these values of x back into **(1)** to find the corresponding values of x:

$$y = 2: \ x^2 + 2 = 2 \Rightarrow x = 0$$
$$y = -1: \ x^2 - 1 = 2 \Rightarrow x^2 = 3 \Rightarrow x = \pm\sqrt{3}$$

So, the solutions are $\boxed{\left(\sqrt{3},-1\right), \left(-\sqrt{3},-1\right), \text{ and } \left(0,2\right)}$.

59. Solve the system $\begin{cases} x^2 + y^2 = 5 & \textbf{(1)} \\ 2x^2 - y = 0 & \textbf{(2)} \end{cases}$

Solve **(2)** for y: $y = 2x^2$ **(3)**

Substitute **(3)** into **(1)** to get an equation in x:

$$x^2 + (2x^2)^2 = 5$$
$$x^2 + 4x^4 = 5$$
$$4x^4 + x^2 - 5 = 0$$
$$\left(4x^2 + 5\right)\left(x^2 - 1\right) = 0$$
$$x = \pm 1$$

Now, substitute each value of x back into **(2)** to find the corresponding values of y:

$$x = 1: \quad y = 2(1)^2 = 2$$
$$x = -1: \ y = 2(-1)^2 = 2$$

So, the solutions are $\boxed{(1,2) \text{ and } (-1,2)}$.

60. Solve the system $\begin{cases} x^2+y^2=16 & \textbf{(1)} \\ 6x^2+y^2=16 & \textbf{(2)} \end{cases}$

Multiply **(1)** by -1, and then add to **(2)**:

$$5x^2=0$$

$$x=0 \quad \textbf{(3)}$$

Substitute **(3)** into **(1)** to find the corresponding values of y:

$$0^2+y^2=16$$

$$y=\pm 4$$

So, the solutions are $\boxed{(0,4) \text{ and } (0,-4)}$.

61. Solve the system $\begin{cases} x+y=3 & \textbf{(1)} \\ x^2+y^2=4 & \textbf{(2)} \end{cases}$

Solve **(1)** for y: $y=3-x$ **(3)**

Substitute **(3)** into **(2)** to get an equation in terms of y:

$$x^2+(3-x)^2=4$$

$$x^2+9-6x+x^2=4$$

$$2x^2-6x+5=0$$

$$x=\frac{6\pm\sqrt{36-40}}{2(2)} \text{ which are not real numbers}$$

Hence, the system has $\boxed{\text{no solution}}$.

62. Solve the system $\begin{cases} xy = 4 & \textbf{(1)} \\ x^2 + y^2 = 16 & \textbf{(2)} \end{cases}$

Solve **(1)** for y: $y = \frac{4}{x}$ **(3)**

Substitute **(3)** into **(2)** to get an equation in terms of x:

$$x^2 + \left(\tfrac{4}{x}\right)^2 = 16$$

$$x^2 + \tfrac{16}{x^2} = 16$$

$$\frac{x^4 + 16}{x^2} = 16$$

$$\frac{x^4 + 16}{x^2} - \frac{16x^2}{x^2} = 0$$

$$\frac{x^4 - 16x^2 + 16}{x^2} = 0$$

$$x^4 - 16x^2 + 16 = 0 \quad \textbf{(4)}$$

Let $u = x^2$ **(5)**. Then, **(4)** can be written in the equivalent form $u^2 - 16u + 16 = 0$. Solve this equation using the quadratic formula:

$$u = \tfrac{16 \pm \sqrt{256 - 4(16)}}{2} = \tfrac{16 \pm \sqrt{192}}{2} = \tfrac{16 \pm 8\sqrt{3}}{2} = 8 \pm 4\sqrt{3}$$

Substituting both of these values into **(5)** now leads to the following two equations in x:

$$x^2 = 8 + 4\sqrt{3} \quad \text{and} \quad x^2 = 8 - 4\sqrt{3}$$

$$x \cong \pm 3.864 \qquad\qquad x \cong \pm 1.035$$

Now, substitute each of these values of x back into **(3)** to find the corresponding value of y:

$$x = 3.864: \quad y = \tfrac{4}{3.864} \cong 1.035$$

$$x = -3.864: \quad y = \tfrac{4}{-3.864} \cong -1.035$$

$$x = 1.035: \quad y = \tfrac{4}{1.035} \cong 3.865$$

$$x = -1.035: \quad y = \tfrac{4}{-1.035} \cong -3.865$$

So, the approximate solutions are $\boxed{(3.864, 1.035),\ (-3.864, -1.035),\ (1.035, 3.865), \text{ and } (-1.035, -3.865)}$.

63. Solve the system $\begin{cases} x^2 + xy + y^2 = -12 & \textbf{(1)} \\ x - y = 2 & \textbf{(2)} \end{cases}$

Solve **(2)** for x: $x = y + 2$ **(3)**

Substitute **(3)** into **(1)** to get an equation in terms of y:

$$(y+2)^2 + (y+2)y + y^2 = -12$$

$$y^2 + 4y + 4 + y^2 + 2y + y^2 = -12$$

$3y^2 + 6y + 16 = 0$ so that $y = \frac{-6 \pm \sqrt{36 - 4(3)(16)}}{2}$ which are not real numbers

Hence, the system has $\boxed{\text{no solution}}$.

64. Solve the system $\begin{cases} x - y^2 = -9 & \textbf{(1)} \\ 3x + y = 3 & \textbf{(2)} \end{cases}$

Solve **(1)** for x: $x = y^2 - 9$ **(3)**

Substitute **(3)** into **(1)** to get an equation in terms of y:

$$3(y^2 - 9) + y = 3$$

$$3y^2 + y - 30 = 0$$

$$(3y + 10)(y - 3) = 0 \text{ so that } y = -\tfrac{10}{3},\ 3$$

Now, substitute these values of y back into **(3)** to find the corresponding values of x:

$$y = -\tfrac{10}{3}:\ x = \left(-\tfrac{10}{3}\right)^2 - 9 = \tfrac{19}{9}$$

$$y = 3:\ x = (3)^2 - 9 = 0$$

So, the solutions are $\boxed{(0,3) \text{ and } (-\tfrac{10}{3}, \tfrac{19}{9})}$.

65. Solve the system $\begin{cases} x^3 - y^3 = -19 & \textbf{(1)} \\ x - y = -1 & \textbf{(2)} \end{cases}$

Solve **(2)** for x: $x = y - 1$ **(3)**

Substitute **(3)** into **(1)** to get an equation in terms of x:

$$\underbrace{(y-1)^3}_{(y-1)(y^2-2y+1)} - y^3 = -19$$

$$(y^3 - 2y^2 + y - y^2 + 2y - 1) - y^3 + 19 = 0$$

$$-3y^2 + 3y + 18 = 0$$

$$y^2 - y - 6 = 0$$

$$(y+2)(y-3) = 0 \text{ so that } y = -2,\ 3$$

Now, substitute these values of y back into **(3)** to find the corresponding values of x:

$$y = -2:\ x = -2 - 1 = -3, \quad y = 3: \quad x = 3 - 1 = 2$$

So, the solutions are $\boxed{(2,3) \text{ and } (-3,-2)}$.

66. Solve the system $\begin{cases} 2x^2+4xy=9 & \textbf{(1)} \\ x^2-2xy=0 & \textbf{(2)} \end{cases}$

Multiply **(2)** by 2 and then add to **(1)** to get an equation in terms of x:

$$4x^2=9$$
$$x^2=\tfrac{9}{4}$$
$$x=\pm\tfrac{3}{2}$$

Now, substitute these values of x back into **(2)** to find the corresponding values of y:

$$x=\tfrac{3}{2}:\quad (\tfrac{3}{2})^2-2(\tfrac{3}{2})(y)=0$$
$$3y=\tfrac{9}{4} \text{ so that } y=\tfrac{3}{4}$$
$$x=-\tfrac{3}{2}:\quad (-\tfrac{3}{2})^2-2(-\tfrac{3}{2})(y)=0$$
$$3y=-\tfrac{9}{4} \text{ so that } y=-\tfrac{3}{4}$$

So, the solutions are $\boxed{(\tfrac{3}{2},\tfrac{3}{4}) \text{ and } (-\tfrac{3}{2},-\tfrac{3}{4})}$.

67. Solve the system $\begin{cases} \dfrac{2}{x^2}+\dfrac{1}{y^2}=15 & \textbf{(1)} \\ \dfrac{1}{x^2}-\dfrac{1}{y^2}=-3 & \textbf{(2)} \end{cases}$

Add **(1)** and **(2)** to get an equation in terms of x:

$$\frac{3}{x^2}=12$$
$$12x^2=3$$
$$x^2=\tfrac{1}{4} \text{ so that } x=\pm\tfrac{1}{2}.$$

Now, substitute these values of x back into **(2)** to find the corresponding value of y:

$$x=\tfrac{1}{2}:\quad \frac{1}{(\frac{1}{2})^2}-\frac{1}{y^2}=-3 \qquad\qquad x=-\tfrac{1}{2}:\quad \frac{1}{(-\frac{1}{2})^2}-\frac{1}{y^2}=-3$$
$$4-\frac{1}{y^2}=-3 \qquad\qquad 4-\frac{1}{y^2}=-3$$
$$7=\frac{1}{y^2} \qquad\qquad 7=\frac{1}{y^2}$$
$$7y^2=1 \qquad\qquad 7y^2=1$$
$$y^2=\tfrac{1}{7} \qquad\qquad y^2=\tfrac{1}{7}$$
$$y=\pm\sqrt{\tfrac{1}{7}} \qquad\qquad y=\pm\sqrt{\tfrac{1}{7}}$$

So, the solutions are $\boxed{(\tfrac{1}{2},\tfrac{1}{\sqrt{7}}),\ (-\tfrac{1}{2},\tfrac{1}{\sqrt{7}}),\ (\tfrac{1}{2},-\tfrac{1}{\sqrt{7}}), \text{ and } (-\tfrac{1}{2},-\tfrac{1}{\sqrt{7}})}$.

68. Solve the system $\begin{cases} x^2 + y^2 = 2 & \textbf{(1)} \\ x^2 + y^2 = 4 & \textbf{(2)} \end{cases}$

Add **(1)** and **(2)** to obtain the false statement 0 = 6. Hence, there are no real values of x and y that can satisfy both equations simultaneously. Hence, there is $\boxed{\text{no solution}}$ of this system.

69.

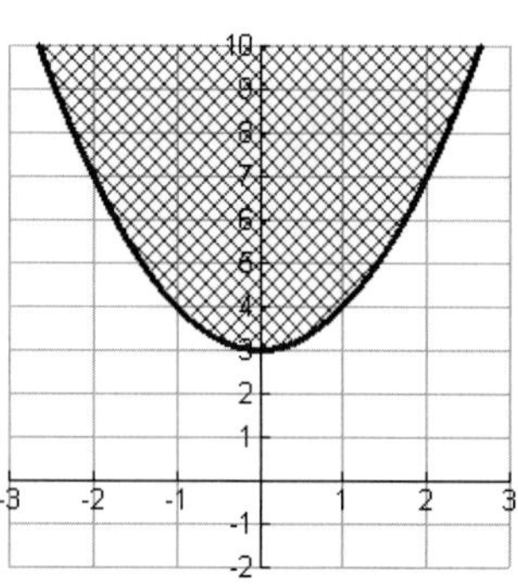

70.

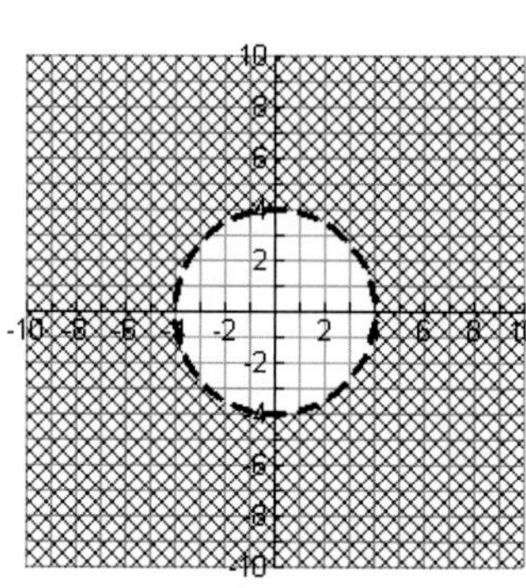

71.

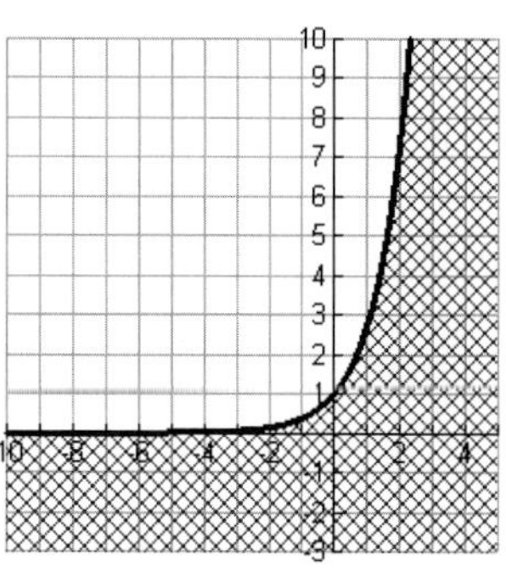

72.

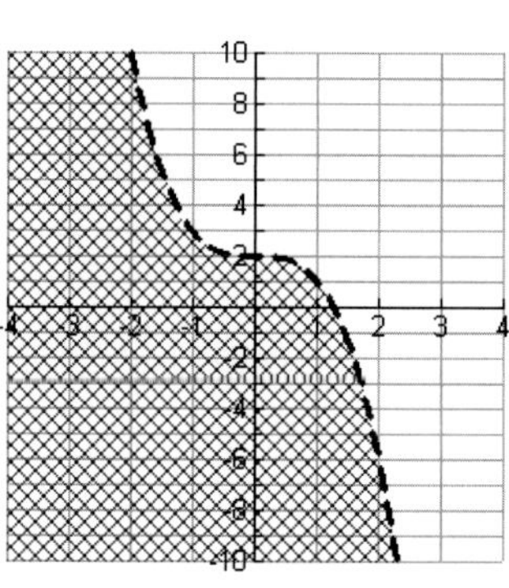

73.

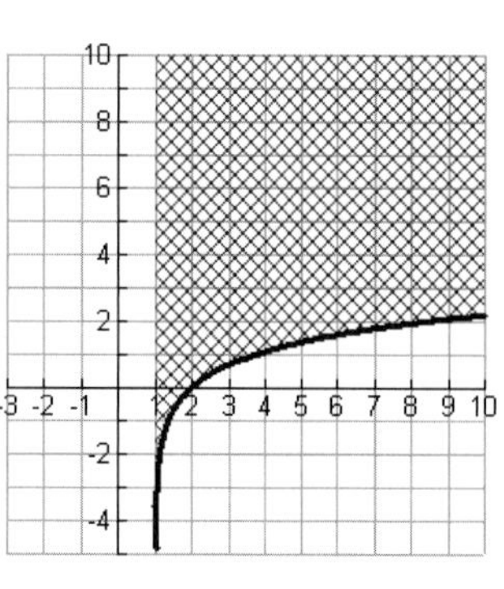

74.

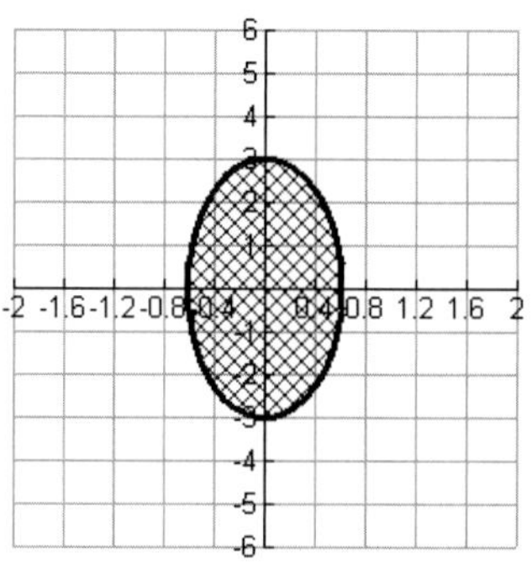

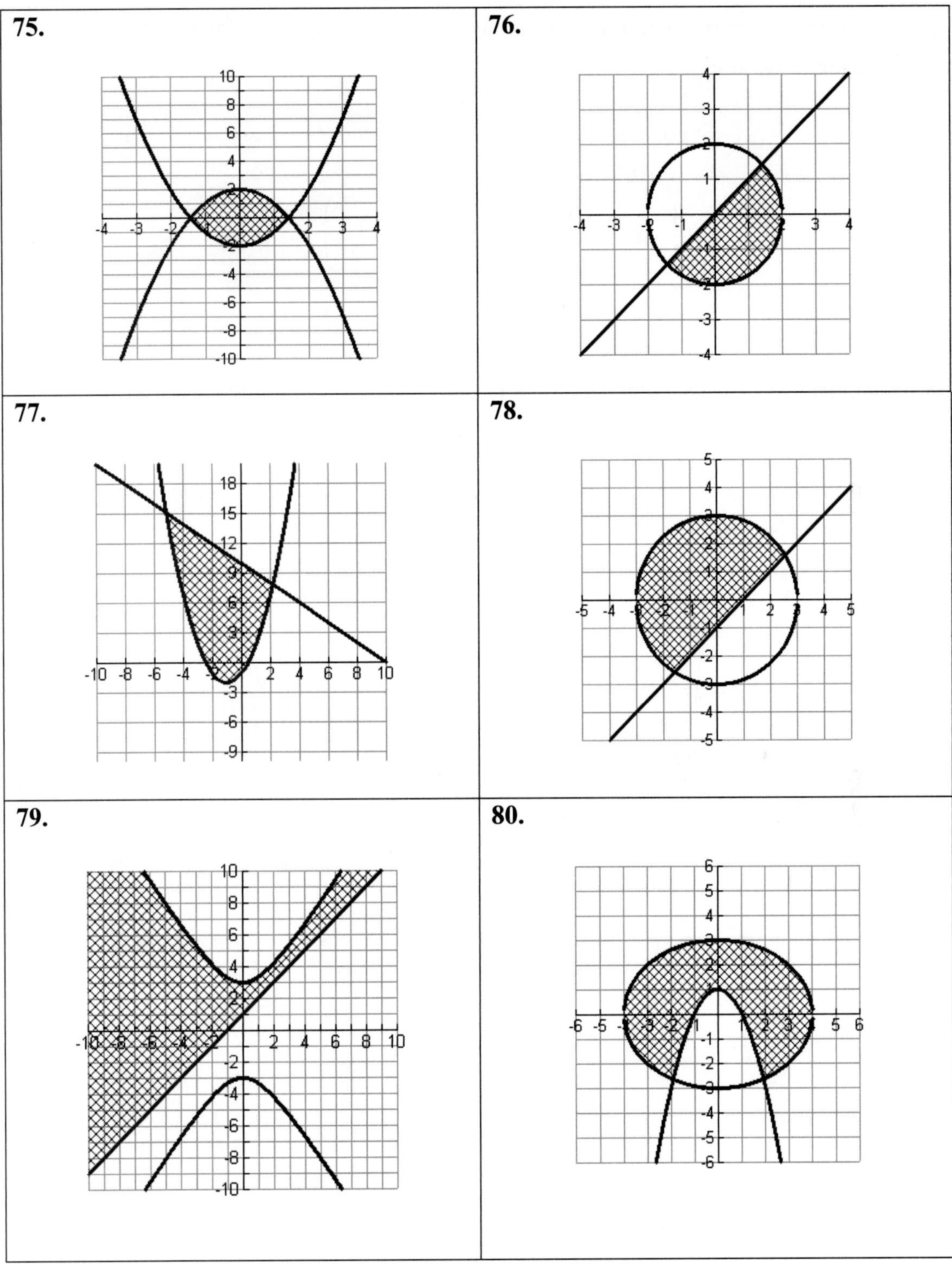
75.
76.
77.
78.
79.
80.

81. The coordinates of the point obtained by rotating $(-3,2)$ by an angle of 60° counterclockwise are:

$$X = x\cos\theta + y\sin\theta = -3\cos 60^\circ + 2\sin 60^\circ = -\frac{3}{2} + 2\left(\frac{\sqrt{3}}{2}\right) = -\frac{3}{2} + \sqrt{3}$$

$$Y = -x\sin\theta + y\cos\theta = 3\sin 60^\circ + 2\cos 60^\circ = 3\left(\frac{\sqrt{3}}{2}\right) + 2\left(\frac{1}{2}\right) = \frac{3\sqrt{3}}{2} + 1$$

So, the corresponding coordinates of this point in the *XY*-system are $\left(-\frac{3}{2} + \sqrt{3}, \frac{3\sqrt{3}}{2} + 1\right)$.

82. The coordinates of the point obtained by rotating $(4,-3)$ by an angle of 45° counterclockwise are:

$$X = x\cos\theta + y\sin\theta = 4\cos 45^\circ - 3\sin 45^\circ = 4\left(\frac{\sqrt{2}}{2}\right) - 3\left(\frac{\sqrt{2}}{2}\right) = \frac{\sqrt{2}}{2}$$

$$Y = -x\sin\theta + y\cos\theta = -4\sin 45^\circ - 3\cos 45^\circ = -4\left(\frac{\sqrt{2}}{2}\right) - 3\left(\frac{\sqrt{2}}{2}\right) = -\frac{7\sqrt{2}}{2}$$

So, the corresponding coordinates of this point in the *XY*-system are $\left(\frac{\sqrt{2}}{2}, -\frac{7\sqrt{2}}{2}\right)$.

83. Consider the equation $2x^2+4\sqrt{3}xy-2y^2-16=0$.
In order to obtain the equation of this curve in the XY-system, we make the following substitution into the original equation with $\theta=30^\circ$:

$$\begin{cases} x = X\cos\theta - Y\sin\theta = \left(\sqrt{3}X-Y\right)\left(\frac{1}{2}\right) \\ y = X\sin\theta + Y\cos\theta = \left(X+\sqrt{3}Y\right)\left(\frac{1}{2}\right) \end{cases}$$

Doing so yields the following, after collecting like terms and simplifying:

$$2\left(\sqrt{3}X-Y\right)\left(\frac{1}{2}\right)^2+4\sqrt{3}\left(X+\sqrt{3}Y\right)\left(\sqrt{3}X-Y\right)\left(\frac{1}{2}\right)^2-2\left(X+\sqrt{3}Y\right)^2\left(\frac{1}{2}\right)^2-16=0$$
$$2\left(3X^2-2\sqrt{3}XY+Y^2\right)+4\sqrt{3}\left(\sqrt{3}X^2-XY+3XY-\sqrt{3}Y^2\right)-2\left(X^2+2\sqrt{3}XY+3Y^2\right)-64=0$$
$$16X^2-16Y^2-64=0$$
$$\frac{X^2}{4}-\frac{Y^2}{4}=1$$

Now, the graph of $2x^2+4\sqrt{3}xy-2y^2-16=0$, along with the new axes $y=\left(\tan 30^\circ\right)x=\left(\frac{1}{\sqrt{3}}\right)x,\quad y=-\sqrt{3}x$ is as follows:

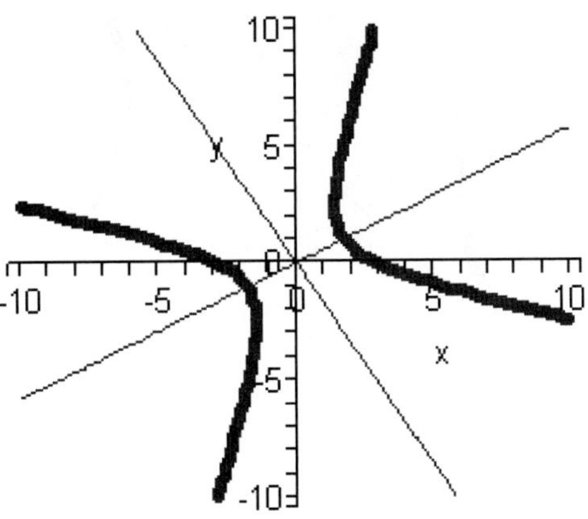

84. Consider the equation $25x^2 + 14xy - 25y^2 - 288 = 0$.
In order to obtain the equation of this curve in the XY-system, we make the following substitution into the original equation with $\theta = 45^\circ$:

$$\begin{cases} x = X\cos\theta - Y\sin\theta = (X-Y)\left(\dfrac{\sqrt{2}}{2}\right) \\ y = X\sin\theta + Y\cos\theta = (X+Y)\left(\dfrac{\sqrt{2}}{2}\right) \end{cases}$$

Doing so yields the following, after collecting like terms and simplifying:

$$25(X-Y)^2\left(\frac{\sqrt{2}}{2}\right)^2 + 14(X-Y)(X+Y)\left(\frac{\sqrt{2}}{2}\right)^2 - 25(X+Y)^2\left(\frac{\sqrt{2}}{2}\right)^2 - 288 = 0$$

$$14X^2 - 14Y^2 - 576 = 0$$

$$\frac{7X^2}{288} - \frac{7Y^2}{288} = 1$$

(c) The graph of $25x^2 + 14xy - 25y^2 - 288 = 0$, along with the new axes $y = (\tan 45^\circ)x = x,\ y = -x$ is as follows:

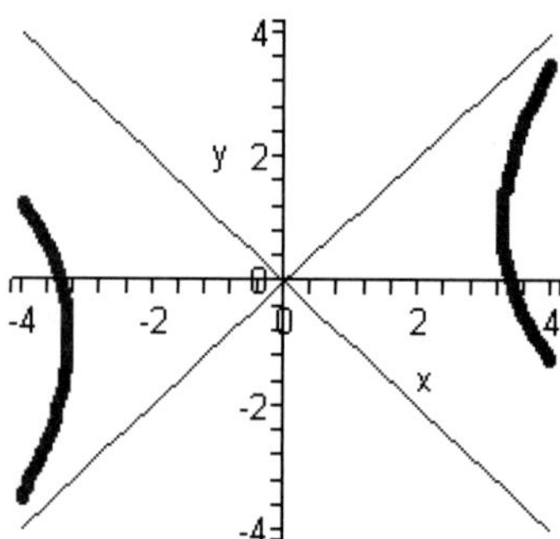

85. To find the amount of rotation needed to transform the given equation into one in the XY-system with no XY term, we seek the solution θ of the equation $\cot 2\theta = \frac{A-C}{B}$ in $[0,\pi]$. In this case, $A=4,\ B=2\sqrt{3},\ C=6$, so that this equation becomes $\cot 2\theta = -\frac{1}{\sqrt{3}}$. Hence, we need to find θ in $[0,\pi]$ such that

$$\cos 2\theta = -\tfrac{1}{2} \text{ and } \sin 2\theta = \tfrac{\sqrt{3}}{2}.$$

But, this occurs precisely when $2\theta = 120^\circ$, so that $\boxed{\theta = 60^\circ}$.

86. To find the amount of rotation needed to transform the given equation into one in the XY-system with no XY term, we seek the solution θ of the equation $\cot 2\theta = \frac{A-C}{B}$ in $[0,\pi]$. In this case, $A=4,\ B=5,\ C=4$, so that solving this equation yields:

$$\cot 2\theta = 0 \Rightarrow \cos 2\theta = 0 \Rightarrow 2\theta = 90^\circ \Rightarrow \boxed{\theta = 45^\circ}.$$

87. Consider the equation $x^2+2xy+y^2+\sqrt{2}x-\sqrt{2}y+8=0$.

Here, $A=1,\ B=2,\ C=1$, so that $B^2-4AC=0$. Thus, the curve is a *parabola*. In order to obtain its graph, follow these steps:

Step 1: Find the angle of counterclockwise rotation necessary to transform the equation into the *XY*-system with no *XY* term.

We seek the solution θ of the equation $\cot 2\theta=\frac{A-C}{B}=0$ in $[0,\pi]$. Hence, we need to find θ in $[0,\pi]$ such that $\cos 2\theta=0$ and $\sin 2\theta=1$.

But, this occurs precisely when $2\theta=90^\circ$, so that $\theta=45^\circ$.

Step 2: Now, determine the equation of this curve in the *XY*-system making use of the following substitution into the original equation with $\theta=45^\circ$:

$$x=X\cos\theta-Y\sin\theta=(X-Y)\left(\frac{\sqrt{2}}{2}\right) \qquad y=X\sin\theta+Y\cos\theta=(X+Y)\left(\frac{\sqrt{2}}{2}\right)$$

Doing so yields the following, after collecting like terms and simplifying:

$$(X-Y)^2\left(\frac{\sqrt{2}}{2}\right)^2+2(X-Y)(X+Y)\left(\frac{\sqrt{2}}{2}\right)^2+(X+Y)^2\left(\frac{\sqrt{2}}{2}\right)^2$$
$$+\sqrt{2}(X-Y)\left(\frac{\sqrt{2}}{2}\right)-\sqrt{2}(X+Y)\left(\frac{\sqrt{2}}{2}\right)+8=0$$
$$\left(X^2-2XY+Y^2\right)+2\left(X^2-Y^2\right)+\left(X^2+2XY+Y^2\right)+2(X-Y)-2(X+Y)+16=0$$
$$X^2+4=Y$$

Step 3: The graph of $5x^2+2xy+y^2+\sqrt{2}x-\sqrt{2}y+8=0$, along with the new axes $y=(\tan 45^\circ)x=x,\ y=-x$ is as follows:

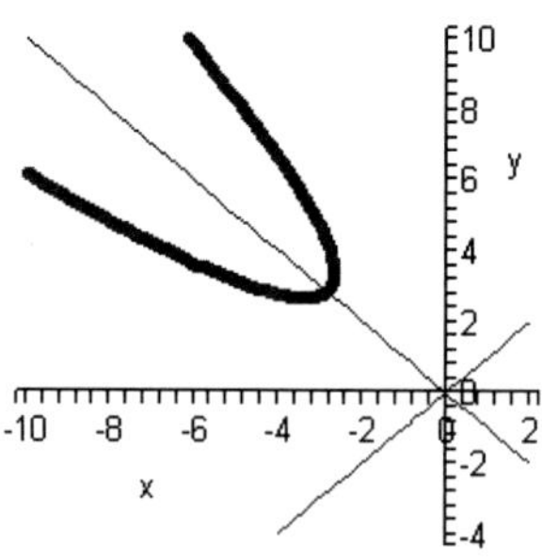

88. Consider the equation $76x^2 + 48\sqrt{3}xy + 28y^2 - 100 = 0$.

Here, $A = 76,\ B = 48\sqrt{3},\ C = 28$, so that $B^2 - 4AC = -1600 < 0$. Thus, the curve is an *ellipse* In order to obtain its graph, follow these steps:

Step 1: Find the angle of counterclockwise rotation necessary to transform the equation into the *XY*-system with no *XY* term.

We seek the solution θ of the equation $\cot 2\theta = \frac{A-C}{B} = \frac{1}{\sqrt{3}}$ in $[0,\pi]$. Hence, we need to find θ in $[0,\pi]$ such that $\cos 2\theta = \frac{1}{2}$ and $\sin 2\theta = \frac{\sqrt{3}}{2}$.

But, this occurs precisely when $2\theta = 60^\circ$, so that $\theta = 30^\circ$.

Step 2: Now, determine the equation of this curve in the *XY*-system making use of the following substitution into the original equation with $\theta = 30^\circ$:

$$x = X\cos\theta - Y\sin\theta = \left(\sqrt{3}X - Y\right)\left(\frac{1}{2}\right) \quad y = X\sin\theta + Y\cos\theta = \left(X + \sqrt{3}Y\right)\left(\frac{1}{2}\right)$$

Doing so yields the following, after collecting like terms and simplifying:

$$76\left(\sqrt{3}X - Y\right)^2\left(\frac{1}{2}\right)^2 + 48\sqrt{3}\left(\sqrt{3}X - Y\right)\left(X + \sqrt{3}Y\right)\left(\frac{1}{2}\right)^2 + 28\left(X + \sqrt{3}Y\right)^2\left(\frac{1}{2}\right)^2 - 100 = 0$$

$$76\left(3X^2 - 2\sqrt{3}XY + Y^2\right) + 48\sqrt{3}\left(\sqrt{3}X^2 + 3XY - XY - \sqrt{3}Y^2\right) + 28\left(X^2 + 2\sqrt{3}XY + 3Y^2\right) - 400 = 0$$

$$400X^2 + 16Y^2 - 400 = 0$$

$$\frac{X^2}{1} + \frac{Y^2}{25} = 1$$

Step 3: The graph of $76x^2 + 48\sqrt{3}xy + 28y^2 - 100 = 0$, along with the new axes $y = \left(\tan 30^\circ\right)x = \left(\frac{1}{\sqrt{3}}\right)x,\ y = -\sqrt{3}x$ is as follows:

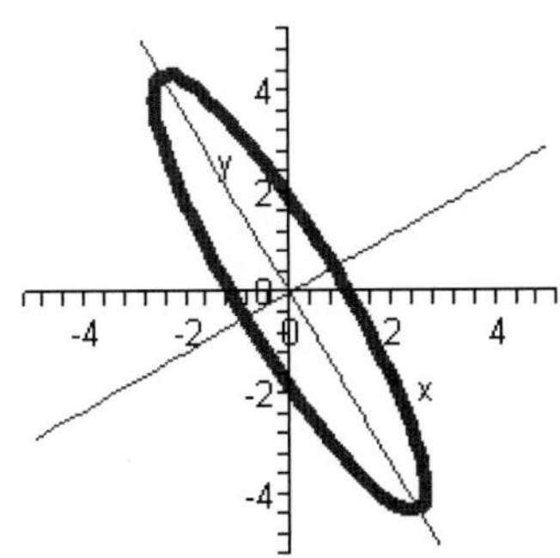

89. The directrix is horizontal and 7 units below the pole, so the equation is $r=\dfrac{\left(\frac{3}{7}\right)(7)}{1-\left(\frac{3}{7}\right)\sin\theta}=\dfrac{3}{1-\frac{3}{7}\sin\theta}$, which simplifies to $\boxed{r=\dfrac{21}{7-3\sin\theta}}$.
90. Since the curve is a parabola, $e=1$. Further, since the directrix is vertical and 2 units to the right of the pole, the equation is $r=\dfrac{(1)(2)}{1+(1)\cos\theta}=\boxed{\dfrac{2}{1+\cos\theta}}$.
91. Observe that $r=\dfrac{6}{4-5\cos\theta}=\dfrac{6}{4\left(1-\frac{5}{4}\cos\theta\right)}=\dfrac{\frac{3}{2}}{1-\frac{5}{4}\sin\theta}$, so that $e=\frac{5}{4}>1$. Hence, this conic is a *hyperbola.*
92. Observe that $r=\dfrac{2}{5+3\sin\theta}=\dfrac{2}{5\left(1+\frac{3}{5}\sin\theta\right)}=\dfrac{\frac{2}{5}}{1+\frac{3}{5}\sin\theta}$, so that $e=\frac{3}{5}<1$. Hence, this conic is an *ellipse.*

93. Consider the equation $r = \dfrac{4}{2+\cos\theta}$.

(a) Since $r = \dfrac{4}{2+\cos\theta} = \dfrac{4}{2\left(1+\frac{1}{2}\cos\theta\right)} = \dfrac{4\left(\frac{1}{2}\right)}{1+\frac{1}{2}\cos\theta}$, $e = \frac{1}{2} < 1$. So, this conic is an *ellipse*.

(b) From (a), we see that $e = \frac{1}{2}$ and $p = 4$. From the general form of the equation, we know the directrix is vertical and is 4 units to the right of the origin. Since the directrix is perpendicular to the major axis, we know that the major axis lies along the x-axis. Hence, the vertices occur when $\theta = 0, \pi$. Observe that $r(0) = \frac{4}{3}$ and $r(\pi) = 4$. So, in rectangular coordinates, the vertices are $\left(\frac{4}{3}, 0\right)$ and $(-4, 0)$.

(c) In order to obtain the graph, we need to find the center and y-intercepts.
Center = midpoint of vertices = $\left(\dfrac{\frac{4}{3}-4}{2}, 0\right) = \left(-\dfrac{4}{3}, 0\right)$

The length of the major axis = $2a = \frac{4}{3} - (-4) = \frac{16}{3}$, so that $a = \frac{8}{3}$. Now, using $e = \dfrac{c}{a}$, we see that $c = ae = \left(\frac{8}{3}\right)\left(\frac{1}{2}\right) = \frac{4}{3}$. Since $b^2 = c^2 - a^2$, we further find that $b^2 = \frac{16}{3}$, so $b = \frac{4\sqrt{3}}{3}$. Hence, the length of the minor axis (which is perpendicular to the major axis through the center) = $2b$. So, the y-intercepts are $\left(-\frac{4}{3} \pm \frac{4\sqrt{3}}{3}, 0\right)$. The graph is as follows:

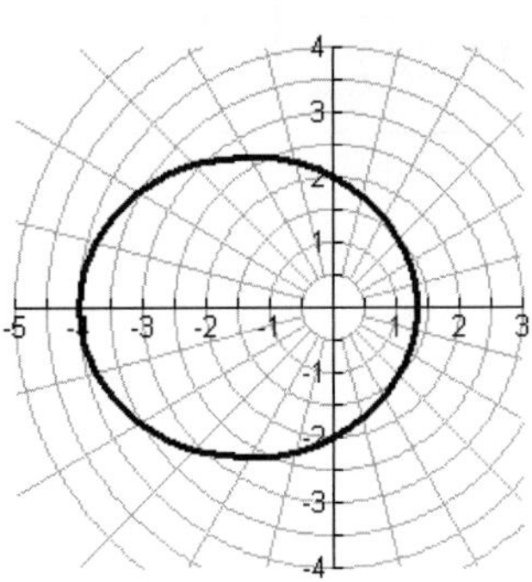

94. Consider the equation $r = \dfrac{6}{1-\sin\theta}$.

(a) Observe that $e = 1$. So, this conic is a *parabola.*

(b) Inspecting the equation in (a), we see that $e = 1$ and $p = 6$.
From the general form of the equation, we know the directrix is horizontal and is below the origin. As such, since the parabola opens up, the vertex must lie on the *y*-axis. Further, since the right-side of the equation is undefined at $\theta = \frac{\pi}{2}$, the vertex must occur at $\theta = \frac{3\pi}{2}$. Evaluating the equation at $\theta = \frac{3\pi}{2}$ yields $r = 3$. So, the vertex is rectangular coordinates is $(0,-3)$.

(c) Consider the following table of points:

θ	$r = \dfrac{6}{1-\sin\theta}$	(r,θ)
0	6	$(6,0)$
$\frac{\pi}{2}$	undefined	---
π	6	$(6,\pi)$
$\frac{3\pi}{2}$	3	$(3,\frac{3\pi}{2})$
2π	6	$(6,2\pi)$

The graph is as follows:

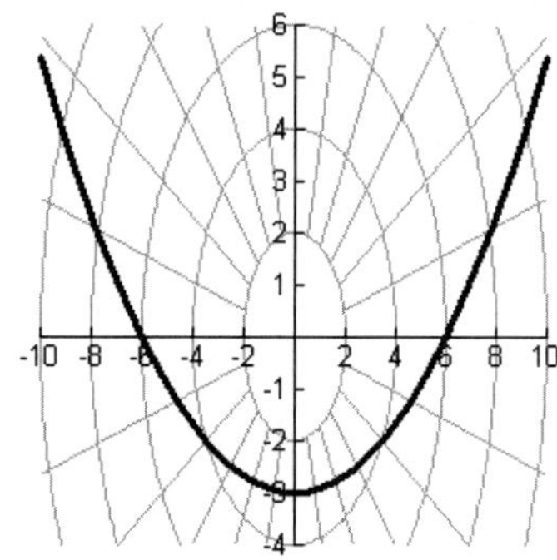

95. In order to graph the curve defined parametrically by

$$x = \sin t,\ \ y = 4\cos t,\ \ t \text{ in } [-\pi, \pi],$$

consider the following sequence of tabulated points:

t	$x = \sin t$	$y = 4\cos t$
$-\pi$	0	-4
$-\pi/2$	-1	0
0	0	4
$\pi/2$	1	0
π	0	-4

Eliminating the parameter reveals that the graph is an ellipse. Indeed, observe that

$$x^2 = \sin^2 t,\ \ \left(\frac{y}{4}\right)^2 = \cos^2 t$$

so that summing then yields $x^2 + \frac{y^2}{16} = 1$.

The graph is as follows:

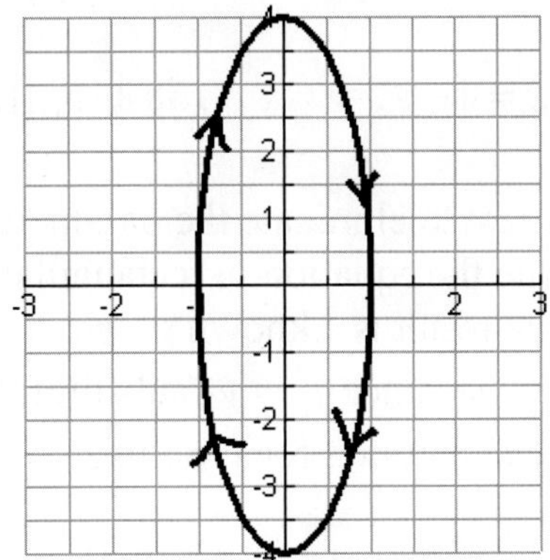

96. In order to graph the curve defined parametrically by

$$x = 5\sin^2 t,\ \ y = 2\cos^2 t,\ \ t \text{ in } [-\pi, \pi],$$

it is easiest to eliminate the parameter t and write the equation in rectangular form since the result is a known graph. Indeed,

$$\frac{x}{5} = \sin^2 t,\ \ \frac{y}{2} = \cos^2 t$$

so that adding the two equations yields

$$\frac{x}{5} + \frac{y}{2} = 1,\ \ 0 \le x \le 5,\ \ 0 \le y \le 2,$$

which is a line segment.

The graph is as follows:

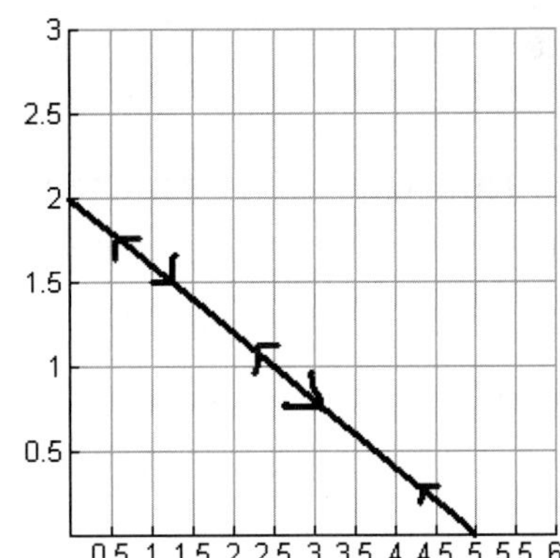

97. In order to graph the curve defined parametrically by

$$x = 4 - t^2,\ y = t^2,\ t \text{ in } [-3,3],$$

it is easiest to eliminate the parameter t and write the equation in rectangular form since the result is a known graph. Indeed, observe that since $t^2 = y$, substituting this into the expression for x yields

$$x = 4 - y \quad x \text{ in } [-5,4].$$

The graph is as follows:

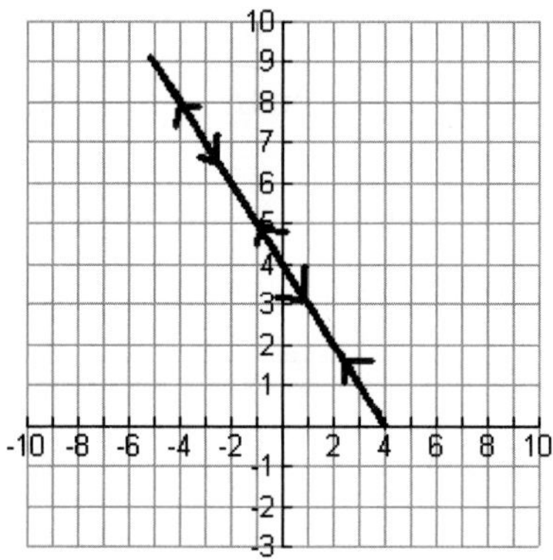

98. In order to graph the curve defined parametrically by

$$x = t + 3,\ y = 4,\ t \text{ in } [-4,4],$$

observe that since the y-coordinate is always 4, the path is a horizontal line segment starting at the point (-1, 4) and ends at (7, 4).

The graph is as follows:

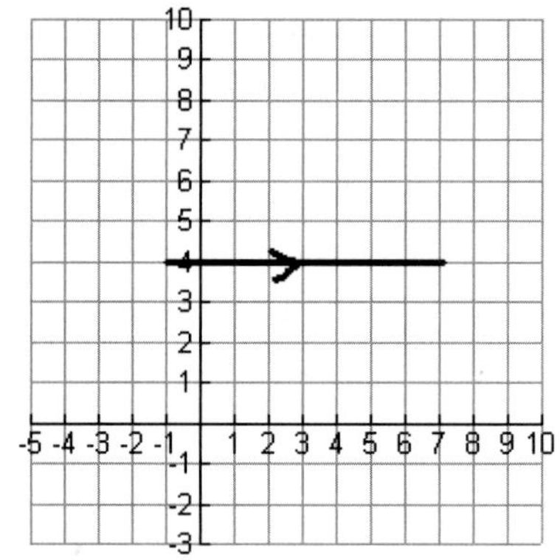

99. In order to express the parametrically-defined curve given by $x = 4 - t^2,\ y = t$ as a rectangular equation, we eliminate the parameter t by simply substituting $y = t$ into the equation for x. So, the equation is $\boxed{x = 4 - y^2}$.

100. In order to express the parametrically-defined curve given by $x = 5\sin^2 t,\ y = 2\cos^2 t$ as a rectangular equation, we eliminate the parameter t. Indeed, observe that $\frac{x}{5} = \sin^2 t,\ \frac{y}{2} = \cos^2 t$, so that adding the two equations yields $\frac{x}{5} + \frac{y}{2} = 1$, which is equivalent to $\boxed{2x + 5y = 10}$.

101. In order to express the parametrically-defined curve given by $x = 2\tan^2 t,\ y = 4\sec^2 t$ as a rectangular equation, we eliminate the parameter t in the following sequence of steps:

$$x = 2\tan^2 t = 2\left(\frac{\sin^2 t}{\cos^2 t}\right) = 2\left(\frac{1-\cos^2 t}{\cos^2 t}\right) = 2\left(\frac{1}{\cos^2 t}\right) - 2$$
$$= 2\left(\sec^2 t\right) - 2 = \frac{2}{4}\left(4\sec^2 t\right) - 2 = \frac{1}{2}y - 2$$

So, the equation is $\boxed{y = 2x + 4}$.

102. In order to express the parametrically-defined curve given by $x = 3t^2 + 4,\ y = 3t^2 - 5$ as a rectangular equation, we eliminate the parameter t in the following sequence of steps:

$$3t^2 = x - 4 \text{ so that } y = (x-4) - 5 = x - 9.$$

So, the equation is $\boxed{y = x - 9}$.

103. The vertex is at (0.6, -1.2), and it should open down. The graph below confirms this:

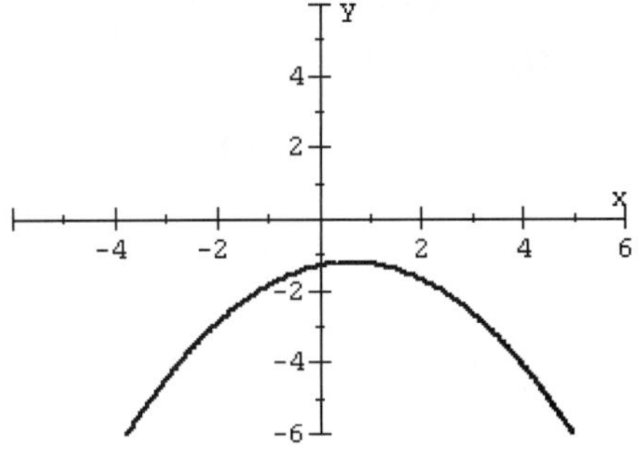

104. The vertex is at (2.8, 0.2) and it should open to the right. The graph below confirms this:

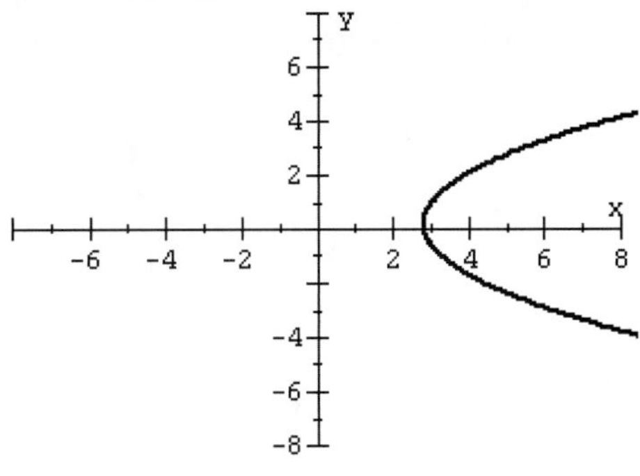

105.

a. Completing the square yields

$$y^2 + 2.8y = -3x + 6.85$$
$$y^2 + 2.8y + 1.4^2 = -3x + 6.85 + 1.4^2$$
$$(y+1.4)^2 = -3x + 8.81$$
$$y + 1.4 = \pm\sqrt{-3x+8.81}$$
$$y = -1.4 \pm \sqrt{-3x+8.81}$$

The graph is as follows:

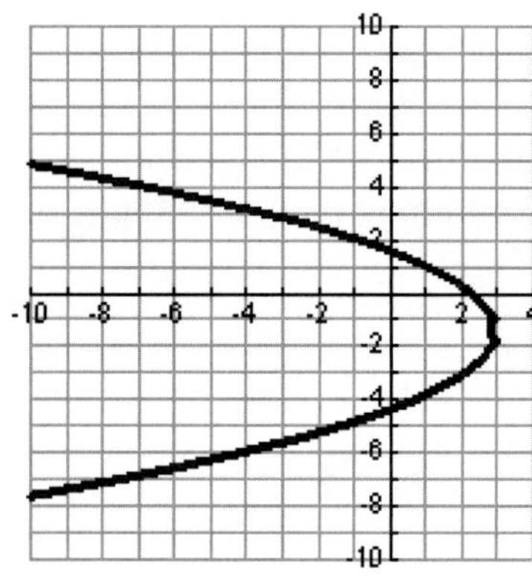

b. From **a**, we obtain

$$y^2 + 2.8y = -3x + 6.85$$
$$y^2 + 2.8y + 1.4^2 = -3x + 6.85 + 1.4^2$$
$$(y+1.4)^2 = -3x + 8.81$$
$$(y+1.4)^2 = -3(x - 2.937)$$
$$(y-(-1.4))^2 = 4\left(-\tfrac{3}{4}\right)(x-2.937)$$

The vertex is (2.937, -1.4). The parabola opens to the left.

c. Yes, parts **a** and **b** agree.

106.

a. The equation is equivalent to

$$y = x^2 - 10.2x + 24.8.$$

The graph is as follows:

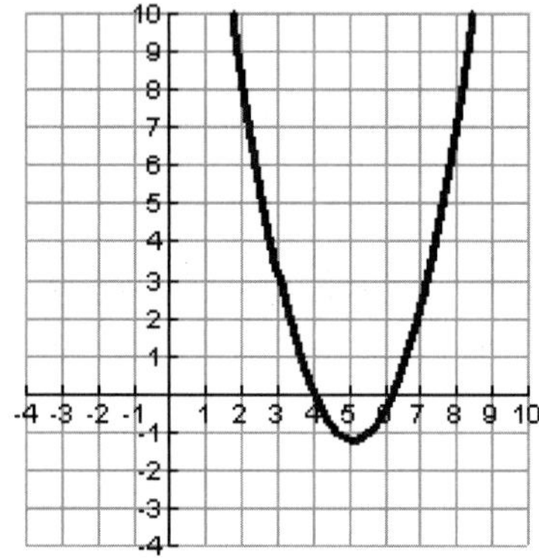

b. Completing the square yields

$$x^2 - 10.2x + 24.8 = y$$
$$x^2 - 10.2x = y - 24.8$$
$$x^2 - 10.2x + 5.1^2 = y - 24.8 + 5.1^2$$
$$(x-5.1)^2 = y + 1.21$$
$$(x-5.1)^2 = 4\left(\tfrac{1}{4}\right)(y-(-1.21))$$

The vertex is (5.1, -1.21). The parabola opens up.

c. Yes, parts **a** and **b** agree.

107. From the graphs, we see that as c increases the minor axis (along the x-axis) of the ellipse whose equation is given by $4(cx)^2 + y^2 = 1$ decreases.

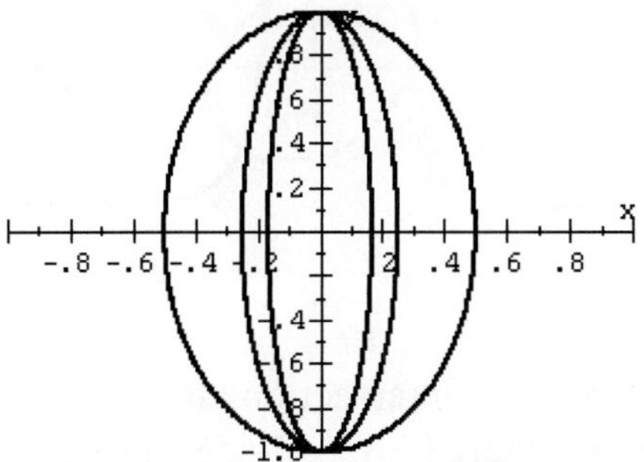

108. From the graphs, we see that as c increases the minor axis (along the y-axis) of the ellipse whose equation is given by $x^2 + 4(cy)^2 = 1$ decreases.

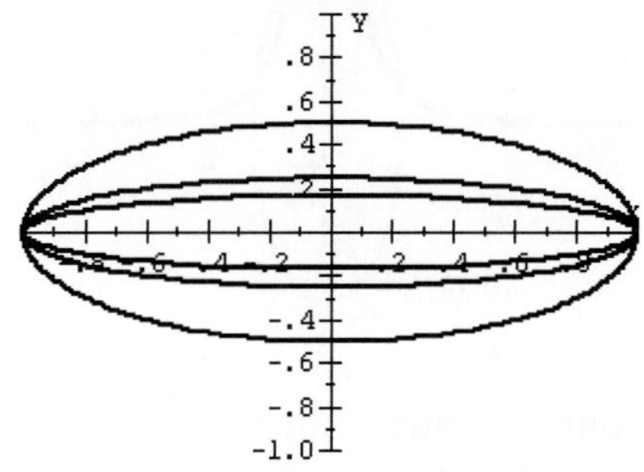

109. From the graphs, we see that as c increases, the vertices of the hyperbola whose equation is given by $4(cx)^2 - y^2 = 1$ are at $\left(\pm\frac{1}{2c}, 0\right)$ are moving towards the origin:

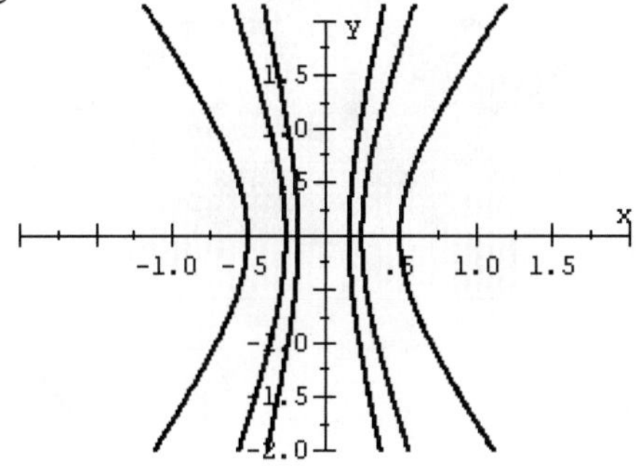

110. From the graphs, we see that as c increases, the vertices of the hyperbola whose equation is given by $x^2 - 4(cy)^2 = 1$ remain at $(\pm 1, 0)$, but the graphs open more narrowly:

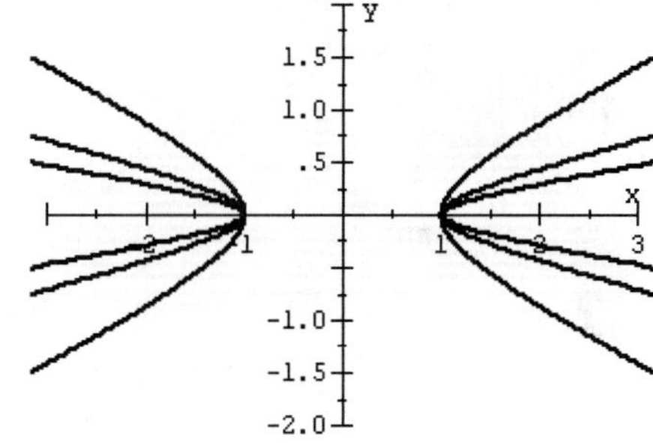

111. The graph of this nonlinear system is as follows:

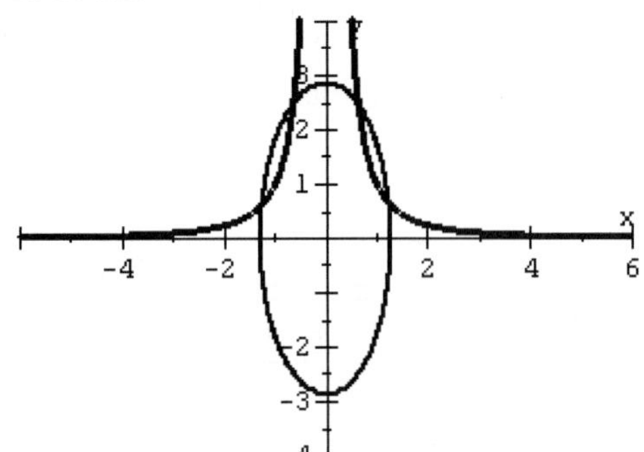

The points of intersection are approximately (0.635, 2.480), (–0.635, 2.480), (–1.245, 0.645), and (1.245, 0.645).

112. The graph of this nonlinear system is as follows:

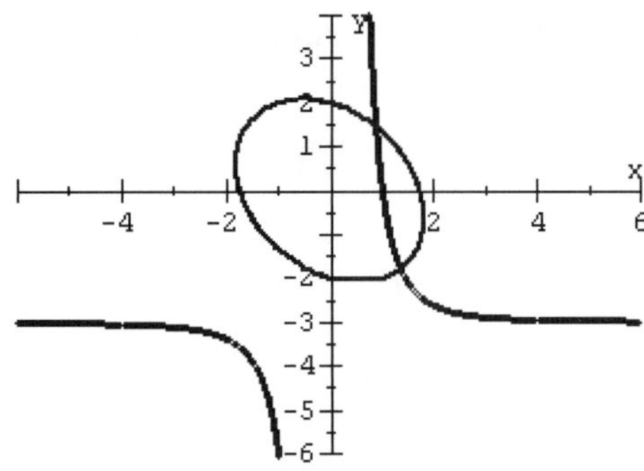

The points of intersection are approximately (0.876, 1.458) and (1.350, –1.781).

113. The darker region below is the one desired:

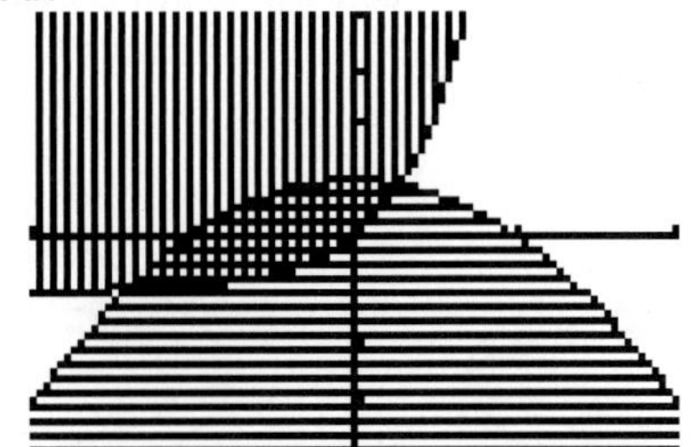

114. The darker region below is the one desired:

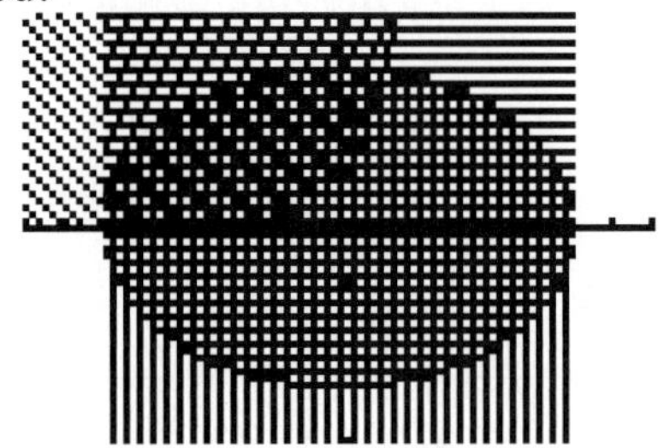

115. a.

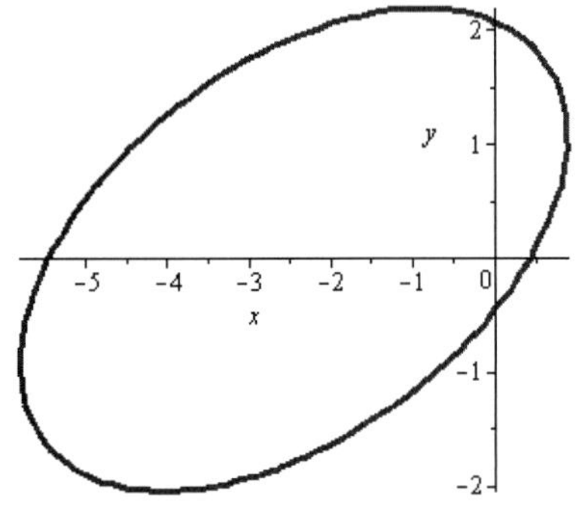

The angle of rotation is computed by

$$\cot 2\theta = \frac{A-C}{B} = \frac{2-5}{-3} = -1$$

$$2\theta = \frac{3\pi}{4}$$

$$\theta = \frac{3\pi}{8} \approx 1.2 \text{ rad}$$

b.

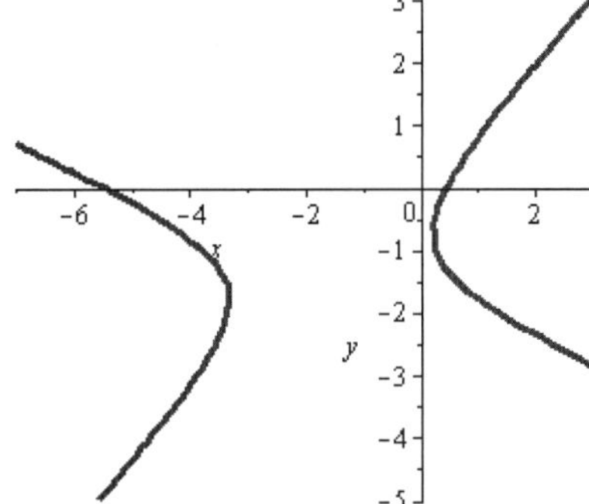

The angle of rotation is computed by

$$\cot 2\theta = \frac{A-C}{B} = \frac{2-(-5)}{3} = \frac{7}{3}$$

$$2\theta = \cot^{-1}\left(\frac{7}{3}\right)$$

$$\theta \approx 0.2 \text{ rad}$$

116. a.

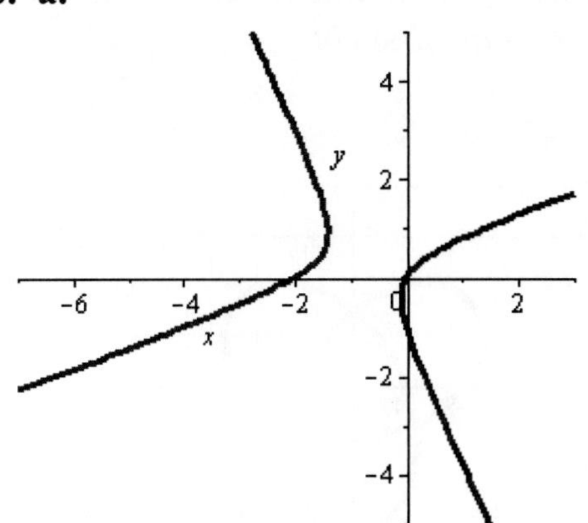

The angle of rotation is computed by

$$\cot 2\theta = \frac{A-C}{B} = \frac{1-(-1)}{-2} = -1$$
$$2\theta = \frac{3\pi}{4}$$
$$\theta = \frac{3\pi}{8} \approx 1.2 \text{ rad}$$

b.

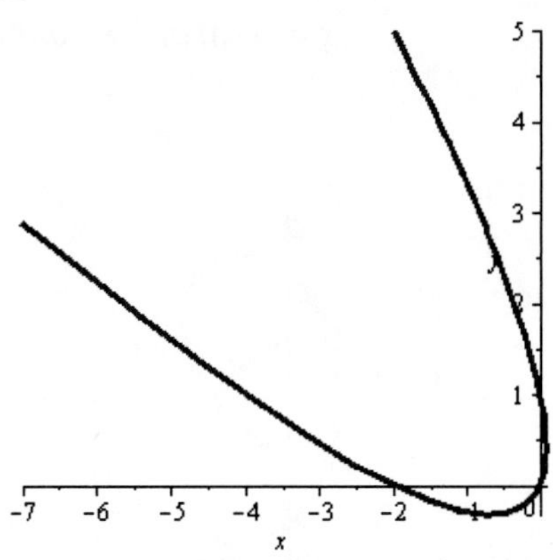

The angle of rotation is computed by

$$\cot 2\theta = \frac{A-C}{B} = \frac{1-1}{2} = 0$$
$$2\theta = \frac{\pi}{2}$$
$$\theta = \frac{\pi}{4} \approx 0.8 \text{ rad}$$

117. The calculator is sampling using to wide a step size, and is jumping over parts of the graph. Using more points yields a better approximation to the graph of the curve.

118. The calculator is sampling using to wide a step size, and is jumping over parts of the graph. Using more points yields a better approximation to the graph of the curve.

119. First, we consider the curve defined by the following parametric equations:

$$x = a\sin at + b\cos bt, \quad y = a\cos at + b\sin bt$$

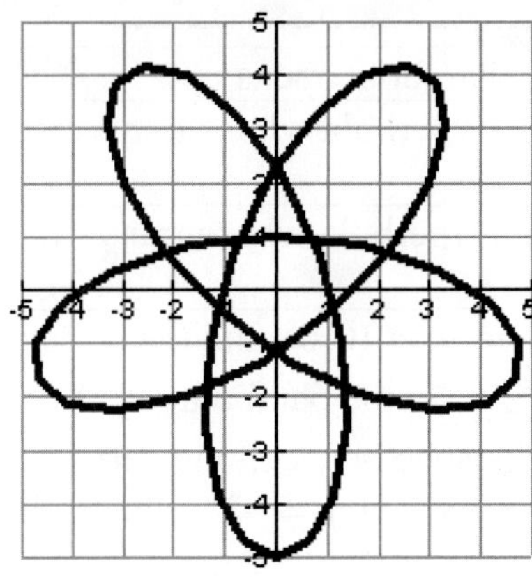

Graph for $a = 2$, $b=3$

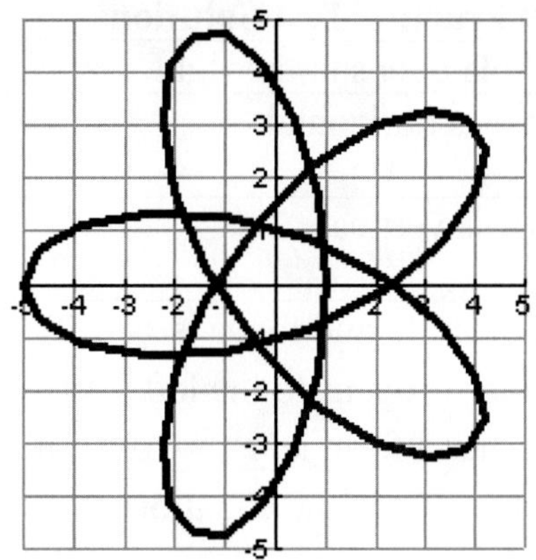

Graph for $a = 3$, $b=2$

All curves are generated in 2π units of time.

120. First, we consider the curve defined by the following parametric equations:

$$x = a\sin at - b\cos bt, \quad y = a\cos at - b\sin bt$$

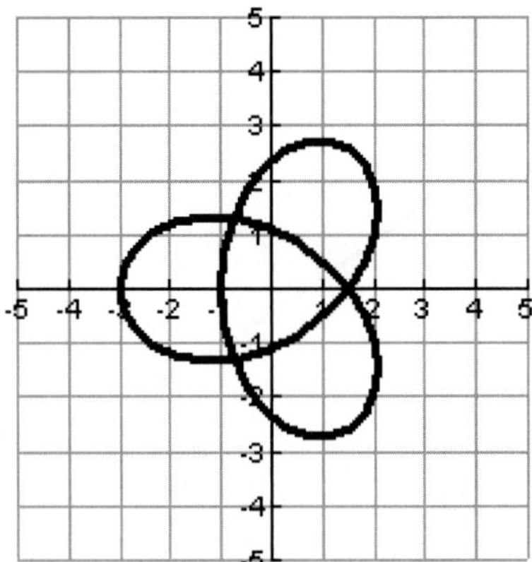

Graph for $a = 1$, b=2

Graph for $a = 2$, b=1

All curves are generated in 2π units of time.

Chapter 9 Practice Test Solutions---

1. c Parabola opens to the right	**2. b** Parabola opens up
3. d Ellipse is horizontal	**4. e** Hyperbola whose transverse axis is the x-axis
5. f Ellipse is vertical	**6. a** Hyperbola whose transverse axis is y-axis.
7. Vertex is $(0,0)$ and focus is $(-4,0)$. So, the parabola opens to the left. As such, the general form is $y^2 = 4px$. In this case, $p = -4$, so the equation is $\boxed{y^2 = -16x}$.	**8.** Vertex is $(0,0)$ and the directrix is $y = 2$. So, the parabola opens down. Since the distance between the vertex and directrix is 2, we know $p = -2$. Hence, the equation is $\boxed{x^2 = 4(-2)y = -8y}$.

9. Vertex is $(-1,5)$ and focus is $(-1,2)$. So, the parabola opens down. As such, the general form is $(x+1)^2 = 4p(y-5)$. In this case, $p=-3$, so the equation is $\boxed{(x+1)^2 = -12(y-5)}$.	**10.** Vertex is $(2,-3)$ and the directrix is $x=0$. So, the parabola opens to the right. Since the distance between the vertex and directrix is 2, we know $p=2$. Hence, the equation is $\boxed{(y+3)^2 = 4(2)(x-2) = 8(x-2)}$.

11. Since the foci are $(0,-3), (0,3)$ and the vertices are $(0,-4), (0,4)$, we know:

i) the ellipse is vertical with center at $(0,0)$,

ii) $c=3$,

iii) the length of the major axis is $4-(-4)=8$. Hence, $a=4$.

Now, since $c^2 = a^2 - b^2$, $9 = 16 - b^2$ so that $b^2 = 7$. Hence, the equation of the ellipse is $\boxed{\frac{x^2}{7} + \frac{y^2}{16} = 1}$.

12. Since the foci are $(-1,0), (1,0)$ and the vertices are $(-3,0), (3,0)$, we know:

i) the ellipse is horizontal with center at $(0,0)$,

ii) $c=1$,

iii) the length of the major axis is $3-(-3)=6$. Hence, $a=3$.

Now, since $c^2 = a^2 - b^2$, $1 = 9 - b^2$ so that $b^2 = 8$. Hence, the equation of the ellipse is $\boxed{\frac{x^2}{9} + \frac{y^2}{8} = 1}$.

13. Since the foci are $(2,-4), (2,4)$ and the vertices are $(2,-6), (2,6)$, we know:

i) the ellipse is vertical with center at $(2, \frac{-6+6}{2}) = (2,0)$,

ii) $c=4$,

iii) the length of the major axis is $6-(-6)=12$. Hence, $a=6$.

Now, since $c^2 = a^2 - b^2$, $16 = 36 - b^2$ so that $b^2 = 20$. Hence, the equation of the ellipse is $\boxed{\frac{(x-2)^2}{20} + \frac{y^2}{36} = 1}$.

14. Since the foci are $(-6,-3), (-5,-3)$ and the vertices are $(-7,-3), (-4,-3)$, we know:

i) the ellipse is horizontal with center at $(\frac{-7+-4}{2}, -3) = (-\frac{11}{2}, -3)$,

ii) $c = \frac{1}{2}$,

iii) the length of the major axis is $-4-(-7)=3$. Hence, $a = \frac{3}{2}$.

Now, since $c^2 = a^2 - b^2$, $(\frac{1}{2})^2 = (\frac{3}{2})^2 - b^2$ so that $b^2 = 2$. Hence, the equation of the ellipse is $\boxed{\frac{(x+\frac{11}{2})^2}{2.25} + \frac{(y+3)^2}{2} = 1}$.

15. Since the vertices are $(-1,0)$, $(1,0)$, and the center is the midpoint of the segment connecting them, the center must be (0,0). Moreover, since the form of the vertices is $(h \pm a, k)$ (which in this case is $(0 \pm a, 0)$) we see that $a = 1$. It remains to find *b*. At this point, we need to use the fact that we are given that the equations of the asymptotes are $y = \pm 2x$. Since the transverse axis is parallel to the *x*-axis, we know that the slopes of the asymptotes for such a hyperbola are $\pm\frac{b}{a}$. Thus, $\pm\frac{b}{1} = \pm 2$, which implies that $b = \pm 2$. Hence, the equation of the hyperbola must be $\boxed{x^2 - \frac{y^2}{4} = 1}$.

16. Since the vertices are $(0,-1)$, $(0,1)$, and the center is the midpoint of the segment connecting them, the center must be (0,0). Moreover, since the form of the vertices is $(h \pm a, k)$ (which in this case is $(0 \pm a, 0)$) we see that $a = 1$. It remains to find *b*. At this point, we need to use the fact that we are given that the equations of the asymptotes are $y = \pm\frac{1}{3}x$. Since the transverse axis is parallel to the *y*-axis, we know that the slopes of the asymptotes for such a hyperbola are $\pm\frac{a}{b}$. Thus, $\pm\frac{a}{b} = \pm\frac{1}{3}$. Since $a = 1$, this implies that $b = \pm 3$. Hence, the equation of the hyperbola must be $\boxed{y^2 - \frac{x^2}{9} = 1}$.

17. Since the vertices are $(2,-4)$, $(2,4)$, we know that the center is $(2,0)$ and the transverse axis is parallel to the *y*-axis. Hence, $0 - a = -4$ so that $a = 4$. Also, since the foci are $(2,-6)$, $(2,6)$, we know that $0 - c = -6$ so that $c = 6$. Now, to find *b*, we substitute the values of *c* and *a* obtained above into $c^2 = a^2 + b^2$ to see that $36 = 16 + b^2$ so that $b^2 = 20$. Hence, the equation of the hyperbola is $\boxed{\frac{y^2}{16} - \frac{(x-2)^2}{20} = 1}$.

18. Since the vertices are $(-6,-3)$, $(-5,-3)$, we know that the center is $(-5.5,-3)$ and the transverse axis is parallel to the *x*-axis. Hence, $-5.5 - a = -6$ so that $a = 0.5$. Also, since the foci are $(-7,-3)$, $(-4,-3)$, we know that $-5.5 - c = -7$ so that $c = 1.5$. Now, to find *b*, we substitute the values of *c* and *a* obtained above into $c^2 = a^2 + b^2$ to see that $(1.5)^2 = (0.5)^2 + b^2$ so that $b^2 = 2$. Hence, the equation of the hyperbola is

$\boxed{\frac{(x+5.5)^2}{0.25} - \frac{(y+3)^2}{2} = 1}$.

19.

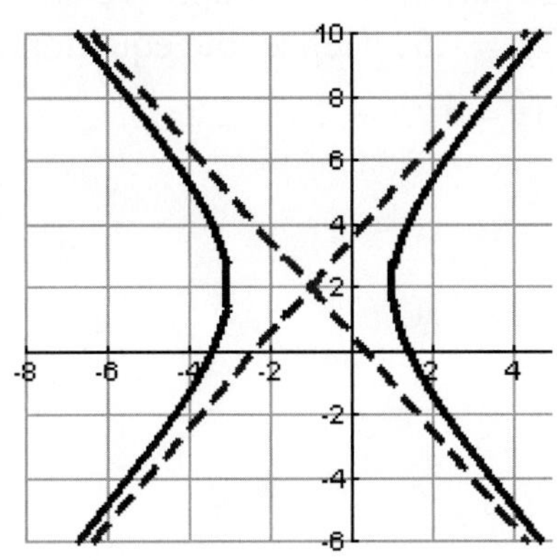

Notes on the Graph:
First, write the equation in standard form by completing the square:

$$9\left(x^2+2x\right)-4\left(y^2-4y\right)=43$$
$$9\left(x^2+2x+1\right)-4\left(y^2-4y+4\right)=43+9-16$$
$$9(x+1)^2-4(y-2)^2=36$$
$$\frac{(x+1)^2}{4}-\frac{(y-2)^2}{9}=1$$

The equations of the asymptotes are $y=\pm\frac{3}{2}(x+1)+2$.

20.

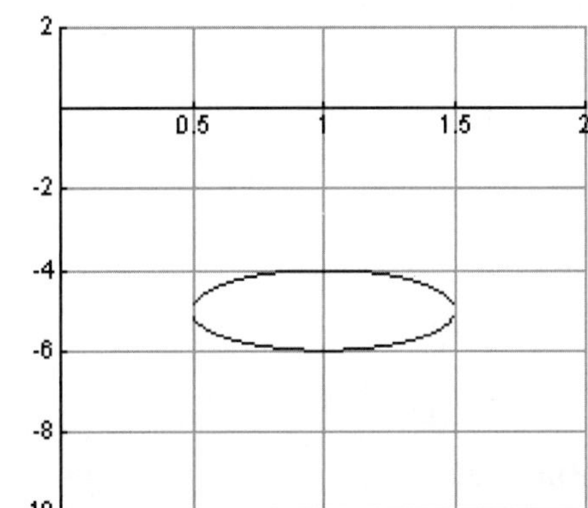

First, write the equation in standard form by completing the square:

$$4(x^2-2x)+(y^2+10y)=-28$$
$$4(x^2-2x+1)+(y^2+10y+25)=-28+4+25$$
$$4(x-1)^2+(y+5)^2=1$$
$$\frac{(x-1)^2}{\frac{1}{4}}+(y+5)^2=1$$

21.

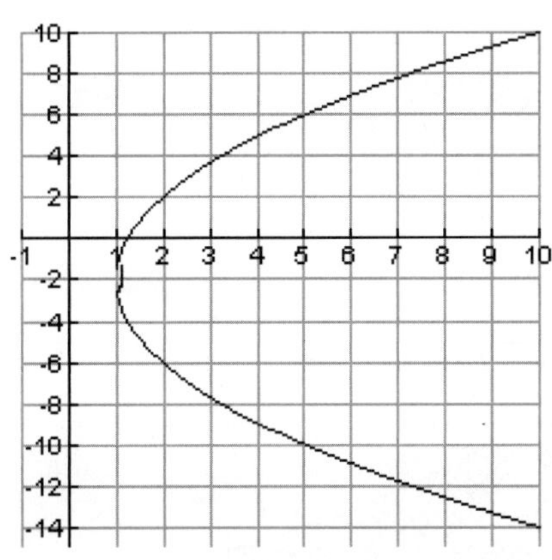

Equation: Completing the square yields:

$$y^2+4y=16x-20$$
$$(y^2+4y+4)=16x-20+4$$
$$(y+2)^2=16x-16$$
$$(y+2)^2=16(x-1)$$
$$(y+2)^2=4(4)(x-1)\quad \textbf{(1)}$$

So, $p=4$ and opens to the right and its vertex is $(1,-2)$.

22.

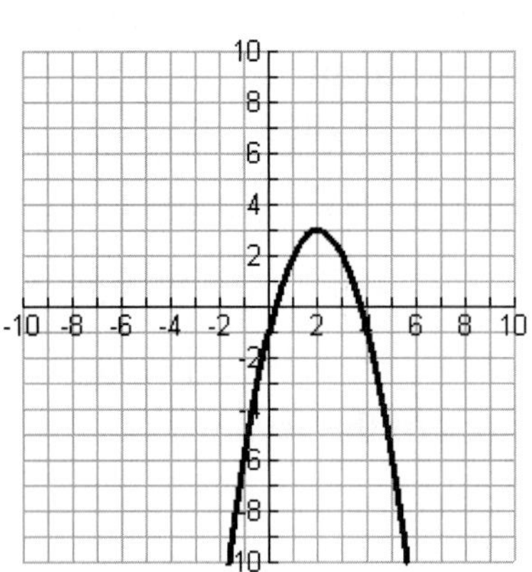

23. Assume that the vertex is at the origin and that the parabola opens up. We are given that $p = 1.5$. Hence, the equation is $x^2 = 4(1.5)y = 6y, \quad -2 \le x \le 2$.

24. Assume that the sun is at the origin. Then, the vertices of Uranus's horizontal elliptical trajectory are: $(-2.739\times10^9, 0)$, $(3.003\times10^9, 0)$

From this, we know that:

i) The length of the major axis is $3.003\times10^9 - (-2.739\times10^9) = 5.742\times10^9$, and so the value of a is half of this, namely 2.871×10^9;

ii) The center of the ellipse is $\left(\frac{3.003\times10^9 + (-2.739\times10^9)}{2}, 0\right) = (1.32\times10^8, 0)$;

iii) The value of c is 1.32×10^8.

Now, since $c^2 = a^2 - b^2$, we have $\left(1.32\times10^8\right)^2 = \left(2.871\times10^9\right)^2 - b^2$, so that $b^2 = 8.225\times10^{18}$. Hence, the equation of the ellipse must be

$$\boxed{\frac{\left(x-(1.32\times10^8)\right)^2}{8.24\times10^{18}} + \frac{y^2}{8.225\times10^{18}} = 1}.$$

25.

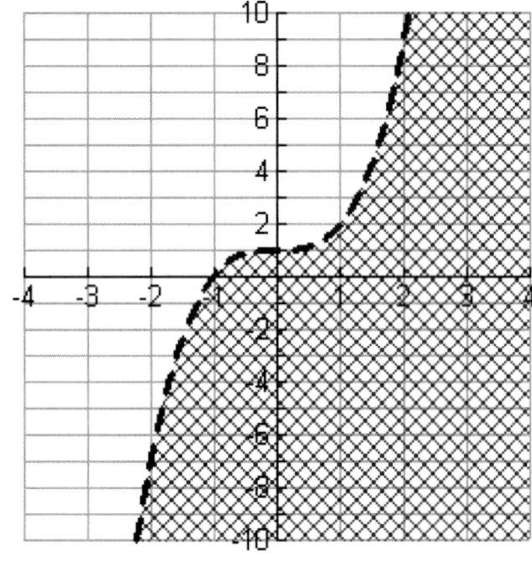

26.

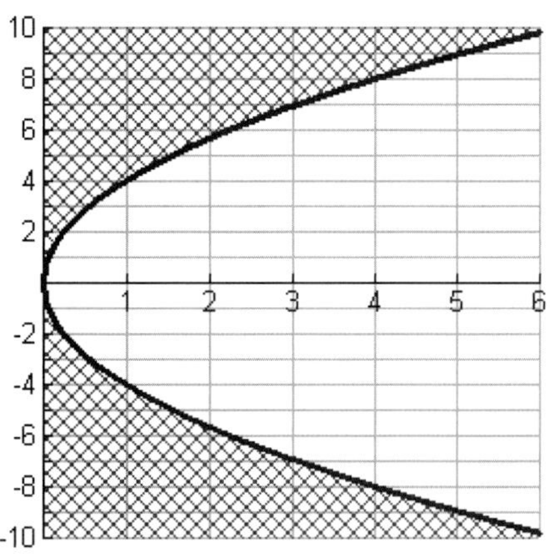

27.

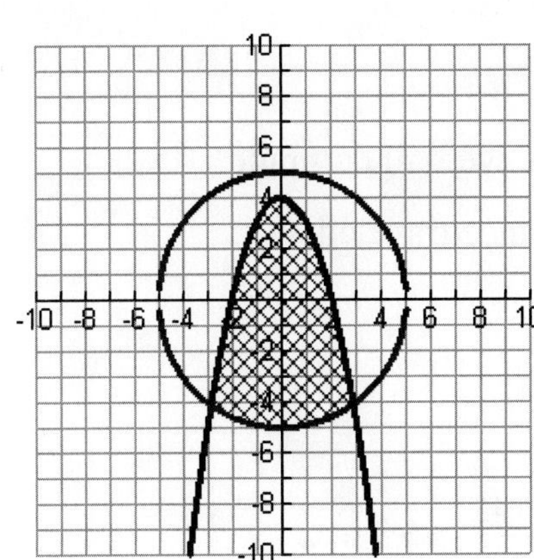

28.

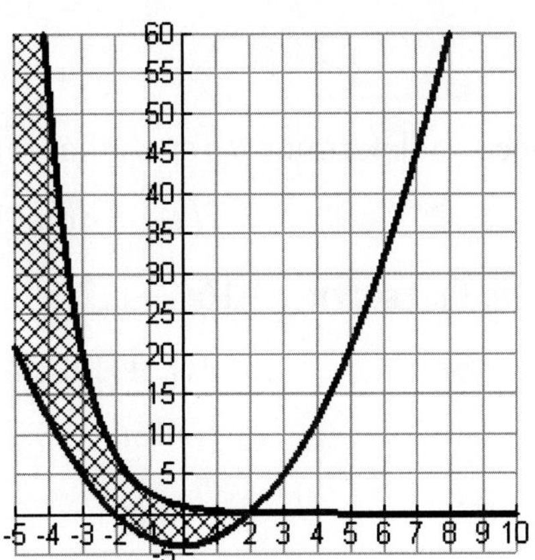

29. Observe that $r = \dfrac{12}{3+2\sin\theta} = \dfrac{12}{3\left(1+\frac{2}{3}\sin\theta\right)} = \dfrac{4}{1+\frac{2}{3}\sin\theta}$, so that $e = \frac{2}{3} < 1$. Hence, this conic is an *ellipse*.

30. Consider the equation $6\sqrt{3}x^2 + 6xy + 4\sqrt{3}y^2 - 21\sqrt{3} = 0$.

Here, $A = 6\sqrt{3},\ B = 6,\ C = 4\sqrt{3}$. In order to obtain its graph, follow these steps:

<u>Step 1</u>: Find the angle of counterclockwise rotation necessary to transform the equation into the *XY*-system with no *XY* term.

We seek the solution θ of the equation $\cot 2\theta = \dfrac{A-C}{B} = \dfrac{1}{\sqrt{3}}$ in $[0,\pi]$. Hence, we need to find θ in $[0,\pi]$ such that $\cos 2\theta = \frac{1}{2}$ and $\sin 2\theta = \frac{\sqrt{3}}{2}$.

But, this occurs precisely when $2\theta = 60°$, so that $\theta = 30°$.

<u>Step 2</u>: Now, determine the equation of this curve in the *XY*-system making use of the following substitution into the original equation with $\theta = 30°$:

$$\begin{cases} x = X\cos\theta - Y\sin\theta = \left(\sqrt{3}X - Y\right)\left(\dfrac{1}{2}\right) \\ y = X\sin\theta + Y\cos\theta = \left(X + \sqrt{3}Y\right)\left(\dfrac{1}{2}\right) \end{cases}$$

Doing so yields the following, after collecting like terms and simplifying:

$$6\sqrt{3}\left(\sqrt{3}X - Y\right)^2\left(\frac{1}{2}\right)^2 + 6\left(\sqrt{3}X - Y\right)\left(X + \sqrt{3}Y\right)\left(\frac{1}{2}\right)^2 + 4\sqrt{3}\left(X + \sqrt{3}Y\right)^2\left(\frac{1}{2}\right)^2 - 21\sqrt{3} = 0$$

$$6\sqrt{3}\left(3X^2 - 2\sqrt{3}XY + Y^2\right) + 6\left(\sqrt{3}X^2 + 3XY - XY - \sqrt{3}Y^2\right) + 4\sqrt{3}\left(X^2 + 2\sqrt{3}XY + 3Y^2\right) - 84\sqrt{3} = 0$$

$$28\sqrt{3}X^2 + 12\sqrt{3}Y^2 + 84\sqrt{3} = 0$$

$$\frac{X^2}{3} + \frac{Y^2}{7} = 1$$

31. From the given information, we know that the parametric equations are:

$$x = \left(120\cos 45^\circ\right)t$$
$$y = -16t^2 + \left(120\sin 45^\circ\right)t + 0$$

We need to determine t_0 such that $y(t_0) = 0$. Then, we will compute

$$x(t_0) = \text{ total horizontal distance traveled}$$

Observe that solving $-16t^2 + \left(120\sin 45^\circ\right)t = 0$ yields

$$-16t^2 + \left(120\sin 45^\circ\right)t = 0$$
$$t\left(-16t + 120\sin 45^\circ\right) = 0$$
$$t = 0, \frac{120\sin 45^\circ}{16} \approx 5.3 \text{ sec.}$$

Now, we conclude that $\boxed{x(5.3) = 450 \text{ ft.} = \text{ total horizontal distance traveled}}$

32. Given the parametric equations $x = \sqrt{1-t},\ y = \sqrt{t},\ t \text{ in } [0,1]$,

observe that since $t = y^2$, we have

$x = \sqrt{1-y^2}$. Now, since $y = \sqrt{t}$ is always ≥ 0 and $x = \sqrt{1-y^2}$ is also always ≥ 0, this generates a $\boxed{\text{quarter circle in QI}}$.

33.

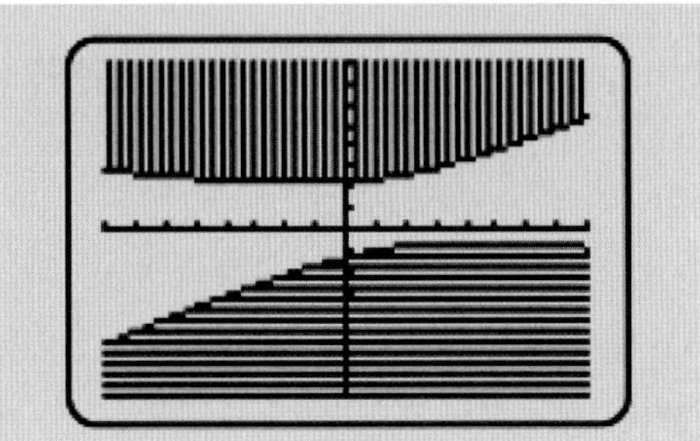

34.

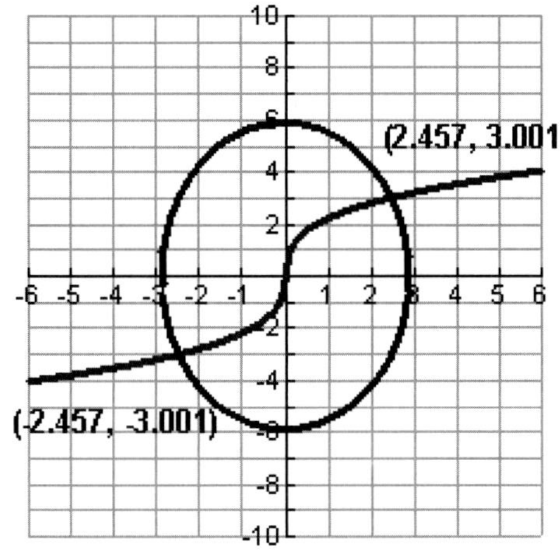

So, the approximate solutions of this system are (2.457, 3.001) and (-2.457, -3.001).

35. a. The equation is equivalent to

$$y = x^2 + 4.2x + 5.61.$$

The graph is as follows:

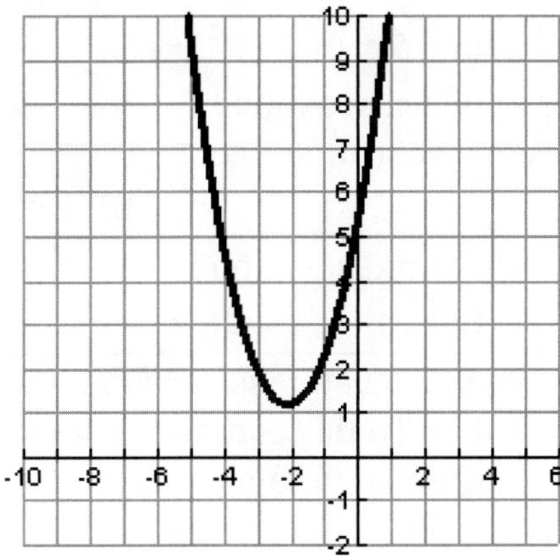

b. Completing the square yields

$$x^2 + 4.2x = y - 5.61$$

$$x^2 + 4.2x + 2.1^2 = y - 5.61 + 2.1^2$$

$$(x + 2.1)^2 = y - 1.2$$

$$(x - (-1.2))^2 = 4\left(\tfrac{1}{4}\right)(y - 1.2)$$

The vertex is (-1.2, 1.2). The parabola opens up.

c. Yes, parts **a** and **b** agree.

36. a.

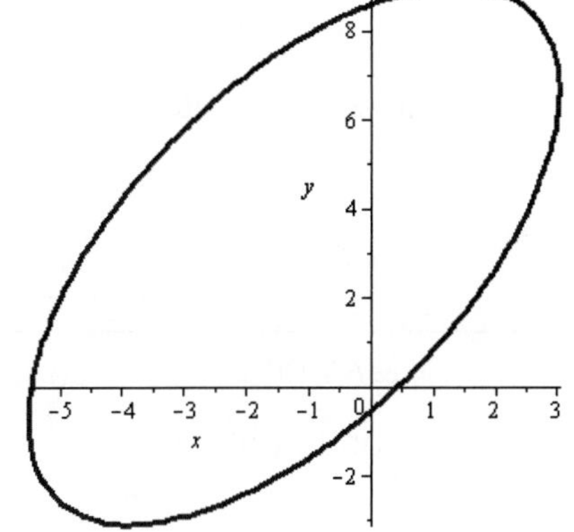

The angle of rotation is computed by

$$\cot 2\theta = \tfrac{A-C}{B} = \tfrac{2-1}{-\sqrt{3}} = -\tfrac{1}{\sqrt{3}}$$

$$2\theta = \cot^{-1}\left(-\tfrac{1}{\sqrt{3}}\right)$$

$$\theta \approx .052 \text{ rad}$$

b.

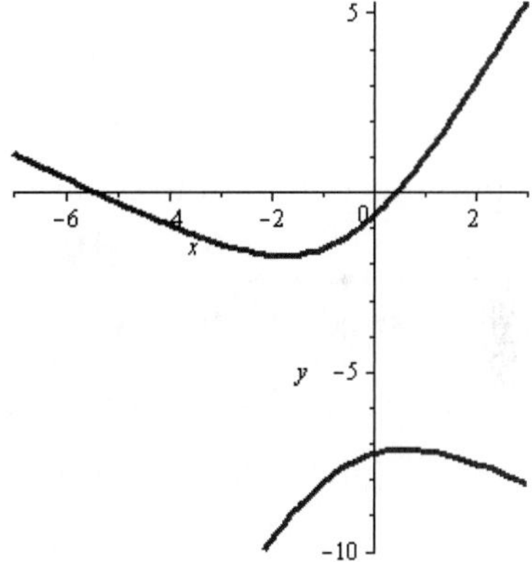

The angle of rotation is computed by

$$\cot 2\theta = \tfrac{A-C}{B} = \tfrac{2-(-1)}{\sqrt{3}} = \tfrac{3}{\sqrt{3}}$$

$$2\theta = \cot^{-1}\left(\tfrac{3}{\sqrt{3}}\right)$$

$$\theta \approx .26 \text{ rad}$$

Chapter 9 Cumulative Test----- ---

1.

$$(x+2)^2-(x+2)-20=0$$
$$\big((x+2)-5\big)\big((x+2)+4\big)=0$$
$$(x-3)(x+6)=0$$
$$x=\boxed{-6,3}$$

2. The equation is of the general form
$$(x-h)^2+(y-k)^2=r^2.$$
Since $(h,k)=(5,1)$, this becomes
$$(x-5)^2+(y-1)^2=r^2.$$
In order to find *r*, substitute the point (6,-2) into the above equation to obtain
$$(6-5)^2+(-2-1)^2=1+9=10=r^2.$$
Hence, the desired equation is
$$(x-5)^2+(y-1)^2=10.$$

3.

$$\frac{(8-7(x+h))-(8-7x)}{h}=\frac{8-7x-7h-8+7x}{h}$$
$$=\frac{-7h}{h}=\boxed{-7}$$

4. The general equation is $I=kPt$. We need to find *k*. To this end, we substitute in the given information to obtain
$$90=k(1500)(2),$$
so that $k=\frac{3}{100}$. Hence, the equation is
$$I=\tfrac{3}{100}Pt.$$

5. The general form is $y=a(x-h)^2+k$. We know that the vertex $(h,k)=(7,7)$, so that this equation becomes $y=a(x-7)^2+7$. In order to find *a*, substitute the point (10,10) into this equation to obtain
$$10=a(10-7)^2+7 \;\Rightarrow\; 3=9a \;\Rightarrow\; a=\tfrac{1}{3}.$$
Hence, the desired equation is $y=\frac{1}{3}(x-7)^2+7$.

6. Use $A=P\left(1+\frac{r}{n}\right)^{nt}$. Here, $r=0.047,\ n=52,\ t=17,\ A=65{,}000$. We need to determine *P*. Substitute all of this information into the equation to obtain
$$65{,}000=P\left(1+\tfrac{0.047}{52}\right)^{52(17)}$$
$$P=\frac{65{,}000}{\left(1+\frac{0.047}{52}\right)^{52(17)}}\approx 29{,}246.16$$
So, you should invest \$29,246.16.

7.

$$\log x^2-\log 16=0 \;\Rightarrow\; 2\log x=\log 16 \;\Rightarrow\; \log x=\tfrac{1}{2}\log 16=\log 4 \;\Rightarrow\; x=10^{\log 4}=\boxed{4}$$

8. Let x = the shortest leg in a $30^\circ-60^\circ-90^\circ$ triangle. If $x=8$ in., then the other leg has length $8\sqrt{3}$ in. and the hypotenuse has length 16 in.

9. $\cot(-27^\circ)=\dfrac{1}{\tan(-27^\circ)}\approx\boxed{1.9626}$

10. Amplitude is 0.007, and frequency is $\frac{2\pi}{850\pi}=\frac{1}{425}$.

11.

$$\tan\theta(\csc\theta+\cos\theta)=\frac{\sin\theta}{\cos\theta}\left(\frac{1}{\sin\theta}+\cos\theta\right)=\boxed{\frac{1}{\cos\theta}+\sin\theta}$$

12.

$$\cos\left(-\tfrac{11\pi}{12}\right)=\cos\left(-\frac{\frac{11\pi}{6}}{2}\right)=\cos\left(\frac{\frac{11\pi}{6}}{2}\right)=-\sqrt{\frac{1+\cos\left(\frac{11\pi}{6}\right)}{2}}=-\sqrt{\frac{1+\frac{\sqrt{3}}{2}}{2}}=\boxed{-\frac{\sqrt{2+\sqrt{3}}}{2}}$$

13. Observe that

$$\begin{aligned}4\cos^2 x+4\cos 2x+1&=0\\4\cos^2 x+4\left(2\cos^2 x-1\right)+1&=0\\12\cos^2 x-3&=0\\\cos^2 x&=\tfrac{1}{4}\\\cos x&=\pm\tfrac{1}{2}\end{aligned}$$

This holds when $x=\boxed{\frac{\pi}{3},\frac{2\pi}{3},\frac{4\pi}{3},\frac{5\pi}{3}}$

14. First, note that at the end of two hours, one plane flew due north for 900 miles, and the other plane flew for 750 miles at an angle 135° clockwise from due north. Consider the following diagram:

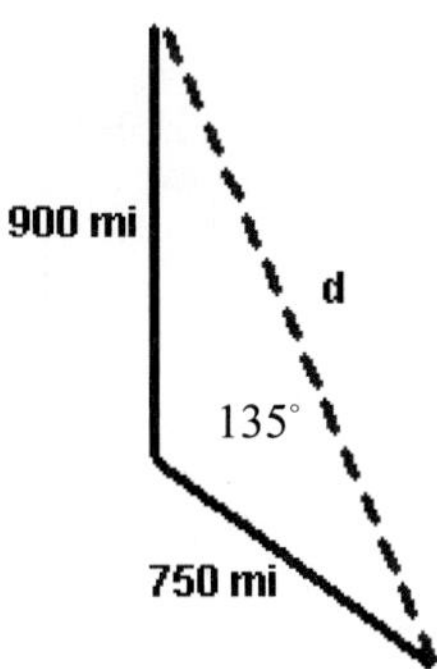

Using the Law of Cosines yields

$$d^2=900^2+750^2-2(900)(750)\cos\left(135^\circ\right)\approx 2327094.155$$

$$d\approx 1525.48$$

So, the two planes are approximately 1525 miles apart after two hours.

15. $u = \langle |u|\cos\theta, |u|\sin\theta \rangle = \langle 15\cos 110^\circ, 15\sin 110^\circ \rangle \approx \langle -5.13, 14.10 \rangle$

16. Observe that

$$z_1 = 5e^{(15^\circ)i},\ z_2 = 2e^{(75^\circ)i}$$

Then, we have

$$\begin{aligned} z_1 \cdot z_2 &= 5e^{(15^\circ)i} \cdot 2e^{(75^\circ)i} \\ &= (2)(5)e^{(15^\circ+75^\circ)i} \\ &= 10e^{(90^\circ)i} \\ &= \boxed{10i} \end{aligned}$$

17. Let x = cost of a soda and
y = cost of a soft pretzel.

Solve the system:

$$\begin{cases} 3x+2y=6.77 & \textbf{(1)} \\ 5x+4y=12.25 & \textbf{(2)} \end{cases}$$

Multiply **(1)** by -5 and **(2)** by 3, and add:

$$\begin{aligned} -15x-10y &= -33.85 \\ +\ 15x+12y &= 36.75 \\ \hline 2y &= 2.90 \\ y &= 1.45 \end{aligned}$$

Substitute this value of y into **(1)** to solve for x:

$$3x+2(1.45)=6.77 \Rightarrow x=1.29$$

So, a soda costs \$1.29 and a soft pretzel costs \$1.45.

18. The partial fraction decomposition is of the form

$$\frac{3x+5}{(x-3)(x^2+5)} = \frac{A}{x-3} + \frac{Bx+C}{x^2+5}$$

Multiply by the LCD to obtain

$$3x+5 = A(x^2+5)+(Bx+C)(x-3) = (A+B)x^2+(C-3B)x+(5A-3C).$$

We must solve the following system:

$$\begin{cases} A+B=0 & \textbf{(1)} \\ C-3B=3 & \textbf{(2)} \\ 5A-3C=5 & \textbf{(3)} \end{cases}$$

Solve **(1)** for B: $B=-A$ **(4)**

Substitute **(4)** into **(2):** $C+3A=3$ **(5)**

Solve the 2×2 system:

$$\begin{cases} 5A-3C=5 & \textbf{(3)} \\ 3A+C=3 & \textbf{(5)} \end{cases}$$

Multiply **(5)** by 3 and add to **(3)** to see that $A=1$.
Substitute this back into **(5)** to obtain $C=0$.
Finally, substitute the value of A into **(1)** to see that $B=-1$.
Hence, the partial fraction decomposition is

$$\frac{3x+5}{(x-3)(x^2+5)} = \frac{1}{x-3} + \frac{-x}{x^2+5}$$

19.

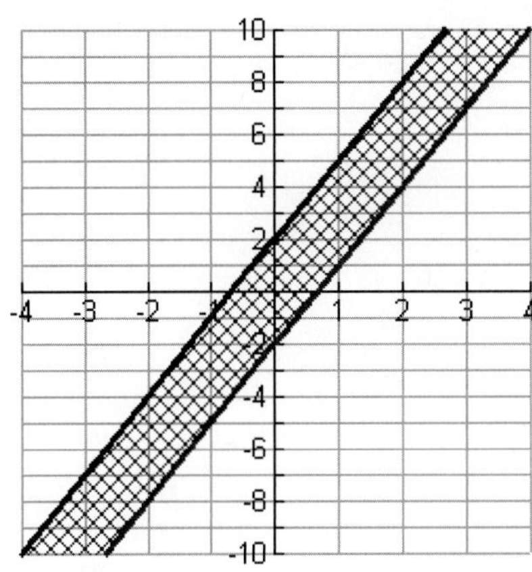

20.

$$\left[\begin{array}{ccc|c} 1 & -2 & 1 & 7 \\ -3 & 1 & 2 & -11 \end{array}\right] \xrightarrow{R_2+3R_1 \to R_2} \left[\begin{array}{ccc|c} 1 & -2 & 1 & 7 \\ 0 & -5 & 5 & 10 \end{array}\right]$$

$$\xrightarrow{-\frac{1}{5}R_2 \to R_2} \left[\begin{array}{ccc|c} 1 & -2 & 1 & 7 \\ 0 & 1 & -1 & -2 \end{array}\right]$$

Let $\boxed{y=a}$. Then, $\boxed{x=a-2}$ and by substituting these two values into the equation corresponding to the first row of the last matrix, we see that $(a-2)-2a+z=7$, so that $\boxed{z=a+9}$.

21.

$$2B-3A=\begin{bmatrix} 16 & -4 & 12 \\ 18 & 0 & -2 \end{bmatrix}-\begin{bmatrix} 9 & 12 & -21 \\ 0 & 3 & 15 \end{bmatrix}$$

$$=\begin{bmatrix} 7 & -16 & 33 \\ 18 & -3 & -17 \end{bmatrix}$$

22. Write the system in matrix form as

$$\begin{bmatrix} 25 & 40 \\ 75 & -105 \end{bmatrix}\begin{bmatrix} x \\ y \end{bmatrix}=\begin{bmatrix} -12 \\ 69 \end{bmatrix}$$

Observe that

$$D=\begin{vmatrix} 25 & 40 \\ 75 & -105 \end{vmatrix}=-5,625$$

$$D_x=\begin{vmatrix} -12 & 40 \\ 69 & -105 \end{vmatrix}=-1,500$$

$$D_y=\begin{vmatrix} 25 & -12 \\ 75 & 69 \end{vmatrix}=2,625$$

So,

$$x=\tfrac{1500}{5625}=\tfrac{4}{15},\quad y=-\tfrac{2625}{5625}=-\tfrac{7}{15}.$$

23. Since the vertices are (6,3) and (6,-7), the center is $\left(6,\frac{3-7}{2}\right)=(6,-2)$.

Using the foci, we see that

$$-2+c=2 \text{ and } -2-c=-6,$$

so that $c=4$. Also, we have

$$-2+a=3 \text{ and } -2-a=-7,$$

so that $a=5$.

As such, $c^2=a^2-b^2 \Rightarrow b=3$.

Thus, the equation of the ellipse is

$$\tfrac{(x-6)^2}{9}+\tfrac{(y+2)^2}{25}=1.$$

24. Since the vertices are (5,-2) and (5,0), the center of the hyperbola is (5,-1). Also, we have $a = 1$. Thus, the equation so far is

$$(y+1)^2-\frac{(x-5)^2}{b^2}=1.$$

To find b, we use the fact that $c = 2$, so that

$$c^2=a^2+b^2 \Rightarrow b=\sqrt{3}$$

Thus, the desired equation is

$$(y+1)^2-\frac{(x-5)^2}{3}=1$$

25. Solve the system:

$$\begin{cases} x+y=6 & \textbf{(1)} \\ x^2+y^2=20 & \textbf{(2)} \end{cases}$$

Solve **(1)** for y: $y=6-x$ **(3)**

Substitute **(3)** into **(2)** and solve for x:

$$x^2+(6-x)^2=20$$
$$2x^2-12x+36=20$$
$$x^2-6x+8=0$$
$$(x-4)(x-2)=0$$
$$x=2,4$$

Substitute each value into **(1)** to find the corresponding y-value. This yields two solutions, namely (2,4) and (4,2).

26. The graph from the calculator is as follows:

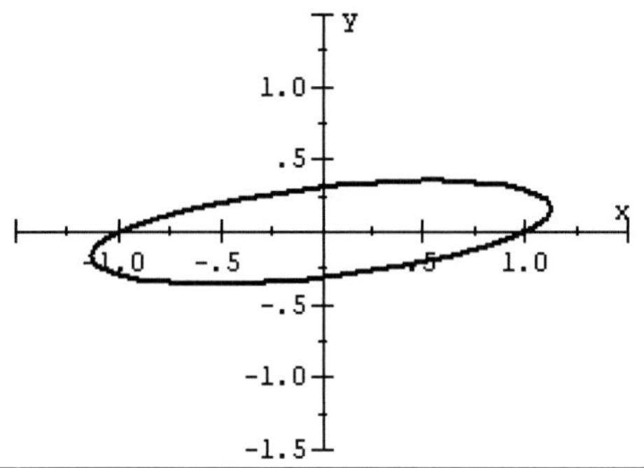

27. The dark region is the one desired:

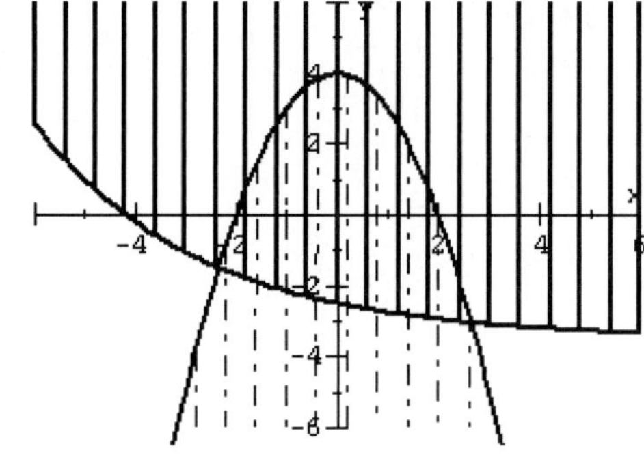

CHAPTER 10

Section 10.1 Solutions--

1. $1, 2, 3, 4$	**2.** $1, 4, 9, 16$
3. $1, 3, 5, 7$	**4.** x, x^2, x^3, x^4
5. $\frac{1}{2}, \frac{2}{3}, \frac{3}{4}, \frac{4}{5}$	**6.** $2, \frac{3}{2}, \frac{4}{3}, \frac{5}{4}$
7. $2, \underbrace{\frac{4}{2!}}_{=\frac{4}{2\cdot 1}}, \underbrace{\frac{8}{3!}}_{=\frac{8}{3\cdot 2\cdot 1}}, \underbrace{\frac{16}{4!}}_{=\frac{16}{4\cdot 3\cdot 2\cdot 1}},$ which is equivalent to $2, 2, \frac{4}{3}, \frac{2}{3}$	**8.** $\frac{n!}{(n+1)!} = \frac{\cancel{n!}}{(n+1)\cancel{n!}} = \frac{1}{n+1},$ so the terms are $\frac{1}{2}, \frac{1}{3}, \frac{1}{4}, \frac{1}{5}$
9. $-x^2, x^3, -x^4, x^5$	**10.** $1, -4, 9, -16$
11. $\frac{-1}{2\cdot 3}, \frac{1}{3\cdot 4}, \frac{-1}{4\cdot 5}, \frac{1}{5\cdot 6}$ which is equivalent to $\frac{-1}{6}, \frac{1}{12}, \frac{-1}{20}, \frac{1}{30}$	**12.** $0, \frac{1}{9}, \underbrace{\frac{4}{16}}_{=\frac{1}{4}}, \frac{9}{25}$
13. $\left(\frac{1}{2}\right)^9 = \frac{1}{512}$	**14.** $\frac{15}{(16)^2} = \frac{15}{256}$
15. $\frac{(-1)^{19}\, 19!}{(21)!} = \frac{-1}{21\cdot 20} = \frac{-1}{420}$	**16.** $\frac{(-1)^{14}(12)(15)}{13} = \frac{180}{13}$
17. $\left(1 + \frac{1}{100}\right)^2 = \left(\frac{101}{100}\right)^2 = \frac{10{,}201}{10{,}000}$	**18.** $1 - \frac{1}{10^2} = \frac{99}{100}$
19. $\log 10^{23} = 23 \log 10 = 23$	**20.** $e^{\ln 49} = 49$
21. $a_n = 2n, n \geq 1$	**22.** $a_n = 3n, n \geq 1$
23. $a_n = \frac{1}{(n+1)n}, n \geq 1$	**24.** $a_n = \left(\frac{1}{2}\right)^n, n \geq 1$
25. $a_n = (-1)^n \left(\frac{2}{3}\right)^n = \frac{(-1)^n 2^n}{3^n},\ n \geq 1$	**26.** $a_n = \frac{3^{n-1}}{2^n}, n \geq 1$

27. $(-1)^{n+1}, n \geq 1$	**28.** $(-1)^{n+1} \dfrac{n}{n+2}, n \geq 1$
29. $\dfrac{9!}{7!} = \dfrac{9 \cdot 8 \cdot \cancel{7!}}{\cancel{7!}} = 72$	**30.** $\dfrac{4!}{6!} = \dfrac{\cancel{4!}}{6 \cdot 5 \cdot \cancel{4!}} = \dfrac{1}{30}$
31. $\dfrac{29!}{27!} = \dfrac{29 \cdot 28 \cdot \cancel{27!}}{\cancel{27!}} = 812$	**32.** $\dfrac{32!}{30!} = \dfrac{32 \cdot 31 \cdot \cancel{30!}}{\cancel{30!}} = 992$
33. $\dfrac{75!}{77!} = \dfrac{\cancel{75!}}{77 \cdot 76 \cdot \cancel{75!}} = \dfrac{1}{5852}$	**34.** $\dfrac{100!}{103!} = \dfrac{\cancel{100!}}{103 \cdot 102 \cdot 101 \cdot \cancel{100!}} = \dfrac{1}{1,061,106}$
35. $\dfrac{97!}{93!} = \dfrac{97 \cdot 96 \cdot 95 \cdot 94 \cdot \cancel{93!}}{\cancel{93!}} = 83,156,160$	**36.** $\dfrac{101!}{98!} = \dfrac{101 \cdot 100 \cdot 99 \cdot \cancel{98!}}{\cancel{98!}} = 999,900$
37. $\dfrac{(n-1)!}{(n+1)!} = \dfrac{\cancel{(n-1)!}}{(n+1)n\cancel{(n-1)!}} = \dfrac{1}{(n+1)n}$	**38.** $\dfrac{(n+2)!}{n!} = \dfrac{(n+2)(n+1)\cancel{n!}}{\cancel{n!}} = (n+2)(n+1)$
39. $\dfrac{(2n+3)!}{(2n+1)!} = \dfrac{(2n+3)(2n+2)\cancel{(2n+1)!}}{\cancel{(2n+1)!}}$ $= (2n+3)(2n+2)$	**40.** $\dfrac{(2n+2)!}{(2n-1)!} = \dfrac{(2n+2)(2n+1)(2n)\cancel{(2n-1)!}}{\cancel{(2n-1)!}}$ $= (2n+2)(2n+1)(2n)$
41. $7, 10, 13, 16$	**42.** $2, 3, 4, 5$
43. $1, 2, 6, 24$	**44.** $2, 6, 24, 120$
45. $100, \underbrace{\dfrac{100}{2!}}_{=50}, \underbrace{\dfrac{50}{3!}}_{=\frac{50}{3\cdot 2}=\frac{25}{3}}, \underbrace{\dfrac{\frac{25}{3}}{4!}}_{=\frac{25}{3\cdot 4\cdot 3\cdot 2}=\frac{25}{72}}$ which is equivalent to 100, 50, $\dfrac{25}{3}, \dfrac{25}{72}$.	**46.** $20, \underbrace{\dfrac{20}{2^2}}_{=5}, \dfrac{5}{3^2}, \underbrace{\dfrac{\frac{5}{9}}{4^2}}_{=\frac{5}{144}},$ which is equivalent to 20, 5, $\dfrac{5}{9}, \dfrac{5}{144}$
47. $1, 2, 2, 4$ (Note that $a_3 = a_2 \cdot a_1$, $a_4 = a_3 \cdot a_2$)	**48.** 1, 2, $\dfrac{1}{2}$, 4 (Note that $a_3 = \dfrac{a_1}{a_2}$, $a_4 = \dfrac{a_2}{a_3}$)
49. $1, -1, -\underbrace{\left[(1)^2 + (-1)^2\right]}_{=-2}, \underbrace{\left[(-1)^2 + (-2)^2\right]}_{=5}$ which is equivalent to $1, -1, -2, 5$	**50.** $1, -1, \underbrace{2(-1)+1}_{=-1}, \underbrace{3(-1)+2(-1)}_{=-5}$ which is equivalent to 1, -1, -1, -5 (Note that $a_3 = 2a_2 + 1a_1$, $a_4 = 3a_3 + 2a_2$.)
51. $2 \cdot 5 = 10$	**52.** $7 \cdot 5 = 35$

53. $0^2+1^2+2^2+3^2+4^2=30$	**54.** $1+\frac{1}{2}+\frac{1}{3}+\frac{1}{4}=\frac{25}{12}$
55. $\sum_{n=1}^{6}(2n-1)=1+3+5+7+9+11=36$	**56.** $\sum_{n=1}^{6}(n+1)=2+3+4+5+6+7=27$
57. $1^n=1$, for all n. So, $\sum_{n=0}^{4}1^n=\sum_{n=0}^{4}1=5(1)=5$.	**58.** $2^0+2^1+2^2+2^3+2^4=1+2+4+8+16$ $=31$
59. $1-x+x^2-x^3$	**60.** $-x+x^2-x^3+x^4$
61. $\frac{2^0}{0!}+\frac{2^1}{1!}+\underbrace{\frac{2^2}{2!}}_{=\frac{4}{2\cdot1}}+\underbrace{\frac{2^3}{3!}}_{=\frac{8}{3\cdot2}}+\underbrace{\frac{2^4}{4!}}_{=\frac{16}{4\cdot3\cdot2}}+\underbrace{\frac{2^5}{5!}}_{=\frac{32}{5\cdot4\cdot3\cdot2}}=$ $1+2+2+\frac{4}{3}+\frac{2}{3}+\frac{4}{15}=$ $7+\frac{4}{15}=\frac{109}{15}$	**62.** $\frac{1}{0!}-\frac{1}{1!}+\frac{1}{2!}-\frac{1}{3!}+\frac{1}{4!}-\frac{1}{5!}=$ $\cancel{1}-\cancel{1}+\frac{1}{2}-\frac{1}{6}+\frac{1}{24}-\frac{1}{120}=\frac{11}{30}$
63. $\frac{1}{0!}+\frac{x}{1!}+\frac{x^2}{2!}+\frac{x^3}{3!}+\frac{x^4}{4!}=$ $1+x+\frac{x^2}{2}+\frac{x^3}{6}+\frac{x^4}{24}$	**64.** $\frac{1}{0!}-\frac{(1)x}{1!}-\frac{(2)x^2}{2!}-\frac{(3)x^3}{3!}-\frac{(4)x^4}{4!}=$ $1-x-x^2-\frac{x^3}{2}-\frac{x^4}{6}$
65. $2\frac{(0.1)^0}{1-0.1}=\frac{2}{0.9}=2.\overline{2}=\frac{20}{9}$	**66.** $5\cdot\frac{\left(\frac{1}{10}\right)^0}{1-\frac{1}{10}}=\frac{5}{0.9}=5.\overline{5}=\frac{50}{9}$
67. Not possible. (The result is infinite.)	**68.** Not possible. (The result is infinite.)
69. $\sum_{n=0}^{6}(-1)^n\frac{1}{2^n}$	**70.** $\sum_{n=0}^{\infty}\frac{1}{2^n}$
71. $\sum_{n=1}^{\infty}(-1)^{n-1}n$	**72.** $\sum_{n=1}^{23}n$
73. $\sum_{n=1}^{6}\frac{(n+1)!}{(n-1)!}=\sum_{n=1}^{5}n(n+1)$ (Remember that $0!=1!=1$.)	**74.** $\sum_{n=1}^{\infty}\frac{2^{n-1}}{(n-1)!}=\sum_{n=0}^{\infty}\frac{2^n}{n!}$

75. $\sum_{n=1}^{\infty}(-1)^{n-1}\frac{x^{n-1}}{(n-1)!}=\sum_{n=0}^{\infty}(-1)^{n}\frac{x^{n}}{n!}$	**76.** $\sum_{n=1}^{6}\frac{x^{n}}{(n-1)!}$
77. $A_{72}=20{,}000\left(1+\frac{0.06}{12}\right)^{72}\approx \$28{,}640.89.$ So, she has approximately \$28,640.89 in her account after 6 years (or 72 months).	**78.** $A_{12}=7{,}000\left(1+\frac{0.05}{4}\right)^{12}\approx \$8{,}125.28$ So, she has approximately \$8,125.28 in her account after 3 years (or 36 months).
79. Let $n=$number of years experience. Then, the salary per hour is given by $a_n=20+2n,\ n\geq 0.$ So, $a_{20}=20+2(20)=60.$ Thus, a paralegal with 20 years experience would make \$60 per hour.	**80.** Here, $a_n=275{,}000+75{,}000(n-1),\ n\geq 1.$ So, $\sum_{n=1}^{3}a_n=$ salary for a 3 year career.
81. Let $n=$ number of years on the job. Then, the salary is given recursively by $a_0=30{,}000$ $a_n=\underbrace{a_{n-1}}_{\text{previous year}}+\underbrace{0.03a_{n-1}}_{\text{raise}}=1.03a_{n-1}$	**82.** Let $n=\#20$ minute periods. Then, the number of E.coli cells is described by: $a_0=2$ $a_n=2\cdot a_{n-1}=2^{n+1},\ n\geq 1.$ After 12 hours, 36 20-minute periods have passed and so, a_{36} E.coli cells exist, which equals $2^{37}\approx 1.374\times 10^{11}$. After 48 hours, 144 20-minute periods have passed and a_{144} E.coli cells exist, which equals $2^{145}\approx 4.46\times 10^{43}$.
83. Let $n=$number of years. Then, the number of T-cells in body is given by: $a_1=1000$ $a_{n+1}=1000-75n,\ n\geq 1.$ We must find n such that $a_{n+1}\leq 200.$ To do so, note that the above formula can be expressed explicitly as $a_{n+1}=1000-75n.$ So, we must solve $1000-75n=200$: $75n=800$ $n\approx 10.7$ As such, after approximately 10.7 years, the person would have full blown AIDS.	**84.** $a_3=3.8+1.6(3)=8.6$ billion in sales $a_4=3.8+1.6(4)=10.2$ billion in sales So, $\frac{1}{2}\sum_{n=3}^{4}a_n=\frac{1}{2}(8.6+10.2)=9.4$ represents the average sales for the years 2003 and 2004.

85. Observe that

$A_1 = 100{,}000(1.001-1) = 100$

$A_2 = 100{,}000\left((1.001)^2-1\right) = 200.10$

$A_3 = 100{,}000\left((1.001)^3-1\right) \approx 300.30$

$A_4 = 100{,}000\left((1.001)^4-1\right) \approx 400.60$

$A_{36} = 100{,}000\left((1.001)^{36}-1\right) \approx 3663.72$

86. Observe that

$A_1 = 50{,}000(1.001-1) = 50$

$A_2 = 50{,}000\left((1.001)^2-1\right) = 100.05$

$A_3 = 50{,}000\left((1.001)^3-1\right) \approx 150.15$

$A_4 = 50{,}000\left((1.001)^4-1\right) \approx 200.30$

$A_{48} = 50{,}000\left((1.001)^{48}-1\right) \approx 2457.27$

87.

	At $x = 2$
1	1
$1+x$	3
$1+x+\frac{x^2}{2}$	5
$1+x+\frac{x^2}{2}+\frac{x^3}{3!}$	≈ 6.33
$1+x+\frac{x^2}{2}+\frac{x^3}{3!}+\frac{x^4}{4!}$	≈ 7.0
	Approx. of $e^2 \approx 7.38906$

88.

$$A_6 = 195{,}000(1.035)^6 = 239{,}704.7886$$

So, approximately \$239,705.

89.

	At $x = 1.1$
$(x-1)$	0.1
$(x-1)-\frac{(x-1)^2}{2}$	0.0950
$(x-1)-\frac{(x-1)^2}{2}+\frac{(x-1)^3}{3}$	0.0953
$(x-1)-\frac{(x-1)^2}{2}+\frac{(x-1)^3}{3}-\frac{(x-1)^4}{4}$	0.09531
$(x-1)-\frac{(x-1)^2}{2}+\frac{(x-1)^3}{3}-\frac{(x-1)^4}{4}+\frac{(x-1)^5}{5}$	0.095306 Approx of ln(1.1)

90. $FV = 5000\left[\dfrac{(1.06)^n-1}{0.06}\right]$

n	***FV***
1	5,000
2	10,300
3	15,918
4	21,873.08
5	28,185.4648

91. The mistake is that $6! \neq 3!2!$, but rather $6! = 6 \cdot 5 \cdot 4 \cdot 3 \cdot 2 \cdot 1$.	**92.** The mistake is that $(2n-2)! \neq (2n-2)(2n-4)(2n-6)\cdot \cdots \cdot (2)$ The terms should be consecutive and decrease by 1 (not 2). Therefore, it should be $(2n-2)! = (2n-2)(2n-3)(2n-4)\cdot \cdots \cdot (1)$. The same error is made when computing $(2n+2)!$.
93. $(-1)^{n+1} = \begin{cases} 1, & n = 1,3,5,\ldots \\ -1, & n = 2,4,6,\ldots \end{cases}$ So, the terms should all be the opposite sign.	**94.** Same error as 93. **95.** True **96.** True
97. False. Let $a_k = 1 = b_k$ and $n = 5$. Then, observe that $\sum_{k=1}^{n} a_k b_k = \sum_{k=1}^{5} 1 = 1 \cdot 5 = 5$ whereas $\left(\sum_{k=1}^{n} a_k\right) \cdot \left(\sum_{k=1}^{n} b_k\right) = \left(\sum_{k=1}^{5} 1\right) \cdot \left(\sum_{k=1}^{5} 1\right) = 25.$	**98.** False. Let $a = 3, b = 2$. Then, observe that $(a!)(b!) = (3!)(2!) = (3 \cdot 2 \cdot 1)(2 \cdot 1) = 12$ whereas $(ab)! = (2 \cdot 3)! = 6! = 720.$
99. False. Not necessarily. Suppose that $a_1 = 1,\ a_2 = 2$, and $a_k = 0$, for $k \geq 3$. Then, $\sum_{k=1}^{n} a_k = 3$.	**100.** True. If $m = n$, then $m! = n!$, not $m! < n!$. If $m > n$, then $m = n + p$, for some positive integer p. Thus, $m! = (n+p)! = (n+p)\cdots(n+1)n! > n!$ So, it must be the case that $m < n$.
101. $a_1 = C$ $a_2 = (C) + D$ $a_3 = (C+D) + D = C + 2D$ $a_4 = (C+2D) + D = C + 3D$	**102.** $a_1 = C$ $a_2 = D(C)$ $a_3 = D(DC) = D^2C$ $a_4 = D(D^2C) = D^3C$

103.

$$F_1 = \frac{\left(1+\sqrt{5}\right)-\left(1-\sqrt{5}\right)}{2\sqrt{5}} = \frac{2\sqrt{5}}{2\sqrt{5}} = 1$$

$$F_2 = \frac{\left(1+\sqrt{5}\right)^2-\left(1-\sqrt{5}\right)^2}{2^2\sqrt{5}}$$

$$= \frac{1+2\sqrt{5}+5-1+2\sqrt{5}-5}{4\sqrt{5}} = 1$$

104.

n	$a_n = \sqrt{a_{n-1}},\ n \geq 2$
1	**7**
2	$\sqrt{7} = 7^{1/2}$
3	$\sqrt{\sqrt{7}} = 7^{1/4}$
4	$\sqrt{\sqrt{\sqrt{7}}} = 7^{1/8}$
	$\sqrt{\sqrt{\sqrt{\sqrt{7}}}} = 7^{1/16}$

In general, $a_n = 7^{\frac{1}{2^{n-1}}},\ n \geq 2$.

105. Using a calculator yields the following table of values:

n	$\left(1+\frac{1}{n}\right)^n$
100	$2.704813 \approx 2.705$
1000	$2.716923 \approx 2.717$
10000	$2.718146 \approx 2.718$
100000	2.718268
1000000	2.718280
10000000	2.718281693
↓	↓
∞	e

106.

n	a_n	a_{n+1}	$\frac{a_{n+1}}{a_n}$
1	1	1	1
2	1	2	2
3	2	3	1.5
4	3	5	$1.\overline{66}$
5	5	8	1.6
6	8	13	1.625
7	13	20	1.53846
8	20	33	1.65
9	33	53	1.60606

10	53	86	1.62264
11	86	139	1.616279
12	139	228	1.618705
13	225	364	$1.6177\overline{7}$
14	364	589	1.618131
15	589	953	1.617996
16	953	1542	1.618048
17	1542	2495	1.618028
18	2498	4037	1.618036
19	4037		

107. The calculator gives $\frac{109}{15}$, which agrees with Exercise 61.

108. The calculator gives $\frac{11}{30}$, which agrees with Exercise 62.

109. Observe that $a_n = 4 - \frac{20}{n+5}$. Then,

$$a_{n+1} = 4 - \frac{20}{(n+1)+5} = \underbrace{4 - \frac{20}{n+6} > 4 - \frac{20}{n+5}}_{\text{since subtracting a smaller fraction.}} = a_n .$$

So, the sequence is monotonic; in fact, increasing.

110. Observe that the terms of this sequence repeat every 8 terms because sine has period 2π. Thus, the sequence is not monotonic.

111. The terms oscillate toward zero because of the term $(-1)^n$ in the numerator. Hence, the sequence is not monotonic.

112. Observe that long division enables us to express a_n in the equivalent form

$$a_n = \frac{3}{5} - \frac{3/5}{5n^2+1}.$$

Since $\frac{3/5}{5n^2+1}$ gets smaller as n gets larger, we are subtracting off less each time. So, the sequence is monotone nondecreasing.

Section 10.2 Solutions---

1. Yes, $d=3$.	**2.** Yes, $d=-3$.
3. No.	**4.** No.
5. Yes, $d=-0.03$.	**6.** Yes, $d=0.5$.
7. Yes, $d=2/3$.	**8.** Yes, $d=1/3$.
9. No.	**10.** No.
11. First four terms: $3,1,-1,-3$ So, yes and $d=-2$.	**12.** First four terms: $-7,-4,-1,2$ So, yes and $d=3$.
13. First four terms: $1,4,9,16$ So, no.	**14.** First four terms are: $1,\underbrace{\frac{4}{2!}}_{\frac{4}{2}=2},\underbrace{\frac{9}{3!}}_{\frac{9}{3\cdot 2}=\frac{3}{2}},\underbrace{\frac{16}{4!}}_{\frac{16}{4\cdot 3\cdot 2}=\frac{2}{3}}$ which is equivalent to $1,2,\frac{3}{2},\frac{2}{3}$. So, no.
15. First four terms: $2,7,12,17$ So, yes and $d=5$.	**16.** First four terms: $1,-3,-7,-11$So, yes and $d=-4$.
17. First four terms: $0,10,20,30$ So, yes and $d=10$.	**18.** First four terms: $4,12,20,28$ So, yes and $d=8$.
19. First four terms: $-1,2,-3,4$ So, no.	**20.** First four terms: $2,-4,6,-8$ So, no.
21. $a_n=11+(n-1)5=5n+6$	**22.** $a_n=5+(n-1)11=11n-6$
23. $a_n=-4+(n-1)(2)=-6+2n$	**24.** $a_n=2+(n-1)(-4)=-4n+6$
25. $a_n=0+(n-1)\frac{2}{3}=\frac{2}{3}n-\frac{2}{3}$	**26.** $a_n=-1+(n-1)\left(\frac{-3}{4}\right)=\frac{-3}{4}n-\frac{1}{4}$
27. $a_n=0+(n-1)e=en-e$	**28.** $a_n=1.1+(n-1)(-0.3)=-0.3n+1.4$
29. Here, $a_1=7,\ d=13$. So, $a_n=7+(n-1)13\ =13n-6$. Thus, $a_{10}=130-6=124$.	**30.** Here, $a_1=7, d=-6$. So, $a_n=7\ +(n-1)(-6)=-6n+13$ Thus, $a_{19}=-6(19)+13=-101$.
31. Here, $a_1=9,\ d=-7$. So, $a_n=9+(n-1)(-7)\ =-7n+16$ Thus, $a_{100}=-700+16=-684$.	**32.** Here, $a_1=13,\ d=6$. So, $a_n=13+(n-1)(6)\ =6n+7$ Thus, $a_{90}=6(90)+7=547$.

33. $a_5=44$, $a_{17}=152$

Find $d: a_{17}=a_5+12d$
$152=44+12d$
$9=d$

Find $a_1: a_5=a_1+4(9)$
$44=a_1+36$
$8=a_1$

So, $a_n=8+(n-1)9$
$=9n-1$

34. $a_9=-19$, $a_{21}=-55$. Find d: $a_{21}=a_9+12d$; $-55=-19+12d$; $-36=12d$; $-3=d$. Find a_1: $a_9=a_1+8(-3)$; $-19=a_1-24$; $5=a_1$. So, $a_n=5+(n-1)(-3)$ $=-3n+8$	
35. $a_7=-1$, $a_{17}=-41$. Find d: $a_{17}=a_7+10d$; $-41=-1+10d$; $-40=10d$; $-4=d$. Find a_1: $a_7=a_1+6(-4)$; $-1=a_1-24$; $23=a_1$. So, $a_n=23+(n-1)(-4)$ $=-4n+27$	
36. $a_8=47$, $a_{21}=112$. Find d: $a_{21}=a_8+13d$; $112=47+13d$; $65=13d$; $d=5$. Find a_1: $a_8=a_1+7(5)$; $47=a_1+35$; $12=a_1$. So, $a_n=12+(n-1)(5)$ $=5n+7$	
37. $a_4=3$, $a_{22}=15$. Find d: $a_{22}=a_4+18d$; $15=3+18d$; $12=18d$; $\frac{2}{3}=d$. Find a_1: $a_4=a_1+3\left(\frac{2}{3}\right)$; $3=a_1+2$; $1=a_1$. So, $a_n=1+(n-1)\frac{2}{3}$ $=\frac{2}{3}n+\frac{1}{3}$	
38. $a_{11}=-3$, $a_{31}=-13$. Find d: $a_{31}=a_1+20d$; $-13=-3+20d$; $-10=20d$; $d=-\frac{1}{2}$. Find a_1: $a_{11}=a_1+10\left(-\frac{1}{2}\right)$; $-3=a_1-5$; $2=a_1$. So, $a_n=2+(n-1)\left(\frac{-1}{2}\right)$ $=-\frac{1}{2}n+\frac{5}{2}$	
39. $\sum_{k=1}^{23}2k=2\sum_{k=1}^{23}k=\frac{23}{2}[2+46]=552$	**40.** $\sum_{k=0}^{20}5k=5\sum_{k=1}^{20}k=\frac{20}{2}[5+100]=1050$ ↑ Note that the 1st term when $k=0$ is 0.
41. $\sum_{n=1}^{30}(-2n+5)=-2\sum_{n=1}^{30}n+\sum_{n=1}^{30}5$ $=\frac{30}{2}[3-55]=-930+150=-780$	**42.** $\sum_{n=0}^{17}(3n-10)=-10+\sum_{n=1}^{17}(3n-10)$ $=-10+\frac{17}{2}[-7+41]$ $=-7+3\left(\frac{17(18)}{2}\right)-10(17)=279$

43. $\sum_{j=3}^{14} 0.5j = \left[\sum_{j=1}^{14} 0.5j - \sum_{j=1}^{2} 0.5j\right]$

$= \frac{14}{2}[0.5+7] - \frac{3}{2} = 0.5[102] = 51$

44. $\sum_{j=1}^{33} \frac{j}{4} = \frac{33}{2}\left[\frac{1}{4}+\frac{33}{4}\right] = 140.25$

45. $\underbrace{2+7+12+\ldots+62}_{\text{arithmetic}}$. We need to find n such that $a_n = 62$. To this end, observe that $a_1 = 2$ and $d = 5$. Thus, since the sequence is arithmetic, $a_n = 2+(n-1)5 = 5n-3$. So, we need to solve $a_n = 5n-3 = 62$: $5n = 65$, so that $n = 13$. Therefore, this sum equals $\frac{13}{2}[2+62] = 416$.

46. $\underbrace{1-3-7-\ldots-75}_{\text{arithmetic}}$. We need to find n such that $a_n = -75$. To this end, observe that $a_1 = 1$ and $d = -4$. Thus, since the sequence is arithmetic, $a_n = 1+(n-1)(-4) = -4n+5$. So, we need to solve $a_n = -4n+5 = -75$: $-4n = -80$, so that $n = 20$. Therefore, this sum equals $\frac{20}{2}[1-75] = -740$.

47. $\underbrace{4+7+10+\ldots+151}_{\text{arithmetic}}$. We need to find n such that $a_n = 151$. To this end, observe that $a_1 = 4$ and $d = 3$. Thus, since the sequence is arithmetic, $a_n = 4+(n-1)3 = 3n+1$. So, we need to solve $a_n = 3n+1 = 151$: $3n = 150$, so that $n = 50$. Therefore, this sum equals $= \frac{50}{2}(4+151) = 3875$.

48. $\underbrace{2+0-2-\ldots-56}_{\text{arithmetic}}$. We need to find n such that $a_n = -56$. To this end, observe that $a_1 = 2$ and $d = -2$. Thus, since the sequence is arithmetic,

$$a_n = 2+(n-1)(-2) = -2n+4.$$

So, we need to solve $a_n = -2n+4 = -56$: $-2n = -60$, so that $n = 30$. Therefore, this sum equals $\frac{30}{2}(2-56) = -810$.

49. $\underbrace{\frac{1}{6}-\frac{1}{6}-\frac{1}{2}-\ldots-\frac{13}{2}}_{\text{arithmetic}}$. We need to find n such that $a_n=-\frac{13}{2}$. To this end, observe that $a_1=\frac{1}{6}$ and $d=-\frac{1}{3}$. Thus, since the sequence is arithmetic,

$a_n=\frac{1}{6}+(n-1)\left(\frac{-1}{3}\right)=-\frac{1}{3}n+\frac{1}{2}$. So, we need to solve $a_n=-\frac{1}{3}n+\frac{1}{2}=-\frac{13}{2}$:

$\frac{-1}{3}n=-7$, so that $n=21$. Therefore, this sum equals $\frac{21}{2}\left[\frac{1}{6}-\frac{13}{2}\right]=\frac{21}{2}\left(\frac{1-39}{6}\right)=-\frac{133}{2}$.

50. $\underbrace{\frac{11}{12}+\frac{7}{6}+\frac{17}{12}+\ldots+\frac{14}{3}}_{\text{arithmetic}}$. We need to find n such that $a_n=\frac{14}{3}$. To this end, observe that $a_1=\frac{11}{12}$ and $d=\frac{7}{6}-\frac{11}{12}=\frac{3}{12}=\frac{1}{4}$. Thus, since the sequence is arithmetic,

$$a_n=\frac{11}{12}+(n-1)\left(\frac{1}{4}\right)=\frac{1}{4}n+\frac{2}{3}.$$

So, we need to solve $a_n=\frac{1}{4}n+\frac{2}{3}=\frac{14}{3}$: $\frac{1}{4}n=\frac{12}{3}=4$, so that $n=16$.

Therefore, this sum equals $=\frac{16}{2}\left[\frac{11}{12}+\frac{14}{3}\right]=8\left(\frac{111-56}{12}\right)=\frac{134}{3}$.

51. Here, $d=4$ and $a_1=1$. Hence, $a_n=1+(n-1)(4)=4n-3$. So, $S_n=\frac{n}{2}(a_1+a_n)=\frac{n}{2}(1+4n-3)=n(2n-1)$. As such, $S_{18}=18(35)=630$.

52. Here, $d=3$ and $a_1=2$. Hence, $a_n=2+(n-1)(3)=3n-1$. So, $S_n=\frac{n}{2}(a_1+a_n)=\frac{n}{2}(2+3n-1)=\frac{n}{2}(3n+1)$. As such, $S_{21}=\frac{21}{2}(64)=672$.

53. Here, $d=-\frac{1}{2}$ and $a_1=1$. Hence, $a_n=1+(n-1)\left(-\frac{1}{2}\right)=-\frac{1}{2}n+\frac{3}{2}$. So, $S_n=\frac{n}{2}(a_1+a_n)=\frac{n}{2}\left(1-\frac{1}{2}n+\frac{3}{2}\right)=\frac{n}{2}\left(-\frac{1}{2}n+\frac{5}{2}\right)$. As such, $S_{43}=\frac{43}{4}(5-43)=-\frac{817}{2}$.

54. Here, $d=-\frac{3}{2}$ and $a_1=3$. Hence, $a_n=3+(n-1)\left(-\frac{3}{2}\right)=-\frac{3}{2}n+\frac{9}{2}$. So, $S_n=\frac{n}{2}(a_1+a_n)=\frac{n}{2}\left(3-\frac{3}{2}n+\frac{9}{2}\right)=\frac{n}{2}\left(-\frac{3}{2}n+\frac{15}{2}\right)$. As such, $S_{37}=\frac{37}{4}(15-3(37))=-888$.

55. Here, $d=10$ and $a_1=-9$. Hence, $a_n=-9+(n-1)(10)=10n-19$. So, $S_n=\frac{n}{2}(a_1+a_n)=\frac{n}{2}(-9+10n-19)=n(5n-14)$. As such, $S_{18}=18(5(18)-14)=1368$.

56. Here, $d=10$ and $a_1=-2$. Hence, $a_n=-2+(n-1)(10)=10n-12$. So, $S_n=\frac{n}{2}(a_1+a_n)=\frac{n}{2}(-2+10n-12)=\frac{n}{2}(10n-14)$. As such, $S_{21}=21(5(21)-7)=2058$.

57. Colin

$a_1 = 28,000$

$d = 1500$

So $a_n = 28,000 + (n-1)(1500)$

$= 26,500 + 1500n$

Thus, $a_{10} = 41,500$.

After 10 years, Colin will have accumulated $\frac{10}{2}(28,000 + 41,500)$ = \$347,500.

Camden

$a_1 = 25,000$

$d = 2000$

So, $a_n = 25,000 + (n-1)2000$

$= 23,000 + 2000n$

So, $a_{10} = 43,000$

After 10 years, Camden will have accumulated $\frac{10}{2}(25,000 + 43,000)$ = \$340,000.

58. Jasmine

$a_1 = 80,000$

$d = 2000$

So, $a_n = 80,000 + (n-1)(2000)$

$= 78,000 + 2000n$

Thus, $a_{15} = 108,000$.

After 15 years, Jasmine will have accumulated $\frac{15}{2}(80,000 + 108,000)$ = \$1,410,000.

Megan

$a_1 = 90,000$

$d = 5000$

So, $a_n = 90,000 + (n-1)5000$

$= 85,000 + 5000n$

Thus, $a_{15} = 160,000$.

After 15 years, Megan will have accumulated $\frac{15}{2}(90,000 + 160,000)$ = \$1,875,000.

59. We are given that $a_1 = 22$. We seek $\sum_{i=1}^{25} a_i$. Observe that $d = 1$. So, $a_n = 22 + (n-1)(1) = 21 + n$. Thus, $a_{25} = 21 + 25 = 46$. As such, using the formula $S_n = \frac{n}{2}(a_1 + a_n)$, we conclude that $\sum_{i=1}^{25} a_i = = \frac{25}{2}[22 + 46] = 850$.

60. We are given that $a_1 = 1$. We seek $\sum_{i=1}^{20} a_i$. Observe that $d = 1$. So, $a_n = 1 + (n-1)(1) = n$. Thus, $a_{20} = 20$. As such, using the formula $S_n = \frac{n}{2}(a_1 + a_n)$, we conclude that $\sum_{i=1}^{20} a_i = = \frac{20}{2}[1 + 20] = 210$. So, there are 210 tulips in each delta.

61. We are given that $a_1=1$ and $n=56$. We need to find a_{56} and d. Well, note that $\sum_{i=1}^{56} a_i = 30,856 = \frac{56}{2}(a_1+a_{56}) = \underbrace{\frac{56}{2}}_{28}(1+a_{56})$. Hence,

$$30,856 = 28(1+a_{56}), \text{ so that } 1101 = \frac{30,828}{28} = a_{56}.$$

Now, since $a_n = a_1 + (n-1)d$, we can substitute $n=56,\ a_1=1$ in to find d:

$$1101 = 1+(56-1)d, \text{ so that } 1100 = 55d \text{ and hence, } d = 20.$$

So, there are 1101 glasses on the bottom row and each row had 20 fewer glasses than the one before.

62. From the information given in the problem, we infer that $a_1=1$, $a_{25}=25$, and $d=1$.

Since the sequence is arithmetic, the sum must be $\frac{25}{2}[1+25] = 325$ logs in the pile.

63. We are given that

$a_1 = 16, d = 32, n = 10$. Hence,

$$a_n = 16 + (n-1)(32) = -16 + 32n.$$

So, $a_{10} = -16 + 320 = 304$. Now, observe that $\sum_{i=1}^{10} a_i = \frac{10}{2}[16+304]$

$= 1600$ feet in 10 seconds

64. We are given that

$a_i = 4.9$, $d = 9.8$, $n = 10$. Hence,

$a_n = 4.9 + (n-1)(9.8) = -4.9 + 9.8n$

So, $a_{10} = 93.1$. Now, observe that

$\sum_{i=1}^{10} a_i = \frac{10}{2}[4.9 + 93.1]$

$= 490$ meters in 10 seconds

65. The sum is 20 + 19 + 18 + . . . + 1.

This is arithmetic with $a_1 = 20,\ d = -1$.

So, $a_n = 20 + (n-1)(-1) = 21 - n$. Thus,

$S_n = \frac{n}{2}(20 + 21 - n) = \frac{n}{2}(41 - n)$. So,

$S_{20} = 10(21) = 210$.

There are 210 oranges in the display.

66. The sum is 45000 + 46500 + … + a_n. This is arithmetic with

$$a_1 = 45000,\ d = 1500.$$

So,

$a_n = 45000 + (n-1)(1500) = 1500n + 43500$

a. $a_{35} = 96,000$ -- salary in year 35

b. Total earnings in 35 years = S_{35}

$= \frac{35}{2}(45000 + 1500(35) - 43500) = 2,467,500$

67. $a_1 + a_2 + a_3 + 26 + 27 + 28 + ... + a_n$

Here, $d = 1$, so that $a_1 = 23$.

a. There are 23 seats in the first row.

b. $a_n = 23 + (n-1)(1) = n + 22$.

So, $S_n = \frac{n}{2}(23 + n + 22)$, and hence,

$S_{30} = 1125$ seats in the theater.

68. Here $d = \frac{2}{e}$ and $a_1 = \frac{1}{e}$. Hence,

$a_n = \frac{1}{e} + (n-1)(2) = -\frac{1}{e} + \frac{2}{e}n$. So,

$S_n = \frac{12}{2}\left(\frac{1}{e} + \frac{23}{e}\right) = \frac{144}{e}$. Since there are 12 terms all told, the total is $S_{12} = 132 + \frac{12}{e}$.

69. $a_n = a_1 + (n-1)d$, not $a_1 + nd$

70. Here, $d = -2$ since consecutive terms decrease by 2

71. There are 11 terms, not 10. So, $n = 11$, and thus, $S_{11} = \frac{11}{2}(1+21) = 121$.

72. $S_n = \frac{n}{2}(a_n + a_1)$, not $S_n = \frac{n}{2}(a_n - a_1)$

73. False. In a series you are adding terms, while in a sequence you are not.

74. False. For instance, $\sum_{i=1}^{\infty}(2i+1)$ doesn't exist. All <u>FINITE</u> arithmetic series can be computed, however.

75. True, since d must be constant. If a sequence alternates, the difference between consecutive terms would need to change sign.

76. False. If the terms are decreasing, d would be negative.

77.

$$a + (a+b) + \ldots + (a+nb) = \sum_{k=1}^{n+1}(a+(k-1)b) = \frac{n+1}{2}(a+a+nb) = \frac{(n+1)(2a+nb)}{2}$$

78. First, observe that $\sum_{k=-29}^{30} \ln e^k = \sum_{k=-29}^{30} k = -29 - 28 - \ldots - 1 + 0 + 1 + \ldots + 30$.

Now, we simplify as follows:

$$\begin{array}{r} = \quad -\cancel{29} - \cancel{28} - \ldots - \cancel{2} - \cancel{1}\ 0 \\ 30 + \cancel{29} + \cancel{28} + \ldots + \cancel{2} + \cancel{1} \\ \hline 30 \end{array}$$

So, the sum is 30.

79. As n gets larger, $\frac{1}{n^2}$ goes to zero. So, $v = R\left(\frac{1}{k^2} - \frac{1}{n^2}\right)$ gets closer to $\frac{R}{k^2}$, which in this case is 27,419.5.

80. Here, $a_1 = -4,\ d = 6$. So, $a_n = -4 + (n-1)(6) = 6n - 10$. Thus, $S_n = \frac{n}{2}(-4 + 6n - 10) = n(3n-7)$.

Solve $n(3n-7) = 570$:

$$3n^2 - 7n - 570 = 0$$

$$n = \frac{7 \pm \sqrt{49 + 4(3)(570)}}{6} = 15 \text{ or } \cancel{-\tfrac{38}{3}}$$

81. Compute $\sum_{i=1}^{100} i$ - you should get 5050.

82. Compute $\sum_{i=1}^{50} 2i$ - you should get 2550.

83. Compute $\sum_{i=1}^{50}(2i-1)$ - you should get 2500.

84. Same answer.

85. 18,850	**86.** 12,320
87. $\sum_{i=1}^{100}(2i)(0.1)=0.2\sum_{i=1}^{100}i=0.2\left(\frac{100(101)}{2}\right)=1010$	**88.** $\sum_{i=1}^{50}(4i-2)(0.01)=0.04\sum_{i=1}^{50}i-\sum_{i=1}^{50}0.02$ $=0.04\left(\frac{50(51)}{2}\right)-0.02(50)$ $=50$
89. $\sum_{i=1}^{43}(6+i)(0.001)=\sum_{i=1}^{43}0.006+0.001\sum_{i=1}^{43}i$ $=0.258+0.001\left(\frac{43(44)}{2}\right)$ $=1.204$	**90.** $\sum_{i=1}^{85}(6-7i)(0.2)=\sum_{i=1}^{85}1.2-1.4\sum_{i=1}^{85}i$ $=1.2(85)-1.4\left(\frac{85(86)}{2}\right)$ $=-5015$

Section 10.3 Solutions---

1. Yes, $r=3$	**2.** Yes, $r=2$
3. No, $\frac{9}{4}\neq\frac{16}{9}$ for instance	**4.** No, $\frac{\frac{1}{9}}{\frac{1}{4}}\neq\frac{\frac{1}{16}}{\frac{1}{9}}$ for instance
5. Yes, $r=\frac{1}{2}$	**6.** Yes, $r=-\frac{1}{2}$
7. Yes, $r=1.7$	**8.** Yes, $r=2.2$
9. $6,18,54,162,486$	**10.** $17,34,68,136,272$
11. $1,-4,16,-64,256$	**12.** $-3,6,-12,24,-48$
13. $10{,}000,\ 10{,}600,\ 11{,}236,$ $11{,}910.16,\ 12{,}624.77$	**14.** 10000, 8000, 6400, 5120, 4096
15. $\frac{2}{3},\frac{1}{3},\frac{1}{6},\frac{1}{12},\frac{1}{24}$	**16.** $\frac{1}{10},-\frac{1}{50},\frac{1}{250},-\frac{1}{1250},\frac{1}{6250}$
17. $a_n=5(2)^{n-1}$	**18.** $a_n=12(3)^{n-1}$
19. $a_n=1(-3)^{n-1}$	**20.** $a_n=-4(-2)^{n-1}$
21. $a_n=1000(1.07)^{n-1}$	**22.** $a_n=1000(0.5)^{n-1}$

23. $a_n = \frac{16}{3}\left(-\frac{1}{4}\right)^{n-1}$	**24.** $a_n = \frac{1}{200}(5)^{n-1}$
25. Since $a_1 = -2,\ r = -2$, we have $$a_n = -2(-2)^{n-1}.$$ In particular, $a_7 = -2(-2)^{7-1} = -128$.	**26.** Since $a_1 = 1,\ r = -5$, we have $$a_n = (-5)^{n-1}$$ In particular, $a_{10} = (-5)^{10-1} = -1{,}953{,}125$.
27. Since $a_1 = \frac{1}{3},\ r = 2$, we have $$a_n = \frac{1}{3}(2)^{n-1}$$ In particular, $a_{13} = \frac{1}{3}(2)^{13-1} = \frac{4096}{3}$.	**28.** Since $a_1 = 100,\ r = \frac{1}{5}$, we have $$a_n = 100\left(\frac{1}{5}\right)^{n-1}$$ In particular, $a_9 = 100\left(\frac{1}{5}\right)^{9-1} = 2.56\times10^{-4}$.
29. Since $a_1 = 1000,\ r = \frac{1}{20}$, we have $$a_n = 1000\left(\frac{1}{20}\right)^{n-1}$$ In particular, $$a_{15} = 1000\left(\frac{1}{20}\right)^{15-1} \approx 6.10\times10^{-16}.$$	**30.** Since $a_1 = 1000,\ r = -\frac{4}{5}$, we have $$a_n = 1000\left(-\frac{4}{5}\right)^{n-1}$$ In particular, $$a_8 = 1000\left(-\frac{4}{5}\right)^{8-1} = -209.7152.$$
31. Use the formula $S_n = a_1\frac{(1-r^n)}{1-r}$. Since $a_1 = \frac{1}{3},\ n = 13,\ r = 2$, the given sum is $S_{13} = \frac{1}{3}\left(\frac{1-2^{13}}{1-2}\right) = \frac{8191}{3}$.	**32.** Use the formula $S_n = a_1\frac{(1-r^n)}{1-r}$. Since $a_1 = 1,\ n = 11,\ r = \frac{1}{3}$, the given sum is $S_{11} = 1\left(\frac{1-\left(\frac{1}{3}\right)^{11}}{1-\frac{1}{3}}\right) \approx 1.5$.

33. Use the formula $S_n = a_1 \frac{(1-r^n)}{1-r}$. Since $a_1 = 2,\ n = 10,\ r = 3$, the given sum is $$S_{10} = 2\left(\frac{1-3^{10}}{1-3}\right) = 59{,}048.$$	**34.** Use the formula $S_n = a_1 \frac{(1-r^n)}{1-r}$. Since $a_1 = 1,\ r = 4,\ n = 10$, the given sum is $S_{10} = 1\left(\frac{1-4^{10}}{1-4}\right) = 349{,}525$.
35. Use the formula $S_n = a_1 \frac{(1-r^n)}{1-r}$. Since $a_1 = 2,\ r = 0.1,\ n = 11$, the given sum is $$S_{11} = 2\left(\frac{1-(0.1)^{10}}{1-0.1}\right) = 2.\overline{2}.$$ (Note: The sum starts with $n = 0$, so there are 11 terms.)	**36.** Use the formula $S_n = a_1 \frac{(1-r^n)}{1-r}$. Since $a_1 = 3,\ r = 0.2,\ n = 12$, the given sum is $$S_{12} = 3\left(\frac{1-(0.2)^{12}}{1-0.2}\right) \approx 3.75$$ (Note: The sum starts with $n = 0$, so there are 12 terms.)
37. Use the formula $S_n = a_1 \frac{(1-r^n)}{1-r}$. Since $a_1 = 2,\ r = 3,\ n = 8$, the given sum is $$S_8 = 2\left(\frac{1-3^8}{1-3}\right) = 6560.$$	**38.** Use the formula $S_n = a_1 \frac{(1-r^n)}{1-r}$. Since $a_1 = \frac{2}{3},\ r = 5,\ n = 9$, the given sum is $S_9 = \frac{2}{3}\left(\frac{1-5^9}{1-5}\right) \approx 325{,}520.6667$.
39. Use the formula $S_n = a_1 \frac{(1-r^n)}{1-r}$. Since $a_1 = 1,\ r = 2,\ n = 14$, the given sum is $S_{14} = 1\left(\frac{1-2^{14}}{1-2}\right) = 16{,}383$.	**40.** Use the formula $S_n = a_1 \frac{(1-r^n)}{1-r}$. Since $a_1 = 1,\ r = \frac{1}{2},\ n = 14$, the given sum is $S_{14} = 1\left(\frac{1-\left(\frac{1}{2}\right)^{14}}{1-\frac{1}{2}}\right) \approx 2.0$.

41. Use the formula $S_\infty = \dfrac{a_1}{1-r}$.

Since $a_1 = 1$, $r = \dfrac{1}{2}$, the given sum is

$$S_\infty = \frac{1}{1-\frac{1}{2}} = 2.$$

42. Use the formula $S_\infty = \dfrac{a_1}{1-r}$.

Since $a_1 = \dfrac{1}{3}$, $r = \dfrac{1}{3}$, the given sum is

$$S_\infty = \frac{\frac{1}{3}}{1-\frac{1}{3}} = \frac{1}{2}.$$

43. Use the formula $S_\infty = \dfrac{a_1}{1-r}$.

Since $a_1 = -\dfrac{1}{3}$, $r = -\dfrac{1}{3}$, the given sum is

$$S_\infty = \frac{-\frac{1}{3}}{1-\left(-\frac{1}{3}\right)} = -\frac{1}{4}.$$

44. Use the formula $S_\infty = \dfrac{a_1}{1-r}$.

Since $a_1 = 1$, $r = -\dfrac{1}{2}$, the given sum is

$$S_\infty = \frac{1}{1-\left(-\frac{1}{2}\right)} = \frac{2}{3}.$$

45. Not possible - diverges.

46. Not possible - diverges.

47. Use the formula $S_\infty = \dfrac{a_1}{1-r}$.

Since $a_1 = -9$, $r = \dfrac{1}{3}$, the given sum is

$$S_\infty = \frac{-9}{1-\frac{1}{3}} = -\frac{27}{2}.$$

48. Use the formula $S_\infty = \dfrac{a_1}{1-r}$.

Since $a_1 = -8$, $r = -\dfrac{1}{2}$, the given sum is

$$S_\infty = \frac{-8}{1-\left(-\frac{1}{2}\right)} = -\frac{16}{3}.$$

49. Use the formula $S_\infty = \dfrac{a_1}{1-r}$.

Since $a_1 = 10{,}000$, $r = 0.05$, the given sum is $S_\infty = \dfrac{10{,}000}{1-0.05} \approx 10{,}526$.

50. Use the formula $S_\infty = \dfrac{a_1}{1-r}$.

Since $a_1 = 200$, $r = 0.04$, the given sum is $S_\infty = \dfrac{200}{1-0.04} = \dfrac{625}{3}$.

51. The sum is $\dfrac{0.4}{1-0.4} = \dfrac{2}{3}$.

52. The sum can be written as $\sum_{n=1}^{\infty} 3\left(\frac{1}{10}\right)^n$, which equals $\dfrac{3\left(\frac{1}{10}\right)}{1-\frac{1}{10}} = \dfrac{1}{3}$.

53. The sum is $\dfrac{(0.99)^0}{1-0.99} = 100$.

54. Diverges – infinite sum

55. Observe that $a_1 = 34,000,\ r = 1.025$. We need to find a_{12}. Observe that the *n*th term is given by $a_n = a_1 r^{n-1} = 34,000(1.025)^{n-1}$. Hence,

$$a_{12} = 34000(1.025)^{11} \approx 44,610.95$$

So, the salary after 12 years is approximately $44,610.95.

56. Observe that $a_1 = 22,000,\ r = 1.15$. We need to find a_{10}. Observe that the *n*th term is given by $a_n = 22,000(1.15)^{n-1}$. Hence,

$$a_{10} = 22,000(1.15)^9 \approx 77,393.28$$

So, the salary after 10 years is approximately $77,393.28.

57. Since $a_1 = 2000,\ r = 0.5$, we see that

$$a_n = \underbrace{2000(0.5)^n}_{\text{Value of laptop after } n \text{ years}}.$$

In particular, we have

$$a_4 = \underbrace{2000(0.5)^4 = 125}_{\text{Worth when graduating from college.}}$$

$$a_7 = \underbrace{2000(0.5)^7 = 16}_{\text{Worth when graduating from graduate school.}}$$

58.

BMW	Honda
$a_1 = 35,000$	$a_1 = 25,000$
$r = 0.8$	$r = 0.9$
$a_n = 35,000(0.8)^{n-1}$	$a_n = 25,000(0.9)^{n-1}$
$a_{10} = 35,000(0.8)^9$	$a_{10} = 25,000(0.9)^9$
$= 4697.62$	$= 9685.51$

So, the Honda will be worth (much) more in 10 years.

59. Since $a_1 = 100,\ r = 0.7$, we see that

$$a_n = 100(0.7)^n.$$

We need to find a_5 and the value of *n* such that $a_n = 0$. Indeed, we see that

$$a_5 = 100(0.7)^5 \approx 17 \text{ feet}.$$

And to answer the second part, we note that $a_n \neq 0$, for all values of *n*. As such, the object would never come to a rest, theoretically.

60. Since $a_1 = 200,\ r = 0.65$, we see that

$$a_n = 200(0.65)^{n-1}.$$

We need to find a_8 and the value of *n* such that $a_n = 0$. Indeed, we see that

$$a_8 = 200(0.65)^7 \approx 9.80 \text{ feet}.$$

And to answer the second part, we note that $a_n \neq 0$, for all values of *n*. As such, the object would never come to a rest, theoretically.

61. Since $a_1 = 36,000,\ r = \dfrac{37,800}{36,000} = 1.05$, we see that $a_n = 36,000(1.05)^{n-1}$. So, in particular, $a_{11} = 36,000(1.05)^{10} \approx 58,640$. As such, in the year 2010, there will be approximately 58,640 students.

62. Since $a_1 = 20000,\ r = 1.05$, we see that $a_n = 20,000(1.05)^{n-1}$. So, in particular, we have

$$a_{52} = 20,000(1.05)^{51} \approx 240,815.$$

So, there will be about 240,815 hits per week one year from now.

63. Since $a_1 = 1000,\ r = 0.90$, we know that $a_n = 1000(0.9)^{n-1}$. First, find the value of n_0 such that $a_{n_0} \le 1$. Since

$$1000(0.9)^{n-1} < 1 \Rightarrow (0.9)^{n-1} < 0.001$$
$$\Rightarrow n > \tfrac{\ln(0.001)}{\ln(0.9)} \approx 67$$

We see that after 67 days, he will pay less than \$1 per day. Next, the total amount of money he paid in January is given by:

$$\sum_{n=1}^{31} 1000(0.9)^{n-1} = 1000\frac{\left(1-0.9^{31}\right)}{1-0.9} \approx 9618$$

64. Since $a_1 = 0.01,\ r = 2$, we see that

$$a_n = 0.01(2)^{n-1}.$$

Hence, the total paid out in January is given by:

$$\sum_{n=1}^{31} 0.01(2)^{n-1} = 0.01\left(\frac{1-2^{31}}{1-2}\right)$$
$$\cong \$21{,}474{,}836.47$$

65. Use the formula $A = P\left(1+\frac{r}{n}\right)^{nt}$. Note that in the current problem, $r = 0.05,\ n = 12,$ so that the formula becomes:

$$A = P\left(1+\frac{0.05}{12}\right)^{12t} = P(1.0042)^{12t}$$

Let $t = \frac{n}{12}$, where n is the number of months of investment. Also, let $A_n = P(1.0042)^n$.

First deposit of \$100 gains interest for 36 months: $A_{36} = \$100(1.0042)^{36}$

Second deposit of \$100 gains interest for 35 months: $A_{35} = \$100(1.0042)^{35}$

⋮

Last deposit of \$100 gains interest for 1 month: $A_1 = \$100(1.0042)^1$

Now, sum the amount accrued for 36 deposits:

$$A_1 + \ldots + A_{36} = \sum_{n=1}^{36} 100(1.0042)^{n-1}$$

Note that here $a_1 = 100, r = 1.0042$. So,

$$\text{So, } S_{36} = 100\left(\frac{1-(1.0042)^{36}}{1-1.0042}\right)$$
$$\approx 3877.64$$

So, he saved \$3,877.64 in 3 years.

66. Use the formula $A = P\left(1+\frac{r}{n}\right)^{nt}$.

Note that in the current problem, $r = 0.04$, $n = 52$, so the formula becomes:

$$A = P\left(1+\frac{0.04}{52}\right)^{52t} = P(1.00077)^{52t}$$

Let $t = \frac{n}{52}$, where n is number of weeks of investment. Also, let $A_n = P(1.00077)^n$.

First deposit of \$50 gains interest for 52 weeks: $A_{52} = 50(1.00077)^{52}$

Second deposit of \$50 gains interest for 51 weeks: $A_{51} = 50(1.00077)^{51}$

⋮

Last deposit of \$50 gains interest for 1 week: $A_1 = 50(1.00077)^1$

Now, sum the amount accrued for 52 weeks:

$$A_1 + \ldots + A_{52} = \sum_{n=1}^{52} 50(1.00077)^{n-1}$$

Note that here $a_1 = 50,\ r = 1.00077$. So,

$$\text{So, } S_{52} = 50\left(\frac{1-(1.00077)^{52}}{1-1.00077}\right)$$
$$\approx 2651.71$$

So, he saved \$2,651.71 in one year.

67. Use the formula $A = P\left(1+\frac{r}{n}\right)^{nt}$. Note that in the current problem, $r = 0.06,\ n = 52,$ so that the formula becomes: $A = P\left(1+\frac{0.06}{52}\right)^{52t} = P(1.0012)^{52t}$

Let $t = \frac{n}{52}$, where n is the number of weeks of investment. Also, let $A_n = P(1.0012)^n$.

26 week investment:
First deposit of \$500 gains interest for 26 weeks: $A_{26} = 500(1.0012)^{26}$

Second deposit of \$500 gains interest for 25 weeks: $A_{25} = 500(1.0012)^{25}$

$\vdots$

Last deposit of \$500 gains interest for 1 week: $A_1 = 500(1.0012)^1$

Now, the amount accrued for 26 weeks:

$$A_1 + \ldots + A_{26} = \sum_{n=1}^{26} 500(1.0012)^{n-1}$$

Note that here $a_1 = 500,\ r = 1.0012$. So,

$$S_{26} = 500\left(\frac{1-(1.0012)^{26}}{1-1.0012}\right) \approx 13{,}196.88.$$

52 week investment:
First deposit of \$500 gains interest for 52 weeks: $A_{52} = 500(1.0012)^{52}$

Second deposit of \$500 gains interest for 51 weeks: $A_{51} = 500(1.0012)^{51}$

$\vdots$

Last deposit of \$500 gains interest for 1 week: $A_1 = 500(1.0012)^1$

Now, the amount accrued for 52 weeks:

$$A_1 + \ldots + A_{52} = \sum_{n=1}^{52} 500(1.0012)^{n-1}$$

Note that here $a_1 = 500,\ r = 1.0012$. So,

$$S_{52} = 500\left(\frac{1-(1.0012)^{52}}{1-1.0012}\right) \approx 26{,}811.75.$$

68. Use the formula $A = P\left(1+\frac{r}{n}\right)^{nt}$. Note that in the current problem, $r = 0.05,\ n = 12$, so that the formula becomes:

$$A = P\left(1+\frac{0.05}{12}\right)^{12t} = P(1.0042)^{12t}$$

Let $t = \frac{n}{12}$, where n is the number of months of investment. Also, let $A_n = P(1.0042)^n$.
First deposit of \$300 gains interest for 60 months: $A_{36} = \$300(1.0042)^{60}$
Second deposit of \$300 gains interest for 59 months: $A_{35} = \$300(1.0042)^{59}$

$\vdots$

Last deposit of \$300 gains interest for 1 month: $A_1 = \$300(1.0042)^1$
Now, sum the amount accrued for 36 deposits:

$$A_1 + \ldots + A_{60} = \sum_{n=1}^{60} 300(1.0042)^{n-1}$$

Note that here $a_1 = 300, r = 1.0042$. So,

$$\text{So, } S_{60} = 300\left(\frac{1-(1.0042)^{60}}{1-1.0042}\right)$$

$$\approx 20{,}422.66$$

So, he saved \$20,422.66 in 5 years.

69. Here, $a_1 = 195,000$ and $r = 1 + 0.065 = 1.065$. So, $a_n = 195,000(1.065)^{n-1}$, $n \geq 1$. So, its worth after 15 years is $a_{15} = \$501,509$.	**70.** Here, $a_1 = 9$ and $r = \frac{1}{3}$. Each time the ball bounces, it travels up and down by the same distance. So, the total distance traveled is: $9 + \sum_{n=1}^{\infty} 2 \cdot 9\left(\frac{1}{3}\right)^n = 9 + \frac{18\left(\frac{1}{3}\right)}{1-\frac{1}{3}} = 18$ ft.
71. The sum is $\frac{\frac{1}{2}}{1-\frac{1}{2}} = 1$.	**72.** The finite sum is $1 \cdot \frac{1-2^{30}}{1-2} = 2^{30} - 1 = 1,073,741,823$. This is clearly the better deal than a one-time check of \$10 million.
73. Should be $r = -\frac{1}{3}$	**74.** Should be $a_1 = 2$, so that you have $\sum_{k=1}^{n} 2 \cdot 2^{k-1}$.
75. Should use $r = -3$ all the way through the calculation. Also, $a_1 = -12$(not 4).	**76.** Formula for S_∞ only applies if $\lvert r \rvert < 1$.
77. False. $1, -\frac{1}{2}, \frac{1}{4}, -\frac{1}{8}$ is geometric with $a_1 = 1, r = -\frac{1}{2}$	**78.** False. Finite geometric series can always be computed, but infinite geometric series only exist if $\lvert r \rvert < 1$.
79. True. We could have either of the following: $1, \frac{1}{2}, \frac{1}{4}, \cdots$ (Here $r = \frac{1}{2}$.) $1, -\frac{1}{2}, \frac{1}{4}, -\frac{1}{8}, \cdots$ (Here $r = -\frac{1}{2}$.)	**80.** False. Common ratio must be <1 in absolute value in order for an infinite geometric series to exist.
81. $a + a \cdot b + \ldots + a \cdot b^n + \ldots = \sum_{n=1}^{\infty} ab^{n-1}$ This sum exists iff $\lvert b \rvert < 1$. In such case, the sum is $\frac{a}{1-b}$. (Here, $a_1 = a, r = b, S_\infty = \frac{a_1}{1-r}$)	**82.** Since $a_1 = 1$, $r = 2$, $n = 21$, and we can simplify the given series as: $\sum_{k=0}^{20} \log 10^{2^k} = \sum_{k=0}^{20} 2^k$ we have $S_{21} = 1\left(\frac{1-2^{21}}{1-2}\right) = 2,097,151$.

83. This decimal can be represented as

$$\sum_{n=1}^{\infty} 47\left(\tfrac{1}{100}\right)^n = \frac{\frac{47}{100}}{1-\frac{1}{100}} = \frac{47}{99}.$$

84. If $\dfrac{2}{1-x}$ is the sum of a geometric series, then $a_1 = 2$ and $r = x$. Hence, the series is $\sum_{n=1}^{\infty} 2x^{n-1}$.

a. First five terms:

$$2 + 2x + 2x^2 + 2x^3 + 2x^4$$

b. This series converges for $|x| < 1$.

85. Since $a_1 = 1,\ r = -2,\ n = 50$, we have

$$\sum_{k=1}^{50} 1(-2)^{k-1} = 1\left(\frac{1-(-2)^{50}}{1-(-2)}\right)$$

$$= -375{,}299{,}968{,}947{,}541$$

86. Yes, you should get:

$$a_1 = 1, r = \frac{1}{3} \text{ and } S_\infty = \frac{1}{1-\frac{1}{3}} = \frac{3}{2}.$$

87. We expect that $\sum_{n=0}^{\infty} x^n = \dfrac{1}{1-x}$ since as $n \to \infty$, the graph of $y_n = 1 + \ldots + x^n$ gets closer to the graph of $\dfrac{1}{1-x}$.

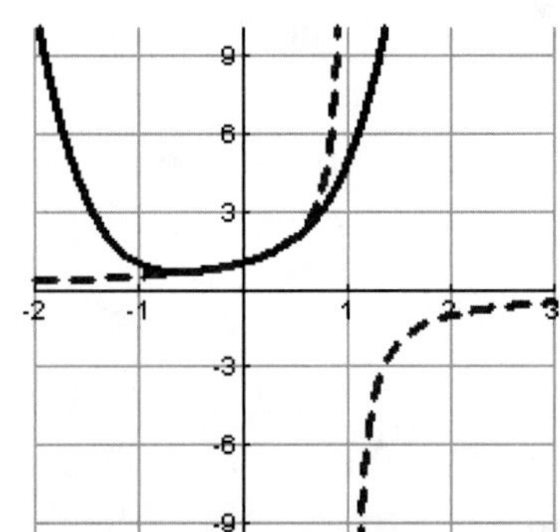

Notes on the graph:
Solid curve: $y_1 = 1 + x + x^2 + x^3 + x^4$

Dashed curve: $y_2 = \dfrac{1}{1-x}$, assuming $|x| < 1$.

88. The plot of the two functions is:

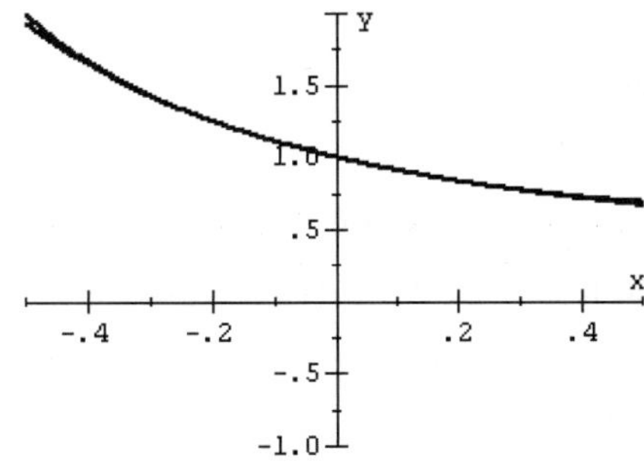

The series will sum to $\dfrac{1}{1+x}$ assuming $|x| < 1$.

89. The series converges since $r = 2/3 < 1$. The sum is $\sum_{n=0}^{\infty}\left(\dfrac{2}{3}\right)^n = \dfrac{1}{1-\frac{2}{3}} = 3$.

90. The series $\sum_{n=0}^{\infty}\left(\frac{5}{4}\right)^n$ diverges since $r = 5/4 > 1$.
91. The series converges since $r = ¼ < 1$. The sum is $\sum_{n=0}^{\infty}\frac{3}{8}\left(\frac{1}{4}\right)^n = \frac{\frac{3}{8}}{1-\frac{1}{4}} = \frac{1}{2}$.
92. The series converges since $\|r\| = 8/9 < 1$. The sum is $\sum_{n=0}^{\infty}\frac{\pi}{3}\left(-\frac{8}{9}\right)^n = \frac{\frac{\pi}{3}}{1-\left(-\frac{8}{9}\right)} = \frac{3\pi}{17}$

Section 10.4 Solutions ---

1. **Claim:** $n^2 \le n^3$, for all $n \ge 1$.

Proof.

Step 1: Show the statement is true for $n = 1$. $1^2 \le 1^3$ is true since $1^2 = 1^3 = 1$.

Step 2: Assume the statement is true for $n = k$: $k^2 \le k^3$

Show the statement is true for $n = k+1$: $(k+1)^2 \le (k+1)^3$

$$\begin{aligned}(k+1)^2 &= k^2 + 2k + 1\\ &\le k^3 + 2k + 1 \text{ (by assumption)}\\ &\le k^3 + 3k + 1 \text{ (since } 2k \le 3k, \text{ for } k > 0)\\ &\le k^3 + 3k^2 + 3k + 1 \text{ (since } 3k^2 > 0)\\ &= (k+1)^3\end{aligned}$$

This completes the proof. ■

2. **Claim:** If $0 < x < 1$, then $0 < x^n < 1$, for all $n \ge 1$.

Proof.

Step 1: Show the statement is true for $n = 1$.

The statement If $0 < x < 1$, then $0 < x < 1$ tautologically true.

Step 2: Assume the statement is true for $n = k$: If $0 < x < 1$, then $0 < x^k < 1$

Show the statement is true for $n = k+1$: If $0 < x < 1$, then $0 < x^{k+1} < 1$

Starting from $0 < x^k < 1$, multiply through the inequality by x (which is fine since $x > 0$. This yields the inequality $0 < \underbrace{x^k \cdot x}_{=x^{k+1}} < \underbrace{1 \cdot x}_{=x}$, which simplifies to $0 < x^{k+1} < x$ **(1)**. Since from Step 1 we know that $x < 1$, we can use this in **(1)** to further conclude that $0 < x^{k+1} < x < 1$, so that $0 < x^{k+1} < 1$, as desired.

This completes the proof. ■

3. Claim: $2n \le 2^n$, for all $n \ge 1$.

Proof.

Step 1: Show the statement is true for $n = 1$.

$$2(1) \le 2^1 \text{ is true since both terms equal 2.}$$

Step 2: Assume the statement is true for $n = k$: $2k \le 2^k$

Show the statement is true for $n = k+1$: $2(k+1) \le 2^{k+1}$

$$\begin{aligned} 2(k+1) = 2k+2 \ &\le 2^k + 2^1 \quad \text{(by assumption)} \\ &\le 2^k + 2^k \quad (\text{since } 2 \le 2^k, \text{ for } k > 0) \\ &= 2(2^k) = 2^{k+1} \end{aligned}$$

This completes the proof. ■

4. Claim: $5^n < 5^{n+1}$, for all $n \ge 1$.

Proof.

Step 1: Show the statement is true for $n = 1$.

$$5^1 < 5^{1+1} = 5^2 = 25 \text{ is clearly true.}$$

Step 2: Assume the statement is true for $n = k$: $5^k < 5^{k+1}$

Show the statement is true for $n = k+1$: $5^{k+1} < 5^{k+2}$

$$5^{k+1} = 5^k \cdot 5 \ < 5^{k+1} \cdot 5 \text{ (by assumption) } = 5^{k+2}$$

This completes the proof.

5. Claim: $n! > 2^n$, for all $n \ge 4$.

Proof.

Step 1: Show the statement is true for $n = 4$.

$$4! > 2^4 \text{ since } 4! = 4 \cdot 3 \cdot 2 \cdot 1 = 24 \text{ and } 2^4 = 16.$$

Step 2: Assume the statement is true for $n = k$: $k! > 2^k$

Show the statement is true for $n = k+1$: $(k+1)! > 2^{k+1}$

$$\begin{aligned} (k+1)! &= (k+1) \cdot k! \\ &> (k+1) \cdot 2^k \quad \text{(by assumption)} \\ &> 2 \cdot 2^k \qquad (\text{since } k+1 > 2 \text{ for } k \ge 4) \\ &= 2^{k+1} \end{aligned}$$

This completes the proof.

6. **<u>Claim</u>:** For any $c \geq 1$, $(1+c)^n \geq nc$, for all $n \geq 1$.

Proof.

<u>Step 1</u>: Show the statement is true for $n = 1$.

$(1+c)^1 \geq 1 \cdot c$ is clearly true.

<u>Step 2</u>: Assume the statement is true for $n = k$: $(1+c)^k \geq kc$

Show the statement is true for $n = k+1$: $(1+c)^{k+1} \geq (k+1)c$

$$\begin{aligned} ck + c &\leq (1+c)^k + c \\ &\leq (1+c)^k + (1+c) \\ &\leq (1+c)^k + (1+c)^k \\ &= 2(1+c)^k \\ &\leq (1+c)(1+c)^k \quad \text{(since } c \geq 1\text{)} \\ &= (1+c)^{k+1} \end{aligned}$$

This completes the proof.

7. **<u>Claim</u>:** $n(n+1)(n-1)$ is divisible by 3, for all $n \geq 1$.

Proof.

<u>Step 1</u>: Show the statement is true for $n = 1$.

$(1)(1+1)(1-1) = 1(2)(0) = 0$, which is clearly divisible by 3.

<u>Step 2</u>: Assume the statement is true for $n = k$: $k(k+1)(k-1)$ is divisible by 3

Show the statement is true for $n = k+1$: $(k+1)(k+2)(k)$ is divisible by 3

First, note that since by assumption $k(k+1)(k-1)$ is divisible by 3, we know that there exists an integer m such that $k(k+1)(k-1) = 3m$ **(1)**.

Now, observe that

$$(k+1)(k+2)(k) = (k^2+3k+2)(k) = k^3+3k^2+2k$$

(At this point, write $2k = 3k - k$ and group the terms as shown.)

$$\begin{aligned} &= (k^3 - k) + (3k^2 + 3k) \\ &= 3m + 3(k^2 + k) \quad \text{(by } \mathbf{(1)}\text{)} \\ &= 3(m + k^2 + k) \end{aligned}$$

Now, choose $p = m + k^2 + k$ to see that you have expressed $(k+1)(k+2)(k)$ as $3p$, thereby showing $(k+1)(k+2)(k)$ is divisible by 3.

This completes the proof.

8. Claim: $n^3 - n$ is divisible by 3, for all $n \geq 1$.

Proof.

Step 1: Show the statement is true for $n = 1$.

$1^3 - 1 = 0$, which is clearly divisible by 3.

Step 2: Assume the statement is true for $n = k$: $k^3 - k$ is divisible by 3

Show the statement is true for $n = k+1$: $(k+1)^3 - (k+1)$ is divisible by 3

First, note that since by assumption $k^3 - k$ is divisible by 3, we know that there exists an integer m such that $k^3 - k = 3m$ **(1)**.

Now, observe that

$$(k+1)^3 - (k+1) = \left(k^3 + 3k^2 + 3k + 1\right) - \left(k + 1\right) = k^3 + 3k^2 + 2k$$

(At this point, write $2k = 3k - k$ and group the terms as shown.)

$$= (k^3 - k) + (3k^2 + 3k)$$

$$= 3m + 3(k^2 + k) \quad \text{(by (1))}$$

$$= 3(m + k^2 + k)$$

Now, choose $p = m + k^2 + k$ to see that you have expressed $(k+1)^3 - (k+1)$ as $3p$, thereby showing $(k+1)^3 - (k+1)$ is divisible by 3.

This completes the proof.

9. Claim: $n^2 + 3n$ is divisible by 2, for all $n \geq 1$.

Proof.

Step 1: Show the statement is true for $n = 1$.

$1^2 + 3(1) = 4 = 2(2)$, which is clearly divisible by 2.

Step 2: Assume the statement is true for $n = k$: $k^2 + 3k$ is divisible by 2

Show the statement is true for $n = k+1$: $(k+1)^2 + 3(k+1)$ is divisible by 2

First, note that since by assumption $k^2 + 3k$ is divisible by 2, we know that there exists an integer m such that $k^2 + 3k = 2m$ **(1)**.

Now, observe that

$$(k+1)^2 + 3(k+1) = \left(k^2 + 2k + 1\right) + 3\left(k + 1\right) = k^2 + 5k + 4$$

(At this point, write $5k = 3k + 2k$ and group the terms as shown.)

$$= (k^2 + 3k) + (2k + 4)$$

$$= 2m + 2(k + 2) \quad \text{(by (1))}$$

$$= 2(m + k + 2)$$

Now, choose $p = m + k + 2$ to see that you have expressed $(k+1)^2 + 3(k+1)$ as $2p$, thereby showing $(k+1)^2 + 3(k+1)$ is divisible by 2.

This completes the proof.

10. <u>Claim</u>: $n(n+1)(n+2)$ is divisible by 6, for all $n \geq 1$.

Proof.

<u>Step 1</u>: Show the statement is true for $n = 1$.

$(1)(1+1)(1+2) = 1(2)(3) = 6$, which is clearly divisible by 6.

<u>Step 2</u>: Assume the statement is true for $n = k$: $k(k+1)(k+2)$ is divisible by 6

Show the statement is true for $n = k+1$: $(k+1)(k+2)(k+3)$ is divisible by 6

First, note that since by assumption $k(k+1)(k+2)$ is divisible by 6, we know that there exists an integer m such that $k(k+1)(k+2) = 6m$ **(1)**.

Now, observe that

$$(k+1)(k+2)(k+3) = (k+1)(k^2+5k+6)$$
$$= k^3 + 5k^2 + 6k + k^2 + 5k + 6$$
$$= k^3 + 6k^2 + 11k + 6$$

(At this point, write $6k^2 = 3k^2 + 3k^2$ and $11k = 2k + 9k$ and group the terms as shown.)

$$= (k^3 + 3k^2 + 2k) + (3k^2 + 9k + 6)$$
$$= 6m + 3(k^2 + 3k + 2) \quad \text{(by } \mathbf{(1)}\text{)}$$
$$= 6m + 3\underbrace{(k+2)(k+1)}$$

For any k, one of these consecutive integers must be even. So, the product must be divisible by 2.

$$= 6m + 3(2s), \text{ for some integer } s$$
$$= 6(m+s)$$

Now, choose $p = m+s$ to see that you've expressed $(k+1)(k+2)(k+3)$ as $6p$, thereby showing $(k+1)(k+2)(k+3)$ is divisible by 6.

This completes the proof. ■

11. <u>Claim</u>: $2+4+...+2n = n(n+1)$, for all $n \geq 1$.

Proof.

<u>Step 1</u>: Show the statement is true for $n = 1$.

$2 = 1(2)$, which is clearly true.

<u>Step 2</u>: Assume the statement is true for $n = k$: $2+4+...+2k = k(k+1)$

Show the statement is true for $n = k+1$: $2+4+...+2(k+1) = (k+1)(k+2)$

Observe that

$$2+4+...+2k+2(k+1) = \left(2+4+...+2k\right) + 2(k+1)$$
$$= k(k+1) + 2(k+1) \quad \text{(by assumption)}$$
$$= (k+2)(k+1)$$

This completes the proof. ■

12. Claim: $1+3+...+(2n-1)=n^2$, for all $n \geq 1$

Proof.

Step 1: Show the statement is true for $n=1$.

$$1=1^2\text{, which is clearly true.}$$

Step 2: Assume the statement is true for $n=k$: $1+3+...+(2k-1)=k^2$

Show the statement is true for $n=k+1$: $1+3+...+(2(k+1)-1)=(k+1)^2$

$$\begin{aligned}1+3+...+(2k-1)+\underbrace{(2(k+1)-1)}_{2k+1}&=\left(1+3+...+(2k-1)\right)+(2k+1)\\&=k^2+(2k+1)\quad\text{(by assumption)}\\&=(k+1)^2\quad\text{(factor)}\end{aligned}$$

This completes the proof. ■

13. Claim: $1+3+...+3^n=\frac{3^{n+1}-1}{2}$, for all $n \geq 1$

Proof.

Step 1: Show the statement is true for $n=1$.

$$1+3^1=\frac{3^2-1}{2}=4\text{ is clearly true.}$$

Step 2: Assume the statement is true for $n=k$: $1+3+...+3^k=\frac{3^{k+1}-1}{2}$

Show the statement is true for $n=k+1$: $1+3+...+3^{k+1}=\frac{3^{k+2}-1}{2}$

$$\begin{aligned}1+3+...+3^k+3^{k+1}&=\left(1+3+...+3^k\right)+3^{k+1}\\&=\frac{3^{k+1}-1}{2}+3^{k+1}\quad\text{(by assumption)}\\&=\frac{3^{k+1}-1+2(3^{k+1})}{2}\\&=\frac{3(3^{k+1})-1}{2}=\frac{3^{k+2}-1}{2}\end{aligned}$$

This completes the proof. ■

14. Claim: $2+4+...+2^n=2^{n+1}-2$, for all $n \geq 1$

Proof.

Step 1: Show the statement is true for $n=1$.

$$2=2^{1+1}-2\text{ is clearly true.}$$

Step 2: Assume the statement is true for $n=k$: $2+4+...+2^k=2^{k+1}-2$

Show the statement is true for $n=k+1$: $2+4+...+2^{k+1}=2^{k+2}-2$

$$\begin{aligned}2+4+...+2^k+2^{k+1}&=\left(2+4+...+2^k\right)+2^{k+1}\\&=\left(2^{k+1}-2\right)+2^{k+1}\quad\text{(by assumption)}\\&=2(2^{k+1})-2\\&=2^{k+2}-2\end{aligned}$$

This completes the proof. ■

15. <u>Claim</u>: $1^2+...+n^2=\frac{n(n+1)(2n+1)}{6}$, for all $n\geq 1$

Proof.

<u>Step 1</u>: Show the statement is true for $n=1$.

$$1^2=\frac{1(1+1)(2(1)+1)}{6}=\frac{6}{6}=1 \text{ is clearly true.}$$

<u>Step 2</u>: Assume the statement is true for $n=k$: $1^2+...+k^2=\frac{k(k+1)(2k+1)}{6}$

Show the statement is true for $n=k+1$: $1^2+...+(k+1)^2=\frac{(k+1)(k+2)\overbrace{(2(k+1)+1)}^{2k+3}}{6}$

$$\begin{aligned}1^2+...+k^2+(k+1)^2&=\left(1^2+...+k^2\right)+(k+1)^2\\&=\frac{k(k+1)(2k+1)}{6}+(k+1)^2 \text{ (by assumption)}\\&=\frac{k(k+1)(2k+1)+6(k+1)^2}{6}\\&=\frac{2k^3+3k^2+k+6k^2+12k+6}{6}\\&=\frac{2k^3+9k^2+13k+6}{6}\\&=\frac{(k+1)(k+2)(2k+3)}{6}\end{aligned}$$

This completes the proof. ■

16. <u>Claim</u>: $1^3+...+n^3=\frac{n^2(n+1)^2}{4}$, for all $n\geq 1$

Proof.

<u>Step 1</u>: Show the statement is true for $n=1$.

$$1^3=\frac{1^2(1+1)^2}{4}=1 \text{ is clearly true.}$$

<u>Step 2</u>: Assume the statement is true for $n=k$: $1^3+...+k^3=\frac{k^2(k+1)^2}{4}$

Show the statement is true for $n=k+1$: $1^3+...+(k+1)^3=\frac{(k+1)^2(k+2)^2}{4}$

$$\begin{aligned}1^3+...+k^3+(k+1)^3&=\left(1^3+...+k^3\right)+(k+1)^3\\&=\frac{k^2(k+1)^2}{4}+(k+1)^3 \text{ (by assumption)}\\&=\frac{k^2(k+1)^2+4(k+1)^3}{4}\\&=\frac{(k+1)^2\left(k^2+4(k+1)\right)}{4}\\&=\frac{(k+1)^2(k+2)^2}{4}\end{aligned}$$

This completes the proof. ■

17. Claim: $\frac{1}{1\cdot 2}+\frac{1}{2\cdot 3}+\ldots+\frac{1}{n(n+1)}=\frac{n}{n+1}$, for all $n\geq 1$

Proof.

Step 1: Show the statement is true for $n=1$.

$\frac{1}{1\cdot 2}=\frac{1}{1+1}=\frac{1}{2}$, which is clearly true.

Step 2: Assume the statement is true for $n=k$: $\frac{1}{1\cdot 2}+\frac{1}{2\cdot 3}+\ldots+\frac{1}{k(k+1)}=\frac{k}{k+1}$

Show the statement is true for $n=k+1$: $\frac{1}{1\cdot 2}+\frac{1}{2\cdot 3}+\ldots+\frac{1}{(k+1)(k+2)}=\frac{k+1}{k+2}$

$$\begin{aligned}\frac{1}{1\cdot 2}+\frac{1}{2\cdot 3}+\ldots+\frac{1}{k(k+1)}+\frac{1}{(k+1)(k+2)}&=\left(\frac{1}{1\cdot 2}+\frac{1}{2\cdot 3}+\ldots+\frac{1}{k(k+1)}\right)+\frac{1}{(k+1)(k+2)}\\&=\frac{k}{k+1}+\frac{1}{(k+1)(k+2)}\quad\text{(by assumption)}\\&=\frac{k(k+2)+1}{(k+1)(k+2)}\\&=\frac{k^2+2k+1}{(k+1)(k+2)}\\&=\frac{(k+1)^2}{(k+1)(k+2)}\\&=\frac{k+1}{(k+2)}\end{aligned}$$

This completes the proof. ■

18. $\frac{1}{2\cdot 3}+\ldots+\frac{1}{(n+1)(n+2)}=\frac{n}{2(n+2)}$, for all $n\geq 1$

Proof.

Step 1: Show the statement is true for $n=1$.

$\frac{1}{2\cdot 3}=\frac{1}{2(1+2)}=\frac{1}{2(3)}$, which is clearly true.

Step 2: Assume the statement is true for $n=k$: $\frac{1}{2\cdot 3}+\ldots+\frac{1}{(k+1)(k+2)}=\frac{k}{2(k+2)}$

Show the statement is true for $n=k+1$: $\frac{1}{2\cdot 3}+\ldots+\frac{1}{(k+2)(k+3)}=\frac{k+1}{2(k+3)}$

$$\begin{aligned}\frac{1}{2\cdot 3}+\ldots+\frac{1}{(k+1)(k+2)}+\frac{1}{(k+2)(k+3)}&=\left(\frac{1}{2\cdot 3}+\ldots+\frac{1}{(k+1)(k+2)}\right)+\frac{1}{(k+2)(k+3)}\\&=\frac{k}{2(k+2)}+\frac{1}{(k+2)(k+3)}\quad\text{(by assumption)}\\&=\frac{k(k+3)+2}{2(k+2)(k+3)}\\&=\frac{k^2+3k+2}{2(k+2)(k+3)}\\&=\frac{\cancel{(k+2)}(k+1)}{2\cancel{(k+2)}(k+3)}\\&=\frac{k+1}{2(k+3)}\end{aligned}$$

This completes the proof. ■

19. Claim: $(1\cdot 2) + (2\cdot 3) + \ldots + n(n+1) = \frac{n(n+1)(n+2)}{3}$, for all $n \geq 1$.

Proof.

Step 1: Show the statement is true for $n=1$.

$$(1\cdot 2) = \frac{1(1+1)(1+2)}{3} = \frac{1(2)(3)}{3} = 2, \text{ which is clearly true.}$$

Step 2: Assume the statement is true for $n=k$:

$$(1\cdot 2) + (2\cdot 3) + \ldots + k(k+1) = \frac{k(k+1)(k+2)}{3}$$

Show the statement is true for $n=k+1$:

$$(1\cdot 2) + (2\cdot 3) + \ldots + (k+1)(k+2) = \frac{(k+1)(k+2)(k+3)}{3}$$

$$\begin{aligned}(1\cdot 2) + (2\cdot 3) + \ldots + k(k+1) + (k+1)(k+2) &= \left((1\cdot 2) + (2\cdot 3) + \ldots + k(k+1)\right) + (k+1)(k+2) \\ &= \frac{k(k+1)(k+2)}{3} + (k+1)(k+2) \quad \text{(by assumption)} \\ &= \frac{k(k+1)(k+2)+3(k+1)(k+2)}{3} \\ &= \frac{(k+1)(k+2)(k+3)}{3} \quad \text{(factor out } (k+1)(k+2))\end{aligned}$$

This completes the proof. ■

20. Claim: $(1\cdot 3) + (2\cdot 4) + \ldots + n(n+2) = \frac{n(n+1)(2n+7)}{6}$, for all $n \geq 1$.

Proof.

Step 1: Show the statement is true for $n=1$.

$$(1\cdot 3) = \frac{1(1+1)(2(1)+7)}{6} = \frac{1(2)(9)}{6} = 3, \text{ which is clearly true.}$$

Step 2: Assume the statement is true for $n=k$:

$$(1\cdot 3) + (2\cdot 4) + \ldots + k(k+2) = \frac{k(k+1)(2k+7)}{6}$$

Show the statement is true for $n=k+1$:

$$(1\cdot 3) + (2\cdot 4) + \ldots + (k+1)(k+3) = \frac{(k+1)(k+2)\overbrace{(2(k+1)+7)}^{2k+9}}{6}$$

$$\begin{aligned}(1\cdot 3) + (2\cdot 4) + \ldots + k(k+2) + (k+1)(k+3) &= \left((1\cdot 3) + (2\cdot 4) + \ldots + k(k+2)\right) + (k+1)(k+3) \\ &= \frac{k(k+1)(2k+7)}{6} + (k+1)(k+3) \quad \text{(by assumption)} \\ &= \frac{k(k+1)(2k+7)+6(k+1)(k+3)}{6} \\ &= \frac{(k+1)[k(2k+7)+6(k+3)]}{6} \quad \text{(factor out } (k+1)) \\ &= \frac{(k+1)\left[2k^2+13k+18\right]}{6} \\ &= \frac{(k+1)(k+2)(2k+9)}{6}\end{aligned}$$

This completes the proof. ■

21. Claim: $1+x+\ldots+x^{n-1}=\frac{1-x^n}{1-x}$, for all $n\geq 1$.

Proof.

Step 1: Show the statement is true for $n=1$.

$1=\frac{1-x^1}{1-x}$, which is clearly true.

Step 2: Assume the statement is true for $n=k$: $1+x+\ldots+x^{k-1}=\frac{1-x^k}{1-x}$

Show the statement is true for $n=k+1$: $1+x+\ldots+x^k=\frac{1-x^{k+1}}{1-x}$

$$\begin{aligned}1+\ldots+x^{k-1}+x^k &= \left(1+\ldots+x^{k-1}\right)+x^k \\ &= \frac{1-x^k}{1-x}+x^k \\ &= \frac{1-x^k+x^k(1-x)}{1-x} \\ &= \frac{1-\cancel{x^k}+\cancel{x^k}-x^k x}{1-x} \\ &= \frac{1-x^{k+1}}{1-x}\end{aligned}$$

This completes the proof. ■

22. Claim: $\frac{1}{2}+\frac{1}{4}+\ldots+\frac{1}{2^n}=1-\frac{1}{2^n}$, for all $n\geq 1$.

Proof.

Step 1: Show the statement is true for $n=1$.

$\frac{1}{2}=1-\frac{1}{2^1}$, which is clearly true.

Step 2: Assume the statement is true for $n=k$: $\frac{1}{2}+\frac{1}{4}+\ldots+\frac{1}{2^k}=1-\frac{1}{2^k}$

Show the statement is true for $n=k+1$: $\frac{1}{2}+\frac{1}{4}+\ldots+\frac{1}{2^{k+1}}=1-\frac{1}{2^{k+1}}$

$$\begin{aligned}\frac{1}{2}+\frac{1}{4}+\ldots+\frac{1}{2^k}+\frac{1}{2^{k+1}} &= \left(\frac{1}{2}+\frac{1}{4}+\ldots+\frac{1}{2^k}\right)+\frac{1}{2^{k+1}} \\ &= \left(1-\frac{1}{2^k}\right)+\frac{1}{2^{k+1}} \\ &= \frac{2^{k+1}-2+1}{2^{k+1}} \\ &= \frac{2^{k+1}-1}{2^{k+1}} \\ &= 1-\frac{1}{2^{k+1}}\end{aligned}$$

This completes the proof. ■

23. Claim: $a_1 + (a_1 + d) + ... + (a_1 + (n-1)d) = \frac{n}{2}\left[2a_1 + (n-1)d\right]$, for all $n \geq 1$.

Proof.

Step 1: Show the statement is true for $n = 1$.

$$a_1 = \frac{1}{2}\left[2a_1 + (1-1)d\right] = \frac{1}{2}\left[2a_1\right], \text{ which is clearly true.}$$

Step 2: Assume the statement is true for $n = k$:

$$a_1 + (a_1 + d) + ... + (a_1 + (k-1)d) = \frac{k}{2}\left[2a_1 + (k-1)d\right]$$

Show the statement is true for $n = k+1$:

$$a_1 + (a_1 + d) + ... + (a_1 + ((k+1)-1)d) = \frac{k+1}{2}\left[2a_1 + (\overbrace{k+1-1}^{k})d\right]$$

Observe that

$$\begin{aligned}
&a_1 + (a_1 + d) + ... + (a_1 + (k-1)d) + (a_1 + kd) \\
&= \left(a_1 + (a_1 + d) + ... + (a_1 + (k-1)d)\right) + (a_1 + kd) \\
&= \tfrac{k}{2}\left[2a_1 + (k-1)d\right] + (a_1 + kd) \\
&= ka_1 + \tfrac{k(k-1)}{2}d + a_1 + kd \\
&= (k+1)a_1 + kd(\tfrac{k-1}{2} + 1) \\
&= (k+1)a_1 + kd(\tfrac{k+1}{2}) \\
&= \tfrac{(k+1)}{2}2a_1 + kd(\tfrac{k+1}{2}) \\
&= \tfrac{(k+1)}{2}\left[2a_1 + kd\right]
\end{aligned}$$

This completes the proof. ■

24. Claim: $a_1 + ra_1 + ... + r^{n-1}a_1 = \left(\frac{1-r^n}{1-r}\right)a_1$, for all $n \geq 1$.

Proof.

Step 1: Show the statement is true for $n = 1$.

$$a_1 = \underbrace{\left(\tfrac{1-r^1}{1-r}\right)}_{=1}a_1, \text{ which is clearly true.}$$

Step 2: Assume the statement is true for $n = k$: $a_1 + ra_1 + ... + r^{k-1}a_1 = \left(\frac{1-r^k}{1-r}\right)a_1$

Show the statement is true for $n = k+1$: $a_1 + ra_1 + ... + r^k a_1 = \left(\frac{1-r^{k+1}}{1-r}\right)a_1$

Observe that

$$\begin{aligned}
a_1 + ra_1 + ... + r^{k-1}a_1 + r^k a_1 &= \left(a_1 + ra_1 + ... + r^{k-1}a_1\right) + r^k a_1 \\
&= \left(\tfrac{1-r^k}{1-r}\right)a_1 + r^k a_1 \\
&= \tfrac{a_1(1-r^k) + r^k a_1(1-r)}{1-r} = \tfrac{a_1 1 - \cancel{a_1 r^k} + \cancel{r^k a_1} - r^{k+1}a_1}{1-r} = \left(\tfrac{1-r^{k+1}}{1-r}\right)a_1
\end{aligned}$$

This completes the proof. ■

25. Label the disks 1, 2, and 3 (smallest = 1 and largest = 3), and label the posts A, B, and C. The following are the moves on would take to solve the problem in the fewest number of step. (<u>Note</u>: The manner in which the disks are stacked (from top to bottom) on each peg form the contents of each cell; a blank cell means that no disk is on that peg in that particular move.)

	Post A	**Post B**	**Post C**
Initial placement	1 2 3		
Move 1	2 3	1	
Move 2	3	1	2
Move 3	3		1 2
Move 4		3	1 2
Move 5	1	3	2
Move 6	1	2 3	
Move 7		1 2 3	

So, the puzzle can be solved in as few as 7 steps. An argument as to why this is actually the fewest number of steps is beyond the scope of the text. (<u>Note</u>: Alternatively, we could have initially placed disk 1 on post C, and proceeded in a similar manner.)

26. Label the disks 1, 2, 3, and 4 (smallest = 1 and largest = 4), and label the posts A, B, and C. The following are the moves on would take to solve the problem in the fewest number of step. (<u>Note</u>: The manner in which the disks are stacked (from top to bottom) on each peg form the contents of each cell; a blank cell means that no disk is on that peg in that particular move.)

	Post A	**Post B**	**Post C**
Initial placement	1 2 3 4		
Move 1	2 3 4	1	

Move 2	3 4	1	2
Move 3	3 4		1 2
Move 4	4	3	1 2
Move 5	1 4	3	2
Move 6	1 4	2 3	
Move 7	4	1 2 3	
Move 8		1 2 3	4
Move 9		2 3	1 4
Move 10	2	3	1 4
Move 11	1 2	3	4
Move 12	1 2		3 4
Move 13	2	1	3 4
Move 14	1		2 3 4
Move 15			1 2 3 4

So, the puzzle can be solved in as few as 15 steps. An argument as to why this is actually the fewest number of steps is beyond the scope of the text. (Note: Alternatively, we could have initially placed disk 1 on post C, and proceeded in a similar manner.)

27. Using the strategy of Problem 25 (and 26), this puzzle can be solved in as few as 31 steps. Have fun trying it!! There are many classical references that discuss this problem, as well as several internet sites.

28. What follows is not a formal proof, but rather an intuitive discussion that hints at what the proof ought to be. To establish a more formal argument, one needs to know how to work with *recurrence relations*, a topic not addressed in the text. But, see if you can follow the spirit of the argument as presented below.
First, note that the general strategy used to solve the problem with n disks is to:

- Move the topmost $n-1$ disks from the left peg to one of the other two,
- Move the bottommost disk from left to right,
- Move the $n-1$ disks from its current location to the right.

And, this process is replicated inductively until you are down to one disks. As such, since this amounts to two steps applied in succession for each of the n disks, this would imply that the number of steps would be 2^n. However, 1 is subtracted since you start off with the problem with an initial configuration. So, truly, the problem can be solved in as few as 2^n-1 moves.

29. If $n = 2$, then the number of wires needed is $\frac{2(2-1)}{2}=1$.
Assume the formula holds for k cities. Then, if k cities are to be connected directly to each other, the number of wires needed is $\frac{k(k-1)}{2}$. Now, if one more city is added, then you must have k additional wires to connect the telephone to this additional city. So, the total wires in such case is $k+\frac{k(k-1)}{2}=\frac{2k+k^2-k}{2}=\frac{k(k+1)}{2}$. Thus, the statement holds for k+1 cities. Hence, we have proven the statement by induction.

30. For $n = 3$, the formula becomes (3-2)(180)=180. The formula is true when n =3 (a triangle), since it is known that the sum of the interior angles of a triangle is 180 degrees. Assume the formula is true for k sided regular polygon. Then, the sum of the interior angles is (k-2)(180). Looking at a few particular examples (for particular values of k) we observe that you can divide a k-sided polygon into (k-2) triangles, and each time we increase the number of sides by one, one additional triangle is formed. As such, the interior angle sum increases by 1. Hence, adding one side to a k-side polygon then has interior angle sum 180 + (k-2)(180) = 180(k-1), thereby showing the statement holds for k+1. Hence, we have proven the statement by induction.

31. False. You first need to show that S_1 is true.

32. False. You must show that if S_k is true, then S_{k+1} is true, <u>for all values</u> of k, not just the first pair of consecutive values of k.

33. <u>Claim</u>: $\sum_{k=1}^{n} k^4 = \frac{n(n+1)(2n+1)(3n^2+3n-1)}{30}$, for all $n \geq 1$.

Proof.

<u>Step 1</u>: Show the statement is true for $n=1$.

$$1^4 = \frac{1(1+1)(2(1)+1)(3(1)^2+3(1)-1)}{30} = 1, \text{ which is clearly true.}$$

<u>Step 2</u>: Assume the statement is true for $n=p$: $\sum_{k=1}^{p} k^4 = \frac{p(p+1)(2p+1)(3p^2+3p-1)}{30}$

Show the statement is true for $n=p+1$: $\sum_{k=1}^{p+1} k^4 = \frac{(p+1)(p+2)(2(p+1)+1)(3(p+1)^2+3(p+1)-1)}{30}$

Observe that

$$\sum_{k=1}^{p+1} k^4 = \sum_{k=1}^{p} k^4 + (p+1)^4$$

$$= \frac{p(p+1)(2p+1)(3p^2+3p-1)}{30} + (p+1)^4 \quad \text{(by assumption)}$$

$$= \frac{p(p+1)(2p+1)(3p^2+3p-1)+30(p+1)^4}{30}$$

$$= \frac{(p+1)\left[p(2p+1)(3p^2+3p-1)+30(p+1)^3\right]}{30}$$

$$= \frac{(p+1)\left[6p^4+39p^3+91p^2+89p+30\right]}{30} \quad \textbf{(1)}$$

Next, note that multiplying out these terms yields

$$\frac{(p+1)(p+2)(2(p+1)+1)(3(p+1)^2+3(p+1)-1)}{30} = \frac{(p^2+3p+2)(2p+3)(3p^2+6p+3+3p+3-1)}{30}$$

$$= \frac{(2p^3+9p^2+13p+6)(3p^2+9p+5)}{30}$$

$$= \frac{6p^5+45p^4+130p^3+180p^2+119p+30}{30} \quad \textbf{(2)}$$

Comparing **(1)** and **(2)**, we see they are, in fact, equal. This is precisely what we needed to show to establish the claim.

This completes the proof.

34. <u>Claim</u>: $\sum_{k=1}^{n} k^5 = \frac{n^2(n+1)^2(2n^2+2n-1)}{12}$, for all $n \geq 1$.

Proof.

<u>Step 1</u>: Show the statement is true for $n = 1$.

$$1^5 = \frac{1^2(1+1)^2(2(1)^2+2(1)-1)}{12} = 1\text{, which is clearly true.}$$

<u>Step 2</u>: Assume the statement is true for $n = p$: $\sum_{k=1}^{p} k^5 = \frac{p^2(p+1)^2(2p^2+2p-1)}{12}$

Show the statement is true for $n = p+1$: $\sum_{k=1}^{p+1} k^5 = \frac{(p+1)^2((p+1)+1)^2\overbrace{(2(p+1)^2+2(p+1)-1)}^{=2p^2+6p+3}}{12}$

Observe that

$$\begin{aligned}\sum_{k=1}^{p+1} k^5 &= \sum_{k=1}^{p} k^5 + (p+1)^5 \\ &= \frac{p^2(p+1)^2(2p^2+2p-1)}{12} + (p+1)^5 \quad \text{(by assumption)} \\ &= \frac{p^2(p+1)^2(2p^2+2p-1)+12(p+1)^5}{12} \\ &= \frac{(p+1)^2\left[p^2(2p^2+2p-1)+12(p+1)^3\right]}{12} \\ &= \frac{(p+1)\left[2p^4+14p^3+35p^2+36p+12\right]}{12} \quad \textbf{(1)}\end{aligned}$$

Next, note that multiplying out these terms yields

$$\begin{aligned}\frac{(p+1)^2((p+1)+1)^2(2(p+1)^2+2(p+1)-1)}{12} &= \frac{(p+1)^2(p+2)^2(2p^2+6p+3)}{12} \\ &= \frac{(p+1)^2(p^2+4p+4)(2p^2+6p+3)}{12} \\ &= \frac{(p+1)^2(2p^4+14p^3+35p^2+36p+12)}{12} \quad \textbf{(2)}\end{aligned}$$

Comparing **(1)** and **(2)**, we see they are, in fact, equal. This is precisely what we needed to show to establish the claim.

This completes the proof. ■

35. **<u>Claim</u>:** $\left(1+\frac{1}{1}\right)\cdot\left(1+\frac{1}{2}\right)\cdot\ldots\cdot\left(1+\frac{1}{n}\right)=n+1$, for all $n\geq 1$.

Proof.

<u>Step 1</u>: Show the statement is true for $n=1$.

$$\left(1+\frac{1}{1}\right)=1+1=2, \text{ which is clearly true.}$$

<u>Step 2</u>: Assume the statement is true for $n=k$: $\left(1+\frac{1}{1}\right)\cdot\left(1+\frac{1}{2}\right)\cdot\ldots\cdot\left(1+\frac{1}{k}\right)=k+1$

Show the statement is true for $n=k+1$: $\left(1+\frac{1}{1}\right)\cdot\left(1+\frac{1}{2}\right)\cdot\ldots\cdot\left(1+\frac{1}{k+1}\right)=\underbrace{(k+1)+1}_{=k+2}$

Observe that

$$\begin{aligned}\left(1+\tfrac{1}{1}\right)\cdot\left(1+\tfrac{1}{2}\right)\cdot\ldots\cdot\left(1+\tfrac{1}{k}\right)\cdot\left(1+\tfrac{1}{k+1}\right)&=\left[\left(1+\tfrac{1}{1}\right)\cdot\left(1+\tfrac{1}{2}\right)\cdot\ldots\cdot\left(1+\tfrac{1}{k}\right)\right]\cdot\left(1+\tfrac{1}{k+1}\right)\\&=(k+1)\cdot\left(1+\tfrac{1}{k+1}\right)\quad\text{(by assumption)}\\&=\cancel{(k+1)}\left(\tfrac{k+1+1}{\cancel{k+1}}\right)\\&=k+2\end{aligned}$$

This completes the proof. ■

36. **<u>Claim</u>:** $x+y$ is a factor of $x^{2n}-y^{2n}$, for all $n\geq 1$.

Proof.

<u>Step 1</u>: Show the statement is true for $n=1$.

$$x^2-y^2=(x+y)(x-y), \text{ so that } x+y \text{ is a factor.}$$

<u>Step 2</u>: Assume the statement is true for $n=k$:

$$x+y \text{ is a factor of } x^{2k}-y^{2k}$$

Show the statement is true for $n=k+1$:

$$x+y \text{ is a factor of } x^{2(k+1)}-y^{2(k+1)}$$

Observe that

$$\begin{aligned}x^{2(k+1)}-y^{2(k+1)}&=x^{2k}x^2-y^{2k}y^2\\&=x^{2k}x^2-y^{2k}x^2+y^{2k}x^2-y^{2k}y^2\quad\text{(Add zero.)}\\&=x^2\left(x^{2k}-y^{2k}\right)+y^{2k}\left(x^2-y^2\right)\\&=x^2\left(x^{2k}-y^{2k}\right)+y^{2k}(x+y)(x-y)\\&=x^2(x+y)p(x,y)+y^{2k}(x+y)(x-y)\quad\text{(by assumption)}\quad\textbf{(1)}\end{aligned}$$

where $p(x,y)$ is some polynomial in x and y. Hence, since $(x+y)$ can be factored out of both terms in **(1)**, we conclude that, in fact, $(x+y)$ is indeed a factor of $x^{2(k+1)}-y^{2(k+1)}$.

This completes the proof. ■

37. $x-y$ is a factor of $x^{2n}-y^{2n}$, for all $n\geq 1$.

Proof.

Step 1: Show the statement is true for $n=1$.

$$x^2-y^2=(x+y)(x-y)\text{, so that } x-y \text{ is a factor.}$$

Step 2: Assume the statement is true for $n=k$:

$$x-y \text{ is a factor of } x^{2k}-y^{2k}$$

Show the statement is true for $n=k+1$:

$$x-y \text{ is a factor of } x^{2(k+1)}-y^{2(k+1)}$$

Observe that

$$\begin{aligned} x^{2(k+1)}-y^{2(k+1)} &= x^{2k}x^2-y^{2k}y^2 \\ &= x^{2k}x^2-y^{2k}x^2+y^{2k}x^2-y^{2k}y^2 \quad \text{(Add zero.)} \\ &= x^2\left(x^{2k}-y^{2k}\right)+y^{2k}\left(x^2-y^2\right) \\ &= x^2\left(x^{2k}-y^{2k}\right)+y^{2k}(x+y)(x-y) \\ &= x^2(x-y)p(x,y)+y^{2k}(x+y)(x-y) \quad \text{(by assumption)} \quad \textbf{(1)} \end{aligned}$$

where $p(x,y)$ is some polynomial in x and y. Hence, since $(x-y)$ can be factored out of both terms in **(1)**, we conclude that, in fact, $(x-y)$ is indeed a factor of $x^{2(k+1)}-y^{2(k+1)}$.

This completes the proof. ■

38. **Claim:** $\ln\left(c_1\cdot c_2\cdot\ldots\cdot c_n\right)=\ln(c_1)+\ldots+\ln(c_n)$, for all $n\geq 1$.

Proof.

Step 1: Show the statement is true for $n=1$.

$$\ln\left(c_1\right)=\ln(c_1) \text{ is clearly true.}$$

Step 2: Assume the statement is true for $n=k$:

$$\ln\left(c_1\cdot c_2\cdot\ldots\cdot c_k\right)=\ln(c_1)+\ldots+\ln(c_k)$$

Show the statement is true for $n=k+1$:

$$\ln\left(c_1\cdot c_2\cdot\ldots\cdot c_{k+1}\right)=\ln(c_1)+\ldots+\ln(c_{k+1})$$

Observe that

$$\begin{aligned} \ln\left(\underbrace{c_1\cdot c_2\cdot\ldots\cdot c_k}_{\text{Treat as a single quantity}}\cdot c_{k+1}\right) &= \ln(c_1\cdot\ldots\cdot c_k)+\ln(c_{k+1}) \\ &= \ln(c_1)+\ldots+\ln(c_k)+\ln(c_{k+1}) \quad \text{(by assumption)} \end{aligned}$$

This completes the proof. ■

39. $\frac{255}{256}$. Yes.	**40.** 2,706,800. Yes.

41. **Claim:** $\sum_{i=1}^{n}(\pi+i)=\frac{n(2\pi+n+1)}{2}$.

Proof.

Step 1: Show the statement is true for $n=2$.

$$\sum_{i=1}^{2}(\pi+i)=(\pi+1)+(\pi+2)=2\pi+3=\frac{2(2\pi+2+1)}{2}$$

Step 2: Assume the statement is true for $n=k$:

$$\sum_{i=1}^{k}(\pi+i)=\frac{k(2\pi+k+1)}{2}$$

Show the statement is true for $n=k+1$:

Observe that

$$\sum_{i=1}^{k+1}(\pi+i)=\sum_{i=1}^{k}(\pi+i)+(\pi+k+1)=\frac{k(2\pi+k+1)}{2}+(\pi+k+1)$$

$$=\frac{k(2\pi+k+1)+2(\pi+k+1)}{2}=\frac{(k+1)(2\pi)+k(k+1)+2(k+1)}{2}$$

$$=\frac{(k+1)(2\pi+k+1+1)}{2}$$

This completes the proof. ■

42. **Claim:** $\sum_{i=1}^{n}(i+i^2)=\frac{n(n+1)(n+2)}{3}$.

Proof.

Step 1: Show the statement is true for $n=2$.

$$\sum_{i=1}^{2}(i+i^2)=(1+1^2)+(2+2^2)=8=\frac{2(2+1)(2+2)}{3}$$

Step 2: Assume the statement is true for $n=k$:

$$\sum_{i=1}^{k}(i+i^2)=\frac{k(k+1)(k+2)}{3}$$

Show the statement is true for $n=k+1$:

Observe that

$$\sum_{i=1}^{k+1}(i+i^2)=\sum_{i=1}^{k}(i+i^2)+\left[(k+1)+(k+1)^2\right]=\frac{k(k+1)(k+2)}{3}+(k+1)+(k+1)^2$$

$$=\frac{k(k+1)(k+2)+3(k+1)\left[1+(k+1)\right]}{3}=\frac{(k+1)(k+2)(k+3)}{3}$$

This completes the proof. ■

43. **Claim**: $\sum_{i=1}^{n}\left(i+i^3\right)=\frac{n(n+1)\left(n^2+n+2\right)}{4}$.

Proof.

Step 1: Show the statement is true for $n=2$.

$$\sum_{i=1}^{2}\left(i+i^3\right)=\left(1+1^3\right)+\left(2+2^3\right)=12=\frac{2(2+1)\left(2^2+2+2\right)}{4}$$

Step 2: Assume the statement is true for $n=k$:

$$\sum_{i=1}^{k}\left(i+i^3\right)=\frac{k(k+1)\left(k^2+k+2\right)}{4}$$

Show the statement is true for $n=k+1$:
Observe that

$$\sum_{i=1}^{k+1}\left(i+i^3\right)=\sum_{i=1}^{k}\left(i+i^3\right)+\left((k+1)+(k+1)^3\right)=\frac{k(k+1)\left(k^2+k+2\right)}{4}+(k+1)+(k+1)^3$$

$$=\frac{(k+1)\left[k\left(k^2+k+2\right)+4+4(k+1)^2\right]}{4}=\frac{(k+1)\left(k^3+5k^2+10k+8\right)}{4}$$

$$=\frac{(k+1)(k+2)(k^2+3k+4)}{4}=\frac{(k+1)(k+2)\left[(k+1)^2+(k+1)+2\right]}{4}$$

This completes the proof. ■

44. <u>**Claim**</u>: $\sum_{i=1}^{n}\left(i^2+i^3\right)=\frac{n(n+1)(n+2)(3n+1)}{12}$.

Proof.

<u>Step 1</u>: Show the statement is true for $n=2$.

$$\sum_{i=1}^{2}\left(i^2+i^3\right)=\left(1^2+1^3\right)+\left(2^2+2^3\right)=14=\frac{2(2+1)(2+2)(3(2)+1)}{12}$$

<u>Step 2</u>: Assume the statement is true for $n=k$:

$$\sum_{i=1}^{k}\left(i^2+i^3\right)=\frac{k(k+1)(k+2)(3k+1)}{12}$$

Show the statement is true for $n=k+1$:
Observe that

$$\begin{aligned}\sum_{i=1}^{k+1}\left(i^2+i^3\right)&=\sum_{i=1}^{k}\left(i^2+i^3\right)+\left((k+1)^2+(k+1)^3\right)\\&=\frac{k(k+1)(k+2)(3k+1)}{12}+(k+1)^2+(k+1)^3\\&=\frac{k(k+1)(k+2)(3k+1)+12(k+1)^2\left[1+(k+1)\right]}{12}\\&=\frac{(k+1)(k+2)\left[k(3k+1)+12(k+1)\right]}{12}\\&=\frac{(k+1)(k+2)\left[3k^2+13k+12\right]}{12}\\&=\frac{(k+1)(k+2)(k+3)(3k+4)}{12}\\&=\frac{(k+1)(k+2)(k+3)(3(k+1)+1)}{12}\end{aligned}$$

This completes the proof. ■

Section 10.5 Solutions ---

1.	$\binom{7}{3} = \frac{7!}{(7-3)!3!} = \frac{7!}{4!3!} = \frac{7\cdot 6\cdot 5\cdot \cancel{4!}}{\cancel{4!}(3\cdot 2\cdot 1)} = 35$
2.	$\binom{8}{2} = \frac{8!}{(8-2)!2!} = \frac{8!}{6!2!} = \frac{8\cdot 7\cdot \cancel{6!}}{\cancel{6!}(2\cdot 1)} = 28$
3.	$\binom{10}{8} = \frac{10!}{(10-8)!8!} = \frac{10!}{2!8!} = \frac{10\cdot 9\cdot \cancel{8!}}{\cancel{8!}(2\cdot 1)} = 45$
4.	$\binom{23}{21} = \frac{23!}{(23-21)!21!} = \frac{23!}{2!21!} = \frac{23\cdot 22\cdot \cancel{21!}}{\cancel{21!}(2\cdot 1)} = 253$
5.	$\binom{17}{0} = \frac{17!}{(17-0)!0!} = \frac{\cancel{17!}}{\cancel{17!}0!} = \frac{1}{0!} = 1$
6.	$\binom{100}{0} = \frac{100!}{(100-0)!0!} = \frac{\cancel{100!}}{\cancel{100!}0!} = \frac{1}{0!} = 1$
7.	$\binom{99}{99} = \frac{99!}{(99-99)!99!} = \frac{\cancel{99!}}{0!\cancel{99!}} = \frac{1}{0!} = 1$
8.	$\binom{52}{52} = \frac{52!}{(52-52)!52!} = \frac{\cancel{52!}}{0!\cancel{52!}} = \frac{1}{0!} = 1$
9.	$\binom{48}{45} = \frac{48!}{(48-45)!45!} = \frac{48!}{3!45!} = \frac{48\cdot 47\cdot 46\cdot \cancel{45!}}{\cancel{45!}(3\cdot 2\cdot 1)} = 17{,}296$
10.	$\binom{29}{26} = \frac{29!}{(29-26)!26!} = \frac{29!}{3!26!} = \frac{29\cdot 28\cdot 27\cdot \cancel{26!}}{\cancel{26!}(3\cdot 2\cdot 1)} = 3654$
11.	$(x+2)^4 = \sum_{k=0}^{4}\binom{4}{k}x^{4-k}2^k$ $= \binom{4}{0}x^4 2^0 + \binom{4}{1}x^3 2^1 + \binom{4}{2}x^2 2^2 + \binom{4}{3}x^1 2^3 + \binom{4}{4}x^0 2^4$ $= \underbrace{\frac{4!}{4!0!}}_{=1}x^4 2^0 + \underbrace{\frac{4!}{3!1!}}_{=4}x^3 2^1 + \underbrace{\frac{4!}{2!2!}}_{=6}x^2 2^2 + \underbrace{\frac{4!}{1!3!}}_{=4}x^1 2^3 + \underbrace{\frac{4!}{0!4!}}_{=1}x^0 2^4$ $= x^4 + 8x^3 + 24x^2 + 32x + 16$

12.

$$(x+3)^5 = \sum_{k=0}^{5} \binom{5}{k} x^{5-k} 3^k$$

$$= \binom{5}{0} x^5 3^0 + \binom{5}{1} x^4 3^1 + \binom{5}{2} x^3 3^2 + \binom{5}{3} x^2 3^3 + \binom{5}{4} x^1 3^4 + \binom{5}{5} x^0 3^5$$

$$= \underbrace{\tfrac{5!}{5!0!}}_{=1} x^5 3^0 + \underbrace{\tfrac{5!}{4!1!}}_{=5} x^4 3^1 + \underbrace{\tfrac{5!}{3!2!}}_{=10} x^3 3^2 + \underbrace{\tfrac{5!}{2!3!}}_{=10} x^2 3^3 + \underbrace{\tfrac{5!}{1!4!}}_{=5} x^1 3^4 + \underbrace{\tfrac{5!}{0!5!}}_{=1} x^0 3^5$$

$$= x^5 + 15x^4 + 90x^3 + 270x^2 + 405x + 243$$

13.

$$(y-3)^5 = \sum_{k=0}^{5} \binom{5}{k} y^{5-k} (-3)^k$$

$$= \binom{5}{0} y^5(-3)^0 + \binom{5}{1} y^4(-3)^1 + \binom{5}{2} y^3(-3)^2 + \binom{5}{3} y^2(-3)^3 + \binom{5}{4} y^1(-3)^4 + \binom{5}{5} y^0(-3)^5$$

$$= \underbrace{\tfrac{5!}{5!0!}}_{=1} y^5(-3)^0 + \underbrace{\tfrac{5!}{4!1!}}_{=5} y^4(-3)^1 + \underbrace{\tfrac{5!}{3!2!}}_{=10} y^3(-3)^2 + \underbrace{\tfrac{5!}{2!3!}}_{=10} y^2(-3)^3 + \underbrace{\tfrac{5!}{1!4!}}_{=5} y^1(-3)^4 + \underbrace{\tfrac{5!}{0!5!}}_{=1} y^0(-3)^5$$

$$= y^5 - 15y^4 + 90y^3 - 270y^2 + 405y - 243$$

14.

$$(y-4)^4 = \sum_{k=0}^{4} \binom{4}{k} y^{4-k} (-4)^k$$

$$= \binom{4}{0} y^4(-4)^0 + \binom{4}{1} y^3(-4)^1 + \binom{4}{2} y^2(-4)^2 + \binom{4}{3} y^1(-4)^3 + \binom{4}{4} y^0(-4)^4$$

$$= \underbrace{\tfrac{4!}{4!0!}}_{=1} y^4(-4)^0 + \underbrace{\tfrac{4!}{4!0!}}_{=4} y^3(-4)^1 + \underbrace{\tfrac{4!}{4!0!}}_{=6} y^2(-4)^2 + \underbrace{\tfrac{4!}{4!0!}}_{=4} y^1(-4)^3 + \underbrace{\tfrac{4!}{4!0!}}_{=1} y^0(-4)^4$$

$$= y^4 - 16y^3 + 96y^2 - 256y + 256$$

15.

$$(x+y)^5 = \sum_{k=0}^{5} \binom{5}{k} x^{5-k} y^k$$

$$= \binom{5}{0} x^5 y^0 + \binom{5}{1} x^4 y^1 + \binom{5}{2} x^3 y^2 + \binom{5}{3} x^2 y^3 + \binom{5}{4} x^1 y^4 + \binom{5}{5} x^0 y^5$$

$$= \underbrace{\tfrac{5!}{5!0!}}_{=1} x^5 y^0 + \underbrace{\tfrac{5!}{4!1!}}_{=5} x^4 y^1 + \underbrace{\tfrac{5!}{3!2!}}_{=10} x^3 y^2 + \underbrace{\tfrac{5!}{2!3!}}_{=10} x^2 y^3 + \underbrace{\tfrac{5!}{1!4!}}_{=5} x^1 y^4 + \underbrace{\tfrac{5!}{0!5!}}_{=1} x^0 y^5$$

$$= x^5 + 5x^4y + 10x^3y^2 + 10x^2y^3 + 5xy^4 + y^5$$

16.

$$(x-y)^6 = \sum_{k=0}^{6}\binom{6}{k}x^{6-k}(-y)^k$$

$$= \binom{6}{0}x^6(-y)^0 + \binom{6}{1}x^5(-y)^1 + \binom{6}{2}x^4(-y)^2 + \binom{6}{3}x^3(-y)^3 + \binom{6}{4}x^2(-y)^4 + \binom{6}{5}x^1(-y)^5 + \binom{6}{6}x^0(-y)^6$$

$$= \underbrace{\tfrac{6!}{6!0!}}_{=1}x^6(-y)^0 + \underbrace{\tfrac{6!}{5!1!}}_{=6}x^5(-y)^1 + \underbrace{\tfrac{6!}{4!2!}}_{=15}x^4(-y)^2 + \underbrace{\tfrac{6!}{3!3!}}_{=20}x^3(-y)^3 + \underbrace{\tfrac{6!}{2!4!}}_{=15}x^2(-y)^4 + \underbrace{\tfrac{6!}{1!5!}}_{=6}x^1(-y)^5 + \underbrace{\tfrac{6!}{0!6!}}_{=1}x^0(-y)^6$$

$$= x^6 - 6x^5y + 15x^4y^2 - 20x^3y^3 + 15x^2y^4 - 6xy^5 + y^6$$

17.

$$(x+3y)^3 = \sum_{k=0}^{3}\binom{3}{k}x^{3-k}(3y)^k$$

$$= \binom{3}{0}x^3(3y)^0 + \binom{3}{1}x^2(3y)^1 + \binom{3}{2}x^1(3y)^2 + \binom{3}{3}x^0(3y)^3$$

$$= \underbrace{\tfrac{3!}{3!0!}}_{=1}x^3(3y)^0 + \underbrace{\tfrac{3!}{2!1!}}_{=3}x^2(3y)^1 + \underbrace{\tfrac{3!}{1!2!}}_{=3}x^1(3y)^2 + \underbrace{\tfrac{3!}{0!3!}}_{=1}x^0(3y)^3$$

$$= x^3 + 9x^2y + 27xy^2 + 27y^3$$

18.

$$(2x-y)^3 = \sum_{k=0}^{3}\binom{3}{k}(2x)^{3-k}(-y)^k$$

$$= \binom{3}{0}(2x)^3(-y)^0 + \binom{3}{1}(2x)^2(-y)^1 + \binom{3}{2}(2x)^1(-y)^2 + \binom{3}{3}(2x)^0(-y)^3$$

$$= \underbrace{\tfrac{3!}{3!0!}}_{=1}(2x)^3(-y)^0 + \underbrace{\tfrac{3!}{2!1!}}_{=3}(2x)^2(-y)^1 + \underbrace{\tfrac{3!}{1!2!}}_{=3}(2x)^1(-y)^2 + \underbrace{\tfrac{3!}{0!3!}}_{=1}(2x)^0(-y)^3$$

$$= 8x^3 - 12x^2y + 6xy^2 - y^3$$

19.

$$(5x-2)^3 = \sum_{k=0}^{3}\binom{3}{k}(5x)^{3-k}(-2)^k$$
$$=\binom{3}{0}(5x)^3(-2)^0+\binom{3}{1}(5x)^2(-2)^1+\binom{3}{2}(5x)^1(-2)^2+\binom{3}{3}(5x)^0(-2)^3$$
$$=\underbrace{\tfrac{3!}{3!0!}}_{=1}(5x)^3(-2)^0+\underbrace{\tfrac{3!}{2!1!}}_{=3}(5x)^2(-2)^1+\underbrace{\tfrac{3!}{1!2!}}_{=3}(5x)^1(-2)^2+\underbrace{\tfrac{3!}{0!3!}}_{=1}(5x)^0(-2)^3$$
$$=125x^3-150x^2+60x-8$$

20.

$$(a-7b)^3 = \sum_{k=0}^{3}\binom{3}{k}(a)^{3-k}(-7b)^k$$
$$=\binom{3}{0}(a)^3(-7b)^0+\binom{3}{1}(a)^2(-7b)^1+\binom{3}{2}(a)^1(-7b)^2+\binom{3}{3}(a)^0(-7b)^3$$
$$=\underbrace{\tfrac{3!}{3!0!}}_{=1}(a)^3(-7b)^0+\underbrace{\tfrac{3!}{2!1!}}_{=3}(a)^2(-7b)^1+\underbrace{\tfrac{3!}{1!2!}}_{=3}(a)^1(-7b)^2+\underbrace{\tfrac{3!}{0!3!}}_{=1}(a)^0(-7b)^3$$
$$=a^3-21a^2b+147ab^2-343b^3$$

21.

$$(\tfrac{1}{x}+5y)^4 = \sum_{k=0}^{4}\binom{4}{k}(\tfrac{1}{x})^{4-k}(5y)^k,\ x\neq 0$$
$$=\binom{4}{0}(\tfrac{1}{x})^4(5y)^0+\binom{4}{1}(\tfrac{1}{x})^3(5y)^1+\binom{4}{2}(\tfrac{1}{x})^2(5y)^2+\binom{4}{3}(\tfrac{1}{x})^1(5y)^3+\binom{4}{4}(\tfrac{1}{x})^0(5y)^4$$
$$=\underbrace{\tfrac{4!}{4!0!}}_{=1}(\tfrac{1}{x})^4(5y)^0+\underbrace{\tfrac{4!}{3!1!}}_{=4}(\tfrac{1}{x})^3(5y)^1+\underbrace{\tfrac{4!}{2!2!}}_{=6}(\tfrac{1}{x})^2(5y)^2+\underbrace{\tfrac{4!}{1!3!}}_{=4}(\tfrac{1}{x})^1(5y)^3+\underbrace{\tfrac{4!}{0!4!}}_{=1}(\tfrac{1}{x})^0(5y)^4$$
$$=\tfrac{1}{x^4}+20\tfrac{y}{x^3}+150\tfrac{y^2}{x^2}+500\tfrac{y^3}{x}+625y^4$$

22.

$$\left(2x+\tfrac{3}{y}\right)^4=\sum_{k=0}^{4}\binom{4}{k}(2x)^{4-k}\left(\tfrac{3}{y}\right)^k,\ \ y\neq 0$$

$$=\binom{4}{0}(2x)^4\left(\tfrac{3}{y}\right)^0+\binom{4}{1}(2x)^3\left(\tfrac{3}{y}\right)^1+\binom{4}{2}(2x)^2\left(\tfrac{3}{y}\right)^2+\binom{4}{3}(2x)^1\left(\tfrac{3}{y}\right)^3+\binom{4}{4}(2x)^0\left(\tfrac{3}{y}\right)^4$$

$$=\underbrace{\tfrac{4!}{4!0!}}_{=1}(2x)^4\left(\tfrac{3}{y}\right)^0+\underbrace{\tfrac{4!}{3!1!}}_{=4}(2x)^3\left(\tfrac{3}{y}\right)^1+\underbrace{\tfrac{4!}{2!2!}}_{=6}(2x)^2\left(\tfrac{3}{y}\right)^2+\underbrace{\tfrac{4!}{1!3!}}_{=4}(2x)^1\left(\tfrac{3}{y}\right)^3+\underbrace{\tfrac{4!}{0!4!}}_{=1}(2x)^0\left(\tfrac{3}{y}\right)^4$$

$$=16x^4+96\tfrac{x^3}{y}+216\tfrac{x^2}{y^2}+216\tfrac{x}{y^3}+81\tfrac{1}{y^4}$$

23.

$$(x^2+y^2)^4=\sum_{k=0}^{4}\binom{4}{k}\left(x^2\right)^{4-k}(y^2)^k$$

$$=\binom{4}{0}\left(x^2\right)^4(y^2)^0+\binom{4}{1}\left(x^2\right)^3(y^2)^1+\binom{4}{2}\left(x^2\right)^2(y^2)^2+\binom{4}{3}\left(x^2\right)^1(y^2)^3+\binom{4}{4}\left(x^2\right)^0(y^2)^4$$

$$=\underbrace{\tfrac{4!}{4!0!}}_{=1}\left(x^2\right)^4(y^2)^0+\underbrace{\tfrac{4!}{3!1!}}_{=4}\left(x^2\right)^3(y^2)^1+\underbrace{\tfrac{4!}{2!2!}}_{=6}\left(x^2\right)^2(y^2)^2+\underbrace{\tfrac{4!}{1!3!}}_{=4}\left(x^2\right)^1(y^2)^3+\underbrace{\tfrac{4!}{0!4!}}_{=1}\left(x^2\right)^0(y^2)^4$$

$$=x^8+4x^6y^2+6x^4y^4+4x^2y^6+y^8$$

24.

$$(r^3-s^3)^3=\sum_{k=0}^{3}\binom{3}{k}(r^3)^{3-k}(-s^3)^k$$

$$=\binom{3}{0}(r^3)^3(-s^3)^0+\binom{3}{1}(r^3)^2(-s^3)^1+\binom{3}{2}(r^3)^1(-s^3)^2+\binom{3}{3}(r^3)^0(-s^3)^3$$

$$=\underbrace{\tfrac{3!}{3!0!}}_{=1}(r^3)^3(-s^3)^0+\underbrace{\tfrac{3!}{2!1!}}_{=3}(r^3)^2(-s^3)^1+\underbrace{\tfrac{3!}{1!2!}}_{=3}(r^3)^1(-s^3)^2+\underbrace{\tfrac{3!}{0!3!}}_{=1}(r^3)^0(-s^3)^3$$

$$=r^9-3r^6s^3+3r^3s^6-s^9$$

25.

$$(ax+by)^5 = \sum_{k=0}^{5}\binom{5}{k}(ax)^{5-k}(by)^k$$

$$= \binom{5}{0}(ax)^5(by)^0 + \binom{5}{1}(ax)^4(by)^1 + \binom{5}{2}(ax)^3(by)^2 + \binom{5}{3}(ax)^2(by)^3$$

$$+ \binom{5}{4}(ax)^1(by)^4 + \binom{5}{5}(ax)^0(by)^5$$

$$= \underbrace{\tfrac{5!}{5!0!}}_{=1}(ax)^5(by)^0 + \underbrace{\tfrac{5!}{4!1!}}_{=5}(ax)^4(by)^1 + \underbrace{\tfrac{5!}{3!2!}}_{=10}(ax)^3(by)^2 + \underbrace{\tfrac{5!}{2!3!}}_{=10}(ax)^2(by)^3$$

$$+ \underbrace{\tfrac{5!}{1!4!}}_{=5}(ax)^1(by)^4 + \underbrace{\tfrac{5!}{0!5!}}_{=1}(ax)^0(by)^5$$

$$= a^5x^5 + 5a^4bx^4y + 10a^3b^2x^3y^2 + 10a^2b^2x^2y^3 + 5ab^4xy^4 + b^5y^5$$

26.

$$(ax-by)^5 = \sum_{k=0}^{5}\binom{5}{k}(ax)^{5-k}(-by)^k$$

$$= \binom{5}{0}(ax)^5(-by)^0 + \binom{5}{1}(ax)^4(-by)^1 + \binom{5}{2}(ax)^3(-by)^2 + \binom{5}{3}(ax)^2(-by)^3$$

$$+ \binom{5}{4}(ax)^1(-by)^4 + \binom{5}{5}(ax)^0(-by)^5$$

$$= \underbrace{\tfrac{5!}{5!0!}}_{=1}(ax)^5(-by)^0 + \underbrace{\tfrac{5!}{4!1!}}_{=5}(ax)^4(-by)^1 + \underbrace{\tfrac{5!}{3!2!}}_{=10}(ax)^3(-by)^2 + \underbrace{\tfrac{5!}{2!3!}}_{=10}(ax)^2(-by)^3$$

$$+ \underbrace{\tfrac{5!}{1!4!}}_{=5}(ax)^1(-by)^4 + \underbrace{\tfrac{5!}{0!5!}}_{=1}(ax)^0(-by)^5$$

$$= a^5x^5 - 5a^4bx^4y + 10a^3b^2x^3y^2 - 10a^2b^2x^2y^3 + 5ab^4xy^4 - b^5y^5$$

27.

$$(\sqrt{x}+2)^6 = \sum_{k=0}^{6}\binom{6}{k}\left(\sqrt{x}\right)^{6-k}2^k,\ \ x \ge 0$$

$$= \binom{6}{0}\left(\sqrt{x}\right)^6 2^0 + \binom{6}{1}\left(\sqrt{x}\right)^5 2^1 + \binom{6}{2}\left(\sqrt{x}\right)^4 2^2 + \binom{6}{3}\left(\sqrt{x}\right)^3 2^3$$

$$+ \binom{6}{4}\left(\sqrt{x}\right)^2 2^4 + \binom{6}{5}\left(\sqrt{x}\right)^1 2^5 + \binom{6}{6}\left(\sqrt{x}\right)^0 2^6$$

$$= \underbrace{\tfrac{6!}{6!0!}}_{=1}\left(\sqrt{x}\right)^6 2^0 + \underbrace{\tfrac{6!}{5!1!}}_{=6}\left(\sqrt{x}\right)^5 2^1 + \underbrace{\tfrac{6!}{4!2!}}_{=15}\left(\sqrt{x}\right)^4 2^2 + \underbrace{\tfrac{6!}{3!3!}}_{=20}\left(\sqrt{x}\right)^3 2^3$$

$$+ \underbrace{\tfrac{6!}{2!4!}}_{=15}\left(\sqrt{x}\right)^2 2^4 + \underbrace{\tfrac{6!}{1!5!}}_{=6}\left(\sqrt{x}\right)^1 2^5 + \underbrace{\tfrac{6!}{0!6!}}_{=1}\left(\sqrt{x}\right)^0 2^6$$

$$= x^3 + 12x^{5/2} + 60x^2 + 160x^{3/2} + 240x + 192x^{1/2} + 64$$

28.

$$(3+\sqrt{y})^4 = \sum_{k=0}^{4}\binom{4}{k}3^{4-k}\left(\sqrt{y}\right)^k,\ \ y \ge 0$$

$$= \binom{4}{0}3^4\left(\sqrt{y}\right)^0 + \binom{4}{1}3^3\left(\sqrt{y}\right)^1 + \binom{4}{2}3^2\left(\sqrt{y}\right)^2 + \binom{4}{3}3^1\left(\sqrt{y}\right)^3 + \binom{4}{4}3^0\left(\sqrt{y}\right)^4$$

$$= \underbrace{\tfrac{4!}{4!0!}}_{=1}3^4\left(\sqrt{y}\right)^0 + \underbrace{\tfrac{4!}{3!1!}}_{=4}3^3\left(\sqrt{y}\right)^1 + \underbrace{\tfrac{4!}{2!2!}}_{=6}3^2\left(\sqrt{y}\right)^2 + \underbrace{\tfrac{4!}{1!3!}}_{=4}3^1\left(\sqrt{y}\right)^3 + \underbrace{\tfrac{4!}{0!4!}}_{=1}3^0\left(\sqrt{y}\right)^4$$

$$= 81 + 108y^{1/2} + 54y + 12y^{3/2} + y^2$$

29.

$$(a^{3/4}+b^{1/4})^4 = \sum_{k=0}^{4}\binom{4}{k}\left(a^{3/4}\right)^{4-k}\left(b^{1/4}\right)^k,\ \ a,b \ge 0$$

$$= \binom{4}{0}\left(a^{3/4}\right)^4\left(b^{1/4}\right)^0 + \binom{4}{1}\left(a^{3/4}\right)^3\left(b^{1/4}\right)^1 + \binom{4}{2}\left(a^{3/4}\right)^2\left(b^{1/4}\right)^2$$

$$+ \binom{4}{3}\left(a^{3/4}\right)^1\left(b^{1/4}\right)^3 + \binom{4}{4}\left(a^{3/4}\right)^0\left(b^{1/4}\right)^4$$

$$= \underbrace{\tfrac{4!}{4!0!}}_{=1}\left(a^{3/4}\right)^4\left(b^{1/4}\right)^0 + \underbrace{\tfrac{4!}{3!1!}}_{=4}\left(a^{3/4}\right)^3\left(b^{1/4}\right)^1 + \underbrace{\tfrac{4!}{2!2!}}_{=6}\left(a^{3/4}\right)^2\left(b^{1/4}\right)^2$$

$$+ \underbrace{\tfrac{4!}{1!3!}}_{=4}\left(a^{3/4}\right)^1\left(b^{1/4}\right)^3 + \underbrace{\tfrac{4!}{0!4!}}_{=1}\left(a^{3/4}\right)^0\left(b^{1/4}\right)^4$$

$$= a^3 + 4a^{9/4}b^{1/4} + 6a^{3/2}b^{1/2} + 4a^{3/4}b^{3/4} + b$$

30.

$$(x^{2/3}+y^{1/3})^3=\sum_{k=0}^{3}\binom{3}{k}\left(x^{2/3}\right)^{3-k}\left(y^{1/3}\right)^{k}$$

$$=\binom{3}{0}\left(x^{2/3}\right)^{3}\left(y^{1/3}\right)^{0}+\binom{3}{1}\left(x^{2/3}\right)^{2}\left(y^{1/3}\right)^{1}+\binom{3}{2}\left(x^{2/3}\right)^{1}\left(y^{1/3}\right)^{2}+\binom{3}{3}\left(x^{2/3}\right)^{0}\left(y^{1/3}\right)^{3}$$

$$=\underbrace{\tfrac{3!}{3!0!}}_{=1}\left(x^{2/3}\right)^{3}\left(y^{1/3}\right)^{0}+\underbrace{\tfrac{3!}{2!1!}}_{=3}\left(x^{2/3}\right)^{2}\left(y^{1/3}\right)^{1}+\underbrace{\tfrac{3!}{1!2!}}_{=3}\left(x^{2/3}\right)^{1}\left(y^{1/3}\right)^{2}+\underbrace{\tfrac{3!}{0!3!}}_{=1}\left(x^{2/3}\right)^{0}\left(y^{1/3}\right)^{3}$$

$$=x^2+3x^{4/3}y^{1/3}+3x^{2/3}y^{2/3}+y$$

31.

$$(x^{1/4}+2\sqrt{y})^4=\sum_{k=0}^{4}\binom{4}{k}\left(x^{1/4}\right)^{4-k}\left(2\sqrt{y}\right)^{k},\ \ x,y\geq 0$$

$$=\binom{4}{0}\left(x^{1/4}\right)^{4}\left(2\sqrt{y}\right)^{0}+\binom{4}{1}\left(x^{1/4}\right)^{3}\left(2\sqrt{y}\right)^{1}+\binom{4}{2}\left(x^{1/4}\right)^{2}\left(2\sqrt{y}\right)^{2}$$

$$+\binom{4}{3}\left(x^{1/4}\right)^{1}\left(2\sqrt{y}\right)^{3}+\binom{4}{4}\left(x^{1/4}\right)^{0}\left(2\sqrt{y}\right)^{4}$$

$$=\underbrace{\tfrac{4!}{4!0!}}_{=1}\left(x^{1/4}\right)^{4}\left(2\sqrt{y}\right)^{0}+\underbrace{\tfrac{4!}{3!1!}}_{=4}\left(x^{1/4}\right)^{3}\left(2\sqrt{y}\right)^{1}+\underbrace{\tfrac{4!}{2!2!}}_{=6}\left(x^{1/4}\right)^{2}\left(2\sqrt{y}\right)^{2}$$

$$+\underbrace{\tfrac{4!}{1!3!}}_{=4}\left(x^{1/4}\right)^{1}\left(2\sqrt{y}\right)^{3}+\underbrace{\tfrac{4!}{0!4!}}_{=1}\left(x^{1/4}\right)^{0}\left(2\sqrt{y}\right)^{4}$$

$$=x+8x^{3/4}y^{1/2}+24x^{1/2}y+32x^{1/4}y^{3/2}+16y^2$$

32.

$$(\sqrt{x}-3y^{1/4})^8=\sum_{k=0}^{8}\binom{8}{k}\left(\sqrt{x}\right)^{8-k}\left(-3y^{1/4}\right)^k,\ \ x,y\ge 0$$

$$=\binom{8}{0}\left(\sqrt{x}\right)^8\left(-3y^{1/4}\right)^0+\binom{8}{1}\left(\sqrt{x}\right)^7\left(-3y^{1/4}\right)^1+\binom{8}{2}\left(\sqrt{x}\right)^6\left(-3y^{1/4}\right)^2$$

$$+\binom{8}{3}\left(\sqrt{x}\right)^5\left(-3y^{1/4}\right)^3+\binom{8}{4}\left(\sqrt{x}\right)^4\left(-3y^{1/4}\right)^4+\binom{8}{5}\left(\sqrt{x}\right)^3\left(-3y^{1/4}\right)^5$$

$$+\binom{8}{6}\left(\sqrt{x}\right)^2\left(-3y^{1/4}\right)^6+\binom{8}{7}\left(\sqrt{x}\right)^1\left(-3y^{1/4}\right)^7+\binom{8}{8}\left(\sqrt{x}\right)^0\left(-3y^{1/4}\right)^8$$

$$=\underbrace{\tfrac{8!}{8!0!}}_{=1}\left(\sqrt{x}\right)^8\left(-3y^{1/4}\right)^0+\underbrace{\tfrac{8!}{7!1!}}_{=8}\left(\sqrt{x}\right)^7\left(-3y^{1/4}\right)^1+\underbrace{\tfrac{8!}{6!2!}}_{=28}\left(\sqrt{x}\right)^6\left(-3y^{1/4}\right)^2$$

$$+\underbrace{\tfrac{8!}{5!3!}}_{=56}\left(\sqrt{x}\right)^5\left(-3y^{1/4}\right)^3+\underbrace{\tfrac{8!}{4!4!}}_{=70}\left(\sqrt{x}\right)^4\left(-3y^{1/4}\right)^4+\underbrace{\tfrac{8!}{3!5!}}_{=56}\left(\sqrt{x}\right)^3\left(-3y^{1/4}\right)^5$$

$$+\underbrace{\tfrac{8!}{2!6!}}_{=28}\left(\sqrt{x}\right)^2\left(-3y^{1/4}\right)^6+\underbrace{\tfrac{8!}{1!7!}}_{=8}\left(\sqrt{x}\right)^1\left(-3y^{1/4}\right)^7+\underbrace{\tfrac{8!}{0!8!}}_{=1}\left(\sqrt{x}\right)^0\left(-3y^{1/4}\right)^8$$

$$=x^4-24x^{7/2}y^{1/4}+252x^3y^{1/2}-1512x^{5/2}y^{3/4}+5670x^2y$$

$$-13{,}608x^{3/2}y^{5/4}+20{,}412xy^{3/2}-17{,}496x^{1/2}y^{7/4}+6561y^2$$

33.

$$(r-s)^4=r^4+4r^3(-s)+6r^2(-s)^2+4r(-s)^3+(-s)^4$$

$$=r^4-4r^3s+6r^2s^2-4rs^3+s^4$$

34.

$$\left(x^2+y^2\right)^7=\left(x^2\right)^7+7\left(x^2\right)^6\left(y^2\right)^1+21\left(x^2\right)^5\left(y^2\right)^2+35\left(x^2\right)^4\left(y^2\right)^3+35\left(x^2\right)^3\left(y^2\right)^4$$

$$+21\left(x^2\right)^2\left(y^2\right)^5+7\left(x^2\right)^1\left(y^2\right)^6+\left(y^2\right)^7$$

$$=x^{14}+7x^{12}y^2+21x^{10}y^4+35x^8y^6+35x^6y^8+21x^4y^{10}+7x^2y^{12}+y^{14}$$

35.

$$\left(ax+by\right)^6=(ax)^6+6(ax)^5(by)+15(ax)^4(by)^2+20(ax)^3(by)^3$$

$$+15(ax)^2(by)^4+6(ax)^1(by)^5+(by)^6$$

$$=a^6x^6+6a^5bx^5y+15a^4b^2x^4y^2+20a^3b^3x^3y^3+15a^2b^4x^2y^4$$

$$+6ab^5xy^5+b^6y^6$$

36.

$$(x+3y)^4 = x^4 + 4x^3(3y)^1 + 6x^2(3y)^2 + 4x^1(3y)^3 + (3y)^4$$
$$= x^4 + 12x^3y + 54x^2y^2 + 108xy^3 + 81y^4$$

37. This term is $\binom{10}{4}x^{10-4}(2)^4 = \frac{10!}{4!6!}(16)x^6 = \frac{10\cdot9\cdot8\cdot7\cdot\cancel{6!}}{(4\cdot3\cdot2\cdot1)\cancel{6!}}x^6 = 3360x^6$

So, the desired coefficient is 3360.

38. This term is $\binom{9}{4}y^{9-4}(3)^4 = \frac{9!}{5!4!}(81)y^5 = \frac{9\cdot8\cdot7\cdot6\cdot\cancel{5!}}{(4\cdot3\cdot2\cdot1)\cancel{5!}}y^5 = 10{,}206y^5$

So, the desired coefficient is 10,206.

39. This term is $\binom{8}{4}y^{8-4}(-3)^4 = \frac{8!}{4!4!}(81)y^4 = \frac{8\cdot7\cdot6\cdot5\cdot\cancel{4!}}{(4\cdot3\cdot2\cdot1)\cancel{4!}}y^4 = 5670y^4$

So, the desired coefficient is 5670.

40. This term is $\binom{12}{7}x^{12-7}(-1)^7 = \frac{12!}{7!5!}(-1)x^5 = \frac{12\cdot11\cdot10\cdot9\cdot8\cdot\cancel{7!}}{(5\cdot4\cdot3\cdot2\cdot1)\cancel{7!}}(-1)x^5 = -792x^5$

So, the desired coefficient is -792.

41. This term is

$$\binom{7}{4}(2x)^{7-4}(3y)^4 = \frac{7!}{4!3!}(8\cdot81)x^3y^4 = \frac{7\cdot6\cdot5\cdot\cancel{4!}}{(3\cdot2\cdot1)\cancel{4!}}(648)x^3y^4 = 22{,}680x^3y^4$$

So, the desired coefficient is 22,680.

42. This term is

$$\binom{9}{7}(3x)^{9-7}(-5y)^7 = \frac{9!}{7!2!}(9)(-78{,}125)x^2y^7 = \frac{9\cdot8\cdot\cancel{7!}}{(2\cdot1)\cancel{7!}}(-703{,}125)x^2y^7 = (-25{,}312{,}500)x^2y^7$$

So, the desired coefficient is $-25{,}312{,}500$.

43. This term is $\binom{8}{4}(x^2)^{8-4}(y)^4 = \frac{8!}{4!4!}x^8y^4 = \frac{8\cdot7\cdot6\cdot5\cdot\cancel{4!}}{(4\cdot3\cdot2\cdot1)\cancel{4!}}x^8y^4 = 70x^8y^4$

So, the desired coefficient is 70.

44. This term is $\binom{10}{4}r^{10-4}(-s^2)^4 = \frac{10!}{4!6!}r^6s^8 = \frac{10\cdot9\cdot8\cdot7\cdot\cancel{6!}}{(4\cdot3\cdot2\cdot1)\cancel{6!}}r^6s^8 = 210r^6s^8$

So, the desired coefficient is 210.

45. Since $\binom{40}{6} = \frac{40!}{6!34!} = \frac{40\cdot39\cdot38\cdot37\cdot36\cdot35\cdot\cancel{34!}}{6!\cancel{34!}} = 3{,}838{,}380$, there are 3,838,380 such 6-number lottery numbers.

46. Since $\binom{60}{6} = \frac{60!}{6!54!} = \frac{60\cdot59\cdot58\cdot57\cdot56\cdot55\cdot\cancel{54!}}{6!\cancel{54!}} = 50{,}063{,}860$, there are 50,063,860 such 6-number lottery numbers.

47. Since $\binom{52}{5}=\frac{52!}{5!47!}=\frac{52\cdot51\cdot50\cdot49\cdot48\cdot\cancel{47!}}{5!\cancel{47!}}=2{,}598{,}960$, there are 2,598,960 possible 5-card poker hands.

48. Since $\binom{108}{11}=\frac{108!}{97!11!}=\frac{108\cdot107\cdot\ldots\cdot98\cdot\cancel{97!}}{11!\cancel{97!}}=344{,}985{,}116{,}800{,}000$, there are $344{,}985{,}116{,}800{,}000$ different 11-card Canasta hands.

49. $\binom{7}{5}\neq\frac{7!}{5!}$, $\binom{7}{5}=\frac{7!}{5!2!}$

50. Should replace powers of y by powers of $2y$. More precisely, the expansion should be

$$(x+2y)^4=x^4+4x^3(2y)+6x^2(2y)^2+4x(2y)^3+(2y)^4=x^4+8x^3y+24x^2y^2+32xy^3+16y^4$$

51. False, there are 11 terms.

52. True

53. True

54. False. This is only defined for $n=0$, and the result in such case is $\binom{0}{0}=1$.

55. True. Observe that $(2x-1)^{12}=\sum_{k=0}^{12}\binom{12}{k}(2x)^{12-k}(-1)^k$. The coefficient of x^8 occurs when $k=4$, which gives

$$\binom{12}{4}(2x)^{12-4}(-1)^4=\frac{12!}{4!8!}(2x)^8=\frac{12(11)(10)(9)8!}{4(3)(2)8!}256x^8=126{,}720x^8.$$

56. False. Observe that $(x^2+y)^{12}=\sum_{k=0}^{10}\binom{10}{k}(x^2)^{10-k}(y)^k$. The 6th term occurs when $k=5$, which gives

$$\binom{10}{5}(x^2)^{10-5}(y)^5=\frac{10!}{5!5!}(x^2)^5y^5=\frac{10(9)(8)(7)(6)5!}{5(4)(3)(2)5!}x^{10}y^5=252x^{10}y^5$$

57. Claim: $\binom{n}{k}=\binom{n}{n-k}$, $0\le k\le n$, for any $n\ge1$.

Proof. Observe that $\binom{n}{k}=\frac{n!}{k!(n-k)!}=\frac{n!}{(n-k)!k!}=\frac{n!}{(n-k)!\underbrace{(n-(n-k))}_{\text{This is } k \text{ written in a different form.}}!}=\binom{n}{n-k}$ ■

58. Observe that

$$2^n=(1+1)^n$$
$$=\binom{n}{0}1^{n-0}1^0+\binom{n}{1}1^{n-1}1^1+\ldots+\binom{n}{k}1^{n-k}1^k+\ldots+\binom{n}{n}1^01^n=\binom{n}{0}+\binom{n}{1}+\ldots+\binom{n}{n}$$

since all powers of 1 equal 1.

59. Note that this is equivalent to $\binom{m+1}{p+1}=\binom{m}{p}+\binom{m}{p+1}$. This can be established using mathematical induction.
60. For a fixed n and k, we have: $$(n-2k)\binom{n}{k}=(n-2k)\frac{n!}{k!(n-k)!}=n\left[\frac{(n-1)!\left[(n-k)-k\right]}{k!(n-k)!}\right]=n\left[\frac{(n-k)-(n-1)!k}{k!(n-k)!}\right]$$ $$=n\left[\frac{(n-1)!}{k!(n-1-k)!}-\frac{(n-1)!}{(k-1)!(n-k)!}\right]=n\left[\binom{n-1}{k}-\binom{n-1}{k-1}\right]$$
61. The binomial expansion of $(1-x)^3$ is $$1-3x+3x^2-x^3.$$ Notes on the Graph: Heavy Solid Curve: $y_1=1-3x+3x^2-x^3$ Heavy Dashed Curve: $y_2=-1+3x-3x^2+x^3$ Heavy Dotted Curve: $y_3=(1-x)^3$ The graphs of y_1 and y_3 are the same, so you only see two graphs. 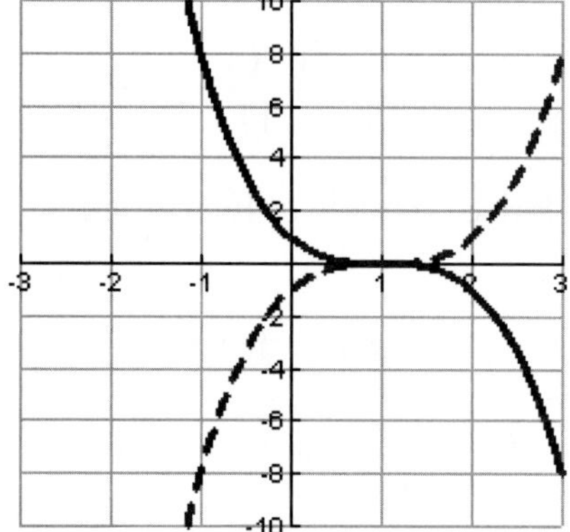
62. The binomial expansion of $(x+3)^4$ is $$x^4+12x^3+54x^2+108x+81.$$ Notes on the Graph: Heavy Solid Curve: $$y_3=x^4+12x^3+54x^2+108x+81$$ Heavy Dashed Curve: $$y_2=x^4+4x^3+6x^2+4x+1$$ Heavy Dotted Curve: $y_1=(x+3)^4$ The graphs of y_1 and y_3 are the same, so you only see two graphs. 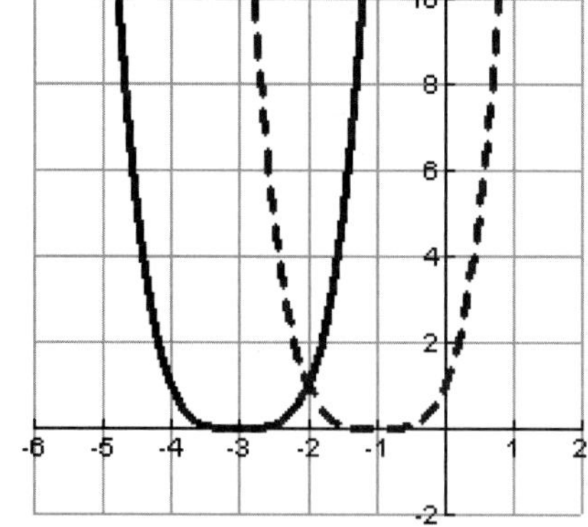

63. We see from the graph that as each term is added, the graphs of the respective functions get closer to the graph of $y_4 = (1-x)^3$ when $1 < x < 2$. However, when $x > 1$, this is no longer true.

Notes on the graph:
Heavy solid curve: y_4
Heavy dashed curve: y_1
Heavy dotted curve: y_2
Thin dashed curve: y_3

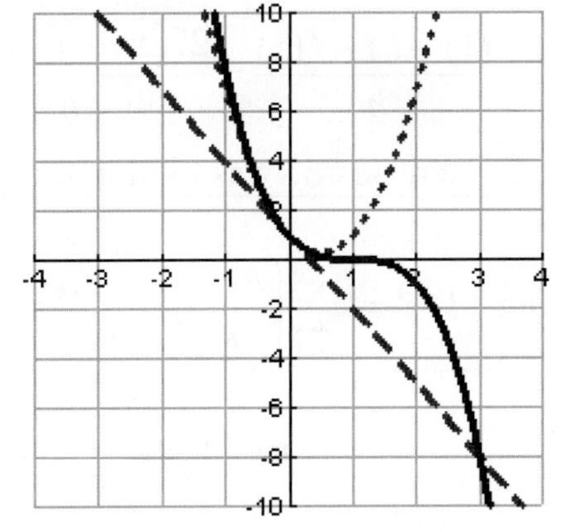

64. We see from the graph that as each term is added, the graphs of the respective functions get closer to the graph of $y_4 = (1-\frac{1}{x})^3$ when $1 < x < 2$. However, when $0 < x < 1$, this is no longer true.

Notes on the graph:
Heavy solid curve: y_4
Heavy dashed curve: y_1
Heavy dotted curve: y_2
Thin dashed curve: y_3

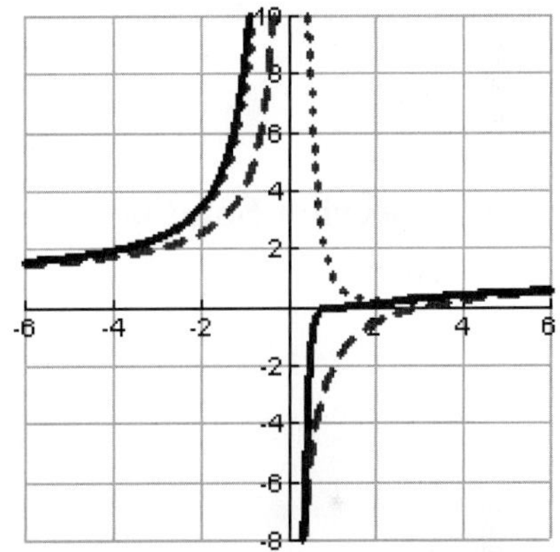

65. As each new term is added, the corresponding graph of the curve is a better approximation to the graph of $y = \left(1+\frac{1}{x}\right)^3$, for $1 < x < 2$. The series of functions does not get closer to this graph if $0 < x < 1$.

66. As each new term is added, the corresponding graph of the curve is a better approximation to the graph of $y = e^x$, for $-1 < x < 1$. The series of functions does not get closer to this graph if $1 < x < 2$.

67.

$$\frac{f(x+h)-f(x)}{h} = \frac{(x+h)^n - x^n}{h} = \frac{\sum_{k=0}^{n}\binom{n}{k}x^{n-k}h^k - x^n}{h} = \frac{\binom{n}{0}x^{n-0}h^0 + \sum_{k=1}^{n}\binom{n}{k}x^{n-k}h^k - x^n}{h}$$

$$= \frac{h\sum_{k=1}^{n}\binom{n}{k}x^{n-k}h^{k-1}}{h} = \sum_{k=1}^{n}\binom{n}{k}x^{n-k}h^{k-1}$$

68. $$\frac{f(x+h)-f(x)}{h}=\frac{2^n(x+h)^n-2^nx^n}{h}=2^n\left[\frac{(x+h)^n-x^n}{h}\right]=2^n\sum_{k=1}^{n}\binom{n}{k}x^{n-k}h^{k-1},$$ where Exercise #67 was used in the last step.
69. $$n(x+1)^{n-1}=n\sum_{k=0}^{n-1}\binom{n-1}{k}x^{n-1-k}1^k=\sum_{k=0}^{n-1}n\left(\frac{(n-1)!}{(n-1-k)!k!}\right)x^{n-1-k}=\sum_{k=0}^{n-1}\frac{n!}{(n-1-k)!k!}x^{n-1-k}$$
70. $$F(x)=\frac{1}{n+1}\sum_{k=0}^{n+1}\binom{n+1}{k}x^{n+1-k}1^k=\sum_{k=0}^{n+1}\frac{(n+1)!}{(n+1)(n+1-k)!k!}x^{n-1-k}=\sum_{k=0}^{n+1}\frac{n!}{(n+1-k)!k!}x^{n+1-k}$$

Chapter 10 Review Solutions --

1. $1,8,27,64$	**2.** $\frac{1!}{1},\frac{2!}{2},\frac{3!}{3},\frac{4!}{4},$ which is equivalent to $1,1,2,6$ (Note that $a_n=\frac{n!}{n}=\frac{\cancel{n}(n-1)!}{\cancel{n}}=(n-1)!$.)
3. $5,8,11,14$	**4.** $(-1)^1x^3,(-1)^2x^4,(-1)^3x^5,(-1)^4x^6$ which is equivalent to $-x^3,x^4,-x^5,x^6$.
5. $a_5=\left(\frac{2}{3}\right)^5=\frac{32}{243}=0.13$	**6.** $a_8=\frac{64}{6561}$
7. $a_{15}=\frac{(-1)^{15}(15-1)!}{15(15+1)!}=-\frac{(14)!}{15(16)!}$ $=-\frac{\cancel{14!}}{15\cdot16\cdot15\cdot\cancel{14!}}=-\frac{1}{3600}$	**8.** $a_{10}=1+\frac{1}{10}=\frac{11}{10}$
9. $a_n=(-1)^{n+1}3n,n\geq1$	**10.** $a_n=\begin{cases}n, & n \text{ odd}\\ \frac{1}{n}, & n \text{ even}\end{cases}$
11. $a_n=(-1)^n,n\geq1$	**12.** $a_n=10^{n-1},n\geq1$

13. $\frac{8!}{6!}=\frac{8\cdot 7\cdot \cancel{6!}}{\cancel{6!}}=56$	**14.** $\frac{20!}{23!}=\frac{\cancel{20!}}{23\cdot 22\cdot 21\cdot \cancel{20!}}=\frac{1}{10{,}626}$
15. $\frac{n(n-1)!}{(n+1)!}=\frac{\cancel{n(n-1)!}}{(n+1)\,\cancel{n(n-1)!}}=\frac{1}{n+1}$	**16.** $\frac{(n-2)!}{n!}=\frac{\cancel{(n-2)!}}{n(n-1)\,\cancel{(n-2)!}}=\frac{1}{n(n-1)}$
17. $5,3,1,-1$	**18.** $1,\underbrace{2^2\cdot 1}_{=4},\underbrace{4^2\cdot 1}_{=16},\underbrace{16^2\cdot 1}_{256}=1,4,16,256$
19. $1,2,\underbrace{(2)^2\cdot(1)}_{=4},\underbrace{4^2\cdot 2}_{=32}$ which is equivalent to $1,2,4,32$.	**20.** $1,2,\underbrace{\frac{1}{(2)^2}}_{=\frac{1}{4}},\underbrace{\frac{2}{\left(\frac{1}{4}\right)^2}}_{=32}$ which is equivalent to $1,2,\frac{1}{4},32$.
21. $\sum_{n=1}^{5}3=3(5)=15$	
22. $\sum_{n=1}^{4}\frac{1}{n^2}=1+\frac{1}{4}+\frac{1}{9}+\frac{1}{16}=\frac{144+36+16+9}{144}=\frac{205}{144}$	
23. $\sum_{n=1}^{6}(3n+1)=\frac{6}{2}(4+19)=69$	
24. $\sum_{k=0}^{5}\frac{2^{k+1}}{k!}=2+\frac{4}{1}+\frac{8}{2!}+\frac{16}{3!}+\frac{32}{4!}+\frac{64}{5!}$ $=2+4+4+\frac{8}{3}+\frac{4}{3}+\frac{8}{15}=\frac{218}{15}$	**25.** $\sum_{n=1}^{7}\frac{(-1)^n}{2^{n-1}}$
26. $\sum_{n=1}^{10}2n$	**27.** $\sum_{n=0}^{\infty}\frac{x^n}{n!}$ (Note: $2!=2,\ 3!=6,\ 4!=24$)
28. $\sum_{n=0}^{\infty}(-1)^n\frac{x^{n+1}}{n!}$	**29.** $A_{60}=30{,}000\left(1+\frac{0.04}{12}\right)^{60}=\$36{,}639.90$ So, the amount in the account after 5 years is \$36,639.90.

30.
Since $a_n = 180,000 + 30,000(n-1), n \geq 1$
we have
$$\sum_{n=1}^{4} a_n = \frac{4}{2}[180,000 + 270,000] = 900,000$$
So, the total salary for 4 years is \$900,000.
Also, the salary in the 4th year is given by
$a_4 = 180,000 + 30,000(3) = 270,000$.

31. Yes, it is arithmetic with $d = -2$.

32. No

33. Yes, it is arithmetic with $d = \frac{1}{2}$.

34. Yes, it is arithmetic with $d = -1$.

35. $a_n = \frac{(n+1)!}{n!} = \frac{(n+1)\cancel{n!}}{\cancel{n!}} = n+1$

Yes, it is arithmetic with $d = 1$.

36. $a_n = 5n - 5$
Yes, it is arithmetic with $d = 5$.

37. $a_n = a_1 + (n-1)d$
$= -4 + (n-1)(5) = 5n - 9$

38. $a_n = 5 + (n-1)(6) = 6n - 1$

39. $a_n = 1 + (n-1)\left(-\frac{2}{3}\right) = -\frac{2}{3}n + \frac{5}{3}$

40. $a_n = 0.001 + (n-1)(0.01) = 0.01n - 0.009$

41. We are given that
$$a_5 = a_i + (5-1)d = 13,$$
$$a_{17} = a_i + (17-1)d = 37.$$
In order to find a_n, we must first find *d*. To do so, subtract the above formulae to eliminate a_1:
$$\begin{array}{rl} & a_1 + 4d = 13 \\ - & \underline{a_1 + 16d = 37} \\ & -12d = -24 \end{array}$$
Hence, $d = 2$. So, $a_n = a_1 + (n-1)(2)$.
To find a_1, observe that
$$a_5 - a_1 = 2(4) = 8$$
So, $13 - a_1 = 8$ so that $a_1 = 5$.
Thus, $a_n = 5 + (n-1)(2) = 2n + 3, n \geq 1$

42. We are given that
$$a_7 = -14 = a_i + (7-1)d,$$
$$a_{10} = -23 = a_i + (10-1)d.$$
In order to find a_n, we must first find *d*. To do so, subtract the above formulae to eliminate a_1:
$$\begin{array}{rl} & a_1 + 6d = -14 \\ - & \underline{a_1 + 9d = -23} \\ & -3d = 9 \end{array}$$
Hence, $d = -3$. So,
$a_n = a_1 + (n-1)(-3)$.
To find a_1, observe that
$$a_7 - a_1 = (-3)(6) = -18$$
So, $-14 - a_1 = -18$, so that $a_1 = 4$.
Thus
$a_n = 4 + (n-1)(-3) = -3n + 7, n \geq 1$

43. We are given that

$$a_8 = 52 = a_1 + (8-1)d,$$
$$a_{21} = 130 = a_1 + (21-1)d.$$

In order to find a_n, we must first find *d*. To do so, subtract the above formulae to eliminate a_1:

$$\begin{array}{r} a_1 + 7d = 52 \\ -\ \ a_1 + 20d = 130 \\ \hline -13d = -78 \end{array}$$

Hence, $d = 6$. So, $a_n = a_1 + (n-1)(6)$.

To find a_1, observe that

$$a_8 - a_1 = 6(7) = 42$$

So, $52 - a_1 = 42$, so that $a_1 = 10$.

Thus, $a_n = 10 + (n-1)6 = 6n + 4,\ n \geq 1$.

44. We are given that

$$a_{11} = -30 = a_1 + (11-1)d,$$
$$a_{21} = -80 = a_1 + (21-1)d.$$

In order to find a_n, we must first find *d*. To do so, subtract the above formulae to eliminate a_1:

$$\begin{array}{r} a_1 + 10d = -30 \\ -\ \ a_1 + 20d = -80 \\ \hline -10d = 50 \end{array}$$

Hence, $d = -5$. So, $a_n = a_1 + (n-1)(-5)$.

To find a_1, observe that

$$a_{11} - a_1 = -5(10) = -50$$

So, $-30 - a_1 = -50$, so that $a_1 = 20$.

Thus, $a_n = 20 + (n-1)(-5) = -5n + 25,\ n \geq 1$.

45. $\sum_{k=1}^{20} 3k = \frac{20}{2}(3+60) = 630$

46. $\sum_{n=1}^{15} (n+5) = \frac{15}{2}(6+20) = 195$

47. The sequence is arithmetic with $n = 12$, $a_1 = 2$, and $a_{12} = 68$. So, the given sum is $\frac{12}{2}(2+68) = 420$.

48. The sequence is arithmetic with $n = 17$, $a_1 = \frac{1}{4}$, and $a_{17} = \frac{-31}{4}$. So, the given sum is

$$\frac{17}{2}\left(\frac{1}{4} - \frac{31}{4}\right) = \frac{17}{2}\left(\frac{-30}{4}\right) = -63.75.$$

49.

Bob

$$a_n = 45{,}000 + (n-1)2000$$

$$\sum_{n=1}^{15} a_n = \frac{15}{2}\left[45{,}000 + \underbrace{73{,}000}_{=a_{15}}\right] = 885{,}000$$

Tania

$$a_n = 38{,}000 + 4000(n-1)$$

$$\sum_{n=1}^{15} a_n = \frac{15}{2}\left[38{,}000 + \underbrace{94{,}000}_{=a_{15}}\right] = 990{,}000$$

So, Bob earns \$885,000 in 15 years, while Tania earns \$990,000 in 15 years.

50. The sequence $16, 48, 80, \ldots,$ is arithmetic with $a_1 = 16$ and $d = 32$. So, the *n*th term is given by $a_n = 16 + (n-1)32 = 32n - 16$. So, $a_5 = 144$. Further,

$\sum_{n=1}^{5} a_n = \frac{5}{2}[16 + 144] = \underbrace{400}$. So, he would have fallen 400 feet in 5 seconds.

51. Yes, it is geometric with $r=-2$.	**52.** No
53. Yes, it is geometric with $r=\frac{1}{2}$.	**54.** Yes, it is geometric with $r=10$.
55. $3,6,12,24,48$	**56.** $10,\frac{10}{4},\frac{10}{16},\frac{10}{64},\frac{10}{256}$
57. $100,\ -400,\ 1600,\ -6400,\ 25,600$	**58.** $-60,\ 30,\ -15,\ \frac{15}{2},\ -\frac{15}{4}$
59. $a_n=a_1r^{n-1}=7\cdot 2^{n-1},\ n\geq 1$	**60.** $a_n=12\left(\frac{1}{3}\right)^{n-1},\ n\geq 1$
61. $a_n=1(-2)^{n-1},\ n\geq 1$	**62.** $a_n=\frac{32}{5}\left(\frac{-1}{4}\right)^{n-1},\ n\geq 1$
63. $a_n=2(2)^{n-1}, n\geq 1$ So, $a_{25}=2(2)^{24}=33,554,432$	**64.** $a_n=\frac{1}{2}(2)^{n-1}, n\geq 1$ $a_{10}=\frac{1}{2}(2)^9=256$
65. $a_n=100\left(\frac{-1}{5}\right)^{n-1},\ n\geq 1$ So, in particular, $a_{12}=100\left(\frac{-1}{5}\right)^{11}=-2.048\times 10^{-6}$	**66.** $a_n=1000\left(\frac{-1}{2}\right)^{n-1},\ n\geq 1$ So, in particular, $a_{11}=1000\left(\frac{-1}{2}\right)^{10}=\frac{1000}{1024}$.
67. $\sum_{n=1}^{9}\left(\frac{1}{2}\right)3^{n-1}=\frac{1}{2}\left[\frac{1-3^9}{1-3}\right]=\frac{19,682}{4}=4920.50$	**68.** $\sum_{n=1}^{11}1\left(\frac{1}{2}\right)^{n-1}=1\left[\frac{1-\left(\frac{1}{2}\right)^{11}}{1-\frac{1}{2}}\right]$ $=2\left(1-\frac{1}{2148}\right)=\frac{4094}{2048}$

69. $\sum_{n=1}^{8} 5(3)^{n-1} = 5\left[\frac{1-3^8}{1-3}\right] = 16,400$

70. First, note that the given sum is not in the standard form since the exponent on the common ratio is n rather than $n-1$. So, you must first simplify the expression to put it into standard form, and then apply the known formula, as follows:

$$\sum_{n=1}^{7} \frac{2}{3}(5)^n = \sum_{n=1}^{7} \frac{10}{3}(5)^{n-1}$$
$$= \frac{10}{3}\left[\frac{1-5^7}{1-5}\right] = \frac{195,310}{3}$$

71. $\frac{1}{1-\frac{2}{3}} = 3$

72. $\frac{\left(\frac{-1}{5}\right)^2}{1+\frac{1}{5}} = \frac{\frac{1}{25}}{\frac{6}{5}} = \frac{1}{30}$

73. $a_n = 48,000(1.02)^{n-1}$, $n \geq 1$

So, in particular,

$$a_{12} = 48000(1.02)^{13-1} = 60,875.61.$$

So, the salary after 12 years is \$60,875.61.

74. $a_n = 15000(0.80)^{n-1}, n \geq 1$

So, in particular,

$$a_3 = 15,000(0.80)^{3-1} = 9600.$$

So, the boat will be worth \$9600 after 3 years.

75. <u>**Claim**</u>: $3n \leq 3^n$, for all $n \geq 1$.

Proof.

<u>Step 1</u>: Show the statement is true for $n = 1$.

$3(1) \leq 3^1$ is true since both terms equal 3.

<u>Step 2</u>: Assume the statement is true for $n = k$: $3k \leq 3^k$

Show the statement is true for $n = k+1$: $3(k+1) \leq 3^{k+1}$

$$\begin{aligned} 3(k+1) &= 3k+3 \\ &\leq 3^k + 3^1 \quad \text{(by assumption)} \\ &\leq 3^k + 3^k \quad (\text{since } 3 \leq 3^k, \text{ for } k \geq 1) \\ &= 2(3^k) < 3(3^k) = 3^{k+1} \end{aligned}$$

This completes the proof. ■

76. Claim: $4^n < 4^{n+1}$, for all $n \geq 1$.

Proof.

Step 1: Show the statement is true for $n=1$.

$$4^1 < 4^{1+1} = 4^2 = 16 \text{ is clearly true.}$$

Step 2: Assume the statement is true for $n=k$: $4^k < 4^{k+1}$

Show the statement is true for $n=k+1$: $4^{k+1} < 4^{k+2}$

$$\begin{aligned} 4^{k+1} &= 4^k \cdot 4 \\ &< 4^{k+1} \cdot 4 \text{ (by assumption)} \\ &= 4^{k+2} \end{aligned}$$

This completes the proof. ■

77. Claim: $2+7+ \ldots + (5n-3) = \frac{n}{2}(5n-1)$, for all $n \geq 1$.

Proof.

Step 1: Show the statement is true for $n=1$.

$$2 = \tfrac{1}{2}(5(1)-1) = \tfrac{1}{2}(4) \text{ is clearly true.}$$

Step 2: Assume the statement is true for $n=k$: $2+7+ \ldots + (5k-3) = \frac{k}{2}(5k-1)$

Show the statement is true for $n=k+1$:

$$2+7+ \ldots + (5(k+1)-3) = \tfrac{k+1}{2}(5(k+1)-1)$$

Observe that

$$\begin{aligned} 2+7+ \ldots + (5k-3)+(5(k+1)-3) &= \left(2+7+ \ldots + (5k-3)\right)+(5(k+1)-3) \\ &= \tfrac{k}{2}(5k-1)+(5(k+1)-3) \text{ (by assumption)} \\ &= \tfrac{k(5k-1)+2(5k+2)}{2} \\ &= \tfrac{5k^2+9k+4}{2} = \tfrac{(k+1)(5k+4)}{2} = \tfrac{k+1}{2}(5(k+1)-1) \end{aligned}$$

This completes the proof. ■

78. Claim: $2n^2 > (n+1)^2$, for all $n \geq 3$.

Proof.

Step 1: Show the statement is true for $n=3$.

$$18 = 2(3)^2 > (3+1)^2 = 16 \text{ is clearly true.}$$

Step 2: Assume the statement is true for $n=k$: $2k^2 > (k+1)^2$

Show the statement is true for $n=k+1$: $2(k+1)^2 > (k+1+1)^2$

$$\begin{aligned} 2(k+1)^2 = 2k^2+4k+1 &> (k+1)^2+4k+1 \\ &= k^2+2k+1+4k+1 \\ &= k^2+4k+4+\underbrace{(2k-2)}_{>1 \text{ since } k\geq 3} \\ &> (k+2)^2 \end{aligned}$$

This completes the proof. ■

79.

$$\binom{11}{8}=\frac{11!}{(11-8)!8!}=\frac{11!}{3!8!}=\frac{11\cdot 10\cdot 9\cdot \cancel{8!}}{\cancel{8!}(3\cdot 2\cdot 1)}=165$$

80.

$$\binom{10}{0}=\frac{10!}{(10-0)!0!}=\frac{\cancel{10!}}{\cancel{10!}0!}=\frac{1}{0!}=1$$

81.

$$\binom{22}{22}=\frac{22!}{(22-22)!22!}=\frac{\cancel{22!}}{\cancel{22!}0!}=\frac{1}{0!}=1$$

82.

$$\binom{47}{45}=\frac{47!}{(47-45)!45!}=\frac{47!}{2!45!}$$

$$=\frac{47\cdot 46\cdot \cancel{45!}}{\cancel{45!}(2\cdot 1)}=1081$$

83.

$$(x-5)^4=\sum_{k=0}^{4}\binom{4}{k}x^{4-k}(-5)^k$$

$$=\binom{4}{0}x^4(-5)^0+\binom{4}{1}x^3(-5)^1+\binom{4}{2}x^2(-5)^2+\binom{4}{3}x^1(-5)^3+\binom{4}{4}x^0(-5)^4$$

$$=\underbrace{\tfrac{4!}{4!0!}}_{=1}x^4(-5)^0+\underbrace{\tfrac{4!}{3!1!}}_{=4}x^3(-5)^1+\underbrace{\tfrac{4!}{2!2!}}_{=6}x^2(-5)^2+\underbrace{\tfrac{4!}{1!3!}}_{=4}x^1(-5)^3+\underbrace{\tfrac{4!}{0!4!}}_{=1}x^0(-5)^4$$

$$=x^4-20x^3+150x^2-500x+625$$

84.

$$(x+y)^5=\sum_{k=0}^{5}\binom{5}{k}x^{5-k}y^k$$

$$=\binom{5}{0}x^5y^0+\binom{5}{1}x^4y^1+\binom{5}{2}x^3y^2+\binom{5}{3}x^2y^3+\binom{5}{4}x^1y^4+\binom{5}{5}x^0y^5$$

$$=\underbrace{\tfrac{5!}{5!0!}}_{=1}x^5y^0+\underbrace{\tfrac{5!}{4!1!}}_{=5}x^4y^1+\underbrace{\tfrac{5!}{3!2!}}_{=10}x^3y^2+\underbrace{\tfrac{5!}{2!3!}}_{=10}x^2y^3+\underbrace{\tfrac{5!}{1!4!}}_{=5}x^1y^4+\underbrace{\tfrac{5!}{0!5!}}_{=1}x^0y^5$$

$$=x^5+5x^4y+10x^3y^2+10x^2y^3+5xy^4+y^5$$

85.

$$(2x-5)^3=\sum_{k=0}^{3}\binom{3}{k}(2x)^{3-k}(-5)^k$$

$$=\binom{3}{0}(2x)^3(-5)^0+\binom{3}{1}(2x)^2(-5)^1+\binom{3}{2}(2x)^1(-5)^2+\binom{3}{3}(2x)^0(-5)^3$$

$$=\underbrace{\tfrac{3!}{3!0!}}_{=1}(2x)^3(-5)^0+\underbrace{\tfrac{3!}{2!1!}}_{=3}(2x)^2(-5)^1+\underbrace{\tfrac{3!}{1!2!}}_{=3}(2x)^1(-5)^2+\underbrace{\tfrac{3!}{0!3!}}_{=1}(2x)^0(-5)^3$$

$$=8x^3-60x^2+150x-125$$

86.

$$(x^2+y^3)^4=\sum_{k=0}^{4}\binom{4}{k}\left(x^2\right)^{4-k}(y^3)^k$$

$$=\binom{4}{0}\left(x^2\right)^4(y^3)^0+\binom{4}{1}\left(x^2\right)^3(y^3)^1+\binom{4}{2}\left(x^2\right)^2(y^3)^2+\binom{4}{3}\left(x^2\right)^1(y^3)^3+\binom{4}{4}\left(x^2\right)^0(y^3)^4$$

$$=\underbrace{\tfrac{4!}{4!0!}}_{=1}\left(x^2\right)^4(y^3)^0+\underbrace{\tfrac{4!}{3!1!}}_{=4}\left(x^2\right)^3(y^3)^1+\underbrace{\tfrac{4!}{2!2!}}_{=6}\left(x^2\right)^2(y^3)^2+\underbrace{\tfrac{4!}{1!3!}}_{=4}\left(x^2\right)^1(y^3)^3+\underbrace{\tfrac{4!}{0!4!}}_{=1}\left(x^2\right)^0(y^3)^4$$

$$=x^8+4x^6y^3+6x^4y^6+4x^2y^9+y^{12}$$

87.

$$(\sqrt{x}+1)^5=\sum_{k=0}^{5}\binom{5}{k}(\sqrt{x})^{5-k}(1)^k,\ \ x\geq 0$$

$$=\binom{5}{0}(\sqrt{x})^5(1)^0+\binom{5}{1}(\sqrt{x})^4(1)^1+\binom{5}{2}(\sqrt{x})^3(1)^2+\binom{5}{3}(\sqrt{x})^2(1)^3$$

$$+\binom{5}{4}(\sqrt{x})^1(1)^4+\binom{5}{5}(\sqrt{x})^0(1)^5$$

$$=\underbrace{\tfrac{5!}{5!0!}}_{=1}(\sqrt{x})^5(1)^0+\underbrace{\tfrac{5!}{4!1!}}_{=5}(\sqrt{x})^4(1)^1+\underbrace{\tfrac{5!}{3!2!}}_{=10}(\sqrt{x})^3(1)^2+\underbrace{\tfrac{5!}{2!3!}}_{=10}(\sqrt{x})^2(1)^3$$

$$+\underbrace{\tfrac{5!}{1!4!}}_{=5}(\sqrt{x})^1(1)^4+\underbrace{\tfrac{5!}{0!5!}}_{=1}(\sqrt{x})^0(1)^5$$

$$=x^{5/2}+5x^2+10x^{3/2}+10x+5x^{1/2}+1$$

88.

$$(x^{2/3}+y^{1/3})^6=\sum_{k=0}^{6}\binom{6}{k}\left(x^{2/3}\right)^{6-k}\left(y^{1/3}\right)^k$$

$$=\binom{6}{0}\left(x^{2/3}\right)^6\left(y^{1/3}\right)^0+\binom{6}{1}\left(x^{2/3}\right)^5\left(y^{1/3}\right)^1+\binom{6}{2}\left(x^{2/3}\right)^4\left(y^{1/3}\right)^2$$

$$+\binom{6}{3}\left(x^{2/3}\right)^3\left(y^{1/3}\right)^3+\binom{6}{4}\left(x^{2/3}\right)^2\left(y^{1/3}\right)^4+\binom{6}{5}\left(x^{2/3}\right)^1\left(y^{1/3}\right)^5+\binom{6}{6}\left(x^{2/3}\right)^0\left(y^{1/3}\right)^6$$

$$=\underbrace{\tfrac{6!}{6!0!}}_{=1}\left(x^{2/3}\right)^6\left(y^{1/3}\right)^0+\underbrace{\tfrac{6!}{5!1!}}_{=6}\left(x^{2/3}\right)^5\left(y^{1/3}\right)^1+\underbrace{\tfrac{6!}{4!2!}}_{=15}\left(x^{2/3}\right)^4\left(y^{1/3}\right)^2+\underbrace{\tfrac{6!}{3!3!}}_{=20}\left(x^{2/3}\right)^3\left(y^{1/3}\right)^3$$

$$+\underbrace{\tfrac{6!}{2!4!}}_{=15}\left(x^{2/3}\right)^2\left(y^{1/3}\right)^4+\underbrace{\tfrac{6!}{1!5!}}_{=6}\left(x^{2/3}\right)^1\left(y^{1/3}\right)^5+\underbrace{\tfrac{6!}{0!6!}}_{=1}\left(x^{2/3}\right)^0\left(y^{1/3}\right)^6$$

$$=x^4+6x^{10/3}y^{1/3}+15x^{8/3}y^{2/3}+20x^2y+15x^{4/3}y^{4/3}+6x^{2/3}y^{5/3}+y^2$$

89.

$$(r-s)^5 = r^5s^0 + 5r^4(-s) + 10r^3(-s)^2 + 10r^2(-s)^3 + 5r(-s)^4 + (-s)^5r^0$$
$$= r^5 - 5r^4s + 10r^3s^2 - 10r^2s^3 + 5rs^4 - s^5$$

90.

$$(ax+by)^4 = (ax)^4(by)^0 + 4(ax)^3(by)^1 + 6(ax)^2(by)^2 + 4(ax)^1(by)^3 + (ax)^0(by)^4$$
$$= a^4x^4 + 4a^3bx^3y + 6a^2b^2x^2y^2 + 4ab^3xy^3 + b^4y^4$$

91. This term is $\binom{8}{2}x^{8-2}(-2)^2 = \frac{8!}{2!6!}(4)x^6 = \frac{8\cdot 7\cdot \cancel{6!}}{(2\cdot 1)\cancel{6!}}(4)x^6 = 112x^6$

So, the desired coefficient is 112.

92. This term is $\binom{7}{3}3^3y^{7-3} = \frac{7!}{3!4!}(27)y^4 = \frac{7\cdot 6\cdot 5\cdot \cancel{4!}}{(3\cdot 2\cdot 1)\cancel{4!}}(27)y^4 = 945y^4$

So, the desired coefficient is 945.

93. This term is

$$\binom{6}{4}(2x)^{6-4}(5y)^4 = \frac{6!}{4!2!}(4\cdot 625)x^2y^4 = \frac{6\cdot 5\cdot \cancel{4!}}{(2\cdot 1)\cancel{4!}}(2500)x^2y^4 = 37{,}500x^2y^4$$

So, the desired coefficient is 37,500.

94. This term is $\binom{8}{4}(r^2)^{8-4}(-s)^4 = \frac{8!}{4!4!}r^8s^4 = \frac{8\cdot 7\cdot 6\cdot 5\cdot \cancel{4!}}{(4\cdot 3\cdot 2\cdot 1)\cancel{4!}}r^8s^4 = 70r^8s^4$

So, the desired coefficient is 70.

95. Since $\binom{53}{6} = \frac{53!}{6!47!} = \frac{53\cdot 52\cdot 51\cdot 50\cdot 49\cdot 48\cdot \cancel{47!}}{6!\cancel{47!}} = 22{,}957{,}480$, there are 22,957,480 different 6-number combinations.

96. Since $\binom{108}{13} = \frac{108!}{95!13!} = \frac{108\cdot 107\cdot \ldots \cdot 96\cdot \cancel{95!}}{13!\cancel{95!}} = 20{,}592{,}957{,}740{,}000{,}000$, there are $20{,}592{,}957{,}740{,}000{,}000$ different 13-card Canasta hands.

97. $\frac{5369}{3600}$	**98.** The sum is infinite.
99. $\frac{34{,}875}{14}$	**100.** 6556

101. The graphs of the two curves are given by: 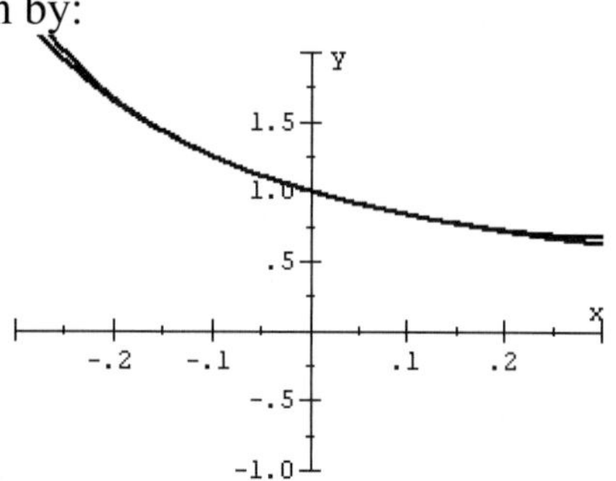 The series will sum to $\frac{1}{1+2x}$.	**102.** Yes, and the sum is approximately 7.4215 since $\sum_{n=0}^{\infty}\left(\frac{e}{\pi}\right)^n = \frac{1}{1-\frac{e}{\pi}} = \frac{\pi}{\pi - e} \approx 7.4215$
103. The sum is 99,900. Yes, it agrees with Exercise 77.	**104.** The graphs are given by Y1=2X2 X=800 Y=1280000 Yes, it confirms the proof in Exercise 78.
105. As each new term is added, the graphs become better approximations of the graph of $y=(1+2x)^4$ for $-0.1<x<0.1$. The series does not get closer to this graph for $0.1<x<1$.	**106.** As each new term is added, the graphs become better approximations of the graph of $y=(1-2x)^4$ for $-0.1<x<0.1$. The series does not get closer to this graph for $0.1<x<1$.

Chapter 10 Practice Test Solutions

1. x^{n-1}	**2.** geometric with $r = x$
3. $S_n = \frac{1-x^n}{1-x}$	**4.** $\sum_{n=0}^{\infty} x^n$
5. We must have $\|x\|<1$ in order for the sum in Problem 4 to exist. In such case, the sum is $\frac{1}{1-x}$.	

6. $\sum_{n=1}^{\infty}\left(\frac{1}{3}\right)^n = \dfrac{\frac{1}{3}}{1-\frac{1}{3}} = \dfrac{1}{2}$

7. $\sum_{n=1}^{10} 3\cdot\left(\frac{1}{4}\right)^n = 3\cdot\left(\frac{1}{4}\right)\cdot\dfrac{1-\left(\frac{1}{4}\right)^{10}}{1-\frac{1}{4}} = 1-\left(\frac{1}{4}\right)^{10} \approx 1$

8. $\frac{50}{2}\big(3+2(50)+1\big) = 25\big(104\big) = 2600$

9. $\sum_{n=1}^{100}\big(5n-3\big) = \dfrac{100}{2}\big(2+497\big) = 24{,}950$

10. **Claim:** $2+4+...+2n = \underbrace{n^2+n}_{=n(n+1)}$, for all $n \ge 1$.

Proof.

Step 1: Show the statement is true for $n=1$.

$$2 = 1+1 = 1^2+1 = 2\text{, which is clearly true.}$$

Step 2: Assume the statement is true for $n=k$: $2+4+...+2k = k(k+1)$

Show the statement is true for $n=k+1$: $2+4+...+2(k+1) = (k+1)(k+2)$

Observe that

$$\begin{aligned} 2+4+...+2k+2(k+1) &= \big(2+4+...+2k\big)+2(k+1) \\ &= k(k+1)+2(k+1) \quad \text{(by assumption)} \\ &= (k+2)(k+1) \end{aligned}$$

This completes the proof. ■

11. $\dfrac{7!}{2!} = \dfrac{7\cdot 6\cdot 5\cdot 4\cdot 3\cdot \cancel{2!}}{\cancel{2!}} = 2520$

12. Note that $(2x+y)^5 = (2x)^5 + \binom{5}{1}(2x)^4 y + \boxed{\binom{5}{2}(2x)^3 y^2} + ...$

Hence, $\binom{5}{2}(2x)^3 y^2 = \dfrac{5!}{2!3!}8x^3y^2 = 80x^3y^2$.

13. $\binom{15}{12} = \dfrac{15!}{12!3!} = \dfrac{15\cdot 14\cdot 13\cdot \cancel{12!}}{3!\cancel{12!}} = 455$

14. $\binom{k}{k} = \dfrac{k!}{(k-k)!k!} = \dfrac{k!}{0!k!} = 1$

15. $\dfrac{42!}{7!37!} = \dfrac{42(41)(40)(39)(38)37!}{7(6)(5)(4)(3)(2)37!} = 20{,}254$

16. $\dfrac{(n+2)!}{(n-2)!} = \dfrac{(n+2)(n-1)n(n-1)(n-2)!}{(n-2)!} = (n+2)(n-1)n(n-1) = n^4+2n^3-n^2-2n$

17.

$$(x^2+\tfrac{1}{x})^5=\sum_{k=0}^{5}\binom{5}{k}\left(x^2\right)^{5-k}\left(\tfrac{1}{x}\right)^k,\ x\neq 0$$

$$=\binom{5}{0}\left(x^2\right)^5\left(\tfrac{1}{x}\right)^0+\binom{5}{1}\left(x^2\right)^4\left(\tfrac{1}{x}\right)^1+\binom{5}{2}\left(x^2\right)^3\left(\tfrac{1}{x}\right)^2$$

$$+\binom{5}{3}\left(x^2\right)^2\left(\tfrac{1}{x}\right)^3+\binom{5}{4}\left(x^2\right)^1\left(\tfrac{1}{x}\right)^4+\binom{5}{5}\left(x^2\right)^0\left(\tfrac{1}{x}\right)^5$$

$$=\underbrace{\tfrac{5!}{5!0!}}_{=1}\left(x^2\right)^5\left(\tfrac{1}{x}\right)^0+\underbrace{\tfrac{5!}{4!1!}}_{=5}\left(x^2\right)^4\left(\tfrac{1}{x}\right)^1+\underbrace{\tfrac{5!}{3!2!}}_{=10}\left(x^2\right)^3\left(\tfrac{1}{x}\right)^2$$

$$+\underbrace{\tfrac{5!}{2!3!}}_{=10}\left(x^2\right)^2\left(\tfrac{1}{x}\right)^3+\underbrace{\tfrac{5!}{1!4!}}_{=5}\left(x^2\right)^1\left(\tfrac{1}{x}\right)^4+\underbrace{\tfrac{5!}{0!5!}}_{=1}\left(x^2\right)^0\left(\tfrac{1}{x}\right)^5$$

$$=x^{10}+5x^7+10x^4+10x+\tfrac{5}{x^2}+\tfrac{1}{x^5}$$

18.

$$(3x-2)^4=(3x)^4(2)^0+\binom{4}{1}(3x)^3(-2)+\binom{4}{2}(3x)^2(-2)^2+\binom{4}{3}(3x)(-2)^3+\binom{4}{4}(3x)^0(-2)^4$$

$$=81x^4-\frac{4!}{3!1!}54x^3+\frac{4!}{2!2!}36x^2-\frac{4!}{1!3!}24+16=81x^4-216x^3+216x^2-96x+16$$

19. **Claim:** $\sum_{i=1}^{n}\left(4i-1\right)=2n^2+n$.

Proof. Step 1: Show the statement is true for $n=2$.

$$\sum_{i=1}^{2}\left(4i-1\right)=(4(1)-1)(4(2)-1)=10=2(2)^2+2$$

Step 2: Assume the statement is true for $n=k$:

$$\sum_{i=1}^{k}\left(4i-1\right)=2k^2+k$$

Show the statement is true for $n=k+1$:

Observe that

$$\sum_{i=1}^{k+1}\left(4i-1\right)=\sum_{i=1}^{k}\left(4i-1\right)+(4(k+1)-1)=2k^2+k+(4(k+1)-1)$$

$$=2k^2+5k+3=\left(2k^2+4k+2\right)+\left(k+1\right)=2\left(k+1\right)^2+\left(k+1\right)$$

This completes the proof. ■

20. **Claim:** $\sum_{i=1}^{n}\left(2^i+i^3\right)=2^{n+1}-2+\frac{n^2(n+1)^2}{4}$.

Proof. Step 1: Show the statement is true for $n=2$.

$$\sum_{i=1}^{2}\left(2^i+i^3\right)=\left(2^1+1^3\right)+\left(2^2+2^3\right)=15=2^{2+1}-2+\frac{2^2(2+1)^2}{4}$$

Step 2: Assume the statement is true for $n=k$:

$$\sum_{i=1}^{k}\left(2^i+i^3\right)=2^{k+1}-2+\frac{k^2(k+1)^2}{4}$$

Show the statement is true for $n=k+1$:

Observe that

$$\sum_{i=1}^{k+1}\left(2^i+i^3\right)=\sum_{i=1}^{k}\left(2^i+i^3\right)+\left(2^{k+1}+(k+1)^3\right)=2^{k+1}-2+\frac{k^2(k+1)^2}{4}+2^{k+1}+(k+1)^3$$

$$=2\left(2^{k+1}\right)-2+\frac{k^2(k+1)^2+4(k+1)^3}{4}=2^{k+2}-2+\frac{(k+1)^2\left[k^2+4(k+1)\right]}{4}$$

$$=2^{k+2}-2+\frac{(k+1)^2(k+2)^2}{4}$$

This completes the proof. ■

21. $\sum_{n=0}^{\infty}\left(\frac{2}{5}\right)^n=\frac{1}{1-\frac{2}{5}}=\frac{5}{3}$	**22.** $\sum_{n=0}^{\infty}\left(\frac{3}{4}\right)^n=\frac{1}{1-\frac{3}{4}}=4$
23. $\sum_{n=1}^{\infty}\frac{5}{2(n+3)}$	**24.** $\sum_{n=1}^{\infty}\frac{2^n+(-1)^n}{2^{n+1}}$

25.

$$(2-3x)^7=\sum_{k=0}^{7}\binom{7}{k}(2)^{7-k}(-3x)^k,\ x\geq 0$$

$$=\binom{7}{0}(2)^7(-3x)^0+\binom{7}{1}(2)^6(-3x)^1+\binom{7}{2}(2)^5(-3x)^2+\binom{7}{3}(2)^4(-3x)^3$$

$$+\binom{7}{4}(2)^3(-3x)^4+\binom{7}{5}(2)^2(-3x)^5+\binom{7}{6}(2)^1(-3x)^6+\binom{7}{7}(2)^0(-3x)^7$$

$$=128-1344x+6048x^2-15120x^3+22680x^4-20412x^5+10206x^6-2187x^7$$

26. $$\left(x^2+y^3\right)^6 = (x^2)^6+6(x^2)^5(y^3)+15(x^2)^4(y^3)^2+20(x^2)^3(y^3)^3 +15(x^2)^2(y^3)^4+6(x^2)^1(y^3)^5+(y^3)^6 = x^{12}+6x^{10}y^3+15x^8y^6+20x^6y^9+15^4y^{12}+6x^2y^{15}+y^{18}$$

27. Expanding this using a calculator reveals that the constant term is 184,756.	**28.** Using a calculator to compute this sum yields $\frac{73,375}{12}$.

Chapter 10 Cumulative Test--

1.

$$\frac{f(x+h)-f(x)}{h} = \frac{\frac{(x+h)^2}{x+h+1}-\frac{x^2}{x+1}}{h} = \frac{\left(x^2+2hx+h^2\right)(x+1)-x^2(x+h+1)}{h(x+1)(x+h+1)}$$

$$= \frac{x^3+2hx^2+h^2x+x^2+2hx+h^2-x^3-x^2h-x^2}{h(x+1)(x+h+1)}$$

$$= \frac{h\left(x^2+hx+2x+h\right)}{h(x+1)(x+h+1)} = \frac{x^2+hx+2x+h}{(x+1)(x+h+1)}$$

2. $f(g(x)) = f\left(\frac{1}{x+3}\right) = \left(\frac{1}{x+3}\right)^2+\frac{1}{x+3} = \frac{1+x+3}{(x+3)^2} = \frac{x+4}{(x+3)^2}$

3. The possible rational zeros are: $\pm1, \pm5$ Using synthetic division yields

5	1	−4	−4	−4	−5
		5	5	5	5
−1	1	1	1	1	0
		−1	0	−1	
	1	0	1	0	

So, $f(x)=(x-5)(x+1)\left(x^2+1\right)=(x-5)(x+1)(x-i)(x+i)$.

4. Solve $y=5x-4$ for x: $x=\frac{y+4}{5}$. We conclude that $f^{-1}(x)=\frac{x+4}{5}$.

5.

$$\begin{array}{r} -6x^3+21x \\ x^2+0x+3\overline{\big)\,-6x^5+0x^4+3x^3+2x^2+0x-7} \\ \underline{-(-6x^5+0x^4-18x^3)} \\ 21x^3+0x^2+0x \\ \underline{-(21x^3+0x^2+63x)} \\ -63x-7 \end{array}$$

So, $\boxed{Q(x)=-6x^3+21x,\quad r(x)=-63x-7}$.

6. $10^{-3}=0.001$

7.

$$\ln(5x-6)=2$$
$$5x-6=e^2$$
$$x=\frac{6+e^2}{5}\approx 2.678$$

8. Using the formula $A=\frac{1}{2}r^2\theta_d\left(\frac{\pi}{180^\circ}\right)$ yields

$$A=\tfrac{1}{2}(21\text{ft.})^2\left(50^\circ\right)\left(\tfrac{\pi}{180^\circ}\right)\approx\boxed{192.4\text{ ft.}^2}.$$

9. $-\frac{\sqrt{2}}{2}$

10. Consider the following triangle:

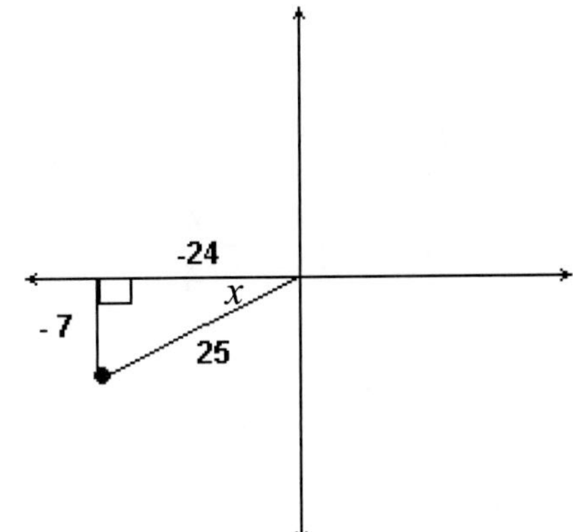

Observe that

$$\cos(2x)=\cos^2 x-\sin^2 x$$
$$=\left(-\tfrac{24}{25}\right)^2-\left(-\tfrac{7}{25}\right)^2=\tfrac{527}{625}$$

11. Factoring yields:

$$\left(2\cos\theta+1\right)\left(\cos\theta-1\right)=0$$
$$\cos\theta=-\tfrac{1}{2},\,1$$
$$\theta=0,\ \tfrac{\pi}{3},\tfrac{2\pi}{3},\ 2\pi$$

12. This is SSA, so use Law of Sines:

Step 1: Determine β.

$$\frac{\sin\gamma}{c}=\frac{\sin\alpha}{a} \Rightarrow \frac{\sin 53^\circ}{17}=\frac{\sin\alpha}{18} \Rightarrow \sin\alpha=\frac{18\sin 53^\circ}{17} \Rightarrow \alpha=\sin^{-1}\left(\frac{18\sin 53^\circ}{17}\right)\approx 57.74^\circ$$

Note that there are two triangles in this case since the angle in QII with the same sine as this value of α is $180^\circ-57.74^\circ=122.26^\circ$ and $\gamma+\alpha<180^\circ$, therefore allowing the formation of a triangle

Step 2: Solve for the triangle I (corresponding to $\alpha=57.74^\circ$).

$$\beta\approx 180^\circ-(53^\circ+57.74^\circ)=69.26^\circ$$

$$\frac{\sin\gamma}{c}=\frac{\sin\beta}{b} \Rightarrow \frac{\sin 53^\circ}{17}=\frac{\sin 69.26^\circ}{b} \Rightarrow b=\frac{17\sin 69.26^\circ}{\sin 53^\circ}\approx 19.91$$

Step 3: Solve for the triangle II (corresponding to $\alpha=122.26^\circ$).

$$\beta\approx 180^\circ-(53^\circ+122.26^\circ)=4.74^\circ$$

$$\frac{\sin\gamma}{c}=\frac{\sin\beta}{b} \Rightarrow \frac{\sin 53^\circ}{17}=\frac{\sin 4.74^\circ}{b} \Rightarrow b=\frac{17\sin 4.74^\circ}{\sin 53^\circ}\approx 1.76$$

13. In order to express $\sqrt{2}-i\sqrt{2}$ in polar form, observe that $x=\sqrt{2},\ y=-\sqrt{2}$, so that the point is in QIV. Now,

$$r=\sqrt{x^2+y^2}=\sqrt{\left(\sqrt{2}\right)^2+\left(-\sqrt{2}\right)^2}=2$$

$$\tan\theta=\frac{y}{x}=\frac{-\sqrt{2}}{\sqrt{2}}=-1,\text{ so that } \theta=\tan^{-1}(-1)=-45^\circ \text{ or } -\frac{\pi}{4}.$$

Since $0\le\theta<2\pi$, we must use the reference angle 315° or $\frac{7\pi}{4}$.

Hence, $\boxed{\sqrt{2}-i\sqrt{2}=2\left[\cos\left(\frac{7\pi}{4}\right)+i\sin\left(\frac{7\pi}{4}\right)\right]=2\left[\cos(315^\circ)+i\sin(315^\circ)\right]}$.

14. Solve the system:

$$\begin{cases} 8x-5y=15 & \textbf{(1)} \\ y=\frac{8}{5}x+10 & \textbf{(2)} \end{cases}$$

Substitute **(2)** into **(1)**:

$$8x-5\left(\tfrac{8}{5}x+10\right)=15 \Rightarrow 0=65$$

Since this results in a false statement, the system has no solution.

15. Using matrix methods, we have

$$\left[\begin{array}{ccc|c} 2 & -1 & 1 & 1 \\ 1 & -1 & 4 & 3 \end{array}\right] \xrightarrow{R_1-2R_2\to R_2} \left[\begin{array}{ccc|c} 2 & -1 & 1 & 1 \\ 0 & 1 & -7 & -5 \end{array}\right]$$

$$\xrightarrow{\frac{1}{2}R_1\to R_1} \left[\begin{array}{ccc|c} 1 & -\frac{1}{2} & \frac{1}{2} & \frac{1}{2} \\ 0 & 1 & -7 & -5 \end{array}\right]$$

Let $y=a$. Then, $x=7a-5$ (using 2nd row). Hence, substituting these into the equation corresponding to the first row, we see that:

$$(7a-5)-\tfrac{1}{2}a+\tfrac{1}{2}z=\tfrac{1}{2} \Rightarrow z=11-13a$$

16. The region is as follows:

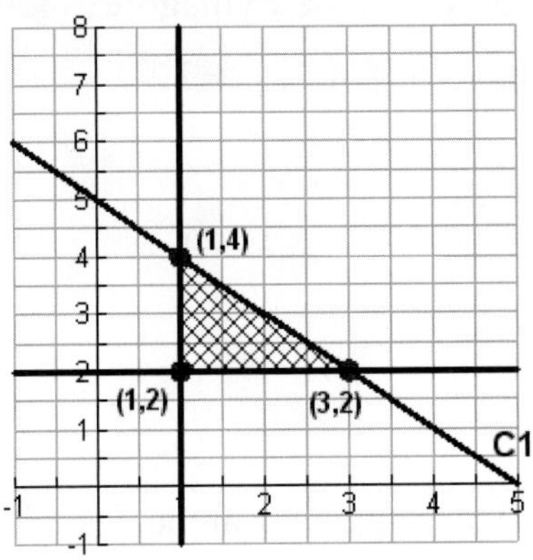

Vertex	$z=4x+5y$
(1,4)	24 MAX
(1,2)	14
(3,2)	22

So, the maximum is $z(1,4) = 24$.

Note on the graph:
C1: $y = 5 - x$

17.

$$\left[\begin{array}{ccc|c} 1 & 5 & -2 & 3 \\ 3 & 1 & 2 & -3 \\ 2 & -4 & 4 & 10 \end{array}\right] \xrightarrow[R_3-2R_1\to R_3]{R_2-3R_1\to R_2} \left[\begin{array}{ccc|c} 1 & 5 & -2 & 3 \\ 0 & -14 & 8 & -12 \\ 0 & -14 & 8 & 4 \end{array}\right]$$

$$\xrightarrow{R_2-R_3\to R_3} \left[\begin{array}{ccc|c} 1 & 5 & -2 & 3 \\ 0 & -14 & 8 & -12 \\ 0 & 0 & 0 & -16 \end{array}\right]$$

The last row suggests the system has no solution.

18.

$$\begin{bmatrix} 9 & 0 \\ 1 & 2 \end{bmatrix}\begin{bmatrix} 11 & 2 & -1 \\ 9 & 1 & 4 \end{bmatrix} = \begin{bmatrix} 99 & 18 & -9 \\ 29 & 4 & 7 \end{bmatrix}$$

19.

$$\begin{vmatrix} 2 & 5 & -1 \\ 1 & 4 & 0 \\ -2 & 1 & 3 \end{vmatrix} = 2\begin{vmatrix} 4 & 0 \\ 1 & 3 \end{vmatrix} - 5\begin{vmatrix} 1 & 0 \\ -2 & 3 \end{vmatrix} - 1\begin{vmatrix} 1 & 4 \\ -2 & 1 \end{vmatrix}$$

$$= 2(12) - 5(3) - 1(9) = 0$$

20. Since (3,5) is the vertex and the directrix is $x = 7$, the parabola opens to the left and has general equation

$$-4p(x-3) = (y-5)^2$$

To find p, note the distance from vertex to directrix is 4 units, so $p = 4$. Hence, the equation is

$$-16(x-3) = (y-5)^2$$

$$x = -\frac{1}{16}(y-5)^2 + 3$$

21.

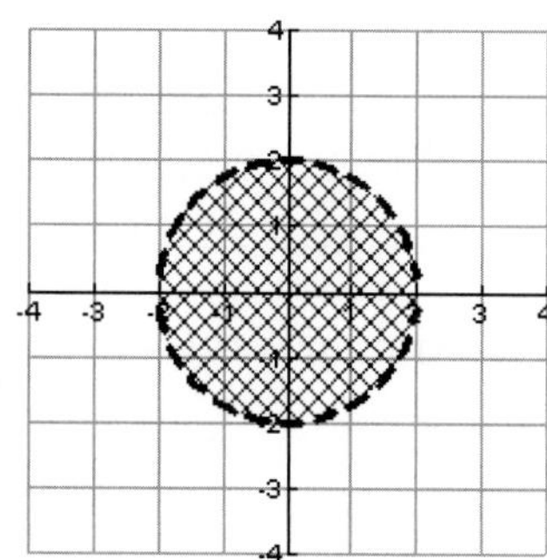

22. Observe that $\frac{x}{2} = \sin t$, $\frac{y}{3} = \cos t$. Hence, using the Pythagorean identity, we have $\left(\frac{x}{2}\right)^2 + \left(\frac{y}{3}\right)^2 = 1$.

23.
$$\frac{2^0}{1!} + \frac{2^1}{2!} + \frac{2^2}{3!} + \frac{2^3}{4!} = 1 + 1 + \tfrac{2}{3} + \tfrac{1}{3} = 3$$

24. The sequence can be described by
$$a_n = 5 \cdot 3^{n-1},\ n \geq 1.$$
So, it is geometric.

25. $\dfrac{7!}{3!4!} = \dfrac{7(6)(5)4!}{3(2)4!} = 35$

26.
$$\begin{aligned}
(x - x^2)^5 &= \sum_{k=0}^{5} \binom{5}{k} x^{5-k}(-x^2)^k \\
&= \binom{5}{0} x^5(-x^2)^0 + \binom{5}{1} x^4(-x^2)^1 + \binom{5}{2} x^3(-x^2)^2 + \binom{5}{3} x^2(-x^2)^3 \\
&\quad + \binom{5}{4} x^1(-x^2)^4 + \binom{5}{5} x^0(-x^2)^5 \\
&= \underbrace{\tfrac{5!}{5!0!}}_{=1} x^5(-x^2)^0 + \underbrace{\tfrac{5!}{4!1!}}_{=5} x^4(-x^2)^1 + \underbrace{\tfrac{5!}{3!2!}}_{=10} x^3(-x^2)^2 + \underbrace{\tfrac{5!}{2!3!}}_{=10} x^2(-x^2)^3 \\
&\quad + \underbrace{\tfrac{5!}{1!4!}}_{=5} x^1(-x^2)^4 + \underbrace{\tfrac{5!}{0!5!}}_{=1} x^0(-x^2)^5 \\
&= -x^{10} + 5x^9 - 10x^8 + 10x^7 - 5x^6 + x^5
\end{aligned}$$

CHAPTER 11

Section 11.1 Solutions--

1. Consider the following table of values:

x	$\frac{x-1}{x^2-1}$
0.9	0.526
0.99	0.5025
0.999	0.50025
0.9999	0.500025
↓	
1	0.5
↑	
1.0001	0.499975
1.001	0.49975
1.01	0.49751
1.1	0.47619

So, $\lim_{x\to 1}\frac{x-1}{x^2-1}=\frac{1}{2}$.

2. Consider the following table of values:

x	$\frac{x+2}{x^2-4}$
-1.9	-0.2564
-1.99	-0.25063
-1.999	-0.250063
-1.9999	-0.2500063
↓	
-2	-0.25
↑	
-2.0001	-0.249994
-2.001	-0.249938
-2.01	-0.24938
-2.1	-0.24390

So, $\lim_{x\to -2}\frac{x+2}{x^2-4}=-\frac{1}{4}$.

3. Consider the following table of values:

x	$\frac{\sqrt{1-x}-2}{x+3}$
-2.9	-0.25158
-2.99	-0.25016
-2.999	-0.250016
-2.9999	-0.2500016
↓	
-3	-0.25
↑	
-3.0001	-0.249918
-3.001	-0.24998
-3.01	-0.24984
-3.1	-0.24846

So, $\lim_{x\to -3}\frac{\sqrt{1-x}-2}{x+3}=-\frac{1}{4}$.

4. Consider the following table of values:

x	$\frac{\sqrt{x}-3}{x-9}$
8.9	0.16713
8.99	0.166713
8.999	0.1666713
8.9999	0.16666713
↓	
9	0.166…
↑	
9.0001	0.1666662
9.001	0.166662
9.01	0.16662
9.1	0.16621

So, $\lim_{x\to 9}\frac{\sqrt{x}-3}{x-9}=\frac{1}{6}$.

5. Consider the following table of values:

x	$\frac{\sin x}{x}$
-0.1	0.99833
-0.01	0.999983
-0.001	0.9999983
-0.0001	0.999999983
↓	
0	1
↑	
0.0001	0.999999983
0.001	0.9999983
0.01	0.999983
0.1	0.99833

So, $\lim_{x\to 0} \frac{\sin x}{x} = 1$.

6. Consider the following table of values:

x	$\frac{\cos(\pi x)+1}{x-1}$
0.9	-0.48943
0.99	-0.04934
0.999	-0.0049348
0.9999	-0.000493
↓	
1	0
↑	
1.0001	0.000493
1.001	0.0049348
1.01	0.04934
1.1	0.48943

So, $\lim_{x\to 1} \frac{\cos(\pi x)+1}{x-1} = 0$.

7. Consider the following table of values:

x	$\frac{e^x-1}{x}$
-0.1	0.951626
-0.01	0.995017
-0.001	0.99950
-0.0001	0.99995
↓	
0	1
↑	
0.0001	1.00005
0.001	1.0005
0.01	1.00501
0.1	1.05171

So, $\lim_{x\to 0} \frac{e^x-1}{x} = 1$.

8. Consider the following table of values:

x	$\frac{1}{e^{1/x}}$
-0.1	22,026.47
-0.01	$2.688\text{x}10^{44}$
-0.001	Large
-0.0001	Even larger
↓	Approaches ∞
0	
↑	Approaches 0
0.0001	$1.135\text{x}10^{-4342}$
0.001	$5.075\text{x}10^{-434}$
0.01	$3.720\text{x}10^{-43}$
0.1	0.000045

So, $\lim_{x\to 0} \frac{1}{e^{1/x}}$ DNE.

9. Consider the following table of values:

x	$x\ln x$
0.1	-0.23026
0.01	-0.04605
0.001	-0.00691
0.0001	-0.00092
↓	
0	0

So, $\lim_{x\to 0^+} x\ln x = 0$.

10. Consider the following table of values:

x	$\frac{\ln(x+1)}{x}$
-0.1	1.0536
-0.01	1.005034
-0.001	1.0005
-0.0001	1.00005
↓	
0	1
↑	
0.0001	0.999950
0.001	0.9995
0.01	0.995
0.1	0.95310

So, $\lim_{x\to 0} \frac{\ln(x+1)}{x} = 1$.

11. 0	**12.** 0
13. DNE (since left and right limits are different)	**14.** DNE (since left and right limits are different)
15. $-\infty$	**16.** 1
17. 0	**18.** 0
19. DNE (since left and right limits are different)	**20.** $-\infty$
21. 0	**22.** DNE
23. DNE (too wild of oscillations near 0 – never settles down to a single real number)	**24.** DNE (too wild of oscillations near 0 – never settles down to a single real number)
25. 5	**26.** 2
27. DNE (since left and right limits are different)	**28.** 5
29. 2	**30.** 2
31. 2	**32.** 2
33. 1	**34.** 4
35. DNE (since left and right limits are different)	**36.** 1
37. 1	**38.** 0
39. DNE (since left and right limits are different)	**40.** 2
41. 0	**42.** 0
43. 0	**44.** 1

45. Consider the following graph:

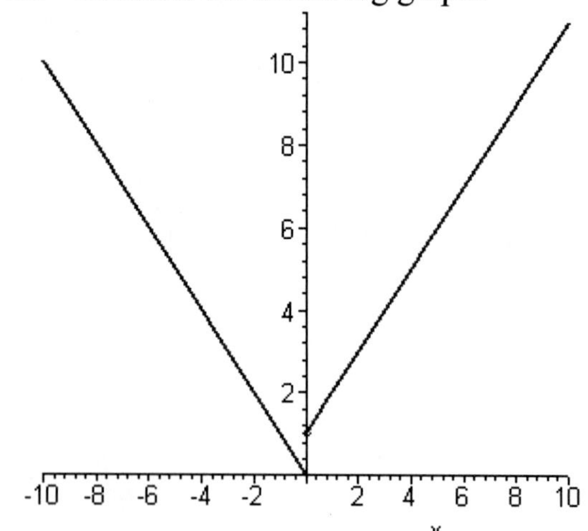

The limit DNE since the left and right limits are different.

46. Consider the following graph:

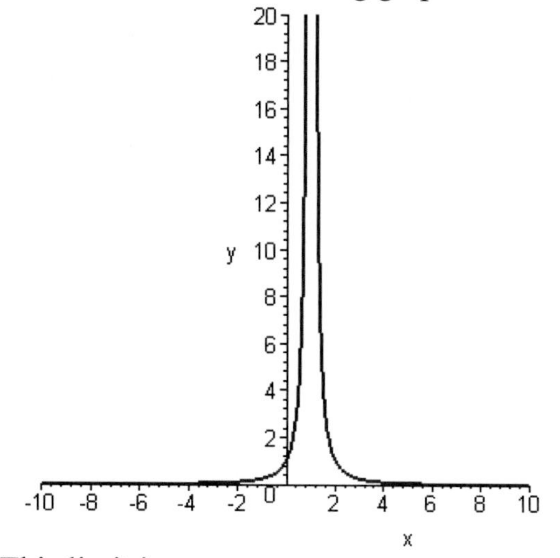

This limit is ∞

47. Consider the following graph:

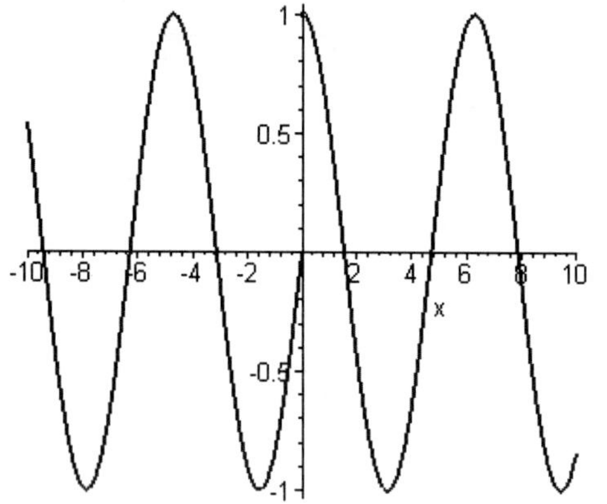

The limit DNE since the left and right limits are different.

48. Consider the following graph:

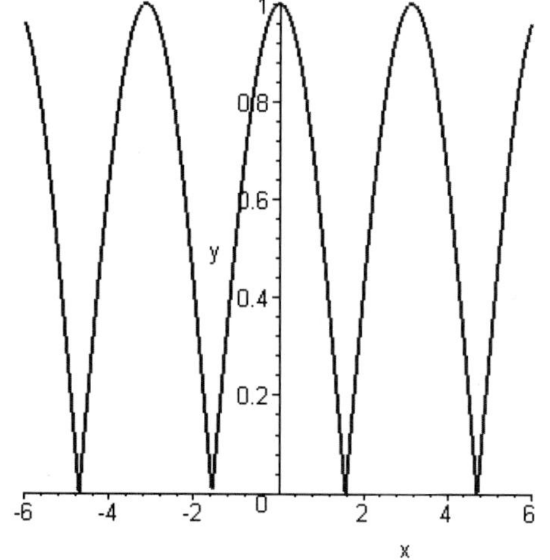

This limit is 1.

49. a) 0
b) 1
c) DNE (since left and right limits are different)
d) 1

50. a) 0 **b)** 1
c) DNE (since left and right limits are different)
d) 1
e) 0

51. The values chosen do not illustrate the oscillatory nature of the function. The function oscillates more rapidly the closer the inputs get to the origin, which means the limit DNE.

52. Need to check values on both sides of 0, not just the functional value at 0. In such case, the left-limit is -1 while the right-limit is 1. Hence, the limit DNE.

53. False. The graph could have an open hole at *a*.	**54.** False. The left and right limits at *a* could still be different, and in such case the limit DNE.
55. True.	**56.** False. They must be equal in order to ensure the two-sided limit exists.
57. The left-limit at *c* is *c*-1, and the right-limit is 1-*c*. They are equal iff $c = 1$.	**58.** The left-limit at *c* is $\sin(\pi c)$ and the right-limit is $\cos(\pi c)$. They are equal if $$\sin(\pi c) = \cos(\pi c).$$ Since by the definition of the function, $0<c<1$, we conclude that $c = ¼$.
59. Consider the graph: 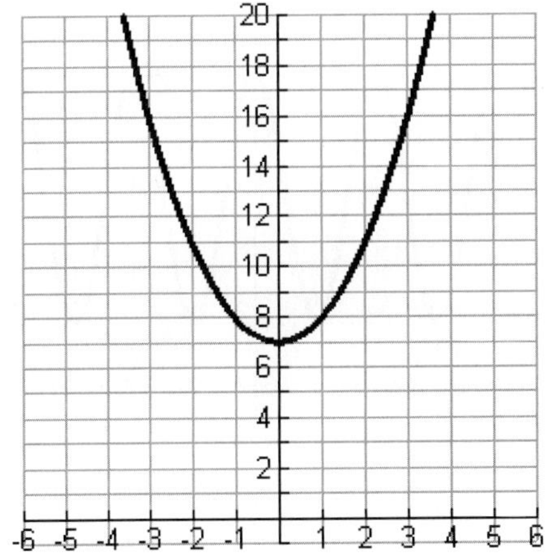 Using synthetic division yields $$\begin{array}{r\|rrrr} 1 & 1 & -1 & 7 & -7 \\ & & 1 & 0 & 7 \\ \hline & 1 & 0 & 7 & 0 \end{array}$$ So, $f(x) = x^2 + 7,\ x \neq 1$. The limit at 1 is 8.	**60.** Consider the graph: 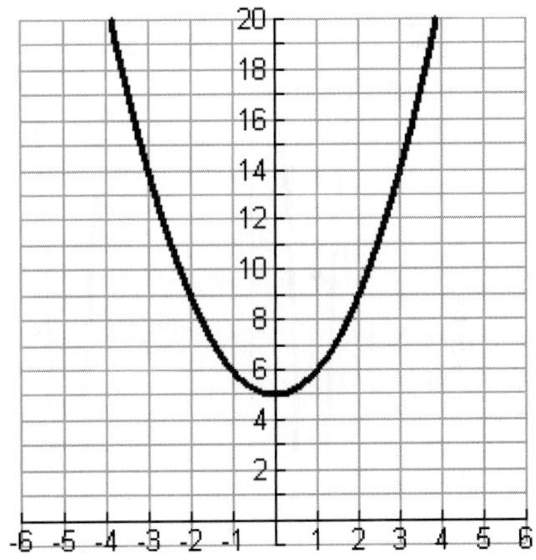 Using synthetic division yields $$\begin{array}{r\|rrrr} 1 & 1 & 1 & 5 & 5 \\ & & -1 & 0 & -5 \\ \hline & 1 & 0 & 5 & 0 \end{array}$$ So, $f(x) = x^2 + 5,\ x \neq -1$. The limit at -1 is 6.

61. Consider the graph: 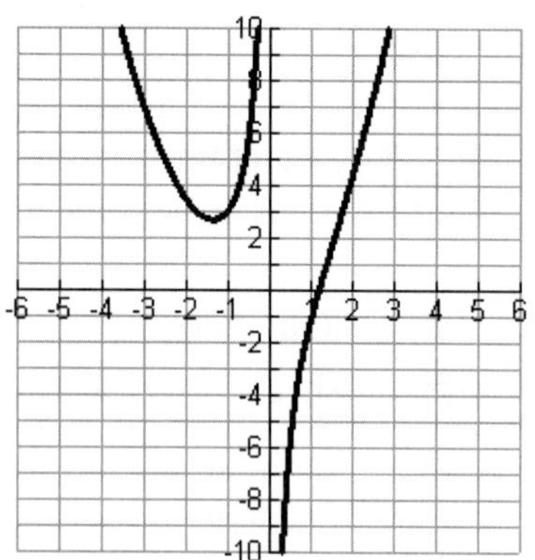 The limit at 0 DNE.	**62.** Consider the graph: 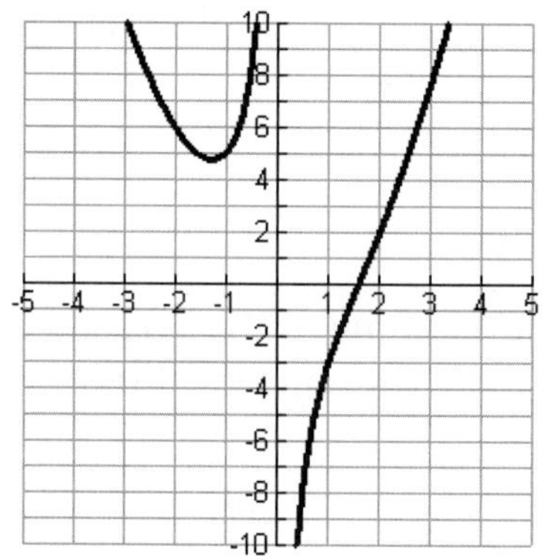 The limit at 0 DNE.
63. Consider the graph: 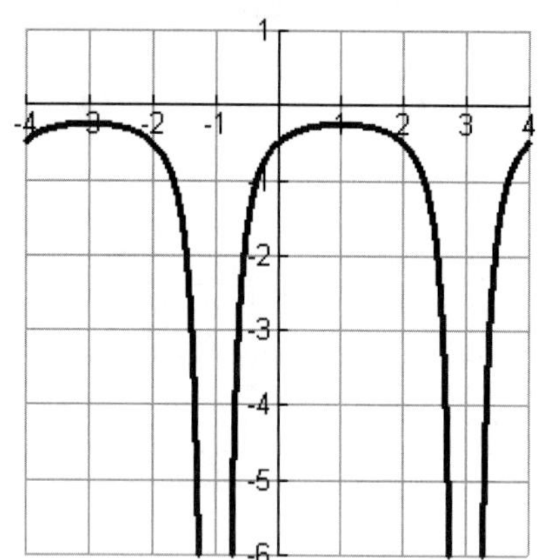 The limit at 1 is 0-.25.	**64.** Consider the graph: 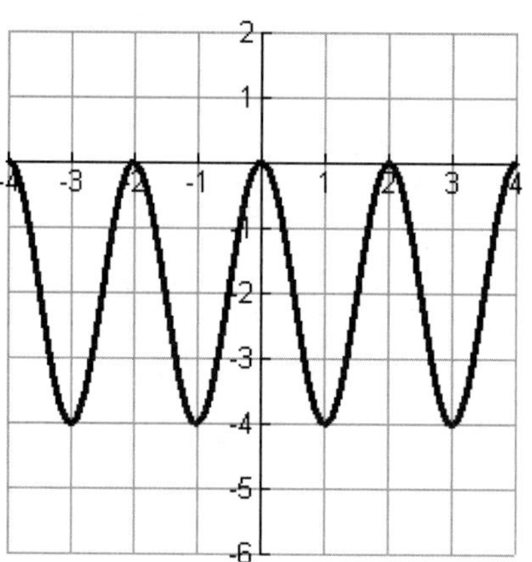 The limit at 1 is -4.

Section 11.2 Solutions---

1. $3\lim_{x\to 0} f(x) + \lim_{x\to 0} g(x) = 3(1) + (1) = 4$
2. $\lim_{x\to 0} f(x) - 3\lim_{x\to 0} g(x) = (1) - 3(1) = -2$
3. $\lim_{x\to -2} f(x) \cdot \lim_{x\to -2} g(x) = (1)(-3) = -3$
4. $\lim_{x\to 1} f(x) \cdot \lim_{x\to 1} g(x) = 0(-2) = 0$

5. $\dfrac{\lim\limits_{x\to 1} f(x)}{\lim\limits_{x\to 1} g(x)} = \dfrac{0}{-2} = 0$
6. $\dfrac{\lim\limits_{x\to -1} f(x)}{\lim\limits_{x\to -1} g(x)}$ **DNE** since the denominator is 0.
7. $\dfrac{\lim\limits_{x\to -2} g(x)}{\lim\limits_{x\to -2} f(x)} = \dfrac{-3}{1} = -3$
8. $\dfrac{\lim\limits_{x\to 0} g(x)}{\lim\limits_{x\to 0} f(x)} = \dfrac{1}{1} = 1$
9. $\left(\lim\limits_{x\to -1} f(x)\right)^2 = (2)^2 = 4$
10. $\left(\lim\limits_{x\to 1} g(x)\right)^2 = (-2)^2 = 4$
11. $\sqrt{\lim\limits_{x\to -2} f(x) - \lim\limits_{x\to -2} g(x)} = \sqrt{1-(-3)} = 2$
12. $\sqrt{3\lim\limits_{x\to 0} f(x) + \lim\limits_{x\to 0} g(x)} = \sqrt{3(1)+1} = 2$
13. $\lim\limits_{x\to 0} f(x) + \lim\limits_{x\to 0} g(x)$ DNE since $\lim\limits_{x\to 0} f(x)$ DNE.
14. $\lim\limits_{x\to -2} g(x) - \lim\limits_{x\to -2} f(x)$ DNE since $\lim\limits_{x\to -2} g(x)$ DNE.
15. $\dfrac{\lim\limits_{x\to -1} f(x)}{\lim\limits_{x\to -1} g(x)} = \dfrac{-3}{1} = -3$
16. $\lim\limits_{x\to 2} f(x) \cdot \lim\limits_{x\to 2} g(x) = 3(2) = 6$
17. $\left(\lim\limits_{x\to 0} f(x)\right)^2 = 3^2 = 9$
18. $\left(\lim\limits_{x\to -2} g(x)\right)^2 = 0^2 = 0$
19. $\sqrt{\lim\limits_{x\to 2} f(x) + \lim\limits_{x\to 2} g(x)} = \sqrt{3+2} = \sqrt{5}$
20. $\sqrt{\lim\limits_{x\to -1} g(x) - \lim\limits_{x\to -1} f(x)} = \sqrt{1-(-3)} = 2$
21. 17 (limit of a constant is the constant)
22. $\left(\lim\limits_{x\to -2} x\right)^3 = (-2)^3 = -8$
23. $\sqrt{\lim\limits_{x\to 9} x} = \sqrt{9} = 3$

24. $\left(\lim_{x\to -1} x\right)^4 - \lim_{x\to -1} x + \lim_{x\to -1} 3 = (-1)^4 - (-1) + 3 = 5$
25. $\dfrac{\lim_{x\to 0}\left(x^2+2\right)}{\lim_{x\to 0}\left(x^2-1\right)} = \dfrac{2}{-1} = -2$
26. $\sqrt{\lim_{x\to 1}\left(x^2+8\right)} = \sqrt{9} = 3$
27. $\left[\lim_{x\to 1}(x-3)\cdot\lim_{x\to 1}(x+2)\right]^2 = \left[(-2)(3)\right]^2 = 36$
28. $\left(\dfrac{\lim_{x\to -1}\left(x^2+2x-1\right)}{\lim_{x\to -1}\left(x^2+2\right)}\right)^2 = \left(\dfrac{-2}{1}\right)^2 = 4$
29. Simply substitute $x = 1$ into the expression to compute the limit in this case. The limit is 1.
30. Simply substitute $x = -5$ into the expression to compute the limit in this case. The limit is -5.
31. Simply substitute $x = 2$ into the expression to compute the limit in this case. The limit is $\frac{31}{12}$.
32. Simply substitute $x = 0$ into the expression to compute the limit in this case. The limit is $\frac{5}{3}$.
33. $\lim_{x\to -2}\dfrac{x^2-x-6}{x+2} = \lim_{x\to -2}\dfrac{(x-3)(x+2)}{x+2} = \lim_{x\to -2}(x-3) = -5$
34. $\lim_{x\to 2}\dfrac{x^2-x-2}{x-2} = \lim_{x\to 2}\dfrac{(x-2)(x+1)}{x-2} = \lim_{x\to 2}(x+1) = 3$
35. $\lim_{x\to 1}\dfrac{x^4-1}{x-1} = \lim_{x\to 1}\dfrac{(x-1)(x+1)\left(x^2+1\right)}{x-1} = \lim_{x\to 1}(x+1)\left(x^2+1\right) = 4$
36. $\lim_{x\to -2}\dfrac{x^4-16}{x+2} = \lim_{x\to -2}\dfrac{(x-2)(x+2)\left(x^2+4\right)}{x+2} = \lim_{x\to -2}(x-2)\left(x^2+4\right) = -32$
37. $\lim_{x\to 0}\dfrac{1-\cos x}{\sin x} = \lim_{x\to 0}\dfrac{\frac{1-\cos x}{x}}{\frac{\sin x}{x}} = \dfrac{\lim_{x\to 0}\frac{1-\cos x}{x}}{\lim_{x\to 0}\frac{\sin x}{x}} = \dfrac{0}{1} = 0$
38. $\lim_{x\to \frac{\pi}{2}}\dfrac{1-\sin x}{\cos x} = \lim_{u\to 0}\dfrac{1-\sin\left(u+\frac{\pi}{2}\right)}{\cos\left(u+\frac{\pi}{2}\right)} = \lim_{u\to 0}\dfrac{1-\cos u}{-\sin u} = \lim_{u\to 0}\dfrac{\frac{1-\cos u}{u}}{\frac{-\sin u}{u}} = \dfrac{\lim_{u\to 0}\frac{1-\cos u}{u}}{-\lim_{u\to 0}\frac{\sin u}{u}} = -\left(\dfrac{0}{1}\right) = 0$
39. $\lim_{x\to 0}\dfrac{\tan x}{\sec x} = \lim_{x\to 0}\left(\dfrac{\sin x}{\cos x}\cdot\dfrac{\cos x}{1}\right) = \sin 0 = 0$

40. $\lim_{x\to 0}\dfrac{\sec x}{\csc x}=\lim_{x\to 0}\dfrac{\sin x}{\cos x}=0$

41. $\lim_{x\to 0}\dfrac{e^{2x}-1}{e^x-1}=\lim_{x\to 0}\dfrac{\left(e^x-1\right)\left(e^x+1\right)}{e^x-1}=\lim_{x\to 0}\left(e^x+1\right)=2$

42. $\lim_{x\to 0}\dfrac{e^{2x}-1}{e^x+1}=\lim_{x\to 0}\dfrac{\left(e^x-1\right)\left(e^x+1\right)}{e^x+1}=\lim_{x\to 0}\left(e^x-1\right)=0$

43. $\lim_{x\to 0}\dfrac{\sqrt{x+1}-1}{x}=\lim_{x\to 0}\dfrac{\left(\sqrt{x+1}-1\right)\left(\sqrt{x+1}+1\right)}{x\left(\sqrt{x+1}+1\right)}=\lim_{x\to 0}\dfrac{1}{\sqrt{x+1}+1}=\dfrac{1}{2}$

44. $\lim_{x\to 0}\dfrac{\sqrt{x+4}-2}{x}=\lim_{x\to 0}\dfrac{\left(\sqrt{x+4}-2\right)\left(\sqrt{x+4}+2\right)}{x\left(\sqrt{x+4}+2\right)}=\lim_{x\to 0}\dfrac{1}{\sqrt{x+4}+2}=\dfrac{1}{4}$

45. $\lim_{x\to 4}\dfrac{2-\sqrt{x}}{x-4}=\lim_{x\to 4}\dfrac{\left(2-\sqrt{x}\right)\left(2+\sqrt{x}\right)}{(x-4)\left(2+\sqrt{x}\right)}=\lim_{x\to 4}\dfrac{4-x}{-(4-x)\left(2+\sqrt{x}\right)}=-\dfrac{1}{4}$

46.

$$\lim_{x\to 1}\frac{\sqrt{x+8}-3}{x-1}=\lim_{x\to 1}\frac{\left(\sqrt{x+8}-3\right)\left(\sqrt{x+8}+3\right)}{(x-1)\left(\sqrt{x+8}+3\right)}=\lim_{x\to 1}\frac{(x-1)}{(x-1)\left(\sqrt{x+8}+3\right)}=\lim_{x\to 1}\frac{1}{\sqrt{x+8}+3}=\frac{1}{6}$$

47. $\lim_{t\to 2}\dfrac{\frac{1}{t}-\frac{1}{2}}{t-2}=\lim_{t\to 2}\dfrac{\frac{2-t}{2t}}{t-2}=\lim_{t\to 2}\dfrac{-(t-2)}{2t(t-2)}=\lim_{t\to 2}\dfrac{-1}{2t}=-\dfrac{1}{4}$

48. $\lim_{t\to -1}\dfrac{\frac{1}{t}+1}{t+1}=\lim_{t\to -1}\dfrac{\frac{1+t}{t}}{t+1}=\lim_{t\to -1}\dfrac{1+t}{t(t+1)}=\lim_{t\to -1}\dfrac{1}{t}=-1$

49. $\lim_{x\to 0}\dfrac{\frac{1}{x+2}-\frac{1}{2}}{x}=\lim_{x\to 0}\dfrac{\frac{2-(x+2)}{2(x+2)}}{x}=\lim_{x\to 0}\dfrac{-x}{2x(x+2)}=\lim_{x\to 0}\dfrac{-1}{2(x+2)}=-\dfrac{1}{4}$

50. $\lim_{x\to 0}\dfrac{\frac{1}{x-1}+1}{x}=\lim_{x\to 0}\dfrac{\frac{1+(x+1)}{x-1}}{x}=\lim_{x\to 0}\dfrac{x}{x(x-1)}=\lim_{x\to 0}\dfrac{1}{x-1}=-1$

51. $\lim_{h\to 0}\dfrac{f(x+h)-f(x)}{h}=\lim_{h\to 0}\dfrac{(5(x+h)+2)-(5x+2)}{h}=\lim_{h\to 0}\dfrac{5h}{h}=5$

52. $\lim_{h\to 0}\dfrac{f(x+h)-f(x)}{h}=\lim_{h\to 0}\dfrac{(3(x+h)+1)-(3x+1)}{h}=\lim_{h\to 0}\dfrac{3h}{h}=3$

53. $\lim_{h\to 0}\dfrac{f(x+h)-f(x)}{h}=\lim_{h\to 0}\dfrac{(-2(x+h)+3)-(-2x+3)}{h}=\lim_{h\to 0}\dfrac{-2h}{h}=-2$

54. $\lim_{h\to 0}\frac{f(x+h)-f(x)}{h}=\lim_{h\to 0}\frac{\left(-3(x+h)-2\right)-\left(-3x-2\right)}{h}=\lim_{h\to 0}\frac{-3h}{h}=-3$

55.

$$\lim_{h\to 0}\frac{f(x+h)-f(x)}{h}=\lim_{h\to 0}\frac{\left((x+h)^2+2\right)-\left(x^2+2\right)}{h}=\lim_{h\to 0}\frac{x^2+2hx+h^2+2-x^2-2}{h}=\lim_{h\to 0}\frac{h\left(2x+h\right)}{h}$$

56.

$$\lim_{h\to 0}\frac{f(x+h)-f(x)}{h}=\lim_{h\to 0}\frac{\left((x+h)^2-3\right)-\left(x^2-3\right)}{h}=\lim_{h\to 0}\frac{x^2+2hx+h^2-3-x^2+3}{h}=\lim_{h\to 0}\frac{h\left(2x+h\right)}{h}=2x$$

57.

$$\lim_{h\to 0}\frac{f(x+h)-f(x)}{h}=\lim_{h\to 0}\frac{\left(-2(x+h)^2+1\right)-\left(-2x^2+1\right)}{h}=\lim_{h\to 0}\frac{-2x^2-4hx-2h^2+1+2x^2-1}{h}$$

$$=\lim_{h\to 0}\frac{-h\left(4x+2h\right)}{h}=-4x$$

58.

$$\lim_{h\to 0}\frac{f(x+h)-f(x)}{h}=\lim_{h\to 0}\frac{\left(-3(x+h)^2+2\right)-\left(-3x^2+2\right)}{h}=\lim_{h\to 0}\frac{-3x^2-6hx-3h^2+2+3x^2-2}{h}$$

$$=\lim_{h\to 0}\frac{-h\left(6x+3h\right)}{h}=-6x$$

59.

$$\lim_{h\to 0}\frac{f(x+h)-f(x)}{h}=\lim_{h\to 0}\frac{\frac{1}{x+h}-\frac{1}{x}}{h}=\lim_{h\to 0}\frac{x-(x+h)}{h(x+h)x}=\lim_{h\to 0}\frac{-h}{h(x+h)x}=\lim_{h\to 0}\frac{-1}{(x+h)x}=-\frac{1}{x^2}$$

60.

$$\lim_{h\to 0}\frac{f(x+h)-f(x)}{h}=\lim_{h\to 0}\frac{\sqrt{x+h}-\sqrt{x}}{h}=\lim_{h\to 0}\frac{\left(\sqrt{x+h}-\sqrt{x}\right)\left(\sqrt{x+h}+\sqrt{x}\right)}{h\left(\sqrt{x+h}+\sqrt{x}\right)}=\lim_{h\to 0}\frac{1}{\sqrt{x+h}+\sqrt{x}}=\frac{1}{2\sqrt{x}}$$

61.

$$\lim_{h\to 0}\frac{f(x+h)-f(x)}{h}=\lim_{h\to 0}\frac{\left(-(x+h)^2+2(x+h)+3\right)-\left(-x^2+2x+3\right)}{h}$$

$$=\lim_{h\to 0}\frac{-x^2-2hx-h^2+2x+2h+3+x^2-2x-3}{h}=\lim_{h\to 0}\frac{h\left(-2x-h+2\right)}{h}=-2x+2$$

62.

$$\lim_{h\to 0}\frac{f(x+h)-f(x)}{h}=\lim_{h\to 0}\frac{\left(-(x+h)^2-3(x+h)+1\right)-\left(-x^2-3x+1\right)}{h}$$

$$=\lim_{h\to 0}\frac{-x^2-2hx-h^2-3x-3h+1+x^2+3x-1}{h}=\lim_{h\to 0}\frac{h\left(-2x-h-3\right)}{h}=-2x-3$$

63. The left and right limits at $x=0$ are both 0. Hence, the limit at $x=0$ is 0.
64. The right-limit at $x=0$ is 2, but the left-limit is 1. Hence, the limit at $x=0$ DNE.
65. The right-limit at $x=1$ is 3, but the left-limit is 0. Hence, the limit at $x=1$ DNE.
66. The right-limit at $x=-1$ is 2, but the left-limit is -1. Hence, the limit at $x=-1$ DNE.
67. The right-limit at $x=0$ is 1, but the left-limit is 0. Hence, the limit at $x=0$ DNE.
68. The right-limit at $x=\frac{\pi}{2}$ is 0, but the left-limit is 1. Hence, the limit at $x=\frac{\pi}{2}$ DNE.
69. The right-limit at $x=3$ is 1, but the left-limit is -1. Hence, the limit at $x=3$ DNE.
70. The right-limit at $x=-4$ is 1, but the left-limit is -1. Hence, the limit at $x=-4$ DNE.

71.

$$\lim_{t\to 3}\frac{h(3)-h(t)}{3-t}=\lim_{t\to 3}\frac{\left[-16(3)^2+32(3)+100\right]-\left[-16t^2+32t+100\right]}{3-t}$$

$$=\lim_{t\to 3}\frac{16\left(t^2-2t-3\right)}{-(t-3)}=\lim_{t\to 3}\frac{16(t-3)(t+1)}{-(t-3)}=\lim_{t\to 3}-16(t+1)=64\,{}^{ft}\!/\!_{sec}$$

72.

$$\lim_{t\to 75}\frac{\left[-16(75)^2+1200(75)\right]-\left[-16t^2+1200t\right]}{75-t}=\lim_{t\to 75}\frac{16t\left(t-75\right)}{-(t-75)}=\lim_{t\to 75}-16t=-16(75)=-1200\,{}^{ft}\!/\!_{sec}$$

73. Did not divide correctly. $\frac{x^3-8}{x}=x^2-\frac{8}{x}$ Then, direct substitution does not work because $\lim_{x\to 0}\frac{8}{x}$ DNE.

74. Did not divide correctly. $\frac{x^2-1}{x}=x-\frac{1}{x}$ Then, direct substitution does not work because $\lim_{x\to 0}\frac{1}{x}$ DNE.

75. Direct substitution cannot be used when the denominator goes to zero. The expression should be simplified first.

76. Limit laws cannot be used because the denominator has limit zero. The expression should be simplified first.

77. False. $\lim_{x\to 0}\frac{1}{x}=\infty$

78. False. $\lim_{x\to -2}\frac{x+2}{-(x+2)}=-1$

79. False. They must be the same value in order for the two-sided limit to exist.

80. True.

81. $\lim_{x \to c^+} f(x) = c^2 - 3c$, $\lim_{x \to c^-} f(x) = 2c + 6$. The two-sided limit exists iff these two are equal. Equating them and solving for c yields: $$c^2 - 3c = 2c + 6 \Rightarrow c^2 - 5c - 6 = 0 \Rightarrow (c-6)(c+1) = 0 \Rightarrow c = -1, 6$$
82. $\lim_{x \to c^+} f(x) = \sin 2c$, $\lim_{x \to c^-} f(x) = \sin c$. The two-sided limit exists iff these two are equal. Equating them and solving for c yields: $$\sin 2c = \sin c \Rightarrow 2\sin c\cos c = \sin c \Rightarrow 1 = 2\cos c \Rightarrow \cos c = \tfrac{1}{2},$$ So, $c = \frac{\pi}{3}$.
83. Consider the graph: 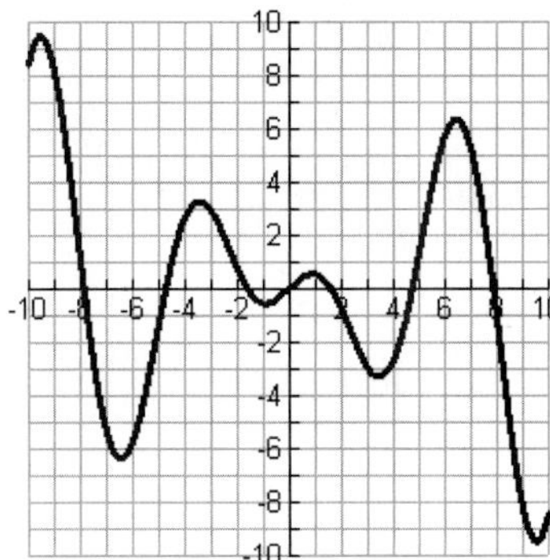 The limit at $x = 0$ is 0.
84. Consider the graph: 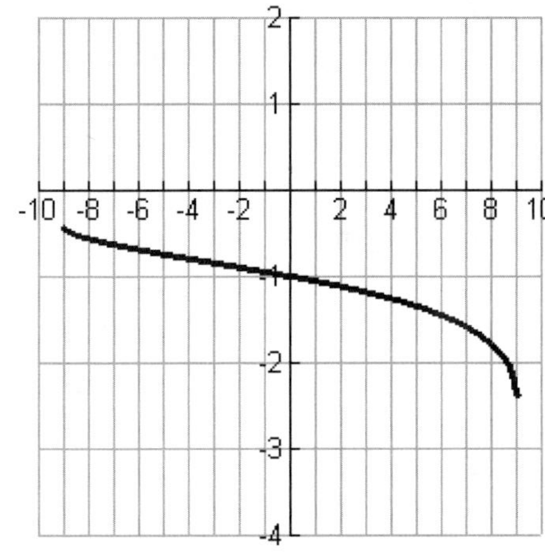 The limit at $x = 0$ is -1.

85. Consider the graph:

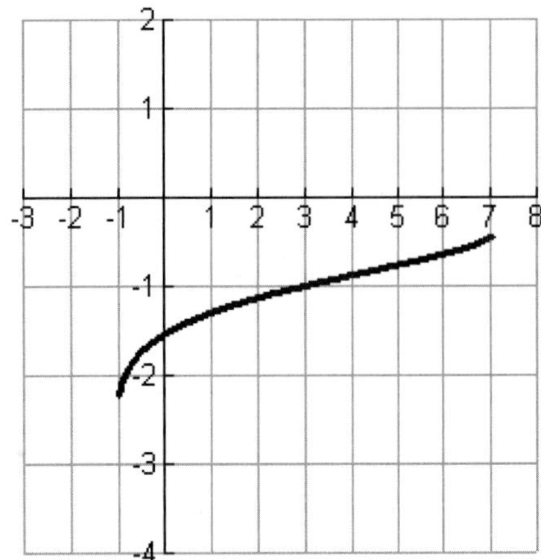

The limit at $x = 3$ is -1.

86. Consider the graph:

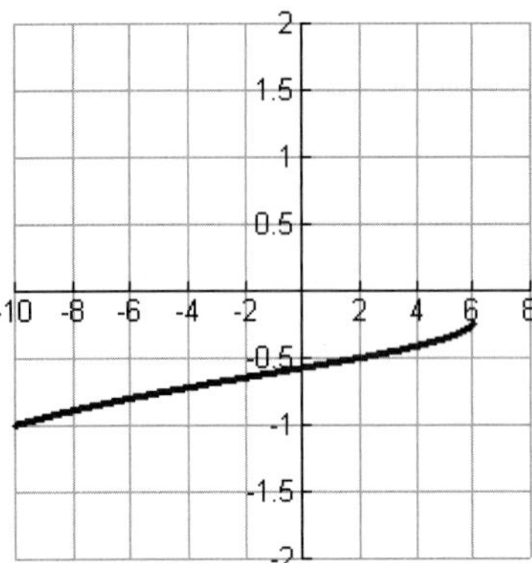

The limit at $x = 2$ is – ½ .

Section 11.3 Solutions--

1. $\lim_{x\to5}\frac{3-3}{x-5}=\lim_{x\to5}0=0$

2. $\lim_{x\to-1}\frac{2-2}{x+1}=\lim_{x\to-1}0=0$

3. $\lim_{x\to0}\frac{(4x-1)-(-1)}{x-0}=\lim_{x\to0}\frac{4x}{x}=\lim_{x\to0}4=4$

4. $\lim_{x\to-3}\frac{(-5x+3)-18}{x-(-3)}=\lim_{x\to-3}\frac{-5(x+3)}{x+3}=\lim_{x\to-3}-5=-5$
5. $\lim_{x\to1}\frac{(-3x^2)-(-3)}{x-1}=\lim_{x\to1}\frac{-3(x^2-1)}{x-1}=\lim_{x\to1}\frac{-3(x-1)(x+1)}{x-1}=\lim_{x\to1}-3(x+1)=-6$
6. $\lim_{x\to1}\frac{(7x^2)-(7)}{x-1}=\lim_{x\to1}\frac{7(x^2-1)}{x-1}=\lim_{x\to1}\frac{7(x-1)(x+1)}{x-1}=\lim_{x\to1}7(x+1)=14$
7. $\lim_{x\to3}\frac{\frac{1}{x+1}-\frac{1}{4}}{x-3}=\lim_{x\to3}\frac{4-(x+1)}{4(x+1)(x-3)}=\lim_{x\to3}\frac{3-x}{-4(x+1)(3-x)}=\lim_{x\to3}\frac{1}{-4(x+1)}=-\frac{1}{16}$
8. $\lim_{x\to-1}\frac{\frac{2}{x^2}-2}{x-(-1)}=\lim_{x\to-1}\frac{2-2x^2}{x^2(x+1)}=\lim_{x\to-1}\frac{2(1-x)(x+1)}{x^2(x+1)}=\lim_{x\to-1}\frac{2(1-x)}{x^2}=4$
9. $\lim_{x\to4}\frac{(\sqrt{x}-1)-(1)}{x-4}=\lim_{x\to4}\frac{\sqrt{x}-2}{x-4}=\lim_{x\to4}\frac{(\sqrt{x}-2)(\sqrt{x}+2)}{(x-4)(\sqrt{x}+2)}=\lim_{x\to4}\frac{x-4}{(x-4)(\sqrt{x}+2)}=\lim_{x\to4}\frac{1}{\sqrt{x}+2}=\frac{1}{4}$
10. $\lim_{x\to5}\frac{\sqrt{x-1}-2}{x-5}=\lim_{x\to5}\frac{(\sqrt{x-1}-2)(\sqrt{x-1}+2)}{(x-5)(\sqrt{x-1}+2)}=\lim_{x\to5}\frac{x-5}{(x-5)(\sqrt{x-1}+2)}=\lim_{x\to5}\frac{1}{\sqrt{x-1}+2}=\frac{1}{4}$
11. Slope: $\lim_{x\to2}\frac{17-17}{x-2}=\lim_{x\to2}0=0$ Point: (2,17) So, the equation of the tangent line is: $y=17$.
12. Slope: $\lim_{x\to5}\frac{-6-(-6)}{x-5}=\lim_{x\to5}0=0$ Point: (5, -6) So, the equation of the tangent line is: $y=-6$
13. Slope: $\lim_{x\to-3}\frac{(-2x+1)-(7)}{x-(-3)}=\lim_{x\to-3}\frac{-2(x+3)}{x+3}=\lim_{x\to-3}-2=-2$ Point: (-3,7) So, the equation of the tangent line is: $y-7=-2(x+3)\Rightarrow y=-2x+1$
14. Slope: $\lim_{x\to4}\frac{(3x-2)-(10)}{x-4}=\lim_{x\to4}\frac{3(x-4)}{x-4}=\lim_{x\to4}3=3$ Point: (4,10) So, the equation of the tangent line is: $y-10=3(x-4)\Rightarrow y=3x-2$

15. Slope:

$$\lim_{x\to -1}\frac{\left(2x^2-3x+1\right)-(6)}{x-(-1)}=\lim_{x\to -1}\frac{2x^2-3x-5}{x+1}=\lim_{x\to -1}\frac{(2x-5)(x+1)}{x+1}=\lim_{x\to -1}(2x-5)=-7$$

Point: (-1,6)

So, the equation of the tangent line is: $y-6=-7(x+1)\ \Rightarrow\ y=-7x-1$

16. Slope:

$$\lim_{x\to 2}\frac{\left(-3x^2+2x+7\right)-(-1)}{x-2}=\lim_{x\to 2}\frac{-3x^2+2x+8}{x-2}=\lim_{x\to 2}\frac{(3x+4)(-x+2)}{x-2}=\lim_{x\to 2}-(3x+4)=-10$$

Point: (2,-1)

So, the equation of the tangent line is: $y+1=-10(x-2)\ \Rightarrow\ y=-10x+19$

17. Slope: $\lim_{x\to 1}\frac{\frac{x}{x+1}-\frac{1}{2}}{x-1}=\lim_{x\to 1}\frac{2x-(x+1)}{2(x+1)(x-1)}=\lim_{x\to 1}\frac{x-1}{2(x+1)(x-1)}=\lim_{x\to 1}\frac{1}{2(x+1)}=\frac{1}{4}$

Point: $\left(1,\frac{1}{2}\right)$

So, the equation of the tangent line is: $y-\frac{1}{2}=\frac{1}{4}(x-1)\ \Rightarrow\ y=\frac{1}{4}x+\frac{1}{4}$

18. Slope: $\lim_{x\to -1}\frac{\frac{-1}{x^2}-(-1)}{x-(-1)}=\lim_{x\to -1}\frac{-1+x^2}{x^2(x+1)}=\lim_{x\to -1}\frac{(x-1)(x+1)}{x^2(x+1)}=\lim_{x\to -1}\frac{x-1}{x^2}=-2$

Point: (-1,-1)

So, the equation of the tangent line is: $y+1=-2(x+1)\ \Rightarrow\ y=-2x-3$

19. Slope:

$$\lim_{x\to 6}\frac{\sqrt{x-2}-2}{x-6}=\lim_{x\to 6}\frac{\left(\sqrt{x-2}-2\right)\left(\sqrt{x-2}+2\right)}{(x-6)\left(\sqrt{x-2}+2\right)}=\lim_{x\to 6}\frac{x-6}{(x-6)\left(\sqrt{x-2}+2\right)}=\lim_{x\to 6}\frac{1}{\sqrt{x-2}+2}=\frac{1}{4}$$

Point: (6,2)

So, the equation of the tangent line is: $y-2=\frac{1}{4}(x-6)\ \Rightarrow\ y=\frac{1}{4}x+\frac{1}{2}$

20. Slope:

$$\lim_{x\to 4}\frac{\left(2-\sqrt{x}\right)-(0)}{x-4}=\lim_{x\to 4}\frac{\left(2-\sqrt{x}\right)\left(2+\sqrt{x}\right)}{(x-4)\left(2+\sqrt{x}\right)}=\lim_{x\to 4}\frac{-(x-4)}{(x-4)\left(\sqrt{x}+2\right)}=\lim_{x\to 4}\frac{-1}{\sqrt{x}+2}=-\frac{1}{4}$$

Point: (4,0)

So, the equation of the tangent line is: $y-0=-\frac{1}{4}(x-4)\ \Rightarrow\ y=-\frac{1}{4}x+1$

21. $f'(2)=\lim_{x\to 2}\frac{\left(1-2x^2\right)-(-7)}{x-2}=\lim_{x\to 2}\frac{8-2x^2}{x-2}=\lim_{x\to 2}\frac{-2(x-2)(x+2)}{x-2}=\lim_{x\to 2}-2(x+2)=-8$

22. $f'(3)=\lim_{x\to 3}\frac{\left(2-x^2\right)-(-7)}{x-3}=\lim_{x\to 3}\frac{9-x^2}{x-3}=\lim_{x\to 3}\frac{-(x-3)(x+3)}{x-3}=\lim_{x\to 3}-(x+3)=-6$

23.	$f'(1)=\lim_{x\to1}\frac{x^3-1}{x-1}=\lim_{x\to1}\frac{(x-1)(x^2+x+1)}{x-1}=\lim_{x\to1}(x^2+x+1)=3$
24.	$f'(-1)=\lim_{x\to-1}\frac{x^4-1}{x+1}=\lim_{x\to-1}\frac{(x+1)(x-1)(x^2+1)}{x+1}=\lim_{x\to1}(x-1)(x^2+1)=-4$
25.	$f'(12)=\lim_{x\to12}\frac{(1-\sqrt{x-3})-(-2)}{x-12}=\lim_{x\to12}\frac{3-\sqrt{x-3}}{x-12}=\lim_{x\to12}\frac{(3-\sqrt{x-3})(3+\sqrt{x-3})}{(x-12)(3+\sqrt{x-3})}$ $=\lim_{x\to12}\frac{-(x-12)}{(x-12)(3+\sqrt{x-3})}=\lim_{x\to12}\frac{-1}{(3+\sqrt{x-3})}=-\frac{1}{6}$
26.	$f'(5)=\lim_{x\to5}\frac{(\sqrt{x-1}+2)-4}{x-5}=\lim_{x\to5}\frac{\sqrt{x-1}-2}{x-5}=\lim_{x\to5}\frac{(\sqrt{x-1}-2)(\sqrt{x-1}+2)}{(x-5)(\sqrt{x-1}+2)}$ $=\lim_{x\to5}\frac{x-5}{(x-5)(\sqrt{x-1}+2)}=\lim_{x\to5}\frac{1}{\sqrt{x-1}+2}=\frac{1}{4}$
27.	$f'(1)=\lim_{x\to1}\frac{\frac{1}{\sqrt{x}}-1}{x-1}=\lim_{x\to1}\frac{1-\sqrt{x}}{\sqrt{x}(x-1)}=\lim_{x\to1}\frac{(1-\sqrt{x})(1+\sqrt{x})}{\sqrt{x}(x-1)(1+\sqrt{x})}=\lim_{x\to1}\frac{1-x}{-\sqrt{x}(1-x)(1+\sqrt{x})}$ $=\lim_{x\to1}\frac{-1}{\sqrt{x}(1+\sqrt{x})}=-\frac{1}{2}$
28.	$f'(1)=\lim_{x\to1}\frac{-\frac{1}{\sqrt{x}}-(-1)}{x-1}=\lim_{x\to1}\frac{-1+\sqrt{x}}{\sqrt{x}(x-1)}=\lim_{x\to1}\frac{(-1+\sqrt{x})(-1-\sqrt{x})}{\sqrt{x}(x-1)(-1-\sqrt{x})}=\lim_{x\to1}\frac{1-x}{-\sqrt{x}(1-x)(-1-\sqrt{x})}$ $=\lim_{x\to1}\frac{-1}{\sqrt{x}(-1-\sqrt{x})}=\frac{1}{2}$
29.	$f'(2)=\lim_{x\to2}\frac{\frac{2}{x-5}-\left(-\frac{2}{3}\right)}{x-2}=\lim_{x\to2}\frac{6+2(x-5)}{3(x-5)(x-2)}=\lim_{x\to2}\frac{2x-4}{3(x-5)(x-2)}=\lim_{x\to2}\frac{2(x-2)}{3(x-5)(x-2)}$ $=\lim_{x\to2}\frac{2}{3(x-5)}=-\frac{2}{9}$
30.	$f'(-3)=\lim_{x\to-3}\frac{\frac{3}{x+2}-(-3)}{x+3}=\lim_{x\to-3}\frac{3+3(x+2)}{(x+2)(x+3)}=\lim_{x\to-3}\frac{3(x+3)}{(x+2)(x+3)}=\lim_{x\to-3}\frac{3}{(x+2)}=-3$
31.	$f'(x)=\lim_{h\to0}\frac{f(x+h)-f(x)}{h}=\lim_{h\to0}\frac{2-2}{h}=0$
32.	$f'(x)=\lim_{h\to0}\frac{f(x+h)-f(x)}{h}=\lim_{h\to0}\frac{\pi-\pi}{h}=0$

33. $f'(x)=\lim_{h\to 0}\frac{f(x+h)-f(x)}{h}=\lim_{h\to 0}\frac{[-7(x+h)+1]-[-7x+1]}{h}=\lim_{h\to 0}\frac{-7h}{h}=-7$

34. $f'(x)=\lim_{h\to 0}\frac{f(x+h)-f(x)}{h}=\lim_{h\to 0}\frac{[9(x+h)+2]-[9x+2]}{h}=\lim_{h\to 0}\frac{9h}{h}=9$

35. $f'(x)=\lim_{h\to 0}\frac{f(x+h)-f(x)}{h}=\lim_{h\to 0}\frac{\left[(x+h)-(x+h)^2\right]-\left[x-x^2\right]}{h}=\lim_{h\to 0}\frac{h(1-2x-h)}{h}$

$=\lim_{h\to 0}(1-2x-h)=1-2x$

36. $f'(x)=\lim_{h\to 0}\frac{f(x+h)-f(x)}{h}=\lim_{h\to 0}\frac{\left[2(x+h)^2-(x+h)\right]-\left[2x^2-x\right]}{h}=\lim_{h\to 0}\frac{h(4x+2h-1)}{h}$

$=\lim_{h\to 0}(4x+2h-1)=4x-1$

37. The domain of f' is $(0,\infty)$.

$$f'(x)=\lim_{h\to 0}\frac{f(x+h)-f(x)}{h}=\lim_{h\to 0}\frac{\frac{1}{1+\sqrt{x+h}}-\frac{1}{1+\sqrt{x}}}{h}=\lim_{h\to 0}\frac{\left(1+\sqrt{x}\right)-\left(1+\sqrt{x+h}\right)}{h\left(1+\sqrt{x}\right)\left(1+\sqrt{x+h}\right)}=\lim_{h\to 0}\frac{\sqrt{x}-\sqrt{x+h}}{h\left(1+\sqrt{x}\right)\left(1+\sqrt{x+h}\right)}$$

$$=\lim_{h\to 0}\frac{\left(\sqrt{x}-\sqrt{x+h}\right)\left(\sqrt{x}+\sqrt{x+h}\right)}{h\left(1+\sqrt{x}\right)\left(1+\sqrt{x+h}\right)\left(\sqrt{x}+\sqrt{x+h}\right)}=\lim_{h\to 0}\frac{x-(x+h)}{h\left(1+\sqrt{x}\right)\left(1+\sqrt{x+h}\right)\left(\sqrt{x}+\sqrt{x+h}\right)}$$

$$=\lim_{h\to 0}\frac{-h}{h\left(1+\sqrt{x}\right)\left(1+\sqrt{x+h}\right)\left(\sqrt{x}+\sqrt{x+h}\right)}=\frac{-1}{2\sqrt{x}\left(1+\sqrt{x}\right)^2}$$

38. The domain of f' is $(1,\infty)$.

$$f'(x)=\lim_{h\to 0}\frac{f(x+h)-f(x)}{h}=\lim_{h\to 0}\frac{\frac{1}{\sqrt{x+h-1}}-\frac{1}{\sqrt{x-1}}}{h}=\lim_{h\to 0}\frac{\sqrt{x-1}-\sqrt{x+h-1}}{h\sqrt{x-1}\sqrt{x+h-1}}$$

$$=\lim_{h\to 0}\frac{\left(\sqrt{x-1}-\sqrt{x+h-1}\right)\left(\sqrt{x-1}+\sqrt{x+h-1}\right)}{h\sqrt{x-1}\sqrt{x+h-1}\left(\sqrt{x-1}+\sqrt{x+h-1}\right)}=\lim_{h\to 0}\frac{-h}{h\sqrt{x-1}\sqrt{x+h-1}\left(\sqrt{x-1}+\sqrt{x+h-1}\right)}$$

$$=\frac{-1}{2\sqrt{x-1}\left(\sqrt{x-1}\right)^2}=\frac{-1}{2(x-1)\sqrt{x-1}}$$

39. The domain of f' is $(-\infty,0)\cup(0,\infty)$.

$$\lim_{h\to 0}\frac{f(x+h)-f(x)}{h}=\lim_{h\to 0}\frac{\frac{1}{(x+h)^4}-\frac{1}{x^4}}{h}=\lim_{h\to 0}\frac{x^4-(x+h)^4}{h(x+h)^4x^4}=\lim_{h\to 0}\frac{-h\left(4x^3+6x^2h+4xh^2+h^3\right)}{h(x+h)^4x^4}$$

$$=-\frac{4x^3}{x^4\left(x^4\right)}=-\frac{4}{x^5}$$

40. The domain of f' is $(-\infty,0)\cup(0,\infty)$.

$$\lim_{h\to 0}\frac{f(x+h)-f(x)}{h}=\lim_{h\to 0}\frac{\frac{1}{(x+h)^3}-\frac{1}{x^3}}{h}=\lim_{h\to 0}\frac{x^3-(x+h)^3}{h(x+h)^3x^3}=\lim_{h\to 0}\frac{-h\left(3x^2+3xh+h^2\right)}{h(x+h)^3x^3}=-\frac{3x^2}{x^3\left(x^3\right)}=-\frac{3}{x^4}$$

41.

$$\lim_{x\to 70}\frac{\left(-0.008(x-50)^2+30\right)-26.8}{x-70}=\lim_{x\to 70}\frac{-0.008x^2+0.8x-16.8}{x-70}=\lim_{x\to 70}\frac{-0.008\left(x^2-100x+2100\right)}{x-70}$$

$$=\lim_{x\to 70}\frac{-0.008(x-70)(x-30)}{x-70}=-0.008(40)=-0.32$$

So, gas mileage is decreasing at a rate of 0.32 mpg/mph when the car is traveling at 70 mph.

42.

$$\lim_{x\to 55}\frac{\left(-0.008(x-50)^2+30\right)-29.8}{x-55}=\lim_{x\to 55}\frac{-0.008x^2+0.8x-19.8}{x-55}=\lim_{x\to 55}\frac{-0.008\left(x^2-100x+2475\right)}{x-55}$$

$$=\lim_{x\to 55}\frac{-0.008(x-55)(x-30)}{x-55}=-0.008(10)=-0.08$$

So, gas mileage is decreasing at a rate of 0.08 mpg/mph when the car is traveling at 55 mph.

43. $\lim_{x\to 40}\frac{\left(-x^2+80x-1000\right)-600}{x-40}=\lim_{x\to 40}\frac{-x^2+80x-1600}{x-40}=\lim_{x\to 40}\frac{-(x-40)^2}{x-40}=0$

The profit is not changing when 40 units are made.

$P'(x)$ is the rate of change of profit with respect to number of units made. The units are dollars/units made.

44.

$$\lim_{x\to 100}\frac{\left(-x^2+80x-1000\right)-\left(-3000\right)}{x-100}=\lim_{x\to 100}\frac{-x^2+80x-2000}{x-100}=\lim_{x\to 100}\frac{-(x-100)(x+20)}{x-100}=-120$$

The profit is decreasing at a rate of 120 dollars per units made.

45. a)

$$\frac{h(40)-h(30)}{40-30}=\frac{\left[-16(40)^2+1200(40)\right]-\left[-16(30)^2+1200(30)\right]}{10}=\frac{800}{10}=80\text{ ft/sec.}$$

b)

$$\lim_{t\to 70}\frac{\left(-16t^2+1200t\right)-5600}{t-70}=\lim_{t\to 70}\frac{-16\left(t^2-75t+350\right)}{t-70}=\lim_{t\to 70}\frac{-16(t-70)(t-5)}{t-70}=-1040$$

So, the velocity is 1040 ft/sec downward when $t = 70$ sec.

Section 11.4 Solutions---

1. 0 since the degree of the numerator is strictly less than the degree of the denominator.
2. 0 since the degree of the numerator is strictly less than the degree of the denominator.
3. 0 since the degree of the numerator is strictly less than the degree of the denominator.
4. 0 since the degree of the numerator is strictly less than the degree of the denominator.
5. ∞ since the degree of the numerator is strictly larger than the degree of the denominator.
6. $-\infty$ since the degree of the numerator is strictly larger than the degree of the denominator.
7. $\lim_{x\to-\infty} \frac{x^2-4x+5}{2x^2+6x-4} \cdot \frac{\frac{1}{2x^2}}{\frac{1}{2x^2}} = \lim_{x\to-\infty} \frac{\frac{1}{2}-\frac{2}{x}+\frac{5}{x^2}}{1+\frac{3}{x}-\frac{2}{x^2}} = \frac{1}{2}$
8. $\lim_{x\to-\infty} \frac{6x^2+6+5}{3x^2-5x-2} \cdot \frac{\frac{1}{3x^2}}{\frac{1}{3x^2}} = \lim_{x\to-\infty} \frac{2+\frac{2}{x}+\frac{1}{3x^2}}{1-\frac{5}{x}-\frac{2}{3x^2}} = 2$
9. DNE since the function is periodic
10. DNE since the function is periodic
11. $\lim_{x\to\infty} 4e^{-x} = \lim_{x\to\infty} \frac{4}{e^x} = 0$
12. $\lim_{x\to-\infty} 5e^x = 0$
13. $\lim_{x\to\infty} \frac{1}{x^3} - \lim_{x\to\infty} \frac{2x+1}{x-5} = 0-2=-2$
14. $\lim_{x\to\infty} \frac{1}{x^2} - \lim_{x\to\infty} \frac{3x+4}{2x-1} = 0+\frac{3}{2}=\frac{3}{2}$
15. ∞
16. ∞
17. 1
18. 2
19. 2
20. $\lim_{n\to\infty} \frac{(n-1)^2}{(n+1)^2} = \lim_{n\to\infty} \left(\frac{n-1}{n+1}\right)^2 = \left(\lim_{n\to\infty} \frac{n-1}{n+1}\right)^2 = 1$
21. $\lim_{n\to\infty} \frac{n!}{(n+1)!} = \lim_{n\to\infty} \frac{n!}{(n+1)n!} = \lim_{n\to\infty} \frac{1}{(n+1)} = 0$
22. $\lim_{n\to\infty} \frac{(n-1)!}{(n+1)!} = \lim_{n\to\infty} \frac{(n-1)!}{(n+1)n(n-1)!} = \lim_{n\to\infty} \frac{1}{(n+1)n} = 0$
23. DNE since it is oscillatory (and the absolute value of the terms do not approach 0)
24. DNE since the *n*th term is equivalent to $(-1)^n$, which is oscillatory
25. DNE since the odd-indexed terms go to $-\infty$ and the even-indexed terms go to ∞
26. DNE since the even-indexed terms go to $-\infty$ and the odd-indexed terms go to ∞
27. 3
28. 2

29. 0	
30. 0	
31. $\lim_{n\to\infty} a_n = 130$ words per minute.	
32. $\lim_{n\to\infty} \frac{600n}{n+20} = 600$ names	
33. $\lim_{x\to\infty} F(x) = 10$	
34. $\lim_{x\to\infty} y(x) = 2800$ cards	
35. $\lim_{t\to\infty} S(t) = 3000$ snakes	
36. $\lim_{t\to\infty} N(t) = 100{,}000$ automobiles	
37. The limit of $2x$ as x gets arbitrarily large is ∞, not 2. Hence, the entire limit is ∞.	
38. DNE since the *n*th term is equivalent to $(-1)^n$ which is oscillatory. Thus, the quotient has no limit.	
39. False. The limit is 0 in such case.	
40. True (and it equals the quotient of the coefficients of the terms of highest degree in the numerator and denominator).	
41. True.	
42. False. The sequence diverges to ∞.	
43. Consider the graph: 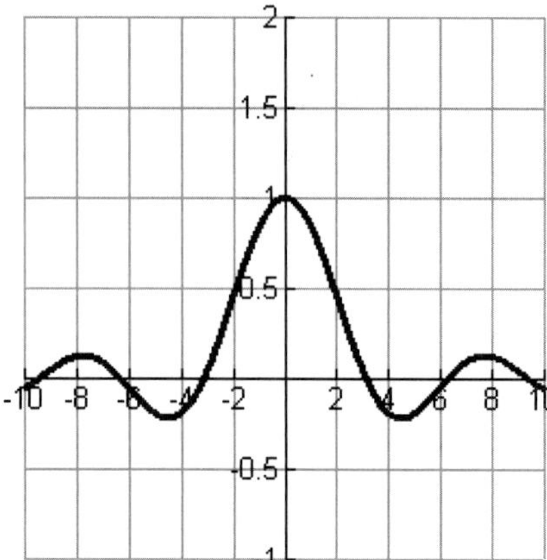	From the graph, it becomes apparent that $\lim_{x\to\infty} \frac{\sin x}{x} = 0$. One can argue this by observing that the numerator is bounded and the denominator becomes arbitrarily large as x goes to infinity, thereby driving the entire quotient to 0.

44. Consider the graph:

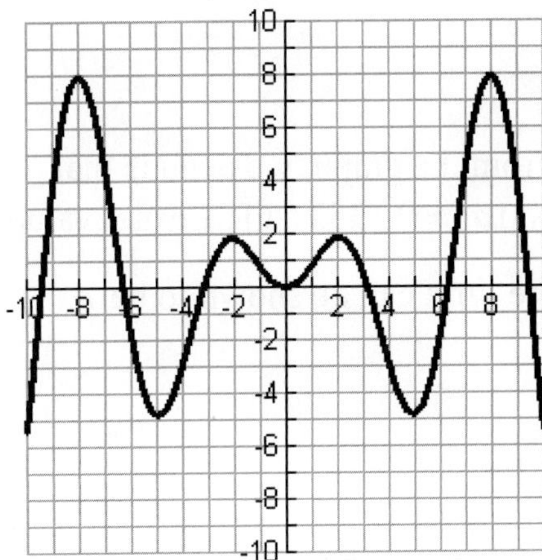

From the graph, it becomes apparent that $\lim_{x\to\infty} x\sin x$ DNE.

45. $\frac{1}{n!}$ goes to zero the fastest of the three since for every fixed *n*, its denominator is the largest and all three terms have the same numerator.

46. $\frac{1}{n^2}$ goes to zero the slowest of the three since its denominator is the smallest and all three terms have the same numerator.

47. Consider the graph:

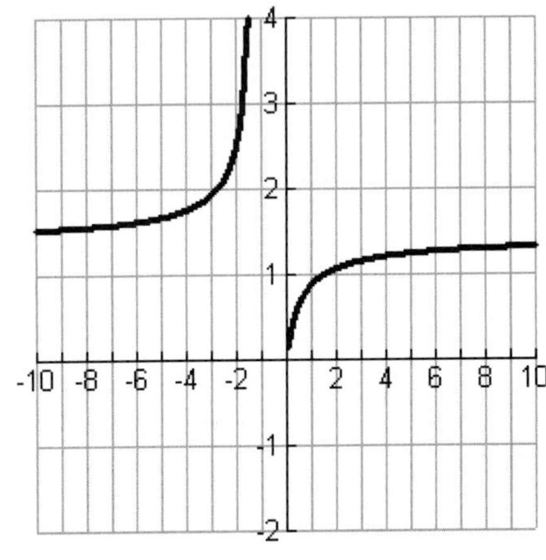

From the graph, we deduce that $\lim_{n\to\infty} a_n \approx 1.4142$. (The exact value of the limit can be shown to be $\sqrt{2}$.)

48. Consider the graph:

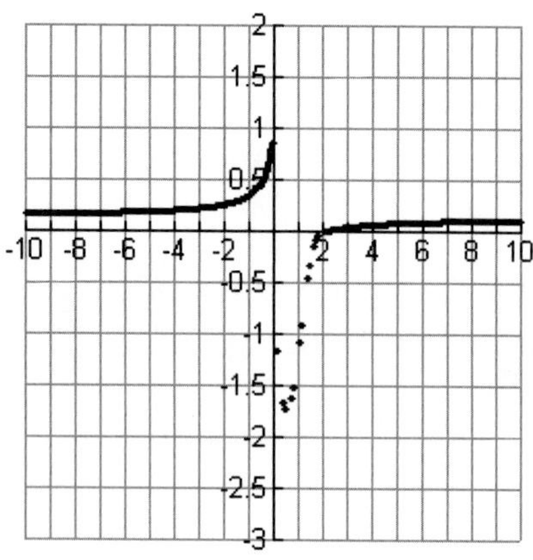

From the graph, we deduce that $\lim_{n\to\infty} a_n \approx 0.1353$. (The exact value of the limit can be shown to be e^{-2}.)

Section 11.5 Solutions---

1. $5(20)=100$

2. $8(30)=240$

3. $3\sum_{k=1}^{25} k = 3\left[\frac{25(26)}{2}\right] = 975$

4. $\sum_{k=1}^{17} 2k^2 = 2\sum_{k=1}^{17} k^2 = 2\left[\frac{(17)(18)(35)}{6}\right] = 3570$

5. $\sum_{k=1}^{n}\left(\frac{3k-4}{n}\right) = \frac{3}{n}\sum_{k=1}^{n} k - \sum_{k=1}^{n}\frac{4}{n} = \frac{3}{n}\left[\frac{n(n+1)}{2}\right] - \frac{4}{n}(n) = \frac{3n-5}{2}$

6. $\frac{2}{n^2}\sum_{k=1}^{n} k - \sum_{k=1}^{n}\frac{5}{n^2} = \frac{2}{n^2}\left[\frac{n(n+1)}{2}\right] - \frac{5}{n^2}(n) = \frac{n-4}{n}$

7. $2\sum_{k=1}^{n} k^3 + \sum_{k=1}^{n} k^2 = 2\left[\frac{n(n+1)}{2}\right]^2 + \frac{n(n+1)(2n+1)}{6} = \frac{3n^4+8n^3+6n^2+n}{6}$

8. $2\sum_{k=1}^{n} k^3 - \sum_{k=1}^{n} k = 2\left[\frac{n(n+1)}{2}\right]^2 - \left[\frac{n(n+1)}{2}\right] = \frac{n(n+1)(n^2+n-1)}{2}$

9.

$$\sum_{k=0}^{n}\left(k^2+2\right) = \left(0^2+2\right) + \sum_{k=1}^{n}\left(k^2+2\right) = 2 + \sum_{k=1}^{n} k^2 + \sum_{k=1}^{n} 2 = 2 + \frac{n(n+1)(2n+1)}{6} + 2n = \frac{2n^3+3n^2+13n+12}{6}$$

10. $\sum_{k=2}^{n}\left(k^2+2\right)=\sum_{k=1}^{n}\left(k^2+2\right)-\left(1^2+2\right)=\sum_{k=1}^{n}k^2-3=\frac{n(n+1)(2n+1)}{6}-3=\frac{2n^3+3n^2+13n-6}{6}$

11. $\lim_{n\to\infty}\sum_{k=1}^{n}\frac{5}{n}=\lim_{n\to\infty}\left[\frac{5}{n}\cdot n\right]=5$

12. $\lim_{n\to\infty}\sum_{k=1}^{n}\frac{8}{n}=\lim_{n\to\infty}\left[\frac{8}{n}\cdot n\right]=8$

13. $\lim_{n\to\infty}\frac{3}{n^2}\sum_{k=1}^{n}k=\lim_{n\to\infty}\left[\frac{3}{n^2}\cdot\frac{n(n+1)}{2}\right]=\lim_{n\to\infty}\frac{3n^2+3n}{2n^2}=\frac{3}{2}$

14. $\lim_{n\to\infty}\frac{2}{n^3}\sum_{k=1}^{n}k^2=\lim_{n\to\infty}\left[\frac{2}{n^3}\cdot\left(\frac{n(n+1)(2n+1)}{6}\right)\right]=\lim_{n\to\infty}\frac{4n^3+6n^2+2n}{6n^3}=\frac{2}{3}$

15. $\lim_{n\to\infty}\left[\frac{2}{n^2}\sum_{k=1}^{n}k-\sum_{k=1}^{n}\frac{3}{n}\right]=\lim_{n\to\infty}\left[\frac{2}{n^2}\cdot\frac{n(n+1)}{2}-\frac{3}{n}\cdot n\right]=\lim_{n\to\infty}\frac{2n^2+2n}{2n^2}-\lim_{n\to\infty}3=1-3=-2$

16. $\lim_{n\to\infty}\left[\sum_{k=1}^{n}\frac{5}{n}-\frac{1}{n^2}\sum_{k=1}^{n}k\right]=\lim_{n\to\infty}\left[\frac{5}{n}\cdot n-\frac{1}{n^2}\cdot\frac{n(n+1)}{2}\right]=\lim_{n\to\infty}\left[5-\frac{n^2+n}{2n^2}\right]=5-\frac{1}{2}=\frac{9}{2}$

17.

$$\lim_{n\to\infty}\left[\frac{2}{n^4}\sum_{k=1}^{n}k^3+\frac{1}{3n^3}\sum_{k=1}^{n}k^2\right]=\lim_{n\to\infty}\left[\frac{2}{n^4}\left(\frac{n(n+1)}{2}\right)^2+\frac{1}{3n^3}\frac{n(n+1)(2n+1)}{6}\right]$$

$$=\lim_{n\to\infty}\frac{2n^4+4n^3+2n^2}{4n^4}+\lim_{n\to\infty}\frac{2n^3+3n^2+n}{18n^3}=\frac{1}{2}+\frac{1}{9}=\frac{11}{18}$$

18.

$$\lim_{n\to\infty}\left[\frac{1}{3n^5}\sum_{k=1}^{n}k^3-\frac{1}{4n^2}\sum_{k=1}^{n}k\right]=\lim_{n\to\infty}\left[\frac{1}{3n^5}\left(\frac{n(n+1)}{2}\right)^2-\frac{1}{4n^2}\frac{n(n+1)}{2}\right]$$

$$=\lim_{n\to\infty}\frac{n^4+2n^3+n^2}{12n^5}-\lim_{n\to\infty}\frac{n^2+n}{8n^2}=0-\frac{1}{8}=-\frac{1}{8}$$

19.

$$\lim_{n\to\infty}\left[\frac{1}{2n^3}\sum_{k=1}^{n}k^2+\sum_{k=1}^{n}\frac{2}{n}-\frac{4}{n^2}\sum_{k=1}^{n}k\right]=\lim_{n\to\infty}\left[\frac{1}{2n^3}\left(\frac{n(n+1)(2n+1)}{6}\right)+\left(\frac{2}{n}\right)(n)-\frac{4}{n^2}\left(\frac{n(n+1)}{2}\right)\right]$$

$$=\lim_{n\to\infty}\left[\frac{2n^3+3n^2+n}{12n^3}+2-\frac{4n^2+4n}{2n^2}\right]=\frac{1}{6}$$

20.

$$\lim_{n\to\infty}\left[\frac{1}{5n^3}\sum_{k=1}^{n}k^2-\sum_{k=1}^{n}\frac{2}{n}+\frac{3}{n^2}\sum_{k=1}^{n}k\right]=\lim_{n\to\infty}\left[\frac{1}{5n^3}\left(\frac{n(n+1)(2n+1)}{6}\right)-\left(\frac{2}{n}\right)(n)+\frac{3}{n^2}\left(\frac{n(n+1)}{2}\right)\right]$$

$$=\lim_{n\to\infty}\left[\frac{2n^3+3n^2+n}{30n^3}-2+\frac{3n^2+3n}{2n^2}\right]=-\frac{13}{30}$$

21. Observe that

Height	Base	Area
$f(0)=0$	1	0
$f(1)=\frac{1}{2}$	1	$\frac{1}{2}$
$f(2)=2$	1	2
$f(3)=\frac{9}{2}$	1	$\frac{9}{2}$

So, the approximate area under the curve is 7 square units.

22. Observe that

Height	Base	Area
$f(1)=\frac{1}{2}$	1	$\frac{1}{2}$
$f(2)=2$	1	2
$f(3)=\frac{9}{2}$	1	$\frac{9}{2}$
$f(4)=8$	1	8

So, the approximate area under the curve is 15 square units.

23. 7 square units < Area under the curve < 15 square units.

24. Observe that

Height	Base	Area
$f(0)=8$	1	8
$f(1)=7$	1	7

So, the approximate area under the curve is 15 square units.

25. Observe that

Height	Base	Area
$f(1)=7$	1	7
$f(2)=0$	1	0

So, the approximate area under the curve is 7 square units.

26. 7 square units < Area under the curve < 15 square units.

27. Observe that

Height	Base	Area
$f(1)=\frac{7}{2}$	1	$\frac{7}{2}$
$f(2)=3$	1	3
$f(3)=\frac{5}{2}$	1	$\frac{5}{2}$
$f(4)=2$	1	2
$f(5)=\frac{3}{2}$	1	$\frac{3}{2}$
$f(6)=1$	1	1
$f(7)=\frac{1}{2}$	1	$\frac{1}{2}$
$f(8)=0$	1	0

So, the approximate area under the curve is 14 square units.

28. Observe that

Height	Base	Area
$f\left(\frac{1}{2}\right)=\frac{5}{4}$	½	$\frac{5}{8}$
$f(1)=2$	½	1
$f\left(\frac{3}{2}\right)=\frac{13}{4}$	½	$\frac{13}{8}$
$f(2)=5$	½	$\frac{5}{2}$

So, the approximate area under the curve is $\frac{23}{4}$ square units.

29. Observe that

Height	Base	Area
$f(0)=0$	1	0
$f(1)=1$	1	1
$f(2)=\sqrt{2}$	1	$\sqrt{2}$
$f(3)=\sqrt{3}$	1	$\sqrt{3}$

So, the approximate area under the curve is approximately 4.15 square units.

30. Observe that

Height	Base	Area
$f(1)=1$	$\frac{1}{8}$	$\frac{1}{8}$
$f\left(\frac{9}{8}\right)=\frac{8}{9}$	$\frac{1}{8}$	$\frac{1}{9}$
$f\left(\frac{10}{8}\right)=\frac{8}{10}$	$\frac{1}{8}$	$\frac{1}{10}$
$f\left(\frac{11}{8}\right)=\frac{8}{11}$	$\frac{1}{8}$	$\frac{1}{11}$
$f\left(\frac{12}{8}\right)=\frac{8}{12}$	$\frac{1}{8}$	$\frac{1}{12}$
$f\left(\frac{13}{8}\right)=\frac{8}{13}$	$\frac{1}{8}$	$\frac{1}{13}$
$f\left(\frac{14}{8}\right)=\frac{8}{14}$	$\frac{1}{8}$	$\frac{1}{14}$
$f\left(\frac{15}{8}\right)=\frac{8}{15}$	$\frac{1}{8}$	$\frac{1}{15}$

So, the approximate area under the curve is 0.725 square units.

31.

$$\lim_{n\to\infty}\sum_{k=1}^{n} f\left(\frac{k}{n}\right)\cdot\left(\frac{1}{n}\right)=\lim_{n\to\infty}\sum_{k=1}^{n}\left[-2\left(\frac{k}{n}\right)+3\right]\cdot\left(\frac{1}{n}\right)=\lim_{n\to\infty}\left[-\frac{2}{n^2}\sum_{k=1}^{n}k+\sum_{k=1}^{n}\frac{3}{n}\right]$$

$$=\lim_{n\to\infty}\left[-\frac{2}{n^2}\left(\frac{n(n+1)}{2}\right)+\frac{3}{n}(n)\right]=\lim_{n\to\infty}\left[\frac{-2n^2-2n}{2n^2}+3\right]=2$$

So, the area is 2 square units.

32.

$$\lim_{n\to\infty}\sum_{k=1}^{n} f\left(\frac{k}{n}\right)\cdot\left(\frac{1}{n}\right)=\lim_{n\to\infty}\sum_{k=1}^{n}\left[\frac{k}{n}-1\right]\cdot\left(\frac{2}{n}\right)=\lim_{n\to\infty}\left[\frac{2}{n^2}\sum_{k=1}^{n}k-\sum_{k=1}^{n}\frac{2}{n}\right]$$

$$=\lim_{n\to\infty}\left[\frac{2}{n^2}\left(\frac{n(n+1)}{2}\right)-\frac{2}{n}(n)\right]=\lim_{n\to\infty}\left[\frac{2n^2+2n}{2n^2}-2\right]=-1$$

So, the area is 2 square units.

33.

$$\lim_{n\to\infty}\sum_{k=1}^{n} f\left(\frac{k}{n}\right)\cdot\left(\frac{1}{n}\right)=\lim_{n\to\infty}\sum_{k=1}^{n}\left[2-\left(\frac{k}{n}\right)^2\right]\cdot\left(\frac{1}{n}\right)=\lim_{n\to\infty}\left[\sum_{k=1}^{n}\frac{2}{n}-\frac{1}{n^3}\sum_{k=1}^{n}k^2\right]$$

$$=\lim_{n\to\infty}\left[\frac{2}{n}(n)-\frac{1}{n^3}\left(\frac{n(n+1)(2n+1)}{6}\right)\right]=\lim_{n\to\infty}\left[2-\frac{2n^3+3n^2+n}{6n^3}\right]=\frac{5}{3}$$

So, the area is $\frac{5}{3}$ square units.

34.

$$\lim_{n\to\infty}\sum_{k=1}^{n} f\left(\frac{2k}{n}\right)\cdot\left(\frac{2}{n}\right)=\lim_{n\to\infty}\sum_{k=1}^{n}\left[5-\left(\frac{2k}{n}\right)^2\right]\cdot\left(\frac{2}{n}\right)=\lim_{n\to\infty}\left[\sum_{k=1}^{n}\frac{10}{n}-\frac{8}{n^3}\sum_{k=1}^{n}k^2\right]$$

$$=\lim_{n\to\infty}\left[\frac{10}{n}(n)-\frac{8}{n^3}\left(\frac{n(n+1)(2n+1)}{6}\right)\right]=\lim_{n\to\infty}\left[10-\frac{16n^3+24n^2+8n}{6n^3}\right]=\frac{22}{3}$$

So, the area is $\frac{22}{3}$ square units.

35.

$$\lim_{n\to\infty}\sum_{k=1}^{n} f\left(\frac{2k}{n}\right)\cdot\left(\frac{2}{n}\right)=\lim_{n\to\infty}\sum_{k=1}^{n}\left[8-\left(\frac{2k}{n}\right)^3\right]\cdot\left(\frac{2}{n}\right)=\lim_{n\to\infty}\left[\sum_{k=1}^{n}\frac{16}{n}-\frac{16}{n^4}\sum_{k=1}^{n}k^3\right]$$

$$=\lim_{n\to\infty}\left[\frac{16}{n}(n)-\frac{16}{n^4}\left(\frac{n(n+1)}{2}\right)^2\right]=\lim_{n\to\infty}\left[16-\frac{16n^4+32n^3+16n^2}{4n^4}\right]=12$$

So, the area is 12 square units.

36.

$$\lim_{n\to\infty}\sum_{k=1}^{n} f\left(\frac{k}{n}\right)\cdot\left(\frac{1}{n}\right)=\lim_{n\to\infty}\sum_{k=1}^{n}\left[\left(\frac{k}{n}\right)^3+1\right]\cdot\left(\frac{1}{n}\right)=\lim_{n\to\infty}\left[\frac{1}{n^4}\sum_{k=1}^{n}k^3+\sum_{k=1}^{n}\frac{1}{n}\right]$$

$$=\lim_{n\to\infty}\left[\frac{1}{n^4}\left(\frac{n(n+1)}{2}\right)^2+\frac{1}{n}(n)\right]=\lim_{n\to\infty}\left[\frac{n^4+2n^3+n^2}{4n^4}+1\right]=\frac{5}{4}$$

So, the area is $\frac{5}{4}$ square units.

37.

$$\lim_{n\to\infty}\sum_{k=1}^{n} f\left(1+\frac{2k}{n}\right)\cdot\left(\frac{2}{n}\right)=\lim_{n\to\infty}\sum_{k=1}^{n}\left[-\left(1+\frac{2k}{n}\right)^2+4\left(1+\frac{2k}{n}\right)\right]\cdot\left(\frac{2}{n}\right)=\lim_{n\to\infty}\left[\sum_{k=1}^{n}\frac{6}{n}+\frac{8}{n^2}\sum_{k=1}^{n}k-\frac{8}{n^3}\sum_{k=1}^{n}k^2\right]$$

$$=\lim_{n\to\infty}\left[\frac{6}{n}(n)+\frac{8}{n^2}\left(\frac{n(n+1)}{2}\right)-\frac{8}{n^3}\left(\frac{n(n+1)(2n+1)}{6}\right)\right]$$

$$=\lim_{n\to\infty}\left[6+\frac{8n^2+8n}{2n^2}-\frac{16n^3+3n^2+4n}{6n^3}\right]=\frac{22}{3}$$

So, the area is $\frac{22}{3}$ square units.

38.

$$\lim_{n\to\infty}\sum_{k=1}^{n} f\left(1+\frac{k}{n}\right)\cdot\left(\frac{1}{n}\right)=\lim_{n\to\infty}\sum_{k=1}^{n}\left[-\left(1+\frac{k}{n}\right)^2+3\left(1+\frac{k}{n}\right)\right]\cdot\left(\frac{1}{n}\right)=\lim_{n\to\infty}\left[\sum_{k=1}^{n}\frac{2}{n}+\frac{1}{n^2}\sum_{k=1}^{n}k-\frac{1}{n^3}\sum_{k=1}^{n}k^2\right]$$

$$=\lim_{n\to\infty}\left[\frac{2}{n}(n)+\frac{1}{n^2}\left(\frac{n(n+1)}{2}\right)-\frac{1}{n^3}\left(\frac{n(n+1)(2n+1)}{6}\right)\right]$$

$$=\lim_{n\to\infty}\left[2+\frac{n^2+n}{2n^2}-\frac{2n^3+3n^2+n}{6n^3}\right]=\frac{13}{6}$$

So, the area is $\frac{13}{6}$ square units.

39.

$$\lim_{n\to\infty}\sum_{k=1}^{n} f\left(2+\frac{k}{n}\right)\cdot\left(\frac{1}{n}\right)=\lim_{n\to\infty}\sum_{k=1}^{n}\left[-\left(2+\frac{k}{n}\right)^2+5\left(2+\frac{k}{n}\right)-4\right]\cdot\left(\frac{1}{n}\right)=\lim_{n\to\infty}\left[\sum_{k=1}^{n}\frac{2}{n}+\frac{1}{n^2}\sum_{k=1}^{n}k-\frac{1}{n^3}\sum_{k=1}^{n}k^2\right]$$

$$=\lim_{n\to\infty}\left[\frac{2}{n}(n)+\frac{1}{n^2}\left(\frac{n(n+1)}{2}\right)-\frac{1}{n^3}\left(\frac{n(n+1)(2n+1)}{6}\right)\right]$$

$$=\lim_{n\to\infty}\left[2+\frac{n^2+n}{2n^2}-\frac{2n^3+3n^2+n}{6n^3}\right]=\frac{13}{6}$$

So, the area is $\frac{13}{6}$ square units.

40.

$$\lim_{n\to\infty}\sum_{k=1}^{n} f\left(3+\frac{k}{n}\right)\cdot\left(\frac{1}{n}\right)=\lim_{n\to\infty}\sum_{k=1}^{n}\left[-\left(3+\frac{k}{n}\right)^2+7\left(3+\frac{k}{n}\right)-10\right]\cdot\left(\frac{1}{n}\right)=\lim_{n\to\infty}\left[\sum_{k=1}^{n}\frac{2}{n}+\frac{1}{n^2}\sum_{k=1}^{n}k-\frac{1}{n^3}\sum_{k=1}^{n}k^2\right]$$

$$=\lim_{n\to\infty}\left[\frac{2}{n}(n)+\frac{1}{n^2}\left(\frac{n(n+1)}{2}\right)-\frac{1}{n^3}\left(\frac{n(n+1)(2n+1)}{6}\right)\right]$$

$$=\lim_{n\to\infty}\left[2+\frac{n^2+n}{2n^2}-\frac{2n^3+3n^2+n}{6n^3}\right]=\frac{13}{6}$$

So, the area is $\frac{13}{6}$ square units.

41.

$$\lim_{n\to\infty}\sum_{k=1}^{n} f\left(\frac{600k}{n}\right)\cdot\left(\frac{600}{n}\right)=\lim_{n\to\infty}\sum_{k=1}^{n}\left[200+\frac{1}{100,000}\left(-\left(\frac{600k}{n}\right)^3+600\left(\frac{600k}{n}\right)^2\right)\right]\cdot\left(\frac{600}{n}\right)$$

$$=\lim_{n\to\infty}\left[\sum_{k=1}^{n}\frac{120,000}{n}-\frac{600^4}{100,000n^4}\sum_{k=1}^{n}k^3+\frac{600^4}{100,000n^3}\sum_{k=1}^{n}k^2\right]$$

$$=\lim_{n\to\infty}\left[\frac{120,000}{n}(n)-\frac{600^4}{100,000n^4}\left(\frac{n(n+1)}{2}\right)^2+\frac{600^4}{100,000n^3}\left(\frac{n(n+1)(2n+1)}{6}\right)\right]$$

$$=228,000$$

So, there are 228,000/43,560 = 5.23 acres.

42.

$$\lim_{n\to\infty}\sum_{k=1}^{n} f\left(\frac{1000k}{n}\right)\cdot\left(\frac{1000}{n}\right)=\lim_{n\to\infty}\sum_{k=1}^{n}\left[-0.006\left(\frac{1000k}{n}\right)^2+5000\right]\cdot\left(\frac{1000}{n}\right)$$

$$=\lim_{n\to\infty}\left[\sum_{k=1}^{n}\frac{5,000,000}{n}-\frac{6000}{n^3}\sum_{k=1}^{n}k^2\right]$$

$$=\lim_{n\to\infty}\left[\frac{5,000,000}{n}(n)-\frac{6000}{n^3}\left(\frac{n(n+1)(2n+1)}{6}\right)\right]$$

$$=4,998,000$$

So, there are 4,998,000/43,560 = 114.7 acres.

43.

$$\lim_{n\to\infty}\sum_{k=1}^{n} f\left(\frac{10k}{n}\right)\cdot\left(\frac{10}{n}\right)=\lim_{n\to\infty}\sum_{k=1}^{n}\left[200-\left(\frac{10k}{n}\right)^2\right]\cdot\left(\frac{10}{n}\right)=\lim_{n\to\infty}\left[\sum_{k=1}^{n}\frac{2000}{n}-\frac{1000}{n^3}\sum_{k=1}^{n}k^2\right]$$

$$=\lim_{n\to\infty}\left[\frac{2000}{n}(n)-\frac{1000}{n^3}\left(\frac{n(n+1)(2n+1)}{6}\right)\right]=1666.7\text{ ft. lbs.}$$

So, about 1666.7 ft. lbs.

44.

$$\lim_{n\to\infty}\sum_{k=1}^{n} f\left(\frac{5k}{n}\right)\cdot\left(\frac{5}{n}\right)=\lim_{n\to\infty}\sum_{k=1}^{n}\left[400-\left(\frac{5k}{n}\right)^2\right]\cdot\left(\frac{5}{n}\right)=\lim_{n\to\infty}\left[\sum_{k=1}^{n}\frac{2000}{n}-\frac{125}{n^3}\sum_{k=1}^{n}k^2\right]$$

$$=\lim_{n\to\infty}\left[\frac{2000}{n}(n)-\frac{125}{n^3}\left(\frac{n(n+1)(2n+1)}{6}\right)\right]=1958.3\text{ ft. lbs.}$$

So, about 1958.3 ft. lbs.

45.

$$\lim_{n\to\infty}\sum_{k=1}^{n} f\left(\frac{10k}{n}\right)\cdot\left(\frac{10}{n}\right)=\lim_{n\to\infty}\sum_{k=1}^{n}\left[10\left(\frac{10k}{n}\right)-\left(\frac{10k}{n}\right)^2\right]\cdot\left(\frac{10}{n}\right)=\lim_{n\to\infty}\left[\frac{1000}{n^2}\sum_{k=1}^{n}k-\frac{1000}{n^3}\sum_{k=1}^{n}k^2\right]$$

$$=\lim_{n\to\infty}\left[\frac{1000}{n}\left(\frac{n(n+1)}{2}\right)-\frac{1000}{n^3}\left(\frac{n(n+1)(2n+1)}{6}\right)\right]=166.7 \text{ ft. lbs.}$$

So, about 1666.7 ft. lbs.

46.

$$\lim_{n\to\infty}\sum_{k=1}^{n} f\left(4+\frac{2k}{n}\right)\cdot\left(\frac{2}{n}\right)=\lim_{n\to\infty}\sum_{k=1}^{n}\left[10\left(4+\frac{2k}{n}\right)-\left(4+\frac{2k}{n}\right)^2\right]\cdot\left(\frac{2}{n}\right)=\lim_{n\to\infty}\left[\sum_{k=1}^{n}\frac{48}{n}+\frac{8}{n^2}\sum_{k=1}^{n}k-\frac{8}{n^3}\sum_{k=1}^{n}k^2\right]$$

$$=\lim_{n\to\infty}\left[\frac{48}{n}(n)+\frac{8}{n^2}\left(\frac{n(n+1)}{2}\right)-\frac{8}{n^3}\left(\frac{n(n+1)(2n+1)}{6}\right)\right]=49.3 \text{ ft. lbs.}$$

So, about 49.3 ft. lbs.

47. The summation should have been evaluated first, then the limit taken.

48. Cannot apply "limit of product is the product of limits" because $\lim_{n\to\infty}\sum_{k=1}^{n}k$ DNE. Evaluate the summation first, then take the limit of the resulting quotient.

49. False. Take n =2. Observe that $\sum_{k=1}^{2}k^2=5$, whereas $\left(\sum_{k=1}^{2}k\right)^2=(1+2)^2=9$.

50. False. $\sum_{k=1}^{\infty}k$ actually diverges.

51. Consider the dashed region:

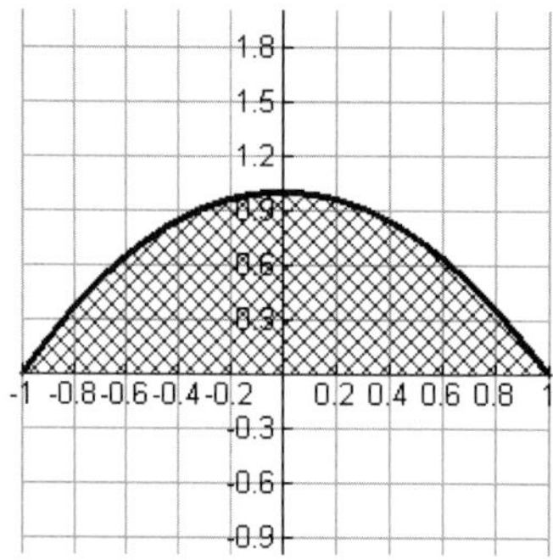

$$\lim_{n\to\infty}\sum_{k=1}^{n} f\left(-1+\frac{2k}{n}\right)\cdot\left(\frac{2}{n}\right)=\lim_{n\to\infty}\sum_{k=1}^{n}\left[-\left(-1+\frac{2k}{n}\right)^2+1\right]\cdot\left(\frac{2}{n}\right)=\lim_{n\to\infty}\left[\frac{8}{n^2}\sum_{k=1}^{n}k-\frac{8}{n^3}\sum_{k=1}^{n}k^2\right]$$

$$=\lim_{n\to\infty}\left[\frac{8}{n^2}\left(\frac{n(n+1)}{2}\right)-\frac{8}{n^3}\left(\frac{n(n+1)(2n+1)}{6}\right)\right]=\frac{4}{3}$$

The area is 4/3 square units.

52. Consider the dashed region:

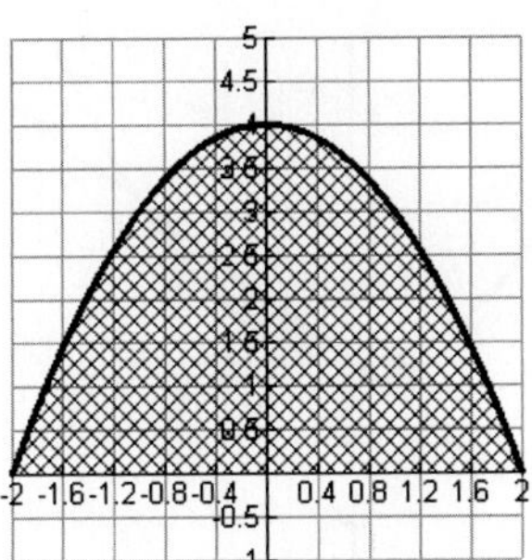

$$\lim_{n\to\infty}\sum_{k=1}^{n} f\left(-2+\frac{4k}{n}\right)\cdot\left(\frac{4}{n}\right)=\lim_{n\to\infty}\sum_{k=1}^{n}\left[-\left(-2+\frac{4k}{n}\right)^2+4\right]\cdot\left(\frac{4}{n}\right)=\lim_{n\to\infty}\left[\frac{64}{n^2}\sum_{k=1}^{n}k-\frac{64}{n^3}\sum_{k=1}^{n}k^2\right]$$

$$=\lim_{n\to\infty}\left[\frac{64}{n^2}\left(\frac{n(n+1)}{2}\right)-\frac{64}{n^3}\left(\frac{n(n+1)(2n+1)}{6}\right)\right]=\frac{32}{3}$$

The area is 32/3 square units.

53. Consider the dashed region:

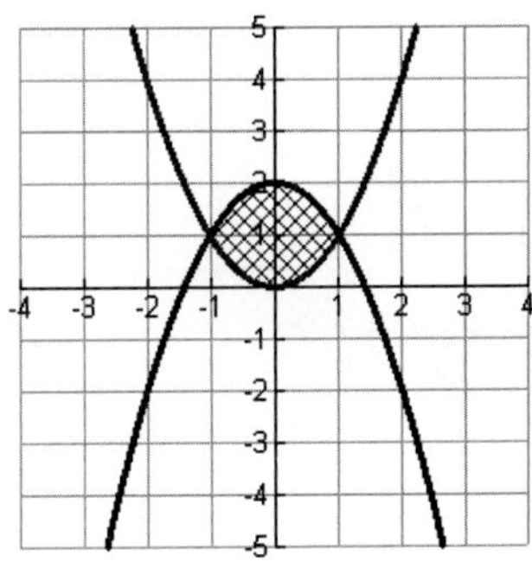

$$\lim_{n\to\infty}\sum_{k=1}^{n}\left[g\left(-1+\frac{2k}{n}\right)-f\left(-1+\frac{2k}{n}\right)\right]\cdot\left(\frac{2}{n}\right)=\lim_{n\to\infty}\sum_{k=1}^{n}\left[\left(-\left(-1+\frac{2k}{n}\right)^2+2\right)-\left(-1+\frac{2k}{n}\right)^2\right]\cdot\left(\frac{2}{n}\right)$$

$$=\lim_{n\to\infty}\sum_{k=1}^{n}\left[-2\left(-1+\frac{2k}{n}\right)^2+2\right]\cdot\left(\frac{2}{n}\right)=\lim_{n\to\infty}\left[\frac{16}{n^2}\sum_{k=1}^{n}k-\frac{16}{n^3}\sum_{k=1}^{n}k^2\right]$$

$$=\lim_{n\to\infty}\left[\frac{16}{n^2}\left(\frac{n(n+1)}{2}\right)-\frac{16}{n^3}\left(\frac{n(n+1)(2n+1)}{6}\right)\right]=\frac{8}{3}$$

The area is 8/3 square units.

54. Consider the dashed region:

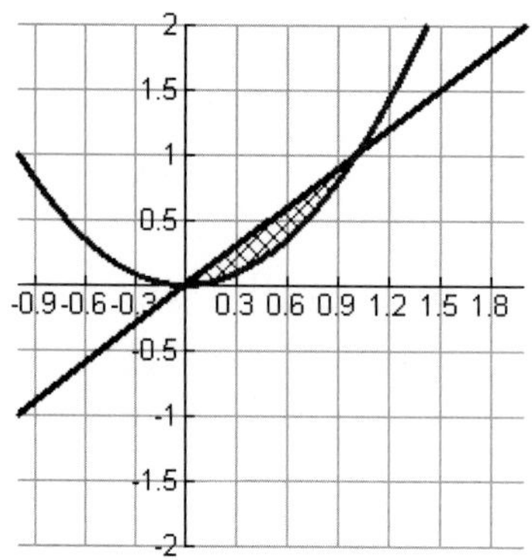

$$\lim_{n\to\infty}\sum_{k=1}^{n}\left[g\left(\frac{k}{n}\right)-f\left(\frac{k}{n}\right)\right]\cdot\left(\frac{1}{n}\right)=\lim_{n\to\infty}\sum_{k=1}^{n}\left[\left(\frac{k}{n}\right)-\left(\frac{k}{n}\right)^2\right]\cdot\left(\frac{1}{n}\right)$$

$$=\lim_{n\to\infty}\left[\frac{1}{n^2}\sum_{k=1}^{n}k-\frac{1}{n^3}\sum_{k=1}^{n}k^2\right]$$

$$=\lim_{n\to\infty}\left[\frac{1}{n^2}\left(\frac{n(n+1)}{2}\right)-\frac{1}{n^3}\left(\frac{n(n+1)(2n+1)}{6}\right)\right]=\frac{1}{6}$$

So, the area is 1/6 square units.

55. Consider the dashed region:

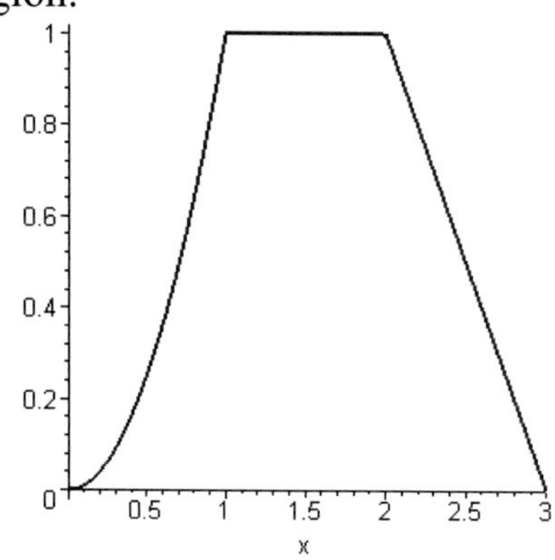

Find the area of each subregion determined by the intervals (0,1), (1,2) and (2,3) separately, and then sum them. For convenience, we label these subregions as I, II, and III.

Region I: $\lim_{n\to\infty}\sum_{k=1}^{n}\left(\frac{k}{n}\right)^2\cdot\left(\frac{1}{n}\right)=\lim_{n\to\infty}\left[\frac{1}{n^3}\sum_{k=1}^{n}k^2\right]=\lim_{n\to\infty}\left[\frac{1}{n^3}\left(\frac{n(n+1)(2n+1)}{6}\right)\right]=\frac{1}{3}$ square units

Region II: It is a rectangle with area 1(1) = 1 square units.

Region III: It is a triangle with area 1/2 (1)(1) = ½ square units.

Thus, the area of the entire region is 11/6 square units.

56. Consider the dashed region:

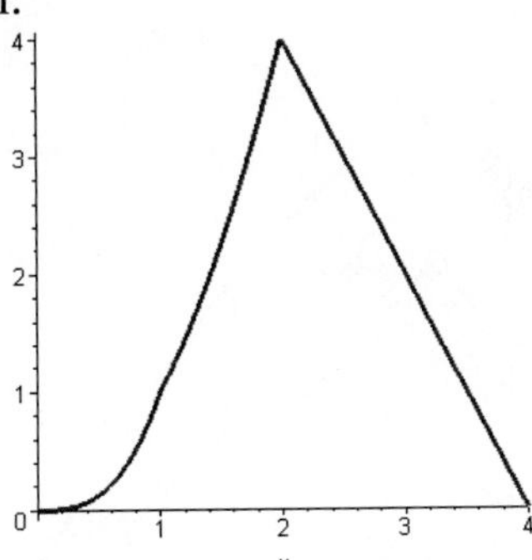

Find the area of each subregion determined by the intervals (0,1), (1,2) and (2,4) separately, and then sum them. For convenience, we label these subregions as I, II, and III.

Region I: $\lim_{n\to\infty}\sum_{k=1}^{n}\left(\frac{k}{n}\right)^3\cdot\left(\frac{1}{n}\right)=\lim_{n\to\infty}\left[\frac{1}{n^4}\sum_{k=1}^{n}k^3\right]=\lim_{n\to\infty}\left[\frac{1}{n^4}\left(\frac{n(n+1)}{2}\right)^2\right]=\frac{1}{4}$ square units

Region II:

$$\lim_{n\to\infty}\sum_{k=1}^{n}\left(1+\frac{k}{n}\right)^2\cdot\left(\frac{1}{n}\right)=\lim_{n\to\infty}\left[\sum_{k=1}^{n}\frac{1}{n}+\frac{2}{n^2}\sum_{k=1}^{n}k+\frac{1}{n^3}\sum_{k=1}^{n}k^2\right]$$

$$=\lim_{n\to\infty}\left[\frac{2}{n}(n)+\frac{1}{n^2}\left(\frac{n(n+1)}{2}\right)+\frac{1}{n^3}\left(\frac{n(n+1)(2n+1)}{6}\right)\right]=\frac{7}{3}\text{ square units}$$

Region III: This is a triangle with area ½ (4)(2) = 4 square units.
Thus, the area of the entire region is $\frac{79}{12}$ square units.

57. Consider the dashed region:

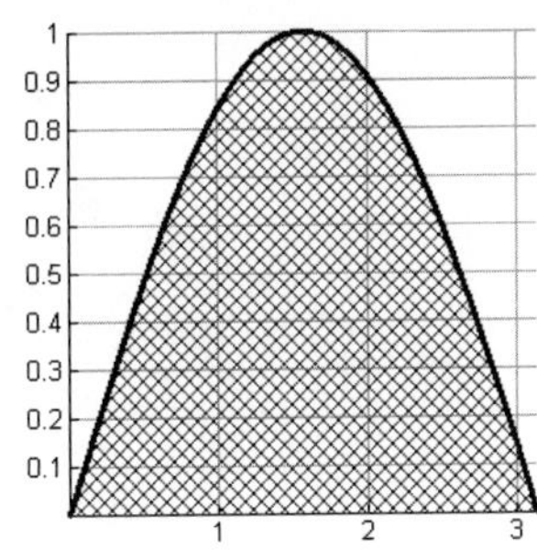

Approximation of the area using 4 rectangles.

$$\sum_{k=1}^{4}f\left(\frac{\pi k}{4}\right)\cdot\left(\frac{\pi}{4}\right)=\frac{\pi}{4}\left[f\left(\frac{\pi}{4}\right)+f\left(\frac{\pi}{2}\right)+f\left(\frac{3\pi}{4}\right)+f(\pi)\right]=\frac{\pi}{4}\left(\sqrt{2}+1\right)\approx 1.8961\text{ square units}$$

Approximation of the area using 10 rectangles.

$$\sum_{k=1}^{10}f\left(\frac{\pi k}{10}\right)\cdot\left(\frac{\pi}{10}\right)\approx 1.9835\text{ square units}$$

58. Consider the dashed region:

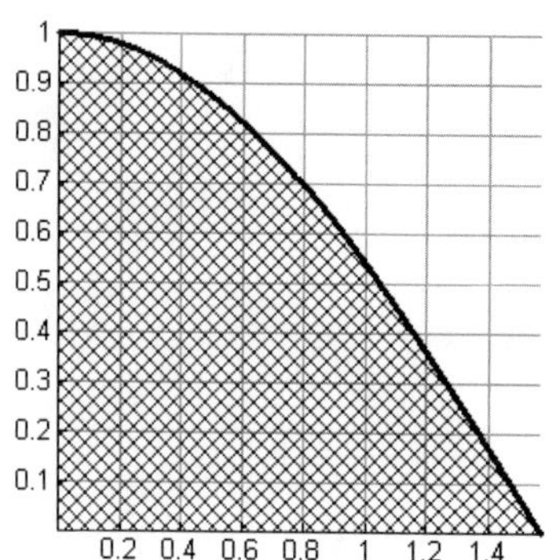

Approximation of the area using 4 rectangles.

$$\sum_{k=1}^{4} f\left(\frac{\frac{\pi}{2}k}{4}\right)\cdot\left(\frac{\frac{\pi}{2}}{4}\right)=\sum_{k=1}^{4} f\left(\frac{\pi k}{8}\right)\cdot\left(\frac{\pi}{8}\right)\text{ square units}$$

Approximation of the area using 10 rectangles.

$$\sum_{k=1}^{10} f\left(\frac{\frac{\pi}{2}k}{10}\right)\cdot\left(\frac{\frac{\pi}{2}}{10}\right)=\sum_{k=1}^{4} f\left(\frac{\pi k}{20}\right)\cdot\left(\frac{\pi}{20}\right)\text{ square units}$$

59. Consider the dashed region:

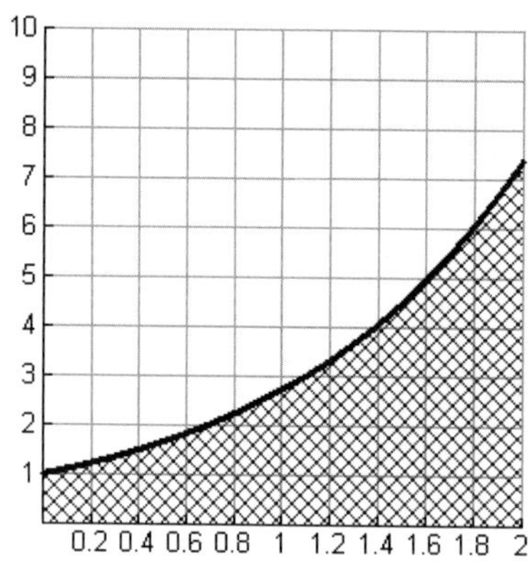

Approximation of the area using n rectangles is given by

$$\sum_{k=1}^{n} f\left(\frac{2k}{n}\right)\cdot\left(\frac{2}{n}\right)=\sum_{k=1}^{n} e^{\left(\frac{2k}{n}\right)}\cdot\left(\frac{2}{n}\right).$$

Using this formula, we see that

If $n = 4$, then the area is approximately 8.1189 square units.
If $n = 100$, then the area is approximately 6.4532 square units.
If $n = 500$, then the area is approximately 6.4018 square units.

60. Consider the dashed region:

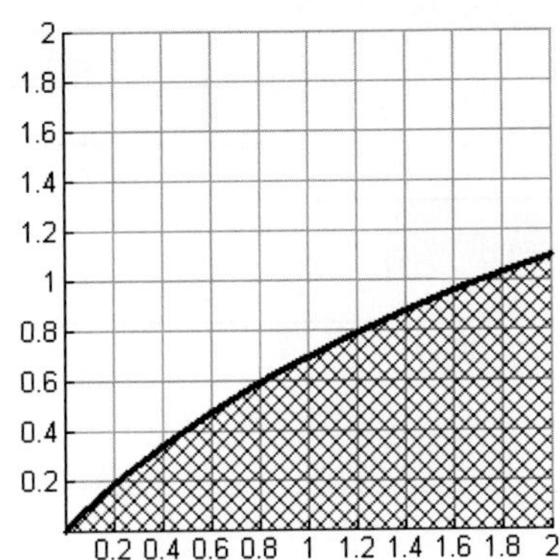

Approximation of the area using n rectangles is given by

$$\sum_{k=1}^{n} f\left(\frac{2k}{n}\right)\cdot\left(\frac{2}{n}\right) = \sum_{k=1}^{n} \ln\left(1+\frac{2k}{n}\right)\cdot\left(\frac{2}{n}\right).$$

Using this formula, we see that

If $n = 4$, then the area is approximately 1.5568 square units.
If $n = 100$, then the area is approximately 1.3068 square units.
If $n = 500$, then the area is approximately 1.2980 square units

Chapter 11 Review Solutions---

1. Consider the following table of values:

x	$\frac{x-3}{x^2-9}$
2.9	0.16949
2.99	0.16694
2.999	0.166694
2.9999	0.1666694
$\downarrow$	
3	1/6
$\uparrow$	
3.0001	0.1666639
3.001	0.166639
3.01	0.16639
3.1	0.16393

So, $\lim_{x\to 3} \frac{x-3}{x^2-3} = \frac{1}{6}$.

2. Consider the following table of values:

x	$\frac{\sqrt{x}-4}{x-16}$
15.9	0.125196
15.99	0.1250196
15.999	0.125002
15.9999	0.1250002
$\downarrow$	
16	1/8
$\uparrow$	
16.0001	0.1249998
16.001	0.124998
16.01	0.12498
16.1	0.124805

So, $\lim_{x\to 16} \frac{\sqrt{x}-4}{x-16} = \frac{1}{8}$.

3. ∞
4. DNE (since left and right limits are different)
5. 1
6. -1
7. 8
8. $\lim_{x\to-1} x^5 = \left(\lim_{x\to-1} x\right)^5 = (-1)^5 = -1$
9. ½ (just substitute in $x = 0$)
10. $\sqrt{\lim_{x\to-2}\left(x^2+5\right)} = \sqrt{9} = 3$
11. $\lim_{x\to-1}\frac{(x-1)(x+1)}{x+1} = -2$
12. $\lim_{x\to9}\frac{\sqrt{x}-3}{x-9} = \lim_{x\to9}\frac{\left(\sqrt{x}-3\right)\left(\sqrt{x}+3\right)}{(x-9)\left(\sqrt{x}+3\right)} = \lim_{x\to9}\frac{x-9}{(x-9)\left(\sqrt{x}+3\right)} = \frac{1}{6}$
13. $\lim_{x\to2\pi}\frac{1-\cos x}{\sin x} = \lim_{x\to2\pi}\frac{\frac{1-\cos x}{x-2\pi}}{\frac{\sin x}{x-2\pi}} = \frac{\lim_{x\to2\pi}\frac{1-\cos x}{x-2\pi}}{\lim_{x\to2\pi}\frac{\sin x}{x-2\pi}} = \frac{0}{1} = 0$
14. $\lim_{x\to\pi}\frac{\sin x}{\cos x}\cdot\cos x = \lim_{x\to\pi}\sin x = 0$
15. $\lim_{h\to0}\frac{f(x+h)-f(x)}{h} = \lim_{h\to0}\frac{\left((x+h)^2-2(x+h)+3\right)-\left(x^2-2x+3\right)}{h}$ $= \lim_{h\to0}\frac{x^2+2hx+h^2-2x-2h+3-x^2+2x-3}{h} = \lim_{h\to0}\frac{h(2x+h-2)}{h} = 2x-2$
16. $\lim_{h\to0}\frac{f(x+h)-f(x)}{h} = \lim_{h\to0}\frac{\left(-7(x+h)+9\right)-\left(-7x+9\right)}{h} = \lim_{h\to0}\frac{-7h}{h} = -7$
17. 0
18. DNE since the left and right limits are different.
19. $\frac{\sqrt{2}}{2}$
20. -1
21. $\lim_{h\to0}\frac{f(0+h)-f(0)}{h} = \lim_{h\to0}\frac{(6h+3)-3}{h} = \lim_{h\to0}\frac{6h}{h} = 6$
22. $\lim_{h\to0}\frac{f(-1+h)-f(-1)}{h} = \lim_{h\to0}\frac{-3(1+h)^2-\left(-3(-1)^2\right)}{h} = \lim_{h\to0}\frac{6h-3h^2}{h} = \lim_{h\to0}\frac{h(6-3h)}{h} = 6$
23. $\lim_{h\to0}\frac{f(-1+h)-f(-1)}{h} = \lim_{h\to0}\frac{\frac{1}{(-1+h)^2}-\frac{1}{(-1)^2}}{h} = \lim_{h\to0}\frac{2h-h^2}{h(-1+h)^2} = \lim_{h\to0}\frac{h(2-h)}{h(-1+h)^2} = 2$

24. $\lim_{h\to 0}\frac{f(3+h)-f(3)}{h}=\lim_{h\to 0}\frac{\sqrt{(3+h)+1}-\sqrt{4}}{h}=\lim_{h\to 0}\frac{\left(\sqrt{4+h}-2\right)\left(\sqrt{4+h}+2\right)}{h\left(\sqrt{4+h}+2\right)}=\lim_{h\to 0}\frac{1}{\sqrt{4+h}+2}=\frac{1}{4}$
25. Slope: $\lim_{h\to 0}\frac{f(2+h)-f(2)}{h}=\lim_{h\to 0}\frac{12-12}{h}=0$ Point: $(2,12)$ So, the equation of the tangent line is: $y=12$.
26. Slope: $\lim_{h\to 0}\frac{f(x+h)-f(x)}{h}=\lim_{h\to 0}\frac{\left(-3(1+h)+5\right)-(2)}{h}=\lim_{h\to 0}\frac{-3h}{h}=-3$ Point: $(1,2)$ So, the equation of the tangent line is: $y-2=-3(x-1)\Rightarrow y=-3x+5$.
27. Slope: $\lim_{h\to 0}\frac{f(0+h)-f(0)}{h}=\lim_{h\to 0}\frac{\left(5(0+h)^2+2\right)-(2)}{h}=\lim_{h\to 0}\frac{5h^2}{h}=0$ Point: $(0,2)$ So, the equation of the tangent line is: $y=2$.
28. Slope: $\lim_{h\to 0}\frac{f(\frac{1}{2}+h)-f(\frac{1}{2})}{h}=\lim_{h\to 0}\frac{\frac{1}{\frac{1}{2}+h}-\frac{1}{\frac{1}{2}}}{h}=\lim_{h\to 0}\frac{-2h}{h(\frac{1}{2}+h)}=-4$ Point: $(\frac{1}{2},2)$ So, the equation of the tangent line is: $y-2=-4(x-\frac{1}{2})\Rightarrow y=-4x+4$.
29. $\lim_{x\to 2}\frac{\left(2-5x^2\right)-(-18)}{x-2}=\lim_{x\to 2}\frac{5\left(4-x^2\right)}{-(2-x)}=\lim_{x\to 2}\frac{5(2-x)(2+x)}{-(2-x)}=\lim_{x\to 2}-5(2+x)=-20$
30. $\lim_{x\to -1}\frac{-x^4-1}{x+1}=\lim_{x\to -1}\frac{-(x+1)(x-1)\left(x^2+1\right)}{x+1}=\lim_{x\to -1}-(x-1)\left(x^2+1\right)=4$
31. $\lim_{x\to 5}\frac{\left(1-x^2\right)-(-24)}{x-5}=\lim_{x\to 5}\frac{25-x^2}{x-5}=\lim_{x\to 5}\frac{(5-x)(5+x)}{-(5-x)}=\lim_{x\to 5}-(5+x)=-10$
32. $\lim_{x\to 0}\frac{\frac{x}{1+x}-0}{x-0}=\lim_{x\to 0}\frac{1}{1+x}=1$
33. $f'(x)=\lim_{h\to 0}\frac{f(x+h)-f(x)}{h}=\lim_{h\to 0}\frac{e-e}{h}=0$
34. $f'(x)=\lim_{h\to 0}\frac{f(x+h)-f(x)}{h}=\lim_{h\to 0}\frac{[6(x+h)+7]-[6x+7]}{h}=\lim_{h\to 0}\frac{6h}{h}=6$

35. $$f'(x)=\lim_{h\to 0}\frac{f(x+h)-f(x)}{h}=\lim_{h\to 0}\frac{\left[3(x+h)^2-2(x+h)+1\right]-\left[3x^2-2x+1\right]}{h}=\lim_{h\to 0}\frac{h(6x+3h-2)}{h}$$ $$=\lim_{h\to 0}(6x+3h-2)=6x-2$$
36. $$f'(x)=\lim_{h\to 0}\frac{f(x+h)-f(x)}{h}=\lim_{h\to 0}\frac{\sqrt{x+h-1}-\sqrt{x-1}}{h}=\lim_{h\to 0}\frac{\left(\sqrt{x+h-1}-\sqrt{x-1}\right)\left(\sqrt{x+h-1}+\sqrt{x-1}\right)}{h\left(\sqrt{x+h-1}+\sqrt{x-1}\right)}$$ $$=\lim_{h\to 0}\frac{h}{h\left(\sqrt{x+h-1}+\sqrt{x-1}\right)}=\frac{1}{2\sqrt{x-1}}$$
37. 0
38. 4/3
39. 0
40. ∞
41. $-\infty$
42. 0 since the degree of the numerator is strictly smaller than the degree of the denominator.
43. The sequence converges to 3.
44. $\lim_{n\to\infty}\frac{(n+1)!}{(n-1)!n(n+1)}=\lim_{n\to\infty}\frac{(n+1)n(n-1)!}{(n+1)n(n-1)!}=\lim_{n\to\infty}1=1$ So, the sequence converges to 1.
45. Since $\sin(\pi n)=0$ for any integer *n*, the sequence converges to 0.
46. DNE since the odd-indexed terms go to $-\infty$ while the even-indexed terms go to ∞.
47. $\sum_{k=1}^{30}3k^2=3\sum_{k=1}^{30}k^2=3\left(\frac{30(31)(61)}{6}\right)=28,365$
48. $\sum_{k=1}^{12}\left(5+k^2\right)=\sum_{k=1}^{12}5+\sum_{k=1}^{12}k^2=5(12)+\frac{12(13)(25)}{6}=710$

49. Consider the region below:

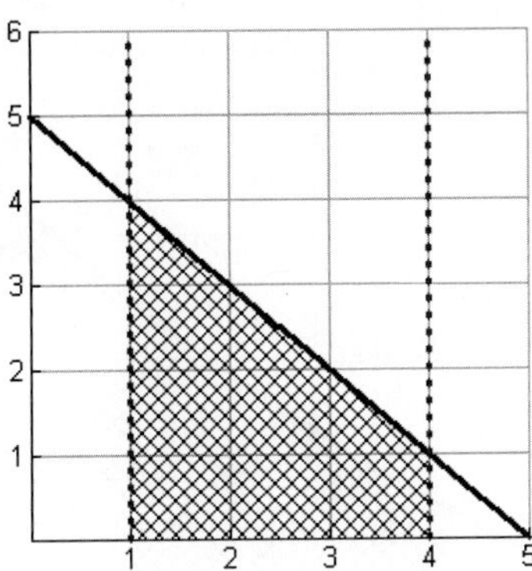

$$\lim_{n\to\infty}\sum_{k=1}^{n} f\left(3+\frac{k}{n}\right)\cdot\left(\frac{3}{n}\right)=\lim_{n\to\infty}\sum_{k=1}^{n}\left[5-\left(1+\frac{3k}{n}\right)\right]\cdot\left(\frac{3}{n}\right)=\lim_{n\to\infty}\left[\sum_{k=1}^{n}\frac{12}{n}-\frac{9}{n^2}\sum_{k=1}^{n}k\right]$$

$$=\lim_{n\to\infty}\left[\frac{12}{n}(n)-\frac{9}{n^2}\left(\frac{n(n+1)}{2}\right)\right]=\frac{15}{2}$$

So, the area is 15/2 square units.

50. Consider the region below:

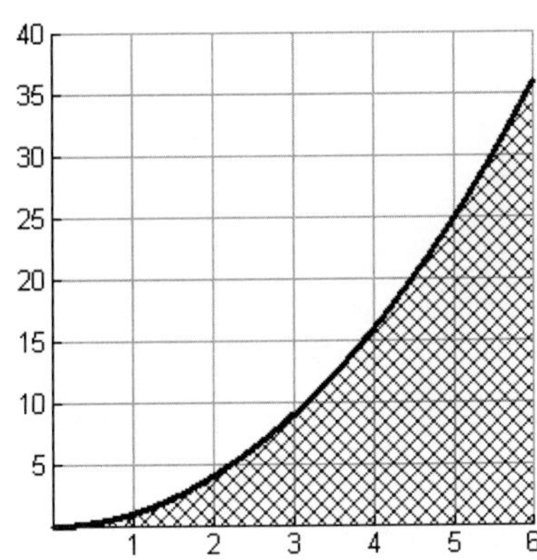

$$\lim_{n\to\infty}\sum_{k=1}^{n} f\left(\frac{6k}{n}\right)\cdot\left(\frac{6}{n}\right)=\lim_{n\to\infty}\sum_{k=1}^{n}\left(\frac{6k}{n}\right)^2\cdot\left(\frac{6}{n}\right)=\lim_{n\to\infty}\left[\frac{216}{n^3}\sum_{k=1}^{n}k^2\right]=\lim_{n\to\infty}\left[\frac{216}{n^3}\left(\frac{n(n+1)(2n+1)}{6}\right)\right]=72$$

So, the area is 72 square units.

51. Consider the region below:

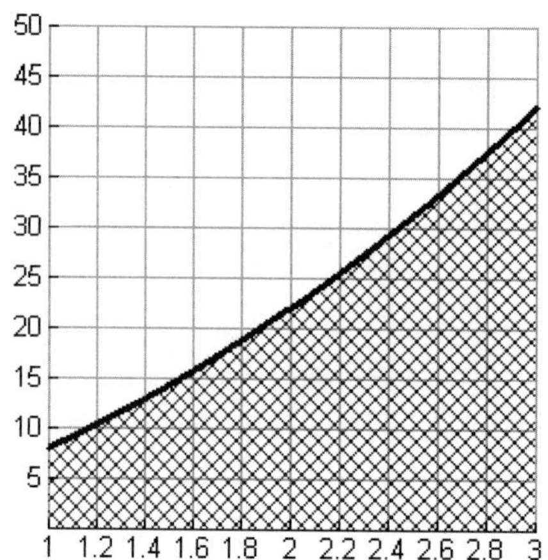

$$\lim_{n\to\infty}\sum_{k=1}^{n} f\left(1+\frac{2k}{n}\right)\cdot\left(\frac{2}{n}\right)=\lim_{n\to\infty}\sum_{k=1}^{n}\left[3\left(1+\frac{2k}{n}\right)^2+5\left(1+\frac{2k}{n}\right)\right]\cdot\left(\frac{2}{n}\right)=\lim_{n\to\infty}\left[\sum_{k=1}^{n}\frac{16}{n}+\frac{44}{n^2}\sum_{k=1}^{n}k+\frac{24}{n^3}\sum_{k=1}^{n}k^2\right]$$

$$=\lim_{n\to\infty}\left[\frac{16}{n}(n)+\frac{44}{n^2}\left(\frac{n(n+1)}{2}\right)+\frac{24}{n^3}\left(\frac{n(n+1)(2n+1)}{6}\right)\right]=46$$

So, the area is 46 square units.

52. Consider the region below:

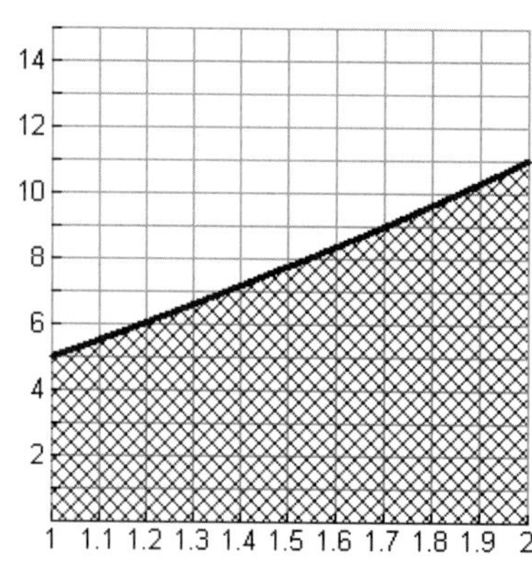

$$\lim_{n\to\infty}\sum_{k=1}^{n} f\left(1+\frac{k}{n}\right)\cdot\left(\frac{1}{n}\right)=\lim_{n\to\infty}\sum_{k=1}^{n}\left[\left(1+\frac{k}{n}\right)^2+3\left(1+\frac{k}{n}\right)+1\right]\cdot\left(\frac{1}{n}\right)=\lim_{n\to\infty}\left[\sum_{k=1}^{n}\frac{5}{n}+\frac{5}{n^2}\sum_{k=1}^{n}k+\frac{1}{n^3}\sum_{k=1}^{n}k^2\right]$$

$$=\lim_{n\to\infty}\left[\frac{5}{n}(n)+\frac{5}{n^2}\left(\frac{n(n+1)}{2}\right)+\frac{1}{n^3}\left(\frac{n(n+1)(2n+1)}{6}\right)\right]=\frac{47}{6}$$

So, the area is 47/6 square units.

53. Consider the following graph near $x = 1$.

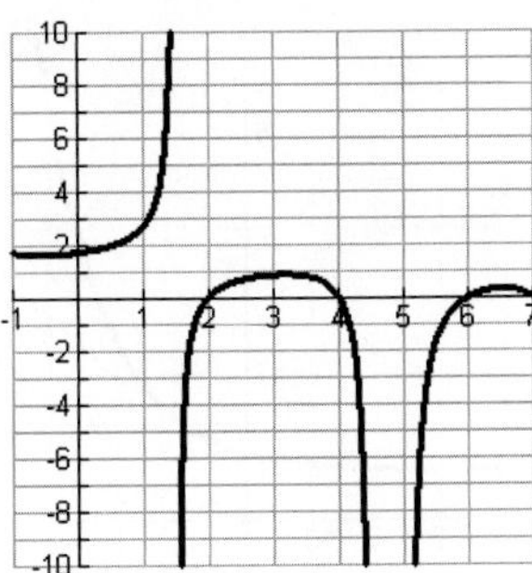

Using the graph, we see that $\lim_{x\to 1}\dfrac{\tan\left(\frac{\pi}{3}x\right)-\sqrt{3}}{\tan\left(\frac{\pi}{4}x\right)-1}=\dfrac{8}{3}$.

54. Consider the following graph near $x = -2$.

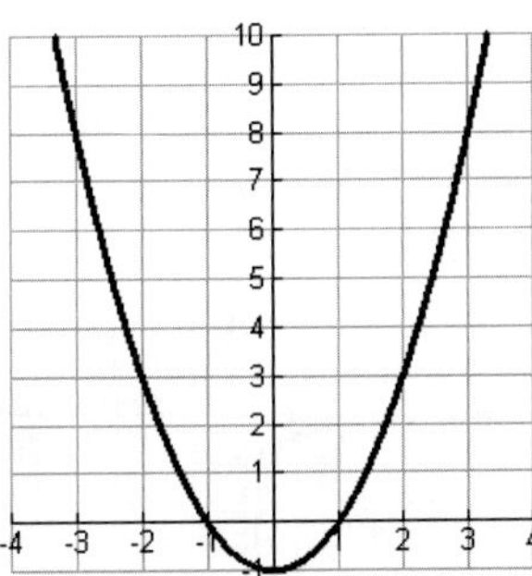

Using the graph, we see that $\lim_{x\to -2}\dfrac{-x^3+2x^2-x-2}{x+2}=3$.

55. Consider the following graph near $x = 3$.

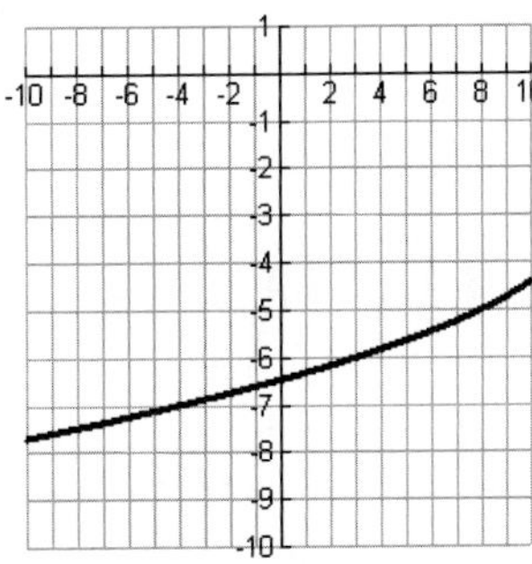

Using the graph, we see that $\lim_{x\to 3}\dfrac{x-3}{\sqrt{12-x}-3}=-6$. (This can be shown algebraically by rationalizing the denominator and then simplifying the resulting expression.)

56. Consider the following graph near $x = 1$.

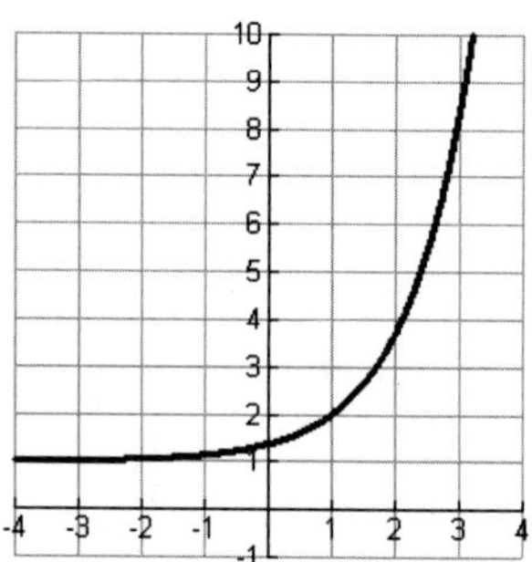

Using the graph, we see that $\lim_{x \to 1} \frac{e^{2(x-1)} - 1}{e^{x-1} - 1} = 2$. (This can be shown algebraically by factoring the numerator as a difference of squares, and then canceling the like factors in the numerator and denominator.)

57. Consider the graphs below.

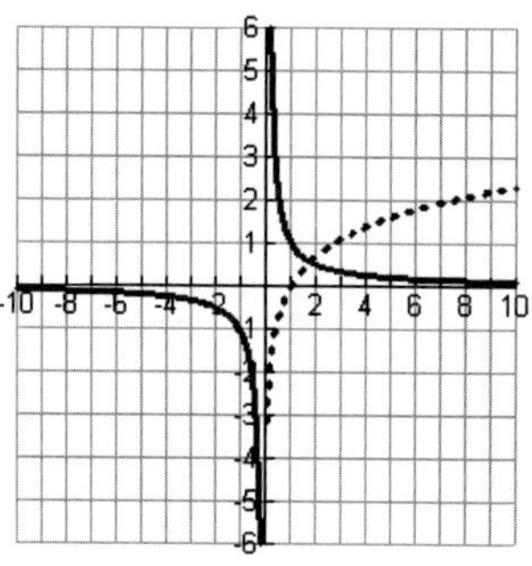

It appears that $f'(x) = \lim_{h \to 0} \frac{f(x+h) - f(x)}{h} = \frac{1}{x}$.

58. Consider the graphs below.

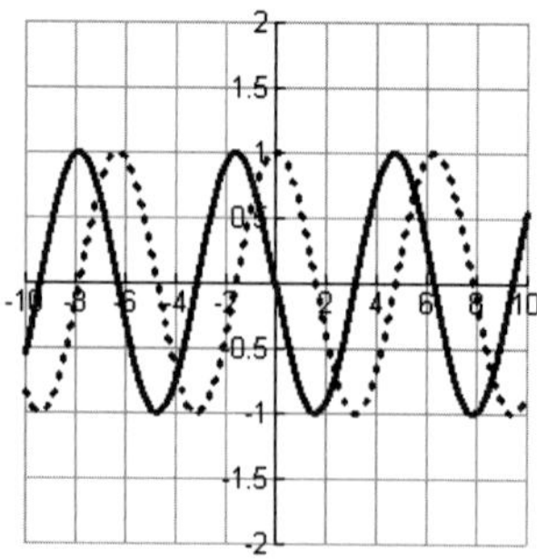

It appears that $f'(x) = \lim_{h \to 0} \frac{f(x+h) - f(x)}{h} = -\sin x$.

59. Consider the graph of $f(x) = x\sin\left(\dfrac{\sqrt{5}}{x}\right)$ below for large x.

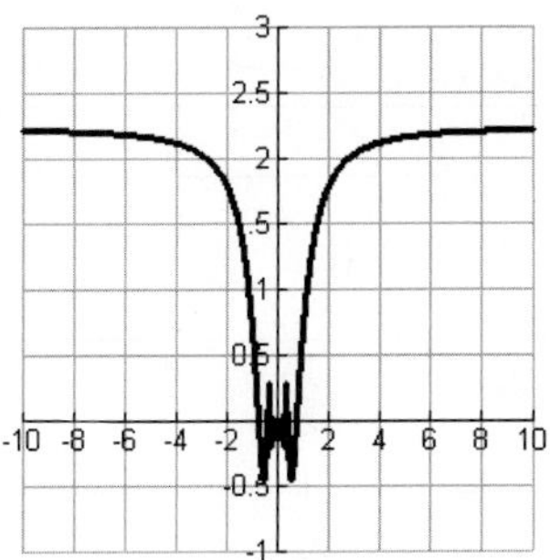

We approximate that the desired limit is 2.2361.

60. Consider the graph of $f(x) = x\left[\cos\left(\frac{1}{x} - 1\right)\right]$ below for large x.

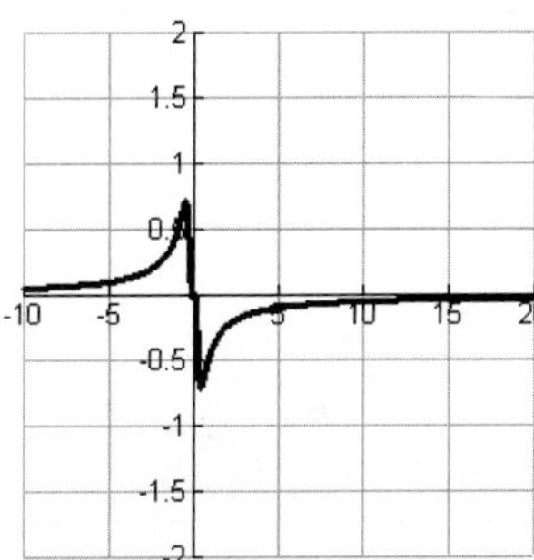

We approximate that the desired limit is 0.

61. Approximation of the area using n rectangles is given by

$$\sum_{k=1}^{n} f\left(\frac{k}{n}\right)\cdot\left(\frac{1}{n}\right) = \sum_{k=1}^{n}\sqrt{1+\frac{2k}{n}}\cdot\left(\frac{1}{n}\right).$$

Using this formula, we see that

If $n = 4$, then the area is approximately 1.4880 square units.
If $n = 100$, then the area is approximately 1.4024 square units.
If $n = 500$, then the area is approximately 1.3994 square units

62. Approximation of the area using n rectangles is given by

$$\sum_{k=1}^{n} f\left(\frac{\frac{\pi}{3}k}{n}\right)\cdot\left(\frac{\frac{\pi}{3}}{n}\right) = \sum_{k=1}^{n}\tan\left(\frac{\pi k}{3n}\right)\cdot\left(\frac{\pi}{3n}\right).$$

Using this formula, we see that

If $n = 4$, then the area is approximately 0.9365 square units.
If $n = 100$, then the area is approximately 0.7022 square units.
If $n = 500$, then the area is approximately 0.6950 square units

Chapter 11 Practice Test Solutions---

1. ∞

2. 0

3. 6

4. 0

5. -1/5

6. $\lim_{x\to -5}\frac{(x-5)(x+5)}{x+5}=\lim_{x\to -5}(x-5)=-10$

7. $\lim_{x\to\pi}\frac{1-\sin x}{\cos x}=\frac{1-\sin\pi}{\cos\pi}=-1$

8. $\lim_{x\to\pi}\left(\frac{1}{\sin x}\right)\left(\frac{\sin x}{\cos x}\right)=\lim_{x\to\pi}\left(\frac{1}{\cos x}\right)=-1$

9. 0

10. 3 (since the degrees of the numerator and denominator are the same)

11. $-\infty$

12. 0 (since the degree of the numerator is strictly less than the degree of the denominator).

13.

$$\lim_{h\to 0}\frac{f(x+h)-f(x)}{h}=\lim_{h\to 0}\frac{\left(-2(x+h)^2+3(x+h)-7\right)-\left(-2x^2+3x-7\right)}{h}$$
$$=\lim_{h\to 0}\frac{-2x^2-4hx-2h^2+3x+3h-7+2x^2-3x+7}{h}=\lim_{h\to 0}\frac{h(-4x-2h+3)}{h}=-4x+3$$

14. 2

15. Slope:

$$\lim_{x\to 1}\frac{\left(-3x^2+2\right)-(-1)}{x-1}=\lim_{x\to 1}\frac{-3x^2+3}{x-1}=\lim_{x\to -1}\frac{-3(x-1)(x+1)}{x-1}=\lim_{x\to 1}-3(x+1)=-6$$

Point: (1,-1)

So, the equation of the tangent line is: $y+1=-6(x-1)\ \Rightarrow\ y=-6x+5$

16. $f'(1)=\lim_{x\to 1}\frac{\left(-x^4+2\right)-1}{x-1}=\lim_{x\to 1}\frac{-(x+1)(x-1)\left(x^2+1\right)}{x-1}=-4$

17. $f'(x)=\lim_{h\to 0}\frac{f(x+h)-f(x)}{h}=\lim_{h\to 0}\frac{[17(x+h)+5]-[17x+5]}{h}=\lim_{h\to 0}\frac{17h}{h}=17$

18.

$$\lim_{h\to 0}\frac{f(x+h)-f(x)}{h}=\lim_{h\to 0}\frac{\left(-2(x+h)^2-3(x+h)+7\right)-\left(-2x^2-3x+7\right)}{h}$$
$$=\lim_{h\to 0}\frac{-2x^2-4hx-2h^2-3x-3h+7+2x^2+3x-7}{h}=\lim_{h\to 0}\frac{h(-4x-2h-3)}{h}=-4x-3$$

19. $\lim_{n\to\infty}\frac{(2n-1)!}{(2n+1)!}=\lim_{n\to\infty}\frac{(2n-1)!}{(2n+1)(2n)(2n-1)!}=\lim_{n\to\infty}\frac{1}{(2n+1)2n}=0$ So, the sequence converges.

20. $\sum_{k=1}^{20}(2k+5)=2\sum_{k=1}^{20}k+\sum_{k=1}^{20}5=2\left(\frac{20(21)}{2}\right)+5(20)=520$

21. Consider the region below.

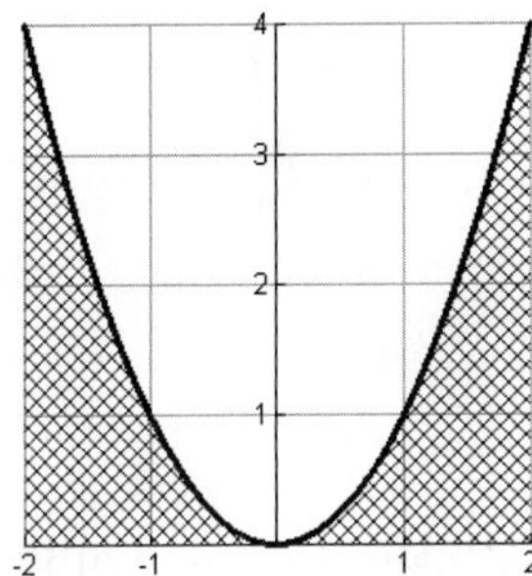

$$\lim_{n\to\infty}\sum_{k=1}^{n}f\left(-2+\frac{4k}{n}\right)\cdot\left(\frac{4}{n}\right)=\lim_{n\to\infty}\sum_{k=1}^{n}\left(-2+\frac{4k}{n}\right)^2\cdot\left(\frac{4}{n}\right)=\lim_{n\to\infty}\left[\sum_{k=1}^{n}\frac{16}{n}-\frac{64}{n^2}\sum_{k=1}^{n}k+\frac{64}{n^3}\sum_{k=1}^{n}k^2\right]$$

$$=\lim_{n\to\infty}\left[\frac{16}{n}(n)-\frac{64}{n^2}\left(\frac{n(n+1)}{2}\right)+\frac{64}{n^3}\left(\frac{n(n+1)(2n+1)}{6}\right)\right]=\frac{16}{3}$$

So, the area is 16/3 square units.

22. Consider the region below.

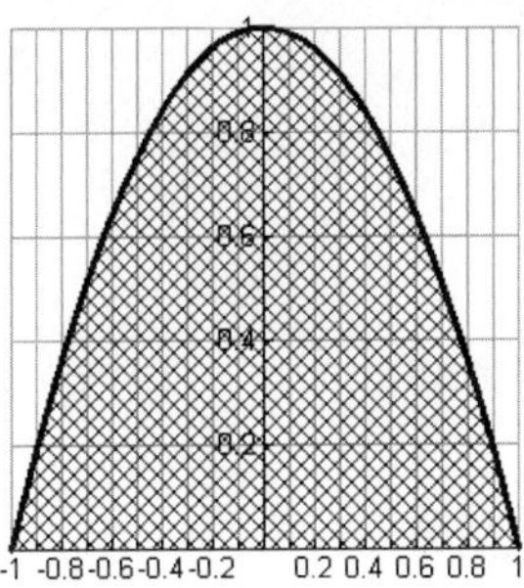

$$\lim_{n\to\infty}\sum_{k=1}^{n}f\left(-1+\frac{2k}{n}\right)\cdot\left(\frac{2}{n}\right)=\lim_{n\to\infty}\sum_{k=1}^{n}\left[-\left(-1+\frac{2k}{n}\right)^2+1\right]\cdot\left(\frac{2}{n}\right)=\lim_{n\to\infty}\left[\frac{8}{n^2}\sum_{k=1}^{n}k-\frac{8}{n^3}\sum_{k=1}^{n}k^2\right]$$

$$=\lim_{n\to\infty}\left[\frac{8}{n^2}\left(\frac{n(n+1)}{2}\right)-\frac{8}{n^3}\left(\frac{n(n+1)(2n+1)}{6}\right)\right]=\frac{4}{3}$$

So, the area is 4/3 square units.

Chapter 11 Cumulative Test --

1.

$$\frac{f(x+h)-f(x)}{h}=\frac{\left(-7(x+h)^2+3(x+h)\right)-\left(-7x^2+3x\right)}{h}=\frac{-7x^2-14hx-7h^2+3x+3h+7x^2-3x}{h}$$

$$=\frac{-h\left(14x+7h-3\right)}{h}=-14x+3-7h$$

2. Solve for x:

$$y=\frac{2}{5+x}$$
$$y(5+x)=2$$
$$5y+xy=2$$
$$xy=2-5y$$
$$x=\frac{2-5y}{y}=\frac{2}{y}-5$$

So, $f^{-1}(x)=\frac{2}{x}-5$.

$$dom(f)=rng(f^{-1})=\left(-\infty,-5\right)\cup\left(-5,\infty\right)$$
$$dom(f^{-1})=rng(f)=\left(-\infty,0\right)\cup\left(0,\infty\right)$$

3. $f(g(x))=\sqrt{\left(x^2+1\right)-1}=\sqrt{x^2}=x$, for any $x\geq 0$. The domain and range of $f\circ g$ are both $\left[0,\infty\right)$.

4. Completing the square yields

$$f(x)=x^2-4x+7$$
$$=\left(x^2-4x+4\right)+3$$
$$=(x-2)^2+3$$

So, the vertex is (2,3).

The graph is as follows:

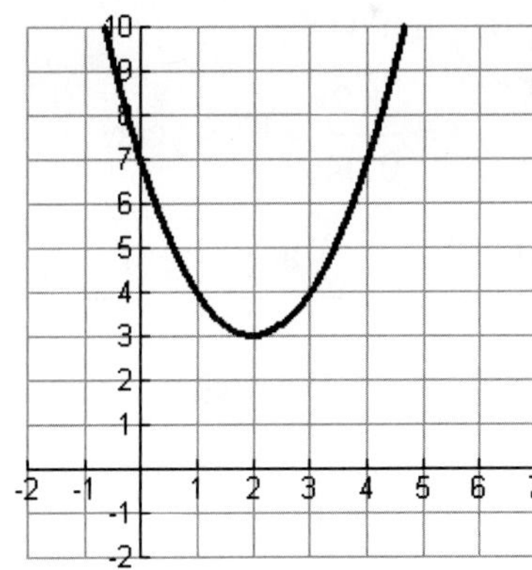

5. The possible rational zeros are $\pm 1, \pm 2, \pm 3, \pm 6, \pm 9, \pm 18$. Using synthetic division yields

$$\begin{array}{r|rrrrr} -1 & 1 & -1 & 7 & -9 & -18 \\ & & -1 & 2 & -9 & 18 \\ \hline 2 & 1 & -2 & 9 & -18 & 0 \\ & & 2 & 0 & 18 & \\ \hline & 1 & 0 & 9 & & \end{array}$$

So, $P(x) = (x-2)(x+1)(x-3i)(x+3i)$.

6. Observe that $f(x) = \dfrac{(3x-2)(x+1)}{(x-2)(x+2)}$. So, the VA are $x = 2$, $x = -2$. The HA is $y = 3$.

7.

$$\log_3(2x+7) - \log_3(x-1) = 1$$
$$\log_3\left(\frac{2x+7}{x-1}\right) = 1$$
$$\frac{2x+7}{x-1} = 3$$
$$2x+7 = 3(x-1)$$
$$x = 10$$

8.

$$e^{2x-1} = 15.7$$
$$2x-1 = \ln 15.7$$
$$x = \frac{1+\ln 15.7}{2} \approx 1.877$$

9. Use $P = P_0 e^{rt}$. Find t such that $P = 2P_0$.

$$2 = e^{0.04t}$$
$$\ln 2 = 0.04t$$
$$t = \frac{\ln 2}{0.04} \approx 17.3 \text{ years}$$

10. $\dfrac{\sqrt{3}}{3}$

11. $dom f = (-\infty, \infty)$ and $rng(f) = [-2, 2]$ since the amplitude is 2 with no vertical shift.

12.

$$-\sin^2 x + \cos x - 1 = 0$$
$$-(1-\cos^2 x) + \cos x - 1 = 0$$
$$\cos^2 x + \cos x - 2 = 0$$
$$(\cos x - 1)(\cos x + 2) = 0$$
$$\underbrace{\cos x = 1}_{x = 2n\pi,\ n \text{ an integer}} \quad \text{or} \quad \underbrace{\cos x = -2}_{\text{No solution.}}$$

So, $x = 2n\pi$, n an integer.

13. Observe that $\tan^{-1}\left(-\dfrac{\sqrt{3}}{3}\right) = -\dfrac{\pi}{6}$. So, $\cos\left(-\dfrac{\pi}{6}\right) = \dfrac{\sqrt{3}}{2}$.

14. Using the Law of Cosines yields

$$a^2 = 3.2^2 + 3.7^2 - 2(3.2)(3.7)\cos 103^\circ \approx 29.25684$$
$$a \approx 5.4 \text{ ft}$$

Now, using the Law of Sines yields

$$\frac{\sin 103^\circ}{5.4} = \frac{\sin \beta}{3.7} \Rightarrow \beta = \sin^{-1}\left(\frac{3.7\sin 103^\circ}{5.4}\right) \approx 42^\circ.$$

Finally, $\gamma = 180^\circ - \left(103^\circ + 42^\circ\right) = 35^\circ$.

15. Solve the system:

$$\begin{cases} 4x+3y-z=-4 & \textbf{(1)} \\ -x+y-z=0 & \textbf{(2)} \\ 3x+2y+5z=25 & \textbf{(3)} \end{cases}$$

First, we eliminate x in two pairs of equations to obtain a 2x2 system in y and z:

Multiply **(2)** by 3 and add to **(3)**: $5y+2z=25$ **(4)**

Multiply **(2)** by 4 and add to **(1)**: $7y-5z=-4$ **(5)**

Now, solve the system **(4) & (5)**:

Multiply **(4)** by 5 and **(5)** by 2 and add the resulting equations: $39y=117 \Rightarrow y=3$

Substitute this into either **(4)** or **(5)** to obtain z= 5.

Substitute into **(2)**: x = -2.

So, the solution is x = -2, y = 3, and z = 5.

16. $\begin{vmatrix} 3 & 0 & 1 \\ 2 & -5 & -1 \\ 1 & 2 & 7 \end{vmatrix} = 3\begin{vmatrix} -5 & -1 \\ 2 & 7 \end{vmatrix} - 0\begin{vmatrix} 2 & -1 \\ 1 & 7 \end{vmatrix} + 1\begin{vmatrix} 2 & -5 \\ 1 & 2 \end{vmatrix} = 3(-35+2)+1(4+5) = -90$

17. $AB = \begin{bmatrix} 13 & 16 \\ 6 & 8 \end{bmatrix}$, $BA = \begin{bmatrix} 10 & 16 & 7 \\ 0 & 4 & 8 \\ 6 & 11 & 7 \end{bmatrix}$

18. Completing the square yields

$$9\left(x^2-2x\right)+4\left(y^2+4y\right)=11$$
$$9\left(x^2-2x+1\right)+4\left(y^2+4y+4\right)=11+9+16$$
$$9(x-1)^2+4(y+2)^2=36$$
$$\frac{(x-1)^2}{4}+\frac{(y+2)^2}{9}=1$$

So, the center is (1, -2).

The graph is given by

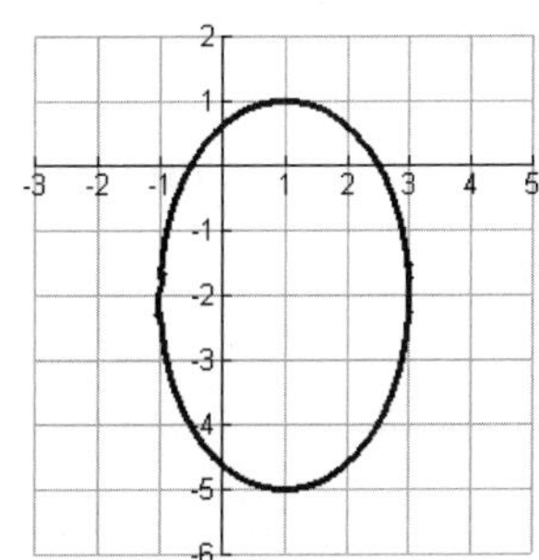

19. Solve the system: $\begin{cases}(x-1)^2+(y+1)^2=9 & \textbf{(1)}\\ x+y=3 & \textbf{(2)}\end{cases}$

Solve **(2)** for *y*: $y=3-x$ **(3)**
Substitute **(3)** into **(1)**:

$$(x-1)^2+(3-x+1)^2=9$$
$$2x^2-10x+8=0$$
$$2(x-4)(x-1)=0$$
$$x=1,4$$

Now, substitute each of these values into **(3)** to find the corresponding *y* values:
$x=1$: $y=2$ and $x=4$: $y=-1$. So, the solutions of the system are (1,2) and (4,-1).

20. $\lim_{x\to 3}\dfrac{x^2-2x-3}{x-3}=\lim_{x\to 3}\dfrac{(x-3)(x+1)}{x-3}=4$

21. ½

22. Consider the region below:

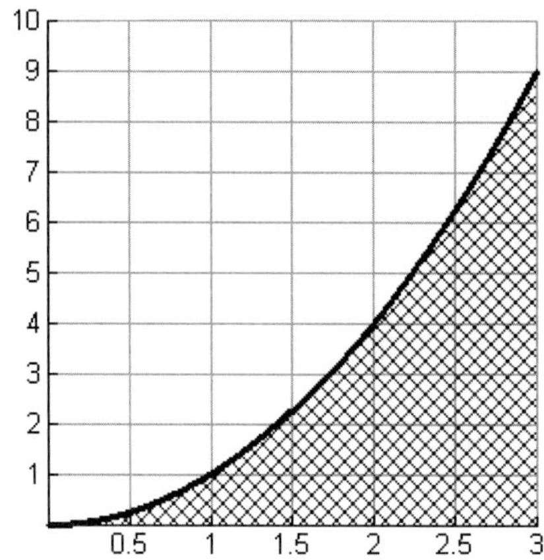

$$\lim_{n\to\infty}\sum_{k=1}^{n} f\left(\frac{3k}{n}\right)\cdot\left(\frac{3}{n}\right)=\lim_{n\to\infty}\sum_{k=1}^{n}\left(\frac{3k}{n}\right)^2\cdot\left(\frac{3}{n}\right)=\lim_{n\to\infty}\left[\frac{27}{n^3}\sum_{k=1}^{n}k^2\right]=\lim_{n\to\infty}\left[\frac{27}{n^3}\left(\frac{n(n+1)(2n+1)}{6}\right)\right]=9$$

So, the area is 9 square units.

APPENDIX

Section A.1 Solutions ---

1. rational (integer/integer)	**2.** rational (integer/integer)
3. irrational (doesn't repeat)	**4.** $\pi = 3.14159...$ irrational (doesn't repeat)
5. rational (repeats)	**6.** rational (repeats)
7. $\sqrt{5} = 2.2360...$ irrational (doesn't repeat)	**8.** $\sqrt{17} = 4.1231...$ irrational (doesn't repeat)

9. a. Rounding: 1 is less than 5, so 7 stays. $\boxed{7.347}$ **b.** Truncating: $7.347\|1$ $\boxed{7.347}$

10. a. Rounding: 9 is greater than 5, so 4 rounds up to 5. $\boxed{9.255}$ **b.** Truncating: $9.254\|9$ $\boxed{9.254}$

11. a. Rounding: 9 is greater than 5, so 4 rounds up to 5. $\boxed{2.995}$ **b.** Truncating: $2.994\|9$ $\boxed{2.994}$

12. a. Rounding: 1 is less than 5, so 5 stays. $\boxed{6.995}$ **b.** Truncating: $6.995\|1$ $\boxed{6.995}$

13. a. Rounding: 4 is less than 5, so 4 stays: $\boxed{0.234}$
b. Truncating: $0.234\|492$ $\boxed{0.234}$

14. a. Rounding: 4 is less than 5, so 4 stays: $\boxed{1.327}$
b. Truncating: $1.327\|491$ $\boxed{1.327}$

15. a. Rounding: 4 is less than 5, so 4 stays: $\boxed{5.238}$
b. Truncating: $5.238\|473$ $\boxed{5.238}$

16. a. Rounding: 4 is less than 5, so 4 stays: $\boxed{2.118}$
b. Truncating: $2.118\|465$ $\boxed{2.118}$

17.
$$5+\underbrace{2\cdot 3}_{6}-7$$
$$\underbrace{5+6}_{11}-7$$
$$11-7=\boxed{4}$$

18.
$$2+\underbrace{5\cdot 4}_{20}+\underbrace{3\cdot 6}_{18}$$
$$\underbrace{2+20}_{22}+18$$
$$22+18=\boxed{40}$$

19.
$$2\cdot\left(5+\underbrace{7\cdot 4}_{28}-20\right)$$
$$2\cdot\left(\underbrace{5+28}_{33}-20\right)$$
$$2\cdot(33-20)$$
$$2\cdot 13=\boxed{26}$$

20.
$$-3\cdot\left(\underbrace{2+7}_{9}\right)+8\cdot\left(7-\underbrace{2\cdot 1}_{2}\right)$$
$$-3\cdot(9)+8\cdot\left(\underbrace{7-2}_{5}\right)$$
$$-3\cdot(9)+8\cdot(5)$$
$$-27+40=\boxed{13}$$

21.
$$\begin{aligned}2-3[4(2\cdot 3+5)] &= 2-3[4(11)]\\ &= 2-3(44)\\ &= \boxed{-130}\end{aligned}$$

22.
$$\begin{aligned}4\cdot 6(5-9) &= 4\cdot 6(-4)\\ &= 24(-4)\\ &= \boxed{-96}\end{aligned}$$

23. $8-(-2)+7=10+7=\boxed{17}$	**24.** $-10-(-9)=-10+9=\boxed{-1}$
25. $-3-(-6)=-3+6=\boxed{3}$	**26.** $-5+2-(-3)=-5+2+3=\boxed{0}$
27. $x-(-y)-z=\boxed{x+y-z}$	**28.** $-a+b-(-c)=\boxed{-a+b+c}$
29. $-(3x+y)=\boxed{-3x-y}$	**30.** $-(4a-2b)=\boxed{-4a+2b}$
31. $\dfrac{-3}{(5)(-1)}=\dfrac{-3}{-5}=\boxed{\dfrac{3}{5}}$	**32.** $\dfrac{-12}{(-3)(-4)}=\dfrac{-12}{12}=\boxed{-1}$
33. $\begin{aligned}-4-6[(5-8)(4)]&=-4-6[(-3)(4)]\\&=-4-6(-12)\\&=-4+72\\&=\boxed{68}\end{aligned}$	**34.** $\dfrac{-14}{5-(-2)}=\dfrac{-14}{7}=\boxed{-2}$
35. $\begin{aligned}-(6x-4y)-(3x+5y)&=-6x+4y-3x-5y\\&=\boxed{-9x-y}\end{aligned}$	**36.** $\dfrac{-4x}{6-(-2)}=\dfrac{-4x}{8}=\boxed{-\dfrac{x}{2}}$
37. $\begin{aligned}-(3-4x)-(4x+7)&=-3+4x-4x-7\\&=\boxed{-10}\end{aligned}$	**38.** $\begin{aligned}2-3[(4x-5)-3x-7]&=2-3(x-12)\\&=2-3x+36\\&=\boxed{-3x+38}\end{aligned}$
39. $\dfrac{-4(5)-5}{-5}=\dfrac{-20-5}{-5}=\boxed{5}$	**40.** $\begin{aligned}&-6(2x+3y)-[3x-(2-5y)]\\&=-6(2x+3y)-[3x-2+5y]\\&=-12x-18y-3x+2-5y\\&=\boxed{-15x-23y+2}\end{aligned}$
41. $\dfrac{1}{3}+\dfrac{5}{4}=\dfrac{1(4)+5(3)}{12}=\dfrac{4+15}{12}=\boxed{\dfrac{19}{12}}$	**42.** $\dfrac{1}{2}-\dfrac{1}{5}=\dfrac{1(5)-1(2)}{10}=\dfrac{5-2}{10}=\boxed{\dfrac{3}{10}}$
43. $\dfrac{5}{6}-\dfrac{1}{3}=\dfrac{5-1\cdot(2)}{6}=\dfrac{5-2}{6}=\dfrac{3}{6}=\boxed{\dfrac{1}{2}}$	**44.** $\dfrac{7}{3}-\dfrac{1}{6}=\dfrac{7\cdot(2)-1}{6}=\dfrac{14-1}{6}=\boxed{\dfrac{13}{6}}$
45. $\frac{3}{2}+\frac{5}{12}=\frac{18}{12}+\frac{5}{12}=\boxed{\frac{23}{12}}$	**46.** $\frac{1}{3}+\frac{5}{9}=\frac{3}{9}+\frac{5}{9}=\boxed{\frac{8}{9}}$
47. $\frac{1}{9}-\frac{2}{27}=\frac{3}{27}-\frac{2}{27}=\boxed{\frac{1}{27}}$	**48.** $\frac{3}{7}-\frac{(-4)}{3}-\frac{5}{6}=\frac{3(6)+4(14)-5(7)}{42}=\boxed{\frac{13}{14}}$

49. $\frac{x}{5}+\frac{x\cdot(2)}{15}=\frac{3\cdot x+2\cdot x}{15}=\frac{5\cdot x}{15}=\boxed{\frac{x}{3}}$	**50.** $\frac{y}{3}-\frac{y}{6}=\frac{y\cdot(2)-y}{6}=\frac{2\cdot y-y}{6}=\boxed{\frac{y}{6}}$
51. $\frac{x}{3}-\frac{2x}{7}=\frac{7x-6x}{21}=\boxed{\frac{x}{21}}$	**52.** $\frac{y}{10}-\frac{y}{15}=\frac{3y-2y}{30}=\boxed{\frac{y}{30}}$
53. $\frac{4y}{15}-\frac{(-3y)}{4}=\frac{16y+(3y)(15)}{60}=\boxed{\frac{61y}{60}}$	**54.** $\frac{6x}{12}-\frac{7x}{20}=\frac{x}{2}-\frac{7x}{20}=\frac{10x-7x}{20}=\boxed{\frac{3x}{20}}$
55. $\frac{3}{40}+\frac{7}{24}=\frac{3(3)+5(7)}{120}=\boxed{\frac{11}{30}}$	**56.** $-\frac{3}{10}-\left(-\frac{7}{12}\right)=-\frac{3}{10}+\frac{7}{12}=\frac{-3(6)+7(5)}{60}=\boxed{\frac{17}{60}}$
57. $\frac{2}{7}\cdot\frac{14}{3}=\frac{2}{1}\cdot\frac{2}{3}=\boxed{\frac{4}{3}}$	**58.** $\frac{2}{3}\cdot\frac{9}{10}=\frac{1}{1}\cdot\frac{3}{5}=\boxed{\frac{3}{5}}$
59. $\frac{2}{7}\div\frac{10}{3}=\frac{2}{7}\cdot\frac{3}{10}=\frac{1}{7}\cdot\frac{3}{5}=\boxed{\frac{3}{35}}$	**60.** $\frac{4}{5}\div\frac{7}{10}=\frac{4}{5}\cdot\frac{10}{7}=\frac{4}{1}\cdot\frac{2}{7}=\boxed{\frac{8}{7}}$
61. $\frac{4b}{9}\div\frac{a}{27}=\frac{4b}{9}\cdot\frac{27}{a}=\boxed{\frac{12b}{a}}$	**62.** $\frac{3a}{7}\div\frac{b}{21}=\frac{3a}{7}\cdot\frac{21}{b}=\frac{3a}{1}\cdot\frac{3}{b}=\boxed{\frac{9a}{b}}$
63. $\frac{3x}{10}\cdot\frac{15}{6x}=\frac{3}{2\cdot 2}=\boxed{\frac{3}{4}}$	**64.** $\frac{21}{5}\cdot\frac{20}{141}=\boxed{\frac{28}{47}}$
65. $\frac{3x}{4}\div\frac{9}{16y}=\frac{3x}{4}\cdot\frac{16y}{9}=\boxed{\frac{4xy}{3}}$	**66.** $\frac{14m}{2}\cdot\frac{4}{7}=\boxed{4m}$
67. $\frac{6x}{7}\cdot\frac{28}{3y}=\boxed{\frac{8x}{y}}$	**68.** $2\frac{1}{3}\cdot 7\frac{5}{6}=\frac{7}{3}\cdot\frac{47}{6}=\boxed{\frac{329}{18}}$
69. $-\frac{(-4)}{2(3)}=\frac{4}{6}=\boxed{\frac{2}{3}}$	**70.** $2(5)+2(10)=\boxed{30}$
71. $\frac{3(4)}{10^2}=\frac{3\cdot 2\cdot 2}{10\cdot 10}=\boxed{\frac{3}{25}}$	**72.** $\frac{100-70}{15}=\frac{30}{15}=\boxed{2}$
73. \$9,176,366,000,000	**74.** 303,818,000
75. $\frac{9{,}176{,}366{,}494{,}947}{303{,}818{,}361}\approx \$30{,}203$	**76.** $\frac{9{,}176{,}366{,}494{,}947}{303{,}818{,}361}\approx \$30{,}203.46$
77. Only look to the right of the digit to round. 13.2<u>7</u>49: the 4 is less than 5, so 7 remains the same. Don't round the 9 first. $\boxed{13.27}$	**78.** Added incorrectly in very first step. Get an LCD. Should be $\frac{2(3)+1}{9}=\frac{7}{9}$.

23. Consider the graph of $f(x) = x\sin\left(\dfrac{\sqrt{3}}{x}\right)$ below for large x.

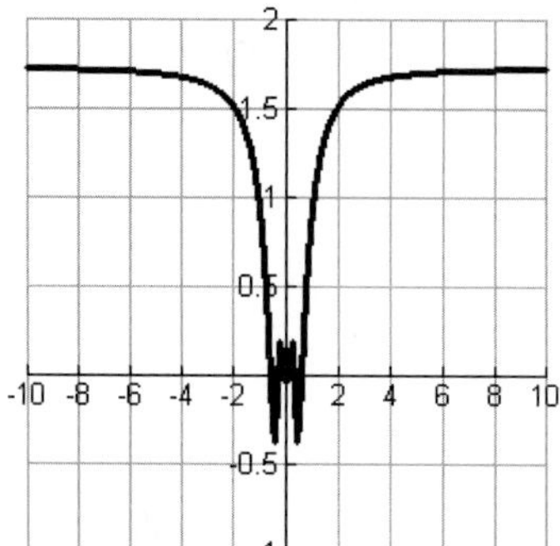

Using the graph, it appears that the desired limit is approximately 1.7321.

24. Consider the dashed region:

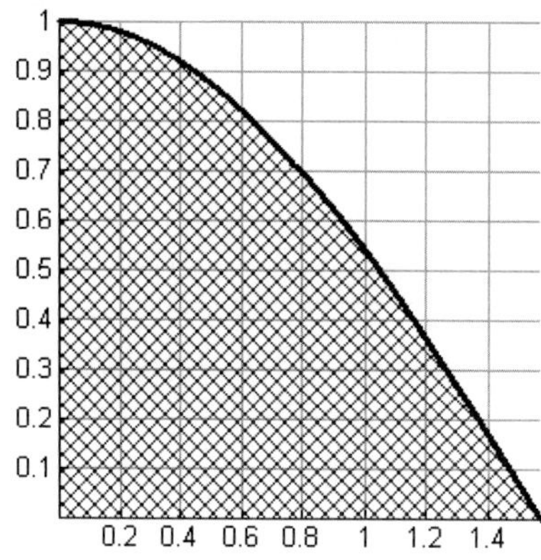

Approximation of the area using n rectangles is given by

$$\sum_{k=1}^{n} f\left(\frac{\frac{\pi}{2}k}{n}\right)\cdot\left(\frac{\frac{\pi}{2}}{n}\right) = \sum_{k=1}^{n}\cos\left(\frac{\pi k}{2n}\right)\cdot\left(\frac{\pi}{2n}\right).$$

Using this formula, we see that

If $n = 4$, then the area is approximately 0.7908 square units.

If $n = 100$, then the area is approximately 0.9921 square units.

79. $3\cdot(x+5)-2\cdot(4+y)=3\cdot x+15-8-2\cdot y$ $=3\cdot x-2\cdot y+7$. Don't forget to distribute the -2 to both terms inside the parenthesis.	**80.** Forgot to distribute "- 1" through the 2nd term in $(1-y)$.
81. False. Not all students are athletes.	**82.** True. To be an active member in a fraternity or sorority, you must also be a student at the university.
83. True. Every integer can be written as $\frac{\text{integer}}{1}$.	**84.** False
85. $x \neq 0$	**86.** $x \neq 0$

87.

$$-2[3(x-2y)+7]+[3(2-5x)+10]-7[-2(x-3)+5]=$$
$$-2[3x-6y+7]+[6-15x+10]-7[-2x+6+5]=$$
$$-6x+12y-14+6-15x+10+14x-42-35=$$
$$\boxed{-7x+12y-75}$$

88.

$$-2\{-5(y-x)-2[3(2x-5)+7(2)-4]+3\}+7=$$
$$-2\{-5(y-x)-2[6x-15+10]+3\}+7=$$
$$-2\{-5y+5x-12x+10+3\}+7=$$
$$-2\{-7x-5y+13\}+7=$$
$$\boxed{14x+10y-19}$$

89. $\sqrt{1260}=35.4964787...$ irrational	**90.** $\sqrt{\frac{144}{25}}=\frac{12}{5}$ rational
91. 67, so rational	**92.** 4.242640687, so appears irrational at this point since no pattern seems to emerge.

Section A.2 Solutions --

1. 256	**2.** 125
3. $(-1)^5 3^5=-243$	**4.** $(-1)^2 4^2=16$

5. -25	**6.** -49
7. $-4\cdot 4=-16$	**8.** $-9\cdot 5=-45$
9. 1	**10.** $-8\cdot 1=-8$
11. $\frac{1}{10}=0.1$	**12.** $\frac{1}{a}$
13. $\frac{1}{8^2}=\frac{1}{64}$	**14.** $\frac{1}{3^4}=\frac{1}{81}$
15. $-6\cdot 25=-150$	**16.** $-2\cdot 16=-32$
17. $\cancel{8}\cdot\frac{1}{\cancel{2^3}}\cdot 5=5$	**18.** $5\cdot\frac{1}{2^4}\cdot 32=5\cdot 2=10$
19. $-6\cdot\frac{1}{3^2}\cdot 81=-6\cdot\frac{1}{9}\cdot 81=-54$	**20.** $6\cdot 4^2\cdot\frac{1}{4^4}=6\cdot\frac{1}{4^2}=\frac{3}{8}$
21. $x^2\cdot x^3=x^{2+3}=\boxed{x^5}$	**22.** y^8
23. $x^2\cdot x^{-3}=x^{2-3}=x^{-1}=\boxed{\frac{1}{x}}$	**24.** $y^{-4}=\frac{1}{y^4}$
25. $\left(x^2\right)^3=x^{2\cdot 3}=\boxed{x^6}$	**26.** $\left(y^3\right)^2=y^{3\cdot 2}=\boxed{y^6}$
27. $(4a)^3=(4)^3\cdot(a)^3=\boxed{64a^3}$	**28.** $\left(4x^2\right)^3=4^3x^6=64x^6$
29. $(-2t)^3=(-2)^3(t)^3=\boxed{-8t^3}$	**30.** $(-3b)^4=(-1)^4 3^4 b^4=81b^4$
31. $\left(5xy^2\right)^2\left(3x^3y\right)=\left(25x^2y^4\right)\left(3x^3y\right)=75x^5y^5$	**32.** $\left(4x^2y\right)\left(2xy^3\right)^2=\left(4x^2y\right)\left(4x^2y^6\right)=16x^4y^7$
33. $\frac{x^5y^3}{x^7y}=\left(x^{5-7}\right)\left(y^{3-1}\right)=x^{-2}\cdot y^2=\boxed{\frac{y^2}{x^2}}$	**34.** $\frac{y^5x^2}{y^{-2}x^{-5}}=x^7y^7$
35. $\frac{(2xy)^2}{(-2xy)^3}=\frac{4x^2y^2}{-8x^3y^3}=\left(\frac{-4}{8}\right)x^{2-3}y^{2-3}=\boxed{-\frac{1}{2xy}}$	**36.** $\frac{-3x^3y}{-4x^6y^9}=\frac{3}{4x^3y^8}$
37. $\left(\frac{b}{2}\right)^{-4}=\left(\frac{2}{b}\right)^4=\frac{2^4}{b^4}=\boxed{\frac{16}{b^4}}$	**38.** $\left(\frac{c}{3}\right)^{-2}=\left(\frac{3}{c}\right)^2=\frac{9}{c^2}$
39. $\left(9a^{-2}b^3\right)^{-2}=(9)^{-2}\left(a^{-2}\right)^{-2}\left(b^3\right)^{-2}$ $=\left(\frac{1}{9^2}\right)a^4b^{-6}=\boxed{\frac{a^4}{81b^6}}$	**40.** $\left(-9x^{-3}y^2\right)^{-4}=\frac{x^{12}}{9^4y^8}=\frac{x^{12}}{6561y^8}$

41. $\dfrac{a^{-2}b^{3}}{a^{4}b^{5}}=\left(\dfrac{a^{-2}}{a^{4}}\right)\left(\dfrac{b^{3}}{b^{5}}\right)=\left(\dfrac{1}{a^{6}}\right)\left(\dfrac{1}{b^{2}}\right)=\boxed{\dfrac{1}{a^{6}b^{2}}}$	**42.** $\dfrac{x^{-3}y^{2}}{y^{-4}x^{5}}=\dfrac{y^{6}}{x^{8}}$
43. $\dfrac{\left(x^{3}y^{-1}\right)^{2}}{\left(xy^{2}\right)^{-2}}=\dfrac{x^{6}y^{-2}}{x^{-2}y^{-4}}=x^{6-(-2)}y^{-2-(-4)}=\boxed{x^{8}y^{2}}$	**44.** $\dfrac{\left(x^{3}y^{-2}\right)^{2}}{\left(x^{4}y^{3}\right)^{-3}}=\dfrac{x^{6}y^{-4}}{x^{-12}y^{-9}}=x^{18}y^{5}$
45. $\dfrac{3\left(x^{2}y\right)^{3}}{12\left(x^{-2}y\right)^{4}}=\dfrac{3x^{6}y^{3}}{12x^{-8}y^{4}}=\dfrac{x^{14}}{4y}$	**46.** $\dfrac{\left(-4x^{-2}\right)^{2}y^{3}z}{\left(2x^{3}\right)^{-2}\left(y^{-1}z\right)^{4}}=\dfrac{16x^{-4}y^{3}z}{2^{-2}x^{-6}y^{-4}z^{4}}=\dfrac{64x^{2}y^{7}}{z^{3}}$
47. $\dfrac{\left(x^{-4}y^{5}\right)^{-2}}{\left[-2\left(x^{3}\right)^{2}y^{-4}\right]^{5}}=\dfrac{x^{8}y^{-10}}{\left(-2x^{6}y^{-4}\right)^{5}}=\dfrac{x^{8}y^{-10}}{-32x^{30}y^{-20}}=\dfrac{y^{10}}{-32x^{22}}$	**48.** $-2x^{2}\left(-2x^{3}\right)^{5}=-2x^{2}\left(-32x^{15}\right)=64x^{17}$
49. $\left[\dfrac{a^{2}(-1)x^{3}y^{12}}{a^{6}x^{4}y^{4}}\right]^{3}=\dfrac{-a^{6}x^{9}y^{36}}{a^{18}x^{12}y^{12}}=-\dfrac{y^{24}}{a^{12}x^{3}}$	**50.** $\left[\dfrac{b^{-3}x^{12}y^{8}(-1)^{4}}{b^{6}y^{2}x^{15}(-1)^{3}}\right]^{5}=\left[\dfrac{-y^{6}}{b^{9}x^{3}}\right]^{5}=-\dfrac{y^{30}}{b^{45}x^{15}}$
51. $2^{8}\cdot16^{3}\cdot64=2^{8}\cdot\left(2^{4}\right)^{3}\cdot2^{6}=2^{26}$	**52.** $3^{9}\cdot81^{5}\cdot9=3^{9}\cdot\left(3^{4}\right)^{5}\cdot3^{2}=3^{31}$

53. $2.76\cdot10^{7}$	**54.** $1.44\cdot10^{11}$	**55.** 9.3×10^{7}	**56.** 1.2345×10^{9}
57. $5.67\cdot10^{-8}$	**58.** $8.28\cdot10^{-6}$	**59.** 1.23×10^{-7}	**60.** 5.0×10^{-9}
61. 47,000,000	**62.** 390,000	**63.** 23,000	**64.** 0.0078
65. 0.000041	**66.** 0.000000092	**67.** 2.0×10^{8} miles	**68.** 1.42×10^{8} miles

69. 0.00000155 meters	**70.** 0.000000693 meters
71. Should be adding exponents here: y^{5}	**72.** Forgot to apply the power to "2".
73. Computed $\left(y^{3}\right)^{2}$ incorrectly. Should be y^{6}.	**74.** Should be subtracting the powers, not dividing them.
75. False. For instance, if $n=2$, then $-2^{2}=-4$, while $(-2)^{2}=(-2)(-2)=4$.	**76.** True

77. False. $x \neq 0$	**78.** False. $x^{-1} + x^{-2} = \frac{1}{x} + \frac{1}{x^2} = \frac{x+1}{x^2}$, $x \neq 0$
79. Multiply all exponents here to get a^{mnk}.	**80.** Multiply all exponents here to get a^{-mnk}.
81. $-(-2)^2 + 2(-2)(3) = -4 - 12 = -16$	**82.** $2(4)^3 - 7(4)^2 = 128 - 112 = 16$
83. $-16(3)^2 + 100(3) = -144 + 300 = 156$	**84.** $\frac{(-2)^3 - 27}{-2-4} = \frac{-8-27}{-6} = \frac{35}{6}$
85. $\frac{(1.5\times10^8)(247)}{6.6\times10^9} = \frac{370.5\times10^8}{6.6\times10^9} \approx 5.6$ acres per person.	**86.** $\frac{(3.79\times10^6)(640)}{3.0\times10^8} = \frac{2425.6\times10^6}{3.0\times10^8} \approx 8.09$ acres per person.
87. $\frac{\left(4\times10^{-23}\right)\left(3\times10^{12}\right)}{6\times10^{-10}} = \frac{12\times10^{-11}}{6\times10^{-10}} = 0.2$	**88.** $\frac{\left(2\times10^{-17}\right)\left(5\times10^{13}\right)}{1\times10^{-6}} = \frac{10\times10^{-4}}{10^{-6}} = 1000$
89. Obtained the same answer using the calculator.	**90.** Obtained the same answer using the calculator.
91. 5.11×10^{14}	**92.** 6.25×10^{23}

Section A.3 Solutions --

1. $-7x^4 - 2x^3 + 5x^2 + 16$ $\boxed{\text{Degree 4}}$	**2.** In standard form. $\boxed{\text{Degree 3}}$
3. $-6x^3 + 4x + 3$ $\boxed{\text{Degree 3}}$	**4.** $5x^5 + 8x^4 - 7x^3 - x^2 + 10$ $\boxed{\text{Degree 5}}$
5. 15 $\boxed{\text{Degree 0}}$	**6.** In standard form. $\boxed{\text{Degree 0}}$
7. $y - 2 = y^1 - 2$ $\boxed{\text{Degree 1}}$	**8.** In standard form. $\boxed{\text{Degree 1}}$
9. $\underline{2x^2} - \underline{\underline{x}} + 7 - \underline{3x^2} + \underline{\underline{6x}} - 2$ $= \boxed{-x^2 + 5x + 5}$	**10.** $\underline{3x^2} + \underline{\underline{5x}} + 2 + \underline{2x^2} - \underline{\underline{4x}} - 9$ $= \boxed{5x^2 + x - 7}$
11. $\underline{\underline{-7x^2}}\underline{-5x} - 8 + \underline{4x} + \underline{\underline{9x^2}} - 10$ $= 2x^2 - x - 18$	**12.** $\underline{\underline{8x^3}}\ \underline{-7x^2} - 10\ \underline{\underline{-7x^3}}\ \underline{-8x^2} + 9x$ $= x^3 - 15x^2 + 9x - 10$
13. $\underline{\underline{2x^4}}\ \underline{-7x^2} + 8\ \underline{-3x^2} + \underline{\underline{2x^4}} - 9$ $= 4x^4 - 10x^2 - 1$	**14.** $\underline{\underline{4x^2}}\ \underline{-9x} - 2 - 5 + \underline{3x} + \underline{\underline{5x^2}}$ $= 9x^2 - 6x - 7$
15. $\underline{7z^2} - 2 - \underline{5z^2} + 2z - 1 = \boxed{2z^2 + 2z - 3}$	**16.** $25y^3 - \underline{7y^2} + \underline{\underline{9y}} - \underline{14y^2} + \underline{\underline{7y}} - 2$ $= \boxed{25y^3 - 21y^2 + 16y - 2}$

17. $\underline{\underline{3y^3}}\ \underline{-7y^2}+\overset{\frown}{8y}-4\ \underline{\underline{-14y^3}}+\overset{\frown}{8y}\ \underline{-9y^2}$ $=-11y^3-16y^2+16y-4$	**18.** $\underline{\underline{2x^2}}+\underline{3xy}\ \underline{\underline{-x^2}}\ \underline{-8xy}+7y^2$ $=x^2-5xy+7y^2$
19. $\underline{\underline{6x}}-2y\underline{\underline{-10x}}+14y=-4x+12y$	**20.** $3a-\left[2a^2-5a+4a^2-3\right]=$ $\underline{\underline{3a}}\underline{-2a^2}+\underline{\underline{5a}}\ \underline{-4a^2}+3=$ $-6a^2+8a+3$
21. $\underline{2x^2}-2-x-1-\underline{x^2}+5=\boxed{x^2-x+2}$	**22.** $3x^3+1-3x^2+1-5x+3$ $=\boxed{3x^3-3x^2-5x+5}$
23. $4t-\underline{\underline{t^2}}-\underline{t^3}-\underline{\underline{3t^2}}+2t-\underline{2t^3}+\underline{3t^3}-1$ $=\boxed{-4t^2+6t-1}$	**24.** $-\underline{z^3}-\underline{\underline{2z^2}}+\underline{\underline{z^2}}-7z+1-\underline{4z^3}-\underline{\underline{3z^2}}+3z-2$ $=\boxed{-5z^3-4z^2-4z-1}$
25. $(5\cdot7)x^{1+1}y^{2+1}=\boxed{35x^2y^3}$	**26.** $(6\cdot4)z^{1+3}=\boxed{24z^4}$
27. $2x^3-2x^4+2x^5=\boxed{2x^5-2x^4+2x^3}$	**28.** $-8z^2-4z^3+4z^4=\boxed{4z^4-4z^3-8z^2}$
29. Distribute and arrange terms in decreasing order (according to degree): $10x^4-2x^3-10x^2$	**30.** Distribute and arrange terms in decreasing order (according to degree): $-2z^3-z^2+5z$
31. Distribute and arrange terms in decreasing order (according to degree): $2x^5+2x^4-4x^3$	**32.** Distribute and arrange terms in decreasing order (according to degree): $3x^5-3x^4+6x^3$
33. $2a^3b^2+4a^2b^3-6ab^4$	**34.** $b^3c^4d^2+bc^4d^5-b^3c^3d^6$
35. $6x^2-8x+3x-4=\boxed{6x^2-5x-4}$	**36.** $12z^2+21z-4z-7=\boxed{12z^2+17z-7}$
37. $x^2+2x-2x-4=\boxed{x^2-4}$	**38.** $y^2+5y-5y-25=\boxed{y^2-25}$
39. $4x^2-6x+6x-9=\boxed{4x^2-9}$	**40.** $25y^2+5y-5y-1=\boxed{25y^2-1}$
41. $2x-4x^2-1+2x=\boxed{-4x^2+4x-1}$	**42.** $16b^2-25y^2$
43. $4x^4-9$	**44.** $16x^2y^2-81$
45. $7y^2-7y^3+7y-2y^3+2y^4-2y^2=$ $\boxed{2y^4-9y^3+5y^2+7y}$	**46.** $24t+4-4t^2-6t^3-t^2+t^4=$ $\boxed{t^4-6t^3-5t^2+24t+4}$

47. $x^3-2x^2+3x+x^2-2x+3=$ $\boxed{x^3-x^2+x+3}$	**48.** $x^3-3x^2+9x+3x^2-9x+27=$ $\boxed{x^3+27}$
49. $(t-2)(t-2)=t^2-2t-2t+4$ $=\boxed{t^2-4t+4}$	**50.** $(t-3)(t-3)=t^2-3t-3t+9$ $=\boxed{t^2-6t+9}$
51. $(z+2)(z+2)=z^2+2z+2z+4$ $=\boxed{z^2+4z+4}$	**52.** $(z+3)(z+3)=z^2+3z+3z+9$ $=\boxed{z^2+6z+9}$
53. $(x+y)^2-6(x+y)+9=$ $\boxed{x^2+2xy+y^2-6x-6y+9}$	**54.** $4x^4+12x^2y+9y^2$
55. $25x^2-20x+4$	**56.** $x^3+x^2+x+x^2+x+1=$ $\boxed{x^3+2x^2+2x+1}$
57. $y\left(6y^2-3y+8y-4\right)=y\left(6y^2+5y-4\right)$ $=\boxed{6y^3+5y^2-4y}$	**58.** $p^2\left(p^2-2p+p-2\right)=p^2\left(p^2-p-2\right)$ $=\boxed{p^4-p^3-2p^2}$
59. $x^4-x^2+x^2-1=\boxed{x^4-1}$	**60.** $\left[(t-5)(t+5)\right]^2=\left[t^2-25\right]^2$ $=\left(t^2-25\right)\left(t^2-25\right)$ $=t^4-25t^2-25t^2+625$ $=\boxed{t^4-50t^2+625}$
61. $(a+2b)(b-3a)(b+3a)=(a+2b)\left(b^2-9a^2\right)=$ $\boxed{ab^2+2b^3-9a^3-18a^2b}$	**62.** $x^3+2x^2y+4xy^2-2x^2y-4xy^2-8y^3$ $=\boxed{x^3-8y^3}$
63. $2x^2-3xy+5xz+2xy-3y^2+5yz-2xz+3yz-5z^2$ $=\boxed{2x^2-3y^2-xy+3xz+8yz-5z^2}$	**64.** $15b^3-5b^4+10b^2-6b^2+2b^3-4b+3b-b^2+2$ $=\boxed{-5b^4+17b^3+3b^2-b+2}$
65. Revenue $=20x$, Cost $=100+9x$. Since Profit $=P=$ Revenue – Cost, we have $P=20x-\left(100+9x\right)=20x-9x-100=\boxed{11x-100}$	
66. $P(x)=25x-75$	

67. Revenue $= -x^2 + 100x$ Cost $= -100x + 7500$	Since Profit $= P =$ Revenue – Cost, we have $P = -x^2 + 100x + 100x - 7500 = \boxed{-x^2 + 200x - 7500}$
68. $P(x) = R(x) - C(x) = \left(-\frac{1}{2}x^2 + 50x\right) - \left(8000 - 150x\right) = -\frac{1}{2}x^2 + 200x - 8000$	
69. Cutting a square with dimensions x from the four corners of the material create sides with lengths $15 - 2x$ and $8 - 2x$, and height x. The resulting volume is $V = (15 - 2x)(8 - 2x)x = 4x^3 - 46x^2 + 120x$.	
70. Cutting a square with dimensions 2 from the four corners of the material create two sides with length $x - 4$, and height 2. The resulting volume is $V(x) = 2(x-4)^2 = 2x^2 - 16x + 32$.	
71. a. The perimeters of the semi-circular pieces are each $\frac{1}{2}(2\pi x) = \pi x$ feet. The perimeter of the exposed rectangular piece is $2(2x+5)$ feet. Thus, the perimeter of the track is $P = (2\pi x + 4x + 10)$ feet. **b.** The areas of the semi-circular pieces are each $\frac{1}{2}(\pi x^2) = \frac{\pi}{2}x^2$ feet^2. The area of the rectangular piece is $(2x)(2x+5)$ feet^2. Thus, the area of the track is $A = (\pi x^2 + 4x^2 + 10x)$ feet^2.	
72. a. The volume of the hemisphere is $\frac{1}{2}\left(\frac{4}{3}\pi r^3\right) = \frac{2}{3}\pi r^3$ units^3. The volume of the cylinder is $\pi r^2(2r) = 2\pi r^3$ units^3. So, the volume of the silo is $\frac{8}{3}\pi r^3$ units^3. **b.** The surface area of the cylinder is $2\pi r(2r) = 4\pi r^2$ units^2. The surface area of the hemispherical top is $\frac{1}{2}(4\pi r^2) = 2\pi r^2$ units^2. The surface area of the bottom of the silo is πr^2 units^2. So, the surface area of the silo is $7\pi r^2$ units^2.	
73. $F = \dfrac{k(x)(3x)}{(10x)^2} = \dfrac{3kx^2}{100x^2} = \dfrac{3k}{100}$	
74. a. $S(t) = -16t^2 + 96t + 192$ **b.** $S(2) = 320$ feet. The ball will NOT hit the ground after 2 seconds since it is still going up at this time.	
75. Forgot to distribute the negative sign through the entire second polynomial. $2x^2 - 5 - 3x + x^2 - 1 = \boxed{3x^2 - 3x - 6}$	**76.** Forgot the middle (inner and outer) terms. $(2+x)(2+x) = 4 + 2x + 2x + x^2$ $= \boxed{x^2 + 4x + 4}$

<table>
<tr><td>77. True.</td><td>78. False. $(3x)(2y)=6xy$
(The product is a monomial.)</td><td>79. False.
$(x+y)^3=(x+y)(x^2+2xy+y^2)$
$=x^3+3xy^2+3x^2y+y^3$</td></tr>
<tr><td colspan="2">80. False $(x-y)(x-y)=x^2-2xy+y^2$</td><td>81. Add the degrees, so $m+n$.</td></tr>
<tr><td colspan="2">82. Take the larger of the two degrees, so m.</td><td>83.
$\left[(7x-4y^2)(7x+4y^2)\right]^2=\left[49x^2-16y^4\right]^2$
$=(49x^2-16y^4)(49x^2-16y^4)$
$=2401x^4-784x^2y^4-784x^2y^4+256y^8$
$=\boxed{2401x^4-1568x^2y^4+256y^8}$</td></tr>
<tr><td colspan="2">84.
$(3x-5y^2)^2(3x+5y^2)^2=\left[(3x-5y^2)(3x+5y^2)\right]^2$
$=\left[9x^2-25y^4\right]^2$
$=81x^4-450x^2y^4+625y^8$</td><td>85.
$(x-a)(x^2+ax+a^2)=$
$x^3+ax^2+a^2x-ax^2-a^2x-a^3=$
x^3-a^3</td></tr>
<tr><td colspan="3">86. $(x+a)(x^2-ax+a^2)=x^3-ax^2+a^2x+ax^2-a^2x+a^3=x^3+a^3$</td></tr>
<tr><td colspan="2">87.
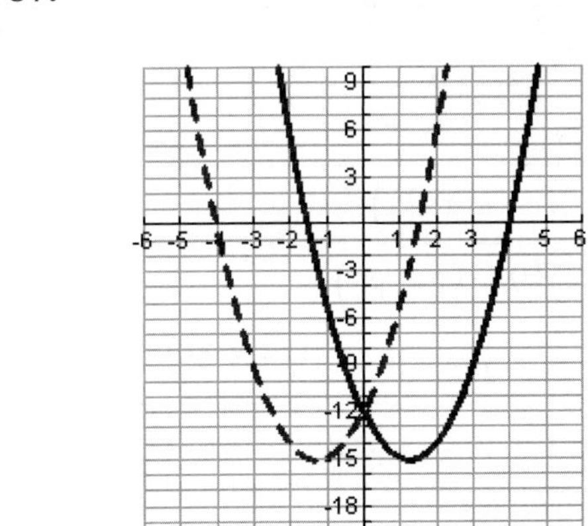

The solid curve represents the graph of both $y=(2x+3)(x-4)$ and $y=2x^2-5x-12$.</td><td>88.
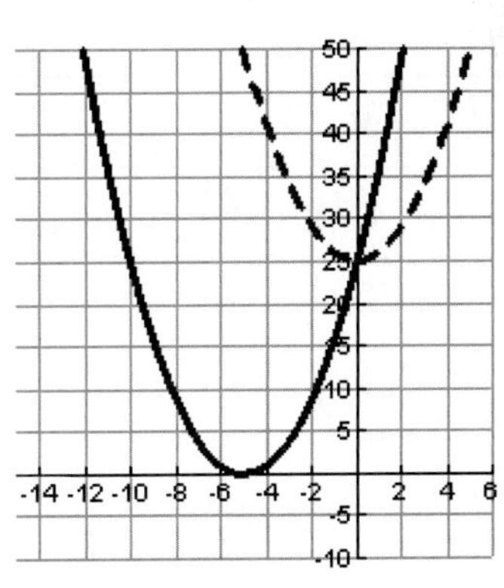

The solid curve represents the graph of both $y=(x+5)^2$ and $y=x^2+10x+25$.</td></tr>
</table>

Section A.4 Solutions ---

1. $5(x+5)$	**2.** $x(x+2)$
3. $2(2t^2-1)$	**4.** $4z(4z-5)$
5. $2x(x^2-25)=2x(x-5)(x+5)$	**6.** $4xy(x-2y+4xy)$
7. $3x(x^2-3x+4)$	**8.** $7x(2x^3-x+3)$
9. $x(x^2-3x-40)=x(x-8)(x+5)$	**10.** $-9y(y-5)$
11. $2xy(2xy^2+3)$	**12.** $3z(z^2-2z+6)$
13. $(x-3)(x+3)$	**14.** $(x-5)(x+5)$
15. $(2x-3)(2x+3)$	**16.** $(1-x^2)(1+x^2)=(1-x)(1+x)(1+x^2)$
17. $2(x^2-49)=2(x-7)(x+7)$	**18.** $(12-9y)(12+9y)=9(4-3y)(4+3y)$
19. $(15x-13y)(15x+13y)$	**20.** $(11y-7x)(11y+7x)$
21. $(x+4)^2$	**22.** $(y-5)^2$
23. $(x^2-2)^2$	**24.** $(1-3y)^2$
25. $(2x+3y)^2$	**26.** $(x-3y)^2$
27. $(x-3)^2$	**28.** $(5x-2y)^2$
29. $(x^2+1)^2$	**30.** $(x^3-3)^2$
31. $(p+q)^2$	**32.** $(p-q)^2$
33. $(t+3)(t^2-3t+9)$	**34.** $(z+4)(z^2-4z+16)$
35. $(y-4)(y^2+4y+16)$	**36.** $(x-1)(x^2+x+1)$
37. $(2-x)(4+2x+x^2)$	**38.** $(3-y)(9+3y+y^2)$
39. $(y+5)(y^2-5y+25)$	**40.** $x(64-x^3)=x(4-x)(16+4x+x^2)$
41. $(3+x)(9-3x+x^2)$	**42.** $(6x-y)(36x^2+6xy+y^2)$
43. $(x-5)(x-1)$	**44.** $(t+1)(t-6)$
45. $(y-3)(y+1)$	**46.** $(y-5)(y+2)$
47. $(2y+1)(y-3)$	**48.** $2(z^2-2z-3)=2(z-3)(z+1)$
49. $(3t+1)(t+2)$	**50.** $2(2x^2-x-6)=2(2x+3)(x-2)$
51. $(-3t+2)(2t+1)$	**52.** $(-2x+1)(3x+10)$

53. $(x^3-3x^2)+(2x-6)=x^2(x-3)+2(x-3)$ $=\boxed{(x^2+2)(x-3)}$	**54.** $(x^5+5x^3)-(3x^2+15)=x^3(x^2+5)-3(x^2+5)$ $=\boxed{(x^3-3)(x^2+5)}$
55. $a^3(a+2)-8(a+2)=(a^3-8)(a+2)$ $=(a+2)(a-2)(a^2+2a+4)$	**56.** $x^4-3x^3-x+3=x^3(x-3)-(x-3)$ $=(x^3-1)(x-3)$ $=(x-1)(x^2+x+1)(x-3)$
57. $3xy+6sy-5rx-10rs=$ $3y(x+2s)-5r(x+2s)=$ $(3y-5r)(x+2s)$	**58.** $2x(3x-5)+(3x-5)=(2x+1)(3x-5)$
59. $4x(5x+2y)-y(5x+2y)=$ $(4x-y)(5x+2y)$	**60.** $x^3(9x^2-a^2)-(9x^2-a^2)=$ $(x^3-1)(9x^2-a^2)=$ $(x-1)(x^2+x+1)(3x-a)(3x+a)$
61. $(x-2y)(x+2y)$	**62.** $(a+3)(a+2)$
63. $(3a+7)(a-2)$	**64.** $ax+bx+b+a=x(a+b)+(a+b)$ $=(x+1)(a+b)$
65. prime	**66.** prime
67. prime	**68.** $\left(\frac{1}{2}\right)^4-b^4=\left[\left(\frac{1}{2}\right)^2-b^2\right]\left[\left(\frac{1}{2}\right)^2+b^2\right]$ $=\left[\frac{1}{2}-b\right]\left[\frac{1}{2}+b\right]\left[\frac{1}{4}+b^2\right]$
69. $2(3x^2+5x+2)=2(3x+2)(x+1)$	**70.** prime
71. $(3x-y)(2x+5y)$	**72.** $15x(1+y)$
73. $9(4s^2-t^2)=9(2s-t)(2s+t)$	**74.** $3x(x^2-36)=3x(x-6)(x+6)$
75. $(ab)^2-(5c)^2=(ab-5c)(ab+5c)$	**76.** $2(x^3+27)=2(x+3)(x^2-3x+9)$
77. $(x-2)(4x+5)$	**78.** $-(x^2-10x+25)=-(x-5)^2$
79. $x(3x^2-5x-2)=\boxed{x(3x+1)(x-2)}$	**80.** $y(2y^2+3y-2)=\boxed{y(2y-1)(y+2)}$

81. $x\left(x^2-9\right)=\boxed{x(x-3)(x+3)}$	**82.** $w\left(w^2-25\right)=\boxed{w(w-5)(w+5)}$
83. $x(y-1)-(y-1)=(x-1)(y-1)$	**84.** $(a+b)+b(a+b)=(1+b)(a+b)$
85. $\left(x^2+3\right)\left(x^2+2\right)$	**86.** $\left(x^3-8\right)\left(x^3+1\right)=$ $(x-2)(x^2+2x+4)(x+1)(x^2-x+1)$
87. $(x-6)(x+4)$	**88.** $(5x+3)^2$
89. $x\left(x^3+125\right)=x(x+5)\left(x^2-5x+25\right)$	**90.** $\left(x^2-1\right)\left(x^2+1\right)=(x-1)(x+1)\left(x^2+1\right)$
91. $\left(x^2-9\right)\left(x^2+9\right)=(x-3)(x+3)\left(x^2+9\right)$	**92.** $(2x-5)(5x-3)$
93. $p=2(2x+4)+2x=6x+8=2(3x+4)$	**94.** $V=x^3+7x^2+12x=x\left(x^2+7x+12\right)$ $=x(x+4)(x+3)$
95. $p=x(2x-15)+2(2x-15)$ $=(x+2)(2x-15)$	**96.** $3x(x+3)-4(x+3)=(3x-4)(x+3)$
97. $-2\left(8t^2+39t-5\right)=-2(8t-1)(t+5)$	**98.** $x(10-x)$
99. Last step. $(x-1)\left(x^2-9\right)=(x-1)(x-3)(x+3)$	**100.** A 2 should have been factored out of both binomials, resulting in a 4. $(2x-4)(2x+10)=2(x-2)2(x+5)=\boxed{4(x-2)(x+5)}$
101. False	**102.** False
103. True	**104.** False. $(x+y)^2=x^2+2xy+y^2$
105. $a^{2n}-b^{2n}=\boxed{\left(a^n-b^n\right)\left(a^n+b^n\right)}$	**106.** $(x+7)(x-2)$ $\boxed{c=5}$; $(x+14)(x-1)$ $\boxed{c=13}$; $(x+2)(x-7)$ $\boxed{c=-5}$; $(x+1)(x-14)$ $\boxed{c=-13}$

107.	**108.**
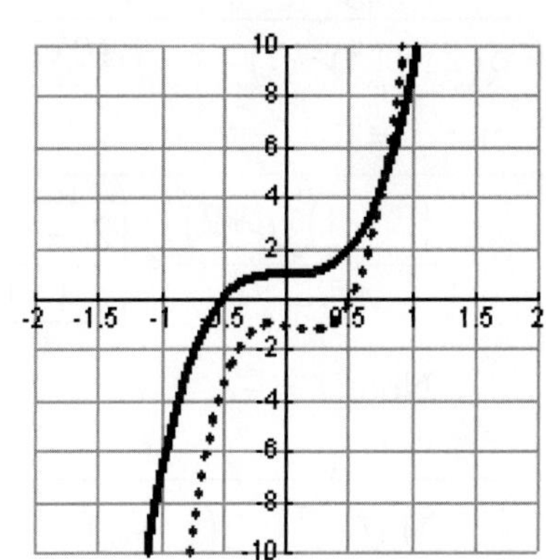 The solid curve represents the graph of both $y = 8x^3 + 1$ and $y = (2x+1)(4x^2 - 2x + 1)$.	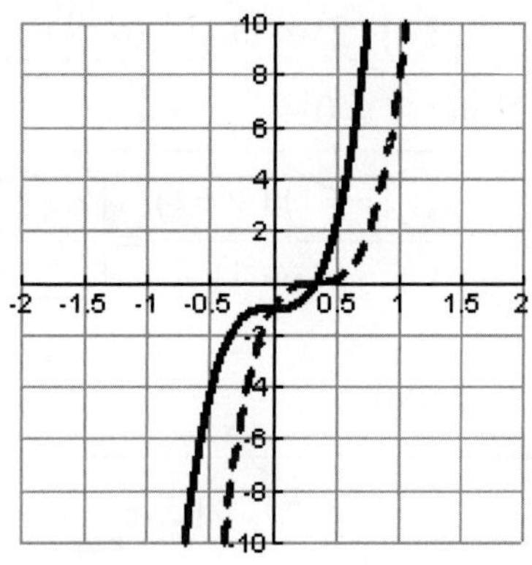 The solid curve represents the graph of both $y = 27x^3 - 1$ and $y = (3x-1)(9x^2 + 3x + 1)$.

Section A.5 Solutions --

1. $\boxed{x \neq 0}$	**2.** $\boxed{x \neq 0}$
3. $x - 1 \neq 0 \Rightarrow \boxed{x \neq 1}$	**4.** $y - 1 \neq 0 \Rightarrow \boxed{y \neq 1}$
5. $x + 1 \neq 0 \Rightarrow \boxed{x \neq -1}$	**6.** $3 - x \neq 0 \Rightarrow \boxed{x \neq 3}$
7. $p^2 - 1 \neq 0 \Rightarrow \boxed{p \neq \pm 1}$	**8.** $t^2 - 9 \neq 0 \Rightarrow \boxed{t \neq \pm 3}$
9. $p^2 + 1$ is never equal to zero for real values of p. So, no restrictions.	**10.** $t^2 + 4$ is never equal to zero for real values of t. So, no real number needs to be eliminated from the domain.
11. $\dfrac{(x-9)\cancel{(x+3)}}{2(x+9)\cancel{(x+3)}} = \dfrac{(x-9)}{2(x+9)}$ Note: $x \neq -9, -3$	**12.** $\dfrac{4y(y-8)\cancel{(y+7)}}{8y(y+8)\cancel{(y+7)}} = \dfrac{(y-8)}{2(y+8)}$ Note: $y \neq -8, -7, 0$
13. $\dfrac{(x-3)\cancel{(x+1)}}{2\cancel{(x+1)}} = \boxed{\dfrac{x-3}{2}}$ Note: $x \neq -1$	**14.** $\dfrac{(2x+1)\cancel{(x-3)}}{3\cancel{(x-3)}} = \boxed{\dfrac{2x+1}{3}}$ Note: $x \neq 3$

15. $\frac{2(3y+1)\cancel{(2y-1)}}{3(3y)\cancel{(2y-1)}}=\frac{2(3y+1)}{9y}$ Note: $y\neq 0,\frac{1}{2}$	**16.** $\frac{7(2y+1)\cancel{(3y-1)}}{5(2y)\cancel{(3y-1)}}=\frac{7(2y+1)}{10y}$ Note: $y\neq 0,\frac{1}{3}$
17. $\frac{\cancel{(5y-1)}(y+1)}{5\cancel{(5y-1)}}=\boxed{\frac{y+1}{5}}$ Note: $y\neq\frac{1}{5}$	**18.** $\frac{(2t-1)\cancel{(t+2)}}{4\cancel{(t+2)}}=\boxed{\frac{2t-1}{4}}$ Note: $t\neq -2$
19. $\frac{(3x+7)\cancel{(x-4)}}{4\cancel{(x-4)}}=\frac{(3x+7)}{4}$ Note: $x\neq 4$	**20.** $\frac{(2t+5)\cancel{(t-7)}}{3\cancel{(t-7)}}=\frac{(2t+5)}{3}$ Note: $t\neq 7$
21. $\frac{\cancel{(x-2)}(x+2)}{\cancel{(x-2)}}=\boxed{x+2}$ Note: $x\neq 2$	**22.** $\frac{t(t^2-1)}{t-1}=\frac{t\cancel{(t-1)}(t+1)}{\cancel{t-1}}=\boxed{t(t+1)}$ Note: $t\neq 1$
23. $\boxed{1}$ Domain: Note $x\neq -7$	**24.** $\frac{2y+9}{2y+9}=1$ Note: $y\neq -\frac{9}{2}$
25. $\boxed{\frac{x^2+9}{2x+9}}$ Note: $x\neq -9/2$	**26.** $\frac{x^2+4}{2(x+2)}$ Note: $x\neq -2$
27. $\frac{(x+3)\cancel{(x+2)}}{(x-5)\cancel{(x+2)}}=\frac{(x+3)}{(x-5)}$ Note: $x\neq -2,5$	**28.** $\frac{(x+15)\cancel{(x+4)}}{(x+4)\cancel{(x+4)}}=\frac{(x+15)}{(x+4)}$ Note: $x\neq -4$
29. $\frac{(3x+1)\cancel{(2x-1)}}{(x+5)\cancel{(2x-1)}}=\frac{(3x+1)}{(x+5)}$ Note: $x\neq -5,\frac{1}{2}$	**30.** $\frac{(3x+1)\cancel{(5x-2)}}{(x+3)\cancel{(5x-2)}}=\frac{(3x+1)}{(x+3)}$ Note: $x\neq -3,\frac{2}{5}$
31. $\frac{3x+5}{x+1}$ Note: $x\neq -1,2$	**32.** $\frac{3x+4}{x-2}$ Note: $x\neq -\frac{5}{4},2$
33. $\frac{2(5x+6)}{5x-6}$ Note: $x\neq 0,\frac{6}{5}$	**34.** $\frac{4(x-2)\cancel{(x+5)}}{\cancel{8x}}\cdot\frac{\cancel{16x}^{2}}{(x-5)\cancel{(x+5)}}=\frac{8(x-2)}{x-5}$ Note: $x\neq 0,\pm 5$

35. $$\frac{2\cancel{(x-1)}}{3x}\cdot\frac{x\cancel{(x+1)}}{\cancel{(x+1)}\cancel{(x-1)}}=\boxed{\frac{2}{3}}$$
Note: $x \neq -1, 0, 1$

36. $$\frac{\cancel{5}\cancel{(x-1)}}{{}^{2}\cancel{10}x}\cdot\frac{\cancel{x}\cancel{(x+1)}}{\cancel{(x-1)}\cancel{(x+1)}}=\frac{1}{2}$$
Note: $x \neq 0, \pm 1$

37. $$\frac{3(x^2-4)}{\cancel{x}}\cdot\frac{\cancel{x}\cancel{(x+5)}}{\cancel{(x+5)}(x-2)}=\frac{3(x+2)\cancel{(x-2)}}{\cancel{(x-2)}}$$
$$=\boxed{3(x+2)}$$
Note: $x \neq -5, 0, 2$

38. $$\frac{4x\cancel{(x-8)}}{\cancel{x}}\cdot\frac{\cancel{x}\cancel{(x+3)}}{\cancel{(x-8)}\cancel{(x+3)}}=4x$$
Note: $x \neq -3, 0, 8$

39. $$\frac{\cancel{t+2}}{3\cancel{(t-3)}}\cdot\frac{\cancel{(t-3)}(t-3)}{\cancel{(t+2)}(t+2)}=\boxed{\frac{t-3}{3(t+2)}}$$
Note: $t \neq -2, 3$

40. $$\frac{\cancel{y+3}}{3\cancel{(y+3)}}\cdot\frac{\cancel{(y-5)}(y-5)}{\cancel{(y-5)}(y+8)}=\frac{y-5}{3(y+8)}$$
Note: $y \neq -8, -3, 5$

41. $$\frac{3t(t^2+4)}{(t-3)(t+2)}$$ Cannot be simplified.
Note: $t \neq -2, 3$

42. $$\frac{\cancel{7}a(a+3)}{14(a-3)\cancel{(a+3)}}\cdot\frac{\cancel{a+3}}{\cancel{7}}=\frac{a(a+3)}{14(a-3)}$$
Note: $a \neq \pm 3$

43. $$\frac{(y-2)\cancel{(y+2)}}{y-3}\cdot\frac{3y}{\cancel{y+2}}=\frac{3y(y-2)}{y-3}$$
Note: $y \neq -2, 3$

44. $$\frac{(t+3)\cancel{(t-2)}}{(t+2)\cancel{(t-2)}}\cdot\frac{\cancel{8t}^{\,4}}{\cancel{2t^2}^{\,t}}=\frac{4(t+3)}{t(t+2)}$$
Note: $t \neq 0, \pm 2$

45. $$\frac{\cancel{3x}\cancel{(x-5)}}{2\cancel{x}(x+5)\cancel{(x-5)}}\cdot\frac{(2x+3)(x-5)}{\cancel{3}x(x+5)}=$$
$$\frac{(2x+3)(x-5)}{2x(x+5)^2}$$ Note: $x \neq 0, \pm 5$

46. $$\frac{5t-1}{4\cancel{t}}\cdot\frac{\cancel{t}\cancel{(4t+3)}}{(4t-3)\cancel{(4t+3)}}=\frac{5t-1}{4(4t-3)}$$
Note: $t \neq 0, \pm\frac{3}{4}$

47. $$\frac{\cancel{(3x+5)}\cancel{(2x-7)}}{(4x+3)\cancel{(2x-7)}}\cdot\frac{(2x-7)(2x+7)}{(3x-5)\cancel{(3x+5)}}=$$
$$\frac{(2x-7)(2x+7)}{(3x-5)(4x+3)}$$ Note: $x \neq -\frac{3}{4}, \frac{7}{2}, \pm\frac{5}{3}$

48. $$\frac{\cancel{x}\cancel{(3x-2)}}{4x^{\cancel{2}}\cancel{(3x-2)}}\cdot\frac{(x+2)\cancel{(x-9)}}{2(x+9)\cancel{(x-9)}}=$$
$$\frac{x+2}{8x(x+9)}$$ Note: $x \neq 0, \frac{2}{3}, \pm 9$

49. $$\frac{3}{x}\cdot\frac{x^2}{12}=\boxed{\frac{x}{4}}$$
Note: $x \neq 0$

50. $$\frac{\cancel{5}}{\cancel{x^2}}\cdot\frac{x^{\cancel{3}}}{\cancel{10}^{\,2}}=\frac{x}{2}$$
Note: $x \neq 0$

51. $$\frac{\cancel{6}}{\cancel{x-2}}\cdot\frac{\cancel{(x-2)}(x+2)}{\cancel{12}^{\,2}}=\frac{x+2}{2}$$
Note: $x \neq \pm 2$

52. $$\frac{5\cancel{(x+6)}}{\cancel{10}(x-6)}\cdot\frac{\cancel{8}}{{}_{5}\cancel{20}\cancel{(x+6)}}=\frac{1}{5(x-6)}$$
Note: $x \neq \pm 6$

53. $\frac{1}{x-1} \div \frac{5}{(x-1)(x+1)} = \frac{1}{x-1}\cdot\frac{(x-1)(x+1)}{5}$

$= \boxed{\frac{x+1}{5}}$ Note: $x \neq -1, 1$

54. $\frac{\cancel{5}}{\cancel{3x-4}}\cdot\frac{\cancel{(3x-4)}(3x+4)}{\cancel{10}^{2}} = \frac{3x+4}{2}$

Note: $x \neq \pm\frac{4}{3}$

55. $\frac{-\cancel{(p-2)}}{(p-1)\cancel{(p+1)}}\cdot\frac{\cancel{(p+1)}}{2\cancel{(p-2)}} = \frac{-1}{2(p-1)}$

$= \boxed{\frac{1}{2(1-p)}}$

Note: $p \neq 2, 1, -1$

56. $\frac{\cancel{4-x}}{\cancel{(x-4)}(x+4)}\cdot\frac{\cancel{x-4}}{3\cancel{(4-x)}} = \frac{1}{3(x+4)}$

Note: $x \neq \pm 4$

57. $\frac{(6-n)\cancel{(6+n)}}{(n-3)\cancel{(n+3)}}\cdot\frac{\cancel{(n+3)}}{\cancel{(n+6)}} = \boxed{\frac{6-n}{n-3}}$

Note: $n \neq -6, -3, 3$

58. $\frac{(7-y)\cancel{(7+y)}}{(y-5)\cancel{(y+5)}}\cdot\frac{2\cancel{(y+5)}}{\cancel{7+y}} =$

$\frac{2(7-y)}{y-5}$ Note: $y \neq -7, \pm 5$

59. $\frac{3t(t-3)(t+1)}{5(t-2)} \div \frac{6(t+1)}{4(t-2)} =$

$\frac{\cancel{3}t(t-3)\cancel{(t+1)}}{5\cancel{(t-2)}}\cdot\frac{{}^{2}\cancel{4}\cancel{(t-2)}}{\cancel{6}\cancel{(t+1)}} = \boxed{\frac{2t(t-3)}{5}}$

Note: $t \neq -1, 2$

60. $\frac{\cancel{x}(x+6)(x+2)}{5\cancel{x}\cancel{(x-2)}}\cdot\frac{\cancel{(x-2)(x+2)}}{4\cancel{(x+2)}} =$

$\frac{(x+6)(x+2)}{20}$ Note: $x \neq 0, \pm 2$

61. $\frac{\cancel{w}(w-1)}{\cancel{w}} \div \frac{\cancel{w}(w^2-1)}{5w^{\cancel{3}2}} = \frac{(w-1)}{1} \div \frac{(w-1)(w+1)}{5w^2}$

$= \frac{\cancel{(w-1)}5w^2}{\cancel{(w-1)}(w+1)}$

$= \boxed{\frac{5w^2}{w+1}}$

Note: $w \neq -1, 1, 0$

62. $\frac{\cancel{y(y-3)}}{\cancel{2}y}\cdot\frac{{}^{4}\cancel{8}\cancel{y}}{\cancel{y^2(y-3)}} = \frac{4}{y}$

Note: $y \neq 0, 3$

63. $\frac{(x-3)\cancel{(x+7)}}{(x-2)\cancel{(x+5)}}\cdot\frac{(x-4)\cancel{(x+5)}}{(x-9)\cancel{(x+7)}} =$

$\frac{(x-3)(x-4)}{(x-2)(x-9)}$ Note: $x \neq -7, -5, 2, 4, 9$

64. $\frac{\cancel{(2y+1)}\cancel{(y-3)}}{\cancel{(2y+1)}\cancel{(y-5)}}\cdot\frac{y\cancel{(y-5)}}{3\cancel{(y-3)}} = \frac{y}{3}$

Note: $y \neq -\frac{1}{2}, 0, 3, 5$

65.

$\frac{\cancel{(5x-2)}\cancel{(4x+1)}}{(5x+2)\cancel{(5x-2)}}\cdot\frac{x\cancel{(3x+5)}}{\cancel{(3x+5)(4x+1)}} =$

$\frac{x}{5x+2}$ Note: $x \neq -\frac{5}{3}, -\frac{1}{4}, 0, \pm\frac{2}{5}$

66. $\frac{(x+3)\cancel{(x-9)}}{(x+7)\cancel{(2x-1)}}\cdot\frac{(x+5)\cancel{(2x-1)}}{(2x+3)\cancel{(x-9)}} =$

$\frac{(x+5)(x+3)}{(x+7)(2x+3)}$ Note: $x \neq -7, -5, -\frac{3}{2}, \frac{1}{2}, 9$

67. $\frac{3(5)-2}{5x}=\frac{15-2}{5x}=\boxed{\frac{13}{5x}}$

Note: $x \neq 0$

68. $\frac{5-3(7)}{7x}=-\frac{16}{7x}$

Note: $x \neq 0$

69. $\frac{3(p+1)+5p(p-2)}{(p-2)(p+1)}=\frac{3p+3+5p^2-10p}{(p-2)(p+1)}$

$=\boxed{\frac{5p^2-7p+3}{(p-2)(p+1)}}$

Note: $p \neq -1, 2$

70. $\frac{4(x-2)-5x(9+x)}{(x+9)(x-2)}=\frac{4x-8-45x-5x^2}{(x+9)(x-2)}$

$=\frac{-(5x^2+41x+8)}{(x+9)(x-2)}$

$=\frac{-(5x+1)(x+8)}{(x+9)(x-2)}$

Note: $x \neq -9, 2$

71. $\frac{2x+1}{5x-1}+\frac{3-2x}{5x-1}=\frac{2x+1+3-2x}{5x-1}=\boxed{\frac{4}{5x-1}}$

Note: $x \neq \frac{1}{5}$

72. $\frac{7}{2x-1}+\frac{5}{2x-1}=\frac{12}{2x-1}$

Note: $x \neq \frac{1}{2}$

73.

$\frac{3y^2(y-1)+(1-2y)(y+1)}{(y+1)(y-1)}=\frac{3y^3-3y^2+y+1-2y^2-2y}{(y+1)(y-1)}$

$=\boxed{\frac{3y^3-5y^2-y+1}{y^2-1}}$

Note: $y \neq \pm 1$

74. $-\frac{3}{x-1}+\frac{4}{x-1}=\frac{1}{x-1}$

Note: $x \neq 1$

75. $\frac{3x}{(x-2)(x+2)}+\frac{(3+x)}{(x+2)}=\frac{3x+(3+x)(x-2)}{(x-2)(x+2)}$

$=\frac{3x+3x-6+x^2-2x}{(x-2)(x+2)}=\frac{x^2+4x-6}{(x-2)(x+2)}$

$=\boxed{\frac{x^2+4x-6}{x^2-4}}$

Note: $x \neq \pm 2$

76. $\frac{x-1}{(2-x)(2+x)}-\frac{(x+1)}{(x+2)}=\frac{(x-1)-(x+1)(2-x)}{(2-x)(x+2)}$

$=\frac{x-1-2x+x^2-2+x}{(2-x)(x+2)}$

$=\boxed{\frac{x^2-3}{(2-x)(x+2)}}$

Note: $x \neq \pm 2$

77. $\frac{(x-1)(x+2)+(x-6)}{(x-2)(x+2)}=\frac{x^2+x-2+x-6}{(x-2)(x+2)}$

$=\frac{x^2+2x-8}{(x-2)(x+2)}=\frac{(x+4)\cancel{(x-2)}}{\cancel{(x-2)}(x+2)}$

$=\frac{x+4}{x+2}$ Note: $x \neq \pm 2$

78. $\frac{2(y+2)+7(y-3)}{(y-3)(y+2)}=\frac{2y+4+7y-21}{(y-3)(y+2)}$

$=\frac{9y-17}{(y-3)(y+2)}$ Note: $y \neq -2, 3$

79. $\dfrac{5a}{(a-b)(a+b)}+\dfrac{7}{a-b}=\dfrac{5a+7(a+b)}{(a-b)(a+b)}$ $=\dfrac{5a+7a+7b}{(a-b)(a+b)}=\dfrac{12a+7b}{a^2-b^2}$ Note: $a \neq \pm b$	**80.** $\dfrac{(y-2)(y+2)+4y-2(y+2)}{y(y-2)(y+2)}=$ $\dfrac{y^2+2y-8}{y(y-2)(y+2)}=\dfrac{(y+4)\cancel{(y-2)}}{y\cancel{(y-2)}(y+2)}$ Note: $y \neq 0, \pm 2$
81. $\dfrac{7(x-3)+1}{x-3}=\dfrac{7x-20}{x-3}$ Note: $x \neq 3$	**82.** $\dfrac{3(y-2)-4(5y+6)+(y^2-y)}{(5y+6)(y-2)}=$ $\dfrac{y^2-18y-30}{(5y+6)(y-2)}$ Note: $y \neq -\frac{6}{5}, 2$
83. $\dfrac{\frac{1}{x}-1}{1-\frac{2}{x}}\cdot\dfrac{x}{x}=\boxed{\dfrac{1-x}{x-2}}$ Note: $x \neq 0, 2$	**84.** $\dfrac{\frac{3-5y}{\cancel{y}}}{\frac{4y-2}{\cancel{y}}}=\dfrac{3-5y}{4y-2}$ Note: $y \neq 0, \frac{1}{2}$
85. $\dfrac{\frac{3x+1}{x}}{\frac{9x^2-1}{x^2}}=\dfrac{\cancel{3x+1}}{\cancel{x}}\cdot\dfrac{x^{\cancel{2}}}{\cancel{(3x+1)}(3x-1)}=\dfrac{x}{3x-1}$ Note: $x \neq 0, \pm\frac{1}{3}$	**86.** $\dfrac{\frac{x+2}{\cancel{x^2}}}{\frac{9x-5}{\cancel{x^2}}}=\dfrac{x+2}{9x-5}$ Note: $x \neq 0, \frac{5}{9}$
87. $\dfrac{\frac{1}{x-1}+1}{1-\frac{1}{x+1}}\cdot\dfrac{(x-1)(x+1)}{(x-1)(x+1)}=\dfrac{x+1+x^2-1}{x^2-1-(x-1)}$ $=\dfrac{x^2+x}{x^2-x}=\dfrac{\cancel{x}(x+1)}{\cancel{x}(x-1)}=\boxed{\dfrac{x+1}{x-1}}$ Note: $x \neq 0, \pm 1$	**88.** $\dfrac{\frac{7}{y+7}}{\frac{y-(y+7)}{y(y+7)}}=\dfrac{\cancel{7}}{\cancel{y+7}}\cdot\dfrac{y\cancel{(y+7)}}{-\cancel{7}}=-y$ Note: $y \neq -7, 0$
89. $\dfrac{\frac{1+x-1}{x-1}}{\frac{1+x+1}{x+1}}=\dfrac{x}{x-1}\cdot\dfrac{x+1}{x+2}=\dfrac{x(x+1)}{(x-1)(x+2)}$ Note: $x \neq \pm 1, -2$	**90.** $\dfrac{\frac{3(x-1)-3(x+1)}{\cancel{(x+1)(x-1)}}}{\frac{5}{\cancel{(x+1)(x-1)}}}=-\dfrac{6}{5}$ Note: $x \neq \pm 1$
91. $A=\dfrac{pi}{\frac{(1+i)^5-1}{(1+i)^5}}=\dfrac{pi(1+i)^5}{(1+i)^5-1}$	**92.** $A=\dfrac{pi}{\frac{(1+i)^{nt}-1}{(1+i)^{nt}}}=\dfrac{pi(1+i)^{nt}}{(1+i)^{nt}-1}$ Substituting in the given information yields $A \approx \$948.10$

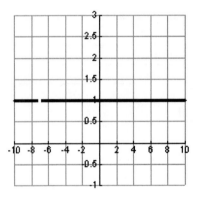

93. $\dfrac{1}{\dfrac{1}{R_1}+\dfrac{1}{R_2}}=\dfrac{1}{\dfrac{R_2+R_1}{R_1R_2}}=\dfrac{R_1R_2}{R_2+R_1}$	**94.** $f=\dfrac{1}{\dfrac{1}{p}+\dfrac{1}{q}}=\dfrac{1}{\dfrac{q+p}{pq}}=\dfrac{pq}{q+p}$
95. Initially, $\frac{x^2+2x+1}{x+1}$ has the restriction $x \neq -1$. $\frac{x^2+2x+1}{x+1}=\boxed{x+1}$	**96.** Cannot cancel the 1's. Must factor first. $\frac{\cancel{x+1}}{\cancel{(x+1)}(x+1)}=\frac{1}{x+1}$
97. False. $\dfrac{x^2-81}{x-9}=x+9\ \boxed{x\neq 9}$	**98.** True.
99. False. $\dfrac{1}{3x}-\dfrac{1}{6x}$ LCD $=6x\left(\text{not } 18x^2\right)$	**100.** False. $\dfrac{x-c}{c-x}=\boxed{-1,\ x\neq c}$
101. $\dfrac{(x+a)}{(x+b)}\div\dfrac{(x+c)}{(x+d)}=\dfrac{(x+a)(x+d)}{(x+b)(x+c)}$ Note: $x\neq -b,-c,-d$	**102.** 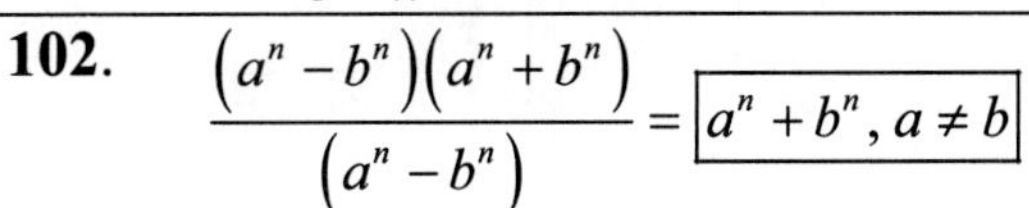 $\dfrac{\left(a^n-b^n\right)\left(a^n+b^n\right)}{\left(a^n-b^n\right)}=\boxed{a^n+b^n,\ a\neq b}$
103. The graph is as follows. There is a hole at $x=-7$. It agrees with Exercise 23. 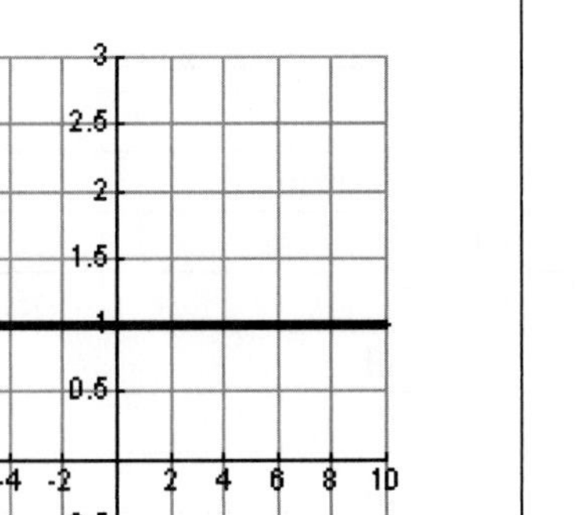 	**104.** The graph is as follows. There is a hole at $x=2$. It agrees with Exercise 21. 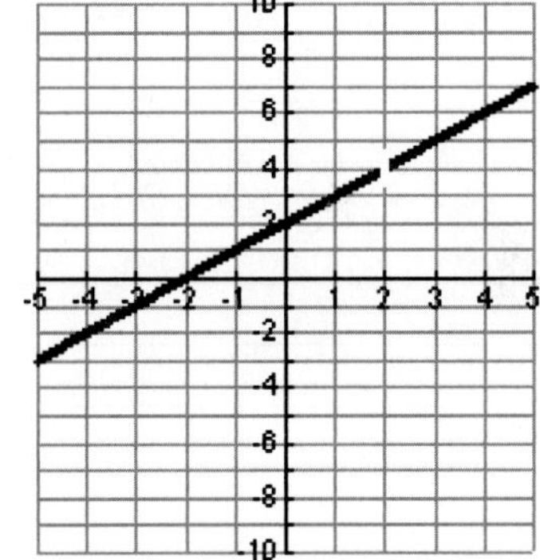

<table>
<tr>
<td>

105. a. $\dfrac{1+\frac{1}{x-2}}{1-\frac{1}{x+2}}=\dfrac{\frac{x-1}{x-2}}{\frac{x+1}{x+2}}=\dfrac{(x-1)(x+2)}{(x+1)(x-2)}$

b.

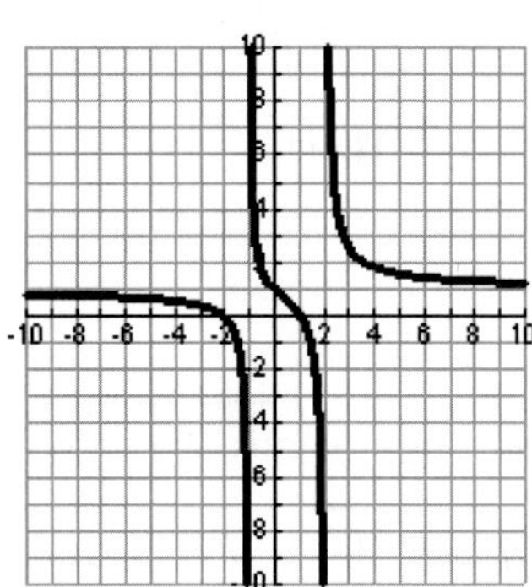

c. The graphs agree as long as $x \neq -1, \pm 2$.

</td>
<td>

106. a. $\dfrac{1-\frac{2}{x+3}}{1+\frac{1}{x-4}}=\dfrac{\frac{x+1}{x+3}}{\frac{x-3}{x-4}}=\dfrac{(x+1)(x-4)}{(x+3)(x-3)}$

b.

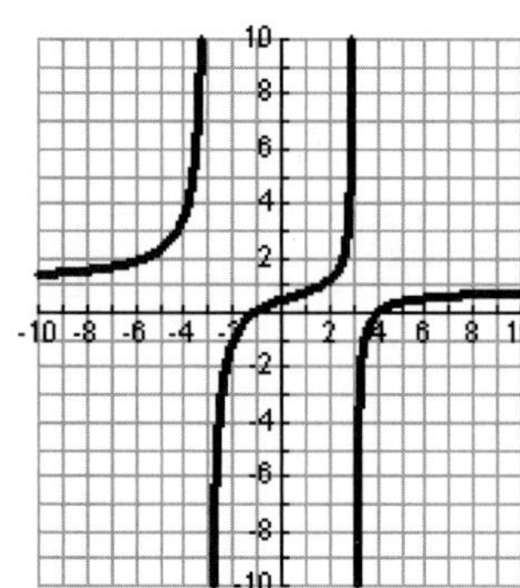

c. The graphs agree as long as $x \neq 4, \pm 3$.

</td>
</tr>
</table>

Section A.6 Solutions ---

1. 10	**2.** 11
3. -12	**4.** not real
5. -6	**6.** -5
7. 7	**8.** 3
9. 1	**10.** -1
11. 0	**12.** 0
13. not real	**14.** -1
15. -3	**16.** -4
17. $\left(8^{1/3}\right)^2=2^2=4$	**18.** $\left((-64)^{1/3}\right)^2=(-4)^2=16$
19. -2	**20.** -3
21. -1	**22.** 1
23. $\left(9^{1/2}\right)^3=3^3=27$	**24.** $27^{2/3}=\left(27^{1/3}\right)^2=3^2=9$
25. $-4\sqrt{2}$	**26.** $-4\sqrt{5}$
27. $8\sqrt{5}$	**28.** $3\sqrt{7}$
29. $2\sqrt{3}\cdot\sqrt{2}=2\sqrt{6}$	**30.** $2\sqrt{5}\cdot 3\left(2\sqrt{10}\right)=2\sqrt{5}\cdot 3\left(2\sqrt{2}\sqrt{5}\right)=60\sqrt{2}$
31. $\sqrt[3]{12}\cdot\sqrt[3]{4}=\sqrt[3]{48}=\sqrt[3]{8\cdot 6}=2\sqrt[3]{6}$	**32.** $\sqrt[4]{8}\cdot\sqrt[4]{4}=\sqrt[4]{32}=2\sqrt[4]{2}$
33. $\sqrt{21}$	**34.** $\sqrt{10}$

35. $8\sqrt{25x^2} = 40\lvert x\rvert$	**36.** $16\left(6y^2\right) = 96y^2$
37. $2\lvert x\rvert\sqrt{y}$	**38.** $4\lvert x\rvert\sqrt{xy}$
39. $-3x^2y^2\sqrt[3]{3y^2}$	**40.** $-2x^2y\sqrt[5]{y^3}$
41. $\frac{1}{\sqrt{3}}\cdot\frac{\sqrt{3}}{\sqrt{3}} = \frac{\sqrt{3}}{3}$	**42.** $\frac{\sqrt{2}}{\sqrt{5}}\cdot\frac{\sqrt{5}}{\sqrt{5}} = \frac{\sqrt{10}}{5}$
43. $\frac{2}{3\sqrt{11}}\cdot\frac{\sqrt{11}}{\sqrt{11}} = \frac{2\sqrt{11}}{33}$	**44.** $\frac{5}{3\sqrt{2}}\cdot\frac{\sqrt{2}}{\sqrt{2}} = \frac{5\sqrt{2}}{6}$
45. $\frac{3}{1-\sqrt{5}}\cdot\frac{1+\sqrt{5}}{1+\sqrt{5}} = \frac{3\left(1+\sqrt{5}\right)}{1-5} = \frac{3+3\sqrt{5}}{-4}$	**46.** $\frac{2}{1+\sqrt{3}}\cdot\frac{1-\sqrt{3}}{1-\sqrt{3}} = \frac{2\left(1-\sqrt{3}\right)}{1-3} = \frac{2-2\sqrt{3}}{-2}$ $= -1+\sqrt{3}$
47. $\frac{1+\sqrt{2}}{1-\sqrt{2}}\cdot\frac{1+\sqrt{2}}{1+\sqrt{2}} = \frac{1+2\sqrt{2}+2}{1-2} = -3-2\sqrt{2}$	**48.** $\frac{3-\sqrt{5}}{3+\sqrt{5}}\cdot\frac{3-\sqrt{5}}{3-\sqrt{5}} = \frac{9-6\sqrt{5}+5}{9-5} = \frac{7-3\sqrt{5}}{2}$
49. $\frac{3}{\sqrt{2}-\sqrt{3}}\cdot\frac{\sqrt{2}+\sqrt{3}}{\sqrt{2}+\sqrt{3}} = \frac{3\left(\sqrt{2}+\sqrt{3}\right)}{2-3} = -3\left(\sqrt{2}+\sqrt{3}\right)$	**50.** $\frac{5}{\sqrt{2}+\sqrt{5}}\cdot\frac{\sqrt{2}-\sqrt{5}}{\sqrt{2}-\sqrt{5}} = \frac{5\left(\sqrt{2}-\sqrt{5}\right)}{2-5} = \frac{5\left(\sqrt{5}-\sqrt{2}\right)}{3}$
51. $\frac{4}{3\sqrt{2}+2\sqrt{3}}\cdot\frac{3\sqrt{2}-2\sqrt{3}}{3\sqrt{2}-2\sqrt{3}} = \frac{4\left(3\sqrt{2}-2\sqrt{3}\right)}{18-12}$ $= \frac{2\left(3\sqrt{2}-2\sqrt{3}\right)}{3}$	**52.** $\frac{7}{2\sqrt{3}+3\sqrt{2}}\cdot\frac{2\sqrt{3}-3\sqrt{2}}{2\sqrt{3}-3\sqrt{2}} = \frac{7\left(2\sqrt{3}-3\sqrt{2}\right)}{12-18}$ $= \frac{7\left(3\sqrt{2}-2\sqrt{3}\right)}{6}$
53. $\frac{4+\sqrt{5}}{3+2\sqrt{5}}\cdot\frac{3-2\sqrt{5}}{3-2\sqrt{5}} = \frac{12+3\sqrt{5}-8\sqrt{5}-10}{9-20}$ $= \frac{-5\sqrt{5}+2}{-11} = \frac{5\sqrt{5}-2}{11}$	**54.** $\frac{6}{3\sqrt{2}+4}\cdot\frac{3\sqrt{2}-4}{3\sqrt{2}-4} = \frac{6\left(3\sqrt{2}-4\right)}{18-16} = 3\left(3\sqrt{2}-4\right)$
55. $\frac{\sqrt{7}+\sqrt{3}}{\sqrt{2}-\sqrt{5}}\cdot\frac{\sqrt{2}+\sqrt{5}}{\sqrt{2}+\sqrt{5}} = \frac{\left(\sqrt{7}+\sqrt{3}\right)\left(\sqrt{2}+\sqrt{5}\right)}{-3}$	**56.** $\frac{\sqrt{y}}{\sqrt{x}-\sqrt{y}}\cdot\frac{\sqrt{x}+\sqrt{y}}{\sqrt{x}+\sqrt{y}} = \frac{\sqrt{xy}+y}{x-y}$
57. x^3y^4	**58.** $y^8\cdot y^3 = y^{11}$
59. $\frac{x^{-1}y^{-3/2}}{x^{-1}y^{1/2}} = \frac{1}{y^2}$	**60.** $\frac{x^{4/3}y^{3/2}}{x^{4/3}y} = y^{1/2}$

61. $x^{7/6}y^2$	**62.** $\dfrac{y^{-9}x^{-8}}{y^6x^{56}}=\dfrac{1}{y^{15}x^{64}}$
63. $\dfrac{8x^2}{16x^{-2/3}}=\dfrac{x^{8/3}}{2}$	**64.** $\dfrac{8x^{-2}}{16x^{-8/3}}=\dfrac{x^{2/3}}{2}$
65. $x^{1/3}\left(x^2-x-2\right)=x^{1/3}(x-2)(x+1)$	**66.** $4x^{1/4}(x+2)$
67. $7x^{3/7}\left(1-2x^{3/7}+3x\right)$	**68.** $7x^{-1/3}\left(1+10x^{4/3}\right)$
69. $\sqrt{\dfrac{1280}{16}}=\sqrt{80}\approx 9$ sec.	**70.** $\sqrt{\dfrac{600}{5}}=\sqrt{120}\approx 11$ sec.
71. $d=(29.46)^{2/3}\approx 9.54$ astronomical units	**72.** $p=2\pi\left(\dfrac{19.6}{9.8}\right)^{1/2}=2\pi\sqrt{2}$ sec.
73. Forgot to square the 4	**74.** Should have multiplied by the conjugate $5+\sqrt{11}$
75. False. Only give principal root	**76.** False. If $x=-1$, then $\sqrt{(-1)^2}=1$, not -1.
77. False. If $a=3, b=4$, then $\sqrt{a^2+b^2}=\sqrt{25}=5$, while $\sqrt{a}+\sqrt{b}=\sqrt{3}+2\neq 5$.	**78.** False. $\sqrt{-4}$ is not a real number.
79. Multiply the exponents to get a^{mnk}.	**80.** Multiply the exponents to get a.
81. $\dfrac{\sqrt{7^2-4(1)(12)}}{2(1)}=\dfrac{1}{2}$	**82.** $\sqrt{7^2-4(1)(12)}=1$
83. $\dfrac{1}{\left(\sqrt{a}+\sqrt{b}\right)^2}=\left(\dfrac{1}{\sqrt{a}+\sqrt{b}}\right)^2=\left(\dfrac{1}{\sqrt{a}+\sqrt{b}}\cdot\dfrac{\sqrt{a}-\sqrt{b}}{\sqrt{a}-\sqrt{b}}\right)^2=\left(\dfrac{\sqrt{a}-\sqrt{b}}{a-b}\right)^2=\dfrac{\left(\sqrt{a}-\sqrt{b}\right)^2}{(a-b)^2}=\dfrac{a-2\sqrt{a}\sqrt{b}+b}{(a-b)^2}=\dfrac{a+b-2\sqrt{ab}}{(a-b)^2}$	**84.** $\dfrac{\sqrt{a+b}-\sqrt{a}}{\sqrt{a+b}+\sqrt{a}}\cdot\dfrac{\sqrt{a+b}-\sqrt{a}}{\sqrt{a+b}-\sqrt{a}}=\dfrac{a+b-2\sqrt{a}\sqrt{a+b}+a}{a+b-a}=\dfrac{2a+b-2\sqrt{a(a+b)}}{b}$
85. 3.317	**86.** 1.913

87.	**88.**
a. $\frac{4}{5\sqrt{2}+4\sqrt{3}} \cdot \frac{5\sqrt{2}-4\sqrt{3}}{5\sqrt{2}-4\sqrt{3}} = \frac{4(5\sqrt{2}-4\sqrt{3})}{50-48}$ $= 10\sqrt{2}-8\sqrt{3}$	**a.** $\frac{2}{4\sqrt{5}-3\sqrt{6}} \cdot \frac{4\sqrt{5}+3\sqrt{6}}{4\sqrt{5}+3\sqrt{6}} = \frac{2(4\sqrt{5}+3\sqrt{6})}{80-54}$ $= \frac{4\sqrt{5}+3\sqrt{6}}{13}$
b. 0.2857291632	**b.** 1.25328778
c. Yes	**c.** Yes

Section A.7 Solutions ---

1. $\sqrt{-16} = \underbrace{\sqrt{16}}_{4} \cdot \underbrace{\sqrt{-1}}_{i} = \boxed{4i}$	**2.** $\sqrt{-100} = \underbrace{\sqrt{100}}_{10} \cdot \underbrace{\sqrt{-1}}_{i} = \boxed{10i}$
3. $\sqrt{-20} = \underbrace{\sqrt{20}}_{2\sqrt{5}} \underbrace{\sqrt{-1}}_{i} = \boxed{2i\sqrt{5}}$	**4.** $\sqrt{-24} = 2\sqrt{6}\underbrace{\sqrt{-1}}_{i} = \boxed{2\sqrt{6} \cdot i}$
5. -4	**6.** -3
7. $8i$	**8.** $3i\sqrt{3}$
9. $3-10i$	**10.** $4-11i$
11. $-10-12i$	**12.** $7-(-5)=12$
13. $3-7i-1-2i = \boxed{2-9i}$	**14.** $1+i+9-3i = \boxed{10-2i}$
15. $10-14i$	**16.** $-5+5i$
17. $(4-5i)-(2-3i) = 4-5i-2+3i = \boxed{2-2i}$	**18.** $(-2+i)-(1-i) = -2+i-1+i = \boxed{-3+2i}$
19. $-1+2i$	**20.** $-1+4i$
21. $12-6i$	**22.** $28-24i$
23. $96-60i$	**24.** $-48-12i$
25. $-48+27i$	**26.** $15-30i$
27. $-102+30i$	**28.** $-96-36i$
29. $(1-i)(3+2i) = 3+2i-3i-2i^2$ $= 3-i+2 = \boxed{5-i}$	**30.** $(-3+2i)(1-3i) = -3+9i+2i-6i^2$ $= -3+11i+6$ $= \boxed{3+11i}$
31. $-15+21i+20i+28 = 13+41i$	**32.** $-32+10i-16i-5 = -37-6i$
33. $42-30i+63i+45 = 87+33i$	**34.** $-21-14i+12i-8 = -29-2i$
35. $-24+36i+12i+18 = -6+48i$	**36.** $16+9=25$
37. $\frac{2}{9}+\frac{8}{9}i-\frac{3}{2}i+6 = \frac{56}{9}-\frac{11}{18}i$	**38.** $-\frac{1}{2}-\frac{1}{3}i+\frac{3}{8}i-\frac{1}{4} = -\frac{3}{4}+\frac{1}{24}i$

39. $-2i+34+51i+3=37+49i$	**40.** $6i+4-9+6i=-5+12i$
41. $\overline{z}=4-7i$ $z\overline{z}=4^2+7^2=65$	**42.** $\overline{z}=2-5i$ $z\overline{z}=2^2+5^2=29$
43. $\overline{z}=2+3i$ $z\overline{z}=2^2+3^2=13$	**44.** $\overline{z}=5+3i$ $z\overline{z}=5^2+3^2=34$
45. $\overline{z}=6-4i$ $z\overline{z}=6^2+4^2=52$	**46.** $\overline{z}=-2-7i$ $z\overline{z}=2^2+7^2=53$
47. $\overline{z}=-2+6i$ $z\overline{z}=2^2+6^2=40$	**48.** $\overline{z}=-3+9i$ $z\overline{z}=3^2+9^2=90$
49. $\frac{2}{i}\cdot\frac{i}{i}=-2i$	**50.** $\frac{3}{i}\cdot\frac{i}{i}=-3i$
51. $\frac{1}{3-i}\cdot\frac{(3+i)}{(3+i)}=\frac{3+i}{9-i^2}=\frac{3+i}{10}=\boxed{\frac{3}{10}+\frac{1}{10}i}$	**52.** $\frac{2}{7-i}\cdot\frac{7+i}{7+i}=\frac{14+2i}{49+1}=\frac{7}{25}+\frac{1}{25}i$
53. $\frac{1}{3+2i}\cdot\frac{(3-2i)}{(3-2i)}=\frac{3-2i}{9-4i^2}=\frac{3-2i}{9+4}$ $=\frac{3-2i}{13}=\boxed{\frac{3}{13}-\frac{2}{13}i}$	**54.** $\frac{1}{4-3i}\cdot\frac{4+3i}{4+3i}=\frac{4+3i}{16+9}=\frac{4}{25}+\frac{3}{25}i$
55. $\frac{2}{7+2i}\cdot\frac{7-2i}{7-2i}=\frac{14-4i}{49+4}=\frac{14}{53}-\frac{4}{53}i$	**56.** $\frac{8}{1+6i}\cdot\frac{1-6i}{1-6i}=\frac{8-48i}{1+36}=\frac{8}{37}-\frac{48}{37}i$
57. $\frac{1-i}{1+i}\cdot\frac{(1-i)}{(1-i)}=\frac{1-2i+i^2}{1-i^2}$ $=\frac{1-2i-1}{1-(-1)}=\frac{-2i}{2}=\boxed{-i}$	**58.** $\frac{3-i}{3+i}\cdot\frac{3-i}{3-i}=\frac{9-6i-1}{9+1}=\frac{8-6i}{10}$ $=\frac{4}{5}-\frac{3}{5}i$
59. $\frac{2+3i}{3-5i}\cdot\frac{3+5i}{3+5i}=\frac{6+9i+10i-15}{9+25}$ $=-\frac{9}{34}+\frac{19}{34}i$	**60.** $\frac{2+i}{3-i}\cdot\frac{3+i}{3+i}=\frac{6+3i+2i-1}{9+1}=\frac{5+5i}{10}$ $=\frac{1}{2}+\frac{1}{2}i$

61. $\frac{4-5i}{7+2i}\cdot\frac{(7-2i)}{(7-2i)}=\frac{28-8i-35i+10i^2}{49-4i^2}$ $=\frac{28-43i-10}{49+4}$ $=\frac{18-43i}{53}=\boxed{\frac{18}{53}-\frac{43}{53}i}$	**62.** $\frac{7+4i}{9-3i}\cdot\frac{9+3i}{9+3i}=\frac{63+36i+21i-12}{81+9}$ $=\frac{51+57i}{90}=\frac{17}{30}+\frac{19}{30}i$
63. $\frac{8+3i}{9-2i}\cdot\frac{9+2i}{9+2i}=\frac{72+27i+16i-6}{81+4}$ $=\frac{66}{85}+\frac{43}{85}i$	**64.** $\frac{10-i}{12+5i}\cdot\frac{12-5i}{12-5i}=\frac{120-12i-50i-5}{144+25}$ $=\frac{115}{169}-\frac{62}{169}i$
65. $i^{15}=i^{12}\cdot i^3=\underbrace{\left(i^4\right)^3}_{1}\cdot i^3=i\left(i^2\right)=i(-1)=\boxed{-i}$	**66.** $i^{99}=i^{96}\cdot i^3=\left(i^4\right)^{24}\cdot i^3$ $=i\left(i^2\right)=i(-1)=\boxed{-i}$
67. $i^{40}=\left(i^4\right)^{10}=(1)^{10}=\boxed{1}$	**68.** $i^{18}=i^{16}\cdot i^2=\left(i^4\right)^4\cdot i^2=(1)^4\cdot i^2=\boxed{-1}$
69. $25-20i-4=21-20i$	**70.** $9-30i-25=-16-30i$
71. $4+12i-9=-5+12i$	**72.** $16-72i-81=-65-72i$
73. $(3+i)^2(3+i)=(9+6i-1)(3+i)$ $=(8+6i)(3+i)$ $=24+18i+8i-6$ $=18+26i$	**74.** $(2+i)^2(2+i)=(4+4i-1)(2+i)$ $=(3+4i)(2+i)$ $=6+8i+3i-4$ $=2+11i$
75. $(1-i)^2(1-i)=(1-2i-1)(1-i)$ $=-2i(1-i)$ $=-2i-2$ $=-2-2i$	**76.** $(4-3i)^2(4-3i)=(16-24i-9)(4-3i)$ $=(7-24i)(4-3i)$ $=28-96i-21i-72$ $=-44-117i$

77. $z=(3-6i)+(5+4i)=8-2i$ ohms	**78.** $\frac{1}{z}=\frac{1}{3-6i}+\frac{1}{5+4i}$ $=\frac{1}{3-6i}\cdot\frac{3+6i}{3+6i}+\frac{1}{5+4i}\cdot\frac{5-4i}{5-4i}$ $=\frac{3+6i}{45}+\frac{5-4i}{41}$ $=\left(\frac{1}{15}+\frac{5}{41}\right)+\left(\frac{2}{15}-\frac{4}{41}\right)i$ $=\frac{116}{615}+\frac{22}{615}i$
79. Should have multiplied by the conjugate $4+i$ (not $4-i$). $\frac{2}{4-i}\cdot\frac{(4+i)}{(4+i)}=\frac{8+2i}{16-i^2}=\frac{8+2i}{17}=\boxed{\frac{8}{17}+\frac{2}{17}i}$	**80.** $10-7i-12i^2=10-7i+12=\boxed{22-7i}$
81. True. **82.** True.	**83.** True. **84.** False.
85. $x^4+2x^2+1=\left(x^2+1\right)^2=(x+i)^2(x-i)^2$	**86.** $\left(x^2+9\right)^2=\left((x+3i)(x-3i)\right)^2$ $=(x+3i)^2(x-3i)^2$
87. $41-38i$	**88.** $-352-936i$
89. $\frac{2}{125}+\frac{11}{125}i$	**90.** $\frac{7}{625}-\frac{24}{625}i$

Appendix Review Solutions --

1. a. 5.22 **b.** 5.21	**2. a.** 7.36 **b.** 7.36
3. $2-10+12=\boxed{4}$	**4.** $-16-49=\boxed{-65}$
5. –2	**6.** $-3x+3y+12x-8y=\boxed{9x-5y}$
7. $\frac{3x}{12}-\frac{4x}{12}=\boxed{-\frac{x}{12}}$	**8.** $\frac{10y}{30}+\frac{6y}{30}-\frac{5y}{30}=\boxed{\frac{11y}{30}}$
9. $\frac{3}{1}\cdot\frac{3}{1}=\boxed{9}$	**10.** $\frac{a^2}{b^3}\cdot\frac{b^2}{2a}=\boxed{\frac{a}{2b}}$
11. $-8z^3$	**12.** $-64z^6$
13. $\frac{9x^6y^4}{2x^8y^4}=\frac{9}{2x^2}$	**14.** $\frac{4x^4y^6}{4^3x^3y^3}=\boxed{\frac{xy^3}{16}}$

15. 2.15×10^{-6}	**16.** $7,200,000,000$
17. $14z^2+3z-2$	**18.** $26y^2-9y+9$
19. $36x^2-4x-5-6x+9x^2-10=$ $45x^2-10x-15$	**20.** $2x-4x^2+7x-3x+2x^2+5x-4=$ $-2x^2+11x-4$
21. $15x^2y^2-20xy^3$	**22.** $2st^3-2s^2t^2+4s^2t^3$
23. $x^2+2x-63$	**24.** $6x^2-x-2$
25. $4x^2-12x+9$	**26.** $25x^2-49$
27. x^4+2x^2+1	**28.** x^4-2x^2+1
29. $2xy^2(7x-5y)$	**30.** $10x^2(3x^2-2x+1)$
31. $\boxed{(x+5)(2x-1)}$	**32.** $\boxed{(3x+1)(2x-7)}$
33. $(4x+5)(4x-5)$	**34.** $(3x-5)(3x-5)=\boxed{(3x-5)^2}$
35. $(x+5)(x^2-5x+25)$	**36.** $(1-2x)(1+2x+4x^2)$
37. $2x(x+5)(x-3)$	**38.** $x(6x^2-5x+1)=\boxed{x(3x-1)(2x-1)}$
39. $(x^2-2)(x+1)$	**40.** $(x^2+3)(2x-1)$
41. $x\neq\pm3$	**42.** x is any real number
43. $x+2,\ x\neq2$	**44.** $1,\ x\neq5$
45. $\dfrac{(t+3)\cancel{(t-2)}}{\cancel{(t-2)}(t+1)}=\dfrac{(t+3)}{(t+1)},\quad t\neq-1,2$	**46.** $\dfrac{\cancel{z}\cancel{(z+1)}(z-1)}{\cancel{z}\cancel{(z+1)}}=z-1,\quad z\neq-1,0$
47. $\dfrac{(x+5)\cancel{(x-2)}}{(x+3)\cancel{(x-1)}}\cdot\dfrac{(x+2)\cancel{(x-1)}}{(x+3)\cancel{(x-2)}}=\boxed{\dfrac{(x+5)(x+2)}{(x+3)^2},x\neq-3,1,2}$	**48.** $\dfrac{(x-2)\cancel{(x+1)}}{x^{\cancel{2}}(x+3)}\cdot\dfrac{x(x+2)}{\cancel{x+1}}=\boxed{\dfrac{(x-2)(x+2)}{x(x+3)},x\neq-3,-2,-1,0}$
49. $\dfrac{x+3}{(x+1)(x+3)}-\dfrac{x+1}{(x+1)(x+3)}=\boxed{\dfrac{2}{(x+1)(x+3)}\ x\neq-1,-3}$	**50.** $\dfrac{(x+1)(x+2)-x(x+2)+x(x+1)}{x(x+1)(x+2)}=\boxed{\dfrac{x^2+2x+2}{x(x+1)(x+2)}\ x\neq0,-1,-2}$
51. $\dfrac{\dfrac{2(x-3)+1}{x-3}}{\dfrac{1+4(5x-15)}{5x-15}}=\dfrac{2x-5}{\cancel{x-3}}\cdot\dfrac{5\cancel{(x-3)}}{20x-59}=\dfrac{10x-25}{20x-59}$ $x\neq3,\frac{59}{20}$	**52.** $\dfrac{\dfrac{x+2}{\cancel{x^2}}}{\dfrac{3x^2-1}{\cancel{x^2}}}=\dfrac{x+2}{3x^2-1},\ x\neq0,\pm\sqrt{\frac{1}{3}}$
53. $\sqrt{4\cdot5}=2\sqrt{5}$	**54.** $\sqrt{16\cdot5}=4\sqrt{5}$

55. $-5xy\sqrt[3]{x^2y}$	**56.** $2\|xy\|\sqrt[4]{2y}$
57. $3\left(2\sqrt{5}\right)+5\left(4\sqrt{5}\right)=26\sqrt{5}$	**58.** $12\sqrt{3x}-16\sqrt{3x}=-4\sqrt{3x}$
59. $2+\sqrt{5}-2\sqrt{5}-5=-3-\sqrt{5}$	**60.** $12+4\sqrt{x}-3\sqrt{x}-x=12-x+\sqrt{x}$
61. $\frac{1}{2-\sqrt{3}}\cdot\frac{2+\sqrt{3}}{2+\sqrt{3}}=\frac{2+\sqrt{3}}{4-3}=2+\sqrt{3}$	**62.** $\frac{1}{3-\sqrt{x}}\cdot\frac{3+\sqrt{x}}{3+\sqrt{x}}=\frac{3+\sqrt{x}}{9-x}$
63. $\frac{9x^{4/3}}{16x^{2/3}}=\frac{9x^{2/3}}{16}$	**64.** $\frac{16x^{3/2}}{4x^{-2/3}}=4x^{13/6}$
65. $5^{\frac{1}{2}-\frac{1}{3}}=5^{\frac{1}{6}}$	**66.** $x^{-8}y^3=\frac{y^3}{x^8}$
67. $13i$	**68.** $\sqrt{-32}=\sqrt{-2\cdot 16}=\boxed{4i\sqrt{2}}$
69. $\left(i^4\right)^5\cdot i^{-1}=\boxed{-i}$	**70.** $\left(i^4\right)^2\cdot i=\boxed{i}$
71. $(2i+2)(3-3i)=\boxed{12}$	**72.** $(6i+1)(1+5i)=\boxed{-29+11i}$
73. $16+56i-49=-33+56i$	**74.** $49-14i-1=48-14i$
75. $\frac{1}{2-i}\cdot\frac{2+i}{2+i}=\frac{2+i}{4+1}=\boxed{\frac{2}{5}+\frac{1}{5}i}$	**76.** $\frac{1}{3+i}\cdot\frac{3-i}{3-i}=\frac{3-i}{9+1}=\boxed{\frac{3}{10}-\frac{1}{10}i}$
77. $\frac{7+2i}{4+5i}\cdot\frac{4-5i}{4-5i}=\frac{28-27i-10i^2}{16+25}=\boxed{\frac{38}{41}-\frac{27}{41}i}$	**78.** $\frac{6-5i}{3-2i}\cdot\frac{3+2i}{3+2i}=\boxed{\frac{28}{13}-\frac{3}{13}i}$
79. $\frac{10}{3i}\cdot\frac{i}{i}=-\frac{10}{3}i$	**80.** $\frac{7}{2i}\cdot\frac{i}{i}=-\frac{7}{2}i$
81. Simplifying the radical using the calculator gives exactly 16.5. So, it is rational.	**82.** Simplifying the radical using the calculator gives the approximation 3.605551275. Since there is no discernable pattern, it seems that the number is irrational.
83. 1.945×10^{-6}	**84.** 1.5625×10^{3}
85. 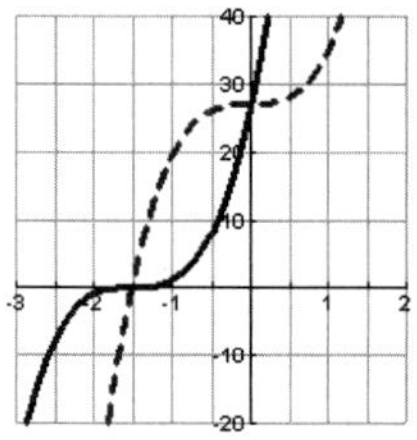 The solid curve represents the graph of both $y=(2x+3)^3$ and $y=8x^3+36x^2+54x+27$.	**86.** 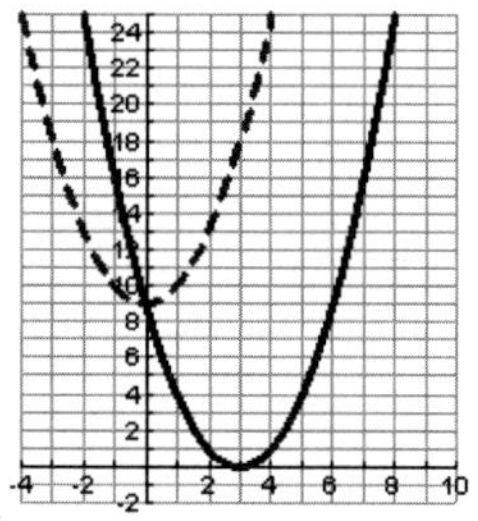 The solid curve represents the graph of both $y=(x-3)^2$ and $y=x^2-6x+9$.

87. 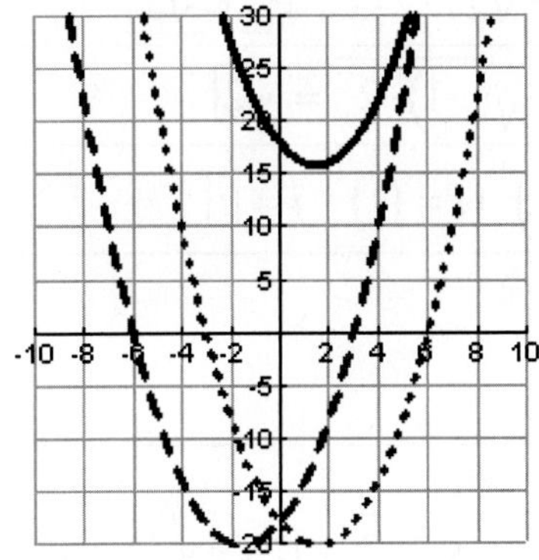 All three graphs are different.	**88.** 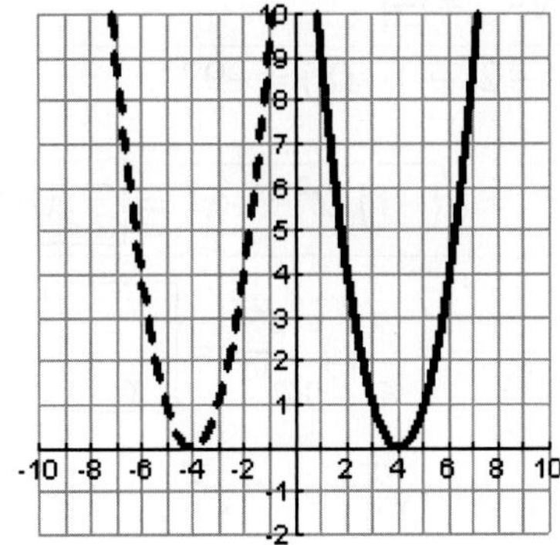 The solid curve represents the graph of both $y = x^2 - 8x + 16$ and $y = (x-4)^2$
89. a. $\frac{1-\frac{4}{x}}{1-\frac{4}{x^2}} = \frac{\frac{x-4}{x}}{\frac{x^2-4}{x^2}} = \frac{x^2(x-4)}{x\left(x^2-4\right)} = \frac{x(x-4)}{(x+2)(x-2)}$ **b.** 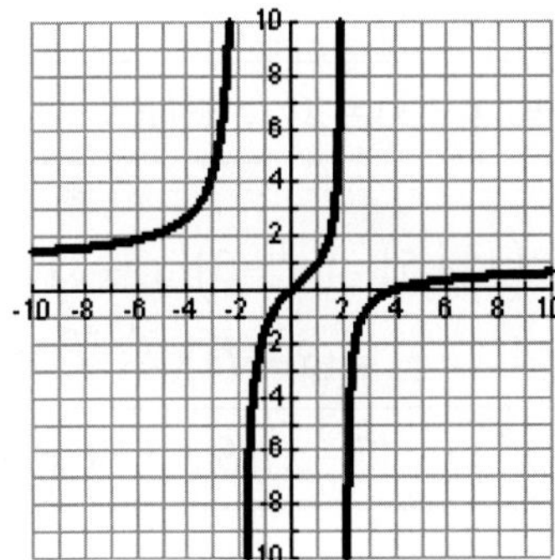 **c.** The graphs agree as long as $x \neq 0, \pm 2$.	**90. a.** $\frac{1-\frac{3}{x}}{1+\frac{9}{x^2}} = \frac{\frac{x-3}{x}}{\frac{x^2+9}{x^2}} = \frac{x^2(x-3)}{x\left(x^2+9\right)} = \frac{x(x-3)}{x^2+9}$ **b.** 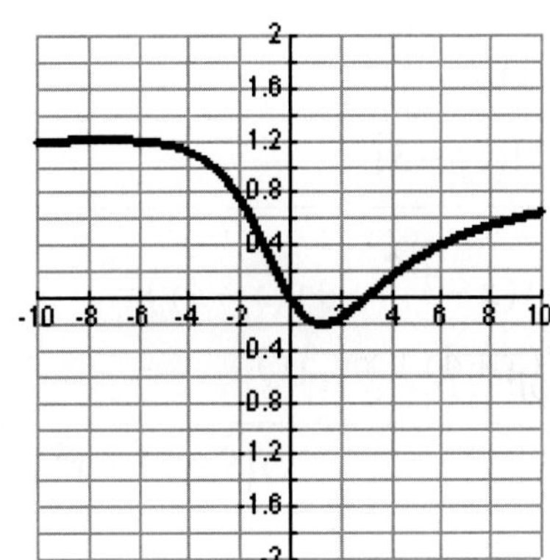 **c.** The graphs agree as long as $x \neq 0$.
91. **a.** $\frac{6}{\sqrt{5}-\sqrt{2}} \cdot \frac{\sqrt{5}+\sqrt{2}}{\sqrt{5}+\sqrt{2}} = \frac{6\left(\sqrt{5}+\sqrt{2}\right)}{5-2}$ $= 2\sqrt{5} + 2\sqrt{2}$ **b.** 7.30056308 **c.** Yes	**92.** **a.** $\frac{11}{2\sqrt{6}+\sqrt{13}} \cdot \frac{2\sqrt{6}-\sqrt{13}}{2\sqrt{6}-\sqrt{13}} = \frac{11\left(2\sqrt{6}-\sqrt{13}\right)}{24-13}$ $= 2\sqrt{6} - \sqrt{13}$ **b.** 1.29342821 **c.** Yes
93. $2868 - 6100i$	**94.** $\frac{7}{2500} + \frac{6}{625}i$
95. 1.6×10^{14}	**96.** $\frac{7}{40{,}000} - \frac{3}{5{,}000}i$

Appendix Practice Test Solutions --

1. $\sqrt{16}=\sqrt{4^2}=\boxed{4}$	**2.** $\sqrt[3]{54x^6}=\sqrt[3]{27\cdot 2\cdot x^6}=\boxed{3x^2\sqrt[3]{2}}$
3. $-3(27)+2(-4)-(8)=-97$	**4.** $\sqrt[5]{-32}=\sqrt[5]{(-1)\cdot 2^5}=\boxed{-2}$
5. $\sqrt{-12x^2}=\sqrt{(-1)\cdot 2^2\cdot 3\cdot x^2}=\boxed{2i\lvert x\rvert\sqrt{3}}$	**6.** $i^{17}=\left(i^4\right)^4\cdot i=(1)^4\cdot i=\boxed{i}$
7. $\dfrac{\left(x^2y^{-3}z^{-1}\right)^{-2}}{\left(x^{-1}y^2z^3\right)^{1/2}}=\dfrac{x^{-4}y^6z^2}{x^{-1/2}yz^{3/2}}=\boxed{\dfrac{y^5z^{1/2}}{x^{7/2}}}$	**8.** $4\sqrt{x}$
9. $3\left(3\sqrt{2}\right)-4\left(4\sqrt{2}\right)=-7\sqrt{2}$	**10.** $30-2\sqrt{2}\sqrt{6}+15\sqrt{6}\sqrt{2}-12$ $=30-4\sqrt{3}+30\sqrt{3}-12$ $=18+26\sqrt{3}$
11. $2y^2-12y+20$	**12.** $10x^2-x-21$
13. $(x-4)(x+4)$	**14.** $3\left(x^2+5x+6\right)=3(x+3)(x+2)$
15. $(2x+3y)^2$	**16.** $\left(x^2-1\right)^2=(x-1)^2(x+1)^2$
17. $(2x+1)(x-1)$	**18.** $(2y-1)(3y+1)$
19. $t(t+1)(2t-3)$	**20.** $x\left(2x^2-5x-3\right)=x(2x+1)(x-3)$
21. $x(x-3y)+4y(x-3y)=$ $(x+4y)(x-3y)$	**22.** $x^2\left(x^2+5\right)-3\left(x^2+5\right)=\left(x^2-3\right)\left(x^2+5\right)$ $=(x-\sqrt{3})(x+\sqrt{3})\left(x^2+5\right)$
23. $3\left(27+x^3\right)=3(3+x)\left(9-3x+x^2\right)$	**24.** $x\left(27-x^3\right)=x(3-x)\left(9+3x+x^2\right)$
25. $\dfrac{2(x-1)+3x}{x(x-1)}=\dfrac{5x-2}{x(x-1)},\ x\neq 0,1$	**26.** $\dfrac{5x}{(x-5)(x-2)}-\dfrac{4}{(x-5)(x+5)}=$ $\dfrac{5x(x+5)-4(x-2)}{(x-5)(x+5)(x-2)}=\dfrac{5x^2+21x+8}{(x-5)(x+5)(x-2)}$ Note: $x\neq 2,\pm 5$

27. $\frac{x-1}{(x-1)(x+1)}\cdot\frac{x^2+x+1}{(x-1)\left(x^2+x+1\right)}=$ $\frac{1}{(x-1)(x+1)}$ Note: $x\neq\pm1$	**28.** $\frac{(2x-3)(2x+3)}{(x-15)(x+4)}\cdot\frac{(x-4)(x+4)}{(2x+3)}=$ $\frac{(2x-3)(x-4)}{(x-15)}$ Note: $x\neq-4,-\frac{3}{2},15$
29. $\frac{x-3}{2x-5}\div\frac{x^2-9}{5-2x}=\frac{x-3}{2x-5}\cdot\frac{-(2x-5)}{(x-3)(x+3)}$ $=-\frac{1}{x+3}$ Note: $x\neq\frac{5}{2},\pm3$	**30.** $\frac{1-t}{3t+1}\cdot\frac{7t+21t^2}{t^2-2t+1}=\frac{1-t}{3t+1}\cdot\frac{7t(1+3t)}{(t-1)(t-1)}$ $=\boxed{\frac{-7t}{t-1},\ t\neq-\frac{1}{3},1}$
31. $(1-3i)(7-5i)=7-26i+15i^2=\boxed{-8-26i}$	**32.** $\frac{2-11i}{4+i}\cdot\frac{4-i}{4-i}=\frac{8-46i-11}{16+1}=\boxed{\frac{-3}{17}-\frac{46}{17}i}$
33. $\frac{7-2\sqrt{3}}{4-5\sqrt{3}}\cdot\frac{4+5\sqrt{3}}{4+5\sqrt{3}}=\frac{28+27\sqrt{3}-10\cdot3}{16-25\cdot3}$ $=\boxed{\frac{2-27\sqrt{3}}{59}}$	**34.** 1.55×10^{-5}
35. $\frac{\frac{x+1-2x}{x(x+1)}}{x-1}=\frac{1-x}{x(x+1)(x-1)}=-\frac{1}{x(x+1)}$ Note: $x\neq0,\pm1$	**36.** **a.** $\frac{1+\frac{5}{x}}{1-\frac{25}{x^2}}=\frac{\frac{x+5}{x}}{\frac{x^2-25}{x^2}}=\frac{x^2(x+5)}{x\left(x^2-25\right)}=\frac{x}{x-5}$ **b.** 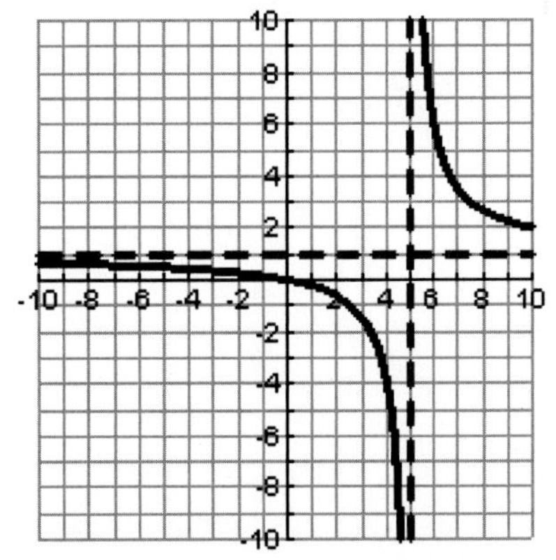 **c.** The graphs agree as long as $x\neq0,\pm5$.
37. 2.330	

NOTES

NOTES

NOTES

NOTES

NOTES

NOTES

NOTES

NOTES

NOTES

NOTES

NOTES

NOTES